Checklist of the British & Irish
BASIDIOMYCOTA

Checklist of the British & Irish
BASIDIOMYCOTA

N.W. Legon and A. Henrici
with P.J. Roberts, B.M. Spooner, and R. Watling

Derived from a database designed by J.A. Cooper and supported by P.M. Kirk

Project Steering Group:

Brian Spooner (Royal Botanic Gardens, Kew–chair), Carl Borges (English Nature), Jerry Cooper (CABI Bioscience), Hubert Fuller (University College, Dublin, representing the Heritage Council of Ireland), Alick Henrici (Royal Botanic Gardens, Kew), Paul Kirk (CABI Bioscience), Nick Legon (Project Officer, Royal Botanic Gardens, Kew), David Mitchel (Environment and Heritage Service, Northern Ireland), Peter Roberts (Royal Botanic Gardens, Kew), Jill Sutcliffe (English Nature), Stephen Ward (Scottish Natural Heritage), Roy Watling (formerly Royal Botanic Garden, Edinburgh, representing the British Mycological Society), Ray Woods (Cyngor Cefn Gwlad Cymru/ Countryside Council for Wales).

Project funding:

British Mycological Society
Cyngor Cefn Gwlad Cymru/Countryside Council for Wales
English Nature
Environment and Heritage Service, Northern Ireland
Fungal Research Trust
Heritage Council of Ireland
Royal Botanic Gardens, Kew
Scottish Natural Heritage

Royal Botanic Gardens, Kew

First published in 2005 by Royal Botanic Gardens, Kew
Richmond, Surrey, TW9 3AB, UK
www.kew.org

ISBN 1 84246 121 4

Cover design by Dan Higham, Media Resources,
Information Services Department,
Royal Botanic Gardens, Kew.

Photographs
Front cover: Nick Legon
Back cover: Brian Spooner, Nick Legon,
Sandra Rickwood, RBG Kew

**For information or to purchase all Kew titles please visit
www.kewbooks.com or email publishing@kew.org**

CONTENTS

INTRODUCTION

This is the first ever comprehensive checklist of the *Basidiomycota* of Great Britain and Ireland. It may seem surprising that such a basic text should have to wait till the 21st century before being published, but until recently the task has seemed overwhelming, given the large numbers of fungal taxa in the British Isles and the small number of taxonomic mycologists.

Partial checklists of various groups of the *Basidiomycota* have, however, previously been published and these earlier works have provided a substantial basis for the preparation of the present checklist. The largest and most important of these was the *New Check List of British Agarics and Boleti* (NCL), published in 1960 by R.W.G. Dennis, P.D. Orton, and F.B. Hora. This provided a critical listing of all the mushrooms, toadstools, and boletes known to occur in the British Isles, clearing away much dead wood and uncertain names, whilst describing many new taxa and providing many new combinations. This revisionary work was itself an update of an earlier checklist of agarics by Pearson & Dennis (1948), which in turn was a radical overhaul of the list of agarics treated by Carleton Rea in his *British Basidiomycetae* (1922). Rea's compilation, updating and extending that by Smith (1908), was a major but uncritical descriptive catalogue of all the British hymenomycetes (a term which excluded rusts, smuts, yeasts and hyphomycetes) recorded up to that time. Important earlier compilations include those by Berkeley (1860, extensively supplemented by Smith, 1891), Cooke (1871, 1883-91), Massee (1892, 1893a,b), and Stevenson (1879, 1886).

More recently, several checklists or monographs of various groups of basidiomycetes have been published. British hypogeous basidiomycetes (the false truffles) were monographed by Pegler, Spooner & Young (1993) and other gasteroid basidiomycetes by Pegler, Læssøe & Spooner (1995). British stipitate hydnoid fungi and chanterelles were monographed by Pegler, Roberts & Spooner (1997). A key to British poroid fungi was published by Pegler (1973) and a list of British aphyllophoroid fungi by Henrici & Cook (1991). A preliminary list of British heterobasidioid fungi was published by Henrici (1990). The British rust fungi, first monographed by Plowright (1889) and later by Grove (1913), were covered in a modern monograph by Wilson & Henderson (1966) with additions by Henderson & Bennell (1979, 1980), and a more recent checklist by Henderson (2000). Mordue & Ainsworth (1984) compiled an account of the smut fungi, which revised the earlier monograph of the British smuts by Ainsworth & Samson (1950).

In addition, compilations of the fungi of the Hebrides (Dennis, 1986) and of South East England (Dennis, 1995) provided comprehensive listings of the fungi known from these regions, the former also including data on all known British fungal genera. Some county mycotas, covering basidiomycetes as well as other groups of fungi, have also been produced, providing details of local species and important data on hosts and distribution. These include most notably those for Yorkshire (resulting from the activities of the Mycological Committee of the Yorkshire Naturalists' Union) first compiled by Massee & Crossland (1905), followed by an extended catalogue by Mason & Grainger (1937), an update by Bramley (1985), and a recent local list for the Scarborough District by Stephenson (2004). Other published county and local mycotas include those of Plowright (1872-73) for Norfolk, Clark (1980) for Warwickshire, Dickson & Leonard (1996) for the New Forest, Bowen (2000) for Dorset, Watling (1994) for Shetland, and Watling *et al.* (1999) for Orkney. Details of many additional regional, county, and local lists can be found in Ainsworth & Waterhouse (1989).

The yeast fungi have tended to be studied quite separately, and it is only with the advent of molecular sequencing that the classification of yeasts has been satisfactorily linked to the classification of filamentous fungi. From this we know that many yeasts are basidiomycetous and should be included in the checklist. However, information on the distribution of these yeasts within the British Isles has been difficult or impossible to find, although the recent compilation by Barnett *et al.* (2000) has been of value.

THE PRESENT CHECKLIST

The present checklist is far more than a simple list of species names. It collates data on some 3670 species or subspecific taxa recorded from the British Isles, together with additional alien and excluded taxa, plus synonyms. It attempts to include all names which have appeared in British literature, a total of over 16,500 names. It has been compiled following an extensive survey of literature references and herbarium collections, undertaken in consultation with a number of specialists. The checklist makes no pretence of being a revisionary work and includes no taxonomic novelties.

The list itself is derived from a database compiled at the Royal Botanic Gardens, Kew since the year 2000. It is intended that this database be maintained and updated at Kew, and be made freely available in a searchable format on the Internet.

This printed edition is just a snapshot of the data currently on file and should not be taken as definitive. Though we have tried to keep them to a minimum, errors and omissions are inevitable (corrections will be welcomed – see below). New research, particularly in the expanding field of molecular sequencing, will also entail the revision of many current species and generic concepts. This checklist is not, therefore, the ultimate guide to the naming of British basidiomycetous fungi. Names are changing right at this moment and will continue to change. But the checklist does attempt to list all the species (and autonomous taxa) which are known to occur in the British Isles, whatever the names they have been given now or in the past.

It is envisaged that the Internet version of the checklist will be regularly updated (with a summary noting any major changes). Corrections and additions should be e-mailed to mycology@rbgkew.org.uk.

WHICH GROUPS OF FUNGI ARE INCLUDED?

The checklist includes entries for all known species within the kingdom *Fungi*, phylum *Basidiomycota*, occurring within the British Isles. Since the higher taxonomy of these fungi is currently in a state of flux, the decision was taken to list species alphabetically by genus within each class, rather than attempt a systematic arrangement by class, order, and family.

The checklist is divided into three main sections, starting with an A-Z listing (by genera and species) of all recognized species and subspecific taxa within the class *Basidiomycetes*. This includes the agarics (mushrooms and toadstools), the cyphelloid fungi, the poroid or bracket fungi, the gasteroid fungi (including puffballs, stinkhorns, and false truffles), the corticioid or patch-forming fungi, the clavarioid or club and coral fungi, the hydnoid or toothed fungi, and many (but not all) of the jelly fungi with their associated yeasts. Altogether, some 3177 species or subspecific taxa of *Basidiomycetes* are recognized as occurring in the British Isles.

The second section is an A-Z listing (by genera and species) of taxa within the class *Urediniomycetes*. This includes the rust fungi, a few of the smut fungi, and the remaining jelly fungi with their associated yeasts. Some 342 species or subspecific taxa are recognized.

The final section is an A-Z listing (by genera and species) of taxa within the class *Ustilaginomycetes*. This contains the majority of the smut fungi, together with a few other plant parasites and yeasts. Some 151 species or subspecific taxa are recognized.

Anamorphs (mainly basidiomycetous yeasts and hyphomycetes) are included under their teleomorph names. Where no teleomorph is known, the anamorph receives a separate entry.

GEOGRAPHICAL COVERAGE

The checklist covers the whole of the British Isles, including the United Kingdom, the Republic of Ireland, the Isle of Man, and the Bailiwicks of Guernsey and Jersey (Channel Islands). Inevitably herbarium collections and published records are not evenly distributed throughout the British Isles, and as a consequence some areas are substantially under-recorded. This is particularly true of the Republic of Ireland and Northern Ireland, though more recording is now being done here than in the past.

CRITERIA FOR INCLUSION

The checklist aims to include at least some reference to all species within the *Basidiomycota* that have ever been reported to occur in the British Isles.

Full entries are given for taxa which are currently recognized as distinct species and whose status as British is supported by at least one herbarium collection. In most (but not all) cases these herbarium collections are in the national collections at the Royal Botanic Gardens, Kew, and/or the Royal Botanic Garden, Edinburgh.

A few subspecific taxa, mainly at the level of 'variety', have also received entries on the same basis. Subspecific taxa at the often rather dubious level of 'form' are generally listed as synonyms, if at all.

EXCLUDED SPECIES

Two rather different categories of species are listed here. Firstly 'good' species that have been claimed to be British, but for which the evidence is lacking (absence of herbarium collections). Secondly, and more numerously, doubtful species, formerly listed as British, but now of uncertain application. These excluded species (and some subspecific taxa) are listed alphabetically by epithet (not by genus) and are split into three sections by class, as in the main list.

ALIEN SPECIES

Alien species receive abbreviated entries in a separate appendix. But defining the criteria for inclusion on this 'alien' list has presented some problems.

There is no clear way to distinguish native British fungi from alien species. We may assume that a species like the bolete *Suillus grevillei*, though originally described from Scotland, is an alien species since it is an exclusive ectomycorrhizal associate of larch, which is an alien tree. But it is less clear (for example) that all the ectomycorrhizal associates of pine, which in the British Isles is only native in Scotland, are alien when found outside Scotland. The distinction is even less clear when considering fungi forming non-specific ectomycorrhizal and saprotrophic associations.

Accordingly, the checklist makes an arbitrary but clear-cut definition of 'alien fungi' as those species which have only ever been reported in the British Isles in an indoor environment, i.e. in hothouses, greenhouses, and other buildings. If a species has been found outdoors, even if only on a compost heap, it is placed in the main part of the checklist. A few rusts recorded once or twice on exotic garden plants have also been relegated to the alien list where the records suggest they have not become established in the British Isles.

The appendix of alien fungi has been included for two reasons. Firstly, some of these alien species have appeared in past accounts or lists of 'British' fungi and should therefore be noted somewhere in the checklist. Secondly, a few species have been described as new to science from British hothouses, some of which (though clearly exotic) are still only known from the British Isles. These too deserve to be noted.

The alien list is arranged alphabetically by name, and is split into three sections by class, in the same way as the main list.

SOURCES OF INFORMATION

Most of the information contained in the checklist has been obtained from the British collections of fungi held in the mycological herbarium at the Royal Botanic Gardens, Kew. These amount to some 250,000 collections from all parts of the British Isles, approximately 100,000 of which belong within the *Basidiomycota*. Substantial additional information has been obtained from the British collections at the Royal Botanic Garden, Edinburgh.

Records in the British Mycological Society's Fungal Records Database (BMSFRD) have been examined and have proved valuable for distribution and frequency data (see below). Much of the literature on the British *Basidiomycota*, both early and recent, has also been examined (see references and literature list below) and many experts in particular groups have been consulted (see acknowledgements below).

GENUS ENTRIES

Accepted genera within the British *Basidiomycota* are listed alphabetically within each class. Each accepted genus receives an entry, giving:
    1) the genus name, with full publication details (author(s), date, and place of publication);
    2) the order and family in which it is placed in the 8th Edition of the *Dictionary of the Fungi* (Hawksworth *et al.*, 1995). This classification is cited rather than that of the 9th Edition (Kirk *et al.*, 2001) which partially incorporates recent revisions based on molecular data. The molecular picture is as yet too fluid and with too many gaps to form a satisfactory basis for a detailed classification;
    3) synonyms (used in the British Isles), with full publication details;
    4) the type species of the accepted genus.

SPECIES ENTRIES

Accepted species (and some subspecific taxa) within the British *Basidiomycota* are listed alphabetically within each genus. Each accepted species receives an entry, giving:
    1) the species name, with full publication details (author(s), date, and place of publication);
    2) synonyms (used in the British Isles), with full publication details;
    3) misapplied names, where widely used;
    4) data on habitat, seasonality, and associated species (see below);
    5) distribution and frequency data (see below);
    6) abbreviated references to relevant descriptions (D) and illustrations (I);
    7) any additional notes.

AUTHOR CITATIONS & PUBLICATION DETAILS

Considerable effort has been made to find the correct author citation and publication details for all taxa in the checklist. This is particularly difficult for older taxa, following the change in the starting point date for fungal nomenclature introduced at the 13th International Botanical Congress in 1981. Previously, the starting point for hymenomycete nomenclature was 1821 (publication date of the first volume of Fries' *Systema mycologicum*), and for gasteromycetes and rusts 1801 (publication date of Persoon's *Synopsis methodica fungorum*). Earlier names were devalidated unless used after the starting point dates, in which case they were credited to the first author to refer back to them. This system was considered unsatisfactory a) because the names of lichenised fungi already started with Linnaeus in 1753, and b) because priority and pre-eminence were inevitably given to any obscure or minor work found to have been the first after the starting point dates to use a devalidated name. Under this system, for example, J.F.B. de St-Amans' little-known *Flore agenaise* (St-Amans, 1821) became one of the most frequently cited mycological works, a reputation it hardly deserved.

Following the 1981 Congress, the starting point date for fungi was put back to 1753 (the date of Linnaeus's *Species Plantarum*) in line with plant and lichen names, except that taxa accepted by Fries (1821, 1822-1823, 1828, & 1829-1832) and Persoon (1801) were 'sanctioned' and could not be replaced by earlier names. More than twenty years after this rule change, the ramifications are still requiring research. Unfamiliar old epithets have been validly resurrected, and a host of author citations and literature references have had to be rechecked. To achieve this, the original literature has been substantially re-examined, facilitated by the extensive mycological library at the Royal Botanic Gardens, Kew.

Following the lead of *The British Ascomycotina: An Annotated Checklist* (Cannon *et al.*, 1985), the sanctioning author has not been included in citations. Thus *Tubaria furfuracea* appears as (Pers.) Gillet rather than (Pers.: Fr.) Gillet. A full list of sanctioned epithets was published by Gams (1984).

SYNONYMS & MISAPPLIED NAMES

The checklist gives full author and publication details for all homotypic and taxonomic synonyms known to have been used in the literature on the British *Basidiomycota*, including some standard continental monographs and field guides. Ideally, taxonomic synonyms would have been credited to the researcher who first published the synonymy (thus answering the question 'Who says it is a synonym?'), but it has not proved feasible to undertake this major additional level of research. Synonyms have thus been accepted in good faith, with comments sometimes added in the notes field if the synonymy is known to be disputed or contentious. Misapplied names are given after the synonymy.

HABITAT & ASSOCIATIONS

Data on habitat (e.g. marshes, dunes, conifer woodland) and associations (e.g. on dead herbaceous stems, with *Salix* and *Populus* spp) have generally been compiled at first hand from notes accompanying collections and records, rather than being copied from the existing literature. In some cases this has revealed interesting differences between data from the British Isles and data from continental Europe. In other cases, it may suggest that collections and records have been misdetermined (and these instances are usually noted).

Associated organisms (the majority of which are tree and shrub species) are given with their Latin names. To save space, familiar native and planted species are referred to by genus only when there is only one main species in the British Isles (thus native beech, *Fagus sylvatica*, is referred to as '*Fagus*'; native holly, *Ilex aquifolium*, is referred to as '*Ilex*'; and so on).

DISTRIBUTION AND FREQUENCY

The entry for each accepted taxon includes basic data on distribution within the British Isles, giving its reported presence within the constituent countries, i.e. England (E), Scotland (S), Wales (W), Northern Ireland (NI), and the Republic of Ireland (ROI). Records from the Channel Islands and the Isle of Man are noted separately under (O). For each of these, frequency is indicated as follows:

(c) common
(o) occasional (or infrequently reported)
(r) rare (or rarely reported)
(!) present, but frequency unknown
(?) reported, but the report is doubtful or uncertain

Since so little is yet known about the distribution of fungi, the categories assigned should be taken as no more than a rough indication of frequency based on available records. In fact, apart from a few species which are visibly common, all taxa in the checklist could easily have been marked 'data deficient'. The problem lies in the lack of fungal recorders (compared to the numbers of people recording birds or plants), the lack of specialists in many difficult or obscure groups, and the lack of knowledge of fruiting patterns for the majority of fungal species. It is not yet possible to tell, for example, whether a 'rare' fungus is genuinely rare, or whether it only produces fruitbodies rarely. Similarly, it is not clear whether a species not seen in the British Isles for many years (e.g. *Gomphus clavatus*) is genuinely 'extinct' (accordingly, all such species are included in the checklist). Additional information may be given in the notes field.

The database, from which this printed checklist has been produced, also contains the herbarium accession number of a voucher collection (where one exists) from each of the constituent parts of the British Isles in which a given taxon has been recorded. This enables anyone interested in following up a report of a taxon in a particular country to check a specific collection (most of which are held at Edinburgh or Kew). For reasons of space, these details have not been included in the printed checklist.

DESCRIPTIONS & ILLUSTRATIONS

For most but not all taxa, abbreviated references are given to relevant descriptions (D) and illustrations (I) (see Bibliography for abbreviations). In the main, the references have been selected from standard works, both British and overseas, currently used for identifying British fungi. Some classic British illustrative works have also been included, together with selected illustrations from recent journals. References to the type description (already given in the publication details for each taxon) are not repeated.

These references are clearly neither comprehensive nor complete, nor should the descriptions and illustrations necessarily be taken as 'recommended' (though most have been checked). They are included simply as a helpful starting point to discover more about the taxon listed.

ADDITIONAL NOTES

The additional notes field has been used for a wide range of brief comments, on the status of a given taxon (especially if uncertain or poorly known in the British Isles), on its distribution (with vice-counties often listed where limited), on collection dates, and so on. In some cases, notes are given on similar taxa or on recognition points. English names are also given, but only where these are familiar and of long usage.

RED DATA LIST SPECIES

The checklist database has provision for recording the Red Data List status of rare and vulnerable British species. At the time of publication of this printed version, however, a new and much revised Red Data List is still in preparation. When this new list is finalised and published, the relevant information will be added to the database and will then appear in any future printed versions of the checklist.

ABBREVIATIONS & LATIN TERMS

Abbreviations of author names follow Brummitt & Powell (1992). Abbreviations of publications follow Bridson (2004) (journals) and Stafleu & Cowan (1976-88) (books) as far as possible; when not included in these works, they are given in full. Abbreviated literature references, mainly found in the descriptions and illustrations section, are given in the bibliography.

Comb. Combination, as in *comb. illegit.*, *comb. inval.* (see *nom. illegit.*, etc.
      below).
f. forma (a taxonomic rank below variety).
Fide On the authority of.
Ined. Unpublished.
Herb. Herbarium. Herb. K is the herbarium of the Royal Botanic Gardens,
      Kew; herb. E the herbarium of the Royal Botanic Garden, Edinburgh.
In litt. In correspondence (i.e. not published).
Mis. Misapplied.
Nom. conf. Nomen confusum. A name with various conflicting interpretations.
Nom. cons. Nomen conservandum. A name specifically conserved by a ruling
      of the International Code of Botanical Nomenclature (Greuter *et al.*,
      2000), usually against a competing name which might otherwise have
      priority.
Nom. dub. Nomen dubium. A name of uncertain application. Typically there

is no type specimen and the original description is inadequate.

Nom. illegit. Nomen illegitimum. A name that is validly published, but contravenes certain articles of the Code. Typically it is a later homonym of an earlier validly published name or a superfluous name.

Nom. inval. Nomen invalidum. A name that is not validly published under the Code.

Nom. nov. Nomen novum. A new name replacing an existing illegitimate name.

Nom. nud. Nomen nudum. A name published without description and thus invalid under the Code.

Nom. rej. Nomen rejiciendum. A name specifically rejected by a ruling of the the Code (usually in favour of a better-known name).

Nom. superfl. Nomen superfluum. A superfluous name, i.e. one with an earlier valid name included by its author as a synonym.

p.p. pro parte. In part.

Pers. comm. Personal communication.

Sensu In the sense of (as interpeted by).

Sensu auct. In the sense of (as interpeted by) various authors, but not in the original sense.

Sensu lato In a broad sense.

Sensu stricto In a narrow sense.

sp. species (singular)

spp species (plural)

ssp. subspecies (a taxonomic rank below species).

var. variety (a taxonomic rank below subspecies).

ACKNOWLEDGEMENTS

The checklist project was jointly funded by the British Mycological Society, The Countryside Council for Wales, English Nature, The Environment & Heritage Service Northern Ireland, The Fungal Research Trust, The Heritage Council of Ireland, The Royal Botanic Gardens, Kew, and Scottish Natural Heritage.

As appointed Project Officer, Nick Legon collated and databased the information on which this checklist is based. Alick Henrici provided additional research. The database was created by Jerry Cooper, with Paul Kirk providing support. Entries were checked and edited by Peter Roberts, Brian Spooner, and Roy Watling.

The project was managed by a Steering Group comprising: Brian Spooner (Royal Botanic Gardens, Kew - chair), Carl Borges (English Nature), Jerry Cooper (CABI Bioscience), Hubert Fuller (University College, Dublin, representing the Heritage Council of Ireland), Alick Henrici (Royal Botanic Gardens, Kew), Paul Kirk (CABI Bioscience), Nick Legon (Project Officer, Royal Botanic Gardens, Kew), David Mitchel (Environment & Heritage Service, Northern Ireland), Peter Roberts (Royal Botanic Gardens, Kew), Jill Sutcliffe (English Nature), Stephen Ward (Scottish Natural Heritage), Roy Watling (formerly Royal Botanic Garden, Edinburgh, representing the British Mycological Society), and Ray Woods (Cyngor Cefn Gwlad Cymru/Countryside Council for Wales).

Thanks should also go to the following for advice on the status and distribution of British species in their various interest groups: Ron Davis, Lynn Davy, Ern Emmett, Alan Hills, Geoffrey Kibby, Pat Leonard, Ludmila Marvanová, Alan Orange, Alan Outen, Derek Schafer, Nigel Stringer, Evelyn Turnbull, and Sheila Wells.

BIBLIOGRAPHY

Short References

A&N1. Antonín, V. & Noordeloos, M.E. (1993). *A Monograph of* Marasmius, Collybia *and related genera in Europe*. Part 1. Marasmius, Setulipes *and* Marasmiellus. *Libri Botanici* 8:1 – 229.

A&N2. Antonín, V. & Noordeloos, M.E. (1997). *A Monograph of* Marasmius, Collybia *and related genera in Europe*. Part 2. Collybia, Gymnopus, Rhodocollybia, Crinipellis, Chaetocalathus *and additions to* Marasmiellus. *Libri Botanici* 17: 1 – 256.

A&N3. Antonín, V. & Noordeloos, M.E. (2004). *A monograph of the genera* Hemimycena. Delicatula, Fayodia, Gamundia, Myxomphalia, Resinomycena, Rickenella *and* Xeromphalina (*Tribus* Mycenae *sensu Singer*, Mycena *excluded) in Europe*. IHW Verlag. 279 pp.

Ainsw. Ainsworth, G.C. & Sampson, K. (1950). *The British Smut Fungi* (Ustilaginales). Kew: Commonwealth Mycological Institute. 137 pp.

Aleur. Núñez, M. & Ryvarden, L. (1997). *The genus* Aleurodiscus *(Basidiomycotina)*. Synopsis Fungorum 12: 1 - 164.

BFF1. Watling, R. (1970). *British Fungus Flora. Agarics and Boleti* 1. Boletaceae: Gomphidiaceae: Paxillaceae. Edinburgh: HMSO. 125 pp.

BFF2. Orton, P.D. & Watling, R. (1979). *British Fungus Flora. Agarics and Boleti* 2. Coprinaceae *Part* 1: Coprinus. Edinburgh: HMSO. 148 pp.

BFF3. Watling, R. (1982). *British Fungus Flora. Agarics and Boleti* 3. Bolbitiaceae: Agrocybe, Bolbitius *and* Conocybe. Edinburgh: HMSO. 138 pp.

BFF4. Orton, P.D. (1984). *British Fungus Flora. Agarics and Boleti* 4. Pluteaceae: Pluteus *and* Volvariella. Edinburgh: HMSO. 98 pp.

BFF5. Watling, R. & Gregory, N.M. (1987). *British Fungus Flora. Agarics and Boleti* 5. Strophariaceae *and* Coprinaceae *p.p.:* Hypholoma, Melanotus, Psilocybe, Stropharia, Lacrymaria *and* Panaeolus. Edinburgh: Royal Botanic Garden. 121 pp.

BFF6. Watling, R. & Gregory, N.M. (1989). *British Fungus Flora. Agarics and Boleti* 6. Crepidotaceae, Pleurotaceae *and other pleurotoid agarics*. Edinburgh: Royal Botanic Garden. 157 pp.

BFF7. Watling, R., Gregory, N.M. & Orton, P.D. (1993). *British Fungus Flora. Agarics and Boleti* 7. Cortinariaceae *p.p.* Galerina, Gymnopilus, Leucocortinarius, Phaeocollybia, Phaeogalera, Phaeolepiota, Phaeomarasmius, Pleuroflammula, Rozites *and* Stagnicola. Edinburgh: Royal Botanic Garden. 131 pp.

BFF8. Watling, R. & Turnbull, E. (1998). *British Fungus Flora. Agarics and Boleti* 8. Cantharellaceae, Gomphaceae *and Amyloid-Spored and Xeruloid Members of* Tricholomataceae (*excl.* Mycena). Edinburgh: Royal Botanic Garden. 189 pp.

B&K2. Breitenbach, J. & Kränzlin, F. (1986). *Fungi of Switzerland* 2. *Non gilled fungi, Heterobasidiomycetes, Aphyllophorales, Gasteromycetes*. Switzerland: Mykologia Luzern. 412 pp.

B&K3. Breitenbach, J. & Kränzlin, F. (1991). *Fungi of Switzerland* 3. *Boletes and agarics, 1st part*. Switzerland: Mykologia Luzern. 361 pp.

B&K4. Breitenbach, J. & Kränzlin, F. (1995). *Fungi of Switzerland* 4. *Agarics, 2nd part*. Switzerland: Mykologia Luzern. 368 pp.

B&K5. Breitenbach, J. & Kränzlin, F. (2000). *Fungi of Switzerland* 5. *Agarics, 3rd Part*. Switzerland: Mykologia Luzern. 338 pp

Bon. Bon, M. (1987). *The Mushrooms and Toadstools of Britain and North-western Europe*. Hodder & Stoughton. 352 pp.

BritChant. Pegler, D.N., Roberts, P.J., & Spooner, B.M. (1997). *British Chanterelles and Tooth Fungi. An account of the British Cantharelloid and Stipitate Hydnoid Fungi*. Kew: Royal Botanic Gardens. 114 pp.

BritPuffb. Pegler, D.N., Læssøe, T., & Spooner, B.M. (1995). *British Puffballs, Earthstars and Stinkhorns. An account of the British Gasteroid Fungi*. Kew: Royal Botanic Gardens. 255 pp.

BritTruff. Pegler, D.N., Spooner, B.M., & Young, T.W.K. (1993). *British Truffles. A Revision of British Hypogeous Fungi*. Kew: Royal Botanic Gardens. 216 pp.

BullBMS. *Bulletin of the British Mycological Society.*

C&D. Courtecuisse, R. & Duhem, B. (1995). *Mushrooms and Toadstools of Britain & Europe*. HarperCollins. 480 pp.

CFP1. Brandrud, T.E. *et al.* (1990). *Cortinarius Flora Photographica* 1 [English version]. Matfors: Cortinarius HB. 40 pp.

CFP2. Brandrud, T.E. *et al.* (1992). *Cortinarius Flora Photographica* 2 [English version]. Matfors: Cortinarius HB. 40 pp.

CFP3. Brandrud, T.E. *et al.* (1994). *Cortinarius Flora Photographica* 3 [English version]. Matfors: Cortinarius HB. 35 pp.

CFP4. Brandrud, T.E. *et al.* (1998). *Cortinarius Flora Photographica* 4 [English version]. Matfors: Cortinarius HB. 31 pp.

CNE2. Eriksson, J. & Ryvarden, L. (1973). *The Corticiaceae of North Europe* 2. Aleurodiscus – Confertobasidium. Oslo: Fungiflora. pp. 60-286

CNE3. Eriksson, J. & Ryvarden, L. (1975). *The Corticiaceae of North Europe* 3. Coronicium – Hyphoderma. Oslo: Fungiflora. pp. 287-546

CNE4. Eriksson, J. & Ryvarden, L. (1976). *The Corticiaceae of North Europe* 4. Hyphodermella – Mycoacia. Oslo: Fungiflora. pp. 547-886

CNE5. Eriksson, J., Hjortstam, K. & Ryvarden, L. (1978). *The Corticiaceae of North Europe* 5. Mycoaciella – Phanerochaete. Oslo: Fungiflora pp. 887-1047

CNE6. Eriksson, J., Hjortstam, K. & Ryvarden, L. (1981). *The Corticiaceae of North Europe* 6. Phlebia – Sarcodontia. Oslo: Fungiflora pp. 1048-1276

CNE7. Eriksson, J., Hjortstam, K. & Ryvarden, L. (1984). *The Corticiaceae of North Europe* 7. Schizopora – Suillosporium. Oslo: Fungiflora pp. 1277-1449

CNE8. Eriksson, J., Hjortstam, K. & Ryvarden, L. (1988). *The Corticiaceae of North Europe* 8. Thanatephorus – Ypsilonidium. Oslo: Fungiflora pp. 1450-1631

Corner. Corner, E.J.H. (1950). *A Monograph of* Clavaria *and Allied Genera*. Ann. Bot., Memoir 1. Oxford University Press. 740 pp.

CRF. Henderson, D.M. (2000). *A Checklist of the Rust Fungi of the British Isles*. Kew: British Mycological Society. 36 pp.

CzM. *Czech Mycology.*

EurPoly1. Ryvarden, L & Gilbertson, R.L. (1993). European Polypores 1: *Abortiporus – Lindtneria. Synopsis Fungorum* 6: 1 – 387.

EurPoly2. Ryvarden, L & Gilbertson, R.L. (1994). European Polypores 2: *Meripilus – Tyromyces. Synopsis Fungorum* 7: 388 – 743.

FAN1. Bas, C., Kuyper, Th.W., Noordeloos, M.E. & Vellinga, E.C. (eds) (1988). *Flora Agaricina Neerlandica* 1. Entolomataceae. Rotterdam: Balkema. 182 pp.

FAN2. Bas, C., Kuyper, Th.W., Noordeloos, M.E. & Vellinga, E.C. (eds) (1990). *Flora Agaricina Neerlandica* 2.

Pleurotaceae, Pluteaceae, and Tricholomataceae (1). Rotterdam: Balkema. 137 pp.

FAN3. Bas, C., Kuyper, Th.W., Noordeloos, M.E. & Vellinga, E.C. (eds) (1995). *Flora Agaricina Neerlandica* 3. Tricholomataceae (2). Rotterdam: Balkema. 183 pp.

FAN4. Bas, C., Kuyper, Th.W., Noordeloos, M.E. & Vellinga, E.C. (eds) (1999). *Flora Agaricina Neerlandica* 4. Strophariaceae, Tricholomataceae (3). Rotterdam: Balkema. 191 pp.

FAN5. Noordeloos, M.E., Kuyper, Th.W., & Vellinga, E.C. (eds) (2001). *Flora Agaricina Neerlandica* 5. Agaricaceae. Rotterdam: Balkema. 169 pp.

FlDan1. Lange, J.E. (1935). *Flora Agaricina Danica.* 1. Copenhagen. 92 pp.

FlDan2. Lange, J.E. (1936). *Flora Agaricina Danica.* 2. Copenhagen. 105 pp.

FlDan3. Lange, J.E. (1938). *Flora Agaricina Danica.* 3. Copenhagen. 96 pp.

FlDan4. Lange, J.E. (1939). *Flora Agaricina Danica.* 4. Copenhagen. 119 pp.

FlDan5. Lange, J.E. (1940). *Flora Agaricina Danica.* 5. Copenhagen. 105 pp.

FM. *Field Mycology.*

FME1 Bon, M. (1990). Les Hygrophores. *Flore Mycologique d'Europe* 1. Amiens: CRDP. 99 pp.

FME2 Bon, M. (1991). Les Tricholomes et Ressembants. *Flore Mycologique d'Europe* 2. Amiens: CRDP. 163 pp.

FME4 Bon, M. (1997). Les Clitocybes, Omphales et Ressemblants. *Flore Mycologique d'Europe* 4. Amiens: CRDP. 181 pp.

FNE1. Boertmann, D. (1995). *The genus* Hygrocybe. *Fungi of Northern Europe* 1. 184pp

FNE2. Heilmann-Clausen, J., Verbeken, A., & Vesterholt, J. (1998). The genus *Lactarius. Fungi of Northern Europe* 2. 287pp

FRIC1. Reid, D.A. (1966). *Fungorum Rariorum Icones Coloratae* 1. Cramer. 32 pp.

FRIC2. Reid, D.A. (1967). *Fungorum Rariorum Icones Coloratae* 2. Cramer. 32 pp.

FRIC3. Reid, D.A. (1968). *Fungorum Rariorum Icones Coloratae* 3. Cramer. 36 pp.

FRIC4. Reid, D.A. (1969). *Fungorum Rariorum Icones Coloratae* 4. Cramer. 32 pp.

FRIC5. Schild, E. (1971). *Fungorum Rariorum Icones Coloratae* 5. Cramer. 44 pp.

FRIC6. Reid, D.A. (1972). *Fungorum Rariorum Icones Coloratae* 6. Cramer. 59 pp.

FRIC7. Moser, M.M. (1978). *Fungorum Rariorum Icones Coloratae* 7. Cramer. 48 pp.

FRIC12. Arnolds, E. & Noordeloos, M.E. (1981). *Fungorum Rariorum Icones Coloratae* 12. Cramer. 35 pp.

FRIC13. Dermek, A. (1984). *Fungorum Rariorum Icones Coloratae* 13. Cramer. 15 pp.

FRIC16. Dermek, A. (1987). *Fungorum Rariorum Icones Coloratae* 16. Cramer. 23 pp.

FRIC17. Dermek, A. (1987). *Fungorum Rariorum Icones Coloratae* 17. Cramer. 23 pp.

FungEur1. Capelli, A. (1984). Agaricus. *Fungi Europaei* 1. 560 pp.

FungEur2. Alessio, C.L. (1985). Boletus. *Fungi Europaei* 2. 712 pp.

FungEur2a. Alessio, C.L. (1991). Boletus *(supplement). Fungi Europaei* 2a. 126 pp.

FungEur3. Riva, A. (1988). Tricholoma. *Fungi Europaei* 3. 618 pp.

FungEur3(suppl.) Riva, A. (2003). Tricholoma. Supplemento. *Fungi Europaei* 3 (suppl.) pp. 619-826.

FungEur4. Candusso, M. & Lanzoni, G. (1990). Lepiota *s.l. Fungi Europaei* 4. 743 pp.

FungEur5. Noordeloos, M.E. (1992). Entoloma *s.l. Fungi Europaei* 5. 760 pp.

FungEur6. Candusso, M. (1997). Hygrophorus *s.l. Fungi Europaei* 6. 784 pp.

FungEur7. Basso, M.T. (1999). Lactarius. *Fungi Europaei* 7. 845 pp.

FungEur8. Ladurner, H. & Simonini, G. (2003). Xerocomus *s.l. Fungi Europaei* 8. 527 pp.

Galli. Galli, R. (2001). *Le Amanite.* Milan: Edinatura. 216 pp.

Galli2. Galli, R. (1996). *Le Russule.* Milan: Edinatura. 480 pp.

Galli3. Galli, R. (1999). *I Tricolomi.* Milan: Edinatura. 271 pp.

H&B. Henderson, D.M. & Bennell, A.P. (1979). British Rust Fungi: Additions and Corrections. *Notes Roy. Bot. Gard. Edinburgh* 37(3): 475-501

Hallenb. Hallenberg, N. (1985). *The* Lachnocladiaceae *and* Coniophoraceae *of North Europe.* Oslo: Fungiflora. 96 pp.

HBF1. Cooke, M.C. (1871). *Handbook of British Fungi.* 2 vols. Macmillan.

HBF2. Cooke, M.C. (1883-91). *Handbook of British Fungi.* Edn 2. London. 398 pp.

Holec. Holec, J. (2001). The Genus *Pholiota* in central and western Europe. *Libri Botanici* 20: 1 - 220.

K&M Konrad, P. & Maublanc, A. (1924-1937). *Icones selectae fungorum.* 6 vols. Paris.

K&R Kühner, R. & Romagnesi, H. (1953). *Flore Analytique des Champignons Supérieurs.* Paris: Masson. 557 pp.

Kits1. Kits van Waveren, E. (1985). The Dutch, French and British species of *Psathyrella. Persoonia Supplement* 2. 300 pp.

Kits2. Kits van Waveren, E. (1987). Additions to our monograph on *Psathyrella. Persoonia* 13(3): 327-368.

Kõljalg. Kõljalg, U. (1995). Tomentella (Basidiomycota) *and related genera in Europe.* Oslo: Fungiflora. 213 pp.

Kriegl1. Krieglsteiner, G.J. (2000). *Die Grosspilze Baden-Württenbergs* - Band 1. Ulmer. 629 pp.

Kriegl2. Krieglsteiner, G.J. (2000). *Die Grosspilze Baden-Württenbergs* - Band 2. Ulmer. 620 pp.

Kriegl3. Krieglsteiner, G.J. (2001). *Die Grosspilze Baden-Württenbergs* - Band 3. Ulmer. 634 pp.

Kuyp. Kuyper, T.W. (1986). A revision of the genus *Inocybe* in Europe. Subgenus *Inosperma* and the smooth-spored species of subgenus *Inocybe. Persoonia Supplement* 3: 1 – 247.

L&E. Lannoy, G. & Estades, A. (1995). *Monographie des* Leccinum *d'Europe.* Féderation Mycologiques d'Auphiné Savoie. 229 pp.

Myc. *Mycologist.*

Nannf. Nannfeldt, J.A. (1981). *Exobasidium:* a taxonomic reassessment applied to the European species. *Symb. Bot. Upsal.* 23(2): 1 – 72.

NBA2. Orton, P.D. (1964). Notes on British Agarics II. *Notes Roy. Bot. Gard. Edinburgh* 26: 43-65.

NBA3. Orton, P.D. (1969). Notes on British Agarics III. *Notes Roy. Bot. Gard. Edinburgh* 29: 75-127.

NBA5. Orton, P.D. (1976). Notes on British Agarics V. *Kew Bull.* 31(3): 709-721.

NBA6. Orton, P.D. (1976). Notes on British Agarics VI. *Notes Roy. Bot. Gard. Edinburgh* 35: 147-154.

NBA7. Orton, P.D. (1980). Notes on British Agarics VII. *Notes Roy. Bot. Gard. Edinburgh* 38: 315-330.

NBA8. Orton, P.D. (1984). Notes on British Agarics VIII. *Notes Roy. Bot. Gard. Edinburgh* 41: 565-624.

NBA9. Orton, P.D. (1988). Notes on British Agarics IX. *Trans. Brit. Mycol. Soc.* 91(4): 543-740.

NCL. Dennis, R.W.G., Orton, P.D., & Hora, F.B. (1960). New Check List of British Agarics and Boleti (Parts I & II). *Trans. Brit. Mycol. Soc. (Supplement).* 225 pp.

NCL3. Dennis, R.W.G., Orton, P.D., & Hora, F.B. (1960). New Check List of British Agarics and Boleti (Part III). *Trans. Brit. Mycol. Soc.* 43(2): 159-459.

NM2. Hansen, I. & Knudsen, H. (1992). *Nordic Macromycetes* 2. Polyporales, Boletales, Agaricales, Russulales. Copenhagen: Nordsvamp. 474 pp.

NM3. Hansen, I. & Knudsen, H. (1997). *Nordic Macromycetes* 3. *Heterobasidioid, Aphyllophoroid and Gasteromycetoid Basidiomycetes.* Copenhagen: Nordsvamp. 444 pp.

Ph. Phillips, R. (1981). *Mushrooms and other fungi of Great Britain & Europe.* London: Pan Books. 288 pp.

Reid4. Reid, D.A. (1970). New or Interesting Records of British hymenomycetes IV. *Trans. Brit. Mycol. Soc.* 55(3): 413-441.

Reid5. Reid, D.A. (1973). New or Interesting Records of British hymenomycetes V. *Persoonia* 7(2): 293-303.

Reid6. Reid, D.A. (1975). New or Interesting Records of British hymenomycetes VI. *Nova Hedwigia Beih.* 51: 199-206.

Reid7. Reid, D.A. (1987). New or Interesting Records of British hymenomycetes VII. *Notes Roy. Bot. Gard. Edinburgh* 44(3): 503-540.

Reid8. Reid, D.A. (1990). New or Interesting Records of British heterobasidiomycetes. *Mycol. Res.* 94(1): 94-108.

Reid84. Reid, D.A. (1984). A revision of the British species of *Naucoria* sensu lato. *Trans. Brit. Mycol. Soc.* 82(2): 191–237.

RivMyc. *Rivista di Micologia.*

Sarn. Sarnari, M. (1998). *Monografia Illustrata del Genere Russula in Europe (Tomo Primo)*. Associazione Micologica Bresadola. 799 pp.

Sow1. Sowerby, J. (1796-1797). *Coloured Figures of English Fungi or Mushrooms* Vol. 1. London. 50 pp [pl. 1-120].

Sow2. Sowerby, J. (1798-1799). *Coloured Figures of English Fungi or Mushrooms* Vol. 2. London. 48 pp [pl. 121-240].

Sow3. Sowerby, J. (1800-1803). *Coloured Figures of English Fungi or Mushrooms* Vol. 3. London. 84 pp [pl. 241-400].

Sow4. Sowerby, J. (1809-1815). *Coloured Figures of English Fungi or Mushrooms* Vol. 4 *(suppl.)*. London. 25 pp [pl. 401-439].

SV. *Svampe.*

TBMS. *Transactions of the British Mycological Society.*

UBI. Mordue, J.E.M. & Ainsworth, G.C. (1984). Ustilaginales of the British Isles. *Mycological Papers* 154: 1 - 96.

Vánky. Vánky, K. (1994). *European Smut Fungi*. Gustav Fischer. 570 pp.

W&H. Wilson, M. & Henderson, D.M. (1966). *British Rust Fungi*. Cambridge University Press. 384 pp.

Standard References

Adamčik, S. & Knudsen, H. (2004). Red-capped species of *Russula* sect. *Xerampelinae* associated with dwarf scrub. *Mycol. Res.* 108: 1463 – 1475.

Ainsworth, G.C. & Waterhouse, G.M. (1989). *British County Foray Lists*. Kew: British Mycological Society. 35 pp.

Ainsworth, G.C. & Sampson, K. (1950). *The British Smut Fungi*. Kew: Commonwealth Mycological Institute. 137 pp.

Barnett, J.A., Payne, R.W., & Yarrow, D. (2000). *Yeasts: Characteristics and identification*. 3rd Edn. Cambridge University Press. 1139 pp.

Berkeley, M.J. (1860). *Outlines of British Fungology*. London: Lovell Reeve. 442 pp.

Boidin, J., Lanquetin, P. & Gilles, G. (1997). Le genre *Gloeocystidiellum* sensu lato. *Bull. Soc. Mycol. France* 113: 1 – 80.

Bolton, J. (1888-89). *An History of Fungusses growing about Halifax*. Huddersfield: J. Brook. 3 vols.

Bon, M. (1987). *The Mushrooms and Toadstools of Britain and North-western Europe*. Hodder & Stoughton. 352 pp.

Bon, M. (1991). Les Tricholomes et Ressemblants. Flore Mycologique d'Europe 2. *Doc. Mycol. Mém. Hors Sér.* 2. 154 pp.

Bourdot, H. & Galzin, A. (1928). *Hyménomycètes de France*. Sociétè Mycologique de France. Sceaux: Marcel Bry. 758 pp.

Bowen, H. (2000). *The flora of Dorset*. Newbury: NatureBureau. 373 pp.

Bramley, W.G. (1985). *A Fungus Flora of Yorkshire*. Yorkshire Naturalists' Union. 277 pp.

Bridson, G.D.R. (2004). *BPH-2. Periodicals with botanical content*. 2 vols. Pittsburgh: Hunt Institute for Botanical Documentation.

Brodie, H.J. (1975). *The Bird's Nest Fungi*. University of Toronto Press. 199 pp.

Brummit, R.K. & Powell, C.E. (1992). *Authors of Plant Names. A list of authors of scientific names of plants, with recommended standard forms of their names, including abbreviations*. Kew: Royal Botanic Gardens. 732 pp.

Bulliard, J.P.F. (1780-1798). *Herbier de la France*. 13 vols. Paris.

Cannon, P.F., Hawksworth, D.L. & Sherwood-Pike, M.A. (1985). *The British Ascomycotina. An Annotated Checklist*. Slough: Commonwealth Mycological Institute. 312 pp.

Clark, M.C. (ed.) (1980). *The Fungus Flora of Warwickshire*. London: British Mycological Society. 272 pp.

Cooke, M.C. (1871). *Handbook of British Fungi*. 2 vols. Macmillan.

Cooke, M.C. (1883-1891). *Handbook of British Fungi*. Edn 2. London. 398 pp.

Cooke, M.C. (1881-1891). *Illustrations of British Fungi (Hymenomycetes)*. 8 vols. London: Williams & Norgate.

Cooke, W.B. (1961) The Cyphellaceous Fungi, a study in the Porotheliaceae. *Sydowia, Beih.* 4: 1-144.

Corner, E.J.H. (1950). *A Monograph of* Clavaria *and Allied Genera*. Ann. Bot., Memoir 1. Oxford University Press. 740 pp.

Corner, E.J.H. (1968). A Monograph of *Thelephora*. *Nova Hedwigia Beih.* 27: 1-110.

Corner, E.J.H. (1970). Supplement to 'A Monograph of *Clavaria* and Allied genera'. *Nova Hedwigia Beih.* 33: 1-299.

Cotton, A.D. & Wakefield, E.M. (1919). A revision of the British *Clavariae*. *Trans. Brit. Mycol. Soc.* 6(2): 164-198.

Dennis, R.W.G. (1948). Some little known British species of Agaricaceae 1. Leucosporae and Rhodosporae. *Trans. Brit. Mycol. Soc.* 31: 191-209.

Dennis, R.W.G. (1986). *Fungi of the Hebrides*. Kew: Royal Botanic Gardens. 383 pp.

Dennis, R.W.G. (1995). *Fungi of South East England*. Kew: Royal Botanic Gardens. 295 pp.

Dickson, G. & Leonard, A. (1996). *Fungi of the New Forest*. London: British Mycological Society. 210 pp.

Donk, M.A. (1966). Check list of European hymenomycetous Heterobasidiae. *Persoonia* 4: 145-335.

Donk, M.A. (1974). Checklist of European Polypores. *Verh. Koninkl. Nederland. Akad. Wettensch. Naturk.* 62: 1–469.

Ellis, M.B. & Ellis, J.P. (1990). *Fungi without Gills (Hymenomycetes and Gasteromycetes). An Identification Handbook*. Chapman Hall. 329 pp.

Favre, J. (1960). Catalogue descriptif des Champignons Supérieurs de la Zone Subalpine du Parc National Suisse. *Ergebn. Wiss. Untersuch. Schweiz. Natn. Parks* 6: 323-610.

Fries, E.M. (1821). *Systema mycologicum*. Vol. 1. Gryphiswald. 520 pp.

Gams, W. (1984). An index to fungal names and epithets sanctioned by Persoon and Fries. *Mycotaxon* 19: 219-270.

Gerhardt, E. (1996). Taxonomische Revision der Gattungen *Panaeolus* und *Panaeolina* (Fungi, Agaricales, Coprinaceae). *Biblioth. Bot.* 147: 1–149.

Greuter *et al.* (2000). International Code of Botanical Nomenclature (Saint Louis Code). *Regnum Veg.* 138. xviii + 474 pp.

Grove, W.B. (1913). *The British Rust Fungi (Uredinales)*. Cambridge University Press. 412 pp.

Grove, W.B. (1922). New or Noteworthy Fungi. *J. Bot.* 60: 167-177.

Guzmán, G. (1983). The Genus *Psilocybe*. *Nova Hedwigia Beih.* 74: 1-439.

Hawksworth, D.L., Kirk, P.M., Sutton, B.C., & Pegler, D.N. (1995). *Ainsworth & Bisby's Dictionary of the Fungi*. Edn 8. Wallingford: CAB International. 616 pp.

Heim, R. (1931). *Le Genre* Inocybe. *Encyclopédie Mycologique* 1: 1-431.

Henderson, D.M. (2000). *A Checklist of the Rust Fungi of the British Isles*. British Mycological Society. 36 pp.

Henderson, D.M. (2004). *The Rust Fungi of The British Isles: Guide to Identification by their Host Plants*. British Mycological Society. 35 pp.

Henderson, D.M. & Bennell, A.P. (1979). British Rust Fungi: Additions and Corrections. *Notes Roy. Bot. Gard. Edinburgh* 37: 475-501.

Henderson, D.M. & Bennell, A.P. (1980). Supplement to British Rust Fungi: Additions and Corrections. *Notes Roy. Bot. Gard. Edinburgh* 38: 184.

Henrici, A. (1990). A preliminary list of British heterobasidiomycetes. *Keys: Newsletter of the BMS Key-Group* 3: 3.2.1 – 3.2.9

Henrici, A. & Cook, P. (1991). A list of British aphyllophoroid fungi. *Keys: Newsletter of the BMS Key-Group* 5: 5.1.1 – 5.1.48

Holec, J. (2001). *The Genus* Pholiota *in central and western Europe. Libri Botanici* 20: 1-220.

Hussey, T.J. (1847-1855). *Illustrations of British Mycology*. 2 vols. London: Lovell Reeve.

Ing, B. (1992). A Provisional Red Data List of British Fungi. *Mycologist* 6(3): 124-128.

Ing, B. (1994). European Exobasidiales and their galls. In Williams, M.A. (ed.) *Plant Galls. Organisms, Interactions, Populations.* Systematics Association Special Vol. 49: 67-76. Oxford: Clarendon Press.

Jülich, W. (1984). Die Nichtblätterpilze, Gallertpilze und Bauchpilze. *Kleine Kryptogamenflora* IIb/1. Stuttgart: Gustav Fischer. 626 pp.

Kauffman, C.H. (1918). *The Agaricaceae of Michigan*. 2 vols. Michigan Geological & Biological Survey Publ. 26.

Kirk, P.M., Cannon, P.F., David, J.C. & Stalpers, J.A. (2001). *Dictionary of the Fungi.* Edn 9. Wallingford: CAB International. 655 pp.

Kühner, R. (1935). *Le Genre* Galera (Fries) Quélet. *Encyclopédie Mycologique* 7: 1-238.

Kühner, R. (1938). *Le Genre* Mycena (Fries). Étude cytologique et systématique des espèces d'Europe et d'Amerique du Nord. *Encyclopédie Mycologique* 10: 1-710.

Kühner, R. & Romagnesi, H. (1953). *Flore Analytique des Champignons Supérieurs*. Paris: Masson. 557 pp.

Kuyper, T.W. (1986). A revision of the genus *Inocybe* in Europe. Subgenus *Inosperma* and the smooth-spored species of subgenus *Inocybe. Persoonia, Suppl.* 3: 1 - 247

Kuyper, T.W. (1996). Notulae ad floram Agaricinam Neerlandicam XXIV-XXVIII. *Persoonia* 16: 225-232.

Maas Geesteranus, R.A. (1992) *Mycenas of the Northern Hemisphere*. North-Holland. 2 vols.

Mason, F.A. & Grainger, J. (1937). *A Catalogue of Yorkshire Fungi*. Hull: Yorkshire Naturalists' Union. 110 pp.

Massee, G. (1892). *British Fungus Flora. A Classified textbook of Mycology.* Vol. 1. London: Bell. 432 pp.

Massee, G. (1893a). *British Fungus Flora. A Classified textbook of Mycology.* Vol. 2. London: Bell. 460 pp.

Massee, G. (1893b). *British Fungus Flora. A Classified textbook of Mycology.* Vol. 3. London: Bell. 512 pp.

Massee, G. (1897). Mycologic Flora of the Royal Botanic Gardens, Kew (*Agaricinae* & *Boletus*). *Bull. Misc. Inform. Kew* 1897: 115-131.

Massee, G. (1899). A revision of the genus *Tilletia. Bull. Misc. Inform. Kew* 1899: 141-149.

Massee, G. (1902). *European Fungus Flora*, Agaricaceae. London: Duckworth. 274 pp.

Massee, G. (1911). *British Fungi with a chapter on lichens*. London: Routledge. 551 pp.

Massee, G. (1913). *Mildews, Rusts and Smuts*. London: Dulau & Co. 229 pp.

Massee, G. & Crossland, C. (1905). The Fungus Flora of Yorkshire. *Bot. Trans. Yorkshire Naturalists Union* 4. London: Brown. 396 pp.

Mordue, J.E.M. & Ainsworth, G.C. (1984). Ustilaginales of the British Isles. *Mycol. Pap.* 154. 96 pp.

Moser, M.M. (1967). Basidiomyceten II. Die Röhrlinge und Blätterpilze (Agaricales). *Kleine Kryptogamenflora* IIb/2. Edn 3. Stuttgart: Gustav Fischer. 443 pp.

Moser, M.M. (1978). Basidiomyceten II. Die Röhrlinge und Blätterpilze (Agaricales). *Kleine Kryptogamenflora* IIb/2. Edn 4. Stuttgart: Gustav Fischer. 532 pp.

Nauta, M.M. (1987). *Revisie van de in Nederland voorkomende soorten van het geschacht* Agrocybe (*Leem hoeden*). Leiden: Rijksherbarium. 103 pp, 52 pl., 11 maps.

Neville, P. & Poumarat, S. (2004). Amaniteae. *Amanita, Limacella* & *Torrendia. Fungi Europaei* 9: 1-1120.

Noordeloos, M.E. (1985). Notulae ad floram Agaricinam Neerlandicam X-XI. *Entoloma. Persoonia* 12(4): 457-462.

Noordeloos, M.E. (1987). *Entoloma* (Agaricales) in Europe. *Nova Hedwigia Beih.* 91: 1-419.

Núñez, M. & Ryvarden, L. (1997). *The genus* Aleurodiscus (Basidiomycotina). Oslo: Fungiflora. 164 pp.

Orton, P.D. (1955). The Genus *Cortinarius* I. *Myxacium* and *Phlegmacium. Naturalist* 1955, supplement. 80 pp.

Orton, P.D. (1958). The Genus *Cortinarius* II. *Inoloma* and *Dermocybe. Naturalist* 1958: 81-149.

Orton, P.D. (1983). Some notes on the genus *Cortinarius* in Britain. *Sydowia* 36: 213-226.

Orton, P.D. (1991). A revised list of the British species of *Entoloma* sensu lato. *Mycologist* 5: 123-138; 172-176.

Palmer, J.T. (1968). A Chronological Catalogue of the Literature to the British Gasteromycetes. *Nova Hedwigia* 15: 65-178.

Pearson, A.A. (1938). New Records and Observations. *Trans. Brit. Mycol. Soc.* 22: 27-46.

Pearson, A.A. (1943). New Records and Observations II. *Trans. Brit. Mycol. Soc.* 26: 36-49.

Pearson, A.A. (1946). New records and observations III. *Trans. Brit. Mycol. Soc.* 29: 191-210.

Pearson, A.A. (1948). The genus *Russula. Naturalist* 1948: 85-108.

Pearson, A.A. (1949). New records and observations IV. *Trans. Brit. Mycol. Soc.* 32: 258-273.

Pearson, A.A. (1950). *Lactarius. Naturalist* 1950: 81-99.

Pearson, A.A. (1952). New records and observations V. *Trans. Brit. Mycol. Soc.* 35: 97-122.

Pearson, A.A. (1954). The Genus *Inocybe. Naturalist* 1954: 117-140.

Pearson, A.A. (1955). *Mycena. Naturalist* 1955: 41-63.

Pearson, A.A. & Dennis, R.W.G. (1948). Revised List of British Agarics and Boleti. *Trans. Brit. Mycol. Soc.* 31: 145-190.

Pegler, D.N. (1973). The Polypores. 2nd Edn. *Bull. Brit. Mycol. Soc.* 7(1), *Suppl.* 43 pp.

Pegler, D.N. (1983). The genus *Lentinus. Kew Bull., Addit. Ser.* 10. HMSO. 281pp.

Pegler, D.N., Læssøe, T. & Spooner, B.M. (1995). British Puffballs, Earthstars and Stinkhorns. An Account of the British Gasteroid Fungi. Kew: Royal Botanic Gardens. 255 pp.

Pegler, D.N., Roberts, P.J. & Spooner, B.M. (1997). *British Chanterelles and Tooth Fungi. An account of the British Cantharelloid and Stipitate Hydnoid Fungi.* Kew: Royal Botanic Gardens. 114 pp.

Pegler, D.N., Spooner, B.M. & Young, T.W.K. (1993). *British Truffles. A Revision of British Hypogeous Fungi.* Kew: Royal Botanic Gardens. 216 pp.

Pegler, D.N. & Young, T.W.K. (1972). Basidiospore form in the British species of *Inocybe. Kew Bull.* 26(3): 499-537.

Persoon, C.H. (1798). *Icones et Descriptiones Fungorum minus cognitorum.* Lipsiae. 60 pp, XIV pl.

Persoon, C.H. (1801). *Synopsis Methodica Fungorum.* Göttingen. 708 pp.

Petersen, R.H. (1975). *Ramaria* subgenus *Lentoramaria* with emphasis on North American Taxa. *Biblioth. Mycol.* 43: 1-161.

Phillips, R. (1981). *Mushrooms and other fungi of Great Britain & Europe*. London: Pan Books. 288 pp.

Pilát, A. (1936). *Atlas des Champignons de l'Europe III*. Polyporaceae. Prague: Musée National. 624 pp.

Pilát, A. (1948). *Atlas des Champignons de l'Europe VI. Monographie des espèces européennes du genre* Crepidotus *Fr.* Prague: Musée National. 84 pp.

Plowright, C.B. (1872-73). *A list of the fungi known to occur in the county of Norfolk*. Norfolk & Norwich Naturalists' Society. 53 pp.

Plowright, C.B. (1889). *A Monograph of the British Uredineae and Ustilagineae*. London: Kegan Paul. 347 pp.

Purton, T. (1821). *An appendix to the Midland flora; comprising also corrections and additions referring to the* two former volumes; and occasional observations tending to elucidate the study of the British fungi. London: Printed by the author.

Ramsbottom, J. (1953). *Mushrooms and Toadstools.* London: Collins. 306 pp.

Rayner, R.W. (1985). *Keys to the British species of* Russula. Edn 3. British Mycological Society. 99 pp.

Rea, C. (1922). *British Basidiomycetae. A Handbook to the Larger British Fungi.* Cambridge University Press. 799 pp.

Rea, C. (1927). Appendix to 'British Basidiomycetae'. *Trans. Brit. Mycol. Soc.* 12(2-3): 205-230.

Rea, C. (1932). Appendix II to 'British Basidiomycetae'. *Trans. Brit. Mycol. Soc.* 17: 35-50.

Redfern, M., Shirley, P. & Bloxham, M. (2002). British Plant Galls. Identification of galls on plants and fungi. *Field Studies* 10: 207-531.

Reid, D.A. (1974). A monograph of the British *Dacrymycetales*. *Trans. Brit. Mycol. Soc.* 62: 433-494.

Reid, D.A. (1985). An annotated list of some fungi from the Channel Islands, mostly from Jersey. *Trans. Brit. Mycol. Soc.* 84: 709-714.

Reid, D.A. (1989). Notes on some Leucocoprinoid fungi from Britain. *Mycol. Res.* 93: 413-423.

Reid, D.A. (1995). Observations on *Leucoagaricus melanotrichus* and similar species. *Mycotaxon* 53: 325-336.

Reid, D.A. (1995a). Three unusual British Fungi. *Mycotaxon* 53: 337-342.

Reid, D.A. & Austwick, P.K.C. (1963). An annotated list of the less common Scottish basidiomycetes. *Glasgow Naturalist* 18(6): 255-336.

Ricken, A. (1915). *Die Blätterpilze* (Agaricaceae) *Deutschlands und der angrenzenden Länder, besonders Oesterreich und der Schweiz.* Leipzig: Weigel. 2 vols.

Romagnesi, H. (1967). *Les Russules d'Europe et d'Afrique du Nord.* Bordas. 989 pp.

Saunders, W. & Smith, W.G. (1871-72). *Mycological Illustrations.* Plates 1-48. London: John van Voorst.

Senn-Irlet, B. (1995). The genus *Crepidotus* (Fr.) Staude in Europe. *Persoonia* 16: 1-80.

Sime, A.D. & Petersen, R.H. (1999). Intercontinental interrelationships among disjunct populations of *Melanotus* (Strophariaceae, Agaricales). *Mycotaxon* 71: 481 – 492.

Singer, R. (1935). Études systematiques sur les *Melanoleuca* d'Europe et clé des espèces observeés en Catalogne. *Cavanillesia* 7: 123-132.

Singer, R. (1986). *The Agaricales in Modern Taxonomy* Edn 4. Koenigstein: Koeltz. 981 pp.

Smith, A.H. (1947). North American species of *Mycena*. *University of Michigan Studies. Science Series* 17: xviii, 1-521.

Smith, A.H. (1949). *Mushrooms in their Natural Habitats.* Portland: Sawyer's. 2 vols.

Smith, A.H. & Singer, R. (1964). *A monograph on the genus Galerina Earle.* Hafner. 384 pp.

Smith, W.G. (1891). *Outlines of British Fungology. Supplement.* London: Reeve & Co. 386 pp.

Smith, W.G. (1908). *Synopsis of the British Basidiomycetes.* London: British Museum. 531 pp.

Spooner, B.M. (1985). *Melanopsichium* (Ustilaginales), a genus new to the British Isles. *Trans. Brit. Mycol. Soc.* 85: 540-544.

Stafleu, F.A. & Cowan, R.S. (1976-1988). *Taxonomic Literature. A selective guide to botanical publications and collections with dates, commentaries and types.* 2nd edn. Utrecht: Bohn, Scheltema & Holkema. 7 vols.

Stalpers, J.A. (1984). A revision of the genus *Sporotrichum*. *Stud. Mycol.* 24: 1-105.

Stalpers, J.A. (1996). The aphyllophoraceous fungi II: keys to the species of the Hericiales. *Stud. Mycol.* 40: 1-185.

St-Amans, J.F.B. de (1821). *Flore agenaise.* Agen. 632 pp.

Stephenson, C.R. (2004). *Fungi of the Scarborough District.* Scarborough Field Naturalists' Society. 198 pp.

Stevenson, J. (1879). *Mycologia Scotica.* Cryptogamic Society of Scotland. 443 pp.

Stevenson, J. (1886). *British Fungi* (*Hymenomycetes*). 2 vols. Blackwood.

Sunhede, S. (1989). *Geastraceae* (*Basidiomycotina*). Morphology, Ecology, and Systematics with special emphasis on the North European species. *Synopsis Fungorum* 1: 1 – 534.

Tofts, R. (2002). The British species of *Lepiota*. 1: Section Stenosporae. *Field Mycology* 3: 124 – 136.

Vellinga, E. (1986). The genus *Flammulaster* in the Netherlands. *Persoonia* 13: 1-26.

Wakefield, E.M. (1917). Notes on British *Thelephoraceae*. *Trans. Brit. Mycol. Soc.* 5(3): 474-480.

Wakefield, E.M. (1952). New or Rare British Hymenomycetes (Aphyllophorales). *Trans. Brit. Mycol. Soc.* 35: 34-65.

Wakefield, E.M. (1969). Tomentelloideae in the British Isles. *Trans. Brit. Mycol. Soc.* 53: 161-206.

Wakefield, E.M. & Bisby, G.R. (1942). List of Hyphomycetes recorded for Britain. *Trans. Brit. Mycol. Soc.* 25: 49-126.

Watling, R. (1973). *Identification of the Larger Fungi.* Amersham: Hulton. 281 pp.

Watling, R. (1986). Observations on the Bolbitiaceae 28: the *Conocybe pubescens* complex. *Bol. Soc. Micol. Madrid* 11: 91-96.

Watling, R. (1987). Larger Arctic-Alpine Fungi in Scotland. In Laursen, G.A. *et al.* (eds). *Arctic & Alpine Mycology* II. pp. 17-45. Plenum Press.

Watling, R. (1994). *The Fungus Flora of Shetland.* Royal Botanic Garden, Edinburgh. 98 pp.

Watling, R., Eggeling, T. & Turnbull, E. (1999). *The Fungus Flora of Orkney.* Royal Botanic Garden, Edinburgh. 124 pp.

Watling, R. & Gregory, N.M. (1981). Census Catalogue of World Members of the *Bolbitiaceae*. *Bibliotheca Mycologica* 82: 1-224.

Watling, R., Kile, G.A. & Gregory, N.M. (1982). The genus *Armillaria* – nomenclature, typification, the identity of *Armillaria mellea* and species differentiation. *Trans. Brit. Mycol. Soc.* 78: 271-285.

Withering, W. (1796). *An arrangement of British plants: according to the latest improvements of the Linnaean system, to which is prefixed, an easy introduction to the study of botany.* 3rd Edn. Birmingham: M. Swinney. 4 vols.

INCLUDED BASIDIOMYCETES

ABORTIPORUS Murrill, *Bull. Torrey Bot. Club* 31: 421 (1904)
Poriales, Coriolaceae
> *Heteroporus* Lázaro Ibiza, *Revista Real. Acad. Ci. Madrid* 15: 119 (1916)
Type: *Boletus distortus* Schwein. (= *Abortiporus biennis*)

biennis (Bull.) Singer, *Mycologia* 36: 68 (1944)
> *Boletus biennis* Bull., *Herb. France*: pl. 449 f. 1 (1790)
> *Sistotrema bienne* (Bull.) Pers., *Syn. meth. fung.*: 550 (1801)
> *Daedalea biennis* (Bull.) Fr., *Syst. mycol.* 1: 332 (1821)
> *Polyporus biennis* (Bull.) Fr., *Epicr. syst. mycol.*: 433 (1838)
> *Heteroporus biennis* (Bull.) Lázaro Ibiza, *Revista Real. Acad. Ci. Madrid* 15: 120 (1916)
> *Daedalea biennis* β *sowerbei* Fr., *Syst. mycol.* 1: 332 (1821)
> *Polyporus biennis* var. *sowerbei* (Fr.) P.W. Graff, *Mycologia* 31: 472 (1939)
> *Boletus distortus* Schwein., *Schriften Naturf. Ges. Leipzig* 1: 97 (1822)
> *Polyporus distortus* (Schwein.) Fr., *Elench. fung.* 1: 79 (1828)
> *Polyporus biennis* var. *distortus* (Schwein.) P.W. Graff, *Mycologia* 31: 480 (1939)
> *Sistotrema rufescens* Pers., *Syn. meth. fung.*: 550 (1801)
> *Polyporus rufescens* (Pers.) Fr., *Syst. mycol.* 1: 351 (1821)
> *Ptychogaster alveolatus* Boud. (anam.), *Bull. Soc. Mycol. France* 4: LV, pl. 3 (1888)
E: c **S:** o **W:** o **NI:** o **ROI:** o **O:** Channel Islands: !
H: On decayed wood of deciduous trees. Often reported 'on the ground', but always attached to dead roots or buried woody debris.
D: EurPoly1: 81, NM3: 220 **D+I:** Ph: 224 (as *Heteroporus biennis*), B&K2: 310 393 **I:** Sow2: 191 (as *Boletus biennis*), Bon: 317, C&D: 141, Kriegl1: 469
Common in southern England, occasional but widespread elsewhere. *Polyporus biennis* var. *distortus* was the name given to abnormal basidiomes in which the tubes show no geotropic orientation and are arranged in sheets over the surface of the basidiomes.

AGARICUS L., *Sp. pl.*: 1171 (1753)
Agaricales, Agaricaceae
> *Fungus* Adans., *Fam. Pl.* 2: 12 (1763)
> *Pratella* (Pers.) Gray, *Nat. arr. Brit. pl.* 1: 626 (1821)
> *Psalliota* (Fr.) P. Kumm., *Führ. Pilzk.*: 23 (1871)
Type: *Agaricus campestris* L.

altipes (F.H. Møller) Pilát, *Sborn. Nár. Mus. Praze* 7B(1): 12 (1951)
> *Psalliota altipes* F.H. Møller, *Friesia* 4: 46 (1950)
> *Psalliota aestivalis* F.H. Møller, *Friesia* 4: 50 (1950)
> *Agaricus aestivalis* (F.H. Møller) Pilát [*nom. illegit.*, non *A. aestivalis* Schumach. (1803)], *Sborn. Nár. Mus. Praze* 7B(1): 50 (1951)
> *Psalliota aestivalis* var. *flavotacta* F.H. Møller, *Friesia* 4: 51 (1950)
> *Agaricus aestivalis* var. *flavotactus* (F.H. Møller) Pilát, *Sborn. Nár. Mus. Praze* 7B(1): 13 (1951)
E: !
H: On soil, associated with conifers. British material reported with *Picea* sp. or *Taxus*.
D: NM2: 208 (as *A. aestivalis*), FAN5: 30 **D+I:** FungEur1: 442 31 **I:** FungEur1: 46 (as *A. aestivalis*)
Known from Surrey (Kew Gardens) and possibly also Yorkshire. Reported from Scotland (Mid-Ebudes: Isle of Mull) but the material in herb. E is labelled *Agaricus* aff. *altipes*.

arvensis Schaeff., *Fung. Bavar. Palat. nasc.* 1: 73 (1762)
> *Psalliota arvensis* (Schaeff.) P. Kumm., *Führ. Pilzk.*: 74 (1871)
> *Psalliota arvensis* f. *obesa* W.G. Sm., *Guide to Worthington Smith's Drawings*: 11 (1910)
> *Psalliota arvensis* f. *epileata* W.G. Sm., *Guide to Worthington Smith's Drawings*: 12 (1910)
> *Psalliota arvensis* var. *hortensis* W.G. Sm., *Guide to Worthington Smith's Drawings*: 11 (1910)
> *Psalliota arvensis* var. *albosquamosa* W.G. Sm., *Guide to Worthington Smith's Drawings*: 11 (1910)
> *Agaricus exquisitus* Vittad., *Descr. fung. mang.*: 146 (1835)
> *Psalliota fissurata* F.H. Møller, *Friesia* 4: 165 (1952)
> *Agaricus fissuratus* (F.H. Møller) F.H. Møller, *Friesia* 4: 204 (1952)
> *Psalliota leucotricha* F.H. Møller, *Friesia* 4: 159 (1952)
> *Agaricus leucotrichus* (F.H. Møller) F.H. Møller, *Friesia* 4: 204 (1952)
E: o **S:** o **W:** ! **NI:** o **ROI:** ! **O:** Channel Islands: !
H: On soil, often amongst grass, in grassland or in open woodland, in gardens and churchyards, parkland, along roadside verges, and on dunes.
D: NM2: 213, FAN5: 43-44 **D+I:** Ph: 166-167, FungEur1: 446 33 **I:** FungEur1: 47 (as *A. leucotrichus*), FungEur1: 34 (as *A. fissuratus*), FungEur1: 160 168
Occasional but widespread and locally common. Possibly declining. English name = 'Horse Mushroom'.

augustus Fr., *Epicr. syst. mycol.*: 212 (1838)
> *Psalliota augusta* (Fr.) Quél., *Mém. Soc. Émul. Montbéliard, Sér. 2*, 5: 255 (1872)
> *Agaricus perrarus* Schulzer, *Verh. zool.-bot. Ges. Wien* 29: 493 (1880)
> *Agaricus augustus* var. *perrarus* (Schulzer) Bon & Cappelli, *Doc. Mycol.* 13(52): 16 (1983)
> *Agaricus peronatus* Massee [*nom. illegit.*, non *A. peronatus* Bolton (1788)], *Brit. fung.-fl.* 1: 415 (1892)
Mis.: *Psalliota subrufescens* sensu Lange (FlDan4: 55 & pl. 136B)
E: o **S:** ! **W:** ! **NI:** ! **ROI:** ! **O:** Channel Islands: !
H: On soil, amongst leaf litter in woodland, or occasionally in parkland, associated with a wide range of deciduous and coniferous trees, including *Taxus*.
D: NM2: 212, FAN5: 44-45 **D+I:** Ph: 164, FungEur1: 480 49, B&K4: 162 169 **I:** FlDan4: 135B (as *Psalliota augusta*), FungEur1: 482 50 (as *Agaricus augustus* var. *perrarus*)
Occasional but widespread, and may be locally frequent on calcareous soils in southern England. English name = 'The Prince'.

benesii (Pilát) Pilát, *Sborn. Nár. Mus. Praze* 7B(1): 56 (1951)
> *Psalliota benesii* Pilát, *Mykologia* 2: 49 (1925)
> *Agaricus carolii* Pilát, *Sborn. Nár. Mus. Praze* 7B(1): 130 (1951)
> *Agaricus squamuliferus* var. *carolii* (Pilát) Pilát [*nom. inval.*], *Naše Houby* 2: 141 (1959)
> *Psalliota squamulifera* F.H. Møller, *Friesia* 4: 21 (1950)
> *Agaricus squamuliferus* (F.H. Møller) Pilát, *Sborn. Nár. Mus. Praze* 7B(1): 7 (1951)
Mis.: *Psalliota exserta* sensu Rea (1922)
E: !
H: On soil in deciduous or coniferous woodland.
D: NM2: 207, FAN5: 41-42 **D+I:** FungEur1: 414 20 **I:** FungEur1: 25 (as *A. squamulifer* var. *carolii*), FungEur1: 24 (as *A. squamulifer*)
A single collection in herb. K from Surrey (Woking, Brookwood Cemetery). Reported in scattered locations from Staffordshire southwards but unsubstantiated with voucher material.

bernardii Quél., *Clavis syn. Hymen. Europ.*: 89 (1878)
> *Psalliota bernardii* (Quél.) Quél., *Bull. Soc. Bot. France* 25: 288 (1879) [1878]
> *Agaricus campestris* ssp. *bernardii* (Quél.) Konrad & Maubl., *Icon. select. fung.* 6: 60 (1937)
> *Agaricus maleolens* F.H. Møller, *Friesia* 4: 203 (1952)
E: ! **S:** ! **W:** ! **NI:** !
H: On soil in coastal grassland, but also inland on roadside verges, spreading due to the de-icing of roads with salt.

D: NM2: 209, FAN5: 33 **D+I:** FungEur1: 370 2, FungEur1: 371 3 (as *Agaricus maleolens*)
Rarely reported in coastal habitat, but increasingly frequent in inland sites.

bisporus (J.E. Lange) Imbach, *Mitt. Naturf. Ges. Luzern* 15: 15 (1946)
    Psalliota hortensis f. *bispora* J.E. Lange, *Dansk. Bot. Ark.* 4(12): 8 (1926)
    Psalliota bispora (J.E. Lange) F.H. Møller & Jul. Schäff., *Ann. Mycol.* 36: 69 (1938)
    Agaricus campestris var. *hortensis* Cooke, *Handb. Brit. fung.*, Edn 1: 138 (1871)
    Psalliota hortensis (Cooke) J.E. Lange, *Dansk. Bot. Ark.* 4(12): 8 (1926)
    Agaricus hortensis (Cooke) Pilát, *Sborn. Nár. Mus. Praze* 7B(1): 37 (1951)
    Agaricus campestris var. *bisporus* Kligman, *Amer. J. Bot.* 3: 746 (1943)
    Mis.: *Agaricus campestris* sensu Cooke [*Ill. Brit. fung.* 4: pl. 527 (1885)]
E: o **S:** ! **W:** ! **NI:** ! **ROI:** ! **O:** Channel Islands: !
H: On soil, amongst grass in parkland, pastures and other grasslands or at the edges of fields. Rarely in woodland. Occasionally on manure and compost heaps or rubbish tips. Extensively cultivated.
D: NM2: 210, FAN5: 34-35 **D+I:** FungEur1: 380 7, B&K4: 164 172
Occasional but widespread. English name = 'Cultivated Mushroom'.

bitorquis (Quél.) Sacc., *Syll. fung.* 5: 998 (1887)
    Psalliota bitorquis Quél., *Compt. Rend. Assoc. Franç. Avancem. Sci.* 12: 500 (1884) [1883]
    Psalliota edulis var. *validus* F.H. Møller, *Friesia* 4: 14 (1950)
    Agaricus bitorquis var. *validus* (F.H. Møller) Bon & Cappelli, *Doc. Mycol.* 13(52): 16 (1983)
    Agaricus campestris var. *edulis* Vittad., *Descr. fung. mang. Italia*: 41 (1832)
    Psalliota edulis (Vittad.) Buchwald, *Spise- og Giftsvampe*: 125 (1937)
    Psalliota peronata Richon & Roze, *Atlas champ.*: 44 (1888)
    Agaricus rodmanii Peck, *Rep. (Annual) New York State Mus. Nat. Hist.* 36: 45 (1883)
    Psalliota rodmanii (Peck) Kauffman, *Michigan Geol. Biol. Surv. Publ., Biol. Ser. 5*, 26: 235 (1918)
E: o **S:** ! **W:** ! **NI:** ! **ROI:** ! **O:** Channel Islands: !
H: On soil (often compacted) on pathsides, in hedgerows, in grassy areas in open woodland, on roadside verges, and in gardens, sometimes pushing through tarmac.
D: NM2: 210, FAN5: 32-33 **D+I:** FungEur1: 368 1, B&K4: 164 173
Occasional but widespread.

bohusii Bon, *Doc. Mycol.* 11(44): 28 (1981)
    Mis.: *Agaricus elvensis* sensu Cooke [*Ill. Brit. fung.* 539 (522) Vol. 4 (1885)] & sensu Rea (1922)
E: ! **S:** !
H: On soil, usually in open deciduous woodland, in parkland and occasionally on roadside verges.
D: FAN5: 37-38 **D+I:** FungEur1: 418 22
Rarely reported, but known from widely scattered localities. A large and distinctive species, unlikely to be overlooked and probably genuinely rare.

bresadolanus Bohus, *Ann. Hist.-Nat. Mus. Natl. Hung.* 61: 154 (1969)
    Agaricus romagnesii Wasser, *Ukrayins'k. Bot. Zhurn.* 34: 305 (1977)
E: ! **W:** ?
H: On soil, in mixed deciduous woodland, parkland and occasionally gardens.
D: FAN5: 60 **D+I:** FRIC6: 6-10 42, FungEur1: 396 14, FungEur1: 526 70 (as *A. romagnesii*)
Rarely reported. Known from Berkshire, East Sussex, South Essex, Middlesex, Oxfordshire, Shropshire and Surrey.

Reported elsewhere in England and from a single location in Wales (Glamorganshire) but unsubstantiated with voucher material.

campestris L., *Sp. pl.*: 1173 (1753)
    Psalliota campestris (L.) Quél., *Mém. Soc. Émul. Montbéliard, Sér. 2*, 5: 140 (1872)
    Psalliota campestris var. *squamulosa* Rea, *Trans. Brit. Mycol. Soc.* 17(1-2): 50 (1932)
    Agaricus campestris var. *squamulosus* (Rea) Pilát, *Sborn. Nár. Mus. Praze* 7B(1): 14 (1951)
    Psalliota campestris var. *fuscopilosella* F.H. Møller, *Friesia* 4: 58 (1950)
    Agaricus campestris var. *fuscopilosellus* (F.H. Møller) Pilát, *Sborn. Nár. Mus. Praze* 7B(1): 14 (1951)
    Psalliota flocculosa Rea, *Trans. Brit. Mycol. Soc.* 17(1-2): 50 (1932)
E: o **S:** o **W:** o **NI:** o **ROI:** o **O:** Channel Islands: ! Isle of Man: !
H: On soil, usually amongst grass in meadows, pasture, playing fields, garden lawns, and cemeteries but occasionally in open woodland, hedgerows or on dunes.
D: NM2: 209, FAN5: 28 **D+I:** Ph: 162, FungEur1: 382 9, B&K4: 164 174 **I:** Sow3: 305, FungEur1: 10 (as *A. campestris* var. *squamulosus*)
Occasional but widespread, sometimes locally abundant but declining in many localities. The epithet '*campestris*' is often applied to any mushroom growing in grassland and many records may refer to *A. bisporus* (and *vice versa*). English name = 'Field Mushroom'.

campestris *var.* **equestris** (F.H. Møller) Pilát, *Sborn. Nár. Mus. Praze* 7B(1): 14 (1951)
    Psalliota campestris var. *equestris* F.H. Møller, *Friesia* 4: 57 (1950)
E: ! **W:** !
H: On soil in grassland, occasionally with shrubs such as *Crataegus monogyna* in such locations.
D: FAN5: 29
Rarely reported. Known from England (South West Yorkshire, Staffordshire, Surrey and West Suffolk) and Wales (Radnorshire).

comtulus Fr., *Epicr. syst. mycol.*: 215 (1838)
    Psalliota comtula (Fr.) Quél., *Mém. Soc. Émul. Montbéliard, Sér. 2*, 5: 140 (1872)
    Agaricus niveolutescens Huijsman, *Persoonia* 1: 321 (1960)
E: o **S:** ! **W:** ?
H: On soil, amongst grass, in grassland, open deciduous woodland, on dunes, in parkland, and occasionally on lawns.
D: NM2: 213, FAN5: 52-53 **D+I:** FungEur1: 488 52, B&K4: 166 175 **I:** FungEur1: 57 (as *A. niveolutescens*)
Occasional but apparently widespread, though frequently misidentified. Few records are substantiated with voucher material.

cupreobrunneus (F.H. Møller) Pilát, *Sborn. Nár. Mus. Praze* 7B(1): 14 (1951)
    Psalliota cupreobrunnea F.H. Møller, *Friesia* 4: 54 (1950)
    Psalliota campestris var. *cupreobrunnea* Jul. Schäff. & Steer [*nom. inval.*], in Michael, *Führer für Pilzfreunde* 1: 147 (1939)
E: ! **S:** ! **NI:** ! **ROI:** ! **O:** Isle of Man: !
H: On soil, often acidic, usually in unimproved grassland, on heath or moorland, and also in pastures and meadows.
D: NM2: 208, FAN5: 31 **D+I:** FungEur1: 390 12, B&K4: 166 176
Rarely reported but apparently widespread.

depauperatus (F.H. Møller) Pilát, *Sborn. Nár. Mus. Praze* 7B(1): 25 (1951)
    Psalliota depauperata F.H. Møller, *Friesia* 4: 24 (1950)
    Agaricus deylii Pilát, *Sborn. Nár. Mus. Praze* 7B(1): 131 (1951)
E: ! **S:** !
H: On soil, amongst grass, in gardens and cemeteries, on roadside verges, and rarely in open woodland.
D: NM2: 207, FAN5: 42-43 **D+I:** FungEur1: 444 32

Rarely reported. Known from East Sussex and Surrey, and from Shetland.

devoniensis P.D. Orton, *Trans. Brit. Mycol. Soc.* 43: 173 (1960)
    Psalliota arenicola Wakef. & A. Pearson, *Trans. Brit. Mycol. Soc.* 29(4): 205 (1946)
E: ! **S:** ! **W:** ! **NI:** ! **O:** Channel Islands: !
H: In sand on dunes, rarely on sandy soil inland.
D: NM2: 209, FAN5: 35-36 **D+I:** FungEur1: 374 4
Rarely reported but apparently widespread.

dulcidulus Schulzer, Kalchbrenner, *Icon. select. Hymenomyc. Hung.*: 29 (1874)
    Psalliota rubella f. *pallens* J.E. Lange, *Dansk. Bot. Ark.* 4(12): 8 (1926)
    Psalliota pallens (J.E. Lange) Rea, *Trans. Brit. Mycol. Soc.* 17(1-2): 37 (1932)
    Psalliota purpurella F.H. Møller, *Friesia* 4: 193 (1952)
    Agaricus purpurellus (F.H. Møller) F.H. Møller, *Friesia* 4: 204 (1952)
    Pratella rubella Gillet, *Hyménomycètes*: 565 (1878)
    Agaricus rubellus (Gillet) Sacc., *Syll. fung.* 5: 1007 (1887)
    Psalliota rubella (Gillet) Rea, *Brit. basidiomyc.*: 90 (1922)
    Mis.: *Psalliota amethystina* sensu Lange (FlDan4: 61 & pl. 135A)
    Mis.: *Psalliota semota* sensu NCL and sensu auct. mult.
E: ! **S:** ! **W:** ! **ROI:** !
H: On humic soil in deciduous woodland and with conifers. Also reported in gardens and on lawns near trees.
D: NM2: 214 (as *A. purpurellus*), FAN5: 50-51 **D+I:** B&K4: 176 191 (as *A. semotus*) **I:** SV34: 9 (as *A. semotus*)
Rarely reported but apparently widespread. Interpreted here in a broad sense for a woodland species that includes both *A. purpurellus* and *A. semota* of NCL but excludes *A. dulcidulus* sensu NCL and Lange (FlDan4: 61 & pl. 135C) which is different and a doubtful member of section *Minores*.

fuscofibrillosus (F.H. Møller) Pilát, *Sborn. Nár. Mus. Praze* 7B(1): 7 (1951)
    Psalliota fuscofibrillosa F.H. Møller, *Friesia* 4: 27 (1950)
E: ! **S:** ! **W:** ! **NI:** ! **ROI:** !
H: On soil in deciduous or coniferous woodland.
D: NM2: 208, FAN5: 41 **D+I:** FungEur1: 416 21, Myc13(4): 156
Rarely reported but apparently widespread.

gennadii (Chatin & Boud.) P.D. Orton, *Trans. Brit. Mycol. Soc.* 43: 174 (1960)
    Chitonia gennadii Chatin & Boud., *J. Bot., Paris* 12: 66 (1898)
    Clarkeinda gennadii (Chatin & Boud.) Sacc. & Syd., *Syll. fung.* 16: 112 (1902)
    Clarkeinda cellaris Bres., *Ann. Mycol.* 3(2): 162 (1905)
    Chitonia cellaris (Bres.) Boud., *Icon. mycol.* 1: pl. 131 (1908)
E: !
H: On soil with conifers, in woodland, cemeteries and on roadside verges. Reported with *Cupressus*, *Larix*, *Pinus* and *Sequoiadendron* spp and *Taxus*.
D: NCL3: 181, FAN5: 34 **D+I:** FungEur1: 376 5
Rarely reported. Known from East Kent, Middlesex and Surrey. Also reported from Cambridgeshire, Oxfordshire and Warwickshire but unsubstantiated with voucher material.

impudicus (Rea) Pilát, *Klíč urč. naš. Hub hrib. bedl.*: 403 (1951)
    Psalliota impudica Rea, *Trans. Brit. Mycol. Soc.* 17(1-2): 50 (1932)
    Psalliota variegata var. *koelerionensis* Bon, *Doc. Mycol.* 3: 12 (1972)
    Agaricus koelerionensis (Bon) Bon, *Doc. Mycol.* 10(37-38): 91 (1980) [1979]
    Agaricus variegans F.H. Møller, *Friesia* 4: 203 (1952)
    Psalliota variegata F.H. Møller, *Friesia* 4: 30 (1950)
E: o **W:** ! **ROI:** ?
H: On soil in deciduous or mixed deciduous and coniferous woodland, also amongst grass in parkland and cemeteries.
D: NM2: 207, FAN5: 39-40 **D+I:** FungEur1: 408 18 & 18bis
Occasional but widespread in England, from Nottinghamshire southwards. Rarely reported elsewhere.

langei (F.H. Møller) F.H. Møller, *Friesia* 4: 203 (1952)
    Psalliota langei F.H. Møller, *Friesia* 4: 28 (1950)
    Psalliota mediofusca F.H. Møller, *Friesia* 4: 32 (1950)
    Agaricus mediofuscus (F.H. Møller) Pilát, *Sborn. Nár. Mus. Praze* 7B(1): 8 (1951)
    Mis.: *Agaricus haemorrhoidarius* sensu Lange (FlDan4: 56 & pl. 137C)
E: o **S:** ! **W:** ! **NI:** !
H: On soil amongst litter, in coniferous or deciduous woodland.
D: NM2: 208, FAN5: 40-41 **D+I:** Ph: 160-161, FungEur1: 430 27 & 27bis, B&K4: 168 179 **I:** FungEur1: 28 (as *A. mediofuscus*)
Occasional but widespread, most frequently reported from England.

lanipes (F.H. Møller & Jul. Schäff.) Singer, *Lilloa* 22: 432 (1951) [1949]
    Psalliota lanipes F.H. Møller & Jul. Schäff., *Ann. Mycol.* 36: 65 (1938)
    Agaricus luteolorufescens P.D. Orton, *Trans. Brit. Mycol. Soc.* 43(2): 182 (1960)
E: ! **S:** ! **NI:** ? **ROI:** ? **O:** Channel Islands: !
H: On soil in deciduous or mixed deciduous and coniferous woodland, also rarely amongst grass in cemeteries and scrub.
D: NM2: 207, FAN5: 59 **D+I:** FungEur1: 420 23
Rarely reported but apparently widespread.

litoralis (Wakef. & A. Pearson) Pilát, *Klíč urč. naš. Hub hrib. bedl.*: 403 (1951)
    Psalliota litoralis Wakef. & A. Pearson, *Trans. Brit. Mycol. Soc.* 29(4): 205 (1946)
    Agaricus maskae Pilát, *Česká Mykol.* 8: 165 (1954)
    Psalliota spissa F.H. Møller, *Friesia* 4: 53 (1950)
    Agaricus spissicaulis F.H. Møller, *Friesia* 4: 203 (1950)
E: ! **W:** !
H: On soil, amongst grass, in pasture, fields, in cemeteries and on lawns, and sometimes (but not always) on dunes or in coastal pasture.
D: NM2: 209, NM2: 209 (as *A. spissicaulis*), FAN5: 58-59 **D+I:** FungEur1: 454 37 (as *A. spissicaulis*), FungEur1: 456 38 (as *A. maskae*)
Rarely reported. Known from widely scattered localities in southern England (East Kent, East Sussex, London, Middlesex, North Devon, West Cornwall and Worcestershire) and Wales (Caernarvonshire, Glamorganshire and Radnorshire).

luteomaculatus (F.H. Møller) F.H. Møller, *Friesia* 4: 204 (1952)
    Psalliota luteomaculata F.H. Møller, *Friesia* 4: 192 (1952)
E: ! **W:** ?
H: On soil, in deciduous or mixed deciduous and coniferous woodland, and occasionally on decayed piles of compost in gardens.
D: FAN5: 51-52 **D+I:** FungEur1: 496 56, B&K4: 170 181
Rarely reported. British collections only from Surrey (Kew: Royal Botanic Gardens). Reported from East and West Cornwall and Wales (Glamorganshire) but unsubstantiated with voucher material.

lutosus (F.H. Møller) F.H. Møller, *Friesia* 4: 204 (1952)
    Psalliota lutosa F.H. Møller, *Friesia* 4: 188 (1952)
E: ! **S:** ! **W:** !
H: On soil, in parkland, on dunes or occasionally with shrubs and trees such as *Crataegus* or *Betula* spp in parkland or grassland. Rarely in scrub woodland.
D+I: FungEur1: 490 53
Rarely reported but apparently widespread. Possibly conspecific with *A. xantholepis* fide Nauta in FAN5.

macrocarpus (F.H. Møller) F.H. Møller, *Friesia* 4: 204 (1952)
    Psalliota macrocarpa F.H. Møller, *Friesia* 4: 153 (1952)
E: ! **S:** o **W:** ! **NI:** !
H: On soil in deciduous woodland, occasionally in grassy places at woodland edges or in cemeteries.
D: FAN5: 48 **D+I:** FungEur1: 470 45, B&K4: 170 182
Rarely reported but apparently widespread; locally frequent in Scotland fide R. Watling (pers. comm.).

moelleri Wasser, *Nov. sist. Niz. Rast.* 13: 77 (1973)
Psalliota meleagris Jul. Schäff., *Z. Pilzk.* 4(2): 28 (1925)
Agaricus meleagris (Jul. Schäff.) Imbach [*nom. illegit.*, non *A. meleagris* With. (1792); non Sowerby (1798)], *Mitt. Naturf. Ges. Luzern* 15: 15 (1946)
Agaricus meleagris var. *terricolor* F.H. Møller [*nom. illegit.*], *Friesia* 4: 208 (1952)
Agaricus praeclaresquamosus var. *terricolor* (F.H. Møller) Bon & Cappelli [*comb. illegit.*], *Doc. Mycol.* 13(52): 16 (1983)
Agaricus praeclaresquamosus A.E. Freeman [*nom. superfl.*], *Mycotaxon* 8: 90 (1979)
Agaricus xanthodermus var. *obscuratus* Maire, *Bull. Soc. Mycol. France* 26: 192 (1910)
Mis.: *Agaricus placomyces* sensu auct. Eur.
E: o **S:** ! **W:** o **NI:** ! **O:** Channel Islands: !
H: On soil (especially calcareous loam) in deciduous woodland, and rarely on dunes.
D: NM2: 211 (as *A. placomyces*), FAN5: 56-57 **D+I:** Ph: 168-169 (as *A. placomyces*), FungEur1: 508 62 (as *A. praeclaresquamosus* var. *praeclaresquamosus*), B&K4: 174 188 (as *A. praeclaresquamosus*) **I:** FungEur1: 63 (as *A. praeclaresquamosus* var. *terricolor*), SV28: 21 (as *A. praeclaresquamosus*)
Occasional but widespread, and may be locally abundant. Better known as *Agaricus placomyces*. N.B. the supposed synonymy of *Psalliota xanthoderma* var. *lepiotoides* with this species (as *Agaricus placomyces*) in NCL is incorrect.

moellerianus Bon, *Doc. Mycol.* 15(60): 6 (1985)
Psalliota campestris var. *floccipes* F.H. Møller, *Friesia* 4: 57 (1950)
Agaricus campestris var. *floccipes* (F.H. Møller) Pilát, *Sborn. Nár. Mus. Praze* 7B(1): 14 (1951)
Agaricus floccipes (F.H. Møller) Bohus [*nom. illegit.*, non *A. floccipes* Fr. (1838)], *Ann. Hist.-Nat. Mus. Natl. Hung.* 70: 107 (1978)
E: !
H: On soil, amongst grass. British collection on roadside verge.
D: FAN5: 29 **D+I:** FungEur1: 388 11 (as *Agaricus floccipes*)
A single collection (West Sussex: Duncton Hill) in 1978 (as *Agaricus campestris* f. *floccipes*).

osecanus Pilát, *Sborn. Nár. Mus. Praze* 7B(1): 133 (1951)
Psalliota nivescens F.H. Møller, *Friesia* 4: 155 (1952)
Agaricus nivescens (F.H. Møller) F.H. Møller, *Friesia* 4: 204 (1952)
Psalliota nivescens var. *parkensis* F.H. Møller, *Friesia* 4: 158 (1952)
Agaricus nivescens var. *parkensis* (F.H. Møller) F.H. Møller, *Friesia* 4: 204 (1952)
Mis.: *Agaricus arvensis* sensu Cooke [Ill. Brit. fung. 540 (523) Vol. 4 (1885)]
E: ! **S:** ! **W:** ! **NI:** ?
H: On soil, amongst grass in pastures, fields, dunes or occasionally under bushes or trees in scrub.
D: FAN5: 45-46 **D+I:** FungEur1: 450 35 (as *A. nivescens*)
Rarely reported but apparently widespread.

pampeanus Speg., *Fungi argentini pugillus* 9: 280 (1880)
E: ! **NI:** !
H: On soil in grassy areas and on roadside verges.
D: FAN5: 29-30 **D+I:** B&K4: 172 186
Rarely reported but likely to be mistaken for *A. campestris*. Known from England (North Essex and Surrey) and Northern Ireland (Fermanagh). *Psalliota flocculosa* Rea, usually considered a synonym of *A. campestris*, may prove to belong here.

phaeolepidotus (F.H. Møller) F.H. Møller, *Friesia* 4: 204 (1952)
Psalliota phaeolepidota F.H. Møller, *Friesia* 4: 170 (1952)
E: ! **W:** ?
H: On soil in deciduous woodland. Reported also with *Taxus* and in grassland on dunes.
D: NM2: 211, FAN5: 57-58 **D+I:** FungEur1: 512 64 **I:** SV28: 21
Rarely reported. Known from Shropshire and Surrey. Reported from North Hampshire, North Somerset and South Devon, and

from Wales (Monmouthshire) but unsubstantiated with voucher material

porphyrizon P.D. Orton, *Trans. Brit. Mycol. Soc.* 43(2): 174 (1960)
Agaricus arvensis var. *purpurascens* Cooke, *Ill. Brit. fung.*: 541 (584) Vol. 4 (1885)
Psalliota arvensis var. *purpurascens* (Cooke) W.G. Sm., *Guide to Worthington Smith's Drawings*: 10 (1910)
Agaricus purpurascens (Cooke) Pilát [*nom. illegit.*, non *A. purpurascens* With. (1792); non Fr. (1818)], *Sborn. Nár. Mus. Praze* 7B(1): 10 (1951)
Psalliota purpurascens (Cooke) F.H. Møller, *Friesia* 4: 187 (1952)
E: ! **S:** ! **W:** ! **O:** Channel Islands: !
H: On soil in deciduous woodland and occasionally with solitary trees in parkland or cemeteries.
D: NM2: 214, FAN5: 53-54 **D+I:** FungEur1: 484 51 & 51bis, B&K4: 174 187
Rarely reported; apparently widespread, although most often reported from southern England.

porphyrocephalus F.H. Møller, *Friesia* 4: 204 (1952)
Psalliota porphyrea F.H. Møller, *Friesia* 4: 53 (1950)
E: ! **S:** ! **W:** !
H: On soil in grassland, often on moors or heaths.
D: NM2: 208, FAN5: 31-32 **D+I:** Ph: 160-161, FungEur1: 394 13
Rarely reported, but apparently widespread.

pseudovillaticus Rauschert, *Nova Hedwigia* 54: 214 (1992)
Agaricus campestris β *vaporarius* Pers., *Syn. meth. fung.*: 418 (1801)
Psalliota vaporaria (Pers.) F.H. Møller & Jul. Schäff., *Ann. Mycol.* 36: 71 (1938)
Agaricus vaporarius (Pers.) Cappelli, *Fungi Europaei: Agaricus* 1: 149 (1984)
Mis.: *Psalliota villatica* sensu auct.
Mis.: *Agaricus campestris* var. *villaticus* sensu Cooke [Ill. Brit. fung. 548 (585) Vol. 4 (1885)]
E: ! **NI:** ! **ROI:** !
H: On soil in woodland or scrub, reported with deciduous and coniferous trees. Occasionally amongst grass at woodland edges or in grassland.
D: NM2: 210 **D+I:** Ph: 160-161, FungEur1: 398 15, B&K4: 180 196
Rarely reported. Better known as *A. vaporarius* and considered by some authorities to be a synonym of *A. subperonatus*.

rufotegulis Nauta, *Persoonia* 17(2): 230 (1999)
E: !
H: On composted grass clippings, soil and decayed woodchip mulch (British material).
D: FAN5: 60-61
Recently collected in Surrey (Esher, Claremont) and London (Sheen Common).

silvaticus Schaeff., *Fung. Bavar. Palat. nasc.* 1: 62 (1762)
Psalliota silvatica (Schaeff.) P. Kumm. [as *P. sylvatica*], *Führ. Pilzk.*: 73 (1871)
Agaricus haemorrhoidarius Schulzer, *Icon. Sel. Hymenomyc. Hung.*: 29 (1874)
Psalliota haemorrhoidaria (Schulzer) Richon & Roze, *Atlas champ.*: 49 (1888)
Agaricus sanguinarius P. Karst., *Bidrag Kännedom Finlands Natur Folk* 37: 232 (1882)
Psalliota sanguinaria (P. Karst.) J.E. Lange, *Dansk. Bot. Ark.* 4: 12 (1926)
Psalliota silvatica var. *pallida* F.H. Møller, *Friesia* 4: 38 (1950)
Agaricus silvaticus var. *pallidus* (F.H. Møller) F.H. Møller, *Friesia* 4: 203 (1952)
Agaricus silvaticus var. *pallens* Pilát, *Sborn. Nár. Mus. Praze* 7B(1): 67 (1951)
Agaricus vinosobrunneus P.D. Orton, *Trans. Brit. Mycol. Soc.* 43(2): 183 (1960)
E: c **S:** c **W:** c **NI:** c **ROI:** c **O:** Channel Islands: ! Isle of Man: !

H: On soil in coniferous and mixed deciduous and coniferous woodland.
D: NM2: 207, FAN5: 38-39 **D+I:** Ph: 160, FungEur1: 404 17 & 17bis, B&K4: 176 192
Common and widespread throughout the British Isles.

silvicola (Vittad.) Peck, *Rep. (Annual) New York State Mus. Nat. Hist.* 23: 97 (1872)
  Agaricus campestris var. *silvicola* Vittad., *Descr. fung. mang.*: 213 (1835)
  Psalliota silvicola (Vittad.) Richon & Roze, *Atlas champ.*: 7 (1885)
  Agaricus essettei Bon, *Doc. Mycol.* 11(44): 28 (1981)
  Pratella flavescens Gillet, *Hyménomycètes*: 564 (1878)
  Mis.: *Agaricus abruptibulbus* sensu auct. Eur.
E: o **S:** o **W:** o **NI:** o
H: On soil in deciduous or coniferous woodland.
D: NM2: 213, FAN5: 47-48 **D+I:** FungEur1: 464 42, FungEur1: 468 44 (as *A. essettei*), B&K4: 178 194
Occasional but widespread. Frequent in southern England, less so northwards.

subfloccosus (J.E. Lange) Pilát, *Sborn. Nár. Mus. Praze* 7B(1): 49 (1951)
  Psalliota hortensis f. *subfloccosa* J.E. Lange, *Dansk. Bot. Ark.* 4(12): 8 (1926)
  Psalliota subfloccosa (J.E. Lange) J.E. Lange, *Fl. agaric. danic.* 5, *Taxonomic Conspectus*: VII (1940)
E: !
H: On soil in woodland, often associated with *Picea* spp in plantations but also known with *Pinus* spp and *Taxus* and reported with deciduous trees such as *Betula* and *Crataegus* spp.
D: NM2: 210, FAN5: 36 **D+I:** FungEur1: 426 26 & 26bis, B&K4: 178 195 **I:** FlDan4: 139D (as *Psalliota subfloccosa*)
Known from widely scattered locations in England, from Cambridgeshire southwards.

subperonatus (J.E. Lange) Singer, *Lilloa* 22: 431 (1951) [1949]
  Psalliota hortensis f. *subperonata* J.E. Lange, *Dansk. Bot. Ark.* 4(12): 8 (1926)
  Psalliota subperonata (J.E. Lange) J.E. Lange, *Fl. agaric. danic.* 5, *Taxonomic Conspectus*: VII (1940)
E: ! **S:** ! **W:** ! **NI:** ! **ROI:** ! **O:** Channel Islands: !
H: On soil in woodland, on roadside verges, edges of paths, parkland and in gardens
D: NM2: 211, FAN5: 37 **D+I:** FungEur1: 402 16 **I:** FlDan3: 140D (as *Psalliota subperonata*), Bon: 275
Rarely reported. Apparently widespread. The description in FAN5: includes *A. pseudovillaticus*, here maintained as distinct.

urinascens (F.H. Møller & Jul. Schäff.) Singer, *Lilloa* 22: 431 (1951) [1949]
  Psalliota urinascens F.H. Møller & Jul. Schäff., *Ann. Mycol.* 36: 79 (1938)
  Agaricus albertii Bon, *Doc. Mycol.* 18(72): 63 (1988)
  Psalliota arvensis ssp. *macrospora* F.H. Møller & Jul. Schäff., *Ann. Mycol.* 36: 78 (1938)
  Psalliota macrospora (F.H. Møller & Jul. Schäff.) F.H. Møller, *Friesia* 4: 181 (1951)
  Agaricus macrosporus (F.H. Møller & Jul. Schäff.) Pilát, *Sborn. Nár. Mus. Praze* 7B(1): 78 (1951)
  Psalliota straminea F.H. Møller & Jul. Schäff., *Ann. Mycol.* 36: 78 (1938)
  Agaricus stramineus (F.H. Møller & Jul. Schäff.) Singer, *Lilloa* 22: 432 (1951) [1949]
  Mis.: *Psalliota villatica* sensu auct.
E: ! **S:** c **NI:** !
H: On soil in grassy areas, also in open woodland.
D: NM2: 212 (as *A. macrosporus*), FAN5: 49-50 **D+I:** FungEur1: 458 39 (as *A. macrosporus*) **I:** FungEur1: 40 (as *A. stramineus*)
Apparently widespread (common in Scotland *fide* R. Watling) and better known as *Agaricus macrosporus*.

urinascens *var.* **excellens** (F.H. Møller) Nauta, *Persoonia* 17: 462 (2001) [2000]

  Psalliota excellens F.H. Møller, *Friesia* 4: 178 (1952)
  Agaricus excellens (F.H. Møller) F.H. Møller, *Friesia* 4: 204 (1952)
E: ! **ROI:** !
H: On soil, amongst grass in meadows, fields and at woodland edges.
D: NM2: 212 (as *A. excellens*), FAN5: 50 **D+I:** FungEur1: 462 41 (as *A. excellens*)
Rarely reported. Known from widely scattered locations in England, from Warwickshire to South Devon, and a single record from Republic of Ireland (Offaly).

xanthodermus Genev., *Bull. Soc. Bot. France* 23: 32 (1876)
  Psalliota xanthoderma (Genev.) Richon & Roze, *Atlas champ.*: 53 (1885)
  Psalliota flavescens Richon & Roze, *Atlas champ.*: 42 (1885)
  Agaricus xanthodermus var. *lepiotoides* Maire, *Bull. Soc. Mycol. France* 24: LVIII (1908)
  Psalliota xanthoderma var. *lepiotoides* (Maire) Rea, *Brit. basidiomyc.*: 85 (1922)
  Psalliota xanthoderma var. *grisea* A. Pearson, *Trans. Brit. Mycol. Soc.* 29(4): 204 (1946)
  Agaricus xanthodermus var. *griseus* (A. Pearson) Bon & Cappelli, *Doc. Mycol.* 13(52): 16 (1983)
E: c **S:** ! **W:** ! **NI:** ! **ROI:** ! **O:** Channel Islands: !
H: On soil in deciduous woodland and amongst grass by roadsides, in cemeteries, gardens, parkland and on dunes.
D: NM2: 211, FAN5: 54-55 **D+I:** Ph: 167, FungEur1: 502 59, FungEur1: 504 60 (as *A. xanthoderma* var. *grisea*), B&K4: 180 197 **I:** FungEur1: 61 (as *A. xanthodermus* var. *lepiotoides*), SV28: 17
Common in England, with numerous records and collections. Fewer records elsewhere but apparently widespread

xantholepis (F.H. Møller) F.H. Møller, *Friesia* 4: 204 (1952)
  Psalliota xantholepis F.H. Møller, *Friesia* 4: 191 (1952)
E: !
H: On soil in deciduous woodland.
D: NM2: 214, FAN5: 53 **D+I:** FungEur1: 500 58
Rarely reported. Known from Surrey (East Molesey) and reported from Staffordshire.

AGROCYBE Fayod, *Ann. Sci. Nat., Bot.*, sér. 7, 9: 358 (1889)
Agaricales, Bolbitiaceae
  Togaria W.G. Sm., *Syn. Brit. Bas.*: 122 (1908)
Type: *Agrocybe praecox* (Pers.) Fayod

arvalis (Fr.) Singer, *Beih. Bot. Centralbl., Abt. B* 56: 167 (1936)
  Agaricus arvalis Fr., *Syst. mycol.* 1: 263 (1821)
  Naucoria arvalis (Fr.) Sacc., *Syll. fung.* 5: 845 (1887)
  Galera arvalis var. *tuberigena* Quél., *Compt. Rend. Assoc. Franç. Avancem. Sci.* 17: t. 15, f. 10 (1889)
  Naucoria arvalis var. *tuberigena* (Quél.) A. Pearson & Dennis, *Trans. Brit. Mycol. Soc.* 31: 174 (1948)
  Naucoria tuberosa Henn., *Hedwigia* 42: 310 (1903)
  Agrocybe tuberosa (Henn.) Singer, *Beih. Bot. Centralbl.* 56: 167 (1936)
E: o **S:** o **W:** ! **ROI:** !
H: On soil in woodland, and in gardens and parks on flowerbeds mulched with woodchips or shredded bark. Occasionally found in cellars in old houses and on decayed straw bales.
D: BFF3: 25-26, NM2: 271 **D+I:** B&K4: 290 361 **I:** FlDan4: 126 D & E (as *Naucoria arvalis*), Bon: 263, C&D: 361
Occasional but widespread. Rather rare in natural habitat but increasing in parks and gardens, at least in southern England, due to use of woodchip mulch on flowerbeds. A distinctive species with diagnostic digitate cheilocystidia, and often arising from a small black sclerotium.

attenuata (Kühner) P.D. Orton, *Trans. Brit. Mycol. Soc.* 43: 174 (1960)
  Agrocybe firma var. *attenuata* Kühner, *Schweiz. Z. Pilzk.* 31: 150 (1953)
E: !
H: On woody debris.

D: NCL3: 184, BFF3: 30
A single collection, from East Norfolk (Surlingham Wood) in 1957. Possibly only a small form of *A. firma* (Peck) Singer, otherwise unknown in Britain.

brunneola (Fr.) Watling, *Biblioth. Mycol.* 82: 29 (1981)
  Agaricus ombrophilus var. *brunneolus* Fr., *Ic. Hymenomyc.* 2: pl. 2 (1884)
  Togaria ombrophila var. *brunneola* (Fr.) W.G. Sm., *Syn. Brit. Bas.*: 122 (1908)
  Pholiota brunneola (Fr.) J.E. Lange, *Fl. agaric. danic.* 5, *Taxonomic Conspectus*: VII (1940)
  Mis.: *Agrocybe ombrophila* sensu auct.Brit.
  Mis.: *Pholiota ombrophila* sensu auct.Brit
E: ? **S:** ?
H: On soil in woodland and shrubberies.
D: NM2: 270 **D+I:** BFF3: 28-29
Rarely recorded and no material traced in herbaria. Possibly only a large-spored form of *Agrocybe erebia* but this remains unresolved. N.B. *Pholiota brunneola* sensu Lange is a different species with small spores, as yet unknown in Britain.

cylindracea (DC.) Maire, *Mem. de la Soc. des Sci. nat. du Maroc* 45: 106 (1938)
  Agaricus cylindraceus DC., *Fl. Fr.* 5: 51 (1815)
  Pholiota cylindracea (DC.) Gillet, *Hyménomycètes*: 439 (1876)
  Agaricus aegerita V. Brig., *Fasc. funghi litograf. Nap.*: t. 1 (1824)
  Pholiota aegerita (V. Brig.) Quél., *Mém. Soc. Émul. Montbéliard, Sér. 2*, 5: 164 (1872)
  Agrocybe aegerita (V. Brig.) Fayod, *Ann. Sci. Nat., Bot., sér. 7*, 9: 358 (1889)
  Agaricus capistratus Cooke, *J. Bot.* 1: 65 (1863)
  Pholiota capistrata (Cooke) Sacc., *Syll. fung.* 5: 743 (1887)
  Agaricus leochromus Cooke, *J. Bot.* 1: 65 (1863)
  Pholiota leochroma (Cooke) Sacc., *Syll. fung.* 5: 742 (1887)
  Mis.: *Pholiota pudica* sensu Rea (1922)
  Mis.: *Agaricus pudicus* sensu Cooke [*Ill. Brit. fung.* 383 (352) Vol. 3 (1884)]
E: o **S:** ! **W:** ! **O:** Channel Islands: !
H: On wood, usually arising from lesions or knot holes on trunks and branches of living deciduous trees, most frequently *Populus nigra* but also *Acer pseudoplatanus*, *Sambucus nigra* and *Ulmus* spp. Rarer hosts include *Acer negundo*, *Aesculus*, *Carpinus*, *Fagus*, *Fraxinus*, *Populus italica*, *Robinia pseudoacacia*, *Quercus* and *Salix* spp.
D: BFF3: 27, NM2: 269 **D+I:** Ph: 170, B&K4: 290 362 **I:** Bon: 263, C&D: 361
Occasional in southern and south western England, rarely reported elsewhere. The epithet *cylindracea* is orthographically correct (as used by De Candolle) rather than the later and incorrect Friesian *cylindrica*.

erebia (Fr.) Singer, *Schweiz. Z. Pilzk.* 17(7): 97 (1939)
  Agaricus erebius Fr., *Syst. mycol.* 1: 246 (1821)
  Pholiota erebia (Fr.) Gillet, *Hyménomycètes*: 86 (1874)
  Togaria erebia (Fr.) W.G. Sm., *Syn. Brit. Bas.*: 122 (1908)
  Agaricus denigratus Pers., *Syn. meth. fung.*: 267 (1801)
  Armillaria denigrata (Pers.) P. Kumm., *Führ. Pilzk.*: 135 (1871)
  Agaricus jecorinus Berk. & Broome [*nom. illegit.*, non *A. jecorinus* With. (1792)], *Ann. Mag. Nat. Hist., Ser. 2* [*Notices of British Fungi no. 328*] 2: 260 (1848)
  Mis.: *Agaricus dissimulans* sensu Cooke (1890) non Rea (1922)
E: o **S:** ! **W:** ! **NI:** ! **ROI:** ! **O:** Isle of Man: !
H: On loam soil in woodland, hedgebanks, and gardens, associated with a wide range of deciduous trees, often abundant in scrub with *Acer pseudoplatanus,* sometimes amongst nettles. Rarely with coniferous trees.
D: BFF3: 29, NM2: 270 **D+I:** Ph: 168-169, B&K4: 292 364 **I:** FlDan3: 105B, Bon: 263, C&D: 360
Occasional but widespread. Usually found singly or in small groups but sometimes occurs in abundance where woodland is felled or scrub cleared.

molesta (Lasch) Singer, *Sydowia* 30: 197 (1978)
  Agaricus molestus Lasch, *Linnaea* 3: 421 (1828)
  Pholiota dura var. *xanthophylla* Bres., *Fungi trident.* 2: 52 (1892)
  Agrocybe dura var. *xanthophylla* (Bres.) P.D. Orton, *Trans. Brit. Mycol. Soc.* 43(2): 174 (1960)
  Agrocybe molesta var. *xanthophylla* (Bres.) Bon & Courtec., *Doc. Mycol.* 18(69): 37 (1987)
  Mis.: *Agrocybe dura* sensu NCL and auct. mult.
E: ! **S:** !
H: On soil in scrub woodland, roadside verges, gardens, grassland and rarely cultivated fields.
D: BFF3: 18-19, NM2: 270 **D+I:** B&K4: 290 363 (as *Agrocybe dura*) **I:** C&D: 361
A spring or summer species, fruiting between May and July, exceptionally later.

paludosa (J.E. Lange) Kühner & Romagn., *Flore Analytique des Champignons Supérieurs*: 341 (1953)
  Pholiota praecox var. *paludosa* J.E. Lange, *Dansk. Bot. Ark.* 2(11): 7 (1921)
E: o **S:** o **ROI:** !
H: On wet soil with *Juncus* and *Carex* spp, in marshy or swampy areas, or with large grasses in wet fields and meadows.
D: BFF3: 14-15, NM2: 270 **D+I:** B&K4: 294 367
Rarely reported, but apparently widespread. Synonymised with *Agrocybe praecox* (or treated as a variety) by some authorities, but differing in habitat and in the more gracile basidiomes.

pediades (Fr.) Fayod, *Ann. Sci. Nat., Bot.*, sér. 7, 9: 358 (1889)
  Agaricus pediades Fr., *Syst. mycol.* 1: 290 (1821)
  Naucoria pediades (Fr.) P. Kumm., *Führ. Pilzk.*: 78 (1871)
  Naucoria arenaria Peck, *Bull. New York State Mus. Nat. Hist.* 157: 29 (1912) [1911]
  Agrocybe arenaria (Peck) Singer, *Nova Hedwigia Beih.* 29: 227 (1969)
  Agaricus arenicola Berk., *J. Bot.* 2: 511 (1843)
  Agrocybe arenicola (Berk.) Singer, *Beih. Bot. Centralbl.* 56: 169 (1936)
  Agaricus pusillus Schaeff. [non *A. pusillus* Fr. (1821)], *Fung. Bav. Palat. Icones*: t. 203 (1800)
  Agaricus semiorbicularis Bull., *Herb. France*: pl. 422 (1789)
  Naucoria semiorbicularis (Bull.) Quél., *Mém. Soc. Émul. Montbéliard, Sér. 2*, 5: 439 (1875)
  Agrocybe semiorbicularis (Bull.) Fayod, *Ann. Sci. Nat., Bot.*, sér. 7, 9: 358 (1889)
  Naucoria subpediades Murrill, *Lloydia* 5: 150 (1942)
  Agrocybe subpediades (Murrill) Watling, *Kew Bull.* 31: 592 (1977)
  Agaricus temulentus Fr., *Syst. mycol.* 1: 268 (1821)
  Naucoria temulenta (Fr.) P. Kumm., *Führ. Pilzk.*: 77 (1871)
  Agrocybe temulenta (Fr.) Singer, *Beih. Bot. Centralbl.* 56: 167 (1936)
  Mis.: *Agrocybe arvalis* sensu auct.
  Mis.: *Naucoria vervacti* sensu Ricken [*Blätterpilze Deutschl.*: 210 (1912)]
E: ! **S:** !
H: On soil amongst grass in fields, grassland, on dunes and in gardens. Several recent records from soil enriched with woodchips and sawdust in parkland.
D: BFF3: 19-22, NM2: 271 **D+I:** B&K4: 294 369 (as *Agrocybe semiorbicularis*) **I:** FlDan4: 126H (as *Naucoria semiorbicularis*), Bon: 263, C&D: 360
British records are uncertain as many are unsubstantiated with voucher material. We follow Nauta (1987: 29) and NM2 in recognising only a single, widespread and variable species covering four species recognised in BFF3.

praecox (Pers.) Fayod, *Ann. Sci. Nat., Bot.*, sér. 7, 9: 358 (1889)
  Agaricus praecox Pers., *Comm.*: 89 (1800)
  Pholiota praecox (Pers.) P. Kumm., *Führ. Pilzk.*: 85 (1871)
  Togaria praecox (Pers.) W.G. Sm., *Syn. Brit. Bas.*: 124 (1908)
  Agaricus gibberosus Fr., *Epicr. syst. mycol.*: 163 (1838)

Agrocybe gibberosa (Fr.) Fayod, *Ann. Sci. Nat., Bot.*, sér. 7,
9: 358 (1889)
    Agaricus togularis Pers. [*nom. illegit.*, non *A. togularis* Bull.
    (1793)], *Syn. meth. fung.*: 262 (1801)
    Mis.: *Pholiota togularis* sensu Gillet (1874)
E: c **S:** o **W:** o **NI:** ! **ROI:** !
H: On soil in grassy areas such as field margins, woodland
edges, cleared areas in woodland, gardens, flowerbeds and
parkland.
D: BFF3: 13-14 (as *Agrocybe gibberosa*), BFF3: 15-17, NM2:
270 **D+I:** Ph: 168-169, B&K4: 294 368 **I:** Bon: 263, C&D: 361
Common and widespread, at least in England, somewhat less so
elsewhere. Morphologically variable and possibly a species
complex. *A. gibberosa*, treated as distinct in BFF3, has been
found to be largely interfertile.

putaminum (Maire) Singer, *Beih. Bot. Centralbl.* 56: 167
(1936)
    Naucoria putaminum Maire, *Ann. Mycol.* 11: 350 (1913)
E: !
H: On soil, usually in parkland or large gardens, on flowerbeds
mulched with woodchips and shredded bark, and recently on
roadsides.
D+I: Myc12(2): 60
A large and distinctive species, until recently known in Britain
only from Kew Gardens where it is an abundant coloniser of
mulched flowerbeds. Now spreading rapidly and found in
several localities in Surrey, West Kent and West Norfolk.
Originally described from France (growing on discarded *Prunus*
stones).

rivulosa Nauta, *Persoonia* 18(2): 272 (2003)
E: !
H: On decayed wood. British collection on a pile of woodchips,
comprised mostly of *Acer pseudoplatanus*, in a garden.
A single collection from Derbyshire (Longsden, near Leek) in
2004.

vervacti (Fr.) Singer, *Beih. Bot. Centralbl.* 56: 167 (1936)
    Agaricus vervacti Fr., *Syst. mycol.* 1: 263 (1821)
    Naucoria vervacti (Fr.) P. Kumm., *Führ. Pilzk.*: 77 (1871)
E: ! **S:** ! **ROI:** !
H: On soil, amongst grass in meadows, fields and on lawns.
D: BFF3: 23-24, NM2: 270 **D+I:** B&K4: 296 371, Myc12(2): 60
  I: FlDan4: 126G (as *Naucoria vervacti*), Bon: 263, C&D: 361
Rarely reported. Often confused with *Agrocybe pediades*
(usually reported as *A. semiorbicularis*), thus records
unsubstantiated with voucher material are dubious.

ALEURODISCUS J. Schröt., in Cohn, *Krypt.-Fl. Schlesien* 3(1): 429 (1888)
Stereales, Aleurodiscaceae
    Acanthophysium (Pilát) G. Cunn., *Bull. New Zealand Dept.
    Sci. Industr. Res.* 145: 150 (1963)
    Acanthobasidium Oberw., *Sydowia* 19: 45 (1965)
Type: *Aleurodiscus amorphus* (Pers.) J. Schröt.
The treatment here follows Núñez & Ryvarden (1997)

amorphus (Pers.) J. Schröt., *Krypt.-Fl. Schlesien* 3(1): 429
(1888)
    Peziza amorpha Pers., *Syn. meth. fung.*: 657 (1801)
    Thelephora amorpha (Pers.) Fr., *Elench. fung.* 1: 183 (1828)
    Corticium amorphum (Pers.) Fr., *Epicr. syst. mycol.*: 559
    (1838)
    Lachnea amorpha (Pers.) Gillet, *Champ. France discomyc.*:
    89 (1881)
    Cyphella amorpha (Pers.) Quél., *Enchir. fung.*: 215 (1886)
    Stereum amorphum (Pers.) E.H.L. Krause, *Arch.
    FreundeNaturgesch. Mecklenburg., Suppl.*: 81 (1929)
E: ! **S:** !
H: On dead and dying attached branches of conifers such as
Abies and *Picea* spp.
D: NM3: 181, Aleur: 39 **D+I:** CNE2: 62-63, B&K2: 78 45 **I:**
C&D: 139, Kriegl1: 145
Rarely reported. Sometimes mistaken for a discomycete,
basidiomes being small and discoid. Collections from

widespread locations in Scotland, and a few from England
(Cumberland, East Kent, Herefordshire and South Devon).
Frequently parasitised by *Tremella simplex*.

apricans Bourdot, *Rev. Sci. Bourbonnais Centre France* 23: 7
(1910)
    Corticium apricans (Bourdot) Sacc. & Trotter, *Syll. fung.* 21:
    403 (1912)
    Acanthophysium apricans (Bourdot) G. Cunn., *Bull. New
    Zealand Dept. Sci. Industr. Res.* 145: 155 (1963)
O: Channel Isles: !
H: On dead stem of *Rubus fruticosus* agg. (British material).
D: Aleur: 45
A single collection from the Channel Islands (Guernsey: Bec du
Nez) in 1948 but not reported since.

aurantius (Pers.) J. Schröt., in Cohn, *Krypt.-Fl. Schlesien* 3(1):
429 (1888)
    Corticium aurantium Pers., *Neues Mag. Bot.* 1: 111 (1794)
    Thelephora aurantia (Pers.) Pers., *Syn. meth. fung.*: 576
    (1801)
    Hypochnus marchandii (Pat.) Pat., *Essai taxon.*: 65 (1900)
    Thelephora rubi Lib., *Pl. Crypt. Ardienn.*: 323 (1837)
    Corticium rubi (Lib.) Cooke, *Grevillea* 20: 12 (1891)
E: o **S:** o **W:** ! **NI:** ! **ROI:** ! **O:** Channel Islands: ! Isle of Man: !
H: On dead woody stems of *Rosa* spp and *Rubus fruticosus*
agg. Rarely on other substrata but reported on *Ulmus* spp,
Picea abies and *Taxus*.
D: NM3: 182, Aleur: 49 **D+I:** CNE2: 64-65
Occasional but widespread. Mostly recorded on brambles,
usually growing at the internodes of dead or dying stems.

botryosus Burt, *Ann. Missouri Bot. Gard.* 5: 198 (1918)
E: !
H: On dead woody stems. British material on *Lonicera
periclymenum* and *Rosa* spp.
D: Aleur: 55
Three collections from South Devon (Torquay) in 2004.

delicatus Wakef., *Trans. Brit. Mycol. Soc.* 35(1): 44 (1952)
    Acanthophysium delicatum (Wakef.) Parmasto, *Eesti N.S.V.
    Tead. Akad. Toimet., Biol.* 16(4): 378 (1967)
    Acanthobasidium delicatum (Wakef.) Jülich, *Persoonia* 10(3):
    335 (1979)
E: !
H: In fens or marshland, on dead and dying culms of *Cladium
mariscus*.
D: Aleur: 69
Described from East Norfolk (Wheatfen Broad) in 1948 but not
recollected in Britain since then. Perhaps overlooked as the
habitat is infrequently visited by specialists in corticioid fungi.

norvegicus J. Erikss. & Ryvarden, *Norweg. J. Bot.* 20: 10
(1973)
    Acanthobasidium norvegicum (J. Erikss. & Ryvarden) Boidin,
    Lanq., Cand., Gilles & Hugueney, *Bull. Soc. Mycol. France*
    101(4): 341 (1986) [1985]
E: ! **S:** ! **O:** Isle of Man: !
H: On dead and dying stems and attached twigs of *Calluna* on
heath and moorland.
D: Aleur: 107, NM3: 181 **D+I:** CNE2: 76-77, Myc15(2): 74
British collections from England (West Sussex: Iping Common),
Scotland (Wigtown) and the Isle of Man.

phragmitis (Boidin, Lanq., Gilles, Cand. & Hugueney) Núñez &
Ryvarden, *Syn. Fungorum* 12: 123 (1997)
    Acanthobasidium phragmitis Boidin, Lanq., Gilles, Cand. &
    Hugueney, *Bull. Soc. Mycol. France* 101(4): 345 (1986)
    [1985]
E: !
H: On dead standing stems of *Phragmites australis* in reedbeds
at, or just above, water level. Also known on dead stems and
culms of *Arundinaria* sp.
D: Myc10(1): 19, Aleur: 123
Known from South Devon (Slapton Ley) and Surrey (Oxshott
Heath).

wakefieldiae Boidin & Beller, *Bull. Soc. Mycol. France* 82: 561 (1966)
  Mis.: *Aleurodiscus oakesii* sensu Wakefield (1952)
E: ! **S:** ! **W:** ! **O:** Isle of Man: !
H: On attached or recently fallen branches of deciduous trees. Reported on *Alnus*, *Fagus* and *Quercus* spp.
D: Aleur: 145
Rarely reported. Known from England (South Devon, South Hampshire and West Cornwall), Scotland (Peebleshire and Kirkudbrightshire), Wales (Caernarvonshire) and the Isle of Man. Probably genuinely rare since the bright pink, orbicular basidiomes would be difficult to overlook.

AMANITA Pers., *Tent. disp. meth. fung.*: 65 (1797)
Agaricales, Amanitaceae
  Vaginata Gray, *Nat. arr. Brit. pl.* 1: 601 (1821)
  Amanitopsis Roze, *Bull. Soc. Bot. France* 23: 50 (1876)
  Lepidella E.-J. Gilbert [*nom. illegit.*, non *Lepidella* Tiegh. (1911) (*Loranthaceae*)], *Bull. Soc. Mycol. France* 41: 303 (1925)
  Aspidella E.-J. Gilbert, in Bresadola, *Icon. mycol.* 27: 73 (1940)
Type: *Amanita muscaria* (L.) Lam.

argentea Huijsman, *Bull. Soc. Mycol. France* 75(1): 14 (1959)
  Amanita mairei var. *argentea* (Huijsman) Bon & Contu, *Doc. Mycol.* 15(59): 53 (1985)
  Mis.: *Amanita mairei* sensu auct. mult.
E: ! **W:** !
H: On soil, associated with *Quercus* spp, in open woodland and occasionally with solitary trees in parkland.
D: Reid7: 507-509 (as *Amanita mairei*) **D+I:** B&K4: 142 141 (as *A. mairei*), Galli: 88-89
Rarely reported. Known from a few widely scattered locations in England (Oxfordshire to South Devon) and Wales (Glamorganshire and Radnorshire).

battarrae (Boud.) Bon, *Doc. Mycol.* 16(61): 16 (1985)
  Amanitopsis battarrae Boud., *Bull. Soc. Mycol. France* 18: 272 (1902)
  Amanita inaurata var. *umbrinolutea* Gillet, *Hyménomycètes*: 42 (1874)
  Amanita umbrinolutea (Gillet) Bataille, *Bull. Soc. Mycol. France* 26: 139 (1910)
  Amanitopsis umbrinolutea (Gillet) E-J. Gilbert, *Bull. Soc. Mycol. France* 44: 164 (1928)
  Amanita umbrinolutea var. *flaccida* D.A. Reid, *Notes Roy. Bot. Gard. Edinburgh* 44(3): 517 (1987)
E: ! **S:** ! **W:** ! **NI:** !
H: On soil, usually with *Quercus* spp, in mixed deciduous woodland or pure oak woodland, but also reported with *Corylus* and *Fraxinus*.
D: Reid7: 515-518 (as *Amanita umbrinolutea*), NM2: 198 **D+I:** Galli: 74-76 (as *A. umbrinolutea*) **I:** Bon: 295, C&D: 275
Rarely reported. Known from England (Oxfordshire, South Devon and Surrey) with single records from Scotland, Wales and Northern Ireland. The illustration and description in B&K4: 138 (pl. 135) is misdetermined *A. submembranacea* [*Persoonia* 16: 262 (1996)].

ceciliae (Berk. & Broome) Bas, *Persoonia* 12(2): 192 (1983)
  Agaricus ceciliae Berk. & Broome, *Ann. Mag. Nat. Hist., Ser. 2* [*Notices of British Fungi no. 663*] 13: 396 (1854)
  Amanita inaurata Gillet, *Hyménomycètes*: 41 (1874)
  Amanitopsis inaurata (Gillet) Fayod, *Ann. Sci. Nat., Bot.*, sér. 7, 9: 317 (1889)
  Mis.: *Amanita strangulata* sensu auct. mult.
E: ! **S:** ! **W:** ! **NI:** ! **ROI:** ! **O:** Isle of Man: !
H: On soil, in open deciduous woodland or scrub, or amongst grass under solitary trees in parkland. Most frequently associated with *Quercus* spp or *Fagus*. Also known (rarely) with *Abies* and *Cedrus* spp.

D: NM2: 197 **D+I:** Ph: 22-23 (as *Amanita inaurata*), B&K4: 140 136, Galli: 96-97 **I:** FlDan1: 7C (as *A. strangulata*), Bon: 295, C&D: 273
Rarely reported but apparently widespread. Most frequently reported from southern England, but nowhere common and often occurring singly.

citrina (Schaeff.) Pers., *Tent. disp. meth. fung.*: 71 (1797)
  Agaricus citrinus Schaeff., *Fung. Bavar. Palat. nasc.* 1: t. 20 (1762)
  Amanita mappa var. *citrina* (Schaeff.) Rea, *Brit. basidiomyc.*: 100 (1922)
  Amanita citrina β *mappalis* Gray, *Nat. arr. Brit. pl.* 1: 599 (1821)
  Agaricus mappus Batsch, *Elench. fung.*: 57 (1783)
  Amanita mappa (Batsch) Bertill., in Dechambre, *Dict. Encyclop. Sci. Méd. Ser. 1*, 3: 500 (1866)
E: c **S:** c **W:** c **NI:** c **ROI:** c **O:** Channel Islands: !
H: On soil, usually associated with *Betula*, *Fagus* and *Quercus* spp, in mixed deciduous woodland. Rarely with conifers.
D: NM2: 195 **D+I:** Ph: 20-21, B&K4: 146 146, Galli: 204-207 **I:** FlDan1: 2B (as *A. mappa*), Bon: 299, C&D: 281, SV48: 21 (as *A. mappa*)
Common and widespread. Cooke 4 (4) Vol. 1 (1881) is atypical. English name = 'False Death Cap'.

citrina var. alba (Pers.) Quél. & Bataille, *Fl. Mon. des Amanites*: 36 (1902)
  Amanita bulbosa var. *alba* Pers., *Traité champ. comest.*: 179 (1818)
  Amanita mappa var. *alba* (Pers.) Rea, *Brit. basidiomyc.*: 100 (1922)
  Agaricus bulbosus Schaeff. [*nom. illegit.*, non *A. bulbosus* Bolton (1791); non Sowerby (1798)], *Fung. Bavar. Palat. nasc.* 3: t. 241 (1770)
E: c **S:** c **W:** c **NI:** c **ROI:** c
H: On soil in mixed deciduous woodland, often growing with *A. citrina*.
D+I: Ph: 20-21, B&K4: 146 147
Common and widespread, but less frequent than *A. citrina*.

crocea (Quél.) Singer, *Lilloa* 22: 386 (1951) [1949]
  Amanita vaginata var. *crocea* Quél., *Rev. Sci. Bourbonnais Centre France* 11(123): 52 (1898)
  Amanitopsis crocea (Quél.) E.-J. Gilbert, *Bull. Soc. Mycol. France* 44(2): 161 (1928)
E: o **S:** o **W:** o **NI:** ! **ROI:** !
H: On acidic soil in mixed deciduous woodland, usually associated with *Betula* spp, but occasionally reported with *Fagus*.
D: NCL3: 184, NM2: 197 **D+I:** Ph: 22-23, B&K4: 140 137, Galli: 70-71 **I:** C&D: 275
Occasional but widespread.

echinocephala (Vittad.) Quél., *Mém. Soc. Émul. Montbéliard, Sér. 2*, 5: 321 (1872)
  Agaricus echinocephalus Vittad., *Descr. fung. mang.*: 346 (1835)
  Lepidella echinocephala (Vittad.) E.-J. Gilbert [*comb. inval.*], *Bull. Soc. Mycol. France* 41: 304 (1925)
  Aspidella echinocephala (Vittad.) E.-J. Gilbert, in Bresadola, *Icon. Mycol.* 27(1): 79 (1940)
  Amanita strobiliformis var. *aculeata* Quél., *Enchir. fung.*: 3 (1886)
  Amanita aculeata (Vittad.) Quél., *Fl. Mycol. Fr.*: 305 (1888)
  Mis.: *Amanita solitaria* sensu auct. mult.
E: !
H: On soil (often calcareous) in open deciduous woodland, usually associated with *Betula* spp or *Fagus*. Occasionally with solitary trees in parkland or open grassland.
D+I: Ph: 21, B&K4: 154 158 (as *A. solitaria*), Galli: 166-169 **I:** Bon: 299, C&D: 279
Most often reported from southern England but confused with *A. strobiliformis*. Both species have often been determined as *A. solitaria* (a *nomen confusum*) thus the true distribution and frequency are uncertain.

eliae Quél., *Mém. Soc. Émul. Montbéliard, Sér. 2,* 5: 230 (1872)
E: ! W: ?
H: On soil in open deciduous woodland.
D: NM2: 196 **D+I:** B&K4: 148 148, Galli: 128-129 **I:** Bon: 297, C&D: 277
Rarely reported. Known from Buckinghamshire, East & West Kent and West Sussex. Reported elsewhere in England and from Wales, but unsubstantiated with voucher material.

excelsa (Fr.) Bertill., in Dechambre, *Dict. Encyclop. Sci. Méd.* Ser. 1, 3: 499 (1866)
  Agaricus excelsus Fr., *Syst. mycol.* 1: 17 (1821)
  Amanita ampla Pers., *Syn. meth. fung.* 2: 255 (1801)
  Agaricus excelsus var. *cariosus* Fr., *Epicr. syst. mycol.*: 8 (1838)
  Amanita cariosa (Fr.) Gillet, *Hyménomycètes*: 48 (1874)
E: ! S: ! W: !
H: In deciduous woodland, mostly on acid soil.
D+I: B&K4: 148 149 **I:** Galli: 143 (lower)
Apparently less common than var. *spissa* with which it has been much confused. It is distinguished by its generally more slender fruitbodies and lack of raphanoid odour.

excelsa var. spissa (Fr.) Neville & Poumerat, *Fungi Europaei* 9, *Amaniteae*: 721 (2004)
  Agaricus spissus Fr., *Epicr. syst. mycol.*: 9 (1838)
  Amanita spissa (Fr.) Opiz, *Seznam*: 114 (1852)
E: c S: c W: c NI: c ROI: c O: Channel Islands: !
H: On soil in mixed deciduous woodland, usually associated with *Quercus* spp, less often with *Betula* spp and *Fagus* and rarely with conifers.
D: NM2: 195 **D+I:** Ph: 17 (as *A. excelsa*), Galli: 140-141 **I:** Bon: 297, C&D: 277
Common and widespread.

flavescens (Gilbert & Lundell) Contu, *Pagine Botaniche* 12: 12 (1988)
  Amanitopsis vaginata var. *flavescens* Gilbert & Lundell, in Gilbert, *Amanites du Sud-Ouest de la France* 2: 217 (1941)
S: !
H: Associated with deciduous trees.
Known from a single collection from Aberdeenshire (Morrone Birchwood) in 2002.

franchetii (Boud.) Fayod, *Ann. Sci. Nat., Bot.,* sér. 7, 9: 316 (1889)
  Amanita aspera var. *franchetii* Boud., *Bull. Soc. Bot. France* 28: 91 (1881)
  Mis.: *Amanita aspera* sensu auct. mult.
E: ! W: !
H: On soil in mixed deciduous woodland, often in open glades and usually associated with *Quercus* spp.
D: NM2: 194 **D+I:** Ph: 18 (as *A. aspera*), B&K4: 148 150, Galli: 144-147 **I:** FlDan1: 6A (as *A. aspera* var. *francheti*), FlDan1: 5B (as *A. aspera*), Bon: 297
Rarely reported. Known from England (Herefordshire and Surrey) and Wales (Denbighshire).

friabilis (P. Karst.) Bas, *Bull. Mens. Soc. Linn. Lyon* 43: 18 (1974)
  Amanitopsis vaginata var. *friabilis* P. Karst., *Bidrag Kännedom Finlands Natur Folk* 32: 547 (1879)
  Amanitopsis friabilis (P. Karst.) Sacc., *Syll. fung.* 5: 22 (1887)
E: ! W: !
H: On soil, always associated with *Alnus* in wet deciduous woodland, marshy areas and fen-carr.
D: Reid7: 503-505, NM2: 196 **D+I:** B&K4: 140 138, Galli: 94-95 **I:** C&D: 273
Rarely reported. Known from England (East Kent) and Wales (Caernarvonshire and Monmouthshire). Reported from South Devon and possibly also this species in West Sussex.

fulva (Schaeff.) Fr., *Observ. mycol.* 1: 2 (1815)
  Agaricus fulvus Schaeff., *Fung. Bavar. Palat. nasc.* 1: 41 (1762)
  Amanitopsis vaginata var. *fulva* (Schaeff.) Sacc., *Syll. fung.* 5: 21 (1887)

  Amanitopsis fulva (Schaeff.) Fayod, *Ann. Sci. Nat., Bot.,* sér. 7, 9: 317 (1889)
E: c S: c W: c NI: c ROI: c O: Channel Islands: !
H: On acidic soil in open deciduous woodland, often associated with *Betula* spp and also reported with *Fagus* and *Quercus* spp. Also on heathland amongst *Calluna* and *Pteridium*.
D: NM2: 197 **D+I:** Ph: 23, B&K4: 142 139, Galli: 72-73 **I:** FlDan1: 7B (as *Amanita vaginata* var. *fulva*), Bon: 295, C&D: 275
Common and widespread. English name = 'Tawny Grisette'.

gemmata (Fr.) Bertill., in Dechambre, *Dict. Encyclop. Sci. Méd. Ser. 1,* 3: 496 (1866)
  Agaricus gemmatus Fr., *Epicr. syst. mycol.*: 12 (1838)
  Amanitopsis gemmata (Fr.) Sacc., *Syll. fung.* 5: 25 (1887)
  Agaricus adnatus W.G. Sm. [*nom. illegit.*, non *A. adnatus* Huds. (1778)], *Myc. Illust.*: pl. 21 (1870)
  Amanitopsis adnata (W.G. Sm.) Sacc., *Syll. fung.* 5: 24 (1887)
  Amanita junquillea Quél., *Bull. Soc. Bot. France* 23: 324 (1876)[1877]
E: ! S: ! NI: ! ROI: ? O: Channel Islands: !
H: On sandy soil in open mixed woodland, often associated with *Pinus sylvestris* and *Betula* spp. Also reported with *Corylus, Fagus, Quercus* spp and *Taxus*.
D: NM2: 196 **D+I:** Ph: 22-23, B&K4: 150 151, Galli: 122-125 (as *A. junquillea*) **I:** FlDan1: 2C (as *A. junquillea*), Bon: 297 (as *A. junquillea*), C&D: 277, Cooke 18 (35) Vol. 1 (1881) (as *Agaricus adnatus*)
Rarely reported, but apparently widespread.

inopinata D.A. Reid & Bas, *Notes Roy. Bot. Gard. Edinburgh* 44(3): 506 (1987)
E: r
H: On soil in man-made habitats such as cemeteries, churchyards or gardens, usually associated with *Chamaecyparis lawsoniana* or *Taxus*. Also reported with 'spruce' and one collection in herb. K supposedly with *Aesculus*.
D: Reid7: 506-507 **I:** FM1(2): 39
Rare and south-eastern. Known from Berkshire, East & West Kent, East Suffolk, East & West Sussex, South Essex and Surrey. A distinctive species, probably an introduction. Also known from the Netherlands and New Zealand, but may also be introduced there.

lividopallescens (Gillet) Seyot, *Les Amanites et la tribu des Amanitées*: 67 (1931)
  Amanita vaginata var. *lividopallescens* Gillet, *Hyménomycètes (Plates)*: pl. 24 (1890)
  Amanitopsis lividopallescens (Gillet) Boud., *Icon. Mycol.* 1: pl. 6 (1905)
E: ! W: !
H: On soil in open mixed deciduous woodland, usually associated with *Betula, Quercus* or *Salix* spp. Rarely in grass with scattered trees in parkland.
D: NM2: 197 **D+I:** B&K4: 142 140 **I:** Bon: 295, C&D: 275
Very rarely reported or collected.

muscaria (L.) Lam., *Encyclop.* 1: 111 (1783)
  Agaricus muscarius L., *Sp. pl.*: 1172 (1753)
  Agaricus nobilis Bolton, *Hist. fung. Halifax* 2: 46 (1788)
  Amanita circinnata Gray, *Nat. arr. Brit. pl.* 1: 600 (1821)
  Amanita muscaria β *minor* Gray, *Nat. arr. Brit. pl.* 1: 600 (1821)
  Agaricus puella Batsch, *Elench. fung.*: 59 (1783)
  Amanita muscaria var. *puella* (Batsch) Pers., *Syn. meth. fung.*: 253 (1801)
E: c S: c W: c NI: c ROI: c O: Channel Islands: c Isle of Man: !
H: On acidic soil in woodland or with scattered trees in parkland or on heathland. Mainly associated with *Betula* spp. Rarely reported with other deciduous trees and conifers.
D: NM2: 196 **D+I:** Ph: 15, B&K4: 150 152, Galli: 112-117 **I:** Sow3: 286 (as *Agaricus muscarius*), FlDan1: 3B, Bon: 297, C&D: 277, Galli: 45, Cooke 5 (117) Vol. 1 (1881).
Very common and widespread. English name = 'Fly Agaric'.

muscaria *var.* aureola (Kalchbr.) Quél., *Enchir. Fung.*: 3 (1886)

Agaricus aureola Kalchbr., *Icon. select. Hymenomyc. Hung.*: 9 (1873)

Amanita aureola (Kalchbr.) Sacc., *Syll. fung.* 5: 12 (1887)

Amanita muscaria f. *aureola* (Kalchbr.) J.E. Lange, *Dansk. Bot. Ark.* 2(3): 9 (1915)

E: ! **S:** ?

H: On soil in woodland. British material associated with *Picea sitchensis*.

D+I: Galli: 120-121 **I:** FlDan1: 3C (as *A. muscaria* f. *aureola*)

A single collection (in herb. K) from Cheviotshire (Holystone Forest) in 2001. Reported from Scotland (Morayshire: Elgin) in 1912 but no material has been traced.

muscaria *var.* formosa (Pers.) Bertill., in Dechambre, *Dict. Encyclop. Sci. Med. Ser. 1*, 3: 496 (1866)

Amanita muscaria β *formosa* Pers., *Observ. mycol.* 2: 37 (1799)

Agaricus muscarius var. *formosus* (Pers.) Fr., *Monogr. Amanit. Sueciae*: 7 (1854)

Amanita muscaria f. *formosa* (Pers.) Veselý, *Ann. Mycol.* 31(4): 253 (1933)

E: !

H: On soil, in similar habitat to *A. muscaria*.

Rarely reported.

nivalis Grev., *Scott. crypt. fl.* 1(4): pl. 18 (1822)

Amanitopsis nivalis (Grev.) Sacc., *Syll. fung.* 5: 22 (1887)

Amanita vaginata var. *nivalis* (Grev.) E.-J. Gilbert, *Le Genre Amanita Persoon*: 141 (1918)

E: ! **S:** ! **W:** !

H: On soil in montane or boreal habitat, always associated with *Salix herbacea*.

D+I: B&K4: 144 142, Galli: 62-63 **I:** C&D: 275

Collections from scattered sites in Scotland, and single locations in England (Cumberland) and Wales (Caernarvonshire). Strictly montane and associated with willows but the name was misapplied by Rea to any white *Amanitopsis* hence there are several old reports from southern England (South Hampshire and West Sussex).

olivaceogrisea Kalamees, *Conspectus florum agaricalium fungorum (Agaricales s.l.) Lithuaniae, Latviae et Estoniae (materies 1778-1984 annorum)*: 45 (1986)

E: ! **W:** !

H: On loam soil in deciduous woodland. English collections associated with *Betula* sp. or *Corylus*.

D+I: FM 2(3): 99-100

Collections from England (South Somerset: Bird's Hill, near Monksilver in 1999 and Surrey: Limpsfield Chart in 2004) and from Wales (Caernarvonshire: Tan-y-Bwlch in 2001).

ovoidea (Bull.) Link, *Handbuch Gewächse* 3: 273 (1833)

Agaricus ovoideus Bull., *Herb. France*: pl. 364 (1788)

E: !

H: On calcareous soil in deciduous woodland. British material is associated with *Quercus* spp.

D: Reid7: 509-511 **D+I:** Galli: 180-183 **I:** Bon: 299, C&D: 279

Known from North Wiltshire and Isle of Wight. Basidiomes are usually large and unlikely to be overlooked, thus probably genuinely rare in Britain.

pachyvolvata (Bon) Krieglst., *Beih. Z. Mykol.* 5: 191 (1984)

Amanitopsis pachyvolvata Bon, *Doc. Mycol.* 8(29): 36 (1978)

E: ! **S:** ?

H: On soil in deciduous woodland.

D: Reid7: 511-513 **D+I:** Galli: 64-65

Known from England (East Gloucestershire and Mid-Lancashire). Reported from Scotland (Berwickshire) but unsubstantiated with voucher material.

pantherina (DC.) Krombh., *Naturgetr. Abbild. Schwämme*: 24 (1836)

Agaricus pantherinus DC., *Fl. Fr.* 5: 52 (1815)

Amanita pantherina f. *robusta* A. Pearson, *Trans. Brit. Mycol. Soc.* 29(4): 191 (1946)

E: o **S:** o **W:** o **NI:** o **ROI:** o

H: On soil in mixed deciduous woodland, usually associated with *Fagus* and *Quercus* spp, rarely with other deciduous trees, or with conifers. Rarely found in open grassland, such as downland or pasture.

D: NM2: 196 **D+I:** Ph: 18, B&K4: 150 153, Galli: 130-133 **I:** FlDan1: 3A, Bon: 297, C&D: 277, Galli: 46, Cooke 6 (6) Vol. 1 (1881) (slender form).

Occasional but widespread.

phalloides (Fr.) Link, *Handbuch Gewächse* 3: 272 (1833)

Agaricus phalloides Fr., *Syst. mycol.* 1: 13 (1821)

Amanita phalloides var. *alba* Costantin & Dufour, *Nouv. fl. champ.*, Edn 2: 256 (1895)

Amanita viridis Pers., *Tent. disp. meth. fung.*: 71 (1797)

E: o **S:** ! **W:** ! **NI:** ! **ROI:** ! **O:** Channel Islands: !

H: On soil, associated with *Fagus* on calcareous loam soils and with *Quercus* spp in similar habitat.

D: NM2: 195 **D+I:** Ph: 18-19, B&K4: 152 154, Galli: 190-195 **I:** FlDan1: 1D, Bon: 299, C&D: 281, Galli: 39

Occasional in England, especially frequent in southern counties. Less frequently reported elsewhere but apparently widespread. Sensu Cooke 2 (2) Vol. 1 (1881) is *Amanita citrina*. English name = 'Death Cap'.

porphyria Alb. & Schwein., *Consp. fung. lusat.*: 142 (1805)

Agaricus recutitus Fr., *Epicr. syst. mycol.*: 6 (1838)

Amanita recutita (Fr.) Gillet, *Hyménomycètes*: 42 (1874)

E: o **S:** ! **W:** o **NI:** ! **ROI:** !

H: On sandy or acidic soils in mixed woodland, usually associated with *Pinus sylvestris* and *Quercus* spp. Also known with *Calluna* on heathland.

D: NM2: 195 **D+I:** Ph: 20-21, B&K4: 152 156, Galli: 208-209 **I:** FlDan1: 1B (as *A. porphyria* var. *recutita*), FlDan1: 1A, Bon: 299, C&D: 281

Occasional to rather rare, but widespread.

rubescens Pers., *Tent. disp. meth. fung.*: 71 (1797)

Amanita rubescens var. *alba* Rea [*nom. illegit., non A. rubescens* var. *alba* Coker (1917)], *Brit. basidiomyc.*: 104 (1922)

E: c **S:** c **W:** c **NI:** c **ROI:** c **O:** Channel Islands: ! Isle of Man: !

H: On soil, most commonly with *Betula* spp, but also frequent with *Fagus* and *Quercus* spp in mixed deciduous woodland. Occasionally reported with conifers such as *Pinus sylvestris*.

D: NM2: 194 **D+I:** Ph: 16-17, Bon: 297, B&K3: 154 157, Galli: 136-139 **I:** FlDan1: 6C, C&D: 277

Very common and widespread. English name = 'The Blusher'.

rubescens *var.* annulosulfurea Gillet, *Hyménomycètes*: 45 (1874)

Mis.: *Amanita rubescens* var. *magnifica* sensu Rea (1922)

E: o **S:** ! **W:** ! **ROI:** ! **O:** Channel Islands: !

H: On soil, in similar habitat to *A. rubescens*.

D: NM2: 194 **I:** Ph: 17

Widespread but much less common than *A. rubescens*.

singeri Bas, *Persoonia* 5(4): 364 (1969)

E: !

H: On soil in wooded parkland or cemeteries, always associated with conifers such as *Cedrus atlantica*, *Pinus nigra* and *Taxus*.

D+I: Galli: 152-153 **I:** C&D: 279

Known from Surrey (Kew, Royal Botanic Gardens and West Molesey). Originally thought to be an introduction from South America, but now known to be widespread and possibly native in the Mediterranean area.

spadicea Pers., *Tent. disp. meth. fung.*: 66 (1797)

Vaginata spadicea (Pers.) Gray, *Nat. arr. Brit. pl.* 1: 601 (1821)

E: ! **S:** !

H: On soil in mixed woodland, associated with *Betula*, *Quercus* and *Picea* spp.

D+I: Galli: 68-69 **I:** FM4(4): 135

Reported by Gray (1821,as *Vaginata spadicea*) from 'fir plantations', but no subsequent records until recently collected from England (Herefordshire: Mortimer Forest) in 2004 and from Scotland (Perthshire: Butterstone Woods) in 2003.

strobiliformis (Vittad.) Bertill., in Dechambre, *Dict. Encyclop. Sci. Med. Ser. 1*, 3: 499 (1866)
   Agaricus strobiliformis Vittad., *Descr. fung. mang.*: 59 (1835)
   Mis.: *Amanita solitaria* sensu NCL
E: ! **S:** ! **W:** ! **ROI:** !
H: On soil (usually calcareous) in open mixed deciduous woodland. Also known amongst grass on downland with *Betula* spp.
D: NM2: 194 **D+I:** Ph: 20-21 (as *A. solitaria*), B&K4: 154 159, Galli: 162-165 **I:** Bon: 299, C&D: 279, Cooke 9 (277) Vol. 1 (1881) (appears slightly too brown)
Rarely reported but apparently widespread.
The true distribution and frequency are unknown due to confusion with *A. echinocephala* and the fact that both species have commonly been reported as *A. solitaria* (a *nomen confusum*).

submembranacea (Bon) Gröger, *Boletus* 3(2): 27 (1979)
   Amanitopsis submembranacea Bon, *Bull. Mens. Soc. Linn. Lyon* 44(6): 176 (1975)
   Amanita submembranacea var. *bispora* D.A. Reid, *Notes Roy. Bot. Gard. Edinburgh* 44(3): 514 (1987)
E: o **S:** ! **NI:** ?
H: On soil, associated with deciduous or coniferous trees, often in mixed woodland.
D: NM2: 197 **D+I:** Myc2(1): 34, B&K4: 144 143, Galli: 98-101 **I:** Bon: 295, C&D: 273
Occasional in England, apparently rare in Scotland and a single record from Northern Ireland, unsubstantiated with voucher material. *Amanita submembranacea* var. *bispora* was based on an immature collection from England (Surrey: Mountain Wood) *fide* Tulloss (*in litt.*).

vaginata (Bull.) Lam., *Encyclop.* 1: 109 (1783)
   Agaricus vaginatus Bull., *Herb. France*: pl. 98 (1782)
   Amanitopsis vaginata (Bull.) Roze, *Bull. Soc. Bot. France* 23: 111 (1876)
   Vaginata livida (Pers.) Gray, *Nat. arr. Brit. pl.* 1: 601 (1821)
   Amanita vaginata var. *livida* (Pers.) Gillet, *Hyménomycètes*: 51 (1874)
   Agaricus plumbeus Schaeff., *Fung. Bavar. Palat. nasc.* 1: 37 (1762)
   Amanita vaginata f. *plumbea* (Schaeff.) Quél., *Fl. mycol. France*: 302 (1888)
   Amanita vaginata var. *plumbea* (Schaeff.) Quél. & Bataille, *Fl. Mon. des Amanites*: 42 (1902)
   Amanita vaginata ssp. *plumbea* (Schaeff.) Konrad & Maubl., *Icon. select. fung.* 6: 33 (1924)
   Amanita vaginata var. *grisea* (DC.) Quél. & Bataille, *Fl. Mon. des Amanites*: 42 (1902)
   Amanita vaginata f. *grisea* (DC.) E.-J. Gilbert, *Le Genre Amanita Persoon*: 139 (1918)
   Amanita violacea Jacz., *Compendium Hymenomycetum* 13: 277 (1923)
   Amanita vaginata f. *violacea* (Jacz.) Veselý, *Ann. Mycol.* 31(4): 280 (1933)
   Amanitopsis vaginata var. *violacea* (Jacz.) E.-J. Gilbert, *Amanitaceae* 1: 75 (1940)
E: o **S:** o **W:** o **NI:** o **ROI:** o **O:** Channel Islands: !
H: On soil in mixed deciduous woodland, usually associated with *Fagus* or *Quercus* spp and less frequently *Corylus* and *Betula* spp.
D: NM2: 197 **D+I:** Ph: 22-23, B&K4: 144 144, Galli: 78-81 **I:** FlDan1: 6B, Bon: 295, C&D: 275
Occasional but widespread (and may be locally frequent) but less common than *A. fulva*. English name = 'Grisette'.

virosa (Fr.) Bertill., in Dechambre, *Dict. Encyclop. Sci. Med. Ser. 1*, 3: 497 (1866)
   Agaricus virosus Fr., *Epicr. syst. mycol.*: 3 (1838)
   Mis.: *Amanita verna* sensu Rea (1922)
E: ! **S:** ! **NI:** ! **ROI:** !

H: On soil in open mixed deciduous woodland, associated with *Castanea*, *Betula*, *Fagus* or *Quercus* spp.
D: NM2: 195 **D+I:** Ph: 21, B&K4: 156 160, Galli: 198--199 **I:** Bon: 299, C&D: 281, Galli: 43
Rarely reported but apparently widespread. English name = 'Destroying Angel'.

vittadinii (Moretti) Vittad., *Tent. mycol.*: 31 (pl. 1) (1826)
   Agaricus vittadinii Moretti, *Giorn. Fis. Chim. Storia Nat. Med. Arti*, Dec. 2, 9: 66 (1826)
   Lepiota vittadinii (Moretti) Quél., *Mém. Soc. Émul. Montbéliard, Sér. 2*, 5: 338 (1873)
   Lepidella vittadinii (Moretti) E.-J. Gilbert, *Bull. Soc. Mycol. France* 41: 304 (1925)
   Aspidella vittadinii (Moretti) E.-J. Gilbert, in Bresadola, *Icon. Mycol.* 27(1): 79 (1940)
E: !
H: On soil in open deciduous woodland or scrub. Also reported from grazed pasture.
D: Reid7: 518-519 **D+I:** Galli: 154-157 **I:** Bon: 299, C&D: 279, Cooke 33 (36) Vol.1 (as *Agaricus (Lepiota) vittadini*).
Rarely reported. First reported from Norfolk (Wymondham) in 1847, with a fine illustration by Hussey (1847: pl. 85) 'such that its authenticity is beyond doubt' (Reid7). Collections in herb. K from Huntingdonshire (Orton Longueville) in 1859, Norfolk (Diss) in 1859 and North Hampshire (Swarraton) in 1901, then unrecorded until found in Oxfordshire (Houlton) in 1976, and West Sussex (Rustington) in 1978.

AMAURODON J. Schröt., in Cohn, *Krypt.-Fl. Schlesien*: 461 (1888)
Thelephorales, Thelephoraceae
   Hypochnopsis P. Karst., *Bidrag Kännedom Finlands Natur Folk* 48: 442 (1889)
   Lazulinospora Burds. & M.J. Larsen, *Mycologia* 66(1): 97 (1974)
   Tomentellago Hjortstam & Ryvarden, *Mycotaxon* 31(1): 40 (1988)
Type: *Amaurodon viridis* (Alb. & Schwein.) J. Schröt.

cyaneus (Wakef.) Kõljalg & K.H. Larss., *Syn. Fungorum* 9: 33 (1996)
   Hypochnus cyaneus Wakef., *Trans. Brit. Mycol. Soc.* 5: 478 (1917)
   Tomentella cyanea (Wakef.) Bourdot & Galzin, *Bull. Soc. Mycol. France* 40: 143 (1924)
   Lazulinospora cyanea (Wakef.) Burds. & M.J. Larsen, *Mycologia* 66: 99 (1974)
E: !
H: On decayed wood of deciduous trees.
D: NM3: 299
Known from Bedfordshire and Surrey.

mustialaënsis (P. Karst.) Kõljalg & K.H. Larss., *Syn. Fungorum* 9: 33 (1996)
   Hypochnus mustialaënsis P. Karst., *Not. Sällsk. Fauna et Fl. Fenn. Förh.* 11: 222 (1871)
   Corticium mustialaënse (P. Karst.) Fr., *Hymenomyc. eur.*: 705 (1874)
   Coniophora mustialaënsis (P. Karst.) Massee, *J. Linn. Soc., Bot.* 25: 139 (1889)
   Hypochnopsis mustialaënsis (P. Karst.) P. Karst., *Bidrag Kännedom Finlands Natur Folk* 48: 442 (1889)
E: !
H: On decayed wood of deciduous tree.
D: NM3: 299
A single collection, from West Sussex (Arundel Park) in 1928.

AMPHINEMA P. Karst., *Bidrag Kännedom Finlands Natur Folk* 51: 228 (1892)
Stereales, Atheliaceae
Type: *Diplonema sordescens* P. Karst. (= *A. byssoides*)

angustispora P. Roberts, *Kew Bull.* 56: 761 (2001)
E: ! **O:** Channel Islands !

H: On fallen bark and decayed wood. British collections all on *Pinus* spp.

Known from Surrey (Kew Gardens) and the Channel Islands (Jersey: St Brelades). Also collected in Belgium.

byssoides (Fr.) J. Erikss., *Symb. Bot. Upsal.* 16(1): 112 (1958)
Thelephora byssoides Fr., *Syst. mycol.* 1: 452 (1821)
Hypochnus byssoides (Fr.) Quél., *Bull. Soc. Mycol. France* 26: 231 (1879)
Coniophora byssoidea (Fr.) P. Karst., *Bidrag Kännedom Finlands Natur Folk* 37: 160 (1882)
Peniophora byssoides (Fr.) Bres., *Jahrb. Westf. Provinz. Verein. Wiss. Kunst.* 26: 130 (1898)
Kneiffia byssoides (Fr.) Herter, *Kryptogamenflora der Mark Brandenburg* 6(1): 107 (1910)
Corticium lacunosum Berk. & Broome, *Ann. Mag. Nat. Hist., Ser. 4 [Notices of British Fungi no. 1371]* 11: 343 (1873)
Corticium strigosum var. *filamentosum* W.G. Sm., *Syn. Brit. Bas.*: 413 (1908)
Kneiffia tomentella Bres., *Ann. Mycol.* 1: 103 (1903)
Peniophora tomentella (Bres.) Pilát, *Bull. Soc. Mycol. France* 42: 115 (1926)
Amphinema tomentellum (Bres.) M.P. Christ., *Dansk. Bot. Ark.* 19(2): 232 (1959)
E: c **S:** c **W:** c **NI:** c **ROI:** c
H: Associated with conifers (and possibly *Betula* spp) in acidic areas on leaf litter, soil, woody debris, and fallen polypore basidiomes.
D: NM3: 207 **D+I:** CNE2: 80-81, B&K2: 98 73 **I:** Kriegl1: 150
Very common and widespread.

AMPULLOCLITOCYBE Redhead, Lutzoni, Moncalvo & Vilgalys, *Mycotaxon* 83: 36 (2002)
Agaricales, Tricholomataceae
Type: *Ampulloclitocybe clavipes* (Pers) Redhead, Lutzoni, Moncalvo & Vilgalys

clavipes (Pers.) Redhead, Lutzoni, Moncalvo & Vilgalys, *Mycotaxon* 83: 36 (2002)
Agaricus clavipes Pers., *Syn. meth. fung.*: 353 (1801)
Clitocybe clavipes (Pers.) P. Kumm., *Führ. Pilzk.*: 124 (1871)
Agaricus comitialis Pers., *Syn. meth. fung.*: 352 (1801)
Clitocybe comitialis (Pers.) P. Kumm., *Führ. Pilzk.*: 124 (1871)
Clitocybe squamulosoides P.D. Orton, *Trans. Brit. Mycol. Soc.* 43(2): 187 (1960)
E: c **S:** c **W:** c **NI:** ! **ROI:** ! **O:** !
H: On acidic to neutral soil, in deciduous and mixed woodland, and with conifers such as *Pinus* spp, also associated with *Pteridium* on heath and moorland.
D: NM2: 109, FAN3: 50 **D+I:** Ph: 48-49, B&K3: 152 154 **I:** FlDan1: 32D, Bon: 135, C&D: 173
Common and widespread, but few records from Ireland. The type of *Clitocybe squamulosoides* has been shown to belong here, although the name was introduced as a *nom. nov.* for *C. trulliformis* sensu Lange, i.e. *C. costata*.

AMYLOSTEREUM Boidin, *Rev. Mycol. (Paris)* 23(3): 345 (1958)
Stereales, Stereaceae
Xerocarpus P. Karst. [*nom. illegit.*, non *Xerocarpus* Guill., Perr. & A. Rich. (1832)(*Papilionaceae*)], *Rev. Mycol. (Toulouse)* 3(9): 22 (1881)
Type: *Amylostereum chailletii* (Pers.) Boidin

areolatum (Chaillet) Boidin, *Rev. Mycol. (Paris)* 23(3): 345 (1958)
Thelephora areolata Chaillet, in Fries, *Elench. fung.* 1: 190 (1828)
E: !
H: On decayed wood (usually old stumps) of *Picea* spp in plantations or parkland.
D: NM3: 185 **D+I:** CNE5: 890-893, B&K2: 178 195, Myc16(3): 124 **I:** SV28: 24, Kriegl1: 151

Known from Berkshire, Buckinghamshire and Surrey but the true distribution and frequency are unknown since it is easily mistaken for *A. chailletii*.

chailletii (Pers.) Boidin, *Rev. Mycol. (Paris)* 23(3): 345 (1958)
Thelephora chailletii Pers., *Mycol. eur.* 1: 125 (1825)
Stereum chailletii (Pers.) Fr., *Epicr. syst. mycol.*: 551 (1838)
Xerocarpus ambiguus P. Karst., *Acta Soc. Fauna Fl. Fenn.* 2(1): 38 (1881)
Peniophora atkinsonii Ellis & Everh., *Proc. Acad. nat. Sci. Philad.*: 324 (1894) [1893]
E: o **S:** o **W:** ? **NI:** ! **ROI:** !
H: On decayed wood (often large fallen trunks) of conifers such as *Abies*, *Cedrus*, *Chamaecyparis*, *Larix*, *Picea* and *Pinus* spp, usually in plantations.
D: NM3: 185 **D+I:** CNE2: 91-93, B&K2: 180 196, Myc10(4): 180-181 **I:** SV28: 24
Occasional but apparently widespread and may be locally common in southern or midland counties of England. A single record from Wales in 1985 but unsubstantiated with voucher material.

laevigatum (Fr.) Boidin, *Rev. Mycol. (Paris)* 23(3): 345 (1958)
Thelephora laevigata Fr., *Elench. fung.* 1: 224 (1828)
Corticium laevigatum (Fr.) Fr., *Epicr. syst. mycol.*: 565 (1838)
Xerocarpus laevigatus (Fr.) P. Karst., *Rev. Mycol. (Toulouse)* 3(9): 22 (1881)
Peniophora laevigata (Fr.) P. Karst., *Meddeland. Soc. Fauna Fl. Fenn.* 6: 12 (1881)
Corticium juniperi P. Karst., *Bidrag Kännedom Finlands Natur Folk* 25: 315 (1876)
Xerocarpus juniperi (P. Karst.) P. Karst., *Rev. Mycol. (Toulouse)* 3(9): 22 (1881)
Hymenochaete stevensonii Berk. & Broome, *Ann. Mag. Nat. Hist., Ser. 5 [Notices of British Fungi no. 1817]* 3: 211 (1879)
E: c **S:** c **W:** c
H: On bark of living or dead conifers, especially on *Taxus*. Less frequent on *Juniperus communis*, *Cupressus macrocarpa*, *Picea* and *Thuja* spp.
D: NM3: 185 **D+I:** CNE2: 94-95, B&K2: 180 197 **I:** Kriegl1: 153
Common and widespread, especially on old yew trees.

ANTRODIA P. Karst., *Meddeland. Soc. Fauna Fl. Fenn.* 5: 40 (1879)
Poriales, Coriolaceae
Coriolellus Murrill, *Bull. Torrey bot. Club* 32: 481 (1905)
Amyloporia Singer, *Mycologia* 36: 67 (1944)
Cartilosoma Kotl. & Pouzar, *Česká Mykol.* 12: 101, 103 (1958)
Fibroporia Parmasto, *Conspectus Systematis Corticiacearum*: 176 (1968)
Type: *Polyporus serpens* Fr. (= *Antrodia albida*)

albida (Fr.) Donk, *Persoonia* 4: 339 (1966)
Daedalea albida Fr., *Observ. mycol.* 1: 107 (1815)
Lenzites albida (Fr.) Fr., *Epicr. syst. mycol.*: 405 (1838)
Coriolellus albidus (Fr.) Bondartsev, *Trutovye griby Evropeiskoi chasti SSSR i Kavkaza*: 504 (1953)
Trametes sepium Berk., *J. Bot.* 6: 322 (1847)
Polyporus serpens Fr., *Observ. mycol.* 2: 265 (1818)
Trametes serpens (Fr.) Fr., *Hymenomyc. eur.*: 586 (1874)
Antrodia serpens (Fr.) P. Karst., *Meddeland. Soc. Fauna Fl. Fenn.* 5: 40 (1880)
Coriolellus serpens (Fr.) Bondartsev, *Trutovye griby Evropeiskoi chasti SSSR i Kavkaza*: 513 (1953)
Polyporus stephensii Berk. & Broome, *Ann. Mag. Nat. Hist., Ser. 2 [Notices of British Fungi no. 356]* 2: 2 (1848)
E: c **S:** ! **W:** o
H: On wood, often small, decorticated fallen branches of deciduous trees and shrubs. Most often reported on *Corylus*, *Fagus* and *Salix* spp. Rarer hosts include *Betula* spp, *Euonymus europaeus*, *Garrya elliptica*, *Ligustrum vulgare*, *Populus* spp, *Quercus* spp, and *Rosa canina*.
D: EurPoly1: 112 **D+I:** Ph: 239 (as *Coriolellus albidus*)

Common in England, especially in woodland on calcareous soils in southern counties, but less often reported elsewhere and seemingly rare in Scotland.

gossypium (Speg.) Ryvarden, *Norweg. J. Bot.* 20: 8 (1973)
  Poria gossypium Speg., *Mus. Nat. Buenos Aires* 6: 169 (1899)
  Leptoporus destructor var. *resupinatus* Bourdot & Galzin, *Hymenomyc. France*: 547 (1927)
  Tyromyces resupinatus (Bourdot & Galzin) Bondartsev & Singer, *Ann. Mycol.* 39: 52 (1941)
  Mis.: *Polyporus destructor* sensu auct.
  Mis.: *Tyromyces destructor* sensu auct.
E: ! S: ?
H: On dead and decayed wood of conifers such as *Pinus* spp; rarely on wood of deciduous trees, e.g. *Fagus*.
D: EurPoly1: 120 (as *Antrodia gossypina*), NM3: 240
Rarely collected in England (Surrey and North Somerset). Reported from Scotland (as *Polyporus destructor*) in 1900 but no material has been traced. Old records should be treated with caution as names have often been misapplied to various other taxa. The epithet is often found as *'gossypina'* which is orthographically incorrect.

macra (Sommerf.) Niemelä, *Karstenia* 25(1): 38 (1985)
  Polyporus macer Sommerf., *Suppl. Fl. lapp.*: 279 (1826)
  Trametes salicina Bres., *Nytt Mag. Naturividensk* 52: 166 (1914)
E: !
H: On dead or decayed wood of *Populus* and *Salix* spp in mixed deciduous woodland.
D: EurPoly1: 126, NM3: 239
A single collection from South Hampshire (New Forest, Millyford Bridge) in 1992. Reported from Norfolk but unsusbtantiated with voucher material.

pseudosinuosa A. Henrici & Ryvarden, *Mycologist* 11: 152 (1997)
E: !
H: On dead and decayed wood of deciduous trees. The type collection was on large, partially fallen and decorticated trunks of *Ulmus* spp in scrub woodland.
D+I: ANT 152-153
Described from Middlesex (Perivale Wood) and now known from Berkshire, North Wiltshire and Surrey.

ramentacea (Berk. & Broome) Donk, *Persoonia* 4: 339 (1966)
  Polyporus ramentaceus Berk. & Broome, *Ann. Mag. Nat. Hist., Ser. 5 [Notices of British Fungi no. 1809]* 3: 210 (1879)
  Poria ramentacea (Berk. & Broome) Cooke, *Grevillea* 14: 112 (1886)
  Trametes subsinuosa Bres., *Ann. Mycol.* 1(1): 82 (1903)
  Cartilosoma subsinuosum (Bres.) Kotlab & Pouzar, *Česká Mykol.* 12: 103 (1958)
E: ! S: !
H: On dead wood, exclusively *Pinus sylvestris* in Scotland, but also known on *Salix* spp in England.
D: EurPoly1: 136 **I:** FM 2(2): 46-49
Rarely reported. Known from widespread localities in the Scottish Highlands, but also collected recently in southern England (Berkshire, South Hampshire and Surrey).

serialis (Fr.) Donk, *Persoonia* 4: 340 (1966)
  Polyporus serialis Fr., *Syst. mycol.* 1: 370 (1821)
  Trametes serialis (Fr.) Fr., *Hymenomyc. eur.*: 584 (1874)
  Coriolellus serialis (Fr.) Murrill, *North American Flora, Ser. 1 (Fungi 3)*, 9(1): 29 (1907)
  Polyporus callosus Fr., *Syst. mycol.* 1: 381 (1821)
  Poria callosa (Fr.) Cooke, *Grevillea* 14: 110 (1886)
E: o S: o W: !
H: On dead wood of conifers, such as *Larix, Picea* and *Pinus* spp. Reported once on *Betula* sp, but unsubstantiated with voucher material. Rarely on structural timber such as window sills and decayed planks.
D: EurPoly1: 137, NM3: 239 **D+I:** B&K2: 276 341 **I:** Kriegl1: 480

Occasional in England and Scotland but rarely reported elsewhere. Basidiomes often cover the cut ends of sawn logs or trunks, especially in stacked timber.

sinuosa (Fr.) P. Karst., *Meddeland. Soc. Fauna Fl. Fenn.* 6: 10 (1881)
  Polyporus sinuosus Fr., *Syst. mycol.* 1: 382 (1821)
  Trametes sinuosa (Fr.) Cooke & Quél., *Clavis syn. Hymen. Europ.*: 190 (1878)
  Mis.: *Poria vaporaria* sensu auct.
  Mis.: *Polyporus vaporarius* sensu auct.
E: ! S: ! NI: ?
H: On dead, decayed and often partially burnt wood of conifers such as *Picea* and *Pinus* spp, often on heathland or moorland. Reported on *Betula* spp and *Populus tremula* but these hosts are not verified.
D: EurPoly1: 138, NM3: 238
Rarely reported. Known from Scotland (Easterness and North Ebudes) and England (Middlesex and Surrey). Reported from Herefordshire, North Somerset and from Northern Ireland but unsubstantiated with voucher material.

vaillantii (DC.) Ryvarden, *Norweg. J. Bot.* 20: 8 (1973)
  Boletus vaillantii DC., *Fl. Fr.* 5: 38 (1815)
  Polyporus vaillantii (DC.) Fr., *Syst. mycol.* 1: 383 (1821)
  Poria vaillantii (DC.) Cooke, *Grevillea* 14: 112 (1886)
  Fibuloporia vaillantii (DC.) Bondartsev & Singer, *Ann. Mycol.* 39: 49 (1941)
  Fibroporia vaillantii (DC.) Parmasto, *Conspectus Systematis Corticiacearum*: 177 (1968)
  Boletus hybridus Sowerby, *Col. fig. Engl. fung.* 3: pl. 289 (1800)
  Polyporus hybridus (Sowerby) Berk. & Broome, *Outl. Brit. fungol.*: xvii (1860)
  Poria hybrida (Sowerby) Cooke, *Grevillea* 14: 109 (1886)
  Mis.: *Polyporus destructor* sensu auct.
E: ! S: ! W: !
H: On decayed structural timber, such as pit props, planks, decayed floorboards or wet plaster walls in old or disused buildings, and in greenhouses on frames and shelving. Rarely outdoors and then always on worked or structural wood. Also recorded (with voucher material) as 'strangling *Lobelias*' in a garden and on 'wet coke' in a coal depot.
D: EurPoly1: 142, NM3: 240
Uncommonly reported but apparently widespread.

xantha (Fr.) Ryvarden, *Norweg. J. Bot.* 20: 8 (1973)
  Polyporus xanthus Fr., *Syst. mycol.* 1: 379 (1821)
  Poria xantha (Fr.) Cooke, *Grevillea* 14: 112 (1886)
  Amyloporia xantha (Fr.) Bondartsev & Singer, *Ann. Mycol.* 39: 50 (1941)
  Antrodia xantha f. *pachymeres* (Fr.) J. Erikss., *Svensk bot. Tidskr.* 43: 3 (1949)
E: o S: o W: o NI: ! O: Channel Islands: !
H: On decayed wood of conifers such as *Abies, Cedrus, Larix, Picea, Pinus* and *Pseudotsuga* spp, usually on large fallen trunks, logs or stumps. Reported on *Betula, Fagus* and *Quercus* spp, but these hosts are not verified.
D: EurPoly1: 145, NM3: 239 **D+I:** B&K2: 276 342 (as *Antrodia xantha* f. *pachymeres*)
Occasional but widespread and may be locally common in southern England. Basidiomes exhibit two distinct colour forms, bright citric-yellow or dull chalky-white, and are usually resupinate but may exhibit a nodular hymenophore (f. *pachymeres*).

ANTRODIELLA Ryvarden & I. Johans., *Preliminary Polypore Flora of East Africa*: 256 (1980)
Poriales, Coriolaceae
Type: *Antrodiella semisupina* (Berk. & M.A. Curtis) Ryvarden

genistae (Bourdot & Galzin) A. David, *Bull. Soc. Mycol. France* 106: 75 (1990)
  Coriolus genistae Bourdot & Galzin, *Bull. Soc. Mycol. France* 41: 145 (1925)
W: !

H: On decayed or dead wood. British material on *Salix* spp in wet woodland (fen-carr).
D: EurPoly1: 156
A single collection from Monmouthshire (Cwm Coed-y-Cerrig) in 1992.

onychoides (Egeland) Niemelä, *Karstenia* 22(1): 11 (1982)
  Polyporus onychoides Egeland, *Nytt Mag. Naturividensk* 22: 11 (1913)
  Tyromyces onychoides (Egeland) Ryvarden, *Blyttia* 25: 25 (1967)
  Antrodia onychoides (Egeland) Ryvarden, *Polyporaceae of North Europe* 1: 85 (1976)
E: !
H: On decayed wood of *Betula*, *Fagus* and *Quercus* spp in old mixed deciduous woodland.
D: EurPoly1: 159, NM3: 208
Known from Dorset, East Kent, Middlesex, Oxfordshire, Leicestershire and Surrey. Reported from Berkshire, but on an odd host (*Taxus*) and unsubstantiated with voucher material.

romellii (Donk) Niemelä, *Karstenia* 22(1): 11 (1982)
  Poria romellii Donk, *Persoonia* 5: 84 (1967)
E: ! **W:** ?
H: On decayed wood of deciduous trees and shrubs, such as *Corylus*, *Fagus* and *Salix* spp, in old deciduous woodland.
D: EurPoly1: 162, NM3: 208
Known from England (Middlesex, Surrey and West Kent) and Wales (Monmouthshire). Closely related to *Antrodiella semisupina* but basidiomes are consistently resupinate.

semisupina (Berk. & M.A. Curtis) Ryvarden, *Preliminary Polypore Flora of East Africa*: 261 (1980)
  Polyporus semisupinus Berk. & M.A. Curtis, *Grevillea* 1: 50 (1872)
  Tyromyces semisupinus (Berk. & M.A. Curtis) Murrill, *North American Flora, Ser. 1 (Fungi 3)*, 9(1): 34 (1907)
E: ! **S:** ! **ROI:** !
H: On decayed wood of *Alnus*, *Betula*, *Corylus*, *Fraxinus* and *Salix* spp.
D: 208, EurPoly1: 164 **D+I:** B&K2: 278 344
Known from England (Derbyshire, Hertfordshire, North Somerset, South Hampshire, Surrey, West Kent and Yorkshire), Scotland (Easterness and Kirkudbrightshire) and Republic of Ireland (North Kerry).

ARMILLARIA (Fr.) Staude, *Schwämme Mitteldeutschl.*: 130 (1857)

Agaricales, Tricholomataceae
  Rhizomorpha Roth (anam.), *Ann. Bot. Usteri* 1: 7 (1791)
  Armillariella (P. Karst.) P. Karst., *Acta Soc. Fauna Fl. Fenn.* 2(1): 4 (1881)
Type: *Armillaria mellea* (Vahl) P. Kumm.

borealis Marxm. & Korhonen, *Bull. Soc. Mycol. France* 98(1): 122 (1982)
E: ? **S:** o
H: Parasitic, then saprotrophic on decayed wood of deciduous trees and conifers. Known on *Betula* spp, *Fraxinus* and *Picea sitchensis*.
D: NM2: 98, FAN3: 36 **I:** C&D: 181
Occasional but widespread in Scotland *fide* R. Watling (pers. comm.). Reported from southern England, but unsubstantiated with voucher material.

ectypa (Fr.) Emel, *Le genre* Armillaria.*(Thesis)*: 50 (1921)
  Agaricus ectypus Fr., *Syst. mycol.* 1: 108 (1821)
  Clitocybe ectypa (Fr.) Sacc., *Syll. fung.* 5: 193 (1887)
  Armillariella ectypa (Fr.) Singer, *Ann. Mycol.* 41: 20 (1943)
E: ! **W:** ! **NI:** !
H: On wet soil or amongst *Sphagnum* spp on moorland and in fens.
D: NM2: 97, FAN3: 38 **I:** Bon: 143, Kriegl3: 123, SV47: 54
Rarely reported. Known from England (Cumberland), Wales (Carmarthenshire) and Northern Ireland (Antrim). Reported elsewhere but unsubstantiated with voucher material.

gallica Marxm. & Romagn., *Bull. Soc. Mycol. France* 103: 152 (1987)
  Mis.: *Armillaria lutea* sensu auct.
  Mis.: *Agaricus melleus* sensu Bolton [Hist. Fung. Halifax: pl. 141 (1791)]
E: c **S:** c **W:** c **NI:** c **ROI:** c
H: Weakly parasitic then saprotrophic on many species of deciduous trees and shrubs. Less frequently on conifers such as *Chamaecyparis*, *Picea*, *Pseudotsuga* and *Thuja* spp.
D: NM2: 98, FAN3: 37 (as *A. lutea*) **I:** Ph: 32-33 (as *A. mellea* chunky form), C&D: 181
Very common and widespread. In southern England, this may be more common than *A. mellea* sensu stricto.

mellea (Vahl) P. Kumm., *Führ. Pilzk.*: 134 (1871)
  Agaricus melleus Vahl, *Fl. Danica*: t. 1013 (1790)
  Armillariella mellea (Vahl) P. Karst., *Acta Soc. Fauna Fl. Fenn.* 2(1): 4 (1881)
  Clitocybe mellea (Vahl) Ricken, *Blätterpilze Deutschl.*: 362 (1914)
  Armillaria mellea var. *sulphurea* (Weinm.) P. Karst., *Bidrag Kännedom Finlands Natur Folk* 32: 22 (1879)
  Armillaria mellea var. *maxima* Barla, *Bull. Soc. Mycol. France* 3: 143 (1887)
  Armillaria mellea var. *minor* Barla, *Bull. Soc. Mycol. France* 3: 143 (1887)
E: c **S:** o **W:** c **NI:** c **ROI:** c **O:** Channel Islands: c Isle of Man: c
H: Virulently parasitic, then saprotrophic. Growing on many species of deciduous trees and shrubs, less frequently on conifers.
D: NM2: 97, FAN3: 35 **D+I:** Ph: 32 (as *Armillaria mellea* slender form), B&K3: 138 134 **I:** Bon: 143, C&D: 181, Kriegl3: 125
Common and widespread, but only occasional in Scotland *fide* R. Watling (pers. comm.). Older records should be viewed with caution, as they may represent other species not previously distinguished. English name = 'Honey Fungus'.

ostoyae (Romagn.) Herink, *Vys. Skola Z. Brné*: 42 (1973)
  Armillariella ostoyae Romagn., *Bull. Soc. Mycol. France* 86(1): 265 (1970)
  Mis.: *Armillaria obscura* sensu auct.
  Mis.: *Agaricus obscurus* sensu auct.
  Mis.: *Armillariella polymyces* sensu auct. Eur.
E: ! **S:** o **W:** ! **NI:** !
H: Parasitic, then saprotrophic on living and dead wood of deciduous and coniferous trees and shrubs, often with a preference for acidic soils.
D: NM2: 97, FAN3: 36 **D+I:** B&K3: 138 135 **I:** C&D: 181, Kriegl3: 127
Apparently widespread and may be locally common, but previously not distinguished from *A. mellea*. The 'commonest *Armillaria* in Scotland after *A. gallica* *fide* Watling (pers. comm.).

tabescens (Scop.) Emel, *Le genre* Armillaria.*(Thesis)*: 50 (1921)
  Agaricus tabescens Scop., *Fl. carniol.* 446 (1772)
  Clitocybe tabescens (Scop.) Bres., *Fungi trident.* 2: 84 (1900)
  Armillaria mellea var. *tabescens* (Scop.) Rea & Ramsb., *Trans. Brit. Mycol. Soc.* 5: 352 (1917)
  Lentinus caespitosus Berk., *J. Bot.* 6: 317 (1847)
  Pleurotus caespitosus (Berk.) Sacc., *Syll. fung.* 5: 352 (1887)
  Mis.: *Omphalia gymnopodia* sensu Quél. [Fl. Mycol. France: 251 (1888)]
  Mis.: *Agaricus gymnopodius* sensu auct. *fide* Pearson & Dennis (1948)
  Mis.: *Clitocybe gymnopodia* sensu K&R
E: o **W:** !
H: Parasitic then saprotrophic, often on roots of deciduous trees in old woodland. Normally associated with *Quercus* spp, less frequently with *Fagus* and *Carpinus*, and rarely with *Betula* spp. Reported with *Cotoneaster* and *Pinus* spp, but these hosts are unverified.
D: FAN3: 38 **D+I:** Ph: 32-33 **I:** Bon: 143, C&D: 181, Kriegl3: 128

Occasional in England, from Lincolnshire southwards, and a single collection from Wales (Glamorganshire).

ARRHENIA Fr., *Summa veg. Scand.*: 312 (1849)
Agaricales, Tricholomataceae
Corniola Gray [*nom. illegit.*, non *Corniola* Adans. (1763)(*Leguminosae*)] *Nat. arr. Brit. pl.* 1: 637 (1821)
Leptotus P. Karst., *Bidrag Kännedom Finlands Natur Folk* 32: 242 (1879)
Leptoglossum P. Karst., *Bidrag Kännedom Finlands Natur Folk* 32: 242 (1879)
Dictyolus Quél., *Enchir. fung.*: 139 (1886)
Phaeotellus Kühner & Lamoure, *Botaniste* 55: 24 (1972)
Type: *Arrhenia auriscalpium* (Fr.) Fr.

acerosa (Fr.) Kühner, *Bull. Mens. Soc. Linn. Lyon* 49(5): 893/992 (1980)
Agaricus acerosus Fr., *Syst. mycol.* 1: 191 (1821)
Pleurotus acerosus (Fr.) Quél., *Mém. Soc. Émul. Montbéliard, Sér. 2,* 5: 246 (1872)
Pleurotellus acerosus (Fr.) Konrad & Maubl., *Icon. select. fung.* 6: 361 (1937)
Leptoglossum acerosum (Fr.) M.M. Moser, *Kleine Kryptogamenflora*, Edn 4: 127 (1978)
Omphalina acerosa (Fr.) M. Lange, *Nordic J. Bot.* 1(5): 695 (1981)
Mis.: *Pleurotus tremulus* sensu Rea (1922)
E: o **S:** o **W:** ! **NI:** o **ROI:** !
H: On soil or decayed wood (stumps and woodchips) associated with mosses, though this is not always evident.
D: BFF6: 31, NM2: 99, FAN3: 87 (as *Omphalina acerosa*) **D+I:** Ph: 185 (as *Pleurotus acerosus*), B&K3: 140 136 **I:** FlDan2: 63B & D (as *P. acerosus*), C&D: 153, C&D: 153
Occasional but widespread.

chlorocyanea (Pat.) Redhead, Lutzoni, Moncalvo & Vilgalys, *Mycotaxon* 83: 46 (2002)
Agaricus chlorocyaneus Pat., *Tab. anal. fung.* 4: 145 (1885)
Omphalia chlorocyanea (Pat.) Sacc., *Syll. fung.* 5: 336 (1887)
Omphalina chlorocyanea (Pat.) Singer, *Lilloa.* 23: 212 (1951) [1949]
Agaricus umbelliferus var. *viridis* Hornem., *Fl. Danica* 10: 10 (1819)
Omphalia umbellifera var. *viridis* (Hornem.) Quél., *Enchir. fung.*: 44 (1886)
Omphalia umbellifera f. *viridis* (Hornem.) Cejp, Kavina & Pilát, *Atlas champ. Eur.* 4: 43 (1936)
Omphalia viridis (Hornem.) Lange, *Dansk. Bot. Ark.* 6(5): 12 (1930)
E: !
H: On sandy soil amongst mosses and lichens.
D: NM2: 172 (as *Omphalina smaragdina*), FAN3: 82 **I:** FlDan1: 60G
Rare. Known from Bedfordshire, Hampshire, Norfolk, Surrey and West Kent. Reported from North Lincolnshire and Shropshire in the late nineteeth century but unsubstantiated with voucher material.

epichysium (Pers.) Redhead, Lutzoni, Moncalvo & Vilgalys, *Mycotaxon* 83: 47 (2002)
Agaricus epichysium Pers., *Syn. Meth. Fung.*: 462 (1801)
Omphalia epichysium (Pers.) P. Kumm., *Führ. Pilzk.*: 107 (1871)
Omphalina epichysium (Pers.) Quél., *Enchir. fung.*: 43 (1886)
S: ?
H: British collections and records are from acidic soil or amongst *Sphagnum* spp, but known on decayed wood elsewhere in Europe.
D: NM2: 172 **D+I:** B&K3: 300 378 **I:** C&D: 185
Reported from four widely scattered sites in Scotland, but only one substantiated with voucher material.

griseopallida (Desm.) Watling, *Notes Roy. Bot. Gard. Edinburgh* 45(3): 554 (1989)
Agaricus griseopallidus Desm., *Pl. Crypt. Nord France* 3: 120 (1826)

Omphalina griseopallida (Desm.) Quél., *Enchir. fung.*: 44 (1886)
Leptoglossum griseopallidum (Desm.) M.M. Moser [*nom. inval.*], *Kleine Kryptogamenflora*, Edn 4: 127 (1978)
Phaeotellus griseopallidus (Desm.) Kühner & Lamoure, *Doc. Mycol.* 15(57-58): 78 (1985) [1984]
E: ! **S:** ! **W:** !
H: On soil amongst grass, usually in short turf, on dunes, downland and other grasslands; several records from bowling greens.
D: NM2: 172, FAN3: 86 **D+I:** Ph: 69, B&K3: 302 379 **I:** Bon: 129, BFF6: 32, C&D: 183
Rarely recorded or collected but apparently widespread.

latispora (J. Favre) Bon & Courtec., *Doc. Mycol.* 18 (69): 37 (1987)
Pleurotellus acerosus var. *latisporus* J. Favre, *Ergebn. Wiss. Untersuch. Schweiz. Natn. Parks* 5: 199 (1955)
Mis.: *Pleurotus acerosus* sensu Rea (1922)
Mis.: *Pleurotellus acerosus* sensu auct.
Mis.: *Pleurotellus tremulus* sensu auct.
Mis.: *Pleurotus tremulus* sensu auct.
E: ! **S:** ! **W:** !
H: On soil, usually amongst mosses in damp areas on moorland, in ditches, open woodland and on dunes.
D: BFF6: 31-32
Rarely reported but apparently widespread. Closely related to *A. acerosa* but differing mainly in the broader spores.

lobata (Pers.) Redhead, *Canad. J. Bot.* 62: 871 (1984)
Merulius lobatus Pers., *Syn. meth. fung.*: 494 (1801)
Corniola lobata (Pers.) Gray, *Nat. arr. Brit. pl.* 1: 596 (1821)
Cantharellus lobatus (Pers.) Fr., *Syst. mycol.* 1: 323 (1821)
Leptotus lobatus (Pers.) P. Karst., *Bidrag Kännedom Finlands Natur Folk* 32: 243 (1879)
Dictyolus lobatus (Pers.) Quél., *Enchir. fung.*: 140 (1886)
Leptoglossum lobatum (Pers.) Ricken, *Blätterpilze Deutschl.*: 6 (1911)
E: ! **S:** ! **NI:** ! **ROI:** !
H: In bogs, associated with mosses and often growing in close proximity to, or actually in, water.
D: BFF6: 28-29, NM2: 99, FAN3: 39-40 **D+I:** B&K3: 140 138 **I:** C&D: 153, C&D: 153, Kriegl3: 129
A distinctive species, supposedly not uncommon *fide* BFF6, and reported from most areas of Britain. However, most specimens named thus in herb. K appear to be misidentified *A. retiruga* collected from habitats other than bogs (old walls, grassland, woodland edges etc).

obatra (Favre) Redhead, Lutzoni, Moncalvo & Vilgalys, *Mycotaxon* 83: 47 (2002)
Omphalia obatra Favre, *Ergebn. Wiss. Untersuch. Schweiz. Natn. Parks* 5: 199 (1955)
Omphalina obatra (J. Favre) P.D. Orton, *Trans. Brit. Mycol. Soc.* 43(2): 180 (1960)
E: !
H: On sandy soil amongst mosses, often in marshy areas.
D+I: B&K3: 302 381 **I:** C&D: 185
Known from Co. Durham and Warwickshire.

obscurata (D.A. Reid) Redhead, Lutzoni, Moncalvo & Vilgalys, *Mycotaxon* 83: 47 (2002)
Omphalina obscurata D.A. Reid, *Trans. Brit. Mycol. Soc.* 41: 419 (1958)
E: ! **S:** !
H: On acidic soil, occasionally amongst *Sphagnum* spp or in grassland; several records from bowling greens, lawns etc.
D: FAN3: 88 **I:** Bon: 129, Kriegl3: 484
Rarely reported.

onisca (Fr.) Redhead, Kutzoni, Moncalvo & Vilgalys, *Mycotaxon* 83: 47 (2002)
Agaricus oniscus Fr., *Observ. mycol.* 2: 209 (1818)
Omphalia oniscus (Fr.) Gillet, *Hyménomycètes*: 297 (1876)
Omphalina oniscus (Fr.) Quél., *Enchir. fung.*: 43 (1886)
E: ! **S:** ! **W:** ! **NI:** ! **ROI:** !

H: In *Sphagnum* spp in bogs on moorland and heathland, rarely in boggy areas in woodland.
D: NM2: 172, FAN3: 85-86 **D+I:** B&K3: 304 382 **I:** Bon: 129
Rarely reported but apparently widespread. Possibly synonymous with *Arrhenia epichysium*.

parvivelutina (Clémençon & Irlet) Redhead, Kutzoni, Moncalvo & Vilgalys, *Mycotaxon* 83: 47 (2002)
  Omphalina parvivelutina Clémençon & Irlet, *Schw. Z. Pilzk.* 1982A, *Sondernummer* 123: 15 (1982)
S: !
H: On soil in montane habitat.
Known from Perthshire and Orkney.

peltigerina (Peck) Redhead, Lutzoni, Moncalvo & Vilgalys, *Mycotaxon* 83: 48 (2002)
  Agaricus peltigerinus Peck, *Rep. (Annual) New York State Mus. Nat. Hist.* 30: 38 (1888) [1887]
  Omphalina peltigerina (Peck) P. Collin, *Bull. Soc. Mycol. France* 110(1): 11 (1994)
  Omphalina cupulatoides P.D. Orton, *Kew Bull.* 31: 712 (1976)
E: ! **S:** ! **W:** !
H: On dead and moribund thalli of the lichen genus *Peltigera*.
Known from Scotland and Wales. Reported from England but unsubstantiated with voucher material. Saprotrophic or weakly parasitic on lichen thalli but apparently not lichenised.

philonotis (Lasch) Redhead, Lutzoni, Moncalvo & Vilgalys, *Mycotaxon* 83: 48 (2002)
  Agaricus philonotis Lasch, *Linnaea* 3: 394 (1828)
  Omphalia philonotis (Lasch) Quél., *Mém. Soc. Emul. Montbéliard, Sér. 2,* 5: 239 (1872)
  Omphalina philonotis (Lasch) Quél., *Enchir. fung.*: 43 (1886)
E: ? **S:** !
H: In boggy areas, usually amongst *Sphagnum* spp and *Philonotis fontana* along small rivulets or around springs in montane habitat.
D: NM2: 172, FAN3: 85
Known from Scotland. Reported from England but the records are unsubstantiated with voucher material.

retiruga (Bull.) Redhead, *Canad. J. Bot.* 62(5): 873 (1984)
  Helvella retiruga Bull., *Herb. France:* pl. 498 (1791)
  Cantharellus retirugus (Bull.) Fr., *Syst. mycol.* 1: 324 (1821)
  Leptoglossum retirugum (Bull.) Ricken, *Blätterpilze Deutschl.*: 6 (1911)
  Helvella membranacea Dicks., *Fasc. pl. crypt. brit.* 1: 21 (1785)
  Merulius membranaceus (Dicks.) With., *Arr. Brit. pl. ed.3* 4: 153 (1796)
  Leptoglossum conchatum Velen., *Mycologia* 2: 46 (1925)
  Mis.: *Cyphella galeata* sensu auct.
E: ! **S:** ! **W:** ! **NI:** ! **ROI:** ! **O:** Isle of Man: !
H: Parasitic on mosses. Known on *Brachythecium* spp, *Calliergon cuspidatum, Camptothecium sericeum, Dicranella subulata, Funaria hygrometrica, Hypnum, Polytrichum* and *Rhytidiadelphus* spp. Often fruiting from late autumn into the spring.
D: BFF6: 29-30 (as *Arrhenia retiruge*), NM2: 99, FAN3: 40 **D+I:** Ph: 264 (as *Leptoglossum retirugum*) **I:** Sow3: 348 (as *Helvella membranacea*)
Rarely reported, but apparently widespread. The correct orthography of the epithet is *retiruga* and not the commonly used *retiruge*.

rickenii (Hora) Watling, *Notes Roy. Bot. Gard. Edinburgh* 45(3): 553 (1989) [1988]
  Omphalina rickenii Hora, *Trans. Brit. Mycol. Soc.* 43(2): 454 (1960)
  Phaeotellus rickenii (Hora) Bon, *Doc. Mycol.* 5(19): 27 (1975)
  Leptoglossum rickenii (Hora) Singer, *Agaricales in Modern Taxonomy,* Edn 4: 269 (1986)
  Cantharellus cupulatus Fr. [*nom. superf.*], *Epicr. syst. mycol.*: 367 (1838)
  Omphalina cupulata (Fr.) P.D. Orton [*comb. inval.*], *Trans. Brit. Mycol. Soc.* 43(2): 179 (1960)

  Leptotus rickenii Singer [*nom. nud.*], *Lilloa* 22: 735 (1951) [1949]
  Mis.: *Cantharellus helvelloides* sensu auct.
  Mis.: *Omphalia muralis* sensu auct.
  Mis.: *Omphalia rustica* sensu K&R
E: ! **W:** ! **ROI:** !
H: On sandy soil and on old dry stone walls, associated with mosses in both habitats.
D: BFF6: 33, FAN3: 86-87 **D+I:** B&K3: 304 384 **I:** Bon: 129, C&D: 183
Rarely reported. Excludes *Omphalina cupulata* sensu NCL, apparently a different species requiring clarification.

rustica (Fr.) Redhead, Lutzoni, Moncalvo & Vilgalys, *Mycotaxon* 83: 48 (2002)
  Agaricus rusticus Fr., *Epicr. syst. mycol.*: 126 (1838)
  Omphalia rustica (Fr.) Gillet, *Hyménomycètes*: 297 (1876)
  Omphalina rustica (Fr.) Quél., *Enchir. fung.*: 43 (1886)
E: ? **S:** ? **ROI:** ?
H: British records on sandy soil amongst mosses and lichens. Variously interpreted in the past, but now lectotypified (Graphis scripta 3: 138-143). British material, however, needs re-examination. Much of it is likely to be *Omphalina wallacei*.

spathulata (Fr.) Redhead, *Canad. J. Bot.* 62(5): 876 (1984)
  Cantharellus spathulatus Fr., *Elench. fung.* 1: 53 (1828)
  Agaricus muscigenus Bull., *Herb. France*: pl. 288 (1786)
  Merulius muscigenus (Bull.) With., *Arr. Brit. pl. ed.3* 4: 153 (1796)
  Cantharellus muscigenus (Bull.) Fr., *Syst. mycol.* 1: 323 (1821)
  Corniola muscigena (Bull.) Gray, *Nat. arr. Brit. pl.* 1: 596 (1821)
  Leptoglossum muscigenum (Bull.) P. Karst., *Bidrag Kännedom Finlands Natur Folk* 32: 242 (1879)
  Dictyolus muscigenus (Bull.) Quél., *Enchir. fung.*: 140 (1886)
  Leptotus muscigenus (Bull.) Maire, *Publ. Inst. Bot. Barcelona* 15: 52 (1933)
E: o **S:** ! **W:** o **ROI:** ! **O:** Channel Islands: !
H: With mosses on heathland and in unimproved grassland.
D: BFF6: 30, NM2: 99, FAN3: 40 **D+I:** B&K3: 142 139 **I:** C&D: 153, C&D: 153, Kriegl3: 131 (as *Arrhenia retiruga* var. *spathulata*)
Occasional but apparently widespread. Said to be closely associated with the moss *Tortula ruralis* but this is not supported by voucher material in herb. K.

sphagnicola (Berk.) Redhead, Lutzoni, Moncalvo & Vilgalys, *Mycotaxon* 83: 48 (2002)
  Agaricus sphagnicola Berk., *Engl. fl.*: 67 (1836)
  Omphalia sphagnicola (Berk.) P. Karst., *Bidrag Kännedom Finlands Natur Folk* 32: 130 (1879)
  Omphalina sphagnicola (Berk.) M.M. Moser, *Kleine Kryptogamenflora,* Edn 3: 72 (1967)
  Omphalina gerardiana (Peck) Singer, *Lilloa* 22: 212 (1951) [1949]
  Omphalina fusconigra P.D. Orton, *Trans. Brit. Mycol. Soc.* 43(2): 335 (1960)
E: ! **S:** ! **NI:** ! **ROI:** !
H: In *Sphagnum* spp on heathland, moorland or montane areas.
D: FAN3: 84 (as *O. gerardiana*) **D+I:** B&K3: 306 386
Described from Britain, but poorly known. Accepted here sensu Redhead *et al.*, but FAN3 considers *A. sphagnicola* sensu stricto a probable synonym of *Lichenomphalia ericetorum*, preferring *O. gerardiana* for the current species.

velutipes (P.D. Orton) Redhead, Lutzoni, Moncalvo & Vilgalys, *Mycotaxon* 83: 48 (2002)
  Omphalina velutipes P.D. Orton, *Trans. Brit. Mycol. Soc.* 43(2): 180 (1960)
E: ! **S:** ! **W:** !
H: On soil amongst grasses, mosses and lichens, on heath and moorland often in upland or montane habitat.
D: FAN3: 88 **D+I:** B&K3: 306 387
Rarely collected or reported. Apparently widespread, but British records mostly from Scotland.

ASEROË Labill., *Voy. rech. Pérouse* 1: 145 (1800)
Phallales, Clathraceae
Type: *Aseroë rubra* Labill.

rubra Labill., *Voy. rech. Pérouse* 1: 145 (1800)
E: !
H: Amongst leaf litter, on sandy, acidic soil in mixed woodland (British material).
D+I: BritPuffb: 188-189 Figs. 151/152
An alien, naturalised in one location in Surrey (Esher Common) and previously known from an old record from Kew Gardens. Native to the southern sub-tropics.

ASTEROPHORA Ditmar, in Link, *J. Bot. (Schrader)* 3: 17 (1809)
Agaricales, Tricholomataceae
Nyctalis Fr., *Syst. orb. veg.* 1: 78 (1825)
Type: *Asterophora lycoperdoides* (Pers.) Ditmar

lycoperdoides (Bull.) Ditmar, *J. Bot. (Schrader)* 3: 56 (1809)
Agaricus lycoperdoides Bull., *Herb. France*: pl. 516 (1791)
Asterophora agaricoides Fr., *Observ. mycol.* 2: 367 (1818)
Nyctalis asterophora Fr., *Epicr. syst. mycol.*: 371 (1838)
E: o **S:** o **W:** o **ROI:** ! **O:** Channel Islands: ! Isle of Man: !
H: On moribund or decayed basidiomes of *Russula* spp, usually *R. nigricans,* rarely *R. adusta.*
D: NM2: 100 **D+I:** Ph: 76 (as *N. asterophora*), Myc4(2): 87, B&K3: 298 375 (as *N. asterophora*) **I:** Sow3: 383 (as *Agaricus lycoperdoides*), FlDan5: 161G (as *N. asterophora* f. *major*), Bon: 169, Kriegl3: 474 (as *N. asterophora*)
Occasional but widespread.

parasitica (Bull.) Singer, *Lilloa* 22: 171 (1951) [1949]
Agaricus parasiticus Bull., *Herb. France*: pl. 574 f. 2 (1792)
Gymnopus parasiticus (Bull.) Gray, *Nat. arr. Brit. pl.* 1: 610 (1821)
Nyctalis parasitica (Bull.) Fr., *Syst. orb. veg.* 1: 69 (1825)
E: o **S:** o **W:** ! **NI:** o **ROI:** ! **O:** Isle of Man: !
H: On moribund or decayed basidiomes of *Russula* spp, usually *R. nigricans,* but also *R. fellea, R. foetens* and *R. densifolia.*
D: NM2: 100 **D+I:** Ph: 76-77, B&K3: 300 376 (as *Nyctalis parasitica*) **I:** Sow3: 343 (as *Agaricus parasiticus*), FlDan5: 162G (as *N. parasitica*), Bon: 169, Kriegl3: 476 (as *N. parasitica*)
Occasional but widespread.

ASTEROSTROMA Massee, *J. Linn. Soc., Bot.* 25: 154 (1889)
Hymenochaetales, Asterostromataceae
Type: *Asterostroma apalum* (Berk. & Broome) Massee

cervicolor (Berk. & M.A. Curtis) Massee, *J. Linn. Soc., Bot.* 25: 155 (1889)
Corticium cervicolor Berk. & M.A. Curtis, *Grevillea* 1: 179 (1873)
Asterostroma medium Bres., *Ann. Mycol.* 18: 49 (1920)
Asterostroma ochroleucum Bres., *Brotéria, Sér. Bot.* 11(1): 82 (1913)
E: ! **S:** ! **W:** ! **ROI:** !
H: Most often on structural timber in buildings, but one collection (in herb. K) on dead stems of *Rumex* spp.
D: NM3: 319 **D+I:** Hallenb: 5-9 Figs 1-3, B&K2: 242 291 (as *Asterostroma ochroleucum*)
Rarely reported but apparently widespread. A major cause of timber decay in buildings in Europe.

laxum Bres., *Bull. Soc. Mycol. France* 36: 46 (1920)
E: ! **S:** !
H: On decayed wood of conifers such as *Pinus* spp. A single collection supposedly on *Quercus* spp, but that host is not verified.
D: NM3: 319 **D+I:** Hallenb: 9-10 Figs 5-7, B&K2: 242 289, Myc11(4): 181

Rarely reported. Known from England (Cheviotshire, Hertfordshire, South Devon, Surrey and West Sussex) and a single collection from Scotland (Perthshire).

ASTRAEUS Morgan, *J. Cincinnati Soc. Nat. Hist.* 12: 20 (1889)
Sclerodermatales, Astraeaceae
Type: *Astraeus hygrometricus* (Pers.) Morgan

hygrometricus (Pers.) Morgan, *J. Cincinnati Soc. Nat. Hist.* 12: 20 (1889)
Geastrum hygrometricum Pers., *Syn. meth. fung.*: 135 (1801)
Geastrum hygrometricum β *anglicum* Pers., *Syn. meth. fung.*: 135 (1801)
Astraeus stellatus (Scop.) E. Fisch., *Nat. Pflanzenfamilien* 1(1): 341 (1900)
Geastrum vulgaris Corda, *Icon. fung.* 5: 64 (1842)
E: ! **W:** ! **NI:** !
H: On sandy soil, associated with deciduous trees in dry woodland.
D+I: Ph: 254-255, BritPuffb: 40-41 Figs. 20/21 **I:** Bon: 303, Kriegl2: 171
Rarely reported. Apparently widespread in England (Leicestershire to West Cornwall) and Wales (Cardiganshire) but mainly southern and south western. Reported from Northern Ireland but only a painting exists, in herb. K, of material collected in Armagh in 1874. English name = 'Barometer Earthstar'.

ATHELIA Pers., *Mycol. eur.* 1: 83 (1822)
Stereales, Atheliaceae
Type: *Athelia epiphylla* Pers.

acrospora Jülich, *Willdenowia. Beih.* 7: 45 (1972)
E: ! **S:** ! **W:** !
H: On decayed leaf litter of deciduous trees, moribund fern fronds and small pieces of woody debris.
D: NM3: 146 **D+I:** CNE2: 99-101
Known from England (Hertfordshire, Surrey and West Kent), Scotland, and Wales (Caernarvonshire).

alnicola (Bourdot & Galzin) Jülich, *Willdenowia. Beih.* 7: 47 (1972)
Corticium centrifugum ssp. *alnicola* Bourdot & Galzin, *Hymenomyc. France*: 198 (1928)
E: !
H: On fallen and decayed leaves of deciduous trees. (British material on *Acer pseudoplatanus*).
A single collection, from Yorkshire (Nun Appleton) in 1939. Considered by some authorities to be a synonym of *A. epiphylla.*

arachnoidea (Berk.) Jülich, *Willdenowia. Beih.* 7: 53 (1972)
Corticium arachnoideum Berk., *Ann. Mag. Nat. Hist., Ser. 1* [*Notices of British Fungi no. 287*] 13: 345 (1844)
Hypochnus bisporus J. Schröt., in Cohn, *Krypt.-Fl. Schlesien* 3: 415 (1889)
Athelia bispora (J. Schröt.) Donk, *Fungus* 27: 12 (1957)
Corticium bisporum (J. Schröt.) Bourdot & Galzin, *Bull. Soc. Mycol. France* 27(2): 241 (1911)
Corticium centrifugum ssp. *bisporum* (J. Schröt.) Bourdot & Galzin, *Hymenomyc. France*: 199 (1928)
Mis.: *Corticium centrifugum* sensu auct. p.p.
E: c **S:** c **W:** c **NI:** ! **ROI:** !
H: On decayed wood of deciduous trees and conifers, on fallen and living leaves of deciduous trees and shrubs, and decayed straw. Occasionally on algae on wood and known to parasitise various species of lichen.
D: NM3: 145 **D+I:** CNE2: 102-103, B&K2: 82 49
Common and widespread.

bombacina (Link) Pers., *Mycol. eur.* 1: 85 (1822)
Sporotrichum bombacinum Link, *Mag. Neuesten Entdeck. Gesammten Naturk. Ges. Naturf. Freunde Berlin* 8: 36 (1816)
E: ! **S:** !
H: On decayed wood, litter (leaves and twigs) and soil (including mushroom compost).
D: NM3: 146 **D+I:** CNE2: 106-109, B&K2: 82 50
Known from England (Buckinghamshire, Oxfordshire, Surrey and West Suffolk) and Scotland. Reported from Staffordshire, South Hampshire.

decipiens (Höhn. & Litsch.) J. Erikss., *Symb. Bot. Upsal.* 16(1): 86 (1958)
Corticium decipiens Höhn. & Litsch., *Sitzungsber. Kaiserl. Akad. Wiss., Wien, Math.-Naturwiss. Cl., Abt. 1* 117: 1116 (1908)
Athelia caucasica Parmasto, *Conspectus Systematis Corticiacearum*: 199 (1968)
E: !
H: On decayed wood of deciduous trees or conifers, fallen leaves, soil, and decayed stems of the grass *Holcus lanatus*.
D: NM3: 146 **D+I:** CNE2: 110-111, B&K2: 82 51
Rarely recorded or collected but apparently widespread in England.

epiphylla Pers., *Mycol. eur.* 1: 84 (1822)
Hypochnus epiphyllus (Pers.) Wallr., *Fl. crypt. Germ., Sect. II*: 310 (1833)
Thelephora epiphylla (Pers.) Fr., *Epicr. syst. mycol.*: 226 (1838)
E: o **S:** o **W:** o **ROI:** !
H: On decayed wood, and leaf litter of deciduous trees. Also reported as a parasite on lichens such as *Lecanora* and *Xanthoria* spp.
D: NM3: 146 **D+I:** CNE2: 112-121, B&K2: 84 52
Occasional but widespread. Probably a species complex.

fibulata M.P. Christ., *Dansk. Bot. Ark.* 19(2): 149 (1960)
W: !
H: British material is from the undersides of clumps of the moss *Thamnobryum alopecurum*.
D: NM3: 146 **D+I:** CNE2: 122-123, B&K2: 84 53
A single collection from Wales (Breconshire: Erwood) in 2001. Reported from Scotland but the specimens in herb. E are redetermined as *Cylindrobasidium laeve*.

neuhoffii (Bres.) Donk, *Fungus* 27: 12 (1957)
Corticium neuhoffii Bres., *Z. Pilzk.* 2(8): 179 (1923)
E: ! **S:** ! **NI:** ?
H: On decayed wood of deciduous trees and conifers, and occasionally on soil.
D: NM3: 146 **D+I:** CNE2: 124-125, B&K2: 84 54
Rarely recorded or collected. Known in widely scattered locations in England (from Yorkshire to South Hampshire) and from Scotland (Perthshire).

nivea Jülich, *Willdenowia. Beih.* 7: 103 (1972)
E: !
H: On decayed wood and fallen bark of deciduous trees.
Rarely reported. Known from widely scattered locations in England (Yorkshire to East Sussex).

pyriformis (M.P. Christ.) Jülich, *Willdenowia. Beih.* 7: 110 (1972)
Xenasma pyriforme M.P. Christ., *Dansk. Bot. Ark.* 19(2): 108 (1960)
E: ! **S:** !
H: On decayed wood of deciduous trees and also known on moribund fronds or decayed litter of *Pteridium*.
D: NM3: 146 **D+I:** CNE2: 126-127, B&K2: 86 55
Rarely recorded or collected, but widespread. Known from widely scattered locations in England, (Yorkshire to Surrey) and Scotland (Wester Ross).

salicum Pers., *Mycol. eur.* 1: 84 (1822)
Corticium centrifugum ssp. *fugax* Bourdot & Galzin, *Hymenomyc. France*: 198 (1928)

Athelia incrustata M.P. Christ., *Dansk. Bot. Ark.* 19(2): 146 (1960)
E: ! **S:** !
H: British material on dead wood of *Quercus* spp, *Sambucus nigra* and *Ulmus glabra* (but also on coniferous substrata in continental Europe).
Known from England (Yorkshire: Pickering area) and an old collection from Scotland.

tenuispora Jülich, *Willdenowia. Beih.* 7: 120 (1972)
E: ! **S:** !
H: On decayed wood of conifers.
Known from England (Surrey) and Scotland (Orkney).

teutoburgensis (Brinkmann) Jülich, *Persoonia* 7(3): 83 (1973)
Corticium teutoburgense Brinkmann, *Jahrb. Westf. Provinz. Verein. Wiss. Kunst.* 44: 38 (1916)
Hyphoderma teutoburgense (Brinkmann) J. Erikss., *Symb. Bot. Upsal.* 16(1): 100 (1958)
Radulomyces teutoburgensis (Brinkmann) Parmasto, *Conspectus Systematis Corticiacearum*: 110 (1968)
Corticium centrifugum var. *macrosporum* Bourdot & Galzin, *Hymenomyc. France*: 198 (1928)
Athelia macrospora (Bourdot & Galzin) M.P. Christ., *Dansk. Bot. Ark.* 19(2): 46 (1968)
Corticium flavescens Bres., *Ann. Mycol.* 3(2): 163 (1905)
E: ! **S:** ? **NI:** ?
H: On decayed wood of deciduous trees. British material on *Fagus*.
Known from England (Warwickshire and West Sussex). Reported from Scotland in 1928 and from Northern Ireland in 1932 but unsubstantiated with voucher material.

ATHELOPSIS Parmasto, *Conspectus Systematis Corticiacearum*: 41 (1968)

Stereales, Atheliaceae
Pteridomyces Jülich, *Persoonia* 10: 331 (1979)
Type: *Athelopsis glaucina* (Bourdot & Galzin) Parmasto

baculifera (Bourdot & Galzin) Jülich, *Willdenowia. Beih.* 6: 217 (1971)
Corticium baculiferum Bourdot & Galzin, *Hymenomyc. France*: 194 (1928)
E: !
H: On decayed wood of conifers. British material on a fallen branch of *Pinus sylvestris*.
A single collection from Cheviotshire (Thrunton Wood) in 2001.

galzinii (Bres.) Hjortstam, *Mycotaxon* 42: 149 (1991)
Epithele galzinii Bres., *Bull. Soc. Mycol. France* 27: 264 (1911)
Pteridomyces galzinii (Bres.) Jülich, *Persoonia* 10(3): 331 (1979)
Pteridomyces bananisporus Boidin & Gilles, *Bull. Soc. Mycol. France* 102: 302 (1986)
Athelopsis bananispora (Boidin & Gilles) Hjortstam, *Mycotaxon* 42: 151 (1991)
E: ! **NI:** ! **O:** Channel Islands: !
H: On decayed debris of ferns such as *Phyllitis scolopendrium* and on decayed wood of conifers.
D+I: Myc9(4): 163 (as *A. bananispora*)
Known from England (South Devon, Surrey and West Cornwall), Northern Ireland (Derry) and the Channel Islands (Guernsey). Note that the description and illustration of the basidiospores in CNE3: 363-364 (as *Epithele galzinii*) are incorrect.

glaucina (Bourdot & Galzin) Parmasto, *Conspectus Systematis Corticiacearum*: 42 (1968)
Corticium glaucinum Bourdot & Galzin, *Hymenomyc. France*: 207 (1928)
Athelia glaucina (Bourdot & Galzin) Donk, *Fungus* 27: 12 (1957)
E: ! **S:** ! **W:** !
H: On decayed wood of deciduous trees, such as *Betula, Carpinus, Corylus, Fagus* and *Populus* spp. Rarely on wood of

conifers such as *Larix* spp, and decayed lianas of *Clematis vitalba*.
D: NM3: 147 **D+I:** CNE2: 136-137
Rarely reported but apparently widespread.

lembospora (Bourdot) Oberw., *Persoonia* 7(1): 3 (1972)
   Corticium lembosporum Bourdot, *Rev. Sci. du Bourb.* 32(1): 10 (1910)
   Luellia lembospora (Bourdot) Jülich, *Persoonia* 8(3): 292 (1975)
   Corticium confusum Bourdot & Galzin, *Bull. Soc. Mycol. France* 27: 250 (1911)
E: ! **S:** ! **W:** !
H: Usually on decayed debris of ferns such as *Dryopteris*, *Polystichum* spp and *Pteridium*. Rarely on decayed wood of *Quercus robur* and *Ulmus* spp, on dying leaves of *Luzula* spp, and on dead lianas of *Clematis vitalba*.
D: NM3: 147
Rarely reported but apparently widespread.

AURANTIPORUS Murrill, *Bull. Torrey Bot. Club* 32: 487 (1905)
Poriales, Coriolaceae
Type: *Polyporus pilotae* Schwein. (= *Aurantiporus croceus* (Pers.) Murrill)

alborubescens (Bourdot & Galzin) H. Jahn, *Westfäl. Pilzbriefe* 9(6-7): 99 (1973)
   Phaeolus albosordescens ssp. *alborubescens* Bourdot & Galzin, *Bull. Soc. Mycol. France* 41: 136 (1925)
   Phaeolus alborubescens (Bourdot & Galzin) Bourdot & Galzin, *Hymenomyc. France*: 556 (1928)
   Tyromyces alborubescens (Bourdot & Galzin) Bondartsev, *Trutovye griby Evropeiskoi chasti SSSR i Kavkaza*: 223 (1953)
E: !
H: On large fallen trunks or stumps in old deciduous woodland. Most collections are on *Fagus* but also known on *Fraxinus*.
D: EurPoly2: 684 (as *Tyromyces alborubescens*) **D+I:** Myc7(1): 15 **I:** SV42: 44
Known from Berkshire, South Hampshire, Surrey and West Kent and apparently increasing in areas such as the New Forest. Possibly a recent colonist, since the large basidiomes and strong, aromatic odour are distinctive, and would be difficult to overlook.

fissilis (Berk. & M.A. Curtis) H. Jahn, *Westfäl. Pilzbriefe* 9: 134 (1973)
   Polyporus fissilis Berk. & M.A. Curtis, *Hooker's J. Bot. Kew Gard. Misc.* 1: 234 (1849)
   Tyromyces fissilis (Berk. & M.A. Curtis) Donk, *Meded. Bot. Mus. Herb. Rijks Univ. Utrecht* 9: 153 (1933)
   Polyporus albosordescens Romell, *Svensk bot. Tidskr.* 6: 637 (1912)
   Leptoporus albosordescens (Romell) Pilát, *Bull. Soc. Mycol. France* 48: 8 (1932)
E: o **W:** !
H: Most frequently on large dead standing trunks of *Fagus* in old woodland on calcareous soil. Rarely on *Malus* spp in gardens and orchards (on old trees) and *Betula*, *Liriodendron*, *Quercus* and *Ulmus* spp.
D: EurPoly2: 687, NM3: 225 (as *Tyromyces fissilis*) **D+I:** B&K2: 312 395 **I:** Kriegl1: 597 (as *T. fissilis*)
Occasional in England as far north as Yorkshire. Single collections from Wales (Glamorganshire) and Northern Ireland (Down).

AUREOBOLETUS Pouzar, *Česká Mykol.* 11: 48 (1957)
Boletales, Boletaceae
Type: *Aureoboletus gentilis* (Quél.) Pouzar

gentilis (Quél.) Pouzar, *Česká Mykol.* 11: 48 (1957)
   Boletus sanguineus var. *gentilis* Quél., *Compt. Rend. Assoc. Franç. Avancem. Sci.* 12: 504 (1884) [1883]

   Ixocomus gentilis (Quél.) Quél., *Fl. mycol. France*: 413 (1888)
   Boletus gentilis (Quél.) Sacc., *Syll. fung.* 6: 8 (1888)
   Xerocomus gentilis (Quél.) Singer, *Rev. Mycol. (Paris)* 5(1): 6 (1940)
   Pulveroboletus gentilis (Quél.) Singer, *Farlowia* 2: 300 (1945)
   Boletus cramesinus Secr. [*nom. inval.*], *Mycogr. Suisse* 3: 39 (1833)
   Pulveroboletus cramesinus Singer, *Pilze Mitteleuropas* 6: 12 (1966)
   Aureoboletus cramesinus (Singer) Watling, *Notes Roy. Bot. Gard. Edinburgh* 29(2): 265 (1969)
   Boletus granulatus var. *tenuipes* Cooke, *Grevillea* 12: 43 (1883)
   Boletus tenuipes (Cooke) Massee, *Brit. fung.-fl.* 1: 281 (1892)
   Mis.: *Boletus auriporus* sensu Kallenbach [*Röhrlinge* p. 96 (1931)]
E: ! **S:** ?
H: On soil, sometimes on old fire sites, in deciduous woodland, or with solitary trees in parkland. Usually associated with *Quercus* spp. Rarely with *Fagus*.
D: NM2: 67 (as *Pulveroboletus gentilis*) **D+I:** Ph: 205, FRIC16: 6-7 121c, B&K3: 76 40 (as *P. gentilis*) **I:** Bon: 45, C&D: 423, Kriegl2: 288 (as *P. gentilis*)
Rarely reported. Known from southern England (Berkshire, Buckinghamshire, East Kent, Middlesex, South Essex, South Hampshire and Surrey). Reported from Scotland (South Aberdeenshire) but unsubstantiated with voucher material.

AURICULARIA Bull., *Hist. champ. France*: 277 (1791)
Auriculariales, Auriculariaceae
Hirneola Fr. [non *Hirneola* Velen. (1939)], *Kongl. Vetensk. Acad. Handl.* 1848: 144 (1849)

auricula-judae (Bull.) Wettst., in Pat., *Cat. pl. cell. Tunisie*: 75 (1897)
   Tremella auricula-judae Bull., *Herb. France*: pl. 427 f. 2 (1789)
   Exidia auricula-judae (Bull.) Fr., *Syst. mycol.* 2(1): 221 (1823)
   Hirneola auricula-judae (Bull.) Berk., *Outl. Brit. fungol.*: 289 (1860)
   Tremella auricula L., *Sp. pl.* II: 1157 (1753)
   Hirneola auricula (L.) H. Karst., *Deut. Fl.*: 93 (1880)
   Auricularia auricula (L.) Underw., *Mem. Torrey Bot. Club* 12: 15 (1902)
   Gyraria auricularis Gray, *Nat. arr. Brit. pl.* 1: 594 (1821)
   Auricularia auricula-judae var. *lactea* Quél., *Enchir. fung.*: 207 (1886)
   Hirneola auricula-judae var. *lactea* (Quél.) D.A. Reid, *Trans. Brit. Mycol. Soc.* 55(3): 440 (1970)
E: c **S:** c **W:** c **NI:** c **ROI:** c **O:** Channel Islands: c Isle of Man: o
H: Saprotrophic or perhaps weakly parasitic. On wood of deciduous trees and shrubs, commonly *Sambucus nigra* but also on *Acer campestre*, *A. palmatum*, *A. pseudoplatanus*, *Berberis* spp, *Buddleja davidii*, *Clematis vitalba*, *Cornus sanguinea*, *Eleagnus* spp, *Euonymus europaeus*, *Fagus*, *Fraxinus*, *Hedera*, *Juglans regia*, *Ilex*, *Laburnum anagyroides*, *Platanus* spp, *Prunus laurocerasus*, *P. spinosa*, *Salix* and *Ulmus* spp. Rarely on conifers such as *Picea* spp.
D: NM3: 97 **D+I:** Ph: 262, B&K2: 54 7 **I:** Bon: 325, C&D: 137, SV33: 5, Kriegl1: 61
Very common and widespread. The var. *lactea* is an unpigmented form, not otherwise distinct. English name = 'Jew's Ear Fungus'.

mesenterica (Dicks.) Pers., *Mycol. eur.* 1: 97 (1822)
   Helvella mesenterica Dicks., *Fasc. pl. crypt. brit.* 1: 20 (1785)
   Stereum mesentericum (Dicks.) Gray, *Nat. arr. Brit. pl.* 1: 653 (1821)
   Auricularia corrugata Sowerby, *Col. fig. Engl. fung.* 3: pl. 290 (1800)
   Auricularia lobata Sommerf., *Mag. Naturvidensk.*: 295 - 299 (1826)

Auricularia tremelloides Bull., *Herb. France*: pl. 290 (1787)
Tremella violacea Relhan, *Fl. Cantab.*: 442 (1785)
Gyraria violacea (Relhan) Gray, *Nat. arr. Brit. pl.* 1: 594 (1821)

E: c **S:** ! **W:** ! **NI:** ! **O:** Channel Islands: !
H: On dead and decayed wood of deciduous trees and shrubs, often old stumps of *Ulmus* spp, or *Fraxinus* but also known on *Acer campestre*, *Betula* spp, *Fagus*, *Ficus carica*, *Malus*, *Quercus*, *Salix* spp, and *Ulex europaeus*. A single collection in herb. K supposedly on *Taxus*.
D: NM3: 97 **D+I:** Ph: 263, B&K2: 54 8 **I:** Bon: 325, C&D: 137, Kriegl1: 63
Common in England. Infrequently reported elsewhere but apparently widespread. English name = 'Tripe Fungus'.

AURISCALPIUM Gray, *Nat. arr. Brit. pl.* 1: 650 (1821)
Hericiales, Auriscalpiaceae
Type: *Auriscalpium vulgare* Gray

vulgare Gray, *Nat. arr. Brit. pl.* 1: 650 (1821)
Hydnum auriscalpium L., *Sp. pl.*: 1178 (1753)
E: o **S:** ! **W:** o **NI:** ! **ROI:** !
H: Only on decayed cones of *Pinus* spp (most often *P. sylvestris*) buried or partially buried in soil and needle litter. Reported on both *Abies* and *Larix* spp, but these hosts are unverified.
D: NM3: 285 **D+I:** Ph: 242, B&K2: 238 283, BritChant: 46-47 **I:** Sow3: 267 (as *Hydnum auriscalpium*), Bon: 313, C&D: 141
Occasional but widespread and may be locally abundant. English name = 'Ear Pick Fungus'.

BAEOSPORA Singer, *Rev. Mycol. (Paris)* 3(6): 193 (1938)
Agaricales, Tricholomataceae
Type: *Baeospora myosuas* (Fr.) Singer

myosura (Fr.) Singer, *Rev. Mycol. (Paris)* 3(6): 193 (1938)
Agaricus myosurus Fr., *Observ. mycol.* 2: 129 (1818)
Collybia myosura (Fr.) Quél., *Mém. Soc. Émul. Montbéliard, Sér. 2*, 5: 95 (1872)
Marasmius myosurus (Fr.) P. Karst., *Bidrag Kännedom Finlands Natur Folk* 48: 102 (1889)
Collybia friesii Bres., *Icon. Mycol.*: t. 214 (1928)
Marasmius friesii (Bres.) Rea, *Trans. Brit. Mycol. Soc.* 17: 46 (1932)
Mis.: *Marasmius conigenus* sensu Rea (1922) p.p.
E: c **S:** c **W:** ! **NI:** !
H: On decayed and often partially buried cones of conifers such as *Abies*, *Cedrus*, *Larix*, *Picea*, *Pinus* and *Pseudotsuga menziesii*.
D: NM2: 100, BFF8: 108, FAN4: 165-166 **D+I:** Ph: 68-69, B&K3: 142 140 **I:** Kriegl3: 134
Common and widespread.

BANKERA Pouzar, *Česká Mykol.* 9: 95 (1955)
Thelephorales, Bankeraceae
Type: *Bankera fuligineoalba* (J.C. Schmidt) Pouzar

fuligineoalba (J.C. Schmidt) Pouzar, *Česká Mykol.* 9: 96 (1955)
Hydnum fuligineoalbum J.C. Schmidt, *Mykol. Hefte* 1: 88 (1817)
Hydnum fragile Fr., *Öfvers. Kongl. Vetensk.-Akad. Förh.* 8: 53 (1852)
Hydnum macrodon Pers., *Syn. meth. fung.*: 560 (1801)
E: ! **S:** o
H: On soil, amongst needle litter of *Pinus sylvestris*.
D: NM3: 309 **D+I:** Ph: 244, B&K2: 230 272, BritChant: 60-61 **I:** Bon: 313, SV27: 4
Occasional but widespread in Scotland. Rare in England and known from Berkshire (Ascot) in 1864 and Yorkshire in 1971.

violascens (Alb. & Schwein.) Pouzar, *Česká Mykol.* 9: 96 (1955)

Hydnum violascens Alb. & Schwein., *Consp. fung. lusat.*: 265 (1805)
S: !
H: On soil, amongst needle litter of *Picea abies*.
D: NM3: 309 **D+I:** B&K2: 230 273, BritChant: 62-63 **I:** SV27: 4
Two recent collections, one from South Aberdeenshire (Inver Wood).

BASIDIODENDRON Rick, *Brotéria, Ci. Nat.* 7: 74 (1938)
Tremellales, Exidiaceae
Type: *Basidiodendron luteogriseum* Rick (= *B. eyrei*)

caesiocinereum (Höhn. & Litsch.) Luck-Allen, *Canad. J. Bot.* 41: 1036 (1963)
Corticium caesiocinereum Höhn. & Litsch., *Sitzungsber. Kaiserl. Akad. Wiss., Wien, Math.-Naturwiss. Cl., Abt. 1* 117: 1116 (1908)
Sebacina caesiocinerea (Höhn. & Litsch.) D.P. Rogers, *Univ. Iowa Stud. Nat. Hist.* 17: 37 (1935)
Bourdotia caesiocinerea (Höhn. & Litsch.) Pilát & Lindtner, *Glasnik (Bull.) Soc. Scient. Skoplje* 18: 175 (1938)
Bourdotia cinerella Bourdot & Galzin, *Bull. Soc. Mycol. France* 36: 71 (1920)
E: ! **S:** ! **W:** !
H: In woodland, often on calcareous soil, on decayed wood of deciduous trees such as *Betula*, *Fraxinus*, *Populus tremula*, *Prunus cerasus* and *Salix* spp.
D: NM3: 98 **D+I:** B&K2: 56 10
Rarely reported but apparently widespread. Mostly reported from southern England, where it would appear to be not infrequent.

cinereum (Bres.) Luck-Allen, *Canad. J. Bot.* 41: 1043 (1963)
Sebacina cinerea Bres., *Fungi trident.* 2(14): 99 (1900)
Bourdotia cinerea (Bres.) Bourdot & Galzin, *Hymenomyc. France*: 49 (1928)
E: ! **S:** ! **NI:** ?
H: British material on dead wood of *Fuchsia magellanica* and also on dead attached leaves of tree ferns.
D: NM3: 98 **D+I:** Reid4: 433-434
Rarely reported. Known from England (Northamptonshire and West Cornwall) and Scotland (North Ebudes). Reported from Northern Ireland but unsubstantiated with voucher material.

cremeum (McNabb) K. Wells & Raitv., *Mycologia* 67: 909 (1975)
Sebacina cremea McNabb, *New Zealand J. Bot.* 7: 249 (1969)
E: !
H: British material is on dead standing stems of *Pteridium*.
A single collection from South Devon (Slapton Wood) in 1992.

eyrei (Wakef.) Luck-Allen, *Canad. J. Bot.* 41: 1034 (1963)
Sebacina eyrei Wakef., *Trans. Brit. Mycol. Soc.* 5(1): 126 (1916)
Bourdotia eyrei (Wakef.) Bourdot & Galzin, *Hymenomyc. France*: 50 (1928)
Gloeocystidium croceotingens Bres., in Bresadola, *Ann. Mycol.* 18: 48 (1920)
E: ! **S:** ! **W:** !
H: In woodland, on decayed wood of deciduous trees such as *Alnus*, *Betula* spp, *Fagus*, *Fraxinus* and *Salix* spp.
D: NM3: 98
Rarely reported but widespread. Known from England (East Gloucestershire, East & West Sussex, Hertfordshire, Middlesex, North & South Hampshire, South Devon, South Essex, Surrey and Yorkshire), Scotland (Mid-Ebudes, Orkney Islands, Westerness and Wester Ross) and Wales (Merionethshire).

pini (H.S. Jacks. & G.W. Martin) Luck-Allen, *Canad. J. Bot.* 41: 1049 (1963)
Sebacina pini H.S. Jacks. & G.W. Martin, *Mycologia* 32: 684 (1940)
E: ! **S:** !
H: On decayed wood of conifers. British collections are on *Picea* spp and *Pinus radiata*.

D: NM3: 98
Known from England (South Devon) and Scotland (South Ebudes).

radians (Rick) P. Roberts, *Kew Bull.* 56: 171 (2001)
   Coniophorella radians Rick, *Brotéria, Ci. Nat.* 3: 167 (1934)
   Basidiodendron nodosum Luck-Allen, *Canad. J. Bot.* 41: 1045 (1963)
E: ! S: ! W: ! O: Isle of Man: !
H: On debris of ferns such as *Dryopteris* spp and *Pteridium*, on decayed wood of *Picea* and *Pinus* spp, and on dead woody stems of *Lonicera periclymenum* and *Ulex europaeus*.
D: NM3: 98 (as *B. nodosum*)
Rarely reported but apparently widespread. Known from England (Berkshire, East Norfolk, Hertfordshire, Shropshire, South Devon and Surrey), Scotland (Kirkudbrightshire), Wales (Caernarvonshire and Monmouthshire), Republic of Ireland (Donegal) and the Isle of Man.

rimosum (H.S. Jacks. & G.W. Martin) Luck-Allen, *Canad. J. Bot.* 41: 1051 (1963)
   Sebacina rimosa H.S. Jacks. & G.W. Martin, *Mycologia* 32: 684 (1940)
E: ! S: !
H: On decayed wood of deciduous trees such as *Crataegus monogyna*, *Quercus* and *Ulmus* spp, and rarely on conifers such as *Juniperus communis* and *Larix* spp.
D: 98
Rarely reported. Known from England (Cornwall, South Devon and Westmorland) and Scotland (South Ebudes).

spinosum (L.S. Olive) Wojewoda, *Mała Flora Grzybów* 2: 91 (1981)
   Sebacina spinosa L.S. Olive, *Bull. Torrey Bot. Club* 85: 27 (1958)
   Bourdotia cinerella var. *trachyspora* Bourdot & Galzin, *Hymenomyc. France*: 50 (1928)
E: o S: ! W: ! O: Channel Islands: !
H: Sapotrophic. In woodland, often on decayed debris of *Pteridium*. but also known on dead lianas of *Clematis vitalba*, and decayed wood of *Corylus* and *Salix* spp, and conifers such as *Larix* and *Pinus* spp.
Occasional in England, especially in areas of calcareous soil in south-eastern England. Rarely reported elsewhere but widespread.

BASIDIORADULUM Nobles, *Mycologia* 59: 192 (1967)
Stereales, Schizoporaceae
Type: *Basidioradulum radula* (Fr.) Nobles

radula (Fr.) Nobles, *Mycologia* 59: 192 (1967)
   Hydnum radula Fr., *Observ. mycol.* 2: 271 (1818)
   Sistotrema radula (Fr.) Pers, *Mycol. eur.* 2: 195 (1825)
   Hyphoderma radula (Fr.) Donk, *Fungus* 27: 15 (1957)
   Radulum corallinum Berk. & Broome, *Ann. Mag. Nat. Hist., Ser. 4 [Notices of British Fungi no. 1441]* 15: 32 (1875)
   Sistotrema digitatum Pers., *Syn. meth. fung.*: 553 (1801)
   Xylodon digitatum (Pers.) Gray, *Nat. arr. Brit. pl.* 1: 649 (1821)
   Radulum epileucum Berk. & Broome, *Ann. Mag. Nat. Hist., Ser. 4 [Notices of British Fungi no. 1442]* 15: 32 (1875)
   Radulum orbiculare Fr., *Syst. orb. veg.* 1: 81 (1825)
   Radulum orbiculare var. *junquillinum* Quél., *Enchir. fung.*: 199 (1886)
   Mis.: *Hydnum spathulatum* sensu Greville [Fl. Edin.: 406 (1824)]
   Mis.: *Radulum tomentosum* sensu auct. Brit.
E: o S: o W: ! NI: ! ROI: !
H: In woodland, on decayed wood and bark of deciduous trees and shrubs. Most often collected on bark of old fallen branches and trunks of *Prunus avium* but also reported on *Betula* spp, *Corylus*, *Fagus*, *Ligustrum vulgare*, *Populus tremula*, *Quercus* spp, *Salix* spp, *Sorbus aucuparia* and *Viburnum lantana*.
D: NM3: 197 D+I: CNE3: 518-521 (as *Hyphoderma radula*), B&K2: 134 128 I: Kriegl1: 224 (as *H. radula*)

Occasional in England and Scotland. Rarely reported elsewhere but apparently widespread.

BATTARRAEA Pers., *Syn. meth. fung.*: 129 (1801)
Tulostomatales, Battarraeaceae
Type: *Battarraea phalloides* (Dicks.) Pers.

phalloides (Dicks.) Pers., *Syn. meth. fung.*: 129 (1801)
   Lycoperdon phalloides Dicks., *Fasc. pl. crypt. brit.* 1: 24 (1785)
   Phallus campanulatus Woodward, *Philos. Trans. Roy. Soc. London* 74: 423 (1784)
E: ! O: Channel Islands: !
H: On dry sandy soil, often amongst suckers of *Ulmus* spp, but also reported with other trees such as *Cupressus macrocarpa*, *Fagus*, *Fraxinus*, *Quercus* spp, and *Taxus*.
D+I: Ph: 250-251, BritPuffb: 53-55 Figs. 30/31 I: Sow3: 390 (as *Lycoperdon phalloides*), Bon: 301
Rarely reported but recently found in several new sites in southern and eastern England and in the Channel Islands. Possibly conspecific with the continental European *Battarraea stevenii*, which would provide an earlier name.

BJERKANDERA P. Karst., *Meddeland. Soc. Fauna Fl. Fenn.* 5: 35 (1879)
Poriales, Coriolaceae
Type: *Bjerkandera adusta* (Willd.) P. Karst.

adusta (Willd.) P. Karst., *Bidrag Kännedom Finlands Natur Folk*: 37: 39 (1882)
   Boletus adustus Willd., *Fl. berol. prodr.*: 392 (1787)
   Polyporus adustus (Willd.) Fr., *Syst. mycol.* 1: 363 (1821)
   Leptoporus adustus (Willd.) Quél., *Enchir. fung.*: 177 (1886)
   Gloeoporus adustus (Willd.) Pilát, *Atlas champ. Eur.* I, 3(1): 157 (1937)
   Polyporus adustus f. *resupinata* Bres., Killerman, *Denkschr. Königl.-Baier. Bot. Ges. Regensburg* 15: 72 (1922)
   Boletus carpineus Sowerby, *Col. fig. Engl. fung.* 2: pl. 231 (1799)
E: c S: c W: c NI: c ROI: c O: Channel Islands: c Isle of Man: c
H: On dead and dying wood of deciduous trees, most frequently *Fagus*. Rarely in wound scars on living trunks, or on decayed wood of conifers such as species of *Araucaria*, *Cupressus*, *Picea*, *Pinus*, *Thuja* and *Taxus*.
D: EurPoly1: 168, NM3: 221 D+I: Myc18(4): 174, Ph: 236, B&K2: 268 329
Very common and widespread.

fumosa (Pers.) P. Karst., *Meddeland. Soc. Fauna Fl. Fenn.* 5: 38 (1879)
   Boletus fumosus Pers., *Syn. meth. fung.*: 530 (1801)
   Polyporus fumosus (Pers.) Fr., *Syst. mycol.* 1: 367 (1821)
   Leptoporus fumosus (Pers.) Quél., *Enchir. fung.*: 177 (1886)
   Gloeoporus fumosus (Pers.) Pilát, *Atlas champ. Eur.* I, 3(1): 161 (1937)
   Tyromyces fumosus (Pers.) Pouzar, *Folia Geobot. Phytotax.* 1: 370 (1966)
   Polyporus fragrans Peck, *Rep. (Annual) New York State Mus. Nat. Hist.* 30: 45 (1878)
   Polyporus fumosus var. *fragrans* (Peck) Rea, *Brit. basidiomyc.*: 587 (1922)
   Polyporus pallescens Fr., *Hymenomyc. eur.*: 546 (1874)
   Daedalea saligna Fr., *Syst. mycol.* 1: 337 (1821)
   Polyporus salignus (Fr.) Fr., *Epicr. syst. mycol.*: 452 (1838)
E: o S: o W: ! NI: ! ROI: !
H: On decayed wood of deciduous trees, most frequently large stumps of *Acer pseudoplatanus* but also known on *Alnus*, *Castanea*, *Betula* spp *Fagus*, *Fraxinus*, *Malus*, *Quercus*, *Salix* and *Ulmus* spp. Rarely reported on conifers such as *Larix* and *Picea* spp, and *Pseudotsuga menziesii* but these hosts are unverified.
D: EurPoly1: 170, NM3: 221 D+I: B&K2: 268 330
Occasional but widespread, sometimes locally common, especially where sycamore scrub is cleared. Basidiomes in active growth often smell strongly of anise.

BOIDINIA Stalpers & Hjortstam, *Mycotaxon* 14: 76 (1982)
Hericiales, Gloeocystidiellaceae
Type: *Boidinia furfuracea* (Bres.) Stalpers & Hjortstam

furfuracea (Bres.) Stalpers & Hjortstam, *Mycotaxon* 14(1): 77 (1982)
    Hypochnus furfuraceus Bres., *Fungi trident.* 2(14): 97 (1900)
    Gloeocystidium furfuraceum (Bres.) Höhn. & Litsch., *Wiesner Festschrift*: 68 (1908)
    Gloeocystidiellum furfuraceum (Bres.) Donk, *Fungus* 26: 9 (1956)
    Corticium granulosum Wakef., *Trans. Brit. Mycol. Soc.* 35(1): 54 (1952)
E: ! **S:** !
H: On decayed wood and fallen bark of conifers such as *Larix*, *Picea* and *Pinus* spp. Rarely on decayed debris of *Pteridium*.
D: NM3: 278 **D+I:** CNE3: 416-417 (as *Gloeocystidiellum furfuraceum*), B&K2: 118 103
Known from England (Hertfordshire, North Hampshire, Shropshire, South Devon, Surrey, West Cornwall, West Norfolk and West Suffolk) and Scotland (East Perthshire).

permixta Boidin, Lanq. & Gilles, *Bull. Soc. Mycol. France* 113(1): 17 (1997)
S: !
H: On decayed wood. British material on a fallen twig of *Betula* spp.
A single collection from Easterness (Coire Adair) in 1997.

peroxydata (Rick) Hjortstam & Ryvarden, *Acta Mycol. Sin.* 7(2): 79 (1988)
    Gloeocystidium peroxydatum Rick, *Brotéria, Ci. Nat.* 3: 46 (1934)
E: !
H: British material on decayed basidiomes of *Stereum hirsutum* and on *Larix decidua* log.
Two British collections from Middlesex (Ealing, Perivale Wood) in 1997 and Surrey (Kew Gardens) in 2003. *Fide* Boidin *et al.* (1997) such European collections are better referred to *B. propinqua* (H.S. Jacks. & Dearn.) Hjortstam & Ryvarden.

BOLBITIUS Fr., *Epicr. syst. mycol.*: 253 (1838)
Agaricales, Bolbitiaceae
    Pluteolus (Fr.) Gillet, *Hyménomycètes*: 549 (1876)
Type: *Bolbitius vitellinus* (Pers.) Fr. (= *B. titubans*)

coprophilus (Peck) Hongo, *Memoirs of the Faculty of Education, Shiga University* 9: 82 (1959)
    Pluteolus coprophilus Peck, *Rep. (Annual) New York State Mus. Nat. Hist.* 46: 59 (1893)
E: !
H: On decayed straw or stable waste, especially when heating up due to fermentation.
D: BFF3: 33 **I:** SV24: 9
Known from Berkshire and Surrey. A distinctive species with pink tints to the pileus this is unlikely to be overlooked and probably genuinely rare in Britain. *Fide* BFF3 'widely distributed under glass in the Netherlands.'

lacteus J.E. Lange, *Fl. agaric. danic.* 5, Taxonomic Conspectus: II (1940)
E: ? **S:** !
H: British material on bare, damp ground, associated with *Salix* sp. by a pond.
A single collection from Stirling (Doune Ponds Reserve) in 1991. Reported from England (North Yorkshire) in BFF3: 36 but based on a single, partially sterile specimen.

reticulatus (Pers.) Ricken, *Blätterpilze Deutschl.*: 68 (1911)
    Agaricus reticulatus Pers.[non *A. reticulatus* With. (1792)], *Icon. descr. fung.*, fasc. 1: 13 (1798)
    Pluteolus reticulatus (Pers.) Gillet, *Hyménomycètes*: 549 (1876)
    Pluteolus aleuriatus var. *reticulatus* (Pers.) J.E. Lange, *Dansk. Bot. Ark.* 9(6): 49 (1938)

    Agaricus aleuriatus Fr., *Observ. mycol.* 1: 49 (1815)
    Pluteolus aleuriatus (Fr.) P. Karst., *Bidrag Kännedom Finlands Natur Folk* 32: 428 (1879)
    Bolbitius aleuriatus (Fr.) Singer, *Lilloa* 22: 490 (1951) [1949]
E: o **S:** ! **W:** !
H: On decayed wood in deciduous woodland especially on calcareous soil. Most often on large, fallen and decayed trunks of *Fagus* or *Ulmus* spp, but also on *Acer pseudoplatanus*, *A. campestre*, *Alnus*, *Corylus*, *Fraxinus*, *Quercus* and *Tilia* spp. Reported on soil or garden peat but then probably arising from small fragments of buried woody material.
D: BFF3: 37, NM2: 271 **D+I:** Ph: 120 (as *Pluteolus aleuriatus*)
I: FlDan4: 131G (as *P. aleuriatus*), C&D: 363, C&D: 363 (as *Bolbitius aleuriatus*)
Rarely reported, most frequent in southern England.

titubans (Bull.) Fr., *Epicr. syst. mycol.*: 254 (1838)
    Agaricus titubans Bull., *Herb. France*: pl. 425 (1789)
    Prunulus titubans (Bull.) Gray, *Nat. arr. Brit. pl.* 1: 632 (1821)
    Bolbitius vitellinus var. *titubans* (Bull.) Bon & Courtec., *Doc. Mycol.* 18(69): 37 (1987)
    Agaricus boltonii Pers., *Syn. meth. fung.*: 414 (1801)
    Prunulus boltonii (Pers.) Gray, *Nat. arr. Brit. pl.* 1: 632 (1821)
    Bolbitius boltonii (Pers.) Fr., *Epicr. syst. mycol.*: 254 (1838)
    Agaricus equestris Bolton [*nom. illegit.*, non *A. equestris* L. (1753)], *Hist. fung. Halifax* 2: 65 (1788)
    Bolbitius flavidus Massee, *Brit. fung.-fl.* 2: 204 (1893)
    Agaricus fragilis L. [*nom. illegit.*, non *A. fragilis* Pers. (1801)], *Sp. pl.* 2: 1175 (1763)
    Bolbitius fragilis (L.) Fr., *Epicr. syst. mycol.*: 254 (1838)
    Agaricus vitellinus Pers., *Syn. meth. fung.*: 402 (1801)
    Bolbitius vitellinus (Pers.) Fr., *Epicr. syst. mycol.*: 254 (1838)
E: c **S:** ! **W:** c **NI:** c **ROI:** c **O:** Channel Islands: c
H: On weathered dung of herbivores, dungy soil, decayed grass leaves or clippings, sawdust, straw and other vegetable debris in woodland and grassland areas, also occasionally as a 'weed' in mushroom beds.
D: BFF3: 33-34, BFF3: 35-36 (as *Bolbitius vitellinus*), NM2: 271 **D+I:** Ph: 154 (as *B. vitellinus*), B&K4: 296 372 (as *B. vitellinus*) **I:** Bon: 261 (as *B. vitellinus*)
Common and widespread. Better known as *B. vitellinus*.

titubans *var.* **olivaceus** (Gillet) Arnolds, *Persoonia* 18(2): 204 (2003)
    Bolbitius vitellinus var. *olivaceus* Rea, *Brit. basidiomyc.*: 497 (1922)
    Bolbitius variicolor G.F. Atk., *Stud. Amer. fungi*: 164 (1900)
    Bolbitius vitellinus var. *variicolor* (G.F. Atk.) Krieglst., *Beitr. Kenntn. Pilze Mitteleurop.* 7: 62 (1991)
E: !
H: On soil in manured grassland, or on old weathered dung.
D: BFF3: 34-35 **D+I:** B&K4: 298 373 (as *Bolbitius vitellinus* var. *variicolor*) **I:** C&D: 363
A single collection from West Suffolk (Lakenheath) in 2003. Possibly an alien, but poorly known in Britain.

BOLETOPSIS Fayod [non *Boletopsis* P. Henn. (1898)], *Malpighia* 3: 72 (1889)
Thelephorales, Thelephoraceae
Type: *Boletopsis leucomelaena* (Pers.) Fayod

leucomelaena (Pers.) Fayod, *Malpighia* 3: 27 (1889)
    Boletus leucomelas Pers., *Syn. meth. fung.*: 515 (1801)
    Polyporus leucomelas (Pers.) Fr., *Mycol. eur.* 2: 40 (1825)
    Caloporus leucomelas (Pers.) Quél., *Fl. mycol. France*: 405 (1888)
S: !
H: In sandy, mineral soils in grassy areas in open conifer woodland.
D: NM3: 309 **D+I:** B&K3: 236 281
Known from Morayshire (Forres), Easterness (Loch an Eilean) and South Aberdeen (Mar Lodge). Reported from Rothiemurchus Forest [TBMS 2: 171 (1907)] but no material traced.

BOLETUS Fr., *Syst. mycol.* 1: 385 (1821)
Boletales, Boletaceae
   Tubiporus P. Karst., *Rev. Mycol. (Toulouse)* 3(9): 16 (1881)
   Dictyopus Quél., *Enchir. fung.*: 159 (1886)
   Xerocomus Quél., Moug. & Ferry, *Fl. Vosges Champ.*: 477
      (1887)
Type: *Boletus edulis* Fr.

aereus Bull., *Herb. France*: pl. 389 (1789)
E: ! **S:** !
H: On soil in mixed deciduous woodland or parkland, usually
   associated with *Quercus* spp, less often *Fagus* or *Tilia* spp.
D: NM2: 57 **I:** Bon: 35, C&D: 431, SV42: 13, Kriegl2: 242
Rarely reported. Most British collections are from Berkshire
   (Windsor Great Park) and South Hampshire (New Forest) but
   also known from Hertfordshire, Surrey and West Kent and
   from Scotland (South Aberdeenshire).

appendiculatus Schaeff., *Fung. Bavar. Palat. nasc.* 2: 130
   (1763)
E: o **S:** ! **W:** ? **ROI:** !
H: On soil in mixed deciduous woodland, usually associated
   with *Quercus* spp. Less often reported with *Fagus* and rarely
   with *Betula* or *Tilia* spp.
D: NM2: 59 **D+I:** Ph: 194-195, B&K3: 52 4 **I:** Bon: 37, C&D:
   429, Kriegl2: 236
Occasional in England, and mostly reported from southern
   counties. Rarely reported elsewhere but apparently
   widespread.

armeniacus Quél., *Ann. Sci. Nat. Bord. Sud-ouest Mém.* 2: 42
   (1884)
   Xerocomus armeniacus (Quél.) Quél., *Fl. mycol. France*: 419
      (1888)
E: !
H: On soil in deciduous woodland. British material associated
   with *Quercus petraea.*
D: NM2: 63 **D+I:** FungEur2: 302 45 **I:** Bon: 43
A single, collection from Oxfordshire (Bagley Wood) in 2004.
   Reported on previous occasions but material is all
   redetermined, usually as *B. declivitatum.*

badius (Fr.) Fr. [non *B. badius* Pers. (1801)], *Syst. mycol.* 1:
   515 (1821)
   Boletus castaneus β *badius* Fr., *Observ. mycol.* 2: 247 (1818)
   Ixocomus badius (Fr.) Quél., *Fl. mycol. France*: 412 (1888)
   Xerocomus badius (Fr.) E.-J. Gilbert, *Les Bolets*: 92 (1931)
E: c **S:** c **W:** c **NI:** c **ROI:** c **O:** Isle of Man: !
H: On acidic soil, in woodland or with isolated trees. Normally
   associated with conifers such as *Pinus sylvestris* but may
   continue to fruit long after conifers have disappeared, and is
   then reported with trees such as *Betula, Carpinus, Castanea,
   Fagus* or *Quercus* spp.
D: NM2: 61 **D+I:** Ph: 196, B&K3: 86 55, FungEur8: 116-121
   86a-86b, 87, 88 (as *Xerocomus badius*) **I:** Bon: 45, C&D: 427
   (as *X. badius*), Kriegl2: 320 (as *X. badius*)
Very common and widespread. English name = 'Bay Bolete'.

betulicola (Vassilkov) Pilát & Dermek, *Hribovite Huby*: 96
   (1974)
   Boletus edulis var. *betulicola* Vassilkov, *Edible and poisonous
      fungi cent. Eur. dist. U.S.S.R.*: 46 (1948)
   Boletus edulis f. *betulicola* (Vassilkov) Vassilkov, *Bekyi Grib*:
      13 (1966)
E: ! **S:** !
H: On soil in open woodland, at woodland edges, and on
   heathland, associated only with *Betula* spp.
D: NM2: 57 **D+I:** FRIC13: 4-5 99 **I:** C&D: 431
Rarely reported.

calopus Pers., *Syn. meth. fung.*: 513 (1801)
   Boletus olivaceus Schaeff., *Fung. Bavar. Palat. nasc.* 2: 105
      (1763)
   Mis.: *Boletus pachypus* sensu NCL & sensu auct. Brit.
E: o **S:** o **W:** o **NI:** o **ROI:** ! **O:** Isle of Man: !

H: On acidic soil in mixed deciduous woodland, usually
   associated with *Fagus* and less often *Quercus* or *Salix* spp.
   Rarely reported with conifers such as *Picea abies.*
D: NM2: 59 **D+I:** Ph: 202 **I:** Bon: 37, C&D: 429, Kriegl2: 232
Occasional but widespread. The illustration in B&K3 may not
   show *Boletus calopus* and is possibly a form of *B. speciosus
   fide* G. Kibby (pers. comm.).

chrysenteron Bull., *Herb. France*: pl. 490 f. 3 (1791)
   Xerocomus chrysenteron (Bull.) Quél., *Fl. mycol. France*: 418
      (1888)
   Boletus communis Bull., *Herb. France*: pl. 393 (1789)
   Xerocomus communis (Bull.) Bon, *Doc. Mycol.* 14 (56): 16
      (1985) [1984]
   Mis.: *Boletus pascuus* sensu auct. Brit.
E: c **S:** c **W:** c **NI:** c **ROI:** c **O:** Channel Islands: !
H: On soil in open woodland, with *Fagus, Larix* and *Cedrus* spp,
   often with solitary trees in parkland and gardens.
D+I: Ph: 204-205 (as *Boletus chrysenteron*), B&K3: 86 56,
   FungEur8: 283-287 264-267 (as *Xerocomus chrysenteron*) **I:**
   Bon: 43, C&D: 427 (as *X. chrysenteron*), Kriegl2: 322 (as *X.
   chrysenteron*)
Widespread. Once considered common but confused with *B.
   cisalpinus* and *B. declivitatum.*

cisalpinus (Simonini, Ladurner & Peintner) Watling & A. Hills,
   Kew Bull. 59(1): 169 (2004)
   Xerocomus cisalpinus Simonini, Ladurner & Peintner, *Mycol.
      Res.* 107(6): 664 (2003)
E: !
H: On soil, in parkland or woodland. British material associated
   with *Fagus, Quercus* spp, *Cedrus* sp. and other conifers.
D+I: FungEur8: 274-277 247-252 (as *Xerocomus cisalpinus*)
Common throughout England and probably elsewhere but only
   recently recognised, previously often misdetermined as *B.
   chrysenteron*, and distribution not yet fully assessed.

declivitatum (C. Martín) Watling, *Edinburgh J. Bot.* 61(1): 43
   (2005)
   Boletus subtomentosus ssp. *declivitatum* C. Martín, *Beitr.
      Kryptogamenfl. Schweiz* 2(1): 18 (1904)
   Mis.: *Boletus communis* sensu auct.
   Mis.: *Xerocomus communis* sensu Bon (*Doc. Mycol.* 14 (56):
      16 (1985) [1984]) & sensu auct.
   Mis.: *Boletus armeniacus* sensu Dennis (1995)
E: c **S:** !
H: On soil in mixed woodland, parkland or gardens, associated
   with *Quercus* spp, but also known with a wide range of other
   deciduous trees.
Apparently widespread and possibly common, but only recently
   recognised as British. Related to (and *fide* some authorities
   synonymous with) *B. rubellus.* Also known as *Xerocomus
   quercinus* H. Engl. & Brückner, an unpublished name.

edulis Bull., *Herb. France*: pl. 60 (1782)
   Leccinum edule (Bull.) Gray, *Nat. arr. Brit. pl.* 1: 647 (1821)
   Boletus edulis var. *laevipes* Massee, *Brit. fung.-fl.* 1: 284
      (1892)
   Boletus edulis f. *arcticus* Vassilkov, *Bekyi Grib.*: 16 (1966)
   Boletus edulis var. *arcticus* (Vassilkov) Hlaváček, *Mykologický
      Sborník* 71(1): 9 (1994)
   Boletus edulis ssp. *trisporus* Duncan & Watling, *Notes Roy.
      Bot. Gard. Edinburgh* 33(2): 326 (1974)
   Boletus solidus Sowerby, *Col. fig. Engl. fung., Suppl.*: pl. 419
      (1809)
E: c **S:** c **W:** c **NI:** c **ROI:** c **O:** Isle of Man: !
H: On soil in woodland, associated with deciduous trees and
   conifers. Also known in grassland on calcareous soil,
   apparently associated with *Helianthemum nummularium.*
D: NM2: 57 **D+I:** Ph: 192-193 **I:** Sow1: 111, Bon: 35, C&D:
   431, SV42: 12, Kriegl2: 247
Common and widespread.

fechtneri Velen., *České Houby* IV-V: 704 (1922)
   Boletus appendiculatus ssp. *pallescens* Konrad, *Bull. Soc.
      Mycol. France* 45: 73 (1929)
   Boletus pallescens (Konrad) Singer, *Ann. Mycol.*: 424 (1936)

E: ! **S:** ? **W:** !
H: On calcareous soil in open, dry, sunny areas in mixed woodland. Usually associated with *Quercus* spp or less frequently *Fagus*.
I: C&D: 429, Kriegl2: 239
Rarely reported. Known from England (Berkshire, Cumberland, Herefordshire, South Hampshire, Surrey and West Sussex) and Wales. Reported from Scotland but unsubstantiated with voucher material.

ferrugineus Schaeff., *Fung. Bavar. Palat. nasc.* 1: 85 (1762)
   Xerocomus ferrugineus (Schaeff.) Bon, *Doc. Mycol.* 14(56): 16 (1985) [1984]
   Boletus citrinovirens Watling, *Notes Roy. Bot. Gard. Edinburgh* 29(2): 266 (1969)
   Boletus spadiceus Fr., *Epicr. syst. mycol.*: 415 (1838)
   Xerocomus spadiceus (Fr.) Quél., *Fl. mycol. France*: 417 (1888)
E: ! **S:** ! **W:** ! **NI:** ! **ROI:** ! **O:** Isle of Man: !
H: On acidic soil in native conifer woodland, or in mixed woodland with deciduous trees and conifers especially where *Fagus* is present.
D+I: FungEur8: 152-155 119-123 (as *Xerocomus ferrugineus*)
I: C&D: 427 (as *X. ferrugineus*)
Rarely reported. Apparently widespread.

fragrans Vittad., *Descr. fung. mang.*: 153 (1835)
E: ! **NI:** ? **O:** Channel Islands: !
H: On soil, usually amongst grass in open deciduous woodland.
D+I: Ph: 196-197 **I:** Bon: 35, C&D: 433
Rarely reported and the majority of records are unsubstantiated with voucher material. Known with certainty from Oxfordshire, Warwickshire and the Channel Islands.

immutatus (Pegler & A.E. Hills) A.E. Hills & Watling, *Edinburgh J. Bot.* 61(1): 44 (2005)
   Boletus luridiformis var. *immutatus* Pegler & A.E. Hills, *Mycologist* 10(2): 80 (1996)
E: r
H: On acidic soil in parkland, associated with *Fagus,* usually amongst short grass under solitary trees.
Described from Britain. At present known only from Berkshire.

impolitus Fr., *Epicr. syst. mycol.*: 421 (1838)
   Boletus suspectus Krombh., *Naturgetr. Abbild. Schwämme* 5: f.7 (1836)
E: o **S:** ! **W:** ! **NI:** !
H: On soil in open woodland, on roadside verges, in parkland with solitary trees and occasionally in gardens. Usually associated with *Quercus* spp, but also reported with *Fagus* and *Tilia* spp.
D: NM2: 61 (as *Boletus suspectus*) **D+I:** Ph: 196-197, B&K3: 56 11, FungEur8: 98-100 60-64 (as *Xerocomus impolitus*) **I:** Bon: 37, C&D: 433
Occasional in England, from Yorkshire southwards. Rarely reported elsewhere but apparently widespread.

legaliae Pilát, *Rev. Mycol. (Paris)* 33(1): 124 (1968)
   Boletus splendidus C. Martín, *Bull. Soc. Bot. Genève* 7: 190 (1894)
   Mis.: *Boletus satanoides* sensu auct. mult.
E: o **NI:** ?
H: On soil in mixed deciduous woodland, usually associated with *Quercus* spp, but also known with *Castanea* and *Fagus*.
D: NM2: 58 **I:** SV26: 8, C&D: 437, SV45: 8
Rarely reported but may be locally frequent in England, from Norfolk southwards. Reported from Northern Ireland (as *B. satanoides*) but unsubstantiated with voucher material.

luridiformis Rostk., *Deutschl. Fl.,* Edn 3: 105 (1844)
   Mis.: *Boletus erythropus* sensu auct. mult.
E: c **S:** c **W:** c **NI:** c **ROI:** c **O:** Channel Islands: ! Isle of Man: !
H: On acidic soil in mixed woodland or parkland. Usually associated with *Quercus* spp, but known with other deciduous trees such as *Betula* spp, *Castanea* and *Fagus* and occasionally with conifers such as *Pinus sylvestris*.

D: NM2: 60 **I:** FRIC3: 17b (as *Boletus discolor*), C&D: 433 (as *B. erythropus*), SV45: 3
Common and widespread.

luridiformis var. discolor (Quél.) Krieglst., *Beitr. Kenntn. Pilze Mitteleurop.* 7: 63 (1991)
   Dictyopus luridus var. *discolor* Quél., *Fl. mycol. France*: 422 (1888)
   Boletus discolor (Quél.) Costantin & L.M. Dufour, *Nouv. fl. champ.*, Edn 3: 296 (1901)
   Boletus erythropus ssp. *discolor* (Quél.) Dermek, Kuthan & Singer, *Česká Mykol.* 30: 1 (1976)
   Boletus queletii var. *discolor* (Quél.) Alessio, *Fungi Europaei: Boletus* 1: 193 (1985)
   Boletus luridiformis ssp. *discolor* (Quél.) Rauschert, *Nova Hedwigia* 45(3-4): 502 (1987)
   Mis.: *Boletus junquilleus* sensu NCL and sensu BFF1 p. 25.
E: !
H: On soil in parkland, often amongst grass and usually associated with *Quercus* spp.
Rarely reported.

luridus Schaeff., *Fung. Bavar. Palat. nasc.* 2: 107 (1763)
   Leccinum luridum (Schaeff.) Gray, *Nat. arr. Brit. pl.* 1: 648 (1821)
   Boletus rubeolarius Bull., *Herb. France*: pl. 490 (1791)
   Leccinum rubeolarium (Bull.) Gray, *Nat. arr. Brit. pl.* 1: 648 (1821)
E: o **S:** o **W:** o **NI:** o **ROI:** !
H: On calcareous soil in mixed deciduous woodland or parkland, and often along roadside verges. Usually associated with *Fagus* and *Quercus* spp, less frequently with *Betula* and *Tilia* spp. Also known in grassland on calcareous soils apparently associated with *Helianthemum nummularium*.
D: NM2: 58 **D+I:** Ph: 198-199, B&K3: 56 12 **I:** Bon: 39, C&D: 435, Kriegl2: 214, SV45: 5
Occasional but widespread.

luridus var. rubriceps (Maire) Dermek, *Fungorum Rar. Icon. Color.* 16: 14 (1987)
   Tubiporus luridus var. *rubriceps* Maire, *Fungi Catalaunici*: 45 (1937)
E: !
H: On calcareous soil in open deciduous woodland. The single British collection was a group of 10 associated with *Fagus*.
Known from Berkshire. Distinguished from the typical variety by the intensely copper-red pileus and strongly red stipe.

moravicus Vacek, *Stud. Bot. Čechoslov.*: 36 (1946)
   Xerocomus moravicus (Vacek) Herink, *Česká Mykol.* 18: 193 (1964)
   Boletus leonis D.A. Reid, *Fungorum Rar. Icon. Color.* 1: 7 (1966)
   Xerocomus leonis (D.A. Reid) Bon, *Doc. Mycol.* 14(56): 16 (1985) [1984]
   Mis.: *Boletus leoninus* sensu Pearson & Dennis (1948) and sensu auct. mult. [non sensu Persoon (1825) nor Krombholz (1846)]
E: ! **S:** !
H: On soil in open deciduous woodland, often amongst grass in glades. Also known with isolated trees in parkland and rarely with conifers in mixed woodland. Associated with *Fagus, Fraxinus, Quercus* spp, and the conifers *Cupressus* and *Pinus* spp.
D: NM2: 62 **D+I:** FRIC1: 7-11 3 (as *Boletus leonis*), FRIC13: 3-4 98, FRIC16: 9-10 122b (as *Xerocomus leonis*), FungEur8: 125128 9, 22, 96a-c (as *X. moravicus*) **I:** C&D: 427 (as *X. leonis*)
Rarely reported. Known from England (Hertfordshire, South Hampshire, Surrey and West Kent) and Scotland (Westerness).

pinophilus Pilát & Dermek, *Česká Mykol.* 27(1): 6 (1973)
   Boletus edulis f. *pinicola* (Vittad.) Vassilkov, *Bekyi Grib*: 14 (1966)
   Boletus pinicola (Vittad.) A. Venturi [*nom. illegit.*], *I mic. ag. Bresc.*: 39 (1863)

E: o **S:** c **W:** !
H: On acidic soil in woodland or with solitary trees, associated with *Pinus sylvestris*.
D: NM2: 58 **I:** Bon: 35, C&D: 431, SV42: 13, Kriegl2: 245
Common in Scotland, occasional in southern England and rarely reported elsewhere.

porosporus G. Moreno & Bon, *Doc. Mycol.* 7(27-28): 6 (1977)
  Xerocomus porosporus (G. Moreno & Bon) Contu, *Bolm Soc. broteriana* 63: 385 (1990)
E: o **S:** ! **W:** ! **NI:** o
H: On soil in mixed woodland with deciduous trees or conifers, in parkland and in gardens and shrubberies.
D: NM2: 62 **D+I:** Ph: 202-203, FungEur8: 194-197 157-161 (as *Xerocomus porosporus*) **I:** Bon: 43, C&D: 427 (as *X. porosporus*)
Occasional to rather rare but apparently widespread.

pruinatus Fr. & Hök, *Boleti fung. gen.*: 9 (1835)
  Xerocomus pruinatus (Fr. & Hök) Quél., *Fl. mycol. France*: 420 (1888)
E: o **S:** o **W:** ! **NI:** !
H: On soil in deciduous woodland.
D: NM2: 62 **D+I:** FungEur8: 259-265 226-235 (as *Xerocomus pruinatus*) **I:** Kriegl2: 333
Occasional to common, and widespread.

pseudoregius (Hubert) Estadès, *Bull. Trimestriel Féd. Mycol. Dauphiné-Savoie* 27(108): 7 (1988)
  Boletus appendiculatus ssp. *pseudoregius* Hubert, *Z. Pilzk.* 17(3-4): 87 (1938)
E: !
H: On soil in parkland, associated only with *Quercus* spp (and a single dubious record where said to be with *Cedrus deodara*).
I: C&D: 429, Kriegl2: 240, FM1(3): 95
Rarely reported. Known from a few scattered sites in Berkshire, East Sussex, Herefordshire and South Hampshire. Reported from East Gloucestershire but unsubstantiated with voucher material.

pseudosulphureus Kallenb., *Z. Pilzk.* 2(10-12): 225 (1923)
  Mis.: *Boletus junquilleus* sensu D.A. Reid [Fung. Rar. Icon. Col. Vol. 3] & sensu auct. mult.
E: ! **S:** !
H: On soil in parkland or open mixed woodland. Usually associated with *Quercus* and *Tilia* spp, but also reported with *Fagus* and *Pinus* spp.
D+I: FRIC3: 1-5 17a (as *Boletus junquilleus*), FRIC16: 10-11 123 (as *B. junquilleus*) **I:** C&D: 433 (as *B. junquilleus*)
Known from England (South Hampshire) and Scotland (East Perthshire and Easter Ross). Reported form Yorkshire but unsubstantiated with voucher material.

pulverulentus Opat., *Arch. Naturgesch.* 2: 27 (1836)
  Xerocomus pulverulentus (Opat.) E.-J. Gilbert, *Les Bolets*: 116 (1931)
  Mis.: *Boletus radicans* sensu Rea (1922) & sensu auct. mult.
E: o **S:** ! **W:** ! **NI:** !
H: On soil in open deciduous woodland, in parkland and gardens and occasionally on heathland. Usually associated with *Quercus* spp, but reported rarely with other trees such as *Betula* sp. and *Carpinus*.
D: NM2: 60 **D+I:** Ph: 199, B&K3: 58 14 **I:** Bon: 37, C&D: 429, Kriegl2: 225
Occasional in England. Rarely reported elsewhere but apparently widespread.

queletii Schulzer, *Hedwigia* 24: 143 (1885)
  Boletus queletii var. *rubicundus* Maire, *Bull. Soc. Mycol. France* 26: 195 (1910)
  Boletus queletii var. *lateritius* (Bres. & R. Schulz) E.-J. Gilbert, *Les Bolets*: 118 (1931)
  Mis.: *Boletus erythropus* sensu Persoon [Syn. Met. Fung. p. 513 (1801)]
E: o **W:** ! **NI:** ! **ROI:** !

H: On soil in parkland and open deciduous woodland, usually associated with *Fagus* and *Tilia* spp, but also reported with *Carpinus*.
D: NM2: 60 **D+I:** FRIC1: 3-5 1, Ph: 201, FRIC16: 11-13 124, B&K3: 58 15 **I:** Bon: 39, C&D: 433, Kriegl2: 212, SV45: 3
Occasional in England and mostly reported from southern counties. Rarely reported elsewhere but apparently widespread.

radicans Pers., *Syn. meth. fung.*: 507 (1801)
  Boletus albidus Roques [*nom. illegit.*, non *B. albidus* Pers. (1801)], *Hist. Champ. comest.*: 70 (1832)
  Boletus amarus Pers., *Syn. meth. fung.*: 511 (1801)
  Boletus pachypus var. *amarus* (Pers.) Fr., *Epicr. syst. mycol.*: 417 (1838)
  Boletus pachypus Fr., *Observ. mycol.* 1: 119 (1815)
  Mis.: *Boletus candicans* sensu auct.
E: o **W:** ! **NI:** ! **O:** Channel Islands: !
H: On soil in open deciduous woodland, or with solitary trees in parkland. Usually associated with *Quercus* spp, but also known with *Betula* spp.
D: NM2: 59 **D+I:** Ph: 196-197 (as *Boletus albidus*), B&K3: 60 16 **I:** Bon: 37, C&D: 429, Kriegl2: 230
Occasional in England, but may be locally common in southern counties. Rarely reported elsewhere but apparently widespread. Sensu Rea (1922) and sensu auct. mult. is *B. pulverulentus*.

regius Krombh., *Naturgetr. Abbild. Schwämme* 2: 3 (1832)
E: r
H: On soil in deciduous woodland. British material was reported with *Quercus* sp., but was possibly with the *Castanea* known at this site.
I: Bon: 37, C&D: 429, Kriegl2: 233
Known from a single location in South Hampshire.

reticulatus Schaeff. [non *B. reticulatus* (Hoffm.) Pers. (1801)], *Fung. Bavar. Palat. nasc.* 2: 108 (1763)
  Boletus edulis f. *reticulatus* (Schaeff.) Vassilkov, *Bekyi Grib*: 18 (1966)
  Boletus aestivalis (Paulet) Fr., *Epicr. syst. mycol.*: 422 (1838)
E: ! **S:** ! **W:** ! **NI:** ! **ROI:** ! **O:** Channel Islands: !
H: On soil in mixed deciduous woodland or with solitary trees in parkland. Usually associated with *Quercus* spp, but also known with *Castanea* and *Fagus*.
D: NM2: 57 **D+I:** B&K3: 60 17 **I:** Bon: 35, C&D: 431, SV42: 12, Kriegl2: 249 (as *Boletus aereus*)
Rarely reported but apparently widespread.

rhodopurpureus Smotl., *Mykologický Sborník* 29(1-3): 31 (1952)
  Boletus rhodopurpureus f. *xanthopurpureus* Smotl., *Cas. Ceska Houb.* 29: 31 (1952)
  Boletus xanthopurpureus (Smotl.) Hlaváček, *Mykologický Sborník* 63(5): 132 (1986)
  Mis.: *Boletus purpureus* sensu NCL & sensu auct. Brit
E: !
H: On soil in open mixed deciduous woodland, or in parkland with solitary trees. Usually associated with *Quercus* spp.
D+I: B&K3: 60 18 **I:** Bon: 39, C&D: 435, SV45: 8 (as *Boletus purpureus*)
Known from Berkshire and South Hampshire.

ripariellus (Redeuilh) Watling & A.E. Hills, *Edinburgh J. Bot.* 61(1): 45 (2005) [2004]
  Xerocomus ripariellus Redeuilh, *Doc. Mycol.* 26(104): 30 (1997)
E: ! **W:** !
H: On soil in open deciduous woodland or parkland with *Quercus* spp, also in grassland and in wet or damp places where it may be associated with *Alnus* or *Salix* spp.
D+I: FungEur8: 245-249 217-222 (as *Xerocomus ripariellus*)
Only recently recognised as distinct and easily confused with similar taxa such as *B. chrysenteron* or *B. rubellus*.

rubellus Krombh., *Naturgetr. Abbild. Schwämme* 5: 12 (1836)
   Xerocomus rubellus (Krombh.) Quél., *Compt. Rend. Assoc.*
     Franç. Avancem. Sci. 24: 620 (1896) [1895]
   Boletus sanguineus With. [*nom. illegit.*, non *B. sanguineus* L.
     (1763)], *Arr. Brit. pl. ed. 3* 4: 319 (1796)
   Boletus versicolor Rostk. [*nom. illegit.*, non *B. versicolor* Gray
     (1821)], *Deutschl. Flora*, Edn 3: 55 (1844)
   Xerocomus versicolor (Rostk.) E.-J. Gilbert, *Les Bolets*: 138
     (1931)
E: o **W:** ! **NI:** ! **ROI:** ! **O:** Channel Islands: !
H: On soil in open mixed deciduous woodland or amongst grass
   under solitary trees in parkland, usually associated with
   Quercus spp.
D: NM2: 62 **D+I:** B&K3: 88 59, FungEur8: 217-224 174-185 (as
   Xerocomus rubellus) **I:** Bon: 43, C&D: 427 (as *X. rubellus*),
   Kriegl2: 324 (as *X. rubellus*)
Occasional in England and most frequent in southern counties.
   Rarely reported elsewhere but apparently widespread.

satanas Lenz, *Schwämme Mitteldeutschl.:* 67 (1831)
E: o **S:** ? **NI:** ?
H: On calcareous soil in open mixed deciduous woodland and
   occasionally with solitary trees in parkland. Usually associated
   with *Fagus* and rarely *Quercus* spp.
D: NM2: 59 **D+I:** Ph: 202-203, B&K3: 62 20 **I:** Bon: 39, SV31:
   3, C&D: 437, Kriegl2: 217, FM1(3): 95, SV45: 7
Occasional but may be locally common in southern England.
   Reports from Scotland and Northern Ireland lack voucher
   material.

subappendiculatus Dermek, Lazedniček & Veselský,
   Fungorum Rar. Icon. Color. 15: 13 (1979)
S: !
H: on soil in open situations with *Abies* & *Picea* spp.
D+I: B&K3: 64 24, FRIC 15 pl 57a
Known from three sites in Perthshire since 1998.

subtomentosus L., *Sp. pl.*: 1178 (1753)
   Leccinum subtomentosum (L.) Gray, *Nat. arr. Brit. pl.* 1: 647
     (1821)
   Xerocomus subtomentosus (L.) Quél., *Fl. mycol. France*: 418
     (1888)
   Boletus lanatus Rostk., *Deutschl. Flora*, Edn 3: 77 (1844)
   Xerocomus lanatus (Rostk.) E.-J. Gilbert, *Les Bolets*: 143
     (1931)
   Boletus leguei Boud., *Bull. Soc. Mycol. France* 11: 417 (1894)
   Xerocomus leguei (Boud.) Bon, *Doc. Mycol.* 14(56): 16
     (1985) [1984]
   Boletus striipes Fr., *Hymenomyc. eur.*: 502 (1874)
   Boletus subtomentosus var. *marginalis* Boud., *Icon. Mycol.*:
     72 (1907)
   Xerocomus subtomentosus f. *xanthus* E.-J. Gilbert, *Les
     Bolets*: 142 (1931)
   Boletus xanthus (E.-J. Gilbert) Merlo [*comb. inval.*], *I Nostri
     Funghi*: 50 (1980)
   Xerocomus xanthus (E.-J. Gilbert) Contu, *Pagine Botaniche*
     14: 29 (1989)
E: c **S:** c **W:** c **NI:** c **ROI:** c **O:** Channel Islands: !
H: On soil in open deciduous woodland.
D: NM2: 61 **D+I:** B&K3: 88 60, FungEur8: 140-144 105-112 (as
   Xerocomus subtomentosus) **I:** Bon: 43, C&D: 427 (as *X.
   subtomentosus*), Kriegl2: 328 (as *X. subtomentosus*)
Common and widespread.

torosus Fr. & Hök, *Boleti fung. gen.*: 10 (1835)
E: !
H: On soil in deciduous woodland. British material with *Quercus*
   sp.
D+I: B&K3: 66 25 **I:** C&D: 435
Known from southern England (Isle of Wight and South Devon).

xanthocyaneus (Ramain) Romagn., *Bull. Soc. Mycol. France*
   92: 305 (1976)
   Boletus purpureus var. *xanthocyaneus* Ramain, *Bull. Soc.
     Naturalistes Oyonnax* 2: 56 (1948)
E: !
H: On soil in woodland or parkland with *Quercus* spp.

I: C&D: 435
British collections only from Berkshire and South Hampshire.

BOTRYOBASIDIUM Donk, *Meded. Ned. Mycol.*
Ver. 18 - 20: 116 (1931)
Stereales, Botryobasidiaceae
   Haplotrichum Link. (anam.)[non *Haplotrichum* Eschw.
     (1824)], in Willd., *Sp. pl.*, Edn 4, 6(1): 52 (1824)
   Oidium Link (anam.) [*nom. rej.*; non *Oidium* Link, 1824, *nom.*
     cons.], *Mag. Neuesten Entdeck. Gesammten Naturk. Ges.*
     Naturf. Freunde Berlin 3: 18 (1809)
   Acladium Link (anam.), *Mag. Neuesten Entdeck. Gesammten*
     Naturk. Ges. Naturf. Freunde Berlin 3: 11 (1809)
   Alysidium Kunze (anam.), *Mykol. Hefte* 1: 11 (1817)
Type: *Botryobasidum subcoronatum* (Höhn. & Litsch.) Donk

aureum Parmasto, *Eesti N.S.V. Tead. Akad. Toimet., Biol.*
   14(2): 220 (1965)
   Oidium aureum Link (anam.), *Mag. Neuesten Entdeck.*
     Gesammten Naturk. Ges. Naturf. Freunde Berlin 3: 18
     (1809)
   Haplotrichum aureum (Link) Hol.-Jech. (anam.), *Česká*
     Mykol. 30(1): 4 (1976)
   Trichoderma dubium Pers. (anam.), *Observ. mycol.* 1: 99
     (1796)
   Alysidium dubium (Pers.) M.B. Ellis (anam.), *Dematiaceous*
     Hyphomycetes: 89 (1971)
   Alysidium fulvum Kunze & J.C. Schmidt (anam.), *Mykol. Hefte*
     1: 11 (1817)
E: c **S:** c **W:** c **ROI:** !
H: On decayed wood of deciduous trees. Most frequently on
   large, fallen trunks, logs or old stumps of *Fagus*.
D: NM3: 120 **D+I:** CNE2: 150-152 **I:** FM1(1): 14
Common and widespread. Closely related to *B. conspersum* and
   difficult to distinguish if the characteristic golden-yellow
   conidial stage is not present.

candicans J. Erikss., *Svensk bot. Tidskr.* 52(1): 6 (1958)
   Acladium capitatum Link (anam.), *Mag. Neuesten Entdeck.*
     Gesammten Naturk. Ges. Naturf. Freunde Berlin 3: 10
     (1809)
   Haplotrichum capitatum (Link) Willd. (anam.), *Sp. pl.* 6(1):
     52 (1824)
   Rhinotrichum bloxamii Berk. & Broome (anam.), *Ann. Mag.*
     Nat. Hist., Ser. 2 7: 177 (1851)
   Monilia candicans Sacc. (anam.), *Nuovo Giorn. Bot. Ital.* 8:
     195 (1876)
   Oidium candicans (Sacc.) Linder (anam.), *Lloydia* 5(2): 183
     (1942)
   Mis.: *Pellicularia pruinata* sensu Wakefield (1952)
E: c **S:** ! **W:** c
H: On decayed wood of deciduous trees or conifers.
D: NM3: 120 **D+I:** CNE2: 156-157, B&K2: 90 62
Common and widespread in old woodland.

conspersum J. Erikss., *Symb. Bot. Upsal.* 16(1): 133 (1958)
   Acladium conspersum Link (anam.), *Mag. Neuesten Entdeck.*
     Gesammten Naturk. Ges. Naturf. Freunde Berlin 3: 11
     (1809)
   Sporotrichum conspersum (Link) Fr. (anam.), *Syst. mycol.* 1:
     419 (1821)
   Oidium conspersum (Link) Linder (anam.), *Lloydia* 5 (2): 176
     (1942)
   Haplotrichum conspersum (Link) Hol.-Jech. (anam.), *Česká*
     Mykol. 30(1): 4 (1976)
E: c **S:** c **W:** c **NI:** c **ROI:** c
H: On decayed wood of deciduous trees or conifers.
D: NM3: 120 **D+I:** CNE2: 158-161
Common and widespread, especially in old woodland.

danicum J. Erikss. & Hjortstam, *Friesia* 9(1): 11 (1969)
E: o **S:** o **W:** o
H: On decayed wood and on debris of *Pteridium*.
D: NM3: 120 **D+I:** CNE2: 162-163
Occasional but apparently widespread.

ellipsosporum Hol.-Jech., *Česká Mykol.* 23(4): 211 (1969)
  Oidium ellipsosporum Hol.-Jech. (anam.), *Česká Mykol.* 23:
    211 (1969)
  Haplotrichum ellipsosporum (Hol.-Jech.) Hol.-Jech. (anam.),
    Česká Mykol. 30(1): 4 (1976)
E: ! S: !
H: On decayed wood of conifers such as *Pinus sylvestris* and
  rarely on wood of deciduous trees such as *Fraxinus, Sorbus
  aucuparia* and *Ulmus* spp.
D: NM3: 120
Rarely reported. Known from widely scattered localities in
Scotland, and a single location in England (Yorkshire: Forest of
Bowland).

intertextum (Schwein.) Jülich & Stalpers, *Verhandelingen der
  Koninklijke Nederlandse Akademie van Wetenschappen* 2(74):
  56 (1980)
  Sporotrichum intertextum Schwein., *Trans. Am. phil. Soc.* 4:
    271 (1832)
  Pellicularia angustispora Boidin, *Publ. Mus. Natl. Hist. Nat.*
    17: 119 (1957)
  Botryobasidium angustisporum (Boidin) P.H.B. Talbot,
    Persoonia 3(4): 395 (1965)
S: !
H: On decayed wood or bark of conifers such as *Pinus
  sylvestris.*
D: NM3: 119 **D+I:** CNE2: 148-149 (as *Botryobasidium
  angustisporum*)
Not well known in Britain. A few recent records from Scotland
(Mid-Perthshire: Black Wood of Rannoch), and possibly also in
England (North Hampshire).

laeve (J. Erikss.) Parmasto, *Eesti N.S.V. Tead. Akad. Toimet.,
  Biol.* 14(4): 220 (1965)
  Botryobasidium pruinatum var. *laeve* J. Erikss., *Svensk bot.
    Tidskr.* 51(1): 10 (1958)
E: ! S: ! W: ! ROI: ?
H: Usually on decayed wood of deciduous trees such as *Alnus,
  Betula* spp and *Corylus*. Rarely on conifers such as *Pinus
  sylvestris.*
D: NM3: 119 **D+I:** CNE2: 164-165, B&K2: 90 63
Rarely reported. Many of the collections are from southern
England.

obtusisporum J. Erikss., *Symb. Bot. Upsal.* 16(1): 57 (1958)
E: ! S: ?
H: On decayed wood of *Fagus* and *Fraxinus*. Reported on *Pinus
  sylvestris* but the host is not verified.
D: NM3: 120 **D+I:** CNE2: 168-169, B&K2: 92 65
Known from North and South Hampshire and Surrey. Reported
from Scotland in 1964 (Dunbartonshire) but the collection in
herb. K is annotated 'compare with' *B. obtusisporum.*

pruinatum (Bres.) J. Erikss., *Svensk bot. Tidskr.* 52: 8 (1958)
  Corticium pruinatum Bres., *Ann. Mycol.* 1(1): 98 (1903)
  Botryobasidium botryoideum (Overh.) Parmasto, *Eesti N.S.V.
    Tead. Akad. Toimet., Biol.* 14(2): 220 (1965)
  Hypochnus coronatus J. Schröt., in Cohn, *Krypt.-Fl. Schlesien*
    3: 418 (1888)
  Corticium coronatum (J. Schröt.) Sacc., *Syll. fung.* 6: 654
    (1888)
  Botryobasidium coronatum (J. Schröt.) Donk, *Meded. Ned.
    Mycol. Ver.* 18 - 20: 117 (1931)
E: ! S: ! W: ! ROI: !
H: On decayed wood of deciduous trees such as *Acer
  pseudoplatanus, Alnus* spp, *Crataegus* spp and *Fagus*. Also
  known on dead, fallen basidiomes of polypores and on debris
  of *Pteridium.*
D: NM3: 119 **D+I:** CNE2: 170-171, B&K2: 92 66
Rarely reported but apparently widespread.

subcoronatum (Höhn. & Litsch.) Donk, *Meded. Ned. Mycol.
  Ver.* 18 - 20: 117 (1931)
  Corticium subcoronatum Höhn. & Litsch., *Sitzungsber.
    Kaiserl. Akad. Wiss., Wien, Math.-Naturwiss. Cl., Abt. 1*
    116: 822 (1907)
  Pellicularia subcoronata (Höhn. & Litsch.) D.P. Rogers,
    Farlowia 1: 104 (1943)
E: c **S:** c **W:** c **NI:** c **ROI:** c **O:** Isle of Man: !
H: On decayed wood of deciduous trees and conifers in old
  woodland. Also known on decayed basidiomes of *Phlebia
  tremellosa* and *Schizopora paradoxa.*
D: NM3: 119 **D+I:** CNE2: 172-173, B&K2: 94 67
Very common and widespread.

vagum (Berk. & M.A. Curtis) D.P. Rogers, *Univ. Iowa Stud. Nat.
  Hist.* 17(1): 17 (1935)
  Corticium vagum Berk. & M.A. Curtis, *Grevillea* 1: 179 (1873)
  Pellicularia vaga (Berk. & M.A. Curtis) Linder, *Lloydia* 5: 170
    (1942)
  Corticium botryosum Bres., *Ann. Mycol.* 1(1): 99 (1903)
  Botryobasidium botryosum (Bres.) J. Erikss., *Symb. Bot.
    Upsal.* 16(1): 53 (1958)
E: o **S:** o **W:** o **NI:** !
H: On decayed wood of both deciduous trees and conifers, in
  old woodland. Also known on fallen and decayed leaves of
  Arundinaria spp.
D: NM3: 120 (as *Botryobasidium botryosum*) **D+I:** CNE2: 154-
155 (as *B. botryosum*), B&K2: 90 61 (as *B. botryosum*)
Occasional but widespread.

BOTRYOHYPOCHNUS Donk, *Meded. Ned. Mycol. Ver.* 18-20: 118 (1931)
Stereales, Botryobasidiaceae
Type: *Botryohypochnus isabellinus* (Fr.) J. Erikss.

isabellinus (Fr.) J. Erikss., *Svensk bot. Tidskr.* 52(1): 2 (1958)
  Hypochnus isabellinus Fr., *Observ. mycol.* 2: 281 (1818)
  Thelephora isabellina (Fr.) Fr., *Summa veg. Scand.* 2: 337
    (1849)
  Corticium isabellinum (Fr.) Fr., *Hymenomyc. eur.*: 660 (1874)
  Tomentella isabellina (Fr.) Höhn. & Litsch., *Sitzungsber.
    Kaiserl. Akad. Wiss., Wien, Math.-Naturwiss. Cl. Abt. 1*
    (115): 1570 (1906)
  Botryobasidium isabellinum (Fr.) D.P. Rogers, *Univ. Iowa
    Stud. Nat. Hist.* 17(1): 11 (1935)
  Hypochnus argillaceus P. Karst., *Meddeland. Soc. Fauna Fl.
    Fenn.* 6: 13 (1881)
  Zygodesmus argillaceus (P. Karst.) P. Karst., in Rabenhorst,
    Fungi europaei V: 3188 (1884)
E: o **S:** ! **NI:** o **O:** !
H: Usually on decayed wood of deciduous trees such as *Betula*
  spp, *Fagus* and *Fraxinus*. Rarely on wood of conifers, but
  known on *Larix* and *Pinus sylvestris*. Also on old polypore
  basidiomes.
D: NM3: 120 **D+I:** CNE2: 178-179, B&K2: 94 69 **I:** Kriegl1: 171
Occasional but apparently widespread.

BOVISTA Pers., *Syn. meth. fung.*: 136 (1801)
Lycoperdales, Lycoperdaceae
Type: *Bovista plumbea* Pers.

aestivalis (Bonord.) Demoulin, *Sydowia, Beih.* 8: 143 (1979)
  Lycoperdon aestivale Bonord., *Handb. Mykol.*: 251 (1851)
  Lycoperdon cepaeforme Bull., *Champ. France,
    Hymenomycetes*: 156 (1791)
  Lycoperdon furfuraceum Schaeff., *Fung. Bavar. Palat. nasc.*
    3: 294 (1770)
  Lycoperdon polymorphum Vittad., *Monogr. Lycoperd.* 5: 183
    (1843)
  Bovista polymorpha (Vittad.) Kreisel, *Repert. Nov. Spec.
    Regn. veg.* 69: 201 (1964)
E: ! S: ! W: ! ROI: ! O: Isle of Man: !
H: On sandy, often calcareous soils, usually in dunes but also
  known from unimproved grassland (downland) and in open
  glades in deciduous woodland.
D: NM3: 333 **D+I:** BritPuffb: 130-131 Figs. 97/98 **I:** Kriegl2:
124
Rarely reported but apparently widespread.

dermoxantha (Vittad.) De Toni, in Sacc., *Syll. fung.* 7: 100
  (1888)
  Lycoperdon dermoxanthum Vittad., *Monog. Lycoperd.*: 178
    (1843)
  Utraria dermoxantha Quél., *Enchir. fung.*: 241 (1886)
  Lycoperdon ericetorum Pers., *J. Bot., Paris* 2: 17 (1809)
  Mis.: *Bovista pusilla* sensu auct. mult.
  Mis.: *Lycoperdon pusillum* sensu auct. mult.
E: ! **S:** ! **W:** ? **NI:** ? **ROI:** ? **O:** Channel Islands: ! Isle of Man: ?
H: On humic soil in warm grassland areas, also on roadside
  verges and occasionally along woodland edges.
D: NM3: 333 (as *Bovista pusilla*) **D+I:** BritPuffb: 132-133 Figs.
  99/100
Numerous records but few collections. Apparently widespread
  and often recorded as *Bovista pusilla* (a *nomen ambiguum*).

limosa Rostr., *Medd. Grønl.* 18: 52 (1894)
E: ! **W:** !
H: On dry sandy soils, sometimes amongst mosses in exposed
  locations. British material on dunes.
D: NM3: 334 **D+I:** BritPuffb: 136-137 Figs. 103/104 **I:** Kriegl2:
  126
Rarely reported. Known from England (South Lancashire and
  Westmorland) and Wales (Anglesey, Carmarthenshire and
  Glamorganshire).

nigrescens Pers., Roemer, *Neues Mag. Bot.* 1: 86 (1794)
E: c **S:** c **W:** c **NI:** c **ROI:** c **O:** Channel Islands: o
H: On soil, in grassland such as meadows, pasture, heathland
  and downland, in open grassy areas on dunes and in
  deciduous woodland. Rarely in cultivated fields and on old
  sewage beds.
D: NM3: 334 **D+I:** Ph: 249, B&K2: 386 507, BritPuffb: 138-139
  Figs. 105/106 **I:** Kriegl2: 127
Common and widespread. Tolerant of grassland fertilisation.

paludosa Lév., *Ann. Sci. Nat., Bot.*, sér. 3, 5: 163 (1846)
  Bovistella paludosa (Lév.) Pat., in Lloyd, *Mycol. not.* 1 (9): 88
    (1902)
E: !
H: On soil (often calcareous) in fens, often amongst mosses
  such as *Aulacomnium palustre*.
D: NM3: 334 **D+I:** BritPuffb: 134-135 Figs. 101/102 **I:** Kriegl2:
  128
Known from East Norfolk (Buxton Heath, and Brundall near
  Norwich) and Yorkshire (Osmotherley).

plumbea Pers., *Observ. mycol.* 1: 5 (1796)
  Lycoperdon bovista Sowerby, *Col. fig. Engl. fung.* 3: pl. 331
    (1803)
  Lycoperdon plumbeum Vittad., *Monog. Lycoperd.*: 174
    (1842)
  Bovista ovalispora Cooke & Massee, *Grevillea* 16: 46 (1887)
E: c **S:** c **W:** c **NI:** c **ROI:** c **O:** Channel Islands: c
H: On soil amongst short grass, often in coastal turf and on
  dunes but also in meadows, pastures, on golf links, lawns,
  downland and roadside verges. Rarely in open woodland.
D: NM3: 334 **D+I:** Ph: 249, B&K2: 388 508, BritPuffb: 140-141
  Figs. 107/108 **I:** Bon: 305, Kriegl2: 129
Very common and widespread.

BOVISTELLA Morgan, *J. Cincinnati Soc. Nat. Hist.* 14: 141 (1892)
Lycoperdales, Lycoperdaceae
Type: *Bovistella ohiensis* (Ellis & Morgan) Morgan

radicata (Durieu & Mont.) Pat., *Bull. Soc. Mycol. France* 15: 55
  (1899)
  Lycoperdon radicatum Durieu & Mont., *Expl. Sci. Algérie [Bot.
    I]* 1: 383 (1848)
E: !
H: On dry acidic soil, amongst short grass.
D: NM3: 335 **D+I:** BritPuffb: 120-121 Figs. 90/91
Known from Surrey (Richmond Park) and North Wiltshire (Spye
  Park) but not seen since 1952 and possibly extinct.

BREVICELLICIUM K.H. Larss. & Hjortstam, *Mycotaxon* 7(1): 117 (1978)
Stereales, Sistotremataceae
Type: *Brevicellicium exile* (H.S. Jacks.) K.H. Larss. & Hjortstam

exile (H.S. Jacks.) K.H. Larss. & Hjortstam, *Mycotaxon* 7(1):
  118 (1978)
  Corticium exile H.S. Jacks., *Canad. J. Res., Sect. C, Bot. Sci.*
    28: 721 (1950)
E: !
H: On decayed wood. British material on *Acer pseudoplatanus*
  and *Taxus*.
D: NM3: 124 **D+I:** CNE8: 1450-1453
Known from Surrey (Kew Gardens) and South Devon (Torquay,
  Lincombe Slopes).

olivascens (Bres.) K.H. Larss. & Hjortstam, *Mycotaxon* 7(1):
  119 (1978)
  Odontia olivascens Bres., *Fungi trident.* 2(8-10): 36 (1892)
  Hydnum granulatum var. *mutabilis* Pers., *Mycol. eur.* 2: 184
    (1825)
  Grandinia mutabilis (Pers.) Bourdot & Galzin, *Bull. Soc. Mycol.
    France* 30(1): 52 (1914)
  Cristella mutabilis (Pers.) Parmasto, *Eesti N.S.V. Tead. Akad.
    Toimet., Biol.* 14(4): 223 (1965)
  Trechispora mutabilis (Pers.) Liberta, *Taxon* 15(8): 319
    (1966)
  Mis.: *Grandinia granulosa* sensu auct.
E: c **S:** c **W:** c **NI:** ! **ROI:** o **O:** Channel Islands: o Isle of Man: o
H: On decayed wood of deciduous trees and shrubs and rather
  frequent on old decayed lianas of *Clematis vitalba*. Rarely on
  wood of conifers such as *Larix* and *Taxus*.
D: NM3: 124 **D+I:** B&K2: 126 116, CNE8: 1453-1455 **I:** Kriegl1:
  172
Common and widespread.

BUCHWALDOBOLETUS Pilát, *Friesia* 9: 217 (1969)
Boletales, Boletaceae
Type: *Buchwaldoboletus lignicola* (Kallenb.) Pilát

lignicola (Kallenb.) Pilát, *Friesia* 9: 217 (1969)
  Boletus lignicola Kallenb., *Pilze Mitteleuropas* 1(9): 57 (1929)
    [1928]
  Phlebopus lignicola (Kallenb.) J.W. Groves, *Mycologia* 54(3):
    320 (1962)
  Pulveroboletus lignicola (Kallenb.) E.A. Dick & Snell,
    Mycologia 57(3): 451 (1965)
E: ! **S:** ! **W:** ! **NI:** ! **ROI:** ?
H: On decayed wood (usually large stumps) of conifers such as
  Larix, Picea abies, Pinus pinaster, Pseudotsuga, Sequoia and
  Sequoiadendron spp. Rarely on wood of deciduous trees but
  known on *Prunus avium*. Often growing on wood decayed by
  Phaeolus schweinitzii and possibly parasitic on its mycelium.
D: NM2: 67 **D+I:** FRIC1: 6-7 2, FRIC13: 1-3 97, B&K3: 76 41 **I:**
  C&D: 423 (as *Pulveroboletus lignicola*), Kriegl2: 290 (as *P.
  lignicola*)
Rarely reported but apparently widespread. Known from
  England (Berkshire, Cumberland, East Kent, Herefordshire,
  Oxfordshire, South Hampshire, South Somerset, Surrey,
  Warwickshire and Worcestershire), Scotland (Clyde Isles, Main
  Argyll and Wester Ross), Wales (Breconshire, Cardiganshire,
  Caernarvonshire, Flintshire and Denbighshire) and Northern
  Ireland (Down). Reported from the Republic of Ireland but
  unsubstantiated with voucher material.

sphaerocephalus (Barla) Watling & T.H. Li, *Edinburgh J. Bot.*
  61(1): 46 (2005)
  Boletus sphaerocephalus Barla, *Champ. Prov. Nice*: t. 36
    (1859)
  Boletus sulphureus Fr. [*nom. illegit.*], *Epicr. syst. mycol.*: 413
    (1838)
  Phlebopus sulphureus (Fr.) Singer [*nom. illegit.*], *Amer. Midl.
    Naturalist* 37: 3 (1947)
  Mis.: *Buchwaldoboletus hemichrysus* sensu auct. Brit.

E: ! S: ?

H: On decayed wood (including sawdust) of conifers.
Collections in herb. K from Berkshire (as *Boletus sulphureus*) in
1970, Hertfordshire in 1947and 1992 and West Sussex
(Stedham Common) in 2004. Reported from Scotland but
lacking voucher material.

BULBILLOMYCES Jülich, *Persoonia* 8: 69 (1974)

Stereales, Hyphodermataceae
Aegerita Pers. (anam.), *Syn. meth. fung.*: 684 (1801)
Type: *Bulbillomyces farinosus* (Bres.) Jülich

farinosus (Bres.) Jülich, *Persoonia* 8(1): 69 (1974)
Kneiffia farinosa Bres., *Ann. Mycol.* 1(1): 105 (1903)
Peniophora farinosa (Bres.) Höhn. & Litsch., *Sitzungsber.
Kaiserl. Akad. Wiss., Wien, Math.-Naturwiss. Cl., Abt. 1*
1(117): 1095 (1908)
Metulodontia farinosa (Bres.) Parmasto, *Conspectus
Systematis Corticiacearum*: 118 (1968)
Peniophora aegerita Höhn. & Litsch., *Sitzungsber. Kaiserl.
Akad. Wiss., Wien, Math.-Naturwiss. Cl., Abt. 1* 1(116):
814 (1907)
Peniophora candida Lyman, *Proc. Boston Soc. nat. Hist.* 33:
168 (1907)
Sclerotium aegerita Hoffm. (anam.), *Deutschl. Fl.*: t. 9 (1795)
Aegerita candida Pers. (anam.), *Syn. meth. fung.*: 685 (1801)
E: c **S:** c **W:** o **NI:** ! **ROI:** o
H: On sodden and decayed wood of deciduous trees in swampy
areas in woodland, or in periodically inundated areas such as
stream edges, ditches, pond margins etc.
D: NM3: 198 **D+I:** CNE4: 554-557, B&K2: 128 119 **I:** Kriegl1:
173
Common and widespread. Usually collected in the anamorphic
Aegerita candida state, which is ecologically adapted to very
wet or periodically flooded habitat. The teleomorph is less
often recorded but can often be found by searching for grey,
effused areas amongst the anamorph.

BULLERA Derx, *Ann. Mycol.* 28: 11 (1930)

Tremellales, Tremellaceae
A genus of anamorphic yeasts with teleomorphs in the genus
Bulleromyces. The two species listed below, however, have no
known teleomorphs.
Type: *Bullera alba* (W.F. Hanna) Derx

armeniaca Buhagiar, *J. Gen. Microbiol.* 129(10): 3152 (1983)
E: !
H: Isolated ex *Brassica oleracea*.
D+I: YST 112

crocea Buhagiar, *J. Gen. Microbiol.* 129(10): 3150 (1983)
E: !
H: Isolated ex *Brassica oleracea* and strawberries (holotype).
D+I: YST 114

BULLEROMYCES Boekhout & A. Fonseca, *Ant. Van. Leeuw.* 59: 90 (1991)

Tremellales, Tremellaceae
Type: *Bulleromyces albus* Boekhout & A. Fonseca

albus Boekhout & A. Fonseca, *Ant. Van Leeuw.* 59: 90 (1991)
Sporobolomyces albus W.F. Hanna (anam.), *Fungi of
Manitoba and Saskatchewan*: 80 (1929)
Bullera alba (W.F. Hanna) Derx (anam.), *Ann. Mycol.* 28: 11
(1930)
E: ! **S:** !
H: Saprotrophs on living leaves. Also isolated ex sausages.
D+I: YST 127 (yeast only)
Widespread, at least in England; often with *Erysiphales* and
Uredinales fide Dennis (1986). Known in Britain only in its
anamorphic yeast stage, *Bullera alba*.

BYSSOCORTICIUM Bondartsev & Singer, *Mycologia* 36: 69 (1944)

Stereales, Atheliaceae
Byssoporia M.J. Larsen & Zak, *Canad. J. Bot.* 56: 1123 (1978)
Type: *Byssocorticium atrovirens* (Fr.) Bondartsev & Singer

atrovirens (Fr.) Bondartsev & Singer, *Mycologia* 36(1): 69
(1944)
Thelephora atrovirens Fr., *Elench. fung.* 1: 202 (1828)
Corticium atrovirens (Fr.) Fr., *Epicr. syst. mycol.*: 562 (1838)
Coniophora atrovirens (Fr.) Cooke, *Grevillea* 20: 13 (1891)
E: ! **S:** ! **NI:** ?
H: On decayed wood of deciduous trees and conifers, usually
when fallen amongst leaf litter.
D: NM3: 152 **D+I:** CNE2: 180-181
Known from England (Hertfordshire and North Hampshire) and
Scotland (Clyde Isles: Isle of Arran). A single record from
Northern Ireland but unsubstantiated with voucher material.

efibulatum Hjortstam & Ryvarden, *Mycotaxon* 7(2): 410 (1978)
E: ! **W:** !
H: On decayed wood of deciduous trees such as *Betula*,
Castanea and *Quercus* spp, usually on large logs or stumps.
D: NM3: 153 **D+I:** Myc8(3): 116
Known from England (East Kent, South Devon, West Cornwall
and West Kent) and Wales (Breconshire and Caernarvonshire).
The bright blue basidiomes and lack of clamps are diagnostic.

pulchrum (S. Lundell) M.P. Christ., *Dansk. Bot. Ark.* 19(2): 158
(1960)
Corticium pulchrum S. Lundell, *Fungi Exsiccati Suecici* 21 -
22: 23 (1941)
E: ! **S:** ! **W:** !
H: On decayed wood and litter of deciduous trees.
D: NM3: 152 **D+I:** CNE2: 182-183, B&K2: 96 70
Known from England (Cheviotshire, East Sussex and Surrey),
Scotland and Wales (Caernarvonshire and Radnorshire).

terrestre (DC.) Bondartsev & Singer, *Ann. Mycol.* 39: 48 (1941)
Boletus terrestris DC., *Fl. Fr.* 5: 39 (1815)
Polyporus terrestris (DC.) Fr., *Syst. mycol.* 1: 383 (1821)
Poria terrestris (DC.) Cooke, *Grevillea* 14: 115 (1886)
Byssoporia terrestris (DC.) M.J. Larsen & Zak, *Canad. J. Bot.*
56: 1123 (1978)
Poria mollicula Bourdot, in Lloyd, *Mycol. not.* 4 (40): 543
(1916)
Byssocorticium molliculum (Bourdot) Jülich, *Persoonia* 8(3):
295 (1975)
E: ! **S:** ? **W:** ?
H: On soil in woodland or on cubically rotted wood.
D: NM3: 152 (as *Byssocorticium molliculum*) **D+I:** CNE2: 186-
187
Known from Hertfordshire (Frithsden Beeches) and Derbyshire
(Baslow). Reported from Scotland and Wales but lacking
voucher material. Basidiomes are variable in colour, often
partially yellow, bluish or greenish, and the species has been
split into several varieties based upon these colour forms.

BYSSOMERULIUS Parmasto, *Eesti N.S.V. Tead. Akad. Toimrt., Biol.* 16(4): 383 (1967)

Stereales, Meruliaceae
Type: *Byssomerulius corium* (Pers.) Parmasto

corium (Pers.) Parmasto, *Eesti N.S.V. Tead. Akad. Toimet.,
Biol.* 16(4): 383 (1967)
Thelephora corium Pers., *Syn. meth. fung.*: 574 (1801)
Merulius corium (Pers.) Fr., *Elench. fung.* 1: 58 (1828)
Meruliopsis corium (Pers.) Ginns, *Canad. J. Bot.* 54(1-2): 126
(1976)
Merulius aurantiacus Klotzsch [*nom. illegit.*, non *M.
aurantiacus* (Wulfen) Pers. (1801)], in Berkeley, *Engl. fl.*
5(2): 128 (1836)
Merulius confluens Schwein., *Schriften Naturf. Ges. Leipzig* 1:
92 (1822)
Auricularia papyrina Bull., *Herb. France*: pl. 147 (1783)
Merulius papyrinus (Bull.) Quél., *Fl. mycol. France*: 32 (1888)
Boletus purpurascens DC. [non *B. purpurascens* Rostk.
(1844)], *Fl. Fr.* 6: 41 (1815)

E: c **S:** c **W:** c **NI:** c **ROI:** c **O:** Channel Islands: c Isle of Man: c
H: On fallen and decayed wood of deciduous trees and shrubs, most frequently *Fraxinus* but also *Acer pseudoplatanus*, *Fagus* and *Salix* spp, and on *Ulex europaeus*. Rarer hosts include *Betula* sp., *Buddleja davidii*, *Crataegus* sp., *Hedera*, *Ligustrum ovalifolium*, *Pittosporum*, *Populus*, *Prunus*, *Quercus*, *Rosa*, *Syringa*, *Tilia* spp and *Viburnum lantana*. Rarely on conifers such as *Pinus* spp.
D: NM3: 156 (as *Meruliopsis corium*) **D+I:** CNE2: 190-192, B&K2: 144 144 (as *M. corium*) **I:** C&D: 139, Kriegl1: 260 (as *M. corium*)
Very common and widespread. Old or dead basidiomes are frequently a dull purplish-red due to colonisation by the ascomycete *Hypomyces rosellus*.

CALOCERA Fr., *Syst. mycol.* 1: 485 (1821)
Dacrymycetales, Dacrymycetaceae
Corynoides Gray, *Nat. arr. Brit. pl.* 1: 654 (1821)
Dacryomitra Tul. & C. Tul., *Ann. Sci. Nat., Bot.*, sér. 5, 15: 217 (1872)
Type: *Calocera viscosa* (Pers.) Fr.

cornea (Batsch) Fr., *Stirp. agri femsion.* 5: 67 (1827)
Clavaria cornea Batsch, *Elench. fung.*: 139 & t. 28 f. 161 (1783)
Corynoides cornea (Batsch) Gray, *Nat. arr. Brit. pl.* 1: 654 (1821)
Tremella palmata Schumach., *Enum. Plant.* 2: 442 (1803)
Calocera palmata (Schumach.) Fr., *Epicr. syst. mycol.*: 581 (1838)
E: c **S:** c **W:** c **NI:** c **ROI:** c **O:** Channel Islands: ! Isle of Man: !
H: On decayed wood, often large fallen trunks, most frequently *Fagus*. Common on both deciduous and conifer wood, but few British collections on the latter.
D: NM3: 91 **D+I:** Ph: 263, B&K2: 50 1 **I:** C&D: 137, Kriegl1: 70
Very common and widespread.

furcata (Fr.) Fr., *Stirp. agri femsion.* 5: 67 (1827)
Clavaria furcata Fr., *Syst. mycol.* 1: 486 (1821)
E: ? **S:** ! **ROI:** ?
H: On decayed coniferous wood.
D: NM3: 91 **I:** Kriegl1: 71
Rarely reported, but known from Scotland (Reid, 1974).

glossoides (Pers.) Fr., *Stirp. agri femsion.* 5: 67 (1827)
Clavaria glossoides Pers., *Comment. Fungis Clavaeform.*: 68 (1797)
Tremella glossoides (Pers.) Pers., *Mycol. eur.* 1: 107 (1822)
Dacryomitra glossoides (Pers.) Bref., *Unters. Gesammtgeb. Mykol.* 7: 162 (1888)
Dacryomitra pusilla Tul. & C. Tul., *Ann. Sci. Nat., Bot.*, sér. 5, 15: 217 (1872)
Dacryomyces pusilla (Tul. & C. Tul.) Lapl., *Dict. Icon, Paris*: 131 (1894)
E: ! **S:** ? **W:** ? **NI:** ? **O:** Isle of Man: ?
H: On decayed wood.
D: NM3: 91 **D+I:** Ph: 263
Known from England (Hertfordshire and West Sussex). Often recorded but rarely collected, and the majority of these collections are misidentified *C. cornea*. Records unsubstantiated with voucher material should therefore be treated with caution.

pallidospathulata D.A. Reid, *Trans. Brit. Mycol. Soc.* 62: 445 (1974)
E: c **S:** c **W:** c **NI:** c **ROI:** c **O:** Isle of Man: c
H: Initially recorded only on decayed wood of conifers, usually *Picea* or *Pinus* spp, but also on *Larix* spp and *Taxus*. More recently noted on wood of deciduous trees such as *Betula*, *Castanea*, *Corylus*, *Fagus*, *Quercus*, *Rhododendron* and *Salix* spp.
D+I: Myc4(1): 34
Common and widespread, but localised in the British Isles and not collected before 1969. Not or little-known in continental Europe. Possibly a naturalised alien species.

viscosa (Pers.) Fr., *Stirp. agri femsion.* 5: 67 (1827)
Clavaria viscosa Pers., *Neues Mag. Bot.* 1: 117 (1794)
Calocera flammea Wallroth, *Fl. crypt. Germ., Sect. II*: 535 (1833)
Calocera cavarae Bres., in Cavara, *Staz. Sperim. Arg. Ital.* 29: 14 (1896)
Calocera viscosa var. *cavarae* (Bres.) McNabb, *New Zealand J. Bot.* 3: 40 (1965)
E: c **S:** c **W:** c **NI:** c **ROI:** c **O:** Channel Islands: ! Isle of Man: c
H: On dead wood of conifers, usually large. Dubiously reported on deciduous wood (*Fagus*, *Fraxinus*, *Quercus* spp, *Hedera* and *Sambucus nigra*).
D: NM3: 91 **D+I:** Ph: 263, B&K2: 50 2 **I:** Bon: 325, C&D: 137, Kriegl1: 73
Very common and widespread. The var. *cavarae* is simply an unpigmented (albino) form, not otherwise distinct.

CALOCYBE Donk, *Nova Hedwigia Beih.* 5: 42 (1962)
Agaricales, Tricholomataceae
Tricholomella Kalamees, *Persoonia* 14(4): 446 (1992)
Type: *Calocybe gambosa* (Fr.) Donk

carnea (Bull.) Donk, *Nova Hedwigia Beih.* 5: 42 (1962)
Agaricus carneus Bull., *Herb. France*: pl. 533 (1792)
Tricholoma carneum (Bull.) P. Kumm., *Führ. Pilzk.*: 132 (1871)
Lyophyllum carneum (Bull.) Kühner & Romagn., *Flore Analytique des Champignons Supérieurs*: 162 (1953)
E: o **S:** o **W:** o **NI:** o **ROI:** ! **O:** Isle of Man: !
H: On soil, amongst short turf on lawns or downland, parkland, or heathland etc. and also known from grassy areas in woodland
D: NM2: 102 **D+I:** Ph: 43, FRIC17: 7-8 131a, B&K3: 144 142 **I:** FlDan1: 24C (as *Tricholoma carneum*), Bon: 167, SV30: 13, Kriegl3: 138, RivMyc: 148 (as *Lyophyllum carneum*)
Occasional but widespread, and often abundant wherever it occurs.

chrysenteron (Bull.) Singer, *Sydowia* 15: 47 (1962) [1961]
Agaricus chrysenteron Bull., *Herb. France*: pl. 556 (1792)
Tricholoma chrysenteron (Bull.) P. Kumm. [as *chrysenterum*], *Führ. Pilzk.*: 133 (1871)
Agaricus cerinus Pers., *Neues Mag. Bot.* 1: 101 (1794)
Tricholoma cerinum (Pers.) P. Kumm., *Führ. Pilzk.*: 133 (1871)
Calocybe cerina (Pers.) Donk, *Nova Hedwigia Beih.* 5: 43 (1962)
E: ! **S:** !
H: On soil, often associated with conifers such as *Picea* spp, and occasionally with *Fagus* in woodland or plantations.
D: NM2: 103 **D+I:** FRIC17: 10-11 131d (as *Calocybe cerina*), B&K3: 144 143 **I:** Bon: 167, Kriegl3: 139
Known from England (Berkshire, Buckinghamshire, North Somerset and South Devon) and a single record from Scotland.
Calocybe cerina and *C. chrysenteron* are considered to be separate taxa by some authorities.

constricta (Fr.) Kühner, *Sydowia* 15: 47 (1962) [1961]
Agaricus constrictus Fr. [non *A constrictus* With. (1796)], *Syst. mycol.* 1: 28 (1821)
Armillaria constricta (Fr.) Gillet, *Hyménomycètes*: 78 (1874)
Tricholoma constrictum (Fr.) Ricken, *Blätterpilze Deutschl.*: 329 (1914)
Lepiota constricta (Fr.) Rea, *Brit. basidiomyc.*: 73 (1922)
Lyophyllum constrictum (Fr.) Singer, *Ann. Mycol.* 41: 100 (1943)
Tricholomella constricta (Fr.) Kalamees, *Persoonia* 14(4): 446 (1992)
Agaricus leucocephalus Bull., *Herb. France*: pl. 536 (1792)
Tricholoma leucocephalum (Bull.) Sacc., in Quélet, *Mém. Soc. Émul. Montbéliard, Sér. 2*, 5: 316 (1872)
Lyophyllum leucocephalum (Bull.) Singer, *Ann. Mycol.* 41: 100 (1943)

Calocybe leucocephala (Bull.) Bon & Courtec., *Doc. Mycol.*
18(69): 37 (1987)
E: ! S: ! W: ! NI: !
H: On soil, often amongst grass in scrub woodland or at
woodland edges. Reported with both deciduous trees and
conifers.
D: NM2: 102 **D+I:** Ph: 41 (as *Tricholoma leucocephalum*) **I:**
FlDan1: 24A (as *T. leucocephalum*)
Rarely reported.

gambosa (Fr.) Donk, *Nova Hedwigia Beih.* 5: 43 (1962)
Agaricus gambosus Fr., *Syst. mycol.* 1: 50 (1821)
Tricholoma gambosum (Fr.) P. Kumm., *Führ. Pilzk.*: 131
(1871)
Lyophyllum gambosum (Fr.) Singer, *Ann. Mycol.* 41: 96
(1943)
Agaricus georgii L. [non *A. georgii* Sowerby (1801)], *Sp. pl.*:
1173 (1753)
Tricholoma georgii (L.) Quél., *Mém. Soc. Émul. Montbéliard*,
Sér. 2, 5: 44 (1872)
Calocybe georgii (L.) Kühner [*nom. inval.*], *Bull. Mens. Soc.
Linn. Lyon* 7: 211 (1938)
E: c S: o W: o NI: o ROI: !
H: On soil, often amongst grass at the edges of woodland or
scrub, associated with a wide variety of deciduous trees and
shrubs, and with conifers. Also on heathland, at the edges of
paths, and rather frequently on roadside verges. A spring
species, fruiting mainly from early April to late June, but also
recorded as early as February and as late as November.
D: NM2: 102 **D+I:** Ph: 41 (as *Tricholoma gambosum*), B&K3:
144 144 **I:** Bon: 167, Kriegl3: 142, RivMyc: 142 (as *Lyophyllum
gambosum*)
Common and widespread at least throughout England, and
probably elsewhere. English name = 'St. George's Mushroom'.

ionides (Bull.) Donk, *Nova Hedwigia Beih.* 5: 43 (1962)
Agaricus ionides Bull., *Herb. France*: pl. 533 (1792)
Tricholoma ionides (Bull.) P. Kumm., *Führ. Pilzk.*: 132 (1871)
Lyophyllum ionides (Bull.) Kühner & Romagn. [*comb. inval.*],
Flore Analytique des Champignons Supérieurs: 162 (1953)
E: o W: ! ROI: !
H: On calcareous loam soils, usually with *Fagus* but also known
with *Acer pseudoplatanus*, *Alnus*, *Fraxinus*, *Quercus* sp. and
Larix.
D: NM2: 102 **D+I:** FRIC17: 8-9 131b, B&K3: 146 145 **I:** FlDan1:
25D (as *Tricholoma ionides*), Bon: 167, FungEur3: 63, RivMyc:
148 (as *Lyophyllum ionides*)
Occasional to rather rare. Most frequently collected from
woodland on calcareous soils in south eastern England where
it may be locally abundant. Less frequent elsewhere and
absent from many areas.

obscurissima (A. Pearson) M.M. Moser, *Kleine
Kryptogamenflora*, Edn 3: 101 (1967)
Tricholoma ionides var. *obscurissima* A. Pearson, *Trans. Brit.
Mycol. Soc.* 29(4): 192 (1946)
Tricholoma obscurissimum (A. Pearson) Hora, *Trans. Brit.
Mycol. Soc.* 43(2): 459 (1960)
Tricholoma conicosporum Métrod [*nom. nud.*], *Rev. Mycol.
(Paris)* 4(3-4): 107 (1939)
E: ! S: ! W: !
H: On soil, often sandy or calcareous, usually associated with
conifers such as *Cupressus*, *Larix*, *Pinus* and *Sequoia* spp, but
also known with deciduous trees such as *Alnus*, *Carpinus*,
Corylus, *Fagus*, *Fraxinus* and *Populus* spp.
I: SV32: 12, Kriegl3: 145 (as *Calocybe ionides* var.
obscurissima), RivMyc: 145 (as *Lyophyllum obscurissimum*)
Rarely reported but apparently widespread. Considered by some
authorities as just a dark-coloured variant of *C. ionides*, and
since rings of basidiomes can be found gradating from typical
C. obscurissima into typical *C. ionides*, this may prove to be so.

onychina (Fr.) Donk, *Nova Hedwigia Beih.* 5: 43 (1962)
Agaricus onychinus Fr., *Epicr. syst. mycol.*: 41 (1838)
Tricholoma onychinum (Fr.) Gillet, *Hyménomycètes*: 113
(1874)

S: !
H: On soil, amongst needle litter of *Pinus* in Caledonian
pinewoods.
D: NM2: 102, NBA2: 65 (as *T. onychinum*) **D+I:** FRIC17: 9-10
131c
Known from Perthshire (Loch Rannoch).

persicolor (Fr.) Singer, *Sydowia* 15: 48 (1962) [1961]
Agaricus ionides var. *persicolor* Fr., *Ic. Hymenomyc.*: 35
(1870)
Tricholoma ionides var. *persicolor* (Fr.) P. Karst., *Bidrag
Kännedom Finlands Natur Folk* 32: 44 (1879)
Tricholoma persicolor (Fr.) Sacc., *Syll. fung.* 5: 117 (1887)
Tricholoma carneum var. *persicolor* (Fr.) Park.-Rhodes,
Trans. Brit. Mycol. Soc. 34(3): 363 (1951)
Agaricus carneolus Fr., *Hymenomyc. eur.*: 65 (1874)
E: ! S: ! NI: !
H: On soil, amongst short turf, usually in unimproved grassland
but also on lawns etc.
D: NM2: 102 **I:** FlDan1: 24G (as *Tricholoma persicolor*), RivMyc:
149 (as *Lyophyllum persicolor*)
Rarely reported. Possibly mistaken for large specimens of
Calocybe carnea which it somewhat resembles. This is
probably '*Tricholoma carneolum*' [see TBMS 3 (5): 376 (1911)]
an unpublished combination.

CALVATIA Fr. [*nom. cons.*], *Summa veg. Scand.*
2: 442 (1849)
Lycoperdales, Lycoperdaceae
Langermannia Rostk. [*nom. rej.*], *Deutschl. Fl.*, Edn 3: 23
(1839)
Type: *Calvatia craniiformis* (Schwein.) De Toni

gigantea (Batsch) Lloyd, *Mycol. not.* 1 (16): 166 (1904)
Lycoperdon giganteum Batsch, *Elench. fung. (Continuatio
Prima)*: 237 & t. 29 f. 165 (1786)
Bovista gigantea (Batsch) Gray, *Nat. arr. Brit. pl.* 1: 583
(1821)
Langermannia gigantea (Batsch) Rostk., *Deutschl. Fl.*, Edn 3:
23 (1839)
E: o S: o W: o NI: o ROI: o O: Channel Islands: o
H: On soil in parks, gardens, grassland, in and around cultivated
fields (especially amongst beds of *Urtica dioica* on highly
nitrogenous soil), on compost heaps, and occasionally in open
woodland.
D: NM3: 337 (as *Langermannia gigantea*) **D+I:** Ph: 247 (as *L.
gigantea*), B&K2: 390 511 (as *L. gigantea*), BritPuffb: 116-117
Figs. 86/87 **I:** Bon: 305 (as *L. gigantea*), Kriegl2: 133
Occasional but widespread. An unmistakable species, fruiting
from early summer.

CALYPTELLA Quél., *Enchir. fung.*: 216 (1886)
Agaricales, Tricholomataceae
Type: *Calyptella capula* (Holmsk.) Quél.

campanula (Nees) W.B. Cooke, *Sydowia, Beih.* 4: 32 (1961)
Peziza campanula Nees, *Syst. Pilze*: 71, f. 295 (1817)
Peziza sulphurea Batsch, *Elench. fung.*: 121 (1783)
Cyphella sulphurea (Batsch) Fr., *Hymenomyc. eur.*: 665
(1874)
E: ! S: !
H: On dead and decayed stems of herbaceous plants. Reported
on *Cirsium*, *Lotus uliginosus*, *Lycopersicum esculentum*, *Urtica*
spp and various umbelliferous herbs.
D+I: B&K2: 198 225
Accepted, but doubtfully distinct from *Calyptella capula*. This is
the causal agent of 'Calyptella Root Rot' of tomatoes but is
rarely reported.

capula (Holmsk.) Quél., *Fl. mycol. France*: 25 (1886)
Peziza capula Holmsk., *Beata ruris* 1: 41 f.21 (1790)
Cyphella capula (Holmsk.) Fr., *Epicr. syst. mycol.*: 568 (1838)
Cyphella capula var. *flavescens* Pat., *Tab. anal. fung.*: 56
(1883)

Cyphella laeta Fr., *Epicr. syst. mycol.*: 568 (1838)
Calyptella laeta (Fr.) W.B. Cooke, *Sydowia, Beih.*: 40 (1961)
Cyphella pimii W. Phillips, *Grevillea* 13: 49 (1884)
E: c **S:** c **W:** c **NI:** ! **ROI:** ! **O:** Channel Islands: ! Isle of Man: !
H: On a wide variety of living, dead and dying herbaceous
plants, especially *Urtica dioica*. Also known on dead twigs of
woody shrubs and a single record (with voucher material) on
old and weathered horse dung.
D: NM2: 103 **D+I:** B&K2: 200 226 **I:** Kriegl3: 592
Common and widespread.

CAMAROPHYLLOPSIS Herink, *Acta Musei Bohemiae septentr. Liberecensis* 1: 61 (1959)
Agaricales, Hygrophoraceae
Hygrotrama Singer, *Sydowia* 12: 221 (1959) [1958]
Type: *Camarophyllopsis schulzeri* (Bres.) Herink

atropuncta (Pers.) Arnolds, *Mycotaxon* 25(2): 642 (1986)
Agaricus atropunctus Pers., *Syn. meth. fung.*: 353 (1801)
Eccilia atropuncta (Pers.) P. Kumm. (as *E. atropunctata*),
Führ. Pilzk.: 94 (1871)
Omphalia atropuncta (Pers.) Quél., *Bull. Soc. Bot. France* 24:
319 (1877)
Hygrophorus atropunctus (Pers.) A.H. Sm. & Hesler, *Lloydia*
5: 15 (1942)
Hygrocybe atropuncta (Pers.) P.D. Orton & Watling, *Notes
Roy. Bot. Gard. Edinburgh* 29(1): 134 (1969)
Hygrotrama atropunctum (Pers.) Singer, *Sydowia, Beih.* 7: 3
(1973)
E: ! **S:** ! **NI:** !
H: On soil, often heavy loam or black soils, under deciduous
trees or amongst short grass in woodland and thickets. Known
with *Carpinus*, *Corylus*, *Fagus* and *Tilia* spp.
D: FAN2: 113, NM2: 75 **I:** FlDan5: 166A (as *Camarophyllus
atropuctus*), FRIC7: 7-8 49d (as *Hygrotrama atropuncta*), Bon:
171, C&D: 233
Rarely reported.

foetens (W. Phillips) Arnolds, *Mycotaxon* 25(2): 643 (1986)
Hygrophorus foetens W. Phillips, *Grevillea* 7(42): 74 (1878)
Camarophyllus foetens (W. Phillips) J.E. Lange, *Dansk. Bot.
Ark.* 4(4): 18 (1923)
Hygrocybe foetens (W. Phillips) P.D. Orton & Watling, *Notes
Roy. Bot. Gard. Edinburgh* 29(1): 134 (1969)
Hygrotrama foetens (W. Phillips) Singer, *Sydowia, Beih.* 7: 3
(1973)
Agaricus abhorrens Berk. & Broome, *Ann. Mag. Nat. Hist.,
Ser. 5 [Notices of British Fungi no. 1751]* 3: 204 (1879)
Omphalia abhorrens (Berk. & Broome) Sacc., *Syll. fung.* 5:
325 (1887)
E: ! **S:** ! **W:** ! **NI:** !
H: On soil, under deciduous trees and shrubs in woodland and
thickets, rarely amongst grass or mosses in open grassland.
D: FAN2: 112-113, NM2: 75 **I:** Bon: 171, FM3(4): 118
Rarely reported. Basidiomes have a pungent smell, reminiscent
of a mixture of coal gas, skatol and napthalene.

hymenocephala (A.H. Sm. & Hesler) Arnolds, *Mycotaxon*
25(2): 643 (1986)
Hygrophorus hymenocephalus A.H. Sm. & Hesler, *Lloydia* 5:
14 (1942)
Armillariella hymenocephala (A.H. Sm. & Hesler) Singer,
Lilloa 22: 217 (1951) [1949]
Hygrotrama hymenocephalum (A.H. Sm. & Hesler) Singer,
Sydowia 12: 222 (1959) [1958]
Hygrocybe hymenocephala (A.H. Sm. & Hesler) P.D. Orton &
Watling, *Notes Roy. Bot. Gard. Edinburgh* 29(1): 134
(1969)
E: !
H: On loam soil, amongst grass under deciduous trees.
D: NM2: 76
Known from Cambridgeshire and Surrey.

micacea (Berk. & Broome) Arnolds, *Persoonia* 13(3): 386
(1987)

Hygrophorus micaceus Berk. & Broome, *Ann. Mag. Nat. Hist.,
Ser. 5 [Notices of British Fungi no. 1779]* 3: 207 (1879)
Hygrocybe micacea (Berk. & Broome) P.D. Orton & Watling,
Notes Roy. Bot. Gard. Edinburgh 29(1): 134 (1969)
Hygrophorus phaeoxanthus Romagn., *Bull. Soc. Mycol.
France* 86(4): 873 (1971) [1970]
E: ! **S:** ! **W:** !
H: On calcareous loam soil in thickets or woodland, also in
unimproved grassland.
D: FAN2: 114-115, NM2: 76
Rarely reported. Apparently widespread.

schulzeri (Bres.) Herink, *Acta Musei Bohemiae septentr.
Liberecensis* 1: 62 (1959)
Hygrophorus schulzeri Bres., *Fungi trident.* 1: 57 (1881)
Camarophyllus schulzeri (Bres.) Ricken, *Vadem. Pilzfr.*: 198
(1920)
Hygrocybe schulzeri (Bres.) Joss., *Bull. Soc. Mycol. France*
53: 206 (1937)
E: ! **S:** ! **W:** !
H: On soil in grassland, occasionally in open spaces in
woodland.
D: NCL3: 264 (as *Hygrophorus schulzeri*), FAN2: 111-112, NM2:
75-76 **I:** FRIC7: 6-7 49c (as *Hygrotrama schulzeri*)
Rarely reported.

CAMPANELLA Henn., *Bot. Jahrb. Syst.* 22: 95 (1895)
Agaricales, Tricholomataceae
Type: *Campanella buettneri* Henn.

caesia Romagn., *Bull. Soc. Mycol. France* 96: 428 (1981)
[1980]
Campanella europaea Singer [*nom. prov.*], *Nova Hedwigia*
26(4): 873 (1975)
E: !
H: On dead and dying culms and stems of various grasses.
D: FAN3: 105 **D+I:** Reid (1995a)
Known from Kent, South Essex, Surrey, and West
Gloucestershire. Reported elsewhere but unsubstantiated with
voucher material.

CANDELABROCHAETE Boidin, *Cah. Maboké* 8: 24 (1970)
Stereales, Botryobasidiaceae
Type: *Candelabrochaete africana* Boidin

septocystidia (Burt) Burds., *Mycotaxon* 19: 392 (1984)
Peniophora septocystidia Burt, *Ann. Missouri Bot. Gard.* 13:
260 (1926)
Phanerochaete septocystidia (Burt) J. Erikss. & Ryvarden,
Corticiaceae of North Europe 5: 1021 (1978)
Scopuloides septocystidia (Burt) Jülich, *Persoonia* 11(4): 422
(1982)
E: !
H: In old woodland, on decayed wood of deciduous trees.
British material on large fallen logs and trunks of *Fagus*.
D: NM3: 167 (as *Phanerochaete septocystidia*) **D+I:** CNE5:
1021-1023, B&K2: 160 167 (as *Scopuloides septocystidia*),
Myc10(4): 156-157
Known from South Hampshire, Surrey and West Sussex.

CANTHARELLULA Singer, *Rev. Mycol. (Paris)* 1(6): 281 (1936)
Agaricales, Tricholomataceae
Type: *Cantharellula umbonata* (J.F. Gmel.) Singer

umbonata (J.F. Gmel.) Singer, *Rev. Mycol. (Paris)* 1(6): 281
(1936)
Merulius umbonatus J.F. Gmel., *Syst. nat.*: 1430 (1792)
Cantharellus umbonatus (J.F. Gmel.) Fr., *Syst. mycol.* 1: 317
(1821)
Clitocybe umbonata (J.F. Gmel.) Konrad, *Bull. Soc. Mycol.
France* 47: 146 (1931)

Hygrophoropsis umbonata (J.F. Gmel.) Kühner & Romagn. [*nom. inval.*], *Flore Analytique des Champignons Supérieurs*: 130 (1953)

E: ! **S:** ! **W:** !

H: On acidic soil, often amongst grass, large mosses such as *Polytrichum* spp or *Calluna* in grassland, open woodland and heathland.

D: NM2: 103, FAN3: 41, BFF8: 48-49 **D+I:** B&K3: 146 146 **I:** FlDan5: 198H (as *Cantharellus umbonatus*), C&D: 183, Kriegl3: 146

Rarely reported, but widespread.

CANTHARELLUS Fr., *Syst. mycol.* 1: 316 (1821)
Cantharellales, Cantharellaceae

Type: *Agaricus cantharellus* L. (= *Cantharellus cibarius*)

amethysteus (Quél.) Sacc., *Syll. fung.* 5: 482 (1887)
Cantharellus cibarius var. *amethysteus* Quél., *Compt. Rend. Assoc. Franç. Avancem. Sci.* 11: 397 (1883)
Craterellus amethysteus (Quél.) Quél., *Fl. mycol. France*: 37 (1888)

E: ! **S:** ! **W:** !

H: On soil, usually associated with deciduous trees such as *Fagus* or *Quercus* spp, but also occasionally reported with conifers.

D: NM3: 261 (as *Cantharellus cibarius* var. *amethysteus*), BFF8: 18 **D+I:** BritChant: 26-27 **I:** SV38: 5

Rarely reported but apparently widespread and may be locally frequent.

aurora (Batsch) Kuyper, *Rivista Micol.* 33(3): 249 (1991) [1990]
Agaricus aurora Batsch, *Elench. fung.*: 93 & t. 9 f. 36 (1783)
Merulius lutescens Pers., *Syn. meth. fung.*: 489 (1801)
Craterellus lutescens (Pers.) Fr., *Epicr. syst. mycol.*: 532 (1838)
Cantharellus xanthopus (Pers.) Duby, *Bot. gall.*: 799 (1830)
Mis.: *Cantharellus lutescens* sensu Fr. (1821)

E: ! **S:** ! **W:** ! **NI:** ! **ROI:** !

H: On soil amongst needle litter under coniferous trees, usually *Pinus sylvestris* and *Picea* spp.

D: NM3: 262, BFF8: 23-24 **D+I:** BritChant: 32-33 **I:** FM7: 30

Rarely reported. Apparently not infrequent in upland areas of Scotland, Northern Ireland, and Republic of Ireland but far less common (and probably often misidentified *Cantharellus tubiformis* var. *lutescens*) in England and Wales.

cibarius Fr., *Syst. mycol.* 1: 318 (1821)
Craterellus cibarius (Fr.) Quél., *Fl. mycol. France*: 37 (1888)
Agaricus cantharellus L. [*nom. illegit.*, non *A. cantharellus* Schwein. (1822)], *Sp. pl.*: 1639 (1753)
Merulius cantharellus (L.) Scop., *Fl. carniol.* 461 (1772)
Agaricus chantarellus Bolton, *Hist. fung. Halifax* 2: 62 (1788)
Cantharellus cibarius f. *neglectus* Souché, *Bull. Soc. Mycol. France* 20: 39 (1904)
Cantharellus cibarius f. *pallidus* R. Schulz, Michael & Schulz, *Führer für Pilzfreunde* 1: pl. 82 (1924)
Cantharellus cibarius var. *albidus* Maire, *Publ. Inst. Bot. Barcelona* 3: 49 (1937)
Cantharellus pallens Pilát, *Acad. Republ. Pop. Romine*: 600 (1959)
Cantharellus vulgaris Gray, *Nat. arr. Brit. pl.* 1: 636 (1821)

E: c **S:** c **W:** c **NI:** c **ROI:** c **O:** Channel Islands: o Isle of Man: o

H: On soil amongst leaf litter in deciduous woodland. Associated with *Betula, Fagus* and *Quercus* spp. Occasionally reported in plantations 'with conifers'.

D: NM3: 261, BFF8: 18-20 **D+I:** Ph: 191, B&K2: 370 481, BritChant: 28-29 **I:** Sow1: 46 (as *Agaricus cantharellus*), Bon: 307, C&D: 145, SV38: 4, Kriegl2: 10

Common and widespread. English name = 'Chanterelle'.

cinereus (Pers.) Fr., *Syst. mycol.* 1: 320 (1821)
Merulius cinereus Pers., *Syn. meth. fung.*: 490 (1801)
Craterellus cinereus (Pers.) Quél., *Fl. mycol. France*: 36 (1888)
Pseudocraterellus cinereus (Pers.) Kalamees, *Tartu Riik. Ülik.*

Toim. 136: 90 (1963).

E: ! **S:** ! **W:** ! **ROI:** !

H: On soil, amongst litter, usually with *Fagus* in woodland.

D: NM3: 262, BFF8: 25-26 **I:** FlDan5: 197A, Bon: 307, Kriegl2: 13

Rarely reported but apparently widespread. Fries [*Syst. mycol.* 1: 320 (1821)] gives *Agaricus infundibuliformis* Bolton as a synonym.

ferruginascens P.D. Orton, *Notes Roy. Bot. Gard. Edinburgh* 29: 83 (1969)
Cantharellus cibarius var. *ferruginascens* (P.D. Orton) Courtec., *Doc. Mycol.* 23(91): 3 (1993)
Mis.: *Cantharellus pallens* sensu Pegler *et al.* (BritChant)

E: ! **S:** ! **W:** !

H: On soil amongst leaf litter in deciduous woodland. Reported with *Betula, Fagus* and *Quercus* spp.

Rarely reported but apparently widespread.

friesii Quél., *Champs Jura Vosges* 1: 191 (1869)

E: ! **S:** ! **O:** Channel Islands: !

H: On soil in mixed woodland.

D: NM3: 262, BFF8: 21, NBA2: 43 **D+I:** B&K2: 370 482, BritChant: 34-35 **I:** Kriegl2: 15

Rarely reported and frequently misidentified. Records scattered throughout England, a few old records from Scotland and a single one from the Channel Islands, mostly unsubstantiated with voucher material.

melanoxeros Desm., *Bot. gall.*: 799 (1830)

E: ! **S:** !

H: On soil amongst leaf litter or mosses, in deciduous woodland. British material is mostly associated with *Fagus* and *Quercus* spp, but also reported with *Corylus* and *Crataegus* sp.

D: NM3: 262, BFF8: 21-22 **D+I:** BritChant: 30-31 **I:** Bon: 307, SV38: 5, Kriegl2: 16

Rarely reported. Known from a few, widely scattered locations in southern England (South Devon, South Hampshire and Surrey) and from Scotland (Westerness). Reported from North Wiltshire and South Essex but unsubstantiated with voucher material. This excludes *C. melanoxeros* sensu B&K2, which is a distinct, non-British species, correctly known as *C. ianthinoxanthus*.

tubaeformis (Bull.) Fr., *Syst. mycol.* 1: 319 (1821)
Helvella tubaeformis Bull., *Herb. France*: pl. 461 (1790)
Helvella cantharelloides Bull., *Herb. France*: pl. 473 (1790)
Agaricus cantharelloides Sowerby, *Col. fig. Engl. fung.* 1: pl. 47 (1796)
Cantharellus infundibuliformis (Scop.) Fr., *Epicr. syst. mycol.*: 366 (1838)
Cantharellus infundibuliformis var. *subramosus* Bres., *Fungi trident.* 1: t. 97 (1881)
Cantharellus tubaeformis var. *lutescens* (Pers.) Fr., *Epicr. syst. mycol.*: 366 (1838)
Peziza undulata Bolton, *Hist. fung. Halifax* 3: 105 (1789)

E: c **S:** c **W:** c **NI:** c **ROI:** c **O:** Channel Islands: ! Isle of Man: !

H: On acidic soil, usually with deciduous trees, in woodland. Often with *Fagus* but also known with *Betula, Quercus* and *Salix* spp, and in mixed woodland with conifers such as *Pinus* and *Picea* spp.

D: NM3: 262 (as *Cantharellus tubaeformis* var. *tubaeformis*), BFF8: 24-25 **D+I:** Ph: 190-191 (as *C. infundibuliformis*), B&K2: 372 485 (as *C. tubaeformis*), BritChant: 36-38 **I:** Sow1: 47 (as *Agaricus cantharelloides*), FlDan5: 198I, Bon: 307, C&D: 145, Kriegl1: 18 (as *C. tubaeformis*)

Common and widespread. The epithet *tubaeformis* (trumpet-shaped) is retained in preference to *tubiformis* (tube-shaped).

CELLYPHA Donk, *Persoonia* 1: 84 (1959)
Agaricales, Tricholomataceae

Type: *Cellypha goldbachii* (Weinm.) Donk

goldbachii (Weinm.) Donk, *Persoonia* 1(1): 85 (1959)
Cyphella goldbachii Weinm., *Hymen. Gasteromyc.*: 522 (1836)

Cyphella lactea Bres., *Fungi trident.* 1: 61 (1884)
Cyphella ochroleuca Berk. & Broome, *Ann. Mag. Nat. Hist.,*
   Ser. 2 [Notices of British Fungi no. 719] 13: 405 (1854)
Phaeocyphella ochroleuca (Berk. & Broome) Rea, *Brit.*
   basidiomyc.: 704 (1922)
Cyphella rubi Fuckel, *Jahrb. Nassauischen Vereins Naturk.*
   23-24: 26 (1870)
E: ! **S:** ! **W:** ! **NI:** ! **O:** Channel Islands: ! Isle of Man: !
H: On dead stems and leaves of *Carex* and *Juncus* spp, also on
   dead grass stems.
D: NM2: 104
Rarely reported but apparently widespread.

CENANGIOMYCES Dyko & B. Sutton, *Trans. Brit. Mycol. Soc.* 72(3): 411 (1979)

Cenangiomyces is a monotypic genus based on an anamorphic
   basidiomycete whose teleomorph is not yet known.
Type: *Cenangiomyces luteus* Dyko & B. Sutton

luteus Dyko & B. Sutton, *Trans. Brit. Mycol. Soc.* 72(3): 411
   (1979)
E: !
H: on pine needles.
The species produces small, disc-shaped conidiomata with
   clamped hyphae. It is only known from the type collection
   (Sussex: Graffham Common).

CEPHALOSCYPHA Agerer, *Sydowia* 27: 193 (1975)

Stereales, Cyphellaceae
Type: *Cephaloscypha morlichensis* (W.B. Cooke) Agerer

mairei (Pilát) Agerer, *Mitt. bot. St Samml., München* 19: 303
   (1983)
   Cyphella mairei Pilát, *Ann. Mycol.* 22: 211 (1924)
   Lachnella mairei (Pilát) W.B. Cooke, *Sydowia, Beih.* 4: 73
      (1961)
   Flagelloscypha mairei (Pilát) Knudsen, *Nordic J. Bot.* 11(4):
      478 (1991)
   Flagelloscypha morlichensis W.B. Cooke, *Sydowia, Beih.*: 63
      (1961)
   Cephaloscypha morlichensis (W.B. Cooke) Agerer, *Sydowia*
      27(1-6): 194 (1975) [1973-1974]
S: !
H: On dead stems and fronds of ferns such as *Blechnum spicant*
   and *Dryopteris* spp.
D: NM2: 123 (as *Flagelloscypha mairei*)
First recorded in Britain from Easterness (Loch Morlich) in 1957
   and only collected twice since.

CERACEOMYCES Jülich, *Willdenowia, Beiehfte* 7: 146 (1972)

Stereales, Meruliaceae
   Ceraceomerulius (Parmasto) J. Erikss. & Ryvarden,
      Corticiaceae of North Europe 2: 196 (1973)
Type: *Ceraceomyces tessulatus* (Cooke) Jülich

borealis (Romell) J. Erikss. & Ryvarden, *Corticiaceae of North*
   Europe 2: 205 (1973)
   Merulius borealis Romell, *Ark. Bot.* 11(3): 27 (1911)
E: ! **NI:** !
D: NM3: 148 **D+I:** CNE2: 204-205, Myc11(3): 112 **I:** FM1(1):
   19
Known from England (Cumberland, Hertfordshire, North & South
   Hampshire, North & South Somerset, South Wiltshire and
   Surrey) and Northern Ireland (Tyrone). Surpisingly, since this
   is a common species in Scandinavia, not yet reported from
   Scotland.

crispatus (O.F. Müll.) Rauschert, *Feddes Repert.* 98(11-12):
   657 (1987)
   Merulius crispatus O.F. Müll., *Fl. Danica*: t. 716, f. 2 (1777)

Polyporus collabefactus Berk. & Broome, *Ann. Mag. Nat.*
   Hist., Ser. 4 [Notices of British Fungi no. 1432] 15: 31
   (1875)
Poria collabefacta (Berk. & Broome) Cooke, *Grevillea* 14: 114
   (1886)
Merulius porinoides Fr., *Observ. mycol.*: 237 (1818)
Merulius serpens Tode, *Abh. Naturf. Ges. Halle.* 1: 355
   (1783)
Xylomyzon serpens (Tode) Pers., *Mycol. eur.* 2: 31 (1825)
Byssomerulius serpens (Tode) Parmasto, *Eesti N.S.V. Tead.*
   Akad. Toimet., Biol. 16: 384 (1967)
Ceraceomerulius serpens (Tode) J. Erikss. & Ryvarden,
   Corticiaceae of North Europe 2: 201 (1973)
Ceraceomyces serpens (Tode) Ginns, *Canad. J. Bot.* 54(1-2):
   147 (1976)
E: ! **W:** !
H: On decayed wood of deciduous trees or conifers in old
   woodland. Also on decayed worked wood such as fence posts.
D: NM3: 148 **D+I:** CNE2: 200-201 (as *Ceraceomerulius*
   serpens), B&K2: 108 90 (dark colour form) **I:** Kriegl1: 176 (as
   Ceraceomyces crispatus)
Known from England (East Suffolk, North Somerset and Surrey)
   and Wales (Caernarvonshire and Monmouthshire). Better
   known as *Ceraceomyces serpens*.

sublaevis (Bres.) Jülich, *Willdenowia. Beih.* 7: 147 (1972)
   Corticium sublaeve Bres., *Ann. Mycol.* 1(1): 95 (1903)
   Peniophora sublaevis (Bres.) Höhn. & Litsch., *Sitzungsber.*
      Kaiserl. Akad. Wiss., Wien, Math.-Naturwiss. Cl., Abt. 1
      1(117): 1088 (1908)
   Athelia sublaevis (Bres.) Parmasto, *Eesti N.S.V. Tead. Akad.*
      Toimet., Biol. 16(4): 382 (1967)
   Corticium microsporum Bourdot & Galzin [*nom. illegit., non*
      C. microsporum Bres & Herter (1911)], *Bull. Soc. Mycol.*
      France 27: 241 (1911)
E: ! **S:** ! **W:** ! **ROI:** !
H: On decayed wood of deciduous trees and shrubs, such as
   Corylus and *Quercus* spp. Rarely on conifers such as *Pinus*
   sylvestris.
D: NM3: 148 **D+I:** CNE2: 208-209, B&K2: 110 91
Rarely reported but apparently widespread. The majority of
   records are from southern England.

tessulatus (Cooke) Jülich, *Willdenowia. Beih.* 7: 154 (1972)
   Corticium tessulatum Cooke, *Grevillea* 6: 132 (1878)
   Athelia tessulata (Cooke) Donk, *Fungus* 27: 12 (1957)
   Corticium illaqueatum Bourdot & Galzin, *Bull. Soc. Mycol.*
      France 27: 238 (1911)
E: ! **S:** !
H: On decayed wood and cones of conifers, usually *Pinus*
   sylvestris in woodland or on heathland. Also known on
   Araucaria araucana and *Betula* spp.
D: NM3: 148 **D+I:** CNE2: 210-211
Known from England (East & West Sussex, South Essex, Surrey
   and Warwickshire) and Scotland (Easterness).

CERATELLOPSIS Konrad & Maubl., *Icon. select. fung.* 6: 502 (1937)

Cantharellales, Clavariadelphaceae
   Ceratella Pat. [*nom. illegit., non Ceratella* Hooker f.
      (1844)(*Compositae*)], *Hyménomyc. Eur.*: 157 (1887)
Type: *Ceratellopsis queletii* (Pat.) Konr. & Maubl.

aculeata (Pat.) Corner, *Monograph of Clavaria and Allied*
   Genera: 200 (1950)
   Pistillaria aculeata Pat., *Tab. anal. fung.*: no. 58 (1883)
E: !
H: Only on decayed culms of *Cladium mariscus* in fens or
   marshes.
D: NM3: 283
Rarely recorded or collected but basidiomes are minute and
   easily overlooked. Known from Cambridgeshire and East
   Norfolk but apparently 'common' in the Cambridgeshire fens
   fide Corner (1950).

acuminata (Fuckel) Corner, *Monograph of Clavaria and Allied Genera*: 202 (1950)
   Pistillaria acuminata Fuckel, *Jahrb. Nassauischen Vereins Naturk.* 23-24: 31 (1870)
   Ceratella acuminata (Fuckel) Pat., *Essai taxon.*: 123 (1900)
   Mis.: *Pistillaria pusilla* sensu B&K2: 336 & pl. 430 (1986), non Corner (1950)
S: !
H: On fallen and decayed needle litter of *Pinus sylvestris* and also reported on fallen leaves of *Eucalyptus cordata*.
D: NM3: 283 (as *Ceratellopsis sagittiformis*)
Rarely reported. Known only from scattered locations in Scotland (Morayshire, South Aberdeen, West Sutherland and Wester Ross).

sagittiformis (Pat.) Corner, *Monograph of Clavaria and Allied Genera*: 206 (1950)
   Pistillaria sagittiformis Pat., *Tab. anal. fung.* 26: no. 56 (1883)
   Mis.: *Ceratellopsis acuminata* sensu Bourdot & Galzin (1928)
   Mis.: *Pistillaria pusilla* sensu auct.
E: !
H: On mosses growing on logs in deciduous woodland.
Known from Hertfordshire (Benington, Combs Wood and Northaw Great Wood) and Surrey (Mickleham).

CERATOBASIDIUM D.P. Rogers, *Univ. Iowa Stud. Nat. Hist.* 17: 4 (1935)
Ceratobasidiales, Ceratobasidiaceae
   Ceratorhiza R.T. Moore (anam.), *Mycotaxon* 29: 94 (1987)
Type: *Ceratobasidium calosporum* D.P. Rogers

anceps (Bres. & Syd.) H.S. Jacks., *Canad. J. Res., Sect. C, Bot. Sci.* 27: 243 (1949)
   Tulasnella anceps Bres. & Syd., *Ann. Mycol.* 8: 490 (1910)
   Corticium anceps (Bres. & Syd.) Gregor, *Ann. Mycol.* 30: 464 (1932)
   Thanatephorus anceps (Bres. & Syd.) Parmasto, *Eesti N.S.V. Tead. Akad. Toimet., Biol.* 17: 225 (1968)
   Sclerotium deciduum Davis (anam.), *Trans. Wis. Acad. Sci. Arts Lett.* 19: 689 (1919)
   Ceratorhiza decidua (Davis) P. Roberts (anam.), *Rhizoc .f. fungi*: 32 (1999)
E: ! **S:** ! **NI:** !
H: On living fronds of *Pteridium*.
D: NM3: 113
Rarely reported. Apparently widespread but most of the records are from Scotland or Northern Ireland.

calosporum D.P. Rogers, *Univ. Iowa Stud. Nat. Hist.* 17: 5 (1935)
   Ceratorhiza anacalospora P. Roberts (anam.), *Rhizoc. f. fungi*: 38 (1999)
E: ! **O:** Channel Islands: !
H: On decayed wood and bark of deciduous trees.
Rarely reported.

cornigerum (Bourdot) D.P. Rogers, *Univ. Iowa Stud. Nat. Hist.* 17: 5 (1935)
   Corticium cornigerum Bourdot, *Rev. Sci. du Bourb.* 35: 4 (1922)
   Ceratobasidium cereale D.I. Murray & Burpee, *Trans. Brit. Mycol. Soc.* 82(1): 172 (1984)
   Rhizoctonia ramicola G.F. Weber & D.A. Roberts (anam.), *Phytopathology* 41: 618 (1951)
   Ceratorhiza ramicola (G.F. Weber & D.A. Roberts) R.T. Moore (anam.), *Mycotaxon* 29: 94 (1987)
   Mis.: *Rhizoctonia cerealis* sensu auct. (anam.)
   Mis.: *Ceratorhiza goodyerae-repentis* sensu auct. (anam.)
   Mis.: *Rhizoctonia goodyerae-repentis* sensu auct. (anam.)
E: ! **S:** !
H: On decaying wood and leaf litter of deciduous trees, on dying or moribund fern fronds, and also on living leaves of herbaceous and woody plants. Also isolated from soil and from orchid mycorrhiza.
D: NM3: 112 **D+I:** CNE2: 222-223

Rarely recorded or collected.

pseudocornigerum M.P. Christ., *Dansk. Bot. Ark.* 19: 46 (1959)
E: ! **W:** !
H: On decayed wood and fallen leaves of deciduous trees.
D: NM3: 112 **D+I:** CNE2: 224-225, B&K2: 76 41
Known from England (South Devon) and Wales (Breconshire).

CERATOSEBACINA P. Roberts, *Mycol. Res.* 97(4): 470 (1993)
Tremellales, Exidiaceae
Type: *Ceratosebacina longispora* (Hauerslev) P. Roberts

calospora (Bourdot & Galzin) P. Roberts, *Folia Cryptog. Estonica* 33: 128 (1998)
   Exidiopsis calospora Bourdot & Galzin, *Bull. Soc. Mycol. France* 39: 263 (1924)
   Sebacina calospora (Bourdot & Galzin) Bourdot & Galzin, *Hymenomyc. France*: 46 (1928)
E: !
H: On decayed wood of deciduous trees.
D: NM3: 106 (as *Sebacina calospora*) **D+I:** Reid4: 432-433 (as *S. calospora*)
Only two collections from Cambridge (Madingley Wood) in 1953 and 1955, and not reported since.

longispora (Hauerslev) P. Roberts, *Mycol. Res.* 97(4): 470 (1993)
   Sebacina longispora Hauerslev, *Friesia* 11: 100 (1976)
   Exidiopsis longispora (Hauerslev) Wojewoda, *Mała Flora Grzybów* 2: 107 (1981)
E: !
H: On decayed wood of deciduous trees such as *Quercus* and *Ulmus* spp.
D: NM3: 102
Rarely reported. Known from Hertfordshire, Shropshire, South Devon, Surrey and West Cornwall.

CERIPORIA Donk, *Meded. Bot. Mus. Herb. Rijks Univ. Utrecht* 9: 170 (1933)
Poriales, Coriolaceae
Type: *Ceriporia viridans* (Berk. & Broome) Donk

excelsa (S. Lundell) Parmasto, *Sporovyja rasterija kamtjatki, Griby (Fungi)* 12: 222 (1959)
   Poria excelsa S. Lundell, *Fungi Exsiccati Suecici* 27-28: 14 (1946)
   Mis.: *Poria rhodella* sensu auct.
E: o **S:** ? **W:** ?
H: On large, decayed, fallen trunks of deciduous trees, in old woodland, usually on calcareous or limestone soils. Most often collected on *Acer pseudoplatanus* and *Fagus* but also known on *Betula* sp., *Castanea* and *Fraxinus*. Rarely recorded on conifer wood but the hosts are unverified.
D: EurPoly1: 183, NM3: 170 **D+I:** B&K2: 296 371
Known from England (Buckinghamshire, Derbyshire, North Somerset, Oxfordshire, South Devon, South Hampshire, South Somerset, Surrey, West Kent and Westmorland) with a few old and unsubstantiated records from Scotland and Wales.

metamorphosa (Fuckel) Ryvarden & Gilb., *Syn. Fungorum* 6: 185 (1993)
   Polyporus metamorphosus Fuckel, *Jahrb. Nassauischen Vereins Naturk.* 27-28: 87 (1873)
   Poria metamorphosa (Fuckel) Cooke, *Grevillea* 14: 112 (1886)
   Sporotrichum aurantiacum Fr. (anam.), *Syst. mycol.* 3: 423 (1829)
   Sporotrichum aureum Link (anam.), *Mag. Neuesten Entdeck. Gesammten Naturk. Ges. Naturf. Freunde Berlin* 3: 13 (1809)
   Sporotrichum laeticolor Cooke & Massee (anam.), *Grevillea* 20: 38 (1891)
E: ! **W:** ?

H: On decayed wood of deciduous trees, usually *Quercus* spp, but also known on *Betula* spp and *Castanea*.
D: EurPoly1: 185
Known from England (Bedfordshire, Berkshire, Cheshire, East Kent, Hertfordshire, Huntingdonshire, South Lancashire, Surrey and Yorkshire). Reported from Wales (Denbighshire) in 1910, but unsubstantiated with voucher material.

purpurea (Fr.) Donk, *Kon. Ned. Akad. Wetensch. Proc. Sect. Sci.* 1: 28 (1971)
  Polyporus purpureus Fr., *Syst. mycol.* 1: 379 (1821)
  Poria purpurea (Fr.) Cooke, *Grevillea* 14: 112 (1886)
  Meruliopsis purpurea (Fr.) Bondartsev, *Eesti N.S.V. Tead. Akad. Toimet., Biol.* 8(4): 274 (1959)
  Mis.: *Poria mellita* sensu auct. Brit.
  Mis.: *Ceriporia mellita* sensu auct. Brit.
  Mis.: *Poria rhodella* sensu auct.
E: o **S:** ! **W:** !
H: On large, decayed, fallen trunks of deciduous trees in old woodland, usually on calcareous soil. Most often on *Fagus* and *Fraxinus* but also known on *Acer pseudoplatanus*, *Betula* sp., *Corylus*, *Crataegus*, *Populus*, *Salix* and *Ulmus* spp.
D: EurPoly1: 186, NM3: 170 **D+I:** B&K2: 296 372 **I:** Kriegl1: 494
Known from England (East Gloucestershire, East & West Norfolk, North Hampshire, Herefordshire, Hertfordshire, South Essex, Surrey, West Sussex and Yorkshire). Single collections from Scotland (South Ebudes) and Wales (Breconshire).

reticulata (Hoffm.) Domański, *Acta Soc. Bot. Poloniae* 32: 732 (1963)
  Mucilago reticulata Hoffm., *Deutschl. Fl.* 3: pl. 12/2 (1795)
  Boletus reticulatus (Hoffm.) Pers. [*nom. illegit.*, non *B. reticulatus* Schaeff. (1774)], *Syn. meth. fung.*: 548 (1801)
  Polyporus reticulatus (Hoffm.) Fr., *Syst. mycol.* 1: 385 (1821)
  Poria reticulata (Hoffm.) Cooke, *Grevillea* 14: 114 (1886)
  Fibuloporia reticulata (Hoffm.) Bondartsev, *Trutovye griby Evropeiskoi chasti SSSR i Kavkaza*: 126 (1953)
  Polyporus farinellus Fr., *Syst. mycol.* 1: 384 (1821)
  Poria farinella (Fr.) Cooke, *Grevillea* 14: 114 (1886)
E: c **S:** ! **W:** o **ROI:** ! **O:** Isle of Man: !
H: On decayed wood of deciduous trees and occasionally on decayed polypore basidiomes.
D: EurPoly1: 187, NM3: 170 **D+I:** B&K2: 298 373 **I:** Kriegl1: 496
Common and widespread but few records from Scotland. Easily recognised by the shallowly reticulate (net like) pores.

viridans (Berk. & Broome) Donk, *Meded. Bot. Mus. Herb. Rijks Univ. Utrecht* 9: 171 (1933)
  Polyporus viridans Berk. & Broome, *Ann. Mag. Nat. Hist., Ser. 3 [Notices of British Fungi no. 937]* 7: 379 (1861)
  Poria viridans (Berk. & Broome) Cooke, *Grevillea* 14: 112 (1886)
  Polyporus blepharistoma Berk. & Broome, *Ann. Mag. Nat. Hist., Ser. 4 [Notices of British Fungi no. 1434]* 15: 31 (1875)
  Poria blepharistoma (Berk. & Broome) Cooke, *Grevillea* 14: 114 (1886)
  Mis.: *Poria rhodella* sensu auct.
E: c **S:** o **W:** ! **ROI:** !
H: On decayed wood of deciduous trees, especially *Betula* spp, *Fagus* and *Fraxinus* but also known on *Acer pseudoplatanus*, *Crataegus*, *Pittosporum*, *Quercus*, *Salix* and *Ulmus* spp. Rarely on conifers such as *Pinus sylvestris* and old decayed basidiomes of polypores.
D: EurPoly1: 189, NM3: 170 **D+I:** B&K2: 3398 374
Common and widespread throughout England, less often reported elsewhere. Basidiomes are macroscopically variable, especially in colour (rarely greenish, as the epithet *viridans* suggests, but frequently bright pink), hence often misidentified.

CERIPORIOPSIS Domański, *Acta Soc. Bot. Poloniae*. 32: 731 (1963)

Poriales, Coriolaceae
  Gelatoporia Niemelä, *Karstenia* 25(1): 22 (1985)
Type: *Ceriporiopsis gilvescens* (Bres.) Domański

aneirina (Sommerf.) Domański, *Acta Soc. Bot. Poloniae.* 32: 732 (1963)
  Polyporus aneirinus Sommerf., *Fl. Lapp.*: 27 (1826)
  Poria aneirina (Sommerf.) Cooke, *Grevillea* 14: 112 (1886)
  Tyromyces aneirinus (Sommerf.) Ryvarden, *Norweg. J. Bot.* 20: 10 (1973)
  Antrodia serena P. Karst., *Meddeland. Soc. Fauna Fl. Fenn.* 6: 10 (1881)
E: ! **S:** !
H: On dead or decayed wood of deciduous trees. British collections on *Populus tremula* and *Quercus* spp.
D: EurPoly1: 193, NM3: 222
Records from England (Cumberland in 1884 and Northamptonshire in 1988) and Scotland (Morayshire in 1870 and 1884, and Easterness in 2000) supported by voucher material. Additionally reported from North Somerset in 1911 and South Hampshire in 1991 but unsubstantiated with voucher material.

gilvescens (Bres.) Domański, *Acta Soc. Bot. Poloniae* 32: 731 (1963)
  Poria gilvescens Bres., *Ann. mycol.* 6: 40 (1908)
  Tyromyces gilvescens (Bres.) Ryvarden, *Norweg. J. Bot.* 20: 10 (1973)
E: o **W:** ? **NI:** ? **ROI:** ? **O:** Channel Islands: !
H: On decayed wood of deciduous trees, especially on large fallen trunks or logs of *Fagus* but also known from *Acer pseudoplatanus*, *Alnus*, *Betula*, *Fraxinus*, *Castanea*, *Quercus* and *Ulmus* spp. Rarely on conifers such as *Pinus sylvestris*. A single collection on dead standing stems of *Heracleum sphondylium*.
D: EurPoly1: 196, NM3: 222 **D+I:** Ph: 237, B&K2: 298 375
Occasional but widespread in England (Derbyshire and Yorkshire southwards to West Cornwall). Rare in northern areas but may be locally common in the south, especially in old woodland such as the New Forest and the Surrey beechwoods. A single collection from the Channel Islands (Sark: Dixcart Valley). Reported rarely elsewhere and unsubstantiated with voucher material.

myceliosa (Peck) Ryvarden & Gilb., *Syn. Fungorum* 6: 198 (1993)
  Poria myceliosa Peck, *Bull. New York State Mus. Nat. Hist.* 54: 952 (1902)
  Anomoporia myceliosa (Peck) Pouzar, *Česká Mykol.* 20: 172 (1966)
E: !
H: On decayed coniferous wood.
D: EurPoly1: 198
Possibly extinct. Two old collections at Kew (1875 and 1878) determined by J.L. Lowe but not reported since.

pannocincta (Romell) Gilb. & Ryvarden, *Mycotaxon* 22(2): 364 (1985)
  Polyporus pannocinctus Romell, *Arch. Bot. (Leipzig)* 11(3): 20 (1911)
  Gloeoporus pannocinctus (Romell) J. Erikss., *Symb. Bot. Upsal.* 1: 136 (1958)
  Tyromyces pannocinctus (Romell) Kotl. & Pouzar, *Česká Mykol.* 18: 65 (1964)
  Gelatoporia pannocincta (Romell) Niemelä, *Karstenia* 25(1): 23 (1985)
E: o **S:** !
H: On decayed wood of deciduous trees in old woodland. Usually on large fallen trunks or logs of *Fagus* but also known on *Acer pseudoplatanus*, *Alnus*, *Betula*, *Fraxinus*, *Quercus*, *Salix* and *Ulmus* spp.
D: EurPoly1: 200, NM3: 221 **D+I:** B&K2: 294 367 (as *Gloeoporus pannocinctus*), Myc6(3): 139 (as *Gelatoporia pannocincta*)
First reported from England (South Hampshire: New Forest) in 1991, this has spread widely throughout southern England

(Berkshire, East Norfolk, Hertfordshire, Middlesex, Oxfordshire, Surrey and West Sussex) and is known from a single site in Scotland (Westerness: Arisaig, Glen Beasdale). Presumably a recent addition to the British mycota.

CERRENA Gray, *Nat. arr. Brit. pl.* 1: 649 (1821)
Poriales, Coriolaceae
Type: *Sistotrema cinereum* Pers. (= *Cerrena unicolor*)

unicolor (Bull.) Murrill, *J. Mycol.* 9: 91 (1903)
    Boletus unicolor Bull., *Herb. France*: pl. 408 (1789)
    Daedalea unicolor (Bull.) Fr., *Syst. mycol.* 1: 336 (1821)
    Trametes unicolor (Bull.) Pilát, *Atlas champ. Eur.* I, 3(1): 279 (1936)
    Sistotrema cinereum Pers., *Neues Mag. Bot.* 1: 109 (1794)
    Daedalea cinerea (Pers.) Fr., *Observ. mycol.* 1: 105 (1815)
    Cerrena cinerea (Pers.) Gray, *Nat. arr. Brit. pl.* 1: 649 (1821)
    Daedalea latissima Fr., *Syst. mycol.* 1: 340 (1821)
E: o **S:** ! **NI:** ? **O:** Channel Islands: !
H: On dead wood of deciduous trees, often old stumps or fallen trunks of *Acer campestre*, *A. pseudoplatanus* and *Fagus*.
D: 227, EurPoly1: 205 **D+I:** B&K2: 278 345 **I:** Sow3: 325 (as *Boletus unicolor*)
Occasional but widespread in England, rarely reported elsewhere. Early collections from Scotland (Angus and Kincardineshire) in herb. K but none since. Reported from Northern Ireland but unsubstantiated with voucher material.

CHAETOCALATHUS Singer, *Lilloa* 8: 518 (1942)
Agaricales, Tricholomataceae
Type: *Chaetocalathus craterellus* (Durrieu & Lév.) Singer

craterellus (Durrieu & Lév.) Singer, *Lilloa* 8: 518 (1942)
    Agaricus craterellus Durrieu & Lév., *Essai. Fl. Mycol. Montp. Et Gard.* 132 (1863)
    Pleurotellus patelloides P.D. Orton, *Trans. Brit. Mycol. Soc.* 43(2): 341 (1960)
E: ! **W:** !
H: On dead and fallen twigs and dead woody stems, fruiting throughout the year. Reported from woodland or scrub vegetation on *Fagus* and *Rubus fruticosus* agg.
D: BFF6: 34
Rarely reported and most records unsubstantiated with voucher material. Easily overlooked, or passed over as a species of *Crepidotus*.

CHAETOTYPHULA Corner, *Monograph of Clavaria and Allied Genera*: 207 (1950)
Cantharellales, Clavariadelphaceae
Type: *Chaetotyphula hyalina* (Jungh.) Corner

actiniceps (Petch) Corner, *Monograph of Clavaria and Allied Genera*: 209 (1950)
    Pistillaria actiniceps Petch, *Rep. Roy. Bot. Gard. Peradeniya* 7: 291 (1922)
E: !
H: On old dead, attached inflorescences of *Buddleja davidii* (British material).
A single collection from South Devon (Torquay) but basidiomes are minute and easily overlooked.

CHALCIPORUS Bataille, *Bull. Soc. Hist. Nat. Doubs.* 15: 39 (1908)
Boletales, Strobilomycetaceae
Type: *Chalciporus piperatus* (Bull.) Bataille

piperatus (Bull.) Bataille, *Bull. Soc. Hist. Nat. Doubs.* 15: 39 (1908)
    Boletus piperatus Bull., *Herb. France*: pl. 451 (1790)
    Leccinum piperatum (Bull.) Gray, *Nat. arr. Brit. pl.* 1: 647 (1821)
    Ixocomus piperatus (Bull.) Quél., *Fl. mycol. France*: 414 (1888)
    Suillus piperatus (Bull.) Kuntze, *Revis. gen. pl.* 3(2): 535 (1898)
E: c **S:** c **W:** c **NI:** c **ROI:** c **O:** Isle of Man: !
H: On soil, usually associated with *Amanita muscaria* near *Betula* spp, but also known with *Fagus* and *Pinus* spp.
D: NM2: 63 **D+I:** Ph: 194 (as *Boletus piperatus*), B&K3: 66 27 **I:** Sow1: 34 (as *B. piperatus*), Bon: 45, C&D: 423, Kriegl2: 251
Common and widespread.

CHAMAEMYCES Earle, *Bull. New York Bot. Gard.* 5: 446 (1909)
Agaricales, Agaricaceae
    Lepiotella (E.-J. Gilbert) Konrad [non *Lepiotella* J. Rick (1938)], *Schweiz. Z. Pilzk.* 12: 117 (1934)
    Drosella Maire [*nom. nud.*], *Bull. Soc. Mycol. France* 50: 15 (1935) [1934]
Type: *Chamaemyces fracidus* (Fr.) Donk

fracidus (Fr.) Donk, *Nova Hedwigia Beih.* 5: 48 (1962)
    Agaricus fracidus Fr., *Epicr. syst. mycol.*: 25 (1838)
    Armillaria fracida (Fr.) Gillet, *Hyménomycètes*: 77 (1874)
    Drosella fracida (Fr.) Singer, *Lilloa* 22: 446 (1951) [1949]
    Agaricus demisannulus Fr., *Hymenomyc. eur.*: 38 (1874)
    Lepiota demisannula (Fr.) Sacc., *Syll. fung.* 5: 69 (1887)
    Lepiota irrorata Quél., *Compt. Rend. Assoc. Franç. Avancem. Sci.* 10: 387 (1882)
    Drosella irrorata (Quél.) Kühner & Maire [*nom. inval.*], *Bull. Soc. Mycol. France* 50(1): 15 (1935) [1934]
    Lepiotella irrorata (Quél.) Singer, *Ann. Mycol.* 34: 338 (1936)
    Armillaria irrorata (Quél.) J.E. Lange, *Fl. agaric. danic. 5, Taxonomic Conspectus*: II (1940)
E: o **S:** ! **W:** ! **NI:** ! **ROI:** !
H: On soil in deciduous woodland, often amongst grass or under cover of *Mercurialis perennis*. Occasionally in grass on dunes and by roadsides.
D: NM2: 215 **D+I:** Ph: 31 (as *Drosella fracida*), FungEur4: 324-327 37, B&K4: 182 200, FAN5: 152-153 **I:** FlDan1: 15B (as *Armillaria irrorata*), Bon: 283, C&D: 241, SV41: 56
Rarely reported but apparently quite frequent in woodland on calcareous soils in southern England.

CHEIMONOPHYLLUM Singer, *Sydowia* 9: 417 (1955)
Agaricales, Tricholomataceae
Type: *Cheimonophyllum candidissimum* (Berk. & M.A. Curtis) Singer

candidissimum (Berk. & M.A. Curtis) Singer, *Sydowia* 9: 417 (1955)
    Agaricus candidissimus Berk. & M.A. Curtis, *Ann. Mag. Nat. Hist., Ser. 3* 4: 288 (1859)
    Pleurotus candidissimus (Berk. & M.A. Curtis) Sacc., *Syll. fung.* 5: 368 (1887)
    Pleurotellus candidissimus (Berk. & M.A. Curtis) Konrad & Maubl., *Icon. select. fung.* 6: 360 (1937)
E: ! **S:** ! **W:** ! **ROI:** !
H: On decayed wood and dead grass stems. Apparently also on mosses.
D: BFF6: 35, NM2: 104
Rarely reported and a poorly known species in Britain. The majority of reports are unsubstantiated with voucher material and collections named this are usually found to be misidentified species of *Crepidotus*.

CHONDROSTEREUM Pouzar, *Česká Mykol.* 13: 17 (1959)
Stereales, Meruliaceae
Type: *Chondrostereum purpureum* (Pers.) Pouzar

purpureum (Pers.) Pouzar, *Česká Mykol.* 13(1): 17 (1959)
    Stereum purpureum Pers., *Neues Mag. Bot.* 1: 110 (1794)
    Thelephora purpurea (Pers.) Pers., *Syn. meth. fung.*: 571 (1801)

Auricularia persistens Sowerby, *Col. fig. Engl. fung.* 3: pl. 388 (1797)

Stereum vorticosum Fr., *Observ. mycol.*: 275 (1796)

Mis.: *Stereum rugosiusculum* sensu Rea [*Brit. basidiomyc.*: 665 (1922)]

E: c **S:** c **W:** c **NI:** c **ROI:** c **O:** Channel Islands: c Isle of Man: !

H: Parasitic then saprotrophic on dead or dying deciduous trees. Rarely reported on conifers such as *Pinus* spp.

D: NM3: 156 **D+I:** CNE2: 236-237, Ph: 236 **I:** B&K2: 180 198, C&D: 139, Kriegl1: 182

Very common and widespread. Of commercial importance as the cause of 'silver leaf' disease of fruit trees. Also used in recent times as a biocontrol agent to remove susceptible thicket-forming shrubs.

CHROMOCYPHELLA De Toni & Levi, *Naturalist (Hull)* 1888: 158 (1888)

Cortinariales, Crepidotaceae

Phaeocyphella Pat. [non *Phaeocyphella* Speg. (1909)], *Essai taxon.*: 57 (1900)

Type: *Cymbella crouani* Pat. & Doass. (= *Chromocyphella muscicola*)

muscicola (Fr.) Donk, *Persoonia* 1(1): 95 (1959)

Cyphella muscicola Fr., *Syst. mycol.* 2(1): 202 (1823)

Phaeocyphella muscicola (Fr.) Rea, *Brit. basidiomyc.*: 704 (1922)

Cyphella fuscospora Cooke, *Grevillea* 20: 9 (1891)

Phaeocyphella fuscospora (Cooke) Rea, *Brit. basidiomyc.*: 704 (1922)

Merulius galeatus Schumach., *Enum. pl.* 2: 371 (1803)

Cyphella galeata (Schumach.) Fr., *Epicr. syst. mycol.*: 567 (1838)

Phaeocyphella galeata (Schumach.) Bourdot & Galzin, *Hymenomyc. France*: 165 (1928)

Chromocyphella galeata (Schumach.) W.B. Cooke, *Sydowia, Beih.*: 136 (1961)

E: ! **S:** o **W:** ! **ROI:** ! **O:** Channel Islands: !

H: Parasitic. On living mosses such as *Dicranum scoparium, Hylocomium brevirostre, H. splendens* and *Hypnum cupressiforme* usually on tree trunks in woodland, occasionally on the ground.

D: NM2: 335

Records only from northern or western areas (rather frequent in Scotland) but basidiomes are small and easily overlooked.

CHROOGOMPHUS (Singer) O.K. Mill., *Mycologia* 51(4): 526 (1964)

Boletales, Gomphidiaceae

Type: *Chroogomphus rutilus* (Schaeff.) O.K. Miller

rutilus (Schaeff.) O.K. Mill., *Mycologia* 51(4): 543 (1964)

Agaricus rutilus Schaeff., *Fung. Bavar. Palat. nasc.* 1: 56 (1762)

Cortinarius rutilus (Schaeff.) Gray, *Nat. arr. Brit. pl.* 1: 629 (1821)

Gomphidius rutilus (Schaeff.) S. Lundell & Nannf., *Fungi Exsiccati Suecici*: 409 (1937)

Chroogomphus britannicus A.Z.M. Khan & Hora, *Trans. Brit. Mycol. Soc.* 70: 155 (1978)

Chroogomphus corallinus O.K. Mill. & Watling, *Notes Roy. Bot. Gard. Edinburgh* 30: 391 (1970)

Chroogomphus rutilus var. *corallinus* (O.K.Mill. & Watling) Watling, *Edinburgh J. Bot.* 61(1): 42 (2005)

Agaricus gomphus Pers., *Icon. descr. fung.*: pl. 51 (1800)

Gomphidius viscidus var. *testaceus* Fr., *Epicr. syst. mycol.*: 319 (1838)

Gomphidius viscidus f. *giganteus* J.E. Lange, *Fl. agaric. danic.* 5: 7 (1940)

Mis.: *Gomphus viscidus* sensu auct.

Mis.: *Gomphidius viscidus* sensu Lange (FlDan5: 6 & pl. 161 D) & sensu auct.

E: o **S:** c **W:** ! **NI:** o **ROI:** ! **O:** Channel Islands: !

H: On soil, associated with *Pinus sylvestris* or other two-needled *Pinus* spp.

D: NM2: 68 **D+I:** Ph: 190-191, B&K3: 96 70 **I:** Sow1: 105 (as *Agaricus rutilus*), FlDan4: 161E (as *Gomphidius viscidus* f. *giganteus*), Bon: 51, C&D: 418, SV33: 17, Kriegl2: 341

Frequent and widespread.

CHRYSOMPHALINA Clémençon, *Z. Mykol.* 48(2): 202 (1982)

Agaricales, Tricholomataceae

Type: *Chrysomphalina chrysophylla* (Fr.) Clémençon

chrysophylla (Fr.) Clémençon, *Z. Mykol.* 48(2): 203 (1982)

Agaricus chrysophyllus Fr., *Syst. mycol.* 1: 167 (1821)

Omphalia chrysophylla (Fr.) P. Kumm., *Führ. Pilzk.*: 107 (1871)

Omphalina chrysophylla (Fr.) Murrill, *North American Flora, Ser. 1 (Fungi 3)*, 9(5): 346 (1916)

Armillariella chrysophylla (Fr.) Singer, *Ann. Mycol.* 41: 20 (1943)

Gerronema chrysophyllum (Fr.) Singer, *Mycologia* 51: 380 (1959)

S: !

H: On decayed wood of *Pinus sylvestris* in Caledonian pinewoods.

D: NM2: 173 (as *Omphalina chrysophylla*) **I:** Bon: 129, C&D: 183 (as *Gerronema chrysophyllum*), Kriegl3: 150, Cooke 1152 (1152) Vol. 8 (1890) (as *Omphalia chrysophila*)

Rare. British records are confined to Scotland.

grossula (Pers.) Norvell, Redhead & Ammirati, *Mycotaxon* 50: 380 (1994)

Agaricus grossulus Pers., *Mycol. eur.* 3: 110 (1828)

Omphalina grossula (Pers.) Singer, *Persoonia* 2: 29 (1961)

Agaricus umbelliferus var. *abiegnus* Berk. & Broome, *Ann. Mag. Nat. Hist., Ser. 4 [Notices of British Fungi no. 1413]* 15: 28 (1875)

Omphalia abiegna (Berk. & Broome) Lange, *Dansk. Bot. Ark.* 6(5): 13 (1930)

Omphalina abiegna (Berk. & Broome) Singer, *Lilloa* 22: 212 (1951) [1949]

Omphalina bibula Quél., *Enchir. fung.*: 44 (1886)

Omphalia umbellifera var. *citrina* Quél., *Enchir. fung.*: 44 (1886)

Omphalia umbellifera var. *chrysoleuca* (Pers.) Rea, *Brit. basidiomyc.*: 429 (1922)

Hygrophorus wynniae Berk. & Broome, *Ann. Mag. Nat. Hist., Ser. 5 [Notices of British Fungi no. 1781]* 3: 208 (1879)

Omphalia wynniae (Berk. & Broome) Quél., *Compt. Rend. Assoc. Franç. Avancem. Sci.* 11: 390 (1883) [1882]

Omphalina wynniae (Berk. & Broome) S. Ito, *Mycol. Fl. Japan* 5(2): 128 (1959)

E: ! **S:** ! **W:** ! **NI:** !

H: On decayed wood of conifers, usually large stumps or logs of *Pinus syvestris*.

D: FAN3: 80 (as *Omphalina grossula*) **D+I:** B&K3: 98 75 (as *Camarophyllus grossulus*) **I:** Kriegl3: 483 (as *O. grossula*)

Rarely reported but supposedly widespread and apparently increasing on bark mulch in Scotland *fide* R. Watling (pers. comm.).

CLATHRUS Pers., *Syn. meth. fung.*: 241 (1801)

Phallales, Clathraceae

Aserophallus Mont. & Lepr., *Ann. Sci. Nat., Bot.*, sér. 3, 4: 360 (1845)

Anthurus Kalchbr. & MacOwan, in Kalchbrenner & Cooke, *Grevillea* 9: 2 (1880)

Mis.: *Lysurus* sensu auct.

Type: *Clathrus ruber* Pers.

archeri (Berk.) Dring, *Kew Bull.* 35: 29 (1980)

Lysurus archeri Berk., *Flora Tasmaniae, Fungi* 2: 264 (1860)

Anthurus archeri (Berk.) E. Fisch., *Jahrb. Bot. Gart. Mus. Berl.* 4: 81 (1886)

Mis.: *Aseroë rubra* sensu auct.

E: ! **O:** Channel Islands: !
H: On garden soil, especially where mixed with decayed woodchips (as on flowerbeds), but also occasionally in parkland or deciduous woodland.
D: NM3: 175 (as *A. archeri*), Dennis & Wakefield [Trans. Brit. Mycol. Soc. 29: 141 (1946)] **D+I:** B&K2: 398 523 (as *A. archeri*), BritPuffb: 182-183 Figs. 145/146 **I:** Bon: 301, Kriegl2: 164
An introduction, currently scarce but apparently spreading in southern and south-western England (reported from Bedfordshire to West Cornwall), probably with the increased use of woodchip mulch in gardens. Native to Australia, New Zealand and Tasmania. English name = 'Devil's Fingers'.

ruber Pers., *Syn. meth. fung.*: 241 (1801)
  Clathrus cancellatus L., *Sp. pl.* 2: 1179 (1753)
E: ! **S:** ! **ROI:** ! **O:** Channel Islands: o
H: On soil in woodland, parkland or gardens, often coastal in distribution.
D+I: B&K2: 398 524, BritPuffb: 184-185 Figs. 147/148 **I:** Bon: 301, Kriegl2: 165, SV41: 49
Rarely reported. Possibly native in the south of Britain, or an old introduction. Slowly spreading in England and reported from Warwickshire to West Cornwall. A single record from the Republic of Ireland (Dublin: Clontarf) and old records from Scotland. Rather frequent in the Channel Islands. English name = 'Red Cage Fungus'.

CLAVARIA L., *Sp. pl.*: 1182 (1753)
Cantharellales, Clavariaceae
Type: *Clavaria fragilis* Holmsk.

acuta Sowerby, *Col. fig. Engl. fung.* 3: pl. 333 (1801)
  Clavaria asterospora Pat., *Tab. anal. fung.*: 568 (1887)
  Clavulinopsis asterospora (Pat.) Corner, *Monograph of Clavaria and Allied Genera*: 357 (1950)
E: c **S:** c **W:** c **NI:** c **O:** Channel Islands: !
H: On soil in basic, unimproved grassland, deciduous woodland, and occasionally under *Taxus*.
D: Corner: 222, NM3: 249 (as *C. falcata*) **D+I:** B&K2: 342 441 (see also plate 442 *C. asterospora*) **I:** Sow3: 333, SV36: 17, Kriegl2: 23 (as *C. falcata*)
Common and widespread. Spores of this species are initially smooth, becoming echinulate as they mature, this being the origin of *Clavaria asterospora*. The name *Clavaria falcata* Pers. is considered by some authorities to pre-date *C. acuta* but no type material exists and the original description is minimal.

argillacea Pers., *Comment. Fungis Clavaeform.*: 74 (1797)
  Clavaria argillacea var. *pusilla* Corner, *Trans. Brit. Mycol. Soc.* 50: 33 (1967)
E: o **S:** c **W:** o **NI:** c **ROI:** !
H: On acidic soil on heathland or moorland, often in short turf and considered mycorrhizal with *Ericaceae* such as *Calluna*.
D: Corner: 225, NM3: 249 **D+I:** Ph: 257
Frequent and widespread, but most often reported from Scotland.

corbierei Bourdot & Galzin, *Hymenomyc. France*: 112 (1928)
E: !
H: On soil, amongst grass in deciduous woodland.
D: Corner: 230
Rarely reported. Known from Warwickshire (Compton Verney) and West Lancashire (Silverdale, Eaves Wood).

crosslandii Cotton, *Naturalist (Hull)* 1912: 86 (1912)
E: !
H: On soil, amongst short grass.
Accepted with doubts. The type, and two other collections in herb. K, need re-evaluation. *Fide* Corner (1950) this is just *C. acuta*. Reported from Warwickshire and Yorkshire, otherwise recorded only from the Faeroes.

fragilis Holmsk., *Beata ruris* 1: 7 (1790)
  Clavaria cylindrica Gray, *Nat. arr. Brit. pl.* 1: 656 (1821)
  Clavaria vermicularis Fr., *Syst. mycol.* 1: 484 (1821)

  Clavaria vermicularis var. *sphaerospora* Bourdot & Galzin, *Hymenomyc. France*: 110 (1928)
  Clavaria vermicularis var. *gracilis* Bourdot & Galzin, *Hymenomyc. France*: 110 (1928)
E: c **S:** c **W:** c **NI:** c **ROI:** c
H: In basic and acidic, unimproved grassland.
D: Corner: 251 (as *C. vermicularis*), NM3: 249 **D+I:** Ph: 257 (as *C. vermicularis*), B&K2: 344 444 (as *C. vermicularis*) **I:** Kriegl2: 25
Common and widespread. Better known as *C. vermicularis*.

fumosa Pers., *Comment. Fungis Clavaeform.*: 76 (1797)
E: c **S:** c **W:** c **NI:** c **ROI:** c **O:** Channel Islands: !
H: In basic and acidic, unimproved grassland such as moorland, coastal turf, downland and occasionally on lawns.
D: Corner: 235, NM3: 249 **D+I:** Ph: 257, B&K2: 244 443 **I:** Kriegl2: 25
Common and widespread.

greletii Boud., *Bull. Soc. Mycol. France* 33: 13 (1917)
E: !
H: On soil in deciduous woodland, and recently collected amongst grass on fixed dunes.
D: Corner: 241, NM3: 248
Rarely reported. Known from Shropshire and South Lancashire. Reported from Oxfordshire and Westmorland but unsubstantiated with voucher material.

guilleminii Bourdot & Galzin, *Hymenomyc. France*: 112 (1928)
E: !
H: On soil, usually on burnt ground such as old fire sites with the moss *Funaria hygrometrica*, and also collected on soil in flowerpots.
D: Corner: 241
Rarely reported and poorly known in Britain. Widely scattered reports from Mid-Lancashire, North East Yorkshire, South Wiltshire and Surrey.

incarnata Weinm., *Hymen. Gasteromyc.*: 510 (1836)
E: ! **S:** ! **W:** ! **ROI:** ?
H: In basic, unimproved grassland, on lawns and in pastures, occasionally in woodland, sometimes near *Taxus*.
D: Corner: 244, NM3: 248
Rarely reported, but apparently widespread.

krieglsteineri Kajan & Grauw., *Beitr. Kenntn. Pilze Mitteleurop.* 3: 358 (1987)
  Mis.: *Clavaria tenuipes* sensu auct.
E: !
H: On soil amongst grass, as in lawns or on playing fields.
D: NM3: 249 (as *Clavaria tenuipes*) **I:** Kriegl2: 26
One verified collection, from Oxfordshire (Milham Ford School) in 2000. Possibly also in Hertfordshire (Bedmond) and West Cornwall (Perranporth: Gear Sands). Many records of *Clavaria tenuipes* may also belong here.

purpurea Fr., *Syst. mycol.* 1: 480 (1821)
E: ! **S:** !
H: On soil, in woodland. British material is often associated with *Taxus*.
D: Corner: 246, NM3: 249
Known from East Gloucestershire, Herefordshire, Mid-Lancashire, and an old, unlocalised Scottish collection (in herb. K).

rosea Fr., *Syst. mycol.* 1: 482 (1821)
  Clavaria rosea var. *subglobosa* (Fr.) Corner, *Monograph of Clavaria and Allied Genera*: 691 (1950)
E: ! **S:** ! **W:** !
H: On soil, usually amongst grass in open areas of woodland, or in grassland.
D: Corner: 248 **D+I:** Ph: 257
Rarely reported, but apparently widespread.

straminea Cotton, *Trans. Brit. Mycol. Soc.* 3(4): 265 (1910)
E: ! **S:** ! **W:** ! **NI:** ? **ROI:** !
H: On acidic or calcareous soil, amongst short turf, or occasionally in areas of scrubby woodland.

D: Corner: 249, NM3: 250 (as *Clavaria flavipes*) **D+I:** FRIC5: 38-42 40b **I:** FM4(2): 66
Rarely collected or reported, but apparently widespread.

tenuipes Berk. & Broome, *Ann. Mag. Nat. Hist., Ser. 2 [Notices of British Fungi no. 369]* 2: 266 (1848)
    Pistillaria tenuipes (Berk. & Broome) Massee, *Brit. fung.-fl.* 1: 91 (1892)
E: ! **S:** ! **W:** ! **ROI:** !
H: On soil in areas of deciduous woodland, often on old fire sites, amongst *Funaria hygrometrica*.
D: Corner: 250 **I:** Kriegl2: 27
Rarely reported. Apparently widespread throughout Britain but confused with *Clavaria krieglsteineri*. Accepted here sensu Schild [*Z. Mycol* 47: 215 (1981)].

zollingeri Lév., *Ann. Sci. Nat., Bot.,* sér. 3, 5: 155 (1846)
    Clavaria lavendula Peck, *Bull. New York State Mus. Nat. Hist.* 139: 47 (1910)
    Mis.: *Clavaria amethystina* sensu auct.
E: ! **S:** ! **W:** ! **NI:** ! **ROI:** !
H: On soil in woodland or rough grassland.
D: Corner: 258, NM3: 247 **D+I:** Ph: 258 (incorrectly as *Clavulina amethystina*), B&K3: 346 445
Rarely reported but apparently widely scattered throughout the British Isles.

CLAVARIADELPHUS Donk, *Revis. Niederl. Homobasidiomyc.* 2: 72 (1933)
Cantharellales, Clavariadelphaceae
Type: *Clavariadelphus pistillaris* (L.) Donk

ligula (Schaeff.) Donk, *Rev. Niederl. Homob. Aphyll.* 2: 73 (1933)
    Clavaria ligula Schaeff., *Fung. Bavar. Palat. nasc.* 2: 171 (1763)
E: ? **S:** !
H: On soil in swampy woodland (British collection associated with *Betula* sp.).
D: Corner: 278, NM3: 269 **D+I:** B&K2: 350 452 **I:** Kriegl2: 37
Apparently rare and only in Scotland (last collected at Aviemore in 1953). Reported several times from England but the single vouchered collection was misidentified, and all other records are unsubstantiated.

pistillaris (L.) Donk, *Revis. Niederl. Homobasidiomyc.* 2: 73 (1933)
    Clavaria pistillaris L., *Sp. pl.*: 1182 (1753)
    Clavaria herculeana Lightf., *Fl. scot.*: 1056 (1777)
E: o **S:** ! **W:** ! **NI:** !
H: On calcareous soil, in woodland, usually associated with *Fagus*.
D: Corner: 279, NM3: 268 **D+I:** Ph: 256-257 **I:** Sow3: 277 (as *Clavaria herculanea*), Kriegl2: 38
Occasional to rare, and possibly declining. Widely scattered throughout mainland Britain, and a single record from Northern Ireland.

truncatus (Quél.) Donk, *Rev. Niederl. Homob. Aphyll.* 2: 73 (1933)
    Clavaria truncata Quél., *Enchir. fung.*: 221 (1886)
E: !
H: On soil in deciduous woodland.
D: Corner: 282, NM3: 268 **D+I:** B&K2: 352 454 **I:** Kriegl2: 40
Known from a single collection by Rea from Worcestershire (Shrawley Wood) in 1924.

CLAVICORONA Doty, *Lloydia* 10: 38 (1947)
Hericiales, Clavicoronaceae
    Artomyces Jülich, *Bibliot. Mycol.* 85: 395 (1982)
Type: *Clavicorona taxophila* (Thom) Doty

pyxidata (Pers.) Doty, *Lloydia* 10: 43 (1947)
    Clavaria pyxidata Pers., *Comment. Fungis Clavaeform.*: 47 (1797)

    Artomyces pyxidatus (Pers.) Fr., *Biblioth. Mycol.* 85: 399 (1982) [1981]
E: !
H: On soil in woodland.
D: Corner: 292
Very rare and unsubstantiated with voucher material There is, however, a good icon by Rea of English material at Kew.

taxophila (Thom) Doty, *Lloydia* 10: 39 (1947)
    Craterellus taxophilus Thom, *Bot. Gaz.*: 3 (1904)
E: ! **NI:** !
H: On soil, amongst litter, usually associated with *Taxus* and occasionally *Chamaecyparis* sp.
D: Corner: 293 **D+I:** Ph: 258-259
Rarely reported. Basidiomes are small and easily overlooked .

CLAVULICIUM Boidin, *Bull. Soc. Hist. Nat. Toulouse.* 92: 280 (1957)
Cantharellales, Clavulinaceae
Type: *Clavulicium pilatii* (Boidin) Boidin

delectabile (H.S. Jacks.) Hjortstam, *Svensk bot. Tidskr.* 67: 107 (1973)
    Corticium delectabile H.S. Jacks., *Canad. J. Res., Sect. C, Bot. Sci.* 26: 145 (1948)
E: !
H: On decayed fern stems and soil, in a conifer plantation (British material).
D+I: CNE2: 246-247
A single collection from South Devon (Kingsteignton) in 2001. Reported from Wales but the collection (in herb. K) has been redetermined as *Phanerochaete martelliana*.

CLAVULINA J. Schröt., in Cohn, *Krypt.-Fl. Schlesien* 3: 442 (1888)
Cantharellales, Clavulinaceae
Type: *Ramaria cristata* Holmsk. (= *Clavulina coralloides*)

cinerea (Bull.) J. Schröt., in Cohn, *Krypt.-Fl. Schlesien* 3: 442 (1888)
    Clavaria cinerea Bull., *Herb. France*: pl. 354 (1788)
    Ramaria cinerea (Bull.) Gray, *Nat. arr. Brit.* pl. 1: 655 (1821)
    Clavaria cinerea var. *gracilis* Rea, *Trans. Brit. Mycol. Soc.* 6(1): 62 (1918)
    Clavulina cinerea var. *gracilis* (Rea) Corner, *Monograph of Clavaria and Allied Genera*: 309 (1950)
    Clavaria fuliginea Pers., *Mycol. eur.* 1: 166 (1822)
    Clavaria grisea Pers., *Comment. Fungis Clavaeform.*: 44 (1797)
E: c **S:** ! **W:** c **NI:** c **ROI:** c
H: On soil in deciduous or mixed woodland.
D: Corner: 308-310, NM3: 254 **D+I:** Ph: 258-259, B&K3: 352 455 **I:** Kriegl2: 54
Very common and widespread. Basidiomes are frequently infected by the ascomycete *Helminthosphaeria clavariorum*, especially around the stipe bases where they become dark greyish.

coralloides (L.) J. Schröt., in Cohn, *Krypt.-Fl. Schlesien* 3: 443 (1888)
    Clavaria coralloides L., *Sp. pl.*: 1182 (1753)
    Ramaria cristata Holmsk., *Beata ruris* 1: 92 (1790)
    Clavaria cristata (Holmsk.) J. Schröt., in Cohn, *Krypt.-Fl. Schlesien* 3: 442 (1888)
    Clavulina cristata var. *lappa* P. Karst., *Bidrag Kännedom Finlands Natur Folk* 37: 168 (1882)
    Clavulina cristata f. *subcinerea* Donk, *Rev. Niederl. Homob. Aphyll.* 2: 19 (1933)
    Clavulina cristata var. *subrugosa* Corner, *Monograph of Clavaria and Allied Genera*: 693 (1950)
    Clavulina cristata var. *coralloides* Corner, *Monograph of Clavaria and Allied Genera*: 693 (1950)
E: c **S:** c **W:** c **NI:** c **ROI:** c **O:** Isle of Man: !
H: On soil in deciduous woodland. Rarely in short turf in grassland, and also known from shingle beaches.

D: Corner: 312 (as *Clavulina cristata*), NM3: 255 **D+I:** Ph: 258-259 (as *C. cristata*), B&K2: 352 456 (as *C. cristata*) **I:** Sow3: 278 (top fig.) (as *Clavaria coralloides*), Kriegl2: 55
Common and widespread. Better known as *Clavulina cristata*.

rugosa (Bull.) J. Schröt., in Cohn, *Krypt.-Fl. Schlesien* 3: 442 (1888)
> *Clavaria rugosa* Bull., *Herb. France*: pl. 448 (1790)
> *Clavaria grossa* Pers., *Comment. Fungis Clavaeform.*: 50 (1797)
> *Clavaria canaliculata* Fr., *Syst. mycol.* 1: 485 (1821)
> *Clavulina rugosa* var. *canaliculata* (Fr.) Corner, *Monograph of Clavaria and Allied Genera*: 338 (1950)
> *Clavulina rugosa* var. *alcyonaria* (Fr.) Corner, *Monograph of Clavaria and Allied Genera*: 337 (1950)
> *Clavaria rugosa* var. *fuliginea* Fr., *Hymenomyc. eur.*: 669 (1874)
> *Clavulina rugosa* var. *fuliginea* (Fr.) Corner, *Monograph of Clavaria and Allied Genera*: 338 (1950)

E: c **S:** c **W:** c **NI:** c **ROI:** ! **O:** Channel Islands: !
H: On rich humic soil, often amongst mosses in woodland. Reported with a wide range of deciduous trees, and rarely with conifers.
D: Corner: 336, NM3: 254 **D+I:** Ph: 258-259, B&K2: 354 457 **I:** Kriegl2: 56
Common and widespread.

CLAVULINOPSIS Overeem, *Bull. Jard. Bot. Buitenzorg, Sér. 3*, 5: 278 (1923)
Cantharellales, Clavariaceae
Ramaria Holmsk. [*nom. rej.*, non *Ramaria* (Fr.) Bonord. (1851)], *Beata ruris* 1: xvii (1790)
Type: *Clavulinopsis sulcata* Overeem

corniculata (Fr.) Corner, *Monograph of Clavaria and Allied Genera*: 362 (1950)
> *Clavaria corniculata* Fr., *Syst. mycol.* 1: 471 (1821)
> *Ramaria corniculata* (Fr.) Gray, *Nat. arr. Brit. pl.* 1: 655 (1821)
> *Clavulinopsis corniculata* f. *bispora* Pilát, *Sborn. Nár. Mus. Praze, II B* 25: f. 18 (1955)
> *Clavaria fastigiata* L., *Sp. pl.*: 1183 (1753)
> *Clavaria muscoides* L., *Sp. pl.*: 1183 (1753)
> *Ramaria pratensis* Gray, *Nat. arr. Brit. pl.* 1: 655 (1821)

E: c **S:** c **W:** c **NI:** c **ROI:** c **O:** Channel Islands: ! Isle of Man: !
H: In grassland or occasionally deciduous woodland.
D: NM3: 250 **D+I:** Ph: 258-259, B&K2: 346 446 **I:** Sow2: 157 (as *Clavaria muscoides*), Kriegl2: 29
Common and widespread. *Clavulinopsis corniculata* f. *bispora* is known (in Britain) only from two recent collections, from England (Oxfordshire: Bix, The Warburg Reserve) and Scotland (Easterness: Kingussie).

fusiformis (Sowerby) Corner, *Monograph of Clavaria and Allied Genera*: 367 (1950)
> *Clavaria fusiformis* Sowerby, *Col. fig. Engl. fung.* 2: pl. 234 (1799)
> *Clavaria ceranoides* Pers., *Syn. meth. fung.*: 594 (1801)
> *Ramaria ceranoides* (Pers.) Gray, *Nat. arr. Brit. pl.* 1: 655 (1821)
> *Clavaria fusiformis* var. *ceranoides* (Pers.) W.G. Sm., *Syn. Brit. Bas.*: 434 (1908)

E: c **S:** c **W:** c **NI:** c **ROI:** c **O:** Channel Islands: !
H: In acidic, unimproved grassland (moorland, coastal grassland, lawns, and heathland).
D: NM3: 251 **D+I:** Ph: 258, B&K2: 346 447 **I:** Kriegl2: 30
Common and widespread.

helvola (Pers.) Corner, *Monograph of Clavaria and Allied Genera*: 372 (1950)
> *Clavaria helvola* Pers., *Comment. Fungis Clavaeform.*: 69 (1797)
> *Clavaria dissipabilis* Britzelm., *Ber. Naturhist. Vereins Augsburg* 29: 289 (1887)
> Mis.: *Clavaria inaequalis* sensu auct.

E: c **S:** c **W:** c **NI:** c **ROI:** c **O:** Channel Islands: c

H: In basic and acidic, unimproved grassland (lawns, heath, moorland, and pasture), occasionally on calcareous loam in deciduous woodland.
D: NM3: 251 **D+I:** Ph: 258-259, B&K2: 348 448 **I:** Kriegl2: 31
Common and widespread.

laeticolor (Berk. & M.A. Curtis) R.H. Petersen, *Mycologia* 57: 522 (1965)
> *Clavaria laeticolor* Berk. & M.A. Curtis, *J. Linn. Soc., Bot.* 10: 338 (1868)
> *Clavaria persimilis* Cotton, *Trans. Brit. Mycol. Soc.* 3: 182 (1909)
> *Clavaria pulchra* Peck, *Rep. (Annual) New York State Mus. Nat. Hist.* 28: 53 (1876)
> *Clavulinopsis pulchra* (Peck) Corner, *Monograph of Clavaria and Allied Genera*: 384 (1950)

E: o **S:** o **W:** o **NI:** ! **O:** Channel Islands: !
H: On soil amongst grass in unimproved grassland, on heaths, moors and in pasture.
D: NM3: 251 **D+I:** B&K2: 348 449 **I:** Kriegl2: 32
Occasional but widespread. Basidiomes are often a distinct peach-yellow colour, but require microscopic examination to distinguish them from *C. helvola* and *C. luteoalba*.

luteoalba (Rea) Corner, *Monograph of Clavaria and Allied Genera*: 374 (1950)
> *Clavaria luteoalba* Rea, *Trans. Brit. Mycol. Soc.* 2: 66 (1903)
> *Clavulinopsis luteoalba* var. *latispora* Corner, *Monograph of Clavaria and Allied Genera*: 375 (1950)

E: c **S:** c **W:** c **NI:** c **ROI:** ! **O:** Channel Islands: !
H: In unimproved grassland, heathland and moorland, rough pasture and meadows, and occasionally on old lawns.
D: NM3: 251 **D+I:** Ph: 258-259
Common and widespread. Often growing with the superficially similar species *C. helvola* and *C. laeticolor*.

luteonana **var. tenuipes** Schild, *Fungorum Rar. Icon. Color.* 5: 29 (1971)
E: !
H: On soil amongst mosses in woodland.
A single collection from South Devon (Slapton) in 1992.

luteo-ochracea (Cavara) Corner, *Monograph of Clavaria and Allied Genera*: 376 (1950)
> *Clavaria luteo-ochracea* Cavara, *Fungi Longobardiae exsiccati II*: no. 64 (1892)

E: !
H: On soil (calcareous loam) in mixed deciduous woodland.
A single collection from Surrey (Mickleham, Norbury Park) in 1974.

rufipes (G.F. Atk.) Corner, *Monograph of Clavaria and Allied Genera*: 386 (1950)
> *Clavaria rufipes* G.F. Atk., *Ann. Mycol.* 6: 57 (1908)
> *Clavulinopsis microspora* Joss., *Bull. Soc. Mycol. France* 64: 29 (1948)

E: !
H: On soil amongst mosses, or on decayed woody debris.
D+I: FRIC4: 1-4 25b
Rarely reported. British collections from widely scattered sites in Berkshire, Cambridgeshire, Mid Lancashire and South Hampshire.

subtilis (Pers.) Corner, *Monograph of Clavaria and Allied Genera*: 391 (1950)
> *Clavaria subtilis* Pers., *Comment. Fungis Clavaeform.*: 51 (1797)
> *Ramariopsis subtilis* (Pers.) R.H. Petersen, *Mycologia* 70(3): 668 (1978)
> *Clavulinopsis dichotoma* (Godey) Corner, *Monograph of Clavaria and Allied Genera*: 365 (1950)

E: ! **S:** ! **NI:** ! **ROI:** !
H: On soil in grassland, lawns, roadside verges, parkland or in deciduous woodland. Rarely on soil in flowerpots.
D: NM3: 253 (as *Ramariopsis subtilis*) **D+I:** B&K2: 348 450
Rarely reported, but apparently widespread.

umbrinella (Sacc.) Corner, *Monograph of Clavaria and Allied Genera*: 393 (1950)
  Clavaria umbrinella Sacc., *Syll. fung.* 6: 695 (1888)
  Clavaria cinereoides G.F. Atk., *Ann. Mycol.* 7: 367 (1909)
  Clavulinopsis cinereoides (G.F. Atk.) Corner, *Monograph of Clavaria and Allied Genera*: 360 (1950)
  Mis.: *Clavaria umbrina* sensu Berkeley (1860)
E: o **S:** ! **W:** o **NI:** !
H: In basic, unimproved grassland and occasionally woodland, sometimes near *Taxus* or *Crataegus monogyna*.
D: NM3: 250 (as *Clavulinopsis cinereoides*) **D+I:** Ph: 258-259 (as *C.cinereoides*) **I:** SV39: 48 (as *C.cinereoides*)
Occasional and widely scattered throughout the British Isles.

CLITOCYBE (Fr.) Staude, *Schwämme Mitteldeutschl.*: 122 (1857)
Agaricales, Tricholomataceae
Type: *Clitocybe nebularis* (Batsch) P. Kumm.

agrestis Harm., *Karstenia* 10: 91 (1969)
  Clitocybe graminicola Bon, *Doc. Mycol.* 9(35): 44 (1979)
  Mis.: *Clitocybe angustissima* sensu NCL, sensu auct.
E: ! **S:** !
H: On soil, usually amongst grass and occasionally on compost.
D: NM2: 112, FAN3: 57 **I:** C&D: 177 (as *C. graminicola*), Kriegl3: 156
Known from England (Cumberland, East Suffolk, London, North Somerset, Staffordshire and West Cornwall) and Scotland. Reported elsewhere but unsubstantiated with voucher material.

albofragrans (Harmaja) Kuyper, *Persoonia* 11(3): 386 (1981)
  Lepista albofragrans Harmaja, *Karstenia* 18: 53 (1978)
  Mis.: *Clitocybe luffii* sensu NCL
E: ! **NI:** ?
H: On soil in deciduous woodland.
D: NM2: 111, FAN3: 47
Known from Derbyshire and South Hampshire. Reported from West Sussex and Northern Ireland (Fermanagh) but unsubstantiated with voucher material.

alexandri (Gillet) Gillet, *Tableaux analytiques des Hyménomycètes*: 28 (1884)
  Paxillus alexandri Gillet, *Bull. Soc. Linn. Normandie* 7: 157 (1873)
  Lepista alexandri (Gillet) Gillet, *Hymenomycètes*: 196 (1876)
E: !
H: On soil in mixed deciduous woodland, usually associated with *Quercus* spp.
D: NM2: 109-110, FAN3: 44 **I:** Bon: 137, C&D: 175
Collections from Herefordshire in 1892, West Suffolk in 1951 and Cambridgeshire in 1960, but not reported since. Reported from Shropshire in the late nineteenth century but lacking details or voucher material. A large species, unlikely to be overlooked and apparently genuinely rare.

amarescens Harmaja, *Karstenia* 10: 98 (1969)
  Clitocybe nitrophila Bon, *Doc. Mycol.* 9(35): 43 (1979)
E: ! **S:** !
H: On soil in parks or gardens, often on flowerbeds mulched with woodchip, and usually in troops or large aggregated clumps.
D: NM2: 113, FAN3: 61 **I:** Bon: 139 (as *Clitocybe nitrophila*), SV28: 57 (as *C. nitrophila*)
Rarely reported and possibly a recent introduction. Known from Surrey and West Kent and reported from Scotland. Apparently spreading with the increased use of woody mulches on flowerbeds.

americana H.E. Bigelow, *Mem. N. Y. bot. Gdn* 28(1): 10 (1976)
E: !
H: On decayed wood of deciduous trees, usually on large fallen and decayed trunks or logs of *Fagus* or *Ulmus* spp, in old areas of woodland. Reported (dubiously) on *Sambucus nigra*.

First reported in Britain from Worcestershire in 1983 [see Bull. B.M.S 18 (2): 139 (1984)] but now also known from Middlesex, Surrey and West Kent.

augeana Mont., *Syll. Crypt.*: 336 (1856)
  Clitocybe ruderalis Harmaja, *Karstenia* 10: 76 (1969)
S: !
H: On soil.
D: FAN3: 48
Known from Scotland where first reported (as *C. ruderalis*) in 1992, from 'an area of reclaimed ground' in West Lothian. There are now several more recent collections in herb. E

barbularum (Romagn.) P.D. Orton, *Trans. Brit. Mycol. Soc.* 43(2): 174 (1960)
  Omphalia barbularum Romagn., *Rev. Mycol. (Paris)* 17(1): 44 (1952)
  Omphalina barbularum (Romagn.) Bon, *Doc. Mycol.* 5(19): 22 (1975)
E: ! **W:** ! **NI:** ! **ROI:** ! **O:** Channel Islands: !
H: On sandy soil, often amongst mosses, in dunes.
D: FAN3: 55 **I:** Bon: 129 (as *Omphalina barbularum*), C&D: 185 (as *O. barbularum*)
Rarely reported but apparently widespread.

brumalis (Fr.) Gillet, *Hyménomycètes*: 148 (1874)
  Agaricus brumalis Fr., *Syst. mycol.* 1: 171 (1821)
E: ! **S:** ! **W:** ! **NI:** ! **ROI:** !
H: On soil in deciduous or mixed coniferous and deciduous woodland. Not confined to winter.
D+I: FlDan1: 83 38D
Rarely reported. *Clitocybe brumalis* is a name of uncertain application but is accepted here, sensu Lange (1935 pl. 38D) and NCL, for a small-spored species. British records may, however, relate to several different species and require revision.

candicans (Pers.) P. Kumm., *Führ. Pilzk.*: 122 (1871)
  Agaricus candicans Pers., *Syn. meth. fung.*: 456 (1801)
  Omphalia candicans (Pers.) Gray, *Nat. arr. Brit. pl.* 1: 613 (1821)
  Agaricus gallinaceus Scop., *Fl. carniol.* 433 (1772)
  Clitocybe gallinacea (Scop.) Gillet, *Hyménomycètes*: 150 (1874)
  Clitocybe tenuissima Romagn., *Bull. Soc. Naturalistes Oyonnax* 8: 74 (1954)
  Agaricus tuba Fr., *Epicr. syst. mycol.*: 72 (1838)
  Clitocybe tuba (Fr.) Gillet, *Hyménomycètes*: 137 (1874)
E: ! **S:** ! **W:** ! **ROI:** !
H: On soil or amongst leaf litter in deciduous or mixed deciduous and conifer woodland.
D: NM2: 108, FAN3: 49 **D+I:** B&K3: 150 152 **I:** C&D: 177
Rarely reported. Apparently widespread.

catinus (Fr.) Quél., *Mém. Soc. Émul. Montbéliard, Sér. 2,* 5: 215 (1872)
  Agaricus catinus Fr., *Hymenomyc. eur.*: 99 (1874)
  Clitocybe infundibuliformis var. *catina* (Fr.) Konrad & Maubl., *Icon. select. fung.* 6: 333 (1937)
S: !
H: On soil in montane conifer woodland.
D: NM2: 106 **D+I:** B&K3: 150 153
Reported from Scotland by Watling (1987). Earlier British records are sensu Rea (1922) and NCL, in litter of *Fagus*, and represent a different species, possibly *C. phyllophila*.

costata Kühner & Romagn., *Bull. Soc. Naturalistes Oyonnax* 8: 73 (1954)
  Mis.: *Clitocybe incilis* sensu NCL
  Mis.: *Clitocybe squamulosoides* sensu NCL
  Mis.: *Clitocybe trulliformis* sensu Lange (FlDan1: 77 & pl. 35E)
E: ! **S:** ! **W:** ! **ROI:** !
H: On soil or in leaf litter, in deciduous or mixed coniferous and deciduous woodland.

D: NM2: 110 (as *Clitocybe squamulosoides*), FAN3: 53 **D+I:** B&K3: 152 155 **I:** FlDan1: 77 35E (as *C. trullaeformis*), Bon: 135, C&D: 173, Kriegl3: 162
Rarely reported but apparently widespread.

diatreta (Fr.) P. Kumm., *Führ. Pilzk.*: 121 (1871)
Agaricus diatretus Fr., *Observ. mycol.* 2: 200 (1818)
Mis.: *Clitocybe pinetorum* sensu auct.
E: ! **S:** ! **ROI:** !
H: On soil in conifer woodland, occasionally in mixed conifer and deciduous woodland.
D: NM2: 112, FAN3: 58 **I:** Bon: 139, C&D: 177, Kriegl3: 164
Rarely reported but apparently widespread.

diosma Einhell., *Ber. Bayer. Bot. Ges.* 44: 24 (1973)
E: !
H: On soil. British collection on a pile of discarded peat compost. A single collection from London in 1998.

ditopa (Fr.) Gillet, *Hyménomycètes*: 166 (1874)
Agaricus ditopus Fr., *Observ. mycol.* 1: 91 (1815)
E: o **S:** o **W:** ! **NI:** o **ROI:** !
H: On soil, usually amongst needle litter of conifers such as *Larix*, *Picea* or *Pinus* spp., and occasionally *Taxus*. Rarely with deciduous trees such as *Quercus* spp.
D: NM2: 112, FAN3: 56 **D+I:** Ph: 50, B&K3: 154 157 **I:** Bon: 141, C&D: 179 (as *Clitocybe ditopa*), Kriegl3: 165 (as *C. ditopa*)
Occasional but widespread.

ericetorum Quél., *Mém. Soc. Émul. Montbéliard, Sér. 2,* 5: 53 (1872)
Agaricus ericetorum Bull. [*nom. illegit.*, non *A. ericetorum* Pers. (1796)], *Herb. France*: pl. 551 (1792)
E: ! **S:** !
H: On sandy soil on heathland or in dunes.
D: NM2: 107 **D+I:** Ph: 46
Known from England (Berkshire, East Norfolk, South Devon and Surrey) and Scotland (East Perthshire).

fragrans (With.) P. Kumm, *Führ. Pilzk.*: 121 (1871)
Agaricus fragrans With., *Bot. Arr. Brit. pl. ed. 2,* 3: 307 (1792)
Omphalia fragrans (With.) Gray, *Nat. arr. Brit. pl.* 1: 613 (1821)
Lepista fragrans (With.) Harmaja, *Karstenia* 15: 14 (1976)
Clitocybe deceptiva H.E. Bigelow, *Nova Hedwigia Beih.* 72: 108 (1982)
Clitocybe fragrans var. *depauperata* J.E. Lange, *Dansk. Bot. Ark.* 6(5): 56 (1930)
Clitocybe depauperata (J.E. Lange) P.D. Orton, *Trans. Brit. Mycol. Soc.* 43(2): 174 (1960)
Agaricus suaveolens Schumach., *Enum. pl.* 2: 337 (1803)
Clitocybe suaveolens (Schumach.) P. Kumm., *Führ. Pilzk.*: 121 (1871)
Mis.: *Clitocybe obsoleta* sensu auct.
E: c **S:** c **W:** c **NI:** c **ROI:** c **O:** Channel Islands: ! Isle of Man: !
H: On soil (often amongst mosses) or on decayed leaf and needle litter in woodland. Occasionally on decayed wood chips or mulch in gardens and then attaining a much larger size than normal. Fruiting throughout the year.
D: NM2: 111, FAN3: 59 **D+I:** Ph: 50, B&K3: 156 160 **I:** Sow4: 10 (as *Agaricus fragrans*), C&D: 179
Common and widespread.

frysica Kuyper, *Persoonia* 16(2): 227 (1996)
Clitocybe sericella Kühner & Romagn. [*nom. nud.*], *Flore Analytique des Champignons Supérieurs*: 138 (1953)
Mis.: *Clitocybe subalutacea* sensu Lange (FlDan1: 75 & pl. 33G) & sensu FAN 3: 51, non NCL
E: ! **S:** !
H: On soil in deciduous woodland.
D: FAN3: 51 (as *Clitocybe subalutacea*)
Collections (originally determined as *Clitocybe subalutacea* sensu Lange) from Northamptonshire and West Kent. Also reported from Scotland but poorly known in Britain.

fuscosquamula J.E. Lange, *Dansk. Bot. Ark.* 6: 48 (1930)
Clitocybe inornata var. *exilis* A. Pearson, *Trans. Brit. Mycol. Soc.* 23(3): 310 (1939)
Mis.: *Clitocybe parilis* sensu Rea (1922)
S: ?
H: On soil in deciduous woodland.
D: NM2: 110
Accepted but with reservations. Listed by Rea (1922) as *Clitocybe parilis* but no material has been traced. Also reported from Scotland (Aviemore) in 1938 but again no material has been traced. The true *Clitocybe parilis* (= *Rhodocybe parilis*) is unknown in Britain.

geotropa (Bull.) Quél., *Mém. Soc. Émul. Montbéliard, Sér. 2,* 5: 89 (1872)
Agaricus geotropus Bull., *Herb. France*: pl. 573 (1792)
Mis.: *Clitocybe maxima* sensu auct. Brit.
Mis.: *Clitocybe subinvoluta* sensu Lange (FlDan: 76 & pl. 34G)
E: o **S:** ! **W:** ! **NI:** o **ROI:** !
H: On soil in deciduous or mixed deciduous and conifer woodland. Most frequently reported with *Fagus* and *Quercus* spp.
D: NM2: 107, FAN3: 51-52 **D+I:** Ph: 46, B&K3: 156 162 **I:** FlDan1: 34D, Bon: 135, C&D: 173, Kriegl3: 170
Occasional but widespread. May be locally common, especially in southern England.

gibba (Pers.) P. Kumm., *Führ. Pilzk.*: 123 (1871)
Agaricus gibbus Pers., *Syn. meth. fung.*: 449 (1801)
Omphalia gibba (Pers.) Gray [non *O. gibba* Pat. (1888)], *Nat. arr. Brit. pl.* 1: 612 (1821)
Agaricus gibbus δ *membranaceus* Fr., *Elench. fung.* 1: 13 (1828)
Clitocybe infundibuliformis var. *membranacea* (Fr.) Massee, *Brit. fung.-fl.* 2: 425 (1893)
Mis.: *Agaricus infundibuliformis* sensu auct.
Mis.: *Clitocybe infundibuliformis* sensu NCL & sensu auct.
E: c **S:** c **W:** c **NI:** c **ROI:** c
H: On soil amongst leaf litter, usually in deciduous woodland. Occasionally found in open plantations, with conifers such as *Picea* sp., and rarely in open grassland. Fruiting from late summer onwards.
D: NM2: 106, FAN3: **D+I:** Ph: 48-49 (as *Clitocybe infundibuliformis*), B&K3: 158 163 **I:** FlDan1: 32C (as *C. infundibuliformis*), Bon: 135, C&D: 173, Kriegl3: 172
Common and widespread. Perhaps better known as *C. infundibuliformis*.

houghtonii (W. Phillips) Dennis, *Kew Bull.* 9: 423 (1954)
Cantharellus houghtonii W. Phillips, *Ann. Mag. Nat. Hist., Ser. 4 [Notices of British Fungi no. 1565]* 17: 135 (1876)
Omphalia roseotincta A. Pearson, *Trans. Brit. Mycol. Soc.* 35(2): 105 (1952)
E: o **S:** ?
H: On decayed leaf litter in woodland, most often associated with *Fagus* in beechwoods on calcareous soil. Usually late fruiting, basidiomes most abundant during November and December.
D: FAN3: 62 **D+I:** FRIC3: 12-13 20, Ph: 50 **I:** Kriegl3: 174
Occasional but widespread. Known from Co. Durham southwards to Herefordshire but most frequent in south-eastern England. A single record from Scotland (Roxburghshire) but unsubstantiated with voucher material.

inornata (Sowerby) Gillet, *Hyménomycètes*: 155 (1874)
Agaricus inornatus Sowerby, *Col. fig. Engl. fung.* 3: pl. 342 (1802)
Paxillus inornatus (Sowerby) Quél., *Fl. mycol. France*: 109 (1888)
Agaricus zygophyllus Cooke & Massee, *Grevillea* 15: 67 (1886)
Clitocybe zygophylla (Cooke & Massee) Sacc., *Syll. fung.* 9: 24 (1891)
Mis.: *Agaricus elixus* sensu Cooke [HBF2: 51 (1883)]
E: ! **S:** ?

H: On soil, usually with *Fagus* but also known with *Larix* and *Pinus* spp in mixed woodland.
D: NM2: 110, FAN3: 44-45 **D+I:** B&K3: 158 165 **I:** Bon: 137, C&D: 175, Kriegl3: 175
Known from Bedfordshire, East Sussex, Hertfordshire, South Devon, South Somerset, South Wiltshire, Surrey, Warwickshire, West Gloucestershire, West Suffolk and Yorkshire. Reported elsewhere and from Scotland but unsubstantiated with voucher material.

metachroa (Fr.) P. Kumm., *Führ. Pilzk.*: 120 (1871)
Agaricus metachrous Fr., *Syst. mycol.* 1: 172 (1821)
Clitocybe decembris Singer, *Sydowia* 15: 48 (1962) [1961]
Agaricus dicolor Pers., *Syn. meth. fung.*: 462 (1801)
Clitocybe dicolor (Pers.) Murrill, *Mycologia* 7: 260 (1915)
E: o **S:** o **W:** o **NI:** o **ROI:** !
H: On soil and decayed leaf or needle litter in deciduous or conifer woodland.
D: NM2: 113, FAN3: 60 (as *Clitocybe metachroa* var. *metachroa*) **D+I:** Ph: 51 (as *C. dicolor*), B&K3: 162 169 **I:** C&D: 179, C&D: 179 (as *C. decembris*), Kriegl3: 178
Occasional but widespread, often locally abundant. Perhaps better known as *C. dicolor*.

nebularis (Batsch) P. Kumm., *Führ. Pilzk.*: 124 (1871)
Agaricus nebularis Batsch, *Elench. fung. (Continuatio Secunda)*: 25 t. 33 f. 193 (1789)
Gymnopus nebularis (Batsch) Gray, *Nat. arr. Brit. pl.* 1: 609 (1821)
Lepista nebularis (Batsch) Harmaja, *Karstenia* 14: 91 (1974)
Clitocybe alba (Bataille) Singer, *Lilloa* 22: 186 (1951) [1949]
Clitocybe nebularis var. *alba* Bataille, *Bull. Soc. Mycol. France* 27: 370 (1911)
E: c **S:** c **W:** c **NI:** c **ROI:** c **O:** Channel Islands: !
H: On soil in woodland, associated with a wide range of deciduous trees and also conifers.
D: NM2: 109, FAN3: 45 **D+I:** Ph: 48 (colour poor), B&K3: 162 170 **I:** FlDan1: 32E, Bon: 137, C&D: 175
Very common and widespread.

odora (Bull.) P. Kumm., *Führ. Pilzk.*: 121 (1871)
Agaricus odorus Bull., *Herb. France*: pl. 176 (1784)
Gymnopus odorus (Bull.) Gray, *Nat. arr. Brit. pl.* 1: 606 (1821)
Clitocybe odora var. *alba* J.E. Lange, *Dansk. Bot. Ark.* 6(5): 45 (1930)
Agaricus trogii Fr., *Epicr. syst. mycol.*: 59 (1838)
Clitocybe trogii (Fr.) Sacc., *Syll. fung.* 5: 153 (1887)
Agaricus virens Scop., *Fl. carniol.*: 437 (1772)
Clitocybe virens (Scop.) Sacc., *Syll. fung.* 5: 152 (1887)
Agaricus viridis With., *Bot. arr. Brit. pl. ed. 2,* 3: 320 (1792)
Clitocybe viridis (With.) Gillet, *Hyménomycètes*: 158 (1874)
Mis.: *Agaricus suaveolens* sensu Fries (1821)
E: c **S:** c **W:** o **NI:** c **ROI:** !
H: On soil in mixed deciduous woodland, usually associated with *Betula*, *Fagus* or *Quercus* spp. Occasionally with conifers in mixed woodland.
D: NM2: 108, FAN3: 46 **D+I:** Ph: 48-49, B&K3: 164 173 (as *Clitocybe odora* var. *alba*), B&K3: 164 172 **I:** Sow1: 42 (as *Agaricus odorus*), FlDan1: 34A, Bon: 137, C&D: 177, Kriegl3: 181
Common and widespread.

ornamentalis Velen., *České Houby* II: 255 (1920)
E: !
H: On soil in deciduous woodland. British material associated with *Fagus*.
D: NBA9: 545 **D+I:** B&K3: 164 174
Known from Gloucestershire and South Hampshire. Reported from West Kent but unsubstantiated with voucher material. Could be regarded as a variant of *Clitocybe phyllophila* with an odour of anise.

phaeophthalma (Pers.) Kuyper, *Persoonia* 11(3): 386 (1981)
Agaricus phaeophthalmus Pers., *Mycol. eur.* 3: 72 (1828)
Agaricus fritilliformis Lasch, in Fries, *Epicr. syst. mycol.*: 74 (1838)

Clitocybe fritilliformis (Lasch) Gillet, *Hyménomycètes*: 146 (1874)
Mis.: *Clitocybe gallinacea* sensu Rea (1922)
Mis.: *Omphalia hydrogramma* sensu auct.
Mis.: *Clitocybe hydrogramma* sensu auct. mult. & sensu NCL
Mis.: *Agaricus hydrogrammus* sensu auct.
E: c **W:** !
H: On decayed leaf litter, usually associated with *Fagus* in beechwoods or mixed deciduous woodland on calcareous soil, and often fruiting late in the season. Less commonly on needle litter of conifers such as *Picea* spp, in plantations.
D: NM2: 112 (as *Clitocybe hydrogramma*), FAN3: 54-55 **D+I:** Ph: 51 (as *C. hydrogramma*), B&K3: 166 175 **I:** Bon: 141, C&D: 177, Kriegl3: 183
Common and widespread in southern England, but few records from elsewhere. Perhaps better known as *C. hydrogramma*. Excludes *C. fritilliformis* sensu Rea (1922) and sensu C&D: p. 179, both doubtful.

phyllophila (Pers.) P. Kumm., *Führ. Pilzk.*: 122 (1871)
Agaricus phyllophilus Pers., *Syn. meth. fung.*: 457 (1801)
Agaricus cerussatus Fr., *Syst. mycol.* 1: 92 (1821)
Clitocybe cerussata (Fr.) P. Kumm, *Führ. Pilzk.*: 122 (1871)
Agaricus pithyophilus Fr., *Epicr. syst. mycol.*: 62 (1838)
Clitocybe pithyophila (Fr.) Gillet, *Hyménomycètes*: 152 (1874)
Clitocybe cerussata var. *pithyophila* (Fr.) J.E. Lange, *Fl. agaric. danic.* 5, *Taxonomic Conspectus*: II (1940)
E: c **S:** o **W:** ! **NI:** ! **ROI:** !
H: On soil in deciduous and conifer woodland. Also known in hedgerows and field borders, in gardens on compost heaps, and on dunes.
D: NM2: 107 (as *Clitocybe cerussata*), NM2: 108, FAN3: 46-47 **D+I:** Ph: 50, B&K3: 166 176 **I:** Bon: 137, C&D: 177 (as *C. cerussata*), C&D: 177, Kriegl3: 185
Occasional but widespread. Some authorities consider *C. cerussata* to be distinct.

pruinosa (Lasch) P. Kumm., *Führ. Pilzk.*: 120 (1871)
Agaricus pruinosus Lasch, *Hymenomyc. eur.*: 101 (1874)
S: !
H: On soil. British material with *Veronica* sp. in a garden. Spring-fruiting.
D: NM2: 109, Myc13(4): 172
A single collection from Midlothian (Balerno) in 1998.

rivulosa (Pers.) P. Kumm., *Führ. Pilzk.*: 122 (1871)
Agaricus rivulosus Pers., *Syn. meth. fung.*: 369 (1801)
Clitocybe dealbata var. *minor* Cooke, *Handb. Brit. fung.*, Edn 2: 50 (1884)
Agaricus rivulosus var. *neptuneus* Berk. & Broome, *Ann. Mag. Nat. Hist., Ser. 5 [Notices of British Fungi no. 1994]* 12: 370 (1883)
Clitocybe rivulosa var. *neptunea* (Berk. & Broome) Massee, *Brit. fung.-fl.* 2: 413 (1893)
Mis.: *Clitocybe dealbata* sensu auct. mult., sensu NCL
E: c **S:** c **W:** c **NI:** ! **ROI:** c **O:** Channel Islands: !
H: On soil, amongst short turf in grassland (including lawns) and on dunes.
D: NM2: 108 (as *Clitocybe dealbata*), FAN3: 48 **D+I:** Ph: 51, B&K3: 152 156 (as *C. dealbata*) **I:** Bon: 137 (as *C. dealbata*), C&D: 177 (as *C. dealbata*), Kriegl3: 188
Common and widespread. Perhaps better known as *C. dealbata*.

sinopica (Fr.) P. Kumm, *Führ. Pilzk.*: 123 (1871)
Agaricus sinopicus Fr., *Syst. mycol.* 1: 83 (1821)
Clitocybe subsinopica Harmaja, *Karstenia* 18(1): 29 (1978)
Mis.: *Clitocybe sinopicoides* sensu auct. Brit.
E: ! **S:** ! **ROI:** !
H: On soil, in deciduous woodland, often on old fire sites, and usually spring-fruiting.
D: NM2: 108, FAN3: 51 **D+I:** Ph: 50 (as *Clitocybe sinopicoides*), B&K3: 166 177 **I:** Bon: 135, C&D: 175, Kriegl3: 190
Rarely reported but apparently widespread. A distinctive species, little known in Britain and probably genuinely rare.

squamulosa (Pers.) P. Kumm., *Führ. Pilzk.*: 123 (1871)
   Agaricus squamulosus Pers., *Syn. meth. fung.*: 449 (1801)
   Clitocybe sinopicoides Peck, *Bull. New York State Mus. Nat. Hist.* 157: 80 (1911)
E: ! **S:** !
H: On soil in leaf litter, associated with deciduous or coniferous trees.
D: NM2: 110, FAN3: 54 **I:** Bon: 135, C&D: 173
Rarely reported and poorly known in Britain. It should be noted that most (if not all) collections named *C. sinopicoides* are in fact *C. sinopica*.

subcordispora Harmaja, *Karstenia* 10: 100 (1969)
   Clitocybe vibecina var. *pseudoobbata* J.E. Lange, *Dansk. Bot. Ark.* 6(5): 55 (1930)
   Clitocybe pseudoobbata (J.E. Lange) Kuyper, *Persoonia* 11(3): 386 (1981)
S: !
H: On soil.
D: FAN3: 62
Known from North Ebudes and Orkney.

subdryadicola Harmaja, *Karstenia* 18(1): 29 (1978)
S: !
H: On soil, associated with *Dryas octopetala* in montane grassland.
A single collection, made during the Nordic Congress Foray (1983).

subspadicea (J.E. Lange) Bon & Chevassut, *Doc. Mycol.* 9: 36 (1973)
   Omphalia umbilicata f. *subspadicea* J.E. Lange, *Fl. agaric. danic.* 2: 54 & pl. 58H (1936)
   Mis.: *Clitocybe umbilicata* sensu NCL, sensu Ricken [*Blätterpilze Deutschl.*: 385 (1915)]
E: ! **S:** !
H: On soil amongst leaf litter of deciduous trees.
D+I: B&K3: 168 178 **I:** Ph: 51 (incorrectly determined as *Clitocybe vibecina*), C&D: 179 (as *C. umbilicata*)
Known from England (North Wiltshire) and Scotland. Reported from Oxfordshire and North Somerset but unsubstantiated with voucher material.

truncicola (Peck) Sacc., *Syll. fung.* 5: 184 (1887)
   Agaricus truncicolus Peck, *Bull. Buffalo Soc. Nat. Sci.* 1: 46 (1873)
E: ! **S:** !
H: On decayed wood of deciduous trees, usually stumps or fallen trunks and occasionally on buried wood. Reported on *Acer pseudoplatanus*, *Fagus* and *Ulmus* spp.
D: NBA9: 546, FAN3: 49
Rarely reported. Most collections are from scattered locations in southern England, but also known from Scotland (Selkirk).

vermicularis (Fr.) Gillet, *Hyménomycètes*: 139 (1874)
   Agaricus vermicularis Fr., *Epicr. syst. mycol.*: 72 (1838)
E: ! **W:** ? **ROI:** ?
H: On soil, usually associated with conifers and fruiting in spring.
D: NM2: 109 **D+I:** B&K3: 168 179
Rarely reported. Known from South Devon. Reported elsewhere but unsubstantiated with voucher material. Records as *C. paropsis* on dunes belong here.

vibecina (Fr.) Quél., *Mém. Soc. Émul. Montbéliard, Sér. 2*, 5: 318 (1872)
   Agaricus vibecinus Fr., *Observ. mycol.* 2: 209 (1818)
   Clitocybe langei Hora, *Trans. Brit. Mycol. Soc.* 43(2): 441 (1960)
   Clitocybe orientalis Harmaja, *Karstenia* 10: 103 (1969)
E: c **S:** c **W:** ! **NI:** ! **ROI:** ! **O:** Channel Islands: !
H: On soil, usually in coniferous or mixed coniferous and deciduous woodland and less often in pure deciduous woodland with *Fagus* or *Quercus* spp.
D: NM2: 113, FAN3: 56-57 **D+I:** B&K3: 168 180 **I:** Bon: 141, C&D: 179 (as *Clitocybe langei*), C&D: 179
Common and widespread.

CLITOCYBULA (Singer) Métrod, *Rev. Mycol. (Paris)* 17(1): 74 (1952)
Agaricales, Tricholomataceae
Type: *Clitocybula lacerata* (Gillet) Métrod

lacerata (Gillet) Métrod, *Rev. Mycol. (Paris)* 17(1): 74 (1952)
   Collybia lacerata Gillet, *Hyménomycètes*: 310 (1876)
   Collybia platyphylla ssp. *lacerata* (Gillet) Konrad & Maubl., *Icon. select. fung.* 6: 242 (1937)
   Fayodia lacerata (Gillet) Singer, *Ann. Mycol.* 43: 63 (1943)
   Agaricus laceratus Lasch [*nom. illegit.*, non *A. laceratus* Bolton (1788)], in Fries, *Epicr. syst. mycol.*: 97 (1838)
E: ! **S:** ?
H: On decayed wood of deciduous trees.
D: BFF8: 109 **D+I:** B&K3: 170 183 **I:** Bon: 131, C&D: 183, Kriegl3: 197
A single collection from Dorset (near Bridport) in herb. K. Reported from South Hampshire and an old record from Scotland, both unsubstantiated with voucher material.

CLITOPILUS P. Kumm., *Führ. Pilzk.*: 23 (1871)
Agaricales, Entolomataceae
   Octojuga Fayod, *Ann. Sci. Nat., Bot.*, sér. 7, 9: 390 (1889)
   Paxillopsis J.E. Lange, *Fl. agaric. danic.* 5, *Taxonomic Conspectus*: VI (1940)
Type: *Clitopilus prunulus* (Scop.) P. Kumm.

daamsii Noordel., *Persoonia* 12(2): 161 (1984)
E: !
H: On decayed wood of decdiuous trees.
D: FAN1: 84, NM2: 342, Myc16(3): 115
Known from Surrey (Betchworth, Dawcombe Nature Reserve). Reported from Scotland but the collection in herb. E is misidentified. Easily mistaken for *C. hobsonii* from which it differs by the markedly larger spores.

hobsonii (Berk.) P.D. Orton, *Trans. Brit. Mycol. Soc.* 43: 174 (1960)
   Agaricus hobsonii Berk., *Outl. Brit. fungol.*: 138 (1860)
   Pleurotus hobsonii (Berk.) Sacc., *Syll. fung.* 5: 382 (1887)
   Octojuga fayodii Konrad & Maubl., *Icon. select. fung.* 6: 234 (1934)
   Octojuga pleurotelloides Kühner, *Botaniste* 17: 158 (1926)
   Clitopilus pleurotelloides (Kühner) Joss., *Bull. Mens. Soc. Linn. Lyon* 10: 14 (1941)
   Mis.: *Clitopilus septicoides* sensu Singer.
E: o **S:** o **W:** ! **ROI:** ! **O:** Isle of Man: !
H: Usually on decayed wood of deciduous trees, most often *Fagus* and *Quercus* spp. Rarely on conifers such as *Larix*, *Picea* and *Pinus* spp, and also *Taxus*. Also known on dead basidiomes of *Ganoderma australe* and *Trametes versicolor*, debris of *Juncus* spp and *Molinia caerulea* and dead stems of *Clematis vitalba* and *Lonicera periclymenum*. Often fruiting into the winter.
D: NCL3: 188, FAN1: 84, BFF6: 113-114, NM2: 342 **D+I:** B&K4: 50 1 **I:** Bon: 189
Occasional but widespread.

passeckerianus (Pilát) Singer, *Farlowia* 2: 560 (1946)
   Pleurotus passeckerianus Pilát, *Atlas champ. Eur.* 2: 49 (1935)
   Mis.: *Clitopilus cretatus* sensu auct.
E: ! **S:** ? **NI:** !
H: On decayed straw and hay, woodchip mulch and wet decayed newpaper. Also occuring as a weed in commercial mushroom beds.
D: BFF6: 114, NM2: 342
Rarely reported. Known from England (East & West Kent, Hertfordshire, Huntingdonshire, Surrey and West Sussex) and Northern Ireland (Antrim).

pinsitus (Fr.) Joss., *Bull. Soc. Mycol. France* 53: 209 (1937)
   Agaricus pinsitus Fr., *Syst. mycol.* 1: 184 (1821)
E: !

H: On decayed wood of deciduous trees, most often standing trunks or large logs of *Fagus* but also reported on *Fraxinus, Quercus* and *Ulmus* spp.
D: BFF6: 115
Rarely reported. Known from Bedfordshire, Hertfordshire, South Hampshire, Surrey and Warwickshire.

prunulus (Scop.) P. Kumm., *Führ. Pilzk.*: 97 (1871)
  Agaricus prunulus Scop., *Fl. carniol.* 437 (1772)
  Paxillopsis prunulus (Scop.) J.E. Lange, *Fl. agaric. danic.* 5, Taxonomic Conspectus: VI (1940)
  Agaricus orcellus Bull., *Herb. France*: pl. 573 f. 1 (1792)
  Pleuropus orcellus (Bull.) Gray, *Nat. arr. Brit. pl.* 1: 615 (1821)
  Clitopilus orcellus (Bull.) P. Kumm., *Führ. Pilzk.*: 97 (1871)
E: c **S:** c **W:** c **NI:** c **ROI:** c **O:** Channel Isles: ! Isle of Man: !
H: On soil, often amongst grass in open deciduous woodland, occasionally reported with conifers.
D: FAN1: 82-83, NM2: 341 **D+I:** Ph: 112-113, B&K4: 50 2 **I:** Bon: 189, C&D: 303, Cooke 342 (322) (as *Agaricus prunulus*) and 344 (323) (as *Agaricus orcella*) Vol. 3 (1885)
Common and widespread. English name = 'The Miller'.

scyphoides (Fr.) Singer, *Farlowia* 2: 554 (1946)
  Agaricus scyphoides Fr., *Syst. mycol.* 1: 163 (1821)
  Omphalia scyphoides (Fr.) P. Kumm., *Führ. Pilzk.*: 106 (1871)
  Clitocybe scyphoides (Fr.) P.D. Orton, *Trans. Brit. Mycol. Soc.* 43: 174 (1960)
  Clitopilus omphaliformis Joss., *Bull. Mens. Soc. Linn. Lyon* 10: 10 (1941)
  Clitopilus scyphoides f. *omphaliformis* (Joss.) Noordel., *Persoonia* 12(2): 159 (1983)
  Agaricus cretatus Berk. & Broome, *Ann. Mag. Nat. Hist., Ser. 3 [Notices of British Fungi no. 903]* 7: 373 (1861)
  Clitopilus cretatus (Berk. & Broome) Sacc., *Syll. fung.* 5: 702 (1887)
  Mis.: *Pleurotus mutilus* sensu Lange (FlDan2: 71 & pl. 79C) & sensu auct. Brit.
E: ! **W:** ! **ROI:** !
H: On soil, often amongst short turf, in grassland, also in woodland, gardens and occasionally on woodchips and woody debris.
D: FAN1: 83 (as *Clitopilus scyphoides* var. *scyphoides*), BFF6: 113, NM2: 342 **D+I:** B&K4: 52 5 (as *C. scyphoides* f. *scyphoides*) **I:** Bon: 189 (as *C. cretatus*)
Rarely reported. British collections mostly from Northumberland southwards. Some authorities e.g. Lange (1936) and NCL have interpreted this epithet as belonging to a white-spored species.

scyphoides *var.* **intermedius** (Romagn.) Noordel., *Persoonia* 12(2): 158 (1983)
  Clitopilus intermedius Romagn., *Bull. Soc. Naturalistes Oyonnax* 8: 74 (1954)
E: !
H: On soil. British material under *Picea* spp or in mixed deciduous woodland.
D: FAN1: 83-84 **D+I:** B&K4: 50 3
Rarely reported. Known from South Hampshire and Surrey.

COLLYBIA (Fr.) Staude, *Schwämme Mitteldeutschl.*: 119 (1857)
Agaricales, Tricholomataceae
  Gymnopus (Pers.) Roussel [non *Gymnopus* Quél. (1887)], *Flora Calvados*, Edn 2: 62 (1806)
  Rhodocollybia Singer, *Schweiz. Z. Pilzk.* 17: 71 (1939)
  Dendrocollybia R.H. Petersen & Redhead, *Mycol. Res.* 105(2): 169 (2001)
Type: *Collybia tuberosa* (Bull.) P. Kumm.

acervata (Fr.) P. Kumm., *Führ. Pilzk.*: 114 (1871)
  Agaricus acervatus Fr., *Syst. mycol.* 1: 122 (1821)
  Marasmius acervatus (Fr.) P. Karst., *Bidrag Kännedom Finlands Natur Folk* 48: 103 (1889)
  Gymnopus acervatus (Fr.) Murrill, *North American Flora, Ser. 1 (Fungi 3)*, 9: 362 (1916)
E: ? **S:** ! **NI:** ? **ROI:** ?

H: In coniferous woodland, on soil and needle litter, on decayed wood (stumps) or amongst mosses.
D: NM2: 114, FAN3: 115 **D+I:** A&N2 57-60 **I:** Kriegl3: 204
Apparently restricted to Caledonian pinewoods, from where it is rarely reported. Records from elsewhere are usually unsubstantiated with voucher material or misidentified *C. confluens* or *C. erythropus*. Cooke 205 (267) Vol. 2 (1882) is probably the latter.

alpina Vilgalys & O.K. Mill., *Trans. Brit. Mycol. Soc.* 88(4): 465 (1987)
  Gymnopus alpinus (Vilgalys & O.K. Mill.) Antonín & Noordel., *Mycotaxon* 63: 363 (1997)
S: !
H: On soil, associated with conifers and *Calluna* on heathland or in montane habitat.
D+I: A&N2 95-96
Rarely reported. Known from Easterness, Orkney, Perthshire and Shetland where it may be common *fide* R. Watling (pers. comm.).

aquosa (Bull.) P. Kumm., *Führ. Pilzk.*: 114 (1871)
  Agaricus aquosus Bull., *Herb. France*: pl. 17 (1780)
  Collybia dryophila var. *aquosa* (Bull.) Quél., *Enchir. fung.*: 31 (1886)
  Marasmius dryophilus var. *aquosus* (Bull.) Rea, *Brit. basidiomyc.*: 337 (1922)
  Gymnopus aquosus (Bull.) Antonín & Noordel., *Mycotaxon* 63: 363 (1997)
E: o **S:** ! **W:** ! **ROI:** ?
H: On soil and humus in deciduous woodland, also reported with conifers, and on roadside verges.
D: FAN3: 117 **D+I:** A&N2 92-94
Rarely reported but possibly misidentified as *C. dryophila* to which it is closely related. The rather pallid basidiomes, often bulbous stipe base and pink rhizomorphs are characteristic of *C. aquosa*.

butyracea (Bull.) P. Kumm., *Führ. Pilzk.*: 117 (1871)
  Agaricus butyraceus Bull., *Herb. France*: pl. 572 (1792)
  Rhodocollybia butyracea (Bull.) Lennox, *Mycotaxon* 9: 218 (1979)
E: c **S:** c **W:** c **NI:** c **ROI:** ! **O:** Channel Islands: ! Isle of Man: !
H: On soil amongst litter under coniferous, and less often deciduous, trees.
D: FAN3: 122 **D+I:** B&K3: 172 186, A&N2 134-136 36 (incorrectly captioned as *Rhodocollybia butyracea* var. *asema*) **I:** Kriegl3: 224
Frequent and widespread, but apparently less so than the var. *asema*.

butyracea *var.* **asema** (Fr.) Quél., *Fl. mycol. France*: 230 (1888)
  Agaricus butyraceus γ *asemus* Fr., *Observ. mycol.* 2: 124 (1818)
  Agaricus asemus (Fr.) Fr., *Syst. mycol.* 1: 121 (1821)
  Collybia asema (Fr.) Gillet, *Hyménomycètes*: 317 (1876)
  Collybia butyracea f. *asema* (Fr.) Singer, *Lilloa* 22: 201 (1951) [1949]
  Rhodocollybia butyracea f. *asema* (Fr.) Antonín, Halling & Noordel., *Mycotaxon* 63: 365 (1997)
  Agaricus bibulosus Massee, *Grevillea* 20: 95 (1891)
  Collybia bibulosa (Massee) Massee, *Brit. fung.-fl.* 3: 125 (1893)
  Collybia butyracea var. *bibulosa* (Massee) Rea, *Brit. basidiomyc.*: 331 (1922)
  Agaricus leiopus Pers., *Tent. disp. meth. fung.*: 21 (1797)
E: c **S:** c **W:** c **NI:** c **ROI:** c **O:** Channel Islands: ! Isle of Man: !
H: On soil and in leaf litter in deciduous woodland and near conifers. Often fruiting well into the winter.
D: NM2: 115, FAN3: 123 **D+I:** Ph: 56-57 (as *Collybia butyracea*), B&K3: 172 185, A&N2 137-138 35 (as *Rhodocollybia butyracea* var. *butyracea*) **I:** FlDan2: 41C (as *C. butyracea*)
Very common and widespread.

cirrhata (Pers.) Quél., *Mém. Soc. Émul. Montbéliard, Sér. 2,* 5: 96 (1872)

   Agaricus amanitae ssp. *cirrhata* Pers., *Observ. mycol.* 2: 53 (1799)

   Agaricus amanitae Batsch, *Elench. fung. (Continuatio Prima)*: 109 & t.18 f.93 (1786)

   Gymnopus tuberosus Gray, *Nat. arr. Brit. pl.* 1: 611 (1821)

E: o S: o W: o NI: ! ROI: !

H: On the mummified or decayed remains of fungal basidiomes. Known on polypores such as *Ganoderma australe* and *Meripilus giganteus* and agarics such as *Lactarius, Pholiota* and *Russula* spp.

D: NM2: 116, FAN3: 108 (as *Collybia amanitae*) **D+I:** B&K3: 182 201, A&N2 15-18 2 **I:** Bon: 181, C&D: 217 (as *C. amanitae*)

Occasional but widespread.

confluens (Pers.) P. Kumm., *Führ. Pilzk.*: 117 (1871)

   Agaricus confluens Pers., *Observ. mycol.* 1: 8 (1796)

   Marasmius confluens (Pers.) P. Karst, *Bidrag Kännedom Finlands Natur Folk* 48: 102 (1889)

   Gymnopus confluens (Pers.) Antonín, Halling & Noordel., *Mycotaxon* 63: 364 (1997)

   Agaricus archyropus Pers., *Mycol. eur.* 3: 135 (1828)

   Marasmius archyropus (Pers.) Fr., *Epicr. syst. mycol.*: 378 (1838)

   Agaricus ingratus Schumach. [non *A. ingratus* (F.H. Møller) Pilát], *Enum. pl.* 2: 304 (1803)

   Collybia ingrata (Schumach.) Quél., *Mém. Soc. Émul. Montbéliard, Sér. 2,* 5: 318 (1872)

   Mis.: *Marasmius hariolorum* sensu Rea (1922)

E: c S: c W: c NI: c ROI: ! O: Channel Islands: ! Isle of Man: !

H: On humus or decayed leaf litter, usually in deciduous woodland, and most often reported with *Fagus*. Occasionally reported with conifers such as *Larix* sp.

D: NM2: 114, FAN3: 110-111 **D+I:** Ph: 54-55, B&K3: 174 187, A&N2 32-37 6 (as *Gymnopus confluens*) **I:** FlDan2: 44G, C&D: 221, Kriegl3: 211

Common and widespread. Sensu Cooke 195 (283) Vol. 2 (1882) (as *Agaricus ingratus*) appears to be *C. erythropus*.

cookei (Bres.) J.D. Arnold, *Mycologia* 27: 413 (1935)

   Collybia cirrhata var. *cookei* Bres., *Icon. Mycol.* 5: 206 (1928)

   Sclerotium fungorum Pers. (anam.), *Syn. meth. fung.*: 120 (1801)

   Mis.: *Collybia cirrhata* sensu Lange (FlDan2: 15 & pl. 44E)

E: c S: c W: c NI: ! ROI: !

H: On decayed or mummified fungal basidiomes, including the polypores *Ganoderma* spp, *Inonotus hispidus* and *Meripilus giganteus* and the agarics *Armillaria mellea, Hypholoma fasciculare* and various species of *Russula*.

D: NM2: 116, FAN3: 109 **D+I:** B&K3: 184 202, A&N2 18-20 3 **I:** C&D: 217, Kriegl3: 200

Common and widespread. The yellow sclerotia are diagnostic.

distorta (Fr.) Quél., *Mém. Soc. Émul. Montbéliard, Sér. 2,* 5: 93 (1872)

   Agaricus distortus Fr., *Epicr. syst. mycol.*: 84 (1838)

   Rhodocollybia prolixa var. *distorta* (Fr.) Antonín, Halling & Noordel., *Mycotaxon* 63: 365 (1997)

E: o S: o W: ! NI: ! ROI: !

H: On acidic, nutrient poor soil, usually amongst leaf litter of deciduous trees such as *Quercus* or *Betula* spp, less frequently *Fagus* and rarely with conifers.

D: NM2: 116, FAN3: 121-122 **D+I:** Ph: 57, B&K3: 174 188, A&N2 125-128 32 (as *Rhodocollybia prolixa* var. *distorta*) **I:** FlDan2: 42B, C&D: 219, Kriegl3: 228 (as *Collybia prolixa* var. *distorta*)

Occasional but widespread.

dryophila (Bull.) P. Kumm., *Führ. Pilzk.*: 115 (1871)

   Agaricus dryophilus Bull., *Herb. France*: pl. 434 (1790)

   Omphalia dryophila (Bull.) Gray, *Nat. arr. Brit. pl.* 1: 612 (1821)

   Marasmius dryophilus (Bull.) P. Karst., *Bidrag Kännedom Finlands Natur Folk* 48: 103 (1889)

   Gymnopus dryophilus (Bull.) Murrill, *North American Flora, Ser. 1 (Fungi 3),* 9: 362 (1916)

   Collybia dryophila var. *aurata* Quél., *Enchir. fung.*: 31 (1886)

   Marasmius dryophilus var. *auratus* (Quél.) Rea, *Brit. basidiomyc.*: 524 (1922)

   Collybia dryophila var. *oedipus* Quél., *Fl. mycol. France*: 227 (1888)

   Marasmius dryophilus var. *oedipus* (Quél.) Rea, *Brit. basidiomyc.*: 525 (1922)

   Collybia dryophila var. *alvearis* Cooke, *Trans. Brit. Mycol. Soc.* 3: 110 (1908)

   Collybia dryophila var. *oedipoides* Singer, *Collect. Bot. (Barcelona)* 1: 233 (1947)

E: c S: c W: c NI: c ROI: c O: Channel Islands: c Isle of Man: c

H: On humus and leaf litter in deciduous and coniferous woodland. Occasionally also found amongst *Calluna* on moorland and amongst *Sphagnum* moss in boggy woodland.

D: NM2: 117, FAN3: 115-116 **D+I:** Ph: 54-55, B&K3: 174 189, A&N2 84-88 20 (as *Gymnopus dryophilus*) **I:** C&D: 219, Kriegl3: 207 (as *Collybia aquosa* var. *dryophila*)

Very common and widespread.

erythropus (Pers.) P. Kumm., *Führ. Pilzk.*: 115 (1871)

   Agaricus erythropus Pers., *Syn. meth. fung.*: 367 (1801)

   Marasmius erythropus (Pers.) Quél., *Mém. Soc. Émul. Montbéliard, Sér. 2,* 5: 221 (1872)

   Gymnopus erythropus (Pers.) Antonín, Halling & Noordel., *Mycotaxon* 63: 364 (1997)

   Collybia bresadolae (Kühner & Romagn.) Singer [*nom. inval.*], *Agaricales in Modern Taxonomy,* Edn 2: 314 (1962)

   Collybia kuehneriana Singer, *Persoonia* 2: 24 (1961)

   Agaricus marasmioides Britzelm., *Ber. Naturhist. Vereins Augsburg* 31: 162 (1894)

   Collybia marasmioides (Britzelm.) Bresinsky & Stangl., *Z. Pilzk.* 35(1-2): 67 (1969)

   Mis.: *Collybia acervata* sensu auct.

   Mis.: *Marasmius acervatus* sensu Pearson & Dennis, TBMS 31: 158 (1948)

E: c S: o W: ! NI: ! ROI: ! O: Isle of Man: !

H: On decayed woody litter, buried wood, and old stumps usually of deciduous trees such as *Betula* spp and *Fagus*. Occasionally in grassland but then growing from buried wood, and also known on old moss-covered stumps of conifers such as *Picea* sp. in plantations.

D: NM2: 117, FAN3: 117-118 **D+I:** Ph: 54-55, B&K3: 180 196 (as *Collybia marasmioides*), A&N2 102-106 26 (as *Gymnopus erythropus*) **I:** FlDan2: 45H, C&D: 219 (as *C. kuehneriana*), Kriegl3: 213

Common to occasional and widespread.

fagiphila Velen., *Novit. mycol.*: 82 (1939)

   Gymnopus fagiphilus (Velen.) Antonín, Halling & Noordel., *Mycotaxon* 63: 364 (1997)

   Mis.: *Collybia fuscopurpurea* sensu auct

   Mis.: *Marasmius fuscopurpureus* sensu auct.

E: !

H: On decayed leaf litter, usually associated with *Fagus* in beechwoods on calcareous soil.

D: FAN3: 118-119 (as *Collybia konradiana*), A&N2 107-109 **D+I:** B&K3: 176 191 (as *C. fuscopurpurea*)

Rarely reported but probably not uncommon in southern and south-western England. The collection illustrated under this name in Mycologist 13 (4):157 (1999) [Profiles of Fungi 109] is incorrect and apparently some other species.

fuscopurpurea (Pers.) P. Kumm., *Führ. Pilzk.*: 116 (1871)

   Agaricus fuscopurpureus Pers. [non *A. fuscopurpureus* With. (1796)], *Icon. descr. fung.*: pl. 12 (1798)

   Marasmius fuscopurpureus (Pers.) Fr., *Epicr. syst. mycol.*: 377 (1838)

   Gymnopus fuscopurpureus (Pers.) Antonín, Halling & Noordel., *Mycotaxon* 63: 364 (1996)

   Collybia obscura J. Favre, *Beitr. Kryptogamenfl. Schweiz* 3: 87 (1948)

   Mis.: *Collybia alkalivirens* sensu auct.

E: ! S: ! W: !

H: On soil in deciduous woodland or on heathland, also on decayed woodchip mulch on flowerbeds, and on lawns, in parks and gardens
D: NM2: 115 (as *Collybia obscura*), NM2: 115, FAN3: 119, A&N2 109-112 **D+I:** Ph: 56-57 (as *C. obscura*)
Previously considered a rarity, this has become commoner, at least in southern England, and there is a single collection from Wales (Denbighshire). There is some doubt that this is the species described in A&N1: 109, since British collections do not show a green reaction with alkali. It has been suggested that British material may represent the North American *Collybia biformis* (Peck) Singer.

fusipes (Bull.) Quél., *Mém. Soc. Émul. Montbéliard, Sér. 2*, 5: 93 (1872)
  Agaricus fusipes Bull., *Herb. France*: pl. 516 (1791)
  Gymnopus fusipes (Bull.) Gray, *Nat. arr. Brit. pl.* 1: 604 (1821)
  Rhodocollybia fusipes (Bull.) Romagn., *Bull. Soc. Mycol. France* 94: 78 (1978)
  Agaricus contortus Bull., *Herb. France*: pl. 36 (1783)
  Collybia fusipes var. *contorta* (Bull.) Gillet, *Hyménomycètes*: 312 (1876)
  Agaricus crassipes Schaeff., *Fung. Bavar. Palat. nasc.* 1: 87 (1762)
  Collybia crassipes (Schaeff.) P. Kumm., *Führ. Pilzk.*: 117 (1871)
  Agaricus fusiformis Bull., *Herb. France*: pl. 76 (1787)
  Agaricus oedematopus Schaeff., *Fung. Bavar. Palat. nasc.* 3: 259 (1770)
  Collybia fusipes var. *oedematopus* (Schaeff.) Gillet, *Hyménomycètes*: 312 (1876)
  Collybia oedematopoda (Schaeff.) Sacc., *Syll. fung.* 5: 206 (1887)
  Agaricus lancipes Fr. *Epicr. syst. mycol.*: 63 (1838)
  Collybia lancipes (Fr.) Gillet, *Hyménomycètes*: 312 (1876)
E: c **S:** c **W:** c **NI:** c **ROI:** c **O:** Channel Islands: c
H: Weakly parasitic then saprotrophic. On roots of living trees, usually *Quercus* spp, less often *Fagus* and *Castanea*. Also reported on *Betula, Carpinus, Corylus, Fraxinus* and *Tilia* spp. Rarely reported on conifers such as *Pinus sylvestris* and a single collection with *Calluna*.
D: Sow2: 129 (as *Agaricus crassipes*), NM2: 115, BFF8: 115 (as *Gymnopus fusipes*) **D+I:** Ph: 54-55, B&K3: 176 192, FAN3: 119-120 (as *Collybia contorta*), A&N2 27-31 5 (as *G. fusipes*) **I:** FlDan2: 43D, C&D: 219, Kriegl3: 216, Cooke 185 (141) Vol. 2 (1881)
Common and widespread.

inodora (Pat.) P.D. Orton, *Notes Roy. Bot. Gard. Edinburgh* 29: 85 (1969)
  Marasmius inodorus Pat., *Tab. anal. fung.*: 13 & pl. 523 (1887)
  Micromphale inodorum (Pat.) Svrček [*nom. illegit.*, non *M. inodorum* Dennis (1961)], *Česká Mykol.* 18: 24 (1964)
  Gymnopus inodorus (Pat.) Antonín & Noordel., *Mycotaxon* 63: 364 (1997)
  Mis.: *Micromphale rufocarneum* sensu auct. Brit.
E: ! **W:** !
H: On decayed wood of deciduous trees, often on mossy logs or apparently on mossy ground, but then arising from buried wood.
D: NBA3: 85-86
Rarely reported or collected. Known from England (East Kent, Hertfordshire, Oxfordshire, Shropshire, South Devon, South Somerset and Surrey) and Wales (Caernarvonshire). Basidiomes are small, cryptically coloured and are often produced during the summer, thus easily overlooked.

luxurians Peck, *Bull. Torrey Bot. Club* 24: 141 (1897)
E: !
H: On soil, on flowerbeds mulched with shredded bark and woodchips and spreading onto nearby lawns.
D: FAN3: 114 (as *Collybia luxurians*) **D+I:** A&N2 45-46 **I:** FM3(4): 142

An alien, previously known only from warm greenhouses at Kew (Royal Botanic Gardens), but now established locally outdoors.

maculata (Alb. & Schwein.) P. Kumm., *Führ. Pilzk.*: 117 (1871)
  Agaricus maculatus Alb. & Schwein., *Consp. fung. lusat.*: 186 (1805)
  Rhodocollybia maculata (Alb. & Schwein.) Singer, *Schweiz. Z. Pilzk.* 17: 71 (1939)
  Agaricus maculatus var. *immaculatus* Cooke, *Ill. Brit. fung.* 187 (221) Vol. 2 (1882)
  Collybia maculata var. *immaculata* (Cooke) Massee, *Brit. fung.-fl.* 3: 124 (1893)
E: c **S:** c **W:** c **NI:** c **ROI:** c **O:** Isle of Man: !
H: On humus and decayed needle litter and especially common in mixed woodland with *Pinus* and *Betula* spp on acidic soils. Also known on moorland with *Calluna*.
D: NM2: 115 **D+I:** Ph: 54, B&K3: 178 195, FAN3: 120, A&N2 117-121 29 (as *Rhodocollybia maculata* var. *maculata*) **I:** FlDan2: 42C1, C&D: 219, Kriegl3: 226
Very common and widespread.

ocior (Pers.) Vilgalys & O.K. Mill., *Trans. Brit. Mycol. Soc.* 88(4): 467 (1987)
  Agaricus ocior Pers., *Mycol. eur.* 3: 115 (1828)
  Gymnopus ocior (Pers.) Antonín & Noordel., *Mycotaxon* 63: 365 (1997)
  Agaricus exculptus Fr., *Epicr. syst. mycol.*: 93 (1838)
  Collybia exculpta (Fr.) Gillet, *Hyménomycètes*: 328 (1876)
  Marasmius exculptus (Fr.) Rea, *Brit. basidiomyc.*: 525 (1922)
  Collybia dryophila ssp. *exculpta* (Fr.) Konrad & Maubl., *Icon. select. fung.* 3: pl. 201 (1934)
  Agaricus dryophilus β *funicularis* Fr., *Syst. mycol.* 1: 125 (1821)
  Marasmius funicularis (Fr.) P. Karst., *Symb. mycol. fenn.* 32: 1 (1894)
  Marasmius dryophilus var. *funicularis* (Fr.) Rea, *Brit. basidiomyc.*: 524 (1922)
  Collybia dryophila var. *funicularis* (Fr.) Halling, *Mycol. Mem., The Genus* Collybia 8: 52 (1983)
  Collybia extuberans (Fr.) Quél., *Mém. Soc. Émul. Montbéliard, Sér. 2*, 5: 237 (1872)
  Collybia luteifolia Gillet, *Hyménomycètes*: 328 (1876)
  Agaricus succineus Fr., *Epicr. syst. mycol.*: 91 (1838)
  Collybia succinea (Fr.) Quél., *Mém. Soc. Émul. Montbéliard, Sér. 2*, 5: 237 (1872)
  Agaricus xanthopus Fr., *Observ. mycol.* 1: 12 (1818)
  Collybia xanthopoda (Fr.) Sacc., *Syll. fung.* 5: 226 (1887)
E: o **S:** ! **W:** !
H: On decayed leaf litter or rich humic soil, and occasionally on decayed wood, in deciduous woodland, often in troops. Rarely in coniferous woodland. Spring-fruiting, most records between April and June.
D: NM2: 117 (as *Collybia exsculpta*), FAN3: 118 **D+I:** B&K3: 178 194 (as *C. luteifolia*), B&K3: 182 200 (as *C. succinea*), A&N2 88-91 21 (as *Gymnopus ocior*) **I:** C&D: 219
Rather frequent in England, less often reported elsewhere.

oreadoides (Pass.) P.D. Orton, *Trans. Brit. Mycol. Soc.* 43(2): 174 (1960)
  Marasmius oreadoides Pass., *Nuovo Giorn. Bot. Ital.* 4: 109 (1872)
  Gymnopus oreadoides (Pass.) Antonín & Noordel., *Mycotaxon* 63: 365 (1997)
E: ! **S:** ! **W:** ! **NI:** !
H: On humic soil, often amongst grass and often near to *Alnus*.
D: A&N2 60-61 **D+I:** Ph: 54
Rarely collected or reported. Known from single collections in England, Wales and Northern Ireland and two from Scotland. Most other records lack voucher material.

peronata (Bolton) P. Kumm., *Führ. Pilzk.*: 116 (1871)
  Agaricus peronatus Bolton [non *A. peronatus* Massee (1892)], *Hist. fung. Halifax* 2: 58 (1788)
  Gymnopus peronatus (Bolton) Gray, *Nat. arr. Brit. pl.* 1: 607 (1821)

Marasmius peronatus (Bolton) Fr., *Sverig Atl. Svamp.*: 52 (1836)

Agaricus urens Bull., *Herb. France*: pl. 528 (1791)

Marasmius urens (Bull.) Fr., *Sverig Atl. Svamp.*: 52 (1836)

Collybia urens (Bull.) P. Kumm., *Führ. Pilzk.*: 115 (1871)

E: c **S:** c **W:** c **NI:** c **ROI:** c **O:** Channel Islands: ! Isle of Man: !

H: Solitary or gregarious on decayed leaf litter of deciduous trees, and occasionally in conifer litter.

D: NM2: 115, FAN3: 111 **D+I:** Ph: 57, B&K3: 180 198, A&N2 41 (as *Gymnopus peronatus*) **I:** Sow1: 37 (as *Agaricus peronatus*), Bon: 179, C&D: 221

Very common and widespread. English name = 'Wood Woolly Foot'.

prolixa (Hornem.) Gillet, *Hyménomycètes*: 317 (1876)

Agaricus prolixus Hornem., *Fl. Danica*: t. 1608 (1818)

Rhodocollybia prolixa (Hornem.) Antonín & Noordel., *Mycotaxon* 63: 365 (1997)

E: ! **S:** !

H: On humus and near to decayed wood in coniferous woodland.

D: NM2: 116 **D+I:** B&K3: 182 199, A&N2 122-125 31 (as *Rhodocollybia prolixa* var. *prolixa*) **I:** Cooke 1139 (950) Vol. 8 (1890, as *Agaricus prolixus*)

Rarely reported. Usually misidentified (as *Collybia distorta*), and records are mostly unsubstantiated with voucher material.

putilla (Fr.) Singer, *Beih. Bot. Centralbl.* 56: 163 (1936)

Agaricus putillus Fr., *Observ. mycol.* 2: 130 (1818)

Gymnopus putillus (Fr.) Antonín, Halling & Noordel., *Mycotaxon* 63: 365 (1997)

E: ? **S:** !

H: On soil, amongst mosses in Caledonian pinewoods.

D: NBA2: 44

Confirmed only from Scotland (see Orton, NBA2). Reported from England but the single collection, from Bedfordshire, in herb. K. is determined as 'compare with' *C. putilla* and a Yorkshire record is unsubstantiated with voucher material.

racemosa (Pers.) Quél., *Mém. Soc. Émul. Montbéliard, Sér. 2,* 5: 342 (1873)

Agaricus racemosus Pers., *Tent. disp. meth. fung.*: 15 (pl. 3) (1797)

Mycena racemosa (Pers.) Gray, *Nat. arr. Brit. pl.* 1: 620 (1821)

Dendrocollybia racemosa (Pers.) R.H. Petersen & Redhead, *Mycol. Res.* 105(2): 169 (2001)

Sclerotium lacunosum Pers. (anam.), *Neues Mag. Bot.* 1: 95 (1794)

Tilachlidiopsis racemosa Keissler (anam.), *Ann. Naturhist. Mus. Wien* 37: 215 (1924)

E: ! **S:** ! **W:** ! **ROI:** !

H: Saprotrophic. On soil, often amongst mosses, in coniferous woodland, on mummified remnants of agarics or where these have decayed.

D: NM2: 116, FAN3: 109 **D+I:** A&N2 20-22 4 **I:** Sow3: 287 (as *Agaricus racemosus*), FlDan2: 45E, SV30: 21, C&D: 217

Rarely reported but widespread.

tergina (Fr.) S. Lundell, *Fungi Exsiccati Suecici* 23-24: 5 (1942)

Agaricus terginus Fr., *Syst. mycol.* 1: 128 (1821)

Marasmius terginus (Fr.) Fr., *Epicr. syst. mycol.*: 377 (1838)

Gymnopus terginus (Fr.) Antonín & Noordel., *Mycotaxon* 63: 365 (1997)

Marasmius gelidus Quél., *Compt. Rend. Assoc. Franç. Avancem. Sci. Assoc. Sci. France* 20: 467 (1892)

E: ? **S:** !

H: On soil, amongst leaf litter in woodland.

D: FAN3: 112 **D+I:** A&N2 48-49 (as *Gymnopus terginus*)

Known from Mid-Perthshire (Dall). Reported elsewhere but unsubstantiated with voucher material or misidentified.

tuberosa (Bull.) P. Kumm., *Führ. Pilzk.*: 116 (1871)

Agaricus tuberosus Bull., *Herb. France*: pl. 552 (1786)

Marasmius sclerotipes Bres., *Fungi trident.* 1(1): 12 (1881)

Sclerotium cornutum Fr. (anam.), *Observ. mycol.*: 205 (1815)

E: c **S:** c **W:** o

H: On decayed remnants of agarics, usually large species of *Russula* and *Lactarius*; also on soil amongst mosses on acidic soil, but then possibly where an agaric has decayed.

D: NM2: 116, FAN3: 107-108 **D+I:** Ph: 56-57, B&K3: 184 203, A&N2 12-14 1 **I:** Bon: 181, C&D: 217

Common and widespread.

COLTRICIA Gray, *Nat. arr. Brit. pl.* 1: 644 (1821)

Hymenochaetales, Hymenochaetaceae

Strilia Gray, *Nat. arr. Brit. pl.* 1: 645 (1821)

Polystictus Fr., *Nova Acta Regiae Soc. Sci. Upsal.* 1: 70 (1851)

Type: *Coltricia connata* Gray (= *C. perennis*)

confluens Keizer, *Persoonia* 16(3): 389 (1997)

E: !

H: On soil. British material on compacted soil mixed with woody debris in dense thickets of *Crataegus monogyna*.

D: Myc17(1): 42 **I:** SV36: 25

Known from one site in Surrey (near Richmond). Described from the Netherlands and there said to be spreading rapidly (Keizer, 1997).

montagnei (Fr.) Murrill, *Mycologia* 12: 13 (1920)

Polyporus montagnei Fr., *Epicr. syst. mycol.*: 434 (1838)

Polystictus montagnei (Fr.) Cooke, *Grevillea* 14: 77 (1886)

E: !

H: On soil associated with deciduous trees. British material with *Tilia* spp.

D: EurPoly1: 215

Accepted but with doubts. A single collection named thus, from Worcestershire (Bromsgrove) in herb. K, but this is in poor condition.

perennis (L.) Murrill, *J. Mycol.* 9: 91 (1903)

Boletus perennis L., *Sp. pl.*: 1177 (1753)

Polyporus perennis (L.) Fr., *Syst. mycol.* 1: 350 (1821)

Polystictus perennis (L.) P. Karst., *Meddeland. Soc. Fauna Fl. Fenn.* 5: 39 (1879)

Coltricia connata Gray, *Nat. arr. Brit. pl.* 1: 644 (1821)

Mis.: *Boletus subtomentosus* sensu Bolton [Hist. Fung. Halifax 2: 87 (1788)]

E: o **S:** o **W:** o **NI:** o

H: Usually on sandy, acidic soil on heathland or moorland. Associated with *Pinus* spp, and often growing on bare or compacted soil by paths and tracksides near to trees. Also reported amongst *Calluna* and *Ulex europaeus* in such habitat. Also rather frequent on old bonfire sites on acid soils. Rarely in deciduous woodland but known with *Betula, Castanea, Fagus* and *Quercus* spp.

D: EurPoly1: 217, NM3: 331 **D+I:** Ph: 218-219, B&K2: 248 298 **I:** Sow2: 192 (as *Boletus perennis*), Bon: 321, C&D: 141, Kriegl1: 430

Occasional but widespread and may be locally common.

CONFERTICIUM Hallenb., *Mycotaxon* 11: 447 (1980)

Hericiales, Gloeocystidiellaceae

Type: *Conferticium insidiosum* (Bourdot & Galzin) Hallenb.

insidiosum (Bourdot & Galzin) Hallenb., *Mycotaxon* 11(2): 448 (1980)

Gloeocystidium insidiosum Bourdot & Galzin, *Bull. Soc. Mycol. France* 28(4): 370 (1913)

Gloeocystidiellum insidiosum (Bourdot & Galzin) Donk, *Fungus* 26: 9 (1956)

W: !

H: On decayed wood of deciduous trees. British material on a stacked pile of logs.

D: Myc15(2): 76

A single collection from Breconshire (Coed Eyrisiog) in 1999.

CONIOPHORA Mérat, *Nouv. fl. env. Paris*, Edn 2, 1: 36 (1821)
Boletales, Coniophoraceae
> *Coniophorella* P. Karst., *Bidrag Kännedom Finlands Natur Folk* 48: 438 (1889)
> *Aldridgea* Massee, *Grevillea* 20: 121 (1892)
Type: *Coniophora membranacea* Mérat (= *C. puteana*)

arida (Fr.) P.Karst, *Not. Sällsk. Fauna et Fl. Fenn. Förh.* 9: 370 (1868)
> *Thelephora arida* Fr., *Elench. fung.* 1: 197 (1828)
> *Coniophora arida* var. *flavobrunnea* Bres., *Ann. Mycol.* 1(1): 110 (1903)
> *Corticium suffocatum* Peck, *Rep. (Annual) New York State Mus. Nat. Hist.* 30: 48 (1878)
> *Coniophora arida* var. *suffocata* (Peck) Ginns, *Opera Bot.* 61: 24 (1982)
> *Coniophora berkeleyi* Massee, *J. Linn. Soc., Bot.* 25: 135 (1890)
> *Coniophora betulae* P. Karst., *Hedwigia* 35: 174 (1896)
> *Coniophora cookei* Massee, *J. Linn. Soc., Bot.* 25: 136 (1889)
> *Corticium subdealbatum* Berk. & Broome, *Ann. Mag. Nat. Hist., Ser. 5 [Notices of British Fungi no. 1823]* 3: 211 (1879)
> *Peniophora subdealbata* (Berk. & Broome) Sacc., *Syll. fung.* 6: 647 (1888)
> *Coniophora subdealbata* (Berk. & Broome) Massee, *J. Linn. Soc., Bot.* 25: 135 (1889)
> *Coniophora sulphurea* var. *ochroidea* Massee, *J. Linn. Soc., Bot.* 25: 133 (1889)
E: c **S:** c **W:** o **NI:** ! **O:** Isle of Man: !
H: Usually on decayed wood or bark of conifers. Uncommon on deciduous trees and shrubs but known on *Acer pseudoplatanus*, *Buxus*, *Fagus*, *Fraxinus*, *Hedera*, *Populus*, *Quercus* and *Salix* spp.
D: NM3: 289 **D+I:** Hallenb: 63 39/40, B&K2: 206 236 **I:** Kriegl1: 361
Common and widespread on wood of conifers. *Coniophora arida* var. *suffocata*, with densely encrusted basal hyphae, is kept separate by some authorities.

fusispora (Cooke & Ellis) Sacc., *Syll. fung.* 6: 650 (1888)
> *Corticium fusisporum* Cooke & Ellis, *Grevillea* 8: 11 (1879)
> *Coniophora bourdotii* Bres., *Ann. Mycol.* 6: 45 (1908)
E: !
H: On decayed wood of conifers.
D: NM3: 289 **D+I:** Hallenb: 65 41/42
Rarely reported. Known from North Essex (Boxted, near Colchester), Surrey (Haslemere and Mickleham: Norbury Park) and West Norfolk (Holkham NNR).

hanoiensis Pat., *Bull. Soc. Mycol. France* 23: 76 (1907)
E: !
H: On decayed wood.
A single old collection, in poor condition, from West Cornwall (Penzance) in herb. K, ex herb. Berkeley.

marmorata Desm., *Cat. pl. omises botanogr. Belgique*: 18 (1823)
E: !
H: On decayed wood, in buildings, damp walls and one record 'on a flower pot'.
Known from Middlesex, South Somerset, Warwickshire and Yorkshire.

olivacea (Pers.) P. Karst., *Bidrag Kännedom Finlands Natur Folk* 37: 162 (1882)
> *Thelephora olivacea* Pers., *Mycol. eur.* 1: 143 (1822)
> *Hypochnus olivaceus* (Pers.) Fr., *Summa veg. Scand.*: 337 (1849)
> *Corticium olivaceum* (Pers.) Fr., *Hymenomyc. eur.*: 660 (1874)
> *Coniophorella olivacea* (Pers.) P. Karst., *Bidrag Kännedom Finlands Natur Folk* 48: 438 (1889)
> *Thelephora sistotremoides* Schwein., *Schriften Naturf. Ges. Leipzig* 1: 109 (1822)

> *Coniophora sistotremoides* (Schwein.) Massee, *J. Linn. Soc., Bot.* 25: 133 (1889)
> *Thelephora umbrina* Alb. & Schwein., *Consp. fung. lusat.*: 281 (1805)
> *Coniophora umbrina* (Alb. & Schwein.) Sacc., *Syll. fung.* 6: 652 (1888)
> *Coniophorella umbrina* (Alb. & Schwein.) Bres., *Ann. Mycol.* 1: 111 (1903)
> *Corticium umbrinum* (Alb. & Schwein.) Fr., *Hymenomyc. eur.*: 658 (1874)
E: ! **S:** !
H: On decayed wood of conifers, usually *Pinus* spp, but also known on *Taxus*. Rarely on wood of deciduous trees such as *Quercus* spp and on old basidiomes of *Hymenochaete* spp.
D+I: Hallenb: 69 43/44, B&K2: 206 237
Rarely reported but probably overlooked. Known from England (Berkshire, Hertfordshire, North Lincolnshire, Shropshire, South Devon, South Essex, South Hampshire, Surrey, Warwickshire and West Kent) and Scotland (Perthshire).

prasinoides (Bourdot & Galzin) Bourdot & Galzin, *Hymenomyc. France*: 361 (1928)
> *Coniophora olivacea* var. *prasinoides* Bourdot & Galzin, *Bull. Soc. Mycol. France* 39: 115 (1923)
E: ! **S:** !
H: British material on decayed bark of *Picea sitchensis* and on old dead stems of *Juncus* sp.
Single collections from Hertfordshire (Little Hormead) in 2004 and South Ebudes (Isle of Jura) in 1993.

puteana (Schumach.) P. Karst., *Not. Sällsk. Fauna et Fl. Fenn. Förh.* 2(6): 370 (1868)
> *Thelephora puteana* Schumach., *Enum. pl.* 1: 379 (1801)
> *Thelephora cerebella* Pers., *Syn. meth. fung.*: 580 (1801)
> *Coniophora cerebella* (Pers.) Pers., *Mycol. eur.* 1: 155 (1822)
> *Aldridgea gelatinosa* Massee, *Grevillea* 20: 121 (1892)
> *Coniophora gelatinosa* (Massee) W.G. Sm., *Syn. Brit. Bas.*: 422 (1908)
> *Coniophora incrustans* Massee, *J. Linn. Soc., Bot.* 25: 132 (1889)
> *Thelephora laxa* Fr. [non *T. laxa* Pers. (1822)], *Elench. fung.* 1: 196 (1828)
> *Coniophora laxa* (Fr.) Quél., *Enchir. fung.*: 212 (1886)
> *Coniophora membranacea* Mérat, *Nouv. fl. env. Paris*, Edn 2: 36 (1821)
> *Coniophora puteana* var. *cellaris* (Pers.) Massee, *J. Linn. Soc., Bot.* 25: 130 (1889)
> *Fibrillaria ramosissima* Sowerby, *Col. fig. Engl. fung.* 3: pl. 387 (1803)
E: c **S:** c **W:** c **NI:** ! **ROI:** ! **O:** Channel Islands: ! Isle of Man: !
H: On decayed wood of deciduous trees or conifers.
D: NM3: 289 **D+I:** Ph: 239, Hallenb: 71 45 & 47
Very common and widespread. The cause of 'Wet Rot' in damp cellars or outhouses. The illustration in B&K2 lacks the typical fleshy appearance of *C. puteana* and may depict *C. arida*.

CONOCYBE Fayod, *Ann. Sci. Nat., Bot., sér. 7*, 9: 357 (1889)
Agaricales, Bolbitiaceae
> *Pholiotina* Fayod, *Ann. Sci. Nat., Bot., sér. 7*, 9: 359 (1889)
Type: *Conocybe tenera* (Schaeff.) Fayod

aeruginosa Romagn., *Bull. Soc. Mycol. France* 84: 368 (1969)
E: !
H: On disturbed soil, usually at the edges of woodland.
D+I: B&K4: 312 395 **I:** C&D: 363 (as *Pholiotina aeruginosa*)
Collections from Dorset, Herefordshire, Huntingdonshire, Northamptonshire, Somerset and Warwickshire.

ambigua Watling, *Notes Roy. Bot. Gard. Edinburgh* 38: 331 (1980)
> *Conocybe siliginea* var. *ambigua* Kühner [*nom. nud.*], *Encycl. Mycol. 7. Le Genre* Galera: 106 (1935)
> *Galera ambigua* (Kühner) J.E. Lange [*nom. nud.*], *Dansk. Bot. Ark.* 9(6): 38 (1938)

E: ! S: ! W: !
H: On soil in grassland (including lawns) and also in woodland clearings.
D: BFF3: 69-70, NM2: 274 **D+I:** B&K4: 298 375
Rarely reported. Apparently widespread but confused with the recently described *C. merdaria*.

anthracophila Watling, *Notes Roy. Bot. Gard. Edinburgh* 40: 540 (1983)
Conocybe siliginea var. *anthracophila* Maire & Kühner [*nom. nudum*], *Encyclop. Mycol. 7, Le Genre* Galera: 97 (1935)
E: !
H: On soil or compost in areas disturbed by human activity such as gardens or waste ground.
D: BFF3: 78
Rarely reported. Several records from scattered sites in northern England and two collections from south eastern England (Surrey: Wisley & West Kent: Sevenoaks).

antipus (Lasch) Fayod, *Ann. Sci. Nat., Bot.,* sér. 7, 9: 357 (1889)
Agaricus antipus Lasch, *Linnaea* 3: 415 (1828)
Galera antipus (Lasch) Quél., *Mém. Soc. Émul. Montbéliard, Sér. 2,* 5: 136 (1872)
E: !
H: On manured soil or decayed mixtures of straw and dung, manure heaps or piles of decayed grass clippings.
D: BFF3: 63-64, NM2: 275
Rarely reported and poorly known in Britain. Scant material in herbaria and few recent collections.

apala (Fr.) Arnolds, *Persoonia* 18(2): 225 (2003)
Agaricus apalus Fr., *Syst. mycol.* 1: 265 (1821)
Galera apala (Fr.) Sacc., *Syll. fung.* 5: 860 (1887)
Bolbitius albipes G.H. Otth, *Mitt. naturf. Ges. Bern* 744: 92 (1871) [1870]
Conocybe albipes (G.H. Otth) Hauskn., *Österr. Z. Pilzk., N.S.* 7: 102 (1998)
Galera lactea J.E. Lange, *Fl. agaric. danic.* 5, *Taxonomic Conspectus*: IV (1940)
Conocybe lactea (J.E. Lange) Métrod, *Bull. Soc. Mycol. France* 56: 46 (1940)
Agaricus tener Sowerby [*nom. illegit.*, non *A. tener* Schaeffer (1762)], *Col. fig. Engl. fung.* 1: pl. 33 (1796)
Bolbitius tener (Sowerby) Berk. & Broome, *Outl. Brit. fungol.*: 183 (1860)
Mycena tenera (Sowerby) Gray, *Nat. arr. Brit. pl.* 1: 620 (1821)
Mis.: *Galera lateritia* sensu Cooke [Ill. Brit. fung. 517(460) (1886)]
Mis.: *Conocybe lateritia* sensu auct.
E: c S: c W: c NI: ! ROI: ! O: Channel Islands: !
H: On soil amongst grass in parkland, meadows, unimproved grassland and gardens, occasionally on dunes, and rarely on flowerbeds mulched with sawdust or woodchips.
D: BFF3: 80 (as *Conocybe lactea*), NM2: 274 (as *C. lactea*) **D+I:** Ph: 154 (as *C. lactea*) **I:** Sow1: 33 (as *Agaricus tener*)
Common and widely distributed. Basidiomes are delicate and fragile, appearing after rain or when lawns are watered, then decaying rapidly. Perhaps better known as *Conocybe lactea*.

aporos Kits van Wav., *Persoonia* 6: 144 (1970)
Pholiotina aporos (Kits van Wav.) Clémençon, *Schweiz. Z. Pilzk.* 54(10): 151 (1976)
E: o S: o W: o
H: On loam soil, associated with a wide range of deciduous trees and shrubs, in herb or moss-rich woodland. Spring-fruiting.
D: BFF3: 97-98, NM2: 272 **D+I:** B&K4: 312 396 **I:** C&D: 363 (as *Pholiotina aporos*)
Occasional but widespread, especially frequent in woodland on calcareous soil in England.

arrhenii (Fr.) Kits van Wav., *Persoonia* 6: 147 (1970)
Agaricus arrhenii Fr., *Epicr. syst. mycol.*: 161 (1838)
Pholiota arrhenii (Fr.) S. Imai, *J. Fac. Agric. Hokkaido Univ.* 43: 185 (1938)

Pholiotina arrhenii (Fr.) Singer, *Sydowia, Beih.* 7: 77 (1973)
Agaricus mesodactylius Berk. & Broome, *Ann. Mag. Nat. Hist., Ser. 2 [Notices of British Fungi no. 329]* 2: 261 (1848)
Mis.: *Pholiota togularis* sensu auct.
E: o S: o W: o NI: o ROI: o
H: On soil in woodland, parkland and along tracks or roadsides.
D: BFF3: 91, NM2: 273 **D+I:** B&K4: 314 398 **I:** C&D: 363 (as *Pholiotina arrhenii*)
Occasional but widespread. Often confused with other annulate species of *Conocybe*.

aurea (Jul. Schäff.) Hongo:, *Jap. J. Bot.* 38: 236 (1963)
Galera aurea Jul. Schäff., *Z. Pilzk.* 9(8-10): 167 (1930)
Conocybe tenera var. *aurea* (Jul. Schäff.) Kühner, *Encycl. Mycol. 7. Le Genre* Galera: 72 (1935)
S: !
H: On soil, in deciduous woodland by an estuary (British material).
D: BFF3: 53 **D+I:** B&K4: 300 376
Known in Britain from one site in south eastern Scotland.

blattaria (Fr.) Kühner, *Encycl. Mycol. 7. Le Genre* Galera: 150 (1935)
Agaricus blattarius Fr., *Syst. mycol.* 1: 246 (1821)
Pholiota blattaria (Fr.) Quél., *Mém. Soc. Émul. Montbéliard, Sér. 2,* 5: 319 (1872)
Pholiotina blattaria (Fr.) Fayod, *Ann. Sci. Nat., Bot.,* sér. 7, 9: 359 (1889)
Togaria blattaria (Fr.) W.G. Sm., *Syn. Brit. Bas.*: 123 (1908)
Pholiota teneroides J.E. Lange, *Dansk. Bot. Ark.* 2(11): 7 (1921)
Conocybe teneroides (J.E. Lange) Kits van Wav., *Persoonia* 6: 160 (1970)
E: ! S: ! NI: !
H: On soil in woodland and scrub, alongside paths and in clearings.
D: BFF3: 92, NM2: 272
Supposedly 'common and widespread' *fide* BFF3, but rarely recorded or collected. Following BFF3, this is a slender two-spored species distinct from the stouter *C. percincta*; but some authorities (e.g. B&K4) use the name for a four-spored species, here listed as *C. vexans*.

brachypodii (Velen.) Hauskn. & Svrček, *Czech Mycol.* 51(1): 43 (1999)
Galera brachypodii Velen., *Opera Bot. Čech.* 4: 67 (1947)
Conocybe macrocephala var. *riedheimensis* Hauskn., *Österr. Z. Pilzk.* 9: 95 (2000)
E: ! W: !
H: On soil.
British records (det. Hausknecht) from East Norfolk (Surlingham), London (Isle of Dogs), Middlesex (Isleworth), South Devon (Rousden), Surrey (Mickleham, Norbury Park) and Merionethshire (Lake Vyrnwy). Often misidentified as *C. brunneola* or *C. mesospora*.

brunnea Watling, *Persoonia* 6: 319 (1971)
Galera brunnea (J.E. Lange & Kühner) J.E. Lange [*nom. nud.*], *Dansk. Bot. Ark.* 9(6): 39 (1938)
Pholiotina brunnea (J.E. Lange & Kühner) Singer [*nom. nud.*], *Sydowia, Beih.* 7: 79 (1973)
E: ! S: !
H: On soil, decayed straw or grass, and on decayed wood
D: BFF3: 98-99, NM2: 273 **I:** C&D: 363 (as *Pholiotina brunnea*)
Rarely recorded or collected but apparently widespread.

brunneola Kühner & Watling, *Notes Roy. Bot. Gard. Edinburgh* 38: 333 (1980)
Conocybe mesospora var. *brunneola* Kühner [*nom. nud.*], *Encycl. Mycol. 7. Le Genre* Galera: 55 (1935)
E: ! S: ! W: ! NI: !
H: On soil in woodland and scrub, also at woodland edges.
D: BFF3: 58-59 **D+I:** B&K4: 300 377
Rarely reported. Apparently widespread.

candida (Cooke & Massee*)* Watling, *Kew Bull.* 31: 593 (1977)
Bolbitius candidus Cooke & Massee, *Grevillea* 21: 37 (1892)

E: !
H: On decayed and fermenting stable waste (manure and straw).
D+I: Hausknecht [*Österr. Z. Pilzk.* 7: 94-95 (1998)]
Collected in Surrey (Windsor Great Park) in 1999. Described from Australia (Victoria: Brighton) in 1892 and then not seen again until this collection.

coprophila (Kühner) Kühner, *Encycl. Mycol. 7. Le Genre Galera*: 125 (1935)
    Galera coprophila Kühner, *Botaniste* 17: 169 (1926)
    Pholiotina coprophila (Kühner) Singer, *Acta Inst. bot. Komarov. Acad. Sci. Plant. Crypt., Ser. 2,* 6: 434 (1950)
E: ! **S:** ! **W:** !
H: On weathered dung of various herbivores (cow and deer) in pastures, dunes and in montane habitat.
D: BFF3: 86, NM2: 276 **D+I:** FRIC7: 23 52c, B&K4: 316 400
Rarely reported. Supposedly 'fairly common and widespread especially northwards' *fide* BFF3.

cyanopus (G.F. Atk.) Kühner, *Encycl. Mycol. 7. Le Genre Galera*: 128 (1935)
    Galerula cyanopus G.F. Atk., *Proc. Amer. Philos. Soc.* 57: 367 (1918)
E: ! **W:** !
H: On soil.
D: BFF3: 85
Accepted with doubt. Mentioned as British in BFF3 (p. 85) and there are records from England (Leicestershire) and Wales (Pembrokeshire) but no material has been traced.

dentatomarginata Watling, *Notes Roy. Bot. Gard. Edinburgh* 38(2): 333 (1980)
    Conocybe appendiculata f. *macrospora* Kühner [*nom. nud.*], *Encycl. Mycol. 7. Le Genre Galera*: 149 (1935)
E: ! **S:** !
H: On soil in deciduous woodland or scrub.
D: BFF3: 101, NM2: 273
Rarely reported.

dumetorum (Velen.) Svrček, *Česká Mykol.* 10: 175 (1956)
    Galera dumetorum Velen., *České Houby* III: 541 (1921)
    Galera laricina Kühner, *Botaniste* 17: 170 (1925)
    Conocybe laricina (Kühner) Kühner, *Encycl. Mycol. 7. Le Genre Galera*: 51 (1935)
E: ! **S:** !
H: On calcareous loam soils, usually under *Fagus* in old woodland, and often attached to old decayed stems of *Mercurialis perennis* or old worm casts.
D: BFF3: 83, NM2: 274
Rarely reported but apparently widespread. Perhaps more frequent in southern and south eastern England but also reported from northern areas. Easily overlooked.

dunensis T.J. Wallace, *Trans. Brit. Mycol. Soc.* 43: 192 (1960)
E: ! **S:** ! **W:** ! **NI:** ! **ROI:** ! **O:** Channel Islands: ! Isle of Man: !
H: On sand or sandy soil in dunes, often associated with *Ammophila arenaria*.
D: NCL3: 192, BFF3: 53-54 **I:** C&D: 365
Widespread, and supposedly 'common' in coastal dunes (*fide* BFF3) but rarely reported.

echinata (Velen.) Singer, *Fieldiana, Bot.* 21: 103 (1989)
    Galera echinata Velen., *Novit. mycol. Novissimae*: 69 (1947)
    Conocybe sordida Kühner & Watling, *Notes Roy. Bot. Gard. Edinburgh* 38(2): 339 (1980)
E: ! **S:** !
H: On soil, often amongst grass, in woodland and copses.
D: BFF3: 62-63 (as *Conocybe sordida*), NM2: 276 (as *C. sordida*)
Rarely reported.

exannulata Kühner & Watling, *Notes Roy. Bot. Gard. Edinburgh* 38: 334 (1980)
    Conocybe blattaria f. *exannulata* Kühner [*nom. nud.*], *Encycl. Mycol. 7. Le Genre Galera*: 153 (1935)
    Pholiotina exannulata (Kühner) M.M. Moser [*nom. nud.*], *Kleine Kryptogamenflora, Edn* 2: 222 (1955)
E: ! **S:** !

H: On soil in deciduous woodland
D: BFF3: 102, NM2: 272
Rarely reported. Known in scattered locations in England (Cumbria southwards) and rarely in Scotland.

excedens *var.* **pseudomesospora** Singer & Hauskn., *Pl. Syst. Evol.* 180(1-2): 95 (1992)
E: !
H: On soil. British collection from a garden.
A single British collection (det. Hausknecht) from Middlesex (Isleworth).

farinacea Watling, *Notes Roy. Bot. Gard. Edinburgh* 25: 311 (1964)
E: ! **S:** !
H: On weathered dung of herbivores (cattle and horses).
D: BFF3: 65, NM2: 275
Rarely collected or reported. Described from Scotland (Mid-Perthshire: Rannoch, Dall Wood). Reported also from southern England but few records accompanied by voucher material.

filaris (Fr.) Kühner, *Encycl. Mycol. 7. Le Genre Galera*: 159 (1935)
    Agaricus togularis var. *filaris* Fr., *Ic. Hymenomyc.* 2: pl. 104 f. 4 (1884)
    Pholiota filaris (Fr.) Peck, *Bull. New York State Mus. Nat. Hist.* 122: 144 (1908)
    Pholiota togularis var. *filaris* (Fr.) J.E. Lange, *Dansk. Bot. Ark.* 2(11): 7 (1921)
    Pholiotina filaris (Fr.) Singer, *Beih. Bot. Centralbl.* 56: 170 (1936)
    Mis.: *Agaricus mycenoides* sensu Cooke [*Ill. Brit. fungi* 405 (1884)]
E: o **S:** ! **W:** ! **NI:** !
H: On soil in deciduous woodland and scrub, and occasionally on mulched flowerbeds in parks and gardens.
D: BFF3: 92-93, NM2: 273 **D+I:** Ph: 154
Rarely reported but widespread. Following BFF3, the relatively robust species illustrated under this name in B&K4: 316, pl. 401 is here distinguished as *Conocybe rugosa*.

fimetaria Watling, *Bol. Soc. Micol. Madrid* 11(1): 92 (1986)
    Mis.: *Conocybe siliginea* var. *neoantipus* sensu Kühner (1935)
E: !
H: On weathered horse dung.
Known only from the type collection from Bedfordshire (Stockgrove).

fuscimarginata (Murrill) Singer, *Nova Hedwigia Beih.* 29: 210 (1969)
    Galera fuscimarginata Murrill, *Lloydia* 5: 148 (1942)
    Mis.: *Conocybe siliginea* sensu auct. Brit.
E: ! **S:** !
H: On manured soil, or rarely dung.
D: BFF3: 71-72 **D+I:** B&K4: 302 379
Rarely reported.

hadrocystis (Kits van Wav.) Watling, *Notes Roy. Bot. Gard. Edinburgh* 38(2): 354 (1980)
    Conocybe arrhenii var. *hadrocystis* Kits van Wav., *Persoonia* 6: 160 (1970)
E: ! **S:** !
H: On bare soil, often amongst grass or herbs in woodland, along woodland edges or in clearings.
D: BFF3: 93-94 **D+I:** Myc7(4): 170
Rarely reported. Previously not differentiated from *Conocybe togularis* or *C. filaris.*

hornana Singer & Hauskn., *Beitr. Kenntn. Pilze Mitteleurop.* 5: 87 (1989)
E: !
H: British material on manured soil on flowerbeds. Originally described from decayed straw and organic debris.
A single British collection from Surrey (Kew, Royal Botanic Gardens) in 1990, originally determined as *C. kuehneriana,* recently redetermined by Hausknecht.

incarnata (Jul. Schäff.) Hauskn. & Arnolds, *Persoonia* 18(2): 246 (2003)
>*Galera incarnata* Jul. Schäff., *Z. Pilzk.* 9: 165 (1930)
>Mis.: *Conocybe fragilis* sensu Watling (BFF3: 76)

E: !
H: On soil in areas disturbed by human activity. British material collected along a roadside.
D: BFF3: 76 (as *Conocybe fragilis*)
Known from Yorkshire (Scarborough).

inocybeoides Watling, *Notes Roy. Bot. Gard. Edinburgh* 38(2): 350 (1980)
E: ! **S:** ?
H: On calcareous or clay soil in deciduous woodland, frequently growing in old wheel ruts.
D: BFF3: 77-78
Known from South Devon, Staffordshire, Surrey and West Sussex. Reported also from Scotland.

intrusa (Peck) Singer, *Sydowia* 4: 133 (1950)
>*Cortinarius intrusus* Peck, *Bull. Torrey Bot. Club* 23: 146 (1896)

H: On compost or composted soil in greenhouses (in botanic gardens) and gardens.
D: BFF3: 81-82, NM2: 274 **D+I:** FRIC3: 23-26 23a, B&K4: 302 380, Moss & Jackson [*Mycologist* 15(4): 155-156 (2001)]
An alien that has succeeded in establishing itself in greenhouses in botanic gardens, but has also recently been collected outdoors on compost.

juniana (Velen.) Hauskn. & Svrček, *Österr. Z. Pilzk., N.S.* 8: 46 (1999)
>*Galera juniana* Velen., *Novit. mycol. Novissimae*: 68 (1947)
>*Conocybe magnicapitata* P.D. Orton, *Trans. Brit. Mycol. Soc.* 43(2): 193 (1960)
>*Conocybe spicula* f. *macrospora* Kühner, *Encycl. Mycol. 7. Le Genre* Galera: 63 (1935)
>*Galera tenera* f. *minor* J.E. Lange, *Dansk. Bot. Ark.* 9(6): 37 (1938)

E: ! **S:** ! **W:** !
H: On soil, often amongst grass in grassland and in woodland glades.
D: BFF3: 61-62 (as *Conocybe magnicapitata*), NM2: 276 (as *C. magnicapitata*) **D+I:** B&K4: 304 382 (as *C. magnicapitata*)
Rarely reported but apparently widespread.

juniana *var.* **sordescens** (P.D. Orton) Hauskn. & Svrček, *Österr. Z. Pilzk., N.S.* 8: 50 (1999)
>*Conocybe sordescens* P.D. Orton, *Trans. Brit. Mycol. Soc.* 91(4): 546 (1988)

E: !
H: On calcareous soil, in deciduous woodland.
Rarely collected or reported. Known from Somerset and South Devon.

lenticulospora Watling, *Notes Roy. Bot. Gard. Edinburgh* 38(2): 351 (1980)
E: ! **S:** ! **W:** !
H: On weathered horse dung and on composted flower beds.
D: BFF3: 76-77
Known from Scotland (several widespread localities), England (South Hampshire: Setley and Surrey: Kew Gardens) and Wales (Carmarthenshire: Tywyn Burrows). Reported from North Somerset and South Hampshire but unsubstantiated with voucher material.

leucopus Watling, *Notes Roy. Bot. Gard. Edinburgh* 40(3): 539 (1983)
>*Conocybe leucopoda* Kühner [*nom. nud.*], *Encycl. Mycol. 7. Le Genre* Galera: 82 (1935)

E: !
H: On nutrient-poor sandy or clay soil, often amongst grass.
D: BFF3: 64-65
Known from Co. Durham and West Sussex and reported elsewhere but unsubstantiated with voucher material.

macrocephala Kühner & Watling, *Notes Roy. Bot. Gard. Edinburgh* 38: 335 (1980)

>*Conocybe tenera* f. *macrocephala* Kühner [*nom. nud.*], *Encycl. Mycol. 7. Le Genre* Galera: 73 (1935)
>*Conocybe abruptibulbosa* Watling, *Notes Roy. Bot. Gard. Edinburgh* 38: 345 (1980)

E: ! **S:** !
H: On soil, often amongst grass, in deciduous woodland, scrub, at the sides of paths or along woodland edges.
D: BFF3: 54-55
Rarely reported but apparently widespread.

macrospora (G.F. Atk.) Hauskn., *Österr. Z. Pilzk., N. S.* 12: 64 (2003)
>*Galerula macrospora* G.F. Atk., *The genus* Galerula *in North America*: 371 (1918)
>*Conocybe rubiginosa* Watling, *Notes Roy. Bot. Gard. Edinburgh* 38(2): 353 (1980)
>*Conocybe tenera* f. *bispora* M. Sass, *Amer. J. Bot.* 16: 692 (1929)

E: ! **S:** !
H: On soil.
D+I: B&K4: 308 389 (as *Conocybe rubiginosa*)
Rarely reported. Perhaps better known as *C. rubiginosa* which was described from Mid-Perthshire (Drummond Hill).

mairei Kühner ex Watling, *Biblioth. Mycol.* 61: 41 (1977)
>*Conocybe mairei* Kühner [*nom. nud.*], *Encycl. Mycol. 7. Le Genre* Galera: 131 (1935)
>*Pholiotina mairei* (Kühner) Singer [*nom. inval.*], *Acta Inst. bot. Komarov. Acad. Sci. Plant. Crypt., Ser. 2*, 6: 435 (1950)
>*Galera mairei* (Kühner) J.E. Lange [*nom. inval.*], *Dansk. Bot. Ark.* 9: 40 (1938)

E: ! **S:** !
H: On soil in deciduous woodland, often in herb rich areas on loam soil.
D: BFF3: 87-88, NM2: 277
Rarely reported but apparently widespread.

merdaria Arnolds & Hauskn., *Persoonia* 18(2): 239 (2003)
>Mis.: *Conocybe ambigua* sensu auct. Brit. p.p.

E: !
H: On nitrogenous soil in scrub or grassy areas in woodland.
Known from Hertfordshire and Surrey. Collections were originally named *C. ambigua* but recently redetermined by Hausknecht.

mesospora Kühner & Watling, *Notes Roy. Bot. Gard. Edinburgh* 38: 336 (1980)
>*Conocybe mesospora* f. *typica* Kühner [*nom. nud.*], *Encycl. Mycol. 7. Le Genre* Galera: 58 (1935)

E: ! **S:** ! **W:** ?
H: On soil in copses, woodland glades or along woodland edges.
D: BFF3: 59
Rarely reported but apparently widespread.

moseri Watling, *Notes Roy. Bot. Gard. Edinburgh* 38(2): 342 (1980)
>*Conocybe kuehneri* Singer [*nom. nud.*], *Collect. Bot. (Barcelona)* 1: 236 (1947)
>Mis.: *Conocybe plumbeitincta* sensu NCL and sensu Moser (1967) non Phillips (Ph)

E: !
H: On soil often amongst grass in fields, gardens and at woodland edges.
D: BFF3: 72-73 **D+I:** B&K4: 304 383 (as *Conocybe moseri* var. *moseri*)
Rarely collected or reported and no voucher material has been traced.

murinacea Watling, *Notes Roy. Bot. Gard. Edinburgh* 38(2): 352 (1980)
E: ! **S:** !
H: On weathered horse dung.
D: BFF3: 79
Known from England (Surrey: Kew Gardens) and Scotland (Mid Ebudes: Isle of Mull).

percincta P.D. Orton, *Trans. Brit. Mycol. Soc.* 43(2): 194 (1960)

Mis.: *Conocybe teneroides* sensu auct.
E: ! **S:** !
H: On soil in deciduous woodland or parkland, also on decayed straw or stable waste, decayed sawdust and composted flowerbeds in gardens.
D: BFF3: 94-95 **D+I:** B&K4: 318 403 (as *Conocybe teneroides*)
Rarely recorded or collected but widespread. Possibly only a stout form of *Conocybe blattaria*, united by some authorities under the name *C. teneroides*.

pilosella (Pers.) Kühner, *Encycl. Mycol. 7. Le Genre* Galera: 92 (1935)
  Agaricus pilosellus Pers., *Syn. meth. fung.*: 387 (1801)
  Agaricus tener β *pilosellus* (Pers.) Fr., *Syst. mycol.* 1: 266 (1821)
  Galera tenera var. *pilosella* (Pers.) P. Kumm., *Führ. Pilzk.*: 75 (1871)
  Galera pilosella (Pers.) Rea, *Brit. basidiomyc.*: 407 (1922)
  Conocybe piloselloides Watling, *Notes Roy. Bot. Gard. Edinburgh* 40(3): 549 (1983)
E: ! **S:** ! **W:** !
H: On soil in deciduous woodland, and also in grassland.
D: BFF3: 70-71 **D+I:** B&K4: 306 385
Rarely reported. Apparently widespread.

pinetorum Watling, Esteve-Rav. & G. Moreno, *Bol. Soc. Micol. Madrid* 11(1): 85 (1986)
E: ! **S:** ! **W:** !
H: British material on wet, decayed sawdust and soil.
Known from England (Surrey: Windsor Great Park) and Scotland (Easterness: Abernethy Forest and Mid-Perthshire: Rannoch, Dall Wood). Close to *Conocybe pubescens* but incompatible in culture.

plicatella (Peck) Kühner, *Encycl. Mycol. 7. Le Genre* Galera: 137 (1935)
  Agaricus plicatellus Peck, *Rep. (Annual) New York State Mus. Nat. Hist.* 29: 66 (1878)
  Galerella plicatella (Peck) Singer, *Lilloa* 22: 490 (1951) [1949]
E: ! **S:** !
H: On soil amongst grass in lawns, parkland and unimproved grassland.
D: BFF3: 84, NM2: 277
Rarely collected or reported.

pubescens (Gillet) Kühner, *Encycl. Mycol. 7. Le Genre* Galera: 85 (1935)
  Galera pubescens Gillet, *Hyménomycètes*: 553 (1876)
  Galera cryptocystis G.F. Atk., *Proc. Amer. Philos. Soc.* 75: 368 (1918)
  Conocybe cryptocystis (G.F. Atk.) Singer, *Sydowia* 8: 125 (1954)
  Mis.: *Conocybe pilosella* sensu Rea (1922)
  Mis.: *Conocybe pinetorum* sensu auct. Brit. p.p.
  Mis.: *Conocybe pseudopilosella* sensu auct. Brit. p.p.
  Mis.: *Conocybe subpubescens* sensu auct.
E: o **S:** o **W:** ! **ROI:** !
H: On weathered dung of herbivores, usually in woodland.
D: BFF3: 66-67, NM2: 275
Occasional but apparently widespread.

pulchella (Velen.) Hauskn. & Svrček, *Czech Mycol.* 51(1): 58 (1999)
  Galera pulchella Velen., *České Houby* III: 543 (1921)
  Galera digitalina Velen., *Novit. mycol. Novissimae*. 70 (1947)
  Conocybe digitalina (Velen.) Singer, *Fieldiana, Bot.* 21: 103 (1989)
  Conocybe pseudopilosella Kühner & Watling, *Notes Roy. Bot. Gard. Edinburgh* 38: 336 (1980)
  Conocybe pubescens var. *pseudopilosella* Kühner [*nom. nud.*], *Encycl. Mycol. 7. Le Genre* Galera: 89 (1935)
  Mis.: *Conocybe subpubescens* sensu auct. Brit. p.p.
  Mis.: *Conocybe tenera* sensu auct. Brit. p.p.
E: ! **S:** o **W:** ! **ROI:** !
H: On soil in woodland, along woodland rides and edges, also in unimproved grassland, meadows and lawns.

D: BFF3: 66 (as *Conocybe pseudopilosella*) **D+I:** B&K4: 306 386 (as *C. pseudopilosella*)
Rarely reported. Apparently widespread but confused with similar taxa. Vouchered records suggest it is not uncommon in Scotland.

pygmaeoaffinis (Fr.) Kühner, *Encycl. Mycol. 7. Le Genre* Galera: 133 (1935)
  Agaricus pygmaeoaffinis Fr., *Monogr. hymenomyc. Suec.* 1: 389 (1857)
  Galera pygmaeoaffinis (Fr.) Quél., *Mém. Soc. Émul. Montbéliard, Sér. 2,* 5: 135 (1872)
  Galerula pygmaeoaffinis (Fr.) Maire, *Publ. Inst. Bot. Barcelona* 15: 94 (1933)
  Pholiotina pygmaeoaffinis (Fr.) Singer, *Acta Inst. bot. Komarov. Acad. Sci. Plant. Crypt., Ser. 2* 6: 435 (1950)
  Conocybe friesii S. Lundell, *Fungi Exsiccati Suecici* 41-42: 2048 (1953)
E: !
H: On soil, in deciduous woodland on calcareous soil, also occasionally in grassland and on lawns.
D: BFF3: 8-89, NM2: 277 **D+I:** B&K4: 316 402
Rarely reported. Known from Bedfordshire, Dorset, Herefordshire, Isle of Wight, Middlesex, North & South Wiltshire, South Essex, Surrey, Warwickshire and West Kent. Easily confused with *C. striaepes*.

rickeniana P.D. Orton, *Trans. Brit. Mycol. Soc.* 43(2): 195 (1960)
  Agaricus teneroides Peck, *Rep. (Annual) New York State Mus. Nat. Hist.* 29: 39 (1898)
  Galera teneroides (Peck) Sacc., *Syll. fung.* 5: 861 (1887)
  Mis.: *Galera spartea* sensu Wakefield & Dennis [*Common British Fungi*: 170 (1950)]
  Mis.: *Galera spicula* sensu Ricken (1915: 226) and sensu Rea (1922)
E: ! **S:** ! **W:** ! **ROI:** !
H: On soil, often amongst grass, in deciduous woodland and copses, at path edges and occasionally in unimproved grassland.
D: BFF3: 59-60, NM2: 276 **D+I:** B&K4: 306 387
Apparently widespread but often confused with similar species of *Conocybe*.

rickenii (Jul. Schäff.) Kühner, *Encycl. Mycol. 7. Le Genre* Galera: 115 (1935)
  Galera rickenii Jul. Schäff., *Z. Pilzk.* 9(11): 171 (1930)
  Mis.: *Galera pygmaeoaffinis* sensu Ricken (1915)
  Mis.: *Conocybe siliginea* sensu auct.
E: ! **S:** ! **ROI:** ! **O:** Isle of Man: !
H: On weathered dung of herbivores, also on mulched soil in parks and gardens, and on loam soils in open deciduous woodland.
D: BFF3: 73-74, NM2: 274 **D+I:** Ph: 154, B&K4: 308 388
Rarely reported but often confused with similar taxa.

rugosa (Peck) Watling, *Biblioth. Mycol.* 82: 133 (1981)
  Pholiota rugosa Peck, *Rep. (Annual) New York State Mus. Nat. Hist.* 50: 102 (1897)
  Pholiotina rugosa (Peck) Singer, *Pap. Michigan Acad. Sci.* 32: 148 (1946)
E: ! **S:** !
H: On soil, or soil mixed with woodchips, mulched flowerbeds and on compost, in parkland, gardens, track and path edges and other disturbed areas.
D: BFF3: 95-96 **D+I:** B&K4: 316 401 (as *Conocybe filaris*)
Occasional but widespread. Considered by some authorities to be only a stout form of *C. filaris*.

sabulicola Hauskn. & Enderle, *Z. Mykol.* 58(2): 203 (1992)
S: !
H: On sand or sandy soil in dunes.
This is the 'unnamed species' noted under *C. dunensis* in BFF3: 54. Apparently 'not infrequent in the right habitat', at least in Scotland *fide* R. Watling where there are records from Midlothian and Orkney.

semiglobata Kühner & Watling, *Notes Roy. Bot. Gard. Edinburgh* 38: 337 (1980)
  Conocybe tenera f. *semiglobata* Kühner [*nom. nud.*], *Encycl. Mycol. 7. Le Genre* Galera: 79 (1935)
  Galera tenera f. *convexa* J.E. Lange, *Dansk. Bot. Ark.* 9(6): 37 (1938)
E: ! **S:** !
H: On soil, amongst grass in grassland, on lawns also on dunes, in copses and along woodland edges.
D: BFF3: 55-56, NM2: 276 **D+I:** B&K4: 308 390
Rarely collected or reported but apparently widespread. Previously confused with *C. tenera* and allied taxa.

siennophylla (Berk. & Broome) Singer, *Sydowia* 9: 402 (1955)
  Agaricus siennophyllus Berk. & Broome, *J. Linn. Soc., Bot.* 11: 545 (1871)
  Naucoria siennophylla (Berk. & Broome) Sacc., *Syll. fung.* 5: 858 (1887)
  Mis.: *Conocybe ochracea* sensu auct.
  Mis.: *Conocybe plumbeitincta* sensu Singer [*Mycologia* 51: 396 (1959)]
E: ! **S:** ! **NI:** ! **ROI:** !
H: On soil, weathered manure and decayed stable waste, old woodchips and sawdust, in woodland and grassland areas, on wasteground and in areas disturbed by human activity.
D: BFF3: 74-75, NM2: 275 **D+I:** B&K4: 310 391
Rarely reported but apparently widespread. Previously confused with similar taxa.

siliginea (Fr.) Kühner, *Encycl. Mycol. 7. Le Genre* Galera: 96 (1935)
  Agaricus siligineus Fr., *Observ. mycol.*: 168 (1818)
  Agaricus tener γ *siligineus* (Fr.) Fr., *Syst. mycol.* 1: 266 (1821)
  Galera tenera var. *siliginea* (Fr.) P. Kumm. [as *G. tenera* var. *salignea*], *Führ. Pilzk.*: 75 (1871)
  Galera siliginea (Fr.) Quél., *Mém. Soc. Émul. Montbéliard, Sér. 2,* 5: 136 (1872)
E: ! **S:** !
H: On soil amongst grass
D: BFF3: 75, NM2: 275
Rarely reported but apparently widespread.

striaepes (Cooke) S. Lundell, *Fungi Exsiccati Suecici* 41-42: 2049 (1953)
  Agaricus striaepes Cooke, *Ill. Brit. fung.* 478 (502) Vol. 4 (1885)
  Naucoria striaepes (Cooke) Sacc., *Syll. fung.* 5: 839 (1887)
  Mis.: *Conocybe pygmaeoaffinis* sensu auct.
E: !
H: On soil in deciduous woodland, grassland, parkland and in gardens.
D: BFF3: 89, NM2: 277 (as *Conocybe striipes*)
Rarely reported.

subovalis Kühner & Watling, *Notes Roy. Bot. Gard. Edinburgh* 38(2): 340 (1980)
  Conocybe tenera var. *subovalis* Kühner [*nom. nud.*], *Encycl. Mycol. 7. Le Genre* Galera: 69 (1935)
  Agaricus ovalis Fr., *Monogr. hymenomyc. Suec.* 1: 389 (1857)
  Galera ovalis (Fr.) Gillet, *Hyménomycètes*: 554 (1876)
  Mis.: *Galera tenera* sensu Lange (FlDan4: 34)
E: ! **S:** ! **W:** ! **NI:** ! **ROI:** !
H: On soil in deciduous woodland, and along tracksides.
D: BFF3: 56-57, NM2: 276 **D+I:** B&K4: 310 393
Frequently reported and apparently widespread yet little voucher material in herbaria.

subpubescens P.D. Orton, *Trans. Brit. Mycol. Soc.* 43(2): 195 (1960)
  Mis.: *Conocybe cryptocystis* sensu auct.
  Mis.: *Conocybe digitalina* sensu auct.
  Mis.: *Conocybe pinetorum* sensu auct. Brit. p.p.
  Mis.: *Conocybe pubescens* sensu auct. Brit. p.p.
  Mis.: *Galera pubescens* sensu Lange (FlDan4: 34)
  Mis.: *Galera tenera* sensu Ricken (1915: 225)

E: ! **S:** !
H: On nitrogenous soil, or weathered dung of herbivores in woodland, on heathland and rarely in grassland.
D: BFF3: 68-69 (as *Conocybe subpubescens*), NM2: 275 (as *C. subpubescens*) **D+I:** B&K4: 300 378 (as *C. subpubescens*)
Described from Surrey (Esher) but rarely reported.

sulcatipes (Peck) Kühner, *Encycl. Mycol. 7. Le Genre* Galera: 127 (1935)
  Agaricus sulcatipes Peck, *Rep. (Annual) New York State Mus. Nat. Hist.* 35: 132 (1884)
  Galera sulcatipes (Peck) Sacc., *Syll. fung.* 5: 866 (1887)
S: ! **W:** !
H: On soil amongst grass and mosses in upland grassland (hill pasture).
D: BFF3: 86-87
Rarely reported.

tenera (Schaeff.) Fayod, *Ann. Sci. Nat., Bot.,* sér. 7, 9: 357 (1889)
  Agaricus tener Schaeff. [non *A. tener* Sowerby (1796)], *Fung. Bavar. Palat. nasc.* 1: 70 (1762)
  Galera tenera (Schaeff.) P. Kumm., *Führ. Pilzk.*: 75 (1871)
  Galera tenera f. *typica* Kühner, *Encycl. Mycol. 7. Le Genre* Galera: 68 (1935)
  Galera tenera f. *tenella* J.E. Lange, *Dansk. Bot. Ark.* 9(6): 37 (1938)
E: ! **S:** ! **W:** ! **NI:** ! **ROI:** ! **O:** Channel Islands: !
H: On soil, often amongst grass, in parkland, woodland and gardens.
D: BFF3: 57-58 **D+I:** B&K4: 312 394
Apparently frequent and widespread. However, the name has been widely misapplied and variously interpreted, having been used for at least six other taxa.

umbonata (Massee) Watling, *Notes Roy. Bot. Gard. Edinburgh* 26: 294 (1966)
  Bolbitius umbonatus Massee, *Bull. Misc. Inform. Kew* 1906: 46 (1906)
E: !
H: Described from tan in a propagating pit.
Known only from the type collection from Surrey (Kew, Royal Botanic Gardens). Perhaps an alien.

utriformis P.D. Orton, *Trans. Brit. Mycol. Soc.* 43(2): 196 (1960)
  Pholiotina subnuda (Kühner) Singer [*nom. nud.*], *Beih. Bot. Centralbl.* 56: 170 (1936)
  Conocybe subnuda Watling, *Notes Roy. Bot. Gard. Edinburgh* 40(3): 553 (1983)
E: !
H: On soil in marshland, partially dried out lakes or ponds and occasionally on damp soil in woodland.
D: BFF3: 89-90 (as *Conocybe utriformis*), NM2: 277 (as *C. utriformis*)
Rarely reported and known only from Middlesex and Norfolk. Reported from Scotland but unsubstantiated by voucher material.

velata (Velen.) Watling, *Kew Bull.* 59(1): 168 (2004)
  Galera velata Velen., *České Houby* III: 547 (1921)
  Pholiotina velata (Velen.) Hauskn., *Česká Mykol.* 51(1): 66 (1999)
  Conocybe appendiculata Watling, *Persoonia* 6: 329 (1971)
  Pholiotina appendiculata (Watling) Courtec. [*nom. nud.*] *Doc. Mycol.* 16 (61): 47 (1985)
  Galera appendiculata (Watling) J.E. Lange [*nom. nud.*], *Dansk. Bot. Ark.* 9: 39 (1938)
  Mis.: *Galera ravida* sensu Ricken (1915)
E: o **S:** !
H: On soil, often calcareous loam, in woodland, copses and scrub.
D: BFF3: 100-101 (as *Conocybe appendiculata*), NM2: 273 (as *C. appendiculata*) **D+I:** B&K4: 314 397 (as *C. appendiculata*) **I:** C&D: 363 (as *Pholiotina appendiculata*)
Rarely reported but apparently widespread. Perhaps better known as *C. appendiculata*.

velutipes (Velen.) Hauskn. & Svrček, *Czech Mycol.* 51(1): 68 (1999)
  Galera velutipes Velen., *Novit. mycol.*: 128 (1910)
  Conocybe kuehneriana Singer, *Nova Hedwigia Beih.* 29: 212 (1969)
  Mis.: *Conocybe ochracea* sensu NCL and sensu Phillips (Ph)
E: ! **S:** ! **NI:** !
H: On soil in woodland glades, along the edges of footpaths and in unimproved grassland, also grassy areas on dunes.
D: BFF3: 72 **I:** Ph: 155 (as *Conocybe ochracea*)
Apparently widespread. Often recorded as *C. ochracea* or *C. kuehneriana*.

vestita (Fr.) Kühner, *Encycl. Mycol. 7. Le Genre* Galera: 155 (1935)
  Galera vestita Fr., *Mém. Soc. Émul. Montbéliard, Sér. 2*, 5: 235 (1872)
  Pholiotina vestita (Fr.) Singer, *Beih. Bot. Centralbl.* 56: 170 (1936)
E: !
H: On loam soils in herb rich woodland.
D: BFF3: 99-100, NM2: 273 **D+I:** B&K4: 318 404 **I:** C&D: 363 (as *Pholiotina vestita*)
Rarely reported.

vexans P.D. Orton, *Trans. Brit. Mycol. Soc.* 43(2): 197 (1960)
  Mis.: *Conocybe blattaria* sensu auct. Eur.
  Mis.: *Pholiota togularis* sensu Ricken (1915: 199)
E: ! **S:** ! **NI:** ! **ROI:** !
H: On loam soil in deciduous woodland, often amongst *Mercurialis perennis* and other herbs.
D: BFF3: 96-97, NM2: 273 **D+I:** Myc7(4): 171
Rarely reported but apparently widespread. Previously confused with *Conocybe blattaria* and similar taxa.

watlingii Hauskn., *Österr. Z. Pilzk.* 5: 193 (1996)
  Mis.: *Conocybe neoantipus* sensu Watling [Bull. Soc. Micol. Madrid 11 (1): 92 (1986)]
E: ! **S:** !
H: On weathered horse dung.
Only three British records (as *C. neoantipus*). This is the 'rooting *C. subpubescens*' noted in BFF3 (p. 69).

CONTUMYCES Redhead, Moncalvo, Vilgalys & Lutzoni, *Mycotaxon* 82: 161 (2002)
Agaricales, Tricholomataceae
Type: *Contumyces rosellus* (M.M. Moser) Redhead, Moncalvo, Vilgalys & Lutzoni

rosellus (M.M. Moser) Redhead, Moncalvo, Vilgalys & Lutzoni, *Mycotaxon* 82: 161 (2002)
  Omphalia rosella J.E. Lange [*nom. illegit.*, non *O. rosella* (Batsch) Gray (1821)], *Dansk. Bot. Ark.* 6(5): 14 (1930)
  Clitocybe rosella M.M. Moser, *Sydowia* 4: 100 (1950)
  Marasmiellus rosellus (M.M. Moser) Kuyper & Noordel., *Proceedings International Symposium on Tricholomataceae, Borgo Taro, 1984*: 100 (1986)
  Omphalina rosella (M.M. Moser) Redhead *et al.*, *Mycologia* 87(6): 880 (1995)
  Mycena carnicolor P.D. Orton, *Trans. Brit. Mycol. Soc.* 43(2): 178 (1960)
E: ! **S:** ! **W:** !
H: On soil amongst grass in pastures, on lawns, dunes and in open woodland.
D: NM2: 142, A&N1 165, FAN3: 126 **I:** Bon: 129 (as *Omphalina rosella*)
Rarely reported but apparently widespread.

COPRINUS Pers., *Tent. disp. meth. fung.*: 62 (1797)
Agaricales, Coprinaceae
  Coprinellus P. Karst., *Bidrag Kännedom Finlands Natur Folk* 32: 542 (1879)
  Coprinopsis P. Karst., *Acta Soc. Fauna Flora fenn.* 2(1): 27 (1881)

  Pseudocoprinus Kühner, *Botaniste* 20: 155 (1928)
  Parasola Redhead, Vilgalys & Hopple, *Taxon* 50: 235 (2001)
Type: *Coprinus comatus* (O.F. Müll.) Pers.

acuminatus (Romagn.) P.D. Orton, *Notes Roy. Bot. Gard. Edinburgh* 29(1): 86 (1969)
  Coprinus atramentarius var. *acuminatus* Romagn., *Rev. Mycol. (Paris)* 16(2): 127 (1951)
E: o **S:** ! **W:** ! **NI:** ! **ROI:** ! **O:** Isle of Man: !
H: On decayed and often buried wood, frequently along the sides or edges of water-filled ditches or wheel ruts in deciduous woodland.
D: BFF2: 32, NBA3: 86-87
Occasional but widespread. Often confused with *C. atramentarius*.

alopecius Lasch, in Fries, *Epicr. syst. mycol.*: 248 (1838)
  Coprinus insignis Peck, *Bull. New York State Mus. Nat. Hist.* 1(2): 54 (1873)
E: !
H: Often caespitose on large decayed stumps or around the roots of living deciduous trees. Usually on *Fagus* but also known on *Acer pseudoplatanus*.
D: BFF2: 33 (as *Coprinus insignis*), NM2: 231 (as *C. alopecia*)
Known from Buckinghamshire, Cambridgeshire, North Essex and Surrey. Easily confused with *C. atramentarius* but distinguished by roughened or warted spores.

ammophilae Courtec., *Doc. Mycol.* 18(72): 76 (1988)
E: ! **S:** ! **W:** ! **NI:** !
H: On sand or sandy soil, associated with *Ammophila arenaria* in dunes.
I: C&D: 262, Myc14(3): 108
Rarely reported but apparently widespread.

amphithallus M. Lange & A.H. Sm., *Mycologia* 45: 774 (1953)
E: !
H: On soil, often amongst grass.
Rarely reported. Known from Berkshire, Buckinghamshire, Oxfordshire and West Sussex.

angulatus Peck, *Bull. New York State Mus. Nat. Hist.* 1(2): 54 (1873)
  Coprinus boudieri Quél., *Bull. Soc. Bot. France* 24: 321 (1877)
E: o **S:** ! **W:** ! **NI:** !
H: Solitary or gregarious, on burnt soil, and burnt wood buried in soil.
D: BFF2: 94, NM2: 229 **D+I:** B&K2: 224 264 **I:** Bon: 273, C&D: 260
Occasional but apparently widespread. Readily distinguished by the habitat and distinctive spore shape.

argenteus P.D. Orton, *Notes Roy. Bot. Gard. Edinburgh* 32: 139 (1972)
E: !
H: On calcareous soil amongst grass.
D: BFF2: 46
Known only from the type collection (Surrey: Mickleham) in 1956.

atramentarius (Bull.) Fr., *Epicr. syst. mycol.*: 243 (1838)
  Agaricus atramentarius Bull., *Herb. France*: pl. 164 (1783)
  Agaricus sobolifer Hoffm., *Nomencl. fung.* 1: 216 (1789)
  Coprinus soboliferus (Hoffm.) Fr., *Epicr. syst. mycol.*: 243 (1838)
  Coprinus atramentarius var. *soboliferus* (Hoffm.) Rea, *Brit. basidiomyc.*: 502 (1922)
  Agaricus luridus Bolton [*nom. illegit.*, non *A. luridus* Schaeff. (1762)], *Hist. fung. Halifax* 1: 25 (1788)
  Coprinus luridus (Bolton) Fr., *Epicr. syst. mycol.*: 243 (1838)
  Agaricus plicatus Pers., *Tent. disp. meth. fung.*: 62 (1797)
  Coprinus plicatus (Pers.) Gray, *Nat. arr. Brit. pl.* 1: 634 (1821)
  Mis.: *Agaricus fimetarius* sensu Sowerby [Col. fig. Engl. fung. 2: pl. 188 (1799)]
E: c **S:** c **W:** c **NI:** c **ROI:** c **O:** Channel Islands: !

H: On decayed wood of deciduous trees, often in large caespitose clusters. Sometimes on buried wood and then appearing as if terrestrial.
D: BFF2: 31, NM2: 231 **D+I:** Ph: 178 (colour poor), B&K4: 226 265 **I:** FlDan4: 157H, Bon: 271, C&D: 263
Common and widespread. Sometimes confused with the superficially similar *C. alopecius* which has roughened or warted spores. English name = 'Common Ink Cap'.

auricomus Pat., *Tab. anal. fung.*: 200 (1886)
   Coprinus hansenii J.E. Lange, *Dansk. Bot. Ark.* 2(3): 48 (1915)
E: c **S:** !
H: Singly or in small groups on soil in woodland and may be abundant on woodchip mulch in parks and gardens.
D: BFF2: 98, NM2: 229 (as *Coprinus hansenii*) **D+I:** B&K4: 226 266 **I:** Bon: 273, C&D: 261
Common and widespread but often misidentified as *C. plicatilis*. Possibly increasing with the use of woodchip mulch on flowerbeds. The presence of thick-walled brownish setae on the pileal disc is diagnostic.

bellulus Uljé, *Persoonia* 13(4): 481 (1988)
E: !
H: On soil amongst grass.
A single collection from Surrey (Kew, Royal Botanic Gardens) in 1993.

bisporiger P.D. Orton, *Notes Roy. Bot. Gard. Edinburgh* 35(1): 147 (1976)
   Mis.: *Coprinus bisporus* sensu Buller in TBMS 6:363 (1920)
E: ! **S:** !
H: On wood (sticks dredged from a pond and deposited on the banks).
D: BFF2: 97-98
Known only from the type material although a collection in herb. K, from West Sussex (Madehurst), labelled as '*C. sclerocystidiosus* 2-spored form', from soil at the base of a trunk of *Fagus*, may be this species.

bisporus J.E. Lange, *Dansk. Bot. Ark.* 2(3): 50 (1915)
E: ! **S:** o **W:** ! **O:** Channel Islands: !
H: On soil, dung or mixtures of both.
D: BFF2: 88 **D+I:** B&K4: 226 267
Rarely reported but apparently widespread. Macroscopically resembles *C. congregatus*.

callinus M. Lange & A.H. Sm., *Mycologia* 45: 770 (1953)
   Mis.: *Coprinus hiascens* sensu Romagnesi (Rev. Mycol. (Paris) 4: 119, 1941)
E: ! **S:** ! **NI:** !
H: On soil, often amongst grass and also on burnt ground.
D: BFF2: 93-94, NM2: 230
Rarely reported. Apparently widespread but easily mistaken for *C. hiascens*.

cinereofloccosus P.D. Orton, *Trans. Brit. Mycol. Soc.* 43(2): 198 (1960)
E: ! **S:** ! **W:** !
H: On soil, often amongst grass, singly or in small groups.
D: BFF2: 73-74 **D+I:** B&K4: 228 268
Rarely reported. Apparently widespread but records are usually unsubstantiated with voucher material.

cinereus (Schaeff.) Gray, *Nat. arr. Brit. pl.* 1: 634 (1821)
   Agaricus cinereus Schaeff., *Fung. Bavar. Palat. nasc.* 1: 100 (1762)
   Coprinus fimetarius var. *cinereus* (Schaeff.) Fr., *Epicr. syst. mycol.*: 246 (1838)
   Coprinus delicatulus Apinis, *Trans. Brit. Mycol. Soc.* 48(4): 653 (1965)
   Agaricus macrorhizus Pers. [non *A. macrorhizus* Lasch (1828)], *Syn. meth. fung.*: 398 (1801)
   Coprinus macrorhizus (Pers.) Rea, *Brit. basidiomyc.*: 503 (1922)
   Hormographiella aspergillata Guarro, Gené & De Vroey (anam.), *Mycotaxon* 45: 182 (1992)
   Mis.: *Coprinus fimetarius* sensu auct. Brit.

E: c **S:** c **W:** c **NI:** !
H: On dung and dungy straw heaps especially those heating up during fermentation. Also on piles of germinating grain and compost heaps.
D: BFF2: 42, NM2: 232 **D+I:** B&K4: 228 269 **I:** Bon: 271, C&D: 263
Common and widespread.

cinnamomeotinctus P.D. Orton, *Trans. Brit. Mycol. Soc.* 91(4): 547 (1988)
E: !
H: On calcareous soil, amongst mosses in woodland.
Known only from Surrey (Boxhill, Ashurst Valley and Mickleham).

comatus (O.F. Müll.) Pers., *Tent. disp. meth. fung.*: 66 (1797)
   Agaricus comatus O.F. Müll., *Fl. Danica*: t. 834 (1767)
   Agaricus ovatus Schaeff., *Fung. Bavar. Palat. nasc.* 1: 7 (1762)
   Coprinus comatus var. *ovatus* (Schaeff.) Fr., *Syst. mycol.* 1: 307 (1821)
   Coprinus ovatus (Schaeff.) Fr., *Epicr. syst. mycol.*: 242 (1838)
   Coprinus comatus var. *caprimammillatus* Bogart, *Mycotaxon* 4(1): 274 (1976)
   Agaricus cylindricus Sowerby, *Col. fig. Engl. fung.* 2: pl. 189 (1799)
   Agaricus fimetarius Bolton, *Hist. fung. Halifax* 1: 44 (1788)
E: c **S:** c **W:** c **NI:** c **ROI:** c **O:** Channel Islands: c Isle of Man: c
H: On soil especially where disturbed by man, or mixed with woody debris, often amongst grass at roadsides, on lawns, and on rubbish tips.
D: BFF2: 29, NM2: 231 **D+I:** Ph: 176-177, B&K4: 228 270 **I:** FlDan4: 156E, Bon: 271, C&D: 263
Very common and widespread. English name = 'Shaggy Ink Cap' or 'Lawyer's Wig'.

congregatus (Bull.) Fr., *Epicr. syst. mycol.*: 249 (1838)
   Agaricus congregatus Bull., *Herb. France*: pl. 94 (1786)
E: ! **S:** !
H: On dungy soil or mixtures of dung and woodchips or straw.
D: BFF2: 88-89 **D+I:** Ph: 179
Rarely reported and poorly known in Britain.

coniophorus Romagn., *Rev. Mycol. (Paris)* 6(3-4): 126 (1941)
E: !
H: On decayed wood, in woodland on calcareous soil. British material on *Fagus*.
Known from Surrey (Mickleham, Norbury Park) and Oxfordshire (Bix, Warburg Reserve). Basidiomes are distinctive in the field, resembling a heavily veiled *C. disseminatus*, the veil fragments so abundant that they readily drop off, littering the area below.

cordisporus Gibbs, *Naturalist (Hull)* 1908: 100 (1908)
   Coprinus volvaceominimus Crossl., *Naturalist (Hull)* 1892: 372 (1892)
   Mis.: *Coprinus patouillardii* sensu NCL
E: c **S:** c **W:** c **O:** Isle of Man: !
H: On dung of various herbivores, often on relatively fresh deposits.
D: BFF2: 66-67, NM2: 233, NM2: 233 **I:** FlDan4: 159E
Rarely collected or recorded, but apparently widespread. The minute, fragile basidiomes are difficult to collect or preserve.

cortinatus J.E. Lange, *Dansk. Bot. Ark.* 2(3): 45 (1915)
E: ! **W:** ! **NI:** ?
H: On soil, often calcareous, and often under *Mercurialis perennis* in deciduous woodland.
D: BFF2: 61-62 **D+I:** B&K4: 230 271 **I:** C&D: 261
Rarely recorded or collected. Apparently widespread but easily overlooked.

cothurnatus Godey, in Gillet, *Hyménomycètes*: 605 (1878)
E: ! **S:** !
H: On weathered cow dung, or dungy straw.
D: BFF2: 66, NBA3: 87-88
Rarely reported. A distinctive species with abundant greyish-rose-pink veil, unlikely to be overlooked and probably genuinely rare.

curtus Kalchbr., *Flora* 59: 424 (1876)
  Coprinus plicatiloides Buller [*nom. inval.*], *Researches Fungi*
  1: 69 (1909)
E: !
H: On weathered horse dung, also on decayed grass clippings.
D: BFF2: 85-86, NM2: 230, NM2: 230 **I:** FlDan4: 160G
Rarely reported. Known from Middlesex and Surrey.

disseminatus (Pers.) Gray, *Nat. arr. Brit. pl.* 1: 634 (1821)
  Agaricus disseminatus Pers., *Comm.*: 87 (1800)
  Coprinarius disseminatus (Pers.) P. Kumm., *Führ. Pilzk.*: 68
  (1871)
  Psathyrella disseminata (Pers.) Quél., *Mém. Soc. Émul.*
  Montbéliard, Sér. 2, 5: 123 (1872)
  Pseudocoprinus disseminatus (Pers.) Kühner, *Botaniste* 20:
  156 (1928)
  Agaricus minutulus Schaeff., *Fung. Bavar. Palat. nasc.* 2: 308
  (1763)
  Mis.: *Agaricus striatus* sensu Sowerby [Col. fig. Engl. fung. 2:
  pl. 166 (1798)] non sensu Schaeff. [Icones pl. 38 (1762)]
E: c **S:** c **W:** c **NI:** c **ROI:** c **O:** Channel Islands: ! Isle of Man: !
H: On decayed wood of deciduous trees (usually large stumps)
often in densely gregarious swarms.
D: BFF2: 82, NM2: 230, NM2: 230 **D+I:** Ph: 181, B&K4: 230
272 **I:** FlDan4: 156A (as *Pseudocoprinus disseminatus*), Bon:
273
Common and widespread. Can be confused with the superficially
similar *Psathyrella pygmaea* which may occur mixed with it.
English name = 'Crumble Cap' or 'Fragile Ink Cap'.

domesticus (Bolton) Gray, *Nat. arr. Brit. pl.* 1: 635 (1821)
  Agaricus domesticus Bolton, *Hist. fung. Halifax* 1: 26 (1788)
E: c **S:** o **W:** o **NI:** ! **ROI:** o
H: On dead and decayed wood of deciduous trees, often *Acer
pseudoplatanus*, *Fagus* and *Fraxinus*. Occasionally on burnt
ground, but then arising from woody debris in soil.
D: BFF2: 56, NM2: 234 **D+I:** Ph: 180, B&K4: 230 273 **I:** Bon:
273, C&D: 261
Common and widespread. Basidiomes arise from a sparse
orange-brown '*Ozonium*' state.

echinosporus Buller, *Trans. Brit. Mycol. Soc.* 6(4): 363 (1920)
[1919]
E: ! **S:** ! **W:** !
H: On wood (decayed twigs or sticks) and one record from a
decayed carpet, discarded in woodland.
D: BFF2: 35 **D+I:** B&K4: 232 275
Rarely reported but apparently widespread. Distinguishable by
the warted spores and small basidiomes resembling those of *C.
lagopus*.

ellisii P.D. Orton, *Trans. Brit. Mycol. Soc.* 43(2): 199 (1960)
E: ! **W:** ?
H: On decayed wood, and on soil around decayed wood such as
stumps or fallen trunks.
D: BFF2: 57
Uncommonly recorded from southern England and a single
record from Wales but doubtfully distinct from *C. domesticus*.

ephemeroides (Bull.) Fr., *Epicr. syst. mycol.*: 250 (1838)
  Agaricus ephemeroides Bull., *Herb. France*: pl. 582 (1793)
  Coprinus bulbillosus Pat., *Tab. anal. fung.*: 60 (1889)
  Agaricus hendersonii Berk., *Engl. fl.* 5(2): 122 (1836)
  Coprinus hendersonii (Berk.) Fr., *Epicr. syst. mycol.*: 250
  (1838)
E: ! **S:** ! **ROI:** !
H: On weathered dung of herbivores, especially frequent on
horse dung.
D: BFF2: 67, NM2: 233 **I:** FlDan4: 159H
Rarely reported. The minute and diaphanous basidiomes are
difficult to collect or preserve. The tiny membranous annulus is
diagnostic if present.

ephemerus (Bull.) Fr., *Epicr. syst. mycol.*: 252 (1838)
  Agaricus ephemerus Bull., *Herb. France*: pl. 542 (1792)
E: ! **S:** ! **O:** Isle of Man: !
H: On weathered dung, soil mixed with dung or decayed straw.

D: BFF2: 87, NM2: 231 **D+I:** B&K4: 232 275 **I:** FlDan4: 160H
Rarely reported. Older records should be treated with caution
due to confusion with related taxa.

episcopalis P.D. Orton, *Trans. Brit. Mycol. Soc.* 40(2): 270
(1957)
E: !
H: On calcareous soil, amongst leaf litter in deciduous
woodland, usually with *Fagus*.
D: BFF2: 37 **D+I:** B&K4: 232 276
Apparently rare and only reported from southern England
(Berkshire, Surrey and West Sussex). Basidiomes resemble
miniature specimens of *Coprinus picaceus*.

erythrocephalus (Lév.) Fr., *Hymenomyc. eur.*: 327 (1874)
  Agaricus erythrocephalus Lév., *Ann. Sci. Nat., Bot.*, sér. 2,
  16: 237 (1841)
  Mis.: *Coprinus dilectus* sensu Lange (FlDan4: 109 & pl. 157A)
E: !
H: On soil, woodchips in soil, decayed mulch on flowerbeds,
rubbish tips and old fire sites.
D: BFF2: 34, NM2: 233 **I:** Bon: 271, C&D: 263
Rarely reported. A distinctive species unlikely to be overlooked,
especially when young and the bright coral to orange-red veil
is fully developed.

filamentifer Kühner, *Bull. Soc. Naturalistes Oyonnax* 10 -11
(Suppl.): 3 (1957)
S: ! **ROI:** !
H: On weathered dung of herbivores (cattle, rabbits and sheep).
D: BFF2: 51
Only four British records from scattered locations in Scotland
and the Republic of Ireland.

flocculosus (DC.) Fr., *Epicr. syst. mycol.*: 245 (1838)
  Agaricus flocculosus DC., *Fl. Fr.* 5: 45 (1815)
  Mis.: *Coprinus rostrupianus* sensu Lange (FlDan4: 109 & pl.
  157E)
E: ! **S:** ! **W:** !
H: On soil, and soil mixed with woody debris.
D: BFF2: 60 **D+I:** B&K4: 234 277 **I:** FM3(4): back cover
Rarely reported but sometimes abundant on mulched
flowerbeds.

foetidellus P.D. Orton, *Notes Roy. Bot. Gard. Edinburgh* 32(1):
139 (1972)
E: ! **W:** !
H: On dung of cattle.
D: BFF2: 76
Known only from the type collection from Somerset and a recent
record from Merionethshire (Harlech).

friesii Quél., *Mém. Soc. Émul. Montbéliard, Sér. 2,* 5: 129
(1872)
  Coprinus rhombisporus P.D. Orton, *Notes Roy. Bot. Gard.
  Edinburgh* 32(1): 145 (1972)
E: ! **S:** ! **W:** !
H: On decayed debris of various grasses, also occasionally on
soil amongst grasses and on decayed culms of *Juncus* spp.
D: BFF2: 45, NM2: 232 **I:** C&D: 263
Rarely reported but apparently widespread.

galericuliformis Watling, *Notes Roy. Bot. Gard. Edinburgh*
28(1): 42 (1967)
E: ! **S:** ! **W:** !
H: On soil in woodland or shaded places in gardens, and a
single record in a cool glasshouse.
D: BFF2: 102, NM2: 229
Rarely reported, but apparently widespread.

gonophyllus Quél., *Ann. Sci. Nat. Bord. Sud-ouest, Suppl.* 14:
5 (1884)
E: ! **S:** ! **O:** Isle of Man: !
H: On burnt soil in woodland, and also known on old plaster in
houses.
D: BFF2: 46 **D+I:** B&K4: 234 278 **I:** Bon: 271
Rarely reported but apparently widespread.

griseofoetidus P.D. Orton, *Trans. Brit. Mycol. Soc.* 91(4): 548 (1988)
E: !
H: On soil in deciduous woodland.
D: NBA9: 548
Only three records, from Gloucestershire and Surrey. Probably misidentified as *C. narcoticus* in the past.

heptemerus M. Lange & A.H. Sm., *Mycologia* 45: 751 (1953)
Coprinus curtus f. *macrosporus* Romagn., *Rev. Mycol. (Paris)* 6(3-4): 126 (1941)
E: ! **S:** ! **W:** ! **O:** Isle of Man: !
H: On weathered dung of various herbivores.
D: BFF2: 86, NM2: 230
Uncommonly reported but apparently widespread.

heptemerus *f.* **parvisporus** J. Breitenb. & F. Kränzl., *Fungi of Switzerland* 4: 236 (1995)
E: !
H: On weathered dung of herbivores.
D: B&K4: 236 280
Known from Oxfordshire, South Hampshire and Surrey.

hercules Uljé & Bas, *Persoonia* 12(4): 483 (1985)
E: !
H: On soil amongst grass, usually on lawns.
Known from London and Surrey. Easily overlooked.

heterosetulosus Watling, *Notes Roy. Bot. Gard. Edinburgh* 35(1): 153 (1976)
Coprinus heterosetulosus Locquin [*nom. nud.*], *Bull. Soc. Mycol. France* 63: 78 (1947).
E: ! **S:** !
H: On weathered dung of herbivores.
D: BFF2: 91 **D+I:** B&K4: 236 281
Rarely recorded or collected but apparently widespread.

hiascens (Fr.) Quél., *Fl. mycol. France*: 42 (1888)
Agaricus hiascens Fr., *Syst. mycol.* 1: 303 (1821)
Psathyrella hiascens (Fr.) Quél., *Mém. Soc. Émul. Montbéliard, Sér. 2*, 5: 123 (1872)
Mis.: *Psathyrella crenata* sensu Rea (1922) & sensu auct.
E: o **S:** ! **W:** ! **NI:** ! **ROI:** !
H: On soil, in woodland, often amongst grass by paths and edges of rides, usually in caespitose clumps, rarely occurring singly.
D: BFF2: 92 **D+I:** Myc8(1): 12
Occasional but apparently widespread.

impatiens (Fr.) Quél., *Fl. mycol. France*: 42 (1888)
Agaricus impatiens Fr., *Syst. mycol.* 1: 302 (1821)
Psathyrella impatiens (Fr.) Gillet, *Hyménomycètes*: 616 (1878)
Pseudocoprinus impatiens (Fr.) Kühner, *Bull. Soc. Mycol. France* 52: 33 (1936)
E: o **S:** !
H: On decayed leaf litter in deciduous woodland, most often associated with *Fagus* on calcareous soil.
D: BFF2: 93 **D+I:** Ph: 179, B&K4: 238 283 **I:** FlDan4: 156B (as *Pseudocoprinus impatiens*), C&D: 261
Occasional but locally abundant in southern England, less frequent northwards.

jonesii Peck, *Bull. Torrey Bot. Club* 22(1): 206 (1895)
Coprinus funariarum Métrod, *Bull. Soc. Mycol. France* 53: 346 (1937)
Coprinus lagopus var. *sphaerosporus* Kühner & Joss. [*nom. inval.*], *Bull. Soc. Mycol. France* 60: 31 (1944)
Mis.: *Coprinus lagopides* sensu auct. mult.
E: c **S:** c **W:** c **NI:** ! **O:** Isle of Man: !
H: On burnt soil and wood, also on charred plaster in houses, frequently subcaespitose and rarely occuring singly.
D: BFF2: 41 (as *Coprinus lagopides*), NM2: 232 (as *C. lagopides*) **D+I:** B&K4: 240 287 (as *C. lagopides*) **I:** Ph: 179 (as *C. lagopides*), C&D: 263 (as *C. lagopides*)
Common and widespread. Better known as *C. lagopides*.

krieglsteineri Bender, *Beitr. Kenntn. Pilze Mitteleurop.* 3: 218 (1987)
E: !
H: On decayed wood (woodchip mulch on flowerbeds in British collections)
Known from Surrey (Royal Botanic Gardens, Kew) since 2000 but occurring in large numbers.

kubickae Pilát & Svrček, *Česká Mykol.* 21: 142 (1967)
E: !
H: On decayed vegetation and grass leaves. British material associated with *Glyceria maxima* and *Phragmites australis*.
Known from Berkshire (Thatcham).

kuehneri Uljé & Bas, *Persoonia* 13(4): 438 (1988)
Coprinus plicatilis var. *microsporus* Kühner & Joss., *Bull. Soc. Mycol. France* 50: 57 (1934)
E: ! **S:** !
H: On soil in deciduous woodland, by the edges of paths and amongst grass.
D+I: B&K4: 285
Rarely recorded but widespread in England. May be confused with *C. plicatilis* (of which it was once considered a small spored form) or *C. leiocephalus*.

laanii Kits van Wav., *Persoonia* 5(2): 146 (1968)
Mis.: *Coprinus cineratus* sensu NCL p.p.
E: ! **S:** ! **W:** !
H: On dead wood, often on the cut ends of stacked logs associated with thick growths of algae. Also collected from 'desalinated wood' on the raised wreck of the Tudor warship 'Mary Rose' in Portsmouth Harbour.
D: BFF2: 71 **D+I:** B&K4: 240 286, C&D: 260
Rarely reported but apparently widespread and distinctive.

lagopides P. Karst., *Bidrag Kännedom Finlands Natur Folk* 32: 535 (1879)
Coprinus phlyctidosporus Romagn., *Rev. Mycol. (Paris)* 10(5-6): 88 (1945)
E: !
H: On burnt soil and wood.
D: BFF2: 36 (as *Coprinus phlyctidiosporus*)
A species with warted spores, apparently rare in Britain (only three collections in herb. K). The commoner smooth-spored species (*C. lagopides* sensu auct.) is *C. jonesii*.

lagopus (Fr.) Fr., *Epicr. syst. mycol.*: 250 (1838)
Agaricus lagopus Fr., *Syst. mycol.* 1: 312 (1821)
Coprinus lagopus f. *macrospermus* Romagn., *Rev. Mycol. (Paris)* 10(5-6): 89 (1945)
E: c **S:** c **W:** c **NI:** c **ROI:** c **O:** Channel Islands: ! Isle of Man: !
H: On soil with woody debris, or in leaf litter in woodland and occasionally in field and gardens.
D: BFF2: 40, NM2: 232 **D+I:** Ph: 179, B&K4: 240 288 **I:** Bon: 271
Common and widespread.

lagopus *var.* **vacillans** Uljé, *Persoonia* 17(3): 468 (2000)
E: ! **S:** !
H: On soil, amongst grass on lawns.
Known from Buckinghamshire (Oving, near Whitchurch) and Perthshire (Kindrogan). Basidiomes are evanescent, usually developing and maturing overnight and disappearing by dawn.

leiocephalus P.D. Orton, *Notes Roy. Bot. Gard. Edinburgh* 29: 88 (1969)
E: c **S:** c **W:** ! **NI:** ! **ROI:** ! **O:** Isle of Man: !
H: On soil in grassy areas in woodland, often arising from small fragments of buried, decayed wood.
D: BFF2: 102, NM2: 229
Common and widespread. Closely resembles *C. plicatilis*. The description in BFF2 also includes *C. kuehneri*, now considered distinct. N.B. The collection illustrated in B&K4 is misdetermined.

lilatinctus Bender & Uljé, *Persoonia* 16(3): 373 (1997)
E: ! **S:** !
H: On decayed woodchip mulch on flowerbeds (British material).

D: Myc16(3): 114
Known from England (Surrey: Kew, Royal Botanic Gardens) and Scotland (Braemar: Manse Wood). A distinctive species with marked lilaceous tints to the pilei, especially in young basidiomes.

luteocephalus Watling, *Notes Roy. Bot. Gard. Edinburgh* 31: 359 (1972)
E: ! **S:** !
H: On weathered horse dung.
D: BFF2: 62
Known from Cumberland and Midlothian.

macrocephalus (Berk.) Berk., *Outl. Brit. fungol.*: 180 (1860)
Agaricus macrocephalus Berk., *Engl. fl.* 5(2): 122 (1836)
E: ! **S:** ! **ROI:** !
H: On dung and straw mixtures, less often on pure dung and occasionally on soil and decayed grain (pheasant feed) in woodland.
D: BFF2: 43 **D+I:** B&K4: 242 291
Rarely reported but apparently widespread.

marculentus Britzelm., *Bot. Centralbl.* 67(13): 440 (1899)
Coprinus hexagonosporus Joss. [*nom. inval.*], *Rev. Mycol. (Paris)* 13(2-3): 82 (1948)
E: ! **S:** !
H: On dung, dungy soil or manured straw.
D: BFF2: 83 (as *Coprinus hexagonosporus*), NM2: 230, NBA3: 88 (as *C. hexagonosporus*) **D+I:** B&K4: 244 292
Rarely reported. The majority of reports and collections are from England, with a single report from Scotland (Edinburgh). The distinctly hexagonal spores (in face view) are diagnostic.

martinii P.D. Orton, *Trans. Brit. Mycol. Soc.* 43(2): 201 (1960)
E: ! **S:** ! **W:** !
H: On decayed debris of *Carex* and *Juncus* spp in wet areas such as fens and marshes, and also on *Scirpus caespitosus* on moorland.
D: BFF2: 80, NM2: 233 **D+I:** B&K4: 244 293
Rarely reported but apparently widespread.

megaspermus P.D. Orton, *Notes Roy. Bot. Gard. Edinburgh* 32: 141 (1972)
E: !
H: On soil.
D: BFF2: 100
Known from Shropshire (Attingham Park). Reported from Northamptonshire (Rockingham Castle) but unsubstantiated with voucher material.

micaceus (Bull.) Fr., *Epicr. syst. mycol.*: 247 (1838)
Agaricus micaceus Bull., *Herb. France*: pl. 246 (1786)
Mis.: *Agaricus congregatus* sensu Sowerby [Col. fig. Engl. fung. 3: pl. 261 (1800)]
E: c **S:** c **W:** c **NI:** c **ROI:** c **O:** Channel Islands: c Isle of Man: c
H: On decayed wood of deciduous trees, and less frequently conifers, usually on large stumps, logs or fallen trunks. Sometimes on buried wood and then appearing as if terrestrial.
D: BFF2: 54, NM2: 234 **D+I:** Ph: 180, B&K4: 244 294, C&D: 260 **I:** Bon: 273
Very common and widespread. Sometimes associated with an orange-brown 'Ozonium' stage, such collections fruiting in spring and possibly representing a distinct species.

miser P. Karst., *Bidrag Kännedom Finlands Natur Folk* 37: 236 (1882)
E: o **S:** o **W:** ! **ROI:** ! **O:** Isle of Man: !
H: On weathered dung of herbivores.
D: BFF2: 104, NM2: 229 **I:** FlDan4: 157B
Occasional but widespread.

narcoticus (Batsch) Fr., *Epicr. syst. mycol.*: 250 (1838)
Agaricus narcoticus Batsch, *Elench. fung. (Continuatio Prima)*: 79 & t. 16 f. 77 (1786)
E: ! **S:** ! **W:** ! **ROI:** !
H: On weathered dung or manured soil.
D: BFF2: 69, NM2: 233 **D+I:** Ph: 179 **I:** Bon: 273

Rarely reported but apparently widespread. Records without voucher material (especially those not on dung) are suspect since other unpleasant-smelling species may have been referred here.

nemoralis Bender, *Persoonia* 15(3): 300 (1993)
E: !
H: On wet soil on a stream bank (British material).
D + I: Reid [Mycologist 9(3): 119–120 (1995)]
A single collection from Hertfordshire (Gobions Wood) in 1994.

niveus (Pers.) Fr., *Epicr. syst. mycol.*: 246 (1838)
Agaricus niveus Pers. [non *A. niveus* Pers., *Syn. Meth. Fung.*: 438 (1801)], *Syn. meth. fung.*: 400 (1801)
Coprinus latisporus P.D. Orton, *Notes Roy. Bot. Gard. Edinburgh* 32: 140 (1972)
E: c **S:** c **W:** c **NI:** c **ROI:** ! **O:** Isle of Man: !
H: On weathered dung of herbivores (especially cattle and horses).
D: BFF2: 63, NM2: 233 **D+I:** Ph: 179, Myc8(1): 12, B&K4: 246 295 **I:** FlDan4: 159I, Bon: 273
Common and widespread.

ochraceolanatus Bas, *Persoonia* 15(3): 362 (1993)
E: !
H: British material on decayed wood-chip of *Tilia* sp.
A single collection from Buckinghamshire (Whitchurch) in 2001.

pachydermus Bogart, *Mycotaxon* 8: 274 (1979)
E: !
H: On decayed woodchips (British material).
Known from Surrey (Kew Gardens), West Sussex (Chichester) and Yorkshire (Thorp Perrow Arboretum). Unknown elsewhere in Europe.

pachyspermus P.D. Orton, *Notes Roy. Bot. Gard. Edinburgh* 32: 144 (1972)
E: ! **S:** !
H: On weathered herbivore dung, usually cow.
D: BFF2: 65 **D+I:** Ph: 178-179
Rarely reported. Single collections from England (Hertfordshire: Hitchin) and Scotland (Perthshire).

patouillardii Quél., *Tab. anal. fung.* 1: 107 (1883)
E: !
H: On decayed straw and hay, weathered dung, vegetable debris and kitchen refuse (including old tea-leaves).
D: BFF2: 68 **I:** FlDan4: 157D (as *Coprinus angulatus*)
Rarely reported. The collection illustrated in B&K4: 246 (pl. 296) is now considered to be either 'poorly depicted or wrongly named' (Anon., Persoonia 16: 262, 1996).

pellucidus P. Karst., *Bidrag Kännedom Finlands Natur Folk* 37: 236 (1882)
E: ! **S:** ! **NI:** ! **ROI:** !
H: On weathered dung (especially cattle dung).
D: BFF2: 89, NM2: 231
Rarely reported but apparently widespread.

phaeosporus P. Karst., *Meddeland. Soc. Fauna Fl. Fenn.* 6: 9 (1882)
Coprinus saichiae D.A. Reid, *Trans. Brit. Mycol. Soc.* 41(4): 430 (1958)
Coprinus xantholepis P.D. Orton, *Notes Roy. Bot. Gard. Edinburgh* 32: 150 (1972)
E: ! **S:** !
H: On soil amongst grass or on decayed grass debris.
D: BFF2: 49 (as *Coprinus rhombisporus*), BFF2: 49 (as *C. xantholepis*), BFF2: 50 (as *C. saichiae*), NM2: 232 **I:** FlDan4: 159F (as *C. phaeosporus* var. *solitarius*)
Rarely reported.

picaceus (Bull.) Gray, *Nat. arr. Brit. pl.* 1: 634 (1821)
Agaricus picaceus Bull., *Herb. France*: pl. 206 (1785)
E: o **S:** ! **W:** ! **NI:** !
H: On soil, amongst leaf litter, most often associated with *Fagus* in beechwoods on calcareous soil but also known with other

deciduous trees. Rarely associated with conifers such as *Larix* spp.
D: BFF2: 36, NM2: 231 **D+I:** B&K4: 246 297 **I:** Sow2: 170 (as *Agaricus picaceus*), FlDan4: 158E, Bon: 271, C&D: 263
Frequent in southern and south-eastern England, sometimes fruiting abundantly, but rare elsewhere. Unlikely to be mistaken for anything other species, although *C. episcopalis* may resemble small basidiomes. English name = 'Magpie Cap'.

plagioporus Romagn., *Rev. Mycol. (Paris)* 6(3-4): 126 (1941)
E: ! **ROI:** !
H: On soil.
D: BFF2: 95
Rarely reported, from widely disjunct areas of the British Isles.

plicatilis (Curtis) Fr., *Epicr. syst. mycol.*: 252 (1838)
    Agaricus plicatilis Curtis, *Fl. londin.* 1: 57, t. 200 (1777)
E: c **S:** c **W:** c **NI:** c **ROI:** c Channel Islands: c Isle of Man: c
H: On soil amongst grass in unimproved grassland, parkland, on lawns and along woodland edges (but not in woodland).
D: BFF2: 101, NM2: 229 **D+I:** B&K4: 248 298 **I:** Bon: 273, C&D: 261
Common and widespread. This is a grassland species and records from other habitats require verification.

poliomallus Romagn., *Rev. Mycol. (Paris)* 10(5-6): 89 (1945)
E: ! **S:** !
H: On weathered dung of cattle, and also reported on bare soil associated with *Crataegus*.
D: BFF2: 60
Rarely reported and poorly known in Britain.

pseudofriesii Pilát & Svrček, *Česká Mykol.* 21(3): 140 (1967)
E: !
H: On soil or soil mixed with decayed woody debris, decayed debris of *Carex pendula* and also apparently on weathered dung of cattle.
Known from Buckinghamshire, London and Surrey.

pseudoniveus Bender & Uljé, *Persoonia* 15(3): 270 (1993)
E: !
H: On weathered dung. British material all on cow dung.
Known from Berkshire, Buckinghamshire and Hertfordshire.

pseudoradiatus Watling, *Notes Roy. Bot. Gard. Edinburgh* 35(1): 154 (1976)
    Mis.: *Coprinus radiatus* sensu Rea (1922)
E: ! **S:** !
H: On weathered dung of herbivores
D: NM2: 232
Rarely reported.

pyrrhanthes Romagn., *Rev. Mycol. (Paris)* 16(2): 128 (1951)
E: !
H: On soil in deciduous woodland.
Known from Surrey (Fetcham Downs). Also recorded from Cumberland (near Ullswater; TBMS 38 (2): 180, 1955), but unsubstantiated with voucher material.

radians (Desm.) Fr., *Epicr. syst. mycol.*: 248 (1838)
    Agaricus radians Desm., *Ann. Sci. Nat., Bot.*, sér. 1, 3: 214 (1828)
    Coprinus hortorum Métrod, *Rev. Mycol. (Paris)* 5(2-3): 80 (1940)
    Coprinus similis Berk. & Broome, *Ann. Mag. Nat. Hist., Ser. 3 [Notices of British Fungi no. 1011]* 15: 317 (1865)
    Ozonium auricomum Link (anam.), *Mag. Neuesten Entdeck. Gesammten Naturk. Ges. Naturf. Freunde Berlin* 3: 21 (1809)
E: o **S:** ! **W:** !
H: On decayed wood of deciduous trees.
D: BFF2: 59 **D+I:** C&D: 260
Occasional but apparently widespread. Basidiomes arise from a dense, rusty-brown '*Ozonium*' state.

radiatus (Bolton) Gray, *Nat. arr. Brit. pl.* 1: 635 (1821)
    Agaricus radiatus Bolton, *Hist. fung. Halifax* 1: 39 (1788)
    Mis.: *Coprinus lagopus* sensu auct.

E: ! **S:** ! **W:** ! **NI:** ! **O:** Isle of Man: !
H: On weathered dung of herbivores, especially of horses.
D: BFF2: 43, NM2: 232 **D+I:** B&K4: 248 299 **I:** Bon: 271
Common and widespread.

radicans Romagn., *Rev. Mycol. (Paris)* 16(2): 127 (1951)
E: ! **S:** !
H: On soil and dung mixtures, also on decayed plant debris (old potatoes and straw).
D: BFF2: 70
Rarely reported. The strongly radicant basidiomes and the markedly stercoraceous smell are useful diagnostic characters.

romagnesianus Singer, *Lilloa* 22: 459 (1951) [1949]
E: ! **S:** ! **W:** ! **NI:** !
H: On and around decayed wood (usually stumps) of deciduous trees. Known on *Betula* and *Salix* spp.
D: BFF2: 33 **D+I:** C&D: 262
Rarely reported but apparently widespread. A large, distinctive agaric, resembling *C. atramentarius* but with pileus and lower half of the stipes covered in bright apricot or rusty veil.

saccharomyces P.D. Orton, *Trans. Brit. Mycol. Soc.* 43(2): 202 (1960)
E: ! **S:** !
H: On soil, often amongst grass.
D: BFF2: 81 **D+I:** B&K4: 248 300
Rarely reported but apparently widespread. Basidiomes have a markedly strong smell, variously described as 'yeast' or 'beer'.

sassii M. Lange & A.H. Sm., *Mycologia* 45: 755 (1953)
E: !
H: On dung. British material on cattle dung.
A single collection, from Oxfordshire (Blenheim Park) in 1973.

schroeteri P. Karst., *Hedwigia* 19(7): 114 (1880)
    Coprinus nudiceps P.D. Orton, *Notes Roy. Bot. Gard. Edinburgh* 32: 142 (1972)
E: ! **S:** ! **NI:** !
H: On soil amongst grass, and apparently also on weathered dung.
D: BFF2: 103 (as *Coprinus nudiceps*)
Rarely reported. The majority of British records are from lawns in the Royal Botanic Gardens, Kew.

sclerocystidiosus M. Lange & A.H. Sm., *Mycologia* 45: 769 (1953)
E: ! **W:** !
H: On soil, also recorded from woodchips buried in soil.
Rarely reported, mostly from England with a single Welsh record (Anglesey).

sclerotiger Watling, *Notes Roy. Bot. Gard. Edinburgh* 32(1): 130 (1972)
S: !
H: On dung heaps and manured straw.
D: BFF2: 78
Described from Mid Ebudes (Isle of Mull) and known only from there.

scobicola P.D. Orton, *Notes Roy. Bot. Gard. Edinburgh* 32: 147 (1972)
E: ! **S:** !
H: On decayed sawdust, also on grass clippings.
D: BFF2: 38
Described from South Devon, otherwise known only from Surrey (Esher Common) and Scotland (Orkney). Material discussed by Kemp under the provisional name '*Coprinus bilanatus*' in TBMS 65 (3): 375-388 (1975) may represent this species (*fide* Uljé in *Persoonia* 17(2): 195)

semitalis P.D. Orton, *Notes Roy. Bot. Gard. Edinburgh* 32(1): 147 (1972)
    Coprinus cinereofloccosus var. *angustisporus* D.A. Reid, *Fungorum Rar. Icon. Color.* 5: 22 & pl. 44d (1972)
    Mis.: *Coprinus cineratus* sensu NCL p.p.
E: ! **S:** ! **NI:** !

H: On soil, often amongst grass, also reported from pine plantation.
D: BFF2: 72 **D+I:** B&K4: 250 301
Rarely reported but apparently widespread.

silvaticus Peck, *Rep. (Annual) New York State Mus. Nat. Hist.* 24: 71 (1872)
  Coprinus tardus P. Karst., *Bidrag Kännedom Finlands Natur Folk* 32: 543 (1879)
  Mis.: *Coprinus tergiversans* sensu Ricken (1915: 63)
E: o **S:** ! **W:** ! **NI:** !
H: On wood (decayed logs or stumps), occasionally 'on soil' (but arising from buried wood).
D: BFF2: 82, NM2: 230 **D+I:** Ph: 180-181, B&K4: 250 302
Rarely reported but apparently widespread. Closely resembles *C. micaceus*.

spelaiophilus Bas & Uljé, *Persoonia* 17(2): 179 (1999)
  Mis.: *Coprinus extinctorius* sensu auct. mult.
E: o **S:** ! **NI:** !
H: On dead or decayed wood, often in small 'pocket' lesions at the bases of living trunks of deciduous trees.
D: BFF2: 39 (as *Coprinus extinctorius*)
Known from England (Buckinghamshire, North Hampshire, North & South Somerset, Oxfordshire, West Kent and West Sussex), Northern Ireland (Down) and Scotland. Perhaps better known as *C. extinctorius*.

stanglianus Enderle, Bender & Gröger, *Z. Mykol.* 54(1): 62 (1988)
E: !
H: On calcareous soil in downland grassland. Fruiting from May to July.
D+I: Myc7(2): 87
First recorded in Britain in 1991 from calcareous grassland in Surrey (Box Hill), but previously collected there by Dennis (herb. K, as *Coprinus* sp.) in 1948. Now known from several places in south-eastern England (Berkshire, Cambridgeshire and Surrey) in similar habitat.

stellatus Buller, in Bisby, Buller & Dearness, *Fungi of Manitoba and Saskatchewan*: 119 (1929)
E: ! **S:** ! **W:** ! **O:** Isle of Man: !
H: On weathered or old dung of various herbivores.
D: BFF2: 90
Rarely reported in the field, but more frequently obtained from dung cultures.

stercoreus Fr., *Epicr. syst. mycol.*: 251 (1838)
  Agaricus stercorarius Bull., *Herb. France*: pl. 542 (1786)
  Coprinus stercorarius (Bull.) Fr., *Epicr. syst. mycol.*: 251 (1838)
  Mis.: *Coprinus velox* sensu auct.
E: c **S:** c **W:** ! **NI:** ! **ROI:** ! **O:** Isle of Man: !
H: On herbivore dung, especially when old or waterlogged.
D: BFF2: 78 **D+I:** B&K4: 250 303 **I:** Sow4: pl. 262 (as *Agaricus stercorarius*)
Common and widespread. Macroscopically resembles *C. cordisporus*.

sterquilinus (Fr.) Fr., *Epicr. syst. mycol.*: 242 (1838)
  Agaricus sterquilinus Fr., *Syst. mycol.* 1: 308 (1821)
  Agaricus oblectus Bolton, *Hist. fung. Halifax* Suppl.: 142 (1791)
  Coprinus oblectus (Bolton) Fr., *Epicr. syst. mycol.*: 243 (1838)
E: ! **S:** ! **W:** ! **NI:** ! **ROI:** !
H: On weathered dung of herbivores, especially that of horses and rabbits
D: BFF2: 30, NM2: 232 **I:** FlDan4: 157F, Bon: 271
Rarely reported but apparently widespread.

subdisseminatus M. Lange & A.H. Sm., *Mycologia* 45: 777 (1953)
E: ! **S:** ! **W:** ! **ROI:** !
H: On leaves, sticks and other plant debris.
D: BFF2: 96

Rarely reported but apparently widespread. Resembles single specimens of *C. disseminatus*.

subimpatiens M. Lange & A.H. Sm., *Mycologia* 45: 772 (1953)
E: ! **S:** !
H: On soil or humus in woodland.
D: BFF2: 97
Rarely reported but apparently widespread.

subpurpureus A.H. Sm., *Mycologia* 40: 684 (1948)
E: !
H: On soil, or decayed woody debris buried in soil.
D: BFF2: 148
Rarely reported. Known from South Devon (Slapton) and a possible collection in K of this species from Surrey.

tigrinellus Boud., *Bull. Soc. Bot. France* 32: 283 (1885)
E: ! **W:** !
H: On decayed debris of *Carex* and *Typha* spp in wet places such as marshes, fens and pond margins.
D: BFF2: 48
Rarely reported. Material illustrated in 252 (pl. 304) is now considered to be either 'poorly depicted or wrongly named' (Persoonia 16: 262, 1996).

trisporus Kemp & Watling, *Notes Roy. Bot. Gard. Edinburgh* 31(1): 128 (1972)
  Coprinus triplex P.D. Orton, *Notes Roy. Bot. Gard. Edinburgh* 35(1): 147 (1976)
E: ! **S:** ! **W:** !
H: On weathered dung. British material on horse dung.
D: BFF2: 75
Rarely reported. Known from England (East Kent, South Hampshire and Warwickshire), Scotland (Edinburgh) and Wales (Swansea).

truncorum (Schaeff.) Fr., *Epicr. syst. mycol.*: 248 (1838)
  Agaricus truncorum Schaeff., *Fung. Bavar. Palat. nasc.* 1: 6 (1762)
  Mis.: *Coprinus micaceus* sensu J. Lange, & sensu auct.
E: !
H: On decayed wood (stumps or logs) of deciduous trees.
D: BFF2: 55, NM2: 234
Rarely reported. Known from Bedfordshire, Buckinghamshire, Cambridgeshire, Northamptonshire, South Essex, South Hampshire, Surrey and West Kent. Closely resembles *C. micaceus* in the field.

tuberosus Quél., *Bull. Soc. Bot. France* 25: 289 (1878)
  Mis.: *Coprinus stercorarius* sensu Lange (FlDan4: 114 & pl. 159A)
E: ! **S:** ! **ROI:** !
H: On weathered dung of herbivores (horse and rabbit), also on decayed 'compost' in gardens and greenhouses.
D: BFF2: 77 **D+I:** B&K4: 252 305 **I:** FlDan4: 159A
Rarely reported but apparently widespread.

urticicola (Berk. & Broome) Buller, *Trans. Brit. Mycol. Soc.* 5: 485 (1917)
  Agaricus urticicola Berk. & Broome, *Ann. Mag. Nat. Hist., Ser. 3 [Notices of British Fungi no. 919]* 7: 376 (1861)
  Psathyra urticicola (Berk. & Broome) Sacc., *Syll. fung.* 5: 1073 (1887)
  Coprinus brassicae Peck, *Rep. (Annual) New York State Mus. Nat. Hist.* 43: 64 (1890) [1889]
E: o **S:** ! **W:** !
H: On decayed grass leaves, often *Holcus lanatus*, in wet areas or woodland edges. Also on debris of *Carex* and *Juncus* spp and *Urtica dioica*. Fruiting especially during the summer.
D: BFF2: 47 **D+I:** B&K4: 252 306
Common and widespread at least in southern England. Seemingly less common northwards. Often fruits under dense layers of decayed grass leaves and may be overlooked.

utrifer Watling, *Notes Roy. Bot. Gard. Edinburgh* 31: 362 (1972)
  Coprinus utrifer Joss. [*nom. nud.*], *Bull. Soc. Mycol. France* 64: 26 (1948).

E: ! S: !
H: On weathered dung of herbivores, especially on sheep dung.
D: BFF2: 53
Rarely reported. Seemingly more frequent in northern England and Scotland, and unrecorded from southern areas.

velatopruinatus Bender, *Beitr. Kenntn. Pilze Mitteleurop.* 5: 80 (1989)
E: !
H: On soil or decayed wood-chips.
Known from Surrey (Kew Gardens) in a warm greenhouse, and from more natural habitat in Buckinghamshire (Whitchurch in 2002 and Weston Turville Reservoir in 2003).

vermiculifer Dennis, *Kew Bull.* 19: 112 (1964)
E: ! S: o
H: On weathered dung of herbivores, especially horse and sheep dung.
D: BFF2: 51
Occasional to rather common in Scotland, especially in upland areas.

xanthothrix Romagn., *Rev. Mycol. (Paris)* 6(3-4): 127 (1941)
Mis.: *Coprinus domesticus* sensu J. Lange
E: ! S: ! W: ! NI: !
H: On woody debris of deciduous trees, also on decayed grass compost in wooded areas, and fallen leaves of *Fagus* in woodland. Two separate records (with voucher material) from damp woodwork of old Morris Traveller cars.
D: BFF2: 58 **D+I:** B&K4: 254 307 (material rather old) **I:** Bon: 273, C&D: 261
Rarely reported, mostly from widely scattered locations in England with single records from Scotland, Wales and Northern Ireland.

xenobius P.D. Orton, *Notes Roy. Bot. Gard. Edinburgh* 35(1): 148 (1976)
E: ! S: !
H: On weathered dung of cattle.
D: BFF2: 52
In Britain known only from two records from Scotland and one from England. Rather doubtfully distinct from *C. luteocephalus*.

CORIOLOPSIS Murrill, *Bull. Torrey Bot. Club* 32: 358 (1905)
Poriales, Coriolaceae
Type: *Coriolopsis occidentalis* (Klotzsch) Murrill

gallica (Fr.) Ryvarden, *Norweg. J. Bot.* 19: 230 (1973)
Polyporus gallicus Fr., *Syst. mycol.* 1: 345 (1821)
Trametes gallica (Fr.) Fr., *Epicr. syst. mycol.*: 489 (1838)
Funalia gallica (Fr.) Bondartsev & Singer, *Ann. Mycol.* 39: 62 (1941)
E: !
H: On dead wood of deciduous trees in old woodland, usually in areas of calcareous soil. Mainly on fallen branches of *Fraxinus* but also known on *Fagus, Quercus* and *Ulmus* spp.
D: EurPoly1: 219, NM3: 227 **D+I:** B&K2: 280 347
Known from Berkshire, East & West Kent, South Essex, Surrey, South Lancashire, West Norfolk and West Sussex.

CORONICIUM J. Erikss. & Ryvarden, *Corticiaceae of North Europe* 3: 295 (1975)
Stereales, Xenasmataceae
Type: *Coronicium gemmiferum* (Bourdot & Galzin) J. Erikss. & Ryvarden

alboglaucum (Bourdot & Galzin) Jülich, *Persoonia* 8(3): 299 (1975)
Corticium alboglaucum Bourdot & Galzin, *Bull. Soc. Mycol. France* 27: 251 (1911)
Xenasma alboglaucum (Bourdot & Galzin) Liberta, *Mycologia* 52(6): 892 (1962)
E: !
H: On decayed wood of deciduous trees. British material on dead wood and bark of *Salix fragilis* in damp woodland.

D: NM3: 124 **D+I:** CNE4: 564-566
Known from England (Middlesex: Perivale Wood)

gemmiferum (Bourdot & Galzin) J. Erikss. & Ryvarden, *Corticiaceae of North Europe* 3: 297 (1975)
Corticium gemmiferum Bourdot & Galzin, *Bull. Soc. Mycol. France* 27: 250 (1911)
Hyphoderma gemmiferum (Bourdot & Galzin) Jülich, *Persoonia* 8(1): 80 (1974)
E: !
H: On decayed wood of deciduous trees. British material on *Fraxinus* and *Salix* spp.
D: NM3: 124 **D+I:** CNE3: 296-297
Known from Hertfordshire (Benington: Combs Wood) and Surrey (Kew: Royal Botanic Gardens)

CORTICIUM Pers., *Neues Mag. Bot.* 1: 110 (1794)
Stereales, Corticiaceae
Mycinema C. Agardh, *Systema Algarum* 26: 32 (1824)
Laeticorticium Donk, *Fungus* 26: 16 (1956)
Type: *Corticium roseum* Pers.

erikssonii Jülich, *Int. J. Mycol. Lichenol.* 1(1): 31 (1982)
Laeticorticium roseum var. *pulverulentum* J. Erikss. & Ryvarden [*nom. inval.*], *Corticiaceae of North Europe* 4: 783 (1976)
Laeticorticium pulverulentum (J. Erikss. & Ryvarden) J. Erikss. & Ryvarden [*nom. inval.*], *Corticiaceae of North Europe* 4: 785 (1976)
Hyphelia rosea Fr. (anam.), *Syst. mycol.* 1: 212 (1821)
E: !
H: On decayed, and often sodden wood of deciduous trees. Known on *Populus tremula* and (apparently) also on *Salix* and *Ulmus* spp.
D: NM3: 178 **D+I:** CNE4: 783-786 (as *Laeticorticium roseum* var. *pulverulentum*)
Known from one teleomorphic collection (Buckinghamshire: Rushbeds Wood) in 2001, and two of the *Hyphelia* anamorph (North Somerset) in 1852 and 1859.

quercicola Jülich, *Int. J. Mycol. Lichenol.* 1(1): 31 (1982)
Laeticorticium quercinum J. Erikss. & Ryvarden, *Corticiaceae of North Europe* 4: 777 (1976) [non *Corticium quercinum* (Pers.) Fr. (1838)]
E: ! O: Isle of Man: !
H: On or near ascocarps of *Colpoma quercinum* on dead attached twigs of *Quercus* spp. Fruiting February to May.
D: NM3: 178 **D+I:** CNE4: 776-779 (as *Laeticorticium quercinum*), Myc10(4): 157 **I:** SV31: 9 (as *L. quercinum*)
First recorded from Berkshire in 1994, but now known from Buckinghamshire, East & West Norfolk, Hertfordshire, North & South Essex, Surrey, and the Isle of Man.

roseum Pers., *Neues Mag. Bot.* 1: 111 (1794)
Thelephora rosea (Pers.) Pers., *Syn. meth. fung.*: 575 (1801)
Hypochnus roseus (Pers.) J. Schröt., in Cohn, *Krypt.-Fl. Schlesien* 3: 140 (1889)
Peniophora rosea (Pers.) Massee, *J. Linn. Soc., Bot.* 25: 146 (1890)
Laeticorticium roseum (Pers.) Donk, *Fungus* 26: 17 (1956)
Corticium roseolum Massee, *J. Linn. Soc., Bot.* 27: 140 (1891)
E: ! W: ! NI: !
H: On dead and decayed wood of deciduous trees, often dead attached branches of *Populus* and *Salix* spp.
D: NM3: 178 **D+I:** CNE4: 780-783 (as *Laeticorticium roseum*), B&K2: 98 75 **I:** Kriegl1: 252
Rarely reported, but widespread in England.

CORTINARIUS (Pers.) Gray, *Nat. arr. Brit. pl.* 1: 627 (1821)
Cortinariales, Cortinariaceae
Myxacium (Fr.) P. Kumm., *Führ. Pilzk.*: 22 (1871)
Hydrocybe (Rabenh.) Wünsche, *Pilze*: 87 (1877)
Inoloma (Fr.) Wünsche, *Pilze*: 87 (1877)

Telamonia (Fr.) Wünsche, *Pilze*: 87 (1877)
Dermocybe (Fr.) Wünsche, *Pilze*: 87 (1877)
Phlegmacium (Fr.) Wünsche, *Pilze*: 87 (1877)
Type: *Cortinarius violaceus* (L.) Gray

acetosus (Velen.) Melot, *Doc. Mycol.* 17(68): 65 (1987)
Hydrocybe acetosa Velen., *České Houby* III: 479 (1921)
Mis.: *Cortinarius rigens* sensu Lange (FlDan3: 44 & pl. 100C)
and sensu auct. p.p.
E: ! **NI:** !
H: On soil, in deciduous woodland.
D: NM2: 301 (as *Cortinarius rigens*)
Rarely reported. Records of *C. rigens* from deciduous woodland
are likely to belong here.

acutus (Pers.) Fr., *Epicr. syst. mycol.*: 314 (1838)
Agaricus acutus Pers., *Syn. meth. fung.*: 316 (1801)
Cortinarius acutostriatulus Rob. Henry, *Bull. Soc. Mycol.
France* 83(4): 1001 (1968) [1967]
Cortinarius striatuloides Rob. Henry, *Bull. Soc. Mycol. France*
83(4): 1000 (1968) [1967]
E: ! **S:** ! **W:** ? **NI:** ! **ROI:** !
H: On acidic soil, amongst litter in mixed deciduous and
coniferous woodland, and in dwarf willow beds in montane
habitat.
D+I: CFP3: C46, B&K5: 234 289 **I:** C&D: 333
Often reported and apparently widespread, but most records
unsubstantiated with voucher material.

alboviolaceus (Pers.) Fr., *Epicr. syst. mycol.*: 280 (1838)
Agaricus alboviolaceus Pers., *Syn. meth. fung.*: 286 (1801)
E: o **S:** o **W:** o **NI:** !
H: On acidic soil, usually associated with *Betula*, *Fagus* or
Quercus spp in mixed deciduous woodland; also known in pure
stands of *Salix* spp.
D: NM2: 298 **D+I:** CFP1: A59, B&K5: 206 248 **I:** FlDan3: 92A,
C&D: 327
Occasional but widespread, and may be locally common.

aleuriosmus Maire, *Bull. Soc. Mycol. France* 3: 180 (1910)
Cortinarius caroviolaceus P.D. Orton, *Trans. Brit. Mycol. Soc.*
43(2): 208 (1960)
Mis.: *Cortinarius rapaceus* sensu auct.
E: !
H: Associated with *Fagus* in old deciduous woodland on
calcareous soil.
I: Ph: 126 (as *Cortinarius caroviolaceus*)
Known from southern England (Berkshire, Oxfordshire and
Surrey) but apparently rarely fruiting. For discussion of
synonymy see Orton (1955).

alnetorum (Velen.) M.M. Moser, *Kleine Kryptogamenflora*, Edn
3: 336 (1967)
Telamonia alnetorum Velen., *České Houby* III: 452 (1921)
Hydrocybe alnetorum (Velen.) M.M. Moser, *Kleine
Kryptogamenflora*: 167 (1953)
Mis.: *Cortinarius iliopodius* sensu Moser (1967)
E: ! **S:** ! **W:** ! **NI:** !
H: On wet soil often around lakes or pond margins, or in wet
areas in deciduous woodland, always associated with *Alnus*
spp.
D: NM2: 306 **D+I:** CFP1: A32, B&K5: 234 291
Rarely reported but apparently widespread.

ammophilus A. Pearson, *Trans. Brit. Mycol. Soc.* 29(4): 198
(1946)
E: ! **ROI:** !
H: On dunes, associated with *Salix repens*.
Rarely reported. Described from North Devon (Braunton
Burrows) and also known from the Republic of Ireland (Co.
Dublin: North Bull Island).

amoenolens P.D. Orton, *Trans. Brit. Mycol. Soc.* 43(2): 206
(1960)
Mis.: *Cortinarius anserinus* sensu B&K5 & sensu CFP
Mis.: *Cortinarius glaucopus* sensu Rea (1922)
E: o **W:** !

H: Associated with *Fagus* in old deciduous woodland, often on
calcareous soil.
D: NM2: 295 **D+I:** B&K5: 166 187 (as *Cortinarius anserinus*) **I:**
SV24: 36, C&D: 339 (as *C. anserinus*)
Occasional but widespread in England, from Yorkshire
southwards. Also known from Wales (Monmouthshire).

angelesianus A.H. Sm., *Lloydia* 7(3): 205 (1944)
Cortinarius strobilaceus M.M. Moser, *Nova Hedwigia* 14(2-4):
516 (1968) [1967]
Mis.: *Cortinarius psammocephalus* sensu auct. [non sensu
Lange (1938) = *C. angelesianus*]
E: !
H: On soil, in conifer woodland.
D+I: CFP3: C33, B&K5: 236 292
A single collection from South Devon (Loddiswell).

anomalus (Fr.) Fr., *Epicr. syst. mycol.*: 286 (1838)
Agaricus anomalus Fr., *Observ. mycol.* 2: 73 (1818)
Dermocybe anomala (Fr.) Ricken, *Blätterpilze Deutschl.*: 157
(1912)
Cortinarius lepidopus Cooke, *Grevillea* 16: 43 (1887)
Cortinarius anomalus var. *lepidopus* (Cooke) J.E. Lange, *Fl.
agaric. danic.* 5 *Taxonomic Conspectus*: II (1940)
Cortinarius anomalus f. *lepidopus* (Cooke) Nespiak, *Fl.
Polska*: 66 (1975)
Cortinarius azureus Fr., *Epicr. syst. mycol.*: 286 (1838)
Dermocybe azurea (Fr.) Ricken, *Blätterpilze Deutschl.*: 157
(1912)
Cortinarius azureovelatus P.D. Orton, *Naturalist (Hull)* 1958
(suppl.): 147 (1958)
Cortinarius epsomiensis P.D. Orton, *Naturalist (Hull)* 1958
(suppl.): 147 (1958)
Mis.: *Cortinarius diabolicus* sensu P.D. Orton [Sydowia 36:
225 (1983)]
Mis.: *Cortinarius myrtillinus* sensu Cooke [Ill. Brit. fung. 769
(817) Vol. 6 (1886)]
E: o **S:** o **W:** o **NI:** o **O:** Channel Islands: !
H: On soil, in deciduous woodland.
D: NM2: 298 **D+I:** B&K5: 206 249, B&K5: 208 252 (as
Cortinarius azureus), B&K5: 208 251 (as *C. azureovelatus*) **I:**
C&D: 327
Occasional but widespread and may be locally common. Perhaps
a species complex requiring taxonomic division. For an
extensive discussion see Orton (1958) and for a more recent
species-rich account see Bidaud *et al.* 1992 [Atlas des
Cortinaires Vol. IV].

anthracinus Fr., *Epicr. syst. mycol.*: 288 (1838)
Dermocybe anthracina (Fr.) Ricken, *Blätterpilze Deutschl.*:
159 (1912)
Cortinarius purpureobadius P. Karst., *Symb. mycol. fenn.* 9:
45 (1883)
Cortinarius subanthracinus Rob. Henry, *Bull. Soc. Mycol.
France* 40: 73 (1944)
Dermocybe subanthracina (Rob. Henry) M.M. Moser, *Kleine
Kryptogamenflora*: 153 (1953)
E: ! **S:** ! **W:** ! **NI:** !
H: On clay or loam soil in deciduous woodland, usually
associated with *Fagus* or *Quercus* spp.
D: NM2: 284 **D+I:** CFP3: C03, B&K5: 236 293 **I:** C&D: 331
Occasional in southern England. Rarely reported elsewhere.
Basidiomes are small, intensely blackish-red and rather easily
overlooked.

aprinus Melot, *Doc. Mycol.* 20(77): 93 (1989)
E: !
H: On soil in deciduous woodland.
A single British collection from West Kent (Darenth Wood) in
2000.

argenteopileatus Nezdoïm., *Shlyapochnye Griby SSSR Rod
Cortinarius Fr.* (Leningrad): 107 (1983)
Cortinarius subargentatus P.D. Orton [*nom. illegit.*, non *C.
subargentatus* Murrill (1939)], *Trans. Brit. Mycol. Soc.*
43(2): 385 (1960)

Cortinarius kauffmanianus Rob. Henry [*nom. illegit.*, non *C. kauffmanianus* A.H. Smith (1933)], *Bull. Soc. Mycol. France* 62: 34 (1947)
Mis.: *Cortinarius argentatus* sensu Kauffmann (1918)
E: ! S: !
H: On soil, in deciduous woodland. British material with *Betula* spp.
D+I: B&K5: 218 265 **I:** C&D: 327 (as *Cortinarius argenteopileatus*)
Known from England (South Essex, Warwickshire and West Kent) and Scotland (Easterness).

armillatus (Fr.) Fr., *Epicr. syst. mycol.*: 295 (1838)
Agaricus armillatus Fr., *Syst. mycol.* 1: 214 (1821)
Mis.: *Cortinarius haematochelis* sensu Cooke [Ill. Brit. fung. 801 (803) Vol. 6 (1888)]
E: o S: o W: o NI: !
H: On soil in woodland, usually with mixed *Betula* spp and *Pinus sylvestris*, but occasionally in pure stands of *Pinus* spp.
D: NM2: 300 **D+I:** CFP2: B09, B&K5: 238 295 **I:** FlDan3: 97E, C&D: 331
Occasional but widespread and may be locally common.

arquatus Fr., Cortinarius:, *Epicr. syst. mycol.*: 265 (1838)
Cortinarius cookianus Rob. Henry, *Rev. Mycol. (Paris)* 8(5-6): 19 (1943)
Cortinarius xanthochrous P.D. Orton, *Trans. Brit. Mycol. Soc.* 43(2): 215 (1960)
Mis.: *Cortinarius pansa* sensu P.D. Orton (1955)
Mis.: *Cortinarius calochrous* sensu Rea (1922) & sensu Cooke [Ill. Brit. fung. 707 (713) Vol. 5 (1887)]
E: !
H: Associated with *Fagus* in old beechwoods on calcareous soil.
Known only from Surrey (Mickleham) (type of *C. xanthochrous*).

arvinaceus Fr., *Epicr. syst. mycol.*: 274 (1838)
E: ! NI: !
H: On soil in deciduous woodland. British material associated with *Fagus, Quercus* and *Populus* spp.
D+I: B&K5: 222 272
Known from England (South Hampshire: New Forest) in 1980 and Northern Ireland (Fermanagh: Crom Estate) in 2000. Reported from Scotland, but not since 1927 and unsubstantiated with voucher material. Sensu Cooke 732 (737) Vol. 5 (1888) and sensu Rea (1922) are doubtful.

atropusillus J. Favre, *Ergebn. Wiss. Untersuch. Schweiz. Natn. Parks* 6: 588 (1960)
E: ! NI: ?
H: On soil, associated with *Alnus* in wet areas such as pond or lake margins or in wet woodland.
A single collection from Shropshire (Colemere Country Park) in 1998 and possibly also this species in Northern Ireland (Fermanagh: Lusty Beg).

atrovirens Kalchbr., *Icon. select. Hymenomyc. Hung.* 2: pl. 19 (1874)
E: ? EI: !
H: With conifers on calcareous soil in continental Europe. One recent Irish report with *Dryas octopetala*.
D+I: CFP2: B27, B&K5: 166 189 **I:** FM 2(1): 38, C&D: 341
A little-known species in the British Isles. Illustrated by Cooke [Ill. Brit. fung. 720 (736) Vol. 5 (1888)] and recently reported by Harrington [FM 2(1): 30-33 (2001)] from the Republic of Ireland (The Burren).

aureomarginatus P.D. Orton, *Notes Roy. Bot. Gard. Edinburgh* 41(3): 566 (1984)
E: ! S: !
H: On acidic soil, usually associated with *Betula, Quercus* and *Pinus* spp, in mixed woodland. Rarely with *Salix* spp in wet deciduous woodland.
Known from England (Dorset, South Devon, South Hampshire and Surrey) and Scotland (Perthshire).

aureopulverulentus M.M. Moser, *Sydowia* 6: 152 (1952)
Mis.: *Cortinarius herpeticus* sensu Cooke [Ill. Brit. fung. 722 (849) Vol. 5 (1888)]

E: !
H: On soil, associated with *Fagus* in old deciduous woodland
A single collection from North Somerset (Leigh Woods) in 1992.

balaustinus Fr., *Epicr. syst. mycol.*: 307 (1838)
E: ! S: ! W: ?
H: On soil, associated with with *Betula, Corylus* and *Quercus* spp in old deciduous woodland.
D+I: CFP2: B40 **I:** C&D: 333
Rarely reported. Apparently widespread but most records are unsubstantiated with voucher material. Known with certainty from England (North Somerset, Surrey and West Sussex) and Scotland (Perthshire).

balteatocumatilis P.D. Orton, *Trans. Brit. Mycol. Soc.* 43(2): 207 (1960)
Cortinarius violaceocinctus P.D. Orton, *Trans. Brit. Mycol. Soc.* 43(2): 213 (1960)
Mis.: *Cortinarius balteatus* sensu Cooke [Ill. Brit. fung. 686 (696) Vol. 5 (1887)] & sensu Lange (FlDan3: 21)
E: ! S: ! W: ! ROI: !
H: On soil, usually associated with *Fagus* in old deciduous woodland but also rarely reported with *Quercus* spp.
D: NM2: 292 **D+I:** B&K5: 168 191 **I:** C&D: 337
Known from England (Isle of Wight, South Hampshire, South Somerset, Surrey and West Kent), Scotland (Easterness and Westerness), Wales (Carmarthenshire) and the Republic of Ireland (South Kerry).

barbatus (Batsch) Melot, *Doc. Mycol.* 20(77): 94 (1989)
Agaricus barbatus Batsch, *Elench. fung.*: 39 t. 3 f. 11 (1783)
Mis.: *Cortinarius crystallinus* sensu NCL
E: ! W: ? NI: !
H: On soil, in deciduous woodland.
D+I: CFP2: B56 **I:** C&D: 343
Known from England (Surrey: Esher Common) and Northern Ireland (Fermanagh: Castle Crom). Reported from Wales in 1950 but unsubstantiated with voucher material.

basililaceus P.D. Orton, *Notes Roy. Bot. Gard. Edinburgh* 41(3): 567 (1984)
E: !
H: On soil in old, open deciduous woodland, often in wet areas amongst grass, usually associated with *Quercus* spp.
D+I: Ph: 138-139 (as *Cortinarius hinnuleus*)
Rarely reported. Known from Cheshire, Cumberland, East Norfolk, North Somerset, South Devon, South Hampshire, Surrey and Warwickshire, and may be locally common in some areas such as the New Forest.

basiroseus P.D. Orton, *Notes Roy. Bot. Gard. Edinburgh* 41(3): 569 (1984)
E: !
H: On soil, in deciduous woodland or with scattered trees in parkland. British material associated with *Fagus* and *Quercus* spp.
Known from Berkshire, Oxfordshire and Surrey. A member of the *C. erythrinus* complex.

bataillei J. Favre, *Ergebn. Wiss. Untersuch. Schweiz. Natn. Parks* 6: 515 (1960)
S: !
H: On soil in conifer woodland.
D: NM2: 281
A single collection from Perthshire (Kindrogan) in 1984.

betuletorum M.M. Moser, *Kleine Kryptogamenflora*, Edn 3: 278 (1967)
Agaricus raphanoides Pers., *Syn. meth. fung.*: 324 (1801)
Cortinarius raphanoides (Pers.) Fr., *Epicr. syst. mycol.*: 290 (1838)
E: ! S: !
H: On soil, usually associated with *Betula* spp, and occasionally reported with *Corylus* and *Quercus* spp, in mixed woodland.
D: NM2: 285 **I:** FlDan3: 96A (as *Cortinarius raphanoides*)
Often reported and apparently widespread throughout England and Scotland. However, only four of the records are substantiated with voucher material.

betulinus J. Favre, *Beitr. Kryptogamenfl. Schweiz* 10(3): 213 (1948)
S: !
H: On acidic soil, associated with *Betula* spp in woodland.
D+I: CFP3: C32, B&K5: 222 273
Rarely reported. Known from various sites in the Scottish Highlands but not further south. A species near to *C. delibutus*.

bibulus Quél., *Compt. Rend. Assoc. Franç. Avancem. Sci.* 9: 666 (1881)
    Cortinarius lilacinopusillus P.D. Orton, *Notes Roy. Bot. Gard. Edinburgh* 38(2): 318 (1980)
    Cortinarius pulchellus J.E. Lange, *Medd. foren. Svampek. frem.* 4: 3 (1926)
E: ! **S:** ! **W:** !
H: On soil, always associated with *Alnus* in wet areas in woodland, around the margins of ponds or lakes and especially in dried-out alder swamps.
D: NM2: 304 **D+I:** CFP2: B25, B&K5: 238 297 **I:** C&D: 331
Rarely reported but widespread. Basidiomes are often small, cryptically coloured and grow in inaccessible habitat thus possibly overlooked.

biformis Fr., *Epicr. syst. mycol.*: 299 (1838)
E: ! **S:** ?
H: On soil, in deciduous woodland. British material associated with *Quercus* spp.
D+I: CFP3: C22, CFP3: C05, B&K5: 240 298
Rarely reported and often misidentified (usually as *C. tabacinus*). Apparently widespread in England and reported from Scotland but known with certainty in recent years only from Cumberland.

bivelus (Fr.) Fr., *Epicr. syst. mycol.*: 292 (1838)
    Agaricus bivelus Fr., *Observ. mycol.* 2: 58 (1818)
E: ! **S:** ! **W:** !
H: On soil, associated with *Betula* spp, in mixed deciduous woodland.
D: NM2: 303 **D+I:** CFP2: B30, B&K5: 240 299 **I:** C&D: 333
Rarely reported but apparently widespread.

bolaris (Pers.) Fr., *Epicr. syst. mycol.*: 282 (1838)
    Agaricus bolaris Pers., *Syn. meth. fung.*: 291 (1801)
E: o **S:** o **W:** o **NI:** o **ROI:** !
H: On acidic or sandy soil in open deciduous woodland, associated with *Fagus* or *Quercus* spp and often with *Vaccinium myrtillus*.
D: NM2: 286 **D+I:** CFP3: C23, B&K5: 152 167 **I:** FlDan3: 93A, C&D: 327, SV33: 24
Occasional but widespread and may be locally common.

bovinus Fr., *Epicr. syst. mycol.*: 297 (1838)
E: ? **S:** !
H: On soil, in woodland.
D: NM2: 303 **D+I:** B&K5: 240 300 **I:** FlDan3: 98C
Known with certainty from Scotland (Clyde Isles). Reported from England but records are unsubstantiated with voucher material.

brunneus (Pers.) Fr., *Epicr. syst. mycol.*: 298 (1838)
    Agaricus brunneus Pers., *Syn. meth. fung.*: 274 (1801)
E: ! **S:** ! **W:** ? **NI:** ! **ROI:** ?
H: On soil, in coniferous woodland. British material associated with *Picea* and *Pinus* spp.
D: NM2: 303 **D+I:** CFP2: B07, B&K5: 242 302 **I:** FlDan3: 99E, C&D: 331
Rarely reported but apparently widespread. Known from England (Isle of Wight, South Devon South Hampshire, West Kent and West Sussex), Scotland (Easterness and Perthshire) and Northern Ireland (Fermanagh), but few recent records and most are unsubstantiated with voucher material.

brunneus *var.* glandicolor (Fr.) H. Lindstr. & Melot, *Cortinarius*: 33 (1992)
    Agaricus gentilis β *glandicolor* Fr., *Syst. mycol.* 1: 213 (1821)
    Cortinarius glandicolor (Fr.) Fr., *Epicr. syst. mycol.*: 298 (1838)
    Cortinarius glandicolor var. *curtus* Rea, *Brit. basidiomyc.*: 176 (1922)
    Mis.: *Cortinarius microcyclus* sensu Cooke [Ill. Brit. fung. 793 (865) Vol. 6 (1886)]
    Mis.: *Cortinarius rubricosus* sensu Lange (FlDan3: 45 & pl. 100F)
E: c **S:** c **W:** c **NI:** !
H: On soil, associated with *Betula* spp in woodland. Reported with other trees such as *Fagus* or *Fraxinus* but birch is always present nearby.
D+I: CFP2: B35, B&K5: 242 303 (as *Cortinarius brunneus* var. *glandicolor* f. *curtus*)
Common and widespread.

bulbosus (Sowerby) Fr., *Epicr. syst. mycol.*: 292 (1838)
    Agaricus bulbosus Sowerby [non *A. bulbosus* Bolton (1791); non Schaeff. (1770)], *Col. fig. Engl. fung.* 2: pl. 130 (1798)
    Mis.: *Cortinarius bovinus* sensu Lange (FlDan3: 37 & pl. 98 C&D)
E: ! **S:** ? **NI:** ?
H: On soil in deciduous woodland.
D+I: Ph: 137, B&K5: 244 304
Poorly known in Britain and reported, in recent times, only from South Hampshire (Stubbs Wood). All other records are unsubstantiated with voucher material.

bulliardii (Pers.) Fr., *Epicr. syst. mycol.*: 282 (1838)
    Agaricus bulliardii Pers., *Observ. mycol.* 2: 43 (1799)
    Cortinarius pseudocolus M.M. Moser, *Schweiz. Z. Pilzk.* 43: 123 (1965)
E: ! **NI:** !
H: On soil, in deciduous woodland. British material usually associated with *Quercus* spp.
D: NM2: 300 **D+I:** B&K5: 244 305 **I:** SV27: 49, C&D: 331
Rarely reported. Known from widely scattered sites from Yorkshire southwards, and reported from Northern Ireland.

caerulescens (Schaeff.) Fr., *Epicr. syst. mycol.*: 265 (1838)
    Agaricus caerulescens Schaeff., *Fung. Bavar. Palat. nasc.* 1: 17 (1762)
    Agaricus cyanus β *caerulescens* (Schaeff.) Pers., *Syn. meth. fung.*: 277 (1801)
    Cortinarius caesiocyaneus Britzelm., *Zur Hymenomyceten-Kunde* 1: 10 (1895)
E: ! **S:** ? **W:** ? **NI:** ?
H: On soil, usually associated with *Fagus* in old deciduous woodland.
D: NM2: 294 **D+I:** CFP2: B51, B&K5: 170 193 **I:** SV24: 36 (as *Cortinarius coerulescens*), C&D: 339
British collections only from scattered sites in southern and south-western England. Records from elsewhere are all old and lack voucher material.

caesiocanescens M.M. Moser, *Sydowia 6:* 21 (1952)
ROI: !
H: On soil, in grassland with *Dryas octopetala*.
D+I: CFP2: B42, B&K5: 170 194
Known from the The Burren.

caesiostramineus Rob. Henry, *Bull. Soc. Mycol. France* 55(1): 73 (1939)
    Cortinarius amarescens M.M. Moser, *Kleine Kryptogamenflora*, Edn 3: 283 (1967)
E: !
H: On soil, associated with *Fagus* in old deciduous woodland.
D: NM2: 292 **D+I:** CFP1: A28 **I:** SV24: 37
A single collections by D.A. Reid (Hertfordshire: Tring) in 1960; a further collection, possibly this species, from Middlesex (Ruislip) in 1945.

cagei Melot, *Doc. Mycol.* 20(80): 58 (1990)
    Cortinarius bicolor Cooke [*nom. inval.*, non *C. bicolor* Gray (1821)], *Grevillea* 16: 45 (1887)
E: ! **S:** !
H: On soil in deciduous or mixed woodland.
D+I: CFP4: D48, B&K5: 246 307

Known from England (Bedfordshire, North Hampshire, West Sussex and Yorkshire) and Scotland (South Aberdeen).

caledoniensis P.D. Orton, *Notes Roy. Bot. Gard. Edinburgh* 26: 44 (1964)
S: !
H: On soil, associated with *Pinus sylvestris* in Caledonian pinewoods.
A poorly known species, rarely reported.

callisteus (Fr.) Fr., *Epicr. syst. mycol.*: 281 (1838)
Agaricus callisteus Fr., *Observ. mycol.* 2: 51 (1818)
E: ? **S:** ! **W:** ! **NI:** ?
H: On soil, associated with conifers in woodland or plantations. British material with *Picea abies* and *Pinus sylvestris*.
D: NM2: 287 **D+I:** B&K5: 152 168 **I:** FlDan3: 92B, C&D: 325
Most collections from scattered sites in Scotland, and a single one from Wales (Denbighshire: Abergele). Reported from England (Herefordshire) in the late nineteenth century, and from Northern Ireland, but unsubstantiated with voucher material.

calochrous (Pers.) Gray, *Nat. arr. Brit. pl.* 1: 629 (1821)
Agaricus calochrous Pers., *Syn. meth. fung.*: 282 (1801)
E: o **NI:** !
H: On calcareous soil, associated with *Fagus* in old deciduous woodland.
D: NCL3: 207, NM2: 295 **D+I:** B&K5: **I:** C&D: 339
Rarely reported. Occasional in England, from Westmorland southwards, and reported from Northern Ireland.

calochrous *var.* **coniferarum** Quadr., *Doc. Mycol.* 16(56): 28 (1986)
Phlegmacium calochroum var. *coniferarum* M.M. Moser [*nom. inval.*], *Die Gattung* Phlegmacium: 353 (1960)
ROI: !
H: On soil, in grassland on limestone, associated with *Dryas octopetala* (British material).
D: NM2: 295 **D+I:** CFP1: A40, B&K5: 172 196
Known from The Burren.

calochrous *var.* **parvus** (Rob. Henry) Brandrud, *Cortinarius*: 33 (1992)
Cortinarius parvus Rob. Henry, *Bull. Soc. Mycol. France* 51: 81 (1935)
E: !
H: On calcareous soil, associated with *Fagus* in deciduous woodland.
D+I: CFP2: B53, B&K5: 170 195
British collections from widely scattered locations in southern England (West Kent to North Somerset).

camphoratus Fr., *Epicr. syst. mycol.*: 280 (1838)
Agaricus camphoratus Fr. [*nom. illegit.*, non *A. camphoratus* Bull. (1791)], *Syst. mycol.* 1: 218 (1821)
Cortinarius hircinus Fr., *Hymenomyc. eur.*: 362 (1874)
S: !
H: On soil in conifer woodland.
D: NM2: 298 **D+I:** CFP1: A12, B&K5: 210 253 **I:** C&D: 327
Known from scattered sites in Scotland. All records from England are with deciduous trees, dubious, and unsubstantiated with voucher material.

camptoros Brandrud & Melot, *Bull. Soc. Mycol. France* 99(2): 219 (1983)
E: !
H: On soil in deciduous woodland. British collection associated with *Fagus*.
A single collection in 2004 from East Sussex (Flatropers Wood).

caninus (Fr.) Fr., *Epicr. syst. mycol.*: 285 (1838)
Agaricus anomalus var. *caninus* Fr., *Syst. mycol.* 1: 221 (1821)
Dermocybe canina (Fr.) Ricken, *Blätterpilze Deutschl.*: 156 (1912)
Cortinarius anomalus ssp. *caninus* (Fr.) Konrad & Maubl. *Icon. select. fung.* 6: 169 (1930)
E: ! **S:** ! **W:** ! **NI:** ! **ROI:** ?

H: On soil in woodland with deciduous trees or conifers (but see comments below).
D: NM2: 298 **D+I:** B&K5: 210 254 **I:** C&D: 327
Often interpreted as a robust montane species associated with conifers, but treated by Orton (1958) as a member of the *C. anomalus* complex from lowland sites (e.g. the New Forest) associated with deciduous trees. Sensu Orton apparently frequent and widespread throughout Britain.

cedretorum Maire, *Bull. Soc. Mycol. France* 30: 210 (1914)
E: !
H: On calcareous soil, associated with *Fagus* in old woodland.
I: C&D: 341, SV40: 23
Known from three sites in Surrey (Mickleham: Norbury Park, West Horsley: Sheepleas, and Polesden Lacey) but fruiting only rarely.

cephalixus Fr., *Epicr. syst. mycol.*: 261 (1838)
E: !
H: On soil in deciduous woodland. British material associated with *Tilia* spp.
D+I: B&K5: 172 198
A single collection from South Essex (Epping Forest) in 2000.

chrysomallus Lamoure, *Travaux Scientifiques du Parc National de la Vanoise* 8: 135 (1977)
E: !
H: On sandy soil.
A single collection from Cumberland (Sandscale Haws NNR) in 1999.

cinnabarinus Fr., *Epicr. syst. mycol.*: 287 (1838)
Dermocybe cinnabarina (Fr.) Wünsche, *Die Pilze*: 125 (1877)
E: ! **S:** ! **W:** ! **NI:** !
H: On soil, in deciduous woodland, usually associated with *Fagus*.
D: NM2: 284 **D+I:** CFP3: C41, B&K5: 248 311 **I:** C&D: 329
Known from widely scattered localities in southern England and single collections from Scotland, Wales and Northern Ireland.

cinnamomeoluteus P.D. Orton, *Trans. Brit. Mycol. Soc.* 43(2): 217 (1960)
Dermocybe cinnamomeolutescens M.M. Moser [*nom. nud.*], *Kleine Kryptogamenflora*: 151 (1953)
Cortinarius cinnamomeolutescens Bon, *Doc. Mycol.* 19(73): 66 (1988)
E: ! **S:** ! **W:** !
H: On acidic or sandy soil, often associated with *Betula* and *Pinus* spp; also reported with *Castanea*, *Fagus*, *Salix* spp and *Picea abies*. Said by Orton [TBMS 43: 217 (1960)] to be 'particularly abundant in damp woods with willows and alders.'
D: NM2: 283 **I:** C&D: 329
Rarely reported but apparently widespread.

cinnamomeus (L.) Gray, *Nat. arr. Brit. pl.* 1: 630 (1821)
Agaricus cinnamomeus L., *Sp. pl.*: 1205 (1753)
E: o **S:** o **W:** ! **NI:** !
H: On soil, in mixed woodland, usually associated with *Betula* and *Pinus* spp. Also reported with *Quercus* spp in oakwoods and with *Salix* spp in wet woodland alongside lakes or large ponds.
D: NM2: 281 **D+I:** CFP2: B39, B&K5: 142 153 **I:** C&D: 329
Occasional but widespread.

cinnamoviolaceus M.M. Moser, *Nova Hedwigia* 14(2-4): 514 (1968) [1967]
S: !
H: On soil, in woodland with conifers and *Betula* spp.
D+I: B&K5: 248 312
A single collection from Perthshire (Tulloch) in 1971.

citrinus P.D. Orton, *Trans. Brit. Mycol. Soc.* 43(2): 208 (1960)
Cortinarius elegantior f. *citrinus* J.E. Lange [*nom. inval.*] *Dansk. Bot. Ark.* 8 (7): 17 (1935)
Cortinarius sulphureus var. *citrinus* (J.E. Lange) J.E. Lange [*nom. inval.*], *Fl. agaric. danic.* 3: 18 (1938)
Cortinarius pseudosulphureus P.D. Orton, *Trans. Brit. Mycol. Soc.* 43(2): 211 (1960)

Mis.: *Cortinarius sulphureus* sensu Orton (1955)
E: !
H: Associated with *Fagus* in old beechwoods on calcareous soil.
D: NM2: 296 **D+I:** CFP4: D05, B&K5: 174 199 **I:** SV40: 18
Known from East Gloucestershire, Surrey and West Sussex.

claricolor (Fr.) Fr., *Epicr. syst. mycol.*: 257 (1838)
  Agaricus multiformis β *claricolor* Fr., *Observ. mycol.* 2: 65
    (1818)
  Phlegmacium subclaricolor M.M. Moser, *Kleine*
    Kryptogamenflora: 181 (1953)
  Cortinarius subclaricolor (M.M. Moser) P.D. Orton, *Trans. Brit.*
    Mycol. Soc. 43(2): 175 (1960)
  Mis.: *Cortinarius turmalis* sensu NCL
E: ? **S:** ? **ROI:** ?
H: On acidic soil, in mixed woodland, associated with *Betula* spp
  or *Pinus sylvestris*.
D: NM2: 293 **D+I:** CFP2: B48, B&K5: 174 200 **I:** C&D: 335,
  SV48: 57
Poorly known in Britain (and possibly not authentically British).
  Most of the records are old and unsubstantiated with voucher
  material. Sensu Orton (1955) is *C. durus* and sensu Lange
  (1938) is *C. vulpinus*.

coerulescentium Rob. Henry, *Bull. Soc. Mycol. France* 67: 282
  (1952) [1951]
E: !
H: On calcareous soil, associated with *Fagus* in old beechwoods.
D+I: CFP2: B17
A single collection from Oxfordshire (Bix, Warburg Reserve) in
  1998.

collinitus (Sowerby) Gray, *Nat. arr. Brit. pl.* 1: 628 (1821)
  Agaricus collinitus Sowerby, *Col. fig. Engl. fung.* 1: pl. 9
    (1797)
  Cortinarius muscigenus Peck, *Rep. (Annual) New York State*
    Mus. Nat. Hist. 41: 71 (1888)
E: ? **S:** ! **W:** ? **NI:** ? **ROI:** ?
H: On acidic soil in conifer woodland, associated with *Pinus*
  sylvestris or *Picea abies*.
D: NM2: 289 (as *Cortinarius muscigenus*) **I:** FlDan3: 88B, C&D:
  343
Rarely reported. Most records are from widely scattered
  locations in Scotland. Reported elsewhere but unsubstantiated
  with voucher material. Records associated with deciduous
  trees are suspect.

conicus (Velen.) Rob. Henry, *Bull. Soc. Mycol. France* 57: 21
  (1942) [1941]
  Telamonia conica Velen., *České Houby* III: 447 (1921)
E: !
H: On soil, in deciduous woodland.
D+I: CFP3: C38
Known from East Sussex and West Kent.

corrosus Fr., *Epicr. syst. mycol.*: 266 (1838)
S: !
H: On soil in woodland, associated with conifers. British material
  with *Pinus sylvestris*.
A single collection from Easternesss (Coylum Bridge) in 1957.

cotoneus Fr., *Epicr. syst. mycol.*: 289 (1838)
  Mis.: *Cortinarius sublanatus* sensu auct.
E: ! **S:** ! **NI:** ?
H: On soil in deciduous woodland.
D: NM2: 286 **D+I:** CFP2: B01, C&D: 325, B&K5: 154 170
Known from England (Dorset and Hertfordshire) and Scotland
  (Easter Ross). Reported elsewhere but unsubstantiated with
  voucher material.

crassus Fr., *Epicr. syst. mycol.*: 257 (1838)
  Cortinarius pseudocrassus P.D. Orton, *Trans. Brit. Mycol. Soc.*
    43(2): 217 (1960)
S: !
H: On soil, associated with *Pinus sylvestris* and *Betula* spp, in
  Caledonian pinewoods.
D+I: B&K5: 202 **I:** CFP2: B41

Rarely reported. English collections at K are misidentified *C.*
  balteatoalbus. Often placed in subgenus *Phlegmacium*.

croceocaeruleus (Pers.) Fr., *Epicr. syst. mycol.*: 269 (1838)
  Agaricus croceocaeruleus Pers., *Icon. descr. fung.*: pl. 2
    (1798)
E: o **S:** ? **W:** ! **NI:** !
H: Associated with *Fagus* in old beechwoods on calcareous soil.
D: NM2: 288 **D+I:** CFP2: B49 **I:** FlDan3: 90C, C&D: 343
Occasional in southern England. Single collections from Wales
  (Denbighshire: Colwyn Bay) and Northern Ireland (Down:
  Tollymore). Reported from Scotland in 1938 in conifer
  woodland on acidic soils, but unsubstantiated with voucher
  material. The epithet is often found as '*croceocoeruleus*' in
  literature.

croceus (Schaeff.) Gray, *Nat. arr. Brit. pl.* 1: 630 (1821)
  Agaricus croceus Schaeff., *Fung. Bavar. Palat. nasc.* 1: 3
    (1762)
  Agaricus cinnamomeus δ *croceus* (Schaeff.) Fr., *Syst. mycol.*
    1: 229 (1821)
  Dermocybe crocea (Schaeff.) Høil., *Opera Bot.* 71: 83 (1983)
  Cortinarius cinnamomeobadius Rob. Henry [*nom. nud.*], *Bull.*
    Soc. Mycol. France 55: 300 (1940) [1939]
  Dermocybe cinnamomeobadia (Rob. Henry) M.M. Moser
    [*nom. nud.*], *Schweiz. Z. Pilzk.* 52(7): 97 (1974)
E: ! **S:** ! **W:** ! **NI:** ! **ROI:** !
H: On acidic soil, in deciduous or mixed woodland, often
  amongst *Polytrichum* moss. Known with *Betula*, *Fagus* and
  Quercus spp and *Picea* and *Pinus* spp.
D: NM2: 282 **D+I:** CFP2: B16, B&K5: 144 155 **I:** C&D: 329
Rarely reported but apparently widespread.

cumatilis Fr., *Epicr. syst. mycol.*: 269 (1838)
E: ? **S:** !
H: On soil, associated with *Betula* spp, in woodland.
D: NM2: 291 **D+I:** CFP1: A47, B&K5: 176 203 **I:** C&D: 335,
  Cooke 723 (726) Vol. 5 (1887)
Rarely reported and only in northern birch woods. Collections
  from England are almost certainly misidentified. Closely related
  to *C. praestans*.

cyanites Fr., *Epicr. syst. mycol.*: 279 (1838)
S: ! **NI:** ?
H: On acidic soil, with *Betula* spp and *Pinus sylvestris*.
D+I: CFP2: B02, B&K5: 210 255 **I:** SV31: 13, C&D: 325
Known from a few scattered sites in Scotland. Reported from
  Northern Ireland but unsubstantiated with voucher material.

cyanopus Fr., *Epicr. syst. mycol.*: 258 (1838)
S: !
H: On soil, associated with *Betula* spp, in the Scottish Highlands.
Poorly known in Britain and confined to Scotland *fide* Orton
  (1986). Retained sensu NCL. Sensu Cooke (1887) is *C. largus*.

damascenus Fr., *Epicr. syst. mycol.*: 304 (1838)
E: ? **S:** ? **NI:** !
H: On soil, in deciduous woodland. British material associated
  with *Betula* spp.
D: NM2: 301 **D+I:** B&K5: 250 314 **I:** FlDan3: 100E
A single collection from Fermanagh (Kilgarrow Woods) in 2000.
  Reported from England in 1945 and Scotland in 1912, but
  these records are dubious and unsubstantiated with voucher
  material.

danicus Høil., *Opera Bot.* 71: 106 (1984) [1983]
E: !
H: On soil, associated with *Fagus* and *Quercus* spp.
D: NM2: 284
Known from South Hampshire (New Forest).

danili Rob. Henry, *Bull. Soc. Mycol. France* 59: 57 (1943)
NI: !
H: On soil in deciduous woodland.
A single collection, from Fermanagh (Castle Coole) in 2000.

decipiens (Pers.) Fr., *Epicr. syst. mycol.*: 312 (1838)
  Agaricus decipiens Pers., *Syn. meth. fung.*: 298 (1801)

E: c **S:** c **W:** c **NI:** c **ROI:** !
H: On soil, associated with *Alnus, Betula, Corylus, Fagus, Quercus, Populus tremula* and *Salix* spp, in mixed deciduous woodland.
D: NM2: 306 **D+I:** CFP3: C02, B&K5: 250 315 **I:** C&D: 333
Common and widespread.

decipiens *var.* **atrocaeruleus** (M.M. Moser) H. Lindstr., *Cortinarius* 4: 11 (1998)
    Hydrocybe atrocaerulea M.M. Moser, *Bull. Soc. Naturalistes Oyonnax* 7: 124 (1953)
    Cortinarius atrocaeruleus (M.M. Moser) M.M. Moser, *Kleine Kryptogamenflora, Edn* 3: 336 (1967)
    Cortinarius sertipes Kühner, *Bull. Mens. Soc. Linn. Lyon* 24: 40 (1955)
E: !
H: On soil, in deciduous woodland.
D+I: CFP4: D44, B&K5: 250 314
Known from West Norfolk.

delibutus Fr., *Epicr. syst. mycol.*: 276 (1838)
E: ! **S:** ! **W:** ? **NI:** !
H: On acidic soil, in deciduous woodland. Often associated with *Betula* spp, *Corylus* and *Fagus* and less frequently with *Castanea* or *Quercus* spp.
D: NM2: 289 **D+I:** CFP3: C11, B&K5: 224 275 **I:** C&D: 343
Occasional but widespread.

depressus Fr., *Epicr. syst. mycol.*: 314 (1838)
    Cortinarius adalbertii J. Favre, *Beitr. Kryptogamenfl. Schweiz* 10(3): 104 (1948)
S: !
H: On acidic soil in conifer woodland. British material amongst *Calluna* under *Pinus sylvestris* in Caledonian pinewoods.
D+I: CFP2: B18
A single collection from Easterness (Rothiemurchus) in 1950 (as *C. adalbertii*).

diabolicoides Moënne-Locc. & Reumaux, *Atlas des Cortinaires* 4: 105 (1992)
NI: !
H: On soil in deciduous woodland.
Two collections from Fermanagh (Castle Coole and Aghagrefin NNR) in 2000. Part of the *C. anomalus* group.

diasemospermus Lamoure, *Travaux Scientifiques du Parc National de la Vanoise* 9: 99 (1978)
E: ! **NI:** !
H: On soil in deciduous woodland.
D+I: CFP4: D47, B&K5: 252 316
Known from London and Fermanagh. A recent segregate from *C. paleaceus*.

dibaphus Fr., *Epicr. syst. mycol.*: 266 (1838)
    Cortinarius nemorosus Rob. Henry, *Bull. Soc. Mycol. France* 52: 168 (1936)
    Cortinarius dibaphus var. *nemorosus* (Rob. Henry) Rob. Henry, *Bull. Soc. Mycol. France* 102(1): 93 (1986)
E: ! **S:** ?
H: On acidic soil, associated with *Fagus* in deciduous woodland.
D+I: CFP3: C26 **I:** C&D: 339
Known from South Devon (Mamhead, near Exeter). Reported from East Gloucestershire, Herefordshire and North Somerset, also from various localities in Scotland, but unsubstantiated with voucher material.

dionysae Rob. Henry, *Doc. Mycol.* 20(77): 69 (1989)
E: ! **NI:** ! **ROI:** !
H: On basic or calcareous soil, in deciduous woodland or associated with *Helianthemum nummularium* in unimproved grassland.
D+I: CFP2: B50, B&K5: 178 206 **I:** C&D: 339, SV43: 24
Known from England (Staffordshire and Derbyshire), Northern Ireland (Fermanagh) and the Republic of Ireland.

disjungendus P. Karst., *Acta Soc. Fauna Fl. Fenn.* 9(1): 6 (1893)

Mis.: *Cortinarius brunneus* sensu Cooke [Ill. Brit. fung. 810 (854) Vol. 6 (1887)]
E: ! **NI:** !
H: On soil, associated with *Betula* spp, in deciduous woodland.
D+I: CFP4: D33
Reported from Fermanagh in 2000 and recorded from East Kent in 2001.

durus P.D. Orton, *Trans. Brit. Mycol. Soc.* 43(2): 209 (1960)
    Cortinarius balteatoclaricolor Jul. Schäff., *Ber. Bayer. Bot. Ges.* 27: 217 (1947)
    Mis.: *Cortinarius claricolor* sensu Orton (1955)
E: ! **S:** !
H: On soil in mixed deciduous woodland or in mixed woodland with conifers.
Known from England (South Hampshire and West Sussex) and Scotland (Easterness and Mid-Perthshire).

elegantior (Fr.) Fr., *Epicr. syst. mycol.*: 267 (1838)
    Agaricus multiformis β *elegantior* Fr., *Observ. mycol.* 2: 64 (1818)
    Mis.: *Cortinarius fulmineus* sensu Rea (1922)
    Mis.: *Cortinarius turbinatus* sensu Cooke [Ill. Brit. fung. 714 (714) Vol. 5 (1887)]
E: !
H: On soil. A conifer associate in continental Europe but British material reported with *Fagus*.
D: NM2: 297 **D+I:** CFP4: D01, B&K5: 178 207
Known from Surrey. Reported from Norfolk and Warwickshire, but unsubstantiated with voucher material.

elegantissimus Rob. Henry, *Doc. Mycol.* 20(77): 69 (1989)
    Phlegmacium aureoturbinatum M.M. Moser [*nom. inval.*], *Die Gattung* Phlegmacium: 294 (1960)
    Cortinarius aurantioturbinatus J.E. Lange [*nom. inval.*], *Fl. agaric. danic.* 3: 19 (1939)
E: !
H: Associated with *Fagus* in old beechwoods on calcareous soil.
D: NM2: 296 **D+I:** CFP3: C54, B&K5: 180 208 **I:** SV24: 40, C&D: 341
Known from Berkshire (Sulham Woods near Reading). Rarely reported elsewhere, from West Kent to Herefordshire, but unsubstantiated with voucher material.

emollitus Fr., *Epicr. syst. mycol.*: 269 (1838)
    Mis.: *Cortinarius crystallinus* sensu Cooke [Ill. Brit. fung. 725 (728) Vol. 5 (1887)] and sensu Rea (1922)
E: ! **S:** ? **NI:** !
H: On soil in deciduous or mixed woodland.
D+I: B&K5: 226 277
Known from England (Hertfordshire and West Kent) and Northern Ireland (Antrim). Reported from Scotland in 1908 but unsubstantiated with voucher material.

emunctus Fr., *Epicr. syst. mycol.*: 275 (1838)
S: !
H: On acidic soil, associated with conifers.
D+I: CFP1: A03
A single collection from Easterness (Rothiemurchus, Lochan Mhor) in 1957. Reported from Wester Ross in 1994 but unsubstantiated with voucher material.

erubescens M.M. Moser, *Nova Hedwigia* 14(2-4): 515 (1968) [1967]
W: !
H: On soil in deciduous woodland.
D+I: CFP4: D51
A single collection from Caernarvonshire (Coed Llyn Mair) in 2001.

erythrinus (Fr.) Fr., *Epicr. syst. mycol.*: 312 (1838)
    Agaricus caesius δ *erythrinus* Fr., *Observ. mycol.* 2: 44 (1818)
E: ! **S:** ? **W:** ?
H: on soil, with conifers.
D: NM2: 306 **I:** FlDan3: 103E
Poorly known in Britain, rarely reported and the majority of records unsubstantiated with voucher material. Retained sensu

Lange (FlDan3: 46) as an autumnal conifer associate. Sensu Ricken (1915) is *C. vernus*, associated with deciduous trees and sometimes fruiting in spring.

evernius (Fr.) Fr., *Epicr. syst. mycol.*: 294 (1838)
  Agaricus evernius Fr., *Observ. mycol.* 2: 79 (1818)
  Cortinarius evernius var. *fragrans* M.M. Moser [*nom. inval.*], *Guida alla Determinazione dei Funghi*: 432 (1986)
E: ? **S:** ! **NI:** ?
H: On acidic soil in mixed woodland of *Betula* spp and *Pinus sylvestris*.
D: NM2: 302 **D+I:** CFP1: A11, B&K5: 254 319 **I:** C&D: 331
Rare and perhaps confined to native woodland in Scotland. Reported from England and Northern Ireland, but unsubstantiated with voucher material.

favrei D.M. Hend., *Notes Roy. Bot. Gard. Edinburgh* 22: 593 (1958)
  Myxacium favrei M.M. Moser [*nom. nud.*], *Kleine Kryptogamenflora*, Edn 2: 195 (1955)
  Mis.: *Cortinarius alpinus* sensu auct.
S: !
H: On soil, with *Salix* spp, in boreal or montane habitat. British material with *Salix herbacea* and rarely *S. repens*.
D+I: B&K5: 226 279 **I:** C&D: 343
Known from Orkney.

fervidus P.D. Orton, *Notes Roy. Bot. Gard. Edinburgh* 26: 47 (1964)
S: !
H: On soil, with *Pinus sylvestris* in Caledonian pinewoods.
D: NM2: 281 **D+I:** CFP1: A14
Known only from the type collection from Easterness (Rothiemurchus) in 1960.

flexipes (Pers.) Fr., *Epicr. syst. mycol.*: 300 (1838)
  Agaricus flexipes Pers., *Syn. meth. fung.*: 275 (1801)
  Cortinarius paleiferus Svrček, *Česká Mykol.* 22(4): 276 (1968)
  Mis.: *Cortinarius paleaceus* sensu auct. p.p.
E: ! **S:** ! **W:** ! **NI:** ! **ROI:** !
H: On soil in deciduous woodland.
D: NM2: 305 (as *Cortinarius paleaceus*) **D+I:** CFP4: D43, B&K5: 256 323
Apparently widespread, but often confused with similar species of *Telamonia*.

flexipes var. **flabellus** (Fr.) Lindst. & Melot, *Cortinarius* 4: 20 (1998)
  Agaricus flabellus Fr., *Syst. mycol.* 1: 231 (1821)
  Cortinarius flabellus (Fr.) Fr., *Epicr. syst. mycol.*: 300 (1838)
  Mis.: *Cortinarius paleaceus* sensu auct. p.p.
E: c **S:** c **W:** c **NI:** c **ROI:** c
H: On soil, associated with *Betula* spp, or *Fagus*, in deciduous woodland or scrub.
D+I: CFP4: D35, CFP4: D34, CFP4: D45, B&K5: 256 322
Common and widespread. Many collections are determined as *C. palaceus*.

fulvescens Fr., *Epicr. syst. mycol.*: 311 (1838)
  Mis.: *Cortinarius fasciatus* sensu Lange (FlDan3: 50 & pl. 104D)
E: ! **S:** !
H: On soil, in woodland, associated with *Pinus sylvestris*.
D: NM2: 305 **D+I:** B&K5: 254 321 (as *Cortinarius fasciatus*), B&K5: 256 324 **I:** FlDan3: 104D (as *C. fasciatus*), C&D: 333 (as *C. fasciatus*)
A few collections from scattered sites in Scotland and a single collection from Surrey in 1945. Reported from South Hampshire and West Sussex, but unsubstantiated with voucher material.

fulvoincarnatus Joachim, *Bull. Soc. linn Seine-Mart.* 24: 62 (1936)
E: !
H: Associated with *Fagus* in old deciduous woodland.
A single collection from Worcestershire (Buckwood) in 1992.

fulvoochraceus var. **cyanophyllus** Rob. Henry, *Bull. Soc. Mycol. France* 82: 154 (1966)
  Cortinarius cyanophyllus (Rob. Henry) M.M. Moser [*comb. inval.*], *Kleine Kryptogamenflora*, Edn 4: 366 (1978)
W: !
H: On soil. British material amongst grass, with *Quercus* spp.
A single collection from Denbighshire (Colwyn Bay) in 1988.

fulvosquamosus P.D. Orton, *Kew Bull.* 31(3): 709 (1977)
E: ! **S:** ! **W:** !
H: On sandy soil, associated with *Betula* spp and *Salix repens*, in deciduous woodland or on dunes.
Known from England (South Devon, South Hampshire and Northumberland), Scotland (Angus) and Wales (Merionethshire).

galeobdolon Melot, *Acta. Bot. Islandica.* 12: 91 (1995)
  Mis.: *Cortinarius causticus* sensu NCL & sensu auct. mult.
E: !
H: Usually associated with *Fagus* in old beechwoods or deciduous woodland on calcareous soil. Rarely reported with *Betula* and *Quercus* spp.
Known in scattered localites from Durham southwards but mostly in southern counties. Widely referred to *C. causticus*, but that is associated with conifers in northern areas.

gausapatus J. Favre, *Ergebn. Wiss. Untersuch. Schweiz. Natn. Parks* 5: 202 (1955)
S: !
H: On soil, associated with *Salix herbacea* in montane habitat.
D+I: B&K5: 258 327, B&K5: 258 325
Known from Cairngorm in 1984.

gentilis (Fr.) Fr., *Epicr. syst. mycol.*: 297 (1838)
  Agaricus gentilis Fr., *Syst. mycol.* 1: 212 (1821)
E: ! **S:** o **NI:** !
H: On acidic soil, usually associated with *Pinus sylvestris* and most often in Caledonian pinewoods.
D: NM2: 287 **D+I:** CFP2: B31, B&K5: 154 171
Occasional but widespread in Scotland. Rare or unreported elsewhere. Reported with *Populus* and *Salix* spp in England but these are misidentifications.

glaucopus (Schaeff.) Fr., *Epicr. syst. mycol.*: 264 (1838)
  Agaricus glaucopus Schaeff., *Fung. Bavar. Palat. nasc.* 1: 53 (1762)
  Mis.: *Cortinarius herpeticus* sensu NCL
E: ! **S:** ?
H: On soil, in deciduous woodland. British material with *Fagus*.
D: NM2: 295 **D+I:** CFP3: C52, CFP3: C30, B&K5: 180 210 **I:** C&D: 339
Known from South Essex and West Sussex. Rarely reported elsewhere but most records are unsubstantiated with voucher material. Many probably refer to *C. amoenolens*.

gracilior (M.M. Moser) M.M. Moser, *Kleine Kryptogamenflora*, Edn 3: 286 (1967)
  Phlegmacium gracilior M.M. Moser, *Die Gattung Phlegmacium*: 350 (1960)
E: !
H: On soil, in deciduous woodland. British material associated with *Corylus*, *Fagus* and *Quercus* spp.
D: NBA9: 548-549
Known from North Somerset and West Gloucestershire.

helvelloides (Fr.) Fr., *Epicr. syst. mycol.*: 297 (1838)
  Agaricus gentilis γ *helvelloides* Fr., *Syst. mycol.* 1: 213 (1821)
E: o **S:** o **W:** o **NI:** !
H: On wet soil, associated with *Alnus* in swampy woodland.
D: NM2: 301 **D+I:** CFP1: A17, B&K5: 260 329 **I:** FlDan3: 97A, C&D: 333
Occasional but widespread. Often fruiting in inaccessible habitat.

hemitrichus (Pers.) Fr., *Epicr. syst. mycol.*: 302 (1838)
  Agaricus hemitrichus Pers., *Syn. meth. fung.*: 296 (1801)
E: c **S:** c **W:** c **NI:** c **ROI:** c
H: On acidic soil, associated with *Betula* spp in birchwoods or mixed woodland. Reported on occasions with other trees such

as *Fagus* and *Quercus* spp but birch always present in the vicinity.
D: NM2: 306 **D+I:** CFP1: A31, B&K5: 262 331 **I:** FlDan3: 00B, C&D: 333
Common and widespread.

herculeolens Bidaud, in Bidaud, Moënne-Loccoz, Reumaux & Henry, *Atlas des Cortinaires* 8: 293 (1996)
E: !
H: On soil in deciduous woodland. British material associated with *Carpinus*.
A single collection from East Kent (Putt Wood) in 2003.

heterosporus Bres., in Henn., *Verh. Bot. Ver. Prov. Brandenburg* 30: 169 (1889)
S: !
H: On acidic, sandy soil associated with conifers.
Known from Easterness. Near to *C. umbrinolens* but spores are narrowly fusiform.

hinnuleus (Sowerby) Fr., *Epicr. syst. mycol.*: 296 (1838)
    Agaricus hinnuleus Sowerby, *Col. fig. Engl. fung.* 2: pl. 173 (1798)
    Lepiota helvola Gray, *Nat. arr. Brit. pl.* 1: 603 (1821)
    Cortinarius hinnuleus var. *radicata* Rob. Henry, *Bull. Soc. Mycol. France* 57: 19 (1941)
E: ! **S:** ! **W:** ! **NI:** ! **ROI:** !
H: On soil, usually associated with *Fagus* or *Quercus* spp, in deciduous woodland. Rarely reported with *Corylus* or *Salix* spp.
D: NM2: 303 **D+I:** CFP1: A19, B&K5: 262 332 **I:** C&D: 331
Apparently widespread and supposedly rather frequent, but the majority of records are unsubstantiated with voucher material.

hinnuloides Rob. Henry, *Bull. Soc. Mycol. France* 57: 21 (1941)
E: ? **S:** ? **NI:** !
H: On soil in deciduous woodland.
Known from Northern Ireland (Fermanagh). Reported from England (South Hampshire) and Scotland (Roxburghshire), but unsubstantiated with voucher material.

hoeftii (Weinm.) Fr., *Epicr. syst. mycol.*: 306 (1838)
    Agaricus hoeftii Weinm., *Hymen. Gasteromyc.*: 178 (1836)
E: ! **NI:** !
H: On soil in deciduous woodland.
Poorly known. Accepted sensu Lange (FlDan3: 48 & pl. 103A) as reported by Pearson (1952). Collections from England (Berkshire, North Devon, South Hampshire and West Sussex) and Northern Ireland (Down).

humicola (Quél.) Maire, *Bull. Soc. Mycol. France* 27(4): 436 (1911)
    Dryophila squarrosa var. *humicola* Quél., *Compt. Rend. Assoc. Franç. Avancem. Sci.* 20(2): 466 (1891)
E: ! **W:** !
H: Associated with *Fagus* in old beechwoods or deciduous woodland.
D: NM2: 287 **D+I:** CFP3: C17, B&K5: 156 172 **I:** FlDan3: 90B, C&D: 325
Known from England (East Gloucestershire and Shropshire) and Wales (Monmouthshire).

huronensis Ammirati & A.H. Sm., *Michigan Botanist* 11(1): 20 (1972)
    Cortinarius huronensis var. *olivaceus* Ammirati & A.H. Sm., *Michigan Botanist* 11(1): 21 (1972)
    Hydrocybe palustris M.M. Moser [*nom. nud.*], *Bull. Soc. Naturalistes Oyonnax* 7: 122 (1953)
    Cortinarius palustris (M.M. Moser) Nezdojm. [*nom. nud.*], *Nov. sist. Niz. Rast.* 17: 55 (1980)
W: !
H: On soil in deciduous woodland.
D: NM2: 282 (as *Cortinarius huronensis* var. *olivaceus*), NM2: 283 **D+I:** B&K5: 146 158 (as *C. huronensis* var. *olivaceus*), B&K5: 146 157
A single record, but British records of *C. aureifolius* may belong here.

illibatus Fr., *Epicr. syst. mycol.*: 276 (1838)

Cortinarius subdelibutus P.D. Orton, *Sydowia* 36: 218 (1983)
S: ! **W:** ?
H: On soil, associated with *Betula* in mixed woodland.
D+I: B&K5: 228 280
A single collection from Mid-Perthshire (Rannoch, Dall Wood) (the type collection of *C. subdelibutus*). There is also a collection from Denbighshire in 1880, but this is in poor condition and possibly misidentified.

illuminus Fr., *Epicr. syst. mycol.*: 305 (1838)
    Mis.: *Cortinarius dilutus* sensu Moser [Mycol. Helv. 1(4): 215 (1984)] non Lange (FlDan3: 43 & pl. 100B)
    Mis.: *Cortinarius saturatus* sensu Lange (FlDan3: 42 & pl. 101A)
E: !
H: On soil in deciduous woodland
D: NM2: 304 (as *Cortinarius dilutus*) **D+I:** CFP2: B15, B&K5: 264 335, B&K5: 264 335
A single collection, from Yorkshire (Pickering, Gundale). Accepted here sensu CFP 2: 19 (1992) with spores subglobose. Sensu Cooke 830 (841) Vol. 6 (1887) may be the same; sensu Rea (1922), quoting spore measurements from Ricken (1915), is different.

imbutus Fr., *Epicr. syst. mycol.*: 306 (1838)
W: !
H: On soil in conifer woodland. British material with *Picea abies*.
D+I: CFP4: D30, CFP4: D60
A single collection from Merionethshire in 2001. Sensu Cooke 834 (870) Vol. 6 (1887) is unknown but apparently different from CFP 4: 16 (1998).

incisus (Pers.) Fr., *Epicr. syst. mycol.*: 301 (1838)
    Agaricus incisus Pers., *Syn. meth. fung.*: 310 (1801)
E: ! **W:** ? **NI:** ?
H: On soil, in deciduous or mixed woodland.
D: NM2: 305 **I:** Cooke 819 (807) Vol. 6 (1886)
Known from Surrey. Reported elsewhere but all records are unsubstantiated with voucher material or have been redetermined. Accepted sensu Cooke (HBF2) and sensu Kühner & Romagnesi (1953), non sensu Moser (1967)(with narrower spores) which is unknown in Britain.

inconspicuus J. Favre, *Ergebn. Wiss. Unters. Schweiz. Natn. Parks* 5: 203 (1955)
S: !
H: On soil, associated with *Salix herbacea* in montane habitat
Known from Mid-Perthshire.

infractus (Pers.) Fr., *Epicr. syst. mycol.*: 261 (1838)
    Agaricus infractus Pers., *Observ. mycol.* 2: 42 (1799)
    Cortinarius anfractus Fr., *Epicr. syst. mycol.*: 262 (1838)
E: o **S:** ! **W:** ! **NI:** !
H: On soil, usually associated with *Fagus* in old deciduous woodland. Also known with *Betula*, *Castanea*, *Quercus* and *Tilia* spp.
D: NM2: 291 **D+I:** CFP1: A09, B&K5: 182 213 **I:** FlDan3: 87C, SV24: 33, C&D: 337
Occasional in England. Rarely reported elsewhere but apparently widespread.

ionophyllus M.M. Moser, *Nova Hedwigia* 14(2-4): 504 (1968) [1967]
    Mis.: *Cortinarius scutulatus* sensu Cooke [Ill. Brit. fung. 796 (820) Vol. 6 (1886)] & sensu auct. mult.
S: ! **W:** ?
H: On acidic soil, associated with *Pinus sylvestris*.
D: NM2: 302 **D+I:** CFP2: B22, B&K5: 284 365 (as *Cortinarius ionophyllus*) **I:** Cooke 796 (820) Vol. 6 (1886) (as *Cortinarius (Telamonia) scutulatus*)
A northern conifer associate, known from South Aberdeen. Reported from Wales (Flintshire) in 1993 but said to be associated with *Fagus sylvatica*.

junghuhnii Fr., *Epicr. syst. mycol.*: 314 (1838)
E: ! **S:** ? **W:** ?
H: On soil, in deciduous or mixed coniferous and deciduous woodland.

D+I: B&K5: 268 340
Accepted sensu Lange (FlDan3: 49 & pl. 104E) e.g. as reported by Pearson (1948) from West Sussex, but poorly known in Britain. The majority of records are unsubstantiated with voucher material and most material so named in herbaria has been redetermined.

laniger Fr., *Epicr. syst. mycol.*: 292 (1838)
E: ? **S:** !
H: On acidic soil, associated with conifers in northern areas.
D: NM2: 303 **D+I:** CFP3: C53, B&K5: 270 343
Known from Easterness and Mid-Perthshire. Reported from West Kent in 1998 but unsubstantiated with voucher material.

largus Fr., *Epicr. syst. mycol.*: 259 (1838)
Agaricus variicolor var. *nemorensis* Fr., *Epicr. syst. mycol.*: 259 (1838)
Cortinarius variicolor var. *nemorensis* (Fr.) Fr., *Hymenomyc. eur.*: 339 (1874)
Cortinarius nemorensis (Fr.) J.E. Lange, *Fl. agaric. danic.* 3: 21 (1938)
Cortinarius lividoviolaceus Rob. Henry, *Doc. Mycol.* 17(68): 27 (1987)
Mis.: *Cortinarius balteatocumatilis* sensu auct. Brit.
Mis.: *Cortinarius cyanopus* sensu Cooke [Ill. Brit. fung. 690 (699) Vol. 5 (1887)]
E: o **S:** ! **W:** ? **NI:** ! **ROI:** !
H: Associated with *Betula, Fagus* or *Quercus* spp in mixed deciduous woodland.
D: NM2: 294 (as *Cortinarius nemorensis*) **D+I:** Ph: 131 (as *C. nemorensis*), CFP2: B59 (as *C. nemorensis* [in error]), CFP4: D22, B&K5: 184 216 (as *C. lividoviolaceus*), B&K5: 184 **I:** SV24: 33 (as *C. nemorensis*), C&D: 337
Occasional in England. Rarely reported elsewhere but apparently widespread. Reported from Wales but unsubstantiated with voucher material.

limonius (Fr.) Fr., *Epicr. syst. mycol.*: 296 (1838)
Agaricus limonius Fr., *Observ. mycol.* 2: 56 (1818)
Mis.: *Cortinarius callisteus* sensu NCL
S: ! **W:** !
H: On soil in coniferous woodland, usually associated with *Pinus sylvestris* and often in wet or boggy areas.
D: NM2: 287 **D+I:** CFP2: B34, B&K5: 156 173
Known from scattered sites in Scotland and recently collected in Wales (Caernarvonshire).

lividoochraceus (Berk.) Berk., *Outl. Brit. fungol.*: 186 (1860)
Agaricus lividoochraceus Berk., *Engl. fung.*: 89 (1836)
Cortinarius elatior Fr., *Epicr. syst. mycol.*: 274 (1838)
Agaricus elatus Pers. [*nom. illegit.*, non *A. elatus* Batsch (1783)], *Syn. meth. fung.*: 332 (1801)
E: o **S:** o **W:** o **NI:** ! **ROI:** ! **O:** Isle of Man: !
H: On acidic soil, usually associated with *Fagus* in beechwoods or mixed deciduous woodland.
D: NM2: 290 **D+I:** CFP1: A41, B&K5: 228 281 **I:** C&D: 343 (as *Cortinarius elatior*)
Occasional but apparently widespread. Older records and those unsubstantiated with voucher material, should be treated with caution as they will include *C. stillatitius*. N.B. This excludes *C. lividoochraceus* sensu Cooke (HBF 2: 250), sensu Orton (1958) and NCL with small spores which is possibly a member of the *C. delibutus* complex.

livor Fr., *Epicr. syst. mycol.*: 306 (1838)
NI: !
H: On soil, in deciduous woodland. British material associated with *Fagus*.
A single collection from Fermanagh (Coaghan) in 2000.

lucorum (Fr.) J.E. Lange, *Fl. agaric. danic.* 3: 36 (1938)
Cortinarius impennis var. *lucorum* Fr., *Epicr. syst. mycol.*: 294 (1838)
E: !
H: Associated with *Castanea, Fagus* or *Quercus* spp in deciduous woodland.
D: NM2: 302 **D+I:** CFP3: C10 **I:** FlDan3: 96E

Known from East Gloucestershire, East Kent, South Devon, Shropshire and Staffordshire. Sensu Cooke 1190 (1192) Vol. 8 (1891) is doubtful.

magicus Eichhorn, *Kleine Kryptogamenflora*, Edn 3: 295 (1967)
E: !
H: On soil in mixed deciduous woodland. Reported with *Betula, Corylus, Fagus* and *Quercus* spp.
D: NBA9: 549
Known from North Somerset and Surrey. Possibly only a variety of *C. glaucopus* associated with deciduous trees.

mairei (M.M. Moser) M.M. Moser, *Kleine Kryptogamenflora*, Edn 3: 298 (1967)
Phlegmacium mairei M.M. Moser, *Die Gattung* Phlegmacium: 355 (1960)
Mis.: *Cortinarius caesiocyaneus* sensu Rea (1922) and sensu Maire (Bull. Soc. Mycol. France 26: 176, 1910)
W: !
H: On soil, in woodland, associated with conifers.
A single collection from Denbighshire (Abergele) in 1981. A conifer associate, paler and with smaller spores than the closely related *C. caesiocyaneus*.

malachius (Fr.) Fr., *Epicr. syst. mycol.*: 280 (1838)
Agaricus malachius Fr., *Observ. mycol.* 2: 71 (1818)
Cortinarius malachioides P.D. Orton, *Naturalist (Hull)* 1958 (suppl.): 148 (1958)
Mis.: *Cortinarius umidicola* sensu Moser (1978)
E: ! **S:** !
H: On soil, in conifer woodland, usually associated with *Pinus sylvestris*.
D: NM2: 299 **D+I:** CFP4: D42, CFP4: D54, B&K5: 214 259 **I:** FlDan3: 91E
Apparently widespread but often confused with similar species. Sensu NCL is *C. quarciticus* and sensu Pearson (1943) is *C. pearsonii*.

malicorius Fr., *Epicr. syst. mycol.*: 289 (1838)
Cortinarius croceifolius Peck, *Bull. New York State Mus. Nat. Hist.* 150: 26 (1910)
E: ! **S:** ! **W:** ! **ROI:** !
H: On soil with conifers such as *Pinus contorta, P. sylvestris* and *Picea abies*. Occasionally reported with deciduous trees such as *Fagus* and *Salix* spp in mixed woodland.
D: NM2: 281 **D+I:** CFP1: A56, B&K5: 146 159 **I:** C&D: 329
Rarely reported. Apparently widespread but most often recorded from Scotland.

microspermus J.E. Lange, *Fl. agaric. danic.* 5, *Taxonomic Conspectus*: III (1940)
Mis.: *Cortinarius vespertinus* sensu Rea (1922) and sensu Ricken (as *Phlegmacium*) (1915)
S: !
H: On acidic soil, with *Pinus sylvestris* in Caledonian pinewoods.
Known from scattered sites in Scotland.

mucifluus Fr., *Epicr. syst. mycol.*: 274 (1838)
Cortinarius pinicola P.D. Orton, *Trans. Brit. Mycol. Soc.* 43(2): 204 (1960)
E: ! **S:** ! **NI:** !
H: On soil in coniferous or mixed woodland.
D: NM2: 290 **I:** FlDan3: 90D, C&D: 343
Supposedly not uncommon, and apparently widespread, but few recent records and little material in herbaria. Most often reported from Scotland (various locations) with a few collections from England (Berkshire, East Kent, Surrey and West Sussex). Last reported from Northern Ireland (Hillsborough) in 1931. Sensu Cooke (1888) and sensu Rea (1922) is *C. trivialis*.

mucosus (Bull.) Cooke, *Ill. Brit. fung.* 734 (739) Vol. 5 (1887)
Agaricus mucosus Bull., *Herb. France*: pl. 549 (1792)
Agaricus collinitus β *mucosus* (Bull.) Fr., *Syst. mycol.* 1: 248 (1821)
Cortinarius collinitus var. *mucosus* (Bull.) Fr., *Epicr. syst. mycol.*: 274 (1838)
E: ! **S:** ! **W:** !

H: On sandy soil, usually associated with *Pinus* spp, in woodland, especially on dunes.
D+I: CFP2: B33, B&K5: 228 282 **I:** C&D: 343
Rarely reported but apparently widespread.

mussivus (Fr.) Melot, *Doc. Mycol.* 17 (68): 67 (1987)
    Agaricus mussivus Fr., *Epicr. syst. mycol.*: 178 (1838)
    Hebeloma mussivum (Fr.) Sacc., *Syll. fung.* 5: 792 (1887)
    Cortinarius russeoides M.M. Moser, *Sydowia* 6: 152 (1952)
    Cortinarius russeus Rob. Henry, *Doc. Mycol.* 16(61): 24 (1985)
    Mis.: *Cortinarius russus* sensu Cooke [Ill. Brit. fung. 696 (751) Vol. 5 (1887)] & sensu NCL
E: ! **ROI:** !
H: On soil, associated with *Dryas octopetala* or *Helianthemum nummularium*.
D+I: CFP4: D20, B&K5: 188 220
Known in Britain from England (Derbyshire) and from the Republic of Ireland. Sensu Saccardo (1887) and Rea (1922) refers to an unknown species of *Hebeloma*.

nanceiensis Maire, *Bull. Soc. Mycol. France* 27: 425 (1911)
E: !
H: On calcareous soil in mixed deciduous woodland.
D+I: CFP2: B21, B&K5: 188 221 **I:** C&D: 337
Known from Westmorland (Roudsea Wood) and probably in Derbyshire (Dovedale).

norvegicus Høil., *Opera Bot.* 71: 85 (1984) [1983]
    Cortinarius croceus ssp. *norvegicus* (Høil.) Brandrud & H. Lindstr., *Nordic J. Bot.* 10(5): 536 (1990)
S: !
H: On sandy soil. British material associated with *Salix repens*.
D: NM2: 282 **D+I:** CFP1: A54 (as *Cortinarius croceus* var. *norvegicus*)
Known from Orkney (Papa Westray) and apparently in the Cairngorms.

nothosaniosus M.M. Moser, *Carinthia II* 76: 28 (1966)
E: ! **S:** !
H: On soil in woodland
Known from England (South Hampshire) in 1985 and from Scotland (Midlothian) in 1960.

obtusus (Fr.) Fr., *Epicr. syst. mycol.*: 313 (1838)
    Agaricus obtusus Fr. [non *A. obtusus* Cooke & Massee (1890)], *Syst. mycol.* 1: 233 (1821)
    Cortinarius pseudostriatulus Rob. Henry, *Atlas des Cortinaires* 2: 27 (1990)
E: ! **S:** ! **W:** ! **NI:** ! **ROI:** ?
H: On soil in deciduous woodland.
D: NM2: 305 **D+I:** CFP3: C57, B&K5: 272 347
Frequently reported but rarely collected and most of the few collections named thus (in herb. K) are misidentified.

occidentalis *var.* **obscurus** (M.M. Moser) Quadr., *Doc. Mycol.* 14(56): 29 (1985) [1984]
    Cortinarius purpurascens var. *obscura* M.M. Moser, *Sydowia* 6: 153 (1960)
E: !
H: On soil, in deciduous woodland.
A single record from West Kent (Orpington, Sparrow Wood) in 2001.

ochroleucus (Schaeff.) Fr., *Epicr. syst. mycol.*: 284 (1838)
    Agaricus ochroleucus Schaeff., *Fung. Bavar. Palat. nasc.* 1: 34 (1762)
E: ! **S:** ! **W:** ? **NI:** ! **ROI:** !
H: On soil, usually associated with *Fagus* or *Quercus* spp in mixed deciduous woodland, but also reported with *Populus* and *Tilia* spp.
I: C&D: 343
Rarely reported. Apparently widespread, but most of the records are from scattered localities in England, from Durham to South Devon.

odorifer Britzelm., *Ber. Naturhist. Vereins Augsburg* 28: 123 (1885)

ROI: !
H: On limestone. British material, rather unusually, associated with *Dryas octopetala*.
D: NM2: 296 **D+I:** CFP3: C15, B&K5: 190 223 **I:** C&D: 341, FM7: 31
Known from The Burren. *Cortinarius orichalceus* sensu Cooke (718) 754 Vol. 5 (1887), sensu Rea (1922) and sensu NCL is probably also this species, although Orton (1955) considered it distinct.

olearioides Rob. Henry, *Bull. Soc. Mycol. France* 73(1): 35 (1957)
    Cortinarius subfulgens P.D. Orton, *Trans. Brit. Mycol. Soc.* 43(2): 212 (1960)
    Mis.: *Cortinarius fulgens* sensu NCL
    Mis.: *Cortinarius fulmineus* sensu NCL & sensu auct. mult.
E: ! **S:** ? **ROI:** ?
H: On soil in deciduous or mixed coniferous and deciduous woodland.
D+I: CFP4: D02 **I:** SV40: 14, Cooke 716 (716) Vol.5 (1888) (as *C. fulgens*)
Rarely reported, but possibly widespread. *Cortinarius fulmineus* sensu Rea (1922), with large spores, may be another species.

olidus J.E. Lange, *Fl. agaric. danic.* 5, *Taxonomic Conspectus*: III (1940)
E: !
H: On calcareous soil, associated with *Fagus* in old woodland.
D+I: B&K5: 190 224 **I:** SV43: 49, SV48: 57
Known from East Gloucestershire, Oxfordshire, South Hampshire, Surrey, Westmorland and West Sussex.

olivaceofuscus Kühner, *Bull. Mens. Soc. Linn. Lyon* 24(2): 39 (1955)
    Cortinarius schaefferi Bres., *Icon. Mycol.* 13: pl. 648 (1930)
    Mis.: *Cortinarius malicorius* sensu Pearson (1952)
E: ! **NI:** ! **O:** Channel Islands: !
H: Associated with *Fagus* and *Quercus* spp in old deciduous woodland, and with *Salix* spp in damp areas.
D: NM2: 281 **D+I:** CFP1: A16 **I:** C&D: 329
Rarely reported but apparently widespread.

orellanus Fr., *Epicr. syst. mycol.*: 288 (1838)
    Dermocybe orellana (Fr.) Ricken, *Blätterpilze Deutschl.*: 160 (1912)
    Cortinarius rutilans Quél., *Compt. Rend. Assoc. Franç. Avancem. Sci. Assoc. Sci. France* 26: 448 (1898)
E: ! **S:** ! **W:** !
H: On soil, usually associated with *Quercus* spp in mixed deciduous woodland but also reported with *Betula* spp in similar habitat.
D: NM2: 286 **D+I:** CFP1: A20, C&D: 325, B&K5: 158 176
Rarely reported. An extremely toxic species.

osmophorus P.D. Orton, *Trans. Brit. Mycol. Soc.* 43(2): 210 (1960)
    Cortinarius evosmus Rob. Henry, *Doc. Mycol.* 20(77): 70 (1989)
E: !
H: Associated with *Fagus* in old beechwoods on calcareous soil.
I: SV33: 61
Known from Berkshire, East Gloucestershire, Surrey and West Kent. The large and strongly scented basidiomes (like a mixture of hyacinths and metal polish or burnt sugar) make this an unmistakable species, probably genuinely rare.

parvannulatus Kühner, *Bull. Mens. Soc. Linn. Lyon* 24(2): 40 (1955)
    Cortinarius cedriolens M.M. Moser, *Kleine Kryptogamenflora,* Edn 3: 337 (1967)
    Mis.: *Cortinarius iliopodius* sensu Cooke [Ill. Brit. fung. 818 (839B) Vol. 6 (1887)]
E: ? **S:** ? **W:** ! **NI:** ?
H: On wet soil, associated with *Alnus* or *Salix* spp, in mixed deciduous woodland.
D+I: CFP1: A60

A single collection from Carmarthenshire in 1994. Reported elsewhere but unsubstantiated with voucher material.

patibilis *var.* **scoticus** Brandrud, *Edinburgh J. Bot.* 54(1): 114 (1997)
Mis.: *Cortinarius largus* sensu auct. Brit.
S: !
H: On soil, associated with *Betula.*
Described from Perthshire (Struan Birchwood) in 1993, and further material from Mid-Perthshire (Glen Lyon and Rannoch). *Fide* Brandrud [Edinburgh J. Bot. 55 (1): 119 (1998)] this is very close to *C. largus.*

pearsonii P.D. Orton, *Naturalist (Hull)* 1958 (suppl.): 148 (1958)
Cortinarius cremeolaniger P.D. Orton, *Sydowia* 36: 220 (1983)
Cortinarius lanigeroides P.D. Orton, *Sydowia* 36: 219 (1983)
Mis.: *Cortinarius malachius* sensu Pearson (1943)
E: !
H: On sandy soil, associated with *Betula* spp, or *Castanea* and *Pinus sylvestris,* often in woodland or wooded areas on heathland.
D+I: FRIC6: 30-31 46
Known from Berkshire, East & West Kent, South Wiltshire, Surrey and West Sussex.

pertristis J. Favre, *Ergebn. Wiss. Untersuch. Schweiz. Natn. Parks* 5: 203 (1955)
S: !
H: On peaty or acidic soil, associated with *Salix herbacea* in montane habitat.
Known from Mid-Perthshire (Ben Lawers).

phaeophyllus P. Karst., *Symb. mycol. fenn.*: 3 (1881)
E: !
H: On soil, in deciduous woodland. British material associated with *Betula* and *Quercus* spp.
A single collection from South Devon (Stover Park) in 1995.

phaeopygmaeus J. Favre, *Ergebn. Wiss. Untersuch. Schweiz. Natn. Parks* 5: 33 (1955)
S: !
H: On peaty soil, in montane habitat, associated with *Salix herbacea.*
D+I: B&K5: 276 352 **I:** C&D: 331
A single collection from Cairngorms (Ben Macdui) in 2002.

pholideus (Fr.) Fr., *Epicr. syst. mycol.*: 282 (1838)
Agaricus pholideus Fr., *Syst. mycol.* 1: 219 (1821)
E: o **S:** o **W:** ! **NI:** !
H: On acidic soil, associated with *Betula* and often *Pinus* spp in mixed woodland.
D: NM2: 297 **D+I:** CFP2: B37, B&K5: 216 262 **I:** FlDan3: 93F, C&D: 327, SV33: 24, FM1(1): 3
Occasional in England and Scotland. Rarely reported elsewhere but apparently widespread.

pluvius (Fr.) Fr., *Epicr. syst. mycol.*: 277 (1838)
Agaricus pluvius Fr., *Syst. mycol.* 1: 236 (1821)
E: ? **S:** ! **W:** ?
H: On acidic soil in montane or boreal habitat, associated with conifers.
D+I: CFP4: D25, CFP4: D23, B&K5: 230 284
Records of this species in lowland habitat in England (South Hampshire and West Sussex) are unsubstantiated with voucher material and probably misidentified.

poecilopus Rob. Henry, *Bull. Soc. Mycol. France* 71: 216 (1956) [1955]
Cortinarius fuscopallens (Fr.) N. Arnold, *Libri Botanici* 7: 117 (1993)
Cortinarius triformis var. *fuscopallens* Fr., *Epicr. syst. mycol.*: 299 (1838)
Mis.: *Cortinarius triformis* sensu auct. mult.
H: On soil in woodland, associated with *Fagus.*
D+I: B&K5: 292 377 (as *Cortinarius triformis*) **I:** C&D: 333
Rarely reported and confused with *C. triformis.*

porphyropus (Alb. & Schwein.) Fr., *Epicr. syst. mycol.*: 271 (1838)
Agaricus porphyropus Alb. & Schwein., *Consp. fung. lusat.*: 153 (1805)
E: ! **S:** ! **W:** ! **NI:** !
H: On soil, in deciduous woodland.
D: NM2: 294 **D+I:** CFP2: B55, B&K5: 194 229 **I:** FlDan3: 87B
Rarely reported but apparently widespread.

praestans (Cordier) Gillet, *Hyménomycètes*: 475 (1876)
Agaricus praestans Cordier, *Champ. France*: 98 (1870)
Cortinarius berkeleyi Cooke, *Handb. Brit. fung.,* Edn 2: 240 (1888)
E: !
H: Associated with *Corylus, Fagus* and *Quercus* spp in mixed deciduous woodland.
D: NM2: 291 **D+I:** CFP1: A42, B&K5: 194 230 **I:** C&D: 335, SV48: 59
Known from Westmorland, Worcestershire and North Somerset. A large and distinctive species unlikely to be overlooked and probably genuinely rare.

pratensis (Bon & Gaugué) Høil., *Opera Bot.* 71: 86 (1984) [1983]
Dermocybe pratensis Bon & Gaugué, *Doc. Mycol.* 5 (17): 14 (1975)
S: ! **ROI:** !
H: On wet soil, associated with *Salix herbacea* in montane habitat.
D: NM2: 283 **I:** C&D: 329
Known from Scotland (Perthshire: The Cairnwell in 1983 and Orkney: Papa Westray in 1992) and from the Republic of Ireland (Laois: Ballyprior) in 1996.

psammocephalus (Bull.) Fr., *Epicr. syst. mycol.*: 301 (1838)
Agaricus psammocephalus Bull., *Herb. France*: pl. 586 (1793)
E: ! **S:** ?
H: On soil in deciduous woodland.
D+I: CFP4: D57
Known from widely scattered sites in England. Reported from Scotland, but in association with conifers and possibly misidentified *C. angelesianus.* Accepted here sensu NCL and sensu CFP 4: 18 (1998) as a species with deciduous trees.

pseudocandelaris (M.M. Moser) M.M. Moser, *Kleine Kryptogamenflora,* Edn 3: 325 (1967)
Hydrocybe pseudocandelaris M.M. Moser, *Sydowia,* Ser. II, *Beih.*1: 240 (1957)
S: !
H: On acidic soil in deciduous woodland.
Known from Mid-Perthshire (Black Wood of Rannoch) in 1967 and reported from Easterness and South Aberdeen. Dubiously distinct from *C. rigens.*

pumilus (Fr.) J.E. Lange, *Fl. agaric. danic.* 3: 23 (1938)
Cortinarius collinitus δ *pumilus* Fr., *Hymenomyc. eur.*: 355 (1874)
E: ! **W:** !
H: On soil in mixed deciduous woodland.
Known from South Hampshire (New Forest) in 1992 and from Glamorganshire (Castell Coch) in 1947.

purpurascens (Fr.) Fr., *Epicr. syst. mycol.*: 265 (1838)
Agaricus purpurascens Fr. [non *A. purpurascens* With. (1792); non (Cooke) Pilát (1951)], *Observ. mycol.* 2: 70 (1818)
E: o **S:** o **W:** ! **NI:** ? **ROI:** ?
H: On soil, usually associated with *Fagus* and less often *Quercus* spp, in mixed deciduous woodland.
D: NM2: 294 **D+I:** B&K5: 196 233 **I:** C&D: 339
Occasional but widespread. Reported from Republic of Ireland (South Kerry) in 1936 and Northern Ireland (Down) in 1948 but unsubstantiated with voucher material.

purpureus (Pers.) Fuckel, *Jahrb. Nassauischen Vereins Naturk.* 15: 114 (1860)
Agaricus purpureus Pers. [non *A. purpureus* Bolton (1788)], *Syn. meth. fung.*: 290 (1801)

Agaricus phoeniceus Vent., in Bulliard, *Hist. champ. France II,* 2: 647 (1812)
  Cortinarius phoeniceus (Vent.) Maire, *Bull. Soc. Mycol. France* 27(4): 434 (1911)
  Dermocybe phoenicea (Vent.) M.M. Moser, *Schweiz. Z. Pilzk.* 52(9): 130 (1974)
  Mis.: *Cortinarius miltinus* sensu Cooke [Ill. Brit. fung. 774 (785) Vol. 6 (1886)] & sensu auct.
E: ! S: ! W: !
H: On acidic soil amongst leaf and needle litter with conifers such as *Picea* or *Pinus* spp, and also known with *Betula* spp.
D: NM2: 284 (as *Cortinarius phoeniceus*) **D+I:** CFP3: C47, B&K5: 148 161 **I:** C&D: 329 (as *C. phoeniceus*)
Rarely reported but apparently widespread. However, many of the records are old and unsubstantiated with voucher material.

quarciticus H. Lindstr., *Cortinarius* 3: 27 (1994)
  Mis.: *Cortinarius malachius* sensu NCL
E: ! S: !
H: On acidic soil in woodland, associated with *Pinus* spp.
D+I: CFP3: C59 **I:** Ph: 131 (as *Cortinarius malachius*)
Rarely reported but apparently more frequent in Britain than true *C. malachius*.

rapaceus Fr., *Epicr. syst. mycol.*: 263 (1838)
S: !
H: On soil in woodland.
I: C&D: 337
A single collection from Morayshire (Darnaway Forest) in 1963. Sensu auct. is *C. aleuriosmus*.

renidens Fr., *Epicr. syst. mycol.*: 308 (1838)
  Mis.: *Cortinarius angulosus* sensu Kühner & Romagnesi (1953)
E: !
H: On soil, in deciduous woodland.
D+I: CFP4: D53, B&K5: 278 356
A single collection from South Devon (Loddiswell) in 1995. Rarely reported elsewhere, but unsubstantiated with voucher material.

rheubarbarinus Rob. Henry, *Bull. Mens. Soc. Linn. Lyon* 71: 229 (1956) [1955]
E: !
H: On soil in deciduous woodland.
D+I: CFP2: B60, CFP3: C19, B&K5: 278 357
A single collection from South Hampshire (New Forest, Millyford Bridge) in 2000. Reported from West Kent but unsubstantiated with voucher material.

rickenianus Maire, *Trabajos del Museo de Ciències Naturales de Barcelona* 3(4): 111 (1937)
  Mis.: *Cortinarius aleuriosmus* sensu Ricken p.p. [Blätterpilze Deutschl.: 136 (1912)]
E: !
H: Associated with *Fagus* in old beechwoods on calcareous soil.
I: SV27: 42
A single collection from West Sussex (Eartham) in 1987.

riederi (Weinm.) Fr., *Epicr. syst. mycol.*: 259 (1838)
  Agaricus riederi Weinm., *Hymen. Gasteromyc.*: 161 (1836)
E: !
H: On soil in deciduous woodland.
D: NM2: 295 **D+I:** B&K5: 196 234 **I:** Cooke 694 (702) Vol. 5 (1887) Poorly known and not reported since the single collection (in herb. K) from South Essex (Epping Forest) by Cooke in 1880, confirmed as *C. riederi* by Moser.

rigens (Pers.) Fr., *Epicr. syst. mycol.*: 311 (1838)
  Agaricus rigens Pers., *Syn. meth. fung.*: 288 (1801)
  Cortinarius duracinus Fr., *Epicr. syst. mycol.*: 304 (1838)
  Mis.: *Cortinarius candelaris* sensu auct. mult.
E: ? S: ? W: ? NI: ?
H: On soil in conifer woodland.
D: NM2: 301 (as *Cortinarius duracinus*) **D+I:** CFP3: C28 (as *C. duracinus*), B&K5: 252 317 (as *C. duracinus*) **I:** FlDan3: 102D (as *C. duracinus*), C&D: 333 (as *C. duracinus*)

Accepted with reservations. British records under three Friesian epithets have been treated as conspecific following Arnolds (1993). See also Kühner & Romagnesi (1953: 314, note 9) for details of numerous efforts to separate them. As treated here *C. rigens* is a mainly northern species associated with conifers. *Cortinarius rigens* sensu Lange (FlDan3: 44) in deciduous woodland is *C. acetosus*.

rigidiusculus Nezdojm., *Shlyapochnye Griby SSSR Rod Cortinarius Fr.* (Leningrad): 142 (1983)
  Mis.: *Cortinarius rigidus* sensu K&R
E: ! S: ! W: ? ROI: ?
H: On soil, in deciduous woodland.
D+I: B&K5: 280 358 (as *Cortinarius rigidus*)
Accepted with reservations. Seemingly less common than *C. umbrinolens* but both have been recorded as *C. rigidus*. There are few collections of either in herbaria and British records are confused.

roseipes (Velen.) Reumaux, *Atlas des Cortinaires* 4: 106 (1992)
  Telamonia roseipes Velen., *České Houby* III: 465 (1921)
E: !
H: On soil in grassland on limestone, associated with *Helianthemum nummularium*.
Known from a single collection from Derbyshire in 2002.

rubellus Cooke, *Grevillea* 16: 44 (1887)
  Cortinarius orellanoides Rob. Henry, *Bull. Soc. Mycol. France* 53(1): 61 (1937)
  Cortinarius speciosissimus Kühner & Romagn. [*nom. illegit.*, non *C. speciosissimus* Earle (1904)], *Flore Analytique des Champignons Supérieurs*: 287 (1953)
  Cortinarius speciosus J. Favre [*nom. illegit.*, non *C. speciosus* Earle (1904)], *Beitr. Kryptogamenfl. Schweiz* 10(3): 213 (1948)
E: ! S: o W: ! NI: !
H: On acidic or sandy soil, usually associated with *Pinus sylvestris* but also reported with *Picea* spp, in woodland and plantations, often growing amongst *Calluna* and *Vaccinium myrtillus*.
D: NM2: 286 (as *Cortinarius orellanoides*), NBA5: 710 (as *C. orellanoides*) **D+I:** CFP1: A58, B&K5: 158 177 **I:** C&D: 325
Occasional in Scotland. Rarely reported elsewhere but apparently widespread. Better known as *C. speciosissimus*. An extremely toxic species.

rubicundulus (Rea) A. Pearson, *Trans. Brit. Mycol. Soc.* 29(4): 197 (1946)
  Agaricus rubicundulus Rea, *Grevillea* 22: 40 (1893)
  Flammula rubicundula (Rea) Rea, *Brit. basidiomyc.*: 318 (1922)
E: ! S: !
H: On soil in deciduous woodland. British material associated with *Quercus* spp.
D: NM2: 286 **D+I:** CFP1: A24, B&K5: 160 178 **I:** C&D: 327
Rarely reported but apparently widespread, perhaps most frequent in Scotland.

rufo-olivaceus (Pers.) Fr., *Epicr. syst. mycol.*: 268 (1838)
  Agaricus rufo-olivaceus Pers., *Syn. meth. fung.*: 285 (1801)
  Cortinarius testaceus Cooke, *Ill. Brit. fung.* 1188 (1190) Vol. 8 (1891)
  Cortinarius vinosus Cooke, *Ill. Brit. fung.* 758 (759) Vol. 6 (1887)
E: !
H: Usually associated with *Fagus* in old beechwoods on calcareous soil, but reported with other trees, such as *Carpinus* or *Quercus* spp, in mixed deciduous woodland.
D+I: CFP2: B23, B&K5: 198 235 **I:** SV27: 43, C&D: 341, SV40:19
Known from England in widely scattered localities from Yorkshire to West Kent and Herefordshire.

rufostriatus J. Favre, *Ergebn. Wiss. Untersuch. Schweiz. Natn. Parks* 5: 204 (1955)
S: !
H: On soil, associated with *Salix herbacea* in montane habitat.

D+I: B&K5: 280 360
Known from Perthshire (The Cairnwell) in 1983 and reported
from South Aberdeenshire.

safranopes Rob. Henry, *Bull. Soc. Mycol. France* 54: 95 (1938)
E: ! **W:** !
H: On soil in mixed deciduous woodland. Associated with *Fagus*,
Quercus or *Salix* spp.
Known from England (Surrey) and Wales (Carmarthenshire).

saginus (Fr.) Fr., *Epicr. syst. mycol.*: 260 (1838)
Agaricus saginus Fr., *Syst. mycol.* 1: 226 (1821)
Cortinarius subtriumphans P.D. Orton, *Trans. Brit. Mycol.
Soc.* 43(2): 212 (1960)
Cortinarius validus J. Favre, *Beitr. Kryptogamenfl. Schweiz*
10(3): 214 (1948)
E: ! **S:** ! **NI:** ?
H: On soil in deciduous woodland.
D: NM2: 293 **D+I:** CFP1: A01, B&K5: 198 236 **I:** SV48: 45
Single collections from England (South Essex: Epping Forest) in
1987, and Scotland (Easterness: Badger Falls; type of *C.
subtriumphans*) in 1956. Reported from Northern Ireland but
unsubstantiated with voucher material. N.B. excludes *C.
saginus* sensu NCL, with small spores, which is doubtful.

salor Fr., *Epicr. syst. mycol.*: 276 (1838)
E: !
H: On soil in woodland.
D+I: CFP1: A02, B&K5: 230 285 **I:** C&D: 343
'Very uncommon' *fide* Orton (1955). A single collection from
Surrey in herb. Orton (*fide* Watling pers. comm.)and one
unlocalised English collection in herb. K in poor condition.
Recorded from Norfolk in 1985 and Yorkshire in 1997, but
unsubstantiated with voucher material.

sanguineus (Wulfen) Fr., *Epicr. syst. mycol.*: 288 (1838)
Agaricus sanguineus Wulfen, in Jacq., *Collect. Bot. Spectantia
(Vindobonae)* 2: t. 15 f. 3 (1788)
Dermocybe sanguinea (Wulfen) Wunsche, *Pilze*: 25 (1877)
Cortinarius puniceus P.D. Orton, *Naturalist (Hull)* 1958
(suppl.): 148 (1958)
Dermocybe punicea (P.D. Orton) M.M. Moser, *Schweiz. Z.
Pilzk.* 52(9): 134 (1974)
E: o **S:** o **W:** ! **NI:** !
H: On acidic soil in deciduous, coniferous or mixed woodland.
D: NM2: 284 **D+I:** CFP1: A57, B&K5: 148 162 **I:** C&D: 329
Occasional but widespread.

saniosus (Fr.) Fr., *Epicr. syst. mycol.*: 213 (1838)
Agaricus saniosus Fr., *Syst. mycol.* 1: 232 (1821)
Flammula saniosa (Fr.) P. Kumm., *Führ. Pilzk.*: 81 (1871)
E: ! **S:** ! **W:** ! **NI:** ! **ROI:** !
H: On soil, most frequently associated with *Betula* or *Salix* spp,
in damp areas in mixed deciduous woodland.
D: NM2: 287 **I:** C&D: 325
Rarely reported but apparently widespread.

saporatus Britzelm., *Zur Hymenomyceten-Kunde* 3: 5 (1897)
Cortinarius turbinatus var. *lutescens* Rea, *Brit. basidiomyc.*:
142 (1922)
Cortinarius lutescens (Rea) Rob. Henry, *Bull. Soc. Mycol.
France* 55: 171 (1939)
Cortinarius subturbinatus P.D. Orton, *Trans. Brit. Mycol. Soc.*
43(2): 213 (1960)
Cortinarius sulphurinus var. *langei* Rob. Henry, *Bull. Soc.
Mycol. France* 55: 169 (1939)
Mis.: *Cortinarius turbinatus* sensu NCL
E: !
H: Associated with *Fagus* in old beechwoods on calcareous soil.
D+I: CFP2: B44
Known from East Gloucestershire, North Somerset, Oxfordshire
and Surrey. Treated as three separate species in NCL but here
as a single, variable species following CFP2.

saturninus (Fr.) Fr., *Epicr. syst. mycol.*: 306 (1838)
Agaricus saturninus Fr., *Syst. mycol.* 1: 219 (1821)
Cortinarius cohabitans P. Karst., *Bidrag Kännedom Finlands
Natur Folk* 32: 388 (1879)

E: ! **S:** ? **W:** ! **NI:** ! **ROI:** ?
H: On soil in deciduous woodland.
D: NM2: 303 **D+I:** CFP3: C09, B&K5: 282 362
Rarely reported but apparently widespread.

scandens Fr., *Epicr. syst. mycol.*: 312 (1838)
Cortinarius obtusus var. *gracilis* Quél., *Grevillea* 8 pl. 129 fig.
1 (1880)
E: ! **S:** ! **NI:** !
H: On soil in deciduous or mixed coniferous and deciduous
woodland.
D+I: B&K5: 282 363
Rarely reported. Apparently widespread, but the majority of
records are unsubstantiated with voucher material.

scaurus (Fr.) Fr., *Epicr. syst. mycol.*: 268 (1838)
Agaricus scaurus Fr., *Observ. mycol.* 2: 75 (1818)
E: ! **S:** !
H: On soil, associated with *Pinus sylvestris* in woodland.
D: NM2: 295 **D+I:** CFP3: C21, B&K5: 198 237 **I:** C&D: 341
Known from England (Mid-Lancashire and Westmorland) and
Scotland (Easterness, Mid-Ebudes and Westerness).

scaurus var. herpeticus (Fr.) Quél., *Enchir. fung.* 76 (1886)
Cortinarius herpeticus Fr., *Epicr. syst. mycol.*: 268 (1838)
Phlegmacium herpeticum var. *fageticola* M.M. Moser, *Die
Gattung* Phlegmacium: 358 (1960)
Cortinarius herpeticus var. *fageticola* (M.M. Moser) Quadr.,
Doc. Mycol. 14(56): 29 (1985) [1984]
Cortinarius subvirentophyllus Rob. Henry [*nom. nud.*], *Bull.
Soc. Mycol. France* 67: 312 (1952)
E: ! **S:** ? **NI:** !
H: On soil in deciduous woodland.
D+I: CFP3: C08
A single collection from England (Cumberland) ex herb. Cooke,
in herb. K, and a collection from Northern Ireland (Fermanagh:
Florence Court) in 2000. Reported from Scotland but
unsubstantiated with voucher material.

scotoides J. Favre, *Ergebn. Wiss. Untersuch. Schweiz. Natn.
Parks* 5: 204 (1955)
S: !
H: On soil in montane habitat.
D+I: B&K5: 284 364
Known from Cairngorm in 1984. A species in the *C. obtusus*
complex.

semisanguineus (Fr.) Gillet, *Hyménomycètes*: 484 (1876)
Agaricus cinnamomeus α *semisanguineus* Fr., *Syst. mycol.* 1:
229 (1821)
Cortinarius cinnamomeus β *semisanguineus* (Fr.) Sacc., *Syll.
fung.* 5: 942 (1887)
Dermocybe semisanguinea (Fr.) M.M. Moser, *Schweiz. Z.
Pilzk.* 52(9): 129 (1974)
E: o **S:** o **W:** o **NI:** o **ROI:** o
H: On acidic soil, usually associated with species of *Pinus* in
conifer woodland. Also reported with *Castanea, Fagus* and
Quercus spp, but conifers usually present nearby.
D: NM2: 284 **D+I:** CFP1: A13, B&K5: 150 163 **I:** C&D: 329
Occasional but widespread.

silvaemonachi (D.A. Reid, Murton & N.J. Westwood) Melot,
Doc. Mycol. 20(77): 96 (1989)
Phlegmacium silvamonachorum D.A. Reid, Murton & N.J.
Westwood, *Trans. Brit. Mycol. Soc.* 68(3): 327 (1977)
E: !
H: On soil in deciduous woodland.
Known only from the type collection (Huntingdonshire: Monks
Wood) but doubtfully distinct from *C. infractus*.

simulatus P.D. Orton, *Naturalist (Hull)* 1958 (suppl.): 148
(1958)
E: ! **S:** ! **W:** !
H: On soil in deciduous woodland. British material associated
with *Betula, Fagus* or *Salix* spp.
Rarely reported but apparently widespread.

sodagnitus Rob. Henry, *Bull. Soc. Mycol. France* 55(1): 78 (1939)

E: o

H: On calcareous soil, associated with *Fagus* in old beechwoods.

D+I: CFP2: B10, B&K5: 200 238 **I:** C&D: 339

Occasional in England, from Westmorland to West Sussex and Herefordshire. The epithet is often incorrectly cited as *sodagnites*.

spadicellus (M.M. Moser) Brandrud, *Edinburgh J. Bot.* 54(1): 114 (1997)

   Phlegmacium spadicellum M.M. Moser, *Die Gattung* Phlegmacium: 248 (1960)

S: !

H: On soil in mixed deciduous woodland. British material associated with *Fagus* and *Quercus* spp.

D+I: CFP4: D10

Known from the Clyde Isles (Arran: Glenashdale and Mull: Glen Aros).

spilomeus (Fr.) Fr., *Monogr. hymenomyc. Suec.* 2: 63 (1863)

   Agaricus spilomeus Fr., *Syst. mycol.* 1: 220 (1821)

E: ! **S:** ! **W:** ? **NI:** ? **ROI:** !

H: On soil, in deciduous or mixed woodland, usually associated with *Betula*, *Quercus* or *Pinus* spp. Recently reported from Republic of Ireland in association with *Dryas octopetala*.

D: NM2: 297 **I:** FlDan3: 96D, FlDan3: 96B (as *Cortinarius spilomeus* var. *depauperatus*)

Apparently widespread but poorly known in Britain. Most records are old and unsubstantiated with voucher material.

splendens Rob. Henry, *Bull. Soc. Mycol. France* 52: 178 (1936)

E: !

H: Associated with *Fagus* in old beechwoods on calcareous soil.

D+I: CFP2: B57, B&K5: 200 240 **I:** C&D: 341, SV40: 15

Known from Berkshire, East Gloucestershire, East & West Kent, Herefordshire, South Hampshire, Surrey and West Sussex.

stillatitius Fr., *Epicr. syst. mycol.*: 277 (1838)

   Cortinarius integerrimus Kühner, *Bull. Mens. Soc. Linn. Lyon* 28: 127 (1959)

   Cortinarius pseudosalor J.E. Lange, *Fl. agaric. danic.* 5, *Taxonomic Conspectus*: III (1940)

E: o **S:** o **W:** o **NI:** !

H: On acidic soil in mixed deciduous woodland, usually associated with *Fagus* or *Quercus* spp, occasionally with *Betula* spp or *Castanea*. Rarely reported with *Pinus* spp.

D: NCL3: 205 (as *Cortinarius pseudosalor*), NM2: 290 **D+I:** CFP1: A33, B&K5: 232 286 **I:** C&D: 343

Occasional but widespread and may be locally frequent. Usually reported (as *C. pseudosalor*) under beech in Britain, but this is a conifer associate in Scandinavia and there may be two taxa involved.

subbalaustinus Rob. Henry, *Doc. Mycol.* 21(81): 46 (1991)

E: ! **S:** ! **NI:** !

H: On soil, usually associated with *Betula* or *Quercus* spp in open mixed deciduous woodland.

D+I: CFP2: B03, B&K5: 286 369

Known from widely scattered sites in England (Warwickshire to South Devon) and single collections from Scotland (Berwickshire) and Northern Ireland (Fermanagh).

subtortus (Pers.) Fr., *Epicr. syst. mycol.*: 273 (1838)

   Agaricus subtortus Pers., *Syn. meth. fung.*: 284 (1801)

E: ? **S:** !

H: On soil in coniferous or mixed woodland. British material associated with *Betula* and *Pinus* spp.

D: NM2: 291 **D+I:** B&K5: 202 243 **I:** C&D: 337

Known from Scotland (Easterness). Reported from England (South Hampshire) but unsubstantiated with voucher material.

subtorvus Lamoure, *Schweiz. Z. Pilzk.* 47: 165 (1969)

E: ! **S:** !

H: On soil, associated with *Salix herbacea* in montane habitat.

D: NM2: 302 **D+I:** CFP1: A04, B&K5: 290 373

Known only from single collections from Cumberland (Ambleside, Crag Hill) in 1992 and Mid-Perthshire (Meall nan Tarmachan) in 2003.

suillus Fr., *Epicr. syst. mycol.*: 281 (1838)

   Mis.: *Cortinarius diabolicus* sensu auct. mult.

E: ! **W:** ! **ROI:** !

H: On soil in deciduous woodland.

D+I: B&K5: 218 266

Known from England (East Gloucestershire, South Essex, Surrey and Warwickshire), Wales (Glamorganshire) and the Republic of Ireland (South Kerry). Accepted sensu NCL and Lange (FlDan3: 27 & pl. 90A) with *Fagus* in deciduous woodland but excluded sensu Favre (1960) with conifers in montane habitat.

tabacinus P.D. Orton, *Notes Roy. Bot. Gard. Edinburgh* 41(3): 570 (1984)

   Mis.: *Cortinarius biformis* sensu Pearson (1952)

E: ! **S:** !

H: On acidic soil, associated with *Pinus sylvestris*.

Rarely reported, but *fide* Orton (NBA8) 'not uncommon in suitable habitat'. Known from England (Dorset and Surrey) and Scotland (Kincardineshire and Perthshire).

tabularis (Fr.) Fr., *Epicr. syst. mycol.*: 284 (1838)

   Agaricus anomalus γ *tabularis* Fr., *Syst. mycol.* 1: 221 (1821)

   Agaricus decoloratus Fr., *Syst. mycol.* 1: 224 (1821)

   Cortinarius decoloratus (Fr.) Fr., *Epicr. syst. mycol.*: 270 (1838)

E: ! **S:** ! **W:** ? **NI:** ? **ROI:** ?

H: On soil, usually associated with *Betula* or *Quercus* spp, in mixed deciduous woodland.

D: NM2: 298 (as *Cortinarius decoloratus*)

Occasional but widespread in England and Scotland. Reported elsewhere but unsubstantiated with voucher material.

talus Fr., *Epicr. syst. mycol.*: 263 (1838)

   Cortinarius melliolens P.D. Orton, *Trans. Brit. Mycol. Soc.* 43(2): 210 (1960)

   Cortinarius ochropallidus Rob. Henry, *Bull. Soc. Mycol. France* 52(2): 153 (1936)

E: ! **S:** !

H: On soil in mixed deciduous woodland or mixed coniferous and deciduous woodland.

D: NM2: 292 **D+I:** CFP2: B47 **I:** C&D: 337

Rarely reported. Accepted sensu CFP2 (p. 28). Excluded sensu Rea (1922) and sensu Cooke 705 (711) Vol. 5 (1887) which are doubtful.

terpsichores Melot, *Doc. Mycol.* 20(77): 96 (1989)

   Mis.: *Cortinarius caerulescens* sensu NCL p.p.

NI: !

H: On soil in deciduous woodland.

I: SV27: 43

A single collection from Fermanagh (Castle Caldwell) in 2000 but not yet well understood and likely to include a proportion of British records of *C. caesiocyaneus* and possibly *C. cyaneus*.

testaceoviolascens Bidaud & Reumaux, *Atlas des Cortinaires* 10(9): 400 (2000)

E: !

H: Associated with *Fagus* in beechwoods on calcareous soil.

A single collection from East Gloucestershire (Cirencester Park) in 1981.

tofaceus Fr., *Epicr. syst. mycol.*: 281 (1838)

   Cortinarius tofaceus var. *redimitus* Fr., *Epicr. syst. mycol.*: 281 (1838)

E: ! **S:** !

H: Associated with *Fagus* in old deciduous woodland.

I: FlDan3: 91D (as *Cortinarius tophaceus*)

Known from England (Berkshire and South Devon) and Scotland (Easterness).

torvus (Fr.) Fr., *Epicr. syst. mycol.*: 293 (1838)

   Agaricus torvus Fr., *Syst. mycol.* 1: 211 (1821)

E: o **S:** ! **W:** o **NI:** ! **ROI:** !

H: Usually with *Fagus* in old beechwoods but also known with *Betula*, *Corylus* and *Quercus* spp in mixed deciduous woodland.
D: NM2: 302 **D+I:** CFP2: B13, B&K5: 292 376 **I:** C&D: 331
Occasional but widespread. Most records are from southern or south western England, less often reported northwards but extending into Scotland.

traganus (Fr.) Fr., *Epicr. syst. mycol.*: 281 (1838)
    Agaricus traganus Fr., *Syst. mycol.* 1: 217 (1821)
    Cortinarius traganus var. *finitimus* Weinm., *Hymen. Gasteromyc.*: 156 (1836)
E: ? **S:** ! **NI:** ?
H: On acidic soil in mixed woodland with *Betula* and *Pinus* spp.
D: NM2: 298 **D+I:** CFP3: C04, B&K5: 218 267 **I:** C&D: 327, SV37: 17
Known from scattered sites in Scotland. Reported from England (Norfolk) and Northern Ireland (Fermanagh) but unsubstantiated with voucher material.

triformis Fr., *Epicr. syst. mycol.*: 299 (1838)
    Mis.: *Cortinarius melleopallens* sensu Lange (FlDan3: 39 & pl. 97F) and sensu auct. Brit.
E: ! **W:** ! **NI:** !
H: On soil, in deciduous woodland.
D+I: B&K5: 292 377
Rarely reported. Apparently widespread but the true distribution and frequency are unknown. Sensu auct. mult. is *C. poecilopus*.

triumphans Fr., *Epicr. syst. mycol.*: 256 (1838)
    Cortinarius crocolitus Quél., *Bull. Soc. Bot. France* 25: 288 (1878)
E: o **S:** o **W:** o **NI:** !
H: On soil, associated with *Betula*, *Fagus* and *Quercus* spp, in mixed deciduous woodland.
D: NM2: 293 **D+I:** Ph: 130 (as *Cortinarius crocolitus*), CFP1: A49, B&K5: 204 244 **I:** C&D: 335, SV48: 45
Occasional but widespread.

trivialis J.E. Lange, *Fl. agaric. danic.* 5, *Taxonomic Conspectus*: III (1940)
    Myxacium collinitum var. *repandum* Ricken, *Blätterpilze Deutschl.*: 124 (1911)
    Mis.: *Cortinarius mucifluus* sensu Rea (1922) and sensu Cooke [Ill. Brit. fung. 735 (740) Vol. 5 (1888)]
E: o **S:** o **W:** ! **NI:** !
H: On soil, usually with species of *Salix* in Britain (including *Salix repens* on dunes) but also reported with *Betula*, *Populus tremula* and *Quercus* spp in mixed deciduous woodland.
D: NM2: 289 **D+I:** CFP1: A36, B&K5: 232 287 **I:** C&D: 343
Occasional in England and Scotland. Rarely reported elsewhere but apparently widespread.

tubarius Ammirati & A.H. Sm., *Michigan Botanist* 11(1): 22 (1972)
    Cortinarius sphagneti P.D. Orton, *Naturalist (Hull)* 1958 (suppl.): 148 (1958)
E: ! **S:** !
H: On wet soil, or amongst bog mosses, usually associated with *Alnus* or *Salix* spp in wet areas in deciduous woodland.
D: NM2: 282 **D+I:** B&K5: 150 165
Rarely reported but often fruiting in inaccessible sites.

turgidus Fr., *Epicr. syst. mycol.*: 278 (1838)
E: !
H: Associated with *Fagus* in old beechwoods on calcareous soil.
D: NM2: 298 **D+I:** CFP2: B58, B&K5: 220 268 **I:** FlDan3: 92C
Known from South Devon, South Somerset, Surrey and West Kent.

uliginosus Berk., *Outl. Brit. fungol.*: 190 (1860)
    Dermocybe uliginosa (Berk.) M.M. Moser, *Schweiz. Z. Pilzk.* 52(7): 103 (1974)
    Cortinarius concinnus P. Karst., *Symb. mycol. fenn.* 3: 178 (1878)
    Cortinarius uliginosus var. *obtusus* J.E. Lange, *Fl. agaric. danic.* 5, *Taxonomic Conspectus*: III (1940)

E: o **S:** o **W:** ! **NI:** o **ROI:** !
H: On wet soil, or amongst *Sphagnum* spp, in woodland or on heathland, usually associated with *Alnus* or *Salix* spp.
D: NM2: 282 **D+I:** CFP4: D50, CFP4: D49, B&K5: 152 166 **I:** C&D: 329
Occasional but widespread.

umbrinolens P.D. Orton, *Notes Roy. Bot. Gard. Edinburgh* 38(2): 319 (1980)
    Mis.: *Cortinarius rigidus* sensu Lange (FlDan3: 41 & pl. 100A)
E: ! **S:** ? **W:** ? **NI:** ? **ROI:** ?
H: On soil in deciduous woodland.
D: NM2: 306 **D+I:** CFP1: A08, B&K5: 292 378 **I:** FlDan3: 100A (as *Cortinarius rigidus*)
Apparently widespread but most of the records are unsubstantiated with voucher material.

urbicus (Fr.) Fr., *Epicr. syst. mycol.*: 293 (1838)
    Agaricus urbicus Fr., *Syst. mycol.* 1: 216 (1821)
E: !
H: On soil, usually associated with *Salix* spp in wet areas in deciduous woodland.
D+I: CFP3: C07, B&K5: 220 269
Known from Bedfordshire (Leighton Buzzard) in 1951 [see TBMS 38(4): 394 (1955)] and Huntingdonshire (Woodwalton Fen) in 1961.

valgus Fr., *Epicr. syst. mycol.*: 290 (1838)
    Cortinarius camurus Fr., *Epicr. syst. mycol.*: 285 (1838)
    Cortinarius valgus ssp. *camurus* (Fr.) Melot, *Doc. Mycol.* 20(80): 70 (1990)
    Cortinarius fuliginosus P.D. Orton, *Notes Roy. Bot. Gard. Edinburgh* 26: 46 (1964)
E: ? **S:** !
H: On soil in deciduous woodland. British material with *Betula* spp, but usually with *Fagus* in continental Europe.
D: NM2: 285 **D+I:** CFP2: B05
Accepted on the basis of a Scottish collection by Orton [as *C. fuliginosus*] in 1960 but other records are dubious and unsubstantiated with voucher material. Sensu Cooke 785 (750) Vol. 6 (1886) [as *C. valgus*] is doubtful.

variegatus Bres., *Fungi trident.* 1: 56 (1884)
    Cortinarius fulvidolilaceus P.D. Orton, *Notes Roy. Bot. Gard. Edinburgh* 29(1): 90 (1969)
    Cortinarius variegatus var. *marginatus* Bres., *Fungi trident.* 1: 56 (1884)
S: ! **ROI:** !
H: On soil in deciduous woodland and reported with *Fagus*.
D+I: CFP3: C58
Records from Ireland referred to *C. variegatus* var. *marginatus* are included here.

variicolor (Pers.) Fr., *Epicr. syst. mycol.*: 259 (1838)
    Agaricus variicolor Pers., *Syn. meth. fung.*: 280 (1801)
    Mis.: *Cortinarius crassus* sensu auct. Brit.
    Mis.: *Cortinarius varius* sensu Cooke [Ill. Brit. fung. 689 (698) Vol. 5 (1887)]
E: ! **S:** ! **NI:** !
H: On acidic soil associated with *Picea* or *Pinus* spp.
D+I: CFP2: B20 (as *Cortinarius variecolor*), B&K5: 204 246 (as *C. variecolor*) **I:** C&D: 337 (as *C. variecolor*)
Known from widely scattered localities in Scotland (Easterness, Mid-Perthshire and Westerness). Reported from England but unsubstantiated with voucher material.

variiformis Malençon, *Fl. Champ. sup. Maroc.* 1: 526 (1970)
    Mis.: *Cortinarius varius* sensu auct. Brit.
E: !
H: On soil, associated with *Quercus* spp in deciduous woodland and also *Carpinus*.
A single collection (in herb. K) from East Kent (Kingstone, Jumping Down). Two additional collections from North Somerset, originally determined as *C. varius*, are mentioned by Brandrud [Edinburgh J. Bot. 53 (3): 357 (1996)].

venetus (Fr.) Fr., *Epicr. syst. mycol.*: 291 (1838)
  Agaricus raphanoides β *venetus* Fr., *Syst. mycol.* 1: 230
    (1821)
E: ! S: ! W: ? NI: ? ROI: !
H: On soil in deciduous or mixed woodland, also on heathland or
  moorland where it may be associated with *Arctostaphyllos uva-ursi*.
D: NM2: 286 **D+I:** CFP1: A15, CFP3: C55, B&K5: 162 181
Rarely reported but apparently widespread. Reported from
  Northern Ireland and Wales but unsubstantiated with voucher
  material.

venustus P. Karst., *Hedwigia* 20: 178 (1881)
  Cortinarius calopus P. Karst., *Hedwigia* 20: 178 (1881)
  Cortinarius traganulus P.D. Orton, *Sydowia* 36: 222 (1983)
S: !
H: On acidic soil, associated with *Pinus sylvestris* in Caledonian
  pinewoods.
D+I: CFP1: A22 (as *Cortinarius calopus*), CFP3: C50
Known from scattered sites in Scotland.

vernus H. Lindstr. & Melot, *Cortinarius* 3: 27 (1994)
  Mis.: *Cortinarius erythrinus* sensu Ricken [as *Hygrocybe
    erythrina* in Blätterpilze Deutschl: 184 (1912)]
  Mis.: *Cortinarius uraceus* sensu Lange (FlDan3: 45 & pl.
    102B) & sensu Pearson (1949)
E: !
H: On soil, in deciduous woodland. A spring species, usually
  asociated with *Fagus* or *Quercus* spp.
D+I: CFP3: C51, B&K5: 294 381
Known from East Sussex, North Somerset, Surrey and South
  Devon, and probably also Berkshire and West Sussex
  (misdetermined as *C. uraceus*).

vibratilis (Fr.) Fr., *Epicr. syst. mycol.*: 277 (1838)
  Agaricus vibratilis Fr., *Syst. mycol.* 1: 227 (1821)
E: ! S: ! NI: ?
H: On soil in coniferous or mixed woodland.
D: NM2: 288 **D+I:** CFP4: D26, B&K5: 232 288 **I:** C&D: 343
Rarely reported but apparently widespread. The single record
  from Northern Ireland (Down) is unsubstantiated with voucher
  material.

violaceofuscus (Cooke & Massee) Massee, *Ann. Bot.* 18: 501
  (1904)
  Agaricus violaceofuscus Cooke & Massee, *Grevillea* 18: 52
    (1890)
  Inocybe violaceofusca (Cooke & Massee) Sacc., *Syll.
    fung.* 9: 96 (1891)
E: !
H: Amongst grass.
Known only from the type collection (Forest of Dean). Confirmed
  as a species of *Cortinarius* by Kuyper (1986), but otherwise
  little-known. Massee assigned it to *Cortinarius* subgenus
  Dermocybe.

violaceovelatus D.A. Reid & Rob. Henry, *Bull. Soc. Mycol.
  France* 77: 89 (1961)
E: !
H: On calcareous soil, in deciduous woodland.
Known only from the type, from Oxfordshire (Blenheim Palace)
  in 1958.

violaceus (L.) Gray, *Nat. arr. Brit. pl.* 1: 217 (1821)
  Agaricus violaceus L., *Fl. Suec.*, Edn 2: 1226 (1755)
E: ! S: o NI: ! ROI: !
H: On acidic or peaty soil, in open deciduous woodland, often
  associated with *Betula* spp, also known with *Fagus, Quercus*
  and *Pinus* spp, and occasionally with *Pteridium* on heathland.
D: NM2: 280 **D+I:** CFP3: C37, B&K5: 142 152 **I:** C&D: 325,
  SV43: 15
Occasional but widespread. Most often collected in Scotland and
  northern areas of England, becoming rare further south.

violilamellatus P.D. Orton, *Notes Roy. Bot. Gard. Edinburgh*
  41(3): 571 (1984)
E: ! S: !
H: On soil, associated with *Pinus sylvestris*.

D+I: CFP4: D39
Known from England (Surrey and West Norfolk) and Scotland
  (Easterness, Kincardineshire and Perthshire).

vulpinus (Velen.) Rob. Henry, *Bull. Soc. Mycol. France* 62(3-4):
  207 (1947)
  Inoloma vulpinum Velen., *České Houby* III: 428 (1921)
  Cortinarius albomarginatus P.D. Orton, *Naturalist (Hull)* 1955:
    58 (1955)
  Cortinarius rufoalbus Kühner, *Bull. Mens. Soc. Linn. Lyon*
    24(2): 45 (1955)
  Mis.: *Cortinarius claricolor* sensu Lange (FlDan3: 20 & pl.
    85A)
E: !
H: On soil in deciduous woodland. British material associated
  with *Carpinus, Fagus* and *Quercus* spp.
D+I: CFP2: B45, B&K5: 222 271 **I:** C&D: 335, SV48: 51
Known from Hertfordshire, Surrey and West Kent.

vulpinus *ssp.* pseudovulpinus (Rob. Henry) Brandrud,
  Cortinarius: 33 (1992)
  Cortinarius pseudovulpinus Rob. Henry, *Bull. Soc. Mycol.
    France* 105(1): 93 (1989)
E: !
H: On soil in mixed deciduous woodland.
D+I: CFP2: B43 **I:** SV48: 53 (as *Cortinarius pseudovulpinus*)
A single collection from South Hampshire (New Forest) in 1999.

xanthocephalus P.D. Orton, *Trans. Brit. Mycol. Soc.* 43(2):
  214 (1960)
  Mis.: *Cortinarius decolorans* sensu auct.
E: ! S: !
H: Usually associated with *Fagus* in old beechwoods.
Known from England (West Kent and West Sussex) and Scotland
  (Perthshire).

xantho-ochraceus P.D. Orton, *Trans. Brit. Mycol. Soc.* 43(2):
  216 (1960)
  Cortinarius langei Rob. Henry, *Doc. Mycol.* 16(61): 22 (1985)
E: !
H: Associated with *Fagus* in old beechwoods on calcareous soil.
Known from Surrey and West Sussex.

xanthophyllus (Cooke) Rob. Henry, *Rev. Mycol. (Paris)* 8(5-6):
  30 (1943)
  Cortinarius dibaphus var. *xanthophyllus* Cooke, *Ill. Brit. fung.*
    713 (753) Vol. 5 (1886)
E: !
H: On soil, usually associated with *Quercus*. spp, and also
  reported with *Betula* spp, in mixed deciduous woodland.
D: NBA9: 550-551 **I:** C&D: 341
Known from Herefordshire, South Devon and South Hampshire.

zosteroides P.D. Orton, *Sydowia* 36: 215 (1983)
E: !
H: Associated with *Quercus* spp in old deciduous woodland.
Known only from South Hampshire.

COTYLIDIA P. Karst., *Rev. Mycol. (Toulouse)* 3(9): 22 (1881)
Stereales, Podoscyphaceae
Type: *Cotylidia undulata* (Fr.) P. Karst.

pannosa (Sowerby) D.A. Reid, *Nova Hedwigia Beih.* 18: 81
  (1965)
  Helvella pannosa Sowerby, *Col. fig. Engl. fung.* 2: pl. 155
    (1799)
  Thelephora pannosa (Sowerby) Fr., *Syst. mycol.* 1: 430
    (1821)
  Thelephora pallida Pers., *Icon. descr. fung.*: pl. 5 (1798)
  Stereum pallidum (Pers.) Cooke, *Cat. Handb. Brit. Basidio.*
    genus no. 88 (1909)
  Thelephora pannosa β *pallida* (Pers.) Fr., *Syst. mycol.* 1: 430
    (1821)
  Thelephora sowerbyi Berk. & Broome, *Outl. Brit. fungol.*: 266
    (1860)

Stereum sowerbyi (Berk. & Broome) Massee, *J. Linn. Soc., Bot.* 25: 164 (1889)
E: ! **W:** ! **NI:** ?
H: On soil in deciduous woodland.
D: NM3: 192 **D+I:** CNE3: 300-301, B&K2: 172 184 **I:** Kriegl1: 186
Known from England (Berkshire, Buckinghamshire, Co. Durham, East Cornwall, East Kent, Herefordshire, North Somerset, Oxfordshire, Shropshire, South Devon, Surrey, West Gloucestershire and West Sussex) and Wales (Glamorganshire and Denbighshire). A single report from Northern Ireland in 1853 is unsubstantiated with voucher material. From published records and collections in herb. K, this was more frequent in the nineteenth and early twentieth centuries than at present.

undulata (Fr.) P. Karst., *Rev. Mycol. (Toulouse)* 3(9): 22 (1881)
Thelephora undulata Fr., *Elench. fung.* 1: 164 (1838)
Stereum undulatum (Fr.) Massee, *Brit. fung.-fl.* 1: 130 (1892)
E: ! **S:** ! **W:** !
H: On soil in deciduous woodland, occasionally on old fire sites in such habitat and rarely amongst mosses on old walls.
D: NM3: 192 **D+I:** CNE3: 302-303, B&K2: 172 185 **I:** SV27: 14
Rarely reported or collected but apparently widespread. Known from England (Shropshire, South Somerset and West Cornwall), Scotland (Angus) and Wales (Denbighshire).

CRATERELLUS Pers., *Mycol. eur.* 2: 4 (1825)
Cantharellales, Craterellaceae
Type: *Craterellus cornucopioides* (L.) Pers.

cornucopioides (L.) Pers., *Mycol. eur.* 2: 5 (1825)
Peziza cornucopioides L., *Sp. pl.*: 1181 (1753)
Merulius cornucopioides (L.) Pers. [non *M. cornucopiodes* (Bolton) With. (1796)], *Syn. meth. fung.*: 491 (1801)
Cantharellus cornucopioides (L.) Fr., *Syst. mycol.* 1: 321 (1821)
Merulius purpureus With., *Bot. Arr. Brit. pl. ed. 2,* 3: 280 (1792)
E: o **S:** o **W:** o **NI:** ! **ROI:** !
H: On soil in deciduous woodland (especially on acidic soil in old beechwoods) often amongst mosses such as *Leucobryum glaucum* or *Polytrichum* spp.
D: NM3: 263, BFF8: 26-27 **D+I:** Ph: 190-191, B&K2: 374 487, BritChant: 16-17 **I:** Sow1: 74 (as *Peziza cornucopioides*), Bon: 307, C&D: 145, Kriegl2: 18
Occasional but widespread.

CREPIDOTUS (Fr.) Staude, *Schwämme Mitteldeutschl.*: 71 (1857)
Cortinariales, Crepidotaceae
Dochmiopus Pat., *Hyménomyc. Eur.*: 113 (1887)
Pleurotellus Fayod, *Ann. Sci. Nat., Bot.,* sér. 7, 9: 339 (1889)
Type: *Crepidotus mollis* (Schaeff.) Staude

applanatus (Pers.) P. Kumm., *Führ. Pilzk.*: 74 (1871)
Agaricus applanatus Pers., *Observ. mycol.* 1: 8 (1796)
Agaricus putrigenus Berk. & M.A. Curtis, *Ann. Mag. Nat. Hist., Ser. 3* 4: 292 (1859)
Crepidotus putrigenus (Berk. & M.A. Curtis) Sacc., *Syll. fung.* 5: 883 (1887)
E: o **S:** o **W:** o **NI:** ! **ROI:** o
H: On decayed wood of deciduous trees, usually on large fallen trunks, branches or logs of *Fagus* but also on *Acer pseudoplatanus, Betula, Quercus* and *Rhododendron* spp. Rarely on coniferous wood (but see var. *subglobiger*).
D: BFF6: 86, NM2: 336 **D+I:** B&K5: 296 384 **I:** C&D: 349
Occasional but widespread and may be locally abundant in old woodland.

applanatus *var.* **subglobiger** Singer, *Nova Hedwigia Beih.* 44: 478 (1973)
E: !
H: On dead wood of conifers. British material on large fallen and decayed trunks of *Pinus* spp.
Known from Surrey.

autochthonus J.E. Lange, *Fl. agaric. danic.* 5, *Taxonomic Conspectus*: III (1940)
Crepidotus fragilis Joss. [*nom. inval.*], *Bull. Soc. Mycol. France* 53(2): 18 (1937)
Mis.: *Crepidotus applanatus* sensu Rea (1922)
E: o **S:** ! **W:** ! **NI:** !
H: On soil in deciduous woodland, usually on calcareous soils, and frequently on bare ground, along forestry tracks or at the edges of footpaths.
D: BFF6: 87-88, NM2: 337
Rarely reported but apparently widespread. Frequent, at least in south-eastern England. Many records, however, are dubious, having been described from woody substrata (such as twigs or sticks) and unsubstantiated with voucher material.

carpaticus Pilát, *Hedwigia* 69: 140 (1929)
Crepidotus wakefieldiae Pilát, *Atlas champ. Eur.* 6: 65 (1948)
E: ! **S:** ! **W:** ! **NI:** ! **ROI:** !
H: On decayed fallen wood (twigs, bark etc.) of deciduous trees.
D: BFF6: 94 (as *Crepidotus wakefieldiae*) **I:** SV33: 61
Rarely reported. Apparently widespread.

cesatii (Rabenh.) Sacc., *Michelia* 1: 2 (1877)
Agaricus cesatii Rabenh., *Flora* 36: 564 (1851)
Agaricus variabilis var. *sphaerosporus* Pat., *Tab. anal. fung.*: 101 (1884)
Crepidotus variabilis var. *sphaerosporus* (Pat.) Quél., *Enchir. fung.*: 108 (1886)
Claudopus sphaerosporus (Pat.) Sacc., *Syll. fung.* 5: 734 (1887)
Crepidotus sphaerosporus (Pat.) J.E. Lange, *Dansk. Bot. Ark.* 9(6): 52 (1938)
E: c **S:** o **W:** o **NI:** o **ROI:** o **O:** Channel Islands: !
H: On decayed wood of deciduous trees, especially in piles of small branches, sticks or twigs. Often on *Acer pseudoplatanus, Fagus* and *Hedera* but known on a wide range of other hosts including conifers such as *Picea* spp, and dead herbaceous or woody stems such as *Chamaerion, Epilobium, Rosa, Rubus, Viburnum* and *Ulex* spp. Basidiomes are frequently produced late in the autumn and continue well into the winter.
D: BFF6: 90, NM2: 337 (as *Crepidotus sphaerosporus*) **D+I:** B&K5: 298 385
Common and widespread, at least throughout England, but frequently misidentified (usually as *C. variabilis*).

cinnabarinus Peck, *Bull. Torrey Bot. Club* 22: 489 (1895)
E: !
H: On fallen and decayed trunks of deciduous trees. British collections on *Fraxinus* and *Populus tremula*.
D: NM2: 336 **D+I:** Myc11(2): 78 **I:** C&D: 349
Known from Hertfordshire and Warwickshire. Probably a recent coloniser as the brightly coloured basidiomes are unlikely to have been previously overlooked.

epibryus (Fr.) Quél., *Fl. mycol. France*: 107 (1888)
Agaricus epibryus Fr., *Syst. mycol.* 1: 275 (1821)
Crepidotus commixtus Bres., *Fungi Saxonici Exsiccati*: 1766 & 1767 (1912)
Pleurotus commixtus (Bres.) Bres., *Icon. Mycol.* 6: 298 (1928)
Dochmiopus commixtus (Bres.) Singer, *Beih. Bot. Centralbl.* 56: 146 (1936)
Pleurotellus graminicola Fayod, *Cens. dei funghi Val. Valdesi*: 12 (1892)
Agaricus herbarum Peck, *Bull. Buffalo Soc. Nat. Sci.* 1: 53 (1873)
Crepidotus herbarum (Peck) Sacc., *Syll. fung.* 5: 888 (1887)
Pleurotellus herbarum (Peck) Singer, *Lilloa* 13: 84 (1947)
Agaricus variabilis var. *hypnophilus* Pers., *Mycol. eur.* 3: 28 (1828)
Pleurotus hypnophilus (Pers.) Sacc., *Syll. fung.* 5: 384 (1887)
Pleurotellus hypnophilus (Pers.) Fayod, *Ann. Sci. Nat., Bot.,* sér. 7, 9: 339 (1889)
Mis.: *Pleurotus perpusillus* sensu auct.
Mis.: *Crepidotus pubescens* sensu K&R
Mis.: *Pleurotellus septicus* sensu auct. Eur.

Mis.: *Pleurotus septicus* sensu auct.
E: c **S:** ! **W:** ! **NI:** ! **O:** Channel Islands: !
H: On fallen wood (twigs etc.) of deciduous trees, dead herbaceous stems, decayed leaf litter, dead and living grass culms and stems. A single collection on a discarded plastic drinking cup. Often fruiting well into the winter or throughout the year.
D: BFF6: 96-97 (as *Pleurotellus graminicola*), NM2: 339 (as *P. hypnophilus*) **D+I:** B&K5: 298 387
Common and widespread. Better known as *Pleurotellus graminicola*.

lundellii Pilát, *Fungi Exsiccati Suecici* 5-6: 10 (1936)
   Crepidotus amygdalosporus Kühner, *Bull. Soc. Naturalistes Oyonnax* 8: 74 (1954)
   Crepidotus subtilis P.D. Orton, *Trans. Brit. Mycol. Soc.* 43(2): 221 (1960)
   Mis.: *Crepidotus inhonestus* sensu NCL, sensu Pegler & Young [Kew Bull. 27:321 (1972)] and sensu BFF6
   Mis.: *Crepidotus sambuci* sensu NCL, and sensu BFF6
E: ! **S:** ! **W:** ! **NI:** ?
H: On decayed wood, bark, small twigs and branches of deciduous trees, most frequently *Acer pseudoplatanus*. Occasionally reported 'on soil' (but then probably on buried wood).
D: NM2: 337 (as *Crepidotus inhonestus*), NM2: 337 (as *C. epibryus*) **D+I:** B&K5: 300 388
Rarely reported but apparently widespread (although often confused with similar taxa).

luteolus (Lambotte) Sacc., *Syll. fung.* 5: 888 (1887)
   Agaricus luteolus Lambotte, *Flore Mycologique de Belge* 1: 181 (1880)
   Mis.: *Crepidotus pubescens* sensu Pearson in TBMS 32: 268 (1949)
E: c **S:** ! **W:** ! **NI:** !
H: On dead woody and herbaceous stems, especially those of *Urtica dioica*, *Rubus fruticosus* and *Pteridium*. Occasionally found on small fallen twigs of deciduous trees, usually *Fraxinus*.
D: BFF6: 90-91, NM2: 337 **D+I:** Ph: 188-189, B&K5: 300 389 **I:** C&D: 349
Common in southern England especially on old nettle stems.

mollis (Schaeff.) Staude, *Schwämme Mitteldeutschl.*: 71 (1857)
   Agaricus mollis Schaeff. [non *A. mollis* Bolton (1788)], *Fung. Bavar. Palat. nasc.* 1: 49 (1762)
   Crepidopus mollis (Schaeff.) Gray, *Nat. arr. Brit. pl.* 1: 616 (1821)
   Agaricus ralfsii Berk. & Broome, *Ann. Mag. Nat. Hist., Ser. 5 [Notices of British Fungi no. 2008]* 12: 372 (1883)
   Crepidotus ralfsii (Berk. & Broome) Sacc., *Syll. fung.* 5: 881 (1881)
E: c **S:** c **W:** c **NI:** c **ROI:** c **O:** Channel Islands: ! Isle of Man: !
H: On decayed wood of deciduous trees, especially common on *Fraxinus* and also frequent on *Fagus* but known on a wide range of other hosts. Rarely reported on coniferous wood but these records are unverified.
D: BFF6: 86, NM2: 336 **D+I:** Ph: 188-189, B&K5: 300 390 **I:** Sow1: 98 (as *Agaricus mollis*)
Common and widespread. A form with a scaly pileus is frequently recorded as *Crepidotus calolepis* or *C. mollis* var. *calolepis*.

muscigenus Velen., *Novit. mycol. Novissimae*: 77 (1947)
W: !
H: On living mosses. British collection on *Rhytidiadelphus squarrosus* in open deciduous woodland.
A single collection, from Caernarvonshire (Coed Gorllwyn) in 1999.

subverrucisporus Pilát, *Atlas champ. Eur.* 6: 51 (1948)
   Crepidotus bickhamensis P.D. Orton, *Notes Roy. Bot. Gard. Edinburgh* 41(3): 573 (1984)
E: !
H: On mossy bark of living trees. British material on *Salix* spp.
D: BFF6: 89 (as *Crepidotus bickhamensis*) **D+I:** B&K5: 302 391

Rarely reported and poorly known in Britain. Records from Somerset and Oxfordshire (as *C. bickhamensis*).

variabilis (Pers.) P. Kumm., *Führ. Pilzk.*: 74 (1871)
   Agaricus variabilis Pers., *Observ. mycol.* 2: 46 (1799)
   Crepidopus variabilis (Pers.) Gray, *Nat. arr. Brit. pl.* 1: 616 (1821)
   Claudopus variabilis (Pers.) Gillet, *Hyménomycètes*: 426 (1876)
E: o **S:** ! **W:** ! **NI:** ! **ROI:** ! **O:** Channel Islands:! Isle of Man: !
H: On fallen woody debris (twigs etc.) or dead attached twigs of deciduous trees, and (apparently) on old weathered herbivore dung.
D: BFF6: 88, NM2: 337 **D+I:** B&K5: 302 392
Frequently reported and widespread. Many collections, however, are redetermined as *C. cesatii* so records unsubstantiated with voucher material are suspect.

versutus (Peck) Sacc., *Syll. fung.* 5: 888 (1887)
   Agaricus versutus Peck, *Rep. (Annual) New York State Mus. Nat. Hist.* 30: 70 (1878)
   Crepidotus bresadolae Pilát [*nom. superf.* for *C. pubescens* Bres.], *Atlas champ. Eur.* 6: 46 (1948)
   Crepidotus pubescens Bres., *Icon. Mycol.* 16: 790 (1930)
E: ! **S:** ! **W:** !
H: On dead wood, usually small fallen twigs and sticks of deciduous trees.
D: NM2: 336 **D+I:** B&K5: 302 393
Rarely reported, but apparently widespread. Said to be 'not uncommon' in BFF6 (as *C. pubescens*). British records as *C. pubescens* are often sensu Pearson (1949) (= *C. luteolus*).

CRINIPELLIS Pat., *J. Bot. (Morot)* 3: 336 (1889)
Agaricales, Tricholomataceae
Type: *Crinipellis scabella* (Alb. & Schwein.) Murrill

scabella (Alb. & Schwein.) Murrill, *North American Flora, Ser. 1 (Fungi 3)*, 4: 287 (1915)
   Agaricus scabellus Alb. & Schwein. [non *A. scabellus* Fr. (1838)], *Consp. fung. lusat.*: 189 (1805)
   Agaricus caulicinalis Bull. [non sensu Sowerby in Col. fig. Eng. fung. 2: pl. 163 (1788)], *Herb. France*: pl. 522 f. 1 (1791)
   Crinipellis caulicinalis (Bull.) Rea, *Trans. Brit. Mycol. Soc.* 5(3): 436 (1917)
   Marasmius epichloë Fr., *Hymenomyc. eur.*: 479 (1874)
   Androsaceus epichloë (Fr.) Rea, *Brit. basidiomyc.*: 533 (1922)
   Agaricus stipitarius Fr., *Syst. mycol.* 1: 138 (1821)
   Collybia stipitaria (Fr.) Gillet, *Hyménomycètes*: 319 (1876)
   Crinipellis stipitaria (Fr.) Pat., *J. Bot. (Morot)* 3: 336 (1889)
E: o **S:** o **W:** o **NI:** ! **ROI:** o **O:** Channel Islands: !
H: On dead and decayed stems and culms of many species of grasses.
D: NM2: 117 (as *Crinipellis scabella*), FAN3: 136 **D+I:** B&K3: 184 204 (as *C. stipitaria*), A&N2 140-144 37 **I:** Bon: 175 (as *C. stipitarius*), C&D: 215, SV32: 5, Kriegl3: 230
Occasional but widespread.

CRISTINIA Parmasto, *Conspectus Systematis Corticiacearum*: 47 (1968)
Stereales, Lindtneriaceae
   Dacryobasidium Jülich, *Biblioth. Mycol.* 85: 396 (1982)
Type: *Cristinia helvetica* (Pers.) Parmasto

coprophila (Wakef.) Hjortstam, *Mycotaxon* 47: 407 (1993)
   Corticium coprophilum Wakef., *Trans. Brit. Mycol. Soc.* 5(3): 480 (1917)
   Athelia coprophila (Wakef.) Jülich, *Willdenowia. Beih.* 7: 66 (1972)
   Byssocorticium coprophilum (Wakef.) J. Erikss. & Ryvarden, *Corticiaceae of North Europe* 2: 186 (1973)
   Dacryobasidium coprophilum (Wakef.) Jülich, *Biblioth. Mycol.* 85: 400 (1982) [1981]
E: !

H: British material on a dead stem of *Glyceria maxima* in a dried-up pond, on weathered horse dung, and on chipboard in grass.
D: NM3: 152 (as *Byssocorticum coprophilum*)
Described from Surrey; also known from London and Middlesex.

gallica (Pilát) Jülich, *Persoonia* 8(3): 298 (1975)
Radulum gallicum Pilát, *Mykol.* 2: 54 (1925)
Mis.: *Cristinia mucida* sensu Erikss. & Ryvarden (1975)
Mis.: *Radulum mucidum* sensu Bourdot & Galzin (1928)
E: ! **S:** ?
H: In woodland, on decayed wood (old stumps etc.) of deciduous trees, also on soil and woody debris in damp areas on decayed leaf litter and old basidiomes of polypores.
D: NM3: 122 **D+I:** CNE3: 310-311 (as *Cristinia mucida*), B&K2: 108 88, Myc8(3): 116 **I:** Kriegl1: 187, SV42: 41
Known from Berkshire, Buckinghamshire, Hertfordshire, North Hampshire, Oxfordshire, South Devon, Surrey, West Kent, West Norfolk, Westmorland and Yorkshire. Reported from Scotland but unsubstantiated with voucher material.

helvetica (Pers.) Parmasto, *Conspectus Systematis Corticiacearum*: 48 (1968)
Hydnum helveticum Pers., *Mycol. eur.* 2: 184 (1825)
Grandinia helvetica (Pers.) Fr., *Hymenomyc. eur.*: 627 (1874)
Corticium helveticum (Pers.) Höhn. & Litsch., *Sitzungsber. Kaiserl. Akad. Wiss., Wien, Math.-Naturwiss. Cl., Abt. 1* 117: 1084 (1908)
E: c **S:** o **W:** o **O:** Channel Islands: !
H: On decayed wood, bark and litter of deciduous trees and conifers, on soil under leaf litter, on dead stems or debris of *Pteridium* and on fallen polypore basidiomes. Rarely on man-made objects, such as a leather handbag and old boots or shoes, discarded in woodland.
D: NM3: 122 **D+I:** CNE3: 307-309, B&K2: 108 89 **I:** Kriegl1: 189
Common and widespread in England, less frequently reported elsewhere.

rhenana Grosse-Brauckm., *Mycotaxon* 47: 407 (1993)
E: !
H: On decayed wood. British material on woodchips of *Populus alba*.
Known from Surrey (Richmond, Ham Lands).

CRUCIBULUM Tul. & C. Tul., *Ann. Sci. Nat., Bot.,* sér. 3, 1: 89 (1844)
Nidulariales, Nidulariaceae
Type: *Crucibulum laeve* (Huds.) Kambly

laeve (Huds.) Kambly, *Gast. Iowa*: 167 (1936)
Peziza laevis Huds., *Fl. angl.*, Edn 2, 2: 634 (1778)
Nidularia laevis (Huds.) Huds., *Fl. Cantab. (Suppl.)*: 529 (1793)
Peziza crucibuliformis Schaeff., *Fung. Bavar. Palat. nasc.* 2: 125 (1763)
Cyathus crucibuliformis (Schaeff.) Hoffm., *Veg. Crypt.* 2: 29 (1790)
Cyathus crucibulum Pers., *Syn. meth. fung.*: 238 (1801)
Nidularia crucibulum (Pers.) Fr., *Syst. mycol.* 2: 299 (1823)
Crucibulum vulgare Tul. & C. Tul., *Ann. Sci. Nat., Bot.,* sér. 3, 1: 90 (1844)
E: c **S:** ! **W:** o **NI:** o **ROI:** o
H: On decayed woody debris, fallen twigs, compost, decayed herbaceous stems and grass culms, in woodland, dunes, parkland and gardens, often abundant on woodchip mulch on flowerbeds.
D: NM3: 194 **D+I:** Ph: 254-255, B&K2: 378 493, BritPuffb: 64-65 Figs. 38/39 **I:** Sow1: 30 (as *Nidularia laevis*), Bon: 301
Common in southern England, occasional elsewhere.

CRUSTODERMA Parmasto, *Conspectus Systematis Corticiacearum*: 87 (1968)
Stereales, Hyphodermataceae
Type: *Crustoderma dryinum* (Berk. & M.A. Curtis) Parmasto

dryinum (Berk. & M.A. Curtis) Parmasto, *Conspectus Systematis Corticiacearum*: 88 (1968)
Corticium dryinum Berk. & M.A. Curtis, *Grevillea* 1: 179 (1873)
E: !
H: On decayed wood. British material on conifer wood and *Acer* sp.
D: NM3: 191 **D+I:** CNE3: 315-317
Known from southern England (Hertfordshire, East Kent and South Hampshire).

CRUSTOMYCES Jülich, *Persoonia* 10: 140 (1978)
Stereales, Corticiaceae
Type: *Crustomyces subabruptus* (Bourdot & Galzin) Jülich

expallens (Bres.) Hjortstam, *Windahlia* 17: 56 (1987)
Corticium expallens Bres., *Ann. mycol.* 6: 43 (1908)
Laeticorticium expallens (Bres.) J. Erikss. & Hjortstam, *Corticiaceae of North Europe* 6: 1069 (1981)
Corticium salicicola M.P. Christ., *Friesia* 5(3 - 5): 209 (1956)
Hyphoderma salicicola (M.P. Christ.) M.P. Christ., *Dansk. Bot. Ark.* 19(2): 210 (1960)
E: ! **S:** !
H: On dead bark or hard, decorticated wood (stumps and large logs) of deciduous trees such as *Acer*, *Salix* and *Ulmus* spp.
D: NM3: 178 (as *Corticium expallens*) **D+I:** CNE6: 1069-1071 (as *Laeticorticium expallens*)
Known from England (Berkshire, London, Middlesex, South Somerset, Surrey and West Sussex) and Scotland (South Ebudes).

subabruptus (Bourdot & Galzin) Jülich, *Persoonia* 10(1): 140 (1978)
Odontia subabrupta Bourdot & Galzin, *Hymenomyc. France*: 430 (1928)
Cystostereum subabruptum (Bourdot & Galzin) J. Erikss. & Ryvarden, *Corticiaceae of North Europe* 3: 327 (1975)
E: ! **S:** !
H: British material on large fallen and decayed trunks of *Fagus* or *Quercus robur* with a single collection on *Pinus sylvestris*.
D: NM3: 205 **D+I:** CNE3: 326-327 (as *Cystostereum subabruptum*), B&K2: 112 95
Known from England (Berkshire, Mid-Lancashire, North Hampshire, Surrey and West Kent) and Scotland (Clyde Isles). Probably genuinely uncommon since basidiomes often cover extensive areas and would be difficult to overlook.

CRYPTOCOCCUS Vuill., *Rev. Gén. Sci. Pures Appl.* 12: 741 (1901)
A genus of anamorphic yeasts with teleomorphs in *Filobasidiella*. The species listed below, however, have no known teleomorphs. Based on molecular research, most do not belong in *Cryptococcus* sensu stricto but are not, as yet, assigned elsewhere.
Tremellales, Filobasidiaceae
Type: *Cryptococcus neoformans* (San Felice) Vuill.

aerius (Saito) Nann., *Com. Fatta R. Accad. Fisiocrit. Siena*: 1 (1927)
Torula aerius Saito, *Jap. J. Bot.* 1(1): 41 (1922)
Torulopsis aeria (Saito) Lodder, *Verh. K. Akad. Wet.* 32: 161 (1934)
Cryptococcus albidus var. *aerius* (Saito) Phaff & Fell, *Yeasts, a taxonomic study*: 1093 (1970)
E: !
H: Isolated ex sausages.
D+I: YST 289
DNA analysis suggests this species should be assigned to the *Filobasidiales*.

albidus (Saito) C.E. Skinner, *Amer. Midl. Naturalist* 43: 249 (1950)
Torula albida Saito, *Jap. J. Bot.* 1(1): 43 (1922)
Torulopsis albida (Saito) Lodder, *Verh. K. Akad. Wet.* 2(32): 163 (1934)

E: !

H: Isolated ex paint, soil in sheep pasture, lamb carcasses, human knee aspirates, and cats.

D+I: YST 291

A member of the *Filobasidiales* reported as the anamorph of *Filobasidium floriforme* but this remains unproven. Possibly part of a species complex.

aquaticus (E.B.G. Jones & Slooff) Rodr. Mir. & Weijman, *Antonie van Leeuwenhoek* 54(6): 550 (1988)

  Candida aquatica E.B.G. Jones & Slooff, *Antonie van Leeuwenhoek* 32: 223 (1966)

  Vanrija aquatica (E.B.G. Jones & Slooff) R.T. Moore, *Bot. mar.* 23(6): 367 (1980)

E: ! S: !

H: Isolated ex scum and foam from lake water. Also known from dead stems of *Equisetum* and on rotting leaves of *Alnus glutinosa* in streams.

D+I: YST 295

Type collection from Yorkshire (Malham Tarn). Collected from other localities in England (Cumberland: Blea Tarn and South Devon: River Teign) and Scotland (Wester Ross: Falls of Maesach). DNA analysis suggests this species belongs in the *Cystofilobasidiales*.

curvatus (Diddens & Lodder) Golubev, *Taksonomiya i identifikatsaya drozhzhevykh gribov roda Cryptococcus (Pushchino)*: 26 (1980)

  Candida heveanensis var. *curvata* Diddens & Lodder, *Die Hefasammlung des 'Centraalbureau voor Schimmelcultures': Beitrage zu einer Monographie der Hefearten. II. Teil. Die anaskosporogenen Hefen. Zweite Halfte*: 486 (1946)

  Candida curvata (Diddens & Lodder) Lodder & Kreger, *The Yeasts*: 576 (1952)

  Vanrija curvata (Diddens & Lodder) R.T. Moore, *Bot. mar.* 23(6): 367 (1980)

E: !

H: Isolated ex uterus of a cow and as a contaminant in culture media.

D+I: YST 300

DNA analysis suggests this species should be assigned to the *Trichosporonales*.

diffluens (Zach) Lodder & Kreger, *The Yeasts*: 391 (1952)

  Torula diffluens Zach, *Archiv Dermatol. Syph. (Berlin)* 170: 690 (1934)

E: !

H: Isolated ex sausages and dogs.

D+I: YST 301

DNA analysis suggests this species should be assigned to the *Filobasidiales*.

gastricus Reiersöl & di Menna, *Leeuwenhoek ned. Tidjdschr.* 24: 28 (1958)

E: !

H: Isolated ex pig carcasses.

D+I: YST 307

DNA analysis suggests this species should be assigned to the *Filobasidiales*.

laurentii (Kuff.) C.E. Skinner, *Amer. Midl. Naturalist* 43: 249 (1950)

  Torula laurentii Kuff., *Ann. Soc. Sci. méd. nat. Brux.* 74: 38 (1920)

  Torulopsis laurentii (Kuff.) Lodder, *Verh. K. Akad. Wet.* 2(32): 160 (1934)

  Torula flavescens Saito, *Jap. J. Bot.* 1: 43 (1922)

  Cryptococcus laurentii var. *flavescens* (Saito) Lodder & Kreger, *The Yeasts*: 381 (1952)

E: !

H: Isolated ex paint, lamb and pig carcasses and (as var. *flavescens*) sausages.

D+I: YST 315-316

macerans (Freder.) Phaff & Fell, *Yeasts, a taxonomic study*: 1127 (1970)

  Rhodotorula macerans Freder., *Friesia* 5: 237 (1956)

E: ! S: !

H: Isolated ex stored wheat, raw meat and sausages.

D+I: YST 318

DNA analysis suggests this species should be assigned to the *Cystofilobasidiales*.

magnus (Lodder & Kreger) Baptist & Kurtzman, *Mycologia* 68(6): 1200 (1976)

  Cryptococcus laurentii var. *magnus* Lodder & Kreger, *The Yeasts*: 381 (1952)

E: !

H: Isolated ex sausages.

D+I: YST 319

Currently assigned to the *Filobasidiales*.

skinneri Phaff & Carmo Souza, *Leeuwenhoek ned. Tidjdschr.* 28: 205 (1962)

E: !

H: Isolated ex sausages.

D+I: YST 322

CYATHUS Haller, *Hist. Stirp. Helv.* 3: 127 (1768)
Nidulariales, Nidulariaceae

  Nidularia Bull. [*nom. rej.*, non *Nidularia* Fr. (1817)], *Hist. champ. France*: 163 (1791)

Type: *Cyathus striatus* (Huds.) Pers.

olla (Batsch) Pers., *Syn. meth. fung.*: 237 (1801)

  Peziza olla Batsch, *Elench. fung.*: 127 (1783)

  Nidularia campanulata With., *Arr. Brit. pl. ed.3* 4: 356 (1796)

  Cyathus olla var. *agrestis* Pers., *Syn. meth. fung.*: 237 (1801)

  Cyathus vernicosus DC., *Fl. Fr.* 2: 270 (1805)

  Cyathus ollaris Gray, *Nat. arr. Brit. pl.* 1: 587 (1821)

E: o S: o W: o NI: ! ROI: ! O: Channel Islands: !

H: On woody debris (twigs, woodchips etc.) and humus in woodland and on compost or woodchip mulch. Also known on soil in plant pots, and on dunes.

D: NM3: 194 D+I: Ph: 254-255, B&K2: 378 494, BritPuffb: 62-63 Figs. 36/37 I: Sow1: pl. 28 (as *Nidularia campanulata*), NCL 301, Kriegl2: 157, FMBC back cover

Occasional but widespread. Increasingly common, and often abundant, on decayed woodchip mulch.

olla f. anglicus (Lloyd) H.J. Brodie, *Mycologia* 44: 417 (1952)

  Cyathus vernicosus f. *anglicus* Lloyd, *Nidulariaceae*: 25 (1906)

E: !

H: On woody debris and humus.

D+I: Brodie (1975)

Distinguished from the typical form by the large, sulcate peridia. Described from England and subsequently reported from the Americas (Brodie, 1975).

stercoreus (Schwein.) De Toni, *Syll. fung.* 7: 40 (1888)

  Nidularia stercorea Schwein., *Trans. Am. phil. Soc.* 4: 253 (1832)

E: ! S: ! W: !

H: On sandy soil in dunes, often near, or attached to, weathered rabbit droppings and occasionally around stem bases of *Ammophila arenaria*.

D: NM3: 194 D+I: B&K2: 378 495, BritPuffb: 60-61 Figs. 34/35

Rarely reported but apparently widespread in coastal areas of Scotland and Wales. Also isolated from soil in South Devon.

striatus (Huds.) Pers., *Syn. meth. fung.*: 237 (1801)

  Peziza striata Huds., *Fl. angl.*, Edn 2, 2: 634 (1778)

  Nidularia striata (Huds.) With., *Bot. arr. Brit. pl. ed.* 2, 3: 446 (1792)

E: c S: o W: ! NI: ! ROI: ! O: Channel Islands: !

H: On fallen woody debris (sticks, twigs etc.) in woodland. Also on flowerbeds mulched with woodchips or shredded bark.

D: NM3: 194 D+I: Ph: 254-255, B&K2: 380 496, Myc2(3): 110, BritPuffb: 58-59 Figs. 32/33 I: Sow1: 29 (as *Nidularia striata*), Bon: 301, Kriegl2: 159

Common and widespread but less so northwards. English name = 'Fluted Bird's Nest'.

CYLINDROBASIDIUM Jülich, *Persoonia* 8: 72 (1974)

Stereales, Hyphodermataceae

Butlerelfia Weresub & Illman, *Canad. J. Bot.* 58(2): 144 (1980)

Type: *Cylindrobasidium evolvens* (Fr.) Jülich (= *C. laeve*)

laeve (Pers.) Chamuris, *Mycotaxon* 20(2): 587 (1984)
  Corticium laeve Pers., *Neues Mag. Bot.* 1: 110 (1794)
  Butlerelfia eustacei Weresub & Illman, *Canad. J. Bot.* 58(2): 145 (1980)
  Thelephora evolvens Fr., *Observ. mycol.* 1: 154 (1815)
  Corticium evolvens (Fr.) Fr., *Epicr. syst. mycol.*: 558 (1838)
  Cylindrobasidium evolvens (Fr.) Jülich, *Persoonia* 8(1): 72 (1974)
  Corticium laeve f. *cucullata* Bourdot & Galzin, *Hymenomyc. France*: 184 (1928)
  Cladoderris minima Berk. & Broome, *Ann. Mag. Nat. Hist., Ser. 5 [Notices of British Fungi no. 1692]* 1: 24 (1878)
E: c **S:** c **W:** c **NI:** c **ROI:** c **O:** Channel Islands: c Isle of Man: c
H: On dead wood and bark of a wide range of deciduous trees, and especially common on brashings of *Acer pseudoplatanus*. Rarely on conifers such as *Picea* and *Pinus* spp., and on stored apples.
D: NM3: 184 **D+I:** CNE4: 568-571 (as *Cylindrobasidium evolvens*), Ph: 237, B&K2: 110 92 (as *C. evolvens*) **I:** Kriegl1: 191
Very common and widespread. Basidiomes are usually resupinate or effuso-reflexed; pileate forms were previously known as 'forma *cucullata*' or *Cladoderris minima*. Causes 'Fish Eye Disease' in stored apples.

parasiticum D.A. Reid, *Trans. Brit. Mycol. Soc.* 91(1): 168 (1988)
E: !
H: On sclerotia of *Typhula incarnata*.
Known only from the type collection.

CYPHELLA Fr., *Syst. mycol.* 2: 201 (1823)

Stereales, Aleurodiscaceae

Type: *Cyphella digitalis* (Alb. & Schwein.) Fr.

ferruginea H. Crouan & P. Crouan, *Florule Finistère*: 61 (1867)
  Mis.: *Phaeosolenia betulae* sensu auct. Brit.
E: ! **S:** ! **O:** Channel Islands: !
H: On mossy bark of living branches of deciduous trees such as *Corylus*, *Prunus padus*, *Quercus* and *Salix* spp.
Poorly known in Britain. There are collections in herb. K from southern England (South Devon: Slapton, South Hampshire: New Forest and Wiltshire: Clarendon Estate), the Channel Islands (Jersey: St. Catherine's Valley) and Scotland (North Ebudes: Isle of Skye). The species probably belongs in *Merismodes* rather than *Cyphella* sensu stricto.

CYPHELLOSTEREUM D.A. Reid, *Nova Hedwigia Beih.* 18: 336 (1965)

Stereales, Podoscyphaceae

Stereophyllum P. Karst. [*nom. illegit.*, non *Stereophyllum* Mitt. (1859)(*Musci*)], *Hedwigia* 28: 190 (1889)

Type: *Cyphellostereum pusiolum* (Berk. & M.A. Curtis) D.A. Reid

laeve (Fr.) D.A. Reid, *Nova Hedwigia Beih.* 18: 337 (1965)
  Cantharellus laevis Fr., *Syst. mycol.* 1: 324 (1821)
  Cyphella laevis (Fr.) S. Lundell, *Fungi Exsiccati Suecici*: 35 (1953)
  Stereophyllum boreale P. Karst., *Meddeland. Soc. Fauna Fl. Fenn.* 16: 104 (1889)
  Thelephora muscigena Pers., *Syn. meth. fung.*: 572 (1801)
  Cyphella muscigena (Pers.) Fr., *Epicr. syst. mycol.*: 567 (1838)
E: ! **S:** ! **W:** ! **O:** Channel Islands: ! Isle of Man: !
H: On acidic soil, amongst or on living mosses such as *Dicranella heteromalla* and *Polytrichum* spp.
D: BFF6: 36-37, NM2: 118, FAN3: 63 **D+I:** CNE3: 320-321, B&K2: 204 233 **I:** Kriegl1: 192

Rarely recorded or collected but widely scattered throughout the British Isles.

CYSTODERMA Fayod, *Ann. Sci. Nat., Bot.*, sér. 7, 9: 350 (1889)

Agaricales, Agaricaceae

Type: *Cystoderma amianthinum* (Scop.) Fayod

amianthinum (Scop.) Fayod, *Ann. Sci. Nat., Bot.*, sér. 7, 9: 351 (1889)
  Agaricus amianthinus Scop., *Fl. carniol.* 434 (1772)
  Lepiota granulosa var. *amianthina* (Scop.) P. Kumm., *Führ. Pilzk.*: 136 (1871)
  Lepiota amianthina (Scop.) P. Karst., *Bidrag Kännedom Finlands Natur Folk* 32: 15 (1879)
  Lepiota amianthina var. *alba* Maire, Rea, *Brit. basidiomyc.*: 76 (1922)
E: c **S:** c **W:** c **NI:** c **ROI:** c **O:** Channel Islands: ! Isle of Man: !
H: On acidic soil in deciduous or conifer woodland, often amongst mosses such as *Polytrichum* spp. Occasionally with other mosses in old pastures, on heathland and moorland and rarely on lawns.
D: NM2: 119, BFF8: 37-38 **D+I:** Ph: 31, B&K4: 184 202 **I:** Bon: 173, C&D: 239
Common and widespread. Specimens with a markedly rugose pileal surface have been referred to 'form' *rugosoreticulatum* but are within the natural variation of the species. Albino forms also occur.

carcharias (Pers.) Fayod, *Ann. Sci. Nat., Bot.*, sér. 7, 9: 351 (1889)
  Agaricus carcharias Pers., *Tent. disp. meth. fung.*: 19 (1797)
  Agaricus granulosus var. *carcharias* (Pers.) Fr., *Epicr. syst. mycol.*: 18 (1838)
  Lepiota granulosa var. *carcharias* (Pers.) P. Kumm., *Führ. Pilzk.*: 136 (1871)
  Lepiota carcharias (Pers.) P. Karst., *Bidrag Kännedom Finlands Natur Folk* 32: 14 (1879)
E: ! **S:** ! **W:** ! **NI:** ! **ROI:** ! **O:** Isle of Man: !
H: On soil in woodland, usually associated with conifers and rarely with deciduous trees. Occasionally with mosses in grassland in upland areas.
D: NM2: 118, BFF8: 40 **D+I:** B&K4: 182 203 **I:** Bon: 173, C&D: 239, Cooke 37(42) Vol. 1 (as *Agaricus (Lepiota) carcharias*)
Rarely reported but apparently widespread.

cinnabarinum (Alb. & Schwein.) Fayod, *Ann. Sci. Nat., Bot.*, sér. 7, 9: 351 (1889)
  Agaricus granulosus var. *cinnabarinus* Alb. & Schwein., *Consp. fung. lusat.*: 147 (1805)
  Agaricus cinnabarinus (Alb. & Schwein.) Fr., *Index. Alphabeticus*: 12 (1832)
  Lepiota cinnabarina (Alb. & Schwein.) Ricken, *Blätterpilze Deutschl.*: 327 (1914)
  Agaricus terrei Berk. & Broome, *Ann. Mag. Nat. Hist., Ser. 4 [Notices of British Fungi no. 1183]* 6: 462 (1870)
  Lepiota terrei (Berk. & Broome) P. Karst., *Bidrag Kännedom Finlands Natur Folk* 32: 14 (1879)
  Cystoderma terrei (Berk. & Broome) Harmaja, *Karstenia* 18: 30 (1978)
E: ! **S:** ! **ROI:** ?
H: On soil in woodland, usually associated with *Pinus* spp.
D: NM2: 119 (as *Cystoderma terrei*), BFF8: 42-43 **D+I:** B&K4: 188 208 **I:** FlDan1: 15F (as *Lepiota cinnabarina*), Bon: 173, C&D: 239 (as *C. terrei*), Kriegl3: 237 (as *C. terrei*)
Rarely reported. Collections from East Norfolk and Surrey, Easterness and Mid-Perthshire. Reports from elsewhere are unsubstantiated with voucher material.

granulosum (Batsch) Fayod, *Ann. Sci. Nat., Bot.*, sér. 7, 9: 351 (1889)
  Agaricus granulosus Batsch, *Elench. fung.*: 79 & t. 6 f. 24 (1783)
  Lepiota granulosa (Batsch) Quél., *Mém. Soc. Émul. Montbéliard, Sér. 2*, 5: 73 (1872)
E: o **S:** o **W:** ! **NI:** ! **ROI:** !

H: On acidic soil in woodland or on heathland, usually associated with *Pinus sylvestris* and *Calluna*.
D: NM2: 119, BFF8: 41-42 **D+I:** B&K4: 184 204 **I:** C&D: 239
Occasional in England and Scotland. Rarely reported elsewhere but apparently widespread. Albino forms also occur.

jasonis (Cooke & Massee) Harm., *Karstenia* 18(1): 29 (1978)
   Agaricus jasonis Cooke & Massee, *Grevillea* 16: 77 (1887)
   Armillaria jasonis (Cooke & Massee) Sacc., *Syll. fung.* 9: 12 (1891)
   Lepiota amianthina var. *longispora* Kühner, *Bull. Soc. Mycol. France* 51: 204 (1936)
   Cystoderma amianthinum var. *longisporum* (Kühner) A.H. Sm. & Singer, *Pap. Michigan Acad. Sci.* 30: 114 (1945) [1944]
E: ! **S:** ! **W:** ! **NI:** ! **ROI:** !
H: On acidic soil, often with mosses such as *Polytrichum* spp, in woodland. Rarely amongst short turf in old, mossy, unimproved grassland.
D: NM2: 119, BFF8: 38-39, NBA8: 574 **D+I:** B&K4: 186 205 **I:** Kriegl3: 235
Apparently widespread but usually mistaken for *C. amianthinum* with which it was once considered synonymous. Records are mostly unsubstantiated with voucher material.

lilacipes Harmaja, *Karstenia* 18(1): 29 (1978)
S: !
H: On soil amongst grass, on a lawn (British collection).
A single collection from South Aberdeen (Balmoral) in 1999.

simulatum P.D. Orton, *Trans. Brit. Mycol. Soc.* 43(2): 222 (1960)
E: ! **W:** ! **NI:** !
H: On decayed wood of deciduous trees or shrubs and on soil mixed with decayed wood.
D: BFF8: 39-40 **D+I:** B&K4: 186 206
Retained here, but possibly just a lignicolous form of *C. amianthinum*.

superbum Huijsman, *Fungus* 26: 39 (1956)
   Mis.: *Cystoderma haematites* sensu auct. mult.
S: !
H: On soil in conifer woodland.
D: BFF8: 40-41 **D+I:** B&K4: 186 207 **I:** C&D: 239
A single collection from Perthshire (Blair Atholl) in 1997. Records of *C. haematites* may also be this species, but all are unsubstantiated with voucher material.

CYSTOFILOBASIDIUM Oberw. & Bandoni, *Syst. Appl. Microbiol.* 4: 116 (1983)
Tremellales, Filobasidiaceae
Type: *Cystofilobasidium bisporidii* (Fell, Hunter & Tallman) Oberw. & Bandoni

capitatum (Fell, I.L. Hunter & Tallman) Oberw. & Bandoni, *Syst. Appl. Microbiol.* 4(1): 116 (1983)
   Rhodosporidium capitatum Fell, I.L. Hunter & Tallman, *Canad. J. Microbiol.* 19: 650 (1973)
E: !
H: Isolated ex rhubarb.
D+I: YST 329

infirmominiatum (Fell, I.L. Hunter & Tallman) Hamam., Sugiy. & Komag., *J. Gen. Appl. Microbiol.* 34(3): 276 (1988)
   Rhodosporidium infirmominiatum Fell, I.L. Hunter & Tallman, *Canad. J. Microbiol.* 19: 656 (1973)
   Torula infirmominiata Okun. (anam.), *Jap. J. Bot.* 5: 319 (1931)
   Cryptococcus infirmominiatus (Okun.) Phaff & Fell (anam.), *Yeasts, a taxonomic study*: 1113 (1970)
E: !
H: Isolated ex soil (from sheep pasture) and ex meat.
D+I: YST 330
Only known in its anamorphic yeast state (*Cryptococcus infirmominiatus*) in the British Isles.

CYSTOLEPIOTA Singer, *Lilloa* 25: 281 (1952) [1951]
Agaricales, Agaricaceae
   Pulverolepiota Bon, *Doc. Mycol.* 22(88): 30 (1993)
Type: *Cystolepiota constricta* Singer

adulterina (F.H. Møller) Bon, *Doc. Mycol.* 7(27-28): 54 (1977)
   Lepiota adulterina F.H. Møller, *Friesia* 6(1-2): 23 (1959) [1957-1958]
   Cystolepiota adulterina f. *reidii* Bon, *Doc. Mycol.* 11(43): 25 (1981)
   Mis.: *Cystolepiota hetieri* sensu Breitenbach & Kränzlin [B&K4:: 188 & pl. 210 (1995)]
   Mis.: *Lepiota hetieri* sensu Lange (FlDan1: 35 & pl. 14J)
E: !
H: On calcareous soil in woodland, usually with *Fagus* in beechwoods, often under dense cover of *Mercurialis perennis*.
D: NM2: 215, FAN5: 157 **D+I:** FRIC6: 10-11 43a (as *Lepiota adulterina*), Ph: 30 (as *L. adulterina*), FungEur4: 85-88 2c, FungEur4: 88-90 2b (as *Cystolepiota adulterina* f. *reidii*), B&K4: 188 210 (as *C. hetieri*), B&K4: 190 212 (as *Cystolepiota* sp.) **I:** SV26: 34, C&D: 241
Rarely reported. Known from East & West Gloucestershire, South Devon, Staffordshire and West Sussex. Reported from Herefordshire, North Somerset and Worcestershire but unsubstantiated with voucher material.

bucknallii (Berk. & Broome) Singer & Clémençon, *Nova Hedwigia* 23: 238 (1972)
   Agaricus bucknallii Berk. & Broome, *Ann. Mag. Nat. Hist.,* Ser. 5 [Notices of British Fungi no. 1836] 7: 124 (1881)
   Lepiota bucknallii (Berk. & Broome) Sacc., *Syll. fung.* 5: 50 (1887)
   Cystoderma bucknallii (Berk. & Broome) Singer, *Schweiz. Z. Pilzk.* 17: 53 (1939)
   Lepiota seminuda var. *lilacina* Quél., *Bull. Soc. Bot. France* 23: 325 (1877)
   Lepiota lilacina (Quél.) Boud., *Bull. Soc. Mycol. France* 9: 6 (1893)
E: o **W:** ! **NI:** ! **ROI:** !
H: On soil, usually calcareous loam, in mixed deciduous woodland or scrub often with *Fagus*, *Corylus* and *Fraxinus*. Rarely with conifers in plantations, or with *Urtica dioica* in nettlebeds.
D: NM2: 216 **D+I:** Ph: 30 (as *Lepiota bucknallii*), FungEur4: 108-110 6b, B&K4: 188 209, FAN5: 158-159 **I:** FlDan1: 13E (as *L. bucknallii*), Bon: 283, SV26: 34, C&D: 241 (unusually pallid)
Most often collected in southern and south-eastern England, but known from Northumberland southwards. Less often reported elsewhere but apparently widespread.

hetieri (Boud.) Singer, *Sydowia, Beih.* 7: 67 (1973)
   Lepiota hetieri Boud., *Bull. Soc. Mycol. France* 18: 137 (1902)
   Agaricus granulosus var. *rufescens* Berk. & Broome, *Ann. Mag. Nat. Hist., Ser. 5 [Notices of British Fungi no. 1834]* 7: 124 (1881)
   Lepiota granulosa var. *rufescens* (Berk. & Broome) Sacc., *Syll. fung.* 5: 47 (1887)
   Lepiota rufescens (Berk. & Broome) J.E. Lange [*nom. illegit.*], *Dansk. Bot. Ark.* 9(6): 65 (1938)
   Lepiota langei Locq. [non sensu Knudsen (1980)], *Bull. Mens. Soc. Linn. Lyon* 14: 87 (1945)
E: o **S:** ! **W:** ! **ROI:** !
H: On soil, often calcareous loam, in deciduous woodland with *Corylus*, *Fagus* and *Fraxinus*, frequently under dense cover of *Mercurialis perennis*.
D: NM2: 215 **D+I:** FungEur4: 94-97 4, FAN5: 156 **I:** FlDan1: 14I, Bon: 283, SV26: 34, C&D: 241, Cooke 40 (213) Vol. 1 (as *Agaricus (Lepiota) granulosus* var. *rufescens*)
Mostly reported from southern counties. Much less often recorded elsewhere but apparently widespread.

moelleri Knudsen, *Bot. Tidsskr.* 73(2): 134 (1978)
   Lepiota rosea Rea, *Trans. Brit. Mycol. Soc.* 6(1): 61 (1918)

Cystolepiota rosea (Rea) Bon [*nom. illegit.*]*, Doc. Mycol.*
6(24): 43 (1976)
E: o W: !
H: On soil, usually calcareous loam, in mixed deciduous
woodland. Usually associated with *Corylus, Fagus* or *Fraxinus*
and often under dense cover of *Mercurialis perennis.*
D: NCL3: 286 (as *Lepiota rosea*), NM2: 216 **D+I:** Ph: 30-31 (as
L. rosea), FungEur4: 98-100 5a, FAN5: 157 **I:** C&D: 241 (as
Cystolepiota rosea)
Local but widespread in England and also known from Wales
(Breconshire and Denbighshire).

pulverulenta (Huijsman) Vellinga, *Persoonia* 14(4): 407 (1992)
Lepiota pulverulenta Huijsman, *Persoonia* 1(3): 328 (1960)
Leucoagaricus pulverulentus (Huijsman) Bon, *Doc. Mycol.*
8(30-31): 70 (1978)
Pulverolepiota pulverulenta (Huijsman) Bon, *Doc. Mycol.*
22(88): 30 (1993)
Mis.: *Lepiota pseudogranulosa* sensu Reid [TBMS 38: 389
(1955)]
E: !
H: On soil, usually calcareous loam, in mixed deciduous
woodland with *Corylus* and *Ulmus* spp. Also collected on
compost in cool greenhouses and in a garden under *Forsythia*
sp.
D+I: FRIC2: 6-7 9 (e-f), FungEur4: 356-260 41 (as
Leucoagaricus pulverulentus), FAN5: 159-160
Rarely reported. Known from East Norfolk, Oxfordshire, South
Devon, South Somerset, South Wiltshire, Surrey, Warwickshire
and West Kent.

seminuda (Lasch) Bon, *Doc. Mycol.* 6(24): 43 (1976)
Agaricus seminudus Lasch, *Linnaea* 3: 157 (1828)
Lepiota seminuda (Lasch) P. Kumm., *Führ. Pilzk.*: 136 (1871)
Cystoderma seminudum (Lasch) Fayod, *Ann. Sci. Nat., Bot.,*
sér. 7, 9: 351 (1889)
Lepiota sororia Huijsman, *Persoonia* 1: 326 (1960)
Cystolepiota sororia (Huijsman) Singer, *Sydowia, Beih.* 7: 67
(1973)
Lepiota seminuda f. *minima* J.E. Lange [*nom. inval.*], *Fl.
agaric. danic.* 1: 36 (1935)
Lepiota sistrata f. *minima* (J.E. Lange) Babos [*nom. inval.*]*,
Ann. Hist.-Nat. Mus. Natl. Hung.* 50: 91 (1958)
Mis.: *Lepiota sistrata* sensu auct. mult.
E: c S: o W: o NI: ! ROI: !
H: On soil, usually calcareous loam, in mixed deciduous
woodland, usually with *Fagus* or *Corylus* and often under
dense cover of *Mercurialis perennis.* Also known with conifers
(especially in old plantations of *Picea* spp) and with *Urtica
dioica* in nettlebeds.
D: NM2: 216 **D+I:** Ph: 30-31 (as *Lepiota sistrata*), FungEur4:
79-84 2a (as *Cystolepiota sistrata*), Myc8(3): 108, B&K4: 190
211, FAN5: 158 **I:** Bon: 283, C&D: 241
Common in England, especially from the midlands southwards.
Occasional elsewhere but widespread.

CYTIDIA Quél., *Fl. mycol. France*: 25 (1888)
Stereales, Corticiaceae
Type: *Cytidia salicina* (Fr.) Burt

salicina (Fr.) Burt, *Rep. (Annual) Missouri Bot. Gard.* 11: 10
(1924)
Thelephora salicina Fr., *Syst. mycol.* 1: 442 (1821)
Corticium salicinum (Fr.) Fr., *Epicr. syst. mycol.*: 558 (1838)
Cytidia rutilans Quél., *Fl. mycol. France*: 25 (1888)
E: ! S: !
H: On living branches of *Salix* spp in woodland. British material
on *S. aurita* and *S. cinerea.*
D: NM3: 179 **D+I:** CNE3: 332-335, B&K2: 114 97, FM1(2): 45-
46 **I:** C&D: 139
Not infrequent in the 19th century, but unrecorded for most of
the 20th, and considered extinct until refound in Cheviotshire
(Kielder Forest) in 1999. More recent collections from Scotland
(Easterness: Aviemore area, Mid-Perthshire: Rannoch, and
Selkirk).

DACRYMYCES Nees, *Syst. Pilze*: 89 (1817)
Dacrymycetales, Dacrymycetaceae
Arrhytidia Berk. & M.A. Curtis, *Hooker's J. Bot. Kew Gard.
Misc.* 1: 235 (1849)
Dacryopsis Massee, *J. Mycol.* 6: 180 (1891)
Type: *Dacrymyces stillatus* Nees

capitatus Schwein., *Trans. Am. phil. Soc.* 4: 186 (1832)
Dacrymyces deliquescens f. *stipitata* Bourdot & Galzin, *Bull.
Soc. Mycol. France* 25: 33 (1909)
Dacrymyces ellisii Coker, *J. Elisha Mitchell Sci. Soc.* 35: 175
(1920)
Dacrymyces deliquescens var. *ellisii* (Coker) L.L. Kenn.,
Mycologia 50(6): 911 (1959)
Ditiola nuda Berk. & Broome, *Ann. Mag. Nat. Hist., Ser. 2
[Notices of British Fungi no. 375]* 2: 267 (1848)
Dacryopsis nuda (Berk. & Broome) Massee, *J. Mycol.* 6: 182
(1891)
Dacryomitra nuda (Berk. & Broome) Pat., *Essai taxon.*: 31
(1900)
Ditiola ulicis Plowr., *Trans. Brit. Mycol. Soc.* 1(2): 55 (1898)
Dacryopsis ulicis (Plowr.) Sacc. & Syd., *Syll. fung.* 16: 223
(1902)
Mis.: *Dacrymyces lutescens* sensu auct.
E: ! S: ! W: ? NI: ? ROI: ?
H: On decayed wood of deciduous trees and conifers.
D+I: B&K2: 50 3 **I:** SV23: 49, Kriegl1: 78
Rarely reported but apparently widespread.

chrysocomus (Bull.) Tul., *Ann. Sci. Nat., Bot.,* sér. 3, 19: 211
(1853)
Peziza chrysocoma Bull., *Herb. France*: pl. 376 f. 2 (1788)
Hymenoscyphus chrysocomus (Bull.) Gray, *Nat. arr. Brit. pl.*
1: 674 (1821)
Calloria chrysocoma (Bull.) Fr., *Summa veg. Scand.* 2: 359
(1849)
Orbilia chrysocoma (Bull.) Sacc., *Syll. fung.* 8: 624 (1889)
Guepiniopsis chrysocoma (Bull.) Brasf., *Amer. Midl. Naturalist*
20: 226 (1938)
E: ? S: !
H: On decayed conifer wood (British material on *Pinus
sylvestris*).
D: NM3: 94
Rarely reported. A few confirmed collections from Scotland and
several old records from England unsubstantiated with voucher
material.

chrysospermus Berk. & M.A. Curtis, *Grevillea* 2: 20 (1873)
Tremella palmata Schwein. [non *T. palmata* Schumach.],
Trans. Am. phil. Soc. 4: 186 (1832)
Dacrymyces palmatus (Schwein.) Bres., *Oesterr. Bot. Z.* 54:
425 (1904)
E: ! S: ?
H: British material on decayed wood of *Pinus sylvestris.*
D: NM3: 94 **I:** Kriegl1: 76
Known from a single collection from Bedfordshire (Woburn).
Reported from East Sutherland (Migdale) but no material has
been traced.

enatus (Berk. & M.A. Curtis) Massee, *J. Mycol.* 6: 182 (1891)
Tremella enata Berk. & M.A. Curtis, *Grevillea* 2: 20 (1873)
E: !
H: On decayed wood of deciduous trees, such as *Fraxinus,
Quercus,* and *Salix* spp, *Sambucus nigra* and *Sorbus aria.*
D: NM3: 93 (as *Dacryomyces enatus*) **I:** SV23: 49
Rarely reported. Known from Hertfordshire, South Devon, Surrey
and West Sussex.

estonicus Raitv., *Eesti N.S.V. Tead. Akad. Toimet., Biol.* 11:
238 (1962)
E: ! S: !
H: On decayed wood of conifers, usually *Pinus sylvestris* but
also known on *Picea* spp and *Taxus.*
D: NM3: 94
Rarely reported. Known from England (Cumberland, East Kent,
Mid-Lancashire, South Devon and Surrey) and Scotland
(Banffshire, Easterness, Mid-Perthshire and Morayshire).

macnabbii D.A. Reid, *Trans. Brit. Mycol. Soc.* 62: 456 (1974)
E: ? **S:** !
H: Typically on decayed wood of *Pinus sylvestris*.
I: SV23: 52
Described from Wester Ross (Kinlochewe). Reported from East Norfolk and Yorkshire but unsubstantiated with voucher material.

minor Peck, *Rep. (Annual) New York State Mus. Nat. Hist.* 30: 49 (1878)
    Dacrymyces deliquescens var. *fagicola* Bourdot & Galzin, *Hymenomyc. France*: 68 (1928)
E: ! **S:** !
H: On decayed wood of *Carpinus, Laburnum anagyroides, Quercus* spp, *Cytisus scoparius* and rarely *Pinus* spp.
D: NM3: 93 **I:** SV23: 52
Known from England (Cheshire, Hertfordshire, Leicestershire, London, Oxfordshire, South Hampshire, Surrey and Yorkshire) and Scotland (East Perthshire, Morayshire and Shetland).

ovisporus Bref., *Unters. Gesammtgeb. Mykol.* 7: 158 (1888)
S: !
H: British material on fallen, decayed wood of *Pinus sylvestris*.
D: NM3: 92
Known from Easterness and South Aberdeen.

punctiformis Neuhoff, *Schweiz. Z. Pilzk.* 12: 81 (1934)
    Dacrymyces romellii Neuhoff, *Schweiz. Z. Pilzk.* 12: 82 (1934)
    Mis.: *Tremella torta* sensu auct.
    Mis.: *Dacrymyces tortus* sensu auct.
E: ! **S:** ! **W:** ?
H: On decayed wood of conifers, usually *Picea* and *Pinus* spp, but also known on *Taxus*.
D: NM3: 93 (as *Dacryomyces tortus*) **I:** SV23: 53 (as *D. tortus*)
Rarely reported but apparently widespread.

stillatus Nees, *Syst. Pilze*: 89 (1817)
    Calloria stillata (Nees) Fr., *Summa veg. Scand.* 2: 359 (1849)
    Tremella abietina Pers., *Observ. mycol.* 1: 78 (1796)
    Dacrymyces abietinus (Pers.) J. Schröt., in Cohn, *Krypt.-Fl. Schlesien* 3: 400 (1888)
    Mis.: *Dacrymyces deliquescens* sensu auct.
    Mis.: *Tremella lacrymalis* sensu auct.
    Mis.: *Dacrymyces lacrymalis* sensu auct.
E: c **S:** c **W:** c **NI:** c **ROI:** c **O:** Channel Islands: c Isle of Man: c
H: On decayed wood of deciduous trees and conifers, and often on worked wood such as window frames, greenhouse slatting, fence posts etc.
D: NM3: 93 **D+I:** Ph: 263, B&K2: 52 4 **I:** SV23: 52 (as *D. lacrymalis*), Kriegl1: 75
Very common and widespread.

variisporus McNabb, *New Zealand J. Bot.* 11: 504 (1973)
E: ! **S:** ! **W:** ?
H: On decayed wood of conifers such as *Pinus nigra, Pinus sylvestris* and *Picea* spp.
D+I: B&K2: 52 5
Rarely reported.

DACRYOBOLUS Fr., *Summa veg. Scand.*: 404 (1849)
Stereales, Meruliaceae
    Gloeocystidium P. Karst., *Bidrag Kännedom Finlands Natur Folk* 48: 429 (1889)
Type: *Dacryobolus sudans* (Alb. & Schwein.) Fr.

karstenii (Bres.) Parmasto, *Conspectus Systematis Corticiacearum*: 98 (1968)
    Stereum karstenii Bres., *Atti Imp. Regia Accad. Roveretana, ser. 3,* 3: 109 (1897)
    Phanerochaete karstenii (Bres.) P. Karst., *Meddeland. Soc. Fauna Fl. Fenn.* 1: 162 (1889)
    Tubulicrinis karstenii (Bres.) Donk, *Fungus* 26: 14 (1956)
    Peniophora crassa Burt, *Rep. (Annual) New York State Mus. Nat. Hist.* 54: 155 (1901)
E: o **S:** !

H: On decayed wood of conifers, usually large, fallen and decorticated trunks. Most often on *Pinus* spp, but also known on *Larix, Picea* spp and *Taxus*.
D: NM3: 154 **D+I:** CNE3: 341-344, Myc10(2): 85
Occasional in England. Rarely reported elsewhere.

sudans (Alb. & Schwein.) Fr., *Summa veg. Scand.* 2: 404 (1849)
    Hydnum sudans Alb. & Schwein., *Consp. fung. lusat.*: 272 (1805)
    Odontia sudans (Alb. & Schwein.) Bres., *Atti Imp. Regia Accad. Roveretana, ser. 3,* 3: 100 (1897)
    Porotheleum confusum Berk. & Broome, *Ann. Mag. Nat. Hist., Ser. 5 [Notices of British Fungi no. 1685]* 1: 24 (1878)
    Porotheleum stevensonii Berk. & Broome, *Ann. Mag. Nat. Hist., Ser. 5 [Notices of British Fungi no. 1683]* 1: 24 (1878)
E: ! **S:** !
H: On decayed conifer wood. Reported on *Pinus contorta* and *P. sylvestris*.
D: NM3: 154 **D+I:** CNE3: 345-349, B&K2: 114 98
Known only from Scotland in recent times, and there rarely reported.

DAEDALEA Pers., *Syn. meth. fung.*: 499 (1801)
Poriales, Coriolaceae
Type: *Daedalea quercina* (L.) Pers.

quercina (L.) Pers., *Syn. meth. fung.*: 500 (1801)
    Agaricus quercinus L., *Sp. pl.*: 1176 (1753)
E: o **S:** o **W:** o **NI:** ! **ROI:** ! **O:** Channel Islands: o
H: Usually on *Quercus* spp and rarely *Castanea*. Reported on *Acer campestre, Aesculus, Betula, Fagus, Fraxinus, Populus* and *Ulmus* spp, but these hosts are unverified. Usually on hard, barely decayed wood.
D: EurPoly1: 222, NM3: 241 **D+I:** Ph: 233, B&K2: 304 383 **I:** Sow2: 181 (as *Agaricus quercinus*), C&D: 143, Kriegl1: 507
Occasional but widespread, and may be locally common in old woodland. Causes a progressive brown rot in heartwood, fruiting after the host dies.

DAEDALEOPSIS J. Schröt., in Cohn, *Krypt.-Fl. Schlesien* 3: 492 (1888)
Poriales, Coriolaceae
Type: *Daedaleopsis confragosa* (Bolton) J. Schröt.

confragosa (Bolton) J. Schröt., in Cohn, *Krypt.-Fl. Schlesien* 3: 492 (1888)
    Boletus confragosus Bolton, *Hist. fung. Halifax* Suppl.: 160 (1791)
    Daedalea confragosa (Bolton) Pers., *Syn. meth. fung.*: 501 (1801)
    Polyporus confragosus (Bolton) P. Kumm., *Führ. Pilzk.*: 59 (1871)
    Trametes confragosa (Bolton) Jörst., *Nytt Mag. Bot.* 15(3): 265 (1968)
    Daedalea rubescens Alb. & Schwein., *Consp. fung. lusat.*: 238 (1805)
    Trametes rubescens (Alb. & Schwein.) Fr., *Epicr. syst. mycol.*: 492 (1838)
E: c **S:** o **W:** c **NI:** c **ROI:** ! **O:** Channel Islands: o Isle of Man: o
H: On dead wood of deciduous trees in parkland and woodland. Especially common on *Betula, Corylus, Fagus, Fraxinus* and *Salix* spp. Also known on *Acer campestre, A. pseudoplatanus, Aesculus, Alnus, Castanea, Ilex, Juglans regia, Platanus, Prunus, Quercus, Sambucus nigra, Sorbus aucuparia, Tilia* and *Ulmus* spp. Reports on conifers such as *Picea* spp and *Taxus* are unverified.
D: EurPoly1: 225, NM3: 228 **D+I:** Ph: 232, B&K2: 304 384
Very common and widespread, but less so northwards and in Scotland. English name = 'Blushing Bracket'.

DATRONIA Donk, *Persoonia* 4: 337 (1967)
Poriales, Coriolaceae
Type: *Datronia mollis* (Sommerf.) Donk

87

mollis (Sommerf.) Donk, *Persoonia* 4: 338 (1966)
  Daedalea mollis Sommerf., *Suppl. Fl. lapp.*: 271 (1826)
  Trametes mollis (Sommerf.) Fr., *Hymenomyc. eur.*: 585 (1874)
  Antrodia mollis (Sommerf.) P. Karst., *Meddeland. Soc. Fauna Fl. Fenn.* 5: 40 (1879)
  Polyporus sommerfeldtii P. Karst., *Meddeland. Soc. Fauna Fl. Fenn.* 5: 53 (1878)
E: c **S:** c **W:** c **NI:** c **ROI:** ! **O:** Channel Islands: ! Isle of Man: !
H: On dead or dying branches of deciduous trees, most often *Acer pseudoplatanus*, *Corylus*, *Fagus* and *Salix* spp but known on many other hosts. Rarely (and dubiously) reported on conifer wood.
D: EurPoly1: 230, NM3: 228 **D+I:** Ph: 234, B&K2: 280 348 **I:** Kriegl1: 513
Common and widespread.

DELICATULA Fayod, *Ann. Sci. Nat., Bot.*, sér. 7, 9: 313 (1889)
Agaricales, Tricholomataceae
Type: *Delicatula integrella* (Pers.) Fayod

integrella (Pers.) Fayod, *Ann. Sci. Nat., Bot.*, sér. 7, 9: 313 (1889)
  Agaricus integrellus Pers., *Icon. descr. fung.*: pl. 54 (1800)
  Mycena integrella (Pers.) Gray, *Nat. arr. Brit. pl.* 1: 621 (1821)
  Omphalia integrella (Pers.) P. Kumm., *Führ. Pilzk.*: 106 (1871)
  Delicatula bagnolensis E.-J. Gilbert, *Bull. Soc. Mycol. France* 42: 62 (1926)
E: o **S:** ! **W:** ! **NI:** ! **ROI:** !
H: On soil, or rarely on decayed wood, in damp places such as stream sides, fen carr, swamps and wet woodland, often amongst liverworts (*Hepaticae*) growing under *Alnus* and *Salix* spp. Sometimes found in vast swarms.
D: NM2: 120, BFF8: 110-111 **D+I:** A&N3 124-130 pl.37&38, B&K3: 186 205 **I:** FlDan2: 62C (as *Omphalia integrella*), Bon: 187, C&D: 221, Kriegl3: 239
Occasional in England, rarely reported elsewhere. Possibly overlooked, since basidiomes are small and often appear during summer in wet areas unfrequented by agaricologists.

DENDROTHELE Höhn. & Litsch., *Sitzungsber. Kaiserl. Akad. Wiss., Wien, Math.-Naturwiss. Cl., Abt. 1* 116: 819 (1907)
Stereales, Corticiaceae
Type: *Dendrothele papillosa* Höhn. & Litsch.

acerina (Pers.) P.A. Lemke, *Persoonia* 3(3): 366 (1965)
  Corticium acerinum Pers. [non *C. acerinum* Velen. (1922)], *Observ. mycol.* 1: 37 (1796)
  Thelephora acerina (Pers.) Pers., *Syn. meth. fung.*: 581 (1801)
  Stereum acerinum (Pers.) Fr., *Epicr. syst. mycol.*: 554 (1838)
  Aleurodiscus acerinus (Pers.) Höhn. & Litsch., *Sitzungsber. Kaiserl. Akad. Wiss., Wien, Math.-Naturwiss. Cl., Abt. 1* 116: 804 (1907)
  Acanthophysium acerinum (Pers.) G. Cunn., *Bull. New Zealand Dept. Sci. Industr. Res.* 145: 166 (1963)
E: c **W:** c
H: Usually on bark of living trunks of *Acer campestre*. Rarely reported on *Acer pseudoplatanus*, *Buxus* and *Ulmus* spp.
D: NM3: 182 **D+I:** CNE3: 352-353 **I:** Kriegl1: 197
Common and widespread in England and Wales, becoming much less frequent northwards (as does the preferred host tree).

amygdalispora Hjortstam, *Windahlia* 17: 56 (1987)
E: !
H: On bark of living trees and shrubs. British material on *Acer campestre*, *Cornus sanguinea*, *Taxus* and *Ulmus* spp.
D: NM3: 182
Known from East Cornwall, South Devon and Surrey.

commixta (Höhn. & Litsch.) J. Eriss. & Ryvarden, *Corticiaceae of North Europe* 3: 355 (1975)
  Corticium commixtum Höhn. & Litsch., *Sitzungsber. Kaiserl. Akad. Wiss., Wien, Math.-Naturwiss. Cl., Abt. 1* 116: 821 (1907)
E: ! **W:** !
H: On bark and dead attached twigs of deciduous trees and shrubs. Known on *Buddleja davidii*, *Buxus*, *Quercus robur*, *Rubus fruticosus*, *Ulex europaeus* and *Ulmus* spp.
D: NM3: 182 **D+I:** CNE3: 355-357
Rarely reported. Known from England (South Devon, Surrey and West Sussex) and Wales (Carmarthenshire). Reported from Scotland but unsubstantiated with voucher material.

griseocana (Bres.) Bourdot & Galzin, *Bull. Soc. Mycol. France* 28: 354 (1913)
  Corticium griseocanum Bres., *Fungi trident.* 2(11-13): 58 (1898)
ROI: ! **O:** Isle of Man: !
H: On stems of living or dead woody shrubs. British material on *Hydrangea* sp. and *Rosa pimpinellifolia*.
D: NM3: 182 **D+I:** CNE3: 358-359
Known from Clare (Paulsallagh) in 1993 and the Isle of Man (Silverdale Glen) in 1994.

sasae Boidin, Cand. & Gilles, *Trans. Mycol. Soc. Japan* 27(4): 466 (1987) [1986]
E: !
H: On dead standing stems of *Phragmites australis* (near water level) in a reedswamp, and dead stems of cultivated bamboos in gardens.
D: Myc16(3): 114
Known from South Devon and West Cornwall.

DERMOLOMA (J.E. Lange) Singer, *Mycologia* 48: 724 (1956)
Agaricales, Tricholomataceae
Type: *Dermoloma pseudocuneifolium* Bon

cuneifolium (Fr.) Bon, *Doc. Mycol.* 17(65): 51 (1986)
  Agaricus cuneifolius Fr., *Observ. mycol.* 2: 99 (1818)
  Tricholoma cuneifolium (Fr.) P. Kumm., *Führ. Pilzk.*: 132 (1871)
  Agaricus atrocinereus Pers., *Syn. meth. fung.*: 348 (1801)
  Tricholoma atrocinereum (Pers.) Quél., *Mém. Soc. Émul. Montbéliard, Sér. 2, 2*: 80 (1872)
  Dermoloma atrocinereum (Pers.) P.D. Orton, *Trans. Brit. Mycol. Soc.* 43(2): 175 (1960)
  Agaricus cinereorimosus Batsch, *Elench. fung. (Continuatio Secunda)*: 63 t. 37 f. 206 (1789)
  Tricholoma cuneifolium var. *cinereorimosum* (Batsch) Cooke, *Ill. Brit. fung.* 92 (261) Vol. 1 (1882)
  Dermoloma fuscobrunneum P.D. Orton, *Notes Roy. Bot. Gard. Edinburgh* 38(2): 326 (1980)
E: o **S:** o **W:** o **NI:** o **ROI:** o **O:** Channel Islands: ! Isle of Man: !
H: On soil, amongst short turf in unimproved grassland.
D: NM2: 121, NM2: 121 (as *Dermoloma atrocinereum*), FAN3: 31 (as *D. cuneifolium* var. *cuneifolium*), BFF8: 89-90, NBA7: 325-326 **D+I:** B&K3: 186 206 **I:** Bon: 171 (as *D. atrocinereum*), C&D: 237, C&D: 237 (as *D. atrocinereum*), Kriegl3: 240
Occasional but widespread and may be locally common. Here treated as a single, variable species with inamyloid spores following Arnolds in FAN3. This excludes *D. cuneifolium* sensu NCL (= *Dermoloma pseudocuneifolium*).

josserandii Dennis & P.D. Orton, *Trans. Brit. Mycol. Soc.* 43(2): 226 (1960)
  Tricholoma hygrophorus Joss. [*nom. nud.*], *Bull. Soc. Mycol. France* 74: 482 (1958)
  Dermoloma pragense Kubička, *Česká Mykol.* 29: 31 (1975)
E: ! **ROI:** ?
H: On soil, amongst short turf in unimproved grassland and in grassy places under trees.
D: BFF8: 90-91

Known from South Hampshire, South Somerset, Staffordshire, Surrey and Yorkshire. Reported from the Republic of Ireland but unsubstantiated with voucher material.

josserandii *var.* phaeopodium (P.D. Orton) Arnolds, *Persoonia* 15(2): 195 (1993)
   Dermoloma phaeopodium P.D. Orton, *Notes Roy. Bot. Gard. Edinburgh* 28: 327 (1980)
E: ! **S:** ?
H: On soil, amongst short turf in unimproved grassland or in grassy places under trees.
D: NM2: 121 (as *Dermoloma phaeopodium*), NM2: 121, BFF8: 91-92 **I:** C&D: 237 (as *D. phaeopodium*)
Rarely reported. Known from London, North Somerset, South Devon, South Hampshire, Surrey, Warwickshire and West Kent. Reported from Scotland but unsubstantiated with voucher material.

magicum Arnolds, *Persoonia* 17(4): 665 (2002)
E: ! **S:** !
H: On soil amongst grass, in old or weakly fertilised grassland.
I: FM4(1): 6
Paratype from Dumfrieshire (Moffat) in 1996. Also recently collected in Surrey (Virginia Water).

pseudocuneifolium Bon, *Doc. Mycol.* 17(65): 52 (1986)
   Mis.: *Dermoloma cuneifolium* sensu NCL
E: ! **S:** ! **W:** !
H: On soil amongst short turf in unimproved grassland.
D: NM2: 120, BFF8: 92, NBA7: 328 **I:** Bon: 171
Known from Nottinghamshire, Oxfordshire, South Devon, South Hampshire and Surrey. Reported from Scotland (East Perthshire) and Wales (Breconshire) but unsubstantiated with voucher material.

DICHOMITUS D.A. Reid, *Revta Biol., Lisboa* 5: 149 (1965)
Poriales, Polyporaceae
Type: *Dichomitus squalens* (P. Karst.) D.A. Reid

campestris (Quél.) Domański & Orlicz, *Acta Soc. Bot. Poloniae.* 35: 627 (1966)
   Trametes campestris Quél., *Mém. Soc. Émul. Montbéliard, Sér. 2*, 5: 286 (1872)
E: ! **S:** ! **NI:** !
D: EurPoly1: 238, NM3: 226 **D+I:** Ph: 234, B&K2: 282 350 **I:** Kriegl1: 609 (as *Polyporus campestris*)
Rarely reported but widespread. The pulvinate basidiomes are characteristic, though microscopically the species is similar to *Polyporus squamosus.*

DICHOSTEREUM Pilát, *Ann. Mycol.* 24: 223 (1926)
Lachnocladiales, Dichostereaceae
Type: *Dichostereum durum* (Bourdot & Galzin) Pilát

durum (Bourdot & Galzin) Pilát, *Ann. Mycol.* 24: 223 (1926)
   Asterostromella dura Bourdot & Galzin, *Bull. Soc. Mycol. France* 36: 74 (1920)
   Vararia dura (Bourdot & Galzin) Boidin, *Bull. Soc. Naturalistes Oyonnax* 5: 78 (1951)
E: !
H: On decayed wood of decdiuous trees.
A single collection from West Sussex (Ebernoe) in 1972.

DICTYONEMA C. Agardh, in Kunth, *Synops. plant. aequinoct. orb. novi* 1: 1 (1822)
Stereales, Meruliaceae
   Rhizonema Thwaites, in Smith & Sowerby, *Engl. Bot., suppl.* 4: 2954 (1849)
   Type: *Dictyonema excentricum* C. Agardh

interruptum (Hook.) Parmasto, *Lichenologist* 11(1): 103 (1979)
   Calothris interruptum Hook., Berkeley, *Engl. fl.* 5: 368 (1833)

   Rhizonema interruptum (Hook.) Thwaites, in Smith & Sowerby, *Engl. Bot., suppl.* 4: 2954 (1849)
   Stigonema interruptum (Hook.) Hassall, *Hist. Brit. freshwater alg.* 2: 229 (1852)
   Scytonema interruptum (Hook.) Cooke, *Brit. fresh-water alg.* 1: 266 (1882)
E: ? **S:** ! **W:** ! **ROI:** !
H: A basidiolichen, forming associations with mosses such as *Thuidium* and *Diplophyllum* spp.
D+ I: Coppins & James (Lichenologist 11: 103-105, 1979) **I:** Cooke (Brit. fresh-water alg. 2: tab. 106, 1884)
Known from Wales (Breconshire, Denbighshire and Montgomeryshire), Scotland (Mid-Ebudes) and the Republic of Ireland. Reported from England (Dorset) but lacking voucher matarial.

DIGITATISPORA Doguet, *Compt. Rend. Hebd. Séances Acad. Sci.* 254(25): 4338 (1962)
Stereales, Amylocorticaceae
Type: *Digitatispora marina* Doguet

marina Doguet, *Compt. Rend. Hebd. Séances Acad. Sci.* 254: 4338 (1962)
O: Isle of Man: !
H: Growing on wood immersed in the sea.
D: NM3: 184
A single record from blocks of beech wood experimentally immersed off the coast of the Isle of Man (Port Erin).

DIPLOMITOPORUS Domański, *Acta Soc. Bot. Poloniae.* 39: 191 (1970)
Poriales, Coriolaceae
Type: *Diplomitoporus flavescens* (Bres.) Domański

flavescens (Bres.) Domański, *Acta Soc. Bot. Poloniae* 39: 191 (1970)
   Trametes flavescens Bres., *Ann. Mycol.* 1(1): 81 (1903)
E: !
H: On fallen and decayed trunks, branches or logs of conifers, and also on treated wood (fence posts). Usually on *Pinus sylvestris* but also known on *Pinus radiata.*
D: EurPoly1: 245, NM3: 209
Rarely reported. Known from Berkshire, South Hampshire, South Devon and Surrey.

lindbladii (Berk.) Gilb. & Ryvarden, *Mycotaxon* 22(2): 364 (1985)
   Polyporus lindbladii Berk., *Grevillea* 1: 54 (1872)
   Poria lindbladii (Berk.) Cooke, *Grevillea* 14: 111 (1886)
   Cinereomyces lindbladii (Berk.) Jülich, *Biblioth. Mycol.* 85: 400 (1982) [1981]
   Antrodia lindbladii (Berk.) Ryvarden, *Mycotaxon* 22: 364 (1985)
   Polyporus cinerascens Bres., *Verh. zool.-bot. Ges. Wien* 1: 54 (1872)
   Poria cinerascens (Bres.) Sacc. & Syd., *Syll. fung.* 16: 161 (1902)
   Tyromyces cinerascens (Bres.) Bondartsev & Singer, *Ann. Mycol.* 39: 52 (1941)
E: o **S:** o **W:** ! **O:** Isle of Man: ?
H: On decayed wood of conifers, often large logs or fallen trunks of *Pinus sylvestris* lying in dense cover of *Calluna* on heathland. Also known on *Larix* and *Picea sitchensis.* Rarely reported on *Betula* and *Quercus* spp but these hosts are unverified.
D: EurPoly1: 245, NM3: 209 **D+I:** B&K2: 280 346 (as *Cinereomyces lindbladii*)
Occasional but apparently widespread. Not infrequent in southern England.

DITIOLA Fr., *Syst. mycol.* 2: 169 (1823)
Dacrymycetales, Dacrymycetaceae
   Femsjonia Fr., *Summa veg. Scand.* 2: 341 (1849)
   Type: *Ditiola radicata* (Alb. & Schwein.) Fr.

peziziformis (Lév.) D.A. Reid, *Trans. Brit. Mycol. Soc.* 62(3): 474 (1974)
  Exidia peziziformis Lév., *Ann. Sci. Nat., Bot.,* sér. 3, 9: 127 (1848)
  Femsjonia peziziformis (Lév.) P. Karst, *Bidrag Kännedom Finlands Natur Folk* 31: 352 (1876)
  Femsjonia luteoalba Fr., *Summa veg. Scand.* 2: 341 (1849)
  Ditiola luteoalba (Fr.) Quél., *Enchir. fung.*: 227 (1886)
E: o **S:** ! **W:** o
H: On decayed wood of deciduous trees, most often *Quercus* spp, less frequently *Betula* spp and *Fagus*, and rarely on conifers such as *Pinus sylvestris*.
D: NM3: 95 (as *Femsjonia pezizaeformis*) **D+I:** B&K2: 52 6 (as *F. pezizaeformis*) **I:** C&D: 137, Kriegl1: 82 (as *F. pezizaeformis*)
Occasional but widespread. Few records from Scotland.

radicata (Alb. & Schwein.) Fr., *Syst. mycol.* 2: 170 (1823)
  Helotium radicatum Alb. & Schwein., *Consp. fung. lusat.*: 348 (1805)
  Dacrymyces deliquescens var. *radicatus* (Alb. & Schwein.) Bourdot & Galzin, *Hymenomyc. France*: 68 (1928)
  Dacrymyces radicatus (Alb. & Schwein.) Donk, *Meded. Ned. Mycol. Ver.* 18-20: 120 (1931)
E: !
H: On decayed conifer wood. British material on worked wood such as railings, planks, railway sleeepers, or driftwood.
D: NM3: 95
Rarely reported. A few collections in the late nineteenth century, then none until 1961 from East Norfolk (Scolt Head) and none since then.

EFIBULOBASIDIUM K. Wells, *Mycologia* 67: 148 (1975)
Tremellales, Exidiaceae
Type: *Efibulobasidium albescens* (Sacc. & Malbr.) K. Wells

albescens (Sacc. & Malbr.) K. Wells, *Mycologia* 67: 149 (1975)
  Epidochium albescens Sacc. & Malbr., *Michelia* 2: 305 (1881)
  Tremella albescens (Sacc. & Malbr.) Sacc., *Syll. fung.* 6: 790 (1888)
  Exidiopsis albescens (Sacc. & Malbr.) D.A. Reid, *Nova Hedwigia Beih.* 51: 199 (1975)
  Tremella fusispora Bourdot & Galzin, *Bull. Soc. Mycol. France* 39: 262 (1923)
  Sebacina fusispora (Bourdot & Galzin) Raitv., *Opr. Getero. Grib. SSSR*: 60 (1967)
E: !
H: On fallen leaves of *Salix* spp and on dead stems of herbaceous plants or grasses such as *Chamaerion angustifolium, Fallopia japonica, Geranium pratense, Glyceria maxima* and *Urtica dioica*.
D+I: Reid6: 199-200 (as *Exidiopsis albescens*)
Known from East Norfolk, Hertfordshire, South Devon, Surrey, Warwickshire and Yorkshire.

EICHLERIELLA Bres., *Ann. Mycol.* 1(1): 115 (1903)
Tremellales, Exidiaceae
Type: *Eichleriella incarnata* Bres.

deglubens (Berk. & Broome) D.A. Reid, *Trans. Brit. Mycol. Soc.* 55: 436 (1970)
  Radulum deglubens Berk. & Broome, *Ann. Mag. Nat. Hist., Ser. 4 [Notices of British Fungi no. 1444]* 15: 32 (1875)
  Radulum kmetii Bres., *Atti. Accad. Sci. Lett. Arti Ag., Ser. III*: 102 (1897)
  Eichleriella kmetii (Bres.) Bres., *Bull. Soc. Mycol. France* 25: 30 (1910) [1909]
  Hirneolina kmetii (Bres.) Sacc. & Trotter, *Syll. fung.* 21: 451 (1912)
  Mis.: *Stereum rufum* sensu auct. Brit.
  Mis.: *Eichleriella spinulosa* sensu auct. Brit.
  Mis.: *Heterochaete spinulosa* sensu auct. Brit.
  Mis.: *Radulum spinulosum* sensu auct. Brit.

E: o **S:** o **W:** ! **NI:** ! **ROI:** !
H: On fallen wood of deciduous trees, most often *Fraxinus* and *Fagus* but also *Acer campestre, Buxus, Corylus, Cotonoeaster, Ilex, Prunus spinosa, Quercus, Sorbus aria* and *Ulmus* spp. Rarely on conifers, but known on *Taxus*.
D: NM3: 99 **D+I:** B&K2: 60 17, Myc13(1): 36-37
Occasional but widespread.

ELAPHOCEPHALA Pouzar, *Česká Mykol.* 37: 206 (1983)
Stereales, Sistotremataceae
Type: *Elaphocephala iocularis* Pouzar

iocularis Pouzar, *Česká Mykol.* 37(4): 206 (1983)
E: !
H: On decayed wood.
D+I: Myc8(3): 115
Known from South Devon, but easily overlooked as the corticioid basidiomes are virtually invisible to the naked eye.

ENDOPERPLEXA P. Roberts, *Mycol. Res.* 97(4): 471 (1993)
Tremellales, Exidiaceae
  Opadorhiza T.F. Andersen & R.T. Moore (anam.), *Rhizoctonia Species*: 25 (1996)
Type: *Endoperplexa dartmorica* P. Roberts

dartmorica P. Roberts, *Mycol. Res.* 97(4): 472 (1993)
E: ! **S:** !
H: On decayed wood of *Picea* spp, and one collection on *Fagus*.
D: NM3: 102
Described from South Devon (Dartmoor) and known from several localities in Devon and also from Scotland (Kirkudbrightshire and Wester Ross).

enodulosa (Hauerslev) P. Roberts, *Mycol. Res.* 97(4): 473 (1993)
  Sebacina enodulosa Hauerslev, *Windahlia* 16: 47 (1986)
  Endoperplexa septocystidiata (Hauerslev) P. Roberts, *Mycol. Res.* 97(4): 472 (1993)
  Rhizoctonia globularis H.K. Saksena & Vaartaja (anam.), *Canad. J. Bot.* 38: 939 (1960)
  Opadorhiza globularis (H.K. Saksena & Vaartaja) T.F. Andersen & R.T. Moore (anam.), *Rhizoctonia Species*: 25 (1996)
E: !
H: On decayed herbaceous and woody plants such as *Acorus calamus* and *Rubus fruticosus*, on debris of *Dryopteris* spp and *Pteridium* and also on decayed wood.
D: NM3: 106
Known from London, South Devon, Surrey and Yorkshire.

subfarinacea (Hauerslev) P. Roberts, *Mycol. Res.* 97(4): 472 (1993)
  Sebacina subfarinacea Hauerslev, *Friesia* 11(2): 104 (1976)
  Exidiopsis subfarinacea (Hauerslev) Wojewoda, *Mała Flora Grzybów* 2: 109 (1981)
E: !
H: British material on dead standing stems of *Chamaerion angustifolium* and *Epilobium* sp.
D: NM3: 102 (as *Exidiopsis subfarinacea*)
Known from Buckinghamshire (Burnham Beeches) and South Devon (Torquay).

ENTOLOMA (Fr.) P. Kumm., *Führ. Pilzk.*: 23 (1871)
Agaricales, Entolomataceae
  Leptonia (Fr.) P. Kumm., *Führ. Pilzk.*: 24 (1871)
  Nolanea (Fr.) P. Kumm., *Führ. Pilzk.*: 24 (1871)
  Eccilia (Fr.) P. Kumm., *Führ. Pilzk.*: 23 (1871)
  Claudopus Gillet, *Hyménomycètes*: 426 (1876)
  Rhodophyllus Quél., *Enchir. fung.*: 57 (1886)
  Pouzaromyces Pilát, *Sborn. Nár. Mus. Praze, B* 9(2): 60 (1953)

Alboleptonia Largent & R.G. Benedict, *Mycologia* 62: 439 (1970)

Pouzarella Mazzer, *Biblioth. Mycol.* 46: 69 (1976)

Inocephalus (Noordel.) P.D. Orton, *Mycologist* 5(3): 130 (1991)

Paraleptonia (Noordel.) P.D. Orton, *Mycologist* 5(4): 174 (1991)

Omphaliopsis (Noordel.) P.D. Orton, *Mycologist* 5(4): 173 (1991)

Trichopilus (Romagn.) P.D. Orton, *Mycologist* 5(4): 175 (1991)

Type: *Entoloma prunuloides* (Fr.) Quél.

acidophilum Arnolds & Noordel., *Persoonia* 10: 285 (1979)
E: ! **S:** !
H: On acidic soil, amongst short turf on heathland and moorland.
D: FAN1: 132, FungEur5: 330 **D+I:** FRIC12: 9-10 91a
A single collection from South Hampshire in 1992; also reported from Scotland.

aethiops (Scop.) G. Stev., *Kew Bull.* 16: 230 (1962)
Agaricus aethiops Scop., *Fl. carniol.* 345 (1772)
Leptonia aethiops (Scop.) Gillet, *Hyménomycètes*: 415 (1876)
E: ! **S:** ! **W:** ! **ROI:** !
H: On peaty or acidic soil with *Betula* or *Salix* spp.
D+I: FungEur5: 506 56c
Rarely reported. Apparently widespread.

albotomentosum Noordel. & Hauskn., *Z. Mykol.* 55(1): 32 (1989)
E: ! **S:** !
H: On decayed leaves and stems of various grasses, in damp woodland or marshy places at the edges of ponds.
D+I: FungEur5: 611 72c
Known from England (London: Buckingham Palace, Berkshire: Windsor Great Park and Surrey: Kew Gardens and Runnymede) and Scotland (East Perthshire: Kindrogan). Basidiomes are small and often hidden under dense mats or tufts of grass.

allospermum Noordel., *Persoonia* 12: 460 (1985)
Leptonia allosperma (Noordel.) P.D. Orton, *Mycologist* 5(3): 130 (1991)
S: !
H: On soil in montane or sub-montane habitat, associated with *Betula* spp in the type locality.
D: FungEur5: 530
Known from Perthshire (Blair Atholl, Struan Birchwood).

ameides (Berk. & Broome) Sacc., *Syll. fung.* 5: 686 (1887)
Agaricus ameides Berk. & Broome, *Ann. Mag. Nat. Hist., Ser. 3 [Notices of British Fungi no. 999]* 15: 315 (1865)
Nolanea ameides (Berk. & Broome) P.D. Orton, *Mycologist* 5(3): 136 (1991)
E: ! **S:** ! **W:** ! **NI:** !
H: On soil amongst grass, in unimproved grassland, meadows and hayfields, occasionally in grassy areas in open woodland.
D: FAN1: 128, NM2: 352 **D+I:** FungEur5: 260 32a **I:** SV29: 36, Cooke 329 (341) Vol. 3 (1884)
Rarely reported but apparently widespread.

anatinum (Lasch) Donk, *Bull. Jard. Bot. Buitenzorg, Sér. 3*, 18: 158 (1949)
Agaricus anatinus Lasch, *Linnaea* 4: 540 (1829)
Leptonia anatina (Lasch) P. Kumm, *Führ. Pilzk.*: 96 (1871)
E: ? **S:** ! **W:** ?
H: On acidic soil amongst grasses and mosses in unimproved grassland.
D: NCL3: 290 (as *Leptonia anatina*), FAN1: 161 **D+I:** FungEur5: 544 63a **I:** C&D: 299
British records with certainty only from Scotland. Reports for Wales (Carmarthenshire) and England (North Somerset and Shropshire) are unsubstantiated with voucher material.

anthracinum (J. Favre) Noordel., *Persoonia* 11(2): 228 (1981)
Rhodophyllus anthracinus J. Favre, *Ergebn. Wiss. Untersuch. Schweiz. Natn. Parks* 5: 200 (1955)

S: !
H: On soil, associated with *Salix herbacea* in boreal or montane habitat.
D: FungEur5: 196
British records only from Cairngorms and Perthshire (near Kindrogan).

aprile (Britzelm.) Sacc., *Syll. fung.* 5: 696 (1887)
Agaricus aprilis Britzelm., *Ber. Naturhist. Vereins Augsburg* 28: 149 (1885)
Mis.: *Entoloma majale* sensu Lange (1940)
E: ! **S:** ! **W:** !
H: On soil, associated with species of *Ulmus* spp in hedgerows and scrub woodland. Spring-fruiting.
D: FAN1: 98-99 **D+I:** Ph: 116, FungEur5: 131 7b **I:** C&D: 291
Rarely reported and even less often collected. Most records are unsubstantiated with voucher material, or are associated with *Crataegus* and probably refer to the more frequent *E. clypeatum*.

araneosum (Quél.) M.M. Moser, *Kleine Kryptogamenflora*, Edn 4: 208 (1978)
Nolanea araneosa Quél., *Bull. Soc. Bot. France* 23: 327 (1877) [1876]
Pouzarella araneosa (Quél.) Mazzer, *Biblioth. Mycol.* 46: 100 (1976)
Pouzaromyces araneosus (Quél.) P.D. Orton, *Mycologist* 5(4): 174 (1991)
Agaricus fulvostrigosus Berk. & Broome, *Ann. Mag. Nat. Hist., Ser. 5 [Notices of British Fungi no. 1650]* 1: 19 (1878)
Nolanea fulvostrigosa (Berk. & Broome) Sacc., *Syll. fung.* 5: 720 (1887)
Leptonia fulvostrigosa (Berk. & Broome) P.D. Orton, *Trans. Brit. Mycol. Soc.* 43(2): 177 (1960)
E: ! **S:** ? **NI:** ! **ROI:** !
H: On soil or calcareous loam, frequently under cover of *Mercurialis perennis* in open deciduous woodland.
D: FAN1: 116-117, NM2: 356 **D+I:** FRIC3: 21-23 19c (as *Nolanea fulvostrigosa*), FungEur5: 362 40a **I:** C&D: 293
Rarely reported but often occuring singly, easily overlooked and apparently widespread. Reported from Scotland but unsubstantiated with voucher material.

argenteostriatum Arnolds & Noordel., *Persoonia* 10: 285 (1979)
S: !
H: On acidic soil in unimproved grassland.
D: FAN1: 132, FungEur5: 332 **D+I:** Myc9(3): 118
A recent collection.

asprellum (Fr.) Fayod, *Ann. Sci. Nat., Bot.,* sér. 7, 9: 383 (1889)
Agaricus asprellus Fr., *Syst. mycol.* 1: 208 (1821)
Leptonia asprella (Fr.) P. Kumm., *Führ. Pilzk.*: 96 (1871)
E: ! **S:** ! **W:** ! **NI:** ! **ROI:** !
H: On acidic to neutral soil in unimproved grassland.
D: NM2: 347 **D+I:** FungEur5: 522 60a **I:** C&D: 299, SV39: 34
Rarely reported but apparently widespread.

atrocoeruleum Noordel., *Nova Hedwigia Beih.* 91: 248 (1987)
Leptonia atrocoerulea (Noordel.) P.D. Orton, *Mycologist* 5(3): 130 (1991)
E: ! **S:** ! **W:** ! **NI:** ! **ROI:** !
H: On acidic soil in unimproved grassland.
D+I: FungEur5: 510 56b
Rarely reported, but apparently widespread.

atromarginatum (Romagn. & J. Favre) Zschiesch., *Wiss. Z. Friedrich-Schiller-Univ. Jena. Math.-Naturwiss. Reihe* 33: 814 (1984)
Rhodophyllus atromarginatus Romagn. & J. Favre, *Rev. Mycol. (Paris)* 3(2-3): 77 (1938)
Leptonia atromarginata (Romagn. & J. Favre) Konrad & Maubl., *Encycl. Mycol.* 14, *Les Agaricales* I: 263 (1948)
E: ! **S:** ! **NI:** !
H: On soil in peat bogs and amongst grass in unimproved grassland.

D+I: FungEur5: 579 69b
Rarely reported.

bisporigerum (P.D. Orton) Noordel., *Persoonia* 11(1): 86
(1980) [1979]
 Eccilia bisporigera P.D. Orton, *Notes Roy. Bot. Gard.*
 Edinburgh 29: 99 (1969)
E: !
H: On soil in damp areas in deciduous woodland.
D: FAN1: 109 **D+I:** FungEur5: 193 20b
Known in Britain only from the type collection from
 Herefordshire (Shobdon) and collections from Norfolk (Herons
 Carr) and West Sussex (Ebernoe Common).

bloxamii (Berk. & Broome) Sacc., *Syll. fung.* 5: 684 (1887)
 Agaricus bloxamii Berk. & Broome, *Ann. Mag. Nat. Hist., Ser.*
 2 [Notices of British Fungi no. 677] 13: 399 (1854)
 Mis.: *Entoloma madidum* sensu NCL and sensu auct. mult.
E: ! **S:** ! **W:** ! **NI:** ! **ROI:** !
H: On soil amongst grass in meadows, unimproved grassland
 and downland.
D: FAN1: 96, NM2: 344 **D+I:** FungEur5: 115 3 **I:** C&D: 291
Rarely reported, but apparently widespread.

brassicolens (D.A. Reid) Noordel., *Persoonia* 11(2): 229 (1981)
 Nolanea brassicolens D.A. Reid, *Trans. Brit. Mycol. Soc.*
 48(4): 518 (1965)
E: ! **NI:** !
H: On soil in deciduous woodland.
D: FungEur5: 161
Described from Northern Ireland and known from two sites in
 Down, also a collection from England (South Devon:
 Membury). Basidiomes smell strongly of rotten cabbage.

byssisedum (Pers.) Donk, *Bull. Jard. Bot. Buitenzorg, Sér. 3,*
18: 158 (1949)
 Agaricus byssisedus Pers., *Syn. meth. fung.*: 482 (1801)
 Crepidotus byssisedus (Pers.) P. Kumm., *Führ. Pilzk.*: 74
 (1871)
 Claudopus byssisedus (Pers.) Gillet, *Hyménomycètes*: 427
 (1876)
E: ! **S:** !
H: On decayed wood and bark in deciduous woodland. Rarely
 on soil and decayed herbaceous debris.
D: FAN1: 172, BFF6: 109-110 (as *Claudopus byssisedus*) **D+I:**
 FungEur5: 605 71b **I:** Bon: 189, C&D: 301
Rarely reported. Possibly overlooked since basidiomes are small
 and occur in cryptic locations such as under decayed bark.
 Known from England (East Norfolk, Northamptonshire,
 Oxfordshire, Staffordshire, Surrey, Warwickshire and West
 Norfolk) and Scotland (South Ebudes: Isle of Islay).

caccabus (Kühner) Noordel., *Persoonia* 11(1): 86 (1980)
[1979]
 Rhodophyllus caccabus Kühner, *Rev. Mycol. (Paris)* 19(1): 3-
 4 (1954)
 Eccilia paludicola P.D. Orton, *Trans. Brit. Mycol. Soc.* 43(2):
 227 (1960)
 Entoloma paludicola (P.D. Orton) Romagn., *Bull. Soc. Mycol.*
 France 103(2): 80 (1987)
 Omphaliopsis paludicola (P.D. Orton) P.D. Orton, *Mycologist*
 5(4): 173 (1991)
E: !
H: On soil in damp woodland, usually associated with *Alnus* or
 Salix spp.
D: FAN1: 108-109, NM2: 351 **D+I:** FungEur5: 191 20c **I:** C&D:
 293
Rarely reported. British records mostly from widely scattered
 locations in southern England, from Surrey to South Devon,
 but also known from Co. Durham.

caeruleoflocculosum Noordel., *Persoonia* 12: 461 (1985)
 Leptonia caeruleoflocculosa (Noordel.) P.D. Orton, *Mycologist*
 5(3): 131 (1991)
S: !
H: On soil amongst grass in open woodland, associated with
 Betula sp. in the type collection.

D: FAN1: 160, FungEur5: 551
Described from Perthshire (Struan Birchwood) but poorly known
 in Britain.

caeruleopolitum Noordel. & Brandt-Ped., *Persoonia* 12(3): 221
(1984)
E: ! **S:** ? **ROI:** !
H: On acidic soil in unimproved grassland and one confirmed
 collection from peaty soil in a dried out swamp, associated
 with *Alnus* and *Salix* spp. **D:** NM2: 346 **D+I:** FungEur5: 200
 21b
Rarely reported, but apparently widespread. Reported from
 Scotland but unsubstantiated with voucher material.

caeruleum (P.D. Orton) Noordel., *Persoonia* 11(4): 470 (1982)
 Leptonia caerulea P.D. Orton, *Trans. Brit. Mycol. Soc.* 43(2):
 290 (1960)
E: ! **S:** ! **W:** ! **NI:** !
H: On soil amongst grass or mosses in unimproved grassland,
 also on dunes.
D: FungEur5: 503
Rarely reported and even less often collected.

caesiocinctum (Kühner) Noordel., *Persoonia* 11(4): 470 (1982)
 Rhodophyllus caesiocinctus Kühner, *Rev. Mycol. (Paris)*
 19(1): 4 (1954)
 Leptonia caesiocincta (Kühner) P.D. Orton, *Trans. Brit. Mycol.*
 Soc. 43(2): 177 (1960)
E: ! **S:** o **W:** o **NI:** ! **ROI:** !
H: On acidic soil amongst grass in unimproved grassland, often
 in upland areas.
D: FAN1: 157-158, NM2: 346 **D+I:** FungEur5: 476 52c **I:** C&D:
 299
Occasional but widespread, especially in upland grassland in
 northern and western Britain, though virtually unknown in
 southern or south eastern England.

calaminare Noordel., *Persoonia* 12(3): 198 (1984)
E: ? **S:** !
H: On soil in deciduous woodland or in grassland.
D: FAN1: 155, FungEur5: 461
Recently collected from Scotland, and reported from England
 (East Norfolk) but unsubstantiated with voucher material.

calthionis Arnolds & Noordel., *Persoonia* 10(2): 285 (1979)
S: !
H: On peaty or acidic soil in unimproved grassland.
D: FAN1: 136 **D+I:** FRIC12: 4-5 89c, FungEur5: 276 78a
A single collection from Orkney (Deerness).

canosericeum (J.E. Lange) Noordel., *Nordic J. Bot.* 2(2): 157
(1982)
 Rhodophyllus canosericeus J.E. Lange, *Fl. agaric. danic.* 5,
 Taxonomic Conspectus: VIII (1940)
 Nolanea canosericea (J.E. Lange) P.D. Orton, *Trans. Brit.*
 Mycol. Soc. 43: 178 (1960)
S: !
H: On soil in deciduous woodland.
D: NM2: 356 **D+I:** FungEur5: 335 38a
Rarely reported and accepted with doubt. Listed in NCL and in
 Orton (1991) but Orton's concept, of a species lacking cystidia,
 is not that of Noordeloos (FungEur5). The Scottish material
 needs re-examination.

carneogriseum (Berk. & Broome) Noordel., *Nova Hedwigia*
 Beih. 91: 216 (1987)
 Agaricus carneogriseus Berk. & Broome, *Ann. Mag. Nat. Hist.,*
 Ser. 3 [Notices of British Fungi no. 1001] 15: 315 (1865)
 Eccilia carneogrisea (Berk. & Broome) Gillet,
 Hyménomycètes: 424 (1876)
 Leptonia carneogrisea (Berk. & Broome) P.D. Orton,
 Mycologist 5(3): 131 (1991)
E: ! **S:** ! **ROI:** ?
H: On soil amongst grass in unimproved grassland.
D+I: FungEur5: 480 52b
Rarely reported. Last record in England (North Devon: Braunton)
 in 1946 and Scotland (South Aberdeen: Aboyne) in 1862 (the

type collection). Sensu Cooke 368 (380) Vol. 3 (1885) is a
different, unknown species possibly *E. mougeotii.*

catalaunicum (Singer) Noordel., *Persoonia* 11(4): 470 (1982)
   Leptonia catalaunica Singer, *Ann. Mycol.* 34: 428 (1936)
E: ! S: !
H: On base rich soil in pastures and unimproved grassland.
D+I: FungEur5: 517 58b
British collections from England (Yorkshire: Arncliffe) and several
localities in Scotland where it is apparently more frequent.
Reported from North Somerset but unsubstantiated with
voucher material.

cephalotrichum (P.D. Orton) Noordel., *Persoonia* 1: 260
   (1979)
   Leptonia cephalotricha P.D. Orton, *Trans. Brit. Mycol. Soc.*
      43(2): 291 (1960)
   Alboleptonia cephalotricha (P.D. Orton) P.D. Orton,
      Mycologist 5(3): 125 (1991)
   Mis.: *Eccilia molluscus* sensu Reid [TBMS 41: 433 (1958)]
E: !
H: On soil in damp deciduous woodland.
D: FAN1: 150, NM2: 344 **D+I:** FungEur5: 422 46d
Rarely reported. British records from widely scattered areas
from Lincolnshire southwards.

cetratum (Fr.) M.M. Moser, *Kleine Kryptogamenflora, Edn 4*:
   206 (1978)
   Agaricus cetratus Fr., *Observ. mycol.* 2: 218 (1818)
   Nolanea cetrata (Fr.) P. Kumm., *Führ. Pilzk.*: 95 (1871)
   Mis.: *Nolanea testacea* sensu NCL
E: c S: c W: o NI: ! ROI: !
H: On acidic, peaty or sandy soil usually in needle litter of
conifers in woodlands or plantations and occasionally under
dense cover of *Calluna* on heath and moorland. Fruitng from
early summer onwards.
D: FAN1: 133-134, NM2: 349 **D+I:** Ph: 117 (as *Nolanea
cetrata*), FRIC12: 2-3 89a, FungEur5: 270 34c **I:** FlDan2: 78F
(as *Rhodophyllus cetratus*), C&D: 925
Common and widespread.

chalybaeum (Pers.) Noordel., *Nordic J. Bot.* 2: 163 (1982)
   Agaricus chalybaeus Pers., *Syn. meth. fung.*: 343 (1801)
   Gymnopus chalybaeus (Pers.) Gray, *Nat. arr. Brit. pl.* 1: 608
      (1821)
   Leptonia chalybaea (Pers.) P. Kumm., *Führ. Pilzk.*: 96 (1871)
E: o S: o W: o NI: ! ROI: ! O: Channel Islands: !
H: On soil, in unimproved grassland.
D: FAN1: 163-164, NM2: 346 **D+I:** FungEur5: 484 53a **I:** C&D:
299, SV38: 28
Occasional but widepread, and in some areas locally common.

chalybaeum *var.* lazulinum (Fr.) Noordel., *Persoonia* 12(3):
   206 (1984)
   Agaricus lazulinus Fr., *Epicr. syst. mycol.*: 153 (1838)
   Leptonia lazulina (Fr.) Quél., *Mém. Soc. Émul. Montbéliard,
      Sér. 2*, 5: 344 (1872)
   Rhodophyllus lazulinus (Fr.) Quél., *Enchir. fung.*: 60 (1886)
   Entoloma lazulinum (Fr.) Noordel., *Nordic J. Bot.* 2: 162
      (1982)
E: o S: o W: o NI: ! ROI: !
H: On soil, amongst grass, in unimproved grassland.
D: FAN1: 164, NM2: 345 **D+I:** Ph: 117 (as *Leptonia lazulina*),
FungEur5: 486 53b
Occasional but widespread, and may be locally abundant.

clandestinum (Fr.) Noordel., *Persoonia* 10(4): 456 (1980)
   Agaricus clandestinus Fr., *Observ. mycol.* 2: 166 (1818)
   Nolanea clandestina (Fr.) P. Kumm., *Führ. Pilzk.*: 95 (1871)
E: ! S: ! ROI: !
H: On soil, amongst grass, in unimproved grassland, heathland
and moorland.
D: FAN1: 122-123, NM2: 350 **D+I:** FRIC12: 14-15 91e,
FungEur5: 230 27b
Rarely reported. Most British records are from Scotland, but also
known from scattered sites in England (South Hampshire and
Yorkshire) and Republic of Ireland (Dublin: North Bull Island).

clypeatum (L.) P. Kumm., *Führ. Pilzk.*: 98 (1871)
   Agaricus clypeatus L. [non *A. clypeatus* Huds. (1778)], *Sp.
      pl.*: 1174 (1753)
   Rhodophyllus clypeatus (L.) Quél., *Enchir. fung.*: 59 (1886)
   Agaricus fertilis Pers., *Syn. meth. fung.*: 328 (1801)
   Entoloma fertile (Pers.) Gillet, *Hyménomycètes*: 405 (1876)
E: c S: ! W: ! NI: ! ROI: !
H: On soil, usually associated with *Crataegus monogyna*, but
occasionally reported with *Prunus spinosa* and *Malus sylvestris*.
Fruiting from April to June.
D: FAN1: 97-98, NM2: 353 **D+I:** Ph: 114-115, FungEur5: 129 6
I: FlDan2: 75C (as *Rhodophyllus clypeatus*), SV27: 52, C&D:
291
Common throughout England, less often reported elsewhere,
but apparently widespread and often occuring in large
numbers.

cocles (Fr.) Noordel., *Persoonia* 11(2): 149 (1981)
   Agaricus cocles Fr., *Epicr. syst. mycol.*: 158 (1838)
   Nolanea cocles (Fr.) Gillet, *Hyménomycètes*: 422 (1876)
E: ! S: !
H: On soil, in unimproved grassland.
D: FAN1: 155-156, NM2: 349, FungEur5: 458
Rarely reported. Known from England (North Somerset) and
from Scotland

conferendum (Britzelm.) Noordel., *Persoonia* 10(4): 446
   (1980)
   Agaricus conferendus Britzelm., *Ber. Naturhist. Vereins
      Augsburg* 26: 140 (1881)
   Nolanea conferenda (Britzelm.) Sacc., *Syll. fung.* 5: 723
      (1887)
   Rhodophyllus rickenii Romagn., *Bull. Soc. Mycol. France* 48:
      320 (1932)
   Entoloma conferendum var. *rickenii* (Romagn.) Bon &
      Courtec., *Doc. Mycol.* 18(69): 38 (1987)
   Nolanea staurospora Bres., *Fungi trident.* 1: 18 (1881)
   Rhodophyllus staurosporus (Bres.) J.E. Lange, *Fl. agaric.
      danic.* 5, *Taxonomic Conspectus*: VII (1940)
   Entoloma staurosporum (Bres.) E. Horak, *Sydowia* 28: 222
      (1976) [1975]
   Mis.: *Entoloma pascuum* sensu auct. mult.
   Mis.: *Nolanea proletaria* sensu Rea (1922)
E: c S: c W: c NI: c ROI: c O: Channel Islands: c
H: On soil, in grassland, including lawns and playing fields and
upland or montane pastures, occasionally in bogs, woodland
glades and woodland margins.
D: FAN1: 120, NM2: 349 **D+I:** Ph: 118 (as *Nolanea
staurospora*), FungEur5: 373 41 **I:** FlDan2: 77A (as
Rhodophyllus staurosporus), C&D: 925
Very common and widespread.

conferendum *var.* pusillum (Velen.) Noordel., *Persoonia*
   10(4): 450 (1980)
   Nolanea pusilla Velen., *České Houby* III: 626 (1921)
   Rhodophyllus xylophilus J.E. Lange, *Dansk. Bot. Ark.* 2(11):
      35 (1921)
   Nolanea xylophila (J.E. Lange) P.D. Orton, *Trans. Brit. Mycol.
      Soc.* 43: 179 (1960)
E: ! S: !
H: On decayed deciduous wood.
D: FAN1: 121, NM2: 349, FungEur5: 376
Rarely reported. Known from England (Berkshire and West
Sussex) and Scotland (Isle of Rhum).

corvinum (Kühner) Noordel., *Nordic J. Bot.* 2(2): 162 (1982)
   Rhodophyllus corvinus Kühner, *Rev. Mycol. (Paris)* 19(1): 4
      (1954)
   Leptonia corvina (Kühner) P.D. Orton, *Trans. Brit. Mycol. Soc.*
      43(2): 177 (1960)
E: o S: o W: ! NI: ! ROI: !
H: On soil, in unimproved grassland
D: FAN1: 165, NM2: 346 **D+I:** FungEur5: 495 55 **I:** C&D: 299
Occasional but widespread.

cruentatum (Quél.) Noordel., *Persoonia* 12(3): 201 (1984)
Nolanea cruentata Quél., *Compt. Rend. Assoc. Franç. Avancem. Sci.* 14: 446 (1886) [1885]
Leptonia cruentata (Quél.) P.D. Orton, *Mycologist* 5(3): 131 (1991)
Rhodophyllus coelestinus var. *cruentatus* (Quél.) Quél., *Enchir. fung.*: 65 (1886)
E: ? **S:** ! **W:** ?
H: On soil, amongst grass, in upland or montane areas.
D: FungEur5: 488
Reported from England and Wales but unsubstantiated with voucher material.

cuneatum (Bres.) M.M. Moser, *Kleine Kryptogamenflora*, Edn 4: 205 (1978)
Nolanea cuneata Bres., *Fungi trident.* 1: 77 (1887)
E: ! **S:** !
H: On soil, in coniferous woodland. Known with *Pinus* and *Picea* spp in Britain.
D: NCL3: 330 (as *Nolanea cuneata*), FAN1: 135 **D+I:** FungEur5: 279 34a
Rarely reported in England, but apparently more widespread in Scotland.

cuspidiferum Noordel., *Persoonia* 10(4): 461 (1980)
Rhodophyllus cuspidifer Kühner & Romagn. [*nom. nud.*], *Flore Analytique des Champignons Supérieurs*: 189 (1953)
Nolanea cuspidifer (Kühner & Romagn.) P.D. Orton [*nom. nud.*], *Trans. Brit. Mycol. Soc.* 43(2): 179 (1960)
Agaricus junceus f. *cuspidatus* Fr., *Ic. Hymenomyc.* 1: pl. 99 f. 2 (1867)
Nolanea juncea var. *cuspidata* (Fr.) J. Favre, *Bull. Soc. Mycol. France* 52(2): 137 (1936)
E: ! **S:** !
H: In *Sphagnum* moss in peat bogs, also in marshy areas amongst grasses, and in partially dried out dune slacks.
D: FAN1: 124, NM2: 349 **D+I:** FungEur5: 241 35a
Rarely reported.

cyaneoviridescens (P.D. Orton) Noordel., *Persoonia* 11(4): 470 (1982)
Leptonia cyaneoviridescens P.D. Orton, *Trans. Brit. Mycol. Soc.* 43(2): 292 (1960)
E: ! **S:** !
H: On acidic soil, amongst short turf, in unimproved grassland.
D: FungEur5: 489
Known from Shropshire, Yorkshire and several sites in Scotland

depluens (Batsch) Hesler, *Nova Hedwigia Beih.* 23: 16 (1967)
Agaricus depluens Batsch, *Elench. fung. (Continuatio Prima)*: 167 & t. 24 f. 122 (1786)
Crepidotus depluens (Batsch) P. Kumm., *Führ. Pilzk.*: 74 (1871)
Claudopus depluens (Batsch) Gillet, *Hyménomycètes*: 427 (1876)
Rhodophyllus depluens (Batsch) Quél., *Enchir. fung.*: 65 (1886)
Agaricus epigaeus Pers., *Observ. mycol.* 2: 47 (1799)
Crepidopus epigaeus (Pers.) Gray, *Nat. arr. Brit. pl.* 1: 616 (1821)
Crepidotus epigaeus (Pers.) Berk. & Broome, *Ann. Mag. Nat. Hist., Ser. 5 [Notices of British Fungi no. 1949]* 9: 179 (1882)
E: ! **S:** !
H: On decayed wood, and also on woody fragments in soil with potted plants.
D: NCL3: 186 (as *Claudopus depluens*), FAN1: 173, BFF6: 110 (as *C. depluens*), FungEur5: 606
Poorly known in Britain, with few records, mostly unsubstantiated with voucher material. Only four specimens in herb. K, three collected between 1881 and 1882 and one from near Aldershot (North Hampshire) in July 2003, and Scottish material in E.

dichrum (Pers.) P. Kumm. (as *E. dichrous*), *Führ. Pilzk.*: 97 (1871)
Agaricus dichrous Pers., *Syn. meth. fung.*: 343 (1801)

Leptonia dichroa (Pers.) P.D. Orton, *Mycologist* 5(3): 132 (1991)
E: !
H: In grass, near *Pinus* sp.
D: FAN1: 152 **D+I:** FungEur5: 427 47b **I:** C&D: 297
Known in Britain from a single site in Shropshire (Attingham Park). Not always distinguished from the similar and more frequent *Entoloma tjallingiorum*.

dysthales (Peck) Sacc., *Syll. fung.* 9: 83 (1891)
Agaricus dysthales Peck, *Rep. (Annual) New York State Mus. Nat. Hist.* 32: 28 (1879)
Pouzarella dysthales (Peck) Mazzer, *Biblioth. Mycol.* 46: 105 (1976)
Pouzaromyces dysthales (Peck) P.D. Orton, *Mycologist* 5(4): 174 (1991)
Inocybe bucknallii Massee, *Ann. Bot.* 18: 473 (1904)
Astrosporina bucknallii (Massee) Rea, *Brit. basidiomyc.*: 213 (1922)
Mis.: *Entoloma babingtonii* sensu auct.
Mis.: *Nolanea babingtonii* sensu Pearson & Dennis (1948)
Mis.: *Leptonia babingtonii* sensu NCL
E: ! **S:** !
H: On calcareous, humic soil in deciduous woodland, often under dense cover of *Mercurialis perennis* or other herbs.
D: FAN1: 113, FungEur5: 343, NM2: 356 **I:** C&D: 293
Rarely reported. Not infrequent in herb rich woodland on calcareous soil in southern England, but known from scattered sites as far north as Perthshire.

dysthaloides Noordel., *Persoonia* 10: 219 (1979)
E: ! **S:** !
H: On soil or humus in deciduous woodland.
D: FAN1: 114, NM2: 356 **D+I:** FungEur5: 347 39a
Rarely reported. British collections from England (North West Yorkshire, Oxfordshire, South Hampshire, South Devon) and Scotland (Kirkudbright).

elodes (Fr.) P. Kumm., *Führ. Pilzk.*: 98 (1871)
Agaricus elodes Fr., *Syst. mycol.* 1: 196 (1821)
Trichopilus elodes (Fr.) P.D. Orton, *Mycologist* 5(4): 175 (1991)
E: ! **S:** o **NI:** ! **ROI:** !
H: On peaty or acidic soil, often amongst species of *Sphagnum* moss.
D: NCL3: 229 (as *Entoloma helodes*), FAN1: 144, NM2: 355 **D+I:** FRIC12: 31-33 94c (as *E. helodes*), FungEur5: 402 43b
Rarely reported. Known from widely scattered locations in Britain, most often from Scotland. The epithet is frequently found in literature as '*helodes*'.

euchroum (Pers.) Donk, *Bull. Jard. Bot. Buitenzorg, Sér. 3,* 18: 157 (1949)
Agaricus euchrous Pers., *Syn. meth. fung.*: 343 (1801)
Leptonia euchroa (Pers.) P. Kumm., *Führ. Pilzk.*: 96 (1871)
E: ! **S:** ! **W:** ! **ROI:** !
H: On decayed wood, usually of *Corylus* but also known on *Alnus*, and *Quercus* spp, and rarely *Castanea, Fagus, Prunus avium* and *Rhamnus carthartica*.
D: FAN1: 151-152, NM2: 345 **D+I:** Ph: 116-117 (as *Leptonia euchroa*), FungEur5: 425 46e **I:** FlDan2: 76A (as *Rhodophyllus euchrous*), C&D: 297
Occasional in southern England, especially in damp woodland or copses on calcareous soil. Rarely reported elsewhere, but apparently widespread. On *Corylus* it is often found fruiting amongst the basal branches of old coppiced trunks.

excentricum Bres., *Fungi trident.* 1: 11 (1881)
E: ! **S:** ! **ROI:** !
H: On calcareous or limestone soil in unimproved grassland.
D: FAN1: 117, NM2: 354 **D+I:** FungEur5: 215 24 **I:** C&D: 293
Rarely reported. The majority of reports are unsubstantiated with voucher material.

exile (Fr.) Hesler, *Nova Hedwigia Beih.* 23: 178 (1967)
Agaricus exilis Fr., *Observ. mycol.* 2: 95 (1818)
Nolanea exilis (Fr.) P. Kumm., *Führ. Pilzk.*: 95 (1871)

Leptonia exilis (Fr.) P.D. Orton, *Trans. Brit. Mycol. Soc.*
43(2): 177 (1960)
Leptonia pyrospila P.D. Orton, *Trans. Brit. Mycol. Soc.* 43(2):
298 (1960)
Entoloma pyrospilum (P.D. Orton) M.M. Moser, *Kleine
Kryptogamenflora*, Edn 4: 200 (1978)
Entoloma exile var. *pyrospilum* (P.D. Orton) Noordel.,
Persoonia 12(4): 461 (1985)
E: ! **S:** o **W:** ! **NI:** ! **ROI:** !
H: On acidic or sandy soil, amongst short grass, in unimproved
grassland and dunes.
D: FAN1: 166, NM2: 357 **D+I:** FungEur5: 555 65 **I:** C&D: 299
(as *Entoloma pyrospilum*)
Rarely reported but apparently widespread. Most records are
from Scotland.

favrei Noordel., *Int. J. Mycol. Lichenol.* 1(1): 56 (1982)
E: !
H: On soil, often amongst mosses and short grass, in
unimproved grassland (British records from cemeteries).
D: FAN1: 125 **D+I:** FungEur5: 250 29c
British records only from Co. Durham (Darlington) and Middlesex
(Hanwell).

fernandae (Romagn.) Noordel., *Persoonia* 10: 250 (1979)
Rhodophyllus fernandae Romagn., *Rev. Mycol. (Paris)* 1(3):
162 (1936)
Nolanea fernandae (Romagn.) P.D. Orton, *Trans. Brit. Mycol.
Soc.* 43(2): 179 (1960)
Entoloma psilopus Arnolds & Noordel., *Persoonia* 10: 293
(1979)
E: ! **S:** !
H: On acidic soil, amongst grass in unimproved grassland, and
also in coniferous woodland.
D: FAN1: 131, NM2: 350 **D+I:** FRIC12: 10-12 91b, FungEur5:
322 35b & 79b
Known from England (North Hampshire: Mixley Wood near
Yateley and Surrey: Elstead) and Scotland (Outer Hebrides:
Barra Isles).

formosum (Fr.) Noordel., *Persoonia* 12(4): 461 (1985)
Agaricus formosus Fr., *Syst. mycol.* 1: 208 (1821)
Leptonia formosa (Fr.) Gillet, *Hyménomycètes*: 414 (1876)
Leptonia fulva P.D. Orton, *Trans. Brit. Mycol. Soc.* 43(2): 293
(1960)
Entoloma fulvum (P.D. Orton) Arnolds, *Biblioth. Mycol.* 90:
331 (1983) [1982]
Rhodophyllus fulvus (P.D. Orton) M.M. Moser, *Kleine
Kryptogamenflora*, Edn 3: 159 (1967)
E: ! **S:** ! **W:** ! **NI:** ! **ROI:** !
H: On acidic soil, amongst grass in unimproved grassland, on
roadside verges, and occasionally in woodland.
D: FAN1: 168, NM2: 357 **D+I:** FungEur5: 560 67a **I:** C&D: 301
Uncommonly reported, but apparently widespread and perhaps
more frequent in northern and western areas.

fuscomarginatum P.D. Orton, *Trans. Brit. Mycol. Soc.* 43(2):
228 (1960)
Trichopilus fuscomarginatus (P.D. Orton) P.D. Orton,
Mycologist 5(4): 176 (1991)
E: ? **S:** o **W:** ! **ROI:** ? **O:** Isle of Man: !
H: On acidic soil in peat bogs.
D: FAN1: 145, FungEur5: 404
Rarely reported, but apparently widespread. Described from
Scotland, and (*fide* R. Watling) not infrequent there.
Noordeloos (FungEur5: 407) says 'common and abundant in
atlantic peat bogs of the United Kingdom' but this is not
supported by the limited number of records.

fuscotomentosum F.H. Møller, *Fungi Faeroes* 1: 251 (1945)
Trichopilus fuscotomentosus (F.H. Møller) P.D. Orton,
Mycologist 5(4): 176 (1991)
S: !
H: On soil, amongst grass, in unimproved grassland.
D: NM2: 355, NM2: 355, FungEur5: 401
Only two British records, from Allt Tarsuinn and Creag Bhaig.

griseocyaneum (Fr.) P. Kumm., *Führ. Pilzk.*: 97 (1871)
Agaricus griseocyaneus Fr., *Observ. mycol.* 2: 96 (1818)
Leptonia griseocyanea (Fr.) P.D. Orton, *Trans. Brit. Mycol.
Soc.* 43(2): 177 (1960)
E: ! **S:** ! **W:** ! **NI:** ! **ROI:** !
H: On soil, often calcareous, amongst grass in unimproved
grassland.
D: FAN1: 162, NM2: 348 **D+I:** FungEur5: 548 63b
Rarely reported, but apparently widespread.

griseorubidum Noordel., *Persoonia* 12: 196 (1984)
Leptonia griseorubida (Noordel.) P.D. Orton, *Mycologist* 5:
132 (1991)
Mis.: *Leptonia griseorubella* sensu NCL
E: ! **S:** !
H: On soil, amongst grass, in unimproved grassland, or
woodland.
D: FAN1: 154, NM2: 357 **D+I:** FungEur5: 455 49a
Rarely reported. Known from England (West Gloucestershire:
Westridge Wood) and Scotland (Mid-Ebudes: Isle of Arran,
Loch Ranza).

hebes (Romagn.) Trimbach, *Doc. Mycol.* 11(44): 6 (1981)
Rhodophyllus hebes Romagn., *Rev. Mycol. (Paris)* 19(1): 4
(1954)
Nolanea hebes (Romagn.) P.D. Orton, *Trans. Brit. Mycol.
Soc.* 43(2): 179 (1960)
Rhodophyllus mammosus var. *obsoletus* Romagn., *Rev.
Mycol. (Paris)* 19(1): 7 (1953)
Nolanea tenuipes P.D. Orton, *Trans. Brit. Mycol. Soc.* 43(2):
334 (1960)
Entoloma leptopus Noordel., *Persoonia* 10(4): 442 (1980)
Mis.: *Nolanea hirtipes* sensu Phillips (Ph)
Mis.: *Nolanea mammosa* sensu auct. Brit. p.p.
E: c **S:** o **W:** o **NI:** ! **ROI:** !
H: On soil or humus in damp deciduous woodland, especially on
calcareous soil with herb-rich ground cover.
D: FAN1: 119-120, NM2: 348 **D+I:** Ph: 117 (as *Nolanea
hirtipes*), FungEur5: 224 26a **I:** C&D: 925
Uncommonly reported, but fairly frequent, especially in southern
counties, and widespread throughout Britain.

henrici E. Horak & Aeberh., *Cryptog. Mycol.* 4(1): 21 (1983)
S: ! **ROI:** !
H: On acidic soil, amongst grass, in unimproved grassland.
D: FungEur5: 396
A single collection, from Glen Nevis in 1983 and material from
ROI reported by Noordeloos (FungEur5).

hirtipes (Schumach.) M.M. Moser, *Kleine Kryptogamenflora*,
Edn 4: 206 (1978)
Agaricus hirtipes Schumach., *Enum. pl.* 2: 272 (1803)
Nolanea hirtipes (Schumach.) P. Kumm., *Führ. Pilzk.*: 95
(1871)
Mis.: *Nolanea majalis* sensu Rea (1922)
E: ? **S:** ? **W:** ? **ROI:** ?
H: On soil in coniferous woodland. Mostly spring-fruiting.
D: FAN1: 118-119, NM2: 351 **D+I:** FungEur5: 220 25
Rarely reported. This is a robust northern species, poorly known
in Britain. Most British records are sensu Phillips (1981) (= *E.
hebes*).

hispidulum (M. Lange) Noordel., *Nordic J. Bot.* 2(2): 159
(1982)
Rhodophyllus hispidulus M. Lange, *Friesia* 3: 210 (1946)
Leptonia hispidula (M. Lange) P.D. Orton, *Mycologist* 5(3):
132 (1991)
Leptonia inocybeoides P.D. Orton, *Trans. Brit. Mycol. Soc.*
43(2): 296 (1960)
Mis.: *Agaricus resutus* sensu Cooke [*Ill. Brit. fung.* 334 (318)
(1884)]
E: ! **ROI:** ?
H: On sandy soil, amongst grass in grassland or on dunes.
D: FAN1: 151, NM2: 354 **D+I:** FungEur5: 450 81b & 82b
Recorded from Nottinghamshire (Worksop) and South Devon
(Dawlish Warren). Reported on several occasions from the
Republic of Ireland, but unsubstantiated with voucher material.

huijsmanii Noordel., *Persoonia* 12: 212 (1984)
  Leptonia huijsmanii (Noordel.) P.D. Orton, *Mycologist* 5(3): 133 (1991)
S: !
H: On soil in grassland (also in deciduous woodland in continental Europe).
D: FAN1: 161, NM2: 347 **D+I:** FungEur5: 527 87f
Recently collected in Scotland. Noordeloos (FungEur5: 560) suggests that *Agaricus formosus* var. *suavis* sensu Cooke 360 (488) Vol. 3 (1885) may also be this species.

incanum (Fr.) Hesler, *Nova Hedwigia Beih.* 23: 147 (1967)
  Agaricus incanus Fr., *Syst. mycol.* 1: 209 (1821)
  Leptonia incana (Fr.) Gillet, *Hyménomycètes*: 414 (1876)
  Agaricus euchlorus Lasch, Fries, *Epicr. syst. mycol.*: 154 (1838)
  Leptonia euchlora (Lasch) P. Kumm., *Führ. Pilzk.*: 96 (1871)
  Leptonia incana var. *citrina* D.A. Reid, *Fungorum Rar. Icon. Color.* 6: 18 & pl. 44a (1972)
  Agaricus murinus Sowerby [*nom. illegit.*, non *A. murinus* Batsch (1787)], *Col. fig. Engl. fung.* 2: pl. 162 (1798)
  Agaricus sowerbei Berk., *Engl. fl.* 5(2): 82 (1836)
E: o **S:** ! **W:** o **NI:** ! **ROI:** !
H: On calcareous soil, in unimproved grassland, meadows, sheep pasture, and downland, also rather commonly in grassy areas on dunes.
D: FAN1: 165-166, NM2: 357 **D+I:** Ph: 116-117 (as *Leptonia incana*), FungEur5: 512 58a **I:** FlDan2: 77C (as *Rhodophyllus euchlorus*), FRIC6: 44b, C&D: 299
Occasional but widespread and may be locally abundant. The distinctively coloured basidiomes and the usually strong smell of 'mouse urine' make this an unmistakable species.

incarnatofuscescens (Britzelm.) Noordel., *Persoonia* 12(4): 461 (1985)
  Agaricus incarnatofuscescens Britzelm., *Hymenomyc. Südbayern* 8: 6 (1891)
  Rhodophyllus leptonipes Kühner & Romagn., *Rev. Mycol. (Paris)* 19(1): 6 (1954)
  Leptonia leptonipes (Kühner & Romagn.) P.D. Orton, *Trans. Brit. Mycol. Soc.* 43(2): 177 (1960)
  Entoloma leptonipes (Kühner & Romagn.) M.M. Moser, *Kleine Kryptogamenflora*, Edn 4: 210 (1978)
  Omphaliopsis leptonipes (Kühner & Romagn.) P.D. Orton, *Mycologist* 5(4): 173 (1991)
E: ! **S:** !
H: On bare soil or loam, often in calcareous sites, usually alongside paths in mixed deciduous woodland, or under dense ground cover of herbs such as *Mercurialis perennis*.
D: FAN1: 175, NM2: 347 **D+I:** FungEur5: 590 85a **I:** Bon: 189 (as *Entoloma leptonipes*), C&D: 301
Rarely reported in Britain. Most records are from southern England but also known as far north as lowland Scotland.

indutoides (P.D. Orton) Noordel., *Persoonia* 12(3): 198 (1984)
  Leptonia indutoides P.D. Orton, *Trans. Brit. Mycol. Soc.* 43(2): 295 (1960)
E: ! **S:** !
H: On acidic soil in unimproved grassland.
D+I: FungEur5: 456 49b
Known only from the type collection from Yorkshire (Ingleton) and a collection from Scotland (Outer Hebrides: Loch Druidibeg). Reported from Malham Tarn but unsubstantiated with voucher material.

indutum Boud., *Bull. Soc. Mycol. France* 16: 194 (1900)
  Leptonia induta (Boud.) P.D. Orton, *Trans. Brit. Mycol. Soc.* 43(2): 177 (1960)
  Pouzaromyces indutus (Boud.) P.D. Orton, *Mycologist* 5(4): 175 (1991)
E: !
H: On calcareous soil in deciduous woodland.
D: FungEur5: 365
A single collection from West Sussex (Duncton Hill) in 1970.

infula (Fr.) Noordel., *Persoonia* 10(4): 503 (1980)
  Agaricus infula Fr., *Spicilegium Pl. neglect.*: 8 (1836)

  Nolanea infula (Fr.) Gillet, *Hyménomycètes*: 421 (1876)
E: ! **S:** ! **W:** ! **NI:** ! **ROI:** !
H: On soil in unimproved grassland and occasionally along roadside verges in woodland.
D: FAN1: 137, NM2: 350 **D+I:** FRIC12: 5-6 90a, FungEur5: 289 36a **I:** C&D: 925
Uncommonly reported but apparently widespread.

infula *var.* **chlorinosum** (Arnolds & Noordel.) Noordel., *Fungi Europaei: Entoloma s.l.* 5: 290 (1992)
  Entoloma chlorinosum Arnolds & Noordel., *Persoonia* 10(5): 287 (1979)
  Nolanea chlorinosa (Arnolds & Noordel.) P.D. Orton, *Mycologist* 5(3): 136 (1991)
E: !
H: On soil in grassland.
D: FAN1: 138 (as *Entoloma chlorinosum*), FungEur5: 290 **D+I:** FRIC12: 8-9 90b (as *E. chlorinosum*)
A single collection from Shropshire (Ludlow, The Patches).

insidiosum Noordel., *Nova Hedwigia Beih.* 91: 182 (1987)
  Leptonia insidiosa (Noordel.) P.D. Orton, *Mycologist* 5(3): 133 (1991)
S: !
H: On acidic soil in unimproved grassland, in boreal or montane habitat.
D: FungEur5: 448
A single British collection (the type) from Westerness (Fort William).

insolitum Noordel., *Nova Hedwigia Beih.* 91: 348 (1987)
S: !
H: On soil, in upland grassland.
D: FungEur5: 602
A single record.

inusitatum Noordel., Enderle & H. Lammers, *Z. Mykol.* 61(2): 192 (1995)
E: !
H: On soil in deciduous woodland.
A single record, from Surrey (Newlands Corner, near Guildford) det. Noordeloos in 1999.

inutile (Britzelm.) Noordel., *Persoonia* 10(4): 512 (1980)
  Nolanea inutilis (Britzelm.) Sacc. & Trotter, *Ber. Naturwiss. Vereins Schwaben Neuberg* 30: 16 (1890)
  Agaricus inutilis Britzelm., *Bot. Centralbl.* 75: 169 (1898)
E: ! **S:** !
H: On acidic sandy soil with *Pinus* spp or other conifers on heathland or moorland, often amongst lichens or mosses.
D: FAN1: 140 **D+I:** FRIC12: 22-23 93e, FungEur5: 310 80a
Rarely reported. Known from England (South Devon) and Scotland (Easterness, Mid-Ebudes and Outer Hebrides).

jennyae Noordel. & Cate [as *E. jennyi*], *Österr. Z. Pilzk.* 3: 29 (1994)
ROI: !
H: On acidic soil in a peat bog, associated with *Calluna, Erica cinerea, Narthecium ossifragum* and *Potentilla erecta*.
Only two collections, including the holotype, from Galway (Kylmore) in 1990 and 1992.

jubatum (Fr.) P. Karst., *Bidrag Kännedom Finlands Natur Folk* 32: 263 (1879)
  Agaricus jubatus Fr., *Syst. mycol.* 1: 196 (1821)
  Trichopilus jubatus (Fr.) P.D. Orton, *Mycologist* 5(4): 176 (1991)
E: o **S:** o **W:** o **NI:** o **ROI:** o **O:** Isle of Man: !
H: On acidic soil, in grassland, also known on dunes and occasionally in woodland clearings.
D: FAN1: 142-143, NM2: 355 **D+I:** FRIC12: 29-31 95b, FungEur5: 399 43a **I:** C&D: 297
Occasional but widespread. Most often collected in northern and western Britain, and almost frequent in Scotland.

juncinum (Kühner & Romagn.) Noordel., *Persoonia* 10: 255 (1979)
  Rhodophyllus juncinus Kühner & Romagn., *Rev. Mycol. (Paris)* 19(1): 5 (1954)
  Nolanea juncina (Kühner & Romagn.) P.D. Orton, *Trans. Brit. Mycol. Soc.* 43(2): 179 (1960)
  Mis.: *Nolanea juncea* sensu auct. mult.
E: ! S: ! ROI: ?
H: On humus or damp soil in grassland or under trees.
D: FAN1: 126, NM2: 350 **D+I:** FungEur5: 243 29a **I:** C&D: 925
Rarely reported.

kervernii (De Guern.) M.M. Moser, *Kleine Kryptogamenflora*, Edn 4: 210 (1978)
  Leptonia kervernii De Guern., in Gillet, *Hyménomycètes*: 413 (1876)
E: ! S: ! NI: !
H: On soil in unimproved grassland.
D+I: FungEur5: 570 76b
Rarely reported. British records from Scotland (various locations), England (Westmorland) and Northern Ireland (Derry).

kuehnerianum Noordel., *Persoonia* 12: 461 (1985)
  Nolanea kuehneriana (Noordel.) P.D. Orton, *Mycologist* 5(3): 138 (1991)
  Mis.: *Nolanea mammosa* sensu NCL p.p. *fide* Orton (1991)
E: ! S: ! W: ! ROI: ?
H: On soil, in unimproved grassland and occasionally on bare soil in open deciduous woodland.
D: FAN1: 119, FungEur5: 222
Rarely reported (usually as *N. mammosa*) but apparently widespread.

lampropus (Fr.) Hesler, *Nova Hedwigia Beih.* 23: 154 (1967)
  Agaricus lampropus Fr., *Observ. mycol.* 1: 19 (1815)
  Leptonia lampropus (Fr.) Quél., *Mém. Soc. Émul. Montbéliard, Sér. 2,* 5: 121 (1872)
E: ! S: ! W: ! NI: ! ROI: !
H: On soil in unimproved grassland, occasionally in grassy areas in woodland and on dunes.
D: NM2: 346, FungEur5: 440 **I:** C&D: 297
Often reported but rarely collected and confused with similar taxa. Most British records are sensu NCL and relate to the more frequent *E. sodale*.

langei Noordel. & T. Borgen, *Persoonia* 12: 292 (1984)
S: ! NI: ? ROI: ?
H: On acidic soil in unimproved grassland.
D: FungEur5: 304-305
British records only from Scotland where it is apparently widespread. Reported from the Republic of Ireland and Northern Ireland but unsubstantiated with voucher material.

lanicum (Romagn.) Noordel., *Persoonia* 11(2): 149 (1981)
  Rhodophyllus lanicus Romagn., *Rev. Mycol. (Paris)* 1(3): 159 (1936)
  Claudopus lanicus (Romagn.) P.D. Orton, *Mycologist* 5(3): 126 (1991)
  Rhodophyllus undatus var. *pusillus* J.E. Lange, *Dansk bot. Ark.* 2(11): 39 (1921)
E: !
H: On calcareous soil in deciduous woodland.
D+I: FungEur5: 615 86g
British collections only from Mid-Lancashire (Gait Barrows NNR) and Surrey (Norbury Park).

lanuginosipes Noordel., *Persoonia* 10: 248 (1979)
E: !
H: On soil, often amongst mosses, in coniferous woodland.
D: FAN1: 135, Myc16(3): 115 **D+I:** FungEur5: 282 34b
British records only from Cumberland (Buttermere, Birtness Wood, Tarn Hows and Thornthwaite Forest).

lividoalbum (Kühner & Romagn.) Kubička, *Česká Mykol.* 29: 27 (1975)
  Rhodophyllus lividoalbus Kühner & Romagn., *Rev. Mycol. (Paris)* 19(1): 6 (1954)

E: ! S: ! NI: !
H: On soil, or heavy calcareous loam, in open deciduous woodland.
D: FAN1: 102, NM2: 353 **D+I:** FungEur5: 145 12 **I:** C&D: 293
Rarely reported.

lividocyanulum Noordel., *Persoonia* 12: 214 (1984)
  Leptonia lividocyanula (Kühner) P.D. Orton [*nom. nud.*], *Trans. Brit. Mycol. Soc.* 43(2): 105 (1960)
  Rhodophyllus lividocyanulus Kühner [*nom. nud.*], *Flora Analytique des Champignons Supérieurs*: 204 (1953)
  Mis.: *Eccilia griseorubella* sensu Bres. (1929) non Lasch (1829)
  Mis.: *Rhodophyllus griseorubellus* sensu J. Lange (1937) p.p. non Lasch (1829)
E: ! S: ! W: ? NI: ? ROI: ?
H: On calcareous or base-rich soil in unimproved grassland.
D: FAN1: 160, NM2: 347 **D+I:** FungEur5: 524 60b **I:** FM3(4): 115
Rarely reported.

longistriatum (Peck) Noordel., *Cryptog. Stud.* 1: 12 (1988)
  Leptonia longistriata Peck, *Rep. (Annual) New York State Mus. Nat. Hist.* 150: 57 (1911)
  Rhodophyllus majusculus Kühner & Romagn., *Rev. Mycol. (Paris)* 19(1): 6 (1954)
  Leptonia majuscula (Kühner & Romagn.) P.D. Orton, *Trans. Brit. Mycol. Soc.* 43(2): 178 (1960)
  Rhodophyllus sarcitulus var. *majusculus* Kühner & Romagn., *Flore Analytique des Champignons Supérieurs*: 204 (1953)
  Entoloma sarcitulum var. *majusculum* (Kühner & Romagn.) Noordel., *Persoonia* 12(4): 462 (1985)
E: ? S: !
H: On soil, in unimproved grassland.
I: SV40: 39
Widely reported from Scotland. Also reported from England (mostly from North Somerset) but unsubstantiated with voucher material.

longistriatum *var.* **sarcitulum** (P.D. Orton) Noordel., *Cryptog. Stud.* 2: 12 (1988)
  Leptonia sarcitula P.D. Orton, *Trans. Brit. Mycol. Soc.* 43(2): 301 (1960)
  Entoloma sarcitulum (P.D. Orton) Arnolds, *Biblioth. Mycol.* 90: 348 (1983) [1982]
  Rhodophyllus sarcitulus var. *spurcifolium* Kühner, *Rev. Mycol. (Paris)* 19(1): 9 (1954)
  Entoloma sarcitulum var. *spurcifolium* (Kühner) Arnolds, *Biblioth. Mycol.* 90: 348 (1983) [1982]
  Mis.: *Leptonia sarcita* sensu Pearson [TBMS 22: 31 (1938)]
E: ! S: ! W: ! ROI: !
H: On soil, in grassland or in open deciduous woodland.
D: FAN1: 170 (as *Entoloma sarcitulum*), NM2: 358 (as *E. sarcitulum*), FungEur5: 574 **I:** C&D: 301 (as *E. sarcitulum*)
Rarely reported but apparently widespread.

lucidum (P.D. Orton) M.M. Moser, *Kleine Kryptogamenflora*, Edn 4: 206 (1978)
  Nolanea lucida P.D. Orton, *Trans. Brit. Mycol. Soc.* 43(2): 331 (1960)
  Mis.: *Nolanea sericea* sensu auct. Brit. p.p.
E: ! S: ! ROI: ?
H: On soil, amongst grass, usually in woodland, occasionally in grassland at woodland edges and rarely in unimproved grassland such as pasture and on dunes.
D: FAN1: 123 **D+I:** Ph: 118 (as *Nolanea lucida*), FungEur5: 232 27c
Rarely reported.

majaloides P.D. Orton, *Trans. Brit. Mycol. Soc.* 43(2): 230 (1960)
E: ? S: !
H: On soil in woodland.
D: FAN1: 105 **D+I:** FungEur5: 177 19
Reported from several sites in Scotland, and a few in England but mostly unsubstantiated with voucher material.

minutum (P. Karst.) Noordel., *Persoonia* 10: 248 (1879)
Nolanea minuta P. Karst., *Meddeland. Soc. Fauna Fl. Fenn.* 5: 10 (1879)
Rhodophyllus minutus (P. Karst.) J.E. Lange, *Dansk. Bot. Ark.* 2(11): 37 (1921)
Rhodophyllus minutus var. *polymorphus* Romagn., *Rev. Mycol. (Paris)* 19(1): 7 (1954)
E: ! **S:** ! **W:** ?
H: On soil, in damp woodland, often associated with *Alnus*, *Betula* and *Salix* spp, frequently under or amongst tussocks of grass.
D: FAN1: 126, NM2: 349 **D+I:** FRIC12: 19-20 93a, FungEur5: 248 29b **I:** C&D: 925
Rarely reported, but minute and easily overlooked.

mougeotii (Fr.) Hesler, *Nova Hedwigia Beih.* 23: 158 (1967)
Eccilia mougeotii Fr., Quélet, *Mém. Soc. Émul. Montbéliard, Sér. 2,* 5: 345 (1873)
Entoloma ardosiacum var. *mougeotii* (Fr.) A. Pearson & Dennis, *Trans. Brit. Mycol. Soc.* 31: 170 (1948)
Leptonia mougeotii (Fr.) P.D. Orton, *Trans. Brit. Mycol. Soc.* 43(2): 290 (1960)
Leptonia serrulata var. *berkeleyi* Maire, *Bull. Soc. Mycol. France* 26: 176 (1910)
E: ! **S:** ! **W:** ! **NI:** ! **ROI:** !
H: On soil, in unimproved grassland.
D: FAN1: 164-165, NM2: 346 **D+I:** FungEur5: 497 56a & 87h **I:** C&D: 299
Rarely reported, but apparently widespread.

mougeotii *var.* **fuscomarginatum** Noordel., *Nova Hedwigia Beih.* 91: 237 (1987)
E: ! **S:** ! **ROI:** !
H: On calcareous soil in unimproved grassland.
D: FungEur5: 500
Rarely reported.

mutabilipes Noordel. & Liiv, *Persoonia* 15(1): 30 (1992)
S: !
D+I: FungEur5: 531 61
Rarely reported.

myrmecophilum (Romagn.) M.M. Moser, *Kleine Kryptogamenflora, Edn 4*: 197 (1978)
Rhodophyllus myrmecophilus Romagn., *Bull. mens. Soc. linn. Lyon*: 386 (1974)
E: !
H: On soil in deciduous woodland.
D: FAN1: 107, NBA9: 552, NM2: 353 **D+I:** FungEur5: 168 17
Rarely reported.

neglectum (Lasch) Arnolds, *Guida alla Determinazione dei Funghi*: 224 (1986)
Agaricus neglectus Lasch, *Linnaea* 4: 401 (1829)
Clitopilus neglectus (Lasch) P. Kumm., *Führ. Pilzk.*: 97 (1871)
Paraleptonia neglecta (Lasch) P.D. Orton, *Mycologist* 5(4): 174 (1991)
Agaricus cancrinus Fr., *Epicr. syst. mycol.*: 150 (1838)
Eccilia cancrina (Fr.) Ricken, *Blätterpilze Deutschl.*: 301 (1913)
Clitopilus cancrinus (Fr.) Quél., *Mém. Soc. Émul. Montbéliard, Sér. 2,* 5: 247 (1872)
E: ! **S:** !
H: On soil, in damp woodland or grassy areas.
D: FAN1: 175, NM2: 357 **D+I:** FungEur5: 596 71c
British records from England (Northamptonshire, North Somerset, South Devon, Staffordshire, Surrey, West Kent, and West Suffolk) and Scotland (North Ebudes: Isle of Rhum and the Shetland Isles).

nigellum (Quél.) Noordel., *Fungi Europaei 5, Entoloma s.l.*: 618 (1992)
Eccilia nigella Quél., *Compt. Rend. Assoc. Franç. Avancem. Sci.* 12(2): 499 (1884) [1883]
Claudopus nigellus (Quél.) P.D. Orton, *Mycologist* 5(3): 126 (1991)
E: ! **S:** ! **W:** !

H: On sandy soil on fixed dunes, associated with *Salix repens*.
D: NCL3: 227 (as *Eccilia nigella*), FungEur5: 618
Rarely reported. British collections from England (South Devon: Dawlish Warren), Wales (Cardiganshire: Ynyslas) and Scotland (East Lothian: Aberlady Bay).

nigroviolaceum (P.D. Orton) Hesler, *Nova Hedwigia Beih.* 23: 50 (1967)
Leptonia nigroviolacea P.D. Orton, *Trans. Brit. Mycol. Soc.* 43(2): 296 (1960)
E: ? **S:** !
H: On acidic soil in unimproved grassland.
D: FungEur5: 505
British records only from three sites in Scotland (Easterness: Tomich, Outer Hebrides: Isle of Mingulay, and South Aberdeen: Braemar). Reported from England (South Devon: Membury) but unsubstantiated with voucher material.

niphoides Noordel., *Persoonia* 12(4): 459 (1985)
Mis.: *Entoloma speculum* sensu Cooke [Ill. Brit. fung. 342 (308) Vol. 3 (1884)]
E: ! **W:** !
H: On soil, associated with various *Rosaceae*, usually *Crataegus* or *Prunus* spp, usually spring-fruiting.
D: FAN1: 99-100, NM2: 343 **D+I:** FungEur5: 134 7a
A distinctive species, known with certainty in Britain only from Huntingdonshire, West Sussex, and Pembrokeshire and probably genuinely rare.

nitens (Velen.) Noordel., *Persoonia* 10: 252 (1979)
Nolanea nitens Velen., *České Houby* III: 627 (1921)
E: ? **S:** ! **NI:** ?
H: On soil in open deciduous woodland, often in grassy areas.
D: FAN1: 127, NM2: 350 **D+I:** FRIC12: 21-22 93d, FungEur5: 246 30b & 78c
In Britain, known with certainty only from Peebleshire (Dawyck Botanic Garden). Reported from Surrey (Windsor Great Park) but the collection in herb. K is annotated 'cf' *E. nitens*. Reported from Northern Ireland but lacking voucher material.

nitidum Quél., *Compt. Rend. Assoc. Franç. Avancem. Sci.* 11: 391 (1883) [1882]
E: ! **S:** ! **NI:** !
H: On soil, often in boggy areas in coniferous or mixed coniferous and deciduous woodland.
D: FAN1: 97, NM2: 344 **D+I:** FungEur5: 116 4 **I:** FlDan2: 74A (as *Rhodophyllus nitidus*), C&D: 291
Rarely reported and less often collected but apparently widespread. Mostly reported from northern areas, although recorded as far south as East Sussex (Ashdown Forest).

occultopigmentatum Noordel. & Arnolds, *Persoonia* 10: 292 (1979)
E: ? **S:** !
H: On soil, in unimproved grassland.
D: FAN1: 135-136 **D+I:** FRIC12: 18-19 92c, FungEur5: 284 80b
In Britain, known with certainty only from Scotland. The single collection from England (Surrey: Englefield Green, Cooper's Hill) is annotated as 'query *E. occultopigmentatum*'.

ochromicaceum Noordel. & Liiv, *Persoonia* 15(1): 27 (1992)
S: !
H: On calcareous soil in unimproved grassland.
D+I: FungEur5: 580 59a
Collected from two sites in Scotland (Wester Ross: Beinn Eighe and Perthshire: Schiehallion) in 2000.

olorinum (Romagn. & J. Favre) Noordel., *Persoonia* 10: 260 (1979)
Rhodophyllus olorinus Romagn. & J. Favre, *Rev. Mycol. (Paris)* 3(2-3): 75 (1938)
W: !
H: On soil, amongst grass and mosses, in unimproved grassland or deciduous woodland.
D: FAN1: 149-150 **D+I:** FungEur5: 419 46b
Known in Britain from a single collection from Coed Rhnd. A record from Yorkshire (Crimsworth Dean) was based on misidentified material.

opacum (Velen.) Noordel., *Nova Hedwigia Beih.* 91: 135 (1987)
 Clitocybe opaca Velen., *České Houby* II: 268 (1920)
E: ! **W:** !
H: On soil in grassland.
D: FungEur5: 390
British records from Warwickshire (Welford on Avon) and Merionethshire (Coed Crafnant).

ortonii Arnolds & Noordel., *Persoonia* 10: 292 (1979)
 Nolanea farinolens P.D. Orton, *Trans. Brit. Mycol. Soc.* 43(2): 330 (1960)
 Entoloma farinolens (P.D. Orton) M.M. Moser [*nom. illegit.*], *Kleine Kryptogamenflora*, Edn 4: 206 (1978)
 Mis.: *Nolanea sericea* sensu auct. Brit. p.p.
E: ! **S:** ! **W:** !
H: On soil in unimproved grassland or on lawns, occasionally in marshy areas in woodland, and in dune slacks.
D: FAN1: 124 **D+I:** FRIC12: 16-18 92b, FungEur5: 240 28a
Rarely reported, but apparently widespread in Britain.

pallens (Maire) Arnolds, *Biblioth. Mycol.* 90: 341 (1983) [1982]
 Eccilia pallens Maire, *Publ. Inst. Bot. Barcelona* 3: 96 (1937)
 Paraleptonia pallens (Maire) P.D. Orton, *Mycologist* 5(4): 174 (1991)
E: !
H: On soil in grassland.
D: FAN1: 176, FungEur5: 598
A single British record from West Sussex (Arundel) in 1976.

papillatum (Bres.) Dennis, *Bull. Soc. Mycol. France* 69: 162 (1953)
 Nolanea papillata Bres., *Fungi trident.* 1: 75 (1887)
E: ! **S:** ! **W:** ! **NI:** ! **ROI:** ! **O:** Channel Islands: !
H: On soil, in unimproved grassland, on roadside verges, and occasionally grassy areas in woodland.
D: FAN1: 122, NM2: 350 **D+I:** FRIC12: 13--14 91d, FungEur5: 228 27a **I:** C&D: 925
Frequent and widespread.

parasiticum (Quél.) Kreisel, *Feddes Repert.* 95: 699 (1984)
 Leptonia parasitica Quél., *Bull. Soc. Bot. France* 25: 287 (1878)
 Claudopus parasiticus (Quél.) Ricken, *Blätterpilze Deutschl.*: 304 (1913)
 Mis.: *Claudopus depluens* sensu Cooke [Ill. Brit. fung. 371 (343) Vol. 3 (1884)] non al.
E: ! **S:** !
H: On decayed wood, herbaceous debris, occasionally on dying and moribund fungi such as *Cantharellus cibarius*, *Polyporus squamosus*, *Trametes versicolor* and possibly various *Corticiaceae*.
D: FAN1: 172, BFF6: 110-111 (as *Claudopus parasiticus*), FungEur5: 609
Rarely reported. British records from widely scattered sites in England (Cumberland, Hertfordshire, North Essex, South Devon, South Hampshire, Staffordshire, Warwickshire and Worcestershire) and Scotland (Easterness).

parkensis (Fr.) Noordel., *Persoonia* 11: 262 (1979)
 Agaricus parkensis Fr., *Kongl. Vetensk. Acad. Handl.* 1851: 45 (1851)
 Eccilia parkensis (Fr.) P. Karst., *Bidrag Kännedom Natur Folk* 32: 289 (1879)
 Omphaliopsis parkensis (Fr.) P.D. Orton, *Mycologist* 5(4): 173 (1991)
E: ! **S:** !
H: On soil, amongst grass, in grassland, also occasionally on humus in open woodland.
D: FungEur5: 594 **D+I:** Myc14(2): 85 (as *Omphaliopsis parkensis*) **I:** C&D: 301 (as *Entoloma parkense*)
British collections from England (Cheshire and East Sussex) and a single collection from Scotland. Reported from North Devon and North Somerset but unsubstantiated with voucher material.

percandidum Noordel., *Nordic J. Bot.* 2(2): 161 (1982)

Rhodophyllus omphaliformis Romagn., *Rev. Mycol. (Paris)* 19(1): 7 (1954)
 Alboleptonia omphaliformis (Romagn.) P.D. Orton, *Mycologist* 5(3): 125 (1991)
E: !
H: On soil, in damp deciduous woodland.
D: NM2: 344 **D+I:** FungEur5: 420 46c
Known from two sites, in East Kent (Gorham) and Yorkshire (near Beverley). Also reported from Somerset, but 'on wood' and possibly misidentified.

phaeocyathus Noordel., *Persoonia* 12(4): 461 (1985)
 Eccilia tristis Bres., Saccardo, *Syll. fung.* 9: 89 (1891)
E: !
H: On sandy soil with *Ammophila arenaria* on dunes.
D: FAN1: 174 **D+I:** FungEur5: 623 72b
Known in Britain with certainty only from East Kent (Sandwich Bay). Reported from Wales but misdetermined. Included in *Eccilia rusticoides* in NCL, but that species lacks cystidia.

plebeioides (Schulzer) Noordel., *Persoonia* 12(4): 462 (1985)
 Agaricus plebeioides Schulzer, *Verh. zool.-bot. Ges. Wien* 26: 428 (1876)
E: !
H: On soil in grassland.
D: FAN1: 147-148 **D+I:** FungEur5: 393 42b
A single British collection, from Warwickshire (Birmingham, Sutton Park) in 1996.

plebejum (Kalchbr.) Noordel., *Persoonia* 12(4): 462 (1985)
 Agaricus plebejus Kalchbr., *Icon. select. Hymenomyc. Hung.*: 22 (1874)
 Inocephalus plebejus (Kalchbr.) P.D. Orton, *Mycologist* 5(3): 130 (1991)
 Agaricus erophilus Fr., *Hymenomyc. eur.*: 190 (1874)
 Entoloma erophilum (Fr.) P. Karst., *Bidrag Kännedom Finlands Natur Folk* 32: 259 (1879)
 Entoloma erophilum var. *pyrenaicum* Quél., *Compt. Rend. Assoc. Franç. Avancem. Sci.* 13: 279 (1885) [1884]
 Entoloma pyrenaicum (Quél.) Sacc., *Syll. fung.* 5: 682 (1887)
E: ! **S:** ! **ROI:** ? **O:** Isle of Man: ?
H: On soil, in damp deciduous woodland, often associated with *Salix* spp. Spring-fruiting.
D: FAN1: 147 **D+I:** FungEur5: 388 42a
British collections from England (Surrey: Englefield Green, Kingswood Hall) and Scotland (Westerness: Loch Teacuis). Reported from Republic of Ireland and the Isle of Man but unsubstantiated with voucher material.

pleopodium (DC.) Noordel., *Persoonia* 12: 459 (1985)
 Agaricus pleopodius DC., *Fl. Fr.* 2: 194 (1805)
 Agaricus icterinus Fr., *Syst. mycol.* 1: 207 (1821)
 Nolanea icterina (Fr.) P. Kumm., *Führ. Pilzk.*: 95 (1871)
 Entoloma icterinum (Fr.) M.M. Moser, *Kleine Kryptogamenflora*, Edn 4: 205 (1978)
 Rhodophyllus icterinus f. *gracillimus* J.E. Lange, *Dansk. Bot. Ark.* 2(11): 37 (1921)
E: ! **S:** ! **W:** ! **NI:** ! **O:** Channel Islands: !
H: On soil, in open deciduous woodland, especially on nitrogenous soil amongst or near to *Urticia dioica* or decayed vegetation.
D: FAN1: 141, NM2: 348 **D+I:** FungEur5: 315 37a **I:** FlDan2: 78A (as *Rhodophyllus icterinus*), C&D: 295 (as *Entoloma icterinum*)
Uncommonly reported but apparently widespread. A distinctive species with a strong and characteristic smell of amyl acetate (pear drops), and unlikely to be misidentified.

poliopus (Romagn.) Noordel., *Persoonia* 10: 262 (1979)
 Rhodophyllus poliopus Romagn., *Rev. Mycol. (Paris)* 19(1): 8 (1954)
 Leptonia poliopus (Romagn.) P.D. Orton, *Mycologist* 5(3): 134 (1991)
E: ! **S:** ! **W:** ! **NI:** ! **ROI:** ?
H: On soil, amongst grass and mosses, in unimproved grassland, also occasionally in grassy places in open woodland, and on dunes.

D: FAN1: 159, NM2: 347 **D+I:** FungEur5: 534 62b **I:** C&D: 299
Rarely reported. Apparently widespread.

poliopus *var.* **discolor** Noordel., *Persoonia* 12(4): 460 (1985)
E: ! **S:** ! **ROI:** ?
H: On soil, amongst grass, in unimproved grassland and
occasionally in open deciduous woodland.
D: FAN1: 159, FungEur5: 537
Rarely reported. Described from Scotland (Perthshire:
Kirkmichael). Recorded from England (North Somerset) on
several occasions, and from Republic of Ireland but most
reports are unsubstantiated with voucher material.

poliopus *var.* **parvisporigerum** Noordel., *Persoonia* 12(4):
460 (1985)
E: ! **S:** ! **W:** ! **ROI:** ?
H: On soil amongst grass, in lawn turf, or unimproved
grassland.
D: FAN1: 159-160, NM2: 347, FungEur5: 538
Rarely reported. Differs from the typical variety in the small
spores.

politum (Pers.) Donk, *Bull. Jard. Bot. Buitenzorg, Sér. 3*, 18:
158 (1949)
Agaricus politus Pers. [non *A. politus* Bolton (1788)], *Syn.
meth. fung.*: 465 (1801)
Eccilia polita (Pers.) P. Kumm., *Führ. Pilzk.*: 95 (1871)
Leptonia polita (Pers.) P.D. Orton, *Trans. Brit. Mycol. Soc.*
43(2): 178 (1960)
Leptonia pernitrosa P.D. Orton, *Trans. Brit. Mycol. Soc.*
43(2): 297 (1960)
Eccilia pernitrosa (P.D. Orton) P.D. Orton, *Notes Roy. Bot.
Gard. Edinburgh* 29(1): 76 (1969)
Entoloma pernitrosum (P.D. Orton) P.D. Orton, *Mycologist*
5(3): 128 (1991)
Entoloma politum f. *pernitrosum* (P.D. Orton) Noordel.,
Persoonia 11(2): 211 (1981)
Mis.: *Entoloma nidorosum* sensu auct. p.p.
E: o **S:** o **W:** ! **NI:** ! **ROI:** ?
H: On soil, often in damp areas in woodland, often associated
with *Alnus* or *Salix* spp.
D: FAN1: 108, NM2: 351 **D+I:** FungEur5: 189 20a
Occasional but widespread and may be locally abundant.

porphyrophaeum (Fr.) P. Karst., *Bidrag Kännedom Finlands
Natur Folk* 32: 259 (1879)
Agaricus porphyrophaeus Fr., *Monogr. hymenomyc. Suec.* 1:
473 (1857)
Rhodophyllus porphyrophaeus (Fr.) J.E. Lange, *Dansk. Bot.
Ark.* 2(11): 28 (1921)
Trichopilus porphyrophaeus (Fr.) P.D. Orton, *Mycologist* 5(4):
176 (1991)
Mis.: *Entoloma jubatum* sensu Cooke [Ill. Brit. fung. 333
(317) (1884)]
E: o **S:** o **W:** o **NI:** o **ROI:** o **O:** Channel Islands: ! Isle of Man:
!
H: On acidic soil, amongst short turf, in unimproved grassland.
D: FAN1: 143-144, NM2: 355 **D+I:** Ph: 116, FRIC12: 33-35 96,
FungEur5: 407 44 **I:** FlDan2: 73D (as *Rhodophyllus
porphyrophaeus*), C&D: 297
Occasional but widespread, and may be locally common in
upland pastures, especially in northern and western Britain.
Infrequent in southern counties.

pratulense Noordel., *Beih. Nova Hedwigia* 91: 89 (1987)
E: !
H: In grassland and open grassy areas in deciduous woodland.
D+I: FungEur5: 301-305
A single collection, from West Gloucestershire (Forest of Dean)
in 1988.

prunuloides (Fr.) Quél., *Mém. Soc. Émul. Montbéliard, Sér. 2*,
5: 117 (1872)
Agaricus prunuloides Fr., *Syst. mycol.* 1: 198 (1821)
Rhodophyllus prunuloides (Fr.) Quél., *Enchir. fung.*: 57
(1886)

Entoloma inocybiforme Bon [*nom. illegit.*], *Doc. Mycol.*
10(37-38): 90 (1980) [1979]
Entoloma inopiliforme Bon, *Doc. Mycol.* 12(46): 32 (1982)
Mis.: *Entoloma repandum* sensu Lange (FlDan2: 95 & pl.
73A) and sensu auct. Brit.
E: o **S:** o **W:** ! **NI:** ! **ROI:** ! **O:** Isle of Man: !
H: On soil, amongst grass, in unimproved or poorly fertilised
sheep pasture.
D: FAN1: 94-95, NM2: 354 **D+I:** FungEur5: 110 1 **I:** FlDan2:
73B (as *Rhodophyllus prunuloides*), Bon: 193 (as *Entoloma
inopiliforme*), C&D: 291
Occasional but widespread.

pseudocoelestinum Arnolds, *Biblioth. Mycol.* 90: 341 (1983)
[1982]
S: !
H: On soil, in grassland, often with *Molinia caerulea*.
D: FAN1: 163 **D+I:** FungEur5: 493 54a
Recently collected in Scotland.

pseudoturci Noordel., *Persoonia* 12(3): 215 (1984)
E: ! **S:** ! **NI:** ? **ROI:** ?
H: On soil, amongst grass, in unimproved grassland.
D: FAN1: 169, NM2: 358 **D+I:** FungEur5: 586 70b
Rarely reported.

pulvereum Rea, *Trans. Brit. Mycol. Soc.* 3: 170 (1907)
Leptonia pulverea (Rea) P.D. Orton, *Trans. Brit. Mycol. Soc.*
43(2): 178 (1960)
Pouzaromyces pulvereus (Rea) P.D. Orton, *Mycologist* 5(4):
175 (1991)
E: !
H: On soil or humus in woodland, associated with *Quercus* spp.
D: FAN1: 115-116 **D+I:** FungEur5: 357 39c
Lacking type material, but accepted by Noordeloos [Persoonia
10: 223 (1979)] on the basis of Rea's description and plate
which agree with later collections from the Netherlands and
England (Banstead: Nork, Park Wood).

queletii (Boud.) Noordel., *Cryptog. Mycol.* 4(1): 37 (1983)
Leptonia queletii Boud., *Bull. Soc. Bot. France* 24: 307 (1877)
E: ! **S:** ? **ROI:** ?
H: On soil, in damp deciduous woodland. Also known on
decayed wood (fallen tree trunks or stumps) in continental
Europe.
D: FAN1: 171-172, NM2: 344 **D+I:** FungEur5: 567 68 **I:** C&D:
301
Rarely reported. Known from Cumberland (Roudsea Woods),
Mid-Lancashire (near Carnforth) and South Devon (Membury).

querquedula (Romagn.) Noordel., *Persoonia* 11(4): 471 (1982)
Rhodophyllus querquedula Romagn., *Rev. Mycol. (Paris)*
19(1): 8 (1954)
Leptonia querquedula (Romagn.) P.D. Orton, *Kew Bull.* 31:
711 (1976)
E: ! **S:** ? **NI:** !
H: On soil in damp areas in mixed deciduous woodland.
D: FungEur5: 478, NBA8: 592-593 **I:** C&D: 299
Rarely reported, the majority of reports unsubstantiated with
voucher material.

reaae (Maire) Noordel., *Fungi Europaei: Entoloma s.l.* 5: 201
(1992)
Leptonia reaae Maire, *Trans. Brit. Mycol. Soc.* 3(3): 170
(1910) [1909]
E: ! **S:** ?
H: On acidic soil, amongst grass, in boggy grassland.
D: FungEur5: 201
Only known from two collections, from Derbyshire (Grindleford)
and West Gloucestershire (Forest of Dean). Dubiously reported
from Scotland (Orkney).

reginae Noordel. & Chrispijn, *Blumea* 41(1): 7 (1996)
E: !
H: On gravelly soil in grassland.
A single record from South Hampshire (New Forest, near
Beaulieu).

rhodocylix (Lasch) M.M. Moser, *Kleine Kryptogamenflora*, Edn
4: 210 (1978)
   Agaricus rhodocylix Lasch, *Linnaea* 4: 542 (1829)
   Eccilia rhodocylix (Lasch) P. Kumm., *Führ. Pilzk.*: 95 (1871)
   Claudopus rhodocylix (Lasch) P.D. Orton, *Mycologist* 5(3):
     126 (1991)
E: ! S: ! W: !
H: On soil, in woodland and also (apparently) on decayed wood.
D: FAN1: 173-174, NM2: 356 **D+I:** FungEur5: 621 81c & 86a **I:**
C&D: 301
Reported from widely scattered sites in England (Berkshire,
Buckinghamshire, East Kent, Lincolnshire, South Hampshire,
Warwickshire and West Sussex), Scotland (Easterness, Isle of
Mull and Perthshire) and also from Wales, but many records
are unsubstantiated with voucher material.

rhodopolium (Fr.) P. Kumm., *Führ. Pilzk.*: 98 (1871)
   Agaricus rhodopolius Fr., *Observ. mycol.* 2: 103 (1818)
   Agaricus nidorosus Fr., *Epicr. syst. mycol.*: 148 (1838)
   Entoloma nidorosum (Fr.) Quél., *Mém. Soc. Émul.*
     Montbéliard, Sér. 2, 5: 119 (1872)
   Entoloma rhodopolium f. *nidorosum* (Fr.) Noordel., *Beitr.*
     Kenntn. Pilze Mitteleurop. 5: 43 (1989)
   Mis.: *Entoloma speculum* sensu Lange (FlDan2: 96 & pl. 75B)
E: o S: o W: ! NI: ! ROI: !
H: On soil in damp deciduous woodland.
D: NM2: 101-102 (as *Entoloma nidorosum*), NM2: 354 **D+I:** Ph:
115 (as *E. nidorosum*), Ph: 115, FungEur5: 141 10 & 11 **I:**
FlDan2: 75A (as *Rhodophyllus rhodopolius*), C&D: 293
Occasional but widespread throughout Britain. The forma
nidorosum, often distinguished as an autonomous species, is
also common, and distinguished only by the slightly nitrous
smell of the basidiomes.

rhombisporum (Kühner & Boursier) E. Horak, *Sydowia* 28: 228
(1976) [1975/1976]
   Leptonia rhombispora Kühner & Boursier, *Bull. Soc. Mycol.*
     France 45: 276 (1929)
   Nolanea rhombispora (Kühner & Boursier) P.D. Orton, *Notes*
     Roy. Bot. Gard. Edinburgh 41(3): 603 (1984)
E: ! S: ! W: ! ROI: ?
H: In marshy areas, often growing amongst *Sphagnum* moss.
D: FAN1: 121-122, NM2: 348, NBA8: 603-604 **D+I:** FungEur5:
379 26b **I:** C&D: 295
Rarely reported. The majority of reports are from other habitats
and unsubstantiated with voucher material.

rhombisporum *var.* **floccipes** Noordel., *Nova Hedwigia Beih.*
91: 77 (1987)
S: !
H: On soil in unimproved grassland.
D: FungEur5: 380
Recently collected in Scotland, and apparently widespread there.

roseum (Longyear) Hesler, *Nova Hedwigia Beih.* 23: 165 (1967)
   Leptonia rosea Longyear, *Rep. (Annual) Michigan Acad. Sci.*
     3: 59 (1902)
   Entoloma griseocyaneum var. *roseum* Maire, *Trans. Brit.*
     Mycol. Soc. 3(3): 170 (1910) [1909]
E: ! S: ! W: ! ROI: !
H: On soil, amongst short turf in unimproved grassland and
occasionally with *Salix repens* on dunes or in dune slacks.
D: NCL3: 300 (as *Leptonia rosea*), FAN1: 171, NM2: 344 **D+I:**
Myc3(3): 143 (as *L. rosea*), FungEur5: 516 58c **I:** C&D: 299
Rarely reported but apparently widespread.

rufocarneum (Berk.) Noordel., *Persoonia* 12(4): 462 (1985)
   Agaricus rufocarneus Berk., *Engl. fl.* 5(2): 82 (1836)
   Nolanea rufocarnea (Berk.) Sacc., *Syll. fung.* 5: 720 (1887)
   Leptonia rufocarnea (Berk.) P.D. Orton, *Mycologist* 5(3): 135
     (1991)
   Leptonia andrianae Bres., *Ann. Mycol.* 18: 3 (1920)
E: ! S: ? W: ? NI: ?
H: On soil, in deciduous woodland.
D+I: FungEur5: 566 76a
Lacking type material but accepted by Noordeloos (1985) on the
   basis of French collections agreeing with the type description

and the illustation by Cooke 364 (378) Vol. 3 (1885). Old
collections in herb. K from Scotland and Wales need re-
examination. Reported from Northern Ireland but
unsubstantiated with voucher material.

rugosum (Malençon) Bon, *Doc. Mycol.* 13(49): 45 (1983)
   Eccilia rugosa Malençon, *Bull. Soc. Mycol. France* 58: 40
     (1942)
E: !
H: On soil in woodland.
D+I: FungEur5: 628 72b **I:** C&D: 301
British collections only from Mid-Lancashire (Gait Barrows NNR),
Surrey (Esher, Claremont), and West Norfolk (Thompson
Common).

rusticoides (Gillet) Noordel., *Persoonia* 11(2): 150 (1981)
   Eccilia rusticoides Gillet, *Hyménomycètes.* 425 (1876)
   Claudopus rusticoides (Gillet) P.D. Orton, *Mycologist* 5(3):
     126 (1991)
E: ! S: ! W: ! NI: !
H: On soil in gardens, parks, in woodland and on dunes.
D: FAN1: 174, NM2: 356 **D+I:** FungEur5: 622 72a **I:** C&D: 301
Accepted with reservations. Some of the few British records may
   be *Entoloma phaeocyathus*. Sensu NCL covered both species.

sacchariolens (Romagn.) Noordel., *Persoonia* 10(4): 474
(1980)
   Rhodophyllus sacchariolens Romagn., *Bull. mens. Soc. linn.*
     Lyon 43 (No. spec.): 385 (1974)
   Nolanea sacchariolens (Romagn.) P.D. Orton, *Mycologist*
     5(4): 172 (1991)
S: !
H: On acidic soil in unimproved grassland.
D: FAN1: 128 **D+I:** FungEur5: 262 32b
Collected from several widely scattered localities in 1998 and
2000.

saepium (Noulet & Dass.) Richon & Roze, *Atlas champ.*: 92
(1888)
   Agaricus saepius Noulet & Dass., *Traité champ.*: 155 (1838)
   Mis.: *Entoloma saundersii* sensu Phillips (1981)
E: !
H: On soil or loam, associated with various *Rosaceae* in
woodland and hedgerows. Fruiting in spring and summer.
D: FAN1: 100, NM2: 354 **D+I:** FungEur5: 136 8 (as *Entoloma*
sepium) **I:** C&D: 291 (as *E. sepium*)
Rarely reported and poorly known in Britain. Possibly confused
with *E. clypeatum*.

saundersii (Fr.) Sacc., *Syll. fung.* 5: 689 (1887)
   Agaricus saundersii Fr., *J. Bot.* 11: 205 (1873)
   Mis.: *Agaricus majalis* sensu Saunders *et al.* [Mycol. Illus. pl.
     46 (1872)]
E: !
H: On loam or clay soil, spring-fruiting and associated with
species of *Ulmus* in woodland or hedgerows.
D: FAN1: 100-101 **D+I:** FungEur5: 138 9 & 77 **I:** C&D: 291
Rarely reported. The material illustrated in Phillips (Ph: 115) is
not this but *E. saepium fide* Noordeloos (FungEur5: 138).

scabiosum (Fr.) Quél., *Compt. Rend. Assoc. Franç. Avancem.*
Sci. 14(2): 445 (1886) [1885]
   Agaricus scabiosus Fr., *Spicilegium Pl. neglect.*: 3 (1836)
   Trichopilus scabiosus (Fr.) P.D. Orton, *Mycologist* 5(4): 176
     (1991)
E: ! S: !
H: On soil or humus in deciduous woodland.
D: FAN1: 146, NM2: 355 **D+I:** FungEur5: 409 45
Rarely reported. A distinctive species, unlikely to be overlooked.

sericatum (Britzelm.) Sacc., *Syll. fung.* 11: 45 (1895)
   Agaricus sericatus Britzelm., *Bot. Centralbl.* 15-17: 8 (1893)
   Mis.: *Entoloma rhodopolium* sensu auct. Brit.
E: ! S: ! W: ! NI: ! ROI: !
H: On soil in damp deciduous woodland.
D: FAN1: 105-106, NM2: 353 **D+I:** FungEur5: 174 18
Rarely reported but apparently widespread.

sericellum (Fr.) P. Kumm., *Führ. Pilzk.*: 97 (1871)
Agaricus sericeus β *sericellus* Fr., *Observ. mycol.* 2: 145 (1818)
Leptonia sericella (Fr.) Barbier, *Bull. Soc. Mycol. France* 27: 178 (1911)
Aboleptonia sericella (Fr.) Largent & R.G. Benedict, *Mycologia* 62: 446 (1970)
Agaricus carneoalbus With., *Arr. Brit. pl. ed.3* 4: 170 (1796)
Eccilia carneoalba (With.) Quél., *Bull. Soc. Bot. France* 26: 49 (1880) [1879]
Rhodophyllus carneoalbus (With.) Quél., *Enchir. fung.*: 62 (1886)
Clitopilus carneoalbus (With.) Sacc., *Syll. fung.* 5: 704 (1887)
Agaricus molluscus Lasch, *Linnaea* 3: 398 (1828)
Eccilia mollusca (Lasch) D.A. Reid [non sensu Reid], *Trans. Brit. Mycol. Soc.* 41(4): 433 (1958)
Aboleptonia rubellotincta Largent & Watling, *Mycologia* 78(1): 132 (1986)
Entoloma sericellum var. *decurrens* Boud., *Icon. Mycol.* 49: pl. 94 (1904)
Leptonia sericella var. *decurrens* (Boud.) Rea, *Brit. basidiomyc.*: 347 (1922)
Agaricus sericellus var. *lutescens* Fr., *Ic. Hymenomyc.* 1: pl. 106 (1867)
E: c **S:** c **W:** c **NI:** c **ROI:** c **O:** Isle of Man: !
H: On soil, often amongst mosses, in grassland, in open deciduous woodland and on dunes.
D: FAN1: 149, NM2: 344 **D+I:** Ph: 116-117 (as *Leptonia sericella*), FungEur5: 415 46a **I:** Bon: 189, C&D: 301
Common and widespread.

sericeoides (J.E. Lange) Noordel., *Persoonia* 10(4): 483 (1980)
Rhodophyllus sericeoides J.E. Lange, *Fl. agaric. danic.* 5, *Taxonomic Conspectus*: VIII (1940)
Leptonia sericeoides (J.E. Lange) P.D. Orton, *Trans. Brit. Mycol. Soc.* 43(2): 178 (1960)
Nolanea sericeoides (J.E. Lange) P.D. Orton, *Mycologist* 5(4): 172 (1991)
E: ! **NI:** ?
H: On soil or loam in open deciduous woodland and scrub.
D: FAN1: 130, NM2: 352 **D+I:** FungEur5: 259 28b
British collections from Buckinghamshire (Wootton Underwood, Rushbeds Wood) and South Devon (Membury). Reported elsewhere but unsubstantiated with voucher material.

sericeonitens (P.D. Orton) Noordel., *Persoonia* 10(4): 459 (1980)
Nolanea sericeonitens P.D. Orton, *Trans. Brit. Mycol. Soc.* 43(2): 333 (1960)
E: !
H: On soil in grasslands, including lawns and on dunes.
D: FAN1: 123-124, FungEur5: 234
Only two British records, from South Devon (Dawlish Warren) and Surrey (Richmond).

sericeum (Bull.) Quél., *Mém. Soc. Émul. Montbéliard, Sér. 2,* 5: 119 (1872)
Agaricus sericeus Bull., *Herb. France*: pl. 413 (1789)
Rhodophyllus sericeus (Bull.) Quél., *Enchir. fung.*: 59 (1886)
Nolanea sericea (Bull.) P.D. Orton, *Trans. Brit. Mycol. Soc.* 43(2): 179 (1960)
Rhodophyllus sericeus var. *nolaniformis* Kühner & Romagn., *Rev. Mycol. (Paris)* 19(1): 9 (1954)
E: c **S:** c **W:** c **NI:** c **ROI:** c **O:** Channel Islands: !
H: On soil in grassland, including fertilised pasture, lawns and playing fields.
D: FAN1: 129, NM2: 352 **D+I:** FungEur5: 256 31 **I:** C&D: 925
Common and widespread.

sericeum *var.* **cinereo-opacum** Noordel., *Persoonia* 10(4): 482 (1980)
E: ! **S:** ! **W:** !
H: On soil in grassland, including lawns.
D: FungEur5: 258

Rarely reported, the majority of reports unsubstantiated with voucher material. This may include *Nolanea radiata* sensu NCL fide Orton (1991).

serrulatum (Fr.) Hesler, *Nova Hedwigia Beih.* 21: 140 (1967)
Agaricus serrulatus Fr., *Observ. mycol.* 2: 216 (1818)
Leptonia serrulata (Fr.) P. Kumm., *Führ. Pilzk.*: 96 (1871)
Agaricus atrides Lasch, *Linnaea* 4: 539 (1829)
Eccilia atrides (Lasch) P. Kumm., *Führ. Pilzk.*: 94 (1871)
Leptonia serrulata var. *laevipes* Maire, *Bull. Soc. Mycol. France* 26: 174 (1910)
E: o **S:** c **W:** o **NI:** o **ROI:** o
H: On soil, amongst grass, in grassland also occasionally in open deciduous woodland.
D: FAN1: 156, NM2: 345 **D+I:** Ph: 117 (as *Leptonia serrulata*), FungEur5: 472 52a **I:** C&D: 299
Common in Scotland, occasional to locally frequent and widespread elsewhere.

sinuatum (Pers.) P. Kumm., *Führ. Pilzk.*: 98 (1871)
Agaricus sinuatus Pers., *Syn. meth. fung.*: 329 (1801)
Rhodophyllus sinuatus (Pers.) Quél., *Enchir. fung.*: 179 (1886)
Entoloma eulividum Noordel., *Persoonia* 12(4): 457 (1985)
Agaricus lividus Bull., *Herb. France*: pl. 382 (1788)
Entoloma lividum (Bull.) Quél., *Mém. Soc. Émul. Montbéliard, Sér. 2,* 5: 116 (1872)
Mis.: *Agaricus fertilis* sensu Berkeley [*Outl. Brit. fungol.*: 142 (1860)]
E: o **S:** ! **W:** ! **NI:** ! **ROI:** !
H: On soil, often associated with *Fagus* or *Quercus* spp, in open deciduous woodland, and occasionally with *Betula* spp. Mainly autumnal but sometimes fruiting in spring.
D: FAN1: 95, NM2: 354 (as *Entoloma eulividum*), NBA8: 575-576 (as *E. lividum*) **D+I:** Ph: 114-115, FungEur5: 111 2 **I:** FlDan2: 74C (as *Rhodophyllus lividus*), C&D: 291 (as *E. lividum*)
Uncommonly reported but apparently widespread. May be locally abundant in certain areas, e.g. the New Forest.

sodale Noordel., *Int. J. Mycol. Lichenol.* 1(1): 58 (1982)
Rhodophyllus sodalis Kühner & Romagn. [*nom. inval.*], *Flore Analytique des Champignons Supérieurs*: 205 (1953)
Leptonia sodalis (Kühner & Romagn.) P.D. Orton, *Mycologist* 5(3): 135 (1991)
Mis.: *Leptonia lampropus* sensu NCL non sensu Fr.
S: o **W:** !
H: On calcareous soil in unimproved upland grassland.
D: FAN1: 158, NM2: 347 **D+I:** FungEur5: 532 62a
Rarely reported. Known from several sites in Scotland and from Wales (Glamorgan: Whiteford Burrows). Reported on several occasions from England but all records are unsubstantiated with voucher material.

solstitiale (Fr.) Noordel., *Persoonia* 10(4): 505 (1980)
Agaricus solstitialis Fr., *Epicr. syst. mycol.*: 152 (1838)
Leptonia solstitialis (Fr.) Gillet, *Hyménomycètes*: 416 (1876)
Nolanea solstitialis (Fr.) P.D. Orton, *Trans. Brit. Mycol. Soc.* 43(2): 179 (1960)
E: ! **S:** ! **W:** ! **NI:** ! **ROI:** !
H: On soil, in unimproved grassland.
D: FAN1: 137-138 **D+I:** FRIC12: 6-8 90c, FungEur5: 293 36b **I:** SV40: 43, Cooke 354 (332) Vol. 3 (1885)
Known from widely scattered locations, but most reports are unsubstantiated with voucher material.

sordidulum (Kühner & Romagn.) P.D. Orton, *Trans. Brit. Mycol. Soc.* 43(2): 175 (1960)
Rhodophyllus sordidulus Kühner & Romagn., *Rev. Mycol. (Paris)* 19(1): 10 (1954)
E: c **W:** !
H: On soil or humus in open deciduous woodland.
D: FAN1: 104-105, NM2: 353 **D+I:** FungEur5: 180 15a & 85b **I:** C&D: 293
Rather frequent (often abundant) and widespread in England.

sphagneti Naveau, *Natuurwet. Tijdschr.* 5: 75 (1932)
E: ! **S:** ? **NI:** ?
H: On peaty soil, and amongst dead and dying species of *Sphagnum* moss.
D: FAN1: 104, NBA9: 552, FungEur5: 166 16 **D+I:** FRIC12: 28 95a
Rarely reported.

sphagnorum (Romagn. & J. Favre) Bon & Courtec., *Doc. Mycol.* 18(69): 38 (1987)
    Rhodophyllus sphagnorum Romagn. & J. Favre, *Rev. Mycol. (Paris)* 3(2-3): 63 (1938)
E: !
H: On peaty soil, amongst *Sphagnum* moss, in peat bogs.
D+I: FungEur5: 564 67b
A single collection, from Cumberland (Tarn Hows).

strigosissimum (Rea) Noordel., *Persoonia* 10: 211 (1979)
    Nolanea strigosissima Rea, *Trans. Brit. Mycol. Soc.* 6(4): 325 (1920)
    Leptonia strigosissima (Rea) P.D. Orton, *Trans. Brit. Mycol. Soc.* 43(2): 178 (1960)
    Pouzaromyces strigosissimus (Rea) E. Horak, *Beitr. Kryptogamenfl. Schweiz* 13: 502 (1968)
E: !
H: On soil or humus and also on decayed wood.
D: FAN1: 112, NM2: 355 **D+I:** FungEur5: 340 39b
British collections only from Middlesex (Horsenden Hill) and Surrey (Weybridge).

subradiatum (Kühner & Romagn.) M.M. Moser, *Kleine Kryptogamenflora*, Edn 4: 197 (1978)
    Rhodophyllus subradiatus Kühner & Romagn., *Rev. Mycol. (Paris)* 19(1): 10 (1954)
E: !
H: On soil or loam in deciduous woodland.
D: FAN1: 103-104 **D+I:** FungEur5: 154 15b **I:** C&D: 293
Rarely reported. Most records are dubious and unsubstantiated with voucher material.

tenellum (J. Favre) Noordel., *Persoonia* 10: 256 (1979)
    Rhodophyllus tenellus J. Favre, *Mat. fl. crypt. Suisse*: 212 (1948)
    Nolanea tenella (J. Favre) P.D. Orton, *Trans. Brit. Mycol. Soc.* 43(2): 179 (1960)
E: ! **S:** ! **W:** ! **NI:** !
H: On soil in damp woodland.
D: FungEur5: 252, NM2: 348
Rarely reported. Apparently widespread but the majority of records are unsubstantiated with voucher material.

testaceum (Bres.) Noordel., *Nova Hedwigia Beih.* 91: 83 (1987)
    Nolanea cetrata var. *testacea* Bres., *Fungi trident.* 1: 77 (1881)
    Nolanea testacea (Bres.) P.D. Orton, *Trans. Brit. Mycol. Soc.* 43(2): 179 (1960)
E: ? **S:** !
H: On soil in woodland, associated with *Pinus* and *Betula* spp.
D: NCL 335 (as *Nolanea testacea*) **D+I:** FungEur5: 286 82b
In Britain, known with certainty only from Scotland. This is a boreal and subalpine species thus English records (e.g. South Hampshire: New Forest), unsubstantiated with voucher material, are doubtful. They may represent *Nolanea testacea* sensu NCL, i.e. a form of *E. cetratum*.

tjallingiorum Noordel., *Persoonia* 11: 465 (1982)
    Leptonia tjallingiorum (Noordel.) P.D. Orton, *Mycologist* 5(3): 135 (1991)
    Mis.: *Entoloma dichroum* sensu auct. mult.
E: ! **ROI:** !
H: On decayed or buried wood of deciduous trees.
D: FAN1: 153, NM2: 345 **D+I:** FungEur5: 429 47a **I:** C&D: 296
British records only from England (Shropshire: Dudmaston Estate and West Sussex: near Pulborough) and Republic of Ireland (Galway: Garryland Wood).

transvenosum Noordel., *Nordic J. Bot.* 2(2): 155 (1982)

    Mis.: *Entoloma costatum* sensu NCL
E: !
H: On soil amongst grass in woodland.
Known with certainty from Gloucestershire (Forest of Dean). However, Orton (1991) suggests that some British records of *E. costatum* may belong here.

turbidum (Fr.) Quél., *Mém. Soc. Émul. Montbéliard, Sér. 2,* 5: 119 (1872)
    Agaricus turbidus Fr., *Syst. mycol.* 1: 205 (1821)
    Entoloma costatum var. *cordae* P. Karst., *Bidrag Kännedom Finlands Natur Folk* 32: 268 (1879)
    Entoloma cordae (P. Karst.) P. Karst., *Meddeland. Soc. Fauna Fl. Fenn.* 5: 9 (1879)
E: ! **S:** ! **W:** ! **NI:** ! **ROI:** ! **O:** Isle of Man: !
H: On sandy soil, usually associated with *Pinus* and *Betula* spp, on heathland, and occasionally in dunes with *Salix repens*.
D: NCL3: 231, FAN1: 109-110, NM2: 353 **D+I:** FRIC12: 26-27 94d, FungEur5: 206 22 **I:** C&D: 293
Rarely reported but apparently widespread, perhaps more frequent in northern areas.

turci (Bres.) M.M. Moser, *Kleine Kryptogamenflora*, Edn 4: 200 (1978)
    Leptonia turci Bres., *Fungi trident.* 1: 47 (1881)
E: ! **S:** ! **W:** ! **NI:** ! **ROI:** !
H: On soil, amongst grass, in unimproved grassland, meadows and open areas in woodland, also occasionally amongst *Sphagnum* spp in swampy woodland.
D: NCL3: 302 (as *Leptonia turci*), FAN1: 168, NM2: 358 **D+I:** FungEur5: 584 70a **I:** C&D: 301
Rarely reported and the majority of records are unsubstantiated with voucher material.

undatum (Gillet) M.M. Moser, *Kleine Kryptogamenflora*, Edn 4: 211 (1978)
    Clitopilus undatus Gillet, *Hyménomycètes*: 407 (1876)
    Eccilia undata (Gillet) Quél., *Bull. Soc. Amis Sci. Nat. Rouen.* 15: 157 (1880)
    Rhodophyllus undatus (Gillet) Quél., *Enchir. fung.*: 62 (1886)
    Agaricus undatus Fr. [*nom. illegit.*, non *A. undatus* Berk. (1836)], *Epicr. syst. mycol.*: 149 (1838)
    Eccilia sericeonitida P.D. Orton, *Trans. Brit. Mycol. Soc.* 43(2): 175 (1960)
    Entoloma sericeonitidum (P.D. Orton) Arnolds, *Biblioth. Mycol.* 90(3): 350 (1983) [1982]
    Claudopus sericeonitidus (P.D. Orton) P.D. Orton, *Mycologist* 5(3): 126 (1991)
    Agaricus undatus var. *viarum* Fr., *Hymenomyc. eur.*: 199 (1874)
    Clitopilus undatus var. *viarum* (Fr.) Sacc., *Syll. fung.* 5: 702 (1887)
    Mis.: *Nolanea proletaria* sensu auct.
E: ! **S:** ! **W:** ! **NI:** ! **ROI:** !
H: On soil or decayed wood (small moss covered stumps) in open, herb-rich deciduous woodland. Rarely (and dubiously) reported from grassland.
D: FAN1: 173, NM2: 356 **D+I:** FungEur5: 613 71a **I:** Bon: 189, C&D: 301
Rarely reported but apparently widespread.

vernum S. Lundell, *Svensk bot. Tidskr.* 31: 193 (1937)
    Nolanea verna (S. Lundell) Kotl. & Pouzar, *Česká Mykol.* 26: 221 (1972)
    Rhodophyllus cucullatus J. Favre, *Ergebn. Wiss. Untersuch. Schweiz. Natn. Parks* 5: 62 (1955)
    Nolanea cucullata (J. Favre) P.D. Orton, *Trans. Brit. Mycol. Soc.* 43(2): 179 (1960)
    Mis.: *Nolanea pascua* sensu Rea (1922), sensu Bres. [*Icon. mycol.* 12 (1929)]
E: ! **S:** ! **W:** !
H: On soil, in coniferous woodland, or amongst grass in unimproved grassland, and on heathland.
D: NM2: 129, NM2: 351 **D+I:** FungEur5: 265 33 **I:** C&D: 925
Rarely reported but apparently widespread.

versatile (Gillet) M.M. Moser, *Kleine Kryptogamenflora*, Edn 4: 209 (1978)
  Nolanea versatilis Gillet, *Hyménomycètes*: 414 (1876)
  Inopilus versatilis (Gillet) Pegler, *Kew. Bull. Addit. Ser.* 9: 345 (1983)
  Pouzaromyces versatilis (Gillet) P.D. Orton, *Mycologist* 5(4): 175 (1991)
  Agaricus versatilis Fr. [*nom. illegit.*, non *A. versatilis* Bertero & Mont. 1837], *Monogr. hymenomyc. Suec.* 2: 297 (1863)
E: ! **S:** !
H: On soil or humus in deciduous woodland.
D: FAN1: 116, FAN1: 116, NM2: 355 **D+I:** FungEur5: 360 40b **I:** C&D: 293
British collections from England (Yorkshire: Crimsworth Dean) and Scotland (Isle of Arran: Kingscross Point and Kirkudbrightshire: Carstramon Wood). Reported elsewhere but unsubstantiated with voucher material.

vinaceum (Scop.) Arnolds & Noordel., *Persoonia* 10: 298 (1979)
  Agaricus vinaceus Scop., *Fl. carniol.* 444 (1772)
  Nolanea vinacea (Scop.) P. Kumm., *Führ. Pilzk.*: 95 (1871)
  Entoloma vinaceum var. *fumosipes* Arnolds & Noordel., *Persoonia* 10(2): 296 (1979)
E: ! **NI:** ! **ROI:** ?
H: On acidic soil in grassland.
D: FAN1: 110, NM2: 352 **D+I:** FRIC12: 24-25 94a, FungEur5: 208 23 & 88e
Accepted dubiously. British records from England (West Kent: Ide Hill) and Northern Ireland (Fermanagh: Inishmakill), though the latter was said to be 'in woodland'. Reported from the Republic of Ireland but unsubstantiated with voucher material.

viridans (Fr.) P. Karst., *Bidrag Kännedom Finlands Natur Folk* 32: 262 (1879)
  Agaricus viridans Fr., *Monogr. hymenomyc. Suec.* 2: 345 (1863)
W: !
H: On soil in deciduous woodland. British material associated with *Quercus* spp.
D+I: FungEur5: 122 84b
A single collection from Merionethshire (Lake Vyrnwy). A rare species otherwise known only from Sweden.

wynnei (Berk. & Broome) Sacc., *Syll. fung.* 5: 696 (1887)
  Agaricus wynnei Berk. & Broome, *Ann. Mag. Nat. Hist., Ser. 4* [*Notices of British Fungi no. 1342*] 11: 340 (1873)
  Leptonia wynnei (Berk. & Broome) P.D. Orton, *Mycologist* 5(3): 136 (1991)
W: !
H: On soil in coniferous woodland.
D+I: FungEur5: 437 83 **I:** Cooke 339 (329) Vol. 3 (1884)
Only known from the type collection from Denbighshire (Coed Coch) in 1872.

xanthocaulon Arnolds & Noordel., *Persoonia* 10: 299 (1979)
S: !
H: On acidic soil in unimproved grassland.
D: FAN1: 133 **D+I:** FRIC12: 12-13 91c, FungEur5: 327 79c
Recently collected in Scotland.

xanthochroum (P.D. Orton) Noordel., *Persoonia* 12(4): 462 (1985)
  Leptonia xanthochroa P.D. Orton, *Notes Roy. Bot. Gard. Edinburgh* 26: 54 (1964)
S: ! **NI:** !
H: On acidic soil in unimproved grassland.
D: FAN1: 167, NM2: 357, FungEur5: 562 **I:** C&D: 301
Known from Scotland and Northern Ireland. Reported elsewhere but records are dubious and unsubstantiated with voucher material.

EPISPHAERIA Donk, *Persoonia* 2: 336 (1962)
Cortinariales, Crepidotaceae
Type: *Episphaeria fraxinicola* (Berk. & Broome) Donk

fraxinicola (Berk. & Broome) Donk, *Persoonia* 2(3): 336 (1962)
  Cyphella fraxinicola Berk. & Broome, *Ann. Mag. Nat. Hist., Ser. 4* [*Notices of British Fungi no. 1446*] 15: 32 (1875)
  Phaeocyphella fraxinicola (Berk. & Broome) Rea, *Brit. basidiomyc.*: 704 (1922)
  Cyphella episphaeria Quél., *Mém. Soc. Émul. Montbéliard, Sér. 2*, 5: 537 (1875)
  Phaeocyphella episphaeria (Quél.) Rea, *Trans. Brit. Mycol. Soc.* 12(2-3): 225 (1927)
E: !
H: On bark of living deciduous trees in woodland, usually on *Fraxinus* but also known on *Acer campestre* and *Fagus* and often (but not always) associated with effete fruitbodies of pyrenomycetes.
D: NM2: 338
Known from Cambridgeshire, Cumberland, Hertfordshire, Middlesex, North Somerset, South Wiltshire, Surrey and Yorkshire. Minute and easily overlooked.

EPITHELE (Pat.) Pat., *Essai taxon.*: 60 (1900)
Stereales, Epitheliaceae
Type: *Epithele typhae* (Pers.) Pat.

typhae (Pers.) Pat., *Essai taxon.*: 60 (1900)
  Athelia typhae Pers., *Mycol. eur.* 1: 84 (1822)
  Thelephora typhae (Pers.) Fr., *Elench. fung.* 1: 226 (1828)
  Corticium typhae (Pers.) Desm., *Pl. crypt. N. France*, Edn 1: no. 2161 (1851)
  Hypochnus typhae (Pers.) Pat., *Tab. anal. fung.* 6(4): 31 (1887)
E: ! **S:** !
H: On dead stem and leaf bases of *Carex acutiformis* and *C. riparia* and rarely *Typha latifolia* and *Glyceria maxima* in marshland, fens, and reedswamps.
D: NM3: 183 **D+I:** CNE3: 364-365, B&K2: 116 102
Rarely reported, but occurring in little-studied habitats. Known from widely scattered sites in England, from Westmorland to South Hampshire, most frequently from East Anglia (Norfolk Broads). A single collection from Scotland (Ayrshire).

ERYTHRICIUM J. Erikss. & Hjortstam, *Svensk bot. Tidskr.* 64(2): 165 (1970)
Stereales, Hyphodermataceae
Type: *Erythricium laetum* (P. Karst.) J. Erikss. & Hjortstam

laetum (P. Karst.) J. Erikss. & Hjortstam, *Svensk bot. Tidskr.* 64(2): 166 (1970)
  Hyphoderma laetum P. Karst., *Rev. Mycol. (Toulouse)* 11(41): 206 (1889)
  Corticium laetum (P. Karst.) Bres., *Ann. Mycol.* 1(1): 94 (1903)
  Mis.: *Corticium anthochroum* sensu Massee (1892), sensu Rea (1922)
E: ! **ROI:** ?
H: On bare soil, large dead herbaceous stems such as *Arctium lappa* and *Digitalis purpurea* and decayed wood of *Acer pseudoplatanus* and *Fraxinus*.
D: NM3: 166 **D+I:** CNE3: 370-371, B&K2: 130 121
Known from Co. Durham, Hertfordshire, Herefordshire, Surrey, Warwickshire, West Norfolk and Yorkshire. A single collection in herb. K., from the Republic of Ireland, is annotated as 'compare with' *E. laetum*. Probably genuinely rare, the rose-pink basidiomes conspicuous and unlikely to be overlooked.

EXIDIA Fr., *Syst. mycol.* 2: 220 (1823)
Tremellales, Exidiaceae
  Myxarium Wallr., *Fl. crypt. Germ., Sect. II*: 260 (1833)
  Ulocolla Bref., *Unters. Gesammtgeb. Mykol.* 7: 95 (1888)
Type: *Exidia glandulosa* (Bull.) Fr.

glandulosa (Bull.) Fr., *Syst. mycol.* 2: 224 (1823)
  Tremella glandulosa Bull., *Herb. France*: pl. 420 f. 1 (1789)
  Tremella atra O.F. Müll., *Fl. Danica*: t. 884 (1782)
  Tremella spiculosa Pers., *Observ. mycol.* 2: 99 (1796)

Gyraria spiculosa (Pers.) Gray, *Nat. arr. Brit. pl.* 1: 594 (1821)
  Exidia spiculosa (Pers.) Sommerf., *Suppl. Fl. lapp.*: 307 (1826)
  Exidia truncata Fr., *Syst. mycol.* 2: 224 (1823)
  Mis.: *Tremella arborea* sensu auct.
  Mis.: *Exidia intumescens* sensu auct.
  Mis.: *Tremella intumescens* sensu auct.
E: c **S:** c **W:** c **NI:** c **ROI:** c **O:** Channel Islands: ! Isle of Man: !
H: On dead attached or fallen wood of deciduous trees, most commonly *Corylus*, *Quercus* spp and *Fagus*.
D+I: Ph: 263, B&K2: 62 21, B&K2: 66 25 (as *Exidia truncata*) **I:** C&D: 137, Kriegl1: 98, Kriegl1: 99 (as *E. plana*)
Common and widespread. Basidiomes are turbinate, unlike those of *E. plana* which are resupinate and cerebriform. *Exidia glandulosa* sensu Fries (1823) is the latter species.

nucleata (Schwein.) Burt, *Rep. (Annual) Missouri Bot. Gard.* 8: 371 (1921)
  Tremella nucleata Schwein., *Schriften Naturf. Ges. Leipzig* 1: 115 (1822)
  Naematelia nucleata (Schwein.) Fr., *Syst. mycol.* 2(1): 228 (1823)
  Myxarium nucleatum Wallr., *Fl. crypt. Germ., Sect. II*: 26 (1833)
  Tremella gemmata Lév., in Demidoff, *Voyag. Russ. Mérid. Crimée* 2: 96 (1842)
  Exidia gemmata (Lév.) Bourdot & Maire, *Bull. Soc. Mycol. France* 36: 69 (1920)
  Mis.: *Tremella albida* sensu auct. p.p.
  Mis.: *Myxarium hyalinum* sensu auct.
E: c **S:** c **W:** c **NI:** c **ROI:** c **O:** Channel Islands: o Isle of Man: o
H: On fallen, usually decorticated wood of deciduous trees. Especially common on *Acer pseudoplatanus*, slightly less so on *Fagus* and *Fraxinus*. Also known on *Corylus*, *Hedera*, *Ilex*, *Laburnum anagyroides*, *Platanus*, *Quercus*, *Salix*, *Tilia* and *Ulmus* spp. Occasionally on large dead stems of herbaceous plants such as *Arctium* spp, and dead lianas of *Clematis vitalba*.
D: NM3: 109 (as *Myxarium nucleatum*) **D+I:** Reid4: 421-423 (as *M. nucleatum*) **I:** C&D: 137 (as *M. nucleatum*), Kriegl1: 112 (as *M. nucleatum*)
Very common and widespread.

plana (F.H. Wigg.) Donk, *Persoonia* 4(2): 228 (1966)
  Tremella plana F.H. Wigg., *Prim. fl. holsat.*: 95 (1780)
  Mis.: *Exidia glandulosa* sensu auct. p.p.
E: c **S:** c **W:** c **NI:** ! **ROI:** !
H: On dead attached or fallen wood of deciduous trees, often large fallen trunks of *Fagus*. Also known on *Acer pseudoplatanus*, *A. campestre*, *Alnus*, *Betula*, *Fraxinus*, *Ilex*, *Prunus avium*, *Salix* and *Tilia* spp. Reported on other hosts but these are unverified.
D: NM3: 100 (as *Exidia glandulosa*)
Common and widespread.

recisa (Ditmar) Fr., *Syst. mycol.* 2: 223 (1823)
  Tremella recisa Ditmar, *Deutschl. Fl.*, Edn 2, 3: 27 (1813)
E: ! **S:** ! **W:** ! **NI:** !
H: On dead attached twigs of *Salix* spp.
D: NM3: 101 **D+I:** B&K2: 64 23 **I:** Kriegl1: 101
Rarely reported but apparently widespread.

repanda Fr., *Syst. mycol.* 2: 225 (1823)
  Tremella repanda (Fr.) Spreng., *Syst. Veg.*: 536 (1827)
  Ulocolla repanda (Fr.) Bres., *Icon. Mycol.*: 23 (1932)
S: ! **NI:** !
H: On dead attached twigs and branches of *Betula* spp, and a single record on *Corylus*.
D: NM3: 101
Known from a few widely scattered localities in Scotland, and a single collection from Northern Ireland (Fermanagh: Florence Court) in 2000.

saccharina (Alb. & Schwein.) Fr., *Syst. mycol.* 2: 225 (1823)

Tremella spiculosa var. *saccharina* Alb. & Schwein., *Consp. fung. lusat.*: 302 (1805)
  Tremella saccharina (Alb. & Schwein.) Bonord., *Handb. Mykol.*: 151 (1851)
  Ulocolla saccharina (Alb. & Schwein.) Bref., *Unters. Gesammtgeb. Mykol.* 7: 95 (1888)
E: ! **S:** ! **W:** ?
H: British material on dead wood (fallen or attached branches) of *Pinus radiata* and *P. sylvestris*.
D: NM3: 101 **I:** Kriegl1: 104
Rarely reported. Known from several widely scattered locations in Scotland and two sites recently in England (South Devon: Bovey Tracey and South Hampshire: New Forest, Markway Inclosure). Recorded for Wales but unsubstantiated with voucher material.

thuretiana (Lév.) Fr., *Hymenomyc. eur.*: 694 (1874)
  Tremella thuretiana Lév., *Ann. Sci. Nat., Bot.*, sér. 3, 9: 127 (1848)
  Mis.: *Tremella albida* sensu auct. p.p.
  Mis.: *Exidia albida* sensu auct. p.p.
  Mis.: *Tremella cerebrina* sensu auct. p.p.
  Mis.: *Tremella hyalina* sensu auct. p.p.
E: o **S:** o **W:** o **NI:** o **ROI:** o **O:** Isle of Man: !
H: On fallen wood of deciduous trees, most frequently on *Fagus* and *Fraxinus* but also known on *Acer pseudoplatanus*, *Aesculus*, *Betula* and *Corylus* spp, *Frangula alnus*, *Hedera*, *Ilex*, *Populus gileadensis*, *Salix* and *Ulmus* spp.
D: NM3: 100 **D+I:** B&K2: 64 24, Reid4: 418-421
Occasional but widespread.

EXIDIOPSIS (Bref.) Møller, *Bot. Mitt. Tropen* 8: 167 (1895)
Tremellales, Exidiaceae
  Bourdotia (Bres.) Bres. & Torrend, *Brotéria, Sér. Bot.* 11: 88 (1913)
Type: *Exidiopsis effusa* (Sacc.) Møller

calcea (Pers.) K. Wells, *Mycologia* 4: 348 (1962)
  Thelephora calcea Pers., *Syn. meth. fung.*: 581 (1801)
  Corticium calceum (Pers.) Fr., *Epicr. syst. mycol.*: 562 (1838)
  Sebacina calcea (Pers.) Bres., *Fungi trident.* 2: 64 (1898)
E: ! **S:** ! **NI:** ! **ROI:** ! **O:** Channel Islands: !
H: On decayed wood of deciduous trees such as *Corylus*, *Fagus*, *Fraxinus*, *Prunus spinosa*, *Salix* spp, *Sambucus nigra*, *Tilia* and *Ulmus* spp; also on dead woody stems of *Rubus idaeus* and lianas of *Clematis vitalba*.
D: NM3: 103 **D+I:** B&K2: 56 11
Rarely reported but apparently widespread.

effusa (Sacc.) Møller, *Bot. Mitt. Tropen* 8: 82 (1895)
  Thelephora effusa Sacc., *Syll. fung.* 6: 541 (1888)
  Sebacina effusa (Sacc.) Maire, *Bull. Soc. Mycol. France* 18: 67 (1902)
  Sebacina laccata Bourdot & Galzin, *Bull. Soc. Mycol. France* 39: 262 (1923)
  Mis.: *Sebacina atra* sensu auct. Brit.
  Mis.: *Sebacina mesomorpha* sensu auct.
  Mis.: *Tremella viscosa* sensu auct. p.p.
E: o **S:** o **W:** o **NI:** ! **ROI:** !
H: On decayed wood of deciduous trees, often on decorticated fallen trunks of *Ulmus* spp, and less frequently *Fraxinus* and *Fagus*. Rarely on *Alnus*, *Buxus*, *Corylus* and *Quercus* spp, or on conifers such as *Pinus radiata*.
D: NM3: 104 **D+I:** B&K2: 56 12, Reid4: 429-430 **I:** Kriegl1: 109 (as *Exidiopsis grisea* var. *effusa*)
Occasional but widespread.

galzinii (Bres.) Killerm., *Die Natür. Pflanzenf.* 6: 113 (1928)
  Sebacina galzinii Bres., *Ann. Mycol.* 6: 46 (1908)
  Bourdotia galzinii (Bres.) Bres. & Torrend, *Brotéria, Sér. Bot.* 11: 88 (1913)
E: !
H: On decayed wood of deciduous trees such as *Fagus*, *Fraxinus* and *Salix* spp.

Known from Cambridgeshire, Hertfordshire, Oxfordshire, Surrey, West Gloucestershire and West Sussex.

grisea (Pers.) Bourdot & Maire, *Bull. Soc. Mycol. France* 36: 71 (1920)
> *Thelephora grisea* Pers., *Mycol. eur.* 1: 149 (1822)
> *Sebacina grisea* (Pers.) Bres., *Ann. Mycol.* 6: 45 (1908)

E: ! **S:** ! **ROI:** ?
H: On decayed bark or wood of conifers such as *Abies grandis* and *Picea* spp.
D: NM3: 103 **D+I:** B&K2: 58 13
Rarely reported. Known from England (South Devon) and Scotland (South Aberdeenshire). Collections from elsewhere on deciduous hosts have been redetermined as *E. effusa*.

molybdea (McGuire) Ervin, *Mycologia* 49: 123 (1957)
> *Sebacina molybdea* McGuire, *Lloydia* 4(1): 17 (1941)

S: !
H: On decayed wood.
Accepted on the basis of a single collection from Scotland (Wester Ross: Loch Maree) [see Mycotaxon 49: 224 (1993)].

opalea (Bourdot & Galzin) D.A. Reid, *Trans. Brit. Mycol. Soc.* 55: 431 (1970)
> *Sebacina opalea* Bourdot & Galzin, *Bull. Soc. Mycol. France* 39: 262 (1924)
> Mis.: *Exidiopsis molybdea* sensu Reid (Reid4)

E: ! **S:** !
H: On decayed wood of deciduous trees such as *Quercus* and *Ulmus* spp.
D: NM3: 103 **D+I:** Reid4: 431-432
Known from England (Hertfordshire, Middlesex and South Devon) and Scotland (Wester Ross).

FAERBERIA Pouzar, *Česká Mykol.* 35: 187 (1981)
Poriales, Lentinaceae
> *Geopetalum* Pat. [*nom. illegit.*], *Hyménomyc. Eur.*: 127 (1887)

Type: *Faerberia carbonaria* (Alb. & Schwein.) Pouzar

carbonaria (Alb. & Schwein.) Pouzar, *Česká Mykol.* 35(4): 187 (1981)
> *Merulius carbonarius* Alb. & Schwein., *Consp. fung. lusat.*: 375 (1805)
> *Cantharellus carbonarius* (Alb. & Schwein.) Fr., *Hymenomyc. eur.*: 456 (1874)
> *Geopetalum carbonarium* (Alb. & Schwein.) Pat., *Hyménomyc. Eur.*: 127 (1887)
> *Merulius leucophaeus* Pers., *Mycol. eur.* 2: 15 (1825)
> *Cantharellus leucophaeus* (Pers.) Nouel, *Mém. Soc. Sci. Agric. Lille*: t. 1 (1831)
> *Cantharellus radicosus* Berk. & Broome, *Ann. Mag. Nat. Hist., Ser. 3 [Notices of British Fungi no. 1134]* 18: 54 (1866)
> Mis.: *Xerotus degener* sensu auct.

E: o **W:** ! **NI:** !
H: On burnt soil, most frequently on old bonfire sites in mixed deciduous woodland.
D: BFF6: 11, NM2: 46 **D+I:** FAN2: 30, B&K3: 186 207, Myc9(2): 60 (as *Geopetalum carbonarium*) **I:** Bon: 125, Kriegl3: 12
Occasional in England, rarely reported elsewhere. Most records are from areas of calcareous soil in south-eastern England.

FAYODIA Kühner, *Bull. Mens. Soc. Linn. Lyon* 9: 68 (1930)
Agaricales, Tricholomataceae
Type: *Fayodia bisphaerigera* (J.E. Lange) Singer

bisphaerigera (J.E. Lange) Singer, *Rev. Mycol. (Paris)* 1(6): 279 (1936)
> *Omphalia bisphaerigera* J.E. Lange, *Dansk. Bot. Ark.* 6(5): 9 (1938)
> *Mycena bisphaerigera* (J.E. Lange) A.H. Sm., *North American Species of Mycena*: 449 (1947)
> Mis.: *Fayodia gracilipes* sensu BFF8 & sensu B&K3: 189 & pl. 208

E: ! **S:** ! **NI:** ?
H: On soil or decayed wood. Associated with conifers such as *Picea* and *Pinus* spp.
D: NM2: 122 (as *Fayodia gracilipes*), FAN3: 154, BFF8: 93-94 (as *F. gracilipes*) **D+I:** A&N3 131-135 pl.39 **I:** FlDan2: 59H (as *Omphalia bisphaerigera*), Bon: 131 (as *F. gracilipes*), C&D: 187 (as *F. gracillipes*)
Rarely reported. Most records and collections are from scattered localities in Scotland. Also known from England (Berkshire and Yorkshire) and reported from Northern Ireland (Fermanagh) but unsubstantiated with voucher material.

FIBRICIELLUM J. Erikss. & Ryvarden, *Corticiaceae of North Europe* 3: 373 (1975)
Stereales, Sistotremataceae
Type: *Fibriciellum silvae-ryae* J. Erikss. & Ryvarden

silvae-ryae J. Erikss. & Ryvarden, *Corticiaceae of North Europe* 3: 375 (1975)

E: !
H: On decayed wood of deciduous trees such as *Fagus*, *Quercus* and *Ulmus* spp, and occasionally on soil mixed with decayed woody debris.
D: NM3: 132 **D+I:** CNE3: 372-375
Known from Berkshire, North & South Hampshire, North Wiltshire, Oxfordshire, South Devon, Surrey and West Sussex. A *Trechispora* species *fide* K.H. Larsson (Ph.D. Thesis, 1992), but no valid combination has yet been made.

FIBULOMYCES Jülich, *Willdenowia. Beih.* 7: 178 (1972)
Stereales, Atheliaceae
Fibulomyces is considered a synonym of *Leptosporomyces*, but one species has not yet been transferred.
Type: *Fibulomyces mutabilis* (Bres.) Jülich (= *Leptosporomyces mutabilis*)

crucelliger Stalpers & Marvanová, *Trans. Brit. Mycol. Soc.* 89(4): 496 (1987)
> *Taeniospora descalsii* Marvanová & Stalpers (anam.), *Trans. Brit. Mycol. Soc.* 89(4): 494 (1987)

E: ! **S:** ! **W:** !
H: Isolated in pure culture from foam samples from streams.
British records from England (South Hampshire: New Forest and South Devon: River Teign), Scotland (Ledard Burn) and Wales (Caernarvonshire: Bryn Gefeila). The teleomorph is only known in culture.

FILOBASIDIELLA Kwon-Chung, *Mycologia* 67: 1199 (1975)
Tremellales, Filobasidiaceae
Type: *Filobasidiella neoformans* Kwon-Chung

depauperata (Petch) Samson, Stalpers & Weijman, *Antonie van Leeuwenhoek* 49: 454 (1983)
> *Aspergillus depauperatus* Petch, *Trans. Brit. Mycol. Soc.* 16: 245 (1932)

E: !
H: Parasitic, probably on ascomycetous moulds. British material on a scale insect (*Lepidosaphes ulmi*).
D+I: YST 380 (yeast only)
Known only from West Norfolk (Hunstanton).

lutea P. Roberts, *Mycotaxon* 63: 198 (1997)

E: !
H: Parasitic. On and in the hymenium of *Hypochnicium vellereum*.
Known from Middlesex, Surrey, South Devon, Surrey and Westmorland. Colonies appear as distinctive pale primrose-yellow areas on the chalk-white to rose-pink hymenium of the host.

neoformans Kwon-Chung, *Mycologia* 67: 1199 (1975)
  Saccharomyces neoformans San Felice (anam.), *Ann. Ig. Sperim.* 5: 241 (1895)
  Cryptococcus neoformans (San Felice) Vuill. (anam.), *Rev. Gén. Sci. Pures Appl.* 12: 747 (1901)
E: ! **S:** !
H: Isolated ex pigeon droppings and ex human mycoses.
D+I: YST 381-382 (yeast only)
In Britain known only as the anamorphic yeast state (*Cryptococcus neoformans*), causing cryptococcosis in humans and other mammals.

FILOBASIDIUM L.S. Olive, *J. Elisha Mitchell Sci. Soc.* 84: 261 (1968)
Tremellales, Filobasidiaceae
Type: *Filobasidium floriforme* L.S. Olive

capsuligenum (Fell, Statzell, I.L. Hunter & Phaff) Rodr. Mir., *Antonie van Leeuwenhoek* 38(1): 96 (1972)
  Leucosporidium capsuligenum Fell, Statzell, I.L. Hunter & Phaff, *Antonie van Leeuwenhoek* 35(4): 444 (1970) [1969]
  Torulopsis capsuligena van der Walt & Kerken (anam.), *Leeuwenhoek ned. Tidjdschr.* 27: 211 (1961)
  Candida japonica Diddens & Lodder (anam.), *Die Hefasammlung des 'Centraalbureau voor Schimmelcultures': Beitrage zu einer Monographie der Hefearten. II. Teil. Die anaskosporogenen Hefen. Zweite Halfte*: 362 (1942)
E: !
H: Isolated ex cider and ex sausages.
D+I: YST 383 (yeast only)
Only known in the British Isles in its anamorphic yeast state.

uniguttulatum Kwon-Chung, *Int. J. Mycol. Lichenol.* 27(3): 293 (1977)
  Eutorulopsis uniguttulata Wolfram & Zach (anam.), *Archiv Dermatol. Syph. (Berlin)* 152: 688 (1934)
  Cryptococcus uniguttulatus (Wolfram & Zach) Phaff & Fell (anam.), *Yeasts, a taxonomic study*: 1140 (1970)
S: !
H: Isolated ex sputum and ex sausages.
D+I: YST 387 (yeast only)
Only known in the British Isles in its anamorphic yeast state.

FISTULINA Bull., *Hist. champ. France*: 313 (1791)
Fistulinales, Fistulinaceae
Type: *Fistulina hepatica* (Schaeff.) With.

hepatica (Schaeff.) With., *Nat. arr. Brit. pl.* 2: 405 (1792)
  Boletus hepaticus Schaeff., *Fung. Bavar. Palat. nasc.* 2: 116 (1763)
E: c **S:** c **W:** c **NI:** c **ROI:** c **O:** Channel Islands: ! Isle of Man: !
H: Weakly parasitic then saprotrophic. On deciduous trees, most often *Quercus* spp, occasionally *Castanea* and rarely *Fagus*. A single collection (in herb. K) in 2003 from the trunk of a fifty year old climbing *Rosa* cultivar.
D: EurPoly1: 249, NM3: 294 **D+I:** Ph: 224, B&K2: 334 428 **I:** Sow1: 58 (as *Boletus hepaticus*), Kriegl1: 517
Very common and widespread. Although basidiomes appear poroid, they are actually composed of multiple individual tubes on a common stroma and are better thought of as cyphelloid. English name = 'Beefsteak Fungus'.

FLAGELLOSCYPHA Donk, *Lilloa* 22: 312 (1951) [1949]
Agaricales, Tricholomataceae
Type: *Flagelloscypha minutissima* (Burt) Donk

faginea (Lib.) W.B. Cooke, *Sydowia, Beih.*: 60 (1961)
  Cyphella faginea Lib., *Pl. Crypt. Ardienn.*: 331 (1837)
E: ! **S:** !
H: On dead woody substrata such as small dead twigs and branches of deciduous trees.
Accepted dubiously. Rarely reported and collections require re-examination, as some may be *Flagelloscypha minutissima*.

kavinae (Pilát) W.B. Cooke, *Sydowia, Beih.*: 62 (1961)
  Cyphella kavinae Pilát, *Ann. Mycol.* 23: 157 (1925)
S: !
H: On a dead stem of *Equisetum* sp. in a ditch (British material).
D: NM2: 123
A single collection from Orkney Isles (Hoy), but basidiomes are minute and easily overlooked.

minutissima (Burt) Donk, in Singer, *Lilloa* 22: 312 (1951) [1949]
  Cyphella minutissima Burt, *Ann. Missouri Bot. Gard.* 1: 367 (1914)
  Cyphella citrispora Pilát, *Ann. Mycol.* 22: 209 (1924)
  Flagelloscypha citrispora (Pilát) D.A. Reid, *Persoonia* 3(1): 98 (1964)
  Cyphella jancheni Pilát, *Ann. Mycol.* 22: 210 (1924)
  Mis.: *Flagelloscypha faginea* sensu Cooke [Beih. Sydowia 4: 60 (1961)]
E: ! **S:** ! **W:** ! **NI:** ! **ROI:** !
H: On woody debris, sticks, twigs and decayed bark of deciduous trees. Occasionally on herbaceous stems or fallen leaves.
D: NM2: 123 **D+I:** B&K2: 200 228
Rarely reported but widespread.

orthospora (Bourdot & Galzin) Berthier & Malençon, *Acta Phytotax. Barcinon.* 19: 30 (1977) [1976]
  Cyphella orthospora Bourdot & Galzin, *Hymenomyc. France*: 160 (1928)
E: ! **S:** !
H: On dead woody stems. British collections are mostly on *Ulex europaeus*.
Rarely reported. Much of the material in herb. K is from widely scattered sites in Scotland, and much of it determined as 'compare with' *F. orthospora*.

pilatii Agerer, *Sydowia* 27(1-6): 239 (1975) [1973-1974]
  Cyphella villosa var. *stenospora* Bourdot & Galzin, *Bull. Soc. Mycol. France* 26: 225 (1910)
  Cyphella punctiformis var. *stenospora* (Bourdot & Galzin) Bourdot & Galzin, *Hyménomyc. France*: 160 (1928)
E: ! **S:** ?
H: On dead grass stems and leaves in meadows and rough pasture, and also on dead fern stems *fide* Rea (1922).
Rarely reported. Known from Northamptonshire, Surrey and Yorkshire. Reported from Scotland but unsubstantiated with voucher material.

punctiformis (Fr.) Agerer, *Sydowia* 27(1-6): 246 (1975) [1973-1974]
  Peziza punctiformis Fr., *Syst. mycol.* 2: 105 (1823)
E: ! **S:** !
H: On herbaceous debris. British material on fallen leaves of *Fraxinus* and dead fronds of *Dryopteris* sp.
D: NM2: 123
Rarely reported. Known from England (Surrey: Mickleham Downs) and Scotland (North Ebudes, Isle of Skye).

FLAMMULASTER Earle, *Bull. New York Bot. Gard.* 5: 435 (1909)
Cortinariales, Cortinariaceae
  Flocculina P.D. Orton, *Trans. Brit. Mycol. Soc.* 43: 168 (1960)
Type: *Flammulaster carpophilus* (Fr.) Earle
The treatment here follows Vellinga in Persoonia 13 (1): 1-26 (1986)

carpophilus (Fr.) Earle, *Bull. New York Bot. Gard.* 5: 435 (1909) [1906]
  Agaricus carpophilus Fr., *Observ. mycol.* 1: 45 (1815)
  Naucoria carpophila (Fr.) Quél., *Mém. Soc. Émul. Montbéliard, Sér. 2,* 5: 134 (1872)
  Galera carpophila (Fr.) Quél., *Fl. mycol. France*: 81 (1888)
  Phaeomarasmius carpophilus (Fr.) Singer, *Sydowia* 2: 37 (1948)
  Flocculina carpophila (Fr.) P.D. Orton, *Trans. Brit. Mycol. Soc.* 43(2): 175 (1960)

Naucoria carpophiloides Kühner, *Bull. Soc. naturalistes Oyonnax* 10 -11 (Suppl.): 5 (1957)

Flocculina carpophiloides (Kühner) P.D. Orton, *Trans. Brit. Mycol. Soc.* 43(2): 168 (1960)

Phaeomarasmius carpophiloides (Kühner) M.M. Moser [*nom. inval.*], *Kleine Kryptogamenflora*, Edn 3: 273 (1967)

Flammulaster carpophiloides (Kühner) Watling, *Notes Roy. Bot. Gard. Edinburgh* 28(1): 65 (1967)

Flammulaster carpophilus var. *autochtonoides* Bon, *Doc. Mycol.* 17 (67): 3 (1987)

E: ! **S:** ! **W:** ! **NI:** ?

H: On fallen litter (usually decayed leaves or mast) of *Fagus* in old beechwoods or mixed deciduous woodland.

D: NM2: 255 **D+I:** B&K4: 318 405 **I:** C&D: 349

Known from scattered locations in England, from Oxfordshire southward, rarely in Scotland, and once in Wales (Breconshire: Cwm Clydach). Reported from Northern Ireland but unsubstantiated with voucher material.

carpophilus *var.* rhombisporus (G.F. Atk.) Vellinga, *Persoonia* 13(1): 10 (1986)

Lepiota rhombispora G.F. Atk., *Proc. Amer. Philos. Soc.* 57: 356 (1918)

Naucoria rhombispora (G.F. Atk.) J. Favre, *Mat. fl. crypt. Suisse* 10(3): 144 (1948)

Flocculina rhombispora (G.F. Atk.) P.D. Orton, *Trans. Brit. Mycol. Soc.* 43: 175 (1960)

Flammulaster rhombisporus (G.F. Atk.) Watling, *Notes Roy. Bot. Gard. Edinburgh* 28(1): 67 (1967)

S: !

H: On decayed leaf litter. British material associated with *Salix cinerea.*

D: NM2: 255 (as *Flammulaster rhombisporus*)

A single record from Morayshire (Culbin Sands) in 1964.

carpophilus *var.* subincarnatus (Joss. & Kühner) Vellinga, *Persoonia* 13(1): 12 (1986)

Naucoria subincarnata Joss. & Kühner, *Bull. Soc. naturalistes Oyonnax* 10 -11 (Suppl.): 6 (1957)

Flocculina subincarnata (Joss. & Kühner) P.D. Orton, *Nov. Actorum Acad. Caes. Leop.-Carol. Nat. Cur.* 43(2): 176 (1960)

Flammulaster subincarnatus (Joss. & Kühner) Watling, *Notes Roy. Bot. Gard. Edinburgh* 28(1): 67 (1967)

E: ! **S:** ! **ROI:** !

H: On fallen litter of *Fagus* (usually on decayed mast) in old mixed deciduous woodland or beechwoods.

D: NM2: 255 (as *Flammulaster subincarnatus*) **D+I:** B&K4: 320 406

Rarely reported but apparently widespread.

ferrugineus (Maire) Watling, *Notes Roy. Bot. Gard. Edinburgh* 28(1): 66 (1967)

Naucoria ferruginea Maire, *Bull. Soc. naturalistes Oyonnax* 10 -11 (Suppl.): 5 (1957)

Tubaria ferruginea (Maire) Maire, *Bull. Soc. Naturalistes Oyonnax* 10 -11 (Suppl.): 6 (1957)

Flocculina ferruginea (Maire) P.D. Orton, *Trans. Brit. Mycol. Soc.* 43(2): 175 (1960)

Mis.: *Naucoria siparia* sensu Lange (FlDan4: 19 & pl. 124E)

E: ! **S:** !

H: On soil or small bits of decayed woody debris, in deciduous woodland.

D: NM2: 255 **I:** C&D: 349

Rarely reported and few of the records are substantiated with voucher material.

fusisporus (P.D. Orton) Watling, *Notes Roy. Bot. Gard. Edinburgh* 28(1): 66 (1967)

Flocculina fusispora P.D. Orton, *Trans. Brit. Mycol. Soc.* 43(2): 235 (1960)

E: ! **W:** !

H: On decayed leaf litter in damp deciduous woodland.

Rarely reported and poorly known in Britain. Possibly only a bisporic form of *Flammulaster carpophilus.*

granulosus (J.E. Lange) Watling, *Notes Roy. Bot. Gard. Edinburgh* 28(1): 66 (1967)

Naucoria granulosa J.E. Lange, *Fl. agaric. danic.* 5, *Taxonomic Conspectus*: VI (1940)

Phaeomarasmius granulosus (J.E. Lange) Singer, *Lilloa* 22: 577 (1951) [1949]

Flocculina granulosa (J.E. Lange) P.D. Orton, *Trans. Brit. Mycol. Soc.* 43(2): 175 (1960)

E: ! **S:** c **W:** ! **NI:** ! **O:** Channel Islands: !

H: On loam soil in deciduous woodland or scrub. Often associated with *Acer pseudoplatanus* or *Fagus* and rarely with *Castanea* and *Salix* spp.

D: NM2: 255 **D+I:** B&K4: 320 407 **I:** Ph: 158 (as *Flocculina granulosa*), Bon: 247

Rarely reported but apparently widespread. Common in Scotland *fide* R. Watling (pers. comm.). Most British records of *Naucoria siparia* probably belong here.

limulatus (Fr.) Watling, *Notes Roy. Bot. Gard. Edinburgh* 28(1): 66 (1967)

Agaricus limulatus Fr., *Observ. mycol.* 2: 28 (1818)

Flammula limulata (Fr.) P. Karst., *Bidrag Kännedom Finlands Natur Folk* 32: 410 (1879)

Dryophila limulata (Fr.) Quél., *Enchir. fung.*: 72 (1886)

Fulvidula limulata (Fr.) Romagn. [*nom. inval.*], *Rev. Mycol. (Paris)* 2(5): 191 (1937)

Flocculina limulata (Fr.) P.D. Orton, *Trans. Brit. Mycol. Soc.* 43(2): 175 (1960)

Phaeomarasmius limulatus (Fr.) Singer, *Sydowia* 15: 75 (1962) [1961]

Naucoria limulata (Fr.) Kühner & Romagn. [*nom.inval.*], *Flore Analytique des Champignons Supérieurs*: 242 (1953)

Flammulaster limulatoides P.D. Orton, *Notes Roy. Bot. Gard. Edinburgh* 41(3): 580 (1984)

E: ! **S:** ! **W:** !

H: On fallen woody debris, usually *Fagus* in beechwoods or mixed deciduous woodland.

D: NM2: 254 (as *Flammulaster limulatoides*)

Rarely reported. Apparently widespread.

limulatus *var.* litus Vellinga, *Persoonia* 13(1): 17 (1986)

Mis.: *Flammulaster limulatus* sensu P.D. Orton [Notes Roy. Bot. Gard. Edinburgh 41: 579 (1984)]

E: ? **S:** !

H: On decayed wood of conifers and apparently deciduous trees.

D: NM2: 254 (as *Flammulaster limulatus*), NBA8: 579-580 (as *F. limulata*)

Described from Perthshire (Rannoch). Reported from England (New Forest: Denny Wood) but this collection (on wood of *Fagus*) needs re-assessment.

limulatus *var.* novasilvensis (P.D. Orton) Vellinga, *Persoonia* 13(1): 18 (1986)

Flammulaster novasilvensis P.D. Orton, *Notes Roy. Bot. Gard. Edinburgh* 41(3): 582 (1984)

E: !

H: On decayed fallen wood.

British collections are all from the type locality in South Hampshire (New Forest: Park Pale).

muricatus (Fr.) Watling, *Notes Roy. Bot. Gard. Edinburgh* 28(1): 66 (1967)

Agaricus muricatus Fr., *Observ. mycol.* 2: 12 (1815)

Pholiota muricata (Fr.) P. Kumm., *Führ. Pilzk.*: 83 (1871)

Dryophila muricata (Fr.) Quél., *Enchir. fung.*: 69 (1886)

Naucoria muricata (Fr.) Kühner & Romagn., *Bull. Soc. Mycol. France* 58: 133 (1942)

Phaeomarasmius muricatus (Fr.) Singer, *Lilloa* 25: 387 (1951) [1961]

Flocculina muricata (Fr.) P.D. Orton, *Trans. Brit. Mycol. Soc.* 43(2): 175 (1960)

Flammulaster denticulatus P.D. Orton, *Notes Roy. Bot. Gard. Edinburgh* 41(3): 577 (1984)

Mis.: *Flocculina erinaceella* sensu NCL

Mis.: *Flammulaster erinaceellus* sensu auct. Brit.

E: ! S: ! W: ! ROI: !
H: In deciduous woodland, on decayed fallen wood of trees such as *Fagus*. Rarely reported from peat soil.
D: NM2: 254 **I:** SV42: 39
Rarely reported. Few of the records are substantiated with voucher material.

speireoides (Romagn.) Watling, *Notes Roy. Bot. Gard. Edinburgh* 28(1): 68 (1967)
  Naucoria speireoides Romagn., *Bull. Soc. naturalistes Oyonnax* 10 -11 (Supl.): 6 (1957)
  Flocculina pusillima P.D. Orton, *Trans. Brit. Mycol. Soc.* 43(2): 236 (1960)
  Flammulaster pusillimus (P.D. Orton) Watling, *Notes Roy. Bot. Gard. Edinburgh* 28(1): 67 (1967)
E: !
H: On soil in deciduous woodland.
D: NM2: 255
Known only from Yorkshire.

FLAMMULINA P. Karst., *Meddeland. Soc. Fauna Fl. Fenn.* 18: 62 (1891)
Agaricales, Tricholomataceae
  Myxocollybia Singer, *Beih. Bot. Centralbl.*: 162 (1936)
Type: *Flammulina velutipes* (Curtis) Singer

velutipes (Curtis) Singer, *Lilloa* 22: 307 (1951) [1949]
  Agaricus velutipes Curtis, *Fl. londin.* 1: 212 (1777)
  Gymnopus velutipes (Curtis) Gray, *Nat. arr. Brit. pl.* 1: 605 (1821)
  Collybia velutipes (Curtis) P. Kumm, *Führ. Pilzk.*: 116 (1871)
  Pleurotus velutipes (Curtis) Quél., *Fl. mycol. France*: 334 (1886)
  Myxocollybia velutipes (Curtis) Singer [*nom. inval.*], *Schweiz. Z. Pilzk.* 17: 72 (1939)
  Collybia eriocephala Rea, *Trans. Brit. Mycol. Soc.* 3: 46 (1908)
  Collybia veluticeps Rea [*nom. illegit.*, non *C. veluticeps* (Cooke & Massee) Sacc. (1891)], *Trans. Brit. Mycol. Soc.* 1(4): 157 (1900)
E: c S: c W: c NI: c ROI: ! O: Channel Islands: !
H: Saprotrophic or weakly parasitic. On dead or dying wood of deciduous trees or shrubs. Very common on *Ulmus* spp, and common on *Acer pseudoplatanus*, *Fraxinus* and *Ulex europaeus*. Also known on *Betula* spp, *Castanea*, *Crataegus monogyna*, *Cystisus scoparius*, *Eleagnus* sp., *Euonymus europaeus*, *Fagus*, *Hedera*, *Ilex*, *Laburnum anagyroides*, *Lavatera arborea*, *Malus* sp., *Pittosporum* sp., *Populus tremula*, *Salix*, *Sambucus nigra* and *Tilia* spp.
D: NM2: 123, FAN3: 172-173 **D+I:** Ph: 58, B&K3: 188 210 **I:** Sow3: pl. 263 (as *Agaricus velutipes*), FlDan2: 43B & B1, Bon: 171, C&D: 235, Kriegl3: 246
Very common and widespread. Fruiting mainly during the winter. English name = 'Velvet Shank'.

velutipes *var.* **lactea** (Quél.) Bas, *Persoonia* 12(1): 63 (1983)
  Pleurotus velutipes var. *lacteus* Quél., *Compt. Rend. Assoc. Franç. Avancem. Sci.* 9: 663 (1881)
  Collybia velutipes var. *lactea* (Quél.) Rea, *Brit. basidiomyc.*: 332 (1922)
E: !
H: On living wood of *Crataegus monogyna* and *Fraxinus* and on decayed hardwoods.
D: FAN3: 173 **D+I:** Myc9(4): 172
Rarely reported. Perhaps just an unpigmented (albino) form. Known from Cambridgeshire, Co. Durham and North East Yorkshire.

FLAVIPORUS Murrill, *Bull. Torrey Bot. Club* 32: 360 (1905)
Poriales, Coriolaceae
Type: *Polyporus rufoflavus* Berk. & M.A. Curtis (= *Flaviporus brownei*)

brownei (Humb.) Donk, *Persoonia* 1: 189 (1960)

  Boletus brownei Humb., *Fl. Friberg. Spec.*: 101 (1793)
  Polyporus brownei (Humb.) Pers., *Mycol. eur.* 2: 121 (1825)
  Polyporus braunii Rabenh., *Fungi europaei* IV: 2005 (1876)
  Polyporus rufoflavus Berk. & M.A. Curtis, *J. Linn. Soc. Bot.* 10: 310 (1868)
  Flaviporus rufoflavus (Berk. & M.A. Curtis) Murrill, *Bull. Torrey Bot. Club* 32: 360 (1905)
  Leptoporus rufoflavus (Berk. & M.A. Curtis) Pilát, *Atlas champ. Eur.* 3(1): 220 (1936)
E: ! O: Channel Islands: !
H: In Britain, mostly on structural timbers in mines but one recent verified collection on an old log of *Ulmus* sp. in deciduous woodland.
D: EurPoly1: 252
Rarely reported. A common and widespread tropical species, but in temperate regions usually found in places with a constant temperature and humidity, such as mines.

FOMES (Fr.) Fr., *Summa veg. Scand.* 2: 321 (1849)
Poriales, Coriolaceae
Type: *Fomes fomentarius* (L.) J. Kickx f.

fomentarius (L.) J. Kickx f., *Fl. Crypt. Flandres* 2: 237 (1867)
  Boletus fomentarius L., *Sp. pl.*: 1176 (1753)
  Polyporus fomentarius (L.) Fr., *Syst. mycol.* 1: 374 (1821)
E: ! S: c W: ? ROI: !
H: Parasitic then saprotrophic on deciduous trees. Most frequent on *Betula* spp, but also known on *Acer pseudoplatanus*, *Aesculus*, *Fagus*, *Quercus robur* and *Populus nigra*. Reported on other deciduous hosts and on conifers such as *Picea* and *Pinus* spp, but these are not verified.
D: EurPoly1: 254, NM3: 230 **D+I:** B&K2: 306 386 **I:** Sow2: 133, Kriegl1: 521
Common in Scotland and now increasingly collected in England (Berkshire, Buckinghamshire, Cumberland, Lincolnshire, Middlesex, North & South Hampshire, Nottinghamshire, Surrey, West Kent and Yorkshire). A single collection from the Republic of Ireland (Leitrim). Reported from Wales in 1910 but unsubstantiated with voucher material. Records from *Salix* spp, are suspect, mostly proving to be misidentified *Phellinus igniarius*.

FOMITOPSIS P. Karst., *Meddeland. Soc. Fauna Fl. Fenn.* 6: 9 (1881)
Poriales, Coriolaceae
Type: *Fomitopsis pinicola* (Sw.) P. Karst.

pinicola (Sw.) P. Karst., *Meddeland. Soc. Fauna Fl. Fenn.* 6: 9 (1881)
  Boletus pinicola Sw., *Kongl. Vetensk. Acad. Nya Handl.* 31: 88 (1810)
  Polyporus pinicola (Sw.) Fr., *Syst. mycol.* 1: 372 (1821)
E: ! S: !
H: On dead wood of conifers such as *Pinus sylvestris* and *Picea sitchensis*, often on dead standing trunks or large logs and also known on dead trunks of *Betula* spp.
D: EurPoly1: 263-264, NM3: 241 **D+I:** Ph: 228-229, B&K2: 306 387 **I:** Bon: 321, Kriegl1: 523, SV41: Front Cover
Rarely reported. Known from England (Cumberland, South Hampshire, Surrey and West Gloucestershire) and Scotland (South Aberdeenshire).

GALERINA Earle, *Bull. New York Bot. Gard.* 5: 423 (1909)
Cortinariales, Cortinariaceae
  Galera (Fr.) P. Kumm. [*nom. illegit.*, non *Galera* Blume (1825) (*Orchidaceae*)], *Führ. Pilzk.*: 21 (1871)
  Velomycena Pilát, *Schweiz. Z. Pilzk.* 31: 175 (1953)
Type: *Galerina vittiformis* (Fr.) Singer

allospora A.H. Sm. & Singer, *Mycologia* 47: 585 (1955)
  Galerina luteofulva P.D. Orton, *Trans. Brit. Mycol. Soc.* 43(2): 240 (1960)

E: ! **S:** ! **W:** ! **ROI:** !
H: On acidic, sandy or peaty soil. Occasionally on burnt heathland and rarely on bare peat.
D: NM2: 312, BFF7: 47-48 (as *Galerina luteofulva*)
Rarely reported but apparently widespread.

ampullaceocystis P.D. Orton, *Trans. Brit. Mycol. Soc.* 43(2): 236 (1960)
Mis.: *Galerina pseudocamerina* sensu auct. p.p.
E: ! **S:** !
H: On decayed wood of conifers such as *Pinus* spp, also on debris of *Pteridium* and apparently on soil.
D: BFF7: 46-47 **D+I:** B&K5: 312 406
Rarely reported.

atkinsoniana A.H. Sm., *Mycologia* 45: 894 (1953)
E: ! **S:** ! **W:** !
H: On soil or with mosses such as *Sphagnum* spp, and also reported 'amongst grass'.
D: NM2: 310, BFF7: 26-27 **D+I:** B&K5: 312 407
Rarely reported. Apparently widespread.

badipes (Fr.) Kühner, *Encycl. Mycol. 7. Le Genre* Galera: 222 (1935)
Agaricus badipes Fr., *Epicr. syst. mycol.*: 196 (1838)
Naucoria badipes (Fr.) P. Kumm. (as '*badipus*'), *Führ. Pilzk.*: 77 (1871)
Galera badipes (Fr.) Ricken, *Blätterpilze Deutschl.*: 228 (1912)
Galera cedretorum Maire, *Bull. Soc. Mycol. France* 44: 49 (1928)
Galerina cedretorum (Maire) Singer, *Acta Inst. bot. Komarov. Acad. Sci. Plant. Crypt., Ser. 2,* 6: 470 (1950)
E: ! **S:** ! **W:** ! **ROI:** !
H: On soil and litter of conifers.
D: NM2: 308, BFF7: 50 **D+I:** B&K5: 314 410
Rarely reported. Apparently widespread.

calyptrata P.D. Orton, *Trans. Brit. Mycol. Soc.* 43(2): 237 (1960)
Galerina cerina var. *calyptrata* (P.D. Orton) Arnolds, *Biblioth. Mycol.* 90: 360 (1982)
E: ! **S:** ! **W:** ! **ROI:** !
H: On soil, associated with mosses.
D: NM2: 312, BFF7: 27-28 **D+I:** B&K5: 314 411
Rarely reported. Apparently widespread but often mis-identified or confused with *Galerina hypnorum* and allied taxa.

camerina (Fr.) Kühner, *Encycl. Mycol. 7. Le Genre* Galera: 212 (1955)
Agaricus camerinus Fr., *Epicr. syst. mycol.*: 196 (1838)
Naucoria camerina (Fr.) Sacc., *Syll. fung.* 5: 841 (1887)
Galera camerina (Fr.) Ricken, *Blätterpilze Deutschl.*: 228 (1912)
Galerina pseudobadipes Joss. [*nom. nud.*], *Bull. Soc. Mycol. France* 71: 109 (1955)
E: ! **S:** !
H: On decayed wood.
D: NM2: 309, BFF7: 48-49 **D+I:** B&K5: 316 412
Rarely reported.

cephalotricha Kühner, *Bull. Soc. Mycol. France* 88(2): 152 (1973) [1972]
E: !
H: On soil, associated with mosses.
Rarely reported.

cerina A.H. Sm. & Singer, *Mycologia* 47: 563 (1955)
E: ! **S:** !
H: On soil, associated with mosses.
D: BFF7: 28
Reported from South Essex, South Devon, South Hampshire, Staffordshire and Warwickshire but most records are unsubstantiated with voucher material. A single record from Scotland in E.

cerina *var.* **longicystis** A.H. Sm. & Singer, *Mycologia* 47: 567 (1955)

S: !
H: On peaty soil in upland or montane habitat.
A single collection from Shetlands (Fair Isle).

cinctula P.D. Orton, *Trans. Brit. Mycol. Soc.* 43(2): 239 (1960)
E: ! **S:** !
H: On soil, usually associated with *Alnus* in damp woodland. Also known on debris of *Calluna*, *Pteridium* and on decayed wood.
D: BFF7: 47 **D+I:** B&K5: 316 413
Known from England (East Norfolk and North Lancashire) and Scotland (Galloway). Reported elsewhere but most records are unsubstantiated with voucher material.

clavata (Velen.) Kühner, *Encycl. Mycol. 7. Le Genre* Galera: 171 (1935)
Galera fragilis var. *clavata* Velen., *České Houby* III: 548 (1921)
Galera clavata (Velen.) J.E. Lange, *Fl. agaric. danic.* 4: 40 (1939)
Mis.: *Galerina heterocystis* sensu auct. Eur.
E: o **S:** o **W:** o **NI:** ! **ROI:** !
H: On soil amongst mosses in grassland, in grassy areas in woodland, and rather frequently on old lawns.
D: NM2: 311, BFF7: 22-23 **D+I:** Myc9(4): 156, B&K5: 316 414
I: C&D: 347 (as *Galerina heterocystis*)
Occasional but widespread. May occur in large numbers on old garden lawns, especially after heavy rain.

clavus Romagn., *Bull. Soc. Mycol. France* 58: 149 (1942)
E: !
H: On soil amongst grass.
D: BFF7: 57
A single collection in herb. K. from South Wiltshire (Swallowcliffe) in 1970.

embolus (Fr.) P.D. Orton, *Trans. Brit. Mycol. Soc.* 43(2): 176 (1960)
Agaricus embolus Fr., *Epicr. syst. mycol.*: 206 (1838)
Tubaria embolus (Fr.) P. Karst., *Bidrag Kännedom Finlands Natur Folk* 32: 446 (1879)
Galera embolus (Fr.) Quél., *Enchir. fung.*: 107 (1886)
E: ! **S:** ! **ROI:** !
H: On sandy soil, associated with mosses and lichens, on heathland or dunes.
D: NCL3: 239, NM2: 314, BFF7: 37 **I:** FlDan4: 127B (as *Tubaria embola*), C&D: 347
Rarely reported, but apparently widespread. Unsubstantiated records from habitats other than those listed should be treated with caution.

harrisonii (Dennis) Bas & Vellinga, *Persoonia* 13(1): 24 (1986)
Phaeomarasmius harrisonii Dennis, *Kew Bull.* 19(1): 113 (1964)
Flammulaster harrisonii (Dennis) Watling, *Notes Roy. Bot. Gard. Edinburgh* 28(1): 68 (1967)
Galerina antheliae Gulden, *Norweg. J. Bot.* 27(4): 245 (1980)
Mis.: *Galerina tundrae* sensu auct.
S: !
H: On soil, with *Oligotrichum*, in snow beds in montane habitat.
D: NM2: 313 (as *Galerina antheliae*), BFF7: 36
Described from North Ebudes (Rhum) and also known from Mid-Perthshire.

heimansii Reijnders, *Persoonia* 1(1): 165 (1959)
E: ! **S:** ! **W:** !
H: On soil amongst litter in fens, swamps or marshy places. Usually associated with *Alnus* and *Salix* spp but also reported with *Pinus* spp.
D: BFF7: 56
Rarely reported. Apparently widespread.

hypnorum (Schrank) Kühner, *Encycl. Mycol. 7. Le Genre* Galera: 194 (1935)
Agaricus hypnorum Schrank, *Baier. Fl.* 2: 605 (1789)
Galera hypnorum (Schrank) P. Kumm., *Führ. Pilzk.*: 75 (1871)
Galerina calyptrospora Kühner, *Botaniste* 17: 172 (1926)

E: ! **S:** ! **W:** ! **NI:** ! **ROI:** ! **O:** Channel Islands: ! Isle of Man: !
H: On soil, associated with mosses in woodland, on heathland and occasionally in grassland.
D: NM2: 313, BFF7: 29 **D+I:** B&K5: 322 423 (as *Galerina sahleri*) **I:** C&D: 347
Commonly recorded but often confused with similar species. Excludes *G. hypnorum* sensu Smith & Singer (1964) and B&K5: 318 pl. 425 which is *G. sahleri*, as yet unknown in Britain.

hypophaea Kühner, *Bull. Soc. Mycol. France* 88(2): 152 (1973) [1972]
S: ! **W:** !
H: On peaty soil in upland or montane grassland. British material is often associated with *Salix herbacea*.
D: BFF7: 41
Known from Scotland (Stirlingshire and Orkney) and Wales (Caernarvonshire).

jaapii A.H. Sm. & Singer, *Mycologia* 47: 574 (1955)
   Mis.: *Galerina mycenoides* sensu NCL and sensu auct. mult.
E: ! **S:** ! **W:** ! **NI:** ! **O:** Isle of Man: !
H: On soil, associated with mosses in boggy or marshy areas, at the margins of pools and ditches and often amongst *Carex* or *Juncus* spp.
D: NM2: 310, BFF7: 41-42
Rarely reported. Apparently widespread.

laevis (Pers.) Singer, *Persoonia* 2(1): 31 (1961)
   Agaricus laevis Pers., *Mycol. eur.* 3: 164 (1828)
   Galera graminea Velen., *České Houby* III: 548 (1921)
   Galerina graminea (Velen.) Kühner, *Encycl. Mycol. 7. Le Genre Galera*: 168 (1935)
E: c **S:** c **W:** c **NI:** c **ROI:** c
H: On soil amongst grass and mosses, in grassland, grassy areas in woodland and especially old, mossy, unimproved lawns.
D: NM2: 310, BFF7: 21-22 **D+I:** B&K5: 318 416 **I:** C&D: 347
Common and widespread.

marginata (Batsch) Kühner, *Encycl. Mycol. 7. Le Genre Galera*: 225 (1935)
   Agaricus marginatus Batsch, *Elench. fung. (Continuatio Secunda)*: 65 & t.37 f.207 (1789)
   Pholiota marginata (Batsch) Quél., *Mém. Soc. Émul. Montbéliard, Sér. 2*, 5: 127 (1872)
   Galera marginata (Batsch) Kühner & Romagn. [*nom. inval.*], *Flore Analytique des Champignons Supérieurs*: 321 (1953)
   Agaricus autumnalis Peck, *Rep. (Annual) New York State Mus. Nat. Hist.* 23: 92 (1872)
   Galerina autumnalis (Peck) A.H. Sm. & Singer, *Monogr. Galerina*: 246 (1964)
   Agaricus unicolor Vahl, *Fl. Danica*: t. 1071 f. 1 (1792)
   Pholiota unicolor (Vahl) Gillet, *Hyménomycètes*: 436 (1876)
   Galerina unicolor (Vahl) Singer, *Beih. Bot. Centralbl., Abt. 1*, 56: 170 (1936)
E: c **S:** c **W:** c **NI:** c **ROI:** c
H: On decayed wood and sawdust of deciduous trees and conifers. Also known on debris of *Pteridium*.
D: NM2: 308 (as *Galerina unicolor*), BFF7: 51, BFF7: 52 (as *G. autumnalis*), NBA5: 711 (as *G. autumnalis*) **D+I:** B&K5: 314 409 (as *G. autumnalis* var. *autumnalis*), B&K5: 318 417 **I:** C&D: 347
Common and widespread. An albino form has been found in England (Buckinghamshire: Rushbeds Wood and North Hampshire: Stockbridge, Leckford Estate).

mniophila (Lasch) Kühner, *Encycl. Mycol. 7. Le Genre Galera*: 192 (1935)
   Agaricus mniophilus Lasch, *Linnaea* 3: 410 (1828)
   Galera hypnorum var. *mniophila* (Lasch) P. Kumm., *Führ. Pilzk.*: 75 (1871)
   Mis.: *Naucoria melinoides* sensu auct. Brit.
   Mis.: *Galera mycenopsis* sensu Rea (1922)
E: o **S:** o **W:** o **ROI:** o
H: On soil, associated with mosses in grasslands and occasionally in grassy areas in woodland or on old lawns.
D: NM2: 313, BFF7: 39-40 **I:** C&D: 347

Occasional but widespread.

nana (Petri) Kühner, *Encycl. Mycol. 7. Le Genre Galera*: 219 (1935)
   Naucoria nana Petri, *Ann. Mycol.* 2(1): 9 (1904)
   Galera nana (Petri) J.E. Lange, *Fl. agaric. danic.* 4: 41 (1939)
   Galerula velenovskyi Kühner, *Bull. Soc. Mycol. France* 50: 78 (1934)
E: ! **S:** ! **W:** !
H: On woody litter of deciduous trees, often on buried wood and then appearing as if terrestrial.
D: NM2: 309, BFF7: 55-56 **I:** C&D: 347
Rarely reported. Apparently widespread.

paludinella P.D. Orton, *Trans. Brit. Mycol. Soc.* 91(4): 553 (1988)
E: ! **W:** !
H: On wet soil. British material associated with the moss *Acrocladium cuspidatum* and *Carex riparia* growing under or near *Salix* spp.
D: NBA9: 553, BFF7: 54-55
Known from England (East Suffolk) and Wales (Merionethshire).

paludosa (Fr.) Kühner, *Encycl. Mycol. 7. Le Genre Galera*: 184 (1935)
   Agaricus paludosus Fr., *Epicr. syst. mycol.*: 209 (1838)
   Galera paludosa (Fr.) P. Kumm., *Führ. Pilzk.*: 75 (1871)
   Tubaria paludosa (Fr.) P. Karst., *Bidrag Kännedom Finlands Natur Folk* 32: 445 (1879)
   Pholiota paludosa (Fr.) Pat., *Hyménomyc. Eur.*: 116 (1887)
E: o **S:** c **W:** o **NI:** ! **ROI:** !
H: Amongst *Sphagnum* spp in bogs and marshes. Often fruiting May to August.
D: NM2: 310, BFF7: 39 **D+I:** B&K5: 320 419 **I:** C&D: 347
Common in Scotland. Occasional but widespread elsewhere.

permixta (P.D. Orton) Pegler & T.W.K. Young, *Kew Bull.* 30(2): 239 (1975)
   Naucoria permixta P.D. Orton, *Trans. Brit. Mycol. Soc.* 43(2): 317 (1960)
   Naucoria cephalescens T.J. Wallace, *Trans. Brit. Mycol. Soc.* 43(2): 315 (1960)
   Galerina cephalescens (T.J. Wallace) Pegler & T.W.K. Young, *Kew Bull.* 30(2): 238 (1975)
E: ! **S:** !
H: On damp soil, associated with *Salix* spp, and occasionally on leaf debris of various species of *Carex*.
D: Reid1: 233, BFF7: 34
Rarely reported. Most records are dubious and unsubstantiated with voucher material.

phillipsii D.A. Reid, *Trans. Brit. Mycol. Soc.* 82(2): 235 (1984)
E: ! **S:** !
H: On heathland. Type material from a large burnt area, amongst *Polytrichum* spp, on wet peaty soil.
D: BFF7: 30-31 **D+I:** Ph: 156-157
Rarely reported but apparently widespread. Described from England (Surrey).

praticola (F.H. Møller) P.D. Orton, *Trans. Brit. Mycol. Soc.* 43(2): 176 (1960)
   Pholiota praticola F.H. Møller, *Fungi Faeroes* 1: 231 (1945)
   Pholiota pumila var. *subferruginea* F.H. Møller & J.E. Lange, *Fl. agaric. danic.* 5, *Taxonomic Conspectus*: VII (1940)
   Mis.: *Pholiota mycenoides* sensu Rea (1922)
   Mis.: *Galerina unicolor* sensu auct. mult.
E: ! **S:** o **W:** ! **NI:** ! **ROI:** !
H: On soil amongst grass or mosses.
D: NCL3: 241-242, NM2: 308 (as *Galerina unicolor*), BFF7: 54 **D+I:** B&K5: 328 431 (as *G. unicolor*)
Occasional in Scotland. Rarely reported elsewhere but apparently widespread. Barely distinct from *Galerina marginata* except in its ecology.

pseudocerina A.H. Sm. & Singer, *Mycologia* 50: 483 (1958)
S: !
H: On soil in montane habitat, often amongst mosses in base rich flushes.

D: BFF7: 45 **D+I:** B&K5: 322 422
Known only from Scotland but increasingly reported there.

pseudomniophila Kühner, *Bull. Soc. Mycol. France* 88(2): 152 (1973) [1972]
E: ? **S:** ! **W:** !
H: On soil amongst mosses in montane habitat.
D: BFF7: 40
Reported from England but unsubstantiated with voucher material.

pseudomycenopsis Pilát, *Friesia* 5: 19 (1954)
Galerina moelleri Bas, *Persoonia* 1(3): 310 (1960)
Galerina pseudopumila P.D. Orton, *Trans. Brit. Mycol. Soc.* 43(2): 176 (1960)
Mis.: *Pholiota pumila* sensu Rea (1922)
E: ! **S:** !
H: On soil associated with mosses, most frequently in boreal and montane habitat but also known on heathland, in upland pasture, and in marshy areas.
D: NM2: 308, BFF7: 52-53 **D+I:** B&K5: 320 418 (as *Galerina moelleri*)
Rarely reported but apparently widespread. Common in Scotland *fide* Watling (pers. comm.), and a gastroid form has been described from Orkney (Watling & Martin, 2002). Older records may refer to other annulate species of *Galerina*.

pseudotundrae Kühner, *Bull. Soc. Mycol. France* 88(2): 152 (1973) [1972]
S: !
H: On peaty soil in montane and upland habitat.
D: NM2: 312, BFF7: 35-36
Known only from Orkney and Shetland.

pumila (Pers.) Singer, *Persoonia* 2(1): 41 (1961)
Agaricus pumilus Pers., *Syn. meth. fung.*: 317 (1801)
Pholiota pumila (Pers.) Gillet, *Hyménomycètes*: 432 (1876)
Agaricus mycenopsis Fr., *Observ. mycol.* 2: 38 (1818)
Galera mycenopsis (Fr.) Quél., *Mém. Soc. Émul. Montbéliard, Sér. 2,* 5: 236 (1872)
Galerina mycenopsis (Fr.) Kühner, *Encycl. Mycol. 7. Le Genre* Galera: 190 (1935)
E: ! **S:** ! **NI:** ! **ROI:** ! **O:** Channels Islands: !
H: On soil amongst mosses.
D: NM2: 313, BFF7: 32 **I:** Ph: 156 (as *Galerina mycenopsis*)
Apparently frequent and widespread.

salicicola P.D. Orton, *Trans. Brit. Mycol. Soc.* 43(2): 242 (1960)
E: ! **W:** !
H: On decayed wood of *Alnus* and *Salix* spp in wet areas of woodland and around the margins of ponds.
D: BFF7: 31 **I:** C&D: 347
Rarely reported. Apparently widespread.

septentrionalis A.H. Sm., *Monogr. Galerina*: 152 (1964)
S: !
H: In boggy areas, usually amongst *Sphagnum* spp. British material with *Pinus sylvestris* in Caledonian pinewoods.
D: BFF7: 38
Known only from Easterness.

sideroides (Bull.) Kühner, *Encycl. Mycol. 7. Le Genre* Galera: 215 (1935)
Agaricus sideroides Bull., *Herb. France*: pl. 588 (1793)
Naucoria sideroides (Bull.) Quél., *Mém. Soc. Émul. Montbéliard, Sér. 2,* 5: 131 (1872)
Galera sideroides (Bull.) Kühner & Romagn., *Flore Analytique des Champignons Supérieurs*: 320 (1953)
E: ! **S:** !
H: On woody debris (stumps, logs, fallen twigs, branches) of conifers.
D: BFF7: 45-46 **D+I:** B&K5: 324 424
Rarely reported but apparently widespread. Older unsubstantiated records may refer to similar taxa.

sphagnorum (Pers.) Kühner, *Encycl. Mycol. 7. Le Genre* Galera: 179 (1935)

Agaricus hypnorum γ *sphagnorum* Pers., *Syn. meth. fung.*: 386 (1801)
Galera sphagnorum (Pers.) P. Karst., *Bidrag Kännedom Finlands Natur Folk* 32: 441 (1879)
E: ! **S:** ! **W:** ! **ROI:** !
H: Amongst *Sphagnum* spp in bogs on heathland, moorland and occasionally in woodland. May be found in standing water in such habitat.
D: NM2: 312, BFF7: 37-38 **D+I:** B&K5: 324 425
Rarely reported. Apparently widespread.

stordalii A.H. Sm., *Monogr. Galerina*: 203 (1964)
Mis.: *Galerina dimorphocystis* sensu Kühner [Bull. Soc. Mycol. France 88 (1) 41-118 (1972)] & sensu auct.
S: !
H: Associated with *Sphagnum* spp in boggy areas in upland and montane habitat.
D: NM2: 311, BFF7: 24
Known only from Orkney and Shetland.

stylifera (G.F. Atk.) A.H. Sm. & Singer, *Sydowia* 11: 449 (1958) [1957]
Galerula stylifera G.F. Atk., *Proc. Amer. Philos. Soc.* 57: 365 (1918)
Naucoria sideroides var. *indusiata* J.E. Lange, *Fl. agaric. danic.* 5, Taxonomic Conspectus: VI (1940)
E: ! **S:** o
H: On woody debris of conifers.
D: NM2: 308, BFF7: 49 **D+I:** B&K5: 324 426
Occasional in Scotland *fide* R. Watling (pers. comm.) but rarely reported elsewhere. Recent records associated with *Pteridium* are dubious and unsubstantiated with voucher material. Not distinguished from *Galerina sideroides* in NCL.

subannulata (Singer) Singer, *Monograph of the genus* Galerina Earle: 293 (1964)
Galerina vittiformis var. *subannulata* Singer, *Sydowia* 7: 246 (1953)
S: !
H: On soil, associated with mosses.
Reported from Shetland by Watling (1994).

subcerina A.H. Sm. & Singer, *Mycologia* 50: 485 (1958)
Galerina subcerina var. *anglica* A.H. Sm., *Monogr. Galerina*: 89 (1964)
E: ! **S:** !
H: On soil, associated with mosses.
D: BFF7: 30
Known from Scotland (Mid-Ebudes) and England (North Somerset).

subclavata Kühner, *Bull. Soc. Mycol. France* 88(2): 152 (1973) [1972]
E: ! **S:** ! **NI:** ?
H: On soil amongst grass or mosses, in parkland, open grassland or heathland.
D: NM2: 311, BFF7: 23 **D+I:** B&K5: 326 427
Rarely reported but apparently widespread.

terrestris V.L. Wells & Kempton, *Lloydia* 32: 385 (1969)
S: !
H: On soil in montane habitat.
D: NM2: 310, BFF7: 26 **D+I:** B&K5: 326 428
Few records, and true distribution unknown.

tibiicystis (G.F. Atk.) Kühner, *Encycl. Mycol. 7. Le Genre* Galera: 176 (1935)
Galerula tibiicystis G.F. Atk., *Proc. Amer. Philos. Soc.* 57: 365 (1918)
Galera tibiicystis (G.F. Atk.) A. Pearson, *Trans. Brit. Mycol. Soc.* 35: 113 (1952)
Mis.: *Galera sphagnorum* sensu Lange (FlDan4: pl. 130E)
E: o **S:** c **W:** ! **ROI:** !
H: Amongst *Sphagnum* spp in woodland, on heathland and moorland, and also in boggy pasture in upland and montane habitat.
D: NM2: 311, BFF7: 42-43 **D+I:** B&K5: 326 429

Rather common in Scotland. Occasional or rarely reported elsewhere but apparently widespread.

triscopa (Fr.) Kühner, *Encycl. Mycol. 7. Le Genre* Galera: 206 (1935)
> *Agaricus triscopus* Fr., *Monogr. hymenomyc. Suec.* 1: 375 (1857)
> *Galera triscopa* (Fr.) Quél., *Enchir. fung.*: 107 (1886)
> *Naucoria triscopoda* (Fr.) Sacc., *Syll. fung.* 5: 841 (1887)
E: ! S: ! NI: ?
H: On acidic peaty soil or old decayed and moss-covered wood. Also known from plant pots in cool greenhouses.
D: NM2: 309, BFF7: 43
Rarely reported. The majority of reports are unsubstantiated with voucher material and records 'on calcareous soils' are suspect.

uncialis (Britzelm.) Kühner, *Encycl. Mycol. 7. Le Genre* Galera: 217 (1935)
> *Agaricus uncialis* Britzelm., *Ber. Naturhist. Vereins Augsburg* 30: 21 (1890)
> *Naucoria uncialis* (Britzelm.) Sacc., *Syll. fung.* 11: 59 (1895)
E: ! S: ! W: ?
H: On sandy soil, often on dunes, associated with mosses.
D: BFF7: 44 **D+I:** B&K5: 328 430 **I:** C&D: 347
Known from England and Scotland but rarely reported. Reported from Wales but unsubstantiated with voucher material.

viscidula P.D. Orton, *Trans. Brit. Mycol. Soc.* 91(4): 554 (1988)
E: ? S: !
H: On soil, usually associated with mosses in woodland, mainly with *Pinus* and *Betula* spp.
D: NBA9: 554, BFF7: 33 **I:** Ph: 157 (misdetermined as *Galerina cinctula*)
Reported once from England but without voucher material.

vittiformis (Fr.) Singer, *Acta Inst. Bot. Acad. Sci. USSR Plant. Crypt., Ser. 2,* 6: 472 (1950)
> *Agaricus vittiformis* Fr., *Epicr. syst. mycol.*: 207 (1838)
> *Galera vittiformis* (Fr.) P. Kumm, *Führ. Pilzk.*: 75 (1871)
> *Galerula muricellospora* G.F. Atk., *Proc. Amer. Philos. Soc.* 57: 360 (1918)
> *Galerina muricellospora* (G.F. Atk.) Kühner, *Encycl. Mycol. 7. Le Genre* Galera: 203 (1935)
> *Agaricus rubiginosus* Pers. [non *A. rubiginosa* With. (1796)], *Syn. meth. fung.*: 385 (1801)
> *Galera hypnorum* var. *rubiginosa* (Pers.) P. Kumm., *Führ. Pilzk.*: 75 (1871)
> *Galera rubiginosa* (Pers.) P. Karst., *Bidrag Kännedom Finlands Natur Folk* 32: 440 (1879)
> *Galerina rubiginosa* (Pers.) Kühner, *Encycl. Mycol. 7. Le Genre* Galera: 200 (1935)
> *Galerina vittiformis* f. *tetraspora* Arnolds, *Biblioth. Mycol.* 90: 381 (1983) [1982]
E: c S: c W: c NI: c ROI: c O: Isle of Man: !
H: On soil amongst grass and mosses in grassland, heathland, woodland, parkland and gardens.
D: NM2: 310 (as *Galerina vittiformis* var. *vittiformis* f. *tetraspora*), BFF7: 25 **D+I:** B&K5: 330 434 (as *G. vittiformis* var. *vitiformis* f. *tetraspora*) **I:** C&D: 347
Common and widespread. The tetrasporic form is less common in Britain than the bisporic form.

GALZINIA Bourdot, *Compt. Rend. Assoc. Franç. Avancem. Sci.* 45: 577 (1922)
Stereales, Sistotremataceaeç
Type: *Galzinia pedicellata* Bourdot

forcipata Pouzar, *Česká Mykol.* 37(4): 209 (1983)
E: !
H: On decayed wood of deciduous trees (British material on *Quercus* sp).
A single collection from South Devon (Orley Common) in 1991.

incrustans Parmasto, *Eesti N.S.V. Tead. Akad. Toimet., Biol.* 14(2): 225 (1965)

> *Corticium incrustans* Höhn. & Litsch. [*nom. illegit.*, non *C. incrustans* Pers. (1796)], *Sitzungsber. Kaiserl. Akad. Wiss., Wien, Math.-Naturwiss. Cl., Abt. 1* 115: 1602 (1906)
E: !
H: On decayed wood of deciduous trees.
D: NM3: 125, Myc14(2): 52 **D+I:** CNE3: 394-395, B&K2: 172 186
Rarely recorded or collected. Known from Berkshire, Herefordshire, Hertfordshire, London, South Devon and Surrey.

pedicellata Bourdot, *Compt. Rend. Assoc. Franç. Avancem. Sci.* 45: 578 (1922)
E: !
H: On decayed wood of deciduous trees. British material on *Salix* spp.
D: NM3: 125 **D+I:** CNE3: 396-397, Myc10(2): 50
Two collections, from South Devon (Bovey Tracey, Great Plantation).

GAMUNDIA Raithelh., *Metrodiana* 8: 34 (1979)
Agaricales, Tricholomataceae
> *Stachyomphalina* H.E. Bigelow, *Mycotaxon* 9: 41 (1979)
Type: *Gamundia pseudoclusilis* (Joss. & Konrad) Raithelh. (= *G. striatula*)

striatula (Kühner) Raithelh., *Metrodiana, Sonderheft* 2: 9 (1983)
> *Rhodocybe striatula* Kühner, *Bull. Mens. Soc. Linn. Lyon* 2: 139 (1928)
> *Omphalia striatula* (Kühner) Kühner & Romagn., *Flore Analytique des Champignons Supérieurs*: 127 (1953)
> *Clitocybe striatula* (Kühner) P.D. Orton, *Trans. Brit. Mycol. Soc.* 43(2): 174 (1960)
> *Stachyomphalina striatula* (Kühner) H.E. Bigelow, *Mycotaxon* 9(1): 42 (1979)
> *Omphalia fuscoalba* F.H. Møller, *Fungi Faeroes*: 263 (1945)
> *Collybia pseudoclusilis* Joss. & Konrad, *Bull. Bi-mens. Soc. Linn. Lyon* 10: 21 (1931)
> *Clitocybe pseudoclusilis* (Joss. & Konrad) P.D. Orton, *Trans. Brit. Mycol. Soc.* 43(2): 174 (1960)
> *Fayodia pseudoclusilis* (Joss. & Konrad) Singer, *Sydowia* 15: 66 (1962) [1961]
> *Gamundia pseudoclusilis* (Joss. & Konrad) Raithelh., *Metrodiana* 9(2): 48 (1980)
> Mis.: *Collybia cessans* sensu Pearson [TBMS 32:263 (1949)]
> Mis.: *Omphalia clusilis* sensu Kühner & Romagnesi (1953)
> Mis.: *Fayodia leucophylla* sensu auct. Brit.
> Mis.: *Omphalia leucophylla* sensu Lange (FlDan2: pl. 59F)
E: ! S: ! W: ! ROI: !
H: On soil in woodland, with deciduous trees or conifers, amongst grass on lawns and also on dunes.
D: NM2: 122 (as *Fayodia pseudoclusilis*), FAN3: 155-156, BFF8: 95 (as *Gamundia pseudoclusilis*), BFF8: 96 **D+I:** A&N3 143-148 pl.40, B&K3: 188 209 (as *F. pseudoclusilis*) **I:** C&D: 187 (as *F. pseudoclusilis*), C&D: 187 (as *F. leucophylla*)
Rarely reported but apparently widespread. Most records are unsubstantiated with voucher material. Some authorities recognise *Gamundia pseudoclusilis* as distinct.

GANODERMA P. Karst., *Rev. Mycol. (Toulouse)* 3(9): 17 (1881)
Ganodermatales, Ganodermataceae
Type: *Ganoderma lucidum* (Curtis) P. Karst.

applanatum (Pers.) Pat., *Hyménomyc. Eur.*: 143 (1887)
> *Boletus applanatus* Pers., *Observ. mycol.* 2: 2 (1799)
> *Polyporus applanatus* (Pers.) Wallr., *Fl. crypt. Germ., Sect. II*: 591 (1833)
> *Fomes applanatus* (Pers.) Gillet, *Hyménomycètes*: 685 (1878)
> Mis.: *Ganoderma lipsiense* sensu auct.
E: o S: o W: o NI: o ROI: o O: Channel Islands: ! Isle of Man: !
H: Parasitic, then saprotrophic. Usually on *Fagus* but also known from *Acer pseudoplatanus*, *Betula*, *Fraxinus*, *Quercus*, *Salix*, *Tilia* and *Ulmus* spp. More unusual hosts include *Chrysolepis*

chrysophila, *Juglans regia*, *Laburnum anagyroides*, *Laurus nobilis*, *Malus* sp, *Morus nigra*, *Prunus amygdaloides* and *Rhus typhina*. Rarely on conifers such as *Abies*, *Araucaria*, *Cedrus*, *Picea* and *Pinus* spp.
D: EurPoly1: 269, NM3: 246 (as *Ganoderma lipsiense*), NM3: 246 (as *G. lipsiense*) **D+I:** Ph: 226-227, B&K2: 332 425 **I:** Kriegl1: 423 (as *G. lipsiense*)
Occasional but widespread. Much less frequent than *G. australe* with which it has been much confused. Basidiomes may be attacked by a dipteran, *Agathomyia wankowiczii*, the larvae of which induce large nipple-like galls on the pore surface.

australe (Fr.) Pat., *Bull. Soc. Mycol. France* 4: 1712 (1887)
  Polyporus australis Fr., *Elench. fung.* 1: 108 (1828)
  Fomes australis (Fr.) Cooke, *Grevillea* 14: 18 (1885)
  Polyporus adspersus Schulzer, *Flora* 61: 11 (1878)
  Ganoderma adspersum (Schulzer) Donk, *Proc. K. Ned. Akad. Wet., Ser. C* 72: 273 (1969)
  Ganoderma europaeum Steyaert, *Bull. Jard. bot. Brux.* 31: 70 (1961)
E: c **S:** c **W:** c **NI:** c **ROI:** c **O:** Channel Islands: c Isle of Man: c
H: Parasitic, then saprotrophic. Usually on *Fagus* but known on a wide range of deciduous trees in woodland, parkland and urban areas. Less common hosts include *Buddleja variabilis*, *Gingko biloba*, *Laurus nobilis*, *Prunus spinosa*, *Pterocarya fraxinifolia*, *Rhododendron ponticum*, *Robinia pseudoacacia*, *Rosa longicuspis*, *Spiraea hypericifolia* and *Ulex europaeus*. Also known on *Araucaria araucana* but reports from conifers are otherwise unverified. **D:** EurPoly1: 271, NM3: 246 (as *Ganoderma adspersum*) **D+I:** Ph: 226-227 (as *G. adspersum*), B&K2: 332 424 (as *G. adspersum*)
Very common and widespread. Perhaps better known as *G. adspersum*. Often confused with *G. applanatum* thus records unsubstantiated with voucher material should be viewed with caution. English name = 'Artist's Fungus'.

carnosum Pat., *Bull. Soc. Mycol. France* 5: 66 (1889)
  Ganoderma atkinsonii H. Jahn, Kotl. & Pouzar, *Westfäl. Pilzbriefe* 11(6): 98 (1980)
  Mis.: *Ganoderma valesiacum* sensu Pegler (1973)
E: ! **S:** !
H: Parasitic. On living trunks of *Taxus* usually in areas of mixed woodland on calcareous soil.
D: EurPoly1: 273 **D+I:** B&K2: 332 426 (? typical)
Known from East Gloucestershire, East Sussex, Hertfordshsire, Middlesex, North & South Hampshire, Surrey, West Kent and West Sussex, and from Stirlingshire. Usually reported as *Ganoderma valesiacum* but this is a central European species, unknown in Britain, and restricted to *Larix* spp *fide* Ryvarden and Gilbertson (EurPoly1).

lucidum (Curtis.) P.Karst, *Rev. Mycol. (Toulouse)* 3(9): 17 (1881)
  Boletus lucidus Curtis, *Fl. londin.* 1: t. 224 (1777)
  Grifola lucida (Curtis) Gray, *Nat. arr. Brit. pl.* 1: 644 (1821)
  Polyporus lucidus (Curtis) Fr., *Syst. mycol.* 1: 353 (1821)
  Fomes lucidus (Curtis.) Fr., *Nov. Symb. Myc.*: 61 (1851)
  Boletus laccatus Timm, *Fl. Megapol. Prodr.*: 269 (1788)
  Polyporus laccatus (Timm) Pers., *Mycol. eur.* 2: 54 (1825)
  Ganoderma laccatum (Timm) Pat., *Ann. Jard. Bot. Buitenzorg* 8: 11 (1897)
E: o **S:** ! **W:** !
H: Weakly parasitic or saprotrophic. On roots or stumps of deciduous trees, most often *Quercus* spp, and *Carpinus*. Also known on *Corylus*, *Fagus*, *Malus* sp., *Prunus domesticus*, *Robinia pseudoacacia*, *Salix* and *Ulmus* spp, and a single collection on *Ulex europaeus*. Rarely on conifers such as *Cedrus libani*, *Picea abies*, *Pseudotsuga menziesii* and *Taxus*.
D: EurPoly1: 275, NM3: 245 **D+I:** Ph: 224-225 **I:** Sow2: 134 (as *Boletus lucidus*), Bon: 321, C&D: 141, Kriegl1: 425
Occasional but widespread in England, from Yorkshire southwards, although mostly recorded from southern or south-western areas. Rarely reported elsewhere. Originally described on *Corylus avellana* from Peckham (South London).

pfeifferi Bres., *Bull. Soc. Mycol. France* 5: 70 (1889)

E: o **S:** !
H: Parasitic, then saprotrophic. Usually on trunks or stumps of *Fagus* in woodland or parkland. Rarely reported on *Quercus* spp, but known on planted *Q. coccinea*.
D: EurPoly1: 277, NM3: 246 **D+I:** Myc9(2): 61 **I:** Kriegl1: 427
Occasional but widespread in England (Bedfordshire, Berkshire, East Gloucestershire, East & West Kent, Herefordshire, Hertfordshire, North & South Hampshire, North Wiltshire, Surrey, Warwickshire, West Sussex and Yorkshire); also known from Scotland (Dumfrieshire).

resinaceum Boud., *Bull. Soc. Mycol. France* 5: 72 (1889)
  Fomes resinaceus (Boud.) Sacc., *Syll. fung.* 9: 179 (1891)
E: o **S:** ! **W:** ! **NI:** o
H: Parasitic. On trunks of deciduous trees, usually *Fagus* or *Quercus* spp. Also known on *Aesculus*, *Betula pendula*, *Carpinus* and *Salix fragilis*.
D: EurPoly1: 279, NM3: 245 **D+I:** B&K2: 334 427
Occasional in England, rarely reported elsewhere but widespread.

GASTROSPORIUM Mattir., *Mem. R. Accad. Torino., Ser. 2* 53: 361 (1903)
Hymenogastrales, Gastrosporiaceae
Type: *Gastrosporium simplex* Mattir.

simplex Mattir., *Mem. Reale Accad. Sci. Torino, Series 2* 53: 361 (1903)
E: !
H: On soil. British material in a garden, associated with *Solidago* sp. Associated with grasses in central Europe.
D: NM3: 315 **D+I:** BritTruff: 163-164 10 A
A single collection, from Hertfordshire (Harpenden) in 1973. English name = 'Steppe Truffle'.

GAUTIERIA Vittad., *Monogr. Tuberac.*: 25 (1831)
Cortinariales, Gautieriaceae
Type: *Gautieria morchelliformis* Vittad.

morchelliformis Vittad., *Monogr. Tuberac.*: 26 (1831)
E: !
H: Epigeous, amongst leaf litter, associated with *Fagus*.
D+I: BritTruff: 153-154 9 A
Known from a restricted area in West Gloucestershire (Wotton under Edge).

GEASTRUM Pers., *Neues Mag. Bot.* 1: 85 (1794)
Lycoperdales, Geastraceae
  Geaster Fr., *Syst. mycol.* 3: 8 (1829)
Type: *Geastrum multifidum* Pers.

berkeleyi Massee [as *Geaster berkleyi*], *Ann. Bot.* 4: 79 (1889)
E: ! **S:** !
H: On calcareous soil in deciduous woodland (associated with *Fraxinus* or *Ulmus* spp in two of the locations) or on open ground.
D: NM3: 342 **D+I:** BritPuffb: 84-85 Figs. 56/57 **I:** FM2(1): 34
Sporadically reported from scattered sites in England (Berkshire, North Lincolnshire, East Norfolk, Northamptonshire, Nottinghamshire, and Surrey) in the period 1880 and 1952. Considered extinct until recollected in Herefordshire 1996 and Worcestershire in 2000. A single collection from Scotland (East Lothian) in 1909.

campestre Morgan [as *Geaster campestre*], *Amer. Naturalist.* 21: 1027 (1887)
E: !
H: On calcareous soil, in parkland, roadsides and churchyards. The most recent collections were associated with *Cedrus* spp on a roadside verge. Also reported with *Taxus*.
D: NM3: 342 **D+I:** BritPuffb: 82-83 Figs. 54/55 **I:** SV31: 13, Kriegl2: 115 (as *Geastrum pedicellatum*)
Known from Hertfordshire (Ickleford) in 1958, Surrey (Kew Gardens) in 1926 and several collections from Pyrford (near Woking) in 2000-2001. Possibly an introduction.

corollinum (Batsch) Hollós, *Gasterom. Ung.*: 57 (1903)
  Lycoperdon corollinum Batsch, *Elench. fung.*: 151 (1783)
  Geastrum mammosum Chevall., *Fl. gén. env. Paris,* Edn 1:
    359 (1826)
  Lycoperdon recolligens With., *Bot. arr. Brit. pl. ed. 2,* 3: 462
    (1792)
  Geastrum recolligens (With.) Desv., *J. Bot. (Paris)* 2: 102
    (1809)
E: ! S: ! W: !
H: On calcareous soil in deciduous woodland or hedgerows.
D: NM3: 343 **D+I:** BritPuffb: 106-107 Figs. 78/79 **I:** Sow1: 80
  (as *Lycoperdon recolligens*), Kriegl2: 108
Apparently rare. Recent collections only from England
(Berkshire, East Kent, East Norfolk, West Suffolk and
Yorkshire). Reported from Cheshire, Co. Durham,
Hertfordshire, Isle of Wight, Leicestershire, Staffordshire, West
Kent, West Norfolk and Worcestershire but unsubstantiated
with voucher material. A single Scottish record (East Lothian).
Collected in Wales (Anglesey) in 1998.

coronatum Pers., *Syn. meth. fung.*: 132 (1801)
  Geastrum limbatum Fr. [as *Geaster limbatus*], *Syst. mycol.* 3:
    15 (1829)
E: ! S: ! W: !
H: On calcareous loam soils in woodland, on dunes, in
hedgerows and in graveyards. Associated with *Crataegus,
Pinus, Quercus, Thuja* and *Ulmus* spp sometimes amongst
dense ground cover of *Hedera.*
D: NM3: 344 **D+I:** Ph: 252-253, BritPuffb: 104-105 Figs. 76/77
  I: Kriegl2: 108
Rarely reported. Known from widely scattered locations in
England, from Co. Durham southwards, from Wales (Anglesey,
Cardiganshire, Denbighshire and Flintshire) and a few old
records from Scotland (Edinburgh and Haddington).

elegans Vittad. [as *Geaster elegans*], *Monogr. Lycoperd.*: 15
  (1842)
  Geastrum badium Pers., *J. Bot., Paris* 2: 31 (1809)
E: ! S: ! W: !
H: On sandy soil, usually in coastal areas in grassland or dunes,
occasionally inland.
D: NM3: 342 **D+I:** BritPuffb: 80-81 Figs. 52/53 **I:** Kriegl2: 109
Rarely reported. Known from England (East Gloucestershire,
East Suffolk, South Lancashire, West Cornwall and West
Norfolk), Scotland (East Lothian) and Wales (Anglesey and
Cardiganshire).

fimbriatum Fr. [as *Geaster fimbriatus*], *Syst. mycol.* 3: 16
  (1829)
  Geastrum rufescens var. *minor* Pers., *Syn. meth. fung.*: 134
    (1801)
  Lycoperdon sessile Sowerby, *Col. fig. Engl. fung., Suppl.*: pl.
    401 (1809)
  Geastrum sessile (Sowerby) Pouzar, *Folia Geobot. Phytotax.*
    6: 95 (1971)
  Geastrum tunicatum Vittad., *Monogr. Lycoperd.*: 18 (1842)
E: o S: o W: o NI: ! ROI: o O: Channel Islands: !
H: On soil (often calcareous) in deciduous woodland, associated
with *Acer pseudoplatanus, Alnus, Carpinus, Corylus, Crataegus,
Fagus, Fraxinus, Populus tremula* and *Prunus avium* or with
conifers such as *Cedrus, Cupressus, Larix, Picea, Pinus,
Sequoia* spp and *Taxus.* Rarely in coastal scrub, with
Hippophae rhamnoides, on dunes, in hedgerows, gardens and
rather frequently with *Urtica dioica* in woodland nettlebeds.
D: NM3: 344 **D+I:** Ph: 252 (as *Geastrum sessile*), B&K2: 382
  501 (as *G. sessile*), BritPuffb: 94-95 Figs. 66/67 **I:** Bon: 303
  (as *G. sessile*), Kriegl2: 110
Occasional but widespread.

floriforme Vittad., *Monogr. Lycoperd.*: 23 (1842)
E: ! W: ? O: Channel Islands: !
H: On sandy soil, often with *Cupressus macrocarpa* but also
known with *Larix* spp and *Quercus ilex.*
D: NM3: 343 **D+I:** BritPuffb: 92-93 Figs. 64/65 **I:** Kriegl2: 111
Known from Co. Durham, East Suffolk, East Sussex,
  Huntingdonshire, South Lancashire, North Essex, West Suffolk,

Surrey, West Cornwall and West Norfolk and the Channel
Islands (Guernsey and Jersey). Reported from Wales but
unsubstantiated with voucher material.

fornicatum (Huds.) Hook., in Curtis, *Fl. londin.*, Edn 2, 4: 575
  (1821)
  Lycoperdon fornicatum Huds., *Fl. angl.*: 502 (1762)
E: o W: ! ROI: ? O: Channel Islands: !
H: On soil in woodland or in hedgerows with deciduous or
coniferous trees. Known with *Corylus, Fagus, Larix, Pinus* spp
and *Taxus.* Rarely reported with *Urtica dioica* in nettlebeds and
single records with *Alnus* in a fen, and *Syringa vulgaris* in a
garden.
D: NM3: 343 **D+I:** Ph: 254, BritPuffb: 96-97 Figs. 68/69 **I:**
  Sow2: 198 (as *Lycoperdon fornicatum*)
Occasional in England. Rarely reported elsewhere.

lageniforme Vittad., *Monogr. Lycoperd.*: 16 (1842)
  Mis.: *Geastrum saccatum* Fr. sensu auct. Brit.
E: ! W: !
H: On soil amongst litter in woodland, gardens and in
hedgerows. Reported with deciduous trees e.g. *Betula* and
Quercus spp, coniferous trees e.g. *Larix* and also in nettlebeds.
D+I: BritPuffb: 110-111 Figs. 82/83 **I:** Kriegl2: 112
Known from England (East Norfolk, Herefordshire,
Northamptonshire, South Devon, Yorkshire, West
Gloucestershire and West Sussex) and Wales
(Glamorganshire). Closely related to *G. triplex* and may be
misidentified as that species. Some authorities [e.g. Sunhede
(1989)] keep *G. saccatum* (described from Brazil) as a
separate species, not confirmed from Britain.

minimum Schwein., *Schriften Naturf. Ges. Leipzig* 1: 58 (1822)
E: ! O: Isle of Man: ?
H: On sandy, calcareous soil in dunes, sometimes amongst
Ammophila arenaria and often near to planted *Pinus nigra.*
D: NM3: 344 **D+I:** BritPuffb: 100-101 Figs. 72/73 **I:** Kriegl2:
  113
Known from Cumberland (near Ravenglass) and West Norfolk
(Holkham Gap). Reported from the Isle of Man (Blue Point) but
unsubstantiated with voucher material.

pectinatum Pers., *Syn. meth. fung.*: 132 (1801)
  Geastrum plicatum Berk. [as *Geaster plicatum*], *Ann. Mag.
    Nat. Hist., Ser. 1* 3: 399 (1839)
  Geastrum tenuipes Berk. [as *Geaster tenuipes*], *London J.
    Bot.* 7: 576 (1848)
E: o S: ! W: ! NI: ! O: Channel Islands: ! Isle of Man: ?
H: On soil in woodland, parkland, hedgerows, gardens, dunes,
cemeteries and reported from old spoil heaps. Usually
associated with conifers such as *Cedrus, Chamaecyparis
lawsoniana, Cupressus macrocarpa, Juniperus, Pinus, Picea*
spp and *Taxus* but also with deciduous trees such as
Crataegus, Fagus and *Fraxinus.*
D: NM3: 343 **D+I:** 254, B&K2: 382 499, BritPuffb: 90-91 Figs.
  62/63 **I:** Kriegl2: 115
Occasional but apparently widespread in England, rarely
reported elsewhere.

quadrifidum Pers., *Neues Mag. Bot.* 1: 86 (1794)
E: !
H: On calcareous soil, usually associated with *Fagus* in
beechwoods, and rarely *Crataegus* and *Quercus* spp. Also
rarely with conifers such as *Pinus sylvestris* and *Taxus.*
D: NM3: 343 **D+I:** Ph: 252-253, B&K2: 382 500, BritPuffb: 98-
  99 Figs. 70/71 **I:** Kriegl2: 117
Known from Bedfordshire, Co. Durham, East Kent, East & West
Norfolk, North & South Essex, North & South Wiltshire, North
Somerset, Shropshire, South Hampshire, and West
Gloucestershire.

rufescens Pers., *Neues Mag. Bot.* 1: 86 (1794)
  Geastrum schaefferi Vittad. [as *Geaster schaefferi*], *Monogr.
    Lycoperd.*: 22 (1842)
  Geastrum vulgatum Vittad. [as *Geaster vulgatum*], *Monogr.
    Lycoperd.*: 20 (1842)
E: o S: ! W: o NI: !

H: On soil (usually calcareous or sandy) in deciduous woodland or on heathland. Occasionally on dunes, roadside verges, and unimproved upland grassland.
D: NM3: 344 **D+I:** Ph: 252-253, B&K2: 384 503 (as *Geastrum vulgatum*), BritPuffb: 102-103 Figs. 74/75 **I:** Kriegl2: 117
Occasional but widespread, most frequent in southern England.

schmidelii Vittad. [as *Geaster schmidelii*], *Monogr. Lycoperd.*: 13 (1842)
    Geastrum nanum Pers. [*nom. illegit.*], *J. Bot. (Morot)* 2: 27 (1809)
E: o **W:** ! **O:** Channel Islands: ! Isle of Man: !
H: On calcareous or sandy soil, frequently in dunes, often in dried out slacks, or in needle litter of *Pinus* spp. Only occasionally from inland sites.
D: NM3: 342 **D+I:** Ph: 252-253 (as *Geastrum nanum*), BritPuffb: 86-87 Figs. 58/59 **I:** Kriegl2: 118
Occasional but apparently widespread and may be locally common. Better known as *Geastrum nanum*.

striatum DC., *Fl. Fr.,* Edn 3, 2: 267 (1805)
    Geastrum bryantii Berk. (as *Geaster*), *Engl. fl.* 5(2): 300 (1836)
E: o **S:** ! **W:** ! **NI:** !
H: In parkland, gardens, dunes and cemeteries. Often on calcareous loam soil (but also on acidic soil), with conifers such as *Cedrus*, *Larix* and *Pinus* spp, and *Taxus* but also known with deciduous trees and shrubs such as *Crataegus*, *Fagus*, *Ligustrum vulgare* and *Ulmus* spp. There are also several collections from 'compost' and 'cool greenhouses'.
D: NM3: 342 **D+I:** BritPuffb: 88-89 Figs. 60/61 **I:** Kriegl2: 119
Occasional but widespread.

triplex Jungh. [as *Geaster triplex*], *Tijdschr. Natuurl. Gesch. Physiol.* 7: 287 (1840)
    Geastrum michelianum W.G. Sm. [as *Geaster michelianum*], *Gard. Chron.* 1873: 608 (1873)
E: c **S:** o **W:** o **NI:** o **ROI:** c **O:** Channel Islands: o Isle of Man: !
H: On soil or humus and amongst leaf litter, usually in deciduous woodland and reported with numerous trees and shrubs. Also known from dunes, gardens and roadside verges. Occasionally under conifers such as *Pinus nigra* and also on heathland in debris of *Pteridium*.
D: NM3: 344 **D+I:** Ph: 252-253, B&K2: 384 502, BritPuffb: 108-109 Figs. 80/81 **I:** Bon: 303, SV39: 8, Kriegl2: 121
Common and widespread especially in southern and eastern England, slightly less frequent northwards.

GERRONEMA Singer, *Mycologia* 43: 599 (1951)
Agaricales, Tricholomataceae
Type: *Gerronema melanomphax* Singer

prescotii (Weinm.) Redhead, *Canad. J. Bot.* 62(5): 884 (1984)
    Cantharellus prescotii Weinm., *Obs. quaed. myc. Floram. Petrop. Spec. (Flora IX)*: 452 (1832)
    Cantharellopsis prescotii (Weinm.) Kuyper, *La Famiglia delle Tricholomataceae* 6: 99 (1986)
    Mis.: *Hygrophoropsis albida* sensu NCL p.p.
    Mis.: *Clitocybe albida* sensu auct.
    Mis.: *Cantharellus albidus* sensu Rea (1922)
E: ? **S:** ! **ROI:** ?
H: On soil in woodland, often amongst mosses. Also apparently growing in fields.
D: NM2: 173, BFF8: 113
Poorly known in Britain. Most records lack voucher material.

stevensonii (Berk. & Broome) Watling, *Edinburgh J. Bot.* 55(1): 157 (1998)
    Cantharellus stevensonii Berk. & Broome, *Ann. Mag. Nat. Hist., Ser. 4 [Notices of British Fungi no. 1422]* 15: 29 (1875)
    Hygrophoropsis stevensonii (Berk. & Broome) Corner, *Monogr. Cantharelloid Fungi*: 135 (1966)
    Mis.: *Hygrophoropsis albida* sensu NCL p.p.
E: ? **S:** !

H: On soil amongst moss and decayed woody fragments.
D: BFF8: 113-114
A poorly known species, described from Angus (Glamis). Reported from England (East Anglia) but lacking voucher material.

GLOBULICIUM Hjortstam, *Svensk bot. Tidskr.* 67: 108 (1973)
Stereales, Hyphodermataceae
Type: *Globulicium hiemale* (Laurila) Hjortstam

hiemale (Laurila) Hjortstam, *Svensk bot. Tidskr.* 67: 109 (1973)
    Corticium hiemale Laurila, *Ann. Bot. Soc. Zool.-Bot. Fenn. 'Vanamo'* 10(4): 4 (1939)
    Aleurodiscus hiemalis (Laurila) J. Erikss., *Symb. Bot. Upsal.* 16(1): 78 (1958)
    Cerocorticium hiemale (Laurila) Jülich & Stalpers, *Verh. Kon. Ned. Akad. Wetensch., Afd. Natuurk., Sect. 2,* 74: 73 (1980)
E: ! **W:** ! **NI:** !
H: On decayed wood of conifers such as *Picea sitchensis* and also reported on decayed lianas of *Clematis vitalba*.
D+I: CNE3: 401-403
A northern species in Europe, rarely recorded in the British Isles. Known from England (South Devon), Wales (Breconshire) and Northern Ireland (Tyrone).

GLOEOCYSTIDIELLUM Donk, *Meded. Ned. Mycol. Ver.* 18-20: 156 (1931)
Hericiales, Gloeocystidiellaceae
Type: *Gloeocystidiellum porosum* (Berk. & M.A. Curtis) Donk

clavuligerum (Höhn. & Litsch.) Nakasone, *Mycotaxon* 14(1): 320 (1982)
    Gloeocystidium clavuligerum Höhn. & Litsch., *Sitzungsber. Kaiserl. Akad. Wiss., Wien, Math.-Naturwiss. Cl., Abt. 1* 115: 1603 (1906)
E: !
H: On decayed wood of deciduous trees. British material on *Acer pseudoplatanus* and *Ulmus* sp.
Known only from Hertfordshire and Surrey.

porosum (Berk. & M.A. Curtis) Donk, *Meded. Ned. Mycol. Ver.* 18 - 20: 156 (1931)
    Corticium porosum Berk. & M.A. Curtis, *Ann. Mag. Nat. Hist., Ser. 3 [Notices of British Fungi no. 1821]* 3: 211 (1879)
    Gloeocystidium porosum (Berk. & M.A. Curtis) Wakef., in Bourdot & Galzin, *Hyménomyc. France.*: 253 (1928)
    Corticium stramineum Bres., *Hedwigia* 39: 221 (1900)
E: c **S:** c **W:** c **NI:** ! **ROI:** ! **O:** Isle of Man: !
H: On decayed wood, most frequently *Fagus* and *Corylus*.
D: NM3: 279 **D+I:** CNE3: 438-441, B&K2: 118 105
Common and widespread.

GLOEOHYPOCHNICIUM (Parmasto) Hjortstam, *Mycotaxon* 28: 30 (1987)
Stereales, incertae sedis
Type: *Gloeohypochnicium analogum* (Bourdot & Galzin) Hjortstam

analogum (Bourdot & Galzin) Hjortstam, *Mycotaxon* 28(1): 38 (1987)
    Gloeocystidium analogum Bourdot & Galzin, *Bull. Soc. Mycol. France.* 28(4): 366 (1913)
    Corticium analogum (Bourdot & Galzin) Wakef., *Trans. Brit. Mycol. Soc.* 35(1): 53 (1952)
    Hypochnicium analogum (Bourdot & Galzin) J. Erikss., *Symb. Bot. Upsal.* 16(1): 101 (1958)
E: !
H: On decayed wood of deciduous trees, especially large logs and fallen trunks in old woodland. British material on *Acer pseudoplatanus* and *Fagus*.
D: NM3: 198 **D+I:** CNE4: 692-695 **I:** SV42: 41

Uncommon in England and possibly restricted to ancient woodlands, especially beech. Several recent collections from South Hampshire (New Forest: Denny Wood area); single collections from Berkshire (Windsor Forest) and South Essex (Epping Forest). Reported from Yorkshire but unsubstantiated with voucher material.

GLOEOPHYLLUM P. Karst., *Bidrag Kännedom Finlands Natur Folk* 37: 79 (1882)
Poriales, Coriolaceae
Osmoporus Singer, *Mycologia* 36: 67 (1944)
Phaeocoriolellus Kotl. & Pouzar, *Česká Mykol.* 11: 162 (1957)
Type: *Gloeophyllum sepiarium* (Wulfen) P. Karst.

odoratum (Wulfen) Imazeki, *Bull. Tokyo Sci. Mus.* 6: 75 (1943)
Boletus odoratus Wulfen, *Collect. Bot. Spectantia (Vindobonae)* 2: 150 (1788)
Polyporus odoratus (Wulfen) Fr., *Syst. mycol.* 1: 37 (1821)
Trametes odorata (Wulfen) Fr., *Epicr. syst. mycol.*: 489 (1838)
Osmoporus odoratus (Wulfen) Singer, *Mycologia* 36: 67 (1944)
S: ! **W:** ?
H: On decayed wood of conifers.
D: EurPoly1: 286, NM3: 241 **D+I:** B&K2: 310 392 **I:** Kriegl1: 527
A single collection in herb. E from South Aberdeen in 2000. Reported from North Wales in 1987, but unsubstantiated with voucher material. A distinctive species with a strong smell of anise when fresh.

sepiarium (Wulfen) P. Karst., *Bidrag Kännedom Finlands Natur Folk* 37: 79 (1882)
Agaricus sepiarius Wulfen, *Jacq. Misc. Austr.* 1: 347 (1786)
Daedalea sepiaria (Wulfen) Fr., *Syst. mycol.* 1: 333 (1821)
Lenzites sepiarius (Wulfen) Fr., *Epicr. syst. mycol.*: 407 (1838)
Agaricus boletiformis Sowerby, *Col. fig. Engl. fung., Suppl.*: pl. 418 (1809)
E: ! **S:** c **O:** Isle of Man: !
H: On dead wood of conifers such as *Picea, Pinus* and *Pseudotsuga* spp, usually on large logs or stumps. Reports on deciduous hosts such as *Alnus* and *Betula* spp are unverified.
D: EurPoly1: 290, NM3: 242 **D+I:** Ph: 234, B&K2: 308 390
Common and widespread throughout Scotland, much less frequent southwards.

trabeum (Pers.) Murrill, *North American Flora* 9(2): 129 (1908)
Agaricus trabeus Pers., *Syn. meth. fung.*: 29 (1801)
Daedalea trabea (Pers.) Fr., *Syst. mycol.* 1: 335 (1821)
Lenzites trabeus (Pers.) Bres., *Atti Imp. Regia Accad. Roveretana, ser. 3,* 3: 91 (1897)
Polyporus trabeus (Pers.) Rea, *Brit. basidiomyc.*: 589 (1922)
Phaeocoriolellus trabeus (Pers.) Kotl. & Pouzar, *Česká Mykol.* 11: 162 (1957)
E: ! **S:** ! **W:** !
H: On structural timbers in buildings and rarely on decayed wood of conifers such as *Pinus* spp and *Taxus* in natural habitat.
D: EurPoly1: 292, NM3: 242 **D+I:** B&K2: 310 391
Known from England (Cheviotshire, Cumberland, North Lincolnshire, South Devon and Surrey) and Wales (Glamorganshire). Known in Scotland only from three collections between 1874 and 1878.

GLOEOPORUS Mont., in de la Sagra, *Hist. de l'Ile Cuba, Botan* 9: 385 (1842)
Stereales, Meruliaceae
Caloporus P. Karst. [non *Caloporus* Quél. (1886)], *Rev. Mycol. (Toulouse)* 3(9): 18 (1881)
Meruliopsis Bondartsev, *Izv. Akad. Nauk Estonsk. SSR, Ser. Biol.* 8: 274 (1959)
Type: *Gloeoporus conchoides* Mont.

dichrous (Fr.) Bres., *Hedwigia* 53: 75 (1912)

Polyporus dichrous Fr., *Syst. mycol.* 1: 364 (1821)
Bjerkandera dichroa (Fr.) P. Karst., *Bidrag Kännedom Finland Natur Folk* 37: 39 (1882)
Caloporus dichrous (Fr.) Ryvarden, *Polyporaceae of North Europe* 1: 109 (1976)
E: ! **S:** !
H: On decayed wood of deciduous trees. British material on *Betula* and *Quercus* spp.
D: EurPoly1: 295, NM3: 157 **D+I:** B&K2: 292 366 **I:** Kriegl1: 530
Collections from Scotland (South Aberdeen and Morayshire) in the 1960's and from England (Berkshire) in 2001.

taxicola (Pers.) Gilb. & Ryvarden, *Mycotaxon* 22(2): 364 (1985)
Xylomyzon taxicola Pers., *Mycol. eur.* 2: 32 (1825)
Merulius taxicola (Pers.) Duby, *Bot. gall.*: 796 (1830)
Poria taxicola (Pers.) Bres., *Atti Imp. Regia Accad. Roveretana, ser. 3,* 3: 80 (1897)
Merulioporia taxicola (Pers.) Bondartsev & Singer, *Ann. Mycol.* 39(1): 48 (1941)
Meruliopsis taxicola (Pers.) Bondartsev & Singer, *Eesti N.S.V. Tead. Akad. Toimet., Biol.* 8(4): 274 (1959)
Caloporus taxicola (Pers.) Ryvarden, *Norweg. J. Bot.* 20(1): 9 (1973)
Poria rufa (Schrad.) Cooke, *Grevillea* 14: 110 (1886)
Boletus rufus Schrad., in Gmelin, *Syst. nat.*: 1435 (1792)
Polyporus rufus (Schrad.) Fr., *Syst. mycol.* 1: 379 (1821)
E: ! **S:** !
H: On dead wood of conifers, most often *Pinus sylvestris* and less frequently *Picea* spp. Rare on wood of deciduous trees or on woody shrubs but recorded on *Betula* spp and *Ulex europaeus*.
D: Myc10(2): 84-85, NM3: 157 **D+I:** Ph: 239 (as *Meruliopsis taxicola*), B&K2: 154 157 (colour poor) (as *M. taxicola*), Myc10(2): 85 **I:** Kriegl1: 262
Rarely reported but apparently widespread.

GLOIOTHELE Bres., *Ann. Mycol.* 18: 44 (1920)
Hericiales, Gloeocystidiellaceae
Type: *Gloiothele lamellosa* (P. Henn.) Bres.

lactescens (Berk.) Hjortstam, *Windahlia* 17: 58 (1987)
Thelephora lactescens Berk., *Engl. fl.* 5(2): 169 (1836)
Corticium lactescens (Berk.) Berk., *Outl. Brit. fungol.*: 274 (1860)
Gloeocystidium lactescens (Berk.) Höhn. & Litsch., *Wiesner Festschrift*: 68 (1908)
Gloeocystidiellum lactescens (Berk.) Boidin, *Compt. Rend. Hebd. Séances Acad. Sci.* 233: 1668 (1951)
Megalocystidium lactescens (Berk.) Jülich, *Persoonia* 10(1): 140 (1978)
E: c **S:** ! **W:** ! **NI:** ! **ROI:** c **O:** Channel Islands: !
H: On decayed wood, usually large stumps, fallen trunks or logs, of deciduous trees. Most frequent on *Fagus* and *Ulmus* spp but also known on *Castanea* and *Fraxinus*.
D: NM3: 280 **D+I:** CNE3: 422-425 (as *Gloeocystidiellum lactescens*), B&K2: 120 107 (as *'Megalocystidiellum' lactescens*), Myc16(4): 177
Frequent and widespread, at least in southern England. Often found covered with orbicular purplish-violaceous lesions, caused by an undescribed mould (*Verticillium* species). In active growth or in damp weather, basidiomes may exude a latex when cut, with taste and odour like that of *Lactarius quietus*.

GOMPHIDIUS Fr., *Fl. Scan.*: 339 (1835)
Boletales, Gomphidiaceae
Type: *Gomphidius glutinosus* (Schaeff.) Fr.

glutinosus (Schaeff.) Fr., *Epicr. syst. mycol.*: 319 (1838)
Agaricus glutinosus Schaeff., *Fung. Bavar. Palat. nasc.* 1: 26 (1762)
Gomphus glutinosus (Schaeff.) P. Kumm., *Führ. Pilzk.*: 93 (1871)
E: ! **S:** ! **W:** ! **NI:** ! **ROI:** ! **O:** Isle of Man: ?

H: On acidic soil, in coniferous or mixed woodland. Reported with *Picea abies*, *Pinus sylvestris* and *Pseudotsuga menziesii*.
D: NM2: 69 **D+I:** Ph: 189, B&K3: 96 71 **I:** Sow1: 7 (as *Agaricus glutinosus*), FlDan5: 161F, Bon: 51, C&D: 418, SV33: 17, Kriegl2: 337
Rarely reported. Known from England (Bedfordshire, Berkshire, East Kent, Northumberland, South Hampshire, South Wiltshire and Yorkshire), Scotland (Angus, Mid-Perthshire and Westerness) and Wales (Denbighshire). Recorded elsewhere but unsubstantiated with voucher material.

maculatus Fr., *Epicr. syst. mycol.*: 319 (1838)
  Gomphidius gracilis Berk. & Broome, *Ann. Mag. Nat. Hist., Ser. 2 [Notices of British Fungi no. 698]* 13: 402 (1854)
  Gomphidius maculatus f. *gracilis* (Berk. & Broome) Kavina, *Mykologia* 1(9-10): 142 (1924)
  Gomphidius maculatus var. *cookei* Massee, *Brit. fung.-fl.* 1: 348 (1892)
E: ! **S:** ! **W:** ! **NI:** ! **ROI:** !
H: On soil, associated with *Larix* spp. **D:** NM2: 69 **D+I:** Ph: 191, B&K3: 98 73 **I:** FlDan5: 161C (as *Gomphidius gracilis*), C&D: 418, Kriegl2: 339
Rarely reported but apparently widespread.

roseus (Fr.) Fr., *Epicr. syst. mycol.*: 319 (1838)
  Agaricus glutinosus β *roseus* Fr., *Syst. Mycol.* 1: 315 (1821)
E: o **S:** o **NI:** ! **ROI:** !
H: On acidic soil in coniferous woodland, associated with *Pinus* spp and *Suillus bovinus* with which there may be a mutualistic relationship.
D: NM2: 69 **D+I:** Ph: 189, B&K3: 98 74 **I:** FlDan5: 161B, C&D: 418, Kriegl2: 335
Occasional but widespread, and may be locally frequent.

GOMPHUS (Pers.) Pers., *Tent. disp. meth. fung.*: 74 (1797)
Gomphales, Gomphaceae
  Nevrophyllum Pat., in Doass. & Pat., *Rev. Mycol. (Toulouse)* 8(29): 26 (1886)
Type: *Gomphus clavatus* (Pers.) Gray

clavatus (Pers.) Gray, *Nat. arr. Brit. pl.* 1: 638 (1821)
  Merulius clavatus Pers., *Observ. mycol.* 1: 21 (1795)
  Cantharellus clavatus (Pers.) Fr., *Syst. mycol.* 1: 322 (1821)
  Craterellus clavatus (Pers.) Fr., *Epicr. syst. mycol.*: 533 (1838)
  Nevrophyllum clavatum (Pers.) Pat. & Doass., *Rev. Mycol. (Toulouse)* 8(29): 27 (1886)
E: ! **ROI:** !
H: On soil or decayed wood in deciduous woodland.
D: NM3: 270, BFF8: 29-30 **D+I:** B&K2: 368 480 **I:** Bon: 307, C&D: 145, Kriegl2: 62, SV43: 39
Rarely collected and possibly extinct. Only six records during the twentieth century, mostly from scattered localities in southern England. Last recorded from East Gloucestershire in 1927. A single old record from the Republic of Ireland (Offaly), unsubstantiated with voucher material.

GRIFOLA Gray, *Nat. arr. Brit. pl.* 1: 643 (1821)
Poriales, Coriolaceae
  Merisma (Fr.) Gillet [*nom. illegit.*, non *Merisma* Pers. (1797)], *Hyménomycètes*: 688 (1878)
Type: *Grifola frondosa* (Dicks.) Gray

frondosa (Dicks.) Gray, *Nat. arr. Brit. pl.* 1: 643 (1821)
  Boletus frondosus Dicks., *Fasc. pl. crypt. brit.* 1: 18 (1785)
  Polyporus frondosus (Dicks.) Fr., *Syst. mycol.* 1: 355 (1821)
  Polyporus intybaceus Fr., *Epicr. syst. mycol.*: 446 (1838)
  Grifola frondosa f. *intybacea* (Fr.) Pilát, *Atlas champ. Eur.* 3(1): 35 (1936)
E: o **S:** ! **W:** ! **NI:** ! **ROI:** ! **O:** Channel Islands: !
H: Weakly parasitic then saprotrophic on roots of deciduous trees, usually *Quercus* spp, but also reported on *Castanea*, *Corylus*, *Fagus*, *Fraxinus*, *Ilex*, *Rhododendron ponticum*,

Sorbus, *Pyrus* and *Ulmus* spp. A single collection in herb. K on roots of *Cedrus* sp.
D: EurPoly1: 299, NM3: 223 **D+I:** Ph: 220, B&K2: 312 394 **I:** Sow1: 87 (as *Boletus frondosus*), Kriegl1: 531, SV47: 45
Occasional but widespread, but with a distinctly southern distribution in England. Rarely reported elsewhere. The large compound basidiomes, formed at the base of trees or stumps, often smell strongly of 'mouse urine' and are fancifully said to resemble a nesting bird, hence the English name 'Hen of the Woods'.

GUEPINIA Fr. [non *Guepinia* Bastard (1812) (*Brassicaceae*)], *Syst. orb. veg.* 1: 92 (1825)
Tremellales, Exidiaceae
  Tremiscus (Pers.) Lév., in Orbigny, *Dict. univ. hist. nat.* 8: 487 (1846)
  Phlogiotis Quél., *Enchir. fung.*: 202 (1886)
Type: *Guepinia helvelloides* (DC.) Fr.

helvelloides (DC.) Fr., *Epicr. syst. mycol.*: 566 (1838)
  Tremella helvelloides DC., *Fl. Fr.* 2: 93 (1805)
  Gyrocephalus helvelloides (DC.) Keissl., *Beih. Bot. Centralbl.* 31: 461 (1914)
  Phlogiotis helvelloides (DC.) G.W. Martin, *Amer. J. Bot.* 23: 628 (1936)
  Tremiscus helvelloides (DC.) Donk, *Taxon* 7: 164 (1958)
  Tremella rufa Jacq., *Misc. Austriaca Bot.* 1: 143 (1788)
  Guepinia rufa (Jacq.) Beck, in M.A. Becker, *Hernstein in Niederosterreich* II, *Flora*: 302 (1884)
  Gyrocephalus rufus (Jacq.) Bref., *Unters. Gesammtgeb. Mykol.* 7: 131 (1888)
E: ! **S:** ! **W:** !
H: On soil amongst grass in woodland glades, and a single collection on a spoil heap by a coal mine.
D: NM3: 111 **D+I:** B&K2: 70 31, Reid5: 297-300, Myc7(2): 86 **I:** Bon: 325, C&D: 137, Kriegl1: 129
Rarely reported. British collections from widely scattered locations in England and Wales with the majority from around the English-Welsh borders. A single collection from Scotland (Stirling: Doune Ponds). Probably genuinely rare in Britain, the distinctive basidiomes making it unlikely to be overlooked.

GUEPINIOPSIS Pat., *Tab. anal. fung.* 1: 27 (1883)
Dacrymycetales, Dacrymycetaceae
Type: *Guepiniopsis tortus* Pat.

alpina (Tracy & Earle) Brasf., *Amer. Midl. Naturalist* 20: 225 (1938)
  Guepinia alpina Tracy & Earle, *Plantae Bakerianae* 1: 23 (1901)
S: !
H: On decayed wood of *Pinus sylvestris*.
D: NM3: 96 (as *Heterotextus alpinus*)
A single collection from Isle of Mull (Glen Aros) in 1971.

buccina (Pers.) L.L. Kenn., *Mycologia* 50: 888 (1958)
  Peziza buccina Pers., *Syn. Meth. Fung.* 2: 659 (1801)
  Guepiniopsis merulina (Pers.) Pat., *Hyménomyc. Eur.*: 159 (1887)
  Ditiola merulina (Pers.) Rea, *Brit. basidiomyc.*: 743 (1922)
  Guepinia peziza Tul., *Ann. Sci. Nat., Bot.*, sér. 3, 19: 224 (1853)
E: ! **S:** ! **W:** ! **ROI:** !
H: On decayed wood (British material on *Fagus*).
Single collections from England (South Devon: Honiton, Offwell) in 1993, Republic of Ireland (Killarney) in 1946, Scotland (Main Argyll: Appin) in the late nineteenth century and Wales Caernarvonshire: Llyn Gwynant) in 1988.

GYMNOMYCES Massee & Rodway, *Bull. Misc. Inform. Kew.* 1898: 125 (1898)
Russulales, Elasmomycetaceae
Type: *Gymnomyces pallidus* Massee & Rodway

xanthosporus (Hawker) A.H. Sm., *Mycologia* 54: 635 (1962)
  Hydnangium carneum var. *xanthosporum* Hawker, *Trans. Brit. Mycol. Soc.* 35: 281 (1952)
E: ! W: !
H: Subhypogeous under leaf litter of deciduous trees such as *Fagus* or with conifers such as *Picea* spp.
D+I: BritTruff: 199-200 12 E
Known only from Herefordshire and Caernarvonshire.

GYMNOPILUS P. Karst., *Bidrag Kännrdom Finlands Natur Folk* 32: 400 (1879)
Cortinariales, Cortinariaceae
  Fulvidula Romagn. [*nom. nud.*], *Rev. Mycol. (Paris)* 1(4): 209 (1936)
Type: *Gymnopilus liquiritiae* (Pers.) P. Karst.

bellulus (Peck) Murrill, *North American Fungi* 10: 200 (1917)
  Agaricus bellulus Peck, *Rep. (Annual) New York State Mus. Nat. Hist.* 26: 58 (1874)
  Naucoria bellula (Peck) Sacc., *Syll. fung.* 5: 841 (1887)
E: ! S: !
H: On decayed wood of conifers such as *Picea* spp and *Taxus*.
D: TBMS52(2): 324, NBA9: 555, BFF7: 67 **D+I:** B&K5: 134 141
First reported from Britain in TBMS 52 (2): 323 (1969). Few records since then, from England (South Wiltshire and Surrey) and Scotland (Isle of Arran).

decipiens (W.G. Sm.) P.D. Orton, *Trans. Brit. Mycol. Soc.* 43(2): 176 (1960)
  Agaricus decipiens W.G. Sm., *J. Bot.* 7: 249 (1869)
  Flammula decipiens (W.G. Sm.) Sacc., *Syll. fung.* 5: 811 (1887)
E: !
H: On burnt wood (logs, stumps etc.) of conifers and also on burnt soil (but then probably growing from buried wood). British material is all on *Pinus* spp.
D: NCL3: 243, BFF7: 69-70 **I:** C&D: 345
Known from South Essex, Surrey and Warwickshire. Reported from Herefordshire and Norfolk but unsubstantiated with voucher material. Possibly not distinct from *G. odini*.

dilepis (Berk. & Broome) Singer, *Lilloa* 22: 560 (1951) [1949]
  Agaricus dilepis Berk. & Broome, *J. Linn. Soc., Bot.* 11: 542 (1871)
  Flammula dilepis (Berk. & Broome) Sacc., *Syll. fung.* 5: 812 (1887)
E: !
H: In greenhouses in botanic gardens. Also on decayed conifer woodchips.
I: FM 3(1) 2002, back cover
An alien, recently established in natural habitat and found on several occasions in scattered locations in East Norfolk (Beeston Common), Surrey (Brentmoor Heath near Bisley and Esher, Claremont) and West Sussex (Iping & Steadham Common near Midhurst).

flavus (Bres.) Singer, *Lilloa* 22: 636 (1951) [1949]
  Naucoria flava Bres., *Ann. Mycol.* 3(2): 162 (1905)
  Flammula flava (Bres.) J.E. Lange, *Fl. agaric. danic.* 4: 12 (1939)
  Fulvidula flava (Bres.) Singer, *Rev. Mycol. (Paris)* 5(1): 12 (1940)
  Flammula dactylidicola J.E. Lange, *Dansk. Bot. Ark.* 5(7): 6 (1928)
E: ! S: ! W: ! NI: ! ROI: !
H: On soil, often (but not always) associated with *Dactylis glomerata* (cocksfoot) in grassland and on wasteground. Rarely with *Ammophila arenaria* on dunes.
D: NM2: 314, BFF7: 67-68 **D+I:** B&K5: 136 142
Rarely collected or recorded but apparently widespread.

fulgens (J. Favre & Maire) Śinger, *Lilloa* 22: 561 (1951) [1949]
  Naucoria fulgens J. Favre & Maire, *Bull. Soc. Mycol. France* 53: 167 (1937)
  Fulvidula fulgens (J. Favre & Maire) Kühner, *Bull. Mens. Soc. Linn. Lyon* 8(2): 44 (1939)

  Flammula fulgens (J. Favre & Maire) S. Lundell, *Fungi Exsiccati Suecici* 55: 9 (1960)
  Mis.: *Naucoria cerodes* sensu Lange (FlDan4: 16)
E: ! S: !
H: On peaty or sandy soil on moorland or heathland, usually amongst lichens or mosses. Also reported on burnt ground in such habitat.
D: NM2: 314, BFF7: 68-69 **D+I:** Ph: 143, B&K5: 136 143 **I:** C&D: 345
Rarely collected or reported.

junonius (Fr.) P.D. Orton, *Trans. Brit. Mycol. Soc.* 43(2): 176 (1960)
  Agaricus junonius Fr., *Syst. mycol.* 1: 244 (1821)
  Pholiota junonia (Fr.) P. Karst., *Bidrag Kännedom Finlands Natur Folk* 32: 301 (1879)
  Pholiota spectabilis var. *junonia* (Fr.) J.E. Lange, *Fl. agaric. danic.* 5: 100 (1940)
  Pholiota grandis Rea, *Trans. Brit. Mycol. Soc.* 2: 37 (1903) [1902]
  Mis.: *Agaricus aureus* sensu Sowerby (Col. fig. Engl. fung. pl. 77 (1797), sensu Cooke [Ill. Brit. fung. 373 (345) Vol. 3 (1895)]
  Mis.: *Gymnopilus spectabilis* sensu Smith (1949) and sensu auct.
E: c S: c W: c NI: c ROI: ! O: Channel Islands: ! Isle of Man: !
H: On decayed wood of deciduous trees such as *Acer pseudoplatanus*, *Betula* spp, *Fagus*, *Fraxinus*, *Quercus*, *Salix*, *Sorbus aucuparia* and *Ulex europaeus*. Rarely reported on conifers such as *Cedrus*, *Pinus* spp and *Taxus*. Also reported as 'terrestrial' but probably then arising from buried wood.
D: NCL3: 244, NM2: 314, BFF7: 71-72 **D+I:** Ph: 144, B&K5: 136 144, Myc14(1): 37 **I:** Sow1: 77 (as *Agaricus aureus*), FlDan3: 108B (as *Pholiota spectabilis*), C&D: 345 (as *Gymnopilus spectabilis*)
Common and widespread.

liquiritiae (Pers.) P. Karst., *Bidrag Kännedom Finlands Natur Folk* 32: 400 (1879)
  Agaricus liquiritiae Pers., *Syn. meth. fung.*: 306 (1801)
  Flammula liquiritiae (Pers.) P. Kumm., *Führ. Pilzk.*: 82 (1871)
E: ! ROI: ?
H: On decayed wood of conifers such as *Pinus* spp.
D: BFF7: 61, Myc15(2): 75
Poorly known in the British Isles. Reported from Killarney by Rea in 1922 (but only a painting survives) and from Surrey in 1999.

odini (Fr.) Kühner & Romagn. [*comb. inval.*], *Flore Analytique des Champignons Supérieurs*: 323 (1953)
  Agaricus odini Fr., *Monogr. hymenomyc. Suec.* 2: 300 (1863)
  Hebeloma odini (Fr.) P. Karst., *Syll. fung.* 5: 808 (1887)
E: !
H: On burnt soil in woodland or scrub, usually on old bonfire sites.
D: NM2: 314, BFF7: 70 **D+I:** B&K5: 138 145
Known from Surrey (Wimbledon Common) and West Gloucestershire (Ruardean Hill).

penetrans (Fr.) Murrill, *Mycologia* 4: 254 (1912)
  Agaricus penetrans Fr., *Observ. mycol.* 1: 23 (1815)
  Flammula penetrans (Fr.) Quél., *Mém. Soc. Émul. Montbéliard, Sér. 2*, 5: 252 (1872)
  Dryophila penetrans (Fr.) Quél., *Enchir. fung.*: 71 (1886)
  Mis.: *Agaricus hybridus* Sowerby sensu Fries (1821)
  Mis.: *Flammula hybrida* (Sowerby) Gillet sensu auct.
  Mis.: *Gymnopilus hybridus* (Sowerby) Maire sensu auct.
  Mis.: *Fulvidula hybrida* (Sowerby) Singer sensu auct.
E: c S: c W: c NI: c ROI: c O: Channel Islands: ! Isle of Man: !
H: On decayed wood of conifers, rarely deciduous trees such as *Betula*, *Fagus* and *Quercus* spp.
D: NM2: 315, BFF7: 62-63 **D+I:** Ph: 143, B&K5: 138 146 **I:** C&D: 345
Very common and widespread but species limits are still unclear. Some authors also include *G. stabilis* and *G. sapineus* in the

synonymy of this species, whilst recent research suggests that *G. hybridus* sensu auct. may be distinct (Holec, pers. comm.).

picreus (Pers.) P. Karst., *Bidrag Kännedom Finlands Natur Folk* 32: 400 (1879)
Agaricus picreus Pers., *Icon. descr. fung.*: pl. 14 (1798)
Flammula picrea (Pers.) P. Kumm., *Führ. Pilzk.*: 82 (1871)
Fulvidula picrea (Pers.) Singer, *Rev. Mycol. (Paris)* 2(6): 239 (1937)
E: ! **W:** ?
H: On decayed wood. British material on a large decayed shank of *Pinus* sp.
D: NM2: 315, BFF7: 62 **D+I:** B&K5: 138 147 **I:** C&D: 345
A single collection from Surrey (Betchworth, Dawcombe Nature Reserve) in 2000. Rarely reported elsewhere in England and from Wales (Carmarthenshire: Llandeilo) but unsubstantiated with voucher material.

sapineus (Fr.) Maire, *Trabajos del Museo de Ciències Naturales de Barcelona* 15(2): 96 (1933)
Agaricus sapineus Fr., *Syst. mycol.* 1: 239 (1821)
Flammula sapinea (Fr.) P. Kumm., *Führ. Pilzk.*: 82 (1871)
Fulvidula sapinea (Fr.) Romagn. *nom. inval.*, *Rev. Mycol. (Paris)* 2(5): 191 (1937)
E: ! **S:** ! **W:** ! **NI:** ! **ROI:** ! **O:** Channel Islands: ! Isle of Man: !
H: On decayed wood of conifers, rarely deciduous trees.
D: BFF7: 65-66 **D+I:** B&K5: 140 149 **I:** C&D: 345
Commonly reported and apparently widespread, but rarely collected. Easily confused with *Gymnopilus penetrans* or *G. hybridus*.

stabilis (Weinm.) Kühner & Romagn., *Flore Analytique des Champignons Supérieurs*: 322 (1953)
Agaricus stabilis Weinm., *Hymen. Gasteromyc.*: 210 (1836)
Flammula stabilis (Weinm.) P. Karst., *Bidrag Kännedom Finlands Natur Folk* 32: 408 (1879)
E: ! **S:** ? **W:** ! **NI:** ?
H: On decayed wood of conifers. British material is on *Pinus sylvestris*.
D: BFF7: 66
Rarely reported. Accepted here following BFF7, but doubtfully distinct from *Gymnopilus penetrans*.

GYRODON Opat., *Arch. Naturgesch.* 2(1): 5 (1836)
Boletales, Gyrodontaceae
Uloporus Quél., *Enchir. fung.*: 162 (1886)
Type: *Gyrodon sistotremoides* (Fr.) Opat. (= *G. lividus*)

lividus (Bull.) Sacc., *Syll. fung.* 6: 52 (1888)
Boletus lividus Bull., *Herb. France*: pl. 490 (1791)
Uloporus lividus (Bull.) Quél., *Enchir. fung.*: 162 (1886)
Boletus brachyporus Pers., *Mycol. eur.* 2: 128 (1825)
Gyrodon sistotrema var. *brachyporus* (Pers.) Rea, *Brit. basidiomyc.*: 557 (1922)
Boletus sistotrema Fr., *Syst. mycol.* 1: 389 (1821)
Gyrodon sistotrema (Fr.) Sacc., *Syll. fung.* 6: 53 (1888)
Boletus sistotremoides Fr., *Observ. mycol.* 1: 120 (1815)
Gyrodon sistotremoides (Fr.) Opat., *Arch. Naturgesch.* 2(1): 5 (1836)
E: ! **W:** ! **ROI:** !
H: On soil, in fens or alder-carr, associated with *Alnus* and often growing amongst grasses.
D: NM2: 53 **D+I:** FRIC6: 44-50 48a, Ph: 206, FRIC16: 3-4 121a, B&K3: 68 28 **I:** Bon: 33, C&D: 421, SV39: 4, Kriegl2: 255
Rarely reported. Known from England (Cambridgeshire, Dorset, Herefordshire, Norfolk, North Hampshire, South Hampshire, Warwickshire, West Gloucestershire, Worcestershire and Yorkshire), Wales (Breconshire and Glamorganshire) and Republic of Ireland (Wexford).

GYROPORUS Quél., *Enchir. fung.*: 161 (1886)
Boletales, Gyrodontaceae
Type: *Gyroporus cyanescens* (Bull.) Quél.

castaneus (Bull.) Quél., *Enchir. fung.*: 161 (1886)
Boletus castaneus Bull., *Herb. France*: pl. 328 (1788)
Boletus testaceus Pers. [*nom. illegit.*, non *B. testaceus* With. (1776)] *Mycol. eur.* 2: 137 (1825)
Boletus fulvidus Fr., *Observ. mycol.* 2: 247 (1818)
E: o **S:** ! **W:** ! **NI:** ! **ROI:** ! **O:** Channel Islands: !
H: On soil, usually associated with *Quercus* spp, in mixed deciduous woodland but also known with *Castanea* and *Fagus*.
D: NM2: 53 **D+I:** Ph: 206-207, B&K3: 68 29 **I:** Bon: 33, C&D: 421, Kriegl2: 257
Occasional but widespread. Most frequently reported from southern and south-eastern England.

cyanescens (Bull.) Quél., *Enchir. fung.*: 161 (1886)
Boletus cyanescens Bull., *Herb. France*: pl. 369 (1788)
Boletus constrictus Pers., *Syn. meth. fung.*: 508 (1801)
Leccinum constrictum (Pers.) Gray, *Nat. arr. Brit. pl.* 1: 647 (1821)
Boletus lacteus Lév., *Ann. Sci. Nat., Bot.*, sér. 3, 9: 124 (1848)
Gyroporus lacteus (Lév.) Quél., *Enchir. fung.*: 161 (1886)
E: ! **S:** ! **W:** ! **NI:** !
H: On mineral or sandy soil, most often with *Betula* spp, in mixed deciduous woodland. Also reported with *Acer pseudoplatanus*, *Castanea*, *Fagus* and *Quercus* spp.
D: NM2: 53 **D+I:** Ph: 206-207, B&K3: 68 30 **I:** Bon: 33, C&D: 421, Kriegl2: 259
Rarely reported. Collections from England (Berkshire, Cumberland, Dorset, East Kent, North Lincolnshire, Shropshire, South Devon, South Hampshire, Surrey, Westmorland, West Gloucestershire, and West Suffolk), Scotland (Angus, Easterness, East Sutherland, East and Mid-Perthshire, and Isle of Mull) and Northern Ireland (Down). Last reported from Wales in 1924.

HAASIELLA Kotl. & Pouzar, *Česká Mykol.* 20: 135 (1966)
Agaricales, Tricholomataceae
Type: *Haasiella splendidissima* Kotl. & Pouzar

venustissima (Fr.) Kotl. & Pouzar, *Česká Mykol.* 20: 135 (1966)
Agaricus venustissimus Fr., *Monogr. hymenomyc. Suec.* 2: 289 (1863)
Clitocybe venustissima (Fr.) P. Karst., *Bidrag Kännedom Finlands Natur Folk* 32: 62 (1879)
E: ! **S:** ?
H: British collections are on calcareous loam soil or decayed, partially buried sticks in scrub woodland, with *Acer pseudoplatanus*, *Buxus*, *Salix caprea*, *Sambucus nigra* and *Larix*. Typically late fruiting (Nov. – Jan.).
D: NM2: 173 (as *Omphalina venustissima*), FAN3: 64
Known from Surrey (Mickleham, Norbury Park) since 1971, and collected in North Hampshire (Alresford: Northington) in 2004. Cooke 133 (265 lower) Vol. 1 (1883), illustrates what appears to be this species from Scotland (Midlothian: Penicuik) but no material has been traced.

HANDKEA Kreisel, *Nova Hedwigia* 48: 282 (1989)
Lycoperdales, Lycoperdaceae
Type: *Handkea utriformis* (Bull.) Pers.

excipuliformis (Scop.) Kreisel, *Nova Hedwigia* 48(3-4): 283 (1989)
Lycoperdon polymorphum var. *excipuliforme* Scop., *Fl. carniol.* 488 (1772)
Lycoperdon excipuliforme (Scop.) Pers., *Syn. meth. fung.*: 143 (1801)
Calvatia excipuliformis (Scop.) Perdeck, *Blumea.* 6: 490 (1950)
Lycoperdon elatum Massee, *J. Roy. Microscop. Soc. London* 1887: 710 (1887)
Lycoperdon excipuliforme f. *flavescens* Quél., *Mém. Soc. Émul. Montbéliard*, Sér. 2, 5: 368 (1873)

Lycoperdon excipuliforme var. *flavescens* (Quél.) Rea, *Brit. basidiomyc.*: 31 (1922)
Lycoperdon saccatum Vahl, *Fl. Danica*: t. 1139 (1792)
Calvatia saccata (Vahl) Morgan, *Gasterom. Ung.*: 89 (1904)
Mis.: *Lycoperdon cervinum* sensu Bolton [Hist. Fung. Halifax pl. 116 (1789) p.p.]
E: c **S:** c **W:** c **NI:** c **ROI:** c **O:** Isle of Man: !
H: On humic soil in woodland areas, and occasionally amongst short turf in grassland.
D: NM3: 336 (as *Calvatia excipuliformis*) **D+I:** Ph: 246-247 (as *C. excipuliformis*), B&K2: 388 510 (as *C. excipuliformis*), BritPuffb: 126-127 Figs. 95/96 **I:** Kriegl2: 136
Common and widespread. Macroscopically polymorphic, thus has acquired a long list of synonyms (see Kriesel 1989).

utriformis (Bull.) Pers., *Nova Hedwigia* 48(3-4): 288 (1989)
Lycoperdon utriforme Bull., *Champ. France, Hymenomycetes*: 153 (1791)
Calvatia utriformis (Bull.) Jaap, *Verh. Bot. Vereins Prov. Brandenburg* 59: 37 (1918)
Lycoperdon bovista Pers., *Observ. mycol.* 1: 4 (1796)
Lycoperdon caelatum Bull., *Champ. France, Hymenomycetes*: 156 (1791)
Calvatia caelata (Bull.) Morgan, *J. Cincinnati Soc. Nat. Hist.* 12: 169 (1890)
Lycoperdon sinclairii Massee, *J. Roy. Microscop. Soc. London.* 1887: 716 (1887)
E: o **S:** o **W:** o **NI:** o **ROI:** o **O:** Channel Islands: ! Isle of Man: !
H: On neutral to acidic soils, amongst short turf in unimproved grassland.
D: NM3: 336 (as *Calvatia utriformis*) **D+I:** Ph: 246-247 (as *C. utriformis*), B&K2: 388 509 (as *C. utriformis*), BritPuffb: 124-125 Figs. 93/94 **I:** Bon: 305 (as *C. utriformis*), Kriegl2: 138
Occasional but widespread.

HAPALOPILUS P. Karst., *Rev. Mycol. (Toulouse)* 3(9): 18 (1881)
Poriales, Coriolaceae
Type: *Hapalopilus nidulans* (Fr.) P. Karst.

nidulans (Fr.) P. Karst., *Rev. Mycol. (Toulouse)* 3(9): 18 (1881)
Polyporus nidulans Fr., *Syst. mycol.* 1: 362 (1821)
Polystictus nidulans (Fr.) Gillot & Lucand, *Cat. Champ. Marn.*: 348 (1890)
Boletus rutilans Pers., *Icon. descr. fung.*: pl. 18 (1798)
Polyporus rutilans (Pers.) Fr., *Observ. mycol.*: 260 (1818)
Hapalopilus rutilans (Pers.) P. Karst., *Rev. Mycol.* 3(9): 18 (1881)
E: o **S:** o **W:** ! **NI:** ! **ROI:** !
H: On dead and decayed wood of deciduous trees and shrubs. Most commonly reported on *Corylus* and *Fagus*. Also known on *Aesculus, Acer campestre, Alnus, Betula, Castanea, Prunus avium, P. padus, Quercus* spp, *Rhododendron ponticum, Salix, Sambucus nigra, Sorbus aucuparia* and *Ulmus* spp.
D: EurPoly1: 303, NM3: 223 (as *Hapalopilus rutilans*) **D+I:** Ph: 224, B&K2: 312 396 (as *H. rutilans*) **I:** Kriegl1: 533
Occasional in England and Scotland. Rarely reported elsewhere.

salmonicolor (Berk. & M.A. Curtis) Pouzar, *Česká Mykol.* 21: 205 (1967)
Polyporus salmonicolor Berk. & M.A. Curtis, *Hooker's J. Bot. Kew Gard. Misc.* 1: 104 (1849)
Poria salmonicolor (Berk. & M.A. Curtis) Cooke, *Grevillea* 14: 112 (1886)
Leptoporus salmonicolor (Berk. & M.A. Curtis) Pat., *Essai taxon.*: 85 (1900)
Hapalopilus aurantiacus (Berk. & M.A. Curtis) Bondartsev & Singer, *Ann. Mycol.* 39: 53 (1941)
S: !
H: On decayed wood of conifers. British material on the underside of a fallen trunk of *Pinus sylvestris*.
D: EurPoly1: 304, NM3: 223
A single collection from Mid-Perthshire (Black Wood of Rannoch) in 1997. Probably genuinely rare since basidiomes are large,

distinctly coloured (dark salmon pink) and would be difficult to overlook.

HAUERSLEVIA P. Roberts, *Cryptog. Mycol* 19: 277 (1998)
Tremellales, Exidiaceae
Type: *Hauerslevia pulverulenta* (Hauerslev) P. Roberts

pulverulenta (Hauerslev) P. Roberts, *Cryptog. Mycol* 19: 279 (1998)
Sebacina pulverulenta Hauerslev, *Friesia* 11: 99 (1976)
E: ! **W:** ! **O:** Channel Islands: !
H: On decayed wood of deciduous trees, also on dead basidiomes of *Hymenochaete corrugata* and dead herbaceous stems.
D: NM3: 105 (as *Sebacina pulverulenta*)
Rarely reported. Known from widely scattered locations in England and Wales, and a single collection from the Channel Islands (Sark).

HEBELOMA (Fr.) P. Kumm., *Führ. Pilzk.*: 22 (1871)
Cortinariales, Cortinariaceae
Hylophila Quél., *Enchir. fung.*: 98 (1886)
Myxocybe Fayod, *Ann. Sci. Nat., Bot.*, sér. 7, 9: 361 (1889)
Type: *Hebeloma fastibile* (Pers.) P. Kumm.

aestivale Vesterh., *Symb. Bot. Upsal.* 30(3): 131 (1995)
E: !
H: On soil in mixed deciduous woodland. British material with *Betula* spp and *Castanea*.
D+I: FM1: 60 f.4
Part of the *H. crustuliniforme* complex, only recently described. In Britain known from southern England (Buckinghamshire, Oxfordshire and Surrey) but probably more widespread.

alpinum (J. Favre) Bruchet, *Bull. Mens. Soc. Linn. Lyon* 39 (6 [Suppl.]): 68 (1970)
Hebeloma crustuliniforme var. *alpinum* J. Favre, *Ergebn. Wiss. Untersuch. Schweiz. Natn. Parks* 5: 202 (1955)
E: ! **S:** ?
H: On soil in montane habitat.
D: NM2: 318 **D+I:** B&K5: 106 97
A single collection from England in non-montane habitat (East Kent: Bedgebury Pinetum), possibly introduced with non-native conifers. Reported from Scotland but material in herb. E is annotated as 'compare with' *H. alpinum*.

ammophilum Bohus [non *H. ammophilum* (Bon) Bon [*nom. illegit*], *Ann. Hist.-Nat. Mus. Natl. Hung.* 70: 101 (1978)
E: ! **W:** ?
H: On sandy soil, usually with *Salix repens* in dried-out dune slacks.
Known from South Lancashire (Ainsdale Dunes) and Cheviotshire (Holy Island). Reported from Wales (Anglesey: Newborough Warren) but unsubstantiated with voucher material.

anthracophilum Maire, *Bull. Soc. Mycol. France* 24: 57 (1908)
E: ! **S:** ! **W:** ! **NI:** !
H: On burnt soil in mixed deciduous woodland, usually on old bonfire sites.
D: NM2: 319 **D+I:** B&K5: 106 98 **I:** Bon: 229, C&D: 309
Rarely reported but apparently widespread. Most records and collections are from widely scattered sites in southern England.

arenosa Burds., Macfall & M.A. Albers, *Mycologia* 78(5): 861 (1986)
S: !
H: On soil, associated with conifers. British collections with *Pinus contorta*.
An introduced member of the *H. crustuliniforme* complex. Known only from Orkney and Shetland in plantations but apparently spreading *fide* R. Watling (pers. comm.).

atrobrunneum Vesterh., *Nordic J. Bot.* 9(3): 311 (1989)
E: ! **S:** !
H: On wet soil, usually associated with *Salix* and *Sphagnum* spp.
D: NM2: 317
Known only from North West Yorkshire (Malham Tarn) and from Orkney.

birrus (Fr.) Sacc. [as *Hebeloma birrum*], *Syll. fung.* 5: 794 (1887)
   Agaricus birrus Fr., *Epicr. syst. mycol.*: 179 (1838)
   Hebeloma danicum Gröger, *Z. Mykol.* 53(1): 53 (1987)
   Hebeloma pumilum J.E. Lange, *Fl. agaric. danic.* 5,
     Taxonomic Conspectus: IV (1940)
   Mis.: *Hebeloma spoliatum* sensu NCL and sensu auct. mult.
E: ! **S:** !
H: On nutrient-rich soil. British material associated with *Betula* spp, *Fagus* and *Pinus sylvestris*.
D: NM2: 317 (as *Hebeloma pumilum*), NM2: 321 (as *H. danicum*) **D+I:** FRIC3: 26-28 24b (as *H. pumilum*), B&K5: 118 117 (as *H. pumilum*) **I:** FlDan3: 120A (as *H. spoliatum*), SV29: 24 (as *H. spoliatum*), C&D: 307 (as *H. pumilum*)
Rarely reported.

circinans (Quél.) Sacc., *Syll. fung.* 9: 103 (1891)
   Hylophila circinans Quél., *Compt. Rend. Assoc. Franç. Avancem. Sci. Assoc. Sci. France* 16: 587 (1888)
   Mis.: *Hebeloma longicaudum* sensu Kühner & Romagnesi (1953)
E: !
H: On calcareous soil, in mixed deciduous woodland.
D: NM2: 321 **D+I:** B&K5: 108 102 **I:** SV29: 24
A single collection, from South Wiltshire (Swallowcliffe) in 1964.

crustuliniforme (Bull.) Quél., *Mém. Soc. Émul. Montbéliard, Sér. 2*, 5: 95 (1872)
   Agaricus crustuliniformis Bull., *Herb. France*: pl. 308 (1787)
   Agaricus crustuliniformis var. *minor* Cooke, *Ill. Brit. fung.* 457 (414) Vol. 3 (1885)
   Hebeloma crustuliniforme var. *minor* (Cooke) Massee, *Brit. fung.-fl.* 2: 176 (1893)
E: c **S:** c **W:** c **NI:** ! **ROI:** c **O:** Channel Islands: !
H: On soil in mixed deciduous woodland, most frequently with *Betula* spp. Less often with conifers and occasionally on dunes amongst *Salix repens*.
D: NM2: 319 **D+I:** Ph: 147, B&K5: 110 105, FM1: 61 f.1
Common and widespread. English name = 'Poison Pie'.

cylindrosporum Romagn., *Bull. Soc. Mycol. France* 81: 328 (1965)
S: !
H: On sandy soil, associated with *Pinus sylvestris* in Caledonian pinewoods.
D: Orton (Notes Roy Bot Gard. Edinburgh 44: 488, 1987) **I:** Bon: 231, C&D: 307
Two collections from Easterness (Abernethy, Garten Wood) in 1985.

dunense L. Corb. & R. Heim, *Mém. Soc. Natn. Scienc. Nat. Cherbourg* 40: 166 (1929)
W: !
H: On sandy soil on dunes, usually associated with *Salix repens*.
I: C&D: 305
A single collection from Glamorgan (Kenfig Burrows) in 1992.

fragilipes Romagn., *Bull. Soc. Mycol. France* 81(3): 341 (1965)
   Mis.: *Hebeloma hiemale* sensu auct.
E: ! **S:** ! **O:** Channel Islands: !
H: On soil in woodland. British material is associated with *Betula* spp.
D+I: FM1: 61 f. 5
Rarely reported. Most often recorded from Scotland, but apparently widely scattered throughout Britain. Part of the *H. crustuliniforme* complex.

fusisporum Gröger & Zschiesch., *Z. Mykol.* 47(2): 204 (1981)
S: !

H: On damp soil, associated with *Salix* spp. British material with *Salix aurita*.
D: NM2: 320 **I:** SV29: 21
A single collection from Orkney in 1994.

groegeri Bon, *Doc. Mycol.* 31(123): 27 (2002)
   Hebeloma gigaspermum Gröger & Zschiesch. [*nom. illegit.*, non *H. gigaspermum* Cooke & Massee (1890)], *Z. Mykol.* 47(2): 201 (1981)
E: ! **S:** ! **NI:** !
H: On damp soil in mixed deciduous woodland. Reported with *Alnus*, *Betula*, *Fagus*, *Populus* and *Salix* spp.
D: NM2: 320 (as *Hebeloma gigaspermum*) **D+I:** B&K5: 112 107 (as *H. gigaspermum*) **I:** SV29: 20 (as *H. gigaspermum*)
Part of the *Hebeloma sacchariolens* complex, only recently distinguished. Apparently widespread.

helodes J. Favre, *Beitr. Kryptogamenfl. Schweiz* 10(3): 214 (1948)
E: ! **S:** ! **W:** !
H: On soil, in boggy areas of woodland, often with *Betula* spp, but also known with *Alnus* and *Salix* spp.
D: NM2: 319 **D+I:** B&K5: 120 118 (misdetermined as *Hebeloma pusillum*), FM1: 63 f. 3
Rarely reported but apparently widespread. Part of the *H. crustuliniforme* complex.

hetieri Boud., *Bull. Soc. Mycol. France* 33: 8 (1917)
   Hebeloma sacchariolens var. *tomentosum* M.M. Moser, *Z. Pilzk.* 36(1-2): 71 (1970)
   Hebeloma tomentosum (M.M. Moser) Gröger & Zschiesch., *Z. Mykol.* 47(2): 203 (1981)
E: ! **S:** !
H: On soil in mixed deciduous woodland. British material associated with *Betula* and *Salix* spp.
D: NM2: 320 (as *Hebeloma tomentosum*) **I:** SV29: 21 (as *H. tomentosum*)
Known from Scotland (Stirlingshire) and England (Yorkshire). Reported (as *Hebeloma tomentosum*) from Warwickshire and West Kent but unsubstantiated with voucher material.

incarnatulum A.H. Sm., *Sydowia* 37: 280 (1984)
   Mis.: *Hebeloma longicaudum* sensu auct. p.p.
E: ! **S:** ! **W:** !
H: On soil in coniferous woodland.
D: NM2: 318 (as *Hebeloma longicaudum* sensu Lange) **D + I:** FM1(2): 64 + front cover) **I:** SV40: 10
Known from England (Westmorland: Roudsea Wood), Scotland (Easterness: Curr Wood) and Wales (Merionethshire: Tan-y-Bwlch).

leucosarx P.D. Orton, *Trans. Brit. Mycol. Soc.* 43(2): 244 (1960)
   Hebeloma velutipes Bruchet, *Bull. Mens Soc. Linn. Lyon.* 39(6, Suppl.): 80 (1970)
   Mis.: *Hebeloma longicaudum* sensu NCL and sensu auct.
E: o **S:** ! **W:** ! **NI:** !
H: On soil, in boggy woodland, often associated with *Betula* or *Salix* spp.
D: NM2: 318 **D+I:** B&K5: 114 110, B&K5: 114 111 (as *Hebeloma longicaudum*), FM1: 65 f.2 **I:** B&K5: 124 126 (as *H. velutipes*)
Uncommonly reported but apparently widespread and probably the commonest member of the *H. crustuliniforme* complex. N.B. the illustration in B&K5: 114 (pl. 110) lacks the typically swollen base to the stipe. Often recorded as *H. longicaudum*.

marginatulum (J. Favre) Bruchet, *Bull. Mens. Soc. Linn. Lyon* 39 (6 [Suppl.]): 43 (1970)
   Hebeloma versipelle var. *marginatulum* J. Favre, *Ergebn. Wiss. Untersuch. Schweiz. Natn. Parks* 5: 202 (1955)
S: !
H: On soil in montane habitat, associated with *Salix herbacea*.
D: NM2: 318 **D+I:** B&K5: 116 112
A single collection from Perthshire (The Cairnwell).

mesophaeum (Pers.) Quél., *Mém. Soc. Émul. Montbéliard, Sér. 2,* 5: 128 (1872)

> *Agaricus fastibilis* var. *mesophaeus* Pers., *Mycol. eur.* 3: 173 (1828)
> *Agaricus mesophaeus* (Pers.) Fr., *Epicr. syst. mycol.*: 179 (1838)
> *Hebeloma flammuloides* Romagn., *Sydowia* 36: 268 (1983)
> *Agaricus mesophaeus* var. *minor* Cooke, *Ill. Brit. fung.* 453 (412) Vol. 3 (1885)
> *Hebeloma mesophaeum* var. *minor* (Cooke) Massee, *Brit. fung.-fl.* 2: 175 (1893)
> *Agaricus strophosus* Fr., *Epicr. syst. mycol.*: 161 (1838)
> *Hebeloma strophosum* (Fr.) Sacc., *Syll. fung.* 5: 808 (1887)

E: c **S:** c **W:** c **NI:** ! **ROI:** c

H: On acidic or sandy soil, most frequently with *Betula* or *Pinus* spp, but also known with a wide variety of other trees. One collection in K with potted specimens of *Dryas octopetala* in cultivation.

D: NM2: 317, NM2: 317 (as *Hebeloma mesophaeum* var. *strophosum*) **D+I:** Ph: 147, B&K5: 116 113 **I:** Ph: 148 (as *H. strophosum*)

Common and widespread.

mesophaeum *var.* **crassipes** Vesterh., *Nordic J. Bot.* 9(3): 298 (1989)

> Mis.: *Hebeloma fastibile* sensu NCL & sensu Lange (FlDan3: 92 & pl. 118F)
> Mis.: *Agaricus fastibilis* sensu Cooke [Ill. Brit. fung. 446 (406) Vol. 3 (1886)]

E: !

H: On soil, mostly in coniferous woodland.

D+I: B&K5: 112 106 (as *Hebeloma fastibile*)

Poorly known in Britain. Last reported in 1903 (as *Hebeloma fastible*) and prior to that in 1885 and 1867.

pallidoluctuosum Gröger & Zschiesch., *Wiss. Z. Friedrich-Schiller-Univ. Jena. Math.-Naturwiss. Reihe* 33(6): 815 (1984)

> *Hebeloma latifolium* Gröger & Zschiesch. [*nom. illegit.*], *Z. Mykol.* 47(2): 198 (1981)
> Mis.: *Hebeloma sacchariolens* sensu NCL and sensu auct. mult.

E: o **S:** ! **W:** ! **NI:** ! **ROI:** !

H: On soil in deciduous woodland. British material associated with *Corylus* and *Quercus* spp. Also collected amongst *Urtica dioica* and *Carex pendula* in wet areas of woodland.

D: NM2: 320 **D+I:** B&K5: 116 114 **I:** Ph: 147 (as *Hebeloma sacchariolens*), SV29: 20

Part of the *H. sacchariolens* complex and evidently the most frequent member of it. Apparently widespread.

populinum Romagn., *Bull. Soc. Mycol. France* 81: 326 (1965)

E: ! **S:** ! **NI:** !

H: On soil in mixed deciduous woodland with *Betula* spp, and also reported with *Picea sitchensis* in plantations.

D+I: B&K5: 118 115 **I:** C&D: 307

Rarely reported but apparently 'well distributed' in Scotland *fide* R. Watling (pers. comm.).

psammophilum Bon, *Doc. Mycol.* 16(62): 70 (1986)

E: !

H: On sand in dunes. British material with *Salix repens*.

A single collection from Westmorland (Sandscale Haws) in 2003.

pseudoamarescens (Kühner & Romagn.) P. Collin, *Doc. Mycol.* 19 (74): 61 (1988)

> *Alnicola pseudoamarescens* Kühner & Romagn., *Ann. Sci. Franche-Comté* 2: 10 (1947)
> *Naucoria pseudoamarescens* (Kühner & Romagn.) Kühner & Romagn. [*comb. inval.*], *Flore Analytique des Champignons Supérieurs*: 236 (1953)
> *Hebeloma funariophilum* M.M. Moser, *Z. Pilzk.* 36(1-2): 61 (1970)

E: ! **S:** !

H: On burnt soil in mixed deciduous woodland, usually on old bonfire sites.

D: NBA9: 555 (as *Hebeloma funariophilum*), NM2: 331 (as *Naucoria pseudoamarescens*), Orton (NBA9: 555, as *H.*

funariophilum), Reid (Reid84, as *N. pseudoamarescens*) **D+I:** B&K5: 118 116

Known from England (Surrey: Brentmoor Heath, and Box Hill) and Scotland (South Aberdeenshire: Braemar). Reported elsewhere but unsubstantiated with voucher material.

pusillum J.E. Lange, *Fl. agaric. danic.* 5, *Taxonomic Conspectus*: IV (1940)

E: o **W:** ! **NI:** ! **ROI:** !

H: On wet soil in fens, marshes, dried out pond margins and occasionally in dried out dune slacks, usually associated with *Salix* spp.

D: NM2: 319 **D+I:** Ph: 148 **I:** FlDan3: 120C, C&D: 306

Occasional but may be locally frequent in southern England. Infrequently reported elsewhere but apparently widespread.

radicosum (Bull.) Ricken, *Blätterpilze Deutschl.*: 115 (1911)

> *Agaricus radicosus* Bull., *Herb. France*: pl. 160 (1783)
> *Pholiota radicosa* (Bull.) P. Kumm., *Führ. Pilzk.*: 84 (1871)
> *Dryophila radicosa* (Bull.) Quél., *Enchir. fung.*: 67 (1886)

E: ! **S:** ! **NI:** !

H: On soil, arising via a long rhizoid from deserted underground latrines or decayed remains of small mammals such as moles. Also known from rabbit warrens.

D: NM2: 316 **D+I:** B&K5: 120 119 **I:** FlDan3: 109E (as *Pholiota radicosa*), SV29: 17, C&D: 309

Rarely reported but apparently widespread.

sacchariolens Quél., *Bull. Soc. Amis Sci. Nat. Rouen., Série II* 15: 185 (1879)

E: ! **S:** ! **W:** ! **NI:** ! **ROI:** !

H: On soil in mixed deciduous woodland.

D: NM2: 320 **D+I:** B&K5: 120 120, B&K5: 120 120 **I:** SV29: 20

Widely reported but the species is part of a complex not yet fully resolved. Most records may refer to *Hebeloma pallidoluctuosum* or *H. groegeri*.

senescens Sacc., *Syll. fung.* 5: 799 (1887)

> *Hebeloma edurum* Bon, *Doc. Mycol.* 16(61): 16 (1985)
> Mis.: *Hebeloma sinapizans* sensu Lange (FlDan3: 94 & pl. 119D)
> Mis.: *Hebeloma sinuosum* sensu NCL

E: ! **W:** !

H: On soil in mixed deciduous woodland.

D: NM2: 321 (as *Hebeloma edurum*) **D+I:** B&K5: 122 121 **I:** Bon: 229 (as *H. edurum*), SV29: 21 (as *H. edurum*), C&D: 309 (as *H. edurum*), Cooke 447 (407) Vol. 3 (1886)

Rarely reported. Known from England, from Northamptonshire southwards, and a single record from Wales. Many reports may relate to *H. sinapizans*.

sinapizans (Paulet) Gillet, *Hyménomycètes*: 527 (1876)

> *Hypophyllum sinapizans* Paulet, *Traité champ.*: pl. 82 (1793)
> *Agaricus sinapizans* (Paulet) Fr., *Epicr. syst. mycol.*: 180 (1838)

E: ! **S:** ! **W:** ! **NI:** ! **ROI:** !

H: On soil in mixed deciduous woodland, especially with *Fagus* on calcareous soils but also occasionally reported with conifers.

D: NM2: 320 **D+I:** Ph: 146, B&K5: 122 122, FM1: 67 f. 6 **I:** Bon: 229, C&D: 309

Apparently widespread, but easily confused with *H. senescens*.

sordescens Vesterh., *Nordic J. Bot.* 9(3): 307 (1989)

> Mis.: *Hebeloma testaceum* sensu NCL and sensu auct. mult.

E: ! **S:** !

H: On soil in woodland, with deciduous trees.

D: NM2: 318 **D+I:** B&K5: 122 123 **I:** FlDan3: 118E (as *H. testaceum*)

Rarely reported.

theobrominum Quadr., *Mycotaxon* 30: 311 (1987)

> Mis.: *Hebeloma truncatum* sensu auct.

E: ! **S:** ! **W:** !

H: On soil in woodland, usually associated with deciduous trees and rarely with conifers.

D: NM2: 320 (as *Hebeloma truncatum*) **D+I:** B&K5: 124 125 (as *H. truncatum*) **I:** Ph: 148 (as *H. truncatum*), Bon: 229 (as *H. truncatum*), C&D: 309

Rarely reported but apparently widespread. Most collections are determined as *H. truncatum.*

vaccinum Romagn., *Bull. Soc. Mycol. France* 81: 333 (1965)
E: ! **W:** ! **ROI:** ?
H: On soil in dried out dune slacks, associated with *Salix repens.* Also recorded from gardens and in woodland, apparently associated with *Alnus, Betula* spp, and *Populus canescens.*
I: C&D: 309
Rarely reported.

HEBELOMINA Maire, *Bull. soc. Hist. nat. afr. N* 16: 14 (1925)
Cortinariales, Cortinariaceae
Type: *Hebelomina domardiana* Maire

neerlandica Huijsman, *Persoonia* 9(4): 490 (1978)
E: !
H: On soil in deciduous or mixed woodland. British collections with *Pinus sylvestris* and *Betula* spp.
D: Myc7(3): 108
Known only from Surrey (Esher Common and Oxshott Heath). It is worth noting that an albino *Galerina* (probably *G. marginata*) occurs in Britain and has been misidentified as *Hebelomina neerlandica.*

HEMIMYCENA Singer, *Rev. Mycol. (Paris)* 3(6): 194 (1938)
Agaricales, Tricholomataceae
Helotium Tode [*nom. rej.*], *Fung. mecklenb. sel.* 1: 22 (1790)
Type: *Hemimycena lactea* (Pers.) Singer

angustispora (P.D. Orton) Singer, *Sydowia* 15: 62 (1962) [1961]
Mycena angustispora P.D. Orton, *Trans. Brit. Mycol. Soc.* 43(2): 305 (1960)
Delicatula angustispora (Joss.) Kühner & Romagn. [*nom. nud., comb. inval.*], *Flore Analytique des Champignons Supérieurs*: 118 (1953)
Omphalia angustispora Joss. [*nom. nud.*], *Ann. Soc. Linn. Lyon* 25: 120 (1937)
Marasmiellus angustisporus (Joss.) Singer [*nom. nud.*], *Lilloa* 22: 298 (1951) [1949]
E: !
H: On decayed leaf litter in deciduous woodland. British material associated with *Fagus.*
D: BFF8: 125 **D+I:** A&N3 71-74 pl.21
Known (in Britain) only from the type collection from Surrey (Mickleham Downs). Reported from East Kent (Blean Woods) and Worcestershire (Wyre Forest), the former unsubstantiated with voucher material, the latter redetermined as *H. tortuosa.*

candida (Bres.) Singer, *Ann. Mycol.* 41: 121 (1943)
Omphalia candida Bres., *Fungi trident.* 2(14): 87 (1900)
Mycena candida (Bres.) Kühner, *Encycl. Mycol.* 10. *Le Genre Mycena*: 659 (1938)
Delicatula candida (Bres.) Kühner & Romagn. [*comb. inval.*], *Flore Analytique des Champignons Supérieurs*: 119 (1953)
E: !
H: On decayed leaves of *Symphytum* spp, and apparently restricted to this host, in woodland, hedgerows or on wasteground.
D: NM2: 126, BFF8: 125 **D+I:** A&N3 80-83 pl.24 **I:** C&D: 221
Rarely reported and even more rarely collected. The majority of the few British records are unsubstantiated with voucher material. Moreover, most cite hosts other than *Symphytum* spp.

cephalotricha (Redhead) Singer, *The Agaricales in Modern Taxonomy* Edn 4: 397 (1986)
Helotium cephalotrichum Redhead, *Canad. J. Bot.* 60(10): 2004 (1982)
Omphalia cephalotricha Joss. [*nom. inval.*], *Ann. Soc. Linn. Lyon* 80: 85 (1937)

Delicatula cephalotricha (Joss.) Cejp [*nom. inval.*], *Atlas champ. Eur.* 4: 138 (1938)
Mycena cephalotricha (Joss.) Romagn. [*nom. inval.*], *Bull. Soc. Mycol. France* 108(2): 86 (1992)
E: ! **S:** ? **ROI:** ?
H: On decayed wood, often mossy fallen trunks of *Quercus* spp, in deciduous woodland.
D: NM2: 126, BFF8: 126 **D+I:** A&N3 33-36 pl.6 **I:** SV37: 24
Rarely reported. Similar to *H. tortuosa* and in similar habitat but evidently much less common.

crispata (Kühner) Singer, *Ann. Mycol.* 41: 121 (1943)
Mycena crispata Kühner, *Encycl. Mycol.* 10. *Le Genre Mycena*: 655 (1938)
Delicatula crispata (Kühner) Kühner & Romagn. [*comb. inval.*], *Flore Analytique des Champignons Supérieurs*: 117 (1953)
E: ! **W:** !
H: On decayed leaves of grass, also fallen leaves and associated debris of deciduous trees.
D: NM2: 126, BFF8: 122 **D+I:** A&N3 99-102 pl. 29&30
Known from England (Cambridgeshire and Surrey) and Wales (Radnorshire).

crispula (Quél.) Singer, *Ann. Mycol.* 41: 121 (1943)
Omphalia crispula Quél., *Enchir. fung.*: 46 (1886)
Mycena crispula (Quél.) Kühner, *Encycl. Mycol.* 10. *Le Genre Mycena*: 642 (1938)
Delicatula crispula (Quél.) Kühner & Romagn. [*comb. inval.*], *Flore Analytique des Champignons Supérieurs*: 117 (1953)
Marasmiellus crispulus (Quél.) Singer, *Lilloa* 22: 298 (1951) [1949]
Helotium hirsutum Tode, *Fung. mecklenb. sel.* 1: 23 (1790)
Omphalia hirsuta (Tode) Quél., *Compt. Rend. Assoc. Franç. Avancem. Sci.* 18: 509 (1890)
Hemimycena hirsuta (Tode) Singer, The Agaricales in Modern Taxonomy: 397 (1986)
E: ! **S:** ! **W:** ! **NI:** !
H: On decayed herbaceous or woody debris, fallen leaves and stems and occasionally on old and decayed cones of conifers.
D: NM2: 125, BFF8: 123 **D+I:** A&N3 58-63 pl.16 & 17 B&K3: 192 215 **I:** FlDan2: 62D (as *Omphalia crispula*)
Rarely reported. Easily confused with other small white, mycenoid taxa.

cucullata (Pers.) Singer, *Persoonia* 2: 20 (1961)
Agaricus cucullatus Pers., *Syn. meth. fung.*: 372 (1801)
Mycena cucullata (Pers.) Redhead, *Sydowia* 37: 252 (1984)
Agaricus gypseus Fr., *Epicr. syst. mycol.*: 104 (1838)
Mycena gypsea (Fr.) Quél., *Mém. Soc. Émul. Montbéliard, Sér. 2*, 5: 342 (1873)
Hemimycena gypsea (Fr.) Singer, *Ann. Mycol.* 41: 121 (1943)
Marasmiellus gypseus (Fr.) Singer, *Lilloa* 22: 298 (1951) [1949]
Mis.: *Mycena echinipes* sensu Lange (FlDan2: 43 & pl. 53I)
E: o **S:** ! **W:** ! **NI:** ! **ROI:** !
H: On fallen and decayed wood, leaf litter and occasionally on soil with mixed woody debris in scrub or woodland.
D: NM2: 126, BFF8: 127 **D+I:** A&N3 23-28 pl.2, B&K3: 192 216 **I:** Bon: 187, C&D: 221, Kriegl3: 253
Occasional in England, and may be quite common locally in southern counties. Rarely reported elsewhere but apparently widespread.

delectabilis (Peck) Singer, *Ann. Mycol.* 41: 121 (1943)
Agaricus delectabilis Peck, *Rep. (Annual) New York State Mus. Nat. Hist.* 27: 93 (1875)
Mycena delectabilis (Peck) Sacc., *Syll. fung.* 5: 262 (1887)
Marasmiellus delectabilis (Peck) Singer, *Lilloa* 22: 299 (1951) [1949]
Delicatula delectabilis (Peck) Kühner & Romagn. [*comb. inval.*], *Flore Analytique des Champignons Supérieurs*: 118 (1953)
Mis.: *Omphalia gracillima* sensu Lange (FlDan2: 62 & pl. 61A)
E: ! **S:** ! **NI:** ?

H: On dead leaves of *Carex* and *Juncus* spp, and grasses. Also on dead rootstocks of *Fallopia japonica,* dead woody stems of *Rubus fruticosus* and decayed debris of *Pteridium.*
D: NM2: 125, BFF8: 121 **D+I:** A&N3 84-88 pl.25
Rarely reported but apparently widespread. Easily confused with other white mycenoid agarics.

epichloë (Kühner) Singer, *Ann. Mycol.* 41: 121 (1943)
   Mycena epichloë Kühner, *Encycl. Mycol.* 10. *Le Genre Mycena*: 653 (1938)
   Delicatula epichloë (Kühner) Kühner & Romagn. [*comb. inval.*], *Flore Analytique des Champignons Supérieurs*: 117 (1953)
E: ! **S:** !
H: On decayed stems and leaves of grasses in grassland or woodland.
D+I: A&N3 102-105 pl.31
Known from widely scattered locations in England (Middlesex, North Somerset and Surrey) and Scotland (Angus: Barry Links) but easily overlooked.

gypsella (Kühner) Elborne & Laessøe, *Nordic J. Bot.* 11(4): 478 (1991)
   Mycena gypsella Kühner, *Travaux du Laboratoire de 'La Jaysinia' a Samoëns* 4: 67 (1972)
S: !
H: On dead fern stems, possibly confined to species of *Dryopteris.*
D: NM2: 126 **D+I:** A&N3 28-30 pl.4
Known from Perthshire (Black Spout Wood) and recorded from Orkney.

lactea (Pers.) Singer, *Rev. Mycol. (Paris)* 3(6): 195 (1938)
   Agaricus lacteus Pers. [non *A. lacteus* Pers., *Syn. meth. fung.*: 439 (1801)], *Syn. meth. fung.*: 394 (1801)
   Mycena lactea (Pers.) P. Kumm., *Führ. Pilzk.*: 110 (1871)
   Hemimycena delicatella (Peck) Singer, *Agaricales in Modern Taxonomy,* Edn 2: 369 (1962)
   Mycena lactella P.D. Orton, *Trans. Brit. Mycol. Soc.* 43(2): 306 (1960)
   Hemimycena lactea f. *lactella* (P.D. Orton) Courtec., *Doc. Mycol.* 16(62): 26 (1986)
   Hemimycena lactella (P.D. Orton) Watling, *Edinburgh J. Bot.* 55(1): 157 (1998)
   Mis.: *Mycena pithya* sensu auct.
E: c **S:** c **W:** ! **NI:** ! **ROI:** !
H: On debris (needle litter, brash and small twigs) of conifers such as *Chamaecyparis, Cupressus, Picea, Pinus, Pseudotsuga, Sequoia* and *Thuja* spp, and rarely on litter of deciduous trees such as *Fagus.*
D: NM2: 125, BFF8: 128 (as *Hemimycena lactella*), BFF8: 128 **D+I:** A&N3 16-20 pl.1, Ph: 76 (as *Mycena lactea*) **I:** Bon: 187 (as *H. delicatella*)
Common and widespread, and may be abundant where it occurs. The illustration of this species in B&K3: 194 pl. 217 is more likely to be *Hemimycena cucullata.*

mairei (E.-J. Gilbert) Singer, *Ann. Mycol.* 41: 121 (1943)
   Omphalia mairei E.-J. Gilbert, *Bull. Soc. Mycol. France* 42: 63 (1926)
   Mycena mairei (E.-J. Gilbert) Kühner, *Encycl. Mycol.* 10. *Le Genre Mycena*: 661 (1938)
   Marasmiellus mairei (E.-J. Gilbert) Singer, *Lilloa* 22: 298 (1951) [1949]
   Delicatula mairei (E.-J. Gilbert) Kühner & Romagn. [*comb. inval.*], *Flore Analytique des Champignons Supérieurs*: 119 (1953)
E: ! **ROI:** !
H: On decayed debris (leaves, stems etc.) of grasses and herbaceous plants in arable fields, gardens, playing fields and in woodland. Most records are from lawns in late autumn.
D: BFF8: 120 **D+I:** A&N3 39-41 pl.7 **I:** Bon: 187, Kriegl3: 258
Rarely reported. Known from widely scattered sites in England, from Co. Durham southwards (mostly in southern England) and also from the Republic of Ireland (Co. Dublin: North Bull Island).

mauretanica (Maire) Singer, *Ann. Mycol.* 41: 121 (1943)
   Omphalia mauretanica Maire, *Bull. Soc. Mycol. France* 44: 43 (1928)
   Mycena mauretanica (Maire) Kühner, *Encycl. Mycol.* 10. *Le Genre Mycena*: 639 (1938)
   Marasmiellus mauretanicus (Maire) Singer, *Lilloa* 22: 298 (1951) [1949]
   Delicatula mauretanica (Maire) Kühner & Romagn. [*comb. inval.*], *Flore Analytique des Champignons Supérieurs*: 117 (1953)
   Omphalia cuspidata var. *stenospora* J.E. Lange, *Dansk. Bot. Ark.* 6(5): 18 (1930)
E: ! **S:** !
H: On woody debris of *Fagus, Quercus* and *Populus* spp. Also on fallen leaves of *Rubus fruticosus* and debris of *Pinus* spp.
D: NM2: 125, BFF8: 124 **D+I:** A&N3 51-55 pl.13
Rarely reported. Most frequently reported from southern England but known from Northumberland and Warwickshire, and from Scotland (Mid-Ebudes: Isle of Mull).

pseudocrispula (Kühner) Singer, *Ann. Mycol.* 41: 121 (1943)
   Mycena pseudocrispula Kühner, *Encycl. Mycol.* 10. *Le Genre Mycena*: 645 (1938)
   Marasmiellus pseudocrispulus (Kühner) Singer, *Lilloa* 22: 298 (1951) [1949]
   Delicatula pseudocrispula (Kühner) Kühner & Romagn. [*comb. inval.*], *Flore Analytique des Champignons Supérieurs*: 117 (1953)
   Mis.: *Mycena cyphelloides* sensu NCL
   Mis.: *Omphalia gibba* sensu Rea [TBMS 12: 217 (1927)]
E: ! **W:** ?
H: On decayed grass leaves or other herbaceous debris in woodland, scrub or pasture.
D: BFF8: 129 **D+I:** A&N3: 63-66 pl.18, B&K3: 196 220
Known from Cambridgeshire, North Devon and Surrey but easily overlooked or confused with other white, mycenoid taxa.

pseudolactea (Kühner) Singer, *Ann. Mycol.* 41: 121 (1943)
   Mycena pseudolactea Kühner, *Encycl. Mycol.* 10. *Le Genre Mycena*: 632 (1938)
S: !
H: On fallen and decayed litter of conifers. British material on needle litter of *Sequoiadendron* spp
D+I: A&N3 30-31 pl.5
A single collection from Peebleshire (Dawyck Botanic Garden) in 2000.

tortuosa (P.D. Orton) Redhead, *Fungi Canadenses No.177.* f. 84a (1980)
   Mycena tortuosa P.D. Orton, *Trans. Brit. Mycol. Soc.* 43(2): 307 (1960)
E: o **S:** o **W:** o **NI:** o **ROI:** o **O:** Isle of Man: o
H: On wood and bark, often on the undersides of large logs in damp places, occasionally on debris such as twigs and sticks in woodland. Usually on *Corylus, Fagus* and *Fraxinus* but also known on *Alnus, Betula* and *Salix* spp, and the conifers *Larix, Picea* spp, and *Taxus.*
D: BFF8: 130 **D+I:** A&N3 77-80 pl.23 **I:** C&D: 221
Occasional but widespread, and may be locally abundant in southern England, especially in woodland on calcareous soil.

HENNINGSOMYCES Kuntze, *Revis. gen. pl.* 3(3): 483 (1898)
Schizophyllales, Schizophyllaceae
   Solenia Pers. [*nom. illegit.*, non *Solena* Lour. (1790)(*Cucurbitaceae*); non *Solenia* Kuntze (1898)], *Neues Mag. Bot.* 1: 116 (1794)
Type: *Henningsomyces candidus* (Pers.) Kuntze

candidus (Pers.) Kuntze, *Revis. gen. pl.* 3: 483 (1893)
   Solenia candida Pers., *Neues Mag. Bot.* 1: 116 (1794)
   Mis.: *Solenia fasciculata* sensu Massee (*Brit. fung.-fl.* 1: 144, 1892), sensu Rea (1922)
E: c **S:** c **W:** o **NI:** ! **ROI:** ! **O:** Channel Islands: ! Isle of Man: !
H: On decayed wood of deciduous trees, especially *Fagus* and *Ulmus* spp.

D: NM3: 157 **D+I:** B&K2: 202 229
Common and widespread.

puber (Romell) D.A. Reid, *Persoonia* 3(1): 119 (1964)
    Solenia pubera W.B. Cooke, *Sydowia, Beih.* 4: 26 (1961)
E: ! **NI:** !
H: On decayed wood of deciduous trees.
D: NM3: 157
Known from England (Oxfordshire: Warburg Reserve) and
    Northern Ireland (Antrim: O'Neill Estate).

HERICIUM Pers., *Neues Mag. Bot.* 1: 109 (1794)
Hericiales, Hericiaceae
    Friesites P. Karst., *Meddeland. Soc. Fauna Fl. Fenn.* 5: 41
      (1879)
    Creolophus P. Karst., *Meddeland. Soc. Fauna Fl. Fenn.* 5: 41
      (1879)
    Dryodon P. Karst., *Rev. Mycol. (Toulouse)* 3(9): 19 (1881)
Type: *Hericium coralloides* (Scop.) Pers.

cirrhatum (Pers.) Nikol., *Acta Inst. Bot. Acad. Sci. USSR Plant.
    Crypt., Ser. 2,* 6: 343 (1950)
    Hydnum cirrhatum Pers., *Syn. meth. fung.*: 558 (1801)
    Creolophus cirrhatus (Pers.) P. Karst., *Meddeland. Soc. Fauna
      Fl. Fenn.* 5: 42 (1879)
    Hydnum diversidens Fr., *Syst. mycol.* 1: 414 (1821)
    Hericium diversidens (Fr.) Nikol., *Acta Inst. Bot. Acad. Sci.
      USSR Plant. Crypt., Ser. 2,* 6: 222 (1961)
E: o **W:** ! **ROI:** ! **O:** Channel Islands: !
H: Weakly parasitic or saprotrophic. On fallen, decayed wood of
    deciduous trees, usually *Fagus* but also known on *Acer
    pseudoplatanus, Betula* spp, *Fraxinus, Quercus robur* and
    Ulmus spp.
D: NM3: 283 (as *Creolophus cirrhatus*) **D+I:** Ph: 246-247 (as *C.
    cirrhatus*), B&K2: 238 284 (as *C. cirrhatus*), BritChant: 52-53 **I:**
    Bon: 313 (as *C. cirrhatus*), Kriegl2: 96, SV48: 35 (as *C.
    cirrhatus*)
Rather rare, but widespread especially in areas of old deciduous
    woodland. Most frequently reported from southern and south-
    western England.

coralloides (Scop.) Pers., *Neues Mag. Bot.* 1: 109 (1794)
    Hydnum coralloides Scop., *Fl. carniol.* 472 (1772)
    Friesites coralloides (Scop.) P. Karst., *Meddeland. Soc. Fauna
      Fl. Fenn.* 5: 41 (1880)
    Dryodon coralloides (Scop.) P. Karst., *Meddeland. Soc. Fauna
      Fl. Fenn.* 6: 15 (1881)
    Hydnum clathroides Pall., *Reise Prov. russ. Reich* 2: 744
      (1773)
    Hericium clathroides (Pall.) Pers., *Comment. Fungis
      Clavaeform.*: 23 (1797)
    Hydnum ramosum Bull., *Herb. France*: pl. 390 (1789)
    Hericium ramosum (Bull.) Letell., *Hist. Descr. Champ.*: 43
      (1826)
E: !
H: Weakly parasitic or saprotrophic. On living and dead wood,
    usually dead standing trunks of *Fagus* and *Fraxinus* in old
    deciduous woodland. A single record on *Ulmus glabra*.
D: NM3: 284 **D+I:** Ph: 245 (as *Hericium ramosum*), BritChant:
    54-55 **I:** Sow3: 252 (as *Hydnum coralloides*), Bon: 313 (as
    Hericium clathroides)
Rare. Known from widely scattered sites from Yorkshire
    southwards but mainly from southern counties.

erinaceus (Bull.) Pers., *Comment. Fungis Clavaeform.*: 27
    (1797)
    Hydnum erinaceus Bull., *Herb. France*: pl. 34 (1780)
    Hydnum caput-medusae Bull., *Herb. France*: pl. 412 (1789)
    Steccherinum quercinum Gray, *Nat. arr. Brit. pl.* 1: 651
      (1821)
    Hericium unguiculatum Pers., *Mycol. eur.* 2: 153 (1825)
E: ! **W:** ?
H: Weakly parasitic or saprotrophic. On living and dead trunks
    of deciduous trees, usually *Fagus* and rarely on *Carpinus* and
    Quercus spp in old deciduous woodland. Reported on *Fraxinus*
    and *Populus* sp. but these hosts are unverified.

D: NM3: 284 **D+I:** Ph: 245, BritChant: 56-57 **I:** Kriegl2: 98
Rare. Virtually confined to old deciduous woodland in southern
    and south-western England, with single records from Cheshire
    (Greater Manchester) in 1989 and from Wales in 1999. Now
    commercially cultivated.

HETEROACANTHELLA Oberw., *Trans. Mycol. Soc. Japan* 31: 208 (1990)
Ceratobasidiales, Ceratobasidiaceae
Type: *Heteroacanthella variabilis* Oberw. & Langer

acanthophysa (Burds.) Oberw., *Trans. Mycol. Soc. Japan*
    31(2): 211 (1990)
    Platygloea acanthophysa Burds., *Mycotaxon* 27: 500 (1986)
    Acanthellorhiza globulifera P. Roberts (anam.), *Rhizoc. f.
      fungi.*: 131 (1999)
E: !
H: On bark and decorticated wood of dead standing trunks of
    Ulmus spp.
D+I: FM12(2): 146-147
Known from Hertfordshire, South Devon and Surrey.

HETEROBASIDION Bref., *Unters. Gesammtgeb. Mykol.* 8: 154 (1888)
Poriales, Coriolaceae
Type: *Heterobasidion annosum* (Fr.) Bref.

annosum (Fr.) Bref., *Unters. Gesammtgeb. Mykol.* 8: 154
    (1888)
    Polyporus annosus Fr., *Syst. mycol.* 1: 373 (1821)
    Fomitopsis annosa (Fr.) P. Karst., *Rev. Mycol. (Toulouse)*
      3(9): 18 (1881)
    Fomes annosus (Fr.) Cooke, *Grevillea* 14: 20 (1885)
    Boletus cryptarum Bull., *Herb. France*: pl. 478 (1790)
    Poria cryptarum (Bull.) Gray, *Nat. arr. Brit. pl.* 1: 639 (1821)
    Polyporus cryptarum (Bull.) Fr., *Syst. mycol.* 1: 376 (1821)
    Fomes cryptarum (Bull.) Sacc., *Syll. fung.* 6: 205 (1888)
    Polystictus cryptarum (Bull.) W.G. Sm., *Syn. Brit. Bas.*: 353
      (1908)
    Polyporus scoticus Berk., *Engl. fl.*: 142 (1836)
E: c **S:** c **W:** c **NI:** c **ROI:** c **O:** Channel Islands: ! Isle of Man - !
H: Parasitic then saprotrophic. A virulent root pathogen
    especially in conifer plantations. Known on a wide range of
    conifers, but also on deciduous hosts, and on *Erica* and
    Calluna.
D: EurPoly1: 308, NM3: 243 **D+I:** Ph: 226-227, B&K2: 314 397
    I: Kriegl1: 536
Very common and widespread. Recent evidence suggests that
    this is probably a species complex.

HETEROGASTRIDIUM Oberw. & R. Bauer, *Mycologia* 82: 55 (1990)
Heterogastridiales, Heterogastridiaceae
    Hyalopycnis Höhn. (anam.), *Hedwigia* 60: 152 (1918)
Type: *Heterogastridium pycnidioideum* Oberw. & R. Bauer

pycnidioideum Oberw. & R. Bauer, *Mycologia* 82(1): 55 (1990)
    Sphaeronema blepharistoma Berk. (anam.), *Mag. Zool. Bot.*
      1: 512 (1837)
    Hyalopycnis blepharistoma (Berk.) Seeler (anam.), *Farlowia*
      1: 124 (1943)
E: !
H: On old and moribund basidiomes of *Russula adusta*.
Known in Britain only from the type collection of the anamorph
    from Northamptonshire (Kings Cliffe) in 1836.

HETEROMYCOPHAGA P. Roberts, *Mycotaxon* 63: 210 (1997)
Incertae sedis
An anamorphic genus belonging to the *Basidiomycota*, but
whose teleomorphs (and taxonomic position) are not yet known.
Type: *Heteromycophaga glandulosae* P. Roberts

glandulosae P. Roberts, *Mycotaxon* 63: 211 (1997)

E: !

H: Parasitic. Forming galls on the hymenium of *Exidia glandulosa*.

Known from southern England (Dorset, Hertfordshire, South Devon, and Surrey).

HOHENBUEHELIA Schulzer, *Verh. zool.-bot. Ges. Wien* 16: 45 (1866)

Agaricales, Tricholomataceae

Acanthocystis (Fayod) Kühner, *Botaniste* 17: 11 (1926)

Type: *Hohenbuehelia petaloides* (Bull.) Schulzer

atrocaerulea (Fr.) Singer, *Lilloa* 22: 255 (1951) [1949]

Agaricus atrocaeruleus Fr., *Syst. mycol.* 1: 190 (1821)

Pleurotus atrocaeruleus (Fr.) P. Kumm., *Führ. Pilzk.*: 104 (1871)

Pleurotus atrocaeruleus f. *albidotomentosus* Pilát, *Atlas champ. Eur.* 2: 98 (1935)

Mis.: *Pleurotus algidus* sensu auct.

E: o **S:** ? **W:** ! **ROI:** ! **O:** Channel Islands: !

H: On dead and decayed wood of deciduous trees, most often on fallen trunks of *Betula*, *Fagus* and *Ulmus* spp. Also rarely on *Alnus*, *Corylus*, *Laburnum anagyroides*, *Quercus*, *Sambucus nigra*, *Sorbus aucuparia* and *Wisteria sinensis* and on conifers such as *Pinus sylvestris*.

D: BFF6: 44, NM2: 128, FAN3: 164-165 **D+I:** B&K3: 196 222, Myc9(2): 60 **I:** C&D: 153

Occasional throughout England, rarely reported elsewhere but apparently widespread. Possibly increasing.

auriscalpium (Maire) Singer, *Lilloa* 22: 255 (1951) [1949]

Pleurotus auriscalpium Maire, *Bull. Soc. Mycol. France* 46: 220 (1931)

Mis.: *Hohenbuehelia petaloides* sensu NCL & sensu auct.

E: ! **S:** ?

H: On decayed wood of deciduous trees, usually large fallen trunks or old moss covered stumps of *Fagus*.

D: BFF6: 41 (as *Hohenbuehelia petaloides*), FAN3: 160-161 **I:** Bon: 123 (as *H. petaloides*), C&D: 153 (as *H. petaloides*), SV49: 51

Known from southern and south-eastern England (Berkshire, North Wiltshire, South Hampshire, Surrey and West Sussex) but previously included with *H. petaloides*. Basidiomes are relatively small and thin-fleshed, spathulate and lignicolous, unlike those of *H. petaloides* which are large, flabelliform to semi-infundibuliform, and terrestrial.

culmicola Bon, *Doc. Mycol.* 10(37-38): 89 (1979)

Acanthocystis petaloides var. *macrospora* Bon, *Bull. Soc. Mycol. France* 86(1): 163 (1970)

Hohenbuehelia petaloides var. *macrospora* (Bon) Courtec., *Doc. Mycol.* 15(57-58): 30 (1985) [1984]

Mis.: *Pleurotus longipes* sensu P.D. Orton (1959)

Mis.: *Pleurotus petaloides* f. *carbonarius* sensu auct. Brit.

E: ! **S:** ! **W:** ! **ROI:** !

H: On dead or dying stems of *Ammophila arenaria* on dunes, often in the sand, at or around the stem bases. A single record on *Leymus arenarius*.

D: BFF6: 42-43, FAN3: 162 **I:** C&D: 153, SV33: 45

Rarely reported but apparently widespread. Basidiomes are small, often hidden in sand and easily overlooked.

cyphelliformis (Berk.) O.K. Mill., *Mycotaxon* 25: 33 (1986)

Agaricus cyphelliformis Berk., *Mag. Zool. Bot.*, [Notices of British Fungi No. 49] 1: 511 (1837)

Pleurotus cyphelliformis (Berk.) Sacc., *Syll. fung.* 5: 379 (1887)

Acanthocystis cyphelliformis (Berk.) Konrad & Maubl., *Icon. select. fung.* 6: 365 (1937)

Resupinatus cyphelliformis (Berk.) Singer, *Lilloa* 22: 253 (1951) [1949]

Geopetalum cyphelliforme (Berk.) Kühner & Romagn., *Flore Analytique des Champignons Supérieurs*: 68 (1953)

Marasmius broomei Berk. & Broome, *Ann. Mag. Nat. Hist., Ser. 5* [Notices of British Fungi no. 1795] 3: 209 (1879)

Marasmius spodoleucus Berk. & Broome, *Ann. Mag. Nat. Hist., Ser. 3* [Notices of British Fungi no. 803] 3: 358 (1859)

Mis.: *Hohenbuehelia silvana* sensu Flora Agaricina Neerlandica 3: 163 (1995)

E: o **S:** ! **W:** !

H: On dead twigs (often attached) of deciduous trees and shrubs, and on dead herbaceous stems.

D: BFF6: 46-47, FAN3: 163

Occasional in southern England. Rarely reported elsewhere but apparently widespread.

fluxilis (Fr.) P.D. Orton, *Notes Roy. Bot. Gard. Edinburgh* 26: 50 (1964)

Agaricus fluxilis Fr., *Syst. mycol.* 1: 189 (1821)

Pleurotus fluxilis (Fr.) Gillet, *Hyménomycètes*: 335 (1876)

E: ! **S:** ! **W:** !

H: On dead wood of deciduous trees and shrubs. Known on *Betula* spp, *Corylus*, *Fagus*, *Prunus avium*, *Sorbus aucuparia*, *Salix* and *Tilia* spp.

D: BFF6: 44-45, NM2: 128, FAN3: 164, NBA2: 50-51

Rarely reported. Known from England (Co. Durham, Herefordshire, North Lincolnshire, South Lincolnshire, South Wiltshire, Surrey, Warwickshire and West Kent), from Wales (Carmarthenshire) and Scotland (Aberdeenshire).

leightonii (Berk.) Watling, *Brit. fung.-fl.* 6: 47 (1989)

Agaricus leightonii Berk., *Ann. Mag. Nat. Hist., Ser. 1* [Notices of British Fungi no. 270] 13: 343 (**1844**)

Pleurotus leightonii (Berk.) Sacc., *Syll. fung.* 5: 378 (1887)

Resupinatus leightonii (Berk.) P.D. Orton, *Trans. Brit. Mycol. Soc.* 43(2): 181 (1960)

E: !

H: On dead wood of deciduous trees.

D: BFF6: 47 **I:** Cooke 290 (260) Vol. 2 (**1883**)

Known with certainty only from the type collection, and two others from the same location (Shropshire: Montford Bridge). Reported elsewhere but records are unsubstantiated with voucher material.

longipes (Boud.) M.M. Moser, *Kleine Kryptogamenflora*, Edn 3: 108 (1967)

Pleurotus longipes Boud., *Bull. Soc. Mycol. France* 21: t. 3 (1905)

Acanthocystis longipes (Boud.) J. Favre, *Beitr. Kryptogamenfl. Schweiz* 10(3): 59 (1948)

E: !

H: On soil amongst mosses.

D: BFF6: 43, NM2: 127 **I:** C&D: 153

Known from East Kent (Bedgebury Pinetum) and a recently discovered specimen in herb.K, collected in 1948 in Surrey (Mickleham). Supposedly a species of boreal or tundra habitat, so the British records are somewhat aberrant.

mastrucata (Fr.) Singer, *Lilloa* 22: 255 (1951) [1949]

Agaricus mastrucatus Fr., *Syst. mycol.* 1: 190 (1821)

Pleurotus mastrucatus (Fr.) Sacc., *Syll. fung.* 5: 376 (1887)

E: ! **S:** !

H: On decayed wood (usually large decayed logs) of deciduous trees such as *Fagus* or *Corylus*.

D: BFF6: 45-46, NM2: 127, FAN3: 165-166 **D+I:** FRIC17: 4-5 129b **I:** C&D: 153

Rarely reported. Known from England (Berkshire, Huntingdonshire and West Kent) and Scotland. Also reported from Lincolnshire, South Hampshire and West Sussex. Probably genuinely rare, since basidiomes are large and distinctive.

petaloides (Bull.) Schulzer, *Verh. zool.-bot. Ges. Wien* 16: 45 (1866)

Agaricus petaloides Bull., *Herb. France*: pl. 226 (1785)

Pleurotus petaloides (Bull.) Quél., *Mém. Soc. Émul. Montbéliard, Sér. 2*, 5: 226 (1872)

Agaricus geogenius (DC) Fr., *Epicr. syst. mycol.*: 134 (1838)

Pleurotus geogenius (DC.) Gillet, *Hyménomycètes*: 339 (1876)

Acanthocystis geogenia (DC.) Kühner, *Botaniste* 17: 111 (1926)

Pleurotus petaloides var. *geogenius* (DC.) Pilát, *Atlas champ. Eur.* 2: 91 (1935)
Hohenbuehelia geogenia (DC.) Singer, *Lilloa* 22: 255 (1951) [1949]
Geopetalum geogenium (DC.) Pat., *Hyménomyc. Eur.*: 127 (1887)
Geopetalum geogenium var. *queletii* Kühner [*nom. nud.*], *Flore Analytique des Champignons Supérieurs*: 70 (1953)
E: ! **S:** ! **W:** ! **O:** Channel Islands: !
H: On soil, trypically with deciduous trees, usually *Fagus* and rarely *Betula* spp. Also reported from dunes with *Pinus* spp and in greenhouses on potting compost.
D: BFF6: 40 (as *Hohenbuehelia geogenia*), NM2: 127 (as *H. geogenia*), NM2: 127, FAN3: 159-160 **D+I:** B&K3: 198 223 (as *H. geogenia*) **I:** FlDan2: 65E (as *Pleurotus petaloides*), C&D: 153 (as *H. geogenia*), Kriegl3: 264
This is a large and distinctive species, most often collected in south-eastern England where it may occasionally be abundant in beechwoods on calcareous soil. N.B. This is not *H. petaloides* of NCL or BFF8, for which see *H. auriscalpium*.

reniformis (G. Mey.) Singer, *Lilloa* 22: 255 (1951) [1949]
Agaricus reniformis G. Mey., *Primit. Fl. Essequ.*: 302 (1818)
Pleurotus reniformis (G. Mey.) P. Karst., *Bidrag Kännedom Finlands Natur Folk* 37: 90 (1879)
E: ! **W:** !
H: On dead twigs of deciduous trees. One British collection on *Populus tremula*.
D: BFF6: 46, NM2: 128
Rarely reported, and poorly known in Britain. The majority of reports are unsubstantiated with voucher material.

tremula (Schaeff.) Thorn & G.L. Barron, *Mycotaxon* 25(2): 414 (1986)
Agaricus tremulus Schaeff., *Fung. Bavar. Palat. nasc.* 1: 54 (1762)
Pleurotus tremulus (Schaeff.) P. Kumm., *Führ. Pilzk.*: 105 (1871)
Pleurotellus tremulus (Schaeff.) Konrad & Maubl., *Encycl. Mycol.* 14, *Les Agaricales* I: 428 (1948)
Hohenbuehelia rickenii P.D. Orton, *Trans. Brit. Mycol. Soc.* 43(2): 176 (1960)
Mis.: *Hohenbuehelia geogenia* sensu Rea (1922)
E: ! **W:** ! **ROI:** !
H: On soil or soil and wood mixtures, sawdust piles and decayed woodchips in coniferous woodland.
D: BFF6: 41-42 (as *Hohenbuehelia rickenii*), NM2: 127, FAN3: 161-162
Rarely reported. Apparently widespread, but previously confused with other *Hohenbuehelia* species.

unguicularis (Fr.) O.K. Mill., *Mycotaxon* 25(1): 44 (1986)
Agaricus unguicularis Fr., *Elench. fung.* 1: 24 (1828)
Resupinatus unguicularis (Fr.) Singer, *Lilloa* 22: 253 (1951) [1949]
S: ? **W:** !
H: On fallen twigs. Reported on *Pinus* spp, and the single Welsh collection was possibly on *Quercus* sp.
D: BFF6: 48-49, NM2: 178 (as *Resupinatus unguicularis*), FAN3: 162-163
Rarely reported and poorly known in Britain.

HORMOMYCES Bonord., *Handb. Mykol.*: 150 (1851)
Tremellales, Tremellaceae
An anamorphic genus with teleomorphs in the genus *Tremella*. One species, however, has no known teleomorph.
Type: *Hormomyces aurantiacus* Bonord.

peniophorae P. Roberts, *Mycotaxon* 63: 214 (1997)
E: !
H: On hymenium of *Peniophora lycii* on *Ulmus* sp. and *Rubus idaeus*.
Known only from two Devon collections (Slapton and Torquay: Scadson Woods).

HYDNANGIUM Wallr., in Dietrich, *Fl. boruss.* 7: 186 (1839)
Agaricales, Hydnangiaceae
Type: *Hydnangium carneum* Wallr.

carneum Wallr., *Fl. Regn. Boruss.* 7: 186 (1839)
E: ! **S:** ! **O:** Channel Islands: !
H: Hypogeous or subepigeous, always under or near *Eucalyptus* spp.
D: NM3: 294 **D+I:** BritTruff: 140-141 8 A
British records initially from Scotland (Glasgow: Botanic Garden in 1875 and Edinburgh: Royal Botanic Gardens in 1880). Now widespread with *Eucalyptus* spp, though easily overlooked.

HYDNELLUM P. Karst., *Meddeland. Soc. Fauna Fl. Fenn.* 5: 41 (1879)
Thelephorales, Thelephoraceae
Calodon P. Karst., *Rev. Mycol. (Toulouse)* 3(9): 20 (1881)
Phaeodon J. Schröt., in Cohn, *Krypt.-Fl. Schlesien* 3: 458 (1888)
Type: *Hydnellum suaveolens* (Scop.) P. Karst.

aurantiacum (Batsch) P. Karst., *Meddeland. Soc. Fauna Fl. Fenn.* 5: 41 (1880)
Hydnum suberosum var. *aurantiacum* Batsch, *Elench. fung. (Continuatio Secunda)*: 103 t. 40 f. 222 (1789)
Hydnum aurantiacum (Batsch) Alb. & Schwein., *Mycol. eur.* 2: 169 (1825)
E: ? **S:** ! **NI:** ?
H: Associated with *Pinus sylvestris*.
D: NM3: 311 **D+I:** B&K2: 220 257, BritChant: 80-81 **I:** Kriegl1: 373
Virtually restricted to Caledonian pinwoods and there uncommon. Reported from southern England and Northern Ireland but unsubstantiated with voucher material.

auratile (Britzelm.) Maas Geest., *Persoonia* 1(1): 111 (1959)
Hydnum auratile Britzelm., *Hymenomyc. Südbayern* 8: 14 (1891)
S: !
H: Amongst litter under *Pinus sylvestris* in Caledonian pinewoods.
D: NM3: 311 **D+I:** B&K2: 220 258
Two collections from Easterness in 1996 and 1997.

caeruleum (Hornem.) P. Karst., *Meddeland. Soc. Fauna Fl. Fenn.* 5: 41 (1879)
Hydnum caeruleum Hornem., *Fl. Danica*: t. 7 (1808)
Hydnum suaveolens var. *caeruleum* (Hornem.) Fr., *Epicr. syst. mycol.*: 507 (1838)
Calodon caeruleum (Hornem.) P. Karst., *Bidrag Kännedom Finlands Natur Folk* 37: 106 (1882)
S: !
H: With *Pinus sylvestris* in Caledonian pinewoods, and rarely with *Picea* spp in plantations. Also known with *Arctostaphylos* spp.
D: NM3: 310 **D+I:** B&K2: 222 259, BritChant: 76-77 **I:** Kriegl1: 374
Rarely reported. Known from widely scattered localities in Scotland.

concrescens (Pers.) Banker, *Mem. Torrey Bot. Club* 12: 157 (1906)
Hydnum concrescens Pers., *Observ. mycol.* 1: 74 (1796)
Hydnum queletii Fr., Quélet, *Mém. Soc. Émul. Montbéliard, Sér. 2*, 5: 277 (1872)
Hydnum zonatum Batsch, *Elench. fung.*: 111 (1783) [1836]
Hydnellum zonatum (Batsch) P. Karst., *Meddeland. Soc. Fauna Fl. Fenn.* 5: 41 (1880)
Hydnellum velutinum var. *zonatum* (Batsch) Maas Geest., *Fungus* 27: 64 (1957)
E: o **S:** o **W:** ! **NI:** ! **ROI:** ! **O:** Channel Islands: !

H: On acidic soil, in mixed deciduous and coniferous woodland, usually with *Castanea*, *Fagus* or *Quercus* spp, intermixed with *Pinus sylvestris*.
D: NM3: 311 **D+I:** Ph: 244, B&K2: 222 261, BritChant: 82-83 **I:** Kriegl1: 376
Occasional but widespread and generally uncommon but may be locally abundant in old woodland. Most frequently reported from southern England and Scotland.

ferrugineum (Fr.) P. Karst., *Meddeland. Soc. Fauna Fl. Fenn.* 5: 41 (1880)
   Hydnum ferrugineum Fr., *Observ. mycol.* 1: 133 (1815)
   Calodon ferrugineus (Fr.) P. Karst., *Rev. Mycol. (Toulouse)* 3(9): 20 (1881)
   Phaeodon ferrugineus (Fr.) J. Schröt., in Cohn, *Crypt. Fl. Schles.* 3: 459 (1888)
E: ! **S:** ! **W:** ! **ROI:** !
H: On sandy soil under *Pinus sylvestris* or *Picea* spp.
D: NM3: 311 **D+I:** B&K2: 224 262, BritChant: 86-87
Rarely reported but apparently widespread.

peckii Banker, Peck, *Bull. New York State Mus. Nat. Hist.* 157: 28 (1912)
   Hydnum peckii (Banker) Trotter, *Syll. fung.* 23: 470 (1925)
   Hydnum diabolus (Banker) Trotter, *Syll. fung.* 23: 470 (1925)
   Hydnellum diabolus Banker, *Mycologia* 5: 194 (1913)
S: o
H: With *Pinus sylvestris* in Caledonian pinewoods.
D: NM3: 312 **D+I:** Ph: 243, B&K2: 224 264, BritChant: 78-79 **I:** Kriegl1: 379
Occasional to locally abundant in Caledonian pinewoods.

scrobiculatum (Fr.) P. Karst., *Meddeland. Soc. Fauna Fl. Fenn.* 5: 41 (1880)
   Hydnum scrobiculatum Fr., *Observ. mycol.* 1: 143 (1815)
   Hydnellum velutinum var. *scrobiculatum* (Fr.) Maas Geest., *Fungus* 27: 63 (1957)
E: o **S:** o **W:** ! **ROI:** !
H: On acidic soil in deciduous and coniferous woodland, usually associated with *Pinus sylvestris*, *Castanea* and *Quercus* spp.
D: NM3: 311 **D+I:** Ph: 243, B&K2: 226 265, BritChant: 84-85
Occasional but widespread.

spongiosipes (Peck) Pouzar, *Česká Mykol.* 14: 130 (1960)
   Hydnum spongiosipes Peck, *Rep. (Annual) New York State Mus. Nat. Hist.* 50: 111 (1897)
   Hydnellum velutinum var. *spongiosipes* (Peck) Maas Geest., *Fungus* 27: 62 (1957)
E: o **S:** ! **W:** ! **ROI:** ! **O:** Channel Islands: o
H: On soil in deciduous woodland, usually associated with *Quercus* spp, but also known with *Betula* spp, *Castanea* and *Fagus*.
D+I: Ph: 244, B&K2: 226 266, BritChant: 88-89
Most records are from southern England and the Channel Islands (Jersey).

HYDNUM L., *Sp. pl.*: 1178 (1753)
Cantharellales, Hydnaceae
   Dentinum Gray, *Nat. arr. Brit. pl.* 1: 650 (1821)
Type: *Hydnum repandum* L.

repandum L., *Sp. pl.*: 1258 (1753)
   Dentinum repandum (L.) Gray, *Nat. arr. Brit. pl.* 1: 650 (1821)
   Mis.: *Hydnum imbricatum* sensu Bolton [Hist. Fung. Halifax 2: 88 (1788)]
   Mis.: *Hydnum rufescens* sensu auct.
E: c **S:** c **W:** c **NI:** c **ROI:** c **O:** Channel Islands: ! Isle of Man: !
H: On soil, amongst leaf litter in deciduous woodland, usually with *Fagus* or *Quercus* spp, less often *Betula* spp, *Castanea* and *Corylus*. Rarely with conifers such as *Larix*, *Picea* or *Pinus* spp.
D: NM3: 264 **D+I:** Ph: 241, B&K2: 236 282, BritChant: 40-41 **I:** Sow2: 176, C&D: 141, Kriegl2: 94
Common and widespread.

rufescens Pers., *Syn. meth. fung.*: 555 (1801)

   Dentinum rufescens (Pers.) Gray, *Nat. arr. Brit. pl.* 1: 650 (1821)
   Hydnum repandum var. *rufescens* (Pers.) Barla, *Champ. Prov. Nice*: 81 (1859)
E: c **S:** c **W:** c **NI:** o **ROI:** ! **O:** Isle of Man: !
H: On acidic soil in deciduous woodland, and occasionally under conifers in areas of mixed woodland.
D: NM3: 264 **D+I:** Ph: 242, BritChant: 42-43 **I:** Bon: 313 (as *Hydnum repandum* var. *rufescens*), C&D: 141
Common and widespread.

HYDROPUS Singer, *Pap. Michigan Acad. Sci.* 32: 127 (1948) [1946]
Agaricales, Tricholomataceae
Type: *Agaricus fuliginarius* Batsch (= *H. nigrita* (Berk. & M.A. Curtis) Singer)

floccipes (Fr.) Singer, *Sydowia* 15: 66 (1962) [1961]
   Agaricus floccipes Fr. [non *A. floccipes* (F.H. Møller) Bohus (1978)], *Epicr. syst. mycol.*: 87 (1838)
   Collybia floccipes (Fr.) Gillet, *Hyménomycètes*: 319 (1876)
   Mycena floccipes (Fr.) Kühner, *Encycl. Mycol.* 10. *Le Genre Mycena*: 540 (1938)
   Marasmiellus floccipes (Fr.) Singer, *Lilloa* 22: 301 (1951) [1949]
E: ! **W:** ! **NI:** !
H: On decayed wood of deciduous trees, usually large fallen and moss-covered trunks of *Fagus* or *Quercus* spp in old woodland. Also known on fallen trunks of *Alnus* and amongst debris of *Thuja* spp, and rarely on soil with intermixed woody debris.
D: BFF8: 132, FAN4: 167-168 **D+I:** Myc15(2): 60 **I:** C&D: 237
Rarely reported but apparently widespread. British records are mainly from southern or south-western England, with single collections from Wales and Northern Ireland.

scabripes (Murrill) Singer, *Sydowia* 15: 66 (1962) [1961]
   Prunulus scabripes Murrill, *North American Flora* 9(5): 331 (1916)
   Mycena scabripes (Murrill) Murrill, *Mycologia*: 221 (1916)
   Mis.: *Mycena excisa* f. *solitaria* sensu Lange (FlDan2: 39 & pl. 51B)
E: !
H: On decayed wood in soil or on soil amongst leaf and needle litter. Reported with deciduous trees such as *Acer pseudoplatanus* and *Salix caprea* and conifers such as *Larix*.
D: NM2: 128, BFF8: 133, FAN4: 168-169 (as *Hydropus scabripes* var. *scabripes*)
Known from Herefordshire, North Hampshire and Surrey. Reported from Yorkshire but unsubstantiated with voucher material.

subalpinus (Höhn.) Singer, *Sydowia* 15: 66 (1962) [1961]
   Mycena subalpina Höhn., *Sitzungsber. Kaiserl. Akad. Wiss., Wien, Math.-Naturwiss. Cl., Abt. 1* 122: 275 (1913)
   Marasmiellus subalpinus (Höhn.) Singer, *Lilloa* 22: 301 (1951) [1949]
   Collybia pseudoradicata J.E. Lange & F.H. Møller, *Friesia* 1: 295 (1936)
E: ! **S:** !
H: On decayed wood of deciduous trees.
D: NM2: 129, BFF8: 133, FAN4: 169-170 **I:** Kriegl3: 271
Known from England (Somerset, South Hampshire, South Wiltshire and Surrey) and Scotland (Kirkudbrightshire). Pearson (TBMS 26: 40, 1943) suggests that Cooke 198 (205) Vol. 2 (1882) [as *Agaricus (Collybia) collinus*] could also be this species.

trichoderma (Joss.) Singer, *Agaricales in Modern Taxonomy*, Edn 2: 389 (1962)
   Mycena trichoderma Joss., *Encycl. Mycol.* 10. *Le Genre Mycena*: 689 (1938)
E: ! **W:** !
H: On decayed wood of deciduous trees.
D: NBA9: 564-565, BFF8: 134

First British record from England (Surrey: Norbury Park) in 1963
and also known from Wales (Denbighshire: Abergele, Glan-yr-
Afon).

HYGROCYBE (Fr.) P. Kumm., *Führ. Pilzk.*: 26 (1871)
Agaricales, Hygrophoraceae
Camarophyllus (Fr.) P. Kumm., *Führ. Pilzk.*: 26 (1871)
Cuphophyllus (Donk) Bon, *Doc. Mycol.* 14(56): 10 (1985)
Type: *Hygrocybe conica* (Schaeff.) P. Kumm.

aurantia (Murrill) Natarajan & Purosh., *Kavaka* 15(1-2): 75 (1989) [1987]
Hydrocybe aurantia Murrill, *Mycologia* 3: 195 (1911)
Hygrophorus aurantius (Murrill) Murrill, *Mycologia* 4: 332 (1912)
E: ! **W:** !
H: On soil. British material in association with conifers and bryophytes (Hepaticae).
D: FungEur6: 500 (as *Hygrophorus aurantius*) **D+I:** FRIC6: 4-5 41b
Only two British collections, from East Kent (Bedgebury Pinetum) and Carmarthenshire (Pembrey Forest). Described from Jamaica and seemingly unknown elsewhere in Europe.

aurantiosplendens R. Haller Aar., *Schweiz. Z. Pilzk.* 32: 86-89 (1954)
Hygrophorus aurantiosplendens (R. Haller Aar.) P.D. Orton, *Trans. Brit. Mycol. Soc.* 43(2): 176 (1960)
E: o **S:** o **W:** ! **NI:** o **ROI:** o
H: On acidic soil in unimproved grassland.
D: NCL3: 261 (as *Hygrophorus aurantiosplendens*), FAN2: 101-102, NM2: 85 **D+I:** FNE1 130-131, FungEur6: 510-514 p.748 **I:** C&D: 159, Kriegl3: 55
Occasional but widespread.

calciphila Arnolds, *Persoonia* 12(4): 475 (1985)
E: ! **S:** ! **W:** ! **NI:** ! **ROI:** !
H: On calcareous soil, in unimproved grassland (dunes, disused railway trackbeds and heathland).
D: FAN2: 100, NM2: 82 **D+I:** B&K3: 102 79, FNE1 104-105, FungEur6: 547-551 p.753 **I:** C&D: 163, Kriegl3: 56
Rarely reported but apparently widespread.

calyptriformis (Berk.) Fayod, *Ann. Sci. Nat., Bot.*, sér. 7, 9: 309 (1889)
Agaricus calyptriformis Berk., *Ann. Nat. Hist. Ser. 1 [Notices of British Fungi No. 63]* 1: 198 (1838)
Hygrophorus calyptriformis (Berk.) Berk., *Outl. Brit. fungol.*: 202 (1860)
Hygrophorus calyptriformis var. *niveus* Cooke, *Handb. Brit. fung.*, Edn 2: 302 (1883)
Hygrocybe calyptriformis f. *nivea* (Cooke) Bon, *Doc. Mycol.* 19 (75): 55 (1989)
E: o **S:** o **W:** o **NI:** o **ROI:** o **O:** Channel Islands: !
H: On soil in unimproved grassland, including open glades in woodland.
D: FAN2: 89, NM2: 79 **D+I:** Ph: 62-63 (as *Hygrocybe calyptraeformis*), B&K3: 102 80, FNE1 136-137, FungEur6: 454-458 **I:** Bon: 105, C&D: 159, Kriegl3: 43
Occasional but widespread and may be locally frequent.

calyptriformis *var.* **domingensis** Lodge & S.A. Cantrell, *Mycol. Res.* 104(7): 876 (2000)
E: !
H: On soil in grassland (British material from a cemetery).
Described from the Caribbean (Dominican Republic) with a paratype collection from England (West Sussex: Petworth). Differs from the type variety in vinaceous stipe, larger spores and the pileus lacking an ixocutis.

canescens (A.H. Sm. & Hesler) P.D. Orton, *Notes Roy. Bot. Gard. Edinburgh* 44(3): 489 (1987)
Hygrophorus canescens A.H. Sm. & Hesler, *Lloydia* 5: 10 (1942)
S: !

H: On acidic soil in unimproved grassland.
D: NM2: 79 **D+I:** FNE1 44-45, FungEur6: 365-367
Known from two sites in Scotland. This is a North American species. also rare in continental Europe.

cantharellus (Schwein.) Murrill (as *Hydrocybe cantharellus*), *Mycologia* 3: 196 (1911)
Agaricus cantharellus Schwein. [non *A. cantharellus* L. (1753)], *Schriften Naturf. Ges. Leipzig* 1: 88 (1822)
Hygrophorus cantharellus (Schwein.) Fr., *Epicr. syst. mycol.*: 329 (1838)
Hygrocybe lepida Arnolds, *Persoonia* 13(2): 139 (1986)
Hygrophorus turundus var. *lepidus* Boud., *Bull. Soc. Mycol. France* 13: 12 (1897)
E: o **S:** o **W:** o **NI:** o **ROI:** o
H: On soil, amongst grass or mosses, in unimproved grassland, and occasionally on bare soil under trees such as *Alnus* and *Betula* spp.
D: NM2: 83 (as *Hygrocybe lepida*), FungEur6: 556 (as *H. lepida*) **D+I:** Ph: 63, FAN2: 98 (as *H. lepida*), FNE1 110-111, FungEur6: 552-556 p.754 **I:** FlDan5: 167B, Bon: 111 (as *H. lepida*), B&K3: 108 90 (as *H. lepida*), C&D: 163 (as *H. lepida*), Kriegl3: 57
Occasional but widespread and may be locally abundant.

ceracea (Wulfen) P. Kumm., *Führ. Pilzk.*: 112 (1871)
Agaricus ceraceus Wulfen, *Misc. Aust. Bot.* 2: 105 (1781)
Gymnopus ceraceus (Wulfen) Gray, *Nat. arr. Brit. pl.* 1: 607 (1821)
Hygrophorus ceraceus (Wulfen) Fr., *Sverig Atl. Svamp.*: 45 (1836)
Hygrocybe vitellinoides Bon, *Doc. Mycol.* 9(35): 39 (1979)
Hygrocybe ceracea var. *vitellinoides* (Bon) Bon, *Doc. Mycol.* 18(69): 35 (1987)
Mis.: *Hygrocybe citrina* sensu Breitenbach & Kränzlin (1991), and sensu Lange (1940)
E: c **S:** c **W:** c **NI:** o **ROI:** c **O:** Channel Islands: !
H: On acidic or basic soils in unimproved grassland.
D: FAN2: 95-96 **D+I:** Ph: 64, B&K3: 104 82 (as *Hygrocybe citrina*), FNE1 120-121, FungEur6: 485-491 p. 744 **I:** Sow1: 20 (as *Agaricus ceraceus*), Bon: 109, C&D: 161
Common and widespread.

chlorophana (Fr.) Wünsche, *Pilze*: 112 (1877)
Agaricus chlorophanus Fr., *Syst. mycol.* 1: 103 (1821)
Hygrophorus chlorophanus (Fr.) Fr., *Epicr. syst. mycol.*: 332 (1838)
Hygrocybe chlorophana var. *aurantiaca* Bon, *Doc. Mycol.* 6(24): 42 (1976)
Hygrocybe euroflavescens Kühner, *Bull. Soc. Mycol. France* 92: 436 (1976)
Hygrophorus euroflavescens (Kühner) Dennis, *Fungi of the Hebrides*: 47 (1986)
Mis.: *Hygrophorus flavescens* sensu auct.Brit
Mis.: *Hygrocybe flavescens* sensu auct. Eur.
E: c **S:** c **W:** c **NI:** c **ROI:** c **O:** Channel Islands: ! Isle of Man: !
H: On soil in unimproved grassland. Rarely on loam soils in deciduous woodland, often associated with *Corylus* or *Prunus spinosa*.
D: FAN2: 89-90 **D+I:** B&K3: 102 81, B&K3: 106 86 (as *H. flavescens*), FNE1 140-141, FungEur6: 465-470 p.742 **I:** FlDan5: 166B, Bon: 106, Bon: 107 (as *H. euroflavescens*), C&D: 157, C&D: 157 (as *H. euroflavescens*), Kriegl3: 44
Common and widespread.

citrinopallida (A.H. Sm. & Hesler) Kobayasi, *Bull. natn. Sci. Mus., Tokyo* 14(1): 62 (1971)
Hygrophorus citrinopallidus A.H. Sm. & Hesler, *Sydowia* 8: 327 (1954)
E: ? **S:** ! **W:** ? **NI:** !
H: On soil amongst mosses and grass.
D+I: FNE1 64-65, FungEur6: 385-388
Known from Northern Ireland (Antrim) in 1998 and a single collection from Scotland. Reported from England and Wales but unsubstantiated with voucher material.

citrinovirens (J.E. Lange) Jul. Schäff., *Ber. bayer bot. Ges.* 27: 222 (1947)
   Camarophyllus citrinovirens J.E. Lange, *Dansk. Bot. Ark.* 4(4): 20 (1923)
   Hygrocybe brevispora F.H. Møller, *Fungi Faeroes* 1: 142 (1945)
   Hygrophorus brevisporus (F.H. Møller) P.D. Orton, *Trans. Brit. Mycol. Soc.* 43(2): 176 (1960)
   Mis.: *Hygrophorus obrusseus* sensu NCL
E: o **S:** ! **W:** o **NI:** ! **ROI:** !
H: On soil in unimproved grassland on acidic or basic soils, rather often in old cemeteries.
D: NM2: 79 **D+I:** FNE1 138-139, FungEur6: 458-462 p.741 **I:** FlDan5: 165A (as *Camarophyllus citrinovirens*)
Occasional but widespread.

coccinea (Schaeff.) P. Kumm., *Führ. Pilzk.*: 112 (1871)
   Agaricus coccineus Schaeff. [non *A. coccineus* Sowerby (1799)], *Fung. Bavar. Palat. nasc.* 1: 70 (1762)
   Hygrophorus coccineus (Schaeff.) Fr., *Sverig Atl. Svamp.*: 45 (1836)
   Hygrocybe coccinea var. *umbonata* Herink, *Sborník severočeského Musea* 1: 77 (1958)
E: c **S:** c **W:** c **NI:** c **ROI:** c **O:** Channel Islands: !
H: On soil in improved and unimproved grassland, on heathland, lawns, pasture, meadows, coastal turf, old cemeteries and rarely in grassy glades in woodland.
D: FAN2: 91-92 **D+I:** Ph: 62-63, B&K3: 104 83, FNE1 116-117, FungEur6: 491-500 p.745 (as *Hygrocybe coccinea* var. *coccinea*) **I:** Bon: 109, C&D: 161, Kriegl3: 59
Common and widespread.

coccineocrenata (P.D. Orton) M.M. Moser, *Kleine Kryptogamenflora*, Edn 3: 68 (1967)
   Hygrophorus coccineocrenatus P.D. Orton, *Trans. Brit. Mycol. Soc.* 43(2): 262 (1960)
   Hygrophorus miniatus var. *sphagnophilus* Peck, *Rep. (Annual) New York State Mus. Nat. Hist.* 53: 856 (1901)
   Mis.: *Hygrocybe turunda* sensu Lange (FlDan5: 27 & pl. 168H)
E: ! **S:** ! **W:** ? **NI:** ? **ROI:** ?
H: On wet ground, amongst or near to *Sphagnum* moss on moorland, in bogs, fens, or around the edges of lakes.
D: NCL3: 262 (as *Hygrophorus coccineocrenatus*), FAN2:, NM2: 83, FungEur6: 557-560 **D+I:** FNE1 114-115 **I:** Bon: 111, C&D: 163, Kriegl3: 60
Rarely reported but often growing in inaccessible areas of moorland or bogs. Most frequently reported from Scotland.

colemanniana (A. Bloxam) P.D. Orton & Watling, *Notes Roy. Bot. Gard. Edinburgh* 29(1): 131 (1969)
   Hygrophorus colemannianus A. Bloxam, *Ann. Mag. Nat. Hist., Ser. 2 [Notices of British Fungi no. 701]* 13: 403 (1854)
   Camarophyllus colemannianus (A. Bloxam) Ricken, *Vadem. Pilzfr.*: 197 (1920)
   Cuphophyllus colemannianus (A. Bloxam) Bon, *Doc. Mycol.* 14(56): 10 (1984)
   Mis.: *Hygrocybe subradiata* sensu auct.
   Mis.: *Hygrophorus subradiatus* sensu Lange (FlDan5: 18 & pl. 165D)
E: o **S:** ! **W:** ! **NI:** ? **ROI:** ?
H: On soil in unimproved grassland.
D: FAN2: 80-81, NM2: 77 (as *Cuphophyllus colemannianus*) **D+I:** B&K3: 104 84, FNE1 54-55, FungEur6: 329-333 p.725 **I:** Bon: 103 (as *C. colemannianus*), C&D: 155 (as *C. colemannianus*), Kriegl3: 36
Occasional to rather rare but apparently widespread.

conica (Schaeff.) P. Kumm., *Führ. Pilzk.*: 111 (1871)
   Agaricus conicus Schaeff. [non *A. conicus* Scop. (1772)], *Fung. Bavar. Palat. nasc.* 1: 2 (1762)
   Hygrophorus conicus (Schaeff.) Fr., *Epicr. syst. mycol.*: 331 (1838)
   Hygrocybe cinereifolia Courtec. & Priou, *Doc. Mycol.* 22(86): 69 (1992)

Agaricus tristis Pers. [*nom. illegit.*, non *A. tristis* Scop. (1772)], *Observ. mycol.* 2: 349 (1796)
   Hygrophorus tristis (Pers.) Bres., *Icon. Mycol.* 7: 349 (1928)
   Hygrocybe tristis (Pers.) F.H. Møller, *Fungi Faeroes* 1: 140 (1945)
   Hygrocybe conica var. *tristis* (Pers.) Heinem., *Bull. Jard. bot. Brux.* 33: 432 (1963)
   Hygrophorus olivaceoniger P.D. Orton, *Trans. Brit. Mycol. Soc.* 43(2): 263 (1960)
   Hygrocybe olivaceonigra (P.D. Orton) M.M. Moser [as *H. olivaceoniger*], *Kleine Kryptogamenflora*, Edn 3: 66 (1967)
   Hygrocybe conica var. *olivaceonigra* (P.D. Orton) Arnolds, *Taxon.* Hygrocybe *Nederland*: 122 (1974)
   Hygrophorus conicus var. *olivaceoniger* (P.D. Orton) Arnolds, *Persoonia* 8(1): 103 (1974)
   Hygrocybe pseudoconica J.E. Lange [as *Hydrocybe*], *Dansk. Bot. Ark.* 4(4): 24 (1923)
   Hygrocybe conica f. *pseudoconica* (J.E. Lange) Arnolds, *Persoonia* 12(4): 476 (1985)
   Hygrophorus conicus var. *chloroides* Malençon, *Fl. Champ. Maroc* 2: 496 (1975)
   Hygrocybe conica var. *chloroides* (Malençon) Bon, *Doc. Mycol.* 15(59): 52 (1985)
   Mis.: *Hygrophorus conicus* var. *nigrescens* sensu auct.
   Mis.: *Hygrocybe nigrescens* sensu auct.
   Mis.: *Hygrophorus nigrescens* sensu auct.
E: c **S:** c **W:** c **NI:** c **ROI:** c **O:** Channel Islands: !
H: On soil in unimproved grassland.
D: FAN2: 83-84, NM2: 78, FungEur6: 416-417 (as *H. olivaceonigra*) **D+I:** Ph: 60-61, Ph: 60-61 (as *H. nigrescens*), B&K3: 106 85, FNE1 158-159, FungEur6: 404-411 p.732, FungEur6: 414-416 (as *H. conica* var. *chloroides*) **I:** C&D: 157 (as *H. pseudoconica*), C&D: 157 (as *H. conica* var. *chloroides*), C&D: 157 (as *H. cinereifolia*), C&D: 157 (as *H. tristis*), Kriegl3: 46
Common and widespread. Extremely variable in appearance hence the long synonymy. *H. conica* var. *olivaceonigra*, growing in sandy grassland near the sea, may prove distinct.

conicoides (P.D. Orton) P.D. Orton & Watling, *Notes Roy. Bot. Gard. Edinburgh* 29(1): 131 (1969)
   Hygrophorus conicoides P.D. Orton, *Trans. Brit. Mycol. Soc.* 34(2): 262 (1960)
   Hygrocybe conica var. *conicoides* (P.D. Orton) Boertm., *Fungi of Northern Europe* 1: 162 (1995)
E: o **S:** o **W:** o **NI:** ! **ROI:** o **O:** Channel Islands: !
H: On soil in coastal grassland, often on dunes.
D: NCL3: 262 (as *Hygrophorus conicoides*), FAN2: 85, NM2: 78 (as *Hygrocybe conica* var. *conicoides*) **D+I:** FNE1 162-163, FungEur6: 418-422 p.734 **I:** Bon: 105, C&D: 157
Occasional but widespread around the coast.

constrictospora Arnolds, *Persoonia* 12(4): 476 (1985)
ROI: !
H: On sandy soil, amongst grass in dunes.
D: FAN2: 94-95, NM2: 84 **D+I:** FNE1 100-101, FungEur6: 520-524
Known only from two coastal sites.

flavipes (Britzelm.) Arnolds, *Persoonia* 14(1): 43 (1989)
   Hygrophorus flavipes Britzelm., *Hymenomyc. Südbayern* 8: 10 (1891)
   Camarophyllus flavipes (Britzelm.) Clémençon, *Beih. Z. Mykol.* 4: 55 (1982)
   Cuphophyllus flavipes (Britzelm.) Bon, *Doc. Mycol.* 14(56): 11 (1985) [1984]
   Mis.: *Hygrophorus lacmus* sensu NCL, and sensu auct. mult
E: o **S:** o **W:** ! **NI:** o **ROI:** !
H: On acidic soil in unimproved grassland.
D: FAN2: 81-82, NM2: 77 (as *Camarophyllus flavipes*) **D+I:** FNE1 58-59, FungEur6: 333-336 **I:** FlDan5: 165B (as *C. lacmus*), Bon: 103, Kriegl3: 37 (as *Hygrocybe lacmus*)
Occasional but widespread.

fornicata (Fr.) Singer, *Lilloa* 22: 152 (1951) [1949]
   Hygrophorus fornicatus Fr., *Epicr. syst. mycol.*: 327 (1838)

Camarophyllus fornicatus (Fr.) P. Karst., *Bidrag Kännedom Finlands Natur Folk* 37: 227 (1879)
>   *Hygrophorus fornicatus* var. *clivalis* Fr., *Monogr. hymenomyc. Suec.* 2: 134 (1863)
>   *Hygrophorus clivalis* (Fr.) Sacc., *Syll. fung.* 5: 406 (1887)
>   *Hygrocybe clivalis* (Fr.) P.D. Orton & Watling, *Notes Roy. Bot. Gard. Edinburgh* 29(1): 131 (1969)
>   *Hygrophorus streptopus* Fr., *Epicr. syst. mycol.*: 327 (1838)
>   *Hygrocybe streptopus* (Fr.) Bon, *Doc. Mycol.* 6(24): 43 (1976)
>   *Hygrocybe fornicata* var. *streptopus* (Fr.) Arnolds, *Persoonia* 12(4): 476 (1985)
>   *Hygrophorus lepidopus* Rea, *Trans. Brit. Mycol. Soc.* 12(2-3): 214 (1927)
>   *Hygrocybe lepidopus* (Rea) P.D. Orton & Watling, *Notes Roy. Bot. Gard. Edinburgh* 29(1): 131 (1969)

E: o **S:** o **W:** o **NI:** o **ROI:** !
H: On soil, in unimproved grassland such as pasture, heathland, dunes and old cemeteries and rarely in grassy areas in open woodland.
D: FAN2: 103-104, NM2: 82 **D+I:** FNE1 72-73, FungEur6: 640-645 p.765 **I:** FlDan5: 165C (as *Camarophyllus fornicatus*), Bon: 113, C&D: 163, Kriegl3: 62
Occasional but widespread and often locally common. Perhaps most frequent in northern and western areas. Collections with dark pilei were formerly distinguished as *H. streptopus*.

glutinipes (J.E. Lange) R. Haller Aar., *Schweiz. Z. Pilzk.* 34: 179 (1956)
>   *Hygrocybe citrina* var. *glutinipes* J.E. Lange, *Fl. agaric. danic.* 5: 27 (1940)
>   *Hygrophorus glutinipes* (J.E. Lange) P.D. Orton, *Trans. Brit. Mycol. Soc.* 43(2): 176 (1960)
>   *Hygrocybe aurantioviscida* Arnolds, *Biblioth. Mycol.* 90: 384 (1983) [1982]
>   Mis.: *Hygrocybe citrina* sensu auct.

E: o **S:** o **W:** o **NI:** o **ROI:** ! **O:** Channel Islands: !
H: On soil in unimproved, often mossy grassland and rarely amongst mosses in deciduous woodland.
D: FAN2: 110, NM2: 78, FungEur6: 476 (as *Hygrocybe aurantioviscida*), FungEur6: 477 (as *H. glutinipes* var. *rubra*) **D+I:** FNE1 142-143, FungEur6: 471-476 p.743 (as *H. glutinipes* var. *glutinipes*) **I:** FlDan5: 167E (as *H. citrina* var. *glutinipes*), Bon: 109 (as *H. citrina*), C&D: 157, C&D: 160 (as *H. aurantioviscida*), Kriegl3: 49
Occasional but widespread.

glutinipes *var.* **rubra** Bon, *Agarica* 4(8): 74 (1983)
E: !
H: On soil amongst grass.
D+I: FNE1 144-145
Rarely reported. Possibly just a colour form of *H. glutinipes*.

helobia (Arnolds) Bon, *Doc. Mycol.* 6(24): 43 (1976)
>   *Hygrophorus helobius* Arnolds, *Persoonia* 8: 99 (1974)
>   Mis.: *Hygrocybe miniata* sensu Phillips (Ph: 63)

E: ! **S:** ! **W:** ! **NI:** ! **ROI:** ! **O:** Isle of Man: !
H: On soil in unimproved grassland.
D: FAN2: 100-101, NBA8: 583 **D+I:** B&K3: 106 87, FNE1 134-135, FungEur6: 567-571 p.755 **I:** Ph: 63 (misdetermined as *Hygrocybe miniata*)
Rarely reported but apparently widespread.

ingrata J.L. Jensen & F.H. Møller, *Fungi Faeroes*: 136 (1945)
>   Mis.: *Hygrophorus nitiosus* sensu NCL

E: ? **S:** ! **W:** ? **ROI:** ?
H: On soil in unimproved grassland.
D+I: FNE1 76-77, FungEur6: 645-749 p.766 **I:** Ph: 60-61 (as *Hygrocybe nitrata* [*fide* D. Boertmann]), Kriegl3: 63
Rarely reported. Apparently widespread but most reports are unsubstantiated with voucher material.

insipida (J.E. Lange) M.M. Moser, *Kleine Kryptogamenflora*, Edn 3: 65 (1967)
>   *Hygrocybe reai* var. *insipida* J.E. Lange, *Dansk. Bot. Ark.* 4(4): 26 (1923)

Hygrophorus insipidus (J.E. Lange) S. Lundell, *Fungi Exsiccati Suecici*: 2331 (1956)
>   Mis.: *Hygrocybe subminutula* sensu auct. Eur.
>   Mis.: *Hygrophorus subminutulus* sensu NCL

E: c **S:** c **W:** c **NI:** c **ROI:** c
H: On acidic soil, in unimproved grassland.
D: NCL3: 268 (as *Hygrophorus subminutulus*), FAN2: 108, NM2: 84 **D+I:** FNE1 122-123, FungEur6: 624-629 **I:** FlDan5: 168C (as *H. reai* var. *insipida*), Bon: 109, C&D: 161 (as *Hygrocybe subminutula*), Kriegl3: 64
Common and widespread.

intermedia (Pass.) Fayod, *Ann. Sci. Nat., Bot.*, sér. 7, 9: 309 (1889)
>   *Hygrophorus intermedius* Pass., *Nuovo Giorn. Bot. Ital.* 4: 103 (1872)

E: o **S:** o **W:** o **NI:** o **ROI:** !
H: On soil in unimproved grassland
D: FAN2: 88-89, NM2: 80 **D+I:** Ph: 62-63, FNE1 148-149, FungEur6: 434-438 p.737 **I:** Bon: 105, C&D: 159
Occasional but widespread.

irrigata (Pers.) Bon, *Doc. Mycol.* 6(24): 41 (1976)
>   *Agaricus irrigatus* Pers., *Syn. meth. fung.*: 361 (1801)
>   *Hygrophorus irrigatus* (Pers.) Fr. [as *H. irriguus*], *Epicr. syst. mycol.*: 329 (1838)
>   *Agaricus unguinosus* Fr., *Syst. mycol.* 1: 101 (1821)
>   *Hygrophorus unguinosus* (Fr.) Fr., *Epicr. syst. mycol.*: 332 (1838)
>   *Hygrocybe unguinosa* (Fr.) P. Karst. [as *Hydrocybe unguinosa*], *Bidrag Kännedom Finlands Natur Folk* 32: 237 (1879)

E: o **S:** o **W:** o **NI:** o **ROI:** !
H: On soil (often acidic) in unimproved grassland, most frequently in upland areas.
D: FAN2: 107-108, NM2: 81 (as *Hygrocybe unguinosa*) **D+I:** Ph: 60 (as *H. unguinosa*), B&K3: 116 102 (as *H. unguinosa*), FNE1 88-89, FungEur6: 592-597 p. 759 **I:** FlDan5: 168I (a little too dark) (as *H. unguinosa*), Bon: 113 (as *H. unguinosa*), C&D: 163 (as *H. unguinosa*)
Occasional but widespread, and may be locally common.

lacmus (Schumach.) P.D. Orton & Watling, *Notes Roy. Bot. Gard. Edinburgh* 29(1): 131 (1969)
>   *Agaricus lacmus* Schumach., *Enum. pl.* 2: 333 (1803)
>   *Hygrophorus subradiatus* var. *lacmus* (Schumach.) Fr., *Epicr. syst. mycol.*: 329 (1838)
>   *Hygrophorus lacmus* (Schumach.) Kalchbr., *Icon. select. Hymenomyc. Hung.* 1: 42 (1873)
>   *Camarophyllus lacmus* (Schumach.) J.E. Lange, *Dansk. Bot. Ark.* 9(6): 96 (1938)
>   *Hygrophorus subviolaceus* Peck, *Rep. (Annual) New York State Mus. Nat. Hist.* 53: 842 (1900)
>   *Hygrocybe subviolacea* (Peck) P.D. Orton & Watling, *Notes Roy. Bot. Gard. Edinburgh* 29(1): 132 (1969)

E: ! **S:** ! **W:** ! **NI:** ! **ROI:** !
H: On soil in unimproved grassland.
D: NCL3: 260 (as *Hygrophorus subviolaceus*), FAN2: 81, NM2: 77 (as *Camarophyllus lacmus*), NBA2: 51 (as *H. subviolaceus*) **D+I:** B&K3: 108 88, FNE1 56-57, FungEur6: 338-341 **I:** Bon: 103 (as *Cuphophyllus subradiatus*), C&D: 155 (as *Cuphophyllus lacmus*)
Rarely reported but apparently widespread.

laeta (Pers.) P. Kumm., *Führ. Pilzk.*: 112 (1871)
>   *Agaricus laetus* Pers., *Observ. mycol.* 2: 48 (1799)
>   *Hygrophorus laetus* (Pers.) Fr., *Epicr. syst. mycol.*: 329 (1838)
>   *Hygrophorus houghtonii* Berk. & Broome, *Ann. Mag. Nat. Hist., Ser. 4 [Notices of British Fungi no. 1360]* 11: 342 (1873)

E: o **S:** o **W:** o **NI:** o **ROI:** o
H: On acidic soil, often in unimproved grassland but also with *Calluna* and *Pteridium* on heath and moorland.

D: FAN2: 106, NM2: 81 **D+I:** Ph: 64-65, B&K3: 108 89, FNE1 84-85, FungEur6: 597-600 p.760 (as *Hygrocybe laeta* var. *laeta*) **I:** FlDan5: 168F & F1, Bon: 113, C&D: 163, Kriegl3: 65
Occasional but widespread, most frequent (and may be locally abundant) in upland areas of northern and western Britain.

lilacina (P. Karst.) M.M. Moser, *Kleine Kryptogamenflora*, Edn 3: 64 (1967)
   Omphalia lilacina P. Karst., *Bidrag Kännedom Finlands Natur Folk* 32: 133 (1879)
   Hygrophorus lilacinus (P. Karst.) M. Lange, *Medd. Grønl.* 148(2): 63 (1957)
   Hygrophorus violeipes M. Lange, *Medd. Grønl.* 147: 18 (1955)
E: ? **S:** !
H: On soil amongst arctic-alpine vegetation (*Rhacomitrium, Empetrum nigrum* etc.).
D: NM2: 81, C&D: 155 (as *Cuphophyllus lilacinus*) **D+I:** FNE1 68-69, FungEur6: 391-395
Known from the Cairngorms. Reported from Cheshire (Macclesfield) but unsubstantiated with voucher material and in unlikely habitat.

marchii (Bres.) Singer, *Lilloa* 22: 153 (1951) [1949]
   Hygrophorus marchii Bres., *Icon. Mycol.* 7: pl. 343 (1928)
E: ! **S:** ! **W:** ! **NI:** ! **ROI:** !
H: On soil in unimproved grassland.
D: FAN2: 92-93 **D+I:** B&K3: 110 91, FNE1 118-119 **I:** Bon: 109, C&D: 161
Widely reported but scarcely distinct from *H. coccinea* fide Boertmann (FNE1), and many records are probably *H. reidii* (= *H. marchii* sensu Reid (FRIC3: 5, as *Hygrophorus*).

miniata (Fr.) P. Kumm., *Führ. Pilzk.*: 112 (1871)
   Agaricus miniatus Fr., *Syst. mycol.* 1: 105 (1821)
   Hygrophorus miniatus (Fr.) Fr., *Epicr. syst. mycol.*: 330 (1838)
   Hygrophorus strangulatus P.D. Orton, *Trans. Brit. Mycol. Soc.* 43(2): 266 (1960)
   Hygrocybe strangulata (P.D. Orton) Svrček, *Česká Mykol.* 16: 167 (1962)
E: o **S:** o **W:** o **NI:** o **ROI:** o
H: On soil in unimproved grassland.
D: FAN2: 98-99, NM2: 82, FungEur6: 578 (as *H. strangulata*) **D+I:** FRIC3: 6-8 18b (as *Hygrophorus strangulatus*), Ph: 63, B&K3: 110 93 (as *H. miniata* var. *mollis*), B&K3: 110 92, FNE1 102-103, FungEur6: 572-577 p.756 **I:** Bon: 111, C&D: 163, Kriegl3: 67
Occasional but widespread and may be locally common.

mucronella (Fr.) P. Karst., *Bidrag Kännedom Finlands Natur Folk* 32: 235 (1879)
   Hygrophorus mucronellus Fr., *Epicr. syst. mycol.*: 331 (1838)
   Hygrophorus reai Maire, *Trans. Brit. Mycol. Soc.* 3(3): 170 (1910) [1909]
   Hygrocybe reai (Maire) J.E. Lange, *Dansk. Bot. Ark.* 4(4): 25 (1923)
E: o **S:** o **W:** o **NI:** ! **ROI:** ! **O:** Channel Islands: !
H: On soil in unimproved grassland.
D: FAN2: 108-109 (as *Hygrocybe reai*) **D+I:** B&K3: 116 100 (as *H. reai*), FNE1 132-133, FungEur6: 629-632 p.764 **I:** FlDan5: 168A (as *H. reai*)
Occasional but widespread.

nitrata (Pers.) Wünsche, *Pilze*: 112 (1877)
   Agaricus nitratus Pers., *Syn. meth. fung.*: 356 (1801)
   Hygrophorus nitratus (Pers.) Fr., *Hymenomyc. eur.*: 421 (1874)
   Mis.: *Hygrocybe murinacea* sensu Moser (1967) & sensu BKriegl3
E: ! **S:** o **W:** ! **NI:** o **ROI:** !
H: On soil in unimproved grassland.
D: FAN2: 103, NM2: 82 **D+I:** B&K3: 112 94 (as *H. murinacea*), FNE1 74-75 **I:** FlDan5: 165E
Occasional in Scotland and Northern Ireland, rarely reported elsewhere.

ovina (Bull.) Kühner, *Botaniste* 17: 43 (1926)
   Agaricus ovinus Bull., *Herb. France*: pl. 520 (1791)
   Hygrophorus ovinus (Bull.) Fr., *Sverig Atl. Svamp.*: 45 (1836)
   Hygrophorus nitiosus A. Blytt, *Videnskabs-Selskabets Skrifter. I Math.-Naturv. Kl.*: 88 (1905)
   Hygrocybe nitiosa (A. Blytt) M.M. Moser, *Kleine Kryptogamenflora*: 37 (1953)
E: ! **S:** ! **W:** ! **NI:** ! **ROI:** !
H: On soil in unimproved grassland, and occasionally on old lawns.
D: FAN2: 102, NM2: 81-82 **D+I:** Ph: 60-61, B&K3: 112 96, FNE1 78-79, FungEur6: 654-659 (as *Hygrocybe nitiosa*) **I:** FlDan5: 166E (as *Camarophyllus ovinus*), Bon: 113, C&D: 163, Kriegl3: 71
Rarely reported, but apparently widespread. A distinctive species, unlikely to be misidentified, and probably genuinely rare.

persistens (Britzelm.) Singer, *Rev. Mycol. (Paris)* 5(1): 8 (1940)
   Hygrophorus conicus β *persistens* Britzelm., *Ber. Naturhist. Vereins Augsburg* 30: 30 (1890)
   Hygrocybe acutoconica (Clem.) Singer, *Lilloa* 22: 153 (1951) [1949]
   Hygrocybe aurantiolutescens P.D. Orton, *Notes Roy. Bot. Gard. Edinburgh* 29(1): 103 (1969)
   Hygrophorus aurantiolutescens (P.D. Orton) Dennis, *Fungi of the Hebrides*: 47 (1986)
   Hygrocybe aurantiolutescens f. *pseudoconica* Bon, *Doc. Mycol.* 6(24): 42 (1976)
   Hygrocybe aurantiolutescens var. *parapersistens* Bon, *Doc. Mycol.* 19(75): 55 (1989)
   Hygrocybe constans J.E. Lange [*nom. illegit.*, non *Hydrocybe constans* Murrill (1912)], *Dansk. Bot. Ark.* 4(4): 23 (1923)
   Hygrophorus cuspidatus Peck, *Bull. Torrey Bot. Club* 24: 141 (1897)
   Hygrocybe persistens var. *cuspidata* (Peck) Arnolds, *Persoonia* 13(2): 143 (1986)
   Hygrocybe langei Kühner, *Botaniste* 18: 174 (1927)
   Hygrophorus langei (Kühner) A.H. Sm. & Hesler, *Lloydia* 5(1): 55 (1942)
   Hygrocybe persistens var. *langei* (Kühner) Bon, *Doc. Mycol.* 18(69): 35 (1987)
   Hygrophorus rickenii Maire, *Bull. Soc. Mycol. France* 46: 220 (1930)
E: o **S:** o **W:** o **NI:** o **ROI:** o **O:** Isle of Man: !
H: On soil in unimproved grassland.
D: FAN2: 85-86, NM2: 79 **D+I:** Ph: 65 (as *H. langei*), B&K3: 114 97, FNE1 152-153 (as *H. persistens* var. *persistens*), FungEur6: 425-430 p.735 & 736 (as *H. acutoconica*) **I:** Bon: 107 (as *H. persistens* var. *langei*), C&D: 159, Kriegl3: 52
Occasional but widespread and may be locally common.

persistens var. konradii (R. Haller Aar.) Boertm., *Fungi of Northern Europe* 1: 154 (1995)
   Hygrocybe konradii R. Haller Aar., *Schweiz. Z. Pilzk.* 33: 172 (1955)
   Hygrophorus konradii (R. Haller Aar.) P.D. Orton, *Trans. Brit. Mycol. Soc.* 43(2): 176 (1960)
   Hygrophorus subglobisporus P.D. Orton, *Trans. Brit. Mycol. Soc.* 43(2): 267 (1960)
   Hygrocybe subglobispora (P.D. Orton) M.M. Moser, *Kleine Kryptogamenflora*, Edn 3: 67 (1967)
   Hygrocybe persistens f. *subglobispora* (P.D. Orton) Boertm., *Fungi of Northern Europe* 1: 156 (1995)
   Mis.: *Hygrocybe amoena* f. *silvatica* sensu Haller & Métrod (1955)
E: o **S:** ! **W:** o
H: On soil, usually calcareous, in unimproved grassland.
D: FAN2: 87 (as *Hygrocybe konradii*) **D+I:** Ph: 64 (as *H. konradii*), FNE1 154-156, FungEur6: 438-442 p.738 (as *H. konradii*), FungEur6: 447-452 p.740 (as *H. subglobispora*)
Occasional but apparently widespread.

phaeococcinea (Arnolds) Bon, *Doc. Mycol.* 15(60): 38 (1985)
   Hygrophorus phaeococcineus Arnolds, *Persoonia* 9: 247 (1977)

E: ! S: ! W: ! NI: !
H: On soil in unimproved grassland.
D: FAN2: 93, NM2: 83 **D+I:** FNE1 96-97, FungEur6: 525-529
Rarely reported but widespread.

pratensis (Pers.) Murrill (as *Hydrocybe pratensis*), *Mycologia* 6(1): 2 (1914)
   Agaricus pratensis Pers. [non *A. pratensis* Schaeff. (1762)], *Syn. meth. fung.*: 304 (1801)
   Cuphophyllus pratensis (Pers.) Fr., *Doc. Mycol.* 14(56): 10 (1985) [1984]
   Gymnopus pratensis (Pers.) Gray, *Nat. arr. Brit. pl.* 1: 604 (1821)
   Hygrophorus pratensis (Pers.) Fr., *Sverig Atl. Svamp.*: 46 (1836)
   Camarophyllus pratensis (Pers.) P. Kumm., *Führ. Pilzk.*: 117 (1871)
   Mis.: *Agaricus miniatus* sensu Sowerby [Col. fig. Engl. fung. 2: pl. 141 (1798)]
   Mis.: *Agaricus claviformis* sensu Withering (1796)
E: c S: c W: c NI: c ROI: c O: Channel Islands: !
H: On soil in unimproved grassland. Rarely on calcareous loam soils in scrub woodland, often with *Corylus* and *Fraxinus*.
D: FAN2: 76-77, NM2: 77 (as *Camarophyllus pratensis* var. *pratensis*) **D+I:** Ph: 60, B&K3: 100 76 (as *Camarophyllus pratensis*), FNE1 40-41, FungEur6: 343-347 p.726 (as *H. pratensis* var. *pratensis*) **I:** FlDan5: 165F & F1 (as *Camarophyllus pratensis*), Bon: 103 (as *Cuphophyllus pratensis*), C&D: 155 (as *Cuphophyllus pratensis*), Kriegl3: 39
Common and widespread.

pratensis *var.* pallida (Cooke) Arnolds, *Persoonia* 12(4): 477 (1985)
   Hygrophorus pratensis var. *pallidus* Cooke, *Grevillea* 2(20): 118 (1874)
   Camarophyllus pratensis var. *pallidus* (Cooke) J.E. Lange, *Dansk. Bot. Ark.* 4(4): 18 (1923)
   Hygrophorus berkeleyi P.D. Orton [non *H. berkeleyi* Sacc (1887)], *Trans. Brit. Mycol. Soc.* 43(2): 259 (1960)
   Hygrocybe berkeleyi (P.D. Orton) P.D. Orton & Watling, *Notes Roy. Bot. Gard. Edinburgh* 29(1): 131 (1969)
   Hygrocybe ortonii Bon, *Doc. Mycol.* 13(50): 27 (1983)
   Hygrophorus ortonii (Bon) Dennis, *Fungi of the Hebrides*: 48 (1986)
   Mis.: *Hygrocybe angustifolia* sensu auct. Eur
E: o S: o W: o NI: o ROI: o
H: On soil in unimproved grassland.
D: FAN2: 77, NM2: 77 **D+I:** FNE1 42-43, FungEur6: 360-364 (as *Hygrocybe berkeleyi*) **I:** Bon: 103 (as *Cuphophyllus berkeleyi*), C&D: 155 (as *C. berkeleyi*)
Occasional but widespread.

psittacina (Schaeff.) P. Kumm., *Führ. Pilzk.*: 112 (1871)
   Agaricus psittacinus Schaeff., *Fung. Bavar. Palat. nasc.* 1: 70 (1762)
   Hygrophorus psittacinus (Schaeff.) Fr., *Epicr. syst. mycol.*: 332 (1838)
E: c S: c W: c NI: c ROI: c O: Channel Islands: !
H: On soil in unimproved grassland. Rarely on loam soils with deciduous trees in open woodland.
D: FAN2: 105, NM2: 81 **D+I:** Ph: 64-65, B&K3: 114 98, FNE1 80-81, FungEur6: 611-617 p.762 **I:** Sow1: 82 (as *Agaricus psittacinus*), FlDan5: 168D, Bon: 113, C&D: 163, FM2(1): 3
Common and widespread.

psittacina *var.* perplexa (A.H. Sm. & Hesler) Boertm., *Fungi of Northern Europe* 1: 82 (1995)
   Hygrophorus perplexus A.H. Sm. & Hesler, *Sydowia* 8: 328 (1954)
   Hygrocybe perplexa (A.H. Sm. & Hesler) Arnolds, *Persoonia* 12(4): 477 (1985)
   Mis.: *Hygrocybe sciophana* sensu auct.
   Mis.: *Hygrophorus sciophanus* sensu NCL and sensu auct. mult.
E: ! S: ! W: ! ROI: !
H: On soil in unimproved grassland.

D: FAN2: 105-106 (as *Hygrocybe perplexa*), NM2: 81 (as *H. perplexa*) **D+I:** FNE1 82-83, FungEur6: 607-611 p.761 (as *H. perplexa*) **I:** C&D: 163 (as *H. perplexa*)
Rarely reported.

punicea (Fr.) P. Kumm., *Führ. Pilzk.*: 112 (1871)
   Agaricus puniceus Fr. [non *A. puniceus* With. (1796)], *Syst. mycol.* 1: 104 (1821)
   Hygrophorus puniceus (Fr.) Fr., *Sverig Atl. Svamp.*: 45 (1836)
E: o S: c W: c NI: c ROI: o O: Channel Islands: !
H: On soil, in unimproved grassland.
D: FAN2: 101, NM2: 85 **D+I:** Ph: 62-63, B&K3: 114 99, FNE1 128-129, FungEur6: 514-518 p.749 **I:** Bon: 107, C&D: 159, SV43: 50, Kriegl3: 74
Occasional but widespread and may be locally common, especially in northern and western areas.

quieta (Kühner) Singer, *Lilloa* 22: 152 (1951) [1949]
   Hygrophorus quietus Kühner, *Ann. Sci. Franche-Comté* 2: 19 (1947)
   Mis.: *Hygrocybe obrussea* sensu auct.
E: c S: o W: c NI: c ROI: o O: Channel Islands: !
H: On soil in unimproved grassland. Occasionally on base rich loam soils in open deciduous woodland.
D: FAN2: 95 (as *Hygrocybe obrussea*), NM2: 84 **D+I:** B&K3: 112 95 (as *H. obrussea*), FNE1 98-99, FungEur6: 529-534 p.750 (as *H. obrussea*) **I:** FlDan5: 166C (as *H. obrussea*), Bon: 109, C&D: 161, Kriegl3: 70 (as *H. obrussea*)
538 p.751 **I:** Bon: 111, C&D: 163
Common and widespread.

radiata Arnolds, *Persoonia* 14(1): 44 (1989)
E: ! W: !
H: On soil in unimproved grassland.
D: FAN2: 79-80, NM2: 77 **D+I:** FNE1 60-61, FungEur6: 352-354 p.728
Rarely reported and poorly known in Britain. Most records are sensu Boertmann in FNE1.

reidii Kühner, *Bull. Soc. Mycol. France* 92: 463 (1976)
   Mis.: *Hygrocybe marchii* sensu NCL, sensu Reid (FRIC3) & sensu auct. mult.
E: c S: c W: c NI: c ROI: o
H: On soil in unimproved grassland.
D: FAN2: 93-94, NM2: 83-84 **D+I:** FRIC3: 5-6 18a (as *Hygrophorus marchii*), Ph: 63 (as *Hygrocybe marchii*), FNE1 92-93, FungEur6: 534-538 p.751 **I:** Bon: 111, C&D: 163
Common and widespread.

russocoriacea (Berk. & T.K. Mill.) P.D. Orton & Watling, *Notes Roy. Bot. Gard. Edinburgh* 29(1): 131 (1969)
   Hygrophorus russocoriaceus Berk. & T.K. Mill., *Ann. Mag. Nat. Hist., Ser. 2 [Notices of British Fungi no. 332]* 2: 261 (1848)
   Camarophyllus russocoriaceus (Berk. & T.K. Mill.) J.E. Lange, *Dansk. Bot. Ark.* 4(4): 20 (1923)
E: c S: c W: o NI: c ROI: o
H: On soil in unimproved grassland.
D: FAN2: 79, NM2: 76 (as *Camarophyllus russocoriaceus*) **D+I:** Ph: 64-65, B&K3: 100 77 (as *C. russocoriaceus*), FNE1 46-47, FungEur6: 371-375 p.730 **I:** FlDan5: 164B (as *C. russocoriaceus*), C&D: 155 (as *Cuphophyllus russocoriaceus*)
Common to occasional, and widespread.

salicis-herbaceae Kühner, *Bull. Soc. Mycol. France* 92: 462 (1976)
S: !
H: On soil in arctic-alpine habitat with *Salix herbacea*.
D+I: FNE1 126-127, FungEur6: 505-508 **I:** C&D: 161
Known from two sites.

spadicea (Scop.) P. Karst., *Bidrag Kännedom Finlands Natur Folk* 32: 237 (1879)
   Agaricus spadiceus Scop., *Fl. carniol.* 443 (1772)
   Hygrophorus spadiceus (Scop.) Fr., *Epicr. syst. mycol.*: 332 (1838)
E: ! S: ! W: !

H: On soil in unimproved grassland.
D: FAN2: 89, NM2: 80 **D+I:** FNE1 150, FungEur6: 442-446 p.739 **I:** Bon: 105, C&D: 159
Rarely reported.

splendidissima (P.D. Orton) M.M. Moser, *Kleine Kryptogamenflora,* Edn 3: 67 (1967)
  Hygrophorus splendidissimus P.D. Orton, *Trans. Brit. Mycol. Soc.* 43(2): 265 (1960)
  Hygrocybe punicea f. *splendidissima* (P.D. Orton) D.A. Reid, *Fungorum Rar. Icon. Color.* 6: 1 & pl. 41a (1972)
  Hygrocybe punicea var. *splendidissima* (P.D. Orton) Krieglst., *Beitr. Kenntn. Pilze Mitteleurop.* 8: 176 (1992)
E: o **S:** o **W:** o **NI:** o **ROI:** !
H: On acidic soil in unimproved grassland.
D: NM2: 83 **D+I:** FNE1 94-95, FungEur6: 538-543 p.752 **I:** Bon: 107, C&D: 159
Occasional but widespread and may be locally common, especially in upland grassland and moorland in the north and west.

subpapillata Kühner, *Sydowia, Beih.* 8: 248 (1979)
E: ! **S:** !
H: On soil in unimproved grassland.
D: NM2: 79, Myc12(2): 55 **D+I:** FNE1 146-147, FungEur6: 478-481 p.742
Rarely reported.

substrangulata (P.D. Orton) P.D. Orton & Watling, *Notes Roy. Bot. Gard. Edinburgh* 29(1): 131 (1969)
  Hygrophorus substrangulatus P.D. Orton, *Trans. Brit. Mycol. Soc.* 43(2): 269 (1960)
E: ! **S:** ! **NI:** ? **ROI:** ?
H: On soil in unimproved grassland.
D: NM2: 83 **D+I:** FNE1 106-107, FungEur6: 579-583 p.757
Rarely reported but apparently widespread.

turunda (Fr.) P. Karst. [as *Hydrocybe turunda*], *Bidrag Kännedom Finlands Natur Folk* 32: 235 (1879)
  Agaricus turundus Fr., *Observ. mycol.* 2: 199 (1818)
  Hygrophorus turundus (Fr.) Fr., *Epicr. syst. mycol.*: 330 (1838)
E: ! **S:** ! **W:** ! **ROI:** !
H: On soil in unimproved grassland, sometimes with *Sphagnum* spp in boggy areas and occasionally amongst mosses under trees.
D: NCL3: 270 (as *Hygrophorus turundus*), FAN2: 96-97, NM2: 83 **D+I:** FNE1 112-113, FungEur6: 585-590 p.758 **I:** Bon: 111, C&D: 163
Rarely reported but apparently widespread.

viola J. Geesink & Bas, *Persoonia* 12(4): 478 (1985)
E: !
H: On damp clay soil, typically amongst liverworts on raised banks in deciduous woodland, usually under *Fagus*.
D: FAN2: 82, FungEur6: 395 **D+I:** FNE1 70-71
Known only from Middlesex (Ruislip Woods) and West Sussex (Petworth, The Mens and The Cut). Basidiomes are small and easily overlooked.

virginea (Wulfen) P.D. Orton & Watling, *Notes Roy. Bot. Gard. Edinburgh* 29(1): 132 (1969)
  Agaricus virgineus Wulfen, *Jacq. Misc. Austr.* 2: 104 (1781)
  Omphalia virginea (Wulfen) Gray, *Nat. arr. Brit. pl.* 1: 613 (1821)
  Hygrophorus virgineus (Wulfen) Fr., *Epicr. syst. mycol.*: 327 (1838)
  Camarophyllus virgineus (Wulfen) P. Kumm., *Führ. Pilzk.*: 117 (1871)
  Cuphophyllus virgineus (Wulfen) Kovalenko, *Opredelitel' Gribov SSSR*: 37 (1989)
  Agaricus niveus Scop. [non *A. niveus* Sowerby (1797) or Pers. (1801)], *Fl. carniol.* 430 (1772)
  Hygrocybe nivea (Scop.) Murrill (as *Hydrocybe nivea*), *North American Flora* 9(6): 377 (1916)
  Hygrophorus niveus (Scop.) Fr., *Epicr. syst. mycol.*: 327 (1838)

  Camarophyllus niveus (Scop.) Wünsche, *Pilze*: 115 (1877)
  Cuphophyllus niveus (Scop.) Bon, *Doc. Mycol.* 14(56): 11 (1985) [1984]
  Hygrophorus virgineus var. *roseipes* Cooke, *Ill. Brit. fung.* 895 (893) Vol. 6 (1888)
  Cuphophyllus niveus f. *roseipes* (Cooke) Bon, *Doc. Mycol.* 19(76): 73 (1989)
  Agaricus subradiatus Schumach., *Enum. pl.* 2: 333 (1803)
  Hygrophorus subradiatus (Schumach.) Fr., *Epicr. syst. mycol.*: 328 (1838)
  Hygrocybe subradiata (Schumach.) P.D. Orton & Watling, *Notes Roy. Bot. Gard. Edinburgh* 29(1): 131 (1969)
  Cuphophyllus subradiatus (Schumach.) Bon, *Doc. Mycol.* 14(56): 11 (1985) [1984]
E: c **S:** c **W:** c **NI:** c **ROI:** c **O:** Channel Islands: !
H: On soil in unimproved grassland; occasionally on loam soils in woodland.
D: FAN2: 77-78, NM2: 76 (as *Camarophyllus virgineus*) **D+I:** Ph: 64-65 (as *Hygrocybe nivea*), B&K3: 118 103, FNE1 48-50, FungEur6: 376-381 p.731 **I:** FlDan5: 164C (as *C. virgineus*), Bon: (as *Cuphophyllus niveus*), C&D: 155 (as *Cuphophyllus virgineus*), Kriegl3: 41
Common and widespread. Occasionally found with bright lilaceous lamellae caused by the hyphomycete *Paecilomyces marquandii*. Specimens with the stipe base reddish or pinkish (caused by *Fusarium* aff. *graminacearum* fide R. Watling) were previously given the varietal epithet *roseipes*.

virginea var. fuscescens (Bres.) Arnolds, *Persoonia* 12: 477 (1985)
  Hygrophorus niveus var. *fuscescens* Bres., *Icon. Mycol.* 7: pl. 330 (1927)
  Hygrophorus fuscescens (Bres.) Kühner & Romagn. [*comb. inval.*], *Flore Analytique des Champignons Supérieurs*: 54 (1953)
  Hygrocybe fuscescens (Bres.) P.D. Orton & Watling, *Notes Roy. Bot. Gard. Edinburgh* 29(1): 132 (1969)
  Mis.: *Hygrocybe subradiata* sensu auct. Brit.
  Mis.: *Hygrophorus subradiatus* sensu NCL
E: ! **S:** ! **W:** ! **NI:** ! **ROI:** !
H: On soil in unimproved grassland.
D: FAN2: 79, NM2: 77 (as *Camarophyllus fuscescens*) **D+I:** B&K4: 118 104, FNE1 50-51, FungEur6: 367-369 p.729 **I:** C&D: 155 (as *Cuphophyllus fuscescens*), Kriegl3: 42, SV44: 53 (as *Hygrocybe fuscescens*)
Occasional but much less frequent than *H. virginea*.

virginea var. ochraceopallida (P.D. Orton) Boertm., *Fungi of Northern Europe* 1: 52 (1995)
  Hygrocybe ochraceopallida P.D. Orton, *Notes Roy. Bot. Gard. Edinburgh* 38(2): 329 (1980)
E: ! **S:** ! **W:** ! **NI:** o **ROI:** !
H: On soil in unimproved grassland.
D: FungEur6: 382 **D+I:** FNE1 52-53
Occasional but much less frequent than *H. virginea*.

vitellina (Fr.) P. Karst., *Bidrag Kännedom Finlands Natur Folk* 32: 233 (1879)
  Hygrophorus vitellinus Fr., *Monogr. hymenomyc. Suec.* 2: 312 (1863)
  Hygrocybe luteolaeta Arnolds, *Persoonia* 12(4): 477 (1985)
E: ! **S:** ! **W:** ! **NI:** ! **ROI:** !
H: On acidic soil in unimproved grassland.
D: FAN2: 110-111 (as *Hygrocybe luteolaeta*), NM2: 80, NBA2: 51-53 **D+I:** Ph: 64, FNE1 90-91, FungEur6: 617-622 p.763 **I:** C&D: 161
Rarely reported. Apparently widespread but known mainly from upland areas of northern and western Britain.

xanthochroa (P.D. Orton) M.M. Moser, *Kleine Kryptogamenflora,* Edn 3: 64 (1967)
  Hygrophorus xanthochrous P.D. Orton, *Trans. Brit. Mycol. Soc.* 43(2): 271 (1960)
E: ? **S:** ! **NI:** ! **ROI:** ?
H: On acidic soil in unimproved boreal-montane grassland.

D: NM2: 81, FungEur6: 396-399 **D+I:** FNE1 66-67 **I:** C&D: 155
(as *Cuphophyllus xanthochrous*)
Described from Easterness (Rothiemuchus). Reports from
Warwickshire in 1974 and from the Republic of Ireland are
unsubstantiated with voucher material.

HYGROPHOROPSIS (J. Schröt.) Maire, in Martin-Sans, *Empoisinnem. Champ.*: 99 (1929)
Boletales, Hygrophoropsidaceae
Type: *Hygrophoropsis aurantiaca* (Wulfen) Maire

aurantiaca (Wulfen) Maire, in Martin-Sans, *Empoisinnem. Champ.*: 99 (1929)
Agaricus aurantiacus Wulfen, *Jacq. Misc. Austr.* 2: 107 (1781)
Merulius aurantiacus (Wulfen) Pers. [non *M. aurantiacus* Klotzsch (1836)], *Syn. meth. fung.*: 488 (1801)
Cantharellus aurantiacus (Wulfen) Fr., *Syst. mycol.* 1: 318 (1821)
Clitocybe aurantiaca (Wulfen) Stud.-Steinh., *Hedwigia* 39(1): 6 (1900)
Cantharellus aurantiacus β *lacteus* Fr., *Syst. Mycol.:* 318 (1821)
Clitocybe aurantiaca var. *lactea* (Fr.) Rea, *Brit. basidiomyc.*: 273 (1922)
Merulius nigripes Pers., *Syn. meth. fung.*: 489 (1801)
Clitocybe aurantiaca var. *nigripes* (Pers.) Rea, *Brit. basidiomyc.*: 274 (1922)
Hygrophoropsis aurantiaca var. *nigripes* (Pers.) Kühner & Romagn. [*comb. illegit.*], *Flore Analytique des Champignons Supérieurs*. 130 (1953)
Cantharellus aurantiacus var. *pallidus* Cooke, *Ill. Brit. fung.* 1057 (1104b) Vol. 7 (1888)
Hygrophoropsis aurantiaca var. *pallida* (Cooke) Kühner & Romagn. [*comb. illegit.*], *Flore Analytique des Champignons Supérieurs*. 130 (1953)
Clitocybe aurantiaca var. *albida* (Gillet) Rea, *Brit. basidiomyc.*: 273 (1922)
Hygrophoropsis aurantiaca var. *rufa* D.A. Reid, *Fungorum Rar. Icon. Color.* 6: 5 & pl. 41c (1972)
Agaricus subcantharellus Sowerby, *Col. fig. Engl. fung., Suppl.*: pl. 413 (1809)
E: c **S:** c **W:** c **NI:** c **ROI:** c **O:** Channel Islands: c Isle of Man: c
H: On acidic soil, amongst needle litter or less often on decayed wood, in conifer woodland. Rarely in deciduous woodland on calcareous soil, with *Sambucus nigra*.
D: NM2: 53 **D+I:** Ph: 66, B&K3: 90 61 **I:** FlDan5: 196B (as *Cantharellus aurantiacus* f. *pallidus*), FlDan5: 196C (as *C. aurantiacus*), Bon: 51, Kriegl3: 274
Very common and widespread. The varieties are mainly colour forms. Var. *pallida* has pale yellowish lamellae, var. *lactea* is entirely white, and var. *nigripes* is characterised by the presence of minute blackish scales on the pileus.

fuscosquamula P.D. Orton, *Trans. Brit. Mycol. Soc.* 43(2): 245 (1960)
E: !
H: On acidic soil, often amongst *Juncus* spp in damp woodland, fen or alder carr.
D: FAN3: 66
Rarely reported. Possibly an ecotype of *H. aurantiaca* and not a distinct species.

macrospora (D.A. Reid) Kuyper, *Persoonia* 16(2): 231 (1996)
Hygrophoropsis aurantiaca var. *macrospora* D.A. Reid, *Fungorum Rar. Icon. Color.* 6: 6 (1972)
Mis.: *Hygrophoropsis pallida* [*nom. inval.*] sensu auct.
E: ! **S:** ? **W:** ? **NI:** ? **ROI:** ?
H: On acidic soil, amongst grass or *Juncus* spp, often in boggy fields in upland or montane areas.
Frequently reported, but usually unsubstantiated with voucher material.

HYGROPHORUS Fr., *Gen. Hym.*: 8 (1836)
Agaricales, Hygrophoraceae
Limacium (Fr.) P. Kumm., *Führ. Pilzk.*: 25 (1871)

Type: *Hygrophorus eburneus* (Bull.) Fr.

agathosmus (Fr.) Fr., *Epicr. syst. mycol.*: 325 (1838)
Agaricus agathosmus Fr., *Observ. mycol.* 1: 16 (1815)
Hygrophorus agathosmus f. *aureofloccosus* Bres., *Icon. Mycol.* 7: pl. 320 (1928)
Hygrophorus agathosmus var. *aureofloccosus* (Bres.) A. Pearson & Dennis, *Trans. Brit. Mycol. Soc.* 31: 163 (1948)
Agaricus cerasinus Berk., *Engl. fl.* 5(2): 12 (1836)
Hygrophorus cerasinus (Berk.) Berk., *Outl. Brit. fungol.*: 197 (1860)
E: ! **S:** o **NI:** ! **ROI:** !
H: On soil with conifers. Known with *Larix*, *Picea* and *Pinus* spp, and rarely reported with *Fagus*.
D: FAN2: 132-133, NM2: 87 **D+I:** B&K3: 118 105, FungEur6: 99-102 p.693 (as *Hygrophorus agathosmus* f. *agathosmus*) **I:** Bon: 119, C&D: 167, SV33: 20, Kriegl3: 109
Occasional but widespread throughout Scotland, rare elsewhere.

arbustivus Fr., *Sverig Atl. Svamp.*: 46 (1836)
E: ! **W:** !
H: On soil, often calcareous, most frequently with *Fagus* but also with mixed deciduous trees. Fruiting from late November onwards.
D: FAN2: 127, NM2: 90 **D+I:** FungEur6: 265-269 p.715 **I:** Bon: 117, C&D: 165, Kriegl3: 93
Rarely reported. British records only from England (Lincolnshire southwards but most frequently collected in the south east) and Wales (Flintshire: Loggerheads).

camarophyllus (Alb. & Schwein.) Dumée, Grandjean & Maire, *Bull. Soc. Mycol. France* 28: 292 (1912)
Agaricus camarophyllus Alb. & Schwein., *Consp. fung. lusat.*: 177 (1805)
Agaricus caprinus Scop., *Fl. carniol.* 438 (1772)
Hygrophorus caprinus (Scop.) Fr., *Epicr. syst. mycol.*: 326 (1838)
E: ? **S:** ? **NI:** ?
H: On soil in conifer woodland.
D: NM2: 88 **D+I:** B&K3: 120 106, FungEur6: 107-115 p.614 **I:** Sow2: pl. 172 (as *Agaricus elixus*), Bon: 119, C&D: 167, Kriegl3: 111, Cooke 889 (916) Vol. 6 (1888) (as *H. caprinus*).
Accepted with doubt. Not recorded since 1902 and old herbarium specimens need re-examination.

chrysodon (Batsch) Fr., *Epicr. syst. mycol.*: 320 (1838)
Agaricus chrysodon Batsch, *Elench. fung. (Continuatio Secunda)*: 79 t. 39 f. 212 (1789)
Hygrophorus chrysodon var. *leucodon* Alb. & Schwein., *Consp. fung. lusat.*: 182 (1805)
E: o **S:** ! **W:** !
H: On loam soil, under deciduous trees. Known with *Fagus*, *Quercus robur* and *Tilia* sp.
D: FAN2: 118-119, NM2: 85 **D+I:** B&K3: 120 108, FungEur6: 159-163 p.699 **I:** FlDan5: 164G (as *Limacium chrysodon*), Bon: 115, C&D: 171, Kriegl3: 80
Occasional but widespread in England, with a few records from Scotland (Mid Ebudes: Isle of Mull) and Wales (Monmouthshire).

cossus (Sowerby) Fr., *Epicr. syst. mycol.*: 321 (1838)
Agaricus cossus Sowerby, *Col. fig. Engl. fung.* 2: pl. 121 (1798)
Hygrophorus eburneus var. *cossus* (Sowerby) Quél., *Fl. mycol. France*: 482 (1888)
Hygrophorus quercetorum P.D. Orton, *Doc. Mycol.* 14(56): 56 (1985) [1984]
Hygrophorus cossus var. *quercetorum* (P.D. Orton) Bon, *Doc. Mycol.* 19(76): 74 (1989)
Hygrophorus eburneus var. *quercetorum* (P.D. Orton) Arnolds, *Persoonia* 13: 75 (1986)
Hygrophorus quercorum P.D. Orton, *Notes Roy. Bot. Gard. Edinburgh* 41: 585 (1984)
Hygrophorus eburneus var. *quercorum* (P.D. Orton) Arnolds, *Persoonia* 12(4): 477 (1985)
E: !
H: On soil, associated with *Quercus* spp.

D+I: FungEur6: 148-151 p. 697

Rare. The name has been applied to any white, strong smelling *Hygrophorus*. Distinguished from *H. eburneus* by the association with *Quercus* spp, the more robust habit and basidiomes becoming cream to ivory-white with age.

discoxanthus (Fr.) Rea, *Trans. Brit. Mycol. Soc.* 3(1): 45 (1908) [1907]
Agaricus discoxanthus Fr., *Observ. mycol.* 1: 15 (1815)
Hygrophorus chrysaspis Métrod, *Rev. Mycol. (Paris)* 3(4-5): 156 (1938)
Mis.: *Hygrophorus cossus* sensu auct. Brit.
Mis.: *Hygrophorus discoideus* sensu auct. Brit.
E: c **S:** ! **W:** !
H: On soil, amongst leaf litter in deciduous woodland. Usually associated with *Fagus* and rarely with *Betula* and *Quercus* spp.
D: FAN2: 119-120, FungEur6: 138-142 **D+I:** B&K3: 122 110 **I:** Bon: 115, C&D: 168
Common in southern and western England, less so elsewhere, but widespread.

eburneus (Bull.) Fr., *Sverig Atl. Svamp.*: 45 (1836)
Agaricus eburneus Bull. [non *A. eburneus* Bolton (1788)], *Herb. France*: pl. 118 (1782)
Gymnopus eburneus (Bull.) Gray, *Nat. arr. Brit. pl.* 1: 610 (1821)
Mis.: *Hygrophorus cossus* sensu auct. p.p.
E: o **S:** o **W:** o **NI:** o **ROI:** o
H: On soil in deciduous woodland, usually associated with *Fagus* but also known with *Quercus* spp.
D: FAN2: 118-119 (as *Hygrophorus eburneus* var. *eburneus*), NM2: 86 **D+I:** Ph: 58-59, B&K3: 122 111, FungEur6: 144-148 p.696 **I:** Ph: 58 (as *H. cossus*), Bon: 115, C&D: 169
Occasional but widespread and may be locally common.

erubescens (Fr.) Fr., *Epicr. syst. mycol.*: 322 (1838)
Agaricus erubescens Fr., *Syst. mycol.* 1: 32 (1821)
E: !
H: On calcareous soil, associated with *Picea* spp in continental Europe. British material mostly with *Fagus* but Cooke 876 (888) Vol. 6 (1888) is given as 'in fir woods'.
D: FAN2: 121, NM2: 88 **D+I:** B&K3: 124 112, FungEur6: 194-199 p.705 **I:** Bon: 117, C&D: 167, Kriegl3: 90
Possibly extinct in Britain. Known only from a few records from Herefordshire and Northamptonshire and unreported since 1877. A large and distinctive species, so unlikely to be overlooked.

hedrychii (Velen.) K. Kult, *Česká Mykol.* 10: 232 (1956)
Limacium hedrychii Velen., *České Houby* I: 96 (1920)
Mis.: *Hygrophorus melizeus* sensu Rea (1922), sensu auct. p.p.
E: ! **S:** !
H: On soil in woodland. British records with *Betula* spp.
D: FAN2: 120-121, FungEur6: 156-158 **D+I:** B&K3: 124 113 **I:** Kriegl3: 83
Known from several sites in Scotland and two in England (South Hampshire and Westmorland).

hypothejus (Fr.) Fr., *Epicr. syst. mycol.*: 324 (1838)
Agaricus hypothejus Fr., *Observ. mycol.* 2: 10 (1818)
Hygrophorus aureus Arrh., *Monogr. hymenomyc. Suec.* 2: 127 (1863)
Hygrophorus hypothejus var. *aureus* (Arrh.) Imler, *Bull. Soc. Mycol. France* 50: 304 (1935) [1934]
Hygrophorus hypothejus var. *expallens* Boud., *Icon. mycol.* 1: pl. 33 (1905)
E: o **S:** ! **W:** o **NI:** o **ROI:** !
H: On acidic soil with conifers, usually *Pinus* spp and rarely *Larix* spp. Rarely recorded with deciduous trees such as *Betula* spp. Usually fruits at the start of the colder months.
D: FAN2: 127-128, NM2: 89 **D+I:** Ph: 59, B&K3: 126 115, FungEur6: 228-232 p.710 **I:** Bon: 119, C&D: 171, SV33: 20, Kriegl3: 101
Occasional but widespread and may be locally common. English name = 'Herald of Winter'.

lindtneri M.M. Moser, *Z. Pilzk.* 33(1-2): 2 (1967)
Hygrophorus carpini Gröger, *Z. Mykol.* 46(2): 162 (1980)
Mis.: *Hygrophorus leucophaeus* sensu NCL p.p.
E: !
H: On calcareous soil, amongst leaf litter in mixed deciduous woodland.
D+I: FungEur6: 248-255 p.713 **I:** C&D: 169
Known only from southern England (North & South Hampshire, Oxfordshire, Surrey and West Kent).

lucorum Kalchbr., *Icon. select. Hymenomyc. Hung.*: 35 (1874)
Tricholoma luteocitrinum Rea, *Trans. Brit. Mycol. Soc.* 3(2): 125 (1908) [1909]
E: ! **S:** ?
H: On calcareous soil, amongst needle litter of *Larix* spp. Very rarely reported with other conifers such as *Pinus* and *Pseudotsuga* spp.
D: FAN2: 127, NM2: 89 **D+I:** B&K3: 128 118, FungEur6: 232-236 p.713 **I:** FlDan5: 163C (as *Limacium lucorum*), C&D: 171, Kriegl3: 103
Known from southern England (Bedfordshire, Berkshire, East & West Kent, East & West Sussex, Hertfordshire, Oxfordshire, Surrey) and a single record from Scotland (Dunballoch Wood) in 1908 (unsubstantiated with voucher material).

mesotephrus Berk., *Ann. Mag. Nat. Hist., Ser. 2 [Notices of British Fungi no. 699]* 13: 402 (1854)
E: ! **W:** ! **NI:** ? **ROI:** ?
H: On soil in woodland. Reported with *Fagus*, *Quercus* and *Pinus* spp.
D: NCL3: 258, FAN2: 131, NM2: 87 **D+I:** FungEur6: 81-84 p.690 **I:** C&D: 169, Kriegl3: 105
Described from Britain but poorly known. Collections (in herb. K) from England (Buckinghamshire, Derbyshire, North Wiltshire, Oxfordshire, South Essex, South Hampshire, Surrey and West Sussex) and Wales (Breconshire and Cardiganshire). Sensu Cooke 887 (914) Vol. 6 (1888) and sensu auct. is *H. unicolor*.

nemoreus (Pers.) Fr., *Epicr. syst. mycol.*: 326 (1838)
Agaricus nemoreus Pers., *Syn. meth. fung.*: 305 (1801)
Camarophyllus nemoreus (Pers.) P. Kumm., *Führ. Pilzk.*: 118 (1871)
Mis.: *Hygrophorus leporinus* sensu Cooke, sensu Rea (1922)
E: ! **S:** ! **NI:** !
H: On calcareous soil in deciduous woodland.
D: FAN2: 124-125, NM2: 90 **D+I:** B&K3: 130 121, FungEur6: 281-285 p.717 **I:** Bon: 117, C&D: 165, Kriegl3: 97
Rarely reported but widely scattered in England, rare in Scotland and Northern Ireland. When found in woodland *Hygrocybe pratensis* is sometimes mistaken for this species.

olivaceoalbus (Fr.) Fr., *Epicr. syst. mycol.*: 324 (1838)
Agaricus olivaceoalbus Fr., *Observ. mycol.* 1: 5 (1815)
E: !
H: With *Picea* spp.
D: FAN2: 129, NM2: 87 **D+I:** FungEur6: 84-88 p.691 **I:** Bon: 119, C&D: 171, Kriegl3: 106
Accepted with doubt. 'Common' *fide* Rea (1922) but many older records are likely to be *H. persoonii*. In continental Europe this species is strictly associated with native *Picea* spp, but British records are all reported with deciduous trees. Very little material in herbaria, mostly in poor condition.

penarius Fr., *Sverig Atl. Svamp.*: 45 (1836)
E: ! **NI:** ? **ROI:** ?
H: On calcareous soils, under trees. British material associated with *Fagus* and rarely *Taxus*.
D: FAN2: 118, NM2: 86 **D+I:** B&K3: 130 123, FungEur6: 175-179 p.702 **I:** Bon: 114, C&D: 165
Known from East Kent, East Sussex, Herefordshire and Lincolnshire. Single records from Northern Ireland and Republic of Ireland but unsubstantiated with voucher material.

persoonii Arnolds, *Persoonia* 10: 365 (1979)
Agaricus limacinus Schaeff. [*nom. illegit.*, non *A. limacinus* Scop. (1772)], *Fung. Bavar. Palat. nasc.* 1: 74 (1762)
Mis.: *Hygrophorus dichrous* sensu NCL

Mis.: *Hygrophorus limacinus* sensu auct.
Mis.: *Limacium olivaceoalbum* sensu Lange (FlDan5, pl. 162A)
Mis.: *Hygrophorus olivaceoalbus* sensu auct.
E: ! **S:** ! **NI:** !
H: On soil in old deciduous woodland, usually with *Quercus* spp, but also known with *Fagus* and *Corylus*.
D: FAN2: 129-130, NM2: 87 **D+I:** FungEur6: 90-95 p.692 **I:** C&D: 171, Kriegl3: 108
Rarely reported but apparently widespread. Scattered widely in England, with single records from Scotland and Northern Ireland.

piceae Kühner, *Bull. Mens. Soc. Linn. Lyon* 18: 179 (1949)
E: ? **S:** !
H: On soil amongst litter with *Picea* spp.
D: NM2: 86 **D+I:** FungEur6: 182-186 p.703
A boreal species, known with certainty only from Perthshire (Black Craig). An English record from West Sussex (Houghton Forest) is unsubstantiated with voucher material.

pudorinus (Fr.) Fr., *Sverig Atl. Svamp.*: 45 (1836)
Agaricus pudorinus Fr., *Syst. mycol.* 1: 33 (1821)
E: !
H: On soil in litter under conifers.
D: NM2: 89 **D+I:** B&K3: 132 126, FungEur6: 217-222 p.708 **I:** Bon: 115, C&D: 169, Kriegl3: 100
Rarely reported, less often collected and poorly known in Britain. Recorded only from a few widely scattered localities in England.

pustulatus (Pers.) Fr., *Epicr. syst. mycol.*: 325 (1838)
Agaricus pustulatus Pers., *Syn. meth. fung.*: 354 (1801)
Agaricus tephroleucus Pers., *Syn. meth. fung.*: 351 (1801)
Hygrophorus tephroleucus (Pers.) Fr., *Epicr. syst. mycol.*: 325 (1838)
E: ! **S:** !
H: On acidic soil in conifer woodland, associated with *Abies* and *Picea* spp.
D: FAN2: 132, NM2: 88 **D+I:** Ph: 59, B&K3: 134 127, FungEur6: 126-130 **I:** C&D: 167, Kriegl3: 115
Rarely reported. Known from England (Co. Durham, Northumberland and Shropshire) and Scotland (Midlothian and Perthshire).

russula (Fr.) Quél., *Enchir. fung.*: 49 (1886)
Agaricus russula Fr., *Syst. mycol.* 1: 38 (1821)
Gymnopus russulus (Fr.) Gray, *Nat. arr. Brit. pl.* 1: 607 (1821)
Tricholoma russula (Fr.) Gillet, *Hyménomycètes*: 91 (1874)
E: !
H: On soil, often amongst grass, in deciduous woodland.
D: FAN2: 121-122, NM2: 88 **D+I:** B&K3: 134 129, FungEur6: 207-211 p.706 **I:** Bon: 117, C&D: 167, Kriegl3: 92
Probably extinct. Collected and illustrated from Kew Gardens in 1886 [see Cooke 1116 (926) Vol. 8 (1889)] and there is a drawing by Rea in 1903 of material collected near Worcester. Unreported since then and no voucher material exists. A distinctive, large and conspicuous fungus unlikely to be overlooked.

unicolor Gröger, *Z. Mykol.* 46(2): 160 (1980)
Mis.: *Hygrophorus leucophaeus* sensu NCL p.p.
Mis.: *Hygrophorus mesotephrus* sensu Cooke [Ill. Brit. fungi 887(914) Vol. 6 (1888)] & sensu auct.
E: !
H: On calcareous soil, associated with *Fagus*.
D: FAN2: 126, NM2: 90, FungEur6: 255 **D+I:** B&K3: 136 131 **I:** Bon: 117, Kriegl3: 97 (as *Hygrophorus lindtneri* var. *unicolor*)
Rarely reported but widely scattered throughout southern England.

HYMENANGIUM Klotzsch, in Dietrich, *Fl. boruss.* 7: 466 (1839)
Cortinariales, Hymenangiaceae
Type: *Hymenangium album* Klotzsch

album Klotzsch, *Fl. Regn. Boruss.* 7: 466 (1839)
Hymenogaster albus (Klotzsch) Berk., *Ann. Mag. Nat. Hist., Ser. 1 [Notices of British Fungi no. 296]* 13: 349 (1844)
Hymenogaster klotzschii Tul., *Fungi hypogaei*: 64 (1851)
Mis.: *Rhizopogon albus* sensu Berkeley [Engl. fl. 5 (2): 229 (1836)]
E: ! **S:** !
H: In soil, associated with *Eucalyptus* spp.
D+I: BritTruff: 162 10 B, Myc14(4): 160
Rarely reported. Described from Glasgow Botanic Garden in 1830 and recently collected at Kew.

HYMENOCHAETE Lév., *Ann. Sci. Nat., Bot.*, sér. 3, 5: 150 (1846)
Hymenochaetales, Hymenochaetaceae
Type: *Hymenochaete rubiginosa* (Dicks.) Lév.

carpatica Pilát, *Hedwigia* 70: 123 (1931)
E: !
H: On dead wood of deciduous trees, usually fallen or dead attached branches. British material on *Fagus* and *Ulmus* spp, and *Ulex europaeus*.
I: Kriegl1: 207
Reported from Berkshire (Windsor Forest) in 2000 and from South Hampshire (New Forest) in 2003-2004. Basidiomes are inconspicuous, thus easily overlooked.

cinnamomea (Pers.) Bres., *Atti Imp. Regia Accad. Roveretana, ser. 3*, 3: 110 (1897)
Thelephora cinnamomea Pers., *Mycol. eur.* 1: 141 (1822)
Corticium cinnamomeum (Pers.) Fr., *Epicr. syst. mycol.*: 561 (1838)
Coniophora cinnamomea (Pers.) Massee, *J. Linn. Soc., Bot.* 25: 130 (1889)
Hymenochaete arida (P. Karst.) Sacc., *Syll. fung.* 9: 228 (1891)
Hymenochaete laxa P. Karst., *Syll. fung.* 9: 222 (1891)
Hymenochaete spreta Peck, *Rep. (Annual) New York State Mus. Nat. Hist.* 30: 47 (1879)
E: ! **S:** ! **W:** ! **NI:** ?
H: On fallen, or dead attached, wood or bark of deciduous trees and shrubs, such as *Corylus*, *Crataegus*, *Fagus* and *Tilia* spp, and a single collection from *Rhododendron ponticum*.
D: NM3: 323 **D+I:** B&K2: 244 292
Known from England (Berkshire, Dorset, Hertfordshire, North & South Hampshire), Scotland (North Ebudes) and Wales (Breconshire and Caernarvonshire). Reported from Northern Ireland but unsubstantiated with voucher material.

corrugata (Fr.) Lév., *Ann. Sci. Nat., Bot.*, sér. 3, 5: 152 (1846)
Thelephora corrugata Fr., *Observ. mycol.* 1: 154 (1815)
Corticium corrugatum (Fr.) Fr., *Epicr. syst. mycol.*: 565 (1838)
Hymenochaete agglutinans Ellis, *Bull. Torrey Bot. Club* 5: 46 (1874)
Hymenochaete corrugata f. *conglutinans* Bourdot & Galzin, *Hymenomyc. France*: 393 (1928)
Hymenochaete croceoferruginea Massee, *J. Linn. Soc., Bot.* 27: 110 (1890)
E: c **S:** c **W:** c **NI:** c **ROI:** c
H: Saprotrophic or weakly parasitic. On dead and living wood of deciduous trees and shrubs. Mostly on *Corylus* but also known on *Buxus*, *Calluna*, *Fagus*, *Myrica gale* and *Rubus fruticosus*.
D: NM3: 323 **D+I:** B&K2: 244 293 **I:** Kriegl1: 210
Common and widespread. Often bridging small gaps between adjacent branches and gluing them together, allowing vegetative spread where trees or shrubs grow close together (especially in copses of *Corylus*). Causes an intense white rot in infected wood.

cruenta (Pers.) Donk, *Persoonia* 1(1): 51 (1959)
Thelephora cruenta Pers., *Syn. meth. fung.*: 575 (1801)
Corticium cruentum (Pers.) J. Schröt., in Cohn, *Krypt.-Fl. Schlesien* 3: 423 (1888)
Cytidia cruenta (Pers.) Herter, *Kryptogamenflora der Mark Brandenburg* 6(1): 83 (1910)

Thelephora mougeotii Fr., *Elech. fung.* 1: 188 (1828)
Stereum mougeotii (Fr.) Berk., *Hooker's J. Bot. Kew Gard. Misc.* 6: 190 (1854)
Hymenochaete mougeotii (Fr.) Cooke, *Grevillea* 8: 147 (1880)
W: !
H: On fallen wood, usually branches, of conifers. British material on *Abies* sp.
D+I: B&K2: 246 295 **I:** Kriegl1: 211
A single collection from Merionethshire (Plas Tan-y-Bwlch) in 1998.

rubiginosa (Dicks.) Lév., *Ann. Sci. Nat., Bot.,* sér. 3, 5: 151 (1846)
Helvella rubiginosa Dicks., *Fasc. pl. crypt. brit.* 1: 20 (1785)
Stereum rubiginosum (Dicks.) Gray, *Nat. arr. Brit. pl.* 1: 652 (1821)
Auricularia ferruginea Bull., *Herb. France:* pl. 378 (1788)
Hymenochaete ferruginea (Bull.) Massee, *J. Linn. Soc., Bot.* 27: 103 (1890)
Stereum ferrugineum (Bull.) Gray, *Nat. arr. Brit. pl.* 1: 653 (1821)
E: c **S:** c **W:** c **NI:** c **ROI:** c **O:** Channel Islands: !
H: On dead wood of deciduous trees, usually large, hard, decorticated stumps and logs of *Quercus* spp, and less frequently *Castanea*. Reported rarely on other hosts such as *Carpinus*, *Corylus*, *Crataegus*, *Fagus*, *Salix*, *Sambucus* and *Ulmus* spp, but these hosts are not verified.
D: NM3: 323 **D+I:** Ph: 235, B&K2: 246 296 **I:** Sow1: 26 (as *Auricularia ferruginea*), Kriegl1: 213
Very common and widespread.

tabacina (Sowerby) Lév., *Ann. Sci. Nat., Bot.,* sér. 3, 5: 145 (1846)
Auricularia tabacina Sowerby, *Col. fig. Engl. fung.* 1: pl. 25 (1797)
Thelephora tabacina (Sowerby) Pers., *Mycol. eur.* 1: 118 (1822)
Stereum tabacinum (Sowerby) Fr., *Epicr. syst. mycol.:* 550 (1838)
Thelephora avellana Fr., *Syst. mycol.* 1: 442 (1821)
Hymenochaete avellana (Fr.) Lév., *Grevillea* 8: 146 (1880)
Hymenochaete badioferruginea (Mont.) Lév., *Ann. Sci. Nat., Bot.,* sér. 3, 5: 152 (1846)
Stereum crocatum Fr., *Epicr. syst. mycol.:* 550 (1838)
Daedalea lirellosa Pers., *Mycol. eur.* 3: 2 (1828)
Phlebia lirellosa (Pers.) Berk. & Broome, *Ann. Mag. Nat. Hist.,* Ser. 5 [Notices of British Fungi no. 1973] 9: 182 (1882)
Helvella nicotiana Bolton, *Hist. fung. Halifax* Suppl.: 174 (1791)
Hymenochaete nigrescens Massee, *J. Linn. Soc., Bot.* 27: 104 (1890)
E: ! **S:** ! **W:** !
H: On dead wood of deciduous trees, usually reported on *Corylus* or *Salix* spp.
D: NM3: 322 **D+I:** B&K2: 246 297 **I:** Sow1: 25 (as *Auricularia tabacina*), Kriegl1: 215
Rarely reported but apparently widespread.

HYMENOGASTER Vittad., *Monogr. Tuberac.*: 20 (1831)
Hymenogastrales, Hymenogasteraceae
Type: *Hymenogaster bulliardii* Vittad.

arenarius Tul. & C. Tul., *Giorn. Bot. Ital.* 2(1): 55 (1844)
Hymenogaster pusillus Berk. & Broome, *Ann. Mag. Nat. Hist.,* Ser. 1 18: 75 (1846)
E: ! **S:** !
H: Hypogeous in humic soil, in deciduous woodland with *Fagus*, *Quercus* and *Tilia* spp. A single record with *Eucalyptus* sp. in a garden.
D: BritTruff: 175, NM3: 316
Known from England (North Somerset, Northamptonshire, Oxfordshire, Shropshire and West Gloucestershire) and

Scotland (Clackmannanshire). Reported elsewhere but unsubstantiated with voucher material.

citrinus Vittad., *Monogr. Tuberac.*: 21 (1831)
E: ! **W:** !
H: Epigeous or hypogeous in calcareous, humic soil in deciduous or mixed woodland. Usually associated with *Fagus* and rarely with conifers such as *Pinus* or *Picea* spp.
D: NM3: 317 **D+I:** BritTruff: 168-169 10 D
Rarely reported. Known from England (Dorset, North Somerset, North & South Wiltshire, Oxfordshire, Surrey, West Gloucestershire and West Kent) and Wales (Caernarvonshire).

griseus Vittad., *Monogr. Tuberac.*: 23 (1831)
E: !
H: Epigeous or hypogeous, in woodland on humic calcareous soil. Associated with deciduous trees such as *Fagus* or conifers such as *Larix* and *Picea* spp.
D: NM3: 317 **D+I:** BritTruff: 180-181 11 H
Known from North Somerset and West Gloucestershire. Reported from South Essex but unsubstantiated with voucher material.

hessei Soehner, *Z. Pilzk.* 2(7): 158 (1923)
E: ! **S:** ? **W:** !
H: Hypogeous in humic soil, in deciduous woodland with *Fagus* or *Quercus* spp, and reported with conifers such as *Picea* spp.
D+I: BritTruff: 181-182 11 E
First reported by Hawker in 1952 (TBMS 35: 279-284). Known from England (Herefordshire, North Somerset and West Gloucestershire) and Wales (Caernarvonshire). Reported from Scotland by Reid & Austwick (1963) but unsubstantiated with voucher material.

luteus Vittad., *Monogr. Tuberac.*: 22 (1831)
E: o **W:** ! **ROI:** !
H: Epigeous or hypogeous. Usually in deciduous woodland with *Fagus* and *Fraxinus*. Rarely with conifers such as *Pinus* spp and *Taxus*.
D+I: BritTruff: 167-168 10 C
Occasional but widespread in England and a single collection from Wales (Caernarvonshire) in herb. K.

muticus Berk. & Broome, *Ann. Mag. Nat. Hist., Ser. 2 [Notices of British Fungi no. 376]* 2: 67 (1848)
E: !
H: Hypogeous in humic soil.
D: NM3: 317 **D+I:** BritTruff: 179 11 B
Known only from the type collection from West Gloucestershire (Stapleton Grove, near Bristol). Reported from Hertfordshire (Tring) but unsubstantiated with voucher material.

olivaceus Vittad., *Monogr. Tuberac.*: 24 (1831)
Hymenogaster decorus Tul. & C. Tul., *Ann. Sci. Nat., Bot.,* sér. 2, 19: 374 (1843)
Hymenogaster olivaceus var. *modestus* Berk. & Broome, *Ann. Mag. Nat. Hist., Ser. 1* 18: 74 (1846)
Hymenogaster pallidus Berk. & Broome, *Ann. Mag. Nat. Hist., Ser. 1* 18: 74 (1846)
Mis.: *Hymenogaster populetorum* sensu Berkeley [British Fungi Exsiccati, Fasc. 4: 304 (1843)] non sensu Tul. & C. Tul. [*Ann. Sci. Nat., Bot.,* sér. 2, 19: 375 (1843)]
E: ! **S:** ! **W:** !
H: Epigeous or hypogeous in humic soil. With *Fagus*, *Quercus* and *Tilia* spp in mixed deciduous woodland but occasionally reported with conifers such as *Larix* and *Picea* spp.
D: NM3: 317 **D+I:** BritTruff: 169-172 10 E
Rarely reported but apparently widespead.

sulcatus R. Hesse, *Hypog. Deutschl.* 1: 111 (1891)
E: ! **W:** !
H: Hypogeous in humic soil.
D: BritTruff: 172-173
Apparently rare. Known from England (Derbyshire and North Somerset) and Wales (Denbighshire).

tener Berk. & Broome, *Ann. Mag. Nat. Hist., Ser. 1* 13: 349
(1844)
> Mis.: *Hymenogaster lilacinus* sensu Berkeley [British Fungi
> Exsiccati, Fasc. 4: 305 (1843)] non Tul. & C. Tul. [*Ann. Sci.
> Nat., Bot.,* sér. 2, 19: 374 (1843)]

E: o **S:** ! **W:** ! **ROI:** !
H: Epigeous under suface litter (and also shallowly hypogeous)
in woodland, associated with a wide range of deciduous trees
and conifers.
D: NM3: 316 **D+I:** BritTruff: 173-175 10 F
Occasional but widespread.

thwaitesii Berk. & Broome, *Ann. Mag. Nat. Hist., Ser. 1* 18: 75
(1846)
E: !
H: Hypogeous in humic soil in woodland.
D+I: BritTruff: 179-180 11 C
Known only from the type collection from North Somerset
(Portbury) in 1845.

vulgaris Tul. & C. Tul., *Ann. Mag. Nat. Hist., Ser. 1* 18: 74
(1846)
> *Rhizopogon albus* Berk. [non *R. albus* Fr. (1823)], *Engl. fl.*
> 5(2): 229 (1836)

E: o **S:** ! **W:** ! **ROI:** !
H: Hypogeous, usually in deciduous woodland, associated with
Castanea, *Fagus* and *Quercus* spp, but also known with *Tilia*
and *Ulmus* spp, and conifers such as *Larix*, *Picea* spp and
Taxus.
D+I: BritTruff: 176-177 11 A
Occasional in southern England. Rarely reported elsewhere but
apparently widespread.

HYPHODERMA Wallr. [non *Hyphoderma* Fr. (1849)], *Fl. crypt. Germ., Sect. II*: 576 (1833)

Stereales, Hyphodermataceae
> *Kneiffia* Fr., *Gen. Hym.*: 17 (1836)
> *Metulodontia* Parmasto, *Conspectus Systematis
> Corticiacearum*: 117 (1968)
> *Conohypha* Jülich, *Persoonia* 8: 303 (1975)
> Type: *Hyphoderma spiculosum* Wallr. (= *H. setigerum*)

albocremeum (Höhn. & Litsch.) J. Erikss. & Å. Strid,
Corticiaceae of North Europe 3: 453 (1975)
> *Corticium albocremeum* Höhn. & Litsch., *Wiesner Festschrift*:
> 61 (1908)
> *Conohypha albocremea* (Höhn. & Litsch.) Jülich, *Persoonia*
> 8(3): 304 (1975)

E: !
H: On decayed wood of deciduous trees. British material on
Betula spp and *Fagus* in old woodland.
D: NM3: 198 (as *Conohypha albocremea*), Myc17(1): 42 **D+I:**
CNE3: 452-453
Known only from South Hampshire.

argillaceum (Bres.) Donk, *Fungus* 27: 14 (1957)
> *Corticium argillaceum* Bres., *Fungi trident.* 2(11-13): 63
> (1898)
> *Peniophora argillacea* (Bres.) Sacc. & Syd., *Syll. fung.* 16:
> 194 (1902)
> *Kneiffia argillacea* (Bres.) Bres., *Ann. Mycol.* 1(1): 100 (1903)
> *Kneiffia carneola* Bres., *Ann. Mycol.* 1(1): 104 (1903)
> *Peniophora carneola* (Bres.) Höhn. & Litsch., *Wiesner
> Festschrift*: 70 (1908)
> *Peniophora reticulata* Wakef., *Trans. Brit. Mycol. Soc.* 35(1):
> 57 (1952)
> *Hyphoderma reticulatum* (Wakef.) Donk, *Fungus* 27: 15
> (1957)

E: o **S:** ! **W:** ! **ROI:** !
H: On decayed wood, in old woodland.
D: NM3: 200 **D+I:** CNE3: 456-459, B&K2: 130 123
Occasional in England. Rarely reported elsewhere but apparently
widespread.

cremeoalbum (Höhn. & Litsch.) Jülich, *Persoonia* 8(1): 80
(1974)
> *Corticium cremeoalbum* Höhn. & Litsch., *Wiesner Festschrift*:
> 63 (1908)

E: ! **W:** !
H: On decayed wood. British material on *Fagus* and *Picea abies*.
D: NM3: 199 **D+I:** CNE3: 464-466
Known from England (Buckinghamshire and Devon) and Wales
(Monmouthshire). Probably also this species in South
Hampshire and Surrey.

cristulatum (Fr.) Donk, *Fungus* 27: 15 (1957)
> *Hydnum cristulatum* Fr., *Syst. mycol.* 1: 422 (1821)
> *Odontia cristulata* (Fr.) Fr., *Epicr. syst. mycol.*: 529 (1838)

E: !
H: On decayed wood of deciduous trees.
A single collection, from Hertfordshire (Brookmans Park) in
1995. Synonymised with *H. setigerum* by some authorities.

cryptocallimon B. de Vries, *Mycotaxon* 28(1): 77 (1987)
E: !
H: On decayed wood of deciduous trees and conifers.
D+I: CNE3: 542-543 (as *Hyphoderma* sp. de Vries 488)
Known only from Oxfordshire, South Hampshire, Surrey and
Yorkshire.

definitum (H.S. Jacks.) Donk, *Fungus* 27: 15 (1957)
> *Corticium definitum* H.S. Jacks., *Canad. J. Res., Sect. C, Bot.
> Sci.* 26: 149 (1948)

E: ! **S:** !
H: On decayed wood. British material on *Fagus* and *Taxus*.
D: NM3: 201 **D+I:** CNE3: 468-469
Known from England (South Hampshire and Surrey) and
Scotland (Mid-Perthshire).

guttuliferum (P. Karst.) Donk, *Persoonia* 2(2): 223 (1962)
> *Gloeocystidium guttuliferum* P. Karst., *Bidrag Kännedom
> Finlands Natur Folk* 48: 430 (1889)
> *Peniophora guttulifera* (P. Karst.) Sacc., *Syll. fung.* 9: 240
> (1891)

E: !
H: On decayed wood of deciduous trees. British material on
Fagus and *Fraxinus*.
D: NM3: 199 **D+I:** CNE3: 474-477
Known only from Berkshire, South Hampshire and Surrey.

litschaueri (Burt) J. Erikss. & Å. Strid, *Corticiaceae of North
Europe* 3: 481 (1975)
> *Corticium litschaueri* Burt, *Ann. Missouri Bot. Gard.* 13: 259
> (1926)
> *Corticium niveum* Bres., *Ann. Mycol.* 1(1): 98 (1903)

E: !
H: On decayed wood of deciduous trees in old woodland.
British material on large fallen trunks or logs of *Fagus*.
D: NM3: 201 **D+I:** CNE3: 480-481, Myc11(3): 111-112
Known only from South Hampshire, Surrey, West Sussex and
Yorkshire. Rather common in the New Forest.

macedonicum (Litsch.) Donk, *Fungus* 27: 15 (1957)
> *Gloeocystidium macedonicum* Litsch., *Glasnik (Bull.) Soc.
> Scient. Skoplje* 18(6): 181 (1938)

S: !
H: On decayed wood.
D: NM3: 201 **D+I:** CNE3: 482-485
Known only from Dunbartonshire and Mid Ebudes.

medioburiense (Burt) Donk, *Fungus* 27: 15 (1957)
> *Peniophora medioburiensis* Burt, *Ann. Missouri Bot. Gard.* 12:
> 328 (1925)

W: !
H: On decayed wood of deciduous trees. British material on
Alnus and *Salix* spp.
D: NM3: 202 **D+I:** CNE3: 486-488
Known only from Monmouthshire and Merionethshire.

occidentale (D.P. Rogers) Boidin & Gilles, *Cryptog. Mycol* 15:
138 (1994)
> *Galzinia occidentalis* D.P. Rogers, *Mycologia* 36: 102 (1944)
> *Hyphoderma subdefinitum* J. Erikss. & Å. Strid, *Corticiaceae
> of North Europe* 3: 539 (1975)

E: ! **S:** !
H: On decayed wood of deciduous trees. British material on *Fagus*.
D+I: CNE3: 538-539 (as *Hyphoderma subdefinitum*), B&K2: 136 130 (as *H. subdefinitum*)
Known from Surrey and a single collection from Scotland. Reported from Oxfordshire but unsubstantiated with voucher material.

pallidum (Bres.) Donk, *Fungus* 27: 15 (1957)
Corticium pallidum Bres., *Fungi trident.* 2(11-13): 59 (1898)
Gloeocystidium pallidum (Bres.) Höhn & Litsch., *Sitzungsber. Kaiserl. Akad. Wiss., Wien, Math.-Naturwiss. Cl., Abt. 1* 116: 838 (1924)
E: ! **S:** ! **W:** ! **O:** Isle of Man: !
H: On decayed wood of conifers such as *Larix*, *Picea* and *Pinus* spp, and rarely on wood of deciduous trees such as *Fagus* and *Sorbus aucuparia*.
D: NM3: 201 **D+I:** CNE3: 501-503
Rarely reported but apparently widespread.

praetermissum (P. Karst.) J. Erikss. & Å. Strid, *Corticiaceae of North Europe* 3: 505 (1975)
Peniophora praetermissa P. Karst., *Bidrag Kännedom Finlands Natur Folk* 48: 423 (1889)
Corticium praetermissum (P. Karst.) Bres., *Ann. Mycol.* 1(1): 100 (1903)
Gloeocystidium praetermissum (P. Karst.) Höhn. & Litsch., *Sitzungsber. Kaiserl. Akad. Wiss., Wien, Math.-Naturwiss. Cl., Abt. 1* 115: 1565 (1906)
Gloeocystidium tenue ssp. *praetermissum* (P. Karst.) Bourdot & Galzin, *Hymenomyc. France*: 256 (1928)
Corticium pertenue P. Karst., *Hedwigia* 29: 270 (1890)
Corticium tenue Pat., *Rev. Mycol. (Toulouse)* 7(26): 152 (1885)
Gloeocystidium tenue (Pat.) Höhn. & Litsch., *Oest. Cort.*: 70 (1907)
Hyphoderma tenue (Pat.) Donk, *Fungus* 27: 15 (1957)
E: c **S:** c **W:** c **NI:** c **ROI:** c **O:** Channel Islands: c Isle of Man: c
H: On decayed wood of deciduous trees and conifers.
D: NM3: 201 **D+I:** CNE3: 504-511, B&K2: 132 126
A common, widespread, and variable specis. Most likely a species complex as yet unresolved.

puberum (Fr.) Wallr., *Fl. crypt. Germ., Sect. II*: 576 (1833)
Thelephora pubera Fr., *Elench. fung.* 1: 215 (1828)
Corticium puberum (Fr.) Fr., *Epicr. syst. mycol.*: 362 (1838)
Peniophora pubera (Fr.) Sacc., *Syll. fung.* 6: 646 (1888)
Phlebia pubera (Fr.) M.P. Christ., *Dansk. Bot. Ark.* 19(2): 171 (1960)
E: c **S:** c **W:** c **NI:** c **ROI:** c **O:** Channel Islands: !
H: On decayed wood of deciduous trees and conifers.
D: NM3: 199 **D+I:** CNE3: 512-516, B&K2: 134 127
Very common and widespread.

roseocremeum (Bres.) Donk, *Fungus* 27: 15 (1957)
Gloeocystidium roseocremeum (Bres.) Brinkmann, *Westfäl. Pilzbriefe* 2: 47 (1904)
Corticium roseocremeum Bres., *Ann. Mycol.* 3(2): 163 (1905)
E: ! **S:** ! **W:** ! **NI:** !
H: On decayed wood of deciduous trees.
D: NM3: 202 **D+I:** CNE3: 522-525
Rarely reported.

setigerum (Fr.) Donk, *Fungus* 27: 15 (1957)
Thelephora setigera Fr., *Elench. fung.* 1: 208 (1828)
Kneiffia setigera (Fr.) Fr., *Epicr. syst. mycol.*: 529 (1838)
Peniophora setigera (Fr.) Höhn. & Litsch., *Ann. Mycol.* 4: 289 (1906)
Odontia setigera (Fr.) Millar, *Mycologia* 26(1): 19 (1934)
E: c **S:** c **W:** c **NI:** c **ROI:** c **O:** Isle of Man: !
H: On decayed wood of deciduous trees and rarely on conifers.
D: NM3: 199 **D+I:** CNE3: 526-533, B&K2: 134 129 **I:** Kriegl1: 226
Very common and widespread. Macroscopically variable. Possibly a species complex, as yet unresolved.

sibiricum (Parmasto) J. Erikss. & Å. Strid, *Corticiaceae of North Europe* 3: 535 (1975)
Radulomyces sibiricus Parmasto, *Conspectus Systematis Corticiacearum*: 223 (1968)
E: !
H: British material on adecayed fallen trunksof *Larix* sp. and *Fagus* in old woodland, on wet ground.
D: NM3: 199, Myc17(1): 42 **D+I:** CNE3: 534-537
Known only from North Hampshire (Hackwood Park Estate, near Basingstoke) and South Hampshire (New Forest, Mallard Wood).

transiens (Bres.) Parmasto, *Conspectus Systematis Corticiacearum*: 114 (1968)
Odontia transiens Bres., in Torrend, *Brotéria, Sér. Ci. Nat.* 11(1): 72 (1913)
E: !
H: On decayed wood or bark of deciduous trees. British material on *Fagus*.
British collections from North Somerset (near Bristol) and South Hampshire (New Forest: Denny Wood).

tsugae (Burt) J. Erikss. & Å. Strid, *Corticiaceae of North Europe* 3: 541 (1975)
Corticium tsugae Burt, *Ann. Missouri Bot. Gard.* 13: 276 (1926)
E: ! **S:** ?
H: On decayed conifer wood.
D: NM3: 200 **D+I:** CNE3: 540-541
Known from Hertfordshire, and reported from Scotland (South Aberdeen) but unsubstantiated with voucher material.

HYPHODERMELLA J. Erikss. & Ryvarden, *Corticiaceae of North Europe* 4: 579 (1976)
Stereales, Hyphodermataceae
Type: *Hyphodermella corrugata* (Fr.) J. Erikss. & Ryvarden

corrugata (Fr.) J. Erikss. & Ryvarden, *Corticiaceae of North Europe* 4: 579 (1976)
Grandinia corrugata Fr., *Hymenomyc. eur.*: 625 (1874)
Odontia corrugata (Fr.) Bres., *Atti Imp. Regia Accad. Roveretana, ser. 3,* 3(1): 98 (1897)
Odontia junquillea Quél., *Bull. Soc. Bot. France* 25: 290 (1878)
E: o **S:** !
H: On decayed wood of deciduous trees or shrubs such as *Acer pseudoplatanus*, *Buxus*, *Fagus* and *Ulex europaeus*. Also known on dead stems of large woody or herbaceous plants such as *Arctium lappa*, *Fallopia japonica*, *Gunnera macrophylla* and *Rubus fruticosus* agg.
D: NM3: 202 **D+I:** CNE4: 578-581, B&K2: 136 131
Known from North Somerset, South Devon, Surrey and West Norfolk. Widely reported in Scotland but most records unsubstantiated with voucher material.

HYPHODONTIA J. Erikss. [*nom. cons.*], *Symb. Bot. Upsal.* 16(1): 101 (1958)
Stereales, Hyphodermataceae
Grandinia Fr. [*nom. rej.*], *Epicr. syst. mycol.*: 527 (1838)
Lyomyces P. Karst. [*nom. rej.*], *Rev. Mycol. (Toulouse)* 3(9): 23 (1881)
Kneiffiella P. Karst. [*nom. rej.*], *Bidrag Kännedom Finlands Natur Folk* 48: 371 (1889)
Chaetoporellus Bondartsev & Singer [*nom. rej.*], *Mycologia* 36: 67 (1944)
Fibrodontia Parmasto, *Conspectus Systematis Corticiacearum*: 174 (1968)
Lagarobasidium Jülich, *Persoonia* 8(1): 84 (1974)
Rogersella Liberta & A.J. Navas, *Canad. J. Bot.* 56: 1777 (1978)
Type: *Hyphodontia pallidula* (Bres.) J. Erikss.

alutacea (Fr.) J. Erikss., *Symb. Bot. Upsal.* 16(1): 104 (1958)
Hydnum alutaceum Fr., *Syst. mycol.* 1: 417 (1821)
Odontia alutacea (Fr.) Quél., *Fl. mycol. France*: 434 (1888)

Kneiffiella alutacea (Fr.) Jülich & Stalpers, *Verh. Kon. Ned. Akad. Wetensch., Afd. Natuurk., Sect. 2,* 74: 134 (1980)
    Hydnum melleum Berk. & Broome, *Ann. Mag. Nat. Hist., Ser. 4 [Notices of British Fungi no. 1436]* 4: 31 (1875)
    Odontia mellea (Berk. & Broome) Rea, *Brit. basidiomyc.*: 638 (1922)
    Kneiffia stenospora P. Karst., *Hedwigia* 25: 231 (1886)
    Grandinia stenospora (P. Karst.) Jülich, *Int. J. Mycol. Lichenol.* 1(1): 36 (1982)
E: ! S: ! W: ! ROI: !
H: On decayed wood of deciduous trees (usually *Fagus*), rarely on conifers.
D: NM3: 212 **D+I:** CNE4: 598-603, B&K2: 106 85 (as *Grandinia stenospora*)
Rarely reported but apparently widespread. The majority of records are from southern England and Wales

alutaria (Burt) J. Erikss., *Symb. Bot. Upsal.* 16(1): 104 (1958)
    Peniophora alutaria Burt, *Ann. Missouri Bot. Gard.* 13: 332 (1926)
    Odontia alutaria (Burt) Nikol., *Novosti Sistematiki Nizshikh Rasteniĭ.* 168 (1964)
    Kneiffiella alutaria (Burt) Jülich & Stalpers, *Verh. Kon. Ned. Akad. Wetensch., Afd. Natuurk., Sect. 2,* 74: 129 (1980)
    Grandinia alutaria (Burt) Jülich, *Int. J. Mycol. Lichenol.* 1(1): 35 (1982)
E: c S: c W: o NI: ! ROI: o O: Channel Islands: o Isle of Man: o
H: On decayed wood of deciduous trees and conifers.
D: NM3: 211 **D+I:** CNE4: 604-607, B&K2: 100 76 (as *Grandinia alutaria*)
Common and widespread.

arguta (Fr.) J. Erikss., *Symb. Bot. Upsal.* 16(1): 104 (1958)
    Hydnum argutum Fr., *Syst. mycol.* 1: 424 (1821)
    Odontia arguta (Fr.) Quél., *Fl. mycol. France*: 435 (1888)
    Odontia arguta var. *alutacea* (Fr.) Bourdot & Galzin, *Bull. Soc. Mycol. France* 30(3): 265 (1914)
    Kneiffiella arguta (Fr.) Jülich & Stalpers, *Verh. Kon. Ned. Akad. Wetensch., Afd. Natuurk., Sect. 2,* 74: 129 (1980)
    Grandinia arguta (Fr.) Jülich, *Int. J. Mycol. Lichenol.* 1(1): 35 (1982)
    Hydnum stipatum Fr., *Syst. mycol.* 1: 425 (1821)
    Odontia stipata (Fr.) Quél., *Fl. mycol. France*: 435 (1888)
E: o S: ! W: !
H: Usually on decayed wood of deciduous trees such as *Fagus*, *Fraxinus* and *Ulmus* spp, and less frequently on conifers such as *Abies*, *Picea*, *Juniperus communis* and *Taxus*.
D: NM3: 212 **D+I:** CNE4: 608-611, B&K2: 100 77 (as *Grandinia arguta*)
Occasional but widespread in England. Rarely reported elsewhere.

aspera (Fr.) J. Erikss., *Symb. Bot. Upsal.* 16(1): 104 (1958)
    Grandinia aspera Fr., *Hymenomyc. eur.*: 627 (1874)
    Odontia aspera (Fr.) Bourdot & Galzin, *Hymenomyc. France*: 428 (1928)
E: ! W: !
H: On decayed wood of deciduous trees and conifers.
D: NM3: 213 **D+I:** CNE4: 612-617
Rarely reported. Known from widely scattered sites in England (Lincolnshire, Shropshire, South Lancashire and Surrey) and from Wales (Monmouthshire).

barbajovis (Bull.) J. Erikss., *Symb. Bot. Upsal.* 16(1): 104 (1958)
    Hydnum barbajovis Bull., *Herb. France*: pl. 481 f. 2 (1791)
    Sistotrema barbajovis (Bull.) Pers., *Mycol. eur.* 2: 200 (1825)
    Odontia barbajovis (Bull.) Fr., *Epicr. syst. mycol.*: 528 (1838)
    Kneiffiella barbajovis (Bull.) P. Karst., *Bidrag Kännedom Finlands Natur Folk* 48: 371 (1889)
    Grandinia barbajovis (Bull.) Jülich, *Int. J. Mycol. Lichenol.* 1(1): 35 (1982)
E: o S: ! W: o NI: ! ROI: !
H: On decayed wood of deciduous trees, especially *Betula* spp, *Corylus* and *Fagus*, and rarely on conifers such as *Picea abies*.

D: NM3: 211 **D+I:** CNE4: 618-622, B&K2: 100 78 (as *Grandinia barba-jovis*) **I:** Kriegl1: 232
Occasional and widespread, at least in England and Wales.

breviseta (P. Karst.) J. Erikss., *Symb. Bot. Upsal.* 16(1): 140 (1958)
    Kneiffia breviseta P. Karst., *Hedwigia* 25: 232 (1886)
    Odontia breviseta (P. Karst.) J. Erikss., *Fungi Exsiccati Suecici*: 22 (1953)
    Kneiffiella breviseta (P. Karst.) Jülich & Stalpers, *Verh. Kon. Ned. Akad. Wetensch., Afd. Natuurk., Sect. 2,* 74: 133 (1980)
    Grandinia breviseta (P. Karst.) Jülich, *Int. J. Mycol. Lichenol.* 1(1): 35 (1982)
E: ! S: ! W: !
H: Usually on decayed wood of conifers such as *Picea* and *Pinus* spp. Rarely on wood of deciduous trees such as *Acer pseudoplatanus* and also known on dead lianas of *Clematis vitalba*.
D: NM3: 213 **D+I:** CNE4: 624-627, B&K2: 102 79 (as *Grandinia breviseta*)
Rarely reported but apparently widespread.

crustosa (Pers.) J. Erikss., *Symb. Bot. Upsal.* 16(1): 104 (1958)
    Odontia crustosa Pers., *Observ. mycol.* 2: 16 (1799)
    Hydnum crustosum (Pers.) Pers., *Syn. meth. fung.*: 561 (1801)
    Grandinia crustosa (Pers.) Fr., *Epicr. syst. mycol.*: 528 (1838)
    Kneiffiella crustosa (Pers.) Jülich & Stalpers, *Verh. Kon. Ned. Akad. Wetensch., Afd. Natuurk., Sect. 2,* 74: 134 (1980)
E: ! S: ! W: ! NI: ! ROI: !
H: On decayed wood of deciduous trees and shrubs such as *Crataegus* spp, *Corylus*, *Fagus*, *Laburnum anagyroides*, *Salix* spp, *Sambucus nigra* and *Quercus* spp. Also on dead briars of *Rosa* spp, and rarely on conifers such as *Picea abies*.
D: NM3: 212 **D+I:** CNE4: 632-635, B&K2: 102 81 (as *Grandinia crustosa*)
Rarely reported but apparently widespread.

detritica (Bourdot) J. Erikss., *Symb. Bot. Upsal.* 16(1): 104 (1958)
    Peniophora detritica Bourdot, *Rev. Sci. du Bourb.* 23(1): 13 (1910)
    Hypochnicium detriticum (Bourdot) J. Erikss. & Ryvarden, *Corticiaceae of North Europe* 4: 701 (1976)
    Lagarobasidium detriticum (Bourdot) Jülich, *Persoonia* 10(3): 334 (1979)
    Odontia pruinosa Bres., *Ann. Mycol.* 18(1-3): 43 (1920)
    Lagarobasidium pruinosum (Bres.) Jülich, *Persoonia* 8(1): 84 (1974)
E: ! S: ! W: ! NI: !
H: On fallen and decayed woody detritus (bark, twigs etc.) of deciduous trees. Rarely on debris of *Picea abies*, decayed fronds of ferns such as *Pteridium* and on decayed leaves of *Carex* spp.
D: NM3: 204 (as *Hypochnicium detriticum*) **D+I:** CNE4: 700-703 (as *H. detriticum*), B&K2: 140 138 (as *Lagarobasidium detriticum*)
Rarely reported but apparently widespread in England and Wales. Single records from Scotland and Northern Ireland.

gossypina (Parmasto) Hjortstam, *Mycotaxon* 39: 416 (1990)
    Fibrodontia gossypina Parmasto, *Conspectus Systematis Corticiacearum*: 207 (1968)
    Mis.: *Odontia stipata* sensu B&G (1928), non sensu Fr. (1821: 425)
E: o
H: On decayed wood of deciduous trees, usually *Ulmus* spp, but also known on *Fagus*, *Fraxinus*, *Populus gileadensis* and *P. tremula*.
D: NM3: 211 **D+I:** CNE6: 1062-1065 (as *Fibrodontia gossypina*)
Rarely reported. Apparently widespread in southern England.

griseliniae (G. Cunn.) Langer, *Biblioth. Mycol.* 154: 120 (1994)
    Corticium griseliniae G. Cunn., *Bull. New Zealand Dept. Sci. Industr. Res.* 145: 71 (1963)

Rogersella griseliniae (G. Cunn.) Stalpers, *New Zealand J. Bot.* 23(2): 305 (1985)

Rogersella asperula Liberta & A.J. Navas, *Canad. J. Bot.* 56(15): 1780 (1978)

E: !

H: British material on decayed stems and debris of *Pteridium*.

D+I: Myc10(4): 154-155

A single record from South Devon (Bantham Ham) in 1995.

latitans (Bourdot & Galzin) Ginns & M.N.L. Lefebvre, *Mycol. Mem.* 19: 89 (1993)

Poria latitans Bourdot & Galzin, *Bull. Soc. Mycol. France* 41: 226 (1925)

Chaetoporellus latitans (Bourdot & Galzin) Singer, *Mycologia* 36: 67 (1944)

E: ! **S:** !

H: On decayed wood of deciduous trees, such as *Fagus* and *Fraxinus*.

D+I: CNE4: 562-563 (as *Chaetoporellus latitans*)

Known from England (South Hampshire: New Forest, Wood Crates) and Scotland (Outer Hebrides: Isle of Harris).

nespori (Bres.) J. Erikss. & Hjortstam, *Corticiaceae of North Europe* 4: 655 (1976)

Odontia nespori Bres., *Ann. Mycol.* 18(1 - 3): 43 (1928)

Grandinia nespori (Bres.) Cejp, *Mon. Hyd.*: 27 (1928)

Kneiffiella nespori (Bres.) Jülich & Stalpers, *Verh. Kon. Ned. Akad. Wetensch., Afd. Natuurk., Sect. 2,* 74: 134 (1980)

Mis.: *Hyphodontia papillosa* sensu auct.

E: o **S:** o **W:** o **NI:** ! **ROI:** ! **O:** Isle of Man: !

H: On decayed wood of conifers and less frequently deciduous trees.

D: NM3: 212 **D+I:** CNE4: 654-656, B&K2: 104 83 (as *Grandinia nespori*) **I:** Kriegl1: 236

Occasional but widespread.

pallidula (Bres.) J. Erikss., *Symb. Bot. Upsal.* 16(1): 104 (1958)

Gonatobotrys pallidula Bres., *Ann. mycol.* 1(1): 127 (1903)

Peniophora pallidula (Bres.) Bres., *Bull. Soc. Mycol. France* 28(4): 390 (1913)

Kneiffiella pallidula (Bres.) Jülich & Stalpers, *Verh. Kon. Ned. Akad. Wetensch., Afd. Natuurk., Sect. 2,* 74: 131 (1980)

Grandinia pallidula (Bres.) Jülich, *Int. J. Mycol. Lichenol.* 1(1): 36 (1982)

E: c **S:** c **W:** c **NI:** ! **ROI:** ! **O:** Isle of Man: o

H: Usually on decayed wood of conifers such as *Abies, Larix, Picea* and *Pinus* spp, and *Taxus*. Also known on wood of deciduous trees and on decayed polypores such as *Phellinus* spp.

D: NM3: 211 **D+I:** CNE4: 657-659, B&K2: 104 84 (as *Grandinia pallidula*)

Very common and widespread.

pruni (Lasch) Svrček, *Česká Mykol.* 27(4): 204 (1973)

Odontia pruni Lasch, in Rabenhorst, *Klotzschii Herbarium Vivum Mycologicum*: 1915 (1851)

Hyphoderma pruni (Lasch) Jülich, *Persoonia* 8(1): 80 (1974)

Grandinia pruni (Lasch) Jülich, *Int. J. Mycol. Lichenol.* 1(1): 36 (1982)

Hyphodontia bugellensis (Ces.) J. Erikss., *Symb. Bot. Upsal.* 16(1): 104 (1958)

E: ! **S:** ! **W:** !

H: On decayed wood of deciduous trees such as *Aesculus, Fagus* and *Salix* spp. Also known on dead lianas of *Clematis vitalba* and dead wood of *Viburnum opulus*.

D: NM3: 212 **D+I:** CNE4: 663-666

Rarely reported.

quercina (Pers.) J. Erikss., *Symb. Bot. Upsal.* 16(1): 104 (1958)

Odontia quercina Pers., *Observ. mycol.* 2: 17 (1799)

Sistotrema quercinum (Pers.) Pers., *Syn. meth. fung.*: 552 (1808)

Xylodon quercinum (Pers.) Gray, *Nat. arr. Brit. pl.* 1: 649 (1821)

Hydnum quercinum (Pers.) Fr., *Syst. mycol.* 1: 423 (1821)

Radulum quercinum (Pers.) Fr., *Epicr. syst. mycol.*: 525 (1838)

Kneiffiella quercina (Pers.) Jülich & Stalpers, *Verh. Kon. Ned. Akad. Wetensch., Afd. Natuurk., Sect. 2,* 74: 134 (1980)

Grandinia quercina (Pers.) Jülich, *Int. J. Mycol. Lichenol.* 1(1): 36 (1982)

Sistotrema fagineum Pers., *Syn. meth. fung.*: 554 (1801)

Hydnum fagineum (Pers.) Fr., *Syst. mycol.* 1: 423 (1821)

Radulum fagineum (Pers.) Fr., *Elench. fung.* 1: 152 (1828)

E: ! **S:** ! **W:** ! **NI:** !

H: On decayed wood of deciduous trees, often *Quercus* spp, but also on a wide range of other hosts such as *Alnus, Corylus, Fagus* and *Salix* spp.

D: NM3: 212 **D+I:** CNE4: 667-669

Rarely reported but apparently widespread.

rimosissima (Peck) Gilb., *Mycologia* 54: 667 (1962)

Odontia rimosissima Peck, *Rep. (Annual) New York State Mus. Nat. Hist.* 50(1): 114 (1898)

Thelephora papillosa Fr., *Elench. fung.* 1: 212 (1828)

Grandinia papillosa (Fr.) Fr., *Epicr. syst. mycol.*: 528 (1838)

Hyphodontia papillosa (Fr.) J. Erikss., *Symb. Bot. Upsal.* 16(1): 104 (1958)

Hyphodontia verruculosa J. Erikss. & Hjortstam, *Corticiaceae of North Europe* 4: 681 (1976)

Kneiffiella verruculosa (J. Erikss. & Hjortstam) Jülich & Stalpers, *Verh. Kon. Ned. Akad. Wetensch., Afd. Natuurk., Sect. 2,* 74: 133 (1980)

Grandinia verruculosa (J. Erikss. & Hjortstam) Jülich, *Int. J. Mycol. Lichenol.* 1(1): 36 (1982)

E: ! **S:** ! **W:** ! **NI:** !

H: Mostly on decayed wood of deciduous trees and shrubs such as *Buxus, Corylus, Prunus avium, Salix* spp and *Sambucus nigra*. Rarely on conifers such as *Larix, Picea, Pinus* spp and *Taxus*.

D: NM3: 213 **D+I:** CNE4: 680-682 (as *Hyphodontia verruculosa*)

Rarely reported but apparently widespread.

sambuci (Pers.) J. Erikss., *Symb. Bot. Upsal.* 16(1): 104 (1958)

Corticium sambuci Pers., *Neues Mag. Bot.* 1: 111 (1794)

Thelephora calcea var. *sambuci* (Pers.) Pers., *Syn. meth. fung.*: 581 (1801)

Thelephora sambuci (Pers.) Fr., *Syst. mycol.* 1: 152 (1821)

Lyomyces sambuci (Pers.) P. Karst., *Bidrag Kännedom Finlands Natur Folk* 37: 153 (1882)

Peniophora sambuci (Pers.) Burt, *Rep. (Annual) Missouri Bot. Gard.* 12: 233 (1925)

Hyphoderma sambuci (Pers.) Jülich, *Persoonia* 8(1): 80 (1974)

Rogersella sambuci (Pers.) Liberta & Navàs, *Canad. J. Bot.* 56(15): 1781 (1978)

Corticium chrysanthemi Plowr., *Trans. Brit. Mycol. Soc.* 2: 91 (1905) [1904]

Peniophora chrysanthemi (Plowr.) W.G. Sm., *Syn. Brit. Bas.*: 419 (1908)

Thelephora cretacea Fr., *Observ. mycol.* 1: 153 (1815)

Corticium serum (Pers.) Fr., *Hymenomyc. eur.*: 659 (1874)

E: c **S:** c **W:** c **NI:** c **ROI:** c **O:** Channel Islands: c

H: On living or dead wood and bark, often at the bases of living trunks of *Sambucus nigra* but known on a wide range of other deciduous hosts. Rarely on coniferous wood or brash.

D: NM3: 210 **D+I:** CNE4: 574-577 (as *Hyphoderma sambuci*), Ph: 240-241, B&K2: 142 139 (as *Lyomyces sambuci*) **I:** Kriegl1: 240

Very common and widespread.

spathulata (Schrad.) Parmasto, *Conspectus Systematis Corticiacearum*: 123 (1968)

Hydnum spathulatum Schrad. [non *H. spathulatum* (Schwein) Fr. (1828)], *Spicil. Fl. Germ.* 1: 178 (1794)

Irpex spathulatus (Schrad.) Fr., *Elench. fung.* 1: 146 (1828)

Kneiffiella spathulata (Schrad.) Jülich & Stalpers, *Verh. Kon. Ned. Akad. Wetensch., Afd. Natuurk., Sect. 2,* 74: 132 (1980)

Grandinia spathulata (Schrad.) Jülich, *Int. J. Mycol. Lichenol.* 1(1): 36 (1982)

E: !

H: On decayed wood. British material on burnt wood of *Pinus sylvestris* on heathland.

D: NM3: 213, Myc15(3): 134 **D+I:** CNE4: 670-673

A single collection, from South Hampshire (New Forest, Dunces Arch, near Lyndhurst) in 1999. Also listed by Rea (1922, as *Irpex spathulatus*), but no records have been traced.

subalutacea (P. Karst.) J. Erikss., *Symb. Bot. Upsal.* 16(1): 104 (1958)
> *Corticium subalutaceum* P. Karst., *Meddeland. Soc. Fauna Fl. Fenn.* 9: 65 (1882)
> *Kneiffia subalutacea* (P. Karst.) Bres., *Ann. Mycol.* 1(1): 104 (1903)
> *Peniophora subalutacea* (P. Karst.) Höhn. & Litsch., *Sitzungsber. Kaiserl. Akad. Wiss., Wien, Math.-Naturwiss. Cl., Abt. 1* 115: 601 (1906)
> *Kneiffiella subalutacea* (P. Karst.) Jülich & Stalpers, *Verh. Kon. Ned. Akad. Wetensch., Afd. Natuurk., Sect. 2,* 74: 131 (1980)

E: o **S:** ! **W:** !

H: On decayed wood of deciduous trees such as *Alnus, Betula, Fagus* and *Quercus* spp. Rarely on conifers such as *Abies, Larix, Picea* and *Pinus* spp.

D: NM3: 210 **D+I:** CNE4: 674-677, B&K2: 106 86 (as *Grandinia subalutacea*)

Uncommonly reported but apparently widespread.

HYPHOLOMA (Fr.) P. Kumm., *Führ. Pilzk.*: 21 (1871)

Agaricales, Strophariaceae
> *Naematoloma* P. Karst., *Meddeland. Soc. Fauna Fl. Fenn.* 5: 61 (1879)

Type: *Hypholoma fasciculare* (Huds.) P. Kumm.

capnoides (Fr.) P. Kumm., *Führ. Pilzk.*: 72 (1971)
> *Agaricus capnoides* Fr., *Syst. mycol.* 1: 289 (1821)
> *Naematoloma capnoides* (Fr.) P. Karst., *Bidrag Kännedom Finlands Natur Folk* 32: 495 (1879)

E: ! **S:** o **W:** ! **NI:** !

H: On decayed conifer wood.

D: BFF5: 12, NM2: 257, FAN4: 69 (as *Psilocybe capnoides*) **D+I:** Ph: 159, B&K4: 320 408 **I:** FlDan4: 144C, C&D: 358

Frequently reported and supposedly 'common' (BFF5: 12), but few collections have been traced in herbaria. *Hypholoma fasciculare* is common on conifers and may be misdetermined in the field as *H. capnoides*.

elongatum (Pers.) Ricken, *Blätterpilze Deutschl.*: 250 (1912)
> *Agaricus elongatus* Pers., *Syn. meth. fung.*: 383 (1801)
> *Psilocybe uda* var. *elongata* (Pers.) Gillet, *Hyménomycètes*: 586 (1878)
> *Psilocybe elongata* (Pers.) J.E. Lange, *Dansk. Bot. Ark.* 9(11): 30 (1936)
> *Hypholoma elongatipes* (Peck) A.H. Sm., *Mycologia* 33: 5 (1941)
> Mis.: *Psilocybe uda* sensu Rea (1922)

E: o **S:** o **W:** o **NI:** o **ROI:** o **O:** Isle of Man: !

H: Associated with *Sphagnum* in bogs and mires, on moorland and also in woodland.

D: BFF5: 16, NM2: 258, FAN4: 73 (as *Psilocybe elongata*) **D+I:** Ph: 160, B&K4: 322 409 **I:** C&D: 358

Occasional but widespread. Records with *Polytrichum* moss may represent *H. polytrichi*.

epixanthum (Fr.) Quél., *Mém. Soc. Émul. Montbéliard, Sér. 2,* 5: 113 (1872)
> *Agaricus epixanthus* Fr., *Epicr. syst. mycol.*: 222 (1838)

E: ! **NI:** ?

H: On decayed wood of deciduous trees.

D: BFF5: 12-13

Rarely recorded or collected. Known from southern England. Reported from Northern Ireland, but unsubstantiated with voucher material.

ericaeoides P.D. Orton, *Trans. Brit. Mycol. Soc.* 43(2): 273 (1960)

E: o **S:** o

H: On peaty or muddy soil, often with *Alnus* and *Salix* spp on wet heathland, and especially around the margins of dry or drying ponds and pools.

D: BFF5: 19-20, NM2: 258, NM2: 258, FAN4: 77 (as *Psilocybe ericaeoides*) **D+I:** B&K4: 322 410 **I:** C&D: 358

Occasional but widespread.

ericaeum (Pers.) Kühner, *Bull. Soc. Mycol. France* 52: 23 (1936)
> *Agaricus ericaeus* Pers., *Syn. meth. fung.*: 412 (1801)
> *Prunulus ericaeus* (Pers.) Gray, *Nat. arr. Brit. pl.* 1: 631 (1821)
> *Psilocybe ericaea* (Pers.) Quél., *Mém. Soc. Émul. Montbéliard, Sér. 2,* 5: 333 (1873)

E: o **S:** o **W:** ! **ROI:** ?

H: On acidic, peaty soil on heath and moorland and occasionally amongst grass in pastures on wet acidic soil.

D: BFF5: 20, NM2: 258, NM2: 258, FAN4: 76 (as *Psilocybe ericaeum*) **I:** C&D: 358

Occasional but widespread. Possibly under-recorded due to confusion with *H. ericaeoides* and *H. udum*

fasciculare (Huds.) P. Kumm., *Führ. Pilzk.*: 72 (1871)
> *Agaricus fascicularis* Huds., *Fl. angl.*, Edn 2, 2: 615 (1778)
> *Pratella fascicularis* (Huds.) Gray, *Nat. arr. Brit. pl.* 1: 627 (1821)
> *Naematoloma fasciculare* (Huds.) P. Karst., *Bidrag Kännedom Finlands Natur Folk* 32: 496 (1879)
> *Agaricus sadleri* Berk. & Broome, *Ann. Mag. Nat. Hist., Ser. 5 [Notices of British Fungi no. 1734]* 3: 203 (1879)
> *Clitocybe sadleri* (Berk. & Broome) Sacc., *Syll. fung.* 5: 163 (1887)

E: c **S:** c **W:** c **NI:** c **ROI:** c **O:** Channel Islands: c Isle of Man: c

H: On decayed wood of both deciduous and coniferous trees.

D: BFF5: 13, NM2: 256, FAN4: 68 (as *Psilocybe fascicularis* var. *fascicularis*) **D+I:** Ph: 158-159, B&K4: 322 411 **I:** Sow3: 285 (as *Agaricus fascicularis*), FlDan4: 144E

Very common and widespread. *Clitocybe sadleri* was a name given to sterile basidiomes in which the lamellae remain yellow. See Cooke 180 (127) Vol.2 (1881) for an illustration of this 'species'. English name = 'Sulphur Tuft'.

fasciculare *var.* **pusillum** J.E. Lange, *Dansk. Bot. Ark.* 40(4): 4 (1923)

E: ! **W:** !

H: In swarms, usually not caespitose, on decayed wood of deciduous and coniferous trees.

D: FAN4: 69

Previously considered a dwarf form without taxonomic value until work by Lamoure [*Sydowia* 36: 176-182 (1984)] showed these dwarfs to be biologically incompatible with the typical form. Known from England (South Hampshire, West Kent and West Sussex) and Wales (Carmarthenshire).

laeticolor (F.H. Møller) P.D. Orton, *Trans. Brit. Mycol. Soc.* 43(3): 176 (1960)
> *Naematoloma laeticolor* F.H. Møller, *Fungi Faeroes* 1: 192 (1945)

E: ! **S:** !

H: On acidic soils amongst grass in rough pasture.

D: BFF5: 17, FAN4: 74 (as *Psilocybe laeticolor*)

Known from England (Lincolnshire and Yorkshire) and Scotland (Easterness and Perthshire). Closely related to *H. elongatum* and growing in similar situations.

lateritium (Schaeff.) P. Kumm., *Führ. Pilzk.*: 72 (1871)
> *Agaricus lateritius* Schaeff. [non *A. lateritius* Fr. (1821)], *Fung. Bavar. Palat. nasc.* 1: 22 (1762)
> *Agaricus pomposus* Bolton, *Hist. fung. Halifax* 1: 5 (1788)
> *Hypholoma sublateritium* f. *pomposum* (Bolton) Massee, *Brit. fung.-fl.* 1: 381 (1892)
> *Hypholoma sublateritium* var. *pomposum* (Bolton) Rea, *Brit. basidiomyc.*: 261 (1922)
> *Hypholoma sublateritium* (Fr.) Quél., *Mém. Soc. Émul. Montbéliard, Sér. 2,* 5: 113 (1872)

Naematoloma sublateritium (Fr.) P. Karst, *Bidrag Kännedom Finlands Natur Folk* 32: 495 (1879)
 Agaricus sublateritius var. *schaefferi* Berk. & Broome, *Ann. Mag. Nat. Hist., Ser. 5 [Notices of British Fungi no. 1768]* 3: 204 (1879)
 Hypholoma sublateritium var. *schaefferi* (Berk. & Broome) Sacc., *Syll. fung.* 5: 1028 (1887)
 Agaricus sublateritius var. *squamosus* Cooke, *Ill. Brit. fung.* 573 (558) Vol. 4 (1885)
 Hypholoma sublateritium var. *squamosum* (Cooke) Sacc., *Syll. fung.* 5: 1028 (1887)
 Hypholoma sublateritium f. *vulgaris* Massee, *Brit. fung.-fl.* 1: 381 (1892)
E: o **S:** o **W:** o **NI:** o **ROI:** ! **O:** Channel Islands: !
H: On decayed wood, usually of deciduous trees and preferentially on or around large old stumps of *Quercus* spp. Rarely on wood of conifers but known on *Picea* and *Pseudotsuga* spp. Also reported 'on soil' or 'in grassland', but then probably arising from dead tree roots.
D: BFF5: 14 (as *Hypholoma sublateritium*), NM2: 257, FAN4: 70 (as *Psilocybe lateritia*) **D+I:** Ph: 159 (as *H. sublateritium*), B&K4: 326 416 (as *H. sublateritium*) **I:** FlDan4: 145D (as *H. sublateritium*), C&D: 358 (as *H. sublateritium*)
Occasional but widespread. Easily recognised by the bright brick-red (laterite) pilei. Several forms, characterised by the size or amount of veil fragments on the pileus, were at one time recognised but have no taxonomic status. Better known as *H. sublateritium*.

marginatum (Pers.) J. Schröt., in Cohn, *Kryptogamenflora Schlesien* 3: 571 (1889)
 Agaricus marginatus Pers., *Observ. mycol.* 1: 11 (1796)
 Agaricus dispersus Fr., *Epicr. syst. mycol.*: 222 (1838)
 Hypholoma dispersum (Fr.) Quél., *Mém. Soc. Émul. Montbéliard, Sér. 2,* 5: 113 (1872)
E: o **S:** o **W:** o **NI:** ! **ROI:** o
H: On acidic soil and decayed needle litter, and occasionally on decayed sawdust or woodchips in conifer woodland and plantations. Known with *Abies, Picea* and *Pinus* spp.
D: BFF5: 15, NM2: 256, FAN4: 72 (as *Psilocybe marginata*) **D+I:** Ph: 158-159, B&K4: 324 412 **I:** C&D: 358
Occasional but widespread. May be abundant where it does occur.

myosotis (Fr.) M. Lange, *Meddr Grønland* 148: 64 (1955)
 Agaricus myosotis Fr., *Observ. mycol.* 2: 34 (1818)
 Naucoria myosotis (Fr.) P. Kumm., *Führ. Pilzk.*: 77 (1871)
 Pholiota myosotis (Fr.) Singer, *Lilloa* 22: 517 (1951) [1949]
 Hemipholiota myosotis (Fr.) Bon, *Doc. Mycol.* 17(65): 52 (1986)
 Mis.: *Naucoria tenax* sensu Rea (1922)
E: o **S:** c **W:** o **NI:** ! **ROI:** o **O:** Isle of Man: !
H: On wet, peaty, acidic soil on moorland or in wet upland pasture amongst *Juncus* spp. Occasionally in boggy areas in woodland, such as willow or alder swamps.
D: BFF5: 22, NM2: 257, NM2: 257, FAN4: 106-107 (as *Pholiota myosotis*) **D+I:** Ph: 146 (as *P. myosotis*), B&K4: 324 413 **I:** Bon: 255 (as *Hemipholiota myosotis*), C&D: 354 (as *H. myosotis*)
Frequent and widespread in the right habitat, more so northwards and common in montane and island communities in Scotland.

polytrichi (Fr.) Ricken, *Blätterpilze Deutschl.*: 249 (1912)
 Agaricus polytrichi Fr., *Observ. mycol.* 1: 51 (1815)
 Psilocybe uda var. *polytrichi* (Fr.) Gillet, *Hyménomycètes*: 586 (1878)
 Psilocybe polytrichi (Fr.) A. Pearson & Dennis, *Trans. Brit. Mycol. Soc.* 31: 184 (1948)
E: ! **S:** o **W:** ! **ROI:** !
H: On soil amongst mosses such as *Polytrichum* and *Atrichum* spp on heathland, or in coniferous woodland.
D: BFF5: 18, NM2: 257, FAN4: 75 (as *Psilocybe polytrichi*) **D+I:** B&K4: 324 414
Rarely reported but apparently widespread. Most frequent in Scotland.

radicosum J.E. Lange, *Dansk. Bot. Ark.* 4(4): 39 (1923)
 Mis.: *Naematoloma epixanthum* sensu auct. p.p.
 Mis.: *Hypholoma epixanthum* sensu auct. p.p.
 Mis.: *Agaricus epixanthus* sensu auct. p.p
 Mis.: *Flammula inopus* sensu Cooke [Ill. Brit. fung. 484 (446) Vol. 4 (1884)]
E: ! **S:** ! **W:** ! **NI:** !
H: On decayed wood, usually large stumps, of conifers.
D: BFF5: 15, NM2: 256, NM2: 256, FAN4: 71 (as *Psilocybe radicosa*) **D+I:** FRIC2: 23-25 15a, B&K4: 326 415 **I:** FlDan4: 145E, C&D: 358
Rarely reported. A large and distinctive species, unlikely to be overlooked and thus probably genuinely rare. Most often reported from Scotland.

subericaeum (Fr.) Kühner, *Bull. Soc. Mycol. France* 52: 27 (1936)
 Agaricus subericaeus Fr., *Ic. Hymenomyc.* 2: pl. 36 (1884)
 Psilocybe subericaea (Fr.) Sacc., *Syll. fung.* 5: 1045 (1887)
 Hypholoma subericaeum f. *verrucosum* Kuhner, *Bull. Soc. Mycol. France* 52: 27 (1936)
 Mis.: *Psilocybe dichroa* sensu Lange (FlDan4: 82 & pl. 149B)
E: o **S:** o **W:** o **O:** Channel Islands: !
H: On base poor mud or 'black soil' around the edges of woodland pools, especially when partially dried out. Also known from marshes.
D: BFF5: 21, NM2: 258, NM2: 258, FAN4: 78 (as *Psilocybe subericaea*) **D+I:** Myc9(4): 157
Occasional but apparently widespread. Specimens with pronounced marginal veil fragments have been called *Psilocybe verrucosum* (*fide* Watling in BFF5, but the combination in *Psilocybe* has not been traced) but fall within the variation of the species.

udum (Pers.) Kühner, *Bull. Soc. Mycol. France* 52: 23 (1936)
 Agaricus udus Pers., *Syn. meth. fung.*: 414 (1801)
 Psilocybe uda (Pers.) Gillet, *Hyménomycètes*: 586 (1878)
 Hypholoma fulvidulum P.D. Orton, *Kew Bull.* 54(3): 705 (1999)
E: o **S:** c **W:** c **NI:** ! **ROI:** !
H: On wet peaty soil in bogs, on heathland and moorland, usually associated with *Sphagnum* spp.
D: BFF5: 22, NM2: 257, NM2: 257, FAN4: 79 (as *Psilocybe uda*) **D+I:** Ph: 158-159, B&K4: 326 417 **I:** C&D: 358
Occasional but widespread. Most frequent in northern areas.

xanthocephalum P.D. Orton, *Notes Roy. Bot. Gard. Edinburgh* 41(3): 586 (1984)
E: !
H: On soil, amongst grass, in open woodland with *Carpinus* and *Quercus* spp.
D: BFF5: 18, FAN4: 74 (as *Psilocybe xanthocephala*)
Rarely reported. Known only from southern England (Hertfordshire and South Hampshire).

HYPOCHNELLA J. Schröt., in Cohn, *Krypt.-Fl. Schlesien* 3: 402 (1888)
Stereales, Atheliaceae
Type: *Hypochnella violacea* J. Schröt.

violacea J. Schröt., in Cohn, *Krypt.-Fl. Schlesien* 3: 420 (1888)
 Hypochnus violaceus (J. Schröt.) Sacc., *Syll. fung.* 6: 659 (1888)
E: !
H: On decayed wood of deciduous trees, usually *Ulmus* spp, but also known on *Fraxinus*. Very rarely reported on conifers such as *Juniperus* spp and a single collection on a basidiome of *Cyathus striatus*.
D: NM3: 121 **D+I:** CNE4: 686-688, B&K2: 142 141
Rarely reported. Known from Cambridgeshire, Derbyshire, Hertfordshire, Middlesex, Oxfordshire, South Devon, Surrey, West Kent, West Sussex and Yorkshire.

HYPOCHNICIELLUM Hjortstam & Ryvarden, *Mycotaxon* 12: 176 (1980)

Stereales, Amylocorticaceae
Type: *Hypochniciellum ovoideum* (Jülich) Hjortstam & Ryvarden

molle (Fr.) Hjortstam, *Mycotaxon* 13(1): 125 (1981)
  Thelephora mollis Fr., *Syst. mycol.* 1: 443 (1821)
  Hypochnus mollis (Fr.) Fr., *Monogr. hymenomyc. Suec.* 2: 264 (1863)
  Corticium molle (Fr.) Fr., *Hymenomyc. eur.*: 660 (1874)
  Peniophora mollis (Fr.) Bourdot & Galzin, *Bull. Soc. Mycol. France* 28: 389 (1912)
  Leucogyrophana mollis (Fr.) Parmasto, *Eesti N.S.V. Tead. Akad. Toimet., Biol.* 16(4): 385 (1967)
E: ! **S:** !
H: On decayed wood and bark of conifers such as *Pinus* spp, with a predilection for the undersides of old fence planks lying on the ground in woodland.
D: NM3: 290 **D+I:** CNE4: 816-817 (as *Leucogyrophana mollis*), B&K2: 116 101, Myc17(1): 41
Known from widely scattered locations in England (Buckinghamshire, East & West Sussex, Hertfordshire, North Hampshire, South Lancashire, Surrey and West Kent) and a single record from Scotland (Orkney).

ovoideum (Jülich) Hjortstam & Ryvarden, *Mycotaxon* 12(1): 177 (1980)
  Leptosporomyces ovoideus Jülich, *Willdenowia. Beih.* 7: 203 (1972)
E: !
H: On decayed wood and leaves of deciduous trees.
D: NM3: 290 **D+I:** B&K2: 88 59
Described from Cambridgeshire (Madingley Wood) and otherwise known only from a single collection from Bedfordshire (Kings Wood) in 1983.

subillaqueatum (Litsch.) Hjortstam, *Mycotaxon* 13(1): 126 (1981)
  Corticium subillaqueatum Litsch., *Ann. Mycol.* 29(2-3): 128 (1941)
  Leucogyrophana subillaqueata (Litsch.) Jülich, *Persoonia* 8(1): 56 (1974)
E: !
H: On decayed conifer wood. British material on *Picea* and *Pinus* spp and *Taxus*.
D: NM3: 290 **D+I:** CNE4: 824-825 (as *Leucogyrophana subillaqueata*)
Known from East Sussex, South Devon, Surrey and West Cornwall.

HYPOCHNICIUM J. Erikss., *Symb. Bot. Upsal.* 16(1): 100 (1958)
Stereales, Hyphodermataceae
  Granulobasidium Jülich, *Persoonia* 10: 328 (1979)
Type: *Hypochnicium bombycinum* (Sommerf.) J. Erikss.

albostramineum (Bres.) Hallenb., *Mycotaxon* 24: 434 (1985)
  Hypochnus albostramineus Bres., *Ann. Mycol.* 1(1): 109 (1903)
  Corticium albostramineum (Bres.) Bourdot & Galzin, *Trans. Brit. Mycol. Soc.* 4: 118 (1913)
  Mis.: *Hypochnicium eichleri* sensu CNE 4 (1976) p.p.
E: ! **S:** !
H: On decayed wood of deciduous trees. British material on *Fagus*.
D: NM3: 203
Known from Surrey (Esher Common, and Mickleham) and a single collection from Scotland (Kirkudbrightshire). Reported from Northern Ireland (Down) in 1918 but unsubstantiated with voucher material.

bombycinum (Sommerf.) J. Erikss., *Symb. Bot. Upsal.* 16(1): 101 (1958)
  Thelephora bombycina Sommerf., *Suppl. Fl. lapp.*: 284 (1826)
  Hypochnus bombycinus (Sommerf.) Fr., *Monogr. hymenomyc. Suec.* 2: 264 (1863)

  Corticium bombycinum (Sommerf.) P. Karst., *Hedwigia* 32: 120 (1893)
E: o **S:** o **W:** !
H: On decayed wood of deciduous trees such as *Betula*, *Quercus, Salix* and *Ulmus* spp and rarely on *Pinus sylvestris*.
D: NM3: 203 **D+I:** CNE4: 696-699, B&K2: 136 132 **I:** Kriegl1: 243
Rarely reported. Apparently widespread.

cremicolor (Bres.) H. Nilsson & Hallenb., *Mycologia* 95(1): 57 (2003)
  Hypochnus cremicolor Bres., *Ann. Mycol.* 1(1): 109 (1903)
  Mis.: *Hypochnicium punctulatum* sensu CNE 4 (1976) and sensu NM3.
E: !
H: On decayed wood, and on soil in woodland.
Only recently separated as distinct from *H. punctulatum* (as the small-spored species in the complex).

eichleri (Bres.) J. Erikss. & Ryvarden, *Corticiaceae of North Europe* 4: 707 (1976)
  Peniophora eichleri Bres., *Syll. fung.* 16: 194 (1902)
  Gloeocystidium albostramineum ssp. *eichleri* (Bres.) Bres., *Hymenomyc. France*: 263 (1928)
E: o **S:** ?
H: On decayed wood of deciduous trees such as *Betula* spp, *Castanea* and *Fagus* and rarely on conifers such as *Pseudotsuga menziesii*. Occasionally found on damp soil around the bases of trees.
D+I: CNE4: 706-709
Rarely reported. Widespread in England, with only a single record from Scotland which is unsubstantiated with voucher material.

erikssonii Hallenb. & Hjortstam, *Windahlia* 18: 44 (1989) [1990]
  Peniophora sphaerospora sensu auct. p.p.
  Hypochnicium sphaerosporum sensu auct. p.p.
E: c **S:** ! **W:** ! **NI:** ! **ROI:** ! **O:** Channel Islands: !
H: On decayed wood of deciduous trees such as *Betula, Fagus* and *Quercus* spp, and also known on *Ulex europaeus*. Rarely on conifers such as *Pinus sylvestris*. Also known on damp soil at the bases of trees, or around the margins of dried out ponds and pools in woodland.
D: NM3: 204 **D+I:** CNE4: 726-729 (as *Hypochnicium sphaerosporum*), B&K2: 140 136 (as *H. sphaerosporum*)
Common at least in southern England, and widespread. Perhaps better known as *Hypochnicium sphaerosporum* sensu auct. p.p.

geogenium (Bres.) J. Erikss., *Symb. Bot. Upsal.* 16(1): 101 (1958)
  Corticium geogenium Bres., *Ann. Mycol.* 1(1): 99 (1903)
  Gloeocystidium geogenium (Bres.) S. Lundell, *Fungi Exsiccati Suecici.* 24 (1950)
  Corticium inaequale (Höhn. & Litsch.) Sacc. & Trotter, *Sitzungsber. Kaiserl. Akad. Wiss., Wien, Math.-Naturwiss. Cl., Abt. 1* 116: 826 (1907)
E: ! **S:** ! **W:** ! **NI:** !
H: On decayed wood of conifers such as *Pinus sylvestris* and *Picea* spp. Rarely on decayed wood of deciduous trees such as *Fagus* and debris of *Pteridium*.
D: NM3: 203 **D+I:** CNE4: 710-712, B&K2: 138 133
Rarely reported but apparently widespread.

lundellii (Bourdot) J. Erikss., *Symb. Bot. Upsal.* 16(1): 101 (1958)
  Corticium lundellii Bourdot, *Svensk bot. Tidskr.* 43(1): 56 (1949)
E: !
H: British material on soil and charred wood and a single collection on living roots of *Populus* sp.
D: 203 **D+I:** CNE4: 713-715, B&K2: 138 134 (poor !)
Known from London, North Hampshire and Surrey. Reported elsewhere but unsubstantiated with voucher material.

multiforme (Berk. & Broome) Hjortstam, *Windahlia* 23: 2 (1998) [1997]
>*Hydnum multiforme* Berk. & Broome, *Ann. Mag. Nat. Hist., Ser. 5 [Notices of British Fungi no. 1687]* 1: 24 (1878)
>*Odontia multiformis* (Berk. & Broome) Rea, *Brit. basidiomyc.*: 648 (1922)
>*Hypochnicium bombycinum* f. *pinicola* (S. Lundell) Parmasto, *Eesti N.S.V. Tead. Akad. Toimet., Biol.* 16: 384 (1967)
>*Corticium karstenii* Bres., *Ann. Mycol.* 9: 427 (1911)
>*Hypochnicium karstenii* (Bres.) Hallenb., *Mycotaxon* 16(2): 565 (1983)

E: ! **S:** !
H: On decayed wood of conifers such as *Picea* and *Pinus* spp.
D: NM3: 203 (as *Hypochnicium karstenii*)
Described from Scotland (Glamis) but only recently recognised as distinct from *Hypochnicium bombycinum*. Records of this latter species on conifer wood possibly belong here, and it may thus be not uncommon in Britain.

polonense (Bres.) Å. Strid, *Wahlenbergia* 1: 68 (1975)
>*Kneiffia polonensis* Bres., *Ann. Mycol.* 1(1): 102 (1903)
>*Peniophora polonensis* (Bres.) Höhn. & Litsch., *Ann. Mycol.* 4(3): 292 (1906)
>*Hyphoderma polonense* (Bres.) Donk, *Fungus* 27: 15 (1957)
>*Hyphodermopsis polonensis* (Bres.) Jülich, *Int. J. Mycol. Lichenol.* 1(1): 28 (1982)
>*Gyrophanopsis polonensis* (Bres.) Stalpers & P.K. Buchanan, *New Zealand J. Bot.* 29(3): 333 (1991)

E: ! **W:** !
H: On decayed, often sodden, wood of deciduous trees such Alnus, Fagus, Salix fragilis and Ulmus spp in damp woodland.
D: NM3: 203 **D+I:** CNE4: 716-720, B&K2: 142 140 (as *Hyphodermopsis polonensis*)
Rarely reported. Known from England (East Norfolk, Hertfordshire, North Hampshire, Shropshire, Staffordshire, Surrey, West Gloucestershire and West Sussex) and Wales (Breconshire and Monmouthshire).

punctulatum (Cooke) J. Erikss., *Symb. Bot. Upsal.* 16(1): 101 (1958)
>*Corticium punctulatum* Cooke, *Grevillea* 6: 132 (1878)
>*Gloeocystidium punctulatum* (Cooke) S. Lundell, *Fungi Exsiccati Suecici*: 25 (1950)
>*Peniophora sphaerospora* Höhn. & Litsch., *Sitzungsber. Kaiserl. Akad. Wiss., Wien, Math.-Naturwiss. Cl., Abt. 1* 115: 1600 (1906)
>*Hypochnicium sphaerosporum* (Höhn. & Litsch.) J. Erikss., *Symb. Bot. Upsal.* 16(1): 101 (1958)
>Mis.: *Hypochnicium eichleri* sensu CNE 4 (1976) p.p.

E: c **S:** ! **W:** ! **O:** Channel Islands: !
H: On decayed wood of deciduous and coniferous trees, also on woody litter, decayed polypore basidiomes, and dead stems of Calluna.
D: NM3: 203 **D+I:** CNE4: 721-725, B&K2: 138 135
Common and widespread in England. *Hypochnicium sphaerosporum* sensu auct. p.p. is the smooth-spored species, *H. erikssonii*.

subrigescens Boidin, *Cah. Maboké* 9(2): 90 (1971)
E: !
H: On decayed wood of deciduous trees. British material on *Quercus robur* in old woodland.
D: Myc17(1): 43
A single collection from South Hampshire (New Forest, Denny Wood) in 2000.

vellereum (Ellis & Cragin) Parmasto, *Conspectus Systematis Corticiacearum*: 116 (1968)
>*Corticium vellereum* Ellis & Cragin, *Bull. Washburn Coll. Lab. Nat. Hist.* 1: 66 (1885)
>*Granulobasidium vellereum* (Ellis & Cragin) Jülich, *Persoonia* 10(3): 328 (1979)

E: o
H: On wood, or most frequently decayed bark, of fallen trunks of deciduous trees. Most often collected on *Ulmus* spp, and

rarely *Acer pseudoplatanus* and *Quercus* spp. Fruiting mainly during the winter.
D: NM3: 203 **D+I:** CNE4: 730-733, Ph: 240-241, B&K2: 130 122 (as *Granulobasidium vellereum*)
Occasional but widespread in England and apparently increasing. The hymenial surface is often spotted with pale primrose-yellow areas due to colonies of the parasitic heterobasidiomycete *Filobasidiella lutea*.

wakefieldiae (Bres.) J. Erikss., *Symb. Bot. Upsal.* 16(1): 101 (1958)
>*Corticium wakefieldiae* Bres., *Ann. Mycol.* 18(1-3): 48 (1920)

E: !
H: On soil or wood (type material).
D+I: CNE4: 724-725 (as *Hypochnicium caucasicum*)
Described *'ad terram vel ligna - Brittania -* leg. Miss. Wakefield' (no location given).

HYPSIZYGUS Singer, *Mycologia* 39: 77 (1947)
Agaricales, Tricholomataceae
Type: *Hypsizygus tessulatus* (Bull.) Singer

tessulatus (Bull.) Singer, *Mycologia* 37: 77 (1947)
>*Agaricus tessulatus* Bull., *Herb. France*: pl. 513 (1791)
>*Pleurotus tessulatus* (Bull.) Gillet, *Hyménomycètes*: 342 (1876)
>*Pleurocybella tessulata* (Bull.) M.M. Moser, *Kleine Kryptogamenflora*, Edn 2: 62 (1955)
>Mis.: *Pleurotus ulmarius* sensu Pearson (1938)

E: !
H: On dead wood of deciduous trees, usually large standing or fallen trunks of *Ulmus* spp.
D: BFF6: 50 **I:** Cooke 272 (254) Vol.2 (1883)
Known from Herefordshire, Lincolnshire, London and Yorkshire. Not distinguished from *H. ulmarius* by some authorities.

ulmarius (Bull.) Redhead, *Trans. Mycol. Soc. Japan* 25(1): 3 (1984)
>*Agaricus ulmarius* Bull., *Herb. France*: pl. 510 (1791)
>*Pleuropus ulmarius* (Bull.) Gray, *Nat. arr. Brit. pl.* 1: 615 (1821)
>*Pleurotus ulmarius* (Bull.) P. Kumm., *Führ. Pilzk.*: 105 (1871)
>*Lyophyllum ulmarium* (Bull.) Kühner, *Bull. Soc. Linn. de Lyon* 7: 211 (1938)
>Mis.: *Pleurotus pantoleucus* sensu auct. Brit.

E: o **W:** ? **NI:** ? **O:** Channel Islands: !
H: Usually on large, dead, standing or fallen trunks of *Ulmus* spp in hedgerows or by roadsides. Also reported on *Aesculus, Populus nigra* and *Quercus* spp.
D: BFF6: 51, NM2: 129 **D+I:** FRIC17: 5-7 130 (as *Hypsizygus tessulatus*), B&K3: 230 272 **I:** Sow1: 67 (as *Agaricus ulmarius* [lower figs.]), FlDan2: 64C (as *Pleurotus ulmarius*), Bon: 167 (as *Lyophyllum ulmarium*), C&D: 211 (as *L. ulmarium*), Kriegl3: 277
Occasional to rare but widespread in England, apparently rare elsewhere. The illustration by Sowerby [Col. fig. Engl. fungi pl. 67 (1797)] illustrates a large basidiome with a strongly tesselated pileus and may represent *Hypsizygus tessulatus*.

HYSTERANGIUM Vittad., *Monogr. Tuberac.*: 13 (1831)
Phallales, Hysterangiaceae
Type: *Hysterangium clathroides* Vittad.

coriaceum R. Hesse, *Hypog. Deutschl.* 1: 101 (1891)
>Mis.: *Hysterangium nephriticum* sensu Rea (1922)

E: !
H: Hypogeous in sandy soil, associated with both deciduous trees and conifers.
D: NM3: 177 **D+I:** BritTruff: 190-191 12 C
A single collection by Rea from Worcestershire (Shrawley Wood) in 1912.

nephriticum Berk., *Ann. Mag. Nat. Hist., Ser. 1 [Notices of British Fungi no. 298]* 13: 350 (1844)

E: ! **O:** Channel Islands: !
H: Epigeous or shallowly hypogeous, usually around roots of *Fagus*.
D+I: BritTruff: 191-194 12 A
Rarely reported but apparently widespread in England and a single record from the Channel Islands (Guernsey) in 1901.

thwaitesii Berk. & Broome, *Ann. Mag. Nat. Hist., Ser. 2* [*Notices of British Fungi no. 377*] 2: 267 (1848)
E: !
H: Hypogeous in humic soil in deciduous woodland.
D+I: BritTruff: 194-195 12 B
Known from North Somerset, Northamptonshire and West Gloucestershire.

ILEODICTYON Tul., in Raoul, *Ann. Sci. Nat., Bot.,* sér. 3, 2: 114 (1844)
Phallales, Clathraceae
Type: *Ileodictyon cibarium* Tul.

cibarium Tul., in Raoul, *Ann. Sci. Nat., Bot.,* sér. 3, 2: 114 (1844)
Clathrus cibarius (Tul.) E. Fisch., *Jahrb. K. Bot. Gard. Berlin* 4: 74 (1886)
E: !
H: On disturbed soil, usually in gardens, but also (rarely) in clearings in deciduous woodland. Usually fruiting in the winter.
D+I: BritPuffb: 190-191 Figs. 153/154
Presumably an alien, native to Australia and New Zealand, established in a few scattered localities in southern England, from where first reported in 1955. Known from East Suffolk, Middlesex and Surrey.

INOCYBE (Fr.) Fr., *Monogr. hymenomyc. Suec.* 2: 346 (1863)
Cortinariales, Cortinariaceae
Clypeus (Britzelm.) Fayod, *Ann. Sci. Nat., Bot.,* sér. 7, 9: 362 (1889)
Astrosporina J. Schröt., in Cohn, *Krypt.-Fl. Schlesien* 3: 576 (1889)
Type: *Inocybe relicina* (Fr.) Quél.

acuta Boud., Bull. Soc. Mycol. France 33: 8 (1917)
Inocybe acutella Bon, *Doc. Mycol.* 6(24): 45 (1976)
Inocybe striata Bres., *Icon. Mycol.* 16: 1259 (1930)
Mis.: *Inocybe umboninota* sensu auct. Brit. & sensu Lange (FlDan3: 85) non Peck [*Rep.* (Annual) *New York State Mus. Nat. Hist.* 139: 58 (1910)]
E: ! **S:** ! **W:** !
H: On soil, associated with deciduous or coniferous trees.
D: NM2: 323 **D+I:** B&K5: 84 66 **I:** C&D: 321
Widespread but rarely reported. Some authorities restrict *I. acuta* to a small, dark agaric associated with *Salix* spp, and distinguish the more frequent conifer associate as *Inocybe striata*. Others refer these forms to *I. acutella* and *I. acuta* respectively.

adaequata (Britzelm.) Sacc., *Syll. fung.* 5: 767 (1887)
Agaricus adaequatus Britzelm., *Ber. Naturhist. Vereins Augsburg* 27: 154 (1883)
Inocybe jurana (Pat.) Sacc., *Syll. fung.* 5: 778 (1887)
Agaricus juranus Pat., *Tab. anal. fung.*: 23 (no. 551) (1886)
Inocybe rhodiola Bres., *Fungi trident.* 1: 80 (1884)
E: ! **W:** ? **NI:** ? **ROI:** ?
H: On soil (often calcareous loam) amongst litter in deciduous woodland. Usually associated with *Fagus* and *Corylus* but also reported with *Betula, Carpinus, Quercus* and *Tilia* spp.
D: Kuyp: 45-47, NM2: 325, NM2: 325 **D+I:** B&K5: 42 1 **I:** FlDan3: 117E (as *Inocybe jurana*), Ph: 148 (as *I. jurana*), Bon: 237, Bon: 237, Bon: 237 (as *I. rhodiola*), C&D: 311 (as *I. jurana*)
Rarely reported but apparently widespread in England from Lincolnshire and Nottinghamshire southwards. Reported from

Wales and Ireland but unsubstantiated with voucher material. Better known as *Inocybe jurana*.

agardhii (N. Lund) P.D. Orton, *Trans. Brit. Mycol. Soc.* 43(2): 177 (1960)
Agaricus agardhii N. Lund, *Consp. hymenomyc.*: 40 (1845)
Flammula agardhii (N. Lund) P. Karst., *Bidrag Kännedom Finlands Natur Folk* 32: 411 (1879)
Pholiota agardhii (N. Lund) M.M. Moser [*comb. inval.*], *Kleine Kryptogamenflora*: 196 (1953)
Mis.: *Inocybe squamata* sensu Pearson [TBMS 32: 260 (1949)]
E: ! **S:** ! **W:** ! **ROI:** !
H: On sand or sandy soil, usually associated with *Salix* spp in dunes, dune slacks, rarely inland.
D: NCL3: 274, NM2: 324 **I:** FlDan3: 122D (as *Flammula agardhii*), C&D: 311
Rarely collected or reported. Apparently widespread but many records from inland areas are unsubstantiated with voucher material.

albomarginata Velen., *České Houby* II: 379 (1920)
Inocybe ovalispora Kauffman, *North American Flora* 10(4): 248 (1924)
Inocybe reducta J.E. Lange, *Fl. agaric. danic.* 5, *Taxonomic Conspectus*: IV (1940)
E: !
H: On soil in deciduous woodland.
D: Kuyp: 186-188 **I:** C&D: 319
Known only from Buckinghamshire, Northamptonshire and West Sussex. Excludes *Inocybe ovalispora* sensu Kühner & Romagnesi (1953) and sensu British records, these being *Inocybe tjallingiorum*.

albovelutipes Stangl, *Z. Mykol.* 46(2): 166 (1980)
Mis.: *Inocybe subnudipes* sensu auct.
E: !
H: On soil in deciduous woodland.
D: Kuyp: 118-119
Known only from East Kent and West Sussex (Stedham Common).

amethystina Kuyper, *Persoonia* 3: 135 (1986)
E: !
D: Kuyp: 135-138 **D+I:** B&K5: 42 3
A single British collection from Surrey (Witley Common) in 2001. Some authorities consider *I. obscuroides* P.D. Orton to be an earlier name but Kuyper (Kuyp) found Orton's type to represent *I. cincinnata*.

appendiculata Kühner, *Bull. Soc. naturalistes Oyonnax* 9 (Suppl.): 4 (1955)
E: ! **S:** ?
H: On soil in deciduous woodland.
D: Kuyp: 75-77 **D+I:** B&K5: 44 4 **I:** C&D: 315
Accepted in NCL on the basis of a 1956 record from Nottinghamshire (Clumber Park) det. Orton. Reported since from elsewhere in England and Scotland but unsubstantiated with voucher material.

arenicola (R. Heim) Bon, *Doc. Mycol.* 12(48): 44 (1983) [1982]
Inocybe fastigiata f. *arenicola* R. Heim, *Encycl. Mycol.* 1. *Le Genre* Inocybe: 178 (1931)
E: ! **W:** !
H: On dunes with *Ammophila arenaria*.
D: Kuyp: 55-57 **I:** C&D: 311
Known from England (Cumbria) and Wales (Glamorgan) but easily overlooked.

assimilata Britzelm., *Ber. Naturhist. Vereins Augsburg* 26: 137 (1881)
Inocybe umbrina Bres., *Fungi trident.* 2: 50 (1894)
Astrosporina umbrina (Bres.) Rea, *Trans. Brit. Mycol. Soc.* 12(2-3): 211 (1927)
E: o **S:** o **W:** ! **NI:** ! **ROI:** !
H: On soil in woodland, associated with deciduous and coniferous trees.

D: NM2: 323 (as *Inocybe umbrina*) **D+I:** B&K5: 86 68 **I:** Ph: 153 (as *I. umbrina*), Bon: 243 (as *I. umbrina*), C&D: 321 (as *I. umbrina*)
Occasional but widespread.

asterospora Quél., *Bull. Soc. Bot. France* 26: 50 (1880) [1879]
    Astrosporina asterospora (Quél.) Rea, *Brit. basidiomyc.*: 210 (1922)
E: c **S:** o **W:** c **NI:** c **ROI:** o **O:** Channel Islands: !
H: On soil in woodland or parkland, associated with deciduous and coniferous trees.
D: NM2: 322 **D+I:** B&K5: 86 69 **I:** FlDan3: 117G, Bon: 243, C&D: 323
Common and widespread, perhaps slightly less so in northern areas.

bongardii (Weinm.) Quél., *Mém. Soc. Émul. Montbéliard, Sér. 2,* 5: 319 (1872)
    Agaricus bongardii Weinm., *Hymen. Gasteromyc.*: 190 (1836)
E: o **S:** ! **W:** ! **NI:** o
H: On soil (usually calcareous loam) in deciduous woodland. Usually associated with *Corylus* and *Fagus* but also reported with *Betula, Fraxinus, Quercus* and *Tilia* spp.
D: Kuyp: 39-41, NM2: 325 **D+I:** Ph: 149, B&K5: 44 6 **I:** FlDan3: 114E, Bon: 235, C&D: 313
Occasional but apparently widespread. Most records are from southern England.

calamistrata (Fr.) Gillet, *Hyménomycètes*: 513 (1876)
    Agaricus calamistratus Fr., *Syst. mycol.* 1: 256 (1821)
    Agaricus hirsutus Lasch, *Linnaea* 4: 546 (1829)
    Inocybe hirsuta (Lasch) Quél., *Mém. Soc. Émul. Montbéliard, Sér. 2,* 5: 178 (1872)
E: o **S:** ! **W:** ! **NI:** ! **ROI:** ! **O:** Isle of Man: !
H: On acidic soil, usually with conifers such as *Abies, Picea* and *Pinus* spp, and rarely with deciduous trees such as *Acer pseudoplatanus* and *Quercus* spp in mixed woodland.
D: Kuyp: 35-36 **D+I:** Ph: 148, B&K5: 46 8 **I:** Bon: 235, C&D: 313
Rarely reported but widespread. An unmistakable species. Excludes *Inocybe hirsuta* sensu NCL which is *I. cervicolor*.

calida Velen., *České Houby* II: 366 (1920)
    Inocybe brunneorufa Stangl & J. Veselský, *Česká Mykol.* 25(1): 5 (1971)
E: ? **S:** !
H: On soil in deciduous woodland.
D+I: B&K5: 88 71
A single collection in herb. K (originally as *I. brunneorufa*) from Scotland (Wester Ross: Kinlochewe) in 1963. Reported from England but unsubstantiated with voucher material.

calospora Quél., Bresadola, *Fungi trident.* 1: 19 (1881)
    Astrosporina calospora (Quél.) Rea, *Brit. basidiomyc.*: 211 (1922)
    Inocybe gaillardii Gillet, *Rev. Mycol. (Toulouse)* 5(17): 31 (1883)
    Astrosporina gaillardii (Gillet) Rea, *Brit. basidiomyc.*: 211 (1922)
E: ! **S:** ! **NI:** ! **ROI:** ?
H: On soil associated with *Corylus, Betula* and *Salix* spp, in damp areas in deciduous woodland.
D: NM2: 324 **D+I:** B&K5: 88 72 **I:** FlDan3: 115C & C1, Bon: 243, C&D: 321
Rarely reported but apparently widespread.

catalaunica Singer, *Collect. Bot. (Barcelona)* 1: 245 (1947)
    Inocybe leiocephala D.E. Stuntz, *Mycologia* 42: 98 (1950)
E: ? **S:** !
H: On soil, in limestone grassland, associated with *Dryas octopetala* (in the British collection).
D: Kuyp: 190-192 (as *Inocybe leiocephala*) **D+I:** B&K5: 64 34 (as *I. leiocephala*)
A single record of this boreal species from Wester Ross (Inchnadamph). Also reported from North Somerset (Leigh Woods) but unsubstantiated with voucher material.

cervicolor (Pers.) Quél., *Enchir. fung.*: 95 (1886)

    Agaricus cervicolor Pers., *Syn. meth. fung.*: 325 (1801)
    Mis.: *Inocybe hirsuta* sensu J. Lange, sensu NCL
E: ! **S:** ! **W:** ! **NI:** ! **ROI:** !
H: On soil in deciduous or mixed coniferous and deciduous woodland.
D: Kuyp: 38-39, NM2: 325 **D+I:** B&K5: 46 9 **I:** C&D: 313
Rarely reported but apparently widespread. Most of the records are unsubstantiated with voucher material.

cincinnata (Fr.) Quél., *Mém. Soc. Émul. Montbéliard, Sér. 2,* 5: 179 (1872)
    Agaricus cincinnatus Fr., *Syst. mycol.* 1: 256 (1821)
    Inocybe cincinnatula Kühner, *Bull. Soc. Naturalistes Oyonnax* 9 (Suppl.): 4 (1955)
    Inocybe conformata P. Karst., *Bidrag Kännedom Finlands Natur Folk* 48: 465 (1889)
    Inocybe obscuroides P.D. Orton, *Trans. Brit. Mycol. Soc.* 43(2): 276 (1960)
    Agaricus phaeocomis Pers., *Mycol. eur.* 3: 192 (1828)
    Inocybe phaeocomis (Pers.) Kuyper, *Persoonia* 3: 138 (1986)
E: o **S:** o **W:** o **NI:** o **ROI:** !
H: On soil, associated with deciduous trees and rarely with conifers in mixed woodland.
D: Kuyp: 138-139, NM2: 327 (as *Inocybe cincinnata*) **D+I:** B&K5: 48 10 (as *I. cincinnata*) **I:** Ph: 151 (as *I. cincinnata*), Bon: 239, C&D: 315 (as *I. cincinnata*)
Occasional but widespread.

cincinnata var. major (S. Petersen) Kuyper, *Z. Mykol.* 55(1): 114 (1989)
    Inocybe obscura var. major S. Petersen, *Dan. Agaric.*: 329 (1911)
    Inocybe phaeocomis var. major (S. Petersen) Kuyper, *Persoonia* 3: 140 (1986)
    Mis.: *Inocybe obscura* sensu auct.
E: o **S:** o **W:** ! **NI:** ! **ROI:** !
H: On soil (frequently calcareous loam) and decayed leaf litter, associated with deciduous and coniferous trees, but most often reported with *Fagus*.
D: Kuyp: 140-142, NM2: 327 (as *Inocybe obscura*) **D+I:** B&K5: 48 11 (as *I. cincinata* var. *major*) **I:** Bon: 239 (as *I. obscura*), C&D: 315 (as *I. obscura*)
Uncommon but widespread. Perhaps better known as *Inocybe obscura*.

concinnula J. Favre, *Ergebn. Wiss. Untersuch. Schweiz. Natn. Parks* 5: 201 (1955)
S: !
H: On soil in montane habitat.
Known only from Shetland.

cookei Bres., *Fungi trident.* 2(8-10): 17 (1892)
E: o **S:** o **W:** o **NI:** o **ROI:** o
H: On soil in deciduous or mixed coniferous and deciduous woodland.
D: Kuyp: 49-51, NM2: 325 **D+I:** Ph: 149, B&K5: 48 12 **I:** Bon: 235, C&D: 313
Occasional but widespread and may be locally common.

corydalina Quél., *Mém. Soc. Émul. Montbéliard, Sér. 2,* 5: 543 (1875)
E: o **S:** ! **W:** ! **NI:** ! **ROI:** !
H: On soil, usually associated with *Corylus* on calcareous loam soils, but also known with *Fagus* and *Quercus* spp.
D: Kuyp: 82-84, NM2: 326 **D+I:** B&K5: 49 13 **I:** Bon: 237, C&D: 313
Occasional but widespread. Most often recorded from southern England. N.B. The collection illustrated in Phillips (Ph: 149) has been re-examined and represents a different species, perhaps *Inocybe maculata*.

corydalina var. erinaceomorpha (Stangl & J. Veselský) Kuyper, *Persoonia* 12: 481 (1985)
    Inocybe erinaceomorpha Stangl & J. Veselský, *Česká Mykol.* 33(2): 72 (1979)
    Agaricus erinaceus Pers. [*nom. illegit.*, non *A. erinaceus* Fr. (1828)], *Mycol. eur.* 3: 191 (1828)

Inocybe pyriodora var. *scabra* Kühner [*nom. inval.*], *Flore Analytique des Champignons Supérieurs*: 220 (1953)
Mis.: *Inocybe scabra* sensu Lange (FlDan3: 71)
H: On soil in woodland, with *Fagus* and *Quercus* sp. (in the British collection).
D: Kuyp: 84-85 **I:** C&D: 313 (as *Inocybe erinaceomorpha*)
A single collection, from North Somerset (Higher Merridge) in 1957.

cryptocystis D.E. Stuntz, *Pap. Michigan Acad. Sci.* 39: 58 (1954) [1953]
E: !
H: On soil (along a roadside verge, in mixed woodland in the British collection).
D: Kuyp: 144-145, Myc14(1): 31
Known only from West Sussex (Crawley Down).

curreyi (Berk.) Sacc., *Syll. fung.* 5: 775 (1887)
Agaricus curreyi Berk., *Outl. Brit. fungol.*: 155 (1860)
Inocybe fastigiata var. *curreyi* (Berk.) R. Heim, *Encycl. Mycol. 1, Le Genre* Inocybe: 184 (1931)
E: !
H: On acidic soil in damp woodland.
I: C&D: 311, Cooke 428 (398) Vol.3 (1886)
The only British collection is the type from Northamptonshire (Fineshade Abbey) in 1859. Kuyper (1986) considered this a synonym of *Inocybe rimosa*, as earlier suggested by Pearson (1954) and NCL. Now, however, interpreted as a good species following Bon [Doc. Mycol. 27 (105): 35 (1997)].

curvipes P. Karst., *Hedwigia* 29: 176 (1890)
Inocybe boltonii R. Heim, *Encycl. Mycol. 1. Le Genre* Inocybe: 345 (1931)
Astrosporina boltonii (R. Heim) A. Pearson, *Trans. Brit. Mycol. Soc.* 26(1-2): 46 (1943)
Inocybe globocystis Velen., *České Houby* II: 368 (1920)
Astrosporina lanuginella J. Schröt., in Cohn, *Krypt.-Fl. Schlesien* 3: 577 (1889)
Inocybe lanuginella (J. Schröt.) Konrad & Maubl., *Icon. select. fung.* 6: 137 (1937)
Mis.: *Astrosporina decipientoides* sensu Pearson (1938)
Mis.: *Inocybe decipientoides* sensu auct. Eur.
Mis.: *Inocybe rickenii* sensu Heim (1931)
E: o **S:** ! **W:** ! **NI:** ! **ROI:** ! **O:** Isle of Man: !
H: On soil, associated with deciduous and, less frequently, coniferous trees.
D: NM2: 324 **I:** FlDan3: 117A (as *Inocybe lanuginella*), Bon: 243, C&D: 321
Occasional to rare but apparently widespread. This excludes *I. boltonii* sensu NCL and sensu auct. mult., for which see *I. subcarpta*.

decemgibbosa (Kühner) Vauras, *Karstenia* 37(2): 51 (1997)
Inocybe oblectabilis f. *decemgibbosa* Kühner, *Bull. Soc. Mycol. France* 49(1): 116 (1933)
E: !
H: On soil in deciduous woodland.
Rarely reported. Known from South Hampshire, Surrey and West Sussex, though some collections referred to *I. oblectabilis* may belong here.

dulcamara P. Kumm., *Führ. Pilzk.*: 79 (1871)
Agaricus dulcamarus Alb. & Schwein. [*nom. illegit.*, non *A. dulcamarus* Pers. (1801)], *Consp. fung. lusat.*: 489 (1805)
Inocybe delecta P. Karst., *Bidrag Kännedom Finlands Natur Folk* 32: 460 (1879)
Inocybe dulcamara f. *squamosoannulata* J. Favre, *Ergebn. Wiss. Untersuch. Schweiz. Natn. Parks* 5: 200 (1955)
Inocybe dulcamara var. *axantha* Kühner, *Bull. Soc. Mycol. France* 71: 169 (1956) [1955]
Inocybe dulcamara var. *homomorpha* Kühner, *Bull. Soc. Mycol. France* 71: 169 (1956) [1955]
E: o **S:** o **W:** ! **NI:** !
H: On sand or sandy soil in dunes, often with *Salix repens* and *Ammophila arenaria*. Also known from inland sites in damp woodland with *Betula* or *Salix* spp.

D: NM2: 324 **D+I:** Ph: 150-151, B&K5: 50 14 **I:** Bon: 235, C&D: 311
Occasional to rare, most frequent in damp areas in dunes.

dunensis P.D. Orton, *Trans. Brit. Mycol. Soc.* 43(2): 277 (1960)
Inocybe heimiana Bon, *Doc. Mycol.* 14(53): 39 (1984)
Mis.: *Inocybe decipiens* sensu Pearson [Naturalist (Hull) 1954: 132 (1954)]
Mis.: *Inocybe maritima* sensu Heim (1931)
E: ! **S:** ! **W:** !
H: On sand or sandy soil in dunes, often associated with *Salix repens* and *Ammophila arenaria* or in short turf.
D+I: B&K5: 90 73 **I:** C&D: 323
Rarely reported but widely scattered around the coast of Britain.

duriuscula Rea, *Trans. Brit. Mycol. Soc.* 3(1): 44 (1908) [1907]
Astrosporina duriuscula (Rea) Rea, *Brit. basidiomyc.*: 208 (1922)
Inocybe fibrosa var. *trivialis* J.E. Lange, *Dansk. Bot. Ark.* 2(7): 43 (1917)
Mis.: *Inocybe fibrosa* sensu auct. Brit.
Mis.: *Inocybe fibrosoides* sensu Pegler & Young (1972) and sensu NCL p.p.
E: ! **S:** ! **W:** !
H: On soil under deciduous trees, in woodland.
D+I: B&K5: 90 75 (as *Inocybe fibrosa*) **I:** FlDan3: 116B (as *I. fibrosa* var. *trivialis*)
Rarely reported and few of the records are substantiated with voucher material. Not reported from Wales since 1918.

erubescens A. Blytt, *Videnskabs-Selskabets Skrifter. I Math.-Naturv. Kl.* 6: 54 (1905) [1904]
Inocybe patouillardii Bres., *Ann. Mycol.* 3(2): 161 (1905)
Mis.: *Inocybe rubescens* sensu auct.
E: ! **ROI:** !
H: On soil (usually calcareous) in deciduous woodland, most frequently associated with *Fagus*, *Corylus* or *Betula* spp. A summer species, appearing as early as June but rarely lasting into autumn.
D: Kuyp: 44-45, NM2: 325 **D+I:** Ph: 149 (as *Inocybe patouillardii*), B&K5: 50 15 **I:** FlDan3: 115E (as *I. patouillardii*), Bon: 237 (as *I. patouillardii*), C&D: 311 (as *I. patouillardii*)
Rarely reported but widespread in England, from Durham southwards, most often recorded from southern counties, and a single report from Republic of Ireland. A highly toxic species.

fibrosoides Kühner, *Bull. Soc. Mycol. France* 49: 91 (1933)
E: ! **S:** ?
H: On soil in woodland, associated with deciduous trees.
I: C&D: 323
Known from Oxfordshire (Bix). Reported from Scotland but records are unsubstantiated with voucher material, and are possibly *Inocybe duriuscula*.

flavella P. Karst., *Meddeland. Soc. Fauna Fl. Fenn.* 16: 100 (1890)
Inocybe xanthocephala P.D. Orton, *Trans. Brit. Mycol. Soc.* 43(2): 277 (1960)
E: ! **S:** ! **W:** !
H: On soil in woodland, asssociated with deciduous or coniferous trees.
D: Kuyp: 53-55 **I:** C&D: 311
Rarely reported but apparently widespread.

flocculosa Sacc., *Syll. fung.* 5: 768 (1887)
Agaricus flocculosus Berk. [*nom. illegit.*, non *A. flocculosus* DC. (1815)], *Engl. fl.* 5(2): 97 (1836)
Inocybe gausapata Kühner, *Bull. Soc. Naturalistes Oyonnax* 9 (Suppl.): 4 (1955)
Inocybe geraniolens Bon & Beller, *Doc. Mycol.* 6(24): 45 (1976)
Inocybe subtigrina Kühner, *Bull. Soc. Naturalistes Oyonnax* 9 (Suppl.): 6 (1955)
Mis.: *Inocybe abjecta* sensu NCL [non sensu Kuyper (Kuyp)]
Mis.: *Inocybe deglubens* sensu Lange (FlDan3: 77)
Mis.: *Inocybe lucifuga* sensu Heim (1931) & sensu NCL
Mis.: *Inocybe pallidipes* sensu Lange (1938), sensu NCL

Mis.: *Inocybe tigrina* sensu auct.

E: o **S:** o **W:** ! **NI:** o **ROI:** ! **O:** Channel Islands: !

H: On soil in woodland, usually associated with deciduous trees, rarely with conifers

D: Kuyp: 159-163, NM2: 330 **D+I:** B&K5: 52 17 **I:** FlDan3: 113D (as *Inocybe pallidipes*), FlDan3: 111C, Ph: 153 (as *I. tigrina*), Bon: 241, C&D: 317

Occasional, but widespread. Intrepreted here, following Kuyper (Kuyp), as a single rather variable species. Listed under four different names in NCL and at least ten in Bon (1987).

fraudans (Britzelm.) Sacc., *Syll. fung.* 5: 778 (1887)
  Agaricus fraudans Britzelm., *Hymenomyc. Südbayern* 3: 5 (1882)
  Inocybe corydalina var. *albidopallens* J.E. Lange, *Fl. agaric. danic.* 5, *Taxonomic Conspectus*: IV (1940)
  Inocybe incarnata Bres., *Fungi trident.* 1: 49 (1884)
  Inocybe pyriodora ssp. *incarnata* (Bres.) Konrad & Maubl., *Icon. select. fung.* 1: pl. 94 (1927)
  Mis.: *Inocybe pyriodora* sensu auct. mult. & sensu NCL

E: o **S:** ! **W:** ! **NI:** ! **ROI:** !

H: On soil in woodland, usually with *Fagus* on calcareous soil but also known with *Betula*, *Corylus* and *Quercus* spp.

D: Kuyp: 78-80, NM2: 326 (as *Inocybe pyriodora*) **D+I:** Ph: 151 (as *I. pyriodora*), B&K5: 52 18 **I:** FlDan3: 110C (as *I. incarnata*), FlDan3: 110B (as *I. corydalina* var. *albidopallens*), Bon: 237, C&D: 313 (as *I. piriodora*)

Occasional but widespread. Most often recorded in beech woodland in southern England, much less frequently elsewhere.

fuligineoatra Huijsman, *Fungus* 25: 27 (1955)

E: !

H: On soil in deciduous woodland.

A single collection from West Gloucestershire (Forest of Dean) in 1999.

fulvella Bres., *Fungi trident.* 1(8-10): 16 (1881)
  Astrosporina fulvella (Bres.) Rea, *Brit. basidiomyc.*: 213 (1922)
  Inocybe scabella var. *fulvella* (Bres.) R. Heim: 322 (1931)

E: ? **S:** !

H: On soil in montane habitat.

Rarely reported. Known from several sites in Scotland and reported from England but in a lowland site and unsubstantiated with voucher material. Sensu Rea (1922) is probably *Inocybe petiginosa*.

furfurea Kühner, *Bull. Soc. Naturalistes Oyonnax* 9 (Suppl.): 4 (1955)

E: !

H: On soil in deciduous woodland. British material with *Quercus* sp.

D+I: Kuyp: 185-186

A single collection from Surrey (Cut Mill, near Elstead) in 2004.

fuscidula Velen., *České Houby* II: 378 (1920)
  Inocybe descissa var. *brunneoatra* R. Heim, *Encycl. Mycol.* 1. *Le Genre Inocybe*: 234 (1931)
  Inocybe brunneoatra (R. Heim) P.D. Orton, *Trans. Brit. Mycol. Soc.* 43(2): 177 (1960)
  Inocybe virgatula Kühner, *Bull. Soc. Naturalistes Oyonnax* 9 (Suppl.): 7 (1955)

E: ! **S:** ! **W:** ! **NI:** ! **ROI:** !

H: On soil in woodland, with deciduous and less frequently coniferous trees.

D: Kuyp: 153-156, NM2: 329 **D+I:** B&K5: 54 19 **I:** FlDan3: 113A (as *Inocybe descissa* var. *brunneoatra*), Bon: 241, C&D: 319

Occasional but apparently widespread. Here includes *Inocybe fuscidula* sensu stricto (= *I. brunneoatra* of NCL), a small, dark agaric mainly of deciduous woodland on calcareous soil in southern England, and *I. virgatula*, larger and paler, found on acidic soil in coniferous woodland, especially with *Picea* spp, in Scotland. Kuyper (Kuyp) finds that these intergrade in continental Europe.

fuscidula *var.* **bisporigera** Kuyper, *Persoonia, Supplement* 3: 156 (1986)
  Inocybe descissa f. *bisporigera* J.E. Lange [*nom. inval.*], *Fl. agaric. danic.* 5: 101 (1940)

E: !

H: On soil in deciduous woodland.

D: Kuyp: 156-157

Known in Britain only from a paratype collection from West Kent (Shoreham).

fuscomarginata Kühner, *Bull. Soc. Mycol. France* 71: 169 (1956) [1955]

E: !

H: On sandy soil. British material associated with *Salix* sp.

D+I: B&K5: 54 20

A single collection from West Sussex (Iping & Stedham Common) in 2000.

geophylla (Fr.) P. Kumm. [as *I. geophyllus*], *Führ. Pilzk.*: 78 (1871)
  Agaricus geophyllus Fr., *Syst. mycol.* 1: 258 (1821)
  Agaricus clarkii Berk. & Broome, *Ann. Mag. Nat. Hist., Ser. 4 [Notices of British Fungi no. 1345]* 11: 340 (1873)
  Inocybe clarkii (Berk. & Broome) Sacc., *Syll. fung.* 5: 784 (1887)
  Inocybe geophylla var. *alba* Hruby, *Hedwigia* 70: 277 (1930)

E: c **S:** c **W:** c **NI:** c **ROI:** c **O:** Channel Islands: ! Isle of Man: !

H: On soil in deciduous, and mixed deciduous and coniferous woodland, often growing with *Inocybe geophylla* var. *lilacina*.

D: Kuyp: 85-89, NM2: 327 **D+I:** Ph: 150-151, B&K5: 54 21 **I:** Sow2: 124 (as *Agaricus geophyllus*), Bon: 241, C&D: 317, Cooke 439 (429) Vol. 3 (1886) (as *Inocybe clarkii*)

Very common and widespread.

geophylla *var.* **lilacina** (Peck) Gillet, *Hyménomycètes*: 520 (1876)
  Agaricus geophyllus var. *lilacinus* Peck, *Rep. (Annual) New York State Mus. Nat. Hist.* 26: 90 (1874)
  Inocybe lilacina (Peck) Kauffman, *Michigan Geol. Biol. Surv. Publ., Biol. Ser.* 5, 26: 466 (1918)
  Agaricus geophilus Pers., *Syn. meth. fung.*: 340 (1801)
  Gymnopus geophilus (Pers.) Gray, *Nat. arr. Brit. pl.* 1: 608 (1821)
  Agaricus geophilus var. *violaceus* Pat., *Tab. anal. fung.* 6: 21 (1886)
  Inocybe geophylla var. *violacea* (Pat.) Sacc., *Syll. fung.* 5: 785 (1887)
  Mis.: *Agaricus geophyllus* sensu Sowerby (1799) non sensu Fr. (1821)

E: c **S:** c **W:** c **NI:** c **ROI:** c **O:** Channel Islands: ! Isle of Man: !

H: On soil in deciduous, and mixed deciduous and conifer woodland, often growing with *Inocybe geophylla*.

D: NM2: 327, Kuyp: 89-90 **D+I:** Ph: 150-151, B&K5: 56 22 **I:** FlDan3: 112E, Bon: 241, C&D: 317

Very common and widespread.

giacomi Bon, *Rivista Micol.* 35(1): 25 (1992)

S: !

H: On soil in montane habitat.

D+I: B&K5: 92 76

A single collection from Cairngorm.

glabrescens Velen., *České Houby* II: 373 (1920)
  Mis.: *Inocybe abietis* sensu NCL

E: ! **W:** !

H: On sandy soil, associated with coniferous trees.

D: Kuyp: 219-220

Single collections from England (Bedfordshire: Rowney Warren) and Wales (Glamorganshire: Whiteford Burrows).

glabripes Ricken, *Blätterpilze Deutschl.*: 107 (1911)
  Inocybe microspora J.E. Lange, *Dansk. Bot. Ark.* 2(7): 38 (1917)
  Inocybe parvispora Alessio [*nom. illegit.*], *Iconogr. mycol.*: 289 (1980)

E: ! **S:** ! **W:** ! **ROI:** !

H: On soil in deciduous woodland.

D: Kuyp: 124-126 **I:** Bon: 239
Rarely collected or reported but apparently widespread.

glabrodisca P.D. Orton, *Trans. Brit. Mycol. Soc.* 43(2): 279 (1960)
Mis.: *Inocybe oblectabilis* sensu NCL & sensu auct.
E: !
H: On soil, amongst mosses such as *Polytrichum* spp.
Known only from the type collection from Yorkshire (Malham) now in poor condition, and possibly not distinct from *I. proximella*. Excludes *I. glabrodisca* sensu Stangl (1989) which is *I. decemgibbosa*.

godeyi Gillet, *Hyménomycètes*: 517 (1876)
Inocybe godeyi var. *rufescens* Cooke, *Trans. Brit. Mycol. Soc.* 3: 110 (1909) [1908]
Inocybe rickenii Kallenb., *Pilz-und Kräuterfreund* 4: 192 (1921)
Inocybe rubescens Gillet, *Rev. Mycol. (Toulouse)* 5(17): 31 (1883)
E: ! **S:** ! **W:** ! **NI:** ! **ROI:** !
H: On soil, often calcareous loam, in deciduous woodland, often associated with *Fagus, Fraxinus* or *Corylus*.
D: Kuyp: 182-184, NM2: 327 **D+I:** Ph: 148, B&K5: 56 24 **I:** FlDan3: 112C (as *Inocybe rickenii*), Bon: 237, C&D: 319
Rarely reported but widespread. Most often collected in southern England.

grammata Quél., *Bull. Soc. Amis Sci. Nat. Rouen* Série II, 15: 164 (1880)[1879]
Mis.: *Inocybe hiulca* sensu Bresadola [Icon. mycol. 16: pl. 762, 1930)]
E: ! **S:** ! **W:** !
H: On soil in deciduous, or mixed deciduous and coniferous woodland
D+I: Ph: 153, B&K5: 92 77 **I:** C&D: 323
Rarely reported but apparently widespread.

griseolilacina J.E. Lange, *Dansk. Bot. Ark.* 2(7): 33 (1917)
E: o **S:** ! **W:** ! **NI:** ! **ROI:** !
H: On soil, usually in deciduous woodland and occasionally coniferous or mixed woodland, sometimes with *Hedera* in hedgerows, roadsides, or along woodland tracks.
D: Kuyp: 133-134, NM2: 326 **D+I:** Ph: 151, B&K5: 58 25 **I:** FlDan3: 111F, Bon: 239, C&D: 315
Occasional but widespread.

haemacta (Berk. & Cooke) Sacc., *Syll. fung.* 5: 763 (1887)
Agaricus haemactus Berk. & Cooke, *Grevillea* 11: 70 (1882)
Inocybe haemacta var. *rubra* Rea, *Trans. Brit. Mycol. Soc.* 4(1): 187 (1913)
E: ! **S:** ? **W:** !
H: On calcareous loam soil, in deciduous woodland, often associated with *Fagus* or *Corylus* and usually under dense cover of *Mercurialis perennis*.
D: Kuyp: 77-78, NM2: 327 **I:** FlDan3: 112B, Bon: 237, C&D: 313
Rarely reported. Mostly southern, but apparently widespread in England with records as far north as Yorkshire and Derbyshire. Reported from Scotland (Perthshire: Drumnadrochit) in 1908 but apparently not since then. The illustration B&K5: 58 pl. 27 appears to be *I. corydalina*.

heimii Bon, *Doc. Mycol.* 14(53): 11 (1984)
Mis.: *Inocybe caesariata* sensu Heim (1931) and sensu auct. mult.
E: ! **W:** ! **O:** Channel Islands: !
H: In sand or sandy soil, usually associated with *Pinus* spp, on dunes.
I: C&D: 311
Known from England (North Devon: Braunton Burrows), Wales (Anglesey: Newborough Warren) and the Channel Islands (Jersey).

hirtella Bres., *Fungi trident.* 1: 52 (1884)
Mis.: *Inocybe langei* sensu Lange (FlDan3: 78)
E: o **S:** ! **W:** ! **ROI:** !
H: On soil in deciduous woodland. Most often reported with *Fagus* and *Corylus*.

D: Kuyp: 195-197, NM2: 328 **D+I:** Ph: 151 **I:** Bon: 241, C&D: 319
Occasional but widespread in England, rarely reported elsewhere but apparently widespread.

hirtella *var.* **bispora** Kuyper, *Persoonia, Supplement* 3: 198 (1986)
Inocybe langei f. *bispora* J.E. Lange, *Dansk. Bot. Ark.* 9(6): 86 (1938)
E: !
H: On soil in woodland, associated with deciduous trees.
D: Kuyp: 198-199 **D+I:** B&K5: 60 28
Evidently more frequent than the typical (4-spored) variety in Britain, though mixed collections occur and some authorities do not regard this as an independent species.

huijsmanii Kuyper, *Persoonia, Supplement* 3: 134 (1986)
E: !
H: On soil in deciduous woodland. British material with *Fagus*.
D: Kuyp: 134-135
A single collection from West Gloucestershire (Frocester Hall).

humilis J. Favre, *Ergebn. Wiss. Untersuch. Schweiz. Natn. Parks* 6: 587 (1960)
E: !
H: On soil in montane habitat.
A single collection from Cumbria in 1997.

hystrix (Fr.) P. Karst., *Bidrag Kännedom Finlands Natur Folk* 32: 453 (1879)
Agaricus hystrix Fr., *Epicr. syst. mycol.*: 171 (1838)
E: ! **S:** o **W:** ! **NI:** ! **ROI:** !
H: On soil in woodland, associated with deciduous trees, rarely with conifers.
D: Kuyp: 130-132, NM2: 328 **D+I:** Ph: 152, B&K5: 60 29 **I:** FlDan3: (as *Inocybe hystrix* f. *minor*), Bon: 239, C&D: 315, Cooke 406 (424) Vol. 3 (1886)
Rarely reported. Most British records are from Scotland, with sporadic reports elsewhere.

impexa (Lasch) Kuyper, *Persoonia* 3: 106 (1986)
Agaricus impexus Lasch, *Linnaea* 4: 545 (1829)
Agaricus maritimus Fr. [*nom. illegit.*, non *A. maritimus* With. (1796)], *Observ. mycol.* 2: 51 (1818)
Inocybe maritima (Fr.) P. Karst, *Bidrag Kännedom Finlands Natur Folk* 32: 457 (1879)
Astrosporina maritima (Fr.) Rea, *Brit. basidiomyc.*: 212 (1922)
Mis.: *Inocybe halophila* sensu Pearson [TBMS 26: 45 (1943)] & sensu NCL p.p.
E: ? **S:** ! **W:** !
H: On sand or sandy soil on dunes.
D: Kuyp: 106-107, NM2: 328 (as *Inocybe maritima*)
Single records from Wales (Gower: Whiteford Burrows) in 1965 and Scotland (Morayshire: Culbin Sands) in 1939. Reported from England but unsubstantiated with voucher material.

inodora Velen., *České Houby* II: 373 (1920)
Mis.: *Inocybe albidodisca* sensu NCL [non sensu Reid in FRIC 6 (1972)]
E: ! **S:** ? **W:** !
H: On sandy soil, often associated with *Pinus* spp or *Salix repens* in dunes and on sandy heathland.
D: Kuyp: 171-174 **D+I:** B&K5: 62 31
Known from England (North Devon and Westmorland) and Wales (Anglesey). Reported from Scotland but unsubstantiated with voucher material

jacobi Kühner, *Bull. Soc. Mycol. France* 71: 170 (1956) [1955]
Agaricus rufoalbus Pat. & Doass., *Rev. Mycol. (Toulouse)* 8(29): 26 (1886)
Inocybe rufoalba (Pat. & Doass.) Sacc., *Syll. fung.* 5: 787 (1887)
Inocybe petiginosa f. *rufoalba* (Pat. & Doass.) R. Heim, *Encycl. Mycol. 1, Le Genre* Inocybe: 337 (1931)
E: ! **S:** !
H: On soil in woodland, associated with conifers.
D+I: B&K5: 92 78

Only two collections from England (Bedfordshire and West Kent) and a single collection from Scotland.

lacera (Fr.) P. Kumm., *Führ. Pilzk.*: 79 (1871)
   Agaricus lacerus Fr., *Syst. mycol.* 1: 257 (1821)
   Inocybe lacera f. *subsquarrosa* F.H. Møller, *Fungi Faeroes*: 226 (1945)
   Inocybe lacera f. *heterospora* Bon, *Doc. Mycol.* 28(110): 15 (1988)
E: ! **S:** ! **W:** ! **NI:** !
H: On sandy soil, associated with coniferous or deciduous trees, in woodland, heathland and also on dunes.
D: Kuyp: 98-101, NM2: 329 **D+I:** Ph: 152, B&K5: 62 33 **I:** Bon: 239, C&D: 317
Common and widespread.

lacera *var.* **helobia** Kuyper, *Persoonia* 3: 103 (1986)
   Inocybe lacera f. *gracilis* J.E. Lange, *Dansk. Bot. Ark.* 2(7): 32 (1917)
E: ! **S:** !
H: On soil in woodland.
D: Kuyp: 103-105 **D+I:** B&K5: 62 32 **I:** C&D: 317 (as *Inocybe lacera* f. *gracilis*.)
Known from England (Worcestershire) and Scotland (Orkney).

lacera *var.* **rhacodes** (J. Favre) Kuyper, *Persoonia* 3: 102 (1986)
   Inocybe rhacodes J. Favre, *Ergebn. Wiss. Untersuch. Schweiz. Natn. Parks* 5: 201 (1955)
S: ! **W:** !
H: On soil, associated with *Salix herbacea* in montane habitat.
D: Kuyp: 102-103 **D+I:** B&K5: 74 50 (as *Inocybe rhacodes*)
Reported from Scotland (several sites) and Wales.

langei R. Heim, *Encycl. Mycol. 1, Le Genre* Inocybe: 335 (1931)
E: ! **W:** !
H: On soil in woodland, associated with a wide range of deciduous trees.
D: Kuyp: 204-205, NM2: 328 **I:** C&D: 317
Known from England (South Wiltshire, Surrey, West Gloucestershire, West Kent, West Sussex and Worcestershire) and Wales (Breconshire). Reported elsewhere but unsubstantiated with voucher material.

lanuginosa (Bull.) P. Kumm., *Führ. Pilzk.*: 80 (1871)
   Agaricus lanuginosus Bull., *Herb. France*: pl. 370 (1788)
   Astrosporina lanuginosa (Bull.) J. Schröt., in Cohn, *Krypt.-Fl. Schlesien* 3: 577 (1889)
   Inocybe longicystis G.F. Atk., *Amer. J. Bot.* 5: 213 (1918)
   Inocybe lanuginosa var. *longicystis* (G.F. Atk.) Stangl & Enderle, *Z. Mykol.* 49(1): 120 (1983)
E: c **S:** c **W:** ! **NI:** ! **ROI:** ! **O:** Isle of Man: !
H: On soil (often acidic) in woodland, with coniferous and deciduous trees.
D: NM2: 323 **D+I:** B&K5: 94 79 **I:** Ph: 152 (as *Inocybe longicystis*), Bon: 243, C&D: 321
Frequent and widespread. Many older records are likely to refer to var. *ovatocystis* which was listed in NCL as *I. lanuginosa*.

lanuginosa *var.* **ovatocystis** (Boursier & Kühner) Stangl, *Hoppea.* 46: 288 (1989)
   Inocybe ovatocystis Boursier & Kühner, *Bull. Soc. Mycol. France* 44: 181 (1928)
   Mis.: *Inocybe lanuginosa* sensu NCL
   Mis.: *Astrosporina sabuletorum* sensu Rea (1922)
E: ! **S:** !
H: On soil in deciduous or coniferous woodland.
D+I: B&K5: 94 80
Rarely reported. Known from England (South Somerset, Surrey, West Sussex and Yorkshire) and Scotland (North Ebudes and Wester Ross).

leptocystis G.F. Atk., *Am. J. Bot.* 5: 212 (1918)
   Inocybe hygrophila J. Favre [*nom. inval.*], *Ergebn. Wiss. Untersuch. Schweiz. Natn. Parks* 6: 587 (1960)
E: ! **S:** !
H: On soil in montane habitat and also in woodland on limestone pavement.

D: Kuyp: 111-112
Known from Mid-Lancashire (Gait Barrows), Inverness-shire (Fort William) and Shetland.

leptophylla G.F. Atk., *Amer. J. Bot.* 5: 212 (1918)
   Inocybe casimiri Velen., *České Houby* II: 369 (1920)
   Inocybe lanuginosa var. *casimiri* (Velen.) R. Heim, *Encycl. Mycol. 1, Le Genre* Inocybe: 365 (1931)
E: ? **S:** ! **NI:** ?
H: On soil, associated with deciduous or coniferous trees.
D+I: B&K5: 94 81
Known from Scotland (Mid Ebudes: Isle of Mull). Reported from England (West Kent) and Northern Ireland, but unsubstantiated with voucher material.

maculata Boud., *Bull. Soc. Bot. France* 32: 283 (1885)
   Mis.: *Inocybe fastigiella* sensu auct. Eur.
   Mis.: *Inocybe maculata* f. *fastigiella* sensu auct. Eur.
E: o **S:** o **W:** o **NI:** ! **ROI:** o
H: On soil in deciduous woodland.
D: Kuyp: 52-53, NM2: 326 **D+I:** Ph: 150, B&K5: 64 36 **I:** Bon: 235, C&D: 311
Occasional but widespread, and may be locally common.

margaritispora (Cooke) Sacc., *Syll. fung.* 5: 781 (1887)
   Agaricus margaritispora Cooke [incorrectly as *A. margarispora*], *Handb. Brit. fung.*, Edn 2: 157 (1887)
   Astrosporina margaritispora (Cooke) Rea, *Brit. basidiomyc.*: 214 (1922)
   Inocybe phaeosticta Furrer-Ziogas, *Schweiz. Z. Pilzk.* 30: 11 (1952)
   Mis.: *Agaricus fastigiatus* sensu Berkeley (1860)
E: ! **W:** !
H: On soil in deciduous woodland.
D: NM2: 322 **D+I:** B&K5: 96 82 **I:** FlDan3: 198C, C&D: 323
Known from England (Bedfordshire, East Gloucestershire, Hertfordshire, North Somerset, Northamptonshire, Northumberland, Oxfordshire, South Devon, South Wiltshire, West Kent, and West Sussex) and Wales (Denbighshire and Glamorganshire)

melanopus D.E. Stuntz, *Pap. Michigan Acad. Sci.* 39: 68 (1954) [1953]
E: !
H: On soil in woodland, associated with conifers. British material with *Pinus sylvestris*.
D+I: Kuyp: 115-117
A single collection from Surrey (Cut Mill, near Elstead) in 2004.

mimica Massee, *Ann. Bot.* 18: 492 (1904)
E: !
H: On soil in deciduous woodland.
D: Kuyp: 58
Known in Britain only from the type collection, from Yorkshire (Castle Howard) in 1892.

mixtilis (Britzelm.) Sacc., *Syll. fung.* 5: 780 (1887)
   Agaricus mixtilis Britzelm., *Ber. Naturhist. Vereins Augsburg* 28: 152 (1885)
   Mis.: *Inocybe scabella* sensu Cooke (Handbook of British Fungi: 159, 1886) and sensu Pearson & Dennis (1948)
   Mis.: *Inocybe trechispora* sensu auct.
E: ! **S:** ! **W:** ! **ROI:** !
H: On soil with deciduous, or rarely coniferous trees in mixed woodland.
D: NM2: 322 **D+I:** B&K5: 96 83 **I:** Bon: 243, C&D: 323
Rarely reported but apparently widespread.

muricellata Bres., *Ann. Mycol.* 3(2): 160 (1905)
   Mis.: *Inocybe hirtella* sensu NCL p.p., sensu Lange (FlDan3: 78 & pl. 113G)
E: ! **S:** !
H: On soil in woodland, with deciduous or coniferous trees. Reported with *Fagus*, *Quercus* and *Pinus* spp.
D: Kuyp: 199-201
Known from southern England (Oxfordshire, Surrey and West Gloucestershire) and a single collection from Scotland (Peebleshire).

napipes J.E. Lange, *Dansk. Bot. Ark.* 2: 44 (1917)
  Astrosporina napipes (J.E. Lange) A. Pearson, *Trans. Brit. Mycol. Soc.* 22(1-2): 28 (1938)
E: c **S:** c **W:** c **NI:** c **ROI:** !
H: On soil in woodland, associated with deciduous, and less frequently, coniferous trees.
D: NM2: 323 **D+I:** B&K5: 96 84 **I:** Bon: 243, C&D: 321
Common and widespread.

nitidiuscula (Britzelm.) Sacc., *Syll. fung.* 1: 53 (1895)
  Agaricus nitidiuscula Britzelm., *Hymenomyc. Südbayern* 8: 7 (1891)
  Inocybe friesii R. Heim, *Encycl. Mycol. 1, Le Genre* Inocybe: 319 (1931)
E: ! **S:** ! **W:** ! **ROI:** !
H: On soil in deciduous, or mixed deciduous and coniferous woodland.
D: Kuyp: 150-152, NM2: 330 **D+I:** B&K5: 66 39 **I:** Bon: 241, C&D: 317
Rarely reported but apparently widespread. Most records are unsubstantiated with voucher material.

oblectabilis (Britzelm.) Sacc., *Syll. fung.* 1: 54 (1895)
  Agaricus oblectabilis Britzelm., *Ber. Naturhist. Vereins Augsburg* 30: 23 (1890)
  Inocybe hiulca f. *major* Bres., *Icon. Mycol.* 16: t.1263 (1930)
E: ! **S:** ! **W:** !
H: On soil in deciduous, or mixed coniferous and deciduous woodland. Reported with *Fagus, Quercus* and *Picea* spp.
Rarely reported but apparently widespread.

obscurobadia (J. Favre) Grund & D.E. Stuntz, *Mycologia* 69: 407 (1977)
  Inocybe furfurea var. *obscurobadia* J. Favre, *Ergebn. Wiss. Untersuch. Schweiz. Natn. Parks* 5: 200 (1955)
  Inocybe tenuicystidiata E. Horak & Stangl, *Sydowia* 33: 148 (1980)
  Mis.: *Inocybe leptocystis* sensu auct. Eur.
E: ! **S:** !
H: On soil in woodland, with deciduous trees or conifers. Reported with *Castanea, Fagus* and *Salix* spp, and also *Picea* spp and *Taxus*.
D: Kuyp: 112-114 **D+I:** B&K5: 68 40
Known from England (Bedfordshire, Berkshire, Mid & South Lancashire, Surrey, Warwickshire, West Sussex and Yorkshire) and Scotland.

ochroalba Bruyl., *Bull. Soc. Mycol. France* 85: 345 (1970) [1969]
E: ! **S:** ! **W:** ?
H: On sandy soil in mixed deciduous and coniferous woodland.
D: Kuyp: 206-209 **D+I:** B&K5: 68 41
Known from England (Surrey: Esher Common and Oxshott) and Scotland (Perthshire). Reported from Wales (Anglesey: Newborough Warren) but unsubstantiated with voucher material.

pallida Velen., *České Houby* II: 366 (1920)
S: !
H: On soil under conifers. British material amongst mosses (including *Sphagnum* sp.) in a *Picea* plantation.
D+I: B&K5: 98 86
A recent collection from Perthshire (Blackcraig Wood) in 2003.

paludinella (Peck) Sacc., *Syll. fung.* 5: 788 (1887)
  Agaricus paludinellus Peck, *Rep. (Annual) New York State Mus. Nat. Hist.* 31: 34 (1879)
  Mis.: *Inocybe trechispora* sensu Lange (FlDan3: 88 & pl. 118B) & sensu NCL
E: ! **S:** !
H: On soil in deciduous woodland.
D: NM2: 324 **I:** C&D: 321
Known from England (Bedfordshire, Northamptonshire and South Devon) and Scotland (Clyde Isles and Mid-Ebudes) but most records are unsubstantiated with voucher material.

pelargonium Kühner, *Bull. Soc. Naturalistes Oyonnax* 9 (Suppl.): 5 (1955)

E: ! **S:** ! **NI:** ! **ROI:** ?
H: On soil in woodland, with deciduous trees and conifers. Reported with *Fagus* and *Picea* spp.
D: Kuyp: 205-206 **D+I:** B&K5: 68 42
Known from England (Surrey, Warwickshire), Scotland (Perthshire) and Northern Ireland (Antrim). Reported from Republic of Ireland but unsubstantiated with voucher material.

perlata (Cooke) Sacc., *Syll. fung.* 5: 774 (1887)
  Agaricus perlatus Cooke, *Grevillea* 15: 40 (1886)
E: !
H: On soil in deciduous woodland, usually associated with *Corylus, Fagus* and *Fraxinus*.
D: NM2: 326 **D+I:** B&K5: 76 52 (as *Inocybe rimosa* var. *perlata*) **I:** FlDan3: 115F
Rarely reported. Considered by Kuyper (Kuyp) to be synonymous with *Inocybe rimosa*.

petiginosa (Fr.) Gillet, *Hyménomycètes*: 521 (1876)
  Agaricus petiginosus Fr., *Syst. mycol.* 1: 243 (1821)
  Hebeloma petiginosum (Fr.) P. Kumm., *Führ. Pilzk.*: 80 (1871)
  Astrosporina petiginosa (Fr.) Rea, *Brit. basidiomyc.*: 213 (1922)
E: o **S:** o **W:** o **NI:** ! **ROI:** !
H: On soil in deciduous or mixed coniferous and deciduous woodland. Usually reported with *Fagus* but known also with other deciduous trees and rarely with conifers.
D: NM2: 324 **D+I:** Ph: 154, B&K5: 98 87 **I:** FlDan3: 118A, Bon: 243, C&D: 321
Occasional but widespread. Basidiomes are small and somewhat cryptically coloured.

phaeodisca Kühner, *Bull. Soc. Naturalistes Oyonnax* 9 (Suppl.): 5 (1955)
  Mis.: *Inocybe descissa* sensu NCL
E: ! **S:** ! **W:** ! **NI:** ! **ROI:** !
H: On soil in woodland, associated with deciduous and coniferous trees
D: Kuyp: 122-124 **D+I:** B&K5: 70 44 **I:** Bon: 239, C&D: 315
Rarely reported but apparently widespread. The majority of records are unsubstantiated with voucher material.

posterula (Britzelm.) Sacc., *Syll. fung.* 5: 778 (1887)
  Agaricus posterulus Britzelm., *Hymenomyc. Südbayern* 3: 5 (1882)
  Inocybe xanthodisca Kühner, *Bull. Soc. Naturalistes Oyonnax* 9 (Suppl.): 7 (1955)
E: ! **S:** ! **W:** ! **NI:** ! **ROI:** !
H: On soil in woodland, associated with deciduous or coniferous trees.
D: Kuyp: 146-147, NM2: 329 **D+I:** B&K5: 70 45 **I:** FlDan3: 113E, C&D: 319
Rarely reported but apparently widespread.

praetervisa Quél., in Bresadola, *Fungi trident.* 1: 35 (1883)
  Astrosporina praetervisa (Quél.) J. Schröt., in Cohn, *Krypt.-Fl. Schlesien* 3: 576 (1889)
E: ! **S:** ! **W:** ! **ROI:** !
H: On soil in woodland, usually associated with deciduous trees, rarely with conifers. Reported with *Corylus, Fagus, Populus* and *Quercus* spp, and with *Calluna* in pinewoods.
D+I: Ph: 152-153, B&K5: 100 89 **I:** Bon: 243, C&D: 323
Rarely reported but apparently widespread.

proximella P. Karst., *Symb. mycol. fenn.* 9: 44 (1882)
  Astrosporina proximella (P. Karst.) Rea, *Brit. basidiomyc.*: 208 (1922)
E: ! **S:** ! **W:** ! **ROI:** !
H: On soil in woodland, often associated with *Quercus* spp, but also known with *Alnus, Betula* and *Pinus* spp.
D+I: B&K5: 100 90
Rarely recorded or collected. Apparently widespread but many collections are determined as 'compare with' *Inocybe proximella*.

pruinosa R. Heim, *Encycl. Mycol. 1, Le Genre* Inocybe: 245 (1931)

Inocybe albidodisca var. *reidii* Stangl & J. Veselský, *Česká Mykol.* 29: 70 (1975)

Mis.: *Inocybe albidodisca* sensu Reid [F.R.I.C. 6: 28 (1972)]

E: ! S: ! W: !

H: On sandy soil on dunes, often associated with *Salix repens*.

D: Kuyp: 169-171

Rarely reported but apparently widespread.

pseudoasterospora Kühner & Boursier, *Bull. Soc. Mycol. France* 48: 121 (1932)

Inocybe pseudoasterospora var. *microsperma* Kuyper & Keizer, *Persoonia* 14(4): 441 (1992)

E: ! S: ?

H: On soil in woodland, associated with conifers.

Known from England (East Kent: Bedgebury Pinetum and Yorkshire: Hovingham). Reported from Scotland (Dumfriesshire: Forest of Ae) but unsubstantiated with voucher material.

pseudodestricta Stangl & J. Veselský, *Česká Mykol.* 27: 19 (1973)

Mis.: *Inocybe virgatula* sensu NCL p.p.

E: ! NI: ?

H: On soil, in deciduous woodland. British material associated with *Fagus*.

D: Kuyp: 152-153

A single collection from Yorkshire (Chevin Forest Park) in 1996. Reported from Northern Ireland but unsubstantiated with voucher material.

pseudohiulca Kühner & Boursier, *Bull. Soc. Mycol. France* 48: 107 (1933)

E: !

H: On soil in woodland.

A single collection from Hertfordshire (Aldbury) in 1954, determined by Kuyper in 1984.

pseudoreducta Stangl & Glowinski, *Karstenia* 21(1): 30 (1981)

E: !

H: On soil in mixed woodland. British material on calcareous loam, with *Corylus* and *Fagus*.

D: Kuyp: 190 **D+I:** B&K5: 72 46

A single collection, from Surrey (West Horsley, Sheepleas) in 2001.

pusio P. Karst., *Bidrag Kännedom Finlands Natur Folk* 48: 465 (1889)

E: ! S: !

H: On soil, with deciduous trees, rarely amongst grass on lawns and then usually near to trees.

D: Kuyp: 147-150, NM2: 326 **D+I:** B&K5: 72 47 **I:** FlDan3: 112A, Bon: 239, C&D: 315

Rarely reported. Apparently widespread in England and reported from Scotland. The majority of records are unsubstantiated with voucher material.

putilla Bres., *Fungi trident.* 1: 81 (1884)

S: !

H: On soil in woodland. Associated with conifers in Britain, but known with *Corylus* in continental Europe.

D+I: B&K5: 102 91

Known only from Mid-Perthshire (Black Wood of Rannoch).

rennyi (Berk. & Broome) Sacc., *Syll. fung.* 5: 788 (1887)

Agaricus rennyi Berk. & Broome, *Ann. Mag. Nat. Hist., Ser. 5* 3: 205 (1879)

Astrosporina rennyi (Berk. & Broome) Rea, *Brit. basidiomyc.*: 212 (1922)

E: !

H: On soil in deciduous woodland.

Known only from the type collection from Herefordshire (Dinedor) in 1880. This collection contains spores of varied sizes and is possibly a teratological form of some better known species.

rimosa (Bull.) P. Kumm., *Führ. Pilzk.*: 78 (1871)

Agaricus rimosus Bull., *Herb. France*: pl. 388 (1789)

Gymnopus rimosus (Bull.) Gray, *Nat. arr. Brit. pl.* 1: 604 (1821)

Agaricus fastigiatus Schaeff., *Fung. Bavar. Palat. nasc.* 1: 13 (1762)

Inocybe fastigiata (Schaeff.) Quél., *Mém. Soc. Émul. Montbéliard, Sér. 2,* 5: 180 (1872)

Inocybe fastigiata f. *alpina* R. Heim, *Encycl. Mycol. 1, Le Genre Inocybe*: 185 (1931)

Inocybe fastigiata f. *argentata* Kühner, *Bull. Soc. Mycol. France* 71: 169 (1956) [1955]

Inocybe umbrinella Bres., *Ann. Mycol.* 3: 161 (1905)

Inocybe fastigiata var. *umbrinella* (Bres.) R. Heim, *Encycl. Mycol. 1, Le Genre Inocybe*: 188 (1931)

Inocybe obsoleta Romagn., *Bull. Soc. Mycol. France* 74: 145 (1958)

Inocybe rimosa var. *obsoleta* (Romagn.) Quadr. & Lunghini, *Quaderni di Accademia Nazionale dei Lincei* 264: 109 (1990)

Inocybe pseudofastigiata Rea, *Trans. Brit. Mycol. Soc.* 12(2-3): 210 (1927)

Agaricus schistus Cooke & W.G. Sm., *Handb. Brit. fung.*, Edn 2: 154 (1887)

Inocybe schista (Cooke & W.G. Sm.) Sacc., *Syll. fung.* 5: 774 (1887)

E: c **S:** c **W:** c **NI:** c **ROI:** c **O:** Isle of Man: !

H: On soil, in deciduous woodland, usually with *Fagus* and *Quercus* spp and rarely reported with conifers such as *Pinus* spp in mixed woodland.

D: Kuyp: 61-68, NM2: 326 (as *Inocybe fastigiata*) **D+I:** Ph: 150 (as *I. fastigiata*), B&K5: 74 51 (as *I. rimosa* var. *obsoleta*), B&K5: 76 53 **I:** Sow3: 323 (as *Agaricus rimosus*), FlDan3: 114B (as *I. fastigiata*), Bon: 235, C&D: 311

Common and widespread.

salicis Kühner, *Bull. Soc. Mycol. France* 77: 175 (1955)

Mis.: *Inocybe xanthomelas* sensu Lange (FlDan3: 84 & pl. 115B), non sensu NCL

E: ! S: ! NI: ?

H: On soil, usually associated with *Salix* spp in sand dunes and wet woodland, but also known in mixed woodland with *Betula* spp and *Fagus*.

D: NM2: 323 **I:** C&D: 323

Rarely reported but apparently widespread.

sambucina (Fr.) Quél., *Mém. Soc. Émul. Montbéliard, Sér. 2,* 5: 182 (1872)

Agaricus sambucinus Fr., *Syst. mycol.* 1: 257 (1821)

S: ! E: ?

H: On soil and needle litter under *Pinus* spp and *Fagus*.

D: Kuyp: 175-176, NM2: 327 **I:** Cooke 436 (399) Vol. 3 (1886)

Known from Angus (Edzell), Easterness (Curr Wood) and Mid-Perthshire (Black Wood of Rannoch). Reported from England (North Somerset) but unsubstantiated with voucher material. The illustration in B&K5 is doubtful.

serotina Peck, *Bull. New York State Mus. Nat. Hist.* 75: 17 (1904)

Inocybe devoniensis T.J. Wallace, *Trans. Brit. Mycol. Soc.* 43(2): 274 (1960)

Inocybe psammophila Bon, *Doc. Mycol.* 14 (53): 25 (1984)

E: ! W: !

H: On sandy soil, mainly on sand dunes, associated with *Salix repens*.

D: Kuyp: 167-169, NM2: 328, NM2: 328 (as *Inocybe devoniensis*) **I:** FlDan3: 111J, C&D: 317 (as *I. psammophila*)

Rarely reported.

sindonia (Fr.) P. Karst., *Bidrag Kännedom Finlands Natur Folk* 32: 465 (1879)

Agaricus sindonius Fr., *Epicr. syst. mycol.*: 176 (1838)

Inocybe commutabilis Furrer-Ziogas, *Schweiz. Z. Pilzk.* 30: 127 (1952)

Inocybe kuehneri Stangl & J. Veselský, *Česká Mykol.* 28: 199 (1974)

Agaricus muticus Fr., *Monogr. hymenomyc. Suec.* 2: 346 (1863)

Inocybe mutica (Fr.) Sacc., *Syll. fung.* 5: 769 (1887)

Mis.: *Inocybe eutheles* sensu NCL & sensu auct. mult.

E: c **S:** c **W:** c **NI:** c **ROI:** !
H: On soil in woodland, with coniferous and deciduous trees.
D: Kuyp: 177-180, NM2: 329 **D+I:** Ph: 152 (as *Inocybe eutheles*), B&K5: 78 55 **I:** FlDan3: 112F, Bon: 241 (as *I. kuehneri*), C&D: 319 (as *I. kuehneri*), C&D: 319 (as *I. euthetes*)
Common and widespread. Perhaps better known as *Inocybe euthetes*.

soluta Velen., *České Houby* II: 365 (1920)
   Inocybe brevispora Huijsman, *Fungus* 25: 23 (1955)
E: ! **S:** !
H: On soil in woodland, associated with deciduous and coniferous trees.
Rarely reported but apparently widespread. Possibly not distinct from *Inocybe proximella*.

splendens R. Heim, *Encycl. Mycol. 1, Le Genre* Inocybe: 328 (1932)
E: ? **W:** !
H: On soil in woodland. British material associated with deciduous trees in parkland.
D: Kuyp: 215-217 **D+I:** B&K5: 78 57 **I:** C&D: 319
A single collection from Pembrokeshire (Amroth, Colby Woodland Gardens) in 1993. Reported from England but unsubstantiated with voucher material.

splendens var. phaeoleuca (Kühner) Kuyper, *Persoonia* 3: 217 (1986)
   Inocybe phaeoleuca Kühner, *Bull. Soc. Naturalistes Oyonnax* 9 (Suppl.): 5 (1955)
   Mis.: *Inocybe brunnea* sensu NCL p.p.
E: ! **S:** !
H: On soil in deciduous woodland, usually associated with *Fagus* in Britain, but also reported with *Corylus* and *Populus* spp.
D: Kuyp: 217-218 **D+I:** B&K5: 78 56 **I:** Bon: 241 (as *Inocybe phaeoleuca*), C&D: 319 (as *I. phaeoleuca*)
Known from East Gloucestershire, East Norfolk, South Wiltshire, Surrey and West Sussex, and from South Aberdeenshire. Reported elsewhere but unsubstantiated with voucher material.

squamata J.E. Lange, *Dansk. Bot. Ark.* 2(7): 39 (1917)
E: ! **W:** ?
H: On soil usually in deciduous woodland, associated with *Betula*, *Populus*, *Quercus* and *Salix* spp, and rarely with conifers such as *Pinus* spp.
D: Kuyp: 59-60, NM2: 326 **I:** C&D: 311
Known from Bedfordshire, East Kent, London, Nottinghamshire, Surrey and Yorkshire. Reported elsewhere in England and Wales but unsubstantiated with voucher material.

squarrosa Rea, *Trans. Brit. Mycol. Soc.* 5(2): 250 (1915)
E: ! **W:** !
H: On damp soil, associated with *Alnus* and *Salix* spp.
D: Pers 2 132 **I:** Bon: 239, C&D: 315
Known from England (Cheshire, East Norfolk, North Somerset and Worcestershire) Scotland (Dumfriesshire) and Wales (Monmouthshire). Reported from East Sussex and Gloucestershire, but unsubstantiated with voucher material.

striatorimosa P.D. Orton, *Trans. Brit. Mycol. Soc.* 43(2): 279 (1960)
   Mis.: *Inocybe striata* sensu Pearson (1946)
E: !
H: On soil in mixed deciduous woodland.
Known only from Surrey (Hindhead, and Esher Common).

subcarpta Kühner & Boursier, *Bull. Soc. Mycol. France* 48: 137 (1932)
   Mis.: *Inocybe boltonii* sensu NCL & sensu Lange (FlDan3: 87 & pl. 116C)
   Mis.: *Inocybe carpta* sensu auct.
E: ! **S:** ! **W:** ! **NI:** !
H: On soil in woodland, usually reported with conifers such as *Pinus sylvestris*, *Picea* and *Pseudotsuga* spp, but also known with deciduous trees such as *Alnus* and *Salix* spp.
D: NM2: 323 (as *Inocybe boltonii*) **D+I:** B&K5: 102 93

Rarely collected or reported. Apparently widespread, but few records are substantiated with voucher material.

tabacina Furrer-Ziogas, *Schweiz. Z. Pilzk.* 30: 11 (1953)
E: ! **W:** !
H: On soil in woodland, associated with deciduous or coniferous trees.
British *fide* NCL and confirmed by Pegler & Young (1972) based on a collection by Orton from Somerset (Aisholt Wood). Few collections since then.

taxocystis (J. Favre) Senn-Irlet, *Bot. Helv.* 102(1): 55 (1992)
   Inocybe decipientoides var. *taxocystis* J. Favre, *Ergebn. Wiss. Untersuch. Schweiz. Natn. Parks* 5: 202 (1955)
S: !
H: On soil in montane habitat, associated with *Salix herbacea*.
D+I: B&K5: 104 95
Known only from Glen Affric and The Cairnwell in 1988.

tenebrosa Quél., *Compt. Rend. Assoc. Franç. Avancem. Sci.* 13: 279 (1885)
   Inocybe atripes G.F. Atk., *Amer. J. Bot.* 5: 210 (1918)
E: ! **NI:** !
H: On soil in deciduous woodland.
D: Kuyp: 209-210, NM2: 328 (as *Inocybe atripes*) **D+I:** FRIC6: 26-27 45a (as *I. atripes*), B&K5: 80 60 **I:** C&D: 319
Known from England (Bedfordshire, Berkshire, Nottinghamshire, West Kent and West Sussex) and Northern Ireland (Fermanagh: Crom Estate). Reported from Oxfordshire and North Somerset but unsubstantiated with voucher material.

tjallingiorum Kuyper, *Persoonia, Supplement* 3: 192 (1986)
   Mis.: *Inocybe ovalispora* sensu Kuhner & Romagnesi (1953)
E: ! **S:** !
H: On soil in deciduous woodland.
D: Kuyp: 192-194
Known from England (North Somerset: Higher Merridge near Bridgwater and Surrey: Ranmore near Dorking) and Scotland (Orkney Islands: Hoy, Berriedale).

umbratica Quél., *Compt. Rend. Assoc. Franç. Avancem. Sci.* 12: 500 (1884)
   Inocybe commixta Bres., *Fungi trident.* 1: 53 (1884)
   Mis.: *Astrosporina infida* sensu Rea *fide* NCL
E: ! **NI:** ?
H: On soil in woodland, associated with conifers. British material with *Picea* sp.
D+I: B&K5: 104 96 **I:** C&D: 321
A single collection from South Devon (Bellever Forest) in 1990. Reported from West Norfolk, West Sussex and Northern Ireland, but unsubstantiated with voucher material.

vaccina Kühner, *Bull. Soc. Naturalistes Oyonnax* 9 (Suppl.): 7 (1955)
E: !
H: On soil in deciduous woodland.
D: Kuyp: 218-219 **D+I:** B&K5: 84 64 **I:** Bon: 241
Known from Berkshire, Surrey and West Kent.

vulpinella Bruyl., *Bull. Soc. Mycol. France* 85: 341 (1970) [1969]
   Mis.: *Inocybe serotina* sensu M. Lange [Med. Groenl. 148: 17 (1957)]
E: ! **W:** ! **ROI:** ?
H: On sand in dunes, usually in damp areas such as dune slacks.
D: Kuyp: 180-182 **I:** C&D: 319
Known from England (Cheviotshire and North Devon) and Wales (Cardiganshire). Reported from the Republic of Ireland (South Kerry) but unsubstantiated with voucher material.

whitei (Berk. & Broome) Sacc., *Syll. fung.* 5: 790 (1887)
   Agaricus whitei Berk. & Broome, *Ann. Mag. Nat. Hist.*, Ser. 4 [Notices of British Fungi no. 1527] 17: 131 (1876)
   Agaricus geophyllus var. *lateritius* Berk. & Broome, *Ann. Mag. Nat. Hist.*, Ser. 4 [Notices of British Fungi no. 1234] 6: 466 (1870)

Inocybe geophylla var. *lateritia* (Berk. & Broome) W.G. Sm., *Syn. Brit. Bas.*: 141 (1908)
Inocybe pudica Kühner, *Ann. Sci. Franche-Comté* 2: 26 (1947)
Mis.: *Inocybe rubescens* sensu Lange (1938)
E: ! **S:** ! **W:** ! **NI:** ? **ROI:** ?
H: On soil in woodland, amongst needle litter of conifers.
D: Kuyp: 90-93, NM2: 327 (as *Inocybe pudica*) **D+I:** B&K5: 84 65 **I:** FlDan3: 112E (as *I. geophylla* var. *lateritia*), Bon: 237 (as *I. pudica*), C&D: 313 (as *I. pudica*)
Rarely reported but apparently widespread. Known from England (North Somerset, Shropshire, South Essex, South Lancashire, South Somerset, West Cornwall, West Sussex and Yorkshire), Scotland (Main Argyll) and Wales (Caernarvonshire and Monmouthshire). Reported from Ireland but unsubstantiated with voucher material.

xanthomelas Boursier & Kühner, *Bull. Soc. Mycol. France* 49: 84 (1933)
E: ! **S:** ! **NI:** !
H: On soil in deciduous or mixed deciduous and coniferous woodland. Reported with *Castanea* and *Betula* spp, and rarely with *Pinus sylvestris*.
I: C&D: 323
Known from England (Surrey: Kew Gardens and Witley Common, West Sussex: West Dean Park), Scotland (North Ebudes: Isle of Skye) and Northern Ireland (Antrim: O'Neill Estate).

INONOTUS P. Karst., *Meddeland. Soc. Fauna Fl. Fenn.* 5: 39 (1879)
Hymenochaetales, Hymenochaetaceae
Phaeoporus J. Schröt. [non *Phaeoporus* Bataille (1908)], in Cohn, *Krypt.-Fl. Schlesien* 3: 489 (1888)
Onnia P. Karst., *Bidrag Kännedom Finlands Natur Folk* 48: 326 (1889)
Type: *Inonotus hispidus* (Bull.) P. Karst.

cuticularis (Bull.) P. Karst., *Meddeland. Soc. Fauna Fl. Fenn.* 5: 39 (1880)
Boletus cuticularis Bull., *Herb. France*: pl. 462 (1790)
Polyporus cuticularis (Bull.) Fr., *Syst. mycol.* 1: 363 (1821)
E: o **W:** !
H: On fallen and living trunks (often in branch scars) of deciduous trees in old woodland. Usually on *Fagus* but also known on *Acer pseudoplatanus*, *Carpinus*, *Fraxinus*, *Quercus* and *Salix* spp.
D: EurPoly1: 319, NM3: 325 **D+I:** Ph: 230-231, B&K2: 250 301 **I:** Kriegl1: 433
Occasional but widespread in England. Reported from Wales (Caernarvonshire: Swallow Falls) in 1924 but unsubstantiated with voucher material.

dryadeus (Pers.) Murrill, *North American Flora* 9(2): 86 (1908)
Boletus dryadeus Pers., *Observ. mycol.* 2: 3 (1799)
Polyporus dryadeus (Pers.) Fr., *Syst. mycol.* 1: 374 (1821)
E: o **S:** ! **W:** o **NI:** o
H: On *Quercus* spp, usually *Q. petraea* and *Q. robur*, but also known on *Q. ilex* and *Q. frainetto*. Single collections in herb. K. supposedly on *Alnus* and *Fagus* but these hosts are unverified.
D: EurPoly1: 321, NM3: 325 **D+I:** Ph: 230-231, B&K2: 250 302 **I:** SV23: 16, Kriegl1: 434
Occasional but widespread. Basidiomes are annual, and often produced on the same part of the trunk over the course of many years.

hispidus (Bull.) P. Karst., *Meddeland. Soc. Fauna Fl. Fenn.* 5: 39 (1880)
Boletus hispidus Bull., *Herb. France*: pl. 210 (1785)
Polyporus hispidus (Bull.) Fr., *Syst. mycol.* 1: 362 (1821)
Boletus spongiosus Lightf., *Fl. scot.*: 1033 (1777)
Poria spongiosa (Lightf.) Gray, *Nat. arr. Brit. pl.* 1: 640 (1821)
Boletus velutinus With. [non *B. velutinus* Pers. (1801)], *Arr. Brit. pl. ed. 3*, 4: 331 (1796)
Boletus villosus Huds., *Fl. angl.*, Edn 2, 2: 626 (1778)

E: c **S:** o **W:** ! **NI:** ! **ROI:** !
H: Saprotrophic or weakly parasitic. On *Fraxinus* and *Malus* spp including cultivars. Very rarely on *Platanus* spp, *Quercus* spp, *Sophora japonica*, *Sorbus torminalis*, *Tilia* and *Ulmus* spp.
D: EurPoly1: 327, NM3: 325 **D+I:** Ph: 230-231, B&K2: 252 304 **I:** Sow3: 345 (as *Boletus velutinus*), Kriegl1: 437
Common in England, especially so in southern parts, occasional but widespread elsewhere.

nodulosus (Fr.) P. Karst., *Bidrag Kännedom Finlands Natur Folk* 37: 73 (1882)
Polyporus nodulosus Fr., *Epicr. syst. mycol.*: 474 (1838)
Polystictus nodulosus (Fr.) Cooke, *Grevillea* 14: 82 (1886)
Polyporus polymorphus Rostk., *Deutschl. Fl.*, Edn 3: 115 (1838)
Polystictus polymorphus (Rostk.) Cooke, *Grevillea* 14: 87 (1886)
E: ! **S:** !
H: On dead wood (usually fallen branches) of *Fagus* in old woodland.
D: EurPoly1: 333, NM3: 325 **D+I:** B&K2: 252 305 **I:** SV23: 12, Kriegl1: 438
Rarely reported. Known from widely scattered locations in southern England (Buckinghamshire, Hertfordshire, North Wiltshire, South Hampshire and West Kent) and Scotland (Midlothian, Peebleshire, and Stirlingshire). Reports on *Betula* spp, are dubious and unsubstantiated with voucher material. They may refer to resupinate basidomes of *Inonotus radiatus* which is frequent on that host.

obliquus (Fr.) Pilát, *Atlas champ. Eur.* 3(1): 572 (1942)
Polyporus obliquus Fr., *Syst. mycol.* 1: 378 (1821)
Poria obliqua (Fr.) P. Karst., *Rev. Mycol. (Toulouse)* 3: 19 (1881)
E: ! **S:** c
H: On living trunks of *Betula* spp. Very rarely reported on *Alnus*, *Malus* and *Platanus* spp but these hosts are unverified.
D: EurPoly1: 335, NM3: 324 **D+I:** B&K2: 252 306 **I:** SV23: 13, Kriegl1: 439
Common in Scotland, rare in England. Produces large, sterile, rock-hard, black 'conks' on the trunks of infected trees. Basidiomes are formed under peeling bark or in cracks on the surface of the conks, but are rarely encountered, being rapidly destroyed by insects or decay.

radiatus (Sowerby) P. Karst., *Bidrag Kännedom Finlands Natur Folk* 37: 73 (1882)
Boletus radiatus Sowerby, *Col. fig. Engl. fung.* 2: pl. 196 (1799)
Polyporus radiatus (Sowerby) Fr., *Syst. mycol.* 1: 369 (1821)
Polystictus radiatus (Sowerby) Cooke, *Grevillea* 14: 82 (1886)
E: c **S:** c **W:** c **NI:** c **ROI:** ! **O:** Isle of Man: !
H: Usually on dead standing trunks of *Alnus* and sometimes *Betula* spp. Also reported on *Carpinus*, *Corylus*, *Fagus*, *Fraxinus*, *Populus tremula*, *Quercus*, *Salix*, *Sambucus nigra*, *Syringa vulgaris*, *Ulex europaeus* and *Ulmus* spp, but these hosts are unverified.
D: EurPoly1: 337, NM3: 325 **D+I:** Ph: 230-231, B&K2: 254 307 **I:** Sow2: 196 (as *Boletus radiatus*), Bon: 321, Kriegl1: 441
Very common and widespread.

IRPICODON Pouzar, *Folia Geobot. Phytotax.* 1: 371 (1966)
Stereales, Amylocorticaceae
Type: *Irpicodon pendulus* (Alb. & Schwein.) Pouzar

pendulus (Alb. & Schwein.) Pouzar, *Folia Geobot. Phytotax.* 1: 371 (1966)
Sistotrema pendulum Alb. & Schwein., *Consp. fung. lusat.*: 261 (1805)
Hydnum pendulum (Alb. & Schwein.) Fr., *Syst. mycol.* 1: 413 (1821)
Irpex pendulus (Alb. & Schwein.) Fr., *Elench. fung.* 1: 143 (1828)
S: !
H: On dead standing trunks of *Pinus sylvestris*.

D: NM3: 286 **D+I:** CNE4: 739-743
Known from two collections (with voucher material) from
Ayrshire (Castle Semple) in 1831 and possibly now extinct.
Frequent in Scandinavia.

ISCHNODERMA P. Karst., *Meddeland. Soc. Fauna Fl. Fenn.* 5: 38 (1879)
Poriales, Coriolaceae
Type: *Ischnoderma resinosum* (Fr.) P. Karst.

benzoinum (Wahlenb.) P. Karst., *Acta Soc. Fauna Fl. Fenn.* 2: 32 (1881)
Boletus benzoinus Wahlenb., *Fl. Suec.,* Edn 1: 1076 (1826)
Polyporus benzoinus (Wahlenb.) Fr., *Elench. fung.* 1: 100 (1828)
E: o **S:** !
H: In woodland or plantations, on dead and decayed wood of
conifers such as *Abies, Araucaria, Picea, Pinus* and
Pseudotsuga spp, often covering large fallen trunks.
D: EurPoly1: 355, NM3: 224 **D+I:** B&K2: 314 398
Generally uncommon, but increasingly recorded in southern and
south-eastern England. Very similar to *Ischnoderma resinosum*
with few distinct differences. Immature basidiomes have a
soft, sappy consistency (the 'leptoporoid' stage) later becoming
hard and fertile (the 'fomitoid' stage).

resinosum (Fr.) P. Karst., *Meddeland. Soc. Fauna Fl. Fenn.* 5: 38 (1880)
Polyporus resinosus Fr., *Syst. mycol.* 1: 361 (1821)
Polyporus resinosus Schrad. [*nom. illegit.*], *Spicil. Fl. Germ.*: 168 (1794)
Polyporus fuliginosus (Scop.) Fr., *Epicr. syst. mycol.*: 451 (1838)
E: ! **S:** ! **NI:** !
H: Supposedly only on dead wood of deciduous trees such as
Betula spp, *Fagus* and *Prunus avium* but several records and
collections said to be on *Pinus sylvestris.*
D: EurPoly1: 357, NM3: 224
Rarely reported but apparently widespread.

ITERSONILIA Derx, *Bull. Jard. Bot. Buitenzorg, Sér. 3,* 17: 471 (1948)
Tremellales, Filobasidiaceae
An anamorphic genus, with no known teleomorphs.
Type: *Itersonilia perplexans* Derx

pastinacae Channon, *Ann. appl. Biol.* 51: 13 (1963)
E: !
H: Isolated from leaves and roots of parsnips.
Reported from Buckinghamshire, Warwickshire and West Kent.
Causes a root canker.

perplexans Derx, *Bull. Jard. Bot. Buitenzorg, Sér. 3,* 17: 471 (1948)
Itersonilia pyriformis Nyland, *Mycologia* 41: 689 (1949)
E: !
H: On various substrata, e.g. basidiomes of *Dacrymyces
stillatus*, decayed wood of *Fagus*, decayed roots of *Daucus
carota* and flowers of *Dendranthema* (*Chrysanthemum*)
cultivars.

JAAPIA Bres., *Ann. Mycol.* 9: 428 (1911)
Boletales, Coniophoraceae
Type: *Jaapia argillacea* Bres

argillacea Bres., *Ann. Mycol.* 9(4): 428 (1911)
E: !
H: On decayed wood of conifers such as *Pinus* spp; also on
decayed leaves of *Juncus* spp at the margins of lakes and
ponds.
D: NM3: 290 **D+I:** CNE4: 746-748
Known from South Devon, South Hampshire, South Somerset,
Surrey and Yorkshire.

ochroleuca (Bres.) Nannf. & J. Erikss., *Svensk bot. Tidskr.* 47(2): 184 (1953)
Coniophora ochroleuca Bres., *Jahres-Ber. Westfäl. Prov.-Vereins Wiss.* 26: 130 (1898)
Peniophora ochroleuca (Bres.) Höhn. & Litsch., *Sitzungsber. Kaiserl. Akad. Wiss., Wien, Math.-Naturwiss. Cl., Abt. 1* 117: 1107 (1908)
Coniophorella ochroleuca (Bres.) Brinkmann *Jahres-Ber. Westfäl. Prov.- Vereins Wiss.* 44: 41 (1916)
E: ! **S:** !
H: On decayed wood of conifers such as *Pinus sylvestris* and
Picea spp.
D: NM3: 291 **D+I:** CNE4: 749-751, B&K2: 208 239
Known from Cheviotshire, South Devon and South Essex, and a
single collection from Scotland (Easterness).

JUNGHUHNIA Corda, *Anleit. Stud. Mykol.*: 195 (1842)
Stereales, Steccherinaceae
Chaetoporus P. Karst., *Hedwigia* 29: 148 (1890)
Type: *Junghuhnia crustacea* (Jungh.) Ryvarden

lacera (P. Karst.) Niemelä & Kinnunen, *Karstenia* 41: 6 (2001)
Physisporus lacer P. Karst., *Meddeland. Soc. Fauna Fl. Fenn.* 9: 69 (1882)
Chaetoporus separabilimus Pouzar, *Česká Mykol.* 21: 210 (1967)
Junghuhnia separabilima (Pouzar) Ryvarden, *Persoonia* 7(1): 18 (1972)
Steccherinum separabilimum (Pouzar) Vesterh., Knudsen & Hansen, *Nordic J. Bot.* 16(2): 216 (1996)
E: !
H: On decayed wood and litter of *Betula, Fagus* and *Salix* spp.
Known from Middlesex, North Essex and West Suffolk.

nitida (Pers.) Ryvarden, *Persoonia* 7(1): 18 (1972)
Poria nitida Pers., *Observ. mycol.* 2: 15 (1799)
Boletus nitidus (Pers.) Pers., *Syn. meth. fung.*: 547 (1801)
Polyporus nitidus (Pers.) Fr., *Observ. mycol.* 2: 262 (1818)
Steccherinum nitidum (Pers.) Vesterh., *Nordic J. Bot.* 16(2): 216 (1996)
Poria eupora (P. Karst.) Cooke, *Grevillea* 14: 110 (1886)
Chaetoporus euporus (P. Karst.) Bondartsev & Singer, *Ann. Mycol.* 39: 51 (1941)
E: o **S:** ! **W:** ! **NI:** ! **O:** Isle of Man: !
H: On decayed wood of deciduous trees, usually in old
woodland on calcareous soil. Most frequently on *Fagus* and
Fraxinus but also recorded on *Betula, Corylus, Populus
tremula, Quercus* and *Salix* spp.
D: NM3: 219 (as *Steccherinum nitidum*) **I:** Kriegl1: 539
Occasional but widespread in England, rarely reported
elsewhere. Microscopically similar to *Steccherinum ochraceum*
but differing in its poroid basidiomes.

KAVINIA Pilát, *Stud. Bot. Čechoslov.* 1: 3 (1938)
Gomphales, Ramariaceae
Type: *Caldesiella sajanensis* Pilát (= *K. alboviridis*)

alboviridis (Morgan) Gilb. & Budington, *J. Ariz. Acad. Sci.* 6(2): 95 (1970)
Hydnum alboviride Morgan, *J. Cincinnati Soc. Nat. Hist.* 10: 12 (1887)
E: !
H: British collections on decayed wood of *Fraxinus* in sparse
deciduous woodland on calcareous soil.
D: NM3: 270, Myc17(1): 43 **D+I:** CNE4: 754-755, B&K2: 88 60
Known from two collections in south eastern England (Surrey:
Chipstead Valley).

KUEHNEROMYCES Singer & A.H. Sm., *Mycologia* 38: 504 (1946)
Agaricales, Strophariaceae
Type: *Kuehneromyces mutabilis* (Schaeff.) Singer & A.H. Sm.

leucolepidotus (P.D. Orton) Pegler & T.W.K. Young, *Kew Bull.* 27(3): 487 (1972)
> *Galerina leucolepidota* P.D. Orton, *Notes Roy. Bot. Gard. Edinburgh* 29: 100 (1969)

E: ! **NI:** !
H: Caespitose on decayed wood (large stumps and logs) of deciduous trees such as *Betula* spp and *Fraxinus*.
D: BFF7: 19-20 (as *Galerina leucolepidota*)
Apparently widespread (*fide* BFF7) but poorly known. The only collection named thus in herb. K is from South Devon (Rousdon). Reported from Northern Ieland (Fermanagh: Kilgarrow Wood) in 2000.

lignicola (Peck) Redhead, *Sydowia* 37: 247 (1984)
> *Agaricus lignicola* Peck, *Rep. (Annual) New York State Mus. Nat. Hist.* 23: 91 (1872)
> *Pholiota lignicola* (Peck) Jacobsson, *Mycotaxon* 36(1): 138 (1989)
> *Galerina myriadophylla* P.D. Orton, *Notes Roy. Bot. Gard. Edinburgh* 29: 101 (1969)
> *Kuehneromyces myriadophyllus* (P.D. Orton) Pegler & T.W.K. Young, *Kew Bull.* 27(3): 487 (1972)

S: !
H: On decayed wood and sawdust of *Pinus sylvestris* in Caledonian pinewood. Usually spring-fruiting.
D: NM2: 259, BFF7: 20 (as *Galerina myriadophylla*) **D+I:** B&K4: 336 430 (as *Pholiota lignicola*)
Known in Britain only from near Loch Rannoch in Perthshire, often growing in large clumps on sawdust heaps.

mutabilis (Schaeff.) Singer & A.H. Sm., *Mycologia* 38: 505 (1946)
> *Agaricus mutabilis* Schaeff., *Fung. Bavar. Palat. nasc.* 1: 9 (1762)
> *Pholiota mutabilis* (Schaeff.) P. Kumm., *Führ. Pilzk.*: 83 (1871)
> *Dryophila mutabilis* (Schaeff.) Quél., *Enchir. fung.*: 69 (1886)
> *Galerina mutabilis* (Schaeff.) P.D. Orton, *Trans. Brit. Mycol. Soc.* 43(2): 176 (1960)
> *Lepiota caudicina* Gray, *Nat. arr. Brit. pl.* 1: 603 (1821)

E: c **S:** c **W:** c **NI:** c **ROI:** c **O:** Channel Islands: ! Isle of Man: !
H: On decayed wood of deciduous trees, often in large caespitose clusters on large fallen trunks and stumps. Most frequently on *Betula, Fagus, Fraxinus* and *Ulmus* spp but also known on *Acer pseudoplatanus, Aesculus, Alnus* spp, *Castanea, Corylus, Crataegus* spp, *Carpinus, Prunus laurocerasus, Salix* spp, *Sambucus nigra, Sorbus aucuparia*, and *Tilia* spp. Reported on conifers such as *Larix, Picea, Pinus* and *Thuja* spp but these hosts are unverified.
D: NM2: 259, BFF7: 18-19 (as *Galerina mutabilis*), FAN4: 105-106 (as *Pholiota mutabilis*) **D+I:** Ph: 156 (as *G. mutabilis*), B&K4: 338 434 (as *P. mutabilis*) **I:** FlDan3: 110A (as *P. mutabilis*), Bon: 255, C&D: 346
Very common and widespread. English name = 'Velvet Tough-shank'.

LACCARIA Berk. & Broome, *Ann. Mag. Nat. Hist., Ser. 5* 12: 370 (1883)
Agaricales, Tricholomataceae
Type: *Laccaria laccata* (Scop.) Cooke

amethystina Cooke, *Grevillea* 12: 70 (1884)
> *Agaricus amethystinus* Huds. [*nom. illegit.*, non *A. amethystinus* Scop. (1772)], *Fl. angl.*, Edn 2, 2: 612 (1778)
> *Laccaria laccata* var. *amethystina* (Cooke) Rea, *Brit. basidiomyc.*: 290 (1922)
> Mis.: *Agaricus amethysteus* Bull. sensu auct.
> Mis.: *Omphalia amethystea* (Bull.) Gray sensu auct.
> Mis.: *Laccaria amethystea* (Bull.) Murrill sensu auct.
> Mis.: *Laccaria laccata* var. *amethystea* (Bull.) Berk. & Broome sensu auct.

E: c **S:** c **W:** c **NI:** c **ROI:** c **O:** Channel Islands: ! Isle of Man: !
H: On soil in deciduous woodland. Usually associated with *Fagus* but not uncommonly with *Betula, Corylus* and *Quercus* spp.

D: NM2: 130, FAN3: 102 **D+I:** Ph: 52-53 (as *Laccaria amethystea*), B&K3: 200 228 (as *L. amethystea*) **I:** Sow2: 187 (as *Agaricus amethystinus*), Bon: 147, C&D: 187 (colour poor)
Very common and widespread. The epithet 'amethystea' is considered incorrect following Mueller & Vellinga in Mycotaxon 37: 385 (1990). English name = 'Amethyst Deceiver'.

bicolor (Maire) P.D. Orton, *Trans. Brit. Mycol. Soc.* 43(2): 177 (1960)
> *Laccaria laccata* var. *bicolor* Maire, *Publ. Inst. Bot. Barcelona* 3: 84 (1937)
> *Laccaria proxima* var. *bicolor* (Maire) Kühner & Romagn. [*comb. inval.*], *Flore Analytique des Champignons Supérieurs*: 131 (1953)

E: o **S:** o **W:** o **NI:** ! **ROI:** !
H: On acidic soil often amongst grass or litter under deciduous or coniferous trees, or on moorland and heathland.
D: NCL3: 280, NM2: 130, FAN3: 99-100 **D+I:** Ph: 52-53, B&K3: 202 229 **I:** Bon: 147, C&D: 187, Kriegl3: 281
Occasional to rare, but widespread.

fraterna (Sacc.) Pegler, *Austral. J. Bot.* 13: 332 (1956)
> *Naucoria fraterna* Sacc., *Syll. fung.* 9: 110 (1891)
> *Agaricus fraternus* Cooke & Massee [*nom. illegit.*, non *A. fraternus* Lasch (1828)], *Grevillea* 16: 31 (1887)

E: ! **S:** !
H: On soil in parkland and gardens, always with *Eucalyptus* spp.
D: FAN3: 100-101 **D+I:** B&K3: 202 230 **I:** C&D: 187
Rarely reported but possibly increasing. See Last & Watling in Mycologist 12 (4): 152-153 (1998).

laccata (Scop.) Cooke, *Grevillea* 12: 70 (1884)
> *Agaricus laccatus* Scop., *Fl. carniol.* 444 (1772)
> *Clitocybe laccata* (Scop.) P. Kumm., *Führ. Pilzk.*: 122 (1871)
> *Laccaria laccata* var. *affinis* Singer, *Bull. Soc. Mycol. France* 83: 111 (1967)
> *Laccaria affinis* (Singer) Bon, *Doc. Mycol.* 13 (51): 49 (1983)
> *Laccaria laccata* var. *anglica* Singer, *Bull. Soc. Mycol. France* 83: 110 (1967)
> *Laccaria affinis* var. *anglica* (Singer) Bon, *Doc. Mycol.* 13 (51): 50 (1983)
> *Agaricus farinaceus* Huds. [*nom. illegit.*, non *A. farinaceus* Schumach. (1803], *Fl. angl.*, Edn 2, 2: 616 (1778)
> *Omphalia farinacea* (Huds.) Gray, *Nat. arr. Brit. pl.* 1: 616 (1821)
> *Clitocybe laccata* var. *pallidifolia* Peck, *Rep. (Annual) New York State Mus. Nat. Hist.* 43: 38 (1890)
> *Laccaria laccata* var. *pallidifolia* (Peck) Peck, *Rep. (Annual) New York State Mus. Nat. Hist.* 157: 92 (1912)
> *Laccaria laccata* var. *moelleri* Singer, *Sydowia, Beih.* 7: 9 (1973)
> *Agaricus rosellus* Batsch [*nom. illegit.*, non *A. rosellus* Fr. (1821)], *Elench. fung. (Continuatio Prima)*: 121 & t. 19 f. 99 (1786)
> *Omphalia rosella* (Batsch) Gray, *Nat. arr. Brit. pl.* 1: 613 (1821)
> *Laccaria tetraspora* var. *scotica* Singer, *Bull. Soc. Mycol. France* 83: 114 (1967)
> *Laccaria scotica* (Singer) Contu, *Micol. ven.* 1(2): 7 (1985)
> *Agaricus amethysteus* Bull., *Herb. France*: pl. 198 (1785)
> *Omphalia amethystea* (Bull.) Gray, *Nat. arr. Brit. pl.* 1: 614 (1821)
> *Laccaria amethystea* (Bull.) Murrill, *North American Flora* 10(1): 1 (1914) [sensu auct. = *L. amethystina*]
> *Laccaria laccata* var. *amethystea* (Bull.) Berk. & Broome, *J. Linn. Soc., Bot.* 11: 518 (1871) [sensu auct. = *L. amethystina*]

E: c **S:** c **W:** c **NI:** c **ROI:** c **O:** Channel Islands: ! Isle of Man: c
H: On soil in woodland, on moorland or heathland, often associated with *Betula* spp, but also known with many other deciduous trees. Occasionally growing amongst *Sphagnum* moss.
D: NM2: 130, FAN3: 97-98 (as *Laccaria laccata* var. *pallidifolia*) **D+I:** Ph: 52, B&K3: 202 231 (as *L. laccata* var. *laccata*) **I:** Bon: 147, C&D: 187, Kriegl3: 283
Very common and widespread. English name = 'The Deceiver'.

maritima (Theodor.) Huhtinen, *Fungi Canadenses No. 319*: f.
99 (1987)
> *Hygrophorus maritimus* Theodor., *Grzyby w. polsk. wybr.*: 31
> (1936)
> *Laccaria trullisata* ssp. *maritima* (Theodor.) O. Andersson,
> *Bot. Not.* 2: 23 (1950)
> Mis.: *Laccaria trullisata* sensu NCL
> Mis.: *Clitocybe venustissima* sensu Rea (1922) non al.

S: !
H: On sandy soil, amongst grass or in woodland on dunes.
D: NM2: 130, FAN3: 103 **I:** C&D: 187
Known only from Morayshire (Culbin Sands) where first reported
in 1912.

proxima (Boud.) Pat., *Hyménomyc. Eur.*: 97 (1887)
> *Clitocybe proxima* Boud., *Bull. Soc. Mycol. France* 28: 91
> (1881)
> *Laccaria laccata* var. *proxima* (Boud.) Maire, *Bull. Soc. Mycol.*
> *France* 24: LV (1908)
> *Laccaria proximella* Singer, *Mycopathol. Mycol. Appl.* 26: 146
> (1965)

E: c **S:** c **W:** ! **NI:** o **ROI:** ! **O:** Isle of Man: !
H: On acidic soil, in wet areas on moorland or heathland, usually
associated with *Betula* spp, and often growing amongst
Calluna, *Juncus*, *Molinia caerulea* and *Sphagnum* spp.
D: NM2: 130, FAN3: 98-99 **D+I:** Ph: 52-53 **I:** Bon: 147 (as
Laccaria laccata var. *moelleri*), C&D: 187 (as *L. laccata* var.
moelleri)
Common and widespread. Restricted to areas of acidic soil. For
some authorities *L. proximella* is a distinct alpine species,
recorded from several montane sites in Scotland.

pumila Fayod, *Ann. Reale Accad. Agric. Torino* 35: 91 (1893)
[1892]
> *Laccaria altaica* Singer, *Bull. Soc. Mycol. France* 83: 122
> (1967)
> Mis.: *Laccaria striatula* sensu P.D. Orton [TBMS 43: 282
> (1960)] & sensu auct.

E: ? **S:** ! **O:** Isle of Man: ?
H: On soil. British material with *Pinus sylvestris* and *Salix* spp.
D: FAN3: 100-101 **D+I:** B&K3: 200 227 (as *Laccaria altaica*) **I:**
C&D: 187
Known from montane habitat in Scotland. Records from England
(Herefordshire, Yorkshire, South Devon) and the Isle of Man
are all unsubstantiated with voucher material.

purpureobadia D.A. Reid, *Fungorum Rar. Icon. Color.* 1: 14-16
pl. 5 (1966)
E: ! **S:** ! **NI:** ?
H: On soil in dried-out woodland swamps, often amongst
Sphagnum spp under *Alnus*, *Betula* and *Salix* spp.
D: NM2: 130, FAN3: 102-103 **D+I:** Ph: 52-53 **I:** C&D: 187
Rarely reported but widespread throughout England. Few
records elsewhere in the British Isles.

tortilis (Bolton) Cooke, *Grevillea* 12: 70 (1884)
> *Agaricus tortilis* Bolton, *Hist. fung. Halifax* 1: 41 (1788)
> *Omphalia tortilis* (Bolton) Gray, *Nat. arr. Brit. pl.* 1: 613
> (1821)
> *Clitocybe tortilis* (Bolton) Gillet, *Hyménomycètes*: 174 (1874)

E: o **S:** o **W:** ! **NI:** ! **ROI:** !
H: On soil in woodland, often in areas which are periodically
flooded such as streambanks, pools and the sides of ditches,
often associated with *Alnus* and *Salix* spp.
D: NCL3: 281, NM2: 131, FAN3: 101 **D+I:** Ph: 52, B&K3: 204
233 **I:** Bon: 147, C&D: 187, Kriegl3: 286
Occasional and generally rather uncommon, but widespread.

LACHNELLA Fr., *Fl. Scan.*: 343 (1835)
Agaricales, Tricholomataceae
Type: *Lachnella alboviolascens* (Alb. & Schwein.) Fr.

alboviolascens (Alb. & Schwein.) Fr., *Summa veg. Scand.* 2:
365 (1849)
> *Peziza alboviolascens* Alb. & Schwein., *Consp. fung. lusat.*:
> 322 (1805)

> *Cyphella alboviolascens* (Alb. & Schwein.) P. Karst., *Not.*
> *Sällsk. Fauna et Fl. Fenn. Förh.* 11: 221 (1870)
> *Cyphella curreyi* Berk. & Broome, *Ann. Mag. Nat. Hist., Ser. 3*
> *[Notices of British Fungi no. 935]* 7: 379 (1861)
> *Cyphella dochmiospora* Berk. & Broome, *Ann. Mag. Nat.*
> *Hist., Ser. 4 [Notices of British Fungi no. 1373]* 11: 343
> (1873)
> *Cyphella stuppea* Berk. & Broome, *Ann. Mag. Nat. Hist., Ser.*
> *5 [Notices of British Fungi no. 1698]* 1: 25 (1878)

E: o **S:** ! **W:** ! **NI:** ! **ROI:** !
H: Usually on small dead twigs and branches of a wide variety
of deciduous trees and shrubs, especially *Sambucus nigra*. Also
frequent on large dead stems of herbaceous or woody plants
such as *Angelica*, *Arctium Epilobium*, *Equisetum*, *Fallopia*,
Humulus, *Lupinus*, *Rumex*, *Symphytum* and *Vicia* spp.
D: NM2: 131 **D+I:** B&K2: 202 230 **I:** Kriegl3: 598
Occasional in England. Rarely reported elsewhere but apparently
widespread and frequent. Basidiomes are minute, easily
overlooked, and often taken to be a discomycete at first
glance.

villosa (Pers.) Donk, *Lilloa* 22: 345 (1951) [1949]
> *Peziza villosa* Pers., *Syn. meth. fung.*: 655 (1801)
> *Solenia villosa* (Pers.) Fr., *Syst. mycol.* 2: 200 (1823)
> *Cyphella villosa* (Pers.) H. Crouan & P. Crouan, *Florule*
> *Finistère*: 61 (1867)
> *Peziza sessilis* Sowerby, *Col. fig. Engl. fung.* 3: pl. 389 (1803)

E: ! **S:** ! **W:** ! **NI:** ! **ROI:** ! **O:** Channel Islands: ! Isle of Man: !
H: On dead and decayed stems of herbaceous plants such as
Angelica, *Arctium*, *Cirsium*, *Digitalis*, *Eupatorium*, *Heracleum*,
Inula, *Iris*, *Lathyrus*, *Lupinus*, *Mentha*, *Oenanthe*, *Rumex*,
Senecio, *Stachys*, *Symphytum*, *Teucrium* and *Urtica* spp.
Rarely on dead attached twigs of trees or shrubs.
D: NM2: 131
Uncommonly reported but apparently widespread. Probably
overlooked or mistaken for *L. alboviolascens* which it
resembles.

LACRYMARIA Pat., *Hyménomyc. Eur.*: 122 (1887)
Agaricales, Coprinaceae
Type: *Lacrymaria lacrymabunda* (Bull.) Pat.

glareosa (J. Favre) Watling, *Notes Roy. Bot. Gard. Edinburgh*
37(2): 376 (1979)
> *Drosophila glareosa* J. Favre, *Schweiz. Z. Pilzk.* 36: 70 (1958)

S: !
H: On disturbed soil, usually near to human habitation.
D: BFF5: 73, NM2: 237 (as *Psathyrella glareosa*)
Rarely reported.

lacrymabunda (Bull.) Pat., *Hyménomyc. Eur.*: 123 (1887)
> *Agaricus lacrymabundus* Bull., *Herb. France*: pl. 194 (1785)
> *Psathyra lacrymabunda* (Bull.) P. Kumm., *Führ. Pilzk.*: 71
> (1871)
> *Hypholoma lacrymabundum* (Bull.) Sacc., *Syll. fung.* 5: 1033
> (1887)
> *Psathyrella lacrymabunda* (Bull.) A.H. Sm., *Mem. N. Y. bot.*
> *Gdn* 24: 53 (1972)
> *Agaricus areolatus* Klotzsch, in Berkeley, *Engl. fl.* 5: 112
> (1836)
> *Psilocybe areolata* (Klotzsch) Sacc., *Syll. fung.* 5: 1043 (1887)
> *Lacrymaria lacrymabunda* f. *gracillima* J.E. Lange 4: 72
> (1939)
> *Agaricus velutinus* Pers., *Syn. meth. fung.*: 409 (1801)
> *Agaricus lacrymabundus* β *velutinus* (Pers.) Fr., *Syst. mycol.*
> 1: 288 (1821)
> *Hypholoma velutinum* (Pers.) P. Kumm., *Führ. Pilzk.*: 72
> (1871)
> *Lacrymaria velutina* (Pers.) Maire, *Treb. Mus. Ci. Nat.*
> *Barcelona* 15(2): 110 (1933)
> *Psathyrella velutina* (Pers.) Singer, *Lilloa* 22: 446 (1951)
> [1949]

E: c **S:** c **W:** c **NI:** c **ROI:** o **O:** Channel Islands: ! Isle of Man: !
H: On soil (often on disturbed ground) by the edges of paths in
woodland, by roadsides, on lawns, in gardens and occasionally
amongst grass in pastures.

D: BFF5: 73, NM2: 237 (as *Psathyrella lacrymabunda*) **D+I**: Ph: 176 (as *Lacrymaria velutina*), B&K4: 254 308, C&D: 264 **I**: Sow1: 41 (as *Agaricus lacrymabundus*), FlDan4: 144B1, Bon: 269 (as *P. lacrymabunda*)

Common and widespread. *Agaricus areolatus* sensu Cooke 596 (570) Vol. 4 (1886) and 1182 (1177) Vol. 8 (1891) may represent a *Hypholoma* sp., perhaps *H. capnoides fide* BFF5: 99.

pyrotricha (Holmsk.) Konrad & Maubl., *Rév. hyménomyc. France*: 91 (1925)
   Agaricus pyrotrichus Holmsk., *Beata ruris* 1: 63 t. 35 (1790)
   Hypholoma pyrotrichum (Holmsk.) Quél., *Mém. Soc. Émul. Montbéliard, Sér. 2*, 5: 114 (1872)
   Psathyrella pyrotricha (Holmsk.) M.M. Moser, *Kleine Kryptogamenflora, Edn 3*: 218 (1967)
E: o **S:** ! **W:** ! **NI:** ! **ROI:** !
H: On soil in open woodland and also occasionally amongst grass in pastures.
D: BFF5: 74, NM2: 237 (as *Psathyrella pyrotricha*) **D+I:** Ph: 176, C&D: 264 **I:** Bon: 269 (as *P. pyrotricha*)
Occasional in England. Rarely reported elsewhere but apparently widespread.

LACTARIUS Pers. [as *Lactaria*], *Tent. disp. meth. fung.*: 63 (1797)
Russulales, Russulaceae
Type: *Lactarius piperatus* (L.) Pers.

acerrimus Britzelm., *Bot. Centralbl.* 54(4): 98 (1893)
   Mis.: *Lactarius insulsus* sensu Pearson [*Naturalist* (Hull) 1950: 90 (1950)] & sensu auct. Eur.
E: o **S:** ? **W:** ! **NI:** ! **ROI:** !
H: On soil in deciduous woodland, usually associated with *Quercus* spp.
D: NM2: 368 **D+I:** FNE2 124-125, FungEur7: 332-336 **I:** FlDan5: t.173F (as *Lactarius insulsus*), Bon: 83, C&D: 403, Kriegl2: 384
Occasional in England. Widespread but rarely reported elsewhere.

acris (Bolton) Gray, *Nat. arr. Brit. pl.* 1: 625 (1821)
   Agaricus acris Bolton, *Hist. fung. Halifax* 2: 60 (1788)
   Agaricus lactifluus [var.] *acris* (Bolton) Pers., *Syn. meth. fung.*, 437 (1801)
E: ! **S:** ! **NI:** ! **ROI:** !
H: On soil (often calcareous) in deciduous scrub or woodland, usually associated with *Corylus*, *Fagus* or *Quercus* spp.
D: NM2: 367 **D+I:** FNE2 228-229, FungEur7: 640-644 **I:** FlDan5: t.169B, Bon: 97, C&D: 417
Rarely reported but apparently widespread. Often confused with closely related taxa and many of the records are unsubstantiated with voucher material.

aspideus (Fr.) Fr., *Epicr. syst. mycol.*: 336 (1838)
   Agaricus aspideus Fr., *Observ. mycol.* 2: 189 (1818)
E: ! **S:** ! **W:** ! **NI:** !
H: On wet soil in swampy areas in deciduous woodland, usually associated with *Betula* or *Salix* spp.
D: NM2: 364 **D+I:** FNE2 96-97, FungEur7: 221-227 **I:** FlDan5: t.170F, Bon: 87, C&D: 409
Rarely reported but apparently widespread.

aurantiacus (Pers.) Gray, *Nat. arr. Brit. pl.* 1: 624 (1821)
   Agaricus lactifluus-testaceus γ *aurantiacus* Pers., *Syn. meth. fung.*: 432 (1801)
   Lactarius aurantiofulvus J. Blum, *Les Lactaires. Études Mycologiques*: 272 (1976)
   Agaricus mitissimus Fr., *Syst. mycol.* 1: 69 (1821)
   Lactarius mitissimus (Fr.) Fr., *Epicr. syst. mycol.*: 345 (1838)
E: c **S:** c **W:** o **NI:** c **ROI:** o
H: On soil in woodland, associated with deciduous trees or conifers.
D: NM2: 372 (as *Lactarius mitissimus*) **D+I:** FNE2 178-179, FungEur7: 541-549 **I:** FlDan5: t.173D, Bon: 91, Bon: 91 (as *L. aurantiofulvus*), C&D: 411 (as *L. aurantiofulvus*), Kriegl2: 406

Apparently common in Britain and Ireland. Previously recorded as *Lactarius mitissimus*.

azonites (Bull.) Fr., *Epicr. syst. mycol.*: 343 (1838)
   Agaricus azonites Bull., *Herb. France*: pl. 559 f. 1 (1791)
   Lactarius fuliginosus f. *albipes* J.E. Lange, *Dansk. Bot. Ark.* 5(5): 33 (1928)
   Mis.: *Lactarius fuliginosus* sensu auct.
E: ! **S:** ! **W:** !
H: On loam or clay soil in deciduous woodland, usually associated with *Quercus* spp, less frequently with *Fagus* or *Corylus*.
D: NM2: 367 **D+I:** FNE2 238-241, FungEur7: 644-650 (as *Lactarius azonites* f. *azonites*) **I:** C&D: 417 (as *L. fuliginosus* var. *albipes*)
Generally rather uncommon and often confused with *L. fuliginosus*. Mainly reported from southern England, but known as far north as Northumberland. Rarely reported from Scotland and Wales.

bertillonii (Z. Schaef.) Bon, *Doc. Mycol.* 10 (37-38): 92 (1980) [1979]
   Lactarius vellereus var. *bertillonii* Z. Schaef., *Česká Mykol.* 33: 9 (1979)
   Lactarius bertillonii var. *queletii* J. Blum [*nom. nud.*], *Les Lactaires. Études Mycologiques*: 309 (1976)
E: ! **S:** ! **W:** ?
H: On soil in deciduous woodland, usually associated with *Fagus* and *Quercus* spp.
D+I: FNE2 254-255, FungEur7: 709-713 **I:** C&D: 399, SV35: 41
British collections restricted to northern England and Scotland. Records from Gloucestershire and Wales (Breconshire) are unsubstantiated with voucher material. Seldom distinguished from *L. vellereus*.

blennius (Fr.) Fr., *Epicr. syst. mycol.*: 337 (1838)
   Agaricus blennius Fr., *Observ. mycol.* 1: 60 (1815)
   Lactarius blennius f. *virescens* J.E. Lange 5: 37 (1940)
E: c **S:** c **W:** c **NI:** c **ROI:** c **O:** Isle of Man: !
H: On soil in deciduous woodland, usually associated with *Fagus* in pure stands of beech woodland. Rarely reported with other trees.
D: NM2: 370 **D+I:** Ph: 82-83, FNE2 44-47, FungEur7: 79-85 **I:** FlDan5: 172E, FlDan5: t.172E, Bon: 87, C&D: 405, Kriegl2: 387
Very common and widespread.

camphoratus (Bull.) Fr., *Epicr. syst. mycol.*: 346 (1838)
   Agaricus camphoratus Bull. [non *A. camphoratus* Fr. (1821)], *Herb. France*: pl. 567 f. 1 (1792)
   Lactarius terrei Berk. & Broome, *Ann. Mag. Nat. Hist., Ser. 5* [*Notices of British Fungi no. 1673*] 1: 22 (1878)
   Lactarius camphoratus var. *terrei* (Berk. & Broome) Cooke, *Handb. Brit. fung., Edn 2*: 317 (1890)
   Agaricus cimicarius Batsch, *Elench. fung. (Continuatio Prima)*: 59 & t. 15 f. 69 (1786)
   Lactarius cimicarius (Batsch) Gillet, *Hyménomycètes*: 221 (1876)
   Agaricus subdulcis β *camphoratus* Fr., *Syst. mycol.* 1: 70 (1821)
E: c **S:** c **W:** c **NI:** c **ROI:** !
H: On soil (usually acidic) in woodland with deciduous trees or conifers, on heathland or occasionally on moorland. Often associated with *Betula* and *Pinus* spp, but also known with *Fagus* and *Quercus* spp, and reported amongst *Pteridium*.
D: NM2: 373 **D+I:** Ph: 90, FNE2 214-215, FungEur7: 587-593 **I:** Bon: 99, C&D: 413, Kriegl2: 407
Common and widespread but few records from the Republic of Ireland. Excludes *L. cimicarius* sensu NCL and sensu auct. mult., for which see *L. subumbonatus*.

chrysorrheus Fr., *Epicr. syst. mycol.*: 342 (1838)
   Lactarius theiogalus var. *chrysorrheus* (Fr.) Quél., *Enchir. fung.*: 129 (1886)
   Mis.: *Agaricus theiogalus* Bull. sensu auct. p.p.
   Mis.: *Lactarius theiogalus* (Bull.) Gray sensu auct. p.p.
E: c **S:** c **W:** o **NI:** o **ROI:** o **O:** Channel Islands: !
H: On soil in deciduous woodland, associated with *Quercus* spp.

D: NM2: 367 **D+I:** Ph: 78-79, FNE2 122-123, FungEur7: 317-322 **I:** FlDan5: 172A, FlDan5: t.172A, Bon: 93, C&D: 413, Kriegl2: 390
Common and widespread.

circellatus Fr., *Epicr. syst. mycol.*: 338 (1838)
Mis.: *Agaricus zonarius* sensu Sowerby (1799)
E: o **S:** o **W:** ! **NI:** ! **ROI:** !
H: On soil in deciduous woodland, usually associated with *Carpinus* but also known (much less frequently) with *Corylus*.
D: NM2: 370 **D+I:** Ph: 84-85, FNE2 50-51, FungEur7: 85-90 **I:** FlDan5: 172D, Bon: 85, C&D: 405
Occasional but apparently widespread.

citriolens Pouzar, *Česká Mykol.* 22: 20 (1968)
Mis.: *Lactarius cilicioides* sensu auct. mult.
E: ! **W:** ? **NI:** ? **ROI:** ?
H: On soil in deciduous woodland, usually associated with *Betula* spp. Very rarely reported with *Fagus* and *Quercus* spp.
D: NM2: 364 **D+I:** Ph: 79 (as *Lactarius cilicioides*), FNE2 120-121, FungEur7: 421-425 **I:** Bon: 83, C&D: 401
Known from Berkshire, Herefordshire, Mid-Lancashire, North Lincolnshire, North Wiltshire, Surrey, Warwickshire, West Kent and West Sussex. Reported elsewhere but lacking voucher material.

controversus Pers., *Observ. mycol.* 2: 39 (1799)
E: o **S:** ! **W:** ! **NI:** ! **ROI:** !
H: On soil in damp deciduous woodland or in dune slacks, often associated with *Salix* spp (frequently *Salix repens*) and with *Populus* spp.
D: NM2: 368 **D+I:** Ph: 77, FNE2 136-137, FungEur7: 336-342 **I:** FlDan5: t.169C, Bon: 95, C&D: 399, Kriegl2: 391
Occasional in England. Rarely reported elsewhere but apparently widespread.

cyathuliformis Bon, *Doc. Mycol.* 8 (30-31): 69 (1978)
E: ! **S:** !
H: On soil, associated with *Alnus* and *Salix* spp, in wet areas in mixed deciduous woodland, in alder swamps and around the margins of lakes and ponds.
D+I: FNE2 210-211, FungEur7: 613-617 **I:** Bon: 99, C&D: 417 (as *Lactarius obscuratus*)
Probably quite common and widespread but previously confused with similar taxa in similar habitats.

decipiens Quél., *Compt. Rend. Assoc. Franç. Avancem. Sci.* 13: 448 (1885)
Lactarius rubescens Bres., *Fungi trident.* 1: 84 (1881)
E: ! **S:** ! **W:** ! **NI:** ! **ROI:** !
H: On soil in deciduous woodland. Most often reported with *Carpinus*, occasionally with *Corylus, Fagus* or *Quercus* spp.
D+I: Ph: 89, FNE2 190-191, FungEur7: 518-523 **I:** Bon: 93, C&D: 413
Apparently widespread, but uncommon.

deliciosus (L.) Gray, *Nat. arr. Brit. pl.* 1: 624 (1821)
Agaricus deliciosus L. [non *A. deliciosus* Bolton (1788)], *Sp. pl.*: 1172 (1753)
Agaricus lactifluus [var.] *deliciosus* (L.) Pers., *Syn. Meth. fung.*: 433 (1801)
E: c **S:** c **W:** o **NI:** c **ROI:** o **O:** Channel Islands: !
H: On acidic soil in woodland or with solitary trees; strictly associated with *Pinus* spp.
D: NM2: 362 **D+I:** Ph: 80-81, FNE2 140-141, FungEur7: 253-260 **I:** Sow2: 202, Bon: 81, C&D: 407, Kriegl2: 352
Common to occasional, but widespread.

deterrimus Gröger, *Westfäl. Pilzbriefe* 7: 10 (1968)
E: c **S:** c **W:** o **NI:** c **ROI:** o
H: On soil in woodland (usually plantations) and strictly associated with *Picea* spp.
D: NM2: 362 **D+I:** Ph: 80-81, FNE2 150-153, FungEur7: 262-267 **I:** FlDan5: t.177A (as *Lactarius deliciosus* [but shown with *Picea* needle litter]), Bon: 81, C&D: 407, Kriegl2: 353
Common to occasional, but widespread and apparently increasing as spruce plantations mature.

evosmus Kühner & Romagn., *Bull. Soc. Mycol. France* 69: 361 (1954)
Mis.: *Lactarius zonarius* sensu NCL & sensu Phillips (Ph)
E: ! **W:** ?
H: On soil, usually associated with *Populus* or *Quercus* spp in woodland or scrub. Also collected recently in grassland where apparently associated with *Helianthemum nummularium*.
D+I: Ph: 50-51 (as *Lactarius zonarius*), FNE2 126-129, FungEur7: 342-347 **I:** C&D: 403
Rarely reported. British records are mostly from southern England and the Peak District (Derbyshire and Staffordshire). Reported from Wales (Caernarvonshire: Great Ormes Head) but without voucher material.

flavidus Boud., *Bull. Soc. Mycol. France* 3: 145 (1887)
Lactarius aspideus var. *flavidus* (Boud.) Neuhoff, *Pilze Mitteleuropas, Die Milchlinge (Lactarii)*: 114 (1956)
Mis.: *Lactarius aspideus* sensu Pearson [Naturalist (Hull) 1950: 88 (1950)]
E: ! **S:** ! **W:** !
H: On soil (often calcareous) in deciduous woodland, usually associated with *Carpinus, Fagus* or *Quercus* spp.
D+I: Ph: 85, FNE2 92-93, FungEur7: 232-237 **I:** FlDan5: 170F, Bon: 87, C&D: 409
Very rarely reported.

flexuosus (Pers.) Gray, *Nat. arr. Brit. pl.* 1: 624 (1821)
Agaricus lactifluus [var.] *flexuosus* Pers., *Syn. meth. fung.*: 430 (1801)
E: ! **S:** ! **W:** ! **ROI:** ?
H: On soil in woodland, associated with *Betula, Quercus* or *Pinus* spp.
D: NM2: 369, FungEur7: 90-93 (as *Lactarius flexuosus* var. *flexuosus*) **D+I:** FNE2 54-55 **I:** C&D: 405, Kriegl2: 393
Rarely reported but apparently widespread. The majority of reports lack voucher material.

flexuosus var. roseozonatus H. Post, in Fries, *Monogr. hymenomyc. Suec.* 2: 163 (1863)
Lactarius roseozonatus (H. Post) Britzelm., *Ber. Naturhist. Vereins Augsburg* 30: 29 (1890)
E: !
H: On soil in woodland, associated with *Betula* and *Quercus* spp.
D: FungEur7: 95-97 **D+I:** FNE2 56-57 (as *Lactarius roseozonatus*) **I:** FlDan5: 172B (as *L. roseozonatus*), FlDan5: t.172B, Bon: 85 (as *L. roseozonatus*), C&D: 405 (as *L. roseozonatus*)
British collections from two sites in northern England (Cumbria and Westmorland).

fluens Boud., *Bull. Soc. Mycol. France* 15: 49 (1899)
Lactarius blennius f. *albidopallens* J.E. Lange, *Dansk. Bot. Ark.* 5(5): 29 (1928)
E: o **S:** o **W:** ! **NI:** o **ROI:** !
H: On soil (often calcareous clay), usually associated with *Fagus* in beechwoods but also reported with *Carpinus, Corylus* and *Quercus* spp.
D: NM2: 370 **D+I:** Ph: 84-85, FNE2 48-49, FungEur7: 97-102 **I:** FlDan5: 173E (as *Lactarius blennius* f. *albido-pallens*), FlDan5: t.173E (as *L. blennius* f. *albidopallens*), Bon: 87, C&D: 405
Occasional but widespread. May be passed over as pallid basidiomes of *L. blennius*.

fuliginosus (Fr.) Fr., *Epicr. syst. mycol.*: 348 (1838)
Agaricus fuliginosus Fr., *Syst. mycol.* 1: 73 (1821)
Mis.: *Lactarius azonites* sensu auct.
E: ? **S:** ? **W:** ? **NI:** ? **ROI:** ?
H: On soil in deciduous woodland or with mixed conifers and deciduous trees.
D: NM2: 367 **D+I:** FNE2 244-245, FungEur7: 653-657 **I:** Bon: 97, C&D: 417, Kriegl2: 378
Frequently reported and apparently widespread, but commonly misidentified or confused with similar taxa such as *L. azonites* or *L. romagnesii*. Most records lack voucher material.

fulvissimus Romagn., *Bull. Soc. Mycol. France* 69: 362 (1954)
  Lactarius britannicus D.A. Reid, *Fungorum Rar. Icon. Color.*
    4: 16 (1969)
  Lactarius subsericatus Bon, *Doc. Mycol.* 9 (35): 39 (1979)
  Mis.: *Lactarius cremor* sensu Lange (FlDan5: 44 & pl. 175A)
  Mis.: *Lactarius decipiens* sensu Pearson [*Naturalist* (Hull)
    1950: 94 (1950)] p.p.
  Mis.: *Lactarius ichoratus* sensu Fries (1838) and sensu auct.
    Brit.
E: o **S:** ! **NI:** !
H: On soil (often calcareous loam) in mixed deciduous
  woodland. Usually associated with *Corylus* or *Fagus*.
D+I: FNE2 196-197, FungEur7: 549-555 **I:** Bon: 91, C&D: 411,
  Kriegl2: 409
Occasional but may be locally frequent in southern England,
  rarely reported elsewhere. Following FNE2 this includes *L.
  britannicus* (but non sensu Basso, 1999).

glaucescens Crossl., *Naturalist (Hull)* 1900: 5 (1900)
  Mis.: *Lactarius pergamenus* sensu auct.
E: ! **S:** !
H: On soil in deciduous woodland. Associated with *Betula* and
  Quercus spp.
D+I: FNE2 250-251, FungEur7: 723-728 **I:** C&D: 399, Kriegl2:
  360
Rarely reported. Known from Cheviotshire, Herefordshire,
  Westmorland and Yorkshire, and from Angus, East Perthshire,
  Easterness and West Sutherland. Reported elsewhere but
  without voucher material. May be confused with *Lactarius
  piperatus*.

glyciosmus (Fr.) Fr., *Epicr. syst. mycol.*: 348 (1838)
  Agaricus glycyosmus Fr., *Observ. mycol.* 2: 194 (1818)
  Mis.: *Lactarius impolitus* sensu auct.
E: c **S:** c **W:** c **NI:** c **ROI:** ! **O:** Isle of Man: !
H: On acidic or sandy soil, associated with *Betula* spp.
D: NM2: 371 **D+I:** Ph: 85, FNE2 170-171, FungEur7: 464-469
  I: FlDan5: t.171a & a1, Bon: 89, C&D: 411
Common and widespread but few records from Republic of
  Ireland.

helvus (Fr.) Fr., *Epicr. syst. mycol.*: 347 (1838)
  Agaricus helvus Fr., *Syst. mycol.* 1: 72 (1821)
  Lactarius aquifluus Peck, *Rep. (Annual) New York State Mus.
    Nat. Hist.* 28: 50 (1879)
  Agaricus tomentosus Krombh., *Naturgetr. Abbild. Schwämme*
    6: 7 (1841)
  Lactarius tomentosus (Krombh.) Cooke, *Handb. Brit. fung.*,
    Edn 2: 314 (1889)
E: o **S:** c **W:** ! **NI:** ! **ROI:** !
H: On acidic soil in woodland, associated with conifers. In Britain
  usually with *Pinus* spp (but also with *Picea* spp in continental
  Europe) and occasionally with *Betula* spp.
D: NM2: 371 **D+I:** Ph: 86-87, FNE2 174-175, FungEur7: 469-
  474 **I:** FlDan5: t.175F, Bon: 89, C&D: 411, Kriegl2: 412
Occasional but widespread in England, rather common in
  Scotland and rarely reported elsewhere.

hepaticus Plowr., in Boudier, *Icon. mycol.* 4: 28 (1905)
  Mis.: *Lactarius theiogalus* sensu Rea (1922)
E: o **S:** ! **W:** ! **NI:** o **ROI:** o
H: On acidic soil in woodland or on heathland, usually
  associated with *Pinus* spp, and rarely reported with *Picea* spp.
D: NM2: 374 **D+I:** Ph: 88-89, FNE2 188-189, FungEur7: 524-
  528 **I:** Bon: 93, C&D: 415, Kriegl2: 413
Occasional but widespread.

hyphoinflatus R.W. Rayner, *Edinburgh J. Bot.* 60(2): 243
  (2003)
E: !
H: On sandy soil in deciduous woodland, associated with *Betula*
  and *Quercus* spp.
Known only from the type collection from West Sussex
  (Sullington, Heath Common) in 1987.

hysginus (Fr.) Fr., *Epicr. syst. mycol.*: 337 (1838)
  Agaricus hysginus Fr., *Observ. mycol.* 2: 192 (1818)

Mis.: *Lactarius curtus* sensu auct. Eur.
E: ! **S:** ! **W:** ! **NI:** ! **ROI:** !
H: On soil, often associated with *Pinus* spp in mixed deciduous
  and conifer woodland. Occasionally reported with *Picea*, *Tsuga*
  and *Betula* spp.
D: NM2: 369 **D+I:** Ph: 82-83, FNE2 80-81, FungEur7: 138-144
  I: FlDan5: t.157B, C&D: 403 (as *Lactarius curtus*), Kriegl2: 395
Rarely reported but apparently widespread.

lacunarum Hora, *Trans. Brit. Mycol. Soc.* 43(2): 444 (1960)
  Lactarius decipiens var. *lacunarum* Romagn. [*nom. inval.*],
    Bull. Soc. Mycol. France 54: 223 (1938)
  Mis.: *Lactarius sphagneti* sensu Moser
E: o **S:** o **W:** ! **NI:** o
H: On soil in deciduous woodland, often in wet areas under
  Alnus, *Betula*, *Quercus* or *Salix* spp. Often collected from
  periodically flooded hollows or depressions in lake- or pond-
  side woodland, in ditches, wheel-ruts, or drainage channels,
  sometimes with *Sphagnum* moss.
D: NM2: 373 **D+I:** FNE2 198-199, FungEur7: 528-533 **I:**
  FlDan5: t.174E, Bon: 93, C&D: 415, Kriegl2: 413
Occasional but apparently widespread. Many of the records are
  from Scotland and Northern Ireland, but it is also not
  uncommon in South Hampshire (New Forest).

lanceolatus O.K. Mill. & Laursen, *Canad. J. Bot.* 51: 43 (1973)
S: !
H: On soil, associated with *Salix herbacea*.
D+I: FNE2 180-181, FungEur7: 555-558
British collections only from Shetland (Foula) and Fair Isle in
  1998.

lilacinus Lasch, in Fries, *Epicr. syst. mycol.*: 348 (1838)
  Lactarius lateritioroseus P. Karst., *Meddeland. Soc. Fauna Fl.
    Fenn.* 16: 14 (1888)
E: ! **S:** ! **W:** ! **NI:** !
H: On soil in wet woodland, along streamsides, drainage
  channels and by lakesides, always associated with *Alnus
  glutinosa*.
D: NM2: 372 **D+I:** FNE2 168-169, FungEur7: 478-482 **I:**
  FlDan5: 171B, Bon: 89, C&D: 411, Kriegl2: 414
Rarely reported but apparently widespread. A distinctive and
  often brightly coloured species, possibly genuinely rare in
  Britain.

luridus (Pers.) Gray, *Nat. arr. Brit. pl.* 1: 625 (1821)
  Agaricus lactifluus [var.] *luridus* Pers., *Syn. meth. fung.* 1:
    436 (1801)
S: !
H: On soil. British collections associated with *Betula* sp. and
  Salix phylicifolia.
D+I: FNE2 84-85, FungEur7: 194-199 **I:** C&D: 409
Reported only from Easterness (near Aviemore) and the Orkney
  Isles.

mairei Malençon, *Bull. Soc. Mycol. France* 55: 34 (1939)
  Lactarius mairei var. *zonatus* A. Pearson, *Naturalist (Hull)*
    1950: 102 (1950)
  Lactarius mairei f. *zonatus* (A. Pearson) D.A. Reid, *Fungorum
    Rar. Icon. Color.* 4: 14 (1969)
  Lactarius pearsonii Z. Schaef., *Česká Mykol.* 22: 19 (1968)
E: ! **S:** !
H: On soil (often calcareous) in deciduous woodland. Often
  associated with *Quercus* spp, but also reported with *Fagus* and
  Fraxinus.
D: NM2: 365 **D+I:** FRIC4: 13-15 30, FNE2 164-165, FungEur7:
  374-379 **I:** Bon: 83, C&D: 401
This is a small but distinctive species, possibly genuinely rare
  though widespread in Britain. Known from England
  (Bedfordshire, Berkshire, Herefordshire, Hertfordshire,
  Oxfordshire, Surrey, West Kent and West Sussex) and a single
  collection from Scotland (Easterness).

mammosus Fr., *Epicr. syst. mycol.*: 347 (1838)
  Lactarius confusus S. Lundell, *Fungi Exsiccati Suecici*: 8
    (1939)

Lactarius fuscus Rolland, *Bull. Soc. Mycol. France* 15: 76 (1899)

E: ! **S:** ! **W:** ! **NI:** !

H: On soil amongst needle litter in conifer woodland, or in litter with mixed conifers and deciduous trees.

D: NM2: 317 **D+I:** FRIC4: 10-13 29, FNE2 172-173, FungEur7: 482-487

Rarely reported, but apparently widespread.

musteus Fr., *Epicr. syst. mycol.*: 337 (1838)

S: !

H: On soil in Caledonian pinewoods, usually in boggy areas, with or near *Sphagnum* spp, growing under *Pinus sylvestris*.

D: NM2: 368 **D+I:** Ph: 82-83, FNE2 76-77, FungEur7: 171-175 **I:** C&D: 403

Rarely reported but rather frequent in Caledonian pine forests.

obscuratus (Lasch) Fr., *Epicr. syst. mycol.*: 346 (1838)

Agaricus obscuratus Lasch, *Linnaea* 3: 430 (1828)

Agaricus obnubilus Lasch, *Linnaea* 3: 161 (1828)

Lactarius obnubilus (Lasch) Fr., *Hymenomyc. eur.*: 438 (1874)

Lactarius radiatus J.E. Lange, *Fl. agaric. danic.* 5, *Taxonomic Conspectus*: V (1940)

Lactarius obscuratus var. *radiatus* (J.E. Lange) Romagn., *Bull. Soc. Mycol. France* 90: 145 (1974)

Mis.: *Lactarius cyathula* sensu Phillips (Ph)

E: o **S:** o **W:** o **NI:** ! **ROI:** !

H: On loam soils, associated with *Alnus* in wet woodland.

D: NM2: 373 **D+I:** Ph: 90, FNE2 206-209, FungEur7: 617-621 (as *Lactarius obscuratus* var. *obscuratus*) **I:** FlDan5: t.176C (as *L. obnubilis*), Bon: 99, C&D: 417 (as *L. obscuratus* var. *radiatus*), Kriegl2: 417

Occasional but apparently widespread. Previously much confused with similar taxa.

omphaliformis Romagn., *Bull. Soc. Mycol. France* 90: 146 (1974)

Mis.: *Lactarius cyathula* sensu NCL

Mis.: *Lactarius tabidus* sensu Rea (1922)

E: ! **NI:** ! **ROI:** !

H: On soil, associated with *Alnus* in wet woodland.

D: NM2: 373 **D+I:** Ph: 90 (as *Lactarius cyathula*), FNE2 212-213, FungEur7: 628-632 **I:** FlDan5: t.175C (as *L. cyathula*), Bon: 99, C&D: 417

Apparently widespread but previously much confused with similar taxa.

pallidus Pers. (as *Lactaria pallida*), *Tent. disp. meth. fung.*: 68 (1797)

Agaricus pallidus (Pers.) Fr. [non *A. pallidus* Sowerby (1802)], *Syst. mycol.* 1: 67 (1821)

Agaricus lactifluus [var.] *pallidus* (Pers.) Pers., *Syn. Meth. Fung.*: 431 (1801)

E: o **S:** o **W:** o **NI:** o **ROI:** o **O:** Isle of Man: !

H: On soil (often calcareous) in deciduous woodland, usually associated with *Fagus* but also reported with *Betula, Corylus* and *Quercus* spp.

D: NM2: 369 **D+I:** Ph: 82-83, FNE2 74-75, FungEur7: 180 **I:** FlDan5: t.175E, Bon: 85, C&D: 403, Kriegl2: 397

Occasional but widespread. May be locally frequent, but nowhere common.

piperatus (L.) Pers. (as *Lactaria piperata*), *Tent. disp. meth. fung.*: 68 (1797)

Agaricus piperatus L., *Sp. pl.*: 1173 (1753)

Agaricus lactifluus [var.] *piperatus* (L.) Pers., *Syn. meth. Fung.*: 429 (1801)

Mis.: *Lactarius pergamenus* sensu auct.

E: o **S:** o **W:** o **NI:** o **ROI:** o

H: On soil in mixed deciduous woodland, usually associated with *Fagus* and *Quercus* spp

D: NM2: 361 **D+I:** Ph: 77, FNE2 248-249, FungEur7: 729-735 **I:** FlDan5: 171D, Bon: 95, C&D: 399

Occasional but widespread. May be locally common, especially on rich loam soils in old woodland.

porninsis Rolland, *Bull. Soc. Mycol. France* 5: 168 (1889)

Lactarius porninae Sacc., *Syll. fung.* 9: 57 (1891)

S: !

H: On soil, in conifer woodland. British records associated with *Larix* spp in upland areas.

D: NM2: 368 **D+I:** FNE2 134-135, FungEur7: 353-358 **I:** Bon: 83, SV29: 37, C&D: 403

First reported and collected from Easterness (Nethy Bridge) in 1982, and from Kingussie in 2004.

pterosporus Romagn., *Rev. Mycol. (Paris)* 14: 110 (1949)

E: o **S:** ! **W:** ! **ROI:** !

H: On soil (often calcareous) in deciduous woodland, usually associated with *Fagus* and less often *Carpinus, Corylus* or *Quercus* spp.

D: NM2: 367 **D+I:** Ph: 87, FNE2 230-231, FungEur7: 669-674 **I:** Bon: 97, C&D: 417, Kriegl2: 382

Occasional in England (and not infrequent in southern counties). Rarely reported elsewhere but apparently widespread. Easily confused with similar taxa.

pubescens (Fr.) Fr., *Epicr. syst. mycol.*: 335 (1838)

Agaricus pubescens Fr., *Observ. mycol.* 1: 56 (1815)

E: c **S:** c **W:** c **NI:** c **ROI:** c

H: On neutral or calcareous soil, in woodland or in grassland, always associated with *Betula* spp.

D: NM2: 365 **D+I:** Ph: 78-79, FNE2 160-161, FungEur7: 380-383 **I:** C&D: 401, Kriegl2: 370

Common and widespread.

pyrogalus (Bull.) Fr., *Epicr. syst. mycol.*: 339 (1838)

Agaricus pyrogalus Bull., *Herb. France*: pl. 529 f. 1 (1792)

Agaricus lactifluus [var.] *pyrogalus* (Bull.) Pers., *Syn. meth. fung.*: 436 (1801)

E: c **S:** c **W:** c **NI:** c **ROI:** c **O:** Isle of Man: !

H: On loam soil in woodland, usually associated with *Corylus* and infrequently with *Carpinus*.

D: NM2: 370 **D+I:** Ph: 84-85, FNE2 52-53, FungEur7: 117-124 **I:** FlDan5: t.174A, Bon: 85, C&D: 403, Kriegl2: 399

Common and widespread.

quieticolor Romagn., *Rev. Mycol. (Paris)* 23(3): 280 (1958)

Lactarius hemicyaneus Romagn., *Rev. Mycol. (Paris)* 23(3): 280 (1958)

Lactarius quieticolor var. *hemicyaneus* (Romagn.) Basso, *Fungi Europaei* 7, *Lactarius*: 275 (1999)

E: ! **S:** ! **W:** !

H: On acidic soil in conifer woodland, usually associated with *Pinus sylvestris*.

D+I: FNE2 142-143, FungEur7: 271-275 (as *Lactarius quieticolor* var. *quieticolor*), FungEur7: 275-279 (as *L. quieticolor* var. *hemicyanea*) **I:** SV27: 14, C&D: 407, C&D: 407 (as *L. hemicyaneus*)

Rarely reported. Known from England (East Sussex, South Hampshire and West Kent), Scotland (Morayshire and South Aberdeen) and Wales (Breconshire) and reported elsewhere but unsubstantiated with voucher material. Resembles *L. deliciosus* when the cobalt-blue colours of the pileipellis have faded.

quietus (Fr.) Fr., *Epicr. syst. mycol.*: 343 (1838)

Agaricus quietus Fr., *Syst. mycol.* 1: 69 (1821)

E: c **S:** c **W:** c **NI:** c **ROI:** c **O:** Channel Islands: ! Isle of Man: !

H: On soil in deciduous woodland, associated with *Quercus* spp.

D: NM2: 374 **D+I:** Ph: 88, FNE2 192-193, FungEur7: 500-504 **I:** FlDan5: t.176E, Bon: 91, C&D: 413, Kriegl2: 419, SV45: 37

Very common and widespread.

repraesentaneus Britzelm., *Ber. Naturhist. Vereins Augsburg* 28: 136 (1885)

E: ! **S:** r **W:** ?

H: On acidic soil, associated with *Pinus sylvestris* and *Betula* spp.

D: NM2: 364 **D+I:** FNE2 104-105, FungEur7: 237-241 **I:** C&D: 409, Kriegl2: 365

Rare, and virtually confined to Scotland. A single record, with voucher material, from England (Cumberland: Loweswater) and reported from Wales (Caernarvonshire: Llyn Padarn) but unsubstantiated with voucher material.

resimus (Fr.) Fr., *Epicr. syst. mycol.*: 336 (1838)
    Agaricus resimus Fr., *Syst. mycol.* 1: 75 (1821)
    Agaricus intermedius var. *expallens* Fr., *Observ. mycol.* 1: 58 (1815)
E: ? **S:** !
H: On soil, associated with *Betula* spp.
D: NM2: 363 **D+I:** FNE2 116-117, FungEur7: 439-444 **I:** FlDan5: 170E, FlDan5: t.170E, C&D: 399
British collections from Highland birchwoods. Records from England are unsubstantiated with voucher material.

romagnesii Bon, *Doc. Mycol.* 9 (35): 39 (1979)
    Lactarius fuliginosus f. *speciosus* J.E. Lange, *Dansk. Bot. Ark.* 5: 33 (1928)
    Lactarius speciosus (J.E. Lange) Romagn. [*nom. illegit.*], *Bull. Soc. Mycol. France* 72: 329 (1957)
    Mis.: *Lactarius fuliginosus* sensu Cooke [Ill. Brit. fung. 959 (996) Vol. 7 (1888)] & sensu NCL p.p.
E: !
H: On loam soil in mixed deciduous woodland, associated with *Corylus*, *Fagus* and *Quercus* spp.
D+I: FNE2 234-235
Accepted sensu FNE2, but not sensu FungEur7, Bon, or C&D (all of which represent a southern species, unknown in Britain). Recorded from southern England but poorly known in Britain. Some records of *Lactarius fuliginosus* may belong here.

rostratus Heilm.-Claus., *Fungi of Northern Europe* 2: 216 (1998)
    Mis.: *Lactarius cremor* sensu Basso (FungEur7)
E: !
H: On soil in mixed deciduous woodland.
D+I: FNE2 216-219, FungEur7: 593-597 (as *Lactarius cremor*)
A single collection from West Sussex (Ebernoe Common) in 2001.

rubrocinctus Fr., *Monogr. hymenomyc. Suec.* 2: 176 (1863)
    Lactarius iners Kühner, *Bull. Soc. Mycol. France* 69: 362 (1954)
    Lactarius subsericeus Hora, *Trans. Brit. Mycol. Soc.* 43(2): 445 (1960)
    Mis.: *Lactarius tithymalinus* sensu auct. Brit.
E: ! **S:** !
H: On soil in deciduous woodland, usually associated with *Fagus* and *Quercus* spp.
D: NM2: 372 **D+I:** FRIC4: 22-27 32, FNE2 202-205, FungEur7: 574-579 **I:** FlDan5: t.176D, FRIC4: 31a (as *Lactarius subsericeus*), Bon: 91, C&D: 413, Kriegl2: 421
Rarely reported. Most records are from southern England, with a few from Scotland.

rufus (Scop.) Fr., *Epicr. syst. mycol.*: 347 (1838)
    Agaricus rufus Scop., *Fl. carniol.* 451 (1772)
    Lactarius mollis D.A. Reid, *Fungorum Rar. Icon. Color.* 4: 19 & pl. 31b (1969)
    Lactarius rufus var. *exumbonatus* Boud., *Icon. mycol.* 4: 26 / pl. 52 (1905)
E: c **S:** c **W:** c **NI:** c **ROI:** o **O:** Channel Islands: ! Isle of Man: !
H: On acidic soil with conifers such as *Abies*, *Picea* and *Pinus* spp, or mixtures of conifers and *Betula* spp, in native woodland, plantations or on heathland. Very rarely reported with *Fagus*.
D: NM2: 373 **D+I:** Ph: 86-87, FNE2 176-177, FungEur7: 487-492 **I:** FlDan5: t.176A, Bon: 89, C&D: 411, Kriegl2: 422
Common and widespread. *L. mollis* is maintained as distinct by some authorities.

ruginosus Romagn., *Bull. Soc. Mycol. France* 72: 340 (1957)
E: ! **S:** ! **W:** !
H: On soil in deciduous woodland, associated with *Carpinus*, *Fagus* and *Quercus* spp.
D+I: FNE2 236-237, FungEur7: 678-684 **I:** C&D: 417
Accepted with some doubt. The few British collections require re-examination, since some or all may be *L. romagnesii*.

salicis-reticulatae Kühner, *Bull. Soc. Mycol. France* 91: 389 (1975)
    Lactarius aspideoides Kühner [*nom. illegit.*, non *L. aspideoides* Burl. (1907)], *Bull. Soc. Mycol. France* 91: 389 (1975)
S: !
H: On soil in unimproved grassland on limestone. Usually associated with dwarf *Salix* spp, but with *Dryas octopetala* in Britain.
D: NM2: 366 **D+I:** FNE2 98-99, FungEur7: 246-250
An arctic-alpine species, known from West Sutherland (Inchnadamph NNR).

salmonicolor (R. Heim & Leclair) R. Heim & Leclair, *Rev. Mycol. (Paris)* 18(3): 221 (1953)
    Lactarius salmoneus R. Heim & Leclair [*nom. illegit.*, non *L. salmoneus* Peck (1898)], *Rev. Mycol. (Paris)* 15(1): 79 (1950)
    Lactarius subsalmoneus Pouzar, *Česká Mykol.* 8: 44 (1954)
E: ! **S:** !
H: On acidic soil in plantation woodland, associated with *Abies* spp. Also reported with *Picea* spp and *Taxus*.
D+I: FNE2 138-139, FungEur7: 282-287 **I:** Bon: 81, C&D: 407, Kriegl2: 355
Rarely reported. Known from England (East Kent and West Sussex) and Scotland. Reported elsewhere in England, but unsubstantiated with voucher material.

scoticus Berk. & Broome, *Ann. Mag. Nat. Hist., Ser. 5 [Notices of British Fungi no. 1783]* 3: 208 (1879)
    Lactarius pubescens var. *scoticus* (Berk. & Broome) Krieglst., *Beitr. Kenntn. Pilze Mitteleurop.* 7: 69 (1991)
S: !
H: On peaty soil, associated with *Betula* spp.
D: NM2: 365 **D+I:** FNE2 162-163, FungEur7: 385-390 **I:** FlDan5: t.169E (as *Lactarius torminosus* var. *gracillimus*), Cooke 938 (1004b) Vol. 7 (1888)
Known in Britain only from the type collection, which is in poor condition, and a collection from Fife (Corl Den) in 2003.

scrobiculatus (Scop.) Fr., *Epicr. syst. mycol.*: 334 (1838)
    Agaricus scrobiculatus Scop., *Fl. carniol.* 450 (1772)
E: ! **S:** ?
H: On soil in woodland. Strictly associated with *Picea* spp.
D: NM2: 362 **D+I:** FNE2 106-107, FungEur7: 444-449 **I:** Bon: 83, C&D: 401, Kriegl2: 373
Accepted only on the strength of some old collections in herb. K, which are in poor condition. Reported on several further occasions from England and Scotland, but unsubstantiated with voucher material.

semisanguifluus R. Heim & Leclair, *Rev. Mycol. (Paris)* 15(1): 79 (1950)
E: ! **NI:** !
H: On soil in woodland, associated with *Pinus* spp. Several collections from cemeteries.
D+I: FNE2 148-149, FungEur7: 287-291 **I:** Bon: 81, C&D: 407, Kriegl2: 358
Known from England (Middlesex and Surrey) and Northern Ireland (Fermanagh).

serifluus (DC.) Fr., *Epicr. syst. mycol.*: 345 (1838)
    Agaricus serifluus DC., *Fl. Fr.* 5: 45 (1815)
    Mis.: *Lactarius subdulcis* var. *cimicarius* sensu Gray (1821)
E: ! **S:** ! **W:** ! **NI:** ! **ROI:** !
H: On soil (often clay) in woodland, usually associated with *Quercus* spp. Also known from fen-carr habitat, with *Alnus* and *Fraxinus*.
D+I: FNE2 220-221, FungEur7: 598-602 **I:** FlDan5: t.174D, C&D: 415
Widely reported but poorly understood in Britain. Frequently confused with *Lactarius subumbonatus*.

sphagneti (Fr.) Neuhoff, *Pilze Mitteleuropas, Die Milchlinge*: 181 (1956)
    Lactarius subdulcis var. *sphagneti* Fr., in Lindblad, *Monogr. Lactar. Suec.* 30 (1855)

S: !
H: In Caledonian pinewoods, usually in wet or boggy areas amongst *Sphagnum* spp, growing under *Pinus sylvestris*.
D: NM2: 374, NBA2: 53-54 **D+I:** FNE2 186-187, FungEur7: 506-510 **I:** Bon: 93
Rarely reported. In Britain, strictly associated with *Pinus sylvestris*. In continental Europe known only with native *Picea* spp, so two taxa may be involved.

spinosulus Quél., *Bull. Soc. Amis Sci. Nat. Rouen* Sér. 2, 15: 168 (1880) [1879]
Lactarius spinosulus var. *violaceus* Cooke, *Handb. Brit. fung.*, Edn 2: 316 (1889)
Lactarius spinulosus Massee, *Eur. Fung. Flora*: 67 (1902)
Mis.: *Lactarius lilacinus* sensu Rea (1922) & sensu Lange (FlDan5: 43 & pl. 171B)
E: ! **S:** o **W:** ! **NI:** ! **ROI:** !
H: On soil in woodland, usually associated with *Betula* spp. Also reported with *Alnus*, *Corylus* and *Fagus*.
D: NM2: 372 **D+I:** Ph: 78-79, FNE2 166-167, FungEur7: 391-395 **I:** FlDan5: t.171B (as *Lactarius lilacinus*), C&D: 411, Kriegl2: 425
Rarely reported but apparently widespread. Seemingly most frequent in Scotland but often confused with *L. lilacinus* and most records are unsubstantiated with voucher material.

subdulcis (Pers.) Gray, *Nat. arr. Brit. pl.* 1: 625 (1821)
Agaricus lactifluus [var.] *subdulcis* Pers., *Syn. meth. fung.*: 433 (1801)
E: c **S:** c **W:** c **NI:** c **ROI:** c **O:** Channel Islands: ! Isle of Man: !
H: On soil in deciduous woodland, usually associated with *Fagus* but also commonly reported with *Betula* and *Quercus* spp.
D: NM2: 374 **D+I:** FNE2 194-195, FungEur7: 510-516 **I:** FlDan5: 170D, FlDan5: t.170D, Bon: 91, C&D: 413
Very common and widespread.

subumbonatus Lindgr., *Bot. Not.*: 200 (1845)
Mis.: *Lactarius cimicarius* sensu NCL & sensu Lange (FlDan5: 49 & pl. 173B)
Mis.: *Lactarius serifluus* sensu Pearson [Naturalist (Hull) 1950: 90 (1950)] & Ricken [Blätterpilze Deutschl.: 40 (1911)]
E: c **S:** o **W:** ! **NI:** c **ROI:** !
H: On soil in deciduous woodland, usually associated with *Quercus* spp.
D+I: Ph: 90 (as *Lactarius cimicarius*), FNE2 222-225, FungEur7: 603-607 **I:** FlDan5: 173B (as *L. cimicarius*), FlDan5: t. 173B (as *L. cimicarius*), Bon: 99, C&D: 415 (as *L. cimicarius*)
Apparently common in England and Northern Ireland, but rarely reported elsewhere. Often confused with similar taxa.

tabidus Fr., *Epicr. syst. mycol.*: 346 (1838)
Lactarius subdulcis var. *tabidus* (Fr.) Quél., *Enchir. fung.*: 131 (1886)
Mis.: *Lactarius theiogalus* sensu auct. p.p.
E: c **S:** c **W:** c **NI:** c **ROI:** c **O:** Channel Islands: !
H: On soil in deciduous woodland, most often reported with *Betula*, *Fagus* and *Quercus* spp.
D+I: Ph: 89, FNE2 200-201, FungEur7: 632-636 **I:** Bon: 93 (as *Lactarius theiogalus*), Bon: 93, C&D: 415 (as *L. theiogalus*), C&D: 415
Very common and widespread.

torminosus (Schaeff.) Pers. (as *Lactaria torminosa*), *Tent. disp. meth. fung.*: 69 (1797)
Agaricus torminosus Schaeff., *Fung. Bavar. Palat. nasc.* 1: 7 (1762)
Agaricus lactifluus [var.] *torminosus* (Schaeff.) Pers., *Syn. meth. Fung.*: 430 (1801)
Lactarius torminosus var. *sublateritius* Kühner & Romagn., *Bull. Soc. Mycol. France* 69: 364 (1954)
E: c **S:** c **W:** c **NI:** c **ROI:** c
H: On soil in deciduous woodland, always associated with *Betula* spp.
D: NM2: 365 **D+I:** Ph: 78-79, FNE2 156-157, FungEur7: 400-406 **I:** FlDan5: t.169A, Bon: 83, C&D: 401, Kriegl2: 373
Common and widespread.

trivialis (Fr.) Fr., *Epicr. syst. mycol.*: 337 (1838)
Agaricus trivialis Fr., *Observ. mycol.* 1: 61 (1815)
E: ? **S:** ! **NI:** ?
H: On acidic soil in woodland. British material is usually said to be associated with *Betula* spp, but these are often intermixed with conifers such as *Pinus* or *Picea* spp.
D: NM2: 370 **D+I:** FNE2 70-71, FungEur7: 152-158 **I:** Bon: 85, C&D: 405, Kriegl2: 401
British collections only from Scotland. Records from elsewhere are unsubstantiated with voucher material. In continental Europe this is a submontane species associated with *Picea* spp in native woodland.

turpis (Weinm.) Fr., *Epicr. syst. mycol.*: 335 (1838)
Agaricus turpis Weinm., *Syll. Pl. Nov.* 2: 85 (1828)
Mis.: *Lactarius necator* sensu auct.
Mis.: *Lactarius plumbeus* sensu Pearson [Naturalist (Hull) 1950: 86 (1950)] & sensu auct.
E: c **S:** c **W:** c **NI:** c **ROI:** c **O:** Channel Islands: ! Isle of Man: !
H: On soil in woodland or on heathland. Usually associated with *Betula* spp, but also known with conifers such as *Picea* and *Larix* spp
D: NM2: 365 (as *Lactarius necator*) **D+I:** Ph: 83, FNE2 42-43 (as *L. plumbeus*), FungEur7: 66-73 **I:** FlDan5: 169D, Bon: 87 (as *L. plumbeus*), C&D: 405 (as *L. necator*), Kriegl2: 375
Very common and widespread. Nomenclature follows Basso (FungEur7). English name = 'Ugly Milk Cap'.

uvidus (Fr.) Fr., *Epicr. syst. mycol.*: 338 (1838)
Agaricus uvidus Fr., *Observ. mycol.* 2: 191 (1815)
E: ! **S:** o **NI:** ! **ROI:** !
H: On soil, often in wet areas in deciduous woodland. Usually associated with *Betula* spp, but also reported with *Alnus*, *Corylus*, *Quercus* and *Salix* spp in similar habitats.
D: NM2: 366 **D+I:** Ph: 84-85, FNE2 82-83, FungEur7: 209-214 **I:** FlDan5: 170C, Bon: 87, C&D: 409, Kriegl2: 366
Occasional in Scotland. Rarely reported elsewhere but apparently widespread.

vellereus (Fr.) Fr., *Epicr. syst. mycol.*: 340 (1838)
Agaricus vellereus Fr., *Syst. mycol.* 1: 76 (1821)
Lactarius albivellus Romagn., *Bull. Soc. Mycol. France* 96(1): 92 (1980)
Lactarius vellereus var. *velutinus* (Bertill.) Bataille, *Fl. Mon. des Ast., Lact., et Russules.*: 35 (1908)
E: o **S:** o **W:** o **NI:** o **ROI:** !
H: On soil in woodland. Usually associated with *Betula*, *Corylus*, *Fagus* and *Quercus* spp.
D: NM2: 361 **D+I:** Ph: 76-77, FNE2 252-253, FungEur7: 713-718 **I:** FlDan5: t.170B, Bon: 95, C&D: 399, SV35: 41, Kriegl2: 361
Occasional but widespread. May be locally common, especially in old deciduous woodland.

vellereus *var.* **hometii** (Gillet) Boud., *Icon. mycol.* 1: 9, t. 49 (1905)
Lactarius hometii Gillet, *Tableaux analytiques des Hyménomycètes*: 14 (1884)
E: !
H: On soil in deciduous woodland. British collection associated with
Betula sp.
A single collection, from East Sussex (Abbots Wood) in 2004. Two other records from Herefordshire are unsubstantiated
with voucher material.

vietus (Fr.) Fr., *Epicr. syst. mycol.*: 344 (1838)
Agaricus vietus Fr., *Syst. mycol.* 1: 66 (1821)
E: c **S:** c **W:** c **NI:** c **ROI:** o
H: On soil or amongst amongst *Sphagnum* spp in woodland, often in wet areas under *Betula* spp.
D: NM2: 371 **D+I:** Ph: 86-87, FNE2 58-59, FungEur7: 125-130 **I:** FlDan5: t.177D, Bon: 87, C&D: 405, Kriegl2: 402
Common and widespread.

violascens (J. Otto) Fr., *Epicr. syst. mycol.*: 344 (1838)
    Agaricus violascens J. Otto, *Vers. Anordn. Agaric.*: 34 (1816)
    Lactarius uvidus var. *violascens* (J. Otto) Quél., *Fl. mycol.*
      France: 352 (1888)
E: ? **S:** ! **W:** ? **NI:** ?
H: On soil in deciduous woodland, usually reported with *Betula*
  and *Quercus* spp, and rarely with *Fagus*.
D: NM2: 366 **D+I:** FNE2 86-87, FungEur7: 215-219 **I:** FlDan5:
  173C, FlDan5: t.173C, Bon: 87, C&D: 409, Kriegl2: 367
Known from Scotland. Recorded from western England
  (Herefordshire and Shropshire), Wales, and Northern Ireland
  but unsubstantiated with voucher material.

volemus (Fr.) Fr., *Epicr. syst. mycol.*: 344 (1838)
    Agaricus volemus Fr., *Syst. mycol.* 1: 69 (1821)
    Agaricus ichoratus Batsch, *Elench. fung. (Continuatio Prima)*:
      37 (1786)
    Lactarius ichoratus (Batsch) Fr., *Epicr. syst. mycol.*: 345
      (1838)
    Agaricus lactifluus L., *Sp. pl.*: 1641 (1753)
    Agaricus oedematopus Scop., *Fl. carniol.*: 453 (1772)
    Lactarius volemus var. *oedematopus* (Scop.) Fr., *Epicr. syst.*
      mycol.: 345 (1838)
    Lactarius volemus var. *subrugatus* Neuhoff, *Pilze*
      Mitteleuropas, Die Milchlinge: 188 (1956)
E: ! **S:** ! **W:** ! **NI:** ! **ROI:** !
H: On soil in deciduous woodland. Usually reported with *Betula,*
  Fagus or *Quercus* spp.
D: NM2: 372 **D+I:** Ph: 88, FNE2 246-247, FungEur7: 702-707
  I: FlDan5: t.176G, Bon: 93, C&D: 413, Kriegl2: 429, SV43: 15
Apparently widespread but rarely reported in recent years and
  possibly a declining species in the British Isles.

zonarius (Bull.) Fr., *Epicr. syst. mycol.*: 336 (1838)
    Agaricus zonarius Bull., *Herb. France*: pl. 104 (1782)
    Mis.: *Lactarius insulsus* sensu NCL & sensu Phillips (Ph)
E: ! **S:** ! **W:** ! **NI:** !
H: On soil in deciduous woodland, associated with *Quercus* spp.
D: NM2: 368 **D+I:** Ph: 80-81 (as *Lactarius insulsus*), FNE2 130-
  131, FungEur7: 364-370 **I:** Bon: 83, C&D: 403
Rarely reported and possibly genuinely rare. All but the most
  recent records are likely to be sensu NCL or Phillips (Ph) (= *L.*
  evosmus).

LAETIPORUS Murrill, *Bull. Torrey Bot. Club* 31: 607 (1904)
Poriales, Coriolaceae
    Polyporus (Pers.) Gray [*nom. illegit.*, non *Polyporus* Adans.
      (1763)], *Nat. arr. Brit. pl.* 1: 645 (1821)
Type: *Laetiporus sulphureus* (Bull.) Bondartsev & Singer

sulphureus (Bull.) Bondartsev & Singer, *Mycologia* 12: 11
  (1920)
    Boletus sulphureus Bull., *Herb. France*: pl. 429 (1789)
    Polyporus sulphureus (Bull.) Fr., *Syst. mycol.* 1: 357 (1821)
    Grifola sulphurea (Bull.) Pilát, *Beih. Bot. Centralbl., Abt. B*: 39
      (1935)
    Boletus citrinus Planer [non *B. citrinus* With. (1776)], *Prog.*
      Pl. Erfurt.: 26 (1788)
    Boletus coriaceus Huds. [*nom. illegit.*, non *B. coriaceus* Scop.
      (1772); non *B. coriaceus* Bull. (1780)], *Fl. angl.*, Edn 2, 2:
      625 (1778)
    Boletus ramosus Bull., *Herb. France*: 418 (1789)
    Polyporus ramosus (Bull.) Gray, *Nat. arr. Brit. pl.* 1: 645
      (1821)
    Ptychogaster aurantiacus Pat. (anam.), *Rev. Mycol.*
      (Toulouse) 7: 28 (1885)
    Ceriomyces aurantiacus (Pat.) Sacc. (anam.), *Syll. fung.* 6:
      386 (1888)
    Calvatia versispora Lloyd (anam.), *Mycol. not.* 4 (40): 548
      (1916)
    Sporotrichum versisporum (Lloyd) Stalpers (anam.), *Stud.*
      Mycol. 24: 25 (1984)
E: c **S:** c **W:** c **NI:** c **ROI:** c **O:** Channel Islands: c Isle of Man: c

H: Parasitic then saprotrophic. Most frequent on *Quercus* spp,
  usually on old trees in woodland or parkland, but known on
  many other deciduous trees. Also frequent on *Taxus*. Rarely
  reported on other conifers such as *Larix, Pinus* and
  Pseudotsuga spp. Fruiting from early summer to autumn.
D: EurPoly1: 373, NM3: 233, Reid (1985, anamorph) **D+I:** Ph:
  222-223, B&K2: 316 400 **I:** Sow2: 135 (as *Boletus*
  sulphureus), Kriegl1: 541
Common and widespread. *Polyporus ramosus* was applied to
  monstrous, etiolated and branched forms growing in cellars or
  mines. English name = 'Chicken of the Woods' or 'Sulphur
  Polypore'.

LAETISARIA Burds., *Trans. Brit. Mycol. Soc.* 72: 420 (1979)
Stereales, Corticiaceae
Type: *Laetisaria fuciformis* (McAlpine) Burds.

fuciformis (McAlpine) Burds., *Trans. Brit. Mycol. Soc.* 72: 420
  (1979)
    Hypochnus fuciformis McAlpine, *Ann. Mycol.* 4: 549 (1906)
    Corticium fuciforme (McAlpine) Wakef., *Trans. Brit. Mycol.*
      Soc. 5: 481 (1917) [1916]
    Athelia fuciformis (McAlpine) Burds., *Trans. Brit. Mycol. Soc.*
      72: 422 (1979)
    Isaria fuciformis Berk. (anam.), *J. Linn. Soc., Bot.* 13: 175
      (1872)
E: ! **S:** ! **NI:** ! **ROI:** !
H: On soil and a parasite of various grasses.
D: NM3: 179
Uncommonly reported but possibly overlooked. May be
  abundant in utility grassland, especially on badly drained soils.
  The cause of 'Red Thread Disease' of turf grasses.

LAXITEXTUM Lentz, *U.S. Dept. Agric. Monogr.* 24: 18 (1955)
Hericiales, Gloeocystidiellaceae
Type: *Laxitextum bicolor* (Pers.) Lentz

bicolor (Pers.) Lentz, *U.S. Dept. Agric. Monogr.* 24: 19 (1955)
    Thelephora bicolor Pers., *Syn. meth. fung.*: 568 (1801)
    Stereum bicolor (Pers.) Fr., *Epicr. syst. mycol.*: 549 (1838)
    Thelephora fusca Schrad., in Gmelin, *Syst. nat.*: 1441 (1792)
    Stereum fuscum (Schrad.) P. Karst., *Ic. Hymenomyc.* 2: pl. 6
      (1883)
    Stereum pannosum Cooke, *Grevillea* 8: 56 (1879)
E: !
H: On decayed wood in open deciduous woodland. Usually on
  Betula spp, but also reported on *Fagus* and *Quercus* spp.
D: NM3: 280 **D+I:** CNE4: 794-799, B&K2: 120 106 **I:** Kriegl1:
  253
Known from Bedfordshire, Oxfordshire, Shropshire, Surrey,
  Warwickshire, West Kent and West Sussex.

LECCINUM Gray, *Nat. arr. Brit. pl.* 1: 646 (1821)
Boletales, Boletaceae
    Krombholzia P. Karst., *Rev. Mycol. (Toulouse)* 3(9): 17
      (1881)
    Krombholziella Maire, *Publ. Inst. Bot. Barcelona* 3(4): 41
      (1937)
Type: *Leccinum aurantiacum* (Bull.) Gray

atrostipitatum A.H. Sm., Thiers & Watling, *Michigan Botanist*
  5: 155 (1966)
E: ! **S:** ?
H: On soil in deciduous woodland, usually associated with *Betula*
  or *Quercus* spp.
D+I: FRIC13: 12-13 104
Known from East Kent, South Hampshire and Surrey. Reported
  from Scotland but unsubstantiated with voucher material. Old
  specimens could easily be mistaken for *L. versipelle*.

aurantiacum (Bull.) Gray, *Nat. arr. Brit. pl.* 1: 646 (1821)
    Boletus aurantiacus Bull., *Herb. France*: pl. 75 (1781)

Boletus scaber var. *aurantiacus* (Bull.) Opat., *Comm. fam. bolet.*: 34 (1836)
Krombholziella aurantiaca (Bull.) Maire, *Publ. Inst. Bot. Barcelona* 3(4): 46 (1937)
Leccinum populinum M. Korhonen, *Karstenia* 35(2): 55 (1995)
Boletus rufus Schaeff. [*nom. illegit.*, non *B. rufus* Schrad. (1792)], *Fung. Bavar. Palat. nasc.* 2: 103 (1763)
Mis.: *Leccinum quercinum* sensu auct.
E: o **S:** o **W:** o **NI:** o
H: On soil in mixed deciduous woodland. Mainly associated with *Populus* spp, including *P. tremula*, but recently also shown to be associated with *Betula* spp.
D: NM2: 65 **D+I:** Ph: 210-211, B&K3: 72 36 (as *Leccinum rufum*), L&E: 132 34 **I:** Bon: 41 (as *Krombholziella aurantiaca*), Kriegl2: 281
Occasional but widespread. May be locally common and abundant.

crocipodium (Letell.) Watling, *Trans. & Proc. Bot. Soc. Edinburgh* 39(2): 200 (1961)
Boletus crocipodius Letell., *Fig. Champ.*: t. 666 (1838)
Boletus nigrescens Richon & Roze, *Atlas champ.*: 191 (1888)
Leccinum nigrescens (Richon & Roze) Singer [*nom. illegit.*], *Amer. Midl. Naturalist* 37: 116 (1947)
Krombholziella nigrescens (Richon & Roze) Šutara, *Česká Mykol.* 36(2): 81 (1982)
Boletus tessellatus Gillet, *Hyménomycètes*: 636 (1878)
E: !
H: On soil in warm deciduous woodland, always associated with *Quercus* spp.
D: NM2: 64 (as *Leccinum tessellatus*) **D+I:** Ph: 211, B&K3: 72 34 (as *L. nigrescens*), L&E: 148 39 **I:** Bon: 41 (as *Krombholziella crocipodia*), Kriegl2: 274
Occasional but widespread throughout England. Mainly southern but scattered reports as far north as South Lancashire and Yorkshire.

cyaneobasileucum Lannoy & Estadès, *Doc. Mycol.* 21 (81): 23 (1991)
Leccinum brunneogriseolum Lannoy & Estadès, *Doc. Mycol.* 21 (82): 1 (1991)
E: ! **S:** ! **W:** ! **NI:** !
H: On acidic soil on heathland, moorland or in woodland, often amongst *Sphagnum* moss in wet or swampy areas, and usually associated with *Betula* spp.
D+I: L&E: 100 21, L&E: 96 18 (as *L. brunneogriseolum*) **I:** C&D: 441 (as *L. brunneogriseolum*), C&D: 441
Rarely reported but apparently widespread. *Fide* A.E. Hills (pers. comm.) this is a variable species influenced by the amount of available moisture in the habitat.

duriusculum (Kalchbr.) Singer, *Amer. Midl. Naturalist* 37: 122 (1947)
Boletus duriusculus Kalchbr., Fries, *Hymenomyc. eur.*: 515 (1874)
E: o **S:** ! **W:** ! **NI:** ! **ROI:** !
H: On soil, associated only with *Populus* spp, in parkland and cemeteries.
D: NM2: 64 **D+I:** Ph: 212, B&K3: 70 32 **I:** Bon: 41 (as *Krombholziella duriuscula*), Kriegl2: 278
Occasional in England, especially in southern counties. Rarely reported elsewhere but apparently widespread.

holopus (Rostk.) Watling, *Trans. Brit. Mycol. Soc.* 43: 692 (1960)
Boletus holopus Rostk., *Deutschl. Fl.* Edn 3: 131 (1844)
Krombholziella holopus (Rostk.) Šutara, *Česká Mykol.* 36(2): 81 (1982)
Leccinum holopus var. *americanum* A.H. Sm. & Thiers, *The Boletes of Michigan*: 183 (1971)
Leccinum nucatum Lannoy & Estadès, *Doc. Mycol.* 23 (89): 63 (1993)
Mis.: *Leccinum aerugineum* sensu Lannoy & Estadès [*Doc. Mycol.* 21 (81): 23 (1991)]
E: o **S:** o **W:** ! **NI:** ! **ROI:** !

H: On soil in deciduous woodland, often in wet areas amongst *Sphagnum* spp, and usually associated with *Betula* spp.
D+I: Ph: 212-213, B&K3: 70 33, L&E: 76 9, L&E: 82 12B (as *Leccinum nucatum*) **I:** Bon: 41 (as *Krombholziella holopus*), SV32: 48, Kriegl2: 267 (as *L. nucatum*)
Occasional to rather rare, but widespread. A reddening variant (*L. holopus* var. *americanum*) has been reported from England (South Devon: Culm Davy Woods) in 1995. A greenish form has been referred to *L. aerugineum*, but this taxon is based on an old and doubtful name.

palustre M. Korhonen, *Karstenia* 35(2): 63 (1995)
E: ! **NI:** !
H: On soil in wet areas in open moorland, always associated with *Betula* spp.
Little known in the British Isles. Collections only from England (South Devon) and Northern Ireland (Fermanagh: Florence Court).

pseudoscabrum (Kallenb.) Šutara, *Česká Mykol.* 43(1): 6 (1989)
Boletus pseudoscaber Kallenb., *Pilze Mitteleuropas* 1: 117 (1935)
Krombholziella pseudoscabra (Kallenb.) Šutara, *Česká Mykol.* 36(2): 82 (1982)
Boletus scaber var. *carpini* R. Schulz, in Michael & Schulz, *Führer für Pilzfreunde* 1: pl. 95 (1924)
Boletus carpini (R. Schulz) A. Pearson, *Naturalist (Hull)* 1946: 96 (1946)
Leccinum carpini (R. Schulz) D.A. Reid, *Trans. Brit. Mycol. Soc.* 48: 525 (1965)
E: o **S:** ! **W:** ! **NI:** !
H: On soil in deciduous woodland. Usually associated with *Carpinus* but also known with *Corylus*.
D: NM2: 65 **D+I:** L&E: 156 41 (as *Leccinum carpini*) **I:** Kriegl2: 276 (as *L. carpini*)
Occasional in England, and (*fide* A.E. Hills, pers. comm.) most frequent in south-eastern counties. Rarely reported elsewhere but apparently widespread.

quercinum (Pilát) Pilát & Dermek, *Hribovite Huby*: 151 (1974)
Leccinum aurantiacum var. *quercinum* Pilát, *Mushrooms and other Fungi*: t. 6 (1961)
Krombholziella quercina (Pilát) Šutara, *Česká Mykol.* 36(2): 82 (1982)
E: o **S:** ? **NI:** !
H: On soil in old deciduous woodland, often in glades or other open areas, associated only with *Quercus* spp.
D: NM2: 65 **D+I:** Ph: 209 & 211, B&K3: 72 35 **I:** Kriegl2: 283
Occasional in England, from Warwickshire to West Sussex and South Hampshire, and may be locally abundant in certain areas such as the New Forest. Rarely reported elsewhere. A single record from Northern Ireland (Fermanagh). Reported from Scotland, but unsubstantiated with voucher material.

rigidipes P.D. Orton, *Trans. Brit. Mycol. Soc.* 91(4): 560 (1988)
Mis.: *Leccinum oxydabile* sensu auct. Brit.
Mis.: *Leccinum umbrinoides* sensu auct. Brit.
E: c **S:** c **W:** c **NI:** ! **ROI:** !
H: On soil in deciduous woodland and on wet heathland, always associated with *Betula* spp.
Common and widespread. *Fide* A.E. Hills (pers. comm.) this is much confused with *Leccinum scabrum*.

roseofractum Watling, *Notes Roy. Bot. Gard. Edinburgh* 28(3): 313 (1968)
Krombholziella roseofracta (Watling) Šutara, *Česká Mykol.* 36(2): 82 (1982)
Leccinum pulchrum Lannoy & Estadès, *Doc. Mycol.* 21 (82): 3 (1991)
E: o **S:** o **W:** o **NI:** ! **ROI:** !
H: On soil in damp deciduous woodland, always associated with *Betula* spp.
D+I: L&E: 56 1B, L&E: 58 2 (as *Leccinum pulchrum*)
Widespread and fairly common.

roseotinctum Watling, *Notes Roy. Bot. Gard. Edinburgh* 29(2): 267 (1969)

  Krombholziella roseotincta (Watling) Šutara, *Česká Mykol.* 36(2): 82 (1982)

E: ? **S:** ! **NI:** ?

H: On soil in woodland, associated with *Betula* and *Pinus* spp, usually amongst grass in glades or open areas.

D+I: L&E: 108 24

Rarely reported. Known from central and northern Scotland. A single record from northern England and recently reported from Northern Ireland but unsubstantiated with voucher material.

salicola Watling, *Notes Roy. Bot. Gard. Edinburgh* 31: 139 (1971)

  Krombholziella salicola (Watling) Šutara, *Česká Mykol.* 36(2): 82 (1982)

E: ? **S:** !

H: On soil in dunes. Always associated with *Salix repens.*

D+I: L&E: 142

Known from northern Scotland and the Outer Hebrides. Reported from England (South Hampshire) but unsubstantiated with voucher material.

scabrum (Bull.) Gray, *Nat. arr. Brit. pl.* 1: 646 (1821)

  Boletus scaber Bull., *Herb. France* pl. 132 (1783)

  Krombholziella scabra (Bull.) Maire, *Publ. Inst. Bot. Barcelona* 3(4): 46 (1937)

  Boletus scaber var. *melaneus* Smotl., *Cas. Csk. Houb* 28: 69 (1951)

  Leccinum scabrum var. *melaneum* (Smotl.) Dermek, *Fungorum Rar. Icon. Color.* 16: 17 (1987)

  Leccinum melaneum (Smotl.) Pilát & Dermek, *Hribovite Huby.* 145 (1974)

  Krombholziella melanea (Smotl.) Šutara, *Česká Mykol.* 36(2): 81 (1982)

  Boletus avellaneus J. Blum, *Bull. Soc. Mycol. France* 85(4): 560 (1970) [1969]

  Leccinum avellaneum (J. Blum) Bon, *Doc. Mycol.* 9 (35): 41 (1979)

  Krombholziella mollis Bon, *Doc. Mycol.* 56 (14): 22 (1984)

  Leccinum molle (Bon) Lannoy & Estadès, *Doc. Mycol.* 19 (75): 58 (1989)

  Boletus murinaceus J. Blum, *Bull. Soc. Mycol. France* 85(4): 560 (1970) [1969]

  Leccinum murinaceum (J. Blum) Bon, *Doc. Mycol.* 9 (35): 41 (1979)

  Leccinum olivaceosum Lannoy & Estadès, *Doc. Mycol.* 24 (94): 10 (1994)

  Leccinum subcinnamomeum Pilát & Dermek [*nom. inval.*], *Hribovite Huby.* 144 (1974)

E: c **S:** c **W:** c **NI:** c **ROI:** c **O:** Isle of Man: !

H: On soil in mixed deciduous woodland or in pure birchwoods but strictly associated with *Betula* spp.

D: NM2: 65 **D+I:** Ph: 212-213, FRIC16: 17 127a (as *Leccinum scabrum* var. *melaneum*), B&K3: 74 37, L&E: 62 4, L&E: 66 5 (as *L. scabrum* var. *melaneum*), L&E: 68 6 (as *L. murinaceum*), L&E: 72 8 (as *L. avellaneum*), L&E: 78 10 (as *L. olivaceosum*), L&E: 80 11 (as *L. molle*) **I:** Bon: 41 (as *Krombholziella scabra*)

Very common and widespread.

variicolor Watling, *Notes Roy. Bot. Gard. Edinburgh* 24: 268 (1969)

  Krombholziella variicolor (Watling) Šutara, *Česká Mykol.* 36(2): 83 (1982)

  Mis.: *Krombholzia scabra* var. *coloratipes* sensu auct.

E: c **S:** c **W:** c **NI:** c **ROI:** c

H: On soil in damp areas in deciduous woodland or open heathland. Strictly associated with *Betula* spp.

D: NM2: 66 **D+I:** Ph: 212, L&E: 86 13 **I:** Bon: 41 (as *Krombholziella variicolor*), Kriegl2: 271

Common and widespread.

versipelle (Fr. & Hök) Snell, *Lloydia* 7(1): 58 (1944)

  Boletus versipellis Fr. & Hök, *Boleti fung. gen.*: 13 (1835)

  Boletus rufescens Konrad, *Bull. Soc. Linn. de Lyon* 1: 151 (1932)

  Krombholziella rufescens (Konrad) Šutara, *Česká Mykol.* 36(2): 82 (1982)

  Leccinum rufescens (Konrad) Šutara, *Česká Mykol.* 43(1): 7 (1989)

  Boletus testaceoscaber Secr. [*nom. inval.*], *Mycogr. Suisse* 3: 8 (1833)

  Leccinum testaceoscabrum Singer, *Amer. Midl. Naturalist* 37: 123 (1947)

  Mis.: *Leccinum cerinum* sensu auct. Brit.

E: o **S:** o **W:** ! **ROI:** !

H: On acidic soil on heathland and in deciduous woodland and scrub, always associated with *Betula* spp.

D: NM2: 65 **D+I:** Ph: 208 & 211, B&K3: 74 38, L&E: **I:** Bon: 41 (as *Krombholziella versipellis*), Kriegl2: 279

Occasional but widespread. Most frequent (and may be locally common) in northern Britain.

vulpinum Watling, *Trans. & Proc. Bot. Soc. Edinburgh* 39(2): 197 (1961)

  Boletus vulpinus (Watling) Hlaváček, *Mykologický Sborník* 67 (4 - 5): 113 (1990)

S: !

H: On acidic soil in Caledonian pinewoods. Associated with *Pinus sylvestris* often fruiting amongst *Vaccinium myrtillus* and *Arctostaphylos uva-ursi.*

D: NM2: 64 **D+I:** L&E: 138 36

Rarely reported.

LENTARIA Corner, *Monograph of Clavaria and Allied Genera*: 437 (1950)

Gomphales, Lentariaceae

Type: *Lentaria surculus* (Berk.) Corner

afflata (Lagger) Corner, *Monograph of Clavaria and Allied Genera*: 438 (1950)

  Clavaria afflata Lagger, *Flora* 19: 231 (1836)

  Clavaria delicata Fr., *Syst. mycol.* 1: 475 (1821)

  Lentaria delicata (Fr.) Corner, *Monograph of Clavaria and Allied Genera* 1: 441 (1950)

E: !

H: On decayed wood. British collections on a decayed and partially burnt sawdust heap, in mixed deciduous woodland.

D+I: Ph: 260-261 (as *Lentaria delicata*), B&K2: 342 439 (as *L. alboviolacea*)

Known only from Buckinghamshire (Ibstone, Common Hill Wood) in 1978 and 1979.

LENTINELLUS P. Karst., *Bidrag Kännedom Finlands Natur Folk* 32: 246 (1879)

Hericiales, Lentinellaceae

Type: *Lentinellus cochleatus* (Pers.) P. Karst.

cochleatus (Pers.) P. Karst., *Bidrag Kännedom Finlands Natur Folk* 32: 247 (1879)

  Agaricus cochleatus Pers., *Tent. disp. meth. fung.*: 22 (1797)

  Omphalia cochleata (Pers.) Gray, *Nat. arr. Brit. pl.* 1: 612 (1821)

  Lentinus cochleatus (Pers.) Fr., *Syst. orb. veg.* 1: 78 (1825)

  Agaricus cornucopioides Bolton, *Hist. fung. Halifax* 1: 8 (1788)

  Merulius cornucopioides (Bolton) With. [non *M. cornucopiodes* (L.) Pers. (1801)], *Arr. Brit. pl. ed. 3* 4: 151 (1796)

  Omphalia cochleata β *cornucopioides* (Bolton) Gray, *Nat. arr. Brit. pl.* 1: 612 (1821)

  Lentinellus cochleatus ssp. *inolens* Konrad & Maubl., *Icon. select. fung.* 4: pl. 316 (1926)

  Agaricus confluens Sowerby, *Col. fig. Engl. fung.* 2: pl. 168 (1798)

  Mis.: *Lentinus umbellatus* sensu Rea (1922)

E: o **S:** o **W:** o **NI:** o **ROI:** o **O:** -

H: On dead and decayed wood, often stumps, of deciduous trees. Most often reported on *Fraxinus* and occasionally *Fagus* and *Corylus*. Reported on wood of conifers but unsubstantiated with voucher material and the hosts unverified.
D: BFF6: 75, NM3: 287 **D+I:** Ph: 188, B&K3: 204 234 **I:** FlDan5: 197H (as *Lentinus cochleatus*), Bon: 125, C&D: 149, Kriegl3: 7
Occasional but widespread. Basidiomes in active growth have a strong scent of anise, lacking in the var. *inolens* which is otherwise identical.

flabelliformis (Bolton) S. Ito, *Mycol. Fl. Japan* 2: 151 (1959)
    Agaricus flabelliformis Bolton, *Hist. fung. Halifax* Suppl.: 157 (1791)
    Lentinus scoticus Berk. & Broome, *Ann. Mag. Nat. Hist., Ser. 4 [Notices of British Fungi no. 1423]* 15: 30 (1875)
    Lentinus omphalodes var. *scoticus* (Berk. & Broome) Pilát, *Atlas champ. Eur.* 5: 35 (1940)
    Mis.: *Lentinellus omphalodes* sensu Pearson & Dennis (1948)
S: !
H: On decayed wood of deciduous trees or conifers.
D: BFF6: 76, NM3: 287 (as *Lentinellus omphalodes*) **D+I:** B&K3: 206 235 (as *L. omphalodes*), Cooke 1094 (1143) Vol. 7 (1890) (as *Lentinus scoticus*).
Two collections, from Angus (Glamis) by J. Stevenson in 1872 and 1873. Confused with *Lentinellus omphalodes* which is unknown in Britain.

laurocerasi (Berk. & Broome) P.D. Orton, *Trans. Brit. Mycol. Soc.* 43(2): 177 (1960)
    Agaricus laurocerasi Berk. & Broome, *Ann. Mag. Nat. Hist., Ser. 5 [Notices of British Fungi no. 1854]* 7: 126 (1881)
    Pleurotus laurocerasi (Berk. & Broome) Sacc., *Syll. fung.* 5: 367(1887)
W: !
H: On dead wood. British material on *Prunus laurocerasus*.
D: BFF6: 78-79 **I:** Cooke 287 (242) Vol. 2 (1883)
Know only from the type, from Denbighshire (Coed Coch) in 1879. Doubtfully a species of *Lentinellus*, since it has unornamented spores.

tridentinus (Sacc. & P. Syd.) Singer, *Ann. Mycol.* 41: 146 (1946)
    Lentinus tridentinus Sacc. & P. Syd., *Syll. fung.* 16: 65 (1902)
    Lentinus badius Bres., *Fungi trident.* 2: 56 (1898)
S: !
H: On dead woody stems, most often *Chamerion angustifolium*.
D: BFF6: 76-77
British collections only from Peeblesshire, Perthshire and South Aberdeen. Rather doubtfully distinct from *Lentinellus flabelliformis*

ursinus (Fr.) Kühner, *Botaniste* 17: 99 (1926)
    Agaricus ursinus Fr., *Syst. mycol.* 1: 185 (1821)
    Lentinus ursinus (Fr.) Fr., *Epicr. syst. mycol.*: 395 (1838)
E: !
H: On decayed wood of deciduous trees in old areas of woodland. Usually on large fallen trunks or stumps of *Fagus* and rarely reported on *Salix* spp.
D: BFF6: 77-78, NM3: 287, NBA8: 587-588 **D+I:** B&K3: 206 236 **I:** Bon: 125, C&D: 149, Kriegl3: 10
Known from Berkshire, South Essex, South Hampshire, Surrey and West Kent.

vulpinus (Sowerby) Kühner & Maire, *Bull. Soc. Mycol. France* 50: 16 (1934)
    Agaricus vulpinus Sowerby, *Col. fig. Engl. fung.* 3: pl. 361 (1803)
    Lentinus vulpinus (Sowerby) Fr., *Synopsis generis Lentinorum*: 12 (1836)
    Lentinus auricula Fr., *Öfvers. Kongl. Vetensk.-Akad. Förh.* 18: 29 (1862) [1861]
E: ! **S:** ?
H: On decayed wood of deciduous trees, usually *Ulmus* or *Tilia* spp.
D: BFF6: 78, NM3: 287 **I:** Kriegl3: 11

Known from England (Berkshire, Buckinghamshire and Surrey). A single rather dubious collection from Scotland (Easterness: Rothiemurchus Forest) on 'pine sawdust'.

LENTINUS Fr., *Syst. orb. veg.* 1: 77 (1825)
Poriales, Lentinaceae
Type: *Lentinus crinitus* (L.) Fr.

tigrinus (Bull.) Fr., *Syst. orb. veg.* 1: 78 (1825)
    Agaricus tigrinus Bull., *Herb. France*: pl. 70 (1786)
    Omphalia tigrina (Bull.) Gray, *Nat. arr. Brit. pl.* 1: 613 (1821)
    Panus tigrinus (Bull.) Singer, *Lilloa* 22: 275 (1951) [1949]
    Lentinus dunalii (DC.) Fr., *Syst. orb. veg.* 1: 78 (1825)
    Lentinus tigrinus var. *dunalii* (DC.) Rea, *Brit. basidiomyc.*: 537 (1922)
    Lentinus fimbriatus Curr., *Trans. Linn. Soc. London* 24: 151 (1863)
E: o **NI:** ?
H: On living or dead wood, often in wet locations such as woodland or trees at the edges of ponds, lakes, rivers etc., and may occur on logs floating or partially submerged in water. Usually on *Salix* spp (especially *S. fragilis*), less frequently *Alnus* and rarely *Fagus*.
D: BFF6: 14, NM2: 47 **D+I:** Myc3(4): 195, FAN2: 26-27, B&K3: 208 240 **I:** Sow1: 68 (as *Agaricus tigrinus*), Bon: 123, C&D: 149, Kriegl3: 20, Cooke 1089 (1138) Vol.7 (1890)
Generally rather rare, but apparently widespread in England, most frequent in southern counties but reported as far north as Yorkshire. A single old record from Northern Ireland, unsubstantiated with voucher material.

LENZITES Fr., *Fl. Scan.*: 339 (1835)
Poriales, Coriolaceae
Type: *Lenzites betulinus* (L.) Fr.

betulinus (L.) Fr., *Epicr. syst. mycol.*: 405 (1838)
    Agaricus betulinus L., *Sp. pl.*: 1176 (1753)
    Daedalea betulina (L.) Fr., *Syst. mycol.* 1: 333 (1821)
    Lenzites flaccidus Fr., *Epicr. syst. mycol.*: 406 (1838)
    Lenzites betulinus f. *flaccidus* (Fr.) Bres., *Hedwigia* 53: 50 (1912)
    Apus coriaceus Gray, *Nat. arr. Brit. pl.* 1: 617 (1821)
    Daedalea variegata Fr., *Observ. mycol.* 2: 240 (1818)
    Mis.: *Agaricus coriaceus* sensu auct.
E: c **S:** o **W:** c **NI:** c **ROI:** c **O:** Channel Islands: !
H: On dead wood of deciduous trees in woodland and on heathland. Most frequently reported on dead stumps of *Betula* spp, and occasionally on fallen trunks of *Fagus*. Also known on *Alnus*, *Corylus*, *Crataegus*, *Populus tremula*, *Quercus* and *Ulmus* spp.
D: EurPoly1: 377, NM3: 228 **D+I:** Ph: 232-233, B&K2: 284 352 **I:** Sow2: 182 (as *Agaricus betulinus*), Kriegl1: 543
Generally common and widespread, but occasional to rather rare in Scotland *fide* R. Watling (pers. comm.).

LEPIDOMYCES Jülich, *Persoonia* 10: 329 (1979)
Stereales, Xenasmataceae
Type: *Lepidomyces subcalceus* (Litsch.) Jülich

subcalceus (Litsch.) Jülich, *Persoonia* 10(3): 330 (1979)
    Peniophora subcalcea Litsch., *Oesterr. Bot. Z.* 8: 119 (1939)
E: !
H: British material on a decayed stem of *Rubus fruticosus* agg.
D: NM3: 140 **D+I:** CNE8: 1454-1456, Myc9(2): 52
A single collection from South Devon (Watcombe Woods) in 1994.

LEPIOTA (Pers.) Gray, *Nat. arr. Brit. pl.* 1: 601 (1821)
Agaricales, Agaricaceae
    Hiatula (Fr.) Mont., *Ann. Sci. Nat., Bot.*, sér. 4, 1: 107 (1854)
    Echinoderma (Bon) Bon, *Doc. Mycol.* 21 (82): 61 (1991)
Type: *Agaricus colubrinus* Pers. (= *Lepiota clypeolaria*)

aspera (Pers.) Quél., *Enchir. fung.*: 5 (1886)

Agaricus asper Pers., in Hoffmann, *Abb. Schwaemme* 3: pl. 21 (1793)

Amanita aspera (Pers.) Pers., *Observ. mycol.* 2: 38 (1799)

Cystolepiota aspera (Pers.) Knudsen, *Bot. Tidsskr.* 73: 129 (1978)

Echinoderma asperum (Pers.) Bon, *Doc. Mycol.* 21 (82): 62 (1991)

Agaricus acutesquamosus Weinm., *Syll. Pl. Not. Ratisb.* 1: 70 (1824)

Lepiota acutesquamosa (Weinm.) P. Kumm., *Führ. Pilzk.*: 136 (1871)

Lepiota friesii var. *acutesquamosa* (Weinm.) Quél., *Mém. Soc. Émul. Montbéliard, Sér. 2,* 5: 72 (1872)

Lepiota acutesquamosa var. *furcata* Kühner, *Bull. Soc. Mycol. France* 52: 210 (1936)

Agaricus elvensis Berk. & Broome, *Ann. Mag. Nat. Hist., Ser. 3 [Notices of British Fungi no. 1009]* 15: 316 (1865)

Agaricus friesii Lasch, *Linnaea* 3: 155 (1828)

Lepiota friesii (Lasch) Quél., *Mém. Soc. Émul. Montbéliard, Sér. 2,* 5: 72 (1872)

Agaricus mariae Klotzsch, *Linnaea* 7: 196 (1832)

E: c **S:** ! **W:** o **NI:** ! **ROI:** ! **O:** Channel Islands: !

H: On soil in deciduous woodland, usually associated with *Corylus, Fagus* or *Fraxinus.* Rarely on decayed wood such as old stumps or large logs.

D: NM2: 221 **D+I:** Ph: 27 (young basidiomes; as *Lepiota friesii*), FungEur4: 126-130 7, B&K4: 192 214, FAN5: 144-145 **I:** Bon: 283 (as *Cystolepiota aspera*), C&D: 241 (as *Echinoderma asperum*)

Rather common in England, especially so in areas of calcareous loam soil in southern counties. Occasional to rather rare elsewhere but widespread.

bickhamensis P.D. Orton, *Notes Roy. Bot. Gard. Edinburgh* 41(3): 588 (1984)

E: !

H: On soil in deciduous woodland.

Described from Somerset. Part of the *Lepiota echinella* complex, poorly known and requiring clarification.

boertmannii Knudsen, *Bot. Tidsskr.* 75(2-3): 150 (1981)

E: !

H: On soil. British material with *Crataegus* sp., in scrubby woodland on limestone.

D: FungEur4: 130-131, NM2: 222 **D+I:** FAN5: 150 **I:** SV26: 35 (as *Echinoderma boertmannii*)

A single collection, from South Devon (Ipplepen, Orley Common) in 1996.

boudieri Bres., *Fungi trident.* 1: 43 (1882)

Lepiota acerina Peck, *Rep. (Annual) New York State Mus. Nat. Hist.* 51 (Appx. 2): 283 (1898)

Lepiota fulvella Rea, *Trans. Brit. Mycol. Soc.* 6(1): 61 (1918)

E: o **S:** ! **W:** o **NI:** o **ROI:** o

H: On soil (often calcareous loam) in deciduous woodland, frequently in nettlebeds in such places. Rarely on decayed stumps and large fallen trunks of *Fagus.*

D: FungEur4: 195-198, FungEur4: 212 (as *Lepiota fulvella*), NM2: 218 (as *L. fulvella*) **D+I:** Ph: 29 (as *L. fulvella*), B&K4: 192 216, FAN5: 132-133 **I:** C&D: 243

Occasional but widespread. Better known as *Lepiota fulvella.*

brunneoincarnata Chodat & C. Martín, *Bull. Soc. Bot. Genève* 5: 222 (1889)

Lepiota barlae Pat. [*nom. illegit.*, non *L. barlae* Quél. (1898)], *Bull. Soc. Mycol. France* 19: 117 (1905)

Lepiota patouillardii Sacc. & Trotter, *Syll. fung.* 21: 17 (1912)

Mis.: *Lepiota helveola* sensu Rea (1922)

E: ! **W:** ! **O:** Channel Islands: ?

H: On soil, with deciduous trees or conifers in native woodland, plantations or gardens, and also amongst grass on dunes.

D: NM2: 220 **D+I:** Ph: 28-29, FungEur4: 248-252 26a, FAN5: 127 **I:** FlDan1: 13F, Bon: 287, C&D: 245

Rarely reported. Known from England (Buckinghamshire, East Gloucestershire, East Kent, Hertfordshire, London, Middlesex, North & South Essex, Oxfordshire, South Devon, Surrey and West Sussex) and Wales (Denbighshire). Reported from the Channel Islands but unsubstantiated with voucher material.

brunneolilacea Bon & Boiffard, *Bull. Soc. Mycol. France* 88: 18 (1972)

Mis.: *Lepiota helveola* sensu Reid [TBMS 84: 719 (1985)]

O: Channel Islands: !

H: On sand or sandy soil in dunes.

D+I: FungEur4: 252-254 27 **I:** Bon: 287 (as *Lepiota helveola*), C&D: 245

Known from Jersey and recorded several times there since 1977.

calcicola Knudsen, *Bot. Tidsskr.* 75(2-3): 140 (1980)

Mis.: *Lepiota hispida* sensu Lange (FlDan1: 26)

E: !

H: On soil (usually calcareous loam) in deciduous woodland.

D: NM2: 222 **D+I:** FungEur4: 131-134 9a, B&K4: 194 217, FAN5: 146-147 **I:** SV34: 33

A single collection from West Sussex (Wisborough Green) in 1969. Can be confused with *L. hystrix.*

castanea Quél., *Compt. Rend. Assoc. Franç. Avancem. Sci.* 9: 661 (1881)

Lepiota ignicolor Bres., *Fungi trident.* 2: 3 (1892)

Lepiota ignipes Bon, *Doc. Mycol.* 8 (30-31): 70 (1978)

Lepiota rufidula Bres., *Atti Imp. Regia Accad. Roveretana, ser. 3,* 8: 129 (1902)

E: o **S:** ! **W:** o **NI:** ! **ROI:** !

H: On soil in woodland, often with *Corylus* or *Fagus* but known with a wide range of other deciduous species. Less frequently reported with conifers such as *Pinus, Thuja* spp and *Taxus.*

D: NM2: 218 **D+I:** FRIC6: 11-14 43b, Ph: 28-29, FungEur4: 198-203 18, B&K4: 194 219, FAN5: 130-131, FM3(4): 128 + 131 **I:** SV26: 39, C&D: 243

Occasional but widespread and may occur in swarms. Especially frequent on calcareous loam soils in southern England, much less so elsewhere and only a single collection from Scotland.

cingulum Kelderman, *Persoonia* 15(4): 537 (1994)

E: !

H: On soil, usually in scrub woodland. Two of the three British collections were with *Salix* spp.

D+I: FAN5: 131-132, FM3(4): 128 + 131 **I:** SV34: 33, Myc12(1): 37 (as *Lepiota tomentella* f. *rubidella*)

Known from Oxfordshire, Surrey and West Gloucestershire.

clypeolaria (Bull.) P. Kumm., *Führ. Pilzk.*: 137 (1871)

Agaricus clypeolarius Bull., *Herb. France:* pl. 405 (1789)

Lepiota clypeolaria var. *minor* J.E. Lange [*nom. inval.*], *Fl. agaric. danic.* 5, *Taxonomic Conspectus:* V (1940)

Agaricus colubrinus Pers., *Syn. meth. fung.*: 258 (1801)

Lepiota colubrina (Pers.) Gray, *Nat. arr. Brit. pl.* 1: 601 (1821)

Lepiota ochraceosulfurescens Bon, *Doc. Mycol.* 16 (61): 46 (1985)

E: o **S:** ! **W:** ! **NI:** !

H: On decayed leaf litter or soil in deciduous woodland. Rarely with conifers.

D: NM2: 219 **D+I:** FungEur4: 168-172 15, B&K4: 196 220, FAN5: 116-117 **I:** Bon: 285, C&D: 243

Occasional in England, especially on calcareous loam in southern counties. Rarely reported elsewhere but apparently widespread.

cortinarius J.E. Lange, *Dansk. Bot. Ark.* 2(3): 25 (1915)

E: ! **S:** ! **W:** ! **NI:** ! **ROI:** !

H: On soil in deciduous woodland and also reported on mulched flowerbeds.

D: NM2: 217 **D+I:** FungEur4: 204-205, FAN5: 120-121

Rarely reported but apparently widespread.

cortinarius var. audreae D.A. Reid, *Fungorum Rar. Icon. Color.* 3: 8 & pl. 19a (1968)

Lepiota audreae (D.A. Reid) Bon, *Doc. Mycol.* 11 (43): 35 (1981)

E: !

H: On soil in deciduous woodland.

D: FAN5: 121 **D+I:** FungEur4: 205 19b
Rarely reported. Dubiously distinct from the typical var., differing chiefly in the colour of the pileus. Reported from the Republic of Ireland, but the collection (in herb. K.) is typical *L. cortinarius*.

coxheadii P.D. Orton, *Notes Roy. Bot. Gard. Edinburgh* 41(3): 589 (1984)
E: !
H: On soil in deciduous woodland. The type collection was under *Hedera* and *Rubus fruticosus* agg.
Described from Somerset. Part of the *Lepiota echinella* complex, poorly known and requiring clarification.

cristata (Bolton) P. Kumm., *Führ. Pilzk.*: 137 (1871)
Agaricus cristatus Bolton [non *A. cristatus* Scop. (1774)], *Hist. fung. Halifax* 1: 7 (1788)
Lepiota cristata var. *felinoides* Bon, *Doc. Mycol.* 11 (43): 34 (1981)
Lepiota felinoides (Bon) P.D. Orton [*nom. illegit.*, non *L. felinoides* Peck (1900)], *Notes Roy. Bot. Gard. Edinburgh* 41: 591 (1984)
Lepiota cristata var. *pallidior* Bon, *Doc. Mycol.* 11 (43): 34 (1981)
Lepiota subfelinoides Bon & P.D. Orton, *Doc. Mycol.* 14 (56): 56 (1985) [1984]
Mis.: *Agaricus clypeolarius* sensu Sowerby [Col. fig. Engl. fung. 1: pl. 14 (1796)]
E: c **S:** c **W:** c **NI:** c **ROI:** c **O:** Channel Islands: ! Isle of Man: !
H: On soil in deciduous woodland, in scrub, gardens and on wasteground. Rarely reported with conifers.
D: NM2: 217, NBA8: 591 (as *Lepiota felinoides*) **D+I:** Ph: 28-29, FungEur4: 205-210 20a,b,& c, B&K4: 196 222 (as *L. cristata* var. *pallidior*), B&K4: 196 221, FAN5: 138, FM3(4): 128 + 134 **I:** C&D: 243, Cooke 31(29) Vol. 1(1881) (as *Agaricus (Lepiota) cristatus*)
Very common and widespread. Macroscopically variable, and several different varieties have been described.

echinacea J.E. Lange, *Fl. agaric. danic.* 5, *Taxonomic Conspectus*: V (1940)
Cystolepiota echinacea (J.E. Lange) Knudsen, *Bot. Tidsskr.* 73: 127 (1978)
Echinoderma echinaceum (J.E. Lange) Bon, *Doc. Mycol.* 21 (82): 63 (1991)
E: ! **W:** ! **NI:** ?
H: On soil (often calcareous) in deciduous woodland.
D: NM2: 221 **D+I:** FungEur4: 138-142 10a, B&K4: 198 223, FAN5: 148 **I:** FlDan1: 10D, Bon: 283 (as *Cystolepiota echinacea*), C&D: 241 (as *Echinoderma echinaceum*)
Rarely reported but apparently widespread. Confused with closely related taxa.

echinella Quél. & G.E. Bernard, *Bull. Soc. Mycol. France* 4: pl. 1 f. 2 (1888)
Lepiota minuta J.E. Lange [*nom. inval.* non *L. minuta* Vogl. (1896)], *Dansk. Bot. Ark.* 4: 48 (1923)
Lepiota setulosa J.E. Lange, *Fl. agaric. danic.* 1: 34 (1935)
E: ! **W:** !
H: On soil (often calcareous loam) in deciduous woodland or scrub.
D: FungEur4: 257-258, NM2: 220 (as *Lepiota setulosa*) **D+I:** FungEur4: 285-288 32b (as *L. setulosa*), FAN5: 125-126 **I:** FlDan1: 13C (as *L. setulosa*)
Rarely reported and part of a species complex poorly known in Britain. Differences between this and related taxa (i.e. *Lepiota bickhamensis*, *L. coxheadii* and *L. locquinii*) are not clearly resolved.

echinella var. rhodorhiza (P.D. Orton) Hardtke & Rödel., *Mykol. Mitteilungsbl.* 35: 62 (1992)
Lepiota rhodorhiza P.D. Orton, *Trans. Brit. Mycol. Soc.* 43(2): 285 (1960)
Lepiota setulosa var. *rhodorrhiza* Romagn. & Locq. [*nom. nud.*], *Bull. Soc. Mycol. France* 60: 38 (1944)
E: !
H: On soil in deciduous woodland.

D: FAN5: 127 **D+I:** FRIC1: 16-19 6a, FungEur4: 281-285 32a (as *Lepiota rhodorhiza*) **I:** Ph: 29 (misident. as *L. subincarnata*), C&D: 245 (as *L. rhodorhiza*)
Rarely reported. Described from East Norfolk (Surlingham, Tuck's Plantation) and also known from Berkshire, Hertfordshire, South Somerset, Surrey and West Sussex.

erminea (Fr.) P. Kumm., *Führ. Pilzk.*: 136 (1871)
Agaricus ermineus Fr., *Syst. mycol.* 1: 22 (1821)
Lepiota alba (Bres.) Sacc., *Syll. fung.* 5: 37 (1887)
Lepiota clypeolaria var. *alba* Bres., *Fungi trident.* 1: 15 (1882)
E: ! **S:** ! **W:** ! **ROI:** !
H: On sandy soil, often amongst grass; usually on dunes but occasionally reported from inland sites.
D: NM2: 218 **D+I:** FungEur4: 163-168 13b (as *Lepiota alba*), FungEur4: 173-174 15, B&K4: 190 213, FAN5: 118-119 **I:** C&D: 243 (as *L. alba*), Cooke 32 (40) Vol. 1 (1881) (as *Agaricus (Lepiota) ermineus*) Rarely reported but apparently widespread.

felina (Pers.) P. Karst., *Bidrag Kännedom Finlands Natur Folk* 32: 10 (1879)
Agaricus felinus Pers., *Syn. meth. fung.*: 261 (1801)
Lepiota clypeolaria var. *felina* (Pers.) Gillet, *Hyménomycètes*: 62 (1874)
Agaricus clypeolarius var. *felinus* (Pers.) Fr., *Syst. mycol.* 1: 21 (1821)
E: o **S:** ! **W:** ! **NI:** o **O:** Channel Islands: !
H: On soil or needle litter in plantations or in mixed deciduous and conifer woodland.
D: NM2: 220 **D+I:** Ph: 28-29, FungEur4: 260-262 28a, FAN5: 122-123 **I:** FlDan1: 12E, Bon: 287, C&D: 245
Occasional to rather rare, but widespread.

forquignonii Quél., *Compt. Rend. Assoc. Franç. Avancem. Sci.* 13: 277 (1885)
Lepiota olivaceobrunnea P.D. Orton, *Trans. Brit. Mycol. Soc.* 91: 561 (1988)
E: !
H: On soil (usually calcareous loam) in mixed deciduous woodland or scrub.
D+I: FRIC2: 2-6 9 (b-d), FungEur4: 263-266 30, FAN5: 124-125 **I:** C&D: 245
British collections from South Devon, South Somerset and West Sussex.

fuscovinacea F.H. Møller & J.E. Lange, *Fl. agaric. danic.* 5, *Taxonomic Conspectus*: V (1940)
E: ! **W:** ! **ROI:** !
H: On soil (often calcareous loam) in woodland, often under mixtures of deciduous trees and conifers. Also collected with *Atropa bella-donna* and in thickets of *Fallopia cuspidata* away from trees.
D: NM2: 219 **D+I:** FRIC2: 8-9 10a, FungEur4: 296-300 33a, B&K4: 198 224, FAN5: 129-130 **I:** FlDan1: 13H, SV28: 48, C&D: 245
Rarely reported but widespread.

grangei (Eyre) Kühner, *Bull. Soc. Linn. de Lyon* 3(5): 79 (1934)
Schulzeria grangei Eyre, *Trans. Brit. Mycol. Soc.* 2: 37 (1902)
Hiatula grangei (Eyre) W.G. Sm., *Syn. Brit. Bas.*: 27 (1908)
E: ! **ROI:** ?
H: On loam soil or decayed leaf litter in deciduous woodland. Occasionally with *Urtica dioica* in nettlebeds.
D: NM2: 218 **D+I:** FungEur4: 213-216 21, B&K4: 198 225, FAN5: 137-138, FM3(4): 129 + 134 **I:** FlDan1: 10, SV26: 38, C&D: 243
Rarely reported but apparently widespread in England. Reported from the Republic of Ireland but unsubstantiated with voucher material.

griseovirens Maire, *Bull. Soc. Mycol. France* 44: 37 (1928)
Lepiota poliochloodes Vellinga & Huijser, *Persoonia* (1993)
E: !

H: On loam soil in deciduous woodland or scrub, sometimes associated with *Urtica dioica* in nettlebeds. Also a single collection with *Fallopia cuspidata* on wasteground.
D: NM2: 218 **D+I:** FRIC6: 14-16 43c-d, FungEur4: 216-220 22a, FAN5: 136-137 (as *Lepiota poliochloodes*) **I:** Bon: 285, SV26: 38
Known from East Norfolk, Hertfordshire, Mid-Lancashire, North & South Somerset, South Devon, South Wiltshire, Surrey, West Kent, and Westmorland. Nomenclature follows Tofts (2002) and differs from that in B&K4 and Vellinga in FAN5.

hymenoderma D.A. Reid, *Fungorum Rar. Icon. Color.* 1: 24 (1966)
E: !
H: On soil in deciduous woodland. Also rarely in gardens.
D+I: FungEur4: 305-308 35a, FAN5: 140
Described from Oxfordshire (Blenheim Park). Collections also from East Kent, South Devon and South Wiltshire but poorly known in Britain.

hystrix F.H. Møller & J.E. Lange, *Fl. agaric. danic.* 5, *Taxonomic Conspectus*: V (1940)
E: ! **W:** ? **ROI:** ?
H: On soil in deciduous woodland.
D: NM2: 221, FAN5: 146 **D+I:** Ph: 26-27, B&K4: 200 227 **I:** FlDan1: 10E
Known from East Gloucestershire. Reported elsewhere but often confused with *L. calcicola* and the majority of records are unsubstantiated with voucher material.

ignivolvata Bousset & Joss., *Rivista Micol.* 33(1): 123 (1990)
E: ! **W:** !
H: On soil or decayed leaf litter in woodland, with deciduous trees and conifers but most often reported with *Fagus* in beechwoods.
D: NM2: 219 **D+I:** Ph: 28, FungEur4: 155-158 11, B&K4: 202 229, FAN5: 122 **I:** Bon: 285, C&D: 243
Rarely reported. Known from England (Bedfordshire, Berkshire, North Wiltshire, Oxfordshire, South Devon and West Sussex) and Wales (Caernarvonshire and Denbighshire). Reported elsewhere but unsubstantiated with voucher material.

jacobii Vellinga & Knudsen, *Persoonia* 14(4): 407 (1992)
Lepiota langei Knudsen [*nom. illegit.,* non *L. langei* Locquin (1945)], *Bot. Tidsskr.* 75(2-3): 130 (1980)
Mis.: *Lepiota eriophora* sensu NCL p.p. (1960)
E: !
H: On soil in deciduous woodland. Reported with *Alnus, Corylus* and under dense cover of *Mercurialis perennis* with *Fagus* in beechwoods.
D: NM2: 221 (as *Lepiota langei*) **D+I:** B&K4: 202 231 (as *L. langei*), FAN5: 148-149 **I:** FlDan1: 12H (as *L. echinella* var. *eriophora*), SV26: 35 (as *Echinoderma jacobii*)
Known from England (Bedfordshire, Co. Durham, North & South Somerset, South Devon, South Wiltshire and Surrey). Confused with closely related taxa.

lilacea Bres., *Fungi trident.* 2: 3 (1892)
E: !
H: On soil (often calcareous loam) in deciduous woodland and occasionally in hedgerows and cemeteries. Rarely in cool greenhouses in private gardens.
D: NM2: 217 **D+I:** FungEur4: 308-310 36, FAN5: 140-141 **I:** FlDan1: 13G, Bon: 287, SV28: 48, C&D: 245
Known from Buckinghamshire, East Gloucestershire, Hertfordshire, Isle of Wight, South Wiltshire, Surrey and West Kent.

locquinii Bon, *Doc. Mycol.* 16 (61): 20 (1985)
Lepiota heimii Locq. [*nom. nud.*], *Doc. Mycol.* 11 (43): 43 (1981)
E: ! **W:** !
H: On soil in gardens or deciduous woodland.
D: FungEur4: 267-270
Known from England (Mid-Lancashire and Surrey) and Wales (Denbighshire). Part of the *Lepiota echinella* complex, poorly known and requiring clarification.

magnispora Murrill, *Mycologia* 4: 237 (1912)
Lepiota ventriosospora D.A. Reid, *Trans. Brit. Mycol. Soc.* 41(4): 427 (1958)
Mis.: *Lepiota clypeolaria* sensu Rea (1922)
Mis.: *Lepiota metulispora* sensu auct. including Berk. & Broome in Notices no. 1182 (1871)
E: o **S:** o **W:** o **NI:** ! **ROI:** !
H: On soil or decayed leaf litter in woodland, often with mixtures of *Pinus* spp., and *Fagus*.
D: NM2: 219 (as *Lepiota ventriosospora*) **D+I:** Ph: 28 (as *L. ventriosospora*), FungEur4: 188-190 17 (as *L. ventriosospora*), B&K4: 206 235 (as *L. ventriosospora*), FAN5: 117-118 **I:** Bon: 285 (as *L. ventriosospora*), C&D: 243 (as *L. ventriosospora*)
Occasional but widespread. Better known as *Lepiota ventriosospora*.

medullata (Fr.) Quél., *Mém. Soc. Émul. Montbéliard, Sér. 2, 5*: 36 (1872)
Agaricus medullatus Fr., *Epicr. syst. mycol.*: 19 (1838)
Limacella medullata (Fr.) P.D. Orton, *Notes Roy. Bot. Gard. Edinburgh* 29(1): 105 (1969)
E: ! **S:** !
H: On soil in woodland.
D: NBA3: 105-106 (as *Limacella medullata*) **I:** Cooke 44 (44) Vol. 1 (1881)
Poorly known in Britain. *Fide* Pegler & Young [Nova Hedwigia, Beih. 35: 78 (1971)] this is a species of *Leucoagaricus*; *fide* Watling (pers. comm.) possibly a *Limacella* or *Drosella* sp.

obscura (Bon) Bon, *Doc. Mycol.* 23 (91): 33 (1991)
Lepiota griseovirens var. *obscura* Bon, *Doc. Mycol.* 6 (24): 44 (1976)
Mis.: *Lepiota griseovirens* sensu auct.
Mis.: *Lepiota pseudofelina* sensu NCL & sensu auct. mult.
E: !
H: On soil (usually calcareous) in deciduous woodland.
D+I: FungEur4: 218-220 22a (as *Lepiota griseovirens*), FungEur4: 222-225 22b (as *L. pseudofelina*), B&K4: 200 226 (as *L. griseovirens*), FAN5: 135-136 (as *L. griseovirens*) **I:** C&D: 242 (as *L. pseudofelina*)
Known from Mid-Lancashire, North Devon, South Somerset and South Wiltshire. Nomenclature follows Tofts (2002).

ochraceofulva P.D. Orton, *Trans. Brit. Mycol. Soc.* 43(2): 284 (1960)
Lepiota cookei Hora, *Trans. Brit. Mycol. Soc.* 43(2): 446 (1960)
E: !
H: On soil, with conifers such as *Cedrus, Chamaecyparis, Taxus* or *Wellingtonia* often with isolated trees in parkland or gardens. Occasionally reported with beech but then usually in mixed deciduous and conifer woodland.
D+I: FRIC1: 19-20 6b, FungEur4: 312-313 33b, FAN5: 141-142 **I:** Bon: 287, SV28: 48
Known from Berkshire, Co. Durham, East Norfolk, Herefordshire, Huntingdonshire, Middlesex, Shropshire, Surrey, West Kent and Worcestershire.

oreadiformis Velen., *České Houby* II: 215 (1920)
Agaricus clypeolarius var. *pratensis* Fr., *Epicr. syst. mycol.*: 15 (1838)
Lepiota clypeolaria var. *pratensis* (Fr.) Gillet, *Hyménomycètes*: 62 (1874)
Lepiota gracilis var. *laevigata* J.E. Lange, *Dansk. Bot. Ark.* 2(3): 24 (1915)
Lepiota laevigata (J.E. Lange) J.E. Lange, *Dansk. Bot. Ark.* 4(4): 47 (1923)
Lepiota pratensis (Fr.) Bigeard & H. Guill. [*nom. illegit.,* non *L. pratensis* Speg. (1898)], *Fl. Champ. Supér. France* 1: 58 (1909)
E: o **S:** ! **W:** !
H: On acidic or basic soil, in grassland.
D: NM2: 219 **D+I:** Ph: 27, FungEur4: 178-182 16a, B&K4: 204 232, FAN5: 119-120 **I:** FlDan1: 11C, C&D: 243 (as *Lepiota laevigata*)

Occasional in England. Rarely reported elsewhere but apparently widespread. Records from woodlands are dubious and unsubstantiated with voucher material.

parvannulata (Lasch) Gillet, *Hyménomycètes*: 66 (1874)
  Agaricus parvannulatus Lasch, *Linnaea* 3: 156 (1828)
E: !
H: On soil in deciduous woodland.
D: FAN5: 129 **D+I:** FungEur4: 243 25b
Listed by Massee (1910) and included by Rea (1922), but this is scarcely known in Britain. A single collection in herb. K (in poor condition) from South Wiltshire (Donhead St. Mary) det. Pearson in 1944.

perplexa Knudsen, *Bot. Tidsskr.* 75(2-3): 137 (1980)
E: o **NI:** !
H: On soil (usually calcareous loam) in deciduous woodland, often with *Fagus* and *Corylus* but also reported with other deciduous species. Rarely in plantations with *Picea* or *Pseudotsuga* spp, and known with *Taxus*.
D: FungEur4: 147-150, NM2: 222 **D+I:** FAN5: 145-146
Occasional in England, and not uncommon, at least in in southern counties. Rarely reported elsewhere but apparently widespread. Confused with similar taxa.

pseudoasperula (Knudsen) Knudsen, *Bot. Tidsskr.* 75(2-3): 128 (1980)
  Cystolepiota pseudoasperula Knudsen, *Bot. Tidsskr.* 73(2): 125 (1978)
  Mis.: *Lepiota eriophora* sensu D.A. Reid [TBMS 41: 426 (1958)] & sensu NCL p.p. (1960)
E: !
H: On soil (usually calcareous loam) in deciduous woodland.
D: NM2: 221 **D+I:** FungEur4: 151-154 10b, FAN5: 150 **I:** SV26: 35 (as *Echinoderma pseudoasperula*)
Known from Mid-Lancashire, South Devon and Westmorland. Reported from South Somerset, South Wiltshire, Surrey and West Sussex. Confused with similar taxa.

pseudolilacea Huijsman, *Bull. Mens. Soc. Linn. Lyon* 16: 183 (1947)
  Lepiota pseudohelveola Hora, *Trans. Brit. Mycol. Soc.* 43(2): 449 (1960)
E: o **W:** !
H: On soil (often calcareous loam) in deciduous woodland.
D+I: Ph: 28-29 (as *Lepiota pseudohelveola*), FungEur4: 274-277 (as *L. pseudohelveola*), FAN5: 123-124 **I:** C&D: 245 (as *L. pseudohelveola*)
Widespread in England and also known from Wales (Denbighshire). Many of the collections in herb. K were originally determined as *Lepiota pseudohelveola*.

subalba P.D. Orton, *Trans. Brit. Mycol. Soc.* 43(2): 287 (1960)
  Mis.: *Lepiota albosericea* sensu Lange (FlDan1: 32 & pl. 12B) & sensu auct. Brit.
E: o **W:** o
H: On calcareous soil in woodland, often under *Mercurialis perennis* with *Fagus* in beechwoods, but also with other deciduous trees.
D: FungEur4: 229-230, NM2: 217 **D+I:** FAN5: 133-134
Known from England (Bedfordshire, Cheshire, East & West Kent, Nottinghamshire, Oxfordshire, South Devon, South Somerset, South Wiltshire, Staffordshire, Surrey, Warwickshire, West Sussex and Yorkshire) and Wales (Carmarthenshire and Denbighshire).

subgracilis Kühner, *Bull. Soc. Mycol. France* 52: 231 (1936)
  Lepiota clypeolaria var. *gracilis* Quél., *Compt. Rend. Assoc. Franç. Avancem. Sci.* 22: 485 (1894) [1893]
  Lepiota gracilis (Quél.) J.E. Lange [*nom. illegit.*, non *L. gracilis* Peck (1899)], *Dansk. Bot. Ark.* 2(3): 24 (1915)
  Lepiota clypeolaria var. *latispora* Wasser, *Ukrayins'k. Bot. Zhurn.* 35(5): 518 (1978)
  Lepiota latispora (Wasser) Bon, *Doc. Mycol.* 11 (43): 30 (1981)
  Lepiota kuehneriana Locq., *Friesia* 5: 296 (1956)
E: o **S:** ! **W:** !

H: On soil (usually calcareous loam) often with *Fagus* in beechwoods but known with other deciduous trees. Often found under dense cover of *Mercurialis perennis*.
D: NM2: 219 **D+I:** FungEur4: 159-162 12 (as *Lepiota kuehneriana*), B&K4: 202 230 (as *L. kuehneriana*), FAN5: 121-122 **I:** FlDan1: 11F
Occasional in England. Rarely reported elsewhere but apparently widespread.

subincarnata J.E. Lange, *Fl. agaric. danic.* 5, *Taxonomic Conspectus*: V (1940)
  Lepiota josserandii Bon & Boiffard, *Bull. Soc. Mycol. France* 90: 289 (1974)
  Mis.: *Lepiota scobinella* sensu NCL p.p. (1960)
E: o **W:** ! **O:** Channel Islands: !
H: On soil in deciduous woodland, and occasionally (but increasingly) in gardens, on lawns and flowerbeds or other composted areas.
D: NM2: 220 **D+I:** FungEur4: 238-241 24 (as *Lepiota josserandii*), FungEur4: 244-246 25a, FAN5: 127-128 **I:** FlDan1: 13I, SV26: 39
Occasional in England. Rarely reported elsewhere but apparently widespread. The illustration in Phillips (Ph) is not this species but *Lepiota echinella* var. *rhodorhiza*.

tomentella J.E. Lange, *Dansk. Bot. Ark.* 4(4): 48 (1923)
E: ! **W:** ! **NI:** ?
H: On soil in deciduous woodland.
D: NBA9: 562, NM2: 218 **D+I:** FAN5: 134-135 **I:** FlDan1: 14D
Rarely reported. Apparently widespread. The description under this name in FungEur4 (*fide* Vellinga in FAN5) is a related species, *Lepiota pilodes*, as yet unknown in Britain.

xanthophylla P.D. Orton, *Trans. Brit. Mycol. Soc.* 43(2): 289 (1960)
  Mis.: *Lepiota citrophylla* sensu Rea (1922) & sensu auct. mult.
  Mis.: *Agaricus citrophyllus* sensu Cooke [Ill. Brit. fung. 1111 (639) Vol. 8 (1889)]
E: o **W:** !
H: On soil (usually calcareous loam) in deciduous woodland, often with *Fagus* or *Corylus*. Rarely reported with conifers such as *Cedrus* and *Larix* spp in mixed woodland.
D+I: FRIC3: 10-12 19b, FungEur4: 292-295 26b, B&K4: 206 236, FAN5: 124-125 **I:** SV26: 38, C&D: 245
Occasional in England (Leicestershire, Mid-Lancashire, North Hampshire, North & South Somerset, Oxfordshire, Shropshire, South Devon, South Wiltshire, West Gloucestershire and West Sussex) and known from Wales (Denbighshire). Reported elsewhere but unsubstantiated with voucher material.

LEPISTA (Fr.) W.G. Sm., *J. Bot.* 8: 248 (1870)
Agaricales, Tricholomataceae
  Rhodopaxillus Maire, *Ann. Mycol.* 11: 338 (1913)
Type: *Paxillus lepista* Fr. (= *Lepista panaeola*)

caespitosa (Bres.) Singer, *Lilloa* 22: 192 (1951) [1949]
  Tricholoma panaeolum var. *caespitosum* Bres., *Fungi trident.* 2: 48 (1898)
  Rhodopaxillus caespitosus (Bres.) Singer, *Rev. Mycol. (Paris)* 4(1-2): 69 (1939)
E: ! **S:** !
H: On soil or decayed wood in woodland. Rarely reported in grassland.
I: Kriegl3: 288
Known from England (South Devon, Surrey and West Kent) and Scotland (Mid-Perthshire).

flaccida (Sowerby) Pat., *Hyménomyc. Eur.*: 96 (1887)
  Agaricus flaccidus Sowerby, *Col. fig. Engl. fung.* 2: pl. 185 (1799)
  Clitocybe flaccida (Sowerby) P. Kumm., *Führ. Pilzk.*: 124 (1871)
  Agaricus lobatus Sowerby, *Col. fig. Engl. fung.* 2: pl. 186 (1799)

Omphalia lobata (Sowerby) Gray, *Nat. arr. Brit. pl.* 1: 612 (1821)

Clitocybe flaccida var. *lobata* (Sowerby) Romagn. & Bon, *Doc. Mycol.* 17 (67): 11 (1987)

Agaricus gilvus Pers., *Syn. meth. fung.*: 448 (1801)

Omphalia gilva (Pers.) Gray, *Nat. arr. Brit. pl.* 1: 612 (1821)

Clitocybe gilva (Pers.) P. Kumm., *Führ. Pilzk.*: 124 (1871)

Lepista gilva (Pers.) Roze, *Bull. Soc. Bot. France* 23: 110 (1876)

Clitocybe inversa (Scop.) Quél., *Mém. Soc. Émul. Montbéliard, Sér. 2,* 5: 235 (1872)

Lepista inversa (Scop.) Pat., *Hyménomyc. Eur.*: 96 (1887)

Agaricus lentiginosus Fr., *Epicr. syst. mycol.*: 69 (1838)

Agaricus splendens Pers. [non *A. splendens* With. (1812)], *Syn. meth. fung.*: 452 (1801)

Clitocybe splendens (Pers.) Gillet, *Hyménomycètes*: 139 (1874)

Mis.: *Clitocybe infundibuliformis* sensu auct.

E: c **S:** c **W:** c **NI:** c **ROI:** c **O:** Channel Islands: !

H: On loam soil or on leaf and needle litter in deciduous and conifer woodland.

D: NM2: 133 **D+I:** Ph: 48-49 (as *Clitocybe flaccida*), B&K3: 212 244 (as *Lepista inversa*), FAN3: 74-75 **I:** Sow2: 185 (as *Agaricus flaccidus*), FlDan1: 35D & D1 (as *C. flaccida*), Bon: 143 (as *L. inversa*), C&D: 175 (as *L. inversa*), C&D: 175 (as *C. gilva*), SV35: 17 (as *L. gilva*), SV35: 17, Kriegl3: 289, Cooke 161 (137) Vol. 2 (1881) (as *Agaricus lobatus*)

Very common and widespread. Macroscopically a variable species, the variations influenced by temperature and humidity *fide* Kuyper & Noordeloos (FAN3: 75).

glaucocana (Bres.) Singer, *Agaricales in Modern Taxonomy,* Edn 1: 193 (1951) [1949]

Tricholoma glaucocanum Bres., *Fungi trident.* 1: 7 (1881)

Tricholoma nudum var. *glaucocanum* (Bres.) L. Corb., *Mém. Soc. Natn. Scienc. Nat. Cherbourg* 40: 53 (1929)

Rhodopaxillus glaucocanus (Bres.) Métrod, *Rev. Mycol. (Paris)* 7 (Suppl.): 27 (1942)

E: ! **S:** ? **NI:** ?

H: On soil in deciduous and conifer woodland.

D: NM2: 133 **D+I:** B&K3: 210 243 **I:** C&D: 201, SV35: 13

Rarely reported. Known only from Surrey (Hindhead, and Ockley), and reported from Scotland (Easterness and Morayshire) in 1912 and from Northern Ireland (Down) in 1968. Previously regarded as a just a pallid form of *Lepista nuda* with a strong and unpleasant smell (said to resemble menthol).

irina (Fr.) H.E. Bigelow, *Canad. J. Bot.* 37: 775 (1959)

Agaricus irinus Fr., *Epicr. syst. mycol.*: 48 (1838)

Tricholoma irinum (Fr.) P. Kumm., *Führ. Pilzk.*: 132 (1871)

Rhodopaxillus irinus (Fr.) Métrod, *Rev. Mycol. (Paris)* 7 (Suppl.): 25 (1942)

E: o **S:** ! **W:** ! **NI:** ! **ROI:** !

H: On soil in deciduous woodland, often with *Fagus* and *Fraxinus*. Rarely in parkland or at the edges of plantations with conifers such as *Cedrus* and *Larix* spp.

D: NM2: 133, FAN3: 70 **D+I:** Ph: 114, B&K3: 212 245 **I:** Bon: 145, SV35: 16

Occasional in England. Rarely reported elsewhere but apparently widespread. May be confused with *Clitocybe nebularis* and *Lepista glaucocana* if the strong odour (said to resemble violets, 'orris' roots, or 'mock orange blossom') is missed.

multiformis (Romell) Gulden, *Sydowia* 36: 65 (1983)

Tricholoma multiforme Romell, *Arch. Bot. (Leipzig)* 3: 3 (1911)

S: !

H: On sandy soil, often amongst grass, on dunes.

D: NM2: 135 **I:** SV35: 20

Known only from Orkney.

nuda (Fr.) Cooke, *Handb. Brit. fung.,* Edn1: 192 (1871)

Agaricus nudus Fr., *Syst. mycol.* 1: 52 (1821)

Tricholoma nudum (Fr.) P. Kumm., *Führ. Pilzk.*: 132 (1871)

Agaricus bulbosus Bolton [*nom illegit.*, non *A. bulbosus* Sowerby (1798); non Schaeff. (1770)], *Hist. fung. Halifax* Suppl.: 147 (1791)

Lepista nuda var. *pruinosa* (Bon) Bon, *Doc. Mycol.* 13 (51): 44 (1983)

Agaricus nudus var. *majus* Cooke, *Handb. Brit. fung.,* Edn 2: 41 (1884)

Tricholoma nudum var. *major* (Cooke) Massee, *Brit. fung.-fl.* 3: 217 (1893)

Mis.: *Tricholoma personatum* sensu auct.

E: c **S:** c **W:** c **NI:** c **ROI:** c **O:** Channel Islands: ! Isle of Man: !

H: On soil or decayed leaf litter in deciduous or conifer woodland, amongst grass in parkland and gardens, and on woodchips, compost heaps or piles of decayed grass clippings.

D: NM2: 133, FAN3: 72-73 **D+I:** Ph: 113, B&K3: 214 247 **I:** Bon: 145, C&D: 201, SV35: 12, Kriegl3: 294

Very common and widespread. Fruiting throughout the year, and often well into the winter. Macroscopically variable, especially with regard to colour and several 'varieties' have been described. English name = 'Wood Blewit'.

ovispora (J.E. Lange) Gulden, *Sydowia* 36: 67 (1983)

Clitocybe aggregata var. *ovispora* J.E. Lange, *Dansk. Bot. Ark.* 6(5): 58 (1930)

Lyophyllum aggregatum var. *ovisporum* (J.E. Lange) Kühner & Romagn., *Flore Analytique des Champignons Supérieurs*: 164 (1953)

Lyophyllum ovisporum (J.E. Lange) D.A. Reid, *Fungorum Rar. Icon. Color.* 3: 13 & pl. 21a (1968)

E: ! **S:** !

H: On soil or leaf and needle litter, in woodland and also known 'amongst grass' in parkland.

D: NM2: 134, FAN3: 71-72 **I:** SV35: 20

Known from England (East Kent, East Norfolk, North Lincolnshire, South Devon and Warwickshire) and Scotland (Mid-Ebudes: Isle of Mull and the Orkney Isles). Reported elsewhere but unsubstantiated with voucher material.

panaeola (Fr.) P. Karst., *Bidrag Kännedom Finlands Natur Folk* 32: 481 (1879)

Agaricus panaeolus Fr., *Epicr. syst. mycol.*: 49 (1838)

Tricholoma panaeolum (Fr.) Quél., *Mém. Soc. Émul. Montbéliard, Sér. 2,* 5: 82 (1872)

Rhodopaxillus panaeolus (Fr.) Maire, *Ann. Mycol.* 11: 338 (1913)

Agaricus calceolus Fr., *Ic. Hymenomyc.* 1: pl. 73 (1873)

Tricholoma panaeolum var. *calceolum* (Fr.) Sacc., *Syll. fung.* 5: 132 (1887)

Paxillus lepista Fr., *Epicr. syst. mycol.*: 316 (1838)

Mis.: *Clitocybe luscina* sensu auct.

Mis.: *Lepista luscina* sensu auct.

Mis.: *Clitocybe nimbata* sensu auct.

E: o **S:** o **W:** ! **NI:** ! **ROI:** ! **O:** Channel Islands: !

H: On soil amongst grass in meadows, pastures and unimproved grassland, often in upland areas.

D: NM2: 133 (as *Lepista luscina*), FAN3: 68 **D+I:** Ph: 113 (as *L. luscina*), B&K3: 212 246 (as *L. luscina*) **I:** Bon: 145, FungEur3: 53 (as *Rhodopaxillus panaeolus*), C&D: 201, SV35: 16 (as *L. luscina*), Kriegl3: 296

Occasional to rather rare, but widespread.

pseudoectypa (M. Lange) Gulden, *Sydowia* 36: 69 (1983)

Clitocybe pseudoectypa M. Lange, *Medd. Grønl.* 11: 33 (1955)

S: !

H: On soil amongst grass, in unimproved grassland.

D: NM2: 134

Known only from theOrkney Islands (Foula and South Ronaldsay).

rickenii Singer, *Sydowia* 2: 26 (1948)

S: !

H: On sandy soil amongst grass, in unimproved grassland.

Known only from the Orkney Islands.

saeva (Fr.) P.D. Orton, *Trans. Brit. Mycol. Soc.* 43(2): 177 (1960)
  Agaricus personatus β *saevus* Fr., *Epicr. syst. mycol.*: 48 (1838)
  Tricholoma personatum var. *saevum* (Fr.) Dumée, *Nouvelle Atlas des Champignons*: 45 (1905)
  Rhodopaxillus saevus (Fr.) Maire, *Ann. Mycol.* 11: 338 (1913)
  Tricholoma saevum (Fr.) Gillet, *Ann. Mycol.* 18: 65 (1920)
  Agaricus anserinus Fr., *Epicr. syst. mycol.*: 72 (1838)
  Tricholoma personatum var. *anserina* (Fr.) Sacc., *Syll. fung.* 5: 130 (1887)
  Agaricus personatus Fr., *Syst. mycol.* 1: 50 (1821)
  Lepista personata (Fr.) Cooke, *Handb. Brit. fung.*, Edn 1: 193 (1871)
  Tricholoma personatum (Fr.) P. Kumm., *Führ. Pilzk.*: 132 (1871)
  Mis.: *Tricholoma amethystinum* sensu auct.
E: o **S:** ! **W:** ! **NI:** ! **ROI:** !
H: On soil amongst grass in scrub woodland with deciduous trees and shrubs such as *Acer pseudoplatanus* or *Crataegus monogyna*. Also known on heathland, in unimproved grassland, and under planted trees on roadsides or in gardens.
D: B&K3: 214 248 (as *Lepista personata*), NM2: 133 (as *L. personata*), FAN3: 74 **D+I:** Ph: 114 **I:** FlDan1: 28A (as *Tricholoma personatum*), Bon: 145, C&D: 201 (as *L. personata*), SV35: 13, Kriegl3: 299
Occasional in England but may be locally abundant. Rarely reported elsewhere but apparently widespread. Often fruiting throughout the winter. English name = 'Field Blewit' or 'Blue Leg'.

sordida (Fr.) Singer, *Lilloa* 22: 193 (1951) [1949]
  Agaricus sordidus Fr., *Syst. mycol.* 1: 51 (1821)
  Tricholoma sordidum (Fr.) P. Kumm., *Führ. Pilzk.*: 134 (1871)
  Rhodopaxillus sordidus (Fr.) Maire, *Ann. Mycol.* 11: 338 (1913)
  Gyrophila nuda var. *lilacea* Quél., *Fl. mycol. France*: 271 (1888)
  Lepista sordida var. *lilacea* (Quél.) Bon, *Doc. Mycol.* 10 (37-38): 91 (1980) [1979]
  Lepista sordida var. *aianthina* Bon, *Doc. Mycol.* 10 (37-38): 91 (1980) [1979]
  Lepista sordida var. *obscurata* (Bon) Bon, *Doc. Mycol.* 10 (37-38): 91 (1980) [1979]
E: o **S:** o **W:** o **NI:** ! **ROI:** ! **O:** Channel Islands: !
H: On soil, often in grassland (pastures or fields), along woodland edges and roadsides, and rather frequently on compost heaps or piles of decayed grass clippings.
D: NM2: 132, FAN3: 73 **D+I:** Ph: 113, B&K3: 216 250 **I:** Bon: 145, C&D: 201, SV35: 12, Kriegl3: 299
Occasional but widespread and may be locally abundant. Basidiomes are macroscopically variable, especially with regard to colour, and several varieties have been described.

LEPTOPORUS Quél., *Enchir. fung.*: 175 (1886)
Poriales, Coriolaceae
Type: *Leptoporus mollis* (Pers.) Quél.

mollis (Pers.) Quél., *Enchir. fung.*: 176 (1886)
  Boletus mollis Pers., *Observ. mycol.* 1: 22 (1796)
  Polyporus mollis (Pers.) Fr., *Syst. mycol.* 1: 360 (1821)
  Tyromyces mollis (Pers.) Kotl. & Pouzar, *Česká Mykol.* 13: 30 (1959)
  Polyporus erubescens Fr., *Epicr. syst. mycol.*: 461 (1838)
E: ! **S:** !
H: On dead wood of conifers such as *Pinus* spp.
D: EurPoly1: 382, NM3: 233 **D+I:** B&K2: 270 332
Not reported since 1957 and possibly extinct. Collections mostly from Scotland, few from England. Recent records from Lincolnshire and Herefordshire are dubious and unsubstantiated with voucher material.

LEPTOSPOROMYCES Jülich, *Willdenowia. Beih.* 7: 192 (1972)
Stereales, Atheliaceae
Type: *Leptosporomyces galzinii* (Bourdot) Jülich

fuscostratus (Burt) Hjortstam, *Windahlia* 17: 58 (1987)
  Corticium fuscostratum Burt, *Ann. Missouri Bot. Gard.* 13: 299 (1926)
  Athelia fuscostrata (Burt) Donk, *Fungus* 27: 12 (1957)
  Mis.: *Athelia olivaceoalba* sensu Dennis (1995)
  Mis.: *Confertobasidium olivaceoalbum* sensu CNE2, & sensu Jülich (1984)
E: c **S:** ! **W:** ! **NI:** ! **ROI:** !
H: Usually on decayed wood, or fallen bark of conifers such as *Cedrus*, *Larix*, *Picea* and *Pinus* spp. Rarely on wood of deciduous trees such as *Betula pendula*, *Corylus* and *Fagus*.
D: NM3: 149 **D+I:** B&K2: 86 56 (incorrectly as *Confertobasidium olivaceoalbum*)
Common and widespread in England. The pallid orange-brownish rhizomorphs ramifying through the substratum are a useful diagnostic character.

galzinii (Bourdot) Jülich, *Willdenowia. Beih.* 7: 192 (1972)
  Corticium galzinii Bourdot, *Rev. Sci. du Bourb.* 23: 11 (1910)
  Athelia galzinii (Bourdot) Donk, *Fungus* 27: 12 (1957)
  Athelia grisea M.P. Christ., *Dansk. Bot. Ark.* 19(2): 153 (1960)
E: ! **S:** ! **W:** !
H: On decayed wood of conifers such as *Abies*, *Larix*, *Picea* and *Pinus* spp in natural woodland and plantations. Also known on debris of *Pteridium* on heathland, and decayed wood of deciduous trees such as *Betula* spp.
D: NM3: 149 **D+I:** CNE4: 802-805
Rarely reported but apparently widespread. Easily overlooked.

mutabilis (Bres.) Krieglst., *Z. Mykol.* 57(1): 53 (1991)
  Corticium mutabile Bres., *Fungi trident.* 2(11-13): 59 (1898)
  Athelia mutabilis (Bres.) Donk, *Fungus* 27: 12 (1957)
  Fibulomyces mutabilis (Bres.) Jülich, *Willdenowia. Beih.* 7: 182 (1972)
E: ! **S:** !
H: On decayed wood of deciduous trees and rarely on acidic soil in woodland.
D: NM3: 149 (as *Fibulomyces mutabilis*) **D+I:** CNE3: 388-389 (as *F. mutabilis*)
Known from England (East Norfolk, Hertfordshire, Huntingdonshire, North Somerset, South Hampshire, Surrey and West Sussex) and Scotland (Wester Ross).

raunkiaeri (M.P. Christ.) Jülich, *Willdenowia. Beih.* 7: 206 (1972)
  Athelia raunkiaeri M.P. Christ., *Dansk. Bot. Ark.* 19(2): 153 (1960)
E: ! **S:** ! **W:** !
H: On living mosses such as *Plagiothecium recurvatum*.
D: NM3: 150 **D+I:** CNE4: 808-809
Known from England (Cheviotshire), Scotland (North Ebudes) and Wales (Breconshire).

septentrionalis (J. Erikss.) Krieglst., *Z. Mykol.* 57(1): 53 (1991)
  Athelia septentrionalis J. Erikss., *Symb. Bot. Upsal.* 16(1): 88 (1958)
  Fibulomyces septentrionalis (J. Erikss.) Jülich, *Willdenowia. Beih.* 7: 187 (1972)
E: ! **S:** ! **W:** !
H: On decayed wood and bark of *Pinus sylvestris*.
D: NM3: 149 (as *Fibulomyces septentrionalis*) **D+I:** CNE3: 390-391 (as *F. septentrionalis*), B&K2: 88 58 (as *F. septentrionalis*)
Known from England (North Hampshire, South Hampshire and Surrey), Wales (Radnorshire) and Scotland (Easter Ross).

LEUCOAGARICUS (Locq.) Singer, *Sydowia* 2: 35 (1948)
Agaricales, Agaricaceae
  Sericeomyces Heinem., *Bull. Jard. Bot. Nat. Belg.* 48: 401 (1978)
Type: *Leucoagaricus macrorhizus* E. Horak (= *L. barssii*)

americanus (Peck) Vellinga, *Mycotaxon* 76: 433 (2000)
  Agaricus americanus Peck, *Rep. (Annual) New York State Mus. Nat. Hist.* 23: 71 (1872) [1870]
  Lepiota bresadolae Schulzer, *Hedwigia* 24: 132 (1885)
  Leucoagaricus bresadolae (Schulzer) Bon, *Doc. Mycol.* 7 (27-28): 15 (1977)
  Mis.: *Leucocoprinus biornatus* sensu auct. Brit. & sensu auct. Eur.
E: !
H: On soil and decayed composted conifer woodchips (British material).
D: NM2: 224 (as *Leucocoprinus bresadolae*) **D+I:** FungEur4: 415-419 48 (as *Leucoagaricus bresadolae*), FAN5: 92-93 **I:** Bon: 289 (as *Leucoagaricus bresadolae*), C&D: 247 (as *Leucoagaricus bresadolae*)
Known from Surrey (Esher, Claremont) where first collected in 1997. Also recently from Norfolk.

badhamii (Berk. & Broome) Singer, *Lilloa* 22: 419 (1951) [1949]
  Agaricus badhamii Berk. & Broome, *Ann. Mag. Nat. Hist.*, Ser. 2 [Notices of British Fungi no. 664] 13: 397 (1854)
  Lepiota badhamii (Berk. & Broome) Quél., *Mém. Soc. Émul. Montbéliard*, Sér. 2, 5: 231 (1872)
  Leucocoprinus badhamii (Berk. & Broome) Locq., *Bull. Mens. Soc. Linn. Lyon* 12: 15 (1943)
  Lepiota meleagroides Huijsman [*nom. nud.*], *Meded. Ned. Mycol. Ver.* 28: 11 (1943)
E: o W: ! ROI: !
H: On soil in deciduous or mixed woodland on calcareous soil, often with *Taxus* and *Fagus*. Rarely with other trees such as *Cedrus* and *Ulmus* spp. Also on flowerbeds mulched with shredded bark and woodchips.
D: NM2: 224 (as *Leucocoprinus badhamii*) **D+I:** FungEur4: 386-388 44, B&K4: 206 237, FAN5: 95-96 **I:** Bon: 289, C&D: 247, Cooke 25 (25) Vol. 1 (1881) (as *Agaricus badhamii*)
Occasional but widespread in England, much less so elsewhere.

barssii (Zeller) Vellinga, *Mycotaxon* 76: 431 (2000)
  Lepiota barssii Zeller, *Mycologia* 26: 211 (1934)
  Lepiota macrorhiza (Locq.) Kühner & Romagn. [*nom. inval.*], *Flore Analytique des Champignons Supérieurs*: 406 (1953)
  Leucoagaricus macrorhizus E. Horak, *Beitr. Kryptogamenfl. Schweiz* 13: 344 (1968)
  Leucocoprinus macrorhizus (E. Horak) D.A. Reid, *Mycol. Res.* 93: 421 (1989)
  Lepiota pinguipes A. Pearson, *Trans. Brit. Mycol. Soc.* 35(2): 97 (1952)
  Leucoagaricus pinguipes (A. Pearson) Bon, *Doc. Mycol.* 11 (43): 54 (1981)
E: ! W: !
H: On soil in cool greenhouses, also in gardens, urban streets and on dunes.
D: FungEur4: 342-345 (as *Leucoagaricus macrorhizus* var. *pinguipes*) **D+I:** FungEur4: 338-342 39 (as *L. macrorhizus*), FAN5: 88 **I:** Bon: 288 (as *L. macrorhizus*), C&D: 247 (as *L. macrorhizus*)
Known from England (East Norfolk, East Sussex, Middlesex, North Devon, South Essex, West Kent and Yorkshire) and Wales (Caernarvonshire). This is a large and distinctive species which would be difficult to overlook, thus probably genuinely rare. Perhaps better known as *Leucoagaricus macrorhizus*.

brunneocingulatus (P.D. Orton) Bon, *Doc. Mycol.* 6 (24): 44 (1976)
  Lepiota brunneocingulata P.D. Orton, *Trans. Brit. Mycol. Soc.* 43(2): 282 (1960)
E: !
H: On soil in mixed deciduous woodland.
D: FungEur4: 333-334
Known from the type collection from North Somerset (Spaxton near Bridgwater, Kenley Bottom) and two other collections from Somerset and Devon all in 1958. Not reported since.

carneifolius (Gillet) Wasser, *Ukrayins'k. Bot. Zhurn.* 34(3): 307 (1977)

Lepiota carneifolia Gillet, *Hyménomycètes*: 65 (1874)
  Mis.: *Leucoagaricus cinerascens* sensu Reid (1989) and sensu auct. mult.
E: ! O: Channel Islands: !
H: On soil in scrub vegetation, in cool greenhouses or gardens and occasionally in grassland such as pastures or meadows.
D: FungEur4: 435, FAN5: 90-91 (as *Leucoagaricus leucothites* var. *carneifolius*) **D+I:** B&K4: 208 239 (as *L. cinerascens*)
Known from England (Isle of Wight, London, Middlesex, Shropshire, South Essex, Surrey, Warwickshire and West Norfolk) and the Channel Islands (Jersey).

croceovelutinus (Bon & Boiffard) Bon & Boiffard, *Doc. Mycol.* 6 (24): 45 (1976)
  Leucocoprinus croceovelutinus Bon & Boiffard, *Bull. Soc. Mycol. France* 88: 26 (1972)
  Leucocoprinus croceovelutinus var. *diversisporus* D.A. Reid (*nom. inval.*), *Mycol. Res.* 94(5): 658 (1990)
E: ! W: !
H: On soil, usually calcareous loam, in mixed woodland. With *Taxus* in Britain.
D+I: FungEur4: 397-400 46, FAN5: 97
Known from England (Shropshire and Surrey) and Wales (Radnorshire).

georginae (W.G. Sm.) Candusso, *Rivista Micol.* 33(1): 10 (1990)
  Agaricus georginae W.G. Sm., *J. Bot.* 9: 1 (1871)
  Lepiota georginae (W.G. Sm.) Sacc., *Syll. fung.* 5: 71 (1887)
  Leucocoprinus georginae (W.G. Sm.) Wasser, *Agarikovye Griby SSSR*: 108 (1985)
E: !
H: On acidic soil in woodland, with deciduous trees such as *Alnus*, *Betula* spp, and *Castanea* and conifers such as *Larix*, *Picea* and *Pinus* spp. Rarely on decayed debris of *Pteridium*.
D: FungEur4: 400-403 **D+I:** FAN5: 96-97 **I:** C&D: 247, Cooke 47 (132) Vol. 1 (1881) (as *Agaricus georginae*)
Known from Berkshire, Dorset, East Norfolk, East Suffolk, Huntingdonshire, North and South Devon, South Hampshire, Surrey, West Kent and West Sussex.

griseodiscus (Bon) Bon & Migl., *Doc. Mycol.* 21 (81): 55 (1991)
  Leucoagaricus gaugueri var. *griseodiscus* Bon, *Doc. Mycol.* 20 (78): 59 (1990)
E: !
H: On soil in deciduous woodland.
Known only from North Somerset and South Devon.

ionidicolor Bellù & Lanzoni, *Rivista Micol.* 31(3-4): 107 (1988)
  Leucocoprinus caeruleoviolaceus D.A. Reid, *Mycol. Res.* 93(4): 413 (1989)
  Leucoagaricus ionidicolor var. *caeruleoviolaceus* (D.A. Reid) D.A. Reid, *Mycotaxon* 53: 327 (1995)
E: !
H: On soil, or decayed wood in soil, in woodland. British material with *Carpinus* and *Pinus* spp.
D+I: FungEur4: 335-337 38b, FAN5: 99-100 **I:** FM3(3): 75
Known from South Hampshire, West Gloucestershire and West Kent.

leucothites (Vittad.) Wasser, *Ukrayins'k. Bot. Zhurn.* 34(3): 308 (1977)
  Agaricus leucothites Vittad., *Descr. fung. mang.*: 310 (1835)
  Lepiota naucina var. *leucothites* (Vittad.) Sacc., *Syll. fung.* 5: 43 (1887)
  Lepiota leucothites (Vittad.) P.D. Orton, *Trans. Brit. Mycol. Soc.* 43(2): 177 (1960)
  Agaricus holosericeus Fr., *Epicr. syst. mycol.*: 16 (1838)
  Lepiota holosericea (Fr.) Gillet, *Hyménomycètes*: 67 (1874)
  Leucocoprinus holosericeus (Fr.) Locq., *Bull. Mens. Soc. Linn. Lyon* 12: 95 (1943)
  Leucoagaricus holosericeus (Fr.) M.M. Moser, *Kleine Kryptogamenflora*, Edn 3: 185 (1967)
  Agaricus laevis Krombh., *Icon. select. fung.*: t. 26 (1831)
  Annularia laevis (Krombh.) Gillet, *Hyménomycètes*: 389 (1876)
  Agaricus naucinus Fr., *Epicr. syst. mycol.*: 16 (1838)

Lepiota naucina (Fr.) P. Kumm., *Führ. Pilzk.*: 136 (1871)
> *Leucoagaricus naucinus* (Fr.) Singer, *Lilloa* 22: 423 (1951)
> [1949]
> Mis.: *Leucoagaricus cretaceus* sensu Moser [*Kleine Kryptogamenflora*: 115 (1953)]
> Mis.: *Agaricus cretaceus* sensu Cooke [Ill. Brit. fung. 542 (524) Vol. 4 (1885)]
> Mis.: *Leucoagaricus pudicus* sensu auct.

E: o **S:** ! **W:** o **ROI:** ! **O:** Channel Islands: !
H: On soil, usually amongst grass in meadows, lawns, pastures etc., and occasionally in grassy areas in woodland or scrub, and near trees in urban streets.
D+I: Ph: 25 (as *Lepiota leucothites*), FungEur4: 428-433 52 & 53, B&K4: 208 240, FAN5: 89-90 **I:** FlDan1: 9A (as *L. naucina*), Bon: 289, C&D: 247
Occasional but widespread in England and Wales, rarely reported elsewhere.

marriagei (D.A. Reid) Bon, *Doc. Mycol.* 6(24): 44 (1976)
> *Lepiota marriagei* D.A. Reid, *Fungorum Rar. Icon. Color.* 1: 20 (1966)

E: ! **W:** !
H: On calcareous soil in deciduous or mixed woodland, usually with *Corylus* and *Fagus* but also with conifers such as *Chamaecyparis lawsoniana* and *Taxus*.
D+I: FungEur4: 345-348, FAN5: 98-99
Known from England (East Kent, North Somerset, South Somerset, Surrey and West Kent) and Wales (Denbighshire).

medioflavoides Bon, *Doc. Mycol.* 6(24): 44 (1976)
E: !
H: On soil in deciduous woodland.
D: FungEur4: 444-445
A single collection from South Devon (Slapton Wood) in 1995.

melanotrichus (Malençon & Bertault) Trimbach, *Doc. Mycol.* 5(20): 42 (1975)
> *Lepiota melanotricha* Malençon & Bertault, *Fl. Champ. Maroc* 1: 134 (1970)
> *Leucoagaricus melanotrichus* var. *septentrionalis* D.A. Reid, *Mycotaxon* 53: 331 (1995)

E: ! **W:** !
H: On soil (rarely on decayed wood) in deciduous or coniferous woodland, also in fen carr.
D: Reid (1995) **D+I:** FungEur4: 348-351 38c, FAN5: 100 **I:** SV29: 40, C&D: 247
Known from England (Huntingdonshire, North Somerset and South Devon) and Wales (Denbighshire).

melanotrichus var. fuligineobrunneus Bon & Boiffard, *Doc. Mycol.* 8(29): 38 (1978)
S: !
H: On soil.
D: FAN5: 100 **D+I:** FungEur4: 352
A single collection from East Lothian.

meleagris (Sowerby) Singer, *Lilloa* 22: 422 (1951) [1949]
> *Agaricus meleagris* Sowerby [*nom. illegit.*, non *A. meleagris* With. (1792); non (Jul. Schäff.) Imbach (1946)], *Col. fig. Engl. fung.* 2: pl. 171 (1798)
> *Gymnopus meleagris* (Sowerby) Gray, *Nat. arr. Brit. pl.* 1: 609 (1821)
> *Lepiota meleagris* (Sowerby) Sacc., *Syll. fung.* 5: 36 (1887)
> *Leucocoprinus meleagris* (Sowerby) Locq., *Bull. Soc. Linn. de Lyon* 14: 93 (1945)

E: ! **W:** !
H: On soil and piles of woodchips or shavings warmed through fermentation and decay (British material).
D+I: FungEur4: 407-410 45a, FAN5: 94-95 **I:** Cooke 26 (26) Vol. 1 (1881) (as *Agaricus meleagris*)
Described from Britain, but probably a semi-naturalised alien, occasionally collected in greenhouses and recently on piles of fermenting woodchips, e.g. in Surrey (Esher, Claremont).

nympharum (Kalchbr.) Bon, *Doc. Mycol.* 7(27-28): 19 (1977)
> *Agaricus nympharum* Kalchbr., *Icon. select. Hymenomyc. Hung.*: 10 t. 2 (1873)

Lepiota nympharum (Kalchbr.) Kalchbr., *Magyar. Tud. Akad. Értes.*: 7 (1878)
> *Macrolepiota nympharum* (Kalchbr.) Wasser, *Agarikovye Griby SSSR*: 114 (1985)
> *Agaricus rhacodes* var. *puellaris* Fr., *Monogr. hymenomyc. Suec.* 2: 285 (1863)
> *Lepiota rhacodes* var. *puellaris* (Fr.) Sacc., *Syll. fung.* 5: 29 (1887)
> *Lepiota procera* var. *puellaris* (Fr.) Massee, *Brit. fung.-fl.* 3: 235 (1893)
> *Lepiota puellaris* (Fr.) Rea, *Brit. basidiomyc.*: 65 (1922)
> *Macrolepiota puellaris* (Fr.) M.M. Moser, *Kleine Kryptogamenflora, Edn* 3: 184 (1967)

E: ! **S:** ! **W:** !
H: On soil, in deciduous woodland.
D: NM2: 225 (as *Macrolepiota nympharum*) **D+I:** FungEur4: 525-530 66 (as *M. puellaris*), Myc6(4): 187, B&K4: 218 255 (as *M. puellaris*), FAN5: 91-92 **I:** FlDan1: 9B (as *M. rhacodes* var. *puellaris*)
Rarely reported. Excludes *Lepiota puellaris* sensu Rea (1922), described with an ochraceous disc and doubtful.

pilatianus (Demoulin) Bon & Boiffard, *Doc. Mycol.* 6(24): 45 (1976)
> *Lepiota pilatiana* Demoulin, *Lejeunia* 39: 11 (1966)
> *Leucocoprinus pilatianus* (Demoulin) Wasser, *Nov. sist. Niz. Rast.*: 219 (1978) [1977]
> *Leucocoprinus pilatianus* var. *subrubens* (Wichanský) Wasser, *Nov. sist. Niz. Rast.*: 221 (1978) [1977]

E: ! **W:** !
H: On soil, often calcareous loam, in woodland and scrub, usually under deciduous trees and occasionally with *Taxus*.
D+I: FungEur4: 410-414 47, FAN5: 97-98 **I:** Bon: 289, C&D: 247
Known from England (Buckinghamshire, East Kent, South Devon, South Somerset and Surrey) and Wales (Glamorganshire).

purpureolilacinus Huijsman, *Fungus* 25: 34 (1955)
> *Leucoagaricus purpureorimosus* Bon & Boiffard, *Doc. Mycol.* 8(29): 37 (1978)

E: !
H: On soil in woodland and gardens.
D: FungEur4: 374 (as *Leucoagaricus purpureorimosus*), FungEur4: 375-378 **D+I:** FAN5: 103-104 **I:** C&D: 247 (as *L. purpureorimosus*)
Known from North Somerset, West Cornwall and West Kent.

serenus (Fr.) Bon & Boiffard, *Bull. Soc. Mycol. France* 90: 301 (1974)
> *Agaricus serenus* Fr., *Hymenomyc. eur.*: 38 (1874)
> *Lepiota serena* (Fr.) Quél., *Bull. Soc. Bot. France* 26: 45 (1880)
> *Pseudobaeospora serena* (Fr.) Locq., *Bull. Soc. Mycol. France* 68: 169 (1952)
> *Sericeomyces serenus* (Fr.) Heinem., *Bull. Jard. Bot. Nat. Belg.* 48: 403 (1978)

E: ! **W:** !
H: On loam soil in scrub, hedgerows, fen-carr and woodland, often under cover of *Mercurialis perennis* with *Fagus* in beechwoods, but also known with *Acer pseudoplatanus*, *Alnus*, *Corylus*, *Crataegus* spp and *Sambucus nigra*.
D+I: FungEur4: 448-451 56, B&K4: 222 260 (as *Sericeomyces serenus*), FAN5: 105-106 **I:** Bon: 289, C&D: 247 (as *S. serenus*)
Known from England (Berkshire, Huntingdonshire, Isle of Wight, South Devon, South Somerset, South Wiltshire, Surrey, West Kent, West Sussex and Worcestershire) and Wales (Glamorganshire).

sericifer (Locq.) Vellinga, *Persoonia* 17(3): 477 (2000)
> *Pseudobaeospora sericifera* Locq., *Bull. Soc. Mycol. France* 68: 169 (1952)
> *Lepiota sericifera* (Locq.) Locq. [*nom. nud.*], *Friesia* 5: 294 (1956)

Lepiota cristata var. *sericea* Cool, *Meded. Ned. Mycol. Ver.* 12: 23 (1922)

Lepiota sericea (Cool) Huijsman, *Meded. Ned. Mycol. Ver.* 28: 46 (1943)

Leucoagaricus sericeus (Cool) Bon & Boiffard [*nom. inval.*], *Doc. Mycol.* 9(35): 40 (1979)

Lepiota sericata Kühner & Romagn. [*nom. superf.*], *Flore Analytique des Champignons Supérieurs*: 405 (1953)

Lepiota sericatella Malençon, *Fl. Champ. Maroc* 1: 152 (1970)

Leucoagaricus sericatellus (Malençon) Bon, *Doc. Mycol.* 9(35): 40 (1979)

Sericeomyces sericatellus (Malençon) Bon, *Bull. Soc. Mycol. France* 96(2): 172 (1980)

Leucoagaricus sericifer f. *sericatellus* (Malençon) Vellinga, *Persoonia* 17(3): 479 (2000)

Mis.: *Lepiota serena* sensu Lange (FlDan1: 29 & pl. 11B)

E: !

H: On loam soil in deciduous woodland.

D: NM2: 228 (as *Sericeomyces sericifer*) **D+I:** FungEur4: 451-452 55d, FungEur4: 452-453 55a (as *Leucoagaricus sericeus*), FAN5: 104-105 **I:** FlDan1: 11B (as *Lepiota serena*), SV29: 40 (as *S. sericifera*)

Known from East Gloucestershire, Herefordshire, Hertfordshire, Oxfordshire, South Somerset, Surrey, West Kent and West Sussex.

subcretaceus Bon, *Doc. Mycol.* 13(49): 49 (1983)

E: !

H: On soil, amongst grass or on decayed bark or woodchip mulch on flowerbeds.

D+I: FungEur4: 437-439 54, B&K4: 210 243

Known from Berkshire, South Devon, South Essex and Surrey. *Fide* FAN5: this is possibly just a large form of *Leucoagaricus leucothites*.

sublittoralis (Hora) Singer, *Nova Hedwigia Beih.* 29: 163 (1969)

Lepiota sublittoralis Hora, *Trans. Brit. Mycol. Soc.* 43(2): 450 (1960)

Leucocoprinus sublittoralis (Hora) Locq. [*nom. inval.*], *Bull. Mens. Soc. Linn. Lyon* 14: 93 (1945)

E: !

H: On soil, often calcareous loam in scrub woodland and also known in fen vegetation.

D+I: FRIC2: 1-2 9a (as *Lepiota sublittoralis*), FungEur4: 381-382 42b, FAN5: 101-102 **I:** Bon: 289

Known from Hertfordshire, Huntingdonshire, Northamptonshire South Somerset and Surrey. A synonym of *Leucoagaricus wichanskyi* for some authorities, and the two species are much confused in Britain.

tener (P.D. Orton) Bon, *Doc. Mycol.* 7(27-28): 54 (1977)

Lepiota tenera P.D. Orton, *Trans. Brit. Mycol. Soc.* 43(2): 288 (1960)

E: ! **W:** !

H: On soil in scrub and woodland.

D+I: FungEur4: 352-354 38a, FAN5: 100-101

Known from England (East Norfolk, Hertfordshire, Middlesex, North Somerset and South Devon) and Wales (Denbighshire).

wichanskyi (Pilát) Bon & Boiffard, *Bull. Soc. Mycol. France* 90: 303 (1974)

Lepiota wichanskyi Pilát, *Sborn. Nár. Mus. Praze* 9B(2): 4 (1953)

Mis.: *Leucoagaricus sublittoralis* sensu D.A. Reid [F.R.I.C. 2: 1 (1967)]

E: !

H: On soil, often calcareous loam, in deciduous woodland and scrub.

D+I: FRIC6: 16-17 43e (as *Lepiota wichanskyi*), FungEur4: 382-385 43, FAN5: 102-103

Known from East Suffolk, North Somerset, South Devon and Surrey, but much confused in Britain with *Leucoagaricus sublittoralis*.

LEUCOCOPRINUS Pat., *J. Bot. (Morot)* 2: 16 (1888)

Agaricales, Agaricaceae

Type: *Leucocoprinus cepistipes* (Sowerby) Pat.

birnbaumii (Corda) Singer, *Sydowia* 15: 67 (1961)

Agaricus birnbaumii Corda, *Icon. fung.* 3: 48 (1839)

Agaricus flammula Alb. & Schwein., *Consp. fung. lusat.*: 149 (1805)

Lepiota flammula (Alb. & Schwein.) Gillet, *Hyménomycètes*: 63 (1874)

Lepiota aurea Massee, *Bull. Misc. Inform. Kew.* 1912: 189 (1912)

Agaricus luteus Bolton, *Hist. fung. Halifax* 2: 50 (1789)

Lepiota lutea (Bolton) Godfrin, *Bull. Soc. Mycol. France* 13: 33 (1897)

Leucocoprinus luteus (Bolton) Locq., *Bull. Soc. Linn. de Lyon* 14: 93 (1945)

Lepiota pseudolicmophora Rea, *Brit. basidiomyc.*: 74 (1922)

Mis.: *Agaricus cepistipes* sensu Sowerby [Col. fig. Engl. fungi. 1: pl. 2 (1796)] (yellow basidiomes)

E: !

H: On mulch or compost usually indoors but recently collected outside.

D+I: FungEur4: 464-468 58a, B&K4: 212 244, FAN5: 80-81 **I:** Sow1: 2 (as *Agaricus cepistipes* [yellow specimens]), Bon: 289, C&D: 249 (as *Leucocoprinus flos-sulphuris*)

An introduced alien, rather common with potted plants in buildings, but collected outdoors in Middlesex (Highgate Woods) in 2003 on a large heap of decayed and fermenting woodchips made from old, recycled Christmas trees.

brebissonii (Godey) Locq., *Bull. Soc. Linn. de Lyon* 12: 95 (1943)

Lepiota brebissonii Godey, Gillet, *Hyménomycètes*: 64 (1874)

Lepiota cepistipes var. *cretacea* Grev., *Scott. crypt. fl.*: t.333 (1828)

Mis.: *Lepiota felina* sensu Cooke [Ill. Brit. fung. 1108 (943A) Vol. 8 (1889)]

Mis.: *Armillaria subcava* sensu Cooke [Ill. Brit. fung. 57 (47) Vol. 1 (1881)]

E: o **S:** ! **W:** ! **NI:** ! **O:** Channel Islands: !

H: On soil in woodland, often with conifers, usually *Pinus* spp but also with *Picea* spp and *Taxus*. Also known with a wide range of deciduous trees such as *Alnus, Betula, Fagus* and *Quercus* spp. Also known from cool greenhouses in botanic gardens.

D: NM2: 224 **D+I:** Ph: 30-31, FungEur4: 468-471 59c, B&K4: 212 245, FAN5: 81-82 **I:** FlDan1: 14H (as *Lepiota brebissonii*), Bon: 289, C&D: 249

Occasional but widespread and may be locally abundant.

cepistipes (Sowerby) Pat., *Tab. anal. fung.* 7: 45 (1889)

Agaricus cepistipes Sowerby, *Col. fig. Engl. fung.* 1: pl. 2 (1796)

Lepiota cepistipes (Sowerby) P. Kumm., *Führ. Pilzk.*: 136 (1871)

Agaricus rorulentus Panizzi, *Comm. Soc. crittog. Ital.* 3: 172 (1862)

Leucocoprinus cepistipes var. *rorulentus* (Panizzi) Babos, *Ann. Hist.-Nat. Mus. Natl. Hung.* 72: 87 (1980)

Agaricus cheimonoceps Berk. & M.A. Curtis, *J. Linn. Soc., Bot.* 10: 283 (1869)

Lepiota cheimonoceps (Berk. & M.A. Curtis) Sacc., *Syll. fung.* 5: 66 (1887)

E: ! **S:** ! **ROI:** !

H: On soil in greenhouses in botanic gardens, also outdoors on fermenting piles of woodchips, stable-waste, compost or manure.

D: NM2: 224 **D+I:** FungEur4: 472-476 59b, B&K4: 212 246, FAN5: 77-78 **I:** Sow1: 2 (as *Agaricus cepistipes* [white specimens]), FlDan1: 14F (as *Lepiota cepaestipes*)

Known from England (Berkshire, Dorset, Leicestershire, North Hampshire, Nottinghamshire, South Devon, South Essex, Surrey, Warwickshire, West Cornwall and West Kent). Possibly a naturalised alien, apparently increasing.

cretaceus (Bull.) Locq., *Bull. Mens. Soc. Linn. Lyon* 14: 93 (1945)

  Agaricus cretaceus Bull., *Herb. France*: pl. 374 (1788)
  Leucoagaricus cretaceus (Bull.) M.M. Moser, *Kleine Kryptogamenflora*: 115 (1953)
  Lepiota cretata Locq. [*nom. nud.*], *Mitth. Aargauischen Naturf. Ges.* 23: 82 (1950)
  Leucocoprinus cretatus Lanzoni, *XIX Com. Sci. Nat.*: 31 (1986)

E: ! **W:** !
H: On soil in cold frames and greenhouses, and occasionally on manure heaps outside.
D+I: FAN5: 78-79
In similar habitat to *Leucocoprinus cepistipes*, and British records of these taxa appear much confused.

cygneus (J.E. Lange) Bon, *Doc. Mycol.* 8(30): 70 (1978)

  Lepiota cygnea J.E. Lange, *Fl. agaric. danic. 5, Taxonomic Conspectus*: V (1940)

E: !
H: On decayed wood, inside a hollow tree trunk (British collection).
D: NM2: 228 (as *Sericeomyces cygneus*) **D+I:** FungEur4: 480-482 59a, FAN5: 82 **I:** FlDan1: 13A (as *Lepiota cygnea*)
Known only from South Devon and West Norfolk.

straminellus (Bagl.) Narducci & Caroti, *Atti Soc. Tosc. Sci. Nat. Pisa, Mem., ser. B* 102: 49 (1996) [1995]

  Agaricus straminellus Bagl., *Comm. Soc. crittog. Ital.* 2(2): 263 (1865)
  Agaricus denudatus Rabenh., *Hedwigia* 6: 45 (1867)
  Lepiota denudata (Rabenh.) Sacc., *Syll. fung.* 5: 52 (1887)
  Leucocoprinus denudatus (Rabenh.) Singer, *Lilloa* 22: 424 (1951) [1949]
  Lepiota guegueni Sacc. & Traverso, *Syll. fung.* 21: 21 (1912)

E: ! **S:** ! **W:** !
H: On soil. In heated greenhouses in botanic gardens, also in gardens on grass clippings and lawns.
D: NM2: 223 (as *Leucocoprinus denudatus*) **D+I:** FungEur4: 492-495 58b (as *L. denudatus*), FAN5: 83
Probably an alien. Rather frequent in greenhouses, but two recent outdoor records from Surrey (Esher, Claremont, and Virginia Water).

LEUCOCORTINARIUS (J.E. Lange) Singer, *Lloydia* 8: 141 (1945)
Cortinariales, Cortinariaceae
Type: *Leucocortinarius bulbiger* (Alb. & Schwein.) Singer

bulbiger (Alb. & Schwein.) Singer, *Lilloa* 8: 141 (1945)

  Agaricus bulbiger Alb. & Schwein., *Consp. fung. lusat.*: 150 (1805)
  Armillaria bulbigera (Alb. & Schwein.) P. Kumm., *Führ. Pilzk.*: 135 (1871)
  Cortinarius bulbiger (Alb. & Schwein.) J.E. Lange, *Dansk. Bot. Ark.* 8(7): 13 (1935)

E: ! **S:** ! **W:** ?
H: On soil in woodland associated with conifers.
D: NM2: 330, BFF7: 73-74 **D+I:** Ph: 123, B&K5: 296 382 **I:** Bon: 173, C&D: 237
Rarely reported and little known in Britain. Reported from Wales but said to be 'under *Fagus sylvatica*' and unsubstantiated with voucher material.

LEUCOGASTER R. Hesse, *Jahrb. Wiss. Bot.* 13: 189 (1882)
Melanogastrales, Leucogastraceae
Type: *Leucogaster floccosus* R. Hesse (= *L. nudus*)

liosporus R. Hesse, *Jahrb. Wiss. Bot.* 13(2): 189 (1882)
E: !
H: On soil amongst litter. British material collected with *Pinus sylvestris* and *Rhododendron* sp.
A single collection from West Sussex (Graffham Common) in 2002.

nudus (Hazsl.) Hollós, *Ann. Hist.-Nat. Mus. Natl. Hung.* 6: 319 (1908)

  Hydnangium nudum Hazsl., *Verh. zool.-bot. Ges. Wien* 25: 64 (1875)
  Leucogaster floccosus R. Hesse, *Bot. Centralbl.* 40: 3 (1889)

E: ! **S:** !
H: Hypogeous, in calcareous loam soils.
D+I: BritTruff: 142-143 8 B
Reported from England (North Somerset: Batheaston) by Broome in 1860 and a Scottish record (as *Leucogaster floccosus*) listed by Reid & Austwick (1963).

LEUCOGYROPHANA Pouzar, *Česká Mykol.* 12: 32 (1958)
Boletales, Coniophoraceae
Type: *Leucogyrophana mollusca* (Fr.) Pouzar

mollusca (Fr.) Pouzar, *Česká Mykol.* 12(1): 33 (1958)

  Merulius molluscus Fr., *Syst. mycol.* 1: 329 (1821)
  Xylomyzon molluscum (Fr.) Pers., *Mycol. eur.* 2: 30 (1825)
  Serpula mollusca (Fr.) Donk, *Persoonia* 3: 209 (1964)
  Merulius fugax Fr., *Observ. mycol.* 1: 100 (1815)
  Merulius laeticolor Berk. & Broome, *Ann. Mag. Nat. Hist., Ser. 5* [*Notices of British Fungi no. 1681*] 1: 23 (1878)
  Merulius pseudomolluscus Parmasto, *Scripta Bot. Belg.* 2: 212 (1962)
  Leucogyrophana pseudomollusca (Parmasto) Parmasto, *Eesti N.S.V. Tead. Akad. Toimet., Biol.* 16(4): 386 (1967)

E: o **S:** ! **W:** ! **ROI:** !
H: On decayed wood of conifers such as *Picea* and *Pinus* spp. Also known on *Taxus*. Very rarely on wood of deciduous trees such as *Malus* and *Ulmus* spp.
D: NM3: 291 **D+I:** CNE4: 822-823 (as *Leucogyrophana pseudomollusca*), Ph: 238-239 (poor photo), B&K2: 210 241 (as *L. pseudomollusca*) **I:** Kriegl1: 365
Occasional in England, and most frequently reported from southern counties. Rarely reported elsewhere but apparently widespread. *Merulius fugax* may be an earlier name.

pinastri (Fr.) Ginns & Weresub, *Mem. N. Y. bot. Gdn* 28: 96 (1976)

  Hydnum pinastri Fr., *Novit. fl. suec. alt.* 2: 38 (1814)
  Sistotrema pinastri (Fr.) Pers., *Mycol. eur.* 2: 199 (1825)
  Merulius pinastri (Fr.) Burt, *Rep. (Annual) Missouri Bot. Gard.* 4: 356 (1917)
  Gyrophana pinastri (Fr.) Bourdot & Galzin, *Bull. Soc. Mycol. France* 39(2): 109 (1923)
  Serpula pinastri (Fr.) W.B. Cooke, *Mycologia* 49(1): 210 (1957)
  Merulius sclerotiorum Falck, *Hausschwammforsch.* 1: 93 (1912)
  Hydnum sordidum Weinm., in Fries, *Hymenomyc. eur.*: 614 (1874)

E: o **S:** ! **O:** Channel Islands: !
H: On decayed wood of conifers such as *Picea* and *Pinus* spp. Rarely on damp wood in buildings, or on woodchip mulch.
D: NM3: 291 **D+I:** Hallenb: 89 Figs. 58-61, Myc13(1): 36
Known from England (Buckinghamshire, East & West Kent, London, North Lincolnshire, South Somerset, Surrey, West Suffolk and Yorkshire), Scotland (Mid-Ebudes) and a single record from the Channel Islands (Jersey).

pulverulenta (Sowerby) Ginns, *Canad. J. Bot.* 56(16): 1966 (1978)

  Auricularia pulverulenta Sowerby, *Col. fig. Engl. fung.* 2: pl. 214 (1799)
  Merulius pulverulentus (Sowerby) Fr., *Elench. fung.* 1: 60 (1828)
  Merulius lacrymans var. *pulverulenta* (Sowerby) Quél., *Hymenomyc. eur.*: 594 (1874)
  Gyrophana pulverulenta (Sowerby) Bourdot & Galzin, *Bull. Soc. Mycol. France* 39(2): 107 (1923)
  Merulius tignicola Harmsen, *Friesia* 4: 245 (1952)
  Serpula tignicola (Harmsen) M.P. Christ., *Dansk. Bot. Ark.* 19(2): 323 (1960)

E: ! S: !
H: On decayed wood.
D: NM3: 291 **D+I:** Hallenb: 93 Figs. 62/63
Described from Britain but rarely recorded or collected. Collections in herb. K (identified by Ginns) from England (East Sussex and Northamptonshire); also known from Scotland (Mid-Ebudes: Isle of Mull).

romellii Ginns, *Canad. J. Bot.* 56(16): 1968 (1978)
   Mis.: *Leucogyrophana mollusca* sensu J. Erikss. & Ryvarden
E: !
H: On dead wood of conifers such as *Abies* spp (British material). Also known on soil under *Alnus*, from 'shingle' in a garden, on decayed wood of *Quercus* spp, and on dead *Glyceria maxima* leaves at the edge of a dried up pond.
D: NM3: 291 **D+I:** CNE4: 818-821 (as *Leucogyrophana mollusca*)
Known from Hertfordshire, Surrey and West Norfolk.

sororia (Burt) Ginns, *Canad. J. Bot.* 54(1 - 2): 150 (1976)
   Merulius sororius Burt, *Ann. Missouri Bot. Gard.* 4: 329 (1917)
E: ! S: !
H: On decayed wood and bark of conifers such as *Picea* and *Pinus* spp, *Pseudotsuga menziesii* and *Taxus*.
D: NM3: 291 **D+I:** CNE4: 828-829 (as *Leucogyrophana* sp.), Myc6(3): 138
Known from England (Surrey) and Scotland (Wester Ross).

LEUCOPAXILLUS Boursier, *Bull. Soc. Mycol. France* 41: 393 (1925)
Agaricales, Tricholomataceae
   Aspropaxillus Kühner & Maire, *Bull. Soc. Mycol. France* 50: 13 (1934)
Type: *Leucopaxillus paradoxus* (Costantin & L.M. Dufour) Boursier

gentianeus (Quél.) Kotl., *Česká Mykol.* 20: 230 (1966)
   Clitocybe gentianea Quél., *Mém. Soc. Émul. Montbéliard, Sér. 2*, 5: 341 (1873)
   Mis.: *Lepista amara* sensu auct. Brit.
   Mis.: *Clitocybe amara* sensu auct. Brit.
   Mis.: *Tricholoma amarum* sensu Rea (1922)
   Mis.: *Leucopaxillus amarus* sensu NCL
   Mis.: *Agaricus amarus* sensu auct. Brit.
E: !
H: On soil in woodland, associated with deciduous trees and conifers.
D: NM2: 136, BFF8: 52 **I:** Bon: 163, C&D: 207 (as *Leucopaxillus amarus*), Kriegl3: 302
Known only from Surrey and West Kent. This is a large and distinctive species, unlikely to be overlooked and probably genuinely rare.

giganteus (Sibth.) Singer, *Schweiz. Z. Pilzk.* 17: 14 (1939)
   Agaricus giganteus Sibth., *Fl. Oxon.* 420 (1794)
   Clitocybe gigantea (Sibth.) Quél., *Mém. Soc. Émul. Montbéliard, Sér. 2*, 5: 88 (1872)
   Paxillus giganteus (Sibth.) Fr., *Hymenomyc. eur.*: 401 (1874)
   Aspropaxillus giganteus (Sibth.) Kühner & Maire, *Bull. Soc. Mycol. France* 50: 13 (1934)
E: o S: o W: ! NI: o ROI: ! O: Channel Islands: !
H: On soil in woodland and at woodland edges. Also amongst grass in gardens, parkland, and meadows.
D: NM2: 135, FAN3: 76-77, BFF8: 51 **D+I:** Ph: 45-46 **I:** Sow3: 244 (as *Agaricus giganteus*), Bon: 163, C&D: 207
Occasional but widespread.

paradoxus (Costantin & L.M. Dufour) Boursier, *Bull. Soc. Mycol. France* 41: 391 (1925)
   Clitocybe paradoxa Costantin & L.M. Dufour, *Nouv. fl. champ.*, Edn 2: 262 (1895)
   Lepista paradoxa (Costantin & L.M. Dufour) Maire, *Bull. Soc. Mycol. France* 40: 307 (1925)
E: ! S: ! ROI: ?

H: On soil, often amongst grass, under trees in woodland and also in cemeteries. Reported with *Betula* and *Picea* spp.
D: NM2: 136, BFF8: 52-53 **I:** Bon: 163, C&D: 207
Known from England (East Kent, Surrey and West Sussex) and Scotland (Dumfriesshire). Reported from the Republic of Ireland but unsubstantiated with voucher material.

rhodoleucus (Romell) Kühner, *Bull. Mens. Soc. Linn. Lyon* 5: 126 (1926)
   Agaricus rhodoleucus Romell, *Bot. Not.* 1895: 66 (1895)
   Clitocybe rhodoleuca (Romell) Sacc., *Syll. fung.* 14: 74 (1899)
   Lepista rhodoleuca (Romell) Maire, *Bull. Soc. Mycol. France* 40: 305 (1924)
E: ! S: !
H: On soil, often in grassy places near scrub or woodland. Reported with *Ulmus* spp and *Chamaecyparis nootkatensis*.
D: NM2: 135, BFF8: 54 **I:** C&D: 207, Kriegl3: 304, FM2(2): 39
Known from England (East Kent, South Hampshire and West Sussex) and Scotland (Peebleshire).

LICHENOMPHALIA Redhead, Lutzoni, Moncalvo & Vilgalys, *Mycotaxon* 83: 36 (2002)
Agaricales, Tricholomataceae
   Phytoconis Bory [*nom. rej.*], *Mem. Conf. et Byssus*: 52 (1797)
   Botrydina Bréb. (anam.), *Mem. Soc. Acad. Agric. Industr. Falaise*: 36 (1839)
   Coriscium Vain. (anam.), *Acta Soc. Fauna Flora fenn.* 7(2): 188 (1890)
Type: *Lichenomphalia hudsoniana* (H.S. Jenn.) Redhead *et al.*

alpina (Britzelm.) Redhead, Lutzoni, Moncalvo & Vilgalys, *Mycotaxon* 83: 36 (2002)
   Agaricus alpinus Britzelm., *Ber. Naturhist. Vereins Augsburg* 30: 13 (1890)
   Omphalina alpina (Britzelm.) Bresinsky & Stangl, *Z. Pilzk.* 40(1-2): 73 (1974)
   Agaricus umbelliferus f. *flavus* Cooke [*nom. nud.*], *Ill. Brit. fung.* 2: pl. 260 (271) (1883)
   Omphalia umbellifera var. *flava* (Cooke) Rea, *Brit. basidiomyc.*: 429 (1922)
   Omphalia umbellifera f. *flava* (Cooke) Cejp, *Atlas champ. Eur.* 4: 43 (1936)
   Omphalia flava (Cooke) F.H. Møller, *Fungi Faeroes*: 260 (1945)
   Omphalina flava (Cooke) M. Lange, *Medd. Grønl.* 147: 25 (1955)
   Omphalia luteovitellina Pilát & Nannf., *Friesia* 5: 23 (1954)
   Omphalina luteovitellina (Pilát & Nannf.) M. Lange, *Medd. Grønl.* 148(2): 63 (1957)
E: ! S: c W: ! NI: ! ROI: !
H: On soil or bare peat, with the anamorphic *Botrydina* state, usually in wet situations such as under rock overhangs or wet mountain and hillsides.
D+I: Ph: 69, B&K3: 190 211
Rather common in Scotland. Poorly known elsewhere but apparently widespread.

hudsoniana (H.S. Jenn.) Redhead, Lutzoni, Moncalvo & Vilgalys, *Mycotaxon*: 38 (2002)
   Hygrophorus hudsonianus H.S. Jenn., *Mem. Carn. Mus.* 12: 2 (1936)
   Omphalina hudsoniana (H.S. Jenn.) H.E. Bigelow, *Mycologia* 62: 15 (1970)
   Phytoconis hudsoniana (H.S. Jenn.) Redhead & Kuyper, *Mycotaxon* 31: 222 (1988)
   Omphalia luteolilacina J. Favre, *Ergebn. Wiss. Untersuch. Schweiz. Natn. Parks* 5: 199 (1955)
   Omphalina luteolilacina (J. Favre) D.M. Hend., *Notes Roy. Bot. Gard. Edinburgh* 22: 595 (1958)
   Verrucaria laetevirens Borrer (anam.), *Engl. fl., suppl.*: t. 2658 (1830)
   Normandina laetevirens (Borrer) Nyl. (anam.), *Lichenogr. Scand.*: 264 (1861)

Endocarpon viride Ach. (anam.), *Lichenogr. Univ.*: 300 (1810)

Coriscium viride (Ach.) Vain. (anam.), *Acta Soc. Fauna Fl. Fenn.* 7: 189 (1890)

Mis.: *Omphalina alpina* sensu auct.

E: ! S: c W: ! NI: ! ROI: !

H: On acidic peaty soil, with the anamorphic lichen stage *Coriscium viride* on heathland and moorland.

D: NM2: 171, FAN3: 91 **D+I:** B&K3: 190 213

Rather common in Scotland. Rarely reported elsewhere but apparently widespread.

umbellifera (L.) Redhead, Lutzoni, Moncalvo & Vilgalys, *Mycotaxon* 83: 38 (2002)

Agaricus umbelliferus L. [non *A. umbelliferus* Schaeff (1770)] *Sp. pl.*: 1175 (1753)

Omphalia umbellifera (L.) P. Kumm., *Führ. Pilzk.*: 107 (1871)

Omphalina umbellifera (L.) Quél., *Enchir. fung.*: 44 (1886)

Agaricus ericetorum Pers. [non *A. ericetorum* Bull. (1792)], *Observ. mycol.* 1: 50 (1796)

Omphalia ericetorum (Pers.) S. Lundell [*nom. illegit.*], *Fungi Exsiccati Suecici*: 1753 (1949)

Omphalina ericetorum (Pers.) H.E. Bigelow, *Mycologia* 62: 13 (1970)

Gerronema ericetorum (Pers.) Singer, *Sydowia, Beih.* 7: 14 (1973)

Phytoconis ericetorum (Pers.) Redhead & Kuyper, *Mycotaxon* 31(1): 222 (1988)

Merulius turfosus Pers., *Mycol. eur.* 2: 26 (1828)

Omphalia umbellifera var. *nivea* Rea, *Brit. basidiomyc.*: 429 (1922)

Omphalia umbellifera f. *albida* J.E. Lange, *Dansk. Bot. Ark.*: 12 (1930)

Omphalia umbellifera f. *bispora* F.H. Møller, *Fungi Faeroes* 1(1): 258 (1945)

Lepraria botryoides (L.) Ach. (anam.), *Lich. Suec. Prodr.*: 10 (1798)

Botrydina vulgaris Bréb. (anam.) [*nom. superf.*], *Mem. Soc. Acad. Agric. Industr. Falaise*: 36 (1939)

Mis.: *Omphalina pseudoandrosacea* sensu auct.

E: o S: c W: ! NI: c ROI: ! O: Isle of Man: !

H: On acidic soil or peat in heathland and moorland, with the anamorphic lichen state. Also known with mosses, including *Sphagnum* spp, and on decayed wood, usually old stumps.

D: FAN3: 91 **D+I:** Ph: 69, B&K3: 190 212 **I:** Bon: 129, C&D: 187

Common in Scotland *fide* R. Watling (pers. comm.). Occasional in England and mostly reported from western and northern counties. Better known as *Omphalina ericetorum*.

velutina (Quél.) Redhead, Lutzoni, Moncalvo & Vilgalys, *Mycotaxon* 83: 43 (2002)

Omphalia velutina Quél., *Compt. Rend. Assoc. Franç. Avancem. Sci.* 14: 445 (1886)

Omphalina velutina (Quél.) Quél., *Enchir. fung.*: 44 (1886)

Omphalia grisella P. Karst., *Meddeland. Soc. Fauna Fl. Fenn.* 16: 92 (1890)

Omphalina grisella (P. Karst.) M.M. Moser, *Kleine Kryptogamenflora*: 70 (1953)

Omphalina pararustica Clémençon, *Z. Mykol.* 48(2): 215 (1982)

E: ! S: ! W: ! NI: !

H: On peaty soil on moorland or upland and montane habitat, associated with the anamorphic *Botrydina* state.

D: FAN3: 90

Rarely reported but apparently widespread. Probably a species complex.

LIMACELLA Earle, *Bull. New York Bot. Gard.* 5: 447 (1909)

Agaricales, Amanitaceae

Type: *Limacella delicata* (Fr.) Earle

delicata (Fr.) Earle, *Bull. New York Bot. Gard.* 5: 447 (1909)

Agaricus delicatus Fr., *Syst. mycol.* 1: 23 (1821)

Lepiota delicata (Fr.) P. Kumm., *Führ. Pilzk.*: 136 (1871)

Armillaria delicata (Fr.) Boud., *Icon. mycol.* 1: pl. 23 (1904)

E: !

H: On soil (usually calcareous loam) in mixed deciduous woodland. With *Acer pseudoplatanus, Corylus, Fagus, Fraxinus* or *Taxus*. Single collections on a lawn and on compost in a cool greenhouse.

Rarely reported. British collections from East & West Kent, South Devon, South Somerset, Surrey and West Sussex.

delicata *var.* **glioderma** (Fr.) Gminder, *Z. Mykol.* 60(2): 386 (1994)

Agaricus gliodermus Fr., *Öfvers. Kongl. Vetensk.-Akad. Förh.* 8: 43 (1852)

Lepiota glioderma (Fr.) Gillet, *Hyménomycètes*: 73 (1874)

Armillaria glioderma (Fr.) Quél., *Mém. Soc. Émul. Montbéliard, Sér. 2,* 5: 541 (1875)

Limacella glioderma (Fr.) Maire, *Bull. Soc. Mycol. France* 40(4): 294 (1924)

E: ! W: ! O: Channel Islands: !

H: On soil (usually calcareous loam) in mixed deciduous woodland.

D: NM2: 198 (as *Limacella glioderma*) **D+I:** B&K4: 156 161 (as *L. glioderma*) **I:** Bon: 295 (as *L. glioderma*), C&D: 273 (as *L. glioderma*)

Known from England (Cumberland, East Kent, Isle of Wight, Mid-Lancashire, North & South Wiltshire, Oxfordshire, South Devon, South Somerset, Surrey, West Gloucestershire and Westmorland), Wales (Anglesey and Denbighshire) and the Channel Islands (Jersey).

delicata *var.* **vinosorubescens** (Furrer-Ziogas) Gminder, *Z. Mykol.* 60(2): 386 (1994)

Limacella vinosorubescens Furrer-Ziogas, *Schweiz. Z. Pilzk.* 47: 214 (1969)

Mis.: *Limacella roseofloccosa* sensu auct.

E: ! W: !

H: On soil (usually calcareous loam) in mixed deciduous woodland often amongst *Hedera* or *Glechoma hederacea*. Rarely reported with conifers such as *Pseudotsuga* or *Thuja* spp.

D: NBA8: 593-594 **D+I:** B&K4: 158 165 (as *Limacella vinosorubescens*) **I:** SV32: 9

Rarely reported. Known from England (North Hampshire, North Somerset, Oxfordshire, South Devon, South Wiltshire, Surrey and West Sussex) and Wales (Caernarvonshire).

guttata (Pers.) Konrad & Maubl., *Icon. select. fung.* 1: pl. 9 (1924)

Agaricus guttatus Pers. [non *A. guttatus* Schaeff. (1770)], *Syn. meth. fung.*: 265 (1801)

Lepiota guttata (Pers.) Quél., *Bull. Soc. Bot. France* 23: 325 (1877) [1876]

Agaricus lenticularis Lasch, *Linnaea* 3: 157 (1828)

Lepiota lenticularis (Lasch) Gillet, *Hyménomycètes*: 66 (1874)

Limacella lenticularis (Lasch) Maire, *Bull. Soc. Mycol. France* 40: 294 (1924)

Agaricus megalodactylus Berk. & Broome, *Outl. Brit. fungol.*: 91 (1860)

Amanita megalodactyla (Berk. & Broome) Sacc., *Syll. fung.* 5: 20 (1887)

Lepiota lenticularis var. *megalodactylus* (Berk. & Broome) Rea, *Brit. basidiomyc.*: 80 (1922)

E: ! S: ! W: ! NI: ! ROI: !

H: On soil in deciduous or mixed conifer and deciduous woodland.

D: NM2: 198 **D+I:** B&K4: 156 162 **I:** FlDan1: 7A (slightly too pink) (as *Limacella lenticularis*), Bon: 295 (as *Limacella lenticularis*), C&D: 273

Rarely reported but apparently widespread. Cooke 15 (11) Vol. 1 (1881) (as *Agaricus megalodactylus*) is possibly this.

illinita (Fr.) Murrill, *North American Flora* 10(1): 40 (1914)

Agaricus illinitus Fr., *Observ. mycol.* 2: 8 (1818)

Lepiota illinita (Fr.) Quél., *Mém. Soc. Émul. Montbéliard, Sér. 2,* 5: 326 (1873)

E: !

H: On soil in deciduous or mixed conifer and deciduous woodland.

D: NM2: 198 **D+I:** B&K4: 158 163 **I:** C&D: 273

Very rarely reported. Known from South Somerset, South Wiltshire, West Sussex and Yorkshire.

ochraceolutea P.D. Orton, *Notes Roy. Bot. Gard. Edinburgh* 29(1): 106 (1969)

E: !

H: On soil in deciduous woodland, on chalk and limestone.

D+I: B&K4: 158 164

Very rarely reported. Known from Mid-Lancashire, North & South Somerset and South Wiltshire.

LIMONOMYCES Stalpers & Loer., *Canad. J. Bot.* 60: 533 (1982)

Stereales, Sistotremataceae

Type: *Limonomyces roseipellis* Stalpers & Loer.

culmigenus (J. Webster & D.A. Reid) Stalpers & Loer., *Canad. J. Bot.* 60(5): 536 (1982)

Exobasidiellum culmigenum J. Webster & D.A. Reid, *Trans. Brit. Mycol. Soc.* 52(1): 20 (1969)

Galzinia culmigena (J. Webster & D.A. Reid) Johri & Bandoni, *Canad. J. Bot.* 53: 2563 (1975)

E: ! **S:** !

H: Parasitic. On various species of *Poaceae* most frequently *Dactylis glomerata*. Rarely reported on *Carex* spp, and there are single collections on dead stems of *Oenanthe crocata* and an old basidiome of *Auriscalpium vulgare*.

Known from England (Nottinghamshire and Yorkshire) and Scotland (Clyde Isles). Easily overlooked.

roseipellis Stalpers & Loer., *Canad. J. Bot.* 60(5): 534 (1982)

Mis.: *Athelia fuciformis* sensu Burdsall [TBMS 72: 422 (1979)]

E: ! **ROI:** !

H: Parasitic. On various species of *Poaceae*, most frequently *Lolium perenne*, *Festuca ovina* and *Festuca rubra* in utility grassland such as sports fields, cricket pitches and lawns.

Rarely reported. Easily overlooked.

LINDTNERIA Pilát, *Stud. Bot. Čechoslov.* 1: 72 (1938)

Stereales, Lindtneriaceae

Type: *Lindtneria trachyspora* (Bourdot & Galzin) Pilát

leucobryophila (Henn.) Jülich, *Persoonia* 9(3): 418 (1977)

Thelephora leucobryophila Henn., *Verh. Bot. Vereins Prov. Brandenburg* 39: 96 (1898)

Trechispora leucobryophila (Henn.) Liberta, *Taxon* 15(8): 318 (1966)

Sistotrema sulphureum var. *variecolor* Bourdot & Galzin, *Bull. Soc. Mycol. France* 30(3): 274 (1914)

Sistotrema variecolor (Bourdot & Galzin) Bourdot & Galzin, *Hymenomyc. France*: 437 (1928)

Cristella variecolor (Bourdot & Galzin) M.P. Christ., *Dansk. Bot. Ark.* 19(2): 97 (1960)

E: o

H: Known on decayed leaves and stems of *Carex pendula*, fallen lianas of *Clematis vitalba*, dead stems of *Rumex obtusifolius*, leaf and needle litter of *Fagus* and *Cupressus macrocarpa* and on humus and decayed fragments of deciduous wood in soil.

D: NM3: 122 **D+I:** Myc11(2): 61

Occasional and widespread in England.

panphyliensis Bernicchia & M.J. Larsen, *Mycotaxon* 37: 350 (1990)

E: !

H: Collected on fallen and decayed deciduous leaves, on decayed wood and also on the dead stem bases of *Carex pendula*.

D+I: Myc9(2): 52

Possibly just a hydnoid form of *Lindtneria leucobryophila*. Known from Hertfordshire, South Devon and Yorkshire.

trachyspora (Bourdot & Galzin) Pilát, *Stud. Bot. Čechoslov.* 1: 72 (1938)

Poria trachyspora Bourdot & Galzin, *Hymenomyc. France*: 659 (1928)

E: ! **W:** ! **ROI:** !

H: On decayed wood of *Prunus avium*, decayed leaves of grasses and *Typha latifolia*, dead stems of *Pteridium* and also found lining the tunnels in a nest of wood ants.

D: NM3: 122 **D+I:** CNE4: 831-833, B&K2: 144 142

Rarely reported, but inconspicuous and easily overlooked. Known from England (Bedfordshire, Cambridgeshire, Derbyshire, Mid-Lancashire and Surrey), Wales (Glamorganshire) and the Republic of Ireland (North Kerry).

LITSCHAUERELLA Oberw., *Sydowia* 19(1-3): 43 (1965)

Stereales, Tubulicrinaceae

Type: *Litschauerella abietis* (Bourdot & Galzin) Jülich

abietis (Bourdot & Galzin) Jülich, *Persoonia* 10(3): 335 (1979)

Peniophora aegerita ssp. *abietis* Bourdot & Galzin, *Bull. Soc. Mycol. France* 28(4): 383 (1913)

Peniophora abietis (Bourdot & Galzin) Bourdot & Galzin, *Hymenomyc. France*: 286 (1928)

E: !

H: On fallen and decayed wood (usually small twigs and branches) of conifers. British material is on dead twigs of *Cupressus macrocarpus*.

A single collection from South Devon (Slapton) in 1994.

clematidis (Bourdot & Galzin) J. Erikss. & Ryvarden, *Corticiaceae of North Europe* 4: 839 (1976)

Peniophora clematidis Bourdot & Galzin, *Bull. Soc. Mycol. France* 28(4): 383 (1913)

Tubulicium clematidis (Bourdot & Galzin) Oberw., *Ann. Mycol.* 19: 56 (1965)

E: o **ROI:** !

H: In scrub or woodland. Virtually restricted to living or fallen lianas of *Clematis vitalba*, rarely on dead stems of *Pteridium*.

D: NM3: 135 **D+I:** CNE4: 838-841, B&K2: 192 214

Rather common, at least in southern England. A single recent collection from the Republic of Ireland (The Burren).

LORELEIA Redhead, Moncalvo, Vilgalys & Lutzoni, *Mycotaxon* 82: 162 (2002)

Agaricales, Tricholomataceae

Type: *Loreleia postii* (Fr.) Redhead, Moncalvo, Vilgalys & Lutzoni

marchantiae (Singer & Clémençon) Redhead, Moncalvo, Vilgalys & Lutzoni, *Mycotaxon* 82: 162 (2002)

Gerronema marchantiae Singer & Clémençon, *Schweiz. Z. Pilzk.* 49: 119 (1971)

Omphalina marchantiae (Singer & Clémençon) Norvell, Redhead & Ammirati, *Mycotaxon* 50 (1994)

E: ! **S:** !

H: On dead or moribund thalli of the frondose liverworts *Conocephalum* and *Lunnularia* spp, and *Marchantia polymorpha*, or on soil amongst them.

D: NM2: 174 **D+I:** Myc3(1): 42 **I:** Ph: 69 (misdet. as *O. postii*)

Known from Scotland (Shetland) and England (South Devon). Reported from Warwickshire but unsubstantiated with voucher material.

postii (Fr.) Redhead, Moncalvo, Vilgalys & Lutzoni, *Mycotaxon* 82: 162 (2002)

Agaricus postii Fr., *Monogr. hymenomyc. Suec.* 2: 291 (1863)

Omphalia postii (Fr.) P. Karst., *Bidrag Kännedom Finlands Natur Folk* 32: 129 (1879)

Omphalina postii (Fr.) Singer, *Mycologia* 39: 83 (1947)

Gerronema postii (Fr.) Singer, *Sydowia* 15: 50 (1962) [1961]

E: ! **S:** !

H: On burnt soil or soil with mixed ashes, often amongst *Sphagnum* spp, or large frondose liverworts such as *Marchantia polymorpha* or *Lunularia* sp.

D: NM2: 174, C&D: 183 (as *Gerronema postii*), FAN3: 81-82
D+I: Myc1(1): 19 **I:** Kriegl3: 485
Rarely collected or reported. Most records are from flowerpots in gardens or from burnt areas amongst *Sphagnum* spp in boggy places.

LUELLIA K.H. Larss. & Hjortstam, *Svensk bot. Tidskr.* 68: 59 (1974)
Stereales, Atheliaceae
Type: *Luellia recondita* (H.S. Jacks.) K.H. Larss. & Hjortstam

cystidiata Hauerslev, *Friesia* 11(5): 283 (1987) [1979]
E: ! **S:** ! **W:** !
H: On decayed wood of conifers such as *Picea abies* and *Pinus sylvestris,* and rarely on deciduous trees such as *Alnus* and *Salix* spp. Also known on debris of *Pteridium.*
D: NM3: 150
Very rarely reported. Collections from England (South Devon), Scotland (Kirkudbrightshire) and Wales (Caernarvonshire). Basidiomes are small, cryptically coloured and inconspicuous.

recondita (H.S. Jacks.) K.H. Larss. & Hjortstam, *Svensk bot. Tidskr.* 68(1): 60 (1974)
Corticium reconditum H.S. Jacks., *Canad. J. Res., Sect. C, Bot. Sci.* 26: 154 (1948)
Athelopsis recondita (H.S. Jacks.) Parmasto, *Conspectus Systematis Corticiacearum*: 43 (1968)
E: ! **O:** Channel Islands: !
H: Most often reported on dead stems of ferns such as *Dryopteris* spp and *Pteridium.* Rarely on decayed wood of *Betula* spp, and conifers such as *Larix, Pinus* and *Picea* spp.
D: NM3: 150 **D+I:** CNE4: 853-854
Rarely reported. Known from central and southern England and the Channel Islands (Jersey). Basidiomes are small, cryptically coloured and inconspicuous.

LYCOPERDON L., *Gen. Pl.*: 493 (1754)
Lycoperdales, Lycoperdaceae
Utraria Quél., *Mém. Soc. Émul. Montbéliard, Sér. 2,* 5: 366 (1873)
Type: *Lycoperdon perlatum* Pers.

atropurpureum Vittad., *Monogr. Lycoperd.* 2: 42 (1842)
E: !
H: On soil in thermophilic woodlands, usually associated with *Quercus* spp.
D+I: BritPuffb: 158-159 Figs. 123/124
A southern species, known from South Devon. Reported from Dorset, Herefordshire, Isle of Wight, North Somerset, South Wiltshire, Surrey, West Gloucestershire, West Suffolk and Worcestershire and Yorkshire but unsubstantiated with voucher material.

caudatum J. Schröt., in Cohn, *Krypt.-Fl. Schlesien* 3: 698 (1889)
Lycoperdon pedicellatum Peck, *Bull. Buffalo Soc. Nat. Hist.* 1: 63 (1873)
E: ? **S:** !
H: On sandy or calcareous soil, usually in dunes or occasionally in open mixed woodland with *Betula, Fagus* and *Pinus* spp.
D: NM3: 337 **D+I:** Ph: 249 (as *Lycoperdon pedicellatum*), B&K2: 394 517 (as *L. pedicellatum*), BritPuffb: 148-149 Figs. 113/114 **I:** Kriegl2: 141
Known from Dunbartonshire, Midlothian (Edinburgh), Perthshire, Fife & Kinross and Angus. Reported from Yorkshire but unsusbtantiated with voucher material.

decipiens Durieu & Mont., *Expl. Sci. Algérie [Bot. I]* 1: 380 (1848)
Bovista cepiformis Wallr., *Fl. crypt. Germ.* 2: 392 (1833)
E: !
H: On dry calcareous soil in unimproved grassland.
D: NM3: 338 **D+I:** BritPuffb: 162-163 Figs. 127/128
Reported from North Lincolnshire, North Somerset, South Wiltshire, Surrey, West Gloucestershire, West Sussex and Worcestershire but unsubstantiated with voucher material. Represented in herb. K only by dubious collections from Surrey and West Sussex.

echinatum Pers., *Tent. disp. meth. fung.*: 53 (1797)
Lycoperdon hoylei Berk. & Broome, *Ann. Mag. Nat. Hist., Ser. 4 [Notices of British Fungi no. 1307]* 7: 430 (1871)
E: o **S:** ! **W:** ! **NI:** ! **O:** Channel Islands: !
H: On soil (usually calcareous) amongst litter of deciduous trees in woodland. Usually found with *Fagus* in beechwoods and uncommonly with *Betula, Corylus* and *Quercus* spp. Rarely reported with conifers such as *Larix* and *Pinus* spp.
D: NM3: 338 **D+I:** Ph: 246-247, B&K2: 390 512, BritPuffb: 150-151 Figs. 115/116 **I:** Bon: 305, Kriegl2: 143
Occasional in England and most common in the south and south-east. Rarely reported elsewhere but apparently widespread and known as far north as Perthshire.

ericaeum Bonord., *Bot. Zeitung (Berlin)* 15: 596 (1857)
E: ! **S:** !
H: On acidic soil amongst short turf.
D: NM3: 340 **D+I:** BritPuffb: 168-169 Figs. 133/134 **I:** Kriegl2: 144
Very poorly known in Britain. Collections in herb. K from Yorkshire in 1883. Recorded from England (West Norfolk: Holkham Dunes) in 1999 but unsubstantiated with voucher material. Reported from Scotland (Midlothian: Edinburgh) in 1878 and Angus (Barry Links) in 1997.

lambinonii Demoulin, *Lejeunia* 62: 13 (1972)
ROI: !
H: On humic calcareous soil under conifers.
D: NM3: 339 **D+I:** BritPuffb: 166-167 Figs. 131/132
A single record from South Tipperary (Glen of Aherlow) in 1977.

lividum Pers., *J. Bot., Paris* 2: 18 (1809)
Lycoperdon cookei Massee, *J. Roy. Microscop. Soc. London.*: 14 (1887)
Lycoperdon spadiceum Pers. [non *L. spadiceum* (Schaeff) Poir. (1808)], *J. Bot., Paris* 2: 20 (1809)
E: c **S:** c **W:** c **NI:** c **ROI:** o **O:** Channel Islands: o Isle of Man: o
H: On soil amongst grass in pastures, meadows, on dunes and sea cliffs, and in cemeteries.
D: NM3: 339 **D+I:** Ph: 249 (as *Lycoperdon spadiceum*), B&K2: 392 514, BritPuffb: 156-157 Figs. 121/122 **I:** Kriegl2: 147
Common and widespread. Most frequently collected in coastal areas.

mammiforme Pers., *Syn. meth. fung.*: 146 (1801)
Lycoperdon velatum Vittad., *Monogr. Lycoperd.*: 43 (1842)
Mis.: *Lycoperdon candidum* sensu Mason [*Naturalist (Hull)* 1928: 306 (1928)]
Mis.: *Lycoperdon cruciatum* sensu Massee & Crossland (1906)
E: o **S:** ! **W:** ! **NI:** ! **ROI:** !
H: On 'black' soil or humus (often calcareous) in open, mixed deciduous woodland, usually associated with *Corylus* and *Fagus* or *Quercus* spp.
D: NM3: 338 **D+I:** Ph: 247, B&K2: 392 515 (as *Lycoperdon mammaeforme*), BritPuffb: 146-147 Figs. 111/112 **I:** Bon: 303, Kriegl2: 148
Occasional but widespread in England. Last reported from Scotland (Easterness: Rothiemurchus) in 1938 and the Republic of Ireland Wicklow) in 1957.

molle Pers., *Syn. meth. fung.*: 150 (1801)
Mis.: *Lycoperdon atropurpureum* sensu auct.
Mis.: *Lycoperdon umbrinum* sensu auct.
E: c **S:** c **W:** c **NI:** c **ROI:** c
H: On soil (often calcareous) in deciduous, or mixed coniferous and deciduous woodland.
D: NM3: 339 **D+I:** B&K2: 393 516, BritPuffb: 160-161 Figs. 125/126 **I:** Kriegl2: 150
Widespread and fairly common.

nigrescens Pers., *Neues Mag. Bot.*: 1 (1794)
Lycoperdon foetidum Bonord., *Handb. Mykol.*: 253 (1851)

Lycoperdon perlatum var. *nigrescens* (Pers.) Pers., *Syn. meth. fung.*: 146 (1801)

E: c **S:** c **W:** c **NI:** c **ROI:** c **O:** Isle of Man: !

H: On acidic soil and humus, in conifer and mixed woodland, on heathland, in grassland and on old mossy lawns.

D: NM3: 339 **D+I:** Ph: 248-249 (as *Lycoperdon foetidum*), B&K2: 390 513 (as *L. foetidum*), BritPuffb: 154-155 Figs. 119/120 **I:** Kriegl2: 145 (as *L. foetidum*)

Common and widespread. Perhaps better known as *Lycoperdon foetidum*.

perlatum Pers., *Observ. mycol.* 1: 4 (1796)
    Lycoperdon gemmatum Batsch, *Elench. fung.*: 147 (1783)
    Lycoperdon gemmatum var. *perlatum* (Pers.) Fr., *Syst. mycol.* 3: 37 (1829)

E: c **S:** c **W:** c **NI:** c **ROI:** c **O:** Channel Islands: c Isle of Man: !

H: On soil or decayed wood (fallen trunks or stumps) in deciduous woodland and with conifers (but less frequent in the latter). Also on flowerbeds mulched with wood chips or shredded bark, where it may occur in quantity.

D: NM3: 338 **D+I:** Ph: 248-249, B&K2: 394 518, BritPuffb: 152-153 Figs. 117/118 **I:** Kriegl2: 151

Very common and widespread.

pyriforme Schaeff., *Fung. Bavar. Palat. nasc.* 2: 128 (1763)
    Lycoperdon pyriforme var. *tesselatum* Pers., *Syn. meth. fung.*: 148 (1801)

E: c **S:** c **W:** c **NI:** c **ROI:** c **O:** Channel Islands: c Isle of Man: c

H: On decayed or buried wood (fallen branches, trunks, stumps etc.) of deciduous and rarely, coniferous substrata. A single recent collection on dead stems of *Dryopteris filix-mas*.

D: NM3: 338 **D+I:** Ph: 248-249, B&K2: 394 519, BritPuffb: 144-145 Figs. 109/110 **I:** Bon: 305, Kriegl2: 152

Very common and widespread. English name = 'Stump Puffball'.

umbrinum Pers., *Syn. meth. fung.*: 147 (1801)

E: ! **S:** ! **W:** ! **NI:** ! **ROI:** !

H: On acidic soils usually under conifers, frequently in plantations of *Picea* spp.

D: NM3: 339 **D+I:** B&K2: 396 520, BritPuffb: 164-165 Figs. 129/130 **I:** Kriegl2: 153

Rarely reported; apparently widespread but old records are often referable to *Lycoperdon molle*.

LYOPHYLLUM P. Karst., *Acta Soc. Fauna Fl. Fenn.* 2(1): 3 (1881)

Agaricales, Tricholomataceae

Type: *Lyophyllum leucophaeatum* (P. Karst.) P. Karst. (= *L. gangraenosum*)

connatum (Schumach.) Singer, *Schweiz. Z. Pilzk.* 17: 55 (1939)
    Agaricus connatus Schumach., *Enum. pl.* 1: 299 (1801)
    Clitocybe connata (Schumach.) Gillet, *Hyménomycètes*: 164 (1874)
    Tricholoma connatum (Schumach.) Ricken, *Blätterpilze Deutschl.*: 360 (1914)

E: ! **S:** c **W:** ! **NI:** ! **ROI:** !

H: On soil in woodland, or occasionally in grassy areas near to woodland.

D: NM2: 138 **D+I:** Ph: 42-43, B&K3: 220 256 **I:** FlDan1: 38F (as *Clitocybe connata*), Bon: 167, C&D: 211, Kriegl3: 305, RivMyc: 139

Uncommonly reported but more frequent in northern areas, and common in Scotland *fide* R. Watling.

decastes (Fr.) Singer, *Lilloa* 22: 165 (1951) [1949]
    Agaricus decastes Fr., *Observ. mycol.* 2: 105 (1818)
    Clitocybe decastes (Fr.) P. Kumm., *Führ. Pilzk.*: 124 (1871)
    Agaricus aggregatus Schaeff., *Fung. Bavar. Palat. nasc.* 4: 305 (1770)
    Clitocybe aggregata (Schaeff.) Gillet, *Hyménomycètes*: 161 (1874)
    Tricholoma aggregatum (Schaeff.) Costantin & L.M. Dufour, *Nouv. fl. champ.*: 16 (1891)
    Lyophyllum aggregatum (Schaeff.) Kühner, *Bull. Mens. Soc. Linn. Lyon* 7: 211 (1938)

Clitocybe aggregata f. *reducta* J.E. Lange, *Dansk. Bot. Ark.* 6(5): 59 (1930)
    Clitocybe molybdina (Bull.) P. Kumm., *Führ. Pilzk.*: 120 (1871)
    Agaricus subdecastes Cooke & Massee, *Handb. Brit. fung.*, Edn 2: 366 (1890)
    Clitocybe subdecastes (Cooke & Massee) W.G. Sm., *Syn. Brit. Bas.*: 51 (1908)
    Mis.: *Tricholoma amplum* sensu Rea (1922)

E: o **S:** o **W:** o **NI:** o **ROI:** !

H: On soil, usually in deciduous woodland or scrub, occasionally in grassland areas near to trees or on wasteground.

D: NM2: 139 **D+I:** B&K3: 220 257 **I:** Ph: 42-43, Bon: 167, C&D: 211, Kriegl3: 307

Occasional but widespread.

eustygium (Cooke) Clémençon, *Mycotaxon* 15: 73 (1982)
    Agaricus eustygius Cooke, *Grevillea* 19: 40 (1890)
    Collybia eustygia (Cooke) Sacc., *Syll. fung.* 9: 33 (1891)
    Mis.: *Lyophyllum amariusculum* sensu C&D: non sensu Clémençon [Mycotaxon 15: 68 (1982)]
    Mis.: *Lyophyllum crassifolium* sensu Lange (FlDan1: pl. 25C) and sensu NM2:, non sensu Fr. (Hymenomyc. eur.: 61, 1874)
    Mis.: *Tricholoma crassifolium* sensu Lange (FlDan1: 58 & pl. 25C)
    Mis.: *Collybia fumosa* sensu Rea (1922)
    Mis.: *Lyophyllum immundum* sensu NCL

E: ! **S:** ! **W:** !

H: On soil often in grassy areas in woodland, less often in grassland.

D: NM2: 138 (as *Lyophyllum crassifolium*) **I:** C&D: 209 (as *L. amariusculum*), Cooke 1146 (1185) Vol. 8 1890) (as *Agaricus (Collybia) eustygius*)

Rarely reported (usually as *L. immundum*).

favrei (R. Haller Aar. & R. Haller Suhr) R. Haller Aar. & R. Haller Suhr, *Schweiz. Z. Pilzk.* 28(4): 51 (1950)
    Tricholoma favrei R. Haller Aar. & R. Haller Suhr, *Schweiz. Z. Pilzk.* 27(9): 132 (1949)
    Calocybe favrei (R. Haller Aar. & R. Haller Suhr) Bon, *Doc. Mycol.* 29(115): 33 (1999)

E: ! **ROI:** !

H: On calcareous soil, in mixed deciduous and coniferous woodland. With *Acer pseudoplatanus*, *Fagus*, *Fraxinus* and *Taxus*.

D+I: Myc3(3): 142, B&K3: 222 259 **I:** RivMyc: 128

Very rare. This is a large, bluish tricholomatoid agaric with bright primrose-yellow lamellae and could hardly be overlooked. First reported from Surrey (Mickleham, Norbury Park) in 1956. Only known from this area until recently collected in Republic of Ireland (Limerick).

fumosum (Pers.) P.D. Orton, *Trans. Brit. Mycol. Soc.* 43(2): 178 (1960)
    Agaricus fumosus Pers., *Syn. meth. fung.*: 348 (1810)
    Clitocybe fumosa (Pers.) P. Kumm., *Führ. Pilzk.*: 124 (1871)
    Collybia fumosa (Pers.) Bres., *Fungi trident.* 2: 50 (1892)
    Tricholoma fumosum (Pers.) Ricken, *Blätterpilze Deutschl.*: 357 (1914)
    Agaricus cinerascens Bull., *Herb. France*: pl. 428 (1789)
    Tricholoma cinerascens (Bull.) Gillet, *Hyménomycètes*: 121 (1874)
    Clitocybe cinerascens (Bull.) Bres., *Icon. Mycol.* 3: 149 (1938)
    Agaricus conglobatus Vittad., *Descr. fung. mang.*: 349 (1835)
    Clitocybe conglobata (Vittad.) Bres., *Fungi trident.* 1: 27 (1881)
    Tricholoma conglobatum (Vittad.) Sacc., *Syll. fung.* 5: 126 (1887)
    Lyophyllum conglobatum (Vittad.) M.M. Moser, *Kleine Kryptogamenflora*: 43 (1953)
    Agaricus tumulosus Kalchbr., *Icon. select. Hymenomyc. Hung.*: 13 t. 5 (1874)
    Clitocybe tumulosa (Kalchbr.) Sacc., *Syll. fung.* 5: 162 (1887)

Mis.: *Tricholoma pescaprae* var. *multiforme* sensu Cooke [HBF2: 365] & sensu Rea (1922)
Mis.: *Tricholoma pescaprae* sensu Cooke [HBF2: 365] & sensu Rea (1922)
E: ! **S:** ! **W:** ! **NI:** ! **ROI:** !
H: On soil in woodland, occasionally in grassland near shrubs or trees.
D: NM2: 139 **D+I:** B&K3: 222 260 **I:** C&D: 211
Rarely reported but apparently widespread. Possibly a species complex.

gangraenosum (Fr.) Gulden, *Nordic J. Bot.* 11(4): 478 (1991)
Agaricus gangraenosus Fr., *Epicr. syst. mycol.*: 56 (1838)
Clitocybe gangraenosa (Fr.) Gillet, *Hyménomycètes*: 153 (1874)
Agaricus fumatofoetens Secr. [*nom. inval.*], *Mycogr. Suisse*: 641 (1833)
Lyophyllum fumatofoetens Jul. Schäff. [*nom. inval.*], *Ber. Bayer. Bot. Ges.* 27: 202 (1947)
Agaricus leucophaeatus P. Karst., *Not. Sällsk. Fauna et Fl. Fenn. Förh.*: 336 (1868)
Collybia leucophaeata (P. Karst.) Sacc., *Syll. fung.* 5: 205 (1887)
Lyophyllum leucophaeatum (P. Karst.) P. Karst., *Meddeland. Soc. Fauna Fl. Fenn.* 2: 3 (1881)
E: ! **S:** ! **W:** ! **NI:** !
H: On soil in deciduous woodland, usually with *Fagus* or *Quercus* spp and rather often under trees by roadsides.
D: NM2: 138 **D+I:** Ph: 42-43 (as *Lyophyllum fumatofoetens*), B&K3: 222 261 (as *L. leucopheatum*) **I:** FlDan1: 32B (as *Clitocybe gangraenosa*), FlDan1: 25E (as *Tricholoma leucophaeatum*), Bon: 167, Kriegl3: 311, RivMyc: 103 (as *L. leucophaeatum*)
Rarely reported but apparently widespread.

infumatum (Bres.) Kühner, *Bull. Mens. Soc. Linn. Lyon* 7: 211 (1938)
Clitocybe ectypa var. *infumata* Bres., *Fungi trident.* 2: 49 (1892)
Clitocybe infumata (Bres.) Kauffman, *Pap. Michigan Acad. Sci.* 8: 202 (1927)
Tricholoma infumatum (Bres.) A. Pearson & Dennis, *Trans. Brit. Mycol. Soc.* 31: 151 (1948)
Collybia infumata (Bres.) Rea, *Trans. Brit. Mycol. Soc.* 12(2-3): 215 (1927)
E: ! **S:** ! **ROI:** ?
H: On soil in woodland.
I: RivMyc: 113
Known from England (Berkshire and Northumberland) and Scotland (Edinburgh: Dalmahoy). Reported from the Republic of Ireland (Wicklow) but unsubstantiated with voucher material.

konradianum (Maire) Konrad & Maubl., *Encycl. Mycol.* 14, *Les Agaricales* I: 368 (1948)
Clitocybe konradianum Maire, *Bull. Soc. Hist. Nat. Afrique N.* 36: 32 (1945)
E: !
H: On soil in woodland, reported with deciduous trees and *Taxus*.
I: RivMyc: 121
Known only from North Somerset and South Wiltshire.

loricatum (Fr.) Kalamees, *Z. Mykol.* 60(1): 14 (1994)
Agaricus loricatus Fr., *Epicr. syst. mycol.*: 37 (1838)
Tricholoma loricatum (Fr.) Gillet, *Hyménomycètes*: 108 (1874)
Mis.: *Clitocybe cartilaginea* sensu auct.
Mis.: *Tricholoma cartilagineum* sensu auct.
E: ! **S:** ! **W:** ! **NI:** ! **ROI:** !
H: On soil in deciduous woodland and occasionally reported in grassland near to trees.
D: NM2: 139 **D+I:** Ph: 42-43, B&K3: 224 262 **I:** RivMyc: 135
Rarely reported but apparently widespread.

semitale (Fr.) Kalamees, *Z. Mykol.* 60(1): 16 (1994)
Agaricus semitalis Fr., *Syst. mycol.* 1: 117 (1821)

Collybia semitalis (Fr.) Quél., *Mém. Soc. Émul. Montbéliard, Sér. 2,* 5: 56 (1872)
Tricholoma semitale (Fr.) Ricken, *Blätterpilze Deutschl.*: 358 (1914)
E: ! **S:** ! **W:** !
H: On soil in deciduous woodland, also reported with conifers and in grassland (downland) on calcareous soil.
D: NM2: 137 **D+I:** B&K3: 228 268 **I:** RivMyc: 123
Rarely reported but apparently widespread.

LYSURUS Fr., *Syst. mycol.* 2: 285 (1823)
Phallales, Clathraceae
Type: *Lysurus mokusin* (L.) Fr.

cruciatus (Lepr. & Mont.) Lloyd, *Mycol. not.* 3: 40 (1909)
Aserophallus cruciatus Lepr. & Mont., *Ann. Sci. Nat., Bot.,* sér 3, 4: 36 (1845)
Lysurus australiensis Cooke & Massee, *Grevillea* 18: 6 (1889)
Mis.: *Lysurus gardneri* sensu Ramsbottom (1953) & Palmer (1968)
E: ! **NI:** !
H: On decayed stable manure, straw (in stubble fields) or manured soil.
D: NM3: 175 **D+I:** BritPuffb: 186-187 Figs. 149/150
Rarely reported and presumably an introduction. First British collection by Rea [TBMS 2 (2): 57 & pl. 2 (1903)] from Worcestershire (near Kidderminster) and also known from Middlesex, South Lancashire, West Sussex, Yorkshire and Northern Ireland (Fermanagh).

MACROCYSTIDIA Joss., *Bull. Soc. Mycol. France* 49: 373 (1934)
Agaricales, Tricholomataceae
Type: *Macrocystidia cucumis* (Pers.) Joss.

cucumis (Pers.) Joss., *Bull. Soc. Mycol. France* 39: 373 (1934)
Agaricus cucumis Pers., *Observ. mycol.* 1: 45 (1796)
Naucoria cucumis (Pers.) P. Kumm., *Führ. Pilzk.*: 78 (1871)
Naucoria cucumis var. *leucospora* J.E. Lange, *Fl. agaric. danic.* 5, *Taxonomic Conspectus*: VI (1940)
Agaricus nigripes Trog, *Verzeich. Geg v.Thun vor. Schwämme (Flora, XV)* 15: 527 (1834)
Nolanea nigripes (Trog) Gillet, *Hyménomycètes*: 420 (1876)
Nolanea picea (Kalchbr.) Gillet, *Hyménomycètes*: 421 (1876)
Nolanea pisciodora (Ces.) Gillet, *Hyménomycètes*: 420 (1876)
E: o **S:** o **W:** o **NI:** ! **ROI:** ! **O:** Channel Islands: !
H: On rich humic or nitrogenous soil, usually in deciduous woodland (but also known from conifer plantations) and often in nettlebeds with *Urtica dioica* in such habitat. May also fruit abundantly on flowerbeds mulched with shredded woodchips or bark.
D: NM2: 142, FAN3: 174 **D+I:** B&K3: 230 273, Myc15(3): 133 **I:** FlDan4: 126B (as *Naucoria cucumis*), Bon: 197, C&D: 305, Kriegl3: 330
Occasional but widespread, and apparently increasing in artificial habitat such as parks and gardens.

MACROLEPIOTA Singer, *Pap. Michigan Acad. Sci.* 32: 141 (1948) [1946]
Agaricales, Agaricaceae
Type: *Macrolepiota procera* (Scop.) Singer
The treatment here largely follows Vellinga in FAN5: 64-73 (2001)

excoriata (Schaeff.) Wasser, *Ukrayins'k. Bot. Zhurn.* 35(5): 516 (1978)
Agaricus excoriatus Schaeff., *Fung. Bavar. Palat. nasc.* 1: 18 (1762)
Lepiota procera β *excoriata* (Schaeff.) Gray, *Nat. arr. Brit. pl.* 1: 601 (1821)
Lepiota excoriata (Schaeff.) P. Kumm., *Führ. Pilzk.*: 135 (1871)
Leucocoprinus excoriatus (Schaeff.) Pat., *Essai taxon.*: 171 (1900)

Leucoagaricus excoriatus (Schaeff.) Singer, *Sydowia* 2: 35 (1948)

Macrolepiota heimii Bon, *Boll. Gruppo Micol. G. Bresadola* 27(1-2): 18 (1984)

E: o **S:** o **W:** o **NI:** ! **ROI:** ! **O:** Channel Islands: !

H: On soil in grassland such as meadows, pastures etc. Also found in grassy areas in woodland glades and often in short turf on dunes.

D: NM2: 226 **D+I:** Ph: 26-27 (as *Lepiota excoriata*), FungEur4: 549-552 72 (as *M. heimii*), FungEur4: 572-576 76, B&K4: 214 249 (as *M. heimii*), FAN5: 70-71 **I:** FlDan1: 8 (as *L. excoriata*), Bon: 291, C&D: 249, Cooke 21 (23) Vol. 1 (1881)

Occasional but widespread.

fuligineosquarrosa Malençon, *Sydowia, Beih.* 8: 26 (1979)

E: ! **W:** !

H: On soil in woodland or scrub, and also known from dunes.

D+I: FungEur4: 581-583 78 **I:** C&D: 249

A single collection from Wales (Anglesey: Newborough Warren) in 1996 [see Boll. Gruppo Micol. G. Bresadola 40: 399-404 (1997)] and recently collected in England (West Kent: Dartford, Beacon Country Park).

fuliginosa (Barla) Bon, *Doc. Mycol.* 11(43): 75 (1981)

Lepiota procera var. *fuliginosa* Barla, *Champ. Alp Marit.*: 21 (1888)

Lepiota rhodosperma P.D. Orton, *Notes Roy. Bot. Gard. Edinburgh* 41(3): 591 (1984)

Macrolepiota rhodosperma (P.D. Orton) Migl., *Boll. Gruppo Micol. G. Bresadola* 38: 140 (1995)

Mis.: *Macrolepiota konradii* sensu B&K4: 216 pl. 250

Mis.: *Lepiota permixta* sensu auct.

Mis.: *Macrolepiota permixta* sensu auct.

E: !

H: On soil, usually in deciduous woodland and occasionally with conifers in mixed woodland.

D+I: FungEur4: 514-517 64, FAN5: 67-68

Rarely reported. The concept used here is that of Vellinga in FAN5. British records of *Macrolepiota permixta* probably belong here.

konradii (P.D. Orton) M.M. Moser, *Kleine Kryptogamenflora*, Edn 3: 185 (1967)

Lepiota konradii P.D. Orton, *Trans. Brit. Mycol. Soc.* 43(2): 283 (1960)

Lepiota excoriata var. *konradii* Huijsman [*nom. nud.*], *Meded. Ned. Mycol. Ver.* 28: 18 (1943)

Mis.: *Lepiota gracilenta* sensu Rea (1922)

E: o **NI:** ! **O:** Channel Islands: ! Isle of Man: !

H: On soil in woodland and copses, and occasionally amongst grass at woodland edges. Reported in grassland and on dunes but see comment below.

D: NM2: 226 **D+I:** Ph: 26-27 (as *Lepiota konradii*), FungEur4: 584-588 79 **I:** Bon: 291

Occasional but widespread. This is a woodland species and records from other habitats, especially those on dunes, should be treated with caution. Considered synonymous with *Macrolepiota mastoidea* in FAN5, but in Britain normally recognised as distinct. Sensu B&K4: 216 & pl. 250 is not this species but *M. fuliginosa*.

mastoidea (Fr.) Singer, *Lilloa* 22: 417 (1951) [1949]

Agaricus mastoideus Fr., *Syst. mycol.* 1: 20 (1821)

Lepiota mastoidea (Fr.) P. Kumm., *Führ. Pilzk.*: 135 (1871)

Agaricus umbonatus Schumach. [*nom. illegit.*, non *A. umbonatus* With. (1792)], *Enum. pl.* 1: 252 (1801)

Lepiota umbonata (Schumach.) J. Schröt., in Cohn, *Krypt.-Fl. Schlesien* 3: 675 (1889)

Agaricus gracilentus Krombh., *Naturgetr. Abbild. Schwämme* 4: 8 (1836)

Lepiota gracilenta (Krombh.) Quél., *Mém. Soc. Émul. Montbéliard, Sér. 2*, 5: 71 (1872)

Macrolepiota gracilenta (Krombh.) Wasser, *Ukrayins'k. Bot. Zhurn.* 35: 516 (1978)

Lepiota rickenii Velen., *Novit. mycol.*: 47 (1939)

Macrolepiota rickenii (Velen.) Bellù & Lanzoni, *Beitr. Kenntn. Pilze Mitteleurop.* 3: 196 (1987)

E: o **S:** ! **W:** ! **NI:** ! **ROI:** !

H: On soil in deciduous woodland.

D: NM2: 226 **D+I:** FungEur4: 552-556 73, FungEur4: 563-568 74 (as *M. rickenii*), Myc6(4): 186, B&K4: 214 248 (as *M. excoriata*), B&K4: 216 251 (as *M. rickenii*), FAN5: 68-69 **I:** FlDan1: 8C (as *Lepiota umbonata*), Bon: 291 (as *M. gracilenta*), C&D: 249, C&D: 249 (as *M. rickenii*) **I:** Cooke 22 (28) Vol. 1 (1881) (as *Agaricus gracilentus*) and 23 (24) (as *A. mastoideus*)

Occasional in England, and most frequently in southern counties. Rarely reported elsewhere but apparently widespread. Some authors consider *M. gracilenta* a distinct species.

olivieri (Barla) Wasser, *Flora Gribov Ukrainy*: 298 (1980)

Lepiota olivieri Barla, *Bull. Soc. Mycol. France* 2: 113 (1886)

Chlorophyllum olivieri (Barla) Vellinga, *Mycotaxon* 83: 416 (2002)

E: !

H: On soil in woodland, often with conifers.

D: FAN5: 73 **D+I:** Ph: 25 (as *Lepiota rhacodes*) **I:** FungEur4: 67 (as *Macrolepiota rhacodes*)

Poorly known in Britain. First reported by Rea in TBMS 17: 35 (1932), and the plate of *Macrolepiota rhacodes* in Phillips (1981) is also this *fide* Vellinga in FAN5. Differing from *M. rhacodes* in the smaller spores and the velar patches on the pileus not markedly contrasting with the background. Recently collected in England (Derbyshire: Chatsworth House and Surrey: Kew, Royal Botanic Gardens), but probably often misidentified as *M. rhacodes*.

procera (Scop.) Singer, *Pap. Michigan Acad. Sci.* 32: 141 (1948) [1946]

Agaricus procerus Scop., *Fl. carniol.*: 418 (1772)

Lepiota procera (Scop.) Gray, *Nat. arr. Brit. pl.* 1: 601 (1821)

E: c **S:** c **W:** c **NI:** c **ROI:** c **O:** Channel Islands: ! Isle of Man: !

H: On soil amongst grass, usually in meadows, pasture, unimproved grassland and parkland. Occasionally in grassy areas in deciduous woodland.

D: NM2: 226 **D+I:** Ph: 24-25 (as *Lepiota procera*), FungEur4: 510-514 63, B&K4: 218 254, FAN5: 65-67 **I:** FlDan1: 8B (as *L. procera*), Bon: 291, C&D: 249

Common and widespread. Cooke 19 (21) Vol. 1 (1881) is atypical.

procera var. **pseudoolivascens** Bellù & Lanzoni, *Beitr. Kenntn. Pilze Mitteleurop.* 3: 190 (1987)

E: !

H: On soil in woodland. British material associated with *Taxus*.

D: FungEur4: 521-524, FAN5: 67 (brief description)

A single British collection, from Surrey (Kew, Royal Botanic Gardens) in 1998.

rhacodes (Vittad.) Singer, *Lilloa* 22: 417 (1951) [1949]

Agaricus rhacodes Vittad., *Descr. fung. mang.*: 158 (1835)

Lepiota rhacodes (Vittad.) Quél., *Mém. Soc. Émul. Montbéliard, Sér. 2*, 5: 70 (1872)

Lepiota procera var. *rhacodes* (Vittad.) Massee, *Brit. fung.-fl.* 3: 234 (1893)

Chlorophyllum rhacodes (Vittad.) Vellinga, *Mycotaxon* 83: 416 (2002)

Mis.: *Agaricus procerus* sensu Sowerby [Col. fig. Engl. fung. 2: pl. 190 (1799)]

E: c **S:** c **W:** c **NI:** c **ROI:** c **O:** Channel Islands: ! Isle of Man: !

H: On soil in woodland, frequently with conifers and less often deciduous trees. Often abundant in plantations. Reported from grassland but mostly unsubstantiated with voucher material.

D: NM2: 225 **D+I:** FungEur4: 530-535 67, FAN5: 71-72 **I:** FlDan1: 9C (as *Lepiota rhacodes*), Bon: 291 (as *L. rhacodes*), C&D: 249

Common and widespread. N.B. originally described as 'rachodes', but this is considered an orthographic error.

rhacodes var. **bohemica** (Wichanský) Bellù & Lanzoni, *Beitr. Kenntn. Pilze Mitteleurop.* 3: 191 (1987)

Lepiota bohemica Wichanský, *Casopsis ceskoslov. houb.* 38: 103 (1961)

Lepiota rhacodes var. *hortensis* Pilát [*nom. nud.*], *Klíč urč. naš. Hub hrib. bedl.*: 242 (1951)

Macrolepiota rhacodes var. *hortensis* (Pilát) Wasser [*nom. inval.*], *Flora Gribov Ukrainy*: 298 (1980)

E: !

H: On soil in nitrogen-enriched habitat such as compost heaps and manured flowerbeds.

D: NM2: 225 **D+I:** Ph: 25 (as *Lepiota rhacodes* var. *hortensis*), FungEur4: 536-542 69, B&K4: 220 257 (as *M. rachodes* var. *hortensis*)

Rarely reported.

MACROTYPHULA R.H. Petersen, *Mycologia* 64: 140 (1972)

Cantharellales, Clavariaceae

Type: *Macrotyphula fistulosa* (Holmsk.) R.H. Petersen

fistulosa (Holmsk.) R.H. Petersen, *Mycologia* 64(1): 140 (1972)

Clavaria fistulosa Holmsk., *Beata ruris* 1: 15 (1790)

Clavariadelphus fistulosus (Holmsk.) Corner, *Monograph of Clavaria and Allied Genera*: 272 (1950)

Clavaria ardenia Sowerby, *Col. fig. Engl. fung.* 2: pl. 215 (1799)

Tremella ferruginea Schumach. [non *T. ferruginea* Sm. (1805)], *Fl. Danica*: t. 1852 (1825)

E: c **S:** ! **W:** ! **NI:** !

H: On soil amongst litter in deciduous woodland.

D: Corner: 272 (as *Clavariadelphus fistulosus*), NM3: 268 (as *C. fistulosus*) **D+I:** B&K2: 340 438, NM3: 256-257 (as *C. fistulosus*) **I:** Kriegl2: 42

Common in England. Rarely reported elsewhere but apparently widespread.

fistulosa var. **contorta** (Holmsk.) Nannf. & L. Holm, *Publ. Herb. Univ. Uppsala* 17: 8 (1985)

Clavaria contorta Holmsk., *Beata ruris* 1: 29 (1790)

Clavariadelphus fistulosus var. *contortus* (Holmsk.) Corner, *Monograph of Clavaria and Allied Genera*: 273 (1950)

E: o **S:** !

H: On dead wood of deciduous trees and shrubs, usually small dead attached branches or twigs. Most frequent on *Corylus* but also known on *Betula*, *Fagus* and *Tilia* spp.

Occasional in England. Rarely reported elsewhere.

juncea (Fr.) Berthier, *Bull. Mens. Soc. Linn. Lyon* 43(6): 186 (1974)

Clavaria juncea Fr., *Syst. mycol.* 1: 479 (1821)

Clavariadelphus junceus (Fr.) Corner, *Monograph of Clavaria and Allied Genera*: 275 (1950)

Clavaria juncea var. *vivipara* Fr., *Syst. mycol.* 1: 426 (1821)

E: c **S:** o **W:** o **NI:** ! **O:** Isle of Man: !

H: On decayed litter, often growing on petioles of fallen leaves, in damp deciduous woodland. Most frequently on *Fraxinus* but also known on *Acer pseudoplatanus* and *Fagus*.

D: Corner: 275, NM3: 269 (as *Clavariadelphus junceus*) **D+I:** Ph: 257, B&K2: 340 437 **I:** Kriegl2: 41 (as *Macrotyphula filiformis*)

Common and widespread. *Typhula phacorrhiza* is macroscopically similar but basidiomes are always associated with a small lenticular sclerotium.

MARASMIELLUS Murrill, *North American Flora* 9(4): 243 (1915)

Agaricales, Tricholomataceae

Type: *Marasmiellus juniperinus* Murrill

candidus (Bolton) Singer, *Pap. Michigan Acad. Sci.* 32: 129 (1946)

Agaricus candidus Bolton, *Hist. fung. Halifax* 1: 39 (1788)

Marasmius candidus (Bolton) Fr., *Epicr. syst. mycol.*: 381 (1838)

Marasmius albus-corticis Singer [*nom. inval.*], *Lilloa* 22: 300 (1951) [1949]

E: ! **NI:** ? **O:** Channel Islands: ?

H: On dead wood of deciduous trees (usually small fallen twigs or sticks) such as *Acer pseudoplatanus*, *Alnus* spp and *Fagus*. Rarely reported on conifers such as *Cupressus macrocarpus*. Also known on dead woody stems of *Hedera* and *Rubus fruticosus* and debris and dead stems of *Juncus* spp and *Urtica dioica*.

D: NM2: 142, A&N1 169-170 13, FAN3: 126-127 **D+I:** Ph: 66-67 (as *Marasmius candidus*) **I:** Bon: 177, C&D: 217, Kriegl3: 332

Known from England (East Kent, South Devon, Surrey and West Sussex). Reported elsewhere but unsubstantiated with voucher material.

humillimus (Quél.) Singer, *Nova Hedwigia Beih.* 44: 308 (1974)

Collybia humillima Quél., *Compt. Rend. Assoc. Franç. Avancem. Sci.* 11: 389 (1883) [1882]

Marasmius flosculinus Bataille, *Bull. Soc. Hist. Nat. Doubs* 30: 80 (1919)

Androsaceus flosculinus (Bataille) Rea, *Brit. basidiomyc.*: 530 (1922)

Marasmius flosculus Quél. [non *M. flosculus* Berk. (1842)], *Bull. Soc. Bot. France* 25: 289 (1879)

E: !

H: On dead stems of *Brachypodium pinnatum*.

D: A&N1 153-155 **I:** SV40:11

A single collection from England (Surrey: Box Hill) in 1992. Reported by Rea (1922) as *Androsaceus flosculinus* but no material has been traced.

ramealis (Bull.) Singer, *Pap. Michigan Acad. Sci.* 32: 130 (1946)

Agaricus ramealis Bull., *Herb. France*: pl. 336 (1788)

Gymnopus ramealis (Bull.) Gray, *Nat. arr. Brit. pl.* 1: 611 (1821)

Marasmius ramealis (Bull.) Fr., *Epicr. syst. mycol.*: 381 (1838)

Agaricus amadelphus Bull., *Herb. France*: pl. 550 (1792)

Marasmius amadelphus (Bull.) Fr., *Epicr. syst. mycol.*: 380 (1838)

Marasmiellus amadelphus (Bull.) M.M. Moser, *Kleine Kryptogamenflora*, Edn 3: 118 (1967)

E: c **S:** c **W:** c **NI:** o **ROI:** o

H: On small fallen sticks and twigs of deciduous trees and shrubs, especially dead stems of *Rubus fruticosus* agg. Also common on decayed brash of conifers such as *Chamaecyparis*, *Larix*, *Picea*, *Pseudotsuga* and *Thuja* spp.

D: NM2: 143, A&N1 149-152, FAN3: 124-125 **D+I:** Ph: 66-67 (as *Marasmius ramealis*), B&K3: 232 274 **I:** FlDan2: 48C (as *M. ramealis*), Bon: 217, C&D: 177

Common and widespread.

tricolor (Alb. & Schwein.) Singer, *Pap. Michigan Acad. Sci.*: 128 (1946) [1948]

Agaricus tricolor Alb. & Schwein., *Consp. fung. lusat.*: 224 (1805)

Omphalia tricolor (Alb. & Schwein.) P. Kumm., *Führ. Pilzk.*: 106 (1871)

Marasmius tricolor (Alb. & Schwein.) Kühner, *Botaniste* 25: 89 (1933)

Agaricus languidus Lasch, *Linnaea* 3: 385 (1828)

Marasmius languidus (Lasch) Fr., *Epicr. syst. mycol.*: 379 (1838)

Marasmiellus languidus (Lasch) Singer, *Lilloa* 22: 300 (1951) [1949]

Mis.: *Marasmius pruinatus* sensu auct.

E: !

H: On dead roots and debris of various grasses.

D: NM2: 143, A&N1 171-172, FAN3: 127

Known from England (South Somerset, Surrey and Yorkshire).

vaillantii (Pers.) Singer, *Nova Hedwigia Beih.* 44: 313 (1973)

Agaricus ericetorum var. *vaillantii* Pers., *Syn. meth. fung.*: 472 (1801)

Agaricus vaillantii (Pers.) Fr., *Syst. mycol.* 1: 136 (1821)

Marasmius vaillantii (Pers.) Fr., *Epicr. syst. mycol.*: 380 (1838)

Marasmius angulatus (Pers.) Berk. & Broome, *Ann. Mag. Nat. Hist., Ser. 3 [Notices of British Fungi no. 1018]* 15: 318 (1865)

Mis.: *Marasmius albus-corticis* sensu auct. Brit.
Mis.: *Androsaceus calopus* sensu Rea (1922)
Mis.: *Marasmius calopus* sensu NCL
Mis.: *Marasmius candidus* sensu Lange (FlDan2: 25 & pl. 47C)
Mis.: *Marasmius insititius* sensu Rea (1922)

E: c **S:** o **W:** ! **NI:** ! **O:** Channel Islands: !
H: On dead culms and stems of various grasses and debris of *Carex* and *Juncus* spp. Also on dead herbaceous stems, dead stems of *Pteridium* and small, fallen twigs of deciduous trees such as *Corylus*, *Fagus* and *Salix* spp.
D: NCL3: 303 (as *Marasmius calopus*), NM2: 143, A&N1 157-159 11, FAN3: 125 **D+I:** Ph: 67 (as *M. calopus*), B&K3: 232 275 **I:** C&D: 217, Kriegl3: 338
Common in England and occasional in Scotland. Rarely reported elsewhere but apparently widespread.

MARASMIUS Fr., *Fl. Scan.*: 339 (1835)
Agaricales, Tricholomataceae
Androsaceus (Pers.) Pat., *Hyménomyc. Eur.*: 105 (1887)
Setulipes Antonín, *Česká Mykol.* 41: 85 (1987)
Type: *Marasmius rotula* (Scop.) Fr.

alliaceus (Jacq.) Fr., *Epicr. syst. mycol.*: 383 (1838)
Agaricus alliaceus Jacq., *Enum. Stirp. Vindob.*: 299 (1762)
E: ! **S:** ! **W:** ? **ROI:** ?
H: On decayed wood of deciduous trees in old woodland, usually large fallen trunks and branches of *Fagus* but also known on *Acer pseudoplatanus*. Reported on *Picea* spp but the host is not verified.
D: NM2: 146, A&N1 105-109, FAN3: 152-153 **D+I:** Ph: 68, B&K3: 232 276 **I:** FlDan2: 47E, Bon: 175, C&D: 215, Kriegl3: 342
Rare. Often cited as common and may have been much more frequent in the past. Reported from widely scattered sites in England, from Yorkshire southwards, rarely elsewhere and then unsubstantiated with voucher material. All recent records and collections have been from West Sussex, in the area around Petworth.

androsaceus (L.) Fr., *Epicr. syst. mycol.*: 385 (1838)
Agaricus androsaceus L., *Sp. pl.*: 1175 (1753)
Androsaceus androsaceus (L.) Rea, *Brit. basidiomyc.*: 531 (1922)
Setulipes androsaceus (L.) Antonín, *Česká Mykol.* 41(2): 86 (1987)
E: c **S:** c **W:** c **NI:** c **ROI:** ! **O:** Isle of Man: !
H: Most frequently on dead leaves and stems of *Calluna* and other *Ericaceae* on heathland and moorland. Also on fallen and decayed needle litter of conifers, including *Taxus*. Rarely on leaf litter of deciduous trees such as *Betula* spp and *Corylus* and *Quercus* spp.
D: NM2: 145, A&N1 137-141 (as *Setulipes androsaceus*), FAN3: 146 **D+I:** Ph: 66-67, B&K3: 234 277 **I:** FlDan2: 48A, Bon: 175, C&D: 215
Very common and widespread, especially so in Scotland where it is associated with 'Snow Blight' of *Calluna* fide R. Watling (pers. comm.). English name = 'Horse Hair Fungus'.

anomalus Lasch, in Rabenhorst, *Klotzschii Herbarium Vivum Mycologicum*: 1806 (1854)
Marasmius epodius Bres., *Fungi trident.* 1: 88 (1881)
Marasmius litoralis Quél., *Bull. Soc. Amis Sci. Nat. Rouen.* 15: 169 (1880)
E: ! **W:** ? **ROI:** !
H: On dead stems and culms of various grasses, including *Ammophila arenaria* on dunes, also by roadsides and in meadows.
D: A&N1 83-86, FAN3: 150 **I:** C&D: 215
Rarely reported. Known from England (East Kent and West Sussex) and the Republic of Ireland (Co. Dublin). Reported

elsewhere in England and from Wales (Glamorganshire) but no voucher material has been traced.

bulliardii Quél., *Bull. Soc. Bot. France* 24: 323 (1878)
E: o **S:** ! **W:** ! **NI:** ! **ROI:** ?
H: On fallen and decayed leaves of deciduous trees, usually *Fagus* and *Fraxinus* in old woodland. Rarely reported on *Alnus, Betula, Corylus* and *Quercus* spp. Often fruiting late in the autumn.
D: NM2: 144, A&N1 33-35, FAN3: 139-140 **D+I:** B&K3: 234 278 **I:** FlDan2: 48F
Occasional but apparently widespread. Possibly under-recorded since basidiomes are small and cryptically coloured. Reports of this species on conifer litter may represent *M. wettsteinii*, a similar species as yet unknown in Britain.

buxi Fr., in Quélet, *Mém. Soc. Émul. Montbéliard, Sér. 2*, 5: 224 (1872)
Androsaceus buxi (Fr.) Pat., *Essai taxon.*: 141 (1900)
E: !
H: On fallen leaves of *Buxus sempervirens* in natural habitat. Occasionally on dying leaves still attached to the bush, usually on branches that are brushing the ground.
D: A&N1 49-50, FAN3: 145-146 **D+I:** B&K3: 234 279
Known from areas of calcareous soil where *Buxus* is native and not uncommon. Most records are from Surrey but also known from Gloucestershire and North Essex.

caricis P. Karst., *Bidrag Kännedom Finlands Natur Folk* 25: 231 (1876)
S: !
H: On decayed leaves and debris of *Carex* spp, in marshes, swamps and at the margins of ponds and lakes.
D: NM2: 145, A&N1 71-72, FAN3: 144
British collections only known from Scotland. However, basidiomes are minute, grow in inaccessible habitat and are easily overlooked.

cohaerens (Pers.) Cooke & Quél., *Clavis syn. Hymen. Europ.*: 153 (1878)
Agaricus cohaerens Pers., *Syn. meth. fung.*: 306 (1801)
Mycena cohaerens (Pers.) P. Kumm., *Führ. Pilzk.*: 111 (1871)
Agaricus balaninus Berk., *Mag. Zool. Bot., [Notices of British Fungi No. 42]* 1: 509 (1837)
Mycena balanina (Berk.) Sacc., *Syll. fung.* 5: 252 (1887)
Marasmius ceratopus (Pers.) Quél., *Fl. mycol. France*: 319 (1886)
E: o **W:** o **ROI:** o
H: On soil (often calcareous) and amongst woody debris and leaf litter in woodland, usually associated with *Fagus* or *Fraxinus*.
D: NM2: 146, A&N1 95-99, FAN3: 150-151 **D+I:** B&K3: 236 282 **I:** FlDan2: 47F, Bon: 175, C&D: 215, Kriegl3: 358
Occasional but widespread. Perhaps most frequent in south-eastern England.

collinus (Scop.) Singer, *Lloydia* 5: 126 (1942)
Agaricus collinus Scop., *Fl. carniol.*: 52 (1772)
Collybia collina (Scop.) P. Kumm., *Führ. Pilzk.*: 115 (1871)
E: !
H: On soil in grassland.
D: A&N1 125-127
Accepted on the strength of a single collection (in poor condition) labelled thus in herb. K. Also accepted in NCL (part II p. 180). Not mentioned as British in A&N1. Apparently easily mistaken for *M. oreades*. Cooke 198 (205) Vol. 2 (1882) supposedly illustrates this, but Pearson & Dennis (1948) suggested it might represent *Hydropus subalpinus*.

corbariensis (Roum.) Singer, *Agaricales in Modern Taxonomy*, Edn 2: 355 (1962)
Agaricus corbariensis Roum., *Rev. Mycol. (Toulouse)* 2(8): 198 (1881)
ROI: !
H: On dead and fallen leaves. Irish material on *Hedera*.
D: A&N1 44

A single collection from Cork (Millstreet), originally determined as *Marasmius minutus*.

cornelii Laessøe & Noordel., *Persoonia* 13(3): 237 (1987)
   Mis.: *Marasmius menieri* sensu Corner in TBMS 19: 285 (1934), sensu auct. Brit.
E: ! **S:** !
H: On dead leaf sheaths of *Cladium mariscus* just above water level in ponds, lakes and on river banks.
D: NM2: 144, FAN3: 144 **D+I:** A&N1 73-75 2
Known from England (East Norfolk) and Scotland (Easterness). Basidiomes usually grow in inaccessible habitat and are easily overlooked.

curreyi Berk. & Broome, *Ann. Mag. Nat. Hist., Ser. 3 [Notices of British Fungi no. 1794]* 3: 209 (1879)
   Androsaceus curreyi (Berk. & Broome) Rea, *Brit. basidiomyc.*: 532 (1922)
   Mis.: *Marasmius graminum* sensu auct. Brit.
E: ! **W:** ! **O:** Channel Islands: !
H: On dead and decayed stems and leaves of grasses in meadows, pasture, downland, and dune turf. Usually reported on *Holcus lanatus* and *H. mollis*, less frequently *Glyceria maxima*, *Phalaris arundinacea*, *Poa annua* and *Phragmites australis*.
D: FAN3: 140 **D+I:** B&K3: 240 286 (as *Marasmius graminum*), A&N1 41-43 1 **I:** C&D: 215, Kriegl3: 354
Rarely reported but apparently widespread. Often growing under dense mats of decayed leaves .

epiphylloides (Rea) Sacc. & Trotter, *Syll. fung.* 23: 145 (1925)
   Androsaceus epiphylloides Rea, *Trans. Brit. Mycol. Soc.* 3(4): 286 (1911) [1910]
   Androsaceus hederae Kühner, *Bull. Soc. Mycol. France* 43: 114 (1927)
   Agaricus squamula Batsch, *Elench. fung. (Continuatio Prima)*: 95 & t.17 f. 84 (1786)
   Merulius squamula (Batsch) With., *Arr. Brit. pl. ed. 3* 4: 151 (1796)
E: o **S:** ! **W:** o **NI:** c **ROI:** ! **O:** Isle of Man: !
H: On fallen and decayed leaves of *Hedera helix* and, in Ireland, *H. hibernica*.
D: NM2: 145, A&N1 67-69, FAN3: 142-143 **D+I:** B&K3: 238 284 **I:** Sow1: 93 (as *Agaricus squamula*), Kriegl3: 345
Occasional but widespread and with a marked bias toward western areas. Rather common in Northern Ireland, and rare in south east England. Sowerby's plate no. 93 (and description) of *Agaricus squamula* is clearly this species.

epiphyllus (Pers.) Fr., *Epicr. syst. mycol.*: 386 (1838)
   Agaricus epiphyllus Pers., *Syn. meth. fung.*: 468 (1801)
   Micromphale epiphyllum (Pers.) Gray, *Nat. arr. Brit. pl.* 1: 622 (1821)
   Androsaceus epiphyllus (Pers.) Pat., *Hyménomyc. Eur.*: 105 (1887)
E: c **S:** ! **W:** ! **NI:** ! **ROI:** ! **O:** Isle of Man: !
H: On fallen and decayed leaves of deciduous trees and shrubs in woodland or scrub. Most frequently on *Acer pseudoplatanus*, *Fagus* and *Fraxinus*.
D: NM2: 145, A&N1 52-57, FAN3: 141-142 **D+I:** Ph: 66-67, B&K3: 238 285 **I:** FlDan2: 49F, Bon: 175, C&D: 215
Common, at least in England. Rarely reported elsewhere.

favrei Antonín, *Mycol. Helv.* 4(2): 238 (1991)
   Mis.: *Marasmius tremulae* sensu auct. Eur.
E: !
H: On decayed leaves of *Populus* spp in damp woodland.
D: NM2: 145 (as *Marasmius tremulae*), A&N1 58-61
A single collection (originally det. as *M. tremulae*) from England (Worcestershire: Uffmore Wood) in 1986. However, basidiomes are minute, shrivel rapidly in dry weather and are easily overlooked.

hudsonii (Pers.) Fr., *Epicr. syst. mycol.*: 386 (1838)
   Agaricus hudsonii Pers., *Syn. meth. fung.*: 390 (1801)
   Mycena hudsonii (Pers.) Gray, *Nat. arr. Brit. pl.* 1: 620 (1821)
   Androsaceus hudsonii (Pers.) Pat., *Essai taxon.*: 141 (1900)

Agaricus pilosus Huds. [*nom. illegit.*, non *A. pilosus* Schaeff. (1762)], *Fl. angl.*, Edn 2, 2: 622 (1778)
E: o **S:** ! **W:** ! **NI:** o **ROI:** !
H: On fallen and decayed leaves of *Ilex*; rarely on other intermixed leaves.
D: NM2: 145, A&N1 48-48, FAN3: 144-145 **D+I:** B&K3: 240 287 **I:** Sow2: 164 (as *Agaricus pilosus*)
Occasional to rare and with a distinct bias toward western areas of Britain.

limosus Quél., *Bull. Soc. Bot. France* 23: 323 (1877)
E: ! **S:** ! **W:** !
H: On dead and decayed leaves of *Phragmites australis* in swampy areas such as fens, marshes and at pond margins. Single records on dead leaves of *Phalaris arundinacea* and *Juncus* sp.
D: NM2: 144, A&N1 35-37, FAN3: 140 **D+I:** B&K3: 240 288 **I:** FlDan2: 48B
Known from England (Buckinghamshire, Cambridgeshire, Co. Durham, Hertfordshire, South Devon and Worcestershire) and also from Scotland and Wales. Basidiomes are minute, cryptically coloured, often grow in inaccessible habitats and are easily overlooked.

minutus Peck, *Rep. (Annual) New York State Mus. Nat. Hist.* 27: 97 (1873)
   Marasmius capillipes Sacc., *Nuovo Giorn. Bot. Ital.* 8: 162 (1876)
E: !
H: On fallen and decayed leaves of *Salix* spp in fens, marshes, pond margins or stream sides.
D: A&N1 45-48, FAN3: 145 **D+I:** B&K3: 236 280 (as *Marasmius capillipes*)
Known from England (East Suffolk, Surrey and West Cornwall). Possibly overlooked since basidiomes are minute, cryptically coloured and shrivel rapidly in dry conditions.

oreades (Bolton) Fr., *Sverig Atl. Svamp.*: 52 (1836)
   Agaricus oreades Bolton, *Hist. fung. Halifax* Suppl.: 151 (1792)
   Agaricus coriaceus Lightf. [non *A. coriaceus* Scop. (1772)], *Fl. scot.*: 1020 (1777)
   Agaricus pratensis Huds. [non *A. pratensis* Pers. (1801)], *Fl. angl.*, Edn 2, 2: 616 (1778)
E: c **S:** c **W:** c **NI:** c **ROI:** c **O:** Channel Islands: c Isle of Man: c
H: On soil amongst grass in meadows, dunes, downland, on lawns and in rough pasture often forming large 'fairy rings'.
D: NM2: 147, A&N1 129-133, FAN3: 149 **D+I:** Ph: 66, B&K3: 242 289 **I:** Sow3: 247 (as *Agaricus pratensis*), FlDan2: 46F, Bon: 175, C&D: 215, Kriegl3: 350
Very common and widespread. English name = 'Fairy Ring Champignon'.

pseudocaricis Noordel., *Persoonia* 11(3): 373 (1981)
S: !
H: On decayed leaves of *Carex* spp just above water level in a pond.
D: A&N1 77-79
Known only from the type collection from Scotland (East Perthshire: Kindrogan).

quercophilus Pouzar, *Česká Mykol.* 36(1): 1 (1982)
   Setulipes quercophilus (Pouzar) Antonín, *Česká Mykol.* 41(2): 86 (1987)
   Mis.: *Agaricus androsaceus* sensu Bolton [Hist. Fung. Halifax Vol. 1: 32 (1788) & Sowerby [Col. fig. Eng. fung. 1: pl. 94 (1797)]
   Mis.: *Androsaceus splachnoides* sensu Rea (1922)
   Mis.: *Marasmius splachnoides* sensu auct.
E: o **S:** !
H: Usually on fallen and decayed leaves of *Quercus petraea* and *Q. robur*, more rarely *Q. ilex* and *Q. rubra*, in woodland and parkland. Rarely on leaves of *Fagus* when mixed amongst oak leaves.
D: NM2: 146, A&N1 141-143 10 (as *Setulipes quercophilus*), FAN3: 147

Rarely reported but apparently widespread and may be locally abundant. Leaves colonised by mycelium become bleached and easily seen amongst large drifts. *Agaricus androsaceus* sensu Sowerby [Vol. 1 pl. 94 (1797)] clearly represents *M. quercophilus*.

rotula (Scop.) Fr., *Epicr. syst. mycol.*: 385 (1838)
  Agaricus rotula Scop., *Fl. carniol.*: 457 (1772)
  Androsaceus rotula (Scop.) Pat., *Hyménomyc. Eur.*: 477 (1887)
  Micromphale collariatum (With.) Gray, *Nat. arr. Brit. pl.* 1: 622 (1821)
  Merulius collariatus With., *Arr. Brit. pl. ed. 3* 4: 148 (1796)
E: c **S:** c **W:** c **NI:** c **ROI:** c **O:** Channel Islands: ! Isle of Man: c
H: On wood of deciduous trees and shrubs in woodland, usually small buried twigs and sticks, only occasionally on larger logs or stumps. Very rarely on conifers such as *Pinus sylvestris*.
D: NM2: 144, A&N1 25-30, FAN3: 139 **D+I:** Ph: 67, B&K3: 242 291 **I:** Sow1: 95 (as *Agaricus rotula*), Bon: 175, C&D: 215, Kriegl3: 356
Very common and widespread.

saccharinus (Batsch) Fr., *Epicr. syst. mycol.*: 386 (1838)
  Agaricus saccharinus Batsch, *Elench. fung. (Continuatio Prima)*: 93 & t. 17 f. 83 (1786)
  Androsaceus saccharinus (Batsch) Rea, *Brit. basidiomyc.*: 533 (1922)
E: ! **S:** ?
H: On fallen and decayed leaves of *Fagus* in beechwoods on calcareous soil.
D: NCL 304, A&N1 65-66 **I:** Cooke 1087 (1136c) Vol. 7 (1890)
Known from England (Surrey: Mickleham, Box Hill area) in 1954. Reports from Staffordshire and Scotland (Mid-Ebudes: Isle of Mull) are unsubstantiated with voucher material. Basidiomes are minute and easily overlooked.

scorodonius (Fr.) Fr., *Sverig Atl. Svamp.*: 53 (1836)
  Agaricus scorodonius Fr., *Observ. mycol.* 1: 29 (1815)
E: ! **NI:** ?
H: On soil amongst leaf litter in mixed deciduous and coniferous woodland.
D: A&N1 111-113, FAN3: 151-152 **D+I:** B&K3: 244 292 **I:** Bon: 175, C&D: 215
Only two collections in herb. K, from England (North Wiltshire: Bowood in 1853 and West Sussex: Goodwood, and Slindon in 1968. Very few other records, and all unsubstantiated with voucher material.

setosus (Sowerby) Noordel., *Persoonia* 13(3): 241 (1987)
  Agaricus setosus Sowerby, *Col. fig. Engl. fung.* 3: pl. 302 (1801)
  Mycena setosa (Sowerby) Gillet, *Hyménomycètes*: 281 (1876)
  Marasmius eufoliatus Kühner, *Bull. Soc. Mycol. France* 43: 110 (1927)
  Marasmius recubans Quél., *Mém. Soc. Émul. Montbéliard, Sér. 2*, 5: 355 (1873)
  Mis.: *Marasmius saccharinus* sensu auct.
E: o **S:** o **W:** ! **NI:** o
H: On fallen and decayed leaves of *Fagus* in beechwoods, often fruiting on the petioles in large drifts of leaves. Rarely reported on *Corylus*, *Quercus robur* and *Salix* spp.
D: NM2: 145, A&N1 63-65, FAN3: 142 **D+I:** B&K3: 242 290 (as *Marasmius recubans*) **I:** Sow3: pl. 302 (as *Agaricus setosus*), FlDan2: 48E (as *M. recubans*)
Occasional but widespread and may occur in large numbers.

tenuiparietalis Singer, *Nova Hedwigia Beih.* 29: 99 (1969)
E: !
H: On fallen leaves of deciduous trees. British material on *Clematis vitalba*.
D: A&N1 57-58
A single collection (the neotype) from West Gloucestershire (Wotton under Edge) in 1948.

torquescens Quél., *Mém. Soc. Émul. Montbéliard, Sér. 2*, 5: 221 (1872)

Mis.: *Marasmius lupuletorum* sensu auct.
E: o **S:** ! **W:** !
H: On soil and decayed leaves or woody litter in deciduous woodland, often on calcareous soil. Usually with *Fagus* and *Fraxinus* but also reported with *Acer pseudoplatanus*, *Betula* and *Quercus* spp.
D: NM2: 147, A&N1 99-102, FAN3: 148-149 **D+I:** B&K3: 244 294
Occasional in England, especially frequent in southern counties. Rarely reported elsewhere.

undatus (Berk.) Fr., *Summa veg. Scand.* 2: 313 (1849)
  Agaricus undatus Berk. [non *A. undatus* Fr. (1838)], *Engl. fl.* 5(2): 51 (1836)
  Marasmius chordalis Fr., *Epicr. syst. mycol.*: 383 (1838)
  Agaricus vertirugis Cooke, *Handb. Brit. fung.*, Edn 1: 57 (1871)
  Collybia vertirugis (Cooke) Sacc., *Syll. fung.* 5: 216 (1887)
E: o **S:** !
H: On dead and dying rhizomes of *Pteridium* and restricted to this host.
D: NM2: 147, FAN3: 153102-105 **D+I:** Ph: 66-67 7, B&K3: 236 281 (as *Marasmius chordalis*) **I:** C&D: 215
Occasional but widespread in England. Rarely reported elsewhere.

wynnei Berk. & Broome, *Ann. Mag. Nat. Hist., Ser. 3 [Notices of British Fungi no. 802]* 3: 358 (1859)
  Marasmius argyropus var. *suaveolens* Rea, *Trans. Brit. Mycol. Soc.* 2: 129 (1906) [1905]
  Marasmius globularis Fr. in Quélet, *Mém. Soc. Émul. Montbéliard, Sér. 2*, 5: 220 (1872)
  Marasmius globularis var. *carpathicus* (Kalchbr.) Costantin & L.M. Dufour, *Nouv. fl. champ.*: 67 (1891)
  Marasmius suaveolens (Rea) Rea, *Brit. basidiomyc.*: 523 (1922)
E: c **S:** ! **W:** o **NI:** o
H: On decayed leaf litter in deciduous woodland. Especially frequent on large drifts of leaves of *Fagus* in beechwoods on calcareous soil. Rarely reported with conifers such as *Picea* and *Thuja* spp.
D: NM2: 147, A&N1 120-125 8, FAN3: 147-148 **D+I:** Ph: 66-67, B&K3: 246 295 **I:** FlDan2: 46D (as *Marasmius globularis*), Bon: 175, C&D: 215, Kriegl3: 351
Common, at least in southern and south-eastern England. Occasional and less frequently reported elsewhere.

MARCHANDIOMYCES Dieder. & D. Hawksw., *Mycotaxon* 37: 311 (1990)
Ceratobasidiales, Ceratobasidiaceae
A genus of anamorphic, *Rhizoctonia*-like fungi.
Type: *Marchandiomyces corallinus* (Roberge) Diederich & D. Hawksw.

corallinus (Roberge) Diederich & D. Hawksw., *Mycotaxon* 37: 312 (1990)
  Illosporium corallinum Roberge, *Pl. Crypt. Fr.* fasc. 32: no. 1551 (1847)
E: c **S:** c **W:** c
H: Parasitic. Forming coral-coloured fructifications on dying and moribund thalli of many species of lichen.
Common and widespread. Only recently recognised as a basidiomycete.

MEGACOLLYBIA Kotl. & Pouzar, *Česká Mykol.* 26: 220 (1972)
Agaricales, Tricholomataceae
Type: *Megacollybia platyphylla* (Pers.) Kotl. & Pouzar

platyphylla (Pers.) Kotl. & Pouzar, *Česká Mykol.* 26: 220 (1972)
  Agaricus platyphyllus Pers., *Observ. mycol.* 1: 47 (1796)
  Collybia platyphylla (Pers.) P. Kumm., *Führ. Pilzk.*: 117 (1871)

Tricholomopsis platyphylla (Pers.) Singer, *Schweiz. Z. Pilzk.*
17: 13 (1939)
Oudemansiella platyphylla (Pers.) M.M. Moser, *Kleine*
Kryptogamenflora, Edn 4: 155 (1978)
Agaricus grammocephalus Bull., *Herb. France*: pl. 594 (1793)
Collybia grammocephala (Bull.) Quél., *Fl. mycol. France*: 228
(1888)
Agaricus repens Fr., *Observ. mycol.*: 14 (1815)
Agaricus tenuiceps Cooke & Massee, *Ill. Brit. fung.* 1121
(1166) Vol. 8 (1889)
Tricholoma tenuiceps (Cooke & Massee) Massee, *Brit. fung.-*
fl. 3: 197 (1893)
E: c S: o W: c NI: ! ROI: o
H: On decayed wood of deciduous trees, usually *Fagus* and less
frequently *Betula, Corylus* and *Quercus* spp. Rarely on conifers,
such as *Picea* and *Pinus* spp.
D: NM2: 147, NM2: 147, BFF8: 135-136, FAN4: 172-173 **D+I:**
Ph: 44 (as *Tricholomopsis platyphylla*), B&K3: 246 296 **I:** Bon:
177, C&D: 235, Kriegl3: 361
Common and widespread but apparently less so in northern
areas and few records from Northern Ireland. Occasionally
appearing as if terrestrial, but then attached to buried wood.

MEGALOCYSTIDIUM Jülich, *Persoonia* 10: 139 (1978)
Hericiales, Gloeocystidiellaceae
Type: *Megalocystidium leucoxanthum* (Bres.) Jülich

leucoxanthum (Bres.) Jülich, *Persoonia* 10(1): 140 (1978)
Corticium leucoxanthum Bres., *Fungi trident.* 2(11 - 13): 57
(1898)
Gloeocystidium leucoxanthum (Bres.) Höhn. & Litsch.,
Sitzungsber. Kaiserl. Akad. Wiss., Wien, Math.-Naturwiss.
Cl. Abt. 1, 116: 744 (1907)
Gloeocystidiellum leucoxanthum (Bres.) Boidin, *Compt.*
Rendu. Hebd. Séanc Acad. Sci. 233: 825 (1951)
E: ! S: !
H: On decayed wood of deciduous trees, and reported several
times on dead stems of *Ulex europaeus*.
D: NM3: 279 (as *Gloeocystidiellum leucoxanthum*) **D+I:** CNE3:
426-427 (as *G. leucoxanthum*)
Known from England (Bedfordshire, Mid-Lancashire, South
Hampshire, Surrey and West Sussex) and Scotland (North
Ebudes: Isle of Skye and Outer Hebrides: Isle of Lewis).

luridum (Bres.) Jülich, *Persoonia* 10(1): 140 (1978)
Corticium luridum Bres., *Fungi trident.* 2(11-13): 59 (1898)
Gloeocystidium luridum (Bres.) Höhn. & Litsch., *Sitzungsber.*
Kaiserl. Akad. Wiss., Wien, Math.-Naturwiss. Cl., Abt. 1
116: 770 (1907)
Gloeocystidiellum luridum (Bres.) Boidin, *Compt. Rend. Hebd.*
Séances Acad. Sci. 233: 1668 (1951)
E: ! S: !
H: On decayed wood of deciduous trees such as *Acer*
pseudoplatanus, Betula, Fagus, Fraxinus and *Ulmus* spp.
D: NM3: 279 (as *Gloeocystidiellum luridum*) **D+I:** CNE3: 430-
433 (as *G. luridum*), B&K2: 118 104 **I:** Kriegl1: 204
Rarely reported. Known from England (South Hampshire and
South Devon) and Scotland (Easterness and Orkney Isles:
Hoy)

MELANOGASTER Corda [*nom. cons.*], *Deutschl.* *Fl.*, Edn 3: 1 (1837)
Melanogastrales, Melanogastraceae
Octaviania Vittad., *Monogr. Tuberac.*; 15 (1831)
Type: *Melanogaster tuberiformis* Corda

ambiguus (Vittad.) Tul. & C. Tul., *Ann. Sci. Nat., Bot.*, sér. 2,
19: 378 (1843)
Octaviania ambigua Vittad., *Monogr. Tuberac.* 18: pl. 4 / 7
(1831)
E: o S: !

H: Epigeous on humic or sandy soil in woodland, usually
associated with *Fagus*, less often *Castanea* or *Quercus* spp.
Sometimes with structural timber in buildings.
D: NM3: 295 **D+I:** B&K2: 374 489, BritTruff: 147-148 8 D
Occasional in England, perhaps most frequent in southern areas.
Rarely reported elsewhere. Distinguished by the unpleasant
odour when older (of decayed onions or rubber). English name
= 'Stinking Slime Truffle'.

broomeianus Berk., *Ann. Sci. Nat., Bot.*, sér. 2, 19: 377 (1843)
Melanogaster variegatus var. *broomeianus* (Berk.) Tul. & C.
Tul., *Fungi hypogaei* 93 (pl. 4) (1851)
E: o W: ! NI: ! ROI: !
H: Epigeous or subhypogeous in rich calcareous soils, often
under thick layers of leaf litter.
D: NM3: 295 **D+I:** B&K2: 376 490, BritTruff: 145-147 8 C
Occasional but widespread in England and can be locally
'common' in areas of calcareous soil, especially in southern
counties. Formerly sold in markets. English names = 'Bath
Truffle' or 'Red Truffle'.

intermedius (Berk.) Zeller & H.R. Dodge, *Rep. (Annual)*
Missouri Bot. Gard. 23: 645 (1936)
Melanogaster ambiguus var. *intermedius* Berk., *Ann. Mag.*
Nat. Hist., Ser. 1 13: 354 (1844)
E: !
H: Epigeous or hypogeous in rich loam soil in deciduous
woodland.
D: BritTruff: 149
A single collection North Wiltshire (Spye Park) in 1845, with
recent collections from Berkshire, Essex and Surrey.

MELANOLEUCA Pat., *Cat. pl. cell. Tunisie*: 22 (1897)
Agaricales, Tricholomataceae
Melaleuca Pat. [*nom. illegit.*, non *Melaleuca* L.
(1767)(*Myrtaceae*)], *Hyménomyc. Eur.*: 96 (1887)
Type: *Melanoleuca melaleuca* (Pers.) Murrill

albifolia Boekhout, *Persoonia* 13(4): 421 (1988)
Melanoleuca leucophylla Métrod [*nom. nud.*], *Bull. Soc.*
Mycol. France 64: 161 (1948)
E: ! S: ! W: !
H: On soil amongst grass, often near to trees in dunes or on
heathland, and also known in lawn turf.
D: BFF8: 67-68, FAN4: 163-164
Rarely reported but apparently widespread.

atripes Boekhout, *Persoonia* 13(4): 419 (1988)
E: !
H: On soil. British material on sandy soil, amongst wood-chip
mulch on flowerbeds.
D: BFF8: 68, FAN4: 161
Poorly known in Britain. Only two collections, from England
(Surrey: Windsor Great Park, Valley Gardens) in 1991.

brevipes (Bull.) Pat., *Essai taxon.*: 158 (1900)
Agaricus brevipes Bull., *Herb. France*: pl. 521 f. 2 (1791)
Gymnopus brevipes (Bull.) Gray, *Nat. arr. Brit. pl.* 1: 609
(1821)
Tricholoma brevipes (Bull.) P. Kumm. [non sensu Lange
(1935)], *Führ. Pilzk.*: 133 (1871)
E: ! S: ! NI: ! ROI: !
H: On soil, often amongst grass, in deciduous woodland or in
grassland.
D: BFF8: 60, FAN4: 159 **D+I:** B&K3: 246 297 **I:** Bon: 165, C&D:
203
Poorly understood in Britain and taken here sensu BFF8 and
FAN4. Sensu NCL is *Melanoleuca langei*.

cinereifolia (Bon) Bon, *Doc. Mycol.* 9(33): 71 (1978)
Melanoleuca strictipes var. *cinereifolia* Bon, *Bull. Soc. Mycol.*
France 86(1): 155 (1970)
Melanoleuca cinereifolia var. *maritima* Bon, *Doc. Mycol.*
16(61): 46 (1985)
E: ! S: ! W: ! NI: ! ROI: !

H: On sand or sandy soil in dunes. Usually growing with, or near to, *Ammophila arenaria*.
D: BFF8: 69, FAN4: 161-162 **I:** C&D: 205
Rarely reported but apparently widespread.

cognata (Fr.) Konrad & Maubl., *Icon. select. fung.* 3: pl. 271 (1926)
    Agaricus arcuatus var. *cognatus* Fr., *Epicr. syst. mycol.*: 46 (1838)
    Tricholoma cognatum (Fr.) Gillet, *Hyménomycètes*: 124 (1874)
    Tricholoma arcuatum f. *robusta* J.E. Lange, *Dansk. Bot. Ark.* 8(3): 34 (1933)
    Mis.: *Tricholoma arcuatum* sensu Rea (1922)
E: o **S:** ! **W:** ! **NI:** ! **ROI:** !
H: On soil in deciduous woodland, occasionally with conifers in mixed woodland, on compost heaps and flowerbeds in parks and gardens, and also on roadside verges. Often fruiting in the spring and again in the autumn.
D: NM2: 149, BFF8: 66, FAN4: 159-160 **D+I:** Ph: 44-45, B&K3: 248 299 **I:** Bon: 165, C&D: 205
Occasional in England. Rarely reported elsewhere.

cognata *var.* **nauseosa** Boekhout, *Persoonia* 13(4): 416 (1988)
    Mis.: *Melanoleuca adstringens* sensu BFF8: 65
E: ! **S:** !
H: On soil at margins of deciduous woodland, in parkland or occasionally in gardens.
D: BFF8: 65 (as *Melanoleuca adstringens*), FAN4: 160
Rarely reported and the majority of records are unsubstantiated with voucher material.

exscissa (Fr.) Singer, *Cavanillesia* 7: 125 (1935)
    Agaricus exscissus Fr., *Syst. mycol.* 1: 114 (1821)
    Tricholoma exscissum (Fr.) Quél., *Mém. Soc. Émul. Montbéliard, Sér. 2*, 5: 344 (1873)
    Melanoleuca cinerascens D.A. Reid, *Fungorum Rar. Icon. Color.* 2: 16 (1967)
    Melanoleuca kuehneri Bon, *Doc. Mycol.* 18(72): 64 (1988)
E: ! **S:** ! **W:** ! **ROI:** !
H: On soil, with a tendency to occur in sandy areas, often amongst grass, under deciduous trees or conifers.
D: NM2: 149, BFF8: 61-62 (as *Melanoleuca cinerascens*), BFF8: 63, FAN4: 156 **D+I:** B&K3: 248 300 **I:** C&D: 203 (as *M. kuehneri*)
Rarely reported but apparently widespread. Accepted here in a broad sense, as in FAN4. In BFF8 *Melanoleuca cinerascens* is considered a distinct species.

exscissa *var.* **iris** (Kühner) Boekhout, *Persoonia* 13(4): 411 (1988)
    Melanoleuca iris Kühner, *Bull. Mens. Soc. Linn. Lyon* 25: 181 (1956)
E: ! **W:** !
H: On soil amongst grass in pastures, meadows, woodland edges and dune grassland.
D: BFF8: 62 (as *Melanoleuca iris*)
Known from England (Berkshire, South Devon, Warwickshire, West Kent, West Sussex and Yorkshire) and Wales (Anglesey).

grammopodia (Bull.) Pat., *Essai taxon.*: 159 (1900)
    Agaricus grammopodius Bull., *Herb. France*: pl. 548 (1792)
    Tricholoma grammopodium (Bull.) Quél., *Mém. Soc. Émul. Montbéliard, Sér. 2*, 5: 83 (1872)
    Melanoleuca grammopodia var. *subbrevipes* (Métrod) Kühner & Romagn. [*nom. nud.*], *Flore Analytique des Champignons Supérieurs*: 147 (1953)
    Melanoleuca grammopodia f. *macrocarpa* Boekhout, *Persoonia* 13(4): 407 (1988)
    Melanoleuca grammopodia var. *obscura* Bon, *Doc. Mycol.* 20(79): 60 (1990)
    Melanoleuca subbrevipes Métrod [*nom. nud.*], *Rev. Mycol. (Paris)* 7(2-4): 90 (1942)
E: o **S:** ! **W:** ! **NI:** ! **ROI:** !
H: On soil in grassland, at woodland edges, in copses, orchards, and on composted flowerbeds in parks and gardens.

D: NM2: 149 (as *Melanoleuca subbrevipes*), NM2: 150, BFF8: 61, FAN4: 158 **D+I:** Ph: 44-45 **I:** Bon: 165
Occasional in England. Rarely reported elsewhere but apparently widespread.

langei (Boekhout) Bon, *Doc. Mycol.* 20(79): 61 (1990)
    Melanoleuca polioleuca f. *langei* Boekhout, *Persoonia* 13(4): 426 (1988)
    Mis.: *Tricholoma brevipes* sensu Lange (FlDan1: 65 & pl. 29D)
    Mis.: *Melanoleuca brevipes* sensu NCL
E: ? **S:** ?
H: On soil in deciduous woodland or with conifers; occasionally reported in grassland.
D: BFF8: 70 (as *Melanoleuca langei*)
Reported from England and Scotland, often as *Melanoleuca brevipes*.

melaleuca (Pers.) Murrill, *Mycologia* 3: 167 (1911)
    Agaricus melaleucus Pers., *Syn. meth. fung.*: 355 (1801)
    Tricholoma melaleucum (Pers.) P. Kumm., *Führ. Pilzk.*: 133 (1871)
    Mis.: *Melanoleuca graminicola* sensu K&R & sensu auct.
    Mis.: *Melanoleuca vulgaris* sensu auct.
E: ? **S:** ? **W:** ? **NI:** ? **ROI:** ?
H: On soil in deciduous woodland, along woodland edges, also in dunes, in parks and gardens and on wasteground.
D: FAN3: 154, BFF8: 76
Here taken sensu K&R & FAN4: (lacking cystidia). Often recorded but poorly known in Britain and no voucher collections exist. Most material in herbaria is *Melanoleuca polioleuca*.

nivea Boekhout, *Persoonia* 13(4): 417 (1988)
    Mis.: *Melanoleuca subpulverulenta* sensu Cooke [Ill. Brit. fung. 124 (219) Vol. 1 (1883)]
E: !
H: On soil amongst grass or under trees.
D: BFF8: 72
Known in Britain only from an illustration by Cooke (as *Melanoleuca subpulverulenta*) of material gathered in Kew Gardens in 1882 and assigned here by Boekhout, and a collection (in herb. K) from Leicestershire in 1993.

oreina (Fr.) Kühner & Maire, *Bull. Soc. Mycol. France* 50: 18 (1934)
    Agaricus oreinus Fr., *Syst. mycol.* 1: 52 (1821)
    Tricholoma oreinum (Fr.) Gillet, *Hyménomycètes*: 128 (1874)
    Agaricus humilis var. *fragillimus* Fr., *Hymenomyc. eur.*: 75 (1874)
    Tricholoma humile var. *fragillimum* (Fr.) J.E. Lange 1: 65 (1935)
    Melanoleuca humilis var. *fragillima* (Fr.) Bon, *Doc. Mycol.* 9(33): 73 (1978)
    Melanoleuca polioleuca f. *pusilla* Boekhout & Kuyper, *Persoonia* 16(2): 254 (1996)
E: ! **W:** !
H: On soil in grassland, also in deciduous woodland and with conifers.
D: BFF8: 67
Rarely reported but apparently widespread. The correct name for this species may prove to be *Melanoleuca polioleuca* f. *pusilla*.

polioleuca (Fr.) Kühner & Maire, *Bull. Soc. Mycol. France* 50: 18 (1934)
    Agaricus melaleucus γ *polioleucus* Fr., *Syst. mycol.* 1: 115 (1821)
    Tricholoma melaleucum var. *polioleucum* (Fr.) Gillet, *Hyménomycètes*: 128 (1874)
    Tricholoma polioleucum (Fr.) Sacc., *Syll. fung.* 5: 134 (1887)
    Melaleuca vulgaris Pat., *Hyménomyc. Eur.*: 96 (1887)
    Melaleuca vulgaris (Pat.) Pat., *Cat. pl. cell. Tunisie*: 22 (1897)
    Mis.: *Melanoleuca arcuata* sensu Singer (1935), sensu Phillips (1981)
    Mis.: *Melanoleuca melaleuca* sensu NCL & sensu auct.mult.

E: c **S:** c **W:** ! **NI:** ! **ROI:** !
H: On soil, in deciduous woodland and with conifers. Occasionally reported in grassland.
D: NM2: 150, BFF8: 73, FAN4: 164-165 **I:** Ph: 45 (as *Melanoleuca arcuata*), Ph: 45 (as *M. melaleuca*)
Common in England and Scotland. Rarely reported elsewhere but apparently widespread.

politoinaequalipes (Beguet) Bon, *Doc. Mycol.* 9(33): 59 (1978)
    Melanoleuca grammopodia var. *politoinaequalipes* Beguet, *Doc. Mycol.* 2(5): 37 (1972)
E: ! **W:** ?
H: On soil. British material in parkland under *Cedrus* sp.
A single collection, from England (Surrey: Kew, Royal Botanic Gardens) in 2003. Reported from Wales but unsubstantiated with voucher material.

rasilis (Fr.) Singer, *Schweiz. Z. Pilzk.* 17: 56 (1939)
    Agaricus rasilis Fr., *Epicr. syst. mycol.*: 54 (1838)
    Tricholoma rasile (Fr.) Sacc., *Syll. fung.* 5: 140 (1887)
E: ? **W:** ? **NI:** ?
H: On sandy soil in dunes.
D: FAN4: 157
Accepted with reservations. Reported from dunes in England (North Devon: Braunton Burrows) and Wales (various sites) and *fide* BFF8: 59 is 'probably widespread'. However, only one collection has been traced (in herb. E) and that is labelled 'compare with' *M. rasilis*.

schumacheri (Fr.) Singer, *Ann. Mycol.* 41: 55 (1943)
    Agaricus schumacheri Fr., *Syst. mycol.* 1: 87 (1821)
    Tricholoma schumacheri (Fr.) Gillet, *Hyménomycètes*: 125 (1874)
    Tricholoma strictipes var. *schumacheri* (Fr.) J.E. Lange, Fl. agaric. danic. 3: 16 (1938)
E: ! **S:** ! **W:** ! **NI:** !
H: On soil, usually along the edges of woodland, also on waste ground or in grassy areas on dunes.
D: BFF8: 70
Rarely reported. Apparently widespread.

strictipes (P. Karst.) Jul. Schäff., *Agaricina Eur. clav. dichot.*: 157 (1951)
    Tricholoma strictipes P. Karst., *Symb. mycol. fenn.* 8: 7 (1882)
    Mis.: *Tricholoma cnista* var. *evenosum* sensu auct.Brit.
    Mis.: *Tricholoma cnista* sensu auct.Brit.
    Mis.: *Melanoleuca evenosa* sensu auct.Brit.
    Mis.: *Tricholoma evenosum* sensu Rea (1922)
E: ! **S:** ! **W:** ! **NI:** ! **ROI:** !
H: On soil, often amongst grass, in open deciduous woodland, thickets and also in grassland such as pastures and garden lawns.
D: NM2: 148, BFF8: 71 **I:** C&D: 205 (as *Melanoleuca evenosa*)
Rarely reported. Apparently widespread.

stridula (Fr.) Singer, *Ann. Mycol.* 41: 57 (1943)
    Agaricus stridulus Fr., *Ic. Hymenomyc.*: pl. 62 f. 2 (1870)
    Collybia stridula (Fr.) Sacc., *Syll. fung.* 5: 210 (1887)
    Tricholoma stridulum (Fr.) Bres., *Icon. Mycol.*: 123 (1928)
    Mis.: *Melanoleuca graminicola* sensu auct.
E: ! **W:** ? **ROI:** ?
H: On soil, often amongst grass in wooded areas or gardens and also reported from grassy areas on dunes.
D: BFF8: 76-77 **D+I:** B&K3: 252 304
Rarely reported. The majority of records are unsubstantiated with voucher material.

turrita (Fr.) Singer, *Cavanillesia* 7: 128 (1935)
    Agaricus turritus Fr., *Epicr. syst. mycol.*: 51 (1838)
    Tricholoma turritum (Fr.) Sacc., *Syll. fung.* 5: 134 (1887)
E: ! **S:** !
H: On soil in woodland, associated with *Salix* spp. Recently collected on mulched flowerbeds in gardens, and on decayed sawdust and wood-ash in deciduous woodland.

D: BFF8: 75, FAN4: 162-163 **D+I:** B&K3: 250 302 (as *Melanoleuca humilis*)
Known from southern England and Scotland.

verrucipes (Fr.) Singer, *Rev. Mycol. (Paris)* 4(1-2): 68 (1939)
    Armillaria verrucipes Fr. in Quélet, *Mém. Soc. Émul. Montbéliard, Sér. 2*, 5: 317 (1872)
E: !
H: On soil, decayed woodchips and leaf litter.
D: NM2: 148, FAN4: 155 **D+I:** B&K3: 254 307 **I:** Bon: 165, Kriegl3: 380
Unknown in Britain until collected on mulch in Middlesex (Highgate Woods) in 2000. Now also known from Buckinghamshire and Surrey. A distinctive species, unlikely to have been overlooked, and probably an introduction, spreading due to the increased use of woodchips as mulch.

MELANOMPHALIA M.P. Christ., *Friesia* 1: 288 (1936)
Cortinariales, Crepidotaceae
Type: *Melanomphalia nigrescens* M.P. Christ.

nigrescens M.P. Christ., *Friesia* 1: 289 (1936)
E: !
H: On bare soil, amongst herbaceous plants and *Betula* sp.(British material).
D: NM2: 338 **D+I:** Myc3(1): 43 **I:** FlDan5: 161A
A single collection, from England (Yorkshire: Sprotborough) in 1987.

MELANOPHYLLUM Velen., *České Houby* III: 569 (1921)
Agaricales, Agaricaceae
    Chlorospora Massee [*nom. illegit.*, non *Chlorospora* Speg. (1891)], *Bull. Misc. Inform. Kew.* 136 (1898)
    Glaucospora Rea, *Brit. basidiomyc.*: 62 (1922)
Type: *Melanophyllum canali* Velen. (= *M. haematospermum*)

eyrei (Massee) Singer, *Lilloa* 22: 436 (1951) [1949]
    Schulzeria eyrei Massee, *Grevillea* 22: 38 (1893)
    Chlorospora eyrei (Massee) Massee, *Bull. Misc. Inform. Kew.* 1898: 136 (1898)
    Glaucospora eyrei (Massee) Rea, *Brit. basidiomyc.*: 62 (1922)
    Lepiota eyrei (Massee) J.E. Lange, *Fl. agaric. danic.* 1: 36 (1935)
    Cystoderma eyrei (Massee) Singer, *Ann. Mycol.* 41: 170 (1943)
E: ! **W:** ! **NI:** ! **ROI:** !
H: On soil (usually calcareous loam) often associated with *Fagus* and *Corylus* in deciduous woodland and scrub. Also known in mixed deciduous and conifer woodland and with *Urtica dioica* in nettlebeds.
D: FungEur4: 69-72 1b, NM2: 227 **D+I:** Ph: 30-31, FAN5: 162 **I:** FlDan1: 13B (as *Lepiota eyrei*), Bon: 283, C&D: 251
Collections from England (Berkshire, Mid-Lancashire, North Hampshire, South Somerset, South Devon, South Wiltshire, Surrey, West Gloucestershire, West Sussex, Westmorland and Yorkshire), Wales (Monmouthshire) and Northern Ireland (Fermanagh).

haematospermum (Bull.) Kreisel, *Reprium nov. Spec. Regni. veg.* 95(9-10): 700 (1984)
    Agaricus haematospermus Bull., *Herb. France*: pl. 595 (1793)
    Lepiota haematosperma (Bull.) Quél., *Bull. Soc. Mycol. France* 9: 6 (1893)
    Psalliota haematosperma (Bull.) S. Lundell & Nannf., *Fungi Exsiccati Suecici*: 341 (1935)
    Melanophyllum canali Velen., *České Houby* III: 570 (1921)
    Agaricus echinatus Roth [non *A. echinatus* Sowerby (1797)], *Catal. bot.* 2: 255 (1800)
    Lepiota echinata (Roth) Quél., *Enchir. fung.*: 8 (1886)
    Inocybe echinata (Roth) Sacc., *Syll. fung.* 5: 773 (1887)
    Cystoderma echinatum (Roth) Singer, *Ann. Mycol.* 34: 338 (1936)

Melanophyllum echinatum (Roth) Singer, *Lilloa* 22: 436
(1951) [1949]
Agaricus haematophyllus Berk., *Mag. Zool. Bot., [Notices of British Fungi No. 38]* 1: 507 (1837)
Agaricus hookeri Klotzsch, in Berkeley, *Engl. fl.* 5: 97 (1836)
E: o **S:** ! **W:** ! **NI:** ! **ROI:** ! **O:** Channel Islands: !
H: On loam or highly nitrogenous soil, usually in deciduous woodland, often under dense cover of *Mercurialis perennis* or with *Urtica dioica* in nettlebeds. Occasionally in plantations with conifers such as *Picea* and *Thuja* spp, and on compost or mulched soil in greenhouses and private gardens.
D: FungEur4: 72-76 1a, NM2: 227 **D+I:** Ph: 30-31 (as *Melanophyllum echinatum*), B&K4: 220 258, Myc11(4): 180, FAN5: 161-162 **I:** FlDan1: 14C (as *Lepiota haematosperma*), Bon: 283 (as *M. haematosporum*), SV29: 50, C&D: 251
Occasional in England. Rarely reported elsewhere but widespread.

MELANOTUS Pat., *Essai taxon.*: 175 (1900)
Agaricales, Strophariaceae
Type: *Melanotus bambusinus* (Pat.) Pat.

horizontalis (Bull.) P.D. Orton, *Notes Roy. Bot. Gard. Edinburgh* 41(3): 595 (1984)
Agaricus horizontalis Bull., *Herb. France*: pl. 324 (1787)
Naucoria horizontalis (Bull.) Sacc., *Syll. fung.* 5: 833 (1887)
Phaeomarasmius horizontalis (Bull.) Kühner, *Encyclop. Mycol.* 7, *Le genre* Galera: 33 (1935)
Psilocybe horizontalis (Bull.) Vellinga & Noordel., *Persoonia* 16(1): 128 (1995)
Agaricus hepatochrous Berk., *London J. Bot.* 7: 574 (1848)
Melanotus hepatochrous (Berk.) Singer, *Sydowia* 5: 472 (1951)
Agaricus proteus Kalchbr., in Thümen, *Flora* 59: 424 (1876)
Melanotus proteus (Kalchbr.) Singer, *Lloydia* 9: 130 (1946)
Melanotus textilis Redhead & Kroeger, *Mycologia* 76(5): 868 (1984)
E: o **S:** o **W:** !
H: On decayed (and usually decorticated) wood of deciduous trees and conifers, often growing on the cut faces of stacked trunks and logs, in woodland and plantations. Also known on discarded and decayed rope, carpets, matting, worked wood etc.
D: BFF5: 27-28, FAN4: 52 (as *Psilocybe horizontalis*), NBA8: 595-596
Occasional but widespread. Often split into several taxa on the basis of preferred substrata but specimens fall within the variation of a single species. Cultural studies (Sime & Petersen, 1999) have established the synonymy of several species considered distinct in BFF5.

phillipsii (Berk. & Broome) Singer, *Sydowia, Beih.* 7: 84 (1973)
Agaricus phillipsii Berk. & Broome, *Ann. Mag. Nat. Hist., Ser. 5 [Notices of British Fungi no. 1658]* 1: 21 (1878)
Crepidotus phillipsii (Berk. & Broome) Sacc., *Syll. fung.* 5: 878 (1887)
Geophila phillipsii (Berk. & Broome) Kühner & Romagn., *Flore Analytique des Champignons Supérieurs*: 334 (1953)
Psilocybe phillipsii (Berk. & Broome) Vellinga & Noordel., *Persoonia* 16(1): 129 (1995)
Psilocybe caricicola P.D. Orton, *Notes Roy. Bot. Gard. Edinburgh* 29: 119 (1969)
Melanotus caricicola (P.D. Orton) Guzmán, *Mycotaxon* 6: 468 (1978)
Naucoria scutellina Quél., *Bull. Soc. Bot. France* 24: 287 (1878)
Crepidotus scutellinus (Quél.) Quél., *Enchir. fung.*: 109 (1886)
Mis.: *Pleurotus roseolus* sensu Lange [*Dansk. bot. Ark.* 6(6): (1930)]
E: c **S:** ! **W:** ! **ROI:** !
H: On decayed debris of *Carex*, *Juncus* and *Typha* spp, on decayed leaves of grasses such as *Deschampsia caespitosa*, *Glyceria maxima* and *Phleum pratense*, and especially on

Holcus lanatus, often on plants growing in dense mats in rough pasture or at woodland edges.
D: BFF5: 26, NM2: 259, FAN4: 52 (as *Psilocybe phillipsii*) **D+I:** FRIC7: 27-28 53c, B&K4: 328 418 **I:** FlDan2: 65C (as *Pleurotus roseolus*)
Common in southern England. Rarely reported elsewhere but apparently widespread and probably overlooked since basidiomes often occur under old fallen leaves.

MERIPILUS P. Karst., *Bidrag Kännedom Finlands Natur Folk* 37: 33 (1882)
Poriales, Coriolaceae
Type: *Meripilus giganteus* (Pers.) P. Karst.

giganteus (Pers.) P. Karst., *Bidrag Kännedom Finlands Natur Folk* 37: 33 (1882)
Boletus giganteus Pers., *Syn. meth. fung.*: 521 (1801)
Polyporus giganteus (Pers.) Fr., *Syst. mycol.* 1: 356 (1821)
Grifola gigantea (Pers.) Pilát, *Beih. Bot. Centralbl., Abt. B*: 35 (1934)
E: c **S:** c **W:** c **NI:** c **ROI:** c **O:** Channel Islands: ! Isle of Man: !
H: Parasitic, then saprotrophic. Most frequently reported on large dead stumps or trunks of *Fagus*. Also known on *Acer pseudoplatanus*, *Aesculus*, *Betula* spp, *Castanea*, *Crataegus* sp., *Ilex*, *Laurus nobilis*, *Platanus*, *Populus nigra*, *Prunus avium*, *Quercus* spp, *Robinia pseudoacacia*, *Sambucus nigra* and *Ulmus* spp. Very rarely on conifers such as *Cedrus*, *Picea* and *Pinus* spp. A single collection from the floorboards of an old house.
D: EurPoly2: 395, NM3: 171 **D+I:** Ph: 220-221, B&K2: 316 402 **I:** C&D: 141, Kriegl1: 547
Common and widespread.

MERISMODES Earle, *Bull. New York Bot. Gard.* 5: 406 (1909)
Cortinariales, Crepidotaceae
Cyphellopsis Donk, *Meded. Ned. Mycol. Ver.* 18-20: 128 (1931)
Phaeocyphellopsis W.B. Cooke, *Sydowia, Beih.* 4: 119 (1961)
Type: *Merismodes fasciculata* (Schwein.) Earle

anomala (Pers.) Singer, *Agaricales in Modern Taxonomy*, Edn 3: 665 (1975)
Peziza anomala Pers., *Observ. mycol.* 1: 29 (1796)
Solenia anomala (Pers.) Fuckel, *Jahrb. Nassauischen Vereins Naturk.* 25-26: 290 (1871)
Cyphella anomala (Pers.) Pat., *Essai taxon.*: 56 (1900)
Cyphellopsis anomala (Pers.) Donk, *Meded. Ned. Mycol. Ver.* 18-20: 128 (1931)
E: c **S:** ! **W:** ! **ROI:** !
H: On fallen and decayed wood of deciduous trees, especially *Fagus*. Often growing on the cut ends of old sawn logs.
D: NM2: 151 **D+I:** B&K2: 198 224 (N.B. illustration is upside down) **I:** Kriegl3: 602
Common in England, although mostly reported from southern counties. Rarely reported elsewhere but apparently widespread, and easily overlooked.

bresadolae (Grélet) Singer, *Agaricales in Modern Taxonomy*, Edn 3: 665 (1975)
Cyphella bresadolae Grélet, *Bull. Soc. Mycol. France* 38: 174 (1922)
Cyphella leochroma Bres., *Fungi trident.* 1: 99 (1882)
Cyphella monacha Speg., *Michelia* 2: 303 (1880)
Maireina monacha (Speg.) W.B. Cooke, *Sydowia, Beih.* 4: 90 (1961)
Cyphellopsis monacha (Speg.) D.A. Reid, *Kew Bull.* 17(2): 297 (1963)
Cyphella tephroleuca Bres., *Fungi trident.* 2: 57 (1892)
Mis.: *Cyphella fulva* sensu Rea (1922)
E: ! **S:** !
H: On decayed wood, usually on small twigs or sticks of various shrubs. Known on *Berberis* and *Lonicera* spp, and *Ulex europaeus*.

D: NM2: 151

Rarely reported. The majority of records are from scattered localities in Scotland.

fasciculata (Schwein.) Earle, *Bull. New York Bot. Gard.* 5: 406 (1909)

    Cantharellus fasciculatus Schwein., *Trans. Amer. Philos. Soc.* 4: 153 (1832)

    Solenia confusa Bres., *Ann. Mycol.* 1(1): 84 (1903)

    Cyphella confusa (Bres.) Bourdot & Galzin, *Hymenomyc. France*: 164 (1928)

    Cyphellopsis confusa (Bres.) D.A. Reid, *Persoonia* 3(1): 110 (1964)

E: ! **S:** ! **O:** Isle of Man: !

H: On dead wood, often small, dead, attached twigs of deciduous trees. Reported on *Betula, Corylus, Fagus, Prunus avium* and *Salix* spp.

D: NM2: 151 (as *Solenia confusa*)

Rarely reported but apparently widespread.

MERULICIUM J. Erikss. & Ryvarden, *Corticiaceae of North Europe* 4: 859 (1976)

Stereales, Corticiaceae

Type: *Merulicium fusisporum* (Romell) J. Erikss. & Ryvarden

fusisporum (Romell) J. Erikss. & Ryvarden, *Corticiaceae of North Europe* 4: 861 (1976)

    Merulius fusisporus Romell, *Arch. Bot. (Leipzig)* 2(3): 27 (1911)

E: !

H: Saprotrophic. On the undersides of piles of lopped *Picea* sp. brash still bearing decayed needles.

D: NM3: 150 **D+I:** CNE4: 860-862, B&K2: 144 143

Known from England (South Hampshire: New Forest, Little Holmhill Inclosure). Reported from Shropshire but unsubstantiated with voucher material.

MICROMPHALE Gray, *Nat. arr. Brit. pl.* 1: 621 (1821)

Agaricales, Tricholomataceae

Type: *Micromphale venosum* (Pers.) Gray (= *M. foetidum*)

brassicolens (Romagn.) P.D. Orton, *Trans. Brit. Mycol. Soc.* 43(2): 178 (1960)

    Marasmius brassicolens Romagn., *Bull. Soc. Mycol. France* 68: 139 (1952)

    Gymnopus brassicolens (Romagn.) Antonín & Noordel., *Mycotaxon* 63: 363 (1997)

    Marasmius cauvetii Maire & Kühner [*nom. nud.*], *Ann. Soc. Linn. Lyon* 79: 100 (1936)

    Collybia cauvetii (Maire & Kühner) Singer [*nom. nud.*], *Ann. Mycol.* 41: 111 (1943)

    Micromphale cauvetii Hora, *Trans. Brit. Mycol. Soc.* 43(2): 451 (1960)

E: o **NI:** ! **ROI:** ?

H: On decayed leaf litter, usually of *Fagus* in beechwoods but also known on *Acer pseudoplatanus, Betula, Corylus* and *Quercus* spp. Occasionally fruiting on decayed fallen branches and logs.

D: NM2: 152, FAN3: 131 **D+I:** Ph: 68, B&K3: 254 308, A&N2 74-77 (as *Gymnopus brassicolens*) **I:** Bon: 177, C&D: 217, Kriegl3: 209

Occasional in southern England (Berkshire, Buckinghamshire, East Suffolk, North Hampshire, Oxfordshire, South Hampshire, Surrey and West Sussex) but often abundant where it does occur. Reported from the Republic of Ireland but unsubstantiated with voucher material.

brassicolens *var.* **pallidus** (Antonín & Noordel.) [*ined.*]

    Gymnopus brassicolens var. *pallidus* Antonín & Noordel., *Libri Botanici* 17: 77 (1997)

E: !

H: On decayed leaf litter of *Fagus* and *Pinus* spp in a small mixed plantation, on calcareous soil.

D+I: A&N2 77-80

British collections from Surrey (Mickleham: Norbury Park, The Scrubs), otherwise known only from France. A combination in *Micromphale* has not yet been published.

foetidum (Sowerby) Singer, *Lilloa* 22: 305 (1951) [1949]

    Merulius foetidus Sowerby, *Col. fig. Engl. fung.* 1: pl. 21 (1796)

    Marasmius foetidus (Sowerby) Fr., *Epicr. syst. mycol.*: 380 (1838)

    Marasmiellus foetidus (Sowerby) Antonín, Halling & Noordel., *Mycotaxon* 63: 366 (1997)

    Marasmius rufocarneus Velen., *České Houby* I: 180 (1920)

    Micromphale rufocarneum (Velen.) Knudsen, *Nordic J. Bot.* 11(4): 478 (1991)

    Agaricus venosus Pers., *Syn. meth. fung.*: 467 (1801)

    Micromphale venosum (Pers.) Gray, *Nat. arr. Brit. pl.* 1: 622 (1821)

E: o **S:** !

H: On decayed woody debris in deciduous woodland, usually small fallen branches or sticks of *Corylus* and *Fagus* but also known on *Acer pseudoplatanus, Buxus, Crataegus monogyna* and *Fraxinus*. Very rarely on conifers such as *Taxus*.

D: NM2: 152, FAN3: 130-131 **D+I:** Ph: 68, B&K3: 254 309, A&N2 175-178 (as *Marasmiellus foetidus*) **I:** FlDan2: 48G (as *Marasmius foetidus*), Bon: 177, C&D: 217, Kriegl3: 333 (as *Marasmiellus foetidus*)

Occasional but widespread in southern and south-eastern England, in areas on calcareous soil. Rarely reported in Scotland.

impudicum (Fr.) P.D. Orton, *Trans. Brit. Mycol. Soc.* 43: 178 (1960)

    Marasmius impudicus Fr., *Epicr. syst. mycol.*: 377 (1838)

    Collybia impudica (Fr.) Singer, *Ann. Mycol.* 43: 11 (1943)

    Gymnopus impudicus (Fr.) Antonín, Halling & Noordel., *Mycotaxon* 63: 364 (1997)

E: ! **S:** ! **W:** !

H: Typically amongst *Pteridium* or *Calluna*, usually near to conifers. Very rarely reported from woodchip mulch on flowerbeds.

D: NM2: 115 (as *Collybia impudica*), FAN3: 113-114 (as *C. impudica*) **D+I:** A&N2 66-69 (as *Gymnopus impudicus*)

Uncommonly recorded but apparently widespread in England. Also known in Scotland (South Ebudes) and Wales (Denbighshire).

perforans (Hoffm.) Gray, *Nat. arr. Brit. pl.* 1: 622 (1821)

    Agaricus perforans Hoffm., *Nomencl. fung.* 1: 215 (1789)

    Androsaceus perforans (Hoffm.) Pat., *Hyménomyc. Eur.*: 105 (1887)

    Marasmiellus perforans (Hoffm.) Antonín, Halling & Noordel., *Mycotaxon* 63: 366 (1997)

E: ! **S:** ! **W:** !

H: Usually on needle litter of *Picea* spp in plantations. Rarely with *Pinus sylvestris* in natural woodland.

D: NM2: 151, FAN3: 130 **D+I:** B&K3: 256 310, A&N2 167-171 43 (incorrectly captioned as *Marasmiellus omphaliformis*)

Supposedly common, but there are few records of this species from Britain and many of the herbarium collections (in herb. K) are misidentified *Marasmius androsaceus*.

MICROSEBACINA P. Roberts, *Mycol. Res.* 97(4): 473 (1993)

Tremellales, Exidiaceae

Type: *Microsebacina fugacissima* (Bourdot & Galzin) P. Roberts

fugacissima (Bourdot & Galzin) P. Roberts, *Mycol. Res.* 97(4): 473 (1993)

    Sebacina fugacissima Bourdot & Galzin, *Bull. Soc. Mycol. France* 25: 28 (1909)

    Exidiopsis fugacissima (Bourdot & Galzin) Sacc. & Trotter, *Syll. fung.* 21: 452 (1912)

E: ! **S:** ? **ROI:** ?

H: On decayed wood.

Only one recent English collection with voucher material at K. The name was formerly applied to any cryptic, effused species with tremelloid basidia, so most old records are misdetermined. Old British collections in herb. K are various *Exidiopsis* and *Stypella* spp.

microbasidia (M.P. Christ. & Hauerslev) P. Roberts, *Mycol. Res.* 97: 474 (1993)
  Sebacina microbasidia M.P. Christ. & Hauerslev, *Dansk. Bot. Ark.* 19: 30 (1959)
E: !
H: On decayed wood of deciduous trees. British collections on *Alnus* and *Ulmus* spp.
D: NM3: 106 (as *Sebacina microbasidia*)
Known from England (South Hampshire, South Devon and Surrey)

MUCRONELLA Fr., *Hymenomyc. eur.*: 629 (1874)
Hericiales, Hericiaceae
Type: *Mucronella calva* (Alb. & Schwein.) Fr.

calva (Alb. & Schwein.) Fr., *Hymenomyc. eur.*: 629 (1874)
  Hydnum calvum Alb. & Schwein., *Consp. fung. lusat.*: 271 (1805)
  Mucronella aggregata Fr., *Monogr. hymenomyc. Suec.* 2: 280 (1863)
  Mucronella calva var. *aggregata* (Fr.) Pilát, *Sborn. Nár. Mus. Praze* 14B(3-4): 245 (1958)
E: o **S:** ! **W:** !
H: On decayed wood, usually large fallen trunks, logs or stumps of conifers such as *Larix*, *Pinus sylvestris* and *Picea* spp. Also known on *Betula*, *Fagus*, *Quercus* and *Ulmus* spp.
D: NM3: 285 **D+I:** B&K2: 240 288 (as *Mucronella calva* var. *aggregata*), BritChant: 49-50
Occasional but widespread.

MULTICLAVULA R.H. Petersen, *Amer. Midl. Naturalist* 77: 207 (1967)
Cantharellales, Clavariaceae
Type: *Multiclavula corynoides* (Peck) R.H. Petersen

vernalis (Schwein.) R.H. Petersen, *Amer. Midl. Naturalist* 77: 216 (1967)
  Clavaria vernalis Schwein., *Schriften. Naturf. Ges. Leipzig* 1: 112 (1822)
  Clavulinopsis vernalis (Schwein.) Corner, *Monograph of Clavaria and Allied Genera*: 394 (1950)
S: !
H: On bare peat, amongst algae (*Botrydina* sp.).
D: NM3: 252 **D+I:** Myc6(2): 67 (picture upside down)
Known from Scotland (Outer Hebrides: Isle of Harris, North Harris and Shetland (Noss Island).

MUTINUS Fr., *Summa veg. Scand.* 2: 434 (1849)
Phallales, Phallaceae
  Cynophallus (Fr.) Corda, *Icon. fung.* 6: 14 (1854)
  Corynites Berk. & M.A. Curtis, *Trans. Linn. Soc. London* 21: 149 (1855)
Type: *Mutinus caninus* (Huds.) Fr.

caninus (Huds.) Fr., *Summa veg. Scand.* 2: 234 (1849)
  Phallus caninus Huds., *Fl. angl.*, Edn 2, 2: 630 (1778)
  Cynophallus caninus (Huds.) Berk., *Outl. Brit. fungol.*: 298 (1860)
  Phallus inodorus Sow., *Col. fig. Engl. fung.* 3: pl. 330 (1801)
  Ithyphallus inodorus (Sow.) Gray, *Nat. arr. Brit. pl.* 1: 675 (1821)
E: c **S:** c **W:** c **NI:** c **ROI:** c **O:** Channel Islands: !
H: On soil in deciduous and less frequently conifer woodland; commonly on decayed wood of deciduous trees, usually large fallen trunks or decayed piles of sawdust and wood-chips. Increasingly common on flowerbeds mulched with wood chips and shredded bark, in parkland and gardens.
D: NM3: 176 **D+I:** Ph: 256-257, B&K2: 400 526, BritPuffb: 177 Figs. 141/142 **I:** Sow3: 330 (as *Phallus inodorus*), Bon: 301, Kriegl2: 167

Common and widespread, especially in southern England, slightly less frequent northwards. English name = 'Dog Stinkhorn'.

ravenelii (Berk. & M.A. Curtis) E. Fisch., *Syll. fung.* 7: 13 (1888)
  Corynites ravenelii Berk. & M.A. Curtis, *Trans. Linn. Soc. London* 21: 151 (1853)
  Mis.: *Cynophallus bambusinus* sensu Rea (1922)
  Mis.: *Mutinus bambusinus* sensu Cooke [*Grevillea* 17: 17 (1888)]
E: !
H: On soil in damp deciduous woodland, and also known on heaps of spent tan in old railway sidings.
D: NM3: 176 **D+I:** BritPuffb: 178 Figs. 143/144
Known from England (Berkshire, Cheshire, and South Lancashire). Recorded erroneously on several occasions (as *Mutinus bambusinus*) but is apparently spreading in Europe. Distinguished from *M. caninus* by the bright carmine-red receptacle and more strongly foetid odour.

MYCAUREOLA Maire & Chemin, *Compt. Rend. Hebd. Séances Acad. Sci.* 175: 319 (1922)
Incertae sedis
Type: *Mycaureola dilseae* Maire & Chemin

dilseae Maire & Chemin, *Compt. Rend. Hebd. Séances Acad. Sci.* 175: 319 (1922)
E: ! **O:** Channel Islands: !
H: Parasitic. On thalli of the marine alga *Dilsea carnosa* growing in shallow sub-littoral water.
Described (as a pyrenomycete) in 1922 and rarely recorded since. Known from England (Dorset: Swanage and Weymouth) and the Channel Islands (Jersey) in 1985.

MYCENA (Pers.) Roussel, *Flora Calvados*, Edn 2: 64 (1806)
Agaricales, Tricholomataceae
  Prunulus Gray, *Nat. arr. Brit. pl.* 1: 630 (1821)
  Insiticia Earle, *Bull. New York Bot. Gard.* 5: 425 (1909)
  Pseudomycena Cejp, *Spisy Přír. Fak. Karlovy Univ.* 98: 25 (1929)
Type: *Mycena galericulata* (Scop.) Gray

abramsii (Murrill) Murrill, *Mycologia* 8: 220 (1916)
  Prunulus abramsii Murrill, *North American Flora* 9(5): 338 (1916)
  Mycena praecox Velen., *České Houby* II: 316 (1920)
E: ! **S:** ! **W:** ! **NI:** !
H: On decayed wood in deciduous woodland, often on large decayed stumps or fallen branches and often fruiting in spring or summer.
D: NM2: 169 **D+I:** B&K3: 256 311 **I:** Kriegl3: 413
Rarely reported but apparently widespread.

acicula (Schaeff.) P. Kumm., *Führ. Pilzk.*: 109 (1871)
  Agaricus acicula Schaeff., *Fung. Bavar. Palat. nasc.* 1: 52 (1762)
  Trogia acicula (Schaeff.) Corner, *Monogr. Cantharelloid Fungi*: 194 (1966)
E: c **S:** c **W:** c **NI:** c **ROI:** c
H: On rich loam soil in woodland or copses, often on fragments of decayed leaf or stem debris of *Mercurialis perennis*.
D: NM2: 160 **D+I:** Ph: 75, B&K3: 256 312 **I:** Sow3: 282 (as *Agaricus acicula*), FlDan2: 53D, Bon: 183, C&D: 231, Kriegl3: 385
Common and widespread.

aciculata (A.H. Sm.) Desjardin & E. Horak, *Sydowia* 54(2): 148 (2002)
  Mycena codoniceps var. *aciculata* A.H. Sm., *Mycologia* 29: 344 (1937)
  Mis.: *Mycena codoniceps* sensu Kühner [*Botaniste* 17: 86 (1926)]
  Mis.: *Mycena longiseta* sensu auct. Eur.

E: ! **S:** ! **W:** !
H: On decayed woody debris or leaf litter in deciduous
woodland. Rarely reported with conifers.
Rarely reported. Basidiomes are minute, often grow singly and
are easily overlooked.

adonis (Bull.) Gray, *Nat. arr. Brit. pl.* 1: 620 (1821)
    Agaricus adonis Bull., *Hist. champ. France II*: 445 (1792)
    Hemimycena adonis (Bull.) Singer, *Ann. Mycol.* 41: 123
      (1943)
    Agaricus floridulus Fr., *Epicr. syst. mycol.*: 94 (1838)
    Collybia floridula (Fr.) Gillet, *Hyménomycètes*: 329 (1876)
    Mycena floridula (Fr.) Quél., *Bull. Soc. Bot. France* 23: 325
      (1877) [1876]
    Marasmiellus floridulus (Fr.) Singer, *Lilloa* 22: 301 (1951)
      [1949]
    Mycena rubella Quél., *Compt. Rend. Assoc. Franç. Avancem.*
      Sci. 12: 499 (1884)
    Mis.: *Mycena clavus* sensu Rea (1922)
    Mis.: *Collybia clavus* sensu Rea (1922), sensu Cooke [*Ill. Brit.*
      fung. 209 (147) Vol. 2 (1882)] p.p.
E: o **S:** o **W:** ! **NI:** ! **ROI:** !
H: On debris (fallen twigs, needle litter etc.) in conifer or mixed
conifer and deciduous woodland. Occasionally in grassland and
then usually near to trees or with *Sphagnum* spp in marshy or
boggy areas.
D: NM2: 160, NM2: 160 (as *Mycena floridula*) **D+I:** B&K3: 258
313 **I:** FlDan2: 53B (as *M. rubella*), FlDan2: 53A, Bon: 183,
C&D: 231, Kriegl3: 386
Occasional but widespread. *M. floridula* sensu NCL, NM2: and
Kühner (1938) is excluded, being a distinct species or possibly
a variety of *M. flavoalba fide* Maas Geesteranus (1992). Cooke
209 (147) Vol. 2 (1881) (as *Collybia clavus*) illustrates both
Mycena adonis and *M. acicula.*

adonis var. coccinea (Sowerby) Kühner [*nom. illegit.*], *Encycl.*
Mycol. 10, *Le Genre* Mycena: 561 (1938)
    Agaricus coccineus Sowerby [*nom. illegit.*, non *A. coccineus*
      Schaeff. (1762)], *Col. fig. Engl. fung.* 2: pl. 197 (1799)
    Mycena coccinea (Sowerby) Quél. [non *M. coccinea* (Murrill)
      Singer (1962)], *Bull. Soc. Amis Sci. Nat. Rouen.* 15: 155
      (1880) [1879]
    Mis.: *Mycena rubella* sensu Lange (FlDan2: 41 & pl. 53B)
E: ! **S:** ! **W:** !
H: On woody debris, often small fallen sticks or twigs, in
deciduous woodland or with conifers. British records are
usually with *Quercus* spp or *Larix.*
Rarely reported but apparently widespread.

adscendens (Lasch) Maas Geest., *Proc. K. Ned. Akad. Wet.*
84(2): 211 (1981)
    Agaricus adscendens Lasch, *Linnaea* 4: 536 (1829)
    Agaricus tenerrimus Berk., *Engl. fl.* 5(2): 61 (1836)
    Mycena tenerrima (Berk.) Quél., *Mém. Soc. Émul.*
      Montbéliard, Sér. 2, 5: 109 (1872)
E: c **S:** c **W:** c **NI:** c **ROI:** c **O:** Channel Islands: !
H: On decayed woody debris (fallen cones, twigs, sticks, stems,
or nut shells) of deciduous trees or conifers. Also on dead
stems of herbaceous or woody plants such as *Chamerion*
angustifolium, Rubus fruticosus and *Urtica dioica.*
D: NM2: 156 **D+I:** B&K3: 258 314 **I:** FlDan2: 57C (as *Mycena*
tenerrima), C&D: 223 (as *M. tenerrima*)
Common and widespread.

aetites (Fr.) Quél., *Mém. Soc. Émul. Montbéliard, Sér. 2,* 5: 242
(1872)
    Agaricus aetites Fr. [as '*aetitis*'], *Epicr. syst. mycol.*: 110
      (1838)
    Mycena cinerea Massee & Crossl., *Naturalist (Hull)* 28: 1
      (1902)
    Agaricus consimilis Cooke, *Grevillea* 19: 41 (1890)
    Mycena consimilis (Cooke) Sacc., *Syll. fung.* 9: 35 (1891)
    Agaricus umbelliferus Schaeff. [*nom. illegit.*, non *A.*
      umbelliferus L. (1753)], *Fung. Bavar. Palat. nasc.* 3: 73
      (1770)

Mycena umbellifera (Schaeff.) Quél., *Mém. Soc. Émul.*
    Montbéliard, Sér. 2, 5: 242 (1872)
    Mis.: *Mycena ammoniaca* sensu Lange (FlDan2: 39 & pl. 51A)
E: c **S:** c **W:** ! **NI:** ! **ROI:** o **O:** Channel Islands: !
H: On soil amongst grass, usually in short turf such as lawns or
pasture. Rarely in grassy glades in woodland or under isolated
trees.
D: NM2: 168 **D+I:** Ph: 72-73, B&K3: 258 315 **I:** Bon: 185, C&D:
227
Rather common in England and Scotland, less often reported
elsewhere but apparently widespread.

albidolilacea Kühner & Maire, *Encycl. Mycol.* 10, *Le Genre*
Mycena: 419 (1938)
E: !
H: On decayed leaf litter and woody fragments.
Known from England (Surrey and Yorkshire).

amicta (Fr.) Quél., *Mém. Soc. Émul. Montbéliard, Sér. 2,* 5: 243
(1872)
    Agaricus amictus Fr., *Syst. mycol.* 1: 141 (1821)
    Agaricus iris Berk., *Engl. fl.* 5(2): 56 (1836)
    Mycena iris (Berk.) Quél., *Mém. Soc. Émul. Montbéliard, Sér.*
      2, 5: 243 (1872)
    Insiticia amicta var. *iris* (Berk.) Park.-Rhodes, *Trans. Brit.*
      Mycol. Soc. 34(3): 362 (1951)
    Mycena iris var. *caerulea* Rea, *Trans. Brit. Mycol. Soc.* 4(1):
      187 (1913)
    Agaricus mirabilis Cooke & Quél., *Clavis syn. Hymen. Europ.*:
      39 (1878)
    Mycena mirabilis (Cooke & Quél.) Massee, *Brit. fung.-fl.* 3: 93
      (1893)
    Mis.: *Mycena marginella* sensu Rea (1922)
E: o **S:** o **W:** ! **NI:** ! **ROI:** o **O:** Channel Islands: !
H: On acidic soil, often amongst needle litter or on decayed
wood of conifers. Less frequently in litter of deciduous trees
such as *Betula* and *Quercus* spp, and also known on debris of
Pteridium.
D: NM2: 157 **D+I:** B&K3: 260 318 **I:** FlDan2: 50C (as *Mycena*
iris), Bon: 183 (also as *M. amicta* var. *iris*), C&D: 223
Occasional, but widespread. Basidiomes with bright peacock-
blue tints on the pilei were once separated as var. *iris*, but
both colour forms can be found arising from the same
mycelium.

arcangeliana Bres., *Boll. Soc. Bot. Ital.* 1904: 78 (1904)
    Mycena arcangeliana var. *oortiana* Kühner [*nom. inval.*],
      Encycl. Mycol. 10, *Le Genre* Mycena: 297 (1938)
    Mycena oortiana Hora, *Trans. Brit. Mycol. Soc.* 43(2): 452
      (1960)
    Mycena lineata var. *olivascens* Quél., in Lucand, *Champs*
      France: pl. 277 (1889)
    Mycena vitilis var. *olivascens* (Quél.) Kühner, *Encycl. Mycol.*
      10, *Le Genre* Mycena: 305 (1938)
    Mis.: *Agaricus galericulatus* sensu Sowerby [*Col. fig. Eng.*
      fung. 2: pl. 165 (1798)]
    Mis.: *Mycena lineata* sensu auct.
E: c **S:** o **W:** o **NI:** o **ROI:** o
H: On decayed wood, typically *Fagus* or *Fraxinus* in mixed
deciduous woodland. Rarely reported on conifers and on dead
stems of *Pteridium* or *Fallopia japonica.*
D: NM2: 166 **D+I:** Ph: 71 (as *Mycena oortiana*), B&K3: 262 319
I: Bon: 183, C&D: 223
Common in England, especially in southern counties. Occasional
elsewhere but apparently widespread. Records from grassland
are dubious and probably represent *M. flavescens.*

aurantiomarginata (Fr.) Quél., *Mém. Soc. Émul. Montbéliard,*
Sér. 2, 5: 240 (1872)
    Agaricus aurantiomarginatus Fr., *Syst. mycol.* 1: 113 (1821)
    Mis.: *Mycena elegans* sensu Rea (1922) & sensu Kühner
      (1938)
E: ! **S:** ! **W:** ! **NI:** ! **ROI:** !
H: On needle litter in conifer woodland.
D: NM2: 157 **D+I:** B&K3: 262 320 **I:** Bon: 181, C&D: 225,
Kriegl3: 441

Rarely reported, with few recent collections, but apparently widespread.

belliae (Johnst.) P.D. Orton, *Trans. Brit. Mycol. Soc.* 43(2): 178 (1960)
  Agaricus belliae Johnst., *Ann. Mag. Nat. Hist., Ser. 1 [Notices of British Fungi no. 143]* 6: 356 (1841)
  Omphalia belliae (Johnst.) P. Karst., *Bidrag Kännedom Finlands Natur Folk* 32: 139 (1879)
E: ! S: ! ROI: !
H: On dead and dying stems of *Phragmites australis* in reed swamps, usually fruiting just above the water line.
D: NCL3: 306, NM2: 154 **I:** FlDan2: 61F (as *Omphalia belliae*), C&D: 223, Cooke 266 (251) Vol. 2 (1883)
Rarely reported but often in inaccessible habitat and easily overlooked.

bulbosa (Cejp) Kühner, *Encycl. Mycol.* 10, *Le Genre* Mycena: 176 (1938)
  Pseudomycena bulbosa Cejp, *Spisy Přír. Fak. Karlovy Univ.* 104: 149 (1930)
E: ! S: ! W: ! NI: ! ROI: ! O: Isle of Man: !
H: On dead stems or decayed debris of *Carex* and *Juncus* spp in wet places, such as pond and lake margins, marshes or swampy areas in woodland. Also reported rarely on *Luzula* spp.
D: NM2: 156 **D+I:** B&K3: 262 321 **I:** C&D: 223, Kriegl3: 393
Rarely reported but apparently widespread.

capillaripes Peck, *Rep. (Annual) New York State Mus. Nat. Hist.* 41: 63 (1888)
  Mycena langei Maire, *Bull. Soc. Mycol. France* 44: 39 (1928)
  Mycena plicosa var. *marginata* J.E. Lange, *Dansk. Bot. Ark.* 1(5): 18 (1914)
  Mis.: *Mycena rubromarginata* sensu Pearson [*Naturalist* (Hull) 1955: 6 (1955)] & sensu auct. Brit.
E: o S: o W: ! NI: ! ROI: !
H: On acidic soil and needle litter in conifer woodland or plantations, usually with *Pinus* spp, but also known with *Picea* spp and *Tsuga heterophylla*. Very rarely reported in leaf litter of deciduous trees such as *Fagus* and *Betula* spp, and also noted with *Sphagnum* spp in boggy woodland.
D: NM2: 158 **D+I:** B&K3: 264 322 **I:** Bon: 181, C&D: 225
Occasional but widespread and often locally abundant.

capillaris (Schumach.) P. Kumm., *Führ. Pilzk.*: 108 (1871)
  Agaricus capillaris Schumach., *Enum. pl.* 2: 268 (1803)
E: c S: o W: o NI: o O: Isle of Man: !
H: On fallen and decayed leaves of deciduous trees, often in deep drifts of *Fagus* leaves in beechwoods, but also known on *Corylus*, *Fraxinus* and *Quercus* spp, with a single record on *Hedera*.
D: NM2: 162 **D+I:** B&K3: 264 323 **I:** FlDan2: 56B, Bon: 187
Frequently reported and apparently widespread but rarely collected. Probably under-recorded since basidiomes are minute, delicate and diffcult to collect.

chlorantha (Fr.) P. Kumm., *Führ. Pilzk.*: 110 (1871)
  Agaricus chloranthus Fr., *Observ. mycol.* 2: 156 (1818)
  Mis.: *Mycena virens* sensu Rea (1922)
E: ! S: ! NI: ! ROI: ! O: Channel Islands !
H: On soil, or occasionally on decayed debris of grasses, in deciduous woodland. Also on debris of *Ammophila arenaria* and *Leymus arenarius* on dunes.
D: NM2: 157 **I:** C&D: 225
Rarely reported, but apparently widespread.

cinerella (P. Karst.) P. Karst, *Bidrag Kännedom Finlands Natur Folk* 32: 113 (1879)
  Agaricus cinerellus P. Karst., *Hedwigia* 18: 22 (1879)
  Mis.: *Omphalia grisea* sensu auct. Brit. & sensu Ricken [*Blätterpilze Deutschl.*: 399 (1915)]
E: o S: o W: ! NI: ! ROI: !
H: On acidic soil amongst needle litter, usually of *Pinus sylvestris* but known with other conifers and occasionally with deciduous trees such as *Betula*, *Fagus* and *Quercus* spp.
D: NM2: 164 **D+I:** B&K3: 264 324 **I:** FlDan2: 61H, C&D: 229, Kriegl3: 401

Occasional but widespread.

citrinomarginata Gillet, *Hyménomycètes*: 266 (1876)
E: ! S: ! W: ! ROI: !
H: On soil, or woody fragments and occasionally in grass (but then growing from buried wood).
D: NM2: 158
Rarely reported but apparently widespread.

clavicularis (Fr.) Gillet, *Hyménomycètes*: 257 (1876)
  Agaricus clavicularis Fr., *Syst. mycol.* 1: 158 (1821)
E: ! S: ! W: ! NI: !
H: On wet bark of living deciduous trees such as *Acer campestre*, *Betula* spp, and *Fagus* and amongst litter of conifers such as *Pinus sylvestris* and *Picea* spp.
D: NM2: 154 **I:** C&D: 229
Rarely reported but easily overlooked.

clavularis (Batsch) Sacc., *Syll. fung.* 5: 298 (1887)
  Agaricus clavularis Batsch, *Elench. fung. (Continuatio Prima)*: 89 & t. 17 f. 81 (1786)
  Insiticia clavularis (Batsch) Park.-Rhodes, *Trans. Brit. Mycol. Soc.* 34(3): 363 (1951)
E: ! S: ? W: ! NI: ? ROI: ?
H: On wet bark of deciduous trees, often in sheltered woodland or copses, most often on the deeply flanged bark of *Ulmus* saplings. Also reported from debris of *Pteridium*.
D: NM2: 156 **D+I:** Ph: 75 **I:** FlDan2: 54B
Rarely reported and even more rarely collected, but apparently widespread. Basidiomes are minute, fragile and evanescent.

corynephora Maas Geest., *Proc. K. Ned. Akad. Wet.* 86(3): 407 (1983)
E: ! S: ! W: !
H: On damp bark of living deciduous trees, often amongst mosses.
D+I: B&K3: 266 325 **I:** C&D: 223, Kriegl3: 465
Rarely reported but apparently widespread. Basidiomes are minute and easily overlooked, or mistaken for *M. adscendens*.

crocata (Schrad.) P. Kumm., *Führ. Pilzk.*: 108 (1871)
  Agaricus crocatus Schrad., *Spicil. Fl. Germ.*: 127 (1794)
E: o W: !
H: On decayed leaf litter or woody litter (usually small fallen sticks and twigs and occasionally on fallen trunks), usually with *Fagus* or *Fraxinus* in deciduous woodland on calcareous soil.
D: NM2: 155 **D+I:** Ph: 70-71, B&K3: 266 326 **I:** FlDan2: 55D, Bon: 181, C&D: 229
Locally common and often abundant in southern and south-eastern England, but rarely reported elsewhere.

dasypus Maas Geest. & Laessøe, *Persoonia* 15(1): 101 (1992)
E: !
H: On needle litter of *Pinus* spp and leaf litter of *Quercus* spp, also a single collection on dead stems of *Rubus fruticosus* agg.
Described from England (Surrey: Esher Common) but only reported on three occasions since.

diosma Krieglst. & Schwöbel, *Z. Mykol.* 48(1): 32 (1982)
E: ! NI: !
H: On decayed leaf litter, usually with *Fagus* in beechwoods, but also known with *Corylus* and *Quercus* spp in mixed deciduous woodland. Apparently late fruiting, from late autumn to early winter.
D: NM2: 164 **D+I:** B&K3: 268 328 **I:** C&D: 231, Kriegl3: 395
Known from England (Derbyshire, East Sussex, Herefordshire, Oxfordshire, North & South Hampshire and Surrey) and Northern Ireland (Fermanagh).

epipterygia (Scop.) Gray, *Nat. arr. Brit. pl.* 1: 619 (1821)
  Agaricus epipterygius Scop., *Fl. carniol.*: 455 (1772)
  Agaricus citrinellus Pers., *Icon. descr. fung.*: pl. 44 (1800)
  Mycena citrinella (Pers.) P. Kumm., *Führ. Pilzk.*: 109 (1871)
  Agaricus citrinellus var. *candidus* Weinm., *Hymen. Gasteromyc.*: 118 (1836)
  Mycena citrinella var. *candida* (Weinm.) Gillet, *Hyménomycètes*: 258 (1876)

Mycena epipterygia var. *lignicola* A.H. Sm., *North American Species of Mycena*: 428 (1947)

Mycena epipterygia var. *rubescens* L. Remy [*nom. inval.*], *Bull. Soc. Mycol. France* 80: 498 (1965)

Agaricus flavipes Sibth., *Fl. Oxon*: 365 (1794)

Agaricus nutans Sowerby, *Col. fig. Engl. fung.* 1: pl. 92 (1797)

Mycena epipterygia β *nutans* (Sowerby) Gray, *Nat. arr. Brit. pl.* 1: 620 (1821)

Agaricus plicatocrenatus Fr., *Monogr. hymenomyc. Suec.* 2: 294 (1863)

Mycena plicatocrenata (Fr.) Gillet, *Hyménomycètes*: 257 (1876)

Mycena splendidipes Peck, *Bull. New York State Mus. Nat. Hist.* 167: 28 (1913) [1912]

Mycena epipterygia var. *splendidipes* (Peck) Maas Geest., *Proc. K. Ned. Akad. Wet.* 92(1): 105 (1989)

Mycena viscosa Maire, *Bull. Soc. Mycol. France* 26: 162 (1910)

Mycena epipterygia var. *viscosa* (Maire) Ricken, *Blätterpilze Deutschl.*: 419 (1915)

E: c **S:** c **W:** c **NI:** c **ROI:** c **O:** Isle of Man: !

H: On acidic soil, needle litter, small woody fragments such as fallen twigs, sticks and brash, and decayed debris of *Pteridium* on heathland, in woodland and plantations and occasionally in grassland.

D: NM2: 153 (as *Mycena epipterygia* var. *viscosa*), NM2: 154 **D+I:** Ph: 74 (as *M. viscosa*), B&K3: 268 330 (as *M. epipterygia* var. *lignicola*), B&K3: 268 329, B&K3: 270 332 (as *M. epipterygia* var. *splendidipes*) **I:** Bon: 183, C&D: 229

Common and widespread. A variable species.

epipterygioides A. Pearson, *Trans. Brit. Mycol. Soc.* 6(2): 153 (1919) [1918]

Mycena epipterygia var. *epipterygioides* (A. Pearson) Kühner, *Encycl. Mycol.* 10, *Le Genre Mycena*: 353 (1938)

E: ! **NI:** !

H: On decayed conifer wood, usually stumps or large fallen branches of *Pinus sylvestris*. Also known on decayed wood of *Betula* spp.

D: NM2: (*Mycena epipterygia* var. *epipterygioides*) **I:** C&D: 229

Rarely reported. British collections are mostly from heathland in southern England.

erubescens Höhn., *Sitzungsber. Kaiserl. Akad. Wiss., Wien, Math.-Naturwiss. Cl., Abt. 1* 122(1): 267 (1913)

Mycena fellea J.E. Lange, *Dansk. Bot. Ark.* 1(5): 26 (1914)

E: ! **S:** ! **NI:** !

H: On bark of living deciduous trees, usually on large, moss covered trunks.

D: NM2: 155 **D+I:** B&K3: 270 333 **I:** C&D: 229, Kriegl3: 436

Rarely reported but apparently widespread.

filopes (Bull.) P. Kumm., *Führ. Pilzk.*: 110 (1871)

Agaricus filopes Bull., *Herb. France*: pl. 320 (1788)

Agaricus amygdalinus Pers., *Mycol. eur.* 3: 255 (1828)

Mycena amygdalina (Pers.) Singer, *Persoonia* 2: 6 (1961)

Mycena iodiolens S. Lundell, *Kungl. Svenska Vetenskapsakad. Skr. Naturskyddsärenden* 22: 8 (1932)

Mycena lineata f. *pumila* J.E. Lange, *Fl. agaric. danic.* 2: 46 (1936)

Mis.: *Mycena vitilis* sensu Ricken [*Blätterpilze Deutschl.*: 430 (1915)]

E: c **S:** c **W:** c **NI:** c **ROI:** c

H: On soil, decayed woody debris and decayed leaf litter in woodland.

D: NM2: 166, C&D: 223 **D+I:** Ph: 72-73 (as *Mycena sepia*), Ph: 74, B&K3: 272 334 **I:** Bon: 183

Common and widespread.

flavescens Velen., *České Houby* II: 323 (1920)

Mycena luteoalba var. *sulphureomarginata* J.E. Lange, *Dansk. Bot. Ark.* 4(4): 46 (1923)

Mis.: *Mycena elegans* sensu Gillet [*Hymenomycetes*: 265 (1876)]

E: o **S:** ! **W:** ! **NI:** ! **ROI:** ! **O:** Channel Islands: !

H: On decayed leaf or needle litter in deciduous or conifer woodland. Also on decayed woody litter, and on soil in grassland.

D: NM2: 157 **D+I:** Ph: 74, B&K3: 272 335 **I:** C&D: 223

Frequent in England, especially in southern counties. Rarely reported elsewhere, although apparently widespread.

galericulata (Scop.) Gray, *Nat. arr. Brit. pl.* 1: 619 (1821)

Agaricus galericulatus Scop., *Fl. carniol.*: 455 (1772)

Mycena berkeleyi Massee, *Brit. fung.-fl.* 3: 104 (1893)

Mycena galericulata var. *albida* Gillet, *Hyménomycètes*: 276 (1876)

Agaricus radicatellus Peck, *Rep. (Annual) New York State Mus. Nat. Hist.* 31: 32 (1879)

Mycena radicatella (Peck) Sacc., *Syll. fung.* 5: 275 (1887)

Agaricus rugosus Fr., *Epicr. syst. mycol.*: 106 (1838)

Mycena rugosa (Fr.) Quél., *Mém. Soc. Émul. Montbéliard, Sér. 2*, 5: 69 (1872)

Collybia rugulosiceps Kauffman, *Pap. Michigan Acad. Sci.* 5: 126 (1926)

Mycena rugulosiceps (Kauffman) A.H. Sm., *Mycologia* 29: 342 (1937)

E: c **S:** c **W:** c **NI:** c **ROI:** c **O:** Channel Islands: ! Isle of Man: !

H: On decayed woody debris in woodland, usually on large logs or stumps of deciduous trees, rarely reported on conifers.

D: NM2: 165 **D+I:** Ph: 70, B&K3: 274 338 **I:** FlDan2: 56C, Bon: 185, C&D: 229

Very common and widespread. Sensu Sowerby (Col. fig. Engl. fung. 2: pl. 165) is almost certainly *M. arcangeliana*.

galopus (Pers.) P. Kumm., *Führ. Pilzk.*: 108 (1871)

Agaricus galopus Pers., *Observ. mycol.* 2: 56 (1799)

E: c **S:** c **W:** c **NI:** c **ROI:** c **O:** Channel Islands: ! Isle of Man: c

H: On soil, leaf and needle litter and decayed woody debris in deciduous and conifer woodland.

D: NM2: 155 **D+I:** Ph: 70-71, B&K3: 274 339 **I:** FlDan2: 51G (as *Mycena galopoda*), Bon: 181, C&D: 229, Kriegl3: 437

Very common and widespread.

galopus var. **candida** J.E. Lange, *Dansk. Bot. Ark.* 1(5): 20 (1914)

Mycena galopus var. *alba* Rea, *Brit. basidiomyc.*: 395 (1922)

E: o **S:** o **W:** ! **NI:** ! **ROI:** !

H: On decayed woody debris and leaf litter in deciduous and conifer woodland.

D+I: Ph: 70-71 **I:** FlDan2: 51G1 (as *Mycena galopoda* f. *alba*)

Frequent and widespread, but less common than typical *M. galopus*. Basidiomes can often be collected among normally coloured *M. galopus*, thus it may be nothing more than an albino colour form.

galopus var. **nigra** Rea, *Brit. basidiomyc.*: 395 (1922)

Agaricus leucogalus Cooke, *Grevillea* 12: 41 (1883)

Mycena leucogala (Cooke) Sacc., *Syll. fung.* 5: 292 (1887)

Mycena galopus var. *leucogala* (Cooke) J.E. Lange, *Fl. agaric. danic.* 2: 36 (1936)

Mycena fusconigra P.D. Orton, *Trans. Brit. Mycol. Soc.* 91(4): 563 (1988)

E: o **S:** o **NI:** o **ROI:** ! **O:** Isle of Man: !

H: On acidic, often sandy soil, and decayed woody debris or leaf litter in deciduous or coniferous woodland. Also on heathland after fires, on soil, and dead or burnt stems of *Calluna*.

D+I: Ph: 70-71 (as *Mycena leucogala*), B&K3: 280 346 (as *M. leucogala*) **I:** C&D: 229 (as *M. leucogala*)

Occasional but widespread and may be locally common.

haematopus (Pers.) P. Kumm., *Führ. Pilzk.*: 108 (1871)

Agaricus haematopus Pers., *Observ. mycol.* 2: 56 (1799)

Mycena haematopus var. *marginata* J.E. Lange, *Dansk. Bot. Ark.* 1(5): 20 (1914)

E: c **S:** c **W:** c **NI:** c **ROI:** c **O:** Isle of Man: !

H: On fallen and decayed wood of deciduous trees, usually large fallen branches, trunks or stumps of *Fagus* and *Quercus* spp, but also known from a wide range of other broadleaved hosts. Very rarely on decayed wood of conifers such as *Picea* spp.

D: NM2: 155 **D+I:** Ph: 70-71, B&K3: 276 340 **I:** FlDan2: 50G (as *Mycena haematopoda*), Bon: 181, C&D: 229

Common and widespread.

hiemalis (Osbeck) Quél., *Mém. Soc. Émul. Montbéliard, Sér. 2,* 5: 110 (1872)
> *Agaricus hiemalis* Osbeck, *Observ. Bot.* 2: 19 (1791)
> *Hemimycena hiemalis* (Osbeck) Singer, *Rev. Mycol. (Paris)* 3(6): 195 (1938)
> *Marasmiellus hiemalis* (Osbeck) Singer, *Lilloa* 22: 302 (1951) [1949]
> *Mycena epiphloea* (Fr.) Sacc., *Syll. fung.* 5: 282 (1887)
> *Agaricus epiphloeus* Fr., *Hymenomyc. eur.:* 146 (1874)
> Mis.: *Mycena corticola* sensu auct.

E: ! **S:** ! **W:** ! **NI:** ! **ROI:** !
H: On damp bark of living deciduous trees, usually amongst living mosses.
D+I: B&K3: 276 341 **I:** Bon: 187, C&D: 231
Rarely reported but apparently widespread.

inclinata (Fr.) Quél., *Mém. Soc. Émul. Montbéliard, Sér. 2,* 5: 105 (1872)
> *Agaricus inclinatus* Fr., *Epicr. syst. mycol.:* 107 (1838)
> *Agaricus galericulatus* var. *calopus* Fr., *Ic. Hymenomyc.* 1: pl. 80 f. 2 (1873)
> *Mycena galericulata* var. *calopus* (Fr.) P. Karst., *Bidrag Kännedom Finlands Natur Folk* 32: 106 (1879)
> Mis.: *Agaricus alcalinus* sensu Cooke [Ill. Brit. fung. 234 (225) Vol. 2 (1882)]

pl. 8f. 2**E:** c **S:** c **W:** c **NI:** c **ROI:** c **O:** Channel Islands: !
H: On decayed wood of *Quercus* spp in woodland; less often on *Castanea* and rarely on *Fagus.*
D: NM2: 165 **D+I:** Ph: 72, B&K3: 276 342, Myc14(1): 36 **I:** FlDan2: 55E & E1, Bon: 185, C&D: 229, SV39: 8
Very common and widespread. Records on hosts other than those noted are dubious.

kuehneriana A.H. Sm., *North American Species of Mycena:* 190 (1947)
E: ! **W:** !
H: In litter in deciduous woodland.
Accepted but with reservations. Reported from England (Buckinghamshire) and Wales but unsubstantiated with voucher material. Resembles *M. pearsoniana* but with amyloid spores. Poorly known in Europe.

latifolia (Peck) A.H. Sm., *Mycologia* 27: 599 (1935)
> *Agaricus latifolius* Peck, *Rep. (Annual) New York State Mus. Nat. Hist.* 23: 81 (1872)

E: ! **S:** ! **W:** !
H: On soil amongst grass and mosses such as *Polytrichum* spp.
D: NM2: 164 **D+I:** FRIC3: 14-16 21b, B&K3: 278 344
Known from England (East Kent, North Somerset, Staffordshire and Surrey) and recently also collected in Scotland and Wales (Breconshire).

leptocephala (Pers.) Gillet, *Hyménomycètes:* 267 (1876)
> *Agaricus leptocephalus* Pers., *Icon. descr. fung.:* pl. 12 (1800)
> *Agaricus alcalinus* ssp. *leptocephalus* (Pers.) Fr., *Syst. mycol.* 1: 143 (1821)
> *Mycena alcalina* var. *chlorinella* J.E. Lange, *Dansk. Bot. Ark.* 1(5): 21 (1914)
> *Mycena chlorinella* (J.E. Lange) Singer, *Ann. Mycol.* 34: 430 (1936)
> Mis.: *Mycena ammoniaca* sensu Pearson (Naturalist (Hull) 1955: 48 (1955)] & sensu auct. mult.
> Mis.: *Mycena metata* sensu Rea (1922) & sensu auct.

E: c **S:** c **W:** c **NI:** c **ROI:** c **O:** Channel Islands: c Isle of Man: !
H: On soil or decayed leaf and woody litter (twigs etc.) in deciduous and coniferous woodland.
D: NM2: 168 **D+I:** Ph: 72-73, B&K3: 278 345 **I:** Bon: 185, C&D: 227
Common and widespread. This is a woodland species, occasionally reported 'in grassland' but then probably misidentified.

luteoalba (Bolton) Gray, *Nat. arr. Brit. pl.* 1: 620 (1821)
> *Agaricus luteoalbus* Bolton, *Hist. fung. Halifax* 1: 38 (1788)

Agaricus flavoalbus Fr., *Epicr. syst. mycol.:* 103 (1838)
> *Mycena flavoalba* (Fr.) Quél., *Mém. Soc. Émul. Montbéliard, Sér. 2,* 5: 103 (1872)
> *Hemimycena flavoalba* (Fr.) Singer, *Rev. Mycol. (Paris)* 3(6): 195 (1938)
> *Marasmiellus flavoalbus* (Fr.) Singer, *Lilloa* 22: 301 (1951) [1949]

E: c **S:** o **W:** o **NI:** ! **ROI:** ! **O:** Channel Islands: !
H: On soil amongst short grass in pasture, meadows, playing fields, unimproved grassland, on dunes, and occasionally grassy places in woodland. Very common on old lawns.
D: NM2: 159 **D+I:** Ph: 74, B&K3: 272 336 **I:** FlDan2: 53G (as *Mycena flavo-alba*), Bon: 183, C&D: 231, Kriegl3: 388
Common in England. Occasional but widespread elsewhere. Better known as *Mycena flavoalba.*

maculata P. Karst., *Meddeland. Soc. Fauna Fl. Fenn.* 19: 89 (1890)
> Mis.: *Mycena alcalina* sensu Rea (1922)

E: ! **S:** ! **NI:** !
H: On decayed wood, often on large stumps of *Quercus* spp, in deciduous woodland, but also known on *Castanea* and *Fagus.* Rarely reported on conifers such as *Picea* spp (but unsubstanbtiated with voucher material).
D: NM2: 165 **D+I:** Ph: 73, B&K3: 280 347 **I:** C&D: 229
Rarely reported but apparently widespread.

megaspora Kauffman, *Pap. Michigan Acad. Sci.* 17: 182 (1933)
> *Mycena uracea* A. Pearson, *Trans. Brit. Mycol. Soc.* 22(1-2): 32 (1938)
> Mis.: *Mycena dissimulabilis* sensu auct.

E: o **S:** o **W:** ! **ROI:** !
H: On peaty or acidic soil, on heath or moorland, especially after burning, usually growing with *Calluna* or *Erica* spp, and occasionally with *Sphagnum* in boggy areas.
D: NM2: 165 **D+I:** Ph: 72-73 (as *Mycena uracea*)
Occasional but widespread and often abundant where it is found.

meliigena (Berk. & Cooke) Sacc., *Syll. fung.* 5: 302 (1887)
> *Agaricus meliigena* Berk. & Cooke, *Grevillea* 6(40): 129 (1878)
> Mis.: *Mycena corticola* sensu NCL & sensu auct. mult.

E: ! **S:** ! **NI:** ?
H: On bark of living deciduous trees, often amongst mosses.
D: NM2: 161 **D+I:** B&K3: 280 348 **I:** C&D: 223
Rarely reported and even less often collected. Known from England (Surrey) and Scotland (Perthshire). Reported elsewhere and from Northern Ireland but records are unsubstantiated with voucher material.

metata (Fr.) P. Kumm., *Führ. Pilzk.:* 109 (1871)
> *Agaricus metatus* Fr., *Syst. mycol.* 1: 144 (1821)
> *Agaricus collariatus* Fr. [*nom. illegit.,* non *A. collariatus* With. (1792)], *Observ. mycol.* 2: 164 (1818)
> *Mycena collariata* (Fr.) Quél., *Mém. Soc. Émul. Montbéliard, Sér. 2,* 5: 244 (1872)
> *Agaricus phyllogenus* Pers., *Mycol. eur.* 3: 242 (1828)
> *Mycena phyllogena* (Pers.) Singer, *Persoonia* 2: 38 (1961)
> Mis.: *Mycena vitrea* var. *tenella* sensu Kühner [Le Genre *Mycena:* 289 (1938)]

E: o **S:** o **W:** o **NI:** o **ROI:** !
H: On soil and needle litter in conifer woodland, often associated with *Pinus* spp. Rarely in leaf litter of *Alnus* in wet deciduous woodland.
D: NM2: 166 **D+I:** B&K3: 282 349 **I:** FlDan2: 65F, Bon: 183, C&D: 223, Cooke 240 (189) Vol. 2 (1882) (as *Mycena collariata*)
Occasional but widespread.

mirata (Peck) Sacc., *Syll. fung.* 5: 290 (1887)
> *Agaricus miratus* Peck, *Bull. Buffalo Soc. Nat. Sci.* 1: 48 (1873)

E: ! **S:** ! **W:** ! **NI:** !
H: Usually on bark of live deciduous trunks, often amongst mosses but also known on leaf litter and fallen twigs. Very rarely on debris of *Picea* spp.

D: NM2: 161 **D+I:** B&K3: 282 350
Rarely reported but apparently widespread.

mucor (Batsch) Gillet, *Hyménomycètes*: 263 (1876)
Agaricus mucor Batsch, *Elench. fung. (Continuatio Prima)*: 91 & t. 17 f. 82 (1786)
E: ! **W:** ! **ROI:** !
H: On fallen woody debris such as bark or twigs in deciduous or coniferous woodland.
D: NM2: 156 **I:** FlDan2: 56A
Rarely reported. Basidiomes are minute and easily overlooked.

olida Bres., *Fungi trident.* 1: 73 (1887)
Marasmiellus olidus (Bres.) Singer, *Lilloa* 22: 302 (1951) [1949]
Mis.: *Mycena gypsea* [forma] sensu Lange (FlDan2: 41 & pl. 52B)
Mis.: *Mycena hiemalis* sensu Cooke [Ill. Brit. fung. 164 (250) lower figs Vol. 2 (1882)]
E: o **S:** ! **W:** ! **NI:** ! **ROI:** !
H: On bark of living or fallen trees, often amongst mosses and common on exposed roots. Occasionally on soil with woody debris.
D+I: B&K3: 282 351 **I:** Bon: 187, C&D: 231
Occasional in England, rarely reported elsewhere.

olivaceomarginata (Massee) Massee, *Brit. fung.-fl.* 3: 116 (1893)
Agaricus olivaceomarginatus Massee, in Cooke, *Handb. Brit. fung.*, Edn 2: 369 (1890)
Mycena avenacea var. *olivaceomarginata* (Massee) Rea, *Brit. basidiomyc.*: 374 (1922)
Mycena thymicola Velen., *České Houby* II: 304 (1920)
Mycena avenacea var. *thymicola* (Velen.) Kühner, *Encycl. Mycol. 10, Le Genre* Mycena: 416 (1938)
Mycena olivaceomarginata f. *thymicola* (Velen.) Maas Geest., *Proc. K. Ned. Akad. Wet.* 89(3): 298 (1986)
Mycena avenacea var. *roseofusca* Kühner, *Encycl. Mycol. 10, Le Genre* Mycena: 418 (1938)
Mycena roseofusca (Kühner) Bon, *Doc. Mycol.* 3: 22 (1972)
Mycena olivaceomarginata f. *roseofusca* (Kühner) Maas Geest., *Proc. K. Ned. Akad. Wet.* 89(3): 297 (1986)
Mycena brunneomarginata Kühner [*nom. nud.*], *Encycl. Mycol. 10, Le Genre* Mycena: 419 (1938)
Mis.: *Mycena avenacea* sensu Rea (1922), sensu auct. mult.
E: c **S:** c **W:** o **NI:** ! **ROI:** ! **O:** Isle of Man: !
H: On soil amongst short turf in grassland, especially on lawns in late autumn.
D: NM2: 158 **D+I:** Ph: 74 **I:** FlDan2: 49B (as *Mycena avenacea*), C&D: 225
Common to occasional but widespread.

pearsoniana Singer, *Sydowia* 12: 233 (1959) [1958]
Mis.: *Mycena pseudopura* sensu Kühner (1938)
E: ! **S:** ! **W:** ! **NI:** ! **ROI:** !
H: On soil and litter in deciduous and conifer woodland.
D: NM2: 163 **D+I:** Ph: 73 **I:** C&D: 231
Rarely reported but apparently widespread.

pelianthina (Fr.) Quél., *Mém. Soc. Émul. Montbéliard, Sér. 2,* 5: 102 (1872)
Agaricus pelianthinus Fr., *Syst. mycol.* 1: 112 (1821)
Agaricus denticulatus Bolton, *Hist. fung. Halifax* 1: 4 (1788)
Prunulus denticulatus (Bolton) Gray, *Nat. arr. Brit. pl.* 1: 630 (1821)
E: c **S:** o **W:** o **NI:** !
H: On soil and leaf litter in woodland, usually associated with *Fagus* in beechwoods on calcareous soil, rarely with other trees such as *Betula* spp. Very rarely on needle litter of conifers such as *Picea* spp in plantations.
D: NM2: 163 **D+I:** Ph: 72-73, B&K3: 284 352 **I:** FlDan2: 49C, Bon: 181, C&D: 231
Common in southern and south-eastern England. Occasional elsewhere but apparently widespread.

picta (Fr.) Harmaja, *Karstenia* 19: 52 (1979)
Agaricus pictus Fr., *Observ. mycol.* 1: 83 (1815)

Omphalia picta (Fr.) Gillet, *Hyménomycètes*: 299 (1876)
Omphalina picta (Fr.) Quél., *Enchir. fung.*: 45 (1886)
Xeromphalina picta (Fr.) A.H. Sm., *Pap. Michigan Acad. Sci.* 38: 76 (1953)
E: ! **S:** !
H: On decayed woody debris and leaf litter of *Fagus* and *Quercus* spp (British material).
D: NM2: 161 **D+I:** B&K3: 284 353
Rarely reported. This is a small, cryptically coloured species, easily overlooked. Once considered extinct in Britain but collected recently from England (East Sussex, South Hampshire, and Surrey).

polyadelpha (Lasch) Kühner, *Encycl. Mycol. 10, Le Genre* Mycena: 262 (1938)
Agaricus polyadelphus Lasch, *Linnaea* 3: 391 (1828)
Omphalia polyadelpha (Lasch) P. Kumm., *Führ. Pilzk.*: 106 (1871)
Androsaceus polyadelphus (Lasch) Pat., *Hyménomyc. Eur.*: 105 (1887)
Marasmius polyadelphus (Lasch) Cooke, *Ill. Brit. fung.* 1088 (1137b) Vol. 7 (1889)
Delicatula polyadelpha (Lasch) Cejp, *Spisy Přír. Fak. Karlovy Univ.* 100: 83 (1929)
E: ! **ROI:** ? **O:** Channel Islands: !
H: On fallen and decayed leaves, usually of *Quercus* spp, and occasionally on *Fagus*.
D: NM2: 162 **D+I:** B&K3: 284 354 **I:** FlDan2: 62A, Kriegl3: 455
Rarely reported. Basidiomes are minute and easily overlooked.

polygramma (Bull.) Gray, *Nat. arr. Brit. pl.* 1: 619 (1821)
Agaricus polygrammus Bull., *Herb. France*: pl. 395 (1789)
Mycena polygramma f. *pumila* J.E. Lange, *Dansk. Bot. Ark.* 1(5): 23 (1914)
Mycena polygramma f. *candida* J.E. Lange, *Fl. agaric. danic.* 2: 40 (1936)
E: c **S:** c **W:** c **NI:** c **ROI:** c **O:** Channel Islands: !
H: On decayed wood of deciduous trees and shrubs, often on old stumps of *Corylus*. Also known on *Betula, Fagus* and *Quercus* spp. Reported on conifers such as *Picea* spp, but these hosts are unverified.
D: NM2: 168 **D+I:** Ph: 70, B&K3: 286 355 **I:** Sow2: 222 (as *Agaricus polygrammus*), FlDan2: 52F, Bon: 185, C&D: 227, Kriegl3: 419
Common and widespread.

pseudocorticola Kühner, *Encycl. Mycol. 10, Le Genre* Mycena: 687 (1938)
Mis.: *Mycena corticola* sensu Lange (FlDan2: 48 & pl. 57E & E1) and sensu auct.
E: o **S:** o **W:** ! **NI:** o
H: On bark of living deciduous trees such as *Acer pseudoplatanus, Fraxinus, Salix* and *Ulmus* spp, often amongst mosses, in sheltered woodland.
D: NM2: 161 **D+I:** B&K3: 286 356 **I:** Bon: 187, C&D: 223
Rarely reported. Most frequent in the west and apparently rare in the south east. It appears to require sheltered locations and damp misty weather before producing basidiomes.

pterigena (Fr.) P. Kumm., *Führ. Pilzk.*: 108 (1871)
Agaricus pteriginus Fr., *Observ. mycol.* 1: 43 (1815)
Mis.: *Agaricus rosellus* sensu Withering [Bot. arr. Brit. pl. ed. 2, 1: 237 (1787)]
E: ! **S:** ! **W:** ! **NI:** !
H: On decayed debris of ferns such as *Athyrium filix-femina, Dryopteris filix-mas, Dryopteris* spp and *Pteridium*. Also, a single collection on dead leaves of *Luzula sylvatica*.
D: NM2: 158 **D+I:** B&K3: 286 357 **I:** FlDan2: 54H, C&D: 225, Kriegl3: 457
Rarely reported but apparently widespread.

pura (Pers.) P. Kumm., *Führ. Pilzk.*: 110 (1871)
Agaricus purus Pers., *Neues Mag. Bot.* 1: 101 (1794)
Gymnopus purus (Pers.) Gray, *Nat. arr. Brit. pl.* 1: 608 (1821)
Agaricus pseudopurus Cooke, *Grevillea* 10: 147 (1882)

Mycena pseudopura (Cooke) Sacc., *Syll. fung.* 5: 257 (1887)
Mycena pura var. *alba* Gillet, *Hyménomycètes*: 283 (1876)
Mycena pura f. *alba* (Gillet) Kühner, *Encycl. Mycol.* 10, *Le Genre* Mycena: 450 (1938)
Mycena pura var. *lutea* Gillet, *Hyménomycètes*: 283 (1876)
Mycena pura f. *lutea* (Gillet) Kühner, *Encycl. Mycol.* 10, *Le Genre* Mycena: 450 (1938)
Mycena pura var. *purpurea* Gillet, *Hyménomycètes*: 283 (1876)
Mycena pura f. *purpurea* (Gillet) Maas Geest., *Proc. K. Ned. Akad. Wet.* 92(4): 498 (1989)
Mycena pura var. *violacea* Gillet, *Hyménomycètes*: 283 (1876)
Mycena pura f. *violacea* (Gillet) Maas Geest., *Proc. K. Ned. Akad. Wet.* 92(4): 498 (1989)
Mycena pura var. *multicolor* Bres., *Fungi trident.* 2: 9 (1892)
Mycena pura var. *carnea* Rea, *Brit. basidiomyc.*: 377 (1922)
Agaricus purpureus Bolton [*nom. illegit.*, non *A. purpureus* Pers. (1801)], *Hist. fung. Halifax* 1: 41 (1788)
Agaricus purus γ *purpureus* (Bolton) Pers., *Syn. meth. fung.*: 339 (1801)
Gymnopus purus β *purpureus* (Bolton) Gray, *Nat. arr. Brit. pl.* 1: 609 (1821)
E: c **S:** c **W:** c **NI:** c **ROI:** c **O:** Channel Islands: ! Isle of Man: !
H: On soil or litter in deciduous and conifer woodland, on sand dunes, on shingle beaches and in grassland.
D: NM2: 163 **D+I:** Ph: 72, B&K3: 288 358 **I:** FlDan2: 52G (two colour forms), Bon: 181, C&D: 231, Kriegl3: 398
Very common and widespread. Macroscopically variable, especially in colour, and many forms and varieties have been described. Forms ranging through dark purplish, lilaceous, reddish, pink, pale blue, yellow, and white are noted, often growing together. There is even a multicoloured form.

purpureofusca (Peck) Sacc., *Syll. fung.* 5: 255 (1887)
Agaricus purpureofuscus Peck, *Rep. (Annual) New York State Mus. Nat. Hist.* 38: 85 (1885)
E: ! **S:** !
H: On soil and fallen woody debris of conifers.
D: NM2: 159 **I:** C&D: 225
Known from Scotland (Easterness: mainly in the area around Rothiemurchus) and a single collection from England (Yorkshire). Reported from Gloucestershire but unsubstantiated with voucher material.

renati Quél., *Enchir. fung.*: 34 (1886)
Mycena flavipes Quél. [*nom. illegit.*], *Mém. Soc. Émul. Montbéliard, Sér. 2,* 5: 422 (1873)
E: !
H: On decayed wood of deciduous trees.
D: NM2: 158 **D+I:** B&K3: 288 359 **I:** C&D: 225, Kriegl3: 461
Apparently rare. Known from England (West Sussex) with three collections in the area around Arundel between 1974 and 2001. There is also an old record from Co. Durham in 1911, this unsubstantiated with voucher material.

rhenana Maas Geest. & Winterhoff, *Z. Mykol.* 51(2): 247 (1985)
E: !
H: On decayed debris (leaves, twigs and old cones) of *Alnus glutinosa*.
A single collection from England (Huntingdonshire: Woodwalton Fen NNR) in 2004. Previously reported from North Wiltshire: Stanton Fitzwarren, Great Wood in 2000 but unsubstatiated with voucher material.

rorida (Fr.) Quél., *Mém. Soc. Émul. Montbéliard, Sér. 2,* 5: 74 (1872)
Agaricus roridus Fr., *Observ. mycol.* 1: 85 (1815)
E: o **S:** o **W:** o **NI:** ! **ROI:** ! **O:** Isle of Man: !
H: On debris such as small twigs, dead stems etc., of woody plants and deciduous or coniferous trees. Most frequently collected on dead stems of *Rubus fruticosus* agg. in open deciduous woodland.
D: NM2: 154 **D+I:** B&K3: 288 360 **I:** FlDan2: 54D, Bon: 183, C&D: 231, Kriegl3: 405

Occasional but widespread. Many of the records are from western areas.

rosea (Bull.) Gramberg [non *M. rosea* (Pers.) Sacc. (1915)], *Pilze Heimat* 1: 36 (1912)
Agaricus roseus Bull. [non *A. roseus* Pers. (1797)], *Herb. France*: pl. 162 (1783)
Agaricus purus var. *roseus* (Bull.) Pers., *Syn. meth. fung.*: 339 (1801)
Mycena pura var. *rosea* Gillet, *Hyménomycètes*: 283 (1876)
E: c **S:** ! **W:** ! **NI:** ! **ROI:** !
H: On soil or decayed leaf litter, often associated with *Fagus* in beechwoods on calcareous soil but also known with *Acer pseudoplatanus*, *Betula*, *Fraxinus* and *Quercus* spp.
D+I: B&K3: 290 361 **I:** Bon: 181 (as *Mycena pura* var. *rosea*), C&D: 231, Kriegl3: 401
Common at least in southern and south-eastern England, but rarely reported elsewhere. Considered by some authorities as a variety or form of *Mycena pura*.

rosella (Fr.) P. Kumm., *Führ. Pilzk.*: 109 (1871)
Agaricus rosellus Fr. [non *A. rosellus* Batsch (1786)], *Syst. mycol.* 1: 151 (1821)
Omphalia rosella (Fr.) Gray, *Nat. arr. Brit. pl.* 1: 613 (1821)
Agaricus roseus Pers. [*nom. illegit.*, non *A. roseus* Bull (1783)], *Tent. disp. meth. fung.*: 24 (1797)
Mycena rosea (Pers.) Sacc. [*nom. illegit.*, non *M. rosea* (Bull.) Gramberg (1912)], *Fl. ital. crypt., Hymeniales* 1: 256 (1915)
Mis.: *Mycena strobilina* sensu Rea (1922)
E: ! **S:** ! **ROI:** ?
H: On soil and in needle litter of conifers.
D: NM2: 158 **D+I:** B&K3: 290 363 **I:** Bon: 181, C&D: 225, Kriegl3: 443
Apparently rare. Known from a few old collections from North Wiltshire by Broome between 1849 and 1865, and a few more recent collections from Scotland (East Perthshire, Easterness and South Aberdeen) between 1927 and 1988. All other records are unsubstantiated with voucher material.

rubromarginata (Fr.) P. Kumm., *Führ. Pilzk.*: 109 (1871)
Agaricus rubromarginatus Fr., *Observ. mycol.* 1: 42 (1815)
E: ? **S:** ? **W:** ? **NI:** ? **ROI:** ? **O:** Isle of Man: ?
H: On decayed wood of conifers.
D: NM2: 159 **D+I:** B&K3: 290 363 **I:** C&D: 225, Myc14(3): 109, Kriegl3: 463
Not infrequently reported and apparently widespread but seldom collected. *Mycena rubromarginata* sensu Pearson (1955) is *M. capillaripes* and much of the herbarium material is this latter species.

sanguinolenta (Alb. & Schwein.) P. Kumm., *Führ. Pilzk.*: 108 (1871)
Agaricus sanguinolentus Alb. & Schwein., *Consp. fung. lusat.*: 196 (1805)
Agaricus cruentus Fr., *Syst. mycol.* 1: 149 (1821)
Mycena cruenta (Fr.) Quél., *Mém. Soc. Émul. Montbéliard, Sér. 2,* 5: 107 (1872)
E: ! **S:** c **W:** ! **NI:** ! **ROI:** ! **O:** Isle of Man: !
H: On fallen wood, often moss-covered brash, and on the ground amongst needle litter of conifers such as *Larix*, *Picea* and *Pinus* spp. Also attached to stems of *Calluna*.
D: NM2: 155 **D+I:** Ph: 70-71, B&K3: 292 364 **I:** FlDan2: 50A, Bon: 181, C&D: 229, Kriegl3: 440
Apparently common and widespread but rarely collected, with few voucher specimens in herbaria. The majority of records and collections are from Scotland.

septentrionalis Maas Geest., *Proc. K. Ned. Akad. Wet.* 87(4): 442 (1984)
S: !
H: On decayed wood of conifers. British material with *Pinus sylvestris*.
D: NM2: 166
A single collection, from Easterness (Coylum Bridge) in 1960, recently discovered in herb. E., and previously misidentified as *Mycena sepia* fide E.E. Emmett (pers. comm.).

seynesii Quél., *Bull. Soc. Bot. France* 23: 351 (1877) [1876]
E: !
H: On fallen and decayed cones. British material is on *Pinus pinaster* and *P. radiata*.
I: C&D: 225
Known from Berkshire, North Somerset, South Hampshire and Surrey. The orthography is as above (not '*seynii*'), since the species was named after M. de Seynes.

silvae-nigrae Maas Geest. & Schwöbel, *Beitr. Kenntn. Pilze Mitteleurop.* 3: 149 (1987)
Mis.: *Mycena alcalina* sensu NCL p.p. & sensu auct.
E: ! **S:** !
H: On decayed wood of conifers.
D+I: B&K3: 292 365 **I:** C&D: 227
This is a species of conifer wood, but several recent records, mostly unsubstantiated with voucher material, are from unusual substrata such as soil and leaf litter, and may have been misdetermined.

smithiana Kühner, *Encycl. Mycol.* 10, *Le Genre* Mycena: 252 (1938)
Mis.: *Mycena debilis* sensu Lange [Fl. agaric. danic. 2: 48 & pl. 57H (1936)]
E: !
H: On fallen and decayed leaves of *Quercus* spp. One collection apparently on fallen leaves of *Rubus fruticosus* agg.
D: NM2: 162
Known from South Devon, Surrey, Warwickshire, West Kent and Worcestershire. Basidiomes are minute and often in deep drifts of decayed leaf litter, so are easily overlooked.

speirea (Fr.) Gillet, *Hyménomycètes*: 280 (1876)
Agaricus speireus Fr., *Observ. mycol.* 1: 90 (1815)
Omphalia speirea (Fr.) Quél., *Mém. Soc. Émul. Montbéliard, Sér. 2,* 5: 220 (1872)
Hemimycena speirea (Fr.) Singer, *Rev. Mycol. (Paris)* 3(6): 195 (1938)
Agaricus camptophyllus Berk., *Engl. fl.* 5(2): 62 (1836)
Omphalia camptophylla (Berk.) Sacc., *Syll. fung.* 5: 329 (1887)
Mycena speirea var. *camptophylla* (Berk.) Kühner, *Encycl. Mycol.* 10, *Le Genre* Mycena: 587 (1938)
Omphalia tenuistipes J.E. Lange, *Dansk. Bot. Ark.* 6(5): 16 (1930)
Omphalia speirea var. *tenuistipes* (J.E. Lange) J.E. Lange, *Fl. agaric. danic.* 2: 62 (1936)
E: c **S:** o **W:** o **NI:** o **ROI:** o
H: On decayed wood or fallen bark, usually on large, fallen and moss-covered tree trunks in deciduous woodland. Much less often on conifer trunks or small bits of woody debris.
D: NM2: 162 **D+I:** Ph: 75, B&K3: 292 366 **I:** Bon: 187, C&D: 231
Common to occasional but widespread and often abundant where it does occur. Collections with a yellow stipe were previously separated as var. *camptophylla*.

stipata Maas Geest. & Schwöbel, *Beitr. Kenntn. Pilze Mitteleurop.* 3: 147 (1987)
Mis.: *Mycena alcalina* sensu NCL p.p. & sensu auct.
E: ! **S:** ! **W:** ! **NI:** ! **ROI:** ! **O:** Isle of Man: !
H: On decayed conifer wood.
D: NM2: 169 **D+I:** Ph: 71 (as *Mycena alcalina*), B&K3: 294 367 **I:** Bon: 185 (as *M. alcalina*), Kriegl3: 421
Apparently widespread, but the majority of records are reported from deciduous substrata, and virtually all are unsubstantiated with voucher material. Until recently, included with *M. silvae-nigrae* in *M. alcalina*.

stylobates (Pers.) P. Kumm., *Führ. Pilzk.*: 108 (1871)
Agaricus stylobates Pers., *Syn. meth. fung.*: 390 (1801)
Agaricus dilatatus Fr., *Observ. mycol.* 1: 40 (1815)
Mycena dilatata (Fr.) Gillet, *Hyménomycètes*: 261 (1876)
Agaricus torquatus Fr., *Syst. mycol.* 1: 153 (1821)
Marasmius torquatus (Fr.) Massee, *Eur. Fung. Flora*: 60 (1902)
E: c **S:** o **W:** o **NI:** ! **ROI:** !

H: In woodland and scrub, on fallen and decayed leaf and needle litter, woody debris (tiny twigs) and decayed stems and leaves of grasses.
D: NM2: 156 **D+I:** B&K3: 294 369 **I:** FlDan2: 54C, Bon: 183, C&D: 223, Kriegl3: 391
Common in England. Less frequently reported elsewhere.

urania (Fr.) Quél., *Mém. Soc. Émul. Montbéliard, Sér. 2,* 5: 243 (1872)
Agaricus uranius Fr., *Observ. mycol.* 2: 156 (1818)
S: !
H: On soil amongst needle litter of *Pinus sylvestris* in Caledonian pinewoods.
D: NM2: 166, NBA2: 55-56

venustula Quél., *Compt. Rend. Assoc. Franç. Avancem. Sci.* 11: 390 (1883)
S: !
H: On moss-covered living tree trunks. British material on *Quercus* spp.
A single collection from Perthshire (Blair Atholl) in 2000.

viridimarginata P. Karst., *Hedwigia* 31: 218 (1892)
S: !
H: On soil, amongst litter, associated with *Pinus sylvestris* in coniferous woodland.
D: NM2: 158 **D+I:** B&K3: 296 370 **I:** C&D: 227
A single collection from Scotland (Inverey Youth Hostel) in 2001.

vitilis (Fr.) Quél., *Mém. Soc. Émul. Montbéliard, Sér. 2,* 5: 106 (1872)
Agaricus vitilis Fr., *Epicr. syst. mycol.*: 113 (1838)
Mis.: *Mycena filopes* sensu auct.
E: c **S:** c **W:** c **NI:** c **ROI:** c **O:** Channel Islands: ! Isle of Man: !
H: On leaf litter and fallen woody debris such as small twigs, bark, wood fragments etc. in deciduous woodland.
D: NM2: 168 **D+I:** B&K3: 296 371 **I:** Bon: 185, C&D: 227
Common and widespread.

vulgaris (Pers.) P. Kumm., *Führ. Pilzk.*: 108 (1871)
Agaricus vulgaris Pers., *Neues Mag. Bot.* 1: 104 (1794)
E: o **S:** o **W:** ! **NI:** ! **ROI:** !
H: On acidic soil and needle litter of conifers such as *Picea* and *Pinus* spp. Occasionally also on debris of *Pteridium*.
D: NM2: 154 **I:** FlDan2: 58B, Bon: 183, C&D: 229
Occasional but widespread. From the records this is most frequent in Scotland.

zephirus (Fr.) P. Kumm., *Führ. Pilzk.*: 110 (1871)
Agaricus zephirus Fr., *Observ. mycol.* 2: 161 (1818)
E: ? **S:** !
H: On soil in deciduous or mixed coniferous and deciduous woodland.
D: NM2: 167 **I:** FlDan2: 52A, Bon: 185, C&D: 227 (as *Mycena zephyrus*), Kriegl3: 426
A single collection from South Aberdeenshire (Braemar) in 1961. Reported from Yorkshire but unsubstantiated with voucher material.

MYCENELLA (J.E. Lange) Singer, *Notulae Syst. Sect. Crypt. Inst. Bot. Acad. Sci. U.S.S.R.* 4(10-12): 9 (1938)
Agaricales, Tricholomataceae
Type: *Mycenella margaritispora* (J.E. Lange) Singer

bryophila (Voglino) Singer, *Lilloa* 22: 291 (1951) [1949]
Mycena bryophila Voglino, *Atti Inst. Veneto Sci. lett., ed Arti, Serie 4* 4: 617 (1886)
Mis.: *Mycena lasiosperma* sensu Rea (1922) & sensu Lange (FlDan2: 50, pl. 57F)
E: !
H: On soil or decayed woody debris. Frequently found near to, or on, tree roots in deciduous woodland.
D: NM2: 169, BFF8: 138, FAN4: 176-177 **D+I:** B&K3: 298 373 **I:** Bon: 187, C&D: 233, Kriegl3: 471
Known from Bedfordshire, Co. Durham, Huntingdonshire, Mid-Lancashire, North Somerset, South Hampshire, Warwickshire

and West Sussex. The statement in BFF8 that this has 'widespread Scottish records' is incorrect; the species is actually unknown there *fide* R. Watling (pers. comm.).

margaritispora (J.E. Lange) Singer, *Lilloa* 22: 291 (1951) [1949]
Mycena margaritispora J.E. Lange, *Dansk. Bot. Ark.* 1(5): 37 (1914)
Mis.: *Mycena lasiosperma* sensu Kühner (1938), sensu auct. Brit. p.p.
E: ! S: !
H: On humic soil, or on decayed wood, in deciduous woodland or scrub. Very rarely reported with conifers such as *Picea* spp.
D: BFF8: 139, FAN4: 175
Known from England (Bedfordshire, Cambridgeshire, Hertfordshire, Mid-Lancashire, Middlesex, North Lincolnshire, North Somerset and Surrey) and Scotland (Kirkudbrightshire and Mid-Perthshire).

rubropunctata Boekhout, *Persoonia* 12(4): 433 (1985)
E: !
H: On soil, or decayed leaf litter in deciduous woodland.
D: BFF8: 139-140, FAN4: 175-176
Known from West Sussex (Chichester and Eartham) in 1967 and 2000 respectively.

salicina (Velen.) Singer, *Lilloa* 22: 291 (1951) [1949]
Mycena salicina Velen., *České Houby* II: 306 (1920)
E: ! W: !
H: On soil, in deciduous or coniferous woodland. Apparently (*fide* BFF8) also on decayed wood.
D: NM2: 169, BFF8: 140, FAN4: 173 **I:** C&D: 233
Rarely reported. The collection from Banstead in Surrey mentioned in BFF8: 140 (1998) as 'recent' was made in 1951. Collected since that time from a few widely scattered sites in England (Oxfordshire, South Somerset and West Gloucestershire) and Wales (Cardiganshire).

MYCOACIA Donk, *Meded. Ned. Mycol. Ver.* 18-20: 150 (1931)
Stereales, Meruliaceae
Acia P. Karst. [*nom. illegit.*, non *Acia* Schreb. (1791)(*Rosaceae*)], *Meddeland. Soc. Fauna Fl. Fenn.* 5: 42 (1879)
Type: *Mycoacia fuscoatra* (Fr.) Donk

aurea (Fr.) J. Erikss. & Ryvarden, *Corticiaceae of North Europe* 4: 877 (1976)
Hydnum aureum Fr., *Elench. fung.* 1: 137 (1828)
Odontia aurea (Fr.) Quél., *Compt. Rend. Assoc. Franç. Avancem. Sci.* 14: 450 (1886) [1885]
Mycoleptodon microcystidius M.P. Christ., *Friesia* 4: 329 (1953)
Hydnum stenodon Pers., *Mycol. eur.* 2: 188 (1825)
Acia stenodon (Pers.) Bourdot & Galzin [*nom. inval.*], *Bull. Soc. Mycol. France* 30(2): 256 (1914)
Mycoacia stenodon (Pers.) Donk, *Meded. Ned. Mycol. Ver.* 18- 20: 151 (1931)
Hydnum nodulosum Fr., *Hymenomyc. eur.*: 616 (1874)
Acia stenodon var. *nodulosa* (Fr.) Rea, *Brit. basidiomyc.*: 642 (1922)
E: o S: o W: o NI: ! ROI: o O: Channel Islands: o Isle of Man: !
H: On fallen and decayed wood of deciduous trees, especially *Acer pseudoplatanus*, *Fagus* and *Fraxinus* in old woodland.
D: NM3: 158 **D+I:** CNE4: 876-878, B&K2: 160 168
Occasional but widespread and rather frequent in woodland on calcareous soil in southern and south-eastern England.

fuscoatra (Fr.) Donk, *Meded. Ned. Mycol. Ver.* 18-20: 151 (1931)
Hydnum fuscoatrum Fr., *Novit. fl. suec. alt.* 2: 39 (1814)
Acia fuscoatra (Fr.) P. Karst. [*nom. illegit.*], *Meddeland. Soc. Fauna Fl. Fenn.* 5: 42 (1879)
Odontia fuscoatra (Fr.) Bres., *Atti Imp. Regia Accad. Roveretana, ser. 3,* 3(3): 95 (1897)

Mycoleptodon fuscoatrum (Fr.) Pilát, *Bull. Soc. Mycol. France* 51(3 - 4): 401 (1936) [1935]
Hydnum weinmannii Fr., *Elench. fung.* 1: 136 (1828)
E: c S: ! W: ! ROI: !
H: On decayed wood of deciduous trees such as *Alnus*, *Betula*, *Fagus* and *Salix* spp, often in wet places such as fens, marshes, and woodland pools or ponds, especially when these have partially dried out.
D: NM3: 158 **D+I:** CNE4: 880-882, B&K2: 162 169, Myc12(1): 36
Common, at least in southern England, but rarely reported elsewhere.

nothofagi (G. Cunn.) Ryvarden, in Hjortstam, Tellería, Ryvarden & Calogne, *Nova Hedwigia* 34(3 - 4): 534 (1981)
Odontia nothofagi G. Cunn., *Trans. Roy. Soc. New Zealand* 86(1 - 2): 88 (1959)
E: !
H: On decayed wood of deciduous trees, usually on large fallen trunks in mixed deciduous woodland or in beechwoods. British collections mostly on *Fagus* but also known on *Betula* spp and *Carpinus*.
I: FM3(2): 39-40, SV49: 50 (as *Phlebia nothofagi*)
A naturalised alien, possibly native to New Zealand. In Britain this was formerly known only near Haslemere in Surrey and Petworth in West Sussex, but there are recent collections from Berkshire, East Sussex, Hertfordshire, South Hampshire and West Kent.

uda (Fr.) Donk, *Meded. Ned. Mycol. Ver.* 18-20: 151 (1931)
Hydnum udum Fr., *Syst. mycol.* 1: 422 (1821)
Acia uda (Fr.) Bourdot & Galzin [*nom. illegit.*], *Bull. Soc. Mycol. France* 30(2): 255 (1914)
Hydnum sepultum Berk. & Broome, *Ann. Mag. Nat. Hist., Ser. 5 [Notices of British Fungi no. 1813]* 3: 210 (1879)
Odontia sepulta (Berk. & Broome) Rea, *Brit. basidiomyc.*: 650 (1922)
E: c S: c W: c NI: c ROI: c O: Isle of Man: !
H: On decayed wood of deciduous trees, most frequently on large fallen branches or logs of *Acer pseudoplatanus*, *Fagus* and *Fraxinus* in old woodland.
D: NM3: 158 **D+I:** CNE4: 884-886, Ph: 240-241, B&K2: 162 170, Myc12(1): 36-37 **I:** Kriegl1: 267
Common and widespread. The rarer, but macroscopically similar, *Mycoaciella bispora* has a dimitic hyphal system.

MYCOACIELLA J. Erikss. & Ryvarden, *Corticiaceae of North Europe* 5: 901 (1978)
Stereales, Steccherinaceae
Type: *Mycoaciella bispora* (Stalpers) J. Erikss.

bispora (Stalpers) J. Erikss., *Corticiaceae of North Europe* 5: 902 (1978)
Resinicium bisporum Stalpers, *Persoonia* 9(1): 145 (1976)
Mis.: *Acia denticulata* sensu auct.
Mis.: *Hydnum denticulatum* sensu auct.
Mis.: *Hydnum squalinum* sensu Rea (1922)
E: ! S: ! ROI: ?
H: On fallen and decayed wood of deciduous trees such as *Acer pseudoplatanus*, *Alnus*, *Fagus*, *Quercus*, *Salix* and *Ulmus* spp.
D: NM3: 159 **D+I:** CNE5: 902-903
Known from England (Berkshire, Cambridgeshire, North Hampshire, Nottinghamshire, Oxfordshire, Surrey, Warwickshire, West Sussex and Worcestershire) and Scotland (Mid-Perthshire). Reported from the Republic of Ireland (Offaly) but unsubstantiated with voucher material.

MYCOCALIA J.T. Palmer, *Taxon* 10: 58 (1961)
Nidulariales, Nidulariaceae
Type: *Mycocalia denudata* (Fr. & Nordholm) J.T. Palmer

denudata (Fr. & Nordholm) J.T. Palmer, *Taxon* 10: 58 (1961)
Nidularia denudata Fr. & Nordholm, *Symb. gasteromyc.*: 2 (1817)

Granularia denudata (Fr. & Nordholm) Kuntze, *Rev. gen. Bot.*
2: 855 (1891)
Nidularia fusispora Massee, *Bull. Misc. Inform. Kew* 1898:
125 (1898)
E: o **S:** o **W:** o **ROI:** o
H: On decayed wood, or dead leaves of *Juncus effusus* and
Carex spp, in wet habitat such as moorland, on acidic soil. Also
known on weathered dung of rabbits and sheep in similar
habitats.
D: NM3: 195 **D+I:** BritPuffb: 71 Figs. 46/47
Occasional but widespread.

duriaeana (Tul. & C. Tul.) J.T. Palmer, *Taxon* 10: 58 (1961)
Nidularia duriaeana Tul. & C. Tul., *Ann. Sci. Nat., Bot., sér. 3,*
1: 99 (1844)
Granularia duriaeana (Tul. & C. Tul.) Kuntze, *Rev. gen. Bot.*
2: 855 (1891)
E: !
H: On dunes, attached to dead culms or stems of *Ammophila
arenaria,* and on decayed wood, fallen conifer needles, and
weathered dung of rabbits.
D+I: BritPuffb: 73 Figs. 48/49
Known from England (South Lancashire: Formby and Ainsdale
area).

minutissima (J.T. Palmer) J.T. Palmer, *Taxon* 10: 58 (1961)
Nidularia minutissima J.T. Palmer, *Naturalist (Hull)* 1957: 4
(1957)
E: ! **S:** ! **W:** ! **ROI:** !
H: In moist conditions on various substrata, such as dead leaves
of *Juncus* spp, dead grass culms, mosses and decayed leaves
of *Betula* spp, sometimes fruiting when fully submerged.
D: NM3: 195 **D+I:** BritPuffb: 70 Figs. 44/45
Rarely reported but apparently widespread. Basidiomes are
small (reduced to a single peridiole) and easily overlooked.

sphagneti J.T. Palmer, *Česká Mykol.* 17: 122 (1963)
E: !
H: On dead culms of *Juncus* spp, on fallen leaves of *Ilex* and
Quercus spp, and old leaves of *Eriophorum angustifolium* in
bogs or boggy areas in woodland.
D: NM3: 195 **D+I:** BritPuffb: 69 Figs. 42/43
Known from Derbyshire and South Hampshire.

MYRIOSTOMA Desv., *J. Bot. (Paris)* 2: 103-104
(1809)
Lycoperdales, Geastraceae
Polystoma Gray, *Nat. arr. Brit. pl.* 1: 586 (1821)
Type: *Myriostoma anglicum* Desv. (= *M. coliforme*)

coliforme (With.) Corda, *Anleit. Stud. Mykol.*: 131 (1842)
Lycoperdon coliforme With., *Bot. arr. veg.* 2: 783 (1776)
Geastrum coliformis (With.) Pers., *Syn. meth. fung.*: 131
(1801)
Polystoma coliforme (With.) Gray, *Nat. arr. Brit. pl.* 1: 586
(1821)
Myriostoma anglicum Desv., *J. Bot. (Paris)* 2: 104 (1809)
E: ! **O:** Channel Islands: !
H: On soil, often amongst *Urtica dioica,* on roadside banks and
in hedgerows.
D: NM3: 345 **D+I:** Ph: 252 (photograph of an old specimen),
BritPuffb: 112-113 Figs. 84/85 **I:** Sow3: 313 (as *Lycoperdon
coliforme*)
Rare in the nineteenth century, this is now apparently extinct in
Britain (last collected from Hillington in Norfolk in 1880) but
still present in the Channel Islands. English name = 'Pepper
Pot Fungus'.

MYXOMPHALIA Hora, *Trans. Brit. Mycol. Soc.*
43(2): 453 (1960)
Agaricales, Tricholomataceae
Type: *Myxomphalia maura* (Fr.) Hora

maura (Fr.) Hora, *Trans. Brit. Mycol. Soc.* 43: 453 (1960)
Agaricus maurus Fr., *Syst. mycol.* 1: 168 (1821)

Omphalia maura (Fr.) Gillet, *Hyménomycètes*: 290 (1876)
Omphalina maura (Fr.) Quél., *Enchir. fung.*: 42 (1886)
Fayodia maura (Fr.) Singer, *Ann. Mycol.* 34: 331 (1936)
Mycena maura (Fr.) Kühner, *Encycl. Mycol.* 10, *Le Genre*
Mycena: 535 (1938)
E: o **S:** ! **W:** ! **NI:** !
H: On soil, or in needle litter, especially on burnt areas and old
fire sites in conifer woodland or on heathland.
D: NM2: 121, NM2: 121 (as *Fayodia maura*), FAN3: 156, BFF8:
97-98 **D+I:** A&N3 156-162 42 & 43, Ph: 68-69, B&K3: 298
374 **I:** FlDan2: 59J (as *Omphalia maura*), Bon: 131, Kriegl3:
473
Occasional in England, rare or unreported elsewhere. An albino
form exists, as reported by Hora in TBMS 42: 12 (1959).

NAUCORIA (Fr.) P. Kumm., *Führ. Pilzk.*: 76 (1871)
Cortinariales, Cortinariaceae
Alnicola Kühner, *Botaniste* 17: 175 (1926)
Type: *Naucoria escharioides* (Fr.) P. Kumm.

albotomentosa D.A. Reid, *Trans. Brit. Mycol. Soc.* 82(2): 195
(1984)
E: !
H: On soil and decayed woodchips in deciduous woodland
(associated with *Quercus* sp. in the type collection).
Known from Buckinghamshire and Warwickshire, but poorly
understood.

alnetorum (Maire) Kühner & Romagn., *Flore Analytique des
Champignons Supérieurs*: 238 (1953)
Naucoria submelinoides var. *alnetorum* Maire, in Kühner,
Bull. Soc. Mycol. France 47: 243 (1931)
Alnicola alnetorum (Maire) Kühner, *Bull. Soc. Mycol. France*
47: 241 (1931)
Naucoria celluloderma P.D. Orton, *Trans. Brit. Mycol. Soc.*
43(2): 314 (1960)
E: ! **S:** ! **ROI:** !
H: On damp soil or partially dried mud, associated with *Alnus* in
damp woodland, fen and alder carr.
D: Reid84: 196-197, NM2: 332 (as *Naucoria celluloderma*) **D+I:**
B&K5: 126 127 (as *Alnicola alnetorum*)
Apparently widespread but few of the records are accompanied
by voucher material.

amarescens Quél., *Compt. Rend. Assoc. Franç. Avancem. Sci.*
11: 393 (1883) [1882]
Alnicola amarescens (Quél.) R. Heim & Romagn., *Rev. Mycol.
(Paris)* 2(5): 194 (1937)
E: ! **S:** !
H: On soil, and occasionally on burnt ground or amongst
mosses, in deciduous woodland.
D: Reid84: 197-199, NM2: 332 **D+I:** B&K5: 126 128 (as *Alnicola
amarescens*)
Reported from a few widely scattered locations in southern and
south-eastern England, and from Scotland. Easily confused
with *Hebeloma pseudoamarescens,* however, and most records
are unsubstantiated with voucher material.

bohemica Velen., *České Houby* III: 527 (1921)
Alnicola bohemica (Velen.) Kühner & Maire, *Rev. Mycol.
(Paris)* 2(5): 195 (1937)
Naucoria rubriceps P.D. Orton, *Notes Roy. Bot. Gard.
Edinburgh* 41(3): 600 (1984)
Mis.: *Naucoria scorpioides* sensu Lange (FlDan4: 22 & pl.
125A)
E: ! **S:** ! **W:** ! **NI:** ! **ROI:** !
H: On soil in damp deciduous woodland, usually associated with
Salix spp. Less often reported with *Alnus* and *Betula* spp.
D: Reid84: 199, NM2: 331 **D+I:** Ph: 157, B&K5: 126 129 (as
Alnicola bohemica)
Rarely reported but apparently widespread.

clavuligeroides P.D. Orton, *Notes Roy. Bot. Gard. Edinburgh*
41(3): 599 (1984)
E: !

H: On damp soil in woodland, associated with *Alnus* and *Salix* spp.
Known only from East Norfolk and South Hampshire.

escharioides (Fr.) P. Kumm., *Führ. Pilzk.*: 76 (1871)
    Agaricus escharioides Fr., *Syst. mycol.* 1: 260 (1821)
    Alnicola escharioides (Fr.) Romagn., *Bull. Soc. Mycol. France*
      58: 126 (1942)
    Mis.: *Naucoria melinoides* sensu auct.mult.
    Mis.: *Alnicola melinoides* sensu auct.mult.
E: c **S:** c **W:** c **NI:** c **ROI:** c **O:** Isle of Man: !
H: On damp or wet soil, always associated with *Alnus* in swamps, fens, marshes and damp areas in mixed deciduous woodland.
D: Reid84: 199-200, NM2: 332 **D+I:** Ph: 157, B&K5: 128 132 (as *Alnicola melinoides*)
Common and widespread. Often incorrectly written as 'escharoides'.

luteolofibrillosa (Kühner) Kühner & Romagn., *Flore Analytique des Champignons Supérieurs*: 237 (1953)
    Alnicola luteolofibrillosa Kühner, *Botaniste* 17: 176 (1926)
E: ! **S:** ! **ROI:** !
H: On damp soil in woodland, associated with *Alnus*.
D: Reid84: 200, NM2: 332 **D+I:** B&K5: 128 131 (as *Alnicola luteolofibrillosa*)
Rarely reported. Apparently widespread, but confused with other members of the genus.

salicetorum D.A. Reid, *Trans. Brit. Mycol. Soc.* 82(2): 202 (1984)
E: ! **S:** ! **W:** !
H: On soil, in wet areas such as fens or boggy areas in deciduous woodland, usually associated with *Alnus* or *Salix* spp.
Rarely reported. Apparently widespread.

salicis P.D. Orton, *Trans. Brit. Mycol. Soc.* 43(2): 318 (1960)
    Naucoria badiolateritia P.D. Orton, *Notes Roy. Bot. Gard. Edinburgh* 41(3): 598 (1984)
    Naucoria langei Kühner [*nom. inval.*], *Bull. Soc. Naturalistes Oyonnax* 10 -11 (Suppl.): 5 (1957)
    Naucoria macrospora J.E. Lange [*nom. illegit.*, non *N. macrospora* Pat. & Doass. (1880)], *Fl. agaric. danic.* 5, *Taxonomic Conspectus*: VI (1940)
E: ! **S:** !
H: On soil in damp deciduous woodland, usually associated with *Salix* spp.
D: Reid84: 203-204, NM2: 331 **D+I:** B&K5: 130 134 (as *Alnicola salicis*)
Rarely reported. Apparently widespread.

scolecina (Fr.) Quél., *Mém. Soc. Émul. Montbéliard, Sér. 2*, 5: 438 (1875)
    Agaricus scolecinus Fr., *Epicr. syst. mycol.*: 194 (1838)
    Alnicola scolecina (Fr.) Romagn., *Bull. Soc. Mycol. France* 58: 121 (1944) [1942]
    Alnicola badia Kühner, *Botaniste* 17: 176 (1926)
    Naucoria phaea Kühner & Maire, *Publ. Inst. Bot. Barcelona* 3: 101 (1937)
    Naucoria scolecina f. *gracillima* J.E. Lange, *Dansk. Bot. Ark.* 9(6): 20 (1938)
E: ! **S:** ! **W:** ! **O:** Channel Islands: ?
H: On wet soil in damp deciduous woodland, swamps or fen-carr, always associated with *Alnus*.
D: Reid84: 205, NM2: 333 **D+I:** B&K5: 130 135 (as *Alnicola scolecina*)
Taken here in the narrow sense of Reid [TBMS 82(2): 205-206 (1984)]. Apparently rare; often reported, but mostly lacking voucher material, and collections frequently misidentified *Naucoria striatula* or *N. subconspersa*.

silvaenovae D.A. Reid, *Trans. Brit. Mycol. Soc.* 82(2): 206 (1984)
E: !
H: On wet soil in woodland, under *Alnus*.

Known only from the type collection from South Hampshire (New Forest, Denny Wood) in 1982. Possibly only a bisporic form of *Naucoria luteolofibrillosa*.

spadicea D.A. Reid, *Trans. Brit. Mycol. Soc.* 82(2): 206 (1984)
    Naucoria macrospora f. *tetraspora* J.E. Lange [*nom. inval.*, non *N. macrospora* f. *tetraspora* Pat. & Doass (1880)], *Dansk. Bot. Ark.* 9(6): 21 (1938)
E: !
H: On soil, associated with *Alnus*.
D: NM2: 332 **I:** FlDan4: 125d (as *Naucoria macrospora* f. *tetraspora*)
Rarely reported and usually unsubstantiated with voucher material. Possibly only a tetrasporic form of *Naucoria salicis*.

sphagneti P.D. Orton, *Trans. Brit. Mycol. Soc.* 43(2): 320 (1960)
    Naucoria conspersa var. *uliginosa* (Fr.) P. Karst., *Bidrag Kännedom Finlands Natur Folk* 32: 433 (1879)
    Agaricus conspersus β *uliginosus* Fr., *Syst. mycol.* 1: 261 (1821)
E: ! **S:** !
H: Associated with *Sphagnum* spp, in bogs and on moorland.
D: Reid84: 208 **D+I:** B&K5: 132 136 (as *Alnicola sphagneti*)
Rarely reported. The majority of British collections are from scattered sites in Scotland.

striatula P.D. Orton, *Trans. Brit. Mycol. Soc.* 43(2): 322 (1960)
E: o **S:** o **W:** o **NI:** ! **ROI:** !
H: On soil, usually associated with *Alnus* in alder-swamps, fens, marshes, and wet areas in mixed deciduous woodland.
D: Reid84: 209-210, NM2: 332 **D+I:** B&K5: 130 133 (as *Alnicola paludosa*)
Occasional but widespread. *Alnicola paludosa* (Peck) Singer shows marked similarities and may provide an earlier name.

subconspersa P.D. Orton, *Trans. Brit. Mycol. Soc.* 43(2): 323 (1960)
    Alnicola subconspersa (P.D. Orton) Bon, *Doc. Mycol.* 9(35): 41 (1979)
    Mis.: *Naucoria conspersa* sensu Lange (FlDan4: 21 & pl. 125G)
E: c **S:** c **W:** c **NI:** !
H: On damp soil, in swamps, fen-carr and boggy areas in deciduous woodland, always associated with *Alnus*.
D: Reid84: 210-211, NM2: 333 **D+I:** B&K5: 132 138 (as *Alnicola subconspersa*) **I:** Ph: 157 (misdetermined as *Naucoria scolecina*)
Common and widespread. The majority of British records of *Naucoria scolecina* belong here *fide* Reid (Reid84).

tantilla J. Favre, *Ergebn. Wiss. Untersuch. Schweiz. Natn. Parks* 5: 202 (1955)
E: ! **S:** ? **W:** ?
H: British material on damp sandy soil, with *Salix repens* amongst coastal dunes.
D: Reid84: 212, NM2: 333, NBA8: 603 **D+I:** B&K5: 134 140 (as *Alnicola tantilla*)
A single collection, from England (North Devon). Reported from Scotland (Orkney Isles) and Wales (Anglesey) but unsubstantiated with voucher material.

NEOLENTINUS Redhead & Ginns, *Trans. Mycol. Soc. Japan* 26(3): 357 (1985)
Poriales, Lentinaceae
    Digitellus Paulet (anam.), *Traité champ.* 2: 420 (1793)
Type: *Neolentinus kauffmanii* (A.H. Sm.) Redhead & Ginns

adhaerens (Alb. & Schwein.) Redhead & Ginns, *Trans. Mycol. Soc. Japan* 26(3): 357 (1985)
    Agaricus adhaerens Alb. & Schwein., *Consp. fung. lusat.*: 186 (1805)
    Lentinus adhaerens (Alb. & Schwein.) Fr., *Synopsis generis Lentinorum*: 9 (1836)
    Agaricus adhaesivus With., *Arr. Brit. pl. ed. 3*, 4:160 (1796)
    Agaricus resinaceus Trog, *Flora* 15: 525 (1832)

Lentinus resinaceus (Trog) Fr., *Epicr. syst. mycol.*: 391
(1838)
E: ! **S:** ?
H: On decayed wood of conifers. British material on *Pinus sylvestris.*
D: BFF6: 16 (as *Lentinus adhaerens*), NM2: 47 **D+I:** FAN2: 27-28 (as *L. adhaerens*), B&K3: 206 237 (as *L. adhaerens*), FM 4(1): 25-27 **I:** C&D: 149, Kriegl3: 14 (as *L. adhaerens*)
Known from Berkshire and Cheshire and an old, rather dubious collection by Berkeley from Scotland. Withering (1796) reported it from Edgbaston (Warwickshire).

lepideus (Fr.) Redhead & Ginns, *Trans. Mycol. Soc. Japan* 26(3): 357 (1985)
Agaricus lepideus Fr., *Observ. mycol.* 1: 21 (1815)
Lentinus lepideus (Fr.) Fr., *Syst. orb. veg.* 1: 78 (1825)
Panus lepideus (Fr.) Corner, *Nova Hedwigia Beih.* 69: 64 (1981)
Lentinus contiguus Fr., *Epicr. syst. mycol.*: 390 (1838)
Lentinus lepideus var. *contiguus* (Fr.) Rea, *Brit. basidiomyc.*: 538 (1922)
Lentinus lepideus var. *hibernicus* McArdle, *J. Bot.* 47: 444 (1909)
Lentinus squamosus (Schaeff.) Quél., *Fl. mycol. France*: 328 (1888)
Agaricus suffrutescens Brot., *Lus.*: 466 (1804)
Lentinus suffrutescens (Brot.) Fr., *Epicr. syst. mycol.*: 393 (1838)
Agaricus tubaeformis Schaeff., *Fung. Bavar. Palat. nasc.* 3: 248 (1770)
E: o **S:** o **W:** o
H: Weakly parasitic then saprotrophic. On wood of conifers such as *Pinus* spp, and rarely *Taxus,* and on worked wood such as railway sleepers, fence posts etc., especially when treated with creosote. Reported on *Sambucus nigra* but the host is unlikely and not verified.
D: BFF6: 18 (as *Lentinus lepideus*), NM2: 47 **D+I:** Bon: 123, FAN2: 28-29 (as *L. lepideus*), B&K3: 208 238 (as *L. lepideus*) **I:** Sow3: 382 (as *Agaricus tubaeformis*), FlDan5: 197G (as *L. lepideus*), Bon: 123, C&D: 149, Kriegl3: 17 (as *L. lepideus*)
Occasional but widespread. Previously an important cause of decay in treated timber (railway sleepers, fence posts, telegraph poles etc.). The species has a high tolerance to creosote. Also well known for forming 'monstrous' basidiomes, often taking the form of sterile antler-like growths, in dark or low light conditions (as in mines, cellars etc.). The genus *Digitellus* was erected for such growths. English name = 'The Train Wrecker'.

schaefferi (Weinm.) Redhead & Ginns, *Trans. Mycol. Soc. Japan* 26(3): 357 (1985)
Agaricus schaefferi Weinm., *Hymen. Gasteromyc.*: 665 (1836)
Agaricus cyathiformis Schaeff., *Fung. Bavar. Palat. nasc.* 3: 252 (1770)
Panus cyathiformis (Schaeff.) Fr., *Epicr. syst. mycol.*: 397 (1838)
Lentinus cyathiformis (Schaeff.) Bres., *Iconographia Mycologica* 11: Tab. 511 (1929)
Lentinus degener Kalchbr., in Fries, *Hymenomyc. eur.*: 482 (1874)
Lentinus leontopodius Kalchbr., in Fries, *Hymenomyc. eur.*: 482 (1874)
Mis.: *Xerotus degener* sensu auct., non Fries (1838)
E: !
H: On decayed wood of deciduous trees (usually large stumps or fallen trunks) and rarely on worked wood.
D: BFF6: 17 (as *Lentinus cyathiformis*), NM2: 48 **D+I:** FAN2: 29-30 (as *L. cyathiformis*) **I:** Bon: 123, Bon: 123 (as *L. cyathiformis*), SV28: 8, C&D: 149, Kriegl3: 15 (as *L. cyathiformis*)
A single collection, from Co. Durham (Darlington) in 1954, on worked wood in a water cooling tower.

NIA R.T. Moore & Meyers, *Mycologia* 51: 874 (1961)
Melanogastrales, Niaceae
Type: *Nia vibrissa* R.T. Moore & Meyers

vibrissa R.T. Moore & Meyers, *Mycologia* 51(6): 874 (1961)
E: ! **O:** Isle of Man: !
H: On strandline driftwood, floating wood etc., in sea water. Collected from the timbers of Henry VIII's flagship 'Mary Rose'.
D: NM3: 296
One of the few obligate marine basidiomycetes.

NIDULARIA Fr. [*nom. cons.,* non *Nidularia* Bull. (1791)], in Fries & Nordholm, *Symb. gasteromyc.* 1: 2 (1817)
Nidulariales, Nidulariaceae
Granularia Roth [non *Granularia* Sowerby (1815)], *Catal. bot.*: 231 (1797)
Type: *Nidularia radicata* Fr. (= *N. deformis*)

deformis (Willd.) Fr., *Symb. gasteromyc.*: 3 (1817)
Cyathus deformis Willd., *Bot. Mag.* 2(4): 14 (1788)
Nidularia berkeleyi Massee, *Ann. Bot.* 4: 59 (1889)
Nidularia confluens Fr., *Symb. gasteromyc.*: 3 (1817)
Cyathus farctus Roth, *Cat. bot.*: 237 (1797)
Nidularia farcta (Roth) Fr., *Syst. mycol.* 2: 301 (1823)
Granularia pisiformis Roth, *Catal. bot.*: 231 (1797)
Nidularia pisiformis (Roth) Tul. & C. Tul., *Ann. Sci. Nat., Bot.,* sér. 3, 1: 100 (1844)
Nidularia pisiformis var. *broomei* Massee, *Ann. Bot.* 4: 58 (1889)
E: ! **S:** ! **W:** ! **ROI:** !
H: On fallen and decayed wood, usually small sticks, twigs or fallen branches in open deciduous woodland, and also on decayed woodchip mulch on flowerbeds.
D: NM3: 195 **D+I:** Ph: 255 (as *Nidularia 'farcata'*), B&K2: 380 479 (as *N. farcta*), BritPuffb: 66-67 Figs. 40/41 **I:** Kriegl2: 161
Rarely reported but apparently widespread.

OCTAVIANINA Kuntze, *Revis. gen. pl.* 3(3): 501 (1893)
Hymenogastrales, Octavianinaceae
Mis.: *Octaviania* sensu Tul. & C. Tul., *Ann. Sci. Nat., Bot.,* sér. 2, 19: 376 (1843)
Type: *Octavianina asterosperma* (Vittad.) Kuntze

asterosperma (Vittad.) Kuntze, *Revis. gen. pl.* 3(2): 501 (1898)
Octaviania asterosperma Vittad., *Monogr. Tuberac.*: 17 (1831)
Arcangeliella asterosperma (Vittad.) Zeller & H.R. Dodge, *Rep. (Annual) Missouri Bot. Gard.* 22: 266 (1935)
E: ! **S:** ! **W:** !
H: Epigeous or shallowly hypogeous in humic, calcareous soil in deciduous woodland with *Fagus.*
D+I: BritTruff: 185-186 11F
Rarely reported. Known from England (East & West Gloucestershire and South Hampshire), Scotland (Wigtownshire) and Wales (Caernarvonshire). A single record from the Republic of Ireland (Co. Dublin) in 1884 but unsubstantiated with voucher material.

ODONTICIUM Parmasto, *Conspectus Systematis Corticiacearum*: 126 (1968)
Stereales, Hyphodermataceae
Type: *Odonticium romellii* (S. Lundell) Parmasto

romellii (S. Lundell) Parmasto, *Conspectus Systematis Corticiacearum*: 126 (1968)
Odontia romellii S. Lundell, *Symb. Bot. Upsal.* 16(1): 124 (1958)
S: !
H: On decayed wood of *Pinus sylvestris.*

D: NM3: 214 **D+I:** CNE5: 904-906
A single collection, from Scotland (South Aberdeenshire: Mar Lodge Estate) in 1997.

OLIVEONIA Donk, *Fungus* 28: 20 (1958)
Ceratobasidiales, Ceratobasidiaceae
Sebacinella Hauerslev, *Friesia* 11: 95 (1976)
Oliveorhiza P. Roberts (anam.), *Folia Cryptog. Estonica* 33: 128 (1998)
Type: *Oliveonia fibrillosa* (Burt) Donk

citrispora (Hauerslev) P. Roberts, *Folia Cryptog. Estonica* 33: 129 (1998)
Sebacinella citrispora Hauerslev, *Friesia* 11: 96 (1976)
E: !
H: British material on dead attached twigs of *Salix* spp.
D: NM3: 114 (as *Sebacinella citrispora*)
Known from Buckinghamshire, South Devon and Surrey.

fibrillosa (Burt) Donk, *Fungus* 28: 20 (1958)
Sebacina fibrillosa Burt, *Ann. Missouri Bot. Gard.* 13: 335 (1926)
E: ! **W:** ! **O:** Channel Islands: !
H: On decayed deciduous wood and dead stems of *Rubus fruticosus*.
Known from England (Herefordshire and South Devon), Wales (Breconshire) and the Channel Islands (Sark).

nodosa (Hauerslev) P. Roberts, *Folia Cryptog. Estonica* 33: 129 (1998)
Sebacinella nodosa Hauerslev, *Friesia* 11: 95 (1976)
E: ! **W:** !
H: On decayed wood of conifers such as *Picea* and *Pinus* spp.
D: NM3: 114 (as *Sebacinella nodosa*)
Known from England (Hertfordshire, Shropshire, South Devon and Surrey) and Wales (Monmouthshire).

pauxilla (H.S. Jacks.) Donk, *Fungus* 28: 20 (1958)
Corticium pauxillum H.S. Jacks., *Canad. J. Res., Sect. C, Bot. Sci.* 28: 724 (1950)
Oliveorhiza anapauxilla P. Roberts (anam.), *Folia Cryptog. Estonica* 33: 129 (1998)
E: ! **W:** !
H: On decayed wood, bark and leaf litter of deciduous trees, debris of ferns such as *Dryopteris* spp and *Pteridium*, and on dead stems of *Rubus fruticosus*.
D: NM3: 113 **D+I:** CNE5: 908-909
Known from England (Bedfordshire, Buckinghamshire, South Devon, Surrey, Westmorland and Yorkshire) and Wales (Glamorganshire).

OMPHALIASTER Lamoure, *Svensk bot. Tidskr.* 65: 281 (1971)
Agaricales, Tricholomataceae
Type: *Omphaliaster borealis* (M. Lange & Skifte) Lamoure

asterosporus (J.E. Lange) Lamoure, *Svensk bot. Tidskr.* 65: 282 (1971)
Omphalia asterospora J.E. Lange, *Dansk. Bot. Ark.* 6(5): 10 (1930)
Clitocybe asterospora (J.E. Lange) M.M. Moser [*nom. inval.*], *Kleine Kryptogamenflora*: 52 (1953)
E: ! **S:** !
H: On acidic soil amongst grasses or mosses, sometimes associated with conifers.
D: NM2: 170, FAN3: 78 **D+I:** Ph: 51 (as *Clitocybe asterospora*), B&K3: 300 377 **I:** FlDan2: 59G (as *Omphalia asterospora*), Kriegl3: 272 (as *Hygroaster asterosporus*)
Known from England (East Norfolk, South Hampshire and Surrey) and Scotland (Orkney). Reported elsewhere but unsubstantiated with voucher material.

borealis (M. Lange & Skifte) Lamoure, *Svensk bot. Tidskr.* 65(3): 281 (1971)
Rhodocybe borealis M. Lange & Skifte, *Acta. Boreal.* 23: 45 (1967)

Clitocybe borealis (M. Lange & Skifte) P.D. Orton & Watling, *Notes Roy. Bot. Gard. Edinburgh* 29: 135 (1969)
S: ! **W:** !
H: On soil, amongst grass, in boreal and montane habitat.
D: NM2: 170
This is a boreal-arctic species poorly known in Britain. The single English collection under this name was recently redetermined as *O. asterospora*.

OMPHALINA Quél., *Enchir. fung.*: 42 (1886)
Agaricales, Tricholomataceae
Omphalia (Fr.) Staude [*nom. illegit.*, non *Omphalea* L. (1759) (*Euphorbiaceae, nom. cons.*); non *Omphalia* (Pers.) Gray (1821)], *Schwämme Mitteldeutschl.*: 117 (1857)
Type: *Omphalina pyxidata* (Pers.) Quél.

demissa (Fr.) Quél., *Enchir. fung.*: 44 (1886)
Agaricus demissus Fr., *Syst. mycol.* 1: 157 (1821)
Omphalia demissa (Fr.) P. Karst., *Bidrag Kännedom Finlands Natur Folk* 32: 132 (1879)
Agaricus rufulus Berk., *Ann. Mag. Nat. Hist., Ser. 2 [Notices of British Fungi no. 325]* 2: 260 (1848)
S: ? **W:** ? **ROI:** ?
H: On soil under deciduous trees.
D: FAN3: 81-82 **I:** C&D: 185
Accepted with reservations. Collections named thus from Wales (Dyfed) and the Republic of Ireland (Co. Dublin) in herb K require re-examination. They may be sensu Lange (FlDan2 pl. 60F) and thus *O. pyxidata*. Reported from Scotland (Perthshire) but unsusbtantiated with voucher material.

favrei Watling, *Astarte* 10(2): 69 (1977)
Mis.: *Omphalina brownii* sensu NCL & sensu auct. mult.
E: ? **S:** !
H: On soil, in moorland.
Known from Scotland (Shetland) on sandy, calcareous soil. A dubious collection from England (Surrey: Gomshall), in woodland on calcareous soil, requires re-examination.

fulvopallens P.D. Orton, *Notes Roy. Bot. Gard. Edinburgh* 41(3): 605 (1984)
Mis.: *Omphalina pseudoandrosacea* sensu auct.
E: ! **S:** !
H: Amongst *Sphagnum* spp, in boggy areas, in upland or montane habitat.
Known from a few widely scattered locations in Scotland and Yorkshire.

galericolor (Romagn.) Bon, *Doc. Mycol.* 5(19): 22 (1975)
Omphalia galericolor Romagn., *Rev. Mycol. (Paris)* 17(1): 45 (1952)
E: ! **S:** ! **W:** !
H: On sandy soil in grassland or on dunes with mosses such as *Tortula ruralis*.
D: FAN3: 84 **I:** C&D: 185
Rarely reported but apparently widespread. Poorly known in Britain.

galericolor var. lilacinicolor (Bon) Kuyper, *Persoonia* 16: 231 (1996)
Omphalina lilacinicolor Bon, *Doc. Mycol.* 10(37-38): 91 (1980) [1979]
W: ! **NI:** ! **ROI:** !
H: On sandy soil with grasses and amongst mosses such as *Tortula ruralis* on dunes.
D: FAN3: 84 **I:** Bon: 129 (as *Omphalina lilacinicolor*), C&D: 184 (as *O. lilacinicolor*)
Known from Wales (Breconshire) and Northern Ireland (Derry). Reported from the Republic of Ireland but unsubstantiated with voucher material.

mutila (Fr.) P.D. Orton, *Trans. Brit. Mycol. Soc.* 43(2): 180 (1960)
Agaricus mutilus Fr., *Syst. mycol.* 1: 191 (1821)
Pleurotus mutilus (Fr.) Gillet, *Hyménomycètes*: 344 (1876)
Omphalia mutila (Fr.) P. Karst., *Bidrag Kännedom Finlands Natur Folk* 32: 128 (1879)

Pleurotellus mutilus (Fr.) Konrad & Maubl., *Encycl. Mycol.* 14,
Les Agaricales I: 427 (1948)
Omphalina josserandii Singer, *Lilloa* 22: 213 (1951) [1949]
E: ! **S:** ! **W:** !
H: On wet acidic soil or peat, on heathland or in unimproved
grassland, occasionally in wet grassy areas in woodland or
associated with decayed debris of *Juncus* spp.
D: BFF6: 52, FAN3: 83 **I:** FIDan2: 79C (as *Pleurotus mutilus*)
Rarely reported but apparently widespread.

pyxidata (Pers.) Quél., *Enchir. fung.*: 43 (1886)
Agaricus pyxidatus Pers., *Syn. meth. fung.*: 471 (1801)
Omphalia pyxidata (Pers.) P. Kumm., *Führ. Pilzk.*: 107 (1871)
Mis.: *Omphalina demissa* sensu Lange [Dansk. Bot. Ark 6(5):
13 (1930)]
E: o **S:** o **W:** o **NI:** ! **ROI:** o **O:** Channel Islands: !
H: On burnt soil, usually on heathland but occasionally on old
bonfire sites in woodland, and not uncommon on dunes,
usually amongst grass with lichens, mosses and liverworts.
D: NM2: 173, FAN3: 83 **D+I:** B&K3: 304 383 **I:** Bon: 129, C&D:
185, Kriegl3: 486
Occasional but widespread.

subhepatica (Batsch) Murrill, *North American Flora* 9(5): 346
(1916)
Agaricus subhepaticus Batsch, *Elench. fung.* (*Continuatio
Secunda*): 77 t. 38 f. 211 (1789)
Agaricus hepaticus Fr., *Observ. mycol.* 1: 86 (1815)
Omphalia hepatica (Fr.) Gillet, *Hyménomycètes*: 294 (1876)
Omphalina hepatica (Fr.) P.D. Orton, *Trans. Brit. Mycol. Soc.*
43(2): 180 (1960)
E: ! **S:** ! **W:** ! **ROI:** !
H: On soil amongst mosses, lichens or grass in short grassland.
D: NCL3: 336 (as *Omphalina hepatica*) **D+I:** B&K3: 302 380 (as
O. hepatica) **I:** C&D: 185
Rarely collected or reported but apparently widespread.

wallacei P.D. Orton, *Notes Roy. Bot. Gard. Edinburgh* 41(3):
606 (1984)
Mis.: *Omphalina rustica* sensu NCL non *Agaricus rusticus* Fr.
[Epicr. syst. mycol.: 124 (1838)]
E: ! **S:** !
H: On sandy soil amongst mosses and lichens.
D+I: Myc14(2): 84
Known from England (North Devon, Warwickshire and West
Cornwall) and Scotland (Rannoch).

OMPHALOTUS Fayod, *Ann. Sci. Nat., Bot.*, sér. 7, 9: 338 (1889)
Boletales, Paxillaceae
Type: *Omphalotus olearius* (DC) Fayod

illudens (Schwein.) Bresinsky & Besl, *Sydowia, Beih.* 8: 106
(1979)
Agaricus illudens Schwein., *Schriften Naturf. Ges. Leipzig* 1:
81 (1822)
Clitocybe illudens (Schwein.) Sacc., *Syll. fung.* 5: 162 (1887)
Mis.: *Omphalotus olearius* sensu auct. Brit.
E: !
H: On decayed wood of deciduous trees such as *Quercus* spp
and *Castanea*, usually on large stumps or fallen trunks.
D: BFF6: 53-54 (as *Omphalotus olearius*), FAN3: 89 **D+I:** Ph:
186 (as *O. olearius*) **I:** Bon: 51
Rarely reported. In Britain, known only from southern and south
eastern England. Supposedly luminescent when actively in
growth. Perhaps better known in Britain as *O. olearius*, but this
morphologically similar species is now considered to be mainly
Mediterranean in distribution, where it is common on olive
trees. Nonetheless, a recent DNA study [Kirchmair *et al.*,
Mycologia 96: 1253-1260 (2004)] has suggested that a
collection from Oxfordshire belongs in *O. olearius* sensu stricto,
making it unclear to which of the two species British collections
should be referred.

OSSICAULIS Redhead & Ginns, *Trans. Mycol. Soc. Japan* 26(3): 362 (1985)
Agaricales, Tricholomataceae
Type: *Ossicaulis lignatilis* (Pers.) Redhead & Ginns

lignatilis (Pers.) Redhead & Ginns, *Trans. Mycol. Soc. Japan*
26(3): 362 (1985)
Agaricus lignatilis Pers. [non *A. lignatilis* Bull. (1792)], *Syn.
meth. fung.*: 368 (1801)
Pleurotus lignatilis (Pers.) P. Kumm., *Führ. Pilzk.*: 105 (1871)
Clitocybe lignatilis (Pers.) P. Karst., *Bidrag Kännedom
Finlands Natur Folk* 32: 86 (1879)
Pleurocybella lignatilis (Pers.) Singer, *Lilloa* 22: 203 (1951)
[1949]
Nothopanus lignatilis (Pers.) Bon, *Doc. Mycol.* 17 (65): 53
(1986)
Agaricus circinatus Fr., *Epicr. syst. mycol.*: 132 (1838)
Pleurotus circinatus (Fr.) Gillet, *Hyménomycètes*: 341 (1876)
Pleurotus lignatilis var. *tephrocephalus* Sacc., *Syll. fung.* 5:
344 (1887)
E: o **S:** ! **W:** ! **ROI:** !
H: On decayed wood of deciduous trees, usually large fallen
trunks or inside large hollow stumps of *Ulmus* spp and *Fagus*.
Rarely on *Acer campestre, A. pseudoplatanus, Crataegus,
Fraxinus* and *Populus nigra*.
D: BFF6: 55-56, NM2: 106 (as *Clitocybe lignatilis*), FAN3: 131-
132 **D+I:** Ph: 182 (as *Pleurotus lignatilis*), B&K3: 160 168 (as
C. lignatilis) **I:** FIDan2: 62F (as *P. lignatilis*), Bon: 121 (as
Nothopanus lignatilis), C&D: 151
Occasional to rather rare but widespread.

OUDEMANSIELLA Speg., *Anales Soc. Ci. Argent.* 12: 24 (1881)
Agaricales, Tricholomataceae
Mucidula Pat., *Hyménomyc. Eur.*: 95 (1887)
Type: *Oudemansiella platensis* Speg.

mucida (Schrad.) Höhn., *Sitzungsber. Kaiserl. Akad. Wiss.,
Wien, Math.-Naturwiss. Cl., Abt. 1* 118: 276 (1909)
Agaricus mucidus Schrad., *Spicil. Fl. Germ.*: 116 (1794)
Armillaria mucida (Schrad.) P. Kumm., *Führ. Pilzk.*: 135
(1871)
Mucidula mucida (Schrad.) Pat., *Hyménomyc. 'Eur.*: 96 (1887)
E: c **S:** c **W:** c **NI:** c **ROI:** c **O:** Channel Islands: c Isle of Man: c
H: Possibly a weak parasite, then saprotrophic. On dying and
dead deciduous trees, typically on *Fagus*. Rarely on *Acer
pseudoplatanus* and *Quercus* spp; also reported on *Acer
campestre, Alnus, Betula, Fraxinus, Prunus* and *Tilia* spp.
D: NM2: 175, BFF8: 141-142, FAN4: 177 **D+I:** Ph: 33, B&K3:
308 388 **I:** FIDan2: 41E, Bon: 171, C&D: 235
Common and widespread. English names = 'Porcelain Fungus' or
'Poached Egg Fungus'.

OXYPORUS (Bourdot & Galzin) Donk, *Meded. Bot. Mus. Herb. Rijks Univ. Utrecht* 9: 202 (1933)
Poriales, Coriolaceae
Type: *Oxyporus connatus* (Weinm.) Kühner (= *O. populinus*)

corticola (Fr.) Ryvarden, *Persoonia* 7(1): 19 (1972)
Polyporus corticola Fr., *Syst. mycol.* 1: 385 (1821)
Poria corticola (Fr.) Cooke, *Grevillea* 14: 113 (1886)
Chaetoporus corticola (Fr.) Bondartsev & Singer, *Ann. Mycol.*
39: 51 (1941)
Polyporus ravidus Fr., *Epicr. syst. mycol.*: 475 (1838)
Polystictus ravidus (Fr.) Cooke, *Grevillea* 14: 81 (1886)
Mis.: *Polyporus aneirinus* sensu Fries (1828)
Mis.: *Polyporus heteroclitus* sensu auct.
E: ! **S:** !
H: On decayed wood of deciduous trees such as *Fagus, Populus
tremula* and *Ulmus* spp. A single collection in herb. K
supposedly on *Pinus* spp, but the host is not verified.
D: EurPoly2: 440, NM3: 172

Rarely reported. Known from a few sites in England (Berkshire, Cumberland, North Hampshire, Surrey and Warwickshire) and a single collection from Scotland (Linlithgow) in 1840.

latemarginatus (Durieu & Mont.) Donk, *Persoonia* 4: 342 (1966)
  Polyporus latemarginatus Durieu & Mont., *Syll. Crypt.*: 163 (1856)
  Poria latemarginata (Durieu & Mont.) Cooke, *Grevillea* 14: 112 (1886)
  Poria ambigua Bres., *Atti Imp. Regia Accad. Roveretana, ser. 3,* 3: 84 (1897)
  Chaetoporus ambiguus (Bres.) Bondartsev & Singer, *Ann. Mycol.* 39: 51 (1941)
E: !
H: On large, fallen and usually decayed trunks of deciduous trees in old woodland. Most often on *Fagus* but also known on *Acer pseudoplatanus, Aesculus, Fraxinus, Salix fragilis, Tilia* and *Ulmus* spp. A single collection in herb. K supposedly on conifer wood, but the host is not verified.
D: EurPoly2: 442 **D+I:** Myc9(4): 156
Rarely reported, but apparently increasing. Known from Bedfordshire, Berkshire, Cambridgeshire, East & West Kent, East Norfolk, Middlesex, Oxfordshire, South Hampshire, Surrey and West Sussex. Basidiomes are up to one centimetre thick, resupinate and often cover large areas, thus unlikely to have been previously overlooked.

obducens (Pers.) Donk, *Meded. Bot. Mus. Herb. Rijks Univ. Utrecht* 9: 202 (1933)
  Polyporus obducens Pers., *Mycol. eur.* 2: 104 (1825)
  Poria obducens (Pers.) Cooke, *Grevillea* 14: 110 (1886)
E: ! **S:** ! **W:** ! **NI:** ! **ROI:** !
H: On decayed wood of deciduous trees. Most frequently on *Ulmus* spp, but also known on *Acer pseudoplatanus, Betula, Fagus, Fraxinus, Quercus* and *Salix* spp.
D: EurPoly2: 443, NM3: 172
Rarely reported but apparently widespread. Collections from England (Bedfordshire, Berkshire, Herefordshire, Hertfordshire, Middlesex, North Hampshire, North Somerset, Surrey, West Kent, West Norfolk, West Sussex and Yorkshire), Scotland (Easterness) and Wales (Anglesey and Flintshire).

populinus (Schumach.) Donk, *Meded. Bot. Mus. Herb. Rijks Univ. Utrecht* 9: 204 (1933)
  Boletus populinus Schumach., *Enum. pl.* 2: 384 (1803)
  Polyporus populinus (Schumach.) Fr., *Syst. mycol.* 1: 367 (1821)
  Fomes populinus (Schumach.) Cooke, *Grevillea* 14: 20 (1886)
  Rigidoporus populinus (Schumach.) Pouzar, *Folia Geobot. Phytotax.*1: 368 (1966)
  Polyporus connatus Weinm., *Hymen. Gasteromyc.*: 332 (1836)
  Fomes connatus (Weinm.) Gillet, *Hyménomycètes*: 684 (1878)
E: o **S:** o **W:** o **NI:** ! **ROI:** ! **O:** Isle of Man: !
H: Weakly parasitic then often saprotrophic. On wood of deciduous trees, with a preference for young trunks of *Acer pseudoplatanus* where it grows in knotholes and old branch scars. Also known on *Acer campestre, Carpinus, Crataegus* spp, *Fraxinus, Malus* spp, *Populus* spp, *Pyrus communis, Salix* spp, *Sambucus nigra, Sorbus aucuparia* and *Ulmus* spp.
D: EurPoly2: 446, NM3: 172 **D+I:** Ph: 231, B&K2: 302 379
Occasional but widespread.

PANAEOLINA Maire, *Treb. Mus. Ci. Nat. Barcelona* 3(2): 109 (1933)
Agaricales, Coprinaceae
Type: *Panaeolina foenisecii* (Pers.) Maire

foenisecii (Pers.) Maire, *Treb. Mus. Ci. Nat. Barcelona* 3(2): 109 (1933)
  Agaricus foenisecii Pers., *Icon. descr. fung.*: pl. 42 (1800)
  Prunulus foenisecii (Pers.) Gray, *Nat. arr. Brit. pl.* 1: 631 (1821)

  Psilocybe foenisecii (Pers.) Quél., *Mém. Soc. Émul. Montbéliard, Sér. 2,* 5: 147 (1872)
  Psathyra foenisecii (Pers.) G. Bertrand, *Bull. Soc. Mycol. France* 17: 227 (1913)
  Panaeolus foenisecii (Pers.) Kühner, *Botaniste* 17: 187 (1926)
  Psathyrella foenisecii (Pers.) A.H. Sm., *Mem. New York Bot. Gard.* 24: 32 (1972)
E: c **S:** c **W:** c **NI:** c **ROI:** c **O:** Channel Islands: ! Isle of Man: !
H: On soil amongst grass, especially common in lawns and roadside verges, less often in open glades in woodland.
D: BFF5: 80 (as *Panaeolus foenisecii*), NM2: 235 (as *P. foenisecii*) **D+I:** Ph: 182, B&K4: 258 313, C&D: 366 **I:** Bon: 265
Very common and widespread. English name = 'Brown Hay Cap'.

PANAEOLUS (Fr.) Quél. [*nom. cons.*], *Mém. Soc. Émul. Montbéliard, Sér. 2,* 5: 151 (1872)
Agaricales, Strophariaceae
  Coprinarius (Fr.) P. Kumm. [*nom. rej.*], *Führ. Pilzk.*: 68 (1871)
  Anellaria P. Karst., *Bidrag Kännedom Finlands Natur Folk* 32: 27, 517 (1879)
  Copelandia Bres., *Hedwigia* 53: 51 (1913)
Type: *Panaeolus papilionaceus* (Bull.) Quél.
Treatment follows Gerhardt (1996).

acuminatus (Schaeff.) Gillet, *Hyménomycètes*: 621 (1878)
  Agaricus acuminatus Schaeff., *Fung. Bavar. Palat. nasc.* 1: 44 (1762)
  Agaricus caliginosus Jungh., *Linnaea* 5: 405 (1830)
  Panaeolus caliginosus (Jungh.) Gillet, *Hyménomycètes*: 623 (1878)
  Panaeolus rickenii Hora, *Trans. Brit. Mycol. Soc.* 43(2): 454 (1960)
E: c **S:** c **W:** c **NI:** ! **ROI:** o **O:** Channel Islands: ! Isle of Man: !
H: On soil amongst grass in meadows and pastures.
D: BFF5: 87-88 (also as *Panaeolus rickenii*), NM2: 235 **D+I:** Ph: 181 (as *P. rickenii*), B&K4: 256 310 (as *P. caliginosus*), C&D: 366 (also as *P. rickenii*)
Frequent and widespread. Perhaps better known as *Panaeolus rickenii*.

atrobalteatus Pegler & A. Henrici, *Folia Cryptog. Estonica* 33: 105 (1998)
E: !
H: In large troops on flowerbeds mulched with woodchips.
D: Myc14(2): 53
Recently described from London (grounds of Buckingham Palace).

cinctulus (Bolton) Sacc., *Syll. fung.* 5: 1124 (1887)
  Agaricus cinctulus Bolton, *Hist. fung. Halifax* Suppl.: 152 (1791)
  Agaricus fimicola var. *cinctulus* (Bolton) Cooke, *Handb. Brit. fung.,* Edn 2: 221 (1888)
  Panaeolus fimicola var. *cinctulus* (Bolton) Rea, *Brit. basidiomyc.*: 372 (1922)
  Panaeolus dunensis Bon & Courtec., *Doc. Mycol.* 13(50): 28 (1983)
  Agaricus subbalteatus Berk. & Broome, *Ann. Mag. Nat. Hist., Ser. 3 [Notices of British Fungi no. 923]* 7: 378 (1861)
  Panaeolus subbalteatus (Berk. & Broome) Sacc., *Syll. fung.* 5: 1124 (1887)
E: o **S:** o **W:** ! **NI:** ! **O:** Channel Islands: !
H: On manured soil, weathered dung, stable waste, decayed leaf litter, decayed hay and straw, and increasingly on decayed woodchip mulch on flowerbeds.
D: BFF5: 90 (as *P. subbalteatus*), NM2: 235 (as *P. subbalteatus*) **D+I:** Ph: 182 (as *P. subbalteatus*), B&K4: 256 311 **I:** FlDan4: 149H (as *P. subalteatus*)
Occasional but widespread in parks and gardens. Better known as *Panaeolus subbalteatus*.

fimicola (Pers.) Gillet, *Hyménomycètes*: 621 (1878)
  Agaricus fimicola Pers., *Syn. meth. fung.*: 412 (1801)

Panaeolus fimicola var. *ater* J.E. Lange, *Fl. agaric. danic.* 5, *Taxonomic Conspectus*: VI (1940)
Panaeolus ater (J.E. Lange) Bon, *Doc. Mycol.* 16(61): 46 (1985)
Panaeolus obliquoporus Bon, *Doc. Mycol.* 13(50): 28 (1983)
Agaricus varius Bolton, *Hist. fung. Halifax* 2: 66 (1788)
Prunulus varius (Bolton) Gray, *Nat. arr. Brit. pl.* 1: 631 (1821)
E: c **S:** c **W:** c **NI:** ! **ROI:** ! **O:** Isle of Man: !
H: On soil amongst grass, in pastures, meadows, along field margins, in woodland glades, and on lawns.
D: BFF5: 81 (as *Panaeolus ater*), BFF5: 83, NM2: 235 **D+I:** Ph: 182 (as *P. ater*), B&K4: 254 309 (as *P. ater*), C&D: 366 (as *P. ater*)
Common and widespread. Few records from Ireland. Probably better known as *Panaeolus ater*.

olivaceus F.H. Møller, *Fungi Faeroes* 1: 171 (1945)
Mis.: *Panaeolus castaneifolius* sensu auct. Eur.
E: ! **S:** ! **W:** !
H: On soil amongst grass on lawns or in unimproved grassland.
D: BFF5: 84 **D+I:** B&K4: 258 315
Rarely reported but apparently widespread.

papilionaceus (Bull.) Quél., *Mém. Soc. Émul. Montbéliard, Sér. 2*, 5: 152 (1872)
Agaricus papilionaceus Bull., *Herb. France*: pl. 58 (1781)
Agaricus campanulatus Bull. [non *A. campanulatus* Bolton (1788)], *Herb. de. La France*, t.58 (1781)
Panaeolus campanulatus (Bull.) Quél., *Mém. Soc. Émul. Montbéliard, Sér. 2*, 5: 151 (1872)
Agaricus sphinctrinus Fr., *Epicr. syst. mycol.*: 235 (1838)
Panaeolus sphinctrinus (Fr.) Quél., *Mém. Soc. Émul. Montbéliard, Sér. 2*, 5: 151 (1872)
Panaeolus campanulatus var. *sphinctrinus* (Fr.) Quél., *Fl. mycol. France*: 54 (1888)
Agaricus retirugis Fr., *Epicr. syst. mycol.*: 235 (1838)
Panaeolus retirugis (Fr.) Gillet, *Hyménomycètes*: 621 (1878)
E: c **S:** c **W:** c **NI:** c **ROI:** c **O:** Channel Islands: ! Isle of Man: !
H: On weathered dung of herbivores such as horses or cows, also on soil enriched with dung in pastures and along tracks in woodland.
D: BFF5: 85, NM2: 235 (as *P. sphinctrinus*), NM2: 236 **D+I:** Ph: 182 (as *P. sphinctrinus*), B&K4: 260 316 **I:** Bon: 265 (as *P. sphinctrinus*), C&D: 367 (as *P. sphinctrinus*)
Common and widespread. Perhaps better known as *P. sphinctrinus*.

papilionaceus var. **parvisporus** Ew. Gerhardt, *Biblioth. Bot.* 147: 58 (1996)
Mis.: *Panaeolus campanulatus* sensu auct. Brit.
Mis.: *Panaeolus retirugis* sensu auct.
E: c **S:** c **W:** c **NI:** c **ROI:** o **O:** Channel Islands: !
H: On weathered dung of herbivores, especially on horse dung, or occasionally on manured soil in pastures, meadows and other grassland.
D: BFF5: 84 (as *P. campanulatus*), BFF5: 86 (as *P. retirugis*), NM2: 235 (as *P. retirugis*) **D+I:** Ph: 181 (as *P. campanulatus*)
Frequent and widespread. British collections named *P. campanulatus* and *P. retirugis* have substantially smaller spores than typical *P. papilionaceus* and agree better with var. *parvisporus*.

semiovatus (Sowerby) S. Lundell, *Fungi Exsiccati Suecici* 11-12: 537 (1938)
Agaricus semiovatus Sowerby, *Col. fig. Engl. fung.* 2: pl. 131 (1798)
Anellaria semiovata (Sowerby) A. Pearson & Dennis, *Trans. Brit. Mycol. Soc.* 31(3/4): 185 (1948)
Agaricus ciliaris Bolton, *Hist. fung. Halifax* 2: 53 (1788)
Agaricus separatus L., *Sp. pl.*: 1175 (1753)
Panaeolus separatus (L.) Gillet, *Hyménomycètes*: 620 (1878)
Anellaria separata (L.) P. Karst., *Bidrag Kännedom Finlands Natur Folk* 32: 517 (1879)
Anellaria separata var. *minor* Sacc., *Syll. fung.* 5: 1126 (1887)
Mis.: *Panaeolus fimiputris* sensu auct. mult.

Mis.: *Anellaria fimiputris* sensu auct. mult.
E: c **S:** c **W:** c **NI:** c **ROI:** o **O:** Channel Islands: ! Isle of Man: !
H: On weathered dung of herbivores, usually horse and cow dung, especially in upland or montane habitat.
D: BFF5: 80, NM2: 234 (as *Panaeolus fimiputris*) **D+I:** Ph: 180, B&K4: 260 318, C&D: 366 (as *Anellaria semiovata*) **I:** Sow2: 131 (as *Agaricus semiovatus*), FlDan4: 142F (as *Stropharia separata*), Bon: 265
Common and widespread. Basidiomes are variable in size, dependent on the amount of nutrient present in the substratum.

semiovatus var. **phalaenarum** (Fr.) Ew. Gerhardt, *Biblioth. Bot.* 147: 24 (1996)
Agaricus phalaenarum Fr., *Epicr. syst. mycol.*: 235 (1838)
Panaeolus phalaenarum (Fr.) Quél., *Mém. Soc. Émul. Montbéliard, Sér. 2*, 5: 151 (1872)
Agaricus egregius Massee, *Grevillea* 13: 91 (1885)
Panaeolus egregius (Massee) Sacc., *Syll. fung.* 5: 1119 (1887)
Mis.: *Panaeolus antillarum* sensu BFF5
E: ! **S:** !
H: On weathered dung of herbivores, usually sheep or horse.
D: BFF5: 82 (as *P. antillarum*) **I:** Bon: 265 (as *P. phalaenarum*)
Rarely reported.

subfirmus P. Karst., *Hedwigia* 28: 365 (1889)
Panaeolus speciosus P.D. Orton, *Notes Roy. Bot. Gard. Edinburgh* 29: 108 (1969)
E: ! **S:** ! **W:** ! **NI:** ! **ROI:** !
H: On weathered dung of herbivores, usually of sheep or horses, in grassy areas such as upland pasture or moorland.
D: BFF5: 89 (as *P. speciosus*) **D+I:** Ph: 181 (as *P. speciosus*)
Rarely reported but apparently widespread. Most records are from the north and west, but the majority of reports are unsubstantiated with voucher material.

PANELLUS P. Karst., *Bidrag Kännedom Finlands Natur Folk* 14: 96 (1879)
Agaricales, Tricholomataceae
Tectella Earle, *Bull. New York Bot. Gard.* 5: 433 (1909)
Type: *Panellus stipticus* (Bull.) P. Karst.

mitis (Pers.) Singer, *Ann. Mycol.* 34: 334 (1936)
Agaricus mitis Pers., *Observ. mycol.* 1: 54 (1796)
Pleurotus mitis (Pers.) Quél., *Mém. Soc. Émul. Montbéliard, Sér. 2*, 5: 245 (1872)
E: o **S:** o **ROI:** !
H: On fallen wood of conifers such *Larix*, *Picea* and *Pinus* spp. Usually on brash or small fallen trunks and branches.
D: BFF6: 59, NM2: 175, FAN3: 169 **D+I:** Ph: 188-189, B&K3: 308 389 **I:** FlDan2: 65G (as *Pleurotus mitis*), C&D: 149
Occasional but widespread. Strictly confined to conifers, so records on deciduous wood are misdetermined.

ringens (Fr.) Romagn., *Bull. Soc. Mycol. France* 61: 38 (1945)
Agaricus ringens Fr., *Elench. fung.* 1: 25 (1828)
Lentinus ringens (Fr.) Fr., *Synopsis generis Llentinorum*: 14 (1836)
Panus ringens (Fr.) Fr., *Hymenomyc. eur.*: 490 (1874)
S: !
H: On wood. British material on fallen twigs of *Prunus* sp.
D: BFF6: 58, NM2: 175 **D+I:** B&K3: 308 390 **I:** C&D: 149
A single collection in herb. K from Morayshire (Forres) in 1887, previously determined as *Panus patellaris*.

serotinus (Pers.) Kühner, *Compt. Rend. Hebd. Séances Acad. Sci.* 230: 1889 (1950)
Agaricus serotinus Pers., in Hoffmann, *Abb. Schwaemme* 3: 4 pl. 30 (1793)
Pleurotus serotinus (Pers.) P. Kumm., *Führ. Pilzk.*: 104 (1871)
Acanthocystis serotina (Pers.) Konrad & Maubl., *Icon. select. fung.* 6: 364 (1937)
Hohenbuehelia serotina (Pers.) Singer, *Lilloa* 22: 254 (1951) [1949]

Sarcomyxa serotina (Pers.) P. Karst., *Meddn Soc. Fauna Fl. Fenn.* 18: 62 (1891)
Agaricus almeni Fr., *Hymenomyc. eur.*: 176 (1874)
Pleurotus serotinus var. *almeni* (Fr.) Bigeard & H. Guill., *Fl. Champ. Supér. France* 2: 120 (1913)
Pleurotus serotinus var. *flaccidus* J.E. Lange, *Dansk. Bot. Ark.* 6(5): 27 (1930)
E: c **S:** ! **W:** ! **NI:** o
H: On decayed wood in deciduous woodland, usually on large fallen trunks or branches of *Fagus* and *Betula* spp. Also known on *Alnus*, *Fraxinus*, *Quercus*, *Salix*, *Sambucus nigra* and *Ulmus* spp. Reported on conifer wood but unsubstantiated with voucher material. The majority of collections are made between late November and February.
D: BFF6: 59-60, NM2: 176, FAN3: 169-170 **D+I:** Ph: 188-189, B&K3: 318 403 (as *Sarcomyxa serotina*), Myc14(3): 132 **I:** FlDan2: 64A (as *Pleurotus serotinus*), Bon: 125, C&D: 125, Kriegl3: 515 (as *S. serotina*)
Rather common in England, less so elsewhere but widespread.

stipticus (Bull.) P. Karst., *Bidrag Kännedom Finlands Natur Folk* 32: 96 (1879)
Agaricus stipticus Bull., *Herb. France*: pl. 140 (1783)
Crepidopus stypticus (Bull.) Gray, *Nat. arr. Brit. pl.* 1: 616 (1821)
Panus stipticus (Bull.) Fr., *Epicr. syst. mycol.*: 399 (1838)
Pleurotus stipticus (Bull.) P. Kumm., *Führ. Pilzk.*: 105 (1871)
Agaricus semipetiolatus Lightf., *Fl. scot.*: 1030 (1777)
Panus farinaceus var. *albidotomentosus* Cooke & Massee, *Grevillea* 15: 107 (1886)
Panus stipticus var. *albidotomentosus* (Cooke & Massee) Rea, *Brit. basidiomyc.*: 536 (1922)
Mis.: *Agaricus betulinus* sensu Bolton (Hist. Fung. Halifax Vol. 2 pl. 72)
Mis.: *Agaricus flabelliformis* sensu. Sowerby [Col. fig. Engl. fung. 1: pl. 109 (1797)]
E: c **S:** c **W:** c **NI:** c **ROI:** c **O:** Channel Islands: ! Isle of Man: !
H: Saprotrophic on deciduous trees, usually on dead wood but occasionally in wounds on living trunks. Often on *Quercus* spp, but also known on *Acer pseudoplatanus*, *A. palmatum*, *Alnus*, *Betula*, *Fagus*, *Fraxinus* and *Sambucus nigra*. Reported on conifers but unsubstantiated with voucher material.
D: BFF6: 58-59, NM2: 176, FAN3: 168-169 **D+I:** Ph: 188-189, B&K3: 310 391 **I:** FlDan2: 67A, Bon: 125, Kriegl3: 494
Very common and widespread.

PANUS Fr. [*nom. cons.*], *Epicr. syst. mycol.*: 396 (1838)
Poriales, Lentinaceae
Pleuropus (Pers.) Gray [*nom. rej.*], *Nat. arr. Brit. pl.* 1: 615 (1821)
Type: *Panus conchatus* (Bull.) Fr.

conchatus (Bull.) Fr., *Epicr. syst. mycol.*: 398 (1838)
Agaricus conchatus Bull., *Herb. France*: pl. 298 (1786)
Lentinus conchatus (Bull.) J. Schröt., in Cohn, *Krypt.-Fl. Schlesien* 3: 555 (1889)
Agaricus carneotomentosus Batsch, *Elench. fung.*: 89 t. 8 f. 33 (1783)
Agaricus carnosus Bolton [*nom. illegit.*, non *A. carnosus* Curtis (1777)], *Hist. fung. Halifax* Suppl.: 146 (1791)
Panus flabelliformis (Schaeff.) Quél., *Fl. mycol. France*: 325 (1888)
Agaricus fornicatus Pers., *Syn. meth. fung.*: 475 (1801)
Pleuropus fornicatus (Pers.) Gray, *Nat. arr. Brit. pl.* 1: 615 (1821)
Agaricus inconstans Pers., *Syn. meth. fung.*: 475 (1801)
Pleuropus inconstans (Pers.) Gray, *Nat. arr. Brit. pl.* 1: 616 (1821)
Agaricus torulosus Pers., *Syn. meth. fung.*: 475 (1801)
Panus torulosus (Pers.) Fr., *Epicr. syst. mycol.*: 397 (1838)
Lentinus torulosus (Pers.) Lloyd, *Mycol. not.* 4, *Letter* 47: 13 (1913)
Mis.: *Panus hirtus* sensu Rea (1922), sensu auct.
E: o **S:** o **W:** o **NI:** ! **ROI:** !

H: On decayed wood in woodland or parkland, usually on large stumps or fallen branches of deciduous trees. Most often reported on *Betula* spp, *Fagus* and *Fraxinus* but also known on *Acer pseudoplatanus*, *Aesculus*, *Populus tremula*, *Quercus* and *Salix* spp.
D: BFF6: 15-16 (as *Lentinus torulosus*), NM2: 47 (as *L. conchatus*) **D+I:** Ph: 186-187 (as *Panus torulosus*), FAN2: 27 (as *L. conchatus*), B&K3: 210 241 (as *L. torulosus*) **I:** FlDan2: 67D, Kriegl3: 21 (as *L. torulosus*)
Occasional but widespread. Young basidiomes are often a dark lilac-purplish but lose this with age. When older they may resemble old specimens of *Pleurotus ostreatus* and be passed over as such. Perhaps better known as *Lentinus torulosus*.

PARVOBASIDIUM Jülich, *Persoonia* 8: 302 (1975)
Stereales, Corticiaceae
Type: *Parvobasidium cretatum* (Bourdot & Galzin) Jülich

cretatum (Bourdot & Galzin) Jülich, *Persoonia* 8(3): 302 (1975)
Gloeocystidium cretatum Bourdot & Galzin, *Bull. Soc. Mycol. France* 28(4): 371 (1913)
E: ! **W:** ! **NI:** ! **ROI:** !
H: On dead stems and debris of ferns such as *Dryopteris* spp, and *Polystichum setiferum*.
D+I: Myc9(4): 163
Rarely reported but apparently widespread. Known from England (Shropshire, South Devon and West Sussex), Wales (Breconshire, Carmarthenshire and Monmouthshire), Northern Ireland (Derry) and the Republic of Ireland (Clare).

PAULLICORTICIUM J. Erikss., *Symb. Bot. Upsal.* 16(1): 66 (1958)
Stereales, Sistotremataceae
Type: *Paullicorticium pearsonii* (Bourdot) J. Erikss.

delicatissimum (H.S. Jacks.) Liberta, *Brittonia* 14(2): 222 (1962)
Corticium delicatissimum H.S. Jacks., *Canad. J. Res., Sect. C, Bot. Sci.* 28: 722 (1950)
E: ! **W:** !
H: British collections on decayed wood of conifers such as *Picea* spp and *Taxus* and one on decayed wood of *Salix* sp.
D: NM3: 125 **D+I:** CNE5: 914
Rarely reported. Known from England (South Devon and Surrey) and Wales (Monmouthshire). Basidiomes are thin, evanescent and easily overlooked.

pearsonii (Bourdot) J. Erikss., *Symb. Bot. Upsal.* 16(1): 67 (1958)
Corticium pearsonii Bourdot, *Trans. Brit. Mycol. Soc.* 7(1 - 2): 52 (1921)
Ceratobasidium pearsonii (Bourdot) M.P. Christ., *Dansk. Bot. Ark.* 19(1): 46 (1959)
E: ! **W:** ?
H: On decayed wood of conifers such as *Cedrus*, *Picea* and *Pinus* spp.
D: NM3: 125 **D+I:** CNE5: 915
Known from England (Buckinghamshire, Cambridgeshire, South Devon, South Wiltshire, Surrey, and Worcestershire) and an old record from Wales (Monmouthshire) in 1925, unsubstantiated with voucher material.

PAXILLUS Fr., *Fl. Scan.*: 339 (1835)
Boletales, Paxillaceae
Type: *Paxillus involutus* (Batsch) Fr.

involutus (Batsch) Fr., *Epicr. syst. mycol.*: 317 (1838)
Agaricus involutus Batsch, *Elench. fung.* (*Continuatio Prima*): 39 & t. 13 f. 61 (1786)
Omphalia involuta (Batsch) Gray, *Nat. arr. Brit. pl.* 1: 611 (1821)
Agaricus adscendibus Bolton, *Hist. fung. Halifax* 2: 55 (1788)
Agaricus contiguus Bull., *Herb. France*: pl. 240 (1785)
Paxillus involutus var. *excentricus* Massee, *Brit. fung.-fl.* 2: 10 (1893)

E: c **S:** c **W:** c **NI:** c **ROI:** c **O:** Channel Islands: !
H: Most frequently associated with *Betula* spp, in deciduous woodland or on heathland. Also with *Castanea*, *Fagus* or *Quercus* spp, and with conifers such as *Larix*, *Picea* and *Pinus* spp. Also known from downland on calcareous soil, associated with *Helianthemum nummularium*.
D: NM2: 54 **D+I:** Ph: 142, B&K3: 92 64 **I:** Sow1: 56 (as *Agaricus contiguus*), FlDan4: 134D, Bon: 51
Very common and widespread.

rubicundulus P.D. Orton, *Notes Roy. Bot. Gard. Edinburgh* 29: 110 (1969)
    Mis.: *Paxillus filamentosus* sensu auct.
E: ! **S:** c **W:** ! **NI:** !
H: On soil, associated with *Alnus glutinosa* on wet or swampy ground.
D+I: Myc2(1): 35, B&K3: 92 65 **I:** Kriegl2: 345 (as *Paxillus filamentosus*)
Common in Scotland. Less frequently reported southwards.

PELLIDISCUS Donk, *Persoonia* 1: 89 (1959)
Cortinariales, Crepidotaceae
Type: *Pellidiscus pallidus* (Berk. & Broome) Donk

pallidus (Berk. & Broome) Donk, *Persoonia* 1(1): 90 (1959)
    Cyphella pallida Berk. & Broome, in Rabenhorst, *Fungi europaei* IV: 1415 (1871)
    Cyphella bloxamii Berk. & W. Phillips, *Ann. Mag. Nat. Hist.*, Ser. 5 [Notices of British Fungi no. 1894] 7: 129 (1881)
E: ! **S:** ! **W:** !
H: In mixed deciduous woodland, on fallen and decayed leaves, often on *Ilex* or *Fagus* and *Quercus* spp. Also on dead branches of *Cytisus scoparius* and *Ulex europaeus*, on dead lianas of *Clematis vitalba* and on old stems of *Urtica dioica*. Rarely reported on decayed wood or needle litter of conifers such as *Pinus* spp.
D: NM2: 338
Rarely reported but apparently widespread.

PENIOPHORA Cooke, *Grevillea* 8: 20 (1879)
Stereales, Peniophoraceae
    Sterellum P. Karst., *Bidrag Kännedom Finlands Natur Folk* 48: 405 (1889)
Type: *Peniophora quercina* (Pers.) Cooke

boidinii D.A. Reid, *Revta Biol., Lisboa* 5(1 - 2): 146 (1965)
    Mis.: *Peniophora versicolor* sensu auct.
E: ! **S:** ! **NI:** ! **ROI:** ! **O:** Isle of Man: !
H: On dead wood of deciduous trees or shrubs such as *Buxus*, *Crataegus monogyna*, *Fraxinus*, *Rosa canina*, *Sambucus nigra* and *Tilia* spp.
Rarely reported but apparently widespread.

cinerea (Pers.) Cooke, *Grevillea* 8: 20 (1879)
    Corticium cinereum Pers., *Neues Mag. Bot.* 1: 111 (1794)
    Thelephora cinerea (Pers.) Pers., *Syn. meth. fung.*: 579 (1801)
E: c **S:** c **W:** c **NI:** c **ROI:** c **O:** Isle of Man: !
H: On dead wood of deciduous trees and shrubs. Most frequent on *Acer pseudoplatanus* and *Fagus* but also known on *Hedera*, *Ligustrum vulgare* and *Syringa vulgaris*. Rarely on conifers such as *Juniperus communis* and *Pinus sylvestris*.
D: NM3: 188 **D+I:** CNE5: 935-939, B&K2: 152 156
Very common and widespread.

erikssonii Boidin, *Bull. Soc. Hist. Nat. Toulouse* 92: 286 (1957)
    Mis.: *Peniophora aurantiaca* sensu auct. Brit.
E: ! **S:** ! **W:** !
H: On dead attached branches of *Alnus*. Also reported on *Ulmus montana*.
D: NM3: 186 **D+I:** CNE5: 923-925
Rarely reported but apparently widespread. An alder associate, so the record on elm from Scotland (Isle of Skye) needs reinvestigating.

incarnata (Pers.) P. Karst., *Hedwigia* 28: 27 (1889)

Thelephora incarnata Pers., *Syn. meth. fung.*: 573 (1801)
    Corticium incarnatum (Pers.) Fr., *Epicr. syst. mycol.*: 564 (1838)
E: c **S:** c **W:** c **NI:** o **ROI:** ! **O:** Channel Islands: o Isle of Man: o
H: On dead wood of deciduous (rarely coniferous) trees and shrubs, usually small attached branches or twigs. Common on *Ulex europaeus* and *Hedera*. Rarely reported on *Myrica gale*, *Prunus laurocerasus*, *P. padus*, *Rosa canina* and *Cytisus scoparius*. Single collections from a decayed cone of *Pinus sylvestris* and old dead stems of *Fallopia japonica*.
D: NM3: 186 **D+I:** CNE5: 926-928, Ph: 240-241, B&K2: 146 147 **I:** Kriegl1: 271
Common and widespread.

laeta (Fr.) Donk, *Fungus* 27: 17 (1957)
    Radulum laetum Fr., *Elench. fung.* 1: 152 (1828)
    Thelephora hydnoidea (Pers.) Pers., *Syn. meth. fung.*: 576 (1801)
    Peniophora incarnata var. *hydnoidea* (Pers.) Bourdot & Galzin, *Hyménomyc. France*: 332 (1928)
E: ! **S:** ! **O:** Channel Islands: !
H: On dead attached branches of *Carpinus* and apparently restricted to this host.
D: NM3: 186 (as *Peniophora hydnoidea*) **D+I:** CNE5: 930-933 **I:** Kriegl1: 272
Rarely reported. Known from England (Bedfordshire, Cheshire, Cumberland, Hertfordshire, Middlesex, South Essex, Staffordshire, Surrey and West Kent), Scotland (Mid-Lothian) and the Channel Islands (Jersey).

lilacea Bourdot & Galzin, *Bull. Soc. Mycol. France* 28(4): 403 (1912)
E: !
H: On dead attached twigs and branches of *Ulmus* spp.
D: Myc8(3): 118, NM3: 187 **D+I:** CNE5: 942-945
Known from Berkshire (Windsor Great Park, and near Ascot), North Hampshire (North Wanborough Common) and South Devon (Slapton).

limitata (Chaillet) Cooke, *Grevillea* 8: 21 (1879)
    Thelephora limitata Chaillet, in Fries, *Elench. fung.* 1: 222 (1828)
    Corticium limitatum (Chaillet) Fr., *Epicr. syst. mycol.*: 565 (1838)
    Thelephora fraxinea Pers., *Mycol. eur.* 1: 145 (1822)
    Peniophora fraxinea (Pers.) S. Lundell, *Fungi Exsiccati Suecici* 1(1 - 2): 29 (1934)
E: c **S:** c **W:** c **NI:** ! **ROI:** c **O:** Isle of Man: c
H: On dead branches of *Fraxinus* in deciduous woodland. Virtually restricted to that host, but there is also a single collection on old stems of *Rosa canina*.
D: NM3: 188 **D+I:** CNE5: 947-949, B&K2: 148 148 **I:** Kriegl1: 276
Common and widespread.

lycii (Pers.) Höhn. & Litsch., *Sitzungsber. Kaiserl. Akad. Wiss., Wien, Math.-Naturwiss. Cl., Abt. 1* 1(116): 747 (1907)
    Thelephora lycii Pers., *Mycol. eur.* 1: 148 (1822)
    Corticium lycii (Pers.) Cooke, *Grevillea* 9: 97 (1881)
    Corticium caesium Bres. [nom. illegit., non *C. caesium* Pers. (1796)], *Fungi Trid.* 2: 145 (1892)
    Peniophora caesia (Bres.) Bourdot & Galzin, *Bull. Soc. Mycol. France* 28(4): 406 (1913)
E: c **S:** c **W:** c **NI:** c **ROI:** c **O:** Channel Islands: c Isle of Man: c
H: On dead wood of deciduous trees, often on small attached branches but also on fallen sticks etc. Most often reported on *Acer pseudoplatanus* but also frequent on *Acer campestre*, *Euonymus europaeus*, *Fagus*, *Fraxinus*, *Hedera*, *Rhamnus cathartica* and *Ulex europaeus*. Rarely on conifers such as *Pinus sylvestris*.
D: NM3: 187 **D+I:** CNE5: 950-952
Very common and widespread.

nuda (Fr.) Bres., *Atti Imp. Regia Accad. Roveretana, ser. 3, 3*: 114 (1897)
    Thelephora nuda Fr., *Syst. mycol.* 1: 447 (1821)
    Corticium nudum (Fr.) Fr., *Epicr. syst. mycol.*: 564 (1838)

E: ! W: ! NI: !
H: On dead attached branches of deciduous trees. Known on *Acer pseudoplatanus, Aesculus, Alnus, Corylus, Populus tremula* and *Syringa vulgaris*. Rarely on conifers such as *Taxus*.
D+I: CNE5: 953-955, B&K2: 148 149
Occasional but widespread, generally rather rare. The majority of records are from southern England.

pini (Schleich. & DC.) Boidin, *Rev. Mycol. (Paris)* 21: 123 (1956)
Thelephora pini Schleich. & DC., *Fl. Fr.* 5: 31 (1815)
Stereum pini (Schleich. & DC.) Fr., *Epicr. syst. mycol.*: 553 (1838)
Sterellum pini (Schleich. & DC.) P. Karst., *Bidrag Kännedom Finlands Natur Folk* 48: 405 (1889)
E: ! S: !
H: Usually on dead attached branches and twigs of *Pinus* spp, most frequently *P. sylvestris*.
D: NM3: 186 D+I: CNE5: 959-961, B&K2: 148 150
Known from England (South Hampshire, Surrey, West Suffolk and West Sussex) and Scotland (Angus, Easterness, Morayshire, South Aberdeen and West Sutherland).

polygonia (Pers.) Bourdot & Galzin, *Hyménomyc. France*: 328 (1928)
Corticium polygonium Pers., *Neues Mag. Bot.* 1: 110 (1794)
Thelephora polygonia (Pers.) Pers., *Syn. meth. fung.*: 574 (1801)
Thelephora maculiformis Fr., *Observ. mycol.* 1: 150 (1815)
Corticium maculiforme (Fr.) Fr., *Hymenomyc. eur.*: 656 (1874)
Peniophora maculiformis (Fr.) Bourdot & Galzin, *Hyménomyc. France*: 324 (1928)
Peniophora nuda var. *maculiformis* (Fr.) Rea, *Brit. basidiomyc.*: 695 (1922)
E: o S: o W: !
H: On dead wood, usually attached branches of *Populus alba* and *Populus tremula*. Occasionally on *Betula pendula, Quercus robur* and *Salix* spp.
D: NM3: 187 D+I: CNE5: 965-967, B&K2: 150 152
Occasional but widespread, and may be locally common.

proxima Bres., *Bull. Soc. Mycol. France* 28(4): 402 (1912)
E: !
H: On living and dead trunks of *Buxus* and restricted to this host.
Rarely reported but locally frequent where the host is native. Known from Berkshire, Buckinghamshire, Gloucestershire, North Wiltshire, and Surrey.

quercina (Pers.) Cooke, *Grevillea* 8: 20 (1879)
Thelephora quercina Pers., *Syn. meth. fung.*: 573 (1801)
Corticium quercinum (Pers.) Fr., *Epicr. syst. mycol.*: 563 (1838)
Auricularia corticalis Bull., *Herb. France*: pl. 436 (1786)
Peniophora pezizoides Massee, *J. Linn. Soc., Bot.* 25: 141 (1890)
E: c S: c W: c NI: c ROI: c O: Channel Islands: ! Isle of Man: !
H: On dead attached branches, usually of *Quercus* spp, rarely on *Fagus* and rarely on *Betula* spp.
D: NM3: 188 D+I: CNE5: 968-971, Ph: 240-241, B&K2: 150 153 I: Kriegl1: 281
Very common and widespread.

reidii Boidin & Lanq., *Trans. Brit. Mycol. Soc.* 81(2): 279 (1983)
E: !
H: On dead attached wood of deciduous trees. British material on *Castanea*.
One collection from East Norfolk (Sheringham) in herb. K.

rufa (Fr.) Boidin, *Bull. Soc. Mycol. France* 74(4): 443 (1958)
Thelephora rufa Fr., *Elench. fung.* 1: 187 (1828)
Stereum rufum (Fr.) Fr., *Epicr. syst. mycol.*: 552 (1838)
Sterellum rufum (Fr.) J. Erikss., *Symb. Bot. Upsal.* 16(1): 128 (1958)
E: !
H: On decayed wood of deciduous trees.

D: NM3: 187 D+I: CNE5: 971-975
A single collection from Surrey (Esher, Arbrook Common) in 1999.

rufomarginata (Pers.) Bourdot & Galzin, *Bull. Soc. Mycol. France* 28: 408 (1912)
Thelephora rufomarginata Pers., *Mycol. eur.* 1: 124 (1822)
Stereum rufomarginatum (Pers.) Quél., *Fl. mycol. France*: 13 (1888)
E: o S: o NI: o
H: Usually on dead attached branches and twigs of *Tilia* spp. Rarely on *Alnus* and *Salix* spp.
D: NM3: 188 D+I: CNE5: 976-979, B&K2: 152 154
Occasional but apparently widespread and may be locally frequent.

tamaricicola Boidin & Malençon, *Rev. Mycol. (Paris)* 26(2): 153 (1961)
E: !
H: Weakly parasitic then saprotrophic. On living and dead branches of *Tamarix gallica* in coastal areas.
Rarely reported but may be locally abundant. Known from South Hampshire and West Sussex.

versicolor (Bres.) Sacc. & Syd., *Syll. fung.* 16: 193 (1902)
Corticium versicolor Bres., *Fungi trident.* 2(11-13): 61 (1898)
Peniophora incarnata ssp. *versicolor* (Bres.) Bourdot & Galzin, *Hyménomyc. France*: 322 (1928)
E: ! NI: ?
H: On dead attached twigs. British material on *Prunus domestica*.
A single collection, from East Norfolk in 1950. Recorded from North Somerset and from Northern Ireland, but unsubstantiated with voucher material.

violaceolivida (Sommerf.) Massee, *J. Linn. Soc., Bot.* 25: 152 (1890)
Thelephora violaceolivida Sommerf., *Suppl. Fl. lapp.*: 283 (1826)
Corticium violaceolividum (Sommerf.) Fr., *Epicr. syst. mycol.*: 564 (1838)
Peniophora nuda ssp. *violaceolivida* (Sommerf.) Bourdot & Galzin, *Hyménomyc. France*: 324 (1928)
E: ! S: ! W: !
H: On wood and bark of deciduous trees and shrubs. Known on *Fraxinus, Fuchsia magellanica, Hedera, Populus tremula, Prunus avium, Rubus fruticosus, Salix* spp, and *Ulex europaeus*.
D: NM3: 188 D+I: CNE5: 984-986, B&K2: 152 155
Rarely reported. Apparently widespread but most often reported from Scotland.

PERENNIPORIA Murrill, *Mycologia* 34: 595 (1942)
Poriales, Coriolaceae
Poria Pers. [*nom. illegit.*, non *Poria* Adans. (1763)], *Neues Mag. Bot.* 1: 109 (1794)
Physisporus Chevall. [*nom. illegit.*], *Fl. gén. env. Paris* 1: 261 (1826)
Type: *Perenniporia unita* (Pers.) Murrill

fraxinea (Bull.) Ryvarden, *Polyporaceae of North Europe* 2: 307 (1978)
Boletus fraxineus Bull., *Herb. France*: pl. 433 (1789)
Fomes fraxineus (Bull.) Fr., *Syst. mycol.* 1: 374 (1821)
Polyporus fraxineus (Bull.) Fr., *Syst. mycol.* 1: 374 (1821)
Vanderbylia fraxinea (Bull.) D.A. Reid, *S. African J. Bot.* 39(2): 166 (1973)
Fomes cytisinus Berk., *Engl. fl.* 5(2): 142 (1836)
Fomitopsis cytisina (Berk.) Bondartsev & Singer, *Ann. Mycol.* 39: 55 (1941)
E: o S: ! O: Channel Islands: !
H: Weakly parasitic then saprotrophic. On wood, usually at the base of living deciduous trees, most frequently *Fraxinus* and less often *Fagus, Juglans regia, Platanus* spp, *Populus canescens, P. monilifera, Quercus* spp, *Robinia pseudoacacia, Salix* spp, *Sophora japonica, Sorbus, Tilia* and *Ulmus* spp.

D: EurPoly2: 453, NM3: 244 **D+I:** B&K2: 296 370
Widespread in England, but with a southern bias. A single old collection by Berkeley from Scotland (Kincardineshire) and also known from the Channel Islands (Jersey).

medulla-panis (Jacq.) Donk, *Persoonia* 5: 76 (1967)
Boletus medulla-panis Jacq., *Misc. Austriaca Bot.* 1: 141 (1778)
Polyporus medulla-panis (Jacq.) Fr., *Syst. mycol.* 1: 380 (1821)
Poria medulla-panis (Jacq.) Cooke, *Grevillea* 14: 109 (1886)
E: ! **S:** ? **W:** ?
H: On decayed wood of deciduous trees such as *Castanea* and *Salix* spp, and occasionally reported from old structural timbers in buildings.
D: EurPoly2: 457, NM3: 244 **D+I:** B&K2: 294 369
Known from Hertfordshire, Northamptonshire, North Wiltshire, Nottinghamshire and West Kent. Reported from Scotland (Easterness) in 1927 and Wales (Denbighshire) in 1910 but unsubstantiated with voucher material.

ochroleuca (Berk.) Ryvarden, *Norweg. J. Bot.* 19: 233 (1972)
Polyporus ochroleucus Berk., *London J. Bot.* 4: 53 (1845)
E: ! **W:** ! **O:** Channel Islands: !
H: British material from living trunks of *Malus* sp. and *Quercus ilex*, attached and fallen twigs of *Prunus spinosa*.
D: EurPoly2: 459, Myc15(1): 38
Known from England (East and West Cornwall, South Devon), Wales (Pembrokeshire) and the Channel Islands (Guernsey and Jersey). Frequent in the Mediterranean region and probably at the northern limit of its range here.

PHAEOCOLLYBIA R. Heim, *Encycl. Mycol.* 1, *Le Genre* Inocybe: 70 (1931)
Cortinariales, Cortinariaceae
Type: *Phaeocollybia lugubris* (Fr.) R. Heim

arduennensis Bon, *Doc. Mycol.* 9(35): 42 (1979)
Mis.: *Naucoria jennyae* sensu Pearson [TBMS 35: 112 (1952)] and sensu Lange (FlDan4: 16 & pl. 123A)
Mis.: *Phaeocollybia jennyae* sensu NCL p.p.
E: ? **S:** !
H: On soil in deciduous or conifer woodland
D: NM2: 334, BFF7: 76 **D+I:** B&K5: 306 398 **I:** Bon: 247, C&D: 345
Known from Easterness. Reported from England (as *P. jennyae*) but no voucher material has been traced.

festiva (Fr.) R. Heim, *Encycl. Mycol.* 1. *Le Genre* Inocybe: 70 (1931)
Agaricus festivus Fr., *Epicr. syst. mycol.*: 192 (1838)
Naucoria festiva (Fr.) Bres., *Fungi trident.* 1: 19 (1882)
E: ? **S:** ? **W:** !
H: On soil in acid conifer woods.
D: BFF7: 77 **D+I:** B&K5: 308 400
Known in Britain from a single Welsh collection (Caernarvonshire: Betws-y-Coed) in herb K. Reports from England and Scotland in BFF7 are unsubstantiated with voucher material. Listed by Rea (1922) based on Cooke [170 (966) Vol. 8] from Surrey (Carshalton), but Cooke's plate shows spores too large for any known *Phaeocollybia* sp.

lugubris (Fr.) R. Heim, *Encycl. Mycol.* 1, *Le Genre* Inocybe: 71 (1931)
Agaricus lugubris Fr., *Syst. mycol.* 1: 254 (1821)
Naucoria lugubris (Fr.) Sacc., *Syll. fung.* 5: 828 (1887)
S: ! **W:** ?
H: On soil in conifer woodland.
D: Bon: 247, NM2: 334, BFF7: 78-79 **D+I:** B&K5: 308 402 **I:** C&D: 345
Known from Morayshire. Reported from Wales but unsubstantiated with voucher material.

PHAEOGALERA Kühner, *Bull. Soc. Mycol. France* 88: 151 (1973)
Agaricales, Strophariaceae
Type: *Phaeogalera stagnina* (Fr.) Pegler & T.W.K. Young

dissimulans (Berk. & Broome) Holec, *Sydowia* 55(1): 83 (2003)
Agaricus dissimulans Berk. & Broome, *Ann. Mag. Nat. Hist., Ser. 5 [Notices of British Fungi no. 1940]* 9: 178 (1882)
Pholiota dissimulans (Berk. & Broome) Sacc., *Syll. fung.* 5: 757 (1887)
Agaricus oedipus Cooke, *Grevillea* 14: 1 (1885)
Psathyrella oedipus (Cooke) Konrad & Maubl., *Encyclop. Mycol.* 14. *Les Agaricales* I: 128 (1948)
Hypholoma oedipus (Cooke) Sacc., *Syll. fung.* 5: 1033 (1887)
Pholiota oedipus (Cooke) P.D. Orton, *Trans. Brit. Mycol. Soc.* 43(2): 180 (1960)
Phaeogalera oedipus (Cooke) Romagn., *Bull. Soc. Mycol. France* 96(3): 251 (1980)
Hemipholiota oedipus (Cooke) Bon, *Doc. mycol.* 17(65): 52 (1986)
Dryophila sordida Kühner [*nom. nud.*], *Flore Analytique des Champignons Supérieurs*: 329 (1953)
E: o
H: On fallen, and usually decayed, deciduous leaves or twigs, usually of *Fraxinus* and less often *Alnus*, *Corylus*, *Crataegus*, *Populus tremula* and *Ulmus* spp. Most collections between November and early March, exceptionally later.
I: Cooke 400 (371) Vol. 3 (1885) [as *Agaricus (Pholiota) dissimulans*]; Cooke 579 (587)A Vol. 4 (1885)
A winter-fruiting species, known from Berkshire, Buckinghamshire, Dorset, Leicestershire, Middlesex, North Hampshire, Oxfordshire, South Wiltshire, Staffordshire, Surrey, Warwickshire, West Gloucestershire and West Sussex. Holec [Sydowia 55: 79-85 (2003)] argues that *Agaricus oedipus* is distinct, possibly a *Psathyrella* sp., but in the absence of authentic material this is debatable, especially since Cooke's plate shows material collected in February.

stagnina (Fr.) Pegler & T.W.K. Young, *Kew Bull.* 30: 228 (1975)
Agaricus stagninus Fr., *Syst. mycol.* 1: 268 (1821)
Galera stagnina (Fr.) P. Kumm., *Führ. Pilzk.*: 75 (1871)
Tubaria stagnina (Fr.) Gillet, *Hyménomycètes*: 539 (1876)
Galerina stagnina (Fr.) Kühner, *Encycl. Mycol.* 7, *Le Genre* Galera: 187 (1935)
Naucoria stagnina (Fr.) P.D. Orton, *Trans. Brit. Mycol. Soc.* 43(2): 311 (1960)
Naucoria stagninoides P.D. Orton, *Trans. Brit. Mycol. Soc.* 43(2): 321 (1960)
Phaeogalera stagninoides (P.D. Orton) Pegler & T.W.K. Young, *Kew Bull.* 30: 229 (1975)
Naucoria zetlandica P.D. Orton, *Trans. Brit. Mycol. Soc.* 43(2): 326 (1960)
Phaeogalera zetlandica (P.D. Orton) Pegler & T.W.K. Young, *Kew Bull.* 30: 229 (1975)
E: ! **S:** !
H: Amongst *Sphagnum* spp, usually on moorland, also in ditches and on peaty soil in similar habitats.
D: Reid84: 229, Reid84: 231 (as *Phaeogalera stagninoides*), NM2: 309 (as *Galerina stagnina*), BFF7: 81 **D+I:** B&K5: 310 404, B&K5: 310 404 **I:** Cooke 530 (468) Vol. 4 (1885) [as *Agaricus (Tubaria) stagninus*]
Most records from Scotland. Three closely related species were recognised in NCL but these fall within the range of variability of *P. stagnina*.

PHAEOLEPIOTA Maire ex Konrad & Maubl., *Icon. select. fung.* 6: 111 (1928)
Cortinariales, Cortinariaceae
Phaeolepiota Maire (*nom. inval.*), *Bull. Soc. Mycol. France* 27: 441 (1911)
Type: *Phaeolepiota aurea* (Matt.) Maire ex Konrad & Maubl.

aurea (Matt.) Maire ex Konrad & Maubl., *Icon. select. fung.* 6: 112 (1928)
Agaricus aureus Matt., *Enum. stirp. Silesia*: 331 (1779)
Pholiota aurea (Matt.) P. Kumm., *Führ. Pilzk.*: 85 (1871)

Togaria aurea (Matt.) W.G. Sm., *Syn. Brit. Bas.*: 122 (1908)
Cystoderma aureum (Matt.) Kühner & Romagn., *Flore Analytique des Champignons Supérieurs*: 393 (1953)
Agaricus aureus var. *herefordensis* Renny, *Trans. Woolhope Naturalists' Field Club* 1872: 20 (1872)
Pholiota aurea var. *herefordensis* (Renny) Sacc., *Syll. fung.* 5: 736 (1887)
Togaria aurea var. *herefordensis* (Renny) W.G. Sm., *Syn. Brit. Bas.*: 122 (1908)
Agaricus vahlii Schumach., *Enum. pl.* 2: 258 (1803)
Pholiota vahlii (Schumach.) J.E. Lange, *Dansk. Bot. Ark.* 2(11): 5 (1921)
Agaricus aureus var. *vahlii* (Schumach.) Cooke, *Handb. Brit. fung.*, Edn 2: 140 (1886)
Pholiota aurea var. *vahlii* (Schumach.) Sacc., *Syll. fung.* 5: 736 (1887)
Togaria aurea var. *vahlii* (Schumach.) W.G. Sm., *Syn. Brit. Bas.*: 122 (1908)
Lepiota pyrenaea Quél., *Compt. Rend. Assoc. Franç. Avancem. Sci. Assoc. Sci. France* 16: 587 (1888)
Agaricus spectabilis Weinm., *Elench. Fung. hort. imp. Petrop.*: 28 (1824)
Pholiota spectabilis (Weinm.) P. Kumm., *Führ. Pilzk.*: 84 (1871)
Fulvidula spectabilis (Weinm.) Romagn. [*nom. inval.*], *Rev. Mycol. (Paris)* 2(5): 191 (1937)
Gymnopilus spectabilis (Weinm.) A.H. Sm., *Mushrooms in their Natural Habitats*: 471 (1949)
Mis.: *Pholiota caperata* sensu P. Kummer [*Fuhr. Pilzk.*: 84 (1871)]
Mis.: *Togaria caperata* sensu W.G. Smith [*Syn. Brit. Bas.*: 122 (1908)]
Mis.: *Agaricus caperatus* sensu Cooke [*Ill. Brit. fung.* 375 (348) Vol. 3 (1885)]
E: o **S:** ! **W:** ! **NI:** o
H: On nitrogenous soil in parks, gardens, cemeteries, woodland, riverbanks and rarely in grassland.
D: NM2: 176, BFF7: 84-85 **D+I:** FRIC17: 18-19 135, B&K4: 222 259 **I:** FlDan3: 105C (as *Pholiota vahlii*), Bon: 173, C&D: 239, SV31: 37
Rarely reported but widespread. Excludes all British records of *Gymnopilus spectabilis*, these relating to *G. junonius*.

PHAEOLUS (Pat.) Pat., *Essai taxon.*: 86 (1900)
Poriales, Coriolaceae
Type: *Phaeolus schweinitzii* (Fr.) Pat.

schweinitzii (Fr.) Pat., *Essai taxon.*: 86 (1900)
Polyporus schweinitzii Fr., *Syst. mycol.* 1: 531 (1821)
Polyporus herbergii Rostk., *Deutschl. Fl.*, Edn 3: 35 (1848)
Hydnum spadiceum Pers., *Icon. descr. fung.*: pl. 34 t. 9 (1798)
Phaeolus spadiceus (Pers.) Rauschert, *Haussknechtia* 4: 54 (1988)
Polyporus spongia Fr., *Monogr. hymenomyc. Suec.* 2: 268 (1863)
E: c **S:** ! **W:** c **NI:** ! **ROI:** ! **O:** Channel Islands: ! Isle of Man: !
H: Parasitic then saprotrophic. Usually on roots of living conifers. Most frequently on *Pinus sylvestris* but also known on *Cedrus*, *Larix*, and *Picea* spp, *Pinus nigra*, *P. contorta*, *Pseudotsuga*, *Sequoia*, *Taxus* and *Sequoiadendron* spp. Rarely reported on deciduous hosts but known on *Betula*, *Fagus*, *Malus baccata* x *pumila*, *Prunus avium* and *Quercus* spp.
D: EurPoly2: 467, NM3: 234 **D+I:** Ph: 222, B&K2: 318 403
Common and widespread.

PHAEOMARASMIUS Scherff., *Hedwigia* 36: 288 (1897)
Cortinariales, Cortinariaceae
Type: *Phaeomarasmius excentricus* Scherff. (= *P. rimulincola*)

erinaceus (Fr.) Kühner, *Encycl. Mycol. 7, Le Genre Galera*: 33 (1935)
Agaricus erinaceus Fr. [non *A. erinaceus* Pers. (1828)], *Elench. fung.* 1: 33 (1828)
Naucoria erinacea (Fr.) Gillet, *Hyménomycètes*: 543 (1876)
Dryophila erinacea (Fr.) Quél., *Enchir. fung.*: 69 (1886)
Pholiota erinacea (Fr.) Rea, *Brit. basidiomyc.*: 121 (1922)
Agaricus aridus Pers., *Mycol. eur.* 3: 193 (1828)
Phaeomarasmius aridus (Pers.) Singer, *Lilloa* 22: 577 (1951) [1949]
Agaricus lanatus Sowerby, *Col. fig. Engl. fung., Suppl.*: pl. 417 (1814)
E: ! **S:** ! **W:** ! **ROI:** !
H: On wood, usually small, dead, attached branches of *Salix* spp in wet areas such as fen or alder carr, stream banks, pond margins and marshland.
D: NM2: 260, BFF7: 86-87 **D+I:** B&K5: 330 435 **I:** Bon: 247, C&D: 349
Occasional but widespread. Basidiomes are small, cryptically coloured and easily overlooked.

rimulincola (Rabenh.) P.D. Orton, *Trans. Brit. Mycol. Soc.* 43(2): 180 (1960)
Agaricus rimulincola Rabenh., *Deutschl. Krypt. Flora* 1: 676 (1851)
Naucoria rimulincola (Rabenh.) Sacc., *Syll. fung.* 5: 833 (1887)
Phaeomarasmius excentricus Scherff., *Hedwigia* 36: 288 (1897)
Mis.: *Naucoria horizontalis* sensu Rea (1922)
Mis.: *Phaeomarasmius horizontalis* sensu auct.
E: ! **S:** !
H: On fallen twigs, reported on *Malus* and *Ulmus* spp.
D: NM2: 259, BFF7: 87-88 **I:** Cooke 496 (509b) Vol. 4 (1884)
Apparently widespread in England and Scotland but most reports are unsubstantiated with voucher material and several collections are misdetermined *Melanotus horizontalis*.

PHALLUS L., *Sp. pl.*: 1178 (1753)
Phallales, Phallaceae
Dictyophora Desv., *J. Bot., Paris* 2: 92 (1809)
Hymenophallus Nees, *Syst. Pilze*: 251 (1817)
Ithyphallus Gray, *Nat. arr. Brit. pl.* 1: 675 (1821)
Type: *Phallus impudicus* L.

hadriani Vent., *Mem. de l'Instit. Nat. Class. Sci. Math. et. Phys., Dissertation sur le genre* Phallus 1: 517 (1798)
Phallus imperialis Schulzer, in Kalchbrenner, *Icon. select. Hymenomyc. Hung.*: 63, t. 40 f. 1 (1873)
Phallus iosmus Berk., *Engl. fung.*: 227 (1836)
E: o **S:** ! **W:** ! **ROI:** !
H: On sandy soil, usually on dunes.
D: NM3: 176 **D+I:** B&K2: 400 527, BritPuffb: 174-175 Figs. 139/140 **I:** Bon: 301, Kriegl2: 168
Rarely reported but may be locally frequent. Most frequent in southern and western areas. Said to smell less foetid than *Phallus impudicus* with various interpretations of the scent; according to Rea (1922) 'like liquorice' and to Phillips & Plowright [Grevillea 4: 118 (1876)] 'like violets' (hence the epithet *iosmus*). English name = 'Dune Stinkhorn'.

impudicus L., *Sp. pl.*: 1648 (1753)
Ithyphallus impudicus (L.) Fr., *Jahrbuch. D. Kgl. bot Gar. Berlin* 4: 42 (1886)
Phallus foetidus Sowerby, *Col. fig. Engl. fung.* 3: pl. 329 (1801)
E: c **S:** c **W:** c **NI:** c **ROI:** c **O:** Channel Islands: c Isle of Man: c
H: On soil in woodland, associated with both deciduous trees and conifers, also on dunes and in gardens and occasionally under (or inside) houses. Fruiting throughout the year.
D: NM3: 176 **D+I:** Ph: 256-257, B&K2: 400 528, BritPuffb: 172-173 Figs. 135-137 **I:** Sow3: 329 (as *Phallus foetidus*), Bon: 301, Kriegl2: 169
Very common and widespread. English name = 'Stinkhorn'.

impudicus var. **togatus** (Kalchbr.) Costantin & L.M. Dufour, *Nouv. fl. champ.*, Edn 2: 288 (1895)

Hymenophallus togatus Kalchbr., *Ungar. Akad. Wiss.* 13(8): 6 (1884)

Mis.: *Dictyophora duplicata* sensu auct. Brit.

E: ! **S:** ! **W:** ! **NI:** ! **ROI:** !

H: On soil with deciduous trees or conifers in woodland or occasionally on dunes.

D: NM3: 176 (as *Phallus duplicatus*) **D+I:** BritPuffb: 172 Fig. 138

Rarely reported. Often recorded as the tropical species *Phallus duplicatus* (= *Dictyophora duplicata*) which is similar in appearance but unknown in Britain.

PHANEROCHAETE P. Karst., *Bidrag Kännedom Finlands Natur Folk* 48: 426 (1889)

Stereales, Meruliaceae

Type: *Thelephora alnea* Fr. (= *Phanerochaete velutina*)

calotricha (P. Karst.) J. Erikss. & Ryvarden, *Corticiaceae of North Europe* 5: 997 (1978)

Corticium calotrichum P. Karst., *Rev. Mycol. (Toulouse)* 10(37): 73 (1888)

E: ! **S:** !

H: On decayed wood of deciduous trees such as *Betula* spp, *Fagus* and *Fraxinus*. A single collection from a dead stem of *Cirsium arvense*.

D: NM3: 168 **D+I:** CNE5: 995-999, B&K2: 154 159

Known from England (Middlesex, Oxfordshire, South Hampshire, Surrey and West Sussex) and a recent collection from Scotland (Clyde Isles).

deflectens (P. Karst.) Hjortstam, *Windahlia* 17: 58 (1987)

Grandinia deflectens P. Karst., *Bidrag Kännedom Finlands Natur Folk* 37: 239 (1882)

Phlebia deflectens (P. Karst.) Ryvarden, *Rept. Kevo subarct. Res. Stn* 8: 150 (1971)

Corticium umbratum Bourdot & Galzin, *Hyménomyc. France*: 220 (1928)

E: !

H: On decayed bark and wood. British material on *Populus tremula*.

D+I: CNE6: 1107-1109

A single confirmed collection from Buckinghamshire (Rushbeds Wood) in 2002, and possibly also this species in North Hampshire and North Somerset.

jose-ferreirae (D.A. Reid) D.A. Reid, *Acta Bot. Croat.* 34: 135 (1975)

Corticium jose-ferreirae D.A. Reid, *Rev. Biol. gen. theor. appl.* 5(1 - 2): 140 (1965)

W: ! **O:** Isle of Man: !

H: On decayed wood. British material on *Fraxinus* and *Ulex europaeus*.

D: NM3: 167, Myc14(1): 31 **D+I:** CNE6: 1077-1079

Known from Wales (Merionethshire: Morfa Harlech Dunes) and the Isle of Man (South Barrule Plantation).

laevis (Pers.) J. Erikss. & Ryvarden, *Corticiaceae of North Europe* 5: 1007 (1978)

Thelephora laevis Pers., *Syn. meth. fung.*: 575 (1801)

Peniophora laevis (Pers.) Burt, *Bull. New York State Mus. Nat. Hist.* 54: 954 (1902)

Peniophora affinis Burt, *Ann. Missouri Bot. Gard.* 13: 266 (1926)

Phanerochaete affinis (Burt) Parmasto, *Conspectus Systematis Corticiacearum*: 84 (1968)

Thelephora populina Sommerf., *Suppl. Fl. lapp.*: 284 (1826)

Corticium populinum (Sommerf.) Fr., *Epicr. syst. mycol.*: 559 (1838)

E: o **S:** ! **W:** !

H: On decayed wood of deciduous trees such as *Betula* spp, *Carpinus*, *Fagus* and *Ulmus* spp.

D: NM3: 168 **D+I:** CNE5: 1006-1011, B&K2: 154 158 **I:** Kriegl1: 286

Known from England (Buckinghamshire, East & West Kent, East & West Norfolk, North Hampshire, Oxfordshire, Surrey,

Warwickshire and Yorkshire), Scotland (Orkney) and Wales (Monouthshire).

leprosa (Bourdot & Galzin) Jülich, *Persoonia* 10(3): 334 (1979)

Peniophora radicata ssp. *leprosa* Bourdot & Galzin, *Bull. Soc. Mycol. France* 28(4): 394 (1913) [1912]

Peniophora leprosa (Bourdot & Galzin) Bourdot & Galzin, *Hyménomyc. France*: 312 (1928)

Scopuloides leprosa (Bourdot & Galzin) Boidin, Lanq. & Gilles, *Cryptog. Mycol* 14(3): 200 (1993)

E: !

H: On decayed wood of deciduous trees, and also reported 'on soil'.

Rarely reported and not accepted as a good species by many authorities. Known from Cumberland, North Hampshire, North & South Wiltshire, South Essex, Surrey, West Kent, West Sussex and Yorkshire. Reported elsewhere but records are old and unsubstantiated with voucher material.

magnoliae (Berk. & M.A. Curtis) Burds., *Mycol. Mem.* 10: 95 (1985)

Radulum magnoliae Berk. & M.A. Curtis, *Hooker's J. Bot. Kew Gard. Misc.* 1: 236 (1849)

Phanerochaete raduloides J. Erikss. & Ryvarden, *Corticiaceae of North Europe* 5: 1015 (1978)

Mis.: *Corticium eichlerianum* sensu Bourdot & Galzin (1928)

E: !

H: Usually on dead basidiomes of *Datronia mollis* but also reported on decayed wood of *Betula* and *Ulmus* spp.

D: NM3: 167 (as *Phanerochaete raduloides*) **D+I:** CNE5: 1014-1017 (as *P. raduloides*)

Known from Bedfordshire, East Gloucestershire, East & West Kent, Middlesex, Surrey, West Norfolk and West Sussex.

martelliana (Bres.) J. Erikss. & Ryvarden, *Corticiaceae of North Europe* 5: 1011 (1978)

Corticium martellianum Bres., *Nuovo Giorn. Bot. Ital.* 22: 258 (1890)

Peniophora martelliana (Bres.) Sacc., *Syll. fung.* 9: 239 (1891)

Peniophora macrospora Bres., in Bourdot & Galzin, *Bull. Soc. Mycol. France* 28: 396 (1913)

E: ! **W:** !

H: On decayed woody stems of *Clematis vitalba*, *Hedera* and *Rubus fruticosus*, decayed wood of *Salix cinerea* and on dead stems of *Pteridium*.

D+I: CNE5: 1011-1013

Known from Bedfordshire, Gloucestershire, South Devon and Surrey and a single record from Wales.

radicata (Henn.) Nakasone, C.R. Bergman & Burds., *Sydowia* 46(1): 46 (1994)

Corticium radicatum Henn., in Engler, *Pflanzenw. Ost-Afrikas, Basidiomyceten*: 54 (1895)

Mis.: *Phanerochaete filamentosa* sensu auct. Eur.

E: ! **W:** !

H: Usually on decayed wood of conifers such as *Picea abies* and *Pseudotsuga menziesii*. Rarely on wood of deciduous hosts such as *Betula* spp and *Fagus* or spreading onto litter below.

D: NM3: 168 **D+I:** CNE5: 1000-1002 (as *P. filamentosa*), B&K2: 156 160 (as *P. filamentosa*)

Known from England (Bedfordshire, Berkshire, Buckinghamshire, East Sussex, North Hampshire, Oxfordshire, South Hampshire, Surrey, Warwickshire, Westmorland and Yorkshire) and Wales (Caernarvonshire). All British records of *P. filamentosa* belong here.

sanguinea (Fr.) Pouzar, *Česká Mykol.* 27(1): 26 (1973)

Thelephora sanguinea Fr., *Elench. fung.* 1: 203 (1828)

Corticium sanguineum (Fr.) Fr., *Epicr. syst. mycol.*: 561 (1838)

Kneiffia sanguinea (Fr.) Bres., *Ann. Mycol.* 1(1): 101 (1903)

Peniophora sanguinea (Fr.) Höhn. & Litsch., *Sitzungsber. Kaiserl. Akad. Wiss., Wien, Math.-Naturwiss. Cl., Abt. 1* 115: 1588 (1907)

E: ! **S:** ! **W:** !

H: On decayed wood of deciduous trees such as *Fagus* but the
most recent British material was on decayed wood of *Picea* sp.
D: NM3: 167 **D+I:** CNE5: 1018-1021, B&K2: 156 161
Supposedly common but rarely reported. Collected on conifers in
Devon in 2003 but prior to this the last collection in herb. K
was in 1937. Should be easily recognisable as actively growing
basidiomes stain wood bright red.

sordida (P. Karst.) J. Erikss. & Ryvarden, *Corticiaceae of North
Europe* 5: 1023 (1978)
    Corticium sordidum P. Karst., *Meddeland. Soc. Fauna Fl.
    Fenn.* 9: 65 (1882)
    Corticium allescheri Bres., *Fungi trident.* 2(11-13): 62 (1898)
    Peniophora cremea ssp. *allescheri* (Bres.) Bourdot & Galzin,
    Hyménomyc. France: 304 (1928)
    Corticium cremeum Bres., *Fungi trident.* 2(11-13): 63 (1898)
    Peniophora cremea (Bres.) Sacc. & Syd., *Syll. fung.* 16: 195
    (1902)
    Phanerochaete cremea (Bres.) Parmasto, *Conspectus
    Systematis Corticiacearum*: 84 (1968)
    Corticium eichlerianum Bres., *Ann. Mycol.* 1(1): 95 (1903)
E: c **S:** c **W:** c **NI:** ! **ROI:** c **O:** Channel Islands: o
H: On decayed wood of deciduous trees, most often on *Fagus*
but also known on *Acer pseudoplatanus, Alnus, Fraxinus,
Juglans regia, Populus gileadensis, Betula, Quercus* and *Salix*
spp. Rarely on conifers such as *Larix* and *Picea abies,* on dead
stems of *Ulex europaeus* and on dead grass stems.
D: 168 **D+I:** CNE5: 1023-1031, B&K2: 156 162
Very common and widespread.

tuberculata (P. Karst.) Parmasto, *Conspectus Systematis
Corticiacearum*: 83 (1968)
    Corticium tuberculatum P. Karst., *Hedwigia* 35(1): 45 (1896)
    Corticium lacteum var. *tuberculatum* (P. Karst.) Bourdot &
    Galzin, *Hyménomyc. France*: 188 (1928)
    Mis.: *Thelephora lactea* sensu auct.
    Mis.: *Corticium lacteum* sensu auct.
E: !
H: On decayed wood of deciduous trees, usually *Fagus* but also
known on *Acer pseudoplatanus, Alnus, Salix* and *Ulmus* spp.
D: NM3: 167 **D+I:** CNE5: 1033-1035, B&K2: 158 163 **I:** Kriegl1:
289
Known from Bedfordshire, Cambridgeshire, Derbyshire,
Oxfordshire, Surrey, West Sussex, Worcestershire and
Yorkshire.

velutina (DC.) Fr., *Krit. öfvers. Finl. basidsvamp.,* Till. 3: 33
(1898)
    Thelephora velutina DC., *Fl. Fr.* 5: 33 (1815)
    Corticium velutinum (DC.) Fr., *Epicr. syst. mycol.*: 561 (1838)
    Peniophora velutina (DC.) Cooke, *Grevillea* 8(45): 21 (1879)
    Peniophora scotica Massee, *J. Linn. Soc., Bot.* 25: 152 (1889)
E: c **S:** c **W:** c **NI:** o **ROI:** !
H: On decayed wood of deciduous trees such as *Acer
pseudoplatanus, Betula* spp, *Fagus, Populus tremula, Quercus
robur* and *Sorbus aucuparia.* Rarely on conifer wood and pit-
props in mines.
D: NM3: 168 **D+I:** CNE5: 1035-1038, B&K2: 158 164 **I:** Kriegl1:
290
Very common and widespread.

PHELLINUS Quél., *Enchir. fung.*: 172 (1886)
Hymenochaetales, Hymenochaetaceae
Type: *Phellinus igniarius* (L.) Quél.

cavicola Kotl. & Pouzar, *Česká Mykol.* 48(2): 155 (1995)
    Mis.: *Phellinus umbrinellus* sensu Ryvarden & Gilbertson
    (EurPoly2)
E: !
H: On dead wood of deciduous trees in old woodland. Usually
on *Fagus* but also known on *Quercus* spp, often inside large
hollow stumps or fallen trunks and also on dead attached
branches.
D+I: Myc10(1): 36 (as *Phellinus umbrinellus*)

Rarely reported. Known from Berkshire, Buckinghamshire,
Middlesex and South Hampshire. Originally reported as the
widespread tropical species *Phellinus umbrinellus.*

conchatus (Pers.) Quél., *Enchir. fung.*: 173 (1886)
    Boletus conchatus Pers., *Observ. mycol.* 1: 24 (1796)
    Polyporus conchatus (Pers.) Fr., *Syst. mycol.* 1: 376 (1821)
    Fomes conchatus (Pers.) Gillet, *Hyménomycètes*: 685 (1878)
    Porodaedalea conchata (Pers.) Fiasson & Niemalä, *Karstenia*
    24: 25 (1984)
    Boletus salicinus Pers., *Syn. meth. fung.*: 543 (1801)
    Polyporus salicinus (Pers.) Fr., *Syst. mycol.* 1: 376 (1821)
    Fomes salicinus (Pers.) Gillet, *Hyménomycètes*: 684 (1878)
E: ! **S:** ? **W:** ?
H: Weakly parasitic, then saprotrophic. On dead or dying wood
of deciduous trees, usually old trunks of *Salix* spp, but also
known on *Betula, Populus* and *Ulmus* spp.
D: EurPoly2: 479, NM3: 331 (as *Porodaedalea conchata*) **D+I:**
B&K2: 256 310 **I:** Kriegl1: 448
Rarely reported, but apparently widespread in England.
Reported from Scotland and Wales but unsubstantiated with
voucher material.

contiguus (Pers.) Pat., *Essai taxon.*: 97 (1900)
    Boletus contiguus Pers., *Syn. meth. fung.*: 544 (1801)
    Polyporus contiguus (Pers.) Fr., *Syst. mycol.* 1: 378 (1821)
    Poria contigua (Pers.) Cooke, *Grevillea* 14: 114 (1886)
    Fuscoporia contigua (Pers.) G. Cunn., *Bull. New Zealand
    Dept. Sci. Industr. Res.* 73: 4 (1948)
    Polyporus cellaris Desm., *Pl. crypt. N. France*: no. 72 (1826)
E: !
H: Most collections are on decayed thatch or timber such as
window frames or sills. Rarely collected on dead branches of
Ulmus spp.
D: EurPoly2: 481, NM3: 328 (as *Fuscoporia contigua*) **D+I:**
B&K2: 256 311
Known from Berkshire, Cambridgeshire, Cumberland, Dorset,
South Hampshire, Surrey, West Norfolk and Yorkshire. Other
records are unsubstantiated with voucher material.

ferreus (Pers.) Bourdot & Galzin, *Hyménomyc. France*: 627
(1928)
    Polyporus ferreus Pers., *Mycol. eur.* 2: 89 (1825)
    Poria ferrea (Pers.) Bourdot & Galzin, *Bull. Soc. Mycol. France*
    41: 247 (1925)
    Fuscoporia ferrea (Pers.) G. Cunn., *Bull. New Zealand Dept.
    Sci. Industr. Res.* 73: 7 (1948)
E: c **S:** c **W:** c **NI:** c **ROI:** c **O:** Channel Islands: ! Isle of Man: !
H: Weakly parasitic then saprotrophic. On fallen or attached
branches, fallen trunks etc., of deciduous trees or shrubs, most
commonly on living trunks of *Corylus* and *Crataegus monogyna*
in scrub woodland. Rarely on conifers but known on *Larix,
Picea, Pinus* and *Pseudotsuga* spp and *Taxus.*
D: EurPoly2: 483, NM3: 328 (as *Fuscoporia ferrea*) **D+I:** Ph:
234
Common and widespread, but many records, unsubstantiated
with voucher material, could refer to *P. ferruginosus.*

ferruginosus (Schrad.) Bourdot & Galzin, *Hyménomyc. France*:
625 (1928)
    Boletus ferruginosus Schrad., *Spicil. Fl. Germ.*: 172 (1794)
    Poria ferruginosa (Schrad.) P. Karst., *Rev. Mycol. (Toulouse)*
    1: 378 (1881)
    Fomes ferruginosus (Schrad.) Massee, *Brit. fung.-fl.* 1: 227
    (1892)
    Fuscoporia ferruginosa (Schrad.) Murrill, *North American
    Flora* 9(1): 5 (1907)
    Polyporus umbrinus Fr. [*nom. illegit.*, non *P. umbrinus* Pers.
    (1825)], *Hyménomyc. eur.*: 571 (1874)
    Poria umbrina (Fr.) Quél., *Enchir. fung.*: 178 (1886)
E: c **S:** c **W:** c **NI:** ! **ROI:** !
H: On dead wood of deciduous trees and shrubs. Most often
reported on *Corylus* and *Fagus* but also known on *Crataegus
monogyna, Ilex, Populus tremula, Prunus avium, Quercus,
Salix* and *Ulmus* spp.

219

D: EurPoly2: 486, NM3: 328 (as *Fuscoporia ferruginosa*) **D+I:** B&K2: 258 313 **I:** Kriegl1: 451
Common and widespread, but many records, unsubstantiated with voucher material, could refer to the macroscopically similar *P. ferreus*.

hippophaeicola H. Jahn, *Mem. N. Y. bot. Gdn* 28: 105 (1978)
  Fomitoporia hippophaeicola (H. Jahn) Fiasson & Niemelä, *Karstenia* 24: 25 (1984)
  Phellinus robustus f. *hippophaes* Donk, *Atlas champ. Eur.* 1: 530 (1942)
E: ! **S:** ! **W:** !
H: Parasitic. On living trunks or branches of *Hippophae rhamnoides* in coastal areas. Restricted to this host in Britain but elsewhere in Europe common on *Eleagnus* spp.
D: EurPoly2: 491, NM3: 327 (as *Fomitoporia hippophaeicola*) **D+I:** B&K2: 258 315
Locally common along the coast of England (East Kent, East Sussex, Isle of Wight, North Lincolnshire, South Lancashire, West Norfolk and Yorkshire) and recently reported from Scotland (East Lothian and Fife) and Wales (Glamorganshire).

igniarius (L.) Quél., *Enchir. fung.*: 172 (1886)
  Boletus igniarius L., *Sp. pl.*: 1176 (1753)
  Polyporus igniarius (L.) Fr., *Syst. mycol.* 1: 375 (1821)
  Fomes igniarius (L.) Cooke, *Grevillea* 14: 18 (1886)
  Fomes trivialis Bres., *Icon. Mycol.* 20: 995 (1931)
  Phellinus trivialis (Bres.) Kreisel, *Repert. Nov. Spec. Regn. veg.* 69: 212 (1964)
E: o **S:** o **W:** o **NI:** o **O:** Isle of Man: !
H: Parasitic. Usually living trunks of *Salix* spp. Reported on *Alnus, Betula, Malus* and (dubiously) *Pinus sylvestris*.
D: EurPoly2: 491, NM3: 329 **D+I:** Ph: 228-229, B&K2: 260 316 **I:** Sow2: 132 (as *Boletus ignarius*), Bon: 321, C&D: 141, Kriegl1: 455
Occasional but widespread. Probably a species complex, not yet fully resolved.

laevigatus (Fr.) Bourdot & Galzin, *Hyménomyc. France*: 624 (1928)
  Polyporus laevigatus Fr., *Hymenomyc. eur.*: 571 (1874)
  Poria laevigata (Fr.) P. Karst., *Meddeland. Soc. Fauna Fl. Fenn.* 6: 10 (1881)
E: ! **S:** !
H: The few British collections are mostly on dead trunks of *Betula* spp with one on *Fagus*.
D: EurPoly2: 495, NM3: 330 **D+I:** B&K2: 260 317 **I:** Kriegl1: 458
Recorded from England (Buckinghamshire, Surrey, Warwickshire and Yorkshire) and Scotland (East & Mid-Perthshire, Easterness and Mid-Ebudes) but most records are unsubstantiated with voucher material.

lundellii Niemelä, *Ann. Bot. Fenn.* 9: 51 (1972)
E: ? **S:** !
H: On decayed wood of *Betula* spp.
D: EurPoly2: 497, NM3: 330 **D+I:** B&K2: 260 318 **I:** Kriegl1: 459
Known from East & Mid-Perthshire, Mid-Ebudes and West Sutherland. Reported from England (North Wiltshire and East Gloucestershire) on *Populus canescens* and *Corylus avellana* but these records are unsubstantiated with voucher material.

nigricans (Fr.) P. Karst., *Finl. basidsvamp.*: 134 (1899)
  Polyporus nigricans Fr. [non *P. nigricans* Lasch (1859)], *Syst. mycol.* 1: 375 (1821)
  Fomes nigricans (Fr.) Gillet, *Hyménomycètes*: 685 (1878)
  Fomes igniarius f. *nigricans* (Fr.) Bondartsev, *Acta Inst. Bot. Acad. Sci. USSR Plant. Crypt., Ser. 2* 6: 495 (1935)
E: ? **S:** ! **W:** ?
H: Parasitic then saprotrophic. On living and dead trunks of *Betula* spp, and also reported on *Salix* spp.
D: EurPoly2: 499, NM3: 330 **D+I:** B&K2: 262 319
A single confirmed collection from Easterness (Craigellachie NNR) in 1997. Recorded from England and Wales but records are old and unsubstantiated with voucher material.

pini (Thore) A. Ames, *Ann. Mycol.* 11(1): 246 (1913)
  Boletus pini Thore [non *B. pini* Broterius (1804)], *Essai Chloris Dept. Land.*: 487 (1803)
  Daedalea pini (Thore) Fr., *Syst. mycol.* 1: 336 (1821)
  Trametes pini (Thore) Fr., *Epicr. syst. mycol.*: 489 (1838)
  Porodaedalea pini (Thore) Murrill, *North American Flora* 9: 111 (1908)
E: ! **S:** ! **NI:** ?
H: Parasitic. British material on living trunks of *Pinus sylvestris* and once on *P. radiata*.
D: EurPoly2: 505, NM3: 331 (as *Porodaedalea pini*) **I:** Kriegl1: 460
Rarely reported. Most collections are from Scotland (Easterness, Mid-Perthshire, Morayshire and South Aberdeen) and from areas of heathland in southern England (Dorset, Surrey, South Hampshire and West Kent). Reported recently from Northern Ireland but unsubstantiated with voucher material.

pomaceus (Pers.) Maire, *Fung. Catal.* 37 (1932)
  Boletus pomaceus Pers., *Observ. mycol.* 2: 5 (1818)
  Fomes pomaceus (Pers.) Bigeard & H. Guill., *Fl. Champ. Supér. France* 2: 355 (1913)
  Fomes pomaceus var. *fulvus* Rea, *Brit. basidiomyc.*: 594 (1922)
  Boletus tuberculosus Baumg., *Flora Lips.*: 635 (1790)
  Phellinus tuberculosus (Baumg.) Niemelä, *Karstenia* 22(1): 12 (1982)
  Mis.: *Boletus igniarius* sensu Bolton [Hist. Fung. Halifax 2: 80 (1788)]
E: o **S:** ! **W:** ! **NI:** o
H: Weakly parasitic, then saprotrophic. Typically on species of *Prunus* in gardens, thickets or hedgerows, most often on *Prunus spinosa*. Rarely on *Crataegus monogyna, Malus* and *Sorbus* spp. Reported (with voucher material) on *Corylus, Salix cinerea* and *Syringa vulgaris* but the host determinations are suspect.
D: EurPoly2: 507, NM3: 329 (as *Phellinus tuberculosus*) **D+I:** B&K2: 266 326 (as *P. tuberculosus*) **I:** Kriegl1: 467 (as *P. tuberculosus*)
Occasional and widespread in England but rarely reported from south-eastern counties or from Scotland or Wales.

punctatus (Fr.) Pilát, *Atlas champ. Eur.* 3(1): 530 (1942)
  Polyporus punctatus Fr., *Hymenomyc. eur.*: 572 (1874)
  Poria punctata (Fr.) P. Karst., *Bidrag Kännedom Finlands Natur Folk* 37: 83 (1882)
E: !
H: On dead or dying wood of deciduous trees. British material on dead standing trunks of *Crataegus monogyna* and *Fraxinus*.
D: EurPoly2: 512, NM3: 327 (as *Fomitoporia punctata*) **D+I:** B&K2: 262 321
Known from West Kent (Cobham) and Worcestershire (Pershore). Reported from Dorset, East Gloucestershire, East Suffolk, East Sussex and South Hampshire but unsubstantiated with voucher material. Resembles *Phellinus ferreus* but distinguished by the lack of setae in the basidiomes.

robustus (P. Karst.) Bourdot & Galzin, *Hyménomyc. France*: 616 (1928)
  Fomes robustus P. Karst., *Bidrag Kännedom Finlands Natur Folk* 48: 467 (1889)
  Fomitiporia robusta (P. Karst.) Fiasson & Niemelä, *Karstenia* 24: 25 (1984)
E: !
H: Parasitic. British material only on veteran oaks in parkland, planted avenues, or old woodland.
D: EurPoly2: 517, NM3: 327 (as *Fomitoporia robusta*) **D+I:** B&K2: 264 323 **I:** Kriegl1: 464
Known from Berkshire, Shropshire and South Hampshire. Probably a species complex.

torulosus (Pers.) Bourdot & Galzin, *Bull. Soc. Mycol. France* 41: 191 (1925)
  Boletus torulosus Pers., *Traité champ. comest.*: 94 (1819)
  Polyporus torulosus (Pers.) Pers., *Mycol. eur.* 2: 79 (1825)
E: ! **ROI:** ?

H: Weakly parasitic then saprotrophic. On trunks (usually at the base) of deciduous trees. British collections on *Quercus cerris, Q. robur, Castanea, Prunus avium* and *Crataegus* spp and reported from the Republic of Ireland on *Arbutus unedo*.
D: EurPoly2: 523
Known from Berkshire, Surrey and West Kent. Reported from East Suffolk in 1956 and West Sussex in 1996, and from the Republic of Ireland [see TBMS 35(1): 35 (1952)] but unsubstantiated with voucher material. Very common in the Mediterranean area, and probably at the northern limit of its range in the British Isles.

tremulae (Bondartsev) Bondartsev & Borissov, *Trutovye griby Evropeiskoi chasti SSSR i Kavkaza*: 358 (1953)
    Fomes igniarius f. *tremulae* Bondartsev, *Fung. Bryansk.*: 22 (1912)
S: !
H: Parasitic. On living trunks of *Populus tremula* in open woodland.
D: EurPoly2: 525, NM3: 329 **D+I:** B&K2: 264 324, Myc15(3): 105-106 **I:** Kriegl1: 466
Only recently recorded from Scotland (Morayshire, Spey Valley) but appears to be 'not infrequent' (*fide* E.E. Emmett) elsewhere in Scotland.

PHELLODON P. Karst., *Rev. Mycol. (Toulouse)* 3(9): 19 (1881)

Thelephorales, Bankeraceae
Type: *Phellodon niger* (Fr.) P. Karst.

atratus K.A. Harrison, *Canad. J. Bot.* 42: 1209 (1964)
S: !
H: On soil with *Pinus sylvestris* in Caledonian pinewoods.
A single collection from Abernethy Forest in 2001.

confluens (Pers.) Pouzar, *Česká Mykol.* 10: 74 (1956)
    Hydnum confluens Pers., *Mycol. eur.* 2: 165 (1825)
    Hydnum amicum Quél., *Grevillea* 8(47): 115 (1880)
    Phellodon amicus (Quél.) Banker, *Mycologia* 5: 62 (1913)
E: ! **S:** ! **W:** !
H: On acidic soil, associated with deciduous trees such as *Betula pendula, Castanea, Fagus* and *Quercus* spp, and occasionally with *Pinus sylvestris*.
D: NM3: 313 **D+I:** 245, B&K2: 228 268, BritChant: 72-73 **I:** SV27: 4
Known from scattered localities in England and Scotland and recently collected in Wales

melaleucus (Schwartz) P. Karst., *Rev. Mycol. (Toulouse)* 3(9): 19 (1881)
    Hydnum melaleucum Schwartz., in Fries, *Observ. mycol.* 1: 141 (1815)
    Hydnum albonigrum Peck, *Rep. (Annual) New York State Mus. Nat. Hist.* 50: 110 (1897)
    Phellodon alboniger (Peck) Banker, *Mem. Torrey Bot. Club* 12: 167 (1902)
    Hydnum leptopus γ *graveolens* Pers., *Mycol. eur.* 2: 171 (1825)
    Hydnum graveolens (Pers.) Fr., *Epicr. syst. mycol.*: 509 (1838)
    Phellodon graveolens (Pers.) P.Karst, *Bidrag Kännedom Finlands Natur Folk* 37: 96 (1882)
E: o **S:** o **W:** ! **NI:** ! **ROI:** o
H: On soil in deciduous and coniferous woodland. Known with *Pinus sylvestris, Betula* spp, *Castanea, Fagus* and *Quercus* spp.
D+I: 244-245, B&K2: 228 269, BritChant: 66-67 **I:** SV27: 5
Occasional but widespread and may be locally frequent.

niger (Fr.) P. Karst., *Rev. Mycol. (Toulouse)* 3(9): 19 (1881)
    Hydnum nigrum Fr., *Observ. mycol.* 1: 134 (1815)
    Hydnellum nigrum (Fr.) P. Karst., *Meddeland. Soc. Fauna Fl. Fenn.* 5: 41 (1880)
    Mis.: *Hydnum cinereum* sensu Rea (1922)
E: ! **S:** ! **W:** ! **ROI:** !

H: On soil in deciduous and coniferous woodland, usually associated with *Castanea, Fagus* and *Quercus* spp, less frequently with conifers.
D: NM3: 313 **D+I:** Ph: 245, B&K2: 228 270, BritChant: 70-71 **I:** Bon: 313, SV27: 5, C&D: 139, Kriegl1: 385
Rarely reported but widespread.

tomentosus (L.) Banker, *Mem. Torrey Bot. Club* 12: 171 (1906)
    Hydnum tomentosum L., *Sp. pl.*: 1178 (1753)
    Hydnum connatum E.S. Schultz, *Prodr. Fl. Starg.*: 491 (1806)
    Hydnum cyathiforme Schaeff., *Fung. Bavar. Palat. nasc.* 2: 139 (1763)
E: o **S:** o **W:** ! **ROI:** !
H: Usually with *Pinus sylvestris* in coniferous woodland.
D: NM3: 313 **D+I:** 244-245, B&K2: 230 271, BritChant: 68-69 **I:** Bon: 313, SV27: 5, Kriegl1: 386
Occasional but widespread.

PHLEBIA Fr., *Syst. mycol.* 1: 426 (1821)

Stereales, Meruliaceae
    Merulius Fr., *Syst. mycol.* 1: 326 (1821)
Type: *Phlebia radiata* Fr.

griseoflavescens (Litsch.) J. Erikss. & Hjortstam, *Corticiaceae of North Europe* 6: 1121 (1981)
    Corticium griseoflavescens Litsch., *Glasnik (Bull.) Soc. Scient. Skoplje* 18(6): 178 (1938)
    Hyphoderma griseoflavescens (Litsch.) Jülich, *Persoonia* 8(1): 80 (1974)
E: !
H: On dead or living wood of deciduous trees. British material on *Salix* spp and *Ulmus procera*.
D: NM3: 161 **D+I:** CNE6: 1121-1122
Known from Surrey (Egham, Langham's Pond and Betchworth: Dawcombe Nature Reserve).

lacteola (Bourdot) M.P. Christ., *Dansk. Bot. Ark.* 19(2): 167 (1960)
    Corticium lacteolum Bourdot, *Rev. Sci. du Bourb.* 22(1): 14 (1922)
E: !
H: On decayed wood of deciduous trees.
A single collection, from Surrey (Dorking, White Down) in 1999.

lilascens (Bourdot) J. Erikss. & Hjortstam, in Eriksson, Hjortstam & Ryvarden, *Corticiaceae of North Europe* 6: 1123 (1981)
    Corticium lilascens Bourdot, *Rev. Sci. du Bourb.* 23(1): 13 (1910)
    Mis.: *Corticium seriale* sensu Rea (1922)
E: ! **S:** ! **ROI:** !
H: On decayed wood of deciduous trees such as *Alnus, Fagus, Fraxinus* and *Quercus robur* and rarely on conifers such as *Pinus sylvestris*. The single collection from Republic of Ireland was on decayed flax straw.
D: NM3: 162 **D+I:** CNE6: 1123-1127, B&K2: 164 174
Rarely reported.

livida (Pers.) Bres., *Epicr. syst. mycol.*: 527 (1838)
    Corticium lividum Pers., *Observ. mycol.* 1: 38 (1796)
    Thelephora livida (Pers.) Fr., *Observ. mycol.* 2: 276 (1818)
    Merulius lividus (Pers.) Park.-Rhodes [*nom. illegit.*, non *M. lividus* Bourdot & Galzin (1928)], *Ann. Bot.* 20(78): 258 (1956)
    Grandinia ocellata Fr., *Epicr. syst. mycol.*: 527 (1838)
    Corticium viscosum Pers., *Observ. mycol.* 2: 18 (1799)
    Thelephora viscosa (Pers.) Pers., *Syn. meth. fung.*: 580 (1801)
    Tremella viscosa (Pers.) Berk. & Broome, *Ann. Mycol.* 2(13): 406 (1854)
    Exidia viscosa (Pers.) Rea, *Brit. basidiomyc.*: 735 (1922)
E: ! **S:** ! **W:** ! **NI:** ?
H: On decayed wood of deciduous trees. Also reported on conifers.

D: NM3: 162 **D+I:** CNE6: 1131-1137, B&K2: 166 175 **I:** Kriegl1: 294

Often reported, but rarely collected and often misidentified. Old collections of *Tremella viscosa* in herb. K represent a variety of taxa, mainly *Exidiopsis effusa, Stypella grilletii* and *Radulomyces confluens.*

radiata Fr., *Syst. mycol.* 1: 427 (1821)
   Auricularia aurantiaca Sowerby, *Col. fig. Engl. fung.* 3: pl. 291 (1800)
   Phlebia aurantiaca (Sowerby) J. Schröt., in Cohn, *Krypt.-Fl. Schlesien* 3: 461 (1888)
   Phlebia aurantiaca var. *radiata* (Fr.) Bourdot & Galzin, *Hyménomyc. France*: 342 (1928)
   Phlebia contorta Fr., *Syst. mycol.* 1: 427 (1821)
   Phlebia radiata f. *contorta* (Fr.) Parmasto, *Eesti N.S.V. Tead. Akad. Toimet., Biol.* 16: 393 (1967)
   Merulius merismoides Fr., *Observ. mycol.* 2: 235 (1818)
   Phlebia merismoides (Fr.) Fr., *Syst. mycol.* 1: 427 (1821)
   Sistotrema carneum Fr., *Observ. mycol.* 2: 268 (1818)
   Irpex carneus (Fr.) Fr., *Elench. fung.*: 148 (1828)
E: c **S:** c **W:** c **NI:** c **ROI:** c **O:** Channel Islands: ! Isle of Man: c
H: On decayed wood of deciduous trees. Especially frequent on *Fagus* and *Prunus avium.*
D: NM3: 160 **D+I:** CNE6: 1152-1157, Ph: 238 (as *Phlebia merismoides*), B&K2: 166 176 **I:** C&D: 139 (as *P. merismoides*), Kriegl1: 295 (as *P. merismoides*)
Very common and widespread. Basidiomes are macroscopically variable in form and colour, usually bright orange but becoming greyish-orange with age or almost white. Forms on *Prunus avium* are often pale lilaceous.

rufa (Pers.) M.P. Christ., *Dansk. Bot. Ark.* 19(2): 164 (1960)
   Merulius rufus Pers., *Syn. meth. fung.*: 498 (1801)
   Xylomyzon rufum (Pers.) Pers., *Mycol. eur.* 2: 31 (1825)
   Phlebia erecta Rea, *Trans. Brit. Mycol. Soc.* 5(2): 252 (1915)
   Merulius lividus Bourdot & Galzin [non *M. lividus* (Pers.) Park.-Rhodes (1956)], *Bull. Soc. Mycol. France* 39(2): 104 (1923)
   Merulius pallens Berk. [non *M. pallens* Schwein. (1832)], *Ann. Mag. Nat. Hist., Ser. 1* 6: 357 (1841)
E: c **S:** o **W:** c **NI:** o **O:** Isle of Man: !
H: On decayed wood of deciduous trees, most frequently on *Quercus* spp, but also on *Acer pseudoplatanus* and *Fagus.*
D: NM3: 160 **D+I:** CNE6: 1157-1159, B&K2: 166 177 **I:** Kriegl1: 297
Common and widespread.

segregata (Bourdot & Galzin) Parmasto, *Eesti N.S.V. Tead. Akad. Toimet., Biol.* 16(4): 393 (1967)
   Peniophora segregata Bourdot & Galzin, *Hyménomyc. France*: 284 (1928)
   Peniophora livida Burt, *Ann. Missouri Bot. Gard.* 12: 239 (1925)
E: ? **S:** !
H: On decayed wood of *Pinus sylvestris* in Caledonian pinewoods.
D: NM3: 162 **D+I:** CNE6: 1160-1163
Known from Scotland (Aberdeenshire, Perthshire and Wester Ross). Reported from England (West Yorkshire: Stubham Wood) but unsubstantiated with voucher material.

subochracea (Bres.) J. Erikss. & Ryvarden, *Corticiaceae of North Europe* 4: 873 (1976)
   Grandinia subochracea Bres., *Hedwigia* 33: 206 (1894)
   Acia subochracea (Bres.) Bourdot & Galzin, *Hymenomyc. France*: 417 (1928)
   Peniophora danica M.P. Christ., *Friesia* 5: 207 (1956)
   Phlebia danica (M.P. Christ.) M.P. Christ., *Dansk. Bot. Ark.* 19(2): 167 (1960)
   Corticium ochraceofulvum Bourdot & Galzin, *Bull. Soc. Mycol. France* 27: 257 (1911)
   Phlebia ochraceofulva (Bourdot & Galzin) Donk, *Fungus* 27: 12 (1957)
E: !

H: On decayed and often sodden wood of deciduous trees, usually *Alnus* or *Salix* spp, often in dried up streams or woodland ponds. Also known on *Fagus* in similar situations.
D: NM3: 161 **D+I:** CNE6: 1169-1171, Myc13(2): 85 **I:** Kriegl1: 297
Known from Bedfordshire, East Norfolk, Herefordshire, Hertfordshire, Middlesex, North Hampshire, Shropshire, Surrey, West Kent and Yorkshire. Perhaps genuinely rare since the bright mustard to sulphur basidiomes would be difficult to overlook.

subserialis (Bourdot & Galzin) Donk, *Fungus* 27: 12 (1957)
   Corticium subseriale Bourdot & Galzin, *Hyménomyc. France*: 219 (1928)
E: ! **S:** !
H: On decayed wood of conifers such as *Pinus sylvestris* and *Picea* spp, and a single collection on *Ulmus* sp.
D: NM3: 163 **D+I:** CNE6: 1172-1173
First collected in 1949 from Scotland (Easterness: Rothiemurchus) then no further records until 1999 and 2001 in England (South Hampshire: New Forest and Surrey: Betchworth: Dawcombe Nature Reserve).

tremellosa (Schrad.) Burds. & Nakasone, *Mycotaxon* 21: 245 (1984)
   Merulius tremellosus Schrad., *Spicil. Fl. Germ.* 1: 139 (1794)
   Merulius imbricatus Balf.-Browne, *Bull. Brit. Mus. (Nat. Hist.), Bot.* 1(7): 192 (1955)
E: c **S:** c **W:** c **NI:** c **ROI:** c
H: Usually on decayed wood of deciduous trees, often large fallen trunks of *Betula* spp, *Fagus* and *Fraxinus*. Rarely on conifers such as *Larix, Picea abies* and *Pinus sylvestris.*
D: NM3: 159 **D+I:** CNE4: 864-867 (as *Merulius tremellosus*), Ph: 239, B&K2: 146 145 (as *M. tremellosus*) **I:** Kriegl1: 264 (as *M. tremellosus*)
Very common and widespread.

PHLEBIELLA P. Karst., *Hedwigia* 29: 271 (1890)
Stereales, Xenasmataceae
   Aphanobasidium Jülich, *Persoonia* 10: 326 (1979)
Type: *Phlebiella sulphurea* (Pers.) Ginns & Lefebvre

albida (Hauerslev) Tellería, Melo & M. Dueñas, *Mycotaxon* 65: 367 (1997)
   Xenasmatella albida Hauerslev, *Friesia* 11(5): 332 (1987)
   Aphanobasidium albidum (Hauerslev) Boidin & Gilles, *Cryptog. Bot.* 1(1): 74 (1989)
   Uncobasidium albidum (Hauerslev) Vesterh., *Nordic J. Bot.* 16(1): 218 (1996)
E: !
H: On living twigs of *Taxus*, fallen twigs of *Platanus orientalis* and decayed wood of *Ulmus procera*. Also on needle litter of *Pinus sylvestris* and dead stems of *Rubus fruticosus* agg.
D: NM3: 143 (as *Uncobasidium albidum*), Myc9(1): 11 (as *Aphanobasidium albidum*)
Known from Hertfordshire, South Devon, Surrey and West Kent.

allantospora (Oberw.) K.H. Larss. & Hjortstam, in Hjortstam & Larsson, *Mycotaxon* 29: 318 (1987)
   Xenasmatella allantospora Oberw., *Sydowia* 19(1 - 3): 37 (1965)
   Aphanobasidium allantosporum (Oberw.) Jülich, *Persoonia* 10(3): 326 (1979)
E: ! **S:** ! **W:** ! **O:** Channel Islands: !
H: On decayed wood of deciduous trees, rarely on conifers or decayed bamboo poles.
D: NM3: 141 **D+I:** B&K2: 194 219 (as *Aphanobasidium allantosporum*), CNE8: 1458-1459
Known from England (Hertfordshire, Middlesex, Shropshire, South Devon and Surrey), Scotland (Mid-Perthshire and South Aberdeen), Wales (Breconshire) and the Channel Islands (Jersey).

aurora (Berk. & Broome) K.H. Larss. & Hjortstam, in Hjortstam & Larsson, *Mycotaxon* 29: 317 (1987)

Corticium aurora Berk. & Broome, *Outl. Brit. fungol.*: 276 (1860)

Aphanobasidium aurora (Berk. & Broome) Boidin & Gilles, *Cryptog. Bot.* 1: 74 (1989)

E: !

H: Type material on decayed leaves and stems of *Carex paniculata*.

Described from North Somerset (Batheaston) in 1851 but never recollected.

boidinii Tellería, Melo & M. Dueñas, *Mycotaxon* 65: 367 (1997)

Aphanobasidium sphaerosporum Boidin & Gilles [non *Phlebiella sphaerospora* (Maire) Bondartsev & Singer (1953)], *Cryptog. Bot.* 1(1): 75 (1989)

E: !

H: On decayed wood of deciduous trees. British material on *Carpinus*.

A single collection from Hertfordshire (Northaw Great Wood) in 2002.

bourdotii (Boidin & Gilles) Tellería, Melo & M. Dueñas, *Mycotaxon* 65: 367 (1997)

Aphanobasidium bourdotii Boidin & Gilles, *Cryptog. Bot.* 1(1): 74 (1989)

E: !

H: On living trunks of *Crataegus monogyna*, on dead stems of *Rubus fruticosus*, on living twigs of *Buddleja davidii* and on dead attached twigs of *Ulmus* spp.

Known from South Devon in scattered localities near the coast.

christiansenii (Parmasto) K.H. Larss. & Hjortstam, in Hjortstam & Larsson, *Mycotaxon* 29: 316 (1987)

Cristella christiansenii Parmasto, *Eesti N.S.V. Tead. Akad. Toimet., Biol.* 14(2): 222 (1965)

Trechispora christiansenii (Parmasto) Liberta, *Taxon* 15(8): 318 (1966)

E: !

H: On litter of *Fagus*, decayed wood of *Picea abies*, decayed bamboo poles, debris of ferns such as *Dryopteris* sp. and *Pteridium* and on bare soil.

D: NM3: 142 **D+I:** B&K2: 122 110 (as *Trechispora christiansenii*), CNE8: 1464-1467

Known from Hertfordshire, Oxfordshire, South Devon and Surrey.

fibrillosa (Hallenb.) K.H. Larss. & Hjortstam, in Hjortstam & Larsson, *Mycotaxon* 29: 316 (1987)

Trechispora fibrillosa Hallenb., *Iranian J. Pl. Pathol.* 14: 75 (1978)

Mis.: *Hypochnus submutabilis* sensu Rea (1922)

Mis.: *Cristella submutabilis* sensu Dennis (1986)

E: ! **S:** ! **W:** !

H: On decayed leaf litter of deciduous trees (especially *Fagus*), dead lianas of *Clematis vitalba*, dead leaves of *Iris pseudacorus* and *Rubus fruticosus* and decayed debris of *Fallopia cuspidata* and *Pteridium*.

D: NM3: 141 **D+I:** CNE8: 1467-1469

Known from England (East Norfolk, Hertfordshire, Northumberland, Surrey, West Suffolk and West Sussex), Scotland (Clyde Isles and Mid-Ebudes) and Wales (Monmouthshire).

filicina (Bourdot) K.H. Larss. & Hjortstam, in Hjortstam & Larsson, *Mycotaxon* 29: 317 (1987)

Corticium filicinum Bourdot, *Rev. Sci. du Bourb.* 23(1): 12 (1918)

Xenasma filicinum (Bourdot) M.P. Christ., *Dansk. Bot. Ark.* 19(2): 106 (1960)

Aphanobasidium filicinum (Bourdot) Jülich, *Persoonia* 10(3): 326 (1979)

E: c **S:** c **W:** c **O:** Channel Islands: ! Isle of Man: !

H: On dead and decayed debris of ferns, most often *Pteridium* but also noted on *Athyrium filix-femina*, *Dicksonia antarctica* and *Dryopteris* spp. Rarely on other substrata such as dead stems of *Rubus fruticosus* agg. and dead leaves of *Luzula* spp.

D: NM3: 141 **D+I:** CNE8: 1470-1471

Very common and widespread. Almost ubiquitous on bracken.

grisella (Bourdot) K.H. Larss. & Hjortstam, in Hjortstam & Larsson, *Mycotaxon* 29: 318 (1987)

Corticium grisellum Bourdot, *Rev. Sci. du Bourb.* 35: 17 (1922)

Xenasma grisellum (Bourdot) Liberta, *Mycologia* 52(6): 911 (1962)

Aphanobasidium grisellum (Bourdot) Jülich, *Persoonia* 10(3): 326 (1979)

Corticium pruina Bourdot & Galzin, *Hyménomyc. France*: 224 (1928)

Aphanobasidium pruina (Bourdot & Galzin) Boidin & Gilles, *Cryptog. Bot.* 1(1): 76 (1989)

E: !

H: On decayed wood of conifers such as *Picea* spp, and of deciduous trees such as *Aesculus* and *Fagus*.

D: NM3: 141 **D+I:** B&K2: 196 220 (as *Aphanobasidium grisellum*), CNE8: 1473-1477

Known from Oxfordshire, Surrey and West Kent.

paludicola Hjortstam & P. Roberts, *Mycologist* 9(4): 161 (1995)

Mis.: *Corticium aurora* sensu Bourdot & Galzin [Hymenomyc. France: 209 (1927)]

E: !

H: In marshes and fens, on decayed basal leaves of *Cladium mariscus* and other vegetation.

D+I: Myc9(4): 161-162

Known from East & West Norfolk, South Devon and West Sussex.

pseudotsugae (Burt) K.H. Larss. & Hjortstam, *Mycotaxon* 29: 317 (1987)

Corticium pseudotsugae Burt, *Ann. Missouri Bot. Gard.* 13: 246 (1926)

Xenasma pseudotsugae (Burt) J. Erikss., *Symb. Bot. Upsal.* 16(1): 65 (1958)

Aphanobasidium pseudotsugae (Burt) Boidin & Gilles, *Cryptog. Bot.* 1(1): 75 (1989)

Corticium asseriphilum Litsch., *Ann. Mycol.* 32(1 - 2): 55 (1934)

E: c **S:** ! **W:** !

H: On decayed wood of conifers such as *Larix*, *Picea abies*, *Pinus sylvestris* and *Taxus*. Rarely on wood of deciduous trees such as *Buxus*, *Castanea*, *Corylus*, *Quercus* and *Ulmus* spp.

D: NM3: 141 **D+I:** CNE8: 1478-1481

Common and widespread, at least throughout England.

sulphurea (Pers.) Ginns & Lefebvre, *Mycol. Mem.* 19: 126 (1993)

Corticium sulphureum Pers., *Observ. mycol.* 1: 38 (1796)

Himantia sulphurea (Pers.) Pers., *Tent. disp. meth. fung.*: 43 (1797)

Thelephora sulphurea (Pers.) Pers., *Syn. meth. fung.*: 579 (1801)

Coniophora sulphurea (Pers.) Quél., *Enchir. fung.*: 212 (1886)

Hypochnus sulphureus (Pers.) J. Schröt., in Cohn, *Krypt.-Fl. Schlesien* 3: 417 (1888)

Cristella sulphurea (Pers.) Donk, *Fungus* 27: 20 (1957)

Trechispora sulphurea (Pers.) Rauschert, *Feddes Repert.* 98(11 - 12): 662 (1987)

Hypochnus fumosus Fr., *Observ. mycol.* 2: 279 (1818)

Corticium fumosum (Fr.) Fr., *Epicr. syst. mycol.*: 562 (1838)

Phlebia vaga Fr., *Syst. mycol.* 1: 428 (1821)

Phlebiella vaga (Fr.) P. Karst., *Hedwigia* 29: 271 (1890)

Trechispora vaga (Fr.) Donk, *Taxon* 15(8): 319 (1966)

E: c **S:** c **W:** c **NI:** !

H: On decayed wood of deciduous trees, especially on *Fagus* in woodland on calcareous soil but also not infrequent on conifers such as *Larix*, *Picea abies*, *Pinus sylvestris* and *Taxus*.

D: NM3: 141 **D+I:** B&K2: 126 117 (as *Trechispora vaga*), CNE8: 1486-1488 (as *Phlebiella vaga*) **I:** Kriegl1: 303 (as *P. vaga*)

Very common and widespread. Possibly better known as *Phlebiella vaga* or *Trechispora vaga*. Basidiomes are distinctive, especially when young and surrounded by sulphur-yellow rhizomorphs.

tulasnelloidea (Höhn. & Litsch.) Ginns & M.N.L. Lefebvre, *Mycol. Mem.* 19: 126 (1993)
  Corticium tulasnelloideum Höhn. & Litsch., *Sitzungsber. Kaiserl. Akad. Wiss., Wien, Math.-Naturwiss. Cl., Abt. 1* 117: 1118 (1908)
  Hypochnus tulasnelloideus (Höhn. & Litsch.) Rea, *Trans. Brit. Mycol. Soc.* 12(2-3): 222 (1927)
  Xenasma tulasnelloideum (Höhn. & Litsch.) Donk, *Fungus* 27: 26 (1957)
  Xenasmatella tulasnelloidea (Höhn. & Litsch.) Jülich, *Persoonia* 10(3): 335 (1979)
E: ! **S:** ! **W:** !
H: On decayed wood of deciduous trees, often in old woodland on calcareous soil. Known on *Carpinus*, *Crataegus monogyna*, *Fagus*, *Fraxinus*, *Salix* and *Ulmus* spp. Rarely on conifers such as *Pinus sylvestris*.
D: NM3: 141 **D+I:** B&K2: 198 223 (as *Xenasmatella tulasnelloidea*), CNE8: 1484-1485
Known from England (Buckinghamshire, Co. Durham, East Sussex, Hertfordshire, Middlesex, Oxfordshire, Shropshire, South Wiltshire, Surrey, West Kent, West Sussex and Yorkshire), Scotland (Peebleshire) and Wales (Breconshire and Monmouthshire).

PHLEBIOPSIS Jülich, *Persoonia* 10: 137 (1978)
Stereales, Meruliaceae
Type: *Phlebiopsis gigantea* (Fr.) Jülich

gigantea (Fr.) Jülich, *Persoonia* 10(1): 137 (1978)
  Thelephora gigantea Fr., *Observ. mycol.* 1: 152 (1815)
  Corticium giganteum (Fr.) Fr., *Epicr. syst. mycol.*: 559 (1838)
  Peniophora gigantea (Fr.) Massee, *J. Linn. Soc., Bot.* 25: 142 (1889)
  Phlebia gigantea (Fr.) Donk, *Fungus* 27: 12 (1957)
  Phanerochaete gigantea (Fr.) S.S. Rattan, *Biblioth. Mycol.* 60: 260 (1977)
  Peniophora crosslandii Massee, *Brit. fung.-fl.* 1: 418 (1892)
  Thelephora pergamenea Pers., *Mycol. eur.* 1: 150 (1822)
E: c **S:** c **W:** o **NI:** ! **O:** Isle of Man: -
H: On decayed wood of conifers, especially on cut stumps or the cut ends of stacked trunks. Known on *Larix*, *Picea abies*, *Pinus nigra*, *Pinus sylvestris* and (rarely) *Taxus*.
D: NM3: 169 **D+I:** CNE6: 1180-1183, B&K2: 158 165
Common and widespread throughout England and Scotland, less often reported elsewhere. Basidiomes are conspicuous, forming thick sheets, the general appearance likened to 'candle wax' poured over the wood. In some plantations, it has been inoculated onto freshly cut stumps as a biocontrol agent to impede growth of the virulently pathogenic *Heterobasidion annosum*.

ravenelii (Cooke) Hjortstam, *Windahlia* 17: 58 (1987)
  Peniophora ravenelii Cooke, *Grevillea* 8: 21 (1879)
  Phanerochaete ravenelii (Cooke) Burds., *Mycol. Mem.* 10: 104 (1985)
  Scopuloides ravenelii (Cooke) Boidin, Lanq. & Gilles, *Cryptog. Mycol.* 14(3): 205 (1993)
  Peniophora molleriana Sacc., *Syll. fung.* 11: 12 (1891)
  Corticium roumeguerii Bres., *Fungi trident.* 2(8-10): 36 (1892)
  Peniophora roumeguerii (Bres.) Höhn. & Litsch., *Sitzungsber. Kaiserl. Akad. Wiss., Wien, Math.-Naturwiss. Cl., Abt. 1* 115: 33 (1906)
  Phlebia roumeguerii (Bres.) Donk, *Fungus* 27: 12 (1957)
  Phlebiopsis roumeguerii (Bres.) Jülich & Stalpers, *Verh. K. Akad. Wet.* 74: 190 (1980)
E: c **S:** o **W:** c **O:** Channel Islands: o
H: On decayed wood of deciduous trees, often decayed fallen trunks of *Fraxinus* but also known from *Buddleja* sp., *Fagus*, *Prunus spinosa*, *Quercus* and *Salix* spp and also on dead stems of *Rubus fruticosus* agg.
D+I: CNE6: 1184-1185 (as *Phlebiopsis roumeguerii*)
Common and widespread (at least in southern England). Wood decayed by this species is often infiltrated with thick, pallid orange-white rhizomorphs.

PHOLIOTA (Fr.) P. Kumm., *Führ. Pilzk.*: 22 (1871)
Agaricales, Strophariaceae
  Flammula (Fr.) P. Kumm., *Führ. Pilzk.*: 22 (1871)
  Dryophila Quél., *Enchir. fung.*: 66 (1886)
  Hemipholiota (Singer) Bon, *Doc. Mycol.* 17(65): 52 (1986)
Type: *Pholiota squarrosa* (Weigel) P. Kumm.

adiposa (Batsch) P. Kumm., *Führ. Pilzk.*: 83 (1871)
  Agaricus adiposus Batsch, *Elench. fung. (Continuatio Prima)*: 147 & t.22 f. 113 (1786)
  Dryophila adiposa (Batsch) Quél., *Enchir. fung.*: 68 (1886)
E: ! **S:** ? **NI:** ? **ROI:** ?
H: On decayed wood of deciduous trees in old woodland. Usually on large decayed logs, fallen trunks or stumps of *Fagus*.
D: FAN4: 84 **D+I:** Ph: 144 **I:** Bon: 257
Apparently a rare species in the British Isles, although often reported and supposedly widespread. Many of the collections in herbaria are misidentified *P. aurivella* or *P. jahnii*.

albocrenulata (Peck) Sacc., *Syll. fung.* 5: 760 (1887)
  Agaricus albocrenulatus Peck, *Bull. Buffalo Soc. Nat. Sci.* 1: 49 (1873)
  Hemipholiota albocrenulata (Peck) Bon, *Doc. Mycol.* 17(65): 52 (1986)
  Pholiota fusca Quél., *Bull. Soc. Bot. France* 23: 327 (1877)
E: !
H: British material on soil in deciduous woodland, associated with *Fagus*.
D: NM2: 260, FAN4: 100 **I:** C&D: 354 (as *Hemipholiota albocrenulata*)
Two collections from Surrey (Box Hill area) in 1954 and 1955. Reported from Herefordshire in 1999 but unsubstantiated with voucher material.

alnicola (Fr.) Singer, *Lilloa* 22: 516 (1951) [1949]
  Agaricus alnicola Fr., *Epicr. syst. mycol.*: 250 (1838)
  Flammula alnicola (Fr.) P. Kumm., *Führ. Pilzk.*: 82 (1871)
  Dryophila alnicola (Fr.) Quél., *Enchir. fung.*: 71 (1886)
  Pholiota apicrea (Fr.) M.M. Moser, *Kleine Kryptogamenflora*, Edn 3: 244 (1967)
  Agaricus apicreus Fr., *Epicr. syst. mycol.*: 188 (1838)
  Pholiota aromatica P.D. Orton, *Trans. Brit. Mycol. Soc.* 43(2): 338 (1960)
  Mis.: *Pholiota flavida* sensu auct. Brit.
E: c **S:** c **W:** ! **NI:** ! **ROI:** !
H: On soil or decayed wood, frequently on wet ground in deciduous woodland. Often associated with *Alnus* but also known with *Betula pendula*, *Quercus* or *Salix* spp.
D: NM2: 263, FAN4: 103 **D+I:** Ph: 145, Holec: 145-152 **I:** FlDan4: 122B & B1 (as *Flammula alnicola*), Bon: 259, C&D: 352
Common and widespread. Macroscopically variable, hence the extensive synonymy.

alnicola var. salicicola (Fr.) Holec, *Libri Botanici* 20: 152 (2001)
  Agaricus alnicola β salicicola Fr., *Epicr. syst. mycol.*: 187 (1838)
  Flammula alnicola var. *salicicola* (Fr.) P. Karst, *Bidrag Kännedom Finlands Natur Folk* 32: 804 (1879)
  Pholiota salicicola (Fr.) Arnolds, *Biblioth. Mycol.* 90: 428 (1983) [1982]
  Agaricus amarus Bull., *Herb. France*: pl. 562 (1792)
  Pholiota amara (Bull.) Singer, *Sydowia, Beih.* 7: 85 (1973)
E: ! **S:** ?
H: On decayed wood of *Salix* spp.
Previously included in *P. alnicola* but differs from this in the host and the bitter taste of the basidiomes, unlike the mild taste of typical *P. alnicola*.

astragalina (Fr.) Singer, *Lilloa* 22: 516 (1951) [1949]
  Agaricus astragalinus Fr., *Observ. mycol.* 2: 32 (1818)
  Flammula astragalina (Fr.) P. Kumm., *Führ. Pilzk.*: 82 (1871)
  Dryophila astragalina (Fr.) Quél., *Enchir. fung.*: 71 (1886)
E: ! **S:** ! **O:** Channel Islands: ?
H: On decayed wood of conifers, usually on large stumps.

D: NM2: 264, FAN4: 87-88 **D+I:** Bon: 96-100, B&K4: 328 420 **I:** FlDan4: 121D (as *Flammula astragalina*), Bon: 259, C&D: 350

Rarely reported and poorly known in Britain. Few records, mostly from the late nineteenth century; most often recorded from Scotland, with two English and a single, rather suspect record from the Channel Islands.

aurivella (Batsch) P. Kumm., *Führ. Pilzk.*: 83 (1871)
   Agaricus aurivellus Batsch, *Elench. fung. (Continuatio Prima)*: 153 & t. 22 f. 115 (1786)
   Lepiota squarrosa β *aurivella* (Batsch) Gray, *Nat. arr. Brit. pl.* 1: 602 (1821)
   Pholiota lilacifolia P.D. Orton, *Kew Bull.* 31(3): 719 (1976)
   Mis.: *Pholiota adiposa* sensu Holec (2001)
   Mis.: *Pholiota aurivella* var. *cerifera* sensu auct.
   Mis.: *Pholiota cerifera* sensu auct. Brit.
E: o **S:** ! **W:** ! **ROI:** !
H: On dead or decayed wood of deciduous trees, usually large logs or fallen trunks of *Fagus*. Also known on planted *Acer rubra* and on native *Alnus*, *Betula*, *Corylus*, *Fraxinus*, *Populus*, *Quercus* and *Ulmus* spp. Reported on *Sambucus nigra* but the host unverified.
D: NM2: 262, FAN4: 83-84 **D+I:** Myc7(3): 116, B&K4: 330 421 (as *Pholiota cerifera*), Holec: 40-47 (as *P. adiposa*) **I:** Bon: 257 (as *P. cerifera*)
Occasional but widespread and rather frequent in southern England.

brunnescens A.H. Sm. & Hesler, *N. Am.* Pholiota: 286 (1968)
E: ! **S:** !
H: British collections on burnt soil, or soil and sawdust mixtures.
D: NBA5: 714-715 **D+I:** Holec: 119-121
Known from England (South Devon, Surrey, and Warwickshire) and Scotland (Perthshire). Not collected elsewhere in Europe. Macroscopically resembles large specimens of *Pholiota highlandensis*.

conissans (Fr.) M.M. Moser, *Kleine Kryptogamenflora*, Edn 3: 224 (1967)
   Agaricus conissans Fr., *Epicr. syst. mycol.*: 187 (1838)
   Flammula conissans (Fr.) Gillet, *Hyménomycètes*: 535 (1876)
   Dryophila conissans (Fr.) Quél., *Enchir. fung.*: 71 (1886)
   Dryophila muricella var. *graminis* Quél., *Compt. Rend. Assoc. Franç. Avancem. Sci. Assoc. Sci. France* 15: 485 (1887)
   Pholiota graminis (Quél.) Singer, *Agaricales in Modern Taxonomy*, Edn 2: 555 (1962)
   Agaricus inauratus W.G. Sm., *J. Bot.* 11: 336 (1873)
   Flammula inaurata (W.G. Sm.) Sacc., *Syll. fung.* 5: 820 (1887)
   Pholiota inaurata (W.G. Sm.) M.M. Moser [*nom. inval.*], *Kleine Kryptogamenflora*: 196 (1953)
   Agaricus juncinus W.G. Sm., *J. Bot.* 11: 336 (1873)
   Flammula juncina (W.G. Sm.) Sacc., *Syll. fung.* 5: 816 (1887)
   Mis.: *Pholiota abstrusa* sensu NCL
   Mis.: *Flammula muricella* sensu auct.
E: ! **S:** ! **W:** ! **NI:** !
H: On wood of *Alnus* and *Salix* spp, and on soil with these trees, often amongst grass, rushes or reeds in wet places such as fen, marsh and dried out pond margins.
D: NM2: 264, FAN4: 89-90 **D+I:** B&K4: 330 422, Holec: 76-84 **I:** FlDan4: 122A (as *Flammula conissans*), Bon: 259 (as *Pholiota graminis*), C&D: 350
Rarely reported but apparently widespread. Excludes *P. graminis* sensu B&K4 which is misdetermined *P. gummosa*.

flammans (Batsch) P. Kumm., *Führ. Pilzk.*: 84 (1871)
   Agaricus flammans Batsch, *Elench. fung.*: 87 & t. 7 f. 30 (1783)
   Dryophila flammans (Batsch) Quél., *Enchir. fung.*: 68 (1886)
E: o **S:** c **W:** o **NI:** o
H: On decayed conifer wood, usually on large stumps of *Pinus* spp, less frequently on *Picea* spp, and rarely *Larix* spp. Rarely on decayed wood of deciduous trees such as *Alnus* and *Betula* spp.

D: NM2: 262, FAN4: 90 **D+I:** Ph: 144, B&K4: 332 424, Holec: 86-91 **I:** FlDan3: 109, Bon: 257, C&D: 350
Common in Caledonian pinewoods, occasional to rare elsewhere but possibly increasing and spreading.

gummosa (Lasch) Singer, *Lilloa* 22: 517 (1951) [1949]
   Agaricus gummosus Lasch [non *A. gummosus* Pers. (1800)], *Linnaea* 3: 406 (1828)
   Flammula gummosa (Lasch) P. Kumm., *Führ. Pilzk.*: 82 (1871)
   Dryophila gummosa (Lasch) Quél., *Enchir. fung.*: 71 (1886)
   Agaricus cookei Fr., *Grevillea* 5: 56 (1876)
   Pholiota cookei (Fr.) Sacc., *Syll. fung.* 5: 757 (1887)
   Agaricus ochrochlorus Fr., *Öfvers. Kongl. Vetensk.-Akad. Förh.* 18: 24 (1862)
   Flammula ochrochlora (Fr.) P. Karst., *Bidrag Kännedom Finlands Natur Folk* 32: 411 (1879)
   Pholiota ochrochlora (Fr.) P.D. Orton, *Trans. Brit. Mycol. Soc.* 43(2): 341 (1960)
   Mis.: *Agaricus terrigenus* sensu Cooke [Ill. Brit. fung. 376 (349) Vol. 3 (1885)]
E: c **S:** c **W:** o **NI:** ! **ROI:** !
H: On decayed wood and occasionally on living roots of deciduous trees. Sometimes reported 'on soil' or 'amongst grass', but then growing from buried wood.
D: NCL3: 341 (as *Pholiota ochrochlora*), NM2: 263, FAN4: 88 **D+I:** Ph: 144 (as *P. ochrochlora*), B&K4: 332 426, B&K4: 3324 425 (misdetermined as *P. graminis*), Myc14(3): 133, Holec: 69-73 **I:** FlDan4: 121F (as *Flammula gummosa*), Bon: 259, C&D: 350
Common and widespread. It should be noted that B&K4: 332 & pl. 425 (1995), as *Pholiota graminis*, is actually *P. gummosa*.

heteroclita (Fr.) Quél., *Mém. Soc. Émul. Montbéliard, Sér. 2*, 5: 125 (1872)
   Agaricus heteroclitus Fr., *Observ. mycol.* 2: 223 (1818)
E: ? **S:** !
H: On dead or decayed wood. British collections on *Betula* and *Salix* spp.
D: NM2: 260, FAN4: 99-100, NBA3: 114-115 **D+I:** Holec: 162-166 **I:** FlDan1: 108 C, FlDan3: 108C, Cooke 389 (366) Vol. 3 (1885)
Rare. Known from Scotland on *Betula* spp. Reported from England but these records lack good voucher material. The illustration of '*Agaricus villosus*' in Bolton [Hist. Fung. Halifax. t. 42 (1788)] supposedly represents this species but this interpretation is open to conjecture.

highlandensis (Peck) Quadr., *Quaderni di Accademia Nazionale dei Lincei* 11: 264 (1990)
   Flammula highlandensis Peck, *Rep. (Annual) New York State Mus. Nat. Hist.* 50: 138 (1879)
   Agaricus carbonarius Fr., *Observ. mycol.* 2: 33 (1818)
   Dryophila carbonaria (Fr.) Quél., *Enchir. fung.*: 70 (1886)
   Flammula carbonaria (Fr.) P. Kumm., *Führ. Pilzk.*: 82 (1871)
   Pholiota carbonaria (Fr.) Singer [*nom. illegit.*, non *P. carbonaria* A.H. Sm. (1944)], *Lilloa* 22: 517 (1951) [1949]
   Flammula carbonaria var. *gigantea* J.E. Lange, *Fl. agaric. danic.* 5, *Taxonomic Conspectus*: IV (1940)
   Pholiota persicina P.D. Orton, *Trans. Brit. Mycol. Soc.* 91(4): 567 (1988)
   Pholiota highlandensis f. *persicina* (P.D. Orton) Holec, *Libri Botanici* 20: 118 (2001)
E: c **S:** c **W:** o **NI:** ! **ROI:** ! **O:** Channel Islands: !
H: On firesites in woodland and on heathland.
D: NM2: 262, FAN4: 96 **D+I:** Ph: 146 (as *Pholiota carbonaria*), Bon: 113-118, B&K4: 334 427 **I:** FlDan4: 121B (as *Flammula carbonaria*), Bon: 259, C&D: 352
Common and widespread. Macroscopically variable especially with regard to size and colour of the basidiomes.

jahnii Tjall.-Beuk. & Bas, *Persoonia* 13: 77 (1986)
   Mis.: *Pholiota adiposa* sensu auct.
   Mis.: *Pholiota muelleri* sensu NCL

E: o **S:** !
H: On living or dead wood of deciduous trees, usually *Fagus* in old woodland, and with a preference for calcareous soils.
D: NM2: 261, FAN4: 85-86 **D+I:** Holec: 58-63 **I:** Bon: 257, C&D: 350
Occasional in southern England, and apparently widespread elsewhere. It should be noted that the illustration referred to this species in B&K4: 334 & pl. 428 (1994) is actually *Pholiota squarrosa*.

lenta (Pers.) Singer, *Lilloa* 22: 516 (1951) [1949]
 Agaricus lentus Pers., *Syn. meth. fung.*: 287 (1801)
 Flammula lenta (Pers.) P. Kumm., *Führ. Pilzk.*: 82 (1871)
 Dryophila lenta (Pers.) Quél., *Enchir. fung.*: 70 (1886)
 Agaricus glutinosus Lindgr., *Bot. Not.*: 199 (1845)
 Hebeloma glutinosum (Lindgr.) Sacc., *Syll. fung.* 5: 793 (1887)
 Mis.: *Hebeloma punctatum* sensu Rea (1922)
E: o **S:** o **W:** o **NI:** ! **ROI:** !
H: On soil or humus in coniferous and deciduous woodland.
D: NM2: 262, FAN4: 93 **D+I:** B&K4: 334 429, Holec: 131-136 **I:** FlDan4: 121E (as *Flammula lenta*), FlDan4: 121 E, Bon: 259, C&D: 352, Cooke 448 (430) Vol. 3 (1886) (as *Hebeloma glutinosum*)
Occasional but widespread, generally rather uncommon. A characteristic species with uniformly pale colours and slimy smooth pileus.

limonella (Peck) Sacc., *Syll. fung.* 5: 753 (1887)
 Agaricus limonellus Peck, *Rep. (Annual) New York State Mus. Nat. Hist.* 31: 33 (1879)
 Pholiota ceriferoides P.D. Orton, *Trans. Brit. Mycol. Soc.* 91(4): 565 (1988)
 Mis.: *Pholiota squarrosoadiposa* sensu auct.
E: o **S:** ! **W:** !
H: On dead wood of deciduous trees, often *Fagus*, and usually at the bases of dead standing trunks, on large felled logs, or rarely in knotholes on living trunks.
D: NM2: 262, FAN4: 86 **D+I:** Holec: 53-58 **I:** FlDan1: 109 D
Rarely reported. Previously not distinguished from *Pholiota aurivella* (which has markedly larger spores).

lubrica (Pers.) Singer, *Lilloa* 22: 516 (1951) [1949]
 Agaricus lubricus Pers., *Syn. meth. fung.*: 307 (1801)
 Flammula lubrica (Pers.) P. Kumm., *Führ. Pilzk.*: 81 (1871)
 Dryophila lubrica (Pers.) Quél., *Enchir. fung.*: 70 (1886)
 Agaricus decussatus Fr., *Epicr. syst. mycol.*: 185 (1838)
 Pholiota decussata (Fr.) M.M. Moser, *Kleine Kryptogamenflora*, Edn 3: 242 (1967)
 Pholiota groenlandica M. Lange, *Medd. Grønl.* 148(2): 7 (1957)
E: ! **S:** ?
H: On decayed, often buried wood (and then appearing as if terrestrial), and woodchips.
D: NM2: 262, FAN4: 92-93 **D+I:** B&K4: 330 423 (as *Pholiota decussata*), B&K4: 336 432, Holec: 122-126 **I:** C&D: 352
Known from West Kent (Limpsfield Chart). Reported from a few other sites in England and Scotland but all records are unsubstantiated with voucher material.

lucifera (Lasch) Quél., *Mém. Soc. Émul. Montbéliard, Sér. 2,* 5: 250 (1872)
 Agaricus lucifer Lasch, *Linnaea* 3: 408 (1828)
 Dryophila lucifera (Lasch) Quél., *Enchir. fung.*: 68 (1886)
E: ! **W:** !
H: On dead roots and buried woody debris. Known on *Fraxinus*.
D: NM2: 261 (as *Pholiota lucifer*), FAN4: 101 **D+I:** B&K4: 338 433, Holec: 172-177 **I:** FlDan1: 107 B
Known from England (Derbyshire: Chatsworth, South Devon: Buckfastleigh and West Cornwall: Heligan Gardens) and from Wales (Glamorganshire: Cardiff).

mixta (Fr.) Kuyper & Tjall.-Beuk., *Persoonia* 13(1): 81 (1986)
 Agaricus mixtus Fr., *Epicr. syst. mycol.*: 185 (1838)
 Flammula filia (Fr.) Massee, *Brit. fung.-fl.* 2: 134 (1893)
 Pholiota filia (Fr.) P.D. Orton, *Kew Bull.* 31(3): 716 (1977)

Pholiota xanthophaea P.D. Orton, *Trans. Brit. Mycol. Soc.* 91(4): 568 (1988)
 Pholiota mixta f. *xanthophaea* (P.D. Orton) Holec, *Libri Botanici* 20: 111 (2001)
E: ! **W:** !
H: On soil near to trees or in open woodland. Known with *Betula, Corylus* and *Quercus* spp, also *Larix* and *Pinus* spp.
D: NM2: 263, FAN4: 95, NBA5: 716-717 (as *P. filia*) **D+I:** Holec: 105-110, Holec: 111-112 (as *Pholiota mixta* f. *xanthophaea*)
Rarely recorded or collected and often confused with similar taxa, especially *P. spumosa*. Known from England (Cheshire, Co. Durham, East Kent, Leicestershire, North Hampshire, North Yorshire and Surrey) and Wales (Flintshire). Excludes *P. mixta* sensu Rea (1922) with very large spores, this possibly referable to *P. albocrenulata* but no material has been traced.

pinicola Jacobsson, *Windahlia* 16: 133 (1986)
S: !
H: On decayed conifer wood in Caledonian pinewoods.
D: FAN4: 104 **D+I:** Holec: 153-156 **I:** C&D: 353
Confined to Scotland where it appears to be rare. Closely related to *Pholiota alnicola*.

populnea (Pers.) Kuyper & Tjall.-Beuk., *Persoonia* 13(1): 81 (1986)
 Agaricus populneus Pers., *Mycol. eur.* 3: 171 (1828)
 Hemipholiota populnea (Pers.) Bon, *Doc. Mycol.* 17(65): 52 (1986)
 Agaricus comosus Fr., *Epicr. syst. mycol.*: 165 (1838)
 Pholiota comosa (Fr.) Quél., *Mém. Soc. Émul. Montbéliard, Sér. 2,* 5: 125 (1872)
 Agaricus destruens Brond., *Rec. Pl. Crypt. Agenais*: pl. 6 (1829)
 Pholiota destruens (Brond.) Gillet, *Hyménomycètes*: 442 (1876)
 Mis.: *Pholiota heteroclita* sensu Cooke [Ill. Brit. fung. 389 (366) Vol. 3 (1885)]
E: !
H: On dead wood of *Populus* (and rarely *Salix*) spp, usually growing from the cut faces of large felled trunks or sawn logs.
D: NM2: 261, FAN4: 98, NBA3: 112-113 (as *P. comosa*) **D+I:** Myc7(3): 117, B&K4: 340 437, Holec: 157-162 **I:** FlDan1: 107 C, FlDan3: 107C (as *Pholiota destruens*), Bon: 255 (as *Hemipholiota populnea*), C&D: 354 (as *H. populnea*), Cooke 388 (600) Vol. 3 (1885) (as *Agaricus comosus*)
Known from East Kent, Herefordshire, Middlesex, North Lincolnshire, Northamptonshire, Oxfordshire, Shropshire, Staffordshire, Surrey, Warwickshire and West Kent. This is a large and distinctive species, unlikely to be overlooked and thus probably genuinely rare.

scamba (Fr.) M.M. Moser, *Kleine Kryptogamenflora*, Edn 3: 244 (1967)
 Agaricus scambus Fr., *Observ. mycol.* 2: 45 (1818)
 Ripartites scambus (Fr.) P. Karst., *Bidrag Kännedom Finlands Natur Folk* 32: 479 (1879)
 Flammula scamba (Fr.) Sacc., *Syll. fung.* 5: 828 (1887)
 Paxillus scambus (Fr.) Quél., *Fl. mycol. France*: 110 (1888)
E: ! **S:** o **W:** ? **NI:** !
H: On decayed fallen wood of conifers such as *Larix, Picea* and *Pinus* spp, and rarely on *Betula* spp. Reported 'on soil' but then probably growing from buried wood.
D: NM2: 263, FAN4: 97 **D+I:** B&K4: 340 438, C&D: 352, Holec: 136-140 **I:** FlDan4: 123F (as *Flammula scamba*), FlDan4: 123 F, Bon: 259
Occasional but widespread in Scotland. Known from England (Cumberland, East Norfolk and South Somerset) and Northern Ireland (Fermanagh) but apparently rare. Reported from Wales in 1910 and 1956 but unsubstantiated with voucher material.

spumosa (Fr.) Singer, *Sydowia* 2: 37 (1948)
 Agaricus spumosus Fr., *Syst. mycol.* 1: 252 (1821)
 Flammula spumosa (Fr.) P. Kumm., *Führ. Pilzk.*: 81 (1871)
 Dryophila spumosa (Fr.) Quél., *Enchir. fung.*: 70 (1886)
 Mis.: *Pholiota graveolens* sensu Orton (NBA5)
E: ! **S:** !

H: On 'soil' or buried wood, frequently amongst grass and often near to coniferous trees.
D: NM2: 263, FAN4: 94, NBA5: 718 (as *P. graveolens*) **D+I:** Bon: 101-105, B&K4: 342 441, C&D: 352, Holec: 101-105
Rarely reported. Many herbarium specimens at K named thus have recently been redetermined by Holec, mostly as *Pholiota mixta*.

squarrosa (Weigel) P. Kumm., *Führ. Pilzk.*: 84 (1871)
    Agaricus squarrosus Weigel, *Observ. Bot.*: 40 (1771)
    Lepiota squarrosa (Weigel) Gray, *Nat. arr. Brit. pl.* 1: 602 (1821)
    Dryophila squarrosa (Weigel) Quél., *Enchir. fung.*: 68 (1886)
    Agaricus floccosus Schaeff., *Fung. Bavar. Palat. nasc.* 1: 61 (1762)
    Agaricus verruculosus Lasch, *Linnaea* 3: 408 (1828)
    Pholiota squarrosa var. *verruculosa* (Lasch) Sacc., *Syll. fung.* 5: 749 (1887)
    Mis.: *Pholiota jahnii* sensu B&K4: 334 & pl. 428 (1995)
E: c **S:** c **W:** c **NI:** c **ROI:** c **O:** Isle of Man: o
H: Parasitic, then saprotrophic. On wood of deciduous trees, often at the base of living trunks or on relatively fresh stumps, causing a white rot. Most often collected on *Fagus* and *Fraxinus* but also known on *Betula* and *Buddleja* spp, *Gleditschia triacanthos*, *Liquidambar styraciflua*, *Malus*, *Platanus*, *Populus*, *Prunus*, and *Salix* spp, *Sophora japonica*, *Sorbus aucuparia* and *Tilia* spp. Rarely reported on conifers such as *Abies*, *Pinus* and *Picea* spp, but those hosts are not verified.
D: NM2: 261, FAN4: 83 **D+I:** Ph: 145, B&K4: 342 440, C&D: 350, Holec: 34-38 **I:** Sow3: 284 (as *Agaricus floccosus*), FlDan1: 110 D, FlDan3: 110D, Bon: 257
Common and widespread.

subochracea (A.H. Sm.) A.H. Sm. & Hesler, *N. Am.* Pholiota: 153 (1968)
    Hypholoma subochracea A.H. Sm., *Mycologia* 36: 250 (1944)
    Dryophila nematolomoides J. Favre, *Schweiz. Z. Pilzk.* 36: 67 (1958)
    Pholiota nematolomoides (J. Favre) M.M. Moser, *Kleine Kryptogamenflora*, Edn 3: 243 (1967)
    Mis.: *Flammula inopus* sensu Rea (1922)
S: !
H: On decayed conifer wood in Caledonian pinewoods.
D: NM2: 264, FAN4: 92 **D+I:** B&K4: 338 435 (as *Pholiota nematolomoides*)
Rarely reported. Confined to Scotland *fide* Orton in Bull. BMS 20: 131 (1986) (using the unpublished combination *Pholiota inopus*).

tuberculosa (Schaeff.) P. Kumm., *Führ. Pilzk.*: 83 (1871)
    Agaricus tuberculosus Schaeff., *Fungi Bavar. Palat. Nasc.* 4: 34 (1774)
    Dryophila tuberculosa (Schaeff.) Quél., *Enchir. fung.*: 68 (1886)
    Pleuroflammula tuberculosa (Schaeff.) E. Horak, *Veröff. geobot. Inst. Zürich* 87: 35 (1986)
    Agaricus curvipes Fr. [*nom. illegit.*, non *A. curvipes* Pers. (1801)], *Hymenomyc. eur.*: 223 (1874)
    ?*Agaricus curvipes* Pers., *Syn. meth. fung.*: 312 (1801)
    ?*Pholiota curvipes* (Pers.) Quél., *Mém. Soc. Émul. Montbéliard*, Sér. 2, 5: 250 (1872)
    ?*Dryophila curvipes* (Pers.) Quél., *Enchir. fung.*: 68 (1886)
E: o **W:** !
H: On decayed wood of deciduous trees. Frequently on fallen trunks of *Prunus avium* but also known on *Alnus*, *Betula*, *Corylus*, *Fagus*, *Fraxinus* and *Quercus* spp. A single collection from a pile of decayed coniferous sawdust.
D: NM2: 261, FAN4: 101 **D+I:** Ph: 145, B&K4: 342 441, C&D: 352, Holec: 167-171 **I:** FlDan1: 108 A, FlDan3: 108A, Bon: 257
Occasional but apparently widespread in England. Also collected from Wales (Denbighshire: Coed Coch, and Monmouthshire: Abergavenny, St. Mary's Vale). The identitiy of *Agaricus curvipes* Pers. is uncertain and more fully discussed by Holec (2001).

PHRAGMOXENIDIUM Oberw., *Syst. Appl. Microbiol.* 13: 187 (1990)
Tremellales, Phragmoxenidiaceae
Type: *Phragmoxenidium mycophilum* Oberw. & Schneller

mycophilum Oberw. & Schneller, *Syst. Appl. Microbiol.* 13: 187 (1990)
E: !
H: Parasitic. In the hymenium of *Thanatephorus fusisporus*. Known from South Devon and Surrey.

PHYLLOPORIA Murrill, *Torreya* 4: 141 (1904)
Hymenochaetales, Hymenochaetaceae
Type: *Phylloporia parasitica* Murrill

ribis (Schumach.) Ryvarden, *Polyporaceae North Europe* 2: 371 (1978)
    Boletus ribis Schumach., *Enum. pl.* 2(2): 386 (1803)
    Polyporus ribis (Schumach.) Fr., *Syst. mycol.* 1: 375 (1821)
    Fomes ribis (Schumach.) Gillet, *Hyménomycètes*: 685 (1878)
    Phellinus ribis (Schumach.) Quél., *Enchir. fung.*: 173 (1886)
    Polyporus ribis var. *euonymi* (Kalchbr.) Quél., *Bull. Soc. Bot. France* 23: 149 (1877)
E: o **W:** ! **ROI:** ! **O:** Channel Islands: !
H: Most frequently on living trunks of *Euonymus europaeus* in scrub woodland on calcareous soil. Rarely on *Ribes rubrum* and *R. uva-crispa* and rarely on *Alnus*, *Carpinus*, *Crataegus* and *Prunus* spp, and *Sambucus nigra*.
D: EurPoly2: 535, NM3: 330 **D+I:** B&K2: 264 322 (as *Phellinus ribis*) **I:** Kriegl1: 462
Frequent in southern England. Rarely reported elsewhere but apparently widespread.

PHYLLOPORUS Quél., *Fl. mycol. France*: 409 (1888)
Boletales, Xerocomaceae
Type: *Phylloporus pelletieri* (Lév.) Quél.

pelletieri (Lév.) Quél., *Fl. mycol. France*: 409 (1888)
    Agaricus pelletieri Lév., in Crouan, *Florule Finistère*: 81 (1867)
    Clitocybe pelletieri (Lév.) Gillet, *Hyménomycètes*: 170 (1874)
    Agaricus paradoxus Kalchbr., *Icon. select. Hymenomyc. Hung.* 2: 27 (1874)
    Flammula paradoxa (Kalchbr.) Sacc., *Syll. fung.* 5: 810 (1887)
    Paxillus paradoxus (Kalchbr.) Cooke, *Grevillea* 5: 6 (1875)
    Phylloporus paradoxus (Kalchbr.) Cleland, *Toadstools and Mushrooms and Other Larger Fungi of South Australia* 1: 178 (1934)
    Mis.: *Phylloporus rhodoxanthus* sensu auct. Eur.
E: ! **S:** ! **W:** ! **NI:** !
H: On soil, frequently on sandy banks in deciduous woodland. Known with *Alnus*, *Betula*, *Castanea*, *Corylus*, *Fagus* and *Quercus* spp.
D: NM2: 66 **D+I:** B&K3: 74 39, FungEur8: 132-135 101 (as *Xerocomus pelletieri*) **I:** FlDan4: 134C (as *Paxillus paradoxus*), Bon: 45, C&D: 429 (as *Phylloporus rhodoxanthus*), Kriegl2: 286
Rarely reported but widespread. Usually occurring singly, and often taken at first sight to be *Boletus chrysenteron*.

PHYLLOTOPSIS Singer, *Beih. Bot. Centralbl.* 56: 143 (1936)
Poriales, Lentinaceae
Type: *Phyllotopsis nidulans* (Pers.) Singer

nidulans (Pers.) Singer, *Rev. Mycol. (Paris)* 1(2): 76 (1936)
    Agaricus nidulans Pers., *Syn. meth. fung.*: 482 (1801)
    Pleurotus nidulans (Pers.) P. Kumm., *Führ. Pilzk.*: 105 (1871)
    Crepidotus nidulans (Pers.) Quél., *Champs Jura Vosges* 3: 114 (1875)
    Panus nidulans (Pers.) Pilát, *Mycologia* 5: 90 (1930)
    Agaricus jonquilla Lév., *Icon. Champ.*: 10 (1855)

Crepidotus jonquilla (Lév.) Quél., *Fl. mycol. France*: 75 (1888)
  Panus stevensonii Berk. & Broome, *Ann. Mag. Nat. Hist., Ser. 5 [Notices of British Fungi no. 1796]* 3: 209 (1879)
E: ! **S:** ! **W:** ?
H: On dead or decayed wood of deciduous trees, usually large fallen trunks of *Fagus* and rarely *Salix* spp.
D: BFF6: 19, FAN2: 24, NM2: 48 **D+I:** B&K3: 310 393 **I:** FIDan2: 65D (as *Pleurotus nidulans*), C&D: 151, Kriegl3: 23, SV44: 52
Known from England (Berkshire, Buckinghamshire, Surrey, West Kent and Yorkshire) and Scotland (Midlothian). Reported from East Gloucestershire, South Hampshire and Wales (Glamorganshire and Denbighshire) but unsubstantiated with voucher material.

PHYSALACRIA Peck, *Bull. Torrey Bot. Club* 9: 2 (1882)
Cantharellales, Physalacriaceae
Type: *Physalacria inflata* Peck

cryptomeriae Berthier & Rogerson, *Mycologia* 73(4): 643 (1981)
E: !
H: On fallen leaf litter and twigs of *Cryptomeria japonica*.
D: NM3: 255 **D+I:** Myc7(4): 162-163, Reid (1995a)
Two collections from Berkshire (Windsor Great Park) and West Gloucestershire (Westonbirt Arboretum).

stilboidea (Cooke) Sacc., *Syll. fung.* 9: 256 (1891)
  Pistillaria stilboidea Cooke, *Grevillea* 19: 2 (1890)
E: !
H: On a decayed fallen leaves. British material possibly on *Hedera*.
D: Corner: 468 **D+I:** Myc7(4): 162-163, Reid (1995a)
Two collections from South Devon (Salcombe, Sharpitor Gardens) in 1979.

PHYSISPORINUS P. Karst., *Bidrag Kännedom Finlands Natur Folk* 48: 324 (1889)
Poriales, Coriolaceae
Type: *Physisporinus vitreus* (Pers.) P. Karst

sanguinolentus (Alb. & Schwein.) Pilát, *Atlas champ. Eur.* 3(1): 247 (1940)
  Boletus sanguinolentus Alb. & Schwein., *Consp. fung. lusat.*: 257 (1805)
  Poria sanguinolenta (Alb. & Schwein.) Cooke, *Bot. Macaronés.* 14: 112 (1886)
  Podoporia sanguinolenta (Alb. & Schwein.) Donk, *Meded. Bot. Mus. Herb. Rijks Univ. Utrecht* 9: 158 (1933)
  Rigidoporus sanguinolentus (Alb. & Schwein.) Donk, *Persoonia* 4(3): 341 (1966)
  Polyporus carmichaelianus Grev., *Scott. crypt. fl.* 4: t. 224 (1826)
  Merulius carmichaelianus (Grev.) Berk., *Engl. fl.* 5(2): 130 (1836)
  Mis.: *Podoporia confluens* sensu auct.
E: c **S:** c **W:** c **NI:** c **ROI:** c **O:** Isle of Man: !
H: On decayed or sodden wood or frequently on soil in damp habitats such as ditches, streambanks or overhangs. Rarely found encrusting large mosses or the basal parts of leaves and stems of *Carex* spp, or the grassses *Molinia caerulea* and *Phalaris arundinacea*.
D: EurPoly2: 539, NM3: 173 **D+I:** B&K2: 300 377
Common and widespread.

vitreus (Pers.) P. Karst., *Bidrag Kännedom Finlands Natur Folk* 48: 324 (1889)
  Poria vitrea Pers., *Observ. mycol.* 1: 15 (1796)
  Boletus vitreus (Pers.) Pers., *Syn. meth. fung.*: 545 (1801)
  Polyporus vitreus (Pers.) Fr., *Observ. mycol.*: 265 (1815)
  Podoporia vitrea (Pers.) Donk, *Meded. Bot. Mus. Herb. Rijks Univ. Utrecht* 9: 159 (1933)
  Rigidoporus vitreus (Pers.) Donk, *Persoonia* 4(3): 341 (1966)

Polyporus adiposus Berk. & Broome, *Ann. Mag. Nat. Hist., Ser. 2 [Notices of British Fungi no. 711]* 13: 404 (1854)
E: o **S:** ! **W:** ! **NI:** ! **ROI:** ! **O:** Isle of Man: !
H: On decayed and frequently sodden wood of deciduous trees, rarely on conifers, often spreading onto surrounding soil or leaf litter in damp habitats.
D: EurPoly2: 541
Occasional but widespread in England, most frequently in southern and south-western counties. Basidiomes have a distinct greenish and semi-translucent 'glassy' appearance (hence the epithet *vitreus*) and do not show any colour changes when bruised or damaged.

PILODERMA Jülich, *Ber. Bayer. Bot. Ges.* 81: 415 (1969)
Stereales, Atheliaceae
Type: *Piloderma bicolor* (Peck) Jülich

bicolor (Peck) Jülich, *Ber. Bayer. Bot. Ges.* 81: 417 (1969)
  Corticium bicolor Peck, *Bull. Buffalo Soc. Nat. Hist.* 1(2): 62 (1873)
  Athelia bicolor (Peck) Parmasto, *Eesti N.S.V. Tead. Akad. Toimet., Biol.* 16(4): 379 (1967)
  Piloderma croceum J. Erikss. & Hjortstam, in Eriksson, Hjortstam & Ryvarden, *Corticiaceae of North Europe* 6: 1201 (1981)
E: ! **S:** !
H: Fruiting on litter and decayed wood, usually with *Pinus sylvestris* on acidic soil. Rarely reported with *Quercus* spp.
D: NM3: 153 (as *Piloderma fallax*) **I:** Kriegl1: 305 (as *P. croceum*)
Rarely reported, but easily recognised by the bright yellow resupinate basidiomes with large yellow-orange rhizomorphs. Perhaps better known as *Piloderma croceum*.

byssinum (P. Karst.) Jülich, *Ber. Bayer. Bot. Ges.* 81: 418 (1969)
  Lyomyces byssinus P. Karst., *Meddeland. Soc. Fauna Fl. Fenn.* 11: 137 (1884)
  Corticium byssinum (P. Karst.) Sacc., *J. Linn. Soc., Bot.* 27: 133 (1890)
E: ! **S:** ! **W:** !
H: Fruiting on decayed wood and litter in mixed woodland.
D: NM3: 153 **D+I:** CNE6: 1192-1198, B&K2: 168 180
Known from England (Herefordshire, South Devon, South Hampshire, Surrey and West Kent) and Wales (Breconshire). Reported from Scotland.

PIPTOPORUS P. Karst., *Meddeland. Soc. Fauna Fl. Fenn.* 6: 9 (1881)
Poriales, Coriolaceae
  Buglossoporus Kotl. & Pouzar, *Česká Mykol.* 20: 82 (1966)
Type: *Piptoporus betulinus* (Bull.) P. Karst.

betulinus (Bull.) P. Karst., *Meddeland. Soc. Fauna Fl. Fenn.* 6: 9 (1881)
  Boletus betulinus Bull., *Herb. France*: pl. 312 (1787)
  Polyporus betulinus (Bull.) Fr., *Syst. mycol.* 1: 358 (1821)
  Boletus suberosus L., *Sp. pl.*: 1176 (1753)
E: c **S:** c **W:** c **NI:** c **ROI:** c **O:** Channel Islands: ! Isle of Man: c
H: Parasitic then saprotrophic. On trunks and logs of *Betula* spp.
D: EurPoly2: 545, NM3: 242 **D+I:** Ph: 227, B&K2: 318 404 **I:** Sow2: 212 (as *Boletus betulinus*), Kriegl1: 600
Very common and widespread. Occurs wherever *Betula* grows and is restricted to this host. English names = 'Birch Bracket' or 'Razor Strop Fungus'.

quercinus (Schrad.) P. Karst., *Meddeland. Soc. Fauna Fl. Fenn.* 6: 9 (1881)
  Boletus quercinus Schrad., *Spicil. Fl. Germ.*: 157 (1794)
  Polyporus quercinus (Schrad.) Fr., *Epicr. syst. mycol.*: 441 (1838)
  Buglossoporus quercinus (Schrad.) Kotl. & Pouzar, *Česká Mykol.* 20: 84 (1966)

Buglossoporus pulvinus (Pers.) Donk, *Proc. K. Ned. Akad. Wet.* 74: 4 (1971) [sensu Donk]
E: r **S:** ?
H: On trunks of old *Quercus* spp, occasionally on stumps or fallen wood, in parkland or oak woodland. A collection on *Fagus* has been re-examined and is misdetermined.
D: EurPoly2: 547, NM3: 232 (as *Buglossoporus quercinus*)
Rarely reported but recent survey work has shown it to be less rare in England than previously thought. One dubious old Scottish record (material not seen).

PISOLITHUS Alb. & Schwein., *Consp. fung. lusat.*: 82 (1805)
Sclerodermatales, Sclerodermataceae
Polysaccum F. Desp. & DC., *Mém. Agric. Soc. Agric. Dép. Seine* 1: 8 (1807)
Type: *Pisolithus arhizus* (Scop.) Rauschert

arhizus (Scop.) Rauschert, *Z. Pilzk.* 25: 51 (1959)
Scleroderma arhizum Scop., *Syn. meth. fung.*: 152 (1801)
Pisolithus arenarius Alb. & Schwein., *Consp. fung. lusat.*: 82 (1805)
Lycoperdon capsuliferum Sowerby, *Col. fig. Engl. fung., Suppl.*: pl. 425 a & b (1809)
Polysaccum olivaceum Fr., *Syst. mycol.* 3: 54 (1829)
Polysaccum pisocarpium Fr., *Syst. mycol.* 3: 54 (1829)
Scleroderma tinctorium Pers., *Syn. meth. fung.*: 152 (1801)
Pisolithus tinctorius (Pers.) Coker & Couch, *Gasteromycetes east. U.S.*: 170 (1928)
E: ! **ROI:** !
H: On warm, dry, sandy soils in open woodland, and by roadsides. With *Pinus* spp in Britain.
D: NM3: 296 **D+I:** Ph: 251, BritPuffb: 36-37 Figs. 18/19 **I:** Bon: 303 (as *Pisolithus tinctorius*), Kriegl2: 173
Known from England (Berkshire, East Norfolk, Middlesex, North Hampshire, South Devon, South Hampshire and West Gloucestershire) and from the Republic of Ireland (South Tipperary). English name = 'Dyeball'.

PISTILLINA Quél., *Compt. Rend. Assoc. Franç. Avancem. Sci.* 9: 671 (1880)
Cantharellales, Typhulaceae
Type: *Pistillina hyalina* Quél.
The genus *Pistillina* is here considered a synonym of *Typhula* (Pers.) Fr., but one British species lacks a combination in *Typhula* and is therefore listed here.

brunneola Pat., *Tab. anal. fung.*: 574 (1887)
Pistillaria brunneola (Pat.) Sacc., *Syll. fung.* 6: 759 (1888)
S: !
H: On dead leaves of *Molinia caerulea*
D: Corner: 498
A single collection, from Scotland (Kirkudbright: Bruce's Stone) in 1993.

PLEUROCYBELLA Singer, *Mycologia* 39: 81 (1947)
Agaricales, Tricholomataceae
Type: *Pleurocybella porrigens* (Pers.) Singer

porrigens (Pers.) Singer, *Mycologia* 39: 81 (1947)
Agaricus porrigens Pers., *Observ. mycol.* 1: 54 (1796)
Pleurotus porrigens (Pers.) P. Kumm., *Führ. Pilzk.*: 104 (1871)
Phyllotus porrigens (Pers.) P. Karst., *Bidrag Kännedom Finlands Natur Folk* 32: 92 (1879)
Pleurotellus porrigens (Pers.) Kühner & Romagn., *Flore Analytique des Champignons Supérieurs*: 74 (1953)
Nothopanus porrigens (Pers.) Singer, *Sydowia, Beih.* 7: 19 (1973)
Pleurotus albolanatus Peck, in Kauffman, *Michigan Geol. Biol. Surv. Publ., Biol. Ser. 5,* 26: 672 (1918)
E: ! **S:** c **W:** ! **ROI:** !

H: On decayed conifer wood, usually *Pinus sylvestris*, but also reported on *Larix, Picea* and *Pseudotsuga* spp.
D: BFF6: 61 (as *Phyllotus porrigens*), NM2: 176 **D+I:** Ph: 186 (as *Pleurotellus porrigens*), B&K3: 312 394 (as *Phyllotus porrigens*) **I:** C&D: 151
Common in Scotland, rare elsewhere but possibly increasing as conifer plantations mature. Strictly confined to conifers thus records on deciduous hosts are misdetermined.

PLEUROFLAMMULA Singer, *Mycologia* 38: 521 (1946)
Agaricales, Strophariaceae
Type: *Pleuroflammula dussii* (Pat.) Singer

ragazziana (Bres.) E. Horak, *Persoonia* 9: 443 (1978)
Crepidotus ragazzianus Bres., *Fung. Sc. Col. Eritr.* 5: 176 (1892)
Crepidotus hibernianus A. Pearson & Dennis, *Trans. Brit. Mycol. Soc.* 32(3-4): 268 (1949)
Pleuroflammula hiberniana (A. Pearson & Dennis) Singer, *Sydowia* 15: 70 (1961)
ROI: !
H: On wood. British material on decayed logs of *Tilia* sp.
D: BFF6: 103-104
A single collection (the type of *Crepidotus hibernianus*) from South Kerry (Muckross Park) in 1948. Known from Portugal and widespread in the tropics, thus possibly an introduction.

PLEUROTUS (Fr.) P. Kumm. [*nom. cons.*], *Führ. Pilzk.*: 24 (1871)
Poriales, Lentinaceae
Crepidopus (Nees) Gray [*nom. rej.*], *Nat. arr. Brit. pl.* 1: 616 (1821)
Dendrosarcos Kuntze [*nom. nud.*], *Revis. gen. pl.* 3(2): 462 (1898)
Type: *Pleurotus ostreatus* (Jacq.) P. Kumm.

cornucopiae (Paulet) Rolland, *Atl. Champ. France*: pl. 44 (1910)
Fungus cornucopiae Paulet, *Traité champ.* 2: 2 (1793)
Agaricus cornucopiae (Paulet) Pers., *Mycol. eur.* 3: 37 (1828)
Dendrosarcos cornucopiae (Paulet) Lév., *Iconogr. champ. Paulet*: pl. 28 (1855)
Pleurotus ostreatus f. *cornucopiae* (Paulet) Quél., *Elench. fung.* 1: 148 (1886)
Agaricus dimidiatus Bull., *Herb. France*: pl. 519 (1791)
Agaricus sapidus Schulzer, in Kalchbrenner, *Icon. select. Hymenomyc. Hung.*: pl. 8 (1873)
Pleurotus sapidus (Schulzer) Sacc., *Syll. fung.* 5: 348 (1887)
Mis.: *Pleurotus ostreatus* sensu Cooke [Ill. Brit. fung. 279 (195) Vol. 2 (1883)]
E: o **S:** o **W:** o **NI:** o **ROI:** !
H: On decayed wood of deciduous trees. Very common following Dutch elm disease on large fallen trunks of *Ulmus* spp, but now declining on that substratum. Occasional on *Fagus* and rarely on *Fraxinus, Quercus* and *Sorbus aucuparia*.
D: BFF6: 24, FAN2: 22-23, NM2: 49 **D+I:** Ph: 184-185, Myc3(4): 194
Occasional but widespread. Old *Pleurotus ostreatus* basidiomes are often misidentified as this species.

dryinus (Pers.) P. Kumm., *Führ. Pilzk.*: 101 (1871)
Agaricus dryinus Pers., *Syn. meth. fung.*: 478 (1801)
Pleurotus acerinus (Fr.) Sacc., *Syll. fung.* 5: 360 (1887)
Agaricus corticatus Fr., *Observ. mycol.*: 92 (1815)
Pleurotus corticatus (Fr.) P. Kumm., *Führ. Pilzk.* 1: 101 (1871)
Pleurotus corticatus var. *tephrotrichus* (Fr.) Gillet, *Hyménomycètes*: 340 (1876)
Pleurotus corticatus var. *albertinii* (Fr.) Rea, *Brit. basidiomyc.*: 441 (1922)
Pleurotus spongiosus (Fr.) Sacc., *Syll. fung.* 5: 340 (1887)
E: o **S:** ! **W:** ! **NI:** ! **O:** Channel Islands: !

H: On living wood of deciduous trees, often in knotholes or branch scars, most frequently on *Fagus* and *Fraxinus* and often *Acer pseudoplatanus* and *Sambucus nigra*. Also known on *Aesculus, Betula, Malus, Populus, Salix, Sorbus aucuparia, Tilia, Ulmus* and *Zelkova* spp. Rarely reported on conifers but a recent, verified collection on *Picea* sp.
D: BFF6: 21-22 **D+I:** FAN2: 23-24, B&K3: 313 395 **I:** FlDan2: 62H (as *Pleurotus corticatus*), FlDan2: 62G (as *P. corticatus* var. *tephrotrichus*), Kriegl3: 24
Occasional in England, rarely reported elsewhere. Basidiomes are easily recognisable by the velutinate pilei and the presence of a thick veil (lacking in other British species of *Pleurotus*).

euosmus (Berk.) Sacc., *Syll. fung.* 5: 358 (1887)
  Agaricus euosmus Berk., *Outl. Brit. fungol.*: 135 (1860)
  Pleurotus ostreatus var. *euosmus* (Berk.) Massee, *Brit. fung.-fl.* 2: 372 (1893)
E: ! **S:** !
H: On decayed wood. British material on large stumps and trunks of *Ulmus* spp.
D: BFF6: 22
Rarely reported. Supposedly distinguished by basidiomes smelling of tarragon, but regarded by many authorities as just a variety or form of *Pleurotus ostreatus*. Cooke 280 (196) Vol. 2 (1883) resembles pallid basidiomes of *P. ostreatus*.

ostreatus (Jacq.) P. Kumm., *Führ. Pilzk.*: 105 (1871)
  Agaricus ostreatus Jacq., *Fl. Aust.* 2: pl. 104 (1775)
  Crepidopus ostreatus (Jacq.) Gray, *Nat. arr. Brit. pl.* 1: 616 (1821)
  Pleurotus columbinus Quél., in Bres., *Fungi trident.* 1: 10 (1881)
  Pleurotus ostreatus var. *columbinus* (Quél.) Quél., *Enchir. fung.*: 148 (1886)
  Crepidopus ostreatus β *atroalbus* Gray, *Nat. arr. Brit. pl.* 1: 616 (1821)
  Pleurotus revolutus (J. Kickx f.) Gillet, *Hyménomycètes*: 347 (1876)
  Agaricus salignus Pers., *Syn. meth. fung.*: 479 (1801)
  Pleurotus salignus (Pers.) P. Kumm., *Führ. Pilzk.*: 105 (1871)
  Mis.: *Pleurotus pulmonarius* sensu auct.
E: c **S:** c **W:** c **NI:** o **ROI:** c **O:** Channel Islands: ! Isle of Man: !
H: Weakly parasitic then saprotrophic. On living, fallen and decayed wood of both deciduous and coniferous trees.
D: BFF6: 22, NM2: 49 **D+I:** Ph: 182-183, FAN2: 20-21, B&K3: 314 397 **I:** Sow3: 241 (as *Agaricus ostreatus*), FlDan2: 63C, Kriegl3: 28
Very common, widespread and morphologically variable. Var. *columbinus* is distinguished by young basidiomes being peacock-blue and occuring on *Salix* spp, but this falls within the normal variation of the species. Sensu Cooke 279 (195) Vol. 2 is typical *Pleurotus cornucopiae*. English name = 'Oyster Mushroom'.

pulmonarius (Fr.) Quél., *Mém. Soc. Émul. Montbéliard, Sér. 2,* 5: 11 (1872)
  Agaricus pulmonarius Fr., *Syst. mycol.* 1: 187 (1821)
  Pleurotus ostreatus f. *pulmonarius* (Fr.) Pilát, *Bull. Soc. Mycol. France* 49: 281 (1933)
E: o **S:** ! **W:** ! **NI:** ! **O:** Isle of Man: !
H: On dead wood of deciduous trees, usually *Fagus* but often on *Betula pendula* and *Ilex*. Rarely *Aesculus* and *Sorbus aucuparia*.
D: BFF6: 23, NM2: 49 **D+I:** Ph: 185, FAN2: 21-22
Often recorded, but frequently confused with *Pleurotus ostreatus*. Spores of *P. pulmonarius* are smaller, as are the basidiomes which also tend to become yellowish when dried.

PLICATURA Peck, *Rep. (Annual) New York State Mus. Nat. Hist.* 24: 75 (1872)
Stereales, Amylocorticaceae
  Plicaturopsis D.A. Reid, *Persoonia* 3: 150 (1964)
Type: *Plicatura alni* Peck (= *P. nivea* (Sommerf.) P. Karst.)

crispa (Pers.) Rea, *Brit. basidiomyc.*: 626 (1922)
  Merulius crispus Pers., *Icon. descr. fung.*: pl. 32 (1800)

  Cantharellus crispus (Pers.) Fr., *Syst. mycol.* 1: 323 (1821)
  Trogia crispa (Pers.) Fr., *Öfvers Kongl. Vetensk.-Akad. Förh.* 1: 30 (1862)
  Plicaturopsis crispa (Pers.) D.A. Reid, *Persoonia* 3(1): 50 (1964)
  Plicatura faginea (Schrad.) P. Karst., *Bidrag Kännedom Finlands Natur Folk* 48: 342 (1889)
E: ! **S:** o
H: On dead wood of deciduous trees, usually *Corylus* and *Fagus* but also known from *Alnus* and (apparently) *Aesculus*.
D: BFF6: 70-71 (as *Plicaturopsis crispa*), NM3: 163 **D+I:** CNE6: 1215-1218 (as *P. crispa*), B&K2: 170 183 **I:** Kriegl1: 306
Rarely reported. Occasional in Scotland, and a few records from northern England (Co. Durham, Northumberland and North Yorkshire).

PLUTEUS Fr., *Fl. Scan.*: 338 (1835)
Agaricales, Pluteaceae
The treatment here follows FAN2 based on Vellinga & Shreurs (1985). As a result many species recognised in BFF4 (1986) are reduced to synonymy.
Type: *Pluteus cervinus* (Schaeff.) P. Kumm.

atromarginatus (Singer) Kühner, *Bull. Mens. Soc. Linn. Lyon* 4(1): 51 (1935)
  Pluteus cervinus var. *atromarginatus* Singer, *Z. Pilzk.* 4(3): 40 (1925)
  Pluteus cervinus ssp. *atromarginatus* (Singer) Kühner, *Bull. Soc. Mycol. France* 43: 148 (1927)
  Pluteus cervinus var. *nigrofloccosus* R. Schulz, *Verh. Bot. Vereins Prov. Brandenburg* 54: 102 (1913)
  Pluteus nigrofloccosus (R. Schulz) J. Favre, *Mat. fl. crypt. Suisse*: 104 (1948)
  Pluteus tricuspidatus Velen., *Novit. mycol.*: 143 (1939)
  Mis.: *Pluteus umbrosus* sensu auct.
E: ! **S:** ! **NI:** ! **ROI:** ?
H: On decayed wood of conifers such as *Pinus* spp.
D: BFF4: 20-21, FAN2: 34-35, NM2: 200 (as *Pluteus tricuspidatus*), NM2: 200 (as *P. tricuspidatus*) **D+I:** B&K4: 124 113 (as *P. nigrofloccosus*) **I:** C&D: 283
Known from England (Surrey and Yorkshire), Scotland (Morayshire), and Northern Ireland (Down). Records on deciduous wood probably refer to *Pluteus luctuosus* or *P. umbrosus*.

aurantiorugosus (Trog) Sacc., *Hedwigia* 35(7): 5 (1896)
  Agaricus aurantiorugosus Trog, *Mitt. naturf. Ges. Bern* 32: 388 (1857)
  Pluteus caloceps G.F. Atk., *Ann. Mycol.* 7: 373 (1909)
  Pluteus leoninus var. *coccineus* Massee, *Brit. fung.-fl.* 2: 291 (1893)
  Pluteus coccineus (Massee) J.E. Lange *Fl. agaric. danic.* 2: 88 (1937)
  Mis.: *Pluteus leoninus* sensu Rea (1922) & sensu Cooke [Ill. Brit. fung. 313(412) Vol. 3 (1884)]
E: !
H: On decayed stumps or fallen trunks of deciduous trees. Most often collected on *Ulmus* spp, more rarely on *Fagus, Fraxinus* and *Populus* spp.
D: BFF4: 50-51, FAN2: 55-56, NM2: 202, NM2: 202 **D+I:** FRIC17: 16-17 134c, Myc14(3): 132 **I:** Bon: 199, C&D: 287
Known from Warwickshire southwards. An unmistakable species, genuinely rare, which became locally common during the epidemic of Dutch elm disease.

cervinus (Schaeff.) P. Kumm., *Führ. Pilzk.*: 99 (1871)
  Agaricus cervinus Schaeff., *Fung. Bavar. Palat. nasc.* 1: 6 (1762)
  Agaricus curtisii Berk. & Broome, *Hooker's J. Bot. Kew Gard. Misc.* 1: 98 (1849)
  Pluteus curtisii (Berk. & Broome) Sacc., *Syll. fung.* 5: 675 (1887)
  Agaricus pluteus Batsch, *Elench. fung.*: 79 (1783)
  Agaricus pluteus β *rigens* Pers., *Syn. meth. fung.*: 357 (1801)
  Pluteus cervinus var. *rigens* (Pers.) Sacc., *Syll. fung.* 5: 665 (1887)

E: c **S:** c **W:** c **NI:** c **ROI:** c **O:** Channel Islands: ! Isle of Man: c
H: On decayed wood of deciduous trees. Occasionally on conifers.
D: BFF4: 19, FAN2: 35-36, NM2: 200 (as *Pluteus atricapillis*)
D+I: Ph: 118-119, B&K4: 118 104 **I:** FlDan2: 69A, Bon: 197, SV29: 51 (as *P. atricapillus*), C&D: 283
Very common and widespread. Collections on coniferous wood should be distinguished from *P. pouzarianus* and *P. primus* (the latter not yet British).

chrysophaeus (Schaeff.) Quél., *Mém. Soc. Émul. Montbéliard, Sér. 2,* 5: 82 (1872)
    Agaricus chrysophaeus Schaeff., *Fung. Bavar. Palat. nasc.* 1: 67 (1762)
    Pluteus galeroides P.D. Orton, *Trans. Brit. Mycol. Soc.* 43(2): 354 (1960)
    Pluteus luteovirens Rea, *Trans. Brit. Mycol. Soc.* 12(2-3): 208 (1927)
    Pluteus xanthophaeus P.D. Orton, *Trans. Brit. Mycol. Soc.* 43(2): 366 (1960)
E: o **S:** ! **W:** ! **ROI:** !
H: On very decayed wood of deciduous trees. Most frequent on *Ulmus* spp, less so on *Fagus* and *Fraxinus*. Also known on *Acer pseudoplatanus, Populus* and *Quercus* spp.
D: BFF4: 47-48 (as *Pluteus galeroides*), BFF4: 48-49 (as *P. xanthophaeus*), BFF4: 49-50 (as *P. luteovirens*), FAN2: 50-51, NM2: 202, NBA8: 612 (as *P. xanthophaeus*), NBA8: 612-613 (as *P. galeroides*) **D+I:** FRIC3: 18-21 22b (as *P. xanthophaeus*), Ph: 120 (as *P. luteovirens*), Myc4(4): 199 (as *P. xanthophaeus*) **I:** C&D: 286 (as *P. luteovirens*)
Occasional but widespread in England, from Yorkshire southwards. Rarely reported elsewhere. Several taxa have been described based on possibly transient colour differences.

cinereofuscus J.E. Lange, *Dansk. Bot. Ark.* 2(7): 9 (1917)
    Pluteus nanus var. *major* Massee, *Brit. fung.-fl.* 2: 288 (1893)
    Pluteus olivaceus P.D. Orton, *Trans. Brit. Mycol. Soc.* 43(2): 359 (1960)
    Mis.: *Pluteus godeyi* sensu Orton (BFF4)
    Mis.: *Pluteus thomsonii* sensu Singer [TBMS 39: 216 (1956)]
E: o **S:** ! **W:** ! **ROI:** !
H: On decayed wood of decdiuous trees, or attached to buried woodchips. Often occurring singly.
D: NCL3: 355 (as *Pluteus godeyi*), BFF4: 43-44, BFF4: 44-45 (as *P. godeyi*), FAN2: 51-51, NM2: 202 **D+I:** B&K4: 118 105 **I:** FlDan2: 71G, Bon: 199, C&D: 287
Rarely reported but apparently widespread.

diettrichii Bres., *Ann. Mycol.* 3(2): 160 (1905)
    Pluteus rimulosus Kühner & Romagn., *Bull. Soc. Mycol. France* 72: 182 (1956)
E: !
H: On soil, or rarely very decayed wood of deciduous trees, including woodchip mulch on flowerbeds.
D: BFF4: 55-56 (as *Pluteus rimulosus*), FAN2: 49 **D+I:** FRIC6: 19-20 44c (as *P. rimulosus*), B&K4: 120 106 **I:** C&D: 287
Known from East Kent, Oxfordshire, South Hampshire, Surrey, Warwickshire and West Sussex.

ephebeus (Fr.) Gillet, *Hyménomycètes*: 392 (1876)
    Agaricus ephebeus Fr., *Observ. mycol.* 2: 87 (1818)
    Pluteus lepiotoides A. Pearson, *Trans. Brit. Mycol. Soc.* 35(2): 109 (1952)
    Pluteus murinus Bres., *Ann. Mycol.* 3(2): 160 (1905)
    Pluteus pearsonii P.D. Orton, *Trans. Brit. Mycol. Soc.* 43(2): 361 (1960)
    Agaricus villosus Bull., *Herb. France*: pl. 214 (1785)
    Pluteus villosus (Bull.) Quél., *Fl. mycol. France*: 187 (1888)
    Mis.: *Pluteus plautus* sensu Pearson [TBMS 35: 108 (1952)]
E: ! **W:** !
H: On soil, buried wood or occasionally on sawdust, especially in woodland on calcareous soils.
D: BFF4: 28 (as *Pluteus murinus*), BFF4: 29 (as *P. pearsonii*), BFF4: 30 (as *P. villosus*), FAN2: 38-39, NM2: 201, NBA8: 613-614 (as *P. villosus*) **I:** C&D: 283, C&D: 283 (as *P. murinus*)

Rarely reported. Known from England (from Yorkshire southwards) and Wales (Flintshire and Monmouthshire). Macroscopically variable and split into several taxa by some authors.

exiguus (Pat.) Sacc., *Syll. fung.* 5: 671 (1887)
    Agaricus exiguus Pat., *Tab. anal. fung.* 1: 190 (1886)
E: ! **S:** ! **W:** !
H: On decayed wood or sawdust of deciduous trees.
D: BFF4: 39-40, FAN2: 41, NM2: 200 **D+I:** B&K4: 120 108
Poorly known in Britain. A few old reports from unlocalised sites and recently reported from Surrey and Norfolk but unsubstantiated with voucher material.

griseoluridus P.D. Orton, *Notes Roy. Bot. Gard. Edinburgh* 41(3): 609 (1984)
E: !
H: On decayed wood or on buried woody debris.
D: BFF4: 41-42
Known from Lincolnshire, Middlesex, Surrey and South Hampshire. Reported elsewhere, but unsubstantiated with voucher material.

hispidulus (Fr.) Gillet, *Hyménomycètes*: 391 (1876)
    Agaricus hispidulus Fr., *Observ. mycol.* 2: 97 (1818)
    Pluteus hispidulus var. *cephalocystis* Schreurs, *Persoonia* 12(4): 348 (1985)
E: o
H: On fallen wood, usually on large, moss covered trunks in deciduous woodland on calcareous soil. Most frequently on *Fagus* but also known on *Acer pseudoplatanus* and *Ulmus* spp.
D: BFF4: 38-39, FAN2: 39-40 (as *Pluteus hispidulus* var. *hispidulus*), NM2: 201 **I:** FlDan2: 70B, C&D: 283
Recorded from Lancashire southwards, and not uncommon in southern or south-eastern counties.

inquilinus Romagn., *Bull. Soc. Mycol. France* 94(4): 375 (1979) [1978]
    Mis.: *Pluteus semibulbosus* sensu NCL & sensu BFF4: 55
E: !
H: On decayed wood of deciduous trees. Reported on *Betula* spp and *Fagus*.
D: BFF4: 55 (as *Pluteus semibulbosus*) **I:** C&D: 287
Known from Leicestershire, Mid-Lancashire, Surrey, Warwickshire and West Kent. Often confused with *P. semibulbosus* sensu Gillet (1874: 395) (= *P. plautus*).

insidiosus Vellinga & Schreurs, *Persoonia* 12(4): 366 (1985)
E: !
H: On decayed wood of deciduous trees. British material on *Fagus* and decayed woodchips.
D: FAN2: 49-50, Myc15(4): 163-164 **D+I:** B&K4: 122 109
Two collections, from Shropshire (Oaks Wood) in 2000 and Surrey (Sheen Common) in 2004.

leoninus (Schaeff.) P. Kumm., *Führ. Pilzk.*: 98 (1871)
    Agaricus leoninus Schaeff., *Fung. Bavar. Palat. nasc.* 1: 21 (1762)
    Agaricus sororiatus P. Karst., *Not. Sällsk. Fauna et Fl. Fenn. Förh.* 9: 339 (1868)
    Pluteus sororiatus (P. Karst.) P. Karst., *Bidrag Kännedom Finlands Natur Folk* 32: 254 (1879)
E: !
H: On very decayed wood of deciduous trees. Mostly reported on *Fagus* but also known on *Betula, Quercus* and *Salix* spp.
D: BFF4: 37, FAN2: 41-42, NM2: 200 **D+I:** Ph: 120, B&K4: 122 110 **I:** C&D: 285
Much confused with other yellow *Pluteus* species and the majority of the records are unsubstantiated with voucher material. The filamentous cap cuticle distinguishes this species from *P. chrysophaeus*.

luctuosus Boud., *Bull. Soc. Mycol. France* 21: 70 (1905)
    Pluteus phlebophorus var. *marginatus* Quél., *Compt. Rend. Assoc. Franç. Avancem. Sci.* 13: 2 (1885)
    Pluteus marginatus (Quél.) Bres., *Icon. Mycol.* 11: pl. 546 (1929)
E: ! **S:** ?

H: On decayed wood of deciduous trees, usually decayed trunks or stumps of *Ulmus* spp, and less often *Fagus* and *Fraxinus*.
D: BFF4: 42, FAN2: 53-54 **I:** C&D: 287
Known from Buckinghamshire, East and West Sussex, Middlesex, South Hampshire, South Devon, South Essex, South Somerset, Surrey, and Yorkshire, and reported from Scotland (Easterness).

nanus (Pers.) P. Kumm., *Führ. Pilzk.*: 98 (1871)
  Agaricus nanus Pers., *Syn. meth. fung.*: 357 (1801)
  Pluteus griseopus P.D. Orton, *Trans. Brit. Mycol. Soc.* 43(2): 356 (1960)
E: o **S:** ! **W:** !
H: On decayed wood of deciduous trees or on buried woodchips, in woodland, parks and on flowerbeds mulched with woodchips.
Occasional in England, rarely reported elsewhere.

pellitus (Pers.) P. Kumm., *Führ. Pilzk.*: 98 (1871)
  Agaricus pellitus Pers., *Syn. meth. fung.*: 366 (1801)
E: ! **S:** ! **W:** ? **ROI:** ?
H: On dead wood of deciduous trees, usually on decayed trunks or stumps of *Fagus* in old woodland. Also reported on *Fraxinus*, *Quercus* and *Ulmus* spp.
D: BFF4: 22, FAN2: 37, NM2: 199, C&D: 283 **I:** FlDan2: 70A
Known from England (East & West Kent, Hertfordshire, Isle of Wight, Leicestershire, Middlesex, South Essex, South Hampshire, South Somerset, Surrey, West Sussex and Worcestershire) and Scotland (Isle of Skye and Orkney). Reported from Wales (Caernarvonshire) in 1977 and the Republic of Ireland (Cork) in 1936 but unsubstantiated with voucher material.

petasatus (Fr.) Gillet, *Hyménomycètes*: 395 (1876)
  Agaricus petasatus Fr., *Epicr. syst. mycol.*: 142 (1838)
  Pluteus cervinus var. *petasatus* (Fr.) Massee, *Brit. fung.-fl.* 2: 285 (1893)
  Agaricus patricius Schulzer, in Kalchbrenner, *Icon. select. Hymenomyc. Hung.*: 20 p. 10 (1873)
  Pluteus cervinus var. *patricius* (Schulzer) Massee, *Brit. fung.-fl.* 2: 284 (1893)
  Pluteus patricius (Schulzer) Boud., *Icon. Mycol.* 1: pl. 87 (1904)
  Mis.: *Pluteus curtisii* sensu auct.
E: ! **S:** !
H: On decayed wood of deciduous trees, usually in old woodland on large fallen trunks or stumps of *Fagus*. Rarely on *Acer pseudoplatanus*, *Betula* spp, *Ilex*, *Quercus*, *Salix* or *Tilia* spp, and also on large piles of decayed sawdust.
D: BFF4: 24-25, FAN2: 37-38, NM2: 199 **D+I:** Myc5(1): 42 **I:** FlDan1: 70C & C1, Bon: 197, C&D: 283
Widespread in England and also known from Scotland (East Perthshire, Lanarkshire and Morayshire). Reported elsewhere but unsubstantiated with voucher material and often confused with large, pale specimens of *P. cervinus*.

phlebophorus (Ditmar) P. Kumm., *Führ. Pilzk.*: 98 (1871)
  Agaricus phlebophorus Ditmar, *Deutschl. Fl.* 3: 31 (1817)
  Mis.: *Pluteus chrysophaeus* sensu auct.
E: o **S:** ! **W:** ! **ROI:** !
H: On decayed wood of deciduous trees usually in old woodland on large, fallen and very decayed trunks of *Fagus* but also on *Acer pseudoplatanus*, *Alnus*, *Betula* spp, *Corylus*, *Fraxinus*, *Ilex*, *Quercus*, *Salix*, *Tilia* and *Ulmus* spp. A single collection on dead stems of *Rubus fruticosus*.
D: BFF4: 45-46, FAN2: 51, NM2: 203 **D+I:** B&K4: 126 116 **I:** C&D: 287
Occasional in England but may be locally frequent, especially in southern counties. Rarely reported elsewhere.

plautus (Weinm.) Gillet, *Hyménomycètes*: 394 (1876)
  Agaricus plautus Weinm., *Hymen. Gasteromyc.*: 136 (1836)
  Pluteus alborugosus Kühner & Romagn. [*nom. nud.*], *Flore Analytique des Champignons Supérieurs*: 423 (1953)
  Pluteus boudieri P.D. Orton, *Trans. Brit. Mycol. Soc.* 43(2): 352 (1960)

Pluteus depauperatus Romagn., *Bull. Soc. Mycol. France* 72: 181 (1956)
  Pluteus dryophiloides P.D. Orton, *Notes Roy. Bot. Gard. Edinburgh* 29(1): 115 (1969)
  Pluteus pellitus var. *gracilis* Bres., *Hedwigia* 24: 134 (1885)
  Pluteus gracilis (Bres.) J.E. Lange, *Dansk. Bot. Ark.* 2: 6 (1917)
  Pluteus granulatus Bres., *Fungi trident.* 1: 10 (1881)
  Pluteus punctipes P.D. Orton, *Trans. Brit. Mycol. Soc.* 43(2): 361 (1960)
  Mis.: *Pluteus hiatulus* sensu auct.
E: o **S:** o **NI:** ! **ROI:** !
H: On decayed wood of deciduous, rarely coniferous, trees in old woodland.
D: BFF4: 26-27 (as *Pluteus punctipes*), BFF4: 27-28, BFF4: 32 (as *P. gracilis*), BFF4: 33 (as *P. granulatus*), BFF4: 33-34 (as *P. dryophiloides*), BFF4: 34-35 (as *P. depauperatus*), BFF4: 35-36 (as *P. boudieri*), FAN2: 45, NM2: 201 (as *P. semibulbosus*), NBA2: 56-57 **D+I:** B&K4: 126 117 (as *P. plautus* [white form]) **I:** Ph: 119 N.B. illustration only (as *P. semibulbosus*), C&D: 284 (as *P. granulatus*), C&D: 285, C&D: 285 (as *P. boudieri*), C&D: 285 (as *P. depauperatus*)
Occasional but widespread throughout much of Britain and Ireland. A variable species, often split into many taxa exhibiting intergrading characters.

podospileus Sacc. & Cub., *Syll. fung.* 5: 672 (1887)
  Pluteus minutissimus Maire, *Publ. Inst. Bot. Barcelona* 3: 94 (1937)
  Leptonia seticeps G.F. Atk., *J. Mycol.*: 116 (1902)
  Pluteus seticeps (G.F. Atk.) Singer, *Lloydia* 21(4): 272 (1959)
  Mis.: *Pluteus spilopus* sensu Berkeley & Broome [*Ann. Mag. Nat. Hist., Ser. 5 [Notices of British Fungi no. 1856]* 7: 514 (1881)]
E: o **W:** ! **NI:** !
H: On decayed wood of deciduous trees in old woodland, usually large, fallen and very decayed trunks of *Fagus* and *Fraxinus*. Rarely on *Alnus*, *Aesculus*, *Betula* spp, *Buxus* and *Ulmus* spp. A single collection on old stems of *Hydrangea* sp. Also reported on soil in dunes and in gardens, but probably misidentified.
D: NCL3: 362, BFF4: 58-59, BFF4: 59-60 (as *Pluteus minutissimus*), FAN2: 45-46, NM2: 203 **D+I:** Ph: 118, B&K4: 128 119, B&K4: 128 118 (as *P. podospileus* f. *minutissimus*) **I:** Bon: 199, C&D: 285 (as *P. seticeps*), C&D: 285
Occasional and may be locally frequent, especially in southern England in old beechwoods on calcareous soil. Rarely reported elsewhere.

pouzarianus Singer, *Sydowia* 36: 283 (1984) [1983]
E: ! **W:** ! **ROI:** ?
H: On decayed wood of conifers, usually on or around stumps or on large piles of decayed sawdust.
D: FAN2: 35 **D+I:** B&K4: 128 120
Known from England (North Hampshire and Surrey) and Wales (Anglesey) but only recently recognised as a separate species. Distinct from *Pluteus cervinus* by the presence of clamped hyphae in the pileipellis. Reported from the Republic of Ireland but unsubstantiated with voucher material.

pseudorobertii M.M. Moser & Stangl, *Z. Pilzk.* 29(2): 39 (1963)
  Mis.: *Pluteus robertii* sensu Ricken [*Blätterpilze Deutschl.*: 277 (1913)]
E: ! **W:** ?
H: On decayed wood of deciduous trees.
D: BFF4: 23-24, NM2: 199
Accepted on the basis of the record in BFF4 and a specimen in herb. E from Somerset. Poorly known in Britain and most records are unsusbtantiated with voucher material.

robertii (Fr.) P. Karst, *Bidrag Kännedom Finlands Natur Folk* 32: 255 (1879)
  Agaricus robertii Fr., *Ic. Hymenomyc.* 1: pl. 101 (1867)
E: !
H: On decayed wood of deciduous trees such as *Fagus* and *Ulmus* spp, in old woodland.

D+I: BFF4: 30-31

Known from Middlesex (Ealing, Perivale Wood). Reported from Berkshire, Huntingdonshire, and South Hampshire but unsubstantiated with voucher material. In FAN2: this is considered a synonym of *P. ephebeus*.

romellii (Britzelm.) Sacc., *Syll. fung.* 11: 44 (1895)
  Agaricus romellii Britzelm., *Hymenomyc. Südbayern* 8: 5 (1891)
  Pluteus lutescens (Fr.) Bres., *Icon. Mycol.* 11: 544 (1929)
  Pluteus nanus var. *lutescens* (Fr.) P. Karst., *Bidrag Kännedom Finlands Natur Folk* 32: 256 (1879)
  Pluteus splendidus A. Pearson, *Trans. Brit. Mycol. Soc.* 35(2): 110 (1952)
E: o **S:** o **W:** o **NI:** ! **ROI:** ! **O:** Isle of Man: !
H: On decayed wood of deciduous trees such as *Betula* spp, *Corylus*, *Fagus* and *Fraxinus* in woodland, on woodchips, sawdust and soil mixed with woody debris including old fire sites.
D: BFF4: 51-52, BFF4: 52 (as *Pluteus splendidus*), FAN2: 48, NM2: 202 **D+I:** Ph: 120 (as *P. lutescens*), B&K4: 130 123, Myc15(2): 60-61 **I:** FlDan2: 72F (as *P. nanus* var. *lutescens*), Bon: 199, C&D: 287
Occasional but widespread.

salicinus (Pers.) P. Kumm., *Führ. Pilzk.*: 99 (1871)
  Agaricus salicinus Pers., *Icon. descr. fung.*: pl. 9 (1798)
  Pluteus salicinus var. *beryllus* Sacc., *Syll. fung.* 5: 668 (1887)
  Mis.: *Pluteus petasatus* sensu Ricken [*Blätterpilze Deutschl.*: 277 (1913)]
E: c **S:** c **W:** c **NI:** c **ROI:** c
H: On decayed wood of deciduous trees in old woodland, often on *Ulmus* spp, *Fagus* and *Fraxinus* and rather uncommonly on *Salix* spp (despite the specific epithet). Also known on *Alnus*, *Betula* spp, *Carpinus*, *Corylus*, *Fraxinus*, *Populus* and *Quercus* spp, and *Sambucus nigra*.
D: BFF4: 21-22, FAN2: 33-34, NM2: 200 **D+I:** Ph: 118-119, B&K4: 132 125 **I:** FlDan2: 69C, Bon: 197, C&D: 283
Common and widespread.

satur Kühner & Romagn., *Bull. Soc. Mycol. France* 72: 182 (1956)
  Pluteus pallescens P.D. Orton, *Trans. Brit. Mycol. Soc.* 43(2): 360 (1960)
E: !
H: On decayed wood of deciduous trees or occasionally on soil containing woody debris.
A very few collections from widely scattered localities (Bedfordshire, Buckinghamshire, East Norfolk, Oxfordshire, Surrey, Warwickshire and West Kent). The concept adopted here follows Orton (BFF4: 54, type examined). In FAN2 this usage was considered a misapplication.

thomsonii (Berk. & Broome) Dennis, *Trans. Brit. Mycol. Soc.* 31(3-4): 206 (1948)
  Agaricus thomsonii Berk. & Broome, *Ann. Mag. Nat. Hist., Ser. 4 [Notices of British Fungi no. 1523]* 17: 131 (1876)
  Entoloma thomsonii (Berk. & Broome) Sacc., *Syll. fung.* 5: 693 (1887)
  Pluteus cinereus Quél., *Ann. Sci. Nat. Bord. Sud-ouest., Suppl.* 14: 3 (1884)
E: o **S:** ! **W:** ! **NI:** ! **ROI:** !
H: On decayed wood of deciduous trees in old woodland, usually *Fagus* and *Fraxinus* and often on small, partially buried fragments. Also known on *Acer pseudoplatanus*, *Betula* spp, *Corylus*, *Crataegus*, *Populus*, *Quercus*, *Salix* and *Ulmus* spp, and a single collection with a potted specimen of *Dracaena* sp. in a house.
D: NCL3: 365, BFF4: 56-57, FAN2: 46-47, NM2: 202 **D+I:** FRIC2: 20-22 14b, B&K4: 132 126, Myc15(2): 61 **I:** Bon: 199, C&D: 287, Cooke 336 (374) Vol. 3 (1885) [as *Agaricus (Entoloma) thomsonii*]
Occasional in southern England, especially in deciduous woodland on calcareous soil, and may be locally abundant. Rarely reported elsewhere.

umbrosus (Pers.) P. Kumm., *Führ. Pilzk.*: 98 (1871)

  Agaricus umbrosus Pers., *Icon. descr. fung.*: pl. 8 (1798)
E: o **S:** ! **W:** ! **NI:** ! **ROI:** !
H: On decayed wood of deciduous trees in old woodland, often large, fallen and very decayed trunks or stumps of *Fagus* and *Ulmus* spp. Also known on *Acer pseudoplatanus*, *Betula*, *Fraxinus*, *Populus* and *Quercus* spp.
D: BFF4: 25-26, FAN2: 43-44, NM2: 201 **D+I:** Ph: 118-119, B&K4: 134 127 **I:** Bon: 197, FP13 3
Occasional in southern England, and for a time common on large fallen trunks and stumps of *Ulmus* spp, in the aftermath of Dutch elm disease. Rarely reported elsewhere, but widespread.

PODOSCYPHA Pat., *Essai taxon.*: 70 (1900)
Stereales, Podoscyphaceae
Type: *Podoscypha surinamensis* (Lév.) Pat.

multizonata (Berk. & Broome) Pat., *Ann. Cryptog. Exot.* 1: 6 (1928)
  Thelephora multizonata Berk. & Broome, *Ann. Mag. Nat. Hist., Ser. 3 [Notices of British Fungi no. 1028]* 15: 321 (1865)
  Stereum multizonatum (Berk. & Broome) Massee, *J. Linn. Soc., Bot.* 27: 167 (1890)
E: o **S:** ?
H: Parasitic. On living roots of *Quercus* spp, rarely *Fagus*, in old open woodland or parkland.
D+I: Ph: 222, Myc10(4): 180
Widespread in southern England. A single record from Scotland (Angus) in 1886 but unsubstantiated with voucher material.

POLYPORUS Adans. [non *Polyporus* Gray (1821)], *Fam. Pl.* 2: 10 (1763)
Poriales, Polyporaceae
  Hexagonia Pollini, *Hort. Veron. Pl. Nov.*: 35 (1816)
  Polyporellus P. Karst., *Meddeland. Soc. Fauna Fl. Fenn.* 5: 37 (1879)
  Dendropolyporus (Pouzar) Jülich, *Biblioth. Mycol.* 85: 397 (1982)
Type: *Polyporus tuberaster* (Jacq.) Fr.

brumalis (Pers.) Fr., *Syst. mycol.* 1: 348 (1821)
  Boletus brumalis Pers., *Neues Mag. Bot.* 1: 107 (1794)
  Polyporus fuscidulus (Schrad.) Fr., *Epicr. syst. mycol.*: 431 (1838)
E: c **S:** c **W:** c **NI:** ! **ROI:** !
H: On dead wood of deciduous trees, usually on small branches, sticks or twigs still attached to fallen trees. Especially common on *Betula* spp and *Fagus*. Also known on *Alnus*, *Corylus*, *Crataegus* spp, *Fraxinus*, *Populus* spp, *Prunus avium*, *Quercus* spp, *Sambucus nigra* and *Ulmus* spp. Fruiting from early winter to early spring.
D: NM2: 50, EurPoly2: 563 **D+I:** B&K2: 326 416 **I:** Bon: 315, Kriegl1: 607
Common and widespread. *Polyporus ciliatus*, a similar but much less frequent species, fruits in summer.

ciliatus Fr., *Observ. mycol.* 1: 123 (1815)
  Polyporus lepideus Fr., *Observ. mycol.* 2: 253 (1818)
  Boletus substrictus Bolton, *Hist. fung. Halifax* Suppl.: 170 (1791)
E: ! **S:** ! **W:** !
H: On dead, decayed wood of deciduous trees in old woodland. Usually reported on *Fagus*. but also said to occur on *Betula*, *Corylus*, *Prunus*, and *Salix* spp, *Sorbus aucuparia* and *Ulmus* spp.
D: NM2: 50, EurPoly2: 565 **D+I:** B&K2: 326 417 **I:** Bon: 315, Kriegl1: 610
Frequently reported but rarely collected. Winter records may represent *Polyporus brumalis*.

durus (Timmerm.) Kreisel, *Boletus* 1: 30 (1984)
  Boletus durus Timmerm., *Fl. Megapol. Prodr.*: 271 (1788)
  Boletus badius Pers. [non *B. badius* (Fr.) Fr. (1821)], *Syn. meth. fung.*: 523 (1801)

Grifola badia (Pers.) Gray, *Nat. arr. Brit. pl.* 1: 644 (1821)
Polyporus badius (Pers.) Schwein., *Trans. Am. phil. Soc.* 4: 155 (1832)
Polyporus picipes Fr., *Epicr. syst. mycol.*: 440 (1838)
Polyporellus picipes (Fr.) P. Karst., *Meddeland. Soc. Fauna Fl. Fenn.* 5: 37 (1880)
E: c **S:** ! **W:** o **NI:** o
H: On dead, decaying wood of deciduous trees, usually large, fallen and very decayed trunks of *Fagus* in old woodland and once common on fallen trunks of *Ulmus* spp. Also known on *Acer pseudoplatanus, Alnus, Betula, Castanea, Crataegus, Fraxinus, Quercus* and *Salix* spp.
D: NM2: 51 (as *Polyporus badius*), EurPoly2: 561 **D+I:** Ph: 218-219, B&K2: 326 415 **I:** SV42: 44 (as *P. badius*), Kriegl1: 605 (as *P. badius*)
Common in southern England. Occasional but widespread elsewhere. Better known as *Polyporus badius*.

leptocephalus (Jacq.) Fr., *Syst. mycol.* 1: 349 (1821)
Boletus leptocephalus Jacq., *Misc. Austriaca Bot.* 1: t. 12 (1778)
Coltricia leptocephala (Jacq.) Gray, *Nat. arr. Brit. pl.* 1: 645 (1821)
Polyporus elegans Bull., *Herb. France*: pl. 46 (1780)
Polyporus varius var. *elegans* (Bull.) Gillot & Lucand, *Cat. Champ. Marn., Suppl.*: 327 (1891)
Boletus nummularius Bull., *Herb. France*: pl. 124 (1783)
Polyporus elegans var. *nummularius* (Bull.) Fr., *Syst. mycol.* 1: 381 (1821)
Polyporus varius var. *nummularius* (Bull.) Fr., *Syst. mycol.* 1: 353 (1821)
Coltricia nummularia (Bull.) Gray, *Nat. arr. Brit. pl.* 1: 644 (1821)
Polyporus nummularius (Bull.) Pers, *Mycol. eur.* 2: 44 (1825)
Boletus nigripes With., *Arr. Brit. pl. ed. 3,* 4: 316 (1796)
Boletus varius Pers., *Syn. meth. fung.*: 85 (1796)
Grifola varia (Pers.) Gray, *Nat. arr. Brit. pl.* 1: 644 (1821)
Polyporus varius (Pers.) Fr., *Syst. mycol.* 1: 352 (1821)
Mis.: *Boletus calceolus* sensu Withering [*Arr. Brit. pl. ed. 3,* 4: 389 (1801)]
E: c **S:** c **W:** c **NI:** c **ROI:** c **O:** Channel Islands: ! Isle of Man: !
H: On dead and decayed wood of deciduous trees, usually large logs, stumps or fallen trunks of *Fagus*. Less often on *Acer pseudoplatanus, Betula, Corylus, Fraxinus, Populus, Quercus, Salix* spp, *Sorbus aucuparia, Tilia* and *Ulmus* spp, and a single collection supposedly on *Taxus*.
D: NM2: 50 (as *Polyporus varius*), EurPoly2: 586 **D+I:** Ph: 218-219, B&K2: 330 423 **I:** Sow1: 89 (as *P. nummularius*), Bon: 315 (as *P. varius*)
Very common and widespread. Perhaps better known as *Polyporus varius*.

melanopus (Pers.) Fr., *Syst. mycol.* 1: 347 (1821)
Boletus melanopus Pers., *Syn. meth. fung.*: 516 (1801)
Polyporellus melanopus (Pers.) P. Karst., *Meddeland. Soc. Fauna Fl. Fenn.* 5: 37 (1880)
E: ! **S:** !
H: Basidiomes appear terrestrial, but arise from dead or decayed tree roots. Usually associated with *Fagus* in beechwoods on calcareous soil, but also reported with *Betula* and *Quercus* spp, and *Sorbus aucuparia*.
D: NM2: 51, EurPoly2: 569 **D+I:** B&K2: 328 418
Known from England (Bedfordshire, Leicestershire, Northumberland, Surrey, West Kent and West Sussex) and Scotland (Easterness). Reported from Co. Durham, Shropshire and Yorkshire but unsubstantiated with voucher material.

squamosus (Huds.) Fr., *Syst. mycol.* 1: 343 (1821)
Boletus squamosus Huds., *Fl. angl.,* Edn 2, 2: 626 (1778)
Boletus testaceus With. [non *B. testaceus* Pers. (1825)], *Bot. arr. veg.* 2: 770 (1776)
Boletus cellulosus Lightf., *Fl. scot.*: 1032 (1777)
Boletus juglandis Schaeff., *Fung. Bavar. Palat. nasc.* 2: 101 (1763)
Boletus rangiferinus Bolton, *Hist. fung. Halifax* 3: 138 (1789)

Grifola platypora Gray, *Nat. arr. Brit. pl.* 1: 643 (1821)
Polyporus rostkovii Fr., *Epicr. syst. mycol.*: 439 (1838)
E: c **S:** c **W:** c **NI:** c **ROI:** c **O:** Channel Islands: o Isle of Man: c
H: A wound parasite, also on dead, dying or decayed wood, usually of deciduous trees. Most common on *Acer pseudoplatanus, Fagus, Fraxinus* and *Ulmus* spp. Also known on wide range of other hosts, and very rarely on conifers including *Juniperus* and *Pinus* spp.
D: NM2: 50, EurPoly2: 578 **D+I:** Ph: 218, B&K2: 330 421, C&D: 144 fig. 105 **I:** Bon: 315, C&D: 145, Kriegl1: 614
Common (perhaps slightly less so in northern areas) and widespread. *Boletus rangiferinus* was applied to abnormal 'stagshorn' forms found in mines and cellars. English name = 'Dryad's Saddle'.

tuberaster (Jacq.) Fr., *Syst. mycol.* 1: 347 (1821)
Boletus tuberaster Jacq., *Collect. Bot. Spectantia (Vindobonae)* 5: pls 8 & 9 (1796)
Polyporus boucheanus (Klotzsch) Fr., *Hymenomyc. eur.*: 533 (1874)
Polyporus coronatus Rostk., *Deutschl. Fl.,* Edn 3: 17 (1848)
Polyporus floccipes Rostk., *Deutschl. Flora,* Edn 3: 25 (1848)
Polyporus forquignonii Quél., *Ann. Mycol.* 18: 67 (1884)
Polyporus lentus Berk., *Engl. fl.* 5(2): 134 (1836)
E: o **S:** ! **W:** ! **NI:** ! **ROI:** ! **O:** Channel Islands: ! Isle of Man: !
H: On fallen branches of deciduous trees in old woodland, usually *Corylus* and *Fagus* and less often *Fraxinus*. Also on *Betula, Buxus, Quercus, Salix, Tilia* spp, and *Ulex europaeus* and known from a single collection on *Rosa* sp.
D: NM2: 50, EurPoly2: 581 **D+I:** Ph: 220 (as *Polyporus floccipes* [small specimens]), B&K2: 330 422 **I:** Bon: 315, Kriegl1: 616
Occasional but may be locally common in England. Rarely reported elsewhere but apparently widespread. Basidiomes are said to arise from a large sclerotium but this is usually not found. Small specimens on thin branches were previously called *Polyporus lentus* but are merely depauperate forms.

umbellatus (Pers.) Fr., *Syst. mycol.* 1: 354 (1821)
Boletus umbellatus Pers., *Syn. meth. fung.*: 519 (1801)
Grifola umbellata (Pers.) Pilát, *Beih. Bot. Centralbl., Abt. A* 52 (1934)
Dendropolyporus umbellatus (Pers.) Jülich, *Biblioth. Mycol.* 85(6): 400 (1982) [1981]
E: ! **S:** ! **W:** !
H: On the ground in deciduous woodland, probably arising from tree roots. Reported with *Betula, Carpinus, Fagus* and *Tilia* spp.
D: NM2: 49, EurPoly2: 583 **D+I:** Ph: 220 (as *Grifola umbellata*), B&K2: 318 405 (as *Dendropolyporus umbellatus*) **I:** Bon: 315, C&D: 145, Kriegl1: 617
Rare. The large basidiomes with multiple, circular, stipitate pilei arising from a common stipe are conspicuous and distinctive. Known from Hertfordshire, Lincolnshire, Northumberland, South Hampshire, West Kent, West Sussex and Worcestershire. Last recorded from Scotland (Perthshire) in 1908 and Wales (Clwyd and Denbighshire) in 1910, but no material has been traced.

POROSTEREUM Pilát, *Bull. Soc. Mycol. Fr.* 52: 330 (1936)
Stereales, Classiculaceae

spadiceum (Pers.) Hjortstam & Ryvarden, *Synopsis Fungorum* 4: 51 (1990)
Thelephora spadicea Pers., *Syn. meth. fung.*: 568 (1801)
Stereum spadiceum (Pers.) Quél., *Fl. mycol. France*: 15 (1888)
Lopharia spadicea (Pers.) Boidin, *Bull. Mens. Soc. Linn. Lyon.* 28(7): 211 (1959)
E: ! **S:** - **W:** - **NI:** - **EI:** - **O:** -
H: On decorticated and decayed wood of deciduous trees. British collection probably on *Fagus*.
D: NM3: 189 **D+I:** CNE4: 847-850 (as *Lopharia spadicea*), B&K2: 208 240 **I:** SV48: 37

A single collection from North Hampshire (Alice Holt Forest, Abbot's Wood) in 2004. Reported previously but unsubstantiated with voucher material or found to be misidentified *Stereum gausapatum*.

POROTHELEUM Fr. [non *Porothelium* Eschw. (1824)], *Observ. mycol.* 2: 272 (1818)
Schizophyllales, Stromatoscyphaceae
Stromatoscypha Donk, *Reinwardtia* 1: 218 (1951)
Type: *Porotheleum fimbriatum* (Pers.) Fr.

fimbriatum (Pers.) Fr., *Observ. mycol.* 2: 272 (1818)
  Poria fimbriata Pers., *Neues Mag. Bot.* 1: 109 (1794)
  Boletus fimbriatus (Pers.) Pers., *Syn. meth. fung.*: 546 (1801)
  Polyporus fimbriatus (Pers.) Fr., *Syst. mycol.* 1: 506 (1821)
  Stromatoscypha fimbriata (Pers.) Donk, *Reinwardtia* 1: 219 (1951)
  Mis.: *Porotheleum friesii* sensu Rea (1922) and sensu auct. Brit.
E: ! S: ?
H: On bare soil, leaf litter and very decayed woody debris on the ground, in old woodland.
D: NM3: 164 **D+I:** B&K2: 204 232 **I:** Kriegl1: 414
Rarely reported. A single collection from West Sussex (Petworth, The Mens and The Cut) in 2001. Reported from South Hampshire but unsubstantiated with voucher material. Also reported from Scotland by Reid & Austwick (1963), but no material has been traced.

PORPHYRELLUS E.-J. Gilbert, *Les Bolets*: 99 (1931)
Boletales, Strobilomycetaceae
Phaeoporus Bataille [*nom. illegit.*, non *Phaeoporus* J. Schröt. (1888)], *Bull. Soc. Hist. Nat. Doubs.* 15: 31 (1908)
Type: *Porphyrellus porphyrosporus* (Fr. & Hök) E.-J. Gilbert

porphyrosporus (Fr. & Hök) E.-J. Gilbert, *Les Bolets*: 99 (1931)
  Boletus porphyrosporus Fr. & Hök, *Boleti fung. gen.*: 13 (1835)
  Phaeoporus porphyrosporus (Fr. & Hök) Bataille, *Les Bolets*: 11 (1908)
  Porphyrellus pseudoscaber Singer, *Farlowia* 2(1): 115 (1945)
E: o S: o W: ! NI: ! ROI: !
H: On soil in old mixed deciduous woodland, usually with *Fagus* or *Quercus* spp. Rarely with conifers such as *Pinus* and *Pseudotsuga* spp.
D: NM2: 72 **D+I:** B&K3: 50 1 **I:** C&D: 421, Kriegl2: 315 (as *Tylopilus porphyrosporus*)
Occasional but widespread. Perhaps most frequent in Scotland. Records show a distinctly northern and western bias in Britain but known as far south as South Devon and West Sussex. Only recently reported from Northern Ireland (Antrim and Down).

PORPOLOMA Singer, *Sydowia* 6: 198 (1952)
Agaricales, Tricholomataceae
Type: *Porpoloma sejunctum* Singer

elytroides (Scop.) Singer, *Sydowia, Beih.* 7: 19 (1973)
  Agaricus elytroides Scop., *Fl. carniol.*: 424 (1772)
  Tricholoma elytroides (Scop.) P. Karst, *Bidrag Kännedom Finlands Natur Folk* 32: 40 (1879)
S: !
H: On soil in conifer woodland, possibly only with *Pinus* spp.
D: Bull. Soc. Mycol. France 65: 132-141 (1949)
A single collection (near Glasgow) in 1951.

metapodium (Fr.) Singer, *Sydowia, Beih.* 7: 19 (1973)
  Agaricus metapodius Fr., *Observ. mycol.* 2: 110 (1818)
  Hygrophorus metapodius (Fr.) Fr., *Epicr. syst. mycol.*: 328 (1838)
  Hygrocybe metapodia (Fr.) M.M. Moser, *Kleine Kryptogamenflora*, Edn 3: 63 (1967)
E: ! S: ! W: ! NI: ! ROI: !

H: On soil amongst grass, in unimproved grassland.
D: NM2: 177, BFF8: 78-79, FAN4: 150 **I:** Bon: 163, FungEur3: 65, SV35: 52
Rarely recorded or collected but apparently widespread. Perhaps slightly more frequent in Scotland and Wales.

spinulosum (Kühner & Romagn.) Singer, *Sydowia* 15: 53 (1961)
  Tricholoma spinulosum Kühner & Romagn., *Bull. Mens. Soc. Linn. Lyon* 16: 134 (1947)
E: !
H: On calcareous soil in deciduous woodland and copses.
D: NM2: 177, BFF8: 79-80, FAN4: 149-150 **D+I:** FRIC2: 11-12 12 **I:** Bon: 163, C&D: 209
Known from Surrey, West Kent and West Sussex.

POSTIA Fr., *Hymenomyc. eur.*: 586 (1874)
Poriales, Coriolaceae
  Oligoporus Bref., *Unters. Gesammtgeb. Mykol.* 8: 114 (1888)
  Osteina Donk, *Schweiz. Z. Pilzk.* 44: 86 (1966)
  Strangulidium Pouzar, *Česká Mykol.* 21: 206 (1967)
  Ceriomyces Corda [non *Ceriomyces* Murrill (1909)] (anam.), in Sturm, *Deutschl. Fl.* Abt. 3, 3: 133 (1837)
  Ptychogaster Corda (anam.), *Icon. fung.* 2: 23 (1838)
Type: *Postia lactea* (Fr.) P. Karst. (= *P. tephroleuca*)

balsamea (Peck) Jülich, *Persoonia* 11(4): 423 (1982)
  Polyporus balsameus Peck, *Rep. (Annual) New York State Mus. Nat. Hist.* 30: 46 (1878)
  Tyromyces balsameus (Peck) Murrill, *North. Polypor.*: 13 (1914)
  Oligoporus balsameus (Peck) Gilb. & Ryvarden, *Mycotaxon* 22(2): 364 (1985)
  Tyromyces kymatodes Donk, *Meded. Bot. Mus. Herb. Rijks Univ. Utrecht* 9: 154 (1933)
  Mis.: *Polyporus kymatodes* sensu auct.
  Mis.: *Leptoporus kymatodes* sensu auct.
E: o W: ! NI: ?
H: On dead, decayed wood of conifers such as *Abies, Cedrus, Cryptomeria, Cupressus, Larix, Picea* and *Pinus* spp, and *Taxus*. Rarely on deciduous hosts, but known on *Acer pseudoplatanus, Alnus, Betula, Crataegus, Fagus, Quercus* and *Ulmus* spp.
D: EurPoly2: 403 (as *Oligoporus balsameus*), NM3: 236
Occasional in England and a single collection from Wales (Caernarvonshire). Reported from Northern Ireland but unsubstantiated with voucher material.

caesia (Schrad.) P. Karst., *Rev. Mycol. (Toulouse)* 3(9): 17 (1881)
  Boletus caesius Schrad., *Spicil. Fl. Germ.*: 167 (1794)
  Polyporus caesius (Schrad.) Fr., *Syst. mycol.* 1: 360 (1821)
  Leptoporus caesius (Schrad.) Quél., *Enchir. fung.*: 176 (1886)
  Tyromyces caesius (Schrad.) Murrill, *North American Flora* 9: 34 (1907)
  Oligoporus caesius (Schrad.) Gilb. & Ryvarden, *Mycotaxon* 22(2): 365 (1985)
E: o S: o W: o NI: o ROI: o O: Channel Islands: ! Isle of Man: !

H: On dead or decayed wood of conifers. Usually on *Pinus sylvestris* but also known on *Larix* and *Picea* spp. Reported on deciduous hosts but unsubstantiated with voucher material. Rarely on worked wood (decayed planks or fence posts) in outdoor habitat.
D: EurPoly2: 404 (as *Oligoporus caesius*), NM3: 234 **D+I:** Ph: 232-233 (as *Tyromyces caesius*), B&K2: 272 334 **I:** Kriegl1: 550 (as *O. caesius*)
Occasional but widespread. Confused with *Postia subcaesia* but less frequent than that species.

ceriflua (Berk. & M.A. Curtis) Jülich, *Persoonia* 11(4): 423 (1982)
  Polyporus cerifluus Berk. & M.A. Curtis, *Grevillea* 1: 50 (1872)
  Tyromyces cerifluus (Berk. & M.A. Curtis) Murrill, *North American Flora* 9: 33 (1907)

Oligoporus cerifluus (Berk. & M.A. Curtis) Ryvarden & Gilb.,
  Mycotaxon 22: 365 (1985)
  Leptoporus minusculoides Pilát, *Atlas champ. Eur.*
    Polyporaceae I 3(1): 193 (1938)
  Polyporus revolutus Bres., *Ann. Mycol.* 18: 35 (1920)
  Tyromyces revolutus (Bres.) Bondartsev & Singer, *Ann.*
    Mycol. 39: 52 (1941)
E: !
H: Usually on dead or decayed wood of conifers, most often
  Pinus sylvestris. Also known on deciduous hosts such as *Acer*
  pseudoplatanus, Betula, Fagus and *Quercus* spp.
D: EurPoly2: 406 (as *Oligoporus cerifluus*), NM3: 234
Widespread in England from Warwickshire southwards.

floriformis (Quél.) Jülich, *Persoonia* 11(4): 423 (1982)
  Polyporus floriformis Quél., in Bres., *Fungi trident.* 1: 61
    (1884)
  Tyromyces floriformis (Quél.) Bondartsev & Singer, *Ann.*
    Mycol. 34: 51 - 52 (1941)
  Oligoporus floriformis (Quél.) Gilb. & Ryvarden, *Mycotaxon*
    22(2): 365 (1985)
E: !
H: Usually on decayed structural timbers but also known in
  natural habitat, on decayed wood of conifers such as *Pinus*
  sylvestris and *Taxus*. Single collections on very decayed fallen
  trunks of *Fagus* and *Prunus avium*.
D: EurPoly2: 408 (as *Oligoporus floriformis*), NM3: 234
Collections from Bedfordshire, Berkshire, Mid-Lancashire,
Oxfordshire, South Devon, Surrey, West Sussex and Yorkshire.

folliculocystidiata (Kotl. & Vampola) Niemelä, *Karstenia* 41: 9
  (2001)
  Oligoporus folliculocystidiatus Kotl. & Vampola, *Czech Mycol.*
    47(1): 59 (1993)
E: !
H: On very decayed wood of conifers. British material probably
  on *Picea* sp.
D: EP2 409
A single collection from Surrey (Sheepleas, Mountain Wood) in
2004.

fragilis (Fr.) Jülich, *Persoonia* 11(4): 423 (1982)
  Polyporus fragilis Fr., *Elench. fung.* 1: 86 (1828)
  Leptoporus fragilis (Fr.) Quel., *Enchir. fung.*: 176 (1886)
  Tyromyces fragilis (Fr.) Donk, *Meded. Bot. Mus. Herb. Rijks*
    Univ. Utrecht 9: 148 (1933)
  Oligoporus fragilis (Fr.) Gilb. & Ryvarden, *Mycotaxon* 22(2):
    365 (1985)
  Polyporus keithii Berk. & Broome, *Ann. Mag. Nat. Hist.,*
    Ser. 4 [Notices of British Fungi no. 1430] 15: 30 (1875)
E: ! **S:** ! **W:** ? **NI:** ? **ROI:** ?
H: On dead, decayed wood of conifers, usually *Pinus sylvestris*
  but also known on *Picea* spp.
D: EurPoly2: 410 (as *Oligoporus fragilis*), NM3: 236 **D+I:** B&K3:
  272 335
Collections from England (Bedfordshire, East Kent, East & West
Sussex, London, North Somerset, South Devon, Surrey, West
Cornwall, Worcestershire and Yorkshire) and Scotland
(Dunbartonshire, East Perthshire and Mid-Ebudes). Often
confused with *P. leucomallella*.

guttulata (Peck) Jülich, *Persoonia* 11(4): 423 (1982)
  Polyporus guttulatus Peck, *Syll. fung.* 6: 106 (1888)
  Tyromyces guttulatus (Peck) Murrill, *North American Flora,*
    Ser.1 (Fungi 3) 9(1): 31 (1907)
  Oligoporus guttulatus (Peck) Gilb. & Ryvarden, *Mycotaxon*
    22(2): 365 (1985)
  Mis.: *Ptychogaster rubescens* (anam.) sensu auct.
E: !
H: Normally associated with decayed wood of conifers but
  British material was (supposedly) on a decayed log of *Fagus*.
D: EurPoly2: 412 (as *Oligoporus guttulatus*), NM3: 236
Accepted with reservations. Only one collection named thus in
  herb. K from West Kent (Westerham, Squerries Court) in 1958,
  but this is in poor condition and on an odd host.

hibernica (Berk. & Broome) Jülich, *Persoonia* 11(4): 423 (1982)
  Polyporus hibernicus Berk. & Broome, *Ann. Mag. Nat. Hist.,*
    Ser. 4 [Notices of British Fungi no. 1291] 7: 428 (1871)
  Poria hibernica (Berk. & Broome) Cooke, *Grevillea* 14: 112
    (1886)
  Tyromyces hibernicus (Berk. & Broome) Ryvarden, *Acta*
    Mycol. Sin. 5(4): 231 (1986)
E: ! **NI:** ? **ROI:** !
H: On fallen and decayed trunks of conifers in woodland or on
  heathland. British material on *Pinus* spp.
D: EurPoly2: 414 (as *Oligoporus hibernicus*), NM3: 235
Described from the Republic of Ireland, but type material has
not been located. Collections from England (South Somerset,
Surrey, West Suffolk and Worcestershire). Reported elsewhere
but unsubstantiated with voucher material.

leucomallella (Murrill) Jülich, *Persoonia* 11(4): 423 (1982)
  Tyromyces leucomallellus Murrill, *Bull. Torrey Bot. Club* 67:
    63 (1940)
  Oligoporus leucomallellus (Murrill) Gilb. & Ryvarden,
    Mycotaxon 22(2): 364 (1985)
  Tyromyces gloeocystidiatus Kotl. & Pouzar, *Westfäl. Pilzbriefe*
    4: 45 - 47 (1963)
  Mis.: *Postia fragilis* sensu auct.
E: o **S:** o **W:** o **NI:** !
H: On very decayed wood (usually large trunks or logs) of
  conifers such as *Larix* spp and *Pinus sylvestris*. Rarely reported
  on deciduous hosts such as *Betula* spp, *Fagus* and *Fraxinus*.
D: EurPoly2: 418 (as *Oligoporus leucomallellus*), NM3: 236
Occasional but widespread. Often mistaken for the far less
frequent *Postia fragilis*. Basidiomes of *P. leucomallella*,
however, do not bruise reddish brown and the large hymenial
gloeocystidia are diagnostic.

lowei (Pilát) Jülich, *Persoonia* 11(4): 423 (1982)
  Leptoporus lowei Pilát, *Atlas champ. Eur. Polyporaceae I*
    3(1): 205 (1938)
E: !
H: British collection on a large fallen trunk of *Fagus* in mixed
  deciduous and conifer woodland. Usually associated with
  conifers such as *Picea* spp in native woodland in continental
  Europe.
D: EurPoly2: 419 (as *Oligoporus lowei*), NM3: 237
Accepted with reservations. A single collection from Surrey
(Chipstead, The Long Plantation) in 1991 (det. L. Ryvarden),
but on an odd host and in an odd habitat. Difficult to
distinguish from *Postia cerifula*.

placenta (Fr.) M.J. Larsen & Lombard, *Mycotaxon* 26: 271
  (1986)
  Polyporus placentus Fr., *Öfvers. Kongl. Vetensk.-Akad. Förh.*
    18: 30 (1862)
  Poria placenta (Fr.) Cooke, *Grevillea* 14: 110 (1886)
  Ceriporiopsis placenta (Fr.) Domański, *Acta Soc. Bot.*
    Poloniae. 32: 732 (1963)
  Oligoporus placentus (Fr.) Gilb. & Ryvarden, *Mycotaxon*
    22(2): 365 (1985)
  Poria incarnata Pers., *Tent. Disp. Meth. fung.*: 75 (1797)
  Boletus incarnatus (Pers.) Pers., *Syn. Meth. Fung.* 2: 546
    (1801)
  Polyporus incarnatus (Pers.) Fr., *Syst. mycol.* 1: 379 (1821)
  Ceriporiopsis incarnata (Pers.) Domański, *Acta Soc. Bot.*
    Poloniae. 32(4): 731 (1963)
E: ! **S:** !
H: On dead and decayed wood of conifers, usually *Pinus*
  sylvestris but also *Larix* sp. and *Pseudotsuga menziesii*.
D: EurPoly2: 424 (as *Oligoporus placentus*) **D+I:** B&K2: 274
  339 (as *Tyromyces placenta*) **I:** Kriegl1: 555 (as *O. placentus*)
Collections from England (Berkshire, Cheshire, North Somerset,
Shropshire, South Hampshire, Surrey and West Kent) and
Scotland (Angus, East Perthshire and Mid-Perthshire).

ptychogaster (F. Ludw.) Vesterh., Knudsen & Hansen, *Nordic*
  J. Bot. 16(2): 213 (1996)
  Polyporus ptychogaster F. Ludw., *Z. Gesammten Naturwiss.*
    (Halle) 3: 424 (1880)

Tyromyces ptychogaster (F. Ludw.) Donk, *Meded. Bot. Mus. Herb. Rijks Univ. Utrecht* 9: 153 (1933)
   Oligoporus ptychogaster (F. Ludw.) Falck & O. Falck, *Hausschwammforschungen*: 12 (1937)
   Ptychogaster albus Corda (anam.), *Icon. fung.* 2: 24 t. 12 (1838)
   Ceriomyces albus (Corda) Sacc. (anam.), *Syll. fung.* 6: 388 (1888)
   Ceriomyces richonii Sacc. (anam.), *Syll. fung.* 6: 388 (1888)
   Ptychogaster fuliginoides (Steud.) Donk (anam.), *Proc. K. Ned. Akad. Wet.* 74: 124 (1972)
   Mis.: *Polyporus destructor* sensu Smith (1908)
E: ! **S:** ! **W:** ! **ROI:** ! **O:** Isle of Man: !
H: Weakly parasitic or saprotrophic. On decayed wood of conifers such as *Chamaecyparis lawsoniana, Picea* and *Pinus* spp, usually on decayed stumps. Occasionally clustered around the stems and roots of living *Calluna* on heathland or moorland.
D: NM3: 235 **I:** Kriegl1: 557 (as *Oligoporus ptychogaster*)
The anamorph (*Ptychogaster albus*) is not infrequent, but the poroid teleomorph (formed on the underside of the anamorph) is more rarely seen.

rancida (Bres.) M.J. Larsen & Lombard, *Mycotaxon* 26: 272 (1986)
   Poria rancida Bres., *Fungi trident.* 2(14): 96 (1900)
   Oligoporus rancidus (Bres.) Gilb. & Ryvarden, *North American Polypores* 2: 482 (1987)
E: !
H: British material on sandy soil and decayed needle litter in a plantation of five-needled *Pinus* sp.
D: EurPoly2: 427 (as *Oligoporus rancidus*)
A single collection, from West Suffolk (Mildenhall Woods) in 1996. One other collection named thus (in herb. K) has been redetermined as sterile *Heterobasidion annosum*.

rennyi (Berk. & Broome) Rajchenb., *Bol. Soc. argent. Bot.* 29(1-2): 117 (1993)
   Polyporus rennyi Berk. & Broome, *Ann. Mag. Nat. Hist., Ser. 4 [Notices of British Fungi no. 1433]* 15: 31 (1875)
   Poria rennyi (Berk. & Broome) Cooke, *Grevillea* 14(72): 112 (1886)
   Strangulidium rennyi (Berk. & Broome) Pouzar, *Česká Mykol.* 21: 206 (1967)
   Oligoporus rennyi (Berk. & Broome) Donk, *Persoonia* 6: 214 (1971)
   Tyromyces rennyi (Berk. & Broome) Ryvarden, *Svensk bot. Tidskr.* 68: 281 (1974)
   Ptychogaster citrinus Boud. (anam.), *J. Bot. (Morot)* 1: 8 (1887)
E: c **S:** !
H: On very decayed wood of conifers. Most frequently on large decayed logs or fallen trunks of *Pinus sylvestris* but also known on *Picea* and *Sequoiadendron* spp. Rarely on decayed structural timber in buildings, on pit props in mines, and on decayed fence posts.
D: NM3: 235
Collections from England (Bedfordshire, Berkshire, Buckinghamshire, Herefordshire, North Somerset, Northamptonshire, Oxfordshire, South Hampshire, Surrey and West Norfolk) and Scotland (Peebleshire). The teleomorph is transient and rarely seen, but the lemon-yellow anamorph (*Ptychogaster citrinus*) is not uncommon.

sericeomollis (Romell) Jülich, *Persoonia* 11(4): 424 (1982)
   Polyporus sericeomollis Romell, *Arch. Bot. (Leipzig)* 11(3): 22 (1911)
   Leptoporus sericeomollis (Romell) Pilát, *Atlas champ. Eur. Polyporaceae I* 3(1): 197 (1936)
   Tyromyces sericeomollis (Romell) Bondartsev, *Ann. Mycol.* 39: 52 (1941)
   Strangulidium sericeomolle (Romell) Pouzar, *Česká Mykol.* 21: 206 (1967)
   Oligoporus sericeomollis (Romell) Pouzar, *Česká Mykol.* 38: 203 (1984)
E: o **W:** ! **O:** Isle of Man: !

H: On decayed wood of conifers, most often on large decayed fallen trunks or logs. Usually on *Pinus sylvestris* but also known on *Cedrus, Larix,* and *Pseudotsuga* spp, and *Taxus*. Very rarely on deciduous trees but there is a confirmed collection on *Fagus* in herb. K.
D: 236, EurPoly2: 431 (as *Oligoporus sericeomollis*)
Widespread in England.

stiptica (Pers.) Jülich, *Persoonia* 11(4): 423 (1982)
   Boletus stipticus Pers., *Syn. meth. fung.*: 525 (1801)
   Polyporus stipticus (Pers.) Fr., *Syst. mycol.* 1: 359 (1821)
   Leptoporus stipticus (Pers.) Quél., *Enchir. fung.*: 176 (1886)
   Tyromyces stipticus (Pers.) Kotl. & Pouzar, *Česká Mykol.* 13: 28 (1959)
   Oligoporus stipticus (Pers.) Gilb. & Ryvarden, *North American Polypores* 2: 485 (1987)
   Polyporus albidus (Schaeff.) Trog in Fr. *Epicr. syst. mycol.*: 475 (1838)
   Bjerkandera colliculosa P. Karst., *Hedwigia* 29: 177 (1890)
E: c **S:** c **W:** c **NI:** o **ROI:** ! **O:** Channel Islands: !
H: On decayed wood of conifers such as *Larix, Picea, Pinus, Pseudotsuga, Tsuga* and *Taxus* spp, usually on the cut ends of large logs or felled trunks. Very rarely on deciduous hosts but known on *Alnus, Crataegus* sp., *Fagus, Fraxinus* and *Ilex*.
D: EurPoly2: 433 (as *Oligoporus stipticus*), NM3: 237 **D+I:** Ph: 232-233 (as *Tyromyces stipticus*), B&K2: 272 336
Common and widespread.

subcaesia (A. David) Jülich, *Persoonia* 11(4): 423 (1982)
   Tyromyces subcaesius A. David, *Bull. Soc. Linn. de Lyon* 43: 120 (1974)
   Oligoporus subcaesius (A. David) Ryvarden & Gilb., *Syn. Fungorum* 7(2): 435 (1994)
E: c **S:** o **W:** o **NI:** c **ROI:** !
H: On fallen and decayed wood of deciduous trees and shrubs, often on fallen sapling trunks or thin branches and sticks in old woodland. Usually on *Acer pseudoplatanus, Corylus, Fagus, Fraxinus* and *Salix* spp, but also known on *Betula* spp, *Buxus, Populus alba, Quercus, Tilia* and *Ulmus* spp.
D: EurPoly2: 435 (as *Oligoporus subcaesius*), NM3: 234 **D+I:** B&K2: 274 337 **I:** SV30: 16
Common in England, especially so in woodland on calcareous soil in southern areas, and in Northern Ireland. Occasional elsewhere. Records on conifer wood are doubtful and possibly all refer to *Postia caesia*.

tephroleuca (Fr.) Jülich, *Persoonia* 11(4): 423 (1982)
   Polyporus tephroleucus Fr., *Syst. mycol.* 1: 360 (1821)
   Tyromyces tephroleucus (Fr.) Donk, *Meded. Bot. Mus. Herb. Rijks Univ. Utrecht* 9: 151 (1933)
   Oligoporus tephroleucus (Fr.) Gilb. & Ryvarden, *Mycotaxon* 22(2): 365 (1985)
   Polyporus lacteus Fr., *Syst. mycol.* 1: 359 (1821)
   Postia lactea (Fr.) P. Karst., *Rev. Mycol. (Toulouse)* 3(9): 17 (1881)
   Leptoporus lacteus (Fr.) Quél., *Enchir. fung.*: 176 (1886)
   Tyromyces lacteus (Fr.) Murrill, *North American Flora, Ser. 1 (Fungi 3),* 9: 36 (1907)
E: c **S:** o **W:** o **NI:** ! **ROI:** ! **O:** Channel Islands: ! Isle of Man: !
H: On decayed wood of deciduous trees, usually large fallen trunks in old woodland. Most frequent on *Fagus* but also known on *Alnus, Crataegus, Corylus, Fraxinus, Malus, Quercus, Populus, Sorbus, Tilia* and *Ulmus* spp. Rarely reported on conifers such as *Picea, Pinus* and *Taxus* spp. A single collection (in herb. K) on *Cytisus scoparius*.
D: EurPoly2: 435 (as *Oligoporus tephroleucus*), NM3: 237 **D+I:** B&K2: 274 338
Common in England. Occasional elsewhere but widespread. Perhaps better known as *Tyromyces lacteus*.

wakefieldiae (Kotl. & Pouzar) Pegler & E.M. Saunders, *Mycologist* 8(1): 28 (1994)
   Tyromyces wakefieldiae Kotl. & Pouzar, *Česká Mykol.* 43(1): 39 (1989)
   Mis.: *Tyromyces ellipsosporus* sensu auct.
   Mis.: *Leptoporus ellipsosporus* sensu auct.

E: o
H: On dead and decayed wood, usually large fallen trunks in old woodland. Known on *Betula, Crataegus, Fagus, Fraxinus, Quercus* and *Ulmus* spp, and *Cupressus macrocarpa, Larix, Picea,* and *Pinus* spp and *Taxus.*
D: EurPoly2: 695 (as *Tyromyces wakefieldiae*)
Known from Warwickshire southwards. Occasional, but can be common in some areas of old woodland, such as the New Forest.

PSATHYRELLA (Fr.) Quél., *Mém. Soc. Émul. Montbéliard, Sér. 2,* 5: 152 (1872)
Agaricales, Coprinaceae
Psathyra (Fr.) P. Kumm. [*nom. illegit.,* non *Psathyra* Spreng. (1818)(*Rubiaceae*)], *Führ. Pilzk.*: 20 (1871)
Pannucia P. Karst., *Bidrag Kännedom Finlands Natur Folk* 32(26): 512 (1879)
Drosophila Quél., *Enchir. fung.*: 115 (1886)
Type: *Psathyrella gracilis* (Pers.) Quél. (= *P. corrugis*)

ammophila (Durieu & Lév.) P.D. Orton, *Trans. Brit. Mycol. Soc.* 34: 180 (1960)
Agaricus ammophilus Durieu & Lév., *Expl. Sci. Algérie [Bot. I]*: 31 (1868)
Psilocybe ammophila (Durieu & Lév.) Gillet, *Hyménomycètes*: 587 (1878)
Drosophila ammophila (Durieu & Lév.) Kühner & Romagn. [*nom. inval.*], *Flore Analytique des Champignons Supérieurs*: 358 (1953)
E: o **S:** o **W:** o **NI:** ! **ROI:** ! **O:** Channel Islands: !
H: On sand or sandy soil, usually associated with *Ammophila arenaria* in dunes.
D: NM2: 240 **D+I:** Kits1: 101-103 **I:** FlDan4: 148B (as *Psilocybe ammophila*), Bon: 267, C&D: 264 f. 789, SV33: 45, Myc14(3): 108
Occasional but widespread.

artemisiae (Pass.) Konrad & Maubl., *Encycl. Mycol.* 14, *Les Agaricales* I: 127 (1948)
Agaricus artemisiae Pass., *Nuovo Giorn. Bot. Ital.* 4: 82 (1872)
Hypholoma artemisiae (Pass.) Massee, *Eur. Fung. Fl. (Agaricaceae)*: 215 (1902)
Mis.: *Psathyrella gossypina* sensu Rea (1922)
Mis.: *Psathyrella squamosa* sensu NCL and sensu auct. mult.
E: o **S:** ! **W:** !
H: In open deciduous woodland (often on acidic soil) attached to small decayed twigs or sticks, usually buried in soil.
D+I: Kits1: 245-248, B&K4: 262 319 **I:** FlDan4: 145C (as *Hypholoma artemisiae*), Ph: 175 (as *Psathyrella squamosa*), Bon: 267, C&D: 270
Widespread in England. Also known from Scotland (Angus and East Sutherland) and Wales (Anglesey).

artemisiae *var.* **microspora** Kits van Wav., *Persoonia* 2: 282 (1985)
W: !
H: On soil in deciduous woodland.
D: Kits1: 248
Described from Merionethshire.

atrolaminata Kits van Wav., *Persoonia* 11(3): 362 (1981)
Mis.: *Psathyrella caudata* sensu Lange (FlDan4: 99 & pl. 155A) & sensu NCL
E: !
H: On soil, usually on buried and decayed woodchips or on bark mulch on flowerbeds. Rarely in natural habitat at the edges of deciduous woodland.
D: NM2: 239 **D+I:** Kits1: 54-56 **I:** FlDan1: 99 pl. 155A (as *Psathyrella caudata*)
Known from Bedfordshire, London, Middlesex, South Somerset, South Wiltshire, Surrey, West Kent and Yorkshire.

bipellis (Quél.) A.H. Sm., *J. Elisha Mitchell Sci. Soc.* 62: 187 (1946)
Psathyra bipellis Quél., *Compt. Rendu Assoc. Franç. Avancem. Sci.* 12: 501 (1884)
Psathyra barlae Bres., *Fungi trident.* 1: 84 (1881)
Psathyrella barlae (Bres.) A.H. Sm., *Contr. Univ. Michigan Herb.* 5: 39 (1941)
Mis.: *Agaricus atrorufus* sensu Cooke [Ill. Brit. fung. 602 (571) (1885)]
E: !
H: On soil along path edges or in open deciduous woodland.
D: NM2: 241 **D+I:** Kits1: 107-111, B&K4: 262 321 **I:** C&D: 266
Known from East & West Norfolk, Staffordshire, Surrey, West Kent and Worcestershire. Reported from other vice-counties but unsubstantiated with voucher material.

borgensis Kits van Wav., *Persoonia* 13(3): 332 (1987)
E: !
H: On clay soil in mixed deciduous woodland. British material associated with *Quercus* sp.
D+I: Kits2: 332-333
A single collection, from Middlesex (Ealing, Perivale Wood) in 1997.

candolleana (Fr.) Maire, in Maire & Werner, *Mem. Soc. Sci. Nat. Maroc* 45: 112 (1937)
Agaricus candolleanus Fr., *Observ. mycol.* 2: 182 (1818)
Hypholoma candolleanum (Fr.) Quél., *Mém. Soc. Émul. Montbéliard, Sér. 2,* 5: 146 (1872)
Drosophila candolleana (Fr.) Quél., *Enchir. fung.*: 115 (1886)
Agaricus appendiculatus Bull., *Herb. France* 9: pl. 392 (1789)
Hypholoma appendiculatum (Bull.) Quél., *Mém. Soc. Émul. Montbéliard, Sér. 2,* 5: 146 (1872)
Psathyrella appendiculata (Bull.) Maire, in Maire & Werner, *Mem. Soc. Sci. Nat. Maroc.* 45: 112 (1937)
Agaricus appendiculatus var. *lanatus* Berk. & Broome, *Ann. Mag. Nat. Hist., Ser. 5 [Notices of British Fungi no. 1876]* 7: 127 (1881)
Agaricus catarius Fr., *Hymenomyc. eur.*: 296 (1874)
Hypholoma catarium (Fr.) Massee, *Brit. fung.-fl.* 1: 393 (1892)
Agaricus egenulus Berk. & Broome, *Ann. Mag. Nat. Hist., Ser. 3 [Notices of British Fungi no. 915]* 7: 375 (1861)
Hypholoma egenulum (Berk. & Broome) Sacc., *Syll. fung.* 5: 1040 (1887)
Psathyrella egenula (Berk. & Broome) M.M. Moser, *Kleine Kryptogamenflora*: 206 (1953)
Agaricus felinus Pass., *Nuovo Giorn. Bot. Ital.* 4: 82 (1872)
Hypholoma felinum (Pass.) Sacc., *Syll. fung.* 5: 1040 (1887)
Psathyrella microlepidota P.D. Orton, *Trans. Brit. Mycol. Soc.* 43(2): 375 (1960)
Agaricus vinosus Corda [non *A. vinosus* Bull. (1786)], in Sturm, *Deutschl. Fl.* Abt. 3, 19-20: 13, pl. 4 (1839)
Psathyra corrugis var. *vinosa* (Corda) Rea, *Brit. basidiomyc.*: 414 (1922)
E: c **S:** c **W:** c **NI:** c **ROI:** c **O:** Channel Islands: ! Isle of Man: !
H: On soil and decayed wood in deciduous woodland, also amongst grass on lawns and on mud or decayed grass debris by dried-out pools or ponds.
D: NM2: 244 **D+I:** Ph: 172-173, B&K4: 264 322 **I:** Sow3: 324 (as *Agaricus appendiculatus*), Bon: 269, C&D: 268
Very common and widespread. A cosmopolitan species, variable macroscopically as reflected in the large number of synonyms [see van Waveren in TBMS 75:429-437 (1980)].

canoceps (Kauffman) A.H. Sm., *Contr. Univ. Michigan Herb.* 5: 43 (1941)
Hypholoma canoceps Kauffman, *Pap. Michigan Acad. Sci.* 5: 132 (1926)
Mis.: *Psathyra pennata* sensu Lange (FlDan4: 94 & pl. 151C)
E: ! **S:** ?
H: In deciduous woodland, on soil and also attached to fragments of buried wood.
D+I: Kits1: 140-141, B&K4: 264 323
Known from Bedfordshire, South Devon, South Wiltshire, Warwickshire, West Norfolk and Worcestershire. Three Scottish records are all old and unsubstantiated with voucher material.

caput-medusae (Fr.) Konrad & Maubl., *Encycl. Mycol.* 14, *Les Agaricales* I: 127 (1948)
 Agaricus caput-medusae Fr., *Epicr. syst. mycol.*: 216 (1838)
 Stropharia caput-medusae (Fr.) P. Karst., *Bidrag Kännedom Finlands Natur Folk* 32: 493 (1879)
 Geophila caput-medusae (Fr.) Quél., *Enchir. fung.*: 112 (1886)
 Agaricus jerdonii Berk. & Broome, *Ann. Mag. Nat. Hist., Ser. 3 [Notices of British Fungi no. 913]* 7: 375 (1861)
 Stropharia jerdonii (Berk. & Broome) Sacc., *Syll. fung.* 5: 1025 (1887)
 Psathyrella jerdonii (Berk. & Broome) Konrad & Maubl., *Encycl. Mycol.* 14, *Les Agaricales* I: 127 (1948)
E: ? **S:** ! **W:** ?
H: On decayed wood of conifers, usually large stumps of *Picea* and *Pinus* spp.
D: NM2: 238 **D+I:** Kits1: 118-121 **I:** FlDan4: 143G (as *Stropharia caput-medusae*), C&D: 266, Cooke 568 (540) [as *Agaricus caput-medusae*] and 569 (541) Vol. 4 (1885) [as *A. jerdoni*]
Known from a few, scattered locations in Scotland (Perthsire & Inverness-shire). Reported from England (Middlesex) but the specimen in herb. K is labelled 'compare with' *C. caput-medusea*. Last reported from Wales in 1910 but unsubstantiated with voucher material. Probably genuinely rare since basidiomes are large, distinctive and unlikely to be overlooked.

cernua (Vahl) G. Hirsch, *Wiss. Z. Friedrich-Schiller-Univ. Jena. Math.-Naturwiss. Reihe* 33: 815 (1984)
 Agaricus cernuus Vahl, *Fl. Danica*: t. 1008 (1790)
 Psathyra cernua (Vahl) P. Kumm., *Führ. Pilzk.*: 70 (1871)
 Psilocybe cernua (Vahl) Quél., *Mém. Soc. Émul. Montbéliard, Sér. 2,* 5: 147 (1872)
 Drosophila cernua (Vahl) Quél., *Enchir. fung.*: 117 (1886)
 Agaricus alneti Schumach., *Enum. pl.* 2: 280 (1803)
 Agaricus farinulentus Schaeff., *Fung. Bavar. Palat. nasc.* 4: 205 (1770)
 Agaricus macer Purton, *Bot. descr. Brit. pl.* 3: 221 (1821)
 Agaricus membranaceus Bolton, *Hist. fung. Halifax* 1: 11 (1788)
 Agaricus papyraceus Pers., *Syn. meth. fung.*: 425 (1801)
 Psilocybe papyracea (Pers.) J.E. Lange, *Dansk. Bot. Ark.* 9(1): 32 (1936)
 Prunulus papyraceus (Pers.) Gray, *Nat. arr. Brit. pl.* 1: 631 (1821)
E: ! **S:** !
H: On decayed wood of deciduous trees, usually *Fagus* in mixed deciduous woodland. Also a single collection growing from an old housebrick discarded in woodland.
D: NM2: 243 **D+I:** Kits1: 159-161, B&K4: 264 324 **I:** FlDan4: 147B (as *Psilocybe papyracea*)
Rare. Known from England (Surrey) and Scotland (Easterness, North Ebudes and Roxburgh). Reported from Cambridgeshire in 1974 and South Devon in 1947, but unsubstantiated with voucher material.

chondroderma (Berk. & Broome) A.H. Sm., *Contr. Univ. Michigan Herb.* 5: 43 (1941)
 Agaricus chondrodermus Berk. & Broome, *Ann. Mag. Nat. Hist., Ser. 4 [Notices of British Fungi no. 1538]* 17: 132 (1876)
 Mis.: *Hypholoma instratum* sensu Massee [*Brit. fung.-fl.* 1: 384 (1892)]
 Mis.: *Agaricus instratus* sensu Cooke [*Grevillea* 18: 53 (1887)]
E: ? **S:** ! **ROI:** ?
H: On conifer wood, usually on large stumps.
D: NM2: 250 **D+I:** Kits1: 191-194 **I:** FlDan4: 147C (as *Hypholoma chondrodermum*), C&D: 268
Described from Scotland (Glamis) and collected since then in Wester Ross, but very poorly known in Britain. Reported from the Republic of Ireland (Kerry) in 1936, and from England on several occasions, but unsubstantiated with voucher material.

clivensis (Berk. & Broome) P.D. Orton, *Trans. Brit. Mycol. Soc.* 43(2): 369 (1960)
 Agaricus clivensis Berk. & Broome, *Ann. Mag. Nat. Hist., Ser. 3 [Notices of British Fungi no. 916]* 7: 367 (1861)
 Psilocybe clivensis (Berk. & Broome) Sacc., *Syll. fung.* 5: 1055 (1887)
E: ! **S:** o
H: On soil, often amongst grass.
D: NCL3: 369, NM2: 250 **D+I:** Kits1: 225-227, B&K4: 266 325
Known from England (North Devon, Surrey and West Gloucestershire) and Scotland.

conopilus (Fr.) A. Pearson & Dennis, *Trans. Brit. Mycol. Soc.* 31: 185 (1948)
 Agaricus conopilus Fr., *Syst. mycol.* 1: 504 (1821)
 Psathyra conopilus (Fr.) P. Kumm., *Führ. Pilzk.*: 70 (1871)
 Drosophila conopilus (Fr.) Quél., *Enchir. fung.*: 116 (1886)
 Agaricus aratus Berk., *Outl. Brit. fungol.*: 176 (1860)
 Psathyrella arata (Berk.) W.G. Sm., *Syn. Brit. Bas.*: 200 (1908)
 Agaricus superbus Jungh., *Linnaea* 5: 388 (1830)
 Agaricus conopilus var. *superbus* (Jungh.) Cooke, *Handb. Brit. fung.*, Edn 2: 378 (1891)
 Agaricus conopilus f. *superbus* (Jungh.) Cooke, *Ill. Brit. fung.* 1185 (1158) Vol. 8 (1891)
 Agaricus subatratus Batsch, *Elench. fung. (Continuatio Prima)*: 103 & t. 18 f. 89 (1786)
 Psathyrella subatrata (Batsch) Gillet, *Hyménomycètes*: 616 (1878)
 Psathyra conopilus var. *subatrata* (Batsch) J.E. Lange, *Dansk. Bot. Ark.* 9(1): 14 (1936)
 Psathyra elata Massee, *Brit. fung.-fl.* 1: 353 (1892)
E: c **S:** c **W:** c **NI:** c **ROI:** c
H: On decayed wood of deciduous trees in woodland. Usually on large logs of *Fagus* but also reported on *Alnus*, *Crataegus* and *Quercus* spp. Frequently appearing as if terrestrial, but then growing from buried wood.
D: NM2: 240 **D+I:** Kits1: 104-107, B&K4: 266 326 **I:** FlDan4: 155E (as *Psathyra conopilea* var. *subatrata*), FlDan4: 155D (as *P. conopilea*), Bon: 267, C&D: 264
Common and widespread. The use of the epithet *conopilea* is orthographically incorrect.

coprophila Watling, *Notes Roy. Bot. Gard. Edinburgh* 31: 146 (1971)
 Psathyrella fimetaria Watling, *Notes Roy. Bot. Gard. Edinburgh* 31: 143 (1971)
E: ? **S:** ! **ROI:** ?
H: On weathered dung of cattle and horses.
D: NM2: 241 **D+I:** Kits1: 99-101
The majority of the few records are from Scotland. Records from England and one from the Republic of Ireland appear to lack voucher collections.

corrugis (Pers.) Konrad & Maubl., *Encycl. Mycol.* 14, *Les Agaricales* I: 123 (1948)
 Agaricus corrugis Pers., *Neues Mag. Bot.* 1: 104 (1794)
 Psathyra corrugis (Pers.) Quél., *Mém. Soc. Émul. Montbéliard, Sér. 2,* 5: 148 (1872)
 Psathyrella gracilis var. *corrugis* (Pers.) A. Pearson & Dennis, *Trans. Brit. Mycol. Soc.* 31: 185 (1948)
 Agaricus caudatus Fr., *Observ. mycol.*: 187 (1818)
 Psathyra caudata (Fr.) Quél., *Mém. Soc. Émul. Montbéliard, Sér. 2,* 5: 258 (1872)
 Psathyrella corrugis var. *vinosa* (Corda) Cooke, *Ill. Brit. fung.* 612 Vol. 4(1885)
 Agaricus gracilis Pers. [non *A. gracilis* With. (1792)] *Syn. meth. fung.* 1: 425 (1801)
 Prunulus gracilis (Pers.) Gray, *Nat. arr. Brit. pl.* 1: 630 (1821)
 Psathyrella gracilis (Pers.) Quél., *Mém. Soc. Émul. Montbéliard, Sér. 2,* 5: 152 (1872)
 Psathyrella gracilis f. *substerilis* Kits van Wav., *Persoonia* 6: 267 (1971)
 Psathyrella gracilis f. *clavigera* Kits van Wav., *Persoonia* 6: 265 (1971)
E: c **S:** c **W:** c **NI:** c **ROI:** c **O:** Channel Islands: c Isle of Man: c

H: On dead wood, frequently attached to buried twigs, sticks or woodchips. Often abundant on flowerbeds mulched with woodchips or shredded bark.
D: NM2: 240, NM2: 240 **D+I:** B&K4: 268 330 (as *Psathyrella gracilis*), B&K4: 268 330 **I:** FlDan4: 153B (as *Psathyra gracilis* var. *corrugis*), C&D: 264 (as *Psathyrella gracilis*), SV38: 29
Often reported and apparently widespread, but uncommonly collected and easily confused with *P. microrhiza*. Many forms and varieties are described, but all fall within the limits of the species. Better known as *Psathyrella gracilis*.

cotonea (Quél.) Konrad & Maubl., *Encycl. Mycol.* 14, *Les Agaricales* I: 126 (1948)
    Stropharia cotonea Quél., *Bull. Soc. Bot. France* 23: 328 (1876) [1877]
    Hypholoma cotonea (Quél.) Lange, *Dansk. Bot. Ark.* 4(4): 41 (1923)
    Agaricus hypoxanthus W. Phillips & Plowr., *Grevillea* 13: 48 (1884)
    Hypholoma hypoxanthum (W. Phillips & Plowr.) Sacc., *Syll. fung.* 5: 1037 (1887)
    Agaricus pseudostorea W.G. Sm., *J. Bot.* 41: 386 (1903)
    Hypholoma pseudostorea (W.G. Sm.) Sacc., *Syll. fung.* 17: 89 (1905)
    Agaricus storea var. *caespitosum* Cooke, *Handb. Brit. fung.*, Edn 2: 204 (1887)
    Hypholoma storea var. *caespitosum* (Cooke) Killerm., *Denkschr. Bayer. Bot. Ges. Regensburg* 20: 64 (1936)
    Mis.: *Hypholoma lacrymabundum* sensu Rea (1922)
E: ! **S:** ? **W:** !
H: On decayed wood, often large stumps, of deciduous trees in woodland, often on acidic soil. Sometimes appearing as if terrestrial but then on buried wood. Associated with *Betula*, *Castanea*, *Fagus*, *Ilex* and *Quercus* spp.
D: NM2: 238 **D+I:** Kits1: 127-130, B&K4: 266 327 **I:** FlDan4: 146C (as *Hypholoma cotoneum*), Bon: 269, C&D: 266 (pileus rather dark)
Known from England (Cumberland, East & West Kent, East Norfolk, East & West Sussex, Hertfordshire, North Lincolnshire, Oxfordshire, Shropshire, South Hampshire, Surrey, Westmorland and Yorkshire) and Wales (Breconshire). Reported from Scotland. A distinctive species, growing in large clumps, with densely scaly-fibrillar pileus and bright yellow mycelium at the stipe bases, difficult to overlook and probably genuinely rare.

dennyensis Kits van Wav., *Persoonia* 13(3): 335 (1987)
E: !
H: On acidic peaty soil on boggy heathland.
Known only from the type collection from South Hampshire (New Forest, Denny Wood) in 1971.

dunensis Kits van Wav., *Persoonia* 2: 281 (1985)
S: !
H: On sandy soil, amongst short turf on dunes.
D+I: Kits1: 254-255
Known from Fife and Kinross, and Shetland.

fatua (Fr.) Konrad & Maubl., *Encycl. Mycol.* 14, *Les Agaricales* I: 125 (1948)
    Agaricus stipatus γ *fatuus* Fr., *Syst. mycol.* 1: 296 (1821)
    Psathyra fatua (Fr.) P. Kumm., *Führ. Pilzk.*: 70 (1871)
    Pannucia fatua (Fr.) P. Karst., *Bidrag Kännedom Finlands Natur Folk* 32: 512 (1879)
    Drosophila fatua (Fr.) Quél., *Enchir. fung.*: 117 (1886)
    Hypholoma fatuum (Fr.) Bigeard & H. Guill., *Fl. Champ. Supér. France* 1: 347 (1909)
E: !
H: On soil in deciduous woodland, usually associated with *Fagus*.
D: NM2: 248 **D+I:** Kits1: 223-225, B&K4: 268 328 **I:** FlDan4: 154D (as *Psathyra fatua*)
Poorly known in Britain. Only five records, mostly unsubstantiated with voucher material or notes. Sensu auct. p.p. is *P. spadiceogrisea*.

flexispora T.J. Wallace & P.D. Orton, *Trans. Brit. Mycol. Soc.* 43(2): 371 (1960)
E: ! **S:** ! **W:** !
H: On soil in dunes, amongst grasses such as *Ammophila arenaria* and *Festuca* spp or *Carex arenaria*.
D+I: Kits1: 242-243
A dubious species close to *P. pennata*. Described from Dorset, based on habitat and a slight suprahilar depression in some (but not all) spores.

friesii Kits van Wav., *Persoonia* 9: 282 (1977)
    Mis.: *Psathyrella fibrillosa* sensu Maire [Mem. Soc. Sci. nat. Maroc 45: 113 (1937)]
    Mis.: *Psathyra fibrillosa* sensu Lange [Dansk. Bot. Ark. 9(1): 9 (1936)]
E: ! **S:** ! **W:** ? **ROI:** ?
H: On soil in grassy places in woodland or amongst *Calluna* on moorland.
D: NM2: 249 **D+I:** Kits1: 250-251 **I:** FlDan4: 152D (as *Psathyra fibrillosa*)
Often recorded and apparently frequent, but few records are accompanied with voucher material.

frustulenta (Fr.) A.H. Sm., *Contr. Univ. Michigan Herb.* 5: 45 (1941)
    Agaricus frustulentus Fr., *Epicr. syst. mycol.*: 209 (1838)
    Psathyrella cortinarioides P.D. Orton, *Trans. Brit. Mycol. Soc.* 43(2): 369 (1960)
E: ! **W:** !
H: In woodland, on very decayed wood of deciduous trees such as *Quercus* and *Ulmus* spp. Reported on large stumps and also on fragments of buried or partially buried woody debris.
D: NM2: 248 **D+I:** Kits1: 189-191 **I:** FlDan4: 151D (as *Psathyra frustulenta*), C&D: 268
Known from East Norfolk, South Hampshire and Surrey. Reported elsewhere in England and a single record from Wales (Caernarvonshire) but unsubstantiated with voucher material.

fulvescens (Romagn.) A.H. Sm., *Mem. N. Y. bot. Gdn* 24: 387 (1972)
    Drosophila fulvescens Romagn., *Bull. mens. Soc. linn. Lyon* 21: 153 (1952)
E: ? **S:** ! **NI:** !
H: On soil in damp woodland.
D: NM2: 251 **D+I:** Kits1: 258-260
Single collections from Scotland (South Ebudes: Isle of Colonsay) and Northern Ireland (Fermanagh: Crom Estate). Dubious records from England, all unsubstantiated with voucher material.

fulvescens var. **brevicystis** Kits van Wav., *Persoonia, Supplement* 2: 281 (1985)
    Psathyrella trivialis Arnolds, *Biblioth. Mycol.* 90: 437 (1983) [1982]
E: ! **S:** !
H: On soil in deciduous woodland.
D+I: Kits1: 260-262
Known from England (Middlesex and Surrey) and Scotland (Orkney).

fusca (Schumach.) A. Pearson, *Trans. Brit. Mycol. Soc.* 35: 120 (1952)
    Agaricus fuscus Schumach., *Enum. pl.* 2: 280 (1803)
E: !
H: On soil or soil mixed with woody debris in woodland.
D: NM2: 252 **D+I:** Kits1: 218-220
Known from South Lincolnshire, South Wiltshire, Surrey and Yorkshire. Reported from a few other vice-counties but unsubstantiated with voucher material.

gordonii (Berk. & Broome) A. Pearson & Dennis, *Trans. Brit. Mycol. Soc.* 31: 184 (1948)
    Agaricus gordonii Berk. & Broome, *Ann. Mag. Nat. Hist., Ser. 3 [Notices of British Fungi no. 922]* 7: 5 (1861)
    Psathyra gordonii (Berk. & Broome) Sacc., *Syll. fung.* 5: 1072 (1887)

A good species *fide* Kits van Waveren (Kits1) but not reported recently and lacks a modern concept. Sensu NCL is *Psathyrella pseudogordonii*.

gossypina (Bull.) A. Pearson & Dennis, *Trans. Brit. Mycol. Soc.* 31: 184 (1948)
> *Agaricus gossypina* (Bull.), *Herb. France*: pl. 425 (1789)
> *Psathyra gossypina* (Bull.) Gillet in Quélet, *Mém. Soc. Émul. Montbéliard, Sér. 2*, 5: 439 (1875)
> *Stropharia spintrigera* var. *semivestita* J.E. Lange, *Dansk. Bot. Ark.* 4(4): 36 (1923)
> *Psathyrella xanthocystis* P.D. Orton, *Trans. Brit. Mycol. Soc.* 43(2): 379 (1960)

E: o **S:** ! **NI:** ! **O:** Isle of Man: !
H: On soil or attached to very decayed, often buried, woody debris.
D: NM2: 245 **D+I:** Kits1: 251-253 **I:** FlDan4: 152G (as *Psathyra gossypina*), C&D: 270
Occasional but apparently widespread, though there is little material in herbaria.

hirta Peck, *Rep. (Annual) New York State Mus. Nat. Hist.* 50: 107 (1897)
> *Psathyrella semivestita* var. *coprobia* J.E. Lange, *Dansk. Bot. Ark.* 9(1): 7 (1936)
> *Psathyra coprobia* (J.E. Lange) J.E. Lange, *Fl. agaric. danic.* 5, *Taxonomic Conspectus*: VII (1940)
> *Psathyrella coprobia* (J.E. Lange) A.H. Sm., *Contr. Univ. Michigan Herb.* 5: 44 (1941)

E: ! **S:** ?
H: On weathered dung of cattle and horses.
D: NM2: 240 **D+I:** Kits1: 96-98, B&K4: 270 331 **I:** FlDan4: 152F (as *Psathyra coprobia*)
Known from South Hampshire and Yorkshire. Reported from Warwickshire and Scotland (Mid-Ebudes) but unsubstantiated with voucher material.

laevissima (Romagn.) Singer, *Nova Hedwigia Beih.* 29: 197 (1969)
> *Drosophila laevissima* Romagn., *Bull. mens. Soc. linn. Lyon* 21: 155 (1952)
> Mis.: *Psathyrella piluliformis* sensu NCL
> Mis.: *Psathyrella subpapillata* sensu Kits van Waveren (Kits1)

E: c **S:** ! **W:** !
H: On decayed wood of deciduous trees. Usually on large fallen and decayed trunks or logs of *Fagus* and *Quercus* spp, and rarely *Fraxinus* in old woodland.
D+I: Kits1: 174-176
Common and widespread in England, sometimes abundant in old woodland such as the New Forest. Rarely reported elsewhere. Macroscopically resembles small specimens of *P. piluliformis* and is often confused with (and recorded as) that species.

leucotephra (Berk. & Broome) P.D. Orton, *Trans. Brit. Mycol. Soc.* 43: 180 (1960)
> *Agaricus leucotephrus* Berk. & Broome, *Ann. Mag. Nat. Hist., Ser. 4 [Notices of British Fungi no. 1256]* 6: 468 (1870)
> *Hypholoma leucotephrum* (Berk. & Broome) Sacc., *Syll. fung.* 5: 1040 (1887)
> *Psathyra leucotephra* (Berk. & Broome) G. Bertrand, *Bull. Soc. Mycol. France* 29: 187 (1913)
> Mis.: *Psathyrella hypsipus* sensu auct.
> Mis.: *Stropharia hypsipus* sensu Lange (FlDan4: 70 & pl. 144D, as *S. hypsipoda*)

E: o **W:** ! **ROI:** !
H: On calcareous soil in mixed deciduous woodland, often fruiting near the base of living tree trunks or large stumps. Usually associated with *Fraxinus* or *Fagus*, frequently under *Mercurialis perennis*.
D: NCL3: 375, NM2: 244 **D+I:** FRIC3: 30-32 24a, Kits1: 145-147, B&K4: 270 332 **I:** FlDan4: 144D (as *Stropharia hypsipoda*), Bon: 269
Occasional in southern England. Rarely reported elsewhere but apparently widespread.

longicauda P. Karst., *Hedwigia* 30: 298 (1891)

E: !
H: On soil in litter or amongst grass, on roadside verges, in gardens or open deciduous woodland.
D+I: Kits1: 76-77
Known from Bedfordshire and Mid-Lancashire. Reported from North Somerset and West Sussex but unsubstantiated with voucher material

lutensis (Romagn.) Bon, *Doc. Mycol.* 12(48): 52 (1982)
> *Drosophila lutensis* Romagn., *Bull. mens. Soc. linn. Lyon* 21: 155 (1952)

E: ! **S:** !
H: On soil or very decayed wood of deciduous trees. Known with *Alnus*, *Betula*, *Corylus*, *Fagus*, *Quercus* and *Ulmus* spp.
D: NM2: 246 **D+I:** Kits1: 201-203, B&K4: 270 333 **I:** C&D: 270
Known from England (Hertfordshire, Oxfordshire, Warwickshire, Westmorland and West Sussex) and Scotland (South Aberdeen). Reported elsewhere but unsubstantiated with voucher material.

maculata (C.S. Parker) A.H. Sm., *Mem. N. Y. bot. Gdn* 24: 56 (1972)
> *Hypholoma maculata* C.S. Parker, *Mycologia* 25: 205 (1933)
> Mis.: *Hypholoma melanthinum* sensu Rea (1922) & sensu Lange (FlDan4: 7 & pl. 146D & D1)
> Mis.: *Psathyrella scobinacea* sensu auct. & sensu NM2:

E: !
H: On decayed deciduous and coniferous wood, usually large stumps.
D: NM2: 238 (as *Psathyrella scobinacea*) **D+I:** Kits1: 124-126, B&K4: 272 334 **I:** Bon: 269, C&D: 266
Known from Warwickshire (Albury Park) in 1968. This is a large and striking species, unlikely to be overlooked and probably genuinely rare in Britain.

marcescibilis (Britzelm.) Singer, *Lilloa* 22: 466 (1951) [1949]
> *Agaricus marcescibilis* Britzelm., *Bot. Centralbl.* 54: 69 (1893)
> *Psathyra fragilissima* Kauffman, *Pap. Michigan Acad. Sci.* 5: 141 (1925)
> *Psathyra lactea* J.E. Lange, *Fl. agaric. danic.* 5, *Taxonomic Conspectus*: VII (1940)
> *Psathyrella lactea* (J.E. Lange) M.M. Moser, *Kleine Kryptogamenflora*: 207 (1953)

E: ! **W:** ! **NI:** ! **ROI:** !
H: On soil and buried woody debris in woodland, associated with deciduous trees, often amongst grass. Fruiting from early summer onwards.
D: NM2: 242 **D+I:** Kits1: 141-143, B&K4: 272 335 **I:** FlDan4: 150A (as *Psathyra lactea*), C&D: 268
Rarely reported but apparently widespread in England. Single records from Wales, Northern Ireland, and Republic of Ireland unsubstantiated with voucher material. Basidiomes resemble *Psathyrella candolleana* and may be passed over as such.

microrrhiza (Lasch) Konrad & Maubl., *Encycl. Mycol.* 14, *Les Agaricales* I: 123 (1948)
> *Agaricus microrrhizus* Lasch, *Linnaea* 3: 426 (1828)
> *Psathyra microrrhiza* (Lasch) P. Kumm., *Führ. Pilzk.*: 70 (1871)
> *Psathyrella badiovestita* P.D. Orton, *Trans. Brit. Mycol. Soc.* 43(2): 368 (1960)
> *Agaricus semivestitus* Berk. & Broome, *Ann. Mag. Nat. Hist., Ser. 3 [Notices of British Fungi no. 920]* 7: 376 (1861)
> *Psathyra semivestita* (Berk. & Broome) Sacc., *Syll. fung.* 5: 1071 (1887)
> *Psathyrella semivestita* (Berk. & Broome) A.H. Sm., *Contr. Univ. Michigan Herb.*: 57 (1941)
> *Psathyrella squamifera* P. Karst., *Meddeland. Soc. Fauna Fl. Fenn.* 5: 60 (1882)

E: o **S:** o **W:** o **NI:** o
H: On soil and decayed woody debris buried in soil, occasionally abundant on woodchip mulch on flowerbeds.
D: NM2: 240 **D+I:** Ph: 175, Kits1: 59-63, B&K4: 272 336 **I:** C&D: 265
Probably common and widespread, but often confused with *P. corrugis*.

mucrocystis A.H. Sm., *Mem. N. Y. bot. Gdn* 24: 373 (1972)
S: !
H: On decayed wood. British material caespitose on decayed stumps.
D: NM2: 247 **D+I:** Kits1: 185-187
A single collection, from Angus (Glenisla, Brewland Estate) cited by Kits van Waveren (Kits1).

multipedata (Peck) A.H. Sm., *Contr. Univ. Michigan Herb.* 5: 33 (1941)
Psathyra multipedata Peck, *Bull. Torrey Bot. Club* 32: 77-81 (1905)
Psathyra stipatissima J.E. Lange, *Dansk. Bot. Ark.* 9(1): 11 (1936)
E: o **S:** ! **W:** ! **NI:** ! **ROI:** !
H: On soil amongst trees in scrub or open woodland, in grassy areas along woodland edges, and on lawns.
D: NM2: 246 **D+I:** Ph: 174, Kits1: 263-264, B&K4: 274 337 **I:** FlDan4: 153E (as *Psathyra stipatissima*), Bon: 267, C&D: 270
Occasional in England, less often reported elsewhere, but apparently widespread.

murcida (Fr.) Kits van Wav., *Persoonia, Supplement* 2: 281 (1985)
Agaricus murcidus Fr., *Syst. mycol.* 1: 299 (1821)
E: !
H: On loam soil in woodland. British material associated with *Fagus*.
D: Kits1: 265-266, NM2: 251 **D+I:** B&K4: 274 338
Known from North Hampshire (Waltham Trinleys, near Popham and Embley Wood, near East Stratton) in 2003.

narcotica Kits van Wav., *Persoonia* 6: 305 (1971)
E: ! **W:** !
H: On decayed vegetable debris (compost heaps) and nitrogen rich soil. Also known from dunes.
D: NM2: 239 **D+I:** Kits1: 41-43, Myc2(4): 171
Known from England (Staffordshire, South Lancashire, West Cornwall and Surrey) and Wales (Monmouthshire). A distinctive species, with a strong stercoraceous smell (skatol).

nolitangere (Fr.) A. Pearson & Dennis, *Trans. Brit. Mycol. Soc.* 31: 184 (1948)
Agaricus nolitangere Fr., *Epicr. syst. mycol.*: 234 (1838)
Psathyra nolitangere (Fr.) Quél., *Mém. Soc. Émul. Montbéliard, Sér. 2,* 5: 150 (1872)
Pannucia nolitangere (Fr.) P. Karst., *Bidrag Kännedom Finlands Natur Folk* 32: 515 (1879)
Drosophila nolitangere (Fr.) Quél., *Enchir. fung.*: 118 (1886)
E: ! **S:** ! **W:** !
H: On soil or decayed vegetation (fallen leaves) in wet areas such as ditches, marshland, fen-carr, pond margins and swampy woodland.
D: NM2: 252 **D+I:** FlDan1: 95 152E, Kits1: 215-217 **I:** FlDan4: 152E (as *Psathyra nolitangere*)
Known from England (Cumberland and South Lincolnshire), Scotland (Main Argyll) and Wales (Pembrokeshire). Rarely reported elsewhere but records are dubious and unsubstantiated with voucher material.

obtusata (Pers.) A.H. Sm., *Contr. Univ. Michigan Herb.* 5: 55 (1941)
Agaricus obtusatus Pers., *Syn. meth. fung.*: 428 (1801)
Psathyra obtusata (Pers.) Gillet, *Hyménomycètes*: 591 (1878)
Drosophila obtusata (Pers.) Quél., *Fl. mycol. France*: 59 (1888)
E: ! **S:** ! **W:** ! **ROI:** ! **O:** Channel Islands: !
H: On soil or very decayed wood in deciduous woodland, and occasionally reported in grassland.
D: NM2: 251 **D+I:** Ph: 174-175, Kits2: 337 **I:** FlDan4: 152A (as *Psathyra obtusata*)
Known from England (Bedfordshire, Buckinghamshire, East Norfolk, Hertfordshire, Middlesex, South Devon, South Essex, South Wiltshire, Surrey, West Sussex, Westmorland and Yorkshire), Scotland (Mid & North Ebudes) and Wales (Anglesey). Reported elsewhere but unsubstantiated with

voucher material. Often recorded in spring, suggesting that some of these records refer to *P. spadiceogrisea*.

obtusata *var.* aberrans Kits van Wav., *Persoonia* 13(3): 340 (1987)
Mis.: *Psathyrella obtusata* var. *utriformis* sensu van Waveren [*Persoonia* (Suppl.) 2: 200 (1985)]
E: !
H: On soil in deciduous woodland.
D+I: Kits2: 338/340
Known from East Norfolk, Hertfordshire, and Middlesex. Differs from the type by the presence of cheilo- and pleurocystidia.

ocellata (Romagn.) M.M. Moser, *Kleine Kryptogamenflora*, Edn 3: 222 (1967)
Drosophila ocellata Romagn., *Bull. mens. Soc. linn. Lyon* 21: 154 (1952)
E: ! **S:** !
H: On soil in marshy places. British material associated with *Alnus*.
D+I: Kits1: 268-269
Accepted with reservation. Collections named thus (in herb. K) from England (South Hampshire) and Scotland (Orkney), but both rather doubtful.

olympiana A.H. Sm., *Contr. Univ. Michigan Herb.* 5: 36 (1941)
Psathyrella amstelodamensis Kits van Wav., *Persoonia* 6: 299 (1971)
Psathyrella olympiana f. *amstelodamensis* (Kits van Wav.) Kits van Wav., *Persoonia* 2: 169 & 281 (1985)
E: !
H: On decayed wood in deciduous woodland. British material on buried wood.
D: NM2: 243 **D+I:** B&K4: 276 341
A single collection under this name from Staffordshire (Piggotts Bottom) and one (of f. *amstelodamensis*) from Surrey (Ockley: Vann Lake). The latter differs from the species in the presence of persistent white veil fragments on the pileus.

panaeoloides (Maire) M.M. Moser, Gams, *Kleine Kryptogamenflora*, Edn 3: 222 (1967)
Psathyra panaeoloides Maire, *Publ. Inst. Bot. Barcelona* 3: 117 (1937)
E: ! **S:** o **ROI:** !
H: On soil, often amongst grass.
D: NM2: 246 **D+I:** Kits1: 212-214, B&K4: 276 342
Known from England (South Essex), Scotland (Main Argyll and Orkney) and the Republic of Ireland (Kerry). Rarely reported elsewhere, and records are dubious and unsubstantiated with voucher material.

pannucioides (J.E. Lange) M.M. Moser, *Kleine Kryptogamenflora*, Edn 3: 220 (1967)
Hypholoma pannucioides J.E. Lange, *Fl. agaric. danic.* 5, *Taxonomic Conspectus*: IV (1940)
E: ! **W:** !
H: On soil or dead wood of deciduous trees, often in large caespitose clumps on or around decayed stumps.
D: NM2: 247 **D+I:** Kits1: 205-207 **I:** FlDan4: 200H (as *Hypholoma pannucioides*)
Known from England (North Hampshire: Waggoners Wells) and Wales (Merionethshire: Lake Vyrnwy). Rarely reported elsewhere, but unsubstantiated with voucher material and possibly misidentified *P. multipedata*.

pellucidipes (Romagn.) M.M. Moser, *Kleine Kryptogamenflora*, Edn 4: 268 (1978)
Drosophila pellucidipes Romagn., *Bull. Soc. Mycol. France* 82: 541 (1967) [1966]
E: !
H: On soil, often amongst grass.
D+I: Ph: 175, Kits1: 56-58, B&K4: 278 343
Known from South Essex (Grays). Rarely reported elsewhere, but unsubstantiated with voucher material.

pennata (Fr.) Konrad & Maubl., *Encycl. Mycol.* 14, *Les Agaricales* I: 125 (1948)
Agaricus pennatus Fr., *Syst. mycol.* 1: 297 (1821)

Psathyra pennata (Fr.) P. Karst., *Meddeland. Soc. Fauna Fl.
Fenn.* 5: 32 (1879)
Drosophila pennata (Fr.) Kühner & Romagn., *Flore Analytique
des Champignons Supérieurs*: 360 (1953)
Psathyra pennata f. *annulata* A. Pearson, *Trans. Brit. Mycol.
Soc.* 26: 49 (1943)
Psathyrella carbonicola A.H. Sm., *Contr. Univ. Michigan Herb.*
5: 31 (1941)
E: o **S:** ! **W:** ! **NI:** !
H: On burnt soil or charred wood in woodland or on heathland.
D: NM2: 245 **D+I:** Kits1: 243-245, Myc5(4): 170, B&K4: 278
344 **I:** C&D: 270
Occasional but widespread in England, rarely reported in
Scotland and single records for Wales and Northern Ireland.

phegophila Romagn., Kits van Waveren, *Persoonia,
Supplement* 2: 282 (1985)
E: !
H: On soil amongst leaf litter. British material with *Fagus* in
deciduous woodland.
D: Myc15(2): 74 **D+I:** Kits1: 227-228, B&K4: 280 346
A single collection from Cumberland (Mell Fell) in 1999.

piluliformis (Bull.) P.D. Orton, *Notes Roy. Bot. Gard. Edinburgh*
29: 116 (1969)
Agaricus piluliformis Bull., *Herb. France*: pl. 112 (1783)
Hypholoma piluliforme (Bull.) Gillet, *Hyménomycètes*: 571
(1878)
Drosophila piluliformis (Bull.) Quél., *Enchir. fung.*: 116 (1886)
Hypholoma hydrophila (Bull.) Quél., *Mém. Soc. Émul.
Montbéliard, Sér. 2,* 5: 146 (1872)
Agaricus hydrophilus Bull., *Herb. France*: pl. 511 (1791)
Drosophila hydrophila (Bull.) Quél., *Enchir. fung.*: 116 (1886)
Psathyrella hydrophila (Bull.) Maire, in Maire & Werner, *Mem.
Soc. Sci. Nat. Maroc.* 45: 113 (1937)
Hypholoma subpapillatum P. Karst., *Meddeland. Soc. Fauna
Fl. Fenn.* 5: 31 (1879)
Psathyrella subpapillata (P. Karst.) Romagn., *Bull. Soc. Mycol.
France* 98: 46 (1982)
E: c **S:** c **W:** c **NI:** o **ROI:** ! **O:** Channel Islands: ! Isle of Man: !
H: On decayed wood of deciduous trees in old woodland.
Usually on large fallen trunks and stumps of *Fagus* or *Quercus*
spp. Rarely on *Betula* spp and a single collection on
Rhododendron sp.
D: NM2: 243, NBA3: 116-118 **D+I:** Ph: 174, Kits1: 180-183,
B&K4: 280 347 **I:** Bon: 267 (as *Psathyrella hydrophila*), C&D:
268
Common and widespread. Better known as *Psathyrella
hydrophila*. N.B. This excludes *P. piluliformis* sensu NCL, for
which see *P. laevissima*. The two taxa are frequently confused.

ploddensis Kits van Wav., *Persoonia* 13(3): 357 (1987)
S: !
H: Amongst bog-mosses under *Betula* spp.
D+I: Kits2: 357-358
Described from Easterness (Plodda Falls).

polycystis (Romagn.) Kits van Wav., *Persoonia* 8: 393 (1976)
Drosophila polycystis Romagn., *Bull. mens. Soc. linn. Lyon*
21: 152 (1952)
E: ! **S:** !
H: On soil in deciduous woodland.
D: NM2: 239 **D+I:** Kits1: 57 / 58-59, B&K4: 280 348
Known from England (Surrey) and Scotland (Orkney).

populina (Britzelm.) Kits van Wav., *Persoonia, Supplement* 2:
282 (1985)
Agaricus populinus Britzelm., *Ber. Naturhist. Vereins
Augsburg* 28: 131 (1885)
Mis.: *Hypholoma sylvestre* sensu Heim & Romagn.[*Bull. Soc.
Mycol. France* 50: 183 (1934)]
Mis.: *Psathyrella sylvestris* sensu Konrad & Maubl. (*Encycl.
Mycol.* 14: 127 (1948)] & sensu Moser [*Kleine
Kryptogamenflora*, Edn 2: 243 (1955)]
E: ! **ROI:** ?
H: On decayed wood of deciduous trees. British material on
Crataegus, Populus and *Ulmus* spp.

D: NM2: 238 **D+I:** Kits1: 121-124 **I:** Bon: 269, C&D: 266
Known from Middlesex, North Essex, South Wiltshire and West
Kent. A single record from the Republic of Ireland, but
unsubstantiated with voucher material.

prona (Fr.) Gillet, *Hyménomycètes*: 618 (1878)
Agaricus pronus Fr., *Epicr. syst. mycol.*: 239 (1838)
Psathyra prona (Fr.) J.E. Lange, *Dansk. Bot. Ark.* 4(9): 16
(1936)
Drosophila albidula Romagn., *Bull. mens. Soc. linn. Lyon* 21:
151 (1952)
Psathyrella albidula (Romagn.) M.M. Moser, *Kleine
Kryptogamenflora*, Edn 3: 215 (1967)
Psathyrella prona f. *albidula* (Romagn.) Kits van Wav.,
Persoonia 7: 43 (1972)
Psathyrella subatomata J.E. Lange, *Dansk. Bot. Ark.* 9(1): 16
(1936)
E: ! **S:** ! **W:** ! **NI:** ! **ROI:** !
H: On soil, often amongst grass, in woodland glades, fields and
gardens.
D: NM2: 242 **D+I:** Kits1: 81-85, C&D: 264, B&K4: 282 351 **I:**
FlDan4: 153C (as *Psathyra subatomata*)
Rarely reported.

prona f. cana Kits van Wav., *Persoonia* 7: 37 (1982)
Mis.: *Psathyrella atomata* sensu auct.
E: ! **S:** ! **W:** ! **NI:** ! **ROI:** !
H: On soil, often amongst grass and woody debris in woodland.
D+I: Kits1: 86-87, B&K4: 282 349
Apparently widespread, but of over 130 records of this species,
only four are accompanied by voucher material.

prona f. orbitarum (Romagn.) Kits van Wav., *Persoonia,
Supplement* 2: 282 (1985)
Drosophila orbitarum Romagn., *Bull. mens. Soc. linn. Lyon*
21: 152 (1952)
Psathyrella orbitarum (Romagn.) M.M. Moser, *Kleine
Kryptogamenflora*, Edn 3: 215 (1967)
E: !
H: On soil, often amongst grass.
D+I: Kits1: 88-89
Known from Surrey and West Sussex.

pseudocasca (Romagn.) Kits van Wav., *Persoonia* 11: 500
(1982)
Drosophila pseudocasca Romagn., *Bull. mens. Soc. linn. Lyon*
21: 154 (1952)
Mis.: *Hypholoma cascum* sensu Lange (FlDan4: 77 & pl.
147A)
E: ! **S:** ?
H: On soil (amongst grass in the British collection).
D+I: Kits1: 194
A single collection in herb. K from East Kent (Bedgebury
Pinetum) in 1969, originally determined as *P. casca* sensu
Lange (1939). Reported from Scotland (Mid Ebudes: Isle of
Mull) but unsubstantiated with voucher material.

pseudocorrugis (Romagn.) Bon, *Doc. Mycol.* 12(48): 52
(1982)
Drosophila pseudocorrugis Romagn., *Bull. mens. Soc. linn.
Lyon* 21: 152 (1952)
E: !
H: On soil in deciduous woodland.
D+I: Kits1: 231-233
A single collection from Oxfordshire (Tubney Woods) in 1969.
Reported from South Essex and Yorkshire but unsubstantiated
with voucher material.

pseudogordonii Kits van Wav., *Persoonia, Supplement* 2: 282
(1985)
Mis.: *Psathyrella gordonii* sensu NCL
E: !
H: On decayed wood of deciduous trees. British material on
Fagus and *Ulmus* spp.
D: NCL3: 372 (as *Psathyrella gordonii*) **D+I:** Kits1: 152-154
Known from Berkshire, Buckinghamshire and Oxfordshire.

pseudogracilis (Romagn.) M.M. Moser, *Kleine Kryptogamenflora*, Edn 3: 214 (1967)
   Drosophila pseudogracilis Romagn., *Bull. mens. Soc. linn. Lyon* 21: 152 (1952)
E: ! **ROI:** !
H: On soil, often amongst grass in rough pasture or along roadsides.
D: NM2: 239 **D+I:** Kits1: 50-51, B&K4: 282 353
Known from England (Bedfordshire, East Kent, Middlesex, Surrey, West Sussex and Yorkshire) and the Republic of Ireland (Offaly). Reported elsewhere but unsubstantiated with voucher material.

pygmaea (Bull.) Singer, *Lilloa* 22: 467 (1951) [1949]
   Agaricus pygmaeus Bull., *Herb. France*: pl. 525 (1790)
   Psathyra pygmaea (Bull.) Quél., *Compt. Rendu Assoc. Franç. Avancem. Sci.* 9: 664 (1881)
   Drosophila pygmaea (Bull.) Quél., *Enchir. fung.*: 117 (1886)
   Psathyrella consimilis Bres. & Henn., *Verh. Bot. Vereins Prov. Brandenburg* 31: 178 (1889)
E: o **W:** ! **NI:** ! **ROI:** !
H: On decayed wood of deciduous trees, usually on large logs or stumps.
D: NM2: 242 **D+I:** Kits1: 155-157, B&K4: 284 354 **I:** FlDan4: 151B (as *Psathyra consimilis*), Bon: 269
Occasional but widespread in England, and rather frequent in southern counties. Rarely reported elsewhere. Occasionally found intermixed with *Coprinus disseminatus*.

romseyensis Kits van Wav., *Persoonia* 13(3): 349 (1987)
E: !
H: On soil amongst moss, on heathland.
D+I: Kits2: 349-351
Known only from the type collection from South Hampshire (Romsey, Ampfield Wood).

rostellata Örstadius, *Windahlia* 16: 156 (1986)
E: !
H: On very decayed wood of deciduous trees. British material on large decayed logs of *Fagus*, *Fraxinus* and *Ulmus* spp.
D: Kits2: 365, NM2: 247, Myc14(1): 30
Known from North Hampshire and Oxfordshire. The markedly bifurcate cheilo- and pleurocystidia are diagnostic.

sacchariolens Enderle, *Beitr. Kenntn. Pilze Mitteleurop.* 1: 35 (1984)
   Psathyrella suavissima Ayer, *Mycol. Helv.* 1(3): 145 (1984)
E: !
H: On soil and decayed woody debris. British material from woodchips partially buried in soil.
D: Kits2: 362 **D+I:** B&K4: 286 355 **I:** SV37: 25, Myc15(4): 157 (misident. as *Psathyrella pervelata*)
Known from South Essex and Surrey. The strong smell of burnt sugar, resembling that of *Hebeloma sacchariolens* s.l., is characteristic.

sarcocephala (Fr.) Singer, *Lilloa* 22: 468 (1951) [1949]
   Agaricus sarcocephalus Fr., *Observ. mycol.* 1: 51 (1815)
   Psilocybe sarcocephala (Fr.) Gillet, *Hyménomycètes*: 586 (1878)
   Drosophila sarcocephala (Fr.) Quél., *Enchir. fung.*: 116 (1886)
   Mis.: *Psilocybe spadicea* sensu Lange (FlDan4: 80 & pl. 148E)
E: ! **S:** ! **W:** ! **ROI:** !
H: On decayed wood of deciduous trees such as *Fagus* or *Ulmus* spp, in old woodland.
D+I: Kits1: 164-166 **I:** FlDan4: 148F (as *Psilocybe sarcocephala*)
Widespread in the British Isles but often confused with the similar, and apparently more frequent, *Psathyrella spadicea*.

solitaria (P. Karst.) Örstadius & Huhtinen, *Österr. Z. Pilzk.* 5: 136 (1996)
   Psathyra solitaria P. Karst., *Medd. Soc. Fauna Flora Fenn.* 16: 102 (1889)
   Psathyrella rannochii Kits van Wav., *Persoonia* 11(4): 501 (1982)

Mis.: *Stropharia spintrigera* sensu auct. Brit.
E: ! **S:** o
H: On woody debris and soil, usually near conifers.
D: NM2: 249 (as *Psathyrella rannochii*) **D+I:** Kits1: 195-196 (as *P. rannochii*)
Occasional but widespread in Scotland. A single English collection (Co. Durham).

spadicea (Schaeff.) Singer, *Lilloa* 22: 468 (1951) [1949]
   Agaricus spadiceus Schaeff., *Fung. Bavar. Palat. nasc.* 1: 60 (1762)
   Psilocybe spadicea (Schaeff.) P. Kumm., *Führ. Pilzk.*: 71 (1871)
   Drosophila spadicea (Schaeff.) Quél., *Enchir. fung.*: 116 (1886)
   Pratella spadicea (Schaeff.) J. Schröt., in Cohn, *Krypt.-Fl. Schlesien* 3: 568 (1889)
   Mis.: *Psathyrella sarcocephala* sensu Lange (FlDan4: 80 & pl. 148 F)
E: o **S:** ! **W:** ! **NI:** ! **ROI:** !
H: On wood, often around the bases of living deciduous trees and very occasionally on decayed trunks or logs.
D+I: Kits1: 161-164, B&K4: 286 356 **I:** FlDan4: 148E (as *Psilocybe spadicea*), C&D: 268
Occasional in England. Rarely reported elsewhere but seemingly widespread.

spadiceogrisea (Schaeff.) G. Bertrand, *Bull. Soc. Mycol. France* 29: 185 (1913)
   Agaricus spadiceogriseus Schaeff., *Fung. Bavar. Palat. nasc.* 1: 59 (1762)
   Psathyra spadiceogrisea (Schaeff.) P. Kumm., *Führ. Pilzk.*: 70 (1871)
   Drosophila spadiceogrisea (Schaeff.) Quél., *Enchir. fung.*: 117 (1886)
   Pratella spadiceogrisea (Schaeff.) Kirchn. & W. Eichler, *Jahresh. Vereins Vaterl. Naturk. Württemberg* 50: 448 (1894)
   Psilocybe spadiceogrisea (Schaeff.) Boud., *Icon. mycol.* 4: 68 (1911)
   Drosophila exalbicans Romagn., *Bull. mens. Soc. linn. Lyon* 21: 155 (1952)
   Psathyrella spadiceogrisea f. *exalbicans* (Romagn.) Kits van Wav., *Persoonia* 2: 282 (1985)
   Drosophila mammifera Romagn., *Bull. Soc. Mycol. France* 92: 194 (1976)
   Psathyrella spadiceogrisea f. *mammifera* (Romagn.) Kits van Wav., *Persoonia, Supplement* 2: 282 (1985)
   Psathyra obtusata var. *vernalis* J.E. Lange, *Fl. agaric. danic.* 5, *Taxonomic Conspectus*: VII (1940)
   Psathyrella spadiceogrisea f. *vernalis* (J.E. Lange) Kits van Wav., *Persoonia, Supplement* 2: 282 (1985)
   Psathyrella vernalis (J.E. Lange) M.M. Moser, *Kleine Kryptogamenflora*, Edn 3: 223 (1967)
   Psathyrella obtusata var. *utriformis* Kits van Wav., *Persoonia* 11(4): 499 (1982)
   Mis.: *Psathyrella fatua* sensu auct.
E: c **S:** o **W:** o **NI:** ! **ROI:** o **O:** Channel Islands: !
H: On soil or decayed organic material, often amongst grass and also at woodland edges. Often fruiting in spring.
D+I: Kits1: 234-240, B&K4: 286 357 **I:** FlDan4: 153D (as *Psathyra spadiceogrisea*), FlDan4: 153A (as *Psathyra obtusata* var. *vernalis*)
Common and widespread. Basidiomes are variable macroscopically, hence the long list of synonyms.

sphaerocystis P.D. Orton, *Notes Roy. Bot. Gard. Edinburgh* 26: 64 (1964)
E: ! **S:** !
H: On weathered dung. British material on horse dung.
D+I: Kits1: 113-114
Described from Scotland (Perthshire: Rannoch, Dall Wood) in 1960. A recent collection (1997) from England (Buckinghamshire: Burnham Beeches) may also belong here.

sphagnicola (Maire) J. Favre, *Bull. Soc. Mycol. France* 53: 282 (1937)
> *Stropharia sphagnicola* Maire, *Bull. Soc. Mycol. France* 26: 192 (1910)
> *Stropharia psathyroides* J.E. Lange, *Dansk. Bot. Ark.* 4(4): 36 (1923)

E: ! **S:** o **ROI:** ?
H: Attached to *Sphagnum* moss in boggy areas in moorland.
D: NM2: 245 **D+I:** Kits1: 203-205, B&K4: 288 358 **I:** FlDan4: 144A (as *Stropharia psathyroides*), C&D: 270
Known from widely scattered localities in Scotland. Single records from England (Cumbria) and the Republic of Ireland (Offaly). Reported from Yorkshire but unsubstantiated with voucher material.

spintrigeroides P.D. Orton, *Trans. Brit. Mycol. Soc.* 43(2): 377 (1960)
E: !
H: On soil in mixed deciduous woodland.
D+I: Kits1: 171-172
Known from the type locality in Surrey. A few subsequent records from widely scattered locations in England, all unsubstantiated with voucher material. Doubtfully distinct from *P. olympiana*.

stercoraria (Kühner & Joss.) Arnolds, *Biblioth. Mycol.* 90: 439 (1983) [1982]
> *Drosophila stercoraria* Kühner & Joss., *Bull. Soc. Naturalistes Oyonnax* 10 -11 (Suppl.): 57 (1957)

E: ! **S:** ? **W:** ?
H: On weathered dung of herbivores. British material on horse dung.
D+I: Kits1: 98-99
Known from England (Middlesex). Reported from Wales (Skokholm) and Scotland but unsubstantiated with voucher material.

tephrophylla (Romagn.) Bon, *Doc. Mycol.* 12(48): 52 (1983)
> *Drosophila tephrophylla* Romagn., *Bull. mens. Soc. linn. Lyon* 21: 154 (1952)

E: ! **S:** !
H: On calcareous loam soil or in very decayed leaf litter, usually associated with *Fagus* in beechwoods.
D: NM2: 251 **D+I:** Kits1: 210-212, B&K4: 288 359 **I:** C&D: 270
Known from England (Bedfordshire, Berkshire, East Gloucestershire, Huntingdonshire, Oxfordshire, South Wiltshire, Surrey, West Suffolk and West Sussex) and Scotland (North Ebudes).

typhae (Kalchbr.) A. Pearson & Dennis, *Trans. Brit. Mycol. Soc.* 31: 185 (1948)
> *Agaricus typhae* Kalchbr., *Math. Term. Közl. Mag. Tudom. Akad.* 2: 160 (1862)

E: o
H: Usually on living, dead or dying stems of *Typha angustifolia* and *T. latifolia* just above the waterline. Also reported on *Glyceria maxima*, *Iris pseudacorus*, *Phalaris arundinacea* and *Phragmites australis*.
D: NM2: 241 **D+I:** Kits1: 134-135, B&K4: 288 360 **I:** C&D: 268
Known from East Norfolk, Hertfordshire, Middlesex and Warwickshire.

umbrina Kits van Wav., *Persoonia* 11(4): 506 (1982)
S: !
H: On soil amongst moss. British material by a footpath in mixed deciduous woodland.
D: Kits1: 187-189
Type collection from Easterness (Tomich) in 1968, but not recorded since.

vinosofulva P.D. Orton, *Trans. Brit. Mycol. Soc.* 43(2): 378 (1960)
> *Psathyrella prona* var. *utriformis* Kits van Wav., *Persoonia* 7: 43 (1972)

E: !
H: On soil amongst woody debris in deciduous woodland.

D: NM2: 242 **D+I:** Kits1: 89-92 (as *Psathyrella prona* var. *utriformis*), B&K4: 284 352 (as *P. prona* var. *utriformis*)
Known from South Devon and Surrey.

vyrnwyensis Kits van Wav., *Persoonia* 13(3): 359 (1987)
W: !
H: On humic soil. British material under *Fagus* in deciduous woodland.
D+I: Kits2: 359-361
Type collection from Merionethshire (Lake Vyrnwy) in 1979. Not recorded since.

PSEUDOBAEOSPORA Singer, *Lloydia* 5: 129 (1942)
Agaricales, Agaricaceae
Type: *Pseudobaeospora oligophylla* (Singer) Singer

albidula Bas, *Persoonia* 18(1): 119 (2002)
E: !
H: On calcareous clay soil in mixed deciduous woodland.
Type collection from Surrey (Mickleham Downs)

celluloderma Bas, *Persoonia* 18(1): 119 (2002)
E: !
H: On calcareous soil, amongst woody debris of *Fagus* and *Fraxinus*
Type collection from Surrey (Mickleham Downs)

dichroa Bas, *Persoonia* 18(1): 120 (2002)
E: !
H: On calcareous or limestone soil in mixed deciduous woodland.
Type collection from South Hampshire (Butser Hill: Queen Elizabeth Country Park) and also known from Mid Lancashire (Silverdale: Gait Barrows NNR)

dichroa f. cystidiata Bas, *Persoonia* 18(1): 120 (2002)
H: On soil, in mixed deciduous woodland on limestone
E: !
Type collection from Mid Lancashire (Silverdale: Waterslack Wood)

laguncularis Bas, *Persoonia* 18(1): 121 (2002)
E: !
H: On soil in litter of *Taxus* in mixed woodland, on limestone.
Type collection from Mid Lancashire (Silverdale: Gait Barrows NNR)

PSEUDOBOLETUS Šutara, *Česká Mykol.* 45(1-2): 2 (1991)
Boletales, Boletaceae
Type: *Pseudoboletus parasiticus* (Bull.) Šutara

parasiticus (Bull.) Šutara, *Česká Mykol.* 45(1-2): 2 (1991)
> *Boletus parasiticus* Bull., *Herb. France*: pl. 451 (1790)
> *Xerocomus parasiticus* (Bull.) Quél., *Fl. mycol. France*: 418 (1888)

E: o **S:** o **W:** o **NI:** ! **ROI:** ! **O:** Channel Islands: !
H: On old basidiomes of *Scleroderma citrinum*.
D: NM2: 60 (as *Boletus parasiticus*) **D+I:** Ph: 204-205 (as *B. parasiticus*), B&K3: 88 58 (as *Xerocomus parasiticus*), FungEur8: 86-88 47a & 47b (as *X. parasiticus*) **I:** Bon: 43 (as *X. parasiticus*), C&D: 427 (as *X. parasiticus*), Kriegl2: 318 (as *X. parasiticus*)
Occasional but widespread, possibly most frequent in southern and south-eastern England. Apparently not parasitic on *Scleroderma* but mutualistic, with interwoven mycelia and an exchange of nutrients *fide* R. Watling (pers. comm.).

PSEUDOCLITOCYBE (Singer) Singer, *Mycologia* 48: 725 (1956)
Agaricales, Tricholomataceae
> *Omphalia* (Pers.) Gray [*nom. illegit.*, non *Omphalea* L. (1759)(*Euphorbiaceae, nom. cons.*); non *Omphalia* (Fr.) Staude (1857)], *Nat. arr. Brit. pl.* 1: 611 (1821)

Type: *Pseudoclitocybe cyathiformis* (Bull.) Singer

cyathiformis (Bull.) Singer, *Mycologia* 48: 725 (1956)
Agaricus cyathiformis Bull., *Herb. France*: pl. 575 (1792)
Clitocybe cyathiformis (Bull.) P. Kumm., *Führ. Pilzk.*: 120 (1871)
Cantharellula cyathiformis (Bull.) Singer, *Ann. Mycol.* 34: 331 (1936)
Omphalia cyathiformis (Bull.) Kühner & Romagn., *Flore Analytique des Champignons Supérieurs*: 129 (1953)
Agaricus sordidus Dicks. [*nom. illegit.*, non *A. sordidus* Fr. (1821)], *Fasc. pl. crypt. brit.* 1: 16 (1785)
Agaricus tardus Pers., *Syn. meth. fung.*: 461 (1801)
Omphalia tarda (Pers.) Gray, *Nat. arr. Brit. pl.* 1: 614 (1821)
Mis.: *Agaricus ectypus* sensu Cooke (1870 & 1882)
E: c **S:** c **W:** ! **NI:** c **ROI:** ! **O:** Channel Islands: !
H: On soil, or less frequently decayed mossy logs and fallen trunks, of deciduous trees and conifers in woodland and plantations. Basidiomes are often produced during winter.
D: NM2: 177, FAN3: 92, BFF8: 81 **D+I:** Ph: 51 (as *Cantharellula cyathiformis*), B&K3: 314 399 **I:** FlDan1: 38E (as *Clitocybe cyathiformis*), Bon: 131, C&D: 181, Kriegl3: 502
Common and widespread. English name = 'The Goblet'.

expallens (Pers.) M.M. Moser, *Kleine Kryptogamenflora*, Edn 3: 106 (1967)
Agaricus expallens Pers., *Syn. meth. fung.*: 461 (1801)
Clitocybe expallens (Pers.) P. Kumm., *Führ. Pilzk.*: 120 (1871)
Omphalia expallens (Pers.) Kühner & Romagn., *Flore Analytique des Champignons Supérieurs*: 129 (1953)
Cantharellula expallens (Pers.) P.D. Orton, *Trans. Brit. Mycol. Soc.* 43(2): 174 (1960)
E: ! **S:** !
H: On soil, often amongst grass, in mixed deciduous woodland.
D: BFF8: 81-82 **I:** Bon: 131
Known from Middlesex, North Lincolnshire, North Somerset, South Wiltshire, Surrey and West Sussex, and also from Scotland.

obbata (Fr.) Singer, *Sydowia* 15: 52 (1962) [1961]
Agaricus obbatus Fr., *Epicr. syst. mycol.*: 74 (1838)
Clitocybe obbata (Fr.) Quél., *Mém. Soc. Émul. Montbéliard, Sér. 2*, 5: 90 (1872)
Cantharellula obbata (Fr.) Bousset, *Bull. Soc. Mycol. France* 55: 123 (1939)
Omphalia obbata (Fr.) Kühner & Romagn., *Flore Analytique des Champignons Supérieurs*: 129 (1953)
E: !
H: On acidic soil amongst grass and moss, in unimproved grassland.
D: NM2: 177, FAN3: 93, BFF8: 82-83 **I:** C&D: 181
A single record from Surrey (Netley Heath) in 1951.

PSEUDOCRATERELLUS Corner, *Sydowia, Beih.* 1: 268 (1957)
Cantharellales, Craterellaceae
Type: *Pseudocraterellus sinuosus* (Fr.) D.A. Reid (= *P. undulatus*)

undulatus (Pers.) Rauschert, *Reprium nov. Spec. Regni. veg.* 98(11-12): 661 (1987)
Merulius undulatus Pers., *Syn. meth. fung.*: 492 (1801)
Cantharellus undulatus (Pers.) Fr., *Syst. mycol.* 1: 321 (1821)
Helvella crispa Bull. [*nom. illegit.*, non *H. crispa* (Scop.) Fr. (1823)], *Herb. France*: pl. 465 (1790)
Craterellus crispus (Bull.) Berk., *Outl. Brit. fungol.*: 266 (1860)
Craterellus sinuosus var. *crispus* (Bull.) Quél., *Fl. mycol. France*: 35 (1888)
Cantharellus pusillus Fr., *Syst. mycol.* 1: 319 (1821)
Craterellus pusillus (Fr.) Fr., *Epicr. syst. mycol.*: 533 (1838)
Cantharellus sinuosus Fr., *Syst. mycol.* 1: 319 (1821)
Craterellus sinuosus (Fr.) Fr., *Epicr. syst. mycol.*: 533 (1838)
Pseudocraterellus sinuosus (Fr.) D.A. Reid, *Persoonia* 2: 122 (1962)

E: o **S:** o **W:** ! **NI:** ! **ROI:** !
H: On soil amongst litter, usually associated with *Corylus*, *Fagus* and *Quercus* spp, in old deciduous woodland.
D: NM3: 263 (as *Pseudocraterellus undulatus*), BFF8: 28 **D+I:** B&K2: 374 488, BritChant: 20-21 **I:** Bon: 307, Kriegl2: 20 (as *P. sinuosus*)
Occasional in England and Scotland, rarely reported elsewhere, but apparently widespread.

PSEUDOHYDNUM P. Karst., *Not. Sällsk. Fauna et Fl. Fenn. Förh.* 9: 374 (1868)
Tremellales, Exidiaceae
Hydnogloea Berk., *Ann. Mag. Nat. Hist., Ser. 4* 7: 429 (1871)
Tremellodon (Pers.) Fr., *Hym. Eur.*: 618 (1874)
Type: *Pseudohydnum gelatinosum* (Scop.) P. Karst.

gelatinosum (Scop.) P. Karst., *Not. Sällsk. Fauna et Fl. Fenn. Förh.* 9: 374 (1868)
Hydnum gelatinosum Scop., *Fl. carniol.*: 472 (1772)
Steccherinum gelatinosum (Scop.) Gray, *Nat. arr. Brit. pl.* 1: 651 (1821)
Hydnogloea gelatinosa (Scop.) Berk., *Grevillea* 1: 101 (1873)
Tremellodon gelatinosum (Scop.) Pers., *Hymenomyc. eur.*: 618 (1874)
E: o **S:** o **W:** o **NI:** ! **ROI:** ! **O:** Isle of Man: !
H: On wood of conifers such as *Picea* and *Pinus* spp, usually on very decayed, large stumps.
D: NM3: 110 **D+I:** Ph: 264, B&K2: 62 19 **I:** C&D: 137, Kriegl1: 114
Occasional but widespread.

PSEUDOLASIOBOLUS Agerer, *Mitt. bot. StSamml., München* 19: 279 (1983)
Agaricales, Tricholomataceae
Type: *Pseudolasiobolus minutissimus* Agerer

minutissimus Agerer, *Mitt. bot. StSamml., München* 19: 279 (1983)
E: !
H: On decayed wood. British material on fallen bark of *Cupressus* sp.
Known from South Devon (Slapton, Strete Gate) in 1994.

PSEUDOMERULIUS Jülich, *Persoonia* 10(3): 350 (1979)
Boletales, Coniophoraceae
Type: *Pseudomerulius aureus* (Fr.) Jülich

aureus (Fr.) Jülich, *Persoonia* 10(3): 330 (1979)
Merulius aureus Fr., *Observ. mycol.* 1: 101 (1815)
Serpula aurea (Fr.) P. Karst., *Bidrag Kännedom Finlands Natur Folk* 48: 344 (1889)
Xylomyzon croceum Pers., *Mycol. eur.* 2: 33 (1825)
Xylomyzon solare Pers., *Mycol. eur.* 2: 29 (1825)
Merulius vastator Fr. [non *M. vastator* Tode (1783)], *Syst. mycol.* 1: 329 (1821)
E: r
H: On very decayed wood of *Pinus sylvestris* on heathland, or rarely in conifer woodland or plantations.
D: NM3: 292 **D+I:** CNE6: 1219-1221, B&K2: 210 243 (old specimen), Myc10(4): 181 **I:** Kriegl1: 366
Known from Surrey, with single collections in Berkshire and North Hampshire. The species may be genuinely rare, since basidiomes are brilliant orange and yellow and would be difficult to overlook.

PSEUDOOMPHALINA (Singer) Singer, *Mycologia* 48: 725 (1956)
Agaricales, Tricholomataceae
Type: *Pseudoomphalina kalchbrenneri* (Bres.) Singer

graveolens (S. Petersen) Singer, *The Agaricales in Modern Taxonomy*: 291 (1986)

Omphalia graveolens S. Petersen, *Dan. Agaric.* 1: 137 (1907)
Cantharellula graveolens (S. Petersen) M.M. Moser, *Kleine Kryptogamenflora*: 77 (1953)
E: ! **S:** ! **W:** !
H: On soil in grassland, copses and parkland.
D: NM2: 178 (as *Pseudoomphalina kalchbrenneri*), BFF8: 84 **I:** Bon: 131 (as *P. compressipes*), C&D: 181 (as *P. compressipes*)
Known from England (West Gloucestershire) and Wales (Monmouthshire and Denbighshire). This is a poorly understood species, considered a synonym of *P. compressipes* by many and of *P. kalchbrenneri* by Knudsen in NM2.

pachyphylla (Fr.) Knudsen, *Nordic J. Bot.* 12(1): 76 (1992)
Agaricus pachyphyllus Fr. [non *A. pachyphyllus* Berk. (1836)], *Observ. mycol.* 1: 76 (1815)
Clitocybe pachyphylla (Fr.) Gillet, *Hyménomycètes*: 169 (1874)
Collybia incomis P. Karst., *Bidrag Kännedom Finlands Natur Folk* 32: 164 (1879)
Clitocybe incomis (P. Karst.) Sacc., *Syll. fung.* 5: 192 (1887)
Mis.: *Clitocybe clusiliformis* sensu NCL
Mis.: *Collybia clusilis* sensu auct.
E: !
H: On soil, often amongst mosses under conifers and occasionally reported amongst grass near trees.
D: NCL3: 186 (as *Clitocybe incomis*), NM2: 178, FAN3: 94, BFF8: 85
Known from East Kent, South Devon, South Hampshire and Surrey. Very poorly known in Britain.

PSEUDOTOMENTELLA Svrček, *Česká Mykol.* 12: 67 (1958)
Thelephorales, Thelephoraceae
Type: *Pseudotomentella mucidula* (P. Karst.) Svrček

flavovirens (Höhn. & Litsch.) Svrček, *Česká Mykol.* 12: 68 (1958)
Tomentella flavovirens Höhn. & Litsch., *Sitzungsber. Kaiserl. Akad. Wiss., Wien, Math.-Naturwiss. Cl., Abt. 1* 116: 831 (1907)
E: !
H: On very decayed wood of conifers.
A single collection in herb. K from Bedfordshire (Little Wavendon Heath) in 1978, previously determined as *Lazulinospora cyanea*.

mucidula (P. Karst.) Svrček, *Česká Mykol.* 12: 68 (1958)
Hypochnus mucidulus P. Karst., *Bidrag Kännedom Finlands Natur Folk* 37: 163 (1882)
Tomentella mucidula (P. Karst.) Höhn. & Litsch., *Sitzungsber. Kaiserl. Akad. Wiss., Wien, Math.-Naturwiss. Cl., Abt. 1* 115: 24 (1906)
Hypochnus roseogriseus Wakef. & A. Pearson, *Trans. Brit. Mycol. Soc.* 6(2): 141 (1919)
Hypochnus roseogriseus var. *lavandulaceus* A. Pearson, *Trans. Brit. Mycol. Soc.* 7: 57 (1921)
E: ! **S:** ! **W:** ! **NI:** !
H: Usually on very decayed wood of conifers such as *Cupressus*, *Picea* and *Pinus* spp. Rarely on decayed wood of deciduous trees such as *Betula* spp, *Castanea* and *Fagus* and on leaf debris of *Cladium mariscus*.
D: NM3: 300 **D+I:** B&K2: 218 253
Rarely recorded or collected, but apparently widespread.

tristis (P. Karst.) M.J. Larsen, *Nova Hedwigia* 22(1 - 2): 613 (1971) [1972]
Hypochnus subfuscus ssp. *tristis* P. Karst., *Meddeland. Soc. Fauna Fl. Fenn.* 9: 71 (1882)
Tomentella tristis (P. Karst.) Höhn. & Litsch., *Sitzungsber. Kaiserl. Akad. Wiss., Wien, Math.-Naturwiss. Cl., Abt. 1* 115: 1572 (1906)
Hypochnus sitnensis Bres., *Atti Imp. Regia Accad. Roveretana, ser. 3*, 3(1): 115 (1897)
Thelephora umbrina Fr., *Elench. fung.* 1: 199 (1828)
Hypochnus umbrinus (Fr.) Fr., *Summa veg. Scand.*: 337 (1849)

Tomentella umbrina (Fr.) Litsch., *Bull. Soc. Mycol. France* 49: 52 (1933)
E: ! **S:** ! **W:** ! **NI:** !
H: On decayed wood of coniferous and deciduous trees, also dead stems of *Calluna*.
D: NM3: 300
Rarely reported. Reid & Austwick (1963) referred to this under the name *Tomentella tristis* f. *sitnensis* (Bres.) Bourdot & Galzin, but this combination has not been traced.

PSILOCYBE (Fr.) P. Kumm., *Führ. Pilzk.*: 21 (1871)
Agaricales, Strophariaceae
Deconica (W.G. Sm.) P. Karst., *Bidrag Kännedom Finlands Natur Folk* 32(26): 515 (1879)
Type: *Psilocybe montana* (Pers.) P. Kumm.
Here taken in the restricted sense of *Psilocybe* subgenus *Psilocybe* as in FAN4.

apelliculosa P.D. Orton, *Notes Roy. Bot. Gard. Edinburgh* 29: 118 (1969)
E: ! **S:** ! **W:** !
H: On soil amongst grass in unimproved grassland.
D: BFF5: 48, FAN4: 39-40
Known from England (Yorkshire) and Wales (Merionethshire). Apparently widespread, but many collections are labelled only as 'compare with' *P. apelliculosa*.

chionophila Lamoure, *Bull. Mens. Soc. Linn. Lyon* 46(7): 215 (1977)
Mis.: *Deconica semistriata* sensu auct.
Mis.: *Psilocybe semistriata* sensu Watling p.p. (in BFF 5)
S: !
H: On peaty soil in montane habitat.
D: BFF5: 47 (as *Psilocybe semistriata*) **D+I:** B&K4: 344 443, FAN4: 35
Known from Scotland (Stirlingshire)

coprophila (Bull.) P. Kumm., *Führ. Pilzk.*: 71 (1871)
Agaricus coprophilus Bull., *Herb. France*: pl. 566 (1792)
Deconica coprophila (Bull.) P. Karst., *Bidrag Kännedom Finlands Natur Folk* 32: 515 (1879)
E: o **S:** o **W:** o **NI:** ! **ROI:** !
H: In small clusters on weathered dung of various herbivores, most frequent in unimproved, upland grassland.
D: BFF5: 39, NM2: 266 **D+I:** Ph: 173 (as *Deconica coprophila*), B&K4: 344 444, FAN4: 43
Occasional but widespread. Many of the records are from northern and western areas.

crobula (Fr.) Singer, *Sydowia* 15: 69 (1961)
Agaricus crobulus Fr., *Epicr. syst. mycol.*: 199 (1838)
Tubaria crobula (Fr.) P. Karst., *Bidrag Kännedom Finlands Natur Folk* 32: 446 (1879)
Naucoria crobula (Fr.) Ricken, *Vadem. Pilzfr.*: 117 (1920)
Deconica crobula (Fr.) Romagn., *Rev. Mycol. (Paris)* 2(6): 244 (1937)
Psilocybe inquilina var. *crobula* (Fr.) Høil., *Norweg. J. Bot.* 25: 120 (1978)
E: ! **S:** ! **W:** ! **ROI:** !
H: On decayed woody debris, herbaceous stems, leaves etc.
D: BFF5: 48, NM2: 266 (as *Psilocybe inquilina* var. *crobulus*) **D+I:** Ph: 173, B&K4: 346 445 (as *P. inquilina* var. *crobula*)
Rarely reported. Apparently widespread, but few of the records are substantiated with voucher material.

cyanescens Wakef., *Trans. Brit. Mycol. Soc.* 29: 141 (1946)
E: o **S:** ! **W:** ! **O:** Channel Islands: !
H: On wood of deciduous trees, especially frequent on woodchip mulch on flowerbeds but spreading into natural habitat where known from fallen and decayed trunks and logs.
D: BFF5: 35-36 **D+I:** Ph: 172-173, Myc12(4): 181, FAN4: 48 **I:** SV29: 37
Locally abundant in southern England, but increasing and now widespread elsewhere. Described from the Royal Botanic

Gardens, Kew from material originally collected there in 1911. Possibly an alien.

fimetaria (P.D. Orton) Watling, *Lloydia* 30: 150 (1967)
Stropharia fimetaria P.D. Orton, *Notes Roy. Bot. Gard. Edinburgh* 30: 150 (1967)
E: ! **S:** !
H: On weathered dung. Usually on horse dung.
D: BFF5: 38 **D+I:** FAN4: 46 **I:** SV30: 24
Rarely reported but often abundant where it does occur. Records and collections mostly from Scotland and South Hampshire (New Forest).

inquilinus (Fr.) Bres., *Icon. Mycol.* 18: pl. 863 (1931)
Agaricus inquilinus Fr., *Syst. mycol.* 1: 264 (1821)
Naucoria inquilinus (Fr.) P. Kumm., *Führ. Pilzk.*: 76 (1871)
Tubaria inquilinus (Fr.) Gillet, *Hyménomycètes*: 538 (1876)
Deconica inquilinus (Fr.) Romagn., *Rev. Mycol. (Paris)* 2(6): 244 (1937)
Deconica muscorum P.D. Orton, *Trans. Brit. Mycol. Soc.* 43(2): 255 (1960)
Psilocybe muscorum (P.D. Orton) M.M. Moser, *Kleine Kryptogamenflora*, Edn 3: 239 (1967)
E: o **S:** o **W:** ! **ROI:** !
H: On soil, often among grass or on decayed grass debris. Less frequently on decayed twigs, leaves, fern fronds, or sawdust in woodland.
D: BFF5: 49, NM2: 266 **D+I:** B&K4: 346 446, FAN4: 38
Occasional but widespread.

magica Svrček, *Česká Mykol.* 43(2): 82 (1989)
S: !
H: On soil amongst grass in upland or montane habitat.
D+I: FAN4: 36
Poorly known in the British Isles.

merdaria (Fr.) Ricken, *Blätterpilze Deutschl.*: 251 (1912)
Agaricus merdarius Fr., *Syst. mycol.* 1: 291 (1821)
Stropharia merdaria (Fr.) Quél., *Mém. Soc. Émul. Montbéliard, Sér. 2,* 5: 111 (1872)
Agaricus merdarius var. *major* Cooke, *Handb. Brit. fung.*, Edn 2: 383 (1890)
Agaricus ventricosus Massee, *Brit. fung.-fl.* 1: 400 (1892)
Stropharia ventricosa (Massee) Sacc., *Syll. fung.* 11: 70 (1895)
E: o **S:** o **W:** o **NI:** ! **ROI:** !
H: On weathered dung of large herbivores. Also sporadically (often in abundance) on dried-out sewage beds. Occasionally on straw or sawdust with dung or stable waste.
D: BFF5: 40, NM2: 265 **D+I:** B&K4: 346 447, FAN4: 41
Occasional but widespread.

merdicola Huijsman, *Persoonia* 2: 93 (1961)
E: ! **S:** !
H: On weathered dung, usually of horse.
D: BFF5: 41 **D+I:** FAN4: 42
Known from Scotland (Clyde Isles). Reported from England (North Somerset) but unsubstantiated with voucher material.

moelleri Guzmán, *Mycotaxon* 7: 245 (1978)
Mis.: *Stropharia merdaria* sensu Rea (1922)
Mis.: *Stropharia merdaria* var. *major* sensu Rea (1922)
S: !
H: On weathered dung. British material from cattle dung in Caledonian pinewoods.
D: BFF5: 42 **D+I:** FAN4: 42
Known from Mid-Perthshire (Black Wood of Rannoch).

montana (Pers.) P. Kumm., *Führ. Pilzk.*: 71 (1871)
Agaricus montanus Pers., *Observ. mycol.* 1: 9 (1796)
Deconica montana (Pers.) P.D. Orton, *Trans. Brit. Mycol. Soc.* 43(2): 175 (1960)
Agaricus physaloides Bull., *Herb. France:* pl. 366 (1788)
Psilocybe physaloides (Bull.) Quél., *Mém. Soc. Émul. Montbéliard, Sér. 2,* 5: 238 (1872)
Deconica physaloides (Bull.) P. Karst., *Bidrag Kännedom Finlands Natur Folk* 32: 516 (1879)
Mis.: *Psilocybe atrorufa* sensu auct.

E: c **S:** o **ROI:** ! **O:** Channel Islands: !
H: On soil, especially amongst mosses such as *Polytrichum* spp, on heathland and moorland, and rarely along field and woodland margins, and by roadsides.
D: BFF5: 43, NM2: 266 **D+I:** B&K4: 348 449 (as *Psilocybe physaloides*), B&K4: 348 448, FAN4: 33
Common to occasional but widespread.

phyllogena (Sacc.) Peck, *Bull. New York State Mus. Nat. Hist.* 157: 99 (1912)
Hypholoma phyllogenum Sacc., *Syll. fung.* 5: 1042 (1887)
Agaricus rhombisporus Britzelm., *Bot. Centralbl.* 15-17: 18 (1893)
Psilocybe rhombispora (Britzelm.) Sacc., *Syll. fung.* 11: 72 (1895)
Deconica rhombispora (Britzelm.) Singer, *Lilloa* 22: 509 (1951) [1949]
E: ! **W:** !
H: On decayed woody debris (sawdust, woodchips etc.) in woodland.
D: BFF5: 46, NM2: 266 (as *Psilocybe rhombispora*), B&K4: 348 450 (as *P. rhombispora*) **D+I:** FAN4: 37 **I:** C&D: 356 (as *P. modesta*)
Rarely reported. Easily confused with *P. crobula*. Most records are unsubstantiated with voucher material.

pratensis P.D. Orton, *Notes Roy. Bot. Gard. Edinburgh* 29: 120 (1969)
E: ! **S:** ! **W:** !
H: On soil, associated with grasses.
D: BFF5: 45-46 **D+I:** FAN4: 36 **I:** C&D: 356
Known from England (West Norfolk and Surrey) and Scotland (East Lothian). Reported from West Sussex and Yorkshire but unsubstantiated with voucher material. Also reported from Wales (Glamorganshire) but the collection in herb. K is determined as 'compare with' *P. pratensis*.

scobicola (Berk. & Broome) Sacc., *Syll. fung.* 5: 1048 (1887)
Agaricus scobicola Berk. & Broome, *Ann. Mag. Nat. Hist., Ser. 5 [Notices of British Fungi no. 1769]* 3: 206 (1879)
Naematoloma scobicola (Berk. & Broome) Guzmán, *Mycotaxon* 6: 474 (1978)
S: !
Known only from the type collection from Angus (Glamis). This appears to be a good species of *Hypholoma* closely related to *H. marginatum* but differing from this by the white pileus. Sensu Cooke 598 (607) Vol. 4 (1886) [as *Agaricus (Psilocybe) scobicola*], with pinkish lamellae, appears to be a species of *Entoloma*.

semilanceata (Fr.) P. Kumm., *Führ. Pilzk.*: 71 (1871)
Agaricus semilanceatus Fr., *Epicr. syst. mycol.*: 231 (1838)
E: c **S:** c **W:** c **NI:** c **ROI:** c **O:** Channel Islands: ! Isle of Man: !
H: On acidic soil, amongst grass in unimproved grassland. May be abundant amongst short turf on acidic soils in upland and montane habitat.
D: BFF5: 37, NM2: 265 **D+I:** Ph: 173, B&K4: 350 451, FAN4: 45 **I:** Bon: 255, C&D: 356
Common and widespread. Most frequent in northern and western areas. English name = 'Liberty Cap'.

strictipes Singer & A.H. Sm., *Mycologia* 50: 141 (1958)
Agaricus semilanceatus var. *coerulescens* Cooke, *Brit. fung.-fl.* 1: 373 (1892)
Psilocybe semilanceata var. *coerulescens* (Cooke) Sacc., *Syll. fung.* 5: 1051 (1887)
Mis.: *Psilocybe callosa* sensu auct.
E: ! **S:** ! **ROI:** !
H: On acidic soil in grassland such as fields, roadside verges and lawns.
D: BFF5: 36 **D+I:** FAN4: 45-46
Known from England (Lancashire, North Hampshire, South Lincolnshire, South Wiltshire and Surrey), Scotland (Easterness) and the Republic of Ireland (Co. Dublin). Often confused with bluing variants of *Psilocybe semilanceata* which it closely resembles.

subcoprophila (Britzelm.) Sacc., *Syll. fung.* 11: 72 (1895)
  Agaricus subcoprophilus Britzelm., *Hymenomyc. Südbayern*
    8: 9 (1891)
  Deconica subcoprophila (Britzelm.) E. Horak, *Darwiniana* 14:
    363 (1967)
E: o **S:** o **W:** o
H: On weathered dung. British material on cow and horse dung.
D: BFF5: 42, NM2: 266, NBA3: 122 **D+I:** B&K4: 350 452, FAN4:
44
Apparently widespread but often confused with *Psilocybe
coprophila*. Most of the records are unsubstantiated with
voucher material.

subviscida *var.* **velata** Noordel. & Verduin, *Persoonia* 17: 256
  (1999)
  Deconica graminicola P.D. Orton, *Notes Roy. Bot. Gard.
    Edinburgh* 26: 49 (1964)
  Psilocybe graminicola (P.D. Orton) P.D. Orton, *Notes Roy.
    Bot. Gard. Edinburgh* 29: 80 (1979)
  Mis.: *Deconica bullacea* sensu auct.mult.
  Mis.: *Psilocybe bullacea* sensu auct.mult.
E: ! **S:** !
H: On soil amongst grass, or on decayed grass stems, culms
etc.
D: BFF5: 51 (as *Psilocybe graminicola*), FAN4: 41 **D+I:** B&K4:
344 442 (as *P. bullacea*)
Rarely reported and most records are unsubstantiated with
voucher material.

turficola J. Favre, *Bull. Soc. Mycol. France* 35: 196 (1939)
  Mis.: *Psilocybe atrobrunnea* sensu Guzmán (1983)
E: ? **S:** !
H: On peaty soil in bogs and on moorland.
D: BFF5: 52, NBA5: 720 **D+I:** FAN4: 51
Poorly known in Britain. Only dubiously recorded from England.

PTERULA Fr., *Syst. orb. veg.*: 90 (1825)
Cantharellales, Pterulaceae
Type: *Clavaria plumosa* Schwein. [= *P. plumosa* (Schwein.) Fr.]

caricis-pendulae Corner, *Nova Hedwigia Beih.* 33: 211 (1970)
E: !
H: On decayed debris of *Carex pendula*, *Juncus* and *Symphytum*
spp, in wet woodland, fens or marshes.
Rarely reported but the minute filiform basidiomes are easily
overlooked. Macroscopically similar to *P. gracilis* but basidia are
tetrasporic.

debilis Corner, *Monograph of Clavaria and Allied Genera*: 508
  (1950)
E: !
H: Gregarious on dead stems of *Juncus* spp.
Known only from the type collection from Cambridgeshire
(Whittlesford Fen).

gracilis (Desm. & Berk.) Corner, *Monograph of Clavaria and
  Allied Genera*: 514 (1950)
  Typhula gracilis Desm. & Berk., *Ann. Mag. Nat. Hist., Ser. 1
    [Notices of British Fungi no. 84]* 1: 202 (1838)
  Clavaria aculina Quél., *C. R. Ass. Fr. Av. Sci.* 9: 670 (1880)
  Pistillaria aculina (Quél.) Pat., *Tab. anal. fung.*: no. 570
    (1883)
E: ! **S:** ! **O:** Isle of Man: !
H: On decayed vegetable matter, leaves, herbaceous stems etc.
in deciduous woodland, fens and marshland.
D: NM3: 268 **I:** B&K2: 367 477 inset
Rarely reported but apparently widespread.

multifida (Chevall.) Fr., *Linnaea* 5: 531 (1830)
  Penicillaria multifida Chevall., *Fl. gén. env. Paris,* Edn 1: 111
    (1826)
E: ! **W:** !
H: On wet soil, and leaf litter in damp woodland. Known on
fallen catkins of *Salix* sp., on needle litter of *Picea* spp, dead
stems of *Juncus subnodulosus*, decayed fronds of ferns such
as *Polystichum* sp., and on dead woody stems of *Rosa* or
Rubus fruticosus agg.

Known from England (Berkshire, East & West Norfolk,
Northamptonshire, North Somerset, South Devon, Surrey, and
Warwickshire) and Wales (Glamorganshire). Reported
elsewhere but unsubstantiated with voucher material.

subulata Fr., *Syst. orb. veg.* 1: 90 (1825)
E: !
H: On soil in damp woodland.
D: NM3: 267 (as *Pterula multifida*), NM3: 267 **D+I:** B&K2: 366
477 (as *P. multifida*) **I:** Kriegl2: 60 (as *P. multifida*)
Known from Bedfordshire, Cambridgeshire and South Somerset.

PYCNOPORUS P. Karst., *Rev. Mycol. (Toulouse)*
3(9): 18 (1881)
Poriales, Coriolaceae
Type: *Pycnoporus cinnabarinus* (Jacq.) P. Karst.

cinnabarinus (Jacq.) P.Karst, *Rev. Mycol. (Toulouse)* 3(9): 18
  (1881)
  Boletus cinnabarinus Jacq., *Fl. Aust.* 4: 2 pl. 304 (1776)
  Polyporus cinnabarinus (Jacq.) Fr., *Syst. mycol.* 1: 371
    (1821)
  Trametes cinnabarina (Jacq.) Fr., *Summa veg. Scand.*: 323
    (1849)
E: ? **S:** !
H: On dead wood of deciduous trees in mixed woodland. British
material on *Betula* spp.
D: EurPoly2: 595, NM3: 229 **D+I:** Ph: 222, B&K2: 284 353 **I:**
Kriegl1: 575
Extremely rare and possibly extinct. Last confirmed collection
(specimen in herb. K) from Scotland (Mid-Perthshire: Murthly)
in 1913. Reported from England (Herefordshire in 1902, South
Lancashire 1981, East Gloucestershire in 1994, and Yorkshire)
but unsubstantiated with voucher material or misdetermined
Daedaleopsis confragosa.

RADULOMYCES M.P. Christ., *Dansk. Bot. Ark.* 19:
230 (1960)
Stereales, Hyphodermataceae
Type: *Radulomyces confluens* (Fr.) M.P. Christ.

confluens (Fr.) M.P. Christ., *Dansk. Bot. Ark.* 19(2): 230 (1960)
  Thelephora confluens Fr., *Observ. mycol.* 1: 152 (1815)
  Corticium confluens (Fr.) Fr., *Epicr. syst. mycol.*: 546 (1838)
  Cerocorticium confluens (Fr.) Jülich & Stalpers, *Verh. Kon.
    Ned. Akad. Wetensch., Afd. Natuurk., Sect. 2,* 74: 73
    (1980)
E: c **S:** c **W:** c **NI:** c **ROI:** c **O:** Channel Islands: ! Isle of Man: !
H: On decayed wood of deciduous trees, most frequently *Fagus*.
Also known on conifers such as *Cedrus, Juniperus, Picea,
Pinus, Sequoia* and *Tsuga* spp, on worked wood in buildings,
and on dead lianas of *Clematis vitalba*.
D: NM3: 205 **D+I:** CNE6: 1238-1241, B&K2: 110 93 (as
Cerocorticium confluens), Myc8(1): 13
Very common and widespread.

molaris (Fr.) M.P. Christ., *Dansk. Bot. Ark.* 19(2): 232 (1960)
  Radulum molare Fr., *Elench. fung.* 1: 151 (1828)
  Cerocorticium molare (Fr.) Jülich & Stalpers, *Verh. Kon. Ned.
    Akad. Wetensch., Afd. Natuurk., Sect. 2,* 74: 72 (1980)
  Hydnum membranaceum Bull., *Herb. France*: pl. 481 f. 1
    (1791)
  Acia membranacea (Bull.) P. Karst., *Meddeland. Soc. Fauna
    Fl. Fenn.* 5: 42 (1880)
  Radulum membranaceum (Bull.) Bres., *Atti Imp. Regia
    Accad. Roveretana, ser. 3,* 3: 103 (1897)
  Sistotrema rude Pers., *Mycol. eur.* 2: 192 (1825)
  Hydnum rude (Pers.) Duby, *Bot. gall.*: 780 (1830)
  Radulum rude (Pers.) S. Lundell, *Fungi Exsiccati Suecici*: 9
    (1947)
E: c **S:** c **W:** c **O:** Channel Islands: !
H: Weakly parasitic. On dead attached branches of *Quercus* spp.
Rarely reported on *Corylus* and *Fraxinus*.
D: NM3: 205 **D+I:** CNE6: 1241-1243, B&K2: 112 94 (as
Cerocorticium molare) **I:** Kriegl1: 180 (as *C. molare*)

Common and widespread. Often reported on *Prunus avium* but many of these collections are *Basidioradulum radula*, basidiomes of which superficially resemble *R. molaris*.

rickii (Bres.) M.P. Christ., *Dansk. Bot. Ark.* 19(2): 128 (1960)
Corticium rickii Bres., *Oesterr. Bot. Z.* 48: 136 (1898)
Cerocorticium rickii (Bres.) Boidin, Gilles & Hugueney, *Cryptog. Mycol* 9(1): 45 (1988)
E: ! **W:** !
H: On a variety of woody substrata, often dead lianas of *Clematis vitalba* and one record from bark of *Taxus*. A single collection from wet and decayed plywood in a garden shed.
Rarely reported. Distinguished from *R. confluens* by the uniformly globose spores.

RAMARIA (Fr.) Bonord. [*nom. cons.*, non *Ramaria* Holmsk. (1790)], *Handb. Mykol.*: 166 (1851)
Gomphales, Ramariaceae
Clavariella P. Karst., *Rev. Mycol. (Toulouse)* 3(9): 21 (1881)
Type: *Ramaria botrytis* (Pers.) Ricken

abietina (Pers.) Quél., *Fl. mycol. France*: 467 (1888)
Clavaria abietina Pers., *Neues Mag. Bot.* 1: 117 (1794)
Clavaria ochraceovirens Jungh., *Linnaea* 5: 407 (1830)
Ramaria ochraceovirens (Jungh.) Donk, *Rev. Niederl. Homob. Aphyll.* 2: 112 (1933)
E: ! **S:** ! **NI:** ! **ROI:** !
H: On soil and needle litter of conifers such as *Abies, Larix, Picea* and *Pinus* spp, in woodland, usually plantations. Rarely on decayed wood of conifers and deciduous trees such as *Fraxinus*.
D: NM3: 275 **D+I:** B&K2: 354 458 **I:** Kriegl2: 65
Rarely reported but apparently widespread.

aurea (Fr.) Quél., *Fl. mycol. France*: 467 (1888)
Clavaria aurea Fr., *Epicr. syst. mycol.*: 574 (1838)
E: ! **S:** ! **W:** !
H: On soil in old deciduous woodland, usually associated with *Fagus*.
D: NM3: 275 **D+I:** B&K2: 354 459 **I:** Kriegl2: 67
Rarely reported, but apparently widespread.

botrytis (Pers.) Ricken, *Vadem. Pilzfr.*: 253 (1918)
Clavaria botrytis Pers., *Comment. Fungis Clavaeform.*: 42 (1797)
Clavaria botrytis var. *alba* A. Pearson, *Trans. Brit. Mycol. Soc.* 29(4): 209 (1946)
E: ! **S:** !
H: On soil in old deciduous woodland, often amongst grass or leaf litter and usually associated with *Fagus*.
D: Corner: 560, NM3: 273 **D+I:** Ph: 260-261, B&K2: 356 461 **I:** Kriegl2: 69
Rarely reported. A very few records from scattered locations in England and a single record from Scotland.

bourdotiana Maire, *Publ. Inst. Bot. Barcelona* 3(4): 32 (1937)
Mis.: *Clavaria stricta* sensu Bourdot & Galzin, *Hyménomyc. France*: 98 (1928)
S: !
H: On soil. British collection amongst leaf litter from shrubs on a disused railway line.
D: Corner: 563-564
A single collection from Midlothian (Edinburgh, Davidson Mains) in 2001.

broomei (Cotton & Wakef.) R.H. Petersen, *Biblioth. Mycol.* 79: 53 (1981)
Clavaria broomei Cotton & Wakef., *Trans. Brit. Mycol. Soc.* 6: 170 (1917)
Ramaria nigrescens (Brinkmann) Donk [*nom. illegit.*], *Rev. Niederl. Homob. Aphyll.* 2: 104 (1933)
E: ! **W:** ?
H: On soil in deciduous woodland.
D: NM3: 272 **I:** Kriegl2: 70
Recently only known from northern England (Westmorland: Meathop Fell and Roudsea Wood), but old collections from

North Somerset (Bath), South Hampshire (New Forest) and Surrey (Mickleham). Reported from Wales (Caernarvonshire: Great Orme) in 1988 but unsubstantiated with voucher material.

decurrens (Pers.) R.H. Petersen, *Biblioth. Mycol.* 79: 124 (1981)
Clavaria decurrens Pers., *Mycol. eur.* 1: 164 (1822)
Mis.: *Clavaria crispula* sensu Bourdot & Galzin (1928) and sensu auct. Brit.
E: !
H: On soil or very decayed wood, under conifers such as *Cupressus, Pinus* and *Taxus* spp. Rarely with deciduous trees such as *Ulmus* sp.
D: NM3: 276
Known from Berkshire, Dorset, East Norfolk, Hertfordshire, Leicestershire, London, South Devon and Surrey.

eumorpha (P. Karst.) Corner, *Monograph of Clavaria and Allied Genera*: 575 (1950)
Clavariella eumorpha P. Karst., *Symb. mycol. fenn.* 9: 55 (1883)
Clavaria invalii Cotton & Wakef., *Trans. Brit. Mycol. Soc.* 6: 176 (1919)
Ramaria invalii (Cotton & Wakef.) Donk, *Meded. Bot. Mus. Herb. Rijks Univ. Utrecht* 9: 113 (1933)
E: ! **S:** !
H: On soil in conifer woodland.
D: NM3: 276
Rarely reported. Most of the reports are unsubstantiated with voucher material.

fennica *var.* **griseolilacina** Schild, *Z. Mykol.* 61(2): 160 (1995)
Clavaria fumigata Peck, *Rep. (Annual) New York State Mus. Nat. Hist.* 31: 38 (1879)
Ramaria fumigata (Peck) Corner, *Monograph of Clavaria and Allied Genera*: 591 (1950)
Ramaria versatilis Quél., *Compt. Rend. Assoc. Franç. Avancem. Sci. Assoc. Sci. France* 21: 6 (1893)
E: !
H: On soil in deciduous woodland.
D+I: B&K2: 360 468 (as *Ramaria fumigata*) **I:** FRIC5: 9-13 35 (as *R. fumigata*), C&D: 147 (as *R. versatilis*)
Known from Herefordshire, North Wiltshire and West Sussex.

flaccida (Fr.) Ricken, *Vadem. Pilzfr.*: 254 (1918)
Clavaria flaccida Fr., *Syst. mycol.* 1: 471 (1821)
E: ! **S:** ! **W:** !
H: On soil and in needle litter in conifer woodland. Rarely reported with deciduous trees.
D: Corner: 576, NM3: 276 **D+I:** B&K2: 358 463 **I:** Kriegl2: 75
Rarely reported but apparently widespread.

flava (Schaeff.) Quél., *Fl. mycol. France*: 466 (1888)
Clavaria flava Schaeff., *Fung. Bavar. Palat. nasc.* 2: 175 (1763)
E: ! **S:** ! **W:** ! **ROI:** !
H: On soil, amongst litter, in deciduous woodland. Usually associated with *Fagus* or *Quercus* spp.
D: Corner: 577, NM3: 274 **D+I:** Ph: 260, B&K2: 358 464 **I:** Kriegl2: 76
Rarely reported but apparently widespread.

formosa (Pers.) Quél., *Fl. mycol. France*: 466 (1888)
Clavaria formosa Pers., *Comment. Fungis Clavaeform.*: 41 (1797)
E: ! **W:** ! **NI:** ! **O:** Channel Islands: !
H: On soil amongst leaf litter in deciduous woodland, usually associated with *Fagus*.
D: Corner: 584, NM3: 275 **D+I:** B&K2: 360 467 **I:** Kriegl2: 78
Rarely reported. Apparently widespread but few of the records are substantiated with voucher material.

gracilis (Pers.) Quél., *Fl. mycol. France*: 463 (1888)
Clavaria gracilis Pers., *Comment. Fungis Clavaeform.*: 50 (1797)
Mis.: *Clavaria palmata* sensu Rea (1922)
E: ! **W:** ? **O:** Channel Islands: !

H: On soil in woodland, usually with conifers, often on calcareous soil.
D: Corner: 594, NM3: 271 **D+I:** B&K2: 366 476 **I:** Kriegl2: 80
Known from England (Berkshire, Cambridgeshire, Mid-Lancashire, Surrey, Warwickshire, West Norfolk and Yorkshire) and the Channel Islands (Jersey). Reported from Wales (Carmarthenshire) but unsubstantiated with voucher material.

stricta (Pers.) Quél., *Fl. mycol. France*: 464 (1888)
  Clavaria stricta Pers., *Ann. Bot. (Usteri)* 15: 33 (1795)
  Clavaria condensata Fr., *Epicr. syst. mycol.*: 575 (1838)
  Clavariella condensata (Fr.) P.Karst, *Bidrag Kännedom Finlands Natur Folk* 37: 187 (1882)
  Ramaria condensata (Fr.) Quél., *Fl. mycol. France*: 467 (1888)
  Clavaria kewensis Massee, *J. Bot.* 34: 153 (1896)
  Clavaria stricta var. *alba* Cotton & Wakef., *Trans. Brit. Mycol. Soc.* 6(2): 174 (1918)
  Mis.: *Ramaria spinulosa* sensu Rea (1922)
E: c **S:** o **W:** o **NI:** o **ROI:** o **O:** Channel Islands: !
H: On decayed wood of deciduous trees (rarely conifers). Sometimes abundant on flowerbeds mulched with woodchips, attaining far greater size than normal.
D: Corner: 565 (as *Ramaria condensata*), NM3: 272 **D+I:** Ph: 260-261, B&K2: 366 475 **I:** Kriegl2: 89
Common to occasional but widespread.

subbotrytis (Coker) Corner, *Monograph of Clavaria and Allied Genera*: 625 (1950)
  Clavaria subbotrytis Coker, *Clav. U.S & Can.* 116 (1923)
E: !
H: On soil under *Taxus*. British material in mixed woodland on limestone.
D+I: FRIC5: 13-17 36 **I:** Kriegl2: 90
Known from West Lancashire (Gait Barrows NNR).

suecica (Fr.) Donk, *Rev. Niederl. Homob. Aphyll.* 2: 105 (1933)
  Clavaria suecica Fr., *Syst. mycol.* 1: 469 (1821)
S: !
H: On soil amongst needle litter under *Pinus sylvestris*.
D: NM3: 273
A single collection from Easterness (Lairig Ghru) in 1963.

RAMARICIUM J. Erikss., *Svensk bot. Tidskr.* 48: 189 (1954)
Gomphales, Ramariaceae
Type: *Ramaricium occultum* J. Erikss.

alboochraceum (Bres.) Jülich, *Persoonia* 9(3): 417 (1977)
  Corticium alboochraceum Bres., *Ann. Mycol.* 1(1): 96 (1903)
S: !
H: British collection on a dead stem of *Ammophila arenaria*.
D: NM3: 277 **D+I:** CNE6: 1244-1246
A single collection (det. K. Hjortstam) from Galloway.

RAMARIOPSIS (Donk) Corner, *Monograph of Clavaria and Allied Genera* 1: 636 (1950)
Cantharellales, Clavariaceae
Type: *Ramariopsis kunzei* (Fr.) Corner

biformis (G.F. Atk.) R.H. Petersen, *Bull. Torrey Bot. Club* 91: 276 (1964)
  Clavaria biformis G.F. Atk., *Ann. Mycol.* 6: 56 (1908)
  Clavulinopsis biformis (G.F. Atk.) Corner, *Monograph of Clavaria and Allied Genera*: 358 (1950)
E: ! **W:** !
H: On soil in woodland or occasionally in grassland.
British collections mostly from Mid-Lancashire (Silverdale, Gait Barrows) but also known from Bedfordshire, Herefordshire, Northumberland, South Hampshire, Surrey, Warwickshire, West Kent, West Sussex and a single collection from Wales (Denbighshire).

crocea (Pers.) Corner, *Monograph of Clavaria and Allied Genera*: 638 (1950)
  Clavaria crocea Pers., *Icon. descr. fung.*: pl. 11 f. 6 (1798)

E: ! **S:** ! **O:** Channel Islands: !
H: On soil in woodland, on peat, on very decayed stumps and small fallen twigs, leaf litter, needle debris of *Taxus* and decayed fronds of *Dryopteris filix-mas*.
D: NM3: 252 **D+I:** FRIC5: 22-25 38a
Known from England (Bedfordshire, Leicestershire, Mid-Lancashire, South Devon, Surrey and West Cornwall) and single collections from Scotland (East Perthshire) and the Channel Islands (Guernsey).

kunzei (Fr.) Corner, *Monograph of Clavaria and Allied Genera*: 640 (1950)
  Clavaria kunzei Fr., *Syst. mycol.* 1: 474 (1821)
  Clavaria chionea Pers., *Mycol. eur.* 1: 167 (1822)
  Clavaria krombholzii Fr., *Epicr. syst. mycol.*: 572 (1838)
  Ramariopsis kunzei var. *subasperata* Corner, *Monograph of Clavaria and Allied Genera*: 642 (1950)
  Ramariopsis kunzei var. *bispora* Schild, *Westfäl. Pilzbriefe* 8(2): 30 (1970)
E: o **S:** o **W:** o **NI:** o **ROI:** o **O:** Channel Islands: !
H: On soil usually in open scrubby woodland and less commonly in grassland. Reported with conifers such as *Cryptomeria japonica, Juniperus* spp and *Taxus*. Also known in deciduous woodland and on debris of *Pteridium*.
D: NM3: 253
Occasional but widespread.

minutula (Bourdot & Galzin) R.H. Petersen, *Mycologia* 58: 202 (1966)
  Clavaria minutula Bourdot & Galzin, *Hyménomyc. France*: 105 (1928)
E: !
H: On soil in deciduous woodland.
Known from Surrey and Warwickshire.

pulchella (Boud.) Corner, *Monograph of Clavaria and Allied Genera*: 645 (1950)
  Clavaria pulchella Boud., *Bull. Soc. Mycol. France* 3: 146, t. 13 (1887)
  Clavaria bizzozeriana Sacc., *Syll. fung.* 6: 693 (1888)
  Clavaria conchyliata W.B. Allen, *Trans. Brit. Mycol. Soc.* 3(2): 92 (1909)
E: ! **S:** !
H: On soil amongst mosses and grass in mixed deciduous woodland or occasionally in grassland on calcareous soil, often in dense undergrowth of *Hedera* or *Rubus fruticosus* and occasionally on bare soil under *Taxus*.
D: NM3: 252 **D+I:** B&K2: 350 451 **I:** Kriegl2: 35
Known from Bedfordshire, East Kent, Herefordshire, Mid-Lancashire, North Wiltshire, South Hampshire, Surrey, Warwickshire, Worcestershire and Yorkshire. Reported from Scotland.

tenuiramosa Corner, *Monograph of Clavaria and Allied Genera*: 646 (1950)
E: !
H: On limestone or calcareous soils. British material associated with *Taxus*.
D: NM3: 253
Known from North Hampshire, West Lancashire and West Sussex.

RECTIPILUS Agerer, *Persoonia* 7: 413 (1973)
Schizophyllales, Schizophyllaceae
Type: *Rectipilus fasciculatus* (Pers.) Agerer

bavaricus Agerer, *Persoonia* 7(4): 4 (1973)
S: !
H: On decayed wood. British material on a log of *Betula* sp.
A single collection from Perthshire (Birks of Aberfeldy) in 2003.

REPETOBASIDIELLUM J. Erikss. & Hjortstam, *Corticiaceae of North Europe* 6: 1247 (1981)
Stereales, Sistotremataceae
Type: *Repetobasidiellum fusisporum* J. Erikss. & Hjortstam

fusisporum J. Erikss. & Hjortstam, *Corticiaceae of North Europe* 6: 1247 (1981)
E: !
H: On decayed fern debris.
D: NM3: 126, Myc14(2): 52 **D+I:** CNE6: 1247-1248
A single collection from Surrey (Esher, West End Common) in 1997.

REPETOBASIDIUM J. Erikss., *Symb. Bot. Upsal.* 16(1): 67 (1958)
Stereales, Sistotremataceae
Type: *Repetobasidium vile* (Bourdot & Galzin) J. Erikss.

mirificum J. Erikss., *Symb. Bot. Upsal.* 16(1): 70 (1958)
S: !
H: On decayed wood of conifers. British material on *Pinus sylvestris* in Caledonian pinewood.
D: CNE6: 1259
A single collection from Perthshire (Black Wood of Rannoch) in 2003.

vile (Bourdot & Galzin) J. Erikss., *Symb. Bot. Upsal.* 16(1): 67 (1958)
 Peniophora vilis Bourdot & Galzin, *Hyménomyc. France*: 282 (1928)
E: !
H: On decayed wood of conifers. British material on *Picea* sp.
D: NM3: 126 **D+I:** CNE6: 1263
A single collection from South Devon (Dartmoor, Bellever Forest) in 1991.

RESINICIUM Parmasto, *Conspectus Systematis Corticiacearum*: 97 (1968)
Stereales, Meruliaceae
Type: *Resinicium bicolor* (Alb. & Schwein.) Parmasto

bicolor (Alb. & Schwein.) Parmasto, *Conspectus Systematis Corticiacearum*: 98 (1968)
 Hydnum bicolor Alb. & Schwein., *Consp. fung. lusat.*: 270 (1805)
 Odontia bicolor (Alb. & Schwein.) Quél., *Enchir. fung.*: 195 (1886)
 Kneiffia subgelatinosa Berk. & Broome, *Ann. Mag. Nat. Hist., Ser. 4 [Notices of British Fungi no. 1440]* 15: 32 (1875)
 Hydnum subtile Fr., *Syst. mycol.* 1: 425 (1821)
 Mis.: *Grandinia mucida* sensu Rea (1922)
E: c **S:** c **W:** c **NI:** ! **ROI:** c **O:** Isle of Man: c
H: On decayed wood of conifers such as *Larix*, *Picea*, *Pseudotsuga* and *Pinus* spp. Rarely on deciduous trees such as *Betula* spp, and known from dead stems of *Heracleum sphondylium*, on worked wood in buildings and a single record from a plywood tea chest.
D: NM3: 164 **D+I:** CNE6: 1265-1267, B&K2: 168 178 **I:** Kriegl1: 309
Very common and widespread.

furfuraceum (Bres.) Parmasto, *Conspectus Systematis Corticiacearum*: 98 (1968)
 Corticium furfuraceum Bres., *Mycologia* 17(2): 69 (1925)
E: !
H: On decayed conifer wood. British material on *Pinus sylvestris*.
D: NM3: 164 **D+I:** CNE6: 1268-1269
A single collection from East Norfolk (Sheringham) in 1984.

RESINOMYCENA Redhead & Singer, *Mycotaxon* 13: 151 (1981)
Agaricales, Tricholomataceae
Type: *Resinomycena rhododendri* (Peck) Redhead & Singer

saccharifera (Berk. & Broome) Redhead, *Canad. J. Bot.* 62(9): 1850 (1984)
 Agaricus sacchariferus Berk. & Broome, *Ann. Mag. Nat. Hist., Ser. 4 [Notices of British Fungi no. 1216]* 6: 465 (1870)
 Mycena saccharifera (Berk. & Broome) Gillet, *Hyménomycètes*: 262 (1876)
 Agaricus electicus Buckn., *Proc. Bristol Naturalist's Soc.* ser. 3, 2: 132 (1882)
 Mycena saccharifera var. *electica* (Buckn.) Massee, *Brit. fung.-fl.* 3: 82 (1893)
 Marasmiellus ornatissimus Barkm. & Noord., *Persoonia* 13: 254 (1987)
 Mycena pudica Hora, *Trans. Brit. Mycol. Soc.* 43(2): 452 (1960)
 Omphalia quisquiliaris Joss. [*nom. inval.*], *Ann. Soc. Linn. Lyon* 60: 88 (1937) [1936]
 Delicatula quisquiliaris (Joss.) Kuhner & Romagn. [*nom. inval.*], *Flore Analytique des Champignons Supérieurs*: 118 (1953)
 Mycena quisquiliaris (Joss.) Kühner [*nom. inval. & comb. illegit.*, non *M. quisquiliaris* (Berk.) Sacc. (1887)], *Encycl. Mycol. 10, Le Genre* Mycena: 388 (1938)
E: o **S:** o **W:** ! **NI:** ! **ROI:** ! **O:** Channel Islands: ! Isle of Man: !
H: On decayed debris of grasses, sedges and rushes in woodland, fen-carr, dunes and boggy fields. Also known on dead stems of *Rubus* sp. Reported on *Molinia caerulea*, *Phragmites australis*, *Carex* and *Juncus* spp.
D: NM2: 163 (as *Mycena saccharifera*), BFF8: 142 **D+I** A&N3 164-167 44 **I:** C&D: 231
Occasional but apparently widespread. Basidiomes are small and easily overlooked.

RESUPINATUS Nees, *Syst. Pilze*: 197 (1817)
Agaricales, Tricholomataceae
 Phyllotus P. Karst., *Bidrag Kännedom Finlands Natur Folk* 32: 94 (1879)
Type: *Resupinatus applicatus* (Batsch) Gray

applicatus (Batsch) Gray, *Nat. arr. Brit. pl.* 1: 617 (1821)
 Agaricus applicatus Batsch, *Elench. fung. (Continuatio Prima)*: 171 & t. 24 f. 125 (1786)
 Pleurotus applicatus (Batsch) P. Kumm., *Führ. Pilzk.*: 105 (1871)
 Acanthocystis applicatus (Batsch) Kühner, *Botaniste* 17: 111 (1926)
 Geopetalum striatulum var. *applicatum* (Batsch) Kühner & Romagn., *Flore Analytique des Champignons Supérieurs*: 68 (1953)
 Agaricus striatulus Pers., *Syn. meth. fung.*: 485 (1801)
 Pleurotus striatulus (Pers.) P. Kumm., *Führ. Pilzk.*: 105 (1871)
 Acanthocystis striatula (Pers.) Kühner, *Botaniste* 17: 112 (1926)
 Geopetalum striatulum (Pers.) Kühner & Romagn., *Flore Analytique des Champignons Supérieurs*: 68 (1953)
 Mis.: *Pleurotus reniformis* sensu Rea (1922), sensu auct.
E: c **S:** o **W:** o **NI:** ! **ROI:** ! **O:** Channel Islands: ! Isle of Man: o
H: On decayed wood of deciduous trees, usually *Fagus* and *Fraxinus* but also known on *Acer pseudoplatanus*, *Alnus*, *Corylus*, *Ilex*, *Populus* spp, *Laburnum anagyroides*, *Malus sylvestris*, *Salix* spp, *Tilia* spp, *Sambucus nigra*, *Ulmus* spp and *Ulex europaeus*. Rarely on dead lianas of *Clematis vitalba* and dead stems of *Rosa* spp and *Rubus fruticosus* agg.
D: BFF6: 62-63, NM2: 179, FAN3: 166-167 **D+I:** Myc4(4): 198 **I:** Bon: 123, C&D: 153, Kriegl3: 506
Frequent and widespread.

kavinii (Pilát) M.M. Moser, *Kleine Kryptogamenflora*, Edn 4: 153 (1978)
 Pleurotus kavinii Pilát, *Mykologia* 8: 23 (1931)
 Resupinatus applicatus f. *kavinii* (Pilát) Pilát, *Micologia Veneta (Padova)*: 66 (1935)
E: !
H: On decayed fallen wood, or under bark on dead standing trunks of *Ulmus* spp.
D: BFF6: 63, FAN3: 167
Known from Surrey and Warwickshire.

trichotis (Pers.) Singer, *Persoonia* 2: 48 (1961)
 Agaricus trichotis Pers., *Mycol. eur.* 3: 18 (1828)

Agaricus rhacodius Berk. & M.A. Curtis, *Ann. Mag. Nat. Hist,* Ser. 3, 4(22): 288 (1859)

Pleurotus rhacodius (Berk. & M.A. Curtis) Sacc., *Syll. fung.* 5: 380 (1887)

Resupinatus rhacodius (Berk. & M.A. Curtis) Singer, *Lilloa* 22: 253 (1951) [1949]

E: o **S:** o **W:** ! **NI:** ! **ROI:** ! **O:** Isle of Man: !

H: On fallen or attached wood, usually small branches of *Acer pseudoplatanus* and *Fagus*. Occasionally reported on *Alnus, Fraxinus, Quercus, Salix, Ulex europaeus* and *Ulmus* spp.

D: BFF6: 64, NM2: 179 **D+I:** B&K3: 316 400 **I:** FlDan2: 66A (as *Pleurotus rhacodium*)

Occasional but widespread. Some authorities consider this a form of *Resupinatus applicatus*.

RHIZOPOGON Fr. & Nordholm, *Symb. gasteromyc.* 1: 5 (1817)

Boletales, Rhizopogonaceae

Splanchnomyces Corda, *Deutschl. Fl.,* Edn 3: 3 (1831)
Hysteromyces Vittad., *Not. Nat. Civil. Lombardia* 1: 340 (1844)

Type: *Rhizopogon luteolus* Fr. & Nordholm

luteolus Fr. & Nordholm, *Symb. gasteromyc.* 1: 5 (1817)
Rhizopogon induratus Cooke, *Grevillea* 8: 59 (1879)

E: o **S:** c **ROI:** !

H: Epigeous or subhypogeous in sandy acidic soil, always associated with *Pinus* spp.

D+I: Ph: 252-253, BritTruff: 158-160 9 B **I:** Bon: 303 (as *R. obtextus*), Kriegl2: 199 (as *R. obtextus*)

Common in Scotland and occasional in England (but may be locally abundant). Rarely reported elsewhere. English name = 'Yellow Beard Truffle'.

ochraceorubens A.H. Sm., *Mem. N. Y. bot. Gdn* 14(2): 93 (1966)
Mis.: *Rhizopogon luteolus* sensu auct. Brit. p.p.
Mis.: *Rhizopogon rubescens* sensu auct. Brit. p.p.

E: ! **S:** !

H: Epigeous or hypogeous, in sandy soil with *Pinus* spp.

Known from Scotland (Culbin Sands) and England (New Forest). Previously misdetermined as either *R. luteolus* or *R. rubescens*.

roseolus (Corda) Th. Fr., *Svensk bot. Tidskr.* 6(3): 238 (1909)
Splanchnomyces roseolus Corda, *Deutschl. Fl.* Edn 3: 3 (1837)
Melanogaster berkeleyanus Broome, *Ann. Mag. Nat. Hist., Ser. 1* 15: 41 (1845)
Rhizopogon luteorubescens A.H. Sm., *Mem. N. Y. bot. Gdn* 14(2): 92 (1966)
Rhizopogon provincialis Tul., *Fungi hypogaei*: 88 (1851)
Hysterangium rubescens Tul. & C. Tul., *Ann. Sci. Nat., Bot.,* sér. 2, 19: 375 (1843)
Rhizopogon rubescens (Tul. & C. Tul.) Tul. & C. Tul., *Giorn. Bot. Ital.* 2(1): 58 (1844)
Hysteromyces vulgaris Vittad., *Not. Nat. Civil. Lombardia* 1: 341 (1844)
Rhizopogon vulgaris (Vittad.) M. Lange, *Dansk. Bot. Ark.* 16: 56 (1956)

E: ! **S:** ! **W:** !

H: Hypogeous (often deeply buried) and always associated with *Pinus* spp.

D: NM2: 67 (as *Rhizopogon rubescens*), NM2: 68 **D+I:** BritTruff: 156-157 9 D, BritTruff: 157 9 E **I:** Kriegl2: 200

Not uncommon and apparently widespread. However some of the records and collections may include *R. ochraceorubens*.

villosulus Zeller, *Mycologia* 33: 196 (1941)
Rhizopogon reticulatus Hawker, *Trans. Brit. Mycol. Soc.* 38(1): 73 (1955)
Rhizopogon hawkerae A.H. Sm., *Mem. N. Y. bot. Gdn* 14(2): 83 (1966)
Rhizopogon colossus A.H. Sm., *Mem. N. Y. bot. Gdn* 14(2): 93 (1966)

E: ! **S:** !

H: Subhypogeous in acidic soil, associated with *Pseudotsuga menziesii* and *Picea* spp.

D+I: BritTruff: 160-161 9 C

Known from England (North Somerset, South Hampshire, West Sussex and Yorkshire) and Scotland (Perthshire).

vinicolor A.H. Sm., *Mem. N. Y. bot. Gdn* 14(2): 67 (1966)

E: !

H: Hypogeous in acid soil. Supposedly a *Pseudotsuga* associate, but English material (det. Martín) under *Pinus* sp.

In Britain known only from the New Forest.

RHODOCYBE Maire, *Bull. Soc. Mycol. France* 40: 298 (1926) [1924]

Agaricales, Entolomataceae

Clitopilopsis Maire, *Publ. Inst. Bot. Barcelona* 3(4): 82 (1937)
Type: *Rhodocybe caelata* (Fr.) Maire

caelata (Fr.) Maire, *Bull. Soc. Mycol. France* 40: 298 (1926) [1924]
Agaricus caelatus Fr., *Epicr. syst. mycol.*: 42 (1838)
Tricholoma caelatum (Fr.) Gillet, *Hyménomycètes*: 114 (1874)
Clitopilus caelatus (Fr.) Kühner & Romagn., *Flore Analytique des Champignons Supérieurs*: 173 (1953)

E: ! **S:** !

H: On soil, amongst grass in open woodland and on roadside verges.

D: FAN1: 78, NM2: 358 **D+I:** B&K4: 114 97 **I:** C&D: 303

Known from England (Bedfordshire) and Scotland (Easterness).

fallax (Quél.) Singer, *Farlowia* 2: 549 (1946)
Omphalia fallax Quél., *Compt. Rend. Assoc. Franç. Avancem. Sci. Assoc. Sci. France* 24(2): 617 (1896)
Clitocybe fallax (Quél.) Sacc. & Trotter, *Syll. fung.* 21: 42 (1912)
Rhodopaxillus fallax (Quél.) Maire, *Bull. Mens. Soc. Linn. Lyon* 6(3): 19 (1927)
Paxillopsis fallax (Quél.) J.E. Lange, *Fl. agaric. danic.* 5, *Taxonomic Conspectus*: VI (1940)
Clitopilopsis fallax (Quél.) Konrad & Maubl., *Encycl. Mycol.* 14, *Les Agaricales* I: 380 (1948)
Clitopilus fallax (Quél.) Kühner & Romagn., *Flore Analytique des Champignons Supérieurs*: 173 (1953)

E: ! **S:** ! **ROI:** !

H: On soil, in woodland or rough pasture.

D: FAN1: 80-81, NM2: 359 **D+I:** B&K4: 114 98 **I:** C&D: 303

Known from England (Bedfordshire, East Kent, Warwickshire and West Kent) and single collections from Scotland (Midlothian) and the Republic of Ireland (Co. Dublin).

gemina (Fr.) Kuyper & Noordel., *Persoonia* 13(3): 379 (1987)
Agaricus geminus Fr., *Epicr. syst. mycol.*: 38 (1838)
Tricholoma geminum (Fr.) S. Petersen, *Dan. Agaric.* 1: 61 (1907)
Rhodocybe truncata var. *subvermicularis* (Maire) Singer, *Farlowia* 2: 286 (1946)
Mis.: *Rhodocybe truncata* sensu auct.

E: ! **S:** ! **W:** ! **NI:** ! **ROI:** !

H: Usually on nitrogenous soil in deciduous woodland and scrub, sometimes among *Urtica dioica*. Rarely with conifers, and very rarely with *Elymus arenarius* in dunes.

D: FAN1: 81-82, NM2: 359 **D+I:** B&K4: 114 99 **I:** Bon: 189 (as *Rhodocybe truncata*), C&D: 303

Rarely reported but apparently widespread. Perhaps better known as *R. truncata* (a *nomen dubium*).

griseospora (A. Pearson) P.D. Orton, *Trans. Brit. Mycol. Soc.* 43(2): 181 (1960)
Collybia griseospora A. Pearson, *Trans. Brit. Mycol. Soc.* 35(2): 102 (1952)

E: ? **W:** !

H: On soil in deciduous woodland. British material associated with *Corylus* and *Mercurialis perennis*.

Accepted, but doubtfully distinct from *R. nitellina* and differing only in the greyish coloured spore print. Known only from the type collection from Caernarvonshire (Bangor). Reported from England (South Somerset: Higher Merridge) but unsubstantiated with voucher material.

hirneola (Fr.) P.D. Orton, *Trans. Brit. Mycol. Soc.* 43: 181 (1960)
  Agaricus hirneolus Fr., *Syst. mycol.* 1: 269 (1821)
  Clitocybe hirneola (Fr.) P. Kumm., *Führ. Pilzk.*: 120 (1871)
  Clitopilopsis hirneola (Fr.) Konrad & Maubl., *Encycl. Mycol.* 14, *Les Agaricales* I: 379 (1948)
  Clitopilus hirneolus (Fr.) Kühner & Romagn., *Flore Analytique des Champignons Supérieurs*: 173 (1953)
  Clitopilopsis arthrocystis Kühner & Maire, *Bull. Soc. Hist. Nat. Afrique N.* 28: 113 (1936)
  Agaricus undulatus Bull., *Herb. France*: pl. 535 (1792)
  Clitocybe hirneola var. *undulata* (Bull.) P. Karst., *Bidrag Kännedom Finlands Natur Folk* 32: 60 (1879)
  Clitocybe hirneola var. *ovispora* J.E. Lange, *Fl. agaric. danic.* 5, *Taxonomic Conspectus*: II (1940)
  Clitocybe xanthophylla Bres., *Fungi trident.* 1: 8 (1881)
E: ! **S:** ? **W:** !
H: On soil in conifer woodland.
D: FAN1: 81, NM2: 359 **I:** C&D: 303
Very poorly known in Britain. Collections from England (Bedfordshire) in 1952 and Wales (Denbighshire) in 1879 in herb. K. Reported from Scotland but unsubstantiated with voucher material.

melleopallens P.D. Orton, *Trans. Brit. Mycol. Soc.* 43(2): 380 (1960)
E: !
H: On soil, often calcareous, in open mixed deciduous woodland, and a single collection from soil in an unheated greenhouse.
D: FAN1: 79, NM2: 359
Rarely reported. Known from Middlesex, Surrey and Yorkshire.

nitellina (Fr.) Singer, *Mycologia* 38: 687 (1946)
  Agaricus nitellinus Fr., *Epicr. syst. mycol.*: 80 (1838)
  Collybia nitellina (Fr.) Quél., *Mém. Soc. Émul. Montbéliard, Sér. 2,* 5: 434 (1875)
E: ! **W:** ! **ROI:** ?
H: On calcareous loam soil in open mixed deciduous woodland.
D: FAN1: 78-79, NM2: 359 **D+I:** FRIC3: 17-18 22a, B&K4: 116 101 **I:** Bon: 189, C&D: 303
Known from England (Oxfordshire, Shropshire, South Hampshire, Surrey and West Sussex) and Wales (Pembrokeshire). Reported from the Republic of Ireland (Co. Dublin) in 1898 but unsubstantiated with voucher material.

popinalis (Fr.) Singer, *Lilloa* 22: 609 (1951) [1949]
  Agaricus popinalis Fr., *Syst. mycol.* 1: 194 (1821)
  Clitopilus popinalis (Fr.) P. Kumm., *Führ. Pilzk.*: 97 (1871)
  Paxillus popinalis (Fr.) Ricken, *Blätterpilze Deutschl.*: 94 (1911)
  Clitocybe popinalis (Fr.) Bres., *Icon. Mycol.* 4: 160 (1928)
  Rhodopaxillus popinalis (Fr.) Konrad & Maubl., *Rév. hyménomyc. France*: 327 (1937)
  Paxillopsis popinalis (Fr.) J.E. Lange, *Fl. agaric. danic.* 5, *Taxonomic Conspectus*: VI (1940)
  Clitopilopsis popinalis (Fr.) Konrad & Maubl., *Encycl. Mycol.* 14, *Les Agaricales* I: 379 (1948)
  Rhodopaxillus lutetianus E.-J. Gilbert, *Bull. Soc. Mycol. France* 42: 66 (1926)
  Agaricus mundulus Lasch, *Linnaea* 4: 527 (1829)
  Clitopilus mundulus (Lasch) P. Kumm., *Führ. Pilzk.*: 97 (1871)
  Rhodopaxillus mundulus (Lasch) Konrad & Maubl., *Icon. select. fung.* 3: pl. 278 (1934)
  Paxillopsis mundula (Lasch) J.E. Lange, *Fl. agaric. danic.* 5, *Taxonomic Conspectus*: VI (1940)
  Clitocybe mundula (Lasch) A. Pearson & Dennis, *Trans. Brit. Mycol. Soc.* 31: 153 (1948)
  Clitopilopsis mundula (Lasch) Konrad & Maubl., *Encycl. Mycol.* 14, *Les Agaricales* I: 379 (1948)
  Rhodocybe mundula (Lasch) Singer, *Lilloa* 22: 609 (1951) [1949]

  Clitopilus mundulus var. *nigrescens* Sacc., *Syll. fung.* 5: 700 (1887)
  Mis.: *Paxillus lepista* sensu Ricken [*Blätterpilze Deutschl.*: 94 (1911)]
  Mis.: *Clitocybe senilis* sensu Rea (1922)
E: ! **S:** ! **W:** ! **NI:** ! **ROI:** ! **O:** Channel Islands: !
H: On soil in deciduous and conifer woodland, also in grassy areas on dunes, and in hill pasture.
D: FAN1: 79-80, NM2: 359, NM2: 359 (as *Rhodocybe mundula*) **D+I:** Ph: 114, B&K4: 116 100 (as *R. mundula*) **I:** FlDan4: 134, FlDan4: 133B (as *Paxillopsis mundulus*), Bon: 189 (as *R. mundula*), Bon: 189, FungEur3: 55, C&D: 303
Rarely reported but apparently widespread.

roseiavellanea (Murrill) Singer, *Lilloa* 22: 609 (1951) [1949]
  Pleuropus roseiavellaneus Murrill, *Mycologia* 30: 367 (1938)
E: !
H: On soil. British material amongst needle litter of *Larix* sp. in a plantation.
D+I: Myc15(4): 151-152 (+ back cover)
A single collection from North Hampshire (Northington) in 1999.

RHODOTUS Maire, *Bull. Soc. Mycol. France* 40: 308 (1926) [1924]
Agaricales, Tricholomataceae
Type: *Rhodotus palmatus* (Bull.) Maire

palmatus (Bull.) Maire, *Bull. Soc. Mycol. France* 40: 308 (1926) [1924]
  Agaricus palmatus Bull., *Herb. France*: pl. 216 (1785)
  Pleuropus palmatus (Bull.) Gray, *Nat. arr. Brit. pl.* 1: 615 (1821)
  Crepidotus palmatus (Bull.) Gillet, *Hyménomycètes*: 558 (1876)
  Pleurotus palmatus (Bull.) Quél., *Compt. Rend. Assoc. Franç. Avancem. Sci.* 11: 390 (1883)
  Entoloma cookei Richon, *Descr. dess. pl. crypt.*: 559 (1879)
  Pleuropus palmatus β *rubescens* Gray, *Nat. arr. Brit. pl.* 1: 615 (1821)
  Agaricus phlebophorus var. *reticulatus* Cooke, *Handb. Brit. fung.*, Edn 2: 118 (1886)
  Agaricus subpalmatus Fr., *Epicr. syst. mycol.*: 131 (1838)
  Pleurotus subpalmatus (Fr.) Gillet, *Hyménomycètes*: 343 (1876)
  Mis.: *Agaricus roseoalbus* sensu Cooke [Ill. Brit. fung. 312 (598) Vol. 3 (1884)]
E: o **S:** ! **W:** ! **ROI:** ! **O:** Channel Islands: !
H: On fallen and decayed wood of deciduous trees. Usually on *Ulmus* spp, less commonly on *Acer campestre, A. pseudoplatanus, Fagus* and *Fraxinus*. Rarely on *Populus* and *Quercus* spp.
D: BFF6: 65, NM2: 179, FAN3: 175-176 **D+I:** Ph: 186-187, Myc5(3): 147 **I:** Sow1: 62 (as *Agaricus palmatus*), Bon: 173, C&D: 239, Kriegl3: 508
Occasional but may be locally common in southern England, with a single recent record from Scotland (Clyde Islands: Chatelherault). Particularly common on elm wood during the epidemic of Dutch elm disease.

RICKENELLA Raithelh., *Metrodiana* 4(4): 67 (1973)
Agaricales, Tricholomataceae
Type: *Rickenella fibula* (Bull.) Raithelh.

fibula (Bull.) Raithelh., *Metrodiana* 4: 67 (1973)
  Agaricus fibula Bull., *Herb. France*: pl. 186 (1783)
  Micromphale fibulare (Bull.) Gray, *Nat. arr. Brit. pl.* 1: 623 (1821)
  Omphalia fibula (Bull.) P. Kumm., *Führ. Pilzk.*: 106 (1871)
  Omphalina fibula (Bull.) Quél., *Enchir. fung.*: 46 (1886)
  Mycena fibula (Bull.) Kühner, *Encycl. Mycol.* 10, *Le Genre Mycena*: 607 (1938)
  Hemimycena fibula (Bull.) Singer, *Ann. Mycol.* 41: 123 (1943)
  Marasmiellus fibula (Bull.) Singer, *Sydowia* 2: 32 (1948)

Agaricus nivalis Vahl, *Fl. Danica*: t. 1072 (1792)
Omphalia fibula var. *nivalis* (Vahl) Rea, *Brit. basidiomyc.*: 433 (1922)
E: c **S:** c **W:** c **NI:** c **ROI:** c **O:** Channel Islands: ! Isle of Man: !
H: On soil, always associated with mosses, in woodland, unimproved grassland, or heathland.
D: NM2: 174, FAN3: 158 **D+I:** A&N3 170-175 45, Ph: 75 (as *Mycena fibula*), B&K3: 316 401 **I:** Sow1: 45 (as *Agaricus fibula*), FlDan2: 61G, Bon: 129, C&D: 183, Kriegl3: 509
Very common and widespread.

pseudogrisella (A.H. Sm.) Gulden, *Arctic and Alpine Fungi* 1: 17 (1985)
Mycena pseudogrisella A.H. Sm., *North American Species of Mycena*: 124 (1947)
S: !
H: On soil. British material with the liverwort *Blasia pusilla* on a streambank.
D&I: A&N3 184-187 48
A single collection from Easterness (Abernethy Forest) in 2003.

swartzii (Fr.) Kuyper, *Persoonia* 12(2): 188 (1983)
Agaricus swartzii Fr., *Observ. mycol.* 1: 90 (1815)
Agaricus fibula var. *swartzii* (Fr.) Fr., *Syst. mycol.* 1: 164 (1821)
Omphalina fibula var. *swartzii* (Fr.) Quél., *Enchir. fung.*: 46 (1886)
Mycena swartzii (Fr.) A.H. Sm., *North American Species of Mycena*: 123 (1947)
Omphalina swartzii (Fr.) Kotl. & Pouzar, *Česká Mykol.* 20: 136 (1966)
Agaricus setipes var. *acrocyaneus* Fr., *Hymenomyc. eur.*: 164 (1874)
Mis.: *Omphalina setipes* sensu auct.
Mis.: *Rickenella setipes* sensu auct.
E: c **S:** c **W:** c **NI:** c **ROI:** ! **O:** Channel Islands: ! Isle of Man: !
H: On soil, always associated with mosses, often in grassland and also in woodland.
D: NM2: 174 (as *Omphalina setipes*), FAN3: 157 **D+I:** A&N3 177-181 46&47, Ph: 75 (as *Mycena swartzii*), B&K3: 316 402 (as *Rickenella setipes*) **I:** FlDan2: 61D (as *O. swartzii*), Bon: 129, C&D: 183, Kriegl3: 510
Very common and widespread.

RIGIDOPORUS Murrill, *Bull. Torrey Bot. Club* 32: 478 (1905)
Poriales, Coriolaceae
Type: *Polyporus micromegas* Mont. (= *Rigidoporus lineatus*)

ulmarius (Sowerby) Imazeki, *Bull. Gov. Forest Exp. Sta., Meguro* 57: 119 (1952)
Boletus ulmarius Sowerby, *Col. fig. Engl. fung.* 1: pl. 88 (1797)
Polyporus ulmarius (Sowerby) Fr., *Syst. mycol.* 1: 365 (1821)
Fomes ulmarius (Sowerby) Gillet, *Hyménomycètes*: 683 (1878)
Fomes geotropus Cooke, *Grevillea* 13: 119 (1885)
E: o **W:** ! **O:** Channel Islands: !
H: Parasitic then saprotrophic. Once common on large trunks or stumps of *Ulmus* spp, especially during the epidemic of Dutch elm disease, but now declining on that host. Frequent on *Aesculus* in parkland, and also known on *Acer negundo*, *A. pseudoplatanus*, *Fagus*, *Fraxinus*, *Platanus*, *Populus nigra*, *P. canescens*, *Robinia pseudoacacia*, *Sambucus nigra*, *S. racemosa* and *Tilia* spp. Also collected from old sewage or drainage pipes in London (made from hollowed-out elm trunks).
D: EurPoly2: 602 **D+I:** Ph: 228, Myc4(3): 146 **I:** Sow1: 88 (as *Boletus ulmarius*)
Occasional but widespread in England, perhaps now decreasing. Most often reported from southern areas but known as far north as Derbyshire and Yorkshire. Two collections from Wales (Flintshire and Glamorganshire) and one from the Channel Islands (Guernsey).

undatus (Pers.) Donk, *Persoonia* 5: 115 (1967)

Polyporus undatus Pers., *Mycol. eur.* 2: 90 (1825)
Poria undata (Pers.) Quél., *Fl. mycol. France*: 380 (1888)
E: ! **O:** Isle of Man: ?
H: On sodden and decayed wood in deciduous woodland, often in wet areas such as ditches or by steams or pools. Usually on *Fagus* and *Fraxinus* but also known on *Betula* and *Salix* spp.
D: EurPoly2: 603 **D+I:** B&K2: 300 378 (misdet. as *Physisporinus vitreus*), Myc17(4): 156
Known from Berkshire, North Hampshire, Oxfordshire, South Devon, South Somerset, Surrey, West Kent and West Sussex. Reported from the Isle of Man but unsubstantiated with voucher material. Possibly under-recorded due to confusion with the macroscopically similar *Physisporinus vitreus*.

RIMBACHIA Pat., *Bull. Soc. Mycol. France* 7: 159 (1891)
Agaricales, Tricholomataceae
Mniopetalum Donk & Singer, *Persoonia* 2: 332 (1962)
Type: *Rimbachia paradoxa* Pat.

arachnoidea (Peck) Redhead, *Canad. J. Bot.* 62(5): 878 (1984)
Cyphella arachnoidea Peck, *Rep. (Annual) New York State Mus. Nat. Hist.* 44: 134 (1891)
Mniopetalum globisporum Donk, *Persoonia* 2: 332 (1962)
E: ! **S:** ! **NI:** !
H: Parasitic on mosses. Most of the collections are on *Mnium hornum* but also known on *Brachythecium* spp and *Sphagnum auriculatum*. Also on debris of *Juncus effusus* growing in *Sphagnum*.
D: NM2: 180, FAN3: 135, CzM54(3-4): 147-149 **D+I:** Myc13(2): 85
Known from England (North Hampshire, Staffordshire, Surrey), Scotland (Dumfriesshire and Kirkudbrightshire) and Northern Ireland (Fermanagh).

bryophila (Pers.) Redhead, *Canad. J. Bot.* 62(5): 878 (1984)
Agaricus bryophilus Pers., *Observ. mycol.* 1: 8 (1796)
E: ! **S:** ! **W:** !
H: Parasitic on mosses. British material on *Mnium hornum* and *Eurynchium* spp.
D: BFF6: 67, NM2: 180, FAN3: 135, CzM54(3-4): 151
Single collections from England (Westmorland: Roudsea Wood), Scotland (Mid Perthshire: Pass of Killiekrankie) and Wales (Caernarvonshire: Coeddyd Aber NNR). Reported also from Cumberland and Kirkudbrightshire but unsubstantiated with voucher material.

neckerae (Fr.) Redhead, *Canad. J. Bot.* 62(5): 879 (1984)
Cyphella muscicola δ *neckerae* Fr., *Syst. mycol.* 2: 203 (1823)
Cyphella neckerae (Fr.) Fr., *Epicr. syst. mycol.*: 568 (1838)
Leptoglossum candidum D.A. Reid, *Trans. Brit. Mycol. Soc.* 48: 514 (1965)
E: ! **W:** ! **NI:** !
H: Parasitic on mosses.
D: BFF6: 68 **D+I:** CzM54(3-4): 152-153
Known from England (Herefordshire and Shropshire), Northern Ireland (Antrim) and Wales (Glamorganshire).

RIPARTITES P. Karst., *Bidrag Kännedom Finlands Natur Folk* 32(24): 477 (1879)
Agaricales, Tricholomataceae
Type: *Ripartites tricholoma* (Alb. & Schwein.) P. Karst.

tricholoma (Alb. & Schwein.) P. Karst., *Bidrag Kännedom Finlands Natur Folk* 32: 477 (1879)
Agaricus tricholoma Alb. & Schwein., *Consp. fung. lusat.*: 118 (1805)
Flammula tricholoma (Alb. & Schwein.) P. Kumm., *Führ. Pilzk.*: 82 (1871)
Inocybe tricholoma (Alb. & Schwein.) Kalchbr., *Icon. select. Hymenomyc. Hung.*: 34 (1874)
Paxillus tricholoma (Alb. & Schwein.) Quél., *Enchir. fung.*: 92 (1886)

Astrosporina tricholoma (Alb. & Schwein.) J. Schröt., in Cohn, *Krypt.-Fl. Schlesien* 3: 577 (1889)
Paxillopsis tricholoma (Alb. & Schwein.) J.E. Lange, *Fl. agaric. danic.* 5, *Taxonomic Conspectus*: VI (1940)
Agaricus helomorphus Fr., *Epicr. syst. mycol.*: 184 (1838)
Flammula helomorpha (Fr.) Quél., *Mém. Soc. Émul. Montbéliard, Sér. 2*, 5: 129 (1872)
Paxillopsis helomorpha (Fr.) J.E. Lange, *Fl. agaric. danic.* 5, *Taxonomic Conspectus*: VI (1940)
Ripartites helomorphus (Fr.) P. Karst., *Bidrag Kännedom Finlands Natur Folk* 32: 479 (1879)
Paxillus helomorphus (Fr.) Quél., *Enchir. fung.*: 92 (1886)
Ripartites metrodii Huijsman, *Persoonia* 1: 337 (1960)
Paxillus panaeolus Fr., *Öfvers. Kongl. Vetensk.-Akad. Förh.* 18: 27 (1862)
Agaricus strigiceps Fr., *Syst. mycol.* 1: 270 (1821)
Flammula strigiceps (Fr.) P. Kumm., *Führ. Pilzk.*: 82 (1871)
Ripartites strigiceps (Fr.) P. Karst., *Bidrag Kännedom Finlands Natur Folk* 32: 478 (1879)
Inocybe strigiceps (Fr.) Sacc., *Syll. fung.* 5: 791 (1887)
E: ! **S:** ! **W:** ! **NI:** ! **ROI:** !
H: On soil amongst leaf litter, usually in conifer woodland and plantations, not infrequently on burnt ground. Rarely reported with deciduous trees.
D: NM2: 180 (as *R. metrodii*), NM2: 180 (as *R. helomorphus*), NM2: 180, FAN3: 95 **D+I:** B&K3: 92 66 (as *R. metrodii*), B&K3: 94 67 **I:** FlDan4: 134F (as *Paxillus helomorphus*), FlDan4: 133D & D1 (as *Paxillopsis tricholoma*), Bon: 147, Kriegl3: 513
Apparently widespread. The treatment here follows Noordeloos (FAN3) who considered all five epithets listed above as belonging to a single variable species.

ROZITES P. Karst., *Bidrag Kännedom Finlands Natur Folk* 32(20): 290 (1879)
Cortinariales, Cortinariaceae
Type: *Rozites caperatus* (Pers.) P. Karst.

caperatus (Pers.) P. Karst., *Bidrag Kännedom Finlands Natur Folk* 32: 290 (1879)
Agaricus caperatus Pers., *Observ. mycol.* 1: 48 (1796)
Cortinarius caperatus (Pers.) Fr., *Epicr. syst. mycol.*: 256 (1838)
Pholiota caperata (Pers.) P. Kumm., *Führ. Pilzk.*: 84 (1871)
Dryophila caperata (Pers.) Quél., *Enchir. fung.*: 66 (1886)
Togaria caperata (Pers.) W.G. Sm., *Syn. Brit. Bas.*: 122 (1908)
E: ! **S:** o **NI:** !
H: On soil, especially in Caledonian pinewoods, but also known with deciduous trees and conifers elsewhere.
D: NM2: 335, BFF7: 91-92 **D+I:** Ph: 141, B&K5: 296 383 **I:** FlDan3: 104F (as *Pholiota caperata*), Bon: 231, C&D: 325, SV32: 36
Occasional in Scotland and may be locally frequent but rarer elsewhere.

RUBINOBOLETUS Pilát & Dermek, *Česká Mykol.* 23: 81 (1969)
Boletales, Strobilomycetaceae
Type: *Rubinboletus rubinus* (W.G. Sm.) Pilát & Dermek

rubinus (W.G. Sm.) Pilát & Dermek, *Česká Mykol.* 23: 81 (1969)
Boletus rubinus W.G. Sm., *J. Bot.* 6: 33 (1868)
Suillus rubinus (W.G. Sm.) Kuntze, *Revis. gen. pl.* 3(2): 536 (1898)
Xerocomus rubinus (W.G. Sm.) A. Pearson, *Naturalist (Hull)* 1946: 96 (1946)
Chalciporus rubinus (W.G. Sm.) Singer, *Persoonia* 7(2): 319 (1973)
E: ! **W:** !
H: On soil, associated with *Quercus* spp, often amongst short grass under isolated trees in parkland, not usually in woodland.

D+I: FRIC1: 12-13 4 (as *Boletus rubinus*), FRIC16: 5-6 121b **I:** Bon: 45 (as *Chalciporus rubinus*), C&D: 423 (as *C. rubinus*)
Uncommonly reported. Apparently widespread from Yorkshire southwards, most frequently in southern England.

RUSSULA Pers., *Observ. mycol.* 1: 100 (1796)
Russulales, Russulaceae
Type: *Russula emetica* (Schaeff.) Pers.

acetolens Rauschert, *Česká Mykol.* 43(4): 195 (1989)
Mis.: *Russula lutea* sensu auct.
E: ! **S:** ! **W:** ! **NI:** ! **ROI:** !
H: On soil, in deciduous woodland.
D+I: Galli2: 294-295 **I:** C&D: 395
Rarely reported but apparently widespread. Many collections and records are as *R. lutea*.

acrifolia Romagn., *Doc. Mycol.* 104: 32 (1997)
Mis.: *Russula densifolia* sensu NCL & Rayner (1985)
E: ! **S:** ! **W:** ?
H: On soil, in deciduous or mixed conifer and deciduous woodland.
D: NM2: 377 **D+I:** Galli2: 48-49, Sarn: 149-153
Rarely reported but apparently widespread. Records of *R. densifolia* will include *R. acrifolia* .

adusta (Pers.) Fr., *Syst. mycol.* 1: 60 (1821)
Agaricus adustus Pers., *Syn. meth. fung.*: 459 (1801)
Omphalia adusta (Pers.) Gray, *Nat. arr. Brit. pl.* 1: 614 (1821)
E: ! **S:** ! **NI:** ! **O:** Channel Islands: !
H: On acidic soil, often in mixed woodland.
D: NM2: 377 **D+I:** Galli2: 56-57, Sarn: 159-163 **I:** SV46: 7
Rarely reported but apparently widespread.

aeruginea Fr., *Monogr. hymenomyc. Suec.* 2: 198 (1863)
Mis.: *Russula graminicolor* sensu Rea (1922)
E: c **S:** c **W:** c **NI:** c **ROI:** c
H: On soil, associated with *Betula* and *Quercus* spp, in deciduous woodland.
D: NM2: 383 **D+I:** Galli2: 108-109, Sarn: 370-374 **I:** C&D: 383, SV46: 9
Common and widespread.

albonigra (Krombh.) Fr., *Hymenomyc. eur.*: 440 (1874)
Agaricus alboniger Krombh., *Naturgetr. Abbild. Schwämme* 9: 27 (1845)
Russula adusta var. *albonigra* (Krombh.) Massee, *Brit. fung.-fl.* 3: 52 (1893)
E: !
H: On acidic soil, usually associated with *Quercus* spp, in mixed deciduous woodland, and also reported with *Pseudotsuga menziesii* in plantation.
D: NM2: 377 **D+I:** Galli2: 46-47, Sarn: 178-183
British collections from South Hampshire (New Forest, Millyford Bridge in 1998) and West Sussex (near Cocking in 2004). Sensu NCL and Rayner (1985) is *Russula anthracina*.

alnetorum Romagn., *Bull. mens. Soc. linn. Lyon* 25: 181 (1956)
Russula pumila Rouzeau & F. Massart, *Act. Soc. linn. Bordeaux* 105(7): 3 (1970) [1968]
E: ! **S:** ! **W:** ! **NI:** !
H: On soil, associated with *Alnus* in swampy or wet areas in mixed deciduous woodland.
D: NM2: 396 **D+I:** Myc1(4): 174, Galli2: 212-213 (as *R. pumila*), Sarn: 531-537 **I:** C&D: 375 (as *R. pumila*)
Rarely reported but apparently widespread.

alutacea (Fr.) Fr., *Epicr. syst. mycol.*: 362 (1838)
Agaricus alutaceus Fr. [non *A. alutaceus* Cooke & Massee (1892)], *Syst. mycol.* 1: 55 (1821)
E: ! **W:** ! **NI:** !
H: On soil, in deciduous woodland.
D: NM2: 391 **D+I:** Galli2: 382-383 **I:** C&D: 397
Rarely reported but apparently widespread.

amarissima Romagn. & E.-J. Gilbert, *Bull. Soc. Mycol. France*
59: 71 (1943)
    Russula lepida var. *amara* Maire [*nom. nud.*], *Bull. Soc.
    Mycol. France* 26: 66 (1910)
E: !
H: On soil, associated with *Fagus* in old beechwoods, or mixed
    deciduous woodland.
D+I: Galli2: 192-193 **I:** C&D: 393
Known from Oxfordshire, South Essex, South Hampshire and
West Sussex.

amethystina Quél., *Compt. Rend. Assoc. Franç. Avancem. Sci.
Assoc. Sci. France* 26: 450 (1898)
E: ! **S:** ! **NI:** ?
H: On soil, often associated with *Betula* spp, and also known
    with conifers in mixed woodland.
D: NM2: 392 **D+I:** Galli2: 2860287 **I:** C&D: 397, SV46: 17
Rarely reported but apparently widespread.

amoenolens Romagn., *Bull. mens. Soc. linn. Lyon* 21: 111
(1952)
E: ! **S:** ? **NI:** ?
H: On soil associated with *Quercus* spp, and less frequently
    Fagus, in mixed deciduous woodland.
D: NM2: 381 **D+I:** Galli2: 152-153, Sarn: 452-456
Rarely reported. Most records are unsubstantiated with voucher
material.

anatina Romagn., *Les Russules d'Europe et d'Afrique du Nord*:
306 (1967)
E: ! **NI:** ! **ROI:** ?
H: On soil, in mixed deciduous woodland.
D: NM2: 382 **D+I:** Galli2: 114-115, Sarn: 336-340
Known from England (Huntingdonshire) and Northern Ireland
(Antrim). Reported from the Republic of Ireland but
unsubstantiated with voucher material.

anthracina Romagn., *Bull. mens. Soc. linn. Lyon* 31: 173
(1962)
    Russula anthracina var. *carneifolia* Romagn., *Bull. mens. Soc.
    linn. Lyon* 31: 173 (1962)
    Mis.: *Russula albonigra* sensu NCL and sensu Rayner (1985)
E: o **S:** o **NI:** o
H: On soil, usually associated with *Quercus* spp and *Fagus* in
    mixed deciduous woodland. Reported with *Betula* spp, and the
    conifers *Cedrus, Larix* and *Pinus* spp.
D: NM2: 377 **D+I:** Galli2: 50-51 **I:** FM2(3): 95
Occasional but widespread.

aquosa Leclair, *Bull. Soc. Mycol. France* 48: 303 (1932)
    Russula fragilis var. *carminea* Jul. Schäff., *Ann. Mycol.* 31:
    461 (1933)
    Russula carminea (Jul. Schäff.) Romagn., *Les Russules
    d'Europe et d'Afrique du Nord*: 447 (1967)
E: ! **S:** ! **W:** !
H: Often amongst *Sphagnum* moss in damp woodland, and also
    noted at pond margins. Usually associated with *Betula* and
    Pinus spp, less often *Quercus* and *Salix* spp.
D: NM2: 396 **D+I:** Galli2: 204-205, Sarn: 499-503 **I:** C&D: 375
Rarely reported but apparently widespread.

atropurpurea (Krombh.) Britzelm. [*nom. inval.*, non *R.
atropurpurea* Peck (1888)], *Bot. Centralbl.* 54: 99 (1893)
    Agaricus atropurpureus Krombh., *Naturgetr. Abbild.
    Schwämme* 9: 6 (1845)
    Russula depallens var. *atropurpurea* (Krombh.) Melzer &
    Zvára, *Arch. Přírodov. Výzk. Čech.* 17(4): 10 (1927)
    Russula atropurpurea var. *krombholzii* Singer [*nom. illegit.*],
    Beih. Bot. Centralbl. 49(2): 301 (1932)
    Russula krombholzii Shaffer, *Lloydia* 33: 82 (1970)
    Russula undulata Velen., *České Houby* I: 131 (1920)
    Mis.: *Russula depallens* sensu Cooke [Ill. Brit. fung. 985
    (1021) Vol.7 (1888)]
    Mis.: *Russula rubra* sensu Cooke [Ill. Brit. fung. 996 (1025)
    Vol. 7 (1888)]
E: c **S:** c **W:** c **NI:** c **ROI:** c **O:** Channel Islands: ! Isle of Man: !
H: On soil, usually associated with *Quercus* spp in oakwoods or
    mixed deciduous woodland but also reported with *Betula* spp
    and *Fagus.*
D: NM2: 396 (as *R. undulata*) **D+I:** Galli2: 208-209 (as *R.
    krombholzii*), Sarn: 492-499 **I:** C&D: 375 (as *R. krombholzii*)
Very common and widespread. N.B. The epithet *atropurpurea* is
invalid, being an illegitimate later homonym of *R. atropurpurea*
Peck [Rep. (Annual) New York State Mus. Nat. Hist. 41: 75
(1888)].

atrorubens Quél., *Compt. Rend. Assoc. Franç. Avancem.
Sci.Assoc. Sci. France* 26: 449 (1898)
S: !
H: On soil, associated with *Salix* spp. British material is with
    Salix aurita and *S. repens.*
D: NM2: 397 **D+I:** Galli2: 216-217, Sarn: 509-513 **I:** SV46: 21
Rarely reported.

aurantiaca Romagn., *Russules d'Europe et d'Afrique du Nord*:
827 (1967)
    Russula integra var. *aurantiaca* Jul. Schäff. [*nom. nud.*], *Ann.
    Mycol.* 21: 404 (1933)
E: ! **S:** ?
H: On soil, associated with *Betula* spp, in mixed deciduous
    woodland.
D: NM2: 385
A single collection, from Berkshire (Windsor Great Park).
Reported from Scotland (Mid-Ebudes) but unsubstantiated with
voucher material.

aurea Pers., *Observ. mycol.* 1: 101 (1796)
    Agaricus auratus With. [non *A. auratus* Paulet (1793)] *Arr.
    Brit. pl. ed. 3,* 4: 184 (1801)
    Russula aurata (With.) Fr., *Epicr. syst. mycol.*: 360 (1838)
E: ! **S:** ! **W:** ! **NI:** ! **O:** Channel Islands: !
H: On soil, in deciduous woodland.
D: NM2: 390 **D+I:** Galli2: 442-443 **I:** C&D: 397
Rarely reported but apparently widespread.

aurora Krombh., *Naturgetr. Abbild. Schwämme* 4: 11 t. 66
(1836)
    Russula lepida var. *aurora* (Krombh.) Rea, *Trans. Brit. Mycol.
    Soc.* 17: 44 (1932)
    Russula lactea var. *incarnata* Cooke, *Handb. Brit. fung.,* Edn
    2: 324 (1893)
    Russula rosea Quél. [*nom. illegit.*, non *R. rosea* Pers. (1796)],
    Fl. mycol. France: 349 (1888)
    Russula velutipes Velen., *České Houby* I: 133 (1920)
    Mis.: *Russula incarnata* sensu Rea (1922)
E: ! **S:** ! **W:** ! **NI:** ! **ROI:** !
H: On soil, in mixed deciduous woodland. Usually reported with
    Betula, Quercus and *Fagus* spp.
D: NM2: 389 (as *R. velutipes*) **D+I:** Galli2: 266-267 **I:** C&D: 393
Occasional but widespread in the British Isles. Perhaps better
known as *R. rosea.*

azurea Bres., *Fungi trident.* 1: 20 (1881)
E: ! **S:** ! **W:** ? **NI:** !
H: On soil in woodland, associated with conifers.
D: NM2: 392 **D+I:** Galli2: 280-281 **I:** C&D: 395
Rarely reported. Apparently widespread but the majority of
records are pre-1940 and mostly unsubstantiated with voucher
material.

badia Quél., *Compt. Rend. Assoc. Franç. Avancem. Sci.* 9: 668
(1881)
    Russula friesii Bres., *Icon. Mycol.*: t. 448 (1929)
E: ! **S:** !
H: On acidic soil in woodland, associated with conifers.
D: NM2: 399 **D+I:** Galli2: 240-241, Sarn: 686-689 **I:** C&D: 379
Known from England (East Kent) and Scotland (Easterness and
Perthshire).

betularum Hora, *Trans. Brit. Mycol. Soc.* 43(2): 456 (1960)
    Russula emetica var. *betularum* (Hora) Romagn., *Les
    Russules d'Europe et d'Afrique du Nord*: 410 (1967)
    Mis.: *Russula emetica* sensu auct. mult. p.p.
    Mis.: *Russula fragilis* sensu Pearson & Dennis (1948)

E: c **S:** c **W:** c **NI:** c **ROI:** !
H: On soil, always associated with *Betula* spp, in pure birch
woodand or mixed deciduous woodland.
D: NM2: 397 **D+I:** C&D: 373, Sarn: 527-531 **I:** Galli2: 180
Common and widespread.

brunneoviolacea Crawshay, *Spore ornamentation of the
Russulas*: 90 (1930)
Russula pseudoviolacea Joachim, *Bull. Soc. Mycol. France* 47:
256 (1931)
E: o **S:** ! **W:** ! **NI:** ! **ROI:** !
H: On soil, usually associated with *Corylus*, *Fagus* and *Quercus*
spp, and frequently in mixed woodland with conifers such as
Pinus spp.
D: NM2: 387 **D+I:** Galli2: 336-337 **I:** C&D: 385
Occasional in England. Rarely reported elsewhere but apparently
widespread.

caerulea (Pers.) Fr., *Epicr. syst. mycol.*: 353 (1838)
Agaricus caeruleus Pers. [non *A. caeruleus* Bolton (1788)],
Syn. meth. fung.: 445 (1801)
Russula amara Kučera, *České Houby* VII(3-4): 50 (1927)
E: o **S:** o **W:** o **NI:** o **ROI:** o
H: On soil, associated with *Pinus* spp (usually *Pinus sylvestris* in
Britain).
D: NM2: 392 (as *Russula coerulea*) **D+I:** Galli2: 392-393 (as *R.
amara*) **I:** C&D: 397, SV46: 17
Occasional but widespread and may be locally frequent.

campestris (Romagn.) Romagn., *Les Russules*: 775 (1967)
Russula integra var. *campestris* Romagn., *Bull. mens. Soc.
linn. Lyon.* 31: 177 (1962)
S: !
H: On soil, amongst litter and mosses in woodland. British
material with *Fagus* annd *Picea* spp in mixed woodland.
A single collection from Perthshire (Blackcraig Forest) in 2003.

carminipes J. Blum, *Bull. Soc. Mycol. France* 69: 449 (1954)
E: !
H: On soil, associated with *Quercus* spp, in oakwoods or mixed
deciduous woodland.
D+I: Galli2: 406-407 **I:** Sarn: 69
Known from East Kent, South Essex and Surrey. Reports from
Cambridgeshire and Oxfordshire lack voucher material.

carpini R. Girard & Heinem., *Bull. Jard. bot. Brux.* 26: 321
(1956)
E: !
H: On soil, associated with *Carpinus* spp.
D: NM2: 385 **D+I:** Galli2: 420-421 **I:** C&D: 389
Two collections from Surrey (Kew, Royal Botanic Gardens) in
1998 and 2001.

cavipes Britzelm., *Hymenomyc. Südbayern* 9: 17 (1893)
Mis.: *Russula violacea* sensu auct. p.p.
NI: !
H: In damp coniferous woodland. With *Abies alba* in Northern
Ireland.
I: Bon: 73, C&D: 375
Listed by Rayner (1985: 80) as possibly British, subsequently
confirmed by Rayner (pers. comm.) and by Kibby from
Northern Ireland (Fermanagh).

cessans A. Pearson, *Naturalist (Hull)* 1950: 101 (1950)
E: ! **S:** ! **W:** ! **NI:** !
H: On soil in woodland, with conifers. British material with *Pinus
sylvestris*.
D: NM2: 387 **D+I:** Galli2: 344-345 **I:** SV46: 13
Rarely reported but apparently widespread.

chloroides (Krombh.) Bres., *Fungi trident.* 2: 89 (1900)
Agaricus chloroides Krombh., *Naturgetr. Abbild. Schwämme*
7: 7 t. 56 (1843)
Russula delica var. *chloroides* (Krombh.) Killerm., *Pilz. Bayern*
6: 7 (1936)
E: ! **S:** !
H: On soil in deciduous woodland. Rarely reported with conifers
in mixed woodland.

D: NM2: 376 **D+I:** Galli2: 68-69, Sarn: 196-199
Rarely reported. Considered a form or variety of *R. delica* by
some authorities.

cicatricata Bon, *Doc. Mycol.* 18(69): 35 (1987)
Russula cicatricata var. *fusca* (Melzer & Zvára) A. Marchand
[*comb. inval.*], *Champignons du Nord et du Midi* 5: 172
(1977)
Russula xerampelina var. *fusca* Melzer & Zvára, *Arch.
Přírodov. Výzk. Čech.* 17(4): 60 (1927)
Mis.: *Russula barlae* sensu Romagnesi [*Les Russules* (1967)]
E: ! **S:** !
H: On soil, usually associated with *Quercus* spp, in oakwoods or
mixed deciduous woodland.
D+I: Galli2: 376-377 **I:** C&D: 391, C&D: 391
Not well distinguished from other members of the *Russula
xerampelina* complex.

claroflava Grove, *Midl. Naturalist (London)* 1888: 265 (1888)
Russula ochroleuca var. *claroflava* (Grove) Cooke, *Handb.
Brit. fung.*, Edn 2: 380 (1890)
Russula constans Britzelm., *Ber. Naturhist. Vereins Augsburg*
28: 141 (1885)
Russula decolorans var. *constans* (Britzelm.) P. Karst.,
Hedwigia 66: 234 (1926)
Russula flava Lindblad, *Ann. Mycol.* 18: 66 (1920)
E: c **S:** c **W:** c **NI:** c **ROI:** !
H: On soil or amongst *Sphagnum* spp under *Betula* spp, in damp
or swampy areas in woodland
D: NM2: 370 **D+I:** Galli2: 388-389 **I:** C&D: 397
Very common and widespread.

clavipes Velen., *České Houby* I: 143 (1920)
Mis.: *Russula elaeodes* sensu auct.
E: !
H: On soil in damp woodland, usually associated with conifers
but also reported with *Alnus* and *Betula* spp.
Rarely reported. Often confused with *R. cicatricata*.

cremeoavellanea Singer, *Rev. Mycol. (Paris)* 1(6): 288 (1936)
E: ! **S:** ! **NI:** ?
H: On soil, in deciduous woodland.
D: NM2: 379 **D+I:** Galli2: 438-439 **I:** C&D: 393
Known from England (South Hampshire: New Forest) and
Scotland (Easterness: Craigellachie NNR). Reported from
Northern Ireland but unsubstantiated with voucher material.

cuprea Krombh., *Naturgetr. Abbild. Schwämme* 9: 11 (1845)
Russula nitida var. *cuprea* (Krombh.) Cooke, *Handb. Brit.
fung.*, Edn 2: 381 (1890)
Russula urens Romell, *Pilz-und Kräuterfreund*: 192 (1921)
Mis.: *Russula firmula* sensu Rayner (1985)
Mis.: *Russula luteoviridans* sensu Pearson (1948)
Mis.: *Russula nitida* sensu Rea (1922) and sensu Pearson
(1948)
E: ! **NI:** !
H: On soil in deciduous woodland.
D: NM2: 400 (as *R. cinnamomicolor*) **D+I:** Galli2: 454-455,
Sarn: 722-727 **I:** C&D: 379, C&D: 379 (as *R. urens*)
Known from England (Berkshire, East Sussex, Lancashire and
West Sussex) and Wales (Denbighshire). Reported from
Northern Ireland but unsubstantiated with voucher material.

curtipes F.H. Møller & Jul. Schäff., *Bull. Soc. Mycol. France* 51:
108 (1935)
E: ! **S:** ! **W:** ! **NI:** !
H: On soil, associated with *Fagus* in old beechwoods or mixed
deciduous woodland.
D: NM2: 386 **D+I:** Galli2: 402-403 **I:** C&D: 389
Rarely reported but apparently widespread.

cyanoxantha (Schaeff.) Fr., *Monogr. hymenomyc. Suec.* 2: 194
(1863)
Agaricus cyanoxanthus Schaeff., *Fung. Bavar. Palat. nasc.* 1:
pl. 93 (1762)
Russula cutefracta Cooke, *Grevillea* 10: 46 (1881)
Russula cyanoxantha var. *cutefracta* (Cooke) Sarnari, *Boll.
Assoc. Micol. Ecol. Romana* 9(27): 38 (1992)

Russula cyanoxantha f. *cutefracta* (Cooke) Sarnari, , *Boll. Assoc. Micol. Ecol. Romana* 10(28): 35 (1993)

Russula cyanoxantha f. *pallida* Singer, *Z. Pilzk.* 2(1): 4 (1923)

Russula cyanoxantha f. *peltereaui* Singer, *Z. Pilzk.* 5(1): 15 (1925)

Mis.: *Russula furcata* sensu auct.

E: c **S:** c **W:** c **NI:** c **ROI:** c **O:** Channel Islands: ! Isle of Man: !

H: On soil, usually associated with *Fagus* and *Quercus* spp, but also not uncommonly *Betula* spp and *Castanea* in mixed deciduous woodland.

D: NM2: 389 **D+I:** Galli2: 74-75, Sarn: 233-237, Sarn: 237-240 (as *R. cyanoxantha* f. *cutefracta*) **I:** C&D: 381 (as *R. cutefracta*), C&D: 381, Galli2: 73 (as *R. cyanoxantha* f. *peltereaui*), Galli2: 77 (as *R. cyanoxatha* f. *peltereaui*)

Very common and widespread. The form *peltereaui* is often recorded as a variety, but the combination has never been made.

decipiens (Singer) Svrček, *Česká Mykol.* 21: 228 (1967)

Russula maculata var. *decipiens* Singer, *Bull. Soc. Mycol. France* 46: 212 (1931) [1930]

E: ! **S:** !

H: On soil in deciduous woodland. British material associated with *Fagus* and *Quercus* spp.

D: NM2: 400 **D+I:** Galli2: 472-473, Sarn: 690-694 **I:** C&D: 379

Rarely reported.

decolorans (Fr.) Fr., *Epicr. syst. mycol.*: 361 (1838)

Agaricus decolorans Fr. [non *A. decolorans* Pers. (1796)], *Syst. mycol.* 1: 56 (1821)

Mis.: *Russula constans* sensu Rea (1922)

E: ? **S:** !

H: On soil, associated with *Pinus sylvestris* in Caledonian pinewoods.

D: NM2: 384 **D+I:** Galli2: 424-425 **I:** C&D: 387, SV46: 16

Known from scattered sites in Scotland but rarely reported. Records from England are dubious and unsubstantiated with voucher material.

delica Fr., *Epicr. syst. mycol.*: 350 (1838)

Lactarius piperatus β *exsuccus* Pers., *Observ. mycol.* 2: 41 (1799)

Lactarius exsuccus (Pers.) W.G. Sm., *J. Bot.* 11: 336 (1873)

E: o **S:** o **W:** o **NI:** o **ROI:** o **O:** Channel Islands: !

H: On soil in mixed deciduous woodland, usually associated with *Corylus*, *Fagus* or *Quercus* spp. Also with *Pseudotsuga*, rarely reported with other conifers in plantations or mixed woodland

D: NM2: 376 **D+I:** Galli2: 64-65 (as *R. flavispora*), Galli2: 66-67, Sarn: 188-193 **I:** SV46: 6

Ocassional but widespread and may be locally abundant.

densifolia Gillet, *Hyménomycètes*: 231 (1876)

E: ! **S:** ! **W:** ! **NI:** ! **ROI:** ! **O:** Channel Islands: !

H: On acidic soil in mixed deciduous woodland, usually associated with *Fagus* or *Quercus* spp. Occasionally reported with *Betula* spp.

D: NM2: 377 **D+I:** Galli2: 54-55, Sarn: 164-168 **I:** FM2(3): back cover, SV46: 7

Occasional but widespread. N.B. non sensu NCL nor Rayner (1985) (= *Russula acrifolia*).

emetica (Schaeff.) Pers., *Observ. mycol.* 1: 100 (1796)

Agaricus emeticus Schaeff., *Fung. Bavar. Palat. nasc.* 1: pl. 15 (1762)

Russula emetica var. *gregaria* Kauffman, *Rep. (Annual) Michigan Acad. Sci.* 11: 79 (1909)

Mis.: *Russula clusii* sensu Cooke [Ill. Brit. fung. 1022 (1031) Vol. 7 (1888)]

E: c **S:** c **W:** c **NI:** c **ROI:** c **O:** Isle of Man: !

H: On acidic soil, on heathland or in woodland, associated with *Pinus sylvestris* or *Picea* spp, and often amongst *Sphagnum* moss in such habitat.

D+I: Galli2: 176-177, Sarn: 554-559 **I:** C&D: 373, SV46: 20

Very common and widespread.

emeticicolor Jul. Schäff., *Ann. Mycol.* 35: 112 (1937)

E: ! **S:** ! **NI:** !

H: On soil, associated with *Fagus* in deciduous woodland.

D: NM2: 389 **D+I:** Galli2: 274-275 **I:** SV34: 37

Known from England (East Sussex and West Sussex), Scotland (Borders) and Northern Ireland (Fermanagh).

exalbicans (Pers.) Melzer & Zvára, Russula, *Arch. Přírodov. Výzk. Čech.* 17(4): 97 (1927)

Agaricus rosaceus β *exalbicans* Pers., *Syn. meth. fung.*: 439 (1801)

Agaricus exalbicans (Pers.) J. Otto, *Vers. Anorden. Agaric.*: 27 (1816)

Russula pulchella I.G. Borshch., *Beitr. Pflanzenk. Russ. Reiches* 9: 58 (1857)

Mis.: *Russula depallens* sensu auct.

E: o **S:** o **W:** o **NI:** o **ROI:** o

H: On soil in woodland, usually with *Betula* spp. Also reported with *Quercus* and *Pinus* spp in mixed woodland.

D: NM2: 398 (as *Russula depallens*) **D+I:** Sarn: 651-656 **I:** C&D: 377

Occasional but widespread. Neotypified by Romagnesi (1967: 467).

faginea Adamčík, *Czech Mycol.* 54(3-4): 185 (2003)

Mis.: *Russula barlae* sensu auct. Brit.

E: ! **S:** ? **W:** ! **NI:** !

H: On soil, associated with *Fagus* in beechwood. Dubiously reported with *Quercus* spp in mixed deciduous woodland.

D: NM2: 393 **D+I:** Ph: 104 105 (upper right) (as *R. xerampelina*), Galli2: 368-369 **I:** C&D: 391, C&D: 391

Rarely reported but apparently widespread.

farinipes Romell, in Britzelm. *Bot. Centralbl.* 15-17: 17 (1893)

Mis.: *Russula simillima* sensu Lange (FlDan5: 66 & pl. 186)

Mis.: *Russula subfoetens* sensu Rea (1922)

E: o **S:** ! **W:** ! **NI:** o **ROI:** !

H: On soil in mixed deciduous woodland, usually associated with *Betula*, *Fagus* or *Quercus* spp, often in grassy glades or clearings or with solitary trees in parkland and avenues.

D: NM2: 381 **D+I:** Galli2: 134-135, Sarn: 411-416 **I:** C&D: 371

Occasional to rather rare, but widespread.

fellea (Fr.) Fr., *Stirp. agri femsion.*: 57 (1825)

Agaricus felleus Fr., *Syst. mycol.* 1: 57 (1821)

Mis.: *Russula ochracea* sensu auct.

E: c **S:** c **W:** c **NI:** c **ROI:** c **O:** Isle of Man: !

H: On soil (often acidic) usually associated with *Fagus* in old beechwoods or mixed deciduous woodland. Rarely reported with *Quercus* spp.

D: NM2: 382 **D+I:** Galli2: 164-165, Sarn: 481-484 **I:** C&D: 373

Very common and widespread.

foetens Pers., *Observ. mycol.* 1: 102 (1796)

E: o **S:** ! **W:** ! **NI:** o **ROI:** ! **O:** Channel Islands: ! Isle of Man: !

H: On soil in old deciduous woodland, usually associated with *Fagus* or *Quercus* spp. Occasionally reported with *Betula* spp, and rarely (and dubiously) *Pinus sylvestris*.

D: NM2: 380 **D+I:** Galli2: 138-139, Sarn: 422-425 **I:** C&D: 371, SV46: 9

Occasional to rather rare, but widespread and may be locally abundant.

font-queri Singer, *Rev. Mycol. (Paris)* 1(2): 81 (1936)

S: !

H: On soil in woodland.

D: NM2: 386 **D+I:** Galli2: 328-329 **I:** C&D: 385

Known from Orkney. Reported elsewhere in Scotland but not substantiated with voucher material.

fragilis (Pers.) Fr., *Stirp. agri femsion.*: 57 (1825)

Agaricus fragilis Pers. [non *A. fragilis* L. (1763)], *Syn. meth. fung.*: 440 (1801)

Russula bataillei Bidaux & Reumaux, *Russules Rares ou Méconnues*: 281 (1996)

Agaricus fallax Schaeff., *Fung. Bavar. Palat. nasc.* 1: 16 (1762)

Russula fallax (Schaeff.) Fr., *Observ. mycol.*: 70 (1815)

Russula fragilis var. *fallax* (Schaeff.) Massee, *Brit. fung.-fl.* 3: 76 (1893)

Russula fragilis var. *violascens* Gillet, *Hyménomycètes*: 245 (1876)

E: c **S:** c **W:** c **NI:** c **ROI:** c **O:** Channel Islands: ! Isle of Man: !
H: On soil in woodland, usually associated with *Betula* spp, but also known with *Fagus* and *Quercus* spp.
D: NM2: 397 **D+I:** Galli2: 214-215, Sarn: 503-509 **I:** C&D: 375, SV46: 20
Common and widespread.

fragilis *var.* **knauthii** (Singer) Kuyper & Vuure, *Persoonia* 12(4): 451 (1985)
    Russula emetica f. *knauthii* Singer, *Hedwigia* 66: 216 (1926)
    Russula knauthii (Singer) Hora, *Trans. Brit. Mycol. Soc.* 43(2): 457 (1960)
E: ! **S:** ! **W:** !
H: On soil, in boggy areas.
D: NM2: 397
Rarely reported.

fragrantissima Romagn., *Les Russules d'Europe et d'Afrique du Nord*: 350 (1967)
E: ! **S:** !
H: On soil, in mixed deciduous woodland.
D: NM2: 381 **D+I:** Sarn: 447-451
Known from England (Hertfordshire and South Hampshire) and Scotland (Outer Hebrides).

fusconigra M.M. Moser, *Sydowia* 31: 97 (1979) [1978]
S: !
H: On soil, in mixed deciduous woodland.
D: NM2: 379 **I:** C&D: 389
A single collection from Easterness (Curr Wood) in 2001.

fuscorubroides Bon, *Bull. Soc. Mycol. France* 91(4): pl. 198 (1976) [1975]
E: !
H: On soil in woodland, with conifers.
I: Bon: 75, Galli2: 244
A single collection, from Herefordshire (Mortimer Forest) in 2004. Reported from North Somerset but the collection, in herb. K, has been redetermined as *R. queletii*.

gigasperma Romagn., *Les Russules d'Europe et d'Afrique du Nord*: 861 (1967)
E: ! **S:** ! **W:** !
H: On soil, in deciduous woodland.
D+I: Sarn: 735-738
Known from England (Cumberland and South Devon), Scotland (Borders and Perthshire) and Wales (Monmouthshire). Reported elsewhere but unsubstantiated with voucher material.

gracillima Jul. Schäff., *Z. Pilzk.* 10(4): 105 (1931)
E: o **S:** o **W:** ! **NI:** ! **O:** Isle of Man: !
H: On soil in mixed deciduous woodland, usually associated with *Betula* spp and occasionally reported with *Fagus*.
D: NM2: 397 **D+I:** Galli2: 234-235, Sarn: 656-661 **I:** C&D: 377
Occasional but widespread.

grata Britzelm., *Bot. Centralbl.* 15-17: 17 (1893)
    Russula subfoetens var. *grata* (Britzelm.) Romagn. [*comb. inval.*], *Les Russules d'Europe et d'Afrique du Nord*: 340 (1967)
    Russula laurocerasi Melzer, *České Houby* II: 243 (1921)
E: ! **S:** ! **W:** ! **NI:** ! **ROI:** ! **O:** Isle of Man: !
H: On soil, usually associated with *Fagus* in old beechwoods but also known with *Betula* and *Quercus* spp in mixed deciduous woodland.
D: NM2: 380 **D+I:** Galli2: 144-145 (as *R. laurocerasi*), Sarn: 437-442 (as *R. laurocerasi*) **I:** C&D: 371 (as *R. laurocerasi*)
Rarely reported but apparently widespread and may be locally frequent in old woodland. Better known as *Russula laurocerasi*.

graveolens Romell, in Britzelm., *Bot. Centralbl.* 15-17: 17 (1893)
    Russula xerampelina var. *graveolens* (Romell) Kühner & Romagn. [*comb. inval.*], *Flora Analytique des Champignons Supérieurs*: 449 (1953)

Russula gilvescens Bon, *Doc. Mycol.* 18(69): 36 (1987)
Russula graveolens var. *megacantha* Bon, *Doc. Mycol.* 18(69): 36 (1987)
Russula xerampelina var. *purpurata* Crawshay, *The Spore Ornamentation of Russula*: 103 (1930)
Russula purpurata (Crawshay) Romagn., *Doc. Mycol.* 13(50): 27 (1983)
E: ! **S:** ! **W:** ! **NI:** !
H: On soil in mixed deciduous woodland, usually associated with *Fagus* or *Quercus* spp.
D: NM2: 393 **D+I:** Ph: 104 105 (lower left), Galli2: 366-367 (as *R. purpurata*), Galli2: 370-371 **I:** C&D: 391 (as *R. purpurata*), C&D: 391, C&D: 391
Apparently widespread. Collections of the *R. xerampelina* group in lowland deciduous woodland (other than *Russula faginea* and *R. cicatricata*) probably belong here.

grisea (Pers.) Fr., *Epicr. syst. mycol.*: 361 (1838)
    Agaricus griseus Pers., *Syn. meth. fung.*: 445 (1801)
    Russula furcata var. *pictipes* Cooke, *Handb. Brit. fung.*, Edn 2: 321 (1889)
    Russula palumbina Quél., *Compt. Rend. Assoc. Franç. Avancem. Sci.* 11: 396 (1883)
E: ! **S:** ! **W:** ! **NI:** ! **ROI:** !
H: On soil, in mixed deciduous woodland.
D: NM2: 384 **D+I:** Galli2: 122-123, Sarn: 289-296 **I:** C&D: 383
Frequently reported and apparently widespread, but rarely collected. Many of the records probably refer to *R. ionochlora*.

helodes Melzer, *Bull. Soc. Mycol. France* 45: 284 (1929)
E: ! **W:** ?
H: On soil, in mixed conifer and deciduous woodland, often in damp areas. British material with mixed *Betula*, *Picea* and *Pinus* spp.
D: NM2: 395 **D+I:** Galli2: 246-247, Sarn: 626-630 **I:** C&D: 377
Known from Westmorland and Yorkshire. Reported from Wales (Glamorganshire) in 1973 but unsubstantiated with voucher material.

heterophylla (Fr.) Fr., *Epicr. syst. mycol.*: 352 (1838)
    Agaricus furcatus β *heterophyllus* Fr., *Syst. mycol.* 1: 59 (1821)
    Agaricus lividus Pers., *Syn. meth. fung.*: 446 (1801)
    Russula heterophylla var. *livida* (Pers.) Gillet, *Hyménomycètes*: 241 (1876)
    Russula livida (Pers.) J. Schröt., in Cohn, *Krypt.-Fl. Schlesien* 3: 546 (1889)
    Mis.: *Russula furcata* sensu auct.
E: o **S:** o **W:** ! **NI:** ! **ROI:** o **O:** Channel Islands: !
H: On soil in mixed deciduous woodland, usually with *Betula*, *Fagus* or *Quercus* spp.
D: NM2: 390 **D+I:** Galli2: 82-83, Sarn: 252-258 **I:** C&D: 381
Occasional but widespread.

illota Romagn., *Bull. mens. Soc. linn. Lyon* 23: 175 (1954)
E: ! **S:** !
H: On soil in woodland with deciduous trees or conifers.
D: NM2: 380 **D+I:** FRIC6: 34-36 47a, Galli2: 142-143, Sarn: 443-447 **I:** C&D: 371
Rarely reported.

inamoena Sarnari, *Bollettino dell'Associazione Micologica ed Ecologica Romana* 33: 10 (1994)
E: !
H: On soil in mixed deciduous woodland. British collections reported with *Carpinus*, *Quercus ilex* and *Tilia* spp.
D+I: Sarn: 433-436
Known from Surrey (Kew, Royal Botanic Gardens). Some collections determined as *R. subfoetens* may be this species.

innocua (Singer) Bon, *Doc. Mycol.* 12(46): 32 (1982)
    Russula smaragdina var. *innocua* Singer, *Ann. Mycol.* 33(5-6): 304 (1935)
    Mis.: *Russula smaragdina* sensu Lange (FlDan5: 73 & pl. 187B) and sensu Pearson (1948)
S: ! **ROI:** !
H: On soil, in deciduous woodland.

D+I: Sarn: 600-603 **I:** SV38: 14
Known from Scotland (Angus) and the Republic of Ireland (North Kerry). Both collections were originally determined as *R. smaragdina*.

insignis (Quél.) Quél., *Compt. Rend. Assoc. Franç. Avancem. Sci. Assoc. Sci. France* 16(2): 588 (1888)
  Russula pectinata var. *insignis* Quél., *Fl. mycol. France*: 346 (1888)
  Russula livescens var. *depauperata* J.E. Lange, *Dansk. Bot. Ark.* 4(12): 35 (1926)
  Mis.: *Russula livescens* sensu Lange (FlDan5: 66 & pl. 185A)
  Mis.: *Russula pectinatoides* sensu NCL and sensu Rayner (1985)
E: !
H: On soil, associated with *Quercus* spp (including *Q. ilex*) in mixed deciduous woodland; also reported with *Pinus radiata*.
D+I: Galli2: 148-149, Sarn: 475-480 **I:** C&D: 371
Known from South Devon, Surrey, West Norfolk and West Sussex. Reported from Herefordshire but unsubstantiated with voucher material.

integra (L.) Fr., *Epicr. syst. mycol.*: 360 (1838)
  Agaricus integer L., *Sp. pl.*: 1171 (1753)
  Russula alutacea f. *purpurella* Singer, *Bot. Centralbl., Beih.* 49: 255 (1952)
  Russula integra f. *purpurella* (Singer) Bon, *Doc. Mycol.* 17(65): 55 (1986)
  Mis.: *Russula polychroma* sensu NCL (1960) and sensu Rayner (1985)
S: !
H: On soil with *Pinus sylvestris* and *Betula* spp in Caledonian pinewoods.
D: NM2: 378 **I:** C&D: 391, SV46: 14
Sensu NCL and sensu Rayner (1985) is *R. melitodes*. Sensu Sowerby (Col. fig. Engl. fung. 2: pl. 201) is a mixture of several different species of *Russula*.

intermedia P. Karst., *Meddeland. Soc. Fauna Fl. Fenn.* 16: 38 (1888)
  Russula lundellii Singer, *Lilloa* 22: 719 (1951) [1949]
  Russula mesospora Singer, *Bull. Soc. Mycol. France* 54: 161 (1938)
E: ! **S:** ! **W:** ! **ROI:** ?
H: On soil in mixed deciduous woodland, usually associated with *Betula* and *Quercus* spp.
D: NM2: 399 (as *R. lundellii*) **D+I:** Galli2: 462-463 (as *R. lundellii*), Sarn: 710-716 (as *R. lundellii*) **I:** C&D: 379 (as *R. lundellii*)
Known from England (Cumberland, South Hampshire and Westmorland), Scotland (Easter Ross and East Perthshire) and Wales (Breconshire). Reported from the Republic of Ireland but unsubstantiated with voucher material.

ionochlora Romagn., *Bull. mens. Soc. linn. Lyon* 21: 110 (1952)
  Russula grisea var. *ionochlora* Kühner & Romagn. [*comb. inval.*], *Flore Analytique des Champignons Supérieurs*: 444 (1953)
  Mis.: *Russula grisea* sensu auct. p.p.
E: o **S:** ! **W:** ! **NI:** ! **ROI:** !
H: On soil in mixed deciduous woodland, often in open grassy glades, usually associated with *Fagus* or *Quercus* spp.
D: NM2: 383 **D+I:** Galli2: 104-105, Sarn: 284-289 **I:** C&D: 383
Locally frequent in southern England and apparently widespread elsewhere.

laccata Huijsman, *Fungus* 25: 40 (1955)
  Russula norvegica D.A. Reid [*nom. inval.*], *Fungorum Rar. Icon. Color.* 6: 36 & pl. 48b (1972)
  Mis.: *Russula atrorubens* sensu Lange (FlDan5: 63 & pl. 180B)
E: ! **S:** ! **W:** ! **NI:** !
H: On soil in deciduous woodland, often in wet or swampy areas such as the margins of ponds or lakes, usually associated with *Salix* spp. Also known in montane habitat with *Salix herbacea*.

D: NM2: 396 (as *R. norvegica*) **D+I:** Galli2: 218-219 (as *R. norvegica*), Sarn: 518-526 **I:** C&D: 375 (as *R. norvegica*)
Rarely reported. Some authorities consider *R. norvegica* distinct.

laeta F.H. Møller & Jul. Schäff., *Russula-Monographie*: 162 (1952)
  Mis.: *Russula borealis* sensu Romagnesi [Les Russules (1967)]
S: ! **W:** ! **NI:** ?
H: On soil in deciduous woodland.
D: NM2: 388 **D+I:** Galli2: 446-447 **I:** C&D: 389, C&D: 393 (as *Russula borealis*)
Accepted with reservations. The interpretation is open to doubt, but is taken here sensu Galli2. Sensu Romagnesi (1967) and Rayner (1985) this includes both *Russula borealis* (not British) and *R. cremeoavellanea*. Known from Scotland and a single collection from Wales (Monmouthshire) in 1977 (originally determined as *R. borealis*). Reported from Northern Ireland but unsubstantiated with voucher material.

langei Bon, *Rev. Mycol. (Paris)* 35(4): 240 (1970)
E: ! **S:** ! **W:** ! **NI:** ?
H: On soil, in deciduous woodland.
D+I: Sarn: 248-249 **I:** C&D: 381
Known from a few scattered locations in England, Scotland and Wales. Reported from Northern Ireland but unsubstantiated with voucher material. Possibly not distinct from *R. cyanoxantha*.

lateritia Quél., *Compt. Rend. Assoc. Franç. Avancem. Sci.* 14: 449 (1886)
E: !
H: On soil in woodland, associated with conifers.
D+I: Galli2: 444-445
Accepted with doubt. A single collection by Kits van Waveren reported by Rayner (1985), but no material has been traced. Reported from West Sussex in 1996 but said to be associated with *Carpinus betulus* and unsubstantiated with voucher material.

lilacea Quél., *Bull. Soc. Bot. France* 23: 330 (1877) [1876]
  Russula carnicolor (Bres.) Rea, *Brit. basidiomyc.*: 477 (1922)
  Russula lilacea var. *carnicolor* Bres., *Fungi trident.* 2: 23 (1892)
E: ! **S:** ! **W:** !
H: On soil, usually with *Quercus* spp, in oakwoods or mixed deciduous woodland.
D: NM2: 388 **D+I:** Galli2: 272-273 **I:** C&D: 393, SV34: 37, Sarn: 73
Rarely reported. Apparently widespread.

luteotacta Rea, *Brit. basidiomyc.*: 469 (1922)
  Mis.: *Russula sardonia* sensu Bresadola [Ic. Mycog. 9: pl. 407 (1929)]
E: ! **S:** ! **W:** ? **NI:** ! **ROI:** ?
H: On soil, with *Betula*, *Fagus* and *Quercus* spp, in damp deciduous woodland, or occasionally in mixed woodland with *Pinus* spp
D: NM2: 394 **D+I:** FRIC4: 9-10 28, Galli2: 194-196, Sarn: 672-677 **I:** C&D: 373
Frequently reported and apparently widespread but rarely collected.

maculata Quél. & Roze, *Bull. Soc. Bot. France* 24: 323 (1878)
  Mis.: *Russula elegans* sensu Cooke [Ill. Brit. fung. 1018 (1027) Vol. 7 (1888)]
E: ! **W:** ? **NI:** ?
H: On soil, with *Fagus*, *Quercus* and *Salix* spp, in mixed deciduous woodland.
D: NM2: 400 **D+I:** Galli2: 466-467, Sarn: 695-699 **I:** C&D: 379
Rarely reported. Most records and collections are from southern England. Reported from Northern Ireland and Wales but records are old and unsubstantiated with voucher material. Some may possibly refer to *R. globispora*.

melitodes Romagn., *Bull. Soc. Mycol. France* 59: 71 (1943)
  Mis.: *Russula integra* sensu NCL and sensu Rayner (1985)
E: ! **NI:** !

H: On soil in mixed deciduous woodland, with *Carpinus*, *Fagus* or *Tilia* spp
D+I: FRIC4: 4-5 25a **I:** C&D: 391
Known from widely scattered locations in southern England and reported from Northern Ireland. This is the lowland counterpart of *Russula integra*.

melliolens Quél., *Compt. Rend. Assoc. Franç. Avancem. Sci. Assoc. Sci. France* 26: 449 (1898)
  Russula rubra var. *sapida* Cooke, *Handb. Brit. fung.*, Edn 2: 326 (1889)
E: ! **NI:** ?
H: On soil, in deciduous woodland.
D: NM2: 384 **D+I:** Galli2: 352-353 **I:** C&D: 387
Known from southern England, but most of the few records are old and unsubstantiated with voucher material.

melzeri Zvára, *Arch. Přír. Výzk. Čech.* XVII: 82 (1927)
E: ! **W:** !
H: On soil in old deciduous woodland, usually with *Fagus* and rarely reported with *Castanea* or *Quercus* spp.
D+I: Galli2: 326-327 **I:** C&D: 385, FM3(4): 111
Known from from Cumberland southward, with a single collection from Wales.

minutula Velen., *České Houby* I: 133 (1920)
S: ! **W:** !
H: On soil, in deciduous woodland.
D: NM2: 389 **D+I:** Galli2: 262-263 **I:** C&D: 393
Known from Scotland (Berwickshire) and Wales (Breconshire).

mustelina Fr., *Epicr. syst. mycol.*: 351 (1838)
E: ? **S:** ? **NI:** ?
H: On soil, with conifers in montane habitat.
D: NM2: 378 **D+I:** Galli2: 80-81, Sarn: 266-271 **I:** C&D: 381, SV46: 10
Accepted with reservations. Reported from England but all records are old and unsubstantiated with voucher material. Records from Scotland were accepted by Rayner (1985) but no voucher material has been traced. Reported from Northern Ireland, but also unsubstantiated with voucher material.

nana Killerm., *Denkschr. Königl.-Baier. Bot. Ges. Regensburg* 20: 38 (1936)
  Mis.: *Russula alpina* sensu auct. Brit.
S: !
H: On soil in montane habitat, with *Salix herbacea* and *Dryas*.
D: NM2: 394 **D+I:** Sarn: 570-574 **I:** C&D: 373, Galli2: 183
Known from a few scattered sites in Scotland.

nauseosa (Pers.) Fr., *Epicr. syst. mycol.*: 363 (1838)
  Agaricus nauseosus Pers., *Syn. meth. fung.*: 446 (1801)
E: ! **S:** ! **W:** ! **NI:** !
H: On soil with conifers such as *Pinus sylvestris* and *Picea abies*.
D: NM2: 387 **D+I:** Galli2: 342-343 **I:** SV46: 13
Rarely reported but apparently widespread.

nigricans (Bull.) Fr., *Epicr. syst. mycol.*: 350 (1838)
  Agaricus nigricans Bull., *Herb. France*: pl. 212 (1785)
  Omphalia adusta β *elephantinus* (Bolton) Gray, *Nat. arr. Brit. pl.* 1: 614 (1821)
  Russula elephantina (Bolton) Fr., *Epicr. syst. mycol.*: 350 (1838)
  Agaricus elephantinus Bolton, *Hist. fung. Halifax* 1: 28 (1788)
  Russula nigrescens Krombh., *Naturgetr. Abbild. Schwämme* 9: 27 (1845)
E: c **S:** c **W:** c **NI:** c **ROI:** c **O:** Isle of Man: !
H: On soil in old deciduous woodland, usually with *Fagus* or *Quercus* spp. Also known with *Salix repens* on cliff tops in Shetland *fide* R. Watling (pers. comm.).
D: NM2: 376 **D+I:** Sarn: 154-158 **I:** Sow1: 36 (as *Agaricus elephantinus*), Galli2: 44-45, SV46: 6
Common and widespread.

nitida (Pers.) Fr., *Epicr. syst. mycol.*: 361 (1838)
  Agaricus nitidus Pers., *Syn. meth. fung.*: 444 (1801)
  Russula venosa Velen., *České Houby* I: 146 (1920)

  Mis.: *Russula roseipes* sensu Cooke [Ill. Brit. fung. 1035 (1081) Vol.7 (1888)]
  Mis.: *Russula sphagnophila* sensu Rea [TBMS 17: 45 (1932)] and sensu auct. mult.
E: c **S:** c **W:** c **NI:** c **ROI:** c
H: On acidic soil in woodland or on heathland, usually with *Betula* spp, and frequently with *Pinus* spp.
D: NM2: 386 **D+I:** Galli2: 334-335 **I:** C&D: 387
Common and widespread.

nobilis Velen., *České Houby* I: 138 (1920)
  Russula fageticola S. Lundell, *Fungi Exsiccati Suecici*: 37 (1956)
  Russula mairei Singer, *Arch. Protistenk.* 65: 306 (1929)
  Russula mairei var. *fageticola* Romagn. [*nom. inval.*], *Bull. Soc. Linn. de Lyon* 31: 174 (1962)
  Mis.: *Russula emetica* sensu auct.
E: c **S:** c **W:** c **NI:** c **ROI:** c **O:** Isle of Man: !
H: On soil in beechwoods or mixed deciduous woodland, with *Fagus*.
D: NM2: 394 (as *Russula mairei*) **D+I:** Galli2: 184-185, Sarn: 563-569 (as *R. mairei*) **I:** C&D: 373
Common and widespread. Better known as *Russula mairei*.

ochroleuca Pers., *Observ. mycol.* 1: 102 (1796)
  Russula citrina Gillet, *Rev. Mycol. (Toulouse)* 3(9): 5 (1881)
  Russula granulosa Cooke, *Grevillea* 17: 40 (1888)
  Russula ochroleuca var. *granulosa* (Cooke) Rea, *Brit. basidiomyc.*: 466 (1922)
E: c **S:** c **W:** c **NI:** c **ROI:** c **O:** Channel Islands: c Isle of Man: c
H: On soil in woodland. With a wide range of deciduous trees but also not uncommon with conifers.
D: NM2: 379 **D+I:** Galli2: 166-167 **I:** C&D: 397, SV46: 11
Very common and widespread.

odorata Romagn., *Bull. mens. Soc. linn. Lyon* 19: 76 (1950)
E: ! **S:** ?
H: On soil in mixed deciduous woodland, usually associated with *Fagus*, *Quercus* and less frequently *Betula* spp.
D: NM2: 388 **D+I:** Galli2: 314-315 **I:** C&D: 385
Known from a few scattered sites in southern England. Reported elsewhere but unsubstantiated with voucher material.

olivacea (Schaeff.) Fr., *Epicr. syst. mycol.*: 356 (1838)
  Agaricus olivaceus Schaeff., *Fung. Bavar. Palat. nasc.* 3: 204 (1770)
E: o **NI:** !
H: On soil in old deciduous woodland, usually reported with *Fagus* and *Quercus* spp.
D: NM2: 391 **D+I:** Galli2: 380-381 **I:** C&D: 397
Occasional but widespread in England. Rare or unreported elsewhere.

paludosa Britzelm., *Bot. Centralbl.* 15-17: 17 (1893)
  Mis.: *Russula linnaei* sensu Bresadola [Icon. mycol. pl. 416 Vol. 9 (1929)]
E: ! **S:** !
H: On soil, in conifer woodland, often in damp or wet areas.
D: NM2: 385 **D+I:** Galli2: 430-431 **I:** C&D: 393, Sarn: 81, SV46: 15
Known from Scotland but rarely reported there. A single confirmed record from England (Surrey).

parazurea Jul. Schäff., *Z. Pilzk.* 10(4): 105 (1931)
  Mis.: *Russula aeruginea* sensu Cooke [Ill. Brit. fung. 1027 (1090) Vol. 7 (1888)]
E: c **S:** ! **W:** ! **NI:** o **ROI:** ! **O:** Channel Islands: !
H: On soil in mixed deciduous woodland, usually associated with *Fagus*, *Quercus* or *Tilia* spp. Also known with *Betula* spp in pure birch woodland.
D: NM2: 384 **D+I:** Galli2: 106-107, Sarn: 279-284 **I:** C&D: 383
Common in England. Occasional or rarely reported elsewhere but apparently widespread.

pascua (F.H. Møller & Jul. Schäff.) Kühner, *Bull. Soc. Mycol. France* 91(3): 331 (1975)
  Russula xerampelina var. *pascua* F.H. Møller & Jul. Schäff., *Ann. Mycol.* 38(2-4): 332 (1940)

Mis.: *Russula oreina* sensu auct. non Singer [*Bull. Soc. Mycol. France* 54: 142 (1938)]
S: ! NI: ?
H: On soil, with *Salix herbacea* and *S. repens*.
D: NM2: 392 (as *Russula oreina*), Adamčik & Knudsen (2004: 1466)
Distribution unclear as it has recently been established (Adamčik & Knudsen, 2004) that the similar *R. subrubens* may also occur in such habitat with dwarf willows.

pectinatoides Peck, *Rep. (Annual) New York State Mus. Nat. Hist.* 116: 43 (1907)
Mis.: *Russula pectinata* sensu NCL and sensu Rayner (1985)
E: ! S: ? W: ! NI: !
H: On soil in deciduous woodland or with solitary trees in parkland and gardens.
D: NM2: 378 **D+I:** Galli2: 150-151
Rarely reported. Apparently widespread. Sensu NCL and sensu Rayner (1985) is *R. insignis*. Sensu auct. is *R. praetervisa*.

pelargonia Niolle, *Ann. Mycol.* 39: 66 (1941)
Mis.: *Russula serotina* sensu Cooke [*Ill. Brit. fung.* 1003 (1042) Vol. 7 (1888)]
E: ! S: ?
H: On soil in deciduous woodland, often in damp or wet areas.
D: NM2: 398 **D+I:** Galli2: 228-229, Sarn: 595-600 **I:** C&D: 375
Rarely reported. Known in a few sites from Bedfordshire southwards. Reported elsewhere but unsubstantiated with voucher material.

persicina Krombh., *Naturgetr. Abbild. Schwämme* 9: 12 (1845)
Russula intactior (Jul. Schäff.) Jul. Schäff., *Ark. Bot.* 15: 54 (1939)
Russula luteotacta ssp. *intactior* Jul. Schäff., *Ann. Mycol.* 36(1): 34 (1937)
Russula persicina var. *intactior* (Jul. Schäff.) Kühner & Romagn. [*comb. inval.*], *Flora Analytique des Champignons Supérieurs*: 461 (1953)
E: ! S: ! W: !
H: On calcareous soil, in deciduous woodland. A large form, usually associated with *Salix repens* occurs on dunes.
D: NM2: 394 **D+I:** Galli2: 198-199, Sarn: 667-672 **I:** C&D: 377
Rarely reported. Excludes *R. intactior* sensu NCL which is *R. renidens*.

poichilochroa Sarnari, *Rivista Micol.* 33(2): 164 (1990)
Russula metachroa Sarnari [non *R. metachroa* Hongo in *J. Jap. Bot.* 30: 219 (1955)], *Rivista Micol.* 33(1): 43 (1990)
E: !
H: On soil in deciduous woodland. British material with *Quercus* sp.
D+I: Galli2: 206-207, Sarn: 513518 (as *Russula poikilochroma*)
Known from Cumberland (Buttermere, Scales Wood) in 1999.

postiana Romell, *Ark. Bot.* 11(3): 5 (1912)
S: !
H: On acidic soil, associated with *Pinus sylvestris* in Caledonian pinewoods
D: NM2: 392 **D+I:** Galli2: 292-293 **I:** C&D: 395
A single collection from Perthshire (Black Wood of Rannoch) in 1997.

praetervisa Sarnari, *Monog. del genere* Russula *in Europa*: 463 (1998)
Mis.: *Russula pectinatoides* sensu auct.
E: ! NI: !
H: On soil in mixed deciduous woodland, often with *Tilia* spp. Also reported with *Betula* spp and *Fagus*.
Rarely reported and usually unsubstantiated with voucher material. Often misidentified as *R. pectinatoides*.

pseudoaeruginea (Romagn.) Kuyper & Vuure, *Persoonia* 12(4): 451 (1985)
Russula aeruginea var. *pseudoaeruginea* Romagn., *Bull. mens. Soc. linn. Lyon* 21: 111 (1952)
Mis.: *Russula galochroa* sensu Cooke [*Ill. Brit. fung.* 1011 (1089) Vol. 7 (1888)]
E: !

H: On soil in deciduous woodland.
D: NM2: 383 **D+I:** Galli: 124-125, Sarn: 299-304
A single collection, from West Sussex (Houghton Forest) in 2001.

pseudoaffinis Migl. & Nicolaj, *Boll. Gruppo Micol. G. Bresadola.* 28(3-4): 107 (1985)
E: !
H: On soil in deciduous woodland. British material with *Quercus* and *Tilia* spp.
D+I: Galli2: 156-157 **I:** C&D: 371
A single collection, from Middlesex (Hampstead Heath) in 2002.

pseudoimpolita Sarnari, *Rivista de Micologia* 30(1-2): 33 (1987)
E: ! S: - W: - NI: - EI: - O: -
H: On soil under *Quercus* spp. British material with *Quercus ilex* and *Q. suber.*
D+I: LR 330-331
Collected twice in 2004, from Surrey (Kew, Royal Botanic Gardens) in grassy areas under planted trees.

pseudointegra Arnould & Goris, *Bull. Soc. Mycol. France* 23: 177 (1907)
E: o S: ! W: ! NI: !
H: On soil in mixed deciduous woodland, usually with *Quercus* spp but occasionally reported with *Fagus*.
D: NM2: 391 **D+I:** Galli2: 300-301 **I:** C&D: 395
Occasional but widespread in England. Rarely reported elsewhere.

puellaris Fr., *Epicr. syst. mycol.*: 362 (1838)
Russula puellaris var. *leprosa* Bres., *Fungi trident.* 1: 58 (1881)
E: o S: o W: ! NI: ! ROI: !
H: On soil in woodland, often with mixed deciduous trees and conifers.
D: NM2: 386 **D+I:** Galli2: 308-309 **I:** C&D: 385, SV46: 11
Occasional but apparently widespread.

puellula Ebbesen, F.H. Møller & Jul. Schäff., *Russula-Monographie*: 185 (1952)
E: !
H: On soil in deciduous woodland.
D: NM2: 388 **D+I:** Galli2: 320-321
Known from Cumberland (Great Mell Fell) and West Sussex (Benges Wood). Possibly also in Leicestershire.

pulchrae-uxoris Reumaux, *Bull. Trimestriel Féd. Mycol. Dauphiné-Savoie* 34(135): 6 (1994)
E: !
H: On soil under *Quercus* spp. One of the British collections was with *Quercus* x *turneri* in a botanic garden.
Two collections, from Surrey (Kew, Royal Botanic Gardens) and West Kent (Lamberhurst, The Owl House), both in 2004.

pungens Beardslee, *J. Elisha Mitchell Sci. Soc.* 33(4): 196 (1918)
Mis.: *Russula rubra* sensu NCL and sensu Rayner (1985)
E: !
H: On basic or calcareous soil in deciduous woodland. British material with *Castanea* and *Quercus* spp.
D: NM2: 395
A single collection from Buckinghamshire (Tring, Dancer's End Nature Reserve).

queletii Fr. in Quélet, *Mém. Soc. Émul. Montbéliard, Sér. 2*, 5: 185 (1872)
Russula drimeia var. *queletii* (Fr.) Rea, *Brit. basidiomyc.*: 467 (1922)
Russula flavovirens J. Bommer & M. Rousseau, *Flora myc. Belg.*: 58 (1887)
Mis.: *Russula sardonia* sensu auct.
E: o S: ! W: ! NI: ! ROI: ?
H: On soil, with *Picea abies* or *Pinus* spp.
D: NM2: 399 **D+I:** Galli2: 242-243, Sarn: 640-644 **I:** C&D: 377, FM2(3): back cover, SV46: 23
Occasional but widespread.

raoultii Quél., *Compt. Rend. Assoc. Franç. Avancem. Sci.* 14: 449 (1886)

Mis.: *Russula fragilis* var. *nivea* sensu Cooke [Ill. Brit. fung. 1029 (1060) Vol. 7 (1888)]

E: ! **S:** ! **W:** !

H: On soil in mixed deciduous woodland, usually with *Fagus* and occasionally *Betula* and *Quercus* spp.

D: NM2: 381 **D+I:** Galli2: 172-173, Sarn: 538-541 **I:** C&D: 373

Rarely reported but apparently widespread.

renidens Ruots., Sarnari & Vauras, in Sarnari, *Monog. del genere* Russula *in Europa*: 661 (1998)

Mis.: *Russula intactior* sensu NCL and sensu Rayner (1985)

Mis.: *Russula persicina* sensu NM2: 394

E: o **S:** ! **NI:** ?

H: On soil with *Betula* spp, in birchwoods.

D+I: Sarn: 661-666

Occasional in northern England. Rarely reported in Scotland. Reported from Northern Ireland but no voucher material has been traced.

risigallina (Batsch) Kuyper & Vuure, *Persoonia* 12(4): 451 (1985)

Agaricus risigallinus Batsch, *Elench. fung. (Continuatio Prima)*: 67 & t. 15 f. 72 (1786)

Agaricus vitellinus Pers., *Syn. meth. fung.*: 442 (1801)

Russula vitellina (Pers.) Gray, *Nat. arr. Brit. pl.* 1: 618 (1821)

Russula chamaeleontina Fr., *Epicr. syst. mycol.*: 363 (1838)

Russula risigallina f. *chamaeleontina* (Fr.) Bon, *Doc. Mycol.* 18(70-71): 108 (1988)

Russula armeniaca Cooke, *Ill. Brit. fung.* 1045 (1064) Vol. 7 (1888)

Russula lutea var. *armeniaca* (Cooke) Rea, *Brit. basidiomyc.*: 478 (1922)

Russula lutea var. *ochracea* Singer, *Ann. Mycol.* 33: 298 (1935)

Mis.: *Russula lutea* sensu auct. mult.

Mis.: *Russula ochracea* sensu auct.

E: o **S:** ! **W:** ! **NI:** ! **ROI:** !

H: On soil in mixed deciduous woodland, with *Betula*, *Fagus* or *Quercus* spp.

D: NM2: 391 (as *Russula lutea*), NM2: 391 **D+I:** Galli2: 298-299 **I:** C&D: 395

Occasional but widespread in England. Rarely reported elsewhere but apparently widespread.

romellii Maire, *Bull. Soc. Mycol. France* 26: 57 (1910)

E: ! **W:** ! **NI:** ! **ROI:** ?

H: On soil in mixed deciduous woodland, usually with *Fagus* and occasionally reported with *Betula* spp.

D: NM2: 385 **D+I:** Galli2: 416-417 **I:** C&D: 389

Rarely reported but apparently widespread.

rosea Pers. [non *R. rosea* Quél. (1888)], *Observ. mycol.* 1: 100 (1796)

Agaricus rosaceus Pers., *Syn. meth. fung.*: 439 (1801)

Russula lactea (Pers.) Fr., *Epicr. syst. mycol.*: 355 (1838)

Agaricus lacteus Pers. [*nom. illegit.*, non *A. lacteus* Pers., *Syn. meth. fung.*: 394 (1801)], *Syn. meth. fung.*: 439 (1801)

Russula lepida Fr., *Sverig Atl. Svamp.*: 50 (1836)

Russula lepida var. *alba* Quél., *Compt. Rend. Assoc. Franç. Avancem. Sci.* 13: 280 (1885)

Mis.: *Russula linnaei* sensu auct.

E: c **S:** ! **W:** ! **NI:** o **ROI:** !

H: On soil in mixed deciduous woodland, usually with *Fagus* and less frequently *Quercus* spp.

D: NM2: 389 **D+I:** Galli2: 189-191 (as *Russula lepida*) **I:** C&D: 393 (as *R. lepida*)

Common in England, apparently less so elsewhere but widespread. Perhaps better known as *Russula lepida*.

ruberrima Romagn., *Bull. Soc. Naturalistes Oyonnax* 4: 56 (1950)

E: !

H: On soil in deciduous woodland.

A single collection, from East Norfolk (Kelling Heath) in 2003.

rubroalba (Singer) Romagn., *Les Russules d'Europe et d'Afrique du Nord*: 780 (1967)

Russula alutacea f. *rubroalba* Singer, *Beiheft Z. Bot. Centralbl.* 2: 254 (1932)

E: !

H: On soil in mixed deciduous woodland. British material with *Quercus* sp.

D+I: Galli2: 414

A single collection, from North Essex (Crowsheath Wood) in 2004.

rutila Romagn., *Bull. mens. Soc. linn. Lyon* 21: 112 (1952)

E: ! **W:** !

H: On soil in mixed deciduous woodland. British material with *Carpinus* or *Quercus* spp.

D: NM2: 400 **D+I:** Galli2: 458-459, Sarn: 777-782

Rarely reported. Apparently widespread.

sanguinaria (Schumach.) Rauschert, *Česká Mykol.* 43: 204 (1989)

Agaricus sanguinarius Schumach., *Enum. pl.* 2: 244 (1803)

Russula sanguinea Fr., *Epicr. syst. mycol.*: 351 (1838)

Russula sanguinea var. *rosacea* (Pers.) J.E. Lange, *Fl. agaric. danic.* 5, *Taxonomic Conspectus*: VIII (1940)

Mis.: *Russula rosacea* sensu Cooke [Ill. Brit. fung. 982 (1020) Vol. 7 (1888)] and sensu Fr. (1838)

E: o **S:** ! **W:** ! **NI:** o **ROI:** !

H: On soil in mixed woodland, usually with *Pinus sylvestris* and *Betula* spp.

D: NM2: 395 (as *Russula sanguinea*) **D+I:** Galli2: 250-251, Sarn: 616-621 (as *R. sanguinea*) **I:** C&D: 377, SV46: 22 (as *R. sanguinea*)

Common and widespread. Perhaps better known as *R. sanguinea*.

sardonia Fr., *Epicr. syst. mycol.*: 353 (1838)

Russula chrysodacryon Singer, *Z. Pilzk.* 2(1): 16 (1923)

Russula drimeia Cooke, *Grevillea* 10: 46 (1881)

Russula chrysodacryon f. *viridis* Singer, *Bot. Centralbl., Beihefte* 49(2): 289 (1932)

Russula drimeia f. *viridis* (Singer) Bon, *Doc. Mycol.* 17(65): 55 (1986)

Russula drimeia var. *flavovirens* Rea, *Trans. Brit. Mycol. Soc.* 17(1-2): 45 (1932)

Russula sardonia var. *mellina* Melzer, *Arch. Přírodov. Výzk. Čech.* 17(4): 96 (1927)

E: c **S:** c **W:** c **NI:** c **ROI:** c **O:** Isle of Man: c

H: On soil, usually with *Pinus sylvestris* but also other conifers such as *Larix* and *Picea* spp.

D: NM2: 399 **D+I:** Galli2: 254-255, Sarn: 631-636 **I:** C&D: 377 (as *Russula drimeia*), FM2(3): 103, SV46: 23

Very common and widespread.

scotica A. Pearson, *Trans. Brit. Mycol. Soc.* 23(3): 309 (1939)

E: ! **S:** !

H: On soil in mixed woodland, with *Pinus sylvestris* and *Betula* spp.

D+I: FM4 113 -117 (2003)

Rarely reported. Perhaps not distinct from *Russula versicolor*.

sericatula Romagn., *Bull. mens. Soc. linn. Lyon* 27: 287 (1958)

E: ! **NI:** !

H: On soil in deciduous woodland.

D+I: Galli2: 394-395 **I:** C&D: 397

Known from England (Surrey and West Sussex) and Northern Ireland (Down).

silvestris (Singer) Reumaux, *Russules Rares ou Méconnues*: 289 (1996)

Russula emetica f. *sylvestris* Singer, *Bot. Centralbl., Beihefte* 49(2): 305 (1932)

Russula emetica var. *silvestris* (Singer) Kühner & Romagn. [*comb. inval.*], *Flora Analytique des Champignons Supérieurs*: 460 (1953)

Russula emeticella (Singer) J. Blum, *Bull. Soc. Mycol. France* 72: 141 (1956)

Russula emetica ssp. *emeticella* (Singer) Singer, *Sydowia* 11: 234 (1957)

Mis.: *Russula emetica* sensu auct. p.p.

Mis.: *Russula fragilis* sensu Cooke [Ill. Brit. fung. 1028 (1091) Vol. 7 (1888)] and sensu Rea (1922)

E: ! S: ! W: ! NI: !

H: On sandy or acidic soil, often in mixed deciduous and conifer woodland and frequently amongst the mosses *Leucobryum glaucum* or *Polytrichum* spp.

D+I: Sarn: 559-562 **I:** C&D: 373 (as *Russula emetica* var. *silvestris*), Galli2: 181 (as *R. emetica* var. *silvestris*)

Rarely reported but apparently widespread.

solaris Ferd. & Winge, *Medd. Svamp Fremme*: 9 (1924) [1922]

E: ! S: ! W: !

H: On soil, in beechwoods or mixed deciduous woodland, with *Fagus*.

D: NM2: 381 **D+I:** Galli2: 170-171, Sarn: 541-545 **I:** C&D: 373

Known in a few widely scattered locations in England, from Norfolk to South Devon, and rarely in Scotland (Mid-Perthshire) and Wales (Carmarthenshire).

sororia Fr., *Epicr. syst. mycol.*: 359 (1838)

Russula consobrina var. *sororia* Gillet, *Hyménomycètes*: 238 (1876)

Russula consobrina var. *intermedia* Cooke, *Handb. Brit. fung.*, Edn 2: 329 (1889)

E: ! S: ! W: ! NI: ! ROI: ? O: Channel Islands: !

H: On soil, with *Quercus* spp, in oakwoods or mixed deciduous woodland.

D: NM2: 381 **D+I:** C&D: 371, Galli2: 154-155, Sarn: 456-460

Commonly reported but rarely collected. Sensu NCL and sensu Rayner (1985) includes *Russula amoenolens* which is probably more frequent than true *R. sororia* .

sphagnophila Kauffman, *Rep. (Annual) Michigan Acad. Sci.* 11: 86 (1909)

E: !

H: On soil, or more usually fruiting amongst *Sphagnum* spp, in mixed deciduous woodland, often in damp or wet areas. British material normally with *Betula* spp.

D: NM2: 386 **D+I:** Galli2: 338-339 **I:** C&D: 387, Galli2: 333

Rarely reported, from Cheshire southwards to South Hampshire and West Gloucestershire. Close to *Russula nitida*.

stenotricha Romagn., *Bull. mens. Soc. linn. Lyon* 31: 174 (1962)

E: !

H: On soil in woodland. British material with *Quercus ilex*.

D: NM2: 383 **D+I:** Sarn: 296-299

A single collection, from Surrey (Kew, Royal Botanic Gardens) in 2002.

subfoetens Wm. G. Sm., *J. Bot.* 11: 337 (1873)

Mis.: *Russula subfoetens* var. *grata* sensu auct.

E: ! S: ! W: !

H: On soil, usually with *Fagus* and *Quercus* spp and less often with *Corylus* in old mixed deciduous woodland.

D: NM2: 380 **D+I:** Galli2: 140-141, Sarn: 428-433

Rarely reported and several of the collections in herb. K are redetermined as *R. foetens*. Apparently widespread. Sensu Rea (1922) is *R. farinipes*.

subrubens (J.E. Lange) Bon, *Doc. Mycol.* 5: 33 (1972)

Russula graveolens var. *subrubens* J.E. Lange, *Fl. agaric. danic.* 5, *Taxonomic Conspectus*: VIII (1940)

Russula chamiteae Kühner, *Bull. Soc. Mycol. France* 91(3): 389 (1975)

E: ! S: !

H: On soil in lowland deciduous woodland with *Salix caprea* and *S. cinerea*, in dunes and in upland areas with dwarf *Salix* spp.

D: NM2: 393, Adamčik & Knudsen (2004: 1464) **D+I:** Galli2: 360-361 (as *Russula chamiteae*) **I:** C&D: 391 (as *R. chamiteae*), C&D: 391 (as *R. subrubens*), FlDan5:190B

Known from England (East Kent and South Northumberland) and Scotland (Angus and South Aberdeen). Some collections

recorded as *R. pascua* under dwarf willows may also belong here.

taeniospora Einhell., *Beitr. Kenntn. Pilze Mitteleurop.* 2: 84 (1986)

Mis.: *Russula carminea* sensu NCL & sensu auct. mult.

E: ! S: !

H: On soil, in deciduous woodland and occasionally reported with conifers.

Rarely reported and less often collected.

torulosa Bres., *Icon. Mycol.* 9: pl. 433 (1929)

S: ! W: !

H: On sandy soil in woodland or plantation on dunes and usually with *Pinus nigra*.

D: NM2: 399 **D+I:** Galli2: 256-257, Sarn: 645-650 **I:** C&D: 376, FM2(3): 102, SV46: 24

Locally abundant in Wales (Anglesey) and Scotland (Morayshire). Reported from England (West Sussex) but from an inland site and unsubstantiated with voucher material.

turci Bres., *Fungi trident.* 1: 22 (1881)

E: o S: o

H: On soil in native woodland, usually associated with *Pinus sylvestris*. Rarely with other *Pinus* spp. in plantations.

D: NM2: 392 **D+I:** Galli2: 288-289 **I:** C&D: 397, Sarn: 84

Occasional in England and Scotland.

unicolor Romagn., *Les Russules d'Europe et d'Afrique du Nord*: 606 (1967)

E: ! S: ?

H: On soil in woodland.

D: NM2: 398 **D+I:** Galli2: 312-313

A single collection from West Kent (Saxtens Wood) in 2003. Reported from the Nordic Congress in Mid-Perthshire (Rannoch) in 1983, but unsubstantiated with voucher material.

velenovskyi Melzer & Zvára, *Arch. Přírodov. Výzk. Čech.*: 92 (1927)

E: o S: ! W: ! NI: ! ROI: ?

H: On soil in mixed deciduous woodland, with *Betula* spp, *Fagus* or *Quercus* spp.

D: NM2: 387 **D+I:** FRIC4: 6-9 27, Galli2: 432-433 **I:** C&D: 393

Occasional in England. Rarely reported elsewhere but apparently widespread.

versicolor Jul. Schäff., *Z. Pilzk.* 10(4): 105 (1931)

E: o S: ! W: ! NI: ! ROI: ! O: Channel Islands: ! Isle of Man: !

H: On soil in mixed deciduous woodland, usually with *Betula* spp, and occasionally with *Salix* spp.

D: NM2: 398 **D+I:** Galli2: 310-311 **I:** C&D: 385

Occasional in England. Rarely reported elsewhere but apparently widespread.

vesca Fr., *Sverig Atl. Svamp.*: 51 (1836)

Russula mitis Rea, *Brit. basidiomyc.*: 463 (1922)

E: c S: c W: c NI: c ROI: c O: Channel Islands: ! Isle of Man: !

H: On soil, usually with *Quercus* spp, in oakwoods or mixed deciduous woodland. Less often reported with *Fagus* or *Betula* spp.

D: NM2: 390 **D+I:** Galli2: 84-85, Sarn: 260-266 **I:** C&D: 381, Galli2: 79

Common and widespread.

veternosa Fr., *Epicr. syst. mycol.*: 354 (1838)

Russula schiffneri Singer, *Bot. Centralbl., Beihefte* 46(2): 88 (1929)

E: ! S: ! W: ! NI: !

H: On soil in deciduous woodland.

D: NM2: 400 **D+I:** Galli2: 460-461, Sarn: 758-762 **I:** C&D: 379

Rarely reported but apparently widespread.

vinosa Lindblad, *Svampbok*: 57 (1901)

Russula decolorans var. *obscura* Romell, *Öfvers. Kongl. Vetensk.-Akad. Förh.* 48: 179 (1891)

Russula obscura (Romell) Peck, *Bull. New York State Mus. Nat. Hist.* 116: 94 (1906)

E: ? S: !

H: On soil in woodland, with conifers.
D: NM2: 390 **D+I:** Galli2: 390-391 **I:** C&D: 397, Galli2: 387, SV46: 16
Known from a few widely scattered sites in Scotland. Reported from England (Cumberland) but unsubstantiated with voucher material.

vinosobrunnea (Bres.) Romagn., *Les Russules d'Europe et d'Afrique du Nord*: 732 (1967)
    Russula alutacea var. *vinosobrunnea* Bres., *Icon. Mycol.* 9: pl. 460 (1929)
E: !
H: On soil, in woodland.
D+I: Galli2: 384-385
A single collection from East Kent (Bedgebury Pinetum) in 1975.

vinosopurpurea Jul. Schäff., *Ann. Mycol.* 36(1): 27 (1938)
E: ! **S:** ! **NI:** ?
H: On soil in deciduous woodland.
D+I: Sarn: 763-767
Known from England (East Kent) and Scotland (Wester Ross). Reported from Northern Ireland but unsubstantiated with voucher material.

violacea Quél., *Compt. Rend. Assoc. Franç. Avancem. Sci.* 11: 397 (1883)
E: ! **W:** ? **NI:** ?
H: On soil in deciduous woodland.
D: NM2: 398 **D+I:** Galli2: 230-231 **I:** C&D: 375
A single confirmed collection from England (Staffordshire: Maer Hills) in 1994. Reported elsewhere in England, Wales and Northern Ireland, but unsubstantiated with voucher material.

violeipes Quél., *Compt. Rend. Assoc. Franç. Avancem. Sci. Assoc. Sci. France* 24: 450 (1898) [1897]
    Russula punctata f. *violeipes* (Quél.) Maire, *Bull. Soc. Mycol. France* 26: 118 (1910)
    Russula heterophylla var. *chlora* Gillet, *Hyménomycètes*: 241 (1876)
    Russula olivascens var. *citrinus* Quél., *Enchir. fung.*: 132 (1886)
    Russula punctata f. *citrina* (Quél.) Maire, *Bull. Soc. Mycol. France* 26: 118 (1910)
    Mis.: *Russula amoena* sensu Pearson [*Naturalist* (Hull) (1948)]
E: o **S:** ! **W:** ! **NI:** ! **ROI:** !
H: On soil in mixed deciduous woodland, usually with *Quercus* spp and *Fagus* and less frequently *Betula* spp.
D: NM2: 379 **D+I:** Galli2: 92-93, Sarn: 398-402 **I:** C&D: 381
Occasional in England, rarely reported elsewhere but apparently widespread.

virescens (Schaeff.) Fr., *Sverig Atl. Svamp.*: 50 (1836)
    Agaricus virescens Schaeff., *Fung. Bavar. Palat. nasc.* 1: 40 (1762)
E: o **S:** ! **W:** o **NI:** !
H: On soil, often with *Quercus* spp, in oakwoods but also said to be frequent with *Fagus* in mixed deciduous woodland.
D: NM2: 389 **D+I:** Galli2: 87-89, Sarn: 375-380 **I:** C&D: 381
Occasional in England. Rarely reported elsewhere but apparently widespread.

viscida Kudřna, *Mykologia* 5: 56 (1928)
    Russula melliolens var. *chrismantiae* Maire, *Bull. Soc. Mycol. France* 26: 110 (1910)
    Russula occidentalis (Singer) Singer, *Lilloa* 22: 705 (1951) [1949]
    Russula vinosa ssp. *occidentalis* Singer, *Pap. Michigan Acad. Sci.* 32: 114 (1946)
E: !
H: On soil, in mixed deciduous or mixed conifer and deciduous woodland.
D: NM2: 384 **D+I:** Galli2: 348-349 **I:** C&D: 387, FM1(3): 75
Known from South Essex (Epping Forest) and West Kent (Shoreham, Meefield Wood).

xenochlora P.D. Orton, *Kew Bull.* 54(3): 707 (1999)
S: !

H: On soil in woodland. British material with *Betula* spp.
Known from Scotland (Easterness)

xerampelina (Schaeff.) Fr., *Epicr. syst. mycol.*: 356 (1838)
    Agaricus xerampelina Schaeff., *Fung. Bavar. Palat. nasc.* 3: 214 (1770)
    Russula alutacea var. *erythropus* Fr., *Hymenomyc. eur.*: 453 (1874)
    Russula erythropus (Fr.) Pelt., *Bull. Soc. Mycol. France* 24: 118 (1908)
    Russula xerampelina var. *erythropus* (Fr.) Kühner & Romagn. [*comb. inval.*], *Flora Analytique des Champignons Supérieurs*: 49 (1953)
    Russula erythropus var. *ochracea* J. Blum, *Bull. Soc. Mycol. France* 77: 162 (1961)
E: c **S:** c **W:** c **NI:** c **ROI:** c **O:** Channel Islands: !
H: On soil in native woodland or plantations, usually with *Pinus* spp.
D: NM2: 393 **D+I:** Ph: 104 (as *Russula erythropus*), Galli2: 364-365 **I:** C&D: 391, C&D: 391, SV46: 19
Common and widespread. Treated in a broad sense in NCL but here restricted to the more or less red species growing with conifers.

zonatula Ebbesen & Jul. Schäff., *Russula-Monographie*: 260 (1952)
E: !
H: On soil in deciduous woodland.
D: NM2: 398, NM2: 398 **D+I:** Galli2: 324-325 **I:** C&D: 385
A single collection from Surrey (Gomshall) in 1958.

zvarae Velen., *České Houby* IV-V: 913 (1920)
E: !
H: On soil in deciduous woodland.
D+I: Galli2: 276-277 **I:** C&D: 395
Known from East Sussex, Hertfordshire South Hampshire and West Kent.

SARCODON P. Karst., *Rev. Mycol. (Toulouse)* 3(9): 20 (1881)
Thelephorales, Thelephoraceae
Type: *Sarcodon imbricatus* (L.) P. Karst.

glaucopus Maas Geest. & Nannf., *Svensk bot. Tidskr.* 63: 407 (1969)
S: !
H: With *Pinus sylvestris* in Caledonian pinewoods.
D: NM3: 314 **D+I:** B&K2: 232 274, BritChant: 100-101
Known from a few collections in Easterness (Abernethy Forest and Rothiemurchus).

regalis Maas Geest., *Verh. K. Akad. Wet.* 65: 109 (1975)
E: !
H: With *Castanea* and *Quercus* spp.
D: BritChant: 96-97
Known only from the type locality in Berkshire (Windsor Great Park) but not seen since 1969, and possibly extinct.

scabrosus (Fr.) P. Karst., *Rev. Mycol. (Toulouse)* 3(9): 20 (1881)
    Hydnum scabrosum Fr., *Sverig Atl. Svamp.*: 62 (1836)
E: ! **S:** ?
H: With *Castanea* and *Quercus* spp in woodland on acidic, sandy soil.
D: NM3: 314 **D+I:** Ph: 243, B&K2: 234 279, BritChant: 98-99
Rare. Most records are from southern England. Reports of this species from Scotland require confirmation as they may all be *Sarcodon glaucopus*

squamosus (Schaeff.) Quél., *Enchir. fung.*: 188 (1886)
    Hydnum squamosum Schaeff., *Fung. Bavar. Palat. nasc.* 1: 99 (1762)
    Mis.: *Sarcodon imbricatus* sensu auct. Brit.
E: ! **S:** ! **ROI:** !
H: With *Pinus sylvestris* on acidic sandy soil.
D+I: Ph: 242 (as *Sarcodon imbricatus*), BritChant: 92-93 (as *S. imbricatus*)

Rare, but the most frequently collected species of *Sarcodon* in Britain. All records of *S. imbricatus* belong here, the true *S. imbricatus* (only recently distinguished) being known only with *Picea* spp in continental Europe.

SARCODONTIA Schulzer, in Schulzer et al., *Verh. zool.-bot. Ges. Wien* 16: 41 (1866)
Stereales, Hyphodermataceae
Type: *Sarcodontia mali* Schulzer (= *S. crocea*)

crocea (Schwein.) Kotl., *Česká Mykol.* 7: 117 (1953)
Sistotrema croceum Schwein., *Schriften Naturf. Ges. Leipzig* 1: 102 (1822)
Hydnum croceum (Schwein.) Fr., *Elench. fung.* 1: 137 (1828)
Sarcodontia mali Schulzer, *Verh. zool.-bot. Ges. Wien* 16: 41 (1866)
Hydnum schiedermayeri Heufl., *Oesterr. Bot. Z.* 20: 37 (1870)
Hydnum setosum Pers., *Mycol. eur.* 2: 213 (1825)
Acia setosa (Pers.) Bourdot & Galzin, *Hyménomyc. France*: 418 (1928)
Mycoacia setosa (Pers.) Donk, *Meded. Ned. Mycol. Ver.*: 152 (1931)
Sarcodontia setosa (Pers.) Donk, *Mycologia* 44: 262 (1952)
Mis.: *Mycoacia squalina* sensu Christiansen [Danish Resupinate Fungi: 2 (1960)]
E: ! **S:** !
H: Parasitic then saprotrophic. On old trunks of *Malus* spp, including cultivars, in gardens and orchards. Reported once on *Fagus* but this host is unverified.
D: NM3: 174 **D+I:** CNE6: 1274-1276, B&K2: 168 179 **I:** Kriegl1: 310
Rare but conspicuous on old trunks, forming large, bright yellow, hydnoid basidiomes with a pungent smell said to resemble 'pineapple'. Apparently spread by horticultural processes (pruning etc.) rather than by spore dispersal.

SCHIZOPHYLLUM Fr., *Syst. mycol.* 1: 330 (1821)
Schizophyllales, Schizophyllaceae
Schizophyllus Fr., *Observ. mycol.* 1: 103 (1815)
Apus Gray, *Nat. arr. Brit. pl.* 1: 617 (1821)
Auriculariopsis Maire, *Rech. cytol. Basid.*: 102 (1902)
Type: *Schizophyllum commune* (Fr.) Fr.

amplum (Lév.) Nakasone, *Mycologia* 88(5): 771 (1996)
Cyphella ampla Lév., *Ann. Sci. Nat., Bot.*, sér. 3, 9: 126 (1848)
Auriculariopsis ampla (Lév.) Maire, *Bull. Soc. Mycol. France* 18: 102 (1902)
Cyphella cyclas Cooke & W. Phillips, *Grevillea* 9: 94 (1881)
Thelephora flocculenta Fr., *Elench. fung.* 1: 184 (1828)
Cytidia flocculenta (Fr.) Höhn. & Litsch., *Sitzungsber. Kaiserl. Akad. Wiss., Wien, Math.-Naturwiss. Cl., Abt. 1* 116: 758 (1907)
Corticium flocculentum (Fr.) Fr., *Epicr. syst. mycol.*: 559 (1838)
E: !
H: On dead and dying attached branches of *Populus* spp (including *P. nigra*) and rarely *Salix fragilis*.
D: NM3: 155 (as *Auriculariopsis ampla*) **D+I:** CNE3: 290-293 (as *A. ampla*) **I:** SV36: 24, Kriegl1: 162 (as *A. ampla*)
Known from Bedfordshire, Cambridgeshire, East Sussex, Hertfordshire, London, Middlesex, Northumberland and Oxfordshire. Perhaps better known as *Auriculariopsis ampla*.

commune (Fr.) Fr., *Syst. mycol.* 1: 330 (1821)
Schizophyllus communis Fr., *Observ. mycol.* 1: 103 (1815)
Agaricus alneus L., *Fl. Suec.*, Edn 2: 1242 (1755)
Apus alneus (L.) Gray, *Nat. arr. Brit. pl.* 1: 617 (1821)
Agaricus multifidus Batsch, *Elench. fung. (Continuatio Prima)*: 173 & t 24 f. 126 (1786)
E: o **S:** ! **W:** ! **NI:** !
H: On fallen wood of deciduous trees, usually *Fagus* but also known on *Acer pseudoplatanus*, *Alnus*, *Ilex*, *Prunus avium*, *Quercus* and *Tilia* spp. Increasingly frequent on bales of hay,

especially those wrapped in black plastic sheeting. Rarely on worked wood such as fence posts. Recorded also from a variety of animal substrata (e.g. whalebone, horn, leather) and as a cause of human mouth ulcers and toe nail infections.
D: BFF6: 72-73, NM3: 165 **D+I:** Ph: 186-187, B&K3: 318 404 **I:** Sow2: 183 (as *Agaricus alneus*), FlDan5: 198F, Bon: 321, Kriegl1: 412
Locally common in south eastern England, occasional to rare elsewhere. Cosmopolitan. English name = 'Split-gill'.

SCHIZOPORA Velen., *České Houby* IV - V: 638 (1922)
Stereales, Hyphodermataceae
Type: *Polyporus laciniatus* Velen. (= *Schizopora paradoxa*)

flavipora (Cooke) Ryvarden, *Mycotaxon* 23: 186 (1985)
Poria flavipora Cooke, *Grevillea* 15: 25 (1886)
Poria carneolutea Rodway & Cleland, *Pap. Proc. R. Soc. Tasm.* 1929: 18 (1930)
Schizopora carneolutea (Rodway & Cleland) Kotl. & Pouzar, *Česká Mykol.* 33: 21 (1979)
E: ! **S:** !
H: On decayed wood of deciduous trees such as *Betula* spp, and on conifers such as *Pinus sylvestris*. Also on decayed boards in buildings and dead stems of *Ulex europaeus*.
D: EurPoly2: 606 **I:** CNE7: 1289-1290 (as *Schizopora carneolutea*), Kriegl1: 312
Known from England (East and West Sussex, London, Surrey, Warwickshire) and Scotland (South Ebudes). Common in southern Europe and in the tropics.

paradoxa (Schrad.) Donk, *Persoonia* 5(1): 104 (1967)
Hydnum paradoxum Schrad., *Spicil. Fl. Germ.* 1: 179 (1794)
Sistotrema paradoxum (Schrad.) Pers., *Syn. meth. fung.*: 225 (1801)
Irpex paradoxus (Schrad.) Fr., *Epicr. syst. mycol.*: 522 (1838)
Hyphodontia paradoxa (Schrad.) Langer & Vesterh., *Nordic J. Bot.* 16(2): 211 (1996)
Hydnum obliquum Schrad., *Spicil. Fl. Germ.* 1: 179 (1794)
Irpex obliquus (Schrad.) Fr., *Elench. fung.* 1: 147 (1828)
Polyporus versiporus Pers., *Mycol. eur.* 2: 105 (1825)
Poria versipora (Pers.) Romell, *Techn. Publ. New York State Coll. Forest.* 90: 63 (1966)
Xylodon versiporus (Pers.) Bondartsev, *Trutovye griby Evropeiskoi chasti SSSR i Kavkaza*: 128 (1953)
Irpex deformis Fr., *Elench. fung.* 1: 147 (1828)
Polyporus laciniatus Velen. [*nom. illegit.*, non *P. laciniatus* Pers. (1825)], *Česka Houby* IV-V: 638 (1922)
Mis.: *Poria mucida* sensu Boudot & Galzin (1928) and sensu Rea (1922)
Mis.: *Poria vaporaria* sensu Rea (1922)
Mis.: *Polyporus vaporarius* sensu Berkeley (1860) and sensu Rea (1922) non sensu Fr. (1821)
E: c **S:** c **W:** c **NI:** c **ROI:** c **O:** Channel Islands: c Isle of Man: c
H: Very common on dead wood of deciduous trees, less frequent on conifers. Occasionally on other substrata such as dead lianas of *Clematis vitalba*, dead stems of *Bambusa* spp, or dead basidiomes of polypores.
D: EurPoly2: 608, NM3: 213 (as *Hyphodontia paradoxa*) **D+I:** Ph: 237, CNE7: 1284-1288, B&K2: 302 380
Very common and widespread. Placed by some authorities in *Hyphodontia*.

radula (Pers.) Hallenb., *Mycotaxon* 18(2): 308 (1983)
Poria radula Pers., *Observ. mycol.* 2: 14 (1799)
Boletus radula (Pers.) Pers., *Syn. meth. fung.*: 547 (1801)
Polyporus radula (Pers.) Fr., *Syst. mycol.* 1: 383 (1821)
Hyphodontia radula (Pers.) Langer & Vesterh., *Nordic J. Bot.* 16(2): 212 (1996)
Poria eyrei Bres., *Trans. Brit. Mycol. Soc.* 3(4): 264 (1910)
E: ! **S:** !
H: On dead wood of deciduous trees in old woodland. Known on *Fagus*, *Fraxinus* and *Salix caprea*.
D: EurPoly2: 609, NM3: 213 (as *Hyphodontia radula*)

Known from England (East Norfolk, Hertfordshire, Middlesex, South Hampshire, Surrey and West Sussex) and Scotland (Midlothian). Easily confused with *Schizopora paradoxa*. Placed by some authorities in *Hyphodontia*.

SCLERODERMA Pers., *Syn. meth. fung.*: 150 (1801)

Sclerodermatales, Sclerodermataceae
Type: *Scleroderma verrucosum* (Bull.) Pers.

areolatum Ehrenb., *Sylv. mycol. berol.* 15: 27 (1818)
E: o **S:** o **W:** o **NI:** o **ROI:** o **O:** Channel Islands: !
H: On soil, often amongst grass at the edges of woodland and fields. Frequently associated with *Quercus* spp, but also reported with a variety of other deciduous trees.
D: NM3: 297 **D+I:** Ph: 250-251, B&K2: 384 504, BritPuffb: 34-35 Figs. 16/17
Occasional but widespread.

bovista Fr., *Syst. mycol.* 3: 48 (1829)
Scleroderma verrucosum var. *bovista* (Fr.) Šebek, *Sydowia* 7: 177 (1953)
Tuber fuscum Corda, *Icon. fung.* 1: 25 (1837)
Scleroderma fuscum (Corda) E. Fisch., in Engl. & Prantl., *Nat. Pflanzenfamilien* 1(1): 336 (1900)
Scleroderma texense Berk., *London J. Bot.* 4: 308 (1845)
E: o **S:** o **W:** o **NI:** o **ROI:** o **O:** Channel Islands: !
H: On sandy or well drained soil, often along roadsides. Frequently with *Tilia* spp.
D: NM3: 297 **D+I:** B&K2: 386 505, BritPuffb: 24-25 Figs. 6/7 **I:** Bon: 303, Kriegl2: 175
Occasional but widespread. Previously regarded as a variety of *Scleroderma verrucosum*, but the presence of clamp connections and a reticulate spore ornament indicates no such relationship. English name = 'Potato Earthball'.

cepa Pers., *Syn. meth. fung.*: 155 (1801)
Scleroderma vulgare var. *cepa* (Pers.) W.G. Sm., *Syn. Brit. Bas.*: 480 (1908)
Scleroderma cepioides Gray, *Nat. arr. Brit. pl.* 1: 582 (1821)
Mis.: *Scleroderma spadiceum* sensu Rea (1922)
E: o **S:** ! **W:** ! **NI:** ! **ROI:** !
H: On sandy soil in woodland, usually with *Quercus* spp. Also reported with *Betula* spp and *Fagus* and there is a single record 'amongst shingle on a beach'. Also reported with *Pteridium*.
D: NM3: 297 **D+I:** BritPuffb: 30-31 Figs. 12/13
Occasional in England, rarely reported elsewhere but apparently widespread. Often regarded as a variety of *Scleroderma verrucosum*. English name = 'Onion Earthball'.

citrinum Pers., *Syn. meth. fung.*: 153 (1801)
Scleroderma vulgare Hornem., *Fl. Danica*: t. 1969 f. 2 (1829)
Mis.: *Lycoperdon aurantium* sensu auct.
Mis.: *Scleroderma aurantium* sensu auct.
E: c **S:** c **W:** c **NI:** c **ROI:** c **O:** Channel Islands: c Isle of Man: c
H: On acidic soils, with deciduous trees such as *Fagus, Quercus* and *Betula* spp, in woodland and on heathland. Fruiting from late summer onwards.
D: NM3: 297 **D+I:** Ph: 250-251, B&K2: 386 506, BritPuffb: 26-27 Figs. 8/9 **I:** Bon: 303
Very common and widespread. Supposedly poisonous and causing gastrointestinal upsets if consumed. English name = 'Common Earthball'.

verrucosum (Bull.) Pers., *Syn. meth. fung.*: 154 (1801)
Lycoperdon verrucosum Bull., *Herb. France*: pl. 24 (1780)
Mis.: *Lycoperdon defossum* sensu Sowerby [Col. fig. Engl. fung. 3: pl. 311 (1801)]
E: c **S:** c **W:** c **NI:** o **ROI:** c **O:** Channel Islands: o Isle of Man: o
H: On humic and sandy soils in a variety of habitats such as heathland, parkland and woodland and occasionally on decayed wood.
D: NM3: 297 **D+I:** Ph: 250-251 Figs. 14/15, BritPuffb: 32-33 **I:** Kriegl2: 178

Common and widespread. English name = 'Scaly Earthball'.

SCLEROGASTER R. Hesse, *Hypog. Deutschl.* 1: 84 (1891)

Hymenogastrales, Octavianinaceae
Type: *Sclerogaster lanatus* R. Hesse

compactus (Tul. & C. Tul.) Sacc., *Syll. fung.* 11: 170 (1895)
Octaviania compacta Tul. & C. Tul., *Giorn. Bot. Ital.* 2(1): 56 (1844)
Sclerogaster broomeianus Zeller & H.R. Dodge, *Rep. (Annual) Missouri Bot. Gard.* 22: 370 (1935)
E: !
H: On soil under leaf litter in deciduous woodland or with conifers.
D: NM3: 318 **D+I:** BritTruff: 184 11 G
Known from North Hampshire, North Somerset, South Devon and West Kent.

hysterangioides (Tul. & C. Tul.) Zeller & H.R. Dodge, *Rep. (Annual) Missouri Bot. Gard.* 22: 370 (1935)
Hydnangium hysterangioides Tul. & C. Tul., *Fungi hypogaei*: 77 (1851)
E: !
H: Hypogeous in deciduous woodland.
A single collection from North Somerset (Cleeve Cliff) in herb. K, originally determined as *Rhizopogon roseolus*.

lanatus R. Hesse, *Hypog. Deutschl.* 1: 85 (pl. 5) (1891)
E: !
H: Hypogeous in calcareous soil in woodland.
A single collection from West Kent (Otford) in 1876, in herb. K, originally determined as *Octaviana compacta*.

SCOPULOIDES (Massee) Höhn. & Litsch., *Wiesner Festschrift*: 57 (1908)

Stereales, Meruliaceae
Type: *Scopuloides hydnoides* (Cooke & Massee) Hjortstam & Ryvarden

hydnoides (Cooke & Massee) Hjortstam & Ryvarden, *Mycotaxon* 9(2): 509 (1979)
Peniophora hydnoides Cooke & Massee, *Grevillea* 16(79): 77 (1888)
Phlebia hydnoides (Cooke & Massee) M.P. Christ., *Dansk. Bot. Ark.* 19(2): 175 (1960)
Peniophora crystallina Höhn. & Litsch., *Sitzungsber. Kaiserl. Akad. Wiss., Wien, Math.-Naturwiss. Cl., Abt. 1* 116: 828 (1907)
E: ! **S:** ! **W:** ! **ROI:** !
H: In woodland, on decayed wood of deciduous trees, often in wet areas such as ditches, periodically flooded hollows or the edges of pools. Often on *Alnus* and *Salix* spp, but also known on a wide variety of other deciduous trees, on dead stems of *Ulex europaeus* and on dead basidiomes of polypores.
D: B&K2: 160 166 (as *Scopuloides rimosa*)
Apparently frequent and widespread but confused with the similar and more frequent *S. rimosa*.

rimosa (Cooke) Jülich, *Persoonia* 11(4): 422 (1982)
Peniophora rimosa Cooke, *Grevillea* 9: 94 (1881)
Phanerochaete rimosa (Cooke) Burds., *Mycol. Mem., St. Paul* 10: 107 (1985)
Odontia conspersa Bres., *Atti Imp. Regia Accad. Roveretana, ser. 3,* 3: 100 (1897)
Peniophora terrestris Massee, *Grevillea* 15: 107 (1887)
E: ! **S:** ! **W:** ! **NI:** !
H: On decayed wood of deciduous trees and occasionally on fallen basidiomes of polypores.
D: NM3: 169 **D+I:** CNE7: 1292-1293 (as *Scopuloides hydnoides*),
Rarely recorded or collected but apparently widespread. Often confused with *S. hydnoides*.

SCOTOMYCES Jülich, *Persoonia* 10: 139 (1978)
Ceratobasidiales, Ceratobasidiaceae
 Hydrabasidium J. Erikss. & Ryvarden, *Corticiaceae of North Europe* 5: 896 (1979)
Type: *Scotomyces fallax* (G. Cunn.) Jülich (= *S. subviolaceus*)

subviolaceus (Peck) Jülich, *Persoonia* 10: 334 (1978)
 Hypochnus subviolaceus Peck, *Rep. (Annual) New York State Mus. Nat. Hist.* 47: 151 (1893)
 Hydrabasidium subviolaceum (Peck) J. Erikss. & Ryvarden, *Corticiaceae of North Europe* 5: 897 (1978)
 Corticium atratum Bres., *Hedwigia* 35: 290 (1896)
 Ceratobasidium atratum (Bres.) D.P. Rogers, *Lloydia* 4: 262 (1941)
 Oliveonia atrata (Bres.) P.H.B. Talbot, *Persoonia* 3: 381 (1965)
E: ! **S:** ! **W:** !
H: On decayed conifer wood, and also known on debris of *Pteridium*.
D: NM3: 113 **D+I:** CNE5: 897-899 (as *Hydrabasidium subviolaceum*), B&K2: 78 43
Known from England (Cumberland, East Sussex, South Devon, Westmorland, and Yorkshire), Scotland (Kirkudbightshire) and Wales (Caernarvonshire). Easily overlooked since basidiomes are thin and cryptically coloured bluish-black.

SCYTINOSTROMA Donk, *Fungus* 26: 19 (1956)
Lachnocladiales, Lachnocladiaceae
Type: *Scytinostroma portentosum* (Berk. & M.A. Curtis) Donk

ochroleucum (Bres. & Torrend) Donk, *Fungus* 26: 20 (1956)
 Gloeocystidium ochroleucum Bres. & Torrend, *Brotéria, Sér. Bot.* 11(1): 81 (1913)
 Corticium lentum Wakef., *Trans. Brit. Mycol. Soc.* 35(1): 54 (1952)
E: !
H: On decayed wood. Known from *Taxus* in woodland on calcareous soils, and a single collection on *Rhododendron* sp.
D+I: Hallenb: 27-31 Figs. 17-19, Myc17(4): 156
First reported from West Sussex (Henley Woods) in 1938 but not again until 1998 when collected in Surrey (Mickleham: Norbury Park, and Kew: Royal Botanic Gardens) and Berkshire (Pangbourne: Basildon Park). Also found again in Surrey (West Horsley: Sheepleas) in 2002.

portentosum (Berk. & M.A. Curtis) Donk, *Fungus* 26: 20 (1956)
 Corticium portentosum Berk. & M.A. Curtis, *Grevillea* 2: 3 (1873)
 Stereum portentosum (Berk. & M.A. Curtis) Höhn. & Litsch., *Sitzungsber. Kaiserl. Akad. Wiss., Wien, Math.-Naturwiss. Cl., Abt. 1* 116: 743 (1907)
 Scytinostroma hemidichophyticum Pouzar, *Česká Mykol.* 20: 217 (1966)
E: ! **S:** !
H: On decorticated wood, or on bark, usually on the undersides of large logs or fallen trunks of deciduous trees. Most frequently collected on *Fraxinus* and also known on *Fagus* and *Salix* spp.
D: NM3: 320 **D+I:** Hallenb: 33-37 Figs. 22-24, B&K2: 116 100, Myc10(2): 84 **I:** Kriegl1: 319
First reported from England (West Kent: Lullingstone Park) in 1991 but apparently spreading and now known from Berkshire, North Hampshire, Surrey and West Norfolk, with a single record from Scotland (South Aberdeen) in 2001. Basidiomes have a pungent odour of naphthalene.

SEBACINA Tul., *J. Linn. Soc., Bot.* 13: 36 (1871)
Tremellales, Exidiaceae
 Cristella Pat., *Hyménomyc. Eur.*: 151 (1887)
 Soppittiella Massee, *Brit. fung.-fl.* 1: 106 (1892)
Type: *Sebacina incrustans* (Pers.)Tul.

dimitica Oberw., *Ber. Bayer. Bot. Ges.* 36: 53 (1963)
E: !

H: On decayed wood.
D: NM3: 105, Myc16(4): 178
A single collection from Oxfordshire (The Warburg Reserve) in 1999.

epigaea (Berk. & Broome) Neuhoff, *Z. Pilzk.* 10(3): 71 (1931)
 Tremella epigaea Berk. & Broome, *Ann. Mag. Nat. Hist., Ser. 2 [Notices of British Fungi no. 373]* 2: 266 (1848)
 Sebacina laciniata f. *epigaea* (Berk. & Broome) Bourdot & Galzin, *Hyménomyc. France*: 39 (1928)
 Sebacina ambigua Bres., *Ann. Mycol.* 1(1): 116 (1903)
 Sebacina livescens Bres., *Fungi trident.* 2(11-13): 64 (1898)
 Sebacina epigaea var. *goniophora* Bourdot & Galzin, *Hyménomyc. France*: 40 (1928)
E: o **S:** ! **W:** o **NI:** ! **ROI:** !
H: On soil, worm casts, decayed leaf litter, and decayed fallen wood of deciduous trees.
D: NM3: 106 **D+I:** B&K2: 58 14
Rarely reported but apparently widespread.

incrustans (Pers.) Tul., *J. Linn. Soc., Bot.* 13: 36 (1871)
 Thelephora incrustans Pers., *Mycol. eur.* 1: 135 (1822)
 Corticium incrustans Pers. [non *C. incrustans* Höhn. & Litsch. (1906)], *Observ. mycol.* 1: 39 (1796)
 Merisma cristatum Pers., *Comment. Fungis Clavaeform.*: 96 (1797)
 Thelephora cristata (Pers.) Fr., *Syst. mycol.* 1: 434 (1821)
 Soppittiella cristata (Pers.) Massee, *Brit. fung.-fl.* 1: 107 (1892)
 Thelephora sebacea Pers., *Syn. meth. fung.*: 577 (1801)
 Soppittiella sebacea (Pers.) Massee, *Brit. fung.-fl.* 1: 106 (1892)
 Corticium sebaceum (Pers.) Massee, *J. Linn. Soc., Bot.* 27: 127 (1890)
E: c **S:** o **W:** o **NI:** ! **ROI:** !
H: On soil or decayed wood and often found creeping up the stems of herbaceous plants or at the base of small trees.
D: NM3: 105 **D+I:** Ph: 262, B&K2: 58 15 **I:** C&D: 137
Common in England. Infrequently reported elsewhere.

SERENDIPITA P. Roberts, *Mycol. Res.* 97(4): 474 (1993)
Tremellales, Exidiaceae
Type: *Serendipita vermifera* (Oberw.) P. Roberts

evanescens (Hauerslev) P. Roberts, *Mycol. Res.* 97(4): 476 (1993)
 Sebacina evanescens Hauerslev, *Friesia* 11: 102 (1976)
 Exidiopsis evanescens (Hauerslev) Wojewoda, *Mała Flora Grzybów* 2: 108 (1981)
E: !
H: British material mostly on fallen and decayed conifer wood, and a single collection on *Salix* sp.
D: NM3: 102 (as *Exidia evanescens*)
Known from scattered sites in South Devon.

orliensis P. Roberts, *Mycol. Res.* 97(4): 477 (1993)
E: !
H: On decayed wood and bark of deciduous trees e.g. *Corylus, Fagus* and *Ulmus* spp.
Known only from South Devon and Surrey.

sigmaspora P. Roberts, *Mycol. Res.* 97(4): 475 (1993)
E: !
H: On decayed wood of conifers such as *Picea* spp, and often in the hymenium of effuse corticioid fungi such as *Tubulicrinis* and *Botryobasidiium* spp, growing on the wood.
Known only from South Devon.

vermifera (Oberw.) P. Roberts, *Mycol. Res.* 97(4): 474 (1993)
 Sebacina vermifera Oberw., *Nova Hedwigia* 7: 495 (1964)
 Exidiopsis vermifera (Oberw.) Wojewoda, *Grzyby (Mycota)* 8: 100 (1977)
E: ! **NI:** !

H: On decayed wood of *Fagus*, *Prunus spinosa* and *Picea* spp, also frequently in the hymenium of effuse corticioid fungi such as *Basidiodendron* and *Botryobasidium* spp.
D: NM3: 101 (as *Exidia vermifera*)
Known from England (South Hampshire and South Devon) and Northern Ireland (Tyrone).

SERPULA (Pers.) Gray, *Nat. arr. Brit. pl.* 1: 637 (1821)
Boletales, Coniophoraceae
Xylomyzon Pers., *Mycol. eur.* 2: 26 (1825)
Gyrophana Pat., *Cat. pl. cel. Tunisie*: 53 (1897)
Type: *Merulius destruens* Pers. (= *S. lacrymans*)

himantioides (Fr.) P.Karst, *Bidrag Kännedom Finlands Natur Folk* 48: 344 (1889)
Merulius himantioides Fr., *Observ. mycol.* 2: 238 (1818)
Gyrophana himantioides (Fr.) Bourdot & Galzin, *Bull. Soc. Mycol. France* 39(2): 108 (1923)
Boletus arboreus Sowerby, *Col. fig. Engl. fung.* 3: pl. 346 (1802)
Merulius papyraceus Fr., *Elench. fung.* 1: 61 (1828)
Merulius silvester O. Falck, in Møller, *Hausschwammforsch.* 6: 53 (1912)
Merulius squalidus Fr., *Elench. fung.* 1: 62 (1828)
Xylomyzon versicolor Pers., *Mycol. eur.* 2: 30 (1825)
E: c **S:** ! **W:** !
H: Usually on decayed wood of conifers such as *Larix*, *Picea*, *Pinus*, *Pseudotsuga* and *Thuja* spp, especially on old stacks of logs in plantations. Much less frequent on wood of deciduous trees, but known on *Aesculus*, *Alnus*, *Betula*, *Corylus*, *Fagus* and *Salix* spp.
D: NM3: 293 **D+I:** Ph: 238-239, Hallenb: 75-77 Figs. 48-51, Myc17(4): 157 **I:** SV42: 51
Common and widespread in England. Rarely reported elsewhere (last reported from Wales in 1915).

SETICYPHELLA Agerer, *Mitt. bot. StSamml., München* 19: 282 (1983)
Stereales, Cyphellaceae
Type: *Seticyphella tenuispora* Agerer

tenuispora Agerer, *Mitt. bot. StSamml., München* 19: 290 (1983)
E: !
H: On fallen and decayed leaves. British material on leaves of *Fagus* and *Salix* sp in mixed deciduous woodland.
A single collection from Surrey (Runfold Wood) in 2000, but basidiomes are minute and easily overlooked .

SIMOCYBE P. Karst., *Bidrag Kännedom Finlands Natur Folk* 32(22): 416 (1879)
Cortinariales, Cortinariaceae
Ramicola Velen., *Mykologia* 6: 76 (1929)
Type: *Simocybe centunculus* (Fr.) P. Karst.

centunculus (Fr.) P. Karst., *Bidrag Kännedom Finlands Natur Folk* 32: 420 (1879)
Agaricus centunculus Fr., *Syst. mycol.* 1: 262 (1821)
Naucoria centunculus (Fr.) P. Kumm., *Führ. Pilzk.*: 78 (1871)
Ramicola centunculus (Fr.) Watling, *Notes Roy. Bot. Gard. Edinburgh* 45(3): 555 (1989) [1988]
E: o **S:** ! **W:** ! **NI:** !
H: In deciduous woodland, usually on large, decayed, fallen trunks, branches or logs of *Fagus*. Also known on *Acer pseudoplatanus*, *Betula*, *Quercus*, *Salix*, *Tilia* and *Ulmus* spp.
D: Reid84: 222, NM2: 339 (as *Ramicola centunculus*) **D+I:** B&K5: 304 394 **I:** C&D: 365 (as *R. centunculus*)
Rarely reported but rather common in southern England. Basidiomes are small and cryptically coloured, often appearing in the late summer.

centunculus *var.* laevigata (J. Favre) Senn-Irlet, *Mycol. Helv.* 7(2): 47 (1995)

Naucoria centunculus var. *laevigata* J. Favre, *Mat. fl. crypt. Suisse* 10(3): 138 (1948)
Naucoria laevigata (J. Favre) Kühner & Romagn. [*comb. inval.*], *Flore Analytique des Champignons Supérieurs*: 236 (1953)
Simocybe laevigata (J. Favre) P.D. Orton, *Notes Roy. Bot. Gard. Edinburgh* 29: 78 (1969)
Ramicola laevigata (J. Favre) Watling, *Notes Roy. Bot. Gard. Edinburgh* 45(3): 555 (1989) [1988]
E: !
H: On decayed debris of *Carex* spp, or dead and dying stems of *Oenanthe crocata* in damp areas such as ditches and partially dried out pond margins.
D: Reid84: 223 (as *Simocybe laevigata*), NM2: 339 (as *Ramicola laevigata*)
Known from East Norfolk, Herefordshire and South Devon.

centunculus *var.* maritima (Bon) Senn-Irlet, *Mycol. Helv.* 7(2): 50 (1995)
Naucoria laevigata var. *maritima* Bon, *Bull. Soc. Mycol. France* 86(1): 127 (1970)
Ramicola maritima (Bon) Bon, *Doc. Mycol.* 21 (83): 38 (1991)
W: !
H: On decayed culms, stems and debris of *Ammophila arenaria* on dunes.
I: C&D: 365 (as *Ramicola maritima*)
Known from Anglesey (Newborough Warren) and Carmarthenshire (Whiteford Burrows).

centunculus *var.* obscura (Romagn.) Singer, *Agaricales in Modern Taxonomy*, Edn 2: 588 (1962)
Naucoria centunculus var. *obscura* Romagn., *Bull. Soc. Mycol. France* 58: 149 (1944)
Simocybe obscura (Romagn.) D.A. Reid, *Trans. Brit. Mycol. Soc.* 82(3): 224 (1984)
E: !
H: On decayed wood. British material 'on the ground in a dense pile of twigs and branches, in woodland' thus may have been on buried wood.
A single collection from Surrey (Dorking, Ranmore Common) in 1980. Reported from South Hampshire (New Forest) in 1987 but unsubstantiated with voucher material.

haustellaris (Fr.) Watling, *Biblioth. Mycol.* 82: 39 (1981)
Agaricus haustellaris Fr., *Observ. mycol.* 2: 232 (1818)
Crepidotus haustellaris (Fr.) P. Kumm., *Führ. Pilzk.*: 74 (1871)
Naucoria haustellaris (Fr.) Kühner & Romagn., *Flore Analytique des Champignons Supérieurs*: 236 (1953)
Ramicola haustellaris (Fr.) Watling, *Notes Roy. Bot. Gard. Edinburgh* 45(3): 555 (1989) [1988]
Naucoria effugiens Quél., *Mém. Soc. Émul. Montbéliard, Sér. 2*, 5: 319 (1872)
Agaricus rubi Berk., *Engl. fl.* 5(2): 102 (1836)
Crepidotus rubi (Berk.) Sacc., *Syll. fung.* 5: 881 (1887)
Naucoria rubi (Berk.) Singer, *Sydowia* 6: 348 (1952)
Simocybe rubi (Berk.) Singer, *Sydowia* 15: 72 (1962) [1961]
Ramicola rubi (Berk.) Watling, *Notes Roy. Bot. Gard. Edinburgh* 45(3): 556 (1989) [1988]
E: ! **S:** ! **W:** !
H: On wood, usually on small fallen twigs of deciduous trees, especially *Acer pseudoplatnus*, *Fagus* and *Quercus* spp. Occasionally on larger decayed logs and fallen branches and also reported on dead stems of *Rubus fruticosus* agg.
D: BFF6: 100 (as *Ramicola haustellaris*), NM2: 339 (as *R. haustellaris*) **D+I:** B&K5: 304 396 (as *S. rubi*) **I:** C&D: 365 (as *R. haustellaris*)
Rarely reported but apparently widespread. Basidiomes are minute, cryptically coloured and easily overlooked. Forms with the stipe reduced or absent and growing on small fallen twigs were previously assigned to *S. rubi*. Normally bisporic, but a tetrasporic form is also known and recorded from Surrey.

sumptuosa (P.D. Orton) Singer, *Sydowia* 15: 74 (1962)
Naucoria sumptuosa P.D. Orton, *Trans. Brit. Mycol. Soc.* 43(2): 324 (1960)

Ramicola sumptuosa (P.D. Orton) Watling, *Notes Roy. Bot. Gard. Edinburgh* 45(3): 556 (1989) [1988]

E: o **W:** !

H: On decayed wood of deciduous trees, usually large fallen trunks or branches of *Fagus* but also *Betula, Castanea, Corylus, Fraxinus* and *Quercus* spp.

D: Reid84: 228, NM2: 340 (as *Ramicola sumptuosa*) **D+I:** FRIC2: 25-26 15b (as *Naucoria sumptuosa*), B&K5: 306 397 **I:** Ph: 157 (as *N. centunculus*), C&D: 365 (as *R. sumptuosa*)

Occasional but may be locally common, especially in woodland on calcareous soil in southern England. Recorded from Yorkshire southwards and also in Wales. A Scottish collection (Angus: Glenisla) in herb. K has been redetermined as *Simocybe centunculus*.

SIROBASIDIUM Lagerh. & Pat., *J. Bot. (Morot)* 6: 468 (1892)

Tremellales, Sirobasidiaceae

Type: *Sirobasidium sanguineum* Lagerh. & Pat. (= *S. rubrofuscum* (Berk.) P. Roberts)

brefeldianum Møller, *Bot. Mitt. Tropen* 8: 65 (1895)
Sirobasidium intermediae Kund. & M.S. Patil, *Indian Phytopath.* 39(3): 357 (1987)

E: !

H: Parasitic. On ascocarps of *Eutypella leprosa* on dead attached twigs and branches of *Acer pseudoplatanus* and *Ulmus* sp.

Known from North Hampshire, South Devon and West Cornwall.

SISTOTREMA Fr., *Syst. mycol.* 1: 426 (1821)

Stereales, Sistotremataceae

Hydnotrema Link, *Handbuch Gewächse* 3: 298 (1833)
Burgoa Goid. (anam.), *Boll. R. Staz. Patalog. Veget. Roma* 17: 359 (1938)

Type: *Sistotrema confluens* Pers.

alboluteum (Bourdot & Galzin) Bondartsev & Singer, *Ann. Mycol.* 39(1): 47 (1941)
Poria alboluteа Bourdot & Galzin, *Bull. Soc. Mycol. France* 41: 217 (1925)
Mis.: *Trechispora onusta* sensu Wakefield (1952)

E: !

H: On decayed wood of deciduous trees.

D: NM3: 127 **D+I:** CNE7: 1312-1315

A single collection, from East Sussex (Tilgate Forest) in 1928.

albopallescens (Bourdot & Galzin) Bondartsev & Singer, *Ann. Mycol.* 39: 47 (1941)
Poria albopallescens Bourdot & Galzin, *Bull. Soc. Mycol. France* 41: 216 (1925)

E: !

H: On decayed conifer wood.

D: Myc17(2): 62

A single collection from West Cornwall (Croft Pascoe) in 2000.

brinkmannii (Bres.) J. Erikss., *Förh. Kungl. Fysiogr. Sällsk. Lund* 18: 134 (1948)
Odontia brinkmannii Bres., *Ann. Mycol.* 1(1): 88 (1903)
Grandinia brinkmannii (Bres.) Bourdot & Galzin, *Bull. Soc. Mycol. France* 30(2): 252 (1914)
Trechispora brinkmannii (Bres.) D.P. Rogers & H.S. Jacks., *Farlowia* 1(2): 288 (1943)

E: c **S:** c **W:** c **NI:** ! **ROI:** c **O:** Channel Islands: c Isle of Man: c

H: On a wide range of substrata including decayed deciduous and conifer wood and bark, old fungal basidiomes, decayed cloth, discarded shoes, coconut matting, fern debris and decayed *Gladiolus* corms.

D: NM3: 129 **D+I:** CNE7: 1317-1323, B&K2: 174 188

Very common and widespread.

citriforme (M.P. Christ.) K.H. Larss. & Hjortstam, *Mycotaxon* 29: 318 (1987)
Uthatobasidium citriforme M.P. Christ., *Dansk. Bot. Ark.* 19(1): 49 (1959)

Sistotrema subangulisporum K.H. Larss. & Hjortstam, *Mycotaxon* 5(2): 479 (1977)

E: !

H: On decayed conifer wood. British material on *Picea abies*.

D: NM3: 129 **D+I:** CNE7: 1365 (as *Sistotrema subangulisporum*)

A single collection from Shropshire (Oswestry) in 1999.

confluens Pers., *Neues Mag. Bot.* 1: 108 (1794)
Hydnotrema confluens (Pers.) Link, *Handbuch Gewächse* 3: 298 (1833)
Irpex confluens (Pers.) P. Kumm., *Führ. Pilzk.*: 49 (1871)
Sistotrema ericetorum (Bourdot & Galzin) Sacc. & Trotter, *Syll. fung.* 23: 479 (1925)
Hydnum sublamellosum Bull., *Herb. France*: pl. 453 (1790)

E: ! **S:** ! **W:** ?

H: On soil, mosses, and leaf litter; one record on dead culms of *Arundinaria* sp.

D: NM3: 127 **D+I:** CNE7: 1324-1326 **I:** Sow1: 112 (as *Hydnum sublamellosum*)

Rarely reported. Known from widely scattered locations in England and Scotland, and an old record for Wales unsubstantiated with voucher material. Apparently more frequent during the late nineteenth and early twentieth centuries.

coroniferum (Höhn. & Litsch.) Donk, *Fungus* 26: 4 (1956)
Gloeocystidium coroniferum Höhn. & Litsch., *Sitzungsber. Kaiserl. Akad. Wiss., Wien, Math.-Naturwiss. Cl., Abt. 1* 116: 825 (1907)
Corticium coroniferum (Höhn. & Litsch.) Sacc. & Trotter, *Syll. fung.* 21: 402 (1912)
Trechispora coronifera (Höhn. & Litsch.) D.P. Rogers & H.S. Jacks., *Farlowia* 1(2): 282 (1943)

E: ! **S:** ! **W:** !

H: On decayed wood of deciduous and coniferous trees, dead stems of *Rubus fruticosus* and dead basidiomes of polypores.

D: NM3: 128 **D+I:** CNE7: 1328-1331, B&K2: 176 190

Rarely reported but apparently widespread.

coronilla (Höhn.) D.P. Rogers, *Univ. Iowa Stud. Nat. Hist.* 17: 23 (1935)
Corticium coronilla Höhn., *Ann. Mycol.* 4: 291 (1906)

E: !

H: On decayed wood or vegetation. British material on the underside of an old and decayed woollen blanket in beechwoods on calcareous soil.

A single collection from Surrey (Mickleham, Norbury Park) in 1994.

dennisii Malençon, *Kew Bull.* 31(3): 490 (1976)

S: !

H: On charred cones of *Pinus sylvestris*.

D: NM3: 127 **D+I:** CNE7: 1331

Known from two collections from the Orkney Isles (Hoy).

diademiferum (Bourdot & Galzin) Donk, *Fungus* 26: 4 (1956)
Corticium diademiferum Bourdot & Galzin, *Bull. Soc. Mycol. France* 27: 244 (1911)

E: ! **S:** !

H: Recorded from decayed wood of deciduous trees such as *Fraxinus*, on litter of conifers such as *Picea* spp and *Pinus sylvestris*, on dead herbaceous stems, decayed leaves of *Iris pseudacorus*, fern debris and on decayed basidiomes of *Stereum gausapatum*.

D: NM3: 129 **D+I:** CNE7: 1333-1334

Rarely reported. The majority of records and collections are from southern England.

efibulatum (J. Erikss.) Hjortstam, *Corticiaceae of North Europe* 7: 1337 (1984)
Sistotrema commune f. *efibulatum* J. Erikss., *Svensk bot. Tidskr.* 43: 314 (1949)

E: !

H: British material on decayed brash of *Picea* sp. and on dead stems of *Rubus fruticosus* agg.

D: NM3: 129 **D+I:** CNE7: 1336-1337, B&K2: 174 189 (as *Sistotrema commune* f. *efibulatum*)
Only two British collections, from Buckinghamshire (Burnham Beeches) and South Devon (Bovey Tracey), both in 2000.

hispanicum M. Dueñas, Ryvarden & Tellería, *Ruizia* 5: 130 (1988)
   Sistotrema quadrisporum Hallenb. & Hjortstam, *Mycotaxon* 31(2): 442 (1988)
E: ! **S:** !
H: On decayed wood of deciduous trees and shrubs such as *Corylus, Fagus* and *Quercus* spp. Also collected from decayed basidiomes of *Datronia mollis* and *Piptoporus betulinus*.
D: NM3: 128 (as *Sistotrema quadrisporum*)
Rarely reported. Widely scattered records in England and a single Scottish collection.

muscicola (Pers.) S. Lundell, *Fungi Exsiccati Suecici* 29-30: 11 (1947)
   Hydnum muscicola Pers., *Mycol. eur.* 2: 181 (1825)
   Grandinia muscicola (Pers.) Bres., *Bull. Soc. Mycol. France* 30(2): 252 (1914)
E: ! **S:** !
H: On decayed wood of conifers, especially when covered in mosses.
D: NM3: 127 **D+I:** CNE7: 1342-1345
Rarely reported. Known from England (Berkshire, Buckinghamshire, South Somerset and Warwickshire) and Scotland (Aberdeenshire).

oblongisporum M.P. Christ. & Hauerslev, *Dansk. Bot. Ark.* 19(2): 82 (1960)
E: ! **S:** ! **W:** !
H: On decayed leaf litter and wood of deciduous trees, on debris of *Pteridium* and on decayed leaves and debris of *Scirpus lacustris*.
D: NM3: 129 **D+I:** CNE7: 1345-1348
Rarely reported but apparently widespread.

octosporum (Höhn. & Litsch.) Hallenb., Eriksson, Hjortstam & Ryvarden, *Corticiaceae of North Europe* 7: 1349 (1984)
   Corticium octosporum Höhn. & Litsch., *Ann. Mycol.* 4(3): 292 (1906)
   Sistotrema commune J. Erikss., *Svensk bot. Tidskr.* 43: 312 (1949)
E: ! **S:** ! **NI:** ! **O:** Channel Islands: !
H: On decayed wood and litter of deciduous and coniferous trees, and dead woody stems of *Rubus fruticosus*. Also known on decayed herbaceous stems of *Lythrum salicaria* and *Oenanthe crocata*. Rarely reported on mosses.
D: NM3: 130 **D+I:** CNE7: 1349-1351, B&K2: 176 191
Rarely reported but apparently widespread.

pistilliferum Hauerslev, *Friesia* 10(4 - 5): 318 (1975)
E: ! **S:** !
H: On decayed conifer wood. British collections on *Pinus sylvestris*.
D: NM3: 128
A single record from South Devon (Bovey Tracey) in 1991 and one from Perthshire (Black Wood of Rannoch) in 2003.

pteriphilum K.H. Larss. & Hjortstam, *Windahlia* 15: 56 (1985)
E: !
H: On decayed leaf litter of deciduous trees.
D: NM3: 129
Known from South Devon and Surrey.

pyrosporum Hauerslev, *Friesia* 10(4 - 5): 319 (1975)
E: ! **W:** !
H: British material on fallen decayed leaves of *Rubus fruticosus* agg.
D: NM3: 129
Known from scattered localities in England (South Devon) and recently also from Wales (Carmarthenshire).

sernanderi (Litsch.) Donk, *Fungus* 26: 4 (1956)
   Gloeocystidium sernanderi Litsch., *Svensk bot. Tidskr.* 25(3): 437 (1931)

E: !
H: On decayed wood of deciduous trees such as *Fagus, Salix* or *Ulmus* spp, and known from decayed basidiomes of *Ganoderma australe*.
D: NM3: 128 **D+I:** CNE7: 1360-1362
Known from Berkshire, Buckinghamshire, Cumberland, Hertfordshire, Middlesex, South Devon, South Hampshire, South Wiltshire, Surrey, Warwickshire, West Kent and West Sussex.

subtrigonospermum D.P. Rogers, *Iowa State Coll. J. Sci.* 17(1): 22 (1935)
E: ! **S:** !
H: British material on decayed culms and stems of *Phragmites australis*, and wood of *Salix viminalis*.
D: NM3: 129 **D+I:** CNE7: 13654-1365, Myc10(2): 50
A single collection from England (Surrey: Esher Common, Black Pond) in 1995 and also known from Scotland (Shetland).

SISTOTREMASTRUM J. Erikss., *Symb. Bot. Upsal.* 16(1): 62 (1958)
Stereales, Sistotremataceae
Type: *Sistotremastrum suecicum* J. Erikss.

niveocremeum (Höhn. & Litsch.) J. Erikss., *Symb. Bot. Upsal.* 16(1): 62 (1958)
   Corticium niveocremeum Höhn. & Litsch., *Sitzungsber. Kaiserl. Akad. Wiss., Wien, Math.-Naturwiss. Cl., Abt. 1* 117: 1117 (1908)
   Sistotrema niveocremeum (Höhn. & Litsch.) Donk, *Fungus* 26: 4 (1956)
   Paullicorticium niveocremeum (Höhn. & Litsch.) Jülich, *Persoonia* 10(3): 335 (1979)
E: ! **S:** ! **W:** ? **NI:** !
H: On decayed wood, usually of deciduous trees and shrubs such as *Corylus, Fagus* and *Laburnum anagyroides*. Rarely on conifers such as *Pinus sylvestris* and also on fallen and decayed fence palings in woodland.
D: NM3: 130 **D+I:** CNE7: 1376-1377, B&K2: 174 187 (as *Paullicorticium niveo-cremeum*)
Known from England (Buckinghamshire, Cambridgeshire, East Norfolk, East & West Sussex, Hertfordshire, North Hampshire, South Devon, Surrey and Warwickshire), Scotland (Isle of Arran and Isle of Mull) and Northern Ireland (Derry). Reported from Wales but unsubstantiated with voucher material.

suecicum J. Erikss., *Symb. Bot. Upsal.* 16(1): 62 (1958)
   Corticium suecicum Litsch. [*nom. nud.*], in Lundell & Nannfeldt, *Fungi Exsiccati Suecici* 9-10: 24 (1937)
E: ! **O:** Channel Islands: !
H: In woodland, or scrub, most frequently collected on decayed lianas of *Clematis vitalba* but also known from dead stems of *Atropa bella-donna*.
D: NM3: 130 **D+I:** CNE7: 1377-1378
Known from England (Surrey: Esher Common and Mickleham: Norbury Park) and the Channel Islands (Guernsey: Cobo, Le Guet).

SISTOTREMELLA Hjortstam, *Corticiaceae of North Europe* 7: 1379 (1984)
Stereales, Sistotremataceae
Type: *Sistotremella perpusilla* Hjortstam

hauerslevii Hjortstam, *Corticiaceae of North Europe* 7: 1379 (1984)
E: !
H: British material on decayed wood of *Fagus* in old mixed deciduous woodland on calcareous soil.
D: NM3: 130, Myc14(2): 53 **D+I:** CNE7: 1379-1380
A single collection from West Kent (Lullingstone Park) in 1999.

perpusilla Hjortstam, *Corticiaceae of North Europe* 7: 1381 (1984)
E: ! **W:** !

H: On decayed wood of deciduous trees such as *Crataegus monogyna*, *Fraxinus* and *Salix* spp. and also known on conifers such as *Taxus*.
D: NM3: 130 **D+I:** CNE7: 1381
Known from England (Surrey) and Wales (Breconshire and Monmouthshire).

SKELETOCUTIS Kotl. & Pouzar, *Česká Mykol.* 12: 103 (1958)
Poriales, Coriolaceae
Incrustoporia Domański, *Acta Soc. Bot. Poloniae* 32: 737 (1963)
Leptotrimitus Pouzar, *Česká Mykol.* 20: 175 (1966)
Type: *Skeletocutis amorpha* (Fr.) Kotl. & Pouzar

alutacea (J. Lowe) Jean Keller, *Persoonia* 10: 353 (1979)
Poria alutacea J. Lowe, *Mycologia* 38: 202 (1946)
Incrustoporia alutacea (J. Lowe) D.A. Reid, *Rev. Mycol. (Paris)* 33(4): 237 (1969)
E: ! **S:** !
H: On decayed wood of deciduous trees or conifers. British collections on *Fagus* and *Pinus sylvestris*.
D: EurPoly2: 620, NM3: 216
Known from England (Surrey: Mickleham, Norbury Park) and Scotland (Easterness: Loch Garten).

amorpha (Fr.) Kotl. & Pouzar, *Česká Mykol.* 12: 103 (1958)
Polyporus amorphus Fr., *Syst. mycol.* 1: 364 (1821)
Tyromyces amorphus (Fr.) Murrill, *Mycologia* 10: 109 (1918)
Gloeoporus amorphus (Fr.) Killerm., *Nat. Pflanzenfamilien* 2(6) (1928)
Polyporus armeniacus Berk., *Engl. fl.* 5(2): 147 (1836)
Poria armeniaca (Berk.) W.G. Sm., *Syn. Brit. Bas.*: 358 (1908)
Polyporus kymatodes Rostk., in Sturm, *Deutschl. Fl.* Abt. 3, 4: 51 (1838)
E: c **S:** c **W:** c **NI:** ! **O:** Isle of Man: !
H: On dead or decayed wood of conifers, usually large fallen trunks, logs or branches in woodland or on heathland. Known on *Larix*, *Picea* and *Pinus* spp., and especially frequent on *Pinus sylvestris*. A single record on *Fagus* but unsubstantiated with voucher material.
D: EurPoly2: 621, NM3: 215 **D+I:** B&K2: 290 362 **I:** Kriegl1: 578
Common and widespread. Easily recognised especially if the pore surface has developed the distinctive apricot-orange colouration.

carneogrisea A. David, *Naturaliste Canad.* 109(2): 245 (1982)
E: o **S:** ?
H: Usually on dead basidiomes of *Trichaptum abietinum* on wood of conifers, rarely on the wood itself. Also reported on wood of deciduous trees such as *Aesculus*, *Crataegus monogyna* and *Fagus* but unsubstantiated with voucher material and the hosts unverified.
D: 215, EurPoly2: 623 **D+I:** B&K2: 290 363, Myc11(2): 60
Occasional but widespread in southern England (Berkshire, North Somerset, Oxfordshire, South Devon, South Hampshire, Surrey, West Kent and West Suffolk). Reported from Scotland but unsubstantiated with voucher material.

kuehneri A. David, *Naturaliste Canad.* 109(2): 248 (1982)
Mis.: *Skeletocutis tschulymica* sensu Pegler (1973)
E: !
H: On decayed wood of conifers. British collections mostly on *Pinus sylvestris* and rarely on *Picea abies*.
D: EurPoly2: 624, NM3: 216
Known from South Devon, South Hampshire, South Somerset, Surrey and Yorkshire.

nivea (Jungh.) Jean Keller, *Persoonia* 10: 353 (1979)
Polyporus niveus Jungh., *Verh. Batav. Genootsch. Kunst. Wet.* 17: 48 (1839)
Incrustoporia nivea (Jungh.) Ryvarden, *Norweg. J. Bot.* 19: 232 (1972)

Polyporus semipileatus Peck, *Rep. (Annual) New York State Mus. Nat. Hist.* 34: 43 (1881)
Incrustoporia semipileata (Peck) Donk, *Proc. Kon. Ned. Akad. Wetensch. C* 74: 39 (1971)
Tyromyces semipileatus (Peck) Murrill, *North American Flora* 9(1): 35 (1907)
Leptotrimitus semipileatus (Peck) Pouzar, *Česká Mykol.* 20: 171 (1966)
E: c **S:** o **W:** c **NI:** c **ROI:** c **O:** Isle of Man: !
H: On small fallen branches or sticks of deciduous trees. Most often reported on *Corylus*, *Fagus*, *Fraxinus* and *Salix* spp. Also known on *Acer pseudoplatanus*, *Alnus*, *Buddleja* sp., *Castanea*, *Crataegus* spp, *Prunus spinosa*, *Quercus* and *Rosa* spp, *Sambucus nigra*, *Sorbus aria*, *Ulex europaeus* and *Ulmus* spp. Rare on conifers; known on *Abies*, *Larix*, *Picea* and *Pinus* spp.
D: EurPoly2: 629, NM3: 215 **D+I:** Ph: 234 (as *Incrustoporia semipileata*), B&K2: 292 365 **I:** Bon: 317, Kriegl1: 582
Very common and widespread.

subincarnata (Peck) Jean Keller, *Persoonia* 10: 535 (1979)
Poria attenuata var. *subincarnata* Peck, *Rep. (Annual) New York State Mus. Nat. Hist.* 18: 118 (1894)
Poria subincarnata (Peck) Murrill, *Mycologia* 13: 86 (1921)
Incrustoporia subincarnata (Peck) Domański, *Acta Soc. Bot. Poloniae* 32: 737 (1963)
S: !
H: British material on decayed wood of *Pinus sylvestris* in Caledonian pinewoods.
D: EurPoly2: 636
Known from two Scottish sites (Easterness: Aviemore and Westerness: Corrieshalloch Gorge).

vulgaris (Fr.) Niemelä & Y.C. Dai, *Ann. Bot. Fenn.* 34(2): 135 (1997)
Polyporus vulgaris Fr., *Syst. mycol.* 1: 381 (1821)
Poria vulgaris (Fr.) Cooke [*nom. illegit.*, non *P. vulgaris* Gray (1821)], *Grevillea* 14: 109 (1886)
Poria calcea (Berk. & Broome) Cooke, *Grevillea* 14: 109 (1886)
Polyporus calceus Berk. & Broome, *J. Linn. Soc., Bot.* 14: 55 (1884)
Mis.: *Skeletocutis lenis* sensu auct. mult.
E: c **S:** ! **W:** ! **NI:** !
H: On decayed wood of deciduous and trees and conifers, in old woodland. Most frequently on large logs or fallen branches of *Fagus* but also on *Betula*, *Fraxinus*, *Populus*, *Quercus*, *Tilia* and *Ulmus* spp, and the conifers *Picea abies* and *Pinus sylvestris*. Often fruiting inside rot-pockets in the wood.
D: EurPoly2: 626-627 (as *Skeletocutis lenis*)
Common and widespread in England. Most collections were previously referred to *Skeletocutis lenis* which is now known to be a boreal species, mainly on conifers, recorded in Europe only from Sweden, Finland and Russia.

SPARASSIS Fr., *Novit. fl. suec. alt.*: 80 (1819)
Cantharellales, Sparassidaceae
Type: *Sparassis crispa* (Wulfen) Fr.

crispa (Wulfen) Fr., *Syst. mycol.* 1: 465 (1821)
Clavaria crispa Wulfen, *Jacq. Misc. Austr.* 2: 100 t. 14 (1781)
E: c **S:** c **W:** c **NI:** ! **ROI:** c **O:** Channel Islands: !
H: Parasitic then saprotrophic on roots of conifers, usually *Pinus* spp, but also known on *Cedrus*, *Picea*, *Pseudotsuga* and *Sequoia* spp. Continuing to fruit after the death of the host.
D: NM3: 255 **D+I:** Ph: 255, B&K2: 368 478
Common and widespread, but few records from Northern Ireland. English name = 'Cauliflower Fungus'.

spathulata (Schwein.) Fr., *Elench. fung.* 1: 227 (1828)
Merisma spathulatum Schwein., *Schriften Naturf. Ges. Leipzig* 1: 110 (1822)
Sparassis brevipes Krombh., *Naturgetr. Abbild. Schwämme* 3: t. 22 (1834)
Sparassis laminosa Fr., *Sverig Atl. Svamp.*: 64 (1836)
Sparassis crispa var. *laminosa* (Fr.) Quél., *Fl. mycol. France*: 15 (1888)

Sparassis simplex D.A. Reid, *Trans. Brit. Mycol. Soc.* 41: 439 (1958)

E: ! S: ? NI: ?

H: Parasitic. On living roots of deciduous trees, such as *Castanea*, *Fagus* and *Quercus* spp.

D+I: B&K2: 368 479 (as *Sparassis laminosa*) **I:** Kriegl2: 57 (as *S. brevipes*)

Poorly known in Britain. Most reports unsubstantiated with voucher material and many said to be 'with conifers'. Last reported from Northern Ireland in 1883 and from Scotland in 1908, both records unsubstantiated with voucher material.

SPHAEROBOLUS Tode, *Fung. mecklenb. sel.* 1: 43 (1790)
Sclerodermatales, Sphaerobolaceae

Carpobolus Micheli ex Paulet, *Abh. Königl. Akad. Wiss. Berlin* 1808: 181 (1808)

Type: *Sphaerobolus stellatus* Tode

stellatus Tode, *Fung. mecklenb. sel.* 1: 43 (1790)
Carpobolus stellatus (Tode) Desm., *Observ. Bot.-zool.*: 9 (1826)
Lycoperdon carpobolus L., *Sp. pl.*: 1654 (1753)
Nidularia dentata With., *Bot. arr. Brit. pl. ed. 2*, 4: 357 (1796)
Sphaerobolus dentatus (With.) W.G. Sm., *J. Bot.* 41: 280 (1903)
Sphaerobolus tubulosus Fr. & Nordholm, *Symb. gasteromyc.*: 1 (1817)

E: c S: c W: c NI: o ROI: o O: Isle of Man: !

H: On decayed plant debris, weathered dung and decayed wood and straw in woodland and on dunes; increasingly frequent in gardens and parks on flowerbeds mulched with woodchips.

D: NM3: 196 **D+I:** Ph: 255, B&K2: 380 498, BritPuffb: 74-75 **I:** Sow1: 22 (as *Lycoperdon carpobolus*), Kriegl2: 162

Common and widespread.

SPHAGNOMPHALIA Redhead, Moncalvo, Vilgalys & Lutzoni, *Mycotaxon* 82: 162 (2002)
Agaricales, Tricholomataceae

Type: *Sphagnomphalia brevibasidiata* (Singer) Redhead *et al.*

brevibasidiata (Singer) Redhead, Moncalvo, Vilgalys & Lutzoni, *Mycotaxon* 82: 163 (2002)
Clitocybe brevibasidiata Singer, *Ann. Mycol.* 41: 45 (1943)
Omphalina brevibasidiata (Singer) Singer, *Mycologia* 34: 84 (1947)
Omphalia cincta J. Favre, *Beitr. Kryptogamenfl. Schweiz* 10(3): 212 (1948)

S: !

H: In *Sphagnum* moss on moorland in montane areas.

D: NM2: 174 (as *Omphalina brevibasidiata*)

A single collection from Argyll (Glen Orchy) in 1985.

SPONGIPELLIS Pat., *Hyménomyc. Eur.*: 140 (1887)
Poriales, Coriolaceae

Type: *Spongipellis spumeus* (Sowerby) Pat.

delectans (Peck) Murrill, *North American Flora* 9(1): 38 (1907)
Polyporus delectans Peck, *Bull. Torrey Bot. Club* 11: 26 (1884)
Leptoporus bredecelensis Pilát, *Atlas champ. Eur. Polyporaceae I* 3(1): 240 (1938)

E: !

H: On dead or decayed wood of deciduous trees in old woodland or parkland. Usually on large fallen trunks or branches of *Fagus* but also known on *Acer pseudoplatanus*, *Aesculus* and *Fraxinus*.

D: EurPoly2: 643, NM3: 224 **I:** SV28: 57, SV48: 35

Known from Berkshire, North Somerset, Oxfordshire, South Hampshire, Surrey, West Kent and West Sussex.

pachyodon (Pers.) Kotl. & Pouzar, *Česká Mykol.* 19: 77 (1965)
Hydnum pachyodon Pers., *Mycol. eur.* 2: 174 (1825)

Sistotrema pachyodon (Pers.) Fr., *Epicr. syst. mycol.*: 520 (1838)
Irpex pachyodon (Pers.) Quél., *Fl. mycol. France*: 377 (1888)

E: ! S: ?

H: On living trunks of deciduous trees. British material on *Fagus* and *Quercus ilex*.

D: EurPoly2: 645, NM3: 224 **D+I:** B&K2: 322 411 **I:** Kriegl1: 583

Known from England (Buckinghamshire, South Hampshire and West Sussex) and a single collection in herb. K from Scotland (Grampian) in 1879.

spumeus (Sowerby) Pat., *Essai taxon.*: 84 (1900)
Boletus spumeus Sowerby, *Col. fig. Engl. fung.* 2: pl. 211 (1799)
Polyporus spumeus (Sowerby) Fr., *Syst. mycol.* 1: 358 (1821)
Leptoporus spumeus (Sowerby) Pilát, *Atlas champ. Eur. Polyporaceae I* 3(1): 237 (1938)
Tyromyces spumeus (Sowerby) Imazeki, *Bull. Tokyo Sci. Mus.* 6: 84 (1943)

E: ! NI: ?

H: Weakly parasitic, then saprotrophic. On living and dead wood of deciduous trees. Mostly reported on *Fagus* in woodland or parkland, with several records on *Malus* spp, or cultivars in gardens. Also known on *Acer grosseri*, *Fraxinus*, *Populus*, *Salix fragilis* and *Ulmus* spp.

D: EurPoly2: 646, NM3: 224 **D+I:** Ph: 234, B&K2: 324 412 **I:** Sow2: pl. 211 (as *Boletus spumeus*)

Known from England (Bedfordshire, Berkshire, Herefordshire, Middlesex, North Somerset, Northamptonshire, South Essex, South Hampshire, South Wiltshire, Surrey, Warwickshire, West Kent, West Norfolk and West Sussex). Reported from Northern Ireland but not since 1884 (Tyrone) and 1885 (Down) and no material has been traced.

SQUAMANITA Imbach, *Mitt. Naturf. Ges. Luzern* 15: 81 (1946)
Agaricales, Agaricaceae

Coolia Huijsman [*nom. nud.*], *Meded. Ned. Mycol. Ver.* 28: 59 (1943)

Type: *Squamanita schreieri* Imbach

contortipes (A.H. Sm. & D.E. Stuntz) Heinem. & Thoen, *Bull. Soc. Mycol. France* 89: 30 (1973)
Cystoderma contortipes A.H. Sm. & D.E. Stuntz, *Sydowia, Beih.* 1: 46 (1957)
Squamanita scotica Bas [*nom. prov.*], *Persoonia* 3(3): 341 (1965)

S: !

H: Parasitic. The type collection was on an indeterminate species of *Galerina*.

D: BFF8: 46

A single fruitbody on an indeterminate host from Easterness (Rothiemurchus Forest) in 1957, provisionally described as *S. scotica*.

paradoxa (A.H. Sm. & Singer) Bas, *Persoonia* 3: 348 (1965)
Cystoderma paradoxum A.H. Sm. & Singer, *Mycologia* 40: 454 (1948)

E: ! S: ! W: ! NI: !

H: Parasitic on *Cystoderma* spp. British material is all on *C. amianthinum*.

D: BFF8: 47 **D+I:** Reid [*Bull. Brit. Mycol. Soc.* 17: 111-113 (1983)] **I:** FM6(1): 11

First collected in Scotland (Mull) in 1969. Five widely scattered collections since.

pearsonii Bas, *Persoonia* 3(3): 345 (1965)
Mis.: *Tricholoma odoratum* sensu Pearson [TBMS 35: 99 (1952)]
Mis.: *Squamanita odorata* sensu NCL

S: !

H: Parasitic. On an unknown agaric host in grass in Caledonian pinewoods.

D: BFF8: 45-46 **D+I:** B&K4: 224 262 **I:** FM6(1): 10

Two collections, from Easterness in 1950 and Aberdeenshire in 2004.

STAGNICOLA Redhead & A.H. Sm., *Can. J. Bot.* 64: 645 (1986)
Cortinariales, Cortinariaceae
Type: *Stagnicola perplexa* (P.D. Orton) Redhead & A.H. Sm.

perplexa (P.D. Orton) Redhead & A.H. Sm., *Canad. J. Bot.* 64(3): 645 (1986)
  Phaeocollybia perplexa P.D. Orton, *Kew Bull.* 31(3): 713 (1976)
  Naucoria cidaris var. *minor* (Fr.) Sacc., *Syll. fung.* 5: 831 (1887)
S: !
H: On woodchips, often deeply buried in soil, in Caledonian pinewoods.
D: BFF7: 93-94
Known from Easterness and Mid-Perthshire.

STECCHERINUM Gray, *Nat. arr. Brit. pl.* 1: 651 (1821)
Stereales, Steccherinaceae
  Odontia Fr. [*nom. illegit.*, non *Odontia* Pers. (1794)], *Fl. Scan.*: 340 (1835)
  Mycoleptodon Pat., *Cat. pl. cell. Tunisie*: 54 (1897)
Type: *Steccherinum ochraceum* (Pers.) Gray

albidum Legon & P. Roberts, *Czech Mycol.* 54(1-2): 7 (2002)
E: !
H: On decayed wood of *Fagus* in old deciduous woodland.
Known only from the type collection from West Sussex (Ebernoe Common).

bourdotii Saliba & A. David, *Cryptog. Mycol* 9(2): 100 (1988)
E: ! **W:** !
H: On decayed wood of deciduous trees. British material on *Fagus* and *Quercus* spp.
I: Kriegl1: 327
Previously treated as a form of *Steccherinum ochraceum* forming small reflexed pilei. Known from England (Bedfordshire: Maulden Woods and South Somerset: Dommett Wood) and Wales (Carmarthenshire: Castell Carreg Cennan). Probably also this species in Buckinghamshire and North Somerset.

fimbriatum (Pers.) J. Erikss., *Symb. Bot. Upsal.* 16: 134 (1958)
  Sistotrema fimbriatum Pers., *Syn. meth. fung.*: 553 (1801)
  Odontia fimbriata (Pers.) Fr., *Observ. mycol.* 2: 16 (1818)
  Hydnum fimbriatum (Pers.) Fr., *Syst. mycol.* 1: 421 (1821)
  Mycoleptodon fimbriatum (Pers.) Bourdot & Galzin, *Hyménomyc. France*: 441 (1928)
E: c **S:** ! **W:** ! **NI:** ! **ROI:** o
H: On fallen and decayed wood of deciduous trees in old woodland. Especially common on *Fraxinus* but also known on *Acer pseudoplatanus, Betula* spp, *Fagus, Populus gileadensis* and *Sorbus aucuparia.* Single collections on fallen lianas of *Clematis vitalba,* dead stems of *Rumex obtusifolius* and fallen wood of *Pinus* sp.
D: NM3: 218 **D+I:** CNE7: 1389-1393, B&K2: 178 193 **I:** Kriegl1: 329
Common and widespread in southern England, especially in areas of calcareous soil. Apparently less common elsewhere.

ochraceum (Pers.) Gray, *Nat. arr. Brit. pl.* 1: 651 (1821)
  Hydnum ochraceum Pers., *Observ. mycol.* 1: 73 (1799)
  Mycoleptodon ochraceum (Pers.) Pat., *Essai taxon.*: 116 (1900)
  Acia denticulata (Pers.) P. Karst., *Meddeland. Soc. Fauna Fl. Fenn.* 5: 42 (1880)
  Hydnum denticulatum Pers., *Mycol. eur.* 2: 181 (1825)
  Hydnum pudorinum Fr., *Elench. fung.* 1: 133 (1828)
E: o **S:** o **W:** o **NI:** ! **ROI:** ! **O:** Channel Islands: ! Isle of Man: !
H: On decayed wood of deciduous trees, usually fallen sticks or twigs, most often on *Fagus* but also known on *Acer*

pseudoplatanus, Betula pendula, Buddleja davidii, Corylus, Quercus robur and *Ulex europaeus.*
D: NM3: 218 **D+I:** CNE7: 1396-1400, B&K2: 178 194 **I:** Kriegl1: 331
Occasional but widespread, and may be locally common in southern England. Microscopically indistinguishable from *Junghuhnia nitida* from which it differs in its hydnoid basidiome.

STEPHANOSPORA Pat., *Bull. Soc. Mycol. France* 30: 349 (1914)
Stereales, Stephanosporaceae
Type: *Stephanospora caroticolor* (Berk.) Pat.

caroticolor (Berk.) Pat., *Bull. Soc. Mycol. France* 30: 349 (1914)
  Hydnangium caroticolor Berk., *Ann. Mag. Nat. Hist., Ser. 1* [*Notices of British Fungi no. 299*] 13: 351 (1844)
  Octaviania caroticolor (Berk.) Corda, *Icon. fung.* 6: 36 (1854)
E: !
H: Epigeous on calcareous loam soils, often associated with *Fagus* or *Taxus* and a single record under *Cupressus macrocarpa.*
D+I: B&K2: 376 491, BritTruff: 150-152 8 E **I:** Kriegl2: 190
Rarely reported. Known as far north as Yorkshire but mostly southern.

STEREOPSIS D.A. Reid, *Nova Hedwigia Beih.* 18: 290 (1965)
Stereales, Podoscyphaceae
Type: *Stereopsis radicans* (Berk.) D.A. Reid

vitellina (Plowr.) D.A. Reid, *Nova Hedwigia Beih.* 18: 326 (1965)
  Thelephora vitellina Plowr., *J. Bot.* 39: 385 (1901)
  Cotylidia vitellina (Plowr.) S. Lundell, *Fungi Exsiccati Suecici*: 52 (1947)
S: !
H: On acidic peaty soil (often with, or near to, *Polytrichum* spp) and rarely on decayed conifer wood. British collections only from Caledonian pinewoods.
D: NM3: 193 **D+I:** CNE7: 1414-1415
Rarely reported.

STEREUM Pers., *Neues Mag. Bot.* 1: 110 (1794)
Stereales, Stereaceae
  Haematostereum Pouzar, *Česká Mykol.* 13: 13 (1959)
Type: *Stereum hirsutum* (Willd.) Gray

gausapatum (Fr.) Fr., *Hymenomyc. eur.*: 638 (1874)
  Thelephora gausapata Fr., *Elench. fung.* 1: 171 (1828)
  Haematostereum gausapatum (Fr.) Pouzar, *Česká Mykol.* 13: 13 (1959)
  Stereum cristulatum Quél., *Mém. Soc. Émul. Montbéliard, Sér. 2,* 5: 15 (1872)
  Stereum hirsutum var. *cristulatum* (Quél.) Massee, *Grevillea* 19: 65 (1891)
  Stereum quercinum Potter, *Trans. Engl. Arbor. Soc.*: 7 (1901)
  Thelephora spadicea Fr., *Elench. fung.* 1: 176 (1828)
  Mis.: *Stereum spadiceum* sensu auct.
E: c **S:** c **W:** c **NI:** c **ROI:** c **O:** Channel Islands: ! Isle of Man: o
H: Weakly parasitic, forming a white pocket rot, then saprotrophic. Almost exclusively on branches of *Quercus* spp. Single collections supposedly on *Alnus, Castanea, Fagus* and *Malus* spp in herb. K but the hosts are unverified.
D: NM3: 189 **D+I:** Ph: 236-237, CNE7: 1418-1420, B&K2: 182 199 **I:** Kriegl1: 333
Very common and widespread.

hirsutum (Willd.) Gray, *Nat. arr. Brit. pl.* 1: 653 (1821)
  Thelephora hirsuta Willd., *Fl. berol. prodr.*: 397 (1787)
  Helvella acaulis Huds., *Fl. angl.,* Edn 2, 2: 633 (1778)
  Auricularia aurantiaca Schumach., *Enum. pl.* 2: 398 (1803)
  Boletus auriformis Bolton, *Hist. fung. Halifax* 2: 82 (1788)

Auricularia reflexa Bull., *Herb. France*: pl. 274 (1787)

E: c **S:** c **W:** c **NI:** c **ROI:** c **O:** Channel Islands: c Isle of Man: c

H: Weakly parasitic then saprotrophic. On living, fallen and decayed wood of many species of deciduous trees and shrubs. Reported on conifers such as *Picea* and *Pinus* spp, and *Taxus* but unsubstantiated with voucher material and the hosts are unverified.

D: NM3: 190 **D+I:** Ph: 236-237, CNE7: 1421-1427, B&K2: 182 200 **I:** Sow1: 27 (as *Auricularia reflexa*)

Very common and widespread.

rameale (Pers.) Burt [*nom. illegit.*], *Rep. (Annual) Missouri Bot. Gard.* 7: 169 (1920)

Thelephora hirsuta δ *ramealis* Pers., *Syn. meth. fung.*: 570 (1801)

Mis.: *Stereum complicatum* sensu NM3 and sensu auct. Brit.

Mis.: *Stereum ochraceoflavum* sensu Jülich (1984) and sensu auct. Brit.

Mis.: *Stereum sulphuratum* sensu auct. Brit.

E: ! **S:** ! **W:** ! **NI:** ! **O:** Channel Islands: ! Isle of Man: !

H: On fallen wood of deciduous trees, most often on thin sticks and twigs of *Quercus* spp, but also reported on *Acer pseudoplatnus, Alnus, Betula, Fagus, Rosa* spp and *Ulex europaeus.*

D: B&K2: 182 281 (as *Stereum ochraceoflavum*)

Often reported but rarely collected and many of the supposed collections appear to be depauperate specimens of *Stereum hirsutum* on a nutrient-poor substratum. N.B. This is an illegitimate later homonym of *Stereum rameale* (Berk.) Massee in J. Linn. Soc. (Bot.) 27:187 (1890) but there is no available legitimate name.

rugosum (Pers.) Fr., *Epicr. syst. mycol.*: 552 (1838)

Thelephora rugosa Pers., *Syn. meth. fung.*: 569 (1801)

Haematostereum rugosum (Pers.) Pouzar, *Česká Mykol.* 13: 13 (1959)

Stereum coryli Pers., *Observ. mycol.* 1: 35 (1796)

Thelephora coryli (Pers.) Pers., *Mycol. eur.* 1: 126 (1822)

Thelephora laurocerasi Berk., *Engl. fl.* 5(2): 173 (1836)

Stereum stratosum Berk. & Broome, *Ann. Mag. Nat. Hist., Ser. 5 [Notices of British Fungi no. 2027]* 12: 374 (1883)

E: c **S:** c **W:** c **NI:** c **ROI:** c **O:** Channel Islands: ! Isle of Man: c

H: Weakly parasitic then saprotrophic. On living and fallen trunks of deciduous trees and shrubs, especially on old coppiced trunks of *Corylus* and *Fagus*. Rarely reported on conifers such as *Larix* and *Pinus* spp, but unsubstantiated with voucher material and the hosts are unverified.

D: NM3: 190 **D+I:** Ph: 236-237, CNE7: 1428-1429, B&K2: 184 202 **I:** Kriegl1: 337

Very common and widespread. Records from coniferous hosts are probably misidentified *S. sanguinolentum.*

sanguinolentum (Alb. & Schwein.) Fr., *Epicr. syst. mycol.*: 549 (1838)

Thelephora sanguinolenta Alb. & Schwein., *Consp. fung. lusat.*: 274 (1805)

Merulius sanguinolentus (Alb. & Schwein.) Spreng., *Syst. Veg.*: 468 (1827)

Haematostereum sanguinolentum (Alb. & Schwein.) Pouzar, *Česká Mykol.* 13: 13 (1959)

E: c **S:** c **W:** c **NI:** c **ROI:** c **O:** Channel Islands: c Isle of Man: c

H: On fallen, decayed wood of conifers such as *Abies, Larix, Picea* and *Pinus* spp. Rarely reported on deciduous hosts such as *Betula, Fagus, Salix* spp and *Sambucus nigra* but these are all unsubstantiated with voucher material and the hosts are unverified.

D: NM3: 189 **D+I:** CNE7: 1430-1433, B&K2: 184 203

Common and widespread. Records on deciduous hosts are probably misidentified *S. rugosum.*

subtomentosum Pouzar, *Česká Mykol.* 18: 147 (1964)

Stereum ochroleucum ssp. *arcticum* Fr., *Hymenomyc. eur.*: 639 (1874)

Mis.: *Stereum insignitum* sensu Rea (1922)

E: c **W:** c **NI:** !

H: On fallen wood of deciduous trees, often large logs, fallen trunks and branches, in old woodland. Common on *Acer pseudoplatanus* and *Fagus* and not infrequent on *Alnus* and *Salix* spp. Also known on *Betula* spp and *Corylus.*

D: TBMS52(2): 328, NM3: 190 **D+I:** CNE7: 1434-1435, B&K2: 184 204 **I:** Kriegl1: 339

Common and widespread throughout England and Wales. Once considered a rarity but has spread rapidly in recent years.

STIGMATOLEMMA Kalchbr., *Grevillea* 10: 104 (1882)

Agaricales, Tricholomataceae

Type: *Stigmatolemma incanum* Kalchbr.

poriiforme (Pers.) W.B. Cooke, *Sydowia, Beih.* 4: 128 (1961)

Peziza anomala γ *poriiformis* Pers., *Syn. meth. fung.*: 656 (1801)

Peziza poriiformis (Pers.) Fr., *Syst. mycol.* 2: 106 (1823)

Solenia poriiformis (Pers.) Fuckel, *Jahrb. Nassauischen Vereins Naturk.* 25-26: 290 (1871)

Cyphella poriiformis (Pers.) Bourdot & Galzin, *Hyménomyc. France*: 163 (1928)

Porotheleum poriiforme (Pers.) W.B. Cooke, *Mycologia* 49: 688 (1957)

Solenia urceolata Fr., *Elench. fung.* 2: 28 (1830)

Stigmatolemma urceolatum (Fr.) Donk, *Persoonia* 2(3): 341 (1962)

Cyphella brunnea W. Phillips, *Grevillea* 13: 49 (1884)

E: ! **S:** ! **W:** !

H: On decayed wood of deciduous trees, usually on large fallen trunks or branches of *Fagus* in old woodland, but also reported on *Corylus* and *Quercus* spp.

D: NM2: 181 (as *Stigmatolemma urceolatum*) **D+I:** B&K2: 206 235 (as *S. urceolatum*)

Rarely reported but apparently widespread. Basidiomes are minute and cryptically coloured (greyish-brown to lead grey) thus easily overlooked, even when crowded together.

STROBILOMYCES Berk., *Hooker's J. Bot. Kew Gard. Misc.* 3: 78 (1851)

Boletales, Strobilomycetaceae

Type: *Strobilomyces strobilaceus* (Scop.) Berk.

strobilaceus (Scop.) Berk., *Hooker's J. Bot. Kew Gard. Misc.* 3: 78 (1851)

Boletus strobilaceus Scop., *Annus Historico-naturalis* 4: 148 (1770)

Boletus strobiliformis Dicks., *Fasc. pl. crypt. brit.* 1: 17 (1785)

Boletus strobiliformis Vill. [*nom. illegit.*, non *B. strobiliformis* Dicks. (1785)], *Hist. pl. Dauphiné* 3: 1039 (1789)

Strobilomyces strobiliformis (Vill.) Beck, *Z. Pilzk.* 2: 148 (1923)

Boletus floccopus Vahl, *Fl. Danica*: t. 1252 (1799)

Strobilomyces floccopus (Vahl) P. Karst., *Bidrag Kännedom Finlands Natur Folk* 37: 16 (1882)

Boletus cinereus Pers., *Syn. meth. fung.*: 504 (1801)

E: ! **S:** ! **W:** ! **NI:** !

H: On soil, usually with *Fagus* and less often *Carpinus* in mixed deciduous woodland, occasionally with conifers such as *Pinus* spp in mixed woodland.

D: NM2: 72 **D+I:** Ph: 206, B&K3: 50 2 **I:** C&D: 421, Kriegl2: 293

Occasional but apparently widespread, and may be locally abundant in areas such as the Welsh borders (Severn Valley). Not reported from Ireland (Down) since 1880. English name = 'Old Man of the Woods'.

STROBILURUS Singer, *Persoonia* 2: 409 (1962)

Agaricales, Tricholomataceae

Type: *Strobilurus conigenoides* (Ellis) Singer

esculentus (Wulfen) Singer, *Persoonia* 2: 414 (1962)

Agaricus esculentus Wulfen, in Jacquin, *Jacq. Misc. Austr.*: pl. 14 (1781)

Collybia esculenta (Wulfen) P. Kumm., *Führ. Pilzk.*: 115 (1871)

Marasmius esculentus (Wulfen) P. Karst, *Bidrag Kännedom Finlands Natur Folk* 48: 103 (1889)

Pseudohiatula esculenta (Wulfen) Singer, *Lilloa* 22: 320 (1951) [1949]

Pseudohiatula conigena var. *esculenta* (Wulfen) M.M. Moser, *Kleine Kryptogamenflora*: 89 (1953)

Mis.: *Collybia tenacella* sensu Lange (FlDan2: 13 & pl. 44F)

E: o **S:** o **W:** ! **NI:** ! **ROI:** !

H: On buried and partially decayed cones of *Picea* spp and rarely on cones of *Pseudotsuga menziesii* in woodland and plantations. Fruiting in spring.

D: NM2: 182, BFF8: 144-145, FAN4: 179 **D+I:** B&K3: 318 405 **I:** Bon: 177, C&D: 233, Kriegl3: 517

Occasional but widespread. Virtually restricted to *Picea* spp. Records on *Pinus* spp are erroneous.

stephanocystis (Hora) Singer, *Persoonia* 2: 409 (1962)

Collybia stephanocystis Kühner & Romagn. [*nom. inval.*], *Flore Analytique des Champignons Supérieurs*: 94 (1953)

Pseudohiatula stephanocystis Hora, *Trans. Brit. Mycol. Soc.* 43(2): 455 (1960)

Marasmius esculentus ssp. *pini* Singer, *Ann. Mycol.* 41: 133 (1943)

Mis.: *Marasmius conigenus* sensu auct.

Mis.: *Marasmius esculentus* sensu Rea (1922)

Mis.: *Agaricus stolonifer* sensu auct.

E: ! **S:** o **W:** ! **NI:** !

H: On buried and partially decayed cones of *Pinus* spp, and a single record (with voucher material of the host) on *Wellingtonia*. Often fruiting in spring.

D: NM2: 182, BFF8: 145-146, FAN4: 179-180 **D+I:** B&K3: 320 406 **I:** C&D: 233

Rarely reported but apparently widespread. Occasional in Scotland *fide* R. Watling (pers. comm.).

tenacellus (Pers.) Singer, *Persoonia* 2: 409 (1962)

Agaricus tenacellus Pers., *Observ. mycol.* 1: 50 (1796)

Collybia tenacella (Pers.) P. Kumm., *Führ. Pilzk.*: 114 (1871)

Marasmius tenacellus (Pers.) J. Favre, *Schweiz. Z. Pilzk.* 17: 166 (1939)

Pseudohiatula tenacella (Pers.) Métrod, *Rev. Mycol. (Paris)* 17(1): 86 (1952)

E: o **S:** o **W:** o **NI:** ! **ROI:** o **O:** Isle of Man: !

H: On buried and partially decayed cones of *Pinus* spp, usually *Pinus sylvestris* in Britain.

D: NM2: 182, BFF8: 146, FAN4: 180-181 **D+I:** Ph: 76-77 (as *Pseudohiatula tenacella*), B&K3: 320 407 **I:** C&D: 233, Kriegl3: 519

Occasional but widespread. Reported on cones of *Picea* spp, but unsubstantiated with voucher material and the host is unverifed.

STROPHARIA (Fr.) Quél., *Mém. Soc. Émul. Montbéliard, Sér. 2,* 5: 141 (1872)

Agaricales, Strophariaceae

Geophila Quél., *Enchir. fung.*: 111 (1886)

Type: *Stropharia inuncta* (Fr.) Quél.

aeruginosa (Curtis) Quél., *Mém. Soc. Émul. Montbéliard, Sér. 2,* 5: 141 (1872)

Agaricus aeruginosus Curtis, *Cat. Pl. London* 2: pl. 309 (1774)

Pratella aeruginosa (Curtis) Gray, *Nat. arr. Brit. pl.* 1: 626 (1821)

E: o **S:** o **W:** o **NI:** o **ROI:** ! **O:** Channel Islands: !

H: On soil in wooded areas, gardens and parkland.

D: BFF5: 61, NM2: 268, FAN4: 53-53 **D+I:** B&K4: 350 453 **I:** Sow3: 264 (as *Agaricus aeruginosus*), C&D: 354

Occasional and apparently widespread, but many records may refer to *Stropharia caerulea*. Records with nettles and unsubstantiated with voucher material are likely to be this latter species.

albonitens (Fr.) P. Karst., *Bidrag Kännedom Finlands Natur Folk* 32: 490 (1879)

Agaricus albonitens Fr., *Monogr. hymenomyc. Suec.* 1: 415 (1857)

E: ! **S:** !

H: On soil in parkland, woodland areas and on waste ground.

D: NBA9: 569, NM2: 268, FAN4: 57 (as *Psilocybe albonitens*) **D+I:** B&K4: 352 454 **I:** FlDan4: 141B

Only four records, from England (South Essex and West Gloucestershire) and Scotland (Perthshire).

aurantiaca (Cooke) M. Imai, *J. Fac. Agric. Hokkaido Imp. Univ.* 43: 267 (1938)

Agaricus squamosus f. *aurantiacus* Cooke, *Handb. Brit. fung.*, Edn 2: 199 (1887)

Stropharia percevalii var. *aurantiaca* (Cooke) Sacc., *Syll. fung.* 5: 1016 (1887)

Stropharia squamosa var. *aurantiaca* (Cooke) Massee, *Brit. fung.-fl.* 1: 402 (1892)

E: o **S:** ! **W:** ! **NI:** o **O:** Channel Islands: !

H: On soil and decayed woodchips or mulch in gardens and parks, also (rarely) in more natural woodland.

D: BFF5: 66, FAN4: 64 (as *Psilocybe aurantiaca*) **D+I:** FRIC1: 29-30 8, Ph: 172-173, Myc12(4): 180 **I:** C&D: 356

Occasional but widespread and increasingly reported. Probably an alien which is now spreading rapidly, facilitated by the increasing use of woodchip mulch in parks and gardens. We here adopt the generally accepted interpretation of this name, but it is likely that the taxon Cooke illustrated was in fact *S. squamosa* var. *thrausta*. There appears to be no appropriate epithet for this species, which may need to be described as new.

caerulea Kreisel, *Sydowia, Beih.* 8: 229 (1980) [1979]

Agaricus politus Bolton [*nom. illegit.*, non *A. politus* Pers. (1801)], *Hist. fung. Halifax* 1: 30 (1788)

Mis.: *Stropharia cyanea* sensu Orton (NBA6)

E: c **S:** o **W:** ! **NI:** o **ROI:** o

H: On rich soils in woodland areas, waste places, dunes and on flowerbeds in parks and gardens, especially where mulched with woodchips or shredded bark. Often associated with *Urtica dioica*.

D: BFF5: 62, NM2: 267 (as *Stropharia cyanea*), FAN4: 54-55, NBA6: 151 (as *S. cyanea*) **D+I:** Myc1(4): 175 (as *S. cyanea*), B&K4: 352 455 **I:** C&D: 354

Occasional but widespread. Previously not distinguished from, and frequently misidentified as, *Stropharia aeruginosa*.

coronilla (Bull.) Quél., *Mém. Soc. Émul. Montbéliard, Sér. 2,* 5: 110 (1872)

Agaricus coronillus Bull., *Herb. France*: pl. 597 f. 1 (1793)

Agaricus obturatus Fr., *Syst. mycol.* 1: 283 (1821)

Stropharia obturata (Fr.) Quél., *Mém. Soc. Émul. Montbéliard, Sér. 2,* 5: 110 (1872)

E: o **S:** ! **W:** ! **ROI:** ! **O:** Channel Islands: !

H: On soil amongst grass in pastures, on heaths and on lawns.

D: BFF5: 59-60, NM2: 268, FAN4: 59-60 **D+I:** Ph: 172-173, B&K4: 352 456 **I:** FlDan4: 141F, C&D: 354

Occasional; widespread but apparently more frequent in southern areas.

halophila Pacioni, *Trans. Brit. Mycol. Soc.* 91(4): 579 (1988)

E: ! **W:** !

H: On sand (often expanding below the surface) in dunes, sometimes with *Ammophila arenaria*.

D: FAN4: 60-61 (as *Psilocybe halophila*)

Known from England (North Devon: Croyde Bay and West Norfolk: Holkham) and Wales (Cardiganshire: Ynyslas).

hornemannii (Fr.) S. Lundell & Nannf., *Fungi Exsiccati Suecici* 1: 7 (1934)

Agaricus hornemannii Fr., *Observ. mycol.* 2: 13 (1818)

Agaricus depilatus Pers., *Syn. meth. fung.*: 408 (1801)

Stropharia depilata (Pers.) Fr., *Hymenomyc. eur.*: 283 (1874)

S: r

H: On decayed wood of *Pinus sylvestris* in Caledonian pinewoods.

D: BFF5: 67, NM2: 268, FAN4: 58 **I:** C&D: 356
This is a large and distinctive species, unlikely to be overlooked and thus probably genuinely rare. Common in Scandinavia.

inuncta (Fr.) Quél., *Mém. Soc. Émul. Montbéliard, Sér. 2,* 5: 110 (1872)
    Agaricus inunctus Fr., *Elench. fung.* 1: 40 (1828)
    Agaricus inunctus f. *upsaliensis* Fr., *Epicr. syst. mycol.*: 219 (1838)
    Stropharia inuncta f. *upsaliensis* (Fr.) Sacc., *Syll. fung.* 5: 1014 (1887)
    Stropharia inuncta var. *upsaliensis* (Fr.) Rea, *Brit. basidiomyc.*: 127 (1922)
    Agaricus inunctus f. *lundensis* Fr., *Epicr. syst. mycol.*: 219 (1838)
    Stropharia inuncta f. *lundensis* (Fr.) Sacc., *Syll. fung.* 5: 1014 (1887)
    Stropharia inuncta var. *lundensis* (Fr.) Rea, *British Basidiomycetae*: 126 (1922)
    Agaricus inunctus var. *pallidus* Berk. & Broome, *Ann. Mag. Nat. Hist., Ser. 5 [Notices of British Fungi no. 1875]* 7 (1881)
    Stropharia inuncta var. *pallida* (Berk. & Broome) W.G. Sm., *Syn. Brit. Bas.*: 176 (1908)
E: o **S:** ! **W:** o **NI:** ! **ROI:** !
H: On soil amongst grass or mosses, on heathland and pasture, also on roadside verges.
D: BFF5: 65, NM2: 268, FAN4: 57 **D+I:** B&K4: 354 457 **I:** FlDan4: 141E, C&D: 354
Occasional but widespread. The comment by Watling (BFF5: 65) that 'it sometimes looks like a viscid *Psathyrella*' is particularly apt.

luteonitens (Fr.) Quél., *Mém. Soc. Émul. Montbéliard, Sér. 2,* 5: 112 (1872)
    Agaricus luteonitens Fr., *Syst. mycol.* 1: 511 (1821)
    Psilocybe luteonitens (Fr.) Park.-Rhodes, *Trans. Brit. Mycol. Soc.* 34(3): 364 (1951)
E: ! **W:** ! **ROI:** !
H: On dungy soil amongst grass.
D: BFF7: 53 (as *Psilocybe luteonitens*), FAN4: 63 (as *P. luteonitens*) **I:** FlDan4: 141A, C&D: 357 (as *P. luteonitens*)
Only recorded six times during the nineteenth and twentieth centuries. Apparently widespread.

melanosperma (Bull.) Gillet, *Hyménomycètes*: 579 (1878)
    Agaricus melanospermus Bull., *Herb. France*: pl. 540 (1792)
E: ! **S:** ? **W:** ?
H: On soil, amongst grass at woodland edges, apparently also in fields, and recently on woodchip mulch in a garden.
D: BFF5: 59, NM2: 268, FAN4: 59 (as *Psilocybe melanosperma*) **D+I:** B&K4: 354 458 **I:** Cooke 559 (536) Vol. 4 (as *Agaricus melaspermus*)
Only seven records since 1887, most of which are unsubstantiated with voucher material. A single confirmed collection (Leicestershire: Coalville) in 2002.

percevalii (Berk. & Broome) Sacc., *Syll. fung.* 5: 1016 (1887)
    Agaricus percevalii Berk. & Broome, *Ann. Mag. Nat. Hist., Ser. 5 [Notices of British Fungi no. 1767]* 3: 206 (1879)
    Psilocybe percevalii (Berk. & Broome) P.D. Orton, *Notes Roy. Bot. Gard. Edinburgh* 29(1): 80 (1969)
    Mis.: *Stropharia magnivelaris* sensu NCL & BFF5
E: r
H: On or near decayed wood (stumps etc.) or woodchip mulch.
D: BFF5: 53-54 (as *Psilocybe percevalii*), NM2: 265, FAN4: 66-67 (as *P. percevalii*) **I:** FM6: 7, Cooke 554 (550) Vol. 4 (1885)
Rare. A large and distinctive agaric unlikely to be overlooked, but only three traceable collections.

pseudocyanea (Desm.) Morgan, *J. Mycol.* 14: 74 (1908)
    Agaricus pseudocyaneus Desm., *Cat. pl. omises botanogr. Belgique*: 22 (1823)
    Agaricus albocyaneus Fr., *Epicr. syst. mycol.*: 219 (1838)
    Stropharia albocyanea (Fr.) Quél., *Mém. Soc. Émul. Montbéliard, Sér. 2,* 5: 255 (1872)
    Agaricus worthingtonii Fr., *J. Bot.* 11: 204 (1873)

Stropharia worthingtonii (Fr.) Sacc., *Syll. fung.* 5: 1017 (1887)
E: o **S:** o **W:** o **NI:** o **ROI:** !
H: On soil, often amongst long grass, especially on acidic soil.
D: BFF5: 63-63, NM2: 268, FAN4: 55-56 (as *Psilocybe pseudocyanea*), NBA6: 152 **I:** FlDan4: 140B (as *Stropharia albocyanea*), C&D: 354
Occasional but widespread. Frequently misidentified as *S. caerulea* but distinguishable by the strong peppery smell of the basidiomes.

rugosoannulata Murrill, *Mycologia* 14: 139 (1922)
    Stropharia ferrii Bres., *Stud. Trent., ser. 2* 7: 4 (1926)
E: r
H: On soil in gardens or on dunes. Originally recorded in Britain from decayed wheat straw in a field.
D: BFF5: 60, NM2: 269, FAN4: 61 (as *Psilocybe rugosoannulata*) **D+I:** B&K4: 354 459 **I:** C&D: 356, Myc14(3): 109, SV41: 38
Rare. A large and distinctive agaric only recorded three times in England since 1945. Possibly introduced from North America. Edible, and grown commercially.

semiglobata (Batsch) Quél., *Mém. Soc. Émul. Montbéliard, Sér. 2,* 5: 112 (1872)
    Agaricus semiglobatus Batsch, *Elench. fung. (Continuatio Prima)*: 141 & t. 21 f. 110 (1786)
    Psilocybe semiglobata (Batsch) Noordel., *Persoonia* 16(1): 129 (1995)
    Agaricus stercorarius Schumach., *Enum. pl.* 2: 386 (1803)
    Stropharia stercoraria (Schumach.) Quél., *Mem. Soc. Émul. Montbéliard* 5: 112 (1872)
    Stropharia semiglobata var. *stercoraria* (Schumach.) Bon, *Bull. Soc. Mycol. France* 86: 113 (1970)
    Agaricus virosus Sowerby, *Col. fig. Engl. fung., Suppl.*: 407 (1809)
E: c **S:** c **W:** c **NI:** c **ROI:** c **O:** Channel Islands: ! Isle of Man: c
H: On weathered dung of herbivores or on dungy soil in pastures, heaths and also in woodland. Also known from clifftop grassland but then on soil enriched with guano from sea birds.
D: BFF5: 68, NM2: 267, FAN4: 62 (as *Psilocybe semiglobata*) **D+I:** Ph: 171, B&K4: 356 460 **I:** FlDan4: 142A, C&D: 354
Very common and widespread. The varietal epithet *stercoraria* was previously given to basidiomes with a strongly umbonate pileus.

squamosa (Pers.) Quél., *Mém. Soc. Émul. Montbéliard, Sér. 2,* 5: 337 (1873)
    Agaricus squamosus Pers., *Syn. meth. fung.*: 409 (1801)
    Psilocybe squamosa (Pers.) P.D. Orton, *Notes Roy. Bot. Gard. Edinburgh* 29: 80 (1969)
E: o **S:** ! **NI:** ! **ROI:** !
H: On buried woody debris, in woodland and increasingly on flowerbeds mulched with woodchips.
D: BFF5: 54 (as *Psilocybe squamosa*), NM2: 265, FAN4: 65-66 (as *P. squamosa* var. *squamosa*) **D+I:** B&K4: 356 461 **I:** FlDan4: 141D, C&D: 356
Occasional in England, rarely reported elsewhere but apparently widespread. Most frequent in southern England.

squamosa var. thrausta (Kalchbr.) Massee, *Brit. fung.-fl.* 1: 402 (1892)
    Agaricus thraustus Kalchbr., *Icon. select. Hymenomyc. Hung.*: 30 (1873)
    Stropharia thrausta (Kalchbr.) Sacc., *Syll. fung.* 5: 1016 (1887)
    Psilocybe squamosa var. *thrausta* (Kalchbr.) Guzmán, *Nova Hedwigia Beih.* 74: 321 (1983)
E: ! **S:** !
H: On soil or soil mixed with woody debris.
D: NM2: 265 (as *Psilocybe thrausta*), FAN4: 66 (as *P. squamosa* var. *thrausta*) **D+I:** FRIC7: 22 52b (as *Stropharia thrausta*)
Rarely reported. An orange-reddish variety probably of no taxonomic status (see Jahnke in Biblioth. Mycol. 96: 83, 1984). *Agaricus squamosus* f. *aurantiacus* Cooke would appear to be a further synonym, but see under *S. aurantiaca*.

squamulosa (Massee) Massee, *Trans. Brit. Mycol. Soc.* 2: 101 (1902)
  Stropharia aeruginosa var. *squamulosa* Massee, *Trans. Brit. Mycol. Soc.* 1(1): 23 (1897)
  Stropharia aeruginosa var. *calolepis* Pilát, *Česká Mykol.* 15: 58 (1961)
E: ! NI: ?
H: On sawdust or soil mixed with sawdust.
D: BFF5: 64, FAN4: 56 (as *Psilocybe squamulosa*) **D+I:** FRIC6: 31-34 47b
Known from England (South Somerset and Staffordshire). Reported from Co. Durham and Northumberland in 1907 and from Northern Ireland but unsubstantiated with voucher material.

STYPELLA Møller, *Bot. Mitt. Tropen* 8(75): 166 (1895)
Tremellales, Exidiaceae
  Protodontia Höhn., *Sitzungsber. Kaiserl. Akad. Wiss., Wien, Math.-Naturwiss. Cl., Abt. 1* 116: 83 (1907)
  Heterochaetella (Bourdot) Bourdot & Galzin, *Hyménomyc. France*: 51 (1928)
  Type: *Stypella papillata* Møller (= *S. vermiformis*)

crystallina (D.A. Reid) P. Roberts, *Mycotaxon* 69: 214 (1998)
  Myxarium crystallinum D.A. Reid, *Persoonia* 7: 293 (1973)
E: ! W: ! ROI: !
H: On decayed wood of deciduous trees and shrubs. Known on *Buxus, Fagus, Fraxinus, Quercus, Salix* and *Ulmus* spp.
D: NM3: 107 (as *Myxarium crystallinum*)
Known from England (Dorset, Hertfordshire, Middlesex, South Devon, Surrey, Warwickshire and West Kent), Wales (Breconshire) and the Republic of Ireland (Offaly).

dubia (Bourdot & Galzin) P. Roberts, *Mycotaxon* 69: 216 (1998)
  Heterochaete dubia Bourdot & Galzin, *Bull. Soc. Mycol. France* 25: 30 (1909)
  Sebacina dubia (Bourdot & Galzin) Bourdot, *Compt. Rend. Assoc. Franç. Avancem. Sci. Assoc. Sci. France* 45: 576 (1922)
  Heterochaetella dubia (Bourdot & Galzin) Bourdot & Galzin, *Hyménomyc. France*: 51 (1928)
  Heterochaetella brachyspora Luck-Allen, *Canad. J. Bot.* 38: 566 (1960)
E: o S: ! W: !
H: On decayed wood of deciduous trees and shrubs, most frequently on *Fagus, Fraxinus* or *Ulmus* spp. Also known on *Acer pseudoplatanus, Betula* spp, *Hedera* and *Salix* spp. Rarely on conifers such as *Picea* spp and *Pinus nigra*. Often late-fruiting, with the majority of collections between December and March.
D: NM3: 107 (as *Heterochaetella dubia*) **D+I:** Reid4: 437-439
Occasional but widespread, rather frequent in southern England. Also known from Scotland (Wigtownshire) and Wales (Breconshire).

glaira (Lloyd) P. Roberts, *Mycotaxon* 69: 219 (1998)
  Tremella glaira Lloyd, *Mycol. not.* 5 (60): 874 (1919)
  Exidiopsis glaira (Lloyd) K. Wells, *Lloydia* 20: 48 (1957)
  Sebacina sphaerospora Bourdot & Galzin, *Bull. Soc. Mycol. France* 39: 263 (1924)
  Myxarium sphaerosporum (Bourdot & Galzin) D.A. Reid, *Persoonia* 7: 297 (1973)
E: ! W: !
H: On decayed wood of deciduous or coniferous trees. Known on *Fagus* and *Pinus* spp.
Known from England (Buckinghamshire, East Sussex, South Devon, South Essex and Yorkshire) and Wales (Breconshire).

grilletii (Boud.) P. Roberts, *Mycotaxon* 69: 223 (1998)
  Tremella grilletii Boud., *Bull. Soc. Bot. France* 32: 284 (1885)
  Myxarium grilletii (Boud.) D.A. Reid, *Persoonia* 7: 297 (1973)
  Tremella glacialis Bourdot & Galzin, *Bull. Soc. Mycol. France* 39: 261 (1923)
  Sebacina podlachica Bres., *Ann. Mycol.* 1(1): 117 (1903)
  Exidiopsis podlachica (Bres.) Ervin, *Mycologia* 49: 123 (1957)

  Myxarium podlachicum (Bres.) Raitv., in Parmasto, *Zhiv. Prir. Dal'nego Vost.*: 113 (1971)
  Mis.: *Sebacina fugacissima* f. *sebacea* sensu auct. Brit.
  Mis.: *Myxarium laccatum* sensu auct.
  Mis.: *Stypella minor* sensu auct.
  Mis.: *Tremella viscosa* sensu auct.
E: c **S:** c **W:** c **ROI:** ! **O:** Channel Islands: c Isle of Man: !
H: On decayed wood of deciduous trees and rarely conifers. Occasionally on dead pyrenomycetes such as *Lasiosphaeria* or *Melanomma* spp, and rarely on dead stems of herbaceous plants such as *Epilobium* spp.
D: NM3: 108 (as *Myxarium grilletii*)
Common and widespread.

legonii P. Roberts, *Mycotaxon* 69: 228 (1998)
  Protodontia ellipsospora D.A. Reid [*nom. inval.*], *Mycol. Res.* 94: 104 (1990)
E: !
H: On (or often inside) soft, decayed wood of deciduous trees and shrubs, usually large logs or fallen trunks and branches of *Ulmus* spp, but also known on *Buxus, Fagus* and *Salix* spp.
Known from Berkshire, Hertfordshire, Middlesex, North Hampshire, South Devon, South Somerset, Staffordshire, Surrey, West Kent and West Sussex.

mirabilis P. Roberts, *Mycotaxon* 69: 232 (1998)
E: !
H: On dead and fallen stems of *Rubus idaeus* in woodland.
Known only from two collections from the same location in South Devon (Torquay, Scadson Woods).

subgelatinosa (P. Karst.) P. Roberts, *Mycotaxon* 69: 232 (1998)
  Hydnum subgelatinosum P. Karst., *Meddeland. Soc. Fauna Fl. Fenn.* 9: 50 (1882)
  Protohydnum subgelatinosum (P. Karst.) S. Lundell, *Fungi Exsiccati Suecici* 29-30: 21 (1947)
  Protodontia subgelatinosa (P. Karst.) Pilát, *Sborn. Nár. Mus. Praze* 13: 200 (1957)
  Protohydnum lividum Bres., *Ann. Mycol.* 1: 117 (1903)
  Protodontia uda Höhn., *Sitzungsber. Kaiserl. Akad. Wiss., Wien, Math.-Naturwiss. Cl., Abt. 1* 116: 83 (1907)
E: !
H: On decayed wood of deciduous trees such as *Betula, Fagus* and *Ulmus* spp.
D: NM3: 109 (as *Protodontia subgelatinosa*) **D+I:** Reid6: 201-203 (as *P. subgelatinosa*), Reid8: 103-104 (as *P. subgelatinosa*)
Known from East Suffolk, South Devon, South Hampshire, Surrey and West Kent.

subhyalina (A. Pearson) P. Roberts, *Mycotaxon* 69: 232 (1998)
  Sebacina subhyalina A. Pearson, *Trans. Brit. Mycol. Soc.* 13(1-2): 71 (1928)
  Myxarium subhyalinum (A. Pearson) D.A. Reid, *Trans. Brit. Mycol. Soc.* 55: 426 (1970)
  Sebacina sublilacina G.W. Martin, *Mycologia* 26: 262 (1934)
  Exidiopsis sublilacina (G.W. Martin) Ervin, *Mycologia* 49: 123 (1957)
  Myxarium sublilacinum (G.W. Martin) Raitv., *Opr. Getero. Grib. SSSR*: 66 (1967)
E: c **S:** c **W:** ! **NI:** ! **ROI:** ! **O:** Channel Islands: !
H: On decayed wood of deciduous trees such as *Betula* spp, *Fagus, Fraxinus, Prunus spinosa* and *Quercus* spp, and occasionally on dead or decayed basidiomes of polypores such as *Ganoderma* or *Phellinus* spp.
D: NM3: 108 (as *Myxarium subhyalinum*) **D+I:** Reid4: 426-428 (as *Myxarium subhyalinum*)
Common and widespread.

vermiformis (Berk. & Broome) D.A. Reid, *Trans. Brit. Mycol. Soc.* 62: 473 (1974)
  Dacrymyces vermiformis Berk. & Broome, *Ann. Mag. Nat. Hist.*, Ser. 5 [*Notices of British Fungi no. 1700*] 1: 25 (1878)
  Sebacina crystallina Bourdot, *Trans. Brit. Mycol. Soc.* 7: 53 (1921)

Heterochaetella crystallina (Bourdot) Bourdot & Galzin, *Hyménomyc. France*: 52 (1928)
    Stypella papillata Møller, *Bot. Mitt. Tropen* 8: 166 (1895)
    Sebacina papillata (Møller) Pat., *Essai taxon.*: 25 (1900)
E: o **S:** o **W:** o **NI:** ! **ROI:** ! **O:** Isle of Man: !
H: On decayed wood of conifers, usually *Pinus sylvestris* but also known on *Larix*, *Picea* spp, and *Taxus*.
D: NM3: 111 (as *Stypella papillata*) **D+I:** B&K2: 60 16, Reid8: 101-102
Occasional but widespread. Basidiomes are inconspicuous and thus probably overlooked and under-recorded.

SUBULICIUM Hjortstam & Ryvarden, *Mycotaxon* 9: 511 (1979)
Stereales, Hyphodermataceae
Type: *Subulicium lautum* (H.S. Jacks.) Hjortstam & Ryvarden

lautum (H.S. Jacks.) Hjortstam & Ryvarden, *Mycotaxon* 9(2): 513 (1979)
    Peniophora lauta H.S. Jacks., *Canad. J. Res., Sect. C, Bot. Sci.* 26: 129 (1948)
E: ! **W:** !
H: On decayed wood of conifers such as *Cupressus macrocarpa*, *Larix*, *Picea* spp and *Pinus sylvestris*. Also on decayed deciduous wood and fern stems.
D: NM3: 135 **D+I:** CNE7: 1438-1439, Myc8(3): 106
Known from England (Bedfordshire, South Devon and Surrey) and Wales (Breconshire).

minus Hjortstam, *Mycotaxon* 19: 511 (1984)
W: !
H: On decayed fallen wood. British material on *Picea* sp.
D: NM3: 135 **D+I:** CNE7: 1440-1441, Myc8(3): 106
A single collection from Monmouthshire (Wentwood Forest) in 1992.

SUBULICYSTIDIUM Parmasto, *Conspectus Systematis Corticiacearum*: 120 (1968)
Stereales, Hyphodermataceae
Type: *Subulicystidium longisporum* (Pat.) Parmasto

longisporum (Pat.) Parmasto, *Conspectus Systematis Corticiacearum*: 121 (1968)
    Hypochnus longisporus Pat., *J. Bot.* 8(12): 221 (1894)
    Kneiffia longispora (Pat.) Bres., *Ann. Mycol.* 1: 105 (1903)
    Peniophora longispora (Pat.) Bourdot & Galzin, *Bull. Soc. Mycol. France* 28: 392 (1912)
    Aegerita tortuosa Bourdot & Galzin (anam.), *Hyménomyc. France*: 298 (1928)
    Aegeritina tortuosa (Bourdot & Galzin) Jülich (anam.), *Int. J. Mycol. Lichenol.* 1(3): 282 (1984)
E: c **S:** c **W:** c **NI:** ! **ROI:** c **O:** Channel Islands: !
H: Usually on decayed wood of deciduous trees, usually *Fagus* and *Ulmus* spp but also reported on dead stems of *Centaurea nigra* and dead culms of *Glyceria maxima*.
D: NM3: 136 **D+I:** CNE7: 1444-1446, B&K2: 186 206
Very common and widespread.

SUILLUS Adans., *Fam. Pl.* 2: 10 (1763)
Boletales, Boletaceae
    Pinuzza Gray, *Nat. arr. Brit. pl.* 1: 646 (1821)
    Boletinus Kalchbr., *Bot. Zeitung (Berlin)* 25: 182 (1867)
    Ixocomus Quél., *Fl. mycol. France*: 411 (1888)
Type: *Suillus luteus* (L.) Roussel

bovinus (L.) Roussel, *Flora Calvados*: 34 (1796)
    Boletus bovinus L., *Fl. Suec.*, Edn 2: 1246 (1755)
    Ixocomus bovinus (L.) Quél., *Fl. mycol. France*: 413 (1888)
E: c **S:** c **W:** c **NI:** c **ROI:** c
H: On acidic soil near conifers, usually associated with *Pinus sylvestris* or less commonly, other two-needled *Pinus* spp.
D: NM2: 70 **D+I:** Ph: 215, B&K3: 76 42 **I:** Bon: 49, C&D: 425, Kriegl2: 310
Common and widespread.

bresadolae *var.* **flavogriseus** Cazzoli & Cons., *Il Fungo, 7th Seminario Internazionale sui Funghi ipogei*: 25 (1998) [1996]
    Suillus nueschii Singer [*nom. inval.*], *Sydowia* 15: 82 (1961)
    Mis.: *Boletus flavus* sensu Bresadola [*Icon. mycol.* 19: pl. 904 (1931)]
E: ! **S:** !
H: On soil, associated with *Larix* spp, usually under solitary trees.
I: C&D: 423 (as *Suillus flavus*)
Known only from three sites in England (East Suffolk: Flatford Mill, South Devon: Dartmoor, Postbridge and Yorkshire: near Sheffield) and one in Scotland (South Aberdeen: Linn of Dee). Perhaps better known as *Suillus nueschii*.

cavipes (Opat.) A.H. Sm. & Thiers, *Contrib. N. Am. Suillus*: 30 (1964)
    Boletus cavipes Opat., *Comm. fam. bolet.* 2(1): 11 (1836)
    Boletinus cavipes (Opat.) Kalchbr., *Bot. Zeitung (Berlin)* 25: 182 (1867)
    Boletinus cavipes f. *aureus* (Rolland) Singer, *Rev. Mycol. (Paris)* 3(4-5): 167 (1938)
    Paxillus porosus Berk., *London J. Bot., Decades of Fungi No. 115* 6: 314 (1847)
E: ! **S:** ! **W:** !
H: On soil, associated with *Larix* spp.
D: NM2: 56 (as *Boletinus cavipes*) **I:** C&D: 421 (as *B. cavipes*)
Known from England (Northumberland and South Somerset), Scotland (Wester Ross) and Wales (Glamorganshire). Reported from Hertfordshire in 1952 but unsubstantiated with voucher material.

collinitus (Fr.) Kuntze, *Revis. gen. pl.* 3(2): 536 (1898)
    Boletus collinitus Fr., *Epicr. syst. mycol.*: 410 (1838)
    Suillus fluryi Huijsman, *Schweiz. Z. Pilzk.* 47(3): 70 (1969)
E: ! **S:** ! **W:** ! **ROI:** !
H: On sandy or calcareous soil, associated with *Pinus* spp, often at the edges of plantations.
D+I: B&K3: 78 43 **I:** Bon: 47, C&D: 425, SV35: 24, Kriegl2: 301 (as *Suillus fluryi*)
Known from England (Berkshire, Buckinghamshire, Cambridgeshire, Huntingdonshire, Northamptonshire, South Lincolnshire, Surrey, Warwickshire, West Kent, West Norfolk and West Suffolk), Scotland (East Lothian), Wales (Anglesey, Carmarthenshire and Glamorganshire) and Republic of Ireland (North Kerry).

flavidus (Fr.) Presley, *Vseobecny Rostlinopsis* 2: 1917 (1846)
    Boletus flavidus Fr., *Observ. mycol.* 1: 110 (1815)
    Ixocomus flavidus (Fr.) Quél., *Fl. mycol. France*: 415 (1888)
    Boletus elegans var. *pulchellus* (Fr.) Rea, *Brit. basidiomyc.*: 559 (1922)
    Boletus pulchellus Fr., *Hymenomyc. eur.*: 497 (1874)
E: ? **S:** o **W:** ?
H: On soil, often amongst mosses and associated with *Pinus sylvestris* or other two-needled *Pinus* spp, in boggy areas in woodland.
D: NM2: 70 **D+I:** Ph: 215, B&K3: 78 44 **I:** Bon: 47, C&D: 425, Kriegl2: 295
Occasional in Scotland, but may be locally common in Caledonian pinewoods. Rarely reported from England or Wales, and unsubstantiated with voucher material.

granulatus (L.) Roussel, *Flora Calvados*: 34 (1796)
    Boletus granulatus L., *Sp. pl.*: 1617 (1762)
    Ixocomus granulatus (L.) Quél., *Fl. mycol. France*: 412 (1888)
    Boletus lactifluus Sowerby, *Col. fig. Engl. fung.* 4: pl. 420 (1814)
    Leccinum lactifluum (Sowerby) Gray, *Nat. arr. Brit. pl.* 1: 647 (1821)
    Suillus lactifluus (Sowerby) A.H. Sm. & Thiers, *Michigan Botanist* 7: 16 (1968)
E: c **S:** c **W:** c **NI:** c **ROI:** c **O:** Channel Islands: !
H: On acidic soil, associated with *Pinus sylvestris* and frequently growing amongst *Calluna* on heathland and less frequently in woodland.

D: NM2: 71 **D+I:** Ph: 216-217, B&K3: 78 45 **I:** Bon: 47, C&D: 425, Kriegl2: 299

Common and widespread. *Suillus lactifluus*, considered distinct by some authorities, is reported from two sites in England but unsubstantiated with voucher material.

grevillei (Klotzsch) Singer, *Farlowia* 2: 259 (1945)
Boletus grevillei Klotzsch, *Linnaea* 7: 198 (1832)
Boletus annularius Bolton [non *B. annularius* sensu Bulliard (1786)], *Hist. fung. Halifax* Suppl.: 169 (1791)
Boletus elegans Schumach. [non *B. elegans* Bolton (1788)], *Enum. pl.* 2: 374 (1803)
Suillus elegans (Schumach.) Snell, *Lloydia* 7: 27 (1944)
Ixocomus flavus var. *elegans* (Schumach.) Quél., *Fl. mycol. France*: 415 (1888)
Ixocomus elegans f. *badius* Singer, *Rev. Mycol. (Paris)* 3(1): 40 (1938)
Suillus grevillei f. *badius* (Singer) Singer, *Pilze Mitteleuropas* 5(1): 66 (1965)
Mis.: *Suillus clintonianus* sensu auct. Brit.
E: c **S:** c **W:** c **NI:** c **ROI:** c **O:** Channel Islands: ! Isle of Man: !
H: On soil, associated with *Larix* spp.
D: NM2: 70 **D+I:** Ph: 216, B&K3: 80 46 **I:** Bon: 47, C&D: 423, Kriegl2: 305
Common and widespread. Forms with a red-brown pileus have been distinguished as var. *badius*, or even assigned to the boreal North American species *Suillus clintonianus*, but are of no taxonomic significance.

lakei (Murrill) A.H. Sm. & Thiers, *Contrib. N. Am. Suillus*: 34 (1964)
Boletus lakei Murrill, *Mycologia* 4: 97 (1912)
Boletus tridentinus var. *landkammeri* Pilát & Svrček, *Sborn. Nár. Mus. Praze* 5B: 1 (1949)
Boletinus landkammeri (Pilát & Svrček) Bon, *Doc. Mycol.* 16(62): 66 (1986)
Suillus lakei var. *landkammeri* (Pilát & Svrček) H. Engel & Klofac, *Schmier- und Filzröhrlinge s.l. in Europa*: 52 (1996)
Mis.: *Suillus amabilis* sensu auct. Eur.
E: !
H: On soil in conifer woodland. British material in a plantation, associated with *Pseudotsuga menziesii*.
D: NM2: 70 (as *Suillus amabilis*) **I:** C&D: 421 (as *Boletinus landkammeri*)
A single record (as *S. amabilis*) from Hertfordshire in 1952.

luteus (L.) Roussel, *Flora Calvados*: 34 (1796)
Boletus luteus L., *Sp. pl.*: 1177 (1753)
Ixocomus luteus (L.) Quél., *Fl. mycol. France*: 414 (1888)
E: c **S:** c **W:** c **ROI:** c
H: On acidic soil, usually associated with *Pinus sylvestris*.
D: NM2: 70 **D+I:** Ph: 214-215, B&K3: 80 47 **I:** Bon: 47, C&D: 425, Kriegl2: 297
Common and widespread.

placidus (Bonord.) Singer, *Farlowia* 2: 42 (1945)
Boletus placidus Bonord., *Beitr. Mykol.* 19: 204 (1861)
E: !
H: On soil with five-needled pines. British material associated with *Pinus strobus* in an arboretum.
D: NM2: 71 **D+I:** B&K3: 80 48 **I:** SV31: 12, C&D: 425, Kriegl2: 302
Known from one site in East Kent (Bedgebury Pinetum) in 1976, 1977 and 1981.

tridentinus (Bres.) Singer, *Farlowia* 2: 260 (1945)
Boletus tridentinus Bres., *Fungi trident.* 1: 13 (1881)
Ixocomus tridentinus (Bres.) Bataille, *Les Bolets*: 23 (1923)
Boletus aurantiporus J. Howse, *Grevillea* 12: 43 (1883)
E: ! **ROI:** !
H: On calcareous soil, associated with *Larix* spp, in mixed woodland or plantations.
D+I: Ph: 216-217, B&K3: 82 51 **I:** Bon: 47, C&D: 423, Kriegl2: 309
Known from England (Berkshire, East & West Kent, Hertfordshire, Northamptonshire, Oxfordshire, South Somerset, Surrey, West Suffolk and West Sussex). Reported elsewhere

but unsubstantiated with voucher material. Last reported from Republic of Ireland (Co. Dublin) in 1898.

variegatus (Sw.) Richon & Roze, *Atlas champ.*: 178 (1888)
Boletus variegatus Sw. [non *B. variegatus* Sowerby (1802)], *Vet. Akad. Handl.* 31: 8 (1810)
Ixocomus variegatus (Sw.) Quél., *Fl. mycol. France*: 414 (1888)
E: o **S:** c **W:** c **NI:** ! **ROI:** !
H: On acidic soil, associated with *Pinus sylvestris* on heathland, often amongst *Calluna* in such habitat, and occasionally on decayed conifer wood such as logs and stumps.
D: NM2: 71 **D+I:** Ph: 216-217, B&K3: 84 52 **I:** Bon: 49, C&D: 425, Kriegl2: 312
Rather common in Scotland and Wales, occasional but widespread in England, and rarely reported in Ireland.

viscidus (L.) Rousel, *Flora Calvados*: 34 (1796)
Boletus viscidus L., *Sp. pl.*: 1177 (1753)
Ixocomus viscidus (L.) Quél., *Fl. mycol. France*: 416 (1888)
Boletus aeruginascens Secr. [*nom. inval.*], *Mycogr. Suisse* 3: 6 (1833)
Suillus aeruginascens Snell, *Lloydia* 7: 25 (1944)
Fuscoboletinus aeruginascens (Snell) Pomerl. & A.H. Sm., *Brittonia*. 14: 167 (1962)
Boletus laricinus Berk., *Engl. fl.* 5(2): 248 (1836)
Suillus laricinus (Berk.) Kuntze, *Rev. Gen. Pl.* 3(2): 535 (1898)
E: o **S:** ! **W:** ! **NI:** ! **ROI:** !
H: On sandy or calcareous soil, associated with *Larix* spp, in mixed woodland or plantations.
D: NM2: 70 (as *Suillus aeruginascens*) **D+I:** Ph: 216-217, B&K3: 84 54 (as *S. viscidus*) **I:** Bon: 47 (as *S. laricinus*), C&D: 423, SV32: 45, Kriegl2: 307
Occasional in England from Westmorland southwards, rarely reported elsewhere but apparently widespread.

SYZYGOSPORA G.W. Martin, *J. Washington Acad. Sci.* 27: 112 (1937)
Tremellales, Syzygosporaceae
Christiansenia Hauerslev, *Friesia* 9: 43 (1969)
Type: *Syzygospora alba* G.W. Martin

bachmannii Diederich & M.S. Christ., *Biblioth. Lichenol.* 61: 30 (1996)
S: ! **W:** !
H: Parasitic. On thalli of the lichen *Cladonia gracilis*.
D: NM3: 86
Known in Britain from a single Scottish site (East Sutherland: Cuthill Links) and two in Wales (Caernarvonshire: Moelydd Bach and Weirglodd Isaf).

pallida (Hauerslev) Ginns, *Mycologia* 78(4): 631 (1986)
Christiansenia pallida Hauerslev, *Friesia* 9: 43 (1969)
E: !
H: Parasitic. In the hymenium of *Phanerochaete sordida*.
D: NM3: 85 **D+I:** CNE2: 242-243 (as *Christiansenia pallida*)
Known from Buckinghamshire, East Norfolk and Shropshire.

physciacearum Diederich, *Biblioth. Lichenol.* 61: 38 (1996)
S: ! **W:** ! **NI:** ? **ROI:** !
H: Parasitic. On thalli of species of the lichen genus *Physcia*.
D: NM3: 86
Rarely reported, but apparently widespread.

sorana Hauerslev, *Opera Bot.* 100: 113 (1989)
E: !
H: Parasitic. In old apothecia of *Ascocoryne cylichnium*.
D: NM3: 85
A single collection from South Devon (Dunsford Woods) in 1992 [see Roberts in Mycologist 7 (2): 62 (1993)].

tumefaciens (Ginns & Sunhede) Ginns, *Mycologia* 78(4): 634 (1986)
Christiansenia tumefaciens Ginns & Sunhede, *Bot. Not.* 131: 168 (1978)
E: r **S:** r

H: Parasitic. On basidiomes of *Collybia dryophila*.
D: NM3: 85 **D+I:** Myc4(1): 35 **I:** C&D: 137
Rare. Probably genuinely so, since affected host basidiomes become distorted and covered in tumour-like growths. Known from a few localities in England (East Gloucestershire, Shropshire, Surrey, West Kent and West Sussex) and Scotland (Stirlingshire).

TAENIOSPORA Marvanová, *Trans. Brit. Mycol. Soc.* 69: 146 (1977)

Stereales, Atheliaceae
A genus of anamorphic aquatic hyphomycetes with teleomorphs, as yet unnamed, in the *Atheliaceae*.
Type: *Taeniospora gracilis* Marvanová

gracilis Marvanová, *Trans. Brit. Mycol. Soc.* 69(1): 146 (1977)
E: ! **W:** !
H: Isolated from foam samples collected in streams.
Known from England (South Hampshire: New Forest) and Wales (Caernarvonshire).

gracilis *var.* **enecta** Marvanová & Stalpers, *Trans. Brit. Mycol. Soc.* 89(4): 492 (1987)
E: ! **S:** ! **W:** !
H: Isolated from foam samples collected in streams.
Known from England (Cumberland, South Devon and South Hampshire), Scotland (various sites) and Wales (Caernarvonshire). The teleomorph was reported as *Leptosporomyces galzinii* by Nawawi *et al.* [TBMS 68: 31-36 (1977)] but this is disproven and it appears to be an unnamed species of *Fibulomyces*.

TAPINELLA E.-J. Gilbert, *Les Bolets*: 67 (1931)

Boletales, Paxillaceae
Type: *Tapinella panuoides* (Fr.) Gilb.

atrotomentosa (Batsch) Šutara, *Česká Mykol.* 46(1-2): 50 (1992)
Agaricus atrotomentosus Batsch, *Elench. fung. (Continuatio Prima)*: 89 & t. 8 f. 32 (1786)
Paxillus atrotomentosus (Batsch) Pers., *Syn. meth. fung.*: 472 (1801)
E: o **S:** c **W:** ! **NI:** ! **ROI:** !
H: On decayed wood of conifers, often *Pinus* spp but also reported on *Picea* spp. Usually on large stumps or fallen trunks and occasionally at the base of living trunks.
D: NM2: 54 (as *Paxillus atrotomentosus*) **D+I:** Ph: 143 (as *P. atrotomentosus*), B&K3: 90 63 (as *P. atrotomentosus*) **I:** FlDan4: 134E (as *P. atrotomentosus*), Bon: 51 (as *P. atrotomentosus*), C&D: 418 (as *P. atrotomentosus*), Kriegl2: 346 (as *P. atrotomentosus*)
Rather common in Scotland where it is often abundant. Occasional but widespread elsewhere. Perhaps better known as *Paxillus atrotomentosus*.

panuoides (Fr.) Gilb., *Les Bolets*: 68 (1931)
Agaricus panuoides Fr., *Observ. mycol.* 2: 227 (1818)
Paxillus panuoides (Fr.) Fr., *Epicr. syst. mycol.*: 318 (1838)
Paxillus fagi Berk. & Broome, *Ann. Mag. Nat. Hist., Ser. 5 [Notices of British Fungi no. 1961]* 9: 181 (1882)
Paxillus panuoides var. *fagi* (Berk. & Broome) Cooke, *Handbook of British Fungi, Edn 2*: 288 (1883)
Paxillus panuoides var. *ionipes* Quél., *Fl. mycol. France*: 111 (1888)
Paxillus panuoides var. *rubrosquamulosus* Svrček & Kubička, *Česká Mykol.* 18: 173 (1964)
E: o **S:** o **W:** ! **NI:** ! **ROI:** !
H: On decayed wood (including sawdust) of conifers, most frequently on fallen trunks of *Pinus sylvestris* and formerly not infrequent on pit props in mines. Rarely on deciduous wood but known on *Betula* spp, *Fagus* and *Prunus avium*.
D: BFF6: 106-107 (as *Paxillus panuoides*), NM2: 55 (as *P. panuoides*) **D+I:** Ph: 142 (as *P. panuoides*), B&K3: 94 68 **I:** FlDan4: 134B (as *P. panuoides*), Bon: 51 (as *P. panuoides*), C&D: 418 (as *P. panuoides*), Kriegl2: 348 (as *P. panuoides*)

Occasional but widespread. Basidiomes with a strong development of violaceous-lilac mycelium at the stipe base have been referred to var. *ionipes* and those with prominent reddish scales on the pileus to var. *rubrosquamulosus*.

TEPHROCYBE Donk, *Nova Hedwigia Beih.* 5: 284 (1962)

Agaricales, Tricholomataceae
Type: *Tephrocybe rancida* (Fr.) Donk

ambusta (Fr.) Donk, *Nova Hedwigia Beih.* 5: 284 (1962)
Agaricus umbratilis β *ambustus* Fr., *Syst. mycol.* 1: 157 (1821)
Collybia ambusta (Fr.) Quél., *Mém. Soc. Émul. Montbéliard, Sér. 2,* 5: 238 (1872)
Lyophyllum ambustum (Fr.) Singer, *Ann. Mycol.* 41: 105 (1943)
Collybia gibberosa Jul. Schäff., *Ann. Mycol.* 40: 124 (1942)
Tephrocybe gibberosa (Jul. Schäff.) P.D. Orton, *Notes Roy. Bot. Gard. Edinburgh* 29(1): 76 (1969)
Lyophyllum gibberosum (Jul. Schäff.) M. Lange, *Rev. Mycol. (Paris)* 19(2): 135 (1954)
E: o **S:** ! **W:** ! **ROI:** !
H: On burnt soil (old bonfire sites etc.) in woodland with deciduous and coniferous trees.
D: NM2: 140 (as *Lyophyllum ambustum*) **D+I:** B&K3: 218 254 **I:** RivMyc155(2): 152 (as *L. ambustum*)
Occasional in England, rarely reported elsewhere but apparently widespread. *T. gibberosa* is kept separate by some authorities

anthracophila (Lasch) P.D. Orton, *Notes Roy. Bot. Gard. Edinburgh* 29(1): 76 (1969)
Agaricus anthracophilus Lasch, *Linnaea* 4: 532 (1829)
Collybia anthracophila (Lasch) P. Kumm., *Führ. Pilzk.*: 114 (1871)
Omphalia carbonaria Velen., *České Houby* II: 288 (1920)
Collybia carbonaria (Velen.) P.D. Orton, *Trans. Brit. Mycol. Soc.* 43(2): 174 (1960)
Tephrocybe carbonaria (Velen.) Donk, *Nova Hedwigia Beih.* 5: 284 (1962)
Lyophyllum carbonarium (Velen.) M.M. Moser, *Kleine Kryptogamenflora*: 40 (1953)
Lyophyllum sphaerosporum Kühner & Romagn., *Flore Analytique des Champignons Supérieurs*: 166 (1953)
Mis.: *Collybia atrata* sensu Rea (1922)
E: o **S:** o **ROI:** !
H: On burnt soil or charcoal (on old bonfire sites etc.) in deciduous and coniferous woodland.
D: NM2: 139 (as *Lyophyllum anthracophilum*) **D+I:** B&K3: 218 255 **I:** C&D: 213, RivMyc155(2): 154 (as *L. anthracophilum*)
Occasional but apparently widespread and may be locally abundant.

atrata (Fr.) Donk, *Nova Hedwigia Beih.* 5: 284 (1962)
Agaricus atratus Fr., *Observ. mycol.* 2: 215 (1818)
Collybia atrata (Fr.) P. Kumm., *Führ. Pilzk.*: 307 (1871)
Lyophyllum atratum (Fr.) Singer, *Ann. Mycol.* 41: 103 (1943)
E: o **S:** ! **W:** ! **NI:** ! **ROI:** !
H: On burnt soil (old fire sites etc.) in woodland or on heathland.
D: NM2: 139 (as *Lyophyllum atratum*) **I:** Bon: 169, RivMyc155(2): 169 (as *L. atratum*)
Occasional but widespread and may be locally abundant.

baeosperma (Romagn.) M.M. Moser, *Kleine Kryptogamenflora, Edn 4*: 132 (1978)
Lyophyllum baeospermum Romagn., *Bull. Soc. Naturalistes Oyonnax* 8: 75 (1954)
E: !
H: On loam soil (often calcareous) in deciduous woodland.
D: NBA8: 616 **I:** RivMyc155(2): 163 (as *Lyophyllum baeospermum*)
Known from Norfolk, North Somerset and Surrey.

boudieri (Kühner & Romagn.) Derbsch, *Z. Pilzk.* 43(2): 186 (1977)

Lyophyllum boudieri Kühner & Romagn., *Bull. Soc. Naturalistes Oyonnax* 8: 75 (1954)

Tephrocybe albofloccosa P.D. Orton, *Notes Roy. Bot. Gard. Edinburgh* 41(3): 614 (1984)

Mis.: *Tephrocybe coracina* sensu P.D. Orton [Bull. BMS 18(2): 114-119 (1984)]

E: ! W: ! S: ! NI: !

H: On loam soil and decayed leaf litter, often with *Betula* spp, *Corylus* and *Fagus* on chalk or limestone.

D: NBA8: 616-617, NBA8: 617-618 (as *T. coracina*) **I:** Bon: 168, C&D: 213, RivMyc155(2): 175 (as *Lyophyllum boudieri*)

Rarely reported but apparently widespread.

cessans (P. Karst.) M.M. Moser, *Kleine Kryptogamenflora,* Edn 3: 114 (1967)

Collybia cessans P. Karst., *Bidrag Kännedom Finlands Natur Folk* 32: 163 (1879)

Lyophyllum cessans (P. Karst.) M.M. Moser, *Kleine Kryptogamenflora*: 41 (1953)

E: !

H: On soil in coniferous woodland. Reported with *Pinus* spp, and *Pseudotsuga menziesii.*

Known from East Sussex, Hertfordshire, North Hampshire and Surrey.

confusa (P.D. Orton) P.D. Orton, *Notes Roy. Bot. Gard. Edinburgh* 29(1): 76 (1969)

Collybia confusa P.D. Orton, *Trans. Brit. Mycol. Soc.* 43(2): 189 (1960)

E: ! ROI: !

H: On acidic soil, often amongst short grass and mosses or near *Pteridium* in open woodland or heathland.

D: NM2: 141 (as *Lyophyllum confusum*) **D+I:** Ph: 56-57 (as *Collybia confusa*) **I:** RivMyc155(2): 165 (as *L. confusum*)

Rarely reported, but widespread in southern and south-western England.

ellisii P.D. Orton, *Trans. Brit. Mycol. Soc.* 91(4): 569 (1988)

E: !

H: On soil, often amongst short grass.

Known from Buckinghamshire, East Norfolk, South Hampshire, Surrey, West Cornwall and West Kent.

ferruginella (A. Pearson) P.D. Orton, *Bull. Brit. Mycol. Soc.* 18(2): 119 (1984)

Collybia ferruginella A. Pearson, *Trans. Brit. Mycol. Soc.* 35(2): 102 (1952)

E: !

H: On soil, often amongst grass in woodland or in grassland.

Known from Norfolk, North Hampshire and West Kent.

fuscipes P.D. Orton, *Notes Roy. Bot. Gard. Edinburgh* 22(1): 122 (1969)

Mis.: *Collybia misera* sensu Ricken [Blätterpilze Deutschl.: 403 (1915)]

S: !

H: On decayed needle litter and cones of *Pinus sylvestris.*

Known only from the type (Perthshire: Rannoch) and three other collections from Scotland.

fusispora (Hora) M.M. Moser, *Kleine Kryptogamenflora,* Edn 3: 115 (1967)

Collybia fusispora Hora, *Trans. Brit. Mycol. Soc.* 43(2): 443 (1960)

E: !

H: In leaf litter of *Fagus* on calcareous soil.

D: NM2: 140 (as *Lyophyllum fusisporum*)

Known (in Britain) only from the type collection from Surrey (Gomshall) in 1953.

impexa (P. Karst.) M.M. Moser, *Kleine Kryptogamenflora,* Edn 3: 115 (1967)

Collybia impexa P. Karst., *Bidrag Kännedom Finlands Natur Folk* 32: 549 (1879)

E: !

H: On burnt soil (old bonfire sites etc.) in woodland, heathland, or in gardens.

D: NBA8: 618

Known from England (Bedfordshire, Berkshire, Buckinghamshire, Co. Durham, Hertfordshire, London, Middlesex, North Somerset, Northumberland, South Devon, Surrey and Warwickshire). Reported elsewhere but unsubstantiated with voucher material.

inolens (Fr.) M.M. Moser, *Kleine Kryptogamenflora,* Edn 3: 114 (1967)

Agaricus inolens Fr., *Epicr. syst. mycol.*: 96 (1838)

Collybia inolens (Fr.) Quél., *Mém. Soc. Émul. Montbéliard, Sér. 2,* 5: 238 (1872)

Lyophyllum inolens (Fr.) Kühner & Romagn., *Ann. Mycol.* 41: 104 (1943)

Mis.: *Collybia mephitica* sensu Rea (1922)

E: ! W: !

H: On soil in deciduous, or mixed coniferous and deciduous woodland. British material with *Fagus* and *Larix.*

D: NM2: 141 (as *Lyophyllum inolens*) **I:** Cooke 211 (154) Vol. 2 (1882) as *Agaricus inolens*

Accepted but poorly distinguished from *Tephrocybe ozes.* Known from England (Co. Durham, Hertfordshire, South Somerset and West Sussex) and Wales (Denbighshire).

langei (J. Favre) Raithelh., *Metrodiana* 19(2): 57 (1992)

Collybia langei J. Favre, *Beitr. Kryptogamenfl. Schweiz* 10(3): 85 (1948)

Clitocybe favrei Bon, *Doc. Mycol.* 26(104): 29 (1997)

Mis.: *Collybia misera* sensu Lange (FlDan2: 16 & pl. 45A, as *C. misera* forma)

E: !

H: On soil in woodland.

D: NM2: 111 (as *Clitocybe favrei*)

Known from England (Berkshire and Surrey) and reported from Lancashire (as *Clitocybe favrei*) but unsubstantiated with voucher material. Possibly equates with *Tricholoma putidum* sensu Rea (1922) and *Tephrocybe admissa* sensu Moser (1967).

mephitica (Fr.) M.M. Moser, *Kleine Kryptogamenflora,* Edn 3: 115 (1967)

Agaricus mephiticus Fr., *Epicr. syst. mycol.*: 96 (1838)

Collybia mephitica (Fr.) Sacc., *Syll. fung.* 5: 244 (1887)

Lyophyllum mephiticum (Fr.) Singer, *Lilloa* 22: 166 (1951) [1949]

E: ! S: !

H: On soil and needle litter in coniferous woodland. Also on decayed basidiomes of *Russula claroflava* and *Lactarius vellereus* (a single record from each).

D: NM2: 140 (as *Lyophyllum mephiticum*) **I:** C&D: 213

Known from England (East & West Norfolk, Hertfordshire, North Hampshire, North Wiltshire, Surrey) and Scotland (Easterness)

murina (Batsch) M.M. Moser, *Kleine Kryptogamenflora,* Edn 3: 116 (1967)

Agaricus murinus Batsch [non *A. murinus* Sowerby (1798)], *Elench. fung.*: 79 & t. 5 f. 19 (1783)

Collybia murina (Batsch) P. Kumm., *Führ. Pilzk.*: 114 (1871)

Mis.: *Tephrocybe misera* sensu NCL & sensu Orton [Bull. BMS 18(2):114-120 (1984)]

E: ! S: !

H: On soil in woodland.

D: NM2: 141 (as *Lyophyllum murinum*)

Accepted here sensu lato to include *T. misera* sensu Orton (NBA8). Apparently rare.

osmophora (E.-J. Gilbert) Bon, *Doc. Mycol.* 28(111): 74 (1998)

Clitocybe osmophora E.-J. Gilbert, *Bull. Soc. Mycol. France* 51: 115 (1935)

E: !

H: On calcareous soil, amongst leaf litter of *Fagus* and also known with *Larix.*

I: RivMyc155(2): 159 (as *Lyophyllum osmophorum*)

Known only from two sites in England (Surrey: Banstead and Mickleham).

palustris (Peck) Donk, *Nova Hedwigia Beih.* 5: 284 (1962)
   Agaricus paluster Peck, *Rep. (Annual) New York State Mus. Nat. Hist.* 23: 82 (1872)
   Collybia palustris (Peck) A.H. Sm., *Pap. Michigan Acad. Sci.* 21: 148 (1936)
   Lyophyllum palustre (Peck) Singer, *Ann. Mycol.* 41: 103 (1943)
   Collybia leucomyosotis Cooke & W.G. Sm., *Handb. Brit. fung.*, Edn 2: 369 (1890)
   Agaricus thelephorus Cooke & Massee, *Grevillea* 18: 51 (1890)
   Collybia thelephora (Cooke & Massee) Sacc., *Syll. fung.* 9: 28 (1891)
E: o **S:** o **W:** o **NI:** ! **ROI:** !
H: Amongst *Sphagnum* spp, in bogs, woodland swamps, alder swamps and fen vegetation.
D: NM2: 139 (as *Lyophyllum palustre*) **D+I:** Ph: 56-57 (as *C. palustris*), B&K3: 226 265 **I:** FlDan2: 45C (as *C. leucomyosotis*), Bon: 169, C&D: 213, Kriegl3: 325 (as *L. palustre*), RivMyc155(2): 170 (as *L. palustre*)
Occasional, but widespread and may be locally abundant

platypus (Kühner) M.M. Moser, *Kleine Kryptogamenflora*, Edn 3: 115 (1967)
   Lyophyllum platypus Kühner, *Bull. Soc. Naturalistes Oyonnax* 8: 75 (1954)
E: !
H: On humic soil and leaf litter in deciduous woodland.
D: NBA9: 570 **D+I:** B&K3: 226 266 (as *Lyophyllum platypum*) **I:** RivMyc155(2): 169 (as *L. platypum*)
Known from England (West Gloucestershire: Forest of Dean, Cannop Pond) in 1987 and possibly also from Mid-Lancashire (Silverdale: Gait Barrows).

putida (Fr.) M.M. Moser, *Kleine Kryptogamenflora*, Edn 3: 116 (1967)
   Collybia putida Fr., *Hymenomyc. eur.*: 125 (1874)
   Agaricus putidus Fr. [*nom. illegit.*, non *A. putidus* Weinm. (1836)], *Epicr. syst. mycol.*: 54 (1838)
   Tricholoma putidum (Fr.) P. Karst. [*nom. illegit.*], *Bidrag Kännedom Finlands Natur Folk* 32: 56 (1879)
   Collybia putidella P.D. Orton [*nom. superfl.*], *Trans. Brit. Mycol. Soc.* 43: 174 (1960)
E: ! **S:** !
H: On soil, amongst grass in woodland.
D: NM2: 140 (as *Lyophyllum putidum*) **I:** Bon: 169, RivMyc155(2): 165 (as *L. putidum*), Cooke 127 (172) Vol. 1 (1883) (as *Agaricus (Tricholoma) putidus*)
Rarely reported and known with certainty only from England (South Wiltshire: Whitmarsh Wood) and Scotland. Reported from Lincolnshire, Oxfordshire and Yorkshire but determined with doubt or unsubstantiated with voucher material.

rancida (Fr.) Donk, *Nova Hedwigia Beih.* 5: 284 (1962)
   Agaricus rancidus Fr., *Syst. mycol.* 1: 141 (1821)
   Collybia rancida (Fr.) Quél., *Mém. Soc. Émul. Montbéliard, Sér. 2,* 5: 99 (1872)
   Lyophyllum rancidum (Fr.) Singer, *Ann. Mycol.* 41: 103 (1943)
E: o **S:** o **W:** ! **NI:** ! **ROI:** !
H: On soil, amongst leaf litter in mixed deciduous woodland.
D: NM2: 139 (as *Lyophyllum rancidum*) **D+I:** Ph: 56-57 (as *Collybia rancida*), B&K3: 226 267 (as *L. rancidum*) **I:** FlDan2: 45I (as *C. rancida*), Bon: 169, FungEur3: 67, C&D: 213, RivMyc155(2): 177 (as *L. rancidum*)
Occasional but widespread. Few records from Ireland.

striipilea (Fr.) Donk, *Nova Hedwigia Beih.* 5: 284 (1962)
   Agaricus striipilea Fr., *Öfvers. Kongl. Vetensk.-Akad. Förh.* 18: 22 (1862)
   Omphalia striipilea (Fr.) Gillet, *Hyménomycètes*: 297 (1876)
   Collybia striipilea (Fr.) P.D. Orton, *Trans. Brit. Mycol. Soc.* 43(2): 175 (1960)
   Lyophyllum striipilea (Fr.) Kühner & Romagn. ex Kalamees, *Z. Mykol.* 60(1): 15 (1994)
E: ! **S:** ! **W:** !

H: On soil under coniferous trees.
D: NM2: 140 (as *Lyophyllum striipileum*) **I:** RivMyc155(2): 160 (as *L. striaepileum*)
Known from England (South Somerset in 1955 and Surrey in 1929). Material named thus (in herb. K), collected in the 1870's from Scotland (Angus) and Wales (Denbighshire) are in poor condition and not determinable with certainty.

tylicolor (Fr.) M.M. Moser, *Kleine Kryptogamenflora*, Edn 4: 131 (1978)
   Agaricus tylicolor Fr., *Syst. mycol.* 1: 132 (1821)
   Collybia tylicolor (Fr.) Gillet, *Hyménomycètes*: 309 (1876)
   Lyophyllum tylicolor (Fr.) M. Lange & Sivertsen, *Bot. Tidsskr.* 62: 205 (1966)
   Agaricus plexipes Fr., *Syst. mycol.* 1: 146 (1821)
   Collybia plexipes (Fr.) P. Kumm., *Führ. Pilzk.*: 114 (1871)
   Lyophyllum plexipes (Fr.) Kühner & Romagn., *Flore Analytique des Champignons Supérieurs*: 167 (1953)
   Agaricus tesquorum Fr., *Öfvers Kongl. Vetensk.-Akad. Förh.* 18: 22 (1862)
   Collybia tesquorum (Fr.) Gillet, *Hyménomycètes*: 309 (1876)
   Lyophyllum tesquorum (Fr.) Singer, *Ann. Mycol.* 41: 105 (1943)
   Tephrocybe tesquorum (Fr.) M.M. Moser, *Kleine Kryptogamenflora*, Edn 3: 116 (1967)
   Mis.: *Collybia erosa* sensu Lange (FlDan2: 17 & pl. 46B)
E: ! **S:** !
H: On acidic soil, often amongst mosses, with deciduous trees.
D: NM2: 141 (as *Lyophyllum tylicolor*) **D+I:** B&K3: 220 258 (as *L. erosa*), B&K3: 228 269 (as *L. tesquorum*)
Rarely reported. Apparently widespread, but the majority of records and collections are from southern England.

TERANA Adans., *Fam. Pl.* 2: 5 (1763)
Stereales, Corticiaceae
   Pulcherricium Parmasto, *Conspectus Systematis Corticiacearum*: 132 (1968)
Type: *Terana caerulea* (Lam.) Kuntze

caerulea (Lam.) Kuntze, *Revis. gen. pl.* 2: 872 (1891)
   Byssus caerulea Lam., *Flore Franc.* 1: 103 (1779)
   Thelephora caerulea (Lam.) DC., *Fl. Fr.* 2: 107 (1805)
   Corticium caeruleum (Lam.) Fr., *Epicr. syst. mycol.*: 562 (1838)
   Pulcherricium caeruleum (Lam.) Parmasto, *Conspectus Systematis Corticiacearum*: 132 (1968)
   Thelephora indigo Schwein., *Schriften Naturf. Ges. Leipzig* 1: 107 (1822)
   Auricularia phosphorea Sowerby, *Col. fig. Engl. fung.* 3: pl. 350 (1815)
E: o **S:** ! **W:** o **NI:** ! **O:** Channel Islands: !
H: On decayed wood of deciduous trees, most frequently *Fraxinus* but also known on *Acer campestre, Crataegus monogyna, Corylus, Euonymus europaeus, Fagus, Hedera, Lonicera nitida, Rubus fruticosus, Salix fragilis* and *Ulex europaeus.*
D: NM3: 179 **D+I:** CNE6: 1226-1227 (as *Pulcherricium caeruleum*), B&K2: 106 87 (as *P. caerulea*) **I:** Kriegl1: 343, SV41: 46 (as *P. caeruleum*)
Occasional in western and south-western England, and Wales, much less frequent elsewhere and rare in southeastern England. Perhaps better known as *Pulcherricium caeruleum.*

TETRAGONIOMYCES Oberw. & Bandoni, *Canad. J. Bot.* 59: 7 (1981)
Tremellales, Tetragoniomycetaceae
Type: *Tetragoniomyces uliginosus* (P. Karst.) Oberw. & Bandoni

uliginosus (P. Karst.) Oberw. & Bandoni, *Canad. J. Bot.* 59: 7 (1981)
   Tremella uliginosa P. Karst., *Meddeland. Soc. Fauna Fl. Fenn.* 9: 111 (1883)
E: !
H: Parasitic. On fungal sclerotia on vegetation in marshes.
D: NM3: 90

Known from East Norfolk and South Devon but basidiomes are cryptic (thin gelatinous films) and easily overlooked.

THANATEPHORUS Donk, *Reinwardtia* 3: 376 (1956)
Ceratobasidiales, Ceratobasidiaceae
Uthatobasidium Donk, *Reinwardtia* 3: 376 (1956)
Cejpomyces Svrček & Pouzar, *Česká Mykol.* 24: 5 (1970)
Ypsilonidium Donk, *Proc. K. Ned. Akad. Wet.* 75: 371 (1972)
Rhizoctonia DC. (anam.), *Mém. Mus. Hist. Nat.* 2: 216 (1815)
Moniliopsis Ruhland (anam.), *Arbeiten Biol. Reichsanst. Land-Forstw.* 6: 76 (1908)
Type: *Hypochnus solani* Prill & Delacr. (= *T. cucumeris*)

amygdalisporus Hauerslev, P. Roberts & Å. Strid, in Knudsen & Hansen, *Nordic J. Bot.* 16(2): 217 (1996)
E: ! O: Channel Islands: !
H: On soil, decayed wood and dead herbaceous stems.
D: NM3: 115
Known from England (Hertfordshire, South Devon, Surrey and West Kent) and the Channel Islands (Guernsey).

cucumeris (A.B. Frank) Donk, *Reinwardtia* 3: 376 (1956)
Hypochnus cucumeris A.B. Frank, *Ber. Deutsch. Bot. Ges.* 1: 62 (1883)
Hypochnus filamentosus Pat., *Bull. Soc. Mycol. France* 7: 163 (1891)
Pellicularia filamentosa (Pat.) D.P. Rogers, *Farlowia* 1: 113 (1943)
Ceratobasidium filamentosum (Pat.) L.S. Olive, *Amer. J. Bot.* 44: 431 (1957)
Corticium praticola Kotila, *Phytopathology* 19: 1065 (1929)
Pellicularia praticola (Kotila) Flentje, *Trans. Brit. Mycol. Soc.* 39: 353 (1956)
Ceratobasidium praticola (Kotila) L.S. Olive, *Amer. J. Bot.* 44: 431 (1957)
Thanatephorus praticola (Kotila) Flentje, *Aust. J. biol. Sci.* 16: 451 (1963)
Hypochnus solani Prill. & Delacr., *Bull. Soc. Mycol. France* 7: 220 (1891)
Corticium solani (Prill. & Delacr.) Bourdot & Galzin, *Hyménomyc. France*: 242 (1928)
Botryobasidium solani (Prill. & Delacr.) Donk, *Meded. Ned. Mycol. Ver.* 18 - 20: 117 (1931)
Ceratobasidium solani (Prill. & Delacr.) Pilát, *Česká Mykol.* 11: 81 (1957)
Rhizoctonia solani J.G. Kühn (anam.), *Die Krankheiten der Kulturgewachsen*: 224 (1858)
Moniliopsis solani (J.G. Kühn) R.T. Moore (anam.), *Mycotaxon* 29: 95 (1987)
Mis.: *Corticium vagum* sensu Rea (1922)
E: c S: !
H: On decayed vegetable debris, wood, leaf litter, soil and also on living leaves and stems of various herbaceous plants.
D: NM3: 115 **D+I:** CNE8: 1489-1491
Common and widespread, at least in England.

fusisporus (J. Schröt.) Hauerslev & P. Roberts, in Knudsen & Hansen, *Nordic J. Bot.* 16(2): 218 (1996)
Hypochnus fusisporus J. Schröt., in Cohn, *Krypt.-Fl. Schlesien* 3: 416 (1888)
Uthatobasidium fusisporum (J. Schröt.) Donk, *Fungus* 28: 22 (1958)
Coniophora vaga Burt, *Ann. Missouri Bot. Gard.* 4: 251 (1917)
Mis.: *Pellicularia flavescens* sensu auct.
Mis.: *Corticium flavescens* sensu auct.
E: c NI: ! O: Channel Islands: o
H: On soil and on decayed wood, herbaceous debris, fungal basidiomes and grass culms.
D: NM3: 114 **D+I:** B&K2: 76 42 (as *Hypochnus fusisporus*), CNE8: 1588-1589 (as *Uthatobasidium fusisporum*)
Common and widespread throughout England (from Yorkshire southwards). Rarely reported elsewhere

ochraceus (Massee) P. Roberts, *Sydowia* 50: 252 (1998)
Coniophora ochracea Massee, *J. Linn. Soc., Bot.* 25: 137 (1889)
Uthatobasidium ochraceum (Massee) Donk, *Fungus* 28: 23 (1958)
Corticium frustulosum Bres., *Ann. Mycol.* 1(1): 98 (1903)
Thanatephorus orchidicola Warcup & P.H.B. Talbot, *Trans. Brit. Mycol. Soc.* 49: 432 (1966)
E: ! S: ! NI: !
H: On decayed wood, also forming endomycorrhizal associations with various *Orchidaceae*.
D: NM3: 114 (as *Thanatephorus orchidicola*) **D+I:** B&K2: 78 44 (as *Uthatobasidium ochraceum*)
Known in Britain mainly from cultures of orchid roots. Recorded from England (Cambridgeshire and Surrey), Scotland (West Sutherland), and Northern Ireland.

sterigmaticus (Bourdot) P.H.B. Talbot, *Persoonia* 3(4): 390 (1965)
Corticium sterigmaticum Bourdot, *Rev. Sci. du Bourb.* 35: 4 (1922)
Ceratobasidium sterigmaticum (Bourdot) D.P. Rogers, *Univ. Iowa Stud. Nat. Hist.* 17: 7 (1935)
Ypsilonidium sterigmaticum (Bourdot) Donk, *Proc. K. Ned. Akad. Wet.* 75: 371 (1972)
Thanatephorus langlei-regis D.A. Reid, *Trans. Brit. Mycol. Soc.* 52: 22 (1969)
Ypsilonidium langlei-regis (D.A. Reid) Donk, *Proc. K. Ned. Akad. Wet.* 75: 371 (1972)
E: !
H: On soil and on stems or leaves of herbaceous plants such as *Plantago lanceolata*.
D: NM3: 114 **D+I:** CNE8: 1610-1611 (as *Ypsilonidium sterigmaticum*)
Rarely recorded. Known in Britain only from South Devon and Surrey.

terrigenus (Bres.) G. Langer, *Biblioth. Mycol.* 158: 324 (1994)
Corticium terrigenum Bres., *Ann. Mycol.* 1(1): 99 (1903)
Ceratobasidium terrigenum (Bres.) Wakef., *Trans. Brit. Mycol. Soc.* 35: 64 (1952)
Cejpomyces terrigenus (Bres.) Svrček & Pouzar, *Česká Mykol.* 24: 5 (1970)
E: !
H: On bare soil.
D: NM3: 115 **D+I:** CNE2: 194-195 (as *Cejpomyces terrigenus*)
Rarely recorded. Known in Britain only from Berkshire and Surrey.

THELEPHORA Willd., *Fl. berol. prodr.*: 396 (1787)
Thelephorales, Thelephoraceae
Merisma Pers. [non *Merisma* Gillet (1878)], *Tent. disp. meth. fung.*: 74 (1797)
Phylacteria (Pers.) Pat., *Hyménomyc. Eur.*: 153 (1887)
Type: *Thelephora terrestris* Ehrh.

anthocephala (Bull.) Fr., *Epicr. syst. mycol.*: 535 (1838)
Clavaria anthocephala Bull., *Herb. France*: pl. 452 (1786)
Thelephora anthocephala var. *clavularis* (Fr.) Quél., *Enchir. fung.*: 203 (1886)
Thelephora clavularis Fr., *Observ. mycol.* 1: 1 (1815)
Phylacteria terrestris var. *digitata* Bourdot & Galzin, *Bull. Soc. Mycol. France* 40: 126 (1924)
Thelephora terrestris var. *digitata* (Bourdot & Galzin) Corner, *Nova Hedwigia Beih.* 27: 37 (1968)
E: o S: ! W: ! NI: ! ROI: !
H: On soil in deciduous and coniferous woodland.
D: NM3: 301 **D+I:** B&K2: 218 254 **I:** Sow2: pl. 156 (as *Clavaria anthocephala*), SV47: 38
Occasional in England. Few records elsewhere.

caryophyllea (Schaeff.) Pers., *Syn. meth. fung.*: 565 (1801)
Helvella caryophyllea Schaeff., *Fung. Bavar. Palat. nasc.* 4: 325 (1770)
Auricularia caryophyllea (Schaeff.) Bull., *Herb. France*: pl. 278 (1786)

Merulius caryophylleus (Schaeff.) With., *Arr. Brit. pl. ed. 3*, 4: 153 (1796)
  Phylacteria caryophyllea (Schaeff.) Pat., *Hyménomyc. Eur.*: 154 (1887)
  Stereum laciniatum Pers., *Observ. mycol.* 1: 36 (1796)
  Thelephora laciniata (Pers.) Pers., *Syn. meth. fung.*: 567 (1801)
E: ! **W:** !
H: On soil, in deciduous and coniferous woodland.
D: NM3: 301 **I:** Kriegl1: 395, SV47: 41
Reported from widely scattered locations in England and Wales from Northumberland southwards but mostly unsubstantiated with voucher material.

cuticularis Berk., Thelephora, *London J. Bot.* 6: 324 (1847)
E: !
H: On bark and mossy branches. British record on soil in cart-track.
D: Corner (1968: 53), Mycol. Helv. 7: 71 (1995)
A North American species with a single record from Somerset in 1958 determined by Corner. A further European record from Italy in 1987.

palmata (Scop.) Fr., *Syst. mycol.* 1: 432 (1821)
  Clavaria palmata Scop., *Fl. carniol.*: 483 (1772)
  Phylacteria palmata (Scop.) Pat., *Essai taxon.*: 119 (1900)
  Merisma foetidum Pers., *Comment. Fungis Clavaeform.*: 92 (1797)
E: ! **S:** ! **W:** ! **NI:** ! **ROI:** !
H: With *Pinus* spp and *Betula pendula* on acidic soil.
D: NM3: 301 **D+I:** Ph: 260-261, B&K2: 218 255 **I:** Kriegl1: 397, SV47: 39
Rarely reported but apparently widespread. Basidiomes in active growth have a strong smell, likened to 'decayed garlic' or 'sewage'.

penicillata Fr., *Syst. mycol.* 1: 434 (1821)
  Thelephora mollissima Pers., *Syn. meth. fung.*: 572 (1801)
  Phylacteria mollissima (Pers.) Rea, *Brit. basidiomyc.*: 653 (1922)
  Thelephora spiculosa Fr., *Syst. mycol.* 1: 434 (1821)
E: ! **S:** ! **W:** ! **NI:** ! **ROI:** !
H: On soil or decayed leaf litter, often in wet areas such as the sides of streams and in swampy woodland.
D+I: NM3: 301 **I:** SV47: 40
Rarely reported but apparently widespread.

terrestris Ehrh., *Pl. Crypt.*: 179 (1786)
  Phylacteria terrestris (Ehrh.) Pat., *Essai taxon.*: 119 (1900)
  Tomentella phylacteris Bourd. & Galzin, *Hyménomyc. France*: 486 (1928)
  Thelephora phylacteris (Bourdot & Galzin) Corner, *Nova Hedwigia Beih.* 27: 78 (1968)
E: c **S:** c **W:** c **NI:** c **ROI:** c **O:** Channel Islands: !
H: On sandy or acidic soil, on heathland or moorland and occasionally in open woodland.
D: NM3: 301 **D+I:** Ph: 260-261, B&K2: 220 256 **I:** C&D: 139, Kriegl1: 398, SV47: 37
Common and widespread.

TOMENTELLA Pat. [*nom. cons.*], *Hyménomyc. Eur.*: 154 (1887)
Thelephorales, Thelephoraceae
  Odontia Pers. [*nom. rej.*], *Neues Mag. Bot.* 1: 110 (1794)
  Hypochnus Fr. [non *Hypochnus* Ehrenb. (1820)], *Observ. mycol.* 2: 278 (1818)
  Caldesiella Sacc. [*nom. rej.*], *Fung. Ital.*: f. 125 (1877)
  Tomentellina Höhn. & Litsch., *Sitzungsber. Kaiserl. Akad. Wiss., Wien, Math.-Naturwiss. Cl., Abt. 1* 115: 1604 (1906)
  Tomentellastrum Svrček, *Česká Mykol.* 12: 68 (1958)
Type: *Tomentella ferruginea* (Pers.) Pat.

asperula (P. Karst.) Höhn. & Litsch., *Sitzungsber. Kaiserl. Akad. Wiss., Wien, Math.-Naturwiss. Cl., Abt. 1* 115: 1570 (1906)
  Hypochnus asperulus P. Karst., *Bidrag Kännedom Finlands Natur Folk* 48: 441 (1889)

Tomentella gibbosa Litsch., *Bull. Soc. Mycol. France* 49: 70 (1933)
E: ? **ROI:** !
H: On decayed wood and bark
D: NM3: 305 **D+I** Kõljalg: 93-96
Known from the Republic of Ireland (North Kerry and West Cork) but not reported since 1936. Dubiously reported from England (West Norfolk) in 1986.

atrovirens (Bres.) Höhn. & Litsch., *Sitzungsber. Kaiserl. Akad. Wiss., Wien, Math.-Naturwiss. Cl., Abt. 1* 116: 831 (1907)
  Hypochnus atrovirens Bres., *Atti Imp. Regia Accad. Roveretana, ser. 3*, 3: 116 (1897)
E: !
H: On decayed wood of deciduous trees. British material on *Fagus* and *Fraxinus* in woodland on calcareous soil.
Known from southern England (Surrey and West Sussex).

badia (Link) Stalpers, *Rev. Mycol. (Paris)* 39(2): 98 (1975)
  Sporotrichum badium Link, *Mag. Neuesten Entdeck. Gesammten Naturk. Ges. Naturf. Freunde Berlin* 3: 12 (1809)
  Tomentellastrum badium (Link) M.J. Larsen, *Nova Hedwigia* 35(1): 5 (1981)
  Tomentella fimbriata M.P. Christ., *Dansk. Bot. Ark.* 19(2): 258 (1960)
  Tomentellastrum floridanum (Ellis & Everh.) M.J. Larsen, *Mycol. Mem.* 4: 116 (1974)
E: ! **W:** !
H: On decayed wood and fallen bark of deciduous trees and rarely on bare sandy or clay soil in woodland.
D: NM3: 304 **D+I** Kõljalg: 145-147
Known from England (East Norfolk, Hertfordshire, Oxfordshire, South Devon and South Hampshire) and from Wales (Caernarvonshire).

bryophila (Pers.) M.J. Larsen, *Mycol. Mem.* 4: 51 (1974)
  Sporotrichum bryophilum Pers., *Mycol. eur.* 1: 78 (1822)
  Tomentella pallidofulva (Peck) Litsch., *Oesterr. Bot. Z.* 88: 131 (1939)
  Tomentella subferruginea (Burt) Skovst., *Compt.-Rend. Trav. Carlsberg Lab., Ser. physiol.* 25: 20 (1950)
E: o **S:** ! **W:** ! **ROI:** !
H: On decayed wood or fallen, decayed bark of deciduous trees. Also known on discarded and decayed artefacts such as leather shoes and wallets.
D+I: B&K2: 212 244, Kõljalg: 157-161
Rarely reported but not infrequent at least in southern England.

calcicola (Bourdot & Galzin) M.J. Larsen, *Taxon* 16: 511 (1967)
  Caldesiella ferruginosa var. *calcicola* Bourdot & Galzin, *Hyménomyc. France*: 471 (1928)
  Caldesiella calcicola (Bourdot & Galzin) M.P. Christ., *Dansk. Bot. Ark.* 19(2): 303 (1960)
E: !
H: On soil.
D+I: Kõljalg: 116-118
Known from Warwickshire (Houndshill, near Stratford) in 1971, from soil at the sides of a disused railway cutting.

cinerascens (P. Karst.) Höhn. & Litsch., *Sitzungsber. Kaiserl. Akad. Wiss., Wien, Math.-Naturwiss. Cl., Abt. 1* 115: 1570 (1906)
  Hypochnus cinerascens P. Karst., *Meddeland. Soc. Fauna Fl. Fenn.* 16: 2 (1888)
  Tomentella subcervina Litsch., *Bull. Soc. Mycol. France* 49: 60 (1933)
E: ! **S:** ! **NI:** ? **ROI:** ?
H: On decayed wood.
D: NM3: 304 **D+I:** Kõljalg: 96-99
Known from England (Shropshire, South Somerset, Surrey and West Norfolk) and Scotland (Easterness). Reported from Northern Ireland in 1948 and the Republic of Ireland in 1936 but unsubstantiated with voucher material.

clavigera Litsch., *Sydowia* 14: 192 (1960)
E: !

H: On a fallen and decayed cupule of *Fagus* (British material).
D+I: Kõljalg: 185-188
A single collection from Hertfordshire (Welwyn, Danesbury Park) in 1998.

coerulea (Bres.) Höhn. & Litsch., *Sitzungsber. Kaiserl. Akad. Wiss., Wien, Math.-Naturwiss. Cl., Abt. 1* 116: 831 (1907)
   Hypochnus coeruleus Bres., *Ann. Mycol.* 1(1): 109 (1903)
   Tomentella jaapii (Bres.) Bourdot & Galzin, *Bull. Soc. Mycol. France* 40: 155 (1924)
   Hypochnus cervinus Burt, *Ann. Missouri Bot. Gard.* 3: 232 (1916)
   Tomentella cervina (Burt) Bourdot & Galzin, *Bull. Soc. Mycol. France* 40: 146 (1924)
   Tomentella sordida Wakef., *Trans. Brit. Mycol. Soc.* 53: 185 (1969)
E: ! **S:** !
H: On decayed wood and occasionally on bare soil.
D: NM3: 307 **D+I:** Kõljalg: 178-182
Known from England (Buckinghamshire, Cumberland, Oxfordshire, South Hampshire, Surrey and West Kent) and Scotland (Mid-Perthshire). Rarely reported elsewhere but unsubstantiated with voucher material. *Tomentella cervina* sensu Wakefield (1969) is doubtfully this species.

crinalis (Fr.) M.J. Larsen, *Taxon* 16: 511 (1967)
   Hydnum crinale Fr., *Epicr. syst. mycol.*: 516 (1838)
   Caldesiella crinalis (Fr.) Rea, *Brit. basidiomyc.*: 651 (1922)
   Hydnum fuscum Pers., *Mycol. eur.* 2: 189 (1825)
   Caldesiella ferruginosa (Pers.) Sacc., *Michelia* 1: 7 (1877)
E: o **S:** ! **W:** ! **NI:** ! **ROI:** ! **O:** Channel Islands: !
H: On decayed wood of deciduous trees, usually *Fagus* or *Ulmus* spp.
D: NM3: 304 **D+I:** B&K2: 212 245, Kõljalg: 118-122
Occasional in southern England. Rarely reported elsewhere, but widespread.

ellisii (Sacc.) Jülich & Stalpers, *Verh. Kon. Ned. Akad. Wetensch., Afd. Natuurk., Sect. 2,* 74: 236 (1980)
   Zygodesmus ellisii Sacc., *Syll. fung.* 4: 808 (1886)
   Tomentella hydrophila (Bourdot & Galzin) Litsch., *Oesterr. Bot. Z.* 88: 128 (1939)
   Tomentella livida Litsch., *Svensk bot. Tidskr.* 26: 450 (1932)
   Tomentella luteomarginata M.P. Christ., *Dansk. Bot. Ark.* 19: 263 (1960)
   Zygodesmus ochraceus Sacc., *Michelia* 2: 565 (1882)
   Tomentella ochracea (Sacc.) M.J. Larsen, *Mycologia Mem.* 4: 73 (1974)
E: c **S:** ! **W:** ! **NI:** ! **O:** Channel Islands: !
H: On decayed wood of deciduous or coniferous trees and rarely reported on decayed debris of *Typha latifolia* and *Phragmites australis.*
D: NM3: 307 **D+I:** Kõljalg: 131-134
Common and widespread at least in England but rarely reported elsewhere.

ferruginea (Pers.) Pat., *Hyménomyc. Eur.*: 154 (1887)
   Corticium ferrugineum Pers., *Observ. mycol.* 2: 18 (1799)
   Thelephora ferruginea (Pers.) Pers., *Syn. meth. fung.* 1: 569 (1801)
   Hypochnus ferrugineus (Pers.) Fr., *Observ. mycol.*: 180 (1818)
   Grandinia coriaria Peck, *Bull. Buffalo Soc. Nat. Sci.* 1(2): 61 (1873)
   Tomentella coriaria (Peck) Bourdot & Galzin, *Bull. Soc. Mycol. France* 40: 159 (1924)
   Hypochnus fuscus P. Karst., *Symb. mycol. fenn.* 8: 13 (1881)
   Tomentella fusca (P. Karst.) J. Schröt., in Cohn, *Krypt.-Fl. Schlesien* 3: 419 (1889)
   Mis.: *Tomentella botryoides* sensu Wakefield (1969)
E: o **S:** ! **W:** ! **NI:** ! **ROI:** !
H: On decayed wood of deciduous trees and rarely on decayed debris of *Juncus* spp.
D: NM3: 304 **D+I:** Kõljalg: 105-109
Occasional in England, rarely reported elsewhere but widespread.

fibrosa (Berk. & M.A. Curtis) Kõljalg, *Syn. Fungorum* 9: 122 (1996)
   Zygodesmus fibrosus Berk. & M.A. Curtis, *Grevillea* 3: 145 (1875)
   Tomentellina fibrosa (Berk. & M.A. Curtis) M.J. Larsen, *Mycol. Mem.* 4: 115 (1974)
   Kneiffiella bombycina P. Karst., *Acta Soc. Fauna Fl. Fenn.* 11: 1 (1895)
   Tomentellina ferruginosa Höhn. & Litsch., *Beitr. Kenn. Cort.* 1: 1604 (1906)
E: ! **S:** ?
H: On decayed wood of deciduous or coniferous trees.
D: NM3: 302 **D+I:** B&K2: 216 251, Kõljalg: 122-126
A single collection from Oxfordshire (Bix, Warburg Reserve) in 1998.

fuscocinerea (Pers.) Donk, *Meded. Bot. Mus. Herb. Rijks Univ. Utrecht* 9: 30 (1933)
   Thelephora fuscocinerea Pers., *Mycol. eur.* 1: 114 (1822)
   Tomentellastrum fuscocinereum (Pers.) Svrček, *Česká Mykol.* 12: 69 (1958)
   Hypochnus alutaceoumbrinus Bres., *Ann. Mycol.* 1(1): 109 (1903)
   Tomentellastrum alutaceoumbrinum (Bres.) M.J. Larsen, *Mycol. Mem.* 4: 107 (1974)
   Tomentella macrospora Höhn. & Litsch., *Sitzungsber. Kaiserl. Akad. Wiss., Wien, Math.-Naturwiss. Cl., Abt. 1* 115: 24 (1906)
   Mis.: *Tomentella caesia* sensu Wakefield (1917)
   Mis.: *Soppittiella caesia* sensu auct. Brit.
   Mis.: *Hypochnus caesius* sensu Rea (1922)
   Mis.: *Thelephora phylacteris* sensu auct. mult.
   Mis.: *Hypochnus phylacteris* sensu Rea in TBMS 12 (2-3): 222 (1927)
E: ! **W:** ! **NI:** ! **ROI:** !
H: On decayed wood, fallen litter, or soil in woodland.
D: NM3: 303 **D+I:** Kõljalg: 150-154
Rarely reported but apparently widespread.

griseoviolacea Litsch., *Ann. Mycol.* 39: 375 (1941)
E: !
H: On decayed wood of deciduous trees.
D+I: B&K2: 214 247
A single collection, from Hertfordshire (Astonbury Wood) in 1999. Reported from Surrey but unsubstantiated with voucher material. Some authorities, e.g. Kõljalg (Kõljalg), consider this a synonym of *T. punicea.*

italica (Sacc.) M.J. Larsen, *Taxon* 16: 511 (1967)
   Caldesiella italica Sacc., *Michelia* 1: 7 (1877)
E: ! **S:** ! **W:** !
H: On decayed wood of deciduous trees, usually large logs and stumps of *Ulmus* spp. Rarely on conifer wood, or on fallen and decayed basidiomes of large woody polypores.
D: NM3: 304 **D+I:** Kõljalg: 89-92
Rarely reported but apparently widespread. Rather frequent in southern England during the epidemic of Dutch elm disease, but now appears to be in decline.

lapidum (Pers.) Stalpers, *Stud. Mycol.* 24: 65 (1984)
   Sporotrichum lapidum Pers., *Mycol. eur.* 1: 78 (1822)
   Zygodesmus ramosissimus Berk. & M.A. Curtis, Grevillea 3: 145 (1875)
   Rhinotrichum ramosissimum Berk. & M.A. Curtis, *Grevillea* 3: 108 (1875)
   Tomentella ramosissima (Berk. & M.A. Curtis) Wakef., Mycologia 52: 927 (1960)
   Thelephora spongiosa Schwein., *Schriften Naturf. Ges. Leipzig* 1: 109 (1822)
   Tomentella spongiosa (Schwein.) Bourdot & Galzin, *Hyménomyc. France*: 503 (1928)
E: o **W:** ! **ROI:** !
H: On decayed wood of deciduous trees.
D: NM3: 305 **D+I:** Kõljalg: 161-164
Known from England (Hertfordshire, Oxfordshire, Surrey and Yorkshire), Wales (Breconshire and Glamorganshire) and the

Republic of Ireland (North Kerry). Reported elsewhere but unsubstantiated with voucher material.

lateritia Pat., *J. Bot.* 8: 221 (1894)
Mis.: *Tomentella punicea* sensu Wakefield (1969)
E: ! S: ! ROI: !
H: On decayed wood.
D+I: Kõljalg: 182-185
Known from England (Hertfordshire, West Kent and Westmorland), Scotland (Easterness) and the Republic of Ireland (North Kerry). Reported elsewhere but unsubstantiated with voucher material and often confused with *T. punicea*.

lilacinogrisea Wakef., *Trans. Brit. Mycol. Soc.* 49: 360 (1966)
Tomentella neobourdotii M.J. Larsen, *Mycologia* 60: 1179 (1968)
Mis.: *Tomentella bourdotii* sensu Wakefield (1969)
E: ! W: !
H: On decayed wood, fallen and decayed leaves and litter of deciduous trees, fern debris and decayed basidiomes of polypores.
D: NM3: 305 **D+I:** Kõljalg: 164-168
Known from England (Hertfordshire, Oxfordshire, South Hampshire and Surrey) and Wales (Breconshire). Reported from a few other sites but the records are old and unsubstantiated with voucher material.

microspora (P. Karst.) Höhn. & Litsch., *Sitzungsber. Kaiserl. Akad. Wiss., Wien, Math.-Naturwiss. Cl., Abt. 1* 115: 1571 (1906)
Hypochnus microsporus P. Karst., *Hedwigia* 35: 174 (1896)
E: ? W: !
H: On decayed wood of deciduous trees. British material on *Alnus*.
A single collection, from Caernarvonshire (Glaslyn Marshes) in 2001. Reported from England (Hertfordshire and Surrey) but these records are old, unsubstantiated with voucher material and probably refer to other species of *Tomentella*.

pilosa (Burt) Bourdot & Galzin, *Bull. Soc. Mycol. France* 40: 151 (1924)
Hypochnus pilosus Burt, *Ann. Missouri Bot. Gard.* 3: 221 (1916)
Tomentella subpilosa Litsch., *Sydowia* 14: 224 (1960)
E: ! S: ! W: ! NI: ! ROI: !
H: On decayed wood of deciduous trees. British material on *Corylus* and *Fagus*.
D: NM3: 303 **D+I:** B&K2: 214 248, Kõljalg: 86-89
Rarely reported but apparently widespread.

puberula Bourdot & Galzin, *Bull. Soc. Mycol. France* 40: 150 (1924)
E: ! NI: !
H: On decayed wood of deciduous trees, also on fallen litter such as leaves and twigs.
Known from England (South Hampshire and Warwickshire) and Northern Ireland (Down). Some authorities, e.g. Kõljalg (Kõljalg), make this a synonym of *T. coerulea* but records are here taken sensu Wakefield (1969), as a distinct species.

punicea (Alb. & Schwein.) J. Schröt., in Cohn, *Krypt.-Fl. Schlesien.* 420 (1889)
Thelephora punicea Alb. & Schwein., *Consp. fung. lusat.*: 278 (1805)
Hypochnus puniceus (Alb. & Schwein.) Sacc., *Syll. fung.* 6: 661 (1888)
Hydnum epiphyllum Schwein., *Trans. Am. Phil. Soc.* 4: 320 (1832)
Tomentella epiphylla (Schwein.) Litsch., *Oest. Bot. Z.* 88: 131 (1939)
Hypochnus epiphyllus (Schwein.) Burt [non *H. epiphyllus* (Pers.) Wallr. (1833)], *Rep. (Annual) Missouri Bot. Gard.* 13: 320 (1926)
Hypochnus granulosus (Peck) Burt, *Rep. (Annual) Missouri Bot. Gard.* 3: 218 (1916)
Hypochnus rubiginosus Bres., *Atti Imp. Regia Accad. Roveretana, ser. 3,* 3: 116 (1897)

Tomentella rubiginosa (Bres.) Maire, *Ann. Mycol.* 4: 335 (1906)
Mis.: *Tomentella lateritia* sensu Wakefield (1969)
E: ! S: ! ROI: !
H: On decayed wood of deciduous trees in old woodland, especially on *Fagus* but also known on *Quercus* and *Salix* spp, and rarely on decayed lianas of *Clematis vitalba*.
D: NM3: 306 **D+I:** Kõljalg: 109-112
Rarely reported but apparently widespread. However, some of the records will refer to *T. lateritia*.

radiosa (P. Karst.) Rick, *Brotéria, Ci. Nat.* 2: 79 (1934)
Hypochnus fuscus var. *radiosus* P. Karst., *Meddeland. Soc. Fauna Fl. Fenn.* 9: 71 (1882)
Tomentella fusca var. *radiosa* (P. Karst.) Höhn. & Litsch., *Sitzungsber. Kaiserl. Akad. Wiss., Wien, Math.-Naturwiss. Cl., Abt. 1* 115: 1571 (1906)
Tomentella cladii Wakef., *Trans. Brit. Mycol. Soc.* 53: 179 (1969)
Tomentella pannosa Bourdot & Galzin, *Bull. Soc. Mycol. France* 40: 150 (1924)
E: ! S: !
H: On decayed wood of deciduous and coniferous trees, also on debris of ferns such as *Blechnum spicant* and *Pteridium*. Rarely on decayed debris of *Cladium mariscus* or *Juncus* spp.
D: NM3: 307 **D+I:** Kõljalg: 137-141
Rarely reported.

stuposa (Link) Stalpers, *Stud. Mycol.* 24: 86 (1984)
Sporotrichum stuposum Link, *Mag. Neuesten Entdeck. Gesammten Naturk. Ges. Naturf. Freunde Berlin* 3: 12 (1809)
Hypochnus bresadolae Brinkmann, *Ann. Mycol.* 1: 108 (1903)
Tomentella bresadolae (Brinkmann) Bourdot & Galzin, *Bull. Soc. Mycol. France* 40: 155 (1925)
Tomentella hoehnelii Skovst., *Compt.-Rend. Trav. Carlsberg Lab., Ser. physiol.* 25: 29 (1950)
Mis.: *Tomentella subfusca* sensu auct.
Mis.: *Hypochnus subfuscus* sensu Rea (1922)
E: ! S: ! NI: ! ROI: ! O: Channel Islands: !
H: On decayed wood of deciduous trees.
D+I: Kõljalg: 168-172
Rarely reported but apparently widespread.

subclavigera Litsch., *Bull. Soc. Mycol. France* 49: 57 (1933)
E: ! W: !
H: On decayed wood of deciduous trees. British material was on *Betula* spp.
D: NM3: 303 **D+I:** B&K2: 214 249, Kõljalg: 191-194
A single confirmed collection from England (Yorkshire: Askham Bog) in 1940, and one from Wales (Pembrokeshire: Tenby) in 1961.

sublilacina (Ellis & Holw.) Wakef., *Mycologia* 52: 931 (1962)
Zygodesmus sublilacinus Ellis & Holw., in Arthur *et al.*, *Bull. Geol. Nat. Hist. Surv. Minnesota* 3: 34 (1887)
Tomentella porulosa f. *albomarginata* Bourdot & Galzin, *Hymenomyc. France*: 505 (1928)
Tomentella albomarginata (Bourdot & Galzin) M.J. Larsen, *Mycologia* 62: 134 (1970)
Tomentella pseudopannosa Wakef., *Trans. Brit. Mycol. Soc.* 53: 189 (1969)
Mis.: *Tomentella fusca* sensu Wakefield (1969)
E: c S: c W: o ROI: ! O: Channel Islands: !
H: On decayed wood, litter, fallen leaves and debris and often covering soil in wet areas in woodland at the margins of lakes or ponds, especially when partially dried-out.
D: NM3: 307 **D+I:** Kõljalg:141-145
Common and widespread.

subtestacea Bourdot & Galzin, *Bull. Soc. Mycol. France* 40: 144 (1924)
E: !
H: On decayed wood of deciduous trees.
D+I: Kõljalg: 194-196

A single collection from Oxfordshire (The Warburg Reserve) in 1998. Reported from Berkshire but unsubstantiated with voucher material.

terrestris (Berk. & Broome) M.J. Larsen, *Mycol. Mem.* 4: 105 (1974)
 Zygodesmus terrestris Berk. & Broome, *Ann. Mag. Nat. Hist., Ser. 5 [Notices of British Fungi no. 1915]* 7: 130 (1881)
E: !
H: On bare chalk (British material).
D+I: Kõljalg: 175-178
The only British record is the type collection from East Kent (Crundale) in 1866.

testaceogilva Bourdot & Galzin, *Bull. Soc. Mycol. France* 40: 149 (1924)
 Mis.: *Tomentella rhodophaea* sensu Jülich (1984)
E: ! **W:** ! **NI:** !
H: On decayed wood and fallen bark of deciduous trees.
D: NM3: 306 **D+I:** Kõljalg: 130-131
Rarely reported but apparently widespread. The last collection was from Middlesex (Ruislip, Mad Bess Woods) in 1953.

umbrinospora M.J. Larsen, *Techn. Publ. New York State Coll. Forest.* 93: 61 (1963)
E: ! **W:** !
H: On decayed wood of deciduous trees. British collections usually on *Fagus* or *Corylus*.
D: NM3: 304 **D+I:** Kõljalg: 126-129
Known from England (Berkshire, Hertfordshire, Mid-Lancashire, Oxfordshire and Surrey) and Wales (Carmarthenshire)

viridula Bourdot & Galzin, *Bull. Soc. Mycol. France* 40: 144 (1924)
E: ! **ROI:** !
H: On decayed wood of deciduous trees such as *Carpinus, Corylus* and *Fagus*.
D: NM3: 303 **D+I:** Kõljalg: 196-199
Known from England (Hertfordshire, South Devon, West Norfolk and West Sussex) and the Republic of Ireland (North Kerry). Reported from a few other locations in England but unsubstantiated with voucher material.

TOMENTELLOPSIS Hjortstam, *Svensk bot. Tidskr.* 64(4): 425 (1970)
Thelephorales, Thelephoraceae
Type: *Tomentellopsis echinospora* (Ellis) Hjortstam

echinospora (Ellis) Hjortstam, *Svensk bot. Tidskr.* 64(4): 426 (1970)
 Corticium echinosporum Ellis, *Bull. Torrey Bot. Club* 8: 64 (1881)
 Hypochnus echinosporus (Ellis) Burt, *Rep. (Annual) Missouri Bot. Gard.* 3: 237 (1916)
 Tomentella echinospora (Ellis) Bourdot & Galzin, *Hyménomyc. France*: 483 (1928)
E: c **S:** ! **W:** ! **NI:** !
H: On decayed wood of deciduous trees, frequently *Ulmus* spp, and occasionally on weathered dung of horses in old woodland such as the New Forest.
D: NM3: 308 **D+I:** B&K2: 216 252, Kõljalg: 69-71
Common, at least in England, but rarely reported elsewhere. A small-spored form has also been collected in Britain and may represent a separate species, but requires further investigation.

pusilla Hjortstam, *Svensk bot. Tidskr.* 68: 53 (1974)
E: !
H: On decayed wood of deciduous trees. British material on *Acer pseudoplatanus*.
D: NM3: 308 **D+I:** Kõljalg: 72-74
A single collection from England (North Somerset: Kingweston) in 2000.

submollis (Svrček) Hjortstam, *Svensk bot. Tidskr.* 68(1): 53 (1974)
 Pseudotomentella submollis Svrček, *Česka Mykol.* 12: 68 (1958)
 Tomentella submollis (Svrček) Wakef., *Techn. Publ. New York State Coll. Forest.* 93: 167 (1968)
E: o **S:** !
H: On decayed wood, mainly of coniferous trees.
D: Wakefield (1969)
Distinguished from *T. echinospora* by its pinkish fruitbodies and mycelia and slightly larger spores.

zygodesmoides (Ellis) Hjortstam, *Svensk bot. Tidskr.* 68: 55 (1974)
 Thelephora zygodesmoides Ellis, *North American Fungi*: no. 715 (1882)
 Tomentella zygodesmoides (Ellis) Höhn. & Litsch., *Sitzungsber. Kaiserl. Akad. Wiss., Wien, Math.-Naturwiss. Cl., Abt. 1* 116: 787 (1907)
 Hypochnus zygodesmoides (Ellis) Burt, *Rep. (Annual) Missouri Bot. Gard.* 3: 236 (1916)
E: ! **S:** ! **W:** !
H: On decayed conifer wood. British material on *Pinus sylvestris*.
D: NM3: 307 **D+I:** Kõljalg: 74-77
Apparently widespread but the majority of the records are old and unsubstantiated with voucher material. Collected recently from South Hampshire (New Forest) and Oxfordshire (Henley).

TRAMETES Fr., *Fl. Scan.*: 339 (1835)
Poriales, Coriolaceae
 Coriolus Quél., *Enchir. fung.*: 175 (1886)
 Pseudotrametes Bondartsev & Singer, *Mycologia* 36: 68 (1944)
Type: *Trametes suaveolens* (L.) Fr.

gibbosa (Pers.) Fr., *Epicr. syst. mycol.*: 492 (1838)
 Merulius gibbosus Pers., *Observ. mycol.* 1: 21 (1796)
 Daedalea gibbosa (Pers.) Pers., *Syn. meth. fung.*: 501 (1801)
 Polyporus gibbosus (Pers.) P. Kumm., *Führ. Pilzk.*: 59 (1871)
 Pseudotrametes gibbosa (Pers.) Bondartsev & Singer, *Mycologia* 36: 68 (1944)
 Trametes gibbosa f. *tenuis* Pilát, *Atlas champ. Eur. Polyporaceae I* 3(1): 290 (1936)
 Mis.: *Daedalea polyzona* sensu auct.
E: c **S:** c **W:** c **NI:** c **ROI:** c **O:** Channel Islands: o
H: On fallen, decayed wood of deciduous trees, especially common on large fallen trunks or stumps of *Acer pseudoplatanus* and *Fagus* in woodland, and not infrequent on *Aesculus* in parkland in southern England. Uncommon on *Betula* spp, *Castanea*, *Corylus*, *Populus* spp, *Prunus avium*, *Rhamnus*, *Rosa*, *Salix*, *Tilia* and *Ulmus* spp, and rare on conifers such as *Larix* and *Pinus* spp.
D: EurPoly2: 656, NM3: 229 **D+I:** Ph: 228-229 (as *Pseudotrametes gibbosa*), B&K2: 284 354 **I:** Kriegl1: 585
Very common and widespread. The form *tenuis* was a name given to thin basidiomes formed on a depauperate substratum, such as small sticks or twigs.

hirsuta (Fr.) Pilát, *Atlas champ. Eur. Polyporaceae I* 3(1): 265 (1939)
 Polyporus hirsutus Fr., *Syst. mycol.* 1: 367 (1821)
 Coriolus hirsutus (Fr.) Quél., *Enchir. fung.*: 175 (1886)
 Polystictus hirsutus (Fr.) Cooke, *Grevillea* 14: 83 (1886)
 Mis.: *Daedalea polyzona* sensu auct. Brit.
E: o **S:** ! **W:** ! **NI:** ! **O:** Channel Islands: !
H: On dead wood of deciduous trees, often on large stumps or fallen branches in old woodland, and often on calcareous soil. Usually on *Betula* spp, and *Fagus* but also known on *Alnus, Corylus, Crataegus, Fraxinus, Prunus, Quercus* and *Salix* spp. Rarely on dead stems of *Ulex europaeus* on heathland.
D: EurPoly2: 657, NM3: 230 **D+I:** Ph: 235 (as *Coriolus hirsutus*), B&K2: 286 355 **I:** Kriegl1: 588
Occasional in England. Rarely reported elsewhere but apparently widespread.

ochracea (Pers.) Gilb. & Ryvarden, *North American Polypores* 2: 752 (1987)
 Boletus ochraceus Pers., *Ann. Bot. Usteri* 11: 29 (1794)

Boletus multicolor Schaeff., *Fung. Bavar. Palat. nasc.* 4: 91 (pl. 269) (1774)

Trametes multicolor (Schaeff.) Jülich, *Persoonia* 11(4): 427 (1982)

Trametes zonatella Ryvarden, *Polyporaceae of North Europe* 2: 436 (1978)

Boletus zonatus Nees, *Syst. Pilze*. 221 (1817)

Polyporus zonatus (Nees) Fr., *Syst. mycol.* 1: 368 (1821)

Polystictus zonatus (Nees) Cooke, *Grevillea* 14: 83 (1886)

Coriolus zonatus (Nees) Quél., *Fl. mycol. France*: 390 (1888)

E: o **S:** ! **W:** ! **NI:** !

H: On dead wood of deciduous trees. Reported on *Acer pseudoplatanus*, *Aesculus*, *Betula* spp, *Corylus*, *Fagus*, *Prunus*, *Salix caprea*, *S. babylonica*, *Syringa* and *Tilia* spp.

D: EurPoly2: 663, NM3: 230 **D+I:** B&K2: 286 356 (as *Trametes multicolor*)

Rarely reported but apparently widespread. Easily confused with *Trametes versicolor*.

pubescens (Schumach.) Pilát, *Atlas champ. Eur. Polyporaceae I* 3(1): 268 (1939)

Boletus pubescens Schumach., *Enum. pl.* 2: 384 (1803)

Polyporus pubescens (Schumach.) Fr., *Syst. mycol.* 1: 367 (1821)

Coriolus pubescens (Schumach.) Quél., *Fl. mycol. France*: 391 (1888)

Polystictus pubescens (Schumach.) Gillot & Lucand, *Hyménomyc. France*: 351 (1890)

Leptoporus pubescens (Schumach.) Pat., *Essai taxon.*: 84 (1900)

Tyromyces pubescens (Schumach.) Imazeki, *Bull. Tokyo Sci. Mus.* 6: 84 (1943)

Boletus velutinus Pers. [non *B. velutinus* With. (1796)], *Syn. Meth. Fung.*: 539 (1801)

Polyporus velutinus (Pers.) Fr., *Syst. mycol.* 1: 368 (1821)

Polystictus velutinus (Pers.) Cooke, *Grevillea* 14: 83 (1886)

E: ! **S:** ! **W:** ! **NI:** !

H: In deciduous woodland, usually on *Betula* spp and *Fagus*. Less often on *Alnus*, *Corylus*, *Fraxinus*, *Ilex*, *Populus tremula*, *Quercus* and *Salix* spp, and on dead or burnt stems of *Ulex europaeus* on heathland.

D: EurPoly2: 664, NM3: 230 **D+I:** B&K2: 286 357

Rarely reported but apparently widespread.

suaveolens (L.) Fr., *Epicr. syst. mycol.*: 491 (1838)

Boletus suaveolens L., *Sp. pl.*: 1177 (1753)

Polyporus suaveolens (L.) Fr., *Syst. mycol.* 1: 366 (1821)

Trametes suaveolens f. *indora* (L.) Pilát, *Bull. Soc. Franc. Mycol. Méd.* 49: 264 (1933)

Trametes bulliardii Fr., *Epicr. syst. mycol.*: 491 (1838)

Trametes odora Fr., *Epicr. syst. mycol.*: 491 (1838)

Mis.: *Haploporus odorus* sensu auct.

E: ! **S:** ? **NI:** ?

H: Parasitic then saprotrophic. On trunks of *Populus* and *Salix* spp, often alongside streams or on riverbanks. Reported on *Acer pseudoplatanus* and *Quercus* spp but the hosts are unverified.

D: EurPoly2: 666, NM3: 229 **D+I:** B&K2: 288 358 **I:** Sow2: 228 (as *Boletus suaveolens*)

Known from England (Bedfordshire, Buckinghamshire, Cheshire, Herefordshire, Oxfordshire, North Somerset, Warwickshire and Worcestershire). Recorded from Northern Ireland and Scotland but unsubstantiated with voucher material.

versicolor (L.) Pilát, *Atlas champ. Eur. Polyporaceae I* 3(1): 261 (1939)

Boletus versicolor L., *Sp. pl.*: 1176 (1753)

Poria versicolor (L.) Scop., *Fl. carniol.*: 468 (1772)

Agaricus versicolor (L.) Lam., *Encycl.* 1: 50 (1783)

Polyporus versicolor (L.) Fr., *Syst. mycol.* 1: 368 (1821)

Coriolus versicolor (L.) Quél., *Enchir. fung.*: 175 (1886)

Polystictus versicolor (L.) Cooke, *Grevillea* 14: 83 (1886)

Polyporus fuscatus Fr., *Observ. mycol.* 2: 259 (1818)

Polystictus versicolor var. *fuscatus* (Fr.) Rea, *Brit. basidiomyc.*: 609 (1922)

Polyporus nigricans Lasch [*nom. illegit.*, non *P. nigricans* Fr. (1821)], in Rabenhorst, *Fungi europaei* IV: 15 (1859)

Polystictus versicolor var. *nigricans* (Lasch) Rea, *Brit. basidiomyc.*: 609 (1922)

E: c **S:** c **W:** c **NI:** c **ROI:** c **O:** Channel Islands: c Isle of Man: c

H: Weakly parasitic then saprotrophic. On wood of deciduous trees and shrubs and rarely on conifers such as *Picea* or *Pinus* spp. Also known on worked timber.

D: EurPoly2: 667, NM3: 230 **D+I:** Ph: 235, B&K2: 288 359 **I:** Sow2: 229 (as *Boletus versicolor*), Kriegl1: 592

Very common and widespread. The basidiomes are variable, especially in the colour and degree of zonation of the pilei, with forms ranging from pale greyish-ochraceous to almost black.

TRECHISPORA P. Karst., *Hedwigia* 29: 147 (1890)

Stereales, Sistotremataceae

Echinotrema Park.-Rhodes, *Trans. Brit. Mycol. Soc.* 38: 367 (1955)

Mis.: *Cristella* sensu Donk [*Reinwardtia* 4: 19 (1957)]

Type: *Trechispora onusta* P. Karst.

alnicola (Bourdot & Galzin) Liberta, *Taxon* 15(8): 318 (1966)

Grandinia alnicola Bourdot & Galzin, *Bull. Soc. Mycol. France* 30(2): 254 (1914)

Cristella alnicola (Bourdot & Galzin) Donk, *Fungus* 27: 19 (1957)

E: !

H: Usually on decayed wood of deciduous trees such as *Betula* spp and *Fagus*. Rarely on conifers such as *Pinus sylvestris*.

D: NM3: 133 **D+I:** CNE8: 1492-1495

Known from North Somerset, South Hampshire, Surrey and West Suffolk.

byssinella (Bourdot) Liberta, *Taxon* 15(8): 318 (1966)

Corticium byssinellum Bourdot, *Rev. Sci. du Bourb.* 23(1): 13 (1910)

Cristella byssinella (Bourdot) Donk, *Fungus* 27: 19 (1957)

E: ! **S:** ! **NI:** !

H: On decayed wood and leaf litter of deciduous trees and on living mosses (*Sphagnum* sp. and *Polytrichum formosum*).

Old collections in herbaria are from a few widely scattered locations in England and Northern Ireland with a single record from Scotland. Two recent collections (2004) on mosses from East Sussex.

caucasica (Parmasto) Liberta, *Taxon* 15: 318 (1966)

Cristella caucasica Parmasto, *Eesti N.S.V. Tead. Akad. Toimet., Biol.* 14(2): 221 (1965)

E: !

H: On fallen wood and litter. British collection on dead stem of *Epilobium* sp.

A single British collection, from East Sussex (Abbot's Wood) in 2004.

clancularis (Park.-Rhodes) K.H. Larss., *Mycol. Res.* 98(10): 1163 (1994)

Echinotrema clanculare Park.-Rhodes, *Trans. Brit. Mycol. Soc.* 38(4): 368 (1955)

E: ! **W:** !

H: On soil, and in decayed needle litter of conifers. British material in puffin burrows and in litter of *Pinus radiata*.

Known in Britain from the type collection from Wales (Pembrokeshire: Skokholm Island) in 1954 and from England (South Devon) in 1994.

cohaerens (Schwein.) Jülich & Stalpers, *Verh. Kon. Ned. Akad. Wetensch., Afd. Natuurk., Sect. 2*, 74: 257 (1980)

Sporotrichum cohaerens Schwein., *Trans. Amer. Phil. Soc.* 4: 272 (1832)

Cristella submicrospora (Litsch.) M.P. Christ., *Dansk. Bot. Ark.* 19(2): 90 (1960)

Corticium submicrosporum Litsch., *Mitt. bot. Inst. tech. Hochsch. Wien* 4(3): 92 - 93 (1927)

E: ! **S:** ! **W:** ! **NI:** !

H: On decayed wood and litter of decidous trees, usually *Fagus,* in old woodland. Also rather common on fallen or decayed basidiomes of *Rigidoporus ulmarius.* Rare on fern debris or decayed wood of conifers.
D: NM3: 132 **D+I:** B&K2: 122 111, CNE8: 1494-1497
Apparently widespread, but confused with *Trechispora confinis* (considered a synonym by some authorities).

confinis (Bourdot & Galzin) Liberta, *Taxon* 15: 318 (1966)
  Corticium confine Bourdot & Galzin, *Bull. Soc. Mycol. France* 27: 260 (1911)
  Cristella confinis (Bourdot & Galzin) Donk, *Fungus* 27: 19 (1957)
E: ! S: ! W: ! NI: !
H: On decayed wood and litter of deciduous trees in old woodland, on fern debris and occasionally overgrowing living mosses.
Apparently widespread, but often confused with *Trechispora cohaerens.*

dimitica Hallenb., *Mycotaxon* 11: 468 - 470 (1980)
E: !
H: On decayed wood of conifers such as *Pinus sylvestris,* on soil, on decayed leaf and woody litter, debris of ferns such as *Dryopteris* spp, and rarely overgrowing living mosses.
D: Myc8(3): 117
British collections from Buckinghamshire, Middlesex, North Hampshire, North Somerset, South Devon and Surrey.

farinacea (Pers.) Liberta, *Taxon* 15: 318 (1966)
  Hydnum farinaceum Pers., *Syn. meth. fung.*: 562 (1801)
  Grandinia farinacea (Pers.) Bourdot & Galzin, *Bull. Soc. Mycol. France* 30: 253 (1914)
  Cristella farinacea (Pers.) Donk, *Fungus* 27: 19 (1957)
E: c S: c W: c NI: c ROI: c O: Channel Islands: c
H: On decayed wood of deciduous trees and conifers, and on fallen and decayed polypore basidiomes. Also known on the basal parts of *Carex pendula* in swampy woodland.
D: NM3: 133
Very common and widespread. The anamorph *Osteomorpha fragilis* Arnaud ex Watl. & W.B. Kend. was referred to *T. farinacea* in CNE8: 1499 but in fact belongs to *T. stevensonii.*

fastidiosa (Pers.) Liberta, *Taxon* 15(8): 318 (1966)
  Merisma fastidiosa Pers., *Comment. Fungis Clavaeform.*: 97 (1797)
  Thelephora fastidiosa (Pers.) Fr., *Syst. mycol.* 1: 435 (1821)
  Corticium fastidiosum (Pers.) P. Karst., *Bidrag Kännedom Finlands Natur Folk* 37: 142 (1882)
  Soppittiella fastidiosa (Pers.) Massee, *Brit. fung.-fl.* 1: 107 (1892)
  Odontia alliacea Weinm., *Hymen. Gasteromyc.*: 370 (1836)
E: ! S: ! W: !
H: On soil and very decayed leaf litter.
D: NM3: 133 **D+I:** B&K2: 124 113, CNE8: 1502-1503
Rarely reported but apparently widespread. A rather distinctive species, basidiomes of which smell strongly of 'decayed garlic'.

hymenocystis (Berk. & Broome) K.H. Larss., *Mycol. Res.* 98(10): 1167 (1994)
  Polyporus hymenocystis Berk. & Broome, *Ann. Mag. Nat. Hist., Ser. 5 [Notices of British Fungi no. 1810]* 3: 210 (1879)
  Poria hymenocystis (Berk. & Broome) Cooke, *Grevillea* 14: 112 (1886)
E: ! S: ! ROI: !
H: On very decayed wood of conifers, usually *Pinus sylvestris* but also known on *Picea* spp.
D: NM3: 131 **I:** CNE8: 1514 (as *Trechispora mollusca*)
Rarely reported but apparently widespread. Macroscopically similar to *Trechispora mollusca* and often confused with that species. Originally described from Scotland (Angus: Glamis) in 1879.

invisitata (H.S. Jacks.) Liberta, *Taxon* 15(8): 318 (1966)
  Corticium invisitatum H.S. Jacks., *Canad. J. Res., Sect. C, Bot. Sci.* 26: 155 (1948)

  Cristella invisitata (H.S. Jacks.) Donk, *Fungus* 27: 20 (1957)
E: !
H: On soil and litter, often under decayed logs in herb-rich, deciduous woodland. Also known on litter and brash of conifers.
D: NM3: 132 **D+I:** CNE8: 1504-1505
Known from Buckinghamshire, Hertfordshire, Oxfordshire, North and South Hampshire, Surrey and Warwickshire.

kavinioides B. de Vries, *Mycotaxon* 28(1): 79 (1987)
E: !
H: On fallen, decayed twigs of conifers. British material on *Larix* and *Pinus* spp.
D+I: Myc10(4): 154
Known from a Berkshire collection in 1863, with further collections from Yorkshire in 1958 and West Kent in 2000.

laevis K.H. Larss., *Nordic J. Bot.* 16(1): 88 (1996)
E: !
H: On decayed wood of deciduous trees and conifers.
Known from Surrey, West Cornwall, Westmorland and West Sussex.

microspora (P. Karst.) Liberta, *Taxon* 15(8): 319 (1966)
  Grandinia microspora P. Karst., *Bidrag Kännedom Finlands Natur Folk* 48: 365 (1889)
E: ! S: ! O: Channel Islands: !
H: On decayed woody debris of deciduous trees and conifers, and also (rarely) on decayed fronds of *Dryopteris filix-mas.*
D: NM3: 133 **D+I:** B&K2: 12 114, CNE8: 1508-1509
Rarely reported but apparently widespread. Mostly reported from south-eastern England with a few Scottish records, and a single record from the Channel Islands (Jersey).

minima K.H. Larss., *Nordic J. Bot.* 16(1): 90 (1996)
E: ! S: ! W: !
H: On decayed leaf litter of deiduous trees, on debris of *Carex* and *Juncus* spp, on debris of ferns such as *Dryopteris* spp, and on decayed wood of conifers such as *Picea* spp, and *Pinus sylvestris.*
Rarely reported but apparently widespread.

mollusca (Pers.) Liberta, *Canad. J. Bot.* 51(10): 1878 (1973)
  Boletus molluscus Pers., *Syn. meth. fung.*: 547 (1801)
  Polyporus molluscus (Pers.) Fr., *Syst. mycol.* 1: 384 (1821)
  Poria mollusca (Pers.) Cooke, *Grevillea* 14: 109 (1886)
  Cristella mollusca (Pers.) Donk, *Persoonia* 5(1): 97 (1967)
  Mis.: *Poria candidissima* sensu auct.
  Mis.: *Trechispora candidissima* sensu auct. Brit.
  Mis.: *Cristella candidissima* sensu auct. Brit.
E: c S: c W: c NI: c ROI: c O: Channel Islands: c
H: On, or inside, decayed wood of deciduous trees and conifers, usually large logs or fallen branches, and often extending onto litter or debris below.
D: NM3: 131 **D+I:** B&K2: 126 115, CNE8: 1510-1513
Common and widespread. Easily confused with *T. hymenocystis.*

nivea (Pers.) K.H. Larss., *Symb. Bot. Upsal.* 30: 110 (1995)
  Odontia nivea Pers., *Neues Mag. Bot.*: 110 (1794)
  Hydnum niveum (Pers.) Fr., *Syst. mycol.* 1: 419 (1821)
  Hydnum hypoleucum Berk. & Broome, *J. Linn. Soc., Bot.* 14: 60 (1873)
E: ! S: ? W: ? O: Channel Islands: ?
H: On decayed wood of deciduous trees such as *Fagus, Prunus avium* and *Quercus* spp.
D: NM3: 134
Rarely reported. All confirmed British records are from southern England. Reported elsewhere but records are old and unsubstantiated by voucher material.

praefocata (Bourdot & Galzin) Liberta, *Taxon* 15(8): 319 (1966)
  Corticium sphaerosporum ssp. *praefocatum* Bourdot & Galzin, *Hyménomyc. France*: 233 (1928)
E: ! ROI: ! O: Channel Islands: !
H: On decayed wood of deciduous trees such as *Fraxinus* and *Ulmus* spp, on litter of conifers such as *Taxus,* on soil mixed

with woody debris, and decayed debris of ferns such as *Dryopteris* spp.
D: NM3: 133 **D+I:** CNE8: 1515-1516, Myc16(3): 61
Known from England (East Norfolk, Hertfordshire, London, Shropshire, South Devon, Surrey, West Cornwall and West Kent), the Channel Islands (Jersey) and a single collection from Republic of Ireland (Cork).

stellulata (Bourdot & Galzin) Liberta, *Taxon* 15(8): 319 (1966)
  Corticium stellulatum Bourdot & Galzin, *Bull. Soc. Mycol. France* 27: 263 (1911)
  Cristella stellulata (Bourdot & Galzin) Donk, *Fungus* 27: 20 (1957)
  Mis.: *Trechispora microspora* sensu B&K2: p. 124 & pl. 114 (1986)
  Mis.: *Trechispora sphaerospora* sensu auct.
E: ! **S:** ! **NI:** !
H: On decayed litter and wood of deciduous trees and conifers. Reported on *Corylus*, *Quercus robur* and *Larix*. Also known on dead stems of ferns such as *Dryopteris* spp and *Pteridium*.
D: NM3: 133 **D+I:** CNE8: 1515 & 1517
Known from England (Middlesex, Surrey, West Norfolk and West Cornwall), Northern Ireland (Derry) and Scotland (Orkney).

stevensonii (Berk. & Broome) K.H. Larss., *Symb. Bot. Upsal.*: 115 (1995)
  Hydnum stevensonii Berk. & Broome, *Ann. Mag. Nat. Hist., Ser. 4 [Notices of British Fungi no. 1437]* 15: 31 (1875)
  Odontia stevensonii (Berk. & Broome) Rea, *Brit. basidiomyc.*: 647 (1922)
  Osteomorpha fragilis Arnaud ex Watling & W.B. Kendr. (anam.), *Naturalist (Hull)* 104 (948): 1 (1979)
E: c **S:** ! **W:** !
H: On decayed wood of deciduous trees such as *Fagus* and *Ulmus procera* and conifers such as *Pinus sylvestris*. Also on large fallen and decayed basidiomes of *Ganoderma* spp.
I: B&K2: 124 112 (as *Trechispora farinacea*), CNE8: 1500 (as *T. farinacea*), Kriegl1: 346 (as *T. farinacea*)
Rather common and widespread in England. Rarely reported elsewhere but easily confused with *T. farinacea*. The only Scottish record is the type collection (Angus: Glamis) in 1874. Closely resembles *T. farinacea* but basidiomes are usually accompanied by white, pulverulent masses of the arthroconidial anamorph.

subsphaerospora (Litsch.) Liberta, *Canad. J. Bot.* 51(10): 1887 (1973)
  Corticium subsphaerosporum Litsch., *Nat. Hist. Juan Fernandez* 2: 549 (1928)
E: !
H: on decayed wood.
D+I: CNE8: 1519
A single British collection, from Kent in 1923 (in UPS, as *Cristella sphaerospora*).

tenuicula (Litsch.) K.H. Larss., *Syn. Fungorum* 17: 53 (2003)
  Corticium tenuiculum Litsch., *Ann. Mycol.* 39: 130 (1941)
E: !
H: British collections in woodland, on decayed litter in sheltered places underneath large logs or piles of brash.
Known from Buckinghamshire (Hell Coppice), Oxfordshire (Chinnor Hill) and Surrey (Runfold Wood).

TREMELLA Pers., *Neues Mag. Bot.* 1: 112 (1794)
Tremellales, Tremellaceae
  Naematelia Fr., *Observ. mycol.* 2: 370 (1818)
  Gyraria Gray, *Nat. arr. Brit. pl.* 1: 593 (1821)
  Epidochium Fr., *Summa veg. Scand.* 2: 471 (1849)
  Phaeotremella Rea, *Trans. Brit. Mycol. Soc.* 4(5): 377 (1912) [1911]
  Hormomyces Bonord. (anam.), *Handb. Mykol.*: 150 (1851)
Type: *Tremella mesenterica* Retz.

aurantia Schwein., *Schriften Naturf. Ges. Leipzig* 1: 114 (1822)
  Mis.: *Tremella frondosa* sensu auct. p.p.
  Mis.: *Tremella rubiformis* sensu auct.

E: o **W:** o **O:** Channel Islands: !
H: Parasitic. On fruitbodies of *Stereum hirsutum* growing on dead branches and trunks of deciduous trees.
D+I: Myc9(3): 110-113, YST 733 (yeast state only)
Rarely reported. Possibly misidentified as the superficially similar *Tremella mesenterica*.

callunicola P. Roberts, *Mycologist* 15(4): 146 (2001)
S: !
H: Parasitic. In the hymenium of *Aleurodiscus norvegicus* on stems of *Calluna*.
Known only from the type collection from the Isle of Arran in 1998. The corticioid host is widespread, but rarely collected in the British Isles.

cetrariicola Diederich & Coppins, *Biblioth. Lichenol.* 61: 57 (1996)
S: !
H: Parasitic. On thalli of the lichen *Tuckermannopsis chlorophylla*.
Described from Easterness (Caennacroc Forest, near Fort Augustus) and also known from Wester Ross (Loch Maree).

cladoniae Diederich & M.S. Christ., *Biblioth. Lichenol.* 61: 65 (1996)
W: !
H: Parasitic. On thalli of the lichen *Cladonia furcata*.
A single collection from Merionethshire (Maentwrog) in 2002.

coppinsii Diederich & G. Marson, *Notes Roy. Bot. Gard. Edinburgh* 45(1): 175 (1988)
S: ! **W:** !
H: Parasitic. On thalli of the lichen *Platismatia glauca*.
D: NM3: 89
Known in Britain from two Scottish collections (Westerness: Greenfield and North Ebudes: Isle of Skye, Loch na Beiste) and one from Wales (Merionethshire: Maesgwm, Coed-y-Brenin).

encephala Pers., *Syn. meth. fung.*: 623 (1801)
  Naematelia encephala (Pers.) Fr., *Observ. mycol.* 2: 370 (1818)
  Mis.: *Naematelia rubiformis* sensu auct.
  Mis.: *Tremella rubiformis* sensu Bourdot & Galzin (1928), sensu auct.
E: o **S:** o **W:** o **NI:** ! **O:** Channel Islands: !
H: Parasitic. On basidiomes of *Stereum sanguinolentum* on conifer wood.
D: NM3: 88 **D+I:** B&K2: 66 26, Myc13(3): 127 (129), YST 737 (yeast state only) **I:** Kriegl1: 120
Occasional but widespread and may be locally frequent.

exigua Desm., *Ann. Sci. Nat., Bot.*, sér. 3, 8: 191 (1847)
  Mis.: *Epidochium atrovirens* sensu auct.
  Mis.: *Tremella atrovirens* sensu auct.
  Mis.: *Dacrymyces virescens* sensu auct.
  Mis.: *Naematelia virescens* sensu auct.
  Mis.: *Tremella virescens* sensu auct.
E: o **S:** o **W:** o **O:** Channel Islands: !
H: Parasitic. On pyrenomycetes (*Diaporthe* spp), usually on dead or dying stems of *Ulex europaeus*.
D: NM3: 87
Occasional but apparently widespread. The small, dark olive basidiomes often grow at the bases of dead or dying gorse stems thus are easily overlooked.

foliacea Pers., *Observ. mycol.* 2: 98 (1799)
  Gyraria foliacea (Pers.) Gray, *Nat. arr. Brit. pl.* 1: 594 (1821)
  Ulocolla foliacea (Pers.) Bref., *Unters. Gesammtgeb. Mykol.* 7: 98 (1888)
  Exidia foliacea (Pers.) P. Karst., *Bidr. Känn. Finl. Natur. Och. Folk* 48: 449 (1889)
  Gyraria ferruginea Gray, *Nat. arr. Brit. pl.* 1: 593 (1821)
  Tremella fimbriata Pers., *Observ. mycol.* 2: 97 (1799)
  Tremella succinea Pers. *Mycol. eur.* 1: 101(1822)
  Tremella foliacea var. *succinea* (Pers.) Neuhoff, *Z. Pilzk.* 10(3): 73 (1931)
  Tremella nigrescens Fr., *Summa veg. Scand.*: 341 (1849)

Phaeotremella pseudofoliacea Rea, *Trans. Brit. Mycol. Soc.* 4(5): 377 (1912) [1911]
Mis.: *Tremella ferruginea* sensu auct.
Mis.: *Tremella frondosa* sensu auct. p.p.
E: c **S:** o **W:** ! **NI:** c **ROI:** ! **O:** Channel Islands: !
H: Parasitic. On the mycelium of *Stereum* spp on deciduous and coniferous trees and shrubs.
D: NM3: 88 **D+I:** B&K2: 66 27, Myc13(3): 129, YST 739 (yeast state only) **I:** C&D: 137, Kriegl1: 122
Common and widespread.

globispora D.A. Reid, *Trans. Brit. Mycol. Soc.* 55: 414 (1970)
Mis.: *Tremella albida* sensu auct. p.p.
Mis.: *Tremella tubercularia* sensu auct.
E: ! **S:** ! **NI:** ?
H: Parasitic. On ascocarps of *Eutypa* and *Diaporthe* spp on dead branches of deciduous trees and shrubs.
D: NM3: 88 **D+I:** YST 742 (yeast state only)
Rarely reported. Known from widely scattered locations in England from Yorkshire to South Devon and in Scotland (Perthshire: Killiecrankie). Reported from Northern Ireland (Antrim and Fermanagh) but unsubstantiated with voucher material. Not clearly distinct from *T. indecorata*.

hypogymniae Diederich & M.S. Christ., *Biblioth. Lichenol.* 61: 90 (1996)
S: ! **W:** !
H: Parasitic. On thalli of the lichen *Hypogymnia physodes*.
D: NM3: 88
Known from Scotland (Mid-Perthshire: Ben Lawers) and Wales (Cardiganshire: Devil's Bridge and Merionethshire: Lake Bala).

indecorata Sommerf., *Suppl. Fl. lapp.*: 306 (1826)
Exidia indecorata (Sommerf.) P. Karst., *Rev. Mycol. (Toulouse)* 12(47): 126 (1890)
E: ! **S:** ! **W:** ? **NI:** ? **ROI:** ?
H: Parasitic. On ascocarps of pyrenomycetes, on twigs and branches of deciduous trees and shrubs.
D: NM3: 87 **D+I:** YST 743 (yeast state only)
Known from widely scattered sites in England and Scotland. Reported from Wales, Northern Ireland, and Republic of Ireland but unsubstantiated with voucher material.

invasa (Hauerslev) Hauerslev, *Nordic J. Bot.* 16: 218 (1996)
Exidiopsis invasa Hauerslev, *Mycotaxon* 49: 221 (1993)
E: !
H: Parasitic. In the hymenium of *Trechispora stellulata* growing on dead fern stems.
D: NM3: 87, Myc15(4): 147
A single collection from Surrey (Oxshott Heath) in 1994.

karstenii Hauerslev, *Mycotaxon* 72: 480 (1999)
E: !
H: Parasitic. On effete ascocarps of *Coccomyces juniperi* growing on branches of *Juniperus communis*.
A single collection from Westmorland (Upper Teesdale, Sun Wood) in 1995.

lichenicola Diederich, *Lejeunia, Nouvelle série* 119: 2 (1986)
S: !
H: Parasitic. On thalli of the lichen *Mycoblastus sterilis*.
D: NM3: 88
A single collection from Northumberland (South Tyne Valley) in 1979.

lobariacearum Diederich & M.S. Christ., *Biblioth. Lichenol.* 61: 103 (1996)
S: !
H: Parasitic. On thalli of the lichen *Lobaria pulmonaria*.
Known from Wester Ross (Kishorn, Rassal).

mesenterica Retz., *Kongl. Vetensk. Acad. Handl.* 30: 249 (1769)
Tremella lutescens Pers., *Mycol. eur.* 1: 100 (1822)
Hormomyces aurantiacus Bonord. (anam.), *Handb. Mykol.*: 150 (1851)
Mis.: *Tremella albida* sensu auct. p.p.
Mis.: *Exidia candida* sensu auct. p.p.

Mis.: *Tremella candida* sensu auct. p.p.
E: c **S:** c **W:** c **NI:** c **ROI:** c **O:** Channel Islands: c Isle of Man: !
H: Parasitic. On the mycelium of *Peniophora* spp, especially on those associated with *Ulex europaeus* and *Corylus*.
D: NM3: 88 **D+I:** Ph: 264, B&K2: 68 29, Myc9(3): 113-114, YST 744 (yeast state only) **I:** C&D: 137, Kriegl1: 125
Very common and widespread. *Hormomyces aurantiacus* refers to asexual fruitbodies only producing conidiophores.

moriformis Sm., *Engl. bot.* 34: 2446 (1812)
Dacrymyces moriformis (Sm.) Fr., *Syst. mycol.* 2: 229 (1823)
E: ! **S:** ? **ROI:** ?
H: Parasitic. On pyrenomycetes on wood of deciduous trees.
D+I: YST 745 (yeast state only)
Known from Berkshire, North Somerset, Shropshire, South Devon and West Norfolk. Reported elsewhere in England, from Scotland in 1872, and Republic of Ireland in 1999 but unsubstantiated with voucher material.

normandinae Diederich *Biblioth. Lichenol.* 61: 120 (1996)
S: !
H: Parasitic. On thalli of the lichen *Normandina pulchella*.
Known from Westerness (Ardnamurchan).

penetrans (Hauerslev) Jülich, *Int. J. Mycol. Lichenol.* 1(2): 196 (1983)
Sebacina penetrans Hauerslev, *Friesia* 11: 272 (1979)
E: ! **W:** !
H: Parasitic. In the hymenium of *Dacrymyces minor* and *D. stillatus*.
D: NM3: 86
Known from England (South Devon and Surrey) and Wales (Monmouthshire: Wentwood Forest).

pertusariae Diederich, *Biblioth. Lichenol.* 61: 133 (1996)
ROI: !
H: Parasitic. On thalli of species of the lichen genus *Pertusaria*.
A single collection, from Kerry (Killarney, Lough Leane, Lamb Island) in 1996.

phaeographidis Diederich, Coppins & Bandoni, *Biblioth. Lichenol.* 61: 136 (1996)
E: !
H: On thalli of species of the lichen genus *Phaeographis*.
Known only from North Devon (Brownsham, near Hartland).

phaeophysciae Diederich & M.S. Christ., *Biblioth. Lichenol.* 61: 142 (1996)
S: ! **W:** ?
H: Parasitic. On thalli of the lichen *Phaeophyscia orbicularis*.
D: NM3: 88
Known from East Lothian and Kincardineshire. Reported from Wales but unsubstantiated with voucher material.

polyporina D.A. Reid, *Trans. Brit. Mycol. Soc.* 55: 416 (1970)
E: ! **S:** ! **O:** Channel Islands: !
H: Parasitic. In the hymenium of *Postia* spp. The majority of records are from *P. lactea*.
D: NM3: 87
Rarely reported, but widespread. The fungus produces basidia amongst those of the poroid host, thus is easily overlooked.

protoparmeliae Diederich & Coppins, *Biblioth. Lichenol.* 61: 147 (1996)
E: ! **S:** !
H: On thalli of the lichen *Protoparmelia badia*.
Rarely reported.

sarniensis P. Roberts, *Mycologist* 15(4): 147-148 (2001)
O: Channel Islands: !
H: Parasitic. In the hymenium of *Phanerochaete sordida*.
Known only from the type collection from Guernsey (Le Guet) in 1993.

simplex H.S. Jacks. & G.W. Martin, *Mycologia* 32: 687 (1940)
E: ! **S:** !
H: Parasitic. On the hymenium of *Aleurodiscus amorphus* on conifer branches.
D: NM3: 87, Myc15(4): 148-149 **D+I:** B&K2: 68 30

Rarely reported. The host is rare in Britain, but a recent survey of specimens in herb. K showed most were parasitised by *T. simplex*.

steidleri (Bres.) Bourdot & Galzin, *Hyménomyc. France*: 21 (1928)

> *Tremella encephala* var. *steidleri* Bres., *Ann. Mycol.* 6: 46 (1908)

E: ! W: !
H: Parasitic. On basidiomes of *Stereum hirsutum* on decayed or dead wood of deciduous trees.
D+I: Myc13(3): 128 (129), Reid8: 99-101
Known from England (Berkshire, South Devon, South Hampshire and West Sussex) and Wales (Caernarvonshire, Carmarthenshire and Glamorganshire).

translucens H.D. Gordon, *Trans. Brit. Mycol. Soc.* 22: 111 (1938)

> *Pseudostypella translucens* (H.D. Gordon) D.A. Reid & Minter, *Trans. Brit. Mycol. Soc.* 72: 347 (1979)

E: ! S: o W: ! NI: !
H: Parasitic. On *Lophodermium conigenum*, *L. pinastri* and *L. seditiosum* growing on needles and decayed cone scales of *Pinus sylvestris*.
D: NM3: 104 (as *Pseudostypella translucens*)
Occasional in Scotland but rarely reported from other parts of the British Isles.

versicolor Berk. & Broome, *Ann. Mag. Nat. Hist., Ser. 2* [*Notices of British Fungi no. 726*] 13: 406 (1854)
E: ! NI: !
H: Parasitic. On the hymenium of *Peniophora* spp, and known from *P. cinerea*, *P. lycii* and *P. quercina*.
D: Myc15(4): 149-150
Rarely recorded or collected and, in fact, 'forgotten' for over a century, with several early and recent records, but none between 1895 and 1996. Basidiomes are small and require a deliberate search.

TREMELLODENDROPSIS (Corner) D.A. Crawford, *Trans. & Proc. Roy. Soc. New Zealand* 82: 618 (1954)
Tremellales, Tremellodendropsidaceae
Type: *Tremellodendropsis tuberosa* (Grev.) D.A. Crawford

tuberosa (Grev.) D.A. Crawford, *Trans. & Proc. Roy. Soc. New Zealand* 82: 619 (1954)

> *Merisma tuberosum* Grev., *Scott. crypt. fl.* 4: t. 178 (1826)
> *Thelephora tuberosa* (Grev.) Fr., *Elench. fung.* 1: 167 (1828)
> *Stereum tuberosum* (Grev.) Massee, *Brit. fung.-fl.* 1: 130 (1892)
> *Aphelaria tuberosa* (Grev.) Corner, *Monograph of Clavaria and Allied Genera*: 192 (1950)
> *Clavaria gigaspora* Cotton, *Naturalist (Hull)* 1907: 97 (1907)

E: ! S: !
H: On soil in woodland, and also on rootstocks of *Filipendula ulmaria* in fen vegetation.
D: NM3: 111 **I:** SV39: 35
Known from England (Berkshire, East Norfolk, Herefordshire, Mid-Lancashire and Yorkshire). Originally described from Scotland (Linlithgow).

TRICHAPTUM Murrill, *Bull. Torrey Bot. Club* 31: 608 (1904)
Poriales, Coriolaceae

> *Hirschioporus* Donk, *Revis. Niederl. Homobasidiomyc.* 2: 168 (1933)

Type: *Trichaptum trichomallum* (Berk. & M.A. Curtis) Murrill

abietinum (Pers.) Ryvarden, *Norweg. J. Bot.* 19: 237 (1972)

> *Boletus abietinus* Pers., in Gmelin, *Syst. nat.*: 1437 (1792)
> *Polyporus abietinus* (Pers.) Fr., *Syst. mycol.* 1: 370 (1821)
> *Polystictus abietinus* (Pers.) Cooke, *Grevillea* 14: 84 (1886)
> *Hirschioporus abietinus* (Pers.) Donk, *Rev. Niederl. Homob. Aphyll.* 2: 168 (1933)

> *Trametes abietina* (Pers.) Pilát, *Atlas champ. Eur. Polyporaceae I:* 273 (1939)
> *Hydnum parasiticum* Willd., *Fl. berol. prodr.*: 396 (1797)

E: c **S:** c **W:** c **NI:** c **ROI:** c **O:** Channel Islands: o Isle of Man: c
H: On fallen and decayed wood of conifers such as *Larix*, *Picea* and *Pinus* spp, usually on fallen or stacked trunks. A single verified collection on *Fagus* from South Hampshire (New Forest: Burley Old Inclosure) in herb. K.
D: EurPoly2: 676, NM3: 219 **D+I:** Ph: 237 (as *Hirschioporus abietinus*), B&K2: 288 360 **I:** Kriegl1: 594
Very common and widespread.

TRICHOLOMA (Fr.) Staude, *Schwämme Mitteldeutschl.*: 125 (1857)
Agaricales, Tricholomataceae

> *Gyrophila* Quél., *Enchir. fung.*: 9 (1886)

Type: *Agaricus flavovirens* Pers. (= *T. equestre*)

acerbum (Bull.) Quél., *Mém. Soc. Émul. Montbéliard, Sér. 2*, 5: 77 (1872)

> *Agaricus acerbus* Bull., *Herb. France*: pl. 571 (1792)

E: o **S:** ! **W:** ! **NI:** ! **ROI:** !
H: On soil, usually associated with *Quercus* spp in old deciduous woodland. Also known with *Fagus* or *Castanea*.
D: NM2: 187, FAN4: 120-121 **D+I:** Ph: 40-41, FungEur3: 329-335 50 [p.527], B&K3: 320 408, Galli3: 186-187 **I:** FlDan1: 23A, Bon: 159, C&D: 197, C&D: 197
Occasional in England. Rarely reported elsewhere but apparently widespread. Most records are from the south and west.

aestuans (Fr.) Gillet, *Hyménomycètes*: 102 (1874)

> *Agaricus aestuans* Fr., *Syst. mycol.* 1: 47 (1821)

E: ? **S:** !
H: On soil, associated with *Pinus sylvestris* in woodland.
D: NM2: 188, FAN4: 139-140, NBA3: 124 **D+I:** FungEur3: 266-268 32 [p.481], Galli3: 136-137 **I:** Bon: 153, C&D: 193, SV42: 3, Kriegl3: 558
Known from Scotland (Mid-Perthshire and Selkirk). Reported from England (South Hampshire) in 1983, but unsubstantiated with voucher material.

album (Schaeff.) P. Kumm., *Führ. Pilzk.*: 131 (1871)

> *Agaricus albus* Schaeff., *Fung. Bavar. Palat. nasc.* 3: 256 (1770)
> *Tricholoma raphanicum* P. Karst., *Symb. mycol. fenn.*: 39 (1883)

E: o **S:** o **W:** o **NI:** o **ROI:** o **O:** Isle of Man: !
H: On soil in woodland, under deciduous trees. Often reported with *Betula* spp but also known with *Fagus* and *Quercus* spp.
D: NM2: 185, FAN4: 143-144 **D+I:** FungEur3: 179-181 8 [p. 431], Galli3: 68-69 **I:** Bon: 151, C&D: 189
Occasional but widespread and may be locally common. Some of the records with *Betula* spp may be misdetermined *Tricholoma stiparophyllum*.

apium Jul. Schäff., *Z. Pilzk.* 5(3-4): 65 (1925)

> *Tricholoma helviodor* Pilát & Svrček, *Stud. Bot. Čechoslov.* 7: 2 (1946)

S: r
H: On soil with conifers. British material with *Pinus sylvestris*.
D: NM2: 186, FAN4: 119-120 **D+I:** FRIC2: 10-11 11 (as *Tricholoma helviodor*), FungEur3: 336-340 52 [p.531], Galli3: 188-189 **I:** FRIC7: 11-12 50b
Known only from a few localities in Scotland. A large and distinctive species unlikely to be overlooked.

arvernense Bon, *Doc. Mycol.* 6(22-23): 168 (1976)

> *Tricholoma sejunctoides* P.D. Orton, *Notes Roy. Bot. Gard. Edinburgh* 44(3): 495 (1987)

S: !
H: On acidic soil with *Pinus sylvestris* in Caledonian pinewoods.
D: FAN4: 114-115 **D+I:** FungEur3: 291-294 40 [p. 497], B&K3: 322 410, Galli3: 144-145 **I:** Bon: 157
Known from Mid-Perthshire.

atrosquamosum (Chevall.) Sacc., *Syll. fung.* 5: 45 (1887)
  Agaricus atrosquamosus Chevall., *Fung. Byss. illustr.* 1: 45 (1837)
  Tricholoma terreum var. *atrosquamosum* (Chevall.) Massee, *Brit. fung.-fl.* 3: 195 (1893)
  Mis.: *Tricholoma ramentaceum* sensu Ricken [Blätterpilze Deutschl.: 338 (1914)]
  Mis.: *Tricholoma 'atromarginatum'* [TBMS 17 (1-2): 8 (1932)]
E: o **S:** ! **NI:** ! **ROI:** !
H: On soil, usually associated with *Fagus* but also in mixed woodland with conifers and *Quercus* spp.
D: NM2: 190, FAN4: 132 (as *Tricholoma atrosquamosum* var. *atrosquamosum*) **D+I:** Ph: 36-37, FungEur3: 224-228 20 [p.457], B&K3: 322 411, Galli3: 132-133 **I:** Bon: 155, C&D: 195
Rarely reported but apparently widespread and may be locally frequent in old beechwoods on calcareous soil in southern and south-western England.

atrosquamosum var. squarrulosum (Bres.) M. Chr. & Noordel., *Persoonia* 17: 312 (1999)
  Tricholoma squarrulosum Bres., *Fungi trident.* 2(11-13): 47 (1892)
  Mis.: *Tricholoma atrosquamosum* sensu Lange (FlDan1: 53 & pl. 22D)
E: !
H: On soil with conifers, or in mixed woodland with deciduous trees and conifers.
D: NM2: 190, FAN4: 133 **D+I:** Ph: 38-39 (as *Tricholoma squarrulosum*), FungEur3: 229-231 21 [p.459] (as *T. squarrulosum*), Galli3: 134-135 (as *T. squarrulosum*) **I:** Bon: 155
Rarely reported but widely scattered in England and apparently quite frequent in Kent.

aurantium (Schaeff.) Ricken, *Blätterpilze Deutschl.*: 332 (1914)
  Agaricus aurantius Schaeff., *Fung. Bavar. Palat. nasc.* 1: 68 (1762)
  Armillaria aurantia (Schaeff.) P. Kumm., *Führ. Pilzk.*: 134 (1871)
S: !
H: On acidic soil under conifers. British material with *Larix* and *Pinus* spp, but also known with *Picea* spp in continental Europe.
D: NM2: 186, FAN4: 127-128 **D+I:** FungEur3: 393-395 66 [p.565], B&K3: 324 412, Galli3: 214-215 **I:** Bon: 161, C&D: 199, Kriegl3: 526
Known from Scotland but not reported in recent years (last record in 1957). The single record from England (Cumberland) in 1996, supposedly with *Picea* sp., is unsubstantiated with voucher material.

batschii M. Chr. & Noordel., *Persoonia* 17(2): 315 (1999)
  Agaricus subannulatus Batsch, *Elench. fung. (Continuatio Prima)*: 75 & t. 16 f. 75 (1786)
  Tricholoma subannulatum (Batsch) Bres. [*nom. illegit.*, non *T. subannulatum* (Peck) Zeller (1922)], *Icon. Mycol.* 2: pl. 63 (1927)
  Mis.: *Tricholoma albobrunneum* sensu Ricken [Blätterpilze Deutschl.: 333 (1914)]
  Mis.: *Tricholoma fracticum* sensu auct. mult.
  Mis.: *Agaricus fracticus* sensu auct.
E: !
H: On soil under conifers. British material on an old spoil heap, associated with *Pinus nigra*, and in a garden, associated with *P. sylvestris*.
D: NM2: 186 (as *Tricholoma fracticum*), Myc9(4): 173 (as *T. fracticum*), FAN4: 124-125 **D+I:** FungEur3: 382-385 63 (p. 557) / 63b (p. 559), B&K3: 328 419 (as *T. fracticum*), Galli3: 218-219 (as *T. fracticum*) **I:** Bon: 161 (as *T. fracticum*), C&D: 199 (as *T. fracticum*), Kriegl3: 531 (as *T. fracticum*)
Two collections, from West Gloucestershire (Forest of Dean) in 1994 and from Cambridgeshire in 2002.

bufonium (Pers.) Gillet, *Hyménomycètes*: 111 (1874)
  Agaricus bufonius Pers., *Syn. meth. fung.*: 359 (1801)

Tricholoma sulphureum var. *bufonium* (Pers.) Quél. (1876)
E: ! **S:** !
H: On soil under deciduous trees.
D: NM2: 184, FAN4: 148 **D+I:** FungEur3: 193-195 12 [p.439], B&K3: 324 414, Galli3: 84-85 **I:** Bon: 153, C&D: 193
Known from Scotland (Perthshire) in recent times but last reported from England (East Norfolk) in 1967. Treated by many authorities as merely a colour form of *Tricholoma sulphureum*

cingulatum (Almfelt) Jacobashch, *Verh. Bot. Vereins Prov. Brandenburg* 33: 55 (1890)
  Agaricus cingulatus Almfelt, in Fries, *Linnaea* 5: 507 (1830)
  Mis.: *Tricholoma ramentaceum* sensu auct.
E: o **S:** ! **W:** o **NI:** !
H: On soil, in damp deciduous woodland, associated with *Salix* spp.
D: NM2: 189, FAN4: 136-137 **D+I:** Ph: 34-35, FungEur3: 24-247 26 [p.469], B&K3: 326 416, Galli3: 116-117 **I:** FlDan1: 23C, Bon: 155, C&D: 195, Kriegl3: 547
Occasional in England and Wales. Rarely reported elsewhere but apparently widespread.

colossus (Fr.) Quél., *Mém. Soc. Émul. Montbéliard, Sér. 2*, 5: 76 (1872)
  Agaricus colossus Fr., *Sverig Atl. Svamp.*: 38 (1836)
  Armillaria colossus (Fr.) Sacc., *Syll. fung.* 16: 20 (1902)
  Megatricholoma colossus (Fr.) G. Kost, *Sydowia, Beih.* 37: 54 (1984)
S: r
H: On acidic soil, amongst needle litter of *Pinus sylvestris* in Caledonian pinewoods.
D: NM2: 185, FAN4: 131-132 **D+I:** FungEur3: 396-399 67 [p.567], Galli3: 216-217 **I:** Bon: 161, C&D: 199, Cooke 66 (75) Vol. 1 (1882)
Known from Scotland but rare and generally little-known in Britain.

columbetta (Fr.) P. Kumm., *Führ. Pilzk.*: 131 (1871)
  Agaricus columbetta Fr., *Syst. mycol.* 1: 44 (1821)
E: o **S:** o **W:** ! **NI:** ! **ROI:** !
H: On soil, usually associated with *Fagus* or *Quercus* spp, in mixed deciduous woodland.
D: NM2: 191, FAN4: 117-118 **D+I:** Ph: 40-41, FungEur3: 278-280 35 [p.488], B&K3: 326 417, Galli3: 152-153 **I:** C&D: 193
Occasional in England and Scotland. Rarely reported elsewhere but apparently widepsread.

equestre (L.) P. Kumm., *Führ. Pilzk.*: 130 (1871)
  Agaricus equestris L. [non *A. equestris* Bolton (1788)], *Sp. pl.*: 1173 (1753)
  Tricholoma auratum (Paulet) Gillet, *Hyménomycètes*: 92 (1874)
  Agaricus auratus Paulet [non *A. auratus* With. (1796)], *Traité champ.* 2: 137 (1793)
  Agaricus flavovirens Pers., *Syn. meth. fung.*: 319 (1801)
  Tricholoma flavovirens (Pers.) S. Lundell & Nannf., *Fungi Exsiccati Suecici*: 1102 (1942)
E: ! **S:** o
H: On acidic sandy soil, often amongst mixtures of *Pinus sylvestris*, *Betula* and *Quercus* spp.
D: NM2: 188 (as *Tricholoma flavovirens*), FAN4: 112-113 (as *T. equestre* var. *equestre*) **D+I:** Ph: 34-35 (as *T. flavovirens*), FungEur3: 311-316 45a/b/c [p.511-517], B&K3: 328 418, Galli3: 164-166 **I:** Bon: 157, C&D: 191, Kriegl3: 560, SV46: 39
Occasional in native woodland in Scotland but rare in England.

focale (Fr.) Ricken, *Blätterpilze Deutschl.*: 332 (1914)
  Agaricus focalis Fr., *Epicr. syst. mycol.*: 20 (1838)
  Armillaria focalis (Fr.) P. Karst., *Bidrag Kännedom Finlands Natur Folk* 32: 18 (1879)
E: ! **S:** ! **W:** !
H: On acidic soil, associated with *Pinus sylvestris* in Caledonian pinewoods.
D: NM2: 185, FAN4: 129-130, NBA2: 62-63 **D+I:** FungEur3: 403-406 69 [p.571] **I:** FlDan1: 16B, Bon: 161, C&D: 199, Kriegl3: 529

Rare and probably extinct in England and Wales. Slightly less rare in Scotland and reported on several occasions in recent years.

fulvum (Bull.) Bigeard & H. Guill., *Fl. Champ. Supér. France* 2: 89 (1913)
   Agaricus fulvus Bull., *Herb. France*: pl. 555 (1792)
   Agaricus flavobrunneus Fr., *Observ. mycol.* 2: 119 (1818)
   Tricholoma flavobrunneum (Fr.) P. Kumm., *Führ. Pilzk.*: 130 (1871)
   Agaricus nictitans Fr., *Syst. mycol.* 1: 38 (1821)
   Tricholoma nictitans (Fr.) Gillet, *Hyménomycètes*: 93 (1874)
E: c **S:** c **W:** c **NI:** c **ROI:** c
H: On acidic soil, usually associated with *Betula* spp, but also frequently with *Fagus* and *Quercus* spp, and in mixed coniferous and deciduous woodland.
D: NM2: 187, FAN4: 128 **D+I:** Ph: 38-39, FungEur3: 352-355 56 [p.539] (as *Tricholoma flavobrunneum*), B&K3: 328 420, Galli3: 200-201 **I:** Bon: 159, C&D: 197, Kriegl3: 532
Very common and widespread.

gausapatum (Fr.) Quél., *Mém. Soc. Émul. Montbéliard, Sér. 2,* 5: 211 (1872)
   Agaricus gausapatus Fr., *Syst. mycol.* 1: 43 (1821)
E: ! **W:** ! **NI:** ! **ROI:** !
H: On soil under deciduous trees and conifers. British material is with *Fagus* and *Pinus sylvestris*.
D+I: FungEur3: 213-216 17 [p.451], Galli3: 112-113 **I:** Bon: 155, C&D: 195
Rarely reported.

imbricatum (Fr.) P. Kumm., *Führ. Pilzk.*: 133 (1871)
   Agaricus imbricatus Fr., *Observ. mycol.* 1: 27 (1815)
E: o **S:** o **W:** ! **NI:** ! **ROI:** !
H: On acidic soil, usually associated with *Pinus* spp in conifer or mixed deciduous and coniferous woodland. Rarely on calcareous soils.
D: NM2: 187, FAN4: 118-119 **D+I:** Ph: 38-39, FungEur3: 348-350 55 [p.537], B&K3: 330 421, Galli3: 192-193 **I:** Bon: 159, C&D: 197, Kriegl3: 534
Occasional in England and Scotland. Rarely reported elsewhere but apparently widespread.

inamoenum (Fr.) Gillet, *Hyménomycètes*: 112 (1874)
   Agaricus inamoenus Fr., *Observ. mycol.* 1: 10 (1815)
   Tricholoma inamoenum var. *insignis* Massee, *Brit. fung.-fl.* 3: 205 (1893)
E: ! **S:** ! **W:** ! **ROI:** !
H: On soil, usually in mixed deciduous and conifer woodland.
D: NM2: 184, FAN4: 142-143 **D+I:** FungEur3: 185-188 10 (p.435), B&K3: 330 422, Galli3: 78-79 **I:** Bon: 153, C&D: 193, SV45: 47
Rarely reported and poorly known in Britain. Herbarium collections are from the late nineteenth century. A few later records are unsubstantiated with voucher material.

inocybeoides A. Pearson, *Trans. Brit. Mycol. Soc.* 22(1-2): 29 (1938)
   Tricholoma argyraceum f. *inocybeoides* (A. Pearson) M. Chr. & Noordel., *Persoonia* 17(2): 309 (1999)
E: !
H: On soil with *Betula* spp, often in mixed woodland with *Pinus* spp.
D: FAN4: 135 (as *Tricholoma argyraceum*) **D+I:** Ph: 37, Galli3: 120-121
Much confused with related species. Excludes *T. inocybeoides* sensu FAN3, which is *T. bonii* Candusso & Basso, as yet unknown in Britain.

lascivum (Fr.) Gillet, *Hyménomycètes*: 111 (1874)
   Agaricus lascivus Fr., *Syst. mycol.* 1: 110 (1821)
   Agaricus lascivus var. *robustus* Cooke, *Handb. Brit. fung.*, Edn 2: 36 (1884)
   Tricholoma lascivum var. *robustum* (Cooke) Sacc., *Syll. fung.* 5: 113 (1887)
E: o **S:** ! **W:** ! **NI:** ! **ROI:** !

H: On soil under deciduous trees, often in areas on calcareous soil.
D: NM2: 185, FAN4: 145 **D+I:** Ph: 40-41, FungEur3: 175-178 7 [p.429], B&K3: 330 423, Galli3: 74-75 **I:** C&D: 189
Occasional in England and may be locally common in southern counties. Rarely reported elsewhere but apparently widespread.

luridum (Schaeff.) P. Kumm., *Führ. Pilzk.*: 132 (1871)
   Agaricus luridus Schaeff. [non *A. luridus* Bolton (1788)], *Fung. Bavar. Palat. nasc.* 1: 30 (1762)
   Gymnopus luridus (Schaeff.) Gray, *Nat. arr. Brit. pl.* 1: 606 (1821)
E: ! **NI:** ?
H: On soil in woodland.
D: NM2: 189, FAN4: 116 **D+I:** FungEur3: 308-310 44 [p.509], B&K3: 332 424, Galli3: 170-171 **I:** C&D: 191, Cooke 75 (214) Vol. 1 (1882)
Accepted but with reservations. A few, old records from England (last reported in 1895 from Lincolnshire) and Northern Ireland. This was 'excluded pending clearer definition' in NCL. Voucher material in herbaria is sparse and in poor condition.

nauseosum (A. Blytt) Kytöv., *Karstenia* 28(2): 69 (1989) [1988]
   Armillaria nauseosa A. Blytt, *Videnskabs-Selskabets Skrifter. I Math.-Naturv. Kl.* 6: 22 (1905)
   Tricholoma caligatum var. *nauseosum* (A. Blytt) Bon, *Doc. Mycol.* 20(78): 38 (1998)
S: r
H: On dry acidic soil, under *Pinus sylvestris* in Caledonian pinewoods.
D: FAN4: 130-131 **D+I:** Galli3: 228-229 **I:** Bon: 161 (as *Tricholoma caligatum*), FungEur3: 68 (as *T. caligatum*)
Rarely reported. The most recent collection in herb. K dates from 1927 but reported from Abernethy Forest in 1993 (both as *Tricholoma caligatum*).

orirubens Quél., *Mém. Soc. Émul. Montbéliard, Sér. 2,* 5: 327 (1873)
   Tricholoma horribile Rea, *Trans. Brit. Mycol. Soc.* 2: 94 (1905) [1904]
E: ! **S:** ! **W:** !
H: On soil amongst leaf litter, usually with *Fagus* on calcareous soils, only rarely on acidic soil.
D: NM2: 190, FAN4: 133-134 **D+I:** Ph: 37 (colour poor), FungEur3: 232-234 22 [p. 461], B&K3: 332 425, Galli3: 130-131 **I:** FlDan1: 22A, Bon: 155, C&D: 195
Rarely reported but apparently widespread and may be locally common. Most frequent in southern and south-eastern England.

pessundatum (Fr.) Quél., *Mém. Soc. Émul. Montbéliard, Sér. 2,* 5: 77 (1872)
   Agaricus pessundatus Fr., *Syst. mycol.* 1: 38 (1821)
E: ! **S:** o
H: On acidic soil with *Pinus sylvestris* mostly in Caledonian pinewoods.
D: NM2: 188, FAN4: 124 **D+I:** FungEur3: 359-364 58 [p.543], Galli3: 206-207 **I:** Bon: 159, C&D: 197
Occasional in Scotland and may be locally frequent in Caledonian pinewoods. Rarely reported elsewhere. Sensu NCL included *Tricholoma stans*.

populinum J.E. Lange, *Dansk. Bot. Ark.* 8: 14 (1933)
E: r
H: On soil, associated with various species of *Populus* (but not *P. tremula*).
D: NM2: 188, FAN4: 125-126 **D+I:** FungEur3: 377-381 62 [p.555], B&K3: 334 427, Galli3: 212-213 **I:** FlDan1: 17D, Bon: 159 (slightly atypical in stature), C&D: 197
Rarely reported and apparently a genuinely rare species in Britain. Only eight known sites, all in southern and south-eastern England.

portentosum (Fr.) Quél., *Mém. Soc. Émul. Montbéliard, Sér. 2,* 5: 338 (1873)

Agaricus portentosus Fr., *Syst. mycol.* 1: 39 (1821)
E: ! **S:** o **NI:** ?
H: On acidic soil, associated with *Pinus sylvestris*.
D: NM2: 191, FAN4: 116-117 **D+I:** Ph: 35, FungEur3: 321-323
47 [p.521], B&K3: 334 428, Galli3: 176-179 **I:** Bon: 157, C&D:
191, SV33: 21, Kriegl3: 563
Occasional in Scotland but may be locally common in Caledonian
pinewoods. Rarely reported in England, and a single report,
unsubstantiated with voucher material, from Northern Ireland.

psammopus (Kalchbr.) Quél., *Mém. Soc. Émul. Montbéliard,
Sér. 2,* 5: 433 (1875)
Agaricus psammopus Kalchbr., *Icon. select. Hymenomyc.
Hung.* 12: pl. 3 (1873)
E: o **S:** o **W:** ! **NI:** ! **O:** Isle of Man: !
H: On soil in woodland, associated with conifers, usually *Larix*
and much less frequently *Pinus* spp.
D: NM2: 187, FAN4: 122 (as *Tricholoma psammopus* var.
psammopus) **D+I:** Ph: 40-41, FungEur3: 326-328 49 [p.525],
B&K3: 334 429, Galli3: 182-183 **I:** Bon: 159, C&D: 197,
Kriegl3: 537
Occasional in England and Scotland. Rarely reported elsewhere
but apparently widespread.

robustum (Alb. & Schwein.) Ricken, *Blätterpilze Deutschl.*: 332
(1914)
Agaricus robustus Alb. & Schwein., *Consp. fung. lusat.*: 147
(1805)
Armillaria robusta (Alb. & Schwein.) P. Kumm., *Führ. Pilzk.*:
135 (1871)
Mis.: *Tricholoma focale* sensu Riva (FungEur3) p.p. & sensu
Galli3
S: !
H: On acidic soil, with *Pinus sylvestris* in Caledonian pinewoods.
D: NM2: 185 **I:** FlDan1: 16C
A single confirmed collection from Mid-Perthshire. Older records
are unconfirmed and may be *T. focale*. Cooke 30 (51) Vol. 1
(1881) [as *Agaricus (Armillaria) focalis* var. *goliath*] might also
be this species or *T. focale*.

saponaceum (Fr.) P. Kumm., *Führ. Pilzk.*: 133 (1871)
Agaricus saponaceus Fr., *Observ. mycol.* 2: 101 (1818)
Agaricus atrovirens Pers., *Syn. meth. fung.*: 319 (1801)
Tricholoma saponaceum var. *atrovirens* (Pers.) P. Karst.,
Bidrag Kännedom Finlands Natur Folk 32: 36 (1879)
Agaricus napipes Krombh., *Naturgetr. Abbild. Schwämme* 4:
22 (1836)
Tricholoma saponaceum var. *napipes* (Krombh.) Barla,
Champ. Alp. Marit. 5: 51 (1890)
E: o **S:** o **W:** o **NI:** o **ROI:** !
H: On acidic soil under deciduous trees, or with conifers in
mixed woodland.
D: NM2: 184, FAN4: 141 (as *Tricholoma saponaceum* var.
saponaceum) **D+I:** Ph: 36-37, FungEur3: 157-161 1 [p.417],
B&K3: 336 430, Galli3: 60-61 **I:** Bon: 151, C&D: 189, Kriegl3:
571
Occasional but widespread.

saponaceum *var.* **lavedanum** Rolland, *Bull. Soc. Mycol.
France* 7: 95 (1891)
Tricholoma boudieri Barla, *Bull. Soc. Mycol. France* 3: 205
(1887)
Tricholoma saponaceum var. *boudieri* (Barla) Bigeard & H.
Guill., *Fl. Champ. Supér. France* 2: 46 (1913)
E: ! **S:** !
H: On soil in woodland, under deciduous trees.
D: FAN4: 141 (as *Tricholoma saponaceum* var. *boudieri*) **D+I:**
FungEur3: 165-166, Galli3: 62 **I:** Ph: 36 (as *T. saponaceum*,
showing rusty tints), C&D: 189 (as *T. boudieri*)
Rarely reported.

saponaceum *var.* **squamosum** (Cooke) Rea, *Brit.
basidiomyc.*: 227 (1922)
Agaricus saponaceus var. *squamosus* Cooke, *Handb. Brit.
fung.*, Edn 2: 33 (1884)
Tricholoma saponaceum var. *ardosiacum* Bres., *Icon. Mycol.*:
t.86 (1927)

E: o **S:** o **W:** ! **NI:** ! **ROI:** !
H: On acidic soil under deciduous trees, or with conifers in
mixed woodland.
D: FAN4: 141 **D+I:** FungEur3: 162-163 2 (p. 419), Galli3: 62-63
I: Ph: 36
Occasional in England and Scotland. Rarely reported elsewhere
but apparently widespread.

scalpturatum (Fr.) Quél., *Mém. Soc. Émul. Montbéliard, Sér. 2,*
5: 232 (1872)
Agaricus scalpturatus Fr., *Epicr. syst. mycol.*: 31 (1838)
Agaricus chrysites Jungh., *Linnaea* 5: 388 (1830)
Tricholoma chrysites (Jungh.) Gillet, *Hyménomycetes*: 98
(1874)
Tricholoma argyraceum var. *chrysites* (Jungh.) Gillet,
Hyménomycetes, Suppl.: 9 (1890)
Mis.: *Tricholoma argyraceum* sensu NCL
E: o **S:** o **W:** o **NI:** ! **ROI:** !
H: On soil, often calcareous, in deciduous woodland. Most
frequently reported with *Fagus*, but also known with *Betula*
spp on heathland.
D: NM2: 190, FAN4: 136 **D+I:** FungEur3: 238-240 24 [p.465],
B&K3: 336 431, Galli3: 118-119 **I:** Ph: 37 (as *Tricholoma
argyraceum*), Bon: 155, C&D: 195
Occasional but widespread and may be locally common. Most
records of '*Tricholoma argyraceum*' are probably this species.

sciodes (Pers.) C. Martín, *Cat. Syst. des Basidiomycetes*: 51
(1919)
Agaricus myomyces var. *sciodes* Pers., *Syn. meth. fung.*: 346
(1810)
Tricholoma virgatum var. *sciodes* (Pers.) Konrad, *Bull. Soc.
Mycol. France* 45: 53 (1929)
Tricholoma sciodellum P.D. Orton, *Kew Bull.* 54(3): 709
(1999)
Mis.: *Tricholoma murinaceum* sensu auct.
Mis.: *Agaricus murinaceus* sensu auct.
E: o **S:** o **W:** o **NI:** !
H: On soil, usually amongst leaf litter under *Fagus* in deciduous
woodland.
D: NM2: 191, FAN4: 138-139 **D+I:** FungEur3: 255-258 29
[p.475], B&K3: 336 432, Galli3: 104-105 **I:** C&D: 193
Occasional but widespread, generally rather uncommon.

sejunctum (Sowerby) Quél., *Mém. Soc. Émul. Montbéliard, Sér.
2,* 5: 76 (1872)
Agaricus sejunctus Sowerby, *Col. fig. Engl. fung.* 2: pl. 126
(1799)
E: o **S:** o **W:** !
H: On soil, often associated with *Quercus* and *Betula* spp, in
mixed deciduous woodland.
D: NM2: 189, FAN4: 113-114 **D+I:** Ph: 34-35, FungEur3: 284-
286 37 (p. 491) B&K3: 338 433, Galli3: 156-157 **I:** Bon: 157,
C&D: 191, Kriegl3: 565
Occasional but widespread, generally rather uncommon.

stans (Fr.) Sacc., *Syll. fung.* 5: 94 (1887)
Agaricus stans Fr., *Hymenomyc. eur.*: 52 (1874)
S: !
H: On acidic soil with *Pinus sylvestris* and *Calluna*.
D: NM2: 188, NBA3: 126-127 **D+I:** FungEur3: 373-376 61 (p.
553), B&K3: 338 434, Galli3: 210-211
Known from Easterness. Excluded 'pending clearer definition'
and 'doubtfully distinct from *Tricholoma pessundatum*' in NCL
but now accepted as a good species.

stiparophyllum (S. Lundell) P. Karst., *Bidrag Kännedom
Finlands Natur Folk* 32: 42 (1879)
Agaricus stiparophyllus S. Lundell, Fries, *Monogr.
hymenomyc. Suec.* 1: 29 (1857)
Tricholoma pseudoalbum Bon, *Bull. Soc. Mycol. France* 85(4):
486 (1970) [1969]
Mis.: *Tricholoma album* sensu NCL p.p. & sensu Lange
(FlDan1: 59 & pl. 27D)
E: ! **S:** ! **W:** ! **NI:** !
H: On soil, associated with *Betula* spp, in mixed deciduous
woodland.

D: FAN4: 144 **D+I:** FungEur3: 182-184 9 [p.433] (as *Tricholoma pseudoalbum*), B&K3: 338 435, Galli3: 72-73 (as *T. pseudoalbum*) **I:** Bon: 151 (as *T. pseudoalbum*), C&D: 189 (as *T. pseudoalbum*)
Rarely reported but apparently widespread. Not always distinguished from *Tricholoma album*.

sulphurescens Bres., *Ann. Mycol.* 3(2): 159 (1905)
 Mis.: *Tricholoma resplendens* sensu auct.
E: ! **S:** ! **NI:** !
H: On acidic soil under deciduous trees.
D: NM2: 184, FAN4: 146, NBA3: 125-126 (as *T. resplendens*) **D+I:** FungEur3: 170-173 6 (p. 427), B&K3: 340 436, Galli3: 70-71 **I:** Bon: 151, C&D: 189, Kriegl3: 573
Only five records, from England (Berkshire and Surrey) and Scotland (Perthshire); also reported from Northern Ireland.

sulphureum (Bull.) P. Kumm., *Führ. Pilzk.*: 133 (1871)
 Agaricus sulphureus Bull., *Herb. France*: pl. 168 (1784)
 Gymnopus sulphureus (Bull.) Gray, *Nat. arr. Brit. pl.* 1: 606 (1821)
E: c **S:** c **W:** o **NI:** c **ROI:** o
H: On soil (often acidic) usually associated with *Quercus* spp in oakwoods, but also not uncommon with *Fagus* in mixed deciduous woodland.
D: NM2: 184, FAN4: 146-147 **D+I:** Ph: 34-35, FungEur3: 189-192 11 [p.437], B&K3: 340 437, Galli3: 80-81 **I:** Sow1: 44 (as *Agaricus sulphureus*), FlDan1: 25F, Bon: 153, C&D: 193, Kriegl3: 574
Common and widespread.

sulphureum *var.* **hemisulphureum** Kühner
 Tricholoma hemisulphureum (Kühner) A. Riva, *Mycol. Helv.* 3(3): 325 (1989)
E: !
H: In grassland on calcareous soil, associated with *Helianthemum nummularium*.
D+I: Galli3: 76-77 (as *Tricholoma hemisulphureum*)
British collections only from Buckinghamshire and Derbyshire.

terreum (Schaeff.) P. Kumm., *Führ. Pilzk.*: 134 (1871)
 Agaricus terreus Schaeff., *Fung. Bavar. Palat. nasc.* 1: 28 (1762)
 Agaricus myomyces Pers., *Tent. disp. meth. fung.*: 20 (1797)
 Gymnopus myomyces (Pers.) Gray, *Nat. arr. Brit. pl.* 1: 608 (1821)
 Tricholoma myomyces (Pers.) J.E. Lange, *Dansk. Bot. Ark.* 8(3): 21 (1933)
 Tricholoma bisporigerum J.E. Lange, *Fl. agaric. danic.* 1: 53 (1935)
E: c **S:** c **W:** c **NI:** c **ROI:** o
H: On soil, associated with *Pinus* spp.
D: NM2: 190, FAN4: 134-135 **D+I:** Ph: 35, FungEur3: 209-212 16 [p. 447]/ 16b [p.449], FungEur3: 217-219 18 [p.453] (as *Tricholoma myomyces*), B&K3: 340 438, Galli3: 110-111, Galli3: 114-115 (as *T. myomyces*) **I:** Sow1: 76 (as *Agaricus terreus*), Bon: 155 (as *T. myomyces*), Bon: 155, C&D: 195
Common. Associated with pines, thus records under deciduous trees are likely to be misdetermined. *Tricholoma myomyces* is accepted by some authorities as a good species.

ustale (Fr.) P. Kumm., *Führ. Pilzk.*: 130 (1871)
 Agaricus ustalis Fr., *Observ. mycol.* 2: 122 (1818)
 Agaricus fulvellus Fr., *Epicr. syst. mycol.*: 28 (1838)
 Tricholoma fulvellum (Fr.) Gillet, *Hyménomycètes*: 93 (1874)
 Mis.: *Tricholoma albobrunneum* sensu auct. Brit.
E: o **S:** o **W:** o **NI:** o **ROI:** !
H: On soil, usually associated with *Fagus* and occasionally *Quercus* spp in beechwoods or in mixed deciduous woodland.
D: NM2: 188, FAN4: 126 **D+I:** FungEur3: 369-371 60 [p. 549], B&K3: 342 439, Galli3: 204-205 **I:** Bon: 159, C&D: 197, SV39: 44, Kriegl3: 540
Occasional but apparently widespread. Often confused with *T. ustaloides* and *T. albobrunneum*.

ustaloides Romagn., *Bull. Soc. Naturalistes Oyonnax* 8: 76 (1954)

 Mis.: *Tricholoma albobrunneum* sensu auct. Brit
E: o **S:** ! **W:** ! **NI:** ! **ROI:** !
H: On soil, usually associated with *Fagus* or *Quercus* spp in mixed deciduous woodland.
D: NM2: 186, FAN4: 127 **D+I:** Ph: 39, FungEur3: 390-392 65 [p. 563], B&K3: 342 440, Galli3: 222-223 **I:** Bon: 161, C&D: 199, SV39: 44, Kriegl3: 541
Occasional in England. Rarely reported elsewhere but apparently widespread. Often confused with *T. ustale* and *T. albobrunneum*.

vaccinum (Schaeff.) P. Kumm., *Führ. Pilzk.*: 133 (1871)
 Agaricus vaccinus Schaeff., *Fung. Bavar. Palat. nasc.* 1: 13 (1762)
E: ! **S:** ! **W:** ?
H: On acidic soil with *Pinus* spp.
D: NM2: 187, FAN3: 121 **D+I:** Ph: 38-39, FungEur3: 341-344 53 [p. 533], B&K3: 342 441, Galli3: 190-191 **I:** Bon: 159, C&D: 197, Kriegl3: 543
Apparently rare. Mostly reported from Scotland, rarely in England, and a single record from Wales by Rea in 1910. Records unsubstantiated with voucher material may be misidentified *T. imbricatum*.

vinaceogriseum P.D. Orton, *Notes Roy. Bot. Gard. Edinburgh* 44(3): 497 (1987)
S: !
H: On soil, associated with *Pinus sylvestris* in Caledonian pinewoods.
Known from the type and two other collections. Near to *T. bresadolanum* and *T. sciodes* but associated with conifers.

virgatum (Fr.) P. Kumm., *Führ. Pilzk.*: 134 (1871)
 Agaricus virgatus Fr., *Observ. mycol.* 2: 113 (1818)
E: o **S:** o **W:** o **NI:** o **ROI:** !
H: On soil in deciduous woodland.
D: NM2: 191, FAN4: 137-138 **D+I:** Ph: 35, FungEur3: 251-254 28 [p. 473], B&K3: 344 442, Galli3: 100-101 **I:** Bon: 153, C&D: 193, Kriegl3: 557
Occasional but widespread and may be locally common.

viridifucatum Bon, *Doc. Mycol.* 6(22-23): 185 (1976)
E: !
H: On loam soil in deciduous woodland. British material with *Quercus* spp.
D+I: FungEur3: 299-302 42 [p.503], B&K3: 344 443, Galli3: 174-175 **I:** C&D: 191
Known from West Sussex. Reported from South Hampshire but unsubstantiated with voucher material.

viridilutescens M.M. Moser, *Fungorum Rar. Icon. Color.* 7: 13 & pl. 50c (1978)
E: ! **S:** o
H: On loam soil in deciduous woodland, often on limestone.
D+I: FungEur3: 288-290 39 [p. 495], Galli3: 160-161 **I:** Bon: 157, C&D: 191
Known from England (Mid-Lancashire: Gait Barrows and South Hampshire: New Forest) and from several sites in Scotland.

TRICHOLOMOPSIS Singer, *Schweiz. Z. Pilzk.* 17: 56 (1939)

Agaricales, Tricholomataceae
Type: *Tricholomopsis rutilans* (Schaeff.) Singer

decora (Fr.) Singer, *Schweiz. Z. Pilzk.* 17: 56 (1939)
 Agaricus decorus Fr., *Syst. mycol.* 1: 108 (1821)
 Clitocybe decora (Fr.) Gillet, *Hyménomycètes*: 171 (1874)
 Tricholoma decorum (Fr.) Quél., *Compt. Rend. Assoc. Franç. Avancem. Sci.* 10: 389 (1882) [1881]
 Pleurotus decorus (Fr.) Sacc., *Syll. fung.* 5: 342 (1887)
E: ! **S:** ! **I:** !
H: On decayed wood of conifers, usually fallen trunks or stumps of *Pinus sylvestris*.
D: NM2: 192, FAN4: 151-152 **D+I:** Ph: 44, B&K3: 344 444 **I:** Bon: 151, C&D: 199, Kriegl3: 577

Occasional in Scotland, and may be locally frequent. Rarely reported elsewhere.

rutilans (Schaeff.) Singer, *Schweiz. Z. Pilzk.* 17: 56 (1939)
  Agaricus rutilans Schaeff., *Fung. Bavar. Palat. nasc.* 3: 219 (1770)
  Gymnopus rutilans (Schaeff.) Gray, *Nat. arr. Brit. pl.* 1: 605 (1821)
  Tricholoma rutilans (Schaeff.) P. Kumm., *Führ. Pilzk.*: 133 (1871)
  Agaricus serratis Bolton, *Hist. fung. Halifax* 1: 14 (1788)
  Agaricus variegatus Scop. [*nom. illegit.*, non *A. variegatus* Pers. (1801)], *Fl. carniol.*: 434 (1772)
  Tricholoma variegatum (Scop.) Sacc., *Syll. fung.* 5: 96 (1887)
  Agaricus xerampelinus Sowerby, *Col. fig. Engl. fung.* 1: pl. 31 (1796)
E: c **S:** c **W:** c **NI:** c **ROI:** c **O:** Channel Islands: ! Isle of Man: c
H: On decayed wood of conifers, usually large stumps or fallen trunks of *Pinus* spp, and rarely collected on decayed debris of *Pteridium*. Reported on wood of deciduous trees such as *Fraxinus* and *Quercus* spp, but unsubstantiated by voucher material and the hosts unverified.
D: NM2: 191, FAN4: 151 **D+I:** Ph: 43, B&K3: 346 445 **I:** Sow1: 31 (as *Agaricus xerampelinus*), Bon: 151, C&D: 199, Kriegl3: 579
Common and widespread. English name = 'Plums and Custard'.

TRICHOSPORON Behrend, (anam.) *Berliner Klin. Wochenschr.* 21: 464 (1890)
Trichosporonales
  Proteomyces Moses & Vianna, *Mem. Inst. Oswaldo Cruz* 5: 192 (1913)
  Geotrichoides Langeron & Talice, *Ann. Parasitol. Humaine Comp.* 10: 62 (1932)
A genus of anamorphic yeasts, many of them dermatotrophs, with no known teleomorphs.
Type: *Trichosporon beigelii* (Küchenm. & Rabenh.) Vuill.

asahii Sugita, Nishikawa & Shinoda, *J. Gen. Appl. Microbiol.* 40: 405 (1994)
E: !
H: Isolated ex Stilton cheese.
D+I: YST 752

cutaneum (Beurm., Gougerot & Vaucher bis) N. Ota, *Ann. Parasitol. Humaine Comp.* 4: 1 (1926)
  Oidium cutaneum Beurm., Gougerot & Vaucher bis, *Bull. Mém. Soc. Méd. Hôp. Paris* sér. 3, 28: 256 (1909)
  Mycoderma cutaneum (Beurm., Gougerot & Vaucher bis) Neveu-Lemaire, *Precis Parasitol.*: 932 (1921)
  Geotrichoides cutanea (Beurm., Gougerot & Vaucher bis) Langeron & Talice, *Ann. Parasitol. Humaine Comp.* 10: 1 (1932)
  Geotrichum cutaneum (Beurm., Gougerot & Vaucher bis) F.P. Almeida, *Ann. Fac. Med. S. Paulo* 9: 78 (1933)
  Monilia cutanea (Beurm., Gougerot & Vaucher bis) Castell. & Chalm., *Manual of tropical medicine*, Edn 2: 830 (1934)
  Proteomyces cutaneus (Beurm., Gougerot & Vaucher bis) C.W. Dodge, *Medical Mycol.*: 210 (1935)
E: !
H: Isolated ex pigs, dogs, turkeys, sausages and skin.
D+I: YST 756-757

dulcitum (Berkhout) Weijman, *Antonie van Leeuwenhoek* 45(1): 126 (1979)
  Oospora dulcita Berkhout, *De Schimmelgeslachten Monilia, Oidium, Oospora en Torula*, Scheveningen. 50 (1923)
S: !
H: Isolated ex sea water.
D+I: YST 759

gracile (Weigmann & A. Wolff) E. Guého & M.T. Sm., *Antonie van Leeuwenhoek* 61(4): 307 (1992)
  Oidium gracile Weigmann & A. Wolff, *Centralbl. Bakteriol. Parasitenk.* Abt. II, 22: 668 (1909)
E: !

H: Isolated ex frozen chicken.
D+I: YST 761

inkin (Oho) Carmo Souza & Uden, *Mycologia* 59: 653 (1967)
  Sarcinomyces inkin Oho, *Jap. J. Bot.* 26: 137 (1926)
  Sarcinosporon inkin (Oho) D.S. King & S.C. Jong, *Mycotaxon* 3(1): 93 (1975)
E: !
H: Isolated ex cheese.
D+I: YST 762

TRIMORPHOMYCES Bandoni & Oberw., *Syst. Appl. Microbiol.* 4: 106 (1983)
Tremellales, Tremellaceae
Type: *Trimorphomyces papilionaceus* Bandoni & Oberw.

papilionaceus Bandoni & Oberw., *Syst. Appl. Microbiol.* 4(1): 106 (1983)
E: !
H: Parasitic. On ascomycetous anamorphic *Arthrinium spp.* British material on a dead stem of *Heraclium sphondylium*.
D+I: Myc8(1): 6-7
A single collection from East Suffolk (Flatford Mill) in 1962.

TUBARIA (W.G. Sm.) Gillet, *Hyménomycètes*: 537 (1876)
Cortinariales, Crepidotaceae
Type: *Tubaria furfuracea* (Pers.) Gillet

albostipitata D.A. Reid, *Fungorum Rar. Icon. Color.* 6: 24 & pl. 44e (1972)
E: !
H: On soil, amongst litter of deciduous trees. British material with *Urtica dioica* under *Salix* spp in wet deciduous woodland.
I: Cooke 528 (483) Vol. 4 (1885)
Known from Bedfordshire (Stockgrove Woods) and possibly from South Wiltshire (Sutton Mandeville).

confragosa (Fr.) Harmaja, *Karstenia* 18: 55 (1978)
  Agaricus confragosus Fr., *Epicr. syst. mycol.*: 169 (1838)
  Pholiota confragosa (Fr.) P. Karst., *Bidrag Kännedom Finlands Natur Folk* 32: 304 (1879)
S: !
H: On decayed wood of deciduous trees, often *Betula* spp, but also known on *Fraxinus* and *Ulmus* spp.
D: NM2: 340, NBA8: 620 **I:** C&D: 349
Known from Scotland (Morayshire, Orkney, Perthshire, Stirling and Westerness). Recorded by Cooke then not seen again until 1975. Reported from England (Norfolk) in 1984 but dubiously determined.

conspersa (Pers.) Fayod, *Ann. Sci. Nat., Bot.,* sér. 7, 9: 355 (1889)
  Agaricus conspersus Pers., *Icon. descr. fung.*: pl. 50 (1800)
  Naucoria conspersa (Pers.) P. Kumm., *Führ. Pilzk.*: 76 (1871)
E: c **S:** c **W:** c **NI:** ! **ROI:** !
H: On soil or litter in woodland or occasionally amongst grass in grassland, throughout the year.
D: NM2: **D+I:** B&K4: 356 462 **I:** Bon: 247, C&D: 349
Common and widespread.

dispersa (Pers.) Singer, *Persoonia* 2(1): 22 (1961)
  Agaricus dispersus Pers., *Mycol. eur.* 3: 161 (1828)
  Agaricus sobrius var. *dispersus* (Pers.) Berk. & Broome, *Ann. Mag. Nat. Hist., Ser. 4 [Notices of British Fungi no. 1348]* 11: 341 (1873)
  Naucoria sobria var. *dispersa* (Pers.) Sacc., *Syll. fung.* 5: 852 (1887)
  Agaricus autochthonus Berk. & Broome, *Ann. Mag. Nat. Hist., Ser. 3 [Notices of British Fungi no. 1121]* 18: 53 (1866)
  Tubaria autochthona (Berk. & Broome) Sacc., *Syll. fung.* 5: 874 (1887)
E: c **S:** ? **W:** o **NI:** ! **ROI:** ! **O:** Isle of Man: !
H: On buried, mummified (and often undetectable) berries of *Crataegus* and occasionally on related genera such as *Cotoneaster*. Rarely on fruits of *Amelanchier canadensis* and

Ilex. Usually fruiting in late spring and then again in early autumn.
D: NM2: 340 **D+I:** Ph: 158 (as *Tubaria autochthona*) **I:** FlDan4: 127C (as *T. autochthona*), C&D: 349 (as *T. autochthona*)
Common and widespread in southern England, rarer northwards. Perhaps better known as *T. autochthona.*

furfuracea (Pers.) Gillet, *Hyménomycètes*: 538 (1876)
   Agaricus furfuraceus Pers., *Syn. meth. fung.*: 454 (1801)
   Naucoria furfuracea (Pers.) P. Kumm., *Führ. Pilzk.*: 77 (1871)
   Tubaria hiemalis Bon, *Doc. Mycol.* 8: 5 (1973)
E: c **S:** c **W:** c **NI:** c **ROI:** c **O:** Channel Islands: !
H: On fallen twigs, sticks, woodchips etc. in deciduous woodland, parks, and gardens.
D: NM2: 341 **D+I:** B&K4: 358 464 (as *Tubaria hiemalis*), B&K4: 358 463 **I:** FlDan4: 127G, Bon: 247, C&D: 349, SV43: 55
Very common and widespread. *T. hiemalis* is sometimes considered a separate species, characterised by fruiting during the winter and by the more broadly capitate cheilocystidia, and has been widely reported from Yorkshire southwards.

minutalis Romagn., *Rev. Mycol. (Paris)* 2(5): 193 (1937)
E: !
H: On soil in damp shady places in woodland. Also in gardens.
I: FlDan1: 197B (as *T. minima*)
Known from North Hampshire, South Wiltshire and Surrey. Distinctive on account of the minute basidiomes and small spores.

pallidispora J.E. Lange, *Fl. agaric. danic.* 5, *Taxonomic Conspectus*: IX (1940)
   Naucoria pallidispora (J.E. Lange) Kühner & Romagn. (as 'pallidospora'), *Flore Analytique des Champignons Supérieurs*: 238 (1953)
E: ! **O:** Isle of Man: ?
H: On soil, in deciduous woodland and on decayed grass culms, usually in grassland.
D+I: FlDan1: 128A1
Rarely reported, usually unsubstantiated with voucher material and poorly known in Britain. Rather close to small forms of *T. conspersa.*

praestans (Romagn.) M.M. Moser, *Kleine Kryptogamenflora*, Edn 4: 305 (1978)
   Naucoria praestans Romagn., *Bull. Soc. Naturalistes Oyonnax* 10 – 11 (Suppl.): 6 (1957)
S: !
H: On soil, in deciduous woodland. British material with *Betula* sp.
A single collection from Berwickshire (Mellerstain House) in 1999 and reported from Wigtownshire (Galloway) in 1986.

romagnesiana Arnolds, *Biblioth. Mycol.* 90: 460 (1983) [1982]
   Mis.: *Tubaria pellucida* sensu auct., non sensu Fries (1821, as *Agaricus pellucidus*)
E: ! **S:** ! **NI:** ? **ROI:** ?
H: On soil, in deciduous woodland. A single record also from weathered cow dung.
D+I: B&K4: 358 465
Little known in Britain and often confused with *T. furfuracea.* Arnolds considered the epithet '*pellucidus*' inadmissible because Fries described the spores as hyaline. Old records of *T. pellucida* are probably *T. furfuracea* s.l.

TUBULICIUM Oberw., *Sydowia* 19: 53 (1965)
Stereales, Tubulicrinaceae
Type: *Tubulicium vermiferum* (Bourdot) Jülich

vermiferum (Bourdot) Jülich, *Persoonia* 10(3): 335 (1979)
   Peniophora vermifera Bourdot, *Rev. Sci. du Bourb.* 23(1): 13 (1910)
   Tubulicrinis vermiferus (Bourdot) M.P. Christ., *Dansk. Bot. Ark.* 19(2): 136 (1960)
E: ! **S:** ! **W:** ! **O:** Channel Islands: !
H: On living trunks and decayed wood of deciduous trees such as *Acer pseudoplatanus, Fraxinus* and *Quercus* spp. Rarely on woody debris of conifers but known on *Picea sitchensis.*

D: NM3: 136 **D+I:** CNE8: 1520-1522
Rarely reported but apparently widespread.

TUBULICRINIS Donk, *Fungus* 26: 13 (1956)
Stereales, Tubulicrinaceae
Type: *Thelephora calcea* var. *glebulosa* Fr. (= *Tubulicrinis gracillimus*)

accedens (Bourdot & Galzin) Donk, *Fungus* 26: 14 (1956)
   Peniophora glebulosa ssp. *accedens* Bourdot & Galzin, *Bull. Soc. Mycol. France* 28(4): 386 (1913)
   Peniophora accedens (Bourdot & Galzin) Wakef. & A. Pearson, *Trans. Brit. Mycol. Soc.* 6: 140 (1919)
E: o **S:** o **W:** o
H: On decayed conifer wood. Known on *Juniperus communis, Pinus radiata, P. sylvestris* and *Taxus.*
D: NM3: 137 **D+I:** B&K2: 186 207, CNE8: 1526-1527
Occasional but widespread.

angustus (D.P. Rogers & Weresub) Donk, *Fungus* 26: 14 (1956)
   Peniophora angusta D.P. Rogers & Weresub, *Canad. J. Bot.* 31: 764 (1953)
S: !
H: British material on decayed wood of *Pinus sylvestris* in Caledonian pinewoods.
D: NM3: 139 **D+I:** B&K2: 188 208, CNE8: 1528-1529
Rarely reported.

chaetophorus (Höhn.) Donk, *Fungus* 26: 14 (1956)
   Hypochnus chaetophorus Höhn., *Sitzungsber. Kaiserl. Akad. Wiss., Wien, Math.-Naturwiss. Cl., Abt. 1* 111: 1007 (1902)
   Peniophora chaetophora (Höhn.) Höhn. & Litsch., *Sitzungsber. Kaiserl. Akad. Wiss., Wien, Math.-Naturwiss. Cl., Abt. 1* 116: 748 (1907)
E: !
H: On decayed conifer wood. British material on *Pinus radiata.*
D: NM3: 137 **D+I:** CNE8: 1538-1539
A single collection from South Devon (Torquay) in 1991.

glebulosus (Fr.) Donk, *Fungus* 26: 14 (1956)
   Thelephora calcea var. *glebulosa* Fr., *Elench. fung.* 1: 215 (1828)
   Corticium glebulosum (Fr.) Bres., *Fungi trident.* 2: 61 (1898)
   Peniophora glebulosa (Fr.) Sacc. & Syd., *Syll. fung.* 16: 195 (1902)
   Peniophora gracillima D.P. Rogers & H.S. Jacks., *Farlowia* 1: 317 (1943)
   Tubulicrinis gracillimus (D.P. Rogers & H.S. Jacks.) G. Cunn., *Bull N. Z. Dept Sci. Ind. Res., Plant Disease Division* 145: 141 (1963)
E: ! **S:** ! **W:** ! **NI:** !
H: On decayed wood of conifers such as *Larix* and *Pinus sylvestris.* Rarely on wood of deciduous trees and shrubs such as *Corylus, Quercus* and *Salix* spp.
D: NM3: 139 **D+I:** B&K2: 188 210, CNE8: 1556-1558 (as *Tubulicrinis gracillimus*)
Rarely reported but apparently widespread. Older British records as *Peniophora glebulosa* include *Tubulicrinis subulatus.*

medius (Bourdot & Galzin) Oberw., *Z. Pilzk.* 31(1-2): 26 (1966)
   Peniophora glebulosa ssp. *media* Bourdot & Galzin, *Bull. Soc. Mycol. France* 28(4): 385 (1913)
E: ! **S:** !
H: On decayed conifer wood. British material on *Pinus sylvestris.*
D: NM3: 139 **D+I:** B&K2: 190 212, CNE8: 1566-1570
Known from scattered localities in Scotland and one in England (South Devon).

propinquus (Bourdot & Galzin) Donk, *Fungus* 26(1-4): 14 (1956)
   Peniophora cretacea ssp. *propinqua* Bourdot & Galzin, *Hyménomyc. France*: 288 (1928)
S: !
H: On decayed wood of conifers. British material on *Pinus sylvestris.*
D: CNE8: 1571

A single collection from South Aberdeen (Inverey Wood) in 2003.

regificus (H.S. Jacks. & Dearden) Donk, *Fungus* 26: 14 (1956)
    Peniophora regifica H.S. Jacks. & Dearden, *Mycologia* 43(1): 57 (1951)
E: ! **ROI:** !
H: On decayed wood of deciduous trees such as *Fraxinus* or conifers such as *Cupressus macrocarpa* and *Pinus* spp. Also on dead branches of *Rhododendron* spp and dead stems of *Pteridium*.
D: NM3: 137 **D+I:** CNE8: 15731577, Myc8(3): 117
Known from England (South Devon) and the Republic of Ireland (Offaly and Clare).

sororius (Bourdot & Galzin) Oberw., *Z. Pilzk.* 31(1-2): 23 (1966)
    Peniophora glebulosa ssp. *sororia* Bourdot & Galzin, *Bull. Soc. Mycol. France* 28(4): 386 (1913)
    Peniophora glebulosa ssp. *juniperina* Bourdot & Galzin, *Bull. Soc. Mycol. France* 28(4): 386 (1913)
    Peniophora juniperina (Bourdot & Galzin) Bourdot & Galzin, *Hyménomyc. France*: 289 (1928)
    Tubulicrinis juniperina (Bourdot & Galzin) Donk, *Fungus* 26: 14 (1956)
E: ! **O:** Isle of Man: !
H: On decayed wood of *Picea abies* and a single record on a living basidiome of *Amylostereum chailletii* growing on *Picea* sp.
D: NM3: 137 **D+I:** CNE8: 1578-1581
Known from England (South Devon: Dartmoor, Bellever Forest) and the Isle of Man (South Barrule Plantation).

subulatus (Bourdot & Galzin) Donk, *Fungus* 26: 14 (1956)
    Peniophora glebulosa ssp. *subulata* Bourdot & Galzin, *Bull. Soc. Mycol. France* 28(4): 385 (1913)
    Peniophora subulata (Bourdot & Galzin) Donk, *Meded. Ned. Mycol. Ver.*: 165 (1931)
    Mis.: *Tubulicrinis glebulosus* sensu auct.
E: o **S:** o **W:** o **O:** Isle of Man: o
H: On decayed conifer wood, often *Pinus sylvestris* but also known on *Abies* spp, *Larix*, *Pinus radiata* and *Juniperus communis*. Rarely on wood of deciduous trees such as *Salix* spp, on dead runners of *Hedera* and on decayed planks in buildings.
D: NM3: 137 **D+I:** B&K2: 190 212, CNE8: 1582-1583
Occasional but apparently widespread. Probably commoner than *Tubulicrinis glebulosus*. Older British records as *Peniophora glebulosa* include both species.

TULASNELLA J. Schröt., in Cohn, *Krypt.-Fl. Schlesien* 3: 397 (1888)
Tulasnellales, Tulasnellaceae
    Prototremella Pat., *J. Bot. (Morot)* 2: 269 (1888)
    Pachysterigma Bref., *Unters. Gesammtgeb. Mykol.* 8: 5 (1889)
    Muciporus Juel, *Bih. Kongl. Svenska Vetensk.-Akad. Handl.* 23: 23 (1897)
    Gloeotulasnella Höhn. & Litsch., *Sitzungsber. Kaiserl. Akad. Wiss., Wien, Math.-Naturwiss. Cl., Abt. 1* 115(1): 1557 (1906)
    Epulorhiza R.T. Moore (anam.), *Mycotaxon* 29: 94 (1987)
Type: *Tulasnella lilacina* J. Schröt. (= *T. violea*)

albida Bourdot & Galzin, *Hyménomyc. France*: 59 (1928)
    Tulasnella intrusa Hauerslev, *Opera Bot.* 100: 114 (1989)
E: !
H: On, or in, the hymenium of corticioid fungi such as *Botryobasidium* spp, *Piloderma byssinum*, and *Sistotrema brinkmannii*. Also on decayed conifer wood and dead basidiomes of polypores such as *Trametes hirsuta*.
D: NM3: 116 (as *Tulasnella intrusa*)
Known from South Devon, Surrey and Warwickshire.

allantospora Wakef. & A. Pearson, *Trans. Brit. Mycol. Soc.* 8: 220 (1923)

E: ! **W:** !
H: On decayed wood of conifers (including old fence posts). Rarely on wood of deciduous trees such as *Fagus* and *Ulmus* spp.
D: NM3: 118
Rarely reported. Known from England in scattered locations from Hertfordshire to South Devon, and recently collected in Wales (Breconshire).

anguifera P. Roberts, *Mycol. Res.* 96(3): 235 (1992)
E: !
H: On decayed wood and decayed basidiomes of *Dacrymyces stillatus*.
Known from South Devon.

bourdotii Jülich, *Int. J. Mycol. Lichenol.* 1(1): 121 (1982)
E: !
H: On decayed conifer wood, and a single collection on *Fagus*.
Known from South Devon and Surrey.

brinkmannii Bres., *Ann. Mycol.* 18: 50 (1920)
E: !
H: British material on a dead stem of *Pteridium* and on fallen twigs of *Zelkova carpinifolia*.
Known from Hertfordshire (Luffenhall) in 1999 and Surrey (Kew) in 2004. Reported from South Devon but the collection has been redetermined as *T. calospora*.

calospora (Boud.) Juel, *Bih. Kongl. Svenska Vetensk.-Akad. Handl.* 23: 23 (1897)
    Prototremella calospora Boud., *J. Bot. (Morot)* 10: 85 (1896)
    Gloeotulasnella calospora (Boud.) D.P. Rogers, *Ann. Mycol.* 31: 201 (1933)
E: ! **W:** ! **NI:** !
H: On decayed wood of deciduous trees and conifers, occasionally in the hymenia of corticioid fungi (e.g. *Botryobasidium conspersum*, *Hyphodontia subalutacea* and *Sistotrema octosporum*).
D: NM3: 117
Known from England (Hertfordshire, Shropshire, South Devon, Surrey and West Kent), Wales (Breconshire) and Northern Ireland (Tyrone).

cystidiophora Höhn. & Litsch., *Sitzungsber. Kaiserl. Akad. Wiss., Wien, Math.-Naturwiss. Cl., Abt. 1* 115: 1557 (1906)
    Gloeotulasnella cystidiophora (Höhn. & Litsch.) Juel, *Arch. Bot. (Leipzig)* 14: 8 (1914)
E: ! **W:** !
H: On decayed wood of deciduous trees such as *Corylus*, *Fagus* and *Quercus* spp.
Known from England (East & West Sussex, Herefordshire, South Devon, South Hampshire, Surrey and West Kent) and Wales (Breconshire and Monmouthshire).

danica Hauerslev, *Friesia* 11(5): 275 (1987)
E: !
H: On decayed wood of deciduous trees.
D: NM3: 117
Known from South Devon and Surrey.

deliquescens (Juel) Juel, *Arch. Bot. (Leipzig)* 14: 8 (1914)
    Muciporus deliquescens Juel, *Bih. Kongl. Svenska Vetensk.-Akad. Handl.* 23: 24 (1897)
    Tulasnella rosella Bourdot & Galzin, *Bull. Soc. Mycol. France* 39: 263 (1923)
    Rhizoctonia repens G.E. Bernard (anam.), *Ann. Sci. Nat., Bot., sér. 9, 9*: 31 (1909)
    Epulorhiza repens (G.E. Bernard) R.T. Moore (anam.), *Mycotaxon* 29: 95 (1987)
E: ! **S:** ! **W:** ! **NI:** !
H: On decayed wood of conifers and rarely deciduous trees. Also known on dead stems of *Petasites* spp and isolated from roots of *Orchis purpurella*.
D: NM3: 117
Rarely reported but apparently widespread.

dissitispora P. Roberts, *Mycol. Res.* 98(11): 1240 (1994)
E: !

H: On decayed wood of deciduous trees.
Known only from South Devon.

echinospora P. Roberts, *Cryptog. Mycol.* 25: 23 (2004)
E: !
H: On decayed wood.
Known in Britain only from the type collection, from North Hampshire (Swarraton), collected in 1903.

eichleriana Bres., *Ann. Mycol.* 1(1): 113 (1903)
 Tulasnella lactea Bourdot & Galzin, *Bull. Soc. Mycol. France* 39: 263 (1923)
 Tulasnella microspora Wakef. & A. Pearson, *Trans. Brit. Mycol. Soc.* 8(4): 220 (1923)
 Tulasnella obscura Bourdot & Galzin, *Bull. Soc. Mycol. France* 39: 265 (1923)
E: c **W:** ! **NI:** ! **O:** Channel Islands: !
H: On decayed wood of deciduous trees or conifers.
D: NM3: 116
Common in England, from Yorkshire southwards; widespread elsewhere.

falcifera P. Roberts, *Mycol. Res.* 96(3): 236 (1992)
E: !
H: On decayed wood of *Salix* sp.
Known only from the type collection (South Hampshire: New Forest, Blackhamsley Cutting) in 1991.

fuscoviolacea Bres., *Fungi trident.* 2(14): 98 (1900)
E: !
H: On decayed wood.
D: NM3: 117
A single collection from Worcestershire (Shrawley Wood) in 1921.

helicospora Raunk., *Bot. Tidsskr.* 36: 205 (1918)
E: !
H: On decayed wood of deciduous trees. British material on *Fraxinus*.
D: NM3: 117
Known only from South Devon (Orley Common and Slapton Ley).

hyalina Höhn. & Litsch., *Sitzungsber. Kaiserl. Akad. Wiss., Wien, Math.-Naturwiss. Cl., Abt. 1* 117: 1114 (1908)
E: !
H: British collection on decayed wood of *Picea* sp.
D: NM3: 118
Known in Britain from a single collection from South Devon. Other reports are doubtful or redetermined.

interrogans P. Roberts, *Mycol. Res.* 96(3): 234 (1992)
E: !
H: On decayed wood of *Corylus*, *Salix* and *Picea* spp.
Known only from South Devon.

pallida Bres., *Ann. Mycol.* 1(1): 122 (1903)
 Tulasnella albolilacea Bourdot & Galzin, *Bull. Soc. Mycol. France* 39: 264 (1923)
 Mis.: *Tulasnella violacea* sensu auct. p.p.
E: !
H: On decayed wood of deciduous trees such as *Betula*, *Fagus* and *Salix* spp and also from dead stems of *Epilobium* sp.
D: NM3: 116
Known from South Devon, Surrey, West Gloucestershire and West Kent.

permacra P. Roberts, *Mycol. Res.* 97(2): 214 (1993)
E: !
H: On decayed wood of deciduous trees and conifers; also known in the hymenium of *Botryobasidium danicum*.
D: NM3: 118
Known from Shropshire and South Devon.

pinicola Bres., *Ann. Mycol.* 1(1): 114 (1903)
 Gloeotulasnella pinicola (Bres.) D.P. Rogers, *Ann. Mycol.* 31: 199 (1933)
 Tulasnella sordida Bourdot & Galzin, *Bull. Soc. Mycol. France* 39: 265 (1923)

 Gloeotulasnella sordida (Bourdot & Galzin) M.P. Christ., *Dansk bot. Ark.* 19(1): 43 (1959)
 Tulasnella tremelloides Wakef. & A. Pearson, *Trans. Brit. Mycol. Soc.* 6(1): 70 (1917)
 Gloeotulasnella tremelloides (Wakef. & A. Pearson) D.P. Rogers, *Ann. Mycol.* 31: 201 (1933)
E: ! **W:** ? **O:** Channel islands: !
H: On decayed wood of deciduous trees and conifers, on old polypore basidiomes and in the hymenium of fungi such as *Dacrymyces stillatus*, *Helicogloea vestita* and *Hymenochaete corrugata*.
D: NM3: 116
Known from England (London, South Devon and Surrey) and the Channel Islands (Sark). Reported from Wales (Merionethshire) but unsubstantiated with voucher material.

pruinosa Bourdot & Galzin, *Bull. Soc. Mycol. France* 39: 264 (1923)
 Tulasnella araneosa Bourdot & Galzin, *Bull. Soc. Mycol. France* 39: 265 (1923)
E: !
H: British collections on decayed wood of *Quercus* and *Picea* spp.
D: NM3: 116
A poorly characterised species, accepted with some doubt. British collections in Herb. K from Herefordshire and South Devon.

quasiflorens P. Roberts, *Mycol. Res.* 98(11): 1238 (1994)
E: !
H: On decayed wood.
Known in Britain only from the type collection (South Devon: Dunsford Woods) in 1992.

saveloides P. Roberts, *Mycol. Res.* 97(2): 217 (1993)
E: !
H: On decayed wood of deciduous trees and conifers, on decayed basidiomes of corticioid fungi, and on dried-out peat.
Known from South Devon, South Hampshire and Surrey.

thelephorea (Juel) Juel, *Arch. Bot. (Leipzig)* 14: 8 (1914)
 Muciporus corticola f. *thelephorea* Juel, *Bih. Kongl. Svenska Vetensk.-Akad. Handl.* 23: 23 (1897)
 Tulasnella cremea Jülich, *Int. J. Mycol. Lichenol.* 1(1): 122 (1982)
 Mis.: *Tulasnella inclusa* sensu auct.
E: ! **W:** !
H: On decayed wood of deciduous trees and conifers, on decayed polypore basidiomes and in the hymenium of fungi such as *Sistotrema brinkmanii* and *Dacrymyces stillatus*.
D: NM3: 118
Known from England (Berkshire, Cambridgeshire, South Devon and Surrey) and Wales (Carmarthenshire and Monmouthshire).

tomaculum P. Roberts, *Mycol. Res.* 97(2): 215 (1993)
E: o **W:** o
H: On decayed wood of deciduous trees and conifers.
D: NM3: 117
Occasional in England and Wales.

violea (Quél.) Bourdot & Galzin, *Bull. Soc. Mycol. France* 25: 31 (1909)
 Hypochnus violeus Quél., *Compt. Rend. Assoc. Franç. Avancem. Sci.* 11: 402 (1883)
 Corticium violeum (Quél.) Costantin & L.M. Dufour, *Nouv. fl. champ.*: 187 (1891)
 Mis.: *Tulasnella incarnata* sensu auct.
E: c **S:** o **W:** o
H: On decayed wood and old polypore basidiomes.
D: NM3: 116 **D+I:** B&K2: 70 33
Common and widespread. The most frequently collected species of the genus, possibly due to the basidiomes often being conspicuous and vivid lilac-pink. Old records may have included *Tulasnella eichleriana* which has much smaller spores.

TULOSTOMA Pers., *Neues Mag. Bot.* 1: 86 (1794)
Tulostomatales, Tulostomataceae
Type: *Tulostoma brumale* Pers.

brumale Pers., *Neues Mag. Bot.* 1: 86 (1794)
Lycoperdon mammosum P. Micheli, *Nova plantarum genera*: 217 (1729)
Tulostoma mammosum (P. Micheli) Fr., *Syst. mycol.* 3: 42 (1829)
Lycoperdon pedunculatum L., *Sp. pl.*: 1654 (1763)
E: o **S:** o **W:** o **NI:** ! **ROI:** ! **O:** Channel Islands: ! Isle of Man: !
H: On sandy or calcareous soil, often amongst mosses in dune slacks; also on calcareous mortar in old, moss-covered walls. Most frequently collected in winter or early spring.
D: NM3: 298 **D+I:** Ph: 251, B&K2: 396 522, BritPuffb: 46-47 Figs. 24/25 **I:** Sow4: 406 (as *Lycoperdon pedunculatum*), Kriegl2: 180
Occasional and widespread in coastal areas, but rarely found inland.

melanocyclum Bres., in Petri, *Ann. Mycol.* 2: 415 (1904)
E: ! **W:** !
H: On dunes, usually in dried-out slacks, often amongst *Salix repens* and mosses such as *Ceratodon purpureus* or *Tortula ruraliformis* and also reported with *Ammophila arenaria*.
D: NM3: 298 **D+I:** BritPuffb: 48-49 Figs. 26/27 **I:** Kriegl2: 183
Rarely reported. British records show a markedly western coastal distribution but this species is also known from West Norfolk.

niveum Kers, *Bot. Not.* 131: 411 (1978)
S: !
H: Amongst mosses on limestone boulders in montane habitat. Basidiomes are produced from early winter to early spring.
D: NM3: 298 **D+I:** BritPuffb: 44-45 Figs. 22/23
Known from South Aberdeen (Braemar) and West Sutherland (Inchnadamph).

TYLOPILUS P. Karst., *Rev. Mycol. (Toulouse)* 3(9): 16 (1881)
Boletales, Strobilomycetaceae
Type: *Tylopilus felleus* (Bull.) P. Karst.

felleus (Bull.) P. Karst., *Rev. Mycol. (Toulouse)* 3(9): 16 (1881)
Boletus felleus Bull., *Herb. France*: pl. 379 (1788)
Boletus alutarius Fr., *Observ. mycol.* 1: 115 (1815)
Tylopilus alutarius (Fr.) Rea, *Brit. basidiomyc.*: 555 (1922)
Tylopilus felleus var. *alutarius* (Fr.) P. Karst., *Bidrag Kännedom Finlands Natur Folk* 37: 2 (1882)
E: o **S:** ! **W:** ! **NI:** ! **ROI:** !
H: On acidic or sandy soil in deciduous woodland, usually associated with *Castanea*, *Fagus* and *Quercus* spp. Less often with conifers such as *Pinus sylvestris* in mixed woodland. Occasionally on decayed large stumps.
D: NM2: 72 **D+I:** Ph: 205, FRIC16: 1920 128a (as *Tylopilus felleus* var. *alutarius*), B&K3: 84 54 **I:** Bon: 49, C&D: 431, SV42: 15, Kriegl2: 314
Occasional but widespread and may be locally frequent.

TYLOSPORA Donk, *Taxon* 9: 220 (1960)
Stereales, Atheliaceae
Tylosperma Donk [*nom. illegit.*, non *Tylosperma* Botsch. (1952)(*Rosaceae*)], *Fungus* 27: 28 (1957)
Type: *Tylospora asterophora* (Bonord.) Donk

asterophora (Bonord.) Donk, *Taxon* 9: 220 (1960)
Hypochnus asterophorus Bonord., *Handb. Mykol.*: 160 (1851)
Tylosperma asterophorum (Bonord.) Donk, *Fungus* 27: 28 (1957)
Corticium trigonospermum Bres., *Ann. Mycol.* 3(2): 163 (1905)
Tomentella trigonosperma (Bres.) Höhn. & Litsch., *Sitzungsber. Kaiserl. Akad. Wiss., Wien, Math.-Naturwiss. Cl., Abt. 1* 117: 1090 (1908)
E: o **S:** o **NI:** !

H: With conifers. Fruiting on litter and soil, especially on acidic soils, and common in *Picea* plantations. Also overgrowing *Sphagnum* moss in swampy woodland.
D: NM3: 151 **D+I:** B&K2: 192 215, CNE8: 1584-1587
Occasional but widespread and often rather common in the right habitat. Apparently absent from calcareous soils and thus rare in south east England.

fibrillosa (Burt) Donk, *Taxon* 9: 220 (1960)
Hypochnus fibrillosus Burt, *Ann. Missouri Bot. Gard.* 3: 238 (1916)
Tomentella fibrillosa (Burt) Bourdot & Galzin, *Hyménomyc. France*: 513 (1928)
Tylosperma fibrillosum (Burt) Donk, *Fungus* 27: 28 (1957)
E: c **S:** c **W:** c **NI:** ! **ROI:** ! **O:** Isle of Man: c
H: With conifers. Fruiting on litter, acidic soil, peat, and decayed wood, especially of *Picea abies*.
D: NM3: 151 **D+I:** B&K2: 192 216, CNE8: 1586-1587
Common and widespread in suitable habitats. Apparently absent from calcareous soils and thus rare in south east England

TYPHULA (Pers.) Fr., *Observ. mycol.* 2: 296 (1818)
Cantharellales, Typhulaceae
Pistillaria Fr., *Syst. mycol.* 1: 496 (1821)
Phacorhiza Pers., *Mycol. eur.* 1: 192 (1822)
Pistillina Quél., *Compt. Rend. Assoc. Franç. Avancem. Sci.* 9: 671 (1880)
Type: *Typhula phacorrhiza* (Reichardt) Fr.

capitata (Pat.) Berthier, *Bull. Mens. Soc. Linn. Lyon* 45: 106 (1976)
Sphaerula capitata Pat., *Tab. anal. fung.* 1: 27 (1883)
Pistillina patouillardii Quél., *Compt. Rend. Assoc. Franç. Avancem. Sci.* 12: 506 (1884)
E: ! **W:** !
H: On decayed leaves of herbaceous plants. British material on *Carex pendula* and *Typha latifolia*.
D: NM3: 257
Single collections from England (Hertfordshire: Bayford) in 2003 and Wales (Merionethshire: Hengwrt) in 1976. Reported by Corner (1950) as 'very common' on dead stems and leaves of *Phragmites australis* in the Cambridgeshire fens, but no material has been traced.

crassipes Fuckel, *Jahrb. Nassauischen Vereins Naturk.* 23-24: 32 (1870)
Pistillaria bulbosa Pat., *Bull. Soc. Mycol. France* 32: 45 (1885)
Typhula bulbosa (Pat.) Corner, *Monograph of Clavaria and Allied Genera*: 666 (1950)
Typhula corallina Quél., *Compt. Rend. Assoc. Franç. Avancem. Sci.* 11: 505 (1883)
Clavaria epiphylla Quél., *Compt. Rend. Assoc. Franç. Avancem. Sci.* 11: 505 (1883)
Pistillaria epiphylla (Quél.) Corner, *Monograph of Clavaria and Allied Genera*: 480 (1950)
E: ! **S:** ! **NI:** ?
H: On fallen and decayed leaves of *Chamerion angustifolium* and of deciduous trees such as *Acer pseudoplatanus* and *Alnus*.
D: NM3: 259
Known from England (Yorkshire: Pickering, Kingthorpe) and Scotland (Berwickshire, Kirkudbrightshire and North Ebudes). Reported elsewhere, but unsubstantiated with voucher material.

culmigena (Mont. & Fr.) Berthier, *Bull. Mens. Soc. Linn. Lyon* 45: 186 (1976)
Pistillaria culmigena Mont. & Fr., *Ann. Sci. Nat., Bot.*, sér. 2, 5: 337 (1836)
Pistillaria cardiospora Quél., *Tab. anal. fung.*: 25 (1883)
E: ! **S:** !
H: On dead leaves of grasses such as *Bromus ramosus*, *Arrhenatherum elatius* and *Triticum* spp, also *Typha* spp and *Caltha palustris*.
D: NM3: 257

Known from England (Buckinghamshire and West Gloucestershire) and Scotland (North Ebudes and Wester Ross). Rarely reported elsewhere but unsubstantiated with voucher material.

erythropus (Pers.) Fr., *Syst. mycol.* 1: 495 (1821)
   Clavaria erythropus Pers., *Comment. Fungis Clavaeform.*: 84 (1797)
E: c **S:** c **W:** ! **NI:** c **O:** Isle of Man: !
H: On herbaceous debris, less often on decayed wood or woody debris such as small fallen twigs. Frequently on fallen leaves of *Acer pseudoplatanus* or *Fraxinus*. Less commonly on *Betula* and *Quercus* spp, on dead grasses and on decayed leaves and stems of *Gunnera macrophylla*
D: Corner: 668, NM3: 256 **D+I:** Ph: 258-259, B&K2: 336 431 **I:** Kriegl2: 47
Common and widespread, but basidiomes are small and easily overlooked.

graminum P. Karst., *Mycol. fenn.* 3: 340 (1876)
E: ? **S:** ?
H: On dead and decayed leaves of the grass *Molinia caerulea* (British material).
D: NM3: 259
Accepted with reservations. Recorded from Kirkudbrightshire (Bruce's Stone) in 1993 but voucher material has not been traced. A single specimen in herb K, from Surrey (Norbury Park) on an old stem of *Atropa bella-donna*, is annotated as 'compare with' *T. graminum*.

hollandii D.A. Reid, *Trans. Brit. Mycol. Soc.* 48: 528 (1965)
ROI: !
H: On moss and 'vegetable debris' accumulated in the fork of a tree branch.
Known only from the type collection from Cork (Glengariff) in 1964.

incarnata Fr., *Epicr. syst. mycol.*: 585 (1838)
   Typhula itoana S. Imai, *Trans. Sapporo Nat. Hist. Soc.* 11: 39 (1929)
E: ! **S:** ! **NI:** !
H: On dead or dying stems and leaves of grasses. Known on *Agrostis stolonifera, A. tenuis, Anthoxanthum odoratum, Dactylis glomerata, Festuca rubra, Holcus lanatus, Lolium perenne* and *Poa trivialis*. Also rather frequently on cereal crops such as *Hordeum* spp where it is the cause of 'Snow Rot Disease'.
D: Corner: 673, NM3: 257
Rarely reported, but basidiomes are minute and easily overlooked.

lutescens Boud., *Bull. Soc. Mycol. France* 16: 197 (1900)
E: ! **S:** ! **NI:** !
H: On decayed leaf litter in woodland. Recorded on *Alnus viridis, Fraxinus* and *Quercus* spp.
D: Corner: 676, NM3: 259
Rarely reported but apparently widespread.

micans (Pers.) Berthier, *Bull. Mens. Soc. Linn. Lyon* 45: 172 (1976)
   Clavaria micans Pers., *Comment. Fungis Clavaeform.*: 85 (1797)
   Pistillaria micans (Pers.) Fr., *Syst. mycol.* 1: 497 (1821)
   Pistillaria fulgida Fr., *Epicr. syst. mycol.*: 587 (1838)
   Pistillaria rosella Fr., *Epicr. syst. mycol.*: 587 (1838)
E: o **S:** o **NI:** ! **O:** Channel Islands: ! Isle of Man: !
H: On dead and dying stems of herbaceous plants. Most often collected on *Cirsium* spp and *Urtica dioica* but known on many other hosts.
D: NM3: 257 **I:** Kriegl2: 48
Occasional but widespread. Basidiomes are a vivid coral-pink, but minute and easily overlooked.

phacorrhiza (Reichardt) Fr., *Syst. mycol.* 1: 495 (1821)
   Clavaria phacorrhiza Reichardt, *Schriften Berlin Ges. Naturf. Freunde* 1: 315 (1780)
   Sclerotium scutellatum Alb. & Schwein. (anam.), *Consp. fung. lusat.*: 289 (1805)

E: o **S:** o **W:** ! **NI:** o
H: On decayed leaf litter of deciduous trees in damp woodland. Especially common on *Fraxinus* and not infrequent on *Acer pseudoplatanus, Alnus* and *Corylus*. Reported on *Gunnera manicata, Ilex* and *Tilia* spp.
D: Corner: 678, NM3: 256 **D+I:** B&K2: 336 432 **I:** Sow2: 233 (as *Clavaria phacorhiza*)
Occasional but widespread and locally abundant. The superficially similar *Macrotyphula juncea* lacks a sclerotium.

quisquiliaris (Fr.) Corner, *Monograph of Clavaria and Allied Genera*: 679 (1950)
   Pistillaria quisquiliaris Fr., *Syst. mycol.* 1: 497 (1821)
   Clavaria obtusa Sowerby, *Col. fig. Engl. fung.* 3: pl. 334 (1801)
   Pistillaria puberula Berk., *Outl. Brit. fungol.*: 286 (1860)
E: c **S:** c **W:** c **NI:** ! **ROI:** ! **O:** Isle of Man: !
H: On dead stems of *Pteridium*. Rarely on dead culms of *Juncus* spp and decayed leaves of deciduous trees.
D: NM3: 258 **D+I:** B&K2: 338 433 **I:** Kriegl2: 50
Common and widespread. Records on other substrata, such as wood, are dubious.

setipes (Grev.) Berthier, *Bull. Mens. Soc. Linn. Lyon* 45: 141 (1976)
   Pistillaria setipes Grev., *Scott. crypt. fl.* 6: 61 (1828)
   Typhula candida Fr., *Öfvers. Kongl. Vetensk.-Akad. Förh.* 18: 32 (1862)
   Typhula grevillei Fr., *Epicr. syst. mycol.*: 585 (1838)
   Clavaria gyrans Batsch, *Elench. fung. (Continuatio Prima)*: 235 & t. 28 f. 164 (1786)
   Typhula gyrans (Batsch) Fr., *Syst. mycol.* 1: 495 (1821)
   Typhula gyrans var. *grevillei* (Fr.) Massee, *Brit. fung.-fl.* 1: 89 (1892)
   Mis.: *Pistillaria ovata* sensu auct.
   Mis.: *Pistillaria pusilla* sensu auct
   Mis.: *Typhula sclerotioides* sensu Corner (1950) & sensu auct. Brit.
E: o **S:** o **W:** o **NI:** !
H: On decayed leaves of deciduous trees, often in wet places such as pondsides or swampy woodland. Known on *Acer pseudoplatanus, Alnus, Fraxinus* and *Salix* spp.
D: Corner: 488 (as *Pistillaria setipes*), NM3: 258 **D+I:** B&K2: 338 435 **I:** Kriegl2: 51
Occasional but apparently widespread.

spathulata (Peck) Berthier, *Bull. Mens. Soc. Linn. Lyon* 45: 152 (1976)
   Clavaria spathulata Peck, *Rep. (Annual) New York State Mus. Nat. Hist.* 27: 100 (1875)
   Pistillaria spathulata (Peck) Corner, *Monograph of Clavaria and Allied Genera*: 492 (1950)
E: ! **W:** ? **NI:** ?
H: On fallen branches, sticks etc. in woodland. British material on woody debris of *Crataegus monogyna* and *Fraxinus*.
D: NM3: 259 **I:** Kriegl2: 52
Known from England (Hertfordshire: Walkern Green, North Hampshire: Alice Holt Forest and West Kent: Pratt's Bottom, Birthday Wood). Reported from Berkshire, Northern Ireland (Fermanagh) and Wales (Monmouthshire) but unsubstantiated with voucher material.

subhyalina Courtec., *Doc. Mycol.* 14(54-55): 84 (1984)
   Pistillina hyalina Quél., *Compt. Rend. Assoc. Franç. Avancem. Sci.* 9: 671 (1881)
   Pistillaria hyalina (Quél.) Sacc., *Syll. fung.* 6: 759 (1886)
   Typhula hyalina (Quél.) Berthier [*nom. illegit.*, non *T. hyalina* Jungh. (1839)], *Bull. Mens. Soc. Linn. Lyon* 45: 79 (1976)
   Typhula hyalinella Nannf. & L. Holm, *Publ. Herb. Univ. Uppsala* 17: 16 (1985)
E: !
H: On dead and dying stems of grasses. British material on *Phragmites australis*.
D: Corner: 498 (as *Pistillina hyalina*), NM3: 257 (as *Typhula hyalinella*)

Known from Cambridgeshire (Wicken Fen) and East Norfolk (Wheatfen Broad).

todei (Fr.) Fr., *Syst. mycol.* 1: 494 (1821)
  Mitrula todei Fr., *Observ. mycol.* 1: 160 (1815)
  Pistillaria todei (Fr.) Corner, *Monograph of Clavaria and Allied Genera*: 493 (1950)
  Typhula athyrii Remsberg, *Mycologia* 32: 91 (1940)
E: ! S: !
H: On decayed debris of ferns such as *Athyrium filix-femina*, *Blechnum spicant* and *Dryopteris* spp. A single record on dead stems of *Chamerion angustifolium*.
D: Corner: 665 (as *Typhula athyrii*), NM3: 258
Known from Scotland (East Perthshire, Isle of Skye and Peebleshire) and England (Derbyshire).

trifolii Rostr., *Ugeskr. Landm.* 35(1): 72 (1890)
S: !
H: On decayed vegetable debris. British material on *Iris pseudacorus*.
D: Corner: 685, NM3: 260
A single collection from Dumfrieshire (Loch of Skene) in 1975.

uncialis (Grev.) Berthier, *Bull. Mens. Soc. Linn. Lyon* 45: 83 (1976)
  Clavaria uncialis Grev., *Scott. crypt. fl.* 2: t. 98 (1824)
  Pistillaria uncialis (Grev.) Costantin & L.M. Dufour, *Fl. Champ. Supér. France*: 177 (1921)
E: ! S: ! W: ! NI: !
H: Usually on dead stems of umbellifers such as *Angelica sylvestris* or *Heracleum sphondylium*, most frequently on the latter. Also reported from *Cirsium* spp and *Mercurialis perennis*. Often fruiting during spring and summer.
D: Corner: 495 (as *Pistillaria uncialis*), NM3: 258 **D+I:** B&K2: 340 436 **I:** Kriegl2: 52
Rarely reported but apparently widespread.

variabilis Riess, *Hedwigia* 1: 21 (1853)
  Typhula intermedia K.R. Appel & Laubert, *Arbeiten Kaiserl. Biol. Anst. Land-Forstw.* 5: 153 (1905)
  Sclerotium semen Fr. (anam.), *Syst. mycol.* 2: 249 (1823)
E: ! S: ?
H: On oil seed rape (*Brassica* sp.).
D: Corner: 675 (as *Typhula intermedia*), Corner: 687, NM3: 260
Rarely reported (though most often by seed-testers) and poorly known in Britain.

TYROMYCES P. Karst., *Rev. Mycol. (Toulouse)* 3(9): 17 (1881)
Poriales, Coriolaceae
Type: *Tyromyces chioneus* (Fr.) P. Karst.

chioneus (Fr.) P. Karst., *Rev. Mycol. (Toulouse)* 3(9): 17 (1881)
  Polyporus chioneus Fr., *Syst. mycol.* 1: 359 (1821)
  Leptoporus chioneus (Fr.) Quél., *Enchir. fung.*: 176 (1886)
  Polyporus albellus Peck, *Rep. (Annual) New York State Mus. Nat. Hist.* 30: 45 (1878)
  Leptoporus albellus (Peck) Bourdot & L. Maire, *Bull. Soc. Mycol. France* 36: 83 (1920)
  Tyromyces albellus (Peck) Bondartsev & Singer, *Ann. Mycol.* 39: 52 (1941)
E: o S: ! W: ! NI: ?
H: On dead wood of deciduous trees in woodland or on heathland. Most often on dead trunks or stacked logs of *Betula* spp, but also known on *Corylus* and *Fagus*.
D: EurPoly2: 686, NM3: 225 **D+I:** Ph: 233 (as *Tyromyces albellus*)
Occasional in England. Rarely reported elsewhere but apparently widespread. Often confused with *Postia tephroleuca*, but dried basidiomes usually acquire a wrinkled, yellowish pellicle on the pileal surface. Perhaps better known as *Tyromyces albellus*.

wynnei (Berk. & Broome) Donk, *Meded. Bot. Mus. Herb. Rijks Univ. Utrecht* 9: 156 (1933)
  Polyporus wynnei Berk. & Broome, *Ann. Mag. Nat. Hist., Ser. 3 [Notices of British Fungi no. 807]* 3: 359 (1859)
  Polystictus wynnei (Berk. & Broome) Cooke, *Grevillea* 14: 84 (1886)
  Leptoporus wynnei (Berk. & Broome) Quél., *Fl. mycol. France*: 385 (1888)
  Fibuloporia wynnei (Berk. & Broome) Bondartsev & Singer, *Ann. Mycol.* 34: 49 (1941)
  Loweomyces wynnei (Berk. & Broome) Jülich, *Persoonia* 11(4): 424 (1982)
E: o W: ! ROI: ?
H: On fallen woody debris, soil, litter and herbaceous stems in woodland. Often on litter under dense cover of *Mercurialis perennis* or in woodland nettlebeds.
D: EurPoly2: 697, NM3: 225 **D+I:** B&K2: 270 333 (as *Loweomyces wynnei*)
Rarely reported but apparently widespread and most frequent in southern England. Last reported from Republic of Ireland (Wicklow) in 1898 but no voucher material has been traced. Basidiomes in active growth have a pungent smell, reminiscent of napthalene.

VARARIA P. Karst., *Krit. öfvers. Finl. basidsvamp.*, Till. 3: 32 (1898)
Lachnocladiales, Lachnocladiaceae
  Asterostromella Höhn. & Litsch., *Sitzungsber. Kaiserl. Akad. Wiss., Wien, Math.-Naturwiss. Cl., Abt. 1*, 116: 773 (1907)
Type: *Vararia investiens* (Schwein.) P. Karst.

gallica (Bourdot & Galzin) Boidin, *Bull. Soc. Naturalistes Oyonnax* 5: 78 (1951)
  Asterostromella epiphylla var. *gallica* Bourdot & Galzin, *Bull. Soc. Mycol. France* 27: 265 (1911)
E: ! W: !
H: On decayed grasses and woody stems. British collections on *Ammophila arenaria*, *Clematis vitalba* and *Rubus fruticosus* agg.
D: NM3: 321 **D+I:** Hallenb: 47 Fig. 31, Myc11(2): 78
Known from England (South Devon and Surrey) and Wales (Carmarthenshire).

ochroleuca (Bourdot & Galzin) Donk, *Ned. kruidk. Archf. Ser. 3*, 40: 79 (1930)
  Asterostromella ochroleuca Bourdot & Galzin, *Bull. Soc. Mycol. France* 27: 266 (1911)
E: !
H: On decayed litter and woody debris of deciduous trees, on soil under decayed logs, on decayed petioles of *Dryopteris filix-mas* and in needle litter of *Larix*.
D: NM3: 321 **D+I:** Hallenb: 53 Figs. 34-35
Known from Surrey (Mickleham Downs) where it was first collected in 1922, with several recent collections from the same locality.

VASCELLUM F. Smarda, *Fl. CSR, Gasteromyc.* 1: 304 (1958)
Lycoperdales, Lycoperdaceae
Type: *Lycoperdon depressum* Bonord. (= *V. pratense*)

pratense (Pers.) Kreisel, *Feddes Repert.* 64: 159 (1962)
  Lycoperdon pratense Pers., *Tent. disp. meth. fung.*: 7 (1797)
  Lycoperdon depressum Bonord., *Bot. Zeitung (Berlin)* 15: 611 (1857)
  Vascellum depressum (Bonord.) F. Šmarda, *Fl. CSR, Gasteromyc.* 1: 305 (1958)
  Lycoperdon hiemale Bull., *Champ. France, Hymenomycetes*: 148 (1781)
E: c S: c W: c NI: c ROI: c O: Channel Islands: c Isle of Man: c
H: On soil amongst short turf on heathland, downland, garden lawns, cricket pitches, golf links, grazed pasture and rarely in open woodland
D: NM3: 340 **D+I:** Ph: 248-249, B&K2: 396 521, BritPuffb: 118-119 Figs. 88/89 **I:** Bon: 305, Kriegl2: 154
Very common and widespread.

VESICULOMYCES (Pers.) E. Hagstr., *Bot. Not.* 130: 53 (1977)

Hericiales, Gloeocystidiellaceae

Type: *Vesiculomyces citrinus* (Pers.) E. Hagstr.

citrinus (Pers.) E. Hagstr., *Bot. Not.* 130: 53 (1977)
Thelephora citrina Pers., *Mycol. eur.* 1: 136 (1822)
Corticium citrinum (Pers.) Fr., *Hymenomyc. eur.*: 655 (1874)
Gloeocystidiellum citrinum (Pers.) Donk, *Fungus* 26: 9 (1956)
Gloiothele citrina (Pers.) Ginns & G.W. Freeman, *Biblioth. Mycol.* 157: 55 (1994)
Thelephora alutacea Schrad., in Gmelin, *Syst. nat.*: 1441 (1792)
Gloeocystidium alutaceum (Schrad.) Bourdot & Galzin, *Bull. Soc. Mycol. France* 28(4): 367 (1913)
Mis.: *Corticium radiosum* sensu Rea (1922)
E: !
H: On decayed wood or bark of conifers such as *Picea* spp, *Pinus sylvestris*, *Sequoia* spp and *Taxus*. Also reported on decayed fence posts.
D: NM3: 281 **D+I:** CNE3: 406-409 (as *Gloeocystidiellum citrinum*), B&K2: 122 109
Rarely reported but may be locally abundant. Collections from Buckinghamshire, North & South Hampshire, Oxfordshire, South Devon, South Somerset, Surrey, and West Norfolk.

VOLVARIELLA Speg. [*nom. cons.*], *Anales Mus. Nac. Hist. Nat. Buenos Aires* 6: 119 (1899)

Agaricales, Pluteaceae

Volvaria (Fr.) P. Kumm. [*nom. illegit.*, non *Volvaria* DC. (1805)], *Führ. Pilzk.*: 23 (1871)
Type: *Volvariella argentina* Speg.

bombycina (Schaeff.) Singer, *Lilloa* 22: 401 (1951) [1949]
Agaricus bombycinus Schaeff., *Fung. Bavar. Palat. nasc.* 1: 42 (1762)
Volvaria bombycina (Schaeff.) P. Kumm., *Führ. Pilzk.*: 99 (1871)
E: o **S:** ! **NI:** !
H: On decayed wood, usually large fallen or standing trunks or stumps, occasionally from knotholes or branch scars on living trees. Usually on *Ulmus* spp, less frequently on *Fagus*. Rarely on *Acer pseudoplatanus*, *Aesculus*, *Betula* spp, *Crataegus monogyna*, *Fraxinus*, *Malus*, *Quercus* and *Populus* spp. Single collections on a large decayed basidiome of *Ganoderma australe* and from an occupied wasps' nest.
D: BFF4: 65-66, FAN2: 57-58, NM2: 204 **D+I:** Ph: 112, B&K4: 134 128 **I:** FlDan2: 68E (as *Volvaria bombycina*), Bon: 199, C&D: 289, SV42: 38
Rather rare, but strongly influenced by the epidemic of Dutch elm disease especially in southern England where for a time it became locally abundant. Single records for Scotland and Northern Ireland.

caesiotincta P.D. Orton, *Bull. Mens. Soc. Linn. Lyon* 43: 319 (1974)
Mis.: *Volvaria murinella* sensu Kühner & Romagnesi (1953)
E: !
H: On decayed wood, usually large stumps and fallen trunks of *Ulmus* spp, less often *Fagus* and *Acer pseudoplatanus*.
D: BFF4: 66-67, FAN2: 59 **D+I:** B&K4: 134 129, Myc12(3): 132 **I:** C&D: 289, SV42: 39
Known in scattered localities from Westmorland southwards.

gloiocephala (DC.) Boekhout & Enderle, *Beitr. Kenntn. Pilze Mitteleurop.* 2: 78 (1986)
Agaricus gloiocephalus DC., *Fl. Fr.* 5: 52 (1815)
Volvaria gloiocephala (DC.) Gillet, *Hyménomycètes*: 388 (1876)
Volvariella speciosa var. *gloiocephala* (DC.) Singer, *Lilloa* 22: 401 (1951) [1949]
Amanita speciosa Fr., *Observ. mycol.* 2: 1 (1818)
Volvaria speciosa (Fr.) P. Kumm., *Führ. Pilzk.*: 99 (1871)
Volvariella speciosa (Fr.) Singer, *Lilloa* 22: 401 (1951) [1949]
Agaricus speciosus (Fr.) Fr., *Syst. mycol.* 1: 278 (1821)

E: c **S:** c **W:** ! **NI:** ! **O:** Channel Islands: !
H: On soil and composted areas, amongst woodchips in flowerbeds, in fields amongst stubble, roadside verges, woodland areas, dunes and grassland.
D: BFF4: 69-70 (as *Volvariella speciosa*), FAN2: 56-57, NM2: 204 **D+I:** Ph: 112 (as *V. speciosa*), B&K4: 136 130, Myc12(3): 133 **I:** FlDan2: 69D & D1 (as *Volvaria speciosa*), Bon: 199 (as *V. speciosa*), C&D: 289
Common and widespread. Increasingly frequent in recent years, due to the extensive use of woodchip mulch on flowerbeds.

hypopithys (Fr.) Shaffer, *Mycologia* 49: 572 (1957)
Agaricus hypopithys Fr., *Hymenomyc. eur.*: 183 (1874)
Volvaria parvula var. *biloba* Massee, *Brit. fung.-fl.* 2: 296 (1893)
Volvaria media var. *biloba* (Massee) A. Pearson & Dennis, *Trans. Brit. Mycol. Soc.* 31: 168 (1948)
Volvaria plumulosa Quél., *Bull. Soc. Bot. France* 24: 320 (1878) [1877]
Agaricus pubescentipes Peck, *Rep. (Annual) New York State Mus. Nat. Hist.* 29: 39 (1878)
Volvariella pubescentipes (Peck) Singer, *Lilloa* 22: 410 (1951) [1949]
E: ! **S:** ! **W:** ! **ROI:** !
H: On soil, often in grassy areas in deciduous woodland or with conifers.
D: NCL3: 384 (as *Volvariella pubescentipes*), BFF4: 72-73, FAN2: 61-62, NM2: 204 **D+I:** B&K4: 136 131 **I:** C&D: 289
Rarely reported but widespread, especially in England from Lancashire southwards.

murinella (Quél.) Courtec., *Bull. Soc. Mycol. France* 34: 19 (1984)
Volvaria murinella Quél., *Compt. Rend. Assoc. Franç. Avancem. Sci.* 11: 391 (1883)
E: ! **S:** ! **W:** !
H: On soil, often amongst grass, occasionally with trees in woodland.
D: BFF4: 68, FAN2: 60-61 (as *Volvariella murinella* f. *murinella*), NM2: 204 **D+I:** B&K4: 136 132 **I:** C&D: 289
Rarely reported but apparently widespread. Old records on wood probably relate to *V. caesiotincta*.

pusilla (Pers.) Singer, *Lilloa* 22: 401 (1951) [1949]
Amanita pusilla Pers., *Observ. mycol.* 2: 36 (1799)
Agaricus parvulus Weinm., *Hymen. Gasteromyc.*: 238 (1836)
Volvaria parvula (Weinm.) P. Kumm., *Führ. Pilzk.*: 99 (1871)
Volvariella parvula (Weinm.) Speg., *Bol. Acad. Nac. Ci. Republ. Argent.* 28: 309 (1926)
E: ! **W:** ! **ROI:** !
H: On soil, often amongst grass on lawns, in gardens or in sparsely wooded areas.
D: BFF4: 70-71 (as *Volvariella parvula*), FAN2: 62-63, NM2: 204 **D+I:** Ph: 112 (as *V. parvula*), B&K4: 138 133 **I:** C&D: 289
Rarely reported but apparently widespread.

reidii Heinem., *Bull. Jard. Bot. Nat. Belg.* 48(1-2): 239 (1978)
Volvariella parvispora D.A. Reid [*nom. inval.*], *Trans. Brit. Mycol. Soc.* 68: 327 (1977)
E: ! **S:** ! **NI:** !
H: On soil in mixed deciduous woodland, possibly with *Crataegus* spp. Also in moss on lawns.
D: BFF4: 71-72 (as *Volvariella parvispora*)
Rarely reported. Mostly southern in England, but recorded as far north as Staffordshire. Recent, single collections from Scotland and Northern Ireland.

surrecta (Knapp) Singer, *Lilloa* 22: 401 (1951) [1949]
Agaricus surrectus Knapp, *Journ. natural.* Edn 1, 1: 363 (1829)
Volvaria surrecta (Knapp) Ramsb., *Trans. Brit. Mycol. Soc.* 25: 328 (1942)
Agaricus loveianus Berk., *Engl. fl.* 5(2): 104 (1836)
Volvaria loveiana (Berk.) Gillet, *Hyménomycètes*: 386 (1876)
E: ! **S:** !
H: On moribund and partially decayed basidiomes of *Clitocybe nebularis*.

D: BFF4: 73-74, FAN2: 59-60, NM2: 204 **D+I:** B&K4: 138 134, Myc12(4): 181 **I:** FlDan2: 68B (as *Volvaria loveiana*), C&D: 289
Rarely reported. A few records from England from Co. Durham southwards and a single record from Scotland. Said to occur on other *Clitocybe* and *Tricholoma* spp, but this is not supported by British records.

taylorii (Berk. & Broome) Singer, *Lilloa* 22: 401 (1951) [1949]
Agaricus taylorii Berk. & Broome, *Ann. Mag. Nat. Hist., Ser. 2 [Notices of British Fungi no. 675]* 13: 398 (1854)
Volvaria taylorii (Berk. & Broome) Gillet, *Hyménomycètes*: 386 (1876)
Volvariella pusilla var. *taylorii* (Berk. & Brome) Boekhout, *Persoonia* 13(2): 207 (1986)
Volvaria murinella var. *umbonata* J.E. Lange, *Fl. agaric. danic.* 5, *Taxonomic Conspectus*: IX (1940)
Mis.: *Volvaria parvula* sensu Kühner & Romagnesi (1953)
Mis.: *Volvaria plumulosa* sensu Lange (1937)
E: !
H: On soil, amongst grass or near to decayed wood.
D: BFF4: 67-68, FAN2: 63 (as *Volvariella pusilla* var. *taylori*), NM2: 204 **I:** C&D: 289
Rarely reported but apparently widespread in England. Records on decayed wood may possibly refer to *V. caesiotincta*.

volvacea (Bull.) Singer, *Lilloa* 22: 401 (1951) [1949]
Agaricus volvaceus Bull., *Herb. France*: pl. 262 (1786)
Volvaria volvacea (Bull.) P. Kumm., *Führ. Pilzk.*: 99 (1871)
Amanita virgata Pers., *Tent. disp. meth. fung.*: 18 (1979)
Vaginata virgata (Pers.) Gray, *Nat. arr. Brit. pl.* 1: 601 (1821)
E: ! **S:** !
H: On soil amongst woodchips or compost, often in greenhouses but also in gardens and parks, and occasionally in woodland.
D: BFF4: 64-65, FAN2: 63-64, NM2: 204 **I:** Sow1: 1 (as *Agaricus volvaceus*)
Rarely reported and possibly an alien that has become established in suitable sites. Recent collections only from south-eastern England (Surrey and West Kent).

VUILLEMINIA Maire, *Bull. Soc. Mycol. France* 18 (Suppl.): 81 (1902)
Stereales, Corticiaceae
Type: *Vuilleminia comedens* (Nees) Maire

alni Boidin, Lanq. & Gilles, *Bull. Soc. Mycol. France* 110(2): 95 (1994)
E: ! **S:** ! **NI:** !
H: Decorticating living and dead attached branches of *Alnus*.
D+I: Myc11(1): 6
Rarely reported but apparently widespread.

comedens (Nees) Maire, *Bull. Soc. Mycol. France* 18 (suppl.): 81 (1902)
Thelephora comedens Nees, *Syst. Pilze*: 239 (1817)
Corticium comedens (Nees) Fr., *Epicr. syst. mycol.*: 565 (1838)
Corticium carlylei Massee, *J. Linn. Soc., Bot.* 27: 146 (1890)
Thelephora nigrescens Schrad., *Spicil. Fl. Germ.* 1: 186 (1794)
Corticium nigrescens (Schrad.) Fr., *Epicr. syst. mycol.*: 565 (1838)
E: c **S:** c **W:** c **NI:** c **ROI:** c **O:** Channel Islands: !
H: Decorticating living and dead attached branches, usually *Quercus* spp, and rarely *Acer pseudoplatanus*, *Alnus*, *Betula* spp, *Castanea*, *Corylus*, *Fagus*, *Prunus laurocerasus*, *P. spinosa*, *Sambucus nigra*, *Tilia* and *Salix* spp.
D: NM3: 180 **D+I:** B&K2: 194 217, CNE8: 1590-1592, Myc11(1): 4 **I:** Kriegl1: 355
Very common and widespread. Old records may refer to other, recently distinguished species.

coryli Boidin, Lanq. & Gilles, *Bull. Soc. Mycol. France* 105(2): 164 (1989)
E: c **S:** ! **W:** ! **NI:** ! **ROI:** !

H: Decorticating living and dead attached branches, usually *Corylus* but single collections on *Acer campestre*, *Salix caprea* and *Sambucus nigra*.
D: NM3: 180 **D+I:** Myc11(1): 5
Common, at least in southern England.

cystidiata Parmasto, *Eesti N.S.V. Tead. Akad. Toimet., Biol.* 14(2): 232 (1965)
E: o **NI:** ! **ROI:** !
H: Decorticating living and dead, attached branches of *Crataegus monogyna* and rarely *Prunus spinosa*.
D: NM3: 180 **D+I:** CNE8: 1592-1593, Myc11(1): 5
Occasional in south-eastern England. Rarely reported elsewhere but apparently widespread.

macrospora (Bres.) Hjortstam, *Windahlia* 17: 58 (1987)
Corticium acerinum var. *macrosporum* Bres., *Ann. Mycol.* 1(1): 96 (1903)
Corticium macrosporum (Bres.) Bres., *Ann. Mycol.* 6: 43 (1908)
Aleurodiscus macrosporus (Bres.) Bres., *Bull. Soc. Mycol. France* 28: 353 (1913)
Dendrothele macrospora (Bres.) P.A. Lemke, *Persoonia* 3(3): 366 (1965)
Laeticorticium macrosporum (Bres.) J. Erikss. & Ryvarden, *Corticiaceae of North Europe* 4: 767 (1976)
Corticium acerinum Velen. [non *C. acerinum* Pers. (1796)], *Česka Houby* IV-V: 756 (1922)
Corticium macrosporopsis Jülich, *Int. J. Mycol. Lichenol.* 1(1): 31 (1982)
E: ! **S:** !
H: British collections on dead and dying stems of *Calluna* in moorland and also on *Tamarix gallica* in coastal scrub.
D: NM3: 180 **D+I:** CNE4: 766-768 (as *Laeticorticium macrosporum*), B&K2: 98 74 (as *Corticium macrosporopsis*)
Known from England (East Suffolk: Minsmere and West Norfolk: Brancaster) and Scotland (Wester Ross: Loch Maree and South Ebudes: Isle of Colonsay).

WAITEA Warcup & P.H.B. Talbot, *Trans. Brit. Mycol. Soc.* 45: 503 (1962)
Ceratobasidiales, Ceratobasidiaceae
Type: *Waitea circinata* Warcup & P.H.B. Talbot

circinata Warcup & P.H.B. Talbot, *Trans. Brit. Mycol. Soc.* 45(4): 503 (1962)
Rhizoctonia oryzae Ryker & Gooch [*nom. inval.*] (anam.), *Phytopathology* 18: 138 (1938)
Moniliopsis oryzae (Ryker & Gooch) R.T. Moore [*nom. inval.*] (anam.), *Mycotaxon* 29: 95 (1987)
Rhizoctonia zeae Voorhees (anam.), *Phytopathology* 24: 1299 (1934)
Moniliopsis zeae (Voorhees) R.T. Moore (anam.), *Mycotaxon* 29: 96 (1987)
E: !
H: Saprotrophic and pathogenic. A soil fungus, causing 'Barley Stunt' disease in Britain. Basidiomes are rarely found, usually on clods of earth, or fallen and decayed wood.
D: Myc17(2): 63
The teleomorph is known in Britain from Hertfordshire (Hitchin, Oughton Head) and the rhizoctonia anamorph from Lincolnshire and Nottinghamshire.

WAKEFIELDIA Corner & Hawker, *Trans. Brit. Mycol. Soc.* 36: 130 (1953)
Hymenogastrales, Octavianinaceae
Type: *Wakefieldia striaespora* Corner & Hawker

macrospora (Hawker) Hawker, *Philos. Trans. Roy. Soc. London* 237: 521 (1954)
Sclerogaster macrosporus Hawker, *Trans. Brit. Mycol. Soc.* 34: 218 (1951)
E: !
H: Epigeous or shallowly hypogeous in calcareous soil with *Fagus*.

D+I: BritTruff: 188 11 H
Known only from West Gloucestershire (near Wotton-under-Edge), where last collected in 1954.

XENASMA Donk, *Fungus* 27: 25 (1957)
Stereales, Xenasmataceae
Type: *Xenasma rimicola* (P. Karst.) Donk

pruinosum (Pat.) Donk, *Fungus* 27: 25 (1957)
  Corticium pruinosum Pat., *Cat. pl. cell. Tunisie*: 60 (1897)
  Peniophora pruinosa (Pat.) H.S. Jacks., *Canad. J. Res., Sect. C, Bot. Sci.* 28: 530 (1950)
  Peniophora chordalis Höhn. & Litsch., *Sitzungsber. Kaiserl. Akad. Wiss., Wien, Math.-Naturwiss. Cl., Abt. 1* 115: 1598 (1906)
  Phanerochaete chordalis (Höhn. & Litsch.) Park.-Rhodes, *Ann. Bot.* 20(78): 258 (1956)
E: ! S: !
H: On decayed wood of deciduous trees, often *Fagus* and rarely *Fraxinus*. A single collection on decayed and fallen basidiomes of *Datronia mollis*.
D: NM3: 143 **D+I:** B&K2: 196 221, CNE8: 1595-1597
Known from England (Shropshire, South Hampshire, Surrey, West Kent and West Sussex) and Scotland (Westerness).

pulverulentum (Litsch.) Donk, *Fungus* 27: 25 (1957)
  Corticium pulverulentum Litsch., *Oesterr. Bot. Z.* 88: 112 (1939)
  Peniophora pulverulenta (Litsch.) H.S. Jacks., *Canad. J. Res., Sect. C, Bot. Sci.* 28: 532 (1950)
E: o W: ! O: Isle of Man: !
H: Usually on decayed wood of *Fagus*, less often *Fraxinus* and rarely *Ulmus* spp in woodland, usually on calcareous soil.
D: NM3: 143 **D+I:** B&K2: 196 222, CNE8: 1598-1599
Occasional in southern England but rarely reported elsewhere. Spores are ornamented with raised striations or ridges (similar to those seen in *Clitopilus* spp), a feature unique amongst corticioid fungi.

XENOLACHNE D.P. Rogers, *Mycologia* 39: 561 (1947)
Tremellales, Sirobasidiaceae
Type: *Xenolachne flagellifera* D.P. Rogers

longicornis Hauerslev, *Friesia* 11: 108 (1976)
E: !
H: Parasitic. On the hymenium of the discomycete *Hymenoscyphus vernus*.
D: NM3: 90
A single collection from Surrey (Esher, Winterdown Wood) in 1981.

XENOSPERMA Oberw., *Sydowia* 19: 45 (1966)
Stereales, Xenasmataceae
Type: *Xenosperma ludibundum* (D.P. Rogers & Liberta) Jülich

ludibundum (D.P. Rogers & Liberta) Jülich., *Persoonia* 10(3): 335 (1979)
  Xenasma ludibundum D.P. Rogers & Liberta, *Mycologia* 52(6): 902 (1962)
E: !
H: On decayed wood of conifers such as *Picea* spp and *Pinus sylvestris*, and a single collection on a dead stem of *Fallopia japonica*.
D: NM3: 143 **D+I:** CNE8: 1602-1604, Myc9(1): 11
Known from South Devon and Surrey.

XEROMPHALINA Kühner & Maire, *Icon. select. fung.* 6: 236 (1934)
Agaricales, Tricholomataceae
Type: *Xeromphalina campanella* (Batsch) Kühner & Maire

campanella (Batsch) Kühner & Maire, *Bull. Soc. Mycol. France* 50: 18 (1934)

  Agaricus campanella Batsch, *Elench. fung.*: 73 (1783)
  Omphalia campanella (Batsch) P. Kumm., *Führ. Pilzk.*: 107 (1871)
  Omphalina campanella (Batsch) Quél., *Enchir. fung.*: 45 (1886)
  Omphalia campanella var. *papillata* Gillet, *Hyménomycètes*: 299 (1876)
E: ! S: ! ROI: ?
H: In large clumps on decayed conifer wood, usually large stumps or fallen branches.
D: NM2: 192, BFF8: 147-148 **D+I:** A&N3 196-202 53, B&K3: 346 446 **I:** Bon: 131, C&D: 233, Kriegl3: 581
Fide BFF8 this is 'not uncommon' in northern England and in Scotland but is virtually unknown in the south. Reported from the Republic of Ireland but unsubstantiated with voucher material.

cauticinalis (Fr.) Kühner & Maire, *Bull. Soc. Mycol. France* 50: 18 (1934)
  Marasmius cauticinalis Fr., *Epicr. syst. mycol.*: 383 (1838)
  Xeromphalina fellea Maire & Malençon, *Bull. Soc. Hist. Nat. Afrique N.* 36: 36 (1945)
  Xeromphalina fulvobulbillosa (R.E. Fr.) Kühner & Romagn. [*nom. superfl.*], *Publ. Inst. Bot. Barcelona* 3(4): 68 (1937)
  Mis.: *Agaricus caulicinalis* sensu Sowerby [Col. fig. Engl. Fung. 2: pl. 163 (1789)]
E: ! S: !
H: Growing singly on soil or needle litter of *Pinus* spp, especially in Caledonian pinewoods.
D: NM2: 192, BFF8: 148-149 **D+I:** A&N3 215-220 56, FRIC7: 35-37 55b **I:** Sow2: 163 (as *Agaricus caulicinalis*), FRIC7: 55c (as *Xeromphalina fellea*), C&D: 233, Kriegl3: 582
Rarely reported. Most often collected in Scotland, rarely in England.

XERULA Maire emend. Dorfelt, *Feddes Repert.* 90: 365 (1979)
Agaricales, Tricholomataceae
Type: *Xerula longipes* (Bull.) Maire (= *X. pudens*)

caussei Maire, *Bull. Soc. Mycol. France* 53: 265 (1937)
  Oudemansiella nigra Dörfelt, *Česká Mykol.* 27: 28 (1973)
  Xerula nigra (Dörfelt) Dörfelt, *Landschaftspflege Naturschutz Thüringen* 14(3): 60 (1977)
E: !
H: On soil, but usually from buried roots or wood of deciduous trees such as *Fagus* and *Quercus* spp.
D: NM2: 193, BFF8: 151-152 (as *Xerula nigra*), FAN4: 183, NBA8: 607 (as *O. nigra*) **I:** Kriegl3: 583
A single collection from Yorkshire. Reported from Oxfordshire in 1991, but unsubstantiated with voucher material.

pudens (Pers.) Singer, *Lilloa* 22: 289 (1951) [1949]
  Agaricus radicatus β *pudens* Pers., *Syn. meth. fung.*: 313 (1801)
  Gymnopus pudens (Pers.) Gray, *Nat. arr. Brit. pl.* 1: 605 (1821)
  Agaricus pudens (Pers.) Pers., *Mycol. eur.* 3: 140 (1828)
  Oudemansiella pudens (Pers.) Pegler & T.W.K. Young, *Trans. Brit. Mycol. Soc.* 87(4): 590 (1987)
  Agaricus longipes Bull. [*nom. illegit.*], *Herb. France*: pl. 232 (1785)
  Collybia longipes P. Kumm., *Führ. Pilzk.*: 117 (1871)
  Mucidula longipes (P. Kumm.) Boursier, *Bull. Soc. Mycol. France* 40: 333 (1924)
  Xerula longipes (P. Kumm.) Maire, *Mus. barcin. Scient. nat. Op.* 15: 66 (1933)
  Oudemansiella longipes (P. Kumm.) M.M. Moser, *Kleine Kryptogamenflora, Edn 5*: 156 (1983)
  Collybia longipes var. *badia* Quél., *Bull. Soc. Amis Sci. Nat. Rouen.* 15: 154 (1880)
  Oudemansiella badia (Quél.) M.M. Moser [*nom. inval.*], *Z. Pilzk.* 19: 11 (1955)
E: ! W: ! NI: ! ROI: !

H: On soil, usually near to or with deciduous trees, and probably arising from buried wood or roots.
D: NM2: 193, BFF8: 150-151, FAN4: 181-182 (as *Xerula longipes*) **D+I:** B&K3: 348 449 **I:** FlDan2: 41A (as *Collybia longipes*), Bon: 171 (as *Oudemansiella longipes*), C&D: 235, SV32: 9, Kriegl3: 585
Rarely reported but apparently widespread. Includes both *Oudemansiella badia* and *O. longipes* of NCL.

radicata (Relhan) Dörfelt, *Veröffentlichungen der Museen der Stadt Gera* 2-3: 67 (1975)
   Agaricus radicatus Relhan, *Fl. Cantab.*: 1040 (1785)
   Gymnopus radicatus (Relhan) Gray, *Nat. arr. Brit. pl.* 1: 605 (1821)
   Collybia radicata (Relhan) Quél., *Mém. Soc. Émul. Montbéliard, Sér. 2,* 5: 92 (1872)
   Mucidula radicata (Relhan) Boursier, *Bull. Soc. Mycol. France* 40: 332 (1924)
   Oudemansiella radicata (Relhan) Singer, *Lilloa* 22: 288 (1951) [1949]
   Collybia radicans P. Kumm., *Führ. Pilzk.*: 117 (1871)
   Mucidula radicata f. *marginata* Konrad & Maubl., *Icon. select. fung.* 2: pl. 199 (1931)
   Oudemansiella radicata var. *marginata* (Konrad & Maubl.) Bon & Dennis, *Doc. Mycol.* 15(59): 51 (1985)
E: c **S:** c **W:** c **NI:** c **ROI:** c **O:** Channel Islands: c
H: On soil under or near to deciduous trees, arising from buried wood. Rarely on decayed wood (stumps) and then rooting deeply. Fruiting from from mid-summer onwards.
D: FlDan2: 41D (as *Collybia radicata*), NM2: 193, BFF8: 152-153, FAN4: 182-183 **D+I:** Ph: 33 (as *Oudemansiella radicata*), B&K3: 348 450 **I:** Sow1: 48 (as *Agaricus radicatus*), Bon: 171, C&D: 235, Kriegl3: 587
Common and widespread. Basidiomes occasionally have lamellae with a brownish emarginate edge and these may possibly represent a distinct species.

xeruloides (Bon) Dörfelt, *Feddes Repert.* 91: 216 (1980)
   Oudemansiella xeruloides Bon, *Doc. Mycol.* 5(17): 31 (1975)
O: Channel Islands: o
H: On sand in dunes.
D: BFF8: 152 **I:** C&D: 235
Locally frequent.

ZELLEROMYCES Singer & A.H. Sm., *Mem. Torrey Bot. Club* 21(3): 18 (1960)
Russulales, Elasmomycetaceae
Type: *Zelleromyces cinnabarinus* Singer & A.H. Sm.

stephensii (Berk.) A.H. Sm., *Mycologia* 54: 635 (1962)
   Hydnangium stephensii Berk., *Ann. Mag. Nat. Hist., Ser. 1 [Notices of British Fungi no. 300]* 13: 352 (1844)
   Octaviania stephensii (Berk.) Tul. & C. Tul., *Fungi hypogaei*: 78 (1851)
E: ! **S:** !
H: Hypogeous in humic soil in deciduous woodland. Associated with *Quercus* and *Tilia* spp in the type locality.
D+I: BritTruff: 196-198 12 D
Rarely reported. Known from England (North Somerset and North Wiltshire) and Scotland (Dunbartonshire and Lanarkshire).

INCLUDED UREDINIOMYCETES

ACHROOMYCES Bonord., *Handb. Mykol.*: 135 (1851)
Platygloeales, Platygloeaceae
Type: *Achroomyces tumidus* Bonord.

henricii P. Roberts, *Mycotaxon* 63: 200 (1997)
E: !
H: Parasitic. On stromata of the pyrenomycete *Diatrype disciformis* growing on decayed wood of *Fagus*.
Described from Surrey (Mickleham Downs) and only known from two collections.

insignis Hauerslev, *Mycotaxon* 49: 218 (1993)
W: ! **O:** Channel Islands !
H: Parasitic. In the hymenium of *Stypella subhyalina*.
D: NM3: 76
Single collections from Wales (Breconshire: Epynt Ranges) and the Channel Islands (Guernsey).

lumbricifer P. Roberts, *Sydowia* 53(1): 152 (2001)
S: !
H: Parasitic. The type collection was in the hymenium of *Hyphodontia subalutacea* growing on *Betula* sp.
Described from Kirkudbrightshire (Glen Trool) and only known in Britain from the type collection.

micrus (Bourdot & Galzin) Wojewoda, *Grzyby* (*Mycota*) 8: 252 (1977)
  Platygloea micra Bourdot & Galzin, *Bull. Soc. Mycol. France* 39: 261 (1923)
  Platygloea microspora McNabb, *Trans. Brit. Mycol. Soc.* 48(2): 191 (1965)
E: ! **S:** ! **O:** Channel Islands !
H: Parasitic. In the hymenium of *Stypella grilletii*.
Rarely reported. Basidiomes are cryptic and easily overlooked.

pachysterigmata P. Roberts, *Windahlia* 22: 17 (1997)
E: !
H: Parasitic. In the hymenium of *Tulasnella violea*.
A single collection from South Devon (Dartmoor, Bellever Forest) in 1994.

ATRACTIELLA Sacc., *Syll. fung.* 4: 578 (1886)
Atractiellales, Hoehnelomycetaceae
  Pilacrella J. Schröt., *Krypt.-Fl. Schlesien*: 384 (1888)
Type: *Atractiella brunaudiana* (Sacc.) Sacc.

solani (Cohn & J. Schröt.) Oberw. & Bandoni, *Canad. J. Bot.* 60(9): 1732 (1982)
  Pilacrella solani Cohn & J. Schröt., *Krypt.-Fl. Schlesien*: 385 (1888)
  Pilacre solani (Cohn & J. Schröt.) Sacc., *Syll. fung.* 10: 686 (1892)
  Ecchyna solani (Cohn & J. Schröt.) Pat., *Essai taxon.*: 17 (1900)
S: !
H: On soil. British collection from a storage area where potato tubers had decayed.
A single collection from the Outer Hebrides (St.Kilda).

BAUERAGO Vánky, *Mycotaxon* 70: 44 (1999)
Microbotryales, ?Ustilentylomataceae
Type: *Bauerago abstrusa* (Malençon) Vánky

vuyckii (Oudem. & Beij.) Vánky, *Mycotaxon* 70: 46 (1999)
  Ustilago vuyckii Oudem. & Beij., *Verslagen Zittingen Wis-Natuurk. Afd., Kon. Ned. Akad. Wetensch.* 3: (55) (1895)
E: !
H: In the ovaries of various species of *Luzula*. In Britain known only from *Luzula campestris*.
D: Vánky: 385 (as *Ustilago vuyckii*)
A single collection from Warwickshire reported in Cecidology 13: 35 (1998).

BENSINGTONIA Ingold, *Trans. Brit. Mycol. Soc.* 86: 325 (1986)
Incertae sedis
Type: *Bensingtonia ciliata* Ingold

ciliata Ingold, *Trans. Brit. Mycol. Soc.* 86(2): 325 (1986)
E: !
H: Type specimen isolated ex basidiomes of *Auricularia auricula judae* var. *lactea*; also known on decayed plant debris.
Described from West Kent (Lullingstone Park).

BIATOROPSIS Räsänen, *Ann. Bot. Soc. Zool.-Bot. Fenn. 'Vanamo'* 5(9): 8 (1934)
Platygloeales, Incertae sedis
Type: *Biatoropsis usnearum* Räsänen

usnearum Räsänen, *Ann. Bot. Soc. Zool.-Bot. Fenn. 'Vanamo'* 5(9): 8 (1934)
E: ! **S:** ! **W:** ! **NI:** ! **ROI:** !
H: Parasitic. On thalli of the lichen genus *Usnea*. British material on *U. subfloridana*.
Widespread.

CHIONOSPHAERA D.E. Cox, *Mycologia* 68: 503 (1976)
Atractiellales, Chionosphaeraceae
Type: *Chionosphaera apobasidialis* D.E. Cox

coppinsii P. Roberts, *Mycotaxon* 63: 195 (1997)
S: !
H: On thalli of the lichen *Parmelia glabratula*.
Described from Wester Ross (Torridon, Inveralligan).

lichenicola Alstrup, B. Sutton & Tønsberg, *Graphis Scripta* 5(2): 97 (1993)
S: !
H: On moribund thalli of the lichen *Lecidella elaeochroma*.
A single collection from West Sutherland (Bettyhill, Invernaver NNR) in 1983.

CHRYSOMYXA Unger, *Beitr. vergl. Pathol.*: 24 (1840)
Uredinales, Coleosporiaceae
Type: *Chrysomyxa abietis* (Wallr.) Unger

abietis (Wallr.) Unger, *Beitr. vergl. Pathol.*: 24 (1840)
  Blennoria abietis Wallr., *Allg. Forst.- Jagd-Zeitung* 17: 65 (1834)
E: ! **S:** ! **W:** ! **ROI:** !
H: III on *Picea abies*, *P. rubens* and *P. sitchensis*.
D: W&H: 58-59
An established introduction, rarely reported, but apparently widespread. The aecidial stage, also on *Picea*, has not been collected in Britain.

empetri Cummins, *Mycologia* 48: 602 (1956)
  Uredo empetri DC., *Fl. Fr.* 5: 87 (1815)
E: ! **S:** ! **W:** ! **NI:** !
H: II on *Empetrum hermaphroditum* and *E. nigrum*.
D: W&H: 60
Rarely reported. The majority of records and collections are from Scotland. The host plant may be dominant in some upland areas but the rust is inconspicuous.

ledi *var.* rhododendri (de Bary) Savile, *Canad. J. Bot.* 33: 491 (1955)
  Chrysomyxa rhododendri de Bary, *Bot. Zeitung (Berlin)* 37: 809 (1879)
E: ! **S:** ! **W:** ? **ROI:** !
H: 0 & I on various species of *Picea*. II & III on numerous species and cultivars of *Rhododendron*.
D: W&H: 62-64 (as *Chrysomyxa rhododendri*)

An alien, apparently established and widespread yet rarely collected. Mostly reported from Scotland. A single collection from the Republic of Ireland (Dublin: Glasnevin) and one from England (West Cornwall: Trewithian). Reported elsewhere in England but unsubstantiated with voucher material.

pirolata G. Winter, *Rabenh. Krypt.-Fl.* 1(1): 250 (1882)
Chrysomyxa pyrolae Rostrup, *Bot. Centralbl.* 5: 127 (1881)
E: ! S: ! W: !
H: II & III on *Pyrola minor, P. rotundifolia* and *P. rotundifolia* ssp. *maritima.* in woodland, also on dunes.
D: W&H: 61
Rarely reported but apparently widespread. None of the hosts is common.

COLACOGLOEA Oberw. & Bandoni, *Canad. J. Bot.* 68: 2532 (1990)
Platygloeales, Platygloeaceae
Type: *Colacogloea peniophorae* (Bourdot & Galzin) Oberw., R. Bauer & Bandoni

bispora (Hauerslev) Oberw. & Bandoni, *Kew Bull.* 54: 764 (1999)
Platygloea bispora Hauerslev, *Friesia.* 11: 331 (1987)
Achroomyces bisporus (Hauerslev) Hauerslev, *Mycotaxon* 49: 218 (1993)
E: !
H: Parasitic. In the hymenium of *Tubulicrinis* spp.
D: NM3: 76 (as *Achroomyces bisporus*)
A single collection from South Devon (Slapton Wood) in 1994.

peniophorae (Bourdot & Galzin) Oberw., R. Bauer & Bandoni, *Canad. J. Bot.* 68(12): 2534 (1990)
Platygloea peniophorae Bourdot & Galzin, *Bull. Soc. Mycol. France.* 25: 17 (1909)
Achroomyces peniophorae (Bourdot & Galzin) Wojewoda, *Grzyby* (*Mycota*) 8: 246 (1977)
Mis.: *Platygloea effusa* sensu auct. p.p.
Mis.: *Achroomyces effusus* sensu auct. p.p.
E: o NI: !
H: Parasitic. Usually in the hymenium of *Hyphoderma praetermissum* but also known with *Resinicium bicolor.*
D: NM3: 75 (as *Achroomyces peniophorae*)
Occasional but widespread in England, with a single collection from Northern Ireland.

COLEOSPORIUM Lév., *Ann. Sci. Nat., Bot.,* sér. 3, 8: 373 (1847)
Uredinales, Coleosporiaceae
Type: *Uredo rhinanthacearum* DC. (= *C. tussilaginis*)

tussilaginis (Pers.) Kleb., in Orbigny, *Dict. univ. hist. nat.* 12: 786 (1848)
Uredo tussilaginis Pers., *Syn. meth. fung.*: 218 (1801)
Peridermium acicolum Cooke, *Grevillea* 6(38): 72 (1877)
Coleosporium cacaliae G.H. Otth, *Mitt. naturf. Ges. Bern.* 179 (1865)
Uredo campanulae Pers., *Syn. meth. fung.*: 217 (1801)
Coleosporium campanulae (Pers.) Cooke, *Microscopic fungi.* 213 (1865)
Caeoma compransor Schltdl., *Fl. berol.* 2: 119 (1824)
Coleosporium euphrasiae G. Winter, *Hedwigia* 19: 54 (1880)
Uredo farinosa β *senecionis* Pers., *Syn. meth. fung.*: 218 (1801)
Uredo melampyri Rebent., *Prodr. fl. neomarch.*: 355 (1804)
Coleosporium melampyri (Rebent.) P. Karst., *Bidrag Kännedom Finlands Natur Folk* 31: 62 (1879)
Coleosporium narcissi Grove [*nom. inval.*], *J. Bot.* 60: 121 (1922)
Uredo petasitidis DC. [*nom. inval.*], *Fl. Fr.* 2: 236 (1805)
Coleosporium petasitis Cooke [*nom. inval.*], *Microscopic fungi.* 213 (1865)
Peridermium plowrightii Kleb., *Z. Pflanzenkrankh.* 2: 268 (1892)

Uredo rhinanthacearum DC., in Lam., *Encyclop.* 8: 229 (1808)
Coleosporium rhinanthacearum J.J. Kickx, *Fl. Crypt. Flandres* 2: 53 (1867)
Coleosporium senecionis J.J. Kickx, *Fl. Crypt. Flandres* 2: 53 (1867)
Uredo tremellosa var. *sonchi* F. Strauss, *Ann. Wetterauischen Ges. Gesammte Naturk.* 2: 90 (1810)
Coleosporium sonchi (F. Strauss) Tul., *Ann. Sci. Nat., Bot.,* sér. 3, 2: 190 (1854)
Uredo tremellosa var. *campanulae* F. Strauss, *Ann. Wetterauischen Ges. Gesammte Naturk.* 2: 90 (1810)
Coleosporium sonchi-arvensis Sacc., *Syll. fung.* 21: 722 (1912)
Coleosporium synantherarum Fr., *Summa veg. Scand.*: 512 (1849)
Uredo tropaeoli Desm., *Ann. Sci. Nat., Bot.,* sér. 2, 6: 243 (1836)
Coleosporium tropaeoli Palm [*nom. nud.*], *Svensk bot. Tidskr.* 1: 271 (1917)
Mis.: *Aecidium pini* sensu auct. p.p.
Mis.: *Trichobasis cichoracearum* sensu auct.
Mis.: *Trichobasis senecionis* sensu auct. p.p.
Mis.: *Uredo sonchi-arvensis* sensu auct. p.p.
E: c S: c W: c NI: c ROI: c O: Channel Islands: c Isle of Man: c
D: W&H: 3-10
H: 0 & I on *Pinus sylvestris, P. nigra* and *P. pinaster.* II & III on a wide range of herbaceous species in *Compositae, Campanulaceae* and *Scrophulariaceae.* Almost ubiquitous on *Tussilago farfara* and common on *Euphrasia* spp, on *Odontites verna, Petasites albus, P. hybridus, Rhinanthus minor, Senecio* and *Sonchus* spp. A full list of hosts will be found in Henderson (2000).
Common and widespread. The rust on *Calanthe reflexa* from Kew Gardens, mentioned in WH, may be *C. bletilae* which is considered distinct by several authorities, but requires further investigation.

CRONARTIUM Fr., *Observ. mycol.* 1: 220 (1815)
Uredinales, Cronartiaceae
Type: *Cronartium asclepiadeum* (Willd.) Fr.

flaccidum (Alb. & Schwein.) G. Winter, *Hedwigia* 19: 55 (1880)
Sphaeria flaccida Alb. & Schwein., *Consp. fung. lusat.*: 31 (1805)
Erineum asclepiadeum Willd., in Funck, *Krypt. Gew. Fichtelgeb.*: 145 (1806)
Cronartium asclepiadeum (Willd.) Fr., *Observ. mycol.* 1: 220 (1815)
Cronartium coleosporioides (Dietel & Holw.) Arthur, *North American Flora* 7: 123 (1907)
Cronartium paeoniae Castagne, *Cat. pl. Marseille* 1: 217 (1845)
Mis.: *Peridermium pini* sensu auct. p.p.
E: ! S: !
H: 0 & I mostly on *Pinus sylvestris* but also *P. nigra* ssp. *laricio* and *P. ponderosa.* II & III on *Paeonia mascula* and *Tropaeolum majus* and on cultivated *Vincetoxicum officinale.*
D: W&H: 51-54
Rarely reported. Stages 0 & I from scattered locations in England and Scotland. Stages II & III from England (Dorset, East Norfolk, East Sussex, Huntingdonshire, North Hampshire, North Somerset, North Wiltshire, South Devon, Surrey, West Norfolk and West Sussex) and Scotland (Midlothian).

quercuum Miyabe, *Bot. Mag. (Tokyo)* 13: 74 (1899)
Uredo quercus Brond., in Duby, *Bot. gall.*: 893 (1830)
E: ! W: ! O: Channel Islands: !
H: II on leaves (especially on sucker shoots) of native and planted species of *Quercus.* Known on *Q. ilex, Q. lobata, Q. macranthera, Q. petraea, Q. reticulata* and *Q. robur.*
D: W&H: 367
Rarely reported but easily overlooked. Known from England (East Sussex, North Devon, South Wiltshire, Suffolk and

Surrey), Wales (Carmarthenshire and Pembrokeshire) and the Channel Islands (Guernsey).

ribicola J.C. Fisch., *Hedwigia* 11: 182 (1872)
 Peridermium strobi Kleb. [*nom. inval.*], *Hedwigia* 27: 119 (1888)
E: ! **S:** ! **W:** ? **NI:** ?
H: 0 & I on five-needled *Pinus* spp. II & III on native *Ribes nigrum, R. rubrum, R. uva-crispa*, and cultivated *R. alpestre, R. odoratum* and *R. sanguineum*.
D: W&H: 54-58
Rarely reported, but apparently widespread.

CUMMINSIELLA Arthur, *Bull. Torrey Bot. Club* 60: 475 (1933)
Uredinales, Pucciniaceae
Type: *Cumminsiella sanguinea* Arthur

mirabilissima (Peck) Nannf., *Fungi Exsiccati Suecici*: 1507a (1947)
 Puccinia mirabilissima Peck, *Bot. Gaz.* 6: 226 (1881)
 Cumminsiella sanguinea Arthur, *Bull. Torrey Bot. Club* 60: 475 (1933)
E: c **S:** c **W:** c **ROI:** o **O:** Isle of Man: !
H: 0, I, II & III on *Mahonia aquifolium* and rarely *M. bealei, M. fortunei*, and *M. undulata*.
D: W&H: 300
Common and widespread. An alien that has spread rapidly throughout the British Isles, and is now found wherever *Mahonia* spp are grown or have escaped from cultivation.

CYSTOBASIDIUM (Lagerh.) Neuhoff, *Bot. Arch.* 8: 274 (1924)
Platygloeales, Platygloeaceae
Type: *Cystobasidium lasioboli* (Lagerh.) Neuhoff (= *C. fimetarium*)

fimetarium (Schumach.) P. Roberts, *Mycologist* 13: 171 (1999)
 Tremella fimetaria Schumach., *Enum. pl.* 2: 440 (1803)
 Achroomyces fimetarius (Schumach.) Wojewoda, *Grzyby* (*Mycota*) 8: 248 (1977)
 Jola lasioboli Lagerh., *Bih. Kongl. Svenska Vetensk.-Akad. Handl.* 24: 15 (1898)
 Cystobasidium lasioboli (Lagerh.) Neuhoff, *Bot. Arch.* 8: 274 (1924)
S: !
H: Parasitic on coprophilous pyrenomycetes. British material on *Thelebolus crustaceus* (cf.) on grouse dung.
D: NM3: 76 (as *Achroomyces fimetarius*) **D+I:** Myc13(4): 171
A single collection from Midlothian (Bonally, near Edinburgh) in 1998.

ENDOCRONARTIUM Y. Hirats., *Canad. J. Bot.* 47: 1493 (1969)
Uredinales, Cronartiaceae
 Peridermium (Link) J.C. Schmidt & J. Kunze (anam.), *Deutschl. Schwämme* 6: 4 (1817)
Type: *Endocronartium harknessii* Y. Hirats.

pini Y. Hirats., *Canad. J. Bot.* 47(9): 1494 (1969)
 Peridermium pini Wallr., *Fl. crypt. Germ., Sect. I*: 262 (1831)
 Mis.: *Aecidium pini* sensu auct. p.p.
E: ! **S:** !
D: W&H: 52
H: 0 & I on *Pinus sylvestris*.
Possibly widespread, but poorly known in Britain.

ENDOPHYLLUM Lév., *Mém. Soc. Linn. Paris* 4: 208 (1825)
Uredinales, Pucciniaceae
Type: *Endophyllum persoonii* Lév.

euphorbiae-sylvaticae (DC.) G. Winter, *Rabenh. Krypt.-Fl.* 1(1): 251 (1882)

 Aecidium euphorbiae-silvaticae DC., *Fl. Fr.* 2: 241 (1805)
 Endophyllum euphorbiae Plowr., *Monograph Brit. Ured.*: 228 (1889)
 Mis.: *Aecidium euphorbiae* sensu Cooke [HBF1: 537 (1871)] *fide* Grove (1913)
E: ! **W:** !
H: 0 & III on *Euphorbia amygdaloides*.
D: W&H: 308
Infrequently reported. Known from England (Derbyshire, East & West Gloucestershire, East & West Kent, Herefordshire, North Hampshire, North Wiltshire, Oxfordshire, South Devon, Worcestershire and Yorkshire) and Wales (Montgomeryshire), but many of these records are old.

sempervivi (Alb. & Schwein.) de Bary, *Ann. Sci. Nat., Bot.*, sér. 4, 20: 86 (1863)
 Uredo sempervivi Alb. & Schwein., *Consp. fung. lusat.*: 26 (1805)
E: ! **W:** ! **ROI:** !
H: 0 & III on species and cultivars of *Sempervivum* in gardens.
D: W&H: 309
Rarely recorded or collected. Known from England (Berkshire, Derbyshire, Middlesex, Oxfordshire and Surrey), Wales (Montgomeryshire) and the Republic or Ireland (Dublin).

EOCRONARTIUM G.F. Atk., *J. Mycol.* 8: 107 (1902)
Platygloeales, Platygloeaceae
Type: *Eocronartium typhuloides* G.F. Atk. (= *E. muscicola*)

muscicola (Pers.) Fitzp., *Phytopathology* 8: 212 (1918)
 Clavaria muscicola Pers., *Observ. mycol.* 2: 60 (1799)
 Pistillaria muscicola (Pers.) Fr., *Syst. mycol.* 1: 498 (1821)
 Typhula muscicola (Pers.) Fr., *Epicr. syst. mycol.*: 585 (1838)
 Eocronartium typhuloides G.F. Atk., *J. Mycol.* 8: 107 (1902)
E: ! **W:** ! **ROI:** !
H: Parasitic on mosses. British material on *Eurynchium, Fissidens, Hyalocomium* and *Leskea* spp.
D: NM3: 77 **D+I:** Reid8: 94-97
Rarely reported. Only fourteen records since 1829, the majority of these unsubstantiated with voucher material.

FROMMEËLLA Cummins & Y. Hirats., *Illustrated Genera of Rust Fungi* Edn 2: 120 (1983)
Uredinales, Phragmidiaceae
 Frommea Arthur, *Bull. Torrey Bot. Club* 44: 503 (1917)
Type: *Frommeëlla tormentillae* (Fuckel) Cummins & Y. Hirats.

tormentillae (Fuckel) Cummins & Y. Hirats., *Illustrated Genera of Rust Fungi* Edn 2: 147 (1983)
 Phragmidium tormentillae Fuckel, *Jahrb. Nassauischen Vereins Naturk.* 23-24: 46 (1870)
 Kuehneola tormentillae (Fuckel) Arthur, *Resultats scientifiques du Congres international de Botanique Wien*: 342 (1904)
 Uredo obtusa F. Strauss, *Ann. Wetterauischen Ges. Gesammte Naturk.* 2: 107 (1810)
 Frommea obtusa (F. Strauss) Arthur, *Bull. Torrey Bot. Club* 44: 503 (1917)
 Mis.: *Phragmidium potentillae* sensu Mason & Grainger [Cat. Yorks. Fungi p. 44 (1937)]
E: o **W:** ! **O:** Channel Islands: !
H: 0, I, II & III usually on *Potentilla erecta* ssp. *erecta* and also rarely on *P. reptans*. Often present late in the autumn on moribund leaves of *P. erecta*.
D: W&H: 110-111 (as *Frommea obtusa*)
Rarely reported but probably overlooked.

GLOMOSPORA D.M. Hend., *Notes Roy. Bot. Gard. Edinburgh* 23(4): 497 (1961)
Incertae sedis
Type: *Glomospora empetri* D.M. Hend.
This anamorphic genus has been placed in the *Ascomycota* by Kirk *et al.* (2001), but was said by its author to be related to

anamorphs of the auricularioid *Herpobasidium deformans* C.J. Gould (now referred to *Insolibasidium* Oberw. & Bandoni).

empetri D.M. Hend. (anam.), *Notes Roy. Bot. Gard. Edinburgh* 23(4): 497 (1961)
S: !
H: On living and moribund leaves of *Empetrum nigrum* and *E. hermaphroditum*.
Described from Scotland, and known from East Lothian, Angus, Morayshire and Perthshire.

GYMNOSPORANGIUM R. Hedw. in DC., *Fl. Fr.* 2: 216 (1805)
Uredinales, Pucciniaceae
Roestelia Rebent., *Prodr. fl. neomarch.*: 350 (1804)
Podisoma Link, *Mag. Neuesten Entdeck. Gesammten Naturk. Ges. Naturf. Freunde Berlin* 3: 9 (1809)
Type: *Gymnosporangium fuscum* DC. (= *G. sabinae*)

clavariiforme (Pers.) DC., *Fl. Fr.* 2: 217 (1805)
Tremella clavariiformis Pers., *Syn. meth. fung.*: 629 (1801)
Tremella juniperina L., *Sp. pl.*: 1625 (1753)
Gyraria juniperina (L.) Gray, *Nat. arr. Brit. pl.* 1: 594 (1821)
Roestelia lacerata Mérat, *Nouv. fl. env. Paris*: 113 (1821)
E: ! **S:** ! **W:** ! **NI:** ! **ROI:** !
H: 0 & I on *Crataegus monogyna* and *C. laevigata* (= *C. oxyacanthoides*). III on *Juniperus communis*.
D: W&H: 116-117 **I:** SV33: 64
Rarely reported but apparently widespread. The conspicuous stage III on *Juniperus* is more often recorded than stages 0 & I on *Crataegus*.

confusum Plowr., *Monograph Brit. Ured.*: 232 (1889)
Aecidium cydoniae Lenorm., *Bot. gall.*: 903 (1830)
Roestelia cydoniae (Lenorm.) Thüm., *Syll. fung.* 7: 834 (1888)
E: !
H: 0 & I on *Crataegus monogyna*, *C. laevigata*, *Cydonia oblonga*, *Mespilus germanicus* and *Pyrus communis*. III on cultivated *Juniperus sabina* and *J. sabina* var. *prostrata*.
D: W&H: 117-118
Rarely reported, although there are many recent records from Warwickshire, where it is apparently not uncommon.

cornutum F. Kern, *Bull. New York Bot. Gard.* 7: 444 (1911)
Aecidium cornutum Pers., in Gmelin, *Syst. nat.*: 1472 (1792)
Roestelia cornuta Fr., *Summa veg. Scand.*: 510 (1849)
Mis.: *Gymnosporangium aurantiacum* sensu auct.
Mis.: *Gymnosporangium juniperi* sensu auct.
Mis.: *Gymnosporangium juniperinum* sensu auct.
E: ! **S:** ! **W:** ! **NI:** ! **ROI:** !
H: 0 & I on *Sorbus aucuparia* and cultivated *Sorbus* spp. III on *Juniperus communis*.
D: W&H: 121 **I:** Sow 3 319 (as *Aecidium cornutum*)
Common in parts of Scotland. Rarely reported elsewhere but apparently widespread.

sabinae (Dicks.) G. Winter, *Hedwigia* 19: 55 (1880)
Tremella sabinae Dicks., *Fasc. pl. crypt. brit.* 1: 14 (1785)
Aecidium cancellatum Pers., in Gmelin, *Syst. nat.*: 1472 (1792)
Roestelia cancellata Rebent., *Prodr. fl. neomarch.*: 350 (1804)
Gymnosporangium fuscum DC. [*nom. illegit.*], *Fl. Fr.* 2: 217 (1805)
Puccinia juniperi Pers., *Tent. disp. meth. fung.*: 38 (1797)
Podisoma juniperi-sabinae Fr., *Syst. mycol.* 3: 508 (1829)
E: ! **S:** ! **W:** !
H: 0 & I on *Pyrus communis*. III on cultivated *Juniperus sabina*.
D: W&H: 119-120 (as *Gymnosporangium fuscum*) **I:** Sow 4 (suppl.) 409 & 410 (as *Aecidium cancellatum*)
An established alien. Apparently widespread and more frequently reported in recent years. Perhaps better known as *Gymnosporangium fuscum*.

HELICOBASIDIUM Pat., *Bull. Soc. Bot. France* 32: 172 (1885)
Platygloeales, Platygloeaceae
Thanatophytum Nees (anam.), *Syst. Pilze*: 148 (1817)
Tuberculina Sacc. (anam.), *Syll. fung.* 4: 653 (1886)
Type: *Helicobasidium purpureum* (Tul.) Pat.

longisporum Wakef., *Bull. Misc. Inform. Kew.*: 310 (1917)
Helicobasidium compactum Boedijn, *Arch. voor de Thee Cultuur* 4: 10 (1930)
Mis.: *Helicobasidium mompa* sensu auct. p.p.
E: !
H: Typically on roots or decayed wood, but the first British collection was on the underside of a concrete block.
I: K1 66 (as *Helicobasidium compactum*)
Known from Buckinghamshire, Hertfordshire, and West Sussex.

purpureum (Tul.) Pat., *Bull. Soc. Bot. France* 32: 172 (1885)
Hypochnus purpureus Tul., *Ann. Sci. Nat., Bot.*, sér. 5, 4: 295 (1865)
Protonema brebissonii Desm. (anam.), *Pl. crypt. N. France*: 651 (1834)
Helicobasidium brebissonii (Desm.) Donk, *Taxon* 7: 164 (1958)
Sclerotium crocorum Pers. (anam.), *Syn. meth. fung.*: 119 (1801)
Rhizoctonia crocorum (Pers.) DC. (anam.), *Fl. Fr.* 5: 111 (1815)
Thanatophytum crocorum (Pers.) Nees (anam.), *Syst. Pilze*: 148 (1817)
Rhizoctonia medicaginis DC. (anam.), *Fl. Fr.* 5: 111 (1815)
E: o **S:** ! **W:** o **NI:** ! **O:** Channel Islands: !
H: On decayed wood or soil, often growing up stems of herbaceous plants such as *Mercurialis perennis* and *Urtica dioica* in open woodland.
D: NM3: 78 (as *H. brebissonii*) **D+I:** B&K2: 54 9 **I:** K1 65 (as *H. brebissonii*)
Occasional but widespread. Collections are commonly sterile (the *Thanatophytum* state) consisting only of thick wefts of hyphae. Some of these may refer to *H. longisporum*. The frequently used combination *H. brebissonii* is based on an anamorph name.

HELICOGLOEA Pat., *Bull. Soc. Mycol. France.* 8: 121 (1892)
Platygloeales, Platygloeaceae
Saccoblastia Møller, *Bot. Mitt. Tropen* 8: 162 (1895)
Infundibura Nag Raj & W.B. Kendr. (anam.), *Canad. J. Bot.* 59(4): 544 (1981)
Type: *Helicogloea lagerheimii* Pat.

angustispora L.S. Olive, *Bull. Torrey Bot. Club* 78: 107 (1951)
Infundibura adhaerens Nag Raj & W.B. Kendr. (anam.), *Canad. J. Bot.* 59(4): 544 (1981)
E: !
H: On decayed leaves of *Fagus* or *Quercus* spp in ponds and also isolated from needle litter of *Pinus sylvestris*.
Known in Britain only in its conidial state, *Infundibura adhaerens*. British collections only from South Devon.

farinacea (Höhn.) D.P. Rogers, *Univ. Iowa Stud. Nat. Hist.* 18(3): 66 (1944)
Helicobasidium farinaceum Höhn., *Sitzungsber. Kaiserl. Akad. Wiss., Wien, Math.-Naturwiss. Cl., Abt. 1* 116: 84 (1907)
Saccoblastia farinacea (Höhn.) Donk, *Persoonia* 4: 217 (1966)
E: ! **O:** Channel Islands: !
H: On decayed wood of deciduous trees such as *Fagus* and *Salix* spp, and a single collection on *Pinus sylvestris*.
D: NM3: 78
Known from South Devon, South Hampshire and Surrey, and the Channel Islands (Jersey).

graminicola (Bres.) Baker, *Rep. (Annual) Missouri Bot. Gard.* 23: 90 (1936)
Saccoblastia graminicola Bres., *Ann. Mycol.* 1(1): 112 (1903)

E: ! **O:** !
H: On fallen and decayed wood of deciduous trees, such as *Salix* spp, and a single collection on dead tree fern stems.
Known from West Cornwall (Trellisick Gardens), Surrey (Kew Gardens), and the Channel Islands (Jersey: St. Peter's Woods).

lagerheimii Pat., *Bull. Soc. Mycol. France* 8: 121 (1892)
Saccoblastia sebacea Bourdot & Galzin, *Bull. Soc. Mycol. France* 25: 15 (1909)
E: ! **S:** ! **W:** !
H: On decayed wood of deciduous trees, usually *Fagus* but also *Betula, Crataegus, Salix* and *Ulmus* spp. Rarely on wood of conifers such as *Pinus sylvestris*.
D: NM3: 78 **D+I:** Reid8: 98-99
Rarely reported but apparently widespread.

vestita (Bourdot & Galzin) P. Roberts, *Windahlia* 22: 19 (1997)
Platygloea vestita Bourdot & Galzin, *Bull. Soc. Mycol. France* 39: 261 (1924)
Achroomyces vestitus (Bourdot & Galzin) Wojewoda, *Grzyby* (*Mycota*) 8: 251 (1977)
E: ! **S:** ! **W:** !
H: On decayed wood of deciduous trees such as *Betula* spp and *Fagus*; also known on dead woody stems of *Rubus fruticosus*.
D: NM3: 76 (as *Achroomyces vestitus*)
Known from England (Hertfordshire, Shropshire, South Devon, Surrey and Yorkshire), Scotland (South Aberdeen) and Wales (Anglesey).

HERPOBASIDIUM Lind, *Ark. Bot.* 7(8): 5 (1908)
Platygloeales, Platygloeaceae
Type: *Herpobasidium filicinum* (Rostr.) Lind

filicinum (Rostr.) Lind, *Arch. Bot. (Leipzig)* 7(8): 7 (1908)
Gloeosporium filicinum Rostr., *Myc. Univ. cent. 16, no. 1570*: 2083 (1881)
E: ! **S:** ! **W:** !
H: Parasitic. On sickly or moribund fern fronds. British material on *Dryopteris filix-mas* and *Thelypteris* spp.
D: NM3: 78 **D+I:** Dennis & Wakefield [*Trans. Brit. Mycol. Soc.* 29: 143-144 (1946)], Reid8: 97-98
Rarely reported. Known from England (Hertfordshire, North Lincolnshire and South Devon), Wales (Montgomeryshire), and Scotland (unlocalised).

HOBSONIA Massee, *Ann. Bot.* 5: 509 (1891)
Platygloeales, Platygloeaceae
Type: *Hobsonia gigaspora* Massee (= *H. mirabilis*)
A genus of anamorphic fungi with no known teleomorphs.

mirabilis (Peck) Linder, *Rep. (Annual) Missouri Bot. Gard.* 16: 340 (1929)
Helicomyces mirabilis Peck, *Rep. (Annual) New York State Mus. Nat. Hist.* 34: 46 (1881)
S: !
H: On decayed wood. British material on *Fagus*.
A single record from Mid Ebudes (Isle of Ulva) in 1968. Much more common in warm temperate and tropical areas. Only recently recognised as a basidiomycete based on molecular research.

HYALOPSORA Magnus, *Ber. Bayer. Bot. Ges.* 19: 582 (1902)
Uredinales, Pucciniastraceae
Type: *Hyalopsora aspidiotus* (Magnus) Magnus

adianti-capilli-veneris Syd., *Ann. Mycol.* 1: 249 (1903)
E: ! **S:** ! **ROI:** !
H: II on *Adiantum capillus-veneris*.
D: W&H: 29-30
Rarely reported, as is the host plant in Britain. British records from Scotland, the Republic of Ireland (Co. Galway) in 1843 and England (Kent) but no material traced in herbaria.

aspidiotus (Magnus) Magnus, *Ber. Deutsch. Bot. Ges.* 19: 582 (1901)
Melampsorella aspidiotis Magnus, *Ber. Deutsch. Bot. Ges.* 13: 288 (1895)
Uredo polypodii J. Schröt., *Krypt.-Fl. Schlesien*: 374 (1887)
Mis.: *Uredo filicum* sensu auct. p.p.
S: !
H: II & III on *Gymnocarpium dryopteris* (= *Thelypteris dryopteris*).
D: W&H: 27
No recent records; last reported and collected in Dundee in 1845.

polypodii (Dietel) Magnus, *Ber. Deutsch. Bot. Ges.* 19: 582 (1901)
Pucciniastrum polypodii Dietel, *Hedwigia* 38: 260 (1899)
Mis.: *Uredo filicum* sensu auct. p.p.
E: ! **S:** ! **W:** ! **ROI:** !
H: II on *Cystopteris fragilis*.
D: W&H: 28
Rarely reported but apparently widespread. The host is not uncommon in the north and west.

KRIEGLSTEINERA Pouzar, *Beitr. Kenntn. Pilze Mitteleurop.* 3: 404 (1987)
Platygloeales, Platygloeaceae
Type: *Krieglsteinera lasiosphaeriae* Pouzar

lasiosphaeriae Pouzar, *Beitr. Kenntn. Pilze Mitteleurop.* 3: 403 (1987)
E: !
H: On ascocarps of *Lasiosphaeria ovina*.
D+I: Myc17(1): 12-13 **I:** SV37: 43
A single collection from Hertfordshire (Waterford Heath) in 2001.

KUEHNEOLA Magnus, *Bot. Centralbl.* 74: 169 (1898)
Uredinales, Phragmidiaceae
Type: *Kuehneola albida* (J.G. Kühn) Magnus (= *K. uredinis*)

uredinis (Link) Arthur, *Résultats scientifiques du Congres international de Botanique Wien*: 342 (1905)
Oidium uredinis Link, in Willd., *Sp. pl.,* Edn 4, 6: 123 (1824)
Chrysomyxa albida J.G. Kühn, *Bot. Centralbl.* 16: 154 (1883)
Phragmidium albidum (J.G. Kühn) Lagerh., *Mitt. Bad. Bot. Vereins* 1888: 44 (1888)
Kuehneola albida (J.G. Kühn) Magnus, *Bot. Centralbl.* 74: 169 (1898)
Uredo muelleri J. Schröt., *Krypt.-Fl. Schlesien*: 375 (1887)
E: c **S:** c **W:** c **NI:** c **ROI:** c **O:** Isle of Man: !
H: 0, I, II & III on *Rubus fruticosus* agg. and cultivated *R. loganobaccus*.
D: W&H: 108-109
Very common and widespread.

LEUCOSPORIDIUM Fell, Statzell, I.L. Hunter & Phaff, *Antonie van Leeuwenhoek* 35(4): 438 (1970) [1969]
Sporidiales, Sporidiobolaceae
Type: *Leucosporidium scottii* Fell, Statzell, I.L. Hunter & Phaff

scottii Fell, Statzell, I.L. Hunter & Phaff, *Antonie van Leeuwenhoek* 35(4): 440 (1970) [1969]
Candida scottii Diddens & Lodder (anam.), *Die Hefasammlung des 'Centraalbureau voor Schimmelcultures': Beitrage zu einer Monographie der Hefearten. II Teil. Die anaskosporogenen Hefen. Zweite Halfte*: 487 (1942)
Vanrija scottii (Diddens & Lodder) R.T. Moore (anam.), *Bot. mar.* 23(6): 369 (1980)
E: !
H: Isolated in its yeast state ex raw meat and ex paint (in Britain). First isolated from Antarctic sea water.
D+I: YST 442 (yeast state only)

MELAMPSORA Castagne, *Observ. Uréd.* 2: 18 (1843)
Uredinales, Melampsoraceae
Type: *Melampsora euphorbiae* (C. Schub.) Castagne

allii-fragilis Kleb., *Jahrb. Wiss. Bot.* 35(4): 674 (1901)
E: ! S: ! ROI: !
H: 0 & I on *Allium ursinum.* II & III on *Salix fragilis* and *S. pentandra* and the hybrid between them.
D: W&H: 92
Rarely reported. What is thought to be the aecidial stage on *Allium ursinum* has been recently collected from several sites in southern England.

allii-populina Kleb., *Z. Pflanzenkrankh.* 12: 25 (1902)
    Caeoma ari-italici F. Rudolphi, *Linnaea* 4: 512 (1829)
E: ! S: ! W: ?
H: 0 & I on *Allium ursinum* and *Arum maculatum* reported, but needs confirmation in Britain. II & III on *Populus nigra* and *P. trichocarpa.*
D: W&H: 71
Rarely reported. Apparently widespread but poorly known in Britain.

amygdalinae Kleb., *Jahrb. Wiss. Bot.* 34: 352 (1900)
E: ! W: ! ROI: !
H: 0, I, II & III on *Salix triandra* and *S. triandra* x *viminalis.*
D: W&H: 90
Rarely reported but apparently widespread.

arctica Rostr., *Medd. Grønl.* 3: 535 (1888)
S: !
H: 0 & I on *Saxifraga hypnoides* and *S. oppositifolia.* II & III on *Salix herbacea.*
Previously included in *Melampsora epitea.*

caprearum Thüm., *Mitt. Forstl. Versuchswesen Österreichs* 2: 34 (1879)
    Lecythea caprearum Berk., *Outl. Brit. fungol.*: 334 (1860)
    Melampsora farinosa J. Schröt., *Krypt.-Fl. Schlesien*: 360 (1887)
    Melampsora larici-capraearum Kleb., *Z. Pflanzenkrankh.* 7: 326 (1897)
E: c S: c W: c ROI: c O: Channel Islands: ! Isle of Man: c
H: 0 & I on *Larix* spp. II & III on *Salix* spp, especially common on *Salix caprea.*
D: W&H: 77-78
Very common and widespread.

epitea Thüm., *Hedwigia.* 18: 77 (1879)
    Melampsora alpina Juel, *Öfvers. Kongl. Vetensk.-Akad. Förh.* 51: 417 (1894)
    Uredo confluens γ *orchidis* Alb. & Schwein., *Consp. fung. lusat.*: 122 (1805)
    Lecythea epitea Lév., *Ann. Sci. Nat., Bot.,* sér. 3, 8: 374 (1847)
    Aecidium euonymi J.F. Gmel., *Syst. nat.*: 1473 (1792)
    Caeoma euonymi (J.F. Gmel.) J. Schröt., *Abh. Schles. Ges. Vaterl. Cult., Abth. Naturwiss. (Naturwiss.-Med. Abth.)* 1869-72: 30 (1870)
    Uredo euonymi H. Mart., *Prodr. fl. mosq.*: 230 (1812)
    Uredo confluens var. *euonymi* Cooke, *Handb. Brit. fung.,* Edn 1: 527 (1871)
    Melampsora euonymi-caprearum Kleb., *Jahrb. Wiss. Bot.* 34: 358 (1900)
    Melampsora hartigii Thüm., *Mitt. Forstl. Versuchswesen Österreichs* 2: 34 (1879)
    Melampsora larici-epitea Kleb., *Z. Pflanzenkrankh.* 9: 88 (1899)
    Melampsora orchidi-repentis Kleb., *Jahrb. Wiss. Bot.* 34: 369 (1900)
    Melampsora repentis Plowr., *Z. Pflanzenkrankh.* 1: 131 (1891)
    Melampsora ribesii-purpureae Kleb., *Jahrb. Wiss. Bot.* 35: 667 (1901)
    Aecidium salicis Sowerby, *Col. fig. Engl. fung.* 3: pl. 398 f. 4 (1803)

E: c S: c W: c NI: c ROI: c O: Isle of Man: !
H: 0 & I on various races on *Euonymus europaeus, Larix* spp, and the orchids *Gymnadenia conopsea, Listera ovata* and *Dactylorrhiza* spp. II & III on *Salix* spp.
D: W&H: 80
II & III very common and widespread on *Salix* spp; I frequent on *Euonymus europaeus* at least in southern England, much less often encountered on the other hosts.

epitea var. reticulatae (A. Blytt) Jørst., *Skrifter norske Vidensk-Akad., Mat.-naturv. Kl.* 6: 31 (1940)
    Melampsora reticulatae A. Blytt, *Forh. Vidensk.-Selsk. Kristiania* 4(6): 65 (1896)
    Lecythea mixta Lév., *Ann. Sci. Nat., Bot.,* sér. 3, 8: 374 (1847)
    Melampsora mixta Thüm., *Mitt. Forstl. Versuchswesen Österreichs* 3: 35 (1879)
    Lecythea saliceti Berk., *Outl. Brit. fungol.*: 334 (1860)
S: !
H: 0 & I on *Saxifraga aizoides.* II & III on *Salix arbuscula, S. lapponicum* and *S. myrsinites.*
D: W&H: 82
Known from South Aberdeen and doubtfully reported from Angus.

euphorbiae (C. Schub.) Castagne, *Observ. Uréd.* 2: 18 (1843)
    Xyloma euphorbiae C. Schub., *Fic. Fl. Dresd.* 2: 31 (1823)
    Lecythea euphorbiae Lév., *Ann. Sci. Nat., Bot.,* sér. 3, 8: 374 (1847)
    Melampsora euphorbiae-dulcis G.H. Otth, *Mitt. naturf. Ges. Bern.* 70 (1868)
    Uredo euphorbiae-helioscopiae Pers., *Syn. meth. fung.*: 215 (1801)
    Melampsora helioscopiae G. Winter, *Rabenh. Krypt.-Fl.* 1(1): 240 (1882)
E: c S: ! W: c NI: ! ROI: c O: Channel Islands: ! Isle of Man: !
H: II & III on *Euphorbia amygdaloides, E. cyparissias, E. dulcis, E. exigua, E. helioscopia, E. hyberna, E. lathyris, E. paralias, E. peplus, E. x pseudovirgata, E. stricta* and recently on cultivated *E. characias* and *E. soongarica.*
D: W&H: 67-68
Common and widespread on *Euphorbia helioscopia* and *E. peplus.* Rather frequent on *E. amygdaloides* in southern England and uncommon on the remaining species.

hypericorum G. Winter, *Rabenh. Krypt.-Fl.* 1(1): 241 (1882)
    Caeoma hypericorum Schltdl., *Fl. berol.* 2: 122 (1804)
    Uredo hypericorum DC. [*nom. inval.*], *Mém. Agric. Soc. Agric. Dép. Seine* 10: 235 (1807)
E: c S: ! W: c NI: ! ROI: c O: Channel Islands: ! Isle of Man: !
H: II & III on *Hypericum* spp and cultivars.
D: W&H: 70-71
Common on *Hypericum androsaemum* and *H. calycinum,* much less so on other species.

larici-pentandrae Kleb., *Forstl.-Naturwiss. Z.* 6: 470 (1897)
S: ! ROI: !
H: II & III on *Salix pentandra* and *S. fragilis* x *pentandra*
D: W&H: 79
Rarely reported, but apparently widespread.

larici-populina Kleb., *Z. Pflanzenkrankh.* 12: 43 (1902)
E: ! S: ! W: !
H: II & III on *Populus* species, hybrids and cultivars, but not on native *P. tremula.* I on *Larix* spp, but not reported from Britain.
D: W&H: 73
Rarely reported, but apparently widespread.

lini (Ehrenb.) Desm., *Pl. crypt. N. France.* 2049 (1850)
    Xyloma lini Ehrenb., *Sylv. mycol. berol.*: 27 (1818)
    Lecythea lini Berk., *Outl. Brit. fungol.*: 334 (1860)
E: o S: ! W: o NI: ! ROI: ! O: Channel Islands: !
H: II & III on *Linum catharticum.*
D: W&H: 64-65
Occasional but widespread.

lini var. liniperda Körn., *Verh. Naturhist. Vereines Preuss. Rheinl. Westphalens* 31: 83 (1874)

Melampsora liniperda (Körn.) Palm, *Svensk bot. Tidskr.* 4: (4) (1910)

    Melampsora lini var. *major* Fuckel, *Jahrb. Nassauischen Vereins Naturk.* 23-24: 44 (1870)

E: ! **S:** ! **ROI:** !

H: 0, I, II & III on *Linum usitatissimum*.

D: W&H: 65-66

Rarely reported.

populnea (Pers.) P. Karst., *Bidrag Kännedom Finlands Natur Folk* 31: 53 (1879)

    Sclerotium populneum Pers., *Syn. meth. fung.*: 125 (1801)

    Melampsora aecidioides Plowr., *Monograph Brit. Ured.*: 241 (1889)

    Uredo confluens Pers., *Observ. mycol.* 1: 98 (1796)

    Melampsora laricis Hartig, *Allg. Forst-u. Jagdztg.* 61: 326 (1885)

    Caeoma mercurialis Link, in Willd., *Sp. pl.*, Edn 4, 2: 35 (1825)

    Melampsora pinitorqua Rostr., *Afb. farl. snyltesv.*: 10 (1889)

    Lecythea populina Lév., *Ann. Sci. Nat., Bot.*, sér. 3, 8: 374 (1847)

    Melampsora rostrupii Wagner [*nom. nud.*], *Oesterr. Bot. Z.* 46: 274 (1896)

    Melampsora tremulae Tul., *Ann. Sci. Nat., Bot.*, sér. 4, 2: 95 (1854)

    Melampsora tremulae f. *laricis* Hartig, *Lehrb. Baumkrankh.*, Edn 2: 14 (1889)

E: c **S:** c **W:** c **O:** Channel Islands: ! Isle of Man: c

H: 0 & I on *Larix*, *Pinus* and *Mercurialis* spp. II & III on *Populus alba* and *P. tremula*.

D: W&H: 74-77

Common and widespread, but no records have been traced from Ireland. The aecidial stage is common on *Mercurialis perennis* but was, until recently, unknown on *M. annua*. It is now reported with increasing frequency on this latter host in southern England (London, Middlesex and Surrey)

ribesii-viminalis Kleb., *Jahrb. Wiss. Bot.* 34: 363 (1900)

E: ! **S:** ! **ROI:** !

H: 0 & I on *Ribes* spp, but not known in Britain. II & III on *Salix viminalis*.

D: W&H: 80

Rarely reported.

salicis-albae Kleb., *Jahrb. Wiss. Bot.* 35: 679 (1901)

    Melampsora allii-salicis-albae Kleb., *Z. Pflanzenkrankh.* 12: 19 (1902)

    Uredo vitellinae DC., *Fl. Fr.* 2: 231 (1805)

    Melampsora vitellinae Thüm., *Hedwigia* 17: 79 (1878)

E: ! **ROI:** !

H: II & III on *Salix alba* and *S. alba* var. *vitellina*.

D: W&H: 91-92

Rarely reported.

vernalis G. Winter, *Rabenh. Krypt.-Fl.* 1(1): 237 (1882)

S: !

H: 0, I & III on *Saxifraga granulata*.

D: W&H: 67

Apparently not collected or reported since the 1950's.

MELAMPSORELLA J. Schröt., *Hedwigia* 13: 85 (1874)

Uredinales, Pucciniastraceae

Type: *Melampsorella caryophyllacearum* J. Schröt.

caryophyllacearum J. Schröt., *Hedwigia* 13: 85 (1874)

    Uredo caryophyllacearum DC., *Fl. Fr.* 2: 85 (1805)

    Melampsora cerastii G. Winter, *Rabenh. Krypt.-Fl.* 1(1): 242 (1882)

    Aecidium elatinum Alb. & Schwein., *Consp. fung. lusat.*: 121 (1805)

    Peridermium elatinum (Alb. & Schwein.) F.C. Holl. & J.C. Schmidt, *Deutschl. Schwämme* 6: 4 (1817)

    Uredo pustulata β *cerastii* Pers., *Syn. meth. fung.*: 219 (1801)

E: ! **S:** ! **W:** ! **NI:** ! **ROI:** ! **O:** Channel Islands: !

H: 0 & I on *Abies* spp. II & III on *Cerastium arcticum*, *C. arvense*, *C. glomeratum*, *C. fontanum* ssp. *holosteoides*, *C. semidecandrum*, *C. tomentosum*, *Stellaria graminea*, *S. holostea* and supposedly *S. media*.

D: W&H: 43-44

Rarely reported, but apparently widespread.

symphyti Bubák, *Ber. Deutsch. Bot. Ges.* 12: 423 (1904)

    Uredo symphyti DC., in Lam., *Encyclop.* 8: 232 (1808)

    Trichobasis symphyti Lév., in Orbigny, *Dict. univ. hist. nat.* 1: 19 (1841)

    Coleosporium symphyti Fuckel, *Jahrb. Nassau. Vereins Naturk.* 23-24: 43 (1870)

E: c **S:** ! **W:** !

H: II & III on leaves of *Symphytum asperum*, *S. officinale*, *S. tuberosum* and *S.* x *uplandicum*.

D: W&H: 46-47

Common and widespread, especially so on *Symphytum* x *uplandicum* in southern England.

MELAMPSORIDIUM Kleb., *Z. Pflanzenkrankh.* 9: 21 (1899)

Uredinales, Pucciniastraceae

Type: *Melampsoridium betulinum* (Pers.) Kleb.

betulinum (Pers.) Kleb., *Z. Pflanzenkrankh.* 9: 21 (1899)

    Uredo populina var. *betulina* Pers., *Syn. meth. fung.*: 219 (1801)

    Melampsora betulina (Pers.) Tul., *Ann. Sci. Nat., Bot.*, sér. 4, 2: 97 (1854)

    Sclerotium betulinum Fr., *Syst. mycol.* 2: 262 (1823)

    Lecythea betulina Plowr., *Monograph Brit. Ured.*: 244 (1889)

    Mis.: *Melampsoridium hiratsukanum* sensu W&H: 47-48

E: c **S:** c **W:** c **NI:** c **ROI:** c **O:** Isle of Man: c

H: 0 & I on *Larix* spp. II & III on *Alnus cordata*, *A. glutinosa*, *A. incana*, *A. rubra* and species or cultivars of *Betula*.

D: W&H: 49-50

Very common and widespread on *Betula*, rare on *Alnus*. Rarely absent from any area with *Betula* spp, and especially damaging on leaves of seedling plants.

hiratsukanum S. Ito, *J. Fac. Agric. Hokkaido Univ.* 21: 10 (1927)

E: ! **S:** ! **W:** !

H: In leaves of *Alnus* spp. Known on *A. glutinosa* and on planted *A. incana* and *A. cordata*.

D: W&H: 47

Rarely reported but apparently widespread. Considered by some authorities to be synonymous with *M. betulinum*.

MICROBOTRYUM Lév., *Ann. Sci. Nat., Bot.*, sér. 3, 8: 372 (1847)

Microbotryales, Microbotryaceae

Type: *Microbotryum antherarum* (DC.) Lév. (= *M. violaceum*)

dianthorum (Liro) H. Scholz & I. Scholz, *Englera* 8: 206 (1988)

    Ustilago dianthorum Liro, *Ann. Acad. Sci. Fenn., Ser. A* 17(1): 35 (1924)

E: !

H: In anthers of cultivated *Dianthus* species and hybrids.

D: ESF 154

Known from Norfolk, Surrey and Sussex but not collected in recent years.

lychnidis-dioicae (Liro) G. Deml & Oberw., *Phytopath. Z.* 104(4): 353 (1982)

    Ustilago lychnidis-dioicae Liro, *Ann. Acad. Sci. Fenn., Ser. A* 17(1): 33 (1924)

E: ! **S:** ! **W:** ! **NI:** ! **ROI:** ! **O:** Channel Islands: ! Isle of Man: !

H: In anthers of *Silene dioica* and *S. latifolia*.

D: ESF 154

Previously included with *Ustilago violacea* s.l. Apparently widespread and common with the hosts.

major (J. Schröt.) G. Deml & Oberw., *Phytopath. Z.* 104(4): 353 (1982)

> *Ustilago major* J. Schröt., *Krypt.-Fl. Schlesien:* 273 (1887)

E: r

H: In anthers and filaments, flowers, and ovaries of *Silene otites.*

D: ESF 155

Rare. Known from Cambridgeshire and West Suffolk. The host has a limited distribution in Britain.

silenes-inflatae (Liro) G. Deml & Oberw., *Phytopath. Z.* 104(4): 354 (1982)

> *Ustilago silenes-inflatae* Liro, *Ann. Acad. Sci. Fenn., Ser. A* 17: 44 (1924)

E: ! **O:** Channel Islands: !

H: In anthers of *Silene inflata* and *S. vulgaris*

D: ESF 154

Previously included in *Ustilago violacea* s.l. Apparently scarce and little collected although the hosts are common in Britain.

stellariae (Sowerby) G. Deml & Oberw., *Phytopath. Z.* 104(4): 354 (1982)

> *Farinaria stellariae* Sowerby, *Col. fig. Engl. fung.* 3: pl. 396 f.1 (1803)
> *Ustilago stellariae* (Sowerby) Liro, *Ann. Acad. Sci. Fenn., Ser. A* 17(1): 39 (1924)
> *Ustilago violacea* var. *stellariae* (Sowerby) Savile, *Canad. J. Bot.* 31: 674 (1953)

E: ! **S:** ! **W:** !

H: In anthers *Myosoton aquaticum, Stellaria graminea, S. holostea* and *S. uliginosa.*

D: ESF 155

Previously included with *Ustilago violacea* s.l. Apparently widespread.

violaceum (Pers.) G. Deml & Oberw., *Phytopath. Z.* 104(4): 353 (1982)

> *Uredo violacea* Pers., *Tent. disp. meth. fung.*: 57 (1797)
> *Ustilago violacea* (Pers.) Roussel, *Flora Calvados,* Edn 2: 47 (1806)
> *Ustilago antherarum* Fr., *Syst. mycol.* 3: 518 (1829)

E: ! **S:** ! **W:** ! **NI:** ! **ROI:** ! **O:** Channel Islands: ! Isle of Man: !

H: In anthers of various *Caryophyllaceae,* especially *Lychnis flos-cuculi.*

D: UB 60 (as *Ustilago violacea* p.p.), ESF 156

Apparently common and widespread, but old collections and reports will include both *M. lychnidis-dioicae* and *M. stellariae.*

MIKRONEGERIA Dietel, *Bot. Jahrb. Syst.* 27: 16 (1899)

Uredinales, Mikronegeriaceae

Type: *Mikronegeria fagi* Dietel & Neger

fagi Dietel & Neger, *Engler's Bot. Jahrb.* 27(1): 16 (1899)

E: !

H: II & III on *Nothofagus* spp.

An alien, established on planted *Nothofagus* in East Gloucestershire (Westonbirt Arboretum) and known here since 1976.

MILESIA F.B. White (anam.), *Scott. Naturalist (Perth)* 4: 162 (1878)

Uredinales, Pucciniastraceae

Type: *Milesia magnusiana* (Jaap) Faull

magnusiana (Jaap) Faull (anam.), *Contr. Arnold Arbor.* 2: 32 (1932)

> *Milesina magnusiana* Jaap, *Verh. Bot. Vereins Prov. Brandenburg* 62: 16 (1915)

E: ! **ROI:** !

H: II & (III?) on *Asplenium adiantum-nigrum.*

D: HB 476

Apparently rare. British records only from Republic of Ireland and south western England.

MILESINA Magnus, *Ber. Bayer. Bot. Ges.* 27: 325 (1909)

Uredinales, Pucciniastraceae

Type: *Milesina kriegeriana* (Magnus) Magnus

blechni (Syd.) Syd., *Ann. Mycol.* 8: 491 (1910)

> *Melampsorella blechni* Syd., *Ann. Mycol.* 1: 537 (1903)
> *Milesia blechni* (Syd.) Arthur, *Bot. Gaz.* 73: 61 (1922)
> Mis.: *Aecidium pseudocolumnare* sensu auct. p.p.
> Mis.: *Uredo scolopendrii* sensu auct. p.p.

E: c **S:** c **W:** c **NI:** ! **ROI:** ! **O:** Isle of Man: c

H: 0 & I on *Abies alba, A. amabilis, A. cephalonica* and *A. pectinata.* II & III on moribund fronds of *Blechnum spicant* but not on fresh growth.

D: W&H: 18-19

Common and widespread on the fern host.

carpatorum Hyl., Jørst. & Nannf., *Nytt Mag. Bot.* 15(3): 259 (1968)

E: ! **W:** ? **ROI:** ?

H: II & III on *Dryopteris filix-mas.*

D: SL3 23; W&H: 23

Rarely reported but easily overlooked. The host is common throughout Britain. Collections known from Herefordshire, North Hampshire, South Devon and Surrey. Reported from the Republic of Ireland (Cavan and Louth) but unsubstantiated with voucher material. The single collection from Wales (in herb. K) has been redetermined as *M. kriegeriana.*

dieteliana (Syd.) Magnus, *Ber. Deutsch. Bot. Ges.* 27: 325 (1909)

> *Melampsorella dieteliana* Syd., *Ann. Mycol.* 1: 537 (1903)
> *Milesia polypodii* F.B. White, *Scott. Naturalist (Perth)* 4: 162 (1877)
> Mis.: *Aecidium pseudocolumnare* sensu auct. p.p.

E: ! **S:** o **W:** o **NI:** ! **ROI:** ! **O:** Channel Islands: ! Isle of Man: !

H: 0 & I on *Abies* spp. II & III on *Polypodium vulgare* agg.

D: W&H: 25-26

Occasional but widespread. Easily overlooked. Records show a western bias, following the main distribution of the host.

kriegeriana (Magnus) Magnus, *Ber. Deutsch. Bot. Ges.* 27: 325 (1909)

> *Melampsorella kriegeriana* Magnus, *Ber. Deutsch. Bot. Ges.* 19: 581 (1901)
> *Milesia kriegeriana* (Magnus) Arthur, *Mycologia* 7: 176 (1915)
> Mis.: *Aecidium pseudocolumnare* sensu auct. p.p.

E: c **S:** c **W:** c **ROI:** c **O:** Channel Islands: ! Isle of Man: !

H: 0 & I on *Abies* spp. II & III usually on *Dryopteris* spp. Less often reported on *Athyrium filix-femina* and rarely *Polystichum aculeatum.*

D: W&H: 21-22

Common and widespread. Sori usually numerous on moribund or overwintered fronds of the hosts, but not on fresh growth.

murariae Syd., *Monogr. Ured.* 3: 477 (1915)

> *Milesia murariae* (Syd.) Faull, *Contr. Arnold Arbor.* 2: 34 (1932)
> *Uredo murariae* Magnus, *Ber. Deutsch. Bot. Ges.* 20: 611 (1902)
> Mis.: *Uredo scolopendrii* sensu auct. p.p.

S: ! **W:** ! **NI:** ! **ROI:** !

H: II & III on *Asplenium ruta-muraria.*

D: W&H: 20

Rarely reported but apparently widespread. Records show a distinctly western bias, following the distribution of the host.

scolopendrii (Faull) D.M. Hend., *Notes Roy. Bot. Gard. Edinburgh* 23: 504 (1961)

> *Milesia scolopendrii* Faull, *Contr. Arnold Arbor.* 2: 113 (1932)
> *Ascospora scolopendrii* Fuckel, *Jahrb. Nassauischen Vereins Naturk.* 27-28: 19 (1873)
> *Uredinopsis scolopendrii* (Fuckel) Rostr., *Bot. Tidsskr.* 21: 42 (1897)
> Mis.: *Uredo scolopendrii* sensu auct. p.p.

E: c **S:** c **W:** c **NI:** ! **ROI:** ! **O:** Channel Islands: ! Isle of Man: ! !

H: 0 & I on *Abies alba*, *A. concolor* and *A. pectinata*. II & III on overwintered and moribund fronds of *Phyllitis scolopendrium*.
D: W&H: 19-20
Common and widespread.

vogesiaca Syd., *Ann. Mycol.* 8: 491 (1910)
Uredo vogesiaca (Syd.) Sacc. & Trotter, *Syll. fung.* 21: 812 (1912)
Milesia vogesiaca (Syd.) Faull, *Contr. Arnold Arbor.* 2: 103 (1932)
E: ! **ROI:** !
H: 0 & I on *Abies alba*. II & III on *Polystichum setiferum* and *P. aculeatum*.
D: W&H: 23-24
Only four British collections in herb. K, from Middlesex, South Somerset, South Hampshire and Surrey. Also reported from the Republic of Ireland (Co. Dublin: Killakee Mountain).

whitei (Faull) Hirats., *Mem. Tottori Agric. Coll.* 4: 123 (1936)
Milesia whitei Faull, *Contr. Arnold Arbor.* 2: 111 (1932)
Milesia polystichi Wineland, in Jackson, *Brooklyn Bot. Gard. Mem.* **1**: 214 (1918)
Milesina polystichi (Wineland) Grove, *J. Bot.* 59: 109 (1921)
E: ! **S:** ! **W:** ! **ROI:** !
H: II & III on *Polytstichum setiferum*.
D: W&H: 24-25
Rarely reported but apparently widespread. The host is not uncommon.

MIYAGIA Miyabe, Syd. & P. Syd., *Ann. Mycol.* 11: 107 (1913)
Uredinales, Pucciniaceae
Type: *Miyagia anaphalidis* Miyabe

pseudosphaeria (Mont.) Jørst., *Nytt Mag. Bot.* 9: 78 (1962) [1961]
Puccinia pseudosphaeria Mont., in Webb & Berthelet, *Hist. nat. Iles Canaries* 2: 89 (1841)
Peristemma pseudosphaeria (Mont.) Jørst., *Friesia* 5: 278 (1956)
Puccinia sonchi Desm., *Ann. Sci. Nat., Bot.,* sér. 3, 11: 274 (1849)
Uromyces sonchi Oudem., in Rabenhorst, *Fungi europaei* IV: 95 (1872)
Peristemma sonchi (Desm.) Syd., *Ann. Mycol.* 19: 175 (1921)
Mis.: *Uredo sonchi-arvensis* sensu auct. p.p.
E: c **S:** o **W:** c **NI:** ! **ROI:** o **O:** Channel Islands: c Isle of Man: !
H: 0, I, II & III on *Sonchus arvensis*, *S. asper*, *S. oleraceus* and *S. palustris*.
D: W&H: 298
Very common and widespread. Most frequent on *S. asper* and *S. oleraceus* and rarely not present on the basal and older leaves of mature plants. Often accompanied by *Coleosporium tussilaginis*.

MYCOGLOEA L.S. Olive, *Mycologia* 42: 835 (1950)
Platygloeales, Platygloeaceae
Type: *Mycogloea carnosa* L.S. Olive

macrospora (Berk. & Broome) McNabb, *Trans. Brit. Mycol. Soc.* 48: 187 (1965)
Dacrymyces macrosporus Berk. & Broome, *Ann. Mag. Nat. Hist, Ser.* 4 [*Notices of British Fungi no. 1374*] 11: 343 (1873)
Fusisporium obtusum Cooke (anam.), *Grevillea* 5(34): 58 (1876)
E: !
H: Parasitic. The few British collections are all associated with the ascomycete *Diatrype stigma*.
Several collections by Broome from North Somerset (Batheaston) in 1862-1872, but not recollected since.

NAIADELLA Marvanová & Bandoni, *Mycologia* 79: 579 (1987)

Classiculales, Classiculaceae
Type: *Naiadella fluitans* Marvanová & Bandoni
A genus of anamorphic fungi with teleomorphs in *Classicula*.

fluitans Marvanová & Bandoni, *Mycologia* 79(4): 579 (1987)
E: ! **S:** ? **W:** !
H: In foam samples collected from streams.
Known from England (South Devon: Manaton, Becka Falls) and Wales (Caernarvonshire: Trefriw, Afon Crafnant). Possibly also in Scotland but this may be misidentified material of *Jaculispora submersa*. *Classicula naiadella* (the auricularioid teleomorph) has not yet been found in the British Isles.

NAOHIDEA Oberw., *Rep. Tottori Mycol. Inst.* 28: 114 (1990)
Platygloeales, Platygloeaceae
Type: *Naohidea sebacea* (Berk. & Broome) Oberw.

sebacea (Berk. & Broome) Oberw., *Rep. Tottori Mycol. Inst.* 28: 114 (1990)
Dacrymyces sebaceus Berk. & Broome, *Ann. Mag. Nat. Hist, Ser.* 4 [*Notices of British Fungi no. 1305*] 7: 430 (1871)
Platygloea sebacea (Berk. & Broome) McNabb, *Trans. Brit. Mycol. Soc.* 48: 188 (1965)
E: !
H: Parasitic. On coelomycetes such as *Diplodia subtecta* or pyrenomycetes such as *Botryosphaeria quercuum* on dead wood.
Rarely reported.

NAOHIDEMYCES S. Sato, Katsuya & Y. Hirats., *Trans. Mycol. Soc. Japan* 34(1): 48 (1993)
Uredinales, Pucciniastraceae
Type: *Naohidemyces vaccinii* (G. Winter) S. Sato, Katsuya & Y. Hirats. (= *N. vacciniorum*)

vacciniorum (J. Schröt.) Spooner, *Vieraea* 27: 175 (1999)
Melampsora vacciniorum J. Schröt., *Krypt.-Fl. Schlesien.* 365 (1885)
Uredo pustulata var. *vaccinii* Alb. & Schwein., *Consp. fung. lusat.:* 126 (1805)
Melampsora vaccinii G. Winter, *Rabenh. Krypt.-Fl.* 1(1): 244 (1882)
Pucciniastrum vaccinii (G. Winter) Jørst., *Skrifter Vidensk.-Akad. Oslo.* 55 (1952) [1951]
Thekopsora vaccinii (G. Winter) Hirats., *Uredin. Studies.* 260 (1955)
Naohidemyces vaccinii (G. Winter) S. Sato, Katsuya & Y. Hirats. [*nom. invalid.*], *Trans. Mycol. Soc. Japan.* 48 (1993)
Uredo vacciniorum Rabenh., *Rabenh. Krypt.-Fl.:* 7 (1844)
Thekopsora vacciniorum P. Karst., *Mycol. fenn.* 4: 58 (1878)
E: c **S:** c **W:** ! **NI:** ! **ROI:** ! **O:** Isle of Man: !
H: II & III on *Vaccinium myrtillus*, *V. oxycoccos*, *V. uliginosum*, *V. vitis-idaea* and *Vaccinium x intermedius*.
D: W&H: 38 (as *Pucciniastrum vaccinii*)
Supposedly scarce *fide* WH but actually common on *Vaccinium myrtillus*. Rarely reported on the other hosts.

NYSSOPSORA Arthur, *Résultats scientifiques du Congres international de Botanique Wien.* 342 (1906)
Uredinales, Sphaerophragmiaceae
Type: *Nyssopsora echinata* (Lév.) Arthur

echinata (Lév.) Arthur, *Résultats scientifiques du Congres international de Botanique Wien.* 342 (1906)
Triphragmium echinatum Lév., *Ann. Sci. Nat., Bot.,* sér. 3, 9: 247 (1848)
S: !
H: III on *Meum athamanticum*.
D: W&H: 114
Known from Perthshire and South Aberdeen.

OCCULTIFUR Oberw., *Rep. Tottori Mycol. Inst.* 28: 119 (1990)

Platygloeales, Platygloeaceae
Type: *Occultifur internus* (L.S. Olive) Oberw.

corticiorum P. Roberts, *Mycotaxon* 63: 202 (1997)
E: !
H: Parasitic. In the hymenium of *Hyphoderma praetermissum*.
A single collection (the type) from South Devon (Torquay, Scadson Woods) in 1994.

internus (L.S. Olive) Oberw., *Rep. Tottori Mycol. Inst.* 28: 120 (1990)
Platygloea peniophorae var. *interna* L.S. Olive, *Bull. Torrey Bot. Club* 81: 331 (1954)
Mis.: *Tremella obscura* sensu Reid [*Trans. Brit. Mycol. Soc.* 62(3): 490 (1974)]) p.p.
E: o **S:** !
H: Parasitic. In the hymenium of *Dacrymyces stillatus*.
Occasional in England, with a single record from Scotland (Wester Ross).

OCHROPSORA Dietel, *Ber. Bayer. Bot. Ges.* 13: 401 (1895)

Uredinales, Chaconiaceae
Type: *Ochropsora sorbi* (Oudem.) Dietel (= *O. ariae*)

ariae (Fuckel) Ramsb., *Trans. Brit. Mycol. Soc.* 4(2): 337 (1914)
Melampsora ariae Fuckel, *Jahrb. Nassauischen Vereins Naturk.* 23-24: 45 (1870)
Endophyllum leucospermum Soppitt, *J. Bot.*: 273 (1893)
Caeoma sorbi Oudem., *Nederl. Kruidkundig. Archief.* 2(1): 177 (1872)
Ochropsora sorbi (Oudem.) Dietel, *Ber. Bayer. Deutsch. Bot. Ges.* 13: 401 (1895)
Mis.: *Aecidium leucospermum* sensu auct. Brit.
E: ! **S:** ! **W:** !
H: 0 & I on *Anemone nemorosa* and possibly *A. blanda*. II & III on *Sorbus aucuparia* and dubiously *S. aria* and *Malus domestica*.
D: W&H: 11-13
Rarely reported. Apparently widespread, but the majority of records are unsubstantiated with voucher material.

PACHNOCYBE Berk., *Engl. fl.* 5(2): 333 (1836)

Atractiellales, Pachnocybaceae
Type: *Pachnocybe ferruginea* Berk.

ferruginea Berk., *English fl.* 5(2): 334 (1836)
Graphium cartwrightii J.F.H. Beyma, *Antonie van Leeuwenhoek* 6: 284 (1940)
E: !
H: British material on decayed deal (*Pinus* sp.) firewood, old timber boards in a school, creosoted deal in a house, and pine floorboards.
A few records on worked timber in the late nineteenth century. Last recorded from West Kent (West Wickham) in 1934, but inconspicuous and easily overlooked.

PHLEOGENA Link, *Handbuch Gewächse* 3: 396 (1833)

Atractiellales, Phleogenaceae
Ecchyna Boud., *Bull. Soc. Mycol. France* 1: 111 (1885)
Mis.: *Pilacre* Fr. sensu auct.
Type: *Phleogena faginea* (Fr.) Link

faginea (Fr.) Link, *Handbuch Gewächse* 3: 396 (1833)
Onygena faginea Fr., *Syst. mycol.* 3: 209 (1829)
Pilacre faginea (Fr.) Berk. & Broome, *Ann. Mag. Nat. Hist, Ser. 2 [Notices of British Fungi no. 380]* 5: 365 (1850)
Ecchyna faginea (Fr.) Boud., *Bull. Soc. Mycol. France* 1: 111 (1885)
Pilacre petersii Berk. & M.A. Curtis, *Ann. Mag. Nat. Hist, Ser. 3 [Notices of British Fungi no. 824]* 3: 362 (1859)
Ecchyna petersii (Berk. & M.A. Curtis) Pat., *Essai taxon.*: 17 (1900)
E: o **S:** !
H: Saprotrophic, and possibly weakly parasitic. On dead woody substrata and living trunks of deciduous trees (rarely on conifer wood).
D: NM3: 79 **D+I:** Myc8(3): 107 **I:** K1 68
Not uncommon in southern England. Basidiomes resemble large fruitbodies of the myxomycete genus *Physarum*. When dried they develop a strong smell of fenugreek or curry powder which lasts for years in herbarium material.

PHRAGMIDIUM Link, *Mag. Neuesten Entdeck. Gesammten Naturk. Ges. Naturf. Freunde Berlin* 7: 30 (1815)

Uredinales, Phragmidiaceae
Aregma Fr., *Syst. mycol.* 3: 495 (1829)
Epitea Fr., *Syst. mycol.* 3: 510 (1829)
Lecythea Lév., *Ann. Sci. Nat., Bot.,* sér. 3, 8: 373 (1847)
Type: *Phragmidium mucronatum* (Pers.) Schltdl.

acuminatum (Fr.) Cooke, *Handb. Brit. fung.,* Edn 1: 490 (1871)
Aregma acuminatum Fr., *Observ. mycol.* 1: 226 (1815)
Phragmidium rubi-saxatilis Liro, *Bidrag Kännedom Finlands Natur Folk* 65: 421 (1908)
S: !
H: I, II & III on *Rubus saxatilis*.
D: W&H: 98
A single collection from Morayshire (Forres) in 1879. .Never recollected and possibly extinct.

bulbosum (F. Strauss) Schltdl., *Fl. berol.* 2: 156 (1824)
Uredo bulbosum F. Strauss, *Ann. Wetterauischen Ges. Gesammte Naturk.* 2: 108 (1810)
Aregma bulbosum (F. Strauss) Fr., *Syst. mycol.* 3: 497 (1829)
Phragmidium cylindricum Bonord., *Kenntn. Coniomyc. Cryptomyc.*: 60 (1860)
Puccinia mucronata β *rubi* Pers., *Syn. meth. fung.*: 230 (1801)
Phragmidium rubi (Pers.) G. Winter, *Rabenh. Krypt.-Fl.* 1(1): 230 (1882)
Puccinia rubi DC. [non *P. rubi* Schum. (1803)], *Fl. Fr.* 2: 218 (1805)
Mis.: *Puccinia ruborum* DC sensu auct. Brit. p.p.
E: o **S:** o **W:** o **NI:** o **ROI:** o **O:** Channel Islands: ! Isle of Man: !
H: 0, I, II & III on *Rubus fruticosus* agg. and *Rubus caesius*.
D: W&H: 95
Occasional but apparently widespread. Much less often reported than *P. violaceum* with which it has been much confused.

fragariae (DC.) Rabenh., *Klotzschii Herbarium Vivum Mycologicum*: 1987 (1855)
Puccinia fragariae DC., in Lam., *Encyclop.* 8: 244 (1808)
Puccinia fragariastri DC., *Fl. Fr.* 5: 55 (1815)
Phragmidium fragariastri (DC.) J. Schröt., *Krypt.-Fl. Schlesien*: 351 (1887)
Phragmidium granulatum Fuckel, *Jahrb. Nassauischen Vereins Naturk.* 23-24: 46 (1870)
Uredo obtusa F. Strauss, *Ann. Wetterauischen Ges. Gesammte Naturk.* 2: 107 (1811) p.p.
Phragmidium obtusum (Fr.) G. Winter, *Hedwigia* 19: 26 (1880)
Aregma obtusatum Fr., *Observ. mycol.* 1: 225 (1815)
Lecythea potentillarum Lév., *Ann. Sci. Nat., Bot.,* sér. 3, 8: 374 (1847)
Mis.: *Uredo potentillarum* sensu auct. p.p.
E: c **S:** c **W:** c **NI:** c **ROI:** c **O:** Channel Islands: ! Isle of Man: !
H: 0, I, II & III on *Potentilla sterilis* and rarely on *P. reptans*.
Possibly also rarely on *Fragaria vesca* - recent collections on this host from North Hampshire (Crab Wood, near Winchester and Embley Wood, near East Stratton) await verification.
D: W&H: 100-101

Very common and widespread on the main host.

fusiforme J. Schröt., *Abh. Schles. Ges. Vaterl. Cult., Abth. Naturwiss. (Naturwiss.-Med. Abth.)* 1869-72: 24 (1870)
Uredo pinguis var. *rosae-alpinae* DC., *Fl. Fr.* 5: 235 (1815)
Phragmidium rosae-alpinae (DC.) G. Winter, *Hedwigia* 19: 227 (1880)
E: !
H: 0, I, II and III on cultivated *Rosa glauca* (= *R. rubrifolia*) and *R. pendulina*.
Very few reports since 1889, and deleted from the British list until refound in a survey of rose rusts in 1973 (see Henderson & Bennell, 1979: 481). Probably an introduction.

mucronatum (Pers.) Schltdl., *Fl. berol.* 2: 156 (1824)
Puccinia mucronata Pers., *Syn. meth. fung.*: 230 (1801)
Aregma mucronatum Fr., *Syst. mycol.* 3: 497 (1829)
Puccinia mucronata α *rosae* Pers., *Syn. meth. fung.*: 230 (1801)
Lecythea rosae Lév., *Ann. Sci. Nat., Bot.,* sér. 3, 8: 374 (1847)
Uredo aurea Purton [non *U. aurea* Sowerby (1801)], *Bot. descr. Brit. pl.* 3: 725 (1821)
Phragmidium bullatum Westend., *Bull. Soc. R. bot. Belg.*: 11 (1863)
Phragmidium disciflorum (Tode) J. James, *Contr. U.S. Natl. Herb.* 3: 276 (1895)
Coleosporium miniatum Bonord., *Kenntn. Coniomyc. Cryptomyc.*: 20 (1860)
Phragmidium subcorticium G. Winter, *Rabenh. Krypt.-Fl.* 1(1): 228 (1882)
Mis.: *Uredo effusa* sensu Berkeley (Engl. fl.: 381, 1836) and sensu auct.
E: c **S:** c **W:** c **NI:** c **ROI:** c **O:** Isle of Man: !
H: 0, I, II and III on *Rosa canina* agg. and also on *Rosa* cultivars.
D: W&H: 104-105
Very common and widespread. Probably a species complex, in need of critical revision.

potentillae (Pers.) Grev., *Scott. crypt. fl.*: 3 (1828)
Puccinia potentillae Pers., *Syn. meth. fung.*: 107 (1801)
Mis.: *Uredo potentillarum* sensu auct. p.p.
E: ? **S:** ? **W:** ! **ROI:** ?
H: 0, I, II and III on *Potentilla anglica*, 'P. argentea', *P. neumanniana* (= *tabernaemontani*) and various *Potentilla* cultivars.
D: W&H: 101-102
Rarely reported. Most of the records are unsubstantiated with voucher material and many may refer to *Fromeëlla tormentillae* on *Potentilla erecta*, the two species indistinguishable in the uredospore stage *fide* WH.

rosae-pimpinellifoliae Dietel, *Hedwigia* 44: 339 (1905)
Phragmidium rosarum f. *rosae-pimpinellifoliae* Rabenh. [*nom. nud.*], *Fungi europaei* IV: 1671 (1873)
Mis.: *Phragmidium disciflorum* sensu Grove (1913)
E: ! **S:** ! **W:** ! **ROI:** ! **O:** Channel Islands: ! Isle of Man: !
H: I, II and III on *Rosa pimpinellifolia* and cultivars.
D: W&H: 103
Rarely reported but apparently widespread. The host is uncommon and rather restricted in its distribution.

rubi-idaei (DC.) P. Karst., *Bidrag Kännedom Finlands Natur Folk* 31: 52 (1879)
Uredo rubi-idaei Pers., *Observ. mycol.* 2: 24 (1799)
Puccinia rubi-idaei DC., *Fl. Fr.* 5: 54 (1815)
Puccinia gracilis Grev., *Fl. edin.*: 428 (1824)
Aregma gracile (Grev.) Berk., *Engl. fl.* 5(2): 358 (1886)
Phragmidium gracilis (Grev.) Cooke, *Handb. Brit. fung.,* Edn 1: 491 (1871)
Uredo gyrosa Rebent., *Prodr. fl. neomarch.*: 356 (1804)
Lecythea gyrosa (Rebent.) Berk. in Cooke, *Microscopic fungi*: 222 (1865)
Puccinia rubi Schumach. [non *P. rubi* DC (1805)], *Enum. pl.* 2: 235 (1803)
Mis.: *Uredo ruborum* sensu auct. Brit. p.p.

E: o **S:** o **W:** ! **NI:** ! **ROI:** !
H: 0, I, II and III on wild *Rubus idaeus* and some cultivars.
D: W&H: 96-97
Occasional but widespread. Stage I is rarely seen.

sanguisorbae (DC.) J. Schröt. in Cohn, *Krypt.-Fl. Schlesien*: 352 (1889)
Puccinia sanguisorbae DC., *Fl. Fr.* 5: 54 (1815)
Mis.: *Phragmidium acuminatum* sensu auct. Brit.
Mis.: *Aregma acuminatum* sensu auct. Brit.
Mis.: *Lecythea potentillarum* sensu auct. Brit. p.p.
E: c **S:** ! **W:** c **NI:** ! **ROI:** ! **O:** Channel Islands: o
H: 0, I, II and III on *Sanguisorba minor* and rarely *Sanguisorba minor* ssp. *muricata*.
D: W&H: 102-103
Common on the main host in southern England and Wales, but much less frequent elsewhere.

tuberculatum J.B. Müll., *Ber. Deutsch. Bot. Ges.* 3: 391 (1885)
E: ! **S:** ! **W:** ! **NI:** ! **O:** Channel Islands: !
H: 0, I, II and III on *Rosa afzeliana, R. rubiginosa, R. rugosa* and various *Rosa* cultivars. Only rarely on *R. canina*.
D: W&H: 106
Rarely reported but apparently widespread.

violaceum (Schultz) G. Winter, *Hedwigia* 19: 54 (1880)
Puccinia violacea Schultz, *Prodr. Fl. Starg.*: 459 (1806)
Lecythea ruborum Lév., *Ann. Sci. Nat., Bot.,* sér. 3, 8: 374 (1847)
Mis.: *Uredo ruborum* sensu auct. Brit. p.p.
E: c **S:** c **W:** c **NI:** c **ROI:** c **O:** Channel Islands: c Isle of Man: c
H: 0, I, II and III on *Rubus fruticosus* agg. and cultivars such as *Rubus loganobaccus* and *R. laciniatus*.
D: W&H: 98-100
Very common and widespread.

PUCCINIA Pers., *Syn. meth. fung.*: 225 (1801)
Uredinales, Pucciniaceae
Dicaeoma Gray, *Nat. arr. Brit. pl.* 1: 541 (1821)
Rostrupia Lagerh., *J. Bot. (Morot)* 3: 188 (1889)
Type: *Puccinia graminis* Pers.

acetosae Körn., *Hedwigia* 15: 184 (1876)
Uredo acetosae Schumach., *Enum. pl.* 2: 231 (1803)
Puccinia vaginalium Link, in Willd., *Sp. pl.,* Edn 4, 6(2): 65 (1825)
Mis.: *Uredo bifrons* sensu auct. Brit.
E: c **S:** c **W:** c **NI:** c **ROI:** c **O:** Channel Islands: c Isle of Man: c
H: II & III usually on *Rumex acetosa* or *R. acetosa* ssp. *biformis*. Rarely on *R. acetosella*.
D: W&H: 159
Common and widespread.

adoxae R. Hedw., in DC., *Fl. Fr.* 2: 220 (1805)
Dicaeoma adoxae (R. Hedw.) Gray, *Nat. arr. Brit. pl.* 1: 543 (1821)
E: c **S:** o **W:** o
H: III on *Adoxa moschatellina*.
D: W&H: 186
Common in southern England. Occasionally reported elsewhere and apparently widespread.

aegopodii Röhl., *Deutschl. Fl.,* Edn 2, 3: 131 (1813)
Uredo aegopodii Schumach., *Enum. pl.* 2: 233 (1803)
Aecidium aegopodii Rebent., *Prodr. fl. neomarch.*: 222 (1804)
E: o **S:** o **W:** o **ROI:** o **O:** Isle of Man: !
H: III on *Aegopodium podagraria*.
D: W&H: 144
Occasional but widespread. The host is common throughout the British Isles.

albescens Plowr., *Monograph Brit. Ured.*: 153 (1889)
Aecidium albescens Grev., *Fl. edin.*: 444 (1824)
E: ! **S:** ! **W:** !
H: 0, I, II & III on *Adoxa moschatellina*.
D: W&H: 188

Rarely reported but apparently widespread. Rather common in southern England, often accompanied by *Puccinia adoxae* on the same plants.

albulensis Magnus, *Ber. Deutsch. Bot. Ges.* 8: 169 (1890)
 Mis.: *Puccinia veronicarum* DC. sensu Grove (1913) p.p.
S: ! **E:** ?
H: III on *Veronica alpina* in montane habitat.
D: W&H: 172
Rarely reported. The host plant is also rare. Reported by Grove (1913, as *P. veronicarum*) on *V. officinalis*, presumably in error.

allii Rudolphi, *Linnaea.* 4: 392 (1829)
 Uredo alliorum DC., *Fl. Fr.* 5: 82 (1815)
 Puccinia mixta Fuckel, *Fungi Rhenani exsiccati. Fasc. IV:* 377 (1863)
 Uredo porri Sowerby, *Col. fig. Engl. fung., Suppl.*: pl. 411 (1810)
E: ! **S:** ! **W:** ! **NI:** ! **ROI:** ! **O:** Channel Islands: !
H: 0 & I on *Allium schoenoprasum.* II & III on *Allium cepa, A. cyaneum, A. fistulosum, A. sativum, A. schoenoprasum, A. scorodoprasum* and *A. vineale.* Usually reported on *Allium vineale* which is common throughout most of the British Isles.
D: W&H: 217 **I:** Sow 4 (suppl.) 411 (as *Uredo porri*)
Rarely reported, but apparently widespread.

angelicae (Schumach.) Fuckel, *Jahrb. Nassauischen Vereins Naturk.* 23-24: 52 (1870)
 Uredo angelicae Schumach., *Enum. pl.* 2: 233 (1803)
 Trichobasis angelicae (Schumach.) Cooke, *Microscopic fungi,* Edn 4: 224 (1878)
 Uredo bullata Pers., *Syn. meth. fung.*: 221 (1801)
 Puccinia bullata (Pers.) J. Schröt. [non *P. bullata* Link (1816)], *Beitr. Biol. Pflanzen* 3: 74 (1879)
 Puccinia silai Fuckel, *Jahrb. Nassauischen Vereins Naturk.* 23-24: 53 (1870)
E: ! **S:** ! **W:** ! **ROI:** !
H: 0, I, II and III on *Angelica sylvestris, Peucedanum palustre, Selinum carvifolia* and *Silaum silaus.*
D: W&H: 145
Rarely reported but apparently widespread.

annularis (F. Strauss) Röhl., *Deutschl. Fl.* Edn 2, 3(3): 134 (1813)
 Uredo annularis F. Strauss, *Ann. Wetterauischen Ges. Gesammte Naturk.* 2: 106 (1810)
 Puccinia scorodoniae Link, in Willd., *Sp. pl.,* Edn 4, 6(2): 72 (1825)
E: c **S:** c **W:** c **NI:** c **ROI:** c **O:** Channel Islands: ! Isle of Man: !
H: III on *Teucrium scorodonia.*
D: W&H: 177
Common and widespread.

antirrhini Dietel & Holw., *Hedwigia* 36: 298 (1897)
E: o **S:** ! **W:** o **ROI:** ! **O:** Channel Islands: ! Isle of Man: !
H: II & III on cultivated *Antirrhinum glutinosum, A. majus* and recently on *A. molle.*
D: W&H: 173-174
A naturalised alien, widespread but nowhere common; less frequent nowadays due to resistant cultivars.

apii Desm., *Cat. pl. omises botanogr. Belgique:* 25 (1823)
 Trichobasis apii (Desm.) Cooke, *Microscopic fungi,* Edn 4: 224 (1878)
E: ! **S:** ! **W:** !
H: 0, I, II and III on *Apium graveolens* and *Apium* cultivars (celery).
D: W&H: 146
Fide Moore [British Parasitic Fungi p.297 (1959)] this was 'not uncommon' in the period 1875 to 1890 when several severe attacks were noted on celery crops, but there are no recent collections.

arenariae (Schumach.) G. Winter, *Hedwigia* 19: 35 (1880)
 Uredo arenariae Schumach., *Enum. pl.* 2: 232 (1803)
 Puccinia dianthi DC., *Fl. Fr.* 2: 220 (1805)

 Puccinia herniariae Unger, *Einfl. Boden. Verth. Gew.*: 218 (1836)
 Puccinia lychnidearum Link, in Willd., *Sp. pl.,* Edn 4, 6(2): 80 (1825)
 Puccinia moehringiae Fuckel, *Jahrb. Nassauischen Vereins Naturk.* 23-24: 51 (1870)
 Puccinia saginae Fuckel, *Jahrb. Nassauischen Vereins Naturk.* 23-24: 51 (1870)
 Puccinia spergulae DC., *Fl. Fr.* 2: 219 (1805)
 Puccinia stellariae Duby, *Bot. gall.*: 887 (1830)
 Mis.: *Uredo caryophyllacearum* sensu auct.
 Mis.: *Trichobasis lychnidearum* sensu Cooke [Microscopic FungI: 224 (1865)]
E: c **S:** c **W:** c **NI:** c **ROI:** c **O:** Channel Islands: c Isle of Man: c
H: III on various *Caryophyllaceae.* Most frequent on *Silene dioica.* Not uncommon on *Moehringia trinervia, S. media* and cultivated *Dianthus barbatus.* Uncommon or rare on the other hosts. Records on *Claytonia sibirica* (*Portulacaceae*) from Scotland require verification.
D: W&H: 127-130
Common and widespread.

asparagi DC., *Fl. Fr.* 2: 595 (1805)
E: ! **S:** !
H: 0, I, II & III on *Asparagus officinalis.*
D: W&H: 219
Rare, with no recent records. The last collection in herb. K is from Norfolk in 1936.

asperulae-cynanchicae Wurth, *Centralbl. Bakteriol. Parasitenk.* Abt. II, 14: 316 (1905)
ROI: !
H: II & III on *Asperula cynanchica.*
Known from a single collection from Galway (in herb. K) collected in 1947 and only recently determined. The host is not uncommon in grassland on calcareous soil.

behenis G.H. Otth, *Mitt. naturf. Ges. Bern* 1870: 113 (1871) [1870]
 Puccinia silenes J. Schröt., in G. Winter, *Rabenh. Krypt.-Fl.* 1(1): 215 (1882)
E: ! **S:** ! **W:** ! **O:** Channel Islands: ! Isle of Man: !
H: II & III on *Silene dioica, S. latifolia, S. uniflora* and *S. vulgaris* and on cultivated *Cucubalus baccifer.*
D: W&H: 131
Rarely reported but apparently widespread. Possibly mistaken for *P. arenariae* on similar hosts.

betonicae (Alb. & Schwein.) DC., *Fl. Fr.* 5: 57 (1815)
 Puccinia anemones var. *betonicae* Alb. & Schwein., *Consp. fung. lusat.*: 131 (1805)
 Uredo betonicae (Alb. & Schwein.) DC., in Lam., *Encyclop.* 8: 247 (1808)
E: ! **S:** ! **W:** !
H: III on *Stachys officinalis.*
D: W&H: 177-178
Rarely reported. *Fide* WH this is 'frequent' in England but the limited number of records, especially recent ones, would suggest otherwise.

bistortae (F. Strauss) DC., *Fl. Fr.* 5: 61 (1815)
 Uredo polygoni var. *bistortae* F. Strauss, *Ann. Wetterauischen Ges. Gesammte Naturk.* 2: 103 (1810)
 Puccinia polygoni var. *bistortae* (F. Strauss) Röhl., *Deutschl. Fl.,* Edn 2, 3(3): 132 (1813)
 Puccinia angelicae-bistortae Kleb., *Z. Pflanzenkrankh.* 12: 142 (1902)
 Puccinia cari-bistortae Kleb., *Z. Pflanzenkrankh.* 9: 157 (1899)
 Puccinia conopodii-bistortae Kleb., *Z. Pflanzenkrankh.* 6: 331 (1869)
 Puccinia polygoni-vivipari P. Karst., *Not. Sällsk. Fauna et Fl. Fenn. Förh.* 8: 221 (1866)
 Mis.: *Aecidium bunii* DC sensu Plowright (1889)
E: ! **S:** ! **W:** !
H: 0 & I on *Angelica sylvestris* and *Conopodium majus.* II & III on *Polygonum bistorta* and *P. viviparum.*

D: W&H: 160
Rarely reported. The aecidial stage on *Angelica sylvestris* has been found only once in Britain.

brachypodii G.H. Otth, *Mitt. naturf. Ges. Bern* 1861: 82 (1861)
  Epitea baryi Berk. & Broome, *Ann. Mag. Nat. Hist., Ser. 2 [Notices of British Fungi no. 755]* 13: 461 (1854)
  Lecythea baryi (Berk. & Broome) Cooke, *Microscopic fungi*. 222 (1865)
  Puccinia baryi G. Winter, *Rabenh. Krypt.-Fl.* 1(1): 178 (1882)
E: c **S:** c **W:** c **NI:** ! **ROI:** ! **O:** Channel Islands: !
H: II & III on *Brachypodium sylvaticum* and dubiously *B. pinnatum*.
D: W&H: 250
Common and widespread. The type host of *Epitea baryi* is given as *B. pinnatum*.

brachypodii *var.* **arrhenatheri** (Kleb.) Cummins & H.C. Greene, *Mycologia* 58: 709 (1966)
  Puccinia perplexans f. *arrhenatheri* Kleb., *Abh. naturw. Ver. Bremen* 12: 366 (1892)
  Puccinia arrhenatheri (Kleb.) Erikss., *Beitr. Biol. Pflanzen* 8: 1 (1898)
  Puccinia airae Mayor & Cruchet, *Bull. Soc. Vaud. Sci. Nat.* 51: 628 (1917)
  Puccinia deschampsiae Arthur, *Bull. Torrey Bot. Club* 37: 570 (1910)
E: c **S:** c **W:** c **ROI:** !
H: 0 & I on *Berberis vulgaris*. II & III on *Arrhenatherum elatius*, *A. elatius* var. *bulbosum* and *Deschampsia caespitosa*.
D: W&H: 244 (as *Puccinia deschampsiae*)
Common and widespread.

brachypodii *var.* **poae-nemoralis** (G.H. Otth) Cummins & H.C. Greene, *Mycologia* 58: 705 (1966)
  Puccinia poae-nemoralis G.H. Otth, *Mitt. naturf. Ges. Bern* 1870: 113 (1871) [1870]
  Uredo anthoxanthina Bubák, *Ann. Mycol.* 3: 223 (1905)
  Puccinia anthoxanthina Gäum., *Ber. Schweiz. Bot. Ges.* 55: 74 (1949)
  Uredo glyceriae Lind, *Dan. fung.*: 343 (1913)
  Puccinia poae-sudeticae Jørst., *Nytt Mag. Naturividensk* 70: 325 (1932)
E: c **S:** ! **W:** ! **NI:** ! **ROI:** ! **O:** Channel Islands: ! Isle of Man: !
H: II & III on *Anthoxanthum odoratum*, *Glyceria fluitans*, *Poa angustifolia*, *P. annua*, *P. compressa*, *P. nemoralis*, *P. pratensis*, *P. trivialis*, *Puccinella maritima* and *P. rupestris*.
D: W&H: 272 (as *Puccinia poae-nemoralis*)
Common, at least in England, especially on *Poa nemoralis*, and apparently widespread.

bulbocastani Fuckel, *Jahrb. Nassauischen Vereins Naturk.* 23-24: 52 (1870)
  Aecidium bulbocastani Cumino, *Atti Reale Accad. Sci. Torino* 1806: 202 (1806)
  Aecidium bunii DC, *Syn. Pl. Gall.*: 51 (1806)
  Puccinia bunii G. Winter, *Rabenh. Krypt.-Fl.* 1(1): 197 (1882)
E: !
H: 0, I, II and III on *Bunium bulbocastanum*.
D: W&H: 147
Known from Bedfordshire and Hertfordshire (the host is restricted to a few southeastern counties). Old records on *Conopodium majus* are unsubstantiated with voucher material and almost certainly referable to *Puccinia tumida*.

bupleuri Rudolphi, *Linnaea.* 4: 514 (1829)
  Aecidium falcariae β *bupleuri-falcati* DC., *Fl. Fr.* 5: 91 (1815)
E: !
H: 0, I, II and III on *Bupleurum tenuissimum*.
D: W&H: 147
Rarely reported;. Known from East Suffolk, North & South Essex and West Sussex.

buxi DC., *Fl. Fr.* 5: 60 (1815)
E: o **S:** ! **W:** ! **NI:** ! **ROI:** !
H: III on *Buxus sempervirens* and *B. balearica*.
D: W&H: 135-136 **I:** Sow 4 (suppl.) pl.439

Occasional in England. Rarely reported elsewhere but apparently widespread. May be locally common where the host is native, but appears to be less frequent on planted specimens. There is a single collection in herb. K from the Republic of Ireland (Wicklow) in 1845, supposedly on planted *Buxus balaearica*.

calcitrapae DC., *Fl. Fr.* 2: 221 (1805)
  Puccinia bardanae Corda, *Icon. fung.* 4: 17 (1840)
  Puccinia calcitrapae var. *centaureae* (DC.) Cummins, *Mycotaxon* 5(2): 402 (1977)
  Puccinia cardui-pycnocephali Syd. & P. Syd., *Monogr. Ured.* 1: 34 (1904)
  Puccinia carduorum Jacky, *Z. Pflanzenkrankh.* 9: 288 (1899)
  Puccinia carlinae Jacky, *Z. Pflanzenkrankh.* 9: 289 (1899)
  Puccinia carthami Corda, *Icon. fung.* 4: 15 (1840)
  Puccinia centaureae DC., *Fl. Fr.* 5: 59 (1815)
  Puccinia centaureae f. *scabiosae* Hasler, *Centralbl. Bakteriol. Parasitenk.* Abt. II, 21: 511 (1908)
  Puccinia cirsii Lasch, in Rabenhorst, *Fungi europaei* IV: 89 (1859)
  Puccinia compositarum Schltdl., *Fl. berol.* 2: 133 (1824)
  Puccinia inquinans-bardanae Wallr., *Fl. crypt. Germ., Sect. II*: 219 (1833)
  Puccinia syngenesiarum Link, in Willd., *Sp. pl.*, Edn 4, 6(2): 74 (1825)
E: c **S:** c **W:** c **NI:** ! **ROI:** ! **O:** Channel Islands: ! Isle of Man: !
H: 0, I, II & III on species of *Arctium, Carduus, Carlina, Carthamus, Centaurea* and *Cirsium*. Most often reported on *Centaurea nigra*.
D: W&H: 191-194
Very common and widespread.

calthae Link, in Willd., *Sp. pl.*, Edn 4, 6(2): 79 (1825)
  Aecidium calthae Grev., *Fl. edin.*: 446 (1824)
E: ! **S:** ! **W:** ! **ROI:** !
H: 0, I, II and III on *Caltha palustris*.
D: W&H: 123
Supposedly frequent *fide* WH but rarely reported and with few recent records. Distribution shows a bias toward northern and western areas, with the majority of records from Scotland and Wales. The host plant is common throughout the British Isles.

calthicola J. Schröt., *Beitr. Biol. Pflanzen* 3: 61 (1879)
  Puccinia zopfii G. Winter, *Hedwigia* 19: 39 (1880)
  Mis.: *Aecidium calthae* sensu auct. p.p.
E: ! **S:** ! **W:** ! **ROI:** !
H: 0, I, II and III on *Caltha palustris*.
D: W&H: 124
Supposedly frequent *fide* WH but rarely collected or reported, especially in recent years. The majority of records are from eastern England (mainly Norfolk and Suffolk) but this species appears to be more evenly scattered throughout Britain than *P. calthae*.

campanulae Carmich., Berkeley, *Engl. fl.* 5(2): 365 (1836)
  Puccinia campanulae-rotundifoliae Gäum. & Jaag, *Hedwigia* 75: 121 (1935)
E: ! **S:** ! **W:** !
H: III on *Campanula rotundifolia*, *C. rapunculus*, and also possibly on *Campanula 'persicifolia'*. A single verified collection on *Jasione montana*.
D: W&H: 182-183
Rarely reported and lacking any recent verifiable records. Last reported from Scotland in 1968 but unsubstantiated with voucher material. Last reliably collected from England in 1912, Scotland in 1892 and Wales in 1866.

cancellata Sacc. & Roum., *Rev. Mycol. (Toulouse)* 3(9): 26 (1881)
  Uredo cancellata Durieu & Mont., *Expl. Sci. Algérie [Bot. I]* 1: 314 (1846)
E: ! **W:** ! **O:** Channel Islands: !
H: (II?) & III on *Juncus acutus* and *J. maritimus*
D: W&H: 221
Rarely reported but easily overlooked. The majority of the few British collections are from the Channel Islands (Guernsey and

Herm) with a single collection from England (North Devon: Braunton), and from Wales.

caricina [sensu lato] DC., *Fl. Fr.* 5: 60 (1815)
  Trichobasis caricina (DC.) Berk., *Outl. Brit. fungol.*: 332 (1860)
  Uredo caricis Schumach. [*nom. illegit.*, non *U. caricis* Pers. (1801)], *Enum. pl.* 2: 231 (1803)
  Puccinia caricis (Schumach.) Rebent. [*nom. illegit.*, non *P. caricis* Rebent. (1804) (= *P. dioicae*)], *Prodr. fl. neomarch.*: 356 (1804)
  Aecidium grossulariae Schumach., *Enum. pl.* 2: 223 (1803)
  Puccinia grossulariae Lagerh., *Tromso Mus. Aarsh.* 17: 60 (1895)
  Trichobasis parnassiae Cooke, *Microscopic fungi.* 210 (1865)
  Puccinia striola Link, in Willd., *Sp. pl.*, Edn 4, 6(2): 68 (1825)
E: ! **S:** ! **W:** ! **NI:** !
H: 0 & I on species of *Ribes*. II and III on various species of *Carex*.
D: W&H: 232 (as *Puccinia caricina*)
Common and widespread. N.B. Stages 0 & I on *Urtica* are now referred to *P. urticata*, with varieties dependent on the uredo- and teleutospore host species of *Carex*.

caricina *var.* caricina DC., *Fl. Fr.* 5: 60 (1815)
  Puccinia ribesii-pseudocyperi Kleb., *Jahrb. Wiss. Bot.* 34: 391 (1900)
E: !
H: II & III on *Carex pseudocyperus*.
D: W&H: 235
Known from East Norfolk, Herefordshire, Hertfordshire, South Hampshire and Surrey.

caricina *var.* magnusii (Kleb.) D.M. Hend., *Notes Roy. Bot. Gard. Edinburgh* 23: 235 (1961)
  Puccinia magnusii Kleb., *Z. Pflanzenkrankh.* 5: 79 (1895)
E: ! **W:** ! **ROI:** ?
H: II & III on *Carex riparia* (0 & I on *Ribes* spp, not yet reported from Britain).
D: W&H: 235
Rarely reported. Known from Buckinghamshire and Carmarthenshire. Apparently widespread.

caricina *var.* pringsheimiana (Kleb.) D.M. Hend., *Notes Roy. Bot. Gard. Edinburgh* 23: 237 (1961)
  Puccinia pringsheimiana Kleb., *Z. Pflanzenkrankh.* 5: 79 (1895)
E: ! **S:** ! **W:** ! **ROI:** !
H: 0 & I on *Ribes nigrum*, *R. sanguineum* and *R. uva-crispa*. II & III on *Carex acuta* and *C. nigra*.
D: W&H: 236
Rarely reported but apparently widespread. Collections of stage I are mostly from Scotland and Wales. Stages II and III are rarely collected in Britain.

caricina *var.* ribesii-pendulae (Hasler) D.M. Hend., *Notes Roy. Bot. Gard. Edinburgh* 23: 237 (1961)
  Puccinia ribesii-pendulae Hasler, *Ber. Schweiz. Bot. Ges.* 55: 15 (1945)
E: c **S:** ! **W:** c **NI:** ! **ROI:** ! **O:** Channel Islands: !
H: II & III on *Carex pendula* (0 & I on *Ribes* spp, not yet reported in Britain).
D: W&H: 237
Apparently widespread. The host is common throughout much of the British Isles.

caricina *var.* ribis-nigri-lasiocarpae (Hasler) D.M. Hend., *Notes Roy. Bot. Gard. Edinburgh* 23: 237 (1961)
  Puccinia ribis-nigri-lasiocarpae Hasler, *Ann. Mycol.* 28: 350 (1930)
S: !
H: II & III on *Carex lasiocarpa*. (0 & I on species of *Ribes* not reported in Britain).
D: W&H: 236
Known from Easterness (Loch Pityoulish). Reported from Wales but on an unlikely host (*Carex pendula*) and unsubstantiated with voucher material.

caricina *var.* ribis-nigri-paniculatae (Kleb.) D.M. Hend., *Notes Roy. Bot. Gard. Edinburgh* 23(3): 237 (1961)
  Puccinia ribis-nigri-paniculatae Kleb., *Jahrb. Wiss. Bot.* 34: 393 (1900)
E: ! **S:** ! **W:** !
H: II & III on *Carex paniculata*.
D: W&H: 236
Rarely reported. Reported 'on leaves of *Phragmites australis*' from England (Cumberland) in 1999 but the record is unsubstantiated with voucher material.

chaerophylli Purton, *Bot. descr. Brit. pl.* 3: 303 (1821)
E: ! **S:** o **W:** ! **ROI:** ! **O:** Isle of Man: !
H: 0, I, II and II on *Anthricus sylvestris*, *Chaerophyllum temulentum* and (most frequently) *Myrrhis odorata*. Reported on *Chaerophyllum aureum* in Scotland, but the host determination is in doubt.
D: W&H: 148
Occasional but widespread in Scotland, rarely reported elsewhere.

chrysosplenii Grev., *Fl. edin.*: 429 (1824)
E: o **S:** o **W:** ! **NI:** ! **ROI:** ! **O:** Channel Islands: !
H: III on *Chrysosplenium alternifolium* and *C. oppositifolium*.
D: W&H: 137
Occasional but widespread in Scotland and in northern and western England. Rarely reported elsewhere and very rarely in eastern, southern and southeastern England.

cicutae Lasch, in Rabenhorst, *Klotzschii Herbarium Vivum Mycologicum*: 787 (1845)
E: !
H: 0, I, II and III on *Cicuta virosa*.
D: W&H: 149
Known from East Norfolk. The host is virtually restricted to this area in Britain.

circaeae Pers., *Tent. disp. meth. fung.*: 40 (1797)
E: c **S:** c **W:** c **NI:** c **ROI:** c
H: III on *Circaea lutetiana*. Rarely reported on *C. intermedia* and dubiously on *C. alpina*.
D: W&H: 141-142
Very common and widespread on *Circaea lutetiana* (usually accompanied by *Pucciniastrum circaeae*).

cladii Ellis & Tracy, *Bull. Torrey Bot. Club* 22: 61 (1895)
E: !
H: II on *Cladium mariscus*.
D: W&H: 229
Known from Cambridgeshire, Norfolk and Suffolk, following the distribution of the host.

clintonii Peck, *Rep. (Annual) New York State Mus. Nat. Hist.* 28: 61 (1876)
  Puccinia clintonii var. *sylvaticae* Savile, *Canad. J. Bot.* 45: 1097 (1967)
  Aecidium pedicularis Link, in Willd., *Sp. pl.*, Edn 4, 6(2): 47 (1825)
S: ! **NI:** ! **ROI:** !
H: III on *Pedicularis palustris* and *P. sylvatica*
D: W&H: 175
Rarely reported. Regarded by some authorities as an endemic, in which case the name *P. clintonii* var. *sylvaticae* would apply to the British species. Also, Savile [*Canad. J. Bot.* 45: 1097 (1967)] noted that the rusts on *Pedicularis palustris* and *P. sylvatica* may be different species.

cnici H. Mart., *Prodr. fl. mosq.*, Edn 2: 226 (1817)
  Puccinia cirsii-eriophori Jacky, *Z. Pflanzenkrankh.* 9: 275 (1899)
  Puccinia cirsii-lanceolati J. Schröt., *Krypt.-Fl. Schlesien*: 317 (1887)
E: c **S:** c **W:** c **NI:** c **ROI:** c **O:** Channel Islands: ! Isle of Man: !
H: 0, I, II & III commonly on *Cirsium vulgare* and rarely on *C. eriophorum*. Reported on *Cirsium arvense*, *C. lanceolatum* and *Carduus crispus* but these hosts are unverified.
D: W&H: 196
Common and widespread.

cnici-oleracei Desm., *Cat. pl. omises botanogr. Belgique*: 24 (1823)
  Puccinia andersonii Berk. & Broome, *Ann. Mag. Nat. Hist*, *Ser. 4 [Notices of British Fungi no. 1464]* 15: 35 (1875)
  Puccinia asteris Duby, *Bot. gall.*: 888 (1830)
  Puccinia asteris β *chrysanthemi-leucanthemi* C. Massal., *Boll. Soc. Bot. Ital.* 1900: 258 (1900)
  Puccinia cardui Plowr., *Monograph Brit. Ured.*: 216 (1889)
  Puccinia cirsiorum var. *cirsii-palustris* Desm., *Cat. pl. omises botanogr. Belgique*: 25 (1823)
  Puccinia cirsii-palustris (Desm.) M. Wilson, *Trans. Brit. Mycol. Soc.* 24: 244 (1940)
  Puccinia lemonnieriana Maire, *Bull. Soc. Mycol. France* 16: 65 (1900)
  Puccinia leucanthemi Pass., *Hedwigia* 13: 47 (1874)
  Puccinia millefolii Fuckel, *Jahrb. Nassauischen Vereins Naturk.* 23-24: 55 (1870)
  Puccinia tripolii Wallr., *Fl. crypt. Germ., Sect. II*: 223 (1833)
E: c **S:** c **W:** c **O:** Channel Islands: !
H: III on *Achillea millefolium*, *A. ptarmica*, *Aster tripolium*, *Carduus acanthoides*, *Chrysanthemum segetum*, *Cirsium heterophyllum*, *C. palustre*, *C. vulgare* and *Leucanthemum vulgare*.
D: W&H: 197
Common and widespread, especially so on *Cirsium palustre*, less so on the other hosts. Recent records referred here in Herb. K on *Carduus*, and on *Cirsium vulgare*, not otherwise recorded as hosts in Britain.

commutata Syd. & P. Syd., *Monogr. Ured.* 1: 201 (1904)
E: ! **S:** !
H: 0, I & III on *Valeriana officinalis* and *V. officinalis* ssp. *sambucifolia*.
D: W&H: 189
Known from England (Mid-Lancashire) and Scotland (Mid-Ebudes and West Perthshire).

conii (F. Strauss) Fuckel, *Jahrb. Nassauischen Vereins Naturk.* 23-24: 33 (1870)
  Uredo conii F. Strauss, *Ann. Wetterauischen Ges. Gesammte Naturk.* 2: 96 (1810)
  Trichobasis conii Cooke, *Microscopic fungi*: 225 (1865)
  Puccinia conii Lagerh., *Tromso Mus. Aarsh.* 17: 54 (1895)
  Mis.: *Puccinia bullaria* sensu auct.
E: c **S:** ! **ROI:** ! **O:** Channel Islands: !
H: II & III on *Conium maculatum*.
D: W&H: 150
Common, at least in southern England. Rarely reported elsewhere but apparently widespread. The host is common (increasingly so alongside roads) and widespread.

coronata Corda, *Icon. fung.* 1: 6 (1837)
  Puccinia calamagrostis Syd., *Ured. Exsic.* Fasc. 13-15: no. 662 (1892)
  Puccinia coronata f.sp. *calamagrostis* Erikss., *Ber. Deutsch. Bot. Ges.* 12: 321 (1894)
  Puccinia coronata var. *calamagrostis* W.P. Fraser & Ledingham, *Sci. Agric.* 13: 316 (1933)
  Puccinia coronata f.sp. *alopecuri* Erikss., *Ber. Deutsch. Bot. Ges.* 12: 321 (1894)
  Puccinia coronata f.sp. *avenae* Erikss., *Ber. Deutsch. Bot. Ges.* 12: 321 (1894)
  Puccinia coronata var. *avenae* W.P. Fraser & Ledingham, *Sci. Agric.* 13: 322 (1933)
  Puccinia coronata var. *bromi* W.P. Fraser & Ledingham, *Sci. Agric.* 13: 322 (1933)
  Puccinia coronata f.sp. *festucae* Erikss., *Ber. Deutsch. Bot. Ges.* 12: 321 (1894)
  Puccinia coronifera Kleb., *Z. Pflanzenkrankh.* 4: 135 (1894)
  Puccinia lolii Nielsen, *Ugesk. Land.*, Edn 4, 9(1): 549 (1875)
  Puccinia lolii f.sp. *holci* Syd. & P. Syd., *Monogr. Ured.* 1: 705 (1904)
  Puccinia coronata f.sp. *lolii* Erikss., *Ber. Deutsch. Bot. Ges.* 12: 321 (1894)
  Aecidium crassum Pers., *Icon. descr. fung.*: pl. 27 (1800)
  Aecidium frangulae Schumach., *Enum. pl.* 2: 225 (1803)

  Aecidium rhamni Pers., in Gmelin, *Syst. nat.*: 1472 (1792) p.p.
E: c **S:** c **W:** c **NI:** c **ROI:** c **O:** Channel Islands: c Isle of Man: c
H: 0 & I on *Frangula alnus* and *Rhamnus carthartica*. II & III on a wide range of grasses. For a full list of host species see Henderson (2000).
D: W&H: 251-255
Very common and widespread. Several of the various 'forma speciales' and 'varieties' of this species commonly cited in British literature appear not to have been validly published.

crepidicola Syd. & P. Syd., *Oesterr. Bot. Z.* 51: 17 (1901)
  Mis.: *Puccinia crepidis* sensu Grove [Brit. Rust Fungi p.156 (1913)]
E: ! **S:** ! **W:** ! **NI:** ! **ROI:** ! **O:** Channel Islands: ! Isle of Man: !
H: II & III on *Crepis biennis*, *C. capillaris* and *C. vesicaria* ssp. *taraxifolia*.
D: W&H: 200
Rarely reported but apparently widespread.

cyani Pass., in Rabenhorst, *Fungi europaei* IV: 1767 (1874)
  Uredo cyani DC., *Syn. Pl. Gall.*: 47 (1806)
E: ! **S:** !
H: 0, I, II and III on *Centaurea cyanus*.
D: W&H: 210
Very rarely reported, with no recent records. The host is almost extinct in natural habitat in Britain.

difformis J. Kunze, *Mykol. Hefte* 1: 71 (1817)
  Aecidium galii β *ambiguum* Alb. & Schwein., *Consp. fung. lusat.*: 116 (1805)
  Puccinia galii Schwein., *Schriften Naturf. Ges. Leipzig* 1: 73 (1822)
E: ! **S:** ! **W:** ! **NI:** !
H: 0, I, II & III on *Galium aparine*.
D: W&H: 184
Apparently widespread but rarely reported despite the abundance of the host.

dioicae Magnus, *Amtl. Ber. 50 Versammt. Deutsch Naturf. Ärtze [München]*: 199 (1877)
  Puccinia caricis Rebent., *Prod. fl. neomarch.*: 356 (1804)
E: ! **S:** ! **W:** ! **NI:** ! **ROI:** !
H: 0 & I on *Cirsium dissectum* and *C. palustre*. II & III on *Carex dioica*.
D: W&H: 241
Rarely reported but apparently widespread.

dioicae *var.* **arenariicola** (Plowr.) D.M. Hend., *Notes Roy. Bot. Gard. Edinburgh* 23: 243 (1961)
  Puccinia arenariicola Plowr., *J. Linn. Soc., Bot.* 24: 90 (1888)
E: !
H: 0 & I on *Centaurea nigra*. II & III on *Carex arenaria*.
D: W&H: 242
Known from East Norfolk (Hemsby, near Yarmouth), where it was last collected in 1890.

dioicae *var.* **extensicola** (Plowr.) D.M. Hend., *Notes Roy. Bot. Gard. Edinburgh* 23(3): 343 (1961)
  Puccinia extensicola Plowr., *Monograph Brit. Ured.*: 181 (1889)
E: ! **W:** !
H: 0 & I on *Aster tripolium*. II & III on *Carex extensa*.
D: W&H: 242
Known from England (North Somerset and West Cornwall) and Wales (Anglesey).

dioicae *var.* **schoeleriana** (Plowr. & Magnus) D.M. Hend., *Notes Roy. Bot. Gard. Edinburgh* 23(3): 244 (1961)
  Puccinia schoeleriana Plowr. & Magnus, *Quart. J. Microscop. Sci.* 25: 170 (1885)
  Aecidium compositarum var. *jacobeae* (Grev.) Cooke, *Handb. Brit. fung.*, Edn 1: 542 (1871)
  Aecidium jacobeae Grev., *Fl. edin.*: 445 (1824)
E: ! **S:** ! **W:** ! **NI:** !
H: 0 & I on *Senecio jacobaea*. II & III on *Carex arenaria*.
D: W&H: 242

Rarely reported. British records are mostly from the eastern coast of England with single records from Wales and Northern Ireland.

dioicae *var.* silvatica (J. Schröt.) D.M. Hend., *Notes Roy. Bot. Gard. Edinburgh* 23(3): 245 (1961)
   Puccinia silvatica J. Schröt., *Beitr. Biol. Pflanzen* 3: 68 (1879)
S: !
H: II & III on *Carex capillaris* (and possibly stage I on *Taraxacum officinale*).
D: W&H: 243
Known from Mid-Perthshire and West Sutherland.

distincta McAlpine, *Agric. Gaz. New South Wales* 6: 4 (1896)
E: c **S:** ! **W:** c **NI:** ! **ROI:** !
H: 0, I & III on wild *Bellis perennis* and *Bellis* cultivars.
An introduction, now common and widespread in England and Wales, and spreading.

elymi Westend., *Bull. Acad. Roy. Sci. Belgique* 18(2): 408 (1851)
   Rostrupia elymi (Westend.) Lagerh., *J. Bot. (Morot)* 3: 188 (1889)
   Uredo ammophilae Syd., *Bot. Not.* 1900: 42 (1900)
   Puccinia ammophilae A.L. Guyot, *Rev. Pathol. Vég. Entomol. Agric. France* 19: 36 (1932)
   Rostrupia ammophilae M. Wilson [*nom. nud.*], *Trans. Bot. Soc. Edinburgh* 33: iv (1940)
E: ! **S:** ! **W:** !
H: II & III on *Ammophila arenaria*, x *Calammophila baltica* and *Leymus arenarius*.
D: W&H: 256
Rarely reported, but apparently widespread.

epilobii DC., *Fl. Fr.* 5: 61 (1815)
E: ! **S:** ! **W:** ! **NI:** ? **ROI:** !
H: III on *Epilobium anagallidifolium*, *E. hirsutum*, *E. montanum*, *E. obscurum* and *E. palustre*.
D: W&H: 142
Rarely reported, but apparently widespread.

eriophori Thüm., *Bull. Soc. Imp. Naturalistes Moscou* 55: 208 (1880)
   Puccinia confinis Syd. & P. Syd., *Ann. Mycol.* 18: 154 (1920)
S: !
H: 0 & I on *Solidago virgaurea*. II & III on *Trichophorum caespitosum* ssp. *germanicum* and a single confirmed collection on *Eleocharis quinqueflora*.
D: W&H: 230
Rarely reported.

eutremae Lindr., *Acta Soc. Fauna Fl. Fenn.* 22(3): 9 (1902)
   Puccinia cochleariae Lindr., *Acta Soc. Fauna Fl. Fenn.* 22(3): 10 (1902)
E: ! **S:** !
H: III on *Cochlearia danica*.
D: W&H: 125
A single collection from Scotland (Wester Ross: Plockton, Loch Carron) in 1957; also reported from England in Henderson (2004).

fergussonii Berk. & Broome, *Ann. Mag. Nat. Hist, Ser. 4 [Notices of British Fungi no. 1465]* 15: 35 (1875)
   Mis.: *Puccinia asarina* sensu Cooke [HBF1: 504 (1871)]
E: ! **S:** ! **W:** ! **ROI:** ?
H: III on *Viola palustris*.
D: W&H: 126
Rarely reported. Known from England (East Suffolk), Scotland (Easterness, Mid-Ebudes and Westerness) and Wales (Caernarvonshire).

festucae Plowr., *Grevillea* 21: 109 (1893)
   Uredo festucae DC., *Fl. Fr.* 5: 82 (1815)
   Aecidium periclymeni Schumach., *Enum. pl.* 2: 225 (1803)
E: ! **S:** ! **W:** ! **ROI:** !
H: 0 & I on *Lonicera periclymenum*. II & III on *Festuca arenaria*, *F. glauca*, *F. longifolia*, *F. ovina* and *F. rubra*.
D: W&H: 258

Rarely reported but apparently widespread.

galii-cruciatae Duby, *Bot. gall.*: 888 (1830)
   Puccinia celakovskyana Bubák, *Sitzungsber. Königl. Böhm. Ges. Wiss. Math.-Naturwiss. Cl.* 28: 11 (1898)
E: ! **W:** !
H: 0, I, II & III on *Cruciata laevipes*.
D: W&H: 183
Rarely reported.

galii-verni Ces., *Bot. Zeitung (Berlin)* 4: 879 (1846)
   Mis.: *Puccinia valantiae* sensu auct.
E: ! **S:** o **W:** ! **NI:** ! **O:** Isle of Man: !
H: III on *Cruciata laevipes*, *Galium saxatile*, *G. uliginosum* and *G. verum*.
D: W&H: 186
Occasional on *Galium saxatile* in Scotland. Rarely reported elsewhere or on the other hosts.

gentianae (F. Strauss) Röhl., *Deutschl. Fl.*, Edn 2, 3(3): 131 (1813)
   Uredo gentianae F. Strauss, *Ann. Wetterauischen Ges. Gesammte Naturk.* 2: 102 (1811)
E: ! **W:** !
H: II & III on cultivated *Gentiana acaulis* and *G. verna*.
D: W&H: 170
An introduction. Known only on cultivated plants. Last reported in 1936 from England (Cheshire) and rarely before then.

glechomatis DC., in Lam., *Encyclop.* 8: 245 (1808)
   Puccinia glechomae DC., *Fl. Fr.* 5: 56 (1815)
E: c **S:** c **W:** c **NI:** c **ROI:** c **O:** Channel Islands: c Isle of Man: c
H: III on *Glechoma hederacea*.
D: W&H: 178
Very common and widespread.

glomerata Grev., *Fl. edin.*: 433 (1824)
   Puccinia expansa Link, in Willd., *Sp. pl.*, Edn 4, 6(2): 75 (1825)
   Puccinia senecionis Lib., *Pl. Crypt. Ardienn.*: 92 (1830)
   Mis.: *Trichobasis senecionis* sensu auct. p.p.
E: ! **S:** o **W:** ! **NI:** ! **ROI:** ! **O:** Isle of Man: !
H: III on *Senecio aquatica* and *S. jacobaea*.
D: W&H: 202
Occasional in Scotland. Rarely reported elsewhere but apparently widespread.

graminis Pers., *Tent. disp. meth. fung.*: 39 (1797)
   Aecidium berberidis Pers., in Gmelin, *Syst. nat.*: 1473 (1792)
   Uredo frumenti Sowerby, *Col. fig. Engl. fung.* 2: pl. 140 (1798)
   Uredo linearis Pers., *Syn. meth. fung.*: 216 (1801)
   Mis.: *Trichobasis linearis* sensu auct.
E: c **S:** c **W:** c **NI:** ! **ROI:** c
H: 0 & I on *Berberis vulgaris*, *Mahonia aquifolium* and *M. bealei*. II & III on *Agrostis canina*, *A. capillaris*, *Arrhenatherum elatius*, *Avena fatua*, *A. sativa*, *A. strigosa*, *Bromus sterilis*, *Elytrigia repens*, *Hordeum vulgare*, *Lolium perenne*, *Secale cereale* and *Trisetum flavescens*.
D: W&H: 259-263 **I:** Sow 2 pl.140 (as *Uredo frumenti*)
Widespread, but records will also include *P. graminis* ssp. *graminicola*. The aecidial stage is now rarely seen due to the deliberate eradication of *Berberis* hedges to prevent outbreaks of the rust in cereal crops.

graminis *ssp.* graminicola Z. Urb., *Česká Mykol.* 21: 14 (1967)
   Puccinia anthoxanthi Fuckel, *Jahrb. Nassauischen Vereins Naturk.* 27-28: 15 (1873)
   Puccinia dactylidis Gäum., *Ber. Schweiz. Bot. Ges.* 55: 79 (1945)
   Puccinia phlei-pratensis Erikss. & Henning, *Z. Pflanzenkrankh.* 4: 140 (1894)
   Puccinia graminis var. *phlei-pratensis* (Erikss. & Henning) Stakman & Piem., *J. Agric. Res.* 10: 433 (1917)
E: ? **S:** ? **W:** ! **NI:** ? **ROI:** ? **O:** ?

H: II & III on *Anthoxanthum odoratum*, *Briza media*, *Dactylis glomerata*, *Deschampsia caespitosa*, *Phleum pratense*, *Poa annua*, *P. pratensis*, *P. trivialis* and *Sesleria caerulea*.
Probably common, but many reports will be as *Puccinia graminis* s.l.

heraclei Grev., *Scott. crypt. fl.* 1: 42 (1823)
  Trichobasis heraclei (Grev.) Berk., *Outl. Brit. fungol.*: 332 (1860)
E: ! **S:** o **W:** ! **O:** Channel Islands: !
H: 0, I, II and III on *Heracleum sphondylium*.
D: W&H: 151
Apparently widespread but with a distinctly northern and western distribution in England. Occasional in Scotland. Rarely reported elsewhere. Records on *Taraxacum officinale* from Yorkshire refer to *P. hieracii*.

hieracii (Röhl.) H. Mart., *Prodr. fl. mosq.*, Edn 2: 226 (1817)
  Puccinia flosculosorum var. *hieracii* Röhl., *Deutschl. Fl.*, Edn 2, 3(3): 131 (1813)
  Uredo cichorii DC., *Fl. Fr.* 5: 74 (1815)
  Puccinia cichorii Bellynck, *Fl. Crypt. Flandres* 2: 65 (1867)
  Puccinia endiviae Pass., *Hedwigia* 12: 114 (1873)
  Uredo hieracii Schumach., *Enum. pl.* 2: 222 (1803)
  Puccinia jaceae G.H. Otth, *Mitt. naturf. Ges. Bern*: 173 (1866) [1865]
  Puccinia leontodontis Jacky, *Z. Pflanzenkrankh.* 9: 339 (1899)
  Puccinia picridis Hazsl., *Math. Term. Közl. Mag. Tudom. Akad.* 14: 152 (1877)
  Puccinia taraxaci Plowr., *Monograph Brit. Ured.*: 186 (1889)
  Puccinia tinctoriae Magnus, *Abh. Naturhist. Ges. Nürnberg.* 13: 37 (1900)
  Puccinia tinctoriicola Magnus, *Oesterr. Bot. Z.* 52: 491 (1902)
E: c **S:** c **W:** c **NI:** c **ROI:** c **O:** Channel Islands: c Isle of Man: !
H: 0, I, II & III on *Centarurea nigra*, *Cichorium endivia*, *C. intybus*, *Hieracium* spp, *Hypochaeris maculata*, *Leontodon* spp, *Picris hieracioides*, *Serratula tinctoria* and *Taraxacum* spp.
D: W&H: 203-206
Common and widespread.

hieracii var. hypochaeridis (Oudem.) Jørst., *K. norske Vidensk. Selsk. Skr.* 38: 27 (1936) [1935]
  Puccinia hypochaeridis Oudem., *Nederl. Kruidkundig. Archief.* 1: 175 (1874)
  Uredo hyposeridis Schumach., *Enum. pl.* 2: 233 (1803)
  Mis.: *Trichobasis cichoracearum* sensu auct.
E: ! **S:** ! **W:** ! **NI:** ! **ROI:** ! **O:** Channel Islands: !
H: 0, I, II & III on *Hypochaeris glabra* and *H. radicata*.
D: W&H: 206
Rarely reported but apparently widespread.

hieracii var. piloselloidarum (Probst) Jørst., *K. norske Vidensk. Selsk. Skr.* 38: 27 (1936) [1935]
  Puccinia piloselloidearum Probst, *Centralbl. Bakteriol. Parasitenk.* Abt. II, 22: 712 (1909)
E: ! **S:** ! **W:** ! **NI:** ! **ROI:** ! **O:** Channel Islands: !
H: 0, I, II & III on *Pilosella aurantiaca* and *P. officinarum*.
D: W&H: 206
Widespread but rarely reported.

hordei G.H. Otth, *Mitt. naturf. Ges. Bern* 1870: 114 (1871) [1870]
  Puccinia anomala Rostr., *Flora* 61: 92 (1878)
  Puccinia fragosoi Bubák, *Hedwigia* 57: 2 (1916)
  Puccinia holcina Erikss., *Ann. Sci. Nat., Bot.*, sér. 8, 9: 274 (1899)
  Puccinia rubigo-vera f. sp. *holcina* (Erikss.) Mains, *Pap. Michigan Acad. Sci.* 17: 381 (1933)
  Puccinia recondita f.sp. *holcina* (Erikss.) D.M. Hend., *Notes Roy. Bot. Gard. Edinburgh* 23: 504 (1961)
  Puccinia loliina Syd., *Ann. mycol.* 19: 247 (1921)
  Puccinia triseti Erikss., *Ann. Sci. Nat., Bot.*, sér. 8, 9: 277 (1899)
  Puccinia rubigo-vera f. sp. *triseti* (Erikss.) Mains, *Pap. Michigan Acad. Sci.* 17: 381 (1933)
  Puccinia recondita f.sp. *triseti* (Erikss.) D.M. Hend., *Notes Roy. Bot. Gard. Edinburgh* 23: 504 (1961)

Puccinia schismi Bubák, *Ann. Naturhist. Mus. Wien* 28: 193 (1914)
  Puccinia straminis var. *simplex* Körn., in Thuemen, *Herbarium Mycologicum Oeçonomicum*: 101 (1873)
  Puccinia simplex (Körn.) Erikss. & Henning [non Peck (1881)], *Meddeland. Kungl. Landtbr.-Akad. Experimentalfält.* 27: 175 (1894)
  Puccinia vulpiae-myuri Mayor & Vienn.-Bourg., *Rev. Mycol. (Paris)* 15(2): 103 (1950)
  Puccinia vulpiana Guyot, *Uredineana* 2: 53 (1947)
E: ! **S:** ! **W:** ! **NI:** ! **ROI:** ! **O:** Channel Islands: ! Isle of Man: !
H: 0 & I on *Ornithogalum pyrenaicum*. II & III on *Bromus racemosus*, *Holcus lanatus*, *H. mollis*, *Hordeum distichon*, *H. murinum*, *H. vulgare*, *Lagurus ovatus*, *Lolium multiflorum*, *Trisetum flavescens*, *Vulpia bromoides* and *V. myuros*.
D: W&H: 264, W&H: 285 (as *Puccinia recondita* f. sp. *holcina*), W&H: 290 (as *Puccinia schismi*)
0 & I are rare in Britain (as is the host plant). II & III are widely recorded on many of the common grass hosts but are poorly represented in herbaria.

horiana Henn., *Beibl. Hedwigia* 40(2): 25 (1901)
E: ! **S:** ! **W:** ! **ROI:** !
H: III on *Dendranthema* (= *Chrysanthemum*) cultivars.
D: W&H: 368
An introduction. Rarely reported but widespread and perhaps increasing.

hydrocotyles (Mont.) Cooke, *Grevillea* 9: 14 (1880)
  Uredo hydrocotyles Mont., in Gay, *Fl. chil.* 8: 50 (1852)
  Trichobasis hydrocotyles Cooke, *J. Bot.* 2: 343 (1864)
  Caeoma hydrocotyles Link, in Willd., *Sp. pl.*, Edn 4, 6(2): 22 (1825)
E: ! **S:** ! **W:** ! **ROI:** ?
H: II & III on *Hydrocotyle vulgaris* in bogs, pond margins or swampy areas in woodland.
D: W&H: 151
Several collections made in the late nineteenth and early part of the twentieth century, from East Norfolk and South Essex. Apparently now rarely seen, the last collection being from North Devon (Braunton) in 1960.

hysterium (F. Strauss) Röhl., *Deutschl. Fl.*, Edn 2, 3(3): 131 (1813)
  Uredo hysterium F. Strauss, *Ann. Wetterauischen Ges. Gesammte Naturk.* 11: 102 (1810)
  Puccinia sparsa Cooke, *Engl. fl.* 5(2): 498 (1836)
  Aecidium tragopogi Pers., *Syn. meth. fung.*: 211 (1801)
  Puccinia tragopogi G. Winter, *Hedwigia* 19: 44 (1880)
  Puccinia tragopogonis Corda, *Icon. fung.* 5: 50 (1842)
  Aecidium tragopogonis Cooke, *Handb. Brit. fung.*, Edn 1: 537 (1871)
E: ! **S:** ! **W:** ! **ROI:** !
H: 0, I & III on *Tragopogon pratensis* and *T. pratensis* ssp. *minor*.
D: W&H: 207-208
Apparently widespread and supposedly frequent *fide* WH but the few records do not support this view.

impatientis J.A. Schubad, *Fl. Geg. Dresd. Krypt.* 2: 252 (1823)
  Trichobasis impatientis (J.A. Schubad) Cooke, *Microscopic fungi*: 225 (1865)
  Aecidium argentatum Schultz, *Prodr. Fl. Starg.*: 454 (1806)
  Puccinia argentata (Schultz) G. Winter, *Hedwigia* 19: 38 (1880)
  Puccinia nolitangere Corda, *Icon. fung.* 4: 16 (1840)
E: !
H: 0 & I on *Adoxa moschatellina*. II & III on *Impatiens capensis*.
D: W&H: 134 (as *Puccinia argentata*)
British collections are confined to a small area of south-eastern England (Surrey: Albury and Shere, at various sites along the Wey Navigation Canal).

iridis Wallr., *Deutschl. Krypt. Flora* 1: 23 (1844)
  Trichobasis iridis (Wallr.) Cooke, *Grevillea* 4(30): 67 (1875)
  Uredo iridis DC., in Lam., *Encyclop.* 8: 224 (1808)

Puccinia truncata Berk. & Broome, *Ann. Mag. Nat. Hist, Ser. 2 [Notices of British Fungi no. 754]* 13: 461 (1854)
E: o **S:** ! **W:** ! **NI:** ! **ROI:** ! **O:** Channel Islands: o
H: II & III on native *Iris foetidissima* and reported (dubiously) on *I. pseudacorus.* Also known on *I. germanica* and *I. missourensis* and various unnamed cultivars.
D: W&H: 226-227
Occasional to frequent in England. Rarely reported elsewhere but apparently widespread. Records and collections show a distinct bias toward western and south-western areas.

kusanoi Dietel, *Bot. Jahrb. Syst.* 27: 568 (1899)
E: !
H: II & III on *Arundinaria fastuosa.*
An alien, native to Japan, China and Taiwan. In Britain, first recorded in 1961 from West Sussex (Wakehurst Place) but now also known from Surrey and West Cornwall.

lagenophorae Cooke, *Grevillea* 13: 6 (1884)
Puccinia erechtitis McAlpine, *Proc. Linn. Soc. N. S. W., Ser. 2* 10: 34 (1895)
Puccinia terrieriana Mayor, *Ber. Schweiz. Bot. Ges.* 72: 266 (1962)
E: c **S:** c **W:** c **NI:** c **ROI:** c **O:** Channel Islands: ! Isle of Man: !
H: I & III common on *Senecio squalidus* and *S. vulgaris.* Also known on *Bellis perennis, Calendula officinalis, Senecio jacobaea* and cultivated *Emilia* spp and *Pericallis hybrida.*
D: W&H: 213
Common and widespread. An alien, originally native to Australia and first reported from East Kent (Dungeness) in 1966 from where it has spread throughout the British Isles.

lapsanae Fuckel, *Jahrb. Nassauischen Vereins Naturk.* 15: 13 (1860)
Trichobasis lapsanae (Fuckel) Cooke, *Microscopic fungi* 224 (1865)
Aecidium lapsanae Schultz, *Prodr. Fl. Starg.*: 454 (1806)
Aecidium compositarum var. *lapsanae* (Schultz) Cooke, *Handb. Brit. fung.,* Edn 1: 543 (1871)
Mis.: *Trichobasis cichoracearum* sensu auct.
E: c **S:** c **W:** c **NI:** ! **ROI:** c **O:** Channel Islands: c Isle of Man: c
H: 0, I, II & III on *Lapsana communis.*
D: W&H: 208
Very common and widespread.

libanotidis Lindr., *Meddeland. Stockholms Högskolas Bot. Inst.* 4(9): 2 (1901)
E: !
H: 0, I, II and III on *Seseli libanotis.*
D: W&H: 153
Known from Cambridgeshire. The host plant is rare in Britain and otherwise known only from West Sussex.

liliacearum Duby, *Bot. gall.*: 891 (1830)
E: ! **S:** !
H: I & III on *Ornithogalum pyrenaicum* and *O. angustifolium* (=*O. umbellatum*).
D: W&H: 220
Scarce. Known from England (East Norfolk, East Suffolk and Surrey) and Scotland (Midlothian).

ljulinica Hinkova & Koeva, *Rev. Roumaine Biol., Sér. Bot.* 11: 109 (1966)
E: !
H: III on *Smyrnium perfoliatum.*
Known from Surrey (Kew Gardens).

longicornis Har. & Pat., *Bull. Soc. Mycol. France* 7: 143 (1891)
E: !
H: II & III on the bamboos *Arundinaria japonica* and *Sasa veitchii.*
An introduced species, native in Asia. Initially reported from West Sussex (Wakehurst Place) and now also known from various locations in Surrey.

longissima J. Schröt., *Beitr. Biol. Pflanzen* 3: 70 (1879)
E: ! **S:** ! **ROI:** ?
H: II & III on *Koeleria macrantha* (=*K. cristata*).

D: W&H: 265-266
Known from England (East Gloucestershire) and Scotland (Angus and South Aberdeenshire). Reported from the Republic of Ireland but unsubstantiated with voucher material.

luzulae Liberta, *Pl. Crypt. Ardienn.*: no.94 (1830)
Caeoma oblongata Link, *Mag. Neuesten Entdeck. Gesammten Naturk. Ges. Naturf. Freunde Berlin* 7: 27 (1815)
Uredo oblongata (Link) Grev., *Fl. edin.*: 437 (1824)
Trichobasis oblongata (Link) Berk., *Outl. Brit. fungol.*: 332 (1860)
Puccinia oblongata G. Winter, *Hedwigia* 19(3): 43 (1880)
E: ! **S:** ! **W:** !
H: II & III on *Luzula pilosa.*
D: W&H: 222
Rarely reported but apparently widespread. Restricted to *Luzula pilosa*, records on other species of *Luzula* being misidentified *Puccinia obscura.*

maculosa (F. Strauss) Röhl., *Deutschl. Fl.,* Edn 2, 3(3): 131 (1813)
Uredo maculosa F. Strauss, *Ann. Wetterauischen Ges. Gesammte Naturk.* 2: 101 (1810)
Puccinia chondrillae Corda, *Icon. fung.* 4: 15 (1840)
Aecidium prenanthis Pers., *Syn. meth. fung.*: 208 (1801)
Uredo prenanthis Schumach., *Enum. pl.* 2: 232 (1803)
Puccinia prenanthis J. Kunze, *Fl. Geg. Dresd. Krypt.* 2: 250 (1823)
Puccinia prenanthis Lindr., *Acta Soc. Fauna Fl. Fenn.* 20(9): 6 (1901)
E: ! **S:** ! **W:** ! **ROI:** !
H: I, II & III on *Mycelis muralis.*
D: W&H: 209
Rarely reported, although the host is not uncommon and the aecidia are conspicuous.

magnusiana Körn., *Hedwigia* 15: 179 (1876)
Puccinia arundinacea β *epicaula* Wallr., *Fl. crypt. Germ., Sect. II.* 225 (1833)
Aecidium ranunculi Schwein., *Schriften Naturf. Ges. Leipzig* 1: 67 (1822)
Aecidium ranunculacearum DC., *Fl. Fr.* 6: 97 (1815) p.p.
Aecidium ranunculacearum var. *linguae* Grove, *British Rust Fungi.* 387 (1913)
E: o **S:** o **W:** o **ROI:** o **O:** Channel Islands: ! Isle of Man: !
H: 0 & I on *Ranunculus bulbosus, R. flammula, R. lingua* and *R. repens.* II & III on *Phragmites australis.*
D: W&H: 266-268
Occasional but widespread and may be locally common.

major (Dietel) Dietel, *Mitth. Thüring. Bot. Vereins* 6: 46 (1894)
Puccinia lapsanae var. *major* Dietel, *Hedwigia* 27: 303 (1888)
Aecidium compositarum var. *prenanthis* Cooke, *Handb. Brit. fung.,* Edn 1: 542 (1871)
E: ! **S:** ! **W:** ! **ROI:** !
H: 0, I, II & III on *Crepis paludosa.*
D: W&H: 210
Rarely reported but apparently widespread. The host is localised and uncommon in the British Isles.

malvacearum Mont., in Gay, *Fl. chil.* 8: 43 (1852)
E: c **S:** c **W:** c **NI:** c **ROI:** c **O:** Channel Islands: ! Isle of Man: !
H: III on various species of *Malvaceae.* Known on *Alcea rosea, Lavatera arborea, L. cretica, Malva moschata, M. neglecta, M. pusilla* and *M. sylvestris.*
D: W&H: 132-134
Introduced from Chile and now common and widespread, especially on hollyhocks (*Alcea rosea* cultivars) and *Malva sylvestris.*

mariana Sacc., *Nuovo Giorn. Bot. Ital.* 22: 29 (1915)
E: !
H: II & III on *Silybum marianum.*
D: Myc11(4): 151
A single collection from West Norfolk (Great Hockham) in 1981.

menthae Pers., *Syn. meth. fung.*: 227 (1801)
Puccinia clinopodii DC., *Fl. Fr.* 5: 57 (1815)

Trichobasis clinopodii (DC.) Cooke, *Microscopic fungi*: 224 (1865)
Uredo labiatarum DC, *Fl. Fr.* 5: 72 (1815)
Trichobasis labiatarum Lév., in Orbigny, *Dict. univ. hist. nat.* 1: 19 (1841)
Aecidium menthae DC., *Fl. Fr.* 5: 95 (1815)
E: c **S:** c **W:** c **NI:** c **ROI:** c **O:** Channel Islands: ! Isle of Man: !
H: 0, I, II & III on numerous species and hybrids of *Mentha* and on *Calamintha ascendens, Clinopodium vulgare, Origanum vulgare* and cultivated *Satureja hortensis*.
D: W&H: 179
Very common and widespread, especially on *Mentha*, less frequent on the other host genera.

microsora Körn., in Fuckel, *Jahrb. Nassauischen Vereins Naturk.* 29-30: 14 (1875)
S: !
H: II & III on *Carex vesicaria*.
D: W&H: 245
A single collection from Easterness (Loch Insh) in 1959.

moliniae Tul., *Ann. Sci. Nat., Bot.*, sér. 4, 2: 141 (1854)
Puccinia brunellarum-moliniae Cruchet, *Centralbl. Bakteriol. Parasitenk.* Abt. II, 13: 96 (1904)
E: ? **S:** !
H: 0 & I on *Prunella vulgaris*. II & III on *Molinia caerulea*.
D: W&H: 268
Rarely reported. Confirmed British collections are only from widely scattered locations in Scotland, but there is also material named thus by Grove and by Rea from England (North Wiltshire: Savernake Forest and North Somerset: Exmoor) in herb. K.

nemoralis Juel, *Öfvers. Kongl. Vetensk.-Akad. Förh.* 51: 506 (1894)
Puccinia aecidii-melampyri Liro, *Acta Soc. Fauna Fl. Fenn.* 29: 55 (1907)
S: ! **W:** !
H: 0 & I on *Melampyrum pratense*. II & III on *Molinia caerulea*.
D: TBMS71(2): 325-326
Rarely reported.

nitida (F. Strauss) Röhl., *Deutschl. Fl.*, Edn 2, 3(3): 130 (1813)
Uredo nitida F. Strauss, *Ann. Wetterauischen Ges. Gesammte Naturk.* 2: 100 (1810)
Puccinia aethusae H. Mart., *Prodr. fl. mosq.*, Edn 2: 225 (1817)
Trichobasis cynapii Cooke, *Microscopic fungi*: 224 (1865)
Uredo petroselini DC., *Fl. Fr.* 2: 597 (1805)
Puccinia petroselini (DC.) Lindr., *Acta Soc. Fauna Fl. Fenn.* 12(1): 84 (1902)
Trichobasis petroselini Berk., *Outl. Brit. fungol.*: 332 (1860)
E: ! **W:** ! **NI:** ! **ROI:** ? **O:** Channel Islands: !
H: 0, I, II & III on *Aethusa cynapium* and rarely on *Petroselinum crispum*.
D: W&H: 153
Rarely reported but apparently widespread.

obscura J. Schröt., *Just's Bot. Jahresber.* 5: 162 (1879)
Puccinia luzulae-maximae Dietel, *Ann. Mycol.* 17: 57 (1919)
E: c **S:** c **W:** c **NI:** ! **ROI:** c **O:** Channel Islands: ! Isle of Man: c
H: 0 & I on *Bellis perennis*. II & III on *Luzula* spp.
D: W&H: 223
Common and widespread. The aecidial stage on *Bellis perennis* may be confused with *Puccinia distincta* or *P. lagenophorae* on the same host.

opizii Bubák, *Centralbl. Bakteriol. Parasitenk.* Abt. II, 9: 925 (1902)
E: ! **W:** ! **ROI:** !
H: 0 & I on *Lactuca sativa* and rarely on *L. virosa*. II & III on *Carex appropinquata, C. divulsa, C. muricata, C. paniculata* and *C. muricata*.
D: W&H: 244
Rarely reported. Virtually all British records of the aecidial stage are from Norfolk, but it has not been reported for many years.

oxalidis Dietel & Ellis, *Hedwigia* 34: 291 (1895)

E: o **S:** ! **W:** ! **ROI:** ! **O:** Channel Islands: ! Isle of Man: !
H: II & III on cultivated species of *Oxalis*, but never on native *O. corniculata*.
Occasional but widespread. An alien, now frequent in southern England, especially in the London area, and apparently spreading elsewhere.

oxyriae Fuckel, *Jahrb. Nassauischen Vereins Naturk.* 29-30: 14 (1875)
S: ! **W:** !
H: II & III on *Oxyria digyna* in montane habitat.
D: W&H: 163
Known from Scotland (Angus: near Brechin) and Wales (unlocalised).

paludosa Plowr., *Monograph Brit. Ured.*: 174 (1889)
Puccinia caricina var. *paludosa* (Plowr.) D.M. Hend., *Notes Roy. Bot. Gard. Edinburgh* 23: 236 (1961)
E: ! **S:** ! **W:** !
H: 0 & I on *Pedicularis palustris*. II & III on *Carex 'bigelowii', C. elata, C. elata* var. *aurea, C. nigra* and *C. panicea*.
D: W&H: 235 (as *Puccinia caricina* var. *paludosa*)
Rarely reported but apparently widespread. *Fide* WH it is frequent on *Carex nigra* and *C. panicea* but much less so on the other hosts.

pazschkei Dietel, *Hedwigia* 30: 103 (1891)
E: ! **S:** !
H: III on cultivated species of *Saxifraga*.
D: W&H: 139
An alien, rarely reported.

pazschkei *var.* **jueliana** (Dietel) Savile, *Canad. J. Bot.* 32: 411 (1954)
Puccinia jueliana Dietel, *Hedwigia* 36: 298 (1897)
S: !
H: III on *Saxifraga aizoides* and *S. oppositifolia*.
D: W&H: 139
Rarely reported. British collections are only from montane habitat in Scotland.

pelargonii-zonalis Doidge, *Bothalia* 2: 98 (1926)
E: ! **O:** Channel Islands: !
H: II & III on cultivated *Pelargonium zonale* and hybrids.
An established alien from South Africa, though rarely reported.

phragmitis (Schumach.) Körn., *Hedwigia* 15: 179 (1876)
Uredo phragmitis Schumach., *Enum. pl.* 2: 231 (1803)
Aecidium rubellum Pers., in Gmelin, *Syst. nat.*: 1473 (1792)
Aecidium rumicis Sowerby, *Col. fig. Engl. fung., Suppl.*: pl. 405 (1809)
Puccinia traillii Plowr., *Monograph Brit. Ured.*: 176 (1889)
Mis.: *Puccinia arundinacea* sensu auct.
E: c **S:** ! **W:** o **NI:** ! **ROI:** ! **O:** Isle of Man: !
H: 0 & I on various species of *Rumex*. II & III on *Phragmites australis*.
D: W&H: 269-271 **I:** Sow 4 (suppl.) 405 (as *Aecidium rumicis*)
Common in England. Widespread but less frequently reported elsewhere.

physospermi Pass., in Rabenhorst, *Fungi europaei* IV: 1969 (1875)
E: !
H: 0 & III on *Physospermum cornubiense*.
D: W&H: 154
Known from Buckinghamshire (Burnham Beeches); limited by the localised distribution of the host.

pimpinellae (F. Strauss) Röhl., *Deutschl. Fl.*, Edn 2, 3(3): 131 (1813)
Uredo pimpinellae F. Strauss, *Ann. Wetterauischen Ges. Gesammte Naturk.* 2: 102 (1810)
Trichobasis pimpinellae (F. Strauss) Cooke, *Microscopic fungi*: 224 (1865)
Lecythea poterii Lév., *Ann. Sci. Nat., Bot.*, sér. 3, 8: 374 (1847)
Aecidium poterii Cooke, *J. Bot.* 2: 39 (1864)

Aecidium bunii var. *poterii* (Cooke) Cooke, *Handb. Brit. fung.*,
Edn 1: 540 (1871)
Mis.: *Aecidium bunii* sensu auct. p.p.
E: ! S: ? W: ?
H: 0, I, II and II on *Pimpinella major* and *P. saxifraga*.
D: W&H: 155
Rarely reported. Noted as 'frequent' by WH but there are few
recent records and many earlier records are unsubstantiated
with voucher material. Old reports, supposedly on '*Heracleum
sphondylium*', are incorrect.

poarum Nielsen, *Bot. Tidsskr.* 2: 34 (1877)
Aecidium compositarum var. *tussilaginis* (Pers.) Cooke,
Handb. Brit. fung., Edn 1: 542 (1871)
Aecidium petasitidis Gray, *Nat. arr. Brit. pl.* 1: 536 (1821)
Aecidium tussilaginis Pers., in Gmelin, *Syst. nat.*: 1473 (1792)
E: c S: c W: c NI: c ROI: c O: Channel islands: ! Isle of Man: !
H: 0 & I on *Tussilago farfara*. II & III on *Poa annua, P. pratensis*
and *P. trivialis*.
D: W&H: 274
Common and widespread.

polemonii Dietel & Holw., *Bot. Gaz.* 18: 255 (1893)
E: !
H: III on *Polemonium caeruleum*.
D: W&H: 170
Known from Derbyshire but not reported in recent years. The
host is also rare and localised in Britain. A record purportedly
on *Angelica sylvestris* from Sheffield in 1958 is incorrect.

polygoni-amphibii Pers., *Syn. meth. fung.*: 227 (1801)
Aecidium sanguinolentum Lindr., *Bot. Not.* 1900: 241 (1900)
E: c S: o W: o NI: ! ROI: ! O: Channel Islands: !
H: II & III on *Persicaria amphibia*.
D: W&H: 164
Frequent and widespread. Occurs only on the terrestrial form of
P. amphibia at the margins of lakes or ponds.

polygoni-amphibii var. convolvuli Arthur, *Manual of the
Rusts in the United States & Canada*: 233 (1934)
Uredo betae β *convolvuli* Alb. & Schwein., *Consp. fung.
lusat.*: 127 (1805)
Puccinia polygoni Alb. & Schwein., *Consp. fung. lusat.*: 132
(1805)
Puccinia polygoni-convolvuli DC., in Lam., *Encyclop.* 8: 251
(1808)
E: ! S: ! W: ! ROI: !
H: 0 & I on *Geranium dissectum*. II & III on *Fallopia
(Polygonum) convolvulus*.
D: W&H: 165
Scarce but apparently widespread.

porri G. Winter, *Rabenh. Krypt.-Fl.* 1(1): 200 (1882)
E: ! S: ! W: !
H: II & III on cultivated *Allium porrum*.
Rarely reported but apparently widespread.

pratensis A. Blytt, *Forh. Vidensk.-Selsk. Kristiania* 4(6): 52
(1896)
E: !
H: II & III on *Helictotrichon pratense*.
D: W&H: 275
Known from North East Yorkshire (Thornton-le-Dale, Dalby
Forest and Allerston Forest) in 1952, 1953 [see Naturalist
(Hull) 1953: 94 (1953)] and last reported in 1959.

primulae Duby, *Bot. gall.*: 891 (1830)
Trichobasis primulae (Duby) Cooke, *Grevillea* 4(30): 67
(1875)
Aecidium primulae DC., *Fl. Fr.* 5: 90 (1815)
Uredo primulae DC., *Fl. Fr.* 5: 68 (1815)
E: o S: o W: ! NI: ! ROI: ! O: Channel Islands: c Isle of Man: !
H: I, II & III on *Primula vulgaris* [and possibly *Primula x veris*].
D: W&H: 167-168
Occasional but widespread. 'Frequent' *fide* WH but uncommonly
reported. Mainly western and south-western in the British
Isles, with numerous collections from the Channel Islands.
Rare in southern and southeastern England.

prostii Duby, *Bot. gall.*: 891 (1830)
E: ! S: !
H: 0 & III on *Tulipa sylvestris* and *T. australis*.
D: W&H: 220
Rarely reported, and most probably an old introduction.

pulverulenta Grev., *Fl. edin.*: 432 (1824)
Aecidium epilobii DC., *Fl. Fr.* 5: 238 (1815)
Puccinia epilobii-tetragoni G. Winter, *Rabenh. Krypt.-Fl.* 1(1):
214 (1882)
E: c S: c W: o NI: ! ROI: ! O: Channel Islands: !
H: 0, I, II and III on *Epilobium adnatum, E. ciliatum, E.
hirsutum, E. montanum, E. parviflorum* and *E. tetragonum*. A
single collection on *Chamaenerion angustifolium*.
D: W&H: 143
Common and widespread, especially so on *E. hirsutum*.

punctata Link, *Mag. Neuesten Entdeck. Gesammten Naturk.
Ges. Naturf. Freunde Berlin* 7: 30 (1815)
Puccinia asperulae-odoratae Wurth, *Centralbl. Bakteriol.
Parasitenk.* Abt. II, 14: 314 (1905)
Trichobasis galii Lév., in Orbigny, *Dict. univ. hist. nat.* 1: 19
(1841)
Puccinia galiorum Link, in Willd., *Sp. pl.*, Edn 4, 6(2): 76
(1825)
E: c S: c W: c NI: ! ROI: ! O: Channel Islands: !
H: 0, I, II & III on *Cruciata laevipes, Galium aparine, G.
mollugo, G. odoratum, G. palustre, G. saxatile, G. sterneri, G.
uliginosum* and *G. verum*.
D: W&H: 185
Common and widespread.

punctiformis (F. Strauss) Röhl., *Deutschl. Fl.*, Edn 2, 3(3): 132
(1813)
Uredo punctiformis F. Strauss, *Ann. Wetterauischen Ges.
Gesammte Naturk.* 2: 103 (1810)
Puccinia obtegens Fuckel, *Jahrb. Nassauischen Vereins
Naturk.* 23-24: 54 (1870)
Uredo suaveolens Pers., *Observ. mycol.* 2: 24 (1799)
Trichobasis suaveolens (Pers.) Lév., in Orbigny, *Dict. univ.
hist. nat.* 1: 19 (1841)
Puccinia suaveolens (Pers.) Rostr., *Förh. Skand. Naturf. Möte*
1: 339 (1874)
E: c S: c W: c NI: c ROI: c O: Channel Islands: c Isle of Man: c
H: 0, I, II & III on *Cirsium arvense*.
D: W&H: 210-212
Very common and widespread.

pygmaea Erikss., *Botanisches Centralblatt* 64: 381 (1895)
E: ! S: ! W: ! O: Channel Islands: !
H: II & III on *Calamagrostis epigejos*.
D: W&H: 276
Rarely reported but apparently widespread.

pygmaea var. ammophilina (Mains) Cummins & H.C. Greene,
Mycologia 58: 714 (1966)
Puccinia ammophilina Mains, *Mycologia* 48: 604 (1956)
Uredo ammophilina Kleb., *Kryptogamenflora der Mark
Brandenburg* 5a: 882 (1914)
E: ! S: ! W: ! ROI: ! O: Channel Islands: ! Isle of Man: !
H: II & III on *Ammophila arenaria*.
D: W&H: 276
Rarely reported. The majority of records are from Scotland.

recondita Desm., *Bull. Soc. Bot. France* 4: 798 (1857)
Puccinia agropyri Ellis & Everh., *J. Mycol.* 7: 131 (1892)
Puccinia agropyrina Erikss., *Ann. Sci. Nat., Bot.*, sér. 8, 9:
273 (1899)
Puccinia recondita f.sp. *agropyrina* (Erikss.) D.M. Hend.,
Notes Roy. Bot. Gard. Edinburgh 23: 504 (1961)
Puccinia agrostidis Plowr. (nom. nud.), *Gard. Chron.* 3: 139
(1890)
Puccinia rubigo-vera f. sp. *agrostidis* (Plowr.) Mains (nom.
illegit.), *Pap. Michigan Acad. Sci.* 17: 352 (1933)
Puccinia agrostidis Oudem., *Rév. Champ.* 1: 528 (1892)
Puccinia recondita f.sp. *agrostidis* (Oudem.) D.M. Hend.,
Notes Roy. Bot. Gard. Edinburgh 23: 504 (1961)

Aecidium anchusae Sacc., *Syll. fung.* 11: 204 (1895)
Aecidium aquilegiae Pers., *Icon. pict. sp. fung.* 4: 58 (1803)
Aecidium asperifolii Pers., *Observ. mycol.* 1: 97 (1799)
Puccinia borealis Juel, *Öfvers. Kongl. Vetensk.-Akad. Förh.* 51: 411 (1894)
Puccinia recondita f.sp. *borealis* (Juel) D.M. Hend., *Notes Roy. Bot. Gard. Edinburgh* 23: 504 (1961)
Puccinia bromina Erikss., *Ann. Sci. Nat., Bot., sér.* 8, 9: 271 (1899)
Puccinia recondita f.sp. *bromina* (Erikss.) D.M. Hend., *Notes Roy. Bot. Gard. Edinburgh* 23: 504 (1961)
Puccinia cerinthes-agropyrina f.sp. *echii-agropyri* Gäum. & Terrier, *Ber. Schweiz. Bot. Ges.* 57: 244 (1947)
Puccinia recondita f.sp. *echii-agropyrina* (Gäum. & Terrier) D.M. Hend., *Notes Roy. Bot. Gard. Edinburgh* 23: 504 (1961)
Puccinia dispersa Erikss. & Henning, *Ber. Deutsch. Bot. Ges.* 12: 315 (1894)
Puccinia dispersa f.sp. *tritici* Erikss. & Henning, *Z. Pflanzenkrankh.* 4: 259 (1894)
Aecidium hellebori E. Fisch., *Ured. der Schweiz.* 526 (1904)
Puccinia perplexans Plowr., *Quart. J. Microscop. Sci.* 25: 164 (1885)
Puccinia rubigo-vera f. sp. *perplexans* (Plowr.) Mains, *Pap. Michigan Acad. Sci.* 17: 357 (1933)
Puccinia recondita f.sp. *perplexans* (Plowr.) D.M. Hend., *Notes Roy. Bot. Gard. Edinburgh* 23: 504 (1961)
Puccinia persistens Plowr., *Monograph Brit. Ured.*: 180 (1889)
Puccinia recondita f.sp. *persistens* (Plowr.) D.M. Hend., *Notes Roy. Bot. Gard. Edinburgh* 23: 504 (1961)
Aecidium ranunculacearum var. *aquilegiae* Cooke, *Handb. Brit. fung.*, Edn 1: 539 (1871)
Aecidium ranunculacearum var. *thalictri* Cooke, *Handb. Brit. fung.*, Edn 1: 540 (1871)
Puccinia recondita f.sp. *triticina* (Erikss. & Henn.) D.M. Hend., *Notes Roy. Bot. Gard. Edinburgh* 23: 504 (1961)
Uredo rubigo-vera DC., *Fl. Fr.* 5: 83 (1815)
Trichobasis rubigo-vera (DC.) Lév., in Orbigny, *Dict. univ. hist. nat.* 1: 19 (1841)
Puccinia rubigo-vera (DC.) G. Winter, *Rabenh. Krypt.-Fl.* 1(1): 217 (1882)
Puccinia secalina Grove, *British Rust Fungi*: 261 (1913)
Puccinia symphyti-bromorum F. Muell., *Beih. Bot. Centralbl.* 10: 201 (1901)
Aecidium thalictri Grev., *Scott. crypt. fl.* 1: 4 (1823)
Puccinia triticina Erikss. & Henning, *Ann. Sci. Nat., Bot., sér.* 8, 9: 270 (1899)
Mis.: *Puccinia straminis* sensu auct. p.p.
E: c **S:** c **W:** c **NI:** c **ROI:** c **O:** Channel Islands: c Isle of Man: c
H: 0 & I on Boraginaceae (*Anchusa*, *Echium* & *Lycopsis* spp), & Ranunculaceae (*Aquilegia*, *Helleborus*, *Ranunculus* & *Thalictrum* spp); II & III on various grasses (*Agrostis*, *Alopecurus*, *Anisantha*, *Bromopsis*, *Bromus* & *Secale* spp). See Henderson (2000) for a full list of host species.
D: W&H: 278-290
The arrangement and synonymy here follows Cummins 'Rust Fungi of Cerals, Grasses and Bamboos' (1971). Common and widespread. The aecidial stage of this group on *Clematis* (*Aecidium clematidis* DC., *Fl. Fr.*, Edn 3 **2**: 243, 1805) listed by Cooke (HBF1: 539 (1871), as *A. ranunculacearum* var. *clematidis*) has not been substantiated from Britain.

ribis DC., *Fl. Fr.* 2: 221 (1805)
S: !
H: III on *Ribes spicatum* [not on *R. rubrum* as stated in many of the records].
D: W&H: 140
Known from Easterness, Morayshire and Perthshire.

rugulosa Tranzschel, *Sitzungsber. St. Petersburg Naturf. Ges.* 1892: 1 (1892)
Puccinia auloderma Lindr., *Meddeland. Stockholms Högskolas Bot. Inst.* 4(9): 2 (1901)
Puccinia carniolica W. Voss, *Oesterr. Bot. Z.* 35: 420 (1885)

Puccinia peucedani-parisiensis (DC.) Lindr., *Acta Soc. Fauna Fl. Fenn.* 22: 79 (1902)
Puccinia umbelliferarum var. *peucedani-parisiensis* DC., *Fl. Fr.* 5: 58 (1815)
E: !
H: 0, I, II and III on *Peucedanum officinale*.
D: W&H: 156
Known from East Kent and North Essex, where it was rediscovered in 1989 [Spooner, Cecidology 5: 47–48 (1990)]. The host is rare and localised.

saniculae Grev., *Fl. edin.*: 431 (1824)
Aecidium saniculae Cooke, *J. Bot.* 2: 39 (1864)
E: o **S:** ! **W:** ! **NI:** o **ROI:** o
H: 0, I, II & III on *Sanicula europaea*.
D: W&H: 157
Occasional but widespread, and may be locally common.

saxifragae Schltdl., *Fl. berol.* 2: 134 (1824)
Puccinia heucherae var. *saxifragae* (Schltdl.) Savile, *Canad. J. Bot.* 32: 408 (1954)
Uredo saxifragarum DC., *Fl. Fr.* 5: 57 (1815)
E: ! **S:** ! **W:** ! **ROI:** !
H: III, on *Saxifraga granulata*, *S. spathulata*, *S. stellaris* and *S. umbrosa*, and recently on *Heuchera* sp.
D: W&H: 138
Rarely reported but apparently widespread.

schroeteri Pass., *Nuovo Giorn. Bot. Ital.* 7: 255 (1875)
E: !
H: III on cultivated *Narcissus jonquilla*, *N. majalis* and *N. pseudonarcissus*.
D: W&H: 224
Very rarely reported and regarded as extinct until a collection from West Gloucestershire in May 2003. A collection from Malvern reported in Bull. Brit. Myc. Soc. 1 (1): 5 (1967) has been redetermined as *Puccinia liliacearum* on leaves of *Ornithogalum umbellatum* fide Henderson & Bennell (1979: 487).

scirpi DC., *Fl. Fr.* 2: 223 (1805)
Aecidium nymphoidis DC., *Fl. Fr.* 2: 597 (1805)
E: ! **W:** ?
H: 0 & I on *Nymphoides peltata*. II & III on *Schoenoplectus lacustris*.
D: W&H: 245-246
Known from England (Herefordshire, Huntingdonshire, Norfolk, Oxfordshire, South Wiltshire and Suffolk) and dubiously reported from Wales (Breconshire).

scorzonerae Jacky, *Composit. Puccin.*: 54 (1899)
Uredo scorzonerae Schumach., *Enum. pl.* 2: 229 (1803)
Puccinia scorzonericola Tranzschel, *Ann. Mycol.* 2: 161 (1904)
E: ! **W:** !
H: On leaves of *Scorzonera humilis*.
Known from England (Dorset) and Wales (Glamorganshire). The form on *Scorzonera humilis* has beeen distinguished by some authors as *Puccinia scorzonericola* Tranzschel.

septentrionalis Juel, *Öfvers. Kongl. Vetensk.-Akad. Förh.* 53(6): 383 (1895)
S: !
H: I on *Thalictrum alpinum*. II & III on *Polygonum viviparum*.
D: W&H: 166
'Frequent' fide W&H, but actually rather rarely reported and only from Scotland.

sessilis J. Schröt., *Abh. Schles. Ges. Vaterl. Cult., Abth. Naturwiss. (Naturwiss.-Med. Abth.)* 1869-72: 19 (1870)
Aecidium allii Grev., *Fl. edin.*: 447 (1824)
Puccinia allii-phalaridis Kleb., *Jahrb. Wiss. Bot.* 34: 399 (1899)
Aecidium ari Desm., *Cat. pl. omises botanogr. Belgique*. 26 (1823)
Puccinia ari-phalaridis Kleb., *Jahrb. Wiss. Bot.* 34: 399 (1899)
Aecidium convallariae Schumach., *Enum. pl.* 2: 224 (1803)
Puccinia digraphidis Soppitt, *J. Bot.* 28: 213 (1890)

Puccinia festucina Syd. & P. Syd., *Ann. Mycol.* 10: 217 (1912)
Puccinia linearis Roberge, *Ann. Sci. Nat., Bot.,* sér. 4, 4: 125 (1855)
Aecidium orchidearum Desm., *Cat. pl. omises botanogr. Belgique:* 26 (1823)
Puccinia orchidearum-phalaridis Kleb., *Z. Pflanzenkrankh.* 7: 33 (1897)
Puccinia paradis Plowr., *Gard. Chron.* 13: 137 (1892)
Puccinia phalaridis Plowr., *J. Linn. Soc., Bot.* 24: 88 (1888)
Puccinia winteriana Magnus, *Hedwigia* 33: 78 (1894)
E: c **S:** c **W:** c **NI:** ! **ROI:** ! **O:** Isle of Man: !
H: 0 & I on *Allium ursinum, Arum italicum* ssp. *neglectum A. maculatum, Convallaria majalis, Dactylorhiza* spp, *Gymnadenia conopsea, Listera ovata* and *Paris quadrifolia.* II & III on *Phalaris arundinacea.*
D: W&H: 291
Common and widespread on *Phalaris arundinacea.* Stage I is common (at least in southern England) on *Arum maculatum,* frequent on *Allium ursinum* and rare on the other hosts. On *Convallaria majalis* it has been collected only from Cumberland (Bowness) and was last reported in 1890. On *Paris quadrifolia* it is known only from East Norfolk (Runhall, near Wymondham) collected in 1952.

smyrnii Biv., *Stirp. Rar. Sic.* 4: 30 (1816)
Aecidium bunii var. *smyrnii-olusatri* DC., *Fl. Fr.* 5: 96 (1815)
Puccinia smyrnii-olusatri Lindr., *Acta Soc. Fauna Fl. Fenn.* 22: 9 (1902)
E: c **S:** ! **W:** c **NI:** ! **ROI:** ! **O:** Channel Islands: c Isle of Man: c
H: 0, I, II & III on *Smyrnium olusatrum.*
D: W&H: 157-158
Common and widespread in many coastal areas, but with a bias toward the west and south-west, and rather rarely noted inland. A recent record on *Ulex europaeus* from Suffolk (Southwold) is an error.

sorghi Schwein., *Synopsis Fung. Amer. bor.* 4: 295 (1832)
Puccinia maydis Berenger, *Atti Ruin. sc. ital. Milano* 6: 475 (1844)
E: ! **W:** !
H: II & III on *Zea mays.* (0 & I on *Oxalis corniculata* by innoculation).
D: W&H: 293
Rarely recorded or collected but may spread with the increasing cultivation of maize.

striiformis Westend., *Bull. Roy. Acad. Belg.* 21: 235 (1854)
Trichobasis glumarum Lév., in Orbigny, *Dict. univ. hist. nat.* 1: 19 (1841)
Puccinia glumarum Erikss. & Henning, *Z. Pflanzenkrankh.* 4: 197 (1894)
Puccinia tritici Oerst., *Om. Sygd. hos planterne:* 95 (1863)
Mis.: *Puccinia rubigo-vera* sensu auct. p.p.
Mis.: *Puccinia straminis* sensu auct. p.p.
E: c **S:** c **W:** ! **NI:** ! **ROI:** !
H: II & III on *Elymus caninus, E. repens, Brachypodium sylvaticum, Anisantha sterilis, Bromus* sp., *Hordeum marinum, H. murinum, H. vulgare, Leymus arenarius, Secale cereale* and *Triticum aestivum.*
D: W&H: 294
Apparently common and widespread but rarely reported.

striiformis *var.* **dactylidis** Manners, *Trans. Brit. Mycol. Soc.* 43: 65 (1960)
Mis.: *Puccinia glumarum* sensu auct. Brit.
E: ! **S:** ! **W:** !
H: II & III on *Dactylis glomerata.*
D: W&H: 297
Rarely reported, though the host is common in Britain.

tanaceti DC., *Fl. Fr.* 2: 222 (1805)
Uredo absinthii DC., in Lam., *Encyclop.* 8: 245 (1808)
Puccinia absinthii DC., *Fl. Fr.* 5: 56 (1815)
Uredo artemisiae Rabenh., *Rabenh. Krypt.-Fl.* 1: 111 (1844)
Trichobasis artemisiae (Rabenh.) Berk., *Outl. Brit. fungol.:* 332 (1860)
Puccinia artemisiae (Rabenh.) Fuckel, *Fungi Rhenani exsiccati. Fasc. IV:* 350 (1863)
Puccinia artemisiella Syd. & P. Syd., *Monogr. Ured.* 1: 14 (1902)
Puccinia chrysanthemi Roze, *Bull. Soc. Mycol. France.* 16: 92 (1900)
Puccinia discoidearum Link, in Willd., *Sp. pl.,* Edn 4, 6(2): 73 (1825)
Puccinia pyrethri Rabenh., *Klotzschii Herbarium Vivum Mycologicum (Editio nova), ser. 2:* 1990 (1880)
E: ! **W:** !
H: II & III on *Artemisia* spp and *Seriphidium maritimum* (= *Artemisia maritima*) and on cultivated *Dendranthema grandiflora, Tanacetum (Chrysanthemum) coccineum* and *T. vulgare.*
D: W&H: 212
Rarely reported. Most frequent on *Artemisia* spp, but easily overlooked.

thesii Duby, *Bot. gall.:* 889 (1830)
Aecidium thesii Desv., *Desv. J. Bot.* 2: 311 (1809)
E: !
H: 0, I, II and III on *Thesium humifusum.*
D: W&H: 144
Rarely reported. British records are only from scattered locations in southern England, within the limited range of the host plant, but the rust is often abundant where it does occur.

thymi (Fuckel) P. Karst., *Bidrag Kännedom Finlands Natur Folk* 9: 44 (1884)
Aecidium thymi Fuckel, *Fungi Rhenani exsiccati (Suppl.). Fasc. VII:* 2113 (1868)
Puccinia caulincola W.G. Schneid., *Jahresber. Schles. Ges. Vaterl. Cult.* 48: 120 (1870)
Puccinia ruebsaameni Magnus, *Ber. Deutsch. Bot. Ges.* 22: 344 (1904)
Puccinia schneideri J. Schröt., *Herb. Schles. Pilze:* no.448 (1879)
E: ! **S:** ! **W:** ! **O:** Isle of Man: !
H: III on *Thymus polytrichus* (= *T. drucei*), *T. pulegioides* and *Origanum vulgare.*
D: W&H: 180
Rarely reported. Causes a distinctive 'witches-broom' gall on affected plants.

tumida Grev., *Fl. edin.:* 430 (1824)
Trichobasis umbellatarum Lév., in Orbigny, *Dict. univ. hist. nat.* 1: 19 (1841)
Puccinia umbelliferarum DC., *Fl. Fr.* 5: 58 (1815)
E: ! **S:** o **W:** ! **NI:** ! **ROI:** ! **O:** Isle of Man: !
H: II & III on *Conopodium majus.*
D: W&H: 158
Occasional in Scotland (where the majority of records are from). Rarely reported but apparently widespread elsewhere.

uliginosa Juel, *Öfvers. Kongl. Vetensk.-Akad. Förh.* 51: 410 (1894)
Puccinia caricina var. *uliginosa* (Juel) Jørst., *Skrifter Vidensk.-Akad. Oslo* 1(2): 30 (1952) [1951]
S: !
H: I on *Parnassia palustris.* II & III on *Carex nigra.*
D: W&H: 237 (as *Puccinia caricina* var. *uliginosa*)
Rarely reported. British collections only from widely scattered sites in Scotland.

umbilici Duby, *Bot. gall.:* 890 (1830)
Puccinia rhodiolae Berk. & Broome, *Ann. Mag. Nat. Hist, Ser. 2* 5: 482 (1850)
E: o **S:** ! **W:** o **NI:** ! **ROI:** ! **O:** Channel Islands: c Isle of Man: !
H: III on *Umbilicus rupestris* and rarely *Sedum rosea.*
D: W&H: 136
Locally frequent, especially in the west and southwest (following the range of the main host), but virtually absent from southern and southeastern parts of England.

urticata [sensu lato] F. Kern, *Mycologia* 9: 214 (1917)
Aecidium urticae DC., *Fl. Fr.* 5: 92 (1815)

E: ! **S:** ! **W:** ! **NI:** ! **ROI:** !
H: 0 & I on *Urtica dioica* and *U. urens*. II & III on various
  species of *Carex*
D: W&H: (as *Puccinia caricina* sensu lat.)
Common and widespread.

urticata *var.* biporula Zwetko, *Biblioth. Mycol.* 153: 81 (1993)
S: !
H: 0 & I on *Urtica dioica*. II & III on *Carex pallescens*.
D: W&H: 240 (as *Puccinia caricina* sensu lato.)
A single collection from Wester Ross (Loch Maree) in 1963.

urticata *var.* urticae-acutae (Kleb.) Zwetko, *Biblioth. Mycol.*
  153: 82 (1993)
    Puccinia urticae-acutae Kleb., *Z. Pflanzenkrankh.* 9: 152
    (1899)
    Puccinia caricina var. *urticae-acutae* (Kleb.) D.M. Hend.,
    Notes Roy. Bot. Gard. Edinburgh 23(3): 238 (1961)
E: ! **W:** !
H: 0 & I on *Urtica dioica* and *U. urens*. II & III on *Carex nigra*
  and *C. elata*.
D: W&H: 238 (as *Puccinia caricina* var. *urticae-acutae*)
Rarely reported, though *Carex nigra* is common in Britain.

urticata *var.* urticae-acutiformis (Kleb.) Zwetko, *Biblioth.*
  Mycol. 153: 83 (1993)
    Puccinia caricis f. *urticae-acutiformis* Kleb., *Z.*
    Pflanzenkrankh. 15: 70 (1905)
    Puccinia caricina var. *urticae-acutiformis* (Kleb.) D.M. Hend.,
    Notes Roy. Bot. Gard. Edinburgh 23(3): 239 (1961)
E: c **S:** ! **W:** ! **NI:** ! **ROI:** !
H: 0 & I on *Urtica dioica* and *U. urens*. II & III on *Carex*
  acutiformis.
D: W&H: 238 (as *Puccinia caricina* var. *urticae-acutiformis*)
Common and widespread in England, less frequently reported
  elsewhere.

urticata *var.* urticae-flaccae (Hasler) Zwetko, *Biblioth. Mycol.*
  153: 85 (1993)
    Puccinia urticae-flaccae Hasler, *Ber. Schweiz. Bot. Ges.* 55: 6
    (1945)
    Puccinia caricina var. *urticae-flaccae* (Hasler) D.M. Hend.,
    Notes Roy. Bot. Gard. Edinburgh 23(3): 239 (1961)
E: ! **S:** ! **W:** ! **ROI:** !
H: 0 & I on *Urtica dioica* and *U. urens*. II & III on *Carex flacca*.
D: W&H: 238 (as *Puccinia caricina* var. *urticae-flaccae*)
Rarely reported, although the host is common and widespread.

urticata *var.* urticae-hirtae (Kleb.) Zwetko, *Biblioth. Mycol.*
  153: 86 (1993)
    Puccinia urticae-hirtae Kleb., *Z. Pflanzenkrankh.* 9: 152
    (1899)
    Puccinia caricina var. *urticae-hirtae* (Kleb.) D.M. Hend., *Notes*
    Roy. Bot. Gard. Edinburgh 23(3): 240 (1961)
E: ! **S:** ! **W:** ! **NI:** ! **ROI:** ! **O:** Isle of Man: !
H: 0 & I on *Urtica dioica* and *U. urens*. II & III on *Carex hirta*.
D: W&H: 239 (as *Puccinia caricina* var. *urticae-hirtae*)
Rarely reported.

urticata *var.* urticae-inflatae (Hasler) Zwetko, *Biblioth.*
  Mycol. 153: 87 (1993)
    Puccinia urticae-inflatae Hasler, *Mitth. Aargauischen Naturf.*
    Ges. 17: 64 (1925)
    Puccinia caricina var. *urticae-inflatae* (Hasler) D.M. Hend.,
    Notes Roy. Bot. Gard. Edinburgh 23(3): 240 (1961)
E: ! **S:** ! **W:** ! **ROI:** !
H: 0 & I on *Urtica dioica* and *U. urens*. II & III on *Carex*
  rostrata.
D: W&H: 239 (as *Puccinia caricina* var. *urticae-inflatae*)
Rarely reported but apparently widespread.

urticata *var.* urticae-paniceae (Mayor) Zwetko, *Biblioth.*
  Mycol. 153: 88 (1993)
    Puccinia urticae-paniceae Mayor, *Bull. Soc. Bot. Suisse* 59:
    274 (1949)
ROI: !
H: II & III on *Carex panicea*
Reported only recently from Donegal.

urticata *var.* urticae-ripariae (Hasler) Zwetko, *Biblioth.*
  Mycol. 153: 89 (1993)
    Puccinia urticae-ripariae Hasler, *Ber. Schweiz. Bot. Ges.* 55:
    10 (1945)
    Puccinia caricina var. *urticae-ripariae* (Hasler) D.M. Hend.,
    Notes Roy. Bot. Gard. Edinburgh 23(3): 240 (1961)
E: !
H: 0 & I on *Urtica dioica* and *U. urens*. II & III on *Carex riparia*.
D: W&H: 239 (as *Puccinia caricina* var. *urticae-ripariae*)
Rarely reported.

urticata *var.* urticae-vesicariae (Kleb.) Zwetko, *Biblioth.*
  Mycol. 153: 90 (1993)
    Puccinia caricis f. *urticae-vesicariae* Kleb., *Z. Pflanzenkrankh.*
    15: 70 (1905)
    Puccinia caricina var. *urticae-vesicariae* (Kleb.) D.M. Hend.,
    Notes Roy. Bot. Gard. Edinburgh 23: 241 (1961)
E: ! **S:** ! **W:** !
H: 0 & I on *Urtica dioica* and *U. urens*. II & III on *Carex*
  vesicaria.
D: W&H: 240 (as *Puccinia caricina* var. *urticae-vesicariae*)
Rarely reported but apparently widespread.

variabilis Grev., *Scott. crypt. fl.* 2: pl. 75 (1824)
    Aecidium grevillei Grove, *J. Bot.* 23: 129 (1885)
    Aecidium taraxaci Grev. [non J. Kunze & J.C. Schmidt 1817),
    Fl. edin.: 444 (1824)
E: ! **S:** ! **W:** ! **NI:** ! **ROI:** ! **O:** Isle of Man: !
H: I, II & III on *Taraxacum officinale* and *T. palustre* agg.
D: W&H: 214
Apparently frequent and widespread.

veronicae J. Schröt., *Beitr. Biol. Pflanzen* 3: 89 (1878)
E: c **S:** ! **W:** c **NI:** ! **ROI:** !
H: III on *Veronica montana*.
D: W&H: 176
Common and widespread in England and Wales. Rarely reported
  elsewhere but apparently widespread.

veronicae-longifoliae Savile, *Canad. J. Bot.* 46: 635 (1968)
    Puccinia veronicae f.sp. *spicatae* Gäum., *Ann. Mycol.* 39: 42
    (1941)
E: ! **W:** !
H: III on *Veronica spicata* var. *hybrida*.
British collections only from England (Lancashire) and Wales
  (Denbighshire).

vincae Berk., *Engl. fl.* 5(2): 364 (1836)
    Trichobasis vincae (Berk.) Berk., *Outl. Brit. fungol.*: 332
    (1860)
    Puccinia berkeleyi Pass., *Hedwigia* 12: 143 (1873)
    Uredo vincae DC., *Fl. Fr.* 5: 70 (1815)
E: c **W:** ! **ROI:** ! **O:** Channel Islands: !
H: 0, I, II & III on *Vinca major* [but not on *Vinca minor*].
D: W&H: 168-169
Common in England, especially so in southern counties. Less
  often reported elsewhere but apparently widespread. Probably
  introduced with the host, and causing distortion and damage
  to affected plants.

violae DC., *Fl. Fr.* 5: 62 (1815)
    Puccinia aegra Grove, *J. Bot.* 21: 274 (1883)
    Aecidium depauperans Vize, *Gard. Chron.* 6(6): 175 (1876)
    Puccinia depauperans Syd., *Monogr. Ured.* 1: 442 (1903)
    Aecidium violae Schumach., *Enum. pl.* 2: 224 (1803)
    Uredo violae Schumach., *Enum. pl.* 2: 233 (1803)
    Puccinia violarum Link, in Willd., *Sp. pl.*, Edn 4, 6(2): 80
    (1825)
    Trichobasis violarum (Link) Berk., *Outl. Brit. fungol.*: 333
    (1860)
E: c **S:** c **W:** c **NI:** c **ROI:** c **O:** Channel Islands: ! Isle of Man: !
H: 0, I, II and III on *Viola canina*, *V. cornuta*, *V. hirta*, *V. lutea*,
  V. odorata, *V. reichenbachiana*, *V. riviniana*, *V. tricolor* and
  Viola tricolor ssp. *curtisii*, also on *Viola* cultivars.
D: W&H: 126
Common and widespread on wild species of *Viola* but rare on
  cultivated forms.

virgae-aureae (DC.) Lib., *Pl. Crypt. Ardienn.* 4: 393 (1837)
  Xyloma virgae-aureae DC., *Syn. Pl. Gall.*: 63 (1806)
  Asteroma atratum Chevall. (anam.), *Fl. gén. env. Paris,* Edn 1: 449 (1826)
  Puccinia clandestina Carmich. in Berkeley, *Engl. fl.* 5(2): 365 (1836)
E: ! **S:** ! **W:** !
H: III on *Solidago virgaurea.*
D: W&H: 216
Rarely reported.

PUCCINIASTRUM G.H. Otth, *Mitt. naturf. Ges. Bern.* 72 (1861)
Uredinales, Pucciniastraceae
  Calyptospora J.G. Kühn, *Hedwigia* 8: 81 (1869)
  Thekopsora Magnus, *Hedwigia* 14: 123 (1875)
Type: *Pucciniastrum epilobii* (Chaillet) G.H. Otth

agrimoniae (Dietel) Tranzschel, *Scripta Bot. Horti Univ. Imper. Petrop.* 4: 301 (1895)
  Thekopsora agrimoniae Dietel, *Hedwigia* 29: 153 (1890)
  Pucciniastrum agrimoniae-eupatoriae Lagerh., *Tromsø Mus. Aarsh.* 17: 92 (1895)
  Coleosporium ochraceum Bonord., *Kenntn. Coniomyc. Cryptomyc.*: 20 (1860)
E: c **S:** ! **W:** ! **ROI:** ! **O:** Channel Islands: !
H: II on *Agrimonia eupatoria* and rarely also on *A. odorata.*
D: W&H: 35
Common in southern England. Rarely reported elsewhere but apparently widespread.

areolatum (Fr.) G.H. Otth, *Mitt. naturf. Ges. Bern* 1863: 85 (1863)
  Xyloma areolatum Fr., *Observ. mycol.* 2: 358 (1817)
  Thekopsora areolata (Fr.) Magnus, *Sitzungsber. Ges. Naturf. Freunde Berlin* 16: 58 (1875)
  Uredo padi J. Kunze & J.C. Schmidt, *Fungi selecti exsiccati* 2: no. 187 (1877)
  Melampsora padi (J. Kunze & J.C. Schmidt) G. Winter, *Rabenh. Krypt.-Fl.* 1(1): 244 (1884)
  Thekopsora padi (J. Kunze & J.C. Schmidt) Kleb., *Jahrb. Wiss. Bot.* 34: 378 (1900)
  Uredo porphyrogenita Chevall., *Fl. gén. env. Paris,* Edn 1: 401 (1826)
  Licea strobilina Alb. & Schwein., *Consp. fung. lusat.*: 109 (1805)
  Aecidium strobilinum (Alb. & Schwein.) Reess, *Abh. Naturf. Ges. Halle* 11: 105 t.2 (1870)
E: ? **S:** !
H: 0 & I on *Picea abies.* II & III on *Prunus padus.*
D: W&H: 35-36
Old collections from widespread localities in Scotland. Reported from northern England but unsubstantiated with voucher material.

circaeae (G. Winter) de Toni, in Saccardo, *Syll. fung.* 7: 763 (1883)
  Phragmospora circaeae G. Winter, *Hedwigia* 18: 171 (1879)
  Uredo circaeae Schumach., *Enum. pl.* 2: 228 (1803)
  Uredo circaeae Alb. & Schwein., *Consp. fung. lusat.*: 124 (1805)
  Melampsora circaeae (Alb. & Schwein.) Thüm. [*nom. inval.*], *Mycotheca universalis*: no. 447 (1876)
E: c **S:** c **W:** c **NI:** c **ROI:** c
H: II & III on *Circaea alpina, C. intermedia* and *C. lutetiana.*
D: W&H: 34-35
Common and widespread, especially on *Circaea lutetiana.*

epilobii G.H. Otth, *Mitt. naturf. Ges. Bern* 1861: 72 (1861)
  Pucciniastrum abieti-chamaenerii Kleb., *Jahrb. Wiss. Bot.* 34: 387 (1900)
  Melampsora epilobii Fuckel, *Jahrb. Nassauischen Vereins Naturk.* 23-24: 44 (1870)
  Uredo fuchsiae Arthur & Holw., *Amer. J. Bot.* 5: 538 (1918)
  Uredo pustulata α *epilobii* Pers., *Syn. meth. fung.*: 219 (1801)

Melampsora pustulata J. Schröt., *Krypt.-Fl. Schlesien.* 364 (1887)
  Pucciniastrum pustulatum Dietel, *Nat. Pflanzenfamilien* 1: 47 (1897)
E: c **S:** c **W:** c **NI:** ! **ROI:** ! **O:** Channel Islands: c
H: 0 & I on *Abies grandis.* II & III on *Chamaerion angustifolium, Epilobium anagallidifolium, E. montanum, E. palustre, E. roseum, Fuchsia magellanica* and cultivars, and cultivated *Clarkia amoena.*
D: W&H: 31-34
Common and widespread, especially on *Chamaerion angustifolium.* The fungus on *Chamaerion* is regarded by some authorities as a separate species, *Pucciniastrum abieti-chamaenerii* Kleb.

goeppertianum (J.G. Kühn) Kleb., *Wirtwechs. Rostpilze*: 391 (1904)
  Calyptospora goeppertiana J.G. Kühn, *Hedwigia* 8: 81 (1869)
  Melampsora goeppertiana (J.G. Kühn) G. Winter, *Rabenh. Krypt.-Fl.* 1(1): 245 (1882)
  Peridermium columnare F.C. Holl & J.C. Schmidt, *Deutschl. Schwämme* 1: no. 10 (1815)
E: ! **S:** ! **NI:** !
H: III on *Vaccinium vitis-idaea* (also on imported *V. corymbosum*).
D: W&H: 40-41
Doubtfully native on *Vaccinium vitis-idaea* and not reported for many years. It should be noted that *Peridermium columnare* sensu Cooke [HBF1: 535] is the aecidial stage of a species of *Milesina.*

goodyerae Arthur, *North American Flora* 7: 105 (1907)
  Uredo goodyerae Tranzschel, *Trudy S.- Petersburgsk. Obshch. Estestvoisp., Otd. Bot.* 23: 27 (1893)
S: !
H: II on *Goodyera repens.*
D: W&H: 42
Known from Morayshire.

guttatum (J. Schröt.) Hyl., Jörst. & Nannf., *Opera Bot.* 1(1): 81 (1953)
  Melampsora guttata J. Schröt., *Abh. Schles. Ges. Vaterl. Cult., Abth. Naturwiss. (Naturwiss.-Med. Abth.)* 1869-72: 26 (1870)
  Thekopsora guttata (J. Schröt.) Syd. & P. Syd., *Monogr. Ured.* 3: 467 (1915)
  Melampsora galii G. Winter, *Rabenh. Krypt.-Fl.* 1(1): 244 (1882)
  Thekopsora galii (G. Winter) de Toni, Saccardo, *Syll. fung.* 7: 765 (1888)
E: c **S:** ! **W:** ! **O:** Isle of Man: !
H: II & III on *Galium odoratum, G. palustre, G. saxatile, G. uliginosum, G. verum* and *Sherardia arvensis.*
D: W&H: 41-42
Common, at least in southern England, especially on old and overwintered leaves of *Galium odoratum* and on *G. saxatile.* Rarely reported elsewhere but apparently widespread.

pyrolae Arthur, *North American Flora* 7: 108 (1907)
  Uredo pyrolae H. Mart., *Prodr. fl. mosq.,* Edn 2: 229 (1817)
  Trichobasis pyrolae Berk., *Outl. Brit. fungol.*: 332 (1860)
E: ! **S:** !
H: II & III on *Orthilia secunda, Pyrola media, P. minor* and possibly on *P. rotundifolia.*
D: W&H: 37-38
Rarely reported.

RHODOTORULA F.C. Harrison, *Proc. & Trans. Roy. Soc. Canada, ser. 3* 21: 349 (1927)
Sporidiales, incertae sedis
  Eutorulopsis Cif., *Atti Ist. Bot. Univ. Pavia* ser. 3, 2: 143 (1925)
Type: *Rhodotorula glutinis* (Fresen.) F.C. Harrison
A genus of anamorphic yeasts.

acheniorum (Buhagiar & J.A. Barnett) Rodr. Mir., *Antonie van Leeuwenhoek* 41: 196 (1975)
 Sterigmatomyces acheniorum Buhagiar & J.A. Barnett, *J. Gen. Microbiol.* 77(1): 78 (1973)
E: !
H: Isolated ex strawberries.
D+I: YST 599

bacarum (Buhagiar) Rodr. Mir. & Weijman, *Antonie van Leeuwenhoek* 54(6): 549 (1988)
 Torulopsis bacarum Buhagiar, *J. Gen. Microbiol.* 86(1): 2 (1975)
E: !
H: Isolated ex blackcurrants.
D+I: YST 605

fragariae (J.A. Barnett & Buhagiar) Rodr. Mir. & Weijman, *Antonie van Leeuwenhoek* 54(6): 549 (1988)
 Torulopsis fragariae J.A. Barnett & Buhagiar, *J. Gen. Microbiol.* 67(2): 237 (1971)
E: !
H: Isolated ex strawberries.
D+I: YST 613

glutinis (Fresen.) F.C. Harrison, *Proc. & Trans. Roy. Soc. Canada, ser. 3* 21(5): 349 (1928)
 Cryptococcus glutinis Fresen., *Beitr. Mykol.* 2: 77 (1852)
E: !
H: Isolated ex blood and pig carcasses.
D+I: YST 616-617

graminis di Menna, *J. Gen. Microbiol.* 18: 270 (1958)
E: !
H: Isolated ex pig carcasses.
D+I: YST 618

marina Phaff, Mrak & O.B. Williams, *Mycologia* 44: 436 (1952)
E: !
H: Isolated ex sausages.
D+I: YST 626

minuta (Saito) F.C. Harrison, *Proc. & Trans. Roy. Soc. Canada, ser. 3* 22(5): 187 (1928)
 Torula minuta Saito, *Jap. J. Bot.* 1(1): 48 (1922)
E: !
H: Isolated ex lamb carcasses, drinking water and sausages.
D+I: YST 627

mucilaginosa (A. Jörg.) F.C. Harrison, *Proc. Trans. Roy. Soc. Canada, ser. 3, 21(5): 349 (1928)
 Torula mucilaginosa A. Jörg., *Die Mikroorg. Gärungsindustrie*: 402 (1909)
E: !
H: Isolated ex cattle, sheep, dogs, cats, turkeys, and blood.
D+I: YST 628-629

pustula (Buhagiar) Rodr. Mir. & Weijman, *Antonie van Leeuwenhoek* 54(6): 548 (1988)
 Torulopsis pustula Buhagiar, *J. Gen. Microbiol.* 86(1): 3 (1975)
E: !
H: Isolated ex blackcurrants.
D+I: YST 635

SPHACELOTHECA de Bary, *Vergl. Morph. Biol. Pilze*: 187 (1884)
Microbotryales, Microbotryaceae
Type: *Sphacelotheca hydropiperis* (Schumach.) de Bary

hydropiperis (Schumach.) de Bary, *Vergl. Morph. Biol. Pilze*: 187 (1884)
 Uredo hydropiperis Schumach., *Enum. pl.* 2: 234 (1803)
 Ustilago hydropiperis (Schumach.) J. Schröt., *Beitr. Biol. Pflanzen* 2: 355 (1877)
 Caeoma utriculosum Nees, *Syst. Pilze.* 1: 14 (1817)
 Ustilago utriculosa (Nees) Tul. & C. Tul., *Ann. Sci. Nat., Bot., sér. 3, 7: 102 (1847)
 Mis.: *Ustilago candollei* sensu auct.

E: c **S:** c **W:** c **NI:** c **ROI:** c
H: In the ovaries of *Persicaria hydropiper, P. maculosa*, and *P. mite.* **D:** UB 33, ESF 193-194
Common and widespread. Most frequent on *P. hydropiper*.

SPICULOGLOEA P. Roberts, *Mycotaxon* 60: 112 (1996)
Platygloeales, Platygloeaceae
Type: *Spiculogloea occulta* P. Roberts

minuta P. Roberts, *Mycotaxon* 63: 204 (1997)
E: !
H: Parasitic. In the hymenium of heterobasidiomycetes such as *Helicogloea lagerheimii, Tulasnella eichleriana, T. saveloides* and *T. violea* and corticioid fungi such as *Tubulicrinis accedens.*
D: Myc15(4): 163
Known from Shropshire, South Devon and Surrey.

occulta P. Roberts, *Mycotaxon* 60: 113 (1996)
E: ! **W:** !
H: Parasitic. British collections usually in the hymenium of *Hyphodontia sambuci* and rarely *H. pallidula.*
Known from England (Northumberland, Surrey and West Kent) and Wales (Breconshire).

subminuta Hauerslev, *Mycotaxon* 72: 474 (1999)
E: !
H: Parasitic. In the hymenium of *Botryobasidium subcoronatum.*
Recorded from Berkshire (Windsor Forest), Surrey (Esher Common) and West Kent (West Farleigh: Quarry Wood).

SPOROBOLOMYCES Kluyver & C.B. Niel, *Centralbl. Bakteriol. Parasitenk.* Abt. II, 63: 19 (1924)
Sporidiales, incertae sedis
Type: *Sporobolomyces salmonicolor* (B. Fisch. & Brebeck) Kluyver & Neil
A genus of anamorphic yeasts.

roseus Kluyver & C.B. Niel, *Centralbl. Bakteriol. Parasitenk. Abt. II, 63: 19 (1924)
E: c **S:** ! **NI:** !
H: Described from beer sediment, but all British records are from leaves of various trees and shrubs. Also on seeds and cereals.
D+I: YST 707-708
Common and widespread.

tsugae (Phaff & Carmo Souza) Nakase & Itoh, *J. Gen. Appl. Microbiol.* 34(6): 501 (1988)
 Bullera tsugae Phaff & Carmo Souza, *Leeuwenhoek ned. Tidjdschr.* 28: 205 (1962)
E: !
H: Isolated ex sausages.
D+I: YST 715

TRACHYSPORA Fuckel, *Bot. Zeitung (Berlin)* 19: 250 (1861)
Uredinales, Phragmidiaceae
Type: *Trachyspora alchemillae* J. Schröt. (= *T. intrusa*)

intrusa (Grev.) Arthur, *Manual of the Rusts in the United States & Canada*: 97 (1934)
 Uredo intrusa Grev., *Fl. edin.*: 436 (1824)
 Uromyces alchemillae Lév., *Ann. Sci. Nat., Bot., sér. 3, 8: 371 (1847)
 Trachyspora alchemillae J. Schröt, *Bot. Zeitung (Berlin)* 19: 250 (1861)
E: ! **S:** o **W:** ! **NI:** ! **ROI:** !
H: 0, I, II & III on *Alchemilla* spp.
D: W&H: 364-366
Rarely reported, but most frequent in Scotland. Widespread elsewhere but virtually absent from southern and southeastern England.

TRANZSCHELIA Arthur, *Résultats scientifiques du Congres international de Botanique Wien 1905*: 340 (1906) [1905]
Uredinales, Uropyxidaceae
Type: *Tranzschelia cohaesa* (Long) Arthur

anemones (Pers.) Nannf., *Fungi Exsiccati Suecici*: 839a (1939)
Puccinia anemones Pers., *Observ. mycol.* 2: 24 (1799)
Puccinia fusca G. Winter, *Rabenh. Krypt.-Fl.* 1(1): 199 (1882)
Tranzschelia fusca (G. Winter) Dietel, *Ann. Mycol.* 20: 31 (1922)
Aecidium fuscum Relhan, *Fl. Cantab.*: 1199 (1785)
Puccinia thalictri Chevall., *Fl. gén. env. Paris,* Edn 1: 417 (1826)
Tranzschelia thalictri (Chevall.) Dietel, *Ann. Mycol.* 20: 31 (1922)
Mis.: *Aecidium leucospermum* sensu auct. & sensu Saccardo [Syll. fung. VII: 669,1888].
E: c **S:** c **W:** ! **ROI:** !
H: II & III common on *Anemone nemorosa* and rarely on cultivated *A. blanda,* rare on native *Thalictrum* spp.
D: W&H: 302 **I:** Sow 1 53 (as *Aecidium fuscum*)
Common on *Anemone nemorosa* in England and Scotland. Much less often reported elsewhere and on other hosts, but apparently widespread.

discolor (Fuckel) Tranzschel & Litv., *Bot. Zhurn. S.S.S.R.* 24: 248 (1939)
Puccinia discolor Fuckel, *Fungi Rhenani exsiccati (Suppl.). Fasc. VII*: 2121 (1868)
Tranzschelia pruni-spinosae var. *discolor* (Fuckel) Dunegan, *Phytopathology* 28: 424 (1938)
Aecidium punctatum Pers., *Ann. Bot. Usteri* 20: 135 (1796)
Aecidium quadrifidum DC., *Fl. Fr.* 5: 239 (1815)
E: ! **S:** ! **W:** !
H: 0 & I on cultivated *Anemone coronaria* and *A. x fulgens*. II & III on *Prunus armeniaca, P. domestica* cultivars, *P. persica* and possibly also on *P. spinosa.*
D: W&H: 304
Frequent and apparently widespread. Records on *Prunus spinosa* are unconfirmed due to confusion with *T. pruni-spinosae.*

pruni-spinosae (Pers.) Dietel, *Ann. Mycol.* 20: 31 (1922)
Puccinia pruni-spinosae Pers., *Syn. meth. fung.*: 226 (1801)
Trichobasis pruni-spinosae (Pers.) Cooke, *Handb. Brit. fung.* 2: 507 (1871)
Puccinia pruni DC., Fl. Fr. 2: 222 (1805)
Puccinia prunorum Link, in Willd., *Sp. pl.,* Edn 4, 6 (2): 82 (1825)
Trichobasis rhamni Cooke, *J. Bot.* 2: 344 (1864)
E: c **W:** ! **NI:** ! **O:** Channel Islands: !
H: II & III on *Prunus spinosa.* **D:** W&H: 307
Common in England. Rarely reported elsewhere but apparently widespread. Confused with *T. discolor* on *Prunus* cultivars.

TRIPHRAGMIUM Link, in Willd., *Sp. pl.,* Edn 4, 6(2): 84 (1825)
Uredinales, Sphaerophragmiaceae
Type: *Triphragmium ulmariae* (DC.) Link

filipendulae Pass., *Nuovo Giorn. Bot. Ital.* 7: 255 (1875)
E: ! **S:** ? **ROI:** ?
H: I, II & III on *Filipendula vulgaris* (never on *F. ulmaria*).
D: W&H: 112
Known from East Gloucestershire, South Devon, Worcestershire and Yorkshire. Apparently rare though the host is locally frequent on calcareous downland in southern England. A report from Scotland in 1998 on *Filipendula ulmaria* is unsubstantiated with voucher material and probably misidentified *T. ulmariae.*

ulmariae (DC.) Link, in Willd., *Sp. pl.,* Edn 4, 6(2): 84 (1825)
Puccinia ulmariae DC., in Lam., *Encyclop.* 8: 245 (1808)
Puccinia spiraeae Purton, *Bot. descr. Brit. pl.* 3: 304 (1821)
Uredo spireae Sowerby, *Col. fig. Engl. fung.* 3: pl. 398 f.7 (1803)
Uredo ulmariae Schumach., *Enum. pl.* 2: 227 (1803)
Uromyces ulmariae Lév., *Ann. Sci. Nat., Bot.,* sér. 3, 8: 371 (1847)
Trichobasis ulmariae Cooke, *Grevillea* 4(30): 67 (1875)
Mis.: *Triphragmium filipendulae* sensu auct. p.p.
E: c **S:** c **W:** c **NI:** c **ROI:** c **O:** Isle of Man: !
H: 0, I, II & III on *Filipendula ulmaria.*
D: W&H: 112-113
Widespread and frequent.

TUBERCULINA Sacc., *Michelia* 2: 34 (1880)
Platygloeales, Platygloeaceae
Anamorphic fungi with teleomorphs in the genus *Helicobasidium,* but not as yet linked to species.
Type: *Tuberculina persicina* (Ditm.) Sacc.

maxima Rostr., *Ust. Dan.*: 46 (1890)
E: ! **S:** !
H: Parasitic on rusts (e.g. *Gymnosporangium confusum, Peridermium pini*).

persicina (Ditm.) Sacc., *Fung. Ital.*: t. 964 (1881)
Tubercularia persicina Ditm., in Sturm, *Deutschl. Fl.* Abt. 3, 1: 99, t. 49 (1817)
E: ! **S:** ! **ROI:** !
H: Parasitic on various rusts (e.g. *Kuehneola uredinis* and *Puccinia* spp.).
At least one British collection of *T. persicina* s.l. has *Helicobasidium purpureum* s.l. as its teleomorph, but collections elsewhere have been assigned to *H. longisporum* [Lutz *et al., Mycol. Res.* 108: 227 - 238 (2004)].

sbrozzii Cavara & Sacc., *Nuovo Giorn. Bot. Ital., Ser. 2*: 326 (1899)
E: o
H: On sori of *Puccinia vincae.*

UREDINOPSIS Magnus, *Atti Congr. Bot. Intern. di Genova* 1892: 167 (1893)
Uredinales, Pucciniastraceae
Type: *Uredinopsis filicina* (Niessl) Magnus

filicina (Niessl) Magnus, *Atti Congr. Bot. Intern. di Genova* 1892: 167 (1893)
Protomyces filicinus Niessl, in Rabenhorst, *Fungi europaei* IV: 1659 (1873)
E: ? **S:** !
H: II & III on *Thelypteris phegopteris.*
D: W&H: 16-17
A single collection from Scotland (Argyll) in 1936. Said to occur in England (Hertfordshire) by Henderson (2004) but this record has not been traced.

UREDO Pers., *Syn. meth. fung.*: 214 (1801)
Uredinales, Incertae sedis
Trichobasis Lév., *Dict. Univ. Hist. Nat.* 12: 785 (1849)
Type: *Uredo betae* Pers.

morvernensis Dennis, *Kew Bull.* 38(2): 203 (1983)
S: ! **W:** ! **ROI:** !
H: II on *Thymus polytrichus* (= *T. drucei*).
Known from Scotland (Westerness) and the Republic of Ireland (Galway), and recently reported from Wales (in litt.).

oncidii Henn., *Hedwigia* 41: 15 (1902)
Trichobasis lynchii Berk., *Gard. Chron.* 8: 242 (1877)
Uredo lynchii (Berk.) Plowr., *Monograph Brit. Ured.*: 259 (1889)
E: !
H: II on the orchids *Oncidium cavendishianum, Cattleya dowiana* (on imported plants in botanic gardens) and on *Spiranthes spiralis.*
D: W&H: 366

A single collection (South Hampshire: East Cosham) on *Spiranthes spiralis* thus possibly native, but otherwise only known on cultivated and imported orchids.

UROMYCES (Link) Unger, *Exanth. Pfl.*: 277 (1832)
Uredinales, Pucciniaceae
Capitularia Rabenh. [*nom. illegit*], *Bot. Zeitung (Berlin)* 9: 449 (1851)
Type: *Uromyces appendiculatus* (Pers.) Unger

acetosae J. Schröt., in Rabenhorst, *Fungi europaei* IV: 2080 (1876)
E: ! **S:** ! **W:** ! **NI:** ! **O:** Channel Islands: !
H: 0, I, II & III on *Rumex acetosa* and (rarely) *R. acetosella*.
D: W&H: 341
'Frequent' *fide* W&H and widespread, but there are few reports or collections. All collections in herb. K are on *R. acetosa*.

aecidiiformis (F. Strauss) R.G. Rees, *Amer. J. Bot.* 4: 369 (1917)
Uredo aecidiiformis F. Strauss, *Ann. Wetterauischen Ges. Gesammte Naturk.* 2: 94 (1811)
Caeoma lilii Link, in Willd., *Sp. pl.,* Edn 4, 6(2): 8 (1825)
Uromyces lilii (Link) Fuckel [non *U. lilii* Clinton (1875)], *Jahrb. Nassauischen Vereins Naturk.* 29-30: 16 (1875)
E: !
H: 0, I, & III on cultivated *Lilium candidum* and a single recent collection on *Fritillaria graeca*.
D: W&H: 352
An introduction, reported from East Norfolk (Norwich) and Surrey (Dorking).

airae-flexuosae Ferd. & Winge, *Bull. Soc. Mycol. France.* 36: 164 (1920)
Uredo airae-flexuosae Liro, *Bidrag Kännedom Finlands Natur Folk* 65: 573 (1908)
E: ! **S:** ! **W:** ! **O:** Isle of Man: !
H: II & III on *Deschampsia flexuosa*.
D: W&H: 359
Rarely reported, but easily overlooked.

ambiguus (DC.) Fuckel, *Jahrb. Nassauischen Vereins Naturk.* 23-24: 64 (1870)
Uredo ambigua DC., *Fl. Fr.* 5: 64 (1815)
E: ! **S:** ! **ROI:** ! **O:** Channel Islands: !
H: II & III on *Allium babingtonii, A. schoenoprasum, A. scorodoprasum* and *A. ursinum*.
D: W&H: 349
Rarely reported, but apparently widespread. Regarded by some authorities as synonymous with *Puccinia allii*.

anthyllidis J. Schröt., *Hedwigia* 14: 162 (1875)
Uromyces hippocrepidis Mayor, *Bull. Soc. Naeuchâteloise Sci. Nat.* 45: 40 (1921)
Uromyces jaapianus Kleb., *Kryptogamenflora der Mark Brandenburg* 5a: 239 (1914)
E: ! **S:** ! **W:** ! **ROI:** ! **O:** Isle of Man: !
H: II & III on *Anthyllis vulneraria, Hippocrepis comosa, Lotus hispidus, Lupinus* spp, *Trifolium dubium* and *T. campestre*.
D: W&H: 319, Myc11(3): 111
Rarely reported, but apparently widespread.

appendiculatus (Pers.) Unger, *Einfl. Boden. Verth. Gew.*: 216 (1836)
Uredo appendiculata Pers., *Syn. meth. fung.*: 222 (1801)
Uredo appendiculata α *phaseoli* Pers., *Syn. meth. fung.*: 222 (1801)
Uromyces phaseoli (Pers.) G. Winter, *Hedwigia* 19: 37 (1880)
Uromyces phaseolorum de Bary, *Ann. Sci. Nat., Bot.,* sér. 4, 20: 80 (1863)
E: ! **W:** !
H: 0, I, II & III on cultivated *Phaseolus coccineus* and *P. vulgaris*.
D: W&H: 321
More frequent since 1952 *fide* W&H, but now rather rarely reported.

armeriae J.J. Kickx, *Fl. Crypt. Flandres* 2: 73 (1867)
Caeoma armeriae Schltdl., *Fl. berol.*: 126 (1824)
Uredo armeriae (Schltdl.) Duby, *Bot. gall.*: 899 (1830)
Mis.: *Uromyces limonii* sensu Plowr. (1889: 122)
Mis.: *Uredo statices* sensu auct.
E: o **S:** o **W:** o **ROI:** ! **O:** Channel Islands: ! Isle of Man: !
H: 0, I, II & III on *Armeria maritima* and occasionally on *Armeria* cultivars in gardens or in montane areas.
D: W&H: 345
Occasional but widespread.

behenis (DC.) Unger, *Einfl. Boden. Verth. Gew.*: 216 (1836)
Uredo behenis DC., *Fl. Fr.* 5: 63 (1815)
E: ! **S:** ! **W:** ! **ROI:** ! **O:** Channel Islands: ! Isle of Man: !
H: 0, I, II & III on *Silene uniflora* (=*S. maritima*) and *S. vulgaris*.
D: W&H: 311-312
Widespread, but rarely reported.

beticola (Bellynck) Boerema, Loer. & Hamers, *Netherlands J. Pl. Pathol.,* Suppl. 93: 17 (1987)
Uredo beticola Bellynck, *Herb. crypt. Belg.* Fasc. 24: no.1170 (1857)
Uredo betae Pers., *Syn. meth. fung.*: 220 (1801)
Trichobasis betae (Pers.) Lév., in Orbigny, *Dict. univ. hist. nat.* 1: 19 (1841)
Uromyces betae (Pers.) Lév., *Ann. Sci. Nat., Bot.,* sér. 3, 8: 375 (1847)
Uredo betaecola Westend., *Bull. Acad. Roy. Sci. Belgique* 11: 650 (1861)
E: o **W:** ! **ROI:** ! **O:** Channel Islands: ! Isle of Man: !
H: 0, I, II & III on *Beta maritima* and cultivars.
D: W&H: 315 (as *Uromyces betae*)
Uncommonly reported.

chenopodii (Duby) J. Schröt., in J. Kunze, *Fungi selecti exsiccati*: no. 214 (1880)
Uredo chenopodii Duby, *Bot. gall.*: 899 (1830)
E: ! **S:** ! **W:** !
H: 0, I, II & III on *Suaeda maritima, S. maritima* var. *flexilis* and *S. vera* (=*S. fruticosa*).
D: W&H: 316
Known from England (Dorset, East Norfolk, East Suffolk, North Lincolnshire, South Essex, and West Sussex) and single locations in Scotland (East Lothian) and Wales (Pembrokeshire).

colchici Massee, *Grevillea* 21: 6 (1892)
E: !
H: III on *Colchicum* sp.
D: W&H: 354
Known with certainty only from the type collection from Surrey (Kew, Royal Botanic Gardens). Dubiously reported from Yorkshire in 1923.

croci Pass., *Uromyces*, in Rabenhorst, *Fungi europaei* IV: 2078 (1876)
S: !
D: W&H: 356
On cultivated *Crocus vernus*.

dactylidis G.H. Otth, *Mitt. naturf. Ges. Bern* 1861: 85 (1861)
Uromyces dactylidis var. *poae* (Rabenh.) Cummins, *Rust Fungi of Cereals, Grasses and Bamboos*: 474 (1971)
Uromyces festucae Syd., *Hedwigia* 39: 117 (1900)
Aecidium ficariae Pers., *Observ. mycol.* 1: 23 (1799)
Uromyces graminum Cooke, *Handb. Brit. fung.,* Edn 1: 520 (1871)
Uredo lycoctoni Kalchbr., *Verz. Zips. Schw.*: n. 900 (1865)
Uromyces lycoctoni (Kalchbr.) Trotter, in Traverso, *Fl. ital. crypt.* 1: 64 (1908)
Uromyces poae Rabenh., in Marcucci, *Unio itineraria*: 38 (1866)
Aecidium ranunculacearum DC., *Fl. Fr.* 5: 97 (1815) p.p.
Uromyces ranunculi-festucae Jaap, *Verh. Bot. Vereins Prov. Brandenburg* 47: 90 (1905)

Mis.: *Aecidium calthae* sensu Berk. & Broome [Ann. Mag. Nat. Hist., ser. 4, 17: 141 (1876)]
E: c **S:** c **W:** c **NI:** c **ROI:** c **O:** Isle of Man: !
H: 0 & I on various species of *Ranunculus*. II & III on *Dactylis glomerata* and species of *Festuca* and *Poa*.
D: W&H: 360
Very common and widespread.

dianthi (Pers.) Niessl, *Verhandl. naturforsch Vereines Brünn.* 10: 162 (1872)
  Uredo dianthi Pers., *Syn. meth. fung.*: 222 (1801)
  Uromyces caryophyllinus (Schrank) J. Schröt., *Rabenh. Krypt.-Fl.* 1(1): 149 (1884)
E: ! **NI:** ! **O:** Isle of Man: !
H: II & III on cultivated *Dianthus barbatus, D. caryophyllinus* and *D. chinensis.*
D: W&H: 312
Rarely reported. Occasionally occurring in epidemic proportions on cultivated species of *Dianthus* in a localised area, and possibly an introduction.

ervi Westend., *Bull. Acad. Roy. Sci. Belgique* 21(2): 234 (1854)
  Aecidium ervi Wallr., *Fl. crypt. Germ., Sect. II*: 247 (1833)
E: ! **S:** ! **W:** ! **O:** Channel Islands: !
H: 0, I, II & III on *Vicia hirsuta.*
D: W&H: 322
Rarely reported although the host is common and widespread.

fallens (Arthur) Barthol., *Handbook of the North American Uredinales*: 61 (1928)
  Nigredo fallens Arthur, *North American Flora* 7: 254 (1912)
  Trichobasis fallens Cooke, *J. Bot.* 2: 344 (1864)
E: ! **S:** ! **W:** ! **NI:** ! **O:** Channel Islands: !
H: 0, I, II & III on *Trifolium pratense* (possibly also on *T. incarnatum* and *T. medium* but these records require confirmation).
D: W&H: 326
Uncommonly reported but apparently widespread.

ficariae (Schumach.) Lév., in Orbigny, *Dict. univ. hist. nat.* 12: 786 (1848)
  Uredo ficariae Schumach., *Enum. pl.* 2: 232 (1803)
  Puccinia ficariae DC., *Fl. Fr.* 2: 225 (1805)
E: c **S:** c **W:** c **NI:** c **ROI:** c **O:** Channel Islands: c Isle of Man: c
H: II & III on *Ranunculus ficaria* ssp. *ficaria* and *R. ficaria* ssp. *bulbifera.*
D: W&H: 310
Very common and widespread.

gageae Beck, *Verh. K. K. Zool.-Bot. Ges. Wien* 30: 26 (1880)
E: !
H: III on *Gagea lutea.*
D: W&H: 351
Known from Co. Durham, Oxfordshire and Yorkshire.

gentianae Arthur, *Bot. Gaz.* 16: 227 (1891)
  Uromyces eugentianae Cummins, *Mycologia* 48: 608 (1956)
E: !
H: II & III on *Gentianella amarella.*
D: W&H: (as *Uromyces eugentianae*)
Known from Dorset, East Kent, North Devon, South Hampshire, West Suffolk, West Sussex and Worcestershire, following the limited distribution of the host.

geranii (DC.) Fr., *Summa veg. Scand.*: 514 (1849)
  Uredo geranii DC., *Syn. Pl. Gall.*: 47 (1806)
  Aecidium geranii DC., *Fl. Fr.* 5: 93 (1815)
  Trichobasis geranii Berk., *Outl. Brit. fungol.*: 333 (1860)
  Uromyces kabatianus Bubák, *Sitzungsber. Königl. Böhm. Ges. Wiss., Math.-Naturwiss. Cl.* 46: 1 (1902)
E: c **S:** o **W:** ! **ROI:** ! **O:** Channel Islands: !
H: 0, I, II & III on *Geranium dissectum, G. molle, G. pratense, G. pusillum, G. pyrenaicum, G. robertianum, G. rotundifolium* and *G. sylvaticum* also on various *Geranium* cultivars
D: W&H:
Common and widespread. Least common on *Geranium robertianum.*

inaequialtus Lasch, in Rabenhorst, *Fungi europaei* IV: 94 (1859)
E: ? **O:** Channel Islands: !
H: II & III on *Silene nutans.*
D: W&H: 313-314
Known from the Channel Islands (Herm). Reported also from West Kent but unsubstantiated with voucher material.

junci (Desm.) Tul., *Ann. Sci. Nat., Bot.,* sér. 4, 2: 146 (1854)
  Puccinia junci Desm., *Pl. crypt. N. France*: no. 81 (1825)
  Aecidium zonale Bréb., in Duby, *Bot. gall.*: 906 (1830)
E: ! **W:** !
H: 0 & I on *Pulicaria dysenterica*. II & III on *Juncus articulatus, J. bufonius, J. effusus, J. inflexus* and *J. subnodulosus.*
D: W&H: 355
Rarely reported with records only from widely scattered locations in southern and eastern England and Wales (N. Stringer, pers. comm.).

limonii (DC.) Berk., *Outl. Brit. fungol.*: 333 (1860)
  Puccinia limonii DC., *Fl. Fr.* 2: 595 (1805)
  Aecidium statices Cooke, *Microscopic fungi*: 197 (1865)
  Mis.: *Uredo statices* sensu auct.
E: ! **S:** ! **W:** !
H: 0, I, II & III on *Limonium* spp.
D: W&H: 346
Rarely reported, occurring mainly along the east coast of England.

lineolatus (Desm.) J. Schröt., in Rabenhorst *Fungi europaei* IV: 2077 (1876)
  Puccinia lineolata Desm., *Ann. Sci. Nat., Bot.,* sér. 3, 11: 273 (1849)
  Aecidium glaucis Dozy & Molk., *Tijdschr. Nat. Gesch. Deel* 12: 16 (1844)
  Uromyces lineolatus f.sp. *glaucis-scirpi* Jaap, *Ann. Mycol.* 3: 397 (1905)
  Uromyces lineolatus f.sp. *scirpi-oenanthi-crocati* Maire, *Compt. Rend. Congr. Natl. Soc. Savantes*: 125 (1912) [1911]
  Uromyces maritimae Plowr., *Gard. Chron.* 9: 188 (1884)
  Uromyces scirpi Burrill, *Bot. Gaz.* 9: 188 (1884)
E: o **S:** o **W:** o
H: 0 & I on *Berula erecta, Glaux maritima, Oenanthe crocata, O fistulosa, O lachenalii* and *Sium latifolium*. III on *Bulboschoenus (Scirpus) lacustris*
D: W&H: 357
Scarce *fide* WH but rather frequently reported and seemingly widespread in Britain.

minor J. Schröt., *Krypt.-Fl. Schlesien*: 310 (1887)
E: ! **W:** ! **ROI:** ? **O:** Channel Islands: !
H: I, II & III on *Trifolium campestre, T. dubium* and *T. incarnatum* ssp. *molinerii.*
D: W&H: 327
Rarely reported but apparently widespread.

muscari (Duby) L. Graves., *Cat. des plantes obs. dept. l'Oise Beuvais*: 280 (1857)
  Uredo muscari Duby, *Bot. gall.*: 898 (1830)
  Uromyces concentricus Lév., *Ann. Sci. Nat., Bot.,* sér. 3, 8: 371 (1847)
  Puccinia scillarum J.W. Baxter, *Stirp. crypt. oxon.*: 40 (1825)
  Uromyces scillarum (J.W. Baxter) G. Winter, *Pilze Deutschl.*: 142 (1884)
  Uredo scillarum Berk., *Engl. fl.* 5(2): 376 (1836)
  Trichobasis scillarum Berk., *Outl. Brit. fungol.*: 332 (1860)
E: c **S:** c **W:** c **NI:** c **ROI:** c **O:** Channel Islands: c Isle of Man: c
H: III on *Hyacinthoides hispanica, H. non-scripta, H. hispanica x non-scripta, Muscari polyanthum, Scilla bifolia* and *S. verna.*
D: W&H: 353
Common and widespread, especially on *Hyacinthoides hispanica.*

pisi-sativi (Pers.) Liro, *Uredineae Fennicae*: 100 (1908)
  Uredo appendiculata β *pisi-sativi* Pers., *Syn. meth. fung.*: 222 (1801)

Uredo appendiculata γ *genistae-tinctoriae* Pers., *Syn. meth. fung.*: 222 (1801)
Uromyces astragali Sacc., *Mycotheca veneta*: 208 (1873)
Aecidium euphorbiae Pers., in Gmelin, *Syst. nat.*: 128 (1791)
Uromyces euphorbiae-astragali Jordi, *Centralbl. Bakteriol. Parasitenk.* Abt. II, 11: 790 (1904)
Uromyces euphorbiae-corniculatae Jordi, *Centralbl. Bakteriol. Parasitenk.* Abt. II, 11: 791 (1904)
Uromyces genistae-tinctoriae (Pers.) G. Winter, *Hedwigia* 19: 36 (1880)
Uromyces genistae-tinctoriae f. *scoparii* MacDon., *Trans. Brit. Mycol. Soc.* 29: 67 (1946)
Uromyces genistae-tinctoriae f. *anglicae* MacDon., *Trans. Brit. Mycol. Soc.* 29: 67 (1946)
Uromyces genistae-tinctoriae f. *ulicis* MacDon., *Trans. Brit. Mycol. Soc.* 29: 67 (1946)
Puccinia laburni DC., *Fl. Fr.* 2: 224 (1805)
Uromyces laburni (DC.) G.H. Otth, *Mitt. naturf. Ges. Bern* 1863: 87 (1863)
Puccinia loti Kirchn., *Lotos* 6: 181 (1856)
Uromyces loti A. Blytt, *Forh. Vidensk.-Selsk. Kristiania* 4(6): 37 (1896)
Uromyces onobrychidis Bubák, *Sitzungsber. Köngl. Böhm. Ges. Wiss., Math.-Naturwiss. Cl.* 46: 7 (1902)
Uromyces phacae Thüm., *Bull. Soc. Imp. Naturalistes Moscou* 53: 218 (1878)
Puccinia pisi DC., *Fl. Fr.* 2: 224 (1805)
Uromyces pisi (DC.) G.H. Otth, *Mitt. naturf. Ges. Bern* 1863: 87 (1863)
Uromyces punctatus J. Schröt., *Abh. Schles. Ges. Vaterl. Cult., Abth. Naturwiss. (Naturwiss.-Med. Abth.)* 1869-72: 10 (1870) [1869]
Uromyces striatus J. Schröt., *Abh. Schles. Ges. Vaterl. Cult., Abth. Naturwiss. (Naturwiss.-Med. Abth.)* 1869-72: 11 (1870) [1869]
E: o **S:** ! **W:** ! **NI:** ! **ROI:** ! **O:** Channel Islands: o
H: 0 & I on *Euphorbia cyparissias*. II & III on various *Fabaceae*.
D: W&H: 330 (as *Uromyces pisi*)
Occasional in England. Rarely reported elsewhere but apparently widespread. Recent collections of uredia on *Galega officinalis* from Surrey (West Molesey) and Middlesex (Shepperton) may belong here but are possibly *U. galegae* (Opiz) Sacc.

polygoni-avicularis (Pers.) P. Karst., *Bidrag Kännedom Finlands Natur Folk* 4: 12 (1879)
Puccinia polygoni-avicularis Pers., *Syn. meth. fung.*: 227 (1801)
Puccinia polygoni Pers., *Tent. disp. meth. fung.*: 40 (1797)
Capitularia polygoni (Pers.) Rabenh., *Bot. Zeitung (Berlin)* 9: 449 (1851)
Uromyces polygoni (Pers.) Fuckel, *Jahrb. Nassauischen Vereins Naturk.* 23-24: 64 (1870)
E: ! **S:** ! **W:** ! **NI:** ! **ROI:** ! **O:** Channel Islands: !
H: 0, I, II & III on *Polygonum avenastrum*, *P. aviculare* and *Rumex acetosella*.
D: W&H: 342
Widespread and frequent, at least on *Polygonum aviculare*.

rumicis (Schumach.) G. Winter, *Hedwigia* 19: 37 (1880)
Uredo rumicis Schumach., *Enum. pl.* 2: 231 (1803)
Uredo bifrons DC., *Fl. Fr.* 5: 229 (1815)
Uredo rumicum DC., *Fl. Fr.* 5: 66 (1815)
E: c **S:** c **W:** o **NI:** ! **ROI:** ! **O:** Channel Islands: c Isle of Man: !
H: 0 & I on *Ranunculus ficaria*. II & III on *Rumex conglomeratus*, *R. crispus*, *R. hydrolapathum*, *R. longifolius*, *R. obtusifolius*, *R. maritimus*, *R. patientia* and *R. sanguineus*
D: W&H: 343
Common and widespread, although the aecidial stage on *Ranunculus ficaria* is however, rarely reported.

salicorniae de Bary, in Rabenhorst, *Fungi europaei* IV: 1386 (1870)
E: ! **W:** ! **O:** Channel Islands: !
H: 0, I, II & III on *Salicornia europaea*, *S. dolichostachya*, *S. perennis* and *S. ramosissima*.
D: W&H: 317

Rarely reported but apparently widespread.

scrophulariae Fuckel, *Fungi Rhenani exsiccati. Fasc. IV*: 395 (1863)
Uromyces concomitans Berk. & Broome, *Ann. Mag. Nat. Hist, Ser. 4 [Notices of British Fungi no. 1470]* 15: 36 (1875)
Aecidium scrophulariae DC., *Fl. Fr.* 5: 91 (1815)
E: ! **W:** ! **NI:** ! **ROI:** ! **O:** Channel Islands: !
H: 0, I & III on *Scrophularia aquatica*, *S. auriculata*, *S. nodosa*, *S. scorodonia* and *S. umbrosa*.
D: W&H: 347
Rarely reported but apparently widespread. Sori are inconspicuous and often formed at on the stems at soil level.

scutellatus (Pers.) Lév., *Ann. Sci. Nat., Bot.*, sér. 3, 8: 371 (1847)
Uredo scutellata Pers., *Syn. meth. fung.*: 220 (1801)
E: ! **S:** !
H: III on *Euphorbia cyparissias* and *Euphorbia* x *pseudovirgata*.
D: W&H: 339
Known from England (East Kent, South Essex and West Suffolk) and Scotland (West Sutherland).

sparsus (J. Kunze & J.C. Schmidt) Cooke, *Microscopic fungi*: 214 (1865)
Uredo sparsa J. Kunze & J.C. Schmidt, *Deutschl. Schwämme* 7: 5 (1817)
E: !
H: 0, I, II & III on *Spergularia marina*, *S. media* and *S. rubra*.
D: W&H: 314
Known mainly from Norfolk and Suffolk, but a few collections from Dorset and North Somerset.

tinctoriicola Magnus, *Verh. zool.-bot. Ges. Wien* 46: 429 (1896)
Uromyces hybernae Liou, *Bull. Soc. Mycol. France* 45: 121 (1929)
ROI: !
H: III on *Euphorbia hyberna*.
D: W&H: 340
Known from North Kerry, South Kerry and West Cork. The host is virtually restricted to the same area.

trifolii (Hedw.) Fuckel, *Jahrb. Nassauischen Vereins Naturk.* 23-24: 63 (1870)
Puccinia trifolii Hedw., in DC., *Fl. Fr.* 2: 225 (1805)
Uredo trifolii (Hedw.) DC., in Lam., *Encyclop.* 8: 223 (1808)
Uromyces flectens Lagerh., *Svensk bot. Tidskr.* 3: 36 (1909)
Puccinia nerviphila Grognot, *Pl. crypt. Saône-et-Loire*: 154 (1863)
Uromyces nerviphilus (Grognot) Hotson, *Publ. Puget Sound Biol. Sta.* 4: 368 (1925)
E: ! **S:** ! **W:** ! **NI:** ! **ROI:** ! **O:** Channel Islands: ! Isle of Man: !
H: III on *Trifolium repens* and *T. fragiferum*.
D: W&H: 328 (as *Uromyces nerviphilus*)
Rarely reported but apparently widespread.

trifolii-repentis Liro, *Acta Soc. Fauna Fl. Fenn.* 29(6): 15 (1906)
Mis.: *Uromyces trifolii* sensu auct.
E: ! **NI:** ! **O:** Channel Islands: !
H: III on *Trifolium repens*.
D: W&H: 337 (as *Uromyces trifolii*)
Rarely reported but apparently widespread.

tuberculatus Fuckel, *Jahrb. Nassauischen Vereins Naturk.* 23-24: 64 (1870)
Uromyces excavatus Cooke, *Grevillea* 2: 161 (1874)
Mis.: *Aecidium euphorbiae* sensu auct.
E: !
H: 0, I, II & III on *Euphorbia exigua*.
D: W&H: 341
Known from East Gloucestershire, Northamptonshire, Oxfordshire, South Hampshire and West Suffolk. Records supposedly on *Euphorbia hyberna* from Ireland are actually *U. tinctoriicola*.

valerianae Fuckel, *Jahrb. Nassauischen Vereins Naturk.* 23-24: 63 (1870)
  Uredo valerianae DC., *Fl. Fr.* 5: 68 (1815)
  Lecythea valerianae Berk., *Outl. Brit. fungol.*: 334 (1860)
  Aecidium valerianearum Duby, *Bot. gall.*: 908 (1830)
  Trichobasis parnassiae Cooke, *J. Bot.* 2: 344 (1864)
E: c **S:** c **W:** c **NI:** ! **ROI:** ! **O:** Isle of Man: !
H: 0, I, II & III on *Valeriana officinalis.*
D: W&H: 348
Common and widespread.

viciae-fabae P. Karst., *Bidrag Kännedom Finlands Natur Folk* 31: 13 (1879)
  Puccinia fabae Grev., *Scott. crypt. fl.* I: t. 29 (1823)
  Trichobasis fabae Lév., in Orbigny. *Dict. univ. hist. nat.* 1: 19 (1841)
  Uromyces fabae (Grev.) de Bary [*nom. nud.*], *Ann. Sci. Nat., Bot.,* sér. 4, 20: 80 (1863)
  Puccinia fallens Cooke, *J. Bot.* 4: 105 (1866)
  Uredo viciae-fabae Pers., *Syn. meth. fung.*: 221 (1801)
E: c **S:** c **W:** c **NI:** c **ROI:** c **O:** Channel Islands: ! Isle of Man: !
H: 0, I, II & III on *Pisum sativum, Vicia angustifolia, V. bithynica, V. cracca, V. faba, V. hirsuta, V. lathyroides, V. lutea, V. sativa* and *V. sepium.*
D: W&H: 323 (as *Uromyces viciae-fabae*)
Common and widespread.

viciae-fabae *var.* **orobi** (Schumach.) Jørst., *K. norske Vidensk. Selsk. Skr.* 38: 46 (1936) [1935]
  Uredo orobi Schumach., *Enum. pl.* 2: 232 (1803)
  Uromyces orobi (Schumach.) Lév., *Ann. Sci. Nat., Bot.,* sér. 3, 8: 371 (1847)
  Aecidium orobi DC., *Fl. Fr.* 5: 95 (1815)
E: ! **S:** ! **W:** ! **ROI:** !
H: II & III on *Lathyrus montanus.*
D: W&H: 326
Rarely reported but apparently widespread.

USTILENTYLOMA Savile, *Canad. J. Bot.* 42: 708 (1964)
Microbotryales, Ustilentylomataceae
Type: *Ustilentyloma pleuropogonis* Savile

brefeldii (Willi Krieg.) Vánky, *Mycotaxon* 41(2): 491 (1991)
  Entyloma brefeldii Willi Krieg., *Hedwigia* 35: 145 (1896)
S: !
H: In leaves of *Arrenatherum elatius.*
A single collection from Isle of Skye (Aird) in 1983.

XENODOCHUS Schltdl., *Linnaea* 1: 237 (1826)
Uredinales, Phragmidiaceae
Type: *Xenodochus carbonarius* Schltdl.

carbonarius Schltdl., *Linnaea* 1: 237 (1826)
  Phragmidium carbonarium G. Winter, *Rabenh. Krypt.-Fl.* 1(1): 227 (1882)
E: o **S:** o **W:** !
H: I, II & III on *Sanguisorba officinalis.*
D: W&H: 109-110
Occasional in northern and western Britain. Reportedly 'frequent' in southern England but the host is rare in this area and the records unsubstantiated with voucher material. Many of these southern records may be misidentified *Phragmidium sanguisorbae* on *Poterium sanguisorba.*

ZAGHOUANIA Pat., *Bull. Soc. Mycol. France* 17: 187 (1901)
Uredinales, Pucciniaceae
Type: *Zaghouania phillyreae* Pat.

phillyreae Pat., *Bull. Soc. Mycol. France* 17: 187 (1901)
  Aecidium crassum var. *phillyreae* Cooke, *Handb. Brit. fung.,* Edn 1: 539 (1871)
  Aecidium phyllyreae DC., *Fl. Fr.* 5: 96 (1815)

  Uredo phyllyreae Cooke [*nom. inval.*], *Fung. Brit. Exs.*: no. 592 (1871)
E: !
H: 0, I, & II on planted *Phyllyrea latifolia* and *P. latifolia* var. *media.*
D: W&H: 13-14
Known from Dorset and West Sussex and possibly an introduction. Unreported since 1907 and probably extinct.

ZYGOGLOEA P. Roberts, *Mycotaxon* 52(1): 241 (1994)
Platygloeales, Incertae sedis
Type: *Zygogloea gemellipara* P. Roberts

gemellipara P. Roberts, *Mycotaxon* 52(1): 243 (1994)
E: !
H: Parasitic. In the hymenium of *Exidia nucleata.*
D+I: Reid4: 424-425 (as conidial *Myxarium nucleatum*)
Known from East Sussex and South Devon.

INCLUDED USTILAGINOMYCETES

ANTHRACOIDEA Bref., *Unters. Gesammtgeb. Mykol.* 12: 144 (1895)
Ustilaginales, Ustilaginaceae
Type: *Anthracoidea caricis* (Pers.) Bref.

arenariae (Syd.) Nannf., *Bot. Not.* 130(4): 365 (1977)
    Cintractia arenariae Syd., *Ann. Mycol.* 22: 289 (1924)
    Mis.: *Ustilago caricis* sensu auct. p.p.
E: ! **W:** ! **O:** Channel Islands: !
H: In ovaries of *Carex arenaria* growing on dunes or beaches, rarely inland.
D: Vánky: 23
Rarely reported but apparently widespread. The host is common in coastal sites.

bigelowii Nannf., *Svensk bot. Tidskr.* 59: 203 (1965)
S: !
H: In the ovaries of *Carex bigelowii.*
D: Vánky: 25
A single collection (undated) from South Aberdeenshire in herb. K.

capillaris Kukkonen, *Ann. Bot. Soc. Zool.-Bot. Fenn. 'Vanamo'* 34: 50 (1963)
S: !
H: In the ovaries of *Carex capillaris* in montane habitat.
D: Vánky: 25
A single collection (as '*Cintractia caricis*') in herb. K from South Aberdeenshire in 1844.

caricis (Pers.) Bref., *Unters. Gesammtgeb. Mykol.* 12: 144 (1895)
    Uredo caricis Pers., *Syn. meth. fung.*: 225 (1801)
    Ustilago caricis (Pers.) Unger, *Einfl. Boden. Verth. Gew.*: 211 (1836)
    Cintractia caricis (Pers.) Magnus, *Verh. Bot. Vereins Prov. Brandenburg* 37: 79 (1896)
    Uredo urceolorum DC. [*nom. nov.* for *Uredo caricis* Pers.], *Fl. Fr.* 5: 78 (1815)
    Caeoma urceolorum (DC.) Schltdl., *Fl. berol.* 2: 130 (1824)
    Ustilago urceolorum (DC.) Tul., *Ann. Sci. Nat., Bot.,* sér. 3, 7: 86 (1847)
E: ! **S:** !
H: In the ovaries of *Carex pallescens, C. panicea, C. pilulifera* and *C. 'trinervis'.*
D: UBI 14 (as *Anthracoidea caricis* p.p.), Vánky: 25-26
Rarely reported but apparently widespread.

heterospora (B. Lindeb.) Kukkonen, *Ann. Bot. Soc. Zool.-Bot. Fenn. 'Vanamo'* 34(3): 63 (1963)
    Cintractia heterospora B. Lindeb., *Svensk bot. Tidskr.* 51: 500 (1957)
    Cintractia carpophila (Schumach.) Liro [*comb. illegit.*], *Ann. Acad. Sci. Fenn., Ser. A,* 42: 27 (1938)
S: !
H: In the ovaries of *Carex echinata.*
D: Vánky: 29
A single collection (in herb. E), from Easterness (Aviemore) in 1955.

karii (Liro) Nannf., *Bot. Not.* 130(4): 368 (1977)
    Cintractia karii Liro, *Mycoth. fenn.* 2: 36 (1935) [1934]
    Mis.: *Cintractia caricis* sensu auct. p.p.
S: !
H: In the ovaries of *Carex echinata.*
D: Vánky: 31
Known from a few localities in Scotland. The host is widespread in Britain.

limosa (Syd.) Kukkonen, *Ann. Bot. Soc. Zool.-Bot. Fenn. 'Vanamo'* 34(3): 91 (1963)
    Cintractia limosa Syd., *Ann. Mycol.* 22: 288 (1924)
S: !

H: In the ovaries of *Carex limosa.*
D: Vánky: 32-33
Known from Mid-Perthshire and West Sutherland.

paniceae Kukkonen, *Ann. Bot. Soc. Zool.-Bot. Fenn. 'Vanamo'* 34(3): 76 (1963)
    Mis.: *Cintractia caricis* sensu auct. p.p.
E: ! **S:** !
H: In the ovaries of *Carex panicea.*
D: Vánky: 34
Rarely reported. Known from scattered sites in northern England and Scotland.

pratensis (Syd.) Boidol & Poelt, *Ber. Bayer. Bot. Ges.* 36: 23 (1963)
    Cintractia pratensis Syd., *Ann. Mycol.* 22: 289 (1924)
    Mis.: *Cintractia caricis* sensu auct. p.p.
E: ! **S:** !
H: In the ovaries of *Carex flacca.*
D: Vánky: 35
Rarely reported. The host is common throughout Britain.

pseudirregularis U. Braun, *Boletus* 6(3): 52 (1982)
S: !
H: In the ovaries of *Carex pallescens.*
D: Vánky: 35
A single collection from the Isle of Mull in 1966.

pulicaris Kukkonen, *Ann. Bot. Soc. Zool.-Bot. Fenn. 'Vanamo'* 34(3): 45 (1963)
E: ! **S:** !
H: In the ovaries of *Carex pulicaris.*
D: Vánky: 35
Known from England (East Norfolk) and Scotland (North Ebudes).

scirpi (J.G. Kühn) Kukkonen, *Ann. Bot. Soc. Zool.-Bot. Fenn. 'Vanamo'* 34(3): 69 (1963)
    Ustilago urceolorum f. *scirpi* J.G. Kühn, in Rabenhorst, *Fungi europaei* IV: 150 (1873)
    Cintractia scirpi (J.G. Kühn) Schellenb., *Beitr. Kryptogamenfl. Schweiz* 3: 77-78 (1911)
S: !
H: In the ovaries of *Trichophorum caespitosum* [= *Scirpus caespitosus*].
D: UBI 16, Vánky: 36
Known only from Scotland (Angus, East Perthshire and Wester Ross), although the host plant is common throughout Britain.

subinclusa (Körn.) Bref., *Unters. Gesammtgeb. Mykol.* 12: 146 (1895)
    Ustilago subinclusa Körn., *Hedwigia* 13: 159 (1874)
    Cintractia subinclusa (Körn.) Magnus, *Verh. Bot. Vereins Prov. Brandenburg* 37: 79 (1896) [1895]
E: ! **S:** ? **ROI:** !
H: In the ovaries of *C. riparia* and *C. rostrata.* (and dubiously *C. lasiocarpa*).
D: UBI 16, Vánky: 37
Rarely reported but apparently widespread. The hosts are common. Reported from Scotland but unsubstantiated with voucher material.

ARCTICOMYCES Savile, *Canad. J. Bot.* 37: 984 (1959)
Exobasidiales, Exobasidiaceae
Type: *Arcticomyces warmingii* (Rostr.) Savile

warmingii (Rostr.) Savile, *Canad. J. Bot.* 37: 984 (1959)
    Exobasidium warmingii Rostr., *Fungi Groenl.*: 530 (1888)
S: !
H: On leaves of *Saxifraga oppositifolia* in montane or boreal habitat.

Collected in Perthshire in 1959 and 1961. The host is not uncommon in montane habitat.

DICELLOMYCES L.S. Olive, *Mycologia* 37: 544 (1945)
Incertae sedis
Type: *Dicellomyces gloeosporus* L.S. Olive

scirpi Raitv., *Eesti N.S.V. Tead. Akad. Toimet., Biol.* 17: 223 (1968)
E: !
H: Parasitic. On living leaves of *Scirpus sylvaticus* in woodland. Known only from Berkshire and Surrey. The host is often common in damp woodland.

DOASSANSIA Cornu, *Ann. Sci. Nat., Bot.*, sér. 6, 15: 280 (1883)
Ustilaginales, Tilletiaceae
Type: *Doassansia alismatis* (Nees) Cornu

alismatis (Nees) Cornu, *Ann. Sci. Nat., Bot.*, sér. 6, 15: 280 (1883)
 Sclerotium alismatis Nees, in Fries, *Syst. mycol.* 2: 257 (1823)
 Perisporium alismatis (Nees) Fr., *Syst. mycol.* 3: 252 (1829)
E: ! **S:** !
H: In living or moribund leaves of *Alisma plantago-aquatica*.
D: UBI 17, Vánky: 61
Known from England (East Norfolk, East Suffolk and Gloucestershire) and Scotland (Midlothian, Perthshire and South Aberdeen).

limosellae (J. Kunze) J. Schröt., in Cohn, *Krypt.-Fl. Schlesien* 3: 287 (1877)
 Protomyces limosellae J. Kunze, in Rabenhorst, *Fungi europaei* IV: 1694 (1873)
 Entyloma limosellae (J. Kunze) G. Winter, *Rabenh. Krypt.-Fl.* 1: 115 (1881)
E: !
H: In living leaves and petioles of *Limosella aquatica*.
D: UBI 17, Vánky: 63
Known only from Warwickshire (Earlswood Reservoir), collected in 1921 (Grove, 1922) and 1929.

sagittariae (Fuckel) C. Fisch, *Ber. Deutsch. Bot. Ges.* 2: 405 (1884)
 Physoderma sagittariae Fuckel, *Fungi Rhenani exsiccati. (Suppl.). Fasc. I*: 1549 (1865)
 Protomyces sagittariae (Fuckel) Fuckel, *Jahrb. Nassauischen Vereins Naturk.* 23-24: 75 (1870)
 Uredo sagittariae Westend. [*nom. nud.*], *Herb. crypt. Belg.* Fasc. 24: no.1177 (1857)
 Aecidium incarceratum Berk. & Broome, *Ann. Mag. Nat. Hist, Ser. 4 [Notices of British Fungi no. 1469]* 15: 36 (1875)
E: !
H: In leaves of *Sagittaria sagittifolia* in rivers, canals and ponds.
D: UBI 18, Vánky: 64
Known from Buckinghamshire, Dorset, East Suffolk, East Sussex, Huntingdonshire, Surrey, Warwickshire, Worcestershire and Yorkshire.

DOASSANSIOPSIS (Setch.) Dietel, *Nat. Pflanzenfamilien*: 21 (1897)
Ustilaginales, Tilletiaceae
Type: *Doassansiopsis deformans* (Setch.) Dietel

hydrophila (A. Dietr.) Lavrov, *Animadv. syst. Herb. Univ. Tomsk* 11: 4 (1937)
 Sphaeria hydrophila A. Dietr., *Arch. Naturk. Liv- Ehst-Kurlands, Ser. 2, Biol. Naturk.* 1: 512 (1859)
 Doassansia hydrophila (A. Dietr.) B. Lindeb., *Symb. Bot. Upsal.* 16(2): 23 (1959)
 Protomyces martianoffianus Thüm., *Byull. Moskovsk. Obshch. Isp. Prir., Otd. Biol.* 53: 207 (1878)

 Doassansia martianoffiana (Thüm.) J. Schröt., in Cohn, *Krypt.-Fl. Schlesien* 3: 287 (1887)
 Ramularia aquatilis Peck (anam.), *Rep. (Annual) New York State Mus. Nat. Hist.* 35: 142 (1884)
 Savulescuella aquatilis (Peck) Cif. (anam.), *Lejeunia Mém.*: 179 (1959)
 Doassansiella aquatilis (Peck) Zambett. (anam.), *Rev. Mycol. (Toulouse)* 35 (1-2): 165 (1970)
E: ! **S:** o **W:** !
H: In floating leaves *Potamogeton* spp, usually in lakes, ponds or streams especially when these have dried out. British material on *Potamogeton natans* and *P. polygonifolius*.
D: UBI 18 (as *Doassansia martianoffiana*), Vánky: 71-72
Rarely reported but apparently widespread. Most frequently collected in Scotland but also known from England (East Norfolk, East Suffolk, Herefordshire and South Hampshire) and Wales (Montgomeryshire).

ENTORRHIZA C.A. Weber, *Bot. Zeitung (Berlin)* 42: 378 (1884)
Ustilaginales, Tilletiaceae
 Schinzia Nägeli, *Linnaea* 16: 281 (1842)
Type: *Entorrhiza cypericola* (Magnus) Webber

aschersoniana (Magnus) Lagerh., *Hedwigia* 27: 262 (1888)
 Schinzia aschersoniana Magnus, *Ber. Deutsch. Bot. Ges.* 6: 103 (1888)
S: !
H: In root nodules of *Juncus bufonius*.
D: UBI 19, Vánky: 73
Rarely reported. Sori are cryptic, easily overlooked, and rarely searched for. The host is common throughout Britain, but collections are only from Scotland (Mid Ebudes and South Aberdeenshire).

casparyana (Magnus) Lagerh., *Hedwigia* 27: 262 (1888)
 Schinzia casparyana Magnus, *Ber. Deutsch. Bot. Ges.* 6: 103 (1888)
 Entorrhiza digitata Lagerh., *Hedwigia* 27: 264 (1888)
S: !
H: In roots nodules of *Juncus articulatus*.
D: UBI 19, Vánky: 74
Known from Scotland (North Ebudes and Wester Ross). Sori are cryptic and easily overlooked. The host is common throughout Britain.

scirpicola (Correns) Sacc. & Syd., *Syll. fung.* 14: 425 (1899)
 Schinzia scirpicola Correns, *Hedwigia* 36: 40 (1897)
E: ! **S:** !
H: In roots of *Eleocharis quinqueflora* in woodland and swampy areas.
D: UBI 20, Vánky: 75
Rarely reported. Known from England (East Norfolk) and Scotland (Perthshire) but sori are cryptic and easily overlooked. The host is widespread in Britain.

ENTYLOMA de Bary, *Bot. Zeitung (Berlin)* 32: 101 (1874)
Ustilaginales, Tilletiaceae
 Entylomella Höhn. (anam.), *Ann. Mycol.* 22: 191 (1924)
Type: *Entyloma ungerianum* de Bary

achilleae Magnus, *Abh. Naturhist. Ges. Nürnberg.* 13: 8 (1900)
 Entylomella microstigma (Sacc.) Cif. (anam.), *Omagiu lui Traian Savulescu*: 177 (1959)
S: !
H: In leaves of *Achillea millefolium*.
D: UBI 21-22, Vánky: 84
Known from Clyde Isles and South Aberdeen but not reported since 1944.

bellidis W. Krieg., *Beibl. Hedwigia* 35: 145 (1896)
 Entyloma calendulae f. *bellidis* (W. Krieg.) Ainsw. & Sampson, *British Smut Fungi (Ustilaginales)*: 104 (1950)

Entylomella bellidis Cif. (anam.), *Omagiu lui Traian Savulescu.* 176 (1959)
E: ! **S:** !
H: In leaves of *Bellis perennis.*
D: UBI 22 (as *Entyloma calendulae* f. *bellidis*), Vánky: 86
Rarely reported.

calendulae (Oudem.) de Bary, *Bot. Zeitung (Berlin)* 32: 102 (1874)
Protomyces calendulae Oudem., *Arch. néerl. Sci.* 8: 384 (1873)
E: ! **S:** ! **W:** ! **NI:** !
H: In leaves of *Calendula offinalis* and other *Calendula* cultivars.
D: UBI 22 (as *E. calendulae* f. *calendulae*), Vánky: 87
Rarely reported but apparently not uncommon in south western England and Scotland. Anamorph present, but as yet un-named.

chrysosplenii (Berk. & Broome) J. Schröt., *Beitr. Biol. Pflanzen* 2: 372 (1877)
Protomyces chrysosplenii Berk. & Broome, *Ann. Mag. Nat. Hist, Ser. 4 [Notices of British Fungi no. 1472]* 15: 36 (1875)
Entylomella chrysospleni Cif. (anam.), *Omagiu lui Traian Savulescu.* 176 (1959)
E: ! **S:** !
H: In leaves of *Chrysosplenium* spp
D: UBI 23, Vánky: 88
Rarely reported. Described from Scotland and also known from Yorkshire.

dactylidis (Pass.) Cif., *Boll. Soc. Bot. Ital.* 1924: 55 (1924)
Thecaphora dactylidis Pass., *Nuovo Giorn. Bot. Ital.* 9: 238 (1877)
Entyloma crastophilum Sacc., *Michelia* 1: 540 (1879)
Entyloma holci Liro [*nom. illegit.*], *Ann. Acad. Sci. Fenn., Ser. A,* 42: 97 (1938)
Entylomella crastophila Cif. (anam.), *Omagiu lui Traian Savulescu.* 176 (1959)
E: ! **S:** !
H: In leaves of *Holcus lanatus* and *H. mollis* and rarely on other grasses.
D: UBI 24, Vánky: 89
Rarely reported but apparently widespread. Probably a species complex in need of revision *fide* Vánky.

dahliae Syd. & P. Syd., *Ann. Mycol.* 10: 36 (1912)
Entyloma calendulae f. *dahliae* Sternon, *Sur une maladie nouvelle de Dahlia*: (non visi) (1918)
Entyloma dahliae (Sternon) Cif. (*nom. illegit.*), *Boll. Soc. Bot. Ital.* 1924: 48 (1924)
Entylomella dahliae Boerema & Hamers (anam.), *Netherlands J. Pl. Pathol.* 96(1): 7 (1990)
E: ! **S:** !
H: In leaves of *Dahlia* cultivars.
D: UBI 22-23 (as *Entyloma calendulae* f. *dahliae*), Vánky: 91
Rarely reported but widespread.

eryngii (Corda) de Bary, *Bot. Zeitung (Berlin)* 32: 105 (1874)
Physoderma eryngii Corda, *Icon. fung.* 3: 3 (1839)
S: ! **W:** !
H: In leaves of *Eryngium maritimum.*
D: UBI 24, Vánky: 92
Known from Scotland (Ayrshire) and Wales (Caernarvonshire). The host plant is quite common around coastal Britain. *Fide* Vánky the fungus on this host from Scotland may actually be *E. eryngii-plani* Cif. [*Boll. Soc. Bot. Ital.* 2: 54 (1924)].

fergussonii (Berk. & Broome) Plowr., *Monograph Brit. Ured.*: 289 (1889)
Protomyces fergussonii Berk. & Broome, *Ann. Mag. Nat. Hist, Ser. 4* 15: 36 (1875)
Entyloma canescens J. Schröt., *Beitr. Biol. Pflanzen* 2: 372 (1877)
Cylindrosporium myosotidis Sacc. (anam.), *Michelia* 1: 533 (1879)
E: ! **S:** !

H: In leaves of *Myosotis arvensis, M. laxa* ssp. *caespitosa* and *M. scorpioides.*
D: UBI 25, Vánky: 92
Rarely reported, though the hosts are all common in Britain.

ficariae Thüm. & A.A. Fisch. Waldh., *Byull. mosk. Obshch. Ispyt. Prir.* 52: 309 (1877)
Entyloma ungerianum f. *ficariae* G. Winter [*nom. nud.*], in Rabenhorst, *Fungi europaei* IV: 1873 (1874)
Protomyces ficariae Cornu & Roze [*nom. nud.*], *Bull. Soc. Bot. France* 21: 161 (1874)
Entyloma ranunculi (Bonord.) J. Schröt., *Beitr. Biol. Pflanzen* 2: 370 (1877)
Cylindrosporium ficariae Berk. (anam.), *British Fungi.* 212 (1837)
Gloeosporium ficariae (Berk.) Cooke, *Handb. Brit. fung.,* Edn 1: 475 (1871)
Entylomella ficariae (Berk.) Höhn., *Ann. Mycol.* 22: 193 (1924)
Fusidium ranunculi Bonord. (anam.), *Handb. Mykol.*: 43 (1851)
Mis.: *Protomyces microsporus* ss Cooke [Grevillea 3: 181 (1875)]
E: c **S:** c **W:** c **NI:** c **ROI:** c **O:** Channel Islands: c Isle of Man: c
H: In leaves of *Ranunculus ficaria* and rarely *R. repens* and *R. sceleratus.*
D: UBI 25, Vánky: 93
Very common and widespread.

fuscum J. Schröt., *Beitr. Biol. Pflanzen* 2: 373 (1877)
Entyloma bicolor Zopf [*nom. nud.*], *Hedwigia.* 17: 88-91 (1878)
Entylomella fusca Cif. (anam.), *Omagiu lui Traian Savulescu.* 177 (1959)
E: !
H: In leaves of *Papaver rhoeas.*
D: UBI 26, Vánky: 94
Known from Surrey and West Norfolk. The host is quite common and widespread.

gaillardianum Vánky, *Mycotaxon* 16(1): 104 (1982)
E: !
H: In leaves of cultivated species of *Gaillardia.*
Only one collection from an unspecified location in England in herb. K.

helosciadii Magnus, *Hedwigia.* 21: 129 (1882)
Cylindrosporium helosciadii-repentis Magnus (anam.), *Verh. Bot. Vereins Prov. Brandenburg* 35: 68 (1893)
Entylomella helosciadii-repentis (Magnus) Höhn., *Oesterr. Bot. Z.* 66: 105 (1916)
E: ! **S:** ! **ROI:** ? **O:** Channel Islands: !
H: In leaves of *Apium nodiflorum* and *Oenanthe crocata* in ditches, ponds and streams.
D: UBI 26, Vánky: 95
Rarely reported but apparently widespread. The hosts are common.

henningsianum Syd., *Hedwigia* 39: 123 (1900)
S: !
H: In leaves of *Samolus valerandi.*
D: UBI 26, Vánky: 95
A single collection, from Main Argyll (Inveraray, Dubh Loch) in 1907.

hieracii Cif., *Boll. Soc. Bot. Ital.* 1924: 50 (1924)
Entyloma calendulae f. *hieracii* J. Schröt., *Beitr. Biol. Pflanzen* 2: 439 (1876)
S: !
H: In leaves of *Hieracium* spp
D: UBI 95-96 (as *Entyloma calendulae* f. *hieracii*)
Known only from North Ebudes and Westerness.

irregulare Johanson, *Öfvers. Kongl. Vetensk.-Akad. Förh.* 41: 159 (1885)
E: !
H: In leaves of *Poa annua.*

A single British record from Gloucestershire, consisting only of the anamorph. Distinguished from *Entyloma dactylidis*, rarely on *Poa* spp, by the agglutinated, irregular spores.

linariae J. Schröt., *Beitr. Biol. Pflanzen* 2: 371 (1877)
E: !
H: In leaves of *Linaria vulgaris*.
D: UBI 27, Vánky: 97
Known only from South Devon. The host is common and widespread. Anamorph present but as yet unnamed.

matricariae Rostr. in Thüm., *Mycotheca universalis*: 2223 (1884)
    Entyloma matricariae Trail [*nom. inval.*], in Plowright, *Monograph Brit. Ured.*: 291 (1889)
    Entyloma trailii Massee, *Brit. fung.*: 192 (1891)
    Entylomella trailii (Massee) Cif. (anam.), *Omagiu lui Traian Savulescu*: 178 (1959)
E: ! **S:** !
H: In leaves of *Tripleurospermum inodorum*.
D: UBI 27, Vánky: 98-99
Known from Scotland (Main Argyll and South Aberdeenshire), and recently collected in England (Middlesex).

microsporum (Unger) J. Schröt., in Rabenhorst, *Fungi europaei* IV: 1872 (1874)
    Protomyces microsporus Unger, *Exanth. Pfl.*: 343 (1833)
    Entyloma ungerianum de Bary, *Bot. Zeitung (Berlin)* 32(7): 101 (1874)
    Cylindrosporium ranunculi var. *microsporum* D. Sacc. (anam.), *Mycotheca ital.*: 1456 (1904)
    Entylomella microsporum (D. Sacc.) Cif., *Fl. ital. crypt.* 17: 197 (1938)
E: ! **S:** ! **W:** ! **NI:** ! **ROI:** ! **O:** Channel Islands: !
H: In leaves of *Ranunculus acris* and *R. repens*.
D: UBI 27-28, Vánky: 99-100
Widespread though not often reported.

ossifragi Rostr., *Bot. Foren. Feskr. Kijöbenh.* 54: 133 (1890)
S: !
H: In leaves of *Narthecium ossifragum*.
D: UBI 28, Vánky: 100
A single collection from West Sutherland (Cnoc Bad na h' Achlaise) in 1954. The host is common in heathland and moorland.

ranunculi-repentis Sternon, *L'heterogenite du genre Ramularia - Thése*: 34 (1925)
    Fusidium eburneum J. Schröt. (anam.), *Beitr. Biol. Pflanzen* 2: 373 (1877)
    Entylomella eburnea (J. Schröt.) Cif., *Omagiu lui Traian Savulescu*: 176 (1959)
    Ramularia gibba Fuckel (anam.), *Fungi Rhenani exsiccati (Suppl.). Fasc. II*: 1636 (1866)
    Entylomella gibba (Fuckel) U. Braun, *A Monograph of Cercosporella, Ramularia and Allied Genera (Phytopathogenic Hyphomycetes)* 2: 298 (1998)
E: ! **S:** ! **NI:** !
H: In leaves of *Ranunculus sceleratus* and a single record on *Ranunculus repens*.
D: Vánky: 102, Myc17(2): 62
Rarely reported, but apparently widespread .

serotinum J. Schröt., *Beitr. Biol. Pflanzen* 2: 437 (1876)
    Entylomella serotina Höhn. (anam.), *Ann. Mycol.* 22: 198 (1924)
E: ! **S:** ! **W:** !
H: In leaves of *Symphytum officinale*.
D: UBI 28, Vánky: 103
Known from England (South Devon), Scotland (Main Argyll) and Wales (Monmouthshire). The host is common and widespread.

veronicae (Halst.) Lagerh., *Bull. Soc. Mycol. France.*: 170 (1891)
    Entyloma veronicicola Lindroth, *Acta Soc. Fauna Fl. Fenn.* 26: 13 (1904)
E: !
H: In leaves of *Veronica serpyllifolia*.

D: Vánky: 106
A single collection from Surrey (Esher Common) in 1997.

EXOBASIDIUM Woronin, *Verh. Naturf. Ges. Freiburg* 4(4): 397 (1867)
Exobasidiales, Exobasidiaceae
Type: *Exobasidium vaccinii* (Fuckel) Woronin

arescens Nannf., *Symb. Bot. Upsal.* 23(2): 40 (1981)
S: ! **W:** !
H: On leaves of *Vaccinium myrtillus* causing inconspicuous, small and hardly thickened, red or yellowish leaf spots.
D: Nannf: 40
Rarely reported, but easily overlooked.

camelliae Shirai, *Bot. Mag. (Tokyo)* 10: 51 (1896)
E: ! **W:** !
H: Causing large fleshy swellings on leaves sepals, petals and peduncles of *Camellia* spp and cultivars.
D+I: Dennis & Wakefield [*Trans. Brit. Mycol. Soc.* 29: 142-143 (1946)]
Rarely reported but apparently increasing. British records confined to southern and south-western England, and Wales.

expansum Nannf., *Symb. Bot. Upsal.* 23(2): 44 (1981)
S: !
H: On shoots of *Vaccinium uliginosum* in montane habitat.
A rare species throughout Europe. Known in Britain from a single collection from Cairngorm in 1994.

japonicum Shirai, *Bot. Mag. (Tokyo)* 10: 52 (1896)
    Exobasidium vaccinii var. *japonicum* (Shirai) McNabb, *Trans. Roy. Soc. New Zealand* 1(20): 267 (1962)
    Mis.: *Exobasidium azaleae* sensu auct. eur. p.p.
    Mis.: *Exobasidium vaccinii* sensu auct. p.p.
E: o **S:** o **W:** o **ROI:** o
H: On leaves of various species of Asiatic *Rhododendron* species causing pallid fleshy galls.
Occasional but widespread.

juelianum Nannf., *Symb. Bot. Upsal.* 23(2): 47 (1981)
S: ! **W:** !
H: On deformed shoots of *Vaccinium vitis-idaea* in montane habitat.
D+I: B&K2: 72 35 **I:** K1 85
Rarely reported. Infected plants are dwarfed, excessively branched and eventually covered in a pinkish hymenium.

karstenii Sacc. & Trotter, *Syll. fung.* 21: 420 (1912)
E: ! **S:** ! **W:** ! **ROI:** !
H: On shoots and leaves of *Andromeda polifolia* which become enlarged, reddish to purplish-black and prematurely dehiscent.
D: Nannf: 48-49 **D+I:** B&K2: 72 36
Distribution is limited by the rarity of the host plant but where it does occur, infection is apparently common *fide* Ing (*Mycologist* 12: 80 - 82, 1998).

myrtilli Siegm., *Mit. ver. Nat. Reich.* 19 (1879)
    Fusidium vaccinii f. *vaccinii-myrtilli* Fuckel, *Fungi Rhenani exsiccati. Fasc. III*: 220 (1863)
    Exobasidium vaccinii-myrtilli (Fuckel) Juel, *Svensk bot. Tidskr.* 6: 364 (1912)
    Mis.: *Exobasidium vaccinii* sensu auct. p.p.
E: c **S:** c **W:** c **NI:** ! **ROI:** c
H: On shoots of *Vaccinium myrtillus* on moorland or heathland.
D: Nannf: 50-51
Common and widespread.

oxycocci Shear, *Bull. Bur. Pl. Industr. U.S.D.A.* 110: 35 (1907)
    Mis.: *Exobasidium vaccinii* sensu auct. p.p.
S: ! **ROI:** !
H: On deformed, dark red shoots of *Vaccinium oxycoccos* in bogs.
D: Nannf: 52-53
Rarely reported.

rhododendri (Fuckel) C.E. Cramer, *Bot. Zeitung (Berlin)* 32: 324 (1874)

Exobasidium vaccinii var. *rhododendri* Fuckel, *Jahrb. Nassauischen Vereins Naturk.* 27-28: 7 (1873)
E: ! **S:** ! **W:** ! **ROI:** !
H: On leaves of *Rhododendron ferrugineum*, *R. hirsutum* and hybrids between them, producing large reddish 'apple galls'.
D: Nannf: 55-56 **D+I:** B&K2: 74 38
In Britain usually found on planted shrubs in gardens, apparently genuinely rare. Common in continental Europe, on wild plants in alpine habitat.

rostrupii Nannf., *Symb. Bot. Upsal.* 23(2): 56 (1981)
E: ! **S:** ! **W:** ! **ROI:** !
H: On leaves of *Vaccinium oxycoccos* and rarely *V. microcarpum* (in Scotland), forming bright red, unthickened leaf spots.
D+I: B&K2: 74 39
Rarely reported but easily overlooked.

sydowianum Nannf., *Symb. Bot. Upsal.* 23(2): 60 (1981)
S: !
H: On leaves of *Arctostaphyllos uva-ursi* on moorland in montane habitat.
Rarely reported. British records all from a few sites in Cairngorm and Wester Ross.

unedonis Maire, *Bull. Stat. Rech. Forest. N. Afrique* 1: 123 (1916)
E: !
H: On leaves of *Arbutus unedo*.
D: Nannf: 61-62
Rarely reported. British records only from North Essex on planted trees.

vaccinii (Fuckel) Woronin, *Ber. Naturf. Ges. Freiburg.* 4: 412 (1867)
Fusidium vaccinii Fuckel, *Bot. Zeitung (Berlin)* 19: 251 (1861)
E: o **S:** c **W:** c **NI:** !
H: Gall-forming on leaves, sometimes also causing hypertrophic shoots and occasionally hypertrophic, malformed flowers of *Vaccinium vitis-idaea* on heathland and moorland.
D: Nannf: 63-64 **D+I:** Ph 240-241, B&K2: 76 40 **I:** K1 88
Frequent and apparently widespread. The name has been misapplied to several other species on *Vaccinium* spp. Most frequently reported from Scotland.

FARYSIA Racib., *Bull. Int. Acad. Sci. Cracovie*: 354 (1909)
Ustilaginales, Ustilaginaceae
Elateromyces Bubák, *Houb. Cesk. Dil* 2: 32 (1912)
Type: *Farysia javanica* Racib.

thuemenii (A.A. Fisch. Waldh.) Nannf., *Symb. Bot. Upsal.* 16(2): 51 (1959)
Ustilago thuemenii A.A. Fisch. Waldh., *Hedwigia* 17: 40 (1878)
Uredo segetum var. *caricis* DC., *Fl. Fr.* 2: 230 (1805)
Farysia caricis (DC.) Liro, *Suomal. Tiedeakat.* 42: 49 (1935)
Uredo olivacea DC., *Fl. Fr.* 5: 78 (1815)
Ustilago olivacea (DC.) Tul. & C. Tul., *Ann. Sci. Nat., Bot.*, sér. 3, 7: 88 (1847)
Elateromyces olivaceus (DC.) Bubák, *Arch. Přírodov. Výzk. Čech.* 15(3): 32 (1912)
Farysia olivacea (DC.) Syd. & P. Syd. [*comb. illegit.*, non (Jaap) Höhn. (1917)], *Ann. Mycol.* 17: 41 (1919)
E: ! **ROI:** !
H: In ovaries and inflorescences of *Carex riparia*. Reported on *C. acutiformis* but this host is unverified.
D: UBI 29, Vánky: 134
Rarely reported, but apparently widespread.

MALASSEZIA Baill., *Traité Bot. Méd. Crypt.*: 234 (1889)
Malasseziales, incertae sedis
Pityrosporum Sabour., *Maladies du cuir chevelu. II Les Maladies Desquamatives*: 296 (1904)
Type: *Malassezia furfur* (C.P. Robin) Baill.

A genus of dermatotrophic yeasts. All species are anamorphic.

furfur (C.P. Robin) Baill., *Traité Bot. Méd. Crypt.*: 234 (1889)
Microsporum furfur C.P. Robin, *Hist. nat. vég. paras.*: 136 (1853)
E: !
H: Isolated ex human ear.
D+I: YST 450

globosa Midgley, E. Guého & J. Guillot, *Antonie van Leeuwenhoek* 69(4): 347 (1996)
E: !
H: Isolated ex human skin.
D+I: YST 451

obtusa Midgley, J. Guillot & E. Guého, *Antonie van Leeuwenhoek* 69(4): 348 (1996)
E: !
H: Isolated ex human skin.
D+I: YST 452

pachydermatis (Weidman) C.W. Dodge, *Medical Mycol.*: 370 (1935)
Pityrosporum pachydermatis Weidman, *Rep. Lab. Mus. Comp. Pathol. Zool. Soc. Philadelphia*: 36 (1925)
E: !
H: Isolated ex dog.
D+I: YST 453

restricta E. Guého, J. Guillot & Midgley, *Antonie van Leeuwenhoek* 69(4): 349 (1996)
E: !
H: Isolated ex human skin.
D+I: YST 454

slooffiae J. Guillot, Midgley & E. Guého, *Antonie van Leeuwenhoek* 69(4): 351 (1996)
E: !
H: Isolated ex human skin (scalp).
D+I: YST 455

sympodialis R.B. Simmons & E. Guého, *Mycol. Res.* 94(8): 1147 (1990)
E: !
H: Isolated ex human skin.
D+I: YST 456

MELANOPSICHIUM Beck, *Ann. K. K. Naturhist. Hofmus.* 9: 122 (1894)
Ustilaginales, Ustilaginaceae
Type: *Melanopsichium austroamericanum* (Speg.) Beck

nepalense (Liro) Zundel, *Ustilaginales of the World*: 46 (1953)
Ustilago nepalensis Liro, *Ann. Acad. Sci. Fenn., Ser. A*, 17(1): 184 (1924)
E: !
H: In galls on stems and inflorescences of *Polygonum aviculare*.
D: Vánky: 140
A single collection from Surrey (Ham, near Richmond) in 1984 (Spooner, 1985).

MELANOTAENIUM de Bary, *Bot. Zeitung (Berlin)* 32: 105 (1874)
Ustilaginales, Tilletiaceae
Type: *Melanotaenium endogenum* (Unger) de Bary

ari (Cooke) Lagerh., *Bull. Soc. Mycol. France* 15: 98 (1899)
Protomyces ari Cooke, *Grevillea* 1: 7 (1872)
E: !
H: In leaves of *Arum maculatum*.
D: Vánky: 144
Known only from Shropshire and Surrey. The host is common and widespread and the sori resemble the blackish pigmented areas naturally present on the leaves.

cingens (Beck) Magnus, *Oesterr. Bot. Z.* 42: 40 (1892)
  Ustilago cingens Beck, *Oesterr. Bot. Z.* 31: 313 (1881)
  Cintractia cingens (Beck) De Toni, in Sacc., *Syll. fung.* 7: 481 (1888)
E: ? **W:** !
H: In stems and leaves of *Linaria vulgaris.*
D: UBI 29, Vánky: 144
Known from Wales (Denbighshire, Flintshire, and Merionethshire) in 1902, 1931 and 1908 respectively. Reported from England (Cambridgeshire) but unsubstantiated with voucher material. Perhaps genuinely rare as the host is common, and infected plants would be difficult to overlook due to the marked distortion of the stems and leaves.

endogenum (Unger) de Bary, *Bot. Zeitung (Berlin)* 32: 106 (1874)
  Protomyces endogenus Unger, *Exanth. Pfl.*: 342 (1833)
  Entyloma endogenum (Unger) Wünsche, *Pilze* 21: 12 (1877)
E: ! **S:** ! **W:** ! **ROI:** ! **O:** Channel Islands: !
H: In stems, shoots and leaves of *G. verum* and rarely *G. mollugo.*
D: UBI 30, Vánky: 145
Rarely reported but apparently widespread. The major host is common and widespread, and the fungus causes a distinct and obvious 'witches broom' in affected plants, hence it would be difficult to overlook and the fungus may be genuinely uncommon or rare.

hypogaeum (Tul. & C. Tul.) Schellenb., *Beitr. Kryptogamenfl. Schweiz* 3: 108 (1911)
  Ustilago hypogaeum Tul. & C. Tul., *Fungi hypogaROI:* 196 (1862)
E: !
H: In rootstocks of *Kickxia spuria* in cornfields.
D: UBI 30, Vánky: 145-146
A single collection, from Isle of Wight (Freshwater) in 1869. Perhaps overlooked as sori are cryptic (occurring in underground parts of the host) and rarely searched for.

jaapii Magnus, *Ber. Deutsch. Bot. Ges.* 29: 457 (1911)
  Melanotaenium lamii R. Beer, *Trans. Brit. Mycol. Soc.* 6(4): 337 (1920)
E: !
H: In the underground parts of the stems and roots of *Lamium album.*
D: UBI 30 (as *Melanotaenium lamii*), Vánky: 146
Known only from Buckinghamshire and West Norfolk.

MICROSTROMA Niessl, *Oesterr. Bot. Z.* 11: 252 (1861)
Exobasidiales, Microstromataceae
Type: *Microstroma album* (Desm.) Sacc.

album (Desm.) Sacc., *Michelia* 1: 273 (1878)
  Fusisporium album Desm., *Ann. Sci. Nat., Bot.,* sér. 2, 1: 309 (1838)
  Torula quercina Opiz, *Lotos* 5: 216 (1855)
  Microstroma quercinum Niessl, *Oesterr. Bot. Z.* 11: 252 (1861)
E: ! **S:** ! **W:** ! **NI:** ! **ROI:** !
H: On the undersurface of living leaves of *Quercus* spp
Known from England (East Sussex, London, Northamptonshire, South Hampshire, Surrey, West Sussex and Worcestershire), Scotland (Angus), Wales (Montgomeryshire) and the Republic of Ireland (Co. Dublin and South Kerry) but easily overlooked.

juglandis (Berenger) Sacc., *Syll. fung.* 4: 9 (1886)
  Fusidium juglandis Berenger, *Il. secc. del Gelso:* 7 f.1 (1847)
  Ascomyces juglandis Berk., *Outl. Brit. fungol.*: 367 (1860)
  Fusisporium pallidum Niessl, *Verh. K. K. Zool.-Bot. Ges. Wien* 8, *Abhandl.*: 329 (1858)
  Microstroma pallidum (Niessl) Niessl, *Oesterr. Bot. Z.* 11: 252 (1861)
  Torula juglandina Opiz, *Seznam:* 147 (1852)
E: !
H: On the undersides of living leaves of *Juglans regia.*

Known from Hertfordshire, Norfolk, South Somerset, Staffordshire, Surrey, West Suffolk and Yorkshire but easily overlooked.

PSEUDOZYMA Bandoni, *J. Linn. Soc., Bot.* 91: 38 (1985)
Ustilaginales, incertae sedis
Type: *Pseudozyma prolifica* Bandoni
A genus of yeasts without known teleomorphs.

fusiformata (Buhagiar) Boekhout, *J. Gen. Appl. Microbiol.* 41(4): 364 (1995)
  Candida fusiformata Buhagiar, *J. Gen. Microbiol.* 110(1): 95 (1979)
E: !
H: Isolated ex cauliflower.
D+I: YST 586

RHAMPHOSPORA D.D. Cunn., *Scientific Mem. by Med. Officers of Army of India*: 32 (1888)
Ustilaginales, Tilletiaceae
Type: *Rhamphospora nymphaeae* D.D. Cunn.

nymphaeae D.D. Cunn., *Scientific Mem. by Med. Officers of Army of India*: 32 (1888)
  Entyloma nymphaeae (D.D. Cunn.) Setch., *Bot. Gaz.* 19: 189 (1894)
E: !
H: In floating leaves of *Nymphaea alba.*
D: Vánky: 184
Known from Berkshire, Oxfordshire and West Sussex .

SCHIZONELLA J. Schröt., *Beitr. Biol. Pflanzen* 2: 362 (1877)
Ustilaginales, Ustilaginaceae
Type: *Schizonella melanogramma* (DC.) J. Schröt.

cocconi (Morini) Liro, *Ann. Acad. Sci. Fenn., Ser. A,* 42: 52 (1938)
  Tolyposporium cocconi Morini, *Mem. Reale Accad. Sci. Ist. Bologna,* Ser. 4, 5: 800 (1884)
S: ! **O:** Channel Islands: !
H: In leaves of *Carex flacca.*
Known from Scotland (South Aberdeenshire) and the Channel Islands (Guernsey). Previously included in *S. melanogramma* s.l.

melanogramma (DC.) J. Schröt., *Beitr. Biol. Pflanzen* 2: 362 (1877)
  Uredo melanogramma DC., *Fl. Fr.* 5: 75 (1815)
E: !
H: In leaves of *Carex ericetorum* (British material).
D: UBI 31, Vánky: 186-187
A single collection from West Suffolk (Icklingham, Foxhole Heath) in 1951.

SPORISORIUM Link, in Willd., *Sp. pl.,* Edn 4, 6(2): 86 (1825)
Ustilaginales, Ustilaginaceae
Type: *Sporisorium sorghi* Link

destruens (Schltdl.) Vánky, *Symb. Bot. Upsal.* 24(2): 115 (1985)
  Caeoma destruens Schltdl., *Fl. berol.* 2: 130 (1830)
  Ustilago destruens (Schltdl.) Rabenh., *Klotzschii Herbarium Vivum Mycologicum (Editio nova)*: 400 (1857)
  Sphacelotheca destruens (Schltdl.) J.A. Stev. & Aar.G. Johnson, *Phytopathology* 34: 613 (1944)
  Ustilago panici-miliacei (Pers.) G. Winter, *Rabenh. Krypt.-Fl.* 1: 89 (1881)
  Sphacelotheca panici-miliacei (Pers.) Bubák, *Arch. Přírodov. Výzk. Čech.* 15: 27 (1912)

Uredo segetum γ *panici-miliacei* Pers., *Syn. meth. fung.*: 224 (1801)

E: !

H: In the inflorescences of millet (*Panicum miliaceum*), often completely destroying them.

D: UBI 32 (as *Sphacelotheca destruens*), Vánky: 210

Known from a few localities in Berkshire, Cambridgeshire, Essex and Northamptonshire where the host is occasionally grown.

sorghi Link, in Willd., *Sp. Plant.*, Edn 4, 6(2): 86 (1825)
Ustilago sorghi (Link) Pass., *Hedwigia* 12: 114 (1873)
Sphacelotheca sorghi (Link) G.P. Clinton, *J. Mycol.* 8: 140 (1902)

E: !

H: In florets of *Sorghum vulgare*.

D: UBI 34 (as *Sphacelotheca sorghi*), Vánky: 206

Occasional where the host is grown in experimental plots. Known from Surrey and reported from Berkshire.

THECAPHORA Fingerh., *Linnaea.* 10: 230 (1836)
Ustilaginales, Ustilaginaceae
Poikilosporium Dietel, *Flora* 83(2): 87 (1897)
Type: *Thecaphora hyalina* Fingerh. (= *T. seminis-convolvuli*)

deformans Durieu & Mont., *Ann. Sci. Nat., Bot.*, sér. 3, 7: 110 (1847)

E: !

H: British material in seeds of *Ulex minor* (typically in seeds of *Medicago* spp).

D: UBI 34-35, Vánky: 225

Known from Dorset and South Hampshire.

lathyri J.G. Kühn, in Rabenhorst, *Fungi europaei* IV: 1797 (1874)

S: !

H: In seeds of *Lathyrus pratensis*.

D: Vánky: 226

Known only from East Lothian and Midlothian.

seminis-convolvuli (Desm.) S. Ito, *Trans. Sapporo Nat. Hist. Soc.* 14: 94 (1935)
Uredo seminis-convolvuli Desm., *Pl. crypt. N. France.* 274 (1827)
Thecaphora hyalina Fingerh., *Linnaea.* 10: 230 (1836)
Gloeosporium antherarum Oudem. (anam.), *Hedwigia* 37: 179 (1898)

E: ! **W:** ! **O:** Channel Islands: !

H: In seeds of *Convolvulus arvensis*, *Calystegia sepium*, *C. silvatica* and *C. soldanella*.

D: UBI 35, Vánky: 228

Rarely reported. The hosts are common throughout Britain.

trailii Cooke, *Grevillea* 11: 155 (1883)
Poikilosporium trailii (Cooke) Vestergr., *Bot. Not.*: no 452 (1902)

S: ! **ROI:** !

H: In flower heads of *Cirsium heterophyllum* and *C. dissectum*.

D: UBI 35, Vánky: 229

Known from widespread sites.

TILLETIA Tul. & C. Tul., *Ann. Sci. Nat., Bot.*, sér. 3, 7: 112 (1847)
Ustilaginales, Tilletiaceae
Type: *Tilletia caries* (DC.) Tul. & C. Tul.

anthoxanthi A. Blytt, *Forh. Vidensk.-Selsk. Kristiania* 4(6): 31 (1896)

S: !

H: In ovaries of *Anthoxanthum odoratum*.

D: Vánky: 243

A single collection from Ayrshire (West Kilbride) in 1921.

caries (DC.) Tul. & C. Tul., *Ann. Sci. Nat., Bot.*, sér. 3, 7: 113 (1847)
Uredo caries DC., *Fl. Fr.* 5: 78 (1815)

Lycoperdon tritici Bjerk. [*nom. nud.*], *Kongl. Vetensk. Acad. Handl.* 36: 326 (1775)
Tilletia tritici G. Winter, *Rabenh. Krypt.-Fl.* 1: 110 (1881)

E: ! **W:** ! **NI:** !

H: In the ovaries of various *Poaceae*.

D: UBI 39 (as *Tilletia tritici*), Vánky: 245

Once a common and widespread pathogen of grasses, especially cereal crops such as *Secale cereale* and species of *Triticum* but, with the advent of improved fungicides and resistant host varieties, is now rarely (if ever) encountered.

holci (Westend.) J. Schröt., *Beitr. Biol. Pflanzen* 2: 365 (1877)
Polycystis holci Westend., *Bull. Acad. Roy. Sci. Belgique* 11: 660 (1861)
Tilletia rauwenhoffii A.A. Fisch. Waldh. [*nom. illegit.*], *Aperçu.* 50 (1877)

E: ! **S:** ! **W:** ! **ROI:** !

H: In the ovaries of *Holcus lanatus* and *H. mollis*.

D: UBI 36, Vánky: 247

Rarely reported but apparently widespread.

menieri Har. & Pat., *Bull. Soc. Mycol. France.* 20: 61 (1904)

E: ! **S:** ! **NI:** ! **ROI:** !

H: In the ovaries of *Phalaris arundinacea*.

D: UBI 37-38, Vánky: 249

Rarely reported but apparently widespread.

olida (Riess) G. Winter, *Beitr. Biol. Pflanzen* 2: 366 (1877)
Uredo olida Riess, *Klotzschii Herbarium Vivum Mycologicum.* 1695 (1852)
Ustilago olida (Riess) Cif., *Fl. ital. crypt.* 17: 296 (1938)
Mis.: *Ustilago macrospora* sensu auct.

E: !

H: In leaves of *Brachypodium pinnatum*.

D: UBI 38, Vánky: 249

Rarely reported. The host is common in southern England.

sphaerococca (Rabenh.) A.A. Fisch. Waldh., *Byull. mosk. Obshch. Ispyt. Prir.* 40: 255 (1867)
Uredo sphaerococca Rabenh., *Deutschl. Krypt. Flora* 1: 4 (1844)
Tilletia decipiens (Pers.) Körn., *Hedwigia* 16: 30 (1877)
Erysibe sphaerococca α *agrostidis* Wallr., *Fl. crypt. Germ.*, Sect. *II.* 213 (1833)
Mis.: *Tilletia separata* sensu Massee [Bull. Misc. Inform. Kew 1899: 159 (1899)]

S: ! **ROI:** !

H: In the ovaries of *Agrostis canina*, *A. capillaris* and *A. stolonifera*.

D: UBI 38, Vánky: 251

Known from Scotland (various locations) and Republic of Ireland (West Mayo: Lough Mask).

TRACYA Syd. & P. Syd., *Beibl. Hedwigia* 40(1): (3) (1901)
Ustilaginales, Tilletiaceae
Type: *Tracya lemnae* (Setch.) Syd. & P. Syd.

hydrocharidis Lagerh., *Bot. Not.*: 175 (1902)

E: ! **NI:** !

H: In leaves of *Hydrocharis morsus-ranae*.

D: Vánky: 278

Single collections from England (Surrey: Runnymede) in 1937 and Northern Ireland (Fermanagh: Sand Lough) in 1948, recently discovered during examination of herbarium specimens of the host plants at Kew.

UROCYSTIS Fuckel, *Jahrb. Nassauischen Vereins Naturk.* 23-24: 41 (1870)
Ustilaginales, Tilletiaceae
Granularia Sowerby [non *Granularia* Roth (1791)], *Col. fig. Engl. fung.*: pl. 440 (1815)
Tuburcinia Fr., *Syst. mycol.* 3: 439 (1829)
Polycystis Lév., *Ann. Sci. Nat., Bot.*, sér. 3, 5: 269 (1846)
Type: *Urocystis occulta* (Wallr.) Fuckel

agropyri (Preuss) J. Schröt., *Bull. Soc. Imp. Naturalistes Moscou* 40: 258 (1867)
> *Uredo agropyri* Preuss, *Deutschl. Fl.*, Edn 3: 1 (1848)
> *Tuburcinia agropyri* (Preuss) Liro, *Ann. Univ. Fenn. Aboen.* 1: 15 (1922)

E: ! **S:** !
H: In leaves of *Agropyron pungens, A. repens, Agrostis tenuis, A. stolonifera, Alopecurus pratensis, Arrhenatherum elatius, Festuca arundinacea, Festuca rubra, Hordelymus europaeus, Lolium perenne* and *Melica uniflora.*
D: UBI 41, Vánky: 283
Rarely reported but widespread.

agropyri-campestris (Massenot) H. Zogg, *Cryptog. Helv.* 16: 112 (1986) [1985]
> *Tuburcinia agropyri-campestris* Massenot, *Rev. Pathol. Vég. Entomol. Agric. France* 34: 193 (1955)

E: !
H: In leaves of *Agropyron pungens.*
Known from Norfolk and Suffolk.

agrostidis (Lavrov) Zundel, *Ustilaginales of the World*: 307 (1953)
> *Tuburcinia agrostidis* Lavrov, *Sist. Zametki Mater. Gerb. Krylova Tomsk. Gosud Univ. Kujbyseva* 11: 2 (1937)

E: !
H: In leaves of *Agrostis stolonifera* and *A. capillaris* (=*A. tenuis*).
Known from South Devon and Surrey.

alopecuri A.B. Frank, *Die Krankheiten der Pflanzen. Pilze*: 440 (1880)
E: !
H: In leaves of *Alopecurus pratensis.*
A single collection from Surrey (Leatherhead) in 1946.

anemones (Pers.) G. Winter, *Hedwigia* 19: 160 (1880)
> *Uredo anemones* Pers., *Tent. disp. meth. fung.*: 56 (1797)
> *Polycystis anemones* (Pers.) Lév., *Ann. Sci. Nat., Bot.*, sér. 3, 8: 372 (1847)
> *Tuburcinia anemones* (Pers.) Liro, *Ann. Univ. Fenn. Aboen.* 1: 55 (1922)
> *Caeoma pompholygodes* Schltdl., *Linnaea.* 1: 248 (1826)
> *Polycystis pompholygodes* (Schltdl.) Lév., *Ann. Sci. Nat., Bot.*, sér. 3, 5: 270 (1846)
> *Urocystis pompholygodes* (Schltdl.) Rabenh., *Fungi europaei* IV: 697 (1864)
> *Uredo pompholygodes* (Schltdl.) Rabenh., *Rabenh. Krypt.-Fl.*: 4 (1881)

E: c **S:** ! **W:** ! **NI:** ! **ROI:** ! **O:** Isle of Man: !
H: In leaves of *Anemone nemorosa* in deciduous woodland, and rarely *Anemone* cultivars in gardens
D: UBI 42, Vánky: 284
Common and widespread, at least in England, but many records are on species of *Ranunculus* and are thus *Urocystis ranunculi.*

avenae-elatioris (Kochman) Zundel, *Ustilaginales of the World*: 311 (1953)
> *Tuburcinia avenae-elatioris* Kochman, *Acta Soc. Bot. Poloniae.* 16: 54 (1939)

E: !
H: In leaves of *Arrhenatherum elatius.*
A single collection, from Norfolk in 1944. The host is common throughout Britain.

bolivarii Bubák & Gonz. Frag., *Bol. Real Soc. Esp. Hist. Nat.* 22: 205 (1922)
E: !
H: In leaves of *Lolium perenne.*
A single collection from Oxfordshire (Henley on Thames) in 1970.

bromi (Lavrov) Zundel, *Ustilaginales of the World*: 312 (1953)
> *Tuburcinia bromi* Lavrov, *Sist. Zametki Mater. Gerb. Krylova Tomsk. Gosud. Univ. Kujbiseva* 11: 2 (1937)

E: !
H: In leaves of species of *Bromus.*
A single collection from England (locality not stated) in 1918.

colchici (Schltdl.) Rabenh., *Fungi europaei* IV: 396 (1861)
> *Caeoma colchici* Schltdl., *Linnaea.* 1: 241 (1826)
> *Uredo colchici* (Schltdl.) Endl., *Fl. Poson.*: 19 (1830)
> *Polycystis colchici* (Schltdl.) F. Strauss, *Deutschl. Fl.*, Edn 3: 45 (1853)
> *Tuburcinia colchici* (Schltdl.) Liro, *Ann. Univ. Fenn. Aboen.* 1: 52 (1922)

E: ! **W:** !
H: In leaves of *Colchicum autumnale* and cultivated *C. vernum.*
D: UBI 43, Vánky: 288
Apparently rare, as is the host plant in Britain. Known from widely scattered sites in England (Buckinghamshire, Herefordshire, Wiltshire and Yorkshire) and Wales (Glamorgan).

eranthidis (Pass.) Ainsw. & Sampson, *British Smut Fungi (Ustilaginales)*: 96 (1950)
> *Urocystis pompholygodes* f. *eranthidis* Pass., *Erb. critt. Ital., Ser. 2*: 549 (1871)
> *Tuburcinia eranthidis* (Pass.) Liro, *Ann. Univ. Fenn. Aboen.* 1: 85 (1922)

E: !
H: In leaves of *Eranthis hyemalis* (winter aconite).
D: UBI 43, Vánky: 290
Known from widely scattered sites in southern England.

ficariae (Liro) Moesz, *Budapest köorny. gomb.*: 137 (1942)
> *Tuburcinia ficariae* Liro, *Ann. Univ. Fenn. Aboen.* 1: 67 (1922)

E: ! **S:** !
H: In leaves or petioles of *Ranunculus ficaria.*
D: UBI 44, Vánky: 291
Known from Scotland, and a single collection in herb. K, probably from Norfolk, misidentified as *U. anemones.* Reported from Yorkshire in 2000, but lacking voucher material.

filipendulae (Tul. & C. Tul.) J. Schröt., *Abh. Schles. Ges. Vaterl. Cult., Abth. Naturwiss. (Naturwiss.-Med. Abth.)* 1869-72: 7 (1870)
> *Polycystis filipendulae* Tul. & C. Tul., *Ann. Sci. Nat., Bot.*, sér. 4, 2: 163 (1854)
> *Tuburcinia filipendulae* (Tul. & C. Tul.) Liro, *Ann. Univ. Fenn. Aboen.* 1: 87 (1922)

E: ! **S:** !
H: In the midribs and petioles of leaves of *Filipendula vulgaris* and rarely on *F. ulmaria.*
D: UBI 44, Vánky: 291
Rarely reported., Perhaps genuinely rare as the sori are large (up to 44mm in length) and obvious.

fischeri G. Winter, *Rabenh. Krypt.-Fl.* 1: 120 (1881)
> *Tuburcinia fischeri* (G. Winter) Liro, *Ann. Univ. Fenn. Aboen.* 1: 29 (1922)

E: ! **S:** ! **ROI:** !
H: In leaves of *Carex flacca.* Also reported dubiously on *C. nigra* and *C. panicea.*
D: UBI 44, Vánky: 291-292
Rarely reported but apparently widespread.

floccosa (Wallr.) D.M. Hend., *Notes Roy. Bot. Gard. Edinburgh* 21: 241 (1955)
> *Erysibe floccosa* Wallr., *Fl. crypt. Germ., Sect. II*: 212 (1833)
> *Tuburcinia floccosa* (Wallr.) Jørst., *Nytt Mag. Naturividensk* 83: 238 (1943)
> *Tuburcinia hellebori-viridis* (DC.) Liro, *Ann. Univ. Fenn. Aboen.* 1: 82 (1922)
> *Urocystis hellebori-viridis* (DC.) Moesz, *Kárpát-Medence Üszöggombái*: 209 (1950)
> *Uredo ranunculacearum* var. *hellebori-viridis* DC., *Fl. Fr.* 5: 75 (1815)

E: !
H: In leaves of *Helleborus viridis.*
D: UBI 45, Vánky: 292
Known only from a few widely scattered locations.

gladiolicola Ainsw., *Trans. Brit. Mycol. Soc.* 32(3-4): 257 (1949)
E: ! **O:** Channel Islands: !

H: In leaves of cultivated *Gladiolus*.
D: UBI 45, Vánky: 292-293
Rarely reported, and almost certainly introduced on corms of the host.

junci Lagerh., *Bot. Not.*: 210 (1888)
E: !
H: In the lower parts of stems of *Juncus acutiflorus* and *J. acutus*.
D: UBI 46, Vánky: 294
Known only from West Norfolk but the sori are inconspicuous.

magica Pass., *Mycotheca universalis*: 223 (1875)
Tuburcinia magica (Pass.) Liro, *Ann. Univ. Fenn. Aboen.* 1: 49 (1922)
Urocystis cepulae Frost, *Ann. Rep. Sec. Mass. St. Bd. Agric.* 24: 175 (1877)
E: ! **S:** ! **ROI:** !
H: In the leaves of *Allium cepa*, *A. porrum*, and *A. vineale*.
D: UBI 42 (as *Urocystis cepulae*), Vánky: 296
Rarely reported. Better known in Britain as *U. cepulae* but see Vánky (1994) for comments on the nomenclature.

melicae (Lagerh. & Liro) Zundel, *Ustilaginales of the World*: 326 (1953)
Tuburcinia melicae Lagerh. & Liro, *Ann. Univ. Fenn. Aboen.* 1(1): 23 (1922)
E: !
H: In leaves of *Melica uniflora*.
A single collection, from Surrey (Mickleham) in 1990. The host is common in open woodland, on calcareous soils in southern England.

occulta (Wallr.) Fuckel, *Jahrb. Nassauischen Vereins Naturk.* 23-24: 41 (1870)
Erysibe occulta Wallr., *Fl. crypt. Germ., Sect. II*: 212 (1833)
Polycystis occulta (Wallr.) Schltdl., *Bot. Zeitung (Berlin)* 10: 602 (1852)
Tuburcinia occulta (Wallr.) Liro, *Ann. Univ. Fenn. Aboen.* 1: 12 (1922)
Uredo parallela Berk., *Engl. fl.* 5(2): 375 (1836)
Polycystis parallela (Berk.) Fr., *Summa veg. Scand.*: 516 (1849)
E: ! **S:** !
H: In leaves, culms and inflorescences of *Secale cereale*.
D: UBI 46, Vánky: 299
Rare in Britain where the host is not a major cereal crop.

poae (Liro) Padwick & A. Khan, *Mycol. Pap.* 10: 2 (1944)
Tuburcinia poae Liro, *Ann. Univ. Fenn. Aboen.* 1: 22 (1922)
S: !
H: In leaves of various species of *Poa*.
A single collection from South Aberdeen (Peterhead) in 1933.

primulae (Rostr.) Vánky, *Symb. Bot. Upsal.* 24(2): 176 (1985)
Sorosporium primulae Rostr., *Vars. Univ. Izv. 1879 (2), Neoffic. Otd* 2: 176 (1879)
Paepalopsis irmischiae J.G. Kühn (anam.), *Hedwigia* 22: 28 (1882)
E: ! **S:** !
H: In the ovaries of *P. vulgaris* and rarely *P. veris*.
D: UBI 47 (as *Urocystis primulicola* p.p.), Vánky: 302
Rarely reported. Previously included in *Urocystis prumulicola* s.l. See Vánky (1994) for comparison with *U. primulicola*.

primulicola Magnus, *Verh. Bot. Vereins Prov. Brandenburg* 20: 53 (1878)
Tuburcinia primulicola (Magnus) Rostr., *Bot. Foren. Feskr. Kijöbenh.* 54: 150 (1890)
E: ! **S:** !
H: In the ovaries of *Primula farinosa* (not in *Primula vulgaris*).
D: UBI 47 (as *Urocystis primulicola* p.p.), Vánky: 302
Previously included with *Urocystis primulae* s.l.

ranunculi (Lib.) Moesz, *Kárpát-Medence Üszöggombái*: 213 (1950)
Sporisorium ranunculi Lib., *Pl. Crypt. Ardienn.*: 195 (1832)

Tuburcinia ranunculi (Lib.) Liro, *Ann. Univ. Fenn. Aboen.* 1: 69 (1922)
E: c **S:** c **W:** c **NI:** c **ROI:** c **O:** Channel Islands: ! Isle of Man: !
H: In leaves, petioles and stems of *Ranunculus* spp, most frequently *R. repens* but also known on *R. acris* and *R. bulbosus*.
D: UBI 42 (as *Urocystis anemones* p.p.), Vánky: 303
Common and widespread.

sorosporioides A.A. Fisch. Waldh., *Aperçu*: 41 (1877)
Tuburcinia sorosporioides (A.A. Fisch. Waldh.) Liro, *Ann. Univ. Fenn. Aboen.* 1: 77 (1922)
E: ! **S:** !
H: In leaflets and leaves of *Thalictrum* and *Aquilegia* spp.
D: UBI 47, Vánky: 305
Rarely reported.

trientalis (Berk. & Broome) B. Lindeb., *Symb. Bot. Upsal.* 16(2): 100 (1959)
Tuburcinia trientalis Berk. & Broome, *Ann. Mag. Nat. Hist, Ser. 2 [Notices of British Fungi no. 488]* 5: 464 (1850)
Sorosporium trientalis (Berk. & Broome) Woronin, *Aperçu*: 32 (1877)
Ascomyces trientalis Berk. (anam.), *Outl. Brit. fungol.*: 376 (1860)
S: !
H: In leaves and stems of *Trientalis europaea*.
D: UBI 48, Vánky: 307
Known from Scotland, where the host plant is most often recorded, but apparently rare there.

trollii Nannf., *Symb. Bot. Upsal.* 16(2): 100 (1959)
E: ! **W:** !
H: In leaves of *Trollius* spp, usually on cultivars but also known on native *Trollius europaeus*.
Two collections from England (Warwickshire: Sutton, on cultivated *Trollius asiaticus* in 1908 and Norfolk: Bressingham, on *Trollius cultorum* in 1986) and one from Wales (Glamorganshire: Cwm Cadlan) in 1998 on native *T. europaeus*.

ulei Magnus, *Hedwigia* 17: 89 (1878)
E: ! **S:** !
H: In leaves of *Festuca* spp British material on *F. arenaria* and *F. arundinacea*.
Known from England (North Lincolnshire, North Wiltshire and South Devon) and Scotland (East Lothian: Gullane).

violae (Sowerby) A.A. Fisch. Waldh., *Byull. mosk. Obshch. Ispyt. Prir.* 40: 258 (1867)
Granularia violae Sowerby, *Col. fig. Engl. fung., Suppl.*: pl. 440 (1815)
Polycystis violae (Sowerby) Berk. & Broome, *Ann. Mag. Nat. Hist, Ser. 2 [Notices of British Fungi no. 487]* 5: 464 (1850)
Tuburcinia violae (Sowerby) Liro, *Ann. Univ. Fenn. Aboen.* 1: 91 (1922)
E: o **S:** ! **W:** ! **NI:** ! **ROI:** !
H: In petioles, roots and leaf veins of *Viola hirta*, *V. odorata*, *V. reichenbachiana* and *V. riviniana*, and also on *Viola* cultivars.
D: UBI 48, Vánky: 309 **D+I:** Fox [*Mycologist* 15(3): 136 (2001)]
Occasional in England. Rarely reported elsewhere but apparently widespread.

USTANCIOSPORIUM Vánky, *Mycotaxon* 70: 31 (1999)
Ustilaginales, Ustilaginaceae
Type: *Ustanciosporium rhynchosporae* Vánky

gigantosporum (Liro) M. Piepenbr. & Begerow, *Nova Hedwigia* 70(3-4): 339 (2000)
Cintractia gigantospora Liro, *Mycoth. fenn.* 26: 11 (1935) [1934]
Ustilago gigantospora (Liro) Lehtola, *Ann. Bot. Soc. Zool.-Bot. Fenn. 'Vanamo'* 17(3): 23 (1942)
Ustilago rhynchosporae Saut., *Bot. Zeitung (Berlin)*: 190 (1854)

E: !
H: In ovaries of *Rhynchospora alba*.
D: Vánky: 373 (as *Ustilago rhynchosporae*)
Known from Berkshire and Cambridgeshire, but lacking recent records.

majus (Desm.) M. Piepenbr., *Nova Hedwigia* 70(3-4): 341 (2000)
> *Ustilago montagnei* var. *major* Desm., *Pl. Crypt. Fr.*: 2126 (1850)
> *Cintractia major* (Desm.) Liro, *Ann. Acad. Sci. Fenn., Ser. A*, 42: 46 (1938)
> *Ustilago intercedens* Lehtola, *Ann. Bot. Soc. Zool.-Bot. Fenn. 'Vanamo'* 17: 23 (1942)

E: ! **ROI:** !
H: In ovaries of *Rhynchospora alba*.
D: Vánky: 363 (as *Ustilago intercedens*)
Only recently recognised as British with collections from England (Dorset) in 1933 and Republic of Ireland (Galway) in 1959.

USTILAGO (Pers.) Roussel, *Flora Calvados*, Edn 2: 47 (1806)
Ustilaginales, Ustilaginaceae
> *Farinaria* Sowerby, *Col. fig. Engl. fung.* 3: pl. 396 (1803)

Type: *Ustilago hordei* (Pers.) Lagerh.

anomala G. Winter, *Rabenh. Krypt.-Fl.* 1: 100 (1881)
E: ! **S:** ! **W:** ! **NI:** !
H: In ovaries and flowers of *Fallopia (Polygonum) convolvulus*.
D: UBI 51, Vánky: 349
Scattered records, mainly from England.

avenae (Pers.) Rostr., *Overs. Kongel. Danske Vidensk. Selsk. Forh. Medlemmers Arbeider.* 13 (1890)
> *Uredo segetum* γ *avenae* Pers., *Syn. meth. fung.*: 224 (1801)
> *Ustilago segetum* var. *avenae* (Pers.) Brunaud, *Act. Soc. linn. Bordeaux* 32: 163 (1878)
> *Ustilago nigra* Tapke, *Phytopathology* 22: 869 (1932)
> *Ustilago perennans* Rostr., *Overs. Kongel. Danske Vidensk. Selsk. Forh. Medlemmers Arbeider.* 15 (1890)

E: c **S:** c **W:** c **NI:** c **ROI:** c
H: In the flower spikes and ovaries of *Arrhenatherum elatius* and less frequently *Avena sativa*.
D: UBI 56 (as *Ustilago segetum* var. *avenae*), Vánky: 350
Very common and widespread and often abundant, especially on *Arrhenatherum elatius*.

bistortarum (DC.) Körn., *Hedwigia* 16: 38 (1877)
> *Uredo bistortarum* DC., *Fl. Fr.* 5: 76 (1815)
> *Uredo bistortarum* γ *ustilaginea* DC., *Fl. Fr.* 5: 76 (1815)
> *Ustilago bistortarum* var. *inflorescentiae* Trel., *Cryptogamic Botany, [Harriman Alaska Expedition]* 5: 35 (1904)
> *Ustilago inflorescentiae* (Trel.) Maire, *Bull. Soc. Bot. France* 54: 49 (1907)
> *Sphacelotheca inflorescentiae* (Trel.) Jaap, *Ann. Mycol.* 6: 194 (1908)
> *Ustilago candollei* Tul. & C. Tul., *Ann. Sci. Nat., Bot.*, sér. 3, 7: 93 (1847)
> *Sphacelotheca polygoni-vivipari* Schellenb., *Ann. Mycol.* 5: 388 (1907)

S: !
H: In flowers and bulbils of *Polygonum bistorta* and *P. viviparum*.
D: UBI 33-34 (as *Sphacelotheca inflorescentiae*), Vánky: 351
Collections on *P. viviparum* are only from Scotland. Reported from Yorkshire on *P. bistorta* in 1987, but unsubstantiated with voucher material.

bullata Berk., *Flora Novae-Zealandiae*: 196 (1855)
> *Ustilago carbo* α *vulgaris* d *bromivora* Tul. & C. Tul., *Ann. Sci. Nat., Bot.*, sér. 3, 7: 81 (1847)
> *Ustilago bromivora* (Tul. & C. Tul.) A.A. Fisch. Waldh., *Bull. Soc. Imp. Naturalistes Moscou.* 40: 252 (1867)
> *Cintractia patagonica* Cooke & Massee, *Grevillea* 18: 34 (1899)

E: !

H: British collections in flower spikelets of various species of *Bromus* but known on other species of *Poaceae* in continental Europe.
D: UBI 51, Vánky: 352-353
Uncommon.

cordae Liro, *Ann. Acad. Sci. Fenn., Ser. A*, 17(1): 12 (1924)
E: ! **S:** ! **NI:** ! **ROI:** !
H: In ovaries of *Persicaria hydropiper* and *P. maculosa*.
Only recently recognised as British. Apparently widespread.

cynodontis (Henn.) Henn., *Bull. Herb. Boissier.* 1: 114 (1893)
> *Ustilago segetum* var. *cynodontis* Henn., *Bot. Jahrb. Syst.* 14: 369 (1892)
> *Ustilago carbo* var. *cynodontis* Pass., *Erb. critt. Ital., Ser. 2*: 450 (1871)

E: !
H: In the inflorescences of *Cynodon dactylon*.
D: UBI 52, Vánky: 356
Known from Dorset (Sandbanks near Bournemouth) and West Cornwall (Penzance).

duriaeana Tul. & C. Tul., *Ann. Sci. Nat., Bot.*, sér. 3, 7: 105 (1847)
E: ! **W:** !
H: In ovaries of *Caryophyllaceae*. British material known from *Cerastium glomeratum* and possibly *Moenchia erecta*.
D: Vánky: 357
Known from South Essex (Chelmsford) and Radnorshire (Newtown).

echinata J. Schröt., *Abh. Schles. Ges. Vaterl. Cult., Abth. Naturwiss. (Naturwiss.-Med. Abth.)* 1869-72: 4 (1870)
E: ! **S:** !
H: On leaf blades and sheaths of *Phalaris arundinacea*.
D: Vánky: 357
Rarely reported (usually as *U. serpens*). Perhaps genuinely uncommon as infected plants are usually conspicuous, grossly distorted and destroyed by the smut.

filiformis (Schrank) Rostr., *Bot. Foren. Feskr. Kijöbenh.* 54: 136 (1890)
> *Lycoperdon filiforme* Schrank, *Hoppe's Botanisches Taschenbuch*: 69 (1793)
> *Uredo longissima* Sowerby, *Col. fig. Engl. fung.* 2: pl. 139 (1798)
> *Caeoma longissimum* (Sowerby) Schltdl., *Fl. berol.* 2: 129 (1824)
> *Ustilago longissima* (Sowerby) Meyen, *Pflanzenpathologie*: 124 (1841)

E: c **S:** c **W:** c **NI:** ! **ROI:** !
H: In leaves of *Glyceria maxima* and less frequently *G. fluitans*.
D: UBI 54 (as *U. longissima*), Vánky: 358
Common and widespread. Better known as *U. longissima*.

flosculorum (DC.) Fr., *Syst. mycol.* 3: 518 (1829)
> *Uredo flosculorum* DC., *Fl. Fr.* 5: 79 (1815)

E: ! **S:** !
H: In the anthers of *Knautia arvensis* and *Succisa pratensis*.
D: UBI 52, Vánky: 358
Known from England (Yorkshire: near Pickering) and Scotland (Fife: Kilconquhar).

grandis Fr., *Syst. mycol.* 3: 518 (1829)
> *Erysibe typhoides* Wallr., *Fl. crypt. Germ., Sect. II*: 215 (1833)
> *Ustilago typhoides* (Wallr.) Berk. & Broome, *Ann. Mag. Nat. Hist, Ser. 2 [Notices of British Fungi no. 480]* 5: 463 (1850)

E: ! **W:** ! **ROI:** !
H: In culms of *Phragmites australis*.
D: UBI 52, Vánky: 359
Rarely reported but apparently widespread.

heufleri Fuckel, *Jahrb. Nassauischen Vereins Naturk.* 23-24: 39 (1870)
S: !
H: On cultivated *Erythronium oregonum* (and on native *Tulipa sylvestris* in continental Europe).

D: UBI 52-53, Vánky: 359-360
Known from Midlothian (Edinburgh).

hordei (Pers.) Lagerh. [non *U. hordei* Bref. (1888)], *Mitt. Bad. Bot. Vereins* 1889: 70 (1889)
   Uredo segetum α *hordei* Pers., *Syn. meth. fung.*: 224 (1801)
   Ustilago segetum var. *hordei* (Pers.) Rabenh., *Klotzschii Herbarium Vivum Mycologicum (Editio nova)*: 397 (1856)
   Ustilago kolleri Wille, *Bot. Not.*: 10 (1893)
   Ustilago segetum Roussel, *Flora Calvados*, Edn 2: 47 (1806)
   Mis.: *Reticularia segetum* sensu auct. p.p.
E: ! **S:** ! **W:** ! **O:** Isle of Man: !
H: In spikelets of various *Poaceae*. Known on *Hordeum vulgare* in Britain.
D: Vánky: 360
Once a common and widespread pathogen, but now rarely encountered with the advent of resistant strains of barley.

hypodytes (Schltdl.) Fr., *Syst. mycol.* 3: 518 (1829)
   Caeoma hypodytes Schltdl., *Fl. berol.*: 129 (1824)
   Uredo hypodytes (Schltdl.) Desm., *Ann. Sci. Nat., Bot.*, sér. 2, 13: 182 (1840)
E: c **S:** c **W:** !
H: In the culms, especially surrounding the internodes, of various *Poaceae*, especially frequent in Britain on *Bromopsis (Bromus) erectus*, *Elytrigia (Agropyron) repens* and *Leymus (Elymus) arenarius*.
D: UBI 53, Vánky: 361-362
Common and widespread.

intermedia J. Schröt., in Rabenhorst, *Fungi europaei* IV: 1696 (1873)
E: ! **S:** !
H: In the anthers of *Scabiosa columbaria*.
D: UBI 53, Vánky: 363
Rarely reported.

kuehneana R. Wolff, *Bot. Zeitung (Berlin)* 32: 815 (1874)
E: ! **S:** ! **O:** Channel Islands: !
H: In ovaries and anthers of *R. acetosella*.
D: UBI 54, Vánky: 364
Rarely reported.

marina Durieu, in Tul., *Ann. Sci. Nat., Bot.*, sér. 5, 5: 134 (1866)
E: !
H: In the rhizomes and culm bases of *Eleocharis parvula* (=*Scirpus parvulus*).
D: UBI 54, Vánky: 366
A single collection, from Dorset (Isle of Purbeck, Little Sea) but easily overlooked. N.B. Possibly not a member of the *Ustilaginales*; may be a *Cladochytrium* fide Liro [Ann. Bot. Soc. Zool.-Bot. Fenn. 'Vanamo' 6: 10 (1935)] or a *Melanotaenium* fide Lindeberg [Symb. Bot. Upsal. 16 (2): 124 (1959)].

maydis (DC.) Corda, *Icon. fung.* 5: 3 (1842)
   Uredo maydis DC., *Fl. Fr.* 5: 77 (1815)
   Ustilago zeae (Link) Unger [*comb. illegit.*], *Einfl. Boden. Verth. Gew.*: 211 (1836)
E: !
H: In the inflorescences and aerial parts of *Zea mays*.
D: UBI 61 (as *Ustilago zeae*), Vánky: 366
Occasional but increasing, wherever the host is cultivated.

ornithogali (J.C. Schmidt & Kunze) Magnus, *Hedwigia* 14: 19 (1875)
   Uredo ornithogali J.C. Schmidt & Kunze, *Deutschl. Schwämme*: 5 (1819)
E: !
H: In leaves and pedicels of *Gagea lutea*.
D: UBI 55, Vánky: 369
Rarely reported (as is the host plant in Britain). Records only from scattered localities in Co. Durham and Yorkshire.

parlatorei A.A. Fisch. Waldh., *Hedwigia* 15: 177 (1876)
E: ! **S:** !
H: In stems, petioles and leaves of *Rumex crispus*.
D: Vánky: 370

Only recently recognised as British. Known from England (Yorkshire: Doncaster, and Marishes near Pickering) and Scotland (South Aberdeen: St. Fergus).

pustulata (DC.) G. Winter, *Hedwigia* 19: 109 (1880)
   Uredo bistortarum α *pustulata* DC., *Fl. Fr.* 5: 76 (1815)
E: !
H: In leaves of *Polygonum bistorta*.
D: Vánky: 372
Previously included in *U. bistortarum* and only recently recognised as British based on a single collection by Grove from an unspecified location in England in 1921, recently located in herb. K.

reticulata Liro, *Ann. Acad. Sci. Fenn., Ser. A,* 17(1): 20 (1924)
   Mis.: *Ustilago utriculosa* sensu Tul. & C. Tul. [Ann. Sci. Nat., Bot., sér. 3, 7: 102 (1847)]
E: !
H: In ovaries of *Persicaria lapathifolia* and *P. maculosa*.
D: UBI 59 (as *Ustilago utriculosa*), Vánky: 373
Known from Cambridgeshire, South Essex, South Hampshire, Warwickshire and Yorkshire but previouly included in *U. bistortarum* s.l.

scabiosae (Sowerby) G. Winter, *Hedwigia* 19: 159 (1880)
   Farinaria scabiosae Sowerby, *Col. fig. Engl. fung.* 3: pl. 396 f. 2 (1803)
E: ! **S:** ! **NI:** ! **ROI:** !
H: In the anthers of *Knautia arvensis*.
D: UBI 55, Vánky: 374
Rarely reported, but apparently widespread. The host is common throughout the British Isles.

serpens (P. Karst.) B. Lindeb., *Symb. Bot. Upsal.* 16(2): 151 (1959)
   Tilletia serpens P. Karst., *Fungi Fennici exsiccati*: 599 (1866)
E: ! **S:** ! **O:** Channel Islands: !
H: In leaves of *Bromus erectus*, *Calamagrostis canescens*, *Elytrigia repens*, *E. farctus* and *Leymus arenarius*.
D: UBI 57, Vánky: 376-376
Rarely reported but apparently widespread.

striiformis (Westend.) Niessl, *Hedwigia* 15: 1 (1876)
   Uredo striiformis Westend., *Bull. Acad. Roy. Sci. Belgique* 18: 406 (1851)
   Tilletia striiformis (Westend.) Sacc., *Malpighia* 1: 8 (1977)
   Ustilago salveii Berk. & Broome, *Ann. Mag. Nat. Hist*, Ser. 2 [Notices of British Fungi no. 482] 5: 463 (1850)
   Uredo salveii (Berk. & Broome) Oudem., *Prodromus florae batavae* 2,4 (Fungi): 180 (1866)
   Tilletia salveii (Berk. & Broome) P. Karst., *Bidrag Kännedom Finlands Natur Folk* 39: 102 (1884)
   Tilletia debaryana A.A. Fisch. Waldh., in Rabenhorst, *Fungi europaei* IV: 1097 (1866)
E: c **S:** c **W:** c **NI:** c **ROI:** c **O:** Channel Islands: c
H: In leaves of various *Poaceae*, frequently *Holcus lanatus* and *H. mollis*, but also known on many other grass species.
D: UBI 58, Vánky: 377-378
Common and widespread. *Ustilago salveii* was shown to be a synonym by Ainsworth & Sampson (1950) and potentially provides an earlier name for the species.

stygia Liro, *Ann. Acad. Sci. Fenn., Ser. A,* 17(1): 25 (1924)
E: ! **S:** ! **W:** ! **O:** Channel Islands: !
H: In flowers of *Rumex acetosa*.
D: Vánky: 379
Rarely reported but apparently widespread.

succisae Magnus, *Hedwigia* 14: 17 (1875)
E: c **S:** c **W:** c **NI:** ! **O:** Isle of Man: !
H: In the anthers of *Succisa pratensis*.
D: UBI 58, Vánky: 380
Common and widespread.

tragopogonis-pratensis (Pers.) Roussel, *Flora Calvados*, Edn 2: 47 (1806)
   Uredo tragopogi-pratensis Pers., *Syn. meth. fung.*: 225 (1801)

Uredo tragopogi Pers., *Tent. disp. meth. fung.*: 57 (1797)
E: ! **S:** !
H: In the inflorescences of *Tragopogon pratensis* and *T. porrifolius.*
D: UBI 59, Vánky: 381
Rarely reported. Possibly genuinely rare since infection is usually obvious, transforming the flower heads into large, blackish-violet spore masses.

tritici (Pers.) Rostr., *Overs. Kongel. Danske Vidensk. Selsk. Forh. Medlemmers Arbeider.* 15 (1890)
Uredo segetum β *tritici* Pers., *Syn. meth. fung.*: 224 (1801)
Ustilago segetum var. *tritici* (Pers.) Brunaud, *Act. Soc. linn. Bordeaux* 32(2): 163 (1878)
Ustilago nuda (J.L. Jensen) Kellerm. & Swingle, *Rep. (Annual) Kansas Agric. Exp. Sta.* 2: 227 (1890)
E: ! **S:** ! **NI:** ! **ROI:** !
H: In the ovaries of *Triticum* and *Hordeum* spp.
D: UBI 57 (as *Ustilago segetum* var. *tritici*), Vánky: 383
Previously a cause of widespread and serious disease ('Loose Smut') of wheat and barley, but with the advent of fungicides and resistant crop varieties now rarely seen.

vaillantii Tul. & C. Tul., *Ann. Sci. Nat., Bot.,* sér. 3, 7: 90 (1847)
Ustilago scillae Cif., *Ann. Mycol.* 29: 24 (1931)
E: ! **S:** ! **W:** ! **O:** Isle of Man: !
H: In the anthers (and less often the ovaries) of various *Liliaceae.* Frequently in *Scilla verna* and also known in cultivated species of *Bellevalia, Chionodoxa* and *Muscari.*
D: UBI 59, Vánky: 384
Rarely reported, but apparently widespread.

vinosa Tul. & C. Tul., *Ann. Sci. Nat., Bot.,* sér. 3, 8: 96 (1847)
E: ! **S:** !
H: In flowers of *Oxyria digyna.*
D: UBI 60, Vánky: 384
Rarely reported.

ALIEN BASIDIOMYCETES

Agaricus endoxanthus Berk. & Broome, *J. Linn. Soc., Bot.*
11: 548 (1871)
Described from Sri Lanka and occasionally found in large troops
on compost in heated greenhouses in botanic gardens.

Amanita nauseosa (Wakef.) D.A. Reid, *Fungorum Rar. Icon.*
Color. 1: 25 (1966)
Lepiota nauseosa Wakef., *Bull. Misc. Inform. Kew* 1918 (2):
230 (1918)
First collected in 1918 from a tropical greenhouse in Kew
Gardens. There are several more recent collections from Kew
in the 1960's and 1970's but it has not been reported since.

Athelia rolfsii (Curzi) C.C. Tu & Kimbr., *Bot. Gaz.* 139(4): 460
(1978)
Corticium rolfsii Curzi, *Boll. R. Staz. Patalog. Veget. Roma*
11(4): 368 (1932)
Recorded from England (South Lancashire) and from Scotland
(Mid Ebudes) on imported decayed sacking.

Bolbitius mexicanus (Murrill) Murrill, *Mycologia* 4: 332 (1912)
Mycena mexicana Murrill, *Mycologia* 4: 73 (1912)
A Central and South American species. The British collection is
from the tropical biome at the Eden Project, Cornwall.

Chlorophyllum molybdites (G. Mey.) Massee, *Bull. Misc.*
Inform. Kew. 1898: 136 (1898)
Agaricus molybdites G. Mey., *Pr. Flor. Esseq.*: 300 (1818)
Rarely collected in beds of ornamental plants in heated
buildings, but may be abundant where it occurs. A common
and widespread pantropical species. Poisonous, and easily
identified by the greenish lamellae.

Clitopilus septicoides (Henn.) Singer, *Lilloa* 22: 606 (1951)
[1949]
Pleurotus septicoides Henn., *Hedwigia* 43: 104 (1904)
Reported from greenhouses in botanic gardens but not noted in
recent years and no voucher material traced in herbaria.

Collybia caldarii (Berk.) Sacc., *Syll. fung.* 5: 251 (1887)
Agaricus caldarii Berk., *Grevillea* 1: 89 (1872)
Collected in *Sphagnum* moss used to cultivate orchids in a
hothouse, in 1879. Never recollected.

Collybia dorotheae (Berk.) Sacc., *Syll. fung.* 5: 219 (1887)
Agaricus dorotheae Berk., *Grevillea* 1: 88 (1872)
Reported from England, 'on a dead tree fern stem from Jamaica'
in a hothouse but never recollected.

Collybia multijuga (Berk. & Broome) Sacc., *Syll. fung.* 5: 231
(1887)
Agaricus multijugus Berk. & Broome, *J. Linn. Soc., Bot.* 11:
519 (1871)
Described from Sri Lanka and collected several times from
compost in greenhouses in botanic gardens.

Collybia nephelodes (Berk. & Broome) Sacc., *Syll. fung.* 5:
208 (1887)
Agaricus nephelodes Berk. & Broome, *J. Linn. Soc., Bot.* 11:
521 (1871)
Described from Sri Lanka, this is occasionally collected on soil or
mulch in tropical greenhouses in botanic gardens.

Collybia purpureogrisea (Petch) Pegler, *Kew Bull.* 12: 127
(1986)
Marasmius purpureogriseus Petch, *Trans. Brit. Mycol. Soc.*
31: 40 (1948)
Occasionally collected on soil or mulch in heated greenhouses in
botanic gardens.

Collybia semiusta (Berk. & M.A. Curtis) Dennis, *Trans. Brit.*
Mycol. Soc. 34: 450 (1951)
Marasmius semiustus Berk. & M.A. Curtis, *J. Linn. Soc., Bot.*
10: 295 (1868)

A single collection from Surrey (Kew, Royal Botanic Gardens) in
1933, on leaves of *Musa textilis* recently imported from
Malaysia.

Conchomyces bursiformis (Berk.) E. Horak, *Sydowia* 34: 110
(1981)
Agaricus bursiformis Berk., *Flora Tasmaniae* 2: 245 (1860)
Hohenbuehelia bursiformis (Berk.) D.A. Reid, *Kew Bull.*
17(2): 304 (1963)
A single collection from Yorkshire (Sheffield, Winter Gardens) in
2003 [Hobart, FM4: 84-87 (2003)] on the trunk of *Dicksonia*
antartica in an unheated greenhouse.

Conocybe crispella (Murrill) Singer, *Sydowia* 4: 132 (1950)
Galerula crispella Murrill, *Lloydia* 5: 148 (1942)
A North American species, recently collected from the tropical
biome at the Eden Project in Cornwall.

Conocybe nivea (Massee) Watling, *Biblioth. Mycol.* 82: 123
(1981)
Bolbitius niveus Massee, *Brit. fung.-fl.* 2: 207 (1893)
Mis.: *Bolbitius conocephalus* sensu Cooke [Ill. Brit. fung.
1186 (1160) Vol. 8 (1891)]
A single collection (the type) 'on soil in palm house at Kew
Gardens' in 1888.

Coprinus brunneofibrillosus Dennis, *Kew Bull.* 15: 118
(1961)
Described from Venezuela. Recorded in Britain only once, from
England (Surrey: Kew, Royal Botanic Gardens, Palm House,
1992).

Coprinus grossii J.A. Schmitt & Watling, *Nova Hedwigia* 67(3-
4): 444 (1998)
A single record from Glasgow, on a damp plaster ceiling in a
house. Also known from Germany.

Coprinus kimurae Hongo & Aoki, *Trans. Mycol. Soc. Japan*
7(1): 16 (1966)
Known only from Yorkshire (Leeds, on damp stuffing of a
hospital couch) and Warwickshire (Atherstone, in a house).
Originally described from decayed rice straw in Japan.

Cyathus berkeleyanus (Tul. & C. Tul.) Lloyd, *Nidulariaceae*:
19 (1906)
Cyathus microsporus var. berkeleyanus Tul. & C. Tul., *Ann.*
Sci. Nat., Bot., sér. 3, 1: 73 (1844)
A single collection from England (Surrey: Kew, Royal Botanic
Gardens, Aroid House) in 1961.

Cystolepiota cystidiosa (A.H. Sm.) Bon, *Doc. mycol.* 11(43):
26 (1981)
Lepiota cystidiosa A.H. Sm., *Pap. Michigan Acad. Sci.* 27: 58
(1942) [1941]
Cystolepiota luteicystidiata (D.A. Reid) Bon, *Doc. mycol.*
6(24): 43 (1976)
Lepiota luteicystidiata D.A. Reid, *Fungorum Rar. Icon. Color.*
2: 9 (1967)
Probably an alien, most often collected from compost in
greenhouses in botanical gardens and not known in natural
habitat in Britain. The type of *Lepiota luteicystidiata* was
collected in the Palm House at Kew Gardens.

Cystolepiota ompnera (Berk. & Broome) Pegler, *Kew Bull.* 12:
284 (1986)
Agaricus ompnerus Berk. & Broome, *J. Linn. Soc., Bot.* 11: 21
(1871)
Occasionally reported in greenhouses in botanic gardens.

Donkioporia expansa (Desm.) Kotl. & Pouzar, *Persoonia* 7:
214 (1973)
Boletus expansus Desm., *Cat. pl. omises botanogr. Belgique*:
pl. 18 (1823)
Polyporus megaloporus Pers., *Myc. Eur.* 2: 88 (1825)
Poria megalopora (Pers.) Cooke, *Grevillea* 14: 115 (1886)
Phellinus megaloporus (Pers.) Bondartsev, *Trutovye griby*

Evropeiskoi chasti SSSR i Kavkaza: 414 (1953)
On worked wood, such as roof-beams, window sills etc. Known from England (Bedfordshire, Dorset, Oxfordshire, South Devon and Warwickshire) and a single collection from Wales (Breconshire).

Eccilia acus (W.G. Sm.) Sacc., *Syll. fung.* 5: 730 (1887)
 Agaricus acus W.G. Sm., *J. Bot.* 13: 97 (1875)
Collected on soil, amongst germinating seeds of *Coffea* sp. in a greenhouse at Kew Gardens. See Cooke 369 (613c) Vol. 3 (1885) for an illustration.

Eccilia flosculus (W.G. Sm.) Sacc., *Syll. fung.* 5: 730 (1887)
 Agaricus flosculus W.G. Sm., *J. Bot.* 13: 97 (1875)
Described from soil and decayed tree fern stems in a greenhouse in England (Middlesex: Chelsea, Veitch's Nursery) in 1870. See Cooke 369 (613b) Vol.3 (1885) for an illustration.

Fomitopsis rosea (Alb. & Schwein.) P. Karst., *Meddeland. Soc. Fauna Fl. Fenn.* 6: 9 (1881)
 Boletus roseus Alb. & Schwein., *Consp. fung. lusat.*: 251 (1805)
 Polyporus roseus (Alb. & Schwein.) Fr., *Syst. mycol.* 1: 372 (1821)
 Fomes roseus (Alb. & Schwein.) Sacc., *Syll. fung.* 6: 189 (1888)
In continental Europe this is a montane species growing mainly on *Abies* and *Picea* spp in native woodland, but in Britain always on softwood timbers in buildings. A few widely scattered collections from England (South Essex, South Lancashire and Surrey) and a single one from Scotland (Angus). Reported from Northern Ireland (Down) in 1997 but unsubstantiated with voucher material.

Galerina steglichii Besl, *Z. Mykol.* 59(2): 216 (1993)
Collected on soil in pots with *Carica hastata* in glasshouses in Scotland (Edinburgh: Royal Botanic Garden) in 1981.

Gymnopilus filiceus (Cooke) Singer, *Sydowia* 9: 411 (1955)
 Agaricus filiceus Cooke, *J. Bot.* 1: 66 (1863)
 Flammula filicea (Cooke) Sacc., *Syll. fung.* 5: 812 (1887)
Collected from decayed stems of tree ferns in a hothouse in 1862 and 1870. See Cooke 491 (450) Vol. 4 (1884) for an illustration.

Gymnopilus purpuratus (Cooke & Massee) Singer, *Sydowia* 9: 411 (1955)
 Agaricus purpuratus Cooke & Massee, *Grevillea* 18: 73 (1890)
 Flammula purpurata (Cooke & Massee) Sacc., *Syll. fung.* 9: 107 (1891)
Described from trunks of old tree ferns in England (Surrey: Kew, Royal Botanic Gardens). Outdoor records may refer to the similar *G. dilepis*. See Cooke 1167 (964) Vol.8 (1890) for an illustration.

Hydropus sphaerosporus (Dennis) Dennis, *Kew Bull.* 3: 47 (1970)
 Pseudohiatula sphaerospora Dennis, *Kew Bull.* 15(1): 91 (1961)
A single collection from England (Surrey, Kew: Royal Botanic Gardens) in 1993, on gravelly soil and decayed wood in a tropical greenhouse.

Hypholoma peregrinum Massee, *Bull. Misc. Inform. Kew* 1907: 239 (1907)
Collected and described from decayed, imported wood (Kew, Royal Botanic Gardens: Tropical Fern House).

Inocybe manukanea (E. Horak) Garrido, *Biblioth. Mycol.* 120: 177 (1988)
 Astrosporina manukanea E. Horak, *New Zealand J. Bot.* 15(4): 734 (1978) [1977]
Collected once in a greenhouse (Kew: Royal Botanic Gardens) in 1995

Inonotus rickii (Pat.) D.A. Reid, Inonotus:, *Kew. Bull.* 12: 141 (1957)
 Xanthochrous rickii Pat., *Bull. Soc. Mycol. France* 24: 6 (1908)

Ptychogaster cubensis Pat. (anam.), *Bull. Soc. Mycol. France* 12: 133 (1896)
A tropical species collected once (in its anamorphic state) on ship's timbers.

Lactocollybia angiospermarum Singer, *Sydowia* 2: 32 (1948)
A single collection from Kew (Royal Botanic Gardens, Palm House) in 1990. Common in the tropics.

Lentinula edodes (Berk.) Pegler, *Kavaka* 3: 20 (1976) [1975]
 Agaricus edodes Berk., *J. Linn. Soc., Bot.* 16: 50 (1878)
This is 'shiitake', long cultivated in China and Japan as an edible fungus. Now widely grown on inoculated logs in Europe, it has occasionally been encountered in the field (possibly on discarded logs).

Lepiota efibulis Knudsen, *Bot. Tidsskr.* 75(2-3): 151 (1980)
Described from a small greenhouse in Surrey (Wallington near Croydon), growing amongst tomato roots. Latterly known only from Kew Gardens (Palm House), associated with the roots of banana species.

Lepiota micropholis (Berk. & Broome) Sacc., *Syll. fung.* 5: 61 (1887)
 Agaricus micropholis Berk. & Broome, *J. Linn. Soc., Bot.* 11: 505 (1871)
Described from Sri Lanka. Possibly a species of *Leucocoprinus*. British material was on decayed coconut fibre in a hothouse. See Cooke 1108 (943) Vol. 8 (1889) for an illustration The name was misapplied by Lange (FlDan1: pl.13D) to a species near to *Lepiota lilacea*.

Lepiota phlyctaenodes (Berk. & Broome) Sacc., *Syll. fung.* 5: 52 (1887)
 Agaricus phlyctaenodes Berk. & Broome, *J. Linn. Soc., Bot.* 11: 501 (1871)
Described from Sri Lanka. British collections from soil in greenhouses in Kew Gardens.

Lepiota rubella Bres., *Verh. Bot. Vereins Prov. Brandenburg* 31: 149 (1890)
Described from South America and collected in large quantities on soil in Kew Gardens (Palm House) in 1990 and again in 1999. For a description and illustration see B&K4: 192 pl. 215 as *Lepiota* aff. *bettinae*. Apparently rather frequent in botanical gardens, and known throughout Europe in such situations *fide* Vellinga (FAN5: 129).

Leucoagaricus gongylophorus (A. Møller) Singer, *Agaricales in Modern Taxonomy*, Edn. 4: 477 (1986)
 Rozites gongylophorus A. Møller, *Bot. Mitt. Tropen* 6: 70, pl. I-II (1893)
Recorded on laboratory colonies of attine ants (*Atta cephalata*) in 1991 (Surrey: Kew Gardens) and 1993 (South Devon: University of Exeter).

Leucoagaricus hortensis (Murrill) Pegler, *Kew Bull.* 9: 414 (1983)
 Lepiota hortensis Murrill, *North American Flora* 10(1): 59 (1914)
Collected on soil in a heated greenhouse in Kew Gardens in 1995.

Leucocoprinus caldariorum D.A. Reid [*nom. inval.*], *Mycol. Res.* 94(5): 652 (1990)
 Mis.: *Agaricus biornatus* sensu Cooke [Ill.Brit.Fung. 27 (37) Vol.1 (1881)]
Described from a heated greenhouse in Kew Gardens (Water Lily House). Possibly conspecific with *Leucoagaricus meleagris*.

Leucocoprinus discoideus (Beeli) Heinem., *Bull. Jard. Bot. Nat. Belg.* 47(1-2): 84 (1979)
 Lepiota discoidea Beeli, *Fl. Icon. champ. Congo* 2: 33 (1936)
Described from Africa. Collected on compost in 1998, in the Princess of Wales Conservatory at Kew.

Leucocoprinus fragilissimus (Ravenel) Pat., *Essai taxon.*: 171 (1900)

Hiatula fragilissima Ravenel, *Ann. Mag. Nat. Hist.* 12: 422 (1853)
Agaricus licmophorus Berk. & Broome, *J. Linn. Soc., Bot.* 11: 500 (1871)
Lepiota licmophora (Berk. & Broome) Sacc., *Syll. fung.* 5: 44 (1887)
Collected on compost in tropical greenhouses in Kew. The single collection in herb. K is however from a fernery in South Lancashire (Bolton) in 1886. See Cooke 1110 (1179) Vol.8 (1889) for an illustration.

Leucocoprinus heinemannii var. melanotrichoides P. Mohr, *Feddes Repert.* 115(1-2): 20 (2004)
A single collection (the type) from a greenhouse in Kew Gardens in 1992.

Leucocoprinus heterosporus (Locq.) Locq., *Bull. Mens. Soc. Linn. Lyon* 14: 92 (1945)
Lepiotophyllum heterosporum Locq., *Bull. Mens. Soc. Linn. Lyon* 11: 46 (1942)
A single British collection from a greenhouse in England (South West Yorkshire: Halifax) in 1955.

Leucocoprinus ianthinus (Cooke ex Sacc.) Locq., *Bull. Mens. Soc. Linn. Lyon* 14: 94 (1945)
Lepiota ianthina Cooke ex Sacc., *Syll. fung.* 9: 10 (1891)
Agaricus ianthinus Cooke [*nom. illegit.*, non *A. ianthinus* Fr. (1821)], *Grevillea* 16: 101 (1888)
Lepiota lilacinogranulosa Henn., *Verh. Bot. Vereins Prov. Brandenburg* 40: 145 (1898)
Leucocoprinus lilacinogranulosus (Henn.) Locq., *Bull. Mens. Soc. Linn. Lyon* 12(6): 95 (1943)
Leucocoprinus lilacinogranulosus var. *subglobisporus* D.A. Reid, *Mycol. Res.* 93(4): 420 (1989)
Occasionally collected from compost in heated greenhouses in botanic gardens, and also known with tropical plants in offices and other heated buildings. See Cooke 1112 (944) Vol. 8 (1889) for a good illustration, also FM3(4): 110 (2002).

Leucocoprinus medioflavus (Boud.) Bon, *Doc. mycol.* 6(24): 45 (1976)
Lepiota medioflava Boud., *Bull. Soc. Mycol. France* 10(1): 59 (1894)
Known from a greenhouse in South West Yorkshire (Hebden Bridge) in 1911, but never recollected.

Leucocoprinus tenellus (Boud.) Locq. [*nom. illegit.*, non *L. tenellus* Pegler (1983)], *Bull. Mens. Soc. linn. Lyon* 12: 95 (1942)
Lepiota tenella Boud., *Icon. mycol.* 1: pl. 18 (1905)
A single British collection from a tropical greenhouse in Surrey (Kew, Royal Botanic Gardens) in herb. K, determined by Locquin.

Leucocoprinus wynniae (Berk. & Broome) Locq., *Bull. Mens. Soc. Linn. Lyon* 12: 94 (1943)
Hiatula wynniae Berk. & Broome, *Ann. Mag. Nat. Hist, Ser. 5* [Notices of British Fungi no. 1772] 3: 206 (1879)
Described from material collected on soil in a hothouse at Kew Gardens in 1872. See Cooke 676 (688) Vol. 5 (1887) for an illustration. Now also known from Australia (Queensland) and Sri Lanka.

Leucocoprinus zeylanicus (Berk.) Boedijn, *Bull. Jard. Bot. Buitenzorg, Sér. 3,* 16: 407 (1940)
Agaricus zeylanicus Berk., *J. Bot.* 6: 480 (1847)
Described from Sri Lanka, and occasionally collected on compost in greenhouses in Kew Gardens.

Macrolepiota abruptibulba (R. Heim) Heinem., *Bull. Jard. Bot. Nat. Belg.* 39: 218 (1969)
Leucocoprinus abruptibulbus R. Heim, *Rev. Mycol. (Paris)* 33(2-3): 213 (1968)
Chlorophyllum abruptibulbum (R. Heim) Vellinga, *Mycotaxon* 83: 416 (2002)
Described from tropical Africa. Collected on soil in a greenhouse in Kew Gardens in 1997 [Mycologist 14(1): 30 (2000)].

Macrometrula rubriceps (Cooke & Massee) Donk & Singer, *Mycologia* 40: 264 (1948)
Agaricus rubriceps Cooke & Massee, *Grevillea* 15: 65 (1887)
Chitonia rubriceps (Cooke & Massee) Sacc., *Syll. fung.* 5: 992 (1887)
Clarkeinda rubriceps (Cooke & Massee) Rea, *Brit. basidiomyc.*: 97 (1922)
Collected from soil in the Aroid House, Kew Gardens in 1886. This is the type (and still the only known collection) of *Macrometrula* Donk & Singer, a volvate relative of *Psathyrella*. See Cooke 1176 (967) Vol. 8 (1891) for an illustration.

Marasmius bambusinus (Fr.) Fr., *Epicr. syst. mycol.*: 385 (1838)
Agaricus bambusinus Fr., *Linnaea* 5: 479 (1830)
Collected in 1895 on dead stems of the bamboo *Phyllostachys nigro-punctata* in a cool greenhouse in Kew Gardens.

Marasmius opalinus Massee, *Bull. Misc. Inform. Kew* 1906: 46 (1906)
Described from recently imported wood from Jamaica, in a hothouse in Kew Gardens in 1906, but the type is now in poor condition.

Microporus xanthopus (Fr.) Kuntze, *Revis. gen. pl.* 3(3): 494 (1898)
Polyporus xanthopus Fr., *Syst. mycol.* 1: 350 (1821)
A common palaeotropical species occasionally seen on damp wood in greenhouses in botanic gardens.

Micropsalliota plumaria Höhn., *Sitzungsber. Kaiserl. Akad. Wiss., Wien, Math.-Naturwiss. Cl., Abt. 1* 123: 79 (1914)
Collected once, on soil in the Palm House in Kew Gardens, in 1993.

Mycena alphitophora (Berk.) Sacc., *Syll. fung.* 5: 290 (1887)
Agaricus alphitophorus Berk., *J. Linn. Soc., Bot.* 15: 48 (1877)
Mycena osmundicola J.E. Lange, *Dansk. Bot. Ark.* 1(5): 35 (1914)
Mycena osmundicola ssp. *imleriana* Kühner [*nom. inval.*], *Encycl. Mycol.* 10. Le Genre Mycena: 210 (1938)
Mycena osmundicola var. *imleriana* (Kühner) A. Pearson [*nom. inval.*], *Trans. Brit. Mycol. Soc.* 35: 100 (1952)
Collected from potting compost or orchid fibre, in greenhouses in botanic gardens.

Mycena codoniceps (Cooke) Sacc., *Syll. fung.* 9: 36 (1891)
Agaricus codoniceps Cooke, *Grevillea* 16: 102 (1888)
Described from tree fern stems in a greenhouse. See Cooke 1149 (952) Vol.8 (1890) for an illustration.

Nolanea rubida (Berk.) Sacc., *Syll. fung.* 5: 728 (1887)
Agaricus rubidus Berk., *Mag. Zool. Bot.*, [Notices of British Fungi No. 3] 1: 44 (1837)
Described from material collected in a hothouse-conservatory. Pearson & Dennis (1948) suggest it is 'not a *Nolanea*, rather *Clitopilus cretatus*', i.e. *Clitopilus scyphoides*. See Cooke 367 (340) Vol. 3 (1885) for an illustration.

Phanerochaete chrysosporium Burds., Burdsall & Eslyn, *Mycotaxon* 1(2): 124 (1974)
A single record 'on a decayed mushroom tray' in England (West Sussex: Littlehampton). Very common in the tropics.

Pleurotus opuntiae (Durieu & Lév.) Sacc., *Syll. fung.* 5: 363 (1887)
Agaricus opuntiae Durieu & Lév., *Expl. Sci. Algérie*: t. 32 (1850)
Occasionally encountered in compost with imported pot plants.

Psilocybe cubensis (Earle) Singer, *Sydowia* 2: 37 (1948)
Stropharia cubensis Earle, *Est. Agron. Cuba.* 1: 240 (1906)
Occasionally cultivated and widely sold in Britain as a recreational drug. Native to the tropics where it usually grows on weathered dung of cows.

Psilocybe stuntzii Guzmán & Ott, *Mycologia* 68: 1261 (1976)
A North American species, recorded from the tropical biome at the Eden Project, Cornwall.

Pycnoporus sanguineus (L.) Murrill, *Bull. Torrey Bot. Club* 31: 421 (1904)
Boletus sanguineus L., *Sp. pl.*: 1646 (1763)
Polyporus sanguineus (L.) Fr., *Syst. mycol.* 1: 371 (1821)
Occasional on decayed wood in warm greenhouses in botanic gardens. Common and widespread throughout the tropics.

Queletia mirabilis Fr., *Öfvers. Kongl. Vetensk.-Akad. Förh.* 28: 171 (1871)
Two British collections, from Surrey (Kew Gardens) in 1863, where waste material from an American collection had been tipped out, and Middlesex (Barnsbury) in 1943 on spent tan. There are few collections worldwide and the origin of this species is not known.

Ramaria filicina (Sacc. & Syd.) Corner, *Monograph of Clavaria and Allied Genera*: 576 (1950)
Clavaria filicina Sacc. & Syd., *Syll. fung.* 14: 238 (1899)
Clavaria cervina W.G. Sm., *J. Bot.* 11: 66 (1873)
Described from Britain 'on and about tree fern stems' in a greenhouse in London (South Kensington: Royal Horticultural Society). Never recollected.

Rigidoporus lineatus (Pers.) Ryvarden, *Norweg. J. Bot.* 19: 236 (1972)
Polyporus lineatus Pers. in Gaudichaud, *Voyage Monde, Uranie Physicienne, Bot.*: 174 (1826)
Rarely reported from decayed wood in warm greenhouses in botanic gardens. Common throughout the tropics.

Rigidoporus microporus (Sw.) Overh., *Icon. Fungorum Malayensium* 5: 1 (1924)
Boletus microporus Sw., *Fl. Ind. Occ.* 3: 1925 (1806)
Polyporus microporus (Sw.) Fr., *Syst. mycol.* 1: 376 (1821)
Occasionally collected on decayed wood in warm greenhouses in botanic gardens. Common throughout the tropics.

Serpula lacrymans (Wulfen) J. Schröt., *Krypt.-Fl. Schlesien* 3(1): 446 (1888)
Boletus lacrymans Wulfen, in Jacq., *Misc. Austriaca Bot.* 2: 111 (1781)
Merulius lacrymans (Wulfen) Schumach., *Enum. pl.* 1: 371 (1801)
Sistotrema cellare Pers., *Syn. meth. fung.*: 554 (1801)
Merulius destruens Pers., *Syn. meth. fung.*: 496 (1801)
Serpula destruens (Pers.) Gray, *Nat. arr. Brit. pl.* 1: 637 (1821)
Merulius domesticus O. Falck, in Møller, *Hausschwammforsch.* 6: 53 (1912)
Merulius giganteus Saut., *Hedwigia* 16: 72 (1877)
Merulius guillemotii Boud., *Bull. Soc. Mycol. France* 10(1): 63 (1894)
Merulius lacrymans var. *guillemotii* (Boud.) Boud., *Icon. Mycol.* 4: 84 (t. 165) (1905)
Merulius lacrymans var. *terrestris* Peck, *Rep. (Annual) New York State Mus. Nat. Hist.* 49: 31 (1897)
Boletus obliquus Bolton, *Hist. fung. Halifax* 1: 74 (1788)
Merulius terrestris (Peck) Burt, *Rep. (Annual) Missouri Bot. Gard.* 4: 346 (1917)
Merulius vastator Tode, *Abh. Naturf. Ges. Halle.* 1: 351 (1783)
Occasional and widespread on structural timber, but much less common than formerly. Basidiomes can develop on any solid surface including glass and metallic objects, and the hyphae frequently fill hollow areas (wall spaces etc.) once the infection has taken hold. English name = 'Dry Rot'.

Stereum ostrea (Blume & Nees) Fr., *Epicr. syst. mycol.*: 547 (1838)
Thelephora ostrea Blume & Nees, *Nov. Actorum Acad. Caes. Leop.-Carol. Nat. Cur.* 13: 13 (1826)
Rarely reported on decayed wood in warm greenhouses in botanic gardens. Also reported from structural timber in a

building in Republic of Ireland (Cork) but unsubstantiated with voucher material. Common throughout the tropics.

Trogia buccinalis (Mont.) Pat., *Tab. Analyt. Fung.*: 57 (1889)
Cantharellus buccinalis Mont., *Ann. Sci. Nat., Bot.,* sér. 4, 1: 108 (1854)
A neotropical species, reported in Britain from soil and mulch in the tropical biome, Eden Project, Cornwall.

Veluticeps abietina (Pers.) Hjortstam & Tellería, *Mycotaxon* 37: 54 (1990)
Thelephora abietina Pers., *Syn. meth. fung.*: 577 (1801)
Stereum abietinum (Pers.) Fr., *Epicr. syst. mycol.*: 553 (1838)
Hymenochaete abietina (Pers.) Massee, *J. Linn. Soc., Bot.* 27: 115 (1890)
Columnocystis abietina (Pers.) Pouzar, *Česká Mykol.* 13: 17 (1959)
Stereum striatum (Schrad.) Fr., *Epicr. syst. mycol.*: 551 (1838)
Thelephora conchata Fr., *Syst. mycol.* 1: 438 (1821)
Stereum conchatum (Fr.) Fr., *Epicr. syst. mycol.*: 549 (1838)
Occasionally found on imported timber or worked wood.

Volvaria temperata (Berk. & Broome) Sacc., *Syll. fung.* 5: 660 (1887)
Agaricus temperatus Berk. & Broome, *Ann. Mag. Nat. Hist, Ser. 5 [Notices of British Fungi no. 1757]* 3: 205 (1879)
Collected by Berkeley in Northamptonshire (Sibbertoft) from soil in a hothouse, and listed by Massee (Bull. Misc. Inform. Kew 1906) from Kew Gardens, also from soil in a hothouse. See Cooke 300 (300) Vol. 3 (1884) for an illustration.

ALIEN UREDINIOMYCETES

Chrysomyxa ledicola (Peck) Lagerh., *Tromsö Mus. Aarsh.* 16: 119 (1893)

> *Uredo ledicola* Peck, *Rep. (Annual) New York State Mus. Nat. Hist.* 25: 90 (1873)

Reported once on imported plants of *Ledum groenlandicum*. The infected bushes were destroyed.

Dietelia codiaei (Syd.) Boerema, *Versl. Medsd. Plziektenk. Dienst Wageningen* 16: 153 (1979) [1978]

> *Aecidium codiaei* Syd., *Ann. Mycol.* 37: 198 (1939)

Reported once from England on imported *Codiaeum variegatum*.

Gymnoconia nitens (Schwein.) F. Kern & Thurst., *Bull. Pennsylvania Agric Exp. Sta.* 239: 16 (1929)

> *Aecidium nitens* Schwein., *Schriften Naturf. Ges. Leipzig* 1: 69 (1822)
>
> *Kunkelia nitens* (Schwein.) Arthur, *Bot. Gaz.* 63: 504 (1917)

Reported once on an imported batch of dewberry (*Rubus* sp.). The infested bushes were destroyed.

Gymnosporangium asiaticum Miyabe ex G. Yamada, *Shokubutse Byorigaku (Pl. Path.) Tokyo Hakubunkwan* 37(9): 304 (1904)

Recorded on imported *Juniperus chinensis* in 1974 and 1979. Never established in natural habitat in Britain.

Gymnosporangium juniperi-virginianae Schwein., *Schriften Naturf. Ges. Leipzig* 1: 74 (1822)

Reported occasionally on imported species of *Malus* [Moore, British Parasitic Fungi: 180 (1959)].

Puccinia belamcandae (Henn.) Dietel, *Ann. Mycol.* 5: 71 (1907)

> *Uredo belamcandae* Henn., *Engler's Bot. Jahrb.* 37: 158 (1905)

A single collection on cultivated *Belamcanda chinensis* from England (Cambridge) in 1983.

Puccinia gladioli (Duby) Castagne, Puccinia, *Observ. Uréd.* 2: 17 (1843)

> *Uredo gladioli* Duby, *Bot. gall.*: 901 (1830)

A single record on cultivated *Gladiolus* sp. from Cornwall in 1924. See W&H: 225.

Puccinia satyrii Syd., *Monogr. Ured.* 1: 594 (1903)

Recorded on dying leaves and tubers of imported specimens of the orchid *Satyrium aureum* in 1929.

Puccinia soldanellae Fuckel, *Jahrb. Nassauischen Vereins Naturk.* 29-30: 14 (1875)

> *Aecidium soldanellae* Hornschusch, *Deutschl. Krypt. Flora* 1: 18 (1844)

Known on cultivated plants in botanic gardens in England (Cambridge) and Scotland (Edinburgh), but not reported recently.

Uredinopsis americana Syd., *Ann. Mycol.* 1: 325 (1903)

> *Uredinopsis mirabilis* Magnus, *Hedwigia* 43: 121 (1904)

A single record on the non-native fern *Onoclea sensibilis* in a London garden.

Uredo behnickiana Henn., *Hedwigia* 44: 169 (1905)

> *Hemileia americana* Massee, *Gard. Chron.* 38: 153 (1905)

Occasionally present on introduced species of *Orchidaceae* in botanic gardens.

Uredo epidendri Henn., *Hedwigia* 35: 254 (1896)

Present on imported plants of *Epidendrum paniculatum* in England (West Sussex: East Grinstead, Plant. Path. Lab.) in 1967.

Uromyces aloes (Cooke) Magnus, *Berl. Deutschl. Bot. Ges.* 10: 48 (1892)

> *Uredo aloes* Cooke, *Grevillea* 20: 16 (1891)
>
> *Uromyces aloicola* Henn., *Engler's Bot. Jahrb.* 14: 370 (1891)

Rarely collected on cultivated *Aloe* spp in botanic gardens.

Uromyces ari-triphylli (Schwein.) Seeler, *Rhodora* 44: 174 (1942)

> *Puccinia ari-triphylli* Schwein., *Trans. Am. phil. Soc.* 4: 297 (1832)
>
> *Aecidium dracontii* Schwein. (as *A. dracontinatum*), *Trans. Am. phil. Soc.* 4: 292 (1832)
>
> *Uromyces ari-virginici* Howe, *Bull. Torrey Bot. Club* 5: 43 (1874)
>
> *Aecidium importatum* Henn., *Verh. Bot. Vereins Prov. Brandenburg* 37: 12 (1895)

Rarely on cultivated plants of *Arisaema triphyllum* and *Peltandra virginica* in botanic gardens.

Uromyces erythronii (DC.) Pass., *Comm. Soc. crittog. Ital.* 2: 452 (1867)

> *Uredo erythronii* DC., *Fl. Fr.* 5: 67 (1815)

On plants of *Erythronium dens-canis* introduced from France to a garden in Dorset. Reported in 1936 but never recollected.

Uromyces holwayi Lagerh., *Hedwigia* 28: 108 (1889)

> *Uromyces lilii* Clinton [non *U. lillii* (Link) Fuckel (1875)], *Rep. (Annual) New York State Mus. Nat. Hist.* 27: 103 (1875)

Collected on cultivated *Lilium* spp in 1936 but never recollected or established.

Uromyces transversalis (Thüm.) G. Winter, *Flora* 42: 263 (1884)

> *Uredo transversalis* Thüm., *Flora*: 570 (1876)

Rarely reported on cut flower spikes of *Gladiolus* cultivars imported from South Africa.

ALIEN USTILAGINOMYCETES

Graphiola phoenicis Poit., *Ann. Sci. Nat. (Paris)* 3: 473 (1824)

Occasionally found on imported palms in conservatories (see Cooke, HBF2: 546).

Tilletia lolii G. Winter, *Rabenh. Krypt.-Fl.* 1: 109 (1881)

Material on *Lolium temulentum* imported from Portugal was noted to have infected native *Lolium* spp in 1937-8 in experimental plots (Moore, British Parasitic Fungi, 1959) but has never established itself in Britain.

Urocystis syncocca (L.A. Kirchn.) B. Lindeb., *Symb. Bot. Upsal.* 16(2): 99 (1959)

> *Uredo syncocca* L.A. Kirchn., *Lotos* 6: 179 (1856)
>
> *Urocystis hepaticae-trilobae* (DC.) Sampson & Moesz, *Kárpát-Medence Üszöggombái*: 208 (1950)

Collected once, from Surrey (Kew Gardens) in 1890.

EXCLUDED BASIDIOMYCETES

abieticola (P. Karst.) W.B. Cooke, Flagelloscypha, *Sydowia, Beih.* 4: 59 (1961)
> *Cyphella abieticola* P. Karst., *Fungi Fennici exsiccati*: no. 718 (1868)
Not authentically British. A collection from Cambridgeshire (Wicken Fen) in herb. K was assigned here by Cooke (1961) but later found to be misdetermined.

abieticola (Bourdot & Galzin) J. Erikss., Hyphodontia, *Symb. Bot. Upsal.* 16(1): 104 (1958)
> *Odontia barbajovis* ssp. *abieticola* Bourdot & Galzin, *Hyménomyc. France*: 426 (1928)
> *Grandinia abieticola* (Bourdot & Galzin) Jülich, *Int. J. Mycol. Lichenol.* 1(1): 35 (1982)
Not authentically British. Reported from England (Gloucestershire, Norfolk and South Devon) and Northern Ireland (Tyrone) but unsubstantiated with voucher material.

abietina (Fuckel) Corner, Typhula, *Monograph of Clavaria and Allied Genera*: 664 (1950)
> *Pistillaria abietina* Fuckel, *Jahrb. Nassauischen Vereins Naturk.* 25-26: 292 (1871)
Not authentically British. The single collection, from England (South Devon: Bellever Forest), in herb. K is annotated as 'compare with' *T. abietina*.

abietinum (Bull.) P. Karst., Gloeophyllum, *Bidrag Kännedom Finlands Natur Folk* 37: 80 (1882)
> *Agaricus abietinus* Bull., *Herb. France*: pl. 442 (1790)
> *Daedalea abietina* (Bull.) Fr., *Syst. mycol.* 1: 334 (1821)
> *Lenzites abietinus* (Bull.) Fr., *Epicr. syst. mycol.*: 407 (1838)
Not authentically British. Most of the records are doubtful, unsubstantiated with voucher material, and probably refer to *G. sepiarium*. Those with voucher material need re-assessment.

abietis Kühner, Inocybe, *Bull. Soc. Naturalistes Oyonnax* 9 (Suppl.): 3 (1955)
Not authentically British. Rejected by Kuyper (1986) as the type was unavailable for study. Sensu NCL is *I. glabrescens*.

abjecta P. Karst., Inocybe, *Bidrag Kännedom Finlands Natur Folk* 32: 456 (1879)
> *Inocybe peronatella* J. Favre, *Ergebn. Wiss. Untersuch. Schweiz. Natn. Parks* 6: 472 (1960)
Sensu Kuyper (1986) is not authentically British. Sensu J.E. Lange [Fl. agaric. danic. 3: 73 (1938)] and sensu NCL is doubtfully distinct from *I. flocculosa*.

abruptibulbus Peck, Agaricus, *Rep. (Annual) New York State Mus. Nat. Hist.* 94: 36 (1905)
A North American species, not authentically British. The name has been misapplied (e.g. Ph 169) to a form of *Agaricus silvicola* with truncately bulbous base (by some authorities distinguished as *A. essettei*).

abstrusa (Fr.) Singer, Pholiota, *Lilloa* 22: 516 (1951)
> *Agaricus abstrusus* Fr., *Hymenomyc. eur.*: 257 (1874)
A *nomen dubium*. Sensu NCL and sensu auct. mult. is *Pholiota conissans*.

acanthoides (Bull.) Fr., Polyporus, *Epicr. syst. mycol.*: 448 (1838)
> *Boletus acanthoides* Bull., *Herb. France*: pl. 486 (1791)
A *nomen dubium* lacking type material. Bulliard's plate of *Boletus acanthoides* appears to depict *Meripilus giganteus* and, *fide* Reid & Austwick (1963), Scottish records are probably also that species.

acariforme Sowerby, Lycoperdon, *Col. fig. Engl. fung.* 2: pl. 146 (1798)
Described from Britain but a *nomen dubium* lacking type material. From the illustration and description, probably not a basidiomycete.

acetabulosa (Sowerby) Sacc., Locellina, *Syll. fung.* 5: 761 (1887)
> *Agaricus acetabulosus* Sowerby, *Col. fig. Engl. fung.* 3: pl. 303 (1801)
> *Acetabularia acetabulosa* (Sowerby) Massee, *Brit. fung.-fl.* 2: 232 (1893)
Described from Britain 'from the banks of the River Thames at Millbank' in 1795, but never recollected and a *nomen dubium* lacking type material. Sowerby illustrates a small volvate, brown-spored agaric which could possibly be a species of *Conocybe fide* Watling (BFF3).

actinophorus (Berk. & Broome) Rea, Androsaceus, *Brit. basidiomyc.*: 533 (1922)
> *Marasmius actinophorus* Berk. & Broome, *J. Linn. Soc., Bot.* 14: 39 (1875)
Described from Britain but a *nomen dubium*. It has been suggested that this may represent *Crinipellis scabellus* or *Marasmius curreyi* but this is unlikely since it was described from 'naked soil'.

acuta Rea, Leptonia, *Trans. Brit. Mycol. Soc.* 17(1-2): 50 (1932)
Described from Britain. Doubtfully distinct from *Entoloma jubatum* according to a type study by Noordeloos (1987) and not *Entoloma sericeum* as suggested by Pearson & Dennis (1948).

admissa (Britzelm.) M.M. Moser, Tephrocybe, *Kleine Kryptogamenflora*, Edn 3: 114 (1967)
> *Collybia admissa* Britzelm., *Ber. Naturhist. Vereins Augsburg* 26: 146 (1881)
Not authentically British. Reported from Oxfordshire (Bix, Warburg Reserve) in 1991 but unsubstantiated with voucher material.

adnatus Huds., Agaricus [non *A. adnatus* W.G. Sm. (1870)], *Fl. angl.*, Edn 2, 2: 619 (1778)
Described from Britain but a *nomen dubium* lacking type material.

adstringens (Pers.) Métrod, Melanoleuca, *Bull. Soc. Mycol. France* 64: 163 (1948)
> *Agaricus adstringens* Pers., *Syn. meth. fung.*: 350 (1801)
> *Tricholoma melaleucum* var. *adstringens* (Pers.) Gillet, *Hyménomycètes*: 128 (1874)
Not authentically British. Sensu BFF8 is *Melanoleuca cognata* var. *nauseosa*.

adulterina (Fr.) Peck, Russula, *Rep. (Annual) New York State Mus. Nat. Hist.* 41: 75 (1887)
> *Russula integra* var. *adulterina* Fr., *Epicr. syst. mycol.*: 360 (1838)
Not authentically British. Reported by Rayner (1985) but later considered by him to be a misdetermination.

adusta var. coerulescens Fr., Russula, *Monogr. hymenomyc. Suec.* 2: 185 (1863)
A *nomen dubium*. Reported from Northern Ireland (Down) in 1931 but unsubstantiated with voucher material.

aelopus (Fr.) Sacc. [as *H. aelopodum*], Hypholoma, *Syll. fung.* 5: 1031 (1887)
> *Agaricus aelopus* Fr., *Öfvers. Kongl. Vetensk.-Akad. Förh.* 30: 6 (1873)
A *nomen dubium*. Apparently a distinctive species, the stipe covered in red squamules, but not reported for many years and no voucher material has been traced. The epithet, originally as '*aelolopus*', was corrected by Fries to *aelopus* in Hym. Eur.: 292 (1874). Saccardo incorrectly changed this to *aelopodum* and Rea (1922) to *aellopum*.

aestivus, With., Agaricus, *Bot. arr. Brit. pl. ed. 2*, 3: 306 (1792)
Described from Britain but a *nomen dubium* lacking type material.

affinis Massee, Bolbitius, *Eur. Fung. Flora*: 147 (1902)
Described from Britain but a *nomen dubium*. Possibly *Conocybe lactea* [i.e. *C. apala*] *fide* Pearson & Dennis (1948).

affinis Pat. & Doass., Clavaria, *Tab. anal. fung.*: no. 470 (1886)
A *nomen dubium*. Recent records are sensu Corner (1950) and may refer to *Clavaria tenuipes*.

affricata (Fr.) Gillet, Omphalia, *Hyménomycètes*: 295 (1876)
   Agaricus affricatus Fr., *Observ. mycol.* 2: 213 (1818)
A *nomen dubium*. Growing in *Sphagnum,* thus possibly *Arrhenia sphagnicola*.

agraria (Fr.) Sacc., Psilocybe, *Syll. fung.* 5: 1047 (1887)
   Agaricus agrarius Fr., *Monogr. hymenomyc. Suec.* 2: 304 (1863)
A *nomen dubium*. Pearson & Dennis (1948) suggest that Cooke 597 (622) Vol. 4 (1886) depicts a species of *Mycena,* but it could just as easily represent young, infertile fruitbodies of a *Psathyrella* in the *corrugis* group. Probably the same as *A. (Psilocybe) 'agnarius'* Fr. sensu Berkeley & Broome [Notices Brit. Fungi No. 1257 (1871)]

agrestis With., Agaricus, *Bot. arr. Brit. pl. ed. 2*, 3: 289 (1792)
Described from Britain but a *nomen dubium*.

alba (Bres.) Kühner, Mycena, *Encycl. Mycol.* 10. *Le Genre Mycena*: 584 (1938)
   Omphalia alba Bres., *Fl. ital. crypt.* 1: 295 (1915)
   Marasmiellus albus (Bres.) Singer, *Lilloa* 22: 302 (1951) [1949]
Not authentically British. Reported on a few occasions but all records are unsubstantiated with voucher material. Easily confused with other small whitish species of *Mycena* such as *M. hiemalis* and *M. olida*.

albellum (Schaeff.) P. Kumm., Tricholoma, *Führ. Pilzk.*: 131 (1871)
   Agaricus albellus Schaeff., *Fung. Bavar. Palat. nasc.* 1: 78 (1762)
   Agaricus fumosus var. *albellus* (Schaeff.) Fr., *Elench. fung.* 1: 15 (1828)
A *nomen dubium* lacking type material. Illustrated by Sowerby (1798) but this was copied in part from Schaeffer. Fries interpreted Schaeffer's plate as representing *Clitopilus prunulus* but Sowerby's plate depicts large clumps of an immature white agaric, possibly young material of *Lyophyllum connatum*.

albida (Huds.) Bref., Exidia, *Unters. Gesammtgeb. Mykol.* 7: 94 (1888)
   Tremella albida Huds., *Fl. angl.,* Edn 2, 2: 565 (1778)
Described from Britain but a *nomen dubium* lacking type material and variously interpreted. Sensu Donk (1966) is *Exidia thuretiana* but old British specimens named thus (in herb. K) are *Tremella mesenterica* and *T. globispora*.

albidodisca Kühner, Inocybe, *Bull. Soc. Naturalistes Oyonnax* 9 (Suppl.): 3 (1955)
Not authentically British. Sensu NCL is *Inocybe inodora* and sensu Reid (FRIC6) is *I. pruinosa*.

albidum (Fr.) Singer, Gerronema, *Sydowia* 15: 49 (1962) [1961]
   Cantharellus albidus Fr., *Syst. mycol.* 1: 319 (1821)
   Hygrophoropsis albida (Fr.) Maire, *Treb. Mus. Ci. Nat. Barcelona* 15(2): 52 (1933)
A *nomen dubium*. Sensu NCL (as *Cantharellus albidus*) includes *Gerronema prescotii* and *G. stevensonii*.

albobrunnea (Romell) Ryvarden, Antrodia, *Norweg. J. Bot.* 20: 8 (1973)
   Polyporus albobrunneus Romell, *Arch. Bot. (Leipzig)* 11(3): 10 (1911)
   Leptoporus albobrunneus (Romell) Pilát, *Atlas champ. Eur.* I 3(1): 178 (1938)
   Tyromyces albobrunneus (Romell) Bondartsev, *Trutovye griby Evropeiskoi chasti SSSR i Kavkaza*: 203 (1953)
Not authentically British. Reported from Scotland in 1964, but misidentified *fide* L. Ryvarden (pers. comm.).

albobrunneum (Pers.) P. Kumm., Tricholoma, *Führ. Pilzk.*: 130 (1871)
   Agaricus albobrunneus Pers., *Syn. meth. fung.*: 293 (1801)
   Mis.: *Tricholoma striatum* sensu auct. mult.
Not authentically British. Reported on numerous occasions and apparently 'widespread' but recent reports are sparse. Virtually all records are unsubstantiated with voucher material, and the species is easily confused with similar taxa. Sensu auct. Brit. is a mixture of *T. ustale* and *T. ustaloides*.

albocarneus Britzelm., Lactarius, *Bot. Centralbl.* 23: 309 (1895)
   Lactarius glutinopallens F.H. Møller & J.E. Lange, *Fl. agaric. danic.* 5, *Taxonomic Conspectus*: IV (1940)
Not authentically British. Reported (as *Lactarius glutinopallens*) from South Hampshire (New Forest) in 1977 but unsubstantiated with voucher material.

albocinerea Rea, Clitocybe, *Trans. Brit. Mycol. Soc.* 4(2): 308 (1913)
Described from Britain but a *nomen dubium*.

alboconicum (J.E. Lange) Clémençon, Tricholoma, *Mycol. helv.* 1: 26 (1983)
   Tricholoma myomyces var. *alboconicum* J.E. Lange, *Fl. agaric. danic.* 5, *Taxonomic Conspectus*: IX (1940)
Listed in NCL and FungEur3 as a synonym of *Tricholoma inocybeoides*, but subsequently considered an independent species (FungEur3a: 708) as yet unknown in Britain.

albocyaneus Fr., Cortinarius, *Monogr. hymenomyc. Suec.* 2: 62 (1863)
A *nomen dubium*. Last reported from England (Cumbria) in 1922 but no material traced. Cooke 771 (748) Vol. 6 (1886) is unidentifiable.

albomarginata Pat., Cyphella, *Tab. anal. fung.*: 164 f.361 (1885)
A *nomen dubium*. A single British collection named thus, from Norfolk, requires re-examination.

albosericea Henn., Lepiota, *Verh. Bot. Vereins Prov. Brandenburg* 40: 143 (1898)
A *nomen dubium*. Described from a hothouse in Berlin Botanic Gardens. Sensu J.E. Lange (1935) and sensu auct. Brit. is *L. subalba*.

album *var.* **caesariatum** Rea, Tricholoma, *Brit. basidiomyc.*: 235 (1922)
Described from Britain but a *nomen dubium*.

albus Bolton, Boletus [*nom. illegit.*, non *B. albus* Hudson (1762)], *Hist. fung. Halifax* 2: 78 (1788)
Described from Britain but a *nomen dubium* lacking type material. Bolton's illustration depicts old, weathered fruitbodies of *Laetiporus sulphureus* but Pilát (1936) places it in the synonymy of *Bjerkandera fumosa*.

albus (Huds.) Fr., Polyporus, *Hymenomyc. eur.*: 549 (1874)
   Boletus albus Huds. [non *B. albus* Bolton (1788)], *Fl. angl.*: 496 (1762)
Described from Britain but a *nomen dubium* lacking type material. Compared by Hudson to *B. suaveolens* L. (= *Trametes suaveolens* (L.) Fr. but lacking smell.

alcalina (Fr.) P. Kumm., Mycena, *Führ. Pilzk.*: 109 (1871)
   Agaricus alcalinus Fr., *Observ. mycol.* 2: 153 (1818)
A *nomen dubium* variously interpreted. Sensu NCL includes both *M. stipata* and *M. silvae-nigrae*.

aldridgei Massee, Flammula, *Grevillea* 20: 25 (1892)
Described from England (North Hampshire: Petersfield). Possibly an alien related to *Phylloporus* spp *fide* NCL, but type material is in poor condition.

alexandri Gillet, Locellina, *Hyménomycètes*: 429 (1876)
A *nomen dubium*. Listed by Rea (1922). Possibly a *Cortinarius* sp. *fide* Singer (1986: 676).

algeriensis (Fr.) Fr., Pilosace, in Quélet, *Mém. Soc. Émul. Montbéliard, Sér. 2,* 5: 351 (1873)

Agaricus algeriensis Fr., *Hymenomyc. eur.*: 283 (1874)
A *nomen dubium*. Fide Pearson & Dennis (1948) 'probably a species of *Agaricus* with the annulus rubbed or fallen off'. Cooke 553 (618) Vol.4 (1885) would appear to be the same.

algidus (Fr.) Gillet, Pleurotus, *Hyménomycètes*: 335 (1876)
Agaricus algidus Fr., *Syst. mycol.* 1: 190 (1821)
A *nomen dubium*. Sensu auct. Brit. is *Hohenbuehelia atrocaerulea*.

alienata (S. Lundell) J. Erikss., Hyphodontia, *Symb. Bot. Upsal.* 16(1): 104 (1958)
Peniophora alienata S. Lundell, *Fungi Exsiccati Suecici* 21-22: 28 (1941)
Not authentically British. Reported from Scotland, but material in herb. E has been redetermined as *Hyphodontia sambuci*.

alkalivirens Singer, Collybia, *Sydowia* 2: 27 (1948)
A North American species not authentically British. Sensu auct. Brit. is possibly *C. fuscopurpurea*.

allenii (Maire) P.D. Orton, Omphalina, *Trans. Brit. Mycol. Soc.* 43(2): 179 (1960)
Omphalia allenii Maire, *Trans. Brit. Mycol. Soc.* 3(3): 169 (1910) [1909]
Listed by Rea (1922), this is a dubious species known only from the type collection, which needs re-examination.

alliacea (Quél.) P.A. Lemke, Dendrothele, *Persoonia* 3(3): 336 (1965)
Corticium alliaceum Quél., *Compt. Rend. Assoc. Franç. Avancem. Sci.* 12: 505 (1884)
Not authentically British. The single record from England (Herefordshire: Humber Marsh) was misidentified *D. acerina*.

alligatus Fr., Polyporus, *Elench. fung.* 1: 78 (1828)
A *nomen dubium*. Listed by Rea (1922) as 'on roots, often wrapping round stipules and grasses'. Donk (1974: 30) cites Bresadola's view (*Rev. Mycol. (Toulouse)* 12: 103, 1890) that this is *Bjerkandera fumosa*.

allutus Fr., Cortinarius, *Epicr. syst. mycol.*: 263 (1838)
A *nomen dubium*. Sensu NCL and sensu auct. mult. is *C. multiformis*.

alnetorum J. Favre, Clitocybe, *Ergebn. Wiss. Untersuch. Schweiz. Natn. Parks* 6: 420 (1960)
Not authentically British. A montane species in continental Europe and strictly associated with *Alnus viridis* which is not native in Britain. Reported from England (East Norfolk and Cheshire) but dubiously identified and unsubstantiated with voucher material.

alpestris (Britzelm.) Singer, Calocybe, *Sydowia* 15: 47 (1962) [1961]
Agaricus alpestris Britzelm., *Hymenomyc. Südbayern* 8: 4 (1891)
Lyophyllum alpestre (Britzelm.) Huijsman, *Fungus* 26: 38 (1956)
Not authentically British. Collections assigned here were found to lack the carminophile basidia typical of the genus *Calocybe* and possibly represent *Limacella delicata* var. *vinosorubescens*.

alpina (A. Blytt) F.H. Møller & Jul. Schäff., Russula, *Russula-Monographie*: 222 (1952)
Russula emetica var. *alpina* A. Blytt, *Videnskabs-Selskabets Skrifter. I Math.-Naturv. Kl.* 6: 105 (1905) [1904]
Not authentically British. Sensu auct. Brit. is *Russula nana*.

alpinus Boud., Cortinarius, *Bull. Soc. Mycol. France* 11: 27 (1895)
Not authentically British. Known only from the European Alps. Sensu auct. Brit. is *C. favrei*.

alternatus (Schumach.) Sacc., Coprinus, *Syll. fung.* 5: 1093 (1887)
Agaricus alternatus Schumach., *Enum. pl.* 2: 1874 (1803)
A *nomen dubium* lacking type material. Cooke 664 (677) Vol. 5 (1886) may be a species of *Psathyrella*.

alumnus Bolton, Agaricus, *Hist. fung. Halifax* Suppl.: 155 (1791)
Described from Britain, but a *nomen dubium* lacking type material. Bolton's illustration and description are of a small agaric growing on old specimens of *Russula* and lacking any 'tuberous root' (sclerotium) thus probably synonymous with *Asterophora parasitica*.

alutacea Sacc., Omphalia, *Syll. fung.* 11: 24 (1895)
Agaricus alutaceus Cooke & Massee [*nom. illegit.*, non *A. alutaceus* Fr. (1821)], *Grevillea* 21: 40 (1892)
Described from Britain but a *nomen dubium*. Possibly *Omphalina pyxidata*.

alutaceus Fr., Polyporus, *Syst. mycol.* 1: 360 (1821)
A *nomen dubium*, variously interpreted [see Donk (1974: 284)]. Listed by Rea (1922).

alveolus (Lasch) P. Kumm., Crepidotus, *Führ. Pilzk.*: 74 (1871)
Agaricus alveolus Lasch, *Linnaea* 4: 547 (1829)
A *nomen dubium* lacking type material. Cooke 534 (499) Vol.4 (1885) appears to be *Crepidotus mollis* or a form of it.

amabilis (Peck) Singer, Suillus, *Mycologia* 58: 159 (1966)
Boletus amabilis Peck, *Bull. Torrey Bot. Club* 27: 612 (1900)
A North American species, not authentically British. Reported from England (Hertfordshire) but this is sensu auct. Eur., i.e. *Suillus lakei*.

amara (Alb. & Schwein.) P. Kumm., Clitocybe, *Führ. Pilzk.*: 121 (1871)
Agaricus rivulosus β *amarus* Alb. & Schwein., *Consp. fung. lusat.*: 185 (1805)
Agaricus amarus (Alb. & Schwein.) Fr., *Syst. mycol.* 1: 87 (1821)
Lepista amara (Alb. & Schwein.) Maire, *Bull. Soc. Mycol. France.* 47: 216 (1930)
Leucopaxillus amarus (Alb. & Schwein.) Kühner, *Ann. Soc. Linn. Lyon* 73: 84 (1927)
A *nomen dubium* lacking type material. Sensu auct. Brit. is *Leucopaxillus gentianeus*.

amarella (Pers.) Sacc., Clitocybe, *Syll. fung.* 5: 151 (1887)
Agaricus amarellus Pers., *Mycol. eur.* 3: 99 (1828)
A *nomen dubium* lacking type material. Revived by Bon [Flore Mycol. d'Eur.4 (1997)] for a species in grassland on acidic soil. British records are all old and non sensu Bon.

amariusculum Clémençon, Lyophyllum, *Mycotaxon* 15: 68 (1982)
Not authentically British. Collections named thus are all *Lyophyllum eustygium*.

ambiguum A.H. Sm. & Thiers, Leccinum, *Boletes of Michigan*: 138 (1971)
Not authentically British. Reported from Scotland and England but records are unsubstantiated with voucher material. This is not the same as *Boletus ambiguus*, a *nom. prov.* of Pearson (see below).

ambiguus A. Pearson [*ined.*], Boletus:
A *nom. prov.* of Pearson, never officially published but used in British literature [see BullBMS 10 (2): 61 (1976) also 15 (2): 93 (1981) and 18 (1): 35 (1983)] and with voucher material named thus in herb. K. Apparently a dark taxon near to *Boletus chrysenteron*.

amblyspora Kühner, Inocybe, *Bull. Soc. Naturalistes Oyonnax.* 9 (Suppl.): 3 (1955)
Not authentically British. A single collection from England in 1957 (North Somerset: Higher Merridge near Bridgwater) in herb. K determined with a query and a further comment by Kuyper saying 'probably not this species'.

americanum J. Erikss. & Hjortstam, Repetobasidium, *Corticiaceae of North Europe* 6: 1251 (1981)
Not authentically British. The single collection named thus in herb. K from England (Surrey: Mickleham Downs) is annotated as 'compare with' *R. americanum*.

amethystina (Battarra) Donk, Clavulina, *Meded. Ned. Mycol. Ver.* 22: 23 (1933)
> *Coralloides amethystina* Battarra, *Fungorum agri Ariminensis historia, Edn 2*: 22 (1759)
> *Clavaria amethystina* (Battarra) Fr., *Syst. mycol.* 1: 472 (1821)
> *Clavaria amethystea* Bull., *Herb. France*: pl. 496 (1791)
> *Ramaria amethystea* (Bull.) Gray, *Nat. arr. Brit. pl.* 1: 655 (1821)

Not authentically British and doubtfully distinct from lilaceous-grey forms of *Clavulina cinerea*. Sensu auct. mult. is *Clavulina zollingeri*.

amethystina (Quél.) J.E. Lange, Psalliota, *Dansk. Bot. Ark.* 4(12): 10 (1926)
> *Psalliota sylvatica* var. *amethystina* Quél., *Compt. Rend. Assoc. Franç. Avancem. Sci.* 13: 280 (1885) [1884]

A *nomen dubium* lacking type material. Sensu auct. Brit. is possibly *Agaricus dulcidulus*.

amethystinum (Scop.) P. Kumm., Tricholoma, *Führ. Pilzk.*: 132 (1871)
> *Agaricus amethystinus* Scop. [non *A. amethystinus* Huds. (1778)], *Fl. carniol.*: 437 (1772)

A *nomen dubium* lacking type material. Sensu Scopoli is a species of *Cortinarius*. Cooke 104 (262) Vol.1 (1882) looks distinctive, possibly *T. sejunctum*, but *fide* Pearson & Dennis (1948) is *Tricholoma personatum* (= *Lepista saeva*).

amianthina (Bourdot & Galzin) Liberta, Trechispora, *Taxon* 15(8): 318 (1966)
> *Corticium amianthinum* Bourdot & Galzin, *Bull. Soc. Mycol. France* 27: 260 (1911)
> *Corticium albo-ochraceum* ssp. *amianthinum* (Bourdot & Galzin) Bourdot & Galzin, *Hyménomyc. France*: 231 (1928)
> *Cristella amianthina* (Bourdot & Galzin) Donk, *Fungus* 27: 19 (1957)

Not authentically British. Reported from England (South Somerset, Surrey and West Kent) but doubtfully distinct from *Trechispora cohaerens*.

amianthina var. **broadwoodiae** (Berk. & Broome) Sacc., Lepiota, *Syll. fung.* 5: 48 (1887)
> *Agaricus amianthinus* var. *broadwoodiae* Berk. & Broome, *Ann. Mag. Nat. Hist., Ser. 5 [Notices of British Fungi no. 1730]* 1: 202 (1878)

A *nomen dubium*. From the original description, this is a distinctive variety of *Cystoderma amianthinum* or possibly an unknown autonomous species.

amica (Fr.) Singer, Melanoleuca, *Ann. Mycol.* 41: 57 (1943)
> *Agaricus amicus* Fr., *Monogr. hymenomyc. Suec.* 1: 86 (1857)
> *Tricholoma amicum* (Fr.) Gillet, *Hyménomycètes*: 109 (1874)

Not authentically British. A doubtful record from England (Oxfordshire: Tetsworth near Thame) reported by Massee (as *Tricholoma amicum*) in Grevillea 22: 39 (1893) but no material has been traced.

ammoniaca (Fr.) Quél., Mycena, *Mém. Soc. Émul. Montbéliard, Sér. 2,* 5: 106 (1872)
> *Agaricus ammoniacus* Fr., *Epicr. syst. mycol.*: 109 (1838)

A *nomen dubium* misapplied to both *Mycena aetites* and *M. leptocephala*. Sensu C&D: 573 is some other species, distinct from both of these.

amoena Quél., Russula, *Compt. Rend. Assoc. Franç. Avancem. Sci.* 9: 668 (1881)
> *Russula punctata* Gillet [*nom. illegit.,* non *R. punctata* Krombh. (1845)], *Hyménomycètes*: 245 (1876)
> Mis.: *Russula mariae* sensu auct.

Not authentically British. British records are assumed to be sensu Pearson (1948) which [*fide* Rayner (1985)] is *Russula violeipes*.

amoenicolor Romagn., Russula, *Les Russules d'Europe et d'Afrique du Nord*: 929 (1967)

Not authentically British, though listed in Rayner (1985). Reported from Gloucestershire in 1984 and Wiltshire in 1995, but unsubstantiated with voucher material.

amorpha W.B. Cooke, Maireina, *Sydowia, Beih.* 4: 84 (1961)
Type material needs re-examination. Described by W.B. Cooke from Wales (Montgomeryshire: Forden) using material collected by M.C. Cooke, but never recollected and otherwise unknown.

amplum (Pers.) Rea, Tricholoma, *Brit. basidiomyc.*: 227 (1922)
> *Agaricus amplus* Pers., *Syn. Meth. Fung.*: 360 (1801)
> *Clitocybe ampla* (Pers.) P. Kumm., *Führ. Pilzk.*: 120 (1871)

Placed in the synonymy of *Lyophyllum decastes* in NCL but requiring further investigation. If this synonymy is correct then *amplum* threatens to displace *decastes* as it is the earlier of the two sanctioned names.

anguinea (Fr.) Sacc., Naucoria, *Syll. fung.* 5: 831 (1887)
> *Agaricus anguineus* Fr., *Epicr. syst. mycol.*: 193 (1838)

A *nomen dubium*. Reported 'on soil in a flower pot' from Norfolk [Grevillea 10: 66 (1881)] but no material has been traced. Cooke 494 (455) Vol. 4 (1884) appears to be *Macrocystidia cucumis*.

angulatus With., Agaricus, *Arr. Brit. pl. ed. 3,* 4: 229 (1796)
Described from Britain but a *nomen dubium*.

angulosus Fr., Cortinarius, *Epicr. syst. mycol.*: 308 (1838)
Not authentically British. Cooke (1192 (1178) Vol.8 (1891) is doubtful. Sensu Kühner & Romagnesi (1953) is *C. renidens*.

angustifolia (Murrill) Candusso, Hygrocybe, *Fungi Europaei* 6: Hygrophorus s.l.: 357 (1997)
> *Camarophyllus angustifolius* Murrill, *North American Flora, Ser. 1 (Fungi 3),* 9(6): 386 (1916)
> *Hygrophorus angustifolius* (Murrill) Hesler & A.H. Sm., *North. Am. Hygroph.*: 60 (1963)

An American species, not authentically British. Sensu auct. Eur. is doubtfully distinct from *H. pratensis* var. *pallida fide* Boertmann (FNE1).

angustifolium Romagn., Hebeloma [*nom. illegit.,* non *H. angustifolium* (Britzelm.) Sacc. (1894)], *Sydowia* 36: 258 (1983)
A member of the *H. crustuliniforme* complex. Two collections in herb. K from England (Mid-Lancashire: Gait Barrows NNR) are assigned here.

angustissima (Lasch) P. Kumm, Clitocybe, *Führ. Pilzk.*: 122 (1871)
> *Agaricus angustissimus* Lasch, *Linnaea* 4: 523 (1829)

A *nomen dubium* lacking type material. Sensu auct. and sensu NCL is *C. agrestis*.

angustus (Pers.) P. Kumm., Clitopilus, *Führ. Pilzk.*: 97 (1871)
> *Agaricus angustus* Pers., *Syn. meth. fung.*: 345 (1801)

A *nomen dubium* lacking type material, but possibly an *Entoloma* sp. There is a single old specimen in poor condition from Wales (Denbighshire: Coed Coch) in herb. K, with a single English record (Yorkshire: Mulgrave Woods) listed in TBMS 3(5): 376 (1911) and Naturalist (Hull) 1911: 16 (1911).

anisata Velen., Clitocybe, *Česke Houby* II: 256 (1920)
Not authentically British. Reported from the Republic of Ireland (Wicklow) in 1989 but the record is unsubstantiated with voucher material.

annae Pilát, Agaricus, *Sborn. Nár. Mus. Praze* 7B: 132 (1951)
Not authentically British. Reported from England (Surrey: Kew Gardens) but the collections, in herb. K, are annotated as 'compare with' *A. annae*. Viewed by some authorities as only doubtfully distinct from *A. silvaticus*.

annulatus Bolton, Agaricus, *Hist. fung. Halifax* 1: 23 (1788)
Described from Britain but a *nomen dubium* lacking type material. Bolton's illustration shows a species of *Macrolepiota*.

anomala (Berk. & Broome) Rea, Odontia, *Brit. basidiomyc.*: 645 (1922)

Hydnum anomalum Berk. & Broome, *Ann. Mag. Nat. Hist.,*
Ser. 4 [Notices of British Fungi no. 1438] 15: 31 (1875)
Described from England (Somerset), but a *nomen dubium*. The
lectotype in herb. K is a poorly preserved specimen of a
species of *Stypella* [Roberts, Mycotaxon 69: 241 (1998)].

anserinus (Velen.) Rob. Henry [*nom. inval.*], Cortinarius, *Bull.*
Soc. Mycol. France 102: 54 (1986)
Phlegmacium anserinum Velen., *České Houby* III: 410 (1921)
Not authentically British. Reported from England (Oxfordshire
and West Kent) but unsubstantiated with voucher material.
Sensu Velenovsky (with a strong smell of roast goose) is
unknown in western Europe. Sensu auct. is *C. amoenolens*.

anthochroum (Pers.) Fr., Corticium, *Hymenomyc. eur.*: 661
(1874)
Thelephora anthochroa Pers., *Syn. meth. fung.*: 576 (1801)
A *nomen dubium*, the type collection not containing an
identifiable fungus *fide* Stalpers [Stud. Mycol. 24: 36 (1984)].
Most old British collections at K are *Helicobasidium purpureum*.

anthracobia (J. Favre) Knudsen, Fayodia, *Nordic J. Bot.* 11(4):
477 (1991)
Fayodia bisphaerigera var. *anthracobia* J. Favre, *Mat. fl.*
crypt. Suisse: 213 (1948)
Not authentically British. Reported from Yorkshire but dubiously
identified and unsubstantiated with voucher material.

anthracophila P. Karst., Tubaria, *Symb. mycol. fenn.* 6: 4
(1881)
A *nomen dubium* doubtfully distinct from *T. furfuracea fide* NCL.
Listed by Rea (1922).

antillarum (Fr.) Dennis, Panaeolus, *Kew Bull.* 15(1): 124
(1960)
Agaricus antillarum Fr., *Elench. fung.* 1: 42 (1828)
Agaricus egregius Massee, *Grevillea* 13: 91 (1885)
Panaeolus egregius (Massee) Sacc., *Sylloge Fungorum* 5:
1119 (1887)
Agaricus phalaenarum Fr., *Epicr. syst. mycol.*: 235 (1838)
Panaeolus phalaenarum (Fr.) Quél., *Mém. Soc. Emul.*
Montbeliard, Sér. 2, 5: 151 (1872)
Panaeolus semiovatus var. *phalaenarum* (Fr.) Ew. Gerhardt,
Biblioth. Bot. 147: 24 (1996)
Not authentically British. Sensu auct. Brit. is *Panaeolus*
semiovatus.

aphthosus Fr., Coprinus, *Epicr. syst. mycol.*: 245 (1838)
A *nomen dubium*. Cooke 653 (666) Vol. 5 (1886) is unknown,
the lamellae being shown as white, thus probably not a
Coprinus but also not *Psathyrella cotonea* as suggested in
BFF2: 111.

aphthosus *var.* **boltonii** Massee, Coprinus, *Brit. fung.-fl.* 1:
311 (1892)
Described from Britain but a *nomen dubium*. Erected by Massee
on the strength of Bolton's illustration and description of
Agaricus domesticus [Hist. Fung. Halifax: 26 (1788)]. This was
copied by Cooke 653 (666) Vol.5 (1886) but there referred to
as *C. aphthosus*.

apicalis W.G. Sm., Bolbitius, Cooke, *Handb. Brit. fung.,* Edn 1:
171 (1871)
Described from Britain but a *nomen dubium*. Probably only
unexpanded basidiomes of *B. vitellinus fide* Watling & Gregory
(1981).

apiculata (Fr.) Donk, Ramaria, *Meded. Bot. Mus. Herb. Rijks*
Univ. Utrecht 9: 105 (1933)
Clavaria apiculata Fr., *Observ. mycol.* 2: 288 (1818)
Not authentically British. Listed by Ing (1992) but no records
have been traced.

apiculata *var.* **brunnea** R.H. Petersen, Ramaria, *Am. J. Bot.*
59: 1042 (1972)
Not authentically British. A collection by Massee from Surrey
(Kew, Royal Botanic Gardens) under the herbarium name
Clavaria ochracea was assigned here by Petersen (1975) but is
now referred to *R. stricta*.

appendiculatum *var.* **flocculosum** Boud., Hypholoma, *Icon.*
mycol. 1: t. 137 (1906)
A *nomen dubium*. This is possibly a form of *Psathyrella*
candolleana. Listed by Rea (1922).

applanatum *var.* **vegetum** (Fr.) Rea, Ganoderma, *Brit.*
basidiomyc.: 597 (1922)
Polyporus vegetus Fr., *Epicr. syst. mycol.*: 464 (1838)
Fomes vegetus (Fr.) Cooke, *Grevillea* 14: 18 (1886)
A *nomen dubium* but from the description either *Ganoderma*
applanatum or *G. australe*. Listed by Rea (1922).

aquatilis Peck, Coprinus, *Rep. (Annual) New York State Mus.*
Nat. Hist. 27: 96 (1875)
A *nomen dubium*. Related to *C. silvaticus fide* BFF2: 11.
Illustrated by Crossland from Yorkshire (Halifax) but no
material has been traced.

aquizonatus Kytöv., Lactarius, *Karstenia* 24(2): 60 (1984)
Not authentically British. A recent collection named thus, from
England (Buckinghamshire), appears intermediate between
this species and *Lactarius resimus*.

aratus Berk. & Broome, Coprinus, *Ann. Mag. Nat. Hist., Ser. 3*
[Notices of British Fungi no. 927] 7: 378 (1861)
Described from Britain, but a *nomen dubium*. Cooke 661 (674)
and 662 (675) Vol. 5 (1886) appear to show two different
fungi under this name, the first a luxuriant form of *Coprinus*
micaceus and the latter a form of *C. domesticus*.

aratus W.G. Sm., Marasmius, *J. Bot.* 11: 66 (1873)
Described from material collected on dead tree ferns in London
(Chelsea: Veitch's Nursery) in 1872. Probably an alien and
never recollected. Allied to *M. fuscopurpureus fide* W.G. Smith
but 'lamellae forming a collarium around the stipe as in *M.*
rotula.'

arborea Huds., Tremella, *Fl. angl.* Edn 2, 2: 563 (1778)
Described from Britain but a *nomen dubium* lacking type
material. Sensu auct. is *Exidia glandulosa*.

arcticaulis With., Agaricus, *Syst. arr. Brit. pl. ed. 4,* 4:238
(1812)
Described from Britain but a *nomen dubium*.

arcuata (Bull.) Singer, Melanoleuca, *Cavanillesia.* 7: 128 (1935)
Agaricus arcuatus Bull., *Herb. France*: pl. 443 (1790)
Tricholoma arcuatum (Bull.) Gillet, *Hyménomycètes*: 125
(1874)
A *nomen dubium* lacking type material. Sensu Rea (1922) [as
Tricholoma arcuatum] is *Melanoleuca cognata*. Sensu Phillips
(1981) is *M. polioleuca*.

arcularius (Batsch) Fr., Polyporus, *Syst. mycol.* 1: 342 (1821)
Boletus arcularius Batsch, *Elench. fung.*: 97 (1783)
Not authentically British. Reported on several occasions, but all
collections so named are misidentified, usually *P. brumalis* and
rarely *P. ciliatus*.

ardosiacum (Bull.) Quél., Entoloma, *Compt. Rend. Assoc.*
Franç. Avancem. Sci. 11: 392 (1883) [1882]
Agaricus ardosiacus Bull., *Herb. France*: pl.348 (1788)
A *nomen dubium* lacking type material and variously interpreted.

ardosiacum Bull., Lycoperdon, *Herb. France* pl. 192 (1784)
A *nomen dubium*. Reported by Withering (1792). From his
description and Bulliard's illustration, probably *Bovista*
plumbea.

arenatus (Pers.) Fr., Cortinarius, *Epicr. syst. mycol.*: 283
(1838)
Agaricus arenatus Pers., *Syn. meth. fung.*: 293 (1801)
A *nomen dubium* lacking type material. Cooke 762 (763) Vol.6
(1886) is possibly *C. pholideus fide* Pearson & Dennis (1948).
See Singer [Persoonia 2: 8 (1961)] for further comments.

argentatus (Pers.) Fr., Cortinarius, *Epicr. syst. mycol.*: 278
(1838)
Agaricus argentatus Pers., *Syn. meth. fung.*: 286 (1801)

A *nomen dubium* lacking type material. Possibly an earlier name for *C. diosmus*. See Cooke 745 (745) Vol. 5 (1888) for an interpretation.

argentatus *var*. **pinetorum** Sacc., Cortinarius, *Syll. fung.* 5: 924 (1887)
A *nomen dubium*. See Cooke 746 (746) Vol. 5 (1888) for an interpretation.

argenteogriseus Rea, Pluteus, *Trans. Brit. Mycol. Soc.* 5(2): 250 (1915)
A *nomen dubium*. Described from England (South Essex: Epping Forest) but never recollected.

argenteus With., Agaricus, *Bot. arr. Brit. pl. ed. 2*, 3: 354 (1792)
Described from Britain but a *nomen dubium*.

argutus Fr., Cortinarius, *Epicr. syst. mycol.*: 278 (1838)
Cortinarius fuscotinctus Rea, *Trans. Brit. Mycol. Soc.* 5(3): 436 (1917)
Mis.: *Cortinarius fraudulosus* sensu NCL
Not authentically British. Associated only with *Populus tremula*. Reported on three occasions during the 20th century, but said to be associated with trees other than *Populus*. Sensu Rea (1922) is doubtful and the few collections named thus (in herb. K) need reappraisal.

argyraceum (Bull.) Gillet, Tricholoma, *Hyménomycètes*: 103 (1874)
Agaricus argyraceus Bull., *Herb. France*: pl. 423 (1789)
Tricholoma scalpturatum var. *argyraceum* (Bull.) Kühner & Romagn. [*nom. inval.*], *Flore Analytique des Champignons Supérieurs*: 154 (1953)
A *nomen dubium* lacking type material. Sensu NCL and Phillips (1981) is *Tricholoma scalpturatum*. Sensu FAN4 is *T. inocybeoides*. Sensu Riva (FungEur3) (as *T. scalpturatum* var. *argyraceum*) is another species, not authentically British.

argyraceum *var*. **virescens** (Cooke) Sacc., Tricholoma, *Syll. fung.* 5: 104 (1887)
Agaricus argyraceus var. *virescens* Cooke, *Ill. Brit. fung.* 1118 (641) Vol. 8 (1886)
Described from Britain but a *nomen dubium*. Cooke's illustration appears to be typical *T. scalpturatum* with old basidiomes showing the characteristic patchy yellowing of the lamellae.

arida (Fr.) Gillet, Lepiota, *Hyménomycètes*: 63 (1874)
Amanita arida Fr., *Epicr. syst. mycol.*: 10 (1838)
A *nomen dubium*, but from the description possibly *Limacella guttata*. Reported by Cooke in Grevillea 20: 37 (1892) from Bedfordshire (Dunstable) but no material survives.

armeniacus (Schaeff.) Fr., Cortinarius, *Epicr. syst. mycol.*: 304 (1838)
Agaricus armeniacus Schaeff., *Fung. Bavar. Palat. nasc.* 1: 81 (1762)
Not authentically British. Not reported since 1927 (from Scotland) and all records are unsubstantiated with voucher material. Collections named thus in herb. K are all misidentified. Illustrated in Phillips (1981: 138) but not clearly from British material.

armeniacus *var*. **falsarius** Fr., Cortinarius, *Hymenomyc. eur.*: 387 (1874)
A *nomen dubium*. Listed by Rea (1922).

aromaticus (Sowerby) Berk., Hygrophorus, *Outl. Brit. fungol.*: 198 (1860)
Agaricus aromaticus Sowerby, *Col. fig. Engl. fung.* 2: pl. 144 (1798)
Described from Britain but a *nomen dubium* lacking type material. Possibly a form of *Hygrocybe laeta fide* Pearson & Dennis (1948). Berkeley (1860) remarks 'not found since the time of Sowerby'.

arvensis *var*. **buchananii** W.G. Sm., Psalliota, *Guide to Worthington Smith's Drawings*: 11 (1910)

Described from Britain but a *nomen dubium*. Probably just a growth form of *Agaricus arvensis*.

astroideus (Fr.) Fr., Coprinus, *Epicr. syst. mycol.*: 247 (1838)
Agaricus astroideus Fr., *Syst. mycol.* 1: 312 (1821)
A *nomen dubium*. Supposedly a variety of *Coprinus niveus* and listed as such in Rea (1922).

aterrimum (Fr.) Fr., Radulum, *Hymenomyc. eur.*: 624 (1874)
Hydnum aterrimum Fr., *Syst. Mycol.* 1: 416 (1821)
An ascomycete, now known as *Xenotypa aterrima* (Fr.) Petrak. Sensu auct., however, is possibly a *Vuilleminia* sp. Listed by Rea (1922).

atomata (Fr.) Quél., Psathyrella, *Mém. Soc. Émul. Montbéliard, Sér. 2*, 5: 153 (1872)
Agaricus atomatus Fr., *Syst. mycol.* 1: 298 (1821)
Psathyra atomata (Fr.) P. Kumm., *Führ. Pilzk.*: 70 (1871)
A *nomen dubium*. Listed by Rea (1922). Sensu NCL is *P. cana* f. *prona*.

atra McGuire, Sebacina, *Lloydia* 4: 27 (1941)
Not authentically British. A North American species reported from Warwickshire in 1980. Re-examination of this specimen shows it to be misidentified *Exidiopsis effusa*.

atra Weinm., Thelephora, *Hymen. Gasteromyc.*: 636 (1836)
Not authentically British. Listed by Rea (1922) but the two old collections named thus (in herb. K) are misdetermined.

atroalba (Bolton) Gillet, Mycena, *Hyménomycètes*: 272 (1876)
Agaricus atroalbus Bolton, *Hist. fung. Halifax* 3: 137 (1789)
Described from Britain but a *nomen dubium* lacking type material.

atrobrunnea (Lasch) Gillet, Psilocybe, *Hyménomycètes*: 586 (1878)
Agaricus atrobrunneus Lasch, *Linnaea* 3: 423 (1828)
A *nomen dubium* lacking type material. There is a single record from England (Somerset) by A.H. Smith, accepted in Guzmán (1983) but the record is unsubstantiated with voucher material and the name has been variously interpreted.

atrocrocea (Massee) W.G. Sm., Lepiota, *Trans. Brit. Mycol. Soc.* 2: 62 (1903)
Agaricus atrocroceus Massee, *J. Bot.* 41: 385 (1903)
Described from England (Somerset) but a *nomen dubium*.

atrocyanea (Batsch) Fr., Mycena, *Observ. mycol.* 1: 147 (1801)
Agaricus atrocyaneus Batsch [non *A. atrocyaneus* Pers. (1801)], *Elench. fung. (Continuatio Prima)*: 101 & t. 18 f. 87 (1786)
A *nomen dubium*, possibly an *Entoloma* sp. *fide* Maas Geesteranus (1992). Cooke 231 (236, lower) Vol. 2 (1882) is an unidentifiable species of *Mycena*. Sensu NCL, referred by Orton (NBA9: 563) to *Mycena fuscoatra*, is here considered a synonym of *M. leucogala*.

atroglauca Einhell., Russula, *Hoppea* 39: 103 (1980)
Not authentically British. Reported from Cumberland, Surrey and West Sussex but unsubstantiated with voucher material.

atromarginata (Lasch) P. Kumm., Mycena, *Führ. Pilzk.*: 109 (1871)
Agaricus atromarginatus Lasch, *Linnaea* 3: 387 (1828)
A *nomen dubium* lacking type material. Possibly *Mycena purpureofusca fide* Maas Geesteranus (1992). Listed by Rea (1922).

atropapillata Kühner & Maire, Mycena, *Encycl. Mycol. 10. Le Genre Mycena*: 589 (1938)
Not authentically British. Reported from England (North Somerset) but unsubstantiated with voucher material.

atrorufa (Schaeff.) Quél., Psilocybe, *Mém. Soc. Émul. Montbéliard, Sér. 2*, 5: 117 (1872)
Agaricus atrorufus Schaeff. [non *A. atrorufus* Bolton (1788)], *Fung. Bavar. Palat. nasc.* 3: 234 (1770)
Agaricus montanus β *atrorufus* (Schaeff.) Fr., *Syst. mycol.* 1: 293 (1821)

A *nomen dubium* lacking type material and variously interpreted. Cooke 602 (571) Vol. 4 (1886) is apparently a species of *Psathyrella* (possibly *P. bipellis*) but the name has also been applied to several species of *Psilocybe* (see NCL Index: 173).

atrorufus Bolton, Agaricus [*nom. illegit.*, non *A. atrorufus* Schaeff. (1763)], *Hist. fung. Halifax* 2: 51 (1788)
Described from Britain but a *nomen dubium* lacking type material.

atrovirens Rea, Mycena, *Trans. Brit. Mycol. Soc.* 6(4): 323 (1920)
Described from Britain but a *nomen dubium*. The type collection (in herb. K) is in poor condition. Not discussed by Maas Geesteranus (1992).

atrovirens (Fr.) Sacc., Tremella [non *T. atrovirens* Bull. (1783)], *Syll. fung.* 6: 790 (1888)
  Agyrium atrovirens Fr., *Syst. mycol.* 2: 232 (1823)
  Epidochium atrovirens (Fr.) Fr., *Summa veg. Scand.*: 471 (1849)
An illegitimate homonym of the earlier *Tremella atrovirens* Bull. (thought to be the cyanobacterium *Nostoc commune*). British specimens under this name in herb. K are all *Tremella exigua*.

aurantiaca (Bres.) Bourdot & Galzin, Peniophora, *Bull. Soc. Mycol. France* 28: 402 (1912)
  Corticium aurantiacum Bres., *Fungi trident.* 2: 37 (1892)
Not authentically British. Restricted to *Alnus viridis* (not native in Britain) and last reported in 1923. Sensu auct. Brit. is probably *P. erikssonii*.

aurantiacum β leucopodium Gray, Leccinum, *Nat. arr. Brit. pl.* 1: 646 (1821)
Described from Britain but a *nomen dubium* lacking type material.

aurantiacum γ rufum Gray, Leccinum, *Nat. arr. Brit. pl.* 1: 646 (1821)
Described from Britain but a *nomen dubium* lacking type material. Possibly *Leccinum quercinum* since Gray cites *Boletus aurantiacus* sensu Sowerby [Col. fig. Engl. fung. pl. 110 (1797)], which appears to be that species, as a synonym.

aurantioferrugineus With., Agaricus, *Syst. arr. Brit. pl. ed. 4*, 4: 295 (1801)
Described from Britain but a *nomen dubium*. The description suggests a species of *Pholiota*.

aurantium (L.) Pers., Scleroderma [*nom. rej.*], *Syn. meth. fung.*: 153 (1801)
  Lycoperdon aurantium L., *Syst. Veg.*: 1019 (1797)
A formally rejected name, variously interpreted. Sensu auct. is *S. citrinum*. The epithet is often incorrectly given as *aurantiacum*.

aurea Fr., Daedalea, *Syst. mycol.* 1: 339 (1821)
A *nomen dubium*. Listed by Rea (1922). Fries' description suggests *Gloeophyllum sepiarium* but the material was supposedly growing on 'oak trunks'.

aureifolius Peck, Cortinarius, *Rep. (Annual) New York State Mus. Nat. Hist.* 38: 89 (1885)
Not authentically British. Reported from England (Warwickshire) and Wales (Caernarvonshire) but unsubstantiated with voucher material. Sensu NCL is probably *C. huronensis*.

auricoma (Batsch) Sacc., Inocybe, *Syll. fung.* 5: 777 (1887)
  Agaricus auricomus Batsch, *Elench. fung.*: 75 & t. 5 f. 21 (1783)
  Inocybe descissa var. *auricomus* (Batsch) Massee, *Brit. fung.-fl.* 2: 197 (1893)
Collections under this name in herb. K prior to 1985 were found by Kuyper to be probably misdetermined. Two collections since then have no accompanying notes and require re-examination.

auriporus Peck, Boletus, *Rep. (Annual) New York State Mus. Nat. Hist.* 23: 133 (1873)
Not authentically British. A North American species. Sensu auct. Brit. is *Aureoboletus gentilis*.

auriscalpium (Fr.) Fr., Arrhenia, *Summa veg. Scand.*: 312 (1849)
  Cantharellus auriscalpium Fr., *Elench. fung.* 1: 54 (1828)
Not authentically British but mentioned in BFF6: 28 and could possibly occur in montane areas of Scotland or Wales.

austera (Fr.) Sacc., Flammula, *Syll. fung.* 5: 821 (1887)
  Agaricus austerus Fr., *Epicr. syst. mycol.*: 188 (1838)
A *nomen dubium*. Listed by Rea (1927). *Fide* Holec (2001) this is possibly *Galerina marginata*.

avellanea (Burt) Bourdot & Galzin, Tomentella, *Bull. Soc. Mycol. France* 40: 153 (1924)
  Hypochnus avellaneus Burt, *Ann. Missouri Bot. Gard.* 3: 225 (1916)
A *nomen dubium*. Sensu Wakefield (1969) is doubtful. Probably a synonym of *T. sublilacina*.

avellaneus With., Agaricus, *Arr. Brit. pl. ed. 3*, 4: 225 (1796)
Described from Britain but a *nomen dubium*.

avenacea (Fr.) Quél., Mycena, *Mém. Soc. Émul. Montbéliard, Sér. 2*, 5: 241 (1872)
  Agaricus avenaceus Fr., *Syst. mycol.* 1: 150 (1821)
A *nomen dubium*. Sensu Rea (1922) and Pearson (1955) is *M. olivaceomarginata*.

azyma (Fr.) Quél., Flammula, *Mem. Soc. Émul. Montbéliard, Sér. 2*, 5: 130 (1872)
  Agaricus azymus Fr., *Epicr. syst. mycol.*: 188 (1838)
A *nomen dubium*. Listed by Rea (1922). Possibly a species of *Tubaria fide* Holec (2001: 182).

babingtonii (A. Bloxam) M.M. Moser, Entoloma, *Kleine Kryptogamenflora, Edn 4*: 207 (1978)
  Agaricus babingtonii A. Bloxam, *Ann. Mag. Nat. Hist., Ser. 2* [Notices of British Fungi no. 680] 13(2): 399 (1854)
  Nolanea babingtonii (A. Bloxam) Sacc., *Syll. fung.* 5: 717 (1887)
  Leptonia babingtonii (A. Bloxam) P.D. Orton, *Trans. Brit. Mycol. Soc.* 43(2): 177 (1960)
Described from Britain but a *nomen dubium*. Type material is in poor condition and unidentifiable. Sensu NCL is *E. dysthales*.

badiophylla (Romagn.) Park.-Rhodes, Psathyrella, *Trans. Brit. Mycol. Soc.* 37: 335 (1954)
  Drosophila badiophylla Romagn., *Bull. Mens. Soc. Linn. Lyon* 21: 155 (1952)
Not authentically British. Reported from England (Yorkshire) and Wales (Pembrokeshire) but unsubstantiated with voucher material.

badiophylla var. neglecta (Romagn.) Kits van Wav., Psathyrella, *Persoonia, Supplement* 2: 280 (1985)
  Drosophila badiophylla var. *neglecta* Romagn., *Bull. Mens. Soc. Linn. Lyon* 21: 155 (1952)
Not authentically British. Reported from Oxfordshire (Bix, The Warburg Reserve) on several occasions but unsubstantiated with voucher material.

badiorufus R. Heim [*nom. inval.*], Boletus, *Rev. Mycol. (Paris)* 25 (3-4): 235 (1960)
  Xerocomus badiorufus (R. Heim) Bon [*nom. inval.*], *Rev. Mycol. (Paris)* 35(4): 231 (1970)
Similar to, or conspecific with *B. badius*. Reported on several occasions from England (Berkshire, South Hampshire and Surrey) but unsubstantiated with voucher material.

badiosanguineus Kühner & Romagn., Lactarius, *Bull. Soc. Mycol. France* 69: 361 (1954)
Not authentically British. Listed as British in FNE2, but the only report traced (Scotland, Easterness: Abernethy) is unsubstantiated with voucher material.

balteatus (Fr.) Fr., Cortinarius, *Epicr. syst. mycol.*: 257 (1838)
  Agaricus balteatus Fr., *Observ. mycol.* 2: 138 (1818)
  Cortinarius subbalteatus Kühner, *Bull. Mens. Soc. Linn. Lyon* 24(2): 40 (1955)

Not authentically British. Records are old, mostly unsubstantiated with voucher material, and probably relate to misidentified material of *C. balteatocumatilis*.

bambusinus (Zoll.) E. Fisch., Mutinus, *Ann. Jard. Bot. Buitenzorg* 6: 30 (1886)
  Phallus bambusinus Zoll., *Syst. Verz. ind. Arch. Jahr.* 1: 11 (1854)
  Cynophallus bambusinus (Zoll.) Rea, *Brit. basidiomyc.*: 23 (1922)
Not authentically British. A south-east Asian species. British records refer to *M. ravenelii, fide* BritPuffb.

barlae Quél., Russula, *Compt. Rend. Assoc. Franc. Avancem. Sci.* 12: 504 (1884)
  Russula xerampelina var. *barlae* (Quél.) Kühner & Romagn. [*comb. inval.*], *Flore Analytique des Champignons Supérieurs*: 449 (1953)
A *nomen dubium* lacking type material and variously interpreted. Sensu Rea (1922) is doubtful. Sensu auct. Brit. is *R. faginea*. Sensu Romagnesi (1967) is *R. cicatricata*.

basirubens (Bon) A. Riva & Bon, Tricholoma, *Rivista Micol.* 31(1-2): 23 (1988)
  Tricholoma orirubens var. *basirubens* Bon, *Doc. mycol.* 5(18): 120 (1975)
Not authentically British. Reported from South Hampshire and Gloucestershire but unsubstantiated with voucher material. This is a Mediterranean and central European species.

bathypora (Rostk.) Cooke, Poria, *Grevillea* 14: 115 (1886)
  Polyporus bathyporus Rostk., *Deutschl. Fl.*, Edn 3: t. 59 (1838)
A *nomen dubium* lacking type material. Sensu Rea (1922), with basidiomes described as 'white', is unknown. Sensu Pilát (1936) is possibly *Inonotus radiatus* f. *nodulosus*.

batschianum (Fr.) Sacc., Entoloma, *Syll. fung.* 5: 684 (1887)
  Agaricus batschianus Fr., *Epicr. syst. mycol.*: 144 (1838)
A *nomen dubium*. Sensu Rea (1922) is possibly *E. turbidum*.

battarrae (Fr.) Sacc., Stropharia, *Syll. fung.* 5: 1024 (1887)
  Agaricus battarrae Fr., *Epicr. syst. mycol.*: 217 (1838)
A *nomen dubium*. Listed by Rea (1922) but no records have been traced.

bella (Pers.) Berk. & Broome, Laccaria, *Ann. Mag. Nat. Hist., Ser. 5 [Notices of British Fungi no. 1994]* 12 (1883)
  Agaricus bellus Pers., *Syn. meth. fung.*: 452 (1801)
  Clitocybe bella (Pers.) P. Kumm., *Führ. Pilzk.*: 123 (1871)
Uncertain. NCL suggests that this is a good species lacking a proper description and type material, awaiting rediscovery. See Cooke 178 (183) Vol. 2 (1881) for an illustration.

bellinii (Inzenga) Watling, Suillus, *Notes Roy. Bot. Gard. Edinburgh* 28(1): 59 (1967)
  Boletus bellinii Inzenga, *Funghi Sicil.*: 25 (1879)
Not authentically British. Reported from Wales (Anglesey) in 1990 but unsubstantiated with voucher material. Included in Ing (1992) but this is a southern European species unlikely in Britain.

berkeleyi (Massee) W.B. Cooke, Cellypha, *Sydowia, Beih.* 4: 52 (1961)
  Cyphella berkeleyi Massee, *Brit. fung.-fl.* 1: 141 (1892)
A *nomen dubium*. This is Massee's concept of *Cyphella griseopallida* sensu Berkeley (1860: 277).

betulae W.B. Cooke, Phaeosolenia, *Sydowia, Beih.* 4: 122 (1961)
Not authentically British. Sensu auct. Brit. is *Merismodes ferruginea*.

biennis DC., Thelephora, *Fl. Fr.* 5: 106 (1815)
A *nomen dubium*. Listed by Rea (1922) as a synonym of *Hypochnus umbrinus* (= *Pseudotomentella tristis*).

bifrons (Berk.) A.H. Sm., Psathyrella, *Contr. Univ. Michigan Herb.* 5: 40 (1941)

Agaricus bifrons Berk., *Engl. fl.* 5(2): 114 (1836)
  Psathyra bifrons (Berk.) Quél., *Bull. Soc. Bot. France* 26: 52 (1880) [1879]
  Agaricus bifrons var. *semitinctus* W. Phillips, in Cooke, *Ill. Brit. fung.* 616 (594b) Vol. 4 (1886)
  Psathyra bifrons var. *semitincta* (W. Phillips) Sacc., *Syll. fung.* 5: 1071 (1887)
A *nomen dubium* variously interpreted. Sensu Rea (1922) (as *Psathyra bifrons*) is unknown, and sensu Kits van Waveren (Kits1) is not authentically British.

bilanatus R.F.O. Kemp, Coprinus [*nom. inval.*], *Trans. Brit. Mycol. Soc.* 65(3): 380 (1975)
A *nom. prov.* for a taxon near to or identical with *C. scobicola*.

binucleospora J. Erikss. & Ryvarden, Athelia, *Corticiaceae of North Europe* 2: 105 (1973)
Not authentically British. All collections named thus in herb. K are determined as 'compare with' *A. binucleospora*.

biornatus (Berk. & Broome) Locq., Leucocoprinus, *Bull. Mens. Soc. Linn. Lyon* 14: 92 (1945)
  Agaricus biornatus Berk. & Broome, *J. Linn. Soc., Bot.* 11: 502 (1871)
  Lepiota biornata (Berk. & Broome) Sacc., *Syll. fung.* 5: 54 (1887)
  Leucoagaricus bresadolae var. *biornatus* (Berk. & Broome) Bon, *Doc. mycol.* 7(27-28): 16 (1977)
Described from Sri Lanka and not authentically British. The name was misapplied by Cooke (HBF2: 13) to *L. caldariorum*. Sensu auct. Eur. is *Leucoagaricus americanus*.

bipellis Romagn., Coprinus, *Bull. Soc. Mycol. France* 92(2): 198 (1976)
Not authentically British. Reported from Hertfordshire but incorrectly determined *fide* D.J. Schafer (pers. comm.).

bisus (Quél.) Kühner & Maire, Lentinellus, *Bull. Soc. Mycol. France* 50: 16 (1934)
  Lentinus bisus Quél., in Bresadola, *Fungi trident.* 1: 12 (1881)
Not authentically British. Mentioned as British in BFF6 (p. 133) but the identity of the collection was in doubt. Possibly only a small form of *Lentinellus flabelliformis*. No voucher material has been traced.

blandus Berk., Agaricus, *Engl. fl.* 5(2): 20 (1836)
Described from Britain, but a *nomen dubium*. Possibly a species of *Melanoleuca*.

boltonii (Fr.) Cooke, Hymenochaete, *Grevillea* 8: 145 (1880)
  Corticium boltonii Fr., *Epicr. syst. mycol.*: 558 (1838)
  Stereum boltonii (Fr.) Sacc., *Michelia* 1: 239 (1878)
A *nomen dubium*. Listed by Rea (1922), but no British material traced.

bombycina (Fr.) Pouzar, Anomoporia, *Česká Mykol.* 20: 172 (1966)
  Polyporus bombycinus Fr., *Elench. fung.* 1: 117 (1828)
  Poria bombycina (Fr.) Cooke, *Grevillea* 14: 112 (1886)
Not authentically British. Reported from England (North Somerset: Portbury) by Berkeley & Broome in 1848 and from Wales (Denbighshire: Coed Coch) in 1880 but there is no material of the former collection and the latter (in herb. K) is sterile and unidentifiable.

bongardii var. **pisciodora** (Donadini & Riousset) Kuyper, Inocybe, *Persoonia, Supplement* 3: 41 (1986)
  Inocybe pisciodora Donadini & Riousset, *Doc. mycol.* 5(20): 5 (1975)
Not authentically British. A single report from North Somerset in 1997, unsubstantiated with voucher material.

boreale (Fr.) Sacc., Tricholoma, *Syll. fung.* 5: 121 (1887)
  Agaricus borealis Fr., *Epicr. syst. mycol.*: 44 (1838)
A *nomen dubium* doubtfully distinct from *Calocybe gambosum fide* NCL. Cooke 1123 (956) Vol. 8 (1889) is certainly not *C. gambosum*. It might be *Calocybe borealis* Riva [Rivista di Micologie 30: 90-94 (1987)] described to provide a valid name

for *Tricholoma boreale* sensu Bresadola [*Icon. mycol.* 3: pl. 105 (1928)].

borealis (Fr.) Kotl. & Pouzar, Climacocystis, *Česká Mykol.* 12: 103 (1958)
Polyporus borealis Fr., *Syst. mycol.* 1: 366 (1821)
Daedalea borealis (Fr.) Quél., *Enchir. fung.*: 184 (1886)
Leptoporus borealis (Fr.) Pilát, *Atlas champ. Eur. I* 3(1): 234 (1938)
Not authentically British. Reported on several occasions from England (East Gloucestershire in 1999, Northumberland in 1907, South Hampshire in 1910 and 1980 and Shropshire in 1995) but all unsubstantiated with voucher material and in lowland habitat. A single collection at Kew from Klotzsch (No. 176), supposedly collected in Scotland, is correct but lacks proof that it was gathered in Britain *fide* Reid & Austwick (1963). Also reported from Scotland (Perthshire in 1908 and 1980) but no material traced in herb. E.

borealis (Peck) Bon, Hygrocybe, *Doc. mycol.* 9: 8 (1973)
Hygrophorus borealis Peck, *Rep. (Annual) New York State Mus. Nat. Hist.* 26: 64 (1874)
Cuphophyllus borealis (Peck) Bon, *Doc. mycol.* 14(56): 19 (1985) [1984]
Not authentically British. Reported from South Hampshire (Roydon Woods) in 1996, but unsubstantiated with voucher material. This is a North American species unlikely to occur here.

borealis Kauffman, Russula, *Rep. (Annual) Michigan Acad. Sci.* 11: 69 (1909)
Not authentically British. A North American species, unknown in Europe. Sensu Romagnesi (1967) and sensu Rayner (1985) is *R. laeta.*

botryoides (Schwein.) Bourdot & Galzin, Tomentella, *Bull. Soc. Mycol. France* 40: 159 (1924)
Thelephora botryoides Schwein., *Schriften Naturf. Ges. Leipzig* 1: 109 (1822)
Hypochnus botryoides (Schwein.) Burt, *Rep. (Annual) Missouri Bot. Gard.* 3: 226 (1916)
Not authentically British (or European). Sensu Kõljalg (Kõljalg) and sensu Wakefield (1969) is *Tomentella ferruginea.*

boudieri Rob. Henry, Cortinarius, *Bull. Soc. Mycol. France* 52: 153 (1936)
Not authentically British. A poorly known species 'unrecorded for Britain' *fide* Orton (1955). A 1951 collection (in herb. K) was determined as this by Pearson, but evidently reassessed by Orton.

bourdotii Svrček, Tomentella, *Česká Mykol.* 12: 76 (1958)
Not authentically British. Sensu Wakefield (1969) and auct. Brit. is *T. lilacinogrisea.*

bresadolae Schulzer, Coprinus, *Hedwigia* 24(4): 136 (1885)
Not authentically British. There is a single record of this doubtful species, unsubstantiated with voucher material.

bresadolae Massee, Inocybe, *Ann. Bot.* 18: 465 (1904)
Mis.: *Astrosporina trinii* sensu Rea (1922)
Not authentically British. Kuyper (1986) considers the description of *Asterosporina trinii* in Rea (1922) fits this species but no British material has been traced.

bresadolana Singer, Clitocybe, *Rev. Mycol. (Paris)* 2(6): 228 (1937)
Mis.: *Clitocybe flaccida* sensu Bresadola [*Icon. mycol.* 4: pl. 169 (1928)]
Not authentically British. Recorded from Northern Ireland in 1999 (apparently associated with *Cupressus macrocarpa*) but unsubstantiated with voucher material and probably misidentified *Lepista flaccida.*

bresadolanum Clémençon, Tricholoma, *Doc. mycol.* 7(27-28): 54 (1977)
Mis.: *Tricholoma murinaceum* sensu Bresadola [*Icon. mycol.* 2, pl. 88 (1927)]

Not authentically British. Included by Cooke 82 (49) Vol.1 (1882) [as *Agaricus (Tricholoma) murinaceus*], and also listed by Rea (1922), apparently in the same sense as Bresadola. Not reported since the 1930's and no voucher material has been traced.

brownii (Berk. & Broome) P.D. Orton, Omphalina, *Trans. Brit. Mycol. Soc.* 43(2): 179 (1960)
Cantharellus brownii Berk. & Broome, *Ann. Mag. Nat. Hist.*, Ser. 3 [*Notices of British Fungi no. 336]* 18 (1866)
Omphalia brownii (Berk. & Broome) J. Favre, *Bull. Soc. Mycol. France.* 55: 212 (1939)
Described from Britain but a *nomen dubium.* The type collection is infertile material of a species of *Agrocybe.* Sensu NCL and sensu auct. mult. is *O. favrei.*

bruchetii Bon, Hebeloma, *Bull. Trimestriel Féd. Mycol. Dauphiné-Savoie* 25 (102): 23 (1986)
Hebeloma repandum Bruchet [*nom. illegit.*], *Bull. Mens. Soc. Linn. Lyon* 39 (6 [Suppl.]): 50 (1970)
Not authentically British. Now considered an exclusively arctic-alpine species, so material reported from England (London: Buckingham Palace) in 1997 is presumably misidentified.

brunnea Quél., Inocybe, *Bull. Soc. Amis Sci. Nat. Rouen.*: 14 (1880)
A *nomen dubium* lacking type material. Sensu NCL (citing K&M) is *Inocybe splendens* var. *phaeoleuca fide* Kuyper (1986). But NCL also cites the concept of Pearson (1954) which appears to be near to *I. lacera.* British material needs re-examination.

brunneofulvus Fr., Cortinarius, *Epicr. syst. mycol.*: 298 (1838)
A *nomen dubium.* British records may refer to *C. brunneus* but no material has been traced.

bryantii *var.* **minor** Massee, Geastrum, *Brit. fung.-fl.* 1: 36 (1892)
Described from Britain but a *nomen dubium.* Probably just small basidiomes of *G. striatum.*

buccinalis (Sowerby) Sacc., Omphalia, *Syll. fung.* 5: 322 (1887)
Agaricus buccinalis Sowerby, *Col. fig. Engl. fung.* 1: pl. 107 (1797)
A *nomen dubium* lacking type material. Cooke 261 (272) Vol. 2 (1883) is suggestive of a species of *Hemimycena.*

bulbifera (Kauffman) Romagn., Conocybe, *Bull. Soc. Mycol. Franc.* 58: 147 (1942)
Galera bulbifera Kauffman, *Michigan Geol. Biol. Surv. Publ., Biol. Ser. 5,* 26: 496 (1918)
Not authentically British. Reported from Surrey (Shere, Coombe Bottom) but the voucher material in herb. K is misidentified *Conocybe inocybeoides fide* Watling (pers. comm.).

bulbigenum (Berk. & Broome) Dennis, Entoloma, *Trans. Brit. Mycol. Soc.* 31(3-4): 205 (1948)
Agaricus bulbigenus Berk. & Broome, *Ann. Mag. Nat. Hist.*, Ser. 5 [*Notices of British Fungi no. 1937]* 9: 177 (1882)
A *nomen dubium,* described from Britain in *Agaricus* (*Entoloma*) and unintentionally combined in *Entoloma* by Dennis, who discussed and discarded the name.

bulbillosus Fr., Bolbitius, *Hymenomyc. eur.*: 334 (1874)
A *nomen dubium.* Rejected by Watling (BFF3).

bulbosa Velen., Armillaria, *Mykologia* 4: 116 (1927)
Armillaria mellea var. *bulbosa* Barla [*nom. illegit.*, non *A. mellea* var. *bulbosa* P. Karst. (1879)], *Bull. Soc. Mycol. France* 3: 143 (1887)
A *nomen dubium,* sometimes considered a synonym of *Armillaria cepistipes,* but widely used in Britain for *A. gallica.*

bullacea (Bull.) P. Kumm., Psilocybe, *Führ. Pilzk.*: 71 (1871)
Agaricus bullaceus Bull., *Herb. France.* pl. 566 (1792)
A *nomen dubium* lacking type material. Sensu BFF5 and sensu auct. mult. this is *Psilocybe subviscida* var. *velata.*

bullula (V. Brig.) Sacc., Omphalia, *Syll. fung.* 5: 334 (1887)

Agaricus bullula V. Brig., *De Fung. rar. Reg Neap. Hist.*: t.16 (1824)
A *nomen dubium* lacking type material. Cooke 267 (252) Vol. 2 (1883) appears to be a species of *Hemimycena.*

byssoides Sowerby, Clavaria, *Col. fig. Engl. fung.* 3: pl. 301 (1801)
A *nomen dubium* lacking type material and probably not a basidiomycete. Sowerby's illustration appears to show the myxomycete *Ceratiomyxa fruticulosa.*

cacuminatus With., Agaricus, *Arr. Brit. pl. ed. 3*, 4: 210 (1796)
Described from Britain but a *nomen dubium.*

caeruleus Bolton, Agaricus [*nom. illegit.*, non *A. caeruleus* Pers. (1801)], *Hist. fung. Halifax* 1: 12 (1788)
Described from Britain but a *nomen dubium* lacking type material. Bolton's illustration and description suggest a species of *Stropharia.*

caesariata (Fr.) P. Karst., Inocybe, *Bidrag Kännedom Finlands Natur Folk* 32: 459 (1879)
  Agaricus caesariatus Fr., *Epicr. syst. mycol.*: 176 (1838)
A *nomen dubium* variously interpreted. Cooke 437 (388) Vol. 3 (1886), in beechwoods, is unknown but possibly *I. cervicolor.* Records from dunes are likely to be *I. heimii.*

caesariata var. fibrillosa (Fr.) Rea, Inocybe, *Brit. basidiomyc.*: 199 (1922)
  Agaricus caesariatus var. *fibrillosus* Fr., *Icones selectae Hymenomycetum* 2: pl. 109, f. 3 (1877)
A *nomen dubium.* Sensu Heim (1931) is *Inocybe heimii.*

caesia (Pers.) Höhn. & Litsch., Tomentella, *Sitzungsber. Kaiserl. Akad. Wiss., Wien, Math.-naturwiss. Cl., Abt. 1* 115: 1570 (1906)
  Corticium caesium Pers., *Observ. mycol.* 1: 15 (1796)
  Thelephora caesia (Pers.) Pers., *Syn. meth. fung.*: 579 (1801)
  Soppittiella caesia (Pers.) Massee, *Brit. fung.-fl.* 1: 107 (1892)
  Hypochnus caesius (Pers.) Bres., *Ann. mycol.* 1: 107 (1903)
  Prillieuxia caesia (Bres.) Park.-Rhodes, *Trans. Brit. Mycol. Soc.* 37: 339 (1954)
A *nomen dubium* lacking type material. Sensu auct. Brit. is *Tomentella fuscocinerea* p.p. Discussed in Wakefield (1969: 204).

caespitosa (Bolton) Cooke, Omphalia, *Handb. Brit. fung.,* Edn 2: 94 (1885)
  Agaricus caespitosus Bolton, *Hist. fung. Halifax* 1: 41 (1788)
Described from Britain but a *nomen dubium* lacking type material. A synonym of *Omphalia unbellifera* var. *flava* Cooke *fide* Pearson & Dennis (1948).

caespitosus (Massee) Sacc., Gyrodon, *Syll. fung.* 11: 81 (1895)
  Boletus caespitosus Massee, *Brit. fung.-fl.* 1: 297 (1892)
Described from Britain but a *nomen dubium,* doubtfully distinct from *Boletus pulverulentus fide* NCL.

calceolus (Bull.) Balb., Polyporus, *Fl. Lyon* 2: 279 (1828)
  Boletus calceolus Bull., *Herb. France*: pl. 360 (1786)
A *nomen dubium* lacking type material. Bulliard's illustration depicts what appears to be young and immature basidiomes of *Polyporus durus.* Listed by Rea (1922).

caligatum (Viv.) Ricken, Tricholoma, *Blätterpilze Deutschl.*: 331 (1914)
  Agaricus caligatus Viv., *Funghi d'Italia*: t. 35 (1834)
  Armillaria caligata (Viv.) Gillet, *Hyménomycètes*: 79 (1874)
Reported from Caledonian pinewoods and last recorded from Abernethy Forest in 1993 but this is now known to be a southern European species, unlikely to occur in Britain. Sensu auct. Brit. is *Tricholoma nauseosum.* Reported also from Wales but the collection has been redetermined as *Lepiota cortinarius.*

caliginosa W.G. Sm., Nyctalis, *J. Bot.* 11: 337 (1873)
Described from Britain but a *nomen dubium.* Possibly a diseased state of a *Clitocybe* sp. *fide* Rea (1922). See Cooke 1067 (1132) Vol. 7 for an illustration.

callosa (Fr.) Gillet, Psilocybe, *Hyménomycètes*: 585 (1878)
  Agaricus callosus Fr., *Observ. mycol.* 2: 180 (1818)
A *nomen dubium* listed by Rea (1922). According to Fries this is the same as *Agaricus confertus* Bolton but that also is a *nomen dubium.*

calolepis (Fr.) P. Karst., Crepidotus, *Bidrag Kännedom Finlands Natur Folk* 32: 414 (1879)
  Agaricus calolepis Fr., *Öfvers. Kongl. Vetensk.-Akad. Förh.* 30: 5 (1873)
  Crepidotus mollis var. *calolepis* (Fr.) Pilát, *Sborn. Nár. Mus. Praze* 2B: 74 (1940)
Not authentically British. There is no conclusive evidence that any of the British records under this name can be distinguished from scaly forms of *C. mollis.*

calopus (Pers.) Fr., Marasmius, *Epicr. syst. mycol.*: 379 (1838)
  Agaricus calopus Pers., *Syn. meth. fung.*: 373 (1801)
A *nomen dubium* lacking type material. Sensu NCL is *Marasmiellus vaillantii.*

calyciformis With., Agaricus, *Arr. Brit. pl. ed. 3*, 4: 209 (1796)
Described from Britain but a *nomen dubium.* Reported by Withering 'in plantations in Edgbaston Park' (Warwickshire) and said to be the same as the species illustrated in Batsch, no. 118 [Elench. Fung. Cont. Prima (1786)]

calyptrosporum Bruchet, Hebeloma, *Bull. Mens. Soc. Linn. Lyon* 39, 6 (Suppl.): 104 (1970)
Excluded pending clearer definition. Possibly a form of *Hebeloma anthracophilum* that does not grow on burnt ground *fide* Vesterholt (pers. comm.). Reported from Surrey (Reigate Heath) in 1985 but requires confirmation.

campanella var. badipus (Fr.) Sacc., Omphalia, *Syll. fung.* 5: 327 (1887)
  Agaricus campanella var. *badipus* Fr., *Hymenomyc. eur.*: 162 (1874)
A *nomen dubium.* Possibly *Xeromphalina cauticinalis fide* NCL. Cooke 263 (273) Vol. 2 (1883) is not helpful.

campanella var. myriadea (Kalchbr.) Sacc., Omphalia, *Syll. fung.* 5: 327 (1887)
  Agaricus campanella var. *myriadea* Kalchbr., in Fries, *Hymenomyc. eur.*: 162 (1874)
A *nomen dubium.* Listed by Rea (1922) but no records have been traced. Possibly a species of *Xeromphalina.*

campanulata Massee, Galera, *Brit. fung.-fl.* 2: 145 (1893)
Described from Britain but a *nomen dubium.* This was Massee's concept of *Agaricus siligineus* Fr. sensu Cooke [1174 (1156) Vol. 8 (1891)], originally collected 'from road scrapings' in Scarborough (Yorkshire) .

campanulatus Bolton, Agaricus [*nom. illegit.*, non *A. campanulatus* Bull. (1781)], *Hist. fung. Halifax* 1: 31 (1788)
Described from Britain but a *nomen dubium* lacking type material. Bolton's illustration depicts a species of *Coprinus.*

canaliculatus With., Agaricus, *Bot. arr. Brit. pl. ed. 2*, 3: 396 (1792)
Described from Britain but a *nomen dubium.*

candelaris Fr., Cortinarius, *Epicr. syst. mycol.*: 305 (1838)
Possibly a synonym of *C. rigens* to which most of the British records should be referred, but lacking an agreed modern interpretation.

candicans Inzenga, Boletus, *Funghi Sicil.* 2: 64 (1869)
A *nomen dubium.* Sensu auct. Brit. is *Boletus radicans.*

candida (Weinm.) Corner, Clavulinopsis, *Monograph of Clavaria and Allied Genera*: 360 (1950)
  Clavaria candida Weinm., *Hymen. Gasteromyc.*: 514 (1836)

A *nomen dubium* lacking type material. Possibly *Clavaria asterospora* (i.e. *C. acuta*) *fide* Corner (1950) but according to the type description the spores are 'pruniform'.

candida (Huds.) Pers., Himantia, *Syn. meth. fung.* 2: 704 (1801)
Byssus candidus Huds., *Fl. angl.*, Edn 2, 2: 607 (1778)
A *nomen dubium* lacking type material.

candida Pers., Tremella, *Syn. meth. fung.*: 624 (1801)
A *nomen dubium*, widely used and variously interpreted. Collections in herbaria under this name (or the unpublished '*Exidia candida*') are a mixture of taxa.

candidissima (Schwein.) Bondartsev & Singer, Trechispora, *Ann. Mycol.* 39: 48 (1941)
Polyporus candidissimus Schwein., *Trans. Amer. Philos. Soc.* 4: 159 (1832)
Poria candidissima (Schwein.) Cooke, *Grevillea* 14: 111 (1886)
Cristella candidissima (Schwein.) Donk, *Mycologia* 35: 228 (1943)
Not authentically British. Sensu auct. Brit. is mainly *Trechispora mollusca* or *T. hymenocystis*.

candidum Pers., Lycoperdon, *Icon. descr. fung.* 2: t. 13 f. 4 (1800)
A *nomen dubium* lacking type material and variously interpreted. Sensu auct. is *L. mammiforme*.

canobrunnea (Batsch) P. Kumm., Psilocybe, *Führ. Pilzk.*: 71 (1871)
Agaricus canobrunneus Batsch, *Elench. fung. (Continuatio Prima)*: 133 & t. 20 f. 105 (1786)
A *nomen dubium*. Reported from Berkshire by Wakefield in 1923 [TBMS 10 (1-2): 5 (1924)] but unsubstantiated with voucher material.

canofaciens (Cooke) Sacc., Psilocybe, *Syll. fung.* 5: 1047 (1887)
Agaricus canofaciens Cooke, *Grevillea* 14: 1 (1885)
Described from Britain but a *nomen dubium*. Cooke 595 (621) Vol. 4 (1886) looks distinctive but is unidentifiable.

capitatum J. Erikss. & Å. Strid, Hyphoderma, *Corticiaceae of North Europe* 3: 461 (1975)
Not authentically British. Collections named thus represent various misidentified corticioid species.

capniocephalum (Bull.) Kühner, Lyophyllum, *Bull. Mens. Soc. Linn. Lyon* 7: 211 (1938)
Agaricus capniocephalus Bull., *Herb. France*: pl. 547 (1792)
Hebeloma capniocephalum (Bull.) Gillet, *Hyménomycètes*: 527 (1876)
A *nomen dubium* lacking type material. Cooke 462 (419) Vol. 3 (1886) appears to be a species of *Hebeloma*.

capsicum Schulzer, Lactarius, Kalchbrenner, *Icon. select. Hymenomyc. Hung.*: 40 (1873)
A *nomen dubium*. Old reports from Scotland (Dumfries) and England (Wiltshire: Savernake Forest) but no voucher material has been traced. See Cooke 939 (977) Vol. 7 (1888) for an illustration of the Scottish material.

carinii Bres., Lepiota, *Icon. Mycol.* 12: pl. 598 (1930) [1929]
Not authentically British. Reported from England (Isle of Wight) in 1999 but unsubstantiated with voucher material. The illustration in B&K4: 204 pl. 234 (as *Lepiota* aff. *pseudoasperula*) is probably of this species *fide* FAN5.

carneosanguinea Rea, Mycena, *Trans. Brit. Mycol. Soc.* 1(4): 157 (1900)
Described from Britain but a *nomen dubium*. From the description almost certainly *Mycena pelianthina*.

carneus Fr., Polyporus, *Epicr. syst. mycol.*: 471 (1838)
A *nomen dubium*. Reported in Grevillea 2: 134 (1874) as 'on fir stumps near Welshpool' but no material survives. Described from Java and thus unlikely in central Wales.

carnosa Massee, Flammula, *British Fungi*: 290 (1911)
Described from Britain but a *nomen dubium*.

carnosus Curtis, Agaricus [non *A. carnosus* Bolton (1791)], *Fl. londin.* 1: t. 71 (1777)
Described from Britain but a *nomen dubium* lacking type material. Sensu Sowerby [Col. fig. Engl. fung. 3: pl. 246 (1800)] is a pure white agaric, bruising or spotting dull red and possibly a stylised interpretation of *Collybia maculata* but the pileal surface is shown as striate.

carnosus Rostk., Boletus, *Deutschl. Fl.*, Edn 3: t. 14 (1848)
A *nomen dubium*, listed in Rea (1922). Possibly *B. badius fide* Pearson & Dennis (1948).

carpta (Scop.) P. Kumm., Inocybe, *Führ. Pilzk.*: 79 (1871)
Agaricus carptus Scop., *Fl. carniol.*: 449 (1772)
A *nomen dubium* lacking type material. British records may be *Inocybe subcarpta*.

cartilaginea S. Lundell & Neuhoff, Exidia, *Pilze Mitteleuropas*, Edn 2, 2: 19 (1935)
Not authentically British. Old records of *Tremella albida* from Yorkshire were placed here by Bramley (1985) and the species was included in Ellis & Ellis (1990), but no voucher material traced.

cartilagineum (Bull.) Quél., Tricholoma, *Mém. Soc. Émul. Montbéliard, Sér. 2, 5*: 328 (1873)
Agaricus cartilagineus Bull., *Herb. France*: pl. 589 (1793)
Clitocybe cartilaginea (Bull.) Bres., *Fungi trident.* 2: 7 (1892)
A *nomen dubium* lacking type material. The original illustration by Bulliard appears to be a species of *Melanoleuca* but sensu auct. mult. is *Lyophyllum loricatum*.

casca (Fr.) Konrad & Maubl., Psathyrella, *Encycl. Mycol.* 14, *Les Agaricales* I: 127 (1948)
Agaricus cascus Fr., *Epicr. syst. mycol.*: 224 (1838)
Hypholoma cascum (Fr.) Gillet, *Hyménomycètes*: 571 (1876)
Not authentically British. The single collection named thus in herb. K is sensu Lange (FlDan4: 77 & pl. 147A) and thus equates to *Psathyrella pseudocasca*. Two additional old records are unsubstantiated with voucher material and excluded as doubtful.

caseus With., Agaricus, *Arr. Brit. pl. ed. 3*, 4: 158 (1796)
Described from Britain but a *nomen dubium*. From Withering's descriptions and comments possibly a large species of *Clitocybe*.

casimiri (Velen.) Huijsman, Cortinarius, *Fungus* 25: 20 (1955)
Telamonia casimiri Velen., *České Houby* III: 464 (1921)
Cortinarius decipiens var. *hoffmannii* Reumaux, *Bull. Trimestriel Féd. Mycol. Dauphiné-Savoie* 28(111): 24 (1988)
Cortinarius subsertipes Romagn., *Bull. Soc. Naturalistes Oyonnax* 6: 61 (1952)
Mis.: *Cortinarius unimodus* sensu Cooke [Ill. Brit. fung. 844 (859) Vol. 6 (1887) & Grevillea 16: 45 (1887)]
Not authentically British. Reported by Cooke (1887) (as *C. unimodus*) and a single record (as *C. decipiens* var. *hoffmanii*) in 1994 from Wales, but unsubstantiated with voucher material.

castaneifolius (Murrill) A.H. Sm., Panaeolus, *Mycologia* 40: 685 (1948)
Psilocybe castaneifolia Murrill, *Mycologia* 15: 17 (1923)
Not authentically British. A North American species. Sensu auct. Brit. (and sensu auct. Eur.) is *P. olivaceus*.

castaneus (Bull.) Fr., Cortinarius, *Epicr. syst. mycol.*: 307 (1838)
Agaricus castaneus Bull., *Herb. France*: pl. 268 (1786)
A *nomen dubium* lacking type material. A small brown *Telamonia* of uncertain identity, considered common by Rea (1922) but probably used for several members of section *Erythrini*.

castaneus (Fr.) Cooke, Fomes, *Grevillea* 14: 20 (1886)
Polyporus castaneus Fr., *Syst. mycol.* 1: 369 (1821)

A *nomen dubium*. Listed by Rea (1922).

castoreus (Fr.) Konrad & Maubl., Lentinellus, *Bull. Soc. Mycol. France* 50: 16 (1934)
Lentinus castoreus Fr., *Epicr. syst. mycol.*: 395 (1838)
Not authentically British; there has been one doubtful British collection *fide* BFF6. Possibly only a form of *Lentinellus ursinus* on conifer wood.

catervata (Massee) P.D. Orton, Psathyrella, *Trans. Brit. Mycol. Soc.* 43(2): 180 (1960)
Psilocybe catervata Massee, *Brit. fung.-fl.* 1: 378 (1892)
Described from Britain but a *nomen dubium*. Possibly a pale form of *Psathyrella marcescibilis* but the descriptions by Massee, and also by Pearson [TBMS 32: 270 (1949)], do not suggest any species currently known. Watling in BFF5: 100 synonymised this with *Psathyrella cernua* but spores of *P. catervata* were said to be significantly larger.

catilla W.G. Sm., Cyphella, *J. Bot.* 11: 337 (1873)
Described from Britain, but a *nomen dubium*. The type description of material 'on leaves in a garden at Kings Lynn' suggests a species of *Arrhenia*.

causticus Fr., Cortinarius, *Epicr. syst. mycol.*: 270 (1838)
Not authentically British. Sensu Fries and sensu CFP4: 10 is a northern conifer associate, unknown in Britain. Sensu NCL and sensu auct. mult. is *C. galeobdolon*.

cavipes Huijsman, Hebeloma, *Persoonia* 2(1): 97 (1961)
Hebeloma lutense Romagn., *Bull. Soc. Mycol. France* 81: 342 (1965)
Not authentically British. A southern European species, doubtfully distinct from *H. helodes fide* Vesterholt (in litt.), recorded from Scotland as *H. lutense,* but requiring confirmation. Material named thus in herb. E is probably *H. senescens fide* R. Watling (pers. comm.).

centrifugum (Lév.) Bres., Corticium [*nom. illegit.*, non *C. centrifugum* (Weinm.) Fr. (1874)], *Ann. Mycol.* 1: 96 (1903)
Rhizoctonia centrifuga Lév., *Ann. Sci. Nat., Bot.*, sér. 2, 20: 225 (1843)
An illegitimate and dubious name applied to various species of *Athelia fide* Jülich [*Willdenowia* 7: 249 (1972)]. Often misapplied to *A. arachnoidea*.

centurio (Kalchbr.) Sacc., Tricholoma, *Syll. fung.* 5: 97 (1887)
Agaricus centurio Kalchbr., *Icon. select. Hymenomyc. Hung.*: t. 4 (1873)
A *nomen dubium*. Doubtful *fide* NCL. Riva (FungEur3: 34) suggests that this is synonymous with *Lyophyllum transforme,* a species unknown in Britain.

cepistipes Velen., Armillaria, *České Houby* II: 283 (1920)
Not authentically British. Reported from a few widely scattered locations in England (Yorkshire to South Devon) but unsubstantiated with voucher material; herbarium material determined only as 'cf' *A. cepistipes.*

cerealis E.P. Hoeven, Rhizoctonia [*nom. inval.*], *Netherlands J. Pl. Pathol.* 83(5): 191 (1977)
The name has been applied to the anamorphic state of *Ceratobasidium cornigerum*.

cerebrina Bull., Tremella, *Herb. France*: pl. 386 (1788)
A *nomen dubium* lacking type material. Considered a good species by Donk (1966), but included in the synonymy of *Exidia thuretiana* by others. Sensu Purton (1821: 176) may be *Tremella aurantia*. A single old British collection under this name in herb. K is *T. globispora.*

cerebrinus Berk. & Broome, Polyporus, *Ann. Mag. Nat. Hist., Ser. 5 [Notices of British Fungi no. 1800]* 3: 202 (1879)
Described from Britain, but a *nomen dubium*. Perhaps a resupinate specimen of *Trametes versicolor fide* Reid & Austwick (1963), but the type (in herb. K) is an *Antrodia* species, in poor condition and not determinable.

cerifera (P. Karst.) P. Karst., Pholiota, *Bidrag Kännedom Finlands Natur Folk* 32: 297 (1879)
Agaricus ceriferus P. Karst., *Bidrag Kännedom Finlands Natur Folk* 25: 369 (1876)
Not authentically British. A boreal species associated with *Salix* and *Betula* spp. Sensu auct. mult. is *Pholiota aurivella*.

cerinum M. Korhonen, Leccinum, *Karstenia* 35(2): 61 (1995)
Not authentically British. Reported from South Hampshire (New Forest) but found to be misidentified material of *L. versipelle*.

cernua (Schumach.) W.B. Cooke, Calyptella, *Sydowia, Beih.* 4: 36 (1961)
Peziza capula β *cernua* Schumach., *Enum. pl.* 2: 421 (1803)
Cyphella cernua (Schumach.) Massee, *Brit. fung.-fl.* 1: 138 (1892)
A *nomen dubium* lacking type material. Reported by Rea (1922). From the description probably just the yellow form of *C. capula* on bark of *Sambucus nigra.*

cerodes (Fr.) P. Kumm., Naucoria, *Führ. Pilzk.*: 77 (1871)
Agaricus cerodes Fr., *Epicr. syst. mycol.*: 195 (1838)
A *nomen dubium*. Cooke 498 (489) Vol. 4 (1884) (lower fig.) is possibly a species of *Agrocybe*. Sensu Lange (FlDan4) is *Gymnopilus fulgens*.

cervinum (L.) Pers., Scleroderma, *Syn. meth. fung.*: 156 (1801)
Lycoperdon cervinum L., *Sp. pl.*: 1053 (1753)
Scleroderma vulgare var. *cervinum* (L.) W.G. Sm., *Syn. Brit. Bas.*: 480 (1908)
Not a basidiomycete. This is a species of the ascomycete genus *Elaphomyces* (probably *E. granulatus*). Sensu Bolton [Hist. Fung. Halifax pl. 116 (1789) as *Lycoperdon cervinum*] appears to be a mixture of various gasteromycete taxa.

cervinus var. bullii (Cooke) Rea, Pluteus, *Brit. basidiomyc.*: 56 (1922)
Agaricus cervinus var. *bullii* Cooke, *Handb. Brit. fung.*, Edn 2: 115 (1886)
Pluteus bullii (Cooke) Rea, *Trans. Brit. Mycol. Soc.* 12(2-3): 208 (1927)
A *nomen dubium*. Probably based on large fruitbodies of *P. cervinus* growing on a nutrient-rich substratum (decayed sawdust). See Cooke 304 (357) Vol. 3 (1884) for an illustration.

cervinus var. eximius (W. Saunders & W.G. Sm.) A. Pearson & Dennis, Pluteus, *Trans. Brit. Mycol. Soc.* 31: 169 (1948)
Agaricus cervinus var. *eximius* W. Saunders & W.G. Sm., *Mycol. Illust.*: t. 38 (1870)
Pluteus eximius (W. Saunders & W.G. Sm.) Sacc., *Syll. fung.* 5: 666 (1887)
A *nomen dubium*. Probably based on overgrown fruitbodies of *P. cervinus* growing on decayed sawdust. See Cooke 303 (302) Vol. 3 (1884) for an illustration.

cesatii var. subsphaerosporus (J.E. Lange) Senn-Irlet, Crepidotus, *Persoonia* 16(1): 53 (1995)
Crepidotus variabilis var. *subsphaerosporus* J.E. Lange, *Fl. agaric. danic.* 5, Taxonomic Conspectus: IV (1940)
Crepidotus subsphaerosporus (J.E. Lange) Hesler & A.H. Sm., *North. Am. Crepidotus*: 121 (1965)
This is a species on conifers which needs confirmation as British. Sensu BFF6 is not worth separating from the type variety.

cheimonophilus (Berk. & Broome) Sacc., Crepidotus, *Syll. fung.* 5: 882 (1887)
Agaricus cheimonophilus Berk. & Broome [as *A. chimonophilus*], *Ann. Mag. Nat. Hist., Ser. 2 [Notices of British Fungi no. 687]* 13: 401 (1854)
A *nomen dubium* variously interpreted. Type material is in poor condition and indeterminable but the original description fits *C. herbarum* (= *C. epibryus*) *fide* NCL. Cooke 536 (515) Vol. 4 (1885) is a small pleurotoid agaric showing pinkish tints on the pileus 'hardly different' from *C. variabilis fide* Pearson & Dennis (1948). Senn-Irlet [*Persoonia* 16 (1): 31 (1995)] suggests this is synonymous with *C. mollis.*

chelidonia (Fr.) Quél., Mycena, *Mém. Soc. Émul. Montbéliard, Sér. 2,* 5: 225 (1872)
  Agaricus chelidonius Fr., *Epicr. syst. mycol.*: 115 (1838)
A *nomen dubium*. 'Probably *Mycena crocata* (or a form of it)' *fide* NCL. Cooke 244 (207a) Vol. 2 (1882) is not that species, however, and Massee's description [Brit. fung.-fl. 3: 88 (1893)] is reminiscent of *Mycena renati*.

chioneus (Pers.) Kühner, Pleurotellus, *Botaniste* 17: 114 (1926)
  Agaricus variabilis var. *chioneus* Pers., *Mycol. eur.* 3: 28 (1828)
  Pleurotus chioneus (Pers.) Gillet, *Hyménomycètes*: 336 (1876)
A *nomen dubium* lacking type material and variously interpreted. Most of the British collections thus named in herb. K are *Clitopilus hobsonii*.

chlorinum Link, Sporotrichum, *Mag. Neuesten Entdeck. Gesammten Naturk. Ges. Naturf. Freunde Berlin* 8: 35 (1816)
A *nomen dubium*, listed as British by Wakefield & Bisby (1942). Formerly thought to be a hyphomycete, but the type is a *Tomentella* sp. *fide* Stalpers (1984), though not *T. chlorina* (Massee) G.H. Cunn.

chloropolium (Fr.) M.M. Moser, Entoloma, *Guida alla Determinazione dei Funghi*: 215 (1980)
  Agaricus chloropolius Fr., *Monogr. hymenomyc. Suec.* 2: 297 (1863)
  Leptonia chloropolia (Fr.) Gillet, *Hyménomycètes*: 414 (1876)
Reported from England, Scotland and Republic of Ireland but all records are prior to 1936 and unsubstantiated with voucher material. Collections named thus in herbaria are misdetermined. There is only a drawing by Rea at Kew but see also Cooke 361 (337) Vol. 3 (1885) for an illustration.

chocolatus With., Agaricus, *Arr. Brit. pl. ed. 3*, 4: 250 (1796)
Described from Britain 'in tufts amongst grass, at Edgbaston' (Warwickshire) but a *nomen dubium*.

chrysenteron var. nanus Massee, Boletus, *Brit. fung.-fl.* 1: 264 (1892)
Described from Britain but a *nomen dubium*.

chrysocreas (Berk. & M.A. Curtis) Burds., Phlebia, *Mycologia* 67(3): 497 (1975)
  Corticium chrysocreas Berk. & M.A. Curtis, *Grevillea* 1 (12): 178 (1873)
Not authentically British. A common pantropical species, unlikely to occur in Europe. A British record on *Fagus* from North Wiltshire (Westbury) is unsubstantiated with voucher material and most likely misidentified.

chrysoloma (Fr.) Donk, Phellinus, *Proc. K. Ned. Akad. Wet.* 74: 39 (1971)
  Polyporus chrysoloma Fr., *Öfvers. Kongl. Vetensk.-Akad. Förh.* 18: 30 (1862)
Not authentically British. Reported from the Nordic Congress Foray at Kindrogan, Scotland (1983) but unsubstantiated with voucher material.

cidaris (Fr.) Romagn., Phaeocollybia, *Bull. Soc. Mycol. France* 58: 127 (1942)
  Agaricus cidaris Fr., *Epicr. syst. mycol.*: 192 (1838)
  Naucoria cidaris (Fr.) Sacc., *Syll. fung.* 5: 830 (1887)
A *nomen dubium*. Cooke 492 (451) Vol. 4 (as *Agaricus cidaris*) and Rea (1922) probably refer to *Phaeocollybia arduennensis*.

cilicioides (Fr.) Fr., Lactarius, *Epicr. syst. mycol.*: 334 (1838)
  Agaricus cilicioides Fr., *Syst. mycol.* 1: 63 (1821)
  Lactarius torminosus ssp. *cilicioides* (Fr.) Konrad, *Bull. Soc. Mycol. France* 51: 164 (1935)
A *nomen dubium* further confused by Fries's own, later re-interpretation. Sensu auct. Brit. is *Lactarius citriolens*.

cincta (Berk.) Cooke, Poria, *Grevillea* 14: 110 (1886)
  Polyporus cinctus Berk., *Outl. Brit. fungol.*: 250 (1860)
Described from Britain 'on decayed deal boards in a greenhouse' but a *nomen dubium*.

cinerascens (Quél.) Bon & Boiffard, Leucoagaricus, *Doc. mycol.* 8(29): 38 (1978)
  Lepiota cinerascens Quél., *Compt. Rend. Assoc. Franç. Avancem. Sci.* 22(2): 484 (1894)
Not authentically British. This is a southern European species unlikely in Britain. Collections named thus are mostly misidentified *Leucoagaricus carneifolius*.

cineratus Quél., Coprinus, *Bull. Soc. Bot. France* 23: 329 (1876)
A *nomen dubium* lacking type material and variously interpreted. Sensu NCL is *C. laani* p.p. and *C. semitalis* p.p.

cinerea Sowerby, Auricularia, *Col. fig. Engl. fung.* 3: pl. 388 f. 3 (1803)
Described from Britain but a *nomen dubium* lacking type material and probably not a basidiomycete. From Sowerby's illustrations this appears to be a species of the ascomycete genus *Hypoxylon*.

cinerea f. sublilascens (Bourdot & Galzin) Bon & Courtec., Clavulina, *Doc. mycol.* 18(69): 37 (1987)
  Clavaria cinerea f. *sublilascens* Bourdot & Galzin, *Hyménomyc. France*: 107 (1928)
A *nomen dubium*. Reported from Cambridgeshire (Chippenham Fen) in 1946 but probably just a grey-lilaceous form of *Clavulina cinerea*.

cinereoumbrina (Bres.) Stalpers, Tomentella, *Stud. Mycol.* 35: 96 (1993)
  Hypochnus cinereoumbrinus Bres., *Stud. Trent., ser. 2* 7: 62 (1926)
  Tomentellastrum caesiocinereum Svrček, *Česká Mykol.* 12: 69 (1958)
  Tomentella litschaueri Svrček, *Česká Mykol.* 12: 75 (1958)
  Tomentellastrum litschaueri (Svrček) M.J. Larsen, *Mycol. Mem.* 4: 119 (1974)
Not authentically British. Collections assigned with doubt to *T. litschaueri* by Wakefield (1969) are not this species. Additional reports are unsubstantiated with voucher material.

cinereum Bull., Hydnum, *Herb. France*: pl. 419 (1789)
A *nomen dubium* lacking type material. Sensu Rea (1922) is *Phellodon niger*.

cinereus (Pers.) Fr., Hygrophorus, *Sverig Atl. Svamp.*: 30 (1861)
  Agaricus pratensis β *cinereus* Pers., *Syn. meth. fung.*: 304 (1801)
  Hygrophorus pratensis var. *cinereus* (Pers.) Sacc., *Syll. fung.* 5: 401 (1887)
  Hygrocybe cinerea (Pers.) P.D. Orton & Watling, *Notes Roy. Bot. Gard. Edinburgh* 29(1): 131 (1969)
A *nomen dubium* lacking type material. Sensu. auct. Brit. is possibly old and dried-out material of *Hygrocybe lacmus* fide Boertmann (FNE1) or, *fide* Watling (pers. comm.), *Hygrocybe canescens*.

cinnamomea (Jacq.) Murrill, Coltricia, *Bull. Torrey Bot. Club* 31: 343 (1904)
  Boletus cinnamomeus Jacq. *Collect. Bot. Spectantia (Vindobonae)* 1: 116 (1787)
  Strilia cinnamomea (Jacq.) Gray, *Nat. arr. Brit. pl.* 1: 645 (1821)
  Polystictus cinnamomeus (Jacq.) Sacc., *Syll. fung.* 6: 210 (1888)
Not authentically British. Reported from England and Scotland but all records are old (last reported in 1958) and herbarium material named thus is misidentified, usually *C. perennis*.

cinnamomeus L., Agaricus, *Sp. pl.*: 1173 (1753)
A *nomen dubium*. The illustration by Bolton (1788, as *A. cinnamoneus*) is indeterminable.

circumseptus Batsch, Agaricus, *Elench. Fung, Cont. Prima*: 119, t. 19, f. 98 (1786)

A *nomen dubium*. Reported by Withering [Arr. Brit. pl. ed. 3, 4: 249 (1796)]. From Batsch's illustration, possibly a species of *Conocybe*.

circumtectum (Cooke & Massee) Sacc., Tricholoma, *Syll. fung.* 11: 8 (1895)
    Agaricus circumtectus Cooke & Massee, *Handb. Brit. fung.,* Edn 2: 382 (1890)
Doubtful species *fide* NCL. Dennis [TBMS 31: 193 (1948)] suggests it may be *Tricholoma scalpturatum*. Cooke 1125 (1182) Vol. 8 (1889) looks distinctive, possibly a form of *T. saponaceum*, but the type is in poor condition and unidentifiable.

citri (Inzenga) Sacc., Armillaria, *Syll. fung.* 5: 83 (1887)
    Agaricus citri Inzenga, *Funghi Sicil.*: 33 pl. 3 (1865)
A *nomen dubium*. Considered a greenish-yellow form of *Armillaria mellea* by Termorshuizen & Arnolds [Mycotaxon 30: 106 (1987)]. Cooke 1115 (1181) Vol. 8 (1889) would seem to support this idea.

citrina (Rea) J.E. Lange, Hygrocybe, *Fl. agaric. danic.* 5: 27 (1940)
    Hygrophorus citrinus Rea, *Trans. Brit. Mycol. Soc.* 3(3): 228 (1910) [1909]
Described from Britain but a *nomen dubium* variously interpreted. Sensu auct. is *H. glutinipes*, sensu Lange (1940) and B&K is *H. ceracea*.

citrinus With., Boletus [non *B. citrinus* Planer (1788)], *Bot. arr. veg.* 2: 769 (1776)
Described from Britain but a *nomen dubium*. Stated to grow 'on trees' and from the description, possibly young material of *Inonotus hispidus* or *Fistulina hepatica*.

citrophylla (Berk. & Broome) Sacc., Lepiota, *Syll. fung.* 5: 57 (1887)
    Agaricus citrophyllus Berk. & Broome, *J. Linn. Soc., Bot.* 11: 499 (1871)
Not authentically British. Described from Sri Lanka and the name misapplied in Europe to both *Lepiota xanthophylla* and *L. elaiophylla*.

civile (Fr.) Sacc., Tricholoma, *Syll. fung.* 5: 130 (1887)
    Agaricus civilis Fr., *Ic. Hymenomyc.* 1: pl. 45 (1867)
Excluded pending clearer definition *fide* NCL. There is a single British record of this species from Scotland (Rothiemurchus) in 1927 [TBMS 13(3-4): 307 (1928)] but unsubstantiated with voucher material. Riva (FungEur3: 34) implies that this is synonymous with *Tricholoma acerbum*.

clarkii (Berk. & Broome) Sacc., Hygrophorus, *Syll. fung.* 5: 406 (1887)
    Agaricus clarkii Berk. & Broome, *Ann. Mag. nat. Hist.,* Ser. 4, 11: 341 (1873)
Described from Britain but a *nomen dubium*.

clavata W.B. Cooke, Cellypha, *Sydowia, Beih.* 4: 53 (1961)
Excluded pending clearer definition. Described from England (Burnham Beeches) but not a *Cellypha* as now understood. The type collection needs re-examination.

clavatus, With., Agaricus, *Arr. Brit. pl. ed. 3,* 4: 200 (1796)
Described from Britain but a *nomen dubium*.

claviceps (Fr.) Sacc., Hebeloma, *Syll. fung.* 5: 794 (1887)
    Agaricus claviceps Fr., *Monogr. hymenomyc. Suec.* 2: 346 (1863)
A *nomen dubium*. Cooke 451 (410) Vol. 3 (1886) is not helpful but possibly represents atypical *Hebeloma mesophaeum*.

claviformis Schaeff., Agaricus, *Fung. Bavar. Palat. nasc.* 3: 307 (1770)
A *nomen dubium*. Sensu Withering (1796) may represent *Hygrocybe pratensis*.

claviformis Gray, Mycena, *Nat. arr. Brit. pl.* 1: 621 (1821)

Described from Britain but a *nomen dubium* lacking type material. Not mentioned by Maas Geesteranus (1992) but the description suggests *M. acicula*.

clavuligerum (Romagn.) P. Collin, Hebeloma, *Doc. mycol.* 19(74): 61 (1988)
    Alnicola clavuligera Romagn., *Bull. Soc. Mycol. Franc.* 58: 148 (1944) [1942]
    Naucoria clavuligera (Romagn.) Kühner & Romagn., *Flore Analytique des Champignons Supérieurs*: 236 (1953)
Not authentically British. Collections from Norfolk by Orton were reassigned to *N. clavuligeroides* (NBA8: 600).

clavulipes Romagn., Hebeloma, *Bull. Soc. Mycol. France* 81(3): 326 (1965)
Not authentically British. Recent possible collections with *Helianthemum* sp. await detailed examination.

clavus (L.) P. Kumm., Collybia, *Führ. Pilzk.*: 114 (1871)
    Agaricus clavus L., *Fl. Suec.,* Edn 2: 1212 (1755)
A *nomen dubium* lacking type material. Several specimens named thus, in herb. K, macroscopically resemble *Rickenella fibula*.

clintonianus (Peck) Kuntze, Suillus, *Revis. gen. pl.* 3(2): 535 (1898)
    Boletus clintonianus Peck, *Rep. (Annual) New York State Mus. Nat. Hist.* 23: 128 (1873) [1872]
Not authentically British. A North American species. Reported from England (FM3: 6) but collections have been redetermined as *Suillus grevilleii* var. *badius fide* R. Watling.

clitopilus (Cooke & W.G. Sm.) Sacc., Flammula, *Syll. fung.* 5: 809 (1887)
    Agaricus clitopilus Cooke & W.G. Sm., *Grevillea* 13: 59 (1885)
Described from Britain but a *nomen dubium*. See Cooke 468 (500) Vol. 4 for an interpretation.

clusiae Syd., Physalacria, *Ann. Mycol.* 28: 35 (1930)
Not authentically British. The single collection named thus, from Gloucestershire (Westonbirt Arboretum), has been redetermined as *Physalacria cryptomeriae*.

clusii Fr., Russula, *Hymenomyc. eur.*: 449 (1874)
    Russula emetica var. *clusii* (Fr.) Cooke & Quél., *Clavis syn. Hymen. Europ.*: 146 (1878)
A *nomen dubium*. Sensu Cooke & Quélet is *R. emetica*. The description in Rea (1922) suggests *R. luteotacta*.

clusiliformis Kühner & Romagn. [*nom. inval.*], Omphalia, *Flore Analytique des Champignons Supérieurs*: 125 (1953)
A *nomem novum* for Collybia clusilis sensu K&M (pl. 207I). Sensu NCL (as *Clitocybe clusiliformis* ined.) is *Pseudoomphalina pachyphylla*.

clusilis (Fr.) Sacc., Collybia, *Syll. fung.* 5: 250 (1887)
    Agaricus clusilis Fr., *Epicr. syst. mycol.*: 98 (1838)
A *nomen dubium*. Antonín & Noordeloos (A&N2) suggest that this may represent either *Pseudoomphalina pachyphylla* or *Tephrocybe palustris*. See Cooke 215 (247) Vol. 2 (1882) for an interpretation.

clypeatus Huds., Agaricus [*nom. illegit.*, non *A. clypeatus* L. (1753)], *Fl. angl.,* Edn 2, 2: 619 (1778)
Described from Britain but a *nomen dubium* lacking type material. The illustration in Bolton [Hist. Fung. Halifax 2: 57 (1788)] shows a mixed collection of unidentifiable agarics.

clypeolarioides Rea, Lepiota, *Brit. basidiomyc.*: 69 (1922)
Described from Britain but a *nomen dubium*. See discussion in FAN5: 126.

cnista (Fr.) Gillet, Tricholoma, *Hyménomycètes*: 121 (1874)
    Agaricus cnista Fr., *Epicr. syst. mycol.*: 50 (1838)
A *nomen dubium*. For some authorities e.g. Rea (1922) this is a white *Melanoleuca*, possibly *M. strictipes*. For others it is a species of *Tricholoma* either a variety of *T. saponaceum* or a separate species, *T. moserianum* Bon, unknown in Britain.

coelestinum (Fr.) Hesler, Entoloma, *Nova Hedwigia Beih.* 23: 111 (1967)
    Agaricus coelestinus Fr., *Epicr. syst. mycol.*: 158 (1838)
    Nolanea coelestina (Fr.) Gillet, *Hyménomycètes*: 422 (1876)
    Leptonia coelestina (Fr.) P.D. Orton, *Trans. Brit. Mycol. Soc.* 43(2): 177 (1960)
Last reported by Rea from England (Shropshire) in 1916, but unsubstantiated with voucher material. Sensu Rea (1922) and sensu Cooke 366 (379) Vol. 3 (1885) growing 'on wood' is doubtful and evidently some other species.

coffeata (Fr.) Gillet, Clitocybe, *Hyménomycètes*: 161 (1874)
    Agaricus coffeatus Fr., *Syst. mycol.* 1: 85 (1821)
A *nomen dubium*. Sensu Lange (FlDan 1: 39) probably *Lyophyllum decastes* or a related species. NCL suggests *L. loricatum*.

collariatum Bruchet, Hebeloma *Bull. Mens. Soc. Linn. Lyon* 39 (6 [Suppl.]): 35 (1970)
Not authentically British. The single British collection in herb. K (England: Surrey, Witley Common) is annotated as 'compare with' *H. collariatum*.

collariatus With., Agaricus [non *A. collariatus* Fr. (1818)], *Bot. arr. Brit. pl. ed. 2*, 3: 375 (1792)
Described from Britain but a *nomen dubium* lacking type material. The description suggests *Marasmius rotula*.

colossus (Fr.) C.F. Baker, Ganoderma, *V Cent. Fungi Malay.* no. 425 (1918)
    Polyporus colossus Fr., *Nova Acta Regiae Soc. Sci. Upsal.* 1: 56 (1851)
Not authentically British. The single record, from Northern Ireland (Down: Warrenpoint), is unsubstantiated with voucher material, and presumed a misidentification. Widely distributed in the tropics.

colossus Huijsman, Hebeloma, *Persoonia* 1: 98 (1961)
Not authentically British. Reported from England (West Sussex: Burpham) and Scotland (Midlothian: Edinburgh) but reliably known only from the French type collection.

colubrina γ pantherina Gray, Lepiota, *Nat. arr. Brit. pl.* 1: 602 (1821)
Described from Britain but a *nomen dubium* lacking type material.

colus With., Agaricus, *Bot. arr. Brit. pl. ed. 2*, 3: 383 (1792)
Described from Britain but a *nomen dubium*.

colus Fr., Cortinarius, *Epicr. syst. mycol.*: 308 (1838)
Not authentically British. Reported from Hertfordshire in the late nineteenth century but from an anonymous source and unsubstantiated with voucher material. There is also (in herb. K) a collection by Pearson, said to be this species, from North Hampshire (Selbourne) in 1946, but it is in poor condition and unidentifiable.

colymbadinus Fr., Cortinarius, *Epicr. syst. mycol.*: 289 (1838)
    Mis.: *Cortinarius isabellinus* sensu NCL & sensu auct. mult.
Not authentically British. A single record (as *C. isabellinus*) from Herefordshire in 1980, unsubstantiated with voucher material and dubious, since this is a northern species associated with conifers on calcareous soil.

compactum (Pers.) P.Karst, Hydnellum, *Meddeland. Soc. Fauna Fl. Fenn.* 5: 41 (1879)
    Hydnum compactum Pers., *Syn. meth. fung.*: 556 (1801)
Not authentically British. Collections named thus are all misidentified (usually *Hydnellum caeruleum*).

compactum (Fr.) P. Karst., Tricholoma, *Bidrag Kännedom Finlands Natur Folk* 32: 36 (1879)
    Agaricus compactus Fr., *Öfvers. Kongl. Vetensk.-Akad. Förh.* 18: 20 (1862)
A *nomen dubium*. Listed by Rea (1922).

complicatum Fr., Stereum, *Epicr. syst. mycol.*: 548 (1838)
Not authentically British. A North American species. Sensu auct. Brit. is *Stereum 'rameale'*.

compressipes (Peck) Singer, Pseudoomphalina, *Agaricales in Modern Taxonomy*, Edn 2: 287 (1962)
    Agaricus compressipes Peck, *Rep. (Annual) New York State Mus. Nat. Hist.* 33: 18 (1880)
    Clitocybe compressipes (Peck) Sacc., *Syll. fung.* 5: 184 (1887)
Not authentically British. A North American species. British material is *Pseudoomphalina graveolens fide* Watling (BFF8).

compressus (With.) Gray, Gymnopus, *Nat. arr. Brit. pl.* 1: 610 (1821)
    Agaricus compressus With., *Arr. Brit. pl. ed. 3*, 4: 243 (1796)
Described from Britain but a *nomen dubium* lacking type material. Sowerby's illustration [as *Agaricus compressus* in Col. fig. Engl. fung. 1: pl. 66 (1797)] is not helpful.

compta (Fr.) Sacc., Psilocybe, *Syll. fung.* 5: 1050 (1887)
    Agaricus comptus Fr., *Hymenomyc. eur.*: 301 (1874)
A *nomen dubium*. Cooke 603 (589) Vol. 4 (1886) depicts an agaric apparently in the *Bolbitiaceae*.

comptulus M.M. Moser, Cortinarius, *Nova Hedwigia* 14 (2-4): 514 (1968) [1967]
    Cortinarius hemitrichus var. *paludosus* (Velen.) Kühner & Romagn. [*comb. inval.*], *Flore Analytique des Champignons Supérieurs*: 308 (1953)
    Telamonia paludosa Velen., *České Houby* III: 455 (1921)
Not authentically British. Reported from Shropshire (Nesscliffe Hill) in 1998 and Surrey (Esher Common) in 1986 but unsubstantiated with voucher material and probabaly misidentified.

concava (Scop.) Gillet, Clitocybe, *Hyménomycètes*: 150 (1874)
    Agaricus concavus Scop., *Fl. carniol.*: 449 (1772)
A *nomen dubium* lacking type material. Reported from Scotland in 1910 and from England (South Hampshire) in 1989 but unsubstantiated with voucher material. Sensu NCL is *C. nitriolens* but this is not authentically British (see notes to that species).

concineus Bolton, Agaricus, *Hist. fung. Halifax* 1: 15 (1788)
Described from Britain but a *nomen dubium* lacking type material. Bolton's illustration shows a rather nondescript and unidentifiable agaric but a written comment by Bolton on the plate that 'this is *Agaricus cervinus* Schaeff.' suggests it is probably a synonym of *Pluteus cervinus*.

concolor (J.E. Lange) Kühner, Mycena, *Encycl. Mycol.* 10. *Le Genre* Mycena: 371 (1938)
    Mycena picta var. *concolor* J.E. Lange, *Fl. agaric. danic.* 2: 61 (1936)
Not authentically British. A single collection named thus has been redetermined as *M. latifolia* and no other records or voucher collections have been traced.

confertus (Bolton) Gray, Prunulus, *Nat. arr. Brit. pl.* 1: 631 (1821)
    Agaricus confertus Bolton, *Hist. fung. Halifax* 1: 18 (1788)
    Galera conferta (Bolton) P. Kumm., *Führ. Pilzk.*: 76 (1871)
Described from Britain but a *nomen dubium* lacking type material. Possibly an alien since collected on soil in 'hothouses and stoves'. Cooke 520 (463) Vol. 4 (1885) is copied directly from Bolton's plate. Apparently a species of *Mycena* but *fide* Pearson & Dennis (1948) and Watling (BFF3) it is a *Conocybe*.

confluens P. Karst., Podoporia, *Beibl. Hedwigia* 31: 297 (1892)
A *nomen dubium*. Sensu Dennis (1986) is *Physisporinus sanguinolentus*.

confragosa var. **angustata** (Sowerby) Rea, Daedalea, *Brit. basidiomyc.*: 619 (1822)
    Boletus angustatus Sowerby, *Col. fig. Engl. fung.* 2: pl. 193 (1799)
    Daedalea angustata (Sowerby) Fr., *Syst. mycol.* 1: 338 (1821)

Described from Britain, but a *nomen dubium* lacking type
material. Said to grow on poplars. Probably a form of
Daedaleopsis confragosa .

congregatus Bolton, Agaricus [*nom. illegit.*, non *A.
congregatus* Bull. (1786)], *Hist. fung. Halifax* Suppl.: 140
(1791)
A *nomen dubium* lacking type material. A synonym of *Armillaria
mellea* s.l. *fide* Termorshuizen & Arnolds [Mycotaxon 30: 106
(1987)] since Bolton referred to *Agaricus annularius* Bull.
(1788) (= *A. mellea* s.l.). Synonymised with *Armillaria ostoyae*
by some authorities, e.g. Watling *et al.* (1982).

conica *var.* conicopalustris Arnolds, Hygrocybe, *Persoonia*
13(2): 143 (1986)
Hygrophorus conicopalustris R. Haller Aar [*nom. inval.*],
Schweiz. Z. Pilzk. 31: 141 (1953)
Hygrophorus conicus var. *conicopalustris* (R. Haller Aar.)
Arnolds [*nom. inval.*], *Persoonia* 8: 103 (1974)
Not authentically British. All records unsubstantiated with
voucher material.

conigena (Pers.) P. Kumm., Collybia, *Führ. Pilzk.*: 116 (1871)
Agaricus conigenus Pers., *Syn. meth. fung.*: 388 (1801)
Marasmius conigenus (Pers.) P. Karst., *Bidrag Kännedom
Finlands Natur Folk* 48: 102 (1889)
A *nomen dubium* lacking type material. Possibly *Baeospora
myosura*, or a species of *Strobilurus fide* NCL.

connata Kits van Wav., Psathyrella, *Persoonia* 8: 363 (1976)
Not authentically British. Reported from South Lincolnshire, West
Kent and Yorkshire but dubious, unsubstantiated with voucher
material and probably misidentified *Psathyrella microrrhiza*.

connatus (P. Karst.) Sacc., Hygrophorus, *Syll. fung.* 11: 28
(1895)
Camarophyllus connatus P. Karst, *Symb. mycol. fenn.* 30: 61
(1891)
A *nomen dubium*. Represented at Kew only by a painting by
Rea.

conocephalus (Bull.) Fr., Bolbitius, *Hymenomyc. eur.*: 334
(1874)
Agaricus conocephalus Bull., *Herb. France*: pl. 563 (1792)
A *nomen dubium* lacking type material. Probably a species of
Conocybe fide Watling (BFF3: 107). Sensu Cooke [1186 (1160)]
Vol. 8 (1891)] from a hothouse at Kew is *Conocybe nivea*, an
alien species.

consobrina (Fr.) Fr., Russula, *Epicr. syst. mycol.*: 359 (1838)
Agaricus consobrinus Fr., *Observ. mycol.* 2: 195 (1818)
Not authentically British. Sensu auct. Brit. is possibly *Russula
sororia* s.l.

constrictus With., Agaricus [*nom. illegit.*, non *A. constrictus* Fr
(1821)], *Arr. Brit. pl. ed. 3*, 4: 199 (1796)
Described from Britain but a *nomen dubium*.

contiguus (P. Karst.) J. Erikss. & Ryvarden, Intextomyces,
Corticiaceae of North Europe 4: 737 (1976)
Corticium calceum ssp. *contiguum* P. Karst., *Acta Soc. Fauna
Fl. Fenn.* 2(1): 39 (1881)
Gloeocystidium contiguum (P. Karst.) Bourdot & Galzin, *Bull.
Soc. Mycol. France* 28: 362 (1913)
Not authentically British. Recorded by Reid (as *Gloeocystidium
contiguum*) from England (Hertfordshire) in 1955 but voucher
material cannot be traced.

contingens With., Agaricus, *Arr. Brit. pl. ed. 3*, 4:285 (1796)
Described from Britain but a *nomen dubium*.

contractus Rob. Henry, Cortinarius, *Doc. mycol.* 16(61): 27
(1985)
Not authentically British. A poorly known member of the *C.
duracinus* complex. A single record from Wales
(Merionethshire: Harlech) in 1976, noted as a 'tentative
identification', is from an anonymous source and
unsubstantiated with voucher material.

cookei Fr., Agaricus, *Grevillea* 5 (34): 56 (1876)
Pholiota cookei (Fr.) Massee, *Brit. fung.-fl.* 2: 228 (1893)
Described by Fries from Derbyshire, but a *nomen dubium*.
Incorrectly listed by Rea (1922) as a synonym of *Inocybe
terrigena*.

cookei Quél., Cortinarius, *Bull. Soc. Bot. France* 25: 288 (1879)
[1878]
A *nomen dubium* lacking type material. See Cooke [821 (840)
Vol. 6 (1887)] for an interpretation.

cookei Z. Schaef., Lactarius, *Česká Mykol.* 14: 236 (1960)
Not authentically British. Described from the Tatra mountains in
the belief that it was *Lactarius picinus* sensu Cooke [960 (997)
Vol. 7 (1888)] but there is no evidence of a distinct British
species corresponding to Schaefer's description.

coracina (Fr.) M.M. Moser, Tephrocybe, *Kleine
Kryptogamenflora*, Edn 3: 116 (1967)
Agaricus coracinus Fr., *Epicr. syst. mycol.*: 95 (1838)
Collybia coracina (Fr.) Gillet, *Hyménomycètes*: 307 (1876)
Lyophyllum coracinum (Fr.) Singer, *Ann. Mycol.* 41: 104
(1943)
A *nomen dubium*. Sensu NCL and sensu Orton (Bull. Brit. Mycol.
Soc. 18, 1984) are doubtfully distinct from *Tephrocybe
boudieri*.

coriaceus Bull., Agaricus [non *A. coriaceus* Lightf. (1777)],
Herb. France: pl. 537 (1792)
A *nomen dubium* lacking type material. Bulliard's plate appears
to show *Daedalea quercina* and *Lenzites betulinus* plus a
species of *Phellinus* and *Trametes versicolor*. Sensu Bolton
[Hist. Fung. Halifax pl. 158 (1791)] is *Lenzites betulinus*.

cornea (Bourdot & Galzin) Parmasto, Phlebia, *Eesti N.S.V. Tead.
Akad. Toimet., Biol.* 16: 390 (1967)
Peniophora gigantea ssp. *cornea* Bourdot & Galzin,
Hyménomyc. France: 318 (1928)
Not authentically British. A record from Dorset was
misdetermined and another from South Lancashire is
unsubstantiated with voucher material.

corneus With., Agaricus, *Arr. Brit. pl. ed. 3*, 4: 228 (1796)
Described from Britain 'on the stump of a fir tree which had
fallen' but a *nomen dubium*.

coronatum β woodwardii (Bryant) Pers., Geastrum, *Syn.
meth. fung.*: 132 (1801)
Lycoperdon woodwardii Bryant, *An historical account of two
species of* Lycoperdon: 58 f. 19 (1782)
Described from Britain but a *nomen dubium* lacking type
material.

coronatus With., Agaricus, *Bot. arr. Brit. pl. ed. 2*, 3: 373
(1792)
Described from Britain but a *nomen dubium*.

corrugata (P. Karst.) Schild, Ramaria, *Schweiz. Z. Pilzk.* 53(9):
130 (1975)
Clavaria corrugata P. Karst., *Mycol. fenn.*: 371 (1876)
Not authentically British. Close to *Ramaria decurrens* and said to
be synonymous with *R. myceliosa* of which there are dubious
British records. No voucher collections traced.

corrugatus With., Agaricus, *Bot. arr. Brit. pl. ed. 2*, 3: 385
(1792)
Described from Britain but a *nomen dubium*.

corticalis (Batsch) Fr., Calocera, *Elench. fung.* 1: 233 (1828)
Clavaria corticalis Batsch, *Elench. fung. (Continuatio Prima)*:
231 & t. 28 f. 162 (1786)
A *nomen dubium* lacking type material and probably not a
Calocera fide McNabb [New Zealand J. Bot. 3(1): 53 (1965)].
Reported from England (West Cornwall: Penzance) in Grevillea
11: 14 (1882) but no material has been traced.

corticola (Quél.) Corner, Lentaria, *Monograph of Clavaria and
Allied Genera*: 440 (1950)
Clavaria corticola Quél., *Bull. Soc. Bot. France* 24: 326 (1877)

Listed as British in Jülich (1984) but no records have been traced.

corticola (Pers.) Gray, Mycena, *Nat. arr. Brit. pl.* 1: 621 (1821)
　　Agaricus corticola Pers., *Syn. meth. fung.*: 394 (1801)
A *nomen dubium* lacking type material and variously interpreted. Sensu NCL is *M. meliigena* but earlier records also included *M. hiemalis* and *M. pseudocorticola*.

coruscans (Fr.) Fr., Cortinarius, *Epicr. syst. mycol.*: 271 (1838)
　　Agaricus coruscans Fr., *Syst. mycol.* 1: 227 (1821)
Not authentically British. Sensu Rea (1922) and sensu Cooke 730 (733) Vol. 5 (1888) is an unknown *Phlegmacium* with bright yellow gills.

coryphaeum (Fr.) Gillet, Tricholoma, *Hyménomycètes*: 95 (1874)
　　Agaricus coryphaeus Fr., *Epicr. syst. mycol.*: 26 (1838)
　　Tricholoma sejunctum var. *coryphaeum* (Fr.) A. Pearson & Dennis, *Trans. Brit. Mycol. Soc.* 31: 151 (1948)
Excluded 'pending clearer definition' in NCL but a good species *fide* Riva (FungEur3). There are a few British records all from the early part of the twentieth century and all unsubstantiated with voucher material.

costatum (Fr.) P. Kumm., Entoloma, *Führ. Pilzk.*: 98 (1871)
　　Agaricus pascuus β *costatus* Fr., *Syst. mycol.* 1: 206 (1821)
Not authentically British. Records are all old and unsubstantiated with voucher material. Sensu auct. Brit. is doubtful, possibly in part *E. transvenosum fide* Orton (1991). See Cooke 340 (320) Vol. 3 (1884) for an interpretation.

cothurnata (Fr.) P. Karst., Stropharia, *Bidrag Kännedom Finlands Natur Folk* 32: 494 (1879)
　　Agaricus cothurnatus Fr., *Epicr. syst. mycol.*: 218 (1838)
A *nomen dubium*. Listed by Rea (1922).

craspedius (Fr.) Gillet, Pleurotus, *Hyménomycètes*: 341 (1876)
　　Agaricus craspedius Fr., *Epicr. syst. mycol.*: 131 (1838)
A *nomen dubium*. Cooke 274 (256) Vol. 2 is possibly *Hypsizygus ulmarius*.

crassa (P. Karst.) Ryvarden, Antrodia, *Norweg. J. Bot.* 20: 8 (1973)
　　Physisporus crassus P. Karst., *Bidrag Kännedom Finlands Natur Folk* 48: 319 (1889)
　　Amyloporia crassa (P. Karst.) Bondartsev & Singer, *Ann. Mycol.* 39: 50 (1941)
Not authentically British. The single record [Isle of Man (1976)] was misidentified *fide* Ryvarden (pers. comm.).

　　crassa Britzelm., Clavaria, *Ber. Naturhist. Vereins Augsburg* 29: 286 (1887)
A *nomen dubium*. Listed by Rea (1922). Possibly a form of *Clavulina rugosa fide* Cotton & Wakefield (1919), but a form of *C. cinerea fide* Corner (1950).

crassifolium (Berk.) Singer, Lyophyllum, *Ann. Mycol.* 41: 99 (1943)
　　Agaricus crassifolius Berk., *Outl. Brit. fungol.*: 100 (1860)
　　Tricholoma crassifolium (Berk.) Sacc., *Syll. fung.* 5: 108 (1887)
　　Agaricus pachyphyllus Berk. [*nom. illegit.*, non *A. pachyphyllus* Fr. (1815)], in Berk., *Engl. fl.* 5(2): 16 (1836)
A *nomen dubium*. The type material is not a *Lyophyllum fide* Clémençon [Mycotaxon 15 (1982)]. A later collection in herb. K is *L. eustygium*.

crassum (Lév.) Hjortstam & Ryvarden, Porostereum, *Syn. Fungorum* 4: 29 (1990)
　　Thelephora crassa Lév., *Ann. Sci. Nat., Bot.*, sér. 3, 2: 209 (1844)
　　Hymenochaete crassa (Lév.) Berk., *Grevillea* 8: 148 (1880)
Listed by Rea (1922), but no British material of this pantropical species has been traced.

crassus Massee, Boletus, *Brit. fung.-fl.* 1: 286 (1892)
Described from Britain but a *nomen dubium* probably based on diseased basidiomes of *Boletus edulis*.

crassus Fr., Paxillus, *Epicr. syst. mycol.*: 318 (1838)
A *nomen dubium*. Possibly *Phylloporus rhodoxanthus* (= *P. pelletieri*) *fide* Pearson & Dennis (1948). Cooke [870 (877) Vol. 6 (1888)], collected from a 'mound of old rifle butts' at Blackheath in West Kent, is unidentifiable but not *P. pelletieri*.

cremor Fr., Lactarius, *Epicr. syst. mycol.*: 343 (1838)
A *nomen dubium* variously interpreted. Sensu Rea (1922) is doubtful. Sensu Basso (FungEur7) is *Lactarius rostratus*.

cremor var. **pauper** P. Karst., Lactarius, *Symb. mycol. fenn.* 10: 58 (1869)
A *nomen dubium*. See Cooke 951 (1008) Vol. 7 (1888) for an illustration of a collection (no longer extant) named thus from England (Cumberland: Carlisle) in 1887.

crenata (Lasch) Gillet, Psathyrella, *Hyménomycètes*: 618 (1878)
　　Agaricus crenatus Lasch, *Linnaea* 3: 465 (1828)
　　Coprinarius crenatus (Lasch) P. Kumm., *Führ. Pilzk.*: 68 (1871)
A *nomen dubium* lacking type material. Sensu Withering (1796) is unknown. Sensu Rea (1922) is *Coprinus hiascens*.

crispa (Longyear) Singer, Conocybe, *Lilloa* 22: 485 (1951) [1949]
　　Galera crispa Longyear, *Bot. Gaz.* 28: 272 (1899)
Not authentically European. The collection referred to in BFF3: 80 has recently been redetermined as a large and crisp-gilled form of *Conocybe rickenii fide* Watling (pers. comm.).

crispula (Fr.) Quél., Ramaria, *Fl. mycol. France*: 464 (1888)
　　Clavaria crispula Fr., *Syst. mycol.* 1: 470 (1821)
A *nomen dubium*. Sensu Rea (1922), as *Clavaria crispula*, and sensu auct. mult. is *R. decurrens*.

cristatus (Fr.) Kotl. & Pouzar, Albatrellus, *Česká Mykol.* 11: 154 (1957)
　　Polyporus cristatus Fr., *Syst. mycol.* 1: 356 (1821)
　　Grifola cristata (Fr.) Gray, *Nat. arr. Brit. pl.* 1: 643 (1821)
　　Boletus cristatus Schaeff. [*nom. illegit.*, non *B. cristatus* Gouan (1765)], *Fungi Bavar. Palat. Nasc.* 4: 93 (1774)
Not authentically British. Listed by Rea (1922), as *Polyporus cristatus*, and reported on several occasions, the last from Lincolnshire in 1938, but all records are unsubstantiated with voucher material.

croceoconus Fr., Cortinarius, *Monogr. hymenomyc. Suec.* 2: 67 (1863)
Not authentically British. Accepted in NCL, but now considered a boreo-nemoral species with native *Picea* spp in swampy woodland in continental Europe. Reported on a few occasions since 1900 but in deciduous woodland (often said to be with *Quercus* spp) and unsubstantiated with voucher material.

croceofulvus (DC.) Fr., Cortinarius, *Epicr. syst. mycol.*: 296 (1838)
　　Agaricus croceofulvus DC., *Fl. Fr.* 5: 49 (1815)
A *nomen dubium*. Cooke 1191 (1193) Vol. 8 (1891) is suggestive of *C. limonius*.

croceus Bolton, Agaricus, *Hist. fung. Halifax* 2: 51 (1788)
Described from Britain but a *nomen dubium* lacking type material. Bolton's plate and description strongly suggest a species of *Cystoderma*, possibly *C. amianthinum* and Sowerby's plate [Col. fig. Engl. fung. 1: pl. 19 (1796)] appears to be the same.

croceus (Pers.) Bondartsev & Singer, Hapalopilus, *Ann. Mycol.* 39: 52 (1921)
　　Boletus croceus Pers., *Observ. mycol.* 1: 87 (1796)
　　Polyporus croceus (Pers.) Fr., *Syst. mycol.* 1: 364 (1821)
Not authentically British. Reported by Bramley [Fungus Flora of Yorkshire (1985)] as a 'guess' for *Boletus heteroclitus* Bolton which is a *nomen dubium*.

crocistipidosum H. Engel & Dermek, Leccinum, *Z. Mykol.* 47(2): 211 (1981)

Not authentically British. Reported once from England (South Hampshire: New Forest) but doubtfully distinct from *Leccinum scabrum* and unsubstantiated with voucher material.

cruciatum Rostk., Lycoperdon, *Deutschl. Fl.,* Edn 3: 19 (1838)
A *nomen dubium* lacking type material. From the description this is probably *Lycoperdon mammiforme*. Sensu Rea [TBMS 2 (1): 39 (1903)] is certainly *L. mammiforme*.

cruentata (Cooke & W.G. Sm.) Sacc., Pholiota, *Syll. fung.* 5: 755 (1887)
   Agaricus cruentatus Cooke & W.G. Sm. [*nom. illegit.*, non *A. cruentatus* With. (1801)], *Grevillea* 13: 58 (1885)
Described from Britain but a *nomen dubium*. Cooke 399 (502) Vol. 3 (1885) may be *Cortinarius rubicundulus fide* Pearson & Dennis (1948).

cruentatus With., Agaricus [non *A. cruentatus* Cooke & W.G. Smith (1885)], *Syst. arr. Brit. pl. ed. 4*, 4: 286 (1801)
Described from Britain but a *nomen dubium*.

cruentus Vent., Boletus, *I micieti del agro Bresciano*: t. 43 f. 3 (1855)
A *nomen dubium*. Doubtful *fide* NCL, but *fide* Pearson and Dennis (1948) this is *B. impolitus*.

crustacea Schum., Thelephora, *Enum. Pl. Saell.* 2: 399 (1803)
   Soppittiella crustacea (Schum.) Massee, *Brit. fung.-fl.* 1: 108 (1892)
   Hypochnus crustaceus (Schum.) P. Karst., *Bidrag Kännedom Finlands Natur Folk* 37: 163 (1882)
Accepted by Corner (1968) for a species near *T. spiculosa*, but considered a *nomen dubium* by Wakefield (1969). Listed by Rea (1922) as *Hypochnus crustaceus*.

cryptarum With., Agaricus [non *A. cryptarum* Letell (1829)], *Arr. Brit. pl. ed. 3*, 4: 265 (1796)
Described from Britain but a *nomen dubium*. Said to grow in clusters and was collected 'under the horizontal wooden door of a wine cellar in Edgbaston' (Warwickshire).

cryptarum (Letell.) C.B. Niel, Clitocybe, *Bull. Soc. Amis Sci. Nat. Rouen*: t. 1 (1896)
   Agaricus cryptarum Letell. [*nom. illegit.*, non *A. cryptarum* With. (1796)], *Fig. Champ.*: t. 611 (1829)
A *nomen dubium*. Listed by Rea (1922).

crystallinus Fr., Cortinarius, *Epicr. syst. mycol.*: 270 (1838)
A *nomen dubium*. Sensu NCL is *C. barbatus*. Sensu Cooke (1887) and sensu Rea (1922) is *C. emollitus*.

culmicola Fuckel, Cyphella, *Jahrb. Nassauischen Vereins Naturk.* 23-24: 25 (1870)
A *nomen dubium*. British material needs reappraisal.

cumulatum K.A. Harrison, Hydnellum, *Canad. J. Bot.* 42: 1225 (1964)
Not authentically British. Reported from South Hampshire (New Forest, Rhinefield) but material in herb. E is labelled as 'compare with' *H. cumulatum*.

cumulatus With., Agaricus, *Bot. arr. Brit. pl. ed. 2*, 3: 292 (1792)
Described from Britain, but a *nomen dubium* lacking type material. *Fide* Termorshuizen & Arnolds [*Mycotaxon* 30: 106 (1987)] represents a species of *Armillaria*.

cuniculorum Arnolds & Noordel., Entoloma, *Persoonia* 10: 289 (1979)
Not authentically British. Reported from Surrey (Esher Common) but the single collection so named has been redetermined as *E. papillatum* by Noordeloos. Reported also from Scotland but no material traced in herb. E.

cupularis (Bull.) Gillet, Tubaria, *Hyménomycètes*: 538 (1876)
   Agaricus cupularis Bull., *Herb. France*: pl. 554 (1792)
A *nomen dubium* lacking type material. Cooke 526 (602) Vol. 4 (1885) is not helpful.

curta (Fr.) Corner, Ramariopsis, *Monograph of Clavaria and Allied Genera*: 639 (1950)
   Clavaria curta Fr., *Monogr. hymenomyc. Suec.* 1: 281 (1857)
A *nomen dubium*. Sensu Corner (1950) is near to *Ramariopsis crocea* but British records may be *Ramaria decurrens*.

curtipes (Fr.) Bon, Melanoleuca, *Doc. mycol.* 3: 38 (1972)
   Agaricus curtipes Fr., *Syst. mycol.* 1: 88 (1821)
   Clitocybe curtipes (Fr.) Gillet, *Hyménomycètes*: 154 (1874)
Not authentically British. The single collection named thus (in herb. K) from Surrey (Cobham) is actually labelled 'cf *M. curtipes*' and no other records of this species are known from Britain. Equated in BFF8 with *Melanoleuca excissa* sensu Lange (FlDan1: 66) but for Bon (1991) is a quite different species lacking cystidia

curtus Britzelm., Lactarius, *Ber. Naturhist. Vereins Augsburg* 28: 137 (1885)
A *nomen dubium*. Sensu auct. mult. is *Lactarius hysginus*.

cuspidatus Bolton, Agaricus, *Hist. fung. Halifax* 2: 66 (1788)
Described from Britain but a *nomen dubium* lacking type material. Bolton's illustration resembles a species of *Conocybe*.

cuticulosa (Dicks.) Berk., Cyphella, *Outl. Brit. fungol.*: 278 (1860)
   Peziza cuticulosa Dicks., *Fasc. pl. crypt. brit.* 3: 22 (1793)
Described from rotten grass in Britain, but a *nomen dubium* lacking type material. See also report in TBMS 1 (2): 55 (1898) 'on dead *Triticum* near King's Lynn', possibly referable to *Cellypha goldbachii fide* Donk [*Persoonia* 1: 85 (1959)].

cyanea (Bolton) Tuom., Stropharia, *Karstenia* 2: 31 (1953)
   Agaricus cyaneus Bolton, *Hist. fung. Halifax* Suppl.: 143 (1791)
A *nomen dubium* lacking type material. Undoubtedly a species of *Stropharia* but the description, and Bolton's illustration, could fit either *S. aeruginosa* or *S. caerulea* equally well.

cyanescens (Berk. & Broome) Sacc., Panaeolus, *Syll. fung.* 5: 1123 (1887)
   Agaricus cyanescens Berk. & Broome, *J. Linn. Soc., Bot.* 11: 557 (1871)
   Copelandia cyanescens (Berk. & Broome) Singer, *Lilloa* 22: 473 (1951) [1949]
Not British. A common hallucinogenic species from the tropics mentioned in BFF5 (as *Copelandia cyanescens*) as likely to occur at some future time in greenhouses or seized consignments of drugs.

cyanophaea (Fr.) Sacc., Clitocybe, *Syll. fung.* 5: 146 (1887)
   Agaricus cyanophaeus Fr., *Hymenomyc. eur.*: 82 (1874)
A *nomen dubium*. The description in Saccardo suggests *Lepista nuda*. Listed by Rea (1922).

cyanulum (Lasch) Noordel., Entoloma, *Persoonia* 12(3): 203 (1984)
   Agaricus cyanulus Lasch, *Linnaea* 4: 540 (1829)
   Leptonia cyanula (Lasch) Sacc., *Syll. fung.* 5: 708 (1887)
   Leptonia lampropus var. *cyanulus* (Lasch) Rea, *Brit. basidiomyc.*: 344 (1922)
Not authentically British. Reported from England and Scotland, but unsubstantiated with voucher material and the single collection named thus in herb. K is annotated 'compare with' *E. cyanulum*.

cyatheae W.G. Sm., Radulum, *J. Bot.* 11: 67 (1873)
A *nomen dubium*. Described from material collected on dead tree fern stems in London (Chelsea: Veitch's Nursery) but never recollected.

cyathiformis Bull., Cellularia, *Herb. France*: pl. 414 (1789)
   Peziza cellularia Sowerby, *Col. fig. Engl. fung.* 1: pl. 91 (1797)
A *nomen dubium* lacking type material and variously interpreted. Bulliard's plate appears to represent a group of young, infertile, and unidentifiable fruitbodies of a species of *Polyporus*. Sensu Sowerby [Col. fig. Engl. fung. 1: pl. 91

(1797)] (as *Peziza cellularia*, but citing Bulliard's species in synonymy) could be interpreted as young, orbicular basidiomes of *Phlebia tremellosa*.

cyathoides Bolton, Agaricus, *Hist. fung. Halifax* Suppl.: 145 (1791)
Described from Britain but a *nomen dubium* lacking type material. Bolton's plate cites Bulliard as the author of this epithet but there is no mention of this in *Herb. France*, and the illustration is ambiguous, resembling *Clitocybe clavipes* or possibly *Pseudoclitocybe cyathiformis*.

cyathula Fr., Lactarius, *Epicr. syst. mycol.*: 344 (1838)
A *nomen dubium*. The name has been used for any of the small brownish *Lactarii* associated with *Alnus glutinosa*.

cylindraceocampanulata (Henn.) Höhn., Omphalia, *Sitzungsber. Kaiserl. Akad. Wiss., Wien, Math.-Naturwiss. Cl., Abt. 1* 118: 289 (1909)
Marasmius cylindraceocampanulatus Henn., *Fung. Monsun.* 1: 53 (1899)
A *nomen dubium*. Described from material gathered in Java, and recorded from Hertfordshire (Whippendell Wood) by Hora [TBMS 39 (2): 282 (1955)]. No material survives but this seems likely to have been *Mycena picta*.

cyphelloides (P.D. Orton) Maas Geest., Hemimycena, *Proc. K. Ned. Akad. Wet.* 84 (4): 437 (1981)
Mycena cyphelloides P.D. Orton, *Trans. Brit. Mycol. Soc.* 43(2): 178 (1960)
Helotium gibbum Alb. & Schwein., *Consp. fung. lusat.*: 19 (1805)
Omphalia gibba (Alb. & Schwein.) Pat. [*nom. illegit.*, non *O. gibba* Gray (1821)], *Tab. anal. fung.*: no. 560 (1888)
Not authentically British. Introduced as a *nom. nov.* for *Omphalia gibba* Alb. & Schwein., mistakenly reported as British by Rea (1927). Also described in BFF8 (p. 121) but no authentic collections are known from Britain.

daulnoyae Quél., Clavaria, *Compt. Rend. Assoc. Franç. Avancem. Sci.* 18: 470 (1892) [1891]
A *nomen dubium* with a single British record from Warwickshire in 1978, doubtfully assigned here.

daviesii Sowerby, Hydnum, *Col. fig. Engl. fung.* 1: pl. 15 (1796)
Described from Britain but a *nomen dubium*. Herbarium material at Kew (probably the type) is in poor condition, and unidentifiable but both it and the Sowerby illustration suggest *Steccherinum bourdotii*.

dealbata (Sowerby) P. Kumm., Clitocybe, *Führ. Pilzk.*: 121 (1871)
Agaricus dealbatus Sowerby, *Col. fig. Engl. fung.* 2: pl. 123 (1799)
Described from Britain but a *nomen dubium* lacking type material. Sensu Sowerby appears to represent *Hemimycena lactea*. Sensu auct. mult. is *C. rivulosa*.

debilis (Fr.) Quél., Mycena, *Mém. Soc. Émul. Montbéliard, Sér. 2*, 5: 107 (1872)
Agaricus debilis Fr., *Epicr. syst. mycol.*: 112 (1838)
A *nomen dubium*. Cooke 240 (189) Vol. 2 (1882) is unidentifiable. Sensu auct. is *M. smithiana*.

decipiens Bres., Inocybe, *Fungi trident.* 2 (8-10): 13 (1892)
Not authentically British. Sensu Pearson (1954) is *Inocybe dunensis*.

decipiens (Singer) Pilát & Dermek, Leccinum, *Hribovite Huby*: 150 (1974)
Leccinum aurantiacum var. *decipiens* Singer, *Pilze Mitteleuropas*: 104 (1966)
Not authentically British. A single record from South Hampshire (New Forest), dubiously identified and unsubstantiated with voucher material.

decipientoides Peck, Inocybe, *Bull. Torrey Bot. Club* 34: 100 (1907)

Astrosporina decipientoides (Peck) A. Pearson, *Trans. Brit. Mycol. Soc.* 22: 28 (1938)
Not authentically British. A North American species. Sensu auct. Eur. is *Inocybe curvipes*.

decumbens (Pers.) Fr., Cortinarius, *Epicr. syst. mycol.*: 284 (1838)
Agaricus decumbens Pers., *Syn. meth. fung.*: 286 (1801)
A *nomen dubium* lacking type material. Cooke 765 (816A) Vol. 6 (1886) appears to depict young basidiomes of *Pholiota gummosa*.

defectus A.H. Sm., Rhizopogon, *Mem. N. Y. bot. Gdn* 14(2): 43 (1966)
Not authentically British. A North American species. Reported from Scotland (Morayshire) in 1975 [Bull. Brit. Mycol. Soc. 10 (2): 63 (1976)] but the specimens (in herb. E) are determined as *Rhizopogon* aff. *defectus*.

defossum Batsch, Lycoperdon, *Elench. fung. (Continuatio Secunda)*: 125 & t. t42 f. 229 (1789)
A *nomen dubium*, possibly *Scleroderma bovista* fide Saccardo [Syll. fung. 7: 135 (1888)]. Sensu auct. Brit. is *Scleroderma verrucosum*.

degener Fr., Xerotus, *Epicr. syst. mycol.*: 400 (1838)
A *nomen dubium* variously interpreted. Cooke 1098 (1150) Vol. 7 (1890) appears to be a species of *Omphalina*. This is not *Lentinus degener* Kalchbr. in Fries [Hym. Eur. p. 482 (1874)] which is a synonym of *Lentinus cyathiformis* [= *Neolentinus schaefferi*].

deglubens (Fr.) Gillet, Inocybe, *Hyménomycètes*: 516 (1876)
Agaricus deglubens Fr., *Epicr. syst. mycol.*: 173 (1838)
A *nomen dubium*. Cooke 420 (394) Vol. 3 (1886) is doubtful. Sensu Pearson (1954) is also doubtful. Sensu auct. is possibly *Inocybe flocculosa*.

delibutus *var.* **elegans** Massee, Cortinarius, *Brit. fung.-fl.* 2: 90 (1893)
Described from Britain but a *nomen dubium*. Possibly just a gracile form of *C. delibutus*.

delicatulus With., Agaricus, *Arr. Brit. pl. ed. 3*, 4: 169 (1796)
Described from Britain but a *nomen dubium*.

deliciosus Bolton, Agaricus [*nom. illegit.*, non *A. deliciosus* L. (1753)], *Hist. fung. Halifax* 1: 9 (1788)
Described from Britain but a *nomen dubium* lacking type material. A species of *Lactarius* but from Bolton's illustration apparently not *L. deliciosus* (L.) Gray.

deliquescens (Bull.) Fr., Coprinus, *Epicr. syst. mycol.*: 249 (1838)
Agaricus deliquescens Bull., *Herb. France*: pl. 558 (1786)
A *nomen dubium* lacking type material. See Cooke 665 (678) Vol. 5 (1886), which seems to depict a small form of *Coprinus atramentarius*.

deliquescens (Bull.) Duby, Dacrymyces, *Bot. gall.*: 729 (1830)
Tremella deliquescens Bull., *Herb. France*: pl. 455 (1790)
A *nomen dubium* lacking type material and variously interpreted. British collections in herb. K are *D. stillatus*.

dentatus With., Agaricus, *Bot. arr. veg.* 2: 756 (1776)
Described from Britain 'on the bottom of gate posts on Hampstead Heath' but a *nomen dubium*.

depallens (Pers.) Fr., Russula, *Epicr. syst. mycol.*: 353 (1838)
Agaricus depallens Pers., *Syn. meth. fung.*: 444 (1801)
Russula atropurpurea var. *depallens* (Pers.) Maire, Rea, *Brit. basidiomyc.*: 469 (1922)
A *nomen dubium* lacking type material and variously interpreted. Sensu Cooke (1888) is *R. atropurpurea*. Sensu auct. is *R. pulchella*.

depexus (Fr.) Fr., Cortinarius, *Epicr. syst. mycol.*: 291 (1838)
Agaricus depexus Fr., *Observ. mycol.* 2: 53 (1818)
A *nomen dubium*. Noted in NCL as 'not British'.

depressus With., Agaricus, *Bot. arr. Brit. pl. ed. 2*, 3: 306 (1792)
Described from Britain but a *nomen dubium*. From the description, a species of *Lactarius*.

descissa (Fr.) Quél., Inocybe, *Mém. Soc. Émul. Montbéliard, Sér. 2*, 5: 154 (1872)
  Agaricus descissus Fr., *Epicr. syst. mycol.*: 174 (1838)
A *nomen superfluum fide* Kuyper (1986: 224). Sensu NCL is *Inocybe phaeodisca*.

destricta (Fr.) Quél., Inocybe, *Mém. Soc. Émul. Montbéliard, Sér. 2*, 5: 181 (1872)
  Agaricus rimosus var. *destrictus* Fr., *Epicr. syst. mycol.*: 174 (1838)
A *nomen dubium*. Sensu Rea (1922) is possibly *Inocybe pseudodestricta*.

destructor (Schrad.) Bondartsev & Singer, Tyromyces, *Ann. Mycol.* 39: 52 (1941)
  Boletus destructor Schrad., *Spicil. Fl. Germ.*: 166 (1794)
  Polyporus destructor (Schrad.) Fr., *Syst. mycol.* 1: 359 (1821)
  Polyporus destructor var. *undulatus* Sacc., *Syll. fung.* 6: 115 (1888)
A *nomen dubium*, discussed by Donk (1974: 274-276). Sensu auct. is *Antrodia gossypium*.

destruens Gray, Poria, *Nat. arr. Brit. pl.* 1: 639 (1821)
Described from Britain but a *nomen novum* for *Polyporus destructor* (above) *fide* Donk (1974).

detonsus (Fr.) Fr., Cortinarius, *Epicr. syst. mycol.*: 313 (1838)
  Agaricus detonsus Fr., *Syst. mycol.* 1: 232 (1821)
Not authentically British. A single record by Stevenson listed in Rea (1922) but unlikely to be in the current sense. The material (in herb. K) is in poor condition and unidentifiable.

detrusa (Fr.) Gillet, Omphalia, *Hyménomycètes*: 290 (1876)
  Agaricus detrusus Fr., *Monogr. hymenomyc. Suec.* 2: 291 (1863)
Excluded 'pending clearer definition' *fide* NCL. According to Pearson & Dennis (1948) this name was incorrectly placed on the British list.

diabolicus (Fr.) Fr., Cortinarius, *Epicr. syst. mycol.*: 285 (1838)
  Agaricus anomalus δ *diabolicus* Fr., *Syst. mycol.* 1: 221 (1821)
A *nomen dubium* variously interpreted. Sensu auct. mult. is *C. suillus* whilst for others (e.g. Orton 1983) it is a member of the *C. anomalus* complex.

diaphana (Schrad.) Gray, Odontia, *Nat. arr. Brit. pl.* 1: 651 (1821)
  Hydnum diaphanum Schrad., *Spicil. Fl. Germ.* 1: 78 (1794)
A *nomen dubium*. Sensu Gray is unknown.

dichotoma With., Tremella, *Bot. arr. veg.* 2: 733 (1776)
Not a basidiomycete. This is the lichen *Collema dichotomum* (With.) Coppins & Laundon.

dichroa (P. Karst.) Sacc., Psilocybe, *Syll. fung.* 5: 1045 (1887)
  Psilocybe ericaea var. *dichroa* P. Karst., *Bidrag Kännedom Finlands Natur Folk* 32: 504 (1879)
A *nomen dubium*. Sensu Lange (1938) and thus sensu auct. Brit. is *Hypholoma subericaeum*.

dichroum (Pers.) Bourdot & Galzin, Mycoleptodon, *Bull. Soc. Mycol. France* 30: 276 (1914)
  Hydnum dichroum Pers., *Mycol. eur.* 2: 213 (1825)
A *nomen dubium* lacking type material. Sensu Bourdot & Galzin is *Steccherinum bourdotii*.

dichrous Kühner & Romagn. [*nom. nud.*], Hygrophorus, *Flore Analytique des Champignons Supérieurs*: 60 (1953)
Sensu NCL is *H. persoonii*.

dicrani (A.E. Jansen) Kits van Wav., Psathyrella, *Persoonia* 2: 281 (1985)

  Psathyrella fulvescens var. *dicrani* A.E. Jansen, *Vegetation and macrofungi of acid oakwoods in the North-East Netherlands.*: 120 (1981)
Not authentically British. Reported from Shropshire in 1998 but unsubstantiated with voucher material.

dictyorrhizus (DC.) Kühner, Pleurotellus, *Botaniste* 17: 114 (1926)
  Agaricus dictyorrhizus DC., *Fl. Fr.* 5: 594 (1815)
  Pleurotus dictyorrhizus (DC.) Gillet, *Hyménomycètes*: 338 (1876)
A *nomen dubium*. Three collections named thus in herb. K need re-evaluating.

diffractum (Fr.) Gillet, Hebeloma, *Hyménomycètes*: 526 (1876)
  Agaricus diffractus Fr., *Epicr. syst. mycol.*: 182 (1838)
Doubtfully distinct from *Hebeloma crustuliniforme fide* NCL. See reports in TBMS 5 (3): 436 (1915) and in Naturalist (Hull) 1917: 131 (1917).

digitalis (Batsch) Fr., Coprinus, *Epicr. syst. mycol.*: 249 (1838)
  Agaricus digitalis Batsch, *Elench. fung.*: 62 t. 1 f. 1a (1783)
A *nomen dubium*. The Batsch plate resembles a single basidiome of *Coprinus micaceus* or a closely related species. Included by Rea (1922).

dilectus Fr., Coprinus, *Epicr. syst. mycol.*: 250 (1838)
Not authentically British. A single British record, from Perthshire (Innerhadden) in herb. E, is determined with a query. *Coprinus dilectus* sensu Lange (Fl.Dan4: 157A) is *C. erythrocephalus*.

dilutus (Pers.) Fr., Cortinarius, *Epicr. syst. mycol.*: 305 (1838)
  Agaricus dilutus Pers., *Syn. meth. fung.*: 300 (1801)
A *nomen dubium* lacking type material. Sensu Rea (1922) and Cooke (832 (810) Vol. 6) possibly equates to *C. illuminus*.

dimorphocystis A.H. Sm. & Singer, Galerina, *Mycologia* 47: 558 (1955)
Not authentically British. A North American species. Sensu auct. Eur. is *Galerina stordalii*.

dipsacoides With., Agaricus, *Bot. arr. veg.* 2: 756 (1776)
Described from Britain but a *nomen dubium*.

directa (Berk. & Broome) Sacc., Omphalia, *Syll. fung.* 5: 331 (1887)
  Agaricus directus Berk. & Broome, *Ann. Mag. Nat. Hist., Ser. 5 [Notices of British Fungi no. 1931]* 9: 177 (1882)
Described from Britain but a *nomen dubium*. See Cooke 266 (251) Vol. 2 (1883) for an illustration which suggests a species of *Mycena* or *Hemimycena*.

disciformis (Vill.) Pat., Aleurodiscus, *Bull. Soc. Mycol. France* 10: 80 (1894)
  Helvella disciformis Vill., *Prosp. Hist. des plantes de Dauph.* 3(2): 1046 (1789)
  Stereum disciforme (Vill.) Fr., *Epicr. syst. mycol.*: 551 (1838)
  Peniophora disciformis (Vill.) Cooke, *Grevillea* 8: 20 (1879)
  Hymenochaete disciformis (Vill.) W.G. Sm., *Syn. Brit. Bas.*: 409 (1908)
Not authentically British. Reported by Massee [Grevillea 19: 66 (1891)] from 'oak trunks' but without a locality and lacking voucher material. Also 'known' to W.G. Smith on the same host in 1908, but not reported since that time.

discoideus Dicks., Boletus, *Fasc. pl. crypt. brit.* 3: 21 (1793)
A *nomen dubium*. This is Dickson's concept of *B. suaveolens* Bull. [non *B. suaveolens* L.], also a *nomen dubium*. Perhaps *Daedaleopsis confragosa* but interpreted as *Trametes suaveolens* by Donk (1974).

discoideus (Pers.) Fr., Hygrophorus, *Epicr. syst. mycol.*: 323 (1838)
  Agaricus discoideus Pers., *Syn. meth. fung.*: 365 (1801)
Listed in NCL but not authentically British. Sensu Rea (1922) is doubtful *fide* NCL or possibly *H. unicolor fide* Arnolds in FAN2. British records are old and usually lack voucher material. Material named thus in herb. K is misidentified *H. discoxanthus*.

discopus (Lév.) Quél., Mycena, *Enchir. fung.*: 40 (1886)
  Agaricus discopus Lév., *Ann. Sci. Nat., Bot.*, sér. 2, 16: 237
  (1841)
A *nomen dubium*. Cooke [249 (192) Vol. 2 (1892)] possibly
represents *Mycena adscendens*.

dispar Batsch, Agaricus, *Elench. fung. (Continuatio Secunda)*:
  75 t. 38 f. 210 (1789)
A *nomen dubium*. Reported by Withering (1796). Batsch's
illustration shows an unidentifiable mycenoid species.

dissiliens (Fr.) Quél., Mycena, *Icon. Mycol.* 5: pl. 241 (1929)
  Agaricus dissiliens Fr., *Epicr. syst. mycol.*: 108 (1838)
A *nomen dubium*. Sensu Cooke [230 (285) Vol. 2 (1882)] and
Rea (1922) is unknown.

dissimulabilis (Britzelm.) Sacc., Mycena, *Syll. fung.* 11: 22
  (1895)
  Agaricus dissimulabilis Britzelm., *Bot. Centralbl.* 15-17: 6
  (1893)
A *nomen dubium* lacking type material and inadequately
described. Sensu auct. and British records is *M. megaspora*.

distans Berk., Hygrophorus, *Outl. Brit. fungol.*: 200 t. 13 (1860)
A *nomen dubium*. The type, and only record, in herb. K requires
re-examination but is possibly *Hygrocybe virginea*. See also
Cooke 899 (902) Vol. 6 (1888) for an illustration.

distinguendum S. Lundell, Tricholoma, *Fungi Exsiccati Suecici*:
  1101 (1942)
Not well known and lacking British material. Listed in NCL where
considered the correct name for *Tricholoma cartilagineum*
sensu Rea (1922).

divaricata (Peck) Corner., Ramaria, *Monograph of Clavaria and
  Allied Genera*: 574 (1950)
  Clavaria divaricata Peck, *Bull. New York State Mus. Nat. Hist.*
  2: 11 (1887)
Excluded pending clearer definition. There is a single specimen
labelled thus, in herb. K, but Corner (1950) regarded this as
'hardly differing' from *Ramaria subbotrytis* var. *intermedia*.

dolabratus Fr., Cortinarius, *Epicr. syst. mycol.*: 311 (1838)
A *nomen dubium*. Cooke 845 (811) Vol. 6 (1887) is doubtful.

dryadophilus Kühner, Lactarius, *Bull. Soc. Mycol. France* 91:
  68 (1975)
Not authentically British. Reported from Sutherland (Ben More
Assynt) but this material has been redetermined as *Lactarius
salicis-reticulatae*.

duplicata (Bosc) E. Fisch., Dictyophora, in Saccardo, *Syll. fung.*
  7: 6 (1888)
  Phallus duplicatus Bosc, *Mag. Neuesten Entdeck. Gesammten
  Naturk. Ges. Naturf. Freunde Berlin* 5: 86 (1811)
Not authentically British. Reported on several occasions, but all
of the records and material refer to *Phallus impudicus* var.
togatus.

duportii W. Phillips, Russula, *Grevillea* 13: 49 (1884)
  Russula vesca var. *duportii* (W. Phillips) Massee, *Brit. fung.-fl.*
  3: 61 (1893)
Described from Britain but a *nomen dubium* lacking type
material. See Cooke 1003 (1042) Vol. 7 (1888) for an
illustration. Described as 'smelling of crab' thus probably a
member of the *R. xerampelina* complex.

dura (Bolton) Singer, Agrocybe, *Beih. Bot. Centralbl., Abt. B* 56:
  165 (1936)
  Agaricus durus Bolton, *Hist. fung. Halifax* 2: 67 (1789)
  Pholiota dura (Bolton) P. Kumm., *Führ. Pilzk.*: 84 (1871)
  Togaria dura (Bolton) W.G. Sm., *Syn. Brit. Bas.*: 123 (1908)
Described from Britain but a *nomen dubium* lacking type
material. British records and collections under this name are
often *A. molesta*.

duracinum (Cooke) Sacc., Tricholoma, *Syll. fung.* 5: 123
  (1887)
  Agaricus duracinus Cooke, *Grevillea* 12: 41 (1883)

A *nomen dubium* lacking type material. Possibly *Tricholoma
cinerascens* fide Pearson & Dennis (1948), i.e. *Lyophyllum
fumosum*. See Cooke 1126 (640) Vol. 8. (1889) for an
illustration of the type collection (no longer extant) from Kew
Gardens.

eburneus Bolton, Agaricus [*nom. illegit.*, non *A. eburneus* Bull.
  (1782)], *Hist. fung. Halifax* 1: 4 (1788)
Described from Britain but a *nomen dubium* lacking type
material. Bolton's illustration suggests a species of
Hemimycena.

eburneus Quél., Coprinus, *Compt. Rend. Assoc. Franç.
  Avancem. Sci.* 12: 501 (1884) [1883]
A *nomen dubium* lacking type material. Listed by Rea (1922) as
'in mountainous pastures' but no records have been traced.

eburneus (Velen.) Bon, Cortinarius, *Doc. mycol.* 15(60): 38
  (1985)
  Phlegmacium eburneum Velen., *České Houby* II: 422 (1920)
A *nomen dubium*. Reported from Wales and the Republic of
Ireland but unsubstantiated with voucher material. The name
has been applied to both *C. barbatus* and *C. emollitus* in
Europe.

echinatus Sowerby, Agaricus, *Col. fig. Engl. fung.* 1: pl. 99
  (1797)
Described from Britain but a *nomen dubium* lacking type
material. Possibly a synonym of *Hohenbuehelia mastrucata*
since Cooke's illustration of *Agaricus mastrucatus* [289 (243)
Vol. 2 (1883)] is almost identical to Sowerby's of *A. echinatus*.

echinipes (Lasch) P. Kumm., Mycena, *Führ. Pilzk.*: 108 (1871)
  Agaricus echinipes Lasch, *Linnaea* 3: 193 (1828)
A *nomen dubium* lacking type material. British records are likely
to be sensu Lange (FlDan2: 43 & pl. 53I) which is *Hemimycena
cucullata*.

echinospora (W.G. Sm.) Sacc., Naucoria, *Syll. fung.* 5: 859
  (1887)
  Agaricus echinosporus W.G. Sm., *J. Bot.* 11: 65 (1873)
Described from London (Chelsea: Bull's Orchid House) but a
nomen dubium. Presumably an alien.

edulis var. elephantinus (With.) Massee, Boletus, *Brit. fung.-
  fl.* 1: 284 (1892)
  Boletus elephantinus With., *Arr. Brit. pl. ed. 3*, 4: 317 (1796)
  Leccinum elephantinum (With.) Gray, *Nat. arr. Brit. pl.* 1: 648
  (1821)
Described from Britain but a *nomen dubium* lacking type
material. Probably just very large fruitbodies of *Boletus edulis*.

effibulata (Ginns & Sunhede) Ginns, Syzygospora, *Mycologia*
  78(4): 626 (1986)
  Christiansenia effibulata Ginns & Sunhede, *Bot. Not.* 131: 168
  (1978)
Not authentically British. Reported (as *Christiansenia* cf.
effibulata) on *Hygrocybe chlorophana* from Scotland (St Kilda)
in Dennis (1986), but *Hygrocybe* is not a previously recorded
host genus for this species, and material in herb. K shows no
evidence of the parasite.

effugiens (Bourdot & Galzin) Oberw., Tubulicrinis, *Z. Pilzk.*
  31(1-2): 35 (1966)
  Peniophora glebulosa ssp. *effugiens* Bourdot & Galzin, *Bull.
  Soc. Mycol. France* 28(4): 386 (1913)
  Peniophora effugiens (Bourdot & Galzin) Bourdot & Galzin,
  Hyménomyc. France: 291 (1928)
Not authentically British. Reported from Mid-Perthshire
(Rannoch) in 1983 on the Nordic Mycological Congress, but
unsubstantiated with voucher material.

egenula J. Favre, Inocybe, *Ergebn. Wiss. Untersuch. Schweiz.
  Natn. Parks* 5: 202 (1955)
Not authentically British. Collections in herb. E, from montane
habitat in Scotland, are all labelled as 'compare with' *I.
egenula*.

elaeodes (Fr.) Gillet, Hypholoma, *Hymenomycètes*: 573 (1878)

Agaricus elaeodes Fr., *Epicr. syst. mycol.*: 222 (1838)
A *nomen dubium*. Possibly *Hypholoma fasciculare* fide NCL.

elasticus Bolton, Agaricus [non *A. elasticus* With. (1792)], *Hist. fung. Halifax* 1: 16 (1788)
Described from Britain but a *nomen dubium* lacking type material. Bolton's plate and description strongly suggest a species of *Armillaria*.

elasticus With., Agaricus [non *A. elasticus* Bolton (1788)], *Bot. arr. Brit. pl. ed. 2*, 3: 313 (1792)
Described from Britain but a *nomen dubium*. However, from Withering's description and comments, probably *Collybia fusipes*.

elatum (Batsch) Gillet, Hebeloma, *Hyménomycètes*: 527 (1876)
Agaricus elatus Batsch [non *A. elatus* Pers. (1801)], *Elench. fung.*: 11 & t. 32 f. 108 (1783)
A *nomen dubium*. Cooke 1165 (962) Vol. 8 (1890) is possibly *H. leucosarx*.

elegans Sowerby, Auricularia, *Col. fig. Engl. fung., Suppl.*: pl. 412 (top) (1809)
Described from Britain but a *nomen dubium* lacking type material. Sowerby's illustration suggests old basidiomes of *Chondrostereum purpureum* but with what appear to be lamellae.

elegans Bolton, Boletus, *Hist. fung. Halifax* 2: 76 (1788)
Described from Britain but a *nomen dubium* lacking type material. Bolton's illustration appears to depict *Grifola frondosa*.

elegans (Pers.) P. Kumm., Mycena, *Führ. Pilzk.*: 109 (1871)
Agaricus elegans Pers., *Syn. meth. fung.*: 391 (1801)
A *nomen dubium* lacking type material and variously interpreted. British records are sensu Rea (1922) and Pearson (1955) and thus *M. aurantiomarginata*.

elegans Bres., Russula, *Fungi trident.* 1: 21 (1881)
A *nomen dubium*. Listed by Rea (1922). Cooke 1018 (1027) Vol. 7 (1888) is *R. maculata*.

elixa (Sowerby) Sacc., Clitocybe, *Syll. fung.* 5: 161 (1887)
Agaricus elixus Sowerby, *Col. fig. Engl. fung.* 2: pl. 172 (1799)
Omphalia elixa (Sowerby) Gray, *Nat. arr. Brit. pl.* 1: 614 (1821)
Described from Britain but a *nomen dubium* lacking type material.

emplastrum (Cooke & Massee) Sacc., Lepiota, *Syll. fung.* 9: 8 (1891)
Agaricus emplastrum Cooke & Massee, *Grevillea* 18: 51 (1889)
A *nomen dubium* lacking type material and represented only by illustrations by Wakefield in herb. K and by Cooke [1106 (1164) Vol. 8 (1889)]. Described from a churchyard in Middlesex (Ealing), it resembles *M. rhacodes* but was said to differ by the much larger spores. Either a good species in need of recollection and a modern description or is, more likely, *Macrolepiota rhacodes* var. *bohemica*.

empyreumatica (Berk. & Broome) Sacc., Psathyrella, *Syll. fung.* 5: 1131 (1887)
Agaricus empyreumaticus Berk. & Broome, *Ann. Mag. Nat. Hist., Ser. 4 [Notices of British Fungi no. 1262]* 6: 469 (1870)
Described from Britain but a *nomen dubium* and inadequately described. Possibly a good species requiring rediscovery and redescription. Cooke (641) 657 Vol. 5 (1886) looks distinctive. For notes on the type collection see Orton in NCL.

epileucus (Fr.) Bondartsev, Coriolus, *Trutovye griby Evropeiskoi chasti SSSR i Kavkaza*: 499 (1953)
Polyporus epileucus Fr., *Epicr. syst. mycol.*: 452 (1838)
A *nomen dubium*. Listed by Rea (1922), with old Yorkshire records cited by Bramley (1985).

epiphyllum With., Lycoperdon, *Bot. arr. Brit. pl. ed. 2*, 3: 468 (1792)
Described from Britain but a *nomen dubium* and probably not a basidiomycete. Withering's description suggests a myxomycete, possibly a species of *Trichia*.

epipoleus Fr., Cortinarius, *Epicr. syst. mycol.*: 277 (1838)
Cortinarius salor ssp. *transiens* Melot, *Doc. mycol.* 20(77): 96 (1989)
Not authentically British. Included in NCL, but all records are old, doubtful and unsubstantiated with voucher material. Possibly a form of *C. emunctus*.

epipterygia *var.* **pelliculosa** (Fr.) Maas Geest., Mycena, *Proc. K. Ned. Akad. Wet.* 83(1): 72 (1980)
Agaricus pelliculosus Fr., *Epicr. syst. mycol.*: 116 (1838)
Mycena pelliculosa (Fr.) Quél., *Mém. Soc. Émul. Montbéliard, Sér. 2*, 5: 343 (1873)
Doubtfully distinct from the type. Cooke [246 (191) Vol. 2 (1882), as *Agaricus pelliculosus*] is, however, some other, unidentifiable species. Sensu NCL is unclear.

ericeus Bull., Agaricus, *Herb. France* pl. 188 (1784)
A *nomen dubium*. Reported by Withering (1796) and possibly a white species of *Hygrocybe*.

erinaceellus (Peck) Watling, Flammulaster, *Notes Roy. Bot. Gard. Edinburgh* 28(1): 65 (1967)
Agaricus erinaceellus Peck, *Rep. (Annual) New York State Mus. Nat. Hist.* 31: 70 (1878)
Phaeomarasmius erinaceellus (Peck) Singer & Digilio, *Lilloa* 22: 577 (1951) [1949]
Flocculina erinaceella (Peck) P.D. Orton, *Trans. Brit. Mycol. Soc.* 43(2): 175 (1960)
Not authentically British. Orton (1984) reassigned British material to *F. denticulatus*, here considered a synonym of *F. muricatus* following Vellinga (1986).

eriophora Peck, Lepiota, *Bull. Torrey Bot. Club* 33: 95 (1903)
Lepiota echinella var. *eriophora* (Peck) J.E. Lange, *Fl. agaric. danic.* 1: 27 (1935)
Cystolepiota eriophora (Peck) Knudsen, *Bot. Tidsskr.* 73(2): 127 (1978)
Sensu Peck is unknown in Europe. Sensu NCL includes both *Lepiota pseudoasperula* and *Lepiota jacobii*.

erosa (Fr.) Sacc., Collybia, *Syll. fung.* 5: 250 (1887)
Agaricus erosus Fr., *Syst. mycol.* 1: 145 (1821)
A *nomen dubium*. Sensu auct. is *Tephrocybe tylicolor*.

eruciformis (Batsch) D.A. Reid, Calathella, *Persoonia* 3(1): 123 (1964)
Peziza eruciformis Batsch, *Elench. fung.*: 125 (1783)
Cyphella eruciformis (Batsch) Fr., *Syst. mycol.* 2: 203 (1823)
Lachnella eruciformis (Batsch) W.B. Cooke, *Sydowia, Beih.* 4: 72 (1961)
Cyphella albocarnea Quél., *Bull. Soc. Bot. France* 25: 290 (1878)
Not authentically British. Recorded from Scotland by Dennis (1986, as *Lachnella eruciformis*) based on records of *Cyphella albocarnea* sensu Wakefield (1952), which is now recognised as a species of *Flagelloscypha* probably *F. pilatii*.

erythropus Pers., Boletus, *Observ. mycol.* 1: 23 (1796)
Boletus luridus β *erythropus* (Pers.) Fr., *Syst. mycol.* 1: 391 (1821)
A *nomen dubium* lacking type material and variously interpreted. The common British species so named is now *B. luridiformis*.

eustriatulus Rob. Henry, Cortinarius, *Doc. mycol.* 16(61): 27 (1985)
Not authentically British. Supposedly reported from Scotland (St. Kilda) in 1967, but material in herb. E is actually determined as *Cortinarius* 'aff.' *striatulus* (not *eustriatulus*).

eutheles (Berk. & Broome) Quél., Inocybe, *Enchir. fung.*: 96 (1886)
Agaricus eutheles Berk. & Broome, *Ann. Mag. Nat. Hist., Ser. 3 [Notices of British Fungi no. 1004]* 15: 5 (1865)

A *nomen dubium*. Usually interpreted as *Inocybe sindonia*.

evenosa (Sacc.) Konrad & Maubl., Melanoleuca, *Icon. select. fung.* 4: pl. 272 (1927)
　Tricholoma cnista var. *evenosum* Sacc., *Syll. fung.* 5: 132 (1887)
　Tricholoma evenosum (Sacc.) Rea, *Trans. Brit. Mycol. Soc.* 17(1-2): 40 (1932)
A *nomen dubium* applied to several white or pale *Melanoleuca* spp. Sensu Rea (1922) is *M. strictipes*.

exaratus With., Agaricus, *Bot. arr. Brit. pl. ed. 2*, 3: 396 (1792)
Described from Britain but a *nomen dubium* lacking type material. Withering cites Bulliard's plate 80 [Herb. France (1782)] thus probably *Coprinus plicatilis* or a similar species.

excisa (Lasch) P. Kumm., Mycena, *Führ. Pilzk.*: 111 (1871)
　Agaricus excisus Lasch, *Linnaea* 4: 534 (1829)
　Mycena excisa f. *solitaria* J.E. Lange, *Fl. agaric. danic.* 2: 39 (1936)
A *nomen dubium* lacking type material and variously interpeted. British records are doubtfully distinct from *M. galericulata*.

exserta (Viv.) Rea, Psalliota, *Trans. Brit. Mycol. Soc.* 3: 285 (1911)
　Agaricus exsertus Viv., *Funghi d' Italia*: 55 (1834)
A *nomen dubium* lacking type material. Sensu Rea (1922) is *Agaricus benesii*.

exstinctorius (Bull.) Fr., Coprinus, *Epicr. syst. mycol.*: 245 (1838)
　Agaricus exstinctorius Bull., *Herb. France*: pl. 437 (1790)
A *nomen dubium*. lacking type material. Bulliard's *Agaricus exstinctorius* refers to a small and ephemeral dung species; *C. exstinctorius* sensu auct. mult. has recently been redescribed as *C. spelaiophilus*.

extenuatus Fr., Paxillus, *Epicr. syst. mycol.*: 316 (1838)
A *nomen dubium*. Cooke 862 (873) Vol. 6 (1888) appears to be *Leucopaxillus gentianeus*.

fagetorum M.M. Moser, Cortinarius, *Kleine Kryptogamenflora*, Edn 3: 337 (1967)
Not authentically British. A single report from Wales (Bettws-y-Coed) in 1988, but doubtfully identified and lacking voucher material.

fagetorum (Fr.) Gillet, Mycena, *Hyménomycètes*: 274 (1876)
　Agaricus excisus ssp. *fagetorum* Fr., *Epicr. syst. mycol.*: 106 (1838)
Not authentically British. Reported on several occasions but mostly lacking voucher material. Vouchered collections in herb. K are in poor condition and unidentifiable.

fagicola R.H. Petersen, Ramaria, *Biblioth. Mycol.* 43: 112 (1975)
A single British record was cited by Petersen (1975) but lacks voucher material. Doubtfully distinct from *R. stricta*.

fagineus With., Boletus [non *B. fagineus* Schrad. (1792), *Bot. arr. veg.* 2: 767 (1776)
Described from Britain but a *nomen dubium*. From the description possibly a species of *Ganoderma*.

falcata Pers., Clavaria, *Comment. Fungis Clavaeform.*: 81 (1797)
A *nomen dubium* lacking type material. Mentioned by Cotton & Wakefield (1919: 197) but no records have been traced. Sensu NM3, an older name for *C. acuta*.

falcata *var.* **citrinipes** Quél., Clavaria, *Tab. anal. fung.*: t. 41 (1883)
A *nomen dubium* lacking type material. Supposedly synonymous with *Clavaria acuta* but Corner (1950) implies that this is doubtful. A collection by Reid from East Kent (Bedgebury Pinetum) is in herb. K.

fallax (Sacc.) Redhead & Singer, Calocybe, *Mycotaxon* 6(3): 501 (1978)
　Tricholoma fallax Sacc., *Syll. fung.* 5: 115 (1887)

　Lyophyllum fallax (Sacc.) Kühner & Romagn. [*comb. inval.*], *Flore Analytique des Champignons Supérieurs*: 162 (1953)
　Agaricus fallax Peck [*nom. illegit.*], *Rep. (Annual) New York State Mus. Nat. Hist.* 25: 74 (1873) [1871]
Not authentically British *fide* NCL p.186 (as *Tricholoma fallax*), but Cooke [1122 (1151) Vol. 8 (1889)] depicts a collection so named from England (North East Yorkshire: Scarborough). No voucher material has been traced.

fallax Quél., Cortinarius, *Bull. Soc. Bot. France* 25: 289 (1879)
A *nomen dubium* lacking type material. No British records traced *fide* Pearson & Dennis (1948).

fallax A.H. Sm., Galerina, *Mycologia* 47: 561 (1955)
Not authentically British. A North American species. Reported from South Hampshire but unsubstantiated with voucher material.

farinacea Rea, Hydnopsis, *Trans. Brit. Mycol. Soc.* 3(2): 127 (1909)
Described from England (North Hampshire: Swarraton) 'growing on decayed litter of *Fagus*' but a *nomen dubium*. Rea's illustration in TBMS is distinctive (somewhat like *Kavinia alboviridis*) but it has never been recollected.

farinacea Quél., Odontia, *Fl. mycol. France*: 435 (1888)
A *nomen dubium* lacking type material but apparently not the same as *Hydnum farinaceum* Pers. (= *Trechispora farinacea*). Some authorities place this in the synonymy of *Steccherinum queletii* which is not British.

farinaceus (Schumach.) P. Karst., Panellus, *Bidrag Kännedom Finlands Natur Folk* 32: 96 (1879)
　Agaricus farinaceus Schumach. [non *A. farinaceus* Huds. (1778)], *Enum. pl.* 2: 365 (1803)
　Panus farinaceus (Schumach.) Fr., *Epicr. syst. mycol.*: 399 (1838)
　Panellus stipticus var. *farinaceus* (Schumach.) Rea, *Brit. basidiomyc.*: 536 (1922)
A *nomen dubium* lacking type material. From the original description probably synonymous with *Panellus stipticus*. Cooke 1097 (1144) Vol. 7 (1890) resembles a species of *Resupinatus*.

farrahi Massee & Crossl., Entoloma, *Naturalist (Hull)* 1904: 1 (1904)
Described from England (Yorkshire) but a *nomen dubium*. Fide Pearson & Dennis (1948) this was smooth-spored thus not an *Entoloma*.

farrea (Lasch) P. Kumm., Mycena, *Führ. Pilzk.*: 109 (1871)
　Agaricus farreus Lasch, Fries, *Epicr. syst. mycol.*: 103 (1838)
A *nomen dubium* lacking type material. Plowright's collection [reported in TBMS 1(2): 53 (1898)] is probably *Cystolepiota seminuda fide* NCL.

fasciata (Cooke & Massee) Sacc., Inocybe, *Syll. fung.* 9: 95 (1891)
　Agaricus fasciatus Cooke & Massee, *Grevillea* 18: 52 (1890)
　Astrosporina fasciata (Cooke & Massee) Rea, *Brit. basidiomyc.*: 210 (1922)
A *nomen dubium*. Described from 'amongst grass' in Kew Gardens but never recollected. Cooke 1164 (1173) Vol. 8 (1890) looks distinctive.

fasciatus Fr., Cortinarius, *Epicr. syst. mycol.*: 315 (1838)
A *nomen dubium*. Sensu Lange (1938) and sensu auct. Brit. is *C. fulvescens*.

fasciculatus (Pers.) Agerer, Rectipilus, *Persoonia* 7: 419 (1973)
　Solenia fasciculata Pers., *Mycol. eur.* 1: 335 (1822)
Not authentically British. Collections named thus are mostly *Henningsomyces candidus*.

fasciculosus With., Agaricus, *Bot. arr. veg.* 2: 757 (1776)
Described from Britain but a *nomen dubium* lacking type material. *Fide* Withering a striking species with yellow-brown pileus and purple stipe.

fastibile (Pers.) P. Kumm., Hebeloma, *Führ. Pilzk.*: 80 (1187)
  Agaricus fastibilis Pers., *Syn. meth. fung.*: 326 (1801)
  Hebeloma fastibile var. *elegans* Massee, *Brit. fung.-fl.* 2: 171 (1893)
The type species of *Hebeloma* but a much disputed name. Sensu NCL is *Hebeloma mesophaeum* var. *crassipes*.

fastibile *var.* **album** (Fr.) Sacc., Hebeloma, *Syll. fung.* 5: 792 (1887)
  Agaricus fastibilis var. *albus* Fr., *Epicr. syst. mycol.*: 178 (1838)
A *nomen dubium*. Listed by Rea (1922).

fastibile *var.* **sulcatum** (Lindbl.) Rea, Hebeloma, *Brit. basidiomyc.*: 253 (1922)
  Agaricus sulcatus Lindbl., *Bot. Not.* 1845(12): 199 (1845)
A *nomen dubium*. Listed by Rea (1922).

fastigiella G.F. Atk., Inocybe, *Amer. J. Bot.* 5: 211 (1918)
  Inocybe maculata f. *fastigiella* (G.F. Atk.) Kühner & Romagn. [*comb. inval.*], *Flore Analytique des Champignons Supérieurs*: 218 (1953)
Not authentically British. A North American species, doubtfully distinct from *Inocybe maculata fide* Kuyper (1986). Reported from Britain but records are unsubstantiated with voucher material and are presumably *I. maculata*.

fennica (P. Karst.) Ricken, Ramaria, *Vadem. Pilzfr.*: 264 (1920)
  Clavaria fennica P. Karst., *Not. Sällsk. Fauna Fl. Fenn. Förh.* 9: 372 (1868)
Not authentically British. This is a conifer associate, and British material named thus, from deciduous woodland, is better referred to *R. fennica* var. *griseolilacina*.

ferruginascens Batsch, Agaricus, *Elench. Fung. Cont. Sec.*: 9 t. 32 f. 187 (1789)
A *nomen dubium*. Reported by Withering (1796). The illustration in Batsch shows an unidentifiable species of *Cortinarius*.

ferruginea Schumach., Daedalea, *Enum. pl.* 2: 373 (1803)
A *nomen dubium* lacking type material, but possibly an *Inonotus* sp. Listed by Rea (1922).

ferruginea Sm., Tremella [*nom. illegit.*, non *T. ferruginea* Schum. (1803)], *Engl. fung.*: pl. 1452 (1805)
Described from Britain but lacking type material. Old specimens named thus in herb. K are *Tremella foliacea*.

fibrillosa (Pers.) Maire, Psathyrella, *Mém. Soc. Sci. Nat. Maroc* 45: 113 (1938)
  Agaricus fibrillosus Pers., *Syn. meth. fung.*: 424 (1801)
  Psathyra fibrillosa (Pers.) Quél., *Mém. Soc. Émul. Montbéliard, Sér.* 2, 5: 150 (1872)
A *nomen dubium* lacking type material and variously interpeted (Kits1: 274). Sensu Rea (1922) is doubtful. Sensu Lange (FlDan4) is *Psathyrella friesii*.

fibrosa (Sowerby) Gillet, Inocybe, *Hyménomycètes*: 517 (1876)
  Agaricus fibrosus Sowerby, *Col. fig. Engl. fung., Suppl.*: pl. 414 (1809)
  Astrosporina fibrosa (Sowerby) Rea, *Brit. basidiomyc.*: 208 (1922)
A *nomen dubium* lacking type material. Sensu Sowerby is possibly *Inocybe duriuscula fide* Kuyper (pers. comm.). Sensu auct. mult. is *Inocybe inedita*, not authentically British.

fibula (Sowerby) Fr., Polyporus, *Hymenomyc. eur.*: 567 (1874)
  Boletus fibula Sowerby, *Col. fig. Engl. fung.* 3: pl. 387 t. 8 (1803)
  Polystictus fibula (Sowerby) Cooke, *Grevillea* 14: 81 (1886)
Described from Britain but a *nomen dubium* lacking type material. Probably *Trametes hirsuta*.

ficoides With., Agaricus, *Bot. arr. Brit. pl. ed. 2*, 3: 400 (1792)
Described from Britain but a *nomen dubium*, lacking type material. Almost certainly *Panellus stipticus* since Withering cites Bolton's plate [Hist. Fung. Halifax pl. 152 f. 1 (1788)] of *Agaricus betulinus*.

filamentosa (Berk. & M.A. Curtis) Burds., Phanerochaete, *Distributional History of the Biota of the Southern Appalachians, 4. Algae and Fungi*: 278 (1976)
  Corticium filamentosum Berk. & M.A. Curtis, *Grevillea* 1: 178 (1873)
  Peniophora filamentosa (Berk. & M.A. Curtis) Moffatt, *J. Elisha Mitchell Sci. Soc.* 36: 162 (1921)
A North American species unknown in Europe. European records all relate to *Phanerochaete radicata*.

filamentosus Fr., Paxillus, *Epicr. syst. mycol.*: 317 (1838)
A *nomen dubium*. Said to occur with *Alnus glutinosa*. Sensu auct. is *P. rubicundulus* (*fide* Watling in BFF1) but the name has been used for another fungus probably not distinct from *P. involutus*. It was reported in this latter sense by Moser from England (North Somerset: near Bristol) in 1955.

filicina (P. Karst.) Agerer, Nochascypha, *Mitt. bot. St Samml., München* 19: 268 (1983)
  Cyphella filicina P. Karst., *Meddeland. Soc. Fauna Fl. Fenn.* 11: 220 (1871)
  Lachnella filicina (P. Karst.) W.B. Cooke, *Sydowia, Beih.* 4: 76 (1961)
Not authentically British. The single collection, from Pembrokeshire, was misdetermined, and no other records are known

filicinus (Velen.) P.D. Orton, Pleurotellus, *Trans. Brit. Mycol. Soc.* 43(2): 180 (1960)
  Pleurotus filicinus Velen., *Mykologia* 4: 30 (1927)
Included in NCL and BFF6 based on the assumption of Pilát (1948) that this is the same as *Pleurotus hypnophilus* sensu Rea (1922). However, the small spores described by Rea could equally be from immature *Crepidotus epibryus*.

filiformis Berk. & Broome, Coprinus, *Ann. Mag. Nat. Hist., Ser. 3 [Notices of British Fungi no. 928]* 7: 379 (1861)
Described from Britain, but a *nomen dubium*. See Cooke 674 (686) Vol. 5 (1886) for an interpretation.

filiformis (Bull.) Fr., Typhula, *Syst. mycol.* 1: 496 (1821)
  Clavaria filiformis Bull., *Herb. France*: pl. 448 f. 1 (1790)
A *nomen dubium* lacking type material. Synonymous with *Macrotyphula juncea fide* Corner (1950) but Rea's (1922) description, copied from Massee, suggests a species of *Pterula*.

filium Bres., Corticium, *Ann. Mycol.* 6: 43 (1908)
A *nomen dubium*. Reported from South Somerset (Porlock) in the late nineteenth century and from Sussex (Broadwater Forest) in 1927 by Wakefield (1952), but no material has been traced. Possibly a species of *Trechispora*.

fimbriatus (Bolton) Sacc., Pleurotus, *Syll. fung.* 5: 344 (1887)
  Agaricus fimbriatus Bolton, *Hist. fung. Halifax* 2: 61 (1788)
  Micromphale fimbriatum (Bolton) Gray, *Nat. arr. Brit. pl.* 1: 622 (1821)
  Clitocybe fimbriata (Bolton) Singer, *Sydowia* 31: 231 (1978) [1978]
Described from Britain but a *nomen dubium* lacking type material. Regarded by some authorities as a synonym of *Ossicaulis lignatilis* and by Fraiture (1993: 75) as synonymous with *Clitocybe gibba*, but Bolton's illustration and description resemble neither.

fimetarius Fr., Coprinus, *Epicr. syst. mycol.*: 245 (1838)
A *nomen dubium*. Listed, pro parte, as a synonym of *C. cinereus* in BFF2.

fimetarius *var.* **pullatus** (Bolton) Sacc., Coprinus, *Syll. fung.* 5: 1087 (1887)
  Coprinus cinereus γ *pullatus* (Bolton) Gray, *Nat. arr. Brit. pl.* 1: 634 (1821)
  Agaricus pullatus Bolton [non *A. pullatus* Berk. & Cooke (1882)], *Hist. fung. Halifax* 1: 20 (1788)
  Coprinus pullatus (Bolton) Fr., *Epicr. syst. mycol.*: 246 (1838)
Described from Britain but a *nomen dubium* lacking type material.

fimiputris (Bull.) Quél., Panaeolus, *Mém. Soc. Émul.*
Montbéliard, Sér. 2, 5: 151 (1872)
   Agaricus fimiputris Bull., *Herb. France*: pl. 66 (1781)
   Anellaria fimiputris (Bull.) P. Karst., *Bidrag Kännedom*
    Finlands Natur Folk 32: 518 (1879)
A *nomen dubium* lacking type material. Listed in BFF5 as a
synonym of *P. semiovatus.*

fingibilis Britzelm., Russula, *Ber. Naturhist. Vereins Augsburg*
28: 140 (1885)
A *nomen dubium* lacking type material. Doubtfully distinct from
Russula ochroleuca. Listed by Rea (1922). See Cooke 1030
(1048) Vol. 7 (1889) for an interpretation.

firmula Jul. Schäff., Russula, *Ann. Mycol.* 38(2-4): 111 (1940)
Not authentically British. Sensu Rayner (1985) is *Russula cuprea.*

firmum (Pers.) Sacc., Hebeloma, *Syll. fung.* 5: 793 (1887)
   Agaricus firmus Pers., *Icon. descr. fung.*: pl.15 (1798)
A *nomen dubium* lacking type material. Listed by Rea (1922).

firmus Fr., Cortinarius, *Epicr. syst. mycol.*: 964 (1838)
Not authentically British. A single collection named thus from
West Sussex (Goodwood) in 1945 (in herb. K) was annotated
as 'not interpreted' by Orton in 1970. Reported once since then
(1989) from South Lancashire but lacking voucher material.

fissus Bolton, Agaricus, *Hist. fung. Halifax* 1: 35 (1788)
Described from Britain but a *nomen dubium* lacking type
material. Bolton's description suggests *Mycena polygramma*
but the illustration resembles a species of *Entoloma.*

flammuloides M.M. Moser [*nom. nud.*], Pholiota, *Kleine
Kryptogamenflora*: 193 (1953)
Recorded from Scotland (Glencoe). Probably not distinct from
Pholiota flammans.

flaveolum Massee, Corticium, *J. Linn. Soc., Bot.* 27: 150 (1890)
A *nomen dubium.* Described from the stem of a tree fern in a
hothouse in Kew Gardens, but never recollected. Presumably
an alien.

flavescens (Kauffman) Singer, Hygrocybe, *Lilloa* 22: 154
(1951) [1949]
   Hygrophorus puniceus var. *flavescens* Kauffman, *Rep.
    (Annual) Michigan Acad. Sci.* 8: 34 (1906)
   Hygrophorus flavescens (Kauffman) A.H. Sm. & Hesler,
    Lloydia 5: 60 (1942)
Not authentically British. A North American species. British
collections named thus are all likely to be *H. chlorophana.*

flavescens (Schaeff.) R.H. Petersen, Ramaria, *Amer. J. Bot.*
61(7): 740 (1974)
   Clavaria flavescens Schaeff., *Fung. Bavar. Palat. nasc.* 3: 205
    (1770)
Not authentically British. A single report from East
Gloucestershire (Buckholt Wood) in 1997, unsubstantiated with
voucher material and probably misidentified.

flavicans With., Agaricus, *Arr. Brit. pl. ed. 3,* 4: 267 (1796)
Described from Britain but a *nomen dubium.*

flavida (Schaeff.) Singer, Pholiota, *Lilloa* 22: 516 (1951) [1949]
   Agaricus flavidus Schaeff., *Fung. Bavar. Palat. nasc.* 3: 295
    (1770)
   Flammula flavida (Schaeff.) P. Kumm., *Führ. Pilzk.*: 82 (1871)
A *nomen dubium fide* Holec (2001) widely used for both *Pholiota
alnicola* and *P. pinicola.* British records could be either. Sensu
Noordeloos in FAN4 is probably a species of *Hypholoma fide*
Holec (2001).

flavidus Bolton, Agaricus [*nom. illegit.*, non *A. flavidus* Schaeff.
(1770)], *Hist. fung. Halifax* Suppl.: 149 (1791)
Described from Britain but a *nomen dubium* lacking type
material. Sensu Sowerby [Col. fig. Engl. fung. 1: pl. 96 (1797)]
is *Bolbitius vitellinus* or *B. titubans.*

flavipes Pers., Clavaria, *Comment. Fungis Clavaeform.*: 75
(1797)

A *nomen dubium* lacking type material. Sensu Corner (1950) is
C. argillacea and sensu NM3 is *C. straminea.*

flavissimum Link, Sporotrichum, *Mag. Neuesten Entdeck.
Gesammten Naturk. Ges. Naturf. Freunde Berlin* 8: 34 (1816)
A *nomen dubium*, listed as British, e.g. by Wakefield & Bisby
(1942). The type is a *Leucogyrophana* sp. *fide* Stalpers (1984).

flavobrunnescens (G.F. Atk.) Corner, Ramaria, *Monograph of
Clavaria and Allied Genera*: 581 (1950)
   Clavaria flavobrunnescens G.F. Atk., *Ann. Mycol.* 7: 367
    (1909)
Not authentically British. Reported from Scotland but
misidentified.

flavovirens Berk. & Ravenel, Polyporus, *Grevillea* 1: 38 (1872)
Accepted as a synonym of *Albatrellus cristatus* (not authentically
British) by some authorities, but sensu Rea (1922) is a
different species (spores much larger than *A. cristatus*) and is
unknown.

flavus With., Boletus, *Arr. Brit. pl. ed. 3,* 4: 320 (1796)
   Pinuzza flava (With.) Gray, *Nat. arr. Brit. pl.* 1: 646 (1821)
   Boletus elegans var. *flavus* (With.) Rea, *Brit. basidiomyc.*:
    559 (1922)
Described from Britain but a *nomen dubium fide* NCL. Sensu
auct. is *Suillus bresadolae* var. *flavogriseus* (=*S. nueschii*).

flexipes P. Karst., Galera, *Bidrag Kännedom Finlands Natur
Folk*: 371 (1876)
A *nomen dubium.* Doubtful *fide* NCL. See Naturalist (Hull) 1914:
145 (1914) and TBMS 5 (2): 251 (1915) where reported from
Yorkshire (Mulgrave Woods).

floccifera (Berk. & Broome) Sacc., Flammula, *Syll. fung.* 5: 811
(1887)
   Agaricus floccifer Berk. & Broome, *Ann. Mag. Nat. Hist.* 7:
    374 (1861)
Described from Britain but a *nomen dubium.* See Cooke 467
(438) Vol. 4 (1884) for an interpretation.

floriformis Schaeff., Helvella, *Fung. Bavar. Palat. nasc.* 3: 278
(1770)
A *nomen dubium* lacking type material. Sowerby's illustration
[Col. fig. Engl. fung. 1: pl. 75 (1797)] resembles *Cotylidia
undulata* but could also represent *Pseudocraterellus sinuosus*
(although not as grey as that species usually is). Fries (1821)
places it in the synonymy of the latter.

fodiens (Kalchbr.) Konrad & Maubl., Collybia, *Icon. select. fung.*
6: 247 (1937)
   Agaricus fodiens Kalchbr., *Icon. select. Hymenomyc. Hung.*:
    62 (1873)
   Rhodocollybia fodiens (Kalchbr.) Antonín & Noordel.,
    Mycotaxon 63: 365 (1997)
Not authentically British. Reported by Cooke from North
Hampshire (Alresford) but unsubstantiated with voucher
material and not reported since. See Cooke 1138 (949) Vol. 8
for an illustration of this collection.

foetens Melot, Clitocybe, *Bull. Soc. Mycol. France* 95: 237
(1979)
Not authentically British. A single dubious record from England
(West Sussex) lacking voucher material and probably
misidentified.

foetidum Berk. & Broome, Corticium, *Ann. Mag. Nat. Hist., Ser.
3 [Notices of British Fungi no. 1824]* 15: 211 (1879)
A *nomen dubium*, possibly *Trechispora fastidiosa*, described from
Wales (Denbighshire: Coed Coch) on decayed sawdust.

foetidus With., Agaricus, *Syst. arr. Brit. pl. ed. 4,* 4: 298 (1801)
Described from Britain but a *nomen dubium.* The description
strongly suggests *Micromphale foetidum.*

formosa var. suavis (Lasch) Sacc., Leptonia, *Syll. fung.* 5: 712
(1887)
   Agaricus formosus var. *suavis* Lasch, *Linnaea.* 3: 285 (1828)

Doubtful *fide* NCL. Cooke 360 (488) Vol. 3 (1885) is a species of *Entoloma*, possibly *E. huijsmannii fide* Noordeloos (1987).

fracticum (Britzelm.) Kreisel, Tricholoma, *Feddes Repert.* 95(9-10): 700 (1984)
    Agaricus fracticus Britzelm., *Bot. Centralbl.* 15-17: 2 (1893)
A *nomen dubium* lacking type material. Britzelmayer's plate could represent any of several species near to *Tricholoma albobrunneum*. Sensu Kreisel (and the only British record) is *T. batschii*.

fragilis (Peck) Singer, Conocybe, *Trudy Bot. Inst. Akad. Nauk S.S.S.R., Ser. 2, Sporov. Rast.* 6: 438 (1950)
    Galera fragilis Peck, *Bull. Torrey Bot. Club* 24: 144 (1897)
Not authentically British. A North American species. Sensu Watling in BFF3 and sensu auct. Eur. is *C. incarnata*.

fragilis var. nivea (Pers.) Cooke, Russula, *Handb. Brit. fung.,* Edn 2: 333 (1889)
    Agaricus niveus Pers. [non *A. niveus* Pers., *Syn. meth. fung.*: 400 (1801)], *Syn. meth. fung.*: 438 (1801)
    Russula nivea (Pers.) Gillet, *Tableaux analytiques des Hyménomycètes*: 47 (1884)
A *nomen dubium* lacking type material. Cooke 1029 (1060) Vol. 7 (1889) is *Russula raoultii*.

fraudulosus Britzelm., Cortinarius, *Ber. Naturhist. Vereins Augsburg* 28: 122 (1885)
Not authentically British. A boreal and montane species. Reported from Herefordshire in 1995 but unsubstantiated with voucher material and most likely misidentified *C. argutus*.

friesii Mont., Porotheleum, *Ann. Sci. Nat., Bot.,* sér. 2, 5: 339: (1836)
A *nomen dubium*. Listed by Rea (1922).

frondosa Fr., Tremella, *Syst. mycol.* 2: 212 (1823)
A *nomen dubium*. Considered a good species by Donk (1966). Old specimens named thus in herb. K are *Tremella foliacea* and *T. aurantia*.

frumentaceum (Bull.) Sacc., Tricholoma, *Syll. fung.* 5: 95 (1887)
    Agaricus frumentaceus Bull., *Herb. France*: pl. 571 (1792)
A *nomen dubium* lacking type material. A possible synonym of *T. scalpturatum fide* Riva (FungEur3) but the Bulliard plate depicts a different agaric. Cooke 330 (470) Vol. 3 (1885) [as *Agaricus (Entoloma) frumentaceus*] was copied from a painting of a *Tricholoma* by W.G. Smith [see J. Bot. 30: 39 (1892)] but with pink spores added.

frustulorum Sacc., Coprinus, *Atti Soc. Veneto-Trentino Sci. Nat. Padova* 2(1): 35 (1873)
A *nomen dubium*. Reported by Rea in TBMS 4 (1): 189 (1913), misspelt 'frustulosum'. Apparently a distinctive species with 'rosy red micaceous meal on the pileal surface' but no material survives.

frustulatus (Pers.) Boidin, Xylobolus, *Rev. Mycol. (Paris)* 23: 341 (1958)
    Thelephora frustulata Pers., *Syn. meth. fung.*: 577 (1801)
    Stereum frustulatum (Pers.) Fr., *Epicr. syst. mycol.*: 552 (1838)
Not authentically British. The single collection named thus, from Surrey (Crystal Palace, near Sydenham) in herb. K, is sterile, but probably *Ditiola radicata*.

fucatum (Fr.) P. Kumm., Tricholoma, *Führ. Pilzk.*: 130 (1871)
    Agaricus fucatus Fr., *Syst. mycol.* 1: 40 (1821)
Not authentically British. Reported from a few scattered sites in England but all records are old, and voucher material in herb. K is in poor condition. Listed in NCL only on the strength of Cooke 62 (73) Vol. 1 (1882) which looks convincing but is unsubstantiated with voucher material.

fugacissima f. sebacea Bourdot & Galzin, Sebacina, *Hyménomyc. France*: 43 (1928)
A *nomen dubium*. Sensu auct. Brit. is *Stypella grilletii*.

fulgens (Alb. & Schwein.) Fr., Cortinarius, *Epicr. syst. mycol.*: 267 (1838)
    Agaricus fulgens Alb. & Schwein., *Consp. fung. lusat.*: 160 (1805)
A *nomen dubium*, lacking type material. Sensu NCL is *C. olearioides*.

fuligineipes Métrod, Clitocybe [*nom. nud.*], *Bull. Soc. Mycol. France* 55: 107 (1939)
Introduced by Métrod for his concept of Ricken's *Agaricus fritilliformis*. Noted in NCL as 'not British'.

fuligineoviolaceus (Kalchbr.) Pat., Sarcodon, *Essai taxon.*: 118 (1900)
    Hydnum fuligineoviolaceum Kalchbr., in Fries, *Hymenomyc. eur.*: 602 (1874)
Not authentically British. Listed by Jülich (1984) but lacking substantiated records. Also listed by Ing (1992).

fuligineus Fr. & Hök, Boletus [*nom. illegit.*, non *B. fuligineus* Pers. (1801)], *Boleti fungorum generis illustratio*: 13 (1836)
    Phaeoporus fuligineus (Fr. & Hök) Bataille, *Bull. Soc. Hist. Nat. Doubs* 15: 11 (1908)
    Phaeoporus porphyrosporus var. *fuligineus* (Fr. & Hök) Rea, *Brit. basidiomyc.*: 555 (1922)
British *fide* Rea (1922) but no material traced and doubtfully distinct from *Phylloporus porphyrosporus*.

fuligineus (Pers.) Gray, Albatrellus, *Nat. arr. Brit. pl.* 1: 645 (1821)
    Boletus fuligineus Pers. [non *B. fuligineus* Fr. & Hök (1836)], *Syn. Meth. Fung.*: 516 (1801)
A *nomen dubium* lacking type material. Reported by Gray (1821) as 'on the ground, at the foot of palings'. Synonymised by Gray with *Boletus polyporus* Bull. which is also a *nomen dubium*.

fuliginosa (Pers.) Lév., Hymenochaete, *Ann. Sci. Nat., Bot.,* sér. 3, 5: 152 (1846)
    Thelephora fuliginosa Pers., *Mycol. eur.* 1: 145 (1822)
    Stereum fuliginosum (Pers.) Fr., *Epicr. syst. mycol.*: 554 (1838)
    Hymenochaete subfuliginosa Bourdot & Galzin, *Bull. Soc. Mycol. France* 38: 184 (1922)
    Hymenochaete rubiginosa ssp. *subfuliginosa* (Bourdot & Galzin) Bourdot & Galzin, *Hyménomyc. France*: 391 (1928)
Not authentically British. Sensu Rea (1922) is *Hymenochaete corrugata*. Reported from North Somerset on two occasions (as *H. subfuliginosa*) but unsubstantiated with voucher material.

fuliginosa Sarnari, Russula, *Rivista Micol.* 36(1): 53 (1993)
Not authentically British. Reported from Cumbria in 1999 but misidentified.

fulmineus Fr., Cortinarius, *Epicr. syst. mycol.*: 267 (1838)
A *nomen dubium*. Sensu NCL and sensu auct. mult. is *C. olearioides*. Sensu Rea (1922) is *C. elegantior*.

fulva Rea, Astrosporina, *Trans. Brit. Mycol. Soc.* 6(4): 326 (1920)
Described from Britain but a *nomen dubium*. Possibly *Inocybe curvipes* var. *ionipes fide* Kuyper (note on the herbarium sheet at Kew) which has not been reported from the British Isles.

fulva Berk. & Ravenel, Cyphella, *Ann. Mag. Nat. Hist., Ser. 3 [Notices of British Fungi no. 936]* 7: 379 (1861)
Described from Britain but a *nomen dubium*. The description of large spores suggests *Merismodes bresadolae*.

fulva Burt, Hymenochaete, *Ann. Missouri. Bot. Gard.* 5: 354 (1918)
Not authentically British. Specimens named thus in herb. K are supposedly from England but this is a tropical species described from Jamaica and unlikely to occur in Britain.

fulvosus Bolton, Agaricus, *Hist. fung. Halifax* 2: 56 (1788)
Described from Britain but a *nomen dubium* lacking type material. Bolton's description and illustration suggest a species of *Hygrocybe*, possibly *H. pratensis*.

fulvus (Fr.) Gillet, Fomes, *Hyménomycètes*: 685 (1878)
  Polyporus fulvus Fr., *Epicr. syst. mycol.*: 466 (1838)
Said to be synonymous with *Inonotus rheades* (not British), but old specimens in herb. K, from Scotland (Outer Hebrides: King Georges Sound), represent an infertile *Phellinus* sp.

furcata W.G. Sm., Pistillaria, in Cooke, *Handb. Brit. fung.*, Edn 1: 343 (1871)
Described from Britain but a *nomen dubium*. Possibly *Clavaria tenuipes fide* Corner (1950).

furcata Pers., Russula, *Observ. mycol.* 1: 102 (1796)
A *nomen dubium* lacking type material and variously interpreted. Sensu auct. refers either to green forms of *Russula cyanoxantha* or to *R. heterophylla*.

furfuracea var. heterosticha (Fr.) Sacc., Tubaria, *Syll. fung.* 5: 873 (1887)
  Agaricus furfuraceus β *heterostichus* Fr., *Observ. mycol.* 2: 25 (1818)
A *nomen dubium* but probably not distinct from the type variety.

furfurosus With., Agaricus, *Arr. Brit. pl. ed. 3*, 4: 183 (1796)
Described from Britain but a *nomen dubium*.

fusa var. superba Massee, Flammula, *Brit. fung.-fl.* 2: 135 (1892)
Described from Britain but a *nomen dubium*. Cooke 478 (434) Vol. 4 (1884) possibly represents large and vividly coloured basidiomes of *Hypholoma lateritium*.

fusca Quél., Russula, *Mém. Soc. Émul. Montbéliard, Sér. 2*, 5: 4 (1872)
A *nomen confusum*. See notes in NCL Index p.189.

fuscella (Sacc.) S. Lundell, Tomentella, *Fungi Exsiccati Suecici* 45-46: 9 (1954)
  Hypochnus fuscellus Sacc., *Syll. fung.* 6: 662 (1886)
A *nomen dubium* lacking type material. Reported from Cheshire and Middlesex in 1988 but unsubstantiated with voucher material. Material named thus (in herb. E) is all labelled with a query.

fuscescens (Schaeff.) Fr., Coprinus, *Epicr. syst. mycol.*: 244 (1838)
  Agaricus fuscescens Schaeff., *Fung. Bavar. Palat. nasc.* 1: 17 (1762)
  Coprinus fuscescens var. *rimososquamosus* Cooke, *Ill. Brit. fung.* 651 (664) Vol. 5 (1886)
A *nomen dubium* lacking type material. See Cooke 650 (663) Vol. 5 (1886) for an interpretation, probably representing a species in section *Micacei*.

fuscipes Sowerby, Agaricus, *Col. fig. Engl. fung.* 3: pl. 344 (1802)
Described from Britain but a *nomen dubium* lacking type material. Possibly *Macrocystidia cucumis*, though the characteristic smell is not mentioned.

fuscoalbum (Sowerby) Lannoy & Estadès, Leccinum, *Doc. mycol.* 24(94): 18 (1994)
  Boletus fuscoalbus Sowerby, *Col. fig. Engl. fung., Suppl.*: pl. 421 (1809)
A *nomen dubium* lacking type material. Some British material named thus is a form of *L. variicolor fide* A.E. Hills (pers. comm.).

fuscoalbus With., Agaricus [non *A. fuscoalbus* Lasch (1828)], *Bot. arr. Brit. pl. ed. 2*, 3: 380 (1792)
Described from Britain but a *nomen dubium*.

fuscoalbus (Lasch) Fr., Hygrophorus, *Epicr. syst. mycol.*: 324 (1838)
  Agaricus fuscoalbus Lasch [*nom. illegit.*, non *A. fuscoalbus* With. (1792)], *Linnaea* 3: 502 (1828)
A *nomen dubium* lacking type material. Sensu NCL is *H. latitabundus*, though this is not authentically British.

fuscocarnea (Pers.) Cooke, Poria, *Grevillea* 14: 110 (1886)
  Polyporus fuscocarneus Pers., *Mycol. eur.* 2: 97 (1825)

A *nomen dubium* lacking type material. Listed by Rea (1922).

fuscoflavus With., Agaricus, *Bot. arr. Brit. pl. ed. 2*, 3: 498 (1792)
Described from Britain but a *nomen dubium.*

fuscopallidus Bolton, Agaricus, *Hist. fung. Halifax* 3: 136 (1789)
Described from Britain but a *nomen dubium* lacking type material. *Fide* Termorshuizen & Arnolds [Mycotaxon 30:107 (1987)] this is a member of the *Armillaria mellea* complex but cannot be identified to species with certainty.

fuscopurpureus With., Agaricus [non *A. fuscopurpureus* Pers. (1798)], *Arr. Brit. pl. ed. 3*, 4: 249 (1796)
Described from Britain but a *nomen dubium*.

fuscoviolaceum (Ehrenb.) Ryvarden, Trichaptum, *Norweg. J. Bot.* 19: 237 (1972)
  Sistotrema fuscoviolaceum Ehrenb., *Sylv. mycol. berol.*: 30 (1818)
  Hydnum fuscoviolaceum (Ehrenb.) Fr., *Syst. mycol.* 1: 421 (1821)
  Irpex fuscoviolaceus (Ehrenb.) Fr., *Elench. fung.* 1: 144 (1828)
  Hirschioporus fuscoviolaceus (Ehrenb.) Donk, *Meddel. Bot. Mus. Herb. Rijks Universit. Utrecht* 9: 169 (1933)
  Irpex candidus (Ehrenb.) Weinm., *Hymen. Gasteromyc.*: 376 (1836)
  Irpex violaceus (Schrad.) Fr., *Elench. fung.* 1: 144 (1828)
  Hydnum hollii (J.C. Schmidt) Fr., *Syst. mycol.* 1: 420 (1821)
  Odontia hollii (J.C. Schmidt) Rea, *Brit. basidiomyc.*: 645 (1922)
  Trichaptum hollii (J.C. Schmidt) Kreisel, *Boletus* 1984 (1): 30 (1984)
Not authentically British. Reported from southern England and Northern Ireland but unsubstantiated with voucher material; the one record with material (South Hampshire: New Forest) is probably *T. abietinum* with the hymenophore slightly more lacerate-dentate than normal.

fuscus With., Merulius [non *M. fuscus* Lloyd (1925)], *Arr. Brit. pl. ed. 3*, 4: 149 (1796)
Described from Britain but a *nomen dubium* lacking type material. Synonymised by Massee (1893b: 395) with *Omphalia myochroa*, itself a *nomen dubium*, but possibly an *Arrhenia* sp.

fusipes Pers., Hydnum, *Mycol. eur.* 2: 162 (1825)
A poorly known species, though there is type material at herb. Leiden. A synonym of *Bankera violascens fide* Jülich (1984) but sensu Smith (1908: 376) is doubtfully Persoon's species.

fusoideus (Jülich) Krieglst., Leptosporomyces, *Z. Mykol.* 57(1): 53 (1991)
  Fibulomyces fusoideus Jülich, *Willdenowia. Beih.* 7: 180 (1972)
Not authentically British. Reported from England (as *Fibulomyces fusoideus*) in Wiltshire Nat. Hist. Mag. 72: 13 (1977) but lacking voucher material.

fusus (Batsch) Singer, Pholiota, *Lilloa* 22: 516 (1951) [1949]
  Agaricus fusus Batsch, *Elench. fung. (Continuatio Secunda)*: 13 t. 32 f. 189 (1789)
  Flammula fusa (Batsch) P. Kumm., *Führ. Pilzk.*: 82 (1871)
A *nomen dubium*. Sensu Rea (1922) (as *Flammula fusa*) is doubtful. Cooke 477 (433) Vol. 4 (1884) looks distinctive but 478 (434), supposedly the same, shows a totally different fungus. Sensu FAN4 is possibly *Hypholoma lateritium fide* Holec (2001).

gadinoides (W.G. Sm.) Sacc., Pleurotus, *Syll. fung.* 5: 364 (1887)
  Agaricus gadinoides W.G. Sm., *J. Bot.* 11: 65 (1873)
A *nomen dubium*. Described from dead tree-fern trunks in a hothouse in London (Chelsea: Veitch's Nursery) thus probably an alien. The description stated 'affinis *P. miti*', but Cooke 286 (276) Vol. 2 is unlike *Panellus mitis*.

galeropsis (Fr.) Sacc., Mycena, *Syll. fung.* 5: 261 (1887)

Agaricus galeropsis Fr., *Ic. Hymenomyc.* 2: pl. 79 (1877)
A *nomen dubium*. Rea (1922) equates it with *Collybia aquosa* of this list.

galochroa (Fr.) Fr., Russula, *Hymenomyc. eur.*: 447 (1874)
Agaricus galochrous Fr., *Observ. mycol.* 1: 65 (1815)
Not authentically British. Reported from East Sussex in 1997 [BMS Newsletter Feb. 98 p. 24] but the material (in herb. K) has been redetermined as a pallid form of *R. heterophylla*. Previously reported from various locations in England but all records are pre-1914 and unsubstantiated with voucher material.

gangraenosus *var.* nigrescens (Lasch) Cooke, Agaricus, *Handb. Brit. fung.*, Edn 2: 46 (1884)
Agaricus nigrescens Lasch, *Linnaea* 4: 521 (1829)
A *nomen dubium* lacking type material. Sensu Cooke appears to be *Lyophyllum gangraenosum* (or near to this) but sensu Saccardo [Syll. fung. 5: 143 (1887)] is *Rhodocybe mundula*.

gardneri Berk., Lysurus, *London J. Bot.* 5: 535 (1846)
Not authentically British. Sensu auct. Brit. is *L. cruciatus*.

gaspesica (Liberta) K.H. Larss. & Hjortstam, Phlebiella, *Mycotaxon* 29: 318 (1987)
Xenasma gaspesicum Liberta, *Mycologia* 58(6): 932 (1967)
Aphanobasidium gaspesicum (Liberta) Jülich, *Persoonia* 10(3): 326 (1979)
Xenasmatella gaspesica (Liberta) Hjortstam, *Mycotaxon* 17: 581 (1983)
Not authentically British. The single record (determined as 'possibly this species') on dead stems of *Dryopteris* sp. from South Devon (Slapton) has not been verified and lacks voucher material.

geophylla *var.* fulva (Pat.) Sacc., Inocybe, *Syll. fung.* 5: 785 (1887)
Agaricus geophilus var. *fulvus* Pat., *Tab. anal. fung.* 6: 40 (1886)
A *nomen dubium*. Possibly *Inocybe obscurobadia fide* Kuyper (1986). Listed by Rea (1922).

georgii Sowerby, Agaricus [*nom. illegit.*, non *A. georgii* L. (1753)], *Col. fig. Engl. fung.* 3: pl. 304 (1801)
Described from Britain, but a *nomen dubium* lacking type material. A true species of *Agaricus*, but specifically unidentifiable.

germanus Fr., Cortinarius, *Epicr. syst. mycol.*: 312 (1838)
A *nomen dubium*. Possibly near to *C. cagei*. See Cooke 851 (844) Vol. 6 (1887) for an interpretation.

gibbosa (Lév.) Quél., Calyptella, *Enchir. fung.*: 216 (1886)
Cyphella gibbosa Lév., *Ann. Sci. Nat., Bot.*, sér. 3, 9: 126 (1848)
Not authentically British. Mentioned on dead stems of *Solanum tuberosum* by W.B. Cooke (1961) but no voucher material traced. Doubtfully distinct from *Calyptella capula*.

gibbosa J. Favre, Galerina, *Bull. Soc. Mycol. France* 53: 140 (1936)
Not authentically British. Reported from South Hampshire and Lancashire but apparently lacking voucher material.

gibbsii Massee & Crossl., Coprinus, *Naturalist (Hull)* 1902: 1 (1902)
Described from Britain but a *nomen dubium*. Possibly *C. miser*.

gilvus (Schwein.) Pat., Phellinus, *Essai taxon.*: 82 (1906)
Boletus gilvus Schwein., *Schriften Naturf. Ges. Leipzig* 1: 96 (1822)
Polyporus gilvus (Schwein.) Fr., *Elench. fung.* 1: 104 (1828)
Not authentically British. Listed by Rea (1922) but no records have been traced. This is a common pantropical species, unlikely to occur in Britain.

glandicalyx With., Agaricus, *Arr. Brit. pl. ed. 3*, 4: 228 (1796)
Described from Britain but a *nomen dubium*.

glandiformis (Cooke) Sacc., Naucoria, *Syll. fung.* 5: 839 (1887)

Agaricus glandiformis Cooke, *Grevillea* 13: 59 (1885)
Described from Britain but a *nomen dubium*. See also illustration in Cooke 500 (490) Vol. 4 (1884).

glareosa (Berk. & Broome) Sacc., Psathyra, *Syll. fung.* 5: 1068 (1887)
Agaricus glareosa Berk. & Broome, *Ann. Mag. Nat. Hist., Ser. 5 [Notices of British Fungi no. 2011]* 7: 372 (1883)
A *nomen dubium*. Cooke 610 (591) Vol. 4 (1886) suggests *Mycena zephirus fide* Pearson & Dennis (1948).

glauca (Batsch) Bon, Arrhenia, *Doc. Mycol.* 18(69): 37 (1987)
Agaricus glaucus Batsch, *Elench. fung. (Continuatio Prima)*: 169 & t. 24 f. 123 (1786)
Cantharellus glaucus (Batsch) Fr., *Hymenomyc. eur.*: 460 (1874)
Dictyolus glaucus (Batsch) Quél., *Enchir. fung.*: 140 (1886)
Not authentically British. Cooke [1065 (1115) B Vol. 7 (1890) as *Cantharellus glaucus*] is doubtful, possibly *Leptoglossum queletii* (though this too is not authentically British). Listed by Rea (1922) as *Dictyolus glaucus*, citing Cooke's illustration.

glaucophylla (Lasch) Gillet, Omphalia, *Hyménomycètes*: 297 (1876)
Agaricus glaucophyllus Lasch, *Linnaea* 3: 393 (1828)
A *nomen dubium* lacking type material. No known *Omphalia* or *Omphalina* has spores as small as given by Rea (1922). Cooke 1153 (959) Vol. 8 (1890) is unidentifiable.

globispora (J. Blum) Bon, Russula, *Doc. mycol.* 17(65): 55 (1986)
Russula maculata var. *globispora* J. Blum, *Bull. Soc. Mycol. France* 68(2): 232 (1952)
Mis.: *Russula maculata* sensu Cooke (1889)
Not authentically British. Sarnari (1998) considers Cooke 983 (1069) Vol. 7 (1889, as *R. maculata*) to represent this species but no material has been traced and there is no other evidence for its presence in Britain.

globosum Bolton, Lycoperdon, *Hist. fung. Halifax* 3: 118 (1789)
Described from Britain but a *nomen dubium* lacking type material. From the description and illustration this is possibly *Bovista nigrescens*.

glutinifer Fr., Hygrophorus, *Epicr. syst. mycol.*: 322 (1838)
Agaricus glutinosus Bull. [*nom. illegit.*, non *A. glutinosus* Schaeff. (1770)], *Herb. France*: pl. 258 (1786)
A *nomen dubium* lacking type material. Sensu Cooke 878 (889) Vol. 6 (1888), illustrating specimens from Berkshire (near Slough), and also sensu Rea (1922) is possibly *Hygrophorus latitabundus*.

godeyi Gillet, Pluteus, *Hyménomycètes*: 395 (1876)
A *nomen dubium*. Sensu Pearson (1946) and sensu K&R is just a form of *P. thomsonii* with a smooth pileus. Sensu BFF4 is *P. cinereofuscus*.

goliath (Fr.) S. Lundell & Nannf., Tricholoma, *Fungi Exsiccati Suecici*: 1102 (1942)
Agaricus focalis var. *goliathus* Fr., *Mon. Arm. Suec.*: 4 (1854)
Armillaria focalis var. *goliathus* (Fr.) P. Karst, *Bidrag Kännedom Finlands Natur Folk* 32: 19 (1879)
Tricholoma focale var. *goliath* (Fr.) Bon, *Doc. mycol.* 6(22-23): 273 (1976)
A *nomen dubium*, possibly *T. robustum*. Cooke 51 (30) Vol. 1 (1882) also appears to be this species. Sensu Lundell & Nannfeldt is *T. nauseosum*.

goodyerae-repentis (Costantin & L.M. Dufour) R.T. Moore, Ceratorhiza, *Mycotaxon* 29: 94 (1987)
Rhizoctonia goodyerae-repentis Costantin & L.M. Dufour, *Rev. Gén. Bot.* 32: 533 (1920)
A *nomen dubium* lacking type material and inadequately described. Sensu auct. is the anamorph of *Ceratobasidium cornigerum*.

gordoniensis Berk. & Broome, Polyporus, *Ann. Mag. Nat. Hist.*, Ser. 3 [Notices of British Fungi no. 1023] 15: 319 (1865)

Poria gordoniensis (Berk. & Broome) Cooke, *Grevillea* 14: 112 (1886)
A *nomen dubium*. Referred to *Poria candidissima* (= *Trechispora* sp.) by Reid & Austwick (1963), but re-examination by K.-H. Larsson of the type collection in herb. K shows this to be a misdetermination. The skeletal hyphae and encrusted hyphal tips suggest a species of *Skeletocutis*.

gossypinum Bull., Lycoperdon, *Herb. France*: t. 435 (1790)
A *nomen dubium* lacking type material. Reported by Gray (1821) as 'growing on rotten trunks of trees' but no material has been traced. The illustration in Bulliard is not helpful.

gossypinus (Moug. & Lév.) Cooke, Polystictus, *Grevillea* 14: 81 (1886)
Polyporus gossypinus Moug. & Lév., *Ann. Sci. Nat., Bot.,* sér. 3, 9: 123 (1843)
A *nomen dubium* but a probable synonym of either *Postia caesia* or *P. subcaesia*. Included by Rea (1922).

gracilipes (Britzelm.) Bresinsky & Stangl, Fayodia, *Z. Pilzk.* 40(1-2): 73 (1974)
Agaricus gracilipes Britzelm., *Bot. Centralbl.* 73: 206 (1898)
A *nomen dubium* lacking type material but widely applied to *F. bisphaerigera* e.g. in BFF8 and B&K3.

gracilis With., Agaricus [*nom. illegit.*, non *A. gracilis* Pers. (1801)], *Bot. arr. Brit. pl. ed. 2*, 3: 313 (1792)
Described from Britain but a *nomen dubium*.

gracilis (Quél.) Watling, Flammulaster, *Notes Roy. Bot. Gard. Edinburgh* 28(1): 68 (1967)
Pholiota muricata var. *gracilis* Quél., *Bull. Soc. Amis Sci. Nat. Rouen.* 9: 40 (1880) [1879]
Naucoria muricata var. *gracilis* (Quél.) Romagn., *Bull. Soc. Mycol. France* 58: 133 (1942)
Not authentically British. Collections named thus in herb. K are all misdetermined.

gracilis (Quél.) Singer, Hemimycena, *Ann. Mycol.* 41: 121 (1943)
Omphalia gracilis Quél., *Compt. Rend. Assoc. Franç. Avancem. Sci.* 9: 662 (1881)
Mycena gracilis (Quél.) Kühner, *Encycl. Mycol.* 10, *Le Genre* Mycena: 650 (1938)
Marasmiellus gracilis (Quél.) Singer, *Lilloa* 22: 299 (1951) [1949]
Delicatula gracilis (Quél.) Kühner & Romagn. [*comb. inval.*], *Flore Analytique des Champignons Supérieurs*: 119 (1953)
Mis.: *Hemimycena pithya* sensu B&K3: 194 & pl. 219 (1991)
Mis.: *Mycena lactea* var. *pithya* sensu Rea (1922)
Not authentically British. Included in BFF8 (p. 119) but the only two collections named thus in herb. K are old and doubtful.

gracilis Burl., Russula, *North American Flora, Ser. 1 (Fungi 3)*, 9(4): 222 (1915)
Not authentically British. A North American species. Recent records and collections so named represent *Russula gracillima* fide G. Kibby (pers. comm.).

gracillima (Weinm.) P. Kumm., Omphalia, *Führ. Pilzk.*: 106 (1871)
Agaricus gracillimus Weinm., *Hymen. Gasteromyc.*: 121 (1836)
A *nomen dubium* lacking type material. Cooke 267 (252) Vol. 2 (1883) appears to be a species of *Hemimycena*.

gracillima Peck, Psathyrella, *Bull. Torrey Bot. Club* 23: 417 (1896)
Not authentically British. A North American species, reported from East Gloucestershire (Lineover Wood) in 1997 but unsubstantiated with voucher material and probably misidentified.

gracillima Berk. & Broome, Typhula, *Ann. Mag. Nat. Hist., Ser. 5 [Notices of British Fungi no. 1699]* 1: 25 (1878)
Described from Britain (Scotland: Rannoch) but a *nomen dubium* now represented only by drawings (in herb. K).

According to Corner (1950) synonymous with *Pistillaria uncialis* (i.e. *Typhula uncialis*).

grallipes Fr., Cortinarius, *Epicr. syst. mycol.*: 275 (1838)
Not authentically British. Last reported from England (Herefordshire) in the late nineteenth century, but this report and Cooke 738 (734) Vol. 5 (1888) are unlikely to be *C. grallipes* as currently understood.

graminea (Cejp) Kühner, Mycena, *Encycl. Mycol.* 10. *Le Genre* Mycena: 190 (1938)
Pseudomycena graminea Cejp, *Spisy Přír. Fak. Karlovy Univ.* 104: 148 (1930) [1929]
A *nomen dubium*. Included in NCL.

gramineus Dicks., Agaricus, *Fasc. pl. crypt. brit.* 3: 20 (1793)
Described from Britain, but a *nomen dubium*.

graminicola (Velen.) Kühner & Maire, Melanoleuca, *Bull. Soc. Mycol. France* 50: 18 (1934)
Tricholoma graminicola Velen., *Novit. mycol.*: 62 (1939)
A *nomen dubium* used for both *Melanoleuca stridula* and *M. melaleuca* sensu stricto. British collections named thus require re-investigation.

graminicola (Nees) P. Kumm., Naucoria, *Führ. Pilzk.*: 76 (1871)
Agaricus graminicola Nees, *Syst. Pilze*: 38, f. 186 (1817)
A *nomen dubium* probably not distinct from *Crinipellis scabellus* fide Pearson and Dennis (1948). See Cooke 515 (513) Vol. 4 (1885) for an illustration.

graminicola Bon, Tephrocybe, *Doc. mycol.* 7(25): 58 (1976)
Not authentically British. Reported from East Norfolk and Warwickshire but unsubstantiated with voucher material.

graminicolor Quél., Russula, *Fl. mycol. France*: 347 (1888)
A *nomen dubium* lacking type material. Sensu Rea (1922) is *R. aeruginea*.

graminum (Liberta) Berk. & Broome, Marasmius, *Outl. Brit. fungol.*: 222 (1860)
Agaricus graminum Liberta, *Pl. Crypt. Ardienn.* Fasc. 2: no. 119 (1832)
Not authentically British. The name is often misapplied to *Marasmius curreyi*. True *M. graminum* is known only from single locations in Belgium and Switzerland fide A&N1.

grandiusculus Cooke & Massee, Bolbitius, *Grevillea* 18: 53 (1890)
Described from Britain but a *nomen dubium*. Possibly a *Conocybe* fide BFF3. See Cooke 1187 (1159) Vol. 8 (1891) for an illustration.

granulosum (Pers.) Boidin & Lanq., Dichostereum, *Mycotaxon* 6(2): 284 (1977)
Thelephora granulosa Pers., *Syn. meth. fung.*: 576 (1801)
Hydnum granulosum (Pers.) Pers., *Mycol. eur.* 2: 184 (1825)
Grandinia granulosa (Pers.) P. Karst., *Rev. Mycol. (Toulouse)* 3(9): 20 (1881)
Asterostromella granulosa (Pers.) Bourdot & Galzin, *Hyménomyc. France*: 396 (1928)
Not authentically British. Included by Rea (1922) as *Grandinia granulosa* but all British collections named thus belong to other species, usually *Brevicellicium olivascens*.

graveolens With., Agaricus [*nom. illegit.*, non *A. graveolens* Pers. (1801)], *Bot. arr. Brit. pl. ed. 2*, 3: 305 (1792)
Described from Britain, but a *nomen dubium* lacking type material.

graveolens (Peck) A.H. Sm. & Hesler, Pholiota, *N. Am. Pholiota*: 296 (1968)
Flammula graveolens Peck, *Bull. New York State Mus. Nat. Hist.* 150: 54 (1911)
Not authentically British. A North American species reported from the Scottish Highlands, but voucher material named thus has been redetermined as *Pholiota spumosa* by Holec (2001).

graveolens (Pers.: Fr.) P. Kumm., Tricholoma, *Führ. Pilzk.*: 131 (1871)

   Agaricus graveolens Pers. [non *A. graveolens* With. (1792)], *Syn. meth. fung.*: 361 (1801)

   Gymnopus graveolens (Pers.) Gray, *Nat. arr. Brit. pl.* 1: 609 (1821)

A *nomen dubium* lacking type material. Treated by some authorities as a synonym of *Calocybe gambosum* but the colour is described as 'fuligineo-cinereo', rather suggesting a species of *Melanoleuca*.

grisea (Peck) Singer, Hohenbuehelia, *Lilloa* 22: 255 (1951) [1949]

   Pleurotus atrocoeruleus var. *griseus* Peck, *Rep. (Annual) New York State Mus. Nat. Hist.* 44: 147 (1891)

   Pleurotus griseus (Peck) Peck, *Rep. (Annual) New York State Mus. Nat. Hist.* 131: 25 (1909)

Not authentically British. A North American species. British material so named is doubtfully distinct from *H. atrocaerulea*.

grisea (Fr.) Quél., Omphalia, *Mém. Soc. Émul. Montbéliard, Sér. 2*, 5: 66 (1872)

   Agaricus griseus Fr., *Epicr. syst. mycol.*: 127 (1838)

A *nomen dubium*. Sensu Fries is unknown. Sensu auct. Brit. is *Mycena cinerella*.

griseofumosa (Secr.) Singer & Clémençon [*nom. inval.*], Melanoleuca, *Nova Hedwigia* 23(2-3): 325 (1973) [1972]

   Agaricus griseofumosus Secr. [*nom. inval.*], *Mycogr. Suisse* 2: 93 (1833)

British material doubtfully assigned to this species needs re-assessment *fide* Watling (BFF8: 75).

griseola (Rea) Corner, Clavulinopsis, *Monograph of Clavaria and Allied Genera*: 370 (1950)

   Clavaria griseola Rea, *Trans. Brit. Mycol. Soc.* 17(1-2): 50 (1932)

Described from Britain but a *nomen dubium*. Possibly near to *C. umbrinella*.

griseolilacinus Britzelm., Cortinarius, *Zur Hymenomyceten-Kunde* 1: 2 (1895)

Not authentically British. Reported from Glamorganshire in 1994 but unsubstantiated with voucher material.

griseoluridum (Kühner) M.M. Moser, Entoloma, *Kleine Kryptogamenflora*, Edn 4: 196 (1978)

   Rhodophyllus griseoluridus Kühner, *Rev. Mycol. (Paris)* 19(1): 4 (1954)

Not authentically British. Reported from England (South Devon: Membury) and Scotland (Isle of Mull) but unsubstantiated with voucher material.

griseopallida (Weinm.) Park.-Rhodes [*comb. inval.*], Calyptella, *Trans. Brit. Mycol. Soc.* 37: 332 (1954)

   Cyphella griseopallida Weinm., *Hymen. Gasteromyc.*: 522 (1836)

   Cellypha griseopallida (Weinm.) W.B. Cooke, *Sydowia* 4: 54 (1961)

A *nomen dubium* lacking type material. British records probably refer to several different species.

griseorubellum (Lasch) Kalamees & Urbonas, Entoloma, *Conspectus florum agaricalium fungorum (Agaricales s.l.) Lithuaniae, Latviae et Estoniae (materies 1778-1984 annorum)*: 39 (1986)

   Agaricus griseorubellus Lasch, *Linnaea* 4: 542 (1829)

   Eccilia griseorubella (Lasch) P. Kumm., *Führ. Pilzk.*: 94 (1871)

   Leptonia griseorubella (Lasch) P.D. Orton, *Trans. Brit. Mycol. Soc.* 43(2): 177 (1960)

A *nomen dubium* lacking type material and variously interpreted. Sensu NCL is *Entoloma griseorubidum*.

griseovelata Kühner, Inocybe, *Bull. Soc. Naturalistes Oyonnax* 9 (Suppl.): 4 (1955)

Not authentically British. A single collection named thus in herb. K, from Kent, is probably misidentified *Inocybe albovelutipes fide* Kuyper (1986).

guttatum P. Kumm., Tricholoma, *Führ. Pilzk.*: 132 (1871)

   Agaricus guttatus Schaeff. [*nom. illegit.*, non *A. guttatus* Pers. (1793): Fr.], *Fung. Bavar. Palat. nasc.* 3: 240 (1770)

   Tricholoma orirubens var. *guttatum* (P. Kumm.) A. Pearson & Dennis, *Trans. Brit. Mycol. Soc.* 31: 151 (1948)

A *nomen dubium* lacking type material. The single collection named thus in herb. K is badly deteriorated and not determinable. Cooke 76 (59) Vol.1 (1882) is distinctive but also not determinable.

guttulatus Bres., Panaeolus, *Fungi trident.* 1: 36 (1881)

Not authentically British. A single old collection from Scotland (Angus: Glamis) in herb. K may represent this species, but is in poor condition and requires confirmation.

gymnopodia (Bull.) A.F.M. Reijnders, Pholiota, *Persoonia* 17(1): 113 (1998)

   Agaricus gymnopodius Bull., *Herb. France*: pl. 601 (1798)

   Flammula gymnopodia (Bull.) Quél., *Mém. Soc. Émul. Montbéliard, Sér. 2*, 5: 346 (1873)

   Clitocybe gymnopodia (Bull.) Gillet, *Hyménomycètes*: 162 (1874)

A *nomen dubium* lacking type material and not authentically British. Sensu Quélet is *Armillaria tabescens fide* Pearson & Dennis (1948). Cooke 465 (431) Vol. 4 (1884) could be this, but *fide* NCL may be a luxuriant form of *Gymnopilus sapineus* (though the illustration shows decurrent lamellae). Sensu Reijnders is also a species of *Gymnopilus* probably *G. sapineus* s.l. *fide* Holec (2001).

gyroflexa (Fr.) P. Kumm., Psathyra, *Führ. Pilzk.*: 70 (1871)

   Agaricus gyroflexus Fr., *Epicr. syst. mycol.*: 232 (1838)

A *nomen dubium*, probably a species of *Coprinus fide* Kits van Waveren (Kits1). Cooke 1184 (970) Vol. 8 (1891) is clearly a species of *Coprinus*.

haematites (Berk. & Broome) Kühner & Maire, Cystoderma, *Icon. select. fung.* 3: 237 (1924)

   Agaricus haematites Berk. & Broome, *Ann. Mag. Nat. Hist., Ser. 5 [Notices of British Fungi no. 1635]* 1: 18 (1878)

   Armillaria haematites (Berk. & Broome) Sacc., *Syll. fung.* 5: 77 (1887)

   Lepiota haematites (Berk. & Broome) Ricken, *Blätterpilze Deutschl.*: 328 (1914)

A name widely used for *Cystoderma superbum* but the type (in herb. K) from Scotland (Angus: Glamis) is probably not a *Cystoderma fide* Orton in NCL p.191.

haematochelis (Bull.) Fr., Cortinarius, *Epicr. syst. mycol.*: 302 (1838)

   Agaricus haematochelis Bull., *Herb. France*: pl. 527 (1791)

A *nomen dubium* lacking type material. Sensu Rea (1922) is *C. armillatus*.

halophila R. Heim, Inocybe, *Encycl. Mycol. 1. Le Genre Inocybe*: 242 (1931)

A *nomen dubium*. Probably *Inocybe pruinosa fide* Kuyper (1986). Sensu Pearson [TBMS 26: 45 (1943)] is *I. impexa*. Sensu NCL probably also includes *I. vulpinella*.

hamadryas (Fr.) Sacc., Naucoria, *Syll. fung.* 5: 830 (1887)

   Agaricus hamadryas Fr., *Öfvers. Kongl. Vetensk.-Akad. Förh.* 8: 47 (1852)

A *nomen dubium*. Possibly *Agrocybe arvalis fide* Pearson & Dennis (1948). Cooke 1172 (965) Vol. 8 (1890) is unknown.

hariolorum (Bull.) Quél., Collybia, *Mém. Soc. Émul. Montbéliard, Sér. 2*, 5: 94 (1872)

   Agaricus hariolorum Bull., *Herb. France*: pl. 58 (1781)

   Marasmius hariolorum (Bull.) Quél., *Fl. mycol. France*: 320 (1888)

   Gymnopus hariolorum (Bull.) Antonín, Halling & Noordel., *Mycotaxon* 63: 364 (1997)

Not authentically British. Frequently reported, but not substantiated with voucher material. The name has also been widely misapplied to *Collybia confluens*.

harmajae Lamoure, Clitocybe, *Travaux Scientifiques du Parc National de la Vanoise*: 132 (1972)
Not authentically British. A species of high montane and boreal habitat, reported from England and the Republic of Ireland from lowland deciduous woodland and unsubstantiated with voucher material. Considered a synonym of *C. amarescens* in FAN3.

hastata (Litsch.) J. Erikss., Hyphodontia, *Symb. Bot. Upsal.* 16(1): 104 (1958)
 Peniophora hastata Litsch., *Oesterr. Bot. Z.* 77(2): 130 (1928)
Not authentically British. Reported from Monmouthshire (Wyndcliff) on dead lianas of *Clematis vitalba* in 1992 but unsubstantiated with voucher material.

hastifer Pouzar, Inonotus, *Česká Mykol.* 35(1): 25 (1981)
Not authentically British. Listed by Ryvarden & Gilbertson (EurPoly1) but no records or voucher material have been traced.

hebes (Fr.) Sacc., Psilocybe, *Syll. fung.* 5: 1054 (1887)
 Agaricus hebes Fr., *Syst. mycol.* 1: 293 (1821)
A *nomen dubium*. Cooke 603 (589) Vol. 4 (1886) is unidentifiable.

helobia (Kalchbr.) Sacc., Psathyra, *Syll. fung.* 5: 1072 (1887)
 Agaricus helobius Kalchbr., *Icon. select. Hymenomyc. Hung.* 2: t.17 (1874)
A *nomen dubium*. Probably not a species of *Psathyrella* and not mentioned in Kits1. See Cooke 619 (579) Vol. 4 (1886) (as *Agaricus helobius*) for an illustration.

helobius Romagn., Cortinarius, *Bull. Soc. Naturalistes Oyonnax* 6: 62 (1952)
Not authentically British. Reported from England and Northern Ireland but all records are dubious and unsubstantiated with voucher material.

helvelloides Henn., Cantharellus, *Hedwigia* 43: 181 (1904)
A *nomen dubium*. Sensu auct. Brit. is *Arrhenia rickenii*.

helvelloides Sowerby, Merulius, *Col. fig. Engl. fung.*, Suppl.: pl. 402 (1809)
Described from Britain but a *nomen dubium* lacking type material. Sowerby's illustration is not helpful.

helveola Bres., Lepiota, *Fungi trident.* 1: 15 (1881)
A *nomen dubium* variously interpreted. Sensu Rea (1922) is *Lepiota brunneoincarnata*. Sensu Reid [TBMS 84:719 (1985)] is *L. brunneolilacea*.

helvolus Fr., Cortinarius, *Epicr. syst. mycol.*: 296 (1838)
A *nomen dubium*. Several old records from England and Scotland and several recent ones from Northern Ireland, but all unsubstantiated with voucher material.

hemerobius Fr., Coprinus, *Epicr. syst. mycol.*: 253 (1838)
A *nomen dubium*. Sensu BFF2 is doubtful, possibly only a form of *C. auricomus* with scarce cap setae. Sensu B&K4: 234, pl. 279 is also doubtful.

hemichrysus (Berk. & M.A. Curtis) Pilát, Buchwaldoboletus, *Friesia* 9: 217 (1969)
 Boletus hemichrysus Berk. & M.A. Curtis, *Ann. Mag. Nat. Hist.* Ser. 2, 12: 429 (1853)
Not authentically British. A North American species. Sensu auct. Brit. is *Buchwaldoboletus sphaerocephalus*.

hemisphaerica Peck, Mycena, *Rep. (Annual) New York State Mus. Nat. Hist.* 46: 104 (1892)
Listed in NCL but unknown to Maas Geesteranus (1992) in Europe. Sensu NCL is probably just a dark form of *Mycena galericulata*. No British records or voucher collections have been traced.

henningsii (Bres.) P.D. Orton, Pholiota, *Trans. Brit. Mycol. Soc.* 43(2): 180 (1960)
 Flammula henningsii Bres., Hennings, *Verh. Bot. Vereins Prov. Brandenburg* 31: 171 (1889)
Not authentically British. A single collection in herb. K named thus, from South Wiltshire (Wardour Wood), was a misidentified species of *Hypholoma*.

henriettae (W.G. Sm.) Sacc., Collybia, *Syll. fung.* 17: 15 (1905)
 Agaricus henriettae W.G. Sm., *J. Bot.* 41: 139 (1903)
'Intermediate between *Agaricus radicatus* and *Agaricus longipes*' *fide* W.G. Smith, and indeed a species of *Xerula* according to Antonín & Noordeloos (A&N2).

heteroclitus (Bolton) Fr., Polyporus, *Syst. mycol.* 1: 344 (1821)
 Boletus heteroclitus Bolton, *Hist. fung. Halifax* Suppl.: 164 (1791)
Described from Britain but a *nomen dubium* lacking type material and variously interpreted. Sensu Bolton is possibly *Grifola frondosa*. Sensu Sowerby [Col. fig. Engl. fung. 3: pl. 367 (1802)], depicting an imbricate polypore on wood, could be *Inonotus cuticularis*.

heterocystis (G.F. Atk.) A.H. Sm. & Singer, Galerina, *Sydowia* 11: 447 (1958) [1957]
 Galerula heterocystis G.F. Atk., *Proc. Amer. Philos. Soc.* 57: 362 (1918)
Not authentically British (nor European). A North American species. Sensu auct. = *G. clavata*.

heteromorpha (Fr.) Donk, Antrodia, *Persoonia* 4: 339 (1961)
 Daedalea heteromorpha Fr., *Syst. mycol.* 1: 340 (1821)
 Lenzites heteromorphus (Fr.) Fr., *Epicr. syst. mycol.*: 407 (1838)
Not authentically British. Material named thus in herb. K is all *Antrodia albida*.

hiatulus Romagn., Pluteus, *Flore Analytique des Champignons Supérieurs*: 421 (1953)
A *nomen confusum*. Probably one of the numerous forms of *P. plautus*. Reported from Derbyshire and Nottinghamshire in 1956, but lacking voucher material.

hiemale Bres., Hebeloma, *Fungi trident.* 2(11-13): 52 (1898)
A *nomen dubium*. British records are, at least in part, *H. fragilipes*.

hilaris (Fr.) Bon, Phaeocollybia, *Doc. mycol.* 21(83): 37 (1991)
 Agaricus hilaris Fr., *Syst. mycol.* 1: 254 (1821)
A *nomen dubium*. Said to occur in Britain *fide* BFF7: 76 but no records traced.

hillieri Rob. Henry, Cortinarius, *Bull. Soc. Mycol. France* 54: 107 (1938)
Not authentically British. A poorly known member of the *C. turgidus* complex. Dubious records from East Kent and Hertfordshire lacking voucher material.

hippopinus With., Agaricus, *Arr. Brit. pl. ed. 3*, 4: 202 (1796)
Described from Britain 'on cones of the Scotch Fir' but a *nomen dubium*. Possibly old basidiomes of *Strobilurus tenacellus* or *Baeospora myosura* but the description (with light brown gills) does not match these well.

hirtus Fr., Panus, *Linnaea* 5: 508 (1830)
Not authentically British. Described from Brazil and a synonym of *Lentinus scleropus*. Sensu auct. Brit. is *Panus conchatus*.

hispida (Lasch) Gillet, Lepiota, *Hyménomycètes*: 60 (1874)
 Agaricus hispidus Lasch, *Linnaea* 4: 518 (1829)
A *nomen dubium* lacking type material, variously applied to several members of *Lepiota* section *Echinatae*.

hiulca (Fr.) Gillet, Inocybe, *Hyménomycètes*: 517 (1876)
 Agaricus hiulcus Fr., *Epicr. syst. mycol.*: 175 (1838)
 Astrosporina hiulca (Fr.) Rea, *Brit. basidiomyc.*: 214 (1922)
A *nomen dubium*. The British record was misidentified *Inocybe godeyi fide* Rea (1922).

holophaeus J.E. Lange, Cortinarius, *Fl. agaric. danic.* 5, *Taxonomic Conspectus*: II (1940)
Not authentically British. Reported from West Sussex by Pearson [TBMS 29 (4): 197 (1946)] but lacking voucher material.

hordum (Fr.) Quél., Tricholoma, *Mém. Soc. Émul. Montbéliard, Sér. 2,* 5: 212 (1872)
Agaricus hordus Fr., *Syst. mycol.* 1: 47 (1821)
Not authentically British. 'Doubtful' *fide* NCL but this is a good species which is apparently rare in Europe.

humile *var.* evectum Grove, Tricholoma, *J. Bot.* 50: 9 (1912)
Described from Britain but a *nomen dubium*. Riva (FungEur3) suggests this is possibly *Melanoleuca verrucipes*. The description in Rea (1922) does not support that idea.

humilis (Pers.) Pat., Melanoleuca, *Essai taxon.*: 159 (1900)
Agaricus humilis Pers., *Syn. meth. fung.*: 360 (1801)
Tricholoma humile (Pers.) Quél., *Mém. Soc. Émul. Montbéliard, Sér. 2,* 5: 2 (1872)
A *nomen dubium* lacking type material. Sensu Rea (1922), NCL and BFF8 is doubtful. Sensu B&K3: pl. 302 is *M. turrita*.

hyalina Pers., Tremella, *Mycol. eur.* 1: 105 (1822)
A *nomen dubium* lacking type material. Considered by Donk (1966) to be an earlier name for *Exidia nucleata* but this was rejected by Reid (Reid4).

hybridus (Sowerby) Maire, Gymnopilus, *Treb. Mus. Ci. Nat. Barcelona,* ser. bot., 15(2): 96 (1933)
Agaricus hybridus Sowerby [*nom. illegit.*, non *A. hybridus* Scop. (1772); non *A. hybridus* Bull. (1788)], *Col. fig. Engl. fung.* 2: pl. 221 (1799)
Agaricus sapineus β *hybridus* (Sowerby) Fr., *Syst. mycol.* 1: 239 (1821)
Flammula hybrida (Sowerby) Gillet, *Hyménomycètes*: 532 (1876)
Fulvidula hybrida (Sowerby) Singer, *Rev. Mycol. (Paris)* 2(6): 239 (1937)
A *nomen dubium* lacking type material. Sensu Fries (1821) and sensu auct. is *Gymnopilus penetrans*.

hydrogramma (Bull.) P. Kumm., Clitocybe, *Führ. Pilzk.*: 122 (1871)
Agaricus hydrogrammus Bull., *Herb. France*: pl. 564 (1792)
Omphalia hydrogramma (Bull.) Gillet, *Hyménomycètes*: 290 (1876)
A *nomen dubium* lacking type material. Bulliard's plate depicts four or five different agarics (apparently all species of *Clitocybe*). Sensu NCL is *C. phaeophthalma*.

hydrophora (Bull.) Gillet, Psathyrella, *Hyménomycètes*: 616 (1878)
Agaricus hydrophorus Bull., *Herb. France*: pl. 558 (1786)
Coprinarius hydrophorus (Bull.) P. Kumm., *Führ. Pilzk.*: 68 (1871)
Coprinus hydrophorus (Bull.) Quél., *Fl. mycol. France*: 47 (1888)
A *nomen dubium* lacking type material. Cooke 638 (655) Vol. 5 (1886) is unknown.

hydrophorum (Bull.) Sowerby, Lycoperdon, *Col. fig. Engl. fung.* 1: pl. 23 (1796)
Peziza hydrophora Bull., *Herb. France*: pl. 410 f. 3 (1789)
Unknown. Related to *Sphaerobolus stellatus* fide Sowerby (Sow1: pl. 22, as *Lycoperdon carpobolus*), but doubtfully a basidiomycete and more likely a species of the ascomycete genus *Nectria*.

hypnophilum (P. Karst.) J. Eriks. & Hjortstam, Erythricium, *Svensk bot. Tidskr.* 64(2): 168 (1970)
Corticium hypnophilum P. Karst., *Rev. Mycol. (Toulouse)* 12(47): 126 (1890)
Not authentically British. The single collection named thus, from the Republic of Ireland, was redetermined as *Erythricium laetum*.

hypnorum (Brond.) Sacc., Cantharellus, *Syll. fung.* 11: 32 (1895)

Agaricus hypnorum Brond., *Flora Agenaise* 4: t. 14 (1828)
Clitocybe hypnorum (Brond.) Rea, *Brit. basidiomyc.*: 274 (1922)
A *nomen dubium* lacking type material. Possibly a form of *Hygrophoropsis aurantiaca* fide NCL.

hypnorum *var.* bryorum P. Kumm., Galera, *Führ. Pilzk.*: 75 (1871)
Agaricus hypnorum β *bryophilus* Pers., *Syn. meth. fung.*: 385 (1801)
Agaricus bryorum (P. Kumm.) Fr., *Hymenomyc. eur.*: 270 (1874)
Galera bryorum (P. Kumm.) Sacc., *Syll. fung.* 5: 868 (1887)
A *nomen dubium* lacking type material. Sensu Rea (1922) is doubtful.

hypogaeus Fuckel, Irpex, *Jahrb. Nassauischen Vereins Naturk.* 27-28: 88 (1873)
A *nomen dubium*. Listed by Rea (1922) as 'encrusting fallen pine leaves, twigs, earth and stones'. Collections thus named in herb. K are *Sebacina incrustans*.

hypsipus (Fr.) Konrad & Maubl. [as *hypsipoda*], Psathyrella, *Encycl. Mycol.* 14, *Les Agaricales* I: 128 (1948)
Agaricus hypsipus Fr., *Epicr. syst. mycol.*: 218 (1838)
Stropharia hypsipus (Fr.) Sacc., *Syll. fung.* 5: 1025 (1887)
A *nomen dubium*. Sensu auct. is *Psathyrella leucotephra*. Cooke 571 (619) Vol. 4 (1886) is distinctive but unidentifiable.

hysginoides M. Korhonen & T. Ulvinen, Lactarius, *Karstenia* 25: 62 (1985)
Not authentically British. The name has been doubtfully attached to the same collections as are here excluded under *L. nanus*. Reported from England (Wyre Forest) but lacking voucher material.

ianthinum (Romagn. & J. Favre) Zerov, Entoloma, *Viznachnik Gribiv Ukraïni* 5 *Basidiomycetes*: 102 (1979)
Rhodophyllus ianthinus Romagn. & J. Favre, *Rev. Mycol. (Paris)* 3(2-3): 76 (1938)
Not authentically British. *Fide* Orton (1991) this was included in NCL by error (as *Leptonia ianthina*).

ianthipes Fr., Cortinarius, *Hymenomyc. eur.*: 397 (1874)
A *nomen dubium*. Listed by Rea (1922) but no British records have been traced.

iliopodius (Bull.) Fr., Cortinarius, *Epicr. syst. mycol.*: 301 (1838)
Agaricus iliopodius Bull., *Herb. France*: pl. 583 (1793)
A *nomen dubium*. Sensu Rea (1922) with pine and beech is unknown. Sensu Moser (1967) = *Cortinarius alnetorum*.

illitus With., Agaricus, *Bot. arr. Brit. pl. ed. 2,* 3: 327 (1792)
Described from Britain but a *nomen dubium*.

imbricatus (Bull.) Fr., Polyporus, *Syst. mycol.* 1: 357 (1821)
Boletus imbricatus Bull., *Herb. France*: pl. 366 (1788)
A *nomen dubium*. Bulliard's illustration appears to depict old material of *Laetiporus sulphureus*. Sensu Sowerby [*Col. fig. Engl. fung.* 1: pl. 86 (1797)] is difficult to interpret, possibly *Meripilus giganteus*.

imbricatus (L.) P. Karst, Sarcodon, *Rev. Mycol. (Toulouse)* 3(9): 20 (1881)
Hydnum imbricatum L., *Sp. pl.*: 1178 (1753)
Not authentically British. The true *S. imbricatus* is known only with native *Picea* spp in continental Europe. British collections are all with *Pinus* spp, and are probably all *S. squamosus*.

immundum (Berk.) Kühner, Lyophyllum, *Bull. Mens. Soc. Linn. Lyon* 7: 211 (1938)
Agaricus immundus Berk., *Outl. Brit. fungol.*: 103 (1860)
Tricholoma immundum (Berk.) Quél., *Mém. Soc. Géol. France,* sér 2, 5: 43 (1872)
A *nomen dubium*. Described from sheep dung thus unlikely to be a *Lyophyllum*. Sensu NCL is *L. eustygium*.

impennis Fr., Cortinarius, *Epicr. syst. mycol.*: 293 (1838)

A *nomen dubium*. Listed by Rea (1922) but no material has been traced.

impexa (Romagn.) Bon, Psathyrella, *Doc. mycol.* 12 (48): 52 (1983) [1982]
Drosophila impexa Romagn., *Bull. Mens. Soc. Linn. Lyon* 21: 153 (1952)
Not authentically British. Reported from England (East Norfolk, North Wiltshire, Shropshire, and South Wiltshire) and the Republic of Ireland (Kildare) but lacking voucher material.

impolita Vellinga & G.M. Muell., Laccaria, *Mycotaxon* 37: 387 (1990)
Not authentically British. Reported with doubt from Scotland (Orkney) associated with *Salix repens.*

impolitum (Lasch) P. Kumm., Tricholoma, *Führ. Pilzk.*: 131 (1871)
Agaricus impolitus Lasch, *Linnaea* 4: 522 (1829)
A *nomen dubium* lacking type material, possibly near to *Tricholoma acerbum.*

impolitus Fr., Lactarius, *Monogr. hymenomyc. Suec.* 2: 175 (1863)
A *nomen dubium*. British records are probably sensu K&R and thus likely to be *Lactarius glyciosmus.*

impuber Sowerby, Boletus, *Col. fig. Engl. fung.* 2: pl. 195 (1799)
Described from Britain but a *nomen dubium* lacking type material. Fries (1874: 548) put this into the synonymy of *Phellinus gilvus*, a common pantropical species unlikely to occur in Britain. It could be a species of *Phellinus* or possibly *Hapalopilus nidulans.*

inaequalis O.F. Müll., Clavaria, *Fl. Danica*: t. 836 (1780)
A *nomen dubium* lacking type material and variously interpreted. Sensu auct. Brit. is *Clavulinopsis helvola.*

incana (Quél.) Rea, Clitocybe, *Trans. Brit. Mycol. Soc.* 4: 186 & pl. 5 (1913)
Omphalia orbiformis var. *incana* Quél., *Compt. Rend. Assoc. Franç. Avancem. Sci.* 15: 485 (1887) [1886]
Included by Rea (1922) but a *nomen dubium* lacking type material.

incarnata Quél., Russula, *Compt. Rend. Assoc. Franç. Avancem. Sci.* 11: 396 (1883)
A *nomen dubium* lacking type material. Sensu Rea (1922) is *R. aurora.*

incarnata (Johan-Olsen) Juel, Tulasnella, *Bih. Kongl. Svenska Vetensk.-Akad. Handl.* 23: 22 (1897)
Pachysterigma incarnatum Johan-Olsen, *Unters. Gesammtgeb. Mykol.* 8: 7 (1889)
A *nomen dubium* lacking type material. British material under this name in herb. K is *T. fuscoviolacea, T. tomaculum,* or *T. violea.*

incilis (Fr.) Gillet, Clitocybe, *Hyménomycètes*: 142 (1874)
Agaricus incilis Fr., *Epicr. syst. mycol.*: 69 (1838)
A *nomen dubium*. Sensu NCL is *C. costata.*

inclusa (M.P. Christ.) Donk, Tulasnella, *Persoonia* 4: 263 (1966)
Gloeotulasnella inclusa M.P. Christ., *Dansk. Bot. Ark.* 19(1): 41 (1959)
A *nomen confusum*. British material under this name in herb. K is *T. pinicola* or *T. thelephorea.*

incomptum Massee, Hypholoma, *Brit. fung.-fl.* 1: 385 (1892)
Described from Britain but a *nomen dubium.*

incrassatus Sowerby, Agaricus, *Col. fig. Engl. fung., Suppl.*: pl. 415 (1809)
Described from Britain but a *nomen dubium* lacking type material. Sowerby's illustration is of a species of *Russula* possibly *R. foetens.*

infida (Peck) Massee, Inocybe, *British Fungi*: 264 (1910)

Hebeloma infidum Peck, *Rep. (Annual) New York State Mus. Nat. Hist.* 27: 95 (1875)
Astrosporina infida (Peck) Rea, *Brit. basidiomyc.*: 209 (1922)
Not authentically British. A North American species. Sensu Massee and sensu Rea (1922) is *Inocybe umbratica.*

infucatus Fr., Cortinarius, *Monogr. hymenomyc. Suec.* 2: 309 (1863)
A *nomen dubium*. Reported by Massee (1911) but lacking voucher material.

infumata (Berk. & Broome) Sacc., Omphalia, *Syll. fung.* 5: 323 (1897)
Agaricus infumatus Berk. & Broome, *Ann. Mag. Nat. Hist., Ser. 5 [Notices of British Fungi no. 1851]* 7: 125 (1881)
A *nomen dubium*. Pearson & Dennis (1948) suggest that this may represent *Omphalia chrysophylla* (= *Chrysomphalina chrysophylla*).

infundibuliformis Quél., Clitocybe, *Mém. Soc. Émul. Montbéliard, Sér. 2,* 5: 88 (1872)
Agaricus infundibuliformis Schaeff. [*nom. illegit.*, non *A. infundibuliformis* Bolton (1788)], *Fung. Bavar. Palat. nasc.* 3: 221 (1770)
A *nomen dubium* lacking type material and variously interpreted. Sensu NCL is *C. gibba*, but the name has also been used for *Lepista flaccida.*

infundibuliformis (Bolton) With., Merulius, *Arr. Brit. pl. ed. 3,* 4: 152 (1796)
Agaricus infundibuliformis Bolton [non *A. infundibuliformis* Schaeff. (1770)], *Hist. fung. Halifax* 1: 34 (1788)
Described from Britain but a *nomen dubium* lacking type material. Fries (1821: 320) gives this as a synonym of *Cantharellus cinereus.*

infundibuliformis Sowerby, Helvella, *Col. fig. Engl. fung.* 2: pl. 153 (1798)
Described from Britain but a *nomen dubium* lacking type material. Sowerby's illustration appears to depict young basidiomes of a polypore (probably *Polyporus durus*) growing from buried wood.

ingratum Bruchet, Hebeloma, *Bull. Mens. Soc. Linn. Lyon* 39: 125 (1970)
Not authentically British. A poorly known species, with a single Scottish collection (under *Fagus*) dubiously assigned here. Described as sub-alpine and especially associated with *Populus* spp.

inhonestus P. Karst., Crepidotus, *Symb. mycol. fenn.* 17: 160 (1886)
A *nomen dubium*. Sensu NCL and BFF6 is *C. lundellii.*

injucundus (Weinm.) Fr., Cortinarius, *Epicr. syst. mycol.*: 298 (1838)
Agaricus injucundus Weinm., *Hymen. Gasteromyc.*: 150 (1836)
A *nomen dubium* lacking type material. Cooke 809 (823) Vol. 6 (1886) is doubtful.

innocua (Lasch) Sacc., Naucoria, *Syll. fung.* 5: 836 (1887)
Agaricus innocuus Lasch, *Linnaea* 3: 398 (1828)
A *nomen dubium* lacking type material. Possibly *Pholiota carbonaria* (= *P. highlandensis*) fide Pearson & Dennis (1948). Cooke 498 (489) Vol. 4 (1884) (as *Agaricus innocuus*) appears not to be that species.

inodermeum (Fr.) Sacc., Tricholoma, *Syll. fung.* 5: 103 (1887)
Agaricus inodermeus Fr., *Öfvers. Kongl. Vetensk.-Akad. Förh.* 8: 43 (1852)
Tricholoma inodermeum var. *amarum* Métrod, *Rev. Mycol. (Paris)* 4(3-4): 101 (1939)
Not authentically British. Records are old and unsubstantiated with voucher material. See TBMS 6 (4): 323 (1920) for a description of a collection from Worcestershire and also Cooke 1120 (945) Vol. 8 (1889) for an illustration of material from Herefordshire in 1885. The collection reported in Clark (1980: 172) has been redetermined as *Tricholoma sciodes.*

inodora Fr., Trametes, *Epicr. syst. mycol.*: 491 (1838)
A *nomen dubium*. Sensu Rea (1922) is doubtful, but possibly an inodorous form of *Trametes suaveolens*.

inopus (Fr.) P. Karst., Flammula, *Bidrag Kännedom Finlands Natur Folk* 32: 407 (1879)
    Agaricus inopus Fr., *Observ. mycol.* 2: 32 (1818)
A *nomen dubium* variously interpreted. Sensu Rea (1922) and Orton [Bull. Brit. Mycol. Soc. 20: 132 (1986)] is *Pholiota subochracea*.

insignitum Quél., Stereum, *Compt. Rend. Assoc. Franç. Avancem. Sci. Assoc. Sci. France* 18: 513 (1889)
Not authentically British. Sensu Rea (1922) is *Stereum subtomentosum*.

insititius Fr., Marasmius, *Epicr. syst. mycol.*: 386 (1838)
    Androsaceus insititius (Fr.) Rea, *Brit. basidiomyc.*: 532 (1922)
    Androsaceus insititius var. *albipes* (Fr.) Rea, *Brit. basidiomyc.*: 532 (1922)
    Marasmius insititius var. *albipes* Fr., *Hymenomyc. eur.*: 479 (1874)
A *nomen dubium*. Cooke 1086 (1135) Vol. 7 (1890) is difficult to interpret. British records are *M. calopus* (now *Marasmiellus vaillantii*) *fide* NCL.

instratum (Britzelm.) Massee, Hypholoma, *Brit. fung.-fl.* 1: 384 (1892)
    Agaricus instratus Britzelm., *Ber. Naturhist. Vereins Augsburg* 27: 171 (1883)
A *nomen dubium* lacking type material, but possibly *Psathyrella bipellis*. Cooke 1181 (1157) Vol.8 (1891), however, is *P. chondroderma fide* Kits van Waveren (Kits1).

insulsus (Fr.) Fr., Lactarius, *Epicr. syst. mycol.*: 336 (1838)
    Agaricus insulsus Fr., *Syst. mycol.* 1: 68 (1821)
A *nomen dubium*. Sensu Pearson (1950) is *Lactarius acerrimus*. Sensu NCL and Phillips (1981) is *L. zonarius*.

integra var. **alba** Cooke, Russula, *Handb. Brit. fung.*, Edn 2: 335 (1889)
Described from Britain but a *nomen dubium*. A doubtful variety but possibly *R. pseudointegra* which can occasionally be white. See Cooke 1038 (1094) Vol. 7 (1889) for an illustration.

intermedia F.H. Møller, Leptonia, *Fungi Faeroes* 1: 241 (1945)
A *nomen dubium*. Accepted in NCL but doubtfully distinct from *E. conferendum fide* Noordeloos (1987).

intermedius Rea, Cortinarius, *Trans. Brit. Mycol. Soc.* 12(2-3): 209 (1927)
Described from Britain but a *nomen dubium* lacking type material. A painting of the type collection (in herb. K) is suggestive of *Cortinarius amoenolens*.

intermedius Berk. & Broome, Lactarius, *Ann. Mag. Nat. Hist.*, Ser. 5 [Notices of British Fungi no. 1887] 7: 128 (1881)
    Lactarius cilicioides ssp. *intermedius* (Berk. & Broome) Sacc., *Syll. fung.* 5: 425 (1887)
    Agaricus intermedius Krombh. [*nom. illegit.*, non *A. intermedius* Scop. (1772)], *Naturgetr. Abbild. Schwämme*: t. 58 (1843)
Not authentically British. Reported from Norfolk in 1879 and Herefordshire in 1880 but no material has been traced. This is a continental montane species associated with *Picea* and *Abies* spp in native woodland, thus unlikely to occur in Britain. Sensu Rea (1922) is probably *Lactarius citriolens*.

interveniens P. Karst., Tricholoma, *Bidrag Kännedom Finlands Natur Folk*: 365 (1876)
A *nomen dubium*. Listed by Rea (1922) but no British records are known, nor voucher material traced. Not mentioned by Riva (FungEur3).

intumescens Sm., Tremella, *Engl. fl.*: t. 1870 (1790)
    Gyraria intumescens (Sm.) Gray, *Nat. arr. Brit. pl.* 1: pl. 1 (1821)
    Exidia intumescens (Sm.) Rea, *Brit. basidiomyc.*: 734 (1922)
Described from Britain but a *nomen dubium* lacking type material. Recognized by Donk (1966) as a good species but old British specimens under this name in herb. K are *Exidia glandulosa*.

intybacea Pers., Thelephora, *Syn. meth. fung.*: 567 (1801)
Synonymous with *Thelephora mollissima fide* Jülich (1984). The British collection named thus and reported in Grevillea 4: 68 (1875) was possibly a form of *Thelephora terrestris* but the material has not survived.

inunctus Krombh., Boletus, *Naturgetr. Abbild. Schwämme* 10: t. 76 (1846)
A *nomen dubium* lacking type material. Reported from Berkshire by Cooke [Grevillea 2: 133 (1874)] and see also Berkeley & Broome (Notices of British Fungi No. 1362).

involuta β **truncigena** Gray, Omphalia, *Nat. arr. Brit. pl.* 1: 611 (1821)
Described from Britain but a *nomen dubium* lacking type material and inadequately described. Synonymised by Gray with *Agaricus cyathiformis* Schaeff. i.e. *Neolentinus schaefferi*, but unlikely to have been that species.

involuta (Romagn.) M.M. Moser, Psathyrella, *Kleine Kryptogamenflora*, Edn 3: 217 (1967)
    Drosophila involuta Romagn., *Bull. Mens. Soc. Linn. Lyon* 21: 156 (1952)
Not authentically British. In the original description suggested to be a robust form of *P. marcescibilis* as also suggested by Kits van Waveren (Kits1). Reported from East Suffolk and South West Yorkshire but lacking voucher material.

involutus Soppitt, Lactarius, *Grevillea* 19: 41 (1890)
Described from Britain but a *nomen dubium*. Described as 'resembling *Lactarius vellereus* but perfectly glabrous, and almost too near to *Lactarius scoticus*'. Cooke 1195 (1194) Vol. 8 (1891) suggests a species of *Clitocybe*.

iris Massee, Cortinarius, *Brit. fung.-fl.* 2: 58 (1893)
Described from Britain but a *nomen dubium*. Said to resemble *C. paleaceus* but with larger spores.

irregulare P. Karst., Tricholoma, *Meddeland. Soc. Fauna Fl. Fenn.*: 2 (1881)
A *nomen dubium*. A synonym of *Tricholoma albobrunneum fide* Pearson & Dennis (1948) but not mentioned by Riva (FungEur3).

irregularis Bolton, Agaricus, *Hist. fung. Halifax* 1: 13 (1788)
Described from Britain but a *nomen dubium* lacking type material. Bolton's illustration vaguely resembles a species of *Entoloma*.

irregularis Sowerby, Boletus, *Col. fig. Engl. fung.*, Suppl.: pl. 423 (1809)
Described from Britain but a *nomen dubium* lacking type material. Assigned by Fries [Linnaea 5: 701 (1830)] to *Skeletocutis amorpha*.

irregularis (Fr.) Fr., Cortinarius, *Epicr. syst. mycol.*: 310 (1838)
    Agaricus caesius γ *irregularis* Fr., *Observ. mycol.* 2: 43 (1818)
Not authentically British. Partially based by Fries on a misinterpretation of a Bolton plate.

irriguus Gillet, Hygrophorus, *Hyménomycètes*: 189 (1876)
A *nomen dubium*. Reported by Cooke [Grevillea 1: 113 (1873)] from Northamptonshire (Laxton) in grassy pasture, but no material survives.

irroratum (P. Karst.) Sacc., Hypholoma, *Syll. fung.* 14: 153 (1899)
    Naematoloma irroratum P. Karst., *Hedwigia* 35: 44 (1896)
A *nomen dubium*. Pearson & Dennis (1948) suggested that British records may represent *Hypholoma radicosum*.

isabellinus Fr., Cortinarius, *Epicr. syst. mycol.*: 308 (1838)
A *nomen dubium*. Listed by Rea (1922). Sensu auct. is *C. colymbadinus*.

ischnostylum (Cooke) Sacc., Hebeloma, *Syll. fung.* 5: 802 (1887)
    Agaricus ischnostylus Cooke, *Grevillea* 12: 98 (1884)
Described from Britain but a *nomen dubium*. Cooke 463 (420) Vol. 3 could well represent a member of the *H. sacchariolens* complex.

jecorinus With., Agaricus [non *A. jecorinus* Berk. & Broome (1848)] *Bot. arr. Brit. pl. ed. 2*, 3: 297 (1792)
Described from Britain but a *nomen dubium*. From Withering's description this appears to be a species of *Lactarius*, possibly *L. hepaticus*.

jennyae (P. Karst.) Romagn., Phaeocollybia, *Bull. Soc. Mycol. France* 58: 127 (1942)
    Naucoria jennyae P. Karst., *Hedwigia* 20: 178 (1881)
Not authentically British. Sensu auct. Brit. is *Phaeocollybia arduennensis*.

johnstonii Berk., Irpex, *Outl. Brit. fungol.*: 262 (1860)
Described from Britain but a *nomen dubium*. Fide Reid & Austwick (1963) this is immature *Steccherinum ochraceum* but the material examined by them (in herb. K) is in poor condition and sterile.

jubarinus Fr., Cortinarius, *Epicr. syst. mycol.*: 309 (1838)
A *nomen dubium*. Reported from England (Cumberland and Surrey), Scotland and Wales (Cardiganshire) but lacking voucher material.

jucundissima (Desm.) Höhn., Cyphella, *Sitzungsber. Kaiserl. Akad. Wiss., Wien, Math.-Naturwiss. Cl. Abt. 1*, 127 (8-9): 549 (1918)
    Peziza jucundissima Desm., *Ann. Sci. Nat., Bot.,* sér. 3, 8: 186 (1847)
A *nomen dubium*, probably an ascomycete. Sensu Reid & Austwick (1963) may be *Seticyphella niveola*, but this latter species is not authentically British.

juncea (Fr.) Gillet, Nolanea, *Hyménomycètes*: 419 (1876)
    Agaricus junceus Fr., *Syst. mycol.* 1: 208 (1821)
A *nomen dubium*. Sensu auct. mult. is *Entoloma juncinum*. Sensu Fries and sensu Rea (1922) is an unknown sphagnicolous member of subgenus *Leptonia* near to *Entoloma asprellum*.

juncicola (R. Heim) Singer, Calocybe, *Sydowia* 15: 47 (1962) [1961]
    Tricholoma chrysenteron var. *juncicola* R. Heim, *Treb. Inst. Bot. Barcelona* 15(3): 101 (1934)
A single unconfirmed report from Scotland (Shetland) in 1985 of a species otherwise known only from the Iberian Peninsula.

juncicola (Fr.) Gillet, Mycena, *Hyménomycètes*: 282 (1876)
    Agaricus juncicola Fr., *Syst. mycol.* 1: 160 (1821)
Not authentically British. Cooke 251 (193) Vol. 2 (1882) resembles *Mycena pterigena*.

junquilleus (Quél.) Costantin & L.M. Dufour, Boletus, *Nouv. fl. champ.*, Edn 3: 298 (1901)
    Dictyopus junquilleus Quél., *Compt. Rend. Assoc. Franç. Avancem. Sci.* 26: 450 (1898) [1897]
    Boletus luridiformis var. *junquilleus* (Quél.) Knudsen, *Nordic J. Bot.* 11(4): 477 (1991)
A *nomen dubium* lacking type material and variously interpreted. Sensu NCL and BFF1 is *Boletus luridiformis* var. *discolor*. Sensu auct. mult. is *Boletus pseudosulphureus*.

karstenii Sacc. & Cub., Hygrophorus, *Syll. fung.* 5: 401 (1887)
Not authentically British. This is a species of boreal coniferous woodland. The only British record (Yorkshire: Egton Bridge in 1902), amongst short grass in open pasture, lacks voucher material and probably refers to *Hygrocybe pratensis*.

keithii Berk. & Broome, Porotheleum, *Ann. Mag. Nat. Hist., Ser. 5 [Notices of British Fungi no. 1684]* 1: 24 (1878)
Described from Britain but a *nomen dubium*. Fide Donk [*Persoonia* 1: 57 (1957)] this may be a papillate form of *Phlebia livida*.

kewensis Massee, Omphalia, *Kew Bull.* 1913: 195 (1913)
A *nomen dubium*, never recollected and possibly an alien, described from dead rhizomes of filmy ferns in a greenhouse at Kew Gardens. See also TBMS 4(2): 309 (1913).

kristiansenii Noordel., Entoloma, *Nova Hedwigia Beih.* 91: 93 (1987)
    Nolanea kristiansenii (Noordel.) P.D. Orton, *Mycologist* 5(3): 138 (1991)
Although listed as British in Orton (1991), no records have been traced.

kuehneri Bruchet, Hebeloma, *Bull. Mens. Soc. Linn. Lyon.* 39 (6 [Suppl.]): 21 (1970)
Not authentically British. Reported from Carmarthenshire (Crymlyn Burrows) but lacking voucher material. Now thought to be an arctic-alpine species.

kuehneri Hora, Lepiota, *Trans. Brit. Mycol. Soc.* 43(2): 448 (1960)
A *nomen dubium*, intended to validate *L. kuehneri* Huijsman. See FAN5: 126 for further discussion.

labrynthiformis With., Agaricus, *Bot. arr. Brit. pl. ed. 2*, 3: 397 (1792)
Described from Britain but a *nomen dubium*.

lacera (Pers.) Fr., Cyphella, *Syst. mycol.* 2: 202 (1823)
    Peziza lacera Pers., *Mycol. eur.* 1: 280 (1822)
A *nomen dubium*. Recorded by Berkeley (1860: 277) but lacking a modern interpretation.

lacerata (Litsch.) J. Erikss. & Ryvarden, Athelopsis, *Corticiaceae of North Europe* 2: 141 (1973)
    Corticium laceratum Litsch., *Ann. Mycol.* 39: 2 (1941)
    Amylocorticium laceratum (Litsch.) Hjortstam & Ryvarden, *Mycotaxon* 10(1): 206 (1979)
Not authentically British. Reported from South Hampshire (as *Amylocorticium laceratum*) but lacking voucher material.

laceratus Bolton, Agaricus [non *A. laceratus* Lasch (1838)], *Hist. fung. Halifax* 2: 68 (1788)
Described from Britain but a *nomen dubium* lacking type material. Bolton's illustration resembles a species of *Hygrocybe*.

laciniatus With., Agaricus, *Bot. arr. Brit. pl. ed. 2*, 3: 360 (1792)
Described from Britain but a *nomen dubium*.

lacrimalis Batsch, Agaricus, *Elench. fung.*: 75 t. 3 f. 7-8 (1783)
A *nomen dubium*. Reported by Withering (1796) citing Batsch Fig. 8., which is also dubious, since the illustration shows two different taxa under the same epithet.

lacrymalis (Pers.) Sommerf., Dacrymyces, *Suppl. Fl. lapp.* (Oslo): no. 1753 (1826)
    Tremella lacrymalis Pers., *Syn. meth. fung.*: 628 (1801)
    Gyraria lacrymalis (Pers.) Gray, *Nat. arr. Brit. pl.* 1: 595 (1821)
A *nomen dubium* lacking type material. Sensu auct. Brit. is probably *D. stillatus*.

lacrymans var. minor Rea, Merulius, *Brit. basidiomyc.*: 622 (1922)
    Merulius minor Falck [*nom. nud.*], in Moeller, *Hausschwammforsch.* 6: 53 (1912)
A *nomen dubium*. Possibly *Serpula himantoides*.

lactea var. pulchella (Fr.) Rea, Mycena, *Brit. basidiomyc.*: 381 (1922)
    Agaricus lacteus f. *pulchellus* Fr., *Hymenomyc. eur.*: 135 (1874)
Sensu Rea (1922) is unknown but sensu Singer is *Hemimycena rickenii* which is not authentically British.

lacteum (Fr.) Fr., Corticium, *Epicr. syst. mycol.*: 560 (1838)
    Thelephora lactea Fr., *Syst. mycol.* 1: 452 (1821)
A *nomen dubium*. Sensu Rea (1922) is *Phanerochaete tuberculata*.

lacteus (Fr.) Fr., Irpex, *Elench. fung.* 1: 145 (1828)
   Sistotrema lacteum Fr., *Observ. mycol.* 2: 266 (1818)
Not authentically British. Reported on numerous occasions but
  all voucher material named thus is misidentified, usually
  Schizopora paradoxa.

lacticaulis With., Agaricus, *Arr. Brit. pl. ed. 3*, 4: 204 (1796)
Described from Britain but a *nomen dubium*.

lactifluus Bolton, Agaricus [*nom. illegit.*, non *A. lactifluus* L.
  (1753)], *Hist. fung. Halifax* 1: 3 (1788)
Described from Britain but a *nomen dubium* lacking type
  material. Bolton's illustration is a species of *Lactarius* but is not
  specifically identifiable.

laestadii (Fr. & Berk.) Cooke, Poria, *Grevillea* 14: 110 (1886)
  Polyporus laestadii Fr. & Berk., *Ann. Mag. Nat. Hist., Ser. 5*
  [*Notices of British Fungi no. 2025*] 12: 372 (1883)
Described from Britain 'on decayed deal boards in a greenhouse'
  but a *nomen dubium*. Probably *Antrodia xantha fide* Donk
  (1974: 165).

laevigata (Lasch) Gillet, Mycena, *Hyménomycètes*: 274 (1876)
  Agaricus laevigatus Lasch, *Linnaea* 3: 388 (1828)
  Mycena galericulata var. *laevigata* (Lasch) P. Kumm, *Führ.*
  Pilzk.: 111 (1871)
Not authentically British. Reported from England (Berkshire:
  Windsor Great Park and South Hampshire: New Forest) but
  unsubstantiated with voucher material. Probably
  misdetermined *M. stipata*.

lagopinus H. Post, Marasmius, in Fries, *Hymenomyc. eur.*: 474
  (1874)
A *nomen dubium*. Listed as British by Mason & Crossland (1905)
  but no records have been traced.

lamellosus Sowerby, Merulius, *Col. fig. Engl. fung., Suppl.*: pl.
  403 (1809)
Described from Britain, but a *nomen dubium* lacking type
  material. Collected from a rotten bin in a winecellar at Charlton
  House in Kent. Sowerby's illustration could be a pallid and
  distorted specimen of *Tapinella panuoides*.

lanaripes (Cooke) Sacc., Hypholoma, *Syll. fung.* 5: 1038 (1887)
  Agaricus lanaripes Cooke, *J. Bot.* 1: 66 (1863)
Described from Britain but a *nomen dubium*. Kits Van Waveren
  (Kits1) gives this as a synonym of *Psathyrella candolleana*
  citing Cooke 585 (545) Vol. 4 (1886), but Cooke's plate bears
  little resemblance to that species.

lappula (Fr.) Quél., Leptonia, *Mém. Soc. Émul. Montbéliard,*
  Sér. 2, 5: 247 (1872)
  Agaricus lappula Fr., *Epicr. syst. mycol.*: 152 (1838)
A *nomen dubium*. Reported from Derbyshire (Baslow) in 1909
  and listed by Rea (1922) but no material has been traced.

laricina (Bolton) Sacc., Armillaria, *Syll. fung.* 5: 81 (1887)
  Agaricus laricinus Bolton, *Hist. fung. Halifax* 1: 19 (1788)
  Armillaria mellea var. *laricina* (Bolton) Barla, *Bull. Soc. Mycol.*
  France 3: 143 (1887)
Described from Britain, but a *nomen dubium* lacking type
  material. Possibly just *Armillaria mellea* but described by Bolton
  as new since it was growing on decayed wood of *Larix* sp. His
  illustration depicts a clump of small, deformed and
  unidentifiable basidiomes.

lasiosperma Bres., Mycena, *Fungi trident.* 1: 33 (1881)
Not authentically British. Material named thus has been
  redetermined as *M. bryophila* or *M. margaritispora*. Listed in
  NCL as combined by Singer in *Mycenella*, but the reference has
  not been traced.

lateralis Bolton, Boletus, *Hist. fung. Halifax* 2: 83 (1788)
Described from Britain but a *nomen dubium* lacking type
  material. The description suggests *Polyporus leptocephalus*.

lateritia (Fr.) P. Kumm, Galera, *Führ. Pilzk.*: 76 (1871)
  Agaricus lateritius Fr. [*nom. illegit.*, non *A. lateritius* Schaeff.
  (1762)], *Syst. mycol.* 1: 265 (1821)

A *nomen dubium*. Probably a *Conocybe*. Cooke 517 (460) Vol. 4
  (1885) is apparently *C. albipes*.

latissima (Cooke) Sacc., Naucoria, *Syll. fung.* 5: 850 (1887)
  Agaricus latissimus Cooke, *Grevillea* 13: 60 (1885)
Described from Britain but a *nomen dubium* inadequately
  described *fide* Pearson & Dennis (1948). Cooke 510 (482) Vol.
  4 (1885) looks distinctive.

latitabundus Britzelm., Hygrophorus, *Bot. Centralbl.* 80: 118
  pl. 437 (1899)
  Hygrophorus olivaceoalbus f. *obesus* Bres., *Fungi trident.* 1:
    84 (1887)
  Hygrophorus olivaceoalbus var. *obesus* (Bres.) Rea, *Brit.*
    basidiomyc.: 296 (1922)
  Mis.: *Hygrophorus limacinus* sensu K&R
  Mis.: *Hygrophorus fuscoalbus* sensu NCL, non Cooke (HBF2),
    non Rea (1922)
Not authentically British. There are several dubious records
  during the early twentieth century, but all are unsubstantiated
  with voucher material.

latus Bolton, Agaricus [*nom. illegit.*, non *A. latus* Pers. (1801)],
  Hist. fung. Halifax 1: 2 (1788)
Described from Britain, but a *nomen dubium* lacking type
  material. From the descripotion, it may be *Pluteus cervinus*.
  Sowerby (Sow1: pl. 108) illustrates a large agaric, with a
  strongly radially fibrillose pileus, growing on the ground,
  possibly *P. ephebeus* or a closely related species.

latus (Pers.) Fr., Cortinarius, *Epicr. syst. mycol.*: 260 (1838)
  Agaricus latus Pers. [non *A. latus* Bolton (1788)], *Syn. meth.*
  fung.: 276 (1801)
A *nomen dubium* lacking type material. Listed by Rea (1922) but
  no records have been traced.

laxipes (Fr.) Gillet, Collybia, *Hyménomycètes*: 319 (1876)
  Agaricus laxipes Fr., *Epicr. syst. mycol.*: 86 (1838)
A *nomen dubium*. *Fide* A&N2 this is possibly a species of
  Flammulina. Cooke 191 (184) Vol. 2 (1882), however, is a
  small white mycenoid-collybioid agaric with a hispid and dull
  brick-red stipe.

lenis (P. Karst.) Niemelä, Skeletocutis, *Karstenia* 31(1): 23
  (1991)
  Physisporus lenis P. Karst., in Rabenhorst, *Fungi europaei* V:
    3527 (1886)
  Poria lenis (P. Karst.) Sacc., *Syll. fung.* 6: 313 (1888)
  Amyloporia lenis (P. Karst.) Bondartsev & Singer, *Trutovye*
    griby Evropeiskoi chasti SSSR i Kavkaza: 149 (1953)
  Antrodia lenis (P. Karst.) Ryvarden, *Norweg. J. Bot.* 20: 8
    (1973)
  Diplomitoporus lenis (P. Karst.) Gilb. & Ryvarden, *Mycotaxon*
    22(2): 364 (1985)
Not authentically British. A boreal montane species, apparently
  rare in, and confined to, northern Europe and Scandinavia. The
  name has been widely used (e.g. in EurPoly2) for *S. vulgaris*.

leonina Berk. & M.A. Curtis, Hymenochaete, *J. Linn. Soc., Bot.*
  10 (46): 334 (1868)
Not authentically British (nor European). A specimen in herb. K
  collected and determined by Massee, labelled 'from
  Scarborough', is correctly identified *fide* Corfixen (*in litt.*), and
  accepted by Bramley (1985), and also by Rea (1922).
  However, it seems likely that this is due to a mix up of British
  and tropical material by Massee as the species is unlikely to
  occur here.

leoninus Krombh., Boletus, *Naturgetr. Abbild. Schwämme* 10: t.
  76 (1846)
A *nomen dubium* lacking type material. Sensu Pearson & Dennis
  (1948) and sensu auct. mult. is *Boletus moravicus*.

leporina (Fr.) P.D. Orton & Watling, Hygrocybe, *Notes Roy.*
  Bot. Gard. Edinburgh 29(1): 132 (1969)
  Hygrophorus leporinus Fr., *Epicr. syst. mycol.*: 323 (1838)

Sensu Fries is doubtful. Specimens in herb. K, collected by Broome in grassland near Bristol, are typical *H. pratensis*. Sensu Rea (1922) is *Hygrophorus nemoreus*.

leptopus Fr., Paxillus, *Monogr. hymenomyc. Suec.* 2: 311 (1863)
A *nomen dubium*. Cooke 868 (929) Vol. 6 (1888) is not helpful.

leucocephala Boud., Inocybe, *Bull. Soc. Bot. France* 32: 282 (1885)
    Astrosporina leucocephala (Boud.) Rea, *Brit. basidiomyc.*: 214 (1922)
A *nomen dubium*. Listed by Rea (1922) but no material has been traced.

leucodiatreta Bon, Clitocybe, *Bull. Soc. Mycol. France* 96(2): 165 (1980)
Not authentically British. Reported from Glamorganshire in 1992, but lacking voucher material. Doubtfully distinct from *Clitocybe diatreta*.

leucophaeus (Scop.) Fr., Hygrophorus, *Epicr. syst. mycol.*: 323 (1838)
    Agaricus leucophaeus Scop., *Fl. carniol.*: 423 (1772)
A *nomen dubium* lacking type material and variously interpreted. Sensu NCL is mostly *Hygrophorus unicolor* but also includes *H. lindtneri*.

leucophanes (Berk. & Broome) Sacc., Panaeolus, *Syll. fung.* 5: 1118 (1887)
    Agaricus leucophanes Berk. & Broome, *Ann. Mag. Nat. Hist., Ser. 3 [Notices of British Fungi no. 1127]* 18: 54 (1866)
Described from Britain but a *nomen dubium*. Sensu Cooke (HBF2) has often been considered to be *P. semiovatus*, but seems unlike that species.

leucophylla (Gillet) M. Lange & Sivertsen, Fayodia, *Bot. Tidsskr.*: 202 (1966)
    Omphalia leucophylla Gillet, *Hyménomycètes*: 296 (1876)
    Agaricus leucophyllus Fr. [*nom. illegit.*, non *A. leucophyllus* Pers. (1801)], *Öfvers. Kongl. Vetensk.-Akad. Förh.* 8: 45 (1852)
A *nomen dubium* lacking type material *fide* Antonín [*Persoonia* 18: 354 (2004)]. Sensu auct. Brit. is *Gamundia striatula*.

leucopus (Bull.) Fr., Cortinarius, *Epicr. syst. mycol.*: 311 (1838)
    Agaricus leucopus Bull., *Herb. France*: pl. 533 f. 2 (1792)
    Cortinarius krombholzii Fr., *Hymenomyc. eur.*: 395 (1874)
Not authentically British. Usually interpreted as a species near to *C. obtusus* but all records are old (last reported in 1951) or doubtful, and material in herb. K is in poor condition.

leucopus (Pers.) Maas Geest. & Nannf., Sarcodon, *Svensk bot. Tidskr.* 63: 415 (1969)
    Hydnum leucopus Pers., *Mycol. eur.* 2: 158 (1822)
    Mis.: *Hydnum laevigatum* sensu Fr. [Monogr. hymenomyc. Suec. 2: 275 (1863)] & auct. mult.
    Mis.: *Sarcodon laevigatum* sensu auct.
Not authentically British. Reported on several occasions (as *S. laevigatum*) from England (Berkshire) and from Scotland (Mid-Perthshire: Rannoch) but not since the early twentieth century. Records are unsubstantiated with voucher material.

levis Sowerby, Auricularia, *Col. fig. Engl. fung.* 3: pl. 388 f. 2 (1803)
Described from Britain but a *nomen dubium* lacking type material. Sowerby's illustration shows an unidentifiable, pale violaceous, non-poroid bracket and in the description he states that 'it may just be the remains of *Auricularia reflexa*' (i.e. *Stereum hirsutum*).

libertiana (Cooke) Agerer, Flagelloscypha, *Persoonia* 10(3): 339 (1979)
    Cyphella libertiana Cooke, *Grevillea* 8: 81 (1880)
    Lachnella libertiana (Cooke) W.B. Cooke, *Sydowia, Beih.* 4: 73 (1961)
Not authentically British. Specimens in M.C. Cooke's herbarium are those distributed by Roumeguère (as Fungi Selecti Gallici

exsiccati no. 604) *fide* W.B. Cooke (1961) and were not collected in Britain.

licinipes *var.* **robustior** Cooke, Cortinarius, *Ill. Brit. fung.* 792 (819) Vol. 6 (1887)
Described from Britain but a *nomen dubium*, possibly *Cortinarius bivelus*.

ligatus With., Agaricus, *Bot. arr. Brit. pl. ed. 2*, 3: 357 (1792)
Described from Britain but a *nomen dubium*.

lignyotus Fr., Lactarius, in Lindblad, *Monogr. Lactar. Suec.*: 25 (1855)
Not authentically British. Listed in NCL but this is a montane, continental species associated with *Picea* spp in native woodland and thus unlikely to occur in Britain. A record from East Norfolk (Foxley Wood) in 1990 is misdetermined and old material named thus (in herb. K) has been variously redetermined.

lilacea Quél., Collybia, *Mém. Soc. Émul. Montbéliard, Sér. 2*, 5: 434 (1875)
A *nomen dubium*. For B&K3, a synonym of *Baeospora myriadophylla* (not yet recorded from Britain). For A&N2, possibly a species of *Pseudobaeospora*. British records are probably sensu K&M (pl. 202) and hence *Cystoderma superbum*.

lilacinus Tul. & C. Tul., Hymenogaster, *Fungi hypogaei*: 66 (1851)
A *nomen dubium*. Sensu Berkeley [British Fungi Exsiccati, Fasc. 4 no. 305 (1843)] is *H. tener*.

limacinus (Scop.) Fr., Hygrophorus, *Epicr. syst. mycol.*: 324 (1838)
    Agaricus limacinus Scop., *Fl. carniol.*: 422 (1772)
    Gymnopus limacinus (Scop.) Gray, *Nat. arr. Brit. pl.* 1: 609 (1821)
A *nomen confusum* variously interpreted, applied to *H. persoonii* and related species.

limonicolor (Berk. & Broome) Quél., Odontia, *Enchir. fung.*: 194 (1886)
    Hydnum limonicolor Berk. & Broome, *Ann. Mag. Nat. Hist., Ser. 5 [Notices of British Fungi no. 1686]* 1: 24 (1878)
A *nomen dubium*. Described as growing 'on stones amongst pine needles' from Scotland (Glamis). Considered a synonym of *Sistotrema muscicola* by Bourdot & Galzin (1928).

limosella P.D. Orton, Nolanea, *Notes Roy. Bot. Gard. Edinburgh* 29: 108 (1969)
Described from Britain but a *nomen dubium*. As described this is a species of *Entoloma* near to *E. cetratum* but both the type packet in herb. K and the isotype at herb. E contain a species of *Hebeloma*.

limpidus (Fr.) Quél., Pleurotus, *Enchir. fung.*: 149 (1886)
    Agaricus limpidus Fr., *Epicr. syst. mycol.*: 135 (1838)
A *nomen dubium*. Cooke 286 (276) Vol. 2 (centre) is unidentifiable.

lineata (Bull.) P. Kumm., Mycena, *Führ. Pilzk.*: 110 (1871)
    Agaricus lineatus Bull., *Herb. France*: pl. 522 (1791)
    Mycena lineata var. *expallens* Fr., *Ic. Hymenomyc.* 1: pl. 78 (1873)
A *nomen dubium* lacking type material. Sensu Bulliard is doubtful. Sensu Fries is probably *Mycena flavoalba*. Sensu Lange (1936) and NCL is possibly *Mycena filopes* *fide* Maas Geesteranus (1992, 1: 171).

linkii (Fr.) P. Kumm., Leptonia, *Führ. Pilzk.*: 96 (1871)
    Agaricus linkii Fr., *Syst. mycol.* 1: 204 (1821)
Sensu Fries, growing on decayed wood, is not British. The single collection named thus, in herb. K, from Scotland (Isle of Yell: West Sandwick) has been redetermined as *Entoloma fuscomarginatum* by Orton.

linnaei Fr., Russula, *Hymenomyc. eur.*: 444 (1874)
A *nomen dubium* variously interpreted. Sensu auct. is *R. rosea*. Sensu Bresadola [Icon. Mycol. 9: pl. 416 (1929)] is *R. paludosa*. Included by Rea (1922).

lipsiense (Batsch) G.F. Atk., Ganoderma, *Ann. Mycol.* 6: 189 (1908)
  Boletus lipsiensis Batsch, *Elench. fung. (Continuatio Prima)*: 183 & t. 25 f. 130 (1786)
A *nomen dubium*. Applied to *G. applanatum* by some British and European authors, e.g. Petersen in NM3.

liquescens (Cooke) Sacc., Entoloma, *Syll. fung.* 5: 688 (1887)
  Agaricus liquescens Cooke, *Handb. Brit. fung.*, Edn 2: 121 (1886)
A *nomen dubium* typified by Cooke 328 (581) Vol. 3 (1884) which appears not to be an *Entoloma* but to represent young material of *Psathyrella spadiceogrisea*. For Noordeloos (1987) 'probably a species of *Lacrymaria*'.

listeri With., Agaricus, *Bot. arr. Brit. pl. ed. 2*, 3: 158 (1792)
Described from Britain, but a *nomen dubium* lacking type material. Sowerby (Sow1, 3) illustrates this on two different plates (104 and 245) but these show two different species. Sowerby himself says that the latter plate may represent *Lactarius acris* or *L. plumbeus* (= *turpis*). Saccardo [Syll. fung. V: 337 (1887)] gives this as a synonym of *L. vellereus* (which Sowerby pl. 104 may well represent) whilst Withering cites pl. 21 in Bolton (1788) which is *Lactarius piperatus*.

lithargyrina Bourdot & Galzin, Peniophora, *Hyménomyc. France*: 323 (1928)
A *nomen dubium*. A single record from East Suffolk (Flatford Mill) in 1956, unsubstantiated with voucher material.

lithocras D.A. Reid, Clavaria, *Trans. Brit. Mycol. Soc.* 41(4): 438 (1958)
Described from Britain but a *nomen dubium* known only from the type collection. Possibly abnormally developed material of some more familiar species.

littoreus With., Agaricus, *Arr. Brit. pl. ed. 3*, 4: 216 (1796)
Described from Britain 'on the green sward, adjoining to the sea shore at Teignmouth' but a *nomen dubium*. The description (gills reddish-grey) however, reads like an immature basidiome of a species of *Agaricus*.

lituus (Fr.) Métrod, Clitocybe, *Bull. Soc. Mycol. France* 62: 42 (1946)
  Agaricus lituus Fr., *Epicr. syst. mycol.*: 121 (1838)
  Omphalia litua (Fr.) Gillet, *Hyménomycètes*: 294 (1876)
Not authentically British. Although listed by Rea (1922) [as *Omphalia litua*] and accepted in NCL all records are unsubstantiated with voucher material. Reported more recently with doubts from Scotland (Shetland Isles).

livescens (Batsch) Bataille, Russula, *Flore monographique des Astérosporés*: 76 (1908)
  Agaricus livescens Batsch, *Elench. fung. (Continuatio Prima)*: 53 & t.14 f.67 (1786)
A *nomen dubium* variously interpreted. Most British records are likely to be *Russula insignis*.

lividoalbus Fr., Hygrophorus, *Epicr. syst. mycol.*: 324 (1838)
A *nomen dubium*. Cooke 888 (915) Vol. 6 (1888) is possibly *Hygrocybe virginea*.

lividorubescens Batsch, Agaricus, *Elench. fung. (Continuatio Secunda)*: 51 t. 36 f. 202 (1789)
A *nomen dubium*. Reported by Withering (1792). From Batsch's illustration this appears to be a species of *Lactarius* with latex becoming violaceous on exposure.

lividum var. roseum (Cooke) Sacc., Entoloma, *Syll. fung.* 5: 680 (1887)
  Agaricus lividus var. *roseus* Cooke, *Ill. Brit. fung.* 318 (469) Vol. 3 (1884)
Described from Britain but a *nomen dubium*. The original illustration suggests a colour form of *Entoloma sinuatum*.

lividus Lambotte, Lactarius, *Flore Mycologique de Belge, Suppl.* 1: (1887)
A *nomen dubium* inadequately described. 'Doubtful' *fide* NCL. Reported in TBMS 5 (2): 249 (1915) from Yorkshire (Mulgrave Woods) and also in Naturalist (Hull) 1914: 382 (1914).

lividus Cooke, Paxillus, *Grevillea* 16: 45 (1887)
Described from Britain but a *nomen dubium*. Fide Pearson & Dennis (1948) this is possibly *Tricholoma cinerascens*, i.e. a species of *Melanoleuca*. Cooke 864 (861) Vol. 6 (1888) appears to be a *Lyophyllum* sp.

lixivium (Fr.) P. Karst., Tricholoma, *Bidrag Kännedom Finlands Natur Folk* 32: 55 (1879)
  Agaricus lixivius Fr., *Epicr. syst. mycol.*: 54 (1838)
A *nomen dubium*. Cooke 126 (120) Vol. 1 (1883) could be a species of *Melanoleuca*. Riva (FungEur3) suggests that this is a synonym of *Lyophyllum semitale*.

longicaudum (Pers.) P. Kumm., Hebeloma, *Führ. Pilzk.*: 80 (1871)
  Agaricus longicaudus Pers., *Syn. meth. fung.*: 332 (1801)
A *nomen dubium* lacking type material and variously interpreted. Sensu NCL, following Lange in Fl. agaric. danic. 3 pl. 119E, is *H. leucosarx*.

longicaulis (Peck) Corner, Ramaria, *Monograph of Clavaria and Allied Genera*: 600 (1950)
  Clavaria longicaulis Peck, *Bull. Torrey Bot. Club* 25: 371 (1898)
Not authentically British. A North American species. Two dubious records (as *Clavaria longicaulis*) from South Hampshire in 1965 and 1967, unsubstantiated with voucher material.

longiseta Höhn., Mycena, *Sitzungsber. Kaiserl. Akad. Wiss., Wien, Math.-Naturwiss. Cl., Abt. 1* 118(1): 282 (1909)
Not authentically British. A tropical species described from Java. British (and European) collections thus named are *M. aciculata*.

longistriatum *var*. microsporum (Noordel.) Noordel., Entoloma, *Cryptog. Stud.* 2: 12 (1988)
  Entoloma sarcitulum var. *microsporum* Noordel., *Persoonia* 12: 461 (1985)
Not authentically British. A few records from England and Republic of Ireland, but all unsubstantiated with voucher material.

loscosii (Rabenh.) Sacc., Psathyra, *Syll. fung.* 5: 1061 (1887)
  Agaricus loscosii Rabenh., *Hedwigia* 2: 82 (1863)
A *nomen dubium*. Listed by Rea (1922). Not mentioned by Kits van Waveren (Kits1).

lucifuga (Fr.) P. Kumm., Inocybe, *Führ. Pilzk.*: 79 (1871)
  Agaricus lucifugus Fr., *Observ. mycol.* 2: 50 (1818)
A *nomen dubium*. Fide Kuyper (1986), Fries's concept is possibly not a species of *Inocybe*. Used in NCL following Heim (1931) for a species in the *Inocybe flocculosa* complex from coniferous woods.

ludia (Fr.) Gillet, Collybia, *Hyménomycètes*: 327 (1876)
  Agaricus ludius Fr., *Epicr. syst. mycol.*: 94 (1838)
A *nomen dubium*. British records are probably *Hemimycena lactea*.

luffii (Massee) Singer, Clitocybe, *Ann. Mycol.* 41: 30 (1943)
  Omphalia luffii Massee, *Trans. Brit. Mycol. Soc.* 1(1): 21 (1897)
A *nomen dubium*. From the type description this is possibly *Clitocybe albofragrans* for which it would provide an earlier name, but type material from the Channel Islands (Guernsey), in herb. K, is in poor condition.

lugens (Lasch) P. Kumm., Hebeloma, *Führ. Pilzk.*: 80 (1871)
  Agaricus lugens Lasch, *Linnaea* 3: 399 (1828)
A *nomen dubium* lacking type material. There is a single collection in poor condition, in herb. K, named thus by Broome in 1868.

lupina (Fr.) P. Karst., Flammula, *Bidrag Kännedom Finlands Natur Folk* 32: 402 (1879)
   Agaricus lupinus Fr., *Epicr. syst. mycol.*: 185 (1838)
A dubious species *fide* Holec (2001: 194). Listed by Rea (1922) but no records have been traced.

lupuletorum (Weinm.) Bres., Marasmius, *Fungi trident.* 2: 24 (1892)
   Agaricus lupuletorum Weinm., *Syll. Pl. Nov.* 2: 88 (1828)
A *nomen dubium* lacking type material. Sensu NCL is not distinct from *M. torquescens fide* Antonín & Noordeloos (A&N1).

luscina (Fr.) Singer, Lepista, *Lilloa* 22: 192 (1951) [1949]
   Agaricus luscinus Fr., *Syst. mycol.* 1: 87 (1821)
   Clitocybe luscina (Fr.) Sacc., *Syll. fung.* 5: 145 (1887)
   Melanoleuca luscina (Fr.) Metrod, *Bull. Soc. Mycol. France* 64: 154 (1948)
A *nomen dubium*. Sensu NCL & auct. mult. is *Lepista panaeolus*.

lustratus Fr., Cortinarius, *Epicr. syst. mycol.*: 258 (1838)
Not authentically British. Listed by Rea (1922). Cooke 688 (799) Vol. 5 (1887) could be *Cortinarius argentatus*.

lutea G.H. Otth, Amanita, *Berner Mitteil.*: 27 (1857)
A *nomen dubium*. Listed by Rea (1922).

lutea Gillet, Armillaria, *Hyménomycètes*: 93 (1874)
A *nomen dubium* [see Marxmüller in *Mycotaxon* 44: 272 (1992)]. Sensu Watling, Kyle & Gregory (TBMS 78: 271,1982) is *Armillaria gallica*.

lutea (Vittad.) Schild, Ramaria, *Persoonia* 9(3): 411 (1977)
   Clavaria lutea Vittad., *Descr. fung. mang.*: 228 (1835)
Not authentically British. Reported from Scotland (Dunbartonshire: Loch Lomond) in 1980, but unsubstantiated with voucher material.

lutea (Huds.) Gray, Russula, *Nat. arr. Brit. pl.* 1: 618 (1821)
   Agaricus luteus Huds., *Fl. angl.*, Edn 2, 2: 611 (1778)
Described from Britain but a *nomen dubium* lacking type material and variously interpreted. British records are mainly *R. risigallina*, but probably include *R. acetolens*.

luteofuscus With., Agaricus, *Arr. Brit. pl. ed. 3*, 4: 226 (1796)
Described from Britain but a *nomen dubium*.

luteolosperma (Britzelm.) Singer, Melanoleuca, *Cavanillesia.*: 127 (1935)
   Agaricus luteolospermus Britzelm., *Ber. Naturhist. Vereins Augsburg* 31: 160 (1894)
Excluded pending clearer definition. There are several collections labelled thus (in herb. E) from England (Northumberland) and Scotland (various sites). Mentioned in BFF8: 66 in the notes to *M. adstringens* but as used there appears to be *M. cognata* var. *nauseosa*. This concept is different from that illustrated in C&D: 205.

luteoviridans C. Martin, Russula, *Bull. Soc. bot. Genève, 2 sér.*: 187 (1894)
   Russula firmula f. *luteoviridans* (C. Martin) Kühner & Romagn., *Flora Analytique des Champignons Supérieurs.*: 464 (1953)
A *nomen dubium*. Reported from England (Hertfordshire) and Scotland (Perthshire) but unsubstantiated with voucher material. Sensu Pearson (1948) is *R. cuprea*.

lutescens Bref., Dacrymyces, *Unters. Gesammtgeb. Mykol.* 7: 152 (1888)
A *nomen dubium* lacking type material. Sensu auct. is *D. capitatus*.

luteus (Redhead & B. Liu) Redhead, Pluteus, *Sydowia* 37: 266 (1984)
   Macrocystidia lutea Redhead & B. Liu, *Canad. J. Bot.* 60: 1485 (1982)
Not authentically British. The single British record of this North American species, from Lincolnshire (Haugham Wood) in 1995, lacks voucher material and is probably a transcription error for *Pluteus lutescens* (= *P. romellii*).

luticola (Lasch) Corner, Clavulinopsis, *Monograph of Clavaria and Allied Genera*: 378 (1950)
   Clavaria luticola Lasch, in Rabenhorst, *Fungi europaei* IV: 1609 (1873)
A *nomen dubium* lacking type material. A collection in herb. K referred here by Rea [TBMS 12: 286 (1927)] needs re-examination.

luxurians (Fr.) Gillet, Pholiota, *Hyménomycètes*: 439 (1876)
   Agaricus luxurians Fr., *Epicr. syst. mycol.*: 164 (1838)
A *nomen dubium*. Cooke (387) 365 Vol. 3 (1885) is possibly a form of *Agrocybe cylindracea*.

lycoperdineus Vittad., Hymenogaster, *Monogr. Tuberac.*: 22 (1831)
Not authentically British. Listed by Rea (1922) but no material has been traced.

lycoperdoides Cooke & Massee, Schulzeria, *Trans. Brit. Mycol. Soc.* 2: 13 (1902)
   Hiatula lycoperdoides (Cooke & Massee) W.G. Sm., *Synops. Brit. Basid.*: 27 (1908)
Described under *Cedrus* sp. in Kew Gardens. A *nomen dubium*, but likely to be *Cystolepiota cystidiosa* or *C. pulverulenta*. The first was independently described as *Lepiota lycoperdoides* Kreisel and occurs in Kew glasshouses. The second is a fairly similar species, recorded outdoors at Kew.

macilenta (Fr.) Gillet, Collybia, *Hyménomycètes*: 329 (1876)
   Agaricus macilentus Fr., *Syst. mycol.* 1: 131 (1821)
A *nomen dubium*. Possibly a species of *Calocybe fide* Antonín & Noordeloos (A&N2). Cooke 208 (268) Vol. 2 (1882) is not helpful.

mackinawensis A.H. Sm., Mycena, *Pap. Michigan Acad. Sci.* 38: 67 (1953)
Not authentically British. A North American species unknown in Europe but included in NCL for a collection from Windsor Great Park by Reid [see TBMS 38(4): 391 (1955)] which requires confirmation. The description suggests *Mycena vulgaris fide* Maas Geesteranus (1992).

macrocephalus (Schulzer) Bohus, Leucopaxillus, *Fragm. Bot. Mus. Hist.-Nat. Hung.* 4: 37 (1966) [*nom. inval.*]
   Agaricus macrocephalus Schulzer, in Kalchbrenner, *Icon. select. Hymenomyc. Hung.*: 11 t. 3 (1874)
   Squamanita macrocephala (Schulzer) M.M. Moser, *Kleine Kryptogamenflora*: 66 (1953)
   Agaricus macrorhizus Lasch [*nom. illegit.*, non *A. macrorhizus* Pers. (1801)], *Linnaea* 3: 240 (1828)
   Tricholoma macrorhizum Sacc., *Syll. fung.* 5: 105 (1887)
Not authentically British. Cooke 87 (278) Vol. 1 (1883) [as *Agaricus (Tricholoma) macrorhizus*] depicts an agaric with a tesselated pileus and thick, clavate, rooting stipe but voucher material does not exist. The name equates to *Leucopaxillus macrocephalus*, a little-known species from eastern Europe.

macropus (Fr.) Fr., Cortinarius, *Epicr. syst. mycol.*: 291 (1838)
   Agaricus macropus Fr., *Syst. mycol.* 1: 215 (1821)
Not authentically British. Rarely reported and unsubstantiated with voucher material. Cooke 787 (788) Vol. 6 (1886) is doubtful.

maculata *var.* **scorzonerea** (Fr.) Gillet, Collybia:, *Hyménomycètes*: 315 (1876)
   Agaricus maculatus var. *scorzonereus* Fr., *Epicr. syst. mycol.*: 84 (1838)
   Rhodocollybia maculata var. *scorzonerea* (Fr.) Lennox, *Mycotaxon* 9: 214 (1979)
Not authentically British. Listed by Rea (1922), but no material traced.

macweeneyi W.G. Sm., Boletus, *Syn. Brit. Bas.*: 331 (1908)
Described from Britain but a *nomen dubium*. From the description this may be *Boletus rubellus* or a species close to it.

madidum (Fr.) Gillet, Entoloma, *Hyménomycètes*: 399 (1876)
   Agaricus madidus Fr., *Spicilegium Pl. neglect.*: 6 (1836)

A *nomen dubium*. Sensu Fries (with a foetid smell) is unknown. Sensu Gillet, NCL and auct. mult. is *E. bloxamii*.

magnifica (Fr.) Sacc., Amanita, *Syll. fung.* 5: 19 (1887)
Agaricus magnificus Fr., *Fl. Danica* 12(36): t. 2146 (1834)
Amanita rubescens var. *magnifica* (Fr.) Rea, *Brit. basidiomyc.*: 104 (1922)
A *nomen dubium*. Sensu Rea (1922) is *Amanita rubescens* var. *annulosulfurea*. Sensu Cooke 14 (34) Vol. 1 (1881) (as *Agaricus magnificus*) is difficult to interpret.

magnimamma (Fr.) Sacc., Hebeloma, *Syll. fung.* 5: 807 (1887)
Agaricus magnimamma Fr., *Monogr. hymenomyc. Suec.* 2: 299 (1863)
A *nomen dubium*, possibly *Inocybe petiginosa*. Cooke 464 (508) Vol. 3 (1886) is a species of *Inocybe*.

magnivelaris Peck, Stropharia, *Harriman Alaska Expedition*: 44 (1904)
Not authentically British. A North American species. Sensu auct. Brit. is *Stropharia percevalii*.

mairei Foley, Amanita, *Mém. Soc. Hist. nat. Afr. Nord.* Hors. série 2: 118 (1949)
Not authentically British. A Mediterranean species. Sensu auct. Brit. is *A. argentea*.

mairei (Battetta) Corner, Clavicorona, *Monograph of Clavaria and Allied Genera*: 291 (1950)
Clavaria mairei Battetta, *Bull. Soc. Mycol. France* 54: 44 (1938)
Not authentically British. A single dubious record from South Hampshire (New Forest, Wilverley Plain) unsubstantiated with voucher material. A doubtful synonym of *Clavicorona taxophila* fide Jülich (1984) and a *nomen dubium* for Stalpers (1996).

mairei Donk, Ramaria, *Rev. Niederl. Homob. Aphyll.* 2: 106 (1933)
British material under this name requires re-examination. Confused with *Ramaria pallida*. For Corner (1950) these are synonymous and clamped. For most other authors they are synonymous and clampless.

mairei Bourdot, Tomentella, *Bull. Soc. Mycol. France* 40: 40 (1924)
A *nomen dubium*. A collection in herb. K, from Surrey (Esher: Fairmile Common) in 1988, was assigned here with doubts by Hjortstam.

majale (Fr.) P. Karst., Entoloma, *Bidrag Kännedom Finlands Natur Folk* 32: 267 (1879)
Agaricus majalis Fr., *Syst. mycol.* 1: 205 (1821)
A *nomen dubium*. Listed by Rea (1922), though 'a doubtful native' fide Cooke (HBF2). Sensu Saunders & Smith (1872) is *E. saundersii*.

malenconii R. Heim, Inocybe, *Encycl. Mycol. 1. Le Genre Inocybe*: 163 (1931)
Not authentically British and excluded pending clearer definition. Part of the unresolved *I. dulcamara* complex, records mostly unsubstantiated with voucher material or only tentatively referred here.

mali-sylvestris With., Agaricus, *Bot. arr. veg.* 2: 766 (1776)
Described from Britain, 'on crab-apple trees' but a *nomen dubium*.

mammillaris (Pass.) Sacc., Inocybe, *Syll. fung.* 5: 785 (1887)
Agaricus mammillaris Pass., *Nuovo Giorn. Bot. Ital.* 4: 76 (1872)
A *nomen dubium*. Sensu Rea (1922) is doubtful.

mammosum (L.) Hesler, Entoloma, *Nova Hedwigia Beih.* 23: 185 (1967)
Agaricus mammosus L., *Sp. pl.*: 1174 (1753)
Nolanea mammosa (L.) Quél., *Mém. Soc. Émul. Montbéliard, Sér. 2*, 5: 122 (1872)

A *nomen confusum* lacking type material and variously interpreted. Sensu NCL is *Entoloma kuehnerianum* fide Orton (1991) but most British records are probably *E. hebes*.

marginatosplendens Reumaux, Cortinarius, *Bull. Soc. Mycol. France* 96(3): 356 (1980)
Not authentically British. Collections named thus in herb. K, from Oxfordshire (Bix, Warburg Reserve), have been redetermined as *C. saturninus*.

marginellus (Pers.) Singer, Hydropus, *Lilloa* 22: 350 (1951) [1949]
Agaricus marginellus Pers., *Syn. meth. fung.*: 309 (1801)
Mycena marginella (Pers.) P. Kumm., *Führ. Pilzk.*: 109 (1871)
Not authentically British. Listed as 'rare in Britain' in Bon (1987: 170), but no British records or herbarium material have been traced. Misapplied by Rea (1922) to *Mycena amicta*.

maritimus With., Agaricus [non *A. maritimus* Fr. (1818)], *Arr. Brit. pl. ed. 3*, 4:182 (1796)
Described from Britain but a *nomen dubium*. The description suggests a species of *Agaricus* in the modern sense.

martialis (Cooke & Massee) Sacc., Lepiota, *Syll. fung.* 9: 10 (1891)
Agaricus martialis Cooke & Massee, *Grevillea* 16: 101 (1888)
Described from Britain but a *nomen dubium* known only from Cooke's plate [1112 (944) Vol. 8 (1889)]. Probably an alien species of *Leucocoprinus*, originally collected in North East Yorkshire (Scarborough) on the decayed trunk of a tree fern in a greenhouse. Reported also by Massee (1897: 117) from the Palm House at Kew Gardens.

mastigera (Berk. & Broome) Sacc., Psathyra, *Syll. fung.* 5: 1060 (1887)
Agaricus mastiger Berk. & Broome, *Ann. Mag. Nat. Hist., Ser. 3 [Notices of British Fungi no. 921]* 7: 377 (1861)
A *nomen dubium* lacking type material. See Cooke 610 (591) Vol. 4 (1886) for an illustration of the type collection (now lost). Regarded by Kits van Waveren (Kits1) as 'an aberrant form of *P. conopilus*'.

maxima (Gaertn. & G. Mey.) P. Kumm., Clitocybe, *Führ. Pilzk.*: 123 (1871)
Agaricus maximus Gaertn. & G. Mey., *Ökon. Tech. Fl. Wett.*: 329 (1799)
Clitocybe geotropa var. *maxima* (Gaertn. & G. Mey.) Nüesch, *Die Trichterlinge*: 121 (1926)
A *nomen dubium* lacking type material. Variously interpreted; British records refer to *C. geotropa*.

maxima (Massee) W.B. Cooke, Maireina, *Sydowia, Beih.* 4: 90 (1961)
Solenia maxima Massee, *Brit. fung.-fl.* 1: 143 (1892)
Cyphellopsis maxima (Massee) Donk, ined.
Described from Britain but a *nomen dubium*. As originally described this is doubtfully distinct from *Henningsomyces candidus*, but W.B. Cooke, based on an examination of Massee's type, described a different species, resembling *Merismodes anomalus*. Collections from Warwickshire [as *Cyphellopsis maxima* (Massee) Donk] need re-examination.

media (Schumach.) Singer, Volvariella, *Lilloa* 22: 401 (1951) [1949]
Agaricus medius Schumach., *Enum. pl.* 2: 227 (1803)
Volvaria media (Schumach.) Gillet, *Hyménomycètes*: 388 (1876)
A *nomen dubium* lacking type material. Sensu BFF4 is probably depauperate *V. gloiocephala*. Sensu Rea (1922) is doubtful but small-spored and possibly *V. pusilla*. Cooke 299 (299) Vol. 3 (1884) is unknown, but *V. cookei* Contu [Mycol. Ital. 27 (3): 37 (1998)] is a recent interpretation of *V. media* sensu Cooke.

medullata Romagn., Russula, *Doc. mycol.* 106: 53 (1997)
Not authentically British. Reported, e.g. from England (Yorkshire) by Rayner (1985) and from Scotland (Fair Isle) by Watling (1994), but no material seen.

megaphyllum Boud., Tricholoma, *Icon. mycol.* 1: pl. 28 (1910)
A *nomen dubium*. Included in Dennis (1995), using the unpublished combination *Melanoleuca megaphylla*, based on a collection from West Sussex, but this material has not been traced. Boudier's illustration has a strong resemblance to *Melanoleuca turrita*.

melanodon (Fr.) Quél., Pluteus, *Enchir. fung.*: 56 (1886)
Agaricus melanodon Fr., *Hymenomyc. eur.*: 187 (1874)
A *nomen dubium*. Listed by Rea (1922) as a dull yellow species with dark lamellar edges, but no records or British collections have been traced.

melanotus Kalchbr., Cortinarius, in Fries, *Hymenomyc. eur.*: 365 (1874)
A southern European species. Reported from England (North Wiltshire: Ravensroost Wood) in 1999, but unsubstantiated with voucher material.

melanthina (Fr.) Kits van Wav., Psathyrella, *Persoonia* Suppl. 2: 281 (1985)
Agaricus melanthinus Fr., *Öfvers. Kongl. Vetensk.-Akad. Förh.* 8: 49 (1852)
Hypholoma melanthinum (Fr.) P. Karst., *Bidrag Kännedom Finlands Natur Folk* 32: 500 (1879)
Not authentically British. Sensu Rea (1922), as *Hypholoma melanthinum*, is *Psathyrella maculata* fide NCL.

meleagris With., Agaricus [*non A. meleagris* Sowerby (1798)], *Bot. arr. Brit. pl. ed. 2*, 3: 379 (1792)
Described from Britain but a *nomen dubium*. Withering's description is suggestive of a species of *Lepiota* or perhaps *Macrolepiota*.

melinoides (Bull.) P. Kumm., Naucoria, *Führ. Pilzk.*: 77 (1871)
Agaricus melinoides Bull., *Herb. France*: pl. 560 (1792)
Alnicola melinoides (Bull.) Kühner, *Botaniste* 17: 175 (1926)
A *nomen dubium fide* Watling [Bibl.Mycol. 82: 167 (1981)]. Sensu auct. Brit is *Galerina mniophila fide* NCL. Cooke (499) 457 Vol. 4 (1884) is also a species of *Galerina*. Sensu auct. mult. is *Naucoria escharioides*.

melizeus Fr., Hygrophorus, *Epicr. syst. mycol.*: 321 (1838)
A *nomen dubium* applied to members of the *H. eburneus* complex. Sensu Rea (1922) is possibly *H. hedrychii*. See TBMS 2: 65 (1907) for the British record, apparently collected in Yorkshire (Rievaulx) on a BMS Foray [also in Naturalist (Hull) 1903: 426 (1903)].

mellea var. glabra Gillet, Armillaria, *Hyménomycètes*: 84 (1874)
A *nomen dubium* inadequately described. Listed by Rea (1922).

mellea var. versicolor (With.) W.G. Sm., Armillaria, *Syn. Brit. Bas.*: 30 (1908)
Agaricus versicolor With., *Bot. arr. Brit. pl. ed. 2*, 3: 93 (1792)
Stropharia versicolor (With.) Sacc., *Syll. fung.* 5: 1013 (1887)
Described from Britain but a *nomen dubium* lacking type material and inadequately described.

mellea var. viridiflava Barla, Armillaria, *Bull. Soc. Mycol. France* 3: 143 (1887)
A *nomen dubium*, probably just a greenish-yellow form of *A. mellea*. Sensu Rea (1922) is certainly that.

melleopallens (Fr.) Britzelm., Cortinarius, *Bot. Centralbl.* 27: 14 (1892)
Cortinarius triformis var. *melleopallens* Fr., *Epicr. syst. mycol.*: 299 (1838)
Not authentically British. Associated with *Picea* spp in native woodland in northern Europe. Collections named thus in herb. K have been redetermined as *C. triformis*.

mellita (Bourdot) Bondartsev & Singer, Ceriporia, *Ann. Mycol.* 34: 50 (1941)
Poria mellita Bourdot, in Lloyd, *Mycol. not.* 4(40): 543 (1916)
Not authentically British. Reported on several occasions but always unsubstantiated with voucher material. Considered a synonym of *Ceriporia purpurea* by Ryvarden & Gilbertson (1993).

menieri Boud., Marasmius, *Bull. Soc. Mycol. France* 10: 61 (1894)
Not authentically British. All known British collections of this species are misdetermined *Marasmius cornelii*.

mesomorpha (Bull.) P. Kumm., Lepiota, *Führ. Pilzk.*: 136 (1871)
Agaricus mesomorphus Bull., *Herb. France*: pl. 506 (1791)
A *nomen dubium* lacking type material. Cooke 42 (85) Vol. 1(1882) looks distinctive, the erect annulus suggesting a *Leucoagaricus* sp.

mesomorpha Bourdot & Galzin, Sebacina, *Bull. Soc. Mycol. France* 39: 262 (1924)
Not authentically British. English and Scottish collections in herb. K have been redetermined either as *Exidiopsis effusa* or *Stypella grilletii*.

mesophaeum var. holophaeum (Fr.) Sacc., Hebeloma, *Syll. fung.* 5: 795 (1887)
Agaricus mesophaeus var. *holophaeus* Fr., *Hymenomyc. eur.*: 240 (1874)
A *nomen dubium*, probably a colour form of *H. mesophaeum*. Listed in Rea (1922) but no material traced.

metrodii Rob. Henry, Cortinarius, *Doc. mycol.* 16(61): 21 (1985)
Mis.: *Cortinarius illibatus* sensu Cooke (1891)
A poorly known species. Listed in Orton (1955) and NCL only on the strength of the spore measurements given by Cooke (HBF2: 250) for his *C. illibatus*. This in turn was based only on a collection of Berkeley's of unknown provenance.

metulispora (Berk. & Broome) Sacc., Lepiota, *Syll. fung.* 5: 38 (1887)
Agaricus metulisporus Berk. & Broome, *J. Linn. Soc., Bot.* 11: 512 (1871)
Not authentically British. Described from Sri Lanka. Sensu Berk. & Broome [Notices of British Fungi No. 1182 (1871)] and sensu auct. Eur. is *Lepiota magnispora*.

micans (Ehrenb.) Fr., Poria, *Grevillea* 14: 112 (1886)
Polyporus micans Ehrenb., *Sylv. mycol. berol.*: 30 (1818)
A *nomen dubium*. The only collection thus named in herb. K is *Ceriporia reticulata fide* Reid & Austwick (1963), but the original description may relate to *Junghuhnia nitida fide* Donk (1974).

michelii (Rea) Corner, Clavulinopsis, *Monograph of Clavaria and Allied Genera*: 378 (1950)
Clavaria michelii Rea, *Trans. Brit. Mycol. Soc.* 2: 39 (1903)
Described from Britain but a *nomen dubium* lacking type material. Only a painting survives. Possibly a yellow variety of *Clavulinopsis vermicularis*.

michelii Fr., Polyporus, *Syst. mycol.* 1: 343 (1821)
A *nomen dubium* lacking type material. Listed by Rea (1922). Probably just a small, pale form of *Polyporus squamosus* on willow *fide* Donk (1974), though he retained it as distinct pending further observations.

microcephala (P. Karst.) Singer, Melanoleuca, *Sydowia* 15: 53 (1962) [1961]
Tricholoma microcephalum P. Karst., *Hedwigia* 20: 177 (1881)
Not authentically British. A single record from Yorkshire in 1996, unsubstantiated with voucher material and probably misidentified.

microcyclus Fr., Cortinarius, *Monogr. hymenomyc. Suec.* 2: 78 (1863)
A *nomen dubium*. Listed by Rea (1922) but no records have been traced. Sensu Cooke (HBF2) is *C. brunneus* var. *glandicolor*.

microsporus Berk. & Broome, Coprinus, *J. Linn. Soc., Bot.* 11(560): 304 (1871)
Not authentically British. Records of this species from Britain refer to *Coprinus plicatilis* var. *microsporus*.The true *C. microsporus* is a tropical species described from Sri Lanka.

militare (Lasch) Gillet, Tricholoma, *Hyménomycètes*: 123 (1874)
  Agaricus militaris Lasch, *Linnaea* 3: 428 (1828)
A *nomen dubium* lacking type material, but near to or synonymous with *Tricholoma acerbum*. Cooke 112 (169) Vol. 1 (1883), however, is not this species. Reported from Wales (Denbighshire) in TBMS 10 (4): 233 (1926).

millavensis Bourdot & Galzin, Poria, *Bull. Soc. Mycol. France* 41: 238 (1925)
A *nomen dubium*, possibly an *Oxyporus* sp. *fide* Donk (1974). A collection from Scotland (Barra) named thus in Dennis (1986) has been redetermined as *Physisporinus sanguinolentus*.

millus (Sowerby) Sacc., Armillaria, *Syll. fung.* 5: 81 (1887)
  Agaricus millus Sowerby, *Col. fig. Engl. fung.* 2: pl. 184 (1799)
Described from Britain but a *nomen dubium* lacking type material, inadequately described and possibly not even a species of *Armillaria*. Sowerby's illustration depicts a distinctive, but unidentifiable, agaric.

miltinus Fr., Cortinarius, *Epicr. syst. mycol.*: 287 (1838)
A *nomen dubium*. Cooke 774 (785) Vol. 6 (1886) is *Cortinarius purpureus*.

milvinus Fr., Cortinarius, *Epicr. syst. mycol.*: 314 (1838)
A *nomen dubium*. Cooke 853 (846) B Vol. 6 (1888) is unidentifiable.

mimica (W.G. Sm.) Sacc., Collybia, *Syll. fung.* 5: 214 (1887)
  Agaricus mimicus W.G. Sm., in Cooke, *Handb. Brit. fung.*, Edn 2: 65 (1884)
Described from Britain but a *nomen dubium*. Possibly a synonym of *Macrocystidia cucumis fide* A&N2. Cooke 192 (129) Vol. 2 (1882) is unrecognisable.

miniata (Berk.) Burt, Peniophora, *Rep. (Annual) Missouri Bot. Gard.* 12: 244 (1925)
  Thelephora miniata Berk., *Engl. fl.* 5(2): 168 (1836)
Described from Britain but a *nomen dubium*. Reported by Rea in TBMS 12 (2-3): 225 (1927) 'on fallen limbs of conifers' in England but no material has been traced. Possibly *Phanerochaete sanguinea*.

minimum Bolton, Hydnum, *Hist. Fung. Halifax* 4 pl.171 (1791)
Described from Britain but a *nomen dubium*, and from Bolton's illustrations, probably not a basidiomycete.

minimus W.G. Sm., Lactarius, *J. Bot.* 11: 205 (1873)
Described from Britain but a *nomen dubium*. Cooke 968 (986) Vol. 7 (1888) is not identifiable.

minus Bruchet, Hebeloma, *Bull. Mens. Soc. Linn. Lyon* 39 (6 [Suppl.]): 93 (1970)
Not authentically British. An arctic-alpine species, reported on several occasions but from lowland woodland sites, and unsubstantiated with voucher material.

minuta (Quél.) Cooke, Galera, *Handb. Brit. fung.*, Edn 2: 186 (1887)
  Agaricus minutus Quél., *Mém. Soc. Émul. Montbéliard, Sér. 2*, 5: 438 (1872)
A *nomen dubium* lacking type material. Cooke 524 (466) Vol. 4 (1885) is unidentifiable.

minutula (Peck) Murrill, Hygrocybe, *North American Flora, Ser. 1 (Fungi 3)*, 9(6): 380 (1916)
  Hygrophorus minutulus Peck, *Bull. New York State Mus. Nat. Hist.* 1: 9 (1888)
Not authentically British (or European). Sensu Bon [Doc. Mycol. 93 (1994)] is probably *H. insipida fide* Boertmann (FNE1). A

record from England (South Devon) has been redetermined as *Hygrocybe insipida*.

minutulus J. Favre, Cortinarius, *Ergebn. Wiss. Untersuch. Schweiz. Natn. Parks* 5: 203 (1955)
Not authentically British. Reported from Scotland (Orkney: Papa Westray) in 1992 but the collection is labelled as '?' *C. minutulus*.

mirabile (Fr.) Wehm., Hydnellum, *Fungi of Maritime Provinces*: 68 (1950)
  Hydnum mirabile Fr., *Monogr. hymenomyc. Suec.* 2: 349 (1863)
  Hydnum acre Quél., *Bull. Soc. Bot. France* 24: 324 (1878)
Not authentically British. Reported on several occasions from Britain but all reports unsubstantiated with voucher material.

misera (Fr.) M.M. Moser, Tephrocybe, *Kleine Kryptogamenflora*, Edn 3: 117 (1967)
  Agaricus miser Fr., *Monogr. hymenomyc. Suec.* 2: 290 (1863)
  Collybia misera (Fr.) Gillet, *Hyménomycètes*: 309 (1876)
A *nomen dubium*. Sensu NCL (as *Collybia misera*) and sensu Orton (1984) is *Tephrocybe murina*.

molle Fr., Hydnum, *Öfvers. Kongl. Vetensk.-Akad. Förh.* 8: 53 (1852)
A *nomen dubium*, possibly a *Phellodon* sp. Listed by Rea (1922).

mollis Bolton, Agaricus [*nom. illegit.*, non *A. mollis* Schaeff. (1774)], *Hist. fung. Halifax* 1: 40 (1788)
Described from Britain, but a *nomen dubium* lacking type material. Bolton's description and illustration strongly suggest *Clitocybe nebularis*.

mollis (Berk. & Broome) M.M. Moser, Hygrocybe, *Kleine Kryptogamenflora*, Edn 3: 69 (1967)
  Hygrophorus turundus var. *mollis* Berk. & Broome, *Ann. Mag. Nat. Hist., Ser. 4 [Notices of British Fungi no. 1279]* 7: 426 (1871)
  Hygrophorus mollis (Berk. & Broome) Kauffman, *Pap. Michigan Acad. Sci.* 5: 130 (1926)
  Hygrocybe miniata var. *mollis* (Berk. & Broome) Arnolds, *Persoonia* 13(2): 148 (1986)
A *nomen dubium*. Sensu Arnolds (with spores constricted) is an orange variety of *Hygrocybe miniata*, considered not worth separating by Boertmann (FNE1). Sensu NCL and B&K3 (as *H. miniata* var. *mollis*, with spores unconstricted) is probably *H. calciphila*. Sensu Moser (1967) is *H. helobia*.

molliscorium (Cooke & Massee) Massee, Pholiota, *Brit. fung.-fl.* 2: 215 (1893)
  Agaricus molliscorium Cooke & Massee, *Grevillea* 17: 1 (1888)
  Togaria molliscorium (Cooke & Massee) W.G. Sm., *Syn. Brit. Bas.*: 123 (1908)
Described from Britain but a *nomen dubium*. Doubtfully distinct from *Agrocybe cylindracea fide* BFF3, but the original description is of an agaric with slender basidiomes growing on soil. Pearson & Dennis (1948) suggested *Agrocybe erebia* but the description does not match this either. Cooke 1161 (1171) Vol. 8 (1890) is unhelpful.

molliusculus Sowerby, Agaricus, *Col. fig. Engl. fung.* 2: pl. 174 (1798)
Described from Britain but a *nomen dubium* lacking type material. From Sowerby's illustration clearly a species of *Pluteus*, possibly *P. cervinus*.

mollusca (Fr.) Rea, Odontia, *Brit. basidiomyc.*: 649 (1922)
  Hydnum molluscum Fr., *Summa veg. Scand.*: 327 (1849)
A *nomen dubium*. Listed by Rea (1922) with a minimal description.

molybdocephalus Bull., Agaricus, *Herb. France*: pl. 523 (1791)
A *nomen dubium* lacking type material but the illustration in Bulliard clearly depicts a *Lyophyllum*, probably *L. decastes* or a related species.

molyoides Fr., Marasmius, *Epicr. syst. mycol.*: 382 (1838)
A *nomen dubium*. Listed by Rea (1922) whose description suggests *M. alliaceus*, a possibility also discussed in A&N1 p. 188.

momentaneus Bull., Agaricus, *Herb. France* pl.128 (1783)
A *nomen dubium* lacking type material. Bulliard's illustration is of a species of *Coprinus*, possibly *C. plicatilis*. It was cited by Withering (1792) where said to 'grow in pastures after continued gentle rain'.

monstrosa (Sowerby) Cooke, Clitocybe, *Handb. Brit. fung.*, Edn 2: 53 (1884)
    Agaricus monstrosus Sowerby, *Col. fig. Engl. fung.* 3: pl. 283 (1800)
Described from Britain but a *nomen dubium* lacking type material. Sowerby's illustration shows a clump of white basidiomes, either diseased or immature, but possibly *Lyophyllum connatum*. Cooke 1134 (648) Vol. 8 (1889) is a different but equally unidentifiable agaric.

morganii (Peck) H.E. Bigelow, Hygrophoropsis, *Nova Hedwigia Beih.* 51: 66 (1975)
    Cantharellus morganii Peck, *Bot. Gaz.* 7(4): 43 (1882)
    Cantharellus olidus Quél., *Clavis syn. Hymen. Europ.*: 148 (1878)
    Hygrophoropsis olida (Quél.) Métrod, *Rev. Mycol. (Paris)* 14(3): 15 (1949)
Not authentically British. Reported on several occasions but material named thus in herb. K is misdetermined *Clitocybe phyllophila*.

mori (Pollini) Fr., Polyporus, *Syst. mycol.* 1: 344 (1821)
    Hexagonia mori Pollini, *Spec. Plant. I*: 35 (1816)
Not authentically British. Collections named thus are misidentified *P. brumalis* or occasionally *P. ciliatus*. Common in southern Europe and unlikely to occur in Britain.

mortuosa (Fr.) Gillet, Clitocybe, *Hyménomycètes*: 168 (1874)
    Agaricus metachrous β *mortuosa* Fr., *Observ. mycol.* 2: 210 (1818)
A *nomen dubium*. The single collection named thus, in herb. K, needs reinvestigation.

morus With., Agaricus, *Bot. arr. Brit. pl. ed. 2*, 3: 373 (1792)
Described from Britain but a *nomen dubium*. From the description probably a species of *Panaeolus*.

mucida (Pers.) Gilb. & Ryvarden, Ceriporiopsis, *Mycotaxon* 22 (2): 364 (1985)
    Poria mucida Pers., *Observ. mycol.* 1: 87 (1796)
    Porpomyces mucidus (Pers.) Jülich, *Persoonia* 11(4): 425 (1982)
    Fibuloporia donkii Domański, *Acta Soc. Bot. Poloniae* 38: 454 (1969)
    Mis.: *Fibuloporia mollusca* sensu auct.
Not authentically British. Repeatedly listed in British literature but without convincing evidence. *Poria mucida* sensu Rea (1922) is a form of *Schizopora paradoxa* on coniferous substrates.

mucida (Pers.) Fr., Grandinia, *Hymenomyc. eur.*: 626 (1874)
    Thelephora mucida Pers., *Mycol. eur.* 1: 135 (1822)
A *nomen dubium*. Sensu Berkeley [Notices No. 1691 (1878)] from Scotland (Glamis) and sensu Wakefield [TBMS 3: 280 (1911)] is *Resinicium bicolor*. Sensu Rea (1922) and later British records is doubtful.

mucida (Pers.) J. Erikss & Ryvarden, Cristinia, *Corticiaceae of N. Europe* 3: 311(1975)
    Hydnum mucidum Pers., in Gmelin, *Syst. nat.*: 1440 (1792)
    Radulum mucidum (Pers.) Bourdot & Galzin, *Bull. Soc. Mycol. France* 30: 247 (1914)
Here considered a *nomen dubium,* but it has also been lectotypified with a collection from Persoon's herbarium that is *Trechispora farinacea*. Sensu Fries (1821) is *Dentipellis fragilis* (not a British species). Sensu Bourdot & Galzin, Rea (1922), Bramley (1985) and CNE3 is *Cristinia gallica*.

mucida (Fr.) R.H. Petersen, Multiclavula, *Am. Midl. Nat.* 77: 212 (1967)
    Lentaria mucida (Fr.) Corner, *Monograph of Clavaria and Allied Genera*: 442 (1950)
    Clavaria mucida Fr., *Syst. mycol.* 1: 476 (1821)
Not authentically British. Reported from Yorkshire in 1995 as *Lentaria mucida* but unsubstantiated with voucher material and probably misidentified.

muelleri (Fr.) P.D. Orton, Pholiota, *Trans. Brit. Mycol. Soc.* 43(2): 180 (1960)
    Agaricus squarrosus var. *muelleri* Fr., *Syst. mycol.* 1: 243 (1821)
A *nomen dubium*. Sensu NCL is *P. jahnii*.

mulgravensis Massee & Crossl., Pluteolus, *Trans. Brit. Mycol. Soc.* 4(1): 188 (1912)
Described from Britain, but a *nomen dubium*. Probably *Bolbitius reticulatus fide* BFF3.

multiformis Fr., Cortinarius, *Epicr. syst. mycol.*: 263 (1838)
    Mis.: *Cortinarius allutus* sensu NCL and sensu auct. mult.
As now interpreted this is a northern conifer associate, not known with certainty from Britain. *C. allutus* sensu NCL may be a form of this species, but is only doubtfully British *fide* Orton (1955) and has not been recorded since. British records of *C. multiformis* from deciduous woodland probably represent *C. polymorphus* or *C. talus.*

multiformis var. flavescens Cooke, Cortinarius, *Handb. Brit. fung.*, Edn 2: 241 (1888)
    Cortinarius flavescens (Cooke) Rob. Henry, *Bull. Soc. Mycol. France* 55: 180 (1939)
Described from Britain but a *nomen dubium*. Cooke 702 (709) Vol. 5 (1887) has been cited in support of various species [see Orton (1955: 74)].

multiplex Fr., Hydnum, *Öfvers. Kongl. Vetensk.-Akad. Förh.* 8: 54 (1852)
A *nomen dubium*, seemingly an *Hericium* sp. Listed by Rea (1922).

muralis (Sowerby) Fr., Omphalia, *Syst. mycol.* 1: 165 (1821)
    Agaricus muralis Sowerby, *Col. fig. Engl. fung.* 3: pl. 322 (1801)
Described from Britain but a *nomen dubium* lacking type material. Sensu auct. mult. is *Arrhenia rickenii* but Sowerby's illustration does not resemble that species.

muricella (Fr.) Bon, Pholiota, *Doc. mycol.* 16(61): 46 (1985)
    Agaricus muricellus Fr., *Monogr. hymenomyc. Suec.* 2: 302 (1863)
    Flammula muricella (Fr.) Sacc., *Syll. fung.* 5: 811 (1887)
A *nomen dubium*. Discussed in NCL, but no British records traced. Sensu Bon is *Pholiota conissans*.

muricinus Fr., Cortinarius, *Epicr. syst. mycol.*: 279 (1838)
A *nomen dubium*. Listed in NCL sensu Ricken (1915) but only on the strength of Cooke 748 (815) Vol. 5 (1888) which is more likely to represent either *C. largus* or *C. variicolor*.

murinaceum (Bull.) Gillet, Tricholoma, *Hyménomycètes*: 100 (1874)
    Agaricus murinaceus Bull., *Herb. France*: pl. 520 (1791)
    Hygrophorus murinaceus (Bull.) Fr., *Epicr. syst. mycol.*: 333 (1838)
    Hygrocybe murinacea (Bull.) P. Kumm., *Führ. Pilzk.*: 112 (1871)
A *nomen confusum*. Possibly *Hygrocybe ovina* but sensu Cooke (HBF2), and sensu Rea (1922) is *Tricholoma bresadolanum fide* Bon (1991: 57). Sensu Sowerby [Col. fig. Engl. fung. 1: pl. 106 (1797)] is an unidentifiable species of *Tricholoma*.

muscaria var. umbrina (Pers.) Sacc., Amanita, *Syll. fung.* 5: 13 (1887)
    Amanita umbrina Pers., *Tent. disp. meth. fung.*: 71 (1797)
A *nomen dubium*. Listed by Rea (1922). Sensu auct. is *Amanita regalis* (not a British species).

muscigena (Schumach.) P. Kumm., Collybia, *Führ. Pilzk.*: 114 (1871)
    Agaricus muscigenus Schumach., *Enum. pl.* 1: 307 (1801)
A *nomen dubium* lacking type material. *Fide* A&N2 this is possibly a species of *Mycena*. Cooke 209 (147) Vol. 2 (1882) is also mycenoid.

muscorum (Hoffm.) Sacc., Tubaria, *Syll. fung.* 5: 874 (1887)
    Agaricus muscorum Hoffm., *Nomencl. fung.*: 181 (1789)
A *nomen dubium*. Listed as British by Massee (Brit. fung.-fl. Vol. 2: 124,1893) who there retracts his plate of this species, reproduced in Cooke [Ill. Brit. fung. 1175 (1175) B Vol. 8 (1891], which he says he wrongly supplied.

mustelina (Fr.) Quél., Pholiota, *Mém. Soc. Émul. Montbéliard, Sér. 2,* 5: 127 (1872)
    Agaricus mustellinus Fr., *Epicr. syst. mycol.*: 169 (1838)
A *nomen dubium* but from the description probably a species of *Galerina*, possibly *G. marginata* or *G. unicolor*. See Cooke 404 (356) Vol. 3 (1885) for an illustration.

mutabilis (J. Favre) M.M. Moser, Tephrocybe, *Kleine Kryptogamenflora,* Edn 3: 116 (1967)
    Lyophyllum mutabile J. Favre, *Ergebn. Wiss. Untersuch. Schweiz. Natn. Parks* 6: 447 (1960)
A montane species, not authentically British. A single collection in herb. K from lowland habitat in Surrey (Mickleham, Norbury Park) is probably misidentified.

mutatum (Peck) Donk, Hyphoderma, *Fungus* 27: 15 (1957)
    Corticium mutatum Peck, *Rep. (Annual) New York State Mus. Nat. Hist.* 43: 69 (1889)
Not authentically British. Reported on several occasions but voucher collections are misdetermined.

myceliosa (Peck) Corner, Ramaria, *Monograph of Clavaria and Allied Genera*: 607 (1950)
    Clavaria myceliosa Peck, *Bull. Torrey Bot. Club* 31: 182 (1904)
Not authentically British. Records are dubious and unsubstantiated with voucher material.

mycenoides (Fr.) Kühner, Galerina, *Encycl. Mycol. 7. Le Genre* Galera: 209 (1935)
    Agaricus mycenoides Fr., *Syst. mycol.* 1: 246 (1821)
    Pholiota mycenoides (Fr.) Quél., *Mém. Soc. Émul. Montbéliard, Sér. 2,* 5: 127 (1872)
    Galera mycenoides (Fr.) Kühner [*comb. inval.*], *Flore Analytique des Champignons Supérieurs*: 320 (1953)
A *nomen dubium* variously interpreted. Sensu Fries is probably a species of *Conocybe*. Sensu Rea (1922) is *Galerina praticola* and sensu auct. mult is *Galerina jaapii*.

mycetophila Peck, Tremella, *Rep. (Annual) New York State Mus. Nat. Hist.* 28: 53 (1876) [1874]
Not authentically British. A North Amercan species, unknown in Europe. The name was once loosely applied to any species of *Syzygospora* parasitic on agarics.

myochroa (Fr.) Sacc., Omphalia, *Syll. fung.* 5: 322 (1887)
    Agaricus umbelliferus var. *myochrous* Fr., *Hymenomyc. eur.*: 161 (1874)
    Omphalina myochroa (Fr.) Quél., *Enchir. fung.*: 44 (1886)
    Omphalia umbellifera var. *myochroa* (Fr.) Massee, *Brit. fung.-fl.* 2: 395 (1893)
    Omphalia umbellifera f. *myochroa* (Fr.) Cejp, Kavina & Pilát, *Atlas champ. Eur.* 4: 42 (1936)
A *nomen dubium*. Listed by Rea (1922).

myodes Schaeff., Agaricus, *Fung. Bavar. Palat. nasc.* 1: 69 (1762)
A *nomen dubium* lacking type material. Sensu Bolton [Hist. Fung. Halifax Suppl. pl. 139 (1789)] was apparently a species of *Amanita*.

myrtillinus Fr., Cortinarius, *Epicr. syst. mycol.*: 285 (1838)
A *nomen dubium*. Retained in NCL as a member of the *C. anomalus* complex, on the strength of Cooke 769 (817) Vol. 6

(1886), but lacking any currently accepted interpretation. For an extensive discussion see Orton (1958: 144).

nana Massee, Laccaria, *Bull. Misc. Inform. Kew* 1913: 195 (1913)
    Clitocybe nana (Massee) Sacc. & Trotter, *Syll. fung.* 23: 62 (1925)
Known only from the type collection, 'under trees' in Kew Gardens. It has been synonymised with *L. tortilis* but is clearly distinct *fide* Dennis [TBMS 31: 196 (1948)] and was excluded 'pending clearer definition' in NCL.

nanus J. Favre, Lactarius, *Ergebn. Wiss. Untersuch. Schweiz. Natn. Parks* 5: 205 (1955)
Not authentically British. Reported from montane habitats, associated with *Salix herbacea*, in England (Cumbria) and Northern Ireland (Mountains of Mourne), but material needs further investigation.

napus Fr., Cortinarius, *Epicr. syst. mycol.*: 263 (1838)
A *nomen dubium* listed in NCL citing Cooke 703 (710) Vol. 5, but not reported since 1900.

nasuta (Kalchbr.) Sacc., Naucoria, *Syll. fung.* 5: 834 (1887)
    Agaricus nasutus Kalchbr., in Rabenhorst, *Deutsch. Krypt. Fl.* 1: 852 (1883)
A *nomen dubium*. Possibly *Psilocybe semilanceata fide* Pearson & Dennis (1948). Cooke 1173 (1172) Vol. 8 (1890) is unidentifiable.

nauseosa *var.* **flavida** Cooke, Russula, *Ill. Brit. fung.* 1102 (1053) Vol. 7 (1889)
Described from Britain but a *nomen dubium*. Possibly *Russula acetolens*.

nauseosum (Cooke) Sacc., Hebeloma, *Syll. fung.* 9: 102 (1891)
    Agaricus nauseosus Cooke, *Grevillea* 16: 43 (1887)
Described from Britain but a *nomen dubium*. From the type description, with large spores and strong smell, possibly *Hebeloma groegeri* and Cooke 1166 (963) Vol. 8 (1890) could well be this.

necator (Bull.) Pers., Lactarius, *Observ. mycol.* 2: 42 (1799)
    Agaricus necator Bull., *Herb. France*: pl. 13 (1780)
A *nomen dubium* lacking type material and variously interpreted. Widely used for *Lactarius turpis*.

nefrens (Fr.) P. Kumm., Leptonia, *Führ. Pilzk.*: 96 (1871)
    Agaricus nefrens Fr., *Syst. mycol.* 1: 209 (1821)
A *nomen dubium*, possibly *Entoloma serrulatum fide* Noordeloos (1987). Listed by Rea (1922).

neglecta Massee, Psathyra, *Brit. fung.-fl.* 1: 356 (1892)
Described from Britain but a *nomen dubium*. Not mentioned by Kits van Waveren (Kits1).

neglecta Singer [*nom. inval.*], Russula, *Mycologia* 34: 186 (1947)
The supposed validation of this name by Hora cited in NCL and in Rayner (1985) appears never to have been published. The description in Rayner (1985) suggests *R. postiana*.

nemophila (Fr.) Sacc., Psilocybe, *Syll. fung.* 5: 1044 (1887)
    Agaricus nemophilus Fr., *Hymenomyc. eur.*: 297 (1874)
A *nomen dubium*; 'inadequately described, no modern records known' (BFF5). Listed by Rea (1922).

neoantipus (G.F. Atk.) Singer, Conocybe, *Ann. Mycol.* 34: 433 (1936)
    Galerula neoantipus G.F. Atk., *Proc. Amer. Philos. Soc.* 57: 371 (1918)
    Conocybe siliginea var. *neoantipus* (G.F. Atk.) Kühner, *Encycl. Mycol. 7. Le Genre* Galera: 98 (1935)
A North American species, not authentically European. Sensu Watling (1986) is *C. watlingii*.

nevillae (Berk.) Sacc., Omphalia, *Syll. fung.* 5: 333 (1887)
    Agaricus nevillae Berk., *Grevillea* 1: 89 (1872)

Described from Britain but a *nomen dubium*, never recollected and possibly an alien. Growing amongst *Sphagnum* moss in orchid pots in a greenhouse.

nicotiana Bolton, Auricularia, *Hist. Fung. Halifax* 4: 174 (1791)
Described from Britain but a *nomen dubium*. Bolton's illustrations and text suggest that this is a species of *Thelephora*, possibly *T. terrestris*.

nigellum Redeuilh, Leccinum, *Bull. Soc. Mycol. France* 111(3): 174 (1995)
A *nomen dubium* inadequately described from immature material supposedly with *Populus* sp. Reported from Scotland, but in association with *Betula* sp.

nigrellus (Pers.) P.D. Orton, Claudopus, *Mycologist* 5(3): 126 (1991)
    Agaricus nigrellus Pers., *Syn. meth. fung.*: 463 (1801)
    Eccilia nigrella (Pers.) Berk. & Broome, *Ann. Mag. Nat. Hist., Ser. 5* 32: 102 (1878)
    Rhodophyllus nigrellus (Pers.) Quél., *Enchir. fung.*: 62 (1886)
    Entoloma nigrellum (Pers.) Noordel. [*nom. inval.*], *Persoonia* 11(2): 150 (1981)
A *nomen dubium* lacking type material. Possibly *Rhodocybe hirneola fide* Noordeloos (1987). British records are sensu Lange (FlDan2: 104 & pl. 80B), an apparently distinct species requiring further investigation.

nigrescens (Quél.) Kühner, Hygrocybe, *Botaniste* 17(1-4): 57 (1926)
    Hygrophorus puniceus var. *nigrescens* Quél., *Champs Jura Vosges* 12: 6 (1883)
    Hygrophorus nigrescens (Quél.) Quél., *Fl. mycol. France*: 254 (1888)
A *nomen dubium* lacking type material. Frequently recorded in the British Isles for a form of *Hygrocybe conica*.

nigridisca Peck, Inocybe, *Rep. (Annual) New York State Mus. Nat. Hist.* 41: 67 (1888) [1887]
Not authentically British. A North American species. Sensu Rea (1922) is doubtful.

nigritula P.D. Orton, Armillaria, *Notes Roy. Bot. Gard. Edinburgh* 38(2): 316 (1980)
Described from Britain but a doubtful species. Near to *Armillaria mellea* with slight spore differences and a strong smell like 'cats' or 'flowering currants'. Probably a synonym *fide* Watling (pers. comm.).

nigrocinnamomeum (Schulzer) Sacc., Entoloma, *Syll. fung.* 5: 694 (1887)
    Agaricus nigrocinnamomeus Schulzer, *Ic. Hymenomyc.* 1: pl. 21 (1873)
A *nomen dubium* lacking type material. Possibly *E. myrmicophilum fide* Noordeloos (1987). See also Cooke 1158 (1153) Vol. 8 (1890) which looks distinct.

nigromarginata Massee, Lepiota, *Eur. Fung. Flora*: 10 (1902)
Described from Britain but a *nomen dubium* lacking type material. The dark gill edge suggests *L. hystrix*.

nimbata (Batsch) Gillet, Clitocybe, *Hyménomycètes*: 154 (1874)
    Agaricus nimbatus Batsch, *Elench. fung. (Continuatio Prima)*: 49 & t. 14 f. 65 (1786)
A *nomen dubium*. Listed by Rea (1922). A synonym of *Lepista panaeolus* (as *L. luscina*) *fide* NCL.

nitens Batsch, Agaricus [non *A. nitens* Schaeff. (1770); non *A. nitens* Cooke & Massee (1891)], *Elench. fung. (Continuatio Secunda)*: 21 t. 33 f. 192 (1789)
A *nomen dubium*. Indexed in NCL as 'not known'.

nitens Schaeff., Agaricus [*nom. illegit.*, non *A. nitens* Batsch (1789); non *A. nitens* Cooke & Massee (1891)], *Fung. Bavar. Palat. nasc.* 3: 238 (1770)
A *nomen dubium* lacking type material. Sensu Sowerby [Col. fig. Engl. fung. 1: pl. 71 (1797)] suggests *Hygrophorus*

discoxanthus, showing the golden brown colours this typically develops with age.

nitens (Cooke & Massee) Sacc., Flammula, *Syll. fung.* 9: 105 (1891)
    Agaricus nitens Cooke & Massee [*nom. illegit.*, non *A. nitens* Schaeff. (1770); non *A. nitens* Batsch (1789)], *Grevillea* 18: 52 (1890)
Described from Britain but a *nomen dubium*. Cooke 1168 (1154) Vol. 8 (1890) is unidentifiable.

nitida Fr., Amanita, *Observ. mycol.* 1: 4 (1818)
A *nomen dubium*. *Fide* NCL British collections are possibly *Amanita strobiliformis* or a form of *A. citrina*. Cooke 12 (70) Vol.1 (1881) is neither.

nitidus (Schaeff.) Fr., Cortinarius, *Epicr. syst. mycol.*: 275 (1838)
    Agaricus nitidus Schaeff., *Fung. Bavar. Palat. nasc.* 1: 97 (1762)
A *nomen dubium* lacking type material. Cooke 1189 (1191) Vol. 8 (1891) is possibly *C. delibutus*.

nitratus *var.* **glauconitens** (Fr.) Fr., Hygrophorus, *Hymenomyc. eur.*: 421 (1874)
    Agaricus glauconitens Fr., *Epicr. syst. mycol.*: 54 (1838)
    Hygrophorus glauconitens (Fr.) Cooke, *Grevillea* 6(40): 121 (1878)
A *nomen dubium*. Reported by Cooke in Grevillea 6: 121 (1878) as in 'woody pastures' and also by Berk. & Broome in Ann. Mag. Nat. Hist. [Notices of British Fungi no. 1671] (1878). Excluded from NCL 'pending clearer definition'.

nitriolens J. Favre, Clitocybe, *Ergebn. Wiss. Untersuch. Schweiz. Natn. Parks* 6: 431 (1960)
Mis.: *Clitocybe concava* sensu NCL
Not authentically British. A montane species listed (as *C. concava*) in NCL with doubts. No material has been traced.

nitrosus Cooke, Cortinarius, *Grevillea* 16: 44 (1887)
A *nomen dubium* described from England (North Somerset: Durdham Down, near Bristol) and illustrated as Cooke 808 (837) Vol. 6 (1887).

nivea Quél., Mycena, *Bull. Soc. Mycol. France* 23: 325 (1877)
A *nomen dubium* lacking type material. No voucher material of British collections has been traced. Reported in TBMS 4 (2): 309 (1913) and also in Naturalist (Hull) 1913: 24 (1913). Sensu Rea (1922) is probably *Hemimycena cucullata*.

nivea (Sommerf.) P. Karst., Plicatura, *Bidrag Kännedom Finlands Natur Folk* 48: 342 (1889)
    Merulius niveus Sommerf., *Suppl. Fl. lapp.*: 268 (1826)
    Merulius rimosus Berk., *Grevillea* 19: 108 (1891)
Not authentically British. The only collection named thus in herb. K (from Scotland) has been redetermined as *Plicatura crispa*.

niveipes (Murrill) Murrill, Mycena, *Mycologia* 8: 221 (1916)
    Prunulus niveipes Murrill, *North American Flora, Ser. 1 (Fungi 3)*, 9(5): 332 (1916)
Not authentically British. A North American species, reported from Somerset in 1998, but unsubstantiated with voucher material.

niveola (Sacc.) Agerer, Seticyphella, *Mitt. bot. St. Samml., Münch.* 19: 284 (1983)
    Cyphella niveola Sacc., *Syll. fung.* 6: 678 (1888)
Not authentically British. A single collection by Dennis in 1949, from Scotland (Aviemore) in herb. K, is infertile and doubtfully assigned here.

niveus Sowerby, Agaricus [non *A. niveus* Scop. (1772) or Pers. (1801)], *Col. fig. Engl. fung.* 1: pl. 97 (1797)
Described from Britain but a *nomen dubium* lacking type material. From the illustration this is apparently a species of *Crepidotus*.

niveus Fr., Boletus, *Observ. mycol.* 1: 111 (1815)

Boletus scaber var. *niveus* (Fr.) Opat., *Comm. fam. bolet.*: 33 (1836)
A *nomen dubium*. Listed by Rea (1922) and cited in NCL in the synonymy of *Leccinum holopus*.

nodi-aurei With., Agaricus, *Bot. arr. veg.* 2: 757 (1776)
Described from Britain but a *nomen dubium*. From the description, possibly *Hypholoma fasciculare*.

nodosus With., Agaricus, *Syst. arr. Brit. pl. ed. 4*, 4: 254 (1801)
Described from Britain but a *nomen dubium*.

nucea (Bolton) Sacc., Naucoria, *Syll. fung.* 5: 839 (1887)
Agaricus nuceus Bolton, *Hist. fung. Halifax* 2: 70 (1788)
Described from Britain but a *nomen dubium* lacking type material. Cooke 500 (490) A Vol. 4 (1884) is copied directly from Bolton's plate.

nuciseda (Fr.) Massee, Psilocybe, *Brit. fung.-fl.* 1: 371 (1892)
Agaricus nucisedus Fr., *Syst. mycol.* 1: 293 (1821)
A *nomen dubium*. Not reported in recent times. Cooke 601 (609) Vol. 4 (1886) is doubtful.

nudipes (Fr.) Sacc., Hebeloma, *Syll. fung.* 5: 801 (1887)
Agaricus nudipes Fr., *Epicr. syst. mycol.*: 181 (1838)
A *nomen dubium*. Listed in NCL in the synonymy of *Hebeloma longicaudum*. British records are probably mostly *H. leucosarx*, but northern records may also include *H. incarnatulum*.

nummularia (Fr.) Gillet, Collybia, *Hyménomycètes*: 325 (1876)
Agaricus nummularius Fr, *Epicr. syst. mycol.*: 91 (1838)
A *nomen dubium*. Possibly *Collybia aquosa* or a species near to *C. dryophila fide* A&N2 but Cooke 203 (151) Vol. 2 (1882) resembles neither.

nycthemerus Fr., Coprinus, *Epicr. syst. mycol.*: 251 (1838)
A *nomen dubium*. Cooke 670 (682) Vol. 5 (1886) is probably *C. heptemerus*.

oakesii (Berk. & M.A. Curtis) Pat., Aleurodiscus, *Rev. Mycol. (Toulouse)* 12(47): 133 (1890)
Corticium oakesii Berk. & M.A. Curtis, *Grevillea* 1: 166 (1873)
A North American species, unknown in Europe. Sensu Wakefield (1952) is *Aleurodiscus wakefieldiae*.

obducta (Berk.) Donk, Osteina, *Schweiz. Z. Pilzk.* 44: 86 (1966)
Polyporus obductus Berk., *London J. Bot.* 4: 304 (1845)
Polyporus osseus Kalchbr., *Math. Term. Közl. Mag. Tudom. Akad.* 3: 217 (1865)
Not authentically British. A central European species. Listed by Smith (1908) and Rea (1922) (as *Polyporus osseus*) with the comment 'rare - on larch stumps' but no voucher material has been located.

obesus (Batsch) Gray, Gymnopus, *Nat. arr. Brit. pl.* 1: 607 (1821)
Agaricus obesus Batsch, *Elench. fung.*: 89 t. 39 f. 216 (1789)
A *nomen dubium*. Sensu Gray (1821) is unknown.

obliquus Fr., Lactarius, *Epicr. syst. mycol.*: 348 (1838)
A *nomen dubium* said to grow caespitose on trunks and banks. Cooke 969 (1014) B Vol. 7 (1888) is unidentifiable.

obrussea (Fr.) Wünsche, Hygrocybe, *Pilze*: 113 (1877)
Agaricus obrusseus Fr., *Syst. mycol.* 1: 104 (1821)
Hygrophorus obrusseus (Fr.) Fr., *Epicr. syst. mycol.*: 331 (1838)
A *nomen confusum* variously interpreted. Sensu NCL is *Hygrocybe citrinovirens* but most British records named thus are *H. quieta*.

obscura (Schaeff.) Herink, Armillaria, *Vys. Skola Z. Brně*: 42 (1973)
Agaricus obscurus Schaeff. [non *A. obscurus* Pers. (1801)], *Fung. Bavar. Palat. nasc.* 1: 32 (1762)
Armillaria mellea var. *obscura* (Schaeff.) Gillet, *Hyménomycètes*: 84 (1874)
A *nomen dubium* lacking type material and inadequately described. Listed by Rea (1922). *Armillariella obscura* (Pers.) Romagnesi (Bull. Soc. Mycol. France 86 (1): 262, 1970) is

different *fide* Termorshuizen & Arnolds [Mycotaxon 30: 110 (1987)].

obscura Gillet, Inocybe, *Hyménomycètes*: 515 (1876)
Agaricus obscurus Pers. [*nom. illegit.*, non *A. obscurus* Schaeff. (1774)], *Syn. meth. fung.*: 347 (1801)
A *nomen dubium* lacking type material. The name has been widely used for *Inocybe cincinnata* var. *major*.

obscura var. rufa (Pat.) Sacc., Inocybe, *Syll. fung.* 5: 770 (1887)
Agaricus obscurus var. *rufus* Pat., *Tab. anal. fung.* 6: 20 (1886)
A *nomen dubium*. Possibly *Inocybe pusio* or *I. amethystina fide* Kuyper (1986).

obscura (L.S. Olive) M.P. Christ., Tremella, *Friesia* 5: 62 (1954)
Tremella mycophaga var. *obscura* L.S. Olive, *Mycologia* 38: 540 (1946)
Not authentically British. A North American species reported as British by Reid [TBMS 62: 490 (1974)] although only conidia and conidiophores were seen. Re-examination shows that these belong to *Occultifur internus*. Subsequent collections named *T. obscura* at K are also *O. internus*.

obscurata Cooke, Clitocybe, *Trans. Brit. Mycol. Soc.* 3: 109 (1909)
Described from Britain but a *nomen dubium* and probably just a form of *Clitocybe clavipes fide* NCL. Synonymised with *Clitocybe trulliformis* (not authentically British) in FAN3 p. 50.

obscuratus (P. Karst.) Rea, Hygrophorus, *Brit. basidiomyc.*: 309 (1922)
Camarophyllus obscuratus P. Karst., *Hedwigia* 28: 364 (1889)
A *nomen dubium*. See TBMS 5 (3): 434 (1916) and Naturalist (Hull) 1917: 13 (1917) for descriptions of material collected in Yorkshire (Buckden). From the original description this seems to be a species near to *Hygrocybe nitrata*.

obsolescens Batsch, Agaricus, *Elench. fung. (Continuatio Prima)*: 127 t. 20 f. 102 (1786)
A *nomen dubium*. Reported by Withering (1796). The illustration in Batsch is not helpful, possibly representing a grey species of *Clitocybe*.

obsoleta (Batsch) Quél., Clitocybe, *Mém. Soc. Émul. Montbéliard, Sér. 2*, 5: 216 (1872)
Agaricus obsoletus Batsch, *Elench. fung. (Continuatio Prima)*: 129 & t. 20 f. 103 (1786)
A *nomen dubium* variously interpreted. Sensu auct. Brit. is *C. fragrans*. Sensu C&D p. 179 (304) and sensu B&K3: 162 (pl. 171) is not British.

obtusa (Cooke & Massee) Sacc., Naucoria, *Syll. fung.* 9: 111 (1891)
Agaricus obtusus Cooke & Massee [*nom. illegit.*, non *A. obtusus* Fr. (1821)], *Grevillea* 18: 52 (1890)
Described from Britain but a *nomen dubium*. Possibly *Psathyrella sarcocephala fide* Pearson & Dennis (1948) but Cooke 1171 (1155) Vol. 8 (1890) is certainly not that species.

obtusa (Schrad.) Gray, Odontia, *Nat. arr. Brit. pl.* 1: 651 (1821)
Hydnum obtusum Schrad., *Spicil. Fl. Germ.*: 178 (1794)
A *nomen dubium*. Said by Donk [Taxon 5: 105 (1956)] to be 'identified with *Radulum quercinum*', i.e. *Hyphodontia quercina* but sensu Gray (1821) is unknown.

obtusifolius Rea, Marasmius, *Trans. Brit. Mycol. Soc.* 6(4): 324 (1920)
Described from Britain but a *nomen dubium*. 'Almost certainly' *Oudemansiella radicata* (=*Xerula radicata*) *fide* NCL. Antonín & Noordeloos (A&N1) suggest it may be *Macrocystidia cucumis* but the spores, as described, are far too large.

obtusiforme J. Erikss. & Å. Strid, Hyphoderma, *Corticiaceae of North Europe* 3: 493 (1975)
Not authentically British. Collections named thus from England (Surrey) are determined as 'compare with' *H. obtusiforme*.

Reported from Scotland during the Nordic Mycological Congress in 1983 but unsubstantiated with voucher material.

obtusum J. Erikss., Hyphoderma, *Symb. Bot. Upsal.* 16(1): 97 (1958)
Not authentically British. Reported from Oxfordshire and Yorkshire but unsubstantiated with voucher material.

occulta (Cooke) Sacc., Clitocybe, *Syll. fung.* 11: 15 (1895)
Agaricus occultus Cooke, *Grevillea* 19: 40 (1890)
Described from Britain but a *nomen dubium*. Cooke 1133 (1184) Vol. 8 (1890) appears to be a species of *Tricholoma*.

ocellata (Fr.) P. Kumm., Collybia, *Führ. Pilzk.*: 114 (1871)
Agaricus ocellatus Fr., *Observ. mycol.* 1: 83 (1815)
A *nomen dubium*, possibly a *Marasmiellus* species *fide* A&N2. Accepted in NCL citing Cooke 209 (147) Vol. 2 (1882) which has a strong resemblance to *Crinipellis scabellus*.

ochracea (Kühner) Singer, Conocybe, *Mycologia* 5: 395 (1959)
Conocybe siliginea var. *ochracea* Kühner [*nom. nud.*], *Encycl. Mycol. 7. Le Genre* Galera: 101 (1935)
A *nomen dubium*. Sensu NCL and Phillips (1981) is *Conocybe kuehneriana*. Sensu auct. is *C. siennophylla*.

ochracea (Hoffm.) D.A. Reid, Merismodes, *Persoonia* 3(1): 116 (1964)
Solenia ochracea Hoffm., *Deutschl. Fl.*: pl. 8 f. 2 (1795)
Solenia anomala var. *ochracea* (Hoffm.) Massee, *British Fungus Flora* 1: 144 (1892)
Phaeocyphellopsis ochracea (Hoffm.) W.B. Cooke, *Sydowia, Beih.* 4: 120 (1961)
A doubtful species. Cooke (1961) assigned fifteen British collections here but none of these fitted Reid's (1964) concept of this species.

ochracea (Alb. & Schwein.) Fr., Russula, *Epicr. syst. mycol.*: 362 (1838)
Agaricus ochraceus Alb. & Schwein. [non *A. ochraceus* Pers. (1828)], *Consp. fung. lusat.*: 213 (1805)
A *nomen dubium* lacking type material and variously interpreted. Sensu auct. is *Russula risigallina* but the name has probably also been used for *R. fellea*.

ochraceoflavum (Schwein.) Sacc., Stereum, *Syll. fung.* 6: 576 (1888)
Thelephora ochraceoflava Schwein., *Trans. Am. phil. Soc.* 4: 167 (1832)
Not authentically British. A North American species. Sensu auct. Brit. is *Stereum 'rameale'*.

ochraceo-velatus R.F.O. Kemp [*nom. inval.*], Coprinus, *Trans. Brit. Mycol. Soc.* 65(3): 380 (1975)
A *nom. prov.* for a member of section *Lanatuli* studied in culture by Kemp, which has appeared in literature but was never validly published.

ochraceum (Fr.) Hallenb., Conferticium, *Mycotaxon* 11(2): 448 (1980)
Thelephora ochracea Fr., *Observ. mycol.* 1: 151 (1815)
Corticium ochraceum (Fr.) Fr., *Epicr. syst. mycol.*: 563 (1838)
Peniophora ochracea (Fr.) Massee, *J. Linn. Soc., Bot.* 25: 50 (1889)
Gloeocystidium ochraceum (Fr.) Höhn. & Litsch., *Hyménomyc. France*: 266 (1928)
Gloeocystidiellum ochraceum (Fr.) Donk, *Fungus* 26: 9 (1956)
Not authentically British. A boreal conifer species reported e.g. from Shetland, but material in herb. K and herb. E is misidentified.

ochroleucum (Fr.) Fr., Stereum, *Hymenomyc. eur.*: 639 (1874)
Thelephora ochroleuca Fr., *Observ. mycol.* 2: 276 (1818)
A *nomen dubium* variously interpreted. Sensu Rea (1922) is doubtful. Two Scottish collections thus named by Berkeley, in herb. K, are misidentified *Phanerochaete sordida* and *Cylindrobasidium laeve*.

ochroviridis (Cooke) Reumaux, Russula, *Russules Rares ou Méconnues*: 286 (1996)
Russula furcata var. *ochroviridis* Cooke, *Handb. Brit. fung.*, Edn 2: 322 (1889)
Described from Britain but a *nomen dubium*. Sarnari (1998) considers Cooke 980 (1100) Vol. 7 (1889) to represent a pale form of *R. atropurpurea*.

octagonus With., Agaricus, *Bot. arr. Brit. pl. ed. 2*, 3: 497 (1792)
Described from Britain but a *nomen dubium*.

oculatum Bruchet, Hebeloma, *Bull. Mens. Soc. Linn. Lyon* 36 (6 [Suppl.]): 63 (1970)
A *nomen dubium*. Reported from Britain but unsubstantiated with voucher material. Possibly near to *Hebeloma pusillum*.

odora (Sacc.) Ginns, Skeletocutis, *Mycotaxon* 21: 332 (1984)
Poria odora Sacc., *Syll. fung.* 6: 294 (1888)
Not authentically British. Shown as British in EurPoly2: 631, based on a misdetermined specimen of *Skeletocutis kuehneri*.

odorata (Cool) Bas, Squamanita, *Persoonia* 3(3): 342 (1965)
Lepiota odorata Cool, *Meded. Ned. Mycol. Ver.* 9: 47-52 (1918)
Coolia odorata (Cool) Huijsman, *Meded. Ned. Mycol. Ver.* 28: 60 (1943)
Tricholoma odoratum (Cool) Konrad & Maubl., *Encycl. Mycol. 14, Les Agaricales* I: 346 (1948)
Not authentically British. Parasitic on *Hebeloma mesophaeum* and known from several European countries. The British collection named thus (under *Pinus* in Scotland) was misdetermined and became instead the type of *Squamanita pearsonii*.

odoratus (M.M. Moser) M.M. Moser, Cortinarius, *Kleine Kryptogamenflora*, Edn 3: 306 (1967)
Phlegmacium odoratum M.M. Moser, *Die Gattung Phlegmacium*: 360 (1960)
Not authentically British. Reported from East Gloucestershire and West Kent but unsubstantiated with voucher material.

odorus (Sommerf.) Singer, Haploporus, *Mycologia* 36: 68 (1944)
Polyporus odorus Sommerf., *Suppl. Fl. lapp.*: 275 (1826)
Not authentically British. Sensu auct. Brit. is *Trametes suaveolens*.

offuciata (Fr.) Gillet, Omphalia, *Hyménomycètes*: 292 (1876)
Agaricus offuciatus Fr., *Epicr. syst. mycol.*: 121 (1838)
A *nomen dubium*. Cooke 253 (287) Vol. 2 (1883) is unknown.

olearius (DC.) Singer, Omphalotus, *Pap. Michigan Acad. Sci.* 32: 123 (1946)
Agaricus olearius DC., *Fl. Fr.* 5: 44 (1815)
Not authentically British. A Mediterranean species restricted to decayed wood of *Olea europaea*. Sensu auct. Brit. is *O. illudens*.

olivacea Cooke & Massee, Bovista, *Grevillea* 16: 77 (1888)
Not authentically British. Described from 'Durdham Down' (ex herb. Broome) and 'Winmera, Victoria (Australia)'. The type collection is Australian and equates to *Calvatia candida* (Rostk.) Hollós *fide* Kreisel [Nova Hedwigia, Beih. 25: 174 (1967)]. No material has been traced of the supposed British collection.

olivaceoalba (Bourdot & Galzin) Ginns & M.N.L. Lefebvre, Scytinostromella, *Mycol. Mem.* 19: 141 (1993)
Corticium olivaceoalbum Bourdot & Galzin, *Bull. Soc. Mycol. France* 27: 239 (1911)
Athelia olivaceoalba (Bourdot & Galzin) Donk, *Fungus* 27: 12 (1957)
Confertobasidium olivaceoalbum (Bourdot & Galzin) Jülich, *Willdenowia. Beih.* 7: 167 (1972)
Not authentically British. Sensu CNE2, B&K2, and others (as *Confertobasidium olivaceoalbum*) is *Leptosporomyces fuscostratus*.

olivaceoviolascens Gillet, Russula, *Hyménomycètes*: pl. 512 (1878)
A *nomen confusum fide* Sarnari (1998). British records are sensu K&R thus are possibly *Russula atrorubens*.

olivascens (Batsch) Fr., Cortinarius, *Epicr. syst. mycol.*: 273 (1838)
    Agaricus olivascens Batsch*, Elench. fung. (Continuatio Secunda)*: 3 t. 31 f. 185 (1789)
A *nomen dubium*. Sensu Rea (1922) is possibly *C. infractus*.

olivascens Singer & M.M. Moser, Macrolepiota, *Schweiz. Z. Pilzk.* 39: 154 (1961)
Not authentically British. The single collection named thus (in herb. K) has been redetermined as *M. procera* var. *pseudoolivascens*.

olivascens Fr., Russula, *Öfvers Kongl. Vetensk.-Akad. Förh.* 18: 28 (1862)
    Russula alutacea var. *olivascens* (Fr.) Rea*, Brit. basidiomyc.*: 475 (1922)
A *nomen dubium*. Sensu Cooke 1001 (1041) Vol. 7 (1888) and sensu Rea (1922) is perhaps *Russula olivacea* but was recorded in 'fir' woods.

ombrophila (Fr.) Konrad & Maubl., Agrocybe, *Encycl. Mycol.* 14, *Les Agaricales* I: 160 (1948)
    Agaricus ombrophilus Fr.*, Hymenomyc. eur.*: 216 (1874)
    Pholiota ombrophila (Fr.) P. Karst.*, Bidrag Kännedom Finlands Natur Folk* 32: 292 (1879)
    Togaria ombrophila (Fr.) W.G. Sm.*, Syn. Brit. Bas.*: 123 (1908)
A *nomen dubium*. British records under this name are probably *Agrocybe brunneola*.

omphalodes (Fr.) P. Karst., Lentinellus, *Bidrag Kännedom Finlands Natur Folk* 32: 248 (1879)
    Lentinus cochleatus var. *omphalodes* Fr.*, Monogr. hymenomyc. Suec.* 2: 235 (1863)
Not authentically British. Following BFF6, this name refers to a species related to *Lentinellus flabelliformis* but with larger spores, perhaps *L. micheneri*. The name has been misapplied in Britain to *L. flabelliformis*.

onusta P. Karst., Trechispora, *Hedwigia* 29: 147 (1890)
A *nomen dubium*, sometimes interpreted as *Trechispora hymenocystis*, but the British collection reported by Wakefield (1952: 39) is *Sistotrema alboluteum*. Fragments of both species and others are in Karsten's surviving type material (CNE7: 1313).

opaca (With.) Gillet, Clitocybe, *Hyménomycètes*: 164 (1874)
    Agaricus opacus With.*, Bot. arr. Brit. pl. ed. 2*, 3: 307 (1792)
Described from Britain but a *nomen dubium* lacking type material. NCL considered this a synonym of *C. tornata*, thus possibly near to *C. candicans*.

opicum (Fr.) Gillet, Tricholoma, *Hyménomycètes*: 107 (1874)
    Agaricus opicus Fr.*, Elench. fung.* 1: 16 (1828)
A *nomen dubium*. A doubtful species *fide* NCL. Riva (FungEur3) suggests that this is a synonym of *Tricholoma gausapatum*.

opimus Fr., Cortinarius, *Epicr. syst. mycol.*: 278 (1838)
    Cortinarius opimus var. *fulvobrunneus* Fr.*, Monogr. hymenomyc. Suec.* 2: 45 (1863)
A *nomen dubium*. Listed in NCL sensu Rea (1922) but no material survives. There is a recent collection (1984) named thus from Surrey (Richmond Park).

opipara (Fr.) P. Kumm., Clitocybe, *Führ. Pilzk.*: 121 (1871)
    Agaricus opiparus Fr.*, Epicr. syst. mycol.*: 59 (1838)
    Tricholoma opiparum (Fr.) Bigeard & H. Guill.*, Fl. Champ. Supér. France* 1: 87 (1909)
A *nomen dubium*. Pearson & Dennis (1948) suggest that this may be *Hygrophorus nemoreus*. Cooke 1128 (1183) Vol. 8 (1889) could possibly represent *Rhodocybe gemina*.

orbiformis (Fr.) Gillet, Clitocybe, *Hyménomycètes*: 166 (1874)
    Agaricus orbiformis Fr.*, Epicr. syst. mycol.*: 76 (1838)

A *nomen dubium*. Reported from Surrey by Pearson [TBMS 35 (2): 100 (1952)] and listed in NCL on the strength of this collection, but no material has been traced.

orcelloides Cooke & Massee, Paxillus, *Grevillea* 16: 46 (1887)
Described from Britain but a *nomen dubium*. Possibly *Clitocybe fallax* (= *Rhodocybe fallax*) *fide* Pearson & Dennis (1948). Cooke 863 (874) Vol. 6 (1888), however, appears to be *Ripartites tricholoma*.

orichalceus (Batsch) Fr., Cortinarius, *Epicr. syst. mycol.*: 267 (1838)
    Agaricus orichalceus Batsch*, Elench. fung. (Continuatio Secunda)*: 1 t.31 f.184 (1789)
Not authentically British A *nomen dubium*, and a *nomen confusum fide* CPF3: 11. Sensu auct. mult. is *C. cupreorufus*, which is unknown in Britain. Sensu Cooke [718 (754) Vol. 5], Rea (1922), and NCL, smelling of 'fennel', is probably *C. odorifer*.

orphanellum (Bourdot & Galzin) Donk, Hyphoderma, *Fungus* 27: 15 (1957)
    Peniophora orphanella Bourdot & Galzin*, Bull. Soc. Mycol. France* 28: 381 (1913)
Not authentically British. Reported from Yorkshire and Suffolk but unsubstantiated with voucher material.

ovata (Pers.) J. Schröt., Typhula, in Cohn, *Krypt.-Fl. Schlesien* 3: 439 (1888)
    Clavaria ovata Pers.*, Comment. Fungis Clavaeform.*: 86 (1797)
    Pistillaria ovata (Pers.) Fr.*, Syst. mycol.* 1: 497 (1821)
A *nomen dubium* lacking type material, but near to *Typhula setipes*. Listed by Rea (1922) but no British material has been traced.

ovinus (Schaeff.) Murrill, Albatrellus, *J. Mycol.* 9: 91 (1903)
    Boletus ovinus Schaeff.*, Fung. Bavar. Palat. nasc.* 1: 83 (1762)
    Boletus albidus Pers. [non *B. albidus* Roques (1832)]*, Syn. meth. fung.*: 515 (1801)
    Albatrellus albidus (Pers.) Gray*, Nat. arr. Brit. pl.* 1: 645 (1821)
Not authentically British. Reported by Gray (1821) as 'in fir plantations, on buried sticks' but this is a species associated with native *Picea* sp. in continental Europe and unlikely in Britain. No material has been traced.

oxydabile (Singer) Singer, Leccinum, *Amer. Midl. Naturalist* 37: 123 (1947)
    Krombholzia oxydabilis Singer*, Rev. Mycol. (Paris)* 3(6): 189 (1938)
    Krombholziella oxydabilis (Singer) Šutara*, Česká Mykol.* 36 (2): 82 (1982)
A *nomen dubium* apparently lacking type material. Most British collections named thus are *L. rigidipes fide* A.E. Hills (pers. comm.).

ozes (Fr.) Bon, Tephrocybe, *Doc. mycol.* 25 (97): 4 (1995)
    Agaricus ozes Fr.*, Epicr. syst. mycol.*: 95 (1838)
    Collybia ozes (Fr.) Sacc.*, Syll. fung.* 5: 244 (1887)
    Lyophyllum ozes (Fr.) Singer*, Ann. Mycol.* 41: 103 (1943)
Not authentically British. Excluded 'pending clearer definition' by NCL and not discussed by Orton (1984). There are two specimens in herb. K, both also annotated as 'cf. inolens'.

padi Pers., Thelephora, *Mycol. eur.* 1: 142 (1822)
A *nomen dubium* lacking type material. Sensu Greville in *Scot. Crypt. Fl.* 4: 233 (1826) is *Hymenochaete corrugata fide* Rea (1922).

paedida (Fr.) Kühner & Maire, Melanoleuca, *Bull. Soc. Mycol. France* 50: 18 (1934)
    Agaricus paedidus Fr.*, Epicr. syst. mycol.*: 53 (1838)
    Tricholoma paedidum (Fr.) Quél.*, Mém. Soc. Émul. Montbéliard, Sér. 2*, 5: 341 (1873)
Sensu Rea (1922) is doubtful. Sensu auct. (e.g. B&K3 pl. 303) is a species near *M. rasilis* but not authentically British.

paeonius Fr., Agaricus, *Epicr. syst. mycol.*: 42 (1838)
A *nomen dubium*. Reported by Cooke (HBF1: 31) from
Buckinghamshire (Burnham Beeches) but no material has been
traced. From the description this is possibly *Calocybe carnea*.

paleaceus (Weinm.) Fr., Cortinarius, *Epicr. syst. mycol.*: 302
(1838)
    Agaricus paleaceus Weinm., *Hymen. Gasteromyc.*: 296
    (1836)
A *nomen dubium* lacking type material and variously applied to
any of the species with a smell of *Pelargonium*. The majority of
British collections so named are *C. flexipes* var. *flabellus*.

pallescens (P. Karst.) Noordel., Entoloma, *Persoonia* 10: 251
(1979)
    Nolanea pascua var. *pallescens* P. Karst., *Bidrag Kännedom*
    Finlands Natur Folk 32: 280 (1879)
Not authentically British. Dubiously reported from England
(Derbyshire) in 2002 and also from Republic of Ireland
(Wicklow) but unsubstantiated with voucher material.

pallida (Pilát) E. Horak & M.M. Moser, Galerina, *Kleine
Kryptogamenflora,* Edn 3: 347 (1967)
    Velomycena pallida Pilát, *Schweiz. Z. Pilzk.* 31: 175 (1953)
Not authentically British. Reported from Surrey but dubious and
unsubstantiated with voucher material.

pallida (Schaeff.) Ricken, Ramaria, *Vadem. Pilzfr.*: 263 (1920)
    Clavaria pallida Schaeff., *Fung. Bavar. Palat. nasc.* 3: 286
    (1770)
Not authentically British. Included by Rea (1927, as *Clavaria
pallida*) but requiring confirmation. See also *R. mairei*.

pallidipes Ellis & Everh., Inocybe, *J. Mycol.*: 24 (1889)
Not authentically British. A North American species, probably a
synonym of *Inocybe auricoma*. Sensu NCL (following FIDan3:
74) is a form of *I. flocculosa*.

pallidispora Kühner & Watling, Conocybe, *Notes Roy. Bot.
Gard. Edinburgh* 40 (3): 540 (1983)
    Conocybe siliginea var. *pallidispora* Kühner [*nom. nud.*],
    Encycl. Mycol. 7. *Le Genre* Galera: 100 (1935)
Not authentically British. A single record (as *Conocybe siliginea*
var. *pallidispora*) but unsubstantiated with voucher material.

pallidolivens (Bourdot & Galzin) Parmasto, Phlebia, *Eesti
N.S.V. Tead. Akad. Toimet., Biol.* 16: 391 (1967)
    Corticium pallidolivens Bourdot & Galzin, *Bull. Soc. Mycol.
    France* 27: 254 (1911)
Not authentically British. Reported from Yorkshire but
unsubstantiated with voucher material.

pallidus Sowerby, Agaricus [*nom. illegit.*, non *A. pallidus* (Pers.)
Fr. (1821)], *Col. fig. Engl. fung.* 3: pl. 365 (1802)
Described from Britain, but a *nomen dubium* lacking type
material. Sowerby's illustration and description are not helpful.

paludosus Massee, Boletus, *Brit. fung.-fl.* 1: 279 (1892)
Described from Britain but a *nomen dubium*. Possibly a slender
ecotype of *Boletus badius* in *Sphagnum* moss.

pansa (Fr.) Fr., Cortinarius, *Epicr. syst. mycol.*: 264 (1838)
    Agaricus pansa Fr., *Observ. mycol.* 2: 67 (1818)
A *nomen dubium*. Sensu Orton (1955) is *C. arquatus*.

pantoleucus (Fr.) Sacc., Pleurotus, *Syll. fung.* 5: 349 (1887)
    Agaricus pantoleucus Fr., *Hymenomyc. eur.*: 172 (1874)
A *nomen dubium*. Cooke 277 (179) and 278 (275) Vol. 2 (1883)
shows two completely different pleurotoid species. Sensu auct.
Brit. is *Hypsizygus ulmarius*.

papillatus (Batsch) Fr., Coprinus, *Epicr. syst. mycol.*: 248
(1838)
    Agaricus papillatus Batsch, *Elench. fung. (Continuatio Prima)*:
    81 & t. 17 f. 78 (1786)
    Coprinus papillatus var. *oxygenus* Fr., *Hymenomyc. eur.*: 327
    (1874)

A *nomen dubium*. The Batsch plate could represent *C.
disseminatus*. Cooke 663 (676) Vol. 5 (1886) is some other
species.

papulosus Fr., Cortinarius, *Epicr. syst. mycol.*: 271 (1838)
    Cortinarius papulosus var. *major* Fr., *Monogr. hymenomyc.
    Suec.* 2: 33 (1863)
Not authentically British. Recorded from Herefordshire and
South Somerset in the nineteeth century, but unsubstantiated
with voucher material. Probably misdetermined as this is a
boreal-montane species.

parabolica (Fr.) Quél., Mycena, *Mém. Soc. Émul. Montbéliard,
Sér. 2,* 5: 242 (1872)
    Agaricus parabolicus Fr., *Epicr. syst. mycol.*: 107 (1838)
A *nomen dubium*. Possibly not a species of *Mycena fide* Maas
Geesteranus (1992). See Cooke 229 (224) Vol. 2 (1882) for an
interpretation.

paraceracea Bon, Hygrocybe, *Doc. mycol.* 19 (75): 56 (1989)
Not authentically British. A single record from Herefordshire
(Fishpool Valley) unsubstantiated with voucher material.
Doubtfully distinct from *Hygrocybe ceracea*.

parafulmineus Rob. Henry [*nom. inval.*], Cortinarius, *Flore
Analytique des Champignons Supérieurs*: 266 (1953)
Not authentically British. Included in NCL based on Rea (1922)
citing large spores for *C. fulmineus*. No British collections have
been traced. *C. fulmineus* sensu Rea (1922) is now thought to
be *C. elegantior*.

paragaudis Fr., Cortinarius, *Epicr. syst. mycol.*: 295 (1838)
Not authentically British. Reported by Orton (NBA9: 51, as *C.
haematochelis*) but no material has been traced. *C. paragaudis*
sensu Rea (1922) is doubtful.

parasiticum With., Lycoperdon, *Bot. arr. Brit. pl. ed. 2,* 3: 464
(1792)
Described from Britain but a *nomen dubium,* and probably not a
basidiomycete. Withering's description is suggestive of a
myxomycete.

pardinum var. unguentatum (Fr.) Bon, Tricholoma, *Doc.
Mycol.* 4 (14): 93 (1974)
    Agaricus unguentatus Fr., *Epicr. syst. mycol.*: 27 (1838)
    Tricholoma unguentatum (Fr.) Sacc., *Syll. fung.* 5:
    103 (1887)
Doubtful *fide* NCL. Reported from Yorkshire in TBMS 5 (3): 434
(1916) and also in Naturalist (Hull) 1917: 131 (1917), but no
British material has been traced.

parherpeticus Rob. Henry, Cortinarius, *Bull. Soc. Mycol. France*
67: 284 (1952)
Not authentically British. Included in NCL, but probably
misdetermined *C. glaucopus fide* Orton (NBA9).

parilis (Fr.) Singer, Rhodocybe, *Agaricales in Modern
Taxonomy,* Edn 2: 678 (1962)
    Agaricus parilis Fr., *Syst. mycol.* 1: 168 (1821)
    Clitocybe parilis (Fr.) Gillet, *Hyménomycètes*: 144 (1874)
Not authentically British. Sensu Rea (1922, as *C. parilis*) is *C.
fuscosquamulosa* which is also dubiously British.

parisianorum Bon, Melanoleuca, *Doc. mycol.* 18 (69): 30
(1987)
Not authentically British. The single collection named thus
(South Devon: Orley Common) is misdetermined.

parisotii (Pat.) Sacc., Crepidotus, *Syll. fung.* 5: 886 (1887)
    Agaricus parisotii Pat., *Tab. anal. fung.*: 158 (1885)
A *nomen dubium*. Sensu Rea (1922) is possibly *C. luteolus*.

paropsis (Fr.) Sacc., Clitocybe, *Syll. fung.* 5: 173 (1887)
    Agaricus paropsis Fr., *Epicr. syst. mycol.*: 72 (1838)
A *nomen dubium*. Recent British collections named thus, from
dunes, are *Clitocybe vermicularis*. Sensu Rea (1922) is
unknown.

partitus With., Agaricus, *Bot. arr. Brit. pl. ed. 2,* 3: 357 (1792)
Described from Britain but a *nomen dubium*.

pascua *var.* **umbonata** Quél., Nolanea, *Enchir. fung.*: 63 (1886)
A *nomen dubium* lacking type material. Listed by Rea (1922) as a synonym of *Nolanea mammosa* sensu Quélet [*Mém. Soc. Émul. Montbéliard, Sér. 2*, 5: 89 (1872)], thus possibly *Entoloma hebes*.

pascuum (Pers.) Donk, Entoloma, *Bull. Jard. Bot. Buitenzorg, Sér. 3*, 18: 158 (1949)
Agaricus pascuus Pers., *Syn. meth. fung.*: 427 (1801)
Nolanea pascua (Pers.) P. Kumm., *Führ. Pilzk.*: 95 (1871)
A *nomen dubium* formerly widely used in Britain for *E. conferendum*. Sensu Rea (1922) is *E. vernum*. Material in the Persoon herbarium, perhaps the type *fide* Singer [*Persoonia* 2 (1): 35 (1961)], appears to be a species of *Cortinarius*.

pascuus (Pers.) Krombh., Boletus, *Naturgetr. Abbild. Schwämme*: t. 76 (1846)
Boletus subtomentosus var. *pascuus* Pers., *Mycol. eur.* 2: 139 (1825)
Xerocomus pascuus (Pers.) E.-J. Gilbert, *Les Bolets*: 139 (1931)
A *nomen dubium* lacking type material. Most likely just a form of *Boletus chrysenteron*. Last reported [TBMS 3 (3): 145 (1910)] from the BMS Baslow Foray in 1909. Considered distinct by some authorities.

patellaris (Fr.) Konrad & Maubl., Panellus, *Icon. select. fung.* 6: 379 (1937)
Panus patellaris Fr., *Epicr. syst. mycol.*: 400 (1838)
Tectella patellaris (Fr.) Murrill, *North American Flora, Ser. 1 (Fungi 3),* 9 (4): 247 (1915)
Not authentically British. Cooke 1097 (1144) Vol. 7 (1890) probably represents *P. ringens* and the single collection (in herb. K) from Scotland has been redetermined as *P. ringens*.

pateriformis Fr., Cortinarius, *Epicr. syst. mycol.*: 310 (1838)
A *nomen dubium*. Sensu Rea (1922) is doubtful.

patouillardii *var.* **lipophilus** R. Heim & Romagn., Coprinus, *Bull. Soc. Mycol. France* 50: 187 (1934)
Not authentically British. Listed in Watling (1973: 257) as *Coprinus lipophilus* (*comb. ined.*) in a section on burnt ground taxa. No British records have been traced.

patulum (Fr.) Quél., Tricholoma, *Mém. Soc. Émul. Montbéliard, Sér. 2*, 5: 338 (1873)
Agaricus patulus Fr., *Epicr. syst. mycol.*: 47 (1838)
A *nomen dubium*. Riva (FungEur3) suggests this as a synonym of *Lyophyllum decastes*. Cooke 108 (279) Vol. 1 (1883) is not helpful. Sensu Rea (1922) is a species of *Melanoleuca*.

pauletii (Fr.) Gillet, Lepiota, *Hyménomycètes*: 71 (1874)
Agaricus pauletii Fr., *Epicr. syst. mycol.*: 17 (1838)
A *nomen dubium*. Sensu Rea [TBMS 12 (2-3): 208 (1927)] is doubtfully distinct from *Lepiota aspera*. Sensu Fries (1838) appears to be *Amanita echinocephala*.

paupercula (Berk.) Sacc., Mycena, *Syll. fung.* 5: 277 (1887)
Agaricus pauperculus Berk., *Engl. fl.* 5(2): 57 (1836)
Described from Britain (near Bristol) but this is a doubtful species and the type is in poor condition. Probably neither a *Mycena* nor a *Hemimycena fide* Maas Geesteranus (1992). Cooke 231 (236) Vol. 2 (1882) is unidentifiable.

paupertina A. Pearson, Collybia, *Trans. Brit. Mycol. Soc.* 35(2): 103 (1952)
A *nomen dubium*. The type (in herb. K) is in poor condition. Described in TBMS 35 (2): 103 (1952) 'on soil with *Urtica dioica* and *Mercurialis perennis* in woodland' in England (North Devon: Braunton) this is possibly a species of *Tephrocybe*.

pausiaca (Fr.) Gillet, Clitocybe, *Hyménomycètes*: 165 (1874)
Agaricus pausiacus Fr., *Epicr. syst. mycol.*: 77 (1838)
A *nomen dubium*. Sensu NCL is possibly *Clitocybe foetens*.

paxillus (Fr.) Gillet, Pholiota, *Hyménomycètes*: 437 (1876)
Agaricus paxillus Fr., *Epicr. syst. mycol.*: 168 (1838)

A *nomen dubium*. Listed in Rea (1922) but no material has been traced. This is probably not a species of *Pholiota* s. str.

pearsonii (Pilát) Komarova, Oxyporus, *Opredelitel' Gribov.* 179 (1964)
Poria pearsonii Pilát, *Trans. Brit. Mycol. Soc.* 19(3): 195 (1935)
A doubtful species, usually included in *Oxyporus obducens*. There is a single collection named thus in herb. K from West Sussex (Barnham) in 1968.

pectinata (Huds.) With., Fistulina, *Bot. Arr. Brit. pl. ed. 2*, 3 406 (1792)
Agaricus pectinatus Huds. [non *A. pectinatus* Bull. (1791)], *Fl. angl.*: 495 (1762)
Described from Britain but a *nomen dubium*. From the description and habitat (in cellars), perhaps a species of *Serpula*, most likely *S. lacrymans*. Sensu Withering, who states that it also grows 'in woods', possibly *S. himantioides*.

pectinata (Bull.) Fr., Russula, *Epicr. syst. mycol.*: 358 (1838)
Agaricus pectinatus Bull. [non *A. pectinatus* Huds. (1762)], *Herb. France*: pl. 509 (1791)
Not authentically British. Sensu NCL and sensu Rayner (1985) is *R. pectinatoides*.

pectinatus Klotzsch, Polyporus, *Linnaea* 8: 485 (1833)
A *nomen dubium* but possibly *Phylloporia ribis*. Reported from West Cornwall [*Grevillea* 11: 13 (1882)] but no material survives.

pellitus *var.* **punctillifer** Quél., Pluteus, *Hedwigia* 24: 133 (1885)
A *nomen dubium* lacking type material. Listed by Rea (1922) but no records have been traced.

pelloporus Bull., Boletus, *Herb. France*: pl. 501 (1791)
A *nomen dubium* lacking type material. Sowerby's illustration and description may represent *Bjerkandera fumosa* but the name is included in the synonymy of *B. adusta* by Pilát (1936).

pellosperma (Bull.) Berk. & Broome, Psathyra, *Ann. Mag. Nat. Hist., Ser. 5* 7: 373 (1883)
Agaricus pellospermus Bull., *Herb. France*: pl. 561 (1792)
A *nomen dubium* lacking type material. Cooke 613 (577) Vol. 4 (1886) suggests a species of *Psathyrella* in the *corrugis* group, as does the Bulliard plate.

pellucida (Bull.) Gillet, Tubaria, *Hyménomycètes*: 539 (1876)
Agaricus pellucidus Bull., *Herb. France*: pl. 550 (1792)
A *nomen dubium* lacking type material and interpreted by Fries (1821) as white-spored. Sensu NCL is a synonym of *Tubaria furfuracea*. Some British records are sensu Romagnesi [*Rev. Mycol. (Paris)* 5: 41 (1940)] and are thus *Tubaria romagnesiana*.

pellucidus With., Boletus, *Bot. arr. Brit. pl. ed. 2*, 3: 406 (1792)
Described from Britain but a *nomen dubium*.

peltata (Fr.) Gillet, Mycena, *Hyménomycètes*: 270 (1876)
Agaricus peltatus Fr., *Epicr. syst. mycol.*: 110 (1838)
A *nomen dubium*. *Fide* Maas Geesteranus (1992) this may represent an immature species of *Entoloma*. Sensu Smith (1947) equates to *Mycena latifolia*. There is an English collection of *Agaricus peltatus* from Broome in 1886, in herb. K, which is also *M. latifolia*.

pengellei (Berk. & Broome) Sacc., Clitocybe, *Syll. fung.* 5: 147 (1887)
Agaricus cyanophaeus var. *pengellei* Berk. & Broome, *Ann. Mag. Nat. Hist., Ser. 5 [Notices of British Fungi no.1993]* 12: 370 (1883)
Described from Britain but a *nomen dubium*. Cooke 131 (264) Vol. 1 (1883) appears to be *Calocybe obscurissima*.

penicillatus Fr., Cortinarius, *Epicr. syst. mycol.*: 283 (1838)
A *nomen dubium*. Cooke 763 (764) Vol. 6 (1886) is doubtful. Sensu Rea (1922) and NCL this may be *C. psammocephalus* [see Orton (1958: 126)].

pequinii (Boud.) Singer, Agaricus, *Bot. Mater. Otd. Sporov. Rast. Bot. Inst. Akad. Nauk S.S.S.R.* 4(10-12): 14 (1938)
> *Chitonia pequinii* Boud., *Bull. Soc. Mycol. France* 17: 26 (1901)
> *Clarkeinda pequinii* (Boud.) Sacc. & Syd., *Syll. fung.* 16: 112 (1902)
> *Agaricus gennadii* ssp. *microsporus* Bohus, *Ann. Hist.-Nat. Mus. Natl. Hung.* 67: 38 (1975)

A southern European species, not authentically British. Most material so named is likely to be *A. gennadii*, fitting the concept of Nauta in FAN5, with spores too large for those of *A. pequinii*.

perbrevis (Weinm.) Gillet, Inocybe, *Hyménomycètes*: 518 (1876)
> *Agaricus perbrevis* Weinm., *Hymen. Gasteromyc.*: 185 (1836)

Not authentically British and excluded pending clearer definition. Poorly understood and part of the unresolved *I. dulcamara* complex.

percandidum (Vassilkov) Watling, Leccinum, *Trans. Brit. Mycol. Soc.* 43: 691 (1960)
> *Boletus percandidus* Vassilkov, *Sovietsk. Bot.* 2: 27 (1944)
> *Boletus versipellis* var. *percandidus* (Vassilkov) Vassilkov, *Edible and poisonous fungi cent.Eur. dist. U.S.S.R.*: 37 (1948)
> *Krombholzia aurantiaca* f. *percandida* (Vassilkov) Vassilkov, *Notulae Syst. Sect. Crypt. Inst. Bot. Acad. Sci. U.S.S.R.* 11: 139 (1956)

Not authentically British. Reported from Scotland and England, but the collections are misidentified *L. roseotinctum* fide A.E. Hills (pers. comm.).

percomis Fr., Cortinarius, *Epicr. syst. mycol.*: 260 (1838)
Awaiting rediscovery in Britain. Reported by Orton (1955) as 'uncommon, recently found under beech and yew' but no voucher material has been traced. Apart from an anonymous report from 'South West England' in 1970 it has not been recorded since.

pergamena (Cooke) Sacc., Clitocybe, *Syll. fung.* 14: 77 (1899)
> *Agaricus pergamenus* Cooke, *Handb. Brit. fung.*, Edn 2: 52 (1884)
> *Tricholoma pergamenum* (Cooke) A. Pearson & Dennis, *Trans. Brit. Mycol. Soc.* 31: 151 (1948)

Described from Britain but a *nomen dubium*. Cooke 1132 (643) Vol. 8 (1889) appears to represent a species of *Lyophyllum*.

pergamenus (Sw.) Fr. [as *L. pargamenus*], Lactarius, *Epicr. syst. mycol.*: 340 (1838)
> *Agaricus pergamenus* Sw., *Svampe Fl. Suec.*: 90 (1809)
> *Lactarius piperatus* f. *pergamenus* (Sw.) S. Imai, *J. Fac. Agric. Hokkaido Imp. Univ.* 43: 316 (1938)
> *Lactarius piperatus* var. *pergamenus* (Sw.) Rea, *Brit. basidiomyc.*: 486 (1922)

A *nomen dubium* lacking type material and variously interpreted as a synonym of *Lactarius piperatus* or *L. glaucescens*, or as a species distinct from both. Sensu Cooke 943 (978) Vol. 7 (1888) and sensu Rea (1922) is doubtful.

periscelis Fr., Cortinarius, *Epicr. syst. mycol.*: 300 (1838)
A *nomen dubium*. Cooke 816 (838) Vol. 6 (1887) is doubtful, but probably a species near to *C. flexipes*.

permixta (Barla) Pacioni, Macrolepiota, *Micologia Italiana* 8(3): 13 (1979)
> *Lepiota permixta* Barla, *Bull. Soc. Mycol. France* 2: 114 (1886)
> *Macrolepiota procera* var. *permixta* (Barla) Candusso, *Fungi Europaei* 4: *Lepiota s.l.*: 518 (1990)

Not authentically British. Following Vellinga in FAN5 this is a largely southern European species. Collections named thus need re-examination but are probably *M. fuliginosa*.

persicinum (Fr.) Gillet, Tricholoma, *Hyménomycètes*: 126 (1874)
> *Agaricus persicinus* Fr., *Hymenomyc. eur.*: 76 (1874)

A *nomen dubium*. Possibly a synonym of *Calocybe ionides* but British material named by Pearson is not a *Calocybe* fide Hora (NCL: 206).

persicinus Beck, Hygrophorus, *Verh. K. K. Zool.-Bot. Ges. Wien* 36: 470 (1886)
A *nomen dubium*. See TBMS 4(2): 310 (1913) where recorded as 'in grass in woodland' with no other details. See also Naturalist (Hull) 1913: 24 (1913), which suggests a form of *Cantharellus cibarius* fide Pearson & Dennis (1948).

persoonianus W. Phillips, Agaricus, *Gard. Chron.* 16: 784 (1881)
Described from Britain but a *nomen dubium*. An *Entoloma*, introduced by Phillips for his concept of *Agaricus sericeus* sensu Persoon (1798, tab. 6 f. 2).

persoonii (DuPort) Sacc., Entoloma, *Syll. fung.* 5: 697 (1887)
> *Agaricus persoonii* Du Port, *Grevillea* 10: 42 (1881)

A *nomen dubium* inadequately described and said to have white spores. The type in herb. K is in poor condition. See notes in TBMS 31: 205 (1948) and Noordeloos (1987: 383). Incorrectly reported as *Entoloma persoonianum* by some authorities.

pervelata Kits van Wav., Psathyrella, *Persoonia* 6: 309 (1971)
Not authentically British. Reported from England but both collections have been redetermined as *Psathyrella sacchariolens*. The illustration (but not the description) in Mycologist 15 (4): 157 (2001) also depicts *P. sacchariolens*.

pescaprae (Fr.) Singer, Porpoloma, *Sydowia* 6: 198 (1952)
> *Agaricus pescaprae* Fr., *Epicr. syst. mycol.*: 45 (1838)
> *Tricholoma pescaprae* (Fr.) Quél., *Mém. Soc. Émul. Montbéliard, Sér. 2*, 5: 328 (1873)

Not authentically British *fide* NCL. Sensu Cooke (1871) and sensu Rea (1922) is *Lyophyllum fumosum*.

pescaprae var. multiforme (Schaeff.) Massee, Tricholoma, *Brit. fung.-fl.* 3: 211 (1893)
> *Agaricus multiformis* Schaeff., *Fung. Bavar. Palat. nasc.* 1: 14 (1762)
> *Agaricus pescaprae* var. *multiforme* (Schaeff.) Cooke, *Handb. Brit. fung.*, Edn 2: 365 (1890)

A *nomen dubium* lacking type material. Sensu Rea (1922) is *Lyophyllum fumosum*.

petalodes Fr., Polyporus, *Hymenomyc. eur.*: 536 (1874)
A *nomen dubium*. Probably equates to the *nummularius* form of *Polyporus varius* [= *P. leptocephalus*] fide Donk (1974). Reported 'on a stump at Sibbertoft' (Northamptonshire) [see Grevillea 12: 68 (1884)] but no material survives.

petaloides f. carbonarius Pilát, Pleurotus, *Atlas champ. Eur.* 2: 176 (1935)
A *nomen dubium*. Sensu auct. Brit. is *Hohenbuehelia culmicola*.

pezizoides (Nees & T. Nees) P. Kumm., Crepidotus, *Führ. Pilzk.*: 74 (1871)
> *Agaricus pezizoides* Nees & T. Nees, *Nov. Actorum Acad. Caes. Leop.-Carol. Nat. Cur.* 9: 249 (1818)

A *nomen dubium* variously interpreted. From the original illustration this is probably synonymous with *Resupinatus applicatus*, but Cooke 537 (516) Vol. 4 (1885) appears to be *Crepidotus epibryus* and material named thus in herb. K is that species.

phaeocephala (Pers.) Sacc., Inocybe, *Syll. fung.* 5: 774 (1887)
> *Agaricus phaeocephalus* Pers., *Syn. meth. fung.*: 302 (1801)

A *nomen dubium* lacking type material and possibly not even a species of *Inocybe* fide Kuyper (1986). Cooke 425 (396) Vol. 3 (1886) is not helpful, but does indeed look like an *Inocybe*.

phaeodisca var. geophylloides Kühner, Inocybe, *Bull. Soc. Naturalistes Oyonnax* 9 (Suppl.): 5 (1955)
Not authentically British. Reported from South Devon but unsubstantiated with voucher material.

phaeopodia (Bull.) Murrill, Melanoleuca, *North American Flora* 10 (1): 20 (1914)

Agaricus phaeopodius Bull., *Herb. France*: pl. 532 (1792)
   Collybia phaeopodia (Bull.) Sacc., *Syll. fung.* 5: 209 (1887)
   Melanoleuca melaleuca var. *phaeopodia* (Bull.) Maire, *Treb. Mus. Ci. Nat. Barcelona* 15(2): 76 (1933)
A *nomen dubium* lacking type material. It is not clear to which species Bulliard's plate refers. *Fide* Wakefield & Dennis [Common British Fungi (Edn 2): 129 (1981)] is 'probably only a colour form of *Melanoleuca melaleuca*' (i.e. *M. polioleuca* of this checklist).

phalerata (Fr.) Quél., Pholiota, *Mém. Soc. Émul. Montbéliard, Sér. 2*, 5: 126 (1872)
   Agaricus phaleratus Fr., *Epicr. syst. mycol.*: 169 (1838)
A *nomen dubium*. Probably not a species of *Pholiota fide* Holec (2001). Listed in Rea (1922) on the basis of a single misdetermined record. Pearson & Dennis (1948) suggested that this may be *Rozites caperatus* but that seems unlikely.

phlebophorus var. albofarinosus Rea, Pluteus, *Trans. Brit. Mycol. Soc.* 6(4): 324 (1920)
Described from Britain but a *nomen dubium*.

phosphorea Sowerby, Clavaria, *Col. fig. Engl. fung.* 1: pl. 100 (1797)
A *nomen dubium* lacking type material. Described from the cellar of a house in London and noted to be luminescent. The illustration by Sowerby shows it to be the rhizomorphs of a species of *Armillaria*.

phrygianus Fr., Cortinarius, *Epicr. syst. mycol.*: 283 (1838)
Not authentically British. A montane conifer associate in continental Europe. Sensu Rea (1922), associated with *Fagus sylvatica* in lowland habitat, is doubtful and not reported in recent years.

phyllophila (Massee) D.P. Rogers & H.S. Jacks., Vararia, *Farlowia* 1(2): 323 (1943)
   Peniophora phyllophila Massee, *J. Linn. Soc., Bot.* 25: 150 (1889)
Not authentically British. A North American species. Reported from Yorkshire (as *Peniophora phyllophila*) but unsubstantiated with voucher material.

piceae Stangl & Schwöbel, Inocybe, *Int. J. Mycol. Lichenol.* 2(1): 59 (1985)
Not authentically British. A collection in poor condition from Herefordshire (Dinedor) in 1881 in herb. K is annotated by Kuyper as 'probably *Inocybe oblectabilis* var. *macrospora*'.

picinus Fr., Lactarius, *Epicr. syst. mycol.*: 348 (1838)
   Lactarius fuliginosus ssp. *picinus* (Fr.) Konrad & Maubl., *Icon. select. fung.* 4: pl. 325 (1927)
Not authentically British. A montane species in continental Europe, associated with *Picea* spp in native woodland. Cooke 960 (997) Vol. 7(1888) shows material collected in Herefordshire (Foxley) in 1879 which may be correct but no material survives. A recent collection, said to be this species, from Scotland, is immature and impossible to name with certainty

pilatianus (Bohus) Bohus, Agaricus, *Ann. Hist.-Nat. Mus. Natl. Hung.* 66: 78 (1974)
   Agaricus xanthodermus var. *pilatianus* Bohus, *Ann. Hist.-Nat. Mus. Natl. Hung.* 63: 80 (1971)
Not authentically British. Reported from Isle of Wight (Bembridge) in 1998 but from an anonymous source and unsubstantiated with voucher material.

pileolarius Bull., Agaricus, *Herb. France*: pl. 400 (1789)
A *nomen dubium* lacking type material. Bulliard illustrates what appears to be *Clitocybe nebularis* but sensu Sowerby (Sow1: 61) is clearly *Clitocybe geotropa*.

pilipes (Sowerby) Gray, Gymnopus, *Nat. arr. Brit. pl.* 1: 611 (1821)
   Agaricus pilipes Sowerby, *Col. fig. Engl. fung.* 3: pl. 249 (1800)
Described from Britain but a *nomen dubium* lacking type material and never recollected. Sowerby illustrates a rather distinctive agaric with a pilose stipe, caespitose on decayed or moribund *Lactarii*. It does not appear to be a species of *Asterophora*.

pillodii (Quél.) Wasser, Pseudobaeospora, *Flora Gribov Ukrainy*: 220 (1980)
   Collybia pillodii Quél., *Compt. Rend. Assoc. Franç. Avancem. Sci.* 18 (2): 509 (1890)
Not authentically British. Reported on several occasions from woodland on calcareous or limestone soils in England but all voucher material has been redetermined as other species of *Pseudobaeospora* by Bas [see Persoonia 18 (1): 115-122 (2002)].

pinetorum Velen., Clitocybe, *České Houby* II: 276 (1920)
A *nomen dubium*. Sensu auct. Brit. is *C. diatreta*.

pistillaris Batsch, Agaricus, *Elench. fung.*: 55 (1783)
A *nomen dubium*. Reported by Withering (1792).

pithya (Fr.) Dörfelt, Hemimycena, *Boletus* 2: 61 (1984)
   Agaricus lacteus var. *pithyus* Fr., *Hymenomyc. eur.* (1874)
   Mycena pithya (Fr.) Sacc., *Syll. fung.* 5: 260 (1887)
   Mycena lactea var. *pithya* (Fr.) J.E. Lange, *Dansk. Bot. Ark.* 1(5): 25 (1914)
A *nomen dubium*. Sensu Rea (1922) and B&K3 (1991) is *H. gracilis*.

pithya (Pers.) J. Erikss., Peniophora, *Symb. Bot. Upsal.* 10(5): 45 (1950)
   Thelephora pithya Pers., *Mycol. eur.* 1: 146 (1822)
   Corticium plumbeum Fr., *Hymenomyc. eur.*: 653 (1874)
Not authentically British. Rarely reported from southern England but unsubstantiated with voucher material and dubiously identified. The single collection in herb. K, from South Devon, is determined as 'compare with' *P. pithya*.

placenta (Batsch) Sacc., Entoloma, *Syll. fung.* 5: 682 (1887)
   Agaricus placenta Batsch, *Elench. fung.*: 79 t. 5 f.18 (1783)
A *nomen dubium*. Sensu Cooke (HBF2) and Rea (1922) is unknown.

placidum (Fr.) Noordel., Entoloma, *Persoonia* 1: 150 (1981)
   Agaricus placidus Fr., *Observ. mycol.* 2: 94 (1818)
   Leptonia placida (Fr.) P. Kumm., *Führ. Pilzk.*: 96 (1871)
   Rhodophyllus placidus var. *gracilis* J.E. Lange, *Fl. agaric. danic.* 5, Taxonomic Conspectus: VIII (1940)
Listed in NCL and by Orton (1991) but not authentically British. Two collections named thus, in herb. K, from Scotland (Killiecrankie) and from Northern Ireland (Antrim: Glen Ariff) are both annotated as dubious. See Cooke 352 (330) Vol. 3 (1885) for an illustration of material collected in 1880 from Herefordshire.

placomyces Peck, Agaricus, *Rep. (Annual) New York State Mus. Nat. Hist.* 29: 40 (1878) [1875]
A North American species, not authentically British (nor European). Frequently used in British literature, misapplied to *Agaricus moelleri*.

plancus (Fr.) Fr., Marasmius, *Epicr. syst. mycol.*: 375 (1838)
   Agaricus plancus Fr., *Syst. mycol.* 1: 127 (1821)
A *nomen dubium*. Listed by Rea (1922). Regarded by Antonín & Noordeloos (A&N1) as probably a species of *Gymnopus* (= *Collybia*). Cooke 1073 (1119) Vol. 7 (1890) would seem to support that view.

planipes (V. Brig.) Sacc., Collybia, *Syll. fung.* 5: 228 (1887)
   Agaricus planipes V. Brig., *Hist. Fung. Reg. Neapol.* t. 16 (1848)
A *nomen dubium* lacking type material *fide* Antonín & Noordeloos (A&N2). Listed by Rea (1922).

planus Sowerby, Agaricus, *Col. fig. Engl. fung.* 3: pl. 362 (1802)
Described from Britain but a *nomen dubium* lacking type material. Sowerby's illustration and description are unhelpful.

platyphylloides Romagn., Rhodophyllus, *Rev. Mycol. (Paris)* 19(1): 8 (1954)

A *nomen dubium*. Possibly aberrant *Entoloma myrmecophyllum* fide Noordeloos (1987). Cooke [338 (342) as *E. rhodopolium*] was considered to represent *R. platyphylloides* by Romagnesi (*Rev. Mycol. (Paris)* 20: 210, 1955).

platypus Berk., Coprinus, in Cooke, *Ill. Brit. fung.* 675 (687b) Vol. 5 (1886)

Described from decayed palm fronds in a greenhouse, this is possibly an alien. Also reported on decayed leaves of *Phalaris arundinacea* by Rea (1922) but lacking voucher material.

plautus var. terrestris Bres., Pluteus, *Fungi trident.* 1: 18 (1881)

A *nomen dubium fide* NCL, possibly *P. pearsonii* which is treated here as a synonym of *P. ephebeus.*

plicosa (Fr.) P. Kumm., Mycena, *Führ. Pilzk.*: 109 (1871)
  Agaricus metatus var. *plicosus* Fr., *Syst. mycol.* 1: 145 (1821)
  Agaricus plicosus (Fr.) Fr., *Epicr. syst. mycol.*: 110 (1838)
A *nomen dubium*. Listed by Rea (1922).

plumbea (Fr.) P. Karst., Mycena, *Bidrag Kännedom Finlands Natur Folk* 32: 114 (1879)
  Agaricus plumbeus Fr., *Hymenomyc. eur.*: 144 (1874)
A *nomen dubium*. Listed by Rea (1922). Probably not a *Mycena*, but possibly a diminutive species in *Entoloma fide* Maas Geesteranus (1992).

plumbeitincta (G.F. Atk.) Singer, Conocybe, *Sydowia* 4: 137 (1950)
  Galerula plumbeitincta G.F. Atk., *Proc. Amer. Philos. Soc.* 57: 372 (1918)
A North American species unknown in Europe. Sensu Phillips (1981) is possibly *Conocybe siennophylla*. Sensu NCL is *Conocybe moseri*.

plumbeus (Bull.) Gray, Lactarius, *Nat. arr. Brit. pl.* 1: 625 (1821)
  Agaricus plumbeus Bull., *Herb. France*: pl. 282 (1786)
A *nomen dubium* lacking type material. Widely used for *Lactarius turpis*.

plumiger Fr., Cortinarius, *Epicr. syst. mycol.*: 294 (1838)
A *nomen dubium*. Listed by Rea (1922), but unsubstantiated with voucher material.

plumosa (Bolton) Quél., Inocybe, *Mém. Soc. Émul. Montbéliard, Sér. 2,* 5: 152 (1872)
  Agaricus plumosus Bolton, *Hist. fung. Halifax* 1: 33 (1788)
  Gymnopus plumosus (Bolton) Gray, *Nat. arr. Brit. pl.* 1: 609 (1821)
  Astrosporina plumosa (Bolton) Rea, *Brit. basidiomyc.*: 214 (1922)
Described from Britain, but a *nomen dubium* lacking type material. Probably not an *Inocybe fide* Pearson & Dennis (1948). Sensu Rea (1922) is doubtful.

plumosa (Duby) Rea, Odontia, *Brit. basidiomyc.*: 647 (1922)
  Hydnum plumosum Duby, *Bot. gall.*: 778 (1830)
A *nomen dubium*. Listed by Rea (1922). Some authorities place this in the synonymy of *Phlebiella sulphurea* but Rea's description does not fit that species.

pluteoides (Fr.) P. Karst., Entoloma, *Bidrag Kännedom Finlands Natur Folk* 32: 265 (1879)
  Agaricus pluteoides Fr., *Monogr. hymenomyc. Suec.* 2: 345 (1863)
A *nomen dubium*. Listed by Rea (1922).

poetarum R. Heim, Hygrophorus, *Bull. Soc. Mycol. France* 63: 127 (1947)
  Mis.: *Limacium pudorinum* sensu J.E. Lange [Fl. agaric. danic. 5: 13 & pl. 163D (1940)]
Not authentically British. Listed in NCL and reported from Scotland, but unsubstantiated with voucher material.

polaris Høil., Cortinarius, *Opera Bot.* 71: 94 (1984) [1983]

Not authentically British. Reported from Scotland (The Cairnwell) in 1984 but material in herb. E is labelled as *Cortinarius* 'aff.' *polaris*.

polia (Fr.) Gillet, Clitocybe, *Hyménomycètes*: 156 (1874)
  Agaricus polius Fr., *Epicr. syst. mycol.*: 57 (1838)
A *nomen dubium*. Listed by Rea (1922).

polychroma Hora, Russula, *Trans. Brit. Mycol. Soc.* 43(2): 457 (1960)

Described from Britain but a *nomen dubium*. Sensu NCL and sensu Rayner (1985) is *Russula integra*.

polymorphus Rob. Henry, Cortinarius, *Doc. mycol.* 16(61): 23 (1985)
  Mis.: *Cortinarius multiformis* sensu NCL and sensu auct. mult.
Not authentically British. Reported (as *C. multiformis*) but no collections have been traced. The true *C. multiformis* is associated with conifers.

polymyces (Pers.) Singer & Clémençon, Armillariella, *Nova Hedwigia* 23: 311 (1972)
  Agaricus polymyces Pers., *Tent. disp. meth. fung.*: 19 (1797)
  Lepiota polymyces (Pers.) Gray, *Nat. arr. Brit. pl.* 1: 603 (1821)
A *nomen dubium* lacking type material and variously interpreted. Sensu auct. Brit. is *Armillaria ostoyae*.

polyporus Bull., Boletus, *Herb. France*: pl. 469 (1790)
A *nomen dubium* lacking type material. Bulliard illustrates a stipitate, poroid basidiome with a dull reddish-brown pileus and grey pore surface. Synonymised by Gray (1821) with *Albatrellus fuligineus* but that is also a *nomen dubium*.

polyrhizum (J.F. Gmel.) Pers., Scleroderma, *Syn. meth. fung.*: 156 (1801)
  Lycoperdon polyrhizum J.F. Gmel., *Syst. nat.*: 1464 (1792)
  Scleroderma geaster Fr., *Syst. mycol.* 3: 46 (1829)
Not authentically British. Collections named thus in herb. K have been re-examined and found to be misidentified *S. citrinum* or *S. verrucosum*.

polysticta (Berk.) Quél., Lepiota, *Mém. Soc. Émul. Montbéliard, Sér. 2,* 5: 541 (1875)
  Agaricus polystictus Berk., *Engl. fl.* 5: 9 (1836)
Described from Britain but a *nomen dubium*. Based on a re-examination of authentic material, Dennis (1948) suggests it may be *L. clypeolarioides*, itself a *nomen dubium*. Cooke 41 (30) Vol. 1 (1881) is evidently another species, possibly *Cystoderma granulosum*.

polyzona Pers., Daedalea, *Mycol. eur.* 3: 8 (1828)
Synonymous with *Coriolopsis polyzona* which has never been reported from Europe, but the name has been used sensu auct. for both *Trametes gibbosa* and *T. hirsuta*.

populicola Pat., Solenia, *Tab. anal. fung.* 4: 201 (1886)
A *nomen dubium*. British collections are *Merismodes anomala*.

porphyrogriseum Noordel., Entoloma, *Nova Hedwigia Beih.* 91: 284 (1987)
Not authentically British. Reported from Scotland but no material traced in herb. E. Known to Noordeloos (FungEur5: 544) only from a few localities in Denmark.

porphyroleucum (Bull.) Sacc., Tricholoma, *Syll. fung.* 5: 134 (1887)
  Agaricus porphyroleucus Bull., *Herb. France*: pl. 443 (1790)
  Agaricus melaleucus β *porphyroleucus* (Bull.) Fr., *Syst. mycol.* 1: 115 (1821)
  Tricholoma melaleucum var. *porphyroleucum* (Bull.) Gillet, *Hyménomycètes*: 128 (1874)
A *nomen dubium* lacking type material. *Fide* NCL this is 'hardly distinct from *Melanoleuca melaleuca*'.

porreus (Pers.) Fr., Marasmius, *Epicr. syst. mycol.*: 466 (1838)
  Agaricus porreus Pers., *Syn. meth. fung.*: 376 (1801)

A *nomen dubium* lacking type material. Sensu Cooke 1071 (1133) Vol. 7 (1890) and sensu NCL is possibly *Marasmius alliaceus*. Sensu Rea (1922) is unknown.

porriginosa (Fr.) P. Karst., Naucoria, *Bidrag Kännedom Finlands Natur Folk*: 431 (1879)
  Agaricus porriginosus Fr., *Epicr. syst. mycol.*: 200 (1838)
A *nomen dubium*. Collections named thus in herb. K are either *Agrocybe praecox* or *A. dura fide* Reid [TBMS 82 (2): 212 (1984)]. Cooke 511 (510) Vol. 4 (1885) shows a distinctive, brightly coloured agaric, but neither a *Naucoria* nor an *Agrocybe*.

postii *var.* **aurea** Massee, Omphalia, *Brit. fung.-fl.* 2: 387 (1893)
Described from Britain but a *nomen dubium* probably not distinct from the type variety (= *Loreleia postii*).

praestigiosus (Fr.) M.M. Moser, Cortinarius, *Schweiz. Z. Pilzk.* 43(8): 131 (1965)
  Cortinarius paragaudis var. *praestigiosus* Fr., *Hymenomyc. eur.*: 379 (1874)
Not authentically British. Sensu Rea (1922) as *C. paragaudis* var. *praestigiosus* is clearly different, but doubtful.

praetextus With., Agaricus, *Syst. arr. Brit. pl. ed. 4*, 4: 192 (1801)
Described from Britain but a *nomen dubium*.

prasinus (Schaeff.) Fr., Cortinarius, *Epicr. syst. mycol.*: 268 (1838)
  Agaricus prasinus Schaeff., *Fung. Bavar. Palat. nasc.* 1: 51 (1762)
Not authentically British. All records are old and are unsubstantiated with voucher material. The only material named thus (in herb. K, as *C. turbinatus*) was collected by Broome in 1842, but is in poor condition and unlikely to be correctly identified.

prasiosmus (Fr.) Fr., Marasmius, *Epicr. syst. mycol.*: 376 (1838)
  Agaricus prasiosmus Fr., *Observ. mycol.* 2: 153 (1818)
A *nomen dubium*. The name has been used in a misapplied sense for *M. quercus* on decayed leaves of *Quercus* spp but this is not authentically British. Fries's original concept of this species differs markedly from his concept in 1838.

pratensis Schaeff., Agaricus [non *A. pratensis* Pers. (1801)], *Fung. Bavar. Palat. nasc.* 1: 96 (1762)
  Psalliota pratensis (Schaeff.) Quél., *Mém. Soc. Émul. Montbéliard, Sér. 2*, 5: 139 (1872)
A *nomen dubium* lacking type material. Listed by Rea (1922) & Cooke [543 (525), Vol. 4 (1885)].

pratensis *β* **vitulinus** Gray, Gymnopus, *Nat. arr. Brit. pl.* 1: 604 (1821)
Described from Britain but a *nomen dubium* lacking type material.

pratensis *var.* **meisneriensis** (Pers.) Fr., Hygrophorus, *Monogr. hymenomyc. Suec.* 2: 132 (1863)
  Agaricus pratensis γ *meisneriensis* Pers., *Syn. meth. fung.*: 305 (1801)
A *nomen dubium* lacking type material. Descriptions indicate that this is either *Hygrocybe lacmus* or *H. radiata* and nothing to do with *H. pratensis*.

pratensis *var.* **umbrinus** Rea, Hygrophorus, *Brit. basidiomyc.*: 299 (1922)
Described from Britain but a *nomen dubium*.

pravum (Lasch) Sacc., Tricholoma, *Syll. fung.* 5: 116 (1887)
  Agaricus pravus Lasch, *Linnaea* 4: 532 (1829)
  Agaricus ionides var. *pravus* (Lasch) Cooke, *Handb. Brit. fung.*, Edn 2: 37 (1884)
A *nomen dubium* lacking type material and possibly an alien. Collected 'from a stove' (greenhouse) at Kew Gardens in 1872 [see Grevillea 1: 72 (1872)] but never recollected. Considered by some authorities to be a variety of *Calocybe ionides*.

primula With., Agaricus, *Arr. Brit. pl. ed. 3*, 4: 239 (1796)
Described from Britain but a *nomen dubium*.

privignoides Rob. Henry, Cortinarius, *Doc. mycol.* 16(61): 24 (1985)
Not authentically British. Reported from West Sussex in 1998 but unsubstantiated with voucher material.

privignus (Fr.) Fr., Cortinarius, *Epicr. syst. mycol.*: 304 (1838)
  Agaricus malachius β *privignus* Fr., *Observ. mycol.* 2: 72 (1863)
Not authentically British. A poorly known species, listed by Rea (1922) and reported from Scotland (Perthshire: Kindrogan) in 1983, but unsubstantiated with voucher material.

proboscideus (Fr.) P. Kumm., Crepidotus, *Führ. Pilzk.*: 74 (1871)
  Agaricus proboscideus Fr., *Observ. mycol.* 2: 232 (1818)
A *nomen dubium*. Possibly *Tectella patellaris fide* Singer [*Persoonia* 2 (1): 40 (1961)] or *Tapinella panuoides fide* Pilát (Atlas 6, 1948). Listed by Rea (1922).

procerus Bolton, Boletus, *Hist. fung. Halifax* 2: 86 (1788)
Described from Britain but a *nomen dubium* lacking type material. Bolton's illustration depicts an unidentifiable species of *Leccinum*.

proletaria (Fr.) Gillet, Nolanea, *Hyménomycètes*: 418 (1876)
  Agaricus proletarius Fr., *Spicilegium Pl. neglect.*: 8 (1836)
A *nomen dubium* variously interpreted. Sensu Rea (1922) is *Entoloma conferendum* and sensu auct. is *E. undatum*.

prolifera (Sowerby) Gillet, Mycena, *Hyménomycètes*: 273 (1876)
  Agaricus proliferus Sowerby, *Col. fig. Engl. fung.* 2: pl. 169 (1798)
Described from Britain but a *nomen dubium* lacking type material. Sowerby's illustration appears to show a sterile *Psathyrella* and certainly not a *Mycena*. Pearson (1948) suggested this was possibly *Mycena inclinata* and Cooke 223 (235) Vol. 2 (1882) could also be this, but this is obviously not Sowerby's original species.

proliferus With., Boletus, *Bot. arr. veg.* 2: 768 (1776)
Described from Britain but a *nomen dubium*. Said to occur 'in mountainous pastures' and 'sometimes to spread over a plot of ground upwards of thirty feet in diameter'.

prominens (Fr.) M.M. Moser, Macrolepiota, *Kleine Kryptogamenflora*, Edn 3: 184 (1967)
  Agaricus prominens Fr., *Hymenomyc. eur.*: 30 (1874)
  Lepiota prominens (Fr.) Sacc., *Syll. fung.* 5: 30 (1887)
Doubtfully distinct from *Macrolepiota mastoidea*. Sensu auct. Brit. is mostly *M. procera*.

prona *var.* **smithii** Massee, Psathyrella, *Brit. fung.-fl.* 1: 344 (1892)
Described from Britain but a *nomen dubium*. Not discussed in Kits1.

proteus Bolton, Boletus, *Hist. fung. Halifax* Suppl.: 166 (1791)
Described from Britain but a *nomen dubium* lacking type material. Bolton's illustration shows an unidentifiable, resupinate, poroid fungus which he took to be an early stage of this 'species' but also shows what appears to be *Stereum rugosum* which he describes as an 'older stage' of *B. proteus*.

proteus Sowerby, Lycoperdon, *Col. fig. Engl. fung.* 3: pl. 332 (1801)
Described from Britain but a *nomen dubium* lacking type material. Sowerby's illustration shows at least three different species of '*Lycoperdon*' under this name.

protracta (Fr.) Gillet, Collybia, *Hyménomycètes*: 311 (1876)
  Agaricus protractus Fr., *Epicr. syst. mycol.*: 97 (1838)
Not authentically British. A species of *Tephrocybe fide* A&N2. Cooke 214 (270) Vol. 2 (1882) was illustrated from specimens collected amongst dead leaves in Surrey (Kew Gardens) but no voucher material has been traced.

pruinatus Rea, Marasmius, *Trans. Brit. Mycol. Soc.* 5(3): 435 (1916)
A *nomen dubium*. An insufficiently known species of *Marasmiellus* according to Antonín & Noordeloos (A&N). Described from material gathered 'on soil amongst moss' at Porlock (South Somerset).

psammicola (Berk. & Broome) Sacc., Mycena, *Syll. fung.* 5: 275 (1887)
    Agaricus psammicola Berk. & Broome, *Ann. Mag. Nat. Hist., Ser. 4 [Notices of British Fungi no. 1518]* 17: 130 (1876)
Doubtful species. The type material, from West Kent (Addington, near Croydon), shows a mixture of spore shapes and colours indicating some kind of contamination. Cooke 225 (186) Vol. 2 (1882) is unidentifiable.

psathyroides (Cooke) Sacc., Collybia, *Syll. fung.* 5: 229 (1887)
    Agaricus psathyroides Cooke, *Handb. Brit. fung.*, Edn 2: 68 (1884)
A *nomen dubium*. Cooke 200 (266) Vol. 2 (1882) is unidentifiable, possibly a species of *Mycena*. Dennis [TBMS 31 (3-4): 198 (1948)] suggests the name could be based on diseased material of a species of *Mycena* and that it should be discarded. Antonín & Noordeloos (A&N2) concur.

pseudoandrosacea (Bull.) M.M. Moser (*comb. inval.*), Omphalina, *Kleine Kryptogamenflora*, Edn 3: 71 (1967)
    Agaricus pseudoandrosaceus Bull., *Herb. France*: pl. 276 (1786)
    Omphalia pseudoandrosacea (Bull.) Gillet, *Hyménomycètes*: 292 (1876)
A *nomen dubium*. lacking type material. Sensu auct. is *Omphalina fulvopallens* and the name has also been used for forms of *O. ericetorum*.

pseudocamerina A.H. Sm. & Singer, Galerina, *Monograph Galerina*: 119 (1964)
    Galera josserandii Kühner [*nom. nud.*], *Bull. Soc. Naturalistes Oyonnax* 10 -11 (Suppl.): 4 (1957)
    Mis.: *Galerina camerina* sensu Kühner (1935) *non al.*
Not authentically British. Collections named thus are *Galerina ampullaceocystis*.

pseudoclypeatus Bolton, Agaricus, *Hist. fung. Halifax* Suppl.: 154 (1791)
Described from Britain but a *nomen dubium* lacking type material. *Fide* Maas Geesteranus (1992) this is synonymous with *M. inclinata* but Bolton's illustration and description cannot be definitely assigned to that species.

pseudoconglobata Rea, Clitocybe, *Trans. Brit. Mycol. Soc.* 12(2-3): 213 (1927)
Described from Britain but a *nomen dubium*. Collected from 'old apple pulp' and *fide* Rea 'near to *Clitocybe conglobata*', thus possibly *Lyophyllum fumosum* or a closely related species.

pseudodelica J.E. Lange, Russula, *Dansk. Bot. Ark.* 4(12): 27 (1926)
A *nomen dubium fide* Sarnari (Sarn). Reported from England and Wales but lacking voucher material.

pseudodirecta W.G. Sm., Omphalia, *Syn. Brit. Bas.*: 88 (1908)
Described from Britain but a *nomen dubium*, never recollected and possibly an alien. Growing on old and decayed cones of a cycad (*Encephalartos* sp.), in a nursery in London (Chelsea).

pseudoduracinus Rob. Henry, Cortinarius, *Bull. Soc. Mycol. France* 54: 41 (1938)
Listed in NCL amongst 'species not yet critically considered' as this is Henry's concept of *C. duracinus* sensu Cooke (HBF2: 275). Not well known and possibly a synonym of *C. microspermus*. Cooke's desciption gives the correct spore measurements for *C. microspermus* but in other respects fits *C. duracinus* as generally understood, i.e. *C. rigens* of this list.

pseudofelina J.E. Lange, Lepiota, *Fl. agaric. danic.* 5, *Taxonomic Conspectus*: V (1940)
A *nomen dubium* possibly not distinct from *L. grangei*. Sensu NCL and sensu auct. mult is *Lepiota obscura*

pseudogracilis (Kuhner) Singer, Hemimycena, *Ann. Mycol.* 41: 121 (1943)
    Mycena pseudogracilis Kühner, *Encycl. Mycol.* 10, *Le Genre Mycena*: 648 (1938)
Not authentically British. A single collection named thus, from Scotland (North Ebudes: Isle of Skye) in herb. K, but in poor condition and requiring confirmation.

pseudogranulosa (Berk. & Broome) Pegler, Cystolepiota, *Kew Bull.* 12: 283 (1986)
    Agaricus pseudogranulosus Berk. & Broome, *J. Linn. Soc., Bot.* 11: 501 (1871)
    Lepiota pseudogranulosa (Berk. & Broome) Sacc., *Syll. fung.* 5: 53 (1887)
Not authentically British. Reported by Reid [TBMS 38: 389 (1955)] from soil under shrubs in a greenhouse (Kew, Royal Botanic Gardens, Aroid House) but subsequently redetermined as *Cystolepiota pulverulenta*.

pseudolagopus R.F.O. Kemp, Coprinus, *Trans. Brit. Mycol. Soc.* 65(3): 380 (1975)
A *nom. prov.* for a member of section *Lanatuli* studied in culture, and mentioned in literature, but never validly published.

pseudoorcella (Fr.) Sacc., Clitopilus, *Syll. fung.* 5: 700 (1887)
    Agaricus pseudoorcella Fr., *Öfvers. Kongl. Vetensk.-Akad. Förh.* 18: 33 (1862)
A *nomen dubium*. From the description this is possibly a species of *Entoloma*.

pseudopicta (J.E. Lange) Kühner, Mycena, *Encycl. Mycol.* 10. *Le Genre Mycena*: 363 (1938)
    Omphalia pseudopicta J.E. Lange, *Dansk. Bot. Ark.* 5(5): 15 (1930)
Not authentically British. The single collection named thus, from West Sussex, is possibly *Mycena cinerella*.

pseudoprivignus Rob. Henry, Cortinarius, *Doc. mycol.* 16(61): 25 (1985)
Not authentically British. A poorly understood species not clearly distinct from from *C. privignus* and *C. privignoides*. The collection illustrated in Phillips (1981) under this name is probably *C. hinnuleus*.

pseudoscolecina D.A. Reid, Naucoria, *Trans. Brit. Mycol. Soc.* 82(2): 202 (1984)
Described from Britain but a dubious species described from a single fruitbody and lacking habitat details. Near to *Naucoria scolecina*.

pseudosubtilis R.H. Petersen, Ramariopsis, *Mycologia* 61: 552 (1969)
Not authentically British (or European). Two collections, in herb. K, are both annotated as 'compare with' *R. pseudosubtilis*.

pseudouvidus Kühner, Lactarius, *Bull. Soc. Mycol. France* 1: 53 (1985)
Not authentically British. Reported from Scotland (Orkney) but material was later redetermined by Watling as *Lactarius luridus*.

pudicus (Bull.) Bon, Leucoagaricus, *Doc. mycol.* 11(43): 64 (1981)
    Agaricus pudicus Bull., *Herb. France*: pl. 597 (1793)
    Pholiota pudica (Bull.) Gillet, *Hyménomycètes*: 439 (1876)
    Lepiota pudica (Bull.) Quél., *Enchir. fung.*: 7 (1886)
A *nomen dubium* lacking type material. Bulliard's description and plate show a large, annulate agaric resembling a species of *Leucoagaricus* but it is unclear which one. British collections named thus are mostly *L. leucothites*.

puellaris *var.* intensior Cooke, Russula, *Handb. Brit. fung.*, Edn 2: 337 (1889)
Described from Britain but a *nomen dubium*. Possibly *R. versicolor*.

pulcherrimum Berk. & M.A. Curtis, Hydnum, *Hooker's J. Bot. Kew Gard. Misc.* 1: 235 (1849)

Not authentically British. A North American species now known as *Climacodon pulcherrimum*. An old collection from Scotland (in herb. K) is *Phaeolus schweinitzii* and a more recent collection named thus from England (South Essex) is a parasitised and unidentifiable species of *Postia*.

pulchralis Britzelm., Russula, *Ber. Naturhist. Vereins Augsburg* 28: 140 (1885)
>*Russula nitida* var. *pulchralis* (Britzelm.) Cooke, *Grevillea* 17: 41 (1888)
>*Russula nauseosa* var. *pulchralis* (Britzelm.) Cooke, *Handb. Brit. fung.*, Edn 2: 336 (1889)

A *nomen dubium* lacking type material. Last reported, from North Somerset (Quantock Hills, Triscombe Stone) in 1911, but unsubstantiated with voucher material.

pulchripes J. Favre, Cortinarius, *Beitr. Kryptogamenfl. Schweiz* 10(3): 213 (1948)
Not authentically British. Reported from England (Surrey) but the few collections in herb. K are annotated as 'compare with' *pulchripes*.

pulla (Schaeff.) Gillet, Collybia, *Hyménomycètes*: 317 (1876)
>*Agaricus pullus* Schaeff., *Fung. Bav. Palat.* 3: 250 (1770)

A *nomen dubium* lacking type material. Antonín & Noordeloos (A&N2) suggest this is possibly *Collybia prolixa* var. *distorta*. See TBMS 2: 13 (1907) for the description and illustration of the British collection (South Essex: Epping Forest).

pullata Sacc., Mycena, *Syll. fung.* 5: 277 (1887)
>*Agaricus pullatus* Berk. & Cooke [*nom. illegit.*, non *A. pullatus* Bolton (1788)], *Grevillea* 11: 69 (1882)

A *nomen dubium*. Possibly *Mycena galopus* var. *nigra* fide Maas Geesteranus (1992). Listed in NCL citing Cooke 232 (237) Vol. 2 (1882) as an illustration.

pulmonarius *var.* **juglandis** (Fr.) Sacc., Pleurotus, *Syll. fung.* 5: 362 (1887)
>*Agaricus pulmonarius* var. *juglandis* Fr., *Ic. Hymenomyc.*: t. 87 (1867)

A *nomen dubium*, lacking type material and, from the description, probably just *Pleurotus ostreatus* growing on *Juglans regia*.

pulverulentus Berk. & Broome, Hygrophorus, *Ann. Mag. Nat. Hist., Ser. 5 [Notices of British Fungi no. 1667]* 1: 22 (1878)
Cooke 875 (895) Vol. 6 (1888) may represent small specimens of *Hygrocybe virginea* with bacterial infection in the stipe bases, i.e. what has commonly been known as *Hygrophorus virgineus* var. *roseipes*. The type (in herb. K) from Scotland (Glamis) requires re-examination.

pulverulentus (Scop.) Fr., Lentinus, *Synopsis generis Lentinorum*: 8 (1836)
>*Agaricus pulverulentus* Scop., *Fl. carniol.*: 391 (1772)

A *nomen dubium* lacking type material. Listed by Rea (1922). Synonymous with *Neolentinus adhaerens* fide Pegler (1983, as *Lentinus adhaerens*).

pulvinatus Bolton, Agaricus, *Hist. fung. Halifax* 2: 49 (1788)
Described from Britain but a *nomen dubium* lacking type material. From the illustrations and descriptions, apparently a species of *Amanita* (*Amanitopsis*).

punctata (Cleland & Cheel) D.A. Reid, Amanita, *Austral. J. Bot., Suppl. Ser.* 8: 50 (1980)
>*Amanitopsis punctata* Cleland & Cheel, *Trans. Proc. Roy. Soc. South Australia* 43: 265 (1919)

Not authentically British. An Australian species, the name having been used in Europe for a doubtful species within the *A. vaginata* group.

punctatum (Fr.) P. Kumm., Hebeloma, *Führ. Pilzk.*: 80 (1871)
>*Agaricus punctatus* Fr., *Elench. fung.* 1: 30 (1828)

A *nomen dubium*. Sensu Rea (1922) is *Pholiota lenta*.

punctatus With., Boletus, *Bot. arr. veg.* 2: 769 (1776)
Described from Britain, 'on the stumps of old elms', but a *nomen dubium*.

punctatus (Pers.) Fr., Cortinarius, *Epicr. syst. mycol.*: 299 (1838)
>*Agaricus punctatus* Pers., *Syn. meth. fung.*: 274 (1801)

Excluded pending clearer definition. Cooke 813 (855) Vol. 6 (1887) is doubtful. Several British records are said to be sensu Lange (FlDan4), which is also doubtful.

punctulata (Kalchbr.) Sacc., Stropharia, *Syll. fung.* 5: 1024 (1887)
>*Agaricus punctulatus* Kalchbr., *Icon. select. Hymenomyc. Hung.*: 25 (1873)
>*Hypholoma punctulata* (Kalchbr.) Massee, *Brit. fung.-fl.* 1: 390 (1892)

A *nomen dubium*. Fide Holec (2002) this is *Pholiota gummosa*. Cooke 579 (587) Vol. 4 (1886) (as *Agaricus (Hypholoma) punctulatus*) looks distinctive, possibly some other species of *Pholiota*.

puniceus With., Agaricus [*nom. illegit.*, non *A. puniceus* Fr. (1821)], *Arr. Brit. pl. ed. 3*, 4:196 (1796)
Described from Britain but a *nomen dubium*.

purpurascens With., Agaricus [*nom. illegit.*, non *A. purpurascens* Fr. (1818); non (Cooke) Pilát (1951)], *Bot. arr. Brit. pl. ed. 2*, 3: 356 (1792)
Described from Britain but a *nomen dubium*.

purpurascens Berk. & Broome, Trametes, *Ann. Mag. Nat. Hist., Ser. 5 [Notices of British Fungi no. 1811]* 3: 210 (1879)
Described from Britain but a *nomen dubium*. From the original description of material gathered on dead trunks of *Salix* sp. in Northamptonshire (Cotterstock) probably just old and reddened *Daedaleopsis confragosa*.

purpurea W.G. Sm., Pistillaria, *J. Bot.* 11: 67 (1873)
Described from Britain but a *nomen dubium*. Said to have globose spores but the accompanying plate suggests *Typhula micans*.

purpureum *var.* **atromarginatum** W.G. Sm., Stereum, *Syn. Brit. Bas.*: 405 (1908)
Described from Britain but a *nomen dubium*. Smith equated this with *Auricularia elegans* Sowerby but that too is a *nomen dubium*.

purpureus Fr. & Hök, Boletus, *Boleti fung. gen.*: 11 (1835)
A *nomen dubium*. Sensu NCL and sensu auct. mult. is *Boletus rhodopurpureus*.

pusilla (Batsch.) Pers., Bovista, *Syn. meth. fung.*: 138 (1801)
>*Lycoperdon pusillum* Batsch., *Elench. fung. (Continuatio Secunda)*: 123 t. 41 f. 228 (1789)

A *nomen dubium* variously interpreted. Sensu auct. is *Bovista dermoxantha* but Batsch's plate apparently shows *B. limosa*.

pusilla Sacc. & Trotter, Eccilia, *I funghi dell'Avellinese*: 10 (1920)
A *nomen dubium*. Reported from Glamorganshire (Welsh St. Donats) in 1973, but lacking voucher material.

pusilla (Pers.) J. Schröt., Typhula, in Cohn, *Krypt.-Fl. Schlesien* 3: 439 (1888)
>*Clavaria pusilla* Pers., *Comment. Fungis Clavaeform.*: 86 (1797)
>*Pistillaria pusilla* (Pers.) Fr., *Syst. mycol.* 1: 498 (1821)

A *nomen dubium* lacking type material. Listed by Rea (1922).

pusio J. Howse, Boletus, *Ann. Mag. Nat. Hist., Ser. 5 [Notices of British Fungi no. 1798]* 3: 209 (1879)
Described from Britain but a *nomen dubium*.

pusiola (Fr.) R. Heim, Agrocybe, *Treb. Mus. Ci. Nat. Barcelona* 3: 129 (1934)
>*Agaricus pusiolus* Fr., *Elench. fung.* 1: 36 (1828)
>*Naucoria pusiola* (Fr.) Gillet, *Hyménomycètes*: 546 (1876)
>*Agaricus pusillus* Fr. [*nom. illegit.*, non *A. pusillus* (Pers.) Fr., non Schaeff. (1800)], *Syst. mycol.* 1: 264 (1821)
>*Naucoria pusilla* (Fr.) P. Kumm., *Führ. Pilzk.*: 77 (1871)
>*Agrocybe pusilla* (Fr.) Watling, *Biblioth. Mycol.* 82: 54 (1981)

Not authentically British. Listed by Rea (1922, as *Naucoria pusiola*), by NCL, and by BFF3 (as *Agrocybe pusilla*) but with no supporting evidence. Recently reported from South Hampshire but unsubstantiated with voucher material.

pyriodora (Pers.) P. Kumm., Inocybe, *Führ. Pilzk.*: 79 (1871)
 Agaricus pyriodorus Pers., *Syn. meth. fung.*: 300 (1801)
A *nomen confusum* lacking type material and variously interpreted. Sensu Persoon (1801: 300) was probably *Inocybe bongardii fide* Kuyper (1986) but most British records are *Inocybe fraudans*.

pyrotricha *var.* egregia Massee, Hypholoma, *Brit. fung.-fl.* 1: 389 (1892)
Described from Britain but a *nomen dubium*. Probably a species of *Lacrymaria*.

quadricolor (Scop.) Fr., Cortinarius, *Epicr. syst. mycol.*: 295 (1838)
 Agaricus quadricolor Scop., *Fl. carniol.*: 446 (1772)
A *nomen dubium* lacking type material. K&R cite Cooke 796 (820B) Vol. 6 (1887, as *C. quadricolor*) as an illustration of *C. bicolor* (= *C. cagei* of this checklist).

queletii Maire & Konrad, Inocybe, *Bull. Soc. Mycol. France* 45: 40 (1929)
Not authentically British. Said to be common by Rea (1922) but no records traced.

queletii (Pilát & Svrček) Corner, Leptoglossum, *Monograph of Cantharelloid Fungi*: 146 (1966)
 Leptotus queletii Pilát & Svrček, *Česka Mykol.* 7: 12 (1953)
Not authentically British. A *nomen novum* for *Arrhenia glauca* sensu Quélet (Enchir. fung.: 140 (1886)], with Cooke [1065 (1115b) Vol. 7 (1890) as *Cantharellus glaucus*] cited as an illustration. Reported on dead stems of *Nepeta* sp. by Parker-Rhodes [TBMS 34(3): 361 (1951)] from Pembrokeshire (Skokholm Island) but no material was kept. *Fide* Watling (BFF6: 30) this is possibly a large-spored form of *Arrhenia spathulata*.

querceti H. Haas & Jul. Schäff., Russula, *Russula-Monographie*: 164 (1952)
A *nomen dubium*. British material named thus (in herb. K) is *Russula melzeri*.

querceus Britzelm., Marasmius, *Bot. Centralbl.* 68: 7 (1896)
 Mis.: *Marasmius prasiosmus* sensu NCL
Not authentically British, though included in Sowerby (Sow1: pl. 81) as 'not uncommon in woodland', growing on fallen oak leaves, and with a smell 'strongly of garlic'. It is not, however, *Marasmius alliaceus*.

quercina Pearson in Hora, Clitocybe, *Trans. Brit. Mycol. Soc.* 43(2): 441 (1960)
Originally intended as a *nomen novum* for *Clitocybe fritilliformis* sensu Ricken (1915: 378). However, Hora's material has not survived, the interpretation of Ricken's species is contentious and the name is best treated as a *nomen dubium*.

quercina β dura Gray, Daedalea, *Nat. arr. Brit. pl.* 1: 638 (1821)
Described from Britain but a *nomen dubium* lacking type material. Probably just abnormal basidiomes of *Daedalea quercina*.

quercinus H. Engel & T. Brückner [*nom. prov.*], Xerocomus, *Schmier- und Filzröhrlinge s.l. in Europa*: 205 (1996)
Used for the species here called *Boletus declivitatum*.

quinquepartitum (L.) Gillet, Tricholoma, *Hyménomycètes*: 96 (1874)
 Agaricus quinquepartitus L., *Sp. pl.*: 1171 (1753)
A *nomen dubium* lacking type material. Cooke 63 (74) Vol. 1 (1882) provides one interpretation.

radiata (J.E. Lange) P.D. Orton, Nolanea, *Trans. Brit. Mycol. Soc.* 43(2): 179 (1960)

Rhodophyllus elaphinus var. *radiatus* J.E. Lange, *Dansk. Bot. Ark.* 2(11): 31 (1921)
A *nomen dubium*. Listed in NCL but omitted by Orton [Mycologist 5 (3): 124 (1991)] 'pending clearer definition'.

radiatum (Peck) Parmasto, Boreostereum, *Conspectus Systematis Corticiacearum*: 187 (1968)
 Stereum radiatum Peck, *Bull. Buffalo Soc. Nat. Sci.* 1: 62 (1873)
Not authentically British. A culture formerly held at the Buildings Research Establishment is from imported timber and this species is not otherwise known from western Europe.

radiatum Sowerby, Lycoperdon, *Col. fig. Engl. fung.* 2: pl. 145 (1798)
Described from Britain but a *nomen dubium* lacking type material. Described from 'the damp plaster wall of a ball-room', this is probably not a basidiomycete.

radicatum (Cooke) Maire, Hebeloma, *Bull. Soc. Mycol. France* 24: LVII (1908)
 Agaricus longicaudus var. *radicatus* Cooke, *Ill. Brit. fung.* 459 (416) Vol. 3 (1885)
 Hebeloma longicaudum var. *radicatum* (Cooke) Massee, *Brit. fung.-fl.* 2: 177 (1893)
Described from Britain but a *nomen dubium*.

radiosum (Fr.) Fr., Corticium, *Epicr. syst. mycol.*: 560 (1838)
 Thelephora radiosa Fr., *Observ. mycol.* 2: 277 (1818)
A *nomen dubium* lacking type material, but sensu Rea (1922) is *Vesiculomyces citrinus*.

raffillii Massee, Marasmius, *Bull. Misc. Inform. Kew.* 1909: 374 (1909)
A *nomen dubium*. Described from 'a stove' in Kew Gardens. Probably an alien and, *fide* Antonín & Noordeloos (A&N1), possibly not a species of *Marasmius*.

ramentaceum (Bull.) Ricken, Tricholoma, *Blätterpilze Deutschl.*: 338 (1914)
 Agaricus ramentaceus Bull., *Herb. France*: pl. 595 (1793)
 Armillaria ramentacea (Bull.) P. Kumm., *Führ. Pilzk.*: 134 (1871)
A *nomen dubium* lacking type material. Sensu Cooke 53 (71) Vol. 1 (1882) and sensu Rea (1922) is possibly *Tricholoma cingulatum*.

ramosoradicatus Bolton, Agaricus, *Hist. fung. Halifax* Suppl.: 148 (1791)
Described from Britain but a *nomen dubium* lacking type material. Bolton's plate shows a collybioid agaric with a long radicant stipe from which arise numerous stipes bearing small pilei. Cited by Fries (1821) in the synonmy of *Agaricus inopus* (= *Flammula inopus*), itself a *nomen dubium*.

rapiolens J. Favre, Mycena, *Bull. Soc. Neuchâteloise Sci. Nat.* 80: 90 (1957)
A montane species, not authentically British. Reported 'on a cone, in grassland' from the Republic of Ireland in 1989, but lacking voucher material.

rasilis *var.* leucophylloides Bon, Melanoleuca, *Doc. mycol.* 3(9): 46 (1973)
Not authentically British. Supposedly recorded from Wales *fide* M. Rotheroe (pers. comm.) but no material has been traced.

rasilis *var.* pseudoluscina (Bon) Boekhout, Melanoleuca, *Persoonia* 13(4): 410 (1988)
 Melanoleuca pseudoluscina Bon, *Doc. mycol.* 10(37-38): 89 (1980) [1979]
Not authentically British. Mentioned in BFF8 (p. 59) but no records have been traced.

ravida (Fr.) Sacc., Galera, *Syll. fung.* 5: 871 (1887)
 Agaricus ravidus Fr., *Observ. mycol.* 2: 132 (1818)
A *nomen dubium* variously interpreted, usually as a *Conocybe* sp. Cooke 525 (467) Vol. 4 (1885) is not helpful.

413

reai Singer, Melanoleuca, *Cavanillesia*. 7: 127 (1935)
Mis.: *Tricholoma subpulverulentum* sensu Rea (1922)
A *nomen dubium*. Introduced as a *nomen novum* for *Tricholoma subpulverulentum* sensu Rea (1922), but without description; there is only an inadequate description in Rea (1922).

redemitus Cooke, Cortinarius, *Handb. Brit. fung.,* Edn 2: 254 (1888)
Described from Britain but a *nomen dubium*.

reducta (Fr.) P. Karst., Simocybe, *Bidrag Kännedom Finlands Natur Folk* 32: 429 (1879)
Agaricus reductus Fr., *Syst. mycol.* 1: 133 (1821)
Naucoria reducta (Fr.) Sacc., *Syll. fung.* 5: 849 (1887)
Not authentically British. Listed by Massee (1902) and by Rea (1922) but no records have been traced.

reedii Berk., Cortinarius, *Outl. Brit. fungol.*: 194 (1860)
Described from Britain but a *nomen dubium*.

regalis (Fr.) Michael, Amanita, *Führer Pilzfr.* 1, Edn 2: 56 (1896)
Agaricus muscarius var. *regalis* Fr., *Hymenomyc. eur.*: 20 (1874)
Amanita muscaria var. *regalis* (Fr.) Sacc., *Syll. fung.* 5: 13 (1887)
Amanita emilii Riel, *Bull. Soc. Mycol. France* 23: 1 (1907)
Not authentically British. Listed in Rea (1922) and reported on a few occasions since then, but all records lack voucher material.

relhani Sowerby, Helvella, *Col. fig. Engl. fung.* 1: pl. 11 (1796)
Described from Britain but a *nomen dubium* lacking type material. Sowerby's illustration appears to depict a species of *Conocybe* showing a slight pubescence on the pileus and a bulbous stipe base.

relicina (Fr.) Quél., Inocybe, *Fl. mycol. France*: 104 (1888)
Agaricus relicinus Fr., *Syst. mycol.* 1: 256 (1821)
Not authentically British. A species with small warty spores, known only from Scandinavia. Sensu Rea (1922) is probably *Inocybe cervicolor*.

rennyi *var.* **major** Massee, Inocybe, *Brit. fung.-fl.* 2: 201 (1893)
Astrosporina rennyi var. *major* (Massee) Rea, *Brit. basidiomyc.*: 212 (1922)
Described from Britain but a *nomen dubium*.

repandum (Bull.) Fr., Entoloma, *Hymenomyc. eur.*: 190 (1874)
Agaricus repandus Bull. [non *A. repandus* Bolton (1788)], *Herb. France*: pl. 423 (1789)
A *nomen dubium* lacking type material. *Fide* Pearson & Dennis (1948) British records are doubtfully distinct from *E. prunuloides*.

repandus Bolton, Agaricus [*nom. illegit.*, non *A. repandus* Bull. (1789)], *Hist. fung. Halifax* 1: 6 (1788)
A *nomen dubium* lacking type material. Bolton's illustration could possibly represent *Entoloma rhodopolium*. It appears not to be the same as *Agaricus repandus* Bull. which is illustrated as a large, stout and strongly umbonate species.

replexus (Fr.) Fr., Cantharellus, *Hymenomyc. eur.*: 459 (1874)
Agaricus replexus Fr., *Syst. mycol.* 1: 158 (1821)
A *nomen dubium*. Listed by Rea (1922).

replexus *var.* **devexus** (Fr.) Fr., Cantharellus, *Hymenomyc. eur.*: 459 (1874)
Agaricus devexus Fr., *Syst. mycol.* 1: 158 (1821)
Cantharellus devexus (Fr.) Sacc., *Syll. fung.* 5: 492 (1887)
A *nomen dubium*, possibly *Mycena cinerella*. Cooke 1098 (1150) Vol. 7 resembles *M. speirea*.

resplendens (Fr.) Quél., Tricholoma, *Bull. Soc. Bot. France* 23: 350 (1876)
Agaricus resplendens Fr., *Monogr. hymenomyc. Suec.* 1: 55 (1857)
A *nomen dubium*. Sensu Orton [Notes RBG Edin. 29: 125 (1969)] is not convincingly distinct from *Tricholoma*

columbetta. Sensu auct. is *T. sulphurescens*. Cooke 64 (55) Vol. 1 (1882) is not helpful.

resupinatus Sowerby, Boletus [*nom. illegit*, non *B. resupinatus* Sw. (1788), non Bolton (1791)] *Col. fig. Engl. fung.,* 4 *(suppl.)*: pl. 424 (1814)
Described from Britain but a *nomen dubium* lacking type material. 'From the plate I do not venture to suggest its identity' (Donk, 1974: 354).

resupinatus (Bolton) Massee, Fomes, *Brit. fung.-fl.* 1: 226 (1892)
Boletus resupinatus Bolton [*nom. illegit.*, non *B. resupinatus* Sw. (1788), non Sowerby (1815)], *Hist. fung. Halifax* Suppl.: 165 (1791)
Poria resupinata (Bolton) W.G. Sm., *Syn. Brit. Bas.*: 356 (1908)
Described from Britain but a *nomen dubium* lacking type material. See Donk (1974: 258) for an extensive discussion.

resutum (Fr.) Quél., Entoloma, *Bull. Soc. Bot. France* 23: 326 (1877)
Agaricus resutus Fr., *Epicr. syst. mycol.*: 145 (1838)
Sensu Rea (1922) is doubtful. Sensu Noordeloos (FAN1: 148) is a species in sugenus *Inocephalus*, growing on dunes and not British.

reticulatus With., Agaricus [*nom. illegit.*, non *A. reticulatus* Pers. (1798)] *Bot. arr. Brit. pl. ed. 2*, 3: 389 (1792)
Described from Britain but a *nomen dubium*. The description (pale flesh-coloured lamellae and raised reticulum on the pileus) suggests *Pluteus thomsonii*.

reticulatus *var.* **pluteoides** (M.M. Moser) Arnolds, Bolbitius, *Persoonia* 18(2): 210 (2003)
Bolbitius pluteoides M.M. Moser, *Fungorum Rar. Icon. Color.* 7: 27 (1978)
Not authentically British. 'Undoubtedly occurs in the British Isles' *fide* BFF3: 37 (as *B. pluteoides*) but no records have been traced.

retigera Bres., Collybia, *Fungi trident.* 1: 8 (1881)
A *nomen dubium*. Possibly a species of *Tephrocybe fide* A&N2. See report in TBMS 1 (4): 157 (1900) where material collected by Eyre from North Hampshire (Alresford, Grange Park) is described as growing 'caespitose at base of *Fraxinus excelsior*'.

retisporus Massee, Lactarius, *Brit. fung.-fl.* 3: 29 (1893)
Described from Britain but a *nomen dubium*. Doubtfully distinct from *L. fuliginosus fide* NCL, but agrees better with *L. romagnesii*.

retosta (Fr.) P.D. Orton, Omphalina, *Trans. Brit. Mycol. Soc.* 43(2): 180 (1960)
Agaricus retostus Fr., *Epicr. syst. mycol.*: 125 (1838)
Omphalia retosta (Fr.) Gillet, *Hyménomycètes*: 294 (1876)
A *nomen dubium*. Described by Massee (1893) as 'near to *Omphalia abhorrens*' (=*Camarophyllopsis foetens*) and may well represent that species. There are three old collections from Wales (Denbighshire) in herb. K all in poor condition. Cooke 261 (272) Vol. 2 (1883) is different and appears to represent an *Entoloma* sp.

revolutus Cooke, Paxillus, *Grevillea* 16: 45 (1887)
Described from Britain but a *nomen dubium*. Possibly a species of *Hygrophorus fide* NCL. Cooke 865 (862) Vol. 6 (1888) is more probably a species of *Hygrocybe* near to *H. virginea*.

revolutus *var.* **anglicus** Massee, Pleurotus, *Brit. fung.-fl.* 2: 373 (1893)
Described from Britain but a *nomen dubium*. Probably just *P. ostreatus*.

revolutus *var.* **subrubens** Bourdot & Galzin, Tyromyces, *Hyménomyc. France*: 549 (1928)
Not authentically British. A single record from Hertfordshire (Rothamsted) as 'compare with' this variety is unsubstantiated with voucher material.

rheoides With., Agaricus, *Arr. Brit. pl. ed. 3*, 4: 213 (1796)
Described from Britain, 'on the stumps of old Hawthorns and rotten Alders', but a *nomen dubium*.

rhodella (Fr.) Cooke, Poria, *Grevillea* 14: 110 (1886)
    Polyporus rhodellus Fr., *Observ. mycol.* 2: 261 (1818)
A *nomen dubium* variously interpreted, the name having been applied to at least three different species of *Ceriporia*.

rhodophaea Höhn. & Litsch., Tomentella, *Sitzungsber. Kaiserl. Akad. Wiss., Wien, Math.-Naturwiss. Cl., Abt. 1* 116: 831 (1907)
A *nomen dubium*. Reported from South Hampshire (New Forest) in 1916. Sensu Jülich (1984) is *T. testaceogilva*.

rhodopus Zvára, Russula, *Arch. Přírodov. Výzk. Čech.* 17(4): 108 (1927)
Illustrated by Rea from South Hampshire in 1936, but otherwise unrecorded from Britain. Accepted by Sarnari (1998) as a good species near to *R. sanguinaria*.

rhodospora (Broome & W.G. Sm.) Sacc. & D. Sacc., Nolanea, *Syll. fung.* 17: 60 (1905)
    Agaricus rhodosporus Broome & W.G. Sm., *J. Bot.* 41: 385 (1903)
A *nomen dubium* described from material collected 'on earth in stoves' from a Royal Horticultural Society orchid house in 1869 and 1870. Probably an introduced alien, never recollected.

rhodoxanthus (Krombh.) Kallenb., Boletus, *Z. Pilzk.* 5 (2): 27 (1925)
    Boletus sanguineus var. *rhodoxanthus* Krombh., *Naturgetr. Abbild. Schwämme* 5 pl. 37 f. 12 – 15 (1836)
Not authentically British. Reported from Worcestershire by Rea in 1895, but unsubstantiated with voucher material. The illustration in Phillips (p. 201) is of material from Corsica.

rhodoxanthus (Schwein.) Bres., Phylloporus, *Fungi trident.* 2: 95 (1900)
    Agaricus rhodoxanthus Schwein., *Schriften Naturf. Ges. Leipzig* 1: 83 (1822)
A North American species, unknown in Europe. The name has been widely used in Europe for *P. pelletieri*.

rhomboideus With., Agaricus, *Arr. Brit. pl. ed. 3*, 4: 298 (1796)
Described from Britain but a *nomen dubium*.

rickenii (A.H. Sm.) Singer, Hemimycena, *Sydowia* 15: 60 (1962) [1961]
    Mycena rickenii A.H. Sm., *North American Species of Mycena*: 158 (1947)
    Marasmiellus rickenii (A.H. Sm.) Singer, *Lilloa* 22: 298 (1951) [1949]
Not authentically British. Listed in NCL as *Mycena rickenii* but there is no evidence that this is British or even European. A single report from Herefordshire in 1992 is unsubstantiated with voucher material.

riculatus Fr., Cortinarius, *Epicr. syst. mycol.*: 284 (1838)
A *nomen dubium*. Sensu Rea (1922) is doubtful.

rigidus Bolton, Agaricus [*nom. illegit.*, non *A. rigidus* Scop. (1772)], *Hist. fung. Halifax* 1: 43 (1788)
Described from Britain but a *nomen dubium* lacking type material. Bolton's illustration could depict *Collybia fusipes*.

rigidus (Scop.) Fr., Cortinarius, *Epicr. syst. mycol.*: 302 (1838)
    Agaricus rigidus Scop. [non *A. rigidus* Bolton (1788)], *Fl. carniol.* 456 (1772)
A *nomen dubium* lacking type material. Listed by Rea (1922) but no material has been traced. Sensu Lange (1938) is *C. umbrinolens* and sensu K&R is *C. rigidiusculus*.

rivulosus Berk. & Broome, Bolbitius, *Ann. Mag. Nat. Hist., Ser. 5 [Notices of British Fungi no. 1773]* 3: 207 (1879)
Described from Britain but a *nomen dubium*. Collected on soil in an 'orchard' (orchid?) house in Middlesex (Chiswick). See Cooke 678 (928) Vol. 5 (1887) for an illustration. Possibly *B.*

variicolor (= *B. titubans* var. *olivaceus*) or an alien *fide* Watling & Gregory (1981).

robiniophila (Murrill) Ryvarden, Perenniporia, *Mycotaxon* 17: 517 (1983)
    Trametes robiniophila Murrill, *North American Flora* 9(1): 42 (1907)
A North American species. Reported by Rea [TBMS 17(1-2): 47 (1932)] as 'uncommon on trunks of *Robinia pseudacacia*'. Rea's description is reminiscent of *Spongipellis spumeus* but no material survives.

roburneus (Fr.) Gillet, Fomes, *Hyménomycètes*: 684 (1878)
    Polyporus roburneus Fr., *Epicr. syst. mycol.*: 464 (1838)
A *nomen dubium*. Lloyds's concept of *Fomes igniarius* var. *roburneus* (Fr.) Lloyd [as quoted by Rea (1922) but never validly published] is based on a specimen in herb. K, referred by Ryvarden to young material of *Phellinus igniarius*. See Donk (1974: 224) for further discussion.

robusta *var.* **minor** Sacc., Armillaria, *Syll. fung.* 5: 75 (1887)
A *nomen dubium*. Cooke 52 (86) Vol. 1 (1881) is not helpful. Possibly a small form of *Tricholoma robustum*.

romagnesii Noordel., Entoloma, *Persoonia* 10(2): 225 (1979)
    Mis.: *Rhodophyllus subnigrellus* sensu Romagnesi [Rev. Mycol. (Paris) 2: 86 (1937)]
Not authentically British. Reported by Orton (1991) as 'possibly British but requiring confirmation'. No records have been traced.

roseipes Bres., Russula, *Fungi trident.* 1: 37 (1881)
Not authentically British. Listed by Rea (1922). Cooke 1035 (1081) Vol. 7 (1888) depicts *R. nitida*.

roseoalbus (Hornem.) P. Kumm., Pluteus, *Führ. Pilzk.*: 98 (1871)
    Agaricus roseoalbus Hornem., *Fl. Danica*: t. 1679 (1819)
A *nomen dubium*. Cooke 312 (598) Vol. 3 (1884) is *Rhodotus palmatus*.

roseofloccosa Hora, Limacella, *Trans. Brit. Mycol. Soc.* 43(2): 450 (1960)
Described from Britain but a *nomen dubium*, lacking type material. Probably *Limacella delicata* var. *vinosorubescens*.

roseolus Quél., Pleurotus, *Bull. Soc. Amis Sci. Nat. Rouen* 15: 155 (1879)
A *nomen dubium* lacking type material. The single record named thus in herb K is *Melanotus phillipsii*.

roseotinctus Rea, Coprinus, *Trans. Brit. Mycol. Soc.* 1(1): 23 (1897)
Described from Worcestershire (Temple Laughern) in 1887 and possibly a good species related to *C. niveus* but never recollected. No type material exists but there is a good illustration by Rea in herb. K.

rostellata (Velen.) Hauskn. & Svrček, Conocybe, *Czech Mycol.* 51(1): 61 (1999)
    Galera rostellata Velen., *Novit. mycol.*: 129 (1940)
Not authentically British. There are two specimens in herb. K, labelled as 'compare with' *Conocybe rostellata* (det. Hausknecht).

rostkovii Fr., Boletus, *Hymenomyc. eur.*: 521 (1874)
A *nomen dubium*. Sensu Rea (1922) is unknown.

rostrupianus E.C. Hansen, Coprinus, *Bot. Zeitung (Berlin)* 55(7): 125 (1897)
A *nomen dubium*. Sensu Lange (FlDan4) and sensu auct. is *C. flocculosus*.

rozei Quél., Entoloma, *Bull. Soc. Bot. France* 23: 326 (1876) [1877]
A *nomen dubium* lacking type material. Included by Rea (1922) but 'not authentically British' *fide* NCL. Probably in subgenus *Trichopilus* or *Inocephalus* *fide* Noordeloos (1987).

rubecundus With., Agaricus, *Arr. Brit. pl. ed. 3*, 4: 259 (1796)
Described from Britain but a *nomen dubium*.

rubellus McWeeney, Gyrodon, *Grevillea* 22: 42 (1893)
Described from Ireland but a *nomen dubium*.

ruber With., Agaricus, *Arr. Brit. pl. ed. 3*, 4: 210 (1796)
Described from Britain but a *nomen dubium*. From the
description probably a species of *Lactarius*.

rubeus Bolton, Agaricus, *Hist. fung. Halifax* 1: 36 (1788)
Described from Britain but a *nomen dubium* lacking type
material. Bolton's illustration and description suggest one of
the colour forms of *Mycena pura* or possibly *M. rosea*.

rubiatus With., Agaricus, *Arr. Brit. pl. ed. 3*, 4: 284 (1796)
Described from Britain but a *nomen dubium*. A distinctive
species, Withering commenting that 'the whole plant is
coloured as though it had been dyed with Madder' (*Rubia
peregrina*).

rubiformis (Fr.) Quél., Tremella, *Fl. mycol. France*: 22 (1888)
Naematelia rubiformis Fr., *Observ. mycol.* 2: 370 (1818)
A *nomen dubium* lacking type material. Sensu auct. = *Tremella
encephala* or *T. aurantia*.

rubiginosus With., Agaricus [*nom. illegit.*, non *A. rubiginosus*
Pers. (1801)], *Arr. Brit. pl. ed. 3*, 4: 258 (1796)
Described from Britain on cow dung, but a *nomen dubium*.

rubiginosus Fr., Boletus, *Observ. mycol.* 2: 245 (1818)
A *nomen dubium* collected only once by Fries in 1815. Sensu
Rea (1922) under beeches is unknown.

rubra (Lam.) Fr. [*nom. illegit.*], Russula, *Epicr. syst. mycol.*: 354
(1838)
Amanita rubra Lam., *Encyclop.*: 104 (1783)
A superfluous name for *Russula emetica*. Sensu Cooke (1888) is
R. atropurpurea. Sensu NCL and sensu Rayner (1985) is *R.
pungens*.

rubricatus (Berk. & Broome) Massee, Marasmius, *Brit. fung.-fl.*
3: 164 (1893)
Agaricus rubricatus Berk. & Broome, *Ann. Mag. Nat. Hist.,
Ser. 5 [Notices of British Fungi no. 1873]* 7: 127 (1881)
Naucoria rubricata (Berk. & Broome) Sacc., *Syll. fung.* 5: 834
(1887)
Described from Britain but a *nomen dubium*. Cooke 496 (509)
Vol. 4 (1885) (as *Agaricus rubricatus*) is *Marasmiellus ramealis*.

rubricosus (Fr.) Fr., Cortinarius, *Epicr. syst. mycol.*: 310 (1838)
Agaricus irregularis var. *rubricosus* Fr., *Observ. mycol.* 2: 44
(1818)
A *nomen dubium*. Old records from Scotland and Wales. A
record from Northern Ireland (Fermanagh) in 2000 is sensu
Lange (FlDan3) and thus *C. brunneus* var. *glandicolor*.

rubromarginata *var.* **fuscopurpurea** (Stev.) Sacc., Mycena,
Syll. fung. 5: 254 (1887)
Agaricus rubromarginatus var. *fuscopurpureus* Stev.,
Mycol. Scot.: 27 (1879)
Agaricus rubromarginatus var. *erosus* Lasch, *Linnaea* 4:
535 (1829)
A *nomen dubium* lacking type material. Listed from Scotland
(Glamis) by Stevenson (1879) and hence by Rea (1922).
Lasch's species grew on willow trunks, but is possibly just old
Mycena sanguinolenta fide Maas Geesteranus (1992).
Stevenson appears to have accidentally created a *nomen
novum* by using the first word of Lasch's description
(fuscopurpureus) rather than his epithet (erosus).

rudis Fr., Panus, *Epicr. syst. mycol.*: 398 (1838)
Panus strigosus Berk. & M.A. Curtis, *Ann. Mag. Nat. Hist.,
Ser. 2, 12 (1853)
Not authentically British. Reported from Lincolnshire (South
Elkington) in TBMS 1(2): 75 (1898) and listed in Rea (1922).
This was probably *Panus conchatus*. Listed in NM2 as a
synonym of *P. strigosus* (not British).

rufa Pers., Clavaria, *Comment. Fungis Clavaeform.*: 71 (1797)

A *nomen dubium* lacking type material. Listed by Rea (1922) as
a synonym of *C. inaequalis* (= *C. helvola*).

rufescens (Schaeff.) Corner, Ramaria, *Monograph of Clavaria
and Allied Genera*: 618 (1950)
Clavaria rufescens Schaeff., *Fung. Bavar. Palat. nasc.* 3: 288
(1770)
A clampless relative of *R. botrytis*, not authentically British.
Reported from South Hampshire (New Forest) by Cooke in
Grevillea (1891) but unsubstantiated with voucher material.

rufidula (Kalchbr.) Sacc., Pholiota, *Syll. fung.* 5: 761 (1887)
Agaricus rufidulus Kalchbr., in Fries, *Hymenomyc. eur.*: 226
(1874)
A *nomen dubium* and not a *Pholiota* fide Holec (2001). Listed by
Rea (1922).

rufipes Morgan, Lepiota, *J. Mycol.* 12: 156 (1906)
Not authentically British, nor European. A poorly described North
American species, given a recent European interpretation (see
Vellinga in FAN5 p. 143). The single British collection named
thus (in herb. K) from South Devon needs re-examination.

rufipes (Massee & W.G. Sm.) Sacc. & D. Sacc., Pleurotus, *Syll.
fung.* 17: 26 (1905)
Agaricus rufipes Massee & W.G. Sm., *J. Bot.* 41: 385 (1903)
Described from Britain but a *nomen dubium*.

rugosiusculum Berk. & M.A. Curtis, Stereum, *Grevillea* 1: 162
(1873)
A North American species, not authentically British. Sensu Rea
(1922) is *Chondrosterum purpureum*.

rugosus Fr. & Hök, Boletus [*nom. illegit.*, non *B. rugosus*
Sowerby (1809)], *Boleti fung. gen.*: 11 (1835)
A *nomen dubium*. Listed by Rea (1922).

rugosus Sowerby, Boletus [non *B. rugosus* Fr. & Hök (1835)],
Col. fig. Engl. fung., Suppl.: pl. 422 (1809)
Described from Britain but a *nomen dubium* lacking type
material. Sowerby's illustration suggests *Abortiporus biennis*.

ruralis With., Agaricus, *Bot. arr. veg.* 2: 761 (1776)
Described from Britain, 'on old ruined cottages', but a *nomen
dubium*.

rusiophyllus Lasch, Agaricus, *Linnaea* 3: 37 (1828)
A *nomen dubium* lacking type material. British material thus
named is usually *Agaricus comtulus*.

russus Fr., Cortinarius, *Epicr. syst. mycol.*: 261 (1838)
A boreal-montane species, not authentically British. Old
Herefordshire records lack voucher material. Sensu Cooke
(1887) and sensu NCL is *C. mussivus*.

ruthae (Berk. & Broome) Sacc., Pleurotus, *Syll. fung.* 5: 345
(1887)
Agaricus ruthae Berk. & Broome, *Ann. Mag. Nat. Hist., Ser. 5
[Notices of British Fungi no. 1754]* 3: 205 (1879)
A *Hohenbuehelia* species, described from Britain and named for
Berkeley's daughter, but a *nomen dubium*. Considered by
some as a synonym of *H. petaloides*. Cooke: 275 (178) Vol. 2
(1883) depicts a pleurotoid agaric with pinkish tints to the
lamellar edges and stipe; but 1154 (634) Vol. 8 (1890) is
different.

rutilus Fr. & Hök, Boletus, *Boleti fung. gen.*: 5 (1835)
A *nomen dubium*. Sensu Rea (1922) 'in oak woods' is unknown.

sabuletorum (Berk. & M.A. Curtis) Sacc., Inocybe, *Syll. fung.*
11: 51 (1895)
Agaricus sabuletorum Berk. & M.A. Curtis, *Grevillea* 19: 103
(1891)
Astrosporina sabuletorum (Berk. & M.A. Curtis) Rea, *Brit.
basidiomyc.*: 212 (1922)
Not authentically British. A North American species. Sensu Rea
(1922) is *Inocybe lanuginosa* var. *ovatocystis*.

saccharatus With., Agaricus, *Bot. arr. Brit. pl. ed. 2*, 3: 377
(1792)

Described from Britain but a *nomen dubium*.

saccharinus Romagn., Coprinus, *Bull. Soc. Mycol. France* 92(2): 203 (1976)
Reported from Buckinghamshire (Aston Clinton) and Oxfordshire (Harpsden Woods and Nettlebed Woods) but a dubious species, not readily distinguished from *Coprinus truncorum*.

sachalinensis (S. Imai) Corner, Clavariadelphus, *Monograph of Clavaria and Allied Genera*: 282 (1950)
Clavaria sachalinensis S. Imai, *Trans. Sapporo Nat. Hist. Soc.* 11: 73 (1930)
Known in Scandinavia, but not authentically British. Reported from the Nordic Mycological Congress (1991) in Perthshire (Kindrogan) but unsubstantiated with voucher material.

sagatus Fr., Agaricus, *Syst. Mycol.* 1: 282 (1821)
A *nomen dubium*. A doubtful species within *Agaricus* section *Minores*. Cooke 1177 (898) Vol. 8 (1891) is possibly *Agaricus dulcidulus*.

sahleri (Quél.) Kühner, Galerina, *Beitr. Kryptogamenfl. Schweiz* 10(3): 136 (1948)
Galera sahleri Quél., *Mém. Soc. Émul. Montbéliard, Sér. 2*, 5: 23 (1872)
A *nomen dubium* lacking type material and variously interpreted. Probably a member of the *Galerina hypnorum* complex. Sensu Rea (1922) is doubtful. Sensu B&K5 is *Galerina hypnorum* of this list.

saliceti P.D. Orton, Naucoria, *Notes Roy. Bot. Gard. Edinburgh* 41(3): 601 (1984)
Described from Britain but unclear, possibly a synonym of *Naucoria salicetorum* if clamped, or *N. spadicea* if not.

saliceticola (Singer) Kühner ex Knudsen & T. Borgen, Russula, *First International Symposium on Arcto-Alpine Mycology*: 224 (1982)
Russula sphagnophila ssp. *saliceticola* Singer, *Ann. Mycol.* 34: 425 (1936)
Not authentically British. Reported from Scotland (Shetland) but requiring confirmation *fide* Watling (pers. comm.).

salicicola P.D. Orton [*nom. prov.*], Cortinarius.
A herbarium name, but included in Dennis (1995).

saliciphilus (J. Favre) Watling, Flammulaster, *Notes Roy. Bot. Gard. Edinburgh* 28(1): 68 (1967)
Naucoria saliciphila J. Favre, *Beitr. Kryptogamenfl. Schweiz* 10(3): 214 (1948)
Flocculina saliciphila (J. Favre) P.D. Orton, *Trans. Brit. Mycol. Soc.* 43(2): 175 (1960)
Not authentically British. Reported from Glamorganshire (Whiteford Burrows) in 1992 but unsubstantiated with voucher material.

salmonifolius M.M. Moser & Lamoure, Leucopaxillus, *Sydowia, Beih.* 8: 268 (1979)
Not authentically British. Reported from Oxfordshire (Bix, The Warburg Reserve) but this collection (in herb. K) has been redetermined as *Clitocybe houghtonii*.

sambuci Velen., Crepidotus, *České Houby* IV-V: 919 (1922)
A *nomen dubium*. Sensu NCL and BFF6 is *C. lundelli*.

sanguifluus (Paulet) Fr., Lactarius, *Epicr. syst. mycol.*: 341 (1838)
Hypophyllum sanguifluum Paulet, *Traité champ.*: 186 t. 81 (1793)
Not authentically British. Reported on four occasions between 1908 and 1927, from England and Scotland, but all records are doubtful and unsubstantiated with voucher material. This is mainly a southern European species, but known from The Netherlands.

sapinea var. terrestris Sacc., Flammula, *Syll. fung.* 5: 824 (1887)
A *nomen dubium*, doubtfully distinct from *Gymnopilus sapineus*. Listed by Rea (1922).

saponaceum var. sulphurinum (Quél.) Rea, Tricholoma, *Brit. basidiomyc.*: 227 (1922)
Gyrophila saponaceum var. *sulphurinum* Quél., *Enchir. fung.*: 13 (1886)
Not authentically British. Listed by Rea (1922) but no records have been traced.

sarcitum (Fr.) Noordel., Entoloma, *Persoonia* 11(2): 150 (1981)
Agaricus sarcitus Fr., *Epicr. syst. mycol.*: 155 (1838)
Leptonia sarcita (Fr.) P. Karst., *Bidrag Kännedom Finlands Natur Folk* 32: 279 (1879)
Reported from North Wiltshire (Stanton Park) in 2000, but unsubstantiated with voucher material and probably *E. longistriatum* var. *sarcitulum*. Sensu Noordeloos is a member of subgenus *Paraleptonia* and is not British.

sarcocephala var. cookei Sacc., Psilocybe, *Syll. fung.* 5: 1043 (1887)
A *nomen dubium*. Cooke (591) 620 Vol. 4 (1886) (cited in the type description) appears to be a *Psathyrella* sp., possibly *P. sarcocephala* or *P. spadicea*.

sarnicus Massee, Clitopilus, *Trans. Brit. Mycol. Soc.* 1(1): 21 (1897)
A *nomen dubium*. Described from material collected 'on soil' in the Channel Islands (Guernsey).

satanoides Smotl., Boletus, *Časopis Československých Houbařů* 2: 29 (1920)
A *nomen dubium*. Sensu BFF1 & auct. mult. is *Boletus legaliae*.

saturatus J.E. Lange, Cortinarius, *Fl. agaric. danic.* 5, *Taxonomic Conspectus*: III (1940)
A *nomen dubium*. Sensu auct. Brit. and sensu FlDan3 is *C. illuminus*.

scabella P. Kumm., Inocybe, *Führ. Pilzk.*: 79 (1871)
Astrosporina scabella (P. Kumm.) J. Schröt., in Cohn, *Krypt.-Fl. Schlesien* 3: 576 (1889)
Agaricus scabellus Fr. [*nom. illegit.*, non *A. scabellus* Alb. & Schwein. (1805)], *Epicr. syst. mycol.*: 177 (1838)
A *nomen dubium* lacking type material. Sensu Cooke (HBF2: 159) is *Inocybe mixtilis*.

scabra (O.F. Müll.) P. Kumm., Inocybe, *Führ. Pilzk.*: 79 (1871)
Agaricus scaber O.F. Müll., *Fl. Danica*: t. 832 f. 3 (1780)
A *nomen dubium* lacking type material. Sensu Rea (1922) is doubtful. Sensu FlDan3 and a collection in herb. K (determined by Kuyper) is *Inocybe corydalina* var. *erinaceomorpha*.

scabra var. firmior (Fr.) Massee [as *I. scaber* var. *firma*], Inocybe, *Brit. fung.-fl.* 2: 186 (1893)
Agaricus scaber var. *firmior* Fr., *Hymenomyc. eur.*: 229 (1874)
A *nomen dubium*, listed in Rea (1922). Not discussed by Kuyper (1986).

scabrosum (Fr.) Noordel., Entoloma, *Persoonia* 12(4): 462 (1985)
Agaricus scabrosus Fr., *Epicr. syst. mycol.*: 154 (1838)
Not authentically British. A species of damp woodlands, but the single collection named thus in herb. K is from unimproved grassland on calcareous soil.

scabrum var. coloratipes (Singer) Singer, Leccinum, *Pilze Mitteleuropas* 6: 95 (1966)
Krombholzia scabra var. *coloratipes* Singer, *Ann. Mycol.* 40: 36 (1942)
A *nomen dubium*. Reported from England (Berkshire, Oxfordshire and Staffordshire) but unsubstantiated with voucher material. *Fide* R. Watling (pers. comm.) this is probably *Leccinum variicolor*.

scariosus With., Agaricus, *Bot. arr. Brit. pl. ed. 2*, 3:370 (1792)
Described from Britain but a *nomen dubium*.

schreieri Imbach, Squamanita, *Mitt. Naturf. Ges. Luzern* 15: 81 (1946)

Not authentically British. Included in error by Ing (1992) and C&D, but there appear to be no records from Britain.

sciophanoides (Rea) P.D. Orton & Watling, Hygrocybe, *Notes Roy. Bot. Gard. Edinburgh* 29: 131 (1969)
 Hygrophorus sciophanoides Rea, *Brit. basidiomyc.*: 303 (1922)
Listed in NCL, but probably only aberrant or dried-out *Hygrocybe psittacina fide* Boertmann (FNE1).

sciophyllus Fr., Cortinarius, *Monogr. hymenomyc. Suec.* 2: 309 (1863)
A *nomen dubium.* Listed by Rea (1922) but no records have been traced.

scitula (Massee) Sacc., Anellaria, *Syll. fung.* 5: 1126 (1887)
 Agaricus scitulus Massee, *Grevillea* 15(75): 65 (1887)
A *nomen dubium* and probably an alien. See Cooke 625 (927) Vol. 5 (1886) for an illustration showing a small, distinctive, volvate, coprinoid species collected in Yorkshire (Scarborough) in a flowerpot in a greenhouse.

sclerotioides (Pers.) Fr., Typhula, *Epicr. syst. mycol.*: 585 (1838)
 Phacorhiza sclerotioides Pers., *Mycol. eur.* 1: 192 (1822)
Nor authentically British. An alpine species unlikely to occur here. British records, following Corner (1950), belong in the *Typhula setipes* complex.

scobinacea (Fr.) Konrad & Maubl., Psathyrella, *Encycl. Mycol.* 14, *Les Agaricales* I: 128 (1948)
 Agaricus scobinaceus Fr., *Epicr. syst. mycol.*: 217 (1838)
 Stropharia scobinacea (Fr.) Sacc., *Syll. fung.* 5: 1024 (1887)
A *nomen dubium.* Sensu NCL and sensu auct. mult. is *Psathyrella maculata.*

scobinella (Fr.) Gillet, Lepiota, *Hyménomycètes*: 69 (1874)
 Agaricus scobinellus Fr., *Epicr. syst. mycol.*: 11 (1838)
A *nomen dubium.* Sensu Rea (1922) (lacking pink tints to the basidiomes) is doubtful. Sensu NCL is, at least in part, *Lepiota subincarnata.*

scocholmica Park.-Rhodes, Deconica, *New Phytol.* 49: 341 (1950)
A *nomen dubium* lacking type material. Supposedly a 'hybrid' between *Psilocybe bullacea* and *P. coprophila.* Described from Wales (Pembrokeshire: Skokholm Island).

scorpioides (Fr.) Sacc., Naucoria, *Syll. fung.* 5: 848 (1887)
 Agaricus scorpioides Fr., *Epicr. syst. mycol.*: 199 (1838)
A *nomen dubium.* There is a single record by Plowright, unsubstantiated with voucher material. Sensu Lange (1939) is *N. bohemica.*

scorteus Fr., Marasmius, *Epicr. syst. mycol.*: 376 (1838)
A *nomen dubium.* Described as resembling *M. oreades.* Cooke 1073 (1119) B Vol. 7 (1890) shows a white, mycenoid agaric, with no resemblance to the original description.

scutellaris (Berk. & M.A. Curtis) Gilb., Athelia, *Fung. Dec. Ponderosa Pine*: 42 (1974)
 Corticium scutellare Berk. & M.A. Curtis, *Grevillea* 2: 4 (1873)
A *nomen dubium,* originally described from Venezuela. Listed by Massee (1892: 121) and by Rea (1922) as *Corticium scutellare* and reported from decayed stumps of *Ulex* sp. in Scotland [*Grevillea* 8: 7 (1879)] but all records are unsubstantiated with voucher material.

scutulatus (Fr.) Fr., Cortinarius, *Epicr. syst. mycol.*: 294 (1838)
 Agaricus scutulatus Fr., *Syst. mycol.* 1: 211 (1821)
A *nomen dubium.* Listed by Rea (1922) but no records have been traced. Sensu Cooke (HBF2) and sensu auct. mult. is *C. ionophyllus.*

scyphiformis (Fr.) Quél., Omphalina, *Fl. mycol. France*: 202 (1888)
 Agaricus scyphiformis Fr., *Observ. mycol.* 2: 221 (1818)
 Omphalia scyphiformis (Fr.) Gillet, *Hyménomycètes*: 295 (1876)

A *nomen dubium.* Listed in NCL but lacking any modern interpretation.

sebaceus Fr., Cortinarius, *Epicr. syst. mycol.*: 258 (1838)
A *nomen dubium.* Records from England (South Hampshire: New Forest) and the Republic of Ireland (Offaly) may represent *C. turmalis* (not otherwise British) but voucher material has not been traced.

semibulbosus (Lasch) Gillet, Pluteus, *Hyménomycètes*: 395 (1876)
 Agaricus semibulbosus Lasch, in Fries, *Epicr. syst. mycol.*: 140 (1838)
A *nomen dubium* lacking type material. Sensu NCL and BFF4 is *P. inquilinus.*

semicircularis With., Boletus, *Bot. arr. veg.* 2: 768 (1776)
Described from Britain but a *nomen dubium.*

semicrema Fr., Russula, *Epicr. syst. mycol.*: 350 (1838)
A *nomen dubium.* See Cooke 974 (1067) Vol. 7 for an illustration.

semiflexa (Berk. & Broome) Sacc., Naucoria, *Syll. fung.* 5: 833 (1887)
 Agaricus semiflexus Berk. & Broome, *Ann. Mag. Nat. Hist., Ser.4 [Notices of British Fungi no.1246]* 6: 467 (1870)
Described from Britain but a *nomen dubium.* Possibly *Psilocybe crobula fide* Pearson & Dennis (1948). Cooke 496 (509) Vol. 4 (1884) is unknown.

semilunatus With., Agaricus, *Arr. Brit. pl. ed. 3,* 4: 217 (1796)
Described from Britain but a *nomen dubium.*

semistriata (Peck) Guzmán, Psilocybe, *Nova Hedwigia Beih.* 74: 193 (1983)
 Deconica semistriata Peck, *Rep. (Annual) New York State Mus. Nat. Hist.* 51: 291 (1898)
Not authentically British. Listed in BFF5 as an earlier name for *Psilocybe chionophila* but only reliably known from North America. Two records were cited, one from Somerset determined with doubt by Gúzman, the other here listed under *P. chionophila.*

semota (Fr.) Ricken, Psalliota, *Blätterpilze Deutschl.*: 238 (1912)
 Agaricus semotus Fr., *Monogr. hymenomyc. Suec.* 2: 347 (1863)
A *nomen dubium,* possibly not even an *Agaricus* sp. Sensu NCL and auct. mult. is *A. dulcidulus.*

senilis (Fr.) Gillet, Clitocybe, *Hyménomycètes*: 143 (1874)
 Agaricus senilis Fr., *Hymenomyc. eur.*: 98 (1874)
A *nomen dubium.* Sensu Rea (1922) is *Rhodocybe popinalis.*

seperina Dupain, Russula, *Bull. Soc. Mycol. France* 29: 181 (1913)
Not authentically British. Listed with doubts in Rayner (1985) based on a collection from West Sussex in 1967.

sepia J.E. Lange, Mycena, *Fl. agaric. danic. 5, Taxonomic Conspectus*: V (1940)
A *nomen dubium.* Most of the collections named thus in herb. K are misidentified *Mycena filopes* or *M. arcangeliana.*

septicus (Fr.) P. Kumm., Pleurotus, *Führ. Pilzk.*: 104 (1871)
 Agaricus septicus Fr., *Syst. mycol.* 1: 192 (1821)
 Pleurotellus septicus (Fr.) Konrad & Maubl., *Icon. select. fung.* 6: 360 (1937)
A *nomen dubium.* Cooke 288 (259) Vol. 2 (1883) is unidentifiable.

serarius Fr., Cortinarius, *Epicr. syst. mycol.*: 269 (1838)
Not authentically British. Included in NCL but the few British records are all old, doubtful and unsubstantiated with voucher material.

serialis (Fr.) Donk, Phlebia, *Fungus* 27: 12 (1957)
 Thelephora serialis Fr., *Syst. mycol.* 1: 445 (1821)
 Corticium seriale (Fr.) Fr., *Epicr. syst. mycol.*: 563 (1838)

Not authentically British. Reported from East Suffolk (Flatford Mill) in 1956, but unsubstantiated with voucher material.

serosus With., Agaricus, *Bot. arr. Brit. pl. ed. 2*, 3: 294 (1792)
Described from England but a *nomen dubium*.

serotina Quél., Russula, *Bull. Soc. Bot. France* 25: 289 (1878)
A *nomen dubium* lacking type material. Cooke 1003 (1042) Vol. 7 (1888) is *Russula pelargonia* or near that species.

serotinus Einhell., Ripartites, *Ber. Bayer. Bot. Ges.* 44: 41 (1973)
Not authentically British. Reported from Republic of Ireland but unsubstantiated with voucher material.

sessilis Bull., Agaricus, *Herb. France* pl. 152 (1783)
A *nomen dubium* lacking type material. Reported by Withering (1792). Bulliard's illustration is clearly a species of *Crepidotus*.

setipes (Fr.) Raithelh., Rickenella, *Metrodiana* 4: 67 (1973)
Agaricus setipes Fr., *Observ. mycol.* 2: 162 (1818)
Omphalia setipes (Fr.) Gillet, *Hyménomycètes*: 300 (1876)
Omphalina setipes (Fr.) Quél., *Enchir. fung.*: 45 (1886)
A *nomen dubium*. The name has been widely misapplied to *Rickenella swartzii* and British records all refer to that species.

siccus (Schwein.) Fr., Marasmius, *Epicr. syst. mycol.*: 382 (1838)
Agaricus siccus Schwein., *Schriften Naturf. Ges. Leipzig* 1: 84 (1822)
Not authentically British. A single record, from West Sussex (Climping) [see TBMS 52 (2): 325 (1969)] but *fide* A&N1 this deviates in many respects from typical *Marasmius siccus* and possibly represents *M. anomalus*.

silaceum (Pers.) Sacc., Hypholoma, *Syll. fung.* 5: 1027 (1887)
Agaricus silaceus Pers., *Syn. meth. fung.*: 421 (1801)
A *nomen dubium* lacking type material. Listed as British by Cooke (HBF2), citing a record from Scotland (Angus: Glamis) but no material has been traced.

silvana (Sacc.) O.K. Mill., Hohenbuehelia, *Famiglia Tricholom., Atti Conv. Internaz. Settembre 1984, Italy*, (Ed. Borghi, E.): 131 (1986) [1984]
Agaricus silvanus Sacc., *Michelia* 1: 1 (1877)
Pleurotus silvanus (Sacc.) Sacc., *Syll. fung.* 5: 379 (1887)
Resupinatus silvanus (Sacc.) Singer, *Lilloa* 22: 253 (1951)
A *nomen dubium*. Sensu BFF6 is near to *Hohenbuehelia unguicularis* but considered a synonym of *H. cyphelliformis* in FAN3.

simillima P. Karst., Mycena, *Hedwigia* 30: 246 (1891)
A *nomen dubium*. See TBMS 4 (2): 309 (1913) where described as 'like *M. galericulata* but fragile and becoming pale'. No voucher material has been traced.

simillima Peck, Russula, *Rep. (Annual) New York State Mus. Nat. Hist.* 24: 75 (1972)
Not authentically British. A North American species. Sensu Lange (FlDan5) and sensu auct. Brit. is *R. farinipes*.

sinopica Romagn., Lepiota, *Bull. Soc. Naturalistes Oyonnax* 10 - 11 (Suppl.): 4 (1957)
A *nomen dubium* known only from the type collection which is in poor condition. No published illustrations are known. There is a single British collection dubiously named thus in herb. K (Buckinghamshire: Burnham Beeches) but also in poor condition.

sinuosa (Sowerby) Gray, Daedalea, *Nat. arr. Brit. pl.* 1: 638 (1821)
Boletus sinuosus Sowerby, *Col. fig. Engl. fung.* 2: pl. 194 (1799)
Described from Britain but a *nomen dubium* lacking type material. Persoon (1801) and Fries (1821) refer this to *Daedalea gibbosa* (= *Trametes gibbosa*) but Sowerby's original illustration is not that species.

sinuosum (Fr.) Quél., Hebeloma, *Mém. Soc. Émul. Montbéliard, Sér. 2*, 5: 345 (1873)

Agaricus sinuosus Fr., *Epicr. syst. mycol.*: 178 (1838)
A *nomen dubium*, and from the original description (with white lamellae when mature) possibly not even a species of *Hebeloma*. Sensu NCL is *H. senescens*.

siparius (Fr.) Watling, Flammulaster, *Notes Roy. Bot. Gard. Edinburgh* 28(1): 67 (1967)
Agaricus siparius Fr., *Syst. mycol.* 1: 261 (1821)
Naucoria siparia (Fr.) Gillet, *Hyménomycètes*: 542 (1876)
Phaeomarasmius siparius (Fr.) Singer, *Lilloa* 22: 577 (1951)
Flocculina siparia (Fr.) P.D. Orton, *Trans. Brit. Mycol. Soc.* 43(2): 236 (1960)
A *nomen dubium*. Cooke 513 (480) B Vol. 4 (1884) is unidentifiable. Sensu auct. Brit. is probably mostly *Flammulaster granulosus*.

sistrata (Fr.) Bon & Bellù, Cystolepiota, *Doc. mycol.* 15(59): 51 (1985)
Agaricus sistratus Fr., *Syst. mycol.* 1: 24 (1821)
Lepiota sistrata (Fr.) Sacc., *Syll. fung.* 5: 50 (1887)
A *nomen dubium*. Sensu NCL and auct. mult. is *Cystolepiota seminuda*.

smaragdina Quél., Russula, *Compt. Rend. Assoc. Franç. Avancem. Sci.* 14: 449 (1886)
Not authentically British. A southern European species. Sensu auct. Brit. is *Russula innocua*.

smithii Massee, Clitopilus, *Brit. fung.-fl.* 2: 248 (1893)
Eccilia smithii (Massee) W.G. Sm., *Syn. Brit. Bas.*: 118 (1908)
Described from Britain but a *nomen dubium*. Massee cites Cooke 350 (599) Vol. 3 (1885). Evidently an *Entoloma* sp., but not discussed by Noordeloos (1987).

sobria (Fr.) P. Kumm., Naucoria, *Führ. Pilzk.*: 77 (1871)
Agaricus furfuraceus γ *sobrius* Fr., *Observ. mycol.* 2: 25 (1818)
Agaricus sobrius (Fr.) Fr., *Hymenomyc. eur.*: 263 (1874)
A *nomen dubium*. Cooke 512 (511) Vol. 4 (1885) shows a small brown agaric, apparently *Tubaria furfuracea*.

socialis (Fr.) Gillet, Clitocybe, *Hyménomycètes*: 159 (1874)
Agaricus socialis Fr., *Hymenomyc. eur.*: 83 (1874)
A *nomen dubium*. Cooke 132 (134) Vol. 1 (1883) is possibly a species of *Calocybe*.

sociatus (Schumach.) Fr., Coprinus, *Epicr. syst. mycol.*: 252 (1838)
Agaricus sociatus Schumach., *Enum. pl.* 2: 353 (1803)
A *nomen dubium* unsubstantiated with voucher material but listed by Rea (1922).

solitaria (Bull.) Fr., Amanita, *Sverig Atl. Svamp.*: 33 (1836)
Agaricus solitarius Bull., *Herb. France*: pl. 593 (1793)
A *nomen dubium* lacking type material and variously interpreted. Widely used for *Amanita strobiliformis* but recently considered by some authorities to be the correct name for *A. echinocephala*.

sordidoflavus With., Agaricus, *Arr. Brit. pl. ed. 3*, 4: 197 (1796)
Described from Britain but a *nomen dubium*. Possibly *Russula foetens* or a similar species.

sororia P. Karst., Pholiota, *Hedwigia* 31: 297 (1892)
A *nomen dubium* inadequately described (Holec 2001). Listed by Rea (1922).

spadicea var. polycephala (Paulet) P. Karst., Psilocybe, *Bidrag Kännedom Finlands Natur Folk* 32: 506 (1879)
Hypophyllum polycephalum Paulet, *Traité champ.*: pl. 111 (1793)
Agaricus spadiceus var. polycephalus (Paulet) Fr., *Epicr. syst. mycol.*: 226 (1838)
A *nomen dubium*. Listed as doubtful in NCL, a verdict endorsed in Kits1: 164.

spadiceum (Schaeff.) Pers., Scleroderma, *Syn. meth. fung.*: 155 (1801)

Lycoperdon spadiceum Schaeff., *Fung. Bavar. Palat. nasc.* 2: 188 (1763)
    Scleroderma vulgare var. *spadiceum* (Schaeff.) W.G. Sm., *Syn. Brit. Bas.*: 480 (1908)
A *nomen dubium* lacking type material. Sensu Rea (1922) is *Scleroderma cepa*.

spartea (Fr.) P. Kumm., Galera, *Führ. Pilzk.*: 75 (1871)
    Agaricus sparteus Fr., *Syst. mycol.* 1: 266 (1821)
A *nomen dubium*. Sensu auct. Brit. is *Conocybe rickeniana*.

spathulata (Schwein.) Rea, Odontia, *Brit. basidiomyc.*: 648 (1922)
    Sistotrema spathulatum Schwein., *Schriften Naturf. Ges. Leipzig* 1: 104 (1822)
    Hydnum spathulatum (Schwein.) Fr. [non *H. spathulatum* Schrad. (1794)], *Elench. fung.* 1: 139 (1828)
A *nomen dubium*. Listed by Rea (1922), with spines 'brick-red or orange'.

speciosus Peck, Hygrophorus, *Rep. (Annual) New York State Mus. Nat. Hist.* 29: 43 (1878)
    Hygrophorus bresadolae Quél., in Bresadola, *Fungi trident.* 1: 11 (1881)
Not authentically British. Strictly associated with *Larix* spp. British records are unsubstantiated with voucher material and are all said to be with *Pinus* spp or unspecified hosts, and may refer to colour forms of *Hygrophorus hypothejus*.

speculum (Fr.) Quél., Entoloma, *Mém. Soc. Émul. Montbéliard, Sér. 2,* 5: 119 (1872)
    Agaricus speculum Fr., *Spicilegium Pl. neglect.*: 4 (1836)
Not authentically British. Sensu Cooke 342 (308) Vol. 3 (1884) is *Entoloma niphoides*. Sensu Lange (FlDan2: pl. 75B) appears to be a slender form of *E. rhodopolium*.

spegazzinii P. Karst., Coprinus, *Bidrag Kännedom Finlands Natur Folk* 32: 550 (1879)
Not authentically British. An old British record from a plant pot was referred here but is unsusbstantiated with voucher material.

spermaticum (Fr.) Gillet, Tricholoma, *Hyménomycètes*: 94 (1874)
    Agaricus spermaticus Fr., *Epicr. syst. mycol.*: 27 (1838)
A *nomen dubium*. Cooke 65 (87) Vol. 1(1882) is not helpful. Last reported from England (Shropshire) in 1937, but unsubstantiated with voucher material. Sensu auct. is *Tricholoma umbonatum*, unknown in Britain.

sphaeroidea Remsberg, Typhula, *Mycologia* 32: 74 (1940)
Excluded pending further investigation. Described from North America on decayed leaves of *Rubus fruticosus*. Reported from Northern Ireland (Fermanagh) but said to be with *Picea abies*.

sphaerospora (Maire) Parmasto, Trechispora, *Conspectus Systematis Corticiacearum*: 46 (1968)
    Hypochnus sphaerosporus Maire, *Bull. Soc. Mycol. France* 21: 164 (1905)
    Corticium sphaerosporum (Maire) Höhn. & Litsch., *Sitzungsber. Kaiserl. Akad. Wiss., Wien, Math.-Naturwiss. Cl., Abt. 1* 117: 1105 (1908)
    Cristella sphaerospora (Maire) Donk, *Fungus* 27: 20 (1957)
Not authentically British. Collections named thus are usually *Trechispora stellulata*.

sphagnicola (G.F. Atk.) A.H. Sm. & Singer, Galerina, *Monog. Galerina*: 62 (1964)
    Galerula sphagnicola G.F. Atk., *Proc. Amer. Philos. Soc.* 57: 362 (1918)
Not authentically British. Reported from West Sussex but said to be growing 'on the ground' and unsubstantiated with voucher material.

sphaleromorpha (Bull.) Fayod, Agrocybe, *Ann. Sci. Nat., Bot., sér. 7,* 9: 358 (1889)
    Agaricus sphaleromorphus Bull., *Herb. France*: pl. 540 (1792)
    Agaricus praecox β *sphaleromorpha* (Bull.) Fr., *Syst. mycol.* 1: 238 (1821)

Pholiota sphaleromorpha (Bull.) Quél., *Mém. Soc. Émul. Montbéliard, Sér. 2,* 5: 124 (1872)
    Togaria sphaleromorpha (Bull.) W.G. Sm., *Syn. Brit. Bas.*: 124 (1908)
A *nomen dubium* lacking type material. Sensu Bon (1987) and BFF3 is doubtfully distinct from *Agrocybe paludosa*.

spicula (Lasch) Kühner, Conocybe, *Encycl. Mycol. 7. Le Genre Galera*: 60 (1935)
    Agaricus spiculus Lasch, *Linnaea* 4: 546 (1829)
    Galera spicula (Lasch) P. Kumm., *Führ. Pilzk.*: 75 (1871)
A *nomen dubium* lacking type material, but widely used, e.g. in Rea (1922), for *Conocybe rickeniana*.

spiculoides Kühner & Watling, Conocybe, *Notes Roy. Bot. Gard. Edinburgh* 38: 339 (1980)
    Conocybe spicula var. *spiculoides* Kühner [*nom. nud.*], *Encycl. Mycol. 7. Le Genre Galera*: 61 (1935)
Not authentically British. The single collection named thus in herb. K has been redetermined as *Conocybe rickeniana*.

spilomaeolus Fr., Paxillus, *Epicr. syst. mycol.*: 317 (1838)
    Paxillus panaeolus var. *spilomaeolus* (Fr.) Sacc., *Syll. fung.* 5: 985 (1887)
A *nomen dubium*. Reported in Grevillea 6: 102 (1877) 'on fir leaves at Stoke Poges' (Buckinghamshire) but no material survives. Probably a form of *Ripartites tricholoma*.

spilopus (Berk. & Broome) Sacc., Pluteus, *Syll. fung.* 5: 669 (1887)
    Agaricus spilopus Berk. & Broome, *Ann. Mag. Nat. Hist., Ser. 5 [Notices of British Fungi no. 1856]* 7: 126 (1881)
Described from Sri Lanka and reported (probably mistakenly) from Britain by Berkeley & Broome. No authentic British records have been traced. Sensu auct. Brit. is *P. podospileus*.

spinipes Sowerby, Agaricus, *Col. fig. Engl. fung.* 2: pl. 206 (1799)
Described from Britain but a *nomen dubium* lacking type material. Sowerby's illustration shows a mycenoid agaric on a cone of *Pinus sylvestris*, apparently not *Baeospora myosura* but possibly *Strobilurus tenacellus*.

spintrigera (Fr.) Konrad & Maubl., Psathyrella, *Encycl. Mycol. 14, Les Agaricales* I: 128 (1948)
    Agaricus spintriger Fr., *Epicr. syst. mycol.*: 217 (1838)
    Stropharia spintrigera (Fr.) Sacc., *Syll. fung.* 5: 1025 (1887)
    Drosophila spintrigera (Fr.) Kühner & Romagn., *Flore Analytique des Champignons Supérieurs*: 368 (1953)
Not authentically British. British collections named thus in herb. K are misdetermined *Psathyrella solitaria*.

spinulosa (Stev. & W.G. Sm.) Sacc., Clitocybe, *Syll. fung.* 5: 171 (1887)
    Agaricus spinulosus Stev. & W.G. Sm., *British Fungi*: 84 (1886)
Described from Britain but a *nomen dubium*.

spinulosa (Berk. & M.A. Curtis) D.A. Reid, Heterochaete, *Trans. Brit. Mycol. Soc.* 55(3): 434 (1970)
    Radulum spinulosum Berk. & M.A. Curtis, *Grevillea* 1: 146 (1873)
    Eichleriella spinulosa (Berk. & M.A. Curtis) Burt, *Ann. Missouri Bot. Gard.* 2: 747 (1915)
A North American species, formerly considered an earlier name for *Eichleriella deglubens* but, following Reid4, now known to be distinct and not authentically European. Sensu auct. Eur. is *E. deglubens*.

spinulosa (Pers.) Quél., Ramaria, *Fl. mycol. France*: 468 (1888)
    Clavaria spinulosa Pers., *Observ. mycol.* 2: 59 (1799)
A *nomen dubium* lacking type material. Sensu Rea (1922) is a form of *Ramaria stricta* growing on soil or sawdust.

splachnoides (Hornem.) Fr., Marasmius, *Epicr. syst. mycol.*: 384 (1838)
    Agaricus splachnoides Hornem., *Fl. Danica*: t. 1678 (1819)
    Androsaceus splachnoides (Hornem.) Rea, *Brit. basidiomyc.*: 533 (1922)

A *nomen dubium*, usually interpreted as *M. quercophilus*. Sensu Rea (1922) appears also to be *M. quercophilus*.

splendens With., Agaricus [*nom. illegit.*, non *A splendens* Pers. (1801)], *Syst. arr. Brit. pl. ed. 4,* 4: 255 (1812)
Described from Britain but a *nomen dubium*. Fide Withering, a distinctive species, initially with white lamellae, these later becoming 'pinky-white', and a pileus 'the colour of tarnished copper, with a metallic lustre'.

splendidissima Kotl. & Pouzar, Haasiella, *Česká Mykol.* 20: 136 (1966)
Not authentically British. Reported from Staffordshire (Lady Hill Chase) in 1974, but unsubstantiated with voucher material.

spoliatum (Fr.) Gillet, Hebeloma, *Hyménomycètes*: 526 (1876)
Agaricus spoliatus Fr., *Epicr. syst. mycol.*: 182 (1838)
A *nomen dubium*. Sensu NCL and sensu auct. mult. is *Hebeloma birrus*.

spongiosus With., Agaricus, *Arr. Brit. pl. ed. 3,* 4: 200 (1796)
Described from Britain but a *nomen dubium*.

spraguei Berk. & M.A. Curtis, Coprinus, *Ann. Mag. Nat. Hist., Ser. 3*: 292 (1859)
Described from North America but a doubtful species. British records are all old and this is not otherwise known in Europe. See Cooke 671(683) Vol. 5 (1886) for an interpretation.

squalens (Fr.) Sacc., Psilocybe, *Syll. fung.* 5: 1054 (1887)
Agaricus squalens Fr., *Epicr. syst. mycol.*: 226 (1838)
A *nomen dubium*. Listed by Rea (1922) on rotten trunks. Apparently a *Psathyrella* sp., but not discussed in Kits1.

squalidus (Krombh.) Fr., Lactarius, *Hymenomyc. eur.*: 428 (1874)
Agaricus squalidus Krombh., *Naturgetr. Abbild. Schwämme* 6: 8 (1884)
A *nomen dubium* lacking type material. Cooke 938 (1004) Vol. 7 (1888), and 1196 (1195) Vol. 8 (1891) illustrate dissimilar taxa under this name. Both are unidentifiable and no material survives.

squalina (Fr.) M.P. Christ., Mycoacia, *Dansk. Bot. Ark.* 19(2): 177 (1960)
Hydnum squalinum Fr., *Syst. mycol.* 1: 420 (1821)
Not authentically British. A synonym of *Dentipellis fragilis* which is not known in Britain. The only collection in herb. K has been redetermined as *Mycoaciella bispora*.

squamosa (P. Karst.) A.H. Sm., Psathyrella, *Mem. N. Y. bot. Gdn* 24: 220 (1972)
Psathyra pennata var. *squamosa* P. Karst., *Meddeland. Soc. Fauna Fl. Fenn.* 5: 32 (1879)
Psathyra squamosa (P. Karst.) P. Karst, *Hedwigia* 32: 59 (1893)
A *nomen dubium*. Sensu NCL and sensu auct. mult. is *Psathyrella artemisiae*. The type has recently been found to be distinct from this and unknown in Britain.

squamosus Morgan, Coprinus, *J. Cincinnati Soc. Nat. Hist.* 6(3): 173 (1885)
A North American species, not authentically British. Supposedly collected from an elm stump in Worcestershire (Hanbury Park) by Rea in 1900 [see TBMS 1(4): 158 (1901)], but unsubstantiated with voucher material.

squamulosus Rea [*nom. illegit.*, non *H. squamulosus* Ellis & Everh. (1893)], Hygrophorus, *Proc. R. Ir. Acad., Clare Island Survey part. XIII* 13: 26 (1912)
A *nomen dubium*. See TBMS (1912) 4 (1): 187 for a description of material collected in the Republic of Ireland (Mayo: Old Deer Park, Mount Browne). Pearson & Dennis (1948) suggest 'possibly a *Tricholoma* near *T. sejunctum*'.

squarrosa *var.* **reflexa** (Pers.) Sacc., Pholiota, *Syll. fung.* 5: 749 (1887)
Agaricus reflexus Pers., *Syn. meth. fung.*: 311 (1801)

Gymnopus reflexus (Pers.) Gray, *Nat. arr. Brit. pl.* 1: 604 (1821)
Agaricus squarrosus var. *reflexus* (Pers.) Fr., *Syst. mycol.* 1: 243 (1821)
A *nomen dubium* lacking type material. Reported by Gray (1821) (as *Gymnopus reflexus*) and listed by Rea (1922), but from the descriptions this is probably just an aberrant form of *Pholiota squarrosa*.

stabularis (Fr.) P. Karst., Coniophora, *Bidrag Kännedom Finlands Natur Folk* 37: 159 (1882)
Thelephora stabularis Fr., *Syst. mycol.* 1: 435 (1821)
Corticium stabulare (Fr.) Fr., *Hymenomyc. eur.*: 658 (1874)
A *nomen dubium* inadequately described *fide* Ginns [Opera Bot. 61: 24 (1982)]. Listed by Rea (1922) but no material traced.

stannea (Fr.) Quél., Mycena, *Mém. Soc. Émul. Montbéliard, Sér. 2,* 5: 242 (1872)
Agaricus stanneus Fr., *Epicr. syst. mycol.*: 111 (1838)
A *nomen dubium*. Cooke 236 (188) Vol. 2 (1882) is unidentifiable.

stellaris Quél., Coprinus, *Bull. Soc. Bot. France* 24: 322 (1872)
A *nomen dubium* lacking type material. Listed by Rea (1922).

stellata (Fr.) P. Kumm., Omphalia, *Führ. Pilzk.*: 106 (1871)
Agaricus stellatus Fr., *Syst. mycol.* 1: 163 (1821)
A *nomen dubium*. Cooke 262 (241) Vol. 2 (1883) appears to be a *Hemimycena* sp. or possibly *Delicatula integrella*.

stellatum (Bull.) Gray, Geastrum, *Nat. arr. Brit. pl.* 1: 585 (1821)
Lycoperdon stellatum Bull., *Herb. France*: pl. 471 (1786)
A *nomen dubium* lacking type material. Fries (1829: 15) cites *G. stellatum* sensu Sowerby (Sow3: pl. 312) as an illustration of *Geastrum coronatum*.

stemmatus Fr., Cortinarius, *Epicr. syst. mycol.*: 309 (1838)
Not authentically British. A dubious report in 1978, and prior to that 1930, both records unsubstantiated with voucher material.

stephensii Berk. & Broome, Marasmius, *Ann. Mag. Nat. Hist., Ser. 2 [Notices of British Fungi no.708]* 13: 403 (1854)
A *nomen dubium*. Fide A&N2 this possibly represents *Marasmius collinus*. Listed by Rea (1922) as a synonym of *M. terginus* (= *Collybia tergina*).

stereoides (Fr.) Ryvarden, Datronia, *Fl. Ov. Kjuk*: 42 (1968)
Polyporus stereoides Fr., *Hymenomyc. eur.*: 569 (1874)
Polystictus stereoides (Fr.) Cooke, *Grevillea* 14: 84 (1886)
Not authentically British. Listed by Rea (1922) as *Polystictus stereoides* and by Ellis & Ellis (1990), but no records have been traced.

stevensonii (Berk. & Broome) Sacc., Collybia, *Syll. fung.* 5: 226 (1887)
Agaricus stevensonii Berk. & Broome, *Ann. Mag. Nat. Hist., Ser.4 [Notices of British Fungi no.1497]* 15: 41 (1875)
A *nomen dubium*. Fide A&N2 it may represent a species near to *Collybia dryophila* but with a viscid pileus. Dennis (1948: 198) states that it could be abnormal material of a *Strobilurus* sp. and that the name should be discarded. Cooke 199 (145) Vol. 2 (1882) appears to support this view.

stilbocephalus (Berk. & Broome) Sacc., Clitopilus, *Syll. fung.* 5: 705 (1887)
Agaricus stilbocephalus Berk. & Broome, *Ann. Mag. Nat. Hist., Ser. 5 [Notices of British Fungi no. 1758]* 3: 205 (1879)
A *nomen dubium*. Cooke 349 (324) Vol. 3 (1885) (as *Agaricus stilbocephalus*) is unidentifiable.

stipitata Fuckel, Solenia, *Jahrb. Nassauischen Vereins Naturk.* 25-26: 290 (1871)
A *nomen dubium*. Old collections so named in herb. K are probably *Merismodes anomalus*.

stipitis With., Agaricus, *Bot. Arr. Brit. pl. ed. 2*, 3: 314 (1792)
Described from Britain, but a *nomen dubium* lacking type material. Apparently a species of *Armillaria*. Sensu Sowerby (Sow1: pl. 101) is possibly *A. gallica*.

stolonifera (Jungh.) P. Kumm., Collybia, *Führ. Pilzk.*: 115 (1871)
    Agaricus stoloniferus Jungh., *Linnaea* 5: 396 (1830)
    Collybia tenacella var. *stolonifera* (Jungh.) Rea, *Brit. basidiomyc.*: 337 (1922)
A *nomen dubium*. A species of *Strobilurus* fide A&N2. Rea (1922) equates it with *Marasmius conigenus*, itself a *nomen dubium* but probably also a species of *Strobilurus*.

straminipes (Massee) Sacc., Clitopilus, *Syll. fung.* 9: 86 (1891)
    Agaricus straminipes Massee, *Grevillea* 16: 43 (1887)
    Eccilia straminipes (Massee) A. Pearson & Dennis, *Trans. Brit. Mycol. Soc.* 31: 171 (1948)
A *nomen dubium*. The description and Cooke's plate of the type [1159 (960) Vol. 8 (1890)] suggest a form of *Entoloma sericellum* fide Noordeloos (1987: 394).

strangulata (Fr.) Quél., Amanita, *Mém. Soc. Émul. Montbéliard, Sér. 2*, 5: 27 (1872)
    Agaricus solitarius var. *strangulatus* Fr., *Epicr. syst. mycol.*: 6 (1838)
    Amanitopsis strangulata (Fr.) Roze, in Karsten, *Bidrag Kännedom Finlands Natur Folk* 32: 7 (1879)
A *nomen dubium*, but widely used for *Amanita ceciliae*. Cooke 17 (13) Vol. 1 (1881) (as *Agaricus strangulatus*) is possibly too brown to be *A. ceciliae*.

striata Pers., Clavaria, *Icon. descr. fung.*: pl. 3 f. 5 (1798)
A *nomen dubium* lacking type material. Cotton & Wakefield (1919) suggest this is discoloured *C. vermicularis* whilst Bourdot & Galzin (1928) suggest a form of *C. fumosa*. Sensu Rea (1922) is unknown.

striatula (Peck) Peck, Laccaria, *Bull. New York State Mus. Nat. Hist.* 157: 93 (1912)
    Clitocybe striatula Peck, *Rep. (Annual) New York State Mus. Nat. Hist.* 48: 274 (1897)
A North American species, not authentically European. Sensu NCL is *L. pumila*.

striatum (Schaeff.) Sacc., Tricholoma, *Fl. ital. crypt., Hymeniales* 1.1: 118 (1915)
    Agaricus striatus Schaeff., *Fung. Bavar. Palat. nasc.* 1: 38 (1762)
A *nomen dubium* lacking type material, but accepted as a good species by some authorities. Sensu auct. mult. is *Tricholoma albobrunneum*.

stricta Fr., Calocera, *Epicr. syst. mycol.*: 581 (1838)
A *nomen dubium* lacking type material *fide* Reid (1974). Considered a synonym of *C. viscosa* by McNabb [New Zealand J. Bot. 3(1): 40 (1965)], but old British collections in herb. K are *C. cornea*.

striipilus J. Favre, Cortinarius, *Beitr. Kryptogamenfl. Schweiz* 10(3): 213 (1948)
Not authentically British. Reported from Scotland (Isle of Skye) but this collection (in herb. K) is determined as 'compare with' *C. striipilus*.

strobilina (Pers.) Gray, Mycena, *Nat. arr. Brit. pl.* 1: 621 (1821)
    Agaricus strobilinus Pers., *Syn. meth. fung.*: 393 (1801)
A *nomen dubium* lacking type material. Sensu Rea (1922) is *M. rosella*. Cooke 218 (131) Vol. 2 (1882) is doubtful.

suavissimus (Fr.) Singer, Panus, *Lilloa* 22: 274 (1951)
    Lentinus suavissimus Fr., *Synopsis generis Lentinorum*: 13 (1836)
Not authentically British. A single collection named thus, from Oxfordshire (Henley on Thames) in 1990, was misdetermined *Panus conchatus*.

subalutacea (Batsch) P. Kumm., Clitocybe, *Führ. Pilzk.*: 124 (1871)
    Agaricus subalutaceus Batsch, *Elench. fung. (Continuatio Secunda)*: 27 t. 33 f. 194 (1789)
A *nomen dubium* lacking type material. Sensu NCL, following Ricken (1915) with very small spores, is unknown. Sensu FlDan1 (pl. 33G) and sensu Kuyper in FAN3 is *Clitocybe frysica*.

subarquatus (M.M. Moser) M.M. Moser, Cortinarius, *Kleine Kryptogamenflora, Edn 3*: 294 (1967)
    Phlegmacium subarquatum M.M. Moser, *Die Gattung Phlegmacium*: 353 (1960)
Not authentically British. Listed in Dennis (1995) from Surrey but no records have been traced. Doubtfully distinct from *C. arquatus*.

subcaeruleus With., Agaricus, *Bot. arr. Brit. pl. ed. 2*, 3: 356 (1792)
Described from Britain but a *nomen dubium*.

subcarnaceum Fr., Hydnum, *Syst. mycol.* 1: 418 (1821)
A *nomen dubium*. Possibly a *Mycoacia* sp.

subcava (Schumach.) Sacc., Armillaria, *Syll. fung.* 5: 84 (1887)
    Agaricus subcavus Schumach., *Fl. Danica*: t. 1843 (1825)
A *nomen dubium* lacking type material. Cooke 57 (47) Vol. 1 (1882) appears to be *Leucocoprinus brebissonii*.

subcollariatum (Berk. & Broome) Sacc., Hebeloma, *Syll. fung.* 5: 798 (1887)
    Agaricus subcollariatus Berk. & Broome, *Ann. Mag. Nat. Hist., Ser. 5 [Notices of British Fungi no. 1942]* 9: 178 (1882)
Doubtfully distinct from *Hebeloma pumilum* (= *H. birrus*) *fide* NCL. See Cooke 454 (506) Vol. 3 (1886) for an illustration.

subcorticalis Pers., Rhizomorpha, *Syn. meth. fung.*: 704 (1801)
A long-defunct name based on rhizomorphs of a species of *Armillaria* (probably *A. mellea*). Nonetheless reported in 1984, 1988 and 2000 from Lancashire and Yorkshire.

subcretacea (Litsch.) M.P. Christ., Phlebia, *Dansk. Bot. Ark.* 19(2): 165 (1960)
    Corticium subcretaceum Litsch., *Oesterr. Bot. Z.* 88: 110 (1939)
Not authentically British. Misreported from Shropshire (Brown Moss Fen), the collection (in herb. K) being correctly labelled as *Phlebia subochracea*. Reported from Scotland (Shetland) but unsubstantiated with voucher material.

subelavata W.G. Sm., Collybia, *Trans. Brit. Mycol. Soc.* 2: 14 (1903)
Described from Britain but a *nomen dubium*. Smith [J. Bot. 41: 139 (1903)] comments on Massee's description (citing Smith as the authority for the name) that 'neither the description nor the name are mine' and 'no such species as *Collybia subelevata* exists'.

subferrugineus (Batsch) Fr., Cortinarius, *Epicr. syst. mycol.*: 303 (1838)
    Agaricus subferrugineus Batsch, *Elench. fung. (Continuatio Secunda)*: 7 t. 31 f. 186 (1789)
Not authentically British. Last reported in 1938, all records unsubstantiated with voucher material.

subfimbriata (Romell) Ginns, Junghuhnia, *Mycotaxon* 21: 327 (1984)
    Mis.: *Poria radula* sensu Cooke [Grevillea 14: 111 (1886)]
Not authentically British. Reported from the Republic of Ireland (Co. Dublin) in 1898, but unsubstantiated with voucher material.

subfusca (P. Karst.) Höhn. & Litsch., Tomentella, *Sitzungsber. Kaiserl. Akad. Wiss., Wien, Math.-Naturwiss. Cl., Abt. 1* 115: 1572 (1906)
    Hypochnus subfuscus P. Karst., *Bidrag Kännedom Finlands Natur Folk* 37: 163 (1882)
A *nomen dubium*. Sensu Rea (1922, as *Hypochnus subfuscus*) is *T. stuposa*.

subfuscoflavida (Rostk.) Cooke, Poria, *Grevillea* 14: 113 (1886)
> *Polyporus subfuscoflavidus* Rostk., *Deutschl. Fl.*, Edn 3: f. 27-28 (1848)

A *nomen dubium.* lacking type material. Sensu auct., possibly including Rea (1922), is *Diplomitoporus lindbladii.*

subgelatinosus Berk. & Broome, Polyporus, *Ann. Mag. Nat. Hist., Ser. 4* 17: 136 (1876)
> *Poria subgelatinosa* (Berk. & Broome) Cooke, *Grevillea* 14: 115 (1886)

Reported from Mid-Perthshire (Rannoch), apparently 'parasitic on *Polyporus (Skeletocutis) amorpha'.* [see Grevillea 5: 8 (1876)]. The type is in poor condition but is possibly *Physisporinus sanguinolentus fide* Reid & Austwick (1963).

subglobosa (Alb. & Schwein.) Sacc., Naucoria, *Syll. fung.* 5: 830 (1887)
> *Agaricus subglobosus* Alb. & Schwein., *Consp. fung. lusat.*: 169 (1805)
> *Nolanea subglobosa* (Alb. & Schwein.) Rea, *Brit. basidiomyc.*: 404 (1922)

A *nomen dubium* lacking type material. Listed twice by Rea (1922), firstly following Ricken (1915) as a species of *Naucoria* and secondly following Cooke 1160 (1170) B Vol. 8 (1890) and HBF2: 372 as a species of *Entoloma*. The Cooke plate resembles *Entoloma pleopodium* but no material survives.

subinconspicua (Litsch.) Jülich, Athelopsis, *Persoonia* 8(3): 292 (1975)
> *Corticium subinconspicuum* Litsch., *Glasnik (Bull.) Soc. Scient. Skoplje* 18: 178 (1938)

A little-known species. British *fide* Jülich (1984) but only on the basis of a misidentified collection.

subinvoluta (Batsch) Sacc., Clitocybe, *Syll. fung.* 5: 170 (1887)
> *Agaricus subinvolutus* Batsch, *Elench. fung. (Continuatio Secunda)*: 57 t. 37 f. 204 (1789)
> *Agaricus geotropus* var. *subinvolutus* (Batsch) Cooke, *Handb. Brit. fung.*, Edn 1: 41 (1871)

A *nomen dubium.* Listed by Rea (1922). From the descriptions and illustrations British records are probably *Clitocybe geotropa.*

subinvolutus W. Saunders & W.G. Sm., Agaricus [*nom. illegit.*, non *A. subinvolutus* Batsch (1789)], *Mycological Illustrations*: t. 36 (1870)

Described from Britain, but a *nomen dubium*. Doubtfully distinct from *Leucopaxillus giganteus fide* NCL.

sublanatus (Sowerby) Fr., Cortinarius, *Epicr. syst. mycol.*: 283 (1838)
> *Agaricus sublanatus* Sowerby, *Col. fig. Engl. fung.* 2: pl. 224 (1799)

Described from Britain but a *nomen dubium* lacking type material. Sensu auct. is *C. cotoneus* but British collections are mostly *C. pholideus fide* Pearson & Dennis (1948).

sublutea (Vahl) Sacc., Pholiota, *Syll. fung.* 5: 757 (1887)
> *Agaricus subluteus* Vahl, *Fl. Danica*: t. 1192 (1797)

A *nomen dubium.* Listed by Rea (1922). *Fide* Holec (2001) this is possibly a species of *Psathyrella.*

submarasmioides (Speg.) Sacc., Lepiota, *Syll. fung.* 5: 68 (1887)
> *Agaricus submarasmioides* Speg., *Ann. Hist.-Nat. Mus. Natl. Hung.* 6: 23 (1899) [1888]

Not authentically British. Described from Argentina. Sensu Rea (1922) is doubtful.

submelinoides (Kühner) Maire, Naucoria, *Bull. Soc. Mycol. France* 46: 225 (1930)
> *Alnicola submelinoides* Kühner, *Botaniste* 17: 175 (1926)

Not authentically British. Two collections from Norfolk in 1981and 1984 require confirmation. Also reported from the Nordic Mycological Congress in Perthshire (Kindrogan) in 1983, but unsubstantiated with voucher material.

subminiata Murrill [as *Hydrocybe*], Hygrocybe, *Mycologia* 3: 198 (1911)

Not authentically British. A North American species with two doubtful collections in poor condition in herb. K.

subminutula (Murrill) Pegler, Hygrocybe, *Kew Bull. Addit. Ser.* 9: 61 (1983)
> *Hydrocybe subminutula* Murrill, *Bull. Torrey Bot. Club* 67: 233 (1940)
> *Hygrophorus subminutulus* (Murrill) P.D. Orton, *Trans,. Brit. Mycol. Soc.* 43: 268 (1960)

Sensu NCL + auct. Eur. is *H. insipida* for which it would supply an earlier name if it could be established that they are truly synonymous.

submutabilis (Höhn. & Litsch.) Donk, Cristella, *Fungus* 27: 20 (1957)
> *Corticium submutabile* Höhn. & Litsch., *Sitzungsber. Kaiserl. Akad. Wiss., Wien, Math.-Naturwiss. Cl., Abt. 1* 116: 822 (1907)
> *Hypochnus submutabilis* (Höhn. & Litsch.) Rea, *Brit. basidiomyc.*: 658 (1922)

A *nomen dubium fide* Larsson [Nordic J. Bot. 16(1): 73-82 (1996)]. Sensu Rea plus material in Herb. K is *Phlebiella fibrillosa.* Sensu auct. is *Trechispora farinacea.*

subnotatus (Pers.) Fr., Cortinarius, *Epicr. syst. mycol.*: 290 (1838)
> *Agaricus subnotatus* Pers., *Syn. meth. fung.*: 296 (1801)

A *nomen dubium* lacking type material. Reported by Eyre from North Hampshire (Alresford) in Grevillea 13: 66 (1887), but no material has been traced. Cooke 784 (832) Vol. 6 (1886) is possibly *C. venetus.*

subnuda (P. Karst.) A.H. Sm., Psathyrella, *Contr. Univ. Michigan Herb.* 5: 61 (1941)
> *Psathyra subnuda* P. Karst., *Symb. mycol. fenn.* 10: 60 (1881)

A *nomen dubium.* British records are sensu Lange (FlDan4: 98) and thus probably *Psathyrella spadiceogrisea.*

subnudipes Kühner, Inocybe, *Bull. Soc. Naturalistes Oyonnax* 6: 6 (1955)

A *nomen dubium fide* Kuyper (1986) who was unable to study the type material. British records are referred to *Inocybe albovelutipes.*

subochracea A.H. Sm., Galerina, *Mycologia* 45: 917 (1953)

A North American species, unknown in Europe. A single record from Scotland is unconfirmed.

subpulverulenta (Pers.) Singer, Melanoleuca, *Schweiz. Z. Pilzk.* 17: 56 (1939)
> *Agaricus subpulverulentus* Pers., *Mycol. eur.* 3: 221 (1828)
> *Tricholoma subpulverulentum* (Pers.) P. Karst., *Bidrag Kännedom Finlands Natur Folk* 32: 54 (1879)

A *nomen dubium* lacking type material. Cooke 124 (219) Vol. 1 (1883) is *Melanoleuca nivea fide* Boekhout in FAN3. Sensu NCL and BFF8 (1998) is doubtful but near to *M. polioleuca.* Sensu Rea (1922) was the basis for *M. reai* which is also doubtful.

subpurpurascens (Batsch) Fr., Cortinarius, *Epicr. syst. mycol.*: 265 (1838)
> *Agaricus subpurpurascens* Batsch, *Elench. fung. (Continuatio Prima)*: 71 & t. 16 f. 74 (1786)

A *nomen dubium.* Listed by Rea (1922).

subpusillum (Pilát) Romagn., Entoloma, *Bull. Soc. Mycol. France* 103: 87 (1987)
> *Eccilia subpusilla* Pilát, *Stud. Bot. Čechoslav.* 12: 36 (1951)

Not authentically British. The single collection named thus in herb. K, from Hertfordshire (Brookmans Park), has been redetermined as *E. rusticoides.*

subrimosa (P. Karst.) Massee, Inocybe, *Brit. fung.-fl.* 2: 200 (1893)
> *Clypeus subrimosus* P. Karst., *Meddeland. Soc. Fauna Fl. Fenn.* 28: 38 (1888)

A *nomen dubium*. Listed by Rea (1922) in the synonymy of *Inocybe asterospora*. Massee, however, considered it a distinct but related species.

subrubiginosa Litsch., Tomentella, *Bull. Soc. Sci. Skoplje* 20: 19 (1939)
A *nomen dubium*. Sensu auct. mult. is *T. punicea*. Sensu Wakefield (1969) is doubtful.

subrufescens (Peck) Kauffman, Psalliota, *Michigan Geol. Biol. Surv. Publ., Biol. Ser. 5*, 26: 239 (1918)
   Agaricus subrufescens Peck, *Rep. (Annual) New York State Mus. Nat. Hist.* 92: 105 (1892)
A North American species, not authentically British. British records are sensu Lange (FlDan4: pl. 136B) and thus refer to *Agaricus augustus*.

subsaponaceum P. Karst., Hebeloma, *Symb. mycol. fenn.* 13: 3 (1884)
A synonym of *Hebeloma syrjense* P. Karst., not authentically British. The two collections named thus in herb. K, from Scotland (Orkney), have been redetermined by Vesterholt as *Hebeloma birrus*. Reported from England (Northumberland and Suffolk) but lacking voucher material.

subsquamosus With., Boletus, *Arr. Brit. pl. ed. 3*, 4: 314 (1796)
Described from Britain 'in upland pasture, amongst heath and furze' but a *nomen dubium*.

subsquarrosa (Locq.) Bon, Macrolepiota, *Doc. mycol.* 11(43): 72 (1981)
   Leucocoprinus subsquarrosus Locq., *Rev. Mycol. (Paris)* 17(1): 54 (1952)
Reported from England (Cumberland, West Kent and Westmorland) but doubtfully distinct from *Macrolepiota mastoidea*.

subsquarrosa (Fr.) Sacc., Pholiota, *Syll. fung.* 5: 750 (1887)
   Agaricus subsquarrosus Fr., *Öfvers. Kongl. Vetensk.-Akad. Förh.* 18: 23 (1862)
A *nomen dubium*. Possibly a good species awaiting rediscovery and a modern description *fide* Holec (2001: 204). Last recorded from England (Derbyshire: Chatsworth) in 1919, and from Scotland (Easterness: Aviemore) in 1938, but lacking voucher material.

substrangulata *var.* **rhodophylla** (Kühner) Boertm., Hygrocybe, *Fungi of Northern Europe* 1: 108 (1995)
   Hygrocybe rhodophylla Kühner, *Bull. Soc. Mycol. France* 92(4): 462 (1977) [1976]
Not authentically British. Reported on two occasions from lowland sites in Yorkshire but unsubstantiated with voucher material. Otherwise known only from Greenland and the French Alps.

subtemulenta (Britzelm.) Sacc., Naucoria, *Syll. fung.* 1: 59 (1895)
   Agaricus subtemulentus Britzelm., *Bot. Centralbl.* 15/17: 12 (1893)
A *nomen dubium* lacking type material and inadequately described. Listed by Rea (1922).

subterfurcata Romagn., Russula, *Les Russules d'Europe et d'Afrique du Nord*: 318 (1967)
Not authentically British. Listed in Rayner (1985) on the strength of a collection from Gloucestershire in 1981 that has not survived. Now considered a rare Mediterranean species, with northern collections mostly reassigned to *R. faustiani* Sarnari, though this too is unknown in the British Isles.

subtomentosus *var.* **radicans** (Pers.) Massee, Boletus, *Brit. fung.-fl.* 1: 266 (1892)
   Boletus radicans Pers., *Syn. Meth. Fung.*: 507 (1801)
A *nomen dubium*.

subulatus W.G. Sm., Marasmius, *J. Bot.* 11: 66 (1873)
Described from Britain but a *nomen dubium*, never recollected and probably an alien. Described from London (Chelsea:

Veitch's Nursery) on dead tree fern stems. *Fide* Smith this was allied to *Marasmius rotula* and *M. androsaceus*.

subuncialis Corner, Pistillaria, *Monograph of Clavaria and Allied Genera*: 492 (1950)
Described from Cambridgeshire (Wicken Fen) and reported from Dorset in 1990 but a *nomen dubium* the type of which is apparently missing. The Dorset record lacks voucher material.

subviolascens Nezdojm., Cortinarius, *Shlyapochnye Griby SSSR Rod Cortinarius Fr.*: 173 (1983)
Not authentically British. Reported from Suffolk but lacking voucher material.

succineus M.M. Moser, Cortinarius, *Nova Hedwigia Beih.* 52: 226 (1975)
Not authentically British. Associated with *Nothofagus* spp in Argentina. Recorded from South Hampshire (New Forest) in 1991 with *Salix repens*, but lacking voucher material and dubious.

sudora (Fr.) Gillet, Mycena, *Hyménomycètes*: 273 (1876)
   Agaricus sudorus Fr., *Epicr. syst. mycol.*: 106 (1838)
A *nomen dubium*. Possibly *Mycena galericulata* var. *albida fide* Maas Geesteranus (1992). See Cooke 226 (206) Vol. 2 (1882) for an illustration.

sudum (Fr.) Quél., Tricholoma, *Mém. Soc. Émul. Montbéliard, Sér. 2*, 5: 340 (1873)
   Agaricus sudus Fr., *Epicr. syst. mycol.*: 38 (1838)
Listed by Rea (1922) but no British records have been traced.

sulphuratum Berk. & Ravenel, Stereum, *J. Linn. Soc., Bot.* 10: 331 (1868)
Not authentically British. A North American species. Sensu auct. Brit. is *Stereum 'rameale'*.

sulphurea (Schwein.) Rea, Odontia, *Brit. basidiomyc.*: 649 (1922)
   Hydnum sulphureum Schwein., *Schriften Naturf. Ges. Leipzig* 1: 104 (1822)
A *nomen dubium*. Listed by Rea (1922).

sulphureoides (Peck) Singer, Tricholomopsis, *Ann. Mycol.* 41: 69 (1943)
   Agaricus sulphureoides Peck, *Rep. (Annual) New York State Mus. Nat. Hist.* 23: 86 (1873)
   Pleurotus sulphureoides (Peck) Sacc., *Syll. fung.* 5: 345 (1887)
A North American species, not authentically British. Reported once from Britain but lacking details and voucher material.

sulphureus (Kauffman) J.E. Lange, Cortinarius [*nom. illegit.*, non *C. sulphureus* Lindgren (1845)], *Fl. agaric. danic.* 5, *Taxonomic Conspectus*: III (1940)
   Cortinarius fulmineus var. *sulphureus* Kauffman, *Michigan Geol. Biol. Surv. Publ., Biol. Ser. 5*, 26: 354 (1918)
A *nomen dubium*. Sensu Orton (1955) = *C. citrinus* in NCL.

sulphureus *var.* **albolabyrinthiporus** Rea, Polyporus, *Trans. Brit. Mycol. Soc.* 4(1): 190 (1913)
Described from Britain, but a *nomen dubium*. From the description this may be old and faded *Laetiporus sulphureus*.

sylvestris (Gillet) Konrad & Maubl., Psathyrella, *Encycl. Mycol.* 14, *Les Agaricales* I: 127 (1948)
   Hypholoma sylvestre Gillet, *Hyménomycètes*: 568 (1878)
A *nomen dubium*. Possibly *Psathyrella cotonea fide* Kits1. Sensu auct. mult. is *P. populina*.

syringinus Z. Schaef., Lactarius, *Česká Mykol.* 10: 171 (1956)
Not authentically British. The single British collection in herb. K named thus has been redetermined as *L. trivialis*.

tabacina (DC.) Konrad & Maubl., Agrocybe, *Encycl. Mycol.* 14, *Les Agaricales* I: 162 (1948)
   Agaricus tabacinus DC., *Fl. Fr.* 5: 46 (1815)
   Naucoria tabacina (DC.) Gillet, *Hyménomycètes*: 574 (1878)
A *nomen dubium* lacking a modern interpretation and included with doubts in BFF3.

tammii (Fr.) Sacc., Flammula, *Syll. fung.* 5: 810 (1887)
> *Agaricus tammii* Fr., *Öfvers. Kongl. Vetensk.-Akad. Förh.* 18: 23 (1862)
A *nomen dubium. Fide* NCL a synonym of *Phylloporus rhodoxanthus* (= *P. pelletieri*). Possibly a terrestrial species of *Gymnopilus fide* Holec (2001).

tectorum With., Agaricus, *Bot. arr. veg.* 2: 763 (1776)
Described from Britain, 'amongst moss on the roofs of houses' but a *nomen dubium.*

tegularis (Schumach.) Sacc., Psilocybe, *Syll. fung.* 5: 1050 (1887)
> *Agaricus tegularis* Schumach., *Enum. pl.* 2: 317 (1803)
A *nomen dubium* lacking type material. Listed by Rea (1922).

telmatiaea (Cooke) Sacc., Omphalia, *Syll. fung.* 5: 314 (1887)
> *Agaricus telmatiaeus* Cooke, *Ill. Brit. fung.* 256 (240) Vol. 2 (1883)
> *Agaricus telmaticus* Cooke, in Berkeley, *Ann. Mag. Nat. Hist., Ser. 5 [Notices of British Fungi no. 1999]* 7: 371 (1883)
A *nomen dubium* typified by Cooke's plate (of specimens collected amongst *Sphagnum* spp at Scarborough). It was also published by Berkeley as *Agaricus telmaticus* Cooke but this cites Cooke's plate and was evidently later. Pearson & Dennis (1948) suggest that this may be *Arrhenia philonotis*. See also Dennis (1948: 201).

tenax Lightf., Boletus, *Fl. scot.*: 1031 (1777)
Described from Britain but a *nomen dubium.* From the description, probably *Laetiporus sulphureus.*

tenax (Fr.) Gillet, Naucoria, *Hyménomycètes*: 549 (1876)
> *Agaricus tenax* Fr., *Syst. mycol.* 1: 290 (1821)
A *nomen dubium.* Cooke 504 (617) Vol. 4 (1885) resembles a species of *Entoloma.* Sensu Rea (1922) is *Pholiota myosotis.*

tenerrima Massee & Crossl., Clavaria, *Naturalist (Hull)* 1904: 2 (1904)
Described from Britain, but a *nomen dubium.* Possibly immature *C. asterospora* (= *C. acuta*) *fide* Corner (1970).

tentaculus Bull., Agaricus, *Herb. France*: pl. 560 f. 3 (1792)
Bulliard's illustrations suggest *Mycena speirea* (especially the middle figures), but the species was not discussed by Maas Geesteranus (1992). Sensu Sowerby (Sow3: pl. 385, f. 1) is unknown.

tenuicula P. Karst., Psathyra, *Bidrag Kännedom Finlands Natur Folk* 32: 511 (1879)
A *nomen dubium.* Listed by Rea (1922). Not mentioned in Kits1.

tenuis (Bolton) Gillet, Mycena, *Hyménomycètes*: 271 (1876)
> *Agaricus tenuis* Bolton, *Hist. fung. Halifax* 1: 37 (1788)
Described from Britain but a *nomen dubium* lacking type material. Cooke 237 (160) Vol. 2 (1882) resembles a species of *Hemimycena.*

tenuis (Sowerby) Fr., Typhula, *Syst. mycol.* 1: 495 (1821)
> *Clavaria tenuis* Sowerby, *Col. fig. Engl. fung.* 3: pl. 386 f. 5 (1803)
> *Pistillaria tenuis* (Sowerby) Corner, *Monograph of Clavaria and Allied Genera*: 493 (1950)
Described from Britain but a *nomen dubium* lacking type material. Doubtfully a basidiomycete *fide* Corner (1950).

tenuivolvatus (F.H. Møller) F.H. Møller, Agaricus, *Friesia* 4: 204 (1952)
> *Psalliota tenuivolvata* F.H. Møller, *Friesia* 4: 149 (1952)
Not authentically British. Included in NCL but no records traced. Doubtfully distinct from *A. silvicola.*

terenopus Romagn., Lactarius, *Bull. Soc. Mycol. France* 72: 340 (1957)
A doubtful species, not authentically British. Reported from Scotland (Easterness, Perthshire, and South Aberdeen) but lacking voucher material.

tergiversans (Fr.) Fr., Coprinus, *Epicr. syst. mycol.*: 247 (1838)
> *Agaricus tergiversans* Fr., *Syst. mycol.* 1: 303 (1821)

A *nomen dubium.* Sensu auct. Brit. is *C. silvaticus.*

terrei Berk. & Broome, Trametes, *Ann. Mag. Nat. Hist., Ser. 4 [Notices of British Fungi no. 1571]* 17: 136 (1876)
Described from Buckinghamshire (Stoke Poges) but a *nomen dubium.* Listed by Rea (1922) as *T. terreyi. Fide* Donk (1974) is possibly just immature *Abortiporus biennis.*

terrestris f. resupinata (Bourdot & Galzin) Donk, Thelephora, *Meded. Ned. Mycol. Ver.* 22: 44 (1933)
> *Phylacteria terrestris* var. *resupinata* Bourdot & Galzin, *Bull. Soc. Mycol. France* 40: 127 (1924)
A *nomen dubium.* Reported a few times, but unsubstantiated with voucher material.

terrigena (Fr.) Kuyper, Inocybe, *Persoonia* 12: 4 (1985)
> *Agaricus terrigenus* Fr., *Öfvers. Kongl. Vetensk.-Akad. Förh.* 8: 46 (1852)
> *Pholiota terrigena* (Fr.) P. Karst., *Bidrag Kännedom Finlands Natur Folk* 32: 292 (1879)
> *Togaria terrigena* (Fr.) W.G. Sm., *Syn. Brit. Bas.*: 122 (1908)
Not authentically British. British collections named thus are misdetermined species of *Pholiota fide* Pegler & Young (1972: 505).

testaceum Quél., Hebeloma, *Mém. Soc. Émul. Montbéliard, Sér. 2,* 5: 250 (1872)
> *Agaricus testaceus* Fr. [*nom. illegit.*, non *A. testaceus* Scop. (1772)], *Epicr. syst. mycol.* 178 (1838),
A *nomen dubium.* Sensu NCL is *Hebeloma sordescens.*

thalassinum Pilát & Dermek, Leccinum, *Hribovite Huby*: 146 (1974)
A *nomen dubium,* reported from Hampshire (New Forest) in 1994. Possibly *Leccinum variicolor fide* auct., but type material appears immature, with spores few and smaller than described *fide* Lannoye & Estades [*Doc. Mycol.* 21 (81): 17, 1991)].

theiogalus (Bull.) Gray, Lactarius, *Nat. arr. Brit. pl.* 1: 624 (1821)
> *Agaricus theiogalus* Bull., *Herb. France*: pl. 567 (1791)
A *nomen dubium.* The name has been variously interpreted, e.g. as *Lactarius chrysorrheus, L. hepaticus* and *L. tabidus.*

tigrina R. Heim, Inocybe, *Encycl. Mycol. 1. Le Genre* Inocybe: 230 (1931)
A *nomen dubium* lacking type material *fide* Kuyper (1986: 236). The name has been used in Britain (e.g. in Phillips 1981: 153) for forms of *Inocybe flocculosa* with dark cap scales.

tigrinum (Schaeff.) Gillet, Tricholoma, *Hyménomycètes*: 118 (1874)
> *Agaricus tigrinus* Schaeff., *Fung. Bavar. Palat. nasc.* 1: 89 (1762)
A *nomen dubium* lacking type material. Sensu auct. is *Tricholoma pardinum,* unknown in Britain. Cooke 106 (64) Vol. 1 (1883) is doubtful.

tintinnabulum (Batsch) Quél., Mycena, *Mém. Soc. Émul. Montbéliard, Sér. 2,* 5: 105 (1872)
> *Agaricus tintinnabulum* Batsch, *Elench. fung.*: 61 (1783)
Not authentically British. Listed in NCL and reported on several occasions but all records lack voucher material. Sensu Rea (1922) is doubtful but apparently not this species.

tithymalinus (Scop.) Fr., Lactarius, *Epicr. syst. mycol.*: 347 (1838)
> *Agaricus tithymalinus* Scop., *Fl. carniol.* 452 (1772)
A *nomen dubium* lacking type material. Sensu auct. Brit. is *Lactarius rubrocinctus.*

togularis (Bull.) Kühner, Conocybe, *Encycl. Mycol. 7. Le Genre* Galera: 161 (1935)
> *Agaricus togularis* Bull. [non *A. togularis* Pers. (1801)], *Herb. France*: pl. 595 (1793)
> *Pholiota togularis* (Bull.) P. Kumm., *Führ. Pilzk.*: 83 (1871)
> *Pholiotina togularis* (Bull.) Fayod, *Ann. Sci. Nat., Bot.,* sér. 7, 9: 359 (1889)

Togaria togularis (Bull.) W.G. Sm., *Syn. Brit. Bas.*: 123 (1908)
A *nomen dubium* lacking type material. The name has been used in Britain for many of the annulate members of *Conocybe* subgenus *Pholiotina*.

tomentella *f*. rubidella Bon, Lepiota, *Doc. mycol.* 22(88): 28 (1993)
Not authentically British. A collection named thus (in herb. K) has been redetermined as *Lepiota cingulum*.

tomentosa (Jungh.) Quél., Inocybe, *Fl. mycol. France*: 106 (1888)
Agaricus tomentosus Jungh., *Linnaea* 5: 403 (1830)
A *nomen dubium*. Sensu Rea (1922) is possibly *Inocybe sindonia* fide Pearson & Dennis (1948).

tomentosum Fr., Radulum, *Epicr. syst. mycol.*: 525 (1838)
A *nomen dubium*. Reported in Grevillea 4: 39 (1875) on *Pyrus aucuparia* (=*Sorbus aucuparia*) from Scotland (Angus: Menmuir), but this collection, in herb. K, has been redetermined as *Basidioradulum radula*.

tomentosus (Bull.) Fr., Coprinus, *Epicr. syst. mycol.*: 246 (1838)
Agaricus tomentosus Bull., *Herb. France*: pl. 138 (1786)
A *nomen dubium* lacking type material. Cooke 659 (672) Vol. 5 (1886) (copied directly from Bolton) is possibly a member of *Coprinus* section *Filamentifer* fide BFF2 p.112.

tomentosus (Fr.) Teng, Inonotus, *Fungi of China*: 761 (1964)
Polyporus tomentosus Fr., *Syst. mycol.* 1: 351 (1821)
Polystictus tomentosus (Fr.) Cooke, *Grevillea* 14: 77 (1886)
Onnia tomentosa (Fr.) P. Karst., *Bidr. Kän. Fin. Nat. Folk* 48: 326 (1889)
Trametes circinatus Fr., *Kongl. Vetensk. Acad. Handl.* 1848: 128 (1849)
Not authentically British. Reported once, from the Channel Islands in 1887, but unsubstantiated with voucher material.

tornata (Fr.) P. Kumm., Clitocybe, *Führ. Pilzk.*: 122 (1871)
Agaricus tornatus Fr., *Syst. mycol.* 1: 91 (1821)
A *nomen dubium*. Accepted in NCL but doubtfully distinct from *Clitocybe candicans*.

tortipes Massee, Entoloma, *Brit. fung.-fl.* 2: 278 (1893)
Described from Britain but a *nomen dubium* and probably not an *Entoloma* fide NCL. Pearson & Dennis (1948) suggest that this may represent *Collybia distorta*.

tortuosus (Fr.) Fr., Cortinarius, *Epicr. syst. mycol.*: 305 (1838)
Agaricus tortuosus Fr., *Syst. mycol.* 1: 235 (1821)
Not authentically British. Reported [Grevillea 16: 45 (1887)] from 'damp pine woods near Scarborough' but unsubstantiated with voucher material. A collection in herb. K from Surrey (Haslemere) in 1945, named thus by Pearson, is marked 'uncertain' by Orton.

tortus (Willd.) Fr., Dacrymyces, *Elench. Fung.* 2: 36 (1828)
Tremella torta Willd., *Bot. Mag. (Römer & Usteri)* 2 (4): 18 (1788)
A *nomen dubium* lacking type material. Considered by Donk (1966) to be an earlier name for *Dacrymyces punctiformis* but this was rejected by Reid (1974). Old British specimens named thus in herb. K represent various species of *Dacrymyces*.

trabutii (Maire) Singer, Marasmiellus, *Lilloa* 22: 300 (1951) [1949]
Marasmius trabutii Maire, *Bull. Soc. Bot. France* 56: 278 (1909)
Marasmiellus caespitosus (Pat.) Singer, *Pap. Michigan Acad. Sci.* 32: 129 (1946) [1948]
Not authentically British. Reported from Oxfordshire (Henley on Thames) on old stems of *Rubus fruticosus* but probably misidentified since normally on *Juncaceae* and *Cyperaceae* in salt marshes.

trachyspora (Rea) Bon, Mycenella, *Doc. mycol.* 9: 28 (1973)

Mycena trachyspora Rea, *Trans. Brit. Mycol. Soc.* 12(2-3): 216 (1927)
A *nomen dubium*. Described from roots of dead *Ulmus* in West Kent (Woolwich). Said by some to be *Mycena olida*, which was collected from the same tree five years later, but there is no type extant.

transforme (Britzelm.) Singer, Lyophyllum, *Ann. Mycol.* 41: 98 (1943)
Agaricus transformis Britzelm., *Hymenomyc. Südbayern* 8: 3 (1891)
Tricholoma transforme (Britzelm.) Sacc., *Syll. fung.* 11: 12 (1895)
Not authentically British. The single collection named thus in herb. K is misdetermined.

transitoria (Britzelm.) Sacc., Inocybe, *Syll. fung.* 5: 788 (1887)
Agaricus transitorius Britzelm., *Ber. Naturhist. Vereins Augsburg* 26: 137 (1879)
A *nomen dubium* lacking type material. Possibly a species near to *Inocybe napipes* but all British records are dubious, and no material has been traced.

translucens Berk. & Broome, Typhula, *Ann. Mag. Nat. Hist., Ser. 4 [Notices of British Fungi no.1589]* 17: 138 (1876)
Described from Britain but 'not a fungus' fide Massee [quoted by Corner (1950)]. Possibly an immature, poorly dried specimen of a myxomycete according to Reid & Austwick (1963).

trechispora (Berk.) P. Karst., Inocybe, *Bidrag Kännedom Finlands Natur Folk* 32: 465 (1879)
Agaricus trechisporus Berk., *Ann. Mag. Nat. Hist., Ser. 1 [Notices of British Fungi no. 71]* 1: 200 (1838)
Astrosporina trechispora (Berk.) Rea, *Brit. basidiomyc.*: 209 (1922)
A *nomen dubium*. Sensu NCL is *I. paludinella*. The name has also been widely used for *I. mixtilis*.

tremulae Velen., Marasmius, *Novit. mycol. Novissimae*: 17 (1947)
A *nomen dubium*. Sensu auct., and a single British collection so named in K, is *M. favrei*.

trepida (Fr.) Gillet, Psathyrella, *Hyménomycètes*: 615 (1878)
Agaricus trepidus Fr., *Epicr. syst. mycol.*: 298 (1838)
Psathyra trepida (Fr.) J.E. Lange, *Dansk. Bot. Ark.* 9(1): 16 (1936)
A *nomen dubium*. Accepted in Kits1 and listed in NCL on the basis of a single old and doubtful record, but lacking a convincing modern interpretation. Also reported from Lincolnshire but without voucher material

tricolor (Pers.) Bondartsev & Singer, Daedaleopsis, *Ann. Mycol.* 31: 64 (1941)
Daedalea tricolor Pers., *Mycol. eur.* 2: 12 (1828)
Daedaleopsis confragosa var. *tricolor* (Pers.) Bondartsev, *Trutovye griby Evropeiskoi chasti SSSR i Kavkaza*: 171 (1953)
Not authentically British. Reported on numerous occasions from widespread localities in England but all records are dubious, probably referring to old, reddened and partially lamellate basidiomes of *D. confragosa*. The single collection so named in herb. K is *D. confragosa*.

tridentinus (Bres.) Kühner & Romagn., Rhodophyllus, *Flora Analytique des Champignons Supérieurs*: 202 (1953)
Nolanea cocles var. *tridentina* Bres., *Fl. ital. crypt.*: 566 (1915)
A *nomen dubium*. A single collection in herb. K reported by Bond [Bull. Brit. Mycol. Soc. 15: 125 (1981), as *Leptonia tridentina ined.*] is misdetermined (spores too small) and its identity is unclear.

triformis *var*. schaefferi Fr., Cortinarius, *Monogr. hymenomyc. Suec.* 2: 73 (1863)
A *nomen dubium*. Reported from North Hampshire (Alresford) by Eyre in Grevillea 14: 38 (1885) but no material has been traced.

trigonophylla (Lasch) Sacc., Tubaria, *Syll. fung.* 5: 873 (1887)
> *Agaricus furfuraceus* var. *trigonophyllus* Lasch, *Linnaea* 3: 390 (1828)
> *Tubaria furfuracea* var. *trigonophylla* (Lasch) Massee, *Brit. fung.-fl.* 2: 123 (1893)

A *nomen dubium* lacking type material. Sensu NCL is *Tubaria albostipitata*.

trilobatus With., Boletus, *Bot. arr. veg.* 2: 796 (1776)
Described from Britain, 'on oak leaves in Madingly Wood near Cambridge' but a *nomen dubium*. Possibly not a basidiomycete.

trilobus Bolton, Agaricus, *Hist. fung. Halifax* 1: 38 (1788)
Described from Britain but a *nomen dubium* lacking type material. Bolton's illustration (with volva somewhat exaggerated) is possibly of *Amanita fulva*.

trinii (Weinm.) Quél., Inocybe, *Mém. Soc. Émul. Montbéliard, Sér. 2*, 5: 154 (1872)
> *Agaricus trinii* Weinm., *Hymen. Gasteromyc.*: 194 (1836)
> *Astrosporina trinii* (Weinm.) Rea, *Brit. basidiomyc.*: 211 (1922)

A *nomen dubium* lacking type material and variously interpreted. Sensu Rea (1922) is possibly *I. bresadolae fide* Kuyper (1986), but no good material survives and this is otherwise unknown in Britain.

trispora *var.* **epilobii** Corner, Pistillaria, *Nova Hedwigia Beih.* 33: 124 (1970)
Described from Warwickshire (Yarningdale Common) but the distinction from *Typhula setipes* is unclear.

triste (Velen.) Noordel., Entoloma, *Persoonia* 10: 254 (1979)
> *Nolanea tristis* Velen., *České Houby* III: 630 (1921)

Not authentically British. Reported from the Republic of Ireland on four occasions, but all lack voucher material.

triste (Scop.) Quél., Tricholoma, *Mém. Soc. Émul. Montbéliard, Sér. 2*, 5: 79 (1872)
> *Agaricus tristis* Scop. [non *A. tristis* Pers. (1796)], *Fl. carniol.* 483 (1772)

A *nomen dubium* lacking type material. Used for several species near to *Tricholoma terreum*. Listed in NCL sensu Bresadola [*Icon. mycol.* 2 pl. 78 (1927)]. British material is old and needs re-examination.

trogii (Berk.) Domański, Coriolopsis, *Mała Flora Grzybów, Kraków* 1: 230 (1974)
> *Trametes trogii* Berk., in Trog, *Verz. Schw. Schwam.* 2: 52 (1850)
> *Funalia trogii* (Berk.) Bondartsev & Singer, *Ann. Mycol.* 39: 62 (1941)

Not authentically British. Reported from the Nordic Congress Foray in Perthshire (Kindrogan) in 1983 but lacking voucher material.

trulliformis (Fr.) Quél., Clitocybe, *Compt. Rend. Assoc. Franç. Avancem. Sci.* 11: 389 (1883)
> *Agaricus trulliformis* Fr., *Syst. mycol.* 1: 174 (1821)
> Mis.: *Clitocybe parilis* sensu J.E. Lange [Fl. agaric. danic.1: pl. 33B (1935)]

Not authentically British. Listed in NCL but lacking voucher material and dubiously identified. At least some records refer to *C. costata*.

trullisata (Ellis) Peck, Laccaria, *Rep. (Annual) New York State Mus. Nat. Hist.* 157: 90 (1912)
> *Agaricus trullisatus* Ellis, *Bull. Torrey Bot. Club* 5: 45 (1874)
> *Clitocybe trullisata* (Ellis) Sacc., *Syll. fung.* 5: 195 (1887)

Not authentically British (or European.). A North American species. Sensu NCL is *L. maritima*.

truncata (Schaeff.) Singer, *Rhodocybe, Mycologia* 38: 687 (1946)
> *Agaricus truncatus* Schaeff., *Fung. Bavar. Palat. nasc.* 3: 251 (1770)
> *Hebeloma truncatum* (Schaeff.) P. Kumm., Hebeloma, *Führ.*

Pilzk.: 80 (1871)
> *Tricholoma truncatum* (Schaeff.) Quél., *Bull. Soc. Amis Sci. Nat. Rouen.* 15: 153 (1880)
> *Clitopilus truncatus* (Schaeff.) Kühner & Romagn. [*comb. inval.*], *Flore Analytique des Champignons Supérieurs*: 172 (1953)

A *nomen dubium* lacking type material and with widely differing interpretations. Sensu NCL and sensu auct. mult. is *Rhodocybe gemina*. Sensu auct. is *Hebeloma theobrominum*.

tschulymica (Pilát) Jean Keller, Skeletocutis, *Persoonia* 10(3): 353 (1979)
> *Poria tschulymica* Pilát, *Bull. Soc. Mycol. France* 48: 35 (1932)

Not authentically British. The single record was based on misidentified material (in herb. K) of *Skeletocutis kuehneri*.

tubaeforme Gillet, Hydnum, *Hyménomycètes*: 717 (1878)
A *nomen dubium*. Sensu Rea [TBMS 17: 47 (1932)] is merely a growth form of *Hydnum repandum* that has become hollow in the centre of the stipe and pileus.

tubercularia Berk., Tremella, *Outl. Brit. fungol.*: 288 (1860)
A synonym of the anamorphic ascomycete *Coryne albida* (Berk.) Korf & Cand. Sensu auct. is *Tremella globispora*.

tuberculatum (Berk. & M.A. Curtis) Hjortstam, Basidioradulum, *Mycotaxon* 54: 183 (1995)
> *Grandinia tuberculata* Berk. & M.A. Curtis, *Hooker's J. Bot. Kew Gard. Misc.* 1: 237 (1849)
> *Radulum pendulum* Fr., *Elench. fung.* 1: 149 (1828)
> *Phlebia albida* H. Post, in Fries, *Monogr. hymenomyc. Suec.* 2: 280 (1863)
> *Stereum subcostatum* P. Karst., *Hedwigia* 20: 178 (1881)
> *Stereum hirsutum* var. *subcostatum* (P. Karst.) Massee, *Grevillea* 19: 65 (1891)
> *Corticium subcostatum* (P. Karst.) Bourdot & Galzin, *Hyménomyc. France*: 311 (1928)

Not authentically British. This is *Phlebia albida* sensu CNE6: 1085 but not sensu auct. Brit. [e.g. Rea (1922: 625)]. The single collection named thus in herb. K was misidentified.

tuberosa (Sowerby) Fr., Calocera, *Syst. mycol.* 1: 486 (1821)
> *Clavaria tuberosa* Sowerby, *Col. fig. Engl. fung.* 2: pl. 199 (1799)

A *nomen dubium* lacking type material. The original illustration suggests young specimens of *Macrotyphula fistulosa* var. *contorta*.

tumidum (Pers.) Ricken, Tricholoma, *Blätterpilze Deutschl.*: 344 (1914)
> *Agaricus tumidus* Pers., *Syn. meth. fung.*: 350 (1801)

Doubtful *fide* NCL but accepted by Riva (FungEur3) as a good species citing Cooke 94 (93) Vol. 1 (1882) as an accepted illustration. The Kew copy has a handwritten note, apparently by Wakefield, saying 'rather *T. portentosum* than *tumidum*'.

tumidum *var.* **keithii** (W. Phillips & Plowr.) Sacc., Tricholoma, *Syll. fung.* 5: 109 (1887)
> *Agaricus tumidus* var. *keithii* W. Phillips & Plowr., *Grevillea* 10: 65 (1881)

Described from Britain but a *nomen dubium*. Described as 'near to *Tricholoma sudum*' by various authorities.

tundrae A.H. Sm. & Singer, Galerina, *Mycologia* 47: 584 (1955)
Not authentically British. The single collection in herb. E (Scotland: St Kilda) is determined as 'compare with' *G. tundrae*. Sensu auct. is *G. harrisonii*.

turbinatus (Bull.) Fr., Cortinarius, *Epicr. syst. mycol.*: 266 (1838)
> *Agaricus turbinatus* Bull., *Herb. France*: pl. 110 (1783)

A *nomen dubium* lacking type material and variously interpreted. Sensu Cooke (HBF2) is *C. elegantior* and sensu NCL is *C. saporatus*.

turfosus Sowerby, Agaricus, *Col. fig. Engl. fung.* 2: pl. 210 (1799)

Described from Britain, but a *nomen dubium* lacking type material. Possibly a species of *Omphalina*, though Massee [Brit. fung.-fl. 2: 307 (1893)] equates Sowerby's plate with *Xerotus degener,* which is also doubtful.

turgidus Grev., Agaricus, *Scott. crypt. fl.*: pl. 9 (1826)
Described from Britain, but a *nomen dubium*. Greville's illustration appears to represent young basidiomes of *Clitocybe nebularis.*

turmalis Fr., Cortinarius, *Epicr. syst. mycol.*: 257 (1838)
No voucher material has been traced. Most British records are probably sensu NCL (= *C. claricolor*).

turundus *var.* **sphaerosporus** Rea, Hygrophorus, *Brit. basidiomyc.*: 305 (1922)
A *nomen dubium*. Described as having warted spores, thus could be a species of *Laccaria* (possibly the little known *Laccaria bella fide* NCL).

typhicola Bourdot & Galzin, Pistillaria, *Hyménomyc. France*: 139 (1928)
A *nomen dubium*. Recorded from Scotland on leaves of *Iris pseudacorus* and on dead stems of *Equisetum* sp., this is probably *Typhula setipes* or close to it.

umbellatus Fr., Lentinus, *Synopsis generis Lentinorum*: 10 (1836)
A *nomen dubium*. Sensu Rea (1922) and sensu auct. is *Lentinellus cochleatus.*

umbellifera *f.* **chrysoleuca** (Pers.) Cejp, Omphalia, Kavina & Pilát, *Atlas champ. Eur.* 4: 42 (1936)
  Agaricus chrysoleucus Pers., *Syn. meth. fung.*: 457 (1801)
  Omphalia chrysoleuca (Pers.) P. Karst., *Bidrag Kännedom Finlands Natur Folk* 32: 125 (1879)
  Omphalina chrysoleuca (Pers.) Quél., *Enchir. fung.*: 42 (1886)
A *nomen dubium*. Sensu Rea (1922) is *Chrysomphalina grossula.*

umbellifera *var.* **pallida** Rea, Omphalia, *Brit. basidiomyc.*: 429 (1922)
Described from Britain but a *nomen dubium*. Possibly just a pale variant of *Omphalina ericetorum.*

umbellifera *var.* **pyriformis** (Pers.) Rea, Omphalia, *Brit. basidiomyc.*: 429 (1922)
  Agaricus pyriformis Pers., *Syn. meth. fung.*: 317 (1801)
A *nomen dubium* lacking type material. Doubtful *fide* NCL. Persoon's fungus was said to grow amongst grass, whereas Rea's was on trunks of *Fagus.*

umbelliferus With., Merulius, *Syst. arr. Brit. pl. ed. 4,* 4:144 (1801)
Described from Britain but a *nomen dubium*. From the description this appears to be *Mycena adscendens* or a closely related species.

umbilicalis With., Agaricus, *Bot. arr. veg.* 2: 756 (1776)
Described from Britain but a *nomen dubium.*

umbilicata (Schaeff.) P. Kumm., Clitocybe, *Führ. Pilzk.*: 123 (1871)
  Agaricus umbilicatus Schaeff. [non *A. umbilicatus* Bolton (1788)], *Fung. Bavar. Palat. nasc.* 3: 207 (1770)
  Omphalia umbilicata (Schaeff.) Gillet, *Hyménomycètes*: 289 (1876)
A *nomen dubium* lacking type material. Sensu NCL is *Clitocybe subspadicea.* For Bon [Flore Mycol. d'Eur. 4 (1997)] *C. umbilicata* occurs with conifers and *C. subspadicea* with deciduous trees, both species placed in the genus *Gerronema.*

umbilicatus Bolton, Agaricus [*nom. illegit.,* non *A. umbilicatus* Schaeff. (1770)] *Hist. Fung. Halifax* 1: pl.17 (1788)
Described from Britain but a *nomen dubium* lacking type material. Apparently a species of *Clitocybe* but Withering [Arr. Brit. pl. ed. 3, 4: 155] cites it as a synonym of *Agaricus cyathiformis* (= *Pseudoclitocybe cyathiformis*).

umbilicatum Fr. [as *Geaster umbilicatus*], Geastrum, *Syst. mycol.* 3: 14 (1829)
A *nomen dubium*. Listed by Rea (1922).

umbilicatum *var.* **smithii** (Lloyd) W.G. Sm. [as *Geaster*], Geastrum, *Syn. Brit. Bas.*: 469 (1908)
  Geastrum smithii Lloyd [as *Geaster*], *Mycol. not.* 2 (23): 287 (1906)
A *nomen dubium*. Listed by Smith (1908) but no records have been traced.

umbonatus With., Agaricus [non *A. umbonatus* Schumach. (1801)], *Bot. arr. Brit. pl. ed. 2,* 3: 290 (1792)
Described from Britain but a *nomen dubium*

umbonatus (Velen.) Rob. Henry, Cortinarius, *Bull. Soc. Mycol. France.* 62: 217 (1947) [1946]
  Hydrocybe umbonata Velen., *České Houby* III: 479 (1921)
Not authentically British. A poorly known species, reported from Caernarvonshire in 1988 but unsubstantiated with voucher material.

umboninota Peck, Inocybe, *Rep. (Annual) New York State Mus. Nat. Hist.* 38: 87 (1885)
  Astrosporina umboninota (Peck) A. Pearson, *Trans. Brit. Mycol. Soc.* 26(1-2): 46 (1943)
Not authentically British (or European). Sensu Pearson and sensu auct. Eur. is *Inocybe acuta.*

umbraculum Batsch, Agaricus, *Elench. fung.*: 77 t. 2 f. 4 (1783)
A *nomen dubium*. Reported 'in a hollow stump' by Withering (1792). Batsch's illustration bears a resemblance to *Xerula radicata.*

umbratilis (Fr.) Redhead, Lutzoni, Moncalvo & Vilgalys, Arrhenia, *Mycotaxon* 83: 48 (2002)
  Agaricus umbratilis Fr., *Syst. mycol.* 1: 157 (1821)
  Omphalia umbratilis (Fr.) Gillet, *Hyménomycètes*: 299 (1876)
  Omphalina umbratilis (Fr.) Quél., *Enchir. fung.*: 45 (1886)
Listed in NCL, but a doubtful and poorly known species in Britain with no recent collections. Last reported from Scotland (Easterness) in 1900. Cooke 265 (274) Vol. 2 (1883) may not represent the current concept of the species.

umbrinellus (Bres.) S. Herrera & Bondartseva, Phellinus, *Mikologiya i Fitopatologiya* 14(1): 8 (1980)
  Poria umbrinella Bres., *Hedwigia* 35: 282 (1896)
Not authentically British. A common and widespread species in the tropics. British material under this name redetermined as *Phellinus cavicola.*

umbrinellus (Sommerf.) Gillet, Pluteus, *Hyménomycètes*: 397 (1876)
  Agaricus umbrinellus Sommerf., *Suppl. Fl. lapp.*: 289 (1826)
A *nomen dubium*. Listed in Rea (1922) and reported in 1936 from the Republic of Ireland but lacking voucher material.

umbrinoides (J. Blum) Lannoy & Estadès, Leccinum, *Doc. mycol.* 21(81): 24 (1991)
  Boletus umbrinoides J. Blum, *Bull. Soc. Mycol. France* 85(4): 560 (1970) [1969]
A *nomen dubium*. Sensu auct. Brit is *L. rigidipes.*

umbrinus Cooke & Massee, Coprinus, *Grevillea* 21: 41 (1892)
Described from Britain but a *nomen dubium*. Doubtfully distinct from *C. sterquilinus fide* NCL index.

umbrinus M.M. Moser, Cortinarius [*nom. illegit.,* non *C. umbrinus* (Pers.) P. Kumm (1871)], *Nova Hedwigia Beih.* 52: 350 (1975)
Not authentically British. Associated with *Nothofagus* spp and described from Argentina. A record from England (Worcestershire: Ankerdine Hill) in 1965 lacks voucher material and was presumably a misprint for some other species.

umbrinus (Paulet) Fr., Lactarius, *Epicr. syst. mycol.*: 339 (1838)
  Hypophyllum umbrinum Paulet, *Traité champ.*: t. 69 (1793)

Agaricus umbrinus (Paulet) Pers., *Syn. meth. fung.*: 435 (1801)
A *nomen dubium* lacking type material. Cooke 942 (1006) Vol. 7 (1888) is unidentifiable.

umidicola Kauffman, Cortinarius, *Bull. Torrey Bot. Club* 32: 322 (1905)
A North American species, not authentically British. Scottish records (Watling, TBMS 90: 18, 1988) are sensu Moser (1978) and have since been referred to *C. malachius.*

undosa (Peck) Jülich, Postia, *Persoonia* 11(4): 424 (1982)
Polyporus undosus Peck, *Rep. (Annual) New York State Mus. Nat. Hist.* 34: 42 (1881)
Not authentically British. The single collection named thus, from Oxfordshire (Bix, The Warburg Reserve), is annotated as 'compare with' *P. undosa.*

unimodus Britzelm., Cortinarius, *Ber. Naturhist. Vereins Augsburg* 28: 131 (1885)
A *nomen dubium* lacking type material. Sensu Cooke (HBF2) is *C. casimiri.*

uraceus Fr., Cortinarius, *Epicr. syst. mycol.*: 309 (1838)
Not authentically British. Most material named thus in herb. K has been redetermined. Sensu Lange (1938) and sensu Pearson (1949) is *C. vernus.*

urticae (Pers.) Mart., Dacrymyces, *Fl. crypt. Erlangensis*: 368 (1817)
Tremella urticae Pers., *Syn. Meth Fung.* 2: 628 (1801)
Not a basidiomycete. This is *Calloria neglecta* (= *Callorina fusarioides*) or its anamorphic stage *Cylindrocolla urticae*, common on old nettle stems.

ustulatus With., Agaricus [non *A. ustulatus* Pegler (1983)], *Bot. arr. veg.* 2: 759 (1776)
Described from Britain but a *nomen dubium.*

ustulata (Bull.) Gray, Gyraria, *Nat. arr. Brit. pl.* 1: 595 (1821)
Tremella ustulata Bull., *Herb. France* 4: pl. 420 (1789)
Sclerotium pyrinum Fr. (anam.), *Syst. mycol.* 2: 258 (1823)
Not a basidiomycete. Reported by Gray (1821) as growing 'on rotten fruits' and illustrated by Bulliard growing on 'old lemons'. Known to Fries as *Sclerotium pyrinum*, this is probably a sclerotial state of a hyphomycete.

utilis (Weinm.) Fr., Lactarius, *Monogr. hymenomyc. Suec.* 2: 159 (1863)
Agaricus utilis Weinm., *Hymen. Gasteromyc.*: 43 (1836)
Not authentically British. Cooke 930 (1084) Vol. 7 (1888) is doubtful, but possibly *Lactarius pallidus.* The single collection named thus in herb. K has been redetermined as *Lepista irina.*

vaccinus Fr., Boletus, *Epicr. syst. mycol.*: 420 (1838)
A *nomen dubium.* Sensu auct. is probably a form of *Boletus badius* growing under deciduous trees (possibly semi-sterile *fide* NCL). *Fide* Pearson & Dennis (1948) this is a form of *Gyroporus castaneus.*

vaginata *var.* alba (De Seynes) Gillet, Amanita, *Hyménomycètes*: 51 (1874)
Agaricus vaginatus var. *albus* De Seynes, *Fl. mycol. Montpellier*: 105 (1863)
Agaricus fungites Batsch, *Elench. fung.*: 51 (1783)
Amanita vaginata var. *fungites* (Batsch) J.E. Lange, *Dansk. Bot. Ark.* 2(3): 11 (1915)
Mis.: *Agaricus vaginatus* var. *nivalis* sensu Cooke (Ill. Brit. fung. 940 (1104) Vol. 8)
Mis.: *Amanitopsis vaginata* var. *nivalis* sensu W.G. Smith (1908)
Not authentically British. Reported on a few occasions recently but records lack voucher material.

valesiacum Boud., Ganoderma, *Bull. Soc. Mycol. France* 11: 28 (1895)
Not authentically British. Now considered a central European species restricted to *Larix* spp. Sensu auct. Brit. (on *Taxus baccata*) has been reassigned to *G. carnosum.*

vaporaria Pers., Poria, *Ann. Bot. Usteri* 11: 30 (1794)
Boletus vaporarius (Pers.) Pers., *Syn. meth. fung.*: 546 (1801)
Polyporus vaporarius (Pers.) Fr., *Observ. mycol.* 2: 260 (1818)
A *nomen dubium* lacking type material and variously interpreted. Sensu auct. is *Antrodia sinuosa* or *Schizopora paradoxa.*

vaporaria *var.* secernibilis Berk. & Broome, Poria, *Ann. Mag. Nat. Hist.*, Ser. 3 *[Notices of British Fungi no. 1022]* 15: 1022 (1865)
Polyporus vaporarius var. *secernibilis* (Berk. & Broome) Rea, *Brit. basidiomyc.*: 602 (1922)
Described from Britain but a *nomen dubium.* Described as growing 'on fir leaves under moss'.

varicosus Fr., Marasmius, *Epicr. syst. mycol.*: 376 (1838)
A *nomen dubium.* Possibly a species of *Mycena*. since a 'red sap' is described in the stipe. Cooke 1075 (1121) Vol. 7 (1890) is difficult to interpret.

variecolor Berk. & Broome, Boletus, *Ann. Mag. Nat. Hist.*, Ser. 3 *[Notices of British Fungi no. 1020]* 15: 318 (1865)
Described from Britain but a *nomen dubium.* The description is suggestive of *Boletus subtomentosus* with a reticulum on the stipe.

variegata (Pers.) Gray, Mycena, *Nat. arr. Brit. pl.* 1: 621 (1821)
Agaricus variegatus Pers. [non *A. variegatus* Scop. (1772)], *Syn. meth. fung.*: 391 (1801)
A *nomen dubium* lacking type material and not discussed by Maas Geesteranus (1992). Gray cites Bulliard (Herb. France pl. 560) and Sowerby (Sow3: pl. 385) as examples of *M. variegata*, but both of these show more than one species, not all of them *Mycena.*

variegatus (Sowerby) Cooke, Fomes, *Grevillea* 14: 21 (1885)
Boletus variegatus Sowerby [non *B. variegatus* Swartz (1810)], *Col. fig. Engl. fung.* 3: pl. 368 (1802)
Polyporus variegatus (Sowerby) Fr., *Hymenomyc. eur.*: 563 (1874)
Described from Britain but a *nomen dubium* lacking type material. Sowerby's illustration may represent *Ganoderma resinaceum.*

variegatus (Vittad.) Tul. & C. Tul., Melanogaster, *Ann. Sci. Nat., Bot.*, sér. 2, 19: 92 (1843)
Octaviania variegata Vittad., *Monogr. Tuberac.*: 16 (1831)
Not authentically British. A southern European species, reported from Northern Ireland in 1931 but lacking voucher material.

varius (Schaeff.) Fr., Cortinarius, *Epicr. syst. mycol.*: 258 (1838)
Agaricus varius Schaeff., *Fung. Bavar. Palat. nasc.* 1: 42 (1762)
Agaricus decolorans Pers. [*nom. illegit.*, non *A. decolorans* Fr. (1821)], *Observ. mycol.* 1: 52 (1796)
Cortinarius decolorans Fr., *Epicr. syst. mycol.*: 271 (1838)
Not authentically British. Records (often as *C. decolorans*) are mostly old (from the mid to late nineteenth century) and lack voucher material. Sensu Cooke (HBF2) is *C. variicolor.*

vatricosa (Fr.) P. Karst., Inocybe, *Bidrag Kännedom Finlands Natur Folk* 32: 465 (1879)
Agaricus vatricosus Fr., *Observ. mycol.* 2: 46 (1818)
A *nomen dubium* lacking type material. Cooke 443 (403) Vol. 3 (1886) 'on wood chips near Watford' is unidentifiable and does not suggest an *Inocybe*, having slightly pinkish lamellae.

velox Godey, Coprinus, in Gillet, *Hyménomycètes*: 614 (1878)
A *nomen dubium* lacking type material. Sensu auct. is *C. stercoreus.*

velutinum (Fr.) P. Karst., Hydnellum, *Meddeland. Soc. Fauna Fl. Fenn.* 5: 41 (1880)
Hydnum velutinum Fr., *Syst. mycol.* 1: 404 (1821)
A *nomen dubium.* British records mainly refer to *H. concrescens.*

velutipes *var.* rubescens (Cooke) Massee, Collybia, *Brit. fung.-fl.* 3: 127 (1893)
Agaricus velutipes var. *rubescens* Cooke, *Ill. Brit. fung.* 1141 (650) Vol. 8 (1890)
A *nomen dubium*. Described from Scotland (Fifeshire: Largo) in 1872. Cooke 1141 (650) Vol. 8 (1890) shows a small reddish-brown agaric, with white lamellae spotting russet brown, and with what appear to be velutinate stipes, growing in clumps on soil with conifers. It looks distinct but is not *Flammulina velutipes*.

venosum Gillet, Entoloma, *Hyménomycètes*: 403 (1876)
A dark species, near to *Entoloma rhodopolium* but not authentically British. Known from montane conifer woods in continental Europe. Listed by Rea (1922) but no material traced.

ventricosa (Bull.) Gillet, Collybia, *Hyménomycètes*: 324 (1876)
Agaricus ventricosus Bull., *Herb. France*: pl. 411 (1789)
A *nomen dubium* lacking type material *fide* Antonín & Noordeloos (A&N2). See Cooke 199 (145) Vol. 2 (1882) for an interpretation.

ventricosum Arnolds & Noordel., Entoloma, *Persoonia* 10: 298 (1979)
Nolanea ventricosa (Arnolds & Noordel.) P.D. Orton, *Mycologist* 5(4): 173 (1991)
Not authentically British. Reported from England and Scotland, but mostly unsubstantiated with voucher material. The few collections in herbaria are all annotated as 'compare with' *E. ventricosum*.

ventricosus Berk. & Broome, Hygrophorus, *Ann. Mag. Nat. Hist., Ser. 5 [Notices of British Fungi no. 1777]* 3: 207 (1879)
Described from Britain but a *nomen dubium*. The type material is in poor condition. Probably just a form of *Hygrocybe virginea fide* Dennis (1948: 147). Cooke 897 (901) Vol. 6 (1888) depicts several deformed basidiomes which could well be this species.

verecundum (Fr.) Noordel., Entoloma, *Persoonia* 10(4): 507 (1980)
Agaricus rubellus var. *verecundus* Fr., *Spicilegium Pl. neglect.*: 6 (1836)
Agaricus verecundus (Fr.) Fr., *Epicr. syst. mycol.*: 158 (1838)
Nolanea verecunda (Fr.) Gillet, *Hyménomycètes*: 422 (1876)
Not authentically British. Recorded from England (East Sussex: Buckhurst Park) and Republic of Ireland (Wicklow: Powerscourt) but lacking voucher material and not reported since 1929. Noordeloos [*Persoonia* 10: 508 (1980)], however, cites Cooke 367 (340) A (1886) as in agreement with his concept of the species.

vermicularis Pers., Daedalea, *Mycol. eur.* 3: 2 (1828)
Boletus resupinatus Sowerby [*nom. illegit.*, non *B. resupinatus* Bolton (1791)], *Col. fig. Engl. fung., Suppl.*: pl. 424 (1814)
A *nomen novum* for Sowerby's species. Described from Britain but a *nomen dubium* lacking type material. Sowerby's illustration is unhelpful but obviously not the same as *B. resupinatus* Bolton.

verna (Bull.) Lam., Amanita, *Encyclop.* 1: 113 (1783)
Agaricus vernus Bull., *Herb. France* 3: pl. 108 (1782)
Amanita phalloides var. *verna* (Bull.) Maire, *Bull. Soc. Hist. Nat. Afrique N.* 7: 145 (1916)
Not authentically British. A southern European species, the name misapplied in Britain to *A. virosa* or a white form of *A. phalloides*. Cooke 3 (3) Vol. 1 (1881) is difficult to interpret, apparently not *A. verna* and most probably *A. citrina* var. *alba*.

verna *var.* grisea Rea, Amanita, *Brit. basidiomyc.*: 98 (1922)
Described from Britain but a *nomen dubium*.

vernalis Bolton, Agaricus, *Hist. fung. Halifax* 2: 48 (1788)
Described from Britain, but a *nomen dubium* lacking type material. Bolton's illustration shows a volvate agaric with a large bulbous base, which could be *Amanita citrina*.

vernalis Weinm., Cyphella, *Hymen. Gasteromyc.*: 522 (1836)
A *nomen dubium* not discussed by W.B. Cooke (1961). A collection by Broome (in herb. K) referred here is indeterminable.

vernicosa (Fr.) Gillet, Clitocybe, *Hyménomycètes*: 159 (1874)
Agaricus cerinus β *vernicosus* Fr., *Syst. mycol.* 1: 90 (1821)
Lepista vernicosa (Fr.) Bon, *Doc. mycol.* 13(51): 41 (1983)
A *nomen dubium*. Doubtfully distinct from *Lepista flaccida* though distinguished in Bon (FME4). Cooke 133 (265) Vol. 1 (1883) shows an unidentifiable agaric with an orange-brownish pileus and strongly decurrent, bright yellow lamellae.

vernicosa Bourdot & Galzin, Tulasnella, *Bull. Soc. Mycol. France* 39: 265 (1923)
Not authentically British. Reported from the Isles of Scilly [News Bull. Brit. Mycol. Soc. 25: 6 (1965), as *Gloeotulasnella vernicosa ined.*], but the record lacks voucher material.

versipelle (Fr.) Gillet, Hebeloma, *Hyménomycètes*: 524 (1876)
Agaricus versipellis Fr., *Epicr. syst. mycol.*: 179 (1838)
A *nomen dubium* variously interpreted. Sensu NCL is *H. collariatum*.

versipellis (Fr.) Nikol., Sarcodon, *Fl. pl. crypt. URSS* 6(2): 283 (1961)
Hydnum versipelle Fr., *Öfvers. Kongl. Vetensk.-Akad. Förh.* 18: 31 (1862)
Not authentically British. Reported from Scotland but the material has been redetermined by R. Watling as *Bankera fuligineoalba*.

vesicaria Bull., Tremella, *Herb. France*: pl. 427 f. 3 (1788)
A *nomen dubium* lacking type material. Originally described as a ground-dwelling species and listed as British by Smith [Engl. Bot. 35: 2451 (1813)] and W.G. Smith (1908), but without voucher material. Possibly ascomycetous.

vespertinus (Fr.) Fr., Cortinarius, *Epicr. syst. mycol.*: 272 (1838)
Agaricus vespertinus Fr., *Syst. mycol.* 1: 233 (1821)
Variously interpreted. Sensu CFP3 is not British. Sensu auct., including Rea (1922), is *C. microspermus*, but Pearson & Dennis (1948) considered the only British record 'not convincing'.

vilis (Fr.) Sacc., Clitopilus, *Syll. fung.* 5: 704 (1887)
Agaricus vilis Fr., *Epicr. syst. mycol.*: 150 (1838)
A *nomen dubium*. Last reported in 1913 from West Sussex (Charlton Forest) but no material survives. See Cooke 351 (487) Vol. 3 (1885) for an interpretation. Listed by Rea (1922) where *Eccilia undata* is considered a synonym, but not discussed by Noordeloos (1987).

villatica (Brond.) Bres., Psalliota, *Fungi trident.* 1: 54 (1884)
Agaricus campestris var. *villaticus* Brond., *Rec. pl. Crypt. Agenais*: pl. 7 (1829)
A *nomen dubium* variously interpreted. Sensu Cooke (HBF2) and Rea (1922) is *Agaricus subperonatus*. Sensu Pearson [TBMS 29: 204 (1946)], following Lange (FlDan4: pl. 139C), is *A. urinascens*.

vinosa (Bull.) Sacc., Flammula, *Syll. fung.* 5: 809 (1887)
Agaricus vinosus Bull. [non *A. vinosus* Corda (1839)], *Herb. France*: pl. 54 (1786)
A *nomen dubium* lacking type material. See Cooke 466 (437) Vol. 4 (1884) for an interpretation. Bulliard's plate could easily represent *Paxillus involutus*.

violacea (Alb. & Schwein.) Cooke, Poria, *Grevillea* 14: 112 (1886)
Boletus nitidus var. *violaceus* Alb. & Schwein., *Consp. fung. lusat.*: 258 (1805)
Polyporus violaceus (Alb. & Schwein.) Fr., *Observ. mycol.* 2: 263 (1818)
A *nomen dubium*, but sensu Cooke, Rea (1922) and material in herb. K is *Gloeoporus taxicola*.

violacea *var*. carneolilacina Bres., Russula, *Icon. Mycol.* 9: pl. 444 (1929)
A *nomen dubium*. Reported by Rea [TBMS 17: 45 (1932)] but no material has been traced.

violacea (Johan-Olsen) Juel, Tulasnella, *Bih. Kongl. Svenska Vetensk.-Akad. Handl.* 23: 22 (1897)
Pachysterigma violaceum Johan-Olsen, *Unters. Gesammtgeb. Mykol.* 8: 6 (1889)
A *nomen dubium* lacking type material. Recorded from England by Wakefield & Pearson [TBMS 8: 219 (1923)], but this collection has been redetermined as *Tulasnella pallida*.

violaceifolia Peck, Inocybe, *Rep. (Annual) New York State Mus. Nat. Hist.* 41: 66 (1888)
A *nomen dubium*. Listed in Rea (1922) but a species of *Cortinarius fide* Heim (1931).

violaceofulvens (Batsch) Singer, Panellus, *Bot. Centralbl., Beih.* 56: 142 (1936)
Agaricus violaceofulvens Batsch, *Elench. fung.*: 95 & t. 9 f. 39 (1783)
Not authentically British. A single record from Aberdeenshire (see BFF8: 100) but the material is missing from herb. E *fide* R. Watling (pers. comm.). The epithet was mistranscribed by Fries (1821) as *violaceofulvus* and has been used incorrectly since.

violaceoides Hora, Russula, *Trans. Brit. Mycol. Soc.* 43(2): 458 (1960)
A *nomen dubium* described from Scotland (Rothiemurchus) in 1958 but not collected or reported since. Type material has not survived. Probably *Russula brunneoviolacea fide* Romagnesi (1967) but the spore ornamentation is atypical for that species.

violaceus Sowerby, Agaricus, *Col. fig. Engl. fung.* 2: pl. 209 (1799)
Described from Britain, but a *nomen dubium* lacking type material. Given as a synonym of *Lepista saeva* (as *Tricholoma personatum*) by Saccardo [Syll. fung. 5: 130 (1887)] but could equally be *Lepista nuda*.

violarius (Massee) Sacc., Pluteus, *Syll. fung.* 5: 670 (1887)
Agaricus violarius Massee, *Grevillea* 13: 89 (1885)
A *nomen dubium*. Originally described 'on rotting wood' from Yorkshire (Scarborough) and, as illustrated in Cooke 311 (518) B Vol. 3 (1884), looks distinctive. Possibly a good species awaiting rediscovery and a modern description *fide* BFF4: 79.

viperina (Fr.) Gillet, Volvaria, *Hyménomycètes*: 389 (1876)
Agaricus viperinus Fr., *Epicr. syst. mycol.*: 139 (1838)
A *nomen dubium*. Listed by Rea (1922). Saccardo (Syll. fung. 5: 662) and Orton (BFF4: 79) both suggest this may be a form of *V. gloiocephala*.

virens Quél., Mycena, *Enchir. fung.*: 220 (1886)
A *nomen dubium* lacking type material. British records are sensu Rea (1922) and thus *M. chlorantha*.

virescens Schumach., Tremella, *Enum. pl.* 2: 439 (1803)
Dacrymyces virescens (Schumach.) Fr., *Syst. mycol.* 2: 229 (1822)
A *nomen dubium* lacking type material. Listed by Rea (1922). Old material named thus, in herb. K, is all *Tremella exigua*.

virescens Corda, Naematelia, *Icon. fung.* 3: 35 (1839)
A *nomen dubium* lacking type material. Listed by Rea (1922) as a synonym of *Tremella virescens*. Sensu auct. is *Tremella exigua*.

virescens (Cooke & Massee) Massee, Psilocybe, *Brit. fung.-fl.* 1: 367 (1892)
Agaricus areolatus var. *virescens* Cooke & Massee, *Handb. Brit. fung.*, Edn 2: 377 (1890)
A *nomen dubium fide* Watling (BFF5: 102). Massee cites Cooke 1182 (1177) *Agaricus* (*Psilocybe*) *areolatus* as an illustration. Possibly related to *P. cyanescens*.

virginea Cooke & Massee, Russula, *Grevillea* 19: 41 (1890)

Russula heterophylla var. *virginea* (Cooke & Massee) A. Pearson & Dennis, *Trans. Brit. Mycol. Soc.* 31: 166 (1948)
Described from Britain but a *nomen dubium*. Perhaps an albino form of *Russula heterophylla*, but described with implausibly small spores. See Cooke 1197 (1197) Vol. 8 for an illustration.

viridis (Alb. & Schwein.) J. Schröt., Amaurodon, in Cohn, *Krypt.-Fl. Schlesien* 3: 461 (1889)
Sistotrema viride Alb. & Schwein., *Consp. fung. lusat.*: 262 (1805)
Hydnum viride (Alb. & Schwein.) Fr., *Syst. mycol.* 1: 421 (1821)
Tomentella chlorina (Massee) G. Cunn., *Proc. Linn. Soc. New South Wales* 77: 279 (1952) [1953]
Caldesiella viridis (Alb. & Schwein.) Pat., *Essai taxon.*: 120 (1900)
Not authentically British. Listed in Rea (1922) as *Caldesiella viridis*, but the one old collection so named in herb. K has been redetermined as *Mycoacia uda*.

viridis (With.) Gray, Gymnopus, *Nat. arr. Brit. pl.* 1: 606 (1821)
Agaricus viridis With., *Bot. arr. Brit. pl. ed. 2*, 3: 320 (1792)
Described from Britain but a *nomen dubium* lacking type material. Gray (1821) treats *Agaricus caeruleus* Bolton as a synonym, which is a species of *Stropharia*.

viscidus (L.) Fr., Gomphidius, *Epicr. syst. mycol.*: 319 (1838)
Agaricus viscidus L., *Sp. pl.*: 1173 (1753)
Cortinarius viscidus (L.) Gray, *Nat. arr. Brit. pl.* 1: 629 (1821)
Gomphus viscidus (L.) P. Kumm., *Führ. Pilzk.*: 93 (1871)
A *nomen dubium* lacking type material. Sensu Fries is *Chroogomphus rutilus* but Linnaeus described the stipe as 'short, white and somwhat swollen' which better fits *Gomphidius glutinosus*.

viscosus Berk., Cantharellus, *J. Bot.* 4: 49 (1845)
Not authentically British. Reported, in 1906, from Northumberland (Morpeth) (with voucher material in herb. K) but this is an Australian species unknown in Europe and the origin of the voucher material is doubtful. Closely allied to *Cantharellus cibarius fide* Corner (1966).

vitilis *var*. amsegetes (Fr.) Rea, Mycena, *Brit. basidiomyc.*: 392 (1922)
Agaricus amsegetes Fr., *Hymenomyc. eur.*: 146 (1874)
A *nomen dubium*. Listed by Rea (1922). Not mentioned by Maas Geesteranus (1992).

vitrea (Fr.) Quél., Mycena, *Mém. Soc. Émul. Montbéliard, Sér. 2*, 5: 243 (1872)
Agaricus vitreus Fr., *Syst. mycol.* 1: 146 (1821)
A *nomen dubium* variously interpreted. Cooke (237) 160 Vol. 2 (1882) is probably *Hydropus subalpinus fide* Pearson (1955).

vitrea *var*. tenella (Schumach.) Kühner, Mycena, *Encycl. Mycol. 10. Le Genre Mycena*: 289 (1938)
Agaricus tenellus Schumach., *Enum. pl.* 2: 302 (1803)
Mycena tenella (Schumach.) Quél., *Mém. Soc. Émul. Montbéliard, Sér. 2*, 5: 343 (1873)
A *nomen dubium* lacking type material. Listed (as *M. tenella*) by Rea (1922).

wieslandri (Fr.) M.M. Moser, Flammulaster, *Kleine Kryptogamenflora*, Edn 4: 302 (1978)
Agaricus wieslandri Fr., *Öfvers. Kongl. Vetensk.-Akad. Förh.* 8: 48 (1852)
Naucoria wieslandri (Fr.) Sacc., *Syll. fung.* 5: 856 (1887)
Phaeomarasmius wieslandri (Fr.) Singer, *Lilloa* 22: 577 (1951) [1949]
Not authentically British *fide* NCL. Reported from Yorkshire (Mulgrave Woods) in Naturalist (Hull) 1914: 145 (1914) and also in TBMS 5 (2): 251 (1915), but these records lack voucher material. A collection named thus from Scotland, in herb. K, was reported in Reid & Austwick (1963) but this was sensu K&R and is *Flammulaster muricatus*.

xantholeuca Kühner, Mycena, *Encycl. Mycol.* 10. *Le Genre*
Mycena: 314 (1938)
Not authentically British. Reported from the Republic of Ireland
in 1991 and Northern Ireland in 2001 but unsubstantiated with
voucher material.

xerampelina *var.* olivascens (Pers.) Zvára, Russula, *Z. Pilzk.*
1: 129 (1923)
Russula olivascens Pers., *Observ. mycol.* 1: 103 (1796)
A *nomen dubium*. British records may refer either to *R. clavipes*
or possibly to *R. pseudoolivascens*.

xerotoides H. Post, Marasmius, in Fries, *Monogr. hymenomyc.*
Suec. 2: 231 (1863)
A *nomen dubium* lacking type material. Listed by Rea (1922).
'Difficult to interpret' *fide* A&N1.

xylopes With., Agaricus, *Bot. arr. Brit. pl. ed. 2*, 3: 384 (1792)
Described from Britain but a *nomen dubium.*

xylophila (Weinm.) Sacc., Collybia, *Syll. fung.* 5: 211 (1887)
Agaricus xylophilus Weinm., *Linnaea* 10: 54 (1835)
A *nomen dubium* lacking type material. Cooke 190 (202) Vol. 2
is not helpful showing a small, collybioid agaric with white
lamellae spotting pinkish-brown. Possibly a species of *Mycena*
fide A&N2.

zylophilus Bull., Agaricus, *Herb. France*. pl. 530 f. 2 (1792)
A *nomen dubium* lacking type material. Sowerby's plate (Sow 2:
pl. 167) is similar to that of Bulliard and shows a rather
distinctive collybioid species resembling *Collybia inodora*.

EXCLUDED UREDINIOMYCETES

aconiti-lycoctoni G. Winter, Uromyces, *Pilze Deutschl.*: 153 (1884)
Not authentically British. Reported by Smith & Ramsbottom [TBMS 4 (2): 329 (1913)] as 'sometimes occurring on cultivated plants (of *Aconitum*)' but no material has been traced.

apiculosa Link, Caeoma, in Willd., *Sp. pl.* Edn 4, 6(2): 32 (1825)
A *nomen dubium* variously applied. Listed by Berkeley [Engl. fl.: 382 (1836)].

arrhytidiae (L.S. Olive) Wojewoda, Achroomyces, *Grzyby (Mycota)* 8: 245 (1977)
    Platygloea arrhytidiae L.S. Olive, *Bull. Torrey Bot. Club* 78: 103 (1951)
Not authentically British. The single collection named thus in herb. K (from Scotland) has been redetermined as *Occultifur internus*.

arundinacea Hedw., Puccinia, in Lamarck, *Encyclop.* 8: 250 (1808)
A *nomen dubium*. Sensu auct. and British records refer to *P. phragmitis*.

asarina Kunze, Puccinia, *Mykol. Hefte* 1: 70 (1817)
Not authentically British. The British record refers to a misdetermined collection of *Puccinia fergussonii* on a misidentified host.

aurantiacum Chevall., Gymnosporangium, *Fl. gén. env. Paris*, Edn 1: 424 (1826)
A *nomen dubium* lacking type material. Sensu auct. and British records refer to *G. cornutum fide* W&H.

aurea Sowerby, Uredo, *Col. fig. Engl. fung.* 3: pl. 320 (1801)
Described from Britain supposedly growing on a fern but a *nomen dubium* lacking type material. Sowerby's illustration is not helpful and this is possibly not even a fungus.

barbareae DC., Aecidium, *Fl. Fr.* 5: 244 (1815)
Not authentically British. The record by Cooke [Grevillea 10: 115 (1882)] was based on material supposedly growing on leaves of '*Barbarea praecox*', but the supporting collection in herb. K is actually of *Aecidium (Puccinia) lapsanae* on a basal leaf of *Lapsana communis*.

behenis DC., Aecidium, *Fl. Fr.* 5: 94 (1815)
A *nomen dubium* variously interpreted. Probably referable to stage I of *Uromyces behenis* on *Silene* sp. Listed by Grove (1913).

bullaria Link, Puccinia, in Willd., *Sp. pl.*, Edn 4, 6(2): 78 (1825)
A *nomen dubium*. Sensu auct. and British records are *P. conii*.

cardui Sowerby, Aecidium, *Col. fig. Engl. fung.* 3: pl. 398 f.5 (1803)
Described from Britain but a *nomen dubium* lacking type material. Sowerby's illustration and description suggest *Puccinia punctiformis*.

carpineus Grove, Achroomyces, *J. Bot., London* 60(714): 170 (1922)
An anamorphic ascomycete, provisionally assigned to the genus *Myxosporium*.

cichoracearum Lév., Trichobasis, *Orbigny Dict. Univ. Hist. Nat.*: 19 (1840)
A *nomen dubium,* variously interpreted. Sensu auct. refers to *Coleosporium tussilaginis*, or *Puccinia hieracii* var. *hypochaeridis*, or *Puccinia lapsanae*.

compositarum Mart., Aecidium, *Fl. Crypt. Erlangensis.* 314 (1817)
A *nomen dubium* variously interpreted. British records refer to several of the rusts on the *Compositae*.

convolvuli Castagne, Puccinia, *Observ. Uréd.* 1: 16 (1842)
Not authentically British. A single report on *Calystegia sepium* in 1889, from an unspecified locality in England and unsubstantiated with voucher material.

corni Sowerby, Aecidium, *Col. fig. Engl. fung.* 3: pl. 397 f. 3 (1803)
Described from Britain but a *nomen dubium* lacking type material and probably not a basidiomycete. There is no known aecidium on *Cornus sanguinea* (= *Thelycrania sanguinea*) on which this was supposedly collected.

crepidis J. Schröt., Puccinia, in Cohn, *Krypt.-Fl. Schlesien* 3: 319 (1889)
Not authentically British. Sensu auct. Brit. is *Puccinia crepidicola*.

curtus Cooke, Xenodochus, *Microscopic fungi.* 201 (1865)
Not a basidiomycete *fide* Grove (1913: 390); probably a hyphomycete.

effusa J. Schröt., Platygloea, in Cohn, *Krypt.-Fl. Schlesien* 3: 384 (1889)
    Achroomyces effusus (J. Schröt.) Mig., *Rabenh. Krypt.-Fl.*, Edn 2, 3: 1 (1912)
A *nomen dubium* lacking type material. Most old British collections in herb. K are *Colacogloea peniophorae*.

effusa F. Strauss, Uredo, *Ann. Wetterauischen Ges. Gesammte Naturk.* 2: 91 (1810)
A *nomen dubium* lacking type material and variously interpreted. Sensu Berkeley [Engl. fl.: 381 (1836)] and sensu auct. is *Phragmidium mucronatum*.

epilobii Berk., Trichobasis, *Outl. Brit. fungol.*: 333 (1860)
Described from Britain but a *nomen dubium*. Considered to refer to *Puccinia pulverulenta* by Plowright (1889) and Grove (1913).

filicum Chevall., Uredo, *Fl. gén. env. Paris*, Edn 1: 408 (1827)
A *nomen dubium* lacking type material and variously interpreted. Sensu auct. and British records refer to *Hyalopsora aspidiotus* and *H. polypodii*.

helianthi Schwein., Puccinia, *Schriften Naturf. Ges. Leipzig* 1: 73 (1822)
Not authentically British *fide* W&H. A doubtful record by Massee (1913), unsubstantiated with voucher material.

intybi (Juel) Syd., Puccinia, *Oesterr. Bot. Z.*: 16 (1901)
    Puccinia variabilis var. *intybi* Juel, *Öfvers. Kongl. Vetensk.-Akad. Förh.* 53: 220 (1896)
Not authentically British. Reported by Smith & Ramsbottom [TBMS 4 (2): 329 (1913)] on leaves of *Crepis praemorsa* but no material has been traced.

juniperi Link, Gymnosporangium, in Willd., *Sp. pl.,* edn 4, 6(2): 127 (1825)
    Podisoma foliicola Berk., *Engl. fl.* 5(2): 362 (1836)
    Podisoma juniperi Link, in Willd., *Sp. pl.,* Edn 4, 6(2): 127 (1825)
A *nomen dubium* variously interpreted. Sensu auct. and British records refer to *G. cornutum fide* W&H.

juniperinum Fr., Gymnosporangium, *Syst. mycol.* 3: 506 (1829)
A *nomen dubium*. Sensu auct. and British records refer to *G. cornutum fide* W&H.

laceratum Sowerby, Aecidium, *Col. fig. Engl. fung.* 3: pl.318 (1801)
Described from Britain but a *nomen dubium* lacking type material. Sowerby's illustration suggests the aecidial stage of either *Gymnosporangium clavariiforme* or *G. confusum*.

leucospermum DC., Aecidium, *Fl. Fr.* 2: 239 (1805)
A *nomen dubium*. Sensu auct. Brit. is *Ochropsora ariae* on *Anemone nemorosa*. Sensu Saccardo [Syll. fung. 7: 669 (1888)] is *Tranzschelia anemones* on the same host.

linearis Lév., Trichobasis, in Orbigny, *Dict. univ. hist. nat.* 1: 19 (1841)
A *nomen dubium*. Sensu auct. and British records refer to *Puccinia graminis*.

lychnidearum Lév., Trichobasis, in Orbigny, *Dict. univ. hist. nat.* 1: 19 (1841)
Not authentically British and variously interpreted. Sensu auct. Brit. is *Puccinia arenariae*.

paliformis Fuckel, Puccinia, *Jahrb. Nassauischen Vereins Naturk.* 23-24: 59 (1870)
Not authentically British. Reported from Scotland but this collection was later redetermined as *Puccinia longissima*.

parnassiae Cooke, Uromyces, *Grevillea* 7: 134 (1879)
Described from Britain but type material not traced. Evidently based on an error in host identification and a 'non existent species' *fide* Grove (1913: 389).

pini Pers., Aecidium, in Gmelin, *Syst. nat.*: 1473 (1792)
A *nomen dubium* lacking type material and variously interpreted. Sensu auct. Brit. is *Coleosporium tussilaginis* and *Endocronartium pini*.

plantaginis Berk. & Broome, Uredo, *Ann. Mag. Nat. Hist, Ser. 5,* 7: 130 (1881)
Not a basidiomycete. Apparently a *Synchytrium* sp. *fide* W&H. The collection from the Isle of Wight mentioned by Grove (1913) is *Uromyces beticola* on *Beta* sp.

polygonorum Schltdl., Puccinia, *Fl. berol.*: 132 (1824)
Trichobasis polygonorum Berk., *Outl. Brit. fungol.*: 332 (1860)
A *nomen dubium* lacking type material.

potentillarum DC., Uredo, *Fl. Fr.* 5: 80 (1815)
A *nomen dubium* variously interpreted. British records refer chiefly to *Phragmidium fragariae fide* Grove (1913).

pseudocolumnare J.G. Kühn, Aecidium, *Hedwigia* 23: 168 (1884)
A *nomen dubium* lacking type material. Applied to the aecidial stage (on *Abies*) of several species of *Milesina*, with Scottish records under this name *fide* W&H.

purpurea Cooke, Puccinia, *Grevillea* 5(33): 15 (1876)
Described from India, and not authentically British. Reported on sorghum in England but no material traced. See Moore [British Parasitic Fungi: 310 (1959)] and Wilson & Bisby [TBMS 37: 74 (1954)].

reiliana (J.G. Kühn) G.P. Clinton, Sphacelotheca, *J. Mycol.* 8: 141 (1902)
Ustilago reiliana J.G. Kühn, in Rabenhorst, *Fungi europaei* IV: 1998 (1875)
Not authentically British. Parasitic on *Zea mays* and mentioned in Cooke [Fungoid pests of cultivated plants (1906)] but no British records have been traced.

rhei Sowerby, Aecidium, *Col. fig. Engl. fung.* 3: pl. 398 f. 6 (1803)
Described from Britain on *Rumex* sp. but a *nomen dubium* lacking type material. From the illustration and brief description, this would appear not to be *Puccinia phragmitis* as suggested by various authors, but possibly the uredo- or teleutospore stage of *Uromyces rumicis*.

ruborum DC., Puccinia, *Fl. Fr.* 5: 234 (1815)
A *nomen dubium*. Applied to various species of *Phragmidium* on *Rubus* spp. British records apparently refer to *P. bulbosum*.

scolopendrii J. Schröt., Uredo, in Cohn, *Krypt.-Fl. Schlesien* 3: 374 (1887)
A *nomen dubium* lacking type material and variously interpreted. British records in Plowright (1889) and Grove (1913) refer to both *Milesina blechni* and *M. scolopendrii*.

senecionis Berk., Trichobasis, *Outl. Brit. fungol.*: 332 (1860)
Described from Britain, but a *nomen dubium* variously interpreted.

sonchi-arvensis Pers., Uredo, *Syn. meth. fung.*: 217 (1801)
A *nomen dubium* lacking type material and variously interpreted. Sensu auct. and British records refer to *Coleosporium tussilaginis* and *Miyagia pseudosphaeria*.

statices Desm., Uredo, *Pl. Crypt.*: 32 (1825)
A *nomen dubium*, variously interpreted. Scottish records by Greville in E are *Uromyces armeriae fide* W&H.

straminis Fuckel, Puccinia, *Jahrb. Nassauischen Vereins Naturk.* 15: 9 (1860)
A *nomen dubium* variously interpreted. Sensu auct. and British records refer to both *P. recondita* and *P. striiformis fide* W&H.

submersa H.J. Huds. & Ingold, Jaculispora, *Trans. Brit. Mycol. Soc.* 43: 475 (1960)
Not authentically British. Reported in foam samples from streams in Scotland, but probably misidentified *Naiadella fluitans fide* L. Marvanová (pers. comm.). *Jaculispora submersa* is a predominantly tropical species unlikely to occur in Britain.

urticae Cooke, Uromyces, *Grevillea* 7: 137 (1879)
Described from Britain but type material not traced. Evidently based on an error in host identification and a 'non existent species' *fide* Grove (1913: 389).

valantiae Pers., Puccinia, *Observ. mycol.* 2: 25 (1799)
A *nomen dubium*. British records refer to *P. gallii-verni fide* W&H.

vulgare Tode, Stilbum, *Fung. mecklenb. sel.* 1: 10 (1790)
This name was once applied loosely to any stilboid fungus, but investigation of old collections at K shows no evidence that *Stilbum vulgare* sensu stricto has ever been found in Britain. Specimens examined represent various anamorphic ascomycetes.

EXCLUDED USTILAGINOMYCETES

aspidii Höhn., Entyloma, *Mitt. bot. Inst. tech. Hochsch. Wien* 4: 103 (1927)
    Septocylindrium aspidii Bres. (anam.), *Hedwigia* 35: 201 (1896)
Based on *Septocylindrium aspidii*, recorded from Scotland, but doubtfully a member of the *Ustilaginales fide* UBI: 61.

azaleae Peck, Exobasidium, *Rep. (Annual) New York State Mus. Nat. Hist.* 26: 72 (1874) [1873]
Not authentically British. A North American species. Sensu auct. Brit. = *Exobasidium japonicum*.

berkeleyi Massee, Tilletia, *Bull. Misc. Inform. Kew.*: 154 (1889)
A *nomen dubium*. The type material from Northamptonshire on *Triticum vulgare* is sterile and indeterminable.

carbonaria Sowerby, Farinaria, *Col. fig. Engl. fung.* 3: pl. 396 f. 4 (1803)
Described from Britain but not fungal and based on an insect gall. See Vánky (as *Cintractia carbonaria*).

cardui A.A. Fisch. Waldh., Ustilago, *Bull. Soc. Imp. Naturalistes Moscou.* 40: 255 (1867)
Not authentically British. Reported by Cooke [Microscopic Fungi: 231 (1878)], Plowright (1889: 282) and Mason [Naturalist (Hull) 1921: 349 (1921)] but no voucher material has been traced.

comari (Berk. & F.B. White) De Toni & Massee, Doassansia, *J. Mycol.*: 18 (1888)
    Protomyces comari Berk. & F.B. White, *Ann. Mag. Nat. Hist, Ser. 5 [Notices of British Fungi no. 1708]* 1: 27 (1878)
Described from Britain. This is not a basidiomycete, but the chytrid *Physoderma comari* (Berk. & F.B. White) Lagerh.

compositarum Farl., Entyloma, *Bot. Gaz.* 8: 275 (1883)
A North American species. Reported British material on cultivated species of *Gaillardia* is *Entyloma galliardiana* Vánky.

cucumis A.B. Griffiths, Ustilago, *Trans. Roy. Soc. Edinburgh* 15: 404 (1888)
Not a basidiomycete. A name given to zoogloea threads in the root nodules of *Cucumis sativus*.

cypericola (Magnus) C.A. Weber, Entorrhiza, *Bot. Zeitung (Berlin)* 42: 378 (1884)
    Schinzia cypericola Magnus, *Verh. Bot. Vereins Prov. Brandenburg* 20: 54 (1878)
Not authentically British. Reported from Perthshire on roots of *Eleocharis quinqueflora* in 1979 but unsubstantiated with voucher material. Probably misidentified *Entorrhiza scirpicola*. *E. cypericola* is restricted to *Cyperus flavescens* which is not British.

delastrina (Tul. & C. Tul.) G. Winter, Schroeteria, *Rabenh. Krypt.-Fl.* 1(1): 117 (1881)
    Thecaphora delastrina Tul. & C. Tul., *Ann. Sci. Nat., Bot.,* sér. 3, 7: 108 (1847)
    Geminella delastrina (Tul. & C. Tul.) J. Schröt., *Beitr. Biol. Pflanzen* 2: 5 (1877)
Not a basidiomycete. Producing sori in seeds of *Veronica arvensis* (in Britain), this has recently been shown, by ultrastructural and germination studies, to be an ascomycete.

ficuum Reichardt, Ustilago, *Verh. zool.-bot. Ges. Wien* 17: 335 (1867)
Not a basidiomycete. The type specimen, growing on dried figs, represents the hyphomycete *Aspergillus niger.*

gladioli W.G. Smith, Urocystis, *Gard. Chron.* 6: 421 (1876)
Not a basidiomycete. Type material is a hyphomycete in the genus *Papulospora.*

grammica Berk. & Broome, Ustilago, *Ann. Mag. Nat. Hist, Ser. 2 [Notices of British Fungi no. 483]* 5: 464 (1850)
Described from Britain but not a basidiomycete *fide* Vánky.

horvathianum (F. Thomas) Nannf., Exobasidium, *Symb. Bot. Upsal.* 23(2): 45 (1981)
    Exobasidium discoideum var. *horvathianum* F. Thomas, *Forstl.-Naturwiss. Z.* 6: 305 (1897)
Not authentically British. Reported on *Rhododendron luteum* in England but lacking voucher material.

laevis J.G. Kühn, Tilletia, in Rabenhorst, *Fungi europaei* IV: 1697 (1873)
Not authentically British. A record exists on 'chicken corn' in Liverpool in 1929, but this was probably imported and no material survives.

macrospora Desm., Ustilago, *Pl. Crypt. Fr.,* edn 3: 356 (1852)
A *nomen dubium*. Sensu auct. is *Tilletia olida*, but British records refer to *Ustilago serpens fide* UB: 57.

montagnei (Tul. & C. Tul.) M. Piep., Begerow & Oberw., Ustanciosporium, *Nova Hedwigia* 70: 289 (2000)
    Ustilago montagnei Tul. & C. Tul., *Ann. Sci. Nat., Bot.,* sér. 3, 7: 88 (1847)
Not authentically British. Revision of material named thus in herb. K reveals a mixture of *Ustanciosporium gigantosporum* and *U. major.*

montiae (Rostr.) Rostr., Tolyposporium, *V. i den danske fl. 2.* 31 (1904)
    Sorosporium montiae Rostr., *Bot. Tidsskr.*: 129 (1896)
Reported on *Montia fontana* by Wakefield & Dennis [TBMS 29: 145 (1946)] but *fide* Vánky [Mycotaxon 38: 276 (1990)] this is an ascomycete.

pachysporum Nannf., Exobasidium, *Symb. Bot. Upsal.* 23 (2): 54 (1981)
A Scandinavian species, erroneously included as British by Ing (1994) followed by Preece & Spooner (in Redfern *et al.,* 2002).

phoenicis Corda, Ustilago, *Icon. fung.* 4: 9 pl.3 (1840)
Not a basidiomycete. Type material, growing on dried dates, is the hyphomycete *Aspergillus niger* var. *phoenicis.*

saponariae F. Rudolphi, Sorosporium, *Linnaea.* 4: 116 (1829)
    Ustilago rudolphii Tul. & C. Tul. [*nom. nov.* & *nom. illegit.*], *Ann. Sci. Nat., Bot.,* sér. 3, 7: 99 (1847)
Not authentically British. Reported (as *Ustilago rudolphi*) in Grevillea 10: 67 (1881) as 'growing in the anthers of *Dianthus deltoides* in a garden in Norwich'. The plants were stated to have been brought back from Switzerland, already infected by the smut. However the description in Grevillea does not agree with that of *S. saponariae* and no material survives.

scabies Berk., Tuburcinia, *Ann. Mag. Nat. Hist, Ser. 2 [Notices of British Fungi no. 489]* 5: 464 (1850)
    Sorosporium scabies (Berk.) A.A. Fisch. Waldh., *Aperçu.* 33 (1877)
Described from Britain but not a basidiomycete. The type collection, on potato tubers, is *Spongospora subterranea* (Wallr.) Lagerh., a member of the Plasmodiophorales.

separata G. Winter, Tilletia, *Rabenh. Krypt.-Fl.* 1: 111 (1881)
Not authentically British. Sensu Massee (1899) is *T. sphaerococca.*

sphagni Navashin, Tilletia, *Bot. Centralbl.* 43: 290 (1890)
Reported in spore capsules of *Sphagnum papillosum* from England (Westmorland: Ambleside, Blelham Tarn) in 1948, but not a basidiomycete. See Redhead & Spicer [Mycologia 73: 906 (1981)].

tritici Körn., Urocystis, *Hedwigia.* 16: 33 (1877)
    Tuburcinia tritici (Körn.) Liro, *Ann. Univ. Fenn. Aboen.* 1: 17 (1922)
Previously included in the synonymy of *Urocystis agropyri.* No British records have been found.

ADDENDUM

INCLUDED BASIDIOMYCETES

Cortinarius balteatoalbus Rob. Henry, *Doc. Mycol.* 16(61): 23 (1985)
Mis.: *Cortinarius crassus* sensu NCL and sensu Lange (FlDan3: 21 & pl. 88A)
E: ! **S:** !
H: On soil, in woodland.
D+I: B&K5: 190 **I:** FlDan3 pl. 88A (as *C. crassus*), CFP D21
Rarely reported or collected. Known from England (Isle of Wight and West Kent) and reported from Scotland.

Russula lepidicolor Romagnesi, *Bull. Mens. Soc. Linn. Lyon* 31: 174 (1962)
E: ! **S:** !
H: On soil, associated with *Fagus* in beechwoods or mixed deciduous woodland.
I: LR 265
Single collections in 2004, from England (Buckinghamshire: Hodgemore Woods) and Scotland (Mid-Perthshire: Rannoch).

Mycena truncosalicicola D.A. Reid, *Czech Mycol.* 48(4): 261 (1996)
E: !
H: On dead standing trunk of *Salix caprea*.
Described from a single collection from Shropshire (Llynclys). Possibly a small form of *Mycena arcangeliana fide* E.E. Emmett (pers. comm.).

EXCLUDED BASIDIOMYCETES

lacunosum Bull., Lycoperdon, *Herb. Fr.*: pl. 52 (1781)
Lycoperdon perlatum var. *lacunosum* (Bull.) Rea, *Brit. Basidio.*: 34 (1922)
A *nomen dubium* lacking type material. Bulliard's illustration shows a species of *Lycoperdon* with an elongated and markedly lacunate stipe, possibly *L. excipuliforme* or a form of it. Sensu Rea is unknown.

INDEX TO TAXA

All species treated in the checklist are indexed below alphabetically by epithet. Infraspecific taxa (varieties, forms) are indexed after their species epithet. Genera are indexed by genus name. Entries for current names of included taxa and aliens are in bold type, with synonyms in italics. Excluded taxa are in plain Roman type, with synonyms in italics. Misapplied names are also in italics.

albostipitata, **Tubaria**, 299
albostramineum, Corticium, 146
albostramineum, Hypochnicium,
146
albostramineum ssp. eichleri,
Gloeocystidium, 146
albostramineus, Hypochnus, 146
albotomentosa, Naucoria, 206
albotomentosum, Entoloma, 91
albovelutipes, Inocybe, 148
alboviolaceus, Agaricus, 64
alboviolaceus, Cortinarius, 64
alboviolascens, Cyphella, 160
alboviolascens, Lachnella, 160
alboviolascens, Peziza, 160
alboviride, Hydnum, 158
alboviridis, Kavinia, 158
albulensis, Puccinia, 321
album, Fusisporium, 345
album, Hymenangium, 138
album, Microstroma, 345
album, Tricholoma, 297
album, Tricholoma, 294
album *var.* caesariatum, Tricholoma,
358
albus, Agaricus, 294
albus, Boletus, 358
albus, Boletus, 358
albus, Bulleromyces, 29
albus, Ceriomyces, 237
albus, Hymenogaster, 138
albus, Marasmiellus, 358
albus, Polyporus, 358
albus, Ptychogaster, 237
albus, Rhizopogon, 138, 140
albus, Sporobolomyces, 29
albus-corticis, Marasmius, 188, 189
alcalina, Mycena, 358
alcalina, Mycena, 201, 204
alcalina var. chlorinella, Mycena, 201
alcalinus, Agaricus, 201, 358
alcalinus ssp. leptocephalus, Agaricus,
201
alchemillae, Trachyspora, 334
alchemillae, Uromyces, 334
Aldridgea, 50
aldridgei, Flammula, 358
aleuriatus, Agaricus, 22
aleuriatus, Bolbitius, 22
aleuriatus, Pluteolus, 22
aleuriatus var. reticulatus, Pluteolus,
22
aleuriosmus, Cortinarius, 75
aleuriosmus, Cortinarius, 64
Aleurodiscus, 7
alexandri, Clitocybe, 42
alexandri, Lepista, 42
alexandri, Locellina, 358
alexandri, Paxillus, 42
algeriensis, Agaricus, 359
algeriensis, Pilosace, 358
algidus, Agaricus, 359
algidus, Pleurotus, 359
algidus, Pleurotus, 127
alienata, Hyphodontia, 359
alienata, Peniophora, 359
alismatis, Doassansia, 341
alismatis, Perisporium, 341
alismatis, Sclerotium, 341
alkalivirens, Collybia, 359
alkalivirens, Collybia, 47
allantospora, Phlebiella, 222
allantospora, Tulasnella, 301
allantospora, Xenasmatella, 222
allantosporum, Aphanobasidium, 222

allenii, Omphalia, 359
allenii, Omphalina, 359
allescheri, Corticium, 219
alliacea, Dendrothele, 359
alliacea, Odontia, 291
alliaceum, Corticium, 359
alliaceus, Agaricus, 189
alliaceus, Marasmius, 189
alligatus, Polyporus, 359
allii, Aecidium, 330
allii, Puccinia, 321
allii-fragilis, Melampsora, 315
allii-phalaridis, Puccinia, 330
allii-populina, Melampsora, 315
allii-salicis-albae, Melampsora, 316
alliorum, Uredo, 321
allosperma, Leptonia, 91
allospermum, Entoloma, 91
allospora, Galerina, 109
allutus, Cortinarius, 399
allutus, Cortinarius, 359
almeni, Agaricus, 213
alneti, Agaricus, 239
alnetorum, Alnicola, 206
alnetorum, Clitocybe, 359
alnetorum, Cortinarius, 64
alnetorum, Hydrocybe, 64
alnetorum, Naucoria, 206
alnetorum, Russula, 256
alnetorum, Telamonia, 64
alneus, Agaricus, 267
alneus, Apus, 267
alni, Vuilleminia, 307
Alnicola, 206
alnicola, Agaricus, 224
alnicola, Athelia, 17
alnicola, Cristella, 290
alnicola, Dryophila, 224
alnicola, Flammula, 224
alnicola, Grandinia, 290
alnicola, Pholiota, 224
alnicola, Trechispora, 290
alnicola var. salicicola, Flammula, 224
alnicola var. salicicola, Pholiota,
224
alnicola β salicicola, Agaricus, 224
aloës, Uredo, 356
aloës, Uromyces, 356
aloicola, Uromyces, 356
alopecius, Coprinus, 56
alopecuri, Urocystis, 347
alpestre, Lyophyllum, 359
alpestris, Agaricus, 359
alpestris, Calocybe, 359
alphitophora, Mycena, 354
alphitophorus, Agaricus, 354
alpina, Collybia, 46
alpina, Guepinia, 118
alpina, Guepiniopsis, 118
alpina, Lichenomphalia, 181
alpina, Melampsora, 315
alpina, Omphalina, 181, 182
alpina, Russula, 262
alpina, Russula, 359
alpinum, Hebeloma, 121
alpinus, Agaricus, 181
alpinus, Cortinarius, 70
alpinus, Cortinarius, 359
alpinus, Gymnopus, 46
altaica, Laccaria, 160
alternatus, Agaricus, 359
alternatus, Coprinus, 359
altipes, Agaricus, 1
altipes, Psalliota, 1
alumnus, Agaricus, 359

alutacea, Hyphodontia, 141
alutacea, Incrustoporia, 273
alutacea, Kneiffiella, 142
alutacea, Odontia, 141
alutacea, Omphalia, 359
alutacea, Poria, 273
alutacea, Russula, 256
alutacea, Skeletocutis, 273
alutacea, Thelephora, 306
alutacea f. purpurella, Russula, 261
alutacea f. rubroalba, Russula, 264
alutacea var. erythropus, Russula, 266
alutacea var. olivascens, Russula, 404
alutacea var. vinosobrunnea, Russula,
266
alutaceoumbrinum, Tomentellastrum,
287
alutaceoumbrinus, Hypochnus, 287
alutaceum, Gloeocystidium, 306
alutaceum, Hydnum, 141
alutaceus Cooke & Massee, Agaricus,
359
alutaceus Fr., Agaricus, 256
alutaceus, Polyporus, 359
alutaria, Grandinia, 142
alutaria, Hyphodontia, 142
alutaria, Kneiffiella, 142
alutaria, Odontia, 142
alutaria, Peniophora, 142
alutarius, Boletus, 303
alutarius, Tylopilus, 303
alveolatus, Ptychogaster, 1
alveolus, Agaricus, 359
alveolus, Crepidotus, 359
Alysidium, 26
amabilis, Boletus, 359
amabilis, Suillus, 281
amabilis, Suillus, 359
amadelphus, Agaricus, 188
amadelphus, Marasmiellus, 188
amadelphus, Marasmius, 188
Amanita, 8
amanitae, Agaricus, 47
amanitae ssp. cirrhata, Agaricus, 47
Amanitopsis, 8
amara, Clitocybe, 359
amara, Clitocybe, 181
amara, Lepista, 181, 359
amara, Pholiota, 224
amara, Russula, 258
amarella, Clitocybe, 359
amarellus, Agaricus, 369
amarescens, Alnicola, 206
amarescens, Clitocybe, 42
amarescens, Cortinarius, 66
amarescens, Naucoria, 206
amarissima, Russula, 257
amariusculum, Lyophyllum, 359
amariusculum, Lyophyllum, 185
amarum, Tricholoma, 181
amarus (Alb. & Schwein.) Fr.,
Agaricus, 359
amarus Bull., Agaricus, 224
amarus sensu auct. Brit., Agaricus,
181
amarus, Boletus, 25
amarus, Leucopaxillus, 181, 359
Amaurodon, 11
ambigua, Conocybe, 53
ambigua, Conocybe, 50
ambigua, Galera, 50
ambigua, Octaviania, 192
ambigua, Poria, 211
ambigua, Sebacina, 269
ambigua, Uredo, 336

atrobrunneus, *Agaricus*, 362
atrocaerulea, Hohenbuehelia, 127
atrocaerulea, Hydrocybe, 69
atrocaeruleus, Agaricus, 127
atrocaeruleus, Cortinarius, 69
atrocaeruleus, Pleurotus, 127
atrocaeruleus f. *albidotomentosus, Pleurotus*, 127
atrocaeruleus var. *griseus, Pleurotus*, 386
atrocinereum, Dermoloma, 88
atrocinereum, Tricholoma, 88
atrocinereus, Agaricus, 88
atrocoerulea, Leptonia, 91
atrocoeruleum, Entoloma, 91
atrocrocea, *Lepiota*, 362
atrocroceus, Agaricus, 362
atrocyanea, *Mycena*, 362
atrocyaneus, Agaricus, 362
atroglauca, *Russula*, 362
atrolaminata, Psathyrella, 238
atromarginata, Leptonia, 91
atromarginata, *Mycena*, 362
atromarginatum, Entoloma, 91
atromarginatum, Tricholoma, 295
atromarginatus, Agaricus, 362
atromarginatus, Pluteus, 230
atromarginatus, Rhodophyllus, 91
atropapillata, *Mycena*, 362
atropuncta, Camarophyllopsis, 32
atropuncta, Eccilia, 32
atropuncta, Hygrocybe, 32
atropuncta, Omphalia, 32
atropunctum, Hygrotrama, 32
atropunctus, Agaricus, 32
atropunctus, Hygrophorus, 32
atropurpurea, Russula, 257
atropurpurea var. *depallens, Russula*, 375
atropurpurea var. *krombholzii, Russula*, 257
atropurpureum, Lycoperdon, 184
atropurpureum, Lycoperdon, 184
atropurpureus, Agaricus, 257
atropusillus, Cortinarius, 65
atrorubens, Russula, 257
atrorubens, Russula, 261
atrorufa, *Psilocybe*, 362
atrorufa, *Psilocybe*, 248
atrorufus Bolton, *Agaricus*, 363
atrorufus Schaeff., *Agaricus*, 362
atrorufus sensu Cooke, *Agaricus*, 238
atrosquamosum, Tricholoma, 295
atrosquamosum, Tricholoma, 295
atrosquamosum *var.* squarrulosum, Tricholoma, 295
atrosquamosus, Agaricus, 295
atrostipitatum, Leccinum, 168
atrotomentosa, Tapinella, 282
atrotomentosus, Agaricus, 282
atrotomentosus, Paxillus, 282
atrovirens, Agaricus, 297
atrovirens, Agyrium, 363
atrovirens, Byssocorticium, 29
atrovirens, Coniophora, 29
atrovirens, Corticium, 29
atrovirens, Cortinarius, 65
atrovirens, Epidochium, 292, 363
atrovirens, Hypochnus, 286
atrovirens, *Mycena*, 363
atrovirens, Thelephora, 29
atrovirens, Tomentella, 286
atrovirens, *Tremella*, 363
atrovirens, Tremella, 292

attenuata, Agrocybe, 5
attenuata var. *subincarnata, Poria*, 273
audreae, *Lepiota*, 171
augeana, Clitocybe, 42
augusta, *Psalliota*, 1
augustus, Agaricus, 1
augustus var. *perrarus, Agaricus*, 1
auloderma, Puccinia, 330
aurantia, *Armillaria*, 295
aurantia, *Hydrocybe*, 130
aurantia, Hygrocybe, 130
aurantia, *Thelephora*, 7
aurantia, Tremella, 292
aurantiaca, Auricularia, 222, 275
aurantiaca, Clitocybe, 136
aurantiaca, Hygrophoropsis, 136
aurantiaca, Krombholziella, 168
aurantiaca, *Peniophora*, 363
aurantiaca, Peniophora, 214
aurantiaca, Phlebia, 222
aurantiaca, Russula, 257
aurantiaca, Stropharia, 277
aurantiaca f. *percandida, Krombholzia*, 407
aurantiaca var. *albida, Clitocybe*, 136
aurantiaca var. *lactea, Clitocybe*, 136
aurantiaca var. *macrospora, Hygrophoropsis*, 136
aurantiaca var. *nigripes, Clitocybe*, 136
aurantiaca var. *nigripes, Hygrophoropsis*, 136
aurantiaca var. *pallida, Hygrophoropsis*, 136
aurantiaca var. *radiata, Phlebia*, 222
aurantiaca var. *rufa, Hygrophoropsis*, 136
aurantiacum, Corticium, 363
aurantiacum, Gymnosporangium, 313
aurantiacum, *Gymnosporangium*, 433
aurantiacum, Hydnellum, 128
aurantiacum, Hydnum, 128
aurantiacum, Leccinum, 168
aurantiacum, Sporotrichum, 35
aurantiacum var. *decipiens, Leccinum*, 375
aurantiacum var. *quercinum, Leccinum*, 168
aurantiacum β leucopodium, *Leccinum*, 363
aurantiacum γ rufum, *Leccinum*, 363
aurantiacus, Agaricus, 136
aurantiacus, Boletus, 168
aurantiacus, Cantharellus, 136
aurantiacus, Ceriomyces, 167
aurantiacus, Hapalopilus, 121
aurantiacus, Hormomyces, 293
aurantiacus, Lactarius, 161
aurantiacus, Merulius, 29, 136
aurantiacus, Ptychogaster, 167
aurantiacus var. *pallidus, Cantharellus*, 136
aurantiacus β *lacteus, Cantharellus*, 136
aurantioferrugineus, *Agaricus*, 363
aurantiofulvus, Lactarius, 161
aurantiolutescens, Hygrocybe, 133
aurantiolutescens, Hygrophorus, 133
aurantiolutescens f. *pseudoconica, Hygrocybe*, 133
aurantiolutescens var. *parapersistens, Hygrocybe*, 133
aurantiomarginata, Mycena, 198
aurantiomarginatus, Agaricus, 198

aurantiorugosus, Agaricus, 230
aurantiorugosus, Pluteus, 230
aurantiosplendens, Hygrocybe, 130
aurantiosplendens, Hygrophorus, 130
aurantioturbinatus, Cortinarius, 69
aurantioviscida, Hygrocybe, 132
Aurantiporus, 19
aurantiporus, Boletus, 281
aurantium, Corticium, 7
aurantium, Lycoperdon, 268, 363
aurantium, *Scleroderma*, 363
aurantium, Scleroderma, 268
aurantium, Tricholoma, 295
aurantius, Agaricus, 295
aurantius, Aleurodiscus, 7
aurantius, Hygrophorus, 130
aurata, *Russula*, 257
auratile, Hydnellum, 128
auratile, Hydnum, 128
auratum, Tricholoma, 295
auratus Paulet, *Agaricus*, 295
auratus With., *Agaricus*, 257
aurea, *Clavaria*, 250
aurea, *Conocybe*, 51
aurea, *Daedalea*, 363
aurea, Galera, 51
aurea, *Lepiota*, 179
aurea, Mycoacia, 205
aurea, Odontia, 205
aurea, Phaeolepiota, 216
aurea, Pholiota, 216
aurea, Ramaria, 250
aurea, Russula, 257
aurea, *Serpula*, 246
aurea, *Togaria*, 217
aurea, *Uredo*, 320
aurea, *Uredo*, 433
aurea var. *herefordensis, Pholiota*, 217
aurea var. *herefordensis, Togaria*, 217
aurea var. *vahlii, Pholiota*, 217
aurea var. *vahlii, Togaria*, 217
aureifolius, *Cortinarius*, 363
Aureoboletus, 19
aureola, Agaricus, 10
aureola, Amanita, 10
aureomarginatus, Cortinarius, 65
aureopulverulentus, Cortinarius, 65
aureoturbinatum, Phlegmacium, 69
aureum, Botryobasidium, 26
aureum, Cystoderma, 217
aureum, Haplotrichum, 26
aureum, Hydnum, 205
aureum, Oidium, 26
aureum, Sporotrichum, 35
aureus, Agaricus, 119, 216
aureus, Hygrophorus, 137
aureus, Merulius, 246
aureus, Pseudomerulius, 246
aureus var. *herefordensis, Agaricus*, 217
aureus var. *vahlii, Agaricus*, 217
auricoma, *Inocybe*, 363
auricomum, Ozonium, 61
auricomus, Agaricus, 363
auricomus, Coprinus, 57
auricula, Auricularia, 19
auricula, Hirneola, 19
auricula, Lentinus, 170
auricula, Tremella, 19
auricula-judae, Auricularia, 19
auricula-judae, Exidia, 19
auricula-judae, Hirneola, 19

bertillonii, Lactarius, 161
bertillonii var. queletii, Lactarius, 161
betae, Trichobasis, 336
betae, Uredo, 336
betae, Uromyces, 336
betae β convolvuli, Uredo, 329
beticola, Uredo, 336
beticola, Uromyces, 336
betonicae, Puccinia, 321
betonicae, Uredo, 321
betulae, Coniophora, 50
betulae, Phaeosolenia, 84
betulae, Phaeosolenia, 364
betularum, Russula, 257
betuletorum, Cortinarius, 65
betulicola, Boletus, 23
betulina, Daedalea, 170
betulina, Lecythea, 316
betulina, Melampsora, 316
betulinum, Melampsoridium, 316
betulinum, Sclerotium, 316
betulinus, Agaricus, 170, 213
betulinus, Boletus, 228
betulinus, Cortinarius, 66
betulinus, Lenzites, 170
betulinus, Piptoporus, 228
betulinus, Polyporus, 228
betulinus f. flaccida, Lenzites, 170
Biatoropsis, 310
bibula, Omphalina, 38
bibulosa, Collybia, 46
bibulosus, Agaricus, 46
bibulus, Cortinarius, 66
bickhamensis, Crepidotus, 81
bickhamensis, Lepiota, 171
bicolor, Athelia, 228
bicolor, Corticium, 228
bicolor, Cortinarius, 66
bicolor, Entyloma, 342
bicolor, Hydnum, 252
bicolor, Laccaria, 159
bicolor, Laxitextum, 167
bicolor, Odontia, 252
bicolor, Piloderma, 228
bicolor, Resinicium, 252
bicolor, Stereum, 167
bicolor, Thelephora, 167
bienne, Sistotrema, 1
biennis, Abortiporus, 1
biennis, Boletus, 1
biennis, Daedalea, 1
biennis, Heteroporus, 1
biennis, Polyporus, 1
biennis, Thelephora, 364
biennis var. distortus, Polyporus, 1
biennis var. sowerbei, Polyporus, 1
biennis β sowerbei, Daedalea, 1
biformis, Clavaria, 251
biformis, Clavulinopsis, 251
biformis, Cortinarius, 77
biformis, Cortinarius, 66
biformis, Ramariopsis, 251
bifrons, Agaricus, 364
bifrons, Psathyra, 364
bifrons, Psathyrella, 364
bifrons, Uredo, 320, 338
bifrons var. semitincta, Psathyra, 364
bifrons var. semitinctus, Agaricus, 364
bigelowii, Anthracoidea, 340
bilanatus, Coprinus, 364
binucleospora, Athelia, 364
biornata, Lepiota, 364
biornatus, Agaricus, 353, 364
biornatus, Leucocoprinus, 177
biornatus, Leucocoprinus, 364

bipellis, Coprinus, 364
bipellis, Psathyra, 238
bipellis, Psathyrella, 238
birnbaumii, Agaricus, 179
birnbaumii, Leucocoprinus, 179
birrus, Agaricus, 122
birrus, Hebeloma, 122
bisphaerigera, Fayodia, 106
bisphaerigera, Mycena, 106
bisphaerigera, Omphalia, 106
bisphaerigera var. anthracobia,
    Fayodia, 361
bispora, Athelia, 17
bispora, Colacogloea, 311
bispora, Mycoaciella, 205
bispora, Platygloea, 311
bispora, Psalliota, 2
bisporiger, Coprinus, 57
bisporigera, Eccilia, 92
bisporigerum, Entoloma, 92
bisporigerum, Tricholoma, 298
bisporum, Corticium, 17
bisporum, Resinicium, 205
bisporus, Achroomyces, 311
bisporus, Agaricus, 2
bisporus, Coprinus, 57
bisporus, Coprinus, 57
bisporus, Hypochnus, 17
bistortae, Puccinia, 321
bistortarum, Uredo, 349
bistortarum, Ustilago, 349
bistortarum var. inflorescentiae,
    Ustilago, 349
bistortarum α pustulata, Uredo, 350
bistortarum γ ustilaginea, Uredo, 349
bisus, Lentinellus, 364
bisus, Lentinus, 364
bitorquis, Agaricus, 2
bitorquis, Psalliota, 2
bitorquis var. validus, Agaricus, 2
bivelus, Agaricus, 66
bivelus, Cortinarius, 66
bizzozeriana, Clavaria, 251
Bjerkandera, 21
blandus, Agaricus, 364
blattaria, Conocybe, 56
blattaria, Conocybe, 51
blattaria, Pholiota, 51
blattaria, Pholiotina, 51
blattaria, Togaria, 51
blattaria f. exannulata, Conocybe, 52
blattarius, Agaricus, 51
blechni, Melampsorella, 317
blechni, Milesia, 317
blechni, Milesina, 317
blennius, Agaricus, 161
blennius, Lactarius, 161
blennius f. albidopallens, Lactarius,
    162
blennius f. virescens, Lactarius, 161
blepharistoma, Hyalopycnis, 126
blepharistoma, Polyporus, 36
blepharistoma, Poria, 36
blepharistoma, Sphaeronaema, 126
bloxamii, Agaricus, 92
bloxamii, Cyphella, 214
bloxamii, Entoloma, 92
bloxamii, Rhinotrichum, 26
boertmannii, Lepiota, 171
bohemica, Alnicola, 206
bohemica, Lepiota, 188
bohemica, Naucoria, 206
bohusii, Agaricus, 2
Boidinia, 22
boidinii, Peniophora, 214

boidinii, Phlebiella, 223
bolaris, Agaricus, 66
bolaris, Cortinarius, 66
Bolbitius, 22
boletiformis, Agaricus, 117
Boletinus, 280
Boletopsis, 22
Boletus, 23
bolivarii, Urocystis, 347
boltonii, Agaricus, 22
boltonii, Astrosporina, 150
boltonii, Bolbitius, 22
boltonii, Corticium, 364
boltonii, Hymenochaete, 364
boltonii, Inocybe, 150, 156
boltonii, Prunulus, 22
boltonii, Stereum, 364
bombacina, Athelia, 18
bombacinum, Sporotrichum, 18
bombycina, Anomoporia, 364
bombycina, Kneiffiella, 287
bombycina, Poria, 364
bombycina, Thelephora, 146
bombycina, Volvaria, 306
bombycina, Volvariella, 306
bombycinum, Corticium, 146
bombycinum, Hypochnicium, 146
bombycinum f. pinicola,
    Hypochnicium, 147
bombycinus, Agaricus, 306
bombycinus, Hypochnus, 146
bombycinus, Polyporus, 364
bongardii, Agaricus, 149
bongardii, Inocybe, 149
bongardii var. pisciodora, Inocybe,
    364
boreale, Stereophyllum, 84
boreale, Tricholoma, 364
borealis, Agaricus, 364
borealis, Armillaria, 14
borealis, Ceraceomyces, 34
borealis, Climacocystis, 365
borealis, Clitocybe, 209
borealis, Cuphophyllus, 365
borealis, Daedalea, 365
borealis, Hygrocybe, 365
borealis, Hygrophorus, 365
borealis, Leptoporus, 365
borealis, Merulius, 34
borealis, Omphaliaster, 209
borealis, Polyporus, 365
borealis, Puccinia, 330
borealis, Rhodocybe, 209
borealis, Russula, 365
borealis, Russula, 261
borgensis, Psathyrella, 238
Botrydina, 181
Botryobasidium, 26
Botryohypochnus, 27
botryoides, Hypochnus, 365
botryoides, Lepraria, 182
botryoides, Thelephora, 365
botryoides, Tomentella, 365
botryoides, Tomentella, 287
botryoideum, Botryobasidium, 27
botryosum, Botryobasidium, 27
botryosum, Corticium, 27
botryosus, Aleurodiscus, 7
botrytis, Clavaria, 250
botrytis, Ramaria, 250
botrytis var. alba, Clavaria, 250
boucheanus, Polyporus, 234
boudieri, Coprinus, 56
boudieri, Cortinarius, 365
boudieri, Lepiota, 171

bursiformis, **Conchomyces**, 352
bursiformis, Hohenbuehelia, 352
Butlerelfia, 84
butyracea, Collybia, 46
butyracea, Rhodocollybia, 46
butyracea f. *asema, Collybia*, 46
butyracea f. *asema, Rhodocollybia*, 46
butyracea *var.* asema, Collybia,
   46
butyracea var. *bibulosa, Collybia*, 46
butyraceus, Agaricus, 46
butyraceus γ *asemus, Agaricus*, 46
buxi, Androsaceus, 189
buxi, Marasmius, 189
buxi, Puccinia, 322
byssinella, Cristella, 290
byssinella, Trechispora, 290
byssinellum, Corticium, 290
byssinum, Corticium, 228
byssinum, Piloderma, 228
byssinus, Lyomyces, 228
byssisedum, Entoloma, 92
byssisedus, Agaricus, 92
byssisedus, Claudopus, 92
byssisedus, Crepidotus, 92
Byssocorticium, 29
byssoidea, Coniophora, 12
byssoides, Amphinema, 12
byssoides, Clavaria, 365
byssoides, Hypochnus, 12
byssoides, Kneiffia, 12
byssoides, Peniophora, 12
byssoides, Thelephora, 12
Byssomerulius, 29
Byssoporia, 29
cacaliae, Coleosporium, 311
caccabus, Entoloma, 92
caccabus, Rhodophyllus, 92
cacuminatus, Agaricus, 366
caelata, Calvatia, 121
caelata, Rhodocybe, 253
caelatum, Lycoperdon, 121
caelatum, Tricholoma, 253
caelatus, Agaricus, 253
caelatus, Clitopilus, 253
caerulea, Byssus, 284
caerulea, Leptonia, 92
caerulea, Russula, 258
caerulea, Stropharia, 277
caerulea, Terana, 284
caerulea, Thelephora, 284
caeruleoflocculosa, Leptonia, 92
caeruleoflocculosum, Entoloma,
   92
caeruleopolitum, Entoloma, 92
caeruleoviolaceus, Leucocoprinus, 177
caerulescens, Agaricus, 66
caerulescens, Cortinarius, 77
caerulescens, Cortinarius, 66
caeruleum, Calodon, 128
caeruleum, Corticium, 284
caeruleum, Entoloma, 92
caeruleum, Hydnellum, 128
caeruleum, Hydnum, 128
caeruleum, Pulcherricium, 284
caeruleus Bolton, Agaricus, 366
caeruleus Pers., *Agaricus*, 258
caesariata, Inocybe, 152
caesariata, Inocybe, 366
caesariata *var.* fibrillosa, Inocybe, 366
caesariatus, Agaricus, 366
caesariatus var. *fibrillosus, Agaricus*,
   366
caesia, Campanella, 32
caesia, Peniophora, 214

caesia, Postia, 235
caesia, Prillieuxia, 366
caesia, Soppittiella, 287, 366
caesia, Thelephora, 366
caesia, Tomentella, 287
caesia, Tomentella, 366
caesiocanescens, Cortinarius, 66
caesiocincta, Leptonia, 92
caesiocinctum, Entoloma, 92
caesiocinctus, Rhodophyllus, 92
caesiocinerea, Bourdotia, 20
caesiocinerea, Sebacina, 20
caesiocinereum, Basidiodendron,
   20
caesiocinereum, Corticium, 20
caesiocinereum, Tomentellastrum,
   369
caesiocyaneus, Cortinarius, 66, 72
caesiostramineus, Cortinarius, 66
caesiotincta, Volvariella, 306
caesium, Corticium, 214, 366
caesius, Boletus, 235
caesius, Hypochnus, 366
caesius, Leptoporus, 235
caesius, Oligoporus, 235
caesius, Polyporus, 235
caesius, Tyromyces, 235
caesius γ *irregularis, Agaricus*, 390
caesius δ *erythrinus, Agaricus*, 69
caespitosa, Lepista, 174
caespitosa, Omphalia, 366
caespitosus, Agaricus, 366
caespitosus, Boletus, 366
caespitosus, Gyrodon, 366
caespitosus, Lentinus, 14
caespitosus, Marasmiellus, 426
caespitosus, Pleurotus, 14
caespitosus, Rhodopaxillus, 174
cagei, Cortinarius, 66
calamagrostidis, Puccinia, 324
calaminare, Entoloma, 92
calamistrata, Inocybe, 149
calamistratus, Agaricus, 149
calcea, Exidiopsis, 105
calcea, Poria, 273
calcea, Sebacina, 105
calcea, Thelephora, 105
calcea var. *glebulosa, Thelephora*, 300
calcea var. *sambuci, Thelephora*, 143
calceolus, Agaricus, 175
calceolus, Boletus, 234, 366
calceolus, Polyporus, 366
calceum, Corticium, 105
calceum ssp. *contiguum, Corticium*,
   372
calceus, Polyporus, 273
calcicola, Caldesiella, 286
calcicola, Lepiota, 171
calcicola, Tomentella, 286
calciphila, Hygrocybe, 130
calcitrapae, Puccinia, 322
calcitrapae var. *centaureae, Puccinia*,
   322
caldarii, Agaricus, 352
caldarii, Collybia, 352
caldariorum, Leucocoprinus, 353
Caldesiella, 286
caledoniensis, Cortinarius, 67
calendulae, Entyloma, 342
calendulae, Protomyces, 342
calendulae f. *bellidis, Entyloma*, 341
calendulae f. *dahliae, Entyloma*, 342
calendulae f. *hieracii, Entyloma*, 342
calida, Inocybe, 149
caligata, Armillaria, 366

caligatum, Tricholoma, 366
caligatum var. *nauseosum,
   Tricholoma*, 296
caligatus, Agaricus, 366
caliginosa, Nyctalis, 366
caliginosus, Agaricus, 211
caliginosus, Panaeolus, 211
callinus, Coprinus, 57
callisteus, Agaricus, 67
callisteus, Cortinarius, 72
callisteus, Cortinarius, 67
callosa, Poria, 13
callosa, Psilocybe, 366
callosa, Psilocybe, 248
callosus, Agaricus, 366
callosus, Polyporus, 13
callunicola, Tremella, 292
caloceps, Pluteus, 230
Calocera, 30
calochroum var. *coniferarum,
   Phlegmacium*, 67
calochrous, Agaricus, 67
calochrous, Cortinarius, 67
calochrous, Cortinarius, 65
**calochrous *var.* coniferarum,
   Cortinarius**, 67
**calochrous *var.* parvus,
   Cortinarius**, 67
Calocybe, 30
Calodon, 128
calolepis, Agaricus, 366
calolepis, Crepidotus, 366
Caloporus, 117
calopus, Agaricus, 366
calopus, Androsaceus, 189
calopus, Boletus, 23
calopus, Cortinarius, 79
calopus, Marasmius, 366
calopus, Marasmius, 189
calospora, Astrosporina, 149
calospora, Ceratosebacina, 35
calospora, Exidiopsis, 35
calospora, Gloeotulasnella, 301
calospora, Inocybe, 149
calospora, Prototremella, 301
calospora, Sebacina, 35
calospora, Tulasnella, 301
calosporum, Ceratobasidium, 35
calotricha, Phanerochaete, 218
calotrichum, Corticium, 218
calthae, Aecidium, 322, 337
calthae, Puccinia, 322
calthicola, Puccinia, 322
calthionis, Entoloma, 92
calva, Mucronella, 197
calva var. *aggregata, Mucronella*, 197
Calvatia, 31
calvum, Hydnum, 197
calyciformis, Agaricus, 366
Calyptella, 31
Calyptospora, 333
calyptrata, Galerina, 110
calyptriformis, Agaricus, 130
calyptriformis, Hygrocybe, 130
calyptriformis, Hygrophorus, 130
calyptriformis f. *nivea, Hygrocybe*, 130
**calyptriformis *var.* domingensis,
   Hygrocybe**, 130
calyptriformis var. *niveus,
   Hygrophorus*, 130
calyptrospora, Galerina, 110
calyptrosporum, Hebeloma, 366
Camarophyllopsis, 32
Camarophyllus, 130
camarophyllus, Agaricus, 136

carbonarius, Merulius, 106
carbonarius, Xenodochus, 339
carbonicola, Psathyrella, 243
carcharias, Cystoderma, 84
carcharias, Lepiota, 84
carcharius, Agaricus, 84
cardiospora, Pistillaria, 303
cardui, Aecidium, 433
cardui, Puccinia, 324
cardui, Ustilago, 435
cardui-pycnocephali, Puccinia, 322
carduorum, Puccinia, 322
cari-bistortae, Puccinia, 321
caricicola, Melanotus, 195
caricicola, Psilocybe, 195
caricina, Puccinia, 323
caricina, Trichobasis, 323
caricina var. caricina, Puccinia, 323
caricina var. magnusii, Puccinia, 323
caricina var. paludosa, Puccinia, 328
caricina var. pringsheimiana, Puccinia, 323
caricina var. ribesii-pendulae, Puccinia, 323
caricina var. ribis-nigri-lasiocarpae, Puccinia, 323
caricina var. ribis-nigri-paniculatae, Puccinia, 323
caricina var. uliginosa, Puccinia, 331
caricina var. urticae-acutae, Puccinia, 332
caricina var. urticae-acutiformis, Puccinia, 332
caricina var. urticae-flaccae, Puccinia, 332
caricina var. urticae-hirtae, Puccinia, 332
caricina var. urticae-inflatae, Puccinia, 332
caricina var. urticae-ripariae, Puccinia, 332
caricina var. urticae-vesicariae, Puccinia, 332
caricis, Anthracoidea, 340
caricis, Cintractia, 340
caricis, Farysia, 344
caricis, Marasmius, 189
caricis, Puccinia, 323, 324
caricis, Uredo, 323, 340
caricis, Ustilago, 340
caricis f. urticae-acutiformis, Puccinia, 332
caricis f. urticae-vesicariae, Puccinia, 332
caricis-pendulae, Pterula, 249
caries, Tilletia, 346
caries, Uredo, 346
carinii, Lepiota, 367
cariosa, Amanita, 9
carlinae, Puccinia, 322
carlylei, Corticium, 307
carmichaelianus, Merulius, 228
carmichaelianus, Polyporus, 228
carminea, Russula, 257, 265
carminipes, Russula, 258
carnea, Calocybe, 30
carneifolia, Lepiota, 177
carneifolius, Leucoagaricus, 177
carneoalba, Eccilia, 102
carneoalbus, Agaricus, 102
carneoalbus, Clitopilus, 102
carneoalbus, Rhodophyllus, 102
carneogrisea, Eccilia, 92

carneogrisea, Leptonia, 92
carneogrisea, Skeletocutis, 273
carneogriseum, Entoloma, 92
carneogriseus, Agaricus, 92
carneola, Kneiffia, 140
carneola, Peniophora, 140
carneolus, Agaricus, 31
carneolutea, Poria, 267
carneolutea, Schizopora, 267
carneosanguinea, Mycena, 367
carneotomentosus, Agaricus, 213
carneum, Hydnangium, 128
carneum, Lyophyllum, 30
carneum, Sistotrema, 222
carneum, Tricholoma, 30
carneum var. persicolor, Tricholoma, 31
carneum var. xanthosporum, Hydnangium, 119
carneus, Agaricus, 30
carneus, Irpex, 222
carneus, Polyporus, 367
carnicolor, Mycena, 56
carnicolor, Russula, 261
carniolica, Puccinia, 330
carnosa, Flammula, 367
carnosum, Ganoderma, 114
carnosus, Agaricus, 213
carnosus, Agaricus, 367
carnosus, Boletus, 367
carolii, Agaricus, 1
caroticola, Stephanospora, 275
caroticolor, Hydnangium, 275
caroticolor, Octaviania, 275
caroviolaceus, Cortinarius, 64
carpatica, Hymenochaete, 138
carpaticus, Crepidotus, 80
carpatorum, Milesina, 317
carpineus, Achroomyces, 433
carpineus, Boletus, 21
carpini, Boletus, 168
carpini, Hygrophorus, 137
carpini, Leccinum, 168
carpini, Russula, 258
Carpobolus, 274
carpobolus, Lycoperdon, 274
carpophila, Cintractia, 340
carpophila, Flocculina, 107
carpophila, Galera, 107
carpophila, Naucoria, 107
carpophiloides, Flammulaster, 108
carpophiloides, Flocculina, 108
carpophiloides, Naucoria, 108
carpophiloides, Phaeomarasmius, 108
carpophilus, Agaricus, 107
carpophilus, Flammulaster, 107
carpophilus, Phaeomarasmius, 107
carpophilus var. autochtonoides, Flammulaster, 108
carpophilus var. rhombisporus, Flammulaster, 108
carpophilus var. subincarnatus, Flammulaster, 108
carpta, Inocybe, 156
carpta, Inocybe, 367
carptus, Agaricus, 367
carthami, Puccinia, 322
cartilaginea, Clitocybe, 186, 367
cartilaginea, Exidia, 367
cartilagineum, Tricholoma, 367
cartilagineum, Tricholoma, 186
cartilagineus, Agaricus, 367
Cartilosoma, 12
cartwrightii, Graphium, 319

caryophyllacearum, Melampsorella, 316
caryophyllacearum, Uredo, 316, 321
caryophyllea, Auricularia, 285
caryophyllea, Helvella, 285
caryophyllea, Phylacteria, 286
caryophyllea, Thelephora, 285
caryophylleus, Merulius, 286
caryophyllinus, Uromyces, 337
casca, Psathyrella, 367
cascum, Hypholoma, 243, 367
cascus, Agaricus, 367
caseus, Agaricus, 367
casimiri, Cortinarius, 367
casimiri, Inocybe, 153
casimiri, Telamonia, 367
casparyana, Entorrhiza, 341
casparyana, Schinzia, 341
castanea, Lepiota, 171
castaneifolia, Psilocybe, 367
castaneifolius, Panaeolus, 367
castaneifolius, Panaeolus, 212
castaneus, Agaricus, 367
castaneus, Boletus, 120
castaneus, Cortinarius, 367
castaneus, Fomes, 367
castaneus, Gyroporus, 120
castaneus, Polyporus, 367
castaneus β badius, Boletus, 23
castoreus, Lentinellus, 368
castoreus, Lentinus, 368
catalaunica, Inocybe, 149
catalaunica, Leptonia, 93
catalaunicum, Entoloma, 93
catarium, Hypholoma, 238
catarius, Agaricus, 238
catervata, Psathyrella, 367
catervata, Psilocybe, 367
catilla, Cyphella, 368
catinus, Agaricus, 42
catinus, Clitocybe, 42
caucasica, Athelia, 18
caucasica, Cristella, 290
caucasica, Trechispora, 290
caudata, Psathyrella, 238, 239
caudatum, Lycoperdon, 184
caudatus, Agaricus, 239
caudicina, Lepiota, 159
caulicinalis, Agaricus, 81, 308
caulicinalis, Crinipellis, 81
caulincola, Puccinia, 331
caussei, Xerula, 308
causticus, Cortinarius, 70
causticus, Cortinarius, 368
cauticinalis, Marasmius, 308
cauticinalis, Xeromphalina, 308
cauvetii, Collybia, 196
cauvetii, Marasmius, 196
cauvetii, Micromphale, 196
cavarae, Calocera, 30
cavicola, Phellinus, 219
cavipes, Boletinus, 280
cavipes, Boletus, 280
cavipes, Hebeloma, 368
cavipes, Russula, 258
cavipes, Suillus, 280
cavipes f. aureus, Boletinus, 280
ceciliae, Agaricus, 8
ceciliae, Amanita, 8
cedretorum, Cortinarius, 67
cedretorum, Galera, 110
cedretorum, Galerina, 110
cedriolens, Cortinarius, 73
Cejpomyces, 285
celakovskyana, Puccinia, 325

cuticulosa, *Peziza*, 374
cyanea, *Lazulinospora*, 11
cyanea, Stropharia, 374
cyanea, *Stropharia*, 277
cyanea, *Tomentella*, 11
cyaneobasileucum, *Leccinum*, 167
cyaneoviridescens, Entoloma, 94
cyaneoviridescens, *Leptonia*, 94
cyanescens, *Agaricus*, 374
cyanescens, *Boletus*, 120
cyanescens, *Copelandia*, 374
cyanescens, Gyroporus, 120
cyanescens, Panaeolus, 374
cyanescens, Psilocybe, 247
cyaneus, *Agaricus*, 374
cyaneus, Amaurodon, 11
cyaneus, *Hypochnus*, 11
cyani, Puccinia, 324
cyani, *Uredo*, 324
cyanites, Cortinarius, 68
cyanophaea, Clitocybe, 374
cyanophaeus, *Agaricus*, 374
cyanophaeus var. *pengellei, Agaricus*,
406
cyanophyllus, *Cortinarius*, 70
cyanopus, Conocybe, 52
cyanopus, Cortinarius, 68
cyanopus, *Cortinarius*, 72
cyanopus, *Galerula*, 52
cyanoxantha, Russula, 258
cyanoxantha f. *cutefracta, Russula*,
259
cyanoxantha f. *pallida, Russula*, 259
cyanoxantha f. *peltereaui, Russula*,
259
cyanoxantha var. *cutefracta, Russula*,
258
cyanoxanthus, *Agaricus*, 258
cyanula, *Leptonia*, 374
cyanulum, Entoloma, 374
cyanulus, *Agaricus*, 374
cyanus β *caerulescens, Agaricus*, 66
cyatheae, Radulum, 374
cyathiforme, *Hydnum*, 221
cyathiformis, *Agaricus*, 208, 246
cyathiformis, *Cantharellula*, 246
cyathiformis, Cellularia, 374
cyathiformis, *Clitocybe*, 246
cyathiformis, *Lentinus*, 208
cyathiformis, *Omphalia*, 246
cyathiformis, *Panus*, 208
cyathiformis, Pseudoclitocybe,
246
cyathoides, Agaricus, 375
cyathula, Lactarius, 375
cyathula, *Lactarius*, 164
cyathuliformis, Lactarius, 162
Cyathus, 83
cyclas, *Cyphella*, 267
cydoniae, *Aecidium*, 313
cydoniae, *Roestelia*, 313
cygnea, *Lepiota*, 180
cygneus, Leucocoprinus, 180
cylindracea, Agrocybe, 6
cylindracea, *Pholiota*, 6
cylindraceocampanulata, Omphalia,
375
cylindraceocampanulatus, Marasmius,
375
cylindraceus, *Agaricus*, 6
cylindrica, *Clavaria*, 39
cylindricum, *Phragmidium*, 319
cylindricus, *Agaricus*, 57
Cylindrobasidium, 84
cylindrosporum, Hebeloma, 122

cynapii, *Trichobasis*, 328
cynodontis, Ustilago, 349
Cynophallus, 197
cypericola, Entorrhiza, 435
cypericola, *Schinzia*, 435
Cyphella, 84
cyphelliforme, *Geopetalum*, 127
cyphelliformis, *Acanthocystis*, 127
cyphelliformis, *Agaricus*, 127
cyphelliformis, Hohenbuehelia,
127
cyphelliformis, *Pleurotus*, 127
cyphelliformis, *Resupinatus*, 127
cyphelloides, Hemimycena, 375
cyphelloides, *Mycena*, 125, 375
Cyphellopsis, 195
Cyphellostereum, 84
cystidiata, Luellia, 184
cystidiata, Vuilleminia, 307
cystidiophora, *Gloeotulasnella*, 301
cystidiophora, Tulasnella, 301
cystidiosa, Cystolepiota, 352
cystidiosa, *Lepiota*, 352
Cystobasidium, 312
Cystoderma, 84
Cystofilobasidium, 85
Cystolepiota, 85
Cytidia, 86
cytisina, *Fomitopsis*, 215
cytisinus, *Fomes*, 215
daamsii, Clitopilus, 45
Dacrymyces, 86
Dacryobasidium, 81
Dacryobolus, 87
Dacryomitra, 30
Dacryopsis, 86
dactylidicola, *Flammula*, 119
dactylidis, Entyloma, 342
dactylidis, *Puccinia*, 325
dactylidis, *Thecaphora*, 342
dactylidis, Uromyces, 336
dactylidis var. *poae, Uromyces*, 336
Daedalea, 87
Daedaleopsis, 87
dahliae, Entyloma, 342
dahliae, *Entylomella*, 342
damascenus, Cortinarius, 68
danica, *Peniophora*, 222
danica, *Phlebia*, 222
danica, Tulasnella, 301
danicum, Botryobasidium, 26
danicum, *Hebeloma*, 122
danicus, Cortinarius, 68
danili, Cortinarius, 68
dartmorica, Endoperplexa, 90
dasypus, Mycena, 199
Datronia, 87
daulnoyae, Clavaria, 375
daviesii, Hydnum, 375
dealbata, *Clitocybe*, 44
dealbata, Clitocybe, 375
dealbata var. *minor, Clitocybe*, 44
dealbatus, Agaricus, 375
debaryana, *Tilletia*, 350
debilis, Agaricus, 375
debilis, *Mycena*, 204
debilis, Mycena, 375
debilis, Pterula, 249
decastes, *Agaricus*, 185
decastes, *Clitocybe*, 185
decastes, Lyophyllum, 185
decembris, *Clitocybe*, 44
decemgibbosa, Inocybe, 150
deceptiva, *Clitocybe*, 43
decidua, *Ceratorhiza*, 35

deciduum, *Sclerotium*, 35
decipiens, *Agaricus*, 68, 119
decipiens, Athelia, 18
decipiens, *Corticium*, 18
decipiens, Cortinarius, 68
decipiens, *Flammula*, 119
decipiens, Gymnopilus, 119
decipiens, Inocybe, 375
decipiens, *Inocybe*, 150
decipiens, *Lactarius*, 163
decipiens, Lactarius, 162
decipiens, Leccinum, 375
decipiens, Lycoperdon, 184
decipiens, Russula, 259
decipiens, *Tilletia*, 346
**decipiens var. atrocaeruleus,
Cortinarius**, 69
decipiens var. *hoffmannii, Cortinarius*,
367
decipiens var. *lacunarum, Lactarius*,
163
decipientoides, Astrosporina, 150, 375
decipientoides, Inocybe, 375
decipientoides, Inocybe, 150
decipientoides var. *taxocystis,
Inocybe*, 156
declivitatum, Boletus, 23
decolorans, *Agaricus*, 259, 429
decolorans, *Cortinarius*, 79, 429
decolorans, Russula, 259
decolorans var. *constans, Russula*,
258
decolorans var. *obscura, Russula*, 265
decoloratus, *Agaricus*, 77
decoloratus, *Cortinarius*, 77
Deconica, 247
decora, *Clitocybe*, 298
decora, Tricholomopsis, 298
decorum, *Tricholoma*, 298
decorus, *Agaricus*, 298
decorus, *Hymenogaster*, 139
decorus, *Pleurotus*, 298
decumbens, *Agaricus*, 375
decumbens, Cortinarius, 375
decurrens, *Clavaria*, 250
decurrens, Ramaria, 250
decussata, *Pholiota*, 226
decussatus, *Agaricus*, 226
defectus, Rhizopogon, 375
definitum, *Corticium*, 140
definitum, Hyphoderma, 140
deflectens, *Grandinia*, 218
deflectens, Phanerochaete, 218
deflectens, *Phlebia*, 218
deformans, Thecaphora, 346
deformis, *Cyathus*, 208
deformis, *Irpex*, 267
deformis, Nidularia, 208
defossum, *Lycoperdon*, 268
defossum, Lycoperdon, 375
degener, *Lentinus*, 208
degener, Xerotus, 375
degener, *Xerotus*, 106, 208
deglubens, *Agaricus*, 375
deglubens, Eichleriella, 90
deglubens, Inocybe, 375
deglubens, *Inocybe*, 150
deglubens, *Radulum*, 90
delastrina, Geminella, 435
delastrina, Schroeteria, 435
delastrina, *Thecaphora*, 435
delecta, *Inocybe*, 150
delectabile, Clavulicium, 40
delectabile, *Corticium*, 40
delectabilis, *Agaricus*, 124

digraphidis, Puccinia, 330
dilatata, Mycena, 204
dilatatus, Agaricus, 204
dilectus, Coprinus, 376
dilectus, Coprinus, 58
dilepis, Agaricus, 119
dilepis, Flammula, 119
dilepis, Gymnopilus, 119
dilseae, Mycaureola, 197
dilutus, Agaricus, 376
dilutus, Cortinarius, 71
dilutus, Cortinarius, 376
dimidiatus, Agaricus, 229
dimitica, Sebacina, 269
dimitica, Trechispora, 291
dimorphocystis, Galerina, 112
dimorphocystis, Galerina, 376
dioicae, Puccinia, 324
**dioicae var. arenariicola,
      Puccinia**, 324
dioicae var. extensicola, Puccinia,
      324
**dioicae var. schoeleriana,
      Puccinia**, 324
dioicae var. silvatica, Puccinia,
      325
dionysae, Cortinarius, 69
diosma, Clitocybe, 43
diosma, Mycena, 199
Diplomitoporus, 89
dipsacoides, Agaricus, 376
directa, Omphalia, 376
directus, Agaricus, 376
disciflorum, Phragmidium, 320
disciforme, Stereum, 376
disciformis, Aleurodiscus, 376
disciformis, Helvella, 376
disciformis, Hymenochaete, 376
disciformis, Peniophora, 376
discoidea, Lepiota, 353
discoidearum, Puccinia, 331
discoideum var. horvathianum,
      Exobasidium, 435
discoideus, Agaricus, 376
discoideus, Boletus, 376
discoideus, Hygrophorus, 137
discoideus, Hygrophorus, 376
discoideus, Leucocoprinus, 353
discolor, Boletus, 24
discolor, Puccinia, 335
discolor, Tranzschelia, 335
discopus, Agaricus, 377
discopus, Mycena, 377
discoxanthus, Agaricus, 137
discoxanthus, Hygrophorus, 137
disjungendus, Cortinarius, 69
dispar, Agaricus, 377
dispersa, Puccinia, 330
dispersa, Tubaria, 299
dispersa f.sp. tritici, Puccinia, 330
dispersum, Hypholoma, 145
dispersus, Agaricus, 145, 299
disseminata, Psathyrella, 58
disseminatus, Agaricus, 58
disseminatus, Coprinarius, 58
disseminatus, Coprinus, 58
disseminatus, Pseudocoprinus, 58
dissiliens, Agaricus, 377
dissiliens, Mycena, 377
dissimulabilis, Agaricus, 377
dissimulabilis, Mycena, 377
dissimulabilis, Mycena, 201
dissimulans, Agaricus, 6, 216
dissimulans, Phaeogalera, 216
dissimulans, Pholiota, 216

dissipabilis, Clavaria, 41
dissitispora, Tulasnella, 301
distans, Hygrophorus, 377
distincta, Puccinia, 325
distinguendum, Tricholoma, 377
distorta, Collybia, 47
distortus, Agaricus, 47
distortus, Boletus, 1
distortus, Polyporus, 1
Ditiola, 89
ditopus, Agaricus, 43
ditopus, Clitocybe, 43
divaricata, Clavaria, 377
divaricata, Ramaria, 377
diversidens, Hericium, 126
diversidens, Hydnum, 126
Doassansia, 341
Doassansiopsis, 341
Dochmiopus, 80
dochmiospora, Cyphella, 160
dolabratus, Cortinarius, 377
domesticus, Agaricus, 58
domesticus, Coprinus, 63
domesticus, Coprinus, 58
domesticus, Merulius, 355
donkii, Fibuloporia, 399
dorotheae, Agaricus, 352
dorotheae, Collybia, 352
dracontii, Aecidium, 356
drimeia, Russula, 264
drimeia f. viridis, Russula, 264
drimeia var. flavovirens, Russula, 264
drimeia var. queletii, Russula, 263
Drosella, 37
Drosophila, 238
dryadeus, Boletus, 157
dryadeus, Inonotus, 157
dryadeus, Polyporus, 157
dryadophilus, Lactarius, 377
dryinum, Corticium, 82
dryinum, Crustoderma, 82
dryinus, Agaricus, 229
dryinus, Pleurotus, 229
Dryodon, 126
Dryophila, 224
dryophila, Collybia, 47
dryophila ssp. Exsculpta, Collybia, 48
dryophila var. alvearis, Collybia, 47
dryophila var. aquosa, Collybia, 46
dryophila var. aurata, Collybia, 47
dryophila var. funicularis, Collybia, 48
dryophila var. oedipoides, Collybia, 46
dryophila var. oedipus, Collybia, 46
dryophiloides, Pluteus, 232
dryophilus, Agaricus, 47
dryophilus, Gymnopus, 47
dryophilus, Marasmius, 47
dryophilus, Omphalia, 47
dryophilus var. aquosus, Marasmius,
      46
dryophilus var. auratus, Marasmius,
      47
dryophilus var. funicularis, Marasmius,
      48
dryophilus var. oedipus, Marasmius,
      46
dryophilus β funicularis, Agaricus, 48
dubia, Heterochaete, 279
dubia, Heterochaetella, 279
dubia, Sebacina, 279
dubia, Stypella, 279
dubium, Alysidium, 26
dubium, Trichoderma, 26
dulcamara, Inocybe, 150

dulcamara f. squamosoannulata,
      Inocybe, 150
dulcamara var. axantha, Inocybe, 150
dulcamara var. homomorpha,
      Inocybe, 150
dulcamarus, Agaricus, 150
dulcidulus, Agaricus, 3
dulcita, Oospora, 299
dulcitum, Trichosporon, 299
dumetorum, Conocybe, 52
dumetorum, Galera, 52
dunalii, Lentinus, 170
dunense, Hebeloma, 122
dunensis, Conocybe, 52
dunensis, Inocybe, 150
dunensis, Panaeolus, 211
dunensis, Psathyrella, 240
duplicata, Dictyophora, 218
duplicata, Dictyophora, 377
duplicatus, Phallus, 377
duportii, Russula, 377
dura, Agrocybe, 6
dura, Agrocybe, 377
dura, Asterostromella, 89
dura, Pholiota, 377
dura, Togaria, 377
dura, Vararia, 89
dura var. xanthophylla, Agrocybe, 6
dura var. xanthophylla, Pholiota, 6
duracinum, Tricholoma, 377
duracinus, Agaricus, 377
duracinus, Cortinarius, 75
duriaeana, Granularia, 206
duriaeana, Mycocalia, 206
duriaeana, Nidularia, 206
duriaeana, Ustilago, 349
duriuscula, Astrosporina, 150
duriuscula, Inocybe, 150
duriusculum, Leccinum, 168
duriusculus, Boletus, 168
durum, Dichostereum, 89
durus, Agaricus, 377
durus, Boletus, 233
durus, Cortinarius, 69
durus, Polyporus, 233
dysthales, Agaricus, 94
dysthales, Entoloma, 94
dysthales, Pouzarella, 94
dysthales, Pouzaromyces, 94
dysthaloides, Entoloma, 94
eburnea, Entylomella, 343
eburneum, Fusidium, 343
eburneum, Phlegmacium, 377
eburneus, Agaricus, 137
eburneus, Agaricus, 377
eburneus, Coprinus, 377
eburneus, Cortinarius, 377
eburneus, Gymnopus, 137
eburneus, Hygrophorus, 137
eburneus var. cossus, Hygrophorus,
      136
eburneus var. quercetorum,
      Hygrophorus, 136
eburneus var. quercorum,
      Hygrophorus, 136
Ecchyna, 319
Eccilia, 90
echinacea, Cystolepiota, 172
echinacea, Lepiota, 172
echinaceum, Echinoderma, 172
echinata, Conocybe, 52
echinata, Galera, 52
echinata, Inocybe, 194
echinata, Lepiota, 194
echinata, Nyssopsora, 318

felina, **Lepiota**, 172
felinoides, Lepiota, 172
felinum, Hypholoma, 238
felinus, Agaricus, 172, 238
fellea, Mycena, 200
fellea, Russula, 259
fellea, Xeromphalina, 308
felleus, Agaricus, 259
felleus, Boletus, 303
felleus, Tylopilus, 303
felleus var. *alutarius, Tylopilus*, 303
Femsjonia, 89
fennica, Clavaria, 380
fennica, Ramaria, 380
**fennica *var*. griseolilacina,
  Ramaria**, 250
fergussonii, Entyloma, 342
fergussonii, Protomyces, 342
fergussonii, Puccinia, 325
fernandae, Entoloma, 95
fernandae, Nolanea, 95
fernandae, Rhodophyllus, 95
ferrea, Fuscoporia, 219
ferrea, Poria, 219
ferreus, Phellinus, 219
ferreus, Polyporus, 219
ferrii, Stropharia, 278
ferruginascens, Agaricus, 380
ferruginascens, Cantharellus, 33
ferruginea, Auricularia, 139
ferruginea, Cyphella, 84
ferruginea, Daedalea, 380
ferruginea, Flocculina, 108
ferruginea, Gyraria, 292
ferruginea, Hymenochaete, 139
ferruginea, Naucoria, 108
ferruginea, Pachnocybe, 319
ferruginea, Thelephora, 287
ferruginea, Tomentella, 287
ferruginea, Tremella, 188, 293
ferruginea, Tremella, 380
ferruginea, Tubaria, 108
ferruginella, Collybia, 283
ferruginella, Tephrocybe, 283
ferrugineum, Corticium, 287
ferrugineum, Hydnellum, 129
ferrugineum, Hydnum, 129
ferrugineum, Stereum, 139
ferrugineus, Boletus, 24
ferrugineus, Calodon, 129
ferrugineus, Flammulaster, 108
ferrugineus, Hypochnus, 287
ferrugineus, Phaeodon, 129
ferrugineus, Xerocomus, 24
ferruginosa, Caldesiella, 287
ferruginosa, Fuscoporia, 219
ferruginosa, Poria, 219
ferruginosa, Tomentellina, 287
ferruginosa var. *calcicola, Caldesiella,*
  286
ferruginosus, Boletus, 219
ferruginosus, Fomes, 219
ferruginosus, Phellinus, 219
fertile, Entoloma, 93
fertilis, Agaricus, 93, 102
fervidus, Cortinarius, 70
festiva, Naucoria, 216
festiva, Phaeocollybia, 216
festivus, Agaricus, 216
festucae, Puccinia, 325
festucae, Uredo, 325
festucae, Uromyces, 336
festucina, Puccinia, 331
Fibriciellum, 106
fibrillosa, Oliveonia, 209

fibrillosa, Phlebiella, 223
fibrillosa, Psathyra, 240, 380
fibrillosa, Psathyrella, 380
fibrillosa, Psathyrella, 240
fibrillosa, Sebacina, 209
fibrillosa, Tomentella, 303
fibrillosa, Trechispora, 223
fibrillosa, Tylospora, 303
fibrillosum, Tylosperma, 303
fibrillosus, Agaricus, 380
fibrillosus, Hypochnus, 303
Fibrodontia, 141
Fibroporia, 12
fibrosa, Astrosporina, 380
fibrosa, Inocybe, 150
fibrosa, Inocybe, 380
fibrosa, Tomentella, 287
fibrosa, Tomentellina, 287
fibrosa var. *trivialis, Inocybe*, 150
fibrosoides, Inocybe, 150
fibrosoides, Inocybe, 150
fibrosus, Agaricus, 380
fibrosus, Zygodesmus, 287
fibula, Agaricus, 254
fibula, Boletus, 380
fibula, Hemimycena, 254
fibula, Marasmiellus, 254
fibula, Mycena, 254
fibula, Omphalia, 254
fibula, Omphalina, 254
fibula, Polyporus, 380
fibula, Polystictus, 380
fibula, Rickenella, 254
fibula var. *nivalis, Omphalia*, 255
fibula var. *swartzii, Agaricus*, 255
fibula var. *swartzii, Omphalina*, 255
fibulare, Micromphale, 254
fibulata, Athelia, 18
Fibulomyces, 106
ficariae, Aecidium, 336
ficariae, Cylindrosporium, 342
ficariae, Entyloma, 342
ficariae, Entylomella, 342
ficariae, Gloeosporium, 342
ficariae, Protomyces, 342
ficariae, Puccinia, 337
ficariae, Tuburcinia, 347
ficariae, Uredo, 337
ficariae, Urocystis, 347
ficariae, Uromyces, 337
ficoides, Agaricus, 380
ficuum, Ustilago, 435
filamentifer, Coprinus, 58
filamentosa, Pellicularia, 285
filamentosa, Peniophora, 380
filamentosa, Phanerochaete, 218
filamentosa, Phanerochaete, 380
filamentosum, Ceratobasidium, 285
filamentosum, Corticium, 380
filamentosus, Hypochnus, 285
filamentosus, Paxillus, 380
filamentosus, Paxillus, 214
filaris, Conocybe, 52
filaris, Pholiota, 52
filaris, Pholiotina, 52
filia, Flammula, 226
filia, Pholiota, 226
filicea, Flammula, 353
filiceus, Agaricus, 353
filiceus, Gymnopilus, 353
filicina, Clavaria, 355
filicina, Cyphella, 380
filicina, Lachnella, 380
filicina, Nochascypha, 380
filicina, Phlebiella, 223

filicina, Ramaria, 355
filicina, Uredinopsis, 335
filicinum, Aphanobasidium, 223
filicinum, Corticium, 223
filicinum, Gloeosporium, 314
filicinum, Herpobasidium, 314
filicinum, Xenasma, 223
filicinus, Pleurotellus, 380
filicinus, Pleurotus, 380
filicinus, Protomyces, 335
filicum, Uredo, 433
filicum, Uredo, 314
filiforme, Lycoperdon, 349
filiformis, Clavaria, 380
filiformis, Coprinus, 380
filiformis, Typhula, 380
filiformis, Ustilago, 349
filipendulae, Polycystis, 347
filipendulae, Triphragmium, 335
filipendulae, Triphragmium, 335
filipendulae, Tuburcinia, 347
filipendulae, Urocystis, 347
filium, Corticium, 380
Filobasidiella, 106
Filobasidium, 107
filopes, Agaricus, 200
filopes, Mycena, 204
filopes, Mycena, 200
fimbriata, Clitocybe, 380
fimbriata, Odontia, 275
fimbriata, Poria, 235
fimbriata, Stromatoscypha, 235
fimbriata, Tomentella, 286
fimbriata, Tremella, 292
fimbriatum, Geastrum, 115
fimbriatum, Hydnum, 275
fimbriatum, Micromphale, 380
fimbriatum, Mycoleptodon, 275
fimbriatum, Porotheleum, 235
fimbriatum, Sistotrema, 275
fimbriatum, Steccherinum, 275
fimbriatus, Agaricus, 380
fimbriatus, Boletus, 235
fimbriatus, Lentinus, 170
fimbriatus, Pleurotus, 380
fimbriatus, Polyporus, 235
fimetaria, Conocybe, 52
fimetaria, Psathyrella, 239
fimetaria, Psilocybe, 248
fimetaria, Stropharia, 248
fimetaria, Tremella, 312
fimetarium, Cystobasidium, 312
fimetarius, Achroomyces, 312
fimetarius, Agaricus, 57
fimetarius, Coprinus, 57
fimetarius, Coprinus, 380
fimetarius var. *cinereus, Coprinus*, 57
fimetarius *var.* pullatus, Coprinus, 380
fimicola, Agaricus, 211
fimicola, Panaeolus, 211
fimicola var. *ater, Panaeolus*, 212
fimicola var. *cinctulus, Agaricus*, 211
fimicola var. *cinctulus, Panaeolus*, 211
fimiputris, Agaricus, 381
fimiputris, Anellaria, 212, 381
fimiputris, Panaeolus, 381
fimiputris, Panaeolus, 212
fingibilis, Russula, 381
firma var. *attenuata, Agrocybe*, 5
firmula, Russula, 258
firmula, Russula, 381
firmula f. *luteoviridans, Russula*, 395
firmum, Hebeloma, 381
firmus, Agaricus, 381
firmus, Cortinarius, 381

fornicata var. *streptopus*, *Hygrocybe*, 132
fornicatum, Geastrum, 115
fornicatum, Lycoperdon, 115
fornicatus, Agaricus, 213
fornicatus, Camarophyllus, 132
fornicatus, Hygrophorus, 131
fornicatus, Pleuropus, 213
fornicatus var. *clivalis, Hygrophorus*, 132
forquignonii, Lepiota, 172
forquignonii, Polyporus, 234
fracida, Armillaria, 37
fracida, Drosella, 37
fracidus, Agaricus, 37
fracidus, Chamaemyces, 37
fracticum, Tricholoma, 382
fracticum, Tricholoma, 295
fracticus, Agaricus, 382, 295
fragariae, Phragmidium, 319
fragariae, Puccinia, 319
fragariae, Rhodotorula, 334
fragariae, Torulopsis, 334
fragariastri, Phragmidium, 319
fragariastri, Puccinia, 319
fragile, Hydnum, 20
fragilipes, Hebeloma, 122
fragilis, Agaricus, 22, 259
fragilis, Bolbitius, 22
fragilis, Clavaria, 39
fragilis, Conocybe, 53
fragilis, Conocybe, 382
fragilis, Crepidotus, 80
fragilis, Galera, 382
fragilis, Leptoporus, 236
fragilis, Oligoporus, 236
fragilis, Osteomorpha, 292
fragilis, Polyporus, 236
fragilis, Postia, 236
fragilis, Postia, 236
fragilis, Russula, 257, 265
fragilis, Russula, 259
fragilis, Tyromyces, 236
fragilis var. *carminea, Russula*, 257
fragilis var. *clavata, Galera*, 110
fragilis var. *fallax, Russula*, 259
fragilis *var*. knauthii, Russula, 260
fragilis var. *nivea, Russula*, 264
fragilis *var*. nivea, Russula, 382
fragilis var. *violascens, Russula*, 260
fragilissima, Hiatula, 354
fragilissima, Psathyra, 241
fragilissimus, Leucocoprinus, 353
fragosoi, Puccinia, 326
fragrans, Agaricus, 43
fragrans, Boletus, 24
fragrans, Clitocybe, 43
fragrans, Lepista, 43
fragrans, Omphalia, 43
fragrans, Polyporus, 21
fragrans var. *depauperata, Clitocybe*, 43
fragrantissima, Russula, 260
franchetii, Amanita, 9
frangulae, Aecidium, 324
fraterna, Laccaria, 159
fraterna, Naucoria, 159
fraternus, Agaricus, 159
fraudans, Agaricus, 151
fraudans, Inocybe, 151
fraudulosus, Cortinarius, 362
fraudulosus, Cortinarius, 382
fraxinea, Peniophora, 214
fraxinea, Perenniporia, 215
fraxinea, Thelephora, 214

fraxinea, Vanderbylia, 215
fraxineus, Boletus, 215
fraxineus, Fomes, 215
fraxineus, Polyporus, 215
fraxinicola, Cyphella, 104
fraxinicola, Episphaeria, 104
fraxinicola, Phaeocyphella, 104
friabilis, Amanita, 9
friabilis, Amanitopsis, 9
friesii, Agaricus, 171
friesii, Cantharellus, 33
friesii, Collybia, 20
friesii, Conocybe, 54
friesii, Coprinus, 58
friesii, Inocybe, 154
friesii, Lepiota, 171
friesii, Marasmius, 20
friesii, Porotheleum, 235
friesii, Porotheleum, 382
friesii, Psathyrella, 240
friesii, Russula, 257
friesii var. *acutesquamosa, Lepiota*, 171
Friesites, 126
fritilliformis, Agaricus, 44
fritilliformis, Clitocybe, 44
Frommea, 312
Frommeëlla, 312
frondosa, Grifola, 118
frondosa, Tremella, 382
frondosa, Tremella, 292, 293
frondosa f. *intybacea, Grifola*, 118
frondosus, Boletus, 118
frondosus, Polyporus, 118
frumentaceum, Tricholoma, 382
frumentaceus, Agaricus, 382
frumenti, Uredo, 325
frustulata, Thelephora, 382
frustulatum, Stereum, 382
frustulatus, Xylobolus, 382
frustulenta, Psathyrella, 240
frustulentus, Agaricus, 240
frustulorum, Coprinus, 382
frustulosum, Corticium, 285
frysica, Clitocybe, 43
fucatum, Tricholoma, 382
fucatus, Agaricus, 382
fuchsiae, Uredo, 333
fuciforme, Corticium, 167
fuciformis, Athelia, 167, 183
fuciformis, Hypochnus, 167
fuciformis, Isaria, 167
fuciformis, Laetisaria, 167
fugacissima, Exidiopsis, 196
fugacissima, Microsebacina, 196
fugacissima, Sebacina, 196
fugacissima f. *sebacea, Sebacina*, 279
fugacissima f. sebacea, Sebacina, 382
fugax, Merulius, 180
fulgens, Agaricus, 382
fulgens, Cortinarius, 73
fulgens, Cortinarius, 382
fulgens, Flammula, 119
fulgens, Fulvidula, 119
fulgens, Gymnopilus, 119
fulgens, Naucoria, 119
fulgida, Pistillaria, 304
fuliginea, Clavaria, 40
fuligineipes, Clitocybe, 382
fuligineoalba, Bankera, 20
fuligineoalbum, Hydnum, 20
fuligineoatra, Inocybe, 151
fuligineosquarrosa, Macrolepiota, 187

fuligineoviolaceum, Hydnum, 382
fuligineoviolaceus, Sarcodon, 382
fuligineus, Albatrellus, 382
fuligineus Fr. & Hök, Boletus, 382
fuligineus Pers., *Boletus*, 382
fuligineus, Phaeoporus, 382
fuliginoides, Ptychogaster, 237
fuliginosa, Hymenochaete, 382
fuliginosa, Macrolepiota, 187
fuliginosa, Russula, 382
fuliginosa, Thelephora, 382
fuliginosum, Stereum, 382
fuliginosus, Agaricus, 162
fuliginosus, Cortinarius, 78
fuliginosus, Lactarius, 161, 165
fuliginosus, Lactarius, 162
fuliginosus, Polyporus, 158
fuliginosus f. *albipes, Lactarius*, 161
fuliginosus f. *speciosus, Lactarius*, 165
fuliginosus ssp. *picinus, Lactarius*, 408
fulmineus, Cortinarius, 69, 73
fulmineus, Cortinarius, 382
fulmineus var. *sulphureus, Cortinarius*, 424
fulva, Amanita, 9
fulva, Amanitopsis, 9
fulva, Astrosporina, 382
fulva, Cyphella, 195
fulva, Cyphella, 382
fulva, Hymenochaete, 382
fulva, Leptonia, 95
fulvella, Astrosporina, 151
fulvella, Inocybe, 151
fulvella, Lepiota, 171
fulvellum, Tricholoma, 298
fulvellus, Agaricus, 298
fulvescens, Cortinarius, 70
fulvescens, Drosophila, 240
fulvescens, Psathyrella, 240
fulvescens *var*. brevicystis, Psathyrella, 240
fulvescens var. *dicrani, Psathyrella*, 376
fulvidolilaceus, Cortinarius, 78
Fulvidula, 119
fulvidulum, Hypholoma, 145
fulvidus, Boletus, 120
fulvissimus, Lactarius, 163
fulvobulbillosa, Xeromphalina, 308
fulvoincarnatus, Cortinarius, 70
fulvo-ochrascens *var*. cyanophyllus, Cortinarius, 70
fulvopallens, Omphalina, 209
fulvosquamosus, Cortinarius, 70
fulvostrigosa, Leptonia, 91
fulvostrigosa, Nolanea, 91
fulvostrigosus, Agaricus, 91
fulvosus, Agaricus, 382
fulvum, Alysidium, 26
fulvum, Entoloma, 95
fulvum, Tricholoma, 296
fulvus, Agaricus, 9, 296
fulvus, Fomes, 383
fulvus, Polyporus, 383
fulvus, Rhodophyllus, 95
fumatofoetens, Agaricus, 186
fumatofoetens, Lyophyllum, 186
fumigata, Clavaria, 250
fumigata, Ramaria, 250
fumosa, Bjerkandera, 21
fumosa, Clavaria, 39
fumosa, Clitocybe, 185
fumosa, Collybia, 185
fumosum, Corticium, 223
fumosum, Lyophyllum, 185

graminum, **Typhula**, 304
graminum, Uromyces, 336
grammata, Inocybe, 152
grammica, Ustilago, 435
grammocephala, Collybia, 192
grammocephalus, Agaricus, 192
grammopodia, Melanoleuca, 193
grammopodia f. macrocarpa, Melanoleuca, 193
grammopodia var. obscura, Melanoleuca, 193
grammopodia var. politoinaequalipes, Melanoleuca, 194
grammopodia var. subbrevipes, Melanoleuca, 193
grammopodium, Tricholoma, 193
grammopodius, Agaricus, 193
Grandinia, 141
grandis, Pholiota, 119
grandis, Ustilago, 349
grandiusculus, Bolbitius, 385
grangei, Hiatula, 172
grangei, Lepiota, 172
grangei, Schulzeria, 172
Granularia, 208, 346
granulatum, Phragmidium, 319
granulatum var. mutabilis, Hydnum, 28
granulatus, Boletus, 280
granulatus, Ixocomus, 280
granulatus, Pluteus, 232
granulatus, Suillus, 280
granulatus var. tenuipes, Boletus, 19
Granulobasidium, 146
granulosa, Asterostromella, 385
granulosa, Flocculina, 108
granulosa, Grandinia, 28, 385
granulosa, Lepiota, 84
granulosa, Naucoria, 108
granulosa, Russula, 262
granulosa, Thelephora, 385
granulosa var. amianthina, Lepiota, 84
granulosa var. carcharias, Lepiota, 84
granulosa var. rufescens, Lepiota, 85
granulosum, Corticium, 22
granulosum, Cystoderma, 84
granulosum, Dichostereum, 385
granulosum, Hydnum, 385
granulosus, Agaricus, 84
granulosus, Flammulaster, 108
granulosus, Hypochnus, 288
granulosus, Phaeomarasmius, 108
granulosus var. carcharias, Agaricus, 84
granulosus var. cinnabarinus, Agaricus, 84
granulosus var. rufescens, Agaricus, 85
grata, Russula, 260
graveolens, Agaricus, 385
graveolens, Agaricus, 385
graveolens, Cantharellula, 247
graveolens, Flammula, 385
graveolens, Gymnopus, 386
graveolens, Hydnum, 221
graveolens, Omphalia, 247
graveolens, Phellodon, 221
graveolens, Pholiota, 226
graveolens, Pholiota, 385
graveolens, Pseudoomphalina, 246
graveolens, Russula, 260
graveolens, Tricholoma, 385
graveolens var. megacantha, Russula, 260

graveolens var. subrubens, Russula, 265
greletii, Clavaria, 39
grevillei, Aecidium, 332
grevillei, Boletus, 281
grevillei, Suillus, 281
grevillei, Typhula, 304
grevillei f. badius, Suillus, 281
Grifola, 118
grilletii, Myxarium, 279
grilletii, Stypella, 279
grilletii, Tremella, 279
grisea, Athelia, 176
grisea, Clavaria, 40
grisea, Exidiopsis, 106
grisea, Hohenbuehelia, 386
grisea, Omphalia, 199
grisea, Omphalia, 386
grisea, Russula, 261
grisea, Russula, 260
grisea, Sebacina, 106
grisea, Thelephora, 106
grisea var. ionochlora, Russula, 261
griseliniae, Corticium, 142
griseliniae, Hyphodontia, 142
griseliniae, Rogersella, 143
grisella, Omphalia, 182
grisella, Omphalina, 182
grisella, Phlebiella, 223
grisellum, Aphanobasidium, 223
grisellum, Corticium, 223
grisellum, Xenasma, 223
griseocana, Dendrothele, 88
griseocanum, Corticium, 88
griseocyanea, Leptonia, 95
griseocyaneum, Entoloma, 95
griseocyaneum var. roseum, Entoloma, 101
griseocyaneus, Agaricus, 95
griseodiscus, Leucoagaricus, 177
griseoflavescens, Corticium, 221
griseoflavescens, Hyphoderma, 221
griseoflavescens, Phlebia, 221
griseofoetidus, Coprinus, 59
griseofumosa, Melanoleuca, 386
griseofumosus, Agaricus, 386
griseola, Clavaria, 386
griseola, Clavulinopsis, 386
griseolilacina, Inocybe, 152
griseolilacinus, Cortinarius, 386
griseoluridum, Entoloma, 386
griseoluridus, Pluteus, 231
griseoluridus, Rhodophyllus, 386
griseopallida, Arrhenia, 15
griseopallida, Calyptella, 386
griseopallida, Cellypha, 386
griseopallida, Cyphella, 386
griseopallida, Omphalina, 15
griseopallidum, Leptoglossum, 15
griseopallidus, Agaricus, 15
griseopallidus, Phaeotellus, 15
griseopus, Pluteus, 232
griseorubella, Eccilia, 97, 386
griseorubella, Leptonia, 95, 386
griseorubellum, Entoloma, 386
griseorubellus, Agaricus, 386
griseorubellus, Rhodophyllus, 97
griseorubida, Leptonia, 95
griseorubidum, Entoloma, 95
griseospora, Collybia, 253
griseospora, Rhodocybe, 253
griseovelata, Inocybe, 386
griseoviolacea, Tomentella, 287
griseovirens, Lepiota, 173
griseovirens, Lepiota, 172

griseovirens var. obscura, Lepiota, 173
griseus, Agaricus, 260, 386
griseus, Hymenogaster, 139
griseus, Pleurotus, 386
groegeri, Hebeloma, 122
groenlandica, Pholiota, 226
grossa, Clavaria, 41
grossii, Coprinus, 352
grossula, Chrysomphalina, 38
grossula, Omphalina, 38
grossulariae, Aecidium, 323
grossulariae, Puccinia, 323
grossulus, Agaricus, 38
guegueni, Lepiota, 180
Guepinia, 118
Guepiniopsis, 118
guilleminii, Clavaria, 39
guillemotii, Merulius, 355
gummosa, Dryophila, 225
gummosa, Flammula, 225
gummosa, Pholiota, 225
gummosus, Agaricus, 225
guttata, Lepiota, 182
guttata, Limacella, 182
guttata, Melampsora, 333
guttata, Thekopsora, 333
guttatum, Pucciniastrum, 333
guttatum, Tricholoma, 386
guttatus, Agaricus, 182, 386
guttulata, Postia, 236
guttulatus, Oligoporus, 236
guttulatus, Panaeolus, 386
guttulatus, Polyporus, 236
guttulatus, Tyromyces, 236
guttulifera, Peniophora, 140
guttuliferum, Gloeocystidium, 140
guttuliferum, Hyphoderma, 140
Gymnomyces, 118
Gymnopilus, 119
gymnopodia, Clitocybe, 14, 386
gymnopodia, Flammula, 386
gymnopodia, Omphalia, 14
gymnopodia, Pholiota, 386
gymnopodius, Agaricus, 14, 386
Gymnopus, 46
Gymnosporangium, 313
gypsea, Hemimycena, 124
gypsea, Mycena, 124, 202
gypsella, Hemimycena, 125
gypsella, Mycena, 125
gypseus, Agaricus, 124
gypseus, Marasmiellus, 124
gyrans, Clavaria, 304
gyrans, Typhula, 304
gyrans var. grevillei, Typhula, 304
Gyraria, 292
Gyrodon, 120
gyroflexa, Psathyra, 386
gyroflexus, Agaricus, 386
Gyrophana, 270
Gyrophila, 294
Gyroporus, 120
gyrosa, Lecythea, 320
gyrosa, Uredo, 320
Haasiella, 120
hadriani, Phallus, 217
hadrocystis, Conocybe, 52
haemacta, Inocybe, 152
haemacta var. rubra, Inocybe, 152
haemactus, Agaricus, 152
haematites, Agaricus, 386
haematites, Armillaria, 386
haematites, Cystoderma, 85
haematites, Cystoderma, 386

lembospora, **Athelopsis**, 19
lembospora, Luellia, 19
lembosporum, Corticium, 19
lemonnieriana, Puccinia, 324
lenis, Amyloporia, 392
lenis, Antrodia, 392
lenis, Diplomitoporus, 392
lenis, Physisporus, 392
lenis, Poria, 392
lenis, Skeletocutis, 273
lenis, Skeletocutis, 392
lenta, Dryophila, 226
lenta, Flammula, 226
lenta, Pholiota, 226
Lentaria, 169
lenticularis, Agaricus, 182
lenticularis, Lepiota, 182
lenticularis, Limacella, 182
lenticularis var. *megalodactylus,
    Lepiota*, 182
lenticulospora, Conocybe, 53
lentiginosus, Agaricus, 175
Lentinellus, 169
Lentinus, 170
lentum, Corticium, 269
lentus, Agaricus, 226
lentus, Polyporus, 234
Lenzites, 170
leochroma, Cyphella, 195
leochroma, Pholiota, 6
leochromus, Agaricus, 6
leonina, Hymenochaete, 392
leoninus, Agaricus, 231
leoninus, Boletus, 24
leoninus, Boletus, 392
leoninus, Pluteus, 230
leoninus, Pluteus, 231
leoninus var. *coccineus, Pluteus*, 230
leonis, Boletus, 24
leonis, Xerocomus, 24
leontodontis, Puccinia, 326
leontopodius, Lentinus, 208
lepida, Hygrocybe, 130
lepida, Russula, 264
lepida var. *alba, Russula*, 264
lepida var. *amara, Russula*, 257
lepida var. *aurora, Russula*, 257
Lepidella, 8
lepideus, Agaricus, 208
lepideus, Lentinus, 208
lepideus, Neolentinus, 208
lepideus, Panus, 208
lepideus, Polyporus, 233
lepideus var. *contiguus, Lentinus*, 208
lepideus var. *hibernicus, Lentinus*, 208
lepidicolor, Russula, Addendum
Lepidomyces, 170
lepidopus, Cortinarius, 64
lepidopus, Hygrocybe, 132
lepidopus, Hygrophorus, 132
Lepiota, 170
Lepiotella, 37
lepiotoides, Pluteus, 231
Lepista, 174
lepista, Paxillus, 175, 254
leporina, Hygrocybe, 392
leporinus, Hygrophorus, 137, 392
leprosa, Peniophora, 218
leprosa, Phanerochaete, 218
leprosa, Scopuloides, 218
leptocephala, Coltricia, 234
leptocephala, Mycena, 201
leptocephalus, Agaricus, 201
leptocephalus, Boletus, 234
leptocephalus, Polyporus, 234

leptocystis, Inocybe, 153
leptocystis, Inocybe, 154
Leptoglossum, 15
Leptonia, 90
leptonipes, Entoloma, 96
leptonipes, Leptonia, 96
leptonipes, Omphaliopsis, 96
leptonipes, Rhodophyllus, 96
leptophylla, Inocybe, 153
Leptoporus, 176
leptopus, Entoloma, 95
leptopus, Paxillus, 392
leptopus γ *graveolens, Hydnum*, 221
Leptosporomyces, 176
Leptotrimitus, 273
Leptotus, 15
leucanthemi, Puccinia, 324
Leucoagaricus, 176
leucobryophila, Lindtneria, 183
leucobryophila, Thelephora, 183
leucobryophila, Trechispora, 183
leucocephala, Astrosporina, 393
leucocephala, Calocybe, 31
leucocephala, Inocybe, 393
leucocephalum, Lyophyllum, 30
leucocephalum, Tricholoma, 30
leucocephalus, Agaricus, 30
Leucocoprinus, 179
Leucocortinarius, 180
leucodiatreta, Clitocybe, 393
leucogala, Mycena, 200
leucogalus, Agaricus, 200
Leucogaster, 180
Leucogyrophana, 180
leucolepidota, Galerina, 159
leucolepidotus, Kuehneromyces,
    159
leucomallella, Postia, 236
leucomallellus, Oligoporus, 236
leucomallellus, Tyromyces, 236
leucomelaena, Boletopsis, 22
leucomelas, Boletus, 22
leucomelas, Caloporus, 22
leucomelas, Polyporus, 22
leucomyosotis, Collybia, 284
Leucopaxillus, 181
leucophaeata, Collybia, 186
leucophaeatum, Lyophyllum, 186
leucophaeatus, Agaricus, 186
leucophaeus, Agaricus, 393
leucophaeus, Cantharellus, 106
leucophaeus, Hygrophorus, 137, 138
leucophaeus, Hygrophorus, 393
leucophaeus, Merulius, 106
leucophanes, Agaricus, 393
leucophanes, Panaeolus, 393
leucophylla, Fayodia, 113
leucophylla, Fayodia, 393
leucophylla, Melanoleuca, 192
leucophylla, Omphalia, 113, 393
leucophyllus, Agaricus, 393
leucopoda, Conocybe, 53
leucopus, Agaricus, 393
leucopus, Conocybe, 53
leucopus, Cortinarius, 393
leucopus, Hydnum, 393
leucopus, Sarcodon, 393
leucosarx, Hebeloma, 122
leucospermum, Aecidium, 433
leucospermum, Aecidium, 319, 335
leucospermum, Endophyllum, 319
Leucosporidium, 314
leucotephra, Psathyra, 241
leucotephra, Psathyrella, 241
leucotephrum, Hypholoma, 241

leucotephrus, Agaricus, 241
leucothites, Agaricus, 177
leucothites, Lepiota, 177
leucothites, Leucoagaricus, 177
leucotricha, Psalliota, 1
leucotrichus, Agaricus, 1
leucoxanthum, Corticium, 192
leucoxanthum, Gloeocystidiellum, 192
leucoxanthum, Gloeocystidium, 192
leucoxanthum, Megalocystidium,
    192
levis, Auricularia, 393
libanotidis, Puccinia, 327
libertiana, Cyphella, 393
libertiana, Flagelloscypha, 393
libertiana, Lachnella, 393
lichenicola, Chionosphaera, 310
lichenicola, Tremella, 293
Lichenomphalia, 181
licinipes var. robustior, Cortinarius,
    393
licmophora, Lepiota, 354
licmophorus, Agaricus, 354
ligatus, Agaricus, 393
lignatilis, Agaricus, 210
lignatilis, Clitocybe, 210
lignatilis, Nothopanus, 210
lignatilis, Ossicaulis, 210
lignatilis, Pleurocybella, 210
lignatilis, Pleurotus, 210
lignatilis var. *tephrocephalus,
    Pleurotus*, 210
lignicola, Agaricus, 159
lignicola, Boletus, 28
lignicola, Buchwaldoboletus, 28
lignicola, Kuehneromyces, 159
lignicola, Phlebopus, 28
lignicola, Pholiota, 159
lignicola, Pulveroboletus, 28
lignyotus, Lactarius, 393
ligula, Clavaria, 40
ligula, Clavariadelphus, 40
lilacea, Collybia, 393
lilacea, Lepiota, 173
lilacea, Peniophora, 214
lilacea, Russula, 261
lilacea var. *carnicolor, Russula*, 261
lilacifolia, Pholiota, 225
lilacina, Hygrocybe, 133
lilacina, Inocybe, 151
lilacina, Lepiota, 85
lilacina, Omphalia, 133
lilacinicolor, Omphalina, 209
lilacinogranulosa, Lepiota, 354
lilacinogranulosus, Leucocoprinus, 354
lilacinogranulosus var. *subglobisporus,
    Leucocoprinus*, 354
lilacinogrisea, Tomentella, 288
lilacinopusillus, Cortinarius, 66
lilacinus, Hygrophorus, 133
lilacinus, Hymenogaster, 393
lilacinus, Hymenogaster, 140
lilacinus, Lactarius, 166
lilacinus, Lactarius, 163
lilacipes, Cystoderma, 85
lilascens, Corticium, 221
lilascens, Phlebia, 221
lilatinctus, Coprinus, 59
liliacearum, Puccinia, 327
lilii, Caeoma, 336
lilii, Uromyces, 336, 356
Limacella, 182
limacinus, Agaricus, 137, 393
limacinus, Gymnopus, 393
limacinus, Hygrophorus, 393

mediaburiensis, Peniophora, 140
medioflava, Lepiota, 354
medioflavoides, Leucoagaricus,
178
medioflavus, Leucocoprinus, 354
mediofusca, Psalliota, 3
mediofuscus, Agaricus, 3
medium, Asterostroma, 17
medius, Agaricus, 396
medius, Tubulicrinis, 300
medulla-panis, Boletus, 216
medulla-panis, Perenniporia, 216
medulla-panis, Polyporus, 216
medulla-panis, Poria, 216
medullata, Lepiota, 173
medullata, Limacella, 173
medullata, Russula, 396
medullatus, Agaricus, 173
Megacollybia, 191
Megalocystidium, 192
megalodactyla, Amanita, 182
megalodactylus, Agaricus, 182
megalopora, Poria, 352
megaloporus, Phellinus, 352
megaloporus, Polyporus, 352
megaphyllum, Tricholoma, 397
megaspermus, Coprinus, 60
megaspora, Mycena, 201
Melaleuca, 192
melaleuca, Melanoleuca, 193
melaleuca, Melanoleuca, 193
melaleuca var. phaeopodia,
Melanoleuca, 408
melaleucum, Hydnum, 221
melaleucum, Tricholoma, 193
melaleucum var. adstringens,
Tricholoma, 357
melaleucum var. polioleucum,
Tricholoma, 193
melaleucum var. porphyroleucum,
Tricholoma, 409
melaleucus, Agaricus, 193
melaleucus, Phellodon, 221
melaleucus β porphyroleucus,
Agaricus, 409
melaleucus γ polioleucus, Agaricus,
193
Melampsora, 315
Melampsorella, 316
Melampsoridium, 316
melampyri, Coleosporium, 311
melampyri, Uredo, 311
melanea, Krombholziella, 169
melaneum, Leccinum, 169
melanocyclum, Tulostoma, 303
melanodon, Agaricus, 397
melanodon, Pluteus, 397
Melanogaster, 192
melanogramma, Schizonella, 345
melanogramma, Uredo, 345
Melanoleuca, 192
Melanomphalia, 194
Melanophyllum, 194
melanopoda, Inocybe, 153
Melanopsichium, 344
melanopus, Boletus, 234
melanopus, Polyporellus, 234
melanopus, Polyporus, 234
melanosperma, Stropharia, 278
melanospermus, Agaricus, 278
Melanotaenium, 344
melanotricha, Lepiota, 178
melanotrichus, Leucoagaricus,
178

melanotrichus var.
fuligineobrunneus,
Leucoagaricus, 178
melanotrichus var. septentrionalis,
Leucoagaricus, 178
Melanotus, 195
melanotus, Cortinarius, 397
melanoxeros, Cantharellus, 33
melanthina, Psathyrella, 397
melanthinum, Hypholoma, 241, 397
melanthinus, Agaricus, 397
meleagris (Jul. Schäff.) Imbach,
Agaricus, 4
meleagris Sowerby, Agaricus, 178
meleagris With., Agaricus, 397
meleagris, Gymnopus, 178
meleagris, Lepiota, 178
meleagris, Leucoagaricus, 178
meleagris, Leucocoprinus, 178
meleagris, Psalliota, 4
meleagris var. terricolor, Agaricus, 4
meleagroides, Lepiota, 177
melicae, Tuburcinia, 348
melicae, Urocystis, 348
meliigena, Agaricus, 201
meliigena, Mycena, 201
melinoides, Agaricus, 397
melinoides, Alnicola, 207, 397
melinoides, Naucoria, 111, 207
melinoides, Naucoria, 397
melitodes, Russula, 261
melizeus, Hygrophorus, 137
melizeus, Hygrophorus, 397
mellea, Armillaria, 14
mellea, Armillariella, 14
mellea, Clitocybe, 14
mellea, Odontia, 142
mellea var. bulbosa, Armillaria, 365
mellea var. glabra, Armillaria, 397
mellea var. laricina, Armillaria, 392
mellea var. maxima, Armillaria, 14
mellea var. minor, Armillaria, 14
mellea var. obscura, Armillaria, 402
mellea var. sulphurea, Armillaria, 14
mellea var. tabescens, Armillaria, 14
mellea var. versicolor, Armillaria, 397
mellea var. viridiflava, Armillaria, 397
melleopallens, Cortinarius, 78
melleopallens, Cortinarius, 397
melleopallens, Rhodocybe, 254
melleum, Hydnum, 142
melleus, Agaricus, 14
melliolens, Cortinarius, 77
melliolens, Russula, 262
melliolens var. chrismantiae, Russula,
266
mellita, Ceriporia, 397
mellita, Ceriporia, 36
mellita, Poria, 36, 397
melzeri, Russula, 262
membranacea, Acia, 249
membranacea, Coniophora, 50
membranacea, Helvella, 16
membranaceum, Hydnum, 249
membranaceum, Radulum, 249
membranaceus, Agaricus, 239
membranaceus, Merulius, 16
menieri, Marasmius, 190
menieri, Marasmius, 397
menieri, Tilletia, 346
menthae, Aecidium, 328
menthae, Puccinia, 327
mephitica, Collybia, 283
mephitica, Tephrocybe, 283
mephiticum, Lyophyllum, 283

mephiticus, Agaricus, 283
mercurialis, Caeoma, 316
merdaria, Conocybe, 53
merdaria, Psilocybe, 248
merdaria, Stropharia, 248
merdaria var. major, Stropharia, 248
merdarius, Agaricus, 248
merdarius var. major, Agaricus, 248
merdicola, Psilocybe, 248
Meripilus, 195
Merisma, 118, 285
Merismodes, 195
merismoides, Merulius, 222
merismoides, Phlebia, 222
Merulicium, 196
merulina, Ditiola, 118
merulina, Guepiniopsis, 118
Meruliopsis, 117
Merulius, 221
mesenterica, Auricularia, 19
mesenterica, Helvella, 19
mesenterica, Tremella, 293
mesentericum, Stereum, 19
mesodactylius, Agaricus, 51
mesomorpha, Lepiota, 397
mesomorpha, Sebacina, 397
mesomorpha, Sebacina, 105
mesomorphus, Agaricus, 397
mesophaeum, Hebeloma, 123
mesophaeum var. crassipes,
Hebeloma, 123
mesophaeum var. holophaeum,
Hebeloma, 397
mesophaeum var. minor, Hebeloma,
123
mesophaeus, Agaricus, 123
mesophaeus var. holophaeus,
Agaricus, 397
mesophaeus var. minor, Agaricus, 123
mesospora, Conocybe, 53
mesospora, Russula, 261
mesospora f. typica, Conocybe, 53
mesospora var. brunneola, Conocybe,
51
mesotephrus, Hygrophorus, 138
mesotephrus, Hygrophorus, 137
metachroa, Clitocybe, 44
metachroa, Russula, 263
metachrous, Agaricus, 44
metachrous β mortuosa, Agaricus,
399
metamorphosa, Ceriporia, 35
metamorphosa, Poria, 35
metamorphosus, Polyporus, 35
metapodia, Hygrocybe, 235
metapodium, Porpoloma, 235
metapodius, Agaricus, 235
metapodius, Hygrophorus, 235
metata, Mycena, 201
metata, Mycena, 201
metatus, Agaricus, 201
metatus var. plicosus, Agaricus, 409
metrodii, Cortinarius, 397
metrodii, Ripartites, 256
metulispora, Lepiota, 173
metulispora, Lepiota, 397
metulisporus, Agaricus, 397
Metulodontia, 140
mexicana, Mycena, 352
mexicanus, Bolbitius, 352
micacea, Camarophyllopsis, 32
micacea, Hygrocybe, 32
micaceus, Agaricus, 60
micaceus, Coprinus, 62
micaceus, Coprinus, 60

mompa, *Helicobasidium*, 313
monacha, *Cyphella*, 195
monacha, *Cyphellopsis*, 195
monacha, *Maireina*, 195
Moniliopsis, 285
monstrosa, *Clitocybe*, 399
monstrosus, *Agaricus*, 399
montagnei, Coltricia, 49
montagnei, *Polyporus*, 49
montagnei, *Polystictus*, 49
montagnei, Ustanciosporium, 435
montagnei, *Ustilago*, 435
montagnei var. *major, Ustilago*, 349
montana, *Deconica*, 248
montana, Psilocybe, 248
montanus, *Agaricus*, 248
montanus β atrorufus, *Agaricus*, 362
montiae, *Sorosporium*, 435
montiae, Tolyposporium, 435
moravicus, Boletus, 24
moravicus, *Xerocomus*, 24
morchelliformis, Gautieria, 114
morganii, *Cantharellus*, 399
morganii, Hygrophoropsis, 399
mori, *Hexagonia*, 399
mori, Polyporus, 399
moriformis, *Dacrymyces*, 293
moriformis, Tremella, 293
morlichensis, *Cephaloscypha*, 34
morlichensis, *Flagelloscypha*, 34
mortuosa, Clitocybe, 399
morus, Agaricus, 399
morvernensis, Uredo, 335
moseri, Conocybe, 53
mougeotii, *Eccilia*, 98
mougeotii, Entoloma, 98
mougeotii, *Hymenochaete*, 139
mougeotii, *Leptonia*, 98
mougeotii, *Stereum*, 139
mougeotii, *Thelephora*, 139
mougeotii *var.* fuscomarginatum,
 Entoloma, 98
mucida, *Armillaria*, 210
mucida, Ceriporiopsis, 399
mucida, *Clavaria*, 399
mucida, *Cristinia*, 82
mucida, Cristinia, 399
mucida, Grandinia, 399
mucida, *Grandinia*, 252
mucida, *Lentaria*, 399
mucida, *Mucidula*, 210
mucida, Multiclavula, 399
mucida, Oudemansiella, 210
mucida, *Poria*, 267, 399
mucida, *Thelephora*, 399
Mucidula, 210
mucidula, Pseudotomentella, 247
mucidula, *Tomentella*, 247
mucidulus, *Hypochnus*, 247
mucidum, *Radulum*, 82, 399
mucidus, *Agaricus*, 210
mucidus, *Porpomyces*, 399
mucifluus, *Cortinarius*, 78
mucifluus, Cortinarius, 72
mucilaginosa, Rhodotorula, 334
mucilaginosa, *Torula*, 334
Muciporus, 301
mucor, *Agaricus*, 202
mucor, Mycena, 202
mucosus, *Agaricus*, 72
mucosus, Cortinarius, 72
mucrocystis, Psathyrella, 242
mucronata, *Puccinia*, 320
mucronata α rosae, *Puccinia*, 320
mucronata β rubi, *Puccinia*, 319

mucronatum, *Aregma*, 320
mucronatum, Phragmidium, 320
Mucronella, 197
mucronella, Hygrocybe, 133
mucronellus, *Hygrophorus*, 133
muelleri, Pholiota, 399
muelleri, *Pholiota*, 225
muelleri, *Uredo*, 314
mulgravensis, Pluteolus, 399
Multiclavula, 197
multicolor, *Boletus*, 290
multicolor, *Trametes*, 290
multifida, *Penicillaria*, 249
multifida, Pterula, 249
multifidus, *Agaricus*, 267
multiforme, *Hydnum*, 147
multiforme, Hypochnicium, 147
multiforme, *Tricholoma*, 175
multiformis, *Agaricus*, 407
multiformis, *Cortinarius*, 409
multiformis, Cortinarius, 399
multiformis, Lepista, 175
multiformis, *Odontia*, 147
multiformis *var.* flavescens,
 Cortinarius, 399
multiformis β claricolor, *Agaricus*, 68
multiformis β elegantior, *Agaricus*, 69
multijuga, Collybia, 352
multijugus, *Agaricus*, 352
multipedata, *Psathyra*, 242
multipedata, Psathyrella, 242
multiplex, Hydnum, 399
multizonata, Podoscypha, 233
multizonata, *Thelephora*, 233
multizonatum, *Stereum*, 233
mundula, *Clitocybe*, 254
mundula, *Clitopilopsis*, 254
mundula, *Paxillopsis*, 254
mundula, *Rhodocybe*, 254
mundulus, *Agaricus*, 254
mundulus, *Clitopilus*, 254
mundulus, *Rhodopaxillus*, 254
mundulus var. *nigrescens, Clitopilus*,
 254
muralis, *Agaricus*, 399
muralis, Omphalia, 399
muralis, *Omphalia*, 16
murariae, *Milesia*, 317
murariae, Milesina, 317
murariae, *Uredo*, 317
murcida, Psathyrella, 242
murcidus, *Agaricus*, 242
muricata, *Dryophila*, 108
muricata, *Flocculina*, 108
muricata, *Naucoria*, 108
muricata, *Pholiota*, 108
muricata var. *gracilis, Naucoria*, 385
muricata var. *gracilis, Pholiota*, 385
muricatus, *Agaricus*, 108
muricatus, Flammulaster, 108
muricatus, *Phaeomarasmius*, 108
muricella, *Flammula*, 225, 399
muricella, Pholiota, 399
muricella var. *graminis, Dryophila*, 225
muricellata, Inocybe, 153
muricellospora, *Galerina*, 113
muricellospora, *Galerula*, 113
muricellus, *Agaricus*, 399
muricinus, Cortinarius, 399
murina, *Collybia*, 283
murina, Tephrocybe, 283
murinacea, Conocybe, 53
murinacea, *Hygrocybe*, 133, 399
murinaceum, *Leccinum*, 169
murinaceum, Tricholoma, 399

murinaceum, *Tricholoma*, 297, 365
murinaceus, *Agaricus*, 297, 399
murinaceus, *Boletus*, 169
murinaceus, *Hygrophorus*, 399
murinella, *Volvaria*, 306
murinella, Volvariella, 306
murinella var. *umbonata, Volvaria*,
 307
murinus, *Agaricus*, 96, 283
murinus, *Pluteus*, 231
muscari, *Uredo*, 337
muscari, Uromyces, 337
muscaria, Amanita, 9
muscaria f. *aureola, Amanita*, 10
muscaria f. *formosa, Amanita*, 10
muscaria *var.* aureola, Amanita,
 10
muscaria *var.* formosa, Amanita,
 10
muscaria var. *puella, Amanita*, 9
muscaria var. *regalis, Amanita*, 414
muscaria var. *umbrina, Amanita*, 399
muscaria β *formosa, Amanita*, 10
muscaria β *minor, Amanita*, 9
muscarius, *Agaricus*, 9
muscarius var. *formosus, Agaricus*, 10
muscarius var. *regalis, Agaricus*, 414
muscicola, Chromocyphella, 38
muscicola, *Clavaria*, 312
muscicola, *Cyphella*, 38
muscicola, Eocronartium, 312
muscicola, *Grandinia*, 272
muscicola, *Hydnum*, 272
muscicola, *Phaeocyphella*, 38
muscicola, *Pistillaria*, 312
muscicola, Sistotrema, 272
muscicola, *Typhula*, 312
muscicola δ *neckerae, Cyphella*, 255
muscigena, Collybia, 400
muscigena, *Corniola*, 16
muscigena, *Cyphella*, 84
muscigena, *Thelephora*, 84
muscigenum, *Leptoglossum*, 16
muscigenus, *Agaricus*, 16, 400
muscigenus, *Cantharellus*, 16
muscigenus, *Cortinarius*, 68
muscigenus, Crepidotus, 81
muscigenus, *Dictyolus*, 16
muscigenus, *Leptotus*, 16
muscigenus, *Merulius*, 16
muscoides, *Clavaria*, 41
muscorum, *Agaricus*, 400
muscorum, *Deconica*, 248
muscorum, *Psilocybe*, 248
muscorum, Tubaria, 400
mussivum, *Hebeloma*, 73
mussivus, *Agaricus*, 73
mussivus, Cortinarius, 73
mustelina, Pholiota, 400
mustelina, Russula, 262
mustellinus, *Agaricus*, 400
musteus, Lactarius, 164
mustialaënse, *Corticium*, 11
mustialaënsis, Amaurodon, 11
mustialaënsis, *Coniophora*, 11
mustialaënsis, *Hypochnopsis*, 11
mustialaënsis, *Hypochnus*, 11
mutabile, *Corticium*, 176
mutabile, *Lyophyllum*, 400
mutabilipes, Entoloma, 98
mutabilis, *Agaricus*, 159
mutabilis, *Athelia*, 176
mutabilis, *Cristella*, 28
mutabilis, *Dryophila*, 159
mutabilis, *Fibulomyces*, 176

nigridisca, Inocybe, 401
nigripes, Agaricus, 186
nigripes, Boletus, 234
nigripes, Merulius, 136
nigripes, Nolanea, 186
nigritula, Armillaria, 401
nigrocinnamomeum, Entoloma, 401
nigrocinnamomeus, Agaricus, 401
nigrofloccosus, Pluteus, 230
nigromarginata, Lepiota, 401
nigroviolacea, Leptonia, 98
nigroviolaceum, Entoloma, 98
nigrum, Hydnellum, 221
nigrum, Hydnum, 221
nimbata, Clitocybe, 401
nimbata, Clitocybe, 175
nimbatus, Agaricus, 401
niphoides, Entoloma, 98
nitellina, Collybia, 254
nitellina, Rhodocybe, 254
nitellinus, Agaricus, 254
nitens, Aecidium, 356
nitens, Agaricus, 401
nitens, Agaricus, 401
nitens, Entoloma, 98
nitens, Flammula, 401
nitens, Gymnoconia, 356
nitens, Kunkelia, 356
nitens, Nolanea, 98
nitida, Amanita, 401
nitida, Junghuhnia, 158
nitida, Poria, 158
nitida, Puccinia, 328
nitida, Russula, 258
nitida, Russula, 262
nitida, Uredo, 328
nitida var. *cuprea, Russula*, 258
nitida var. *pulchralis, Russula*, 412
nitidiuscula, Agaricus, 154
nitidiuscula, Inocybe, 154
nitidum, Entoloma, 98
nitidum, Steccherinum, 158
nitidus, Agaricus, 262, 401
nitidus, Boletus, 158
nitidus, Cortinarius, 401
nitidus, Polyporus, 158
nitidus var. *violaceus, Boletus*, 430
nitiosa, Hygrocybe, 133
nitiosus, Hygrophorus, 132, 133
nitrata, Hygrocybe, 133
nitratus, Agaricus, 133
nitratus, Hygrophorus, 133
nitratus *var.* glauconitens,
   Hygrophorus, 401
nitriolens, Clitocybe, 401
nitrophila, Clitocybe, 42
nitrosus, Cortinarius, 401
nivalis, Agaricus, 255
nivalis, Amanita, 10
nivalis, Amanitopsis, 10
nivea, Athelia, 18
nivea, Conocybe, 352
nivea, Hygrocybe, 135
nivea, Incrustoporia, 273
nivea, Melanoleuca, 193
nivea, Mycena, 401
nivea, Odontia, 291
nivea, Plicatura, 401
nivea, Russula, 382
nivea, Skeletocutis, 273
nivea, Trechispora, 291
niveipes, Mycena, 401
niveipes, Prunulus, 401
niveocremeum, Corticium, 272
niveocremeum, Paullicorticium, 272

niveocremeum, Sistotrema, 272
niveocremeum, Sistotremastrum,
   272
niveola, Cyphella, 401
niveola, Seticyphella, 401
niveolutescens, Agaricus, 2
nivescens, Agaricus, 4
nivescens, Psalliota, 4
nivescens var. *parkensis, Agaricus*, 4
nivescens var. *parkensis, Psalliota*, 4
niveum, Corticium, 140
niveum, Hydnum, 291
niveum, Tulostoma, 303
niveus, Agaricus, 401
niveus, Agaricus, 60, 135, 382
niveus, Bolbitius, 352
niveus, Boletus, 401
niveus, Camarophyllus, 135
niveus, Coprinus, 60
niveus, Cuphophyllus, 135
niveus, Hygrophorus, 135
niveus, Merulius, 401
niveus, Polyporus, 273
niveus f. *roseipes, Cuphophyllus*, 135
niveus var. *fuscescens, Hygrophorus*,
   135
nobilis, Agaricus, 9
nobilis, Russula, 262
nodi-aurei, Agaricus, 402
nodosa, Oliveonia, 209
nodosa, Sebacinella, 209
nodosum, Basidiodendron, 21
nodosus, Agaricus, 402
nodulosum, Hydnum, 205
nodulosus, Inonotus, 157
nodulosus, Polyporus, 157
nodulosus, Polystictus, 157
Nolanea, 90
noli-tangere, Agaricus, 242
noli-tangere, Drosophila, 242
noli-tangere, Pannucia, 242
noli-tangere, Psathyra, 242
noli-tangere, Psathyrella, 242
noli-tangere, Puccinia, 326
normandinae, Tremella, 293
norvegica, Russula, 261
norvegicum, Acanthobasidium, 7
norvegicus, Aleurodiscus, 7
norvegicus, Cortinarius, 73
nothofagi, Mycoacia, 205
nothofagi, Odontia, 205
nothosaniosus, Cortinarius, 73
novasilvensis, Flammulaster, 108
nucatum, Leccinum, 168
nucea, Naucoria, 402
nuceus, Agaricus, 402
nuciseda, Psilocybe, 402
nucisedus, Agaricus, 402
nucleata, Exidia, 105
nucleata, Naematelia, 105
nucleata, Tremella, 105
nucleatum, Myxarium, 105
nuda, Dacryomitra, 86
nuda, Dacryopsis, 86
nuda, Ditiola, 86
nuda, Lepista, 175
nuda, Peniophora, 214
nuda, Thelephora, 214
nuda, Ustilago, 351
nuda ssp. *violaceolivida, Peniophora*,
   215
nuda var. *lilacea, Gyrophila*, 176
nuda var. *maculiformis, Peniophora*,
   215
nuda var. *pruinosa, Lepista*, 175

nudiceps, Coprinus, 61
nudipes, Agaricus, 402
nudipes, Hebeloma, 402
nudum, Corticium, 214
nudum, Hydnangium, 180
nudum, Tricholoma, 175
nudum var. *glaucocanum, Tricholoma*,
   175
nudum var. *majus, Tricholoma*, 175
nudus, Agaricus, 175
nudus, Leucogaster, 180
nudus var. *majus, Agaricus*, 175
nueschii, Suillus, 280
nummularia, Collybia, 402
nummularia, Coltricia, 234
nummularius, Agaricus, 402
nummularius, Boletus, 234
nummularius, Polyporus, 234
nutans, Agaricus, 200
nycthemerus, Coprinus, 402
nymphaeae, Entyloma, 345
nymphaeae, Rhamphospora, 345
nympharum, Agaricus, 178
nympharum, Lepiota, 178
nympharum, Leucoagaricus, 178
nymphoidis, Macrolepiota, 178
nymphoidis, Aecidium, 330
Nyssopsora, 318
oakesii, Aleurodiscus, 402
oakesii, Aleurodiscus, 8
oakesii, Corticium, 402
obatra, Arrhenia, 15
obatra, Omphalia, 15
obatra, Omphalina, 15
obbata, Cantharellula, 246
obbata, Clitocybe, 246
obbata, Omphalia, 246
obbata, Pseudoclitocybe, 246
obbatus, Agaricus, 246
obducens, Oxyporus, 211
obducens, Polyporus, 211
obducens, Poria, 211
obducta, Osteina, 402
obductus, Polyporus, 402
obesus, Agaricus, 402
obesus, Gymnopus, 402
oblectabilis, Agaricus, 154
oblectabilis, Inocybe, 152
oblectabilis, Inocybe, 154
oblectabilis f. *decemgibbosa, Inocybe*,
   150
oblectus, Agaricus, 62
oblectus, Coprinus, 62
obliqua, Poria, 157
obliquoporus, Panaeolus, 212
obliquum, Hydnum, 267
obliquus, Boletus, 355
obliquus, Inonotus, 157
obliquus, Irpex, 267
obliquus, Lactarius, 402
obliquus, Polyporus, 157
oblongata, Caeoma, 327
oblongata, Puccinia, 327
oblongata, Trichobasis, 327
oblongata, Uredo, 327
oblongisporum, Sistotrema, 272
obnubilus, Agaricus, 164
obnubilus, Lactarius, 164
obrussea, Hygrocybe, 134
obrussea, Hygrocybe, 402
obrusseus, Agaricus, 402
obrusseus, Hygrophorus, 131, 402
obscura, Armillaria, 402
obscura, Armillaria, 14
obscura, Collybia, 47

olivaceoalba, *Athelia*, 176, 403
olivaceoalba, Scytinostromella, 403
olivaceoalbum, Confertobasidium,
176, 403
olivaceoalbum, Corticium, 403
olivaceoalbum, Limacium, 138
olivaceoalbus, Agaricus, 137
olivaceoalbus, Hygrophorus, 138
olivaceoalbus f. *obesus, Hygrophorus,*
392
olivaceoalbus var. *obesus,*
Hygrophorus, 392
olivaceobrunnea, Lepiota, 172
olivaceofuscus, Cortinarius, 73
olivaceogrisea, Amanita, 10
olivaceomarginata, Mycena, 202
olivaceomarginata f. *roseofusca,*
Mycena, 202
olivaceomarginata f. *thymicola,*
Mycena, 202
olivaceomarginatus, Agaricus, 202
olivaceoniger, Hygrophorus, 131
olivaceonigra, Hygrocybe, 131
olivaceosum, Leccinum, 169
olivaceoviolascens, Russula, 404
olivaceum, Corticium, 50
olivaceum, Polysaccum, 229
olivaceus, Agaricus, 262
olivaceus, Boletus, 23
olivaceus, Elateromyces, 344
olivaceus, Hymenogaster, 139
olivaceus, Hypochnus, 50
olivaceus, Panaeolus, 212
olivaceus, Pluteus, 231
olivaceus var. *modestus,*
Hymenogaster, 139
olivascens, Agaricus, 404
olivascens, Brevicellicium, 28
olivascens, Cortinarius, 404
olivascens, Macrolepiota, 404
olivascens, Odontia, 28
olivascens Fr., Russula, 404
olivascens Pers., *Russula*, 432
olivascens var. *citrinus, Russula*, 266
Oliveonia, 209
Oliveorhiza, 209
olivieri, Chlorophyllum, 187
olivieri, Lepiota, 187
olivieri, Macrolepiota, 187
olla, Cyathus, 83
olla, Peziza, 83
olla f. *anglicus, Cyathus*, 83
olla var. *agrestis, Cyathus*, 83
ollaris, Cyathus, 83
olorinum, Entoloma, 98
olorinus, Rhodophyllus, 99
olympiana, Psathyrella, 242
olympiana f. *amstelodamensis,*
Psathyrella, 242
ombrophila, Agrocybe, 6
ombrophila, Pholiota, 6, 404
ombrophila, Togaria, 404
ombrophila var. *brunneola, Togaria*, 6
ombrophilus, Agaricus, 404
ombrophilus var. *brunneolus,*
Agaricus, 6
Omphalia, 209, 245
Omphaliaster, 209
omphaliformis, Alboleptonia, 99
omphaliformis, Lactarius, 164
omphaliformis, Rhodophyllus, 99
omphaliiformis, Clitopilus, 46
Omphalina, 209
Omphaliopsis, 91
omphalodes, Lentinellus, 170

omphalodes, Lentinellus, 404
omphalodes var. *scoticus, Lentinus,*
170
Omphalotus, 210
ompnera, Cystolepiota, 352
ompnerus, Agaricus, 352
oncidii, Uredo, 335
onisca, Arrhenia, 15
oniscus, Agaricus, 15
oniscus, Omphalia, 15
oniscus, Omphalina, 15
Onnia, 157
onobrychidis, Uromyces, 338
onusta, Trechispora, 404
onusta, Trechispora, 271
onychina, Calocybe, 31
onychinum, Tricholoma, 31
onychinus, Agaricus, 31
onychoides, Antrodia, 14
onychoides, Antrodiella, 14
onychoides, Polyporus, 14
onychoides, Tyromyces, 14
oortiana, Mycena, 198
opaca, Clitocybe, 404
opaca, Clitocybe, 99
opacum, Entoloma, 99
opacus, Agaricus, 404
Opadorhiza, 90
opalea, Exidiopsis, 106
opalea, Sebacina, 106
opalinus, Marasmius, 354
opicum, Tricholoma, 404
opicus, Agaricus, 404
opimus, Cortinarius, 404
opimus var. *fulvobrunneus,*
Cortinarius, 404
opipara, Clitocybe, 404
opiparum, Tricholoma, 404
opiparus, Agaricus, 404
opizii, Puccinia, 328
opuntiae, Agaricus, 354
opuntiae, Pleurotus, 354
orbiculare, Radulum, 21
orbiculare var. *junquillinum, Radulum,*
21
orbiformis, Agaricus, 404
orbiformis, Clitocybe, 404
orbiformis var. *incana, Omphalia*, 389
orbitarum, Drosophila, 243
orbitarum, Psathyrella, 243
orcelloides, Paxillus, 404
orcellus, Agaricus, 46
orcellus, Clitopilus, 46
orcellus, Pleuropus, 46
orchidearum, Aecidium, 331
orchidearum-phalaridis, Puccinia, 331
orchidicola, Thanatephorus, 285
orchidis-repentis, Melampsora, 315
oreades, Agaricus, 190
oreades, Marasmius, 190
oreadiformis, Lepiota, 173
oreadoides, Collybia, 48
oreadoides, Gymnopus, 48
oreadoides, Marasmius, 48
oreina, Melanoleuca, 193
oreina, Russula, 263
oreinum, Tricholoma, 193
oreinus, Agaricus, 193
orellana, Dermocybe, 73
orellanoides, Cortinarius, 75
orellanus, Cortinarius, 73
orichalceus, Agaricus, 404
orichalceus, Cortinarius, 404
orientalis, Clitocybe, 45
orirubens, Tricholoma, 296

orirubens var. *basirubens, Tricholoma,*
364
orirubens var. *guttatum, Tricholoma,*
386
orliensis, Serendipita, 269
ornamentalis, Clitocybe, 44
ornatissimus, Marasmiellus, 252
ornithogali, Uredo, 350
ornithogali, Ustilago, 350
orobi, Aecidium, 339
orobi, Uredo, 339
orobi, Uromyces, 339
orphanella, Peniophora, 404
orphanellum, Hyphoderma, 404
orthospora, Cyphella, 107
orthospora, Flagelloscypha, 107
ortonii, Entoloma, 99
ortonii, Hygrocybe, 134
ortonii, Hygrophorus, 134
oryzae, Moniliopsis, 307
oryzae, Rhizoctonia, 307
osecanus, Agaricus, 4
osmophora, Clitocybe, 283
osmophora, Tephrocybe, 283
osmophorus, Cortinarius, 73
Osmoporus, 117
osmundicola, Mycena, 354
osmundicola ssp. *imleriana, Mycena,*
354
osmundicola var. *imleriana, Mycena,*
354
osseus, Polyporus, 402
Ossicaulis, 210
ossifragi, Entyloma, 343
Osteina, 235
ostoyae, Armillaria, 14
ostoyae, Armillariella, 14
ostrea, Stereum, 355
ostrea, Thelephora, 355
ostreatus, Agaricus, 230
ostreatus, Crepidopus, 230
ostreatus, Pleurotus, 229
ostreatus, Pleurotus, 230
ostreatus f. *cornucopiae, Pleurotus,*
229
ostreatus f. *pulmonarius, Pleurotus,*
230
ostreatus var. *columbinus, Pleurotus,*
230
ostreatus var. *euosmus, Pleurotus,*
230
ostreatus β *atroalbus, Crepidopus,*
230
Oudemansiella, 210
ovalis, Agaricus, 55
ovalis, Galera, 55
ovalispora, Bovista, 28
ovalispora, Inocybe, 148, 156
ovata, Clavaria, 404
ovata, Pistillaria, 304, 404
ovata, Typhula, 404
ovatocystis, Inocybe, 153
ovatus, Agaricus, 57
ovatus, Coprinus, 57
ovina, Hygrocybe, 133
ovinus, Agaricus, 133
ovinus, Albatrellus, 404
ovinus, Boletus, 404
ovinus, Hygrophorus, 133
ovispora, Lepista, 175
ovisporum, Lyophyllum, 175
ovisporus, Dacrymyces, 87
ovoidea, Amanita, 10
ovoideum, Hypochniciellum, 146
ovoideus, Agaricus, 10

paradoxus, Paxillus, 227
paradoxus, Phylloporus, 227
parafulmineus, Cortinarius, 405
paragaudis, Cortinarius, 405
paragaudis var. praestigiosus,
  Cortinarius, 410
Paraleptonia, 91
parallela, Polycystis, 348
parallela, Uredo, 348
pararustica, Omphalina, 182
parasitica, Asterophora, 17
parasitica, Leptonia, 99
parasitica, Nyctalis, 17
parasiticum, Cylindrobasidium, 84
parasiticum, Entoloma, 99
parasiticum, Hydnum, 294
parasiticum, Lycoperdon, 405
parasiticus, Agaricus, 17
parasiticus, Boletus, 245
parasiticus, Claudopus, 99
parasiticus, Gymnopus, 17
parasiticus, Pseudoboletus, 245
parasiticus, Xerocomus, 245
Parasola, 56
parazurea, Russula, 262
pardinum var. unguentatum,
  Tricholoma, 405
parherpeticus, Cortinarius, 405
parilis, Agaricus, 405
parilis, Clitocybe, 43, 405, 427
parilis, Rhodocybe, 405
parisianorum, Melanoleuca, 405
parisotii, Agaricus, 405
parisotii, Crepidotus, 405
parkensis, Agaricus, 99
parkensis, Eccilia, 99
parkensis, Entoloma, 99
parkensis, Omphaliopsis, 99
parlatorei, Ustilago, 350
parnassiae, Trichobasis, 323, 339
parnassiae, Uromyces, 434
paropsis, Agaricus, 405
paropsis, Clitocybe, 405
partitus, Agaricus, 405
parvannulata, Lepiota, 174
parvannulatus, Agaricus, 174
parvannulatus, Cortinarius, 73
parvispora, Inocybe, 151
parvispora, Volvariella, 306
parvivelutina, Arrhenia, 16
parvivelutina, Omphalina, 16
Parvobasidium, 213
parvula, Volvaria, 306, 307
parvula, Volvariella, 306
parvula var. biloba, Volvaria, 306
parvulus, Agaricus, 306
parvus, Cortinarius, 67
pascua, Nolanea, 103, 406
pascua, Russula, 262
pascua var. pallescens, Nolanea, 405
pascua var. umbonata, Nolanea, 406
pascuum, Entoloma, 93
pascuum, Entoloma, 406
pascuus, Agaricus, 406
pascuus, Boletus, 23
pascuus, Boletus, 406
pascuus, Xerocomus, 406
pascuus β costatus, Agaricus, 373
passeckerianus, Clitopilus, 45
passeckerianus, Pleurotus, 45
pastinacae, Itersonilia, 158
patagonica, Cintractia, 349
patellaris, Panellus, 406
patellaris, Panus, 406
patellaris, Tectella, 406

patelloides, Pleurotellus, 37
pateriformis, Cortinarius, 406
patibilis var. scoticus, Cortinarius,
  74
patouillardii, Coprinus, 60
patouillardii, Coprinus, 57
patouillardii, Inocybe, 150
patouillardii, Lepiota, 171
patouillardii, Pistillina, 303
patouillardii var. lipophilus, Coprinus,
  406
patricius, Agaricus, 232
patricius, Pluteus, 232
patulum, Tricholoma, 406
patulus, Agaricus, 406
pauletii, Agaricus, 406
pauletii, Lepiota, 406
Paullicorticium, 213
paupercula, Mycena, 406
pauperculus, Agaricus, 406
paupertina, Collybia, 406
pausiaca, Clitocybe, 406
pausiacus, Agaricus, 406
pauxilla, Oliveonia, 209
pauxillum, Corticium, 209
Paxillopsis, 45
Paxillus, 213
paxillus, Agaricus, 406
paxillus, Pholiota, 406
pazschkei, Puccinia, 328
pazschkei var. jueliana, Puccinia,
  328
pearsoniana, Mycena, 202
pearsonii, Ceratobasidium, 213
pearsonii, Corticium, 213
pearsonii, Cortinarius, 74
pearsonii, Lactarius, 163
pearsonii, Oxyporus, 406
pearsonii, Paullicorticium, 213
pearsonii, Pluteus, 231
pearsonii, Poria, 406
pearsonii, Squamanita, 274
peckii, Hydnellum, 129
peckii, Hydnum, 129
pectinaceus, Agaricus, 406
pectinata, Fistulina, 406
pectinata, Russula, 263
pectinata, Russula, 406
pectinata var. insignis, Russula, 261
pectinatoides, Russula, 261, 263
pectinatoides, Russula, 263
pectinatum, Geastrum, 115
pectinatus, Agaricus, 406
pectinatus, Polyporus, 406
pediades, Agaricus, 6
pediades, Agrocybe, 6
pediades, Naucoria, 6
pedicellata, Galzinia, 113
pedicellatum, Lycoperdon, 184
pedicularis, Aecidium, 323
pedunculatum, Lycoperdon, 303
pelargonia, Russula, 263
pelargonii-zonalis, Puccinia, 328
pelargonium, Inocybe, 154
pelianthina, Mycena, 202
pelianthinus, Agaricus, 202
pelletieri, Agaricus, 227
pelletieri, Clitocybe, 227
pelletieri, Phylloporus, 227
pelliculosa, Mycena, 378
pelliculosus, Agaricus, 378
Pellidiscus, 214
pellitus, Agaricus, 232
pellitus, Pluteus, 232
pellitus var. gracilis, Pluteus, 232

pellitus var. punctillifer, Pluteus, 406
pelloporus, Boletus, 406
pellosperma, Psathyra, 406
pellospermus, Agaricus, 406
pellucida, Tubaria, 406
pellucida, Tubaria, 300
pellucidipes, Drosophila, 242
pellucidipes, Psathyrella, 242
pellucidus, Agaricus, 406
pellucidus, Boletus, 406
pellucidus, Coprinus, 60
peltata, Mycena, 406
peltatus, Agaricus, 406
peltigerina, Arrhenia, 16
peltigerina, Omphalina, 16
peltigerinus, Agaricus, 16
penarius, Hygrophorus, 137
pendulum, Hydnum, 157
pendulum, Radulum, 427
pendulum, Sistotrema, 157
pendulus, Irpex, 157
pendulus, Irpicodon, 157
penetrans, Agaricus, 119
penetrans, Dryophila, 119
penetrans, Flammula, 119
penetrans, Gymnopilus, 119
penetrans, Sebacina, 293
penetrans, Tremella, 293
pengellei, Clitocybe, 406
penicillata, Thelephora, 286
penicillatus, Cortinarius, 406
Peniophora, 214
peniophorae, Achroomyces, 311
peniophorae, Colacogloea, 311
peniophorae, Hormomyces, 128
peniophorae, Platygloea, 311
peniophorae var. interna, Platygloea,
  319
pennata, Drosophila, 243
pennata, Psathyra, 238, 243
pennata, Psathyrella, 242
pennata f. annulata, Psathyra, 243
pennata var. squamosa, Psathyra, 421
pennatus, Agaricus, 242
pequinii, Agaricus, 407
pequinii, Chitonia, 407
pequinii, Clarkeinda, 407
perbrevis, Agaricus, 407
perbrevis, Inocybe, 407
percandidum, Entoloma, 99
percandidum, Leccinum, 407
percandidus, Boletus, 407
percevalii, Agaricus, 278
percevalii, Psilocybe, 278
percevalii, Stropharia, 278
percevalii var. aurantiaca, Stropharia,
  277
percincta, Conocybe, 53
percomis, Cortinarius, 407
peregrinum, Hypholoma, 353
perennans, Ustilago, 349
Perenniporia, 215
perennis, Boletus, 49
perennis, Coltricia, 49
perennis, Polyporus, 49
perennis, Polystictus, 49
perforans, Agaricus, 196
perforans, Androsaceus, 196
perforans, Marasmiellus, 196
perforans, Micromphale, 196
pergamena, Clitocybe, 407
pergamenea, Thelephora, 224
pergamenum, Tricholoma, 407
pergamenus, Agaricus, 407
pergamenus, Lactarius, 163, 164

phoenicis, Ustilago, 435
pholideus, Agaricus, 74
pholideus, Cortinarius, 74
Pholiota, 224
Pholiotina, 50
phosphorea, Auricularia, 284
phosphorea, Clavaria, 408
Phragmidium, 319
phragmitis, Acanthobasidium, 7
phragmitis, Aleurodiscus, 7
phragmitis, Puccinia, 328
phragmitis, Uredo, 328
Phragmoxenidium, 227
phrygianus, Cortinarius, 408
Phylacteria, 285
phylacteris, Hypochnus, 287
phylacteris, Thelephora, 286, 287
phylacteris, Tomentella, 286
phyllogena, Mycena, 201
phyllogena, Psilocybe, 248
phyllogenum, Hypholoma, 248
phyllogenus, Agaricus, 201
phyllophila, Clitocybe, 44
phyllophila, Peniophora, 408
phyllophila, Vararia, 408
phyllophilus, Agaricus, 44
Phylloporia, 227
Phylloporus, 227
Phyllotopsis, 227
phyllyreae, Aecidium, 339
phyllyreae, Uredo, 339
Physalacria, 228
physaloides, Agaricus, 248
physaloides, Deconica, 248
physaloides, Psilocybe, 248
physciacearum, Syzygospora, 281
Physisporinus, 228
Physisporus, 215
physospermi, Puccinia, 328
Phytoconis, 181
picaceus, Agaricus, 60
picaceus, Coprinus, 60
picea, Nolanea, 186
piceae, Hygrophorus, 138
piceae, Inocybe, 408
picinus, Lactarius, 408
picipes, Polyporellus, 234
picipes, Polyporus, 234
picrea, Flammula, 120
picrea, Fulvidula, 120
picreus, Agaricus, 120
picreus, Gymnopilus, 120
picridis, Puccinia, 326
picta, Mycena, 202
picta, Omphalia, 202
picta, Omphalina, 202
picta, Xeromphalina, 202
picta var. *concolor, Mycena*, 371
pictus, Agaricus, 202
Pilacrella, 310
pilatiana, Lepiota, 178
pilatianus, Agaricus, 408
pilatianus, Leucoagaricus, 178
pilatianus, Leucocoprinus, 178
pilatianus var. *subrubens, Leucocoprinus*, 178
pilatii, Flageloscypha, 107
pileolarius, Agaricus, 408
pilipes, Agaricus, 408
pilipes, Gymnopus, 408
pillodii, Collybia, 408
pillodii, Pseudobaeospora, 408
Piloderma, 228
pilosa, Tomentella, 288
pilosella, Conocybe, 54

pilosella, Conocybe, 54
pilosella, Galera, 54
piloselloidarum, Puccinia, 326
piloselloides, Conocybe, 54
pilosellus, Agaricus, 54
pilosus, Agaricus, 190
pilosus, Hypochnus, 288
piluliforme, Hypholoma, 243
piluliformis, Agaricus, 243
piluliformis, Drosophila, 243
piluliformis, Psathyrella, 243
piluliformis, Psathyrella, 241
pimii, Cyphella, 32
pimpinellae, Puccinia, 328
pimpinellae, Trichobasis, 328
pimpinellae, Uredo, 328
pinastri, Gyrophana, 180
pinastri, Hydnum, 180
pinastri, Leucogyrophana, 180
pinastri, Merulius, 180
pinastri, Serpula, 180
pinastri, Sistotrema, 180
pinetorum, Clitocybe, 43
pinetorum, Clitocybe, 408
pinetorum, Conocybe, 54
pinetorum, Conocybe, 54, 55
pinguipes, Lepiota, 177
pinguipes, Leucoagaricus, 177
pinguis var. *rosae-alpinae, Uredo*, 320
pini, Aecidium, 311, 312
pini, Aecidium, 434
pini, Basidiodendron, 20
pini, Boletus, 220
pini, Daedalea, 220
pini, Endocronartium, 312
pini, Peniophora, 214
pini, Peridermium, 311, 312
pini, Phellinus, 220
pini, Porodaedalea, 220
pini, Sebacina, 20
pini, Sterellum, 215
pini, Stereum, 215
pini, Thelephora, 215
pini, Trametes, 220
pinicola, Boletus, 24, 109
pinicola, Cortinarius, 72
pinicola, Fomitopsis, 109
pinicola, Gloeotulasnella, 302
pinicola, Pholiota, 226
pinicola, Polyporus, 234
pinicola, Tulasnella, 302
pinitorqua, Melampsora, 316
pinophilus, Boletus, 24
pinsitus, Agaricus, 45
pinsitus, Clitopilus, 45
Pinuzza, 280
piperatum, Leccinum, 37
piperatus, Agaricus, 164
piperatus, Boletus, 37
piperatus, Chalciporus, 37
piperatus, Ixocomus, 37
piperatus, Lactarius, 164
piperatus, Suillus, 37
piperatus f. *pergamenus, Lactarius*, 407
piperatus var. *pergamenus, Lactarius*, 407
piperatus β *exsuccus, Lactarius*, 259
Piptoporus, 228
pirolata, Chrysomyxa, 311
pisciodora, Inocybe, 364
pisciodora, Nolanea, 186
pisi, Puccinia, 338
pisi, Uromyces, 338
pisiformis, Granularia, 208

pisiformis, Nidularia, 208
pisiformis var. *broomei, Nidularia*, 208
pisi-sativi, Uromyces, 337
pisocarpium, Polysaccum, 229
Pisolithus, 229
Pistillaria, 303
pistillaris, Agaricus, 408
pistillaris, Clavaria, 40
pistillaris, Clavariadelphus, 40
pistilliferum, Sistotrema, 272
Pistillina, 303
Pistillina, 229
pithya, Hemimycena, 385
pithya, Hemimycena, 408
pithya, Mycena, 125, 408
pithya, Peniophora, 408
pithya, Thelephora, 408
pithyophila, Clitocybe, 44
pithyophilus, Agaricus, 44
Pityrosporum, 344
placenta, Agaricus, 408
placenta, Ceriporiopsis, 236
placenta, Entoloma, 408
placenta, Poria, 236
placenta, Postia, 236
placentus, Oligoporus, 236
placentus, Polyporus, 236
placida, Leptonia, 408
placidum, Entoloma, 408
placidus, Agaricus, 408
placidus, Boletus, 281
placidus, Suillus, 281
placidus var. *gracilis, Rhodophyllus*, 408
placomyces, Agaricus, 408
placomyces, Agaricus, 4
plagioporus, Coprinus, 61
plana, Exidia, 105
plana, Tremella, 105
plancus, Agaricus, 408
plancus, Marasmius, 408
planipes, Agaricus, 408
planipes, Collybia, 408
plantaginis, Uredo, 434
planus, Agaricus, 408
platyphylla, Collybia, 192
platyphylla, Megacollybia, 191
platyphylla, Oudemansiella, 192
platyphylla, Tricholomopsis, 192
platyphylla ssp. *lacerata, Collybia*, 45
platyphylloides, Rhodophyllus, 409
platyphyllus, Agaricus, 191
platypora, Grifola, 234
platypus, Coprinus, 409
platypus, Lyophyllum, 284
platypus, Tephrocybe, 284
plautus, Agaricus, 232
plautus, Pluteus, 231
plautus, Pluteus, 232
plautus var. terrestris, Pluteus, 409
plebeioides, Agaricus, 99
plebeioides, Entoloma, 99
plebejum, Entoloma, 99
plebejus, Agaricus, 99
plebejus, Inocephalus, 99
pleopodium, Entoloma, 99
pleopodius, Agaricus, 99
Pleurocybella, 229
Pleuroflammula, 229
Pleuropus, 213
pleurotelloides, Clitopilus, 45
pleurotelloides, Octojuga, 45
Pleurotellus, 80
Pleurotus, 229
plexipes, Agaricus, 284

porphyrizon, **Agaricus**, 4
porphyrocephalus, Agaricus, 4
porphyrogenita, Uredo, 333
porphyrogriseum, Entoloma, 409
porphyroleucum, Tricholoma, 409
porphyroleucus, Agaricus, 409
porphyrophaeum, Entoloma, 100
porphyrophaeus, Agaricus, 100
porphyrophaeus, Rhodophyllus, 100
porphyrophaeus, Trichopilus, 100
porphyropus, Agaricus, 74
porphyropus, Cortinarius, 74
porphyrosporus, Boletus, 235
porphyrosporus, Phaeoporus, 235
porphyrosporus, Porphyrellus, 235
porphyrosporus var. fuligineus, Phaeoporus, 382
Porpoloma, 235
porreus, Agaricus, 409
porreus, Marasmius, 409
porri, Puccinia, 329
porri, Uredo, 321
porrigens, Agaricus, 229
porrigens, Nothopanus, 229
porrigens, Phyllotus, 229
porrigens, Pleurocybella, 229
porrigens, Pleurotellus, 229
porrigens, Pleurotus, 229
porriginosa, Naucoria, 410
porriginosus, Agaricus, 410
portentosum, Corticium, 269
portentosum, Scytinostroma, 269
portentosum, Stereum, 269
portentosum, Tricholoma, 296
portentosus, Agaricus, 297
porulosa f. albomarginata, Tomentella, 288
posterula, Inocybe, 154
posterulus, Agaricus, 154
Postia, 235
postiana, Russula, 263
postii, Agaricus, 183
postii, Gerronema, 183
postii, Loreleia, 183
postii, Omphalia, 183
postii, Omphalina, 183
postii var. aurea, Omphalia, 410
potentillae, Phragmidium, 312
potentillae, Phragmidium, 320
potentillae, Puccinia, 320
potentillarum, Lecythea, 319, 320
potentillarum, Uredo, 319, 320
potentillarum, Uredo, 434
poterii, Aecidium, 328
poterii, Lecythea, 328
Pouzarella, 91
pouzarianus, Pluteus, 232
Pouzaromyces, 90
praeclaresquamosus, Agaricus, 4
praeclaresquamosus var. terricolor, Agaricus, 4
praecox, Agaricus, 6
praecox, Agrocybe, 6
praecox, Mycena, 197
praecox, Pholiota, 6
praecox, Togaria, 6
praecox var. paludosa, Pholiota, 6
praecox β sphaleromorpha, Agaricus, 420
praefocata, Trechispora, 291
praestans, Agaricus, 74
praestans, Cortinarius, 74
praestans, Naucoria, 300
praestans, Tubaria, 300

praestigiosus, Cortinarius, 410
praetermissa, Peniophora, 141
praetermissum, Corticium, 141
praetermissum, Gloeocystidium, 141
praetermissum, Hyphoderma, 141
praetervisa, Astrosporina, 154
praetervisa, Inocybe, 154
praetervisa, Russula, 263
praetextus, Agaricus, 410
pragense, Dermoloma, 88
prasinoides, Coniophora, 50
prasinus, Agaricus, 410
prasinus, Cortinarius, 410
prasiosmus, Agaricus, 410
prasiosmus, Marasmius, 413
prasiosmus, Marasmius, 410
Pratella, 1
pratense, Lycoperdon, 305
pratense, Vascellum, 305
pratensis, Agaricus, 134, 190
pratensis, Agaricus, 410
pratensis, Anthracoidea, 340
pratensis, Camarophyllus, 134
pratensis, Cintractia, 340
pratensis, Cortinarius, 74
pratensis, Cuphophyllus, 134
pratensis, Dermocybe, 74
pratensis, Gymnopus, 134
pratensis, Hygrocybe, 134
pratensis, Hygrophorus, 134
pratensis, Lepiota, 173
pratensis, Psilocybe, 248
pratensis, Puccinia, 329
pratensis, Ramaria, 41
pratensis var. cinereus, Hygrophorus, 369
pratensis var. meisneriensis, Hygrophorus, 410
pratensis var. pallida, Hygrocybe, 134
pratensis var. pallidus, Camarophyllus, 134
pratensis var. pallidus, Hygrophorus, 134
pratensis var. umbrinus, Hygrophorus, 410
pratensis β cinereus, Agaricus, 369
pratensis β vitulinus, Gymnopus, 410
pratensis γ meisneriensis, Agaricus, 410
praticola, Ceratobasidium, 285
praticola, Corticium, 285
praticola, Galerina, 111
praticola, Pellicularia, 285
praticola, Pholiota, 111
praticola, Thanatephorus, 285
pratulense, Entoloma, 100
pravum, Tricholoma, 410
pravus, Agaricus, 410
prenanthis, Aecidium, 327
prenanthis, Puccinia, 327
prenanthis, Uredo, 327
prescotii, Cantharellopsis, 116
prescotii, Gerronema, 116
prescottii, Cantharellus, 116
primula, Agaricus, 410
primulae, Aecidium, 329
primulae, Puccinia, 329
primulae, Sorosporium, 348
primulae, Trichobasis, 329
primulae, Uredo, 329
primulae, Urocystis, 348
primulicola, Tuburcinia, 348
primulicola, Urocystis, 348

pringsheimiana, Puccinia, 323
privignoides, Cortinarius, 410
privignus, Cortinarius, 410
proboscideus, Agaricus, 410
proboscideus, Crepidotus, 410
procera, Lepiota, 187
procera, Macrolepiota, 187
procera var. fuliginosa, Lepiota, 187
procera var. permixta, Macrolepiota, 407
procera var. pseudo-olivascens, Macrolepiota, 187
procera var. puellaris, Lepiota, 178
procera var. rhacodes, Lepiota, 187
procera β excoriata, Lepiota, 186
procerus, Agaricus, 187
procerus, Boletus, 410
proletaria, Nolanea, 93, 103
proletaria, Nolanea, 410
proletarius, Agaricus, 410
prolifera, Mycena, 410
proliferus, Agaricus, 410
proliferus, Boletus, 410
prolixa, Collybia, 49
prolixa, Rhodocollybia, 49
prolixa var. distorta, Rhodocollybia, 47
prolixus, Agaricus, 49
prominens, Agaricus, 410
prominens, Lepiota, 410
prominens, Macrolepiota, 410
prona, Psathyra, 243
prona, Psathyrella, 243
prona f. albidula, Psathyrella, 243
prona f. cana, Psathyrella, 243
prona f. orbitarum, Psathyrella, 243
prona var. smithii, Psathyrella, 410
prona var. utriformis, Psathyrella, 245
pronus, Agaricus, 243
propinquus, Tubulicrinis, 300
prostii, Puccinia, 329
Proteomyces, 299
proteus, Agaricus, 195
proteus, Boletus, 410
proteus, Lycoperdon, 410
proteus, Melanotus, 195
Protodontia, 279
protoparmeliae, Tremella, 293
Prototremella, 301
protracta, Collybia, 410
protractus, Agaricus, 410
provincialis, Rhizopogon, 253
proxima, Clitocybe, 160
proxima, Laccaria, 160
proxima, Peniophora, 215
proxima var. bicolor, Laccaria, 159
proximella, Astrosporina, 154
proximella, Inocybe, 154
proximella, Laccaria, 160
pruina, Aphanobasidium, 223
pruina, Corticium, 223
pruinata, Pellicularia, 26
pruinatum, Botryobasidium, 27
pruinatum, Corticium, 27
pruinatum var. laeve, Botryobasidium, 27
pruinatus, Boletus, 25
pruinatus, Marasmius, 188
pruinatus, Marasmius, 411
pruinatus, Xerocomus, 25
pruinosa, Clitocybe, 44
pruinosa, Inocybe, 154
pruinosa, Odontia, 142
pruinosa, Peniophora, 308
pruinosa, Tulasnella, 302

pubescens var. *scoticus, Lactarius*, 165
pubescentipes, Agaricus, 306
pubescentipes, Volvariella, 306
Puccinia, 320
Pucciniastrum, 333
pudens, Agaricus, 308
pudens, Gymnopus, 308
pudens, Oudemansiella, 308
pudens, Xerula, 308
pudica, Inocybe, 157
pudica, Lepiota, 411
pudica, Mycena, 252
pudica, Pholiota, 6, 411
pudicus, Agaricus, 6, 411
pudicus, Leucoagaricus, 178
pudicus, Leucoagaricus, 411
pudorinum, Hydnum, 275
pudorinum, Limacium, 409
pudorinus, Agaricus, 138
pudorinus, Hygrophorus, 138
puella, Agaricus, 9
puellaris, Lepiota, 178
puellaris, Macrolepiota, 178
puellaris, Russula, 263
puellaris *var. intensior, Russula*, 411
puellaris var. *leprosa, Russula*, 263
puellula, Russula, 263
pulchella, Clavaria, 251
pulchella, Conocybe, 54
pulchella, Galera, 54
pulchella, Ramariopsis, 251
pulchella, Russula, 259
pulchellus, Boletus, 280
pulchellus, Cortinarius, 66
Pulcherricium, 284
pulcherrimum, Hydnum, 411
pulchra, Clavaria, 41
pulchra, Clavulinopsis, 41
pulchrae-uxoris, Russula, 263
pulchralis, Russula, 412
pulchripes, Cortinarius, 412
pulchrum, Byssocorticium, 29
pulchrum, Corticium, 29
pulchrum, Leccinum, 168
pulicaris, Anthracoidea, 340
pulla, Collybia, 412
pullata, Mycena, 412
pullatus, Agaricus, 380, 412
pullatus, Coprinus, 380
pullus, Agaricus, 412
pulmonarius, Agaricus, 230
pulmonarius, Pleurotus, 230
pulmonarius, Pleurotus, 230
pulmonarius var. *juglandis, Agaricus*, 412
pulmonarius *var. juglandis, Pleurotus*, 412
pulverea, Leptonia, 100
pulvereum, Entoloma, 100
pulvereus, Pouzaromyces, 100
Pulverolepiota, 85
pulverulenta, Auricularia, 180
pulverulenta, Cystolepiota, 86
pulverulenta, Gyrophana, 180
pulverulenta, Hauerslevia, 121
pulverulenta, Lepiota, 86
pulverulenta, Leucogyrophana, 180
pulverulenta, Peniophora, 308
pulverulenta, Puccinia, 329
pulverulenta, Pulverolepiota, 86
pulverulenta, Sebacina, 121
pulverulentum, Corticium, 308
pulverulentum, Laeticorticium, 63

pulverulentum, Xenasma, 308
pulverulentus, Agaricus, 412
pulverulentus, Boletus, 25
pulverulentus, Hygrophorus, 412
pulverulentus, Lentinus, 412
pulverulentus, Leucoagaricus, 86
pulverulentus, Merulius, 180
pulverulentus, Xerocomus, 25
pulvinatus, Agaricus, 412
pulvinus, Buglossoporus, 229
pumila, Galerina, 112
pumila, Laccaria, 160
pumila, Pholiota, 112
pumila, Russula, 256
pumila var. *subferruginea, Pholiota*, 111
pumilum, Hebeloma, 122
pumilus, Agaricus, 112
pumilus, Cortinarius, 74
punctata, Amanita, 412
punctata, Amanitopsis, 412
punctata, Poria, 220
punctata, Puccinia, 329
punctata, Russula, 360
punctata f. *citrina, Russula*, 266
punctata f. *violeipes, Russula*, 266
punctatum, Aecidium, 335
punctatum, Hebeloma, 226
punctatum, Hebeloma, 412
punctatus, Agaricus, 412
punctatus, Boletus, 412
punctatus, Cortinarius, 412
punctatus, Phellinus, 220
punctatus, Polyporus, 220
punctatus, Uromyces, 338
punctiformis, Dacrymyces, 87
punctiformis, Flagelloscypha, 107
punctiformis, Peziza, 107
punctiformis, Puccinia, 329
punctiformis, Uredo, 329
punctiformis var. *stenospora, Cyphella*, 107
punctipes, Pluteus, 232
punctulata, Hypholoma, 412
punctulata, Stropharia, 412
punctulatum, Corticium, 147
punctulatum, Gloeocystidium, 147
punctulatum, Hypochnicium, 146
punctulatum, Hypochnicium, 147
punctulatus, Agaricus, 412
pungens, Russula, 263
punicea, Dermocybe, 76
punicea, Hygrocybe, 134
punicea, Thelephora, 288
punicea, Tomentella, 288
punicea, Tomentella, 288
punicea f. *splendidissima, Hygrocybe*, 135
punicea var. *splendidissima, Hygrocybe*, 135
puniceus Fr., *Agaricus*, 134
puniceus With., Agaricus, 412
puniceus, Cortinarius, 76
puniceus, Hygrophorus, 134
puniceus, Hypochnus, 288
puniceus var. *flavescens, Hygrophorus*, 381
puniceus var. *nigrescens, Hygrophorus*, 401
pura, Mycena, 202
pura f. *alba, Mycena*, 203
pura f. *lutea, Mycena*, 203
pura f. *purpurea, Mycena*, 203
pura f. *violacea, Mycena*, 203
pura var. *alba, Mycena*, 203

pura var. *carnea, Mycena*, 203
pura var. *lutea, Mycena*, 203
pura var. *multicolor, Mycena*, 203
pura var. *purpurea, Mycena*, 203
pura var. *rosea, Mycena*, 203
pura var. *violacea, Mycena*, 203
purpurascens (Cooke) Pilát, *Agaricus*, 4
purpurascens Fr., *Agaricus*, 74
purpurascens, Boletus, 29
purpurascens, Cortinarius, 74
purpurascens, Psalliota, 4
purpurascens, Trametes, 412
purpurascens var. *obscura, Cortinarius*, 73
purpurascens With., Agaricus, 412
purpurata, Flammula, 353
purpurata, Russula, 260
purpuratus, Agaricus, 353
purpuratus, Gymnopilus, 353
purpurea, Ceriporia, 36
purpurea, Clavaria, 39
purpurea, Meruliopsis, 36
purpurea, Pistillaria, 412
purpurea, Poria, 36
purpurea, Puccinia, 434
purpurea, Thelephora, 37
purpurella, Psalliota, 3
purpurellus, Agaricus, 3
purpureobadia, Laccaria, 160
purpureobadius, Cortinarius, 64
purpureofusca, Mycena, 203
purpureofuscus, Agaricus, 203
purpureogrisea, Collybia, 352
purpureogriseus, Marasmius, 352
purpureolilacinus, Leucoagaricus, 178
purpureorimosus, Leucoagaricus, 178
purpureum, Chondrostereum, 37
purpureum, Helicobasidium, 313
purpureum, Stereum, 37
purpureum *var.* atromarginatum, Stereum, 412
purpureus Bolton, *Agaricus*, 203
purpureus Pers., *Agaricus*, 74
purpureus, Boletus, 25
purpureus, Boletus, 412
purpureus, Cortinarius, 74
purpureus, Hypochnus, 313
purpureus, Merulius, 80
purpureus, Polyporus, 36
purpureus var. *xanthocyaneus, Boletus*, 26
purus, Agaricus, 202
purus, Gymnopus, 202
purus var. *roseus, Agaricus*, 203
purus β *purpureus, Gymnopus*, 203
purus γ *purpureus, Agaricus*, 203
pusilla, Agrocybe, 412
pusilla, Amanita, 306
pusilla, Bovista, 412
pusilla, Bovista, 28
pusilla, Clavaria, 412
pusilla, Dacrymyces, 30
pusilla, Dacryomitra, 30
pusilla, Eccilia, 412
pusilla, Naucoria, 412
pusilla, Nolanea, 93
pusilla, Pistillaria, 35, 304, 412
pusilla, Tomentellopsis, 289
pusilla, Typhula, 412
pusilla, Volvariella, 306
pusilla var. *taylorii, Volvariella*, 307
pusillima, Flocculina, 109
pusillimus, Flammulaster, 109

radula, Boletus, 267
radula, Hydnum, 21
radula, Hyphoderma, 21
radula, Hyphodontia, 267
radula, Polyporus, 267
radula, Poria, 267, 422
radula, Schizopora, 267
radula, Sistotrema, 21
raduloides, Phanerochaete, 218
Radulomyces, 249
raffillii, Marasmius, 413
ragazziana, Pleuroflammula, 229
ragazzianus, Crepidotus, 229
ralfsii, Agaricus, 81
ralfsii, Crepidotus, 81
Ramaria, 41
Ramaria, 250
Ramaricium, 251
Ramariopsis, 251
rameale, Stereum, 276
ramealis, Agaricus, 188
ramealis, Gymnopus, 188
ramealis, Marasmiellus, 188
ramealis, Marasmius, 188
ramentacea, Antrodia, 13
ramentacea, Armillaria, 413
ramentacea, Poria, 13
ramentaceum, Tricholoma, 295
ramentaceum, Tricholoma, 413
ramentaceus, Agaricus, 413
ramentaceus, Polyporus, 13
Ramicola, 270
ramicola, Ceratorhiza, 35
ramicola, Rhizoctonia, 35
ramosissima, Fibrillaria, 50
ramosissima, Tomentella, 287
ramosissimum, Rhinotrichum, 287
ramosissimus, Zygodesmus, 287
ramosoradicatus, Agaricus, 413
ramosum, Hericium, 126
ramosum, Hydnum, 126
ramosus, Boletus, 167
ramosus, Polyporus, 167
rancida, Collybia, 284
rancida, Poria, 237
rancida, Postia, 237
rancida, Tephrocybe, 284
rancidum, Lyophyllum, 284
rancidus, Agaricus, 284
rancidus, Oligoporus, 237
rangiferinus, Boletus, 234
rannochii, Psathyrella, 244
ranunculacearum, Aecidium, 327, 336
ranunculacearum var. aquilegiae,
    Aecidium, 330
ranunculacearum var. hellebori-viridis,
    Uredo, 347
ranunculacearum var. linguae,
    Aecidium, 327
ranunculacearum var. thalictri,
    Aecidium, 330
ranunculi, Aecidium, 327
ranunculi, Entyloma, 342
ranunculi, Fusidium, 342
ranunculi, Sporisorium, 348
ranunculi, Tuburcinia, 348
ranunculi, Urocystis, 348
ranunculi var. microsporum,
    Cylindrosporium, 343
ranunculi-festucae, Uromyces, 336
ranunculi-repentis, Entyloma, 343
raoultii, Russula, 264
rapaceus, Cortinarius, 64
rapaceus, Cortinarius, 75
raphanicum, Tricholoma, 294

raphanoides, Agaricus, 65
raphanoides, Cortinarius, 65
raphanoides β venetus, Agaricus, 79
rapiolens, Mycena, 413
rasile, Tricholoma, 194
rasilis, Agaricus, 194
rasilis, Melanoleuca, 194
rasilis var. leucophylloides,
    Melanoleuca, 413
rasilis var. pseudoluscina,
    Melanoleuca, 413
raunkiaeri, Athelia, 176
raunkiaeri, Leptosporomyces, 176
rauwenhoffii, Tilletia, 346
ravenelii, Corynites, 197
ravenelii, Mutinus, 197
ravenelii, Peniophora, 224
ravenelii, Phanerochaete, 224
ravenelii, Phlebiopsis, 224
ravenelii, Scopuloides, 224
ravida, Galera, 413
ravida, Galera, 55
ravidus, Agaricus, 413
ravidus, Polyporus, 210
ravidus, Polystictus, 210
reae, Entoloma, 100
reae, Hygrocybe, 133
reae, Hygrophorus, 133
reae, Leptonia, 100
reae var. insipida, Hygrocybe, 132
reai, Melanoleuca, 414
recisa, Exidia, 105
recisa, Tremella, 105
recolligens, Geastrum, 115
recolligens, Lycoperdon, 115
recondita, Athelopsis, 184
recondita, Luellia, 184
recondita, Puccinia, 329
recondita f.sp. agropyrina, Puccinia,
    329
recondita f.sp. agrostidis, Puccinia,
    329
recondita f.sp. borealis, Puccinia, 330
recondita f.sp. bromina, Puccinia, 330
recondita f.sp. echii-agropyrina,
    Puccinia, 330
recondita f.sp. holcina, Puccinia, 326
recondita f.sp. perplexans, Puccinia,
    330
recondita f.sp. persistens, Puccinia,
    330
recondita f.sp. triseti, Puccinia, 326
recondita f.sp. triticina, Puccinia, 330
reconditum, Corticium, 184
Rectipilus, 251
recubans, Marasmius, 191
recutita, Amanita, 10
recutitus, Agaricus, 10
redemitus, Cortinarius, 414
reducta, Inocybe, 148
reducta, Naucoria, 414
reducta, Simocybe, 414
reductus, Agaricus, 414
reedii, Cortinarius, 414
reflexa, Auricularia, 276
reflexus, Agaricus, 421
reflexus, Gymnopus, 421
regalis, Amanita, 414
regalis, Sarcodon, 266
regifica, Peniophora, 301
regificus, Tubulicrinis, 301
reginae, Entoloma, 100
regius, Boletus, 25
reidii, Hygrocybe, 134
reidii, Peniophora, 215

reidii, Volvariella, 306
reiliana, Sphacelotheca, 434
reiliana, Ustilago, 434
relhanii, Helvella, 414
relicina, Inocybe, 414
relicinus, Agaricus, 414
renati, Mycena, 203
renidens, Cortinarius, 75
renidens, Russula, 264
reniformis, Agaricus, 128
reniformis, Hohenbuehelia, 128
reniformis, Pleurotus, 128, 252
rennyi, Agaricus, 155
rennyi, Astrosporina, 155
rennyi, Inocybe, 155
rennyi, Oligoporus, 237
rennyi, Polyporus, 237
rennyi, Poria, 237
rennyi, Postia, 237
rennyi, Strangulidium, 237
rennyi, Tyromyces, 237
rennyi var. major, Astrosporina, 414
rennyi var. major, Inocybe, 414
repanda, Exidia, 105
repanda, Tremella, 105
repanda, Ulocolla, 105
repandum, Dentinum, 129
repandum, Entoloma, 100
repandum, Entoloma, 414
repandum, Hebeloma, 365
repandum, Hydnum, 129
repandum var. rufescens, Hydnum,
    129
repandus, Agaricus, 414
repandus, Agaricus, 414
repens, Agaricus, 192
repens, Epulorhiza, 301
repens, Rhizoctonia, 301
repentis, Melampsora, 315
Repetobasidiellum, 251
Repetobasidium, 252
replexus, Agaricus, 414
replexus, Cantharellus, 414
replexus var. devexus, Cantharellus,
    414
repraesentaneus, Lactarius, 164
resimus, Agaricus, 165
resimus, Lactarius, 165
resinaceum, Ganoderma, 114
resinaceus, Agaricus, 207
resinaceus, Fomes, 114
resinaceus, Lentinus, 208
Resinicium, 252
Resinomycena, 252
resinosum, Ischnoderma, 158
resinosus Fr., Polyporus, 158
resinosus Schrad., Polyporus, 158
resplendens, Agaricus, 414
resplendens, Tricholoma, 298
resplendens, Tricholoma, 414
restricta, Malassezia, 344
resupinata, Poria, 414
Resupinatus, 252
resupinatus, Boletus, 414
resupinatus, Boletus, 414, 430
resupinatus, Fomes, 414
resupinatus, Tyromyces, 13
resutum, Entoloma, 414
resutus, Agaricus, 95, 414
reticulata, Ceriporia, 36
reticulata, Fibuloporia, 36
reticulata, Mucilago, 36
reticulata, Peniophora, 140
reticulata, Poria, 36
reticulata, Ustilago, 350

Roestelia, 313
Rogersella, 141
rolfsii, Athelia, 352
rolfsii, Corticium, 352
romagnesiana, Tubaria, 300
romagnesianus, Coprinus, 61
romagnesii, Agaricus, 2
romagnesii, Entoloma, 415
romagnesii, Lactarius, 165
romellii, Agaricus, 233
romellii, Antrodiella, 14
romellii, Dacrymyces, 87
romellii, Leucogyrophana, 181
romellii, Odontia, 208
romellii, Odonticium, 208
romellii, Pluteus, 233
romellii, Poria, 14
romellii, Russula, 264
romseyensis, Psathyrella, 244
rorida, Mycena, 203
roridus, Agaricus, 203
rorulentus, Agaricus, 179
rosacea, Russula, 264
rosaceus, Agaricus, 264
rosaceus β exalbicans, Agaricus, 259
rosae, Lecythea, 320
rosae-alpinae, Phragmidium, 320
rosae-pimpinellifoliae,
   Phragmidium, 320
rosarum f. rosae-pimpinellifoliae,
   Phragmidium, 320
rosea, Clavaria, 39
rosea, Cystolepiota, 86
rosea, Fomitopsis, 353
rosea, Hyphelia, 63
rosea, Lepiota, 85
rosea, Leptonia, 101
rosea, Mycena, 203
rosea, Mycena, 203
rosea, Peniophora, 63
rosea, Russula, 264
rosea, Russula, 257
rosea, Thelephora, 63
rosea var. subglobosa, Clavaria, 39
roseiavellanea, Rhodocybe, 254
roseiavellaneus, Pleuropus, 254
roseipellis, Limonomyces, 183
roseipes, Cortinarius, 75
roseipes, Russula, 262
roseipes, Russula, 415
roseipes, Telamonia, 75
rosella, Clitocybe, 56
rosella, Mycena, 203
rosella, Omphalia, 56, 159, 203
rosella, Omphalina, 56
rosella, Pistillaria, 304
rosella, Tulasnella, 301
rosellus, Agaricus, 159, 202, 203
rosellus, Contumyces, 56
rosellus, Marasmiellus, 56
roseoalbus, Agaricus, 254, 415
roseoalbus, Pluteus, 415
roseocremeum, Corticium, 141
roseocremeum, Gloeocystidium, 141
roseocremeum, Hyphoderma, 141
roseofloccosa, Limacella, 182
roseofloccosa, Limacella, 415
roseofracta, Krombholziella, 168
roseofractum, Leccinum, 168
roseofusca, Mycena, 202
roseogriseus, Hypochnus, 247
roseogriseus var. lavandulaceus,
   Hypochnus, 247
roseolum, Corticium, 63
roseolus, Pleurotus, 415

roseolus, Pleurotus, 195
roseolus, Rhizopogon, 253
roseolus, Splanchnomyces, 253
roseotincta, Krombholziella, 169
roseotincta, Omphalia, 43
roseotinctum, Leccinum, 169
roseotinctus, Coprinus, 415
roseozonatus, Lactarius, 162
roseum, Corticium, 63
roseum, Entoloma, 101
roseum, Laeticorticium, 63
roseum var. pulverulentum,
   Laeticorticium, 63
roseus, Agaricus, 203
roseus, Boletus, 353
roseus, Fomes, 353
roseus, Gomphidius, 118
roseus, Hypochnus, 63
roseus, Polyporus, 353
roseus, Sporobolomyces, 334
rostellata, Conocybe, 415
rostellata, Galera, 415
rostellata, Psathyrella, 244
rostkovii, Boletus, 415
rostkovii, Polyporus, 234
rostratus, Lactarius, 165
Rostrupia, 320
rostrupianus, Coprinus, 58
rostrupianus, Coprinus, 415
rostrupii, Exobasidium, 344
rostrupii, Melampsora, 316
rotula, Agaricus, 191
rotula, Androsaceus, 191
rotula, Marasmius, 191
roumeguerei, Corticium, 224
roumeguerei, Peniophora, 224
roumeguerei, Phlebia, 224
roumeguerei, Phlebiopsis, 224
rozei, Entoloma, 415
Rozites, 256
rubecundus, Agaricus, 416
rubella, Lepiota, 353
rubella, Mycena, 198
rubella, Pratella, 3
rubella, Psalliota, 3
rubella f. pallens, Psalliota, 3
rubellotincta, Alboleptonia, 102
rubellum, Aecidium, 328
rubellus, Agaricus, 3
rubellus, Boletus, 26
rubellus, Cortinarius, 75
rubellus, Gyrodon, 416
rubellus, Xerocomus, 26
rubellus var. verecundus, Agaricus,
   430
rubeolarium, Leccinum, 24
rubeolarius, Boletus, 24
ruber, Agaricus, 416
ruber, Clathrus, 39
ruberrima, Russula, 264
rubescens, Amanita, 10
rubescens var. alba, Amanita, 10
rubescens var. annulosulphurea,
   Amanita, 10
rubescens var. magnifica, Amanita,
   10, 395, 396
rubescens, Daedalea, 87
rubescens, Hysterangium, 253
rubescens, Inocybe, 150, 152, 157
rubescens, Lactarius, 162
rubescens, Ptychogaster, 236
rubescens, Rhizopogon, 253
rubescens, Trametes, 87
rubeus, Agaricus, 416
rubi, Agaricus, 270

rubi, Corticium, 7
rubi, Crepidotus, 270
rubi, Cyphella, 34
rubi, Naucoria, 270
rubi, Phragmidium, 319
rubi, Puccinia, 319, 320
rubi, Ramicola, 270
rubi, Simocybe, 270
rubi, Thelephora, 7
rubiatus, Agaricus, 416
rubicundula, Flammula, 75
rubicundulus, Agaricus, 75
rubicundulus, Cortinarius, 75
rubicundulus, Paxillus, 214
rubida, Nolanea, 354
rubidus, Agaricus, 354
rubiformis, Naematelia, 292, 416
rubiformis, Tremella, 416
rubiformis, Tremella, 292
rubiginosa, Conocybe, 53
rubiginosa, Galera, 113
rubiginosa, Galerina, 113
rubiginosa, Helvella, 139
rubiginosa, Hymenochaete, 139
rubiginosa, Tomentella, 288
rubiginosa ssp. subfuliginosa,
   Hymenochaete, 382
rubiginosum, Stereum, 139
rubiginosus Pers., Agaricus, 113
rubiginosus With., Agaricus, 416
rubiginosus, Boletus, 416
rubiginosus, Hypochnus, 288
rubigo-vera, Puccinia, 330, 331
rubigo-vera, Trichobasis, 330
rubigo-vera, Uredo, 330
rubigo-vera f.sp. agrostidis, Puccinia,
   329
rubigo-vera f.sp. holcina, Puccinia,
   326
rubigo-vera f.sp. perplexans, Puccinia,
   330
rubigo-vera f.sp. triseti, Puccinia, 326
rubi-idaei, Phragmidium, 320
rubi-idaei, Puccinia, 320
rubi-idaei, Uredo, 320
Rubinoboletus, 256
rubinus, Boletus, 256
rubinus, Chalciporus, 256
rubinus, Rubinoboletus, 256
rubinus, Suillus, 256
rubinus, Xerocomus, 256
rubi-saxatilis, Phragmidium, 319
ruborum, Lecythea, 320
ruborum, Puccinia, 319
ruborum, Puccinia, 434
ruborum, Uredo, 320
rubra, Amanita, 416
rubra, Aseroë, 17
rubra, Aseroë, 38
rubra, Russula, 416
rubra, Russula, 257, 263
rubra var. sapida, Russula, 262
rubricata, Naucoria, 416
rubricatus, Agaricus, 416
rubricatus, Marasmius, 416
rubriceps, Agaricus, 354
rubriceps, Chitonia, 354
rubriceps, Clarkeinda, 354
rubriceps, Macrometrula, 354
rubriceps, Naucoria, 206
rubricosus, Cortinarius, 66
rubricosus, Cortinarius, 416
rubroalba, Russula, 264
rubrocinctus, Lactarius, 165
rubromarginata, Mycena, 199

salicis-albae, Melampsora, 316
salicis-herbaceae, Hygrocybe, 134
salicis-reticulatae, Lactarius, 165
salicola, Krombholziella, 169
salicola, Leccinum, 169
salicorniae, Uromyces, 338
salicum, Athelia, 18
saligna, Daedalea, 21
salignus, Agaricus, 230
salignus, Pleurotus, 230
salignus, Polyporus, 21
salmoneus, Lactarius, 165
salmonicolor, Hapalopilus, 121
salmonicolor, Lactarius, 165
salmonicolor, Leptoporus, 121
salmonicolor, Polyporus, 121
salmonicolor, Poria, 121
salmonifolius, Leucopaxillus, 417
salor, Cortinarius, 76
salor ssp. transiens, Cortinarius, 378
salvei, Ustilago, 350
salveii, Ustilago, 435
sambuci, Corticium, 143
sambuci, Crepidotus, 81
sambuci, Crepidotus, 417
sambuci, Hyphoderma, 143
sambuci, Hyphodontia, 143
sambuci, Lyomyces, 143
sambuci, Peniophora, 143
sambuci, Rogersella, 143
sambuci, Thelephora, 143
sambucina, Inocybe, 155
sambucinus, Agaricus, 155
sanguifluum, Hypophyllum, 417
sanguifluus, Lactarius, 417
sanguinaria, Psalliota, 4
sanguinaria, Russula, 264
sanguinarius, Agaricus, 4, 264
sanguinea, Cumminsiella, 312
sanguinea, Dermocybe, 76
sanguinea, Kneiffia, 218
sanguinea, Peniophora, 218
sanguinea, Phanerochaete, 218
sanguinea, Russula, 264
sanguinea, Thelephora, 218
sanguinea var. rosacea, Russula, 264
sanguineum, Corticium, 218
sanguineus, Agaricus, 76
sanguineus, Boletus, 26, 355
sanguineus var. gentilis, Boletus, 19
sanguineus var. rhodoxanthus,
    Boletus, 415
sanguineus, Cortinarius, 76
sanguineus, Polyporus, 355
sanguineus, Pycnoporus, 355
sanguinolenta, Mycena, 203
sanguinolenta, Podoporia, 228
sanguinolenta, Poria, 228
sanguinolenta, Thelephora, 276
sanguinolentum, Aecidium, 329
sanguinolentum, Haematostereum,
    276
sanguinolentum, Stereum, 276
sanguinolentus, Agaricus, 203
sanguinolentus, Boletus, 228
sanguinolentus, Merulius, 276
sanguinolentus, Physisporinus,
    228
sanguinolentus, Rigidoporus, 228
sanguisorbae, Phragmidium, 320
sanguisorbae, Puccinia, 320
saniculae, Aecidium, 330
saniculae, Puccinia, 330
saniosa, Flammula, 76
saniosus, Agaricus, 76

saniosus, Cortinarius, 76
sapidus, Agaricus, 229
sapidus, Pleurotus, 229
sapinea, Flammula, 120
sapinea, Fulvidula, 120
sapinea var. terrestris, Flammula, 417
sapineus, Agaricus, 120
sapineus, Gymnopilus, 120
sapineus β hybridus, Agaricus, 388
saponaceum, Tricholoma, 297
saponaceum var. ardosiacum,
    Tricholoma, 297
saponaceum var. atrovirens,
    Tricholoma, 297
saponaceum var. boudieri,
    Tricholoma, 297
saponaceum var. lavedanum,
    Tricholoma, 297
saponaceum var. napipes, Tricholoma,
    297
saponaceum var. squamosum,
    Tricholoma, 297
saponaceum var. sulphurinum,
    Gyrophila, 417
saponaceum var. sulphurinum,
    Tricholoma, 417
saponaceus, Agaricus, 297
saponaceus var. squamosus, Agaricus,
    297
saponariae, Sorosporium, 435
saporatus, Cortinarius, 76
sarcita, Leptonia, 97, 417
sarcitula, Leptonia, 97
sarcitulum, Entoloma, 97
sarcitulum var. majusculum,
    Entoloma, 97
sarcitulum var. microsporum,
    Entoloma, 394
sarcitulum var. spurcifolium,
    Entoloma, 97
sarcitulus var. majusculus,
    Rhodophyllus, 97
sarcitulus var. spurcifolium,
    Rhodophyllus, 97
sarcitum, Entoloma, 417
sarcitus, Agaricus, 417
sarcocephala, Drosophila, 244
sarcocephala, Psathyrella, 244
sarcocephala, Psathyrella, 244
sarcocephala, Psilocybe, 244
sarcocephala var. cookei, Psilocybe,
    417
sarcocephalus, Agaricus, 244
Sarcodon, 266
Sarcodontia, 267
sardonia, Russula, 261, 263
sardonia, Russula, 264
sardonia var. mellina, Russula, 264
sarnicus, Clitopilus, 417
sarniensis, Tremella, 293
sasae, Dendrothele, 88
sassii, Coprinus, 61
satanas, Boletus, 26
satanoides, Boletus, 24
satanoides, Boletus, 417
satur, Pluteus, 233
saturatus, Cortinarius, 417
saturatus, Cortinarius, 71
saturninus, Agaricus, 76
saturninus, Cortinarius, 76
satyrii, Puccinia, 356
saundersii, Agaricus, 101
saundersii, Entoloma, 101
saundersii, Entoloma, 101
saveloides, Tulasnella, 302

saxifragae, Puccinia, 330
saxifragarum, Uredo, 330
sbrozzii, Tuberculina, 335
scabella, Astrosporina, 417
scabella, Crinipellis, 81
scabella, Inocybe, 153
scabella, Inocybe, 417
scabella var. fulvella, Inocybe, 151
scabellus, Agaricus, 81, 417
scaber, Boletus, 169
scaber var. aurantiacus, Boletus, 168
scaber var. carpini, Boletus, 168
scaber var. melaneus, Boletus, 169
scaber var. niveus, Boletus, 402
scabies, Sorosporium, 435
scabies, Tuburcinia, 435
scabiosae, Farinaria, 350
scabiosae, Ustilago, 350
scabiosum, Entoloma, 101
scabiosus, Agaricus, 101
scabiosus, Trichopilus, 101
scabra, Agaricus, 417
scabra, Inocybe, 150
scabra, Inocybe, 417
scabra, Krombholziella, 169
scabra var. coloratipes, Krombholzia,
    169, 417
scabra var. firmior, Agaricus, 417
scabra var. firmior, Inocybe, 417
scabripes, Hydropus, 129
scabripes, Mycena, 129
scabripes, Prunulus, 129
scabrosum, Entoloma, 417
scabrosum, Hydnum, 266
scabrosus, Agaricus, 417
scabrosus, Sarcodon, 266
scabrum, Leccinum, 169
scabrum var. coloratipes, Leccinum,
    417
scabrum var. melaneum, Leccinum,
    169
scalpturatum, Tricholoma, 297
scalpturatum var. argyraceum,
    Tricholoma, 362
scalpturatus, Agaricus, 297
scamba, Flammula, 226
scamba, Pholiota, 226
scambus, Agaricus, 226
scambus, Paxillus, 226
scambus, Ripartites, 226
scandens, Cortinarius, 76
scariosus, Agaricus, 417
scaurus, Agaricus, 76
scaurus, Cortinarius, 76
scaurus var. herpeticus,
    Cortinarius, 76
schaefferi, Agaricus, 208
schaefferi, Cortinarius, 73
schaefferi, Geastrum, 115
schaefferi, Neolentinus, 208
schiedermayeri, Hydnum, 267
schiffneri, Russula, 265
Schinzia, 341
schismi, Puccinia, 326
schista, Inocybe, 155
schistus, Agaricus, 155
Schizonella, 345
Schizophyllum, 267
Schizophyllus, 267
Schizopora, 267
schmidelii, Geastrum, 116
schneideri, Puccinia, 331
schoeleriana, Puccinia, 324
schreieri, Squamanita, 417
schroeteri, Coprinus, 61

separata, Tilletia, 346
separata, Tilletia, 435
separatus, Agaricus, 212
separatus, Panaeolus, 212
seperina, Russula, 418
sepia, Mycena, 418
sepiaria, Daedalea, 117
sepiarium, Gloeophyllum, 117
sepiarius, Agaricus, 117
sepiarius, Lenzites, 117
sepium, Trametes, 12
septentrionalis, Athelia, 176
septentrionalis, Fibulomyces, 176
septentrionalis, Galerina, 112
septentrionalis,
    Leptosporomyces, 176
septentrionalis, Mycena, 203
septentrionalis, Puccinia, 330
septicoides, Clitopilus, 352
septicoides, Clitopilus, 45
septicoides, Pleurotus, 352
septicus, Agaricus, 418
septicus, Pleurotellus, 80, 418
septicus, Pleurotus, 418
septicus, Pleurotus, 81
septocystidia, Candelabrochaete,
    32
septocystidia, Peniophora, 32
septocystidia, Phanerochaete, 32
septocystidia, Scopuloides, 32
septocystidiata, Endoperplexa, 90
sepulta, Odontia, 205
sepultum, Hydnum, 205
serarius, Cortinarius, 418
serena, Antrodia, 36
serena, Lepiota, 178, 179
serena, Pseudobaeospora, 178
Serendipita, 269
serenus, Agaricus, 178
serenus, Leucoagaricus, 178
serenus, Sericeomyces, 178
seriale, Corticium, 221, 418
serialis, Antrodia, 13
serialis, Coriolellus, 13
serialis, Phlebia, 418
serialis, Polyporus, 13
serialis, Thelephora, 418
serialis, Trametes, 13
sericata, Lepiota, 179
sericatella, Lepiota, 179
sericatellus, Leucoagaricus, 179
sericatellus, Sericeomyces, 179
sericatula, Russula, 264
sericatum, Entoloma, 101
sericatus, Agaricus, 101
sericea, Lepiota, 179
sericea, Nolanea, 97, 99, 102
sericella, Alboleptonia, 102
sericella, Clitocybe, 43
sericella, Leptonia, 102
sericella var. decurrens, Leptonia, 102
sericellum, Entoloma, 102
sericellum var. decurrens, Entoloma,
    102
sericellus var. lutescens, Agaricus, 102
sericeoides, Entoloma, 102
sericeoides, Leptonia, 102
sericeoides, Nolanea, 102
sericeoides, Rhodophyllus, 102
sericeomolle, Strangulidium, 237
sericeomollis, Leptoporus, 237
sericeomollis, Oligoporus, 237
sericeomollis, Polyporus, 237
sericeomollis, Postia, 237
sericeomollis, Tyromyces, 237

Sericeomyces, 176
sericeonitens, Entoloma, 102
sericeonitens, Nolanea, 102
sericeonitida, Eccilia, 103
sericeonitidum, Entoloma, 103
sericeonitidus, Claudopus, 103
sericeum, Entoloma, 102
sericeum var. cinereo-opacum,
    Entoloma, 102
sericeus, Agaricus, 102
sericeus, Leucoagaricus, 179
sericeus, Rhodophyllus, 102
sericeus var. nolaniformis,
    Rhodophyllus, 102
sericeus β sericellus, Agaricus, 102
sericifer, Leucoagaricus, 178
sericifer f. sericatellus, Leucoagaricus,
    179
sericifera, Lepiota, 178
sericifera, Pseudobaeospora, 178
serifluus, Agaricus, 165
serifluus, Lactarius, 166
serifluus, Lactarius, 165
sernanderi, Gloeocystidium, 272
sernanderi, Sistotrema, 272
serosus, Agaricus, 419
serotina, Acanthocystis, 212
serotina, Entylomella, 343
serotina, Hohenbuehelia, 212
serotina, Inocybe, 156
serotina, Inocybe, 155
serotina, Russula, 263
serotina, Russula, 419
serotina, Sarcomyxa, 213
serotinum, Entyloma, 343
serotinus, Agaricus, 212
serotinus, Panellus, 212
serotinus, Pleurotus, 212
serotinus, Ripartites, 419
serotinus var. almeni, Pleurotus, 213
serotinus var. flaccidus, Pleurotus, 213
serpens, Antrodia, 12
serpens, Byssomerulius, 34
serpens, Ceraceomerulius, 34
serpens, Ceraceomyces, 34
serpens, Coriolellus, 12
serpens, Merulius, 34
serpens, Polyporus, 12
serpens, Tilletia, 350
serpens, Trametes, 12
serpens, Ustilago, 350
serpens, Xylomyzon, 34
Serpula, 270
serratis, Agaricus, 299
serrulata, Leptonia, 102
serrulata var. berkeleyi, Leptonia, 98
serrulata var. levipes, Leptonia, 102
serrulatum, Entoloma, 102
serrulatus, Agaricus, 102
sertipes, Cortinarius, 69
serum, Corticium, 143
sessile, Geastrum, 115
sessile, Lycoperdon, 115
sessilis, Agaricus, 419
sessilis, Peziza, 160
sessilis, Puccinia, 330
seticeps, Leptonia, 232
seticeps, Pluteus, 232
Seticyphella, 270
setigera, Kneiffia, 141
setigera, Odontia, 141
setigera, Peniophora, 141
setigera, Thelephora, 141
setigerum, Hyphoderma, 141
setipes, Agaricus, 419

setipes, Omphalia, 419
setipes, Omphalina, 255, 419
setipes, Pistillaria, 304
setipes, Rickenella, 419
setipes, Rickenella, 255
setipes, Typhula, 304
setipes var. acrocyaneus, Agaricus,
    255
setosa, Acia, 267
setosa, Mycena, 191
setosa, Mycoacia, 267
setosa, Sarcodontia, 267
setosum, Hydnum, 267
setosus, Agaricus, 191
setosus, Marasmius, 191
Setulipes, 189
setulosa, Lepiota, 172
setulosa var. rhodorhiza, Lepiota, 172
seynesii, Mycena, 204
sibiricum, Hyphoderma, 141
sibiricus, Radulomyces, 141
siccus, Agaricus, 419
siccus, Marasmius, 419
sideroides, Agaricus, 112
sideroides, Galera, 112
sideroides, Galerina, 112
sideroides, Naucoria, 112
sideroides var. indusiata, Naucoria,
    112
siennophylla, Conocybe, 55
siennophylla, Naucoria, 55
siennophyllus, Agaricus, 55
sigmaspora, Serendipita, 269
silaceum, Hypholoma, 419
silaceus, Agaricus, 419
silai, Puccinia, 321
silenes, Puccinia, 321
silenes-inflatae, Microbotryum,
    317
silenes-inflatae, Ustilago, 317
siliginea, Conocybe, 52, 54
siliginea, Conocybe, 55
siliginea, Galera, 55
siliginea var. ambigua, Conocybe, 50
siliginea var. anthracophila, Conocybe,
    51
siliginea var. neoantipus, Conocybe,
    52, 400
siliginea var. ochracea, Conocybe, 403
siliginea var. pallidispora, Conocybe,
    405
siligineus, Agaricus, 55
silvaemonachi, Cortinarius, 76
silvae-nigrae, Mycena, 204
silvaenovae, Naucoria, 207
silvae-ryae, Fibriciellum, 106
silvamonachorum, Phlegmacium, 76
silvana, Hohenbuehelia, 127
silvana, Hohenbuehelia, 419
silvanus, Agaricus, 419
silvanus, Pleurotus, 419
silvanus, Resupinatus, 419
silvatica, Psalliota, 4
silvatica, Puccinia, 325
silvatica var. pallida, Psalliota, 4
silvaticus, Agaricus, 4
silvaticus, Coprinus, 61
silvaticus var. pallens, Agaricus, 4
silvaticus var. pallidus, Agaricus, 4
silvester, Merulius, 270
silvestre, Hypholoma, 424
silvestris, Psathyrella, 424
silvestris, Russula, 264
silvicola, Agaricus, 5
silvicola, Psalliota, 5

strigiceps, *Flammula*, 256
strigiceps, *Inocybe*, 256
strigiceps, *Ripartites*, 256
strigosissima, *Leptonia*, 103
strigosissima, *Nolanea*, 103
strigosissimum, Entoloma, 103
strigosissimus, *Pouzaromyces*, 103
strigosum var. *filamentosum*,
  Corticium, 12
strigosus, *Panus*, 416
striiformis, Puccinia, 331
striiformis, *Tilletia*, 350
striiformis, *Uredo*, 350
striiformis, Ustilago, 350
striiformis *var.* dactylidis,
  Puccinia, 331
striipes, *Boletus*, 26
striipilea, *Agaricus*, 284
striipilea, *Collybia*, 284
striipilea, *Omphalia*, 284
striipilea, Tephrocybe, 284
striipileum, *Lyophyllum*, 284
Strilia, 49
striola, *Puccinia*, 323
strobi, *Peridermium*, 312
strobilaceus, *Boletus*, 276
strobilaceus, *Cortinarius*, 64
strobilaceus, Strobilomyces, 276
strobiliformis, *Agaricus*, 11
strobiliformis, Amanita, 11
strobiliformis, *Boletus*, 276
strobiliformis, *Strobilomyces*, 276
strobiliformis var. *aculeata*, *Amanita*, 8
strobilina, *Licea*, 333
strobilina, *Mycena*, 203
strobilina, *Mycena*, 422
strobilinum, *Aecidium*, 333
strobilinus, *Agaricus*, 422
Strobilomyces, 276
Strobilurus, 276
Stromatoscypha, 235
Stropharia, 277
strophosum, *Hebeloma*, 123
strophosus, *Agaricus*, 123
stuntzii, Psilocybe, 355
stuposa, Tomentella, 288
stuposum, *Sporotrichum*, 288
stuppea, *Cyphella*, 160
stygia, Ustilago, 350
stylifera, Galerina, 112
stylifera, *Galerula*, 112
stylobates, *Agaricus*, 204
stylobates, Mycena, 204
Stypella, 279
stypticus, *Crepidopus*, 213
suaveolens, *Agaricus*, 43, 44
suaveolens, *Boletus*, 290
suaveolens, *Clitocybe*, 43
suaveolens, *Marasmius*, 191
suaveolens, *Polyporus*, 290
suaveolens, *Puccinia*, 329
suaveolens, Trametes, 290
suaveolens, *Trichobasis*, 329
suaveolens, *Uredo*, 329
suaveolens f. *indora*, *Trametes*, 290
suaveolens var. *caeruleum*, *Hydnum*,
  128
suavissima, *Psathyrella*, 244
suavissimus, *Lentinus*, 422
suavissimus, *Panus*, 422
subabrupta, *Odontia*, 82
subabruptum, *Cystostereum*, 82
subabruptus, Crustomyces, 82
subalba, Lepiota, 174
subalpina, *Mycena*, 129

subalpinus, Hydropus, 129
subalpinus, *Marasmiellus*, 129
subalutacea, *Clitocybe*, 422
subalutacea, *Clitocybe*, 43
subalutacea, Hyphodontia, 144
subalutacea, *Kneiffia*, 144
subalutacea, *Kneiffiella*, 144
subalutacea, *Peniophora*, 144
subalutaceum, *Corticium*, 144
subalutaceus, *Agaricus*, 422
subangulisporum, *Sistotrema*, 271
subannulata, Galerina, 112
subannulatum, *Tricholoma*, 295
subannulatus, *Agaricus*, 295
subanthracina, *Dermocybe*, 64
subanthracinus, *Cortinarius*, 64
subappendiculatus, Boletus, 26
subargentatus, *Cortinarius*, 64
subarquatum, *Phlegmacium*, 422
subarquatus, *Cortinarius*, 422
subatomata, *Psathyrella*, 243
subatrata, *Psathyrella*, 239
subatratus, *Agaricus*, 239
subbalaustinus, Cortinarius, 77
subbalteatus, *Agaricus*, 211
subbalteatus, *Cortinarius*, 363
subbalteatus, *Panaeolus*, 211
subbotrytis, *Clavaria*, 251
subbotrytis, Ramaria, 251
subbrevipes, *Melanoleuca*, 193
subcaeruleus, *Agaricus*, 422
subcaesia, Postia, 237
subcaesius, *Oligoporus*, 237
subcaesius, *Tyromyces*, 237
subcalcea, *Peniophora*, 170
subcalceus, Lepidomyces, 170
subcantharellus, *Agaricus*, 136
subcarnaceum, *Hydnum*, 422
subcarpta, Inocybe, 156
subcava, *Armillaria*, 422
subcava, *Armillaria*, 179
subcavus, *Agaricus*, 422
subcerina, Galerina, 112
subcerina var. *anglica*, *Galerina*, 112
subcervina, *Tomentella*, 286
subcinnamomeum, *Leccinum*, 169
subclaricolor, *Cortinarius*, 68
subclaricolor, *Phlegmacium*, 68
subclavata, Galerina, 112
subclavigera, Tomentella, 288
subcollariatum, *Hebeloma*, 422
subcollariatus, *Agaricus*, 422
subconspersa, *Alnicola*, 207
subconspersa, Naucoria, 207
subcoprophila, *Deconica*, 249
subcoprophila, Psilocybe, 249
subcoprophilus, *Agaricus*, 249
subcordispora, Clitocybe, 45
subcoronata, *Pellicularia*, 27
subcoronatum, Botryobasidium,
  27
subcoronatum, *Corticium*, 27
subcorticalis, *Rhizomorpha*, 422
subcorticium, *Phragmidium*, 320
subcostatum, *Corticium*, 427
subcostatum, *Stereum*, 427
subcretacea, *Phlebia*, 422
subcretaceum, *Corticium*, 422
subcretaceus, Leucoagaricus, 179
subdealbata, *Coniophora*, 50
subdealbata, *Peniophora*, 50
subdealbatum, *Corticium*, 50
subdecastes, *Agaricus*, 185
subdecastes, *Clitocybe*, 185
subdefinitum, *Hyphoderma*, 140

subdelibutus, *Cortinarius*, 71
subdisseminatus, Coprinus, 62
subdryadicola, Clitocybe, 45
subdulcis, Lactarius, 166
subdulcis var. *cimicarius*, *Lactarius*,
  165
subdulcis var. *sphagneti*, *Lactarius*,
  166
subdulcis var. *tabidus*, *Lactarius*, 166
subdulcis β *camphoratus*, *Agaricus*,
  161
subelavata, *Collybia*, 422
subericaea, *Psilocybe*, 145
subericaeum, Hypholoma, 145
subericaeum f. *verrucosum*,
  Hypholoma, 145
subericaeus, *Agaricus*, 145
suberosum var. *aurantiacum*,
  Hydnum, 128
suberosus, *Boletus*, 228
subfarinacea, Endoperplexa, 90
subfarinacea, *Exidiopsis*, 90
subfarinacea, *Sebacina*, 90
subfelinoides, *Lepiota*, 172
subferruginea, *Tomentella*, 286
subferrugineus, *Agaricus*, 422
subferrugineus, *Cortinarius*, 422
subfimbriata, *Junghuhnia*, 422
subfirmus, Panaeolus, 212
subfloccosa, *Psalliota*, 5
subfloccosus, Agaricus, 5
subfoetens, *Russula*, 259
subfoetens, Russula, 265
subfoetens var. *grata*, *Russula*, 260,
  265
subfulgens, *Cortinarius*, 73
subfuliginosa, *Hymenochaete*, 382
subfusca, *Tomentella*, 422
subfusca, *Tomentella*, 288
subfuscoflavida, *Poria*, 423
subfuscoflavidus, *Polyporus*, 423
subfuscus, *Hypochnus*, 288, 422
subfuscus ssp. *tristis*, *Hypochnus*, 247
subgelatinosa, *Kneiffia*, 252
subgelatinosa, *Poria*, 423
subgelatinosa, *Protodontia*, 279
subgelatinosa, Stypella, 279
subgelatinosum, *Hydnum*, 279
subgelatinosum, *Protohydnum*, 279
subgelatinosus, *Polyporus*, 423
subglobispora, *Hygrocybe*, 133
subglobisporus, *Hygrophorus*, 133
subglobosa, *Naucoria*, 423
subglobosa, *Nolanea*, 423
subglobosus, *Agaricus*, 423
subgracilis, Lepiota, 174
subhepatica, Omphalina, 210
subhepaticus, *Agaricus*, 210
subhyalina, *Sebacina*, 279
subhyalina, Stypella, 279
subhyalina, Typhula, 304
subhyalinum, *Myxarium*, 279
subillaqueata, *Leucogyrophana*, 146
subillaqueatum, *Corticium*, 146
subillaqueatum, Hypochniciellum,
  146
subimpatiens, Coprinus, 62
subincarnata, *Flocculina*, 108
subincarnata, *Incrustoporia*, 273
subincarnata, Lepiota, 174
subincarnata, *Naucoria*, 108
subincarnata, *Poria*, 273
subincarnata, Skeletocutis, 273
subincarnatus, *Flammulaster*, 108
subinclusa, Anthracoidea, 340

sulphuratum, Stereum, 424
sulphurea, Coniophora, 223
sulphurea, Cristella, 223
sulphurea, Cyphella, 31
sulphurea, Grifola, 167
sulphurea, Himantia, 223
sulphurea, Odontia, 424
sulphurea, Peziza, 31
sulphurea, Phlebiella, 223
sulphurea, Thelephora, 223
sulphurea, Trechispora, 223
sulphurea var. ochroidea, Coniophora, 50
sulphureoides, Agaricus, 424
sulphureoides, Pleurotus, 424
sulphureoides, Tricholomopsis, 424
sulphurescens, Tricholoma, 298
sulphureum, Corticium, 223
sulphureum, Hydnum, 424
sulphureum, Tricholoma, 298
sulphureum var. bufonium, Tricholoma, 295
sulphureum var. hemisulphureum, Tricholoma, 298
sulphureum var. variecolor, Sistotrema, 183
sulphureus, Agaricus, 298
sulphureus, Boletus, 28, 167
sulphureus, Cortinarius, 68
sulphureus, Cortinarius, 424
sulphureus var. citrinus, Cortinarius, 67
sulphurinus var. langei, Cortinarius, 76
sulphureus, Gymnopus, 298
sulphureus, Hypochnus, 223
sulphureus, Laetiporus, 167
sulphureus, Phlebopus, 28
sulphureus, Polyporus, 167
sulphureus var. albolabyrinthiporus, Polyporus, 424
sumptuosa, Naucoria, 270
sumptuosa, Ramicola, 271
sumptuosa, Simocybe, 270
superbum, Cystoderma, 85
superbus, Agaricus, 239
surrecta, Volvaria, 306
surrecta, Volvariella, 306
surrectus, Agaricus, 306
suspectus, Boletus, 24
swartzii, Agaricus, 255
swartzii, Mycena, 255
swartzii, Omphalina, 255
swartzii, Rickenella, 255
sydowianum, Exobasidium, 344
sylvatica var. amethystina, Psalliota, 360
sylvestre, Hypholoma, 243
sylvestris, Psathyrella, 243
symphyti, Coleosporium, 316
symphyti, Melampsorella, 316
symphyti, Trichobasis, 316
symphyti, Uredo, 316
symphyti-bromorum, Puccinia, 330
sympodialis, Malassezia, 344
synantherarum, Coleosporium, 311
syncocca, Uredo, 356
syncocca, Urocystis, 356
syngenesiarum, Puccinia, 322
syringinus, Lactarius, 424
Syzygospora, 281
tabacina, Agrocybe, 424
tabacina, Auricularia, 139
tabacina, Hymenochaete, 139
tabacina, Inocybe, 156

tabacina, Naucoria, 424
tabacina, Thelephora, 139
tabacinum, Stereum, 139
tabacinus, Agaricus, 424
tabacinus, Cortinarius, 77
tabescens, Agaricus, 14
tabescens, Armillaria, 14
tabescens, Clitocybe, 14
tabidus, Lactarius, 164
tabidus, Lactarius, 166
tabularis, Cortinarius, 77
Taeniospora, 282
taeniospora, Russula, 265
talus, Cortinarius, 77
tamaricicola, Peniophora, 215
tammii, Agaricus, 425
tammii, Flammula, 425
tanaceti, Puccinia, 331
tantilla, Naucoria, 207
Tapinella, 282
taraxaci, Aecidium, 332
taraxaci, Puccinia, 326
tarda, Omphalia, 246
tardus, Agaricus, 246
tardus, Coprinus, 62
taxicola, Caloporus, 117
taxicola, Gloeoporus, 117
taxicola, Merulioporia, 117
taxicola, Meruliopsis, 117
taxicola, Merulius, 117
taxicola, Poria, 117
taxicola, Xylomyzon, 117
taxocystis, Inocybe, 156
taxophila, Clavicorona, 40
taxophilus, Craterellus, 40
taylorii, Agaricus, 307
taylorii, Volvaria, 307
taylorii, Volvariella, 307
Tectella, 212
tectorum, Agaricus, 425
tegularis, Agaricus, 425
tegularis, Psilocybe, 425
Telamonia, 64
telmatiaea, Omphalia, 425
telmatiaeus, Agaricus, 425
telmaticus, Agaricus, 425
temperata, Volvaria, 355
temperatus, Agaricus, 355
temulenta, Agrocybe, 6
temulenta, Naucoria, 6
temulentus, Agaricus, 6
tenacella, Collybia, 277
tenacella, Pseudohiatula, 277
tenacella var. stolonifera, Collybia, 422
tenacellus, Agaricus, 277
tenacellus, Marasmius, 277
tenacellus, Strobilurus, 277
tenax, Agaricus, 425
tenax, Boletus, 425
tenax, Naucoria, 425
tenax, Naucoria, 145
tenebrosa, Inocybe, 156
tenella, Lepiota, 354
tenella, Mycena, 431
tenella, Nolanea, 103
tenellum, Entoloma, 103
tenellus, Agaricus, 431
tenellus, Leucocoprinus, 354
tenellus, Rhodophyllus, 103
tener, Agaricus, 51, 55
tener, Bolbitius, 51
tener, Hymenogaster, 140
tener, Leucoagaricus, 179
tener β pilosellus, Agaricus, 54

tener γ siligineus, Agaricus, 55
tenera, Conocybe, 54
tenera, Conocybe, 55
tenera, Galera, 55
tenera, Lepiota, 179
tenera, Mycena, 51
tenera f. bispora, Conocybe, 53
tenera f. convexa, Galera, 55
tenera f. macrocephala, Conocybe, 53
tenera f. minor, Galera, 53
tenera f. semiglobata, Conocybe, 55
tenera f. tenella, Galera, 55
tenera f. typica, Galera, 55
tenera var. aurea, Conocybe, 51
tenera var. pilosella, Galera, 54
tenera var. siliginea, Galera, 55
tenera var. subovalis, Conocybe, 55
teneroides, Agaricus, 54
teneroides, Conocybe, 51, 53
teneroides, Galera, 54
teneroides, Pholiota, 51
tenerrima, Clavaria, 425
tenerrima, Mycena, 198
tenerrimus, Agaricus, 198
tentaculus, Agaricus, 425
tenue, Corticium, 141
tenue, Gloeocystidium, 141
tenue, Hyphoderma, 141
tenue ssp. praetermissum, Gloeocystidium, 141
tenuiceps, Agaricus, 192
tenuiceps, Tricholoma, 192
tenuicula, Psathyra, 425
tenuicula, Trechispora, 292
tenuiculum, Corticium, 292
tenuicystidiata, Inocybe, 154
tenuiparietalis, Marasmius, 191
tenuipes, Boletus, 19
tenuipes, Clavaria, 40
tenuipes, Clavaria, 39
tenuipes, Geastrum, 115
tenuipes, Nolanea, 95
tenuipes, Pistillaria, 40
tenuiramosa, Ramariopsis, 251
tenuis, Agaricus, 425
tenuis, Clavaria, 425
tenuis, Mycena, 425
tenuis, Pistillaria, 425
tenuis, Typhula, 425
tenuispora, Athelia, 18
tenuispora, Seticyphella, 270
tenuissima, Clitocybe, 42
tenuistipes, Omphalia, 204
tenuivolvata, Psalliota, 425
tenuivolvatus, Agaricus, 425
Tephrocybe, 282
tephroleuca, Cyphella, 195
tephroleuca, Postia, 237
tephroleucus, Agaricus, 138
tephroleucus, Hygrophorus, 138
tephroleucus, Oligoporus, 237
tephroleucus, Polyporus, 237
tephroleucus, Tyromyces, 237
tephrophylla, Drosophila, 245
tephrophylla, Psathyrella, 245
Terana, 284
terenopus, Lactarius, 425
tergina, Collybia, 49
terginus, Agaricus, 49
terginus, Gymnopus, 49
terginus, Marasmius, 49
tergiversans, Agaricus, 425
tergiversans, Coprinus, 425
tergiversans, Coprinus, 62
terpsichores, Cortinarius, 77

trachyspora, **Lindtneria**, 183
trachyspora, Mycena, 426
trachyspora, Mycenella, 426
trachyspora, Poria, 183
Tracya, 346
traganulus, Cortinarius, 79
traganus, Agaricus, 78
traganus, Cortinarius, 78
traganus var. *finitimus, Cortinarius*, 78
tragopogi, Aecidium, 326
tragopogi, Puccinia, 326
tragopogi, Uredo, 351
tragopogi-pratensis, Uredo, 350
tragopogonis, Aecidium, 326
tragopogonis, Puccinia, 326
tragopogonis-pratensis, Ustilago, 350
trailii, Entyloma, 343
trailii, Entylomella, 343
trailii, Poikilosporium, 346
trailii, Puccinia, 328
trailii, Thecaphora, 346
Trametes, 289
transforme, Lyophyllum, 426
transforme, Tricholoma, 426
transformis, Agaricus, 426
transiens, Hyphoderma, 141
transiens, Odontia, 141
transitoria, Inocybe, 426
transitorius, Agaricus, 426
translucens, Pseudostypella, 294
translucens, Tremella, 294
translucens, Typhula, 426
transvenosum, Entoloma, 103
transversalis, Uredo, 356
transversalis, Uromyces, 356
Tranzschelia, 335
Trechispora, 290
trechispora, Astrosporina, 426
trechispora, Inocybe, 426
trechispora, Inocybe, 153, 154
trechisporus, Agaricus, 426
Tremella, 292
Tremellodendropsis, 294
Tremellodon, 246
tremelloides, Auricularia, 20
tremelloides, Gloeotulasnella, 302
tremelloides, Tulasnella, 302
tremellosa, Phlebia, 222
tremellosa var. *campanulae, Uredo*, 311
tremellosa var. *sonchi, Uredo*, 311
tremellosus, Merulius, 222
Tremiscus, 118
tremula, Hohenbuehelia, 128
tremulae, Marasmius, 426
tremulae, Marasmius, 190
tremulae, Melampsora, 316
tremulae, Phellinus, 221
tremulae f. *laricis, Melampsora*, 316
tremulus, Agaricus, 128
tremulus, Pleurotellus, 15, 128
tremulus, Pleurotus, 15, 128
trepida, Psathyra, 426
trepida, Psathyrella, 426
trepidus, Agaricus, 426
Trichaptum, 294
Trichobasis, 335
trichoderma, Hydropus, 129
trichoderma, Mycena, 129
Tricholoma, 294
tricholoma, Agaricus, 255
tricholoma, Astrosporina, 256
tricholoma, Flammula, 255
tricholoma, Inocybe, 255

tricholoma, Paxillopsis, 256
tricholoma, Paxillus, 255
tricholoma, Ripartites, 255
Tricholomella, 30
Tricholomopsis, 298
Trichopilus, 91
Trichosporon, 299
trichotis, Agaricus, 252
trichotis, Resupinatus, 252
tricolor, Agaricus, 188
tricolor, Daedalea, 426
tricolor, Daedaleopsis, 426
tricolor, Marasmiellus, 188
tricolor, Marasmius, 188
tricolor, Omphalia, 188
tricuspidatus, Pluteus, 230
tridentinus, Boletus, 281
tridentinus, Ixocomus, 281
tridentinus, Lentinellus, 170
tridentinus, Lentinus, 170
tridentinus, Rhodophyllus, 426
tridentinus, Suillus, 281
tridentinus var. *landkammeri, Boletus*, 281
trientalis, Ascomyces, 348
trientalis, Sorosporium, 348
trientalis, Tuburcinia, 348
trientalis, Urocystis, 348
trifolii, Puccinia, 338
trifolii, Typhula, 305
trifolii, Uredo, 338
trifolii, Uromyces, 338
trifolii, Uromyces, 338
trifolii-repentis, Uromyces, 338
triformis, Cortinarius, 74
triformis, Cortinarius, 78
triformis var. *fuscopallens, Cortinarius*, 74
triformis var. *melleopallens, Cortinarius*, 397
triformis *var.* schaefferi, Cortinarius, 426
trigonophylla, Tubaria, 427
trigonosperma, Tomentella, 303
trigonospermum, Corticium, 303
trilobatus, Boletus, 427
trilobus, Agaricus, 427
Trimorphomyces, 299
trinii, Agaricus, 427
trinii, Astrosporina, 365, 427
trinii, Inocybe, 427
Triphragmium, 335
triplex, Coprinus, 62
triplex, Geastrum, 116
tripolii, Puccinia, 324
triscopa, Galera, 113
triscopa, Galerina, 113
triscopoda, Naucoria, 113
triscopus, Agaricus, 113
triseti, Puccinia, 326
trispora *var.* epilobii, Pistillaria, 427
trisporus, Coprinus, 62
triste, Entoloma, 427
triste, Tricholoma, 427
tristis, Agaricus, 131, 427
tristis, Eccilia, 99
tristis, Hygrocybe, 131
tristis, Hygrophorus, 131
tristis, Nolanea, 427
tristis, Pseudotomentella, 247
tristis, Tomentella, 247
tritici, Lycoperdon, 346
tritici, Puccinia, 331
tritici, Tilletia, 346
tritici, Tuburcinia, 435

tritici, Urocystis, 435
tritici, Ustilago, 351
triticina, Puccinia, 330
triumphans, Cortinarius, 78
trivialis, Agaricus, 166
trivialis, Cortinarius, 78
trivialis, Fomes, 220
trivialis, Lactarius, 166
trivialis, Phellinus, 220
trivialis, Psathyrella, 240
trogii, Agaricus, 44
trogii, Clitocybe, 44
trogii, Coriolopsis, 427
trogii, Funalia, 427
trogii, Trametes, 427
trollii, Urocystis, 348
tropaeoli, Coleosporium, 311
tropaeoli, Uredo, 311
trulliformis, Agaricus, 427
trulliformis, Clitocybe, 42, 427
trullisata, Clitocybe, 427
trullisata, Laccaria, 160
trullisata, Laccaria, 427
trullisata ssp. *maritima, Laccaria*, 160
trullisatus, Agaricus, 427
truncata, Clavaria, 40
truncata, Exidia, 105
truncata, Puccinia, 327
truncata, Rhodocybe, 427
truncata, Rhodocybe, 253
truncata var. *subvermicularis, Rhodocybe*, 253
truncatum, Hebeloma, 123, 427
truncatum, Tricholoma, 427
truncatus, Agaricus, 427
truncatus, Clavariadelphus, 40
truncatus, Clitopilus, 427
truncicola, Clitocybe, 45
truncicolus, Agaricus, 45
truncorum, Agaricus, 62
truncorum, Coprinus, 62
truncosalicicola, Mycena, Addendum
tschulymica, Poria, 427
tschulymica, Skeletocutis, 427
tschulymica, Skeletocutis, 273
tsugae, Bullera, 334
tsugae, Corticium, 141
tsugae, Hyphoderma, 141
tsugae, Sporobolomyces, 334
tuba, Agaricus, 42
tuba, Clitocybe, 42
tubaeforme, Hydnum, 427
tubaeformis, Agaricus, 208
tubaeformis, Cantharellus, 33
tubaeformis, Helvella, 33
tubaeformis var. *lutescens, Cantharellus*, 33
Tubaria, 299
tubarius, Cortinarius, 78
tuberarius, Boletus, 234
tuberaster, Polyporus, 234
tubercularia, Tremella, 427
tubercularia, Tremella, 293
tuberculata, Grandinia, 427
tuberculata, Phanerochaete, 219
tuberculatum, Basidioradulum, 427
tuberculatum, Corticium, 219
tuberculatum, Phragmidium, 320
tuberculatus, Uromyces, 338
Tuberculina, 313
Tuberculina, 335
tuberculosa, Dryophila, 227
tuberculosa, Pholiota, 227
tuberculosa, Pleuroflammula, 227

umbrinellus, Pluteus, 428
umbrinoides, Boletus, 428
umbrinoides, Leccinum, 168
umbrinoides, Leccinum, 428
umbrinolens, Cortinarius, 78
umbrinolutea, Amanita, 8
umbrinolutea, Amanitopsis, 8
umbrinolutea var. flaccida, Amanita, 8
umbrinospora, Tomentella, 289
umbrinum, Corticium, 50
umbrinum, Hypophyllum, 428
umbrinum, Lycoperdon, 185
umbrinum, Lycoperdon, 184
umbrinus, Agaricus, 429
umbrinus, Coprinus, 428
umbrinus, Cortinarius, 428
umbrinus, Hypochnus, 247
umbrinus, Lactarius, 428
umbrinus, Polyporus, 219
umbrosus, Agaricus, 233
umbrosus, Pluteus, 230
umbrosus, Pluteus, 233
umidicola, Cortinarius, 72
umidicola, Cortinarius, 429
uncialis, Agaricus, 113
uncialis, Clavaria, 305
uncialis, Galerina, 113
uncialis, Naucoria, 113
uncialis, Pistillaria, 305
uncialis, Typhula, 305
undata, Eccilia, 103
undata, Poria, 255
undatum, Entoloma, 103
undatus, Agaricus, 103, 191
undatus, Clitopilus, 103
undatus, Marasmius, 191
undatus, Polyporus, 255
undatus, Rhodophyllus, 103
undatus, Rigidoporus, 255
undatus ssp. *viarum, Clitopilus*, 103
undatus var. *pusillus, Rhodophyllus*, 97
undatus var. *viarum, Agaricus*, 103
undosa, Postia, 429
undosus, Polyporus, 429
undulata, Cotylidia, 80
undulata, Peziza, 33
undulata, Russula, 257
undulata, Thelephora, 80
undulatum, Stereum, 80
undulatus, Agaricus, 254
undulatus, Cantharellus, 246
undulatus, Merulius, 246
undulatus, Pseudocraterellus, 246
unedonis, Exobasidium, 344
ungerianum, Entyloma, 343
ungerianum f. *ficariae, Entyloma*, 342
unguentatum, Tricholoma, 405
unguentatus, Agaricus, 405
unguicularis, Agaricus, 128
unguicularis, Hohenbuehelia, 128
unguicularis, Resupinatus, 128
unguiculatum, Hericium, 126
unguinosa, Hygrocybe, 132
unguinosus, Agaricus, 132
unguinosus, Hygrophorus, 132
unicolor, Agaricus, 111
unicolor, Boletus, 37
unicolor, Cerrena, 37
unicolor, Daedalea, 37
unicolor, Galerina, 111
unicolor, Hygrophorus, 138
unicolor, Pholiota, 111
unicolor, Russula, 265
unicolor, Trametes, 37

uniguttulata, Eutorulopsis, 107
uniguttulatum, Filobasidium, 107
uniguttulatus, Cryptococcus, 107
unimodus, Cortinarius, 429
unimodus, Cortinarius, 367
uracea, Mycena, 201
uraceus, Cortinarius, 79
uraceus, Cortinarius, 429
urania, Mycena, 204
uranius, Agaricus, 204
urbicus, Agaricus, 78
urbicus, Cortinarius, 78
urceolata, Solenia, 276
urceolatum, Stigmatolemma, 276
urceolorum, Caeoma, 340
urceolorum, Uredo, 340
urceolorum, Ustilago, 340
urceolorum f. *scirpi, Ustilago*, 340
uredinis, Kuehneola, 314
uredinis, Oidium, 314
Uredinopsis, 335
Uredo, 335
urens, Agaricus, 49
urens, Collybia, 49
urens, Marasmius, 49
urens, Russula, 258
urinascens, Agaricus, 5
urinascens, Psalliota, 5
urinascens *var.* excellens, Agaricus, 5
Urocystis, 346
Uromyces, 336
ursinus, Agaricus, 170
ursinus, Lentinellus, 170
ursinus, Lentinus, 170
urticae, Aecidium, 331
urticae, Dacrymyces, 429
urticae, Tremella, 429
urticae, Uromyces, 434
urticae-acutae, Puccinia, 332
urticae-flaccae, Puccinia, 332
urticae-hirtae, Puccinia, 332
urticae-inflatae, Puccinia, 332
urticae-paniceae, Puccinia, 332
urticae-ripariae, Puccinia, 332
urticata, Puccinia, 331
urticata *var.* biporula, Puccinia, 332
urticata *var.* urticae-acutae, Puccinia, 332
urticata *var.* urticae-acutiformis, Puccinia, 332
urticata *var.* urticae-flaccae, Puccinia, 332
urticata *var.* urticae-hirtae, Puccinia, 332
urticata *var.* urticae-inflatae, Puccinia, 332
urticata *var.* urticae-paniceae, Puccinia, 332
urticata *var.* urticae-ripariae, Puccinia, 332
urticata *var.* urticae-vesicariae, Puccinia, 332
urticicola, Agaricus, 62
urticicola, Coprinus, 62
urticicola, Psathyra, 62
usnearum, Biatoropsis, 310
ustale, Tricholoma, 298
ustalis, Agaricus, 298
ustaloides, Tricholoma, 298
Ustanciosporium, 348
Ustilago, 349
Ustilentyloma, 339
ustulata, Gyraria, 429

ustulata, Tremella, 429
ustulatus, Agaricus, 429
Uthatobasidium, 285
utilis, Agaricus, 429
utilis, Lactarius, 429
Utraria, 184
utriculosa, Ustilago, 334, 350
utriculosum, Caeoma, 334
utrifer, Coprinus, 62
utriforme, Lycoperdon, 121
utriformis, Calvatia, 121
utriformis, Conocybe, 55
utriformis, Handkea, 121
uvidus, Agaricus, 166
uvidus, Lactarius, 166
uvidus var. *violascens, Lactarius*, 167
vaccina, Inocybe, 156
vaccinii, Exobasidium, 343
vaccinii, Exobasidium, 344
vaccinii, Fusidium, 344
vaccinii, Melampsora, 318
vaccinii, Naohidemyces, 318
vaccinii, Pucciniastrum, 318
vaccinii, Thekopsora, 318
vaccinii f. *vaccinii-myrtilli, Fusidium*, 343
vaccinii var. *japonicum, Exobasidium*, 343
vaccinii var. *rhododendri, Exobasidium*, 344
vaccinii-myrtilli, Exobasidium, 343
vacciniorum, Melampsora, 318
vacciniorum, Naohidemyces, 318
vacciniorum, Thekopsora, 318
vacciniorum, Uredo, 318
vaccinum, Hebeloma, 124
vaccinum, Tricholoma, 298
vaccinus, Agaricus, 298
vaccinus, Boletus, 429
vaga, Coniophora, 285
vaga, Pellicularia, 27
vaga, Phlebia, 223
vaga, Phlebiella, 223
vaga, Trechispora, 223
vaginalium, Puccinia, 320
Vaginata, 8
vaginata, Amanita, 11
vaginata, Amanitopsis, 11
vaginata f. *grisea, Amanita*, 11
vaginata f. *plumbea, Amanita*, 11
vaginata f. *violacea, Amanita*, 11
vaginata ssp. *plumbea, Amanita*, 11
vaginata *var.* alba, Amanita, 429
vaginata var. *crocea, Amanita*, 8
vaginata var. *flavescens, Amanita*, 9
vaginata var. *friabilis, Amanitopsis*, 9
vaginata var. *fulva, Amanitopsis*, 9
vaginata var. fungites, Amanita, 429
vaginata var. *grisea, Amanita*, 11
vaginata var. *livida, Amanita*, 11
vaginata var. *lividopallescens, Amanita*, 9
vaginata var. *nivalis, Amanita*, 10
vaginata var. nivalis, Amanitopsis, 429
vaginata var. *plumbea, Amanita*, 11
vaginata var. *violacea, Amanitopsis*, 11
vaginatus, Agaricus, 11
vaginatus var. albus, Agaricus, 429
vaginatus var. nivalis, Agaricus, 429
vagum, Botryobasidium, 27
vagum, Corticium, 27, 285
vahlii, Agaricus, 217
vahlii, Pholiota, 217
vaillantii, Agaricus, 189

verrucosum var. bovista, Scleroderma, 268
verruculosa, Grandinia, 143
verruculosa, Hyphodontia, 143
verruculosa, Kneiffiella, 143
verruculosus, Agaricus, 227
versatile, Entoloma, 104
versatilis, Agaricus, 104
versatilis, Inopilus, 104
versatilis, Nolanea, 104
versatilis, Pouzaromyces, 104
versatilis, Ramaria, 250
versicolor, Agaricus, 290, 397
versicolor, Boletus, 26, 290
versicolor, Coriolus, 290
versicolor, Corticium, 215
versicolor, Peniophora, 215
versicolor, Peniophora, 214
versicolor, Polyporus, 290
versicolor, Polystictus, 290
versicolor, Poria, 290
versicolor, Russula, 265
versicolor, Stropharia, 397
versicolor, Trametes, 290
versicolor, Tremella, 294
versicolor, Xerocomus, 26
versicolor, Xylomyzon, 270
versicolor var. fuscatus, Polystictus, 290
versicolor var. nigricans, Polyporus, 290
versicolor var. nigricans, Polystictus, 290
versipelle, Hebeloma, 430
versipelle, Hydnum, 430
versipelle, Leccinum, 169
versipelle var. marginatulum, Hebeloma, 122
versipellis, Agaricus, 430
versipellis, Boletus, 169
versipellis, Sarcodon, 430
versipellis var. percandidus, Boletus, 407
versipora, Poria, 267
versiporus, Polyporus, 267
versiporus, Xylodon, 267
versispora, Calvatia, 167
versisporum, Sporotrichum, 167
versutus, Agaricus, 81
versutus, Crepidotus, 81
vertirugis, Agaricus, 191
vertirugis, Collybia, 191
vervacti, Agaricus, 7
vervacti, Agrocybe, 7
vervacti, Naucoria, 6, 7
vesca, Russula, 265
vesca var. duportii, Russula, 377
vesicaria, Tremella, 430
Vesiculomyces, 306
vespertinus, Agaricus, 430
vespertinus, Cortinarius, 430
vespertinus, Cortinarius, 72
vestita, Conocybe, 56
vestita, Galera, 56
vestita, Helicogloea, 314
vestita, Pholiotina, 56
vestita, Platygloea, 314
vestitus, Achroomyces, 314
veternosa, Russula, 265
vexans, Conocybe, 56
vibecina, Clitocybe, 45
vibecina var. pseudoobbata, Clitocybe, 45
vibecinus, Agaricus, 45
vibratilis, Agaricus, 79

vibratilis, Cortinarius, 79
vibrissa, Nia, 208
viciae-fabae, Uredo, 339
viciae-fabae, Uromyces, 339
viciae-fabae var. orobi, Uromyces, 339
vietus, Agaricus, 166
vietus, Lactarius, 166
vile, Repetobasidium, 252
vilis, Agaricus, 430
vilis, Clitopilus, 430
vilis, Peniophora, 252
villatica, Psalliota, 4, 5
villatica, Psalliota, 430
villosa, Cyphella, 160
villosa, Lachnella, 160
villosa, Peziza, 160
villosa, Solenia, 160
villosa var. stenospora, Cyphella, 107
villosulus, Rhizopogon, 253
villosus, Agaricus, 231
villosus, Boletus, 157
villosus, Pluteus, 231
vinacea, Nolanea, 104
vinaceogriseum, Tricholoma, 298
vinaceum, Entoloma, 104
vinaceum var. fumosipes, Entoloma, 104
vinaceus, Agaricus, 104
vincae, Puccinia, 332
vincae, Trichobasis, 332
vincae, Uredo, 332
vinicolor, Rhizopogon, 253
vinosa, Flammula, 430
vinosa, Russula, 265
vinosa, Uredo, 351
vinosa, Ustilago, 351
vinosa ssp. occidentalis, Russula, 266
vinosobrunnea, Russula, 266
vinosobrunneus, Agaricus, 4
vinosofulva, Psathyrella, 245
vinosopurpurea, Russula, 266
vinosorubescens, Limacella, 182
vinosus, Agaricus, 238, 430
vinosus, Cortinarius, 75
viola, Hygrocybe, 135
violacea, Amanita, 11
violacea, Gyraria, 20
violacea, Hypochnella, 145
violacea, Poria, 430
violacea, Puccinia, 320
violacea, Russula, 258
violacea, Russula, 266
violacea, Tremella, 20
violacea, Tulasnella, 431
violacea, Tulasnella, 302
violacea, Uredo, 317
violacea, Ustilago, 317
violacea var. carneolilacina, Russula, 431
violacea var. stellariae, Ustilago, 317
violaceifolia, Inocybe, 431
violaceocinctus, Cortinarius, 65
violaceofulvens, Agaricus, 431
violaceofulvens, Panellus, 431
violaceofusca, Inocybe, 79
violaceofuscus, Agaricus, 79
violaceofuscus, Cortinarius, 79
violaceoides, Russula, 431
violaceolivida, Peniophora, 215
violaceolivida, Thelephora, 215
violaceolividum, Corticium, 215
violaceovelatus, Cortinarius, 79
violaceum, Microbotryum, 317
violaceum, Pachysterigma, 431

violaceum, Phragmidium, 320
violaceus, Agaricus, 431
violaceus, Agaricus, 79
violaceus, Cortinarius, 79
violaceus, Hypochnus, 145
violaceus, Irpex, 383
violaceus, Polyporus, 430
violae, Aecidium, 332
violae, Granularia, 348
violae, Polycystis, 348
violae, Puccinia, 332
violae, Tuburcinia, 348
violae, Uredo, 332
violae, Urocystis, 348
violarius, Agaricus, 431
violarius, Pluteus, 431
violarum, Puccinia, 332
violarum, Trichobasis, 332
violascens, Agaricus, 167
violascens, Bankera, 20
violascens, Hydnum, 20
violascens, Lactarius, 167
violea, Tulasnella, 302
violeipes, Hygrophorus, 133
violeipes, Russula, 266
violeum, Corticium, 302
violeus, Hypochnus, 302
violilamellatus, Cortinarius, 79
viperina, Volvaria, 431
viperinus, Agaricus, 431
virens, Agaricus, 44
virens, Clitocybe, 44
virens, Mycena, 199
virens, Mycena, 431
virescens, Agaricus, 266
virescens, Dacrymyces, 292, 431
virescens, Naematelia, 292, 431
virescens, Psilocybe, 431
virescens, Russula, 266
virescens, Tremella, 431
virescens, Tremella, 292
virgae-aureae, Puccinia, 333
virgae-aureae, Xyloma, 333
virgata, Amanita, 307
virgata, Vaginata, 307
virgatula, Inocybe, 151, 155
virgatum, Tricholoma, 298
virgatum var. sciodes, Tricholoma, 297
virgatus, Agaricus, 298
virginea, Hygrocybe, 135
virginea, Omphalia, 135
virginea, Russula, 431
virginea var. fuscescens, Hygrocybe, 135
virginea var. ochraceopallida, Hygrocybe, 135
virgineus, Agaricus, 135
virgineus, Camarophyllus, 135
virgineus, Cuphophyllus, 135
virgineus, Hygrophorus, 135
virgineus var. roseipes, Hygrophorus, 135
viridans, Agaricus, 104
viridans, Ceriporia, 36
viridans, Entoloma, 104
viridans, Polyporus, 36
viridans, Poria, 36
viride, Coriscium, 182
viride, Endocarpon, 182
viride, Hydnum, 431
viride, Sistotrema, 431
viridifucatum, Tricholoma, 298
viridilutescens, Tricholoma, 298
viridimarginata, Mycena, 204

Printed in the United Kingdom by
Lightning Source UK Ltd., Milton Keynes
137446UK00001B/127-132/A